ORGANIC CHEMISTRY

pK_a VALUES (*Continued*)

Acid	Base	pK_a
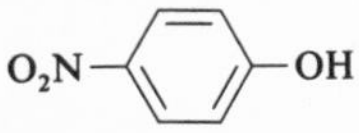	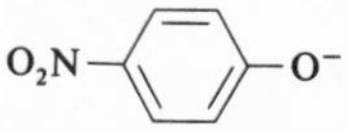	7.2
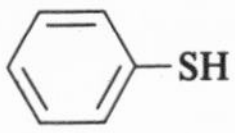	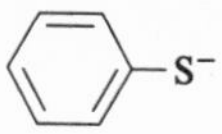	7.8
$CH_3\overset{O}{\overset{\|}{C}}CH_2\overset{O}{\overset{\|}{C}}CH_3$	$CH_3\overset{O}{\overset{\|}{C}}\overset{-}{C}H\overset{O}{\overset{\|}{C}}CH_3$	9.0
HCN	CN^-	9.1
NH_4^+	NH_3	9.4
$(CH_3)_3NH^+$	$(CH_3)_3N$	9.8
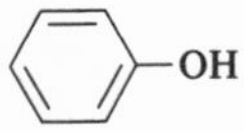	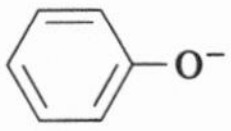	10.0
HCO_3^-	CO_3^{2-}	10.2
CH_3NO_2	$\overset{-}{C}H_2NO_2$	10.2
CH_3CH_2SH	$CH_3CH_2S^-$	10.5
$CH_3NH_3^+$	CH_3NH_2	10.6
$CH_3\overset{O}{\overset{\|}{C}}CH_2\overset{O}{\overset{\|}{C}}OCH_2CH_3$	$CH_3\overset{O}{\overset{\|}{C}}\overset{-}{C}H\overset{O}{\overset{\|}{C}}OCH_2CH_3$	11.0
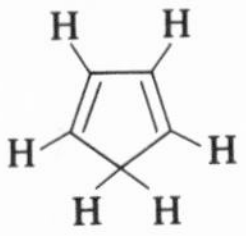	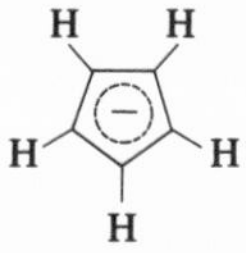	15.0
$CH_3\overset{O}{\overset{\|}{C}}NH_2$	$CH_3\overset{O}{\overset{\|}{C}}\overset{-}{N}H$	15.0
CH_3OH	CH_3O^-	15.5
H_2O	OH^-	15.7
CH_3CH_2OH	$CH_3CH_2O^-$	17
$CH_3-\underset{CH_3}{\underset{\|}{\overset{CH_3}{\overset{\|}{C}}}}-OH$	$CH_3-\underset{CH_3}{\underset{\|}{\overset{CH_3}{\overset{\|}{C}}}}-O^-$	19
$CH_3\overset{O}{\overset{\|}{C}}CH_3$	$CH_3\overset{O}{\overset{\|}{C}}\overset{-}{C}H_2$	19
$CH_3\overset{O}{\overset{\|}{C}}OCH_2CH_3$	$\overset{-}{C}H_2\overset{O}{\overset{\|}{C}}OCH_2CH_3$	23
$CHCl_3$	$\overset{-}{C}Cl_3$	25
$HC\equiv CH$	$HC\equiv C^-$	26
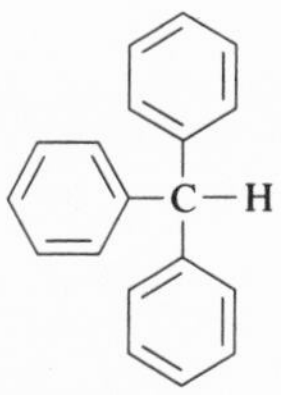	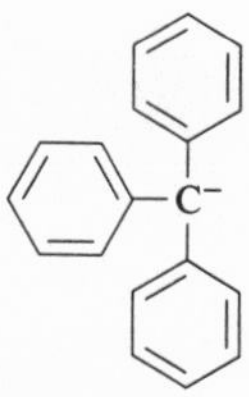	31
NH_3	NH_2^-	36
$CH_2=CH_2$	$CH_2=CH^-$	36
CH_4	CH_3^-	49

ORGANIC CHEMISTRY

SECOND EDITION

SEYHAN N. EĞE
THE UNIVERSITY OF MICHIGAN

D. C. HEATH AND COMPANY
LEXINGTON, MASSACHUSETTS TORONTO

ACKNOWLEDGMENTS

Cover Photograph: Photomicrograph of cyclohexanone oxime by Manfred Kage/ Peter Arnold, Inc.

Mass spectra adapted from *Registry of Mass Spectral Data,* Vol. 1, by E. Stenhagen, S. Abrahamsson, and F. W. McLafferty. Copyright © 1974 by John Wiley & Sons, Inc. Reprinted by permission of John Wiley & Sons, Inc.

Carbon-13 NMR spectra adapted from *Carbon-13 NMR Spectra,* by LeRoy F. Johnson and William C. Jankowski. Copyright © 1972 by John Wiley & Sons, Inc. Reprinted by permission of John Wiley & Sons, Inc.

Ultraviolet spectra adapted from *UV Atlas of Organic Compounds,* Vols. I and II. Copyright © 1966 by Butterworth and Verlag Chemie. Reprinted by permission of Butterworth and Company, Ltd.

Infrared spectra from *The Aldrich Library of FT-IR Spectra,* by Charles J. Pouchert. Copyright © 1985 by Aldrich Chemical Company. Reprinted by permission of Aldrich Chemical Company.

NMR spectra from *The Aldrich Library of NMR Spectra,* 2nd ed., by Charles J. Pouchert. Copyright © 1983 by Aldrich Chemical Company. Reprinted by permission of Aldrich Chemical Company.

Acquisitions Editor: Mary Le Quesne
Production Editor: Cathy Labresh Brooks
Designer: Sally Thompson Steele
Production Coordinator: Michael O'Dea
Photo Researcher: Martha L. Shethar
Text Permissions Editor: Margaret Roll

Published simultaneously in Canada.

Printed in the United States of America.

International Standard Book Number: 0-669-18178-1

Library of Congress Catalog Card Number: 88-080264

10 9 8 7 6 5 4

To my teachers and my students

Preface

The pedagogical aim of *Organic Chemistry,* Second Edition is to educate students to think independently about organic chemistry. The first edition and now, the second edition, have the same philosophy of teaching: Students can truly learn organic chemistry only if they are actively involved in developing a practical understanding of the causes of chemical change, rather than trying to master organic chemistry through memorization. In both editions, I have presented organic chemistry by consistently emphasizing important themes and by returning to fundamentals again and again. In this way I have helped students to think as practicing chemists do in predicting reactivity from structure. Students have told me that they have learned an entirely new way of thinking—of analyzing problems, sorting facts, reasoning by analogy, looking for patterns—and that consequently their approach to all of their other work has changed.

Like the first edition, the second edition of *Organic Chemistry* has been designed to lead students quickly to the concept that structures of organic compounds determine their chemical reactivity. This theme is apparent immediately, even within the first two chapters of the book that introduce students to structure and bonding. Chapter 3, the first chapter devoted to chemical reactivity, uses the reactions of organic compounds as acids and bases to focus the student's attention on two simple reactions, protonation and deprotonation of organic compounds. Through the practice of solving problems on acidity and basicity, students gain confidence in their ability to predict reactivity as chemists do: by looking at structures of organic compounds and applying fundamental concepts such as atomic and ionic sizes, resonance stabilization of species, and pK_a values. In addition, students learn reactions that are important steps in many of the organic transformations they will study later.

Chapter 3 also introduces mechanisms of organic reactions and the convention of using curved arrows for symbolizing the motion of electrons. Coverage of the concepts of nucleophile and electrophile in Chapter 4 is built on the chemistry, and the language and symbolism of chemistry, learned in Chapter 3. In Chapter 4, the reactions of chloromethane with hydroxide ion and hydrogen bromide with propene are used to introduce thermodynamics and kinetics and the idea of reaction pathways.

Chapter 5 presents the nomenclature and conformation of both alkanes and cycloalkanes. Stereoisomerism is discussed in Chapter 6. With these first six chapters, students have most of the concepts they will need to understand the chemistry presented to them in the rest of the book. Extensive cross-referencing allows instructors considerable flexibility in choosing the order of subjects to follow.

NEW IN THE SECOND EDITION

Features

True mastery of organic chemistry requires that students learn to make discriminating use of their eyes and minds. To this end, several new features have been introduced in the second edition.

Visualizing the Reaction
As in the first edition, complete mechanisms are given for each type of reaction. These mechanisms are now highlighted in "Visualizing the Reaction" boxes set apart from the text. Students must practice developing their powers of imagination and following a process with the "inner eye," and these complete mechanisms, which feature the

judicious use of four colors, enhance the process of visualizing. Acidic or electrophilic sites are highlighted as red atoms, blue shading signifies basic or nucleophilic species, and grey shading emphasizes leaving groups. Color is used to indicate whether the reactive species on one side of an equation are converted into new reactive species after reaction, in order to show students the reversibility of many reactions, especially acid-base reactions.

Four Colors

While four-color printing appears primarily in the ''Visualizing the Reaction'' boxes, color is also used to enhance figures and to stress, with consistency, various important structural features. For example, in sections where students are just learning to see stereochemistry, green and red shading highlights stereochemical relationships.

Problem-Solving Skills Sections

In working with my students, I have become convinced that encouraging them to analyze problems systematically is the single most important factor in increasing their overall intellectual skills. These new Problem-Solving Skills sections, unique among organic chemistry texts, offer students a systematic, questioning approach to solving organic chemistry problems. These sections do not simply provide a way for students to learn to plug data into a prelearned formula; rather, students learn to reason their way to a solution.

- In Chapter 1, students are introduced to the idea that the solution to a problem in chemistry requires a step-by-step analysis of the problem. This analysis takes the form of questions that students pose to themselves in a systematic way.
- In Chapter 4, students are shown how to reason backwards in solving problems involving simple syntheses.
- In Chapter 7, students are led through the types of questions that help them to predict the product of a reaction and to transform a given starting material into a desired product. These are complex questions with many types of answers, depending on the particular problem being solved. Not all of the questions are directly applicable to the problem under consideration, but represent steps in the processes of deciding how to use the data given in the problem.
- The same method of questioning is applied in Chapter 9 to writing mechanisms. To reinforce this practice for students, in most chapters through Chapter 23, one or two problems are worked out using the same set of questions.
- The *Study Guide* further reinforces the questioning approach used in the book by applying it to solving some of the problems in and at the end of each chapter.

Concept Maps

Concept maps, which appear in the *Study Guide,* are a fourth innovation in this edition of *Organic Chemistry*. Conceived as a practical way to organize and summarize the material presented in the book, these concept maps present major ideas in outline form. Notes in the margin of the textbook alert students that the concept maps are available. While the concept maps provided can be quite helpful to students, students are encouraged to examine the maps and then create their own, because the process of creating a concept map requires them to give up a purely linear way of thinking about a subject and to explore interrelationships. My students who have used this method to organize their lecture notes and to outline related subjects have found the concept maps not only useful, but fun. Encouraging students to work in this way promotes an actively thinking approach to organic chemistry.

A package of transparency masters containing the complete set of concept maps that appear in the *Study Guide* is available to instructors.

In-Text Summaries
Besides the concept maps in the *Study Guide*, two other forms of summary appear in the textbook itself.

- An end-of-chapter *Summary* offers a concise review of the major concepts covered in the chapter.
- End-of-chapter *Tables* summarize the reactions that appear in the chapter. Organized so as to remind students of how the reactions proceed, they are not made up of general reactions to be memorized, but take students briefly through the stages of the reaction again, reminding them of the types of reagents needed, reactive intermediates involved, and the stereochemistry of the reaction. These tables are particularly helpful to students when they are used together with the concept maps in the *Study Guide*.

Easy-to-Use Cross-Referencing
The second edition is cross-referenced with page numbers to enable both students and teachers to locate related topics quickly.

Reorganization

This revision of *Organic Chemistry* has resulted in considerable rearrangement of topics. Important changes include:

- Early and separate chapters on infrared spectroscopy (Chapter 10) and nuclear magnetic spectroscopy (Chapter 11).
- The introduction to stereochemistry concentrated in a separate chapter (Chapter 6).
- Free radicals in a separate chapter. While free radicals are mentioned briefly in Chapter 5 in connection with the reactivity of alkanes, the separate chapter allows a fuller treatment of their use in synthesis and their biological significance.
- Separation of the contents of each of the longer chapters in the first edition into two shorter chapters. The first chapter in each pair contains the material on that topic commonly covered in most courses; the second chapter may either be omitted or taught in a different order. (Note that this edition retains the integration of biologically interesting examples of chemistry throughout the text.) For example, Chapter 14 concentrates on nucleophilic substitution reactions at the carbonyl group of carboxylic acids and their derivatives. Chapter 15 contains the reactions of carboxylic acids and their derivatives with metal hydrides and organometallic reagents. Chapter 16, which covers the chemistry of enolate anions, follows directly after the chemistry of carboxylic acid derivatives and emphasizes aldol and Claisen condensations. Chapter 18, the second chapter in the pair, involves the chemistry of α,β-unsaturated carbonyl compounds and ylides, and follows a chapter on polyenes (Chapter 17). Chapter 19 covers electrophilic aromatic substitution reactions, and the rest of aromatic chemistry comes in Chapter 23, after students have learned about free radicals (Chapter 20), mass spectrometry (Chapter 21), and diazotization reactions (Chapter 22). Aromatic chemistry is further reinforced by a study of heterocyclic compounds in Chapter 24.

The second edition is unchanged from the first edition in its emphasis on an understanding of reactivity rather than on an encyclopedic knowledge of reactions. Students suffer from an overload of different things to remember, and from not learning to think their way toward predicting the outcome of a reaction they have never seen before. Thinking for themselves is a skill that will be valuable to them as graduate students in chemistry, in other sciences with a strong chemical component, and in medicine. A thorough understanding, with an emphasis on mechanism of a few major reactions, enables students to apply the principles they have learned and gives them the confidence to tackle new situations. I would rather teach students who are confident about their abilities than students who are overwhelmed by a mass of undigested material. Once the fundamental thought processes of chemistry have been learned, teachers find that they can introduce their own favorite reactions to their students.

The length of this book, thus, comes from meticulous explanations and fully detailed mechanisms for selected reactions, mechanisms that were chosen to represent many other reactions of the same type. My experience convinces me that only such an explicit, consistent approach gives students the reinforcement they need to learn a subject they perceive as being difficult.

Does this approach work? Feedback from students has been so positive that it has encouraged me to keep the same approach in this edition as I used in the first edition. I particularly cherish reports from faculty at various schools that students using the book receive better scores than before on the standardized ACS examinations at the end of the year and even complain that the examination was too easy! If the book is used as it is intended, students come to enjoy and be challenged by their intellectual competence. To watch this happen, is, of course, the ultimate satisfaction for a teacher.

SUPPLEMENTS

Study Guide

Roberta W. Kleinman of Lock Haven University, Pennsylvania, and Marjorie L. C. Carter of the University of Michigan are my coauthors for the first and second editions of the *Study Guide*. Both of them have given me invaluable help—especially Roberta Kleinman, who, with her skill at the computer, is responsible for transforming the material into camera-ready copy. As in the first edition, the second edition contains detailed solutions to every problem in the text as well as explanations of the reasoning processes behind the answers for many problems. Some problems in this edition are worked out using the questions developed in the Problem-Solving Skills sections of the text to reinforce students' understanding of this approach.

New concept maps, charts summarizing relationships between key ideas in a section or portion of a section, provide an innovative tool for practice and review. Notes in the margin of the text refer students to relevant concept maps in the *Study Guide*.

Transparency Masters

For classroom use by the instructor, a package of *Transparency Masters* contains the complete set of concept maps that appear in the *Study Guide*.

Transparencies

A complete set of *Transparencies* ($8\frac{1}{2} \times 11''$), many of them full color, is available free to college adopters of the textbook. Reproduced from selected figures and chemical structures in the text, the transparencies include spectra, molecular orbitals, space-filling models, stereochemistry, and reaction mechanisms.

Software

A variety of quality *software* programs are offered by arrangement with COMPress, a division of Queue, Inc. For information and demonstration disks, contact the Marketing Department at D. C. Heath at 1-(800)-235-3565.

ACKNOWLEDGMENTS

Many people have contributed to converting the first edition of *Organic Chemistry* into the second edition. Suggestions and corrections from colleagues and students who have used the book are particularly valuable. I owe special thanks to Brian Coppola, Richard Lawton, and John Wiseman of the University of Michigan; Sally Weersing, Muskegon Community College; Dorothy Goldish, Edwin Harris, and Tom Maricich, California State University, Long Beach; David Reingold, Juniata College; Hans Cerfontain, Henk Hiemstra, and Gerrit-Jan Koomen, University of Amsterdam; and Clarisse Habraken, University of Leiden.

Dr. Alex Aisen of the Department of Radiology at the University of Michigan supplied me with information on magnetic resonance imaging and the photograph that appears on page 416 of the text. Torin Dewey, a student of mine who is now doing graduate work at the University of California, Berkeley, took most of the proton magnetic resonance spectra. Frank Parker and James Windak helped with the spectra illustrating Fourier transform nuclear magnetic resonance and infrared methods.

I very much appreciate the suggestions of the reviewers for the second edition: R. G. Bass, Virginia Commonwealth University; George B. Clemans, Bowling Green State University; James H. Cooley, University of Idaho; William A. Donaldson, Marquette University; Richard R. Doyle, Denison University; Barbara V. Enagonio, Montgomery College; William A. Feld, Wright State University; Dorothy M. Goldish, California State University, Long Beach; Richard Jochman, St. John's University; Ronald H. Kluger, University of Toronto; Robert J. Newland, Glassboro State College; Daniel H. O'Brien, Texas A&M University; Suzanne T. Purrington, North Carolina State University; Harold R. Rogers and Robert Spenger, both at California State University, Fullerton; Joseph J. Tufariello, State University of New York at Buffalo.

No amount of thanks will repay my debt to Marjorie Carter and Roberta Kleinman. Not only have they contributed substantially to the *Study Guide,* but they have helped with the text itself too. All of the three-dimensional figures in the text originated with Roberta Kleinman, who combines artistic talent with an interest in how students visualize and learn. She joins me in struggling to see things as the student sees them and not as we, with years of experience, know them to be. I owe a great deal to her critical eye. Marjorie Carter also brings the viewpoint of the students and a questioning mind to the thankless task of proofreading. Many times she has insisted that Roberta and I try again in drawing structures or explaining our reasoning for greater clarity for the student. I value the help of both of these good friends.

Mary Le Quesne, Senior Acquisitions Editor at D. C. Heath, saw the book through revision with an openness toward my ideas, even when they were unconventional, for which I am grateful. Her knowledge of chemistry and her insight contributed greatly to this edition. Cathy Brooks, Senior Production Editor, has guided me patiently and with endless good humor through the traumatic process of the publication of a technical book in four colors. I am grateful for her expertise and her meticulous attention to detail.

Finally, none of this would have been possible without the encouragement of my family and friends. I thank them for their understanding during all of the times when I could not be with them and for their steady love that supports me.

Seyhan N. Eğe

Contents

1

An Introduction to Structure and Bonding in Organic Compounds

A • L O O K • A H E A D

Organic chemistry was born in 1828 when Friedrich Wöhler attempted to synthesize ammonium cyanate, NH_4CNO, and instead obtained urea,

$$NH_2\overset{\overset{O}{\|}}{C}NH_2$$

Wöhler, who studied to be a doctor of medicine before he decided to become a chemist, discovered that the compound he had made was identical with urea recovered from urine. Up to that time, scientists had thought that the compounds present in living plants and animals could not be synthesized in the laboratory from inorganic reagents. Wöhler recognized the importance of his experiment and wrote to a friend, "I must tell you that I can make urea without the use of kidneys, either man or dog. Ammonium cyanate is urea."

Wöhler's discovery was important because it gave impetus to a long series of experiments in which chemists probed the nature of the chemical substances that exist in living organisms and in petroleum and coal, which are formed from the remains of plants and animals that lived long ago. As early chemists struggled to isolate and purify the components of plants, animals, coal, and petroleum, they quickly realized that the chemistry of carbon was associated with life in a special way that distinguished that element from the others. Compounds containing carbon were called organic compounds to reflect their origin in living systems and to distinguish them from the inorganic compounds, the acids, bases, and salts derived from the other elements in the periodic table.

Organic chemistry, the chemistry of the compounds of carbon, is central to many disciplines. Life processes are supported by the chemical reactions of complex organic compounds such as enzymes, hormones, proteins, carbohydrates, lipids, and nucleic acids. Chemists, in attempts to improve on nature, have created millions of organic compounds that did not exist in nature originally. And the search continues, for example, for a synthetic painkiller that works as well as natural opiates such as morphine without having the undesirable side effect of being addictive. Industrial chemists have developed synthetic rubber, called Neoprene, and synthetic silks, such as rayon and nylon, which improve on the properties of the natural substances and offset shortages of natural supplies. Food additives, dyes, artificial flavorings, artificial sweeteners, preservatives, and pesticides, most of them organic compounds, are regularly in the headlines. Crude petroleum is converted by organic reactions into fuels that supply energy for heat, transportation, and industry. Petroleum is also the chemical basis for giant molecules engineered to have useful properties. The names of these new materials—Teflon, Orlon, Acrilan, polystyrene, polypropylene, polyurethane—have become household words.

The progress of chemistry as a science depends on experimental manipulations of substances in the laboratory, leading to the observation of new phenomena. In thinking about these phenomena, chemists arrive at ideas about the nature of the chemical substances under investigation. These ideas are tested by new experiments and new observations. The range of manipulations and observations available to chemists has expanded enormously in recent years. The power that chemists have to transform chemical compounds into new ones, sometimes useful and sometimes harmful, has also increased.

Chemists have refined the way they visualize and think about the submicroscopic units called atoms and molecules. They have created models that help them picture and understand experimental facts about chemical substances. Scientists are constantly creating and refining models that explain the physical world. Some models are pictorial; others are more abstract and mathematical. Some models are widely adopted because they are useful; others are soon modified or discarded on the basis of new observations. In this first chapter, we will examine some of the models chemists currently find most useful for thinking about experimental observations of structures and properties of organic compounds.

1.1 HOW TO STUDY ORGANIC CHEMISTRY

Learning organic chemistry is like learning a new language, a language that is both verbal and pictorial. Organic chemistry is a highly organized discipline, based on the premise that the structure of a compound determines its reactivity. Organic chemistry is, therefore, a study of the relationship between the structures of molecules and their reactions. As you develop an understanding of this relationship, you will be able to make predictions about molecules and reactions that are new to you. Your eyes will learn to dissect complex structures and to distinguish the pieces you recognize. You will be able to reason by analogy from systems and reactions that you have already learned to new systems and reactions that resemble the earlier ones in important ways.

Your pencil is an indispensable tool in your studies. While you are training your eye to look carefully, you should be training your hand to draw. Professional organic chemists cannot talk to each other without drawing structures. If you expect to learn organic chemistry well, you must draw and redraw the structures of compounds and write out equations as you are studying. The ways structures are illustrated in this book and drawn by your instructor on the blackboard represent the result of years of evolution in thinking about organic compounds. Different kinds of pictures represent different degrees of precision. You must train your hand to produce the correct pictures; *precision in drawing leads to precision in thinking*.

The correct representation of a structure will often give you insight into the correct solution to a problem.

Organic chemists are concerned about the shapes of molecules. You should acquire molecular models* and examine the structures of carbon compounds in three dimensions. Then you will learn to translate three-dimensional structures to the two dimensions of a page. Chemists have developed specific ways of representing three-dimensionality on a flat surface in order to convey the maximum amount of information clearly and consistently. If the rules are not followed exactly, the resulting pictures are meaningless.

To study effectively, you should read the assignments before attending lectures. Spend most of your study time working out the problems. *Work all of the problems,* no matter how simple they seem, in writing and in full detail. There is no other way to develop the skills you will need to work with the more and more complex structures and concepts you will encounter as you go through the book. The important reactions and ideas will come up again and again in different problems throughout the book. If you find that you cannot do a problem, go back to the text and review the reactions and ideas that apply to it; then try the problem once more. This method of studying will ensure that you spend the most time working with the ideas that you find difficult. As you go back and forth between the problems and the text, you will gradually learn the most important facts without making a specific effort to memorize them. You will also learn a way of thinking, of looking for patterns and of recognizing qualitative similarities between seemingly unrelated facts. As an aid to your study, this book has sections on problem-solving skills. They outline the thinking processes that are necessary for the successful solution of the most common types of problems in organic chemistry. The development of such skills will be one of the most important results of your study of organic chemistry, because an ability to think this way can be applied to problems in every area of life.

The answers to the problems in the text are found in the *Study Guide* for this book. In addition, the *Study Guide* contains concept maps, which are summaries of important ideas presented in outline form. Notes have been placed in the margin to alert you that these summaries are available. After looking at some of the concept maps, you may want to practice making your own (which need not look exactly like the ones in the *Study Guide*) to summarize material presented in lectures or to reorganize your study notes. Concept maps may be helpful to you as an alternative way of organizing ideas and presenting relationships.

Gradually, you will learn the names of a large number of organic compounds in order to be able to communicate your thoughts about them. All compounds have formal names that can be assigned by the application of definite rules agreed upon by chemists. Some compounds also have trivial names by which they have been known historically. These names still appear on labels of reagent bottles and in articles in scientific journals. To be literate in organic chemistry, you must be able to recognize both formal and trivial names.

The successful mastery of organic chemistry requires a lot of hard work and consistent studying. It is not a subject that can be crammed. Many students make the mistake of trying to memorize the text. An understanding of the basis of chemical transformations is what is really needed. It is true that facts must be learned, but you will be overwhelmed by them unless you develop an ability to see relationships.

---

*Framework Molecular Models, available from Prentice-Hall, Inc., Englewood Cliffs, New Jersey 07632, and Fieser Molecular Models, available from Aldrich Chemical Co., Inc., Milwaukee, Wisconsin 53233, are recommended.

Although the study of organic chemistry requires a lot of concentrated work, many students find that it is also fun. The thinking that goes into such study is related to the thinking used in solving puzzles. For example, solving problems in organic chemistry requires you to recognize patterns and fill in the missing pieces, much as you do when you put together a jigsaw puzzle. You also learn to be precise in thinking about qualitative concepts, just as you are already precise in quantitative ways in mathematics. You will experience the power of your mind to analyze an unfamiliar problem and to arrive at a correct picture of the disparate facts that must be brought together for a solution. You will come to trust in your ability to think correctly to predict experimental outcomes. Such self-knowledge is exhilarating.

1.2 IONIC AND COVALENT COMPOUNDS

Compounds are divided broadly into two classes, ionic and covalent. **Ionic compounds** are composed of ions, which are units of matter that may be single atoms or groups of atoms, bearing positive or negative charges. In **covalent compounds,** the structural units are molecules having no net charge. Ionic compounds are usually crystalline solids with high melting points. Many of these compounds dissolve in water to form solutions that conduct electricity. Common table salt, NaCl, is a typical example of an ionic compound in which the ions, Na^+ and Cl^-, are single atoms. Magnesium sulfate, $MgSO_4$, commonly known as Epsom salts, and sodium bicarbonate, $NaHCO_3$, which is baking soda, are other familiar ionic compounds. In these compounds, the negatively charged ions (the sulfate anion, SO_4^{2-}, and the bicarbonate anion, HCO_3^-) are composed of atoms held together by covalent bonds. In other ionic compounds, such as ammonium chloride, NH_4Cl, the positively charged ion (the ammonium ion, NH_4^+) contains covalent bonds.

Covalent compounds may be gases, liquids, or solids. Methane, CH_4, is the principal component of natural gas. Carbon tetrachloride, CCl_4, is a typical covalent liquid that was once commonly used in dry cleaning. Naphthalene, $C_{10}H_8$, which used to be the main constituent of mothballs, is a covalent solid. Methane, carbon tetrachloride, and naphthalene do not dissolve in water to any great extent. Such compounds are **nonpolar covalent compounds.**

Other covalent compounds, such as ethanol, CH_3CH_2OH (a liquid), and glucose, $C_6H_{12}O_6$ (a solid), are quite soluble in water and form solutions that do not conduct electricity. Such compounds ionize only slightly in aqueous solutions. Water, ethanol, and glucose are examples of **polar covalent compounds.** In the rest of this chapter, we will look more closely at ideas about the nature of chemical bonding and at the various factors that influence the different physical properties of compounds, such as boiling point, melting point, and solubility.

1.3 IONIC BONDING

Crystalline sodium chloride consists of an arrangement of positively charged sodium ions and negatively charged chloride ions arranged alternately in a three-dimensional array called a **crystal lattice.** Each sodium ion is surrounded by six chloride ions, and each chloride ion by six sodium ions. The ions are held in place by strong electrostatic forces between the positively and negatively charged ions.

Ionic bonding consists of electrostatic attractions between ions of opposite charge. The individual ion is a sphere bearing a symmetrical distribution of charge. For this reason, there is no particular direction to bonding in ionic compounds. In the solid state, there are no individual molecules of sodium chloride composed of one Na^+ and one Cl^-.

Sodium chloride has a high melting point, 801 °C, and a very high boiling point, 1413 °C. These physical properties are an indication of the strength of the electrostatic forces holding the ions together. Large amounts of energy must be applied to the sodium chloride crystal to overcome electrostatic forces and break down the crystal lattice by the process of melting. Even more energy is necessary to vaporize sodium chloride.

Study Guide
Concept Maps 1.1 and 1.2

1.4 COVALENT BONDING

A. Lewis Structures

Early in this century, Gilbert N. Lewis at the University of California at Berkeley proposed that the **covalent bond** be represented as the sharing of a pair of electrons between two atoms. He also proposed that, with a few exceptions, stable molecules or ions have eight electrons, or four pairs of electrons, in the outermost shell, the valence shell, of each atom. This stable configuration of electrons is called an **octet.** His suggestions for drawing structures of covalent compounds have proved enormously useful to organic chemists.

The **Lewis structure** of a covalent molecule shows all the electrons in the valence shell of each atom: the bonds between atoms are shown as shared pairs of electrons. The total number of electrons in the valence shell of each atom can be determined from its group number in the periodic table. The shared electrons are called the **bonding electrons** and may also be represented by a line or lines between the two atoms. The valence electrons that are not being shared are the **nonbonding electrons:** they are shown by dots oriented in a square around the symbol of the atom. The construction of the Lewis structure for a covalent compound, hydrogen chloride, is shown below.

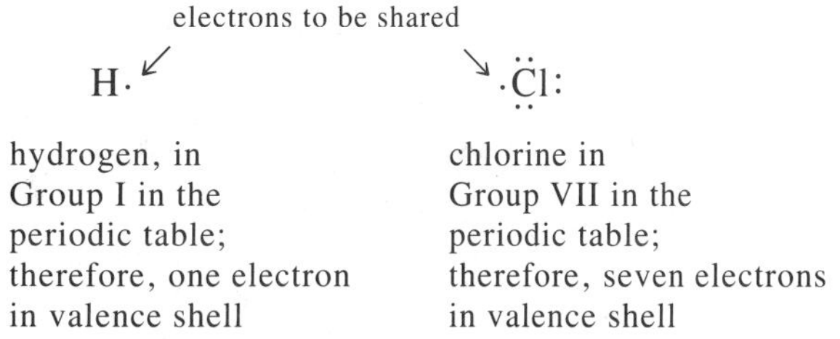
electrons to be shared
H·
·Cl:
hydrogen, in
Group I in the
periodic table;
therefore, one electron
in valence shell
chlorine in
Group VII in the
periodic table;
therefore, seven electrons
in valence shell

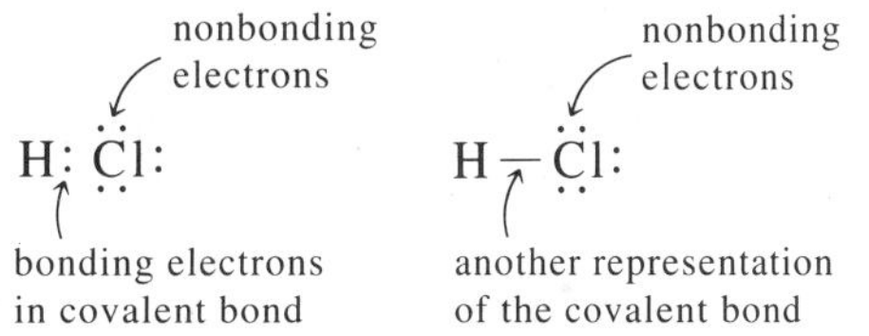
nonbonding
electrons
H: Cl:
bonding electrons
in covalent bond
nonbonding
electrons
H—Cl:
another representation
of the covalent bond

The hydrogen atom in hydrogen chloride shares two electrons, which is all its valence shell can hold. The chlorine atom has eight electrons around it in a stable octet.

TABLE 1.1 Condensed Formulas and Lewis Structures of Some Ions and Compounds

Name	Condensed Formula	Lewis Structures
water	H_2O	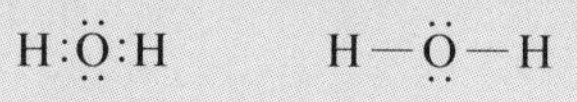
hydronium ion	H_3O^+	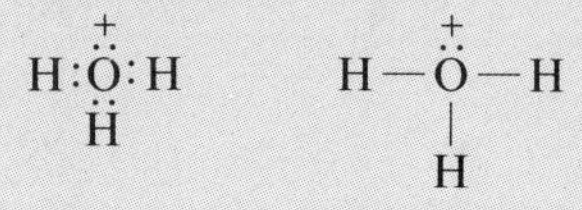
ammonia	NH_3	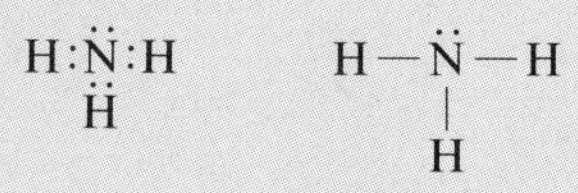
methane	CH_4	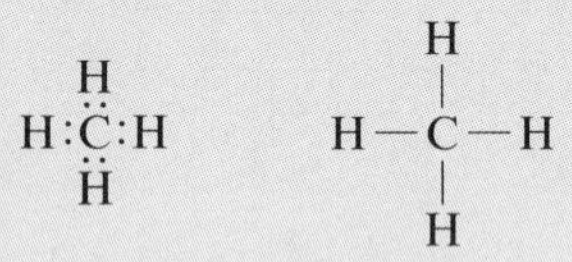
methanol	CH_3OH	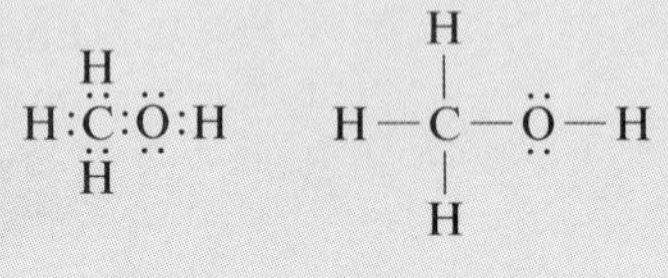
methoxide anion	CH_3O^-	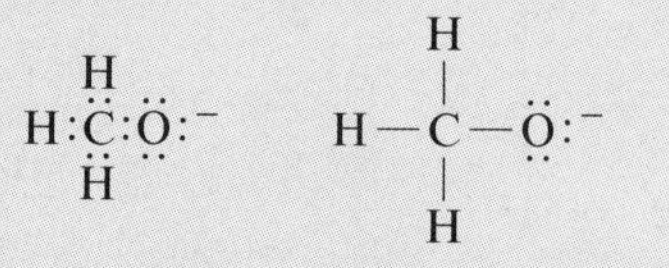
hydroxylamine	H_2NOH	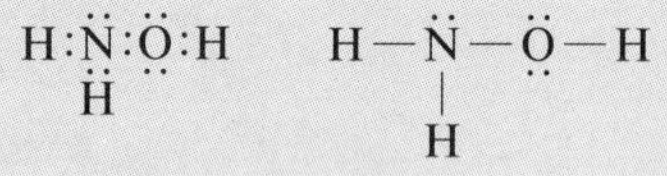

Table 1.1 includes some more examples of Lewis structures of compounds and ions, written first in the abbreviated form known as the condensed formula and then as both the full Lewis structure showing all the valence electrons and the Lewis structure with the covalent bonds represented by lines and only the nonbonding electrons shown as dots.

In all the structures in Table 1.1, the valence electrons are arranged so that the atoms of all elements except hydrogen share eight electrons. The restriction that a hydrogen atom usually cannot share more than two electrons means that we cannot draw structures in which a hydrogen atom has more than one covalent bond. In constructing the water molecule, for example, we see from the periodic table that the oxygen atom has six valence electrons and the hydrogen atoms one each. A total of eight dots representing the electrons can be placed around the oxygen atom. The hydrogen atoms can each share two of these electrons with the oxygen atom, which then has two pairs of unshared, or nonbonding, electrons.

When drawing Lewis structures, we must be careful to keep track of the number of electrons available to form bonds and the location of the electrons. The hydronium ion, for example, is formed when a water molecule accepts a proton (a hydrogen atom without its electron) from an acid. Eight electrons are available to bond one oxygen atom with three hydrogen atoms.

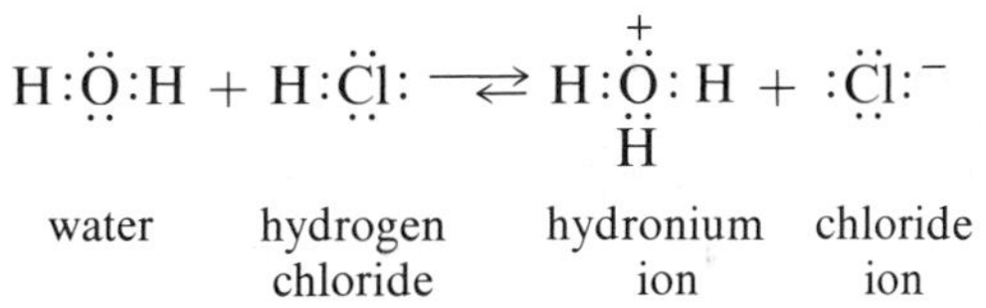
water
hydrogen chloride
hydronium ion
chloride ion

In the equation above, two uncharged species, water and hydrogen chloride, react to give a positively charged hydronium ion and a negatively charged chloride ion. The sum of the charges on one side of the equation, zero in this case, equals the sum of the charges, also zero, on the other side. Charges as well as numbers of atoms must always be balanced in writing equations.

B. Formal Charges

Organic chemists are not satisfied with the simple statement that the hydronium ion is positively charged. They find it useful to locate the charge on a particular atom in the ion. In the equation in the preceding section, the hydronium ion is shown with the positive charge next to the oxygen atom. The decision as to where to put the charge is made by calculating the formal charge for each atom in an ion.

The **formal charge** for an atom may be calculated using this formula:

$$\begin{aligned}\text{Formal charge} = &\ (\text{number of valence electrons})\\ &- (\text{number of nonbonding electrons})\\ &- \tfrac{1}{2}(\text{number of bonding electrons})\end{aligned}$$

To locate the formal charge in the hydronium ion, all of the electrons in the valence shell of each atom are counted.

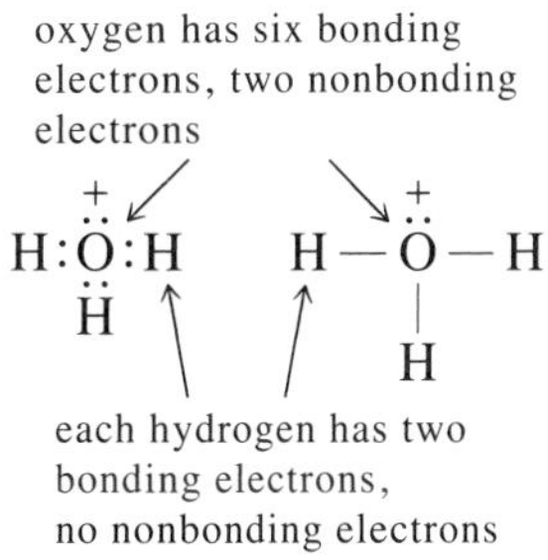
oxygen has six bonding electrons, two nonbonding electrons
each hydrogen has two bonding electrons, no nonbonding electrons

Each hydrogen atom in the hydronium ion has no formal charge because each has one valence electron, no nonbonding electrons, and two bonding electrons.

$$\text{Formal charge for hydrogen} = 1 - 0 - \tfrac{1}{2}(2) = 0$$

In the hydronium ion, oxygen shares six electrons with the hydrogen atoms and therefore has six bonding electrons. It has two nonbonding electrons. Because it is in Group VI of the periodic table, it has six valence electrons.

$$\text{Formal charge for oxygen} = 6 - 2 - \tfrac{1}{2}(6) = 6 - 5 = +1$$

Calculating a formal charge is essentially the same as asking whether an atom in a molecule or ion has in its valence shell more electrons or fewer electrons than are necessary to balance its nuclear charge. An uncharged oxygen atom has to have six electrons in its valence shell. In the hydronium ion, oxygen is bonding with three hydrogen atoms, so only five electrons effectively belong to oxygen, which is one fewer than it needs. Therefore, it bears a formal charge of +1.

In the case of the hydronium ion, the formal charge for oxygen also represents the charge on the ion because no other atoms in the ion are charged. *The sum of the formal charges on the atoms in an uncharged molecule is zero. For an ion, the sum of the formal charges on different atoms should add up to the charge on the ion.*

PROBLEM 1.1

Draw Lewis structures for the species represented by the following condensed formulas. Identify any atoms bearing formal charges.

(a) CCl_4 (b) CH_3Br (c) $CH_3OH_2^+$ (d) NH_2^- (e) $CH_3NH_3^+$

(f) H_2NNH_2 (g) PH_3 (h) H_2S (i) CH_3CH_2OH

(j) $HOCH_2CH_2OH$ (k) $CH_3OCH_3^+$ (with a CH_3 on the O)

PROBLEM 1.2

Decide whether the central atom in each of the following formulas is uncharged, positively charged, or negatively charged. All nonbonding electrons are shown.

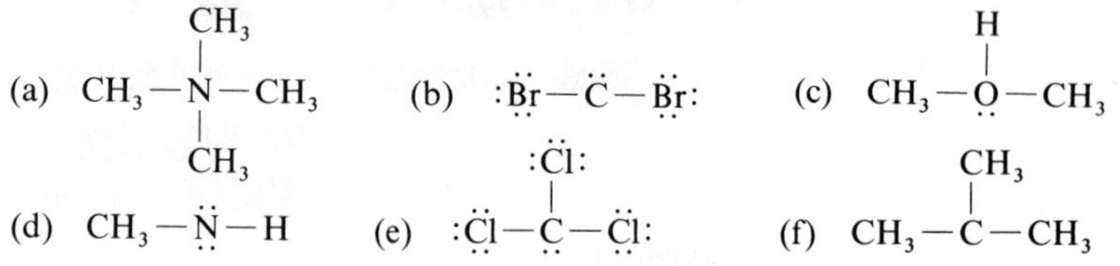

C. Molecules with Open Shells

Sometimes there are not enough electrons in a system to provide an octet around the central atom. Boron trifluoride, BF_3, is such a molecule. Boron is in Goup III of the periodic table and has only three electrons in its valence shell. Adding these three to the twenty-one electrons contributed by the three fluorine atoms gives twenty-four electrons available for bonding. If we put octets around the fluorine atoms, the boron atom ends up with only six electrons.

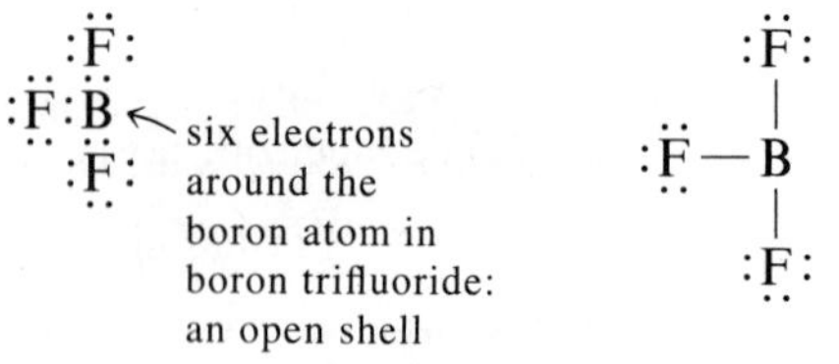
six electrons
around the
boron atom in
boron trifluoride:
an open shell

This leaves the boron atom with an **open shell,** meaning that it can accept another pair of electrons to complete an octet. As a result, boron trifluoride reacts with compounds such as ammonia that have nonbonding electrons.

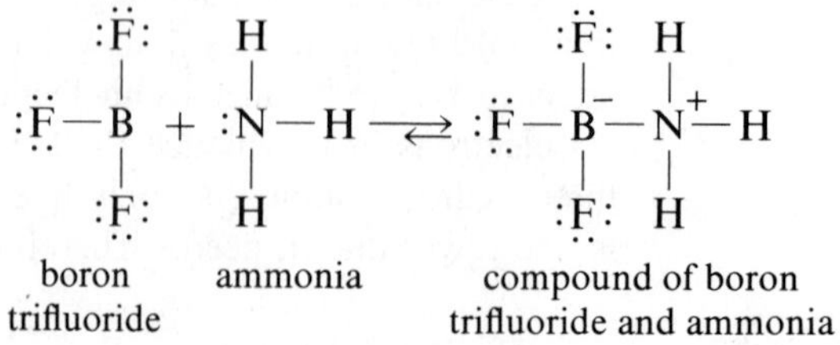
boron
trifluoride
ammonia
compound of boron
trifluoride and ammonia

PROBLEM 1.3

In the equation above, formal charges are shown in the formula for the reaction product of boron trifluoride and ammonia. Make sure you understand how they were obtained.

PROBLEM 1.4

Of the species shown in Table 1.1, pick out the ones that would react with BF_3. Write equations for the reactions. Be sure to include any formal charges that result.

D. Multiple Bonds

Carbon, the central element in organic compounds, usually has a valence of four, meaning that it forms four covalent bonds to other atoms. In some compounds, the carbon atoms are held together by double or triple covalent bonds, which are also called **multiple bonds.** Lewis structures can be written to show the sharing of two or three pairs of electrons by two carbon atoms. For example, the carbon atoms in ethylene, C_2H_4, share two pairs of electrons in a double bond. Acetylene, C_2H_2, is represented as having a triple bond formed by the sharing of three pairs of electrons. In both of these molecules, each carbon atom has eight electrons around it and four bonds.

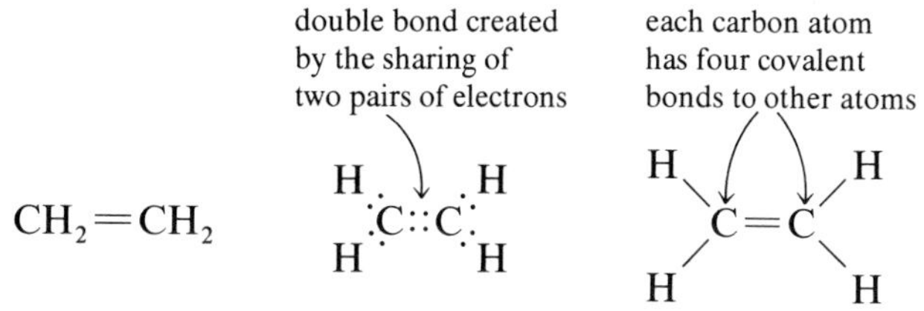

CH₂=CH₂
double bond created by the sharing of two pairs of electrons
each carbon atom has four covalent bonds to other atoms

ethylene

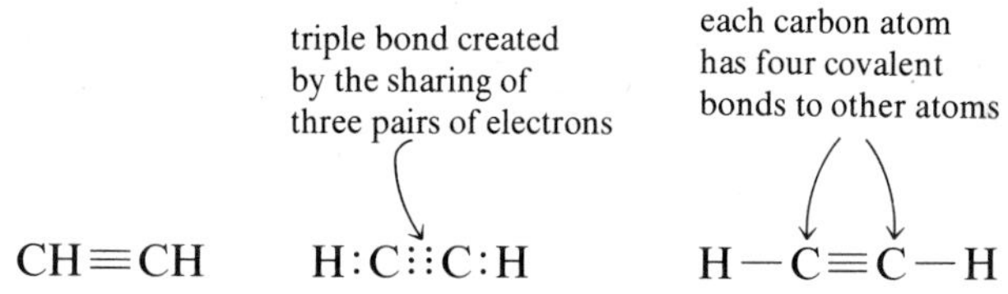

CH≡CH
triple bond created by the sharing of three pairs of electrons
each carbon atom has four covalent bonds to other atoms
H:C⋮⋮C:H
H—C≡C—H

acetylene

Many important organic compounds, such as carbon dioxide, CO_2, and formaldehyde, CH_2O, have double bonds between carbon and oxygen. Triple bonds between carbon and nitrogen are found in cyanides, such as hydrogen cyanide.

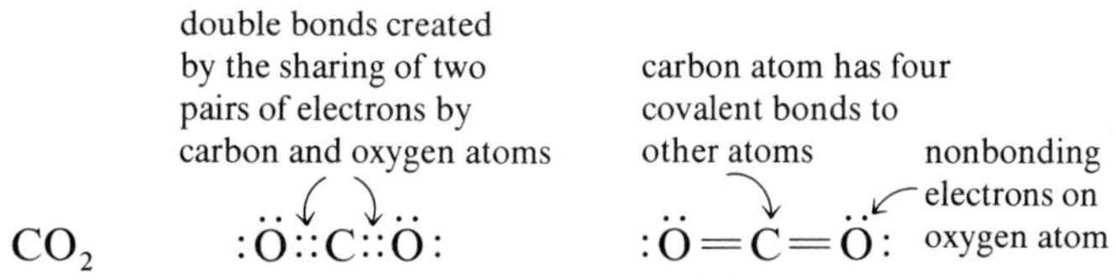

CO₂
double bonds created by the sharing of two pairs of electrons by carbon and oxygen atoms
carbon atom has four covalent bonds to other atoms
nonbonding electrons on oxygen atom

carbon dioxide

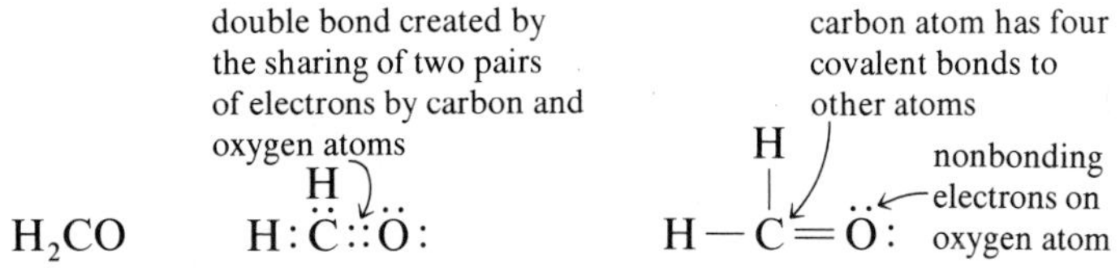

H₂CO
double bond created by the sharing of two pairs of electrons by carbon and oxygen atoms
carbon atom has four covalent bonds to other atoms
nonbonding electrons on oxygen atom

formaldehyde

triple bond created by the sharing of three pairs of electrons by carbon and nitrogen atoms

carbon atom has four covalent bonds to other atoms

nonbonding electrons on nitrogen atom

HCN H:C:::N: H—C≡N:

hydrogen cyanide

PROBLEM 1.5

Draw Lewis structures for the compounds represented by the following formulas.

(a) $CH_3C(=O)CH_3$ (b) $CH_3C{\equiv}CH$ (c) $HC(=O)NH_2$ (d) $HC(=O)OH$

(e) $CH_3C(=O)OCH_3$ (f) $HON{=}O$ (g) $CH_3N{=}NCH_3$ (h) $CH_2{=}CHCl$

E. Resonance

Some covalent molecules and ions cannot be represented satisfactorily by a single Lewis structure. The carbonate ion, CO_3^{2-}, is an example of such a species. The carbonate ion in calcite, $CaCO_3$, is planar, with bond angles of 120° and three equivalent carbon-oxygen bonds 1.29 Å (1.29×10^{-8} cm) long.

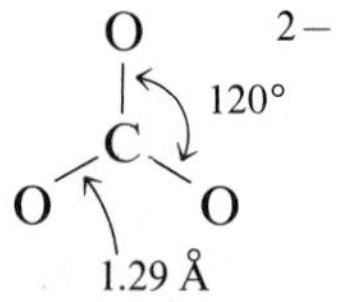
2−
120°
1.29 Å

structure determined experimentally for the carbonate ion, CO_3^{2-}

But, if there is to be an octet around each atom, the Lewis structure of CO_3^{2-} must be drawn with a double bond between carbon and one of the oxygen atoms.

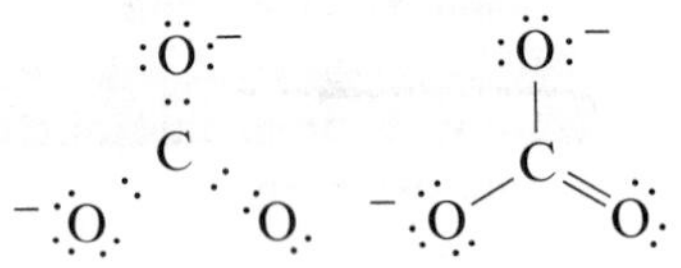

Lewis structures for the carbonate ion, CO_3^{2-}

Two of the oxygen atoms have formal charges of −1, giving a total charge of −2 for the ion. This Lewis structure for the carbonate ion suggests that one of the oxygen atoms is different from the other two. The experimental evidence, however, indicates that all three oxygen atoms are equivalent. For example, the distances from the carbon atom to each oxygen atom are equal. As far as we can tell experimentally, each oxygen atom has some negative charge and an equal probability of reacting with an acid to pick up a proton. The Lewis structure does not adequately depict the experimental reality.

The experimental observations for the carbonate ion are better represented by a picture in which the electrons are equally distributed among all three oxygen atoms. Three structures may be drawn for the carbonate ion, differing only in the location of pairs of electrons. The individual structures are **resonance contributors** to the structure of the carbonate ion; the carbonate ion is pictured as a **resonance hybrid** of these contributors. These structures differ only in the arrangement of electrons, not in the positions of the atoms.

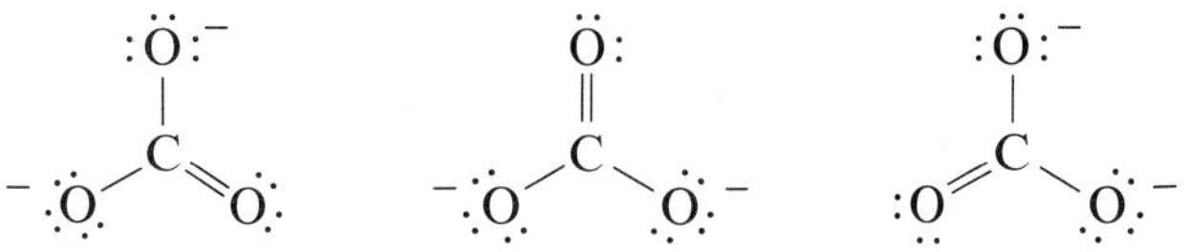

resonance contributors to the structure of the carbonate ion

The actual properties of the carbonate ion cannot be represented by any one of the Lewis structures taken alone. The experimental facts are rationalized by drawing the three resonance contributors. These three taken together indicate that each oxygen atom bears two-thirds of the charge on an electron and that the carbon-oxygen bonds are all the same length, a length between that typical of a carbon-oxygen single bond (1.43 Å) and that typical of a carbon-oxygen double bond (1.22 Å).

Resonance contributors are significant for many other ions. The nitrate ion is another example of a species for which a single Lewis structure is not satisfactory.

NO_3^-

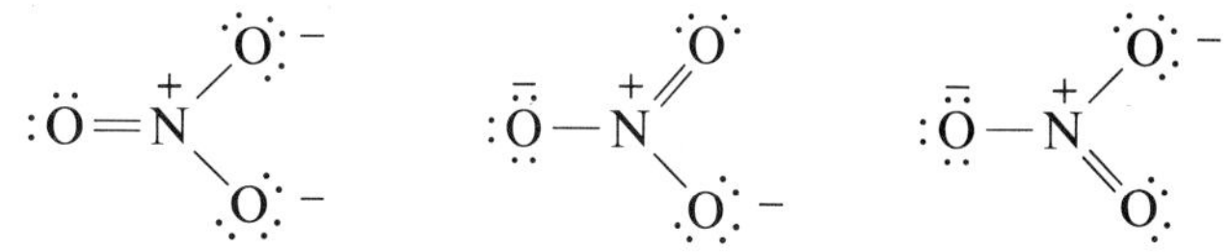

nitrate ion

resonance contributors to the structure of the nitrate ion

Again, these structures differ from each other only in the location of pairs of electrons. In the nitrate ion, two oxygen atoms bear a formal charge of -1, and the nitrogen atom has a formal charge of $+1$. The sum of these formal charges is the charge on the anion. Experimentally, the three oxygen atoms in the nitrate ion are equivalent.

Structures for uncharged molecules may also have resonance contributors. Nitromethane is an example of a molecule for which we can draw more than one Lewis structure.

CH_3NO_2

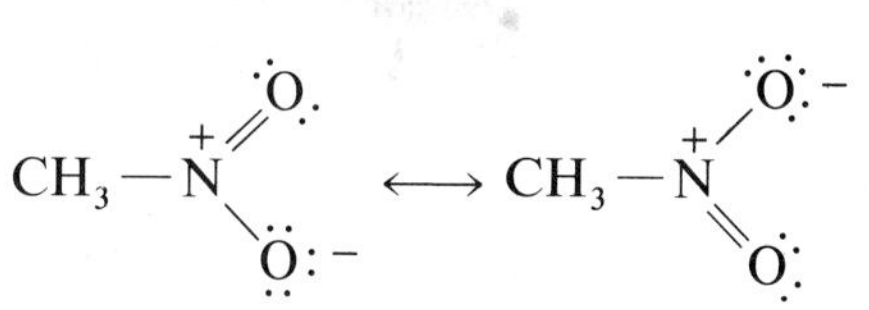

resonance contributors for nitromethane

In nitromethane, the formal charge of $+1$ on the nitrogen atom and the formal charge of -1 on the oxygen atom cancel each other, so the molecule as a whole is not charged.

A double-headed arrow, $\longleftrightarrow$, is used between the resonance contributors of nitromethane to indicate their relationship. This symbol does *not* mean that the two forms are in equilibrium with each other. No reaction is implied by the double-headed resonance arrow. There is only one structure for nitromethane, which is a hybrid of the two Lewis structures we are able to draw. The symbol for equilibrium is two arrows pointing in opposite directions, showing a reversible chemical reaction, for example,

$$H_2CO_3 + H_2O \rightleftharpoons H_3O^+ + HCO_3^-$$

Resonance is an example of a model that was developed to deal with experimental observations that could not be explained in terms of a simpler model, such as a single Lewis structure for a molecular species. It is important to remember that the individual representations of resonance contributors have no reality. The compound, such as nitromethane, for which resonance contributors are written does not exist as a mixture of different forms. The actual molecular species has properties suggested by all the resonance contributors taken together. For example, in nitromethane, the nitrogen atom bears a positive charge, each oxygen atom bears part of a negative charge, and both nitrogen-oxygen bonds are the same length.

The resonance contributors shown for the carbonate ion are equivalent to each other, as are the ones shown for the nitrate ion and nitromethane (p. 11). This is not always the case. For example, three resonance contributors can be written for the formate ion.

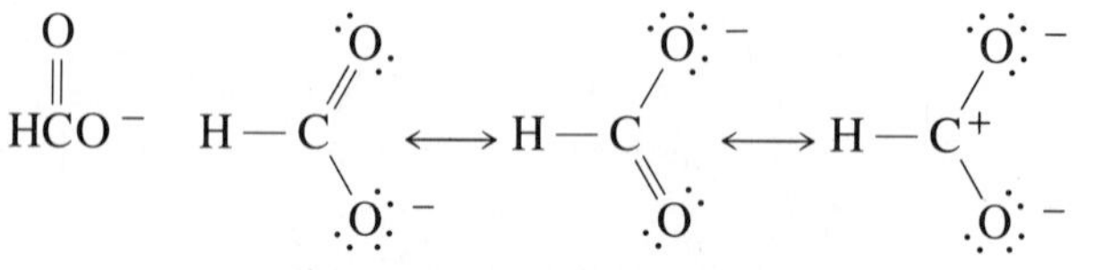
HCO⁻
major contributors
minor contributor

resonance contributors for the formate ion

Two of them are identical, with a double bond to one oxygen atom and a single bond to the other one. Each atom (except hydrogen) has an octet of electrons around it. The third structure is different. The carbon atom has only six electrons around it and bears a positive charge. There is no double bond, and both oxygen atoms are negatively charged. This third structure has a higher energy and is less stable than the other two, because it has fewer bonds and a separation of charge. A **separation of charge,** by which one atom becomes positively charged while another one becomes negatively charged, can be achieved only through an expenditure of energy. The resonance contributors with no separation of charge, with the maximum number of covalent bonds, and with octets around each atom (except hydrogen) contribute the most to the experimentally observed properties of the species being represented. These resonance contributors are thus more important and are known as **major contributors.** Those that have fewer covalent bonds and a separation of charge have less effect on the properties of the species and are often called **minor contributors.**

Minor resonance contributors, in which there is a separation of charge, are frequently used in order to explain chemical reactivity. For example, carbon dioxide reacts readily with hydroxide ion, OH^-. The ease with which this reaction occurs can be rationalized by writing a resonance contributor for carbon dioxide in which the carbon atom has only six electrons around it and bears a positive charge.

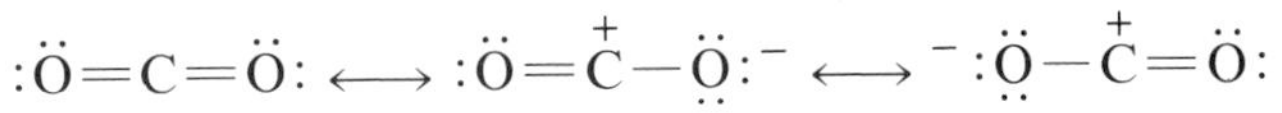

resonance contributors for carbon dioxide, with a separation of charge and six electrons around carbon

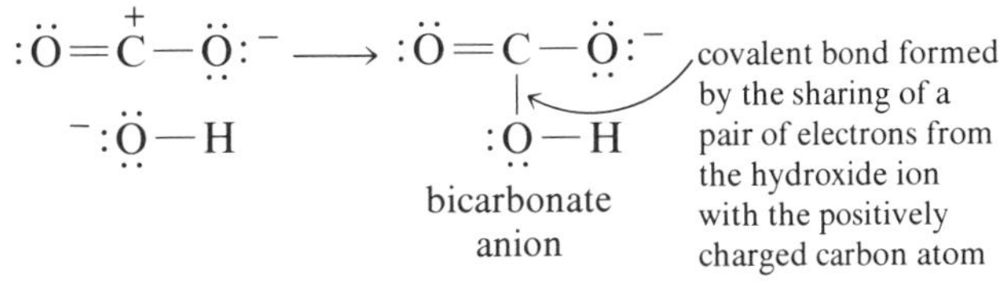
bicarbonate anion
covalent bond formed by the sharing of a pair of electrons from the hydroxide ion with the positively charged carbon atom

These are the rules for writing resonance contributors:

1. Only nonbonding electrons and electrons in multiple bonds change locations from one resonance contributor to another. The electrons in single covalent bonds are not involved.
2. The nuclei of atoms in different resonance contributors are in the same positions.
3. All resonance contributors must have the same numbers of paired and unpaired electrons.
4. Resonance contributors in which atoms of elements from the second period all have eight electrons around them are more important than those in which such atoms have fewer than eight electrons. Similarly, resonance contributors with a greater number of covalent bonds are more important than those with a smaller number. For atoms of elements from periods beyond the second period, such as sulfur and phosphorus, it is possible to write structures with ten or more electrons around a central atom.
5. Resonance contributors in which there is little or no separation of charge are more important than those with a large separation of charge.
6. When structures with a separation of charge are written, the more important resonance contributor has the negative charge on the more electronegative atom. (A review of the concept of electronegativity is found on p. 21.)

The concept of resonance was developed by Linus Pauling of the California Institute of Technology. In 1954, Pauling was awarded the Nobel Prize for his research into the nature of the chemical bond and his application of this knowledge to the determination of the structures of complex substances. Resonance is best understood in the context of different examples. The rationalization of the reactivity of carbon dioxide with the hydroxide ion illustrates the way the idea is most often used. Based on the experimental facts known about a compound, chemists draw resonance contributors for the molecule to explain reactivity. In the case of carbon dioxide, resonance contributors are used to explain the reactivity of the carbon atom toward reagents, such as hydroxide ion, that have pairs of electrons to share with atoms having open shells. Section 2.10 will introduce another important application of the concept of resonance, the idea of stabilization of a species by resonance. You will gradually develop an intuitive understanding of the concept as it is applied to many situations throughout this book.

Study Guide Concept Map 1.3

F. Problem-Solving Skills

When chemists solve a problem, they usually do so by systematically analyzing it and approaching the answer in steps. When they get very good at it, they are not always conscious of all the steps in the process, because they do them rapidly and automatically. Only when they are faced with a difficult and unfamiliar problem do they slow down and become conscious of all the different things they do in order to reach a solution.

For students just starting the study of organic chemistry, all problems are unfamiliar. Therefore, this book has sections on problem-solving skills. These sections ask the kinds of questions and give the kinds of answers chemists do when they have to solve similar problems. Asking yourself these questions and finding the answers to them for different problems will help you develop the skills you need for the successful study of organic chemistry.

Problem: Write resonance contributors for this ion: CH_2CHO^-. Evaluate each resonance contributor in terms of the rules given in Section 1.4E (p. 13) and decide which are major contributors and which minor.

Question: Which atoms are present, and how many electrons are available?

Answer:

2 C (Group IV)	=	2 × 4 electrons	=	8 electrons
3 H (Group I)	=	3 × 1 electron	=	3 electrons
1 O (Group VI)	=	6 electrons	=	6 electrons
1 negative charge	=	1 electron	=	1 electron
				18 electrons

Question: What is a Lewis structure for this ion?

Answer: Start by bonding the carbon and oxygen atoms together and putting in the hydrogen atoms; then use the electrons that are left to complete the octets around the carbon and oxygen atoms.

Stage 1

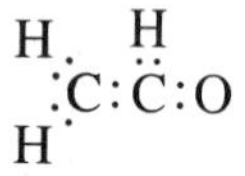
uses 10 electrons

Stage 2

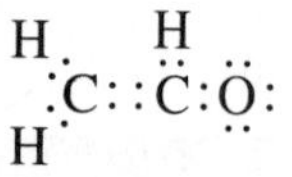
uses 18 electrons

Stage 3 Simplify the structure, and locate any formal charge.

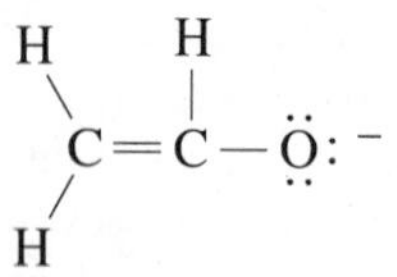

Oxygen has six valence electrons; in this structure, it has six nonbonding electrons and two bonding electrons.

$$\text{Formal charge} = 6 - 6 - \tfrac{1}{2}(2) = -1$$

Note that the task for this part of the problem was made easier because the original problem statement showed the order in which the atoms are bonded to one another. It is important to use all of the information given in a problem.

Question: What resonance contributors are possible?

Answer: Explore ways of moving nonbonding electrons and electrons in the double bond.

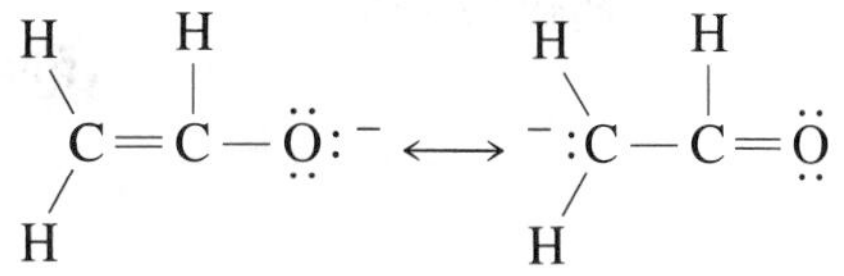

1 2

There seem to be two resonance contributors.

Question: Which one is the more important (major) contributor?

Answer: (Review the rules on p. 13 if necessary.) Both contributors have the same number of covalent bonds and eight electrons around each carbon and oxygen atom. Contributor 1 is the major contributor because the negative charge is on the oxygen atom rather than on the carbon atom, as it is in contributor 2, and oxygen is the more electronegative element.

PROBLEM 1.6

Write resonance contributors for the following ions and molecules, including formal charges where applicable. Evaluate each resonance contributor you write in terms of the rules given in Section 1.4E, and decide which are major contributors and which minor.

(a) HCO_3^- (b) H_2CO (c) NO_2^- (d) NCO^- (e) HNO_3

(f) O_3 (experimental evidence shows the molecule is not cyclic) (g) NO_2^+

PROBLEM 1.7

The reaction between hydroxide ion and carbon dioxide is similar in some ways to the one shown between ammonia and boron trifluoride on p. 8. What similarities do you see?

PROBLEM 1.8

On the basis of the resonance contributors you wrote for formaldehyde, H_2CO (part b of Problem 1.6), and your answer to Problem 1.7, predict whether ammonia will react with formaldehyde.

1.5 ISOMERS

Carbon atoms form strong single bonds and multiple bonds among themselves and with oxygen and nitrogen. The compounds of carbon, therefore, have wide structural variety. The versatility of carbon allows for the creation of the complex structures that are important in living organisms. As the early organic chemists determined the molecular formulas for the compounds they had recovered from natural sources, they discovered that it is possible for two or more compounds with very different properties to have the same molecular formula. In 1830, the Swedish chemist Jakob Berzelius named such compounds "isomeric bodies" from the Greek words *isos,* meaning "equal," and *meros,* meaning "part." We call them isomers.

Isomers are compounds that have identical molecular formulas but differ in the ways in which the atoms are bonded to each other. Isomers may be constitutional isomers or stereoisomers. **Constitutional isomers** differ in the order and the way in which the atoms are bonded together in their molecules. **Stereoisomers** differ only in the arrangement of their atoms in space. They can be distinguished only by exploring their structures in three dimensions. Stereoisomerism will be discussed in Chapter 6.

Two structural formulas may be written for the molecular formula C_4H_{10}.

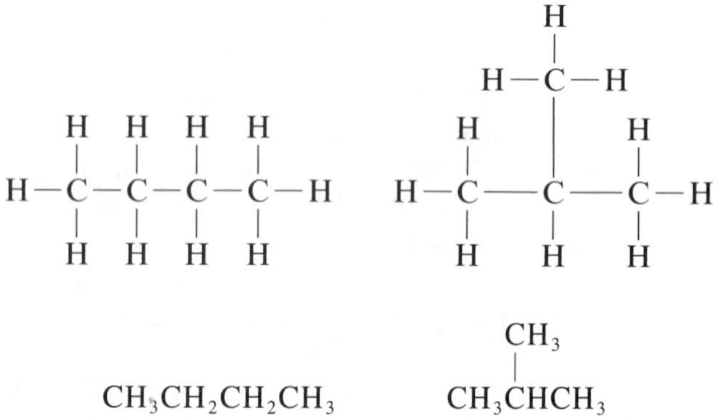

$CH_3CH_2CH_2CH_3$
butane
bp −0.6 °C

2-methylpropane
bp −10.2 °C

Butane and 2-methylpropane are constitutional isomers of each other; they have the same molecular formula but different structural formulas. Their structures differ in the order and the way in which the carbon atoms are bonded to one another. Because their structures differ, the compounds have different physical properties, for example, different boiling points.

Similarly, the molecular formula C_2H_6O corresponds to two structures.

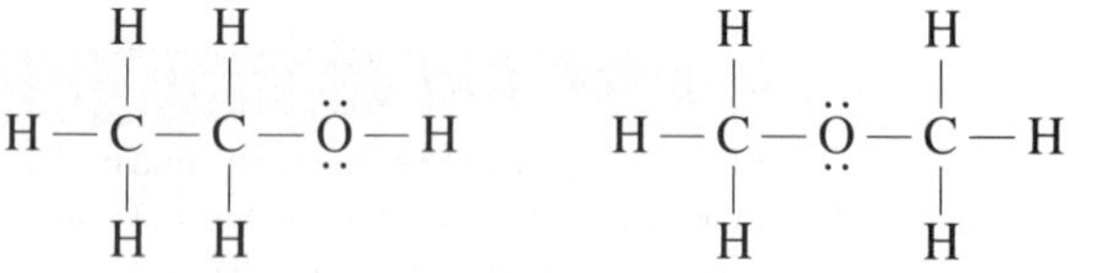

CH_3CH_2OH
ethanol
bp 78 °C

CH_3OCH_3
dimethyl ether
bp −24 °C

Ethanol and dimethyl ether differ from each other more than do butane and 2-methylpropane, which contain only carbon-carbon and carbon-hydrogen bonds. For example, the oxygen atom is bonded to a carbon atom and a hydrogen atom in ethanol, whereas it is bonded only to carbon atoms in dimethyl ether. This difference in structure is reflected in the large difference in boiling point between ethanol and dimethyl ether. Some of the ways in which structure affects physical properties are discussed starting on p. 24.

As the number of atoms in a molecule increases, the number of possible structures increases rapidly. For example, the molecular formula C_3H_6O gives rise to nine constitutional isomers, of which seven are shown below.

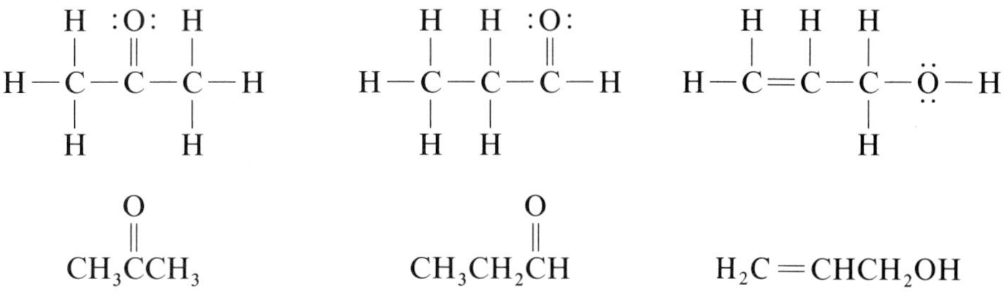

acetone, bp 56.5 °C — propanal, bp 48.8 °C — 2-propen-1-ol (allyl alcohol), bp 97 °C

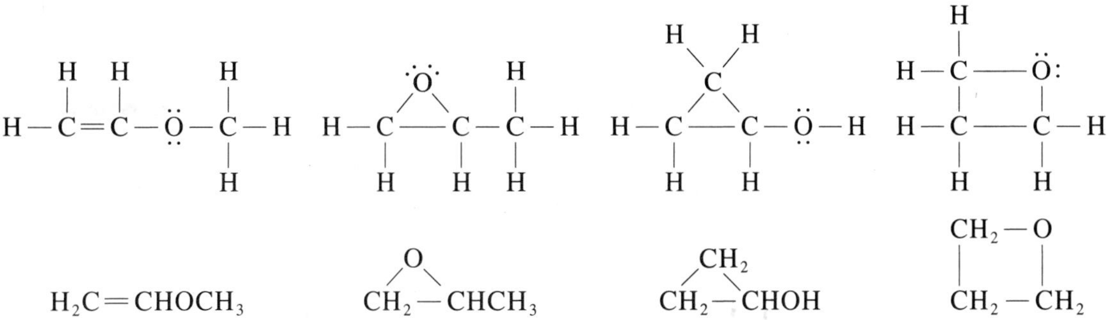

methyl vinyl ether, bp 8 °C — 2-methyloxirane, bp 35 °C — cyclopropanol, bp 100 °C — oxetane, bp 47.9 °C

Two more structural formulas can be written for C_3H_6O. Both of these have a hydroxyl group, —OH, on one of the carbon atoms of a carbon-carbon double bond. These unstable species, called enols, exist in equilibrium with acetone and propanal.

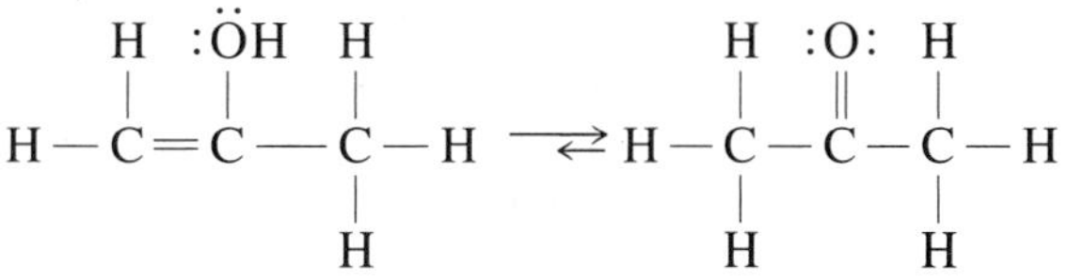

enol of acetone — acetone

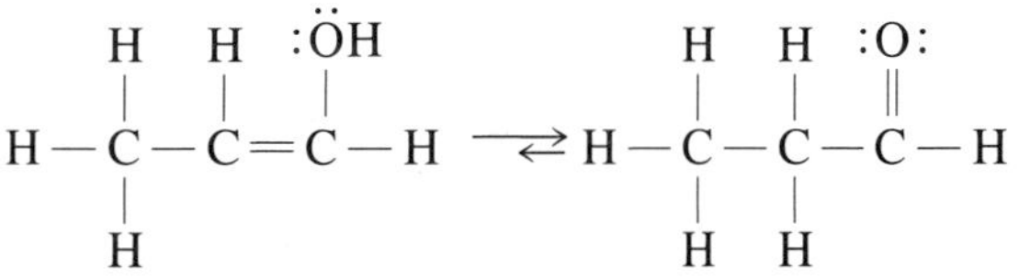

enol of propanal — propanal

We will see later (in Chapter 16, for example) that such species are important in the reactions of compounds such as acetone and propanal even though they cannot be isolated and are usually not counted as constitutional isomers.

The formulas shown above demonstrate the large variety of ways in which carbon atoms bond among themselves and with atoms of another element, oxygen in this case. In determining the number of isomers for a given molecular formula, we must consider structures with single and multiple bonds and cyclic structures.

Study Guide Concept Map 1.4

PROBLEM 1.9

Draw structural formulas for the constitutional isomers of C_3H_7Cl, C_3H_8O, C_5H_{12}, and C_2H_4O.

1.6 SHAPES OF COVALENT MOLECULES

A. Tetrahedral Molecules

The covalent bond, in contrast to the ionic bond (p. 5), has direction in space. If we have more than two atoms covalently bound to each other, we must decide how to arrange them in three dimensions. In 1916, when Lewis postulated that four pairs of electrons form an octet around a central atom, he also suggested that the pairs of electrons are located at the corners of a tetrahedron, as far from each other as possible.

About twenty years after Lewis made his suggestion, experimental values for the distance and angles between atoms in simple covalent molecules were determined using a technique called electron diffraction. In this section, we will consider some of the experimental observations about the shapes of covalent compounds. In Chapter 2, we will study a theory of covalent bonding that explains the molecular geometries that have been observed.

Chemists use two parameters, bond lengths and bond angles, to describe the three-dimensional structure of a molecule. A **bond length** is the average distance between the nuclei of the atoms that are covalently bound together. A **bond angle** is the angle formed by the intersection of two covalent bonds at the atom common to both. Electron diffraction experiments have shown that the four hydrogens bonded to the carbon atom in methane, CH_4, lie at the corners of a regular tetrahedron, with the carbon atom itself at the center of the tetrahedron. A three-dimensional representation of methane, showing carbon-hydrogen bond lengths of 1.09 Å and H—C—H bond angles of 109.5°, the tetrahedral angle, is shown in Figure 1.1.

Figure 1.1 shows a space-filling model and the molecule in perspective. In the perspective drawing, the carbon atom and two of the hydrogen atoms are shown to be lying in the plane of the paper by the use of ordinary solid lines as bonds. The

a space-filling representation

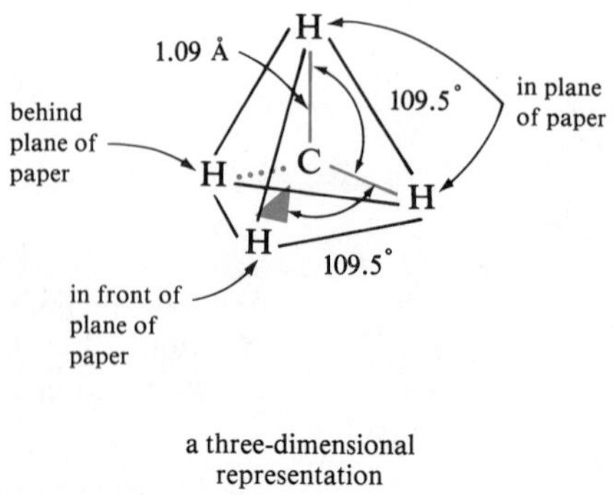
1.09 Å
109.5°
in plane of paper
behind plane of paper
H
C
H
H
H
109.5°
in front of plane of paper

a three-dimensional representation

FIGURE 1.1 Three-dimensional representations of methane.

a space-filling
representation

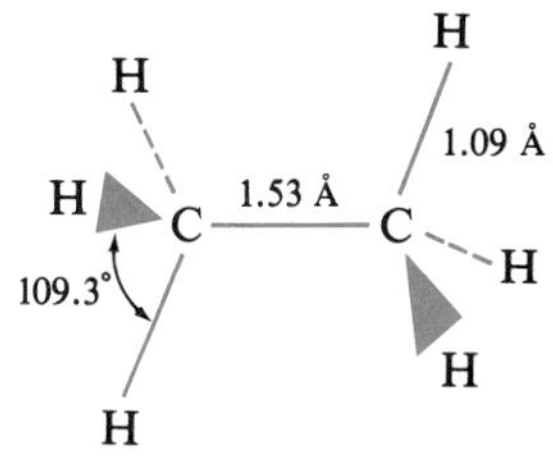

H
H
H
C
1.53 Å
C
1.09 Å
H
H
H
109.3°

a three-dimensional
representation

FIGURE 1.2 Three-dimensional representations of ethane.

solid wedge used to attach one of the hydrogen atoms to the carbon atom indicates that this hydrogen atom is coming out of the plane of the paper toward the viewer. The dashed bond to the fourth hydrogen atom indicates that the atom is behind the plane of the paper. You should look at molecular models of tetrahedral carbon atoms so that you can visualize these directional relationships clearly and draw them.

Ethane, C_2H_6, has carbon-hydrogen bond lengths and H—C—H bond angles similar to those in methane. It also has a carbon-carbon bond that is 1.53 Å long. (See Figure 1.2.)

The ammonium ion, NH_4^+, has the same shape as the methane molecule. The experimentally determined shape of the ammonia molecule is pyramidal, with bond angles that are close to the tetrahedral angle. The experimental methods for determining structure show only the locations of the nuclei of atoms; therefore, in the ammonia molecule, the nonbonding electrons that occupy the fourth corner of the tetrahedron cannot be seen. Similarly, in the water molecule, pairs of nonbonding electrons occupy two corners of the tetrahedron. The experimentally determined shape of the water molecule is triangular, with a bond angle of 104.5°.

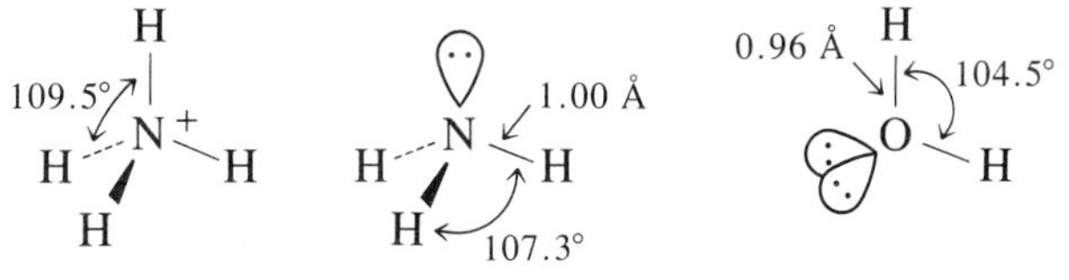

109.5°
1.00 Å
107.3°
0.96 Å
104.5°
ammonium ion
ammonia
water

representations of the ammonium ion, ammonia, and water

In fact, whenever there are four pairs of electrons, either in single bonds or as nonbonding electrons, around a central atom from the second row of the periodic table, the bond angles for the species are close to the tetrahedral angle, 109.5°. This generalization allows us to predict the shapes of many organic molecules and ions.

Bond lengths depend on the identities of the atoms being bonded. For example, a carbon-hydrogen bond is longer than a nitrogen-hydrogen bond, which is longer than an oxygen-hydrogen bond. The length of a carbon-carbon single bond in ethane, however, does not vary much from that of carbon-carbon single bonds in other organic molecules. Similarly, the carbon-hydrogen bond length in methane can be considered representative of carbon-hydrogen bond lengths in other organic compounds having such bonds at tetrahedral carbon atoms.

PROBLEM 1.10

Draw three-dimensional representations of the following compounds.

(a) CCl_4 (b) $CHCl_3$ (c) CH_3Br (d) CH_3CH_2Cl

(e) CH_3OH (f) CH_3NH_2 (g) CH_3OCH_3 (h) CH_3SCH_3

For the cases for which you have the necessary information, show the bond lengths and bond angles you expect to find in these compounds.

B. Planar Molecules

Not all carbon compounds have only tetrahedral carbon atoms in them. Organic molecules containing double or triple bonds have characteristic shapes. Ethylene, $CH_2{=}CH_2$, is a flat molecule with bond angles of approximately 120° around each carbon atom. The length of the carbon-carbon double bond is 1.34 Å, considerably shorter than the carbon-carbon single bond in ethane. The carbon-hydrogen bonds are also a little shorter than the ones in methane and ethane. Two representations of ethylene, one in which the molecule is entirely in the plane of the paper and the other in which the molecule is being viewed from one edge, are shown below.

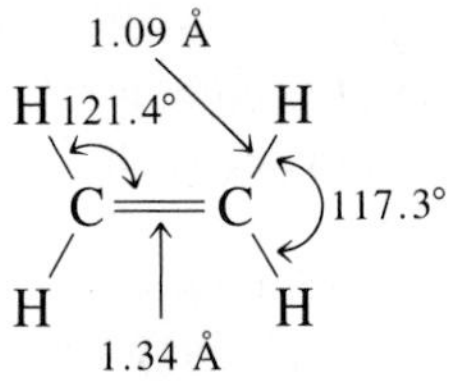
1.09 Å
121.4°
117.3°
1.34 Å

molecule in plane of paper

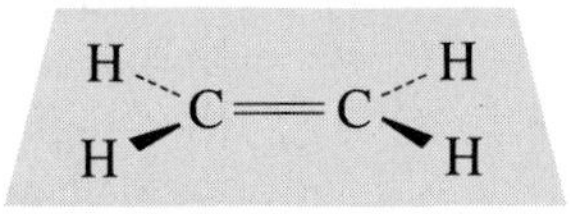

molecule viewed from one edge

ethylene

C. Linear Molecules

The triple bond in acetylene, HC≡CH, is 1.20 Å, shorter than the carbon-carbon bond in ethane or ethylene. The carbon-hydrogen bond in acetylene is only 1.06 Å, which again is shorter than that type of bond in ethane and ethylene. Acetylene is linear, with bond angles of 180°.

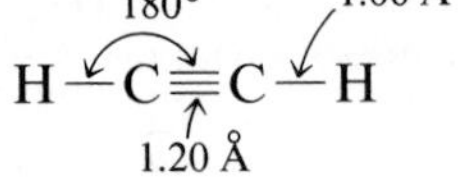
180°
1.06 Å
1.20 Å

acetylene

The bond lengths in ethylene and acetylene are very close to the bond lengths for double and triple carbon-carbon bonds in other organic compounds. The bond angles are even more general. Whenever carbon is bonded by a double bond to one other atom, bond angles of approximately 120° are observed. If carbon is bonded by a triple bond to another atom, the portion of the molecule containing the triple bond is linear.

PROBLEM 1.11

Draw three-dimensional representations of the following compounds. Give approximate bond lengths and bond angles wherever you can.

(a) $CH_2{=}CHCl$ (b) $CH_3C(CH_3){=}CH_2$

(c) $CHCl{=}CHCl$ (Two different arrangements of the atoms are possible.)

(d) $CH_3C{\equiv}CH$ (e) $H_2C{=}O$

1.7 THE POLARITY OF COVALENT MOLECULES

A. Polar Covalent Bonds

Electrons in covalent bonds are shared equally when the two bonded atoms are the same but unequally when different elements participate in bonding. For example, in the hydrogen molecule, where electrons are shared equally by two hydrogen atoms, there is a **nonpolar covalent bond.** In hydrogen chloride, the electrons of the covalent bond are drawn closer to the chlorine atom, which is the more electronegative of the two atoms. A covalent bond in which there is unequal sharing of electrons is a **polar covalent bond.** The presence of such a bond is shown by writing partial positive and partial negative charges, $\delta+$ and $\delta-$, over the bonded atoms, pointing out a permanent and directional distortion of the electrons in the bond. The polarized covalent bond may also be shown as a **bond dipole.** The dipole has a negative pole and a positive pole and is represented by the symbol $\leftarrow\!\!\!+\!\!\rightarrow$, with the point of the arrow drawn toward the more electronegative atom.

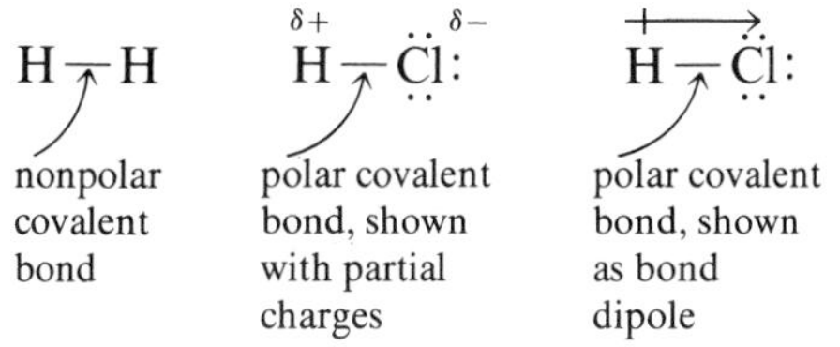
H—H
nonpolar covalent bond
δ+ δ−
H—Cl:
polar covalent bond, shown with partial charges
H—Cl:
polar covalent bond, shown as bond dipole

The **polarity** in a bond arises from the different electronegativities of the two atoms participating in the bond. The **electronegativity** of an element was defined by Pauling, who developed the concept, as "the power of an atom in a molecule to attract electrons to itself." The most electronegative elements are in the upper right-hand corner of the periodic table; electronegativity increases as one moves up in a group or to the right in any period. The electronegativities of some elements of interest in organic chemistry are shown in Table 1.2 (p. 22), arranged according to the groups in the periodic table.

The greater the difference in electronegativity between the bonded atoms, the greater is the polarity of the bond. Compounds formed between the metals at the left-hand side of the table and the nonmetals at the extreme right are ionic. Covalent bonds of all gradations of polarity form between nonmetals and between many metals and nonmetals. For example, the electronegativities of carbon and hydrogen are close enough that carbon-hydrogen bonds do not have much polarity. Bonds of high polarity are those between hydrogen and oxygen, fluorine, chlorine, or nitrogen. Carbon-halogen, carbon-oxygen, and carbon-nitrogen bonds are also polar. Bond polarities contribute significantly to the physical and chemical properties of molecules, and we will refer to them often as we consider chemical reactivity.

TABLE 1.2 Electronegativity Values for Some Elements

I	II		III	IV	V	VI	VII
H 2.1							
Li 1.0			B 2.0	C 2.5	N 3.0	O 3.5	F 4.0
Na 0.9	Mg 1.2		Al 1.5	Si 1.8	P 2.1	S 2.5	Cl 3.0
K 0.8							Br 2.8
							I 2.4

PROBLEM 1.12

Write $\delta+$ and $\delta-$ to predict the direction of polarization of the covalent bonds indicated by the arrows in the following compounds.

(a) 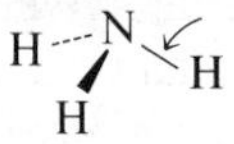
(b) 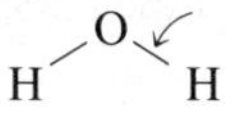
(c) 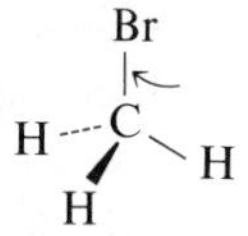
(d) 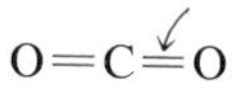

(e) 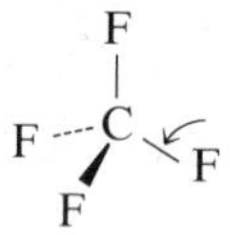
(f) 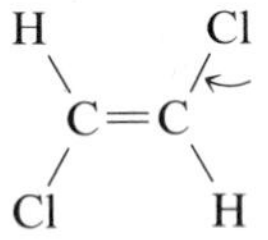
(g) 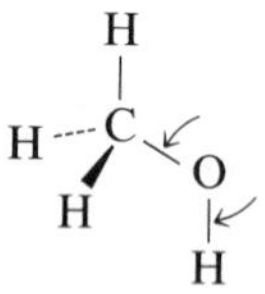

B. Dipole Moments of Covalent Molecules

For diatomic molecules, those containing two atoms, the bond dipole is also the **dipole moment, μ.** The dipole moment results from the separation of the centers of positive and negative charge and is given a unit, D, the debye, that is derived from the magnitude of the overall charge and the distance separating the centers of charge. Diatomic molecules in which both atoms are the same have no dipole moment. For the hydrogen halides, the more electronegative the halogen, the larger is the dipole moment for the molecule.

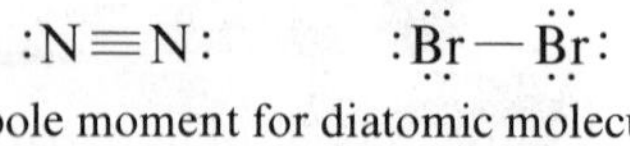

no dipole moment for diatomic molecules in which both atoms are the same

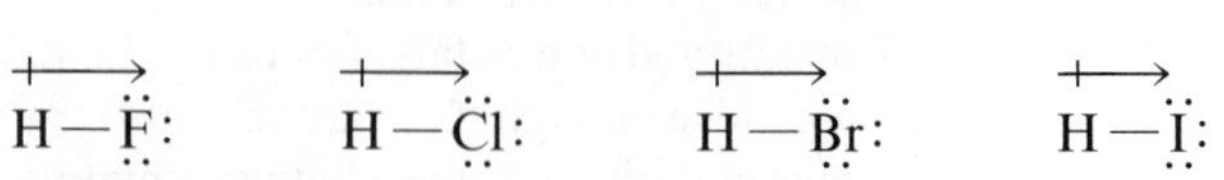

μ, 1.98 D μ, 1.03 D μ, 0.78 D μ, 0.38 D

decreasing dipole moment for hydrogen halides with decreasing electronegativity of the halogen atom

The overall dipole moment of a molecule containing more than two atoms is the vector sum of the individual bond dipole moments. A molecule may contain polar bonds, but have no overall dipole moment if the shape of the molecule is such that the individual bond moments cancel out. In such a case, the average position of the partial positive charges coincides with the average position of the partial negative charges. This is the case for carbon dioxide, CO_2, and carbon tetrachloride, CCl_4.

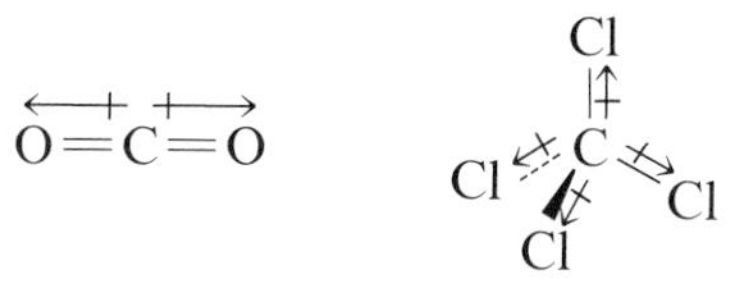

carbon dioxide μ, 0 D — carbon tetrachloride μ, 0 D

cancellation of individual bond moments in symmetrical polyatomic molecules

In carbon dioxide, bond moments of equal magnitude point in exactly opposite directions. The net result is that the molecule as a whole has no dipole moment, even though the individual bond moments are large. The absence of a measurable dipole moment is one of the reasons chemists have assigned a linear structure to carbon dioxide. The cancellation of the bond moments in carbon tetrachloride is harder to visualize. Because of the symmetrical nature of the tetrahedron, the vector sum of the bond moments of any three carbon-chlorine bonds is exactly equal to and opposite in direction to the bond moment for the fourth carbon-chlorine bond. The overall result is that the molecule as a whole has no permanent dipole, though the individual carbon-chlorine bonds are polarized.

In most molecules, the vector sum of the individual bond moments is not zero, and there is a dipole moment. Water, for example, has a dipole moment with the negative pole at the oxygen atom and the positive pole between the two hydrogen atoms. Similarly, the dipole moment in chloromethane, CH_3Cl, has the negative pole at the chlorine atom. Thus, the dipole moment of a molecule is related to the polarities of its individual bonds and to its molecular geometry.

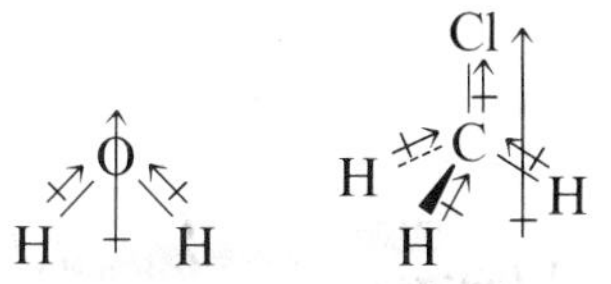

water μ, 1.84 D — chloromethane μ, 1.86 D

molecules with dipole moments resulting from vector sums of individual bond moments

Study Guide Concept Map 1.5

PROBLEM 1.13

Draw three-dimensional representations for the following compounds. For each compound, predict whether it will have a dipole moment and, if so, the direction of the moment.

(a) $CHCl_3$ (b) CH_3OH (c) ICl (d) NH_3

(e) NH_4^+ (f) CH_2Br_2 (g) CH_3OCH_3 (h) $CH_2{=}CCl_2$

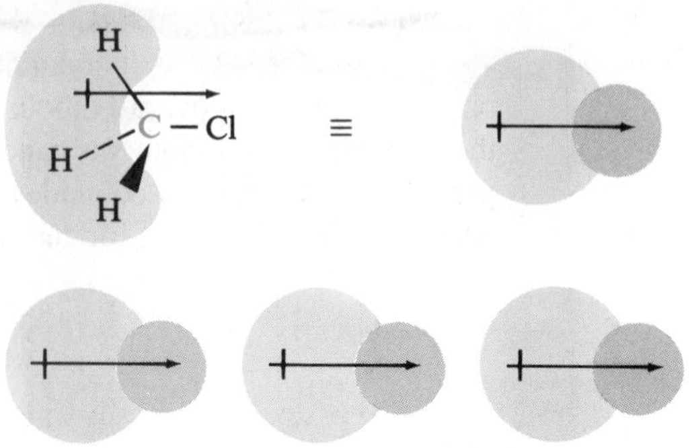
H
C
Cl
H
H
≡

FIGURE 1.3 Schematic representation of dipole-dipole interactions in chloromethane.

1.8 NONBONDING INTERACTIONS BETWEEN MOLECULES

Covalent compounds may be gases, liquids, or solids. Compounds that have low molecular weights and no dipole moments, such as methane and carbon dioxide, are gases. The forces that act between such molecules are very weak, so that condensation to the liquid or solid phase takes place only at low temperatures or high pressures. The forces that act between molecules are called **intermolecular forces,** or **intermolecular nonbonding interactions.** These interactions increase significantly as the molecular weight, and hence the size of the molecules, increases. They also increase with increasing polarity of molecules. Three types of intermolecular forces are important: (1) dipole-dipole interactions, (2) hydrogen bonding, and (3) van der Waals forces.

In the following sections, a number of organic compounds with a variety of structures are introduced. These compounds appear many times in later chapters, and you will gradually become familiar with their structures, their names, and their reactions. For the moment, you should concentrate only on bond polarities and the interactions that are possible between these kinds of molecules. As always, use the problems as a guide to what you should learn from this chapter.

A. Dipole-Dipole Interactions

Molecules with dipole moments tend to orient themselves in the liquid and solid phases so that the negative end of one molecule is facing the positive end of another one. The interactions of the permanent dipoles in different molecules are called **dipole-dipole interactions.** Chloromethane has a dipole moment of 1.86 D, which lies along the carbon-chlorine bond. Chloromethane molecules orient themselves so that the positive end of one dipole is pointed toward the negative end of another dipole, as depicted in the schematic representation in Figure 1.3.

An ordinary covalent bond has bond energy in the range from 30 to 100 kcal/mol (p. 61). Dipole-dipole interactions are much weaker, approximately 1 to 3 kcal/mol.

B. Hydrogen Bonding

When a hydrogen atom is covalently bonded to a strongly electronegative atom, such as oxygen, fluorine, or nitrogen, the bond is very polar. A hydrogen atom in this situation has a large affinity for nonbonding electrons on other oxygen, fluorine, or nitrogen atoms. The strong interaction that results is called a **hydrogen bond.** A hydrogen bond is a particularly strong dipole-dipole interaction.

Significant formation of hydrogen bonds is seen only with hydrogen atoms covalently bonded to oxygen, fluorine, and nitrogen. Even though chlorine is as electronegative as nitrogen, neutral chlorine atoms are not significantly involved in hydrogen bonding. Chlorine, in the third period of the periodic table, has larger atoms than those of fluorine, oxygen, or nitrogen. The partial negative charge on a chlorine atom in a covalent bond is, therefore, more diffuse and does not attract the hydrogen atom as strongly. Furthermore, hydrogen atoms bonded to carbon are not usually involved in the formation of hydrogen bonds.

The strength of hydrogen bonds is reflected in the physical properties of compounds in which such bonding occurs. Water boils at a much higher temperature than hydrogen sulfide, even though it has a lower molecular weight. This difference in boiling points is attributed to the strong hydrogen bonds formed between oxygen and hydrogen atoms.

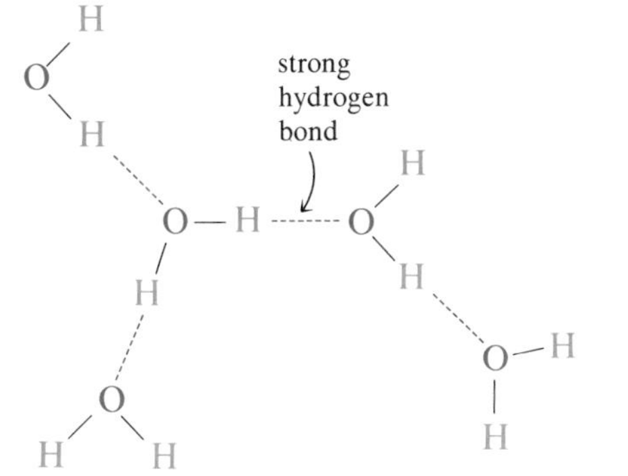
strong
hydrogen
bond

water
molecular weight 18
bp 100 °C

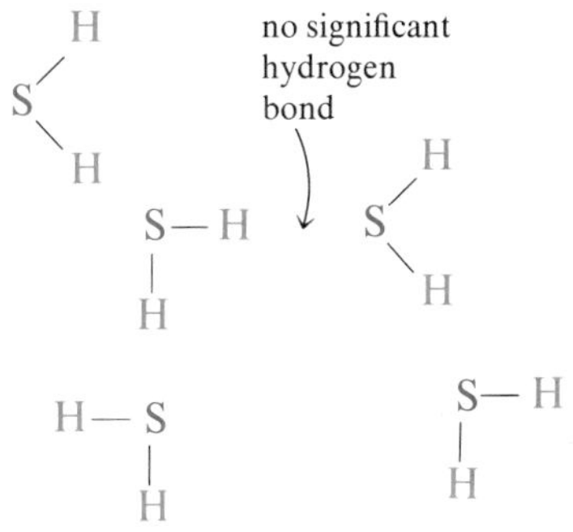
no significant
hydrogen
bond

hydrogen sulfide
molecular weight 34
bp −62 °C

The attractive forces of hydrogen bonding are usually indicated by a dashed line rather than the solid line used for a covalent bond. The strength of a hydrogen bond involving an oxygen, fluorine, or nitrogen atom ranges from 3 to 10 kcal/mol, making hydrogen bonds the strongest known type of intermolecular interaction. Strong as a hydrogen bond is, though, it is still much weaker than a chemical bond (p. 61), ionic or covalent.

The effect of hydrogen bonding is demonstrated dramatically by the difference in the boiling points of ethanol, CH_3CH_2OH, and its isomer dimethyl ether, CH_3OCH_3 (p. 16). Ethanol is an alcohol. An alcohol resembles a water molecule in which one of the hydrogen atoms has been replaced by a nonpolar group containing carbon and hydrogen atoms. The polar part of the molecule is the hydroxyl group (p. 46).

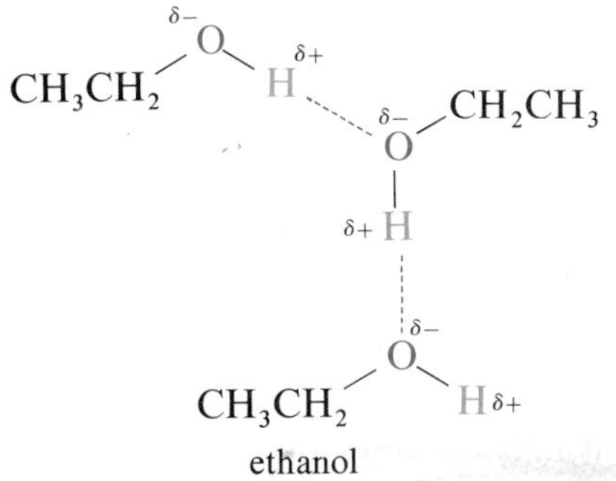

ethanol
molecular weight 46
bp 78 °C

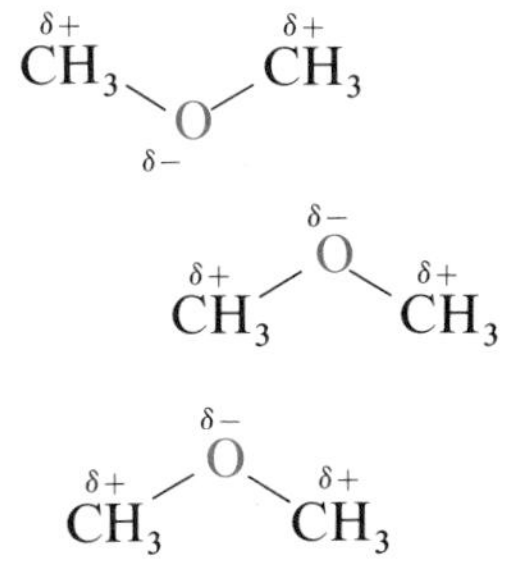

dimethyl ether
molecular weight 46
bp −24 °C

The hydrogen bonding that takes place between molecules of ethanol contributes to a much higher boiling point for this compound than that of dimethyl ether, in which the molecules are held together only by dipole-dipole interactions. In an ether, both hydrogen atoms that are present in water have been replaced by nonpolar groups containing carbon and hydrogen atoms. As a result, an ether does not have a polar hydroxyl group.

PROBLEM 1.14

For each set, predict which compound will have the highest boiling point. Indicate, with drawings, the reasoning behind your conclusions.

(a) $CH_3CH_2CH_2CH_2CH_3$, $CH_3CH_2CH_2CH_2OH$, $CH_3CH_2OCH_2CH_3$

(b) CH_3CH_3, CH_3F, CH_3OH

(c) $CH_3CH_2CH_3$, CH_3SH, CH_3OH

Hydrogen bonding is important in determining the solubility of organic compounds in water. Molecules that can participate in the formation of hydrogen bonds with water will dissolve in water if the nonpolar part of the molecule (the part made up of carbon and hydrogen alone) is not too large. For example, the solubilities in water of three alcohols—ethanol, 1-butanol, and 1-hexanol—vary with the size of the nonpolar portion of their molecules.

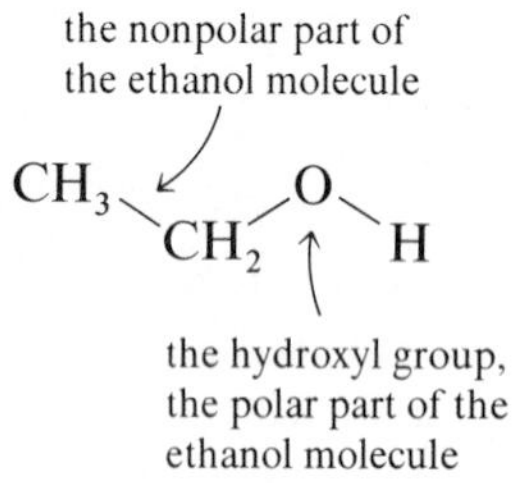
the nonpolar part of the ethanol molecule
CH3
CH2
O
H
the hydroxyl group, the polar part of the ethanol molecule

ethanol mixes with water in all proportions

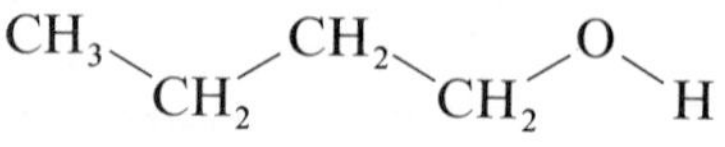
CH3
CH2
CH2
CH2
O
H

1-butanol

7.9 g dissolves in 100 mL of water

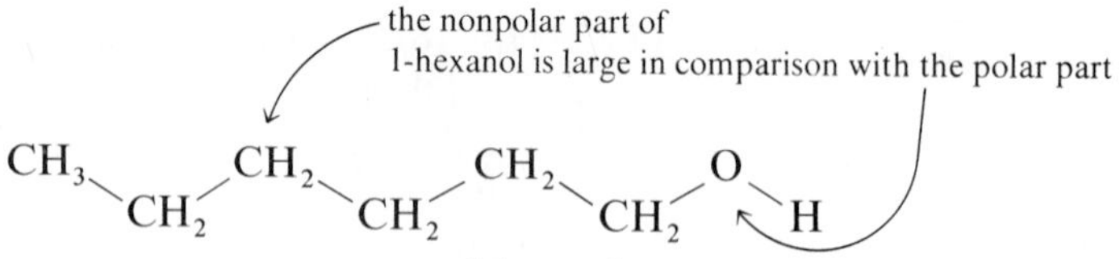
the nonpolar part of 1-hexanol is large in comparison with the polar part
CH3
CH2
CH2
CH2
CH2
CH2
O
H

1-hexanol

0.59 g dissolves in 100 mL of water

Ethanol, in which the polar part of the molecule, the hydroxyl group, is large in proportion to the whole molecule, is completely soluble in water. The hydroxyl group is a much less significant portion of the 1-hexanol molecule, so that compound has low water-solubility. The structure of 1-butanol is intermediate between those of ethanol and 1-hexanol, and so is its solubility in water.

The interaction between a dissolved species and the molecules of a solvent is known as **solvation.** Hydrogen bonding is a particularly effective interaction between solute and solvent. The solvation of ethanol by water is shown in the structure below.

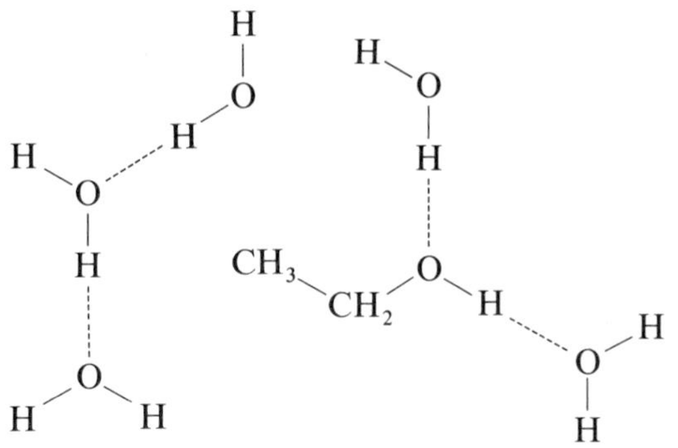
H
H
O
H
O
H
H
O
H
H
CH3
O
CH2
H
H
O
O
H
H
H
H

solvation of ethanol by water

Solvation is especially important in stabilizing ionic species. For example, when magnesium bromide, $MgBr_2$, is dissolved in water, the magnesium ions and the bromide ions are solvated by water molecules. The **dielectric constant, ε,** of a solvent measures the ability of the solvent to separate ionic charges and, therefore, to dissolve ionic compounds. Water has a high dielectric constant, 78.5 at 25 °C. (In contrast, the dielectric constant of carbon tetrachloride is only 2.2.)

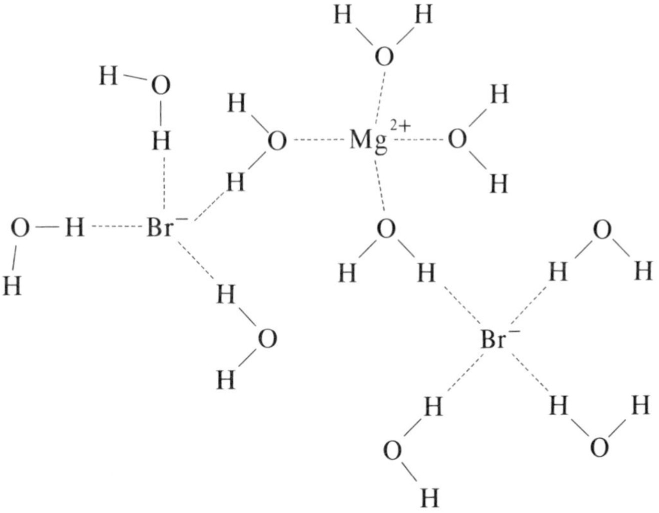
Mg2+
Br−
Br−

solvation of magnesium bromide by water

The interaction between a polar molecule and an ion is a strong one known as **ion-dipole interaction.** The hydroxyl group in water and alcohols, having the large polarity of the oxygen-hydrogen bond and a high concentration of positive charge on the small hydrogen atom, is particularly effective at stabilizing anions.

Compounds that contain oxygen or nitrogen atoms, but lack hydrogen atoms bonded to these electronegative elements, may participate in hydrogen bonding as **hydrogen-bond acceptors** even though they cannot function as **hydrogen-bond donors.** The solubilities in water of diethyl ether, $CH_3CH_2OCH_2CH_3$, and pentane, $CH_3CH_2CH_2CH_2CH_3$, demonstrate this fact.

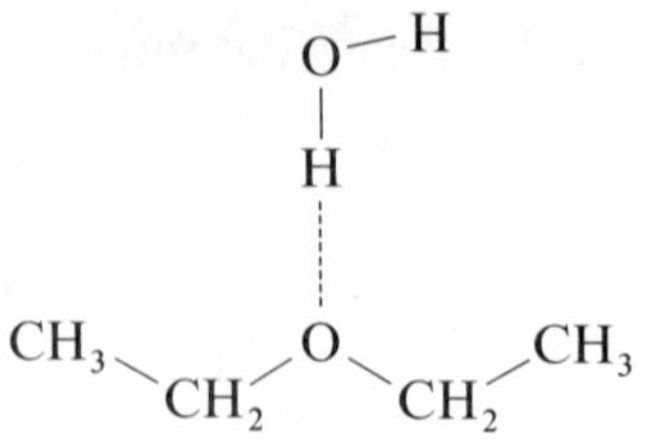

diethyl ether

7.5 g dissolves in 100 mL of water

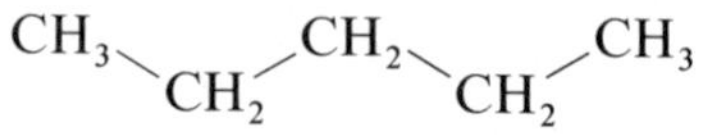

pentane

0.036 g dissolves in 100 mL of water

Diethyl ether is almost as soluble in water as 1-butanol (p. 26) and much more so than pentane, which is a completely nonpolar compound containing only carbon and hydrogen atoms.

PROBLEM 1.15

For each pair of compounds, predict which one will be more soluble in water. Indicate, with drawings, the reasons for your conclusions.

(a) CH_3CH_2Cl or CH_3CH_2OH (b) $CH_3CH_2CH_2OH$ or $CH_3CH_2CH_2SH$

(c) $CH_3CH_2CH_2CH_2CH_2OH$ or $HOCH_2CH_2CH_2CH_2CH_2OH$

(d) $CH_3CH_2\overset{\overset{O}{\|}}{C}OH$ or $CH_3\overset{\overset{O}{\|}}{C}OCH_3$ (e) $CH_3CH_2CH_2Cl$ or $CH_3CH_2CH_2NH_2$

The hydrogen bond plays an important role in chemistry and biochemistry. It is especially important in interactions between different parts of large molecules where many hydrogen bonds are possible. The shape of an enzyme that catalyzes chemical processes in the human body, for example, is determined to a great extent by hydrogen bonding between distant parts of the large molecule. The two strands of the double helix of deoxyribonucleic acid are held together by a precise pattern of hydrogen bonding, believed to be responsible for the transmission of the genetic code. You will learn much more about this when we look at the structures and the chemistry of proteins, carbohydrates, and other giant molecules found in living systems.

C. Van der Waals Forces

Intermolecular forces act to attract even nonpolar molecules to each other, as demonstrated by the physical properties of three nonpolar compounds: methane, CH_4; hexane, C_6H_{14}; and icosane, $C_{20}H_{42}$.

CH_4

methane
molecular weight 16
bp −162 °C
gas at room temperature

$CH_3CH_2CH_2CH_2CH_2CH_3$

hexane
molecular weight 86
bp 69 °C
liquid at room temperature

$CH_3CH_2CH_2CH_2CH_2CH_2CH_2CH_2CH_2CH_2CH_2CH_2CH_2CH_2CH_2CH_2CH_2CH_2CH_2CH_3$

icosane
molecular weight 282
mp 36 °C
solid at room temperature

The forces of attraction between the small molecules of methane are so weak that methane exists as a gas at room temperature. Molecules of hexane are larger than those of methane, and the attractive forces between hexane molecules are increased enough that hexane is a liquid. The still larger molecules of icosane attract each other so strongly that the compound is a solid at room temperature.

The weak forces of attraction that exist between nonpolar molecules are called **van der Waals forces.** These forces are the result of the constant motion of electrons within bonds and molecules, giving rise to effects known as London dispersion forces. The motion of electrons creates small distortions in the distribution of charge in nonpolar molecules. A small and momentary dipole results. This small dipole in one molecule can then create a dipole with the opposite orientation, an **induced dipole,** in a second molecule. Although the induced dipoles are constantly changing, the net result is a slight attraction between molecules. As the number of carbon and hydrogen atoms increases, the additive effect of these weak intermolecular forces becomes more significant, as evidenced by the increase in boiling and melting points from methane to hexane to icosane.

Van der Waals forces can act only through the parts of different molecules that are within a certain distance of each other. The three-dimensional shapes of molecules, therefore, determine to some extent the intermolecular interactions between molecules (see Figure 1.4). For example, the isomers butane and 2-methylpropane, both with the molecular formula C_4H_{10}, have different boiling points (p. 16).

Study Guide Concept Map 1.6

2-Methylpropane is a more compact molecule than butane. If you build models of the two compounds, you can see that 2-methylpropane is almost spherical, whereas butane is elongated. The molecules of butane have a greater surface area for interaction with each other than do the molecules of 2-methylpropane. The stronger interactions that are possible for butane are reflected in its boiling point, which is higher than the boiling point of 2-methylpropane.

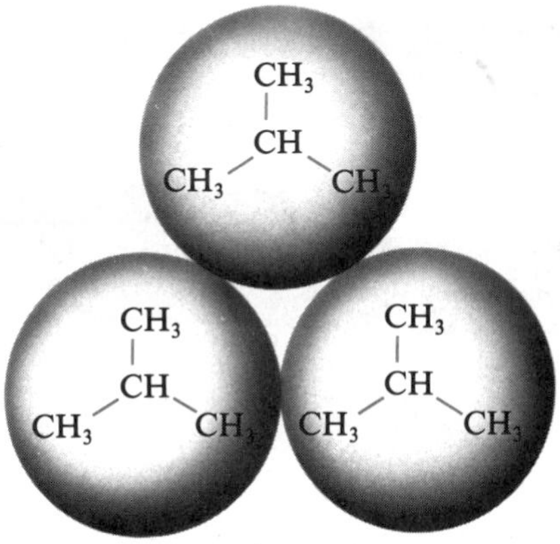

CH3
CH
CH3
CH3
interactions among spherical molecules
2-methylpropane
bp − 10.2 °C

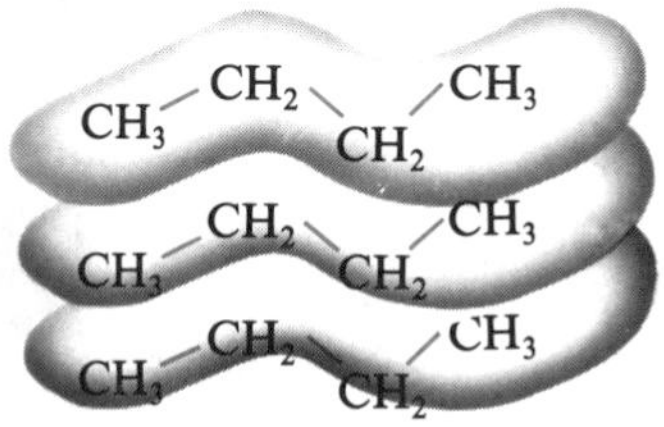

CH3
CH2
CH2
CH3
interactions among elongated molecules
butane
bp − 0.6 °C

FIGURE 1.4 A comparison of intermolecular interactions for 2-methylpropane and butane.

S U M M A R Y

Compounds are classified as ionic compounds or covalent compounds. An ionic bond consists of electrostatic forces holding together positively and negatively charged ions. A covalent bond arises from the sharing of a pair of electrons by two atoms.

Covalent compounds may be represented by Lewis structures, in which lines are used to represent covalent bonds and dots to show nonbonding electrons. Structures are drawn so as to have eight electrons, an octet, around each atom except hydrogen, which can have only two electrons. A few other atoms, such as boron from Group III of the periodic table, which often has an open shell, are also exceptions. Double or triple bonds may be drawn between atoms in order to create structures in which there are octets around atoms such as carbon, oxygen, and nitrogen. In Lewis structures, any of the atoms may bear a formal charge, that is, may have in its valence shell more electrons or fewer electrons than are necessary to balance its nuclear charge.

Not all compounds can be represented satisfactorily by a single Lewis structure. The physical and chemical properties of some species are best accounted for by a hybrid of several Lewis structures, called resonance contributors, that differ from each other in the location of electrons.

Covalent bonds have direction in space. Shapes of covalent molecules are defined by the bond lengths and bond angles. Molecules in which there are four pairs of bonding and/or nonbonding electrons around a central atom have bond angles that approach the tetrahedral angle, 109.5°. Molecules in which the central atom is bonded to three other atoms and there are no nonbonding electrons are planar and have bond angles of 120°. Molecules in which the central atom is bonded to two other atoms are linear with bond angles of 180°.

Covalent bonds may be nonpolar or polar. Nonpolar covalent bonds occur when the atoms that are bonded have similar electronegativities. Polar covalent bonds result from bonding between atoms of differing electronegativities.

A molecule has a dipole moment, μ, when there is a separation of the centers of positive and negative charge in the molecule. Molecules containing only nonpolar covalent bonds are nonpolar and have very small dipole moments or none at all. Molecules containing polar covalent bonds have significant dipole moments unless the shape of the molecule causes the individual bond moments to cancel one another out.

Physical properties such as boiling points, melting points, and solubilities are determined to a large extent by intermolecular nonbonding interactions. Interactions between polar molecules are dipole-dipole interactions, of which hydrogen bonding is the strongest form. Compounds containing hydrogen bonded to oxygen or nitrogen participate in hydrogen bonding both as hydrogen-bond donors and as hydrogen-bond acceptors. Compounds containing oxygen or nitrogen to which no hydrogen is bonded are hydrogen-bond acceptors. The degree of hydrogen bonding possible for a compound strongly influences its boiling point and solubility in water. Interactions between nonpolar molecules are called van der Waals forces. These interactions depend on fluctuating induced dipoles and therefore on the sizes and shapes of covalent molecules.

ADDITIONAL PROBLEMS

1.16 For each of the following compounds, tell whether its bonds are ionic, covalent, or of both kinds. Of the covalent bonds, show which ones have polarity.

(a) MgF_2 (b) SiF_4 (c) NaH (d) ClF (e) SCl_2 (f) OF_2

(g) SiH_4 (h) PH_3 (i) $NaOCH_3$ (j) CH_3Na (k) Na_2CO_3 (l) BrCN

1.17 Draw Lewis structures for the following species, using lines for covalent bonds and showing any nonbonding electrons that are present. Show formal charges where relevant.

(a) NF_3 (b) $AlCl_3$ (c) CH_3SCH_3 (d) CH_3NH_2 (e) $CH_3CHClCH_3$

(f) OH^- (g) $CH_3CH_2CH_2OH$ (h) H_2O_2 (i) CH_3NHOH (j) SO_2

(k) CH_3SH (l) SiH_4 (m) H_2SO_4 (n) HNO_3

1.18 Draw each of the following compounds and ions in three dimensions, showing in each case the geometry you would expect to find around the atom that is shaded with color.

(a) SCl_2 (b) OF_2 (c) $CH_3-\overset{H}{\overset{|}{C}}=CH_2$ (d) $CH_3-C\equiv N$

(e) BF_4^- (f) $CFCl_3$ (g) $CH_3-\underset{CH_3}{\underset{|}{\overset{CH_3}{\overset{|}{N}}}}-CH_3^+$ (h) PH_3

(i) $CH_3OH_2^+$ (j) $CH_3-\underset{CH_3}{\underset{|}{N}}-CH_3$

1.19 Which of the following molecules will have a dipole moment? Show your reasoning by drawing a three-dimensional representation of each molecule, showing the direction of the dipole.

(a) $CH_2=CHCl$ (b) CH_3SCH_3 (c) $CH_3C\equiv CCH_3$ (d) $FCBr_3$

(e) CH_3CH_2OH (f) $HC\equiv CCl$ (g) $Cl_2C=CCl_2$

1.20 For which of the following compounds will hydrogen bonding among its molecules be important?

(a) CH_2F_2 (b) $CH_3CH_2CH_2OH$ (c) $CH_3CH_2OCH_3$

(d) $HOCH_2CH_2CH_2CH_2OH$ (e) $CH_3CH_2CH_2NH_2$ (f) $CH_3CH_2\overset{O}{\overset{\|}{C}}NH_2$

(g) $CH_3CH_2CH_2SH$

1.21 Which of the following compounds will participate in hydrogen bonding with water? For each compound, indicate whether it will be a hydrogen-bond donor, hydrogen-bond acceptor, or both. Illustrate your reasoning with drawings.

(a) $CH_3CH_2OCH_3$ (b) $CH_3CH_2\underset{CH_2CH_3}{\underset{|}{N}}CH_2CH_3$ (c) $CH_3\overset{O}{\overset{\|}{C}}NHCH_3$

(d) $CH_3C\equiv N$ (e) $CH_3CH_2\overset{O}{\overset{\|}{C}}OH$ (f) $CH_3CH_2CH_2Cl$

(g) CH_3NHOH (h) 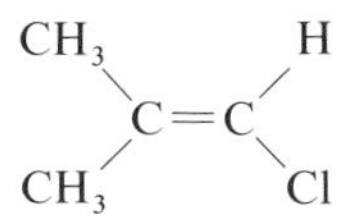 (i) $CH_3\overset{O}{\overset{\|}{S}}CH_3$

1.22 Which compound in each pair do you expect to be more soluble in water?

(a) CH_3Cl or $NaCl$ (b) $H\overset{O}{\overset{\|}{C}}OH$ or $CH_3CH_2CH_2CH_2\overset{O}{\overset{\|}{C}}OH$

(c) $H\overset{\overset{\displaystyle O}{\|}}{C}OH$ or $H\overset{\overset{\displaystyle O}{\|}}{C}OCH_2CH_2CH_2CH_2CH_3$

(d) $CH_3CH_2SCH_2CH_3$ or $CH_3CH_2OCH_2CH_3$

(e) $HO\overset{\overset{\displaystyle O}{\|}}{C}CH_2CH_2CH_3$ or $HO\overset{\overset{\displaystyle O}{\|}}{C}CH_2CH_2\overset{\overset{\displaystyle O}{\|}}{C}OH$

1.23 Draw a segment of the crystal lattice of sodium chloride (p. 4) and explain with diagrams why the compound dissolves easily in water, even though a great deal of energy is required to melt or vaporize it.

1.24 The boiling point of water (100 °C) is higher than the boiling point of hydrogen fluoride (19 °C) or ethanol (78 °C). Use drawings to explain this experimental observation.

1.25 Acetonitrile, $CH_3C{\equiv}N$, has a large dipole moment (greater than 3 D).

(a) Draw a Lewis structure for acetonitrile.
(b) Predict the direction of the dipole moment for acetonitrile.
(c) The large dipole moment for acetonitrile has been interpreted to mean that the molecule has a major resonance contributor that reflects a separation of charge. Draw such a resonance contributor for acetonitrile.

1.26 Draw resonance contributors for each of the following species. Be sure to show any formal charges. Decide which resonance contributors are major and which minor.

(a) Nitrous oxide, N_2O (a linear molecule in which the two nitrogen atoms are bonded to each other)
(b) Sulfate anion, SO_4^{2-} (an ion in which all of the oxygens are equivalent)
(c) Acetic acid, $CH_3\overset{\overset{\displaystyle O}{\|}}{C}OH$

1.27

(a) Carbon monoxide, CO, has a much smaller dipole moment than expected. This observation has puzzled chemists and led to much argument about the nature of the bonding in the molecule. The small dipole moment for CO can be rationalized by concluding that there are three important resonance contributors for the molecule. One of them has no formal charges; the other two are polarized in opposite directions. Write these resonance contributors and analyze each one to see whether (1) each atom has an octet around it, and (2) the formal charges are in accord with the relative electronegativities of carbon and oxygen.
(b) Carbon monoxide is highly toxic because it binds tightly to iron in hemoglobin and thus prevents that molecule from binding to and carrying oxygen in the blood. The carbon atom in carbon monoxide binds to iron(II), Fe^{2+}. What does this experimental fact indicate about the relative importance of the various resonance contributors to the structure of carbon monoxide?

1.28 The azide anion, N_3^-, is linear. The nitrogen-nitrogen bond lengths in the ion are all 1.15 Å. Write resonance contributors for the azide anion that account for these experimental observations. (The typical nitrogen-nitrogen double bond length is 1.20 Å; the typical nitrogen-nitrogen triple bond length is 1.10 Å.)

1.29 Amides are compounds containing the following group of atoms.

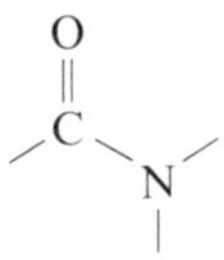

This type of group is important in proteins. In protein chains, amino acids are held together by amide bonds. The properties of an amide bond are best rationalized on the basis of resonance contributors for an amide. Write three resonance contributors for the amide group and discuss their relative importance.

1.30 Decide which of the following sets of structural formulas represent resonance contributors. Also, decide which resonance contributors for a given molecule are more important and which less so. Describe your reasoning.

(a) 

(b) 

(c) 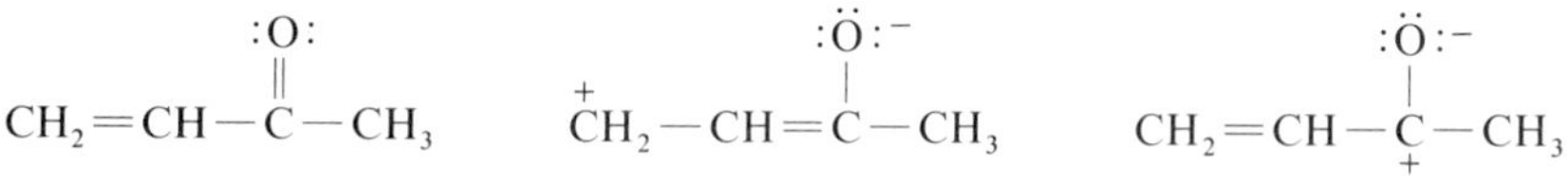

(d) 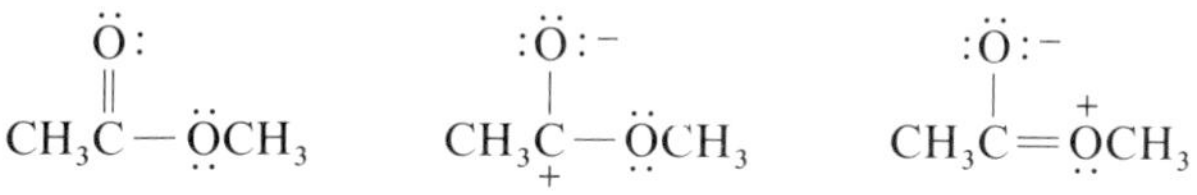

(e) 

(f) 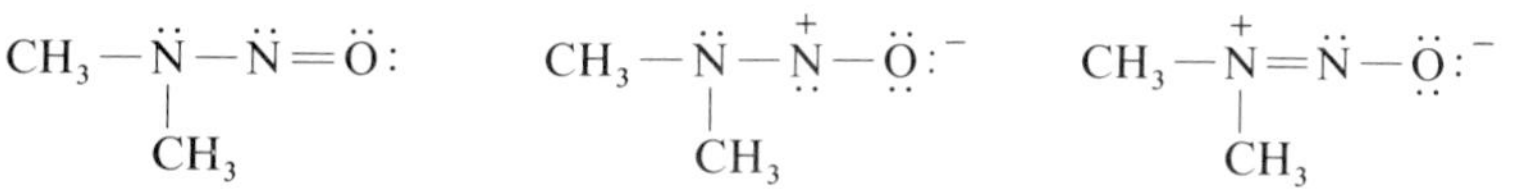

(g) 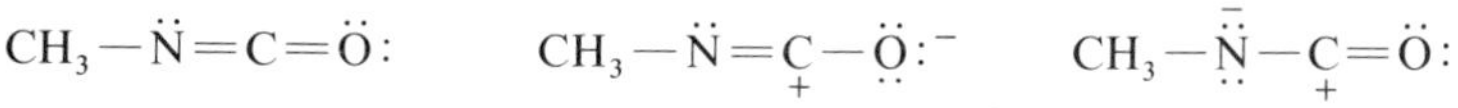

(h) 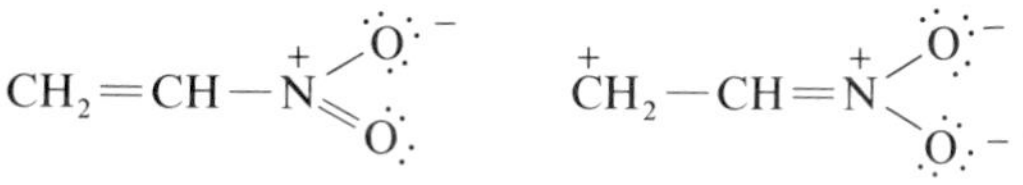

(i) 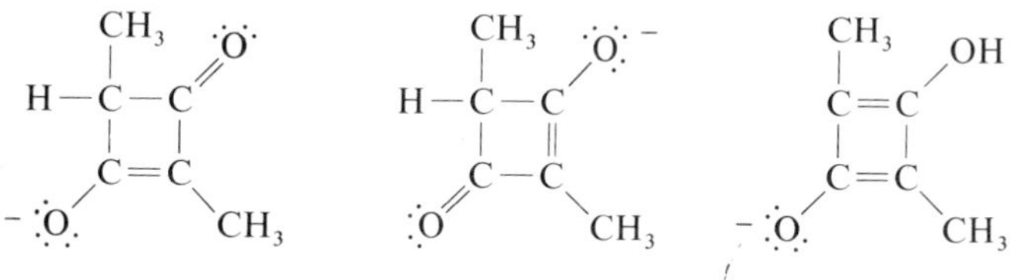

(j) 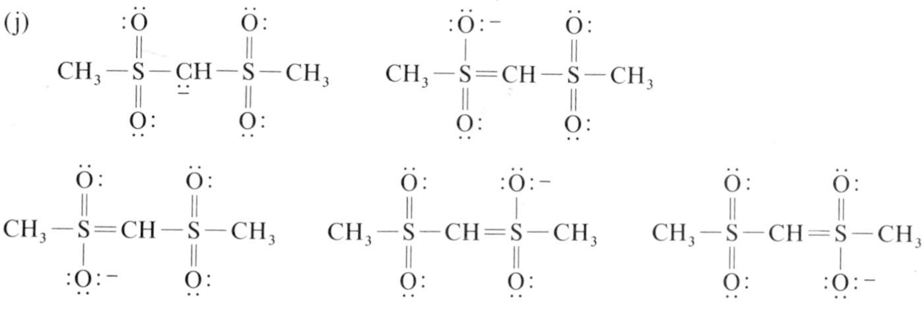

2

Covalent Bonding and Chemical Reactivity

A • L O O K • A H E A D

The complete three-dimensional structures of covalent compounds can be described in terms of bond lengths and bond angles that are experimentally determined. For example, two carbon atoms are bonded to each other and to three hydrogen atoms each in ethane, C_2H_6, a molecule in which the bond angles around the carbon atoms are tetrahedral. In ethylene, C_2H_4, the two carbon atoms are held together by a double bond. All of the atoms in ethylene lie in a plane, with bond angles close to 120°. The two carbon atoms in acetylene, C_2H_2, share a triple bond, and each carbon is bonded to one hydrogen atom in this linear molecule. In general, two carbon atoms can be bonded together in three different ways: with a single bond, a double bond, or a triple bond.

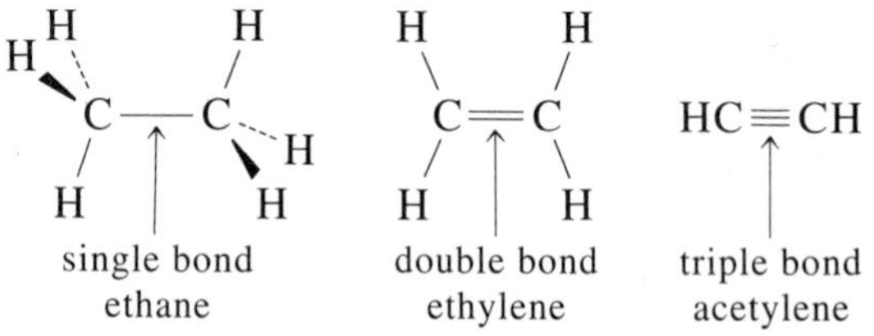

single bond
ethane
double bond
ethylene
triple bond
acetylene

The double bond in ethylene and the triple bond in acetylene are functional groups. A functional group is a structural unit consisting of an atom or a group of atoms that serves as a site of chemical reactivity in a molecule.

How can the electronic structure of carbon be used to explain the different kinds of bonding and the different shapes that are observed for various organic molecules? Scientists have put forth many ideas about the nature of chemical bonding and the reasons for chemical reactivity. In this chapter, we will examine two systems of ideas, called molecular orbital theory and orbital hybridization, that are currently used by organic chemists to rationalize the facts known about the compounds of carbon. We will also learn to recognize important functional groups.

2.1 INTRODUCTION

Chemists' ideas about electrons in molecules and how they participate in bonding are derived from quantum mechanics. In the mathematical equations of quantum mechanics, electrons are treated as if they have the properties of both waves and particles. In 1923, the French physicist Louis de Broglie first introduced the idea that the motion of electrons can be described by equations similar to those associated with waves. This idea was further developed independently by Erwin Schrödinger of Austria and Werner Heisenberg of Germany. In the Schrödinger equation, the motion of the electron is related to a set of allowed energy values by mathematical expressions known as wave equations. Each energy level allowed for an electron corresponds to a particular solution of the wave equation, called a wave function.

Nobel Prizes in Physics were awarded to de Broglie (1929), Heisenberg (1932), and Schrödinger (1933) in recognition of the importance of quantum mechanics to our understanding of the nature of chemical bonding. This way of looking at electrons and bonding is a highly sophisticated and mathematical model; in contrast, the Lewis structure for a covalent compound is a simple and pictorial one. Both systems, however, are human creations; we impose them on the facts of nature as explanations and aids to prediction.

2.2 ATOMIC ORBITALS

A. The Atomic Orbitals of Hydrogen. A Review

According to the prevailing model of atomic structure, the exact location of an electron in an atom cannot be determined. Theories of atomic structure deal with the probability of finding an electron at a given distance and direction from the nucleus. This probability is determined by the wave function for the electron, which is called an **atomic orbital.** An orbital is usually represented by a picture that shows the boundary of the space surrounding the nucleus of an atom within which there is some finite probability (such as 90%) of finding the electron.

The hydrogen atom has only a single proton and a single electron. The two atomic orbitals of lowest energy in the hydrogen atom are the 1*s* and 2*s* orbitals, which are spherically symmetrical (Figure 2.1). There is equal probability of finding the electron at a given distance in all directions from the nucleus. The 2*s* orbital is larger than the 1*s* orbital. An electron in the 2*s* orbital is, on the average, farther away from the nucleus than is one in the 1*s* orbital. The electron in the 2*s* orbital is less attracted by the nucleus and is, therefore, in a higher energy level.

The three 2*p* atomic orbitals of the hydrogen atom are oriented at right angles to each other and are usually given the designation of the coordinate axes $2p_x$, $2p_y$, and $2p_z$. Note that a *p* orbital does *not* have spherical symmetry, but does have symmetry about the axis along which it lies (Figure 2.1). The probability of finding an electron in a 2*p* atomic orbital is greatest in its two lobes, which are on opposite sides of the nucleus. There is zero probability of finding an electron at the nucleus in a *p* orbital. A region where the probability of finding an electron is zero is called a **node.** A nodal plane passes through the nucleus for *p* orbitals. The 2*s* orbital also has a nodal region, the surface of a sphere buried within the boundary surface as seen in Figure 2.1.

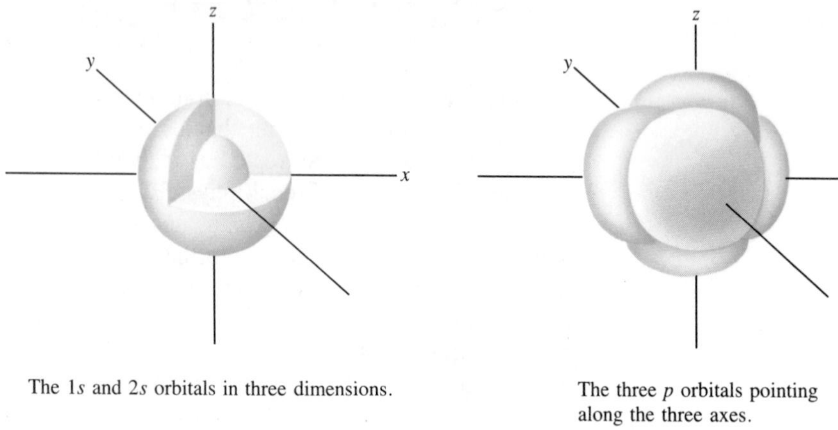
z
y
x
z
y
x
The 1s and 2s orbitals in three dimensions.
The three p orbitals pointing along the three axes.

FIGURE 2.1 The s and p atomic orbitals of the hydrogen atom.

A more precise mathematical way of describing a p atomic orbital is to assign different mathematical signs to the two lobes of the orbital to indicate that the wave function defining the orbital changes sign as it goes through the nodal plane. Unfortunately, positive and negative signs also mean positive and negative charges in the language of chemistry, so pictures with many such signs in them could be confusing. Therefore, in this book the two lobes of a p orbital will be shown with different colored shading to indicate the change in sign that takes place across the nodal plane (Figure 2.2).

It is of no importance which lobe of a p orbital is seen as positive and which negative, or which grey and which colored. Such an assignment is purely arbitrary. The change in sign within such an orbital is important because it determines how the orbital will interact with other orbitals.

The characteristics of the s and p atomic orbitals of the hydrogen atom are used as rough approximations of the properties of similar orbitals for elements such as carbon, oxygen, nitrogen, and the halogens. The relative energies and the exact

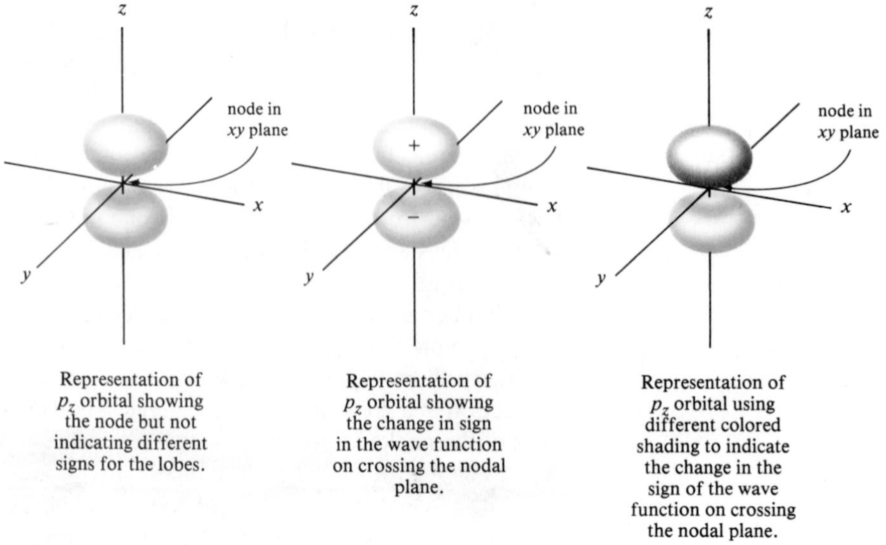
z
node in xy plane
x
y
Representation of p_z orbital showing the node but not indicating different signs for the lobes.
z
node in xy plane
+
−
x
y
Representation of p_z orbital showing the change in sign in the wave function on crossing the nodal plane.
z
node in xy plane
x
y
Representation of p_z orbital using different colored shading to indicate the change in the sign of the wave function on crossing the nodal plane.

FIGURE 2.2 Different ways in which a p orbital may be represented.

sizes of atomic orbitals change with the charge on the nucleus of an atom and with the number of electrons it has. Nevertheless, chemists use the picture developed for the hydrogen atom when making predictions about bonding in more complex atoms.

B. Orbitals of Atoms in the Second Row of the Periodic Table

One of the complications that arises whenever there is more than one electron in an atom is the question of **electron spin.** Electrons can have one of two possible spin orientations, symbolized by arrows pointing up or down, ↑ or ↓. Three rules govern the assignment of the most stable electronic configuration to an atom.

1. The Aufbau Principle states that electrons fill atomic orbitals in order according to increasing energy. This means that the 1*s* orbital is filled first, then the 2*s*, then the 2*p*.
2. The Pauli Exclusion Principle states that two electrons in the same orbital must have opposing spins.
3. Hund's Rule states that when orbitals of equal energy are available, one electron must be assigned to each of those orbitals before any orbital receives two.

Table 2.1 shows the assignment of electrons to the orbitals of the first ten elements in the periodic table. Note that the three 2*p* orbitals are of equal energy. They are said to be **degenerate.**

PROBLEM 2.1

The orbitals that are in the next higher energy level above the 2*p* orbitals are the 3*s* orbitals, followed by the 3*p* orbitals. Assign electronic configurations to the elements sodium, magnesium, aluminum, silicon, phosphorus, sulfur, chlorine, and argon.

TABLE 2.1 Assignment of Electrons to Orbitals

Element	Electronic Configuration	Electrons in Orbitals 1*s*	2*s*	2*p*		
Hydrogen	$1s^1$	↑				
Helium	$1s^2$	⇅				
Lithium	$1s^22s^1$	⇅	↑			
Beryllium	$1s^22s^2$	⇅	⇅			
Boron	$1s^22s^22p^1$	⇅	⇅	↑		
Carbon	$1s^22s^22p^2$	⇅	⇅	↑	↑	
Nitrogen	$1s^22s^22p^3$	⇅	⇅	↑	↑	↑
Oxygen	$1s^22s^22p^4$	⇅	⇅	⇅	↑	↑
Fluorine	$1s^22s^22p^5$	⇅	⇅	⇅	⇅	↑
Neon	$1s^22s^22p^6$	⇅	⇅	⇅	⇅	⇅

A. The Hydrogen Molecule

The Lewis structure of a molecule shows a covalent bond arising from the sharing of a pair of electrons, one from each atom. The orbital picture of the covalent bond shows a molecular orbital formed by the overlap of two atomic orbitals, each containing an electron. The directional properties of atomic orbitals are important in determining the extent of overlap that is possible and the geometry of the molecule that results when an atom forms more than one covalent bond.

The hydrogen molecule, arising from the combination of two hydrogen atoms, is the simplest molecule possible. An orbital picture of bonding requires that we imagine two hydrogen atoms, each with one electron in a 1*s* orbital, approaching each other. If the orbitals have the same mathematical sign, they can interact to reinforce each other. They are said to be **in phase.** The interaction of two orbitals of the same mathematical sign results in the formation of a **bonding molecular orbital,** shown schematically in Figure 2.3. The two electrons, one from each hydrogen atom, occupy the bonding molecular orbital. There is an increase in electron density between the two nuclei.

The hydrogen molecule is more stable than the individual hydrogen atoms because the electron of each atom is attracted to the positively charged nucleus of the other atom as well as to its own nucleus. The interaction shown in Figure 2.3 results in a lowering of the overall energy of the system. The bonding molecular orbital is lower in energy than the two separate atomic orbitals. Note that the bonding molecular orbital of hydrogen is symmetrical about an axis connecting the two nuclei. A bonding molecular orbital with cylindrical symmetry about an internuclear axis is called a **σ molecular orbital** (σ is read "sigma").

The hydrogen molecule is most stable when the nuclei of its atoms are a certain distance apart. If the nuclei are any closer, they repel each other too strongly. If they are farther apart, the atomic orbitals do not overlap enough to form a good covalent bond. The equilibrium distance that allows for the most overlap without excessive nuclear repulsion is called the **bond distance,** which is 0.74 Å, or 0.074 nm (a nanometer is 10^{-9} m), for the hydrogen molecule (Figure 2.3).

There is another possible combination of the two 1*s* atomic orbitals of hydrogen, one in which the orbitals are of opposite mathematical sign and are said to be **out of phase** with each other. In this combination, there is a node, a region of no electron density, between the two nuclei (Figure 2.4). The lack of electron density between the two nuclei results in repulsion between them. This interaction gives

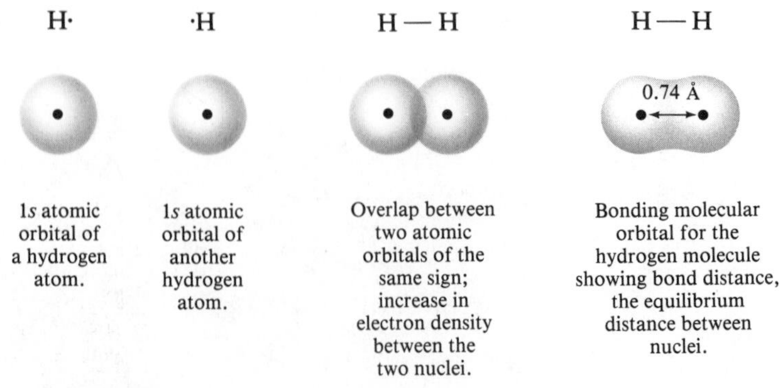
H·
·H
H—H
H—H
0.74 Å
1*s* atomic orbital of a hydrogen atom.
1*s* atomic orbital of another hydrogen atom.
Overlap between two atomic orbitals of the same sign; increase in electron density between the two nuclei.
Bonding molecular orbital for the hydrogen molecule showing bond distance, the equilibrium distance between nuclei.

FIGURE 2.3 A schematic representation of the formation of a bonding molecular orbital in hydrogen by the overlap of two 1*s* atomic orbitals.

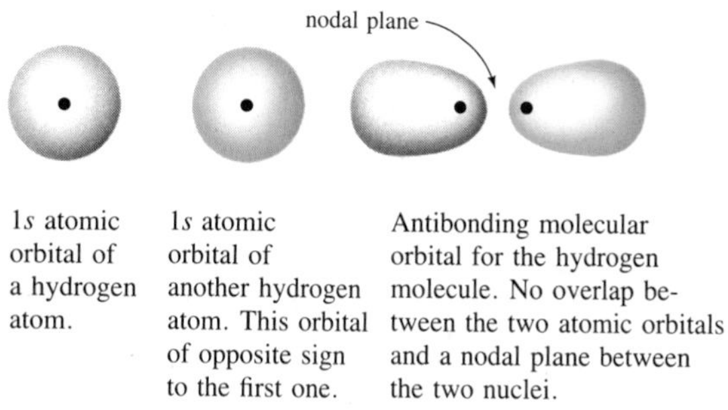
nodal plane
1s atomic orbital of a hydrogen atom.
1s atomic orbital of another hydrogen atom. This orbital of opposite sign to the first one.
Antibonding molecular orbital for the hydrogen molecule. No overlap between the two atomic orbitals and a nodal plane between the two nuclei.

FIGURE 2.4 A schematic representation of the antibonding interaction between 1*s* atomic orbitals of two hydrogen atoms.

rise to an **antibonding molecular orbital,** a molecular orbital that is of higher energy than the two separate atomic orbitals. The antibonding molecular orbital corresponding to the σ bonding molecular orbital is called the **σ* molecular orbital** (σ^* is read "sigma star").

When atomic orbitals are combined to give molecular orbitals, the number of molecular orbitals formed equals the number of atomic orbitals used. Thus, the combination of two atomic orbitals gives rise to two molecular orbitals, a bonding molecular orbital and an antibonding molecular orbital. The bonding and antibonding molecular orbitals of the hydrogen molecule are shown schematically in Figure 2.5.

The two electrons of the hydrogen molecule are usually in the bonding molecular orbital (Figure 2.5). Molecular orbitals, like atomic orbitals, can hold only two electrons of opposing spin. When a bonding molecular orbital is occupied by two electrons, a covalent bond results. The bond between the hydrogen atoms in a molecule of hydrogen is called a σ bond. Electrons in an antibonding molecular orbital do not contribute to bonding between atoms and, in fact, serve to destabilize a molecule.

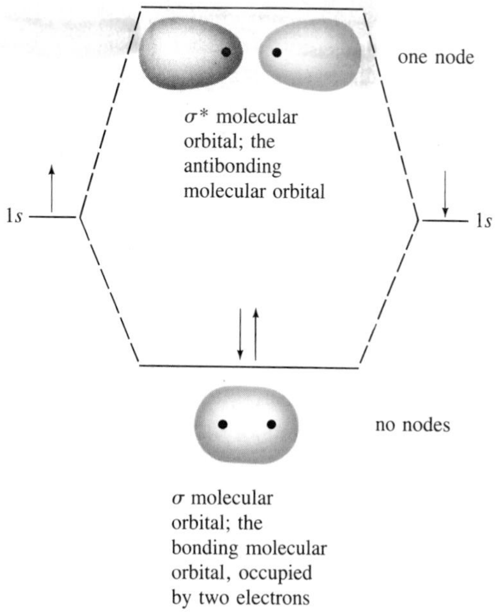
one node
σ* molecular orbital; the antibonding molecular orbital
1s
1s
no nodes
σ molecular orbital; the bonding molecular orbital, occupied by two electrons

FIGURE 2.5 The relative energies and shapes of the σ and σ^* molecular orbitals for the hydrogen molecule.

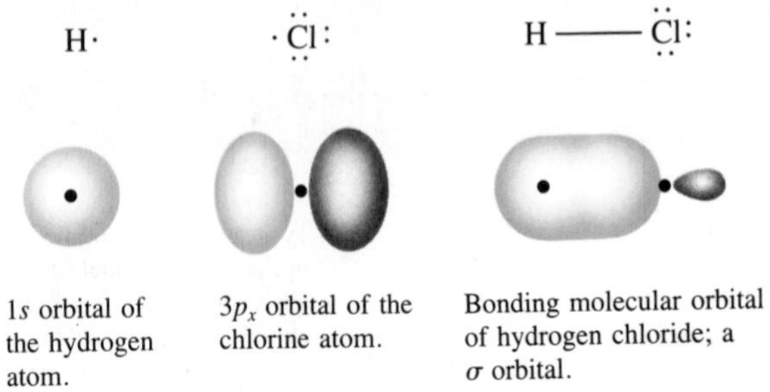
H·
·Cl:
H—Cl:
1s orbital of the hydrogen atom.
$3p_x$ orbital of the chlorine atom.
Bonding molecular orbital of hydrogen chloride; a σ orbital.

FIGURE 2.6 A schematic representation of the formation of a bonding molecular orbital in hydrogen chloride by the overlap of the 1*s* orbital of hydrogen and the $3p_x$ orbital of chlorine.

A hydrogen molecule is more stable than two separate hydrogen atoms by about 104 kcal/mol. This is the amount of energy required to break the bond in the hydrogen molecule and separate the two hydrogen atoms. It is known as the **bond dissociation energy** (p. 59) for hydrogen.

B. Sigma Bonds Involving *p* Orbitals

Different types of atomic orbitals can overlap to create σ bonds. For example, the covalent bond in a molecule of hydrogen chloride is thought to arise from the overlap of a 1*s* orbital of a hydrogen atom with a 3*p* orbital of a chlorine atom. All of the *p* orbitals are equivalent so any one of them may be used for the bond. The bonding molecular orbital that is formed when the spherically symmetrical 1*s* orbital of the hydrogen atom approaches the $3p_x$ orbital of the chlorine atom along the *x* axis is shown schematically in Figure 2.6.

The molecular orbital in hydrogen chloride is shaped differently from the atomic orbitals that go into its formation. In particular, the lobe of the $3p_x$ orbital that is not involved in the bonding is greatly contracted. The greatest electron density in the molecular orbital is observed between the two nuclei. The orbital is cylindrically symmetrical with respect to the axis between the two nuclei and is, therefore, a σ orbital. The presence of two electrons of opposing spin in the orbital gives rise to a single bond, also called a **σ bond,** between the hydrogen and chlorine atoms in hydrogen chloride.

In general, a σ bond is formed by the overlap of two *s* atomic orbitals, an *s* and a *p* atomic orbital, or two *p* atomic orbitals. The bonding molecular orbitals that are formed by the overlap of atomic orbitals are called **σ orbitals** when they are symmetrical around the axis joining the two atomic nuclei. All σ bonds have electron density concentrated between the nuclei of the bonded atoms.

In summary, the overlap of atomic orbitals gives rise to bonding (and antibonding) molecular orbitals. Bonds are formed when bonding molecular orbitals are occupied by pairs of electrons. For the sake of simplicity, we will ignore the antibonding molecular orbitals. In all the compounds we will discuss, the bonding molecular orbitals are filled with electrons, so distinctions between the formation of a bonding orbital and the formation of a bond are not always made.

Study Guide Concept Map 2.1

PROBLEM 2.2

Draw the atomic orbitals involved and show the bonding molecular orbitals of hydrogen fluoride (HF) and the fluorine molecule (F_2).

PROBLEM 2.3

The following drawings show atomic orbitals approaching each other along the x axis for each atom. In each case, decide whether good overlap and bonding will occur for those orientations of the orbitals.

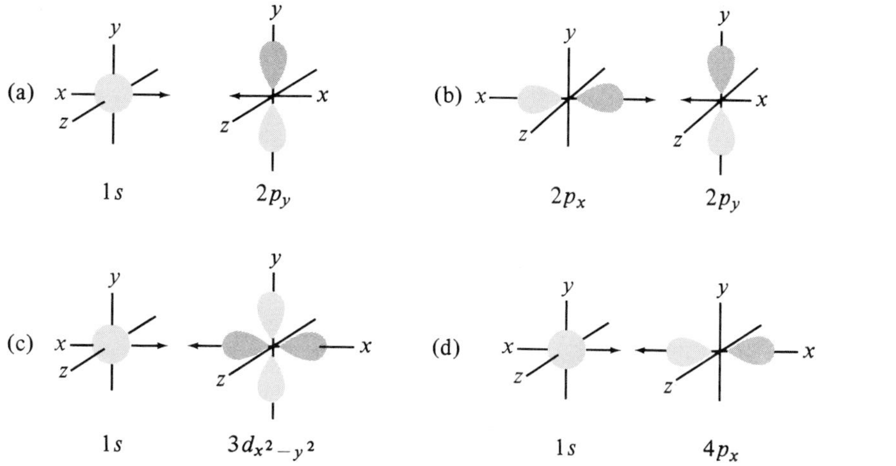
(a)
1s
2p_y
(b)
2p_x
2p_y
(c)
1s
3d_{x^2-y^2}
(d)
1s
4p_x
x
y
z

2.4 HYBRID ORBITALS

A. Tetrahedral Carbon Atoms

The simplest compound of carbon and hydrogen, methane, CH_4, is a symmetrical molecule with four carbon-hydrogen bonds of equal length directed toward the corners of a regular tetrahedron (Figure 1.1, p. 18). A similar tetrahedral arrangement of bonds has been shown experimentally to occur whenever carbon is bonded to four other atoms.

Methane has eight bonding electrons, four from the carbon atom and one from each of the four hydrogen atoms. In the language of molecular orbital theory, methane has four bonding molecular orbitals, each of which holds two of the bonding electrons. The hydrogen atoms (or the electron pairs) repel each other and, therefore, move as far as possible from each other in a tetrahedral arrangement. In this model, more than one molecular orbital contributes to each carbon-hydrogen bond.

Chemists also explain the bonding at a carbon atom that is bonded to four other atoms in a tetrahedral geometry by using the concept of **orbital hybridization.** According to this idea, overlap of hybrid atomic orbitals instead of the pure atomic orbitals of atoms forms molecular orbitals. **Hybrid orbitals** are mathematical combinations of atomic orbitals. For a tetrahedral carbon atom, the wave functions for the four atomic orbitals—$2s$, $2p_x$, $2p_y$, and $2p_z$—are combined to create four new hybrid orbitals, shown in Figure 2.7 on the next page.

The hybrid atomic orbitals are called ***sp^3* hybrid orbitals** to show that they arise from a mathematical combination of one s orbital and three p orbitals, indicated by the superscript 3 on the p. *The number of hybrid orbitals generated is always equal to the number of atomic orbitals combined.*

Methane has four identical σ bonds; therefore, four atomic orbitals of the central carbon atom must be mixed. In other words, once chemists know what the

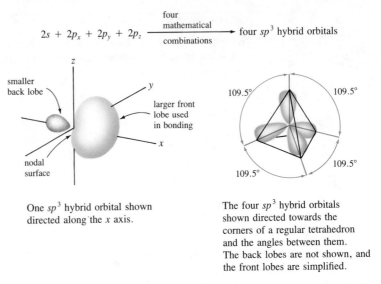

$$2s + 2p_x + 2p_y + 2p_z \xrightarrow[\text{combinations}]{\text{four mathematical}} \text{four } sp^3 \text{ hybrid orbitals}$$

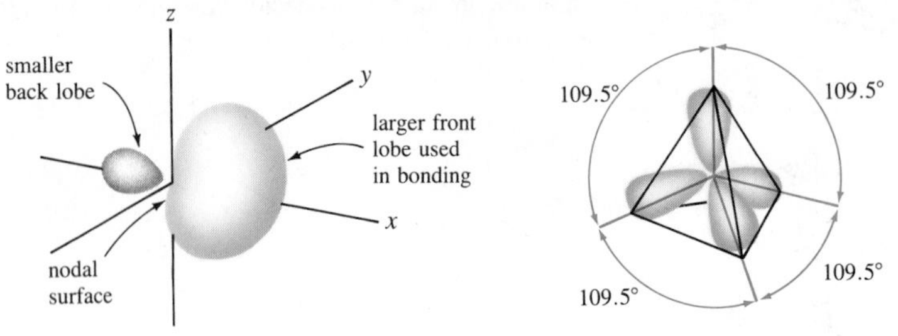
z
y
x
smaller back lobe
larger front lobe used in bonding
nodal surface
109.5°
109.5°
109.5°
109.5°

One sp^3 hybrid orbital shown directed along the x axis.

The four sp^3 hybrid orbitals shown directed towards the corners of a regular tetrahedron and the angles between them. The back lobes are not shown, and the front lobes are simplified.

FIGURE 2.7 Hybridization of the $2s$, $2p_x$, $2p_y$, and $2p_z$ atomic orbitals of carbon to produce sp^3 hybrid orbitals.

structure of a molecule is, they decide which mathematical combination of atomic orbitals will result in hybrid orbitals that have the directional properties necessary for that structure. The mathematical combination of one *s* and three *p* orbitals was chosen deliberately to create four hybrid orbitals directed to the corners of a tetrahedron. Like the pictures of pure atomic orbitals, the pictures of hybrid orbitals represent regions in space within the boundaries of which there is some finite probability of finding an electron.

Note that the overall shape of a single sp^3 hybrid orbital resembles that of a *p* orbital in that it has a node at the nucleus of the carbon atom. An sp^3 orbital, however, does not have two equal lobes, as a pure *p* orbital does. The larger lobe of an sp^3 hybrid orbital extends farther into space from the nucleus of the carbon atom than any of the atomic orbitals do. Thus, the hybrid orbitals of carbon are better able to overlap with orbitals of other atoms and to form stronger covalent bonds than the pure atomic orbitals.

According to the orbital hybridization picture, each of the four carbon-hydrogen bonds in methane is formed by the overlap of a $1s$ orbital of a hydrogen atom and an sp^3 hybrid orbital of a carbon atom (Figure 2.8). In this representation,

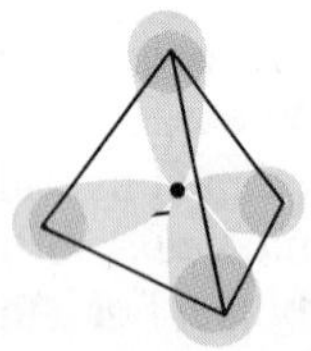

schematic drawing of the overlap between the $1s$ orbitals of the hydrogen atoms and the sp^3 hybrid orbitals of the carbon atom, forming four bonds directed towards the corners of a tetrahedron

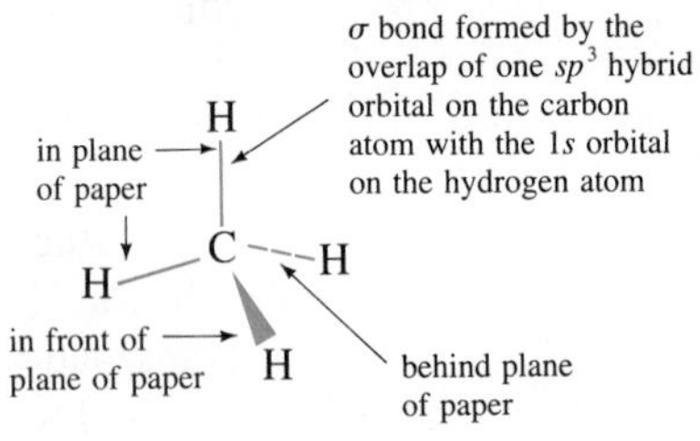
σ bond formed by the overlap of one sp3 hybrid orbital on the carbon atom with the 1s orbital on the hydrogen atom
H
in plane of paper
C
H
H
H
in front of plane of paper
behind plane of paper

FIGURE 2.8 Formation of methane by the overlap of hybrid sp^3 orbitals of a carbon atom with $1s$ orbitals of four hydrogen atoms.

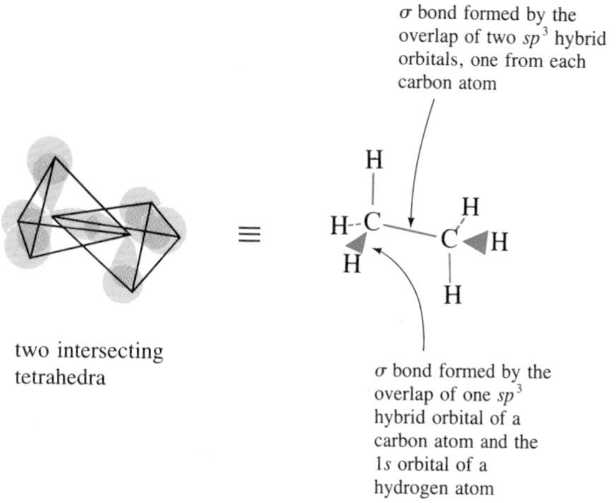
σ bond formed by the overlap of two sp^3 hybrid orbitals, one from each carbon atom
two intersecting tetrahedra
H
H--C
H
H
C
H
H
σ bond formed by the overlap of one sp^3 hybrid orbital of a carbon atom and the 1s orbital of a hydrogen atom

FIGURE 2.9 Orbital picture and structural formula of the ethane molecule.

the eight bonding electrons of methane are localized in four equivalent carbon-hydrogen bonds directed to the corners of a tetrahedron. Methane is pictured as having four equivalent σ bonds with an angle of 109.5°, the tetrahedral angle, between any two.

Organic chemists find the model of orbital hybridization to be useful in accounting for many experimental observations and thus use this way of thinking about bonding in organic compounds. We will use this language in describing the structure and bonding of a number of representative organic compounds in the remainder of this chapter.

A single bond between two carbon atoms is pictured as being formed by the overlap of an sp^3 hybrid orbital from each one. For example, ethane, C_2H_6, has this type of single bond. Tetrahedral geometry is maintained around each carbon atom. In Figure 2.9, the orbital picture of the bonding in ethane is shown, along with a three-dimensional structural formula showing the geometry of the molecule.

Methane and ethane are the simplest members of a family of compounds known as **alkanes.** Alkanes are a subclass of a much larger group of organic compounds called **hydrocarbons.** Hydrocarbons contain only the elements carbon and hydrogen. The alkane family is characterized by the presence of tetrahedral carbon atoms bonded to hydrogen atoms or to other tetrahedral carbon atoms.

A group derived from an alkane by removal of one of its hydrogen atoms is known as an **alkyl group.** For example, CH_3—, from methane, CH_4, is the methyl group; CH_3CH_2—, from ethane, CH_3CH_3, is the ethyl group. The systematic naming of alkanes and alkyl groups is covered in Chapter 5.

PROBLEM 2.4

Chloroform, $CHCl_3$, is another compound that contains a tetrahedral carbon atom. Sketch an orbital picture for it, showing how the bonding in the molecule arises. Also draw a three-dimensional representation of the molecule.

PROBLEM 2.5

Ethyl fluoride, C_2H_5F, is like ethane, except that one of the hydrogen atoms has been replaced by a fluorine atom. Draw an orbital picture and a three-dimensional representation of ethyl fluoride.

B. sp^3-Hybridized Atoms Other than Carbon

The concept of sp^3 hybridization used to describe the bonding of tetrahedral carbon atoms can also be used to describe bonding in ammonia and related organic compounds. Like methane, ammonia has bond angles close to 109.5° (p. 19). Nitrogen, the central atom in ammonia, is bonded to three hydrogen atoms and has a pair of nonbonding electrons. Nitrogen has two electrons in the 2*s* orbital and one electron in each 2*p* orbital, for a total of five electrons in its valence shell. Chemists explain the bonding in ammonia by postulating the hybridization of one 2*s* and three 2*p* orbitals to create the four sp^3 hybrid orbitals of nitrogen. One of the sp^3 hybrid orbitals of nitrogen is filled with a pair of electrons. The other three electrons of the valence shell of the nitrogen atom, along with three electrons from three hydrogen atoms, are in bonding molecular orbitals formed by the overlap of the sp^3 hybrid orbitals of the nitrogen atom with the 1*s* orbitals of the hydrogen atoms (Figure 2.10).

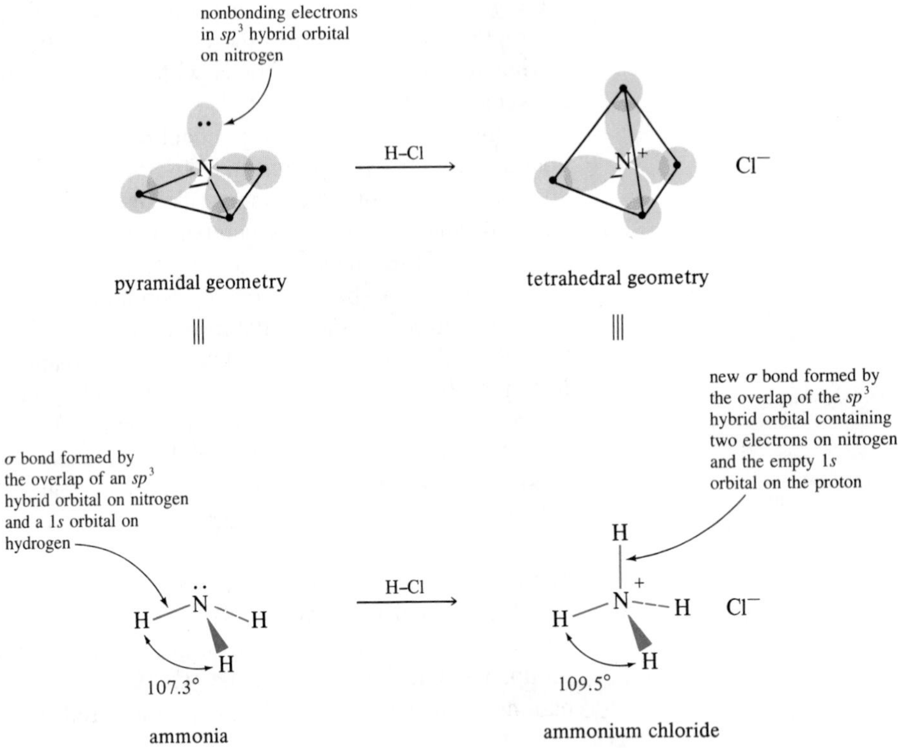
nonbonding electrons
in sp³ hybrid orbital
on nitrogen
N
H–Cl
N+
Cl⁻
pyramidal geometry
tetrahedral geometry
new σ bond formed by
the overlap of the sp³
hybrid orbital containing
two electrons on nitrogen
and the empty 1s
orbital on the proton
σ bond formed by
the overlap of an sp³
hybrid orbital on nitrogen
and a 1s orbital on
hydrogen
H
N
H
H
107.3°
H–Cl
H
N
H
H
H
Cl⁻
109.5°
ammonia
ammonium chloride

FIGURE 2.10 Orbital pictures and three-dimensional structural formulas of ammonia and ammonium chloride.

When ammonia reacts with an acid such as hydrogen chloride, the nonbonding electrons of nitrogen form a covalent bond with a proton, which has no electrons in the 1*s* orbital. The resulting ammonium ion, NH_4^+, contains four equal nitrogen-hydrogen bonds directed to the corners of a tetrahedron (Figure 2.10). The positively charged ammonium ion is held by an ionic bond to the negatively charged chloride ion, which is also formed in the reaction.

PROBLEM 2.6

What is wrong with a picture of the covalent bonding in ammonia that locates the nonbonding electrons of the molecule in the 2*s* orbital of nitrogen and postulates that the three nitrogen-hydrogen single bonds are formed by the overlap of 2*p* atomic orbitals of nitrogen with 1*s* atomic orbitals of hydrogen?

There are many organic compounds that contain a nitrogen atom having tetrahedral geometry of its electron pairs and a pair of nonbonding electrons. These compounds, which react with acids in the way shown in Figure 2.10, are **amines,** organic relatives of ammonia in which one or more of the hydrogen atoms of ammonia have been replaced by a group that contains carbon. The simplest amine is methylamine, CH_3NH_2. The carbon-nitrogen bond in this molecule is formed by the overlap of an sp^3 hybrid orbital of a carbon atom with an sp^3 hybrid orbital of a nitrogen atom. The C—N—H bond angle in methylamine is 107°, which is close to the bond angles in ammonia. The carbon-nitrogen bond length is 1.47 Å, which is shorter than the 1.54 Å of a carbon-carbon σ bond (Figure 2.11). The portion of the methylamine molecule consisting of the nitrogen and two hydrogen atoms is called the amino group. The amino group is a **functional group,** a structural unit that serves as the site of chemical reactivity that is typical of amines.

Another element that is often found in organic compounds is oxygen. The H—O—H bond angle in water has been measured as 104.5°, somewhat smaller than the tetrahedral angle, 109.5°. Oxygen has six valence electrons, two in the 2*s* orbital and four in the three 2*p* orbitals. After hybridization, two of the sp^3 hybrid orbitals of oxygen are filled with pairs of nonbonding electrons. The σ bonds in water are formed when the two unfilled sp^3 hybrid orbitals of the oxygen atom overlap with 1*s* orbitals of two hydrogen atoms. Just as ammonia reacts with an acid to give an ammonium ion (Figure 2.10), water reacts with an acid to give a hydronium ion. A hybrid orbital of the oxygen atom that contains nonbonding electrons overlaps with the empty 1*s* orbital of a proton. The molecular orbital pictures for water and the hydronium ion are shown in Figure 2.12.

The fact that the bond angle in the water molecule deviates from the tetrahedral angle is explained by postulating that the two pairs of nonbonding electrons on

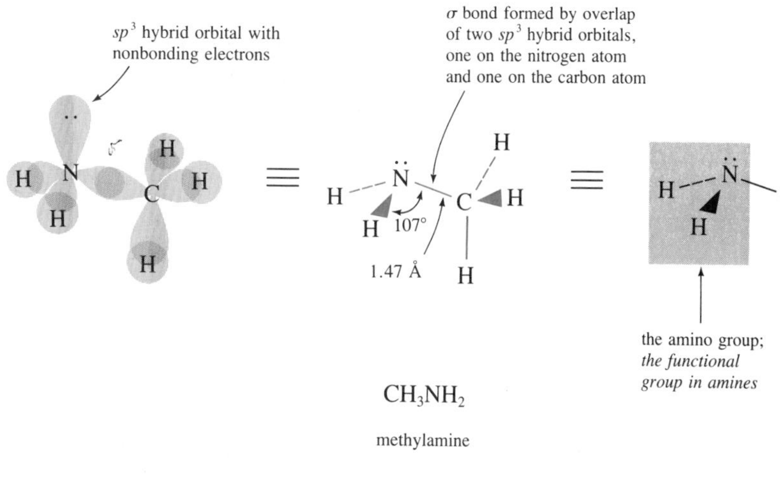

sp^3 hybrid orbital with nonbonding electrons
σ bond formed by overlap of two sp^3 hybrid orbitals, one on the nitrogen atom and one on the carbon atom
H
N
C
107°
1.47 Å
CH_3
the amino group; *the functional group in amines*
CH_3NH_2
methylamine

FIGURE 2.11 Orbital picture and three-dimensional structural formula of methylamine.

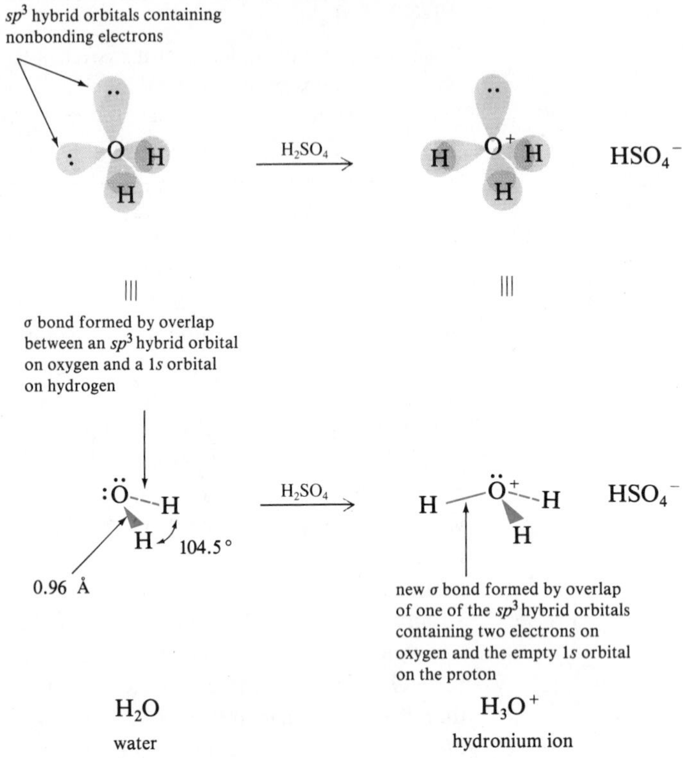
sp^3 hybrid orbitals containing nonbonding electrons
H_2SO_4
HSO_4^-
σ bond formed by overlap between an sp^3 hybrid orbital on oxygen and a 1s orbital on hydrogen
104.5°
0.96 Å
new σ bond formed by overlap of one of the sp^3 hybrid orbitals containing two electrons on oxygen and the empty 1s orbital on the proton
H_2O
water
H_3O^+
hydronium ion

FIGURE 2.12 Orbital pictures and three-dimensional structural formulas for the water molecule and the hydronium ion.

oxygen occupy more space and repel each other more strongly than do the two pairs of electrons in the oxygen-hydrogen bonds because bonding electrons are constrained in space by their interaction with two nuclei.

Alcohols are organic relatives of water. In alcohol molecules, there is the same geometry around the oxygen atom as in the water molecule. In alcohols, one of the hydrogen atoms of water has been replaced by a group containing one or more tetrahedral carbon atoms. The functional group in an alcohol is the hydroxyl group, —OH. The structure of a typical alcohol, methanol, CH_3OH, is shown in Figure 2.13. The carbon-oxygen bond in this compound is formed by the overlap of an sp^3 hybrid orbital from each atom.

Ethers are also organic relatives of water, in which both hydrogens have been replaced by groups containing carbon atoms. The structure of dimethyl ether is shown in Figure 2.13. Ethers, like water and alcohols, contain an sp^3-hybridized oxygen atom with two pairs of nonbonding electrons occupying two of its hybrid orbitals. The oxygen atom and its nonbonding electrons are the functional group in ethers.

In this section, we have examined some compounds that contain atoms of carbon, nitrogen, and oxygen that are covalently bonded to other atoms by single bonds with bond angles that are close to tetrahedral. The compounds fall into four functional group classes. Alkanes contain only tetrahedral carbon atoms bonded to hydrogen atoms. Amines are organic relatives of ammonia. Alcohols and ethers are related to water. The orbital pictures and three-dimensional structures of simple representatives of each of these functional group classes are also characteristic of their more complex members.

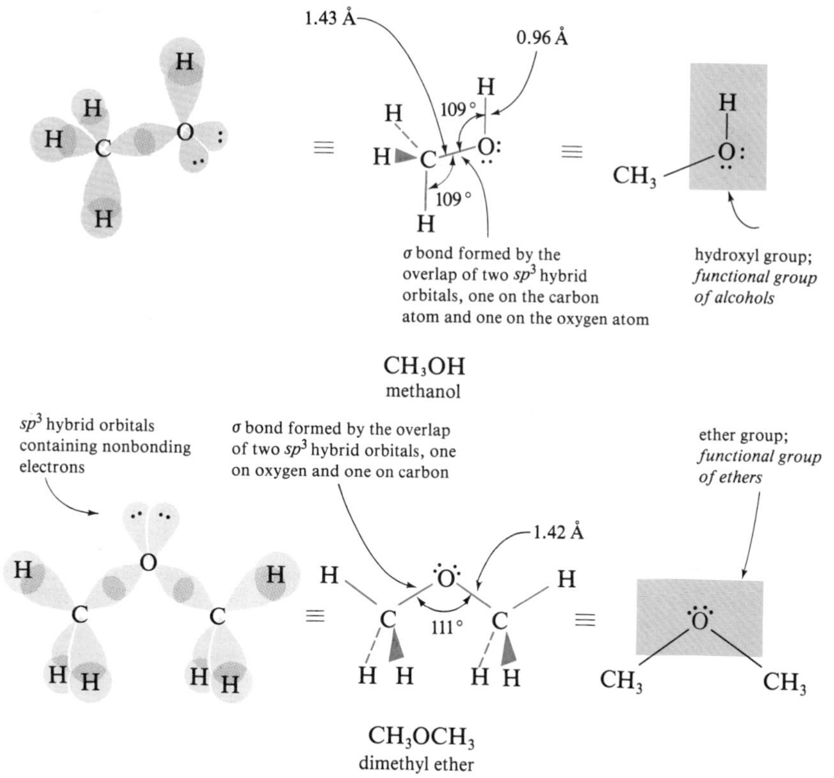

1.43 Å
0.96 Å
109°
109°
σ bond formed by the overlap of two sp^3 hybrid orbitals, one on the carbon atom and one on the oxygen atom
hydroxyl group; *functional group of alcohols*
CH_3OH
methanol
sp^3 hybrid orbitals containing nonbonding electrons
σ bond formed by the overlap of two sp^3 hybrid orbitals, one on oxygen and one on carbon
ether group; *functional group of ethers*
1.42 Å
111°
CH_3OCH_3
dimethyl ether

FIGURE 2.13 Orbital pictures and three-dimensional structural formulas of methanol and dimethyl ether.

The reactions of organic compounds are directly related to their structures. Thus, in many cases we can predict the chemistry of a complex molecule from what we know about the reactivity of a simple one of the same functional group class. The ability to predict reactivity from an inspection of the structure of a compound is important in the study of organic chemistry. By solving the following problems, you can take the first steps in developing this skill.

PROBLEM 2.7

Figure 2.10 illustrates the change in bonding that takes place when ammonia reacts with an acid. The nonbonding electrons on amines also interact with the proton from an acid in the same way. Draw structural formulas showing what happens when methylamine (Figure 2.11) reacts with hydrogen chloride.

PROBLEM 2.8

Alcohols and ethers react with acids the same way water does because of the presence of nonbonding electrons on the oxygen atom. Using Figure 2.12 as your guide, write structural formulas that show what happens when methanol and dimethyl ether (Figure 2.13) react with sulfuric acid.

PROBLEM 2.9

Write three-dimensional structural formulas for the species on the following page. In each case, indicate which types of orbitals overlap to form each bond to the atom that is underlined. For some of the species, you may find it helpful to write Lewis structures first.

(a) $H_2N{-}\underline{N}H_2$ (b) $\underline{B}F_4^-$ (c) $CH_3{-}\overset{CH_3}{\underset{CH_3}{\overset{+}{N}}}{-}\underline{C}H_3$ (d) $CH_3{-}\overset{CH_3}{\overset{+}{\underline{O}}}{-}CH_3$

PROBLEM 2.10

Identify each of the following compounds as belonging to one of these functional group classes: alkanes, amines, alcohols, or ethers.

(a) $CH_3CH_2CH_2CH_2CH_2OH$ (b) $CH_3CH_2CH_2OCH_3$ (c) $CH_3CH_2CH_2CH_3$

(d) $CH_3CH_2NH_2$ (e) $CH_3\underset{OH}{CH}CH_2CH_3$ (f) $CH_3CH_2\overset{H}{N}CH_2CH_3$

PROBLEM 2.11

Draw a structural formula illustrating the important electronic features of the functional group in each of the compounds in Problem 2.10. Predict which compounds will react with sulfuric acid (see Problems 2.7 and 2.8).

2.5 THE ORBITAL PICTURE FOR COMPOUNDS CONTAINING TRIGONAL CARBON ATOMS

A. Covalent Bonding in Alkenes

Not all compounds of carbon contain tetrahedral carbon atoms. Ethylene, $CH_2{=}CH_2$, for example, is a flat molecule in which all six atoms lie in the same plane (p. 20). Experiments show that the H—C—H and H—C—C bond angles are close to 120° and that the carbon-carbon bond length is 1.34 Å, shorter than the carbon-carbon bond in ethane. Ethylene is the simplest member of the family of hydrocarbons called **alkenes.** The functional group of the alkenes is the carbon-carbon double bond.

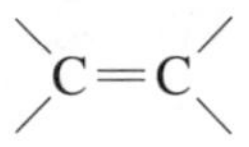

In ethylene, each carbon atom is bonded to three other atoms, all lying in a plane. Such a carbon atom is known as a **trigonal carbon atom.** Other atoms, such as boron in boron trifluoride, BF_3, are also trigonal.

Each carbon atom in ethylene is bonded to only three other atoms. Therefore, to rationalize the bonding in ethylene, chemists say that only two of the 2*p* orbitals of carbon are combined with the 2*s* orbital to create three new ***sp*² hybrid orbitals.** The shape of an sp^2 orbital is similar to that of an sp^3 hybrid orbital, but the spatial orientation is quite different. The three sp^2 hybrid orbitals lie in a plane, are directed to the corners of an equilateral triangle, and have angles of 120° between them. The third 2*p* orbital of an sp^2-hybridized carbon is unhybridized. This orbital retains its shape as an atomic orbital and is perpendicular to the plane defined by the three sp^2 hybrid orbitals (Figure 2.14).

The skeleton of ethylene is made up of the σ bonds between the various atoms. One carbon-carbon bond in ethylene is formed by the overlap of sp^2 hybrid orbitals

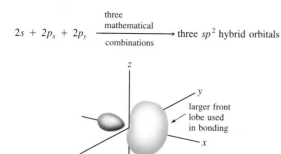

$$2s + 2p_x + 2p_y \xrightarrow[\text{combinations}]{\text{three mathematical}} \text{three } sp^2 \text{ hybrid orbitals}$$

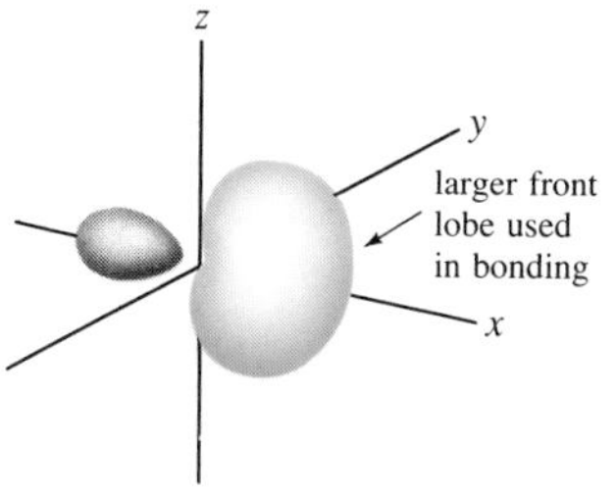

z
y
larger front lobe used in bonding
x

One sp^2 hybrid orbital

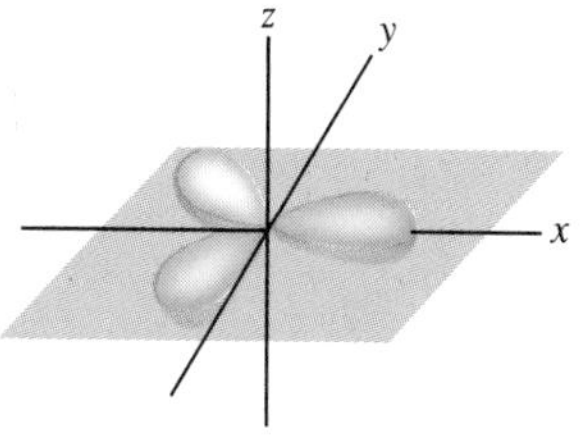

z
y
x

The sp^2 hybrid orbitals lying in the xy plane. Back lobes not shown and front lobes simplified.

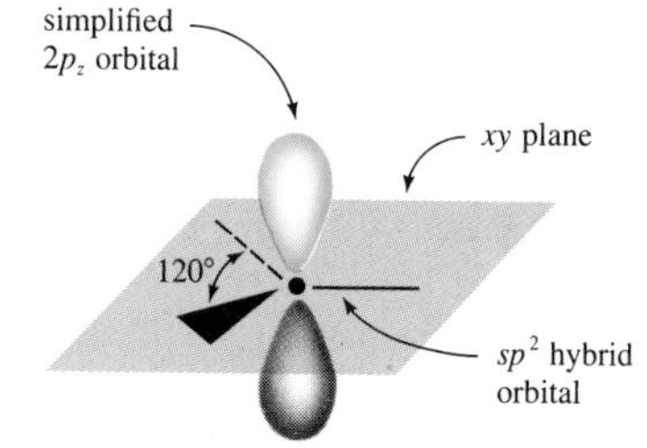

simplified $2p_z$ orbital
xy plane
120°
sp^2 hybrid orbital

p orbital perpendicular to plane defined by sp^2 hybrid orbitals.

FIGURE 2.14 Formation of the sp^2 hybrid orbitals of a trigonal carbon atom.

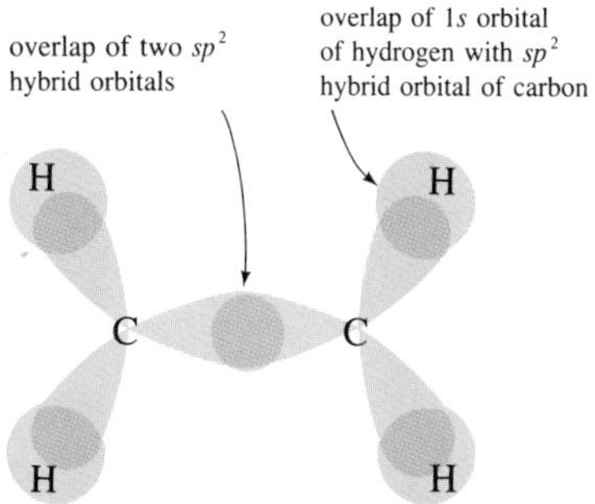

overlap of two sp^2 hybrid orbitals
overlap of $1s$ orbital of hydrogen with sp^2 hybrid orbital of carbon
H
H
C
C
H
H

Schematic drawing of the overlap between the $1s$ orbitals of the hydrogen atoms and the sp^2 hybrid orbitals of the carbon atoms forming the C—H σ bonds of ethylene. The C—C σ bond is formed by overlap of two sp^2 hybrid orbitals.

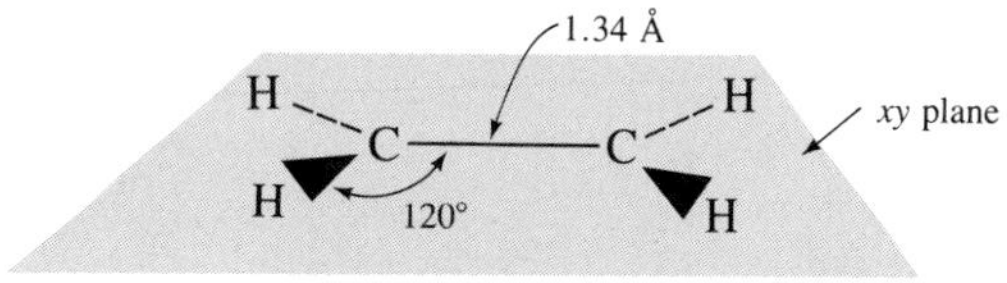

1.34 Å
H
C
C
H
H
120°
H
xy plane

σ bond skeleton of ethylene lying in xy plane.

FIGURE 2.15 Skeleton of σ bonds in ethylene.

from each carbon atom. The carbon-hydrogen bonds are created by the overlap of sp^2 hybrid orbitals of the carbon atoms with $1s$ orbitals of the hydrogen atoms. Each carbon atom contributes three of its four valence electrons to these bonds (Figure 2.15).

The fourth valence electron of each carbon atom is in the unhybridized $2p$ orbital. Two p atomic orbitals on adjacent atoms can interact to create two new molecular orbitals, the π molecular orbitals (π is read ''pi''). Two p orbitals next to each other can be oriented so that lobes of the same sign (same color) are on the same side of the nodal plane. In this case, the lobes that bear the same mathematical sign interact to reinforce each other. The two orbitals are in phase and combine to form a **bonding π molecular orbital.** If, on the other hand, the p orbitals are located so that lobes of opposite sign (differing colors) are on the same side of the nodal plane, they are out of phase and no bonding takes place. In this situation, there is a node (a region of no electron density) between the nuclei of the two atoms and an **antibonding π^* molecular orbital** (π^* is read ''pi star'') results.

When the two carbon atoms in ethylene form a σ bond, their p atomic orbitals are close enough that their parallel lobes interact as shown in Figure 2.16. Two electrons, one from each carbon atom, are in the bonding π molecular orbital, creating a π bond between the two carbon atoms. Thus, the double bond in ethylene and other alkenes consists of a σ bond and a π bond.

Each p atomic orbital has two lobes and a node at the nucleus. The π molecular orbital, because it is created by the side-to-side overlap of p orbitals, also has two lobes and a nodal plane. Thus, in a π bond there is some finite probability of finding

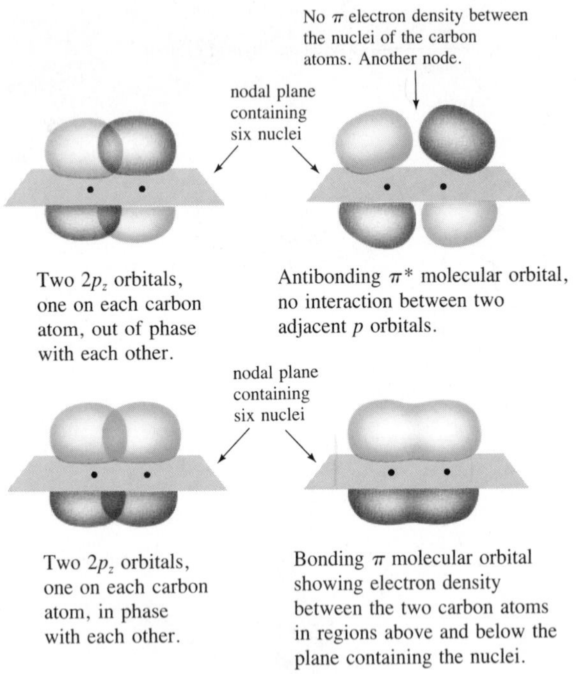
No π electron density between the nuclei of the carbon atoms. Another node.
nodal plane containing six nuclei
Two $2p_z$ orbitals, one on each carbon atom, out of phase with each other.
Antibonding π* molecular orbital, no interaction between two adjacent p orbitals.
nodal plane containing six nuclei
Two $2p_z$ orbitals, one on each carbon atom, in phase with each other.
Bonding π molecular orbital showing electron density between the two carbon atoms in regions above and below the plane containing the nuclei.

FIGURE 2.16 Molecular orbital picture of the π bond in ethylene.

electrons in lobes above and below the plane of the molecule. Figure 2.16 shows that a π bond does not have the cylindrical symmetry of a σ bond.

B. Covalent Bonding in Carbonyl Compounds. Aldehydes and Ketones

A number of organic compounds have π bonds between carbon and oxygen atoms. An important structural unit in such compounds is the carbonyl group, which contains a carbon-oxygen double bond,

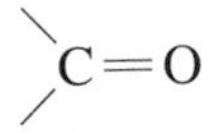

Compounds in which there is a carbonyl group are divided into different functional group classes depending on other groups or atoms that are bonded to the carbon atom of the carbonyl group. In **aldehydes,** for example, the functional group is the carbonyl group bonded to at least one hydrogen atom. The functional group of **ketones** is the carbonyl group bonded to two carbon atoms. The bonding in acetaldehyde, an aldehyde, and in acetone, a ketone, is shown in Figure 2.17.

The carbon atom of the carbonyl group is bonded to three other atoms, all lying in a plane, and therefore is a trigonal carbon atom. The bond angles of the carbonyl group are approximately 120°. The carbon atom of the carbonyl group is sp^2 hybridized. Its three hybrid orbitals form the skeleton of σ bonds for the carbonyl group. The σ bond between carbon and oxygen is formed by overlap of one of these hybrid orbitals with a *p* atomic orbital of oxygen.

Just as with ethylene, we can identify a $2p$ orbital on the carbon atom of the carbonyl group that is perpendicular to the plane defined by the three sp^2 hybrid orbitals. Overlap between this orbital and a $2p$ orbital of oxygen results in the π bond between carbon and oxygen. The π bond in a carbonyl compound has the

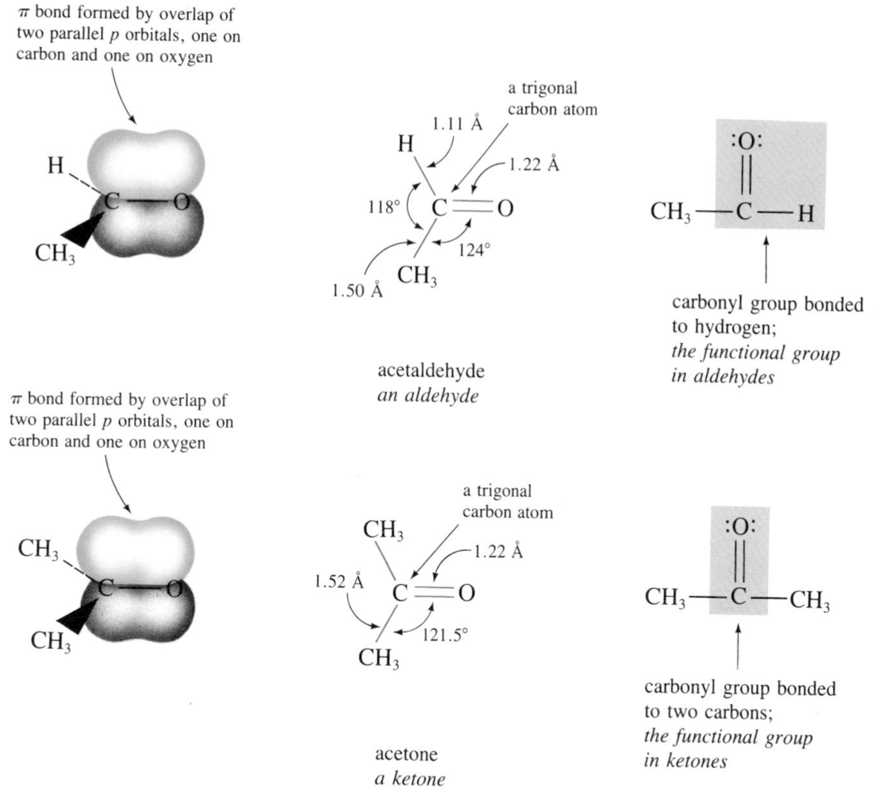
π bond formed by overlap of two parallel p orbitals, one on carbon and one on oxygen
H
C
O
CH3
a trigonal carbon atom
1.11 Å
1.22 Å
118°
124°
1.50 Å
:O:
CH3—C—H
carbonyl group bonded to hydrogen; the functional group in aldehydes
acetaldehyde
an aldehyde
π bond formed by overlap of two parallel p orbitals, one on carbon and one on oxygen
CH3
C
O
CH3
a trigonal carbon atom
1.22 Å
1.52 Å
121.5°
:O:
CH3—C—CH3
carbonyl group bonded to two carbons; the functional group in ketones
acetone
a ketone

FIGURE 2.17 Bonding in acetaldehyde and acetone.

same shape and symmetry as the π bond in ethylene. The nonbonding electrons on oxygen, on the basis of experimental evidence, are placed in the *s* and *p* orbitals of that atom.

PROBLEM 2.12

Construct the skeleton of σ bonds for acetaldehyde (Figure 2.17), showing the different orbitals that must overlap to form the bonds. Also show how the π bond is formed.

C. Carboxylic Acids and Esters

An oxygen atom attached by a single bond to the carbon atom of a carbonyl group is characteristic of two classes of organic compounds, carboxylic acids and esters. In **carboxylic acids,** the functional group is the **carboxyl group,** in which the carbonyl group is bonded to a hydroxyl group. In **esters,** the hydrogen atom on the hydroxyl group of a carboxylic acid has been replaced by a group containing carbon. In other words, both functional groups, carboxyl and ester, contain a carbonyl group bonded to an sp^3-hybridized oxygen atom. Figure 2.18 shows the structures of a carboxylic acid and an ester.

PROBLEM 2.13

Identify the hybridization of all the atoms and the origin of all the bonds in methyl acetate (Figure 2.18).

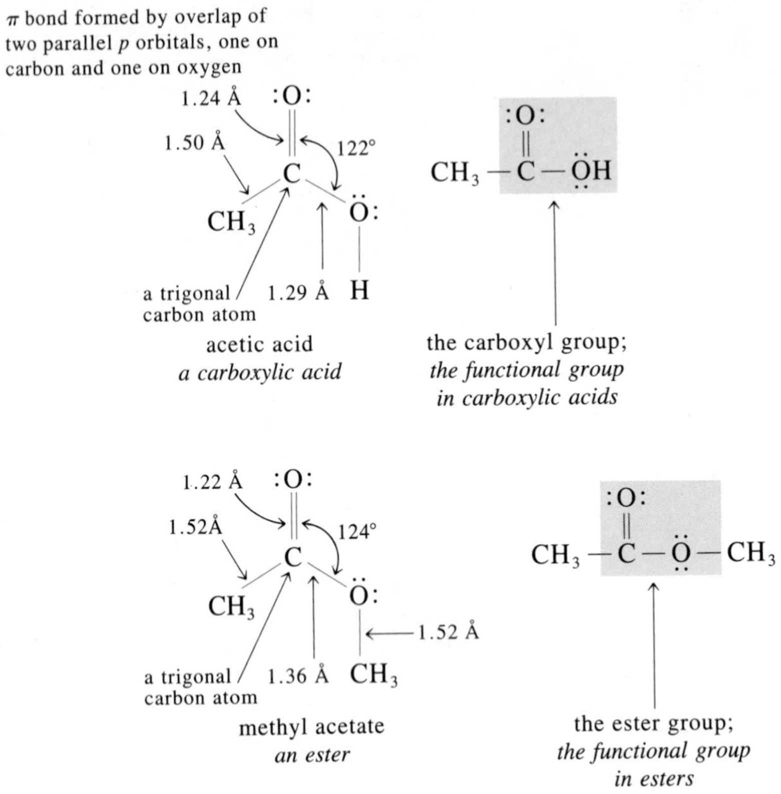
π bond formed by overlap of two parallel p orbitals, one on carbon and one on oxygen
1.24 Å
1.50 Å
122°
1.29 Å
a trigonal carbon atom
acetic acid
a carboxylic acid
the carboxyl group; the functional group in carboxylic acids
1.22 Å
1.52Å
124°
1.52 Å
1.36 Å
a trigonal carbon atom
methyl acetate
an ester
the ester group; the functional group in esters

FIGURE 2.18 Structures of acetic acid and methyl acetate.

PROBLEM 2.14

The functional group classes you have already learned to recognize are alkanes, alkenes, amines, alcohols, ethers, aldehydes, ketones, carboxylic acids, and esters. To which functional group class does each of the following compounds belong?

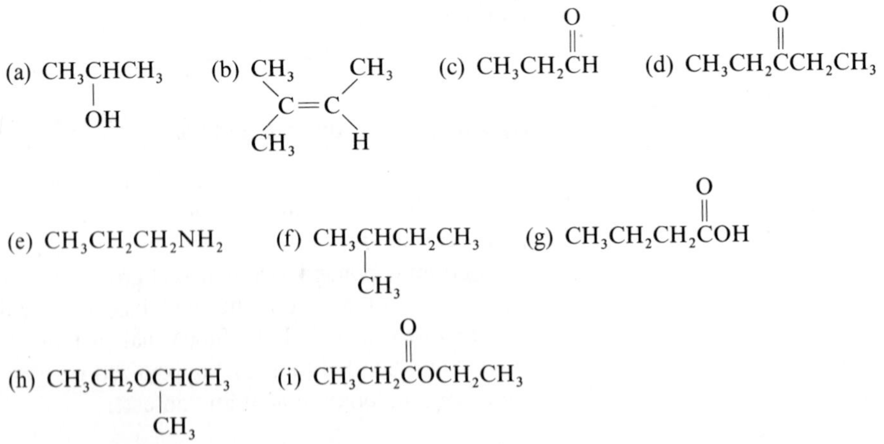
(a)
(b)
(c)
(d)
(e) $CH_3CH_2CH_2NH_2$
(f)
(g)
(h)
(i)

PROBLEM 2.15

Some compounds contain more than one functional group. Identify the functional groups that are present in the following naturally occurring compounds.

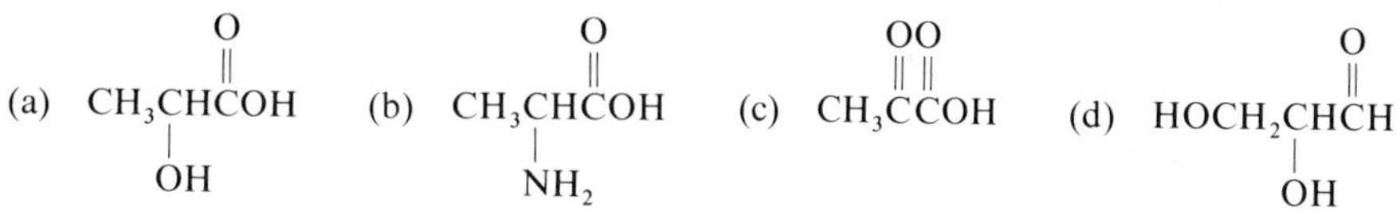

lactic acid — alanine — pyruvic acid — glyceraldehyde

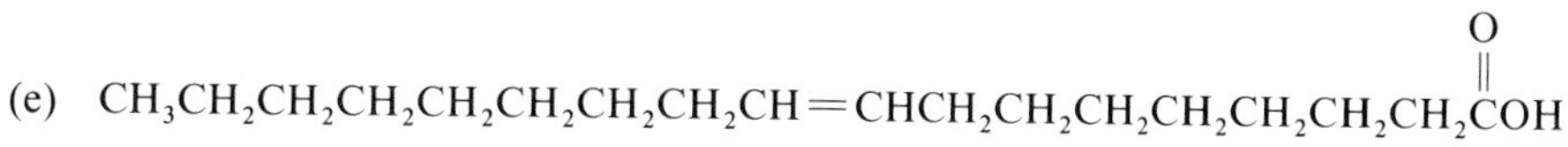

oleic acid

PROBLEM 2.16

What is the hybridization of the atoms shown in color in each of the following compounds?

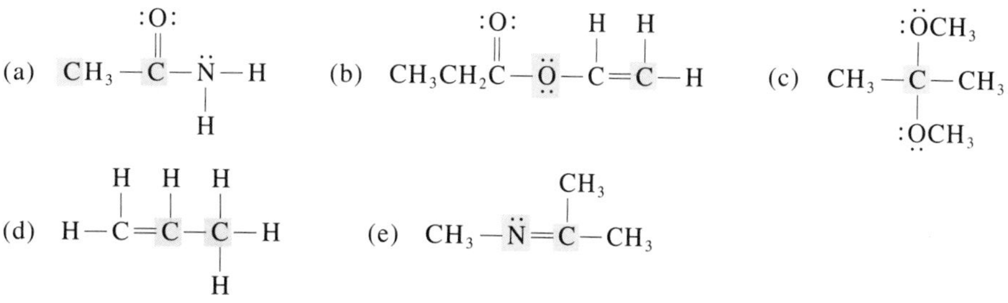

2.6 THE ORBITAL PICTURE FOR LINEAR MOLECULES

A. Bonding in Alkynes

Alkynes are a class of hydrocarbons in which the functional group is a carbon-carbon triple bond, —C≡C—. The simplest member of this class is acetylene, HC≡CH. Experimental measurements show that the four atoms in acetylene lie in a straight line with an H—C—C bond angle of 180° (p. 20). The molecule is linear.

Each carbon atom in acetylene is bonded to only two other atoms. Such a carbon atom is said to be *sp*-hybridized. Two ***sp* hybrid orbitals** are formed by a combination of one 2*s* orbital and one 2*p* orbital. These hybrid orbitals of carbon point away from each other along a straight line. The two carbon atoms of acetylene are joined by the overlap of one *sp* hybrid orbital from each atom. The carbon-hydrogen σ bond results from the overlap of the *sp* hybrid orbital of carbon with the 1*s* orbital of hydrogen (Figure 2.19).

Two of the valence electrons of each carbon atom in acetylene are used to form the σ bonds. Each carbon atom also has two unhybridized 2*p* orbitals at right angles to each other, with one electron in each. As is the case for ethylene, the *p* orbitals of the carbon atoms overlap to form bonding π molecular orbitals. Because there are two sets of *p* orbitals, two sets of π molecular orbitals are formed. Each bonding π molecular orbital has two electrons in it, giving rise to a π bond. Thus, the triple bond in acetylene (and other alkynes) consists of a σ bond and two π bonds. Two views of the π bonds are shown in Figure 2.20. A side view of the molecule shows that there is electron density above and below the line defined by the nuclei, as well as in front of and behind it. A view from the end of the molecule shows that there

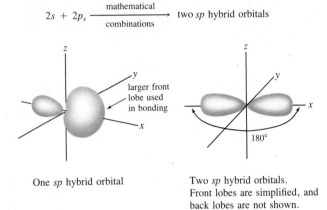

$$2s + 2p_x \xrightarrow[\text{combinations}]{\text{two mathematical}} \text{two } sp \text{ hybrid orbitals}$$

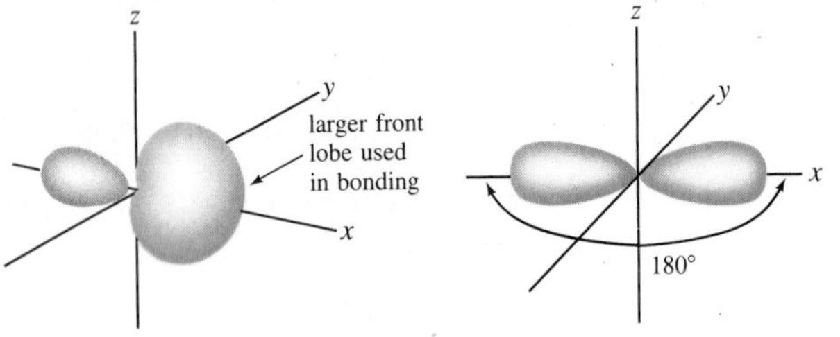
z
y
larger front lobe used in bonding
x
z
y
x
180°

One *sp* hybrid orbital

Two *sp* hybrid orbitals. Front lobes are simplified, and back lobes are not shown.

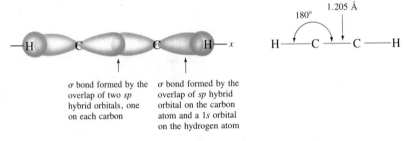

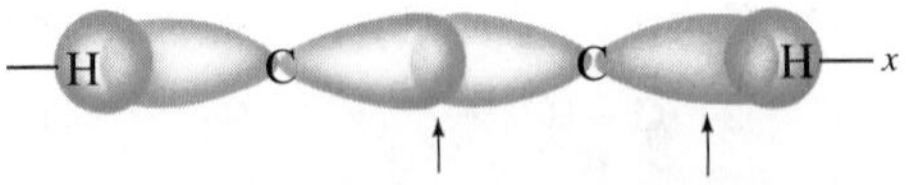
H
C
C
H
x

σ bond formed by the overlap of two *sp* hybrid orbitals, one on each carbon

σ bond formed by the overlap of *sp* hybrid orbital on the carbon atom and a 1*s* orbital on the hydrogen atom

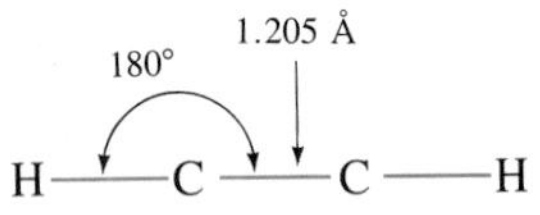
1.205 Å
180°
H
C
C
H

FIGURE 2.19 Orbital hybridization and the σ bond skeleton of acetylene.

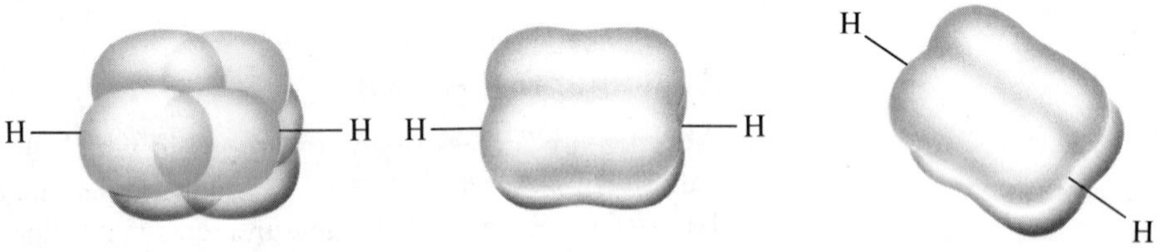
H
H
H
H
H
H

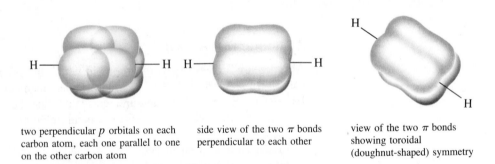

two perpendicular *p* orbitals on each carbon atom, each one parallel to one on the other carbon atom

side view of the two π bonds perpendicular to each other

view of the two π bonds showing toroidal (doughnut-shaped) symmetry

FIGURE 2.20 The π bonds in acetylene.

is electron density all around the axis of the molecule. There is a nodal axis along the line between the nuclei.

B. Bonding in Nitriles

Nitriles are organic compounds that contain a triple bond between a carbon and a nitrogen atom. The functional group in nitriles is the cyano group, —C≡N. The carbon atom of the cyano group is bonded to nitrogen and to one other atom. A typical nitrile is acetonitrile, $CH_3C{\equiv}N$ (Figure 2.21).

The carbon atom and the nitrogen atom in the cyano group are *sp*-hybridized and the carbon-nitrogen triple bond is formed in the same way that the carbon-carbon triple bond in acetylene is. The cyano triple bond consists of a σ bond, formed by the overlap of one *sp* hybrid orbital from carbon and one from nitrogen,

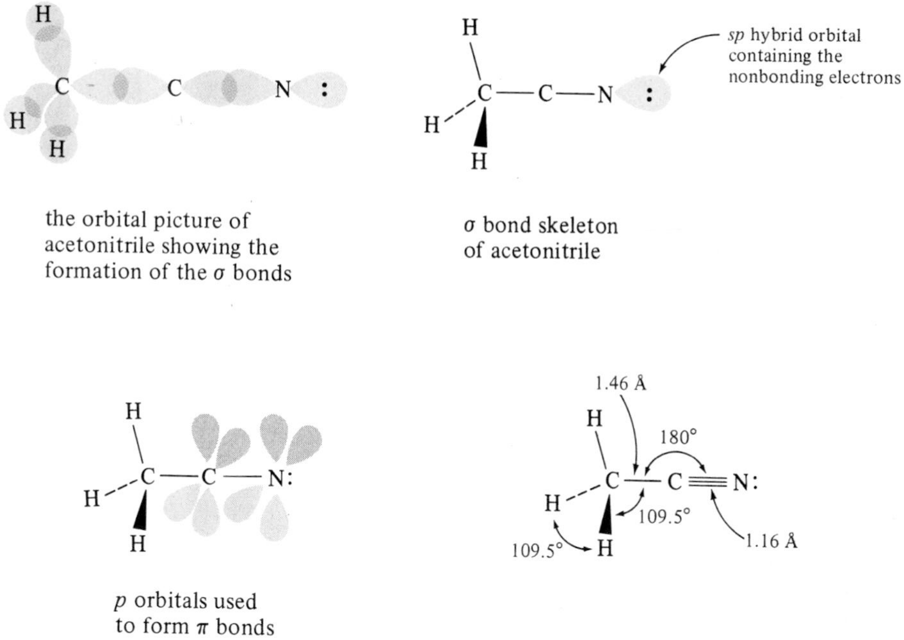
H
C
C
N :
H
H
the orbital picture of acetonitrile showing the formation of the σ bonds
H
C—C—N :
H
H
sp hybrid orbital containing the nonbonding electrons
σ bond skeleton of acetonitrile
H
C—C—N:
H
H
p orbitals used to form π bonds
1.46 Å
H
180°
C—C≡N:
H
109.5°
109.5°
H
1.16 Å

FIGURE 2.21 The orbital picture and structure of acetonitrile.

and two π bonds, formed by the overlap of two *p* orbitals on each atom. One of the *sp* orbitals of carbon overlaps with the sp^3 orbital of the tetrahedral carbon atom to which it is bonded. The nonbonding electrons of nitrogen occupy its second *sp* orbital.

Study Guide Concept Map 2.2

PROBLEM 2.17

Identify the functional group class to which each of the following compounds belongs.

(a) $CH_3CH_2C{\equiv}CH$ (b) $CH_3C{=}CHCH_2CH_2CH_3$ with CH_3 on C-2 (c) $CH_3CH_2\overset{O}{\overset{\|}{C}}CH_2CH_3$

(d) $CH_3CH_2CH_2\overset{O}{\overset{\|}{C}}OCH_3$ (e) $CH_3CH_2CH_2C{\equiv}N$ (f) $CH_3CH_2CH_2\overset{O}{\overset{\|}{C}}H$

(g) $CH_3CH_2CH\overset{O}{\overset{\|}{C}}OH$ with CH_3 on the CH (h) $CH_3CH_2CHCH_2CH_3$ with NH_2 on the CH

PROBLEM 2.18

How many σ and how many π bonds are present in each of the following molecules?

(a) $H{-}\underset{H}{\underset{|}{C}}{=}O$ (b) CH_3CH_2OH (c) $H{-}C{\equiv}C{-}C{\equiv}C{-}H$

(d) $H{-}\underset{H}{\underset{|}{C}}{=}\underset{H}{\underset{|}{C}}{-}C{\equiv}N$ (e) $CH_3{-}\overset{O}{\overset{\|}{C}}{-}OCH_3$ (f) $CH_3CH_2CH_3$

PROBLEM 2.19

Draw a detailed picture of the orbitals involved in the bonding in hydrogen cyanide, HCN.

PROBLEM 2.20

Draw a detailed picture of the orbitals involved in the bonding in propyne, $CH_3C{\equiv}CH$.

2.7 COVALENT BOND LENGTHS AND THEIR RELATION TO ORBITAL HYBRIDIZATION

A. Bond Lengths in Hydrocarbons

Bond lengths and bond angles for a number of typical covalent compounds are given in Chapter 1 and in the earlier sections of this chapter. The values are remarkably constant for each particular kind of bond and are determined not only by the atoms involved but also by the hybridization of orbitals of those atoms.

Electrons in 2*s* atomic orbitals are closer to the nucleus than electrons in 2*p* atomic orbitals are. Orbital hybridization affects bond length because a hybrid orbital with a greater percentage of *s* character does not extend as far from the nucleus as does a hybrid orbital with less *s* character. The percentage of *s* character is the ratio of the number of *s* orbitals to the total number of orbitals used in hybridization. For example, in ethane the sp^3-hybridized carbon atoms have one-fourth, or 25%, *s* character and three-fourths, or 75%, *p* character. For an *sp*-hybridized carbon atom, such as is found in acetylene, the hybrid orbitals have 50% *s* character.

The more *s* character a hybrid orbital has, the closer to the nucleus and the more tightly held are its electrons. An orbital with a large percentage of *s* character, therefore, forms a shorter σ bond than one with a small percentage of *s* character. The lengths of carbon-carbon and carbon-hydrogen bonds in propane, propene, and propyne are compared in Table 2.2.

The carbon-carbon single bond between sp^3-hybridized carbon atoms in propane is longer (1.53 Å) than the carbon-carbon single bond between an sp^3-hybridized carbon atom and an sp^2-hybridized carbon atom in propene (1.50 Å). The comparable bond between sp^3-hybridized and *sp*-hybridized carbon atoms in propyne is the shortest of all (1.46 Å). A similar trend is seen for carbon-hydrogen bonds as the hybridization of the carbon atom changes: sp^3, 1.10 Å; sp^2, 1.08 Å; and *sp*, 1.06 Å.

This series of compounds also demonstrates the effect of π bonding between two carbon atoms. The carbon-carbon double bond in propene is shorter (1.34 Å) than the carbon-carbon single bond in propane (1.53 Å). Propyne, with two π bonds between two of the carbon atoms, has the shortest carbon-carbon bond length of all (1.21 Å). Carbon atoms are held closer together by multiple bonds than by single bonds. The shorter bonds are also stronger bonds. This relationship is presented in detail in Section 2.8 (p. 59).

B. Orbital Hybridization and the Electronegativities of Carbon Atoms

Chemists have rationalized many experimental observations of the properties and reactivity of compounds containing carbon atoms with tetrahedral, trigonal, or linear geometries by suggesting that these carbon atoms have different electro-

TABLE 2.2 Orbital Hybridization and Bond Lengths in Propane, Propene, and Propyne

Compound	Bond	Bond Length, Å	Orbitals Used for Bonding
 propane	C—C	1.53	C, sp^3; C, sp^3
	C—H	1.10	C, sp^3; H, s
 propene	C—C	1.50	C, sp^3; C, sp^2
	C=C	1.34	C, sp^2; C, sp^2
	C—H	1.10	C, sp^3; H, s
	C—H	1.08	C, sp^2; H, s
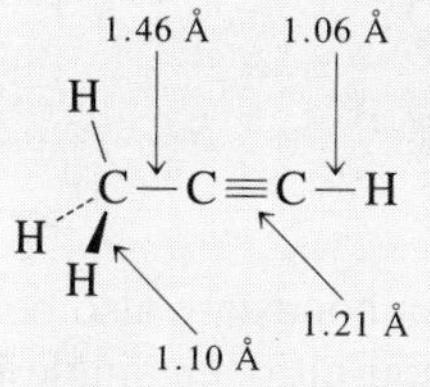 propyne	C—C	1.46	C, sp^3; C, sp
	C≡C	1.21	C, sp; C, sp
	C—H	1.10	C, sp^3; H, s
	C—H	1.06	C, sp; H, s

negativities. This argument is based directly on the differing percentages of *s* character in the hybrid orbitals used in bonding by different types of carbon atoms. An electron in an *s* orbital is, on the average, closer to the nucleus than an electron in a *p* orbital. The hybrid orbitals of the carbon atoms in acetylene, with 50% *s* character, hold the electrons of the bond more tightly to the nucleus of the carbon atom, and are thus more electronegative, than the carbon atoms in ethane, whose hybrid orbitals have only 25% *s* character. This kind of reasoning is used many times in this book to explain the different chemical properties of compounds composed of different types of carbon atoms.

C. Bond Lengths in Organic Compounds Containing Oxygen and Nitrogen

Like carbon-carbon bonds, carbon-oxygen and carbon-nitrogen bonds have lengths related to orbital hybridization and the presence of multiple bonds. Information about the bonding in some organic compounds containing nitrogen and oxygen is summarized in Table 2.3. Carbon-nitrogen bonds are shorter than comparable carbon-carbon bonds and longer than comparable carbon-oxygen bonds. These differences reflect the decrease in atomic size that occurs from left to right in the second row of the periodic table.

TABLE 2.3 Orbital Hybridization and Bond Lengths in Organic Compounds Containing Nitrogen and Oxygen

Compound	Bond	Bond Length, Å	Orbitals Used for Bonding
CH_3NH_2 methylamine	C—N	1.47	C, sp^3; N, sp^3
$(CH_3)_2C{=}NOH$ acetoxime	C=N	1.29	C, sp^2; N, sp^2
	C—C	1.49	C, sp^3; C, sp^2
$CH_3C{\equiv}N$ acetonitrile	C≡N	1.16	C, sp; N, sp
	C—C	1.46	C, sp^3; C, sp
CH_3OH methanol	C—O	1.43	C, sp^3; O, sp^3
	O—H	0.96	O, sp^3; H, s
CH_3OCH_3 dimethyl ether	C—O	1.42	C, sp^3; O, sp^3
$CH_3CH(=O)$ acetaldehyde	C=O	1.22	C, sp^2; O, sp^2
	C—C	1.50	C, sp^3; C, sp^2
$CH_3C(=O)OCH_3$ methyl acetate	C=O	1.22	C, sp^2; O, sp^2
	C—O	1.36	C, sp^2; O, sp^3
	O—C	1.46	C, sp^3; O, sp^3

The trends for carbon-carbon bonds are also observed for carbon-nitrogen and carbon-oxygen bonds. The carbon-nitrogen single bond in methylamine is longer than the carbon-nitrogen triple bond in acetonitrile. Similarly, a carbon-oxygen single bond in either methanol or dimethyl ether is longer than the carbon-oxygen double bond in acetaldehyde. Three different carbon-oxygen bond lengths are seen in methyl acetate. The shortest carbon-oxygen bond (1.22 Å) is the double bond of the carbonyl group. The next longer (1.36 Å) is the bond between the sp^2-hybridized carbon atom of the carbonyl group and the sp^3-hybridized oxygen atom. The longest bond (1.46 Å) is between the sp^3-hybridized oxygen atom and a tetrahedral carbon atom.

It is worthwhile to pay this much attention to the structural details of these few compounds because the types of bonding observed in them are representative of the bonding found in a wide range of organic compounds. The bond lengths for propane are typical of those in all compounds where two or more tetrahedral carbon atoms are bonded to each other. Whenever we see a double or a triple bond between carbon atoms, we can assume that it has properties similar to the bond in propene or propyne, respectively. Similarly, the properties of the functional groups in the simple amines, alcohols, ethers, and carbonyl compounds analyzed here in detail closely resemble the properties of those functional groups in more complex compounds.

PROBLEM 2.21

Using the data in Tables 2.2 and 2.3 as a guide, assign approximate lengths to the bonds indicated by arrows in the following structural formulas.

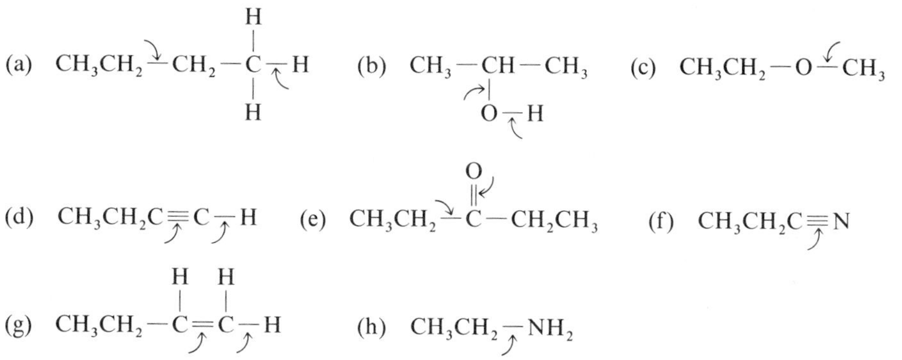

2.8 COVALENT BOND STRENGTHS

A. Bond Dissociation Energies

Bond strengths, as well as bond lengths, are related to orbital hybridization (p. 56). One measure of bond strength is bond dissociation energy. For example, it takes 104 kcal/mol to break the bond in a hydrogen molecule to get two hydrogen atoms.

$$\mathrm{H{-}H(g)} \longrightarrow 2\,\mathrm{H}\cdot\mathrm{(g)} \qquad \Delta H^\circ = 104\ \mathrm{kcal/mol}$$

This reaction is **endothermic;** energy is consumed in breaking the covalent bond. The enthalpy (the heat of reaction) of this endothermic reaction is called the bond dissociation energy, *DH*, for the hydrogen molecule.

The covalent bond in the hydrogen molecule is broken homolytically. In a **homolytic cleavage** of a bond, one electron of the covalent bond being broken goes to each fragment of the molecule. The **standard bond dissociation energy,** DH°, is defined as the change in enthalpy for a reaction in which one specific covalent bond in a molecule is broken homolytically while the reactants and the products are in the standard state (in the gas phase at 1 atm and 25 °C).

Bond dissociation energies have been measured for a variety of bonds involving carbon. Bond dissociation energies for some carbon-hydrogen bonds are shown in Table 2.4. The values for bond dissociation energies for carbon-hydrogen bonds in ethylene and acetylene show the extra strength of such bonds at sp^2- and sp-hybridized carbon atoms as compared to similar bonds at sp^3-hybridized carbon atoms. Also, bond dissociation energies may be correlated with bond lengths of carbon-hydrogen bonds in propane, propene, and propyne (p. 57).

$CH_3CH_2CH_2{-}H$	$CH_3CH{=}CH{-}H$	$CH_3C{\equiv}C{-}H$
carbon-hydrogen bond at sp^3-hybridized carbon atom; 1.10 Å	carbon-hydrogen bond at sp^2-hybridized carbon atom; 1.08 Å	carbon-hydrogen bond at sp-hybridized carbon atom; 1.06 Å
DH° = 98 kcal/mol	DH° = 108 kcal/mol	DH° = 125 kcal/mol
longest carbon-hydrogen bond; weakest bond		*shortest carbon-hydrogen bond; strongest bond*

Bond dissociation energies for some other types of bonds are also given in Table 2.4. The carbon-carbon triple bond in acetylene is shorter (1.21 Å) and

TABLE 2.4 Standard Bond Dissociation Energies for Representative Bonds

R—H	***DH*° (kcal/mol)**
CH_3—H	104
CH_3CH_2—H	98
$CH_3CH_2CH_2$—H	98
CH_3 | CH_3CH—H	95
CH_3 | CH_3CHCH_2—H	98
CH_3 | CH_3C—H | CH_3	91
CH_2=CH—H	108
HC≡C—H	125
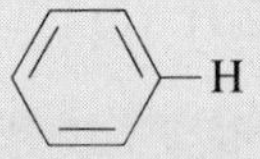	110
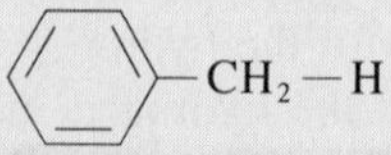	85
CH_2=$CHCH_2$—H	89
CH_3—CH_3	88
CH_2=CH_2	145
$(CH_3)_2C$=O	176
HC≡CH	190

stronger than the carbon-carbon double bond in ethylene (1.34 Å), which in turn is shorter and stronger than the carbon-carbon single bond in ethane (1.54 Å).

B. Average Bond Energies

By making a variety of experimental measurements on large numbers of organic and inorganic compounds, chemists have arrived at values for average bond energies for different types of covalent bonds. These bond energies differ from bond dissociation energies in that they do not refer to a specific bond in a particular molecule. Instead, they represent average values for types of bonds, taken from data for many different molecules. Average bond energies are useful in estimating

TABLE 2.5 Average Bond Energies (kcal/mol)

Single Bonds									
	H	**C**	**N**	**O**	**F**	**Cl**	**Br**	**I**	**Si**
H	104	99	93	111	135	103	87	71	76
C		83	73	86	116	81	68	52	72
N			39	53	65	46			
O				47	45	52	48	56	108
F					37				135
Cl						58			91
Br							46		74
I								36	56
Si									53

Multiple Bonds	
C=C 146	C≡C 200
C=N 147	C≡N 213
C=O (aldehydes) 176	
C=O (ketones) 179	

the changes in energy that occur in chemical reactions, which always involve the breaking and forming of bonds. Table 2.5 gives average bond energies for single bonds between the elements shown and for some multiple bonds.

Using average bond energies, we can predict whether a reaction is exothermic or endothermic, that is, whether it has a negative or positive enthalpy of reaction. We do this by calculating the energy that is consumed in breaking bonds and the energy that is released in forming new bonds in the reaction. If more energy is released than is consumed, the reaction is **exothermic.** The following problem illustrates how this is done.

Problem: Ethylene, an alkene, reacts with hydrogen to give ethane, an alkane. Calculate the enthalpy of the reaction, ΔH_r.

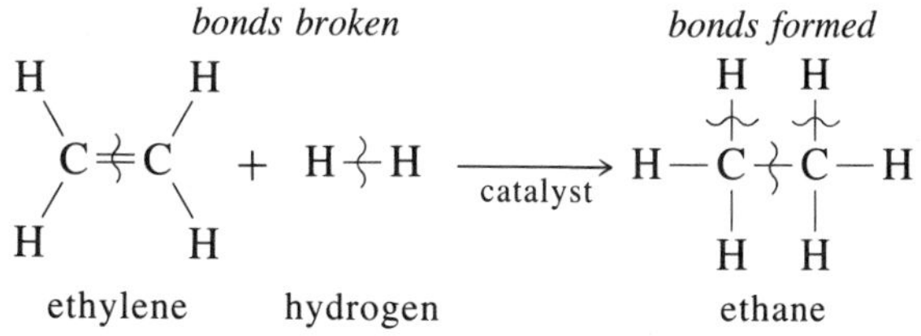
bonds broken
bonds formed
catalyst
ethylene
hydrogen
ethane

Solution: The carbon-carbon double bond and the hydrogen-hydrogen single bond are broken in the reaction, requiring an input of energy. Therefore, the enthalpy changes for the bonds being broken are given as positive numbers. The formation of a carbon-carbon single bond and two carbon-hydrogen single bonds releases energy; the enthalpy changes for the bonds being formed are negative numbers.

Bonds broken:		Bonds formed:	
C=C	+146 kcal/mol	C—C	−83 kcal/mol
H—H	+104 kcal/mol	2 × C—H	2 × −99 kcal/mol
	+250 kcal/mol		−281 kcal/mol

Enthalpy of reaction, ΔH_r = +250 kcal/mol − 281 kcal/mol = −31 kcal/mol

Study Guide Concept Map 2.3

The sum of the changes on both sides of the reaction equation indicates that more energy is released than is consumed, so the reaction is exothermic.

PROBLEM 2.22

Calculate ΔH_r for each of the following reactions using the average bond energies given in Table 2.5.

(a) $CH_3—H + Cl—Cl \longrightarrow CH_3—Cl + H—Cl$

(b) 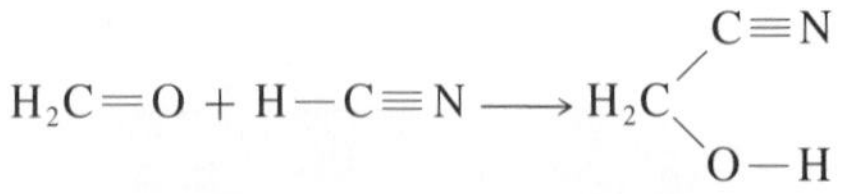

(c) 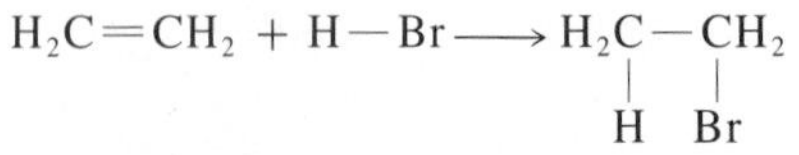

(d) 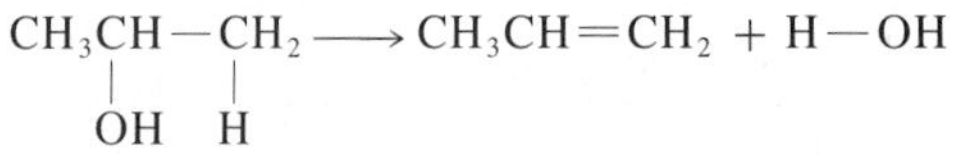

2.9 EFFECTS OF BONDING ON CHEMICAL REACTIVITY

Functional groups of organic compounds are the sites of chemical reactions. Among the hydrocarbons, ethane, a representative of the alkane family, and ethylene, a typical member of the alkene family, behave quite differently in the presence of bromine. When ethane, CH_3CH_3, is treated with bromine in carbon tetrachloride (a solvent) in the dark at room temperature, no reaction takes place. The solution remains red, the color of bromine.

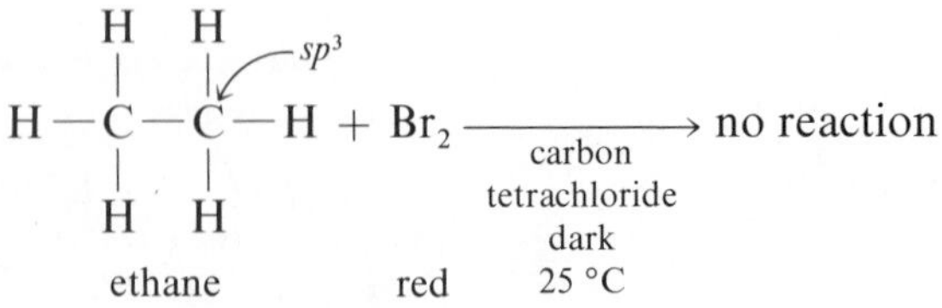
sp³
+ Br₂
no reaction
carbon
tetrachloride
dark
25 °C
ethane
red

an alkane

When ethylene, $H_2C{=}CH_2$, is treated with bromine under the same conditions, the red color disappears rapidly, indicating that a chemical transformation has taken place.

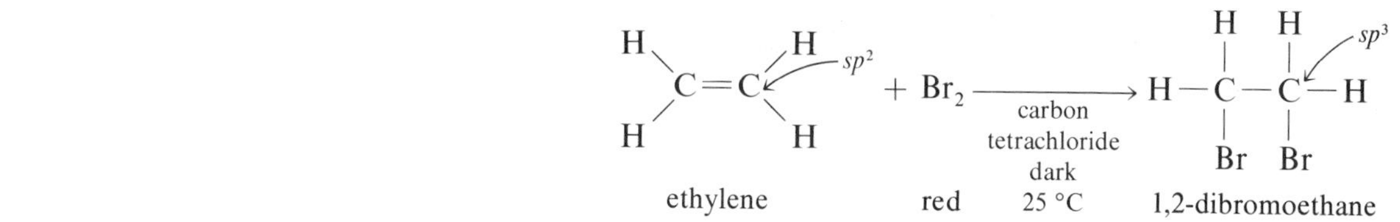
sp²
+ Br₂
carbon
tetrachloride
dark
25 °C
sp³

ethylene — red — 1,2-dibromoethane colorless

an alkene — *an alkyl halide*

an addition reaction

The product of the reaction of bromine with ethylene, 1,2-dibromoethane, is colorless, so the progress of the reaction is easily seen. The reaction of bromine with ethylene is an example of an addition reaction. In an **addition reaction,** the product contains all of the elements of the two reacting species. The reaction of an alkene with hydrogen to give an alkane (p. 61) is another example of an addition reaction.

In ethane, all of the bonding electrons are in σ bonds formed by sp^3-hybridized carbon atoms. Electrons in σ bonds are not readily involved in chemical reactions. The electrons in the π bond of ethylene, in contrast, react easily with bromine. The addition product 1,2-dibromoethane has no π bond, and the sp^2 carbon atoms of ethylene have become rehybridized into sp^3 carbon atoms in this product. In general, electrons in π bonds (called π electrons) are involved more readily in chemical reactions than are electrons in σ bonds. The π molecular orbitals are at a higher energy level than the σ molecular orbitals are. We can visualize π electrons as being more loosely held than are electrons in a σ bond, and thus more available to reagents that are seeking electrons.

Acetylene, HC≡CH, the simplest member of the alkyne family, also undergoes addition of bromine in the dark. Bromine adds successively to each of the two π bonds of the alkyne.

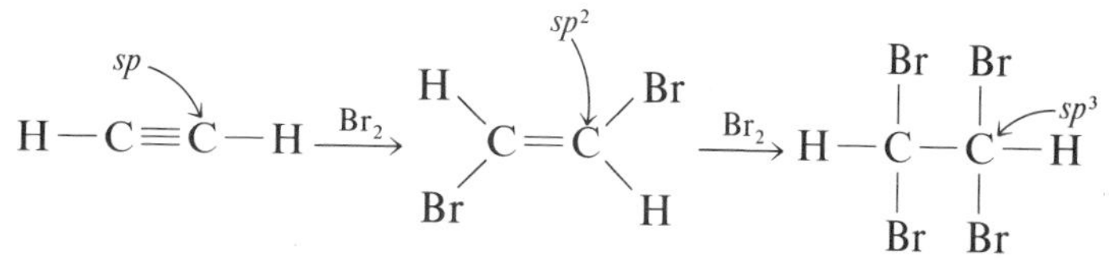
sp
sp²
sp³
Br₂
Br₂

acetylene — 1,2-dibromoethene — 1,1,2,2-tetrabromoethane

an alkyne — *an alkene* — *an alkyl halide*

addition reactions

At the end of the first stage of the reaction, the linear carbon atoms in acetylene have been converted to the trigonal carbon atoms of an alkene, 1,2-dibromoethene. The addition of another molecule of bromine to the π bond of this alkene converts it to 1,1,2,2-tetrabromoethane, which has only tetrahedral carbon atoms.

The addition of bromine is a reaction that is typical of many compounds having π electrons in carbon-carbon double or triple bonds. The final products formed by such addition reactions of bromine are members of a class of organic compounds called alkyl halides. **Alkyl halides** are compounds in which a halogen atom is bonded to a tetrahedral carbon atom. The details of the addition reactions of bromine to π bonds are presented in Section 8.6 (p. 289).

Not all hydrocarbons that contain π bonds react with bromine at room temperature in the dark. Benzene has the molecular formula C_6H_6, which indicates that the carbon atoms in benzene must be linked to each other by multiple bonds. Benzene is the simplest member of a class of hydrocarbons known as **aromatic**

hydrocarbons. These hydrocarbons are pictured as having ring structures in which double bonds alternate with single bonds. Yet benzene does not react with bromine under the conditions that ethylene does.

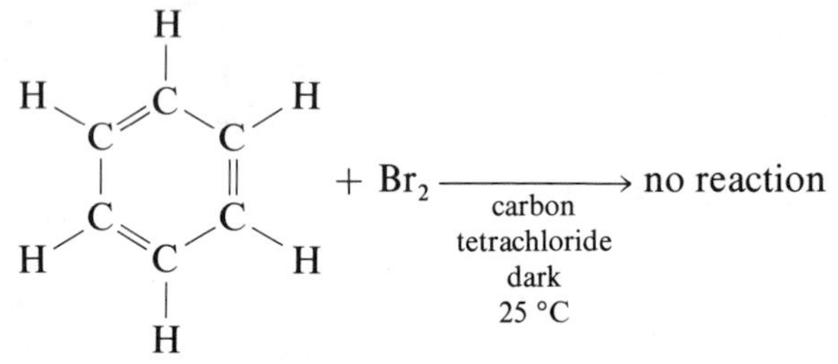
H
C
H
no reaction
carbon
tetrachloride
dark
25 °C

benzene
an aromatic hydrocarbon

How can this significant difference in reactivity between ethylene and benzene be explained? At first glance, we see that both molecules contain π bonds and π electrons, but the bonds of benzene must somehow be different from those of ethylene if the addition reaction with bromine does not occur. The lack of reactivity of benzene and other aromatic hydrocarbons in addition reactions is explored in detail in Chapter 19. The molecular orbital picture of benzene is related to ideas already presented; therefore, it can be examined now.

PROBLEM 2.23

Write equations showing the addition of bromine to the following compounds. How many moles of bromine will react with one mole of each compound before the reaction stops?

(a) $CH_3CH_2CH{=}CHCH_3$ (b) $CH_3C{\equiv}CCH_3$

(c) $HC{\equiv}CCH_2C{\equiv}CH$ (d) $CH_3CH{=}CHCH_2CH{=}CHCH_3$

PROBLEM 2.24

Different classes of compounds containing important functional groups have been introduced throughout this chapter. Review the chapter now, making a list of the different functional group classes and including the structural features that are typical of each one.

2.10 THE STRUCTURE OF BENZENE

The experimental technique of x-ray crystallography provides convincing evidence that the structure of benzene is symmetrical. The six carbon atoms lie at the corners of a regular hexagon. Each carbon atom is bonded to two other carbon atoms and a hydrogen atom, all of which lie in a plane. Each carbon-carbon bond in benzene is 1.39 Å long, and each carbon-hydrogen bond, 1.09 Å. All H—C—C and C—C—C bond angles are 120°, as expected for a regular hexagon. Benzene, in fact, is usually represented by a hexagon from which the symbols for the carbon and hydrogen atoms are omitted.

The carbon atoms in benzene are trigonal. They are bonded to three other atoms in a plane with bond angles of 120°. Such carbon atoms are sp^2-hybridized. Thus,

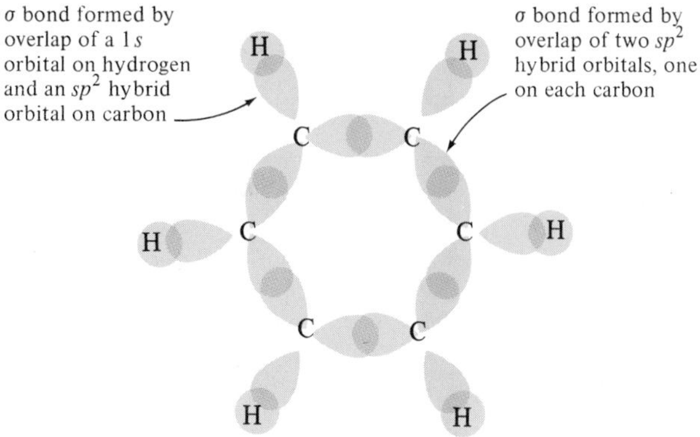
σ bond formed by overlap of a 1s orbital on hydrogen and an sp^2 hybrid orbital on carbon
σ bond formed by overlap of two sp^2 hybrid orbitals, one on each carbon
H
H
C
C
H
C
C
H
C
C
H
H

the orbital picture of the σ bonds in benzene

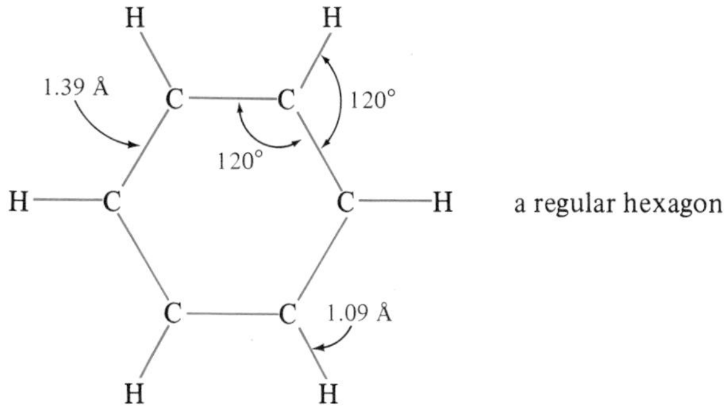
H
H
1.39 Å
C
C
120°
120°
H
C
C
H
a regular hexagon
C
C
1.09 Å
H
H

the experimentally determined bond lengths and bond angles of benzene

FIGURE 2.22 The σ bond skeleton of benzene.

the σ bonds in benzene result from the overlap of sp^2 hybrid orbitals between adjacent carbon atoms and the overlap of sp^2 orbitals of carbon atoms with 1*s* orbitals of hydrogens (Figure 2.22).

Three valence electrons from each carbon atom are involved in the σ bonding in benzene. Each carbon atom also has a 2*p* orbital, which has a lobe above and below the plane of the benzene ring, and an electron in that orbital. All of the six *p* orbitals are perpendicular to the plane of the ring, an orientation that allows for overlap of the orbitals to form π bonds between adjacent carbon atoms. The overlap of six *p* orbitals in benzene gives rise to six π molecular orbitals. Three of them are bonding molecular orbitals and three are antibonding molecular orbitals. The six electrons occupy the three bonding π orbitals of benzene (pictures and relative energies of these orbitals are shown on p. 774). The lowest-energy bonding molecular orbital in benzene is shown in Figure 2.23 on the next page. This molecular orbital has two circular lobes, one above and one below the plane of the ring. The π electrons in benzene are completely **delocalized** over the entire ring. Thus, the molecule is highly symmetrical with all carbon-carbon bonds of equal length (1.39 Å), longer than that of a double bond (1.34 Å) but shorter than that of a single bond (1.53 Å).

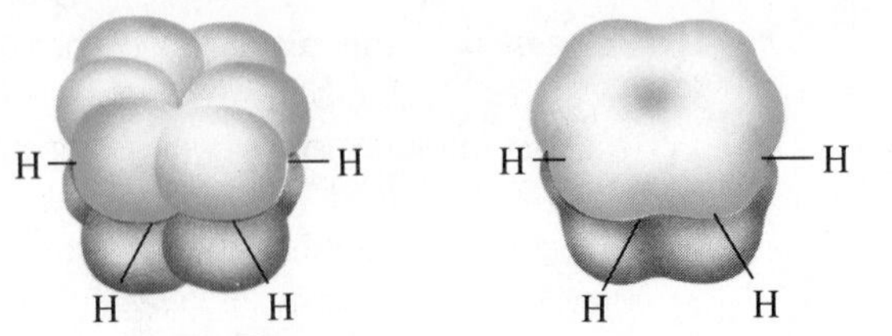
H
H
H
H
H
H
H
H

side view showing the six parallel p orbitals in phase

lowest energy bonding π molecular orbital showing delocalization

FIGURE 2.23 Molecular orbital picture of the π bonds in benzene.

The structural formula for benzene is written with three double bonds in the ring.

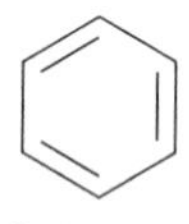

benzene

Such a representation is called the Kekulé formula for benzene, after the German chemist August Kekulé, who first proposed that structure for benzene in 1865. Even that long ago, however, Kekulé recognized that a structure that depicts benzene with single and double bonds is not a satisfactory representation of the properties of the compound. He suggested that the structure of benzene should be represented by two formulas in which the single and double bonds have different positions.

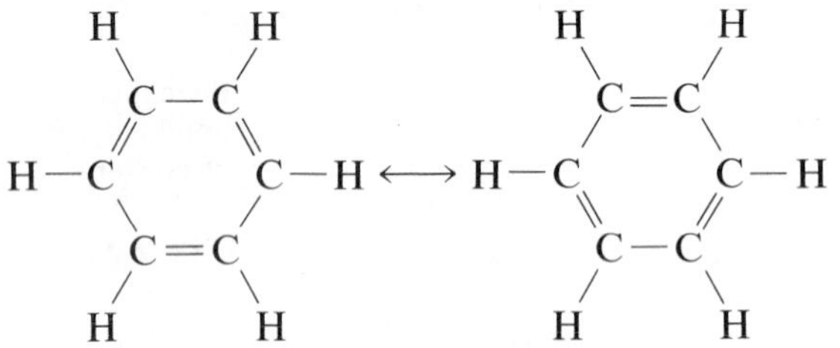

Kekulé formulas for benzene

Kekulé thought that these structural formulas corresponded to different species that were in dynamic equilibrium with each other. It is now recognized that structures that differ from each other only in the location of electrons cannot be distinguished. Such structures are called resonance contributors (p. 11) to the true structure of the compound, which has properties best represented by all of the resonance contributors taken together. The correct structure for the compound is said to be a resonance hybrid of the various resonance contributors that can be written for it.

Each Kekulé formula for benzene suggests that benzene has three double bonds and three single bonds in the six-membered ring. Yet we saw in Section 2.9 that benzene does not react with bromine in the same way as do compounds such as ethylene that have normal carbon-carbon double bonds. Thus, neither Kekulé formula is an accurate picture of the real benzene molecule. The resonance hybrid, the two Kekulé formulas taken together, suggests that the double bonds in benzene are not fixed in position and that the π electrons are delocalized over the entire ring.

As a result, the reactivity of benzene is unlike that of compounds containing localized double bonds.

Aromatic hydrocarbons represent a large class of organic compounds the chemistry of which is both interesting and often different from that of alkanes and alkenes. The chemistry of such compounds is described in detail in Chapters 19 and 23, but aromatic rings appear in many compounds in all parts of the book. Recognition of the special stability of this type of hydrocarbon is therefore important.

PROBLEM 2.25

Before the structure of benzene was finally determined experimentally, one of the many structures that had been proposed for the hydrocarbon with the molecular formula C_6H_6 was the following, called Dewar benzene:

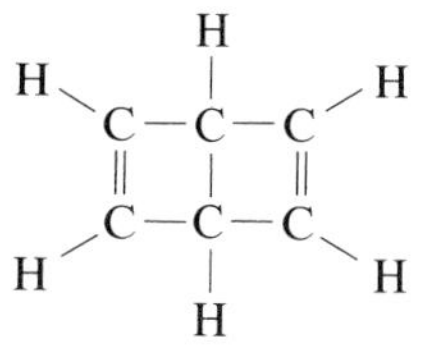

Dewar benzene

What would be the hybridization of each carbon atom in the structure shown? How would you expect this compound to behave if a solution of bromine were added to it?

S U M M A R Y

When two atomic orbitals interact, two molecular orbitals result. In-phase interactions give bonding molecular orbitals, whereas out-of-phase interactions give antibonding molecular orbitals. A covalent bond results when a bonding molecular orbital is occupied by a pair of electrons.

Both σ orbitals and σ bonds have cylindrical symmetry around the axis connecting the two nuclei. Sigma bonds can arise from the overlap of *s* atomic orbitals either with other *s* orbitals or with *p* orbitals or from the overlap of two *p* orbitals. Both π orbitals and π bonds have a nodal plane and arise from the sideways overlap of *p* orbitals.

Organic chemists find the concept of hybrid orbitals useful for rationalizing the different shapes of covalent molecules and the different kinds of bonds found in them. Hybrid orbitals are created by mathematical combinations of the wave functions for *s* and *p* orbitals. When carbon is bonded by single bonds to four other atoms, it is said to be sp^3-hybridized. The sp^3 hybrid orbitals are directed to the corners of a tetrahedron, with angles of 109° between them. Such a carbon is also called a tetrahedral carbon atom.

When a central atom is bonded to three other atoms by single or multiple bonds, it is said to be an sp^2-hybridized atom. The sp^2 hybrid orbitals are directed to the corners of an equilateral triangle, with angles of 120° between them. Such an atom is also called a trigonal atom.

A central atom bonded to two other atoms by single or multiple bonds is an *sp*-hybridized atom. The *sp* hybrid orbitals of an atom point away from each other at an angle of 180°. The portion of a molecule containing *sp* hybrid orbitals is linear.

In the orbital hybridization picture of bonding, σ bonds are formed by the overlap of hybrid orbitals with other hybrid orbitals or with *s* or *p* atomic orbitals.

A carbon atom that is sp^2-hybridized has an unhybridized *p* orbital, which can form a π bond by overlapping side to side with a *p* orbital of an adjacent atom (carbon, oxygen, or nitrogen) to which the carbon is joined by a σ bond. The π bond and σ bond together constitute a double bond. An *sp*-hybridized carbon atom has two unhybridized *p* orbitals and forms a triple bond, consisting of a σ and two π bonds, with another carbon or a nitrogen atom.

The strength of a covalent bond is related to its length, which in turn is related to the sizes of the atoms bonded, their hybridization, and whether the bond is a single, double, or triple bond. The shorter a bond, the stronger it is.

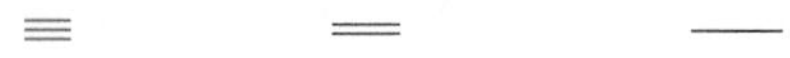

shortest bond *strongest* — longest bond *weakest*

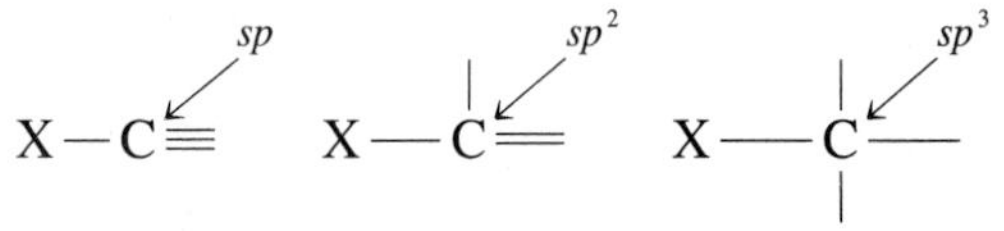
sp
sp²
sp³
X—C
X—C
X—C

shortest bond *strongest* — longest bond *weakest*

Bond strengths may be expressed as bond dissociation energies, *DH* (the energy necessary to break a bond so that one electron of the covalent bond goes to each fragment of the molecule) or as average bond energies (determined from many experimental measurements to be the average covalent bond energy for a variety of typical molecules). Average bond energies are useful for estimating enthalpy changes in reactions.

Carbon atoms bond among themselves and to atoms of other elements with single, double, and triple bonds, giving rise to a variety of functional groups. A functional group is a structural unit consisting of an atom or group of atoms that serves as a site of reactivity in a molecule. Organic compounds that contain the same functional group are classified together into families of compounds. The names of the functional group classes met so far and the structure of the functional group in each are given in Table 2.6. (Table 2.6 also appears in abridged form inside the cover of this book.)

TABLE 2.6 Functional Groups

Functional Group Class	Functional Group	Structural Unit
Alkanes	carbon-carbon and carbon-hydrogen single bonds	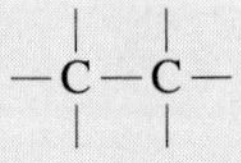 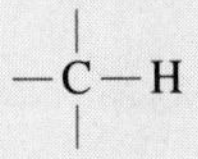
Alkenes	carbon-carbon double bond	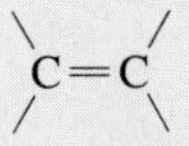

(Continued)

TABLE 2.6 *(Continued)*

Functional Group Class	Functional Group	Structural Unit
Alkynes	carbon-carbon triple bond	−C≡C−
Alkyl halides	carbon-halogen bond	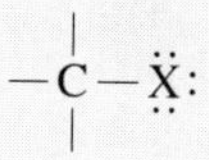 where X = F, Cl, Br, or I
Aromatic hydrocarbons	ring with six delocalized electrons	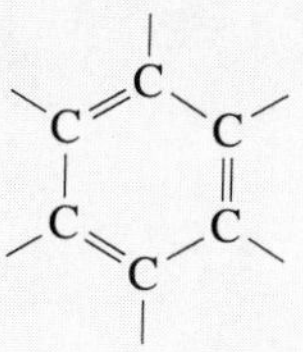
Alcohols	hydroxyl group	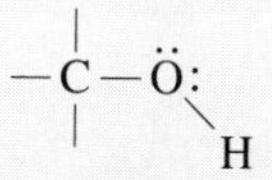
Ethers	oxygen atom bonded to two carbon atoms	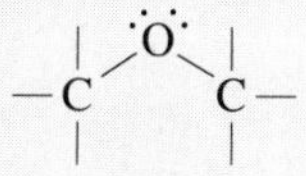
Amines	nitrogen atom	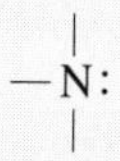 (*not* bonded to a carbonyl group)
Aldehydes	carbonyl group bonded to at least one hydrogen atom	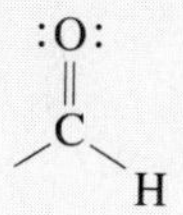
Ketones	carbonyl group bonded to two carbon atoms	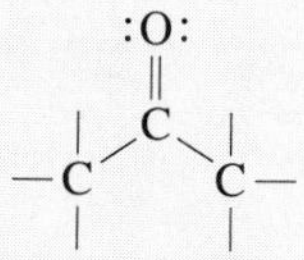
Acids	carboxyl group	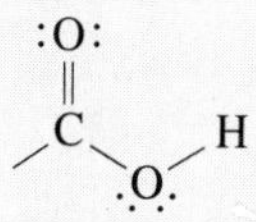
Esters	carboxyl group with carbon replacing the hydrogen atom	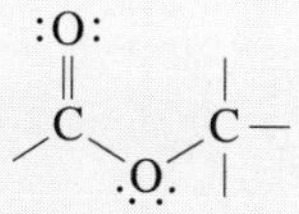
Nitriles	cyano group	−C≡N:

Where no atom is indicated at the end of a bond, that position may be taken by a carbon or a hydrogen.

ADDITIONAL PROBLEMS

2.26 Draw Lewis structures for each of the following species and indicate the hybridization and geometry of the central atom.

(a) BeH_2 (b) CO_2 (c) CH_3^+ (d) CF_4 (e) Cl_2CO (f) CCl_3^-
(g) NF_3 (h) BF_4^-

2.27 How many σ bonds and π bonds are present in each of the following?

(a) $CH_2{=}C{=}CHCH_3$ (b) $CH_2{=}C{=}O$ (c) $CH_3OCH{=}CHCH_3$
(d) $CH_2{=}NOH$

2.28 Indicate the hybridization of all of the carbon and oxygen atoms in each of the following.

(a) $CH_2{=}CHCH_2OH$ (b) 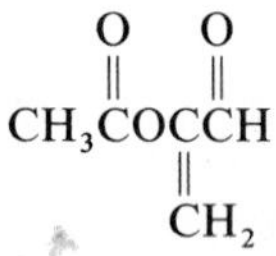 (c) $CH_3CH_2CH_3$

(d) $CH_3C{\equiv}CCH{=}CHCH_3$ (e) 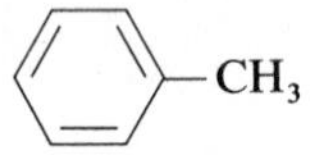

2.29 Rank each of the following groups of compounds according to lengths of the indicated bonds.

(a) 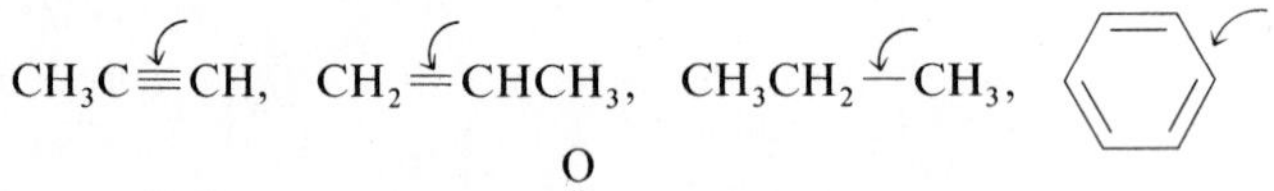

(b) 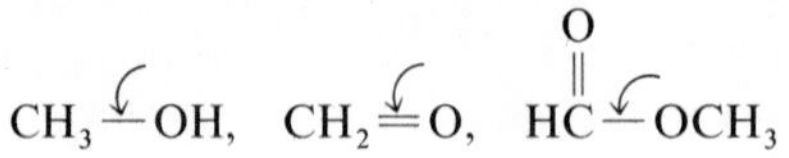

(c) 

2.30 Circle and name all the functional groups in each of the following molecules.

(a) 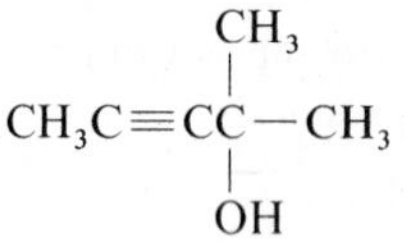 (b) 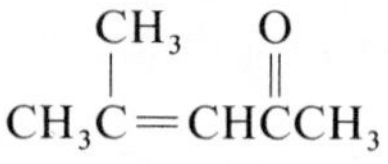 (c) 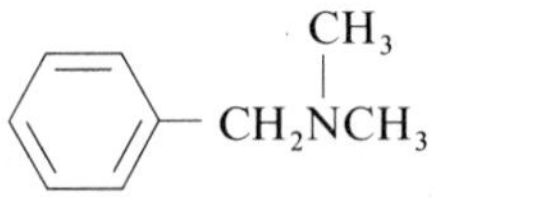

(d) 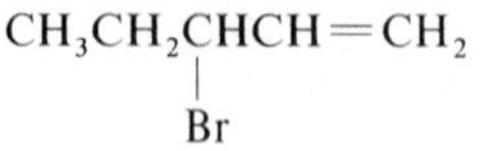 (e) 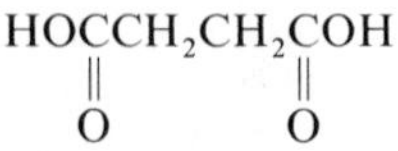 (f) 

(g) 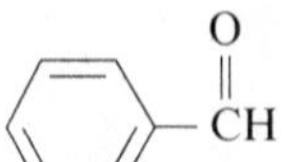 (h) 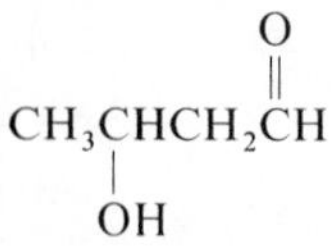 (i) 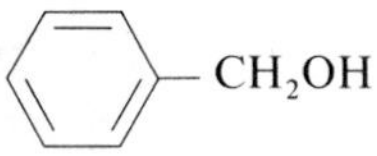

(j) $CH_3CH_2CH_2C{\equiv}N$

2.31 Interesting compounds isolated from nature often have more than one functional group. Identify the functional groups found in each of the natural products shown on the following page.

(a)

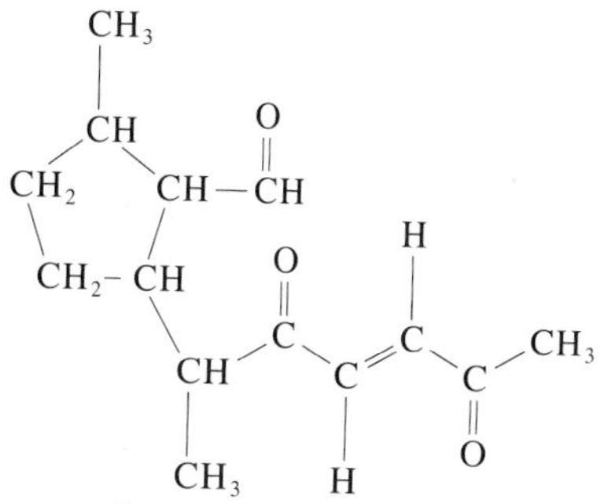

a defensive toxin produced by the whirligig beetle

(b)

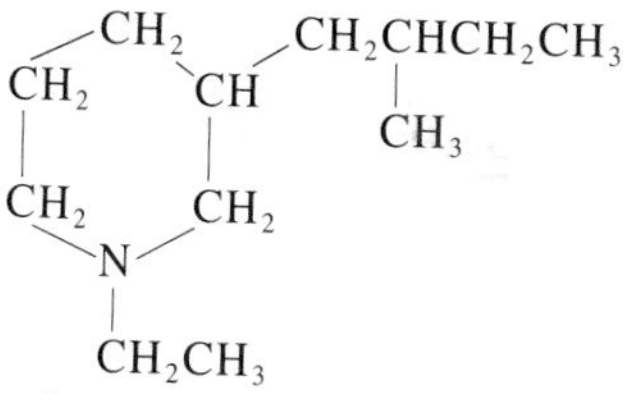

a substance produced by a beetle to help it float on water

(c)

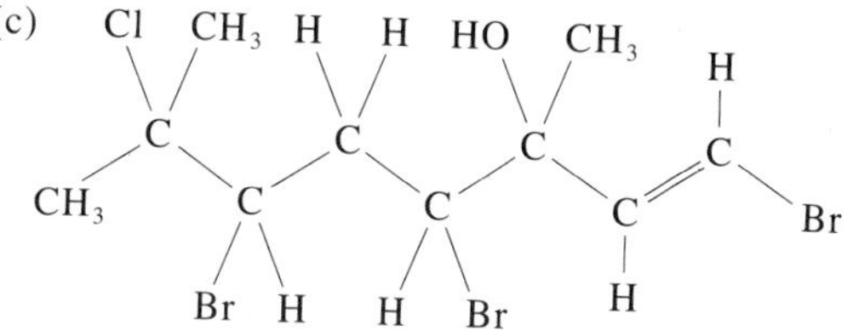

a metabolite of a seaweed that has strong antimicrobial activity

(d)

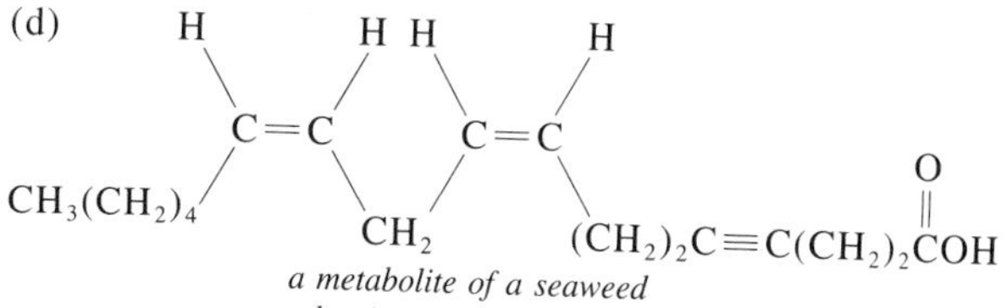

a metabolite of a seaweed that is poisonous to fish

(e)

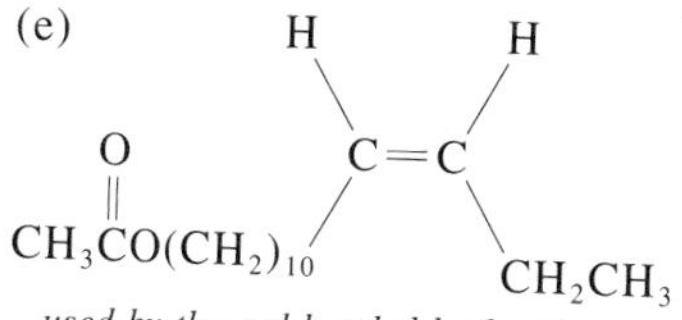

used by the red-banded leaf roller to communicate with others of its species

2.32 For each of the following compounds, assign hybridization to each carbon, oxygen, and nitrogen atom. Show which types of orbitals overlap to create the different kinds of bonds in the molecules.

(a) $CH_2{=}O$ (b) CH_3NHOH (c) $CH_3CH{=}CHCH_3$ (d) $CH_3C{\equiv}CCH_3$

2.33 The formation of a bond between nonbonding electrons of nitrogen and oxygen atoms and the empty 1*s* orbital of a proton was shown in Figures 2.10 and 2.12. Referring to those figures, predict, by writing an equation, what would happen if each of the following compounds were put into sulfuric acid.

(a) $CH_3C{\equiv}N{:}$ (b) $CH_3\overset{\overset{:O:}{\|}}{C}CH_3$ (c) $CH_3\overset{\overset{CH_3}{|}}{C}{=}\ddot{N}CH_3$ (d) $CH_3\overset{\overset{:O:}{\|}}{C}\ddot{O}CH_3$

2.34 Predict, by completing the equations, what would happen if the following reagents were mixed. (A review of Sections 2.9 and 2.10 might be helpful.)

(a) $CH_3CH{=}CHCH_3 + Br_2 \xrightarrow[\text{carbon tetrachloride, dark, 25 °C}]{}$

(b) $CH_3CH_2CH{=}CH_2 + H_2 \xrightarrow[\text{catalyst}]{}$

(c) $CH_3CH_2CH_2CH_3 + Br_2 \xrightarrow[\text{carbon tetrachloride, dark, 25 °C}]{}$

(d) 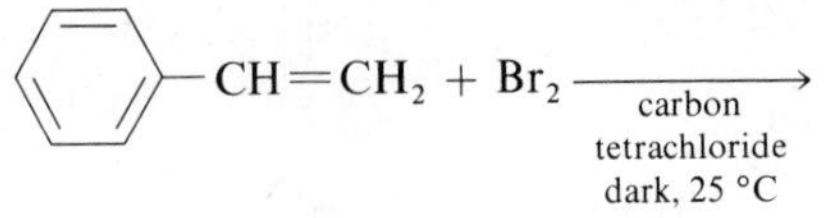

(e) $CH_3C{\equiv}CH + Br_2$ (excess) $\xrightarrow[\text{carbon tetrachloride, dark, 25 °C}]{}$

(f) $CH_3C{\equiv}CH + H_2$ (excess) $\xrightarrow[\text{catalyst}]{}$

2.35 Calculate ΔH_r for each of the following reactions using the average bond energies given in Table 2.5 (p. 61).

(a) $H_2C{=}CH_2 + Br{-}Br \longrightarrow H_2C(Br){-}CH_2(Br)$

(b) $CH_3{-}C(CH_3)_2{-}OH + H{-}Cl \longrightarrow CH_3{-}C(CH_3)_2{-}Cl + H{-}OH$

(c) $CH_3C({=}O)CH_3 + H{-}OH \longrightarrow CH_3C(O{-}H)(OH)CH_3$

2.36 Isomeric compounds with the molecular formula C_3H_6O are shown on p. 17. Analyze the structural formulas to determine what functional groups are present. Assign the compounds to functional group classes.

2.37 Allene has the structural formula shown below. What is the hybridization of each of the carbon atoms? What are the orbitals involved in the formation of the σ bond skeleton and of the π bonds in this molecule? Illustrate your answers with drawings. (Hint: Figures 2.16 and 2.20 may be helpful.)

$$CH_2{=}C{=}CH_2$$

allene

2.38 Methylamine (Figure 2.11) forms a compound with boron trifluoride, just as ammonia does (p. 8). Write an equation for the reaction. Write a Lewis structure for the product and show any formal charges that are present. What are the hybridization and geometry of the nitrogen atom and the boron atom in the product? How would you show the bonding present in this system in a three-dimensional picture? (Hint: Another look at the representations of the structure of ethane in Figure 2.9 may be helpful.)

2.39

(a) Sections 2.5 and 2.6 put forth the ideas that a carbon atom that is trigonal planar is sp^2-hybridized and that a carbon atom that is bonded to two other groups in a linear molecule is sp-hybridized. Similar reasoning can be applied to other elements. For example, water can also be viewed as a planar triangular molecule. What would the hybridization of the oxygen atom be in this model of water? Which orbitals would the two pairs of nonbonding electrons occupy?

(b) Photoelectron emission spectroscopy, a form of spectroscopy that measures how easily electrons can be removed from various energy levels in a molecule, was used in an experiment that indicated that the two pairs of nonbonding electrons in water occupy different energy levels. Which model for the water molecule better fits this experimental fact: the one in which the oxygen atom is sp^3-hybridized or the one in which it is sp^2-hybridized?

(c) Would any other model for the bonding in water account for the observed experimental fact?

(d) Lithium hydroxide, LiOH, is a linear covalent molecule. Propose an orbital picture of the bonding in lithium hydroxide. What kind of hybridization does the oxygen atom have here?

3

Reactions of Organic Compounds as Acids and Bases

A • L O O K • A H E A D

The first step in most chemical reactions is the interaction of a pair of nonbonding electrons in one molecule or ion with a center of electron deficiency (which may be a vacant orbital or a partial positive charge) at an atom in another species. For example, the nonbonding electrons of a molecule of ammonia are donated to a vacant orbital of boron trifluoride (p. 8) to give a new covalent bond. The movement of the pair of electrons from the nitrogen to the boron atom is symbolized by a curved arrow.

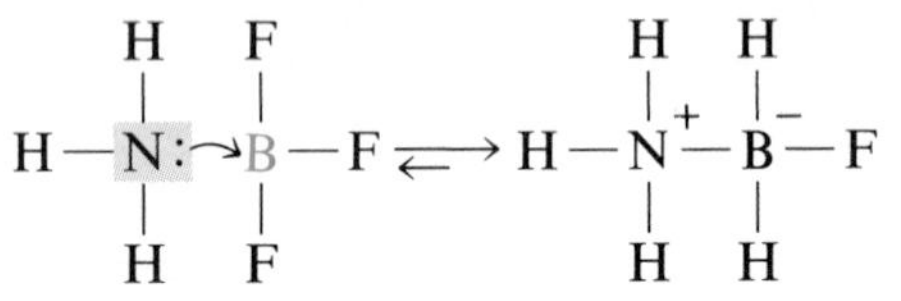

curved arrow symbolizing the donation of a pair of electrons from the nitrogen atom to the boron atom

An understanding of the generality of this way of looking at organic reactions is so fundamental that we will explore it in this chapter using acid-base reactions as examples. In Chapter 1, you acquired some of the skills needed to predict the reactivity of a wide variety of compounds. You developed these skills by locating the nonbonding electrons on Lewis structures, by writing resonance contributors, and by making decisions about the polarities of covalent bonds. The relationship between the structure of a molecule and its chemical reactivity was further explored in Chapter 2. For example, oxygen and nitrogen atoms bearing nonbonding electrons were identified as reactive sites toward acids such as sulfuric acid. Your goal in this chapter is to learn to see the sites of electron density and electron deficiency in the structures of compounds. These sites make the molecules vulnerable to a variety of chemical reagents. A species with electrons to donate is a base. An electron-deficient species that accepts an electron pair is an acid.

You should not attempt, at this point, to memorize structures or names of compounds; you will become familiar with them gradually as you study and write reactions. In this chapter, you should concentrate on learning to predict certain kinds of reactivity from the structural features that are apparent in the formulas and on depicting reactions correctly using the curved-arrow notation.

3.1 BRØNSTED AND LEWIS ACIDS AND BASES

A. Brønsted-Lowry Theory of Acids and Bases

What are acids and bases? Two typical acid-base reactions are shown below.

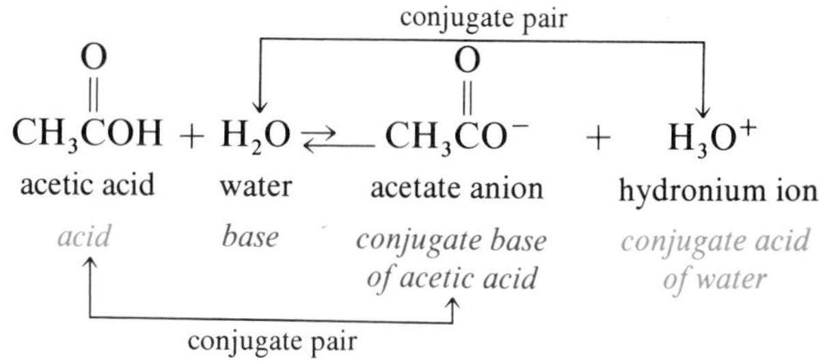
conjugate pair
CH₃COH + H₂O ⇄ CH₃CO⁻ + H₃O⁺
acetic acid
water
acetate anion
hydronium ion
acid
base
conjugate base of acetic acid
conjugate acid of water
conjugate pair

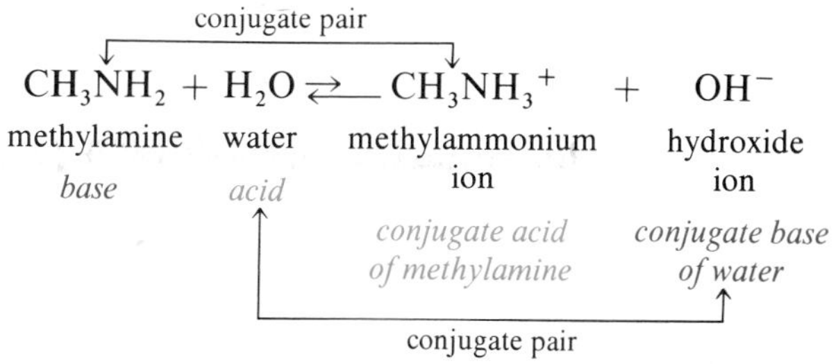
conjugate pair
CH₃NH₂ + H₂O ⇄ CH₃NH₃⁺ + OH⁻
methylamine
water
methylammonium ion
hydroxide ion
base
acid
conjugate acid of methylamine
conjugate base of water
conjugate pair

In each reaction, a proton is transferred from one to the other of the molecules on the left-hand side of the equation. In the **Brønsted-Lowry theory** of acids and bases, an **acid** is defined as a **proton donor** and a **base** as a **proton acceptor.** In the first equation, therefore, acetic acid is the acid (the proton donor), and water is the base (the proton acceptor). Acetate ion, which is formed by the loss of a proton from acetic acid, is a base. It can accept a proton to become acetic acid again. A pair of species that can be interconverted by the loss and gain of a proton are called a **conjugate acid** and **conjugate base.** Brønsted acids are also called **protic acids,** acids that react via the transfer of a proton. Note that water can be an acid or a base. It can gain a proton to become a hydronium ion, H_3O^+, its conjugate acid, or lose a proton to become the hydroxide ion, OH^-, its conjugate base.

Study Guide Concept Map 3.1

PROBLEM 3.1

Identify the conjugate acid and conjugate base in each of the following pairs. Draw Lewis structures for each species and locate any formal charge.

(a) $(CH_3)_2O$, $(CH_3)_2OH^+$ (b) H_2SO_4, HSO_4^- (c) NH_2^-, NH_3
(d) $CH_3OH_2^+$, CH_3OH (e) H_2CO, H_2COH^+ (f) CH_3OH, CH_3O^-

It is not possible to define the acidity of a substance in absolute terms. Acidity and basicity are always described in terms of equilibria. A review of the quantitative treatment of acid-base equilibria is given later in this chapter. First, we will concentrate on achieving a qualitative understanding of the phenomena.

A useful way to picture acid-base reactions is to imagine two bases competing for the same proton. For example, in the reaction of acetic acid with water (p. 75), acetate anion and water are the bases in competition for the proton. Their relative strengths as bases determine where the equilibrium lies. Methylamine and hydroxide ion are the bases that are competing for the proton in an equilibrium process when methylamine reacts with water (p. 75). An amine (p. 45) has basic properties similar to those of ammonia and reacts with acids to give a substituted ammonium ion.

If two bases have quite different basicities, the transfer of a proton from one to the other may be virtually complete. The reaction of hydrogen chloride with dimethylamine, for example, is not written as an equilibrium.

conjugate pair

$$(CH_3)_2NH \; + \; HCl \longrightarrow (CH_3)_2NH_2^+ \; + \; Cl^-$$

$(CH_3)_2NH$	HCl	$(CH_3)_2NH_2^+$	Cl^-
dimethylamine	hydrogen chloride	dimethylammonium ion	chloride ion
base	*acid*	*conjugate acid of dimethylamine*	*conjugate base of hydrogen chloride*

conjugate pair

Chloride ion, the conjugate base of the strong acid hydrogen chloride, is a weak base and does not compete effectively with dimethylamine for the proton. **Strong acids** have weak conjugate bases, **weak acids** have strong conjugate bases.

B. Lewis Theory of Acids and Bases

The proton acts as an acid because it has an empty 1*s* orbital that can accept a pair of nonbonding electrons from some species that is serving as a base. This view of an acid-base reaction is shown in Figures 2.10 and 2.12 (pp. 44 and 46) and leads to another definition of acids and bases. The **Lewis theory** of acids and bases defines an **acid** as an **electron-pair acceptor** and a **base** as an **electron-pair donor.** Thus, a proton is only one of a large number of species that may function as a Lewis acid. Any molecule or ion may be an acid if it has room in its orbitals to accept a pair of electrons. Any molecule or ion with an unshared pair of electrons to donate can be a base. For example, dimethyl ether acts as a Lewis base toward boron trichloride, a Lewis acid.

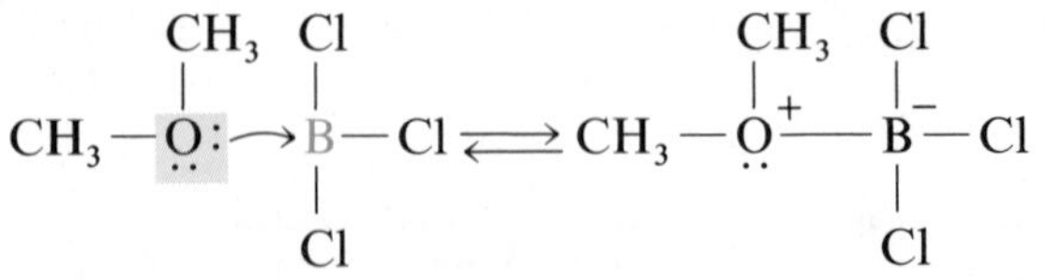

dimethyl ether	boron trichloride	complex of dimethyl ether and boron trichloride
Lewis base	*Lewis acid*	

Ethanol may donate a pair of electrons to a metal ion such as zinc(II).

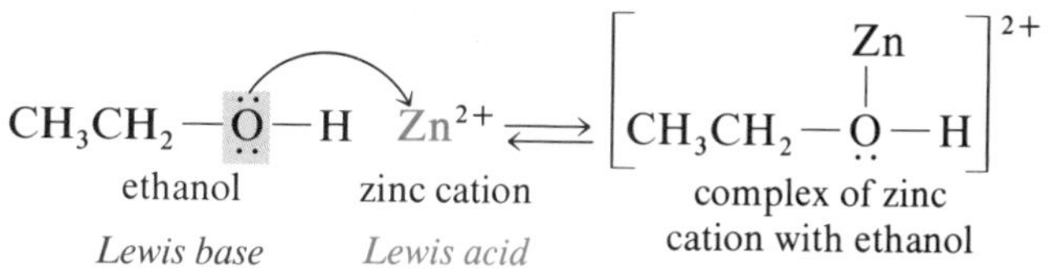

ethanol
Lewis base
zinc cation
Lewis acid
complex of zinc cation with ethanol

An alkyl halide may complex with aluminum chloride.

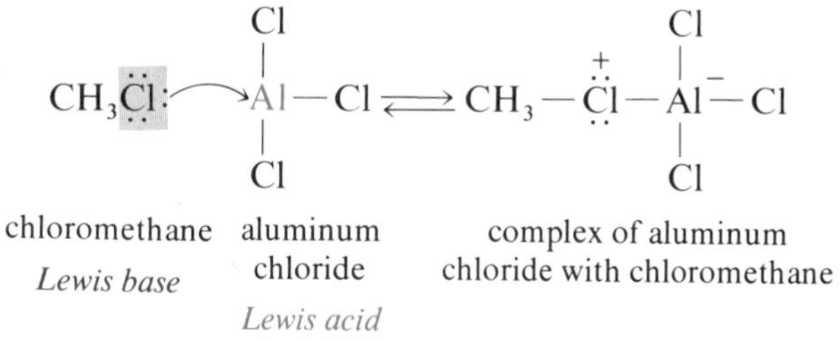

chloromethane
Lewis base
aluminum chloride
Lewis acid
complex of aluminum chloride with chloromethane

Study Guide Concept Map 3.2

In each of these cases, a species with a vacant orbital accepts a pair of electrons from a donor species. Lewis acids are known as **aprotic acids,** compounds that react with bases by accepting pairs of electrons, not by donating protons.

PROBLEM 3.2

Write Lewis structures for the ions or molecules involved in the following reactions, and point out which ones are Lewis acids and which are Lewis bases.

(a) $Cu^{2+} + 4\,NH_3 \rightleftharpoons Cu(NH_3)_4{}^{2+}$

(b) 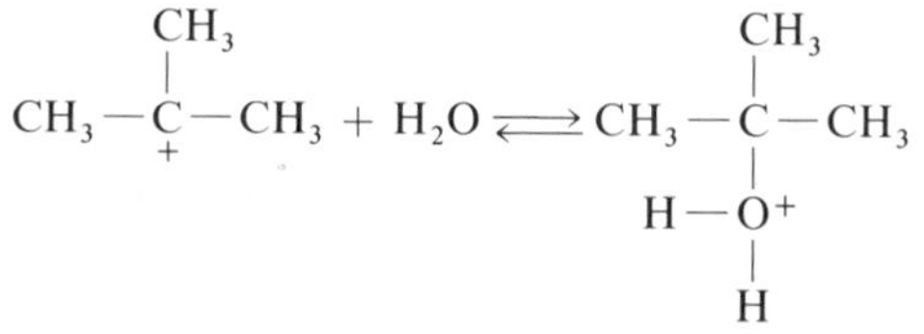

(c) 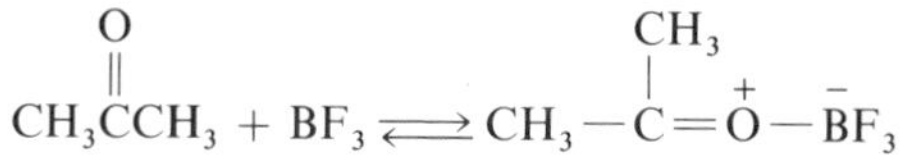

(d) 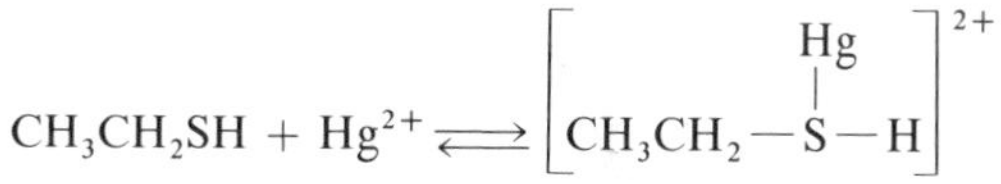

3.2 REACTIONS OF ORGANIC COMPOUNDS AS BASES

Organic compounds with nonbonding electrons on nitrogen, oxygen, or sulfur can react as Lewis bases or Brønsted bases. They react with Lewis acids or Brønsted acids. We have already considered the reactions of nitrogen bases, or amines, with acids (pp. 44 and 76). Just as water is protonated to give a hydronium ion, organic oxygen compounds are protonated to give **oxonium ions.** The reaction of diethyl ether with concentrated hydriodic acid is typical of that of an oxygen base with a protic acid.

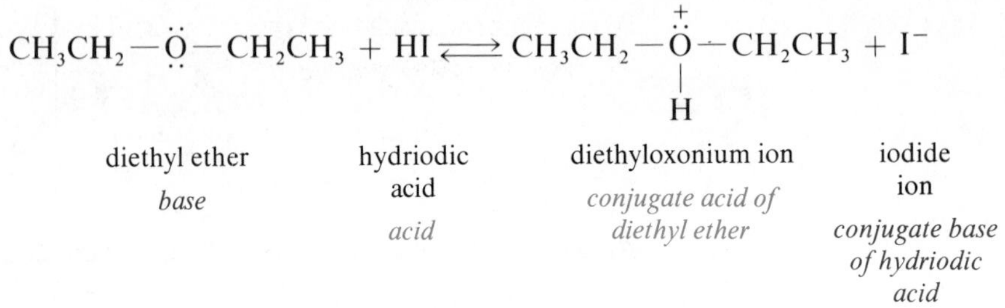

$$CH_3CH_2-\ddot{O}-CH_2CH_3 + HI \rightleftharpoons CH_3CH_2-\overset{+}{\underset{|}{\ddot{O}}}-CH_2CH_3 + I^-$$
$$\underset{H}{}$$

diethyl ether	hydriodic acid	diethyloxonium ion	iodide ion
base	*acid*	*conjugate acid of diethyl ether*	*conjugate base of hydriodic acid*

Alcohols such as ethanol are protonated by acids.

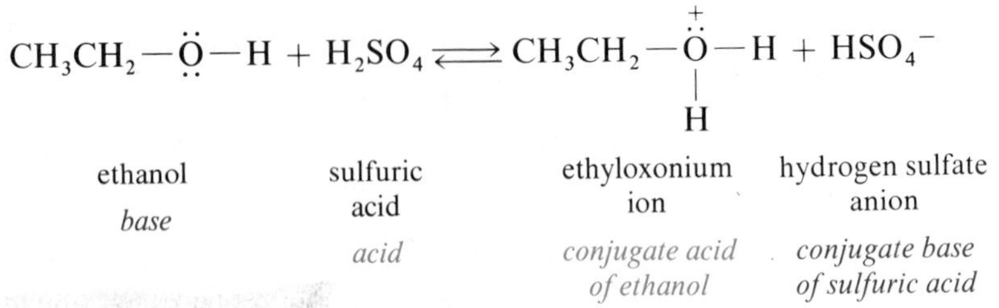

$$CH_3CH_2-\ddot{O}-H + H_2SO_4 \rightleftharpoons CH_3CH_2-\overset{+}{\underset{|}{\ddot{O}}}-H + HSO_4^-$$
$$\underset{H}{}$$

ethanol	sulfuric acid	ethyloxonium ion	hydrogen sulfate anion
base	*acid*	*conjugate acid of ethanol*	*conjugate base of sulfuric acid*

A ketone is another type of oxygen compound that can behave as a base (p. 50). For example, acetone donates electrons to boron trifluoride, a Lewis acid.

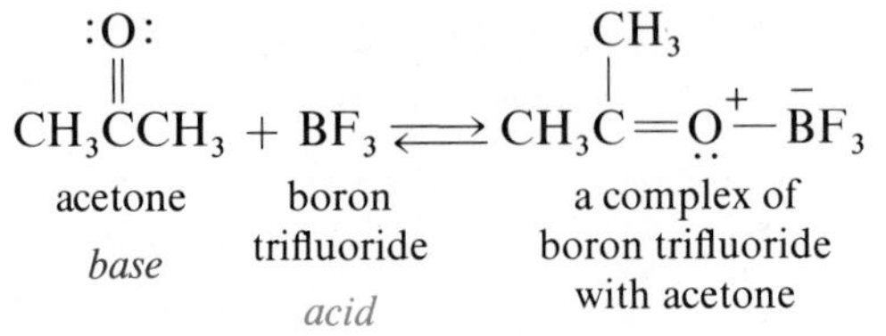

$$\overset{:O:}{\underset{CH_3CCH_3}{\|}} + BF_3 \rightleftharpoons \overset{CH_3}{\underset{CH_3C=\overset{+}{\underset{..}{O}}-\overset{-}{B}F_3}{|}}$$

acetone	boron trifluoride	a complex of boron trifluoride with acetone
base	*acid*	

A **thiol** is a typical organic sulfur compound, formed when an organic group replaces one of the hydrogen atoms in hydrogen sulfide, H_2S. Ethanethiol reacts with the strong protic acid sulfuric acid.

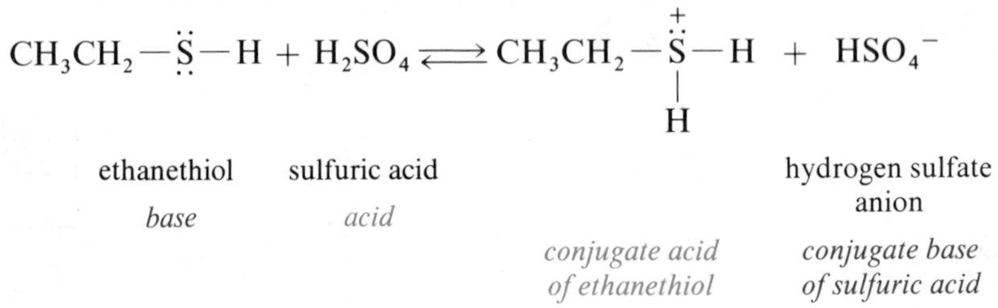

$$CH_3CH_2-\ddot{S}-H + H_2SO_4 \rightleftharpoons CH_3CH_2-\overset{+}{\underset{|}{\ddot{S}}}-H + HSO_4^-$$
$$\underset{H}{}$$

ethanethiol	sulfuric acid		hydrogen sulfate anion
base	*acid*	*conjugate acid of ethanethiol*	*conjugate base of sulfuric acid*

Compounds with π bonds, such as alkenes and alkynes, are also bases. They too react with strong acids.

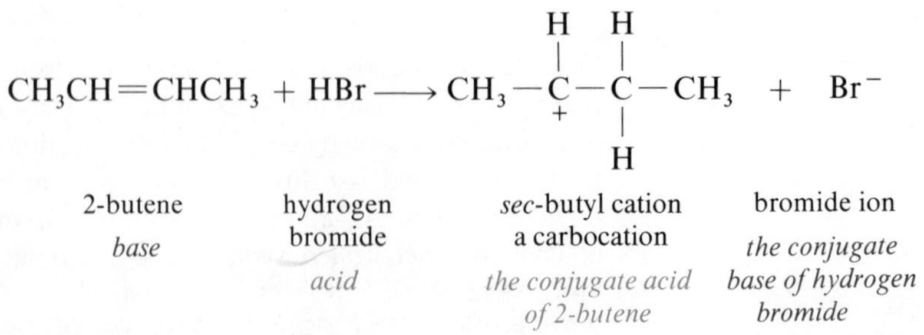

$$CH_3CH=CHCH_3 + HBr \longrightarrow CH_3-\overset{H}{\underset{+}{C}}-\overset{H}{\underset{|}{C}}-CH_3 + Br^-$$
$$\underset{H}{}$$

2-butene	hydrogen bromide	*sec*-butyl cation a carbocation	bromide ion
base	*acid*	*the conjugate acid of 2-butene*	*the conjugate base of hydrogen bromide*

When 2-butene reacts with the strong acid hydrogen bromide, a positively charged carbon atom, a **carbocation,** is formed. The cationic carbon atom is sp^2-hybridized, with an empty *p* orbital (Problem 2.26c). In a subsequent reaction, the cation, as a Lewis acid, can combine with bromide anion, a Lewis base, to give 2-bromobutane.

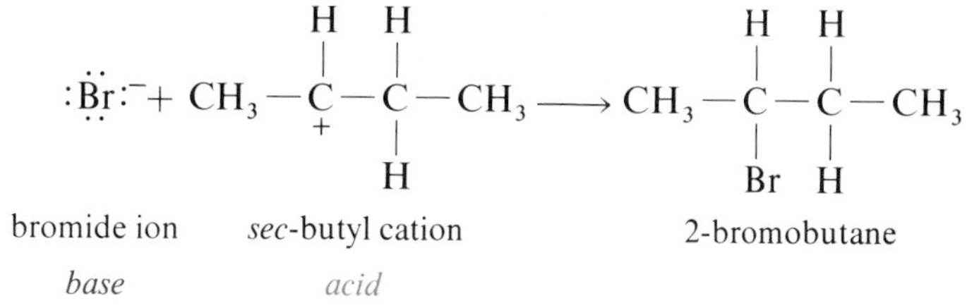
bromide ion
base
sec-butyl cation
acid
2-bromobutane

Generally, nonbonding electrons on nitrogen, oxygen, and sulfur are protonated more easily than the electrons of π bonds.

PROBLEM 3.3

Write equations showing the reactions you would expect for the following compounds with $AlCl_3$ (a Lewis acid) and with H_2SO_4 (a strong protic acid).

(a) CH_3NHCH_3 dimethylamine

(b) HCH with $=O$ (formaldehyde)

(c) $HC{\equiv}CH$ acetylene

(d) $CH_3CH_2PCH_2CH_3$ with CH_2CH_3 on P (triethylphosphine)

(e) CH_3CH_2OH ethanol

(f) $CH_3CH_2SCH_2CH_3$ diethyl sulfide

Some organic compounds have more than one atom with nonbonding electrons; more than one site in such a molecule can react with acids. For example, acetamide has nonbonding electrons on both an oxygen and a nitrogen atom, and either may be protonated.

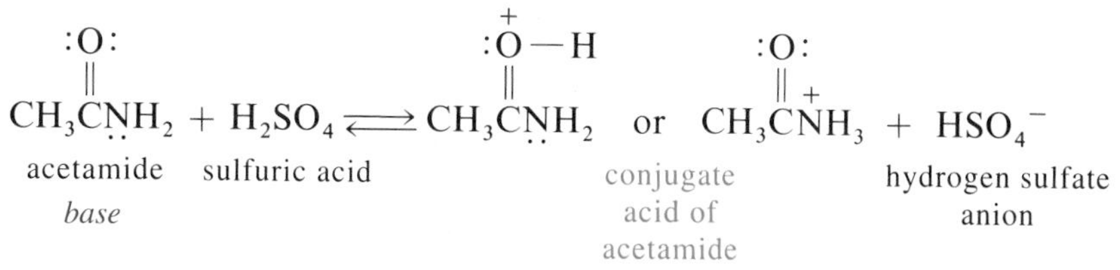
acetamide
base
sulfuric acid
conjugate
acid of
acetamide
hydrogen sulfate
anion

Acetic acid can react as a base at either of two different oxygen atoms.

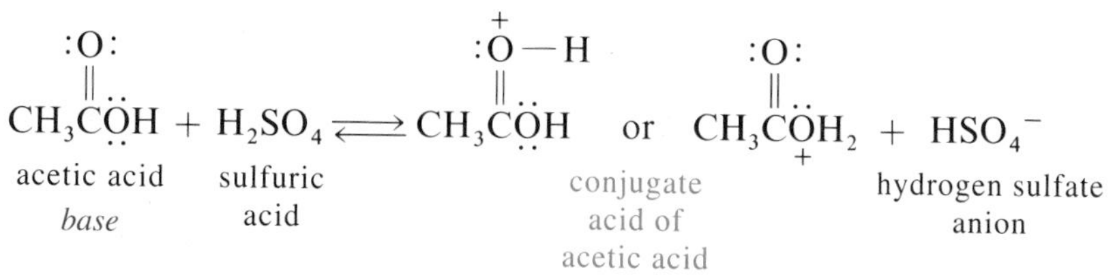
acetic acid
base
sulfuric
acid
conjugate
acid of
acetic acid
hydrogen sulfate
anion

Note that neither compound is protonated twice. Once a molecule has been protonated and bears a positive charge, the availability of the other electrons in the system is greatly reduced. It is possible to protonate the compound again, but only under very strongly acidic conditions. Normally, the reaction stops when one proton is added to the molecule.

Both acetamide and acetic acid are more readily protonated at the oxygen atom of the carbonyl group than at their other basic site. This experimental observation can be explained by writing resonance contributors for each cation. The form of acetamide in which the oxygen atom is protonated, for example, has resonance contributors with the positive charge distributed among an oxygen, a carbon, and a nitrogen atom.

$$\overset{+}{:}\overset{\displaystyle O-H}{\underset{\displaystyle \|}{}} \qquad \overset{:\ddot{O}-H}{\underset{\displaystyle |}{}} \qquad \overset{:\ddot{O}-H}{\underset{\displaystyle |}{}}$$

$$CH_3C-\underset{..}{N}H_2 \longleftrightarrow CH_3\overset{+}{C}-NH_2 \longleftrightarrow CH_3C=\overset{+}{N}H_2$$

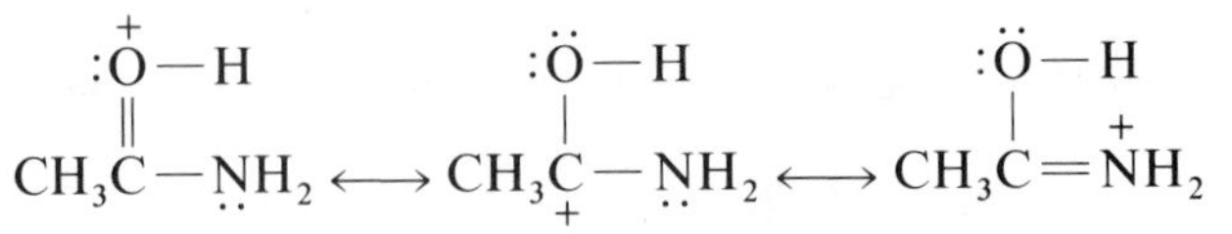

resonance contributors for the cation
resulting from the protonation of acetamide
at the oxygen atom; the cation is
stabilized by delocalization of charge

The spreading of charge to two or more atoms is called **delocalization of charge.** Cations (or anions) for which delocalization of charge is possible are more stable than species in which charge is localized. The cation shown above can be compared with the cation that results if the nitrogen atom in acetamide is protonated.

$$\overset{:O:}{\underset{\displaystyle \|}{}} \qquad \overset{:\ddot{O}:^-}{\underset{\displaystyle |}{}}$$

$$CH_3C-\overset{+}{N}H_3 \longleftrightarrow CH_3\overset{+}{C}-\overset{+}{N}H_3$$

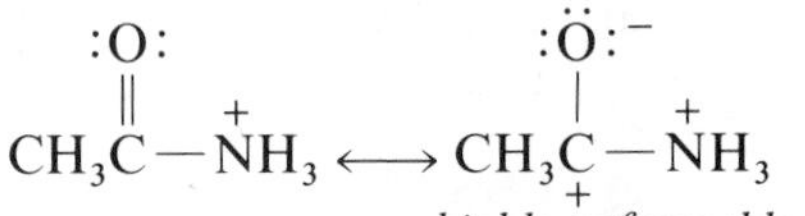

highly unfavorable
resonance contributor;
two positive charges
adjacent to each other

no resonance contributor that delocalizes
the positive charge at the nitrogen atom

This cation from acetamide in which the nitrogen has been protonated has a high concentration of positive charge at a single atom and is, therefore, less stable than the one in which the oxygen has been protonated.

The protonation of the nonbonding electrons on the oxygen atom of a carbonyl or a hydroxyl group is an important first step in the reactions under acidic conditions of compounds such as acetamide, acetic acid, and alcohols. The conjugate acids of these compounds are usually more reactive toward Lewis bases than the unprotonated forms are. Consequently, acids are often used as catalysts to promote reactions of organic compounds.

PROBLEM 3.4

The following compounds all have more than one site where they can react with a strong protic acid such as sulfuric acid. For each compound, draw Lewis structures, and write equations showing the structures of the different products you would expect to obtain from protonation reactions.

$$\text{(a)} \quad CH_3\overset{\displaystyle O}{\underset{\displaystyle \|}{C}}OCH_2CH_3 \qquad\qquad \text{(b)} \quad HOCH_2CH_2\overset{\displaystyle CH_3}{\underset{\displaystyle |}{N}}CH_3$$

(a) 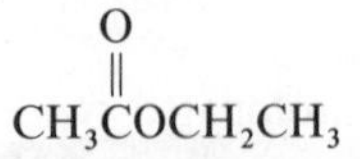 *ethyl acetate*   (b) 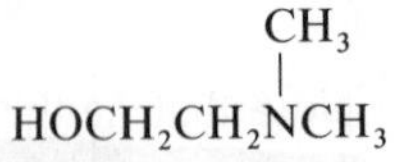 *2-(N,N-dimethylamino)ethanol*

(c) $CH_2{=}CHCH_2OH$ 2-propen-1-ol (d) $CH_3N{=}C{=}O$ methyl isocyanate

PROBLEM 3.5

Give the structural formula of the conjugate acid of each of the following compounds.

(a) 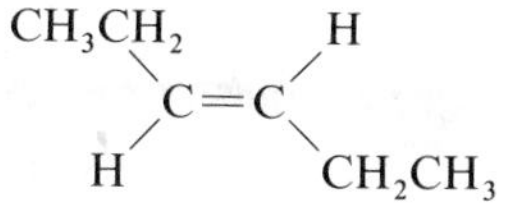
CH3CH2
H
C=C
H
CH2CH3

(b) $CH_3CH_2OCH_2CH_3$

(c) $CH_3CH_2\overset{\displaystyle H}{\overset{|}{N}}CH_2CH_3$ (d) $CH_3CH_2CH_2SCH_3$ (e) $CH_3CH_2\overset{\displaystyle O}{\overset{\|}{C}}H$

(f) $CH_3CH_2CH_2OH$ (g) $CH_3CH_2\overset{\displaystyle O}{\overset{\|}{C}}CH_2CH_3$

3.3 THE USE OF CURVED ARROWS IN WRITING MECHANISMS

Chemists are not satisfied to write equations showing the reactants and the products of reactions. They are also interested in the details of the transformations taking place as reactants are converted to products. In order to study these details, chemists examine the progress of reactions using many experimental techniques. For example, they determine the rate of a chemical reaction and see how the rate and the products vary when experimental conditions are changed. They explore changes in stereochemistry (p. 180) that occur during the reaction. From observations such as these, they postulate the details of the pathway the reactants follow as they are converted into the products. Such a detailed reaction pathway is called the **mechanism** of the reaction. Mechanisms serve as an aid to understanding the reactivity of different types of organic compounds. Subsequent sections of this book will contain many mechanisms proposed for reactions. In this section, we explore the symbolism widely used by organic chemists to trace the details of the changes that occur as reactants are converted to products.

For example, the equation that shows the reactants diethyl ether and hydriodic acid on one side and the products diethyloxonium ion and iodide ion on the other side (p. 78) can be rewritten to show greater detail.

VISUALIZING THE REACTION

Protonation of an ether

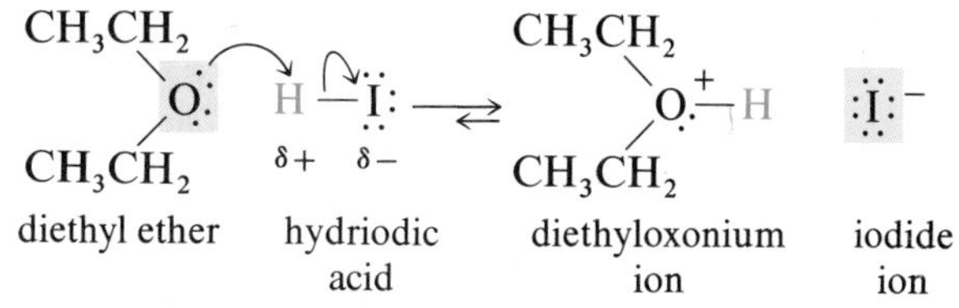
CH3CH2
O:
CH3CH2
H
I:
δ+ δ−
CH3CH2
O
H
CH3CH2
:I:−
diethyl ether
hydriodic acid
diethyloxonium ion
iodide ion

The curved arrows in the equation above show the electronic changes that take place as covalent bonds are formed and broken during the reaction. Such arrows are part of a symbolic language used by many chemists. This kind of symbolism has several important features. First, the structures of both reactants are written to show nonbonding electrons and sites of electron deficiency. Partial positive and negative charges are shown where appropriate, to indicate that bonding is between an atom with unshared electrons and an atom with a deficiency of electrons. A **curved arrow** is drawn from the site of electron density, such as a pair of nonbonding electrons, to the site of electron deficiency, such as an atom with a partial positive charge. This arrow depicts the start of the bonding process. It represents the flow of electrons, not the movement of atoms. If this flow of electrons would result in too many bonds around an atom, such as two covalent bonds to a hydrogen atom, another curved arrow is drawn symbolizing the simultaneous release of bonding electrons to some atom or group of atoms, which then detaches itself. Iodide ion is the species formed by this process in the equation above.

Let's look more closely at another reaction, the reaction of ethanethiol with sulfuric acid.

VISUALIZING THE REACTION

Protonation of a thiol

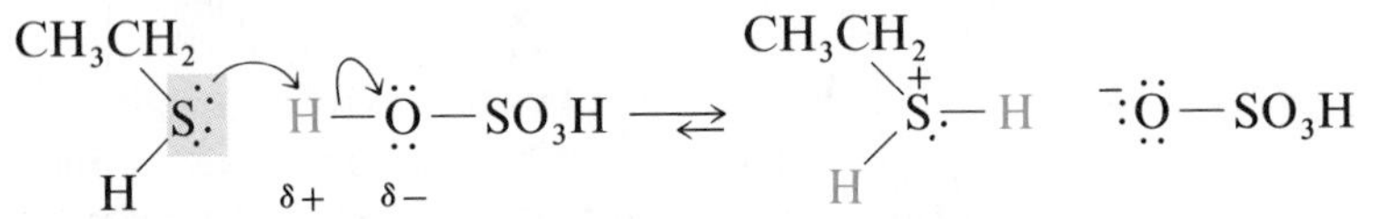
CH₃CH₂
S
H
H—O—SO₃H
δ+ δ−
CH₃CH₂
S—H
H
:O—SO₃H

ethanethiol | sulfuric acid | conjugate acid of ethanethiol | hydrogen sulfate anion

The nonbonding electrons on the sulfur atom are shown moving toward the positively charged hydrogen atom of sulfuric acid. As the proton is transferred to the sulfur atom of ethanethiol, the hydrogen sulfate ion is detached, taking with it the pair of electrons that bonded it to the proton.

The π electrons of a multiple bond may also serve as the site of electron density in a reaction.

VISUALIZING THE REACTION

Protonation of an alkene

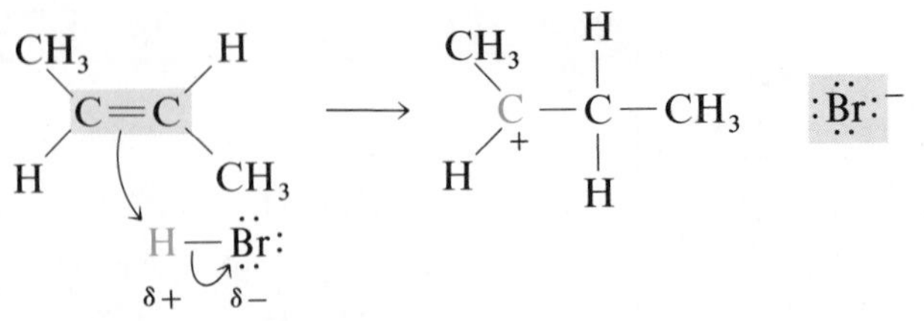
CH₃
H
C=C
H
CH₃
H—Br:
δ+ δ−
CH₃
H
C—C—CH₃
H
H
:Br:

2-butene | hydrobromic acid | *sec*-butyl cation | bromide ion

The reaction of 2-butene with hydrobromic acid starts with the flow of the π electrons toward the positively charged hydrogen atom of the hydrogen halide. The carbon atom that loses its share of the electrons in the π bond becomes positively charged. Bromide ion is the other product of the reaction.

Whenever a mechanism is written for a reaction, it is important to make sure that the net charge is the same on both sides of the equation. Thus, in all the previous examples within this chapter, two neutral reactants with a net charge of zero give products with one positive and one negative charge, also a net charge of zero.

Often the site of electron deficiency is an empty orbital or a positively charged atom. The reaction of the *sec*-butyl cation with bromide ion illustrates this kind of reaction.

VISUALIZING THE REACTION

Reaction of a carbocation with an anion

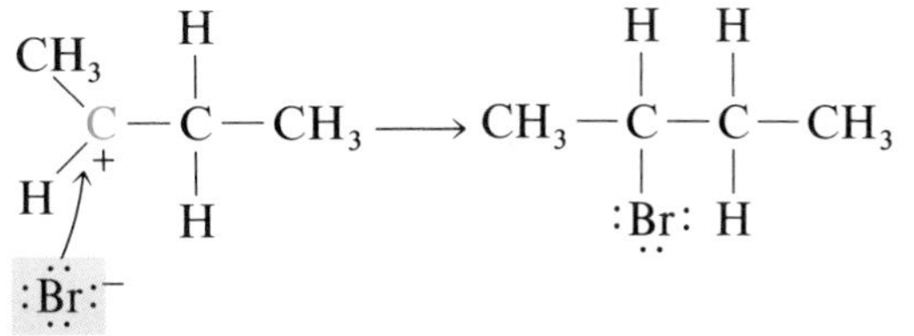

sec-butyl cation and bromide ion — 2-bromobutane

The curved arrow shows the flow of electrons from the bromide ion, the site of electron density, to the positively charged carbon atom, the site of electron deficiency.

The equations and the description given in this section provide an introduction to a language that will be used frequently in the rest of this book. It is important that you start to practice it right away. Observe carefully the way that the curved arrows are drawn, especially where they start and where they end. These arrows, which represent the flow of electrons as bonds are formed and broken, must be used with precision.

PROBLEM 3.6

Rewrite the equations in your answers for Problems 3.3 and 3.4 to show the flow of electrons in the reactions.

PROBLEM 3.7

In the following equations, reactants and arrows showing the flow of electrons are indicated. Supply the products of the reactions.

(a) 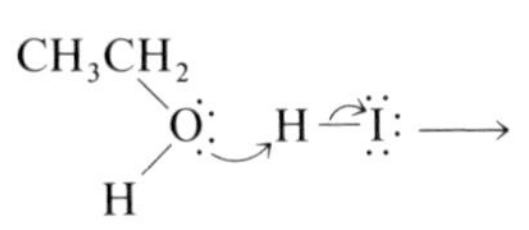

(b) 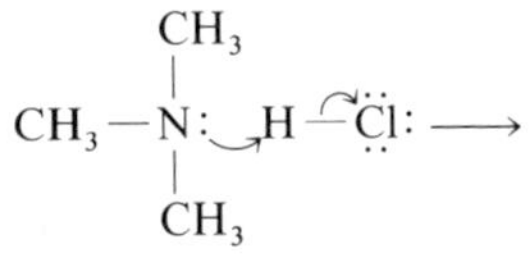

(c) 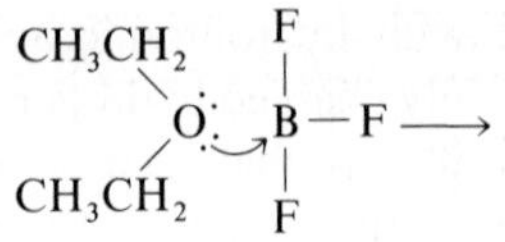

(d) 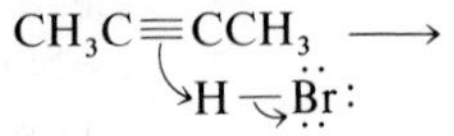

(e) 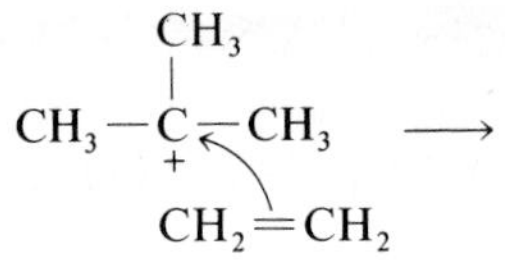

3.4 CARBON, NITROGEN, OXYGEN, SULFUR, AND HALOGEN ACIDS

A. Relative Acidities and Basicities Within Groups in the Periodic Table

The reactions of an organic compound as an acid depend on the ease with which it can lose a proton to a base. Methanol has two kinds of bonds to hydrogen atoms: (1) carbon-hydrogen bonds, which are nonpolar, and (2) an oxygen-hydrogen bond, which is highly polar because of the differing electronegativities of oxygen and hydrogen. When methanol behaves as an acid, it is the hydrogen atom on the oxygen atom that is lost as a proton.

VISUALIZING THE REACTION

Deprotonation of an alcohol

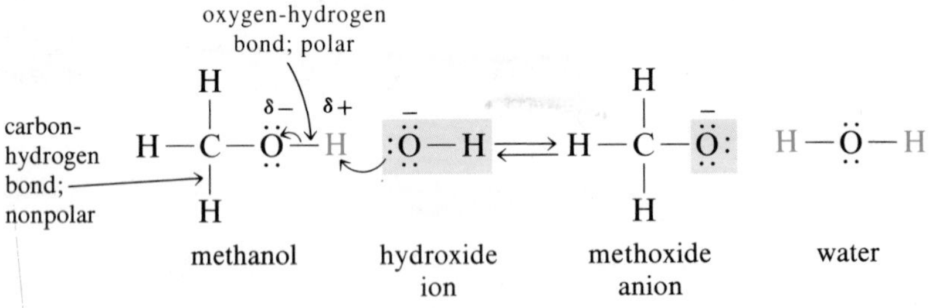
oxygen-hydrogen bond; polar
carbon-hydrogen bond; nonpolar
δ−
δ+
methanol
hydroxide ion
methoxide anion
water

Equilibrium is reached in the reaction of methanol with hydroxide ion when the concentrations of hydroxide ion and methoxide anion are roughly equal, because the acidities of methanol and water are similar. The substitution of a methyl group for a hydrogen atom in the water molecule does not greatly affect the acidity of the hydroxyl group.

In the reaction of methanethiol with hydroxide ion, the equilibrium lies much more to the right.

VISUALIZING THE REACTION

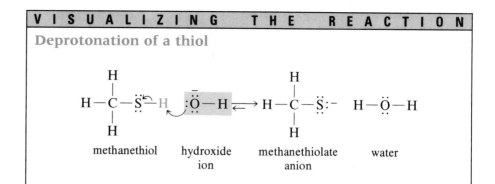

Deprotonation of a thiol

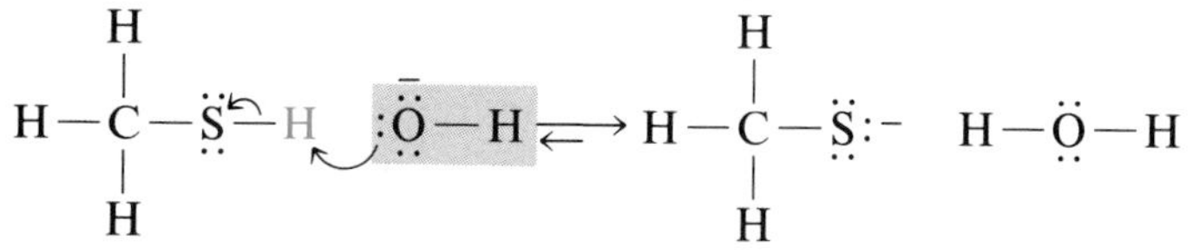

methanethiol hydroxide ion methanethiolate anion water

A hydrogen atom bonded to a sulfur atom is more acidic than one bonded to oxygen and is transferred more completely to hydroxide ion. Another way of describing the same phenomenon is to say that the methanethiolate anion is a weaker base than hydroxide ion (or methoxide ion). Sulfur is below oxygen in the periodic table and is a larger atom. Therefore, a negative charge on sulfur is more spread out, less concentrated at any one point in space, than is a negative charge on oxygen. Consequently, a sulfur anion is less likely than an oxygen anion to attract a proton. The basicity of an anion is weaker the further down the atom is in any group in the periodic table; the acidity of the corresponding conjugate acid increases in the same order.

$$R-\ddot{\underset{..}{O}}:^- \;>\; R-\ddot{\underset{..}{S}}:^-$$

stronger base *weaker base*

$$R-\ddot{\underset{..}{O}}-H \;<\; R-\ddot{\underset{..}{S}}-H$$

weaker acid *stronger acid*

The same trend is characteristic of the halogen acids, which become stronger from hydrogen fluoride to hydrogen iodide.

$$HF < HCl < HBr < HI$$

weakest acid *strongest acid*

$$F^- > Cl^- > Br^- > I^-$$

strongest base *weakest base*

In general, then, a protic acid is stronger the further down the atom to which the proton is bonded is in the periodic table.

PROBLEM 3.8

Which is the stronger acid, H_2O or H_2S?

PROBLEM 3.9

Write an equation showing the reaction you expect between ethanethiol, CH_3CH_2SH, and methoxide anion, CH_3O^-.

B. Relative Acidities and Basicities Across a Period in the Periodic Table

The acidity of a hydrogen atom bound to a central atom increases the further to the right the central atom is in any period of the periodic table. Ethane, in which the hydrogen atoms are bonded to carbon, is a very weak acid. Methylamine does not give up a proton from its nitrogen atom to hydroxide ion, but methanol, in which a hydrogen atom is bonded to oxygen, does give up that proton (p. 84).

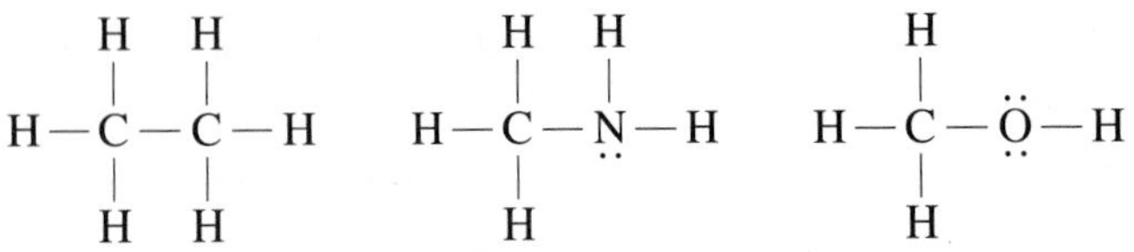

ethane | methylamine | methanol

increasing acidity of hydrogen bonded to carbon, nitrogen, and oxygen

The conjugate bases of ethane, methylamine, and methanol are shown below.

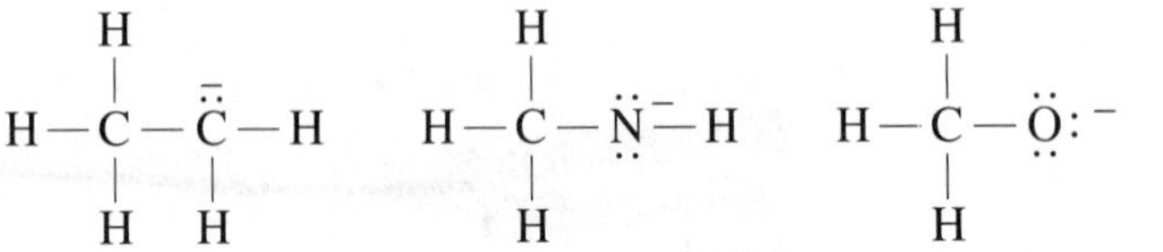

ethyl anion — *a carbanion* — *conjugate base of ethane; strongest base*

methylamide anion — *conjugate base of methylamine*

methoxide anion — *conjugate base of methanol; weakest base*

The weakest acid in the series, ethane, has the strongest conjugate base, and the strongest acid, methanol, has the weakest conjugate base.

The conjugate base of ethane is the ethyl anion, a carbanion. A **carbanion** is a species in which there is a carbon atom with a pair of nonbonding electrons and a negative charge on it. In contrast, a carbocation has only six electrons around a carbon atom that has a positive charge (p. 79). A carbanion is like ammonia in its shape (Problem 2.26f). Most carbanions are strong bases. Section 3.7A (p. 100) will explore some carbanions that are formed easily enough to be important intermediates in organic reactions.

The relative basicities of the ethyl anion, the methylamide anion, and methoxide anion are related to the electronegativities of the carbon, nitrogen, and oxygen atoms. Carbon is less electronegative than nitrogen, which means that it attracts and holds the electrons around it less firmly than nitrogen does. The nonbonding electrons on a nitrogen atom are more available for donation to a proton or a Lewis acid than are the nonbonding electrons on an oxygen atom. Therefore, nitrogen compounds are more basic than similar oxygen compounds. For example, ammonia is more basic than water.

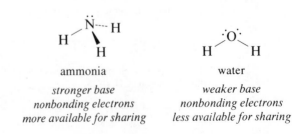

ammonia — *stronger base, nonbonding electrons more available for sharing*

water — *weaker base, nonbonding electrons less available for sharing*

The conjugate base of an amine, a nitrogen anion, is a much stronger base than an oxygen anion.

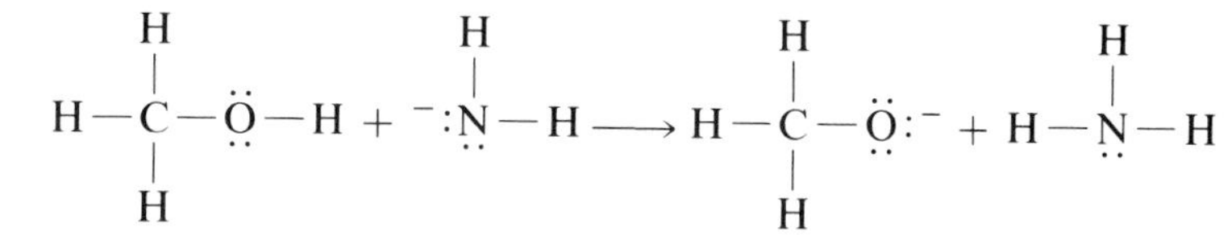

methanol	amide anion	methoxide anion	ammonia
acid	*base*	*conjugate base of methanol*	*conjugate acid of amide anion*

Amide anion, the conjugate base of ammonia, reacts with methanol to give methoxide anion and ammonia. As a base, amide anion is much stronger than methoxide ion—so much stronger that the reaction shown above goes essentially to completion.

Assertions about acidity that depend on the concept of electronegativity can only be made about species containing atoms in the same period of the periodic table. Comparisons based on electronegativity of atoms in different periods do not work. The effects of atomic size, mentioned earlier in comparing the acidity of alcohols and thiols (p. 85), are more important in such cases.

Study Guide Concept Map 3.3

PROBLEM 3.10

Arrange the species in each of the following sets in order of decreasing basicity.

(a) CH_3NHCH_3, $CH_3CH_2CH_3$, CH_3OCH_3

(b) NH_2^-, CH_3^-, F^-, OH^-

(c) $CH_3\bar{N}CH_3$, CH_3O^-

C. Organic Cations as Acids

The protonated forms of organic compounds, which are organic cations, are also acids. The conjugate base of methylammonium ion is methylamine; methylamine is a stable, neutral molecule, therefore methylammonium ion loses a proton relatively easily.

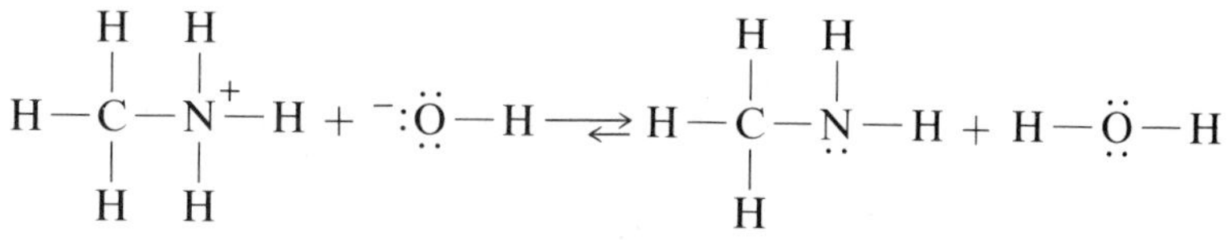

methylammonium ion	hydroxide ion	methylamine	water
acid	*base*	*conjugate base of methylammonium ion*	*conjugate acid of hydroxide ion*

Protonation reactions of alcohols, ethers, and ketones yield oxonium ions, which are acidic. Methyloxonium ion, the conjugate acid of methanol, has an acidity similar to that of the hydronium ion. See the reaction on the next page.

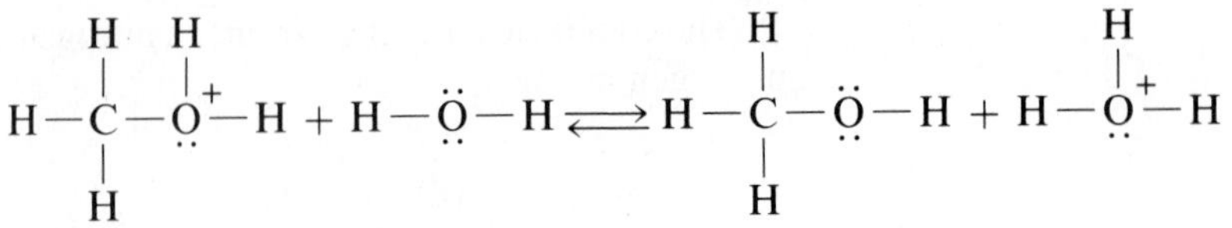

methyloxonium ion	water	methanol	hydronium ion
acid	*base*	*conjugate base of methyloxonium ion*	*conjugate acid of water*

Carbocations are strong Brønsted acids as well as Lewis ones (p. 83). The hydrogen atoms on the carbon atoms adjacent to the cationic carbon are acidic and easily lost to bases, even weak bases such as water.

$$CH_3\overset{+}{C}HCH_3 + H_2O \longrightarrow CH_3CH{=}CH_2 + H_3O^+$$

2-propyl cation	water	propene	hydronium ion
acid	*base*	*conjugate base of 2-propyl cation*	*conjugate acid of water*

PROBLEM 3.11

(a) Water is a Lewis base as well as a Brønsted one. What other reaction besides the one shown above is possible between a carbocation and water? Write a mechanism for the reaction using the curved-arrow notation.

(b) The product of the reaction you wrote in part a is itself an acid. How will it react with excess water?

PROBLEM 3.12

Arrange the species in each of the following sets in order of increasing acidity.

(a) CH_3CH_2OH, $CH_3CH_2CH_3$, $CH_3CH_2NH_2$, CH_3CH_2SH

(b) $CH_3\overset{+}{O}(H)CH_3$, CH_3OCH_3, CH_3NHCH_3

(c) $CH_3\overset{+}{N}(CH_3)(H)CH_3$, $CH_3CH_2CH_3$, $CH_3\overset{+}{O}(H)CH_3$

PROBLEM 3.13

Give the structural formula for the conjugate base of each of the following species.

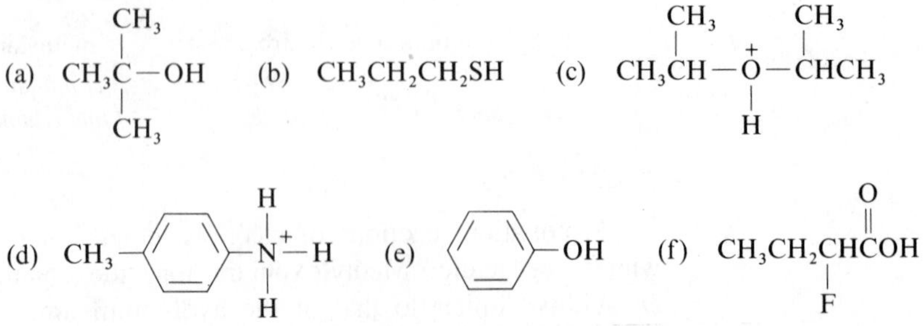

3.5 EQUILIBRIA IN ACID-BASE REACTIONS

A. Acidity Constants and pK_a

Measuring the acidities of organic compounds involves the measurement of equilibrium constants. When acetic acid is dissolved in water, certain concentrations of acetic acid, water, acetate ions, and hydronium ions are present at equilibrium. Acetate ion is more basic than water, so the equilibrium lies on the side of the undissociated acetic acid.

$$\underset{\text{acetic acid}}{CH_3\overset{\overset{\large O}{\|}}{C}OH} + \underset{\text{water}}{H_2O} \rightleftharpoons \underset{\text{acetate anion}}{CH_3\overset{\overset{\large O}{\|}}{C}O^-} + \underset{\text{hydronium ion}}{H_3O^+}$$

The **equilibrium constant** for this reaction, K_{eq}, is written as

$$K_{eq} = \frac{[CH_3CO_2^-][H_3O^+]}{[CH_3CO_2H][H_2O]}$$

For dilute solutions, in which the concentration of water is large and almost constant, another expression, the **acidity constant, K_a,** is used. For acetic acid,

$$K_{eq}[H_2O] = K_a = \frac{[CH_3CO_2^-][H_3O^+]}{[CH_3CO_2H]} = 1.75 \times 10^{-5}$$

The acidity constants for different acids have magnitudes ranging from 10^{14} to 10^{-50}. Such large and small numbers are most conveniently expressed as logarithms. Just as pH is $-\log [H_3O^+]$, so pK_a is defined as $-\log K_a$. The pK_a for acetic acid is 4.76, a value that is typical of many organic acids. A strong inorganic acid such as sulfuric acid has $K_a \sim 10^9$ and p$K_a \sim -9$. The pK_a of hydronium ion is -1.7; that of ammonium ion is 9.4. Hydrocarbons, such as methane, are extraordinarily weak acids, with pK_a values of approximately 50.

Inside the front cover of this book is a table of pK_a values for a variety of acids, shown with their conjugate bases. Note that the strongest acids, with the weakest conjugate bases, are at the top of the table with negative or small positive pK_a values. *The weaker the acid, the stronger its conjugate base and the larger its pK_a value*. The pK_a values for the very weak acids at the bottom of the table can be measured only indirectly, so different values for the same compound can be found in different tables. These pK_a values are rough indications of relative acidities. Note that pK_a values are only given for protic acids. As you use the table, you will acquire some sense of which acids are strong and which are weak. Do not attempt to memorize pK_a values; refer to the table if the solution of a problem requires it.

PROBLEM 3.14

This is formic acid:

$$H\overset{\overset{\large O}{\|}}{C}OH$$

Its K_a is 1.99×10^{-4}. What is its pK_a? Is it a stronger or weaker acid than acetic acid, with a pK_a of 4.76?

PROBLEM 3.15

The imidazole ring is present in the amino acid histidine. This amino acid plays an important role in proton-transfer reactions in the human body. The protonated form of imidazole (imidazolinium ion) has pK_a 7.0. What is K_a for this acid?

imidazole — imidazolinium ion, *the conjugate acid of imidazole*

B. Using the Table of pK_a Values to Predict Acid-Base Reactions

Reactions in which a proton is transferred are fast and reversible. They are the first reactions to occur whenever a strong enough acid is brought into the presence of a base.

The table of pK_a values can be used to predict whether acid-base reactions will take place. In general, an acid will transfer a proton to the conjugate base of any acid that is below it in the table. If the pK_a values for the two acids differ by only one or two units, both acids will be present in substantial quantities at equilibrium. If there is a large difference in acidity, the transfer of the proton will be nearly complete. For example, suppose we wanted to know whether we could use ammonia, NH_3, to prepare free trimethylamine, $(CH_3)_3N$, from trimethylammonium chloride, $(CH_3)_3NH^+Cl^-$. The proton donor in this reaction is the trimethylammonium ion, $(CH_3)_3NH^+$, and the proton acceptor is ammonia.

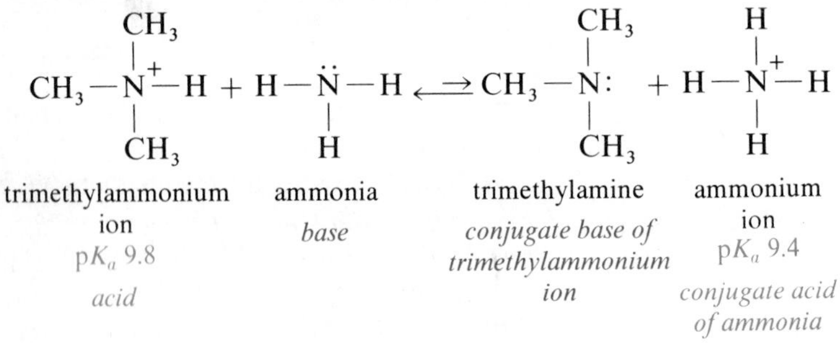
trimethylammonium ion
pKa 9.8
acid
ammonia
base
trimethylamine
conjugate base of trimethylammonium ion
ammonium ion
pKa 9.4
conjugate acid of ammonia

We look up the two acids, trimethylammonium ion and ammonium ion, in the pK_a table and find that their pK_a values are very close to each other, 9.4 for the ammonium ion and 9.8 for the trimethylammonium ion. The larger pK_a value for the trimethylammonium ion indicates that trimethylamine is a stronger base than ammonia. In a competition for a proton, trimethylamine is more likely to retain the proton than to release it to ammonia. The species shown in the equation above are in equilibrium, but the arrow pointing to the left is larger than the one pointing to the right. This reaction could not be used to generate trimethylamine from trimethylammonium ion.

To remove the proton from trimethylammonium ion, we must find a stronger base than ammonia. A search of the table of pK_a values reveals that hydroxide ion, the conjugate base of water, has pK_a 15.7 and thus would be a good reagent to use.

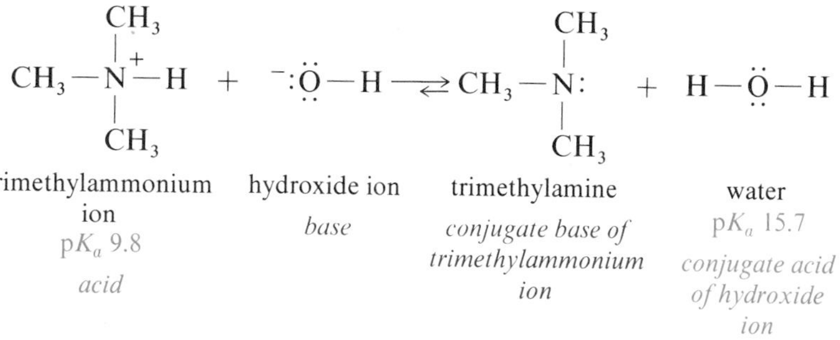
trimethylammonium ion
pKa 9.8
acid
hydroxide ion
base
trimethylamine
conjugate base of trimethylammonium ion
water
pKa 15.7
conjugate acid of hydroxide ion

The difference in the pK_a values of the two acids, trimethylammonium ion and water, is large, indicating that hydroxide ion is a much stronger base than trimethylamine. In the competition between the two bases, the proton is transferred essentially completely from the trimethylammonium ion to the hydroxide ion, freeing the weaker base, trimethylamine.

The table of pK_a values does not include all possible acids and bases. What happens when we need to make a prediction about a compound that does not appear in the table? For instance, we might wish to know whether propyne, $CH_3C{\equiv}CH$, will be deprotonated by amide anion, NH_2^-.

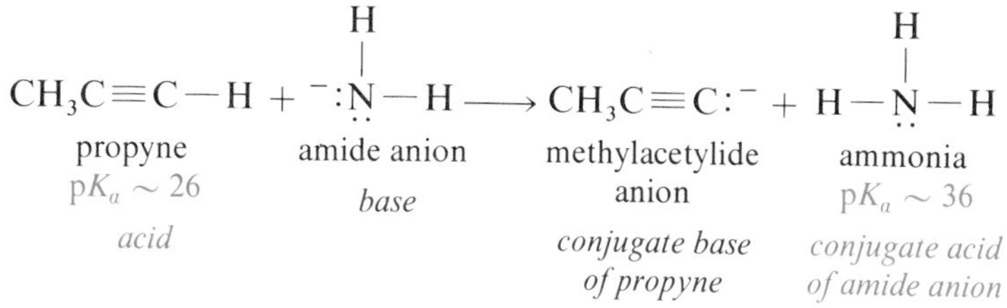
propyne
pKa ~ 26
acid
amide anion
base
methylacetylide anion
conjugate base of propyne
ammonia
pKa ~ 36
conjugate acid of amide anion

Propyne does not appear in the table of pK_a values. Acetylene, $HC{\equiv}CH$, which resembles propyne, is listed with pK_a 26. We can assign the same pK_a value to propyne, reasoning that the acidity of propyne, like that of acetylene, is derived from the hydrogen atom bonded to the *sp*-hybridized carbon atom (p. 53) of the triple bond. (Because there are no other kinds of hydrogen atoms in acetylene, the acidity must come from that hydrogen.) We are also reasoning that the substitution of an alkyl group for a hydrogen atom in acetylene will not dramatically change the acidity of the hydrogen atom on the triple bond. In general, functional groups are affected only in subtle ways by the particular alkyl group attached to them. Therefore, we can generalize further that a series of alkynes such as

$CH_3C{\equiv}CH$	$CH_3CH_2C{\equiv}CH$	$CH_3CH_2CH_2C{\equiv}CH$
propyne	1-butyne	1-pentyne

will have similar pK_a values and similar reactivity. This process of reasoning by analogy to make judgments and estimates about unknown situations is one you should cultivate during your study of organic chemistry. Once we conclude that the pK_a of propyne is approximately 26, we can complete the equation by showing that the alkyne will be deprotonated by amide anion, the conjugate base of ammonia, which has pK_a 36.

The table of pK_a values can be used to make predictions about a large number of reactions. Developing familiarity with the table and skill in using it will help you learn organic chemistry.

PROBLEM 3.16

From the table of pK_a values, select an oxonium ion, an ammonium ion, a carboxylic acid, an alcohol, a thiol, an amine, and an alkane, and arrange them in order of increasing acidity. This series summarizes the trends that were discussed in Section 3.4 and will be useful in problems that follow.

PROBLEM 3.17

A number of acid-base reactions were given in Section 3.4. Rewrite each equation, leaving out the arrows between the reactants and the products. Use the pK_a table to assign pK_a values to the acids that appear in each equation. Make your own judgment in each case about whether the reaction goes essentially to completion as written or whether reactants and products are present together in significant quantities.

PROBLEM 3.18

Use the table of pK_a values to decide whether or not the following reactions will take place. Complete the equations for those that will take place. Write "no reaction" for the others.

(a) $CH_3CH_2OH + KOH \longrightarrow$

(b) $CH_3CH_2\overset{+}{N}H_2(CH_2CH_3)\ Cl^- + NaOH \longrightarrow$

(c) $CH_2F\overset{O}{\overset{\|}{C}}OH + NaHCO_3 \longrightarrow$

(d) 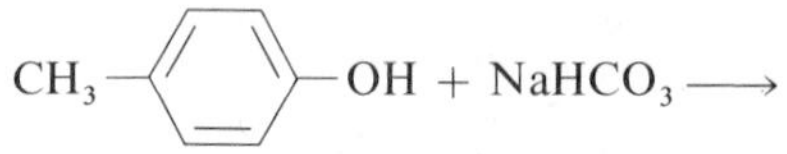

(e) 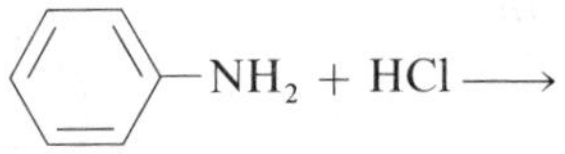

(f) $CH_3CH_2CH_2SH + CH_3CH_2ONa \longrightarrow$

(g) 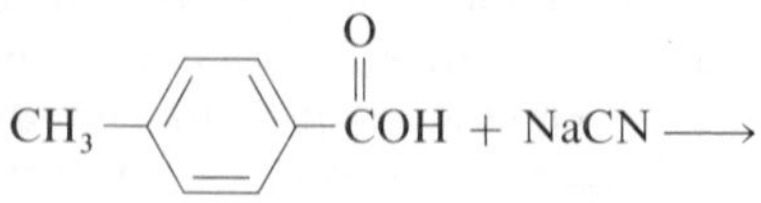

(h) $CH_3\overset{O}{\overset{\|}{C}}CH_2\overset{O}{\overset{\|}{C}}CH_3 + CH_3ONa \longrightarrow$

C. The Importance of Solvation in Acidity

The pK_a of an acid varies somewhat depending on the solvent it is dissolved in and the temperature at which the determination is made. The importance of the solvent has become much clearer since the development of mass spectrometers (Chapter 21), instruments that allow chemists to study reactions in the gas phase.

For example, the proton-transfer reaction of acetic acid with water ι phase has been examined.

$$CH_3\overset{\overset{\displaystyle O}{\|}}{C}OH(g) + H_2O(g) \rightleftharpoons CH_3\overset{\overset{\displaystyle O}{\|}}{C}O^-(g) + H_3\overset{+}{O}(g)$$

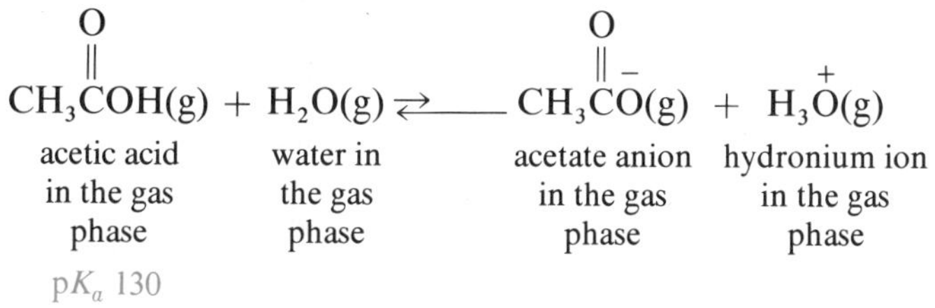

| acetic acid in the gas phase | water in the gas phase | acetate anion in the gas phase | hydronium ion in the gas phase |

pK$_a$ 130

The pK$_a$ of acetic acid under these conditions is 130, indicating that it is enormously difficult to separate the negatively charged acetate ion from the positively charged hydronium ion unless solvent molecules are present to stabilize the charged particles. Clearly, the interaction of the solvent with the undissociated acid and with anions and cations that form on dissociation contributes significantly to the stabilization of the various species and, therefore, to the position of the equilibrium.

Such interactions are particularly important in water, which participates in extensive hydrogen bonding (p. 25). When an acid is dissolved in water, the bonding between the solvent molecules is disrupted, and new hydrogen bonds are formed between water molecules and the solute particles. Experimentation has shown that such interactions may be the most important factor in determining the relative acidities of closely related compounds.

Section 3.6 will discuss the relative acidities of a series of acids and offer rationalizations for the trends observed. The language being used here to talk about acidity is used by chemists to correlate facts and predict reactivity in all areas of organic chemistry. In order to understand the assumptions inherent in this language, let's first review the relationship between equilibrium and the changes in energy that occur during a reaction.

D. Equilibrium, Free Energy, Enthalpy, and Entropy

The equilibrium constant for a reaction is related to the change in standard free energy by the following equation.

$$\Delta G° = -2.303RT \log K_{eq}$$

$\Delta G°$ is the change in free energy for the reaction when the reactants and the products are in their standard states (1 M solution for solutions; 1 atm of pressure for gases); R is the gas constant, 1.987×10^{-3} kcal/deg·mol; T is the absolute temperature; and K_{eq} is the equilibrium constant for the reaction. If there is a decrease in free energy during the reaction, that is, if $\Delta G°$ is negative, the equilibrium constant is greater than 1. In other words, in that case the reaction proceeds so that more products than reactants are present at equilibrium. If $\Delta G°$ is positive, the reverse reaction is favored. For example, chloroacetate anion and acetate anion compete for a proton in the following reaction.

$$ClCH_2\overset{\overset{\displaystyle O}{\|}}{C}OH + CH_3\overset{\overset{\displaystyle O}{\|}}{C}O^- \underset{H_2O}{\rightleftharpoons} ClCH_2\overset{\overset{\displaystyle O}{\|}}{C}O^- + CH_3\overset{\overset{\displaystyle O}{\|}}{C}OH$$

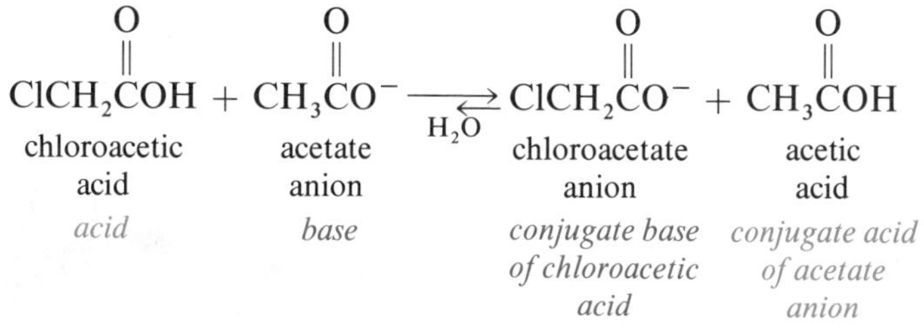

| chloroacetic acid | acetate anion | chloroacetate anion | acetic acid |

acid   *base*   *conjugate base of chloroacetic acid*   *conjugate acid of acetate anion*

$\Delta G°$ for this reaction at 298 K has been determined to be -2.59 kcal/mol. There is a *decrease* in free energy during the reaction; therefore, the equilibrium lies to the right. Acetate anion takes a proton away from chloroacetic acid; it is a stronger base than chloroacetate anion. Therefore, at equilibrium there is more chloroacetate anion present than acetate anion.

The equilibrium constant for the above reaction can be calculated from $\Delta G°$ and is also related to the acidity constants for chloroacetic acid and acetic acid by this equation:

$$K_{eq} = \frac{K_a \text{ of } ClCH_2CO_2H}{K_a \text{ of } CH_3CO_2H}$$

PROBLEM 3.19

Calculate K_{eq} for the reaction of chloroacetic acid and acetate anion and the pK_a for chloroacetic acid from the data given above. K_a for acetic acid is 1.75×10^{-5}.

The change in free energy, ΔG, of the system determines whether a reaction will take place as written. ΔG has two components: (1) the heat of the reaction, or the change in enthalpy, ΔH, and (2) the change in entropy during the reaction, ΔS. Entropy is a measure of disorder or randomness in the system. A restriction on the freedom of motion of or within molecules leads to a decrease in entropy. For example, a reaction in which two molecules combine to give a single one results in a decrease in entropy. A reaction in which a molecule dissociates to give two fragments gives an increase in entropy because each piece is free to move in ways that were not possible when they were bound together.

The relationship among the changes in free energy, enthalpy, and entropy during a reaction is summed up in the following equation.

$$\Delta G = \Delta H - T\,\Delta S$$

If ΔG is negative for a particular reaction, that is, if there is a decrease in free energy during the reaction, the reaction will take place as written. If ΔG is positive, the reaction will not proceed as written; instead, the reverse reaction will occur. If ΔG is zero, the reaction must be at equilibrium, with no net change observed.

Negative values for ΔH and positive values for ΔS contribute to making ΔG negative. In order words, exothermic reactions, reactions in which energy is lost to the surroundings as heat, will occur if there is not a large decrease in entropy. For many organic reactions, entropy changes tend to be small and predictions about reactions can be based on the values of the enthalpy of reaction alone. (Calculating enthalpies of reactions from average bond energies was described on p. 61.) However, although we have an intuitive feeling that exothermic reactions proceed easily and endothermic ones do not, this is not totally justified by the facts.

Study Guide Concept Map 3.4

3.6 THE EFFECTS OF STRUCTURAL CHANGES ON ACIDITY

A. The Resonance Effect

Acetic acid is a typical member of the family of organic compounds known as the carboxylic acids (p. 51). Carboxylic acids, in which a hydroxyl group is bonded to the carbon of a carbonyl group, are much stronger acids than are alcohols, in which the hydroxyl group is bonded to an sp^3-hybridized carbon atom. A simple experimental observation illustrates this difference in acidity. When a carboxylic acid is put into a solution containing bicarbonate ion, bubbles of carbon dioxide are seen.

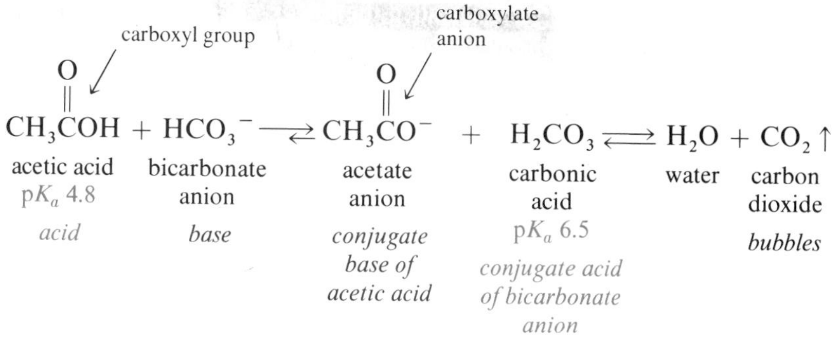
carboxyl group
carboxylate anion
acetic acid
pKa 4.8
acid
bicarbonate anion
base
acetate anion
conjugate base of acetic acid
carbonic acid
pKa 6.5
conjugate acid of bicarbonate anion
water
carbon dioxide
bubbles

When an alcohol is placed in a bicarbonate solution, there is no such evidence of reaction. Ethanol is not a strong enough acid to protonate bicarbonate ion.

$$CH_3CH_2OH + HCO_3^- \longrightarrow \text{no bubbles}$$

ethanol, pK_a 17; bicarbonate anion

Acetic acid, pK_a 4.8, on the other hand, is a stronger acid than carbonic acid, H_2CO_3, pK_a 6.5, and protonates bicarbonate anion, the conjugate base of carbonic acid. The stronger acid, acetic acid, transfers a proton to the conjugate base of the weaker acid, carbonic acid. Carbonic acid is in equilibrium with carbon dioxide dissolved in water; in the absence of external pressure (such as exists in bottles of soft drink), bubbles of carbon dioxide can be seen whenever carbonic acid is formed.

The difference in the acidities of ethanol and acetic acid can be rationalized by comparing the relative basicities of their conjugate bases. The conjugate base of acetic acid, acetate anion, is a weaker base than ethoxide anion, the conjugate base of ethanol. In the acetate anion, the negative charge is distributed equally to the two oxygen atoms. This delocalization of charge (p. 80) by resonance stabilizes the anion. An anion in which the charge is delocalized to two oxygen atoms is a weaker base than one in which the charge is concentrated on a single oxygen atom, as it is in the ethoxide anion.

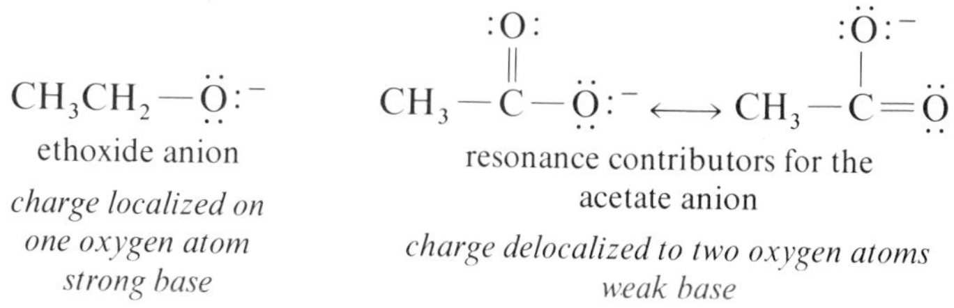
ethoxide anion
charge localized on one oxygen atom
strong base
resonance contributors for the acetate anion
charge delocalized to two oxygen atoms
weak base

In summary, one factor that affects the basicity of an anion and therefore the acidity of its conjugate acid is whether there can be delocalization of charge over several atoms by resonance. The resonance effect is used to account for part of the observed difference of about 12 units between the pK_a of ethanol and that of acetic acid. It is a strong and important effect.

B. Inductive and Field Effects

Other structural features of acids also affect their acidity. For example, one of the hydrogen atoms on the carbon atom attached to the carboxyl group of acetic acid can be replaced by another group or atom. The carbon atom adjacent to a carboxyl group is called the **α-carbon** of the acid, and the hydrogen atoms bonded to it are the **α-hydrogen atoms.**

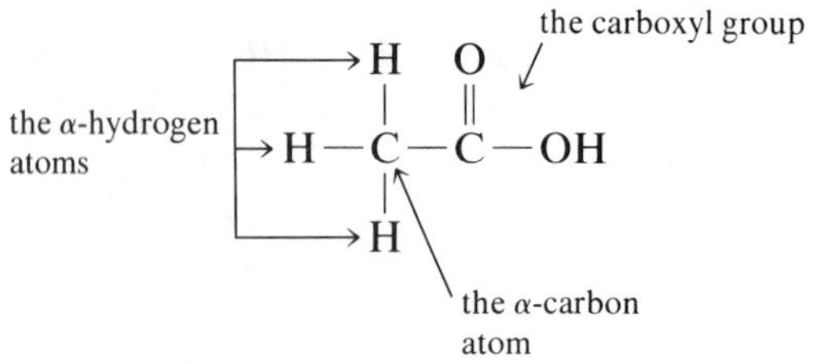
the carboxyl group
the α-hydrogen atoms
H
O
H—C—C—OH
H
the α-carbon atom

In the series of compounds shown in Table 3.1, an α-hydrogen atom in acetic acid has been replaced in turn by a methyl group, a fluorine atom, a chlorine atom, a bromine atom, and an iodine atom. The pK_a values given indicate that acidity is decreased slightly by substitution of a methyl group, but increased by substitution of a halogen atom. Substitution by a fluorine atom at the α-carbon produces the

TABLE 3.1 The Acidity of Substituted Carboxylic Acids Determined in Water at 25 °C

Name	Structure	pK_a
acetic acid	$\underset{\|\atop H}{CH_2}\overset{O\atop \|}{C}OH$	4.76
propanoic acid	$\underset{\|\atop CH_3}{CH_2}\overset{O\atop \|}{C}OH$	4.87
fluoroacetic acid	$\underset{\|\atop F}{CH_2}\overset{O\atop \|}{C}OH$	2.59
chloroacetic acid	$\underset{\|\atop Cl}{CH_2}\overset{O\atop \|}{C}OH$	2.86
bromoacetic acid	$\underset{\|\atop Br}{CH_2}\overset{O\atop \|}{C}OH$	2.90
iodoacetic acid	$\underset{\|\atop I}{CH_2}\overset{O\atop \|}{C}OH$	3.17

TABLE 3.2 The Effect on Acidity (Determined in Water at 25 °C) of the Distance Between the Substituent and the Carboxyl Group

Name	Structure	pK_a
butanoic acid	$CH_3CH_2CH_2\overset{\overset{O}{\|}}{C}OH$	4.82
2-chlorobutanoic acid	$CH_3CH_2\underset{\underset{Cl}{\vert}}{CH}\overset{\overset{O}{\|}}{C}OH$	2.86
3-chlorobutanoic acid	$CH_3\underset{\underset{Cl}{\vert}}{CH}CH_2\overset{\overset{O}{\|}}{C}OH$	4.05
4-chlorobutanoic acid	$\underset{\underset{Cl}{\vert}}{CH_2}CH_2CH_2\overset{\overset{O}{\|}}{C}OH$	4.52

greatest increase in acidity. The acidity decreases as the substituted atom changes from fluorine to chlorine to bromine to iodine.

The pK_a values for another series of acids illustrate the effect of a halogen atom that has been substituted farther and farther away from the carboxyl group (Table 3.2). Butanoic acid has an acidity very close to that of propanoic acid, as shown in Table 3.1.

$$\underset{\text{p}K_a\ 4.82}{\underset{\text{butanoic acid}}{CH_3CH_2CH_2\overset{\overset{O}{\|}}{C}OH}} \qquad \underset{\text{p}K_a\ 4.87}{\underset{\text{propanoic acid}}{CH_3CH_2\overset{\overset{O}{\|}}{C}OH}}$$

Acidity does not change significantly with the size of the alkyl group attached to the carboxyl group. A chlorine atom on the α-carbon atom increases the acidity of butanoic acid, but the effect falls off rapidly as the chlorine atom is moved farther from the carboxyl group in 3-chlorobutanoic acid and still farther in 4-chlorobutanoic acid.

These data suggest that somehow the acidity of an acid increases when electronegative atoms are introduced into the molecule. The greater the electronegativity of the substituent, the greater is the increase in acidity (Table 3.1). This increase in acidity also depends on the distance between the substituent and the carboxyl group (Table 3.2).

Chemists explain these facts in two ways. An electronegative substituent, such as a chlorine atom, draws the electrons of the carbon-chlorine bond toward the chlorine atom, creating an electron deficiency in that carbon atom and in neighboring carbon atoms. This effect results in a withdrawal of electron density from the carboxyl group, which acts to stabilize the carboxylate anion by making its electrons less available to a proton. The carboxylate anion becomes a weaker base; consequently, the corresponding conjugate acid is stronger.

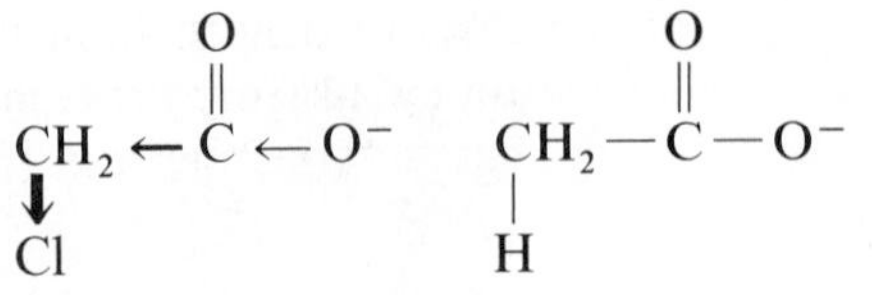

chloroacetate anion

stabilized by the inductive effect, a weaker base than acetate anion

acetate anion

The transmission of the effect of an electron-withdrawing (or electron-donating) group through σ bonds is called an **inductive effect.** It diminishes rapidly as the number of bonds between the substituent and the carboxyl group increases.

The **field effect** arises in the bond dipole moments of a molecule and is transmitted not through the bonds but through the environment around the molecule. For example, chloroacetate ion is stabilized relative to acetate ion by the bond dipole of the carbon-chlorine bond.

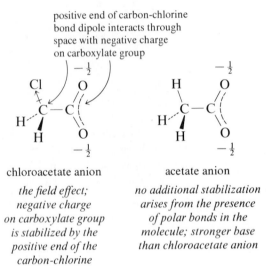

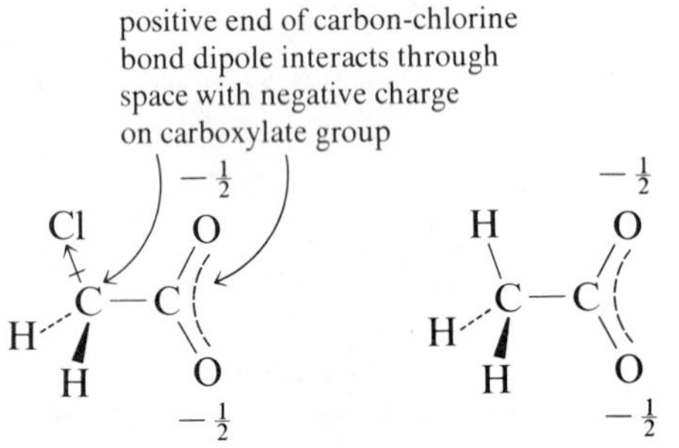
positive end of carbon-chlorine bond dipole interacts through space with negative charge on carboxylate group

chloroacetate anion

the field effect; negative charge on carboxylate group is stabilized by the positive end of the carbon-chlorine bond dipole

acetate anion

no additional stabilization arises from the presence of polar bonds in the molecule; stronger base than chloroacetate anion

It is theorized that the positive end of the dipole of the carbon-chlorine bond in chloroacetate anion interacts directly through space with the negative charge on the carboxylate anion to reduce charge density at the carboxylate group. The effect of the carbon-chlorine bond dipole on the carboxylate anion decreases as the distance between the chlorine atom and the carboxylate group increases (Table 3.2).

The inductive effect and the field effect operate in the same direction. Many ingenious experiments have been devised to measure these effects separately. The most successful experiments indicate that the transmission of the effect through space by interaction of bond dipoles is more important than the transmission of the effect through bonds. Customarily, inductive effect is used to refer to the combination of both of these effects.

In water, propanoic acid is slightly weaker than acetic acid. Chemists have explained the acidity of propanoic acid in water by suggesting that there must be a slight release of electrons from the added methyl group toward the carboxyl group. Such a release of electrons destabilizes the carboxylate anion, making it more basic and more easily protonated; thus, its conjugate acid is correspondingly weaker. Many experimental results can be rationalized if we propose that an alkyl group, especially one bonded to an sp^2- or sp-hybridized carbon atom, is more electron-releasing than a hydrogen atom is. The theoretical explanations given for this effect vary, but the effect itself is clearly seen.

PROBLEM 3.20

Predict whether CF_3CH_2OH is more or less acidic than CH_3CH_2OH. Why?

PROBLEM 3.21

Look up the pK_a values for chloroacetic acid, $CH_2Cl\overset{O}{\overset{\|}{C}}OH$, dichloroacetic acid, $CHCl_2\overset{O}{\overset{\|}{C}}OH$, and trichloroacetic acid, $CCl_3\overset{O}{\overset{\|}{C}}OH$, in the table of p$K_a$ values. How do you explain the data?

Other substituents that increase the acidity of carboxylic acids are the hydroxyl, cyano, and nitro groups (Table 3.3). In hydroxyacetic acid, the electronegative oxygen atom substituted on the α-carbon atom reduces electron density at the carboxylate anion, making it a weaker base in comparison with the unsubstituted acetate ion. Consequently, hydroxyacetic acid is a stronger acid than acetic acid.

The cyano and nitro groups are strongly electron-withdrawing. Note that nitroacetic acid is a stronger acid than fluoroacetic acid (Table 3.1) and cyanoacetic acid is slightly stronger than fluoroacetic acid. The powerful electron-withdrawing effects of these groups are rationalized by pointing to their electronic structures. In each case, the atom attached to the α-carbon atom of the acid either is positively charged or may be positively charged in an important resonance contributor. The presence

TABLE 3.3 The Effects of Electron-Withdrawing Groups on the Acidity of Carboxylic Acids

Name	Structure	pK_a
acetic acid	$\underset{H}{\underset{\vert}{CH_2}}\overset{O}{\overset{\|}{C}}OH$	4.76
hydroxyacetic acid	$\underset{OH}{\underset{\vert}{CH_2}}\overset{O}{\overset{\|}{C}}OH$	3.83
cyanoacetic acid	$\underset{CN}{\underset{\vert}{CH_2}}\overset{O}{\overset{\|}{C}}OH$	2.46
nitroacetic acid	$\underset{NO_2}{\underset{\vert}{CH_2}}\overset{O}{\overset{\|}{C}}OH$	1.68

of this positive charge in the substituent contributes to the stabilization of the carboxylate anion.

PROBLEM 3.22

Draw Lewis structures for cyanoacetic acid and nitroacetic acid (Table 3.3). Draw resonance contributors for them. Explain in each case why the substituent increases the acidity of the acid over that of acetic acid, and also explain why nitroacetic acid is stronger than cyanoacetic acid.

In summary, an electron-withdrawing atom or group substituted on the alkyl chain of a carboxylic acid stabilizes the corresponding carboxylate anion, thereby increasing the acidity of the carboxylic acid. Important electron-withdrawing groups are the halogens, hydroxyl and ether groups, the cyano group, and the nitro group. Alkyl groups are weakly electron-releasing.

A word of caution is necessary. The relative acidities on which these generalizations are based were determined in water. In the gas phase, reversals in the order of related compounds are often seen. For example, in the gas phase, bromoacetic acid is a stronger acid than fluoroacetic acid. The models presented above, which rationalize acidity solely in terms of the electronegativity of the substituents and of the polarization of bonds within molecules, are too simple. Such models are retained, however, because they are enormously useful in rationalizing a wide range of experimental observations and in making predictions about reactivity in many different systems. Resonance effects and inductive effects will be invoked frequently as we try to understand how the structures of organic compounds determine their reactivity.

3.7 THE EFFECTS OF STRUCTURAL CHANGES ON BASICITY

A. Carbanions

Methane is an extremely weak acid and will not react with hydroxide ion, but trichloromethane (chloroform) does react to give low concentrations of trichloromethyl anion.

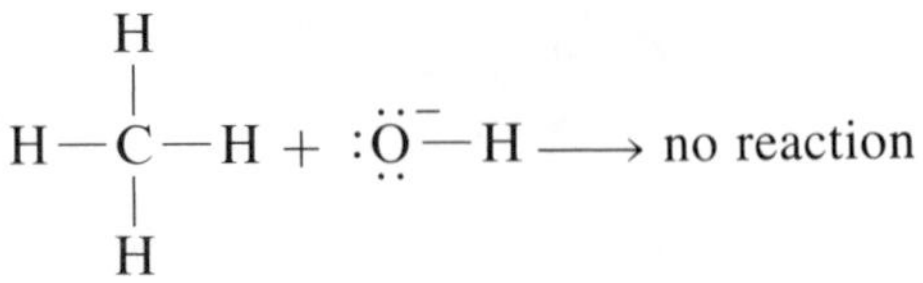
no reaction

methane (pK_a ~50) hydroxide ion

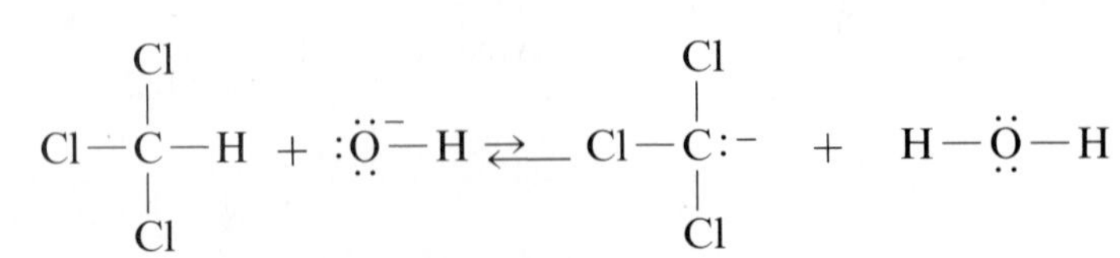

trichloromethane (pK_a ~25, *acid*) hydroxide ion (*base*) trichloromethyl anion (*conjugate base of trichloromethane*) water (pK_a 15.7, *conjugate acid of hydroxide ion*)

The substitution of electron-withdrawing chlorine atoms for three of the hydrogen atoms in methane makes the remaining hydrogen atom acidic enough to react with hydroxide ion. Trichloromethane is a stronger acid than methane because the negative charge on the trichloromethyl anion can be stabilized by the inductive effect due to the chlorine atoms.

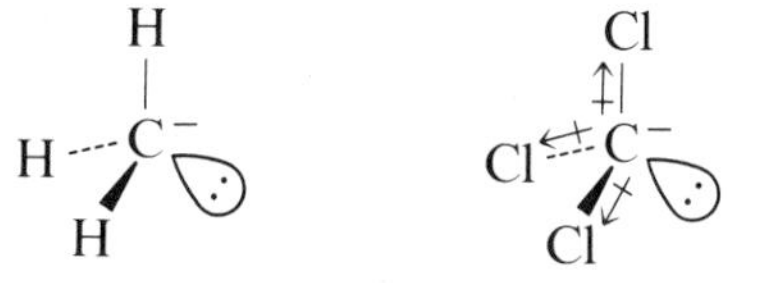

methyl anion

very strong base

trichloromethyl anion

weaker base than methyl anion

The dipoles of the carbon-chlorine bonds act to disperse some of the negative charge of the anionic carbon atom. Thus, the carbanion from trichloromethane is a weaker base than the carbanion derived from methane.

PROBLEM 3.23

Would you expect triiodomethane (iodoform), CHI_3, to be a stronger or weaker acid than trichloromethane, $CHCl_3$? Explain.

Other substituents besides halogens are effective in increasing the acidity of nearby carbon-hydrogen bonds. The nitro group greatly increases the acidity of a carboxylic acid (Table 3.3), and its electron-withdrawing effect is also seen in nitromethane, CH_3NO_2, which, with pK_a 10.2, is a much stronger acid than methane or even trichloromethane. The conjugate base of nitromethane is stabilized by the negative inductive effect of the nitro group and also by resonance.

VISUALIZING THE REACTION

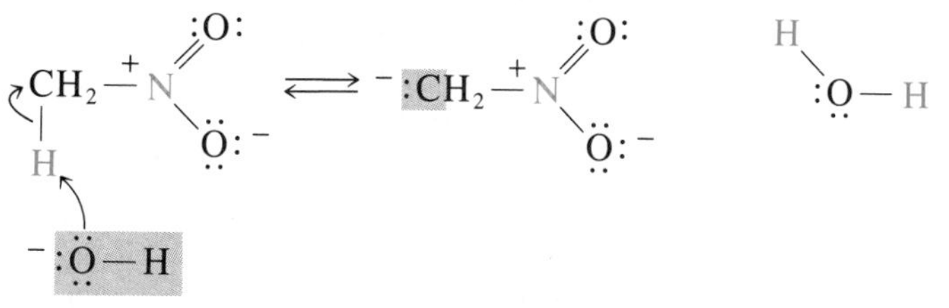

nitromethane and hydroxide ion

carbanion from nitromethane

water

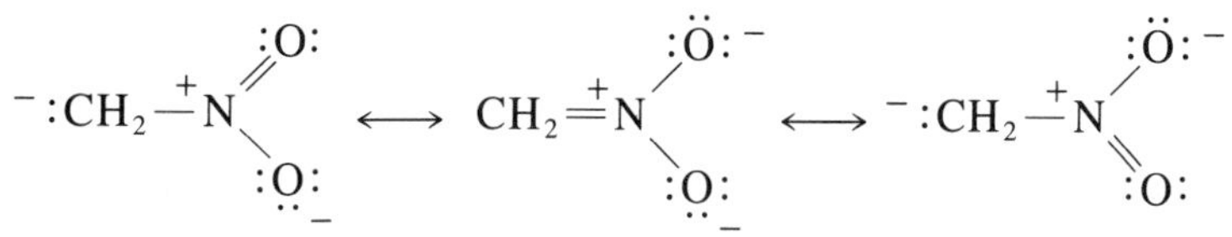

resonance contributors for the carbanion from nitromethane

For the carbanion from nitromethane, we can write resonance contributors that delocalize the negative charge on the carbon atom to the oxygen atoms in the nitro group. The delocalization of charge from the carbon atom to the much more electronegative oxygen atom decreases the basicity of the anion compared to that of a carbanion such as the trichloromethyl anion, where the charge must remain on a carbon atom. Whenever resonance contributors that delocalize charge can be written for an ion, stabilization of the ion is postulated.

Stabilization by resonance is always much more important than stabilization by the inductive effect. Thus, chemists use resonance to explain the low basicities of the carboxylate anion and the carbanion from nitromethane. Subsequent chapters contain many other examples in which we can rationalize the stability of ions by writing resonance contributors that delocalize the charge on the ion.

PROBLEM 3.24

For each of the following compounds, identify the most acidic proton(s) by writing a mechanism, using the curved-arrow convention, for the reaction with hydroxide ion. What factors are important in the stabilization of each organic anion?

(a) $HOCH_2\overset{\overset{\displaystyle O}{\|}}{C}OH$ (b) $CH_3CH_2CH_2NO_2$

(c) $HSCH_2CH_2CH_2OH$ (d) $Br\underset{\underset{\displaystyle Br}{|}}{C}H\overset{\overset{\displaystyle O}{\|}}{C}CH_3$

B. Amines

Amines (p. 45) are the most important organic bases. The basicities of amines are determined by the relative availability of the nonbonding electrons on the nitrogen atom to a proton donor or Lewis acid and by the stabilization of the positively charged nitrogen atom by solvation or, in some special cases, by resonance. On p. 89, the basicities of various species were related to the pK_a values of their conjugate acids. For an amine, too, the larger the pK_a of its conjugate acid, the weaker the acid and the stronger the corresponding base.

Amines in which alkyl groups are substituted on the nitrogen atom have basicities similar to that of ammonia. Their conjugate acids are alkylammonium ions with pK_a values of approximately 10.8 to 9.5. Methylamine, dimethylamine, and trimethylamine are protonated by water just as ammonia is. All amines with low molecular weights, in which the nitrogen atom is a significant portion of the molecule, dissolve in water to give solutions that turn red litmus paper blue.

$$\underset{\text{ammonia}}{NH_3} + \underset{\text{p}K_a\ 15.7}{H_2O} \rightleftharpoons \underset{\substack{\text{ammonium}\\ \text{ion}\\ \text{p}K_a\ 9.2}}{NH_4^+} + OH^-$$

$$\underset{\text{methylamine}}{CH_3NH_2} + \underset{\text{p}K_a\ 15.7}{H_2O} \rightleftharpoons \underset{\substack{\text{methylammonium}\\ \text{ion}\\ \text{p}K_a\ 10.6}}{CH_3NH_3^+} + OH^-$$

$$(CH_3)_2NH + H_2O \rightleftharpoons (CH_3)_2NH_2^+ + OH^-$$

dimethylamine — pK_a 15.7 — dimethylammonium ion, pK_a 10.7

$$(CH_3)_3N + H_2O \rightleftharpoons (CH_3)_3NH^+ + OH^-$$

trimethylamine — pK_a 15.7 — trimethylammonium ion, pK_a 9.8

The one methyl group on the nitrogen atom in methylamine and the two methyl groups in dimethylamine increase basicity. Trimethylamine is slightly more basic than ammonia but is a weaker base than dimethylamine. The three alkyl groups around the nitrogen atom interfere with protonation and with stabilization of the cation by solvation, making the amine less basic (Figure 3.1)

An amine with a nitrogen atom bonded directly to an aromatic ring is a much weaker base than ammonia. For example, the conjugate acid of aniline is slightly more acidic (pK_a 4.6) than acetic acid (pK_a 4.8) and much more acidic than methylammonium ion (pK_a 10.6).

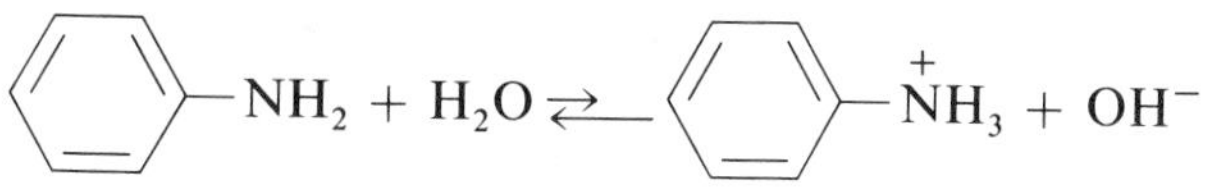

NH2 + H2O
NH3 + OH−

aniline *base* — anilinium ion pK_a 4.6 *conjugate acid*

The nonbonding electrons on the nitrogen atom in aniline are said to be less available for reactions with acids for two reasons. (1) The nitrogen is bonded to an sp^2-hybridized carbon atom of the aromatic ring (p. 65), which is more electronegative (p. 56) than the sp^3-hybridized carbon atom of methylamine. (2) The nonbonding electrons can be delocalized to the aromatic ring. Resonance contributors for aniline indicate that it has decreased electron density at the nitrogen atom and increased electron density at three carbon atoms in the ring.

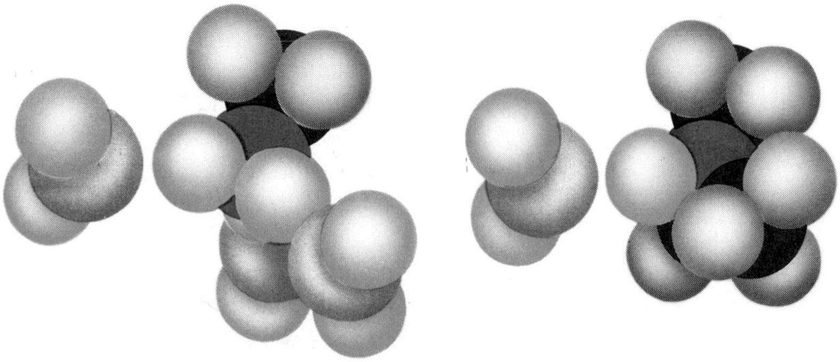

methylammonium ion hydrogen bonding with three water molecules — trimethylammonium ion hydrogen bonding with only one water molecule

FIGURE 3.1 A comparison of the solvation of trimethylammonium and methylammonium ions.

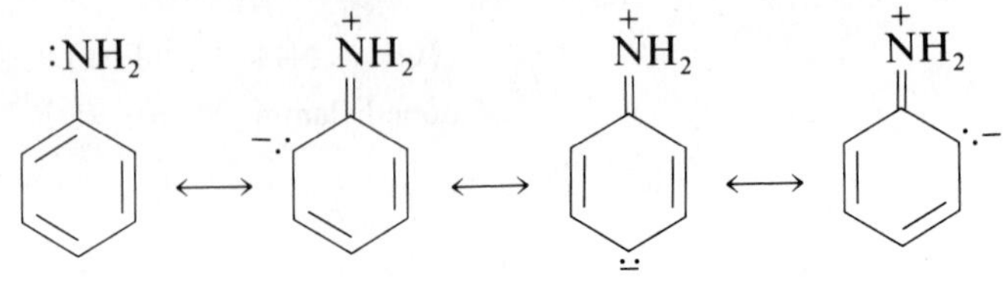

resonance contributors for aniline

The most important resonance contributor is the first one, in which there is no separation of charge and the π electrons of the aromatic ring (p. 65) are undisturbed.

The data given in this section and the preceding one show that resonance effects are more important than inductive effects in influencing acidity or basicity. Aniline, in which the nonbonding electrons on the nitrogen atom can be delocalized by resonance, is a much weaker base than ammonia. The basicity of the methylamines, in which resonance effects are not possible, differs only slightly from that of ammonia.

Study Guide Concept Map 3.5

PROBLEM 3.25

Arrange the following amines in order of increasing basicity.

$(CH_3CH_2)_2NH$, $CH_3CH_2NH_2$, 

PROBLEM 3.26

The pK_a of the conjugate acid of diphenylamine is 0.8. Is diphenylamine a weak or strong base? How would you rationalize this conclusion?

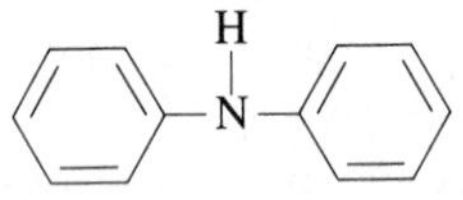

diphenylamine

S U M M A R Y

Many organic reactions can be classified as acid-base reactions. Two ways of defining acids and bases are useful. According to the Brønsted-Lowry definition, an acid is a proton donor, and a base, a proton acceptor. A Brønsted acid is also known as a protic acid. According to the Lewis definition, an acid is an electron-pair acceptor, and a base, an electron-pair donor. In general, any species, including a proton, that can accept a pair of electrons is classified as a Lewis acid. Important Lewis acids include compounds such as boron trifluoride, BF_3, and aluminum chloride, $AlCl_3$. Such acids are known as aprotic acids.

Atoms bearing nonbonding electrons, such as oxygen, nitrogen, and sulfur, are the basic sites in molecules or ions. Protic acids transfer a proton to such atoms, and Lewis acids coordinate with them. The electrons in π bonds also react with protic acids or Lewis acids.

The basicity of a species depends on the position in the periodic table of the atom bearing the nonbonding electrons. Basicity decreases for a series of atoms that go from left to right across any period or down any group. An anion containing a given atom is always more basic than a neutral molecule containing the same atom (for example, OH^- is more basic than H_2O).

The acidities of protic acids are usually expressed as their pK_a values. The loss of a proton converts a protic acid into its conjugate base. Similarly, a base that gains a proton is converted into its conjugate acid. A strong acid has a weak conjugate base, and a weak acid has a strong conjugate base. Bases can be stabilized by delocalization of electron density through resonance or inductive effects. These stabilized bases are weaker (have stronger conjugate acids) than bases in which such effects are not as important.

ADDITIONAL PROBLEMS

3.27 Arrange the following compounds in order of increasing acidity of the hydrogen atoms.

AlH_3, H_2S, HCl, NaH

3.28 Arrange the following compounds in order of increasing acidity.

(a) $Cl^-\ H_3\overset{+}{N}CH_2\overset{\overset{O}{\|}}{C}OH$, $ClCH_2\overset{\overset{O}{\|}}{C}OH$, $CH_3CH_2\overset{\overset{O}{\|}}{C}OH$

(b) $CF_3\overset{\overset{O}{\|}}{C}OH$, $CCl_3\overset{\overset{O}{\|}}{C}OH$, $CH_3\overset{\overset{O}{\|}}{C}OH$

(c) CH_3CH_2OH, FCH_2CH_2OH, $ClCH_2CH_2OH$

(d) CH_3CH_2SH, CH_3CH_2OH, $CH_3CH_2NH_2$

(e) $CH_3\overset{\overset{O}{\|}}{C}OH$, $CH_3\overset{\overset{^+OH}{\|}}{C}OH$, $CH_3\overset{\overset{O}{\|}}{C}OCH_3$

(f) $CH_3\overset{\overset{O}{\|}}{C}OH$, $CH_3\overset{\overset{O}{\|}}{C}NH_2$, $CH_3\overset{\overset{O}{\|}}{C}OCH_3$

3.29 Arrange each group of species in order of increasing basicity.

(a) $Cl_3C\!:^-$, $CH_3\overset{\overset{CH_3}{|}}{\underset{..}{N}}CH_3$, $CH_3\ddot{\underset{..}{O}}CH_3$ (b) $\ddot{N}F_3$, $\ddot{N}H_3$, $\ddot{N}H_2\ddot{\underset{..}{O}}H$

(c) $\ddot{N}H_3$, $:\ddot{N}H_2^-$, NH_4^+ (d) $CH_3\ddot{\underset{..}{S}}CH_3$, $CH_3\overset{\overset{CH_3}{|}}{\underset{..}{P}}CH_3$, $CH_3\overset{\overset{CH_3}{|}}{\underset{\underset{CH_3}{|}}{Si}}CH_3$

3.30 The conjugate acids of the following amines have the pK_a values shown. Explain the observed trend.

	pK_a of conjugate acid
$CH_3CH_2CH_2CH_2NH_2$	10.60
$CH_3OCH_2CH_2CH_2NH_2$	9.92
$CH_3O\underset{\underset{OCH_3}{\vert}}{C}HCH_2NH_2$	8.54
$N{\equiv}CCH_2CH_2NH_2$	7.80

3.31 For each acid-base reaction given below, label acids, bases, conjugate acids, and conjugate bases, and tell whether Brønsted acids or Lewis acids are present.

(a) $F^- + BF_3 \longrightarrow BF_4^-$ (b) $Ag^+ + 2\,NH_3 \longrightarrow Ag(NH_3)_2^+$

(c) $Al(H_2O)_6^{3+} + OH^- \longrightarrow Al(OH)(H_2O)_5^{2+} + H_2O$

(d) $CH_3CH_2\overset{\overset{O}{\|}}{C}OH + OH^- \longrightarrow CH_3CH_2\overset{\overset{O}{\|}}{C}O^- + H_2O$

(e) $CH_3CH_2\overset{\overset{CH_2CH_3}{|}}{N}CH_2CH_3 + CF_3\overset{\overset{O}{\|}}{C}OH \longrightarrow CH_3CH_2\underset{+}{\overset{\overset{CH_2CH_3}{|}}{N}}HCH_2CH_3 + CF_3\overset{\overset{O}{\|}}{C}O^-$

(f) $CH_3CH_2SCH_2CH_3 + BF_3 \longrightarrow CH_3CH_2\underset{+}{\overset{\overset{^-BF_3}{|}}{S}}CH_2CH_3$

(g) $ClCH_2\overset{\overset{O}{\|}}{C}OH + HCO_3^- \longrightarrow ClCH_2\overset{\overset{O}{\|}}{C}O^- + H_2CO_3$

(h) $H_2PO_4^- + OH^- \longrightarrow HPO_4^{2-} + H_2O$

(i) $CH_3CH_2CH_2SH + CH_3CH_2O^- \longrightarrow CH_3CH_2CH_2S^- + CH_3CH_2OH$

3.32 Use the table of pK_a values to decide whether or not the following reactions will take place. Complete the equations for those that will take place. Write "no reaction" for the others.

(a) 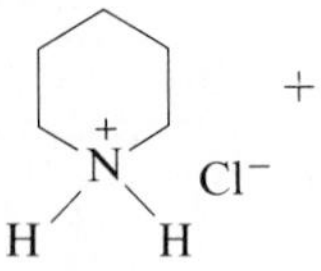 $+ NaOH \longrightarrow$ (b) $CH_3CH_2\underset{+}{\overset{\overset{H}{|}}{O}}CH_2CH_3 + H_2O \longrightarrow$

(c) $CH_3\overset{\overset{O}{\|}}{C}OH + CH_3CH_2SNa \longrightarrow$ (d) $CCl_3\overset{\overset{O}{\|}}{C}OH + NaHCO_3 \longrightarrow$

(e) $CH_3CH_2\overset{\overset{O}{\|}}{C}OH + NH_3 \longrightarrow$ (f) $CH_3NO_2 + CH_3CH_2ONa \longrightarrow$

(g) $CH_3CH_2SH + NaNH_2 \longrightarrow$ (h) $HC{\equiv}CH + NaHCO_3 \longrightarrow$

3.33 Complete the following equations.

(a) 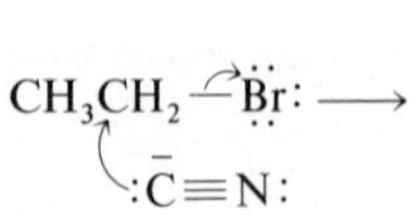

(b) 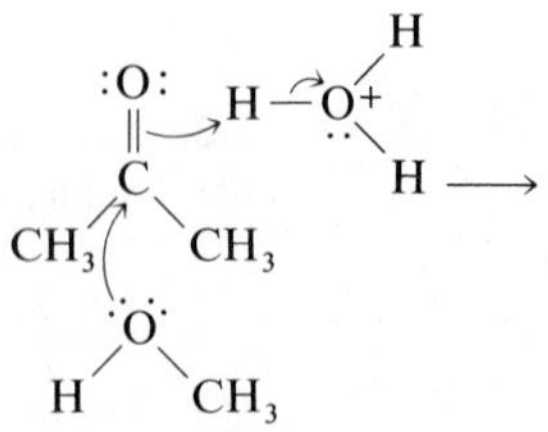

(c) 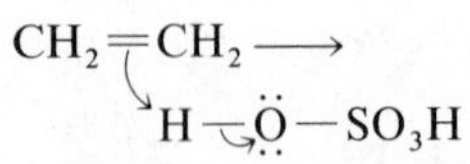

(d) 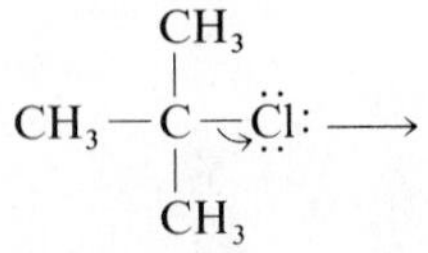

(e) 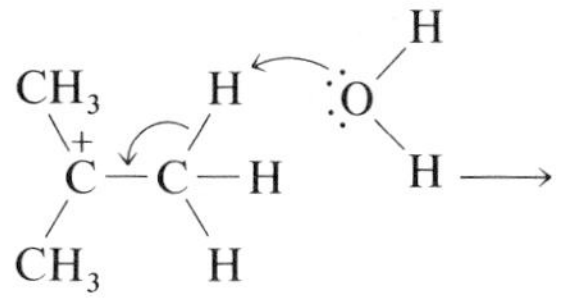

(f) 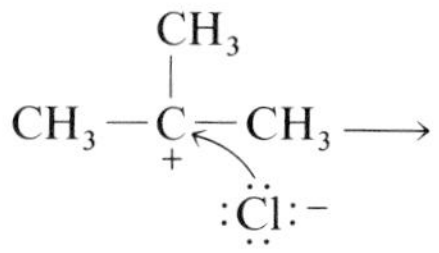

(g) 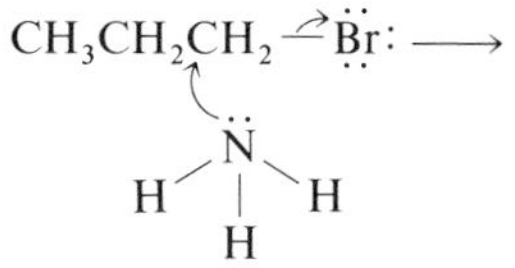

(h) 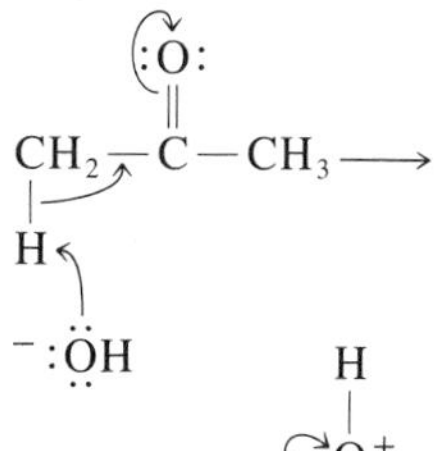

(i) 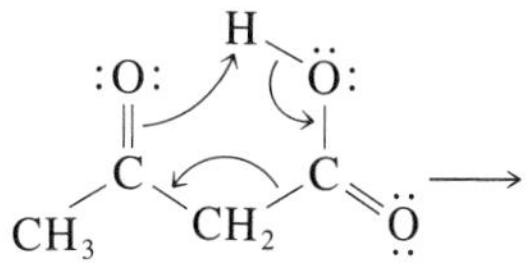

(j) 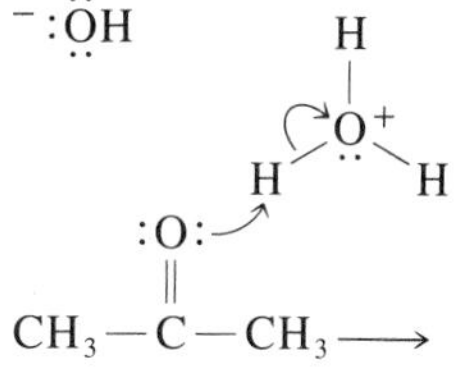

(k) 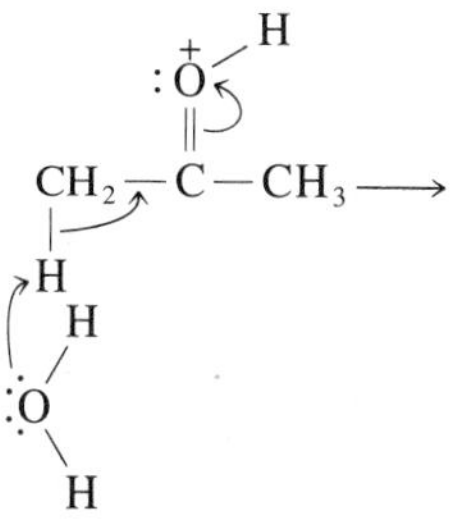

3.34 For the following compounds, pK_a values are given. Calculate the acidity constants.

(a) $CHCl_2\overset{\overset{\displaystyle O}{\|}}{C}OH$, p$K_a$ 1.3 (b) $CH_3NH_3^+$, pK_a 10.4 (c) CCl_3CH_2OH, pK_a 12.2

3.35 The following compounds have the acidity constants shown. What are their pK_a values?

(a) CH_3CH_2OH, $K_a = 10^{-17}$ (b) CH_3CH_2SH, $K_a = 3.16 \times 10^{-11}$

(c) $CH_3CH_2-\overset{\overset{\displaystyle H}{|}}{\underset{+}{O}}-CH_2CH_3$, $K_a = 3.98 \times 10^3$

3.36 Proteins are built up from units called amino acids. The simplest amino acid is glycine, the structure of which can be written in two ways:

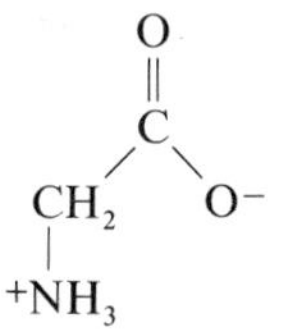 and 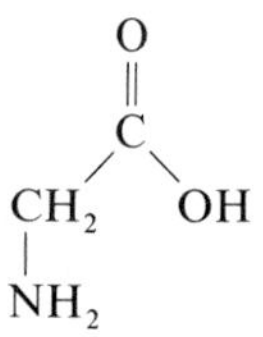

(a) Consult the table of pK_a values and decide which is the better representation of the structure of the compound. (Hint: Which is more basic, an amino group or a carboxylate anion?)

(b) Glycine is a crystalline solid that decomposes at 233 °C as it melts. It has a solubility of 26 g in 100 mL of water. Do these facts fit the structure that you have chosen for glycine? Explain.

3.37 Each of the following compounds has two sites where protonation might occur. For each compound, write a reaction with hydronium ion showing the ionic species that will be formed. (Hint: You may wish to check the table of pK_a values to decide on the relative basicities of some of the sites.)

(a) 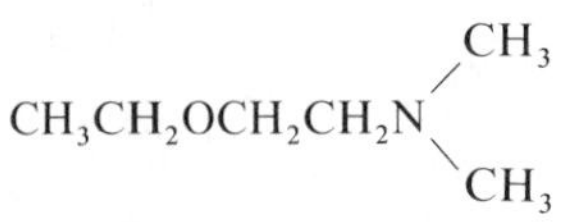

(b) 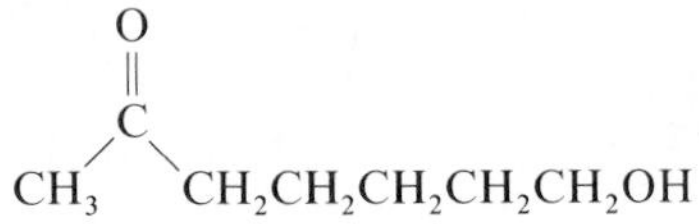

(c) $CH_3C{\equiv}CCH_2CH_2OCH_3$

3.38 The acidity constant for any acid, HA, in water is

$$K_a = \frac{[A^-][H_3O^+]}{[HA]}$$

This expression for K_a can be converted into another one showing the relationship between the pK_a of an acid and the pH of the solution:

$$pH = pK_a + \log\frac{[A^-]}{[HA]}$$

This expression, sometimes called the Henderson-Hasselbach equation, is useful in calculating the relative amounts of dissociated and undissociated acid present at any pH when the pK_a of the acid is known.

(a) Derive the Henderson-Hasselbach equation from the expression for K_a.
(b) What does the Henderson-Hasselbach equation tell you about the concentrations of HA and A^- when pK_a = pH?
(c) In Problem 3.15, you learned that the imidazole ring in the amino acid histidine is important in proton-transfer reactions in the human body. The protonated form of imidazole has pK_a 7.0. The pH of body fluids is ~6.5. What can you say in a qualitative way about the relative quantities of protonated and unprotonated imidazole rings present in the human body?

3.39 A research paper in the *Journal of the American Chemical Society* describes the preparation of trifluoromethanol, CF_3OH, and trifluoromethylamine, CF_3NH_2, two compounds that had not been prepared until recently because of their tendency to lose HF to form multiple bonds. The compounds were found to be reasonably stable at temperatures around 0 °C. The authors of the paper report that "CF_3OH is certainly not a typical alcohol but rather an acid." They also say "CF_3OH is of surprisingly high volatility" (meaning that it has a much lower boiling point than expected). For CF_3NH_2, they report that "the basic character of CF_3NH_2 is lower than the basicity of normal organic bases like $(CH_3)_3N$."

(a) How would you rationalize each of the three observations quoted?
(b) On the basis of the information given above, complete the following equation either by writing "no reaction" or by writing in the products.

$$CF_3NH_3^+\ Cl^- + (CH_3)_3N \longrightarrow$$

3.40 When strong acids or strong bases are dissolved in a solvent such as water, a phenomenon known as the leveling effect is observed. Differences in acidity for strong acids such as sulfuric acid and hydrochloric acid cannot be measured in water. They appear to have the same acidity in that solvent. What is happening? What is the strongest acid that can exist in aqueous solution? What is the strongest base? Write equations illustrating your answers.

4

Reaction Pathways

A • L O O K • A H E A D

In Chapter 3, you learned to make predictions about one important class of reactions, proton-transfer reactions between Brønsted acids and bases. You also learned to recognize Lewis acids, which react by accepting pairs of electrons, and Lewis bases, which are electron-pair donors.

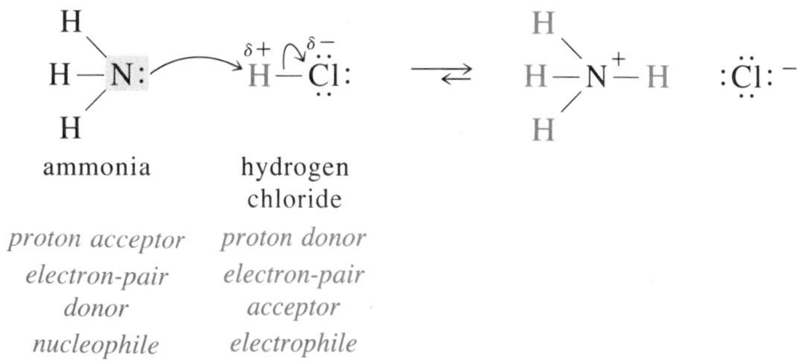
ammonia
hydrogen chloride
proton acceptor
electron-pair donor
nucleophile
proton donor
electron-pair acceptor
electrophile

Hydrogen chloride, an acid, is also an electrophile, a species that is ''electron-seeking'' and can accept a pair of electrons. Ammonia, a base, is a nucleophile, a species ''seeking a nucleus'' to which to donate a pair of electrons.

In this chapter, we will examine two important classes of reactions that result from interactions of nucleophiles and electrophiles. In nucleophilic substitution reactions, a nucleophile displaces a group already present in another molecule.

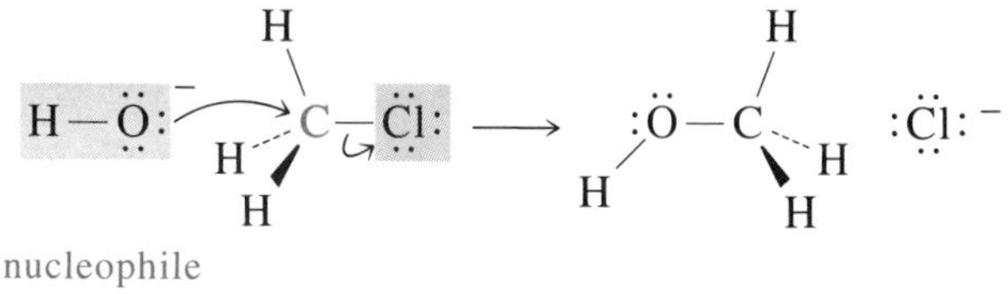
nucleophile

a nucleophilic substitution reaction

In electrophilic addition reactions, an acid adds to a double bond.

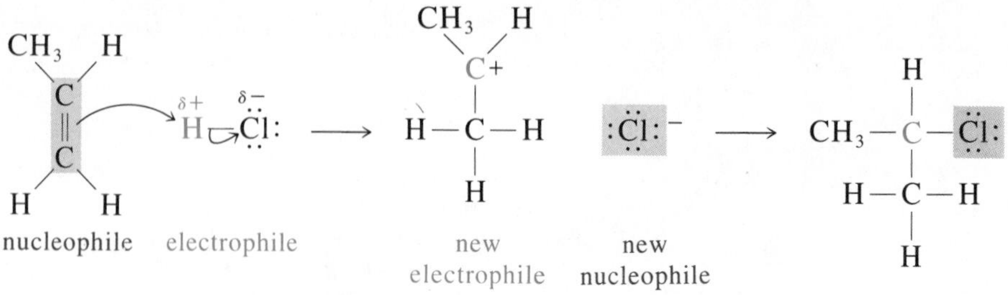
nucleophile
electrophile
new electrophile
new nucleophile

an electrophilic addition reaction

We will study in depth the two reactions shown above and will apply the ideas about chemical reactivity they illustrate to many other reactions.

4.1 INTRODUCTION. ELECTROPHILES AND NUCLEOPHILES

Two acid-base reactions are shown below.

VISUALIZING THE REACTION

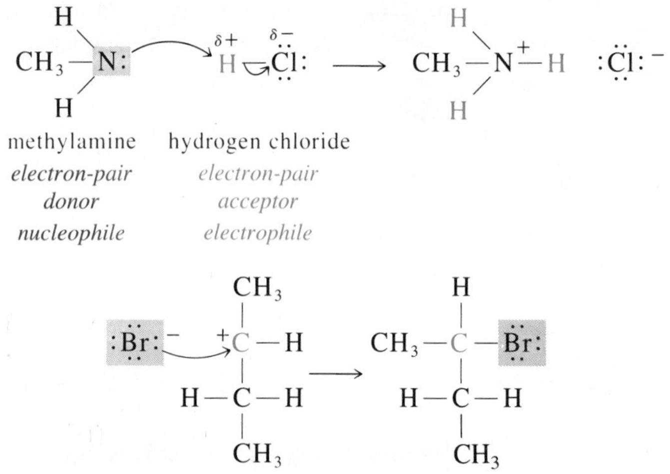
methylamine
electron-pair donor
nucleophile
hydrogen chloride
electron-pair acceptor
electrophile
bromide ion
electron-pair donor
nucleophile
2-butyl cation
electron-pair acceptor
electrophile

The acids in these reactions are examples of **electrophiles,** ''electron-seeking'' reagents that have room in their orbitals to accept a pair of electrons. Electrophiles are shown in red in the equations above. Hydrogen chloride is an electrophile because the electrons of the covalent bond between hydrogen and chlorine are pulled closer to the more electronegative atom, chlorine. Hydrogen has a partial positive charge and is therefore attractive to the nonbonding electrons of methylamine. Carbon in the 2-butyl cation is an electrophile because it has only six electrons in its outermost shell, leaving room in its orbitals to accept a pair of electrons from the bromide ion.

The species that react with electrophiles are called **nucleophiles,** reagents ''seeking a nucleus'' to which to donate a pair of electrons. In the examples shown above, the nucleophiles, methylamine and bromide ion, are identified by blue shaded boxes. Most Lewis and Brønsted bases are nucleophiles.

In predicting whether a reaction between two species is likely, we must consider three factors. First, we examine the structures of the two species to see whether there are sites of electron deficiency and electron density. In other words, there must be an electron acceptor and an electron donor, an electrophile, and a nucleophile, for each reaction. In most cases, the identities of the electrophile and the nucleophile are easy to determine, though sometimes they are not obvious. But in every reaction we will be looking for structural features that allow for interactions between orbitals with easily available electrons and orbitals with room to accept these electrons. Only then is there the possibility for the formation of a new bond.

The second factor we must consider is the position of the equilibrium between the reactants and the products of the reaction in question. For example, pK_a values are used to make predictions about acid-base reactions (p. 90). Equilibrium is determined by thermodynamic considerations, the overall enthalpy and entropy changes that occur during a reaction (p. 94).

Finally, we need to consider the mechanism of the reaction, the details of the bond breaking and bond making that must take place to get from reactants to products (p. 81). A reaction may have an equilibrium constant that favors the products but may occur too slowly to be practical. The rate of a reaction depends on the reaction pathway, the mechanism of the reaction (p. 113). The field of kinetics deals with the factors that affect the rate of a reaction. In the next two sections, we will examine two reactions that represent important classes of reactions. We will identify electrophiles and nucleophiles and examine energy considerations, equilibria, and mechanisms for the two reactions.

Study Guide
Concept Map 4.1

PROBLEM 4.1

Decide which of the following compounds and ions are electrophiles and which ones are nucleophiles. Writing Lewis structures may be helpful.

(a) CH_3O^- (b) PH_3 (c) Cu^{2+} (d) HBr (e) CH_3Cl

(f) CH_3NH_2 (g) H^- (h) $B(CH_3)_3$ (i) $HONH_2$ (j) $HC{\equiv}CH$

(k) $AlCl_3$ (l) 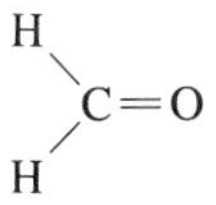 (m) I^- (n) $CH_3CH_2S^-$ (o) Hg^{2+}

4.2 THE REACTION OF CHLOROMETHANE WITH HYDROXIDE ION

A. A Nucleophilic Substitution Reaction

Chloromethane reacts with sodium hydroxide in solution in water to give methanol and sodium chloride.

$$CH_3Cl + Na^+OH^- \xrightarrow[H_2O]{} CH_3OH + Na^+Cl^-$$

The polarization of the carbon-chloride bond in chloromethane is due to the differing electronegativities of carbon and chlorine. The unequal sharing of the electrons in this covalent bond creates a site of electron deficiency at the carbon atom. Chloromethane is, therefore, an electrophile, or, more accurately, it contains an electrophilic center. Hydroxide ion is a good nucleophile. It has pairs of electrons to share and a negative charge, making it a site of electron density.

The reaction starts with the interaction of the nucleophile with the electrophile.

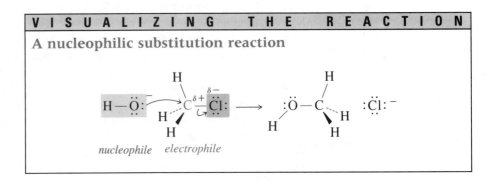

VISUALIZING THE REACTION

A nucleophilic substitution reaction

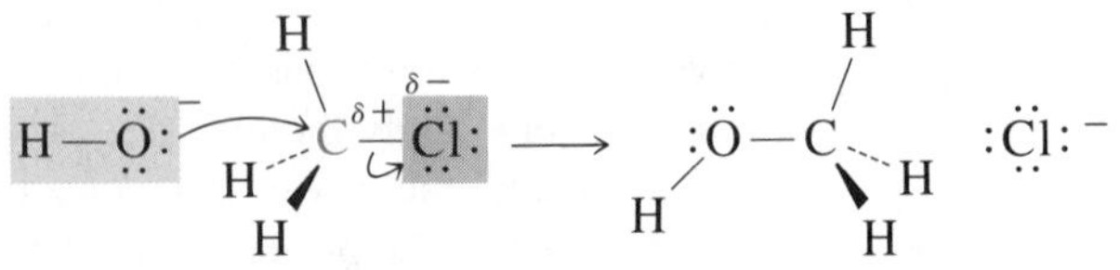
nucleophile
electrophile

Carbon can be involved in only four covalent bonds, so the carbon-chlorine bond breaks as the carbon-oxygen bond forms. The hydroxide ion, the nucleophile in this reaction, displaces the chloride ion. This type of reaction is called a **nucleophilic substitution reaction,** or nucleophilic displacement reaction. Sodium ion, Na^+, appears on both sides of the overall reaction equation shown at the beginning of this section to balance the charges, but does not appear in the mechanism of the reaction.

Study Guide Concept Map 4.2

Nucleophilic substitution reactions are an important class of organic reactions and will be covered in greater detail in Chapter 7.

PROBLEM 4.2

Predict, by using curved arrows to show the movements of electrons, what will happen when each pair of reagents shown below is mixed. Identify the electrophile and the nucleophile in each case.

(a) $CH_3Br + Na^+OH^- \longrightarrow$

(b) $CH_3I + Na^+OH^- \longrightarrow$

(c) $CH_3CH_2Cl + Na^+OH^- \longrightarrow$

(d) $CH_3I + NH_3 \longrightarrow$

PROBLEM 4.3

Sodium ion, Na^+, is not a good electrophile. How would you rationalize this fact? Why are mercury(II) ion, Hg^{2+}, and copper(II) ion, Cu^{2+}, better electrophiles than sodium ion?

B. Energy Changes in the Reaction. Equilibrium

The extent of the reaction of chloromethane with hydroxide ion to give methanol and chloride ion is indicated by the equilibrium constant for the reaction, which can be calculated from the change in standard free energy that occurs when the reaction takes place (p. 93). The change in standard free energy for the reaction is the difference between the standard free energies of formation for the products and those for the reactants. Values for the standard free energies of formation for many covalent and ionic species can be found in handbooks of chemistry and physics. The values for the reaction we are considering are given below. The standard conditions for these values are defined as 1 M solutions in water at a temperature of 298 K. These values are used to calculate the change in standard free energy for the reaction.

$$CH_3Cl(aq) + OH^-(aq) \longrightarrow CH_3OH(aq) + Cl^-(aq)$$

	$CH_3Cl(aq)$	$OH^-(aq)$	$CH_3OH(aq)$	$Cl^-(aq)$
ΔG_f°, kcal/mol	−12.3	−37.6	−41.9	−31.4

ΔG_f°(reactants) = −49.9 kcal/mol ΔG_f°(products) = −73.3 kcal/mol

$$\begin{aligned}\Delta G_r^\circ(\text{reaction}) &= \Delta G_f^\circ(\text{products}) - \Delta G_f^\circ(\text{reactants}) \\ &= -73.3\ \text{kcal/mol} - (-49.9\ \text{kcal/mol}) \\ &= -23.4\ \text{kcal/mol}\end{aligned}$$

There is a decrease in the standard free energy of the system for the reaction as it is written above; ΔG_r° for the reaction has a negative value. This means that the reaction will go as written; that is, the products are favored thermodynamically over the reactants.

The equilibrium constant for the reaction can be calculated from the value for ΔG_r°.

$$\begin{aligned}\Delta G_r^\circ &= -2.303RT \log K_{eq} \\ \log K_{eq} &= \frac{-\Delta G_r^\circ}{2.303RT} \\ &= \frac{-(-23.4\ \text{kcal/mol})}{2.303 \times 1.987 \times 10^{-3}\ \text{kcal/deg·mol} \times 298\ \text{K}} \\ &= \frac{23.4\ \text{kcal/mol}}{1.36\ \text{kcal/mol}} \\ \log K_{eq} &= 17.2 \\ K_{eq} &= 1.61 \times 10^{17}\end{aligned}$$

The value for the equilibrium constant is very large; this reaction is essentially complete at equilibrium. No detectable amount of the reactant chloromethane remains. The equation relating the equilibrium constant to ΔG_r° contains a temperature term because equilibrium constants are dependent on temperature. The one calculated above is for the reaction at room temperature, 25 °C (298 K).

Note that in talking about equilibrium, we are concerned only with the difference in energy between the initial state of the system, the reactants, and the final state of the system when the products have formed. Nothing has been said about how fast the reaction will take place. That is the subject of the next section.

Study Guide Concept Map 4.3

C. The Rate of the Reaction. Kinetics

Questions about how fast a chemical reaction goes are in the realm of kinetics. **Kinetics** deals with the rate of a chemical reaction and the factors that influence that rate.

The rate of the reaction of chloromethane with hydroxide ion has been studied in water solution at 38 °C, with the initial concentration of chloromethane at 0.003 M and that of the hydroxide ion at 0.01 M. The rate was studied by measuring the decrease in the concentration of hydroxide ion or the increase in the concentration of chloride ion as the reaction progressed.

CH_3Cl	+	OH^-	$\longrightarrow$	CH_3OH	+	Cl^-
chloromethane		hydroxide ion		methanol		chloride ion
		concentration decreases as the reaction progresses				*concentration increases as the reaction progresses*

From these quantities, the amount of chloromethane that had reacted at any given time was determined.

The rate of the reaction was found to be dependent on the concentrations of both of the reactants. The rate equation is

$$\text{Rate} = k_r[CH_3Cl][OH^-]$$

where k_r is the rate constant for the reaction. The reaction is a **second-order reaction.** The **order** of a chemical reaction is defined as the sum of the exponents of the concentration terms that appear in the rate equation. In this rate equation, the concentration of chloromethane and the concentration of hydroxide ion each appear once; each exponent is 1. (Of course, exponents that are ones are not generally shown in written equations.) Therefore, the order of the reaction is 1 + 1, or 2. The reaction is first-order with respect to chloromethane, first-order with respect to hydroxide ion, and second-order overall. The order of a reaction is always determined experimentally and cannot be predicted from its equation. The rate constant, k_r, is a constant for any given temperature. On p. 116, we will discuss the effect of temperature on k_r.

For the reaction of chloromethane with hydroxide ion at 38 °C (311 K), the value for k_r is 3.55×10^{-5} L/mol·s. Therefore, for the initial reaction conditions where the concentration of chloromethane is 0.003 M and that of hydroxide ion is 0.01 M, we have

$$\begin{aligned}\text{Initial rate} &= 3.55 \times 10^{-5}\ \text{L/mol·s} \times 3 \times 10^{-3}\ \text{mol/L} \times 1 \times 10^{-2}\ \text{mol/L} \\ &= 1.07 \times 10^{-9}\ \text{mol/L·s}\end{aligned}$$

The reaction is very slow. Only a very small percentage of the chloromethane molecules in this reaction mixture are converted to methanol in a day. Thus, even though the equilibrium is highly favorable for the formation of methanol from chloromethane (p. 113), the reaction proceeds extremely slowly. We will examine why this is so in the next three parts of this section.

Study Guide Concept Map 4.4

PROBLEM 4.4

What would be the initial rate of the reaction of chloromethane with hydroxide ion at 311 K if the concentration of hydroxide ion were increased to 0.05 M? What would happen to the rate if the initial concentration of chloromethane were decreased to 0.001 M?

D. The Transition State

The mechanism for the reaction of chloromethane with hydroxide ion shown on p. 112 shows the two species coming together in a way that allows for bonding to start between the oxygen atom of the hydroxide ion and the carbon atom of chloromethane. The two reagents have to collide with each other and must do so in an orientation that allows bonding to take place. The hydroxide ion could collide with the chlorine end of the chloromethane molecule, and no reaction would take place. Instead, repulsion would occur between the negatively charged hydroxide ion and the partially negatively charged chlorine atom. The two species would bounce off one another, with no chemical change taking place.

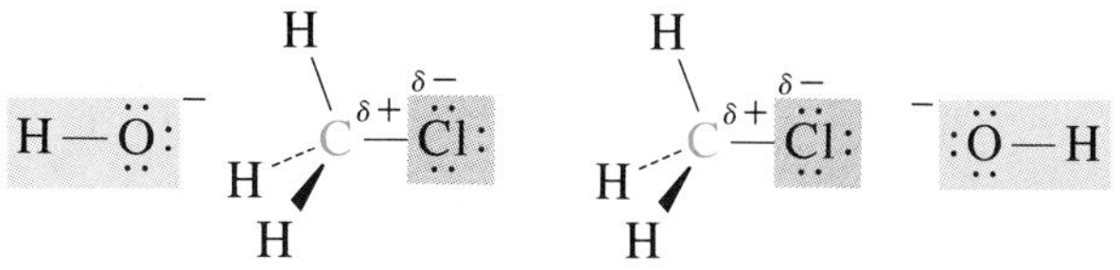

Approach of hydroxide ion that leads to reaction with chloromethane. Attraction between negatively charged hydroxide ion and carbon atom with partial positive charge.

Approach of hydroxide ion to chloromethane that will not lead to chemical reaction. Repulsion between negatively charged hydroxide ion and chlorine atom with partial negative charge.

Therefore, not all encounters between potentially reactive species produce the chemical reaction.

The reaction of chloromethane with hydroxide ion involves the breaking of a bond between carbon and chlorine and the forming of a new bond between oxygen and carbon. At some point during the reaction, the carbon atom is partially bonded to both the hydroxide ion and the chlorine atom. This intermediate state, lying between the reactants and the products, is known as the **transition state.** The molecular complex that exists at the transition state in equilibrium with the reactants is called the **activated complex** and is represented in equations enclosed in square brackets with a symbol called a double dagger (‡) as a superscript.

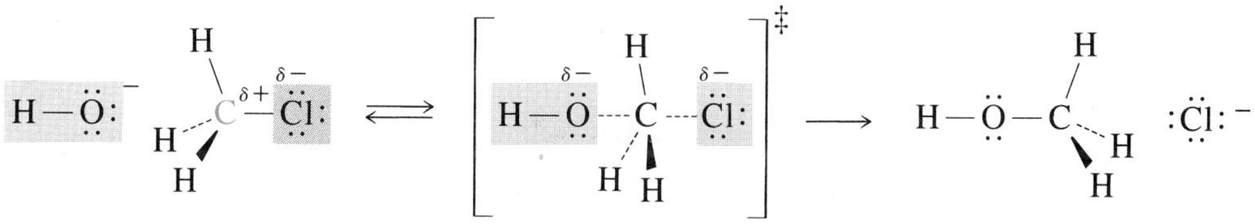

reactants *the activated complex the molecular configuration at the transition state* *products*

The transition state is a high energy state. Energy has been put into the molecule to partially break the carbon-chlorine bond, but the energy from the formation of the full carbon-oxygen bond has not yet been gained.

Once the activated complex is formed, it is converted very rapidly to product, at some absolute rate, k_0. At 25 °C, k_0 is 6.2×10^{12}/s. This number is of the same magnitude as the frequency of the vibration of a covalent bond in a molecule (see p. 357 for a description of some vibrational motions of covalent bonds). Thus, once the reactants acquire enough energy to become an activated complex, bonds break and new bonds form during the time required for a molecular vibration.

E. Free Energy of Activation

The energy relationships between the reactants, the transition state, and the products for the reaction of chloromethane and hydroxide ion can be represented on an **energy diagram** (Figure 4.1). The *y* axis represents energy, in this case, the free energy of the system. The *x* axis is called the **reaction coordinate** and represents the changes that must take place in bond lengths and bond angles within molecules

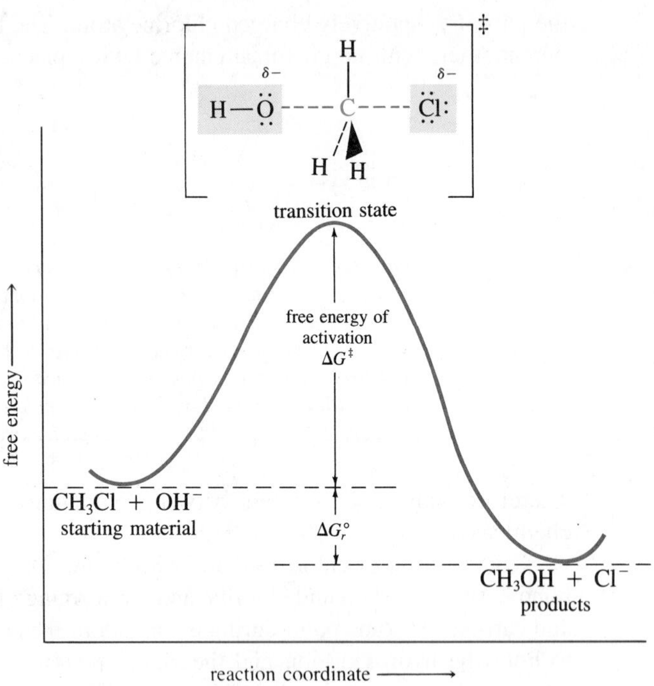
H
δ−
δ−
H—Ö----C----Cl:
H H
‡
transition state
free energy of activation $\Delta G^{\ddagger}$
free energy
CH_3Cl + OH^-
starting material
ΔG_r°
CH_3OH + Cl^-
products
reaction coordinate

FIGURE 4.1 Relative energy levels for reactants, transition state, and products for the reaction of chloromethane with hydroxide ion.

as reactants are converted to products. Moving along the x axis from left to right corresponds to the progress of the reaction from reactants to products.

In going from the reactants to the products, the system has to go over an energy barrier. The height of this energy barrier is the **free energy of activation $\Delta G^{\ddagger}$**, for the reaction, which is the energy that the reactants attain in reaching the transition state. The free energy of activation is a quantity that has to be determined experimentally for each reaction; it cannot be predicted.

The rate constant for a reaction is related to the free energy of activation by the following equation,

$$k_r = k_0 e^{-(\Delta G^{\ddagger}/RT)}$$

in which e is 2.718, the base of natural logarithms, and k_0 is the absolute rate constant mentioned at the end of the last subsection. The negative exponential relationship between the rate constant and $\Delta G^{\ddagger}$ is very important. *As $\Delta G^{\ddagger}$ gets larger, $e^{-(\Delta G^{\ddagger}/RT)}$ must get smaller; therefore, the observed rate of the reaction decreases. And a smaller $\Delta G^{\ddagger}$ means an increased rate of reaction.* Small differences in energies of activation produce very large differences in rates. For example, a doubling of the free energy of activation from 10 kcal/mol to 20 kcal/mol would decrease the rate to a ten-millionth of its value.

Study Guide Concept Map 4.5

F. The Effect of Temperature on the Rate of the Reaction

The rate of a reaction is different at different temperatures. For example, the rate constant, k_r, for the reaction of chloromethane with hydroxide ion has been measured at several temperatures. The results are shown in Table 4.1. The rate of the reaction increases by a factor of about two to four for each rise in temperature of 10°, according to the data in the table.

TABLE 4.1 The Effect of Temperature on the Rate of the Reaction of Chloromethane with Hydroxide Ion

T, K	k_r, L/mol·s
311	3.55×10^{-5}
322	1.46×10^{-4}
329	3.00×10^{-4}
333	4.88×10^{-4}
340	9.97×10^{-4}

We are already familiar with the equation that relates the rate constant with temperature.

$$k_r = k_0 e^{-(\Delta G^{\ddagger}/RT)}$$

Temperature, T, appears in the denominator of the exponent; therefore, an increase in T increases the value of $e^{-(\Delta G^{\ddagger}/RT)}$ and also the value of k_r. We are assuming that $\Delta G^{\ddagger}$ does not vary much with temperature.

From the experimental measurements of the rate of a reaction at different temperatures, an **energy of activation, E_a,** is determined. The relationship between the rate constant and this energy of activation is

$$k_r = Ae^{-E_a/RT}$$

This relationship, known as the Arrhenius equation, is similar in form to the relationship between the rate constant and $\Delta G^{\ddagger}$. The experimentally determined energy of activation for the reaction of chloromethane with hydroxide ion is 24.28 kcal/mol. The A in the Arrhenius equation is known as the **frequency factor** and is also determined experimentally for each reaction. It is similar to the absolute rate constant, k_0 (p. 115), and contains within it the spatial effects that determine whether a collision between molecules will lead to reaction.

What physical model accounts for the dependence of rate on temperature? At any temperature, the molecules in a reaction mixture have a wide distribution of energies, as shown in Figure 4.2 for two temperatures. Most of the molecules have energies close to some average value, but some have much lower energies or much higher ones. The molecules are moving rapidly and collide frequently with one another. As a result of these collisions, energy is transferred from one molecule to another. The kinetic energy of the molecules is also converted to other forms of energy, such as the vibrational energy of bonds within a molecule. Therefore, the amount of energy available to a given molecule is constantly changing.

An increase in temperature increases the kinetic energy of the molecules and thus the number of molecules with higher kinetic energies. If $\Delta G_1^{\ddagger}$ is the free energy of activation for the reaction, more molecules will have enough energy to get to the top of the energy barrier in the reaction pathway (see Figure 4.1) at the higher temperature, T_2, than at the lower temperature, T_1. Thus, an increase in temperature increases the rate of a reaction.

If we compare a reaction with a low free energy of activation ($\Delta G_1^{\ddagger}$ in Figure 4.2) to one with a high free energy of activation ($\Delta G_2^{\ddagger}$), we see that at either

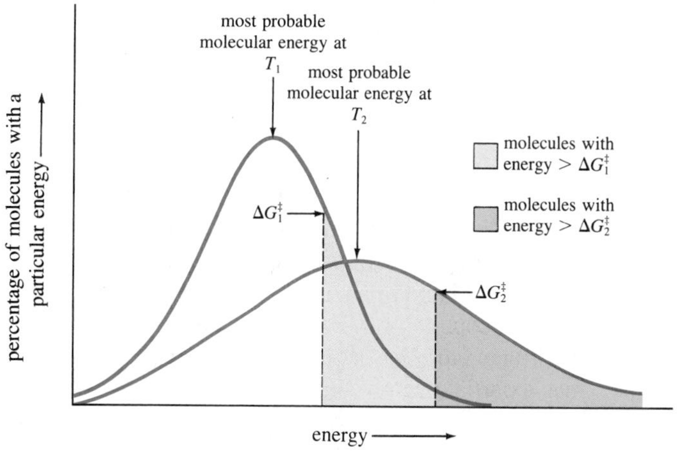
most probable
molecular energy at
T_1
most probable
molecular energy at
T_2
molecules with
energy > $\Delta G_1^{\ddagger}$
molecules with
energy > $\Delta G_2^{\ddagger}$
$\Delta G_1^{\ddagger}$
$\Delta G_2^{\ddagger}$
percentage of molecules with a
particular energy
energy

FIGURE 4.2 Distribution of energy among molecules at temperatures T_1 and T_2.

temperature more molecules will have enough energy to get over the lower energy barrier. *The rate of the reaction with the lower free energy of activation will be greater than the rate of reaction with the higher free energy of activation.*

In summary, the rate of a reaction is affected by four factors:

1. concentration of reagents
2. energy of activation
3. temperature
4. spatial effects that are due to the sizes and shapes of the colliding molecules and that determine which orientations of collisions lead to reactions

Study Guide Concept Map 4.6

In the rate expression that appears on p. 114, all of these factors except concentration are included in the experimental rate constant, k_r.

PROBLEM 4.5

Bromomethane, CH_3Br, also reacts with hydroxide ion in the way that chloromethane does. The second-order rate constant, k_r, for the reaction at 308 K is 5.3×10^{-4} L/mol·s. What is the initial rate of the reaction in a water solution that is 0.001 M in bromomethane and 0.01 M in hydroxide ion?

PROBLEM 4.6

The frequency factor, A, for the reaction of iodomethane, CH_3I, with hydroxide ion was determined to be 1.24×10^{12} L/mol·s. The energy of activation for the reaction is 22.22 kcal/mol. Calculate the second-order rate constant, k_r, at 298 K for this reaction.

4.3 ADDITION OF HYDROGEN BROMIDE TO PROPENE

A. An Electrophilic Addition Reaction

The π electrons in alkenes are protonated by strong acids (p. 78). The vulnerability of a carbon-carbon double bond to acids is illustrated by the reaction of hydrogen bromide and propene.

$$\underset{\text{propene}}{CH_3CH{=}CH_2} + \underset{\text{hydrogen bromide}}{HBr} \xrightarrow[\substack{18\ \text{h} \\ \text{absence} \\ \text{of oxygen}}]{25\ °C} \underset{\substack{\text{2-bromopropane} \\ \text{isopropyl bromide} \\ 95\%}}{\overset{}{CH_3\underset{\displaystyle Br}{\underset{|}{C}}HCH_3}}$$

When the reaction is carried out in the gas phase in the absence of oxygen (air) and with highly purified reagents for eighteen hours at room temperature, a 95% yield of 2-bromopropane, also called isopropyl bromide, is obtained.

The π bond in an alkene is a site of electron density. The alkene, therefore, is the nucleophile in this reaction. The hydrogen bromide molecule is polarized; the hydrogen atom has a partial positive charge and the electronegative bromide atom has a partial negative charge. The hydrogen atom of hydrogen bromide is thus the electrophile. The approach of the polarized hydrogen bromide molecule to the π bond of propene initiates the reaction. The mechanism of the reaction shows the donation of electrons from the nucleophile, specifically, the electrons of the π bond of propene, to the electrophile, the proton from hydrogen bromide.

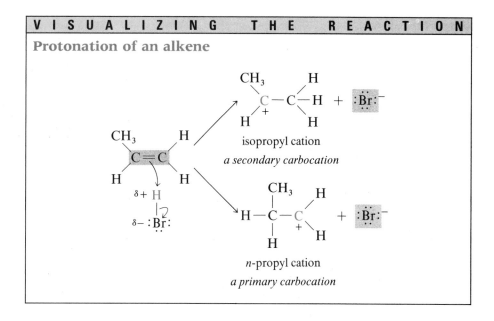

VISUALIZING THE REACTION

Protonation of an alkene

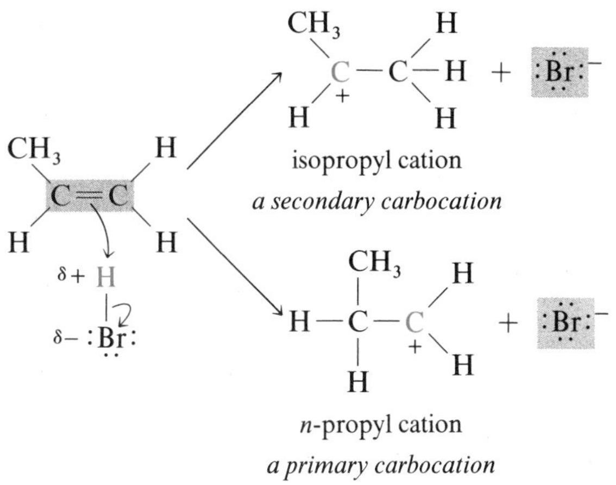

CH3
H
C=C
H
H
δ+ H
δ− :Br:
isopropyl cation
a secondary carbocation
n-propyl cation
a primary carbocation

A bond can form between the proton and the double bond of propene in two ways. The proton can bond to the carbon atom at the end of the propene molecule, leaving the middle carbon atom deficient in electrons. This creates a **secondary carbocation,** a positively charged species in which two alkyl groups are attached to the electron-deficient carbon atom. If bonding occurs between the proton and the middle carbon atom in the chain, the end carbon atom will have a positive charge, creating a **primary carbocation,** a species in which only one alkyl group is attached to the electron-deficient carbon atom.

Carbocations are Lewis acids. They are electron-seeking reagents, electrophiles, and react with nucleophiles that are present in a reaction mixture. A negatively charged bromide ion, which is one of the nucleophiles in the reaction mixture we are considering, combines with an isopropyl cation to give the observed product.

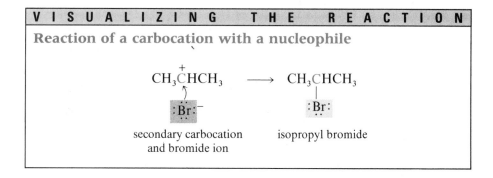

VISUALIZING THE REACTION

Reaction of a carbocation with a nucleophile

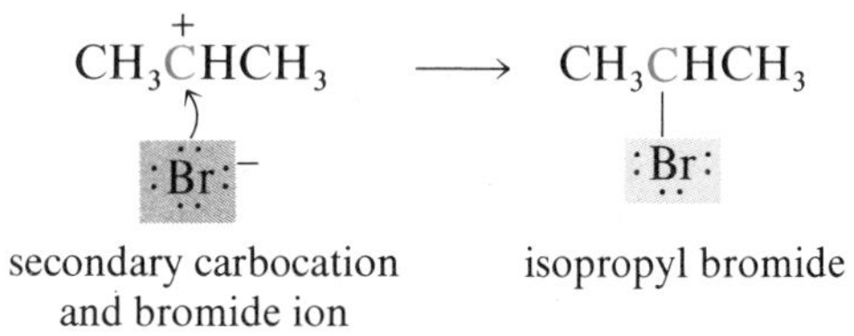

CH3CHCH3
:Br:
secondary carbocation and bromide ion
isopropyl bromide

No n-propyl bromide is obtained, suggesting that little of the primary carbocation is formed in this reaction mixture.

The overall reaction, the addition of hydrogen bromide to the double bond, is an example of a large and important class of reactions known as **electrophilic addition reactions.** Many more examples of this type of reaction will be seen in Chapters 8 and 9.

Study Guide Concept Map 4.7

PROBLEM 4.7

What other nucleophile might be found in the reaction mixture for the reaction of propene with hydrogen bromide? Write a mechanism for the reaction of isopropyl cation with this nucleophile. (Hint: A review of Sections 3.2 and 3.3 may be helpful.)

B. Markovnikov's Rule. Regioselectivity in a Reaction

Propene and hydrogen bromide are both unsymmetrical molecules. The reaction between them could form 1-bromopropane (*n*-propyl bromide) as well as 2-bromopropane, but under the conditions described earlier (p. 118) only the latter compound is produced.

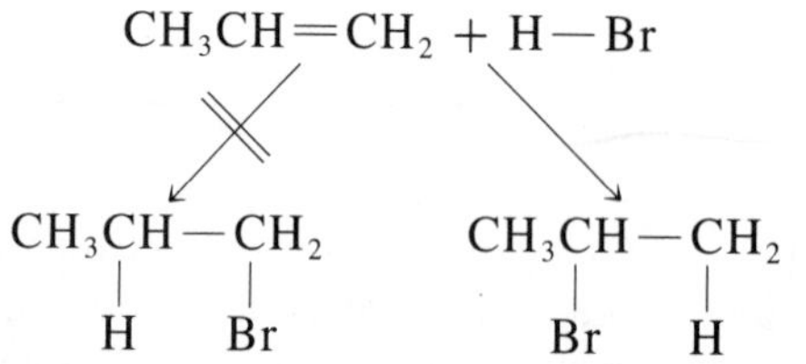
CH3CH=CH2 + H—Br
CH3CH—CH2
H Br
CH3CH—CH2
Br H

1-bromopropane *n*-propyl bromide *not observed* — 2-bromopropane isopropyl bromide *the only product*

Of the two possible ways in which these reagents can react, one reaction pathway leading to one product is favored. Such an observed preference in the direction in which molecules react with each other is called **regioselectivity.** Such selectivity is useful if we wish to synthesize the compound, in this case isopropyl bromide, that is favored in a reaction. On the other hand, it works against us if we want to synthesize the unfavored product of the reaction.

The regioselectivity of the addition of protic acids to alkenes has been recognized for a long time. In 1869, the Russian chemist Vladimir Markovnikov summarized his experimental observations in a rule that bears his name. Markovnikov said, "When a hydrocarbon of unsymmetrical structure combines with a halogen hydracid, the halogen adds itself to the less hydrogenated carbon atom, i.e., to the carbon atom that is more under the influence of other carbon atoms." As Markovnikov's Rule calls for, the bromine atom from hydrogen bromide bonds to the carbon atom of the two involved in the double bond in propene that is bonded to fewer hydrogen atoms, giving rise to isopropyl bromide.

Study Guide Concept Map 4.8

PROBLEM 4.8

Predict the products of the following reactions.

(a) $CH_3\overset{CH_3}{\overset{|}{C}}=CH_2 + HBr \longrightarrow$

(b) $CH_3CH_2CH{=}CH_2 + HBr \longrightarrow$

(c) $CH_3CH_2CH{=}CHCH_3 + HBr \longrightarrow$

C. Relative Stabilities of Carbocations

Carbocations are **reactive intermediates** that are formed as the reaction progresses from the reactants to the products. They are less stable than the reactants and the products and have only a short lifetime in the reaction mixture. Nevertheless, there is good experimental evidence for the existence of such intermediates, including physical measurements that have been made on some solutions. The cationic carbon atom, bonded to three other atoms, is trigonal and is sp^2-hybridized. An empty p orbital is perpendicular to the plane containing the carbon atom and the three atoms to which it is bonded (Figure 4.3).

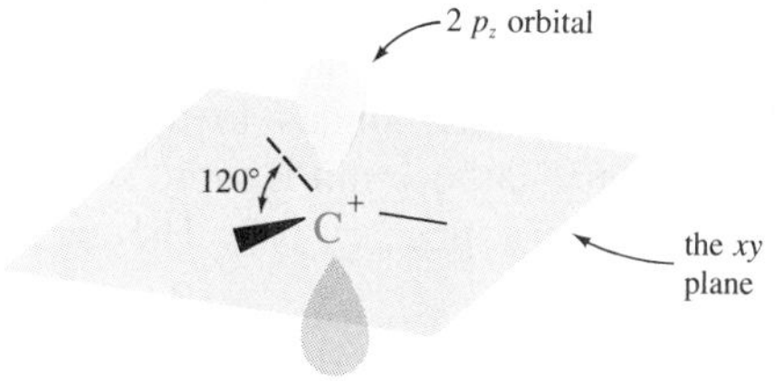
2 p_z orbital
120°
C^+
the xy plane

FIGURE 4.3 The geometry at a carbocation.

Why are secondary carbocations formed more easily than primary cations in the reaction mixture we have been studying? This observation is explained by postulating that the methyl group is slightly electron-releasing in its effect on neighboring atoms (p. 98). A methyl group is more **polarizable** than a hydrogen atom; it has a larger density of electrons that can be drawn toward a site of electron deficiency. This electron-releasing effect is particularly noticeable when the sp^3-hybridized carbon atom of an alkyl group is bonded to a more electronegative sp^2- or sp-hybridized carbon atom (p. 56). The electrons around the carbon atom of the alkyl group are drawn toward the more electronegative carbon atom to which it is bonded. In the case of a carbocation, this effect results in the delocalization of some of the positive charge to the alkyl group, which stabilizes the ion.

In a primary carbocation, one alkyl group and two hydrogen atoms are bonded to the carbon atom bearing the positive charge. In a secondary carbocation, there are two alkyl groups and one hydrogen atom on the electron-deficient carbon atom. A secondary carbocation is more stable than a primary carbocation because the electron deficiency in the secondary carbocation is partially offset by the electron-releasing effect of two methyl groups. Only one alkyl group is present to stabilize a primary carbocation.

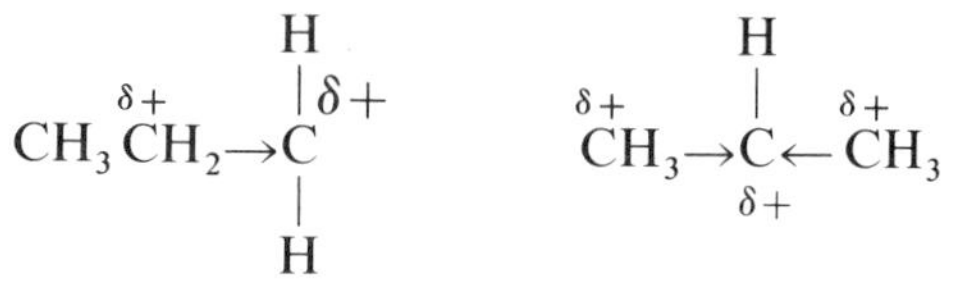
H
δ+
δ+
$CH_3CH_2 \rightarrow C$
H
H
δ+
δ+
$CH_3 \rightarrow C \leftarrow CH_3$
δ+

n-propyl cation
a primary carbocation delocalization of charge to one alkyl group

isopropyl cation
a secondary carbocation delocalization of charge to two alkyl groups

PROBLEM 4.9

Would you expect the methyl cation, CH_3^+, to be more stable or less stable than the *n*-propyl cation, $CH_3CH_2CH_2^+$?

PROBLEM 4.10

Would a tertiary carbocation, $(CH_3)_3C^+$, be more stable or less stable than the isopropyl cation, $(CH_3)_2CH^+$, which is a secondary carbocation? List the four cations given in this problem and Problem 4.9 in order of decreasing stability.

The regioselectivity observed for the addition of hydrogen bromide to propene can be rationalized by saying that the more stable of the two possible cationic intermediates is formed in the reaction. A modern version of Markovnikov's Rule is as follows: *For the addition of a hydrogen halide to a double bond, the major product is that isomer that results from the formation of the more stable cationic intermediate*. Markovnikov himself recognized the importance of the "carbon that is under the influence of other carbon atoms" as the site at which the halogen would attach itself. The stability of a cationic intermediate increases as the number of carbon atoms bonded to the positively charged carbon atom increases. Thus, a tertiary carbocation is more stable than a secondary one, which in turn is more stable than a primary one. The methyl cation is the least stable of all alkyl carbocations.

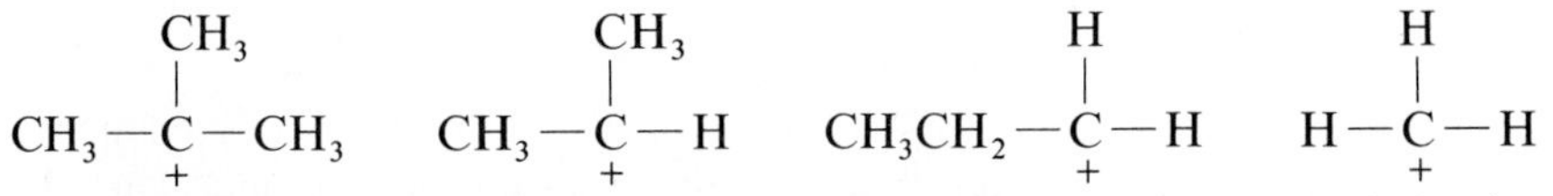

tert-butyl cation — isopropyl cation — *n*-propyl cation — methyl cation

most stable — *least stable*

In recent years, carbocations have been generated by the ionization of alkyl halides in solvents that are highly polar but do not react with carbocations as nucleophiles. In these solvents, the formation of a carbocation can be an exothermic process. For example, in sulfuryl chlorofluoride, SO_2ClF, to which antimony pentafluoride, SbF_5, has been added, the formation of *tert*-butyl chloride ion at −55 °C has an enthalpy of ionization, ΔH_i, of −24.8 kcal/mol. The ionization of isopropyl chloride is less exothermic under the same conditions with a ΔH_i value of −15.3 kcal/mol. The ionization is helped by the interaction of the Lewis acid, SbF_5, with chloride ion, a Lewis base.

$$RCl + SbF_5 \xrightarrow[-55\ ^\circ C]{SO_2ClF} R^+\ SbF_5Cl^-$$

Study Guide
Concept Map 4.9

These experiments have shown that a tertiary carbocation is more easily formed than a secondary carbocation. Primary carbocations have never been observed in these experiments.

D. Equilibria in the Addition of Hydrogen Bromide to Propene

The reaction of propene with hydrogen bromide can give two products, isopropyl bromide and *n*-propyl bromide. The two products do not differ much in energy. The change in standard free energy (p. 113) in going from the reactants to either one of the products is calculated as follows:

$$\begin{array}{llll} & CH_3CH{=}CH_2(g) & + \; HBr(g) \longrightarrow & CH_3CH_2CH_2Br(g) \\ \Delta G_f^\circ & +14.99\ \text{kcal/mol} & -12.73\ \text{kcal/mol} & -5.37\ \text{kcal/mol} \end{array}$$

$$\begin{aligned} \Delta G_r^\circ &= -5.37 - (14.99 - 12.73) \\ &= -7.63\ \text{kcal/mol} \end{aligned}$$

$$\begin{array}{llll} & CH_3CH{=}CH_2(g) & + \; HBr(g) \longrightarrow & CH_3\underset{\underset{\displaystyle Br}{|}}{C}HCH_3 \\ \Delta G_f^\circ & +14.99\ \text{kcal/mol} & -12.73\ \text{kcal/mol} & -6.51\ \text{kcal/mol} \end{array}$$

$$\begin{aligned} \Delta G_r^\circ &= -6.51 - (14.99 - 12.73) \\ &= -8.77\ \text{kcal/mol} \end{aligned}$$

The change in standard free energy for each reaction is a negative value, indicating that both reactions will proceed as written. Note that isopropyl bromide is more stable than *n*-propyl bromide; its ΔG° value is lower by 1.14 kcal/mol.

Equilibrium constants for the two reactions can be calculated using the expression relating free energy changes to the equilibrium constant (p. 113). The equilibrium constant for the formation of *n*-propyl bromide from propene and hydrogen bromide is 3.89×10^5. The equilibrium constant for the formation of isopropyl bromide is 2.63×10^6. The ratio of the two equilibrium constants is 6.76, indicating that the equilibrium mixture should contain 87% isopropyl bromide and 13% *n*-propyl bromide. In actual experiments, however, isopropyl bromide is the only product. The composition of the product of this reaction is *not* determined by the relative stabilities of isopropyl bromide and *n*-propyl bromide.

PROBLEM 4.11

Carry out the calculation of the equilibrium constants for the formation of isopropyl bromide and of *n*-propyl bromide.

E. The Energy Diagram for the Addition of Hydrogen Bromide to Propene

The reaction of propene with hydrogen bromide proceeds by means of a carbocation, a reactive intermediate that is higher in energy than the reactants or the product (p. 121). These energy relationships can be represented on an energy diagram (Figure 4.4). In going from the reactants to the reactive intermediate, the system has to go over an energy barrier that is greater than the difference in energy between propene and isopropyl cation. This high-energy state is the transition state for the formation of the secondary carbocation, and the energy barrier represents the free energy of activation for the reaction, $\Delta G^{\ddagger}$. Note that the peaks labeled as transition states in Figure 4.4 represent the highest points from the reactants to the reactive intermediate and from the intermediate to the product. The molecular configuration at the transition state for the formation of the secondary carbocation is shown in detail below.

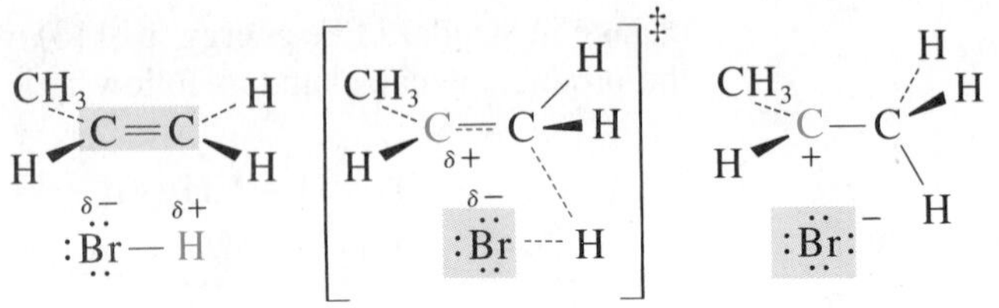
starting materials
activated complex at the transition state
intermediates: isopropyl cation and bromide ion

The two molecules, propene and hydrogen bromide, must first approach each other in the critical orientation with the positive end of the hydrogen bromide molecule pointing toward the π-electron cloud of propene. A flow of electrons starts from the π bond to the hydrogen atom. As this bonding starts, the bond between the hydrogen and bromine atoms becomes weaker as the bonding electrons move closer to bromine, which becomes more like a bromide ion. At the same time, one of the carbon atoms involved in the double bond develops a partial positive charge and begins to look like a carbocation. The hybridization of the carbon atom that is bonding to the hydrogen atom starts to change from sp^2 to sp^3.

The energy diagram in Figure 4.4 indicates that the formation of the secondary carbocation and bromide ion from propene and hydrogen bromide is an endothermic reaction. The molecules of the reagents must absorb energy from their surroundings in order to get to that state. The free energy of activation, the energy necessary to get to the transition state, is even greater than the energy difference between the starting materials and the intermediate. The isopropyl cation is con-

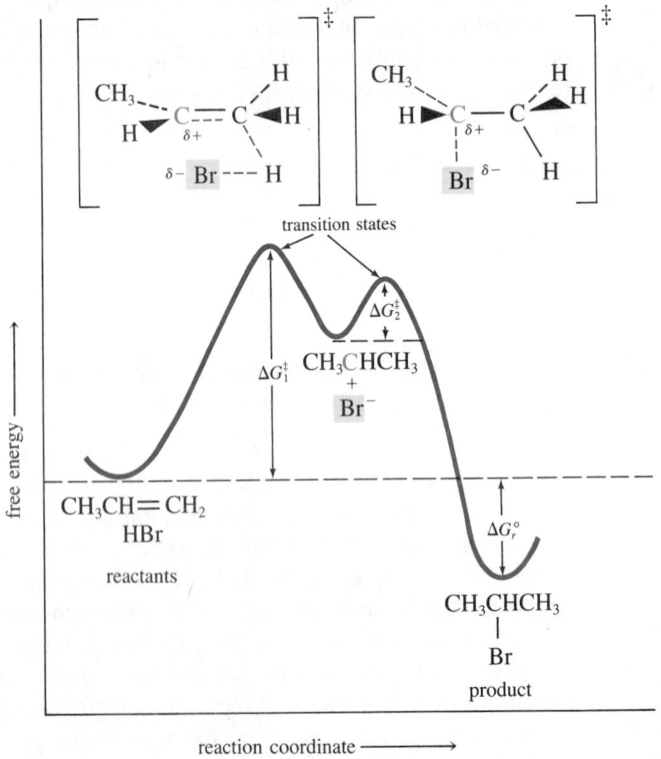
transition states
$\Delta G_2^\ddagger$
$\Delta G_1^\ddagger$
$CH_3\overset{+}{C}HCH_3$
Br^-
$CH_3CH{=}CH_2$
HBr
reactants
ΔG_r°
CH_3CHCH_3
Br
product
free energy
reaction coordinate

FIGURE 4.4 Relative energy levels for reactants, reactive intermediate, and product in the conversion of propene and hydrogen bromide to isopropyl bromide.

verted to isopropyl bromide by a highly exothermic reaction. This reaction also has a transition state and a free energy of activation, though a small one.

F. The Basis for Markovnikov's Rule

Two different carbocations are possible from the protonation of propene. The product observed for the overall reaction comes from only one of these, the secondary carbocation. Two reaction pathways are in competition in this addition reaction. What determines which pathway leads to the major product of the reaction?

Isopropyl bromide is more stable than *n*-propyl bromide by about 1 kcal/mol (p. 123). Indirect measurements show that the isopropyl cation is more stable than the *n*-propyl cation by about 16 kcal/mol. The energy diagram in Figure 4.5 shows the two different pathways that the addition of hydrogen bromide to propene can take. The large difference in the stabilities of the secondary and primary carbocations is reflected in the large difference in the energies of activation for the formation of the two species. The reaction pathway of lower energy proceeds via the more stable secondary carbocation. The rate at which this secondary carbocation is formed is much greater than the rate at which the primary carbocation is formed. Therefore, the product of the reaction is the one from the intermediate the formation of which has the lower energy of activation, even though overall energy considerations (p. 123) allow for the formation of the other product. The products obtained in this reaction are determined by the relative rates at which they are formed and not by their relative stabilities.

Study Guide Concept Map 4.10

PROBLEM 4.12

The reaction of hydrogen iodide with propene in the gas phase at 238.5 °C has been shown to be a second-order reaction.

(a) Write an equation for this reaction. (b) Write a rate equation for the reaction.

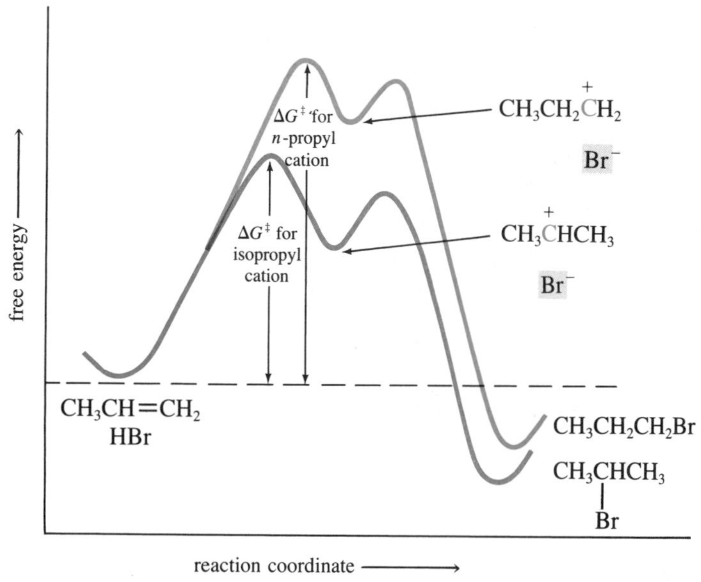

ΔG‡ for n-propyl cation
ΔG‡ for isopropyl cation
CH3CH2CH2 +
Br−
CH3CHCH3 +
Br−
free energy
CH3CH=CH2
HBr
CH3CH2CH2Br
CH3CHCH3
Br
reaction coordinate

FIGURE 4.5 Comparison of the reaction pathways leading from propene and hydrogen bromide to the two possible products.

PROBLEM 4.13

At 238.5 °C when propene and hydrogen iodide are both present at an initial pressure of 45 mm, the second-order rate constant is determined to be 1.66×10^{-6}/atm·s. What is the overall rate of the reaction under these conditions? (760 mm = 1 atm).

PROBLEM 4.14

The energy of activation of the reaction of hydrogen iodide with propene (Problem 4.13) has been experimentally determined to be 22.4 kcal/mol. The observed second-order rate constant, k_r, for the reaction at 490 K is 2.61×10^{-3} L/mol·s. The relationship between k_r and the experimental energy of activation is given by the Arrhenius equation (p. 117). Calculate A, the frequency factor for the reaction.

PROBLEM 4.15

Assume that the energy of activation for the formation of *n*-propyl cation from propene and hydrogen iodide is at least 16 kcal/mol higher than that for the formation of isopropyl cation, or approximately 38 kcal/mol. Using the Arrhenius equation and the value of A calculated in Problem 4.14, calculate k_r for the formation of the *n*-propyl cation at 490 K. How much faster is the reaction that produces the isopropyl cation (Problem 4.14) than the one leading to the *n*-propyl cation?

G. Kinetics of a Reaction with Two Steps. The Rate-Determining Step

The reaction of chloromethane with hydroxide ion occurs in one step (p. 112). No reactive intermediate is formed in the reaction, and the energy diagram for the reaction (Figure 4.1) shows a single transition state (and, therefore, a single free energy of activation) on the path between the reactants and the products. The rate for this reaction is determined by the height of the energy barrier for this single step.

The energy diagram for the reaction of propene with hydrogen bromide (Figure 4.4) has two transition states and two free energies of activation. The reaction occurs in two consecutive steps. These two consecutive steps of a single overall reaction should not be confused with the two competing reactions for the addition of hydrogen bromide to propene illustrated in Figure 4.5.

So far, we have focused on the conversion of propene to isopropyl cation and have not considered the reaction leading from isopropyl cation to isopropyl bromide, the product obtained from the reaction. Isopropyl cation reacts with bromide ion in the second step of the reaction, which has a small energy of activation, goes through a second transition state (see Figure 4.4), and gives isopropyl bromide in an exothermic reaction. Because the energy of activation for this second step is so small in comparison to the energy of activation for the first step, the rate constant for the reaction of isopropyl cation with bromide ion is large. Once the molecules make it to the top of the first hill, they have more than enough energy to make it over the second hill. Thus, although there are actually two different rate constants involved in the overall reaction leading to the formation of isopropyl bromide, experimental measurements are made of only one of them, k_1, the rate constant for the first step.

$$CH_3CH{=}CH_2 + HBr \xrightarrow[\text{(slow step)}]{k_1} CH_3\overset{}{\underset{+}{C}}HCH_3 + Br^- \xrightarrow[\text{(fast step)}]{k_2} CH_3\underset{\underset{Br}{|}}{C}HCH_3$$

The slow step of a reaction sequence, the step with the highest energy of activation, is known as the **rate-determining step** of the reaction. The overall rate of a reaction cannot be any faster than the rate-determining step. For the reaction of propene and hydrogen bromide, the formation of the high-energy reactive intermediate, isopropyl cation, is the rate-determining step. Once the cation is formed, it reacts rapidly with bromide ion to give the product.

4.4 CHEMICAL TRANSFORMATIONS

A. Writing Equations for Organic Reactions

The equations in previous sections of this chapter are attempts to represent accurately the reagents and reaction conditions used in the organic reactions being discussed. The exact details are given because it is important for you to become familiar with the nature of experiments in organic chemistry. It is not necessary, however, for you to memorize all the details that appear in equations. The important reagents will appear many times in problems. If you use the problems as your guide, as you were advised to do in Section 1.1, your attention will be focused on the most important reagents. In order words, you need to learn only the detail that is necessary to answer the problems.

Two types of equations are written for organic reactions. The first type is represented in this chapter by the equations for the reaction of chloromethane with hydroxide ion on page 111 and of hydrogen bromide with propene on page 118. In these equations, all of the reactants are shown on the left side of the arrow and reaction conditions are shown below the arrow. Another way of showing organic reactions is the type of equation in which some reagents are written above the reaction arrow. These are often inorganic reagents, especially acids and bases. For example, equations for the two reactions we have been studying can also be written like this:

$$CH_3Cl \xrightarrow[H_2O]{OH^-} CH_3OH$$

$$CH_3CH{=}CH_2 \xrightarrow[\substack{25\ ^\circ C \\ 18\ h \\ \text{absence of} \\ \text{oxygen}}]{HBr} \underset{\displaystyle Br}{CH_3\overset{}{C}HCH_3}$$

In this type of equation, the emphasis is on the transformation of the organic compounds, and no attempt is made to show a complete and balanced equation. For example, in the top equation, chloride ion, one of the products of the reaction, is not shown. To help you distinguish between the chief reagent and other reaction conditions, the important reagents will appear over the arrow, and any catalysts, solvents, and other conditions, such as temperature and reaction time, will appear below the arrow. Thus, for the second equation, hydrogen bromide is the important reagent and is written over the arrow. It adds to the double bond in propene. The conditions that are necessary for the reaction to give isopropyl bromide with the best yield are shown under the arrow. You are not required to memorize these conditions.

B. Choosing Reagents for Simple Syntheses

Organic chemists are interested in synthesizing organic compounds, transforming simple, readily available reagents into more complex compounds with interesting physical and chemical properties. Syntheses are carried out to make biologically active compounds that were originally isolated from natural sources or to make similar compounds that will have even more useful properties. Much synthetic work is done to produce medicinal products, to find medications having the maximum effectiveness and the minimum number of undesirable side effects.

The first three sections of this chapter covered two important types of reactions, nucleophilic substitution and electrophilic addition reactions. Now we will look at other examples of these types, as we have already done in Problems 4.2 (p. 112) and 4.8 (p. 120). We will also begin to put reactions together to transform one organic compound into another.

Other simple alkyl halides undergo a nucleophilic substitution reaction in the same way that chloromethane does. Many other anions besides hydroxide ion can serve as nucleophiles. Neutral compounds such as ammonia or amines are also nucleophilic (Problem 4.1, p. 111). Other alkenes will add hydrogen halides in the way that propene does. Halogens also add to double bonds (p. 63). Table 4.2 lists some nucleophiles that react readily with alkyl halides as well as some electrophiles, both acids and halogens, that add to alkenes. The following problem provides practice in writing equations with these reagents.

TABLE 4.2 Nucleophilic and Electrophilic Reagents

Nucleophiles	Electrophiles
HO^-, CH_3O^-, $CH_3CH_2O^-$	HCl, HBr, HI
HS^-, CH_3S^-, $CH_3CH_2S^-$	H_2SO_4 $(HOSO_3H)$
CN^-	Cl_2, Br_2
I^-, Br^-	
NH_3, CH_3NH_2	
PH_3	

PROBLEM 4.16

Complete the following equations, showing the major product(s) expected in each case. Identify the electrophile and the nucleophile for each. Write a mechanism for the reactions in parts a–d.

(a) $CH_3CH_2Br + NaI \longrightarrow$

(b) $CH_3\underset{\displaystyle Br}{\underset{|}{C}}HCH_3 + NaCN \longrightarrow$

(c) $CH_3CH_2CH{=}CH_2 + HCl \longrightarrow$

(d) $CH_3CH_2\underset{\displaystyle Br}{\underset{|}{C}}HCH_3 + CH_3SNa \longrightarrow$

(e) $CH_3CH{=}CH_2 + Cl_2 \longrightarrow$

In planning a synthesis, organic chemists analyze the structures of the desired product and the reagents that are available and propose reactions that will convert the starting material to the products. For example, how could the following transformation be carried out?

$$\underset{\text{chloromethane}}{CH_3Cl} \xrightarrow{?} \underset{\text{acetonitrile}}{CH_3CN}$$

An inspection of the two structures reveals that the only change has been the substitution of a cyano group (—CN) for a chlorine atom. Sodium cyanide, a source of the nucleophilic cyanide ion, is the reagent needed to make this transformation.

$$CH_3Cl \xrightarrow{NaCN} CH_3CN$$

What about the following transformation?

$$\underset{\text{2-butene}}{CH_3CH{=}CHCH_3} \xrightarrow{?} \underset{\text{2-iodobutane}}{CH_3CH_2\underset{\displaystyle I}{\underset{|}{C}}HCH_3}$$

The product has an extra hydrogen atom and an iodine atom attached to the carbon atoms that had a π bond between them in the starting material. The π bond is a source of electron density; therefore, an electrophilic reagent incorporating hydrogen and iodine, HI, is needed.

$$CH_3CH{=}CHCH_3 \xrightarrow{HI} CH_3\underset{\displaystyle H}{\underset{|}{C}}H\underset{\displaystyle I}{\underset{|}{C}}HCH_3$$

The starting material is symmetrical; the carbon atoms on both sides of the double bond have the same substituents. Therefore, the reaction will show no regioselectivity.

PROBLEM 4.17

What reagent will bring about the chemical change shown in each of the following equations? Identify each reagent as an electrophile or a nucleophile and explain why you chose it.

(a) $CH_3CH_2I \xrightarrow{?} CH_3CH_2OCH_3$

(b) $CH_3CH_2CH{=}CH_2 \xrightarrow{?} CH_3CH_2\underset{\displaystyle OSO_3H}{\underset{|}{C}}HCH_3$

(c) $CH_3CH_2CH{=}CH_2 \xrightarrow{?} CH_3CH_2\underset{\displaystyle Br}{\underset{|}{C}}H\underset{\displaystyle Br}{\underset{|}{C}}H_2$

(d) $CH_3CH_2CH_2Br \xrightarrow{?} CH_3CH_2CH_2\overset{+}{N}H_3\ Br^-$

Sometimes a chemical transformation requires more than one reaction. For example, how could the following change be carried out?

$$\underset{\text{1-butene}}{CH_3CH_2CH{=}CH_2} \xrightarrow{?} \underset{\textit{sec}\text{-butyl methyl thioether}}{CH_3CH_2\underset{\displaystyle SCH_3}{\underset{|}{C}}HCH_3}$$

There is no reagent that will carry out this change in one reaction. It looks as if H—SCH_3 is being added to the double bond, but that compound is not acidic enough ($pK_a \sim 10.5$ from the table in the front of this book) to serve as an electrophile toward a double bond. The acids listed in Table 4.2 as electrophiles all have negative pK_a values; in other words, they are very strong acids. The —SCH_3 group, however, appears in Table 4.2 as a nucleophile. If a compound that has a halogen atom where the —SCH_3 group is found in the product were available, it could be transformed into the desired product, the thioether. The problem thus requires two reactions.

$$CH_3CH_2CH{=}CH_2 \xrightarrow{?} CH_3CH_2\underset{\displaystyle Br}{\underset{|}{C}}HCH_3 \xrightarrow{CH_3SNa} CH_3CH_2\underset{\displaystyle SCH_3}{\underset{|}{C}}HCH_3$$

The electrophilic reagent HBr will convert the alkene to the alkyl bromide in the first step of this sequence of reactions.

$$CH_3CH_2CH{=}CH_2 \xrightarrow{HBr} CH_3CH_2\underset{\displaystyle Br}{\underset{|}{C}}HCH_3 \xrightarrow{CH_3SNa} CH_3CH_2\underset{\displaystyle SCH_3}{\underset{|}{C}}HCH_3$$

When solving synthesis problems that require more than one step, you should work backwards. Another example will illustrate this thinking process.

$$\underset{\text{propene}}{CH_3CH{=}CH_2} \xrightarrow{?} \underset{\text{isopropyl cyanide}}{CH_3\underset{\displaystyle CN}{\underset{|}{C}}HCH_3}$$

A check of the pK_a table shows that the acid (HCN) that would have to be added to the double bond for a one-step reaction is too weak (pK_a 9.1) to serve as a good electrophile toward a double bond. A direct electrophilic addition reaction is not possible. The structure of the product must be analyzed in order to decide where and what type of reaction is likely.

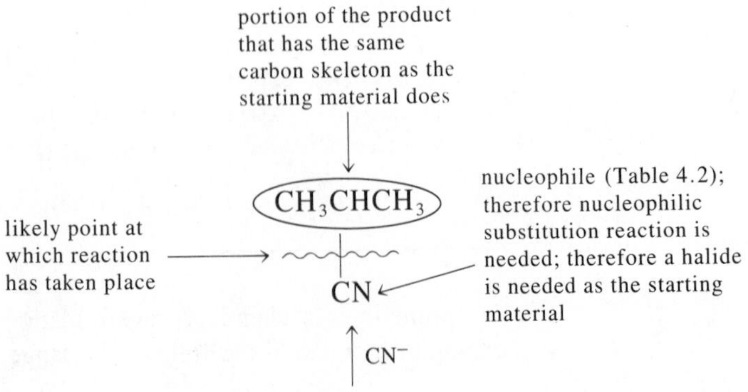
portion of the product that has the same carbon skeleton as the starting material does
CH3CHCH3
likely point at which reaction has taken place
CN
CN−
nucleophile (Table 4.2); therefore nucleophilic substitution reaction is needed; therefore a halide is needed as the starting material

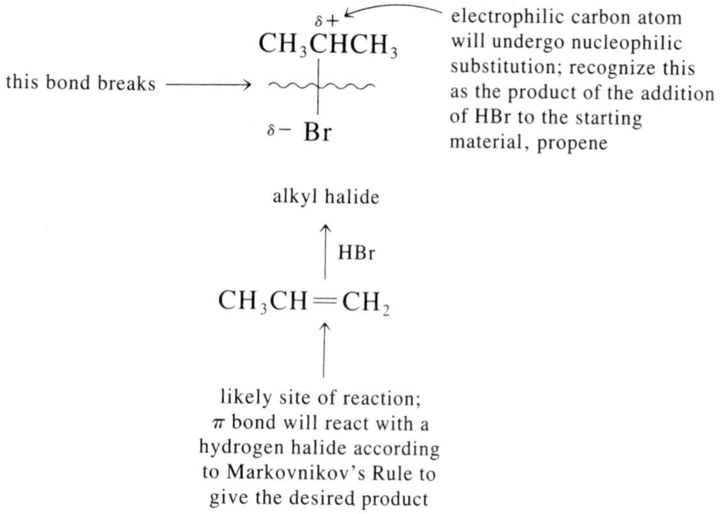

δ+
CH_3CHCH_3
this bond breaks
δ− Br
electrophilic carbon atom will undergo nucleophilic substitution; recognize this as the product of the addition of HBr to the starting material, propene
alkyl halide
HBr
$CH_3CH{=}CH_2$
likely site of reaction; π bond will react with a hydrogen halide according to Markovnikov's Rule to give the desired product

The process of working backwards that is illustrated above is important. It is a good idea to practice it verbally, reasoning aloud to yourself as you work the problems, until this type of analysis becomes second nature.

PROBLEM 4.18

How could each of the following transformations be carried out? Some of them may require more than one reaction. In each case, show as much of your reasoning process as possible.

(a) $CH_3CH_2CH{=}CH_2 \xrightarrow{?} CH_3CH_2\underset{\displaystyle SH}{\underset{|}{C}}HCH_3$

(b) $CH_3CH{=}CHCH_3 \xrightarrow{?} CH_3CH_2\underset{\displaystyle Cl}{\underset{|}{C}}HCH_3$

(c) $CH_2{=}CH_2 \xrightarrow{?} CH_3CH_2\overset{+}{N}H_3\ I^-$

(d) $CH_3\underset{\displaystyle Br}{\underset{|}{C}}HCH_2CH_2CH_3 \xrightarrow{?} CH_3\underset{\displaystyle CN}{\underset{|}{C}}HCH_2CH_2CH_3$

S U M M A R Y

Reagents in chemical reactions can be classified as electrophiles or nucleophiles. Electrophiles are electron-deficient species that are seeking electrons. Nucleophiles are electron-rich species that are seeking a nucleus to donate electrons to.

One example of a nucleophilic substitution reaction is when an alkyl halide, the electrophile, reacts with hydroxide ion, the nucleophile, with the loss of a halide ion.

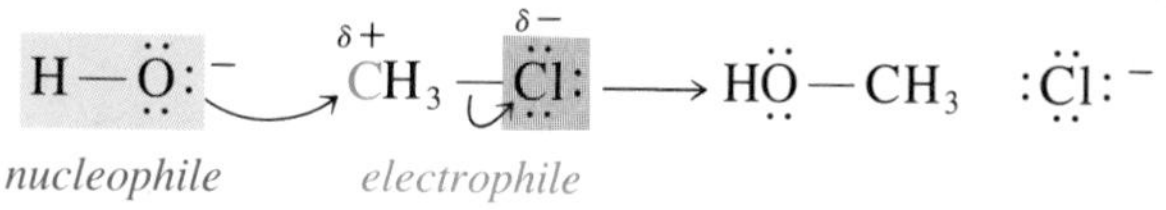

H—O:⁻
δ+
CH₃
δ−
Cl:
HO—CH₃
:Cl:⁻
nucleophile
electrophile

One example of an electrophilic addition reaction is when an alkene, acting as the nucleophile, reacts with a strong acid, acting as the electrophile.

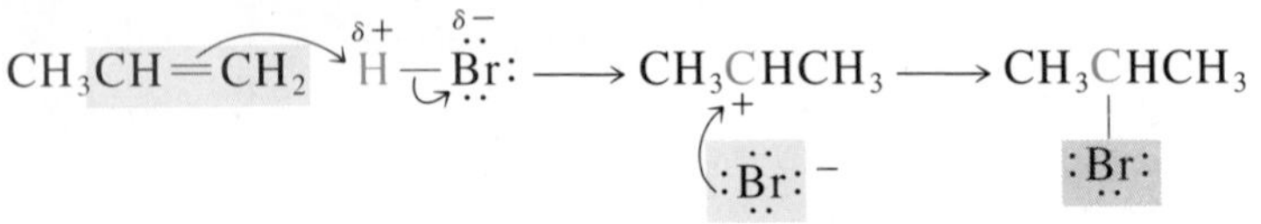
δ+
δ−
CH₃CH=CH₂
H
Br:
CH₃CHCH₃
+
:Br: −
CH₃CHCH₃
:Br:

nucleophile *electrophile*

Chemical reactions occur when there is a decrease in free energy in going from reactants to products. A decrease in free energy corresponds to an equilibrium constant that favors the products over the reactants.

For a reaction to be practical, it must proceed at a reasonable rate. The rate of a reaction is determined by the free energy of activation for the rate-determining step, that is, the energy necessary to reach the transition state on the way to a reactive intermediate or a product. An increase in temperature increases the number of molecules having energy equal to or above the free energy of activation and, therefore, increases the rate of a reaction.

A nucleophilic substitution reaction is a one-step reaction with only one free energy of activation and one transition state. Electrophilic addition to an alkene is a two-step reaction. The first step gives a reactive intermediate, a carbocation; then the second step leads to the product by a combination of the cation with an anion. The overall reaction has two transition states and two free energies of activation. The step with the larger free energy of activation is the rate-determining step.

When a reaction gives as the major product one of several possible products, the reaction is said to be regioselective. The addition of an acid to an unsymmetrical alkene is a regioselective reaction that follows Markovnikov's Rule. The observed regioselectivity is often determined by the relative stabilities of the different reactive intermediates that lead to the products. For the addition of hydrogen bromide to propene, the more stable secondary carbocation is favored over the primary carbocation as the reactive intermediate that leads to the product.

ADDITIONAL PROBLEMS

4.19 Complete the following equations, showing the major product(s) expected in each case. Identify each reagent as an electrophile or a nucleophile. Write the full mechanism for the reactions in parts a–d.

(a) $CH_3CH{=}CH_2 + H_2SO_4 \longrightarrow$

(b) $CH_3CH_2CH_2Br + PH_3 \longrightarrow$

(c) $CH_3CH_2I + CH_3CH_2SNa \longrightarrow$

(d) $(CH_3)_2C{=}CH_2 + HCl \longrightarrow$ (two CH_3 groups on C)

(e) $CH_3CH_2CH{=}CHCH_3 + Br_2 \longrightarrow$

4.20 What reagent will bring about the chemical change shown in each of the following equations? Identify each reagent as an electrophile or a nucleophile and explain why you chose it.

(a) $CH_3Br \xrightarrow{?} CH_3SH$

(b) $CH_3CH_2Br \xrightarrow{?} CH_3CH_2CN$

(c) $CH_2{=}CH_2 \xrightarrow{?} \underset{\displaystyle Cl}{\underset{|}{CH_2}}\underset{\displaystyle Cl}{\underset{|}{CH_2}}$

(d) $CH_3CH_2CH_2Br \xrightarrow{?} CH_3CH_2CH_2OCH_2CH_3$

(e) $CH_3CH_2CH{=}CH_2 \xrightarrow{?} CH_3CH_2\underset{\displaystyle I}{\underset{|}{C}}HCH_3$

(f) $CH_3\underset{\displaystyle Br}{\underset{|}{C}}HCH_3 \xrightarrow{?} CH_3\underset{\displaystyle I}{\underset{|}{C}}HCH_3$

4.21 How could each of the following transformations be carried out? Some of them may require more than one reaction. In each case, show as much of your reasoning process as possible.

(a) $CH_2{=}CH_2 \xrightarrow{?} CH_3CH_2OCH_3$

(b) $CH_3CH{=}CH_2 \xrightarrow{?} CH_3\underset{\displaystyle I}{\underset{|}{C}}HCH_3$

(c) $CH_3CH_2CH{=}CH_2 \xrightarrow{?} CH_3CH_2\underset{\displaystyle OH}{\underset{|}{C}}HCH_3$

(d) $CH_2{=}CH_2 \xrightarrow{?} CH_3CH_2\overset{\displaystyle H}{\overset{|}{\underset{\displaystyle H}{\underset{|}{N}}}}\!\!{}^{+}\!{-}CH_3 \quad Br^-$

4.22 For the following reactions, most of which are as yet unfamiliar, complete the Lewis structures for the pertinent parts of the molecules and identify electrophiles and nucleophiles. Draw arrows showing the movements of electrons that take place to convert the reactants to the products. For example, for

$$N{\equiv}C^- + H{-}\overset{\displaystyle CH_3}{\overset{|}{\underset{\displaystyle H}{\underset{|}{C}}}}{-}Cl \longrightarrow N{\equiv}C{-}\overset{\displaystyle CH_3}{\overset{|}{\underset{\displaystyle H}{\underset{|}{C}}}}{-}H + Cl^-$$

write

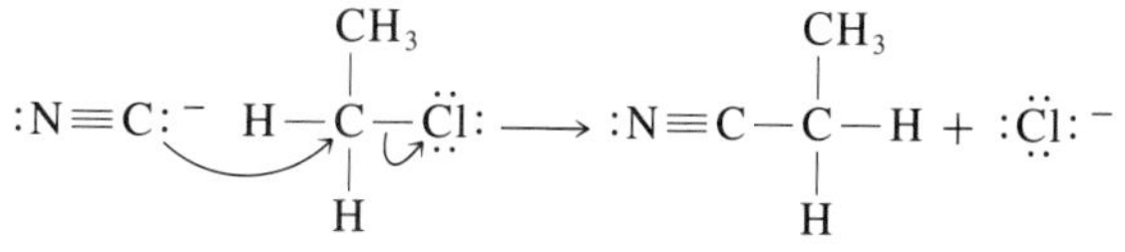

nucleophile electrophile

(a) $I^- + CH_3Cl \longrightarrow ICH_3 + Cl^-$

(b)

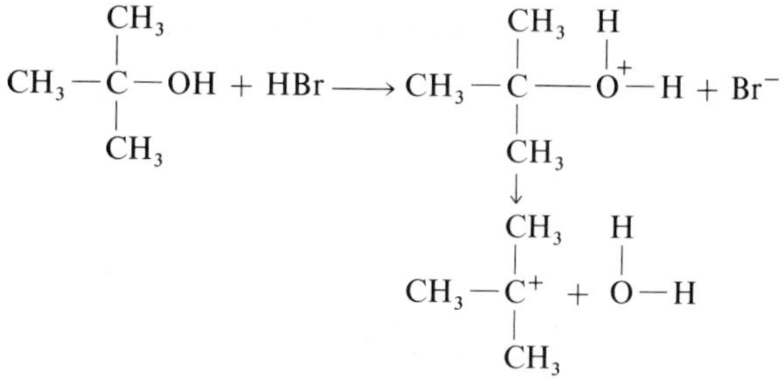

(c) 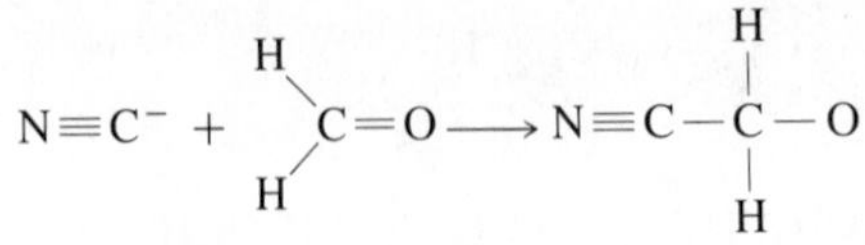

(d) 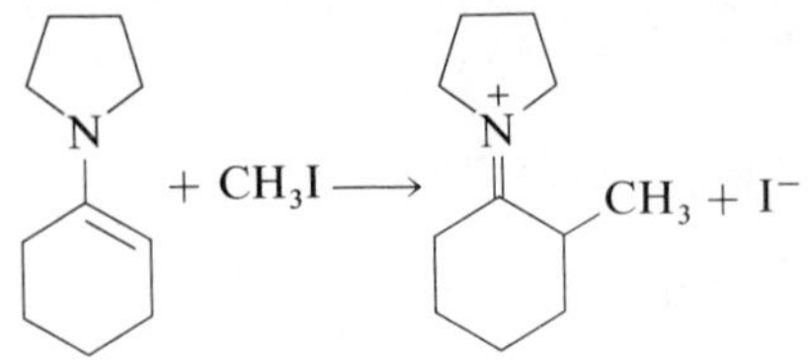

(e) 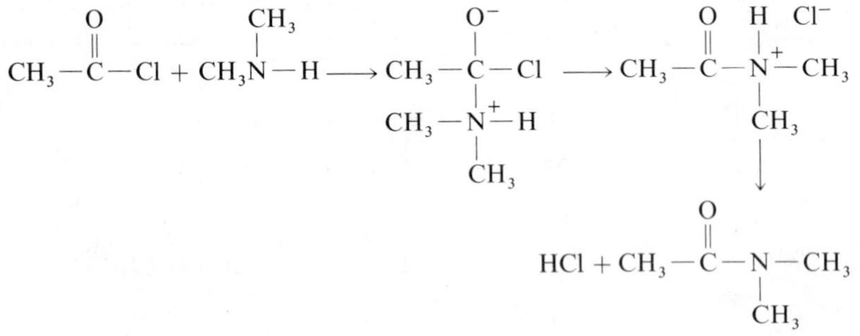

(f) 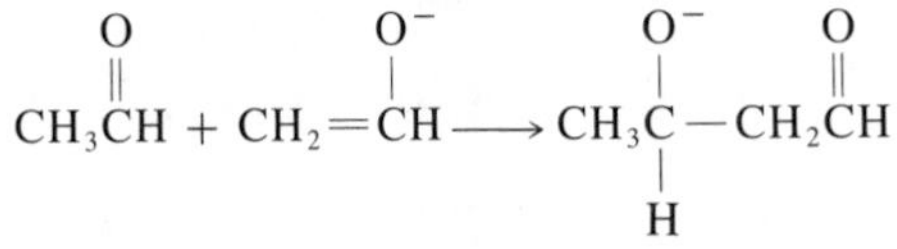

(g) 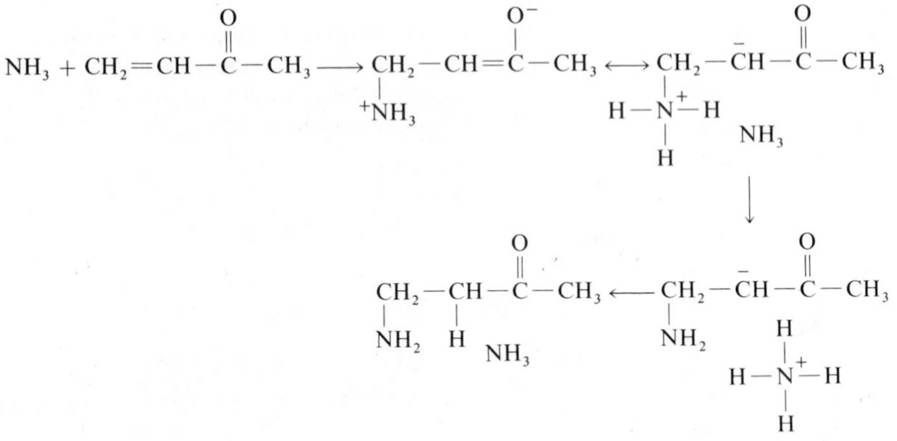

4.23 The initial rate of the reaction of chloromethane with hydroxide ion was calculated on p. 114. Does the rate of the reaction remain the same during the course of the reaction? Explain.

4.24 Refer back to Problem 4.5 (p. 118) and calculate the rate of the reaction of bromomethane with hydroxide ion when the reaction is halfway complete.

4.25 What value of $\Delta G°$ is necessary if a reaction is to go 99% to completion at 298 K?

4.26

(a) Chloromethane reacts with water, as well as with hydroxide ion, to give methanol as the product. The rate constant given for the reaction of hydroxide ion with chloromethane (p. 114) was corrected for the reaction with water. Write an equation for the reaction of chloromethane with water. What is the nucleophile in the reaction? What are the other products of the reaction?

(b) The mechanism for the reaction of chloromethane with water is essentially the same as the one for the reaction with hydroxide ion. Write out a detailed mechanism for the reaction of chloromethane with water.

(c) Given the similarity of the mechanisms for the reactions of chloromethane with hydroxide ion and with water, what is the rate equation for the reaction with water?

(d) When bubbled through water, chloromethane dissolves in it to the extent of about 0.003 mol/L. What is the concentration of water in such a solution? Under these conditions, is it possible to see a change in the concentration of water as the chloromethane reacts? What will the rate equation look like for this experiment?

4.27 The rate of the reaction of propene with hydrogen iodide was measured at 238.5 °C (Problem 4.13, p. 126). The following thermodynamic data are available for the reactants and the products at 500 K. Calculate the equilibrium constant for the reaction at that temperature. A review of pp. 93–94 may be helpful.

	S_f°, cal/mol·deg	ΔH_f°, kcal/mol
propene	73.48	2.80
hydrogen iodide	52.98	−1.35
isopropyl iodide	91.02	−19.65

4.28

(a) The thermodynamic data shown have been determined for the following reaction.

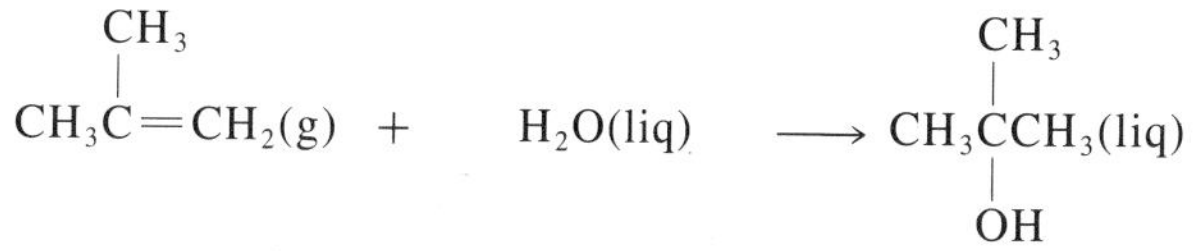

CH3
CH3C=CH2(g) + H2O(liq) → CH3CCH3(liq)
OH

S_f° (at 298 K)	70.17 cal/deg·mol	16.72 cal/deg·mol	46.30 cal/deg·mol
ΔH_f° (at 298 K)	−4.04 kcal/mol	−68.32 kcal/mol	−85.87 kcal/mol

What is the equilibrium constant for this reaction at 298 K? A review of pp. 93–94 may be helpful.

(b) Water alone does not add to the double bond, but when a solution of sulfuric acid in water is used, the alkene is converted into an alcohol. Propose an explanation for this observation based on the pK_a values for the species involved. What is the electrophilic species in a dilute solution of sulfuric acid in water? What is the nucleophile? Why won't pure water add to the double bond?

4.29 When cyclohexanol is heated in the presence of phosphoric acid, it loses water to become cyclohexene. This reaction is known as the dehydration of an alcohol. The acid serves as a catalyst and is regenerated in the reaction. The equation for the reaction and thermodynamic data for the compounds involved are as follows:

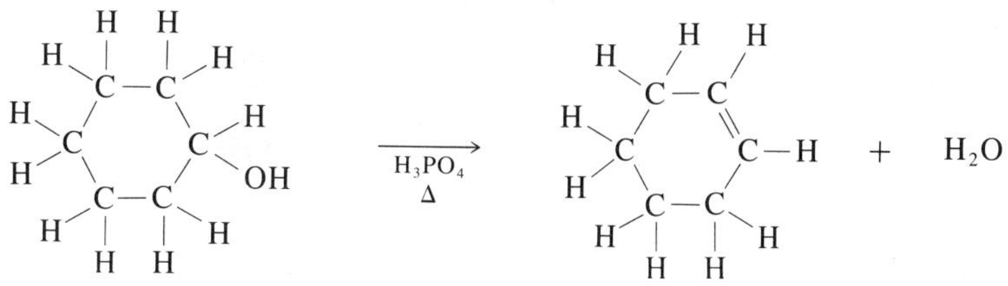

OH
H3PO4
Δ
+ H2O

S_f° (at 298 K)	78.32 cal/deg·mol	74.27 cal/deg·mol	45.11 cal/deg·mol
ΔH_f° (at 298 K)	−70.40 kcal/mol	−1.28 kcal/mol	−57.80 kcal/mol

(a) Does the entropy of the system increase or decrease during this reaction? How can you account for this result? What about the enthalpy of the system?

(b) Calculate the equilibrium constant for this reaction at 298 K. Will products be formed from the starting material at that temperature?

(c) The stages of the reaction (shown using simplified formulas, p. 153, for the cyclic compounds) are believed to be as follows:

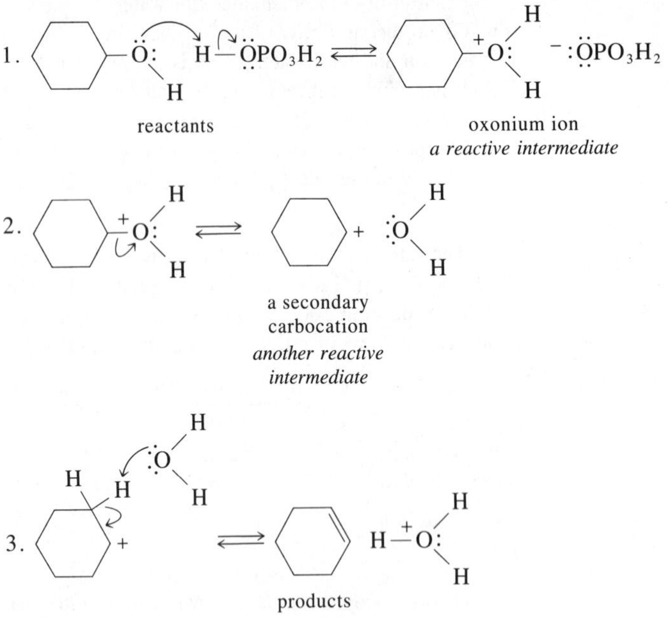
1.
reactants
oxonium ion
a reactive intermediate
2.
a secondary
carbocation
another reactive
intermediate
3.
products

4. $H_3O^+ + {}^-\!:\ddot{O}PO_3H_2 \rightleftharpoons H_2O + HOPO_3H_2$

When cyclohexanol is mixed with phosphoric acid at room temperature, heat is evolved. What does this say about the relative energy levels of all the species involved in the first step of the reaction?

(d) No product is formed at room temperature. The reaction mixture must be heated for cyclohexene to be formed. Which step of this reaction do you think has the highest free energy of activation?

5

Alkanes and Cycloalkanes

A • LOOK • AHEAD

Alkanes are hydrocarbons with the general formula C_nH_{2n+2}. The structures of methane, CH_4, and ethane, C_2H_6, the two smallest members of the family of alkanes, were examined thoroughly on pp. 42 and 43. Removal of a hydrogen atom from an alkane gives rise to an alkyl group (p. 43). Cycloalkanes are hydrocarbons with the general formula C_nH_{2n}. In cycloalkanes, the carbon atoms form a ring.

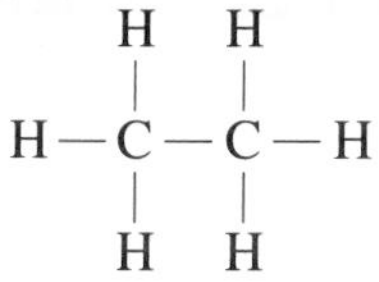

C_2H_6
ethane
an alkane

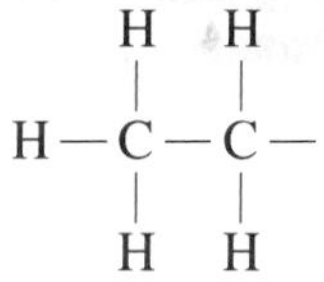

C_2H_5
ethyl group
an alkyl group

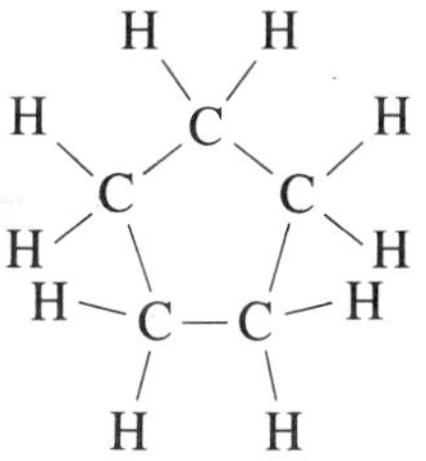

C_5H_{10}
cyclopentane
a cycloalkane

Many organic compounds have alkyl or cycloalkyl groups as part of their structures. Therefore, in this chapter, alkanes and cycloalkanes are used to further examine ideas about structure and isomerism and to introduce systematic ways of naming organic compounds.

5.1 ISOMERISM AND PHYSICAL PROPERTIES

Alkanes have only σ bonds (p. 40) and are described as **saturated hydrocarbons,** meaning that all valence electrons of carbon are involved in single bonds. Compounds that differ from each other in their molecular formulas by the unit $-CH_2-$ are called members of a **homologous series.** Thus, methane and ethane

TABLE 5.1 Boiling Points and Melting Points for Some Alkanes

Molecular Formula	Name	Molecular Weight	bp, °C	mp, °C
CH_4	methane	16	−164	−182.5
C_2H_6	ethane	30	−88.6	−183.3
C_3H_8	propane	44	−42.1	−189.7
C_4H_{10}	butane	58	−0.6	−138.4
C_4H_{10}	2-methylpropane	58	−10.2	−138.3
C_5H_{12}	pentane	72	36.1	−129.7
C_5H_{12}	2-methylbutane	72	27.9	−159.9
C_5H_{12}	2,2-dimethylpropane	72	9.5	−16.6
C_6H_{14}	hexane	86	68.9	−93.5
C_7H_{16}	heptane	100	98.4	−90.6
C_8H_{18}	octane	114	125.7	−56.8
C_9H_{20}	nonane	128	150.8	−51.0
$C_{10}H_{22}$	decane	142	174.1	−29.7
$C_{20}H_{42}$	icosane	282	343	36.8

belong to a homologous series that has many other members, all of which are saturated hydrocarbons. All alkanes have similar chemical properties, but their physical properties vary with molecular weight and the shape of the molecules. Table 5.1 shows the boiling and melting points of some representative alkanes.

The boiling points of the alkanes increase steadily with molecular weight from methane to butane; these first four alkanes are all gases at room temperature. In compounds with four or more carbon atoms, several different arrangements of the carbon chains are possible. The boiling points of different compounds having the same molecular formula and functional groups depend on the compactness of the molecular shape. Compounds with a long carbon chain tend to have higher boiling points than those with a more spherical shape (p. 29).

On the other hand, compact, spherical molecules pack better in crystal lattices, so compounds such as these have higher melting points. These trends are illustrated by the different compounds that have the molecular formula C_5H_{12}, which are shown below.

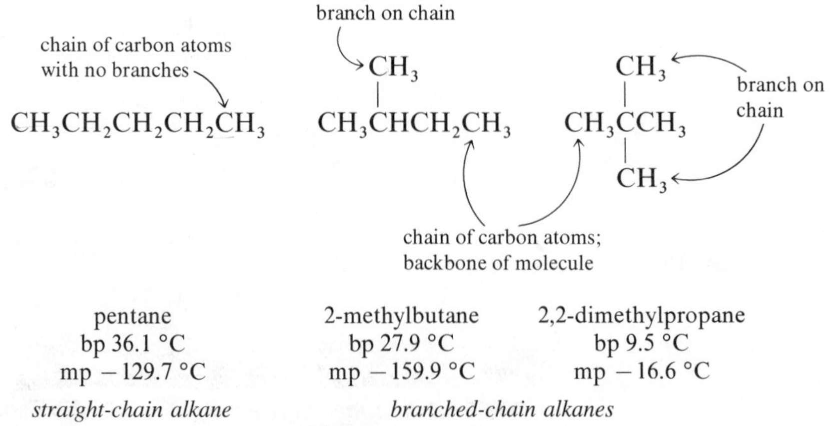
branch on chain
chain of carbon atoms with no branches
branch on chain
chain of carbon atoms; backbone of molecule
pentane
bp 36.1 °C
mp −129.7 °C
straight-chain alkane
2-methylbutane
bp 27.9 °C
mp −159.9 °C
2,2-dimethylpropane
bp 9.5 °C
mp −16.6 °C
branched-chain alkanes

Pentane has the most elongated molecular shape and the highest boiling point of the three compounds. 2,2-Dimethylpropane has the most compact and spherical shape, the lowest boiling point, and by far the highest melting point of the three.

Pentane, 2-methylbutane, and 2,2-dimethylpropane all have the molecular formula C_5H_{12} but exhibit different bonding arrangements among the five carbon atoms and are constitutional isomers (p. 16). Pentane is a **straight-chain hydrocarbon;** 2-methylbutane and 2,2-dimethylpropane are **branched-chain hydrocarbons.** In pentane, all the carbon atoms form one continuous chain. In the other two compounds, some of the carbon atoms form the ''backbone'' of the molecule, but others branch off it. Constitutional isomers have different physical properties, as shown above for the isomers of C_5H_{12}.

The higher the number of carbon atoms in an alkane (and therefore the higher its molecular weight), the higher are its boiling and melting points. The actual values are given in Table 5.1 for the straight-chain hydrocarbons CH_4 through $C_{10}H_{22}$ and $C_{20}H_{42}$. If a linear alkane contains approximately eighteen carbon atoms, it is a solid at room temperature. Icosane, $C_{20}H_{42}$, has a melting point of 36.8 °C, which is very close to body temperature.

5.2 METHANE

Methane, CH_4, is the simplest hydrocarbon. The details of the structure of methane were given on pp. 18 and 42. The molecule is tetrahedral and has four equivalent carbon-hydrogen bonds. It may be represented in three dimensions, as a Lewis structure, or by a condensed formula (Figure 5.1).

Any one of the hydrogen atoms in methane may be replaced by another atom or group to give a new compound. Chloromethane, for example, is a compound in which one of the hydrogen atoms in methane has been substituted by a chlorine atom. Chloromethane is an alkyl halide; the hydrocarbon portion of the molecule

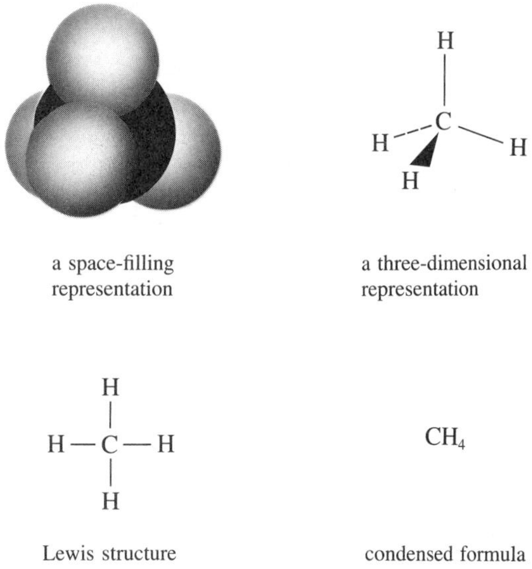
H
C
H
H
H
a space-filling
representation
a three-dimensional
representation
H
H—C—H
H
CH4
Lewis structure
condensed formula

FIGURE 5.1 Different representations of methane.

is called a methyl group. Methanol, an alcohol, is another example of a compound containing a methyl group.

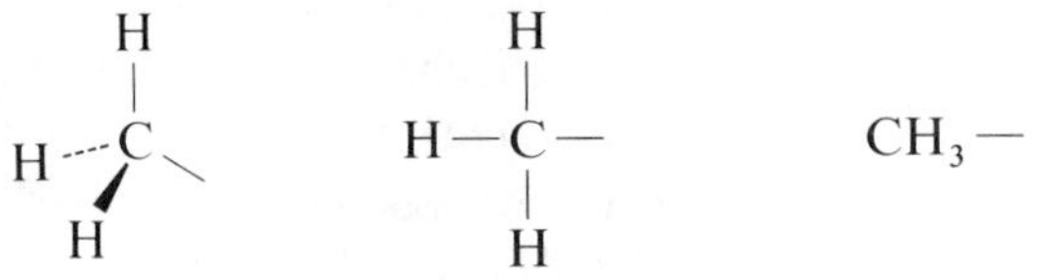

the methyl group
an alkyl group

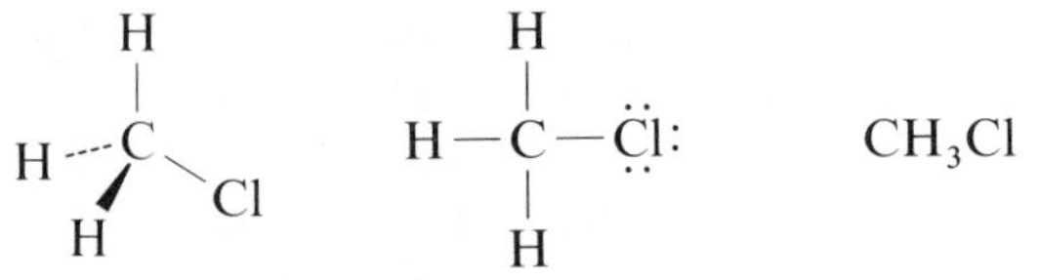

chloromethane
or methyl chloride
an alkyl halide

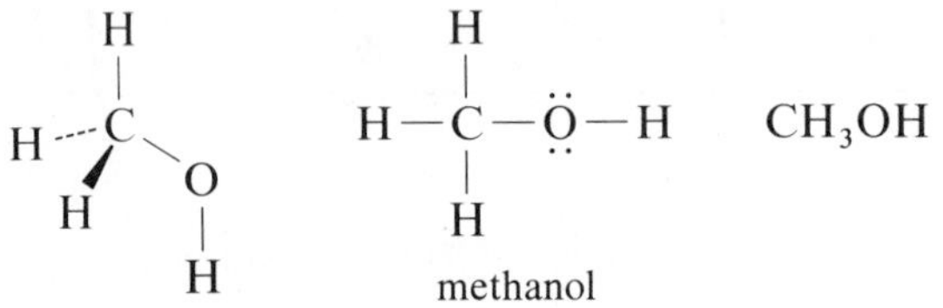

methanol
or methyl alcohol
an alcohol

The first name given for each compound above is its systematic name; the second is another acceptable and widely used name. The **methyl group** is the alkyl group derived from a particular alkane, methane. The name of an alkyl group is obtained by changing the **ane** ending of the name of an alkane to **yl.** Thus, the methyl group is clearly related to methane. An alkane is sometimes represented in tables or equations by the symbol RH; the corresponding alkyl group is symbolized by R. An introduction to the rules for the systematic naming of organic compounds will be given in Section 5.8. In this and the next three sections, you should concentrate on the structures and the names of the first four alkanes and the alkyl groups derived from them.

It is important to recognize that the chlorine atom in chloromethane or the hydroxyl group in methanol could have replaced any one of the hydrogen atoms in methane; in each case, the same compound would have been formed. The structural formulas below, for example, represent a single methanol molecule tumbling in space, not three different molecular species.

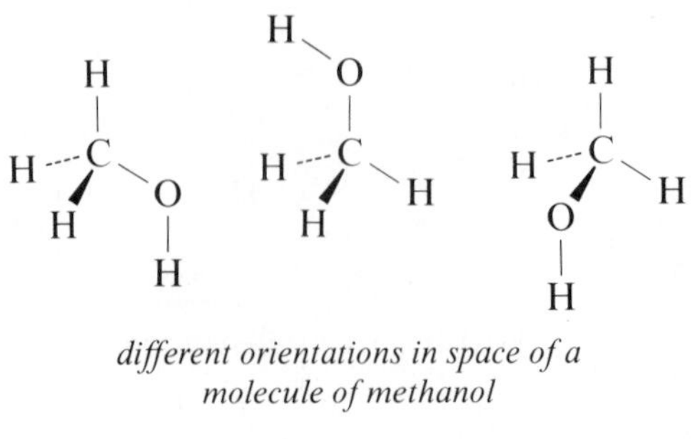

different orientations in space of a molecule of methanol

All of the hydrogen atoms in methane are equivalent. Substitution of a particular group for any one of them results in the same compound.

Similarly, only one compound results if two of the hydrogen atoms of methane are replaced. For example, a molecule of dichloromethane may be represented in a number of ways.

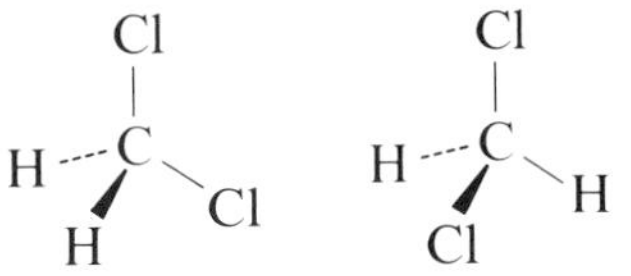

perspective formulas

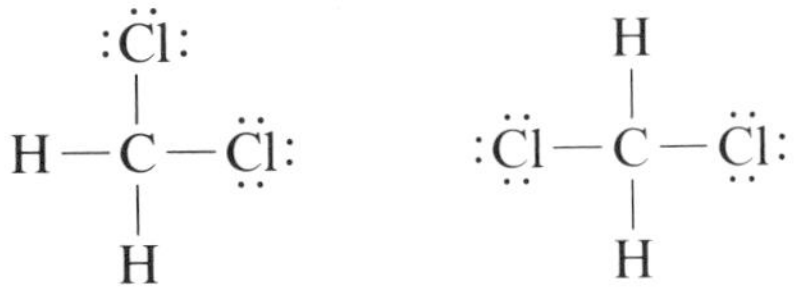

CH_2Cl_2

Lewis structures *condensed formula*

dichloromethane

The representation a chemist chooses to use depends on the particular molecular features we are trying to emphasize.

PROBLEM 5.1

Draw at least two different three-dimensional representations for each of the following compounds. (You may wish to review Sections 1.6 and 2.4 before attempting this.)

(a) CH_2ClBr (b) CH_3SH (c) CH_3OCH_3 (d) $CHCl_3$ (e) CH_3NH_2

5.3 ETHANE

A. The Ethyl Group

Ethane, C_2H_6, follows methane in the series of alkanes. The structure of ethane (considered in detail on pp. 19 and 43) may be represented by a structural formula showing the molecule in three dimensions, by a Lewis structure, or by a condensed formula.

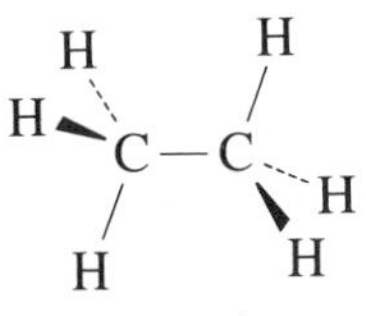

perspective formula

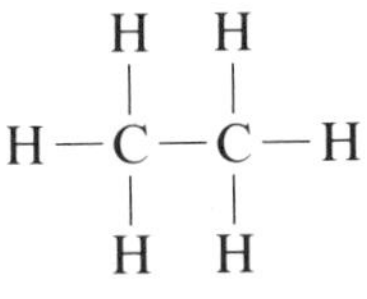

Lewis structure

CH_3CH_3

condensed formula

ethane

All of the hydrogen atoms in ethane are equivalent. Removal of any one of them gives the **ethyl group,** and substitution of another atom or group for any one of them results in the same compound.

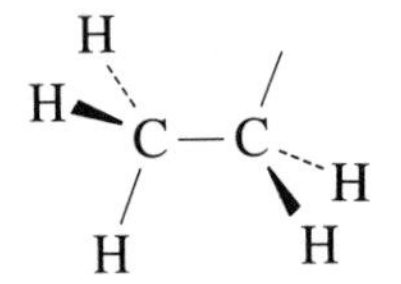

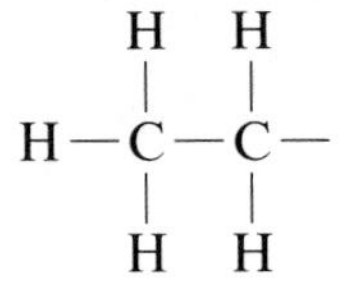

the ethyl group

CH_3CH_2-

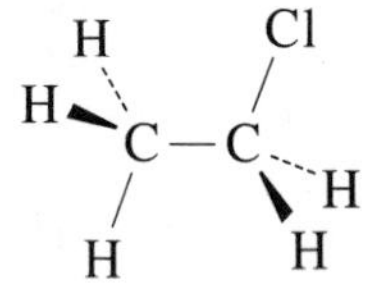

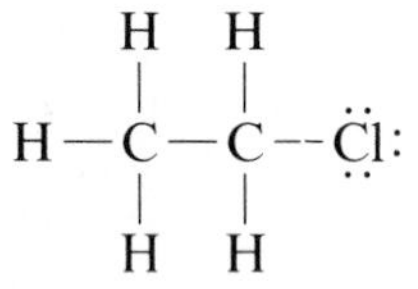

chloroethane
or ethyl chloride

CH_3CH_2Cl

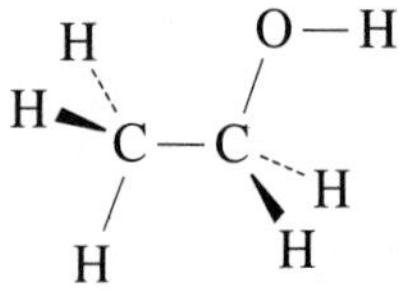

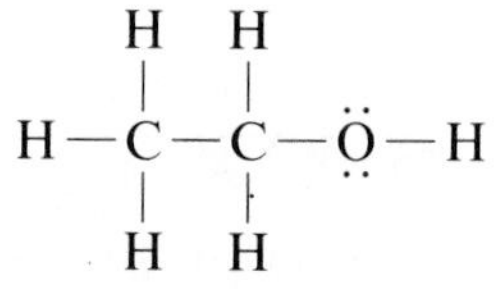

ethanol
or ethyl alcohol

CH_3CH_2OH

perspective formula *Lewis structure* *condensed formula*

The three-dimensional representations of chloroethane shown below are drawn in different ways to represent views of the molecule from different points in space. All four of these structural formulas represent the same molecule.

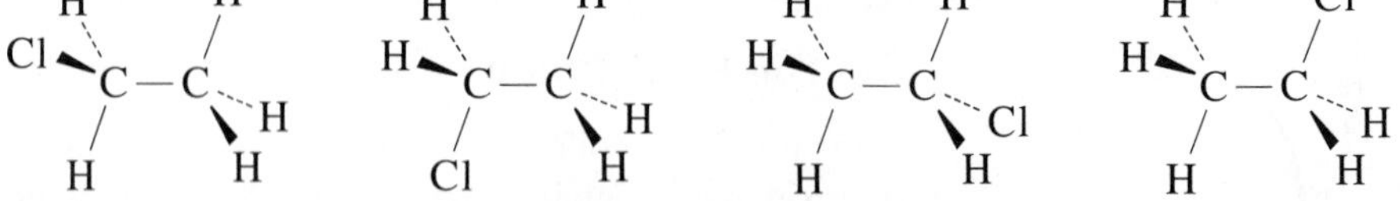

representations of chloroethane, viewed from several different perspectives

B. Isomerism in Disubstituted Ethanes

The replacement of any two hydrogen atoms of methane by two other atoms gives the same compound (p. 141). An inspection of the structure of chloroethane, however, reveals that the five hydrogen atoms are not equivalent to each other. There are two hydrogen atoms on the carbon atom already bearing a chlorine atom and three on the other carbon atom. If one of the remaining hydrogen atoms is replaced by a chlorine atom, the new compound, $C_2H_4Cl_2$, could have one of two possible structures.

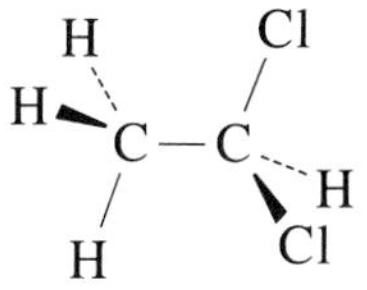

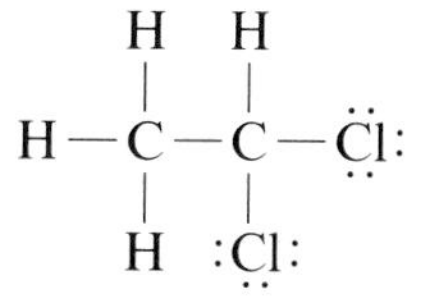

CH_3CHCl_2

bp 57 °C

1,1-dichloroethane

mp −97 °C

$\mu = 1.95$ D

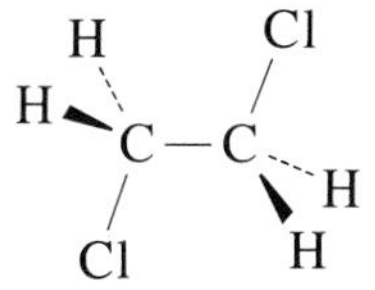

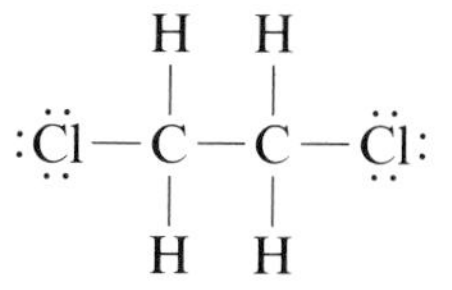

$ClCH_2CH_2Cl$

bp 83 °C

1,2-dichloroethane

mp −35 °C

$\mu = 1.42$ D

isomers of $C_2H_4Cl_2$

1,1-Dichloroethane is derived from chloroethane by replacement of a hydrogen atom on the same carbon atom as the first chlorine atom. 1,2-Dichloroethane is obtained by replacement of one of the three hydrogen atoms on the originally unsubstituted carbon atom. The two compounds are another example of constitutional isomers, compounds having the same molecular formula but different arrangements of the bonds. They have different physical properties, including boiling and melting points and dipole moments. Their names reflect the difference in their structures. The name 1,1-dichloroethane indicates that both chlorine atoms are on the same carbon atom. 1,2-Dichloroethane identifies the positions of the chlorine atoms on adjacent carbon atoms.

PROBLEM 5.2

How many compounds with the molecular formula $C_2H_3Cl_3$ are possible? How about $C_2H_2Cl_4$? Draw condensed formulas, Lewis structures, and three-dimensional representations of these compounds.

5.4 PROPANE

The three carbon atoms and eight hydrogen atoms in propane, C_3H_8, can be combined in only one way. However, propane is the first of the alkane series to contain two different kinds of carbon and hydrogen atoms.

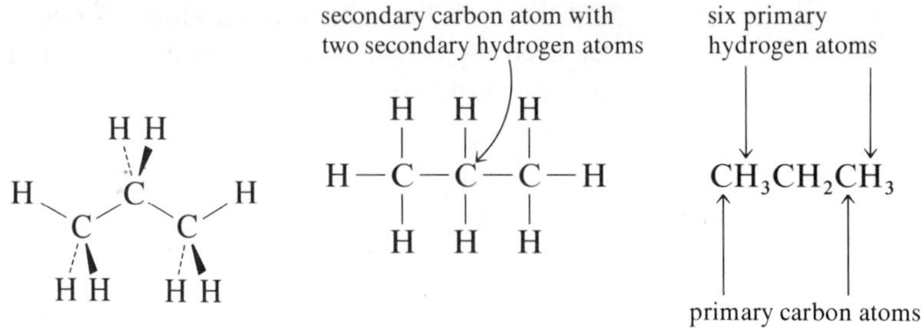
secondary carbon atom with two secondary hydrogen atoms
six primary hydrogen atoms
primary carbon atoms

perspective formula | *Lewis structure* | *condensed formula*

propane

The two end carbon atoms in propane that are each attached to only one other carbon atom are called **primary carbon atoms.** The hydrogen atoms on these carbon atoms are designated as **primary hydrogen atoms.** There are two equivalent primary carbon atoms and six equivalent primary hydrogen atoms in propane. The middle carbon atom in propane is bonded to two carbon atoms and is a **secondary carbon atom,** bearing two **secondary hydrogen atoms.**

Two different propyl groups may be created from propane. Removal of a primary hydrogen atom gives the **propyl,** or ***n*-propyl, group** (*n* stands for normal), a primary alkyl group.

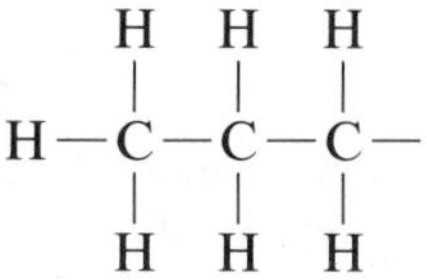

$CH_3CH_2CH_2-$

propyl, or *n*-propyl, group
a primary alkyl group

$CH_3CH_2CH_2Cl$
1-chloropropane
or *n*-propyl chloride
bp 46.6 °C
a primary alkyl halide

$CH_3CH_2CH_2OH$
1-propanol
or *n*-propyl alcohol
bp 97.4 °C
a primary alcohol

If one of the secondary hydrogen atoms is removed, a secondary alkyl group known as the **1-methylethyl,** or **isopropyl, group** is formed.

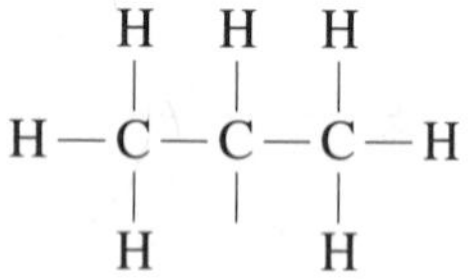

CH_3CHCH_3 (with a free bond on the central carbon)

1-methylethyl, or isopropyl, group
a secondary alkyl group

CH_3CHCH_3 with Cl on the central carbon
2-chloropropane
or isopropyl chloride
bp 35.7 °C
a secondary alkyl halide

CH_3CHCH_3 with OH on the central carbon
2-propanol
or isopropyl alcohol
bp 82.4 °C
a secondary alcohol

1-Chloropropane and 2-chloropropane are constitutional isomers. They have different physical properties, for example, different boiling points. 1-Chloropropane is classified as a **primary alkyl halide** because the halogen is bonded to a primary carbon atom; 2-chloropropane is a **secondary alkyl halide** because the halogen is bonded to a secondary carbon atom. The two compounds have some chemical properties that are similar and some that are different. These properties are treated in more detail in Chapter 7. Likewise, 1-propanol and 2-propanol are isomeric alcohols. They share some chemical properties but differ in that the first is a **primary alcohol** and the other is a **secondary alcohol** (Chapter

12). The kind of carbon atom to which a functional group is bonded often makes a difference in reactivity. Therefore, it is important that you learn to distinguish isomers and to classify them according to their structural features. For example, distinctions were made among primary, secondary, and tertiary carbocations in Chapter 4 (p. 122). Differences in the stabilities of primary and secondary carbocations determine the products formed when hydrogen bromide adds to propene, for example.

PROBLEM 5.3

How many constitutional isomers of $C_3H_6Cl_2$ are possible? Draw them as condensed formulas, Lewis structures, and three-dimensional formulas.

5.5 BUTANES

Alkanes with the molecular formula C_4H_{10} are butanes. There are two ways in which four carbon atoms and ten hydrogen atoms can be bonded together.

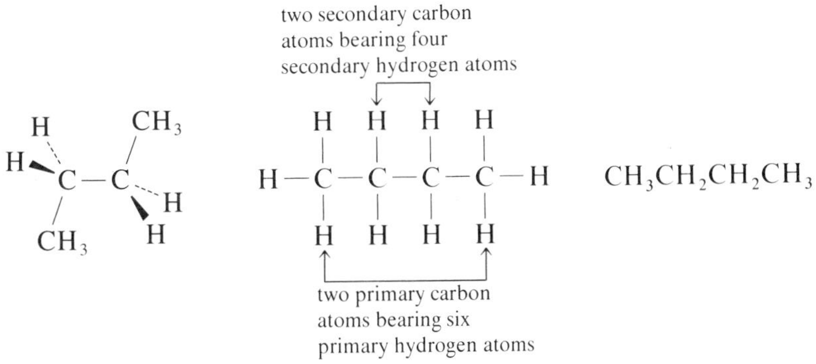
two secondary carbon atoms bearing four secondary hydrogen atoms
two primary carbon atoms bearing six primary hydrogen atoms
CH3CH2CH2CH3

butane or *n*-butane
bp −0.6 °C

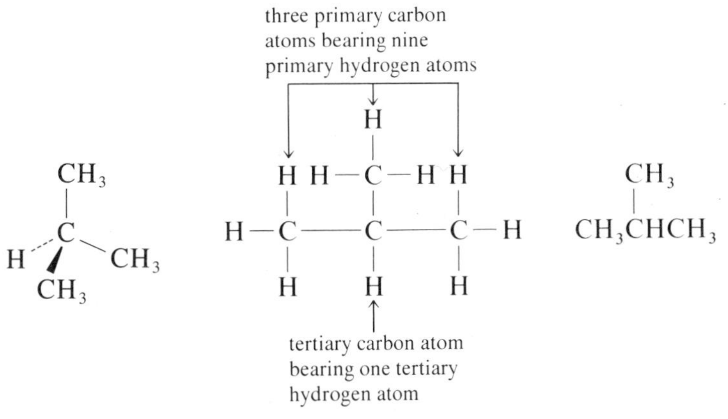
three primary carbon atoms bearing nine primary hydrogen atoms
tertiary carbon atom bearing one tertiary hydrogen atom

2-methylpropane or isobutane
bp −10 °C

One of the butanes is the straight-chain isomer with two equivalent primary carbon atoms and two equivalent secondary ones. The other isomer of C_4H_{10} is a branched-chain alkane. In this isomer, the three carbon atoms that are attached to only one other carbon atom are, by definition, primary carbon atoms. All of the hydrogen atoms on these three carbon atoms are equivalent and are also primary. The fourth carbon atom, which is bonded to three other carbon atoms, is called a **tertiary carbon atom** and bears a **tertiary hydrogen atom.**

The straight-chain isomer of C_4H_{10} is called butane or normal or *n*-butane. The systematic name for the branched-chain isomer is 2-methylpropane, indicating that the compound has a three-carbon chain, propane, on which a methyl group is substituted on the second carbon atom. This compound is also called isobutane.

Four different alkyl groups with the molecular formula C_4H_9 can be created by removing a different type of hydrogen atom from butane or 2-methylpropane. Removal of a primary hydrogen atom from butane creates a primary alkyl group, the **butyl,** or *n*-butyl, group.

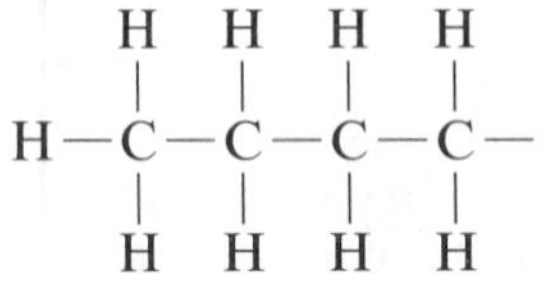

$$H-\overset{\overset{\displaystyle H}{|}}{\underset{\underset{\displaystyle H}{|}}{C}}-\overset{\overset{\displaystyle H}{|}}{\underset{\underset{\displaystyle H}{|}}{C}}-\overset{\overset{\displaystyle H}{|}}{\underset{\underset{\displaystyle H}{|}}{C}}-\overset{\overset{\displaystyle H}{|}}{\underset{\underset{\displaystyle H}{|}}{C}}- \qquad CH_3CH_2CH_2CH_2-$$

butyl, or *n*-butyl, group

a primary alkyl group

$CH_3CH_2CH_2CH_2Cl$	$CH_3CH_2CH_2CH_2OH$
1-chlorobutane	1-butanol
or *n*-butyl chloride	or *n*-butyl alcohol
bp 78.4 °C	bp 117 °C
a primary alkyl halide	*a primary alcohol*

Removal of one of the secondary hydrogen atoms of butane gives the **1-methylpropyl,** or *sec*-butyl, group.

$$H-\overset{\overset{\displaystyle H}{|}}{\underset{\underset{\displaystyle H}{|}}{C}}-\overset{\overset{\displaystyle H}{|}}{\underset{\underset{\displaystyle H}{|}}{C}}-\overset{\overset{\displaystyle H}{|}}{C}-\overset{\overset{\displaystyle H}{|}}{\underset{\underset{\displaystyle H}{|}}{C}}-H \qquad CH_3CH_2CHCH_3$$

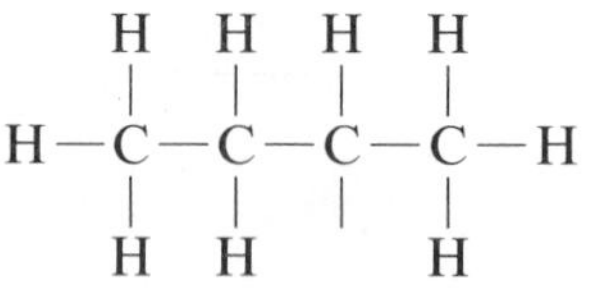

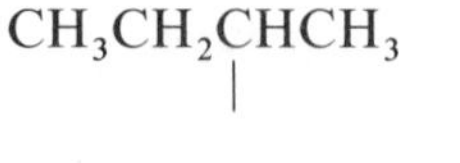

1-methylpropyl, or *sec*-butyl, group

a secondary alkyl group

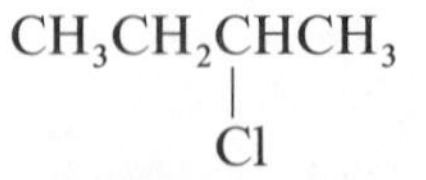

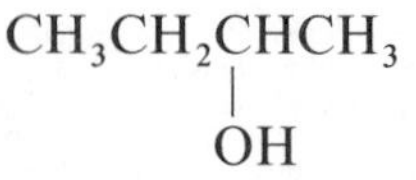

$CH_3CH_2\underset{\underset{\displaystyle Cl}{\vert}}{C}HCH_3$	$CH_3CH_2\underset{\underset{\displaystyle OH}{\vert}}{C}HCH_3$
2-chlorobutane	2-butanol
or *sec*-butyl chloride	or *sec*-butyl alcohol
bp 68.3 °C	bp 99.5 °C
a secondary alkyl halide	*a secondary alcohol*

Two different alkyl groups can be formed from 2-methylpropane. A primary alkyl group, the **2-methylpropyl,** or **isobutyl, group,** results from the removal of any one of the primary hydrogen atoms.

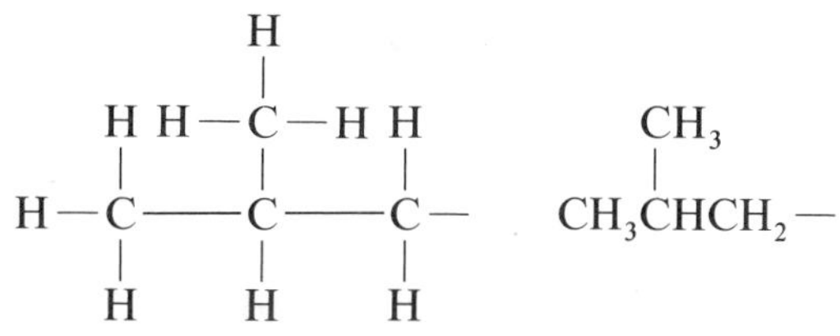

2-methylpropyl, or isobutyl, group

a primary alkyl group

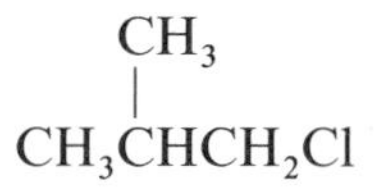

1-chloro-2-methylpropane
or isobutyl chloride
bp 68.9 °C

a primary alkyl halide

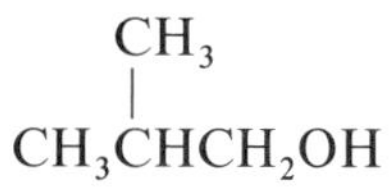

2-methyl-1-propanol
or isobutyl alcohol
bp 108 °C

a primary alcohol

Removal of the tertiary hydrogen atom gives a tertiary alkyl group, usually called the **tertiary butyl** (abbreviated ***tert*-butyl) group.**

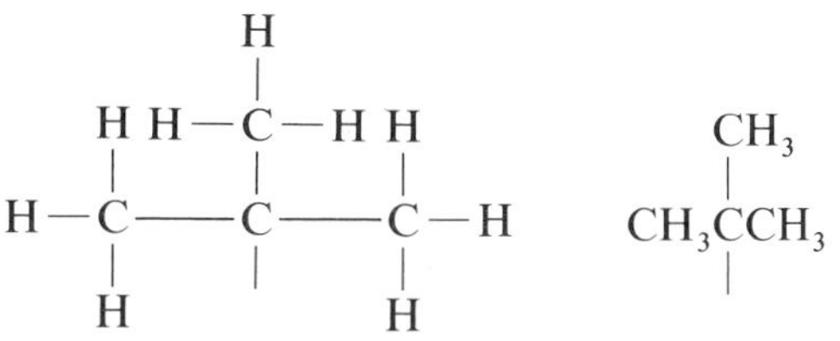

1,1-dimethylethyl or *tert*-butyl group

a tertiary alkyl group

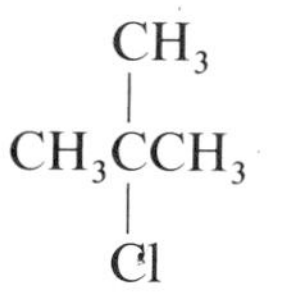

2-chloro-2-methylpropane
tert-butyl chloride
bp 52 °C

a tertiary alkyl halide

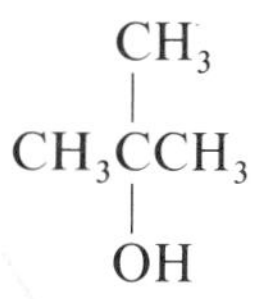

2-methyl-2-propanol
tert-butyl alcohol
bp 82 °C

a tertiary alcohol

The four alkyl halides with the molecular formula C_4H_9Cl shown above are constitutional isomers of one another, as are the four alcohols with the molecular formula $C_4H_{10}O$. Each constitutional isomer has distinctive physical properties; for example, note the different boiling points for the isomeric compounds shown in this section. Thus, various isomers may be separated from each other by methods such as fractional distillation of liquids, recrystallization of solids, and chromatography.

You should learn the structures and the names of the alkyl groups derived from methane, ethane, propane, and the two butanes because they are used in naming more complicated molecules.

PROBLEM 5.4

Write structural formulas for all compounds having the molecular formula $C_4H_8Cl_2$.

PROBLEM 5.5

Many aromatic hydrocarbons have alkyl groups on benzene rings. Identify the alkyl group(s) on the benzene ring in each of the following compounds.

(a) 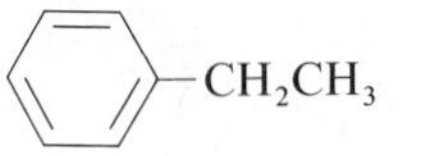

(b) 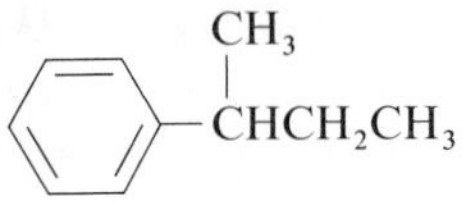

(c) 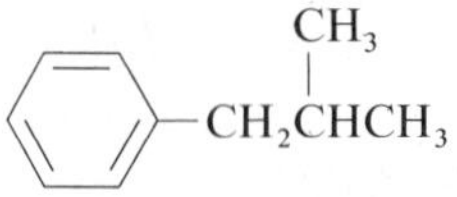

(d) 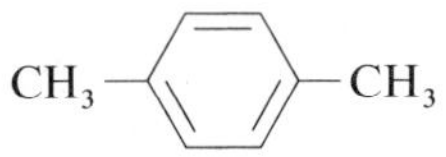

(e) 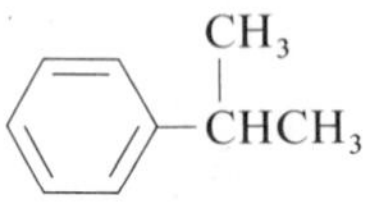

(f) 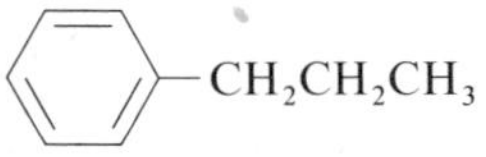

(g) 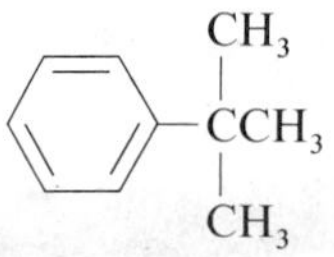

(h) 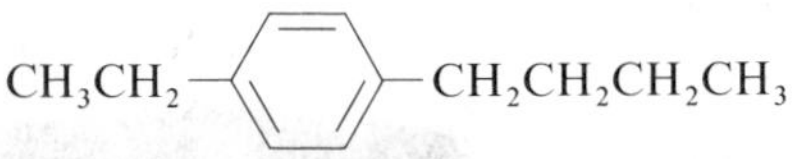

PROBLEM 5.6

Decide whether or not each of the following sets of structural formulas represents constitutional isomers.

(a) $CH_3CH{=}CHCH_2OH$, $CH_3CH{=}CHOCH_3$

(b) $ClCH_2CH_2CH_2CH_3$, $CH_3CH_2CH_2CH_2Cl$

(c) $CH_3CH_2CH_2CH_2NH_2$, $CH_3CH_2NHCH_2CH_3$

(d) $CH_3CH_2CH_2\overset{O}{\overset{\|}{C}}H$, $CH_3CH_2\overset{O}{\overset{\|}{C}}CH_3$

(e) $ClCH_2CH_2CH_2Cl$, $CH_3\underset{Cl}{\underset{|}{\overset{Cl}{\overset{|}{C}}}}CH_3$

(f) $CH_2{=}CHCH_2CH_2OH$, $CH_3CH_2\overset{O}{\overset{\|}{C}}CH_3$

(g) $CH_3\underset{OH}{\underset{|}{C}}HCH_2CH_3$, $CH_3CH_2\underset{OH}{\underset{|}{C}}HCH_3$

5.6 CONFORMATION

A. Conformations of Ethane

The different three-dimensional pictures of chloroethane shown on p. 142 are drawn so that the carbon-chlorine bond bisects the angle made by two of the hydrogen atoms on the adjacent carbon atom. The molecule, however, is not frozen in that form. The carbon atoms are constantly in motion, rotating relative to each other around the single bond that joins them, giving rise to different orientations in space of the hydrogen atoms. The **conformations** of a molecule are different arrangements in space of the atoms within the molecule due to rotations around single bonds. In ethane, the conformation in which the bonding electrons of the carbon-hydrogen bonds on the two carbon atoms are as far apart as possible is called the **staggered conformation.** This is a low-energy conformation. As the two carbon atoms rotate with respect to each other, at some point the carbon-hydrogen bonds on one carbon atom line up paralleling the similar bonds on the other carbon atom, so that they eclipse them if viewed from one end of the molecule. This **eclipsed conformation** corresponds to a maximum energy.

The relative positions of the hydrogen atoms on the two carbon atoms are best seen in a Newman projection, named for Melvin Newman of Ohio State University. A **Newman projection** is a view of a molecule down the axis of a carbon-carbon bond. The carbon atom toward the front is represented by a dot, and the one toward the rear by a circle. The atoms or groups on the carbon atoms are shown as being bonded to the dot or the circle. Newman projections of the staggered and eclipsed conformations of ethane are shown below, with their equivalent perspective formulas.

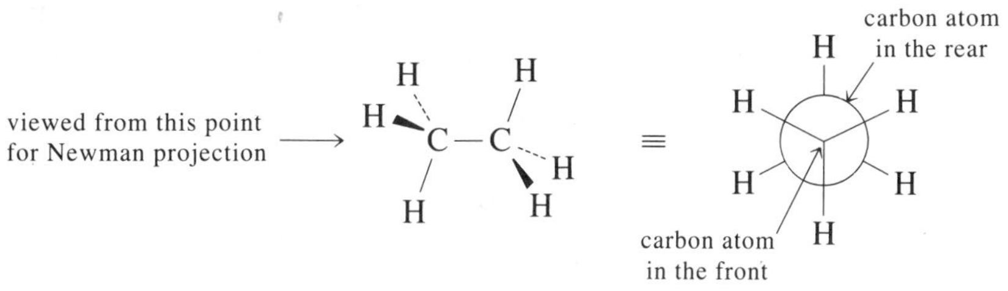
viewed from this point
for Newman projection
carbon atom
in the rear
carbon atom
in the front
staggered conformation of ethane

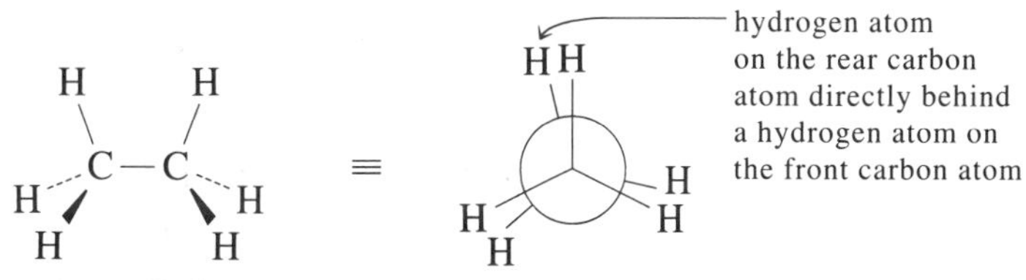
hydrogen atom
on the rear carbon
atom directly behind
a hydrogen atom on
the front carbon atom
eclipsed conformation of ethane

perspective formulas *Newman projections*

The difference in energy between the staggered and eclipsed conformations of ethane is approximately 3 kcal/mol. This energy difference is believed to arise from repulsion between the electrons in the carbon-hydrogen bonds when the bonds are closer together in the eclipsed conformation (Figure 5.2).

The Newman projections of ethane in Figure 5.2 show the rear carbon atom rotating relative to the front one around the carbon-carbon bond. A rotation of 60° converts a staggered conformation into an eclipsed one. Another 60° rotation (or a total rotation of 120°) returns the molecule to a staggered conformation. At room temperature, molecules have enough kinetic energy to get over energy barriers up to approximately 20 kcal/mol. As Figure 5.2 illustrates, the energy required for rotation around the carbon-carbon bond is only 3 kcal/mol. As a result, rotation around the carbon-carbon bond in ethane molecules is constant. Yet the molecules spend most of their time in the energy valleys represented by the staggered conformations.

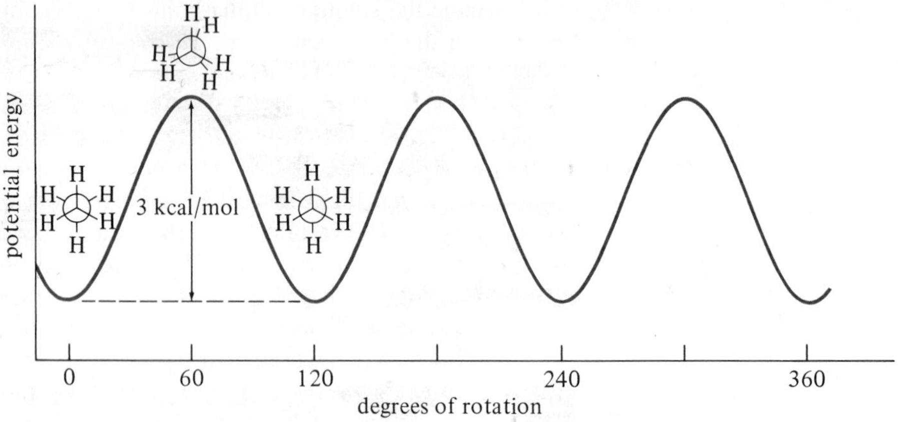
potential energy
3 kcal/mol
0
60
120
240
360
degrees of rotation

FIGURE 5.2 Energy diagram showing the energy difference between the staggered and eclipsed conformations of ethane.

The different conformations of ethane are isomeric. The isomeric species that result from conformational changes are called **conformers.** Usually conformers are being converted so rapidly from one into the other that they cannot be isolated as individual species.

PROBLEM 5.7

Draw perspective formulas and Newman projections for the different conformations of chloroethane.

B. Conformations of Butane

The three-dimensional representation of butane on p. 145 shows the first and last carbon atoms of the chain as far apart from each other as possible. The hydrogen atoms on the second and third carbon atoms are staggered. This arrangement, which is the most stable one, is called the **anti conformation** of butane. There are other conformations also. Shown below and on the following page are perspective formulas and Newman projections for the conformations of butane that result from a rotation of 60° around the bond between the second and third carbon atoms in the chain.

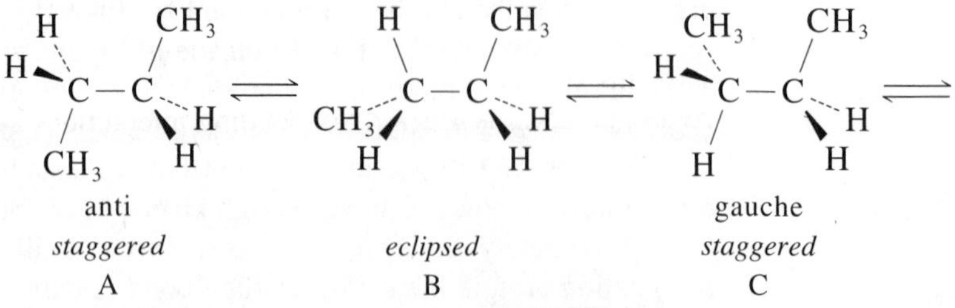
anti
staggered
A
eclipsed
B
gauche
staggered
C

perspective formulas

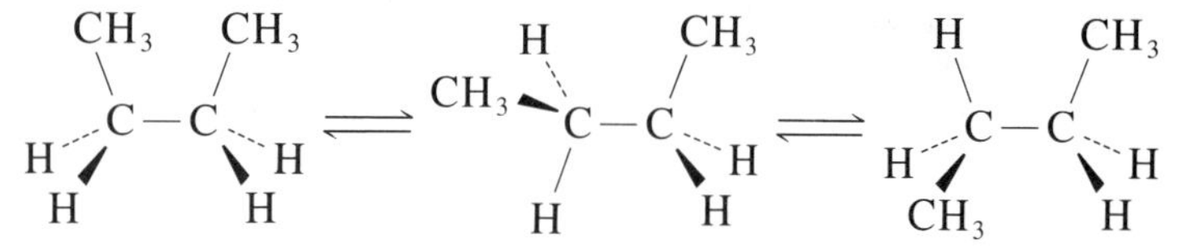

gauche

eclipsed D | *staggered* E | *eclipsed* F

perspective formulas

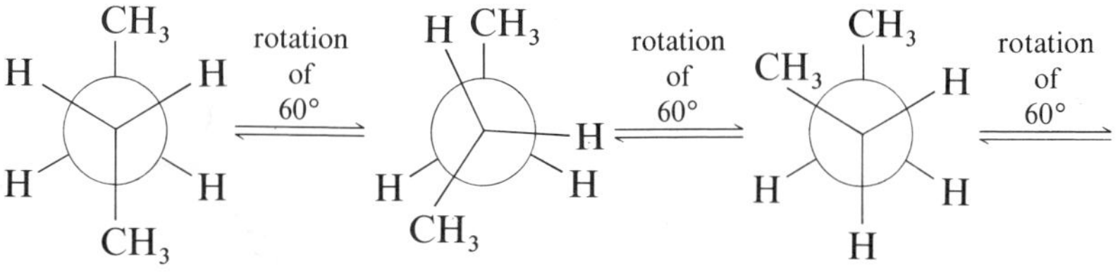
rotation of 60°
rotation of 60°
rotation of 60°

anti *staggered* A | *eclipsed* B | gauche *staggered* C

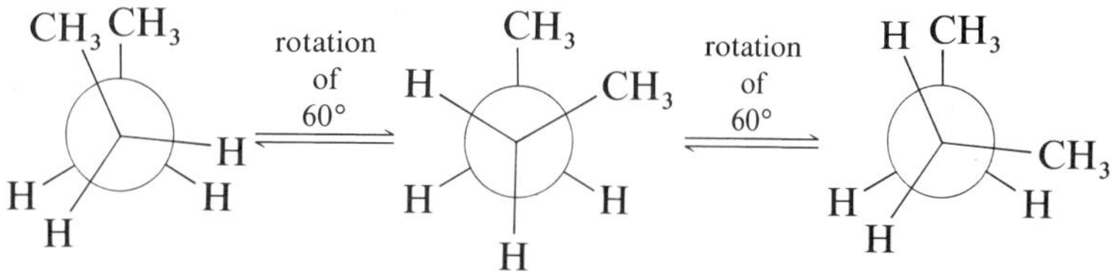
rotation of 60°
rotation of 60°

eclipsed D | gauche *staggered* E | *eclipsed* F

Newman projections

Starting with the anti conformation, rotation of the front carbon atom relative to the rear one gives different conformations of butane. Besides the anti conformation, A, there are two other staggered conformations of butane, C and E; these are called **gauche conformations.** In C and E, the bond between the first and the second carbon atoms and the bond between the third and the fourth carbon atoms are at an angle of 60° with respect to each other in the Newman projection (Figure 5.3). When the methyl groups are that close, nonbonding interactions (p. 24) known as van der Waals repulsions occur. Figure 5.3 (right side) shows that atoms have space-filling properties and exert an influence that is not apparent from looking at molecular formulas. The effective sizes of atoms in molecules are expressed in terms of van der Waals radii for those atoms. The van der Waals radius of a hydrogen atom is 1.2 Å. Nonbonding interactions between molecules or between different parts of the same molecule result in van der Waals attractions (p. 28), as long as the interacting atoms do not get too close to each other. At distances shorter than the van der Waals radii of the atoms, repulsion occurs. In the gauche conformation of butane, the hydrogen atoms of the two methyl groups are close enough that van der Waals repulsions result. Consequently, the gauche conformers are less stable than the anti form by approximately 0.9 kcal/mol.

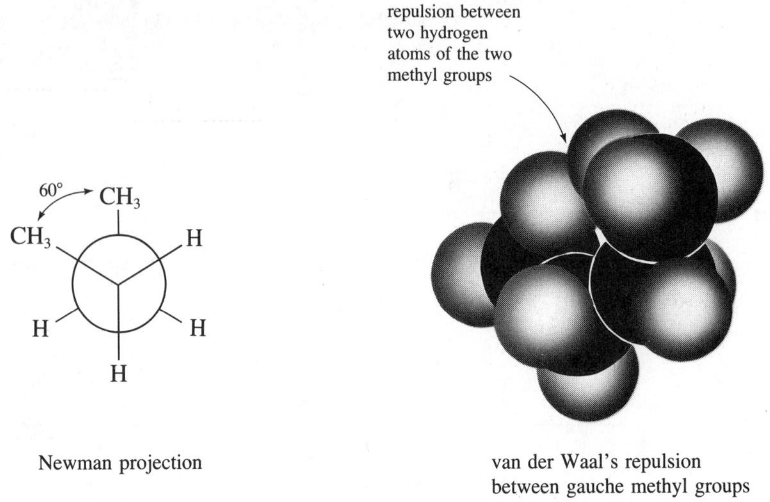
repulsion between
two hydrogen
atoms of the two
methyl groups
60°
CH_3
CH_3
H
H
H
H
Newman projection
van der Waal's repulsion
between gauche methyl groups

FIGURE 5.3 Repulsion between the methyl groups in the gauche conformation of butane.

At room temperature, about 25% of butane molecules are in gauche conformations and 75% in the anti conformation. The energy barrier to rotation around the central bond in butane is small enough, approximately 3.8 kcal/mol, that the different conformers cannot be separated from each other at room temperature; they are rapidly interconverted. The physical properties actually observed for butane represent an average of all the properties that would be expected for all of the different forms present at any one time.

The interconversions of the anti and gauche conformers require rotation through three high-energy conformations, the eclipsed conformations B, D, and F. The highest-energy conformation is D, in which the two methyl groups are eclipsed. The energy relationships of the different conformations of butane are shown in Figure 5.4.

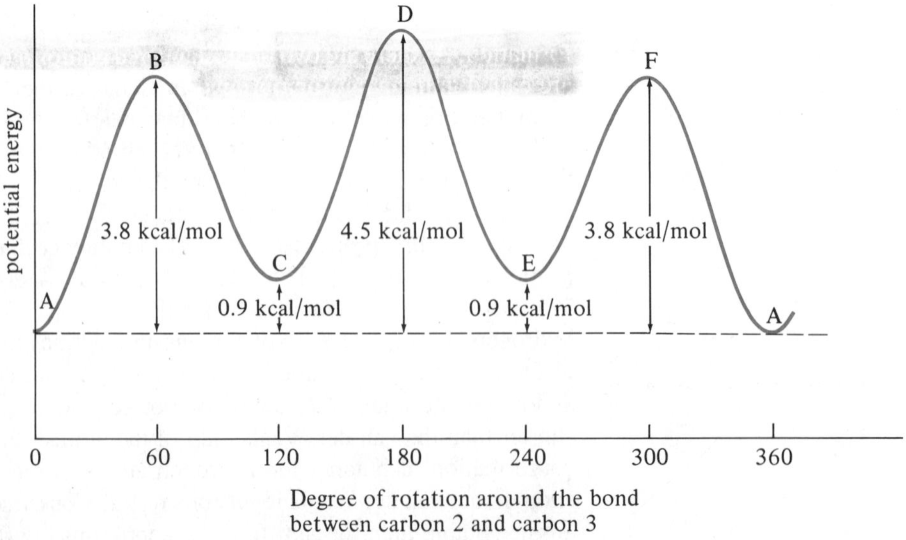
potential energy
A
B
C
D
E
F
A
3.8 kcal/mol
4.5 kcal/mol
3.8 kcal/mol
0.9 kcal/mol
0.9 kcal/mol
0
60
120
180
240
300
360
Degree of rotation around the bond
between carbon 2 and carbon 3

FIGURE 5.4 Energy diagram for the conformations of butane.

For larger alkanes, too, the most stable conformations are generally staggered, with the largest groups anti to each other. The gauche conformations are also present and are only slightly less stable than the anti ones. Thus, even though the structural formulas of alkanes are usually written in a straight line for convenience, it is important to keep in mind the zig-zag nature of hydrocarbon chains, as illustrated below for octane, $CH_3CH_2CH_2CH_2CH_2CH_2CH_2CH_3$.

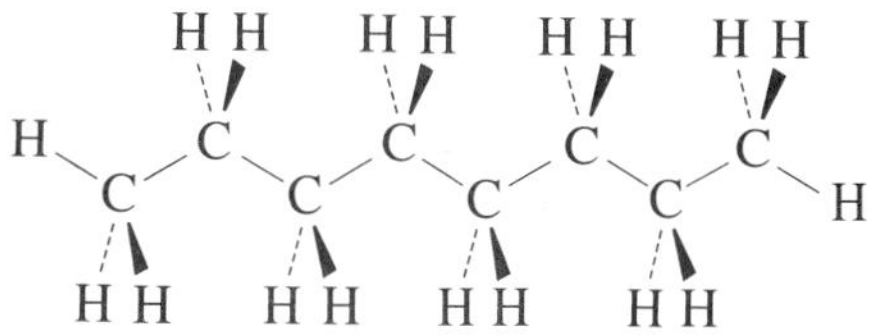

Study Guide
Concept Map 5.1

the most stable conformation of octane;
largest substituents at any one carbon-carbon bond anti to each other

---

PROBLEM 5.8

Draw possible conformations of 1-chloropropane, $CH_3CH_2CH_2Cl$, as perspective formulas and as Newman projections. Draw an energy diagram showing the relative energies of the different conformers.

PROBLEM 5.9

Show the low-energy conformers of 1,2-dichloroethane (p. 143) as perspective formulas and as Newman projections. Predict, from looking at your pictures, whether or not each conformer has a dipole moment. (You may wish to review pp. 22–24.) 1,2-Dichloroethane has a dipole moment of 1.42 D. What does this tell you about the conformational composition of 1,2-dichloroethane at room temperature?

5.7 LINE STRUCTURES FOR ORGANIC COMPOUNDS

Chemists, in drawing complex organic structures, especially those containing rings, often use a simplified way of representing them. In this type of representation, known as **line structures,** or **skeletal structural formulas,** each junction of two straight lines or the end of a straight line represents a carbon atom and the number of hydrogen atoms necessary to give the carbon four bonds, for example:

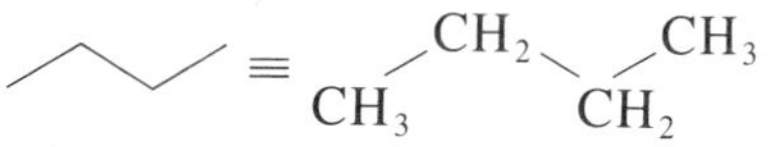

butane

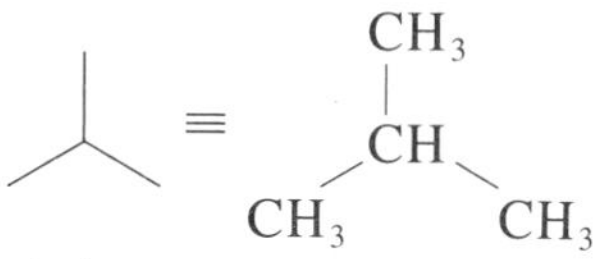

2-methylpropane

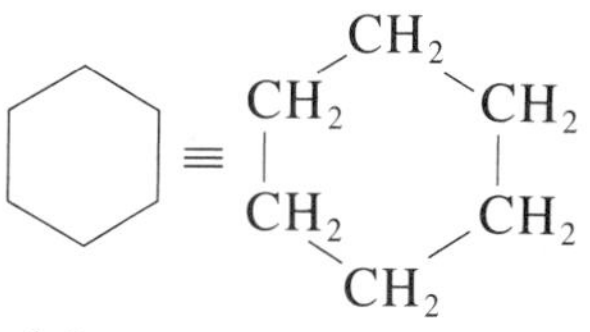

cyclohexane

Note that the end of a line is equivalent to CH_3. The intersection of two lines is CH_2, and the intersection of three lines is CH. Four lines meeting at a point represent a carbon atom with no hydrogen atoms, a **quaternary carbon atom.**

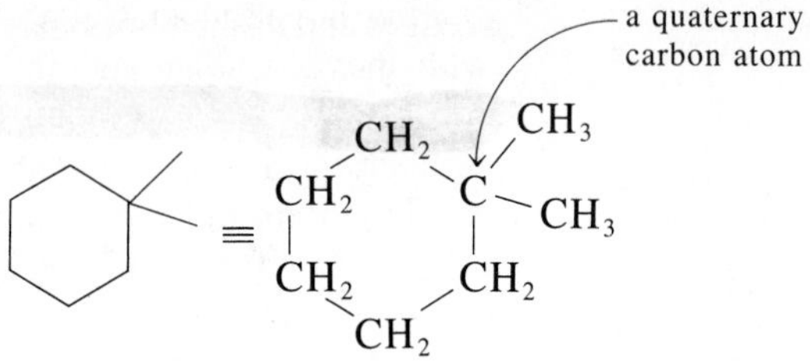
a quaternary carbon atom

Functional groups are represented by their usual symbols. A carbon-carbon double bond is shown as two parallel lines, a triple bond as three.

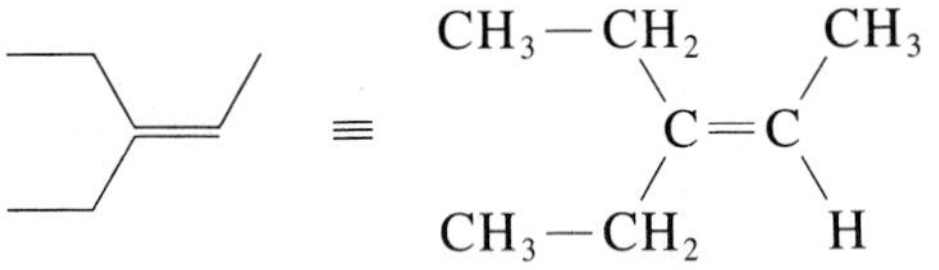

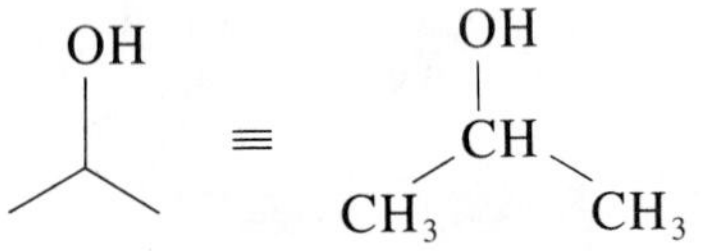

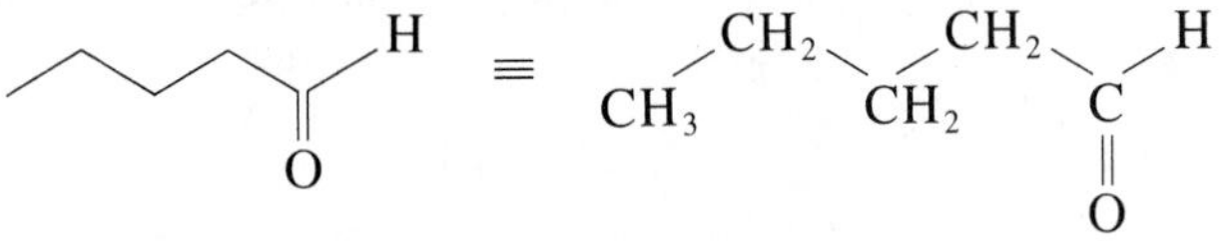

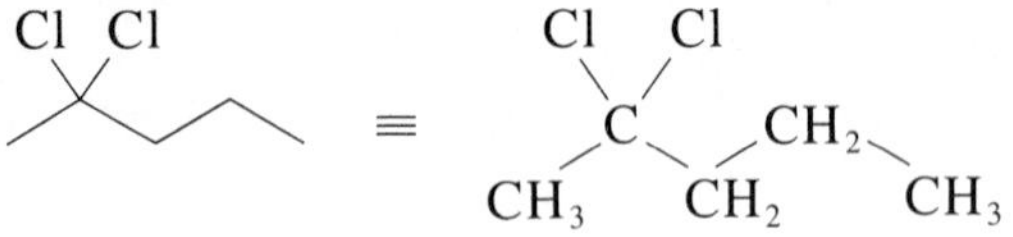

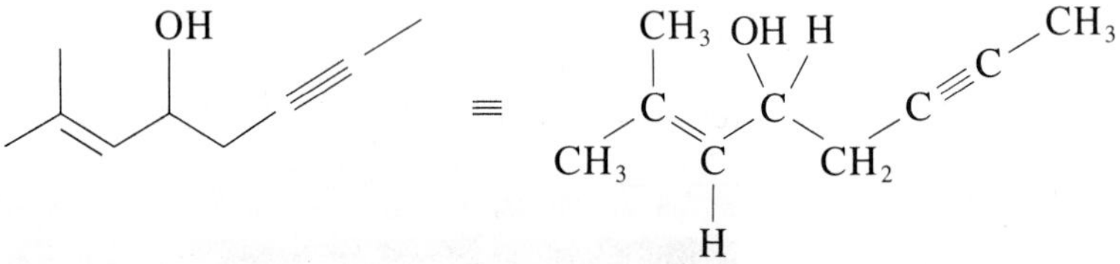

If necessary, nonbonding electrons and formal charges may also be shown on line structures.

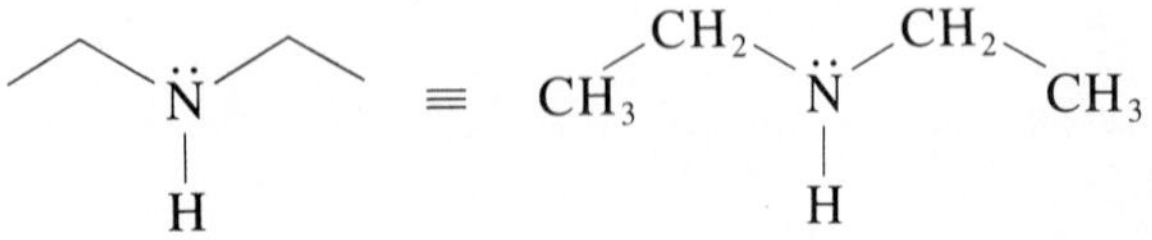

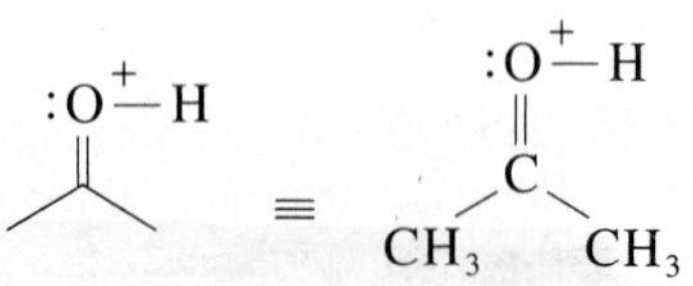

Wedges and dashed lines may be used with line structures to indicate the orientation of groups in space.

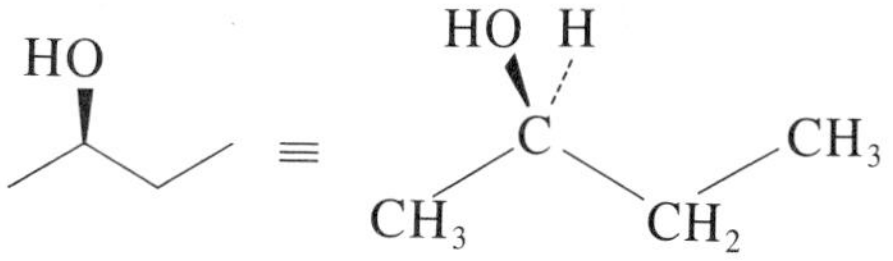

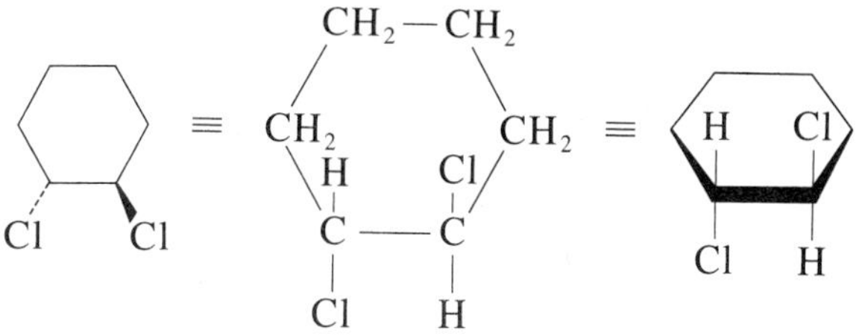

Study Guide
Concept Map 5.2

In this book, line structures will be used for drawing cyclic molecules and some complex molecules. Most of the time, structural formulas showing all of the carbon and hydrogen atoms will be used.

PROBLEM 5.10

Write line structural formulas for the following compounds.

(a) $CH_3CH_2CH_2CH{=}CH_2$ (b) 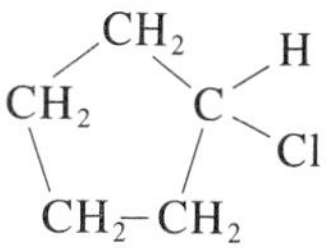

(c) 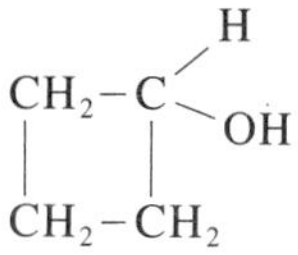 (d) 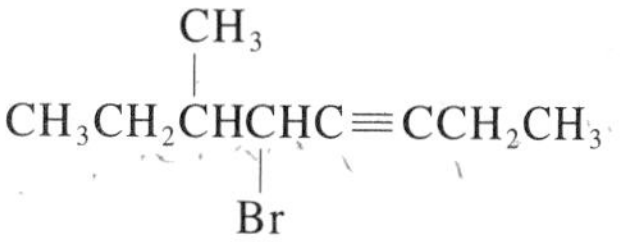

(e) 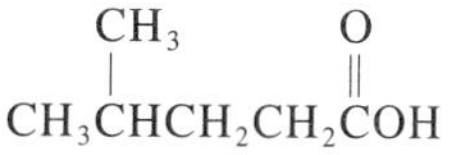

PROBLEM 5.11

Write condensed structural formulas for the following compounds.

(a) 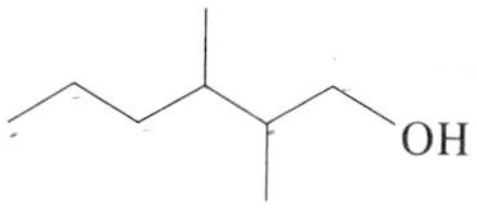 (b) 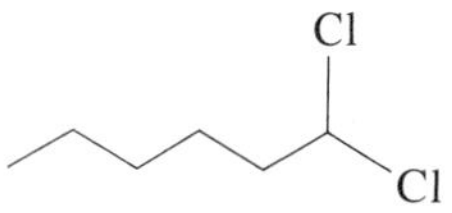

(c) 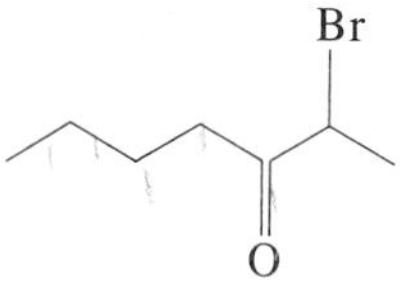 (d) 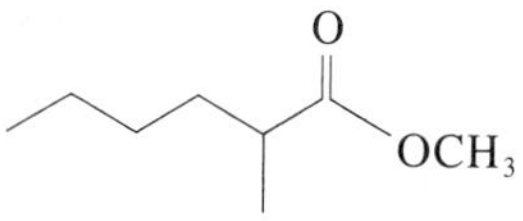

(e) 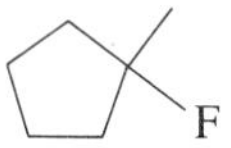

A. Introduction

Historically, compounds were given names that reflected their origin or their properties. For example, the painkiller morphine was named for Morpheus, the Greek god of dreams. Cholesterol, the chief component of gallstones, got its name from the Greek words for bile and solid. This way of naming compounds is descriptive but does not lead to any systematic procedure for assigning names to new and related compounds.

By the end of the nineteenth century, the number of organic compounds that were being synthesized or isolated from natural sources was growing rapidly. Under these conditions, it eventually became impossible for chemists to learn the names randomly assigned to compounds by their discoverers, especially when the names showed no correlation to the structures. Since 1892, chemists from all over the world have met periodically to decide on systematic rules for naming organic compounds. These rules, which are constantly evolving, are called the International Union of Pure and Applied Chemistry (abbreviated as IUPAC) rules. The IUPAC system of nomenclature was developed so that each organic compound would have a unique name that would allow the structural formula to be written from it.

Ideally, a chemist would have to learn only the systematic names of compounds. In the real world, however, there are three complications. First, even the IUPAC rules allow for some variations in the naming of compounds. Second, the correct IUPAC names for some compounds are so complicated and cumbersome that everybody continues to use the common, unsystematic names for them. These names are usually derived from the natural origins of compounds or their molecular shapes, or sometimes even the whimsy of their creators. The structural formulas and systematic names of morphine and cholesterol are shown below, not so you can memorize them but so you will have some idea why the IUPAC names are not generally used to refer to these compounds.

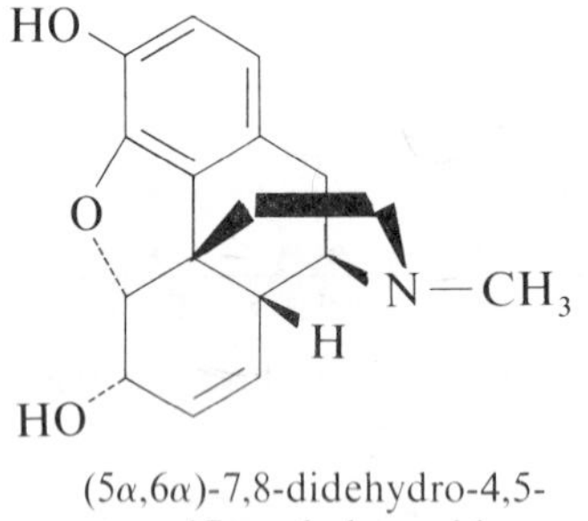
HO
O
N—CH3
H
HO

(5α,6α)-7,8-didehydro-4,5-epoxy-17-methylmorphinan-3,6-diol

morphine

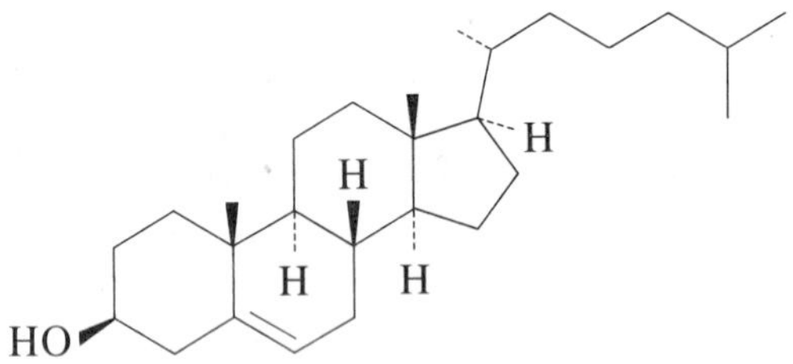
H
H
H
H
HO

(3β)-cholest-5-en-3-ol

cholesterol

Finally, many compounds had names before the IUPAC rules came into being, and these names are still used, especially by chemical supply houses and by industry. To be literate in the laboratory, you must be able to recognize many common names. Therefore, while the emphasis in this book is on the systematic names of compounds, the common names of important compounds are also given and are used in cases where they are the ones overwhelmingly used by chemists in their daily work.

B. Nomenclature of Alkanes

The names of the first four alkanes are methane, ethane, propane, and butane. The systematic nomenclature of the other members of the series is based on a prefix indicating the number of carbon atoms in the chain, followed by the suffix **ane.** The prefixes come from Greek or Latin words for the numbers. The names of the first twelve straight-chain alkanes, along with those of three larger alkanes that form the basis for names of biologically important compounds, are shown in Table 5.2. These names are used in naming all other types of organic compounds derived from alkanes, so you should learn them.

TABLE 5.2 The Names of Some Straight-Chain Alkanes

Molecular Formula	Name	Molecular Formula	Name
CH_4	methane	C_8H_{18}	octane
C_2H_6	ethane	C_9H_{20}	nonane
C_3H_8	propane	$C_{10}H_{22}$	decane
C_4H_{10}	butane	$C_{11}H_{24}$	undecane
C_5H_{12}	pentane	$C_{12}H_{26}$	dodecane
C_6H_{14}	hexane	$C_{16}H_{34}$	hexadecane
C_7H_{16}	heptane	$C_{18}H_{38}$	octadecane
		$C_{20}H_{42}$	icosane

For a compound that is not a straight-chain alkane, the name is derived from the name of the alkane that corresponds to the longest continuous chain of carbon atoms in the molecule. For example, the branched-chain compounds with the molecular formula C_5H_{12} (p. 138) are named 2-methylbutane and 2,2-dimethylpropane. The longest chain in the first one is four carbons long; therefore, the compound is classified as a substituted butane. The longest chain in the second one is three carbons long, so the compound is a substituted propane.

Alkanes with the molecular formula C_6H_{14} exist in five isomeric forms. The straight-chain isomer is called hexane. Two of the branched-chain isomers have five carbon atoms in the longest continuous chain and are named as substituted pentanes.

$CH_3CH_2CH_2CH_2CH_2CH_3$
1 2 3 4 5 6
hexane

straight-chain isomer of C_6H_{14}

$CH_3CH(CH_3)CH_2CH_2CH_3$
1 2 3 4 5
2-methylpentane

$CH_3CH_2CH(CH_3)CH_2CH_3$
1 2 3 4 5
3-methylpentane

two of the branched-chain isomers of C_6H_{14}

The other two isomers of C_6H_{14} have only four carbon atoms in the longest continuous chain and are, therefore, substituted butanes.

$$
\begin{array}{ccc}
& CH_3 & \\
& | & \\
& CH_3CCH_2CH_3 & \\
& 1 \quad |2\ 3 \quad 4 & \\
& CH_3 &
\end{array}
\qquad
\begin{array}{c}
CH_3 \quad CH_3 \\
| \qquad | \\
CH_3CH-CHCH_3 \\
1 \quad 2 \quad\quad 3 \quad 4
\end{array}
$$

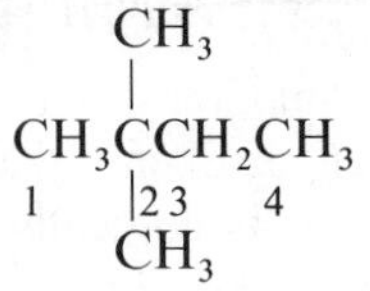

2,2-dimethylbutane

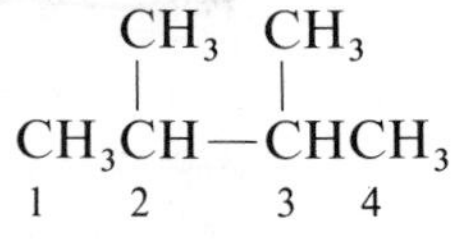

2,3-dimethylbutane

two more of the branched-chain isomers of C_6H_{14}

After the longest continuous chain of carbon atoms in the molecule is identified, the alkyl groups attached to that backbone are named. Each one is then assigned a number indicating its position on the chain. The chain is numbered so that the substituents have the lowest possible numbers. The name of the compound is then created by listing the names of the substituents along with their positions on the chain. For example, the substituents in the branched-chain C_6H_{14} compounds are all methyl groups. The first isomer is called 2-methylpentane to indicate that the methyl group is on the second carbon atom from the end of the five-carbon chain. (The name 4-methylpentane, which is arrived at by numbering the chain in the opposite direction, is incorrect because it violates the rule that substituents must be given the lowest possible numbers.) The second isomer has the methyl group attached to the third carbon atom of the chain and is named 3-methylpentane.

When there is more than one substituent on a chain, the lowest possible number is the number of the first position in the chain at which a choice has to be made, for example:

$$
\begin{array}{c}
CH_3 \qquad CH_3 \\
| \qquad\quad | \quad 3\ \ 2\ \ 1 \\
CH_3CH_2CHCH_2CHCHCH_2CH_3 \\
8 \quad 7 \quad 6 \quad 5 \quad 4 \ | \\
CH_3
\end{array}
$$

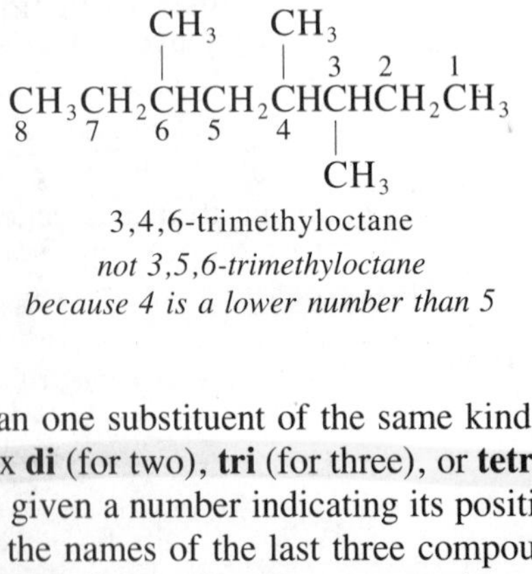

3,4,6-trimethyloctane

not 3,5,6-trimethyloctane because 4 is a lower number than 5

If there is more than one substituent of the same kind, the name of the alkyl group is given the prefix **di** (for two), **tri** (for three), or **tetra** (for four). As before, each substituent is also given a number indicating its position on the chain. These rules are illustrated by the names of the last three compounds shown above. The name 2,2-dimethylbutane indicates that the compound has a chain of four carbon atoms with two methyl groups attached to the second carbon of that chain. Note that the number 2 is repeated for each methyl group so that there is no doubt about the placement of the groups. The name of the isomeric compound 2,3-dimethylbutane differs from 2,2-dimethylbutane in only one number. The name 3,4,6-trimethyloctane also has a number for each methyl group on the chain. Note that the numbers are separated by commas and joined to the name by a hyphen.

An application of these rules to somewhat more complicated examples may be helpful. A $C_{12}H_{26}$ hydrocarbon is shown on the next page.

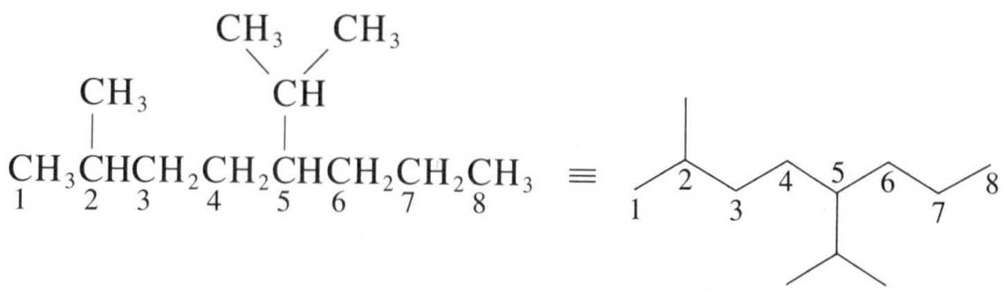

5-isopropyl-2-methyloctane

The longest continuous chain contains eight carbon atoms. The compound is, therefore, named as a substituted octane. The alkyl substituents on the chain are identified as the methyl group (on carbon atom 2) and the isopropyl group (on carbon atom 5). The compound is given the name 5-isopropyl-2-methyloctane. When there are several different kinds of substituents on the chain, they are listed in alphabetical order. Prefixes such as di or tri are not considered when substituents are alphabetized. Neither are italicized prefixes that are followed by hyphens; for example, *tert*-butyl is treated as a butyl group.

Another example further illustrates the rules.

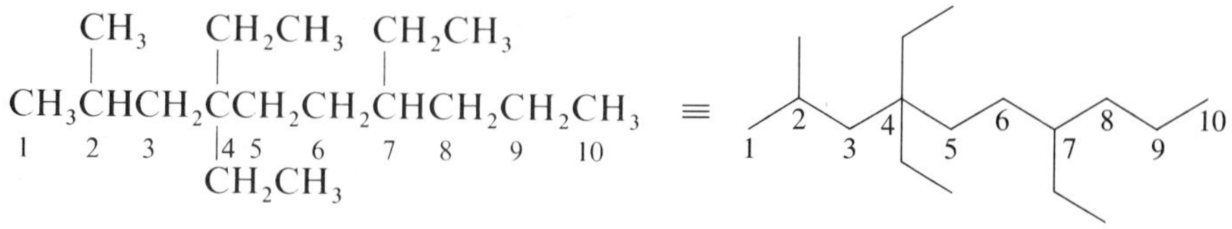

4,4,7-triethyl-2-methyldecane

In this compound, the longest continuous chain contains ten carbon atoms, and, therefore, this is a substituted decane. The backbone of the molecule has three ethyl groups (two at carbon atom 4 and one at carbon atom 7) and a methyl group (at carbon atom 2) attached. The name is 4,4,7-triethyl-2-methyldecane; the ethyl groups are listed before the methyl group because the prefix tri is ignored in alphabetizing. Note again the use of commas to separate a series of numbers and the use of hyphens to join numbers to the substituents. The name of the last substituent listed is merged with that assigned to the backbone.

This list summarizes the rules for the nomenclature of alkanes:

1. Locate the longest continuous straight chain of carbon atoms in the molecule. The name of the straight-chain alkane with the same number of carbon atoms becomes the last part of the name of the compound.
2. Find and name all of the alkyl groups that are branches off the backbone of the molecule. Assign each one a position on the chain, numbering the chain so that the substituents have the lowest possible numbers.
3. If there are several substituents of the same kind, indicate how many by using the prefix di, tri, or tetra, and use a number to assign a position on the chain to each one.
4. Construct the name of the compound by listing all the substituents in alphabetical order, ignoring the prefixes di, tri, and tetra and italicized prefixes such as *n*-, *sec*-, and *tert*-.
5. Use commas to separate numbers that are grouped together. Separate numbers from names of groups by hyphens. Merge the name of the last substituent with the name of the straight chain alkane that is the basis of the name of the compound.

One last example will provide a review of these points.

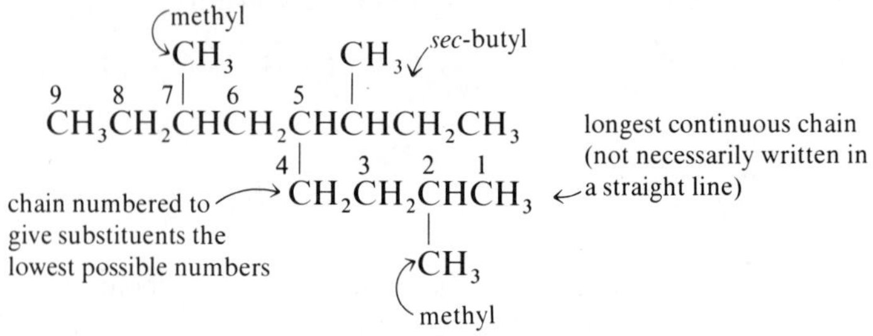
methyl
sec-butyl
longest continuous chain (not necessarily written in a straight line)
chain numbered to give substituents the lowest possible numbers
methyl

5-*sec*-butyl-2,7-dimethylnonane

treated alphabetically as a butyl group

note placement of comma and hyphens in the name

PROBLEM 5.12

Name the following compounds.

(a) 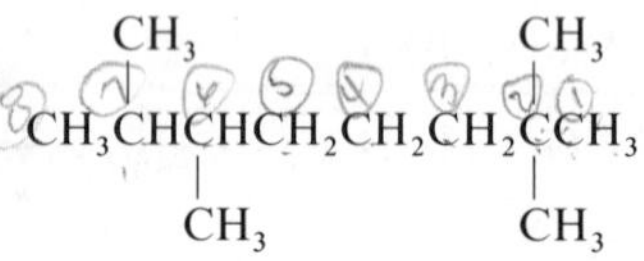

(b) 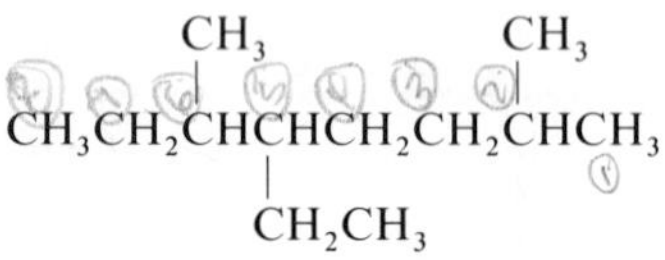

(c) 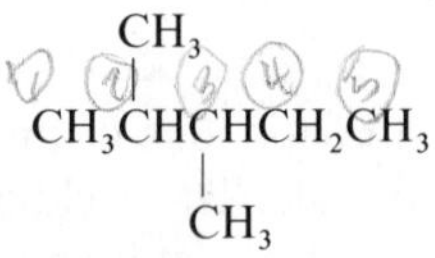

(d) 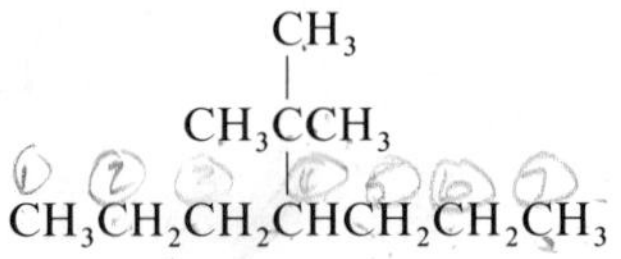

(e) 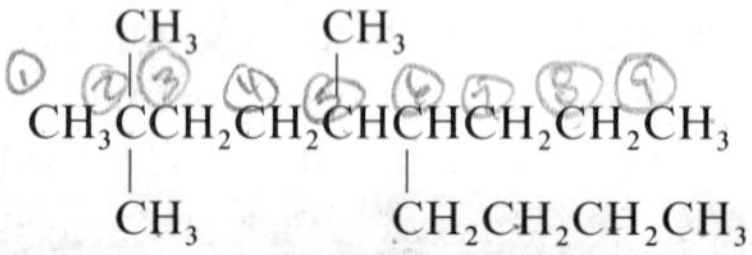

(f) 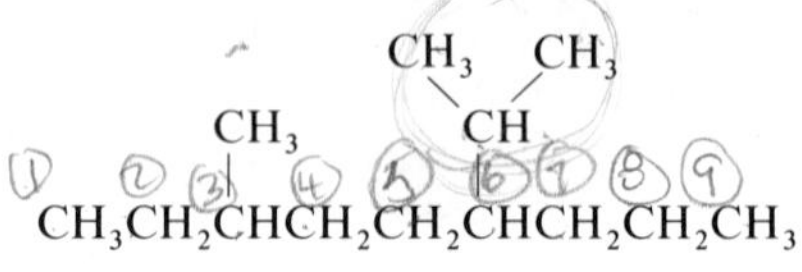

PROBLEM 5.13

Draw structural formulas for the following compounds.

(a) 2,2-dimethylheptane
(b) 6-isobutyl-2-methyldecane
(c) 5,5-diisopropyl-2,8-dimethylnonane
(d) 4-*tert*-butyl-3-ethyloctane
(e) 6-ethyl-2,2,4-trimethyldodecane

PROBLEM 5.14

Draw structural formulas for the nine alkanes having the molecular formula C_7H_{16}. Identify primary, secondary, and tertiary carbon atoms in each one. Show which hydrogen atoms are equivalent. Name each compound.

C. Nomenclature of Alkyl Halides and Alcohols

The systematic names of alkyl halides are assigned in the same way as the names of branched-chain alkanes. The prefixes **fluoro, chloro, bromo,** and **iodo** are used to indicate the presence of halogens, for example.

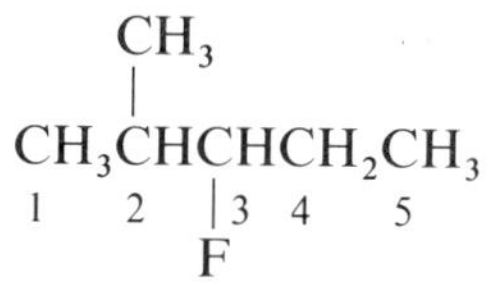

$$\begin{array}{c} CH_3 \\ | \\ CH_3CHCHCH_2CH_3 \\ {\scriptstyle 1\ \ \ 2\ \ |3\ \ 4\ \ 5} \\ F \end{array}$$

3-fluoro-2-methylpentane

$$\begin{array}{c} Cl \\ | \\ CH_3CH_2CCH_3 \\ {\scriptstyle 4\ \ \ 3\ \ 2|\ 1} \\ Cl \end{array}$$

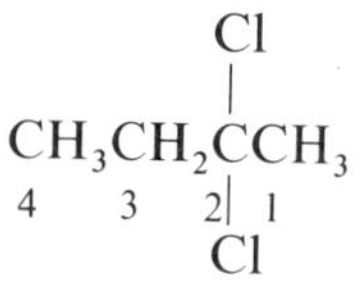

2,2-dichlorobutane

$$\begin{array}{c} CH_3\ \ CH_3 \\ |\ \ \ \ \ | \\ CH_3CH_2C\!\!-\!\!CCH_2CH_3 \\ {\scriptstyle 1\ \ \ 2\ \ |3\ \ 4|\ 5\ \ 6} \\ Br\ \ Br \end{array}$$

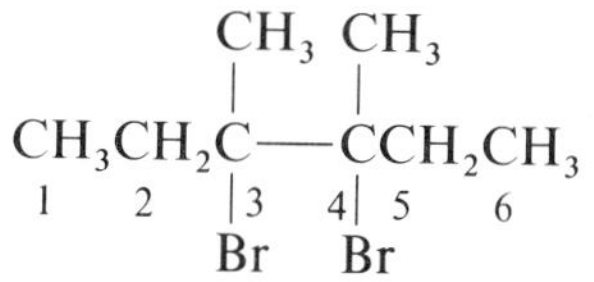

3,4-dibromo-3,4-dimethylhexane

$$\begin{array}{c} CH_2CH_3 \\ | \\ CH_3CH_2CCH_2CH_2CH_2CH_3 \\ {\scriptstyle 1\ \ \ 2\ \ 3|\ 4\ \ 5\ \ \ 6\ \ \ 7} \\ I \end{array}$$

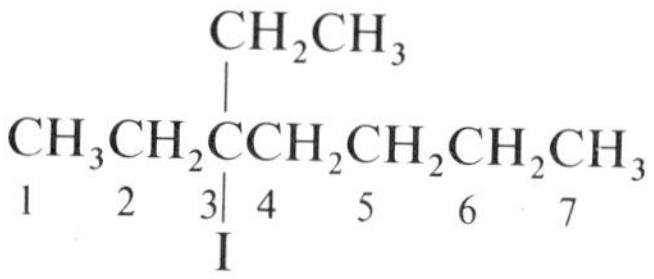

3-ethyl-3-iodoheptane

Some organic compounds are named by changing the suffix of the name of the hydrocarbon chain, instead of adding a prefix to it as is done for an alkyl or halogen substituent. Alcohols are usually named by changing the **e** ending of the name of the alkane to **ol** and using a number to indicate the position of the hydroxyl group, which is not specifically named. An alcohol owes its characteristic properties and reactivity to the hydroxyl group, so in naming an alcohol, the carbon chain is numbered so that the carbon atom bearing the hydroxyl group has the lowest possible number.

$$\begin{array}{c} CH_3 \\ | \\ CH_3CHCH_2CH_2CH_2OH \end{array}$$

4-methyl-1-pentanol

$$\begin{array}{c} CH_3 \\ | \\ CH_3CCH_2OH \\ | \\ CH_3 \end{array}$$

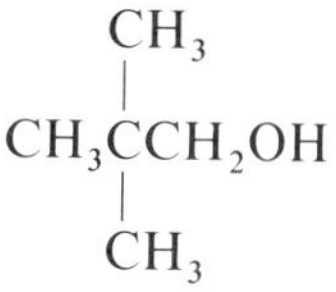

2,2-dimethyl-1-propanol

$$HOCH_2CH_2CH_2OH$$

1,3-propanediol

$$ClCH_2CH_2CH_2CH_2OH$$

4-chloro-1-butanol

PROBLEM 5.15

Name the following compounds.

(a) 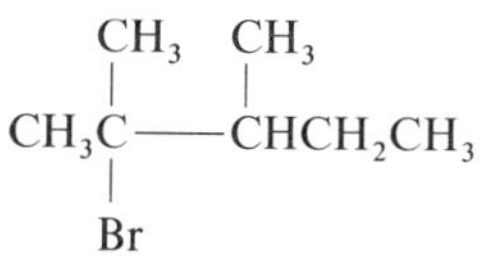
$$\begin{array}{c} CH_3\ \ CH_3 \\ |\ \ \ \ \ | \\ CH_3C\!\!-\!\!CHCH_2CH_3 \\ | \\ Br \end{array}$$

(b) 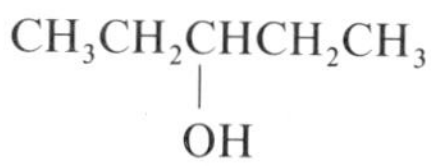
$$\begin{array}{c} CH_3CH_2CHCH_2CH_3 \\ | \\ OH \end{array}$$

(c) 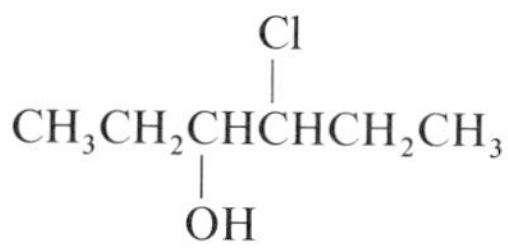
$$\begin{array}{c} Cl \\ | \\ CH_3CH_2CHCHCH_2CH_3 \\ | \\ OH \end{array}$$

(d) 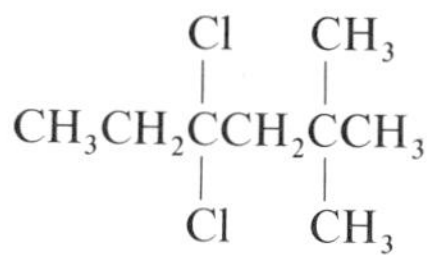
$$\begin{array}{c} Cl\ \ \ CH_3 \\ |\ \ \ \ \ | \\ CH_3CH_2CCH_2CCH_3 \\ |\ \ \ \ \ | \\ Cl\ \ \ CH_3 \end{array}$$

(e) 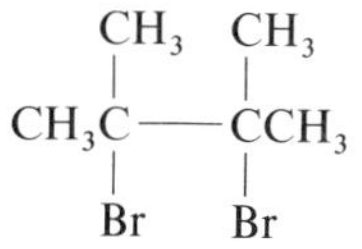
$$\begin{array}{c} CH_3\ \ CH_3 \\ |\ \ \ \ \ | \\ CH_3C\!\!-\!\!CCH_3 \\ |\ \ \ \ \ | \\ Br\ \ Br \end{array}$$

(f) 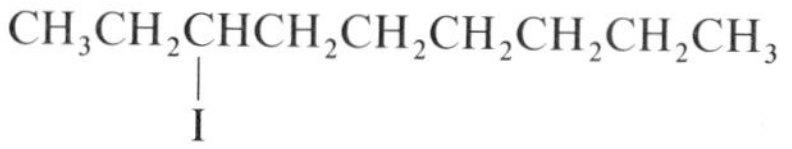
$$\begin{array}{c} CH_3CH_2CHCH_2CH_2CH_2CH_2CH_3 \\ | \\ I \end{array}$$

(g) 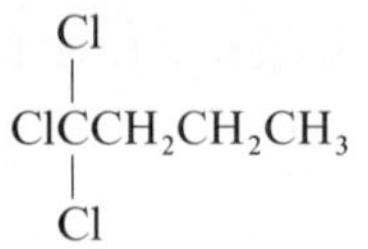 (h) 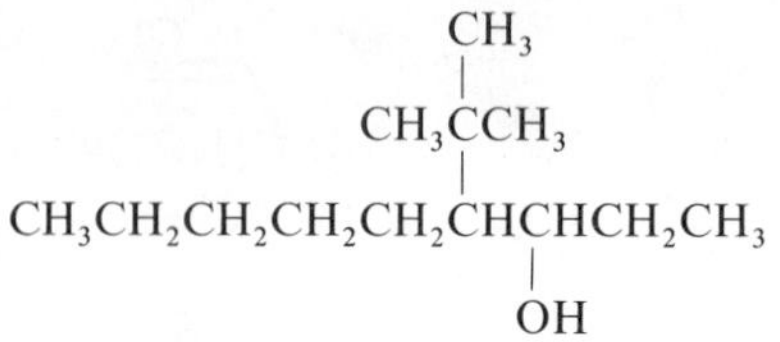

PROBLEM 5.16

Write structural formulas for the following compounds.

(a) 2-iodooctane (b) 3-hexanol (c) 2,2,2-trifluoroethanol
(d) 1-chloro-4,4-dimethyl-2-pentanol (e) 1,2,3-propanetriol
(f) 1-pentanol (g) 1,2-ethanediol (h) 2,3-dichloro-3-ethylheptane
(i) 4-bromo-3,3-dimethylheptane (j) 2-bromo-3-ethylhexane

PROBLEM 5.17

There are seventeen constitutional isomers with the molecular formula $C_6H_{13}Cl$. Write structural formulas for them. Name each one, and tell whether it is a primary, secondary, or tertiary halide. Reviewing Section 5.5 (p. 145) may help you remember the structural features of alkyl halides.

PROBLEM 5.18

Name all the isomeric compounds having the molecular formula $C_4H_8Cl_2$ (Problem 5.4, p. 147).

D. The Phenyl Group

There is one other group that you should learn at this stage. The **phenyl group** is formed by the removal of one of the six equivalent hydrogen atoms on benzene. Groups such as the phenyl group, derived from aromatic hydrocarbons (p. 63), are known as **aryl groups.** An aromatic hydrocarbon is sometimes represented by the symbol ArH, and an aryl group by Ar.

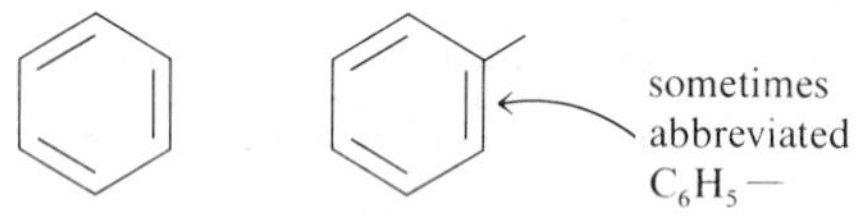
sometimes
abbreviated
C6H5—

benzene phenyl group

The phenyl group is used in much the same way as alkyl groups are used in naming compounds. When the hydrocarbon chain attached to the benzene ring is complex or contains more than five carbon atoms, phenyl is used as a part of the name of the compound. However, benzene rings on which small alkyl groups are substituted are usually named as benzene derivatives.

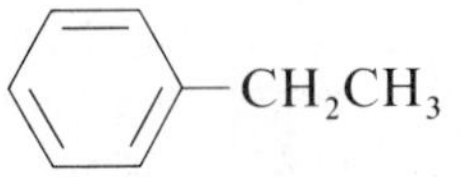
CH2CH3

ethylbenzene

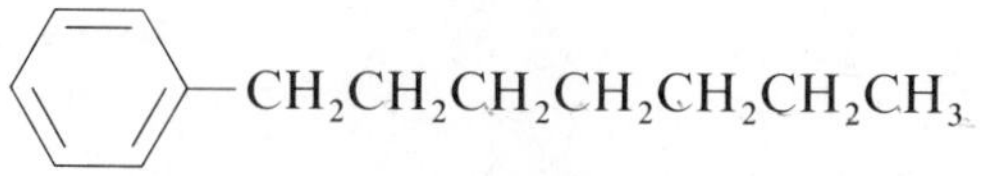
CH2CH2CH2CH2CH2CH2CH3

1-phenylheptane

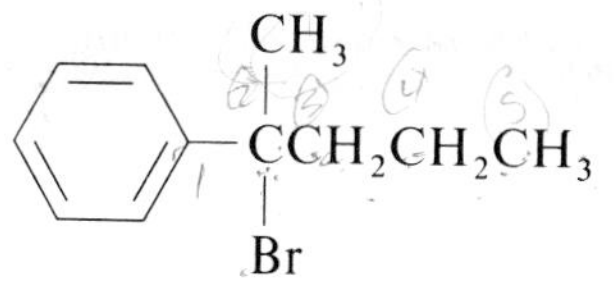
CH3
CCH2CH2CH3
Br

2-bromo-2-phenylpentane

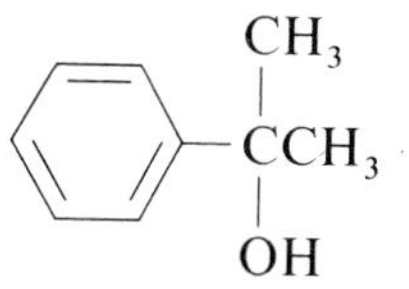
CH3
CCH3
OH

2-phenyl-2-propanol

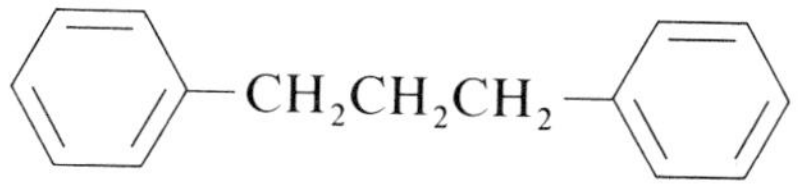
CH2CH2CH2

1,3-diphenylpropane

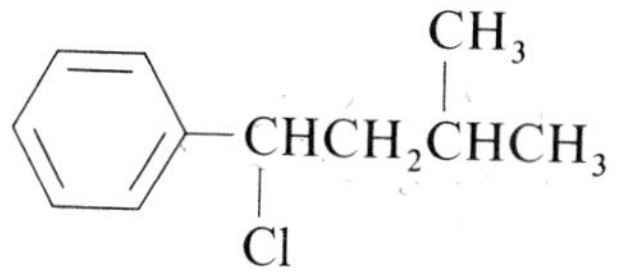
CH3
CHCH2CHCH3
Cl

1-chloro-3-methyl-1-phenylbutane

PROBLEM 5.19

Name the compounds in Problem 5.5 (p. 148). (Hint: Positions on a benzene ring may be numbered too.)

PROBLEM 5.20

Name the following compounds.

(a) 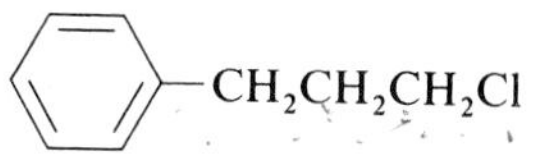
CH2CH2CH2Cl

(b) 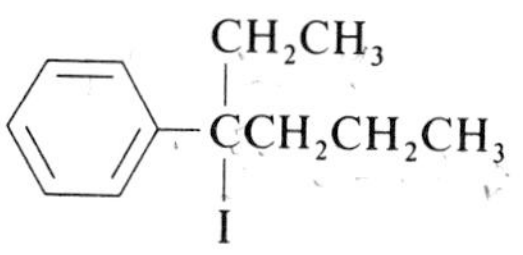
CH2CH3
CCH2CH2CH3
I

(c) 
CHCHCH3
Cl Cl

(d) 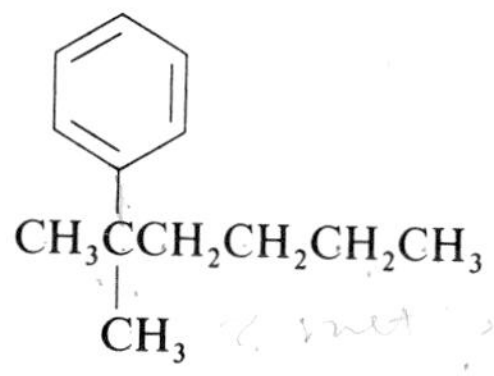
CH3CCH2CH2CH2CH3
CH3

(e) 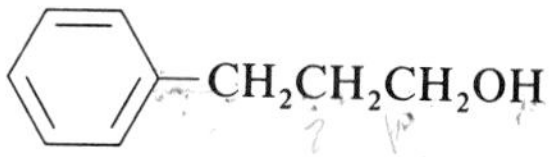
CH2CH2CH2OH

(f) 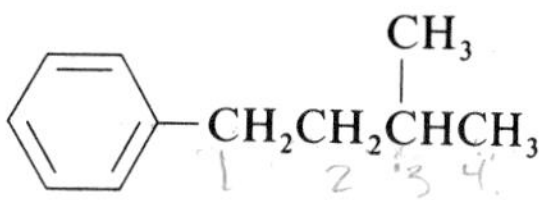
CH3
CH2CH2CHCH3

PROBLEM 5.21

Draw structural formulas for the following compounds.

(a) 1-phenyl-1-pentanol (b) 3-bromo-3-methyl-1-phenylbutane
(c) 4-methyl-2,6-diphenylheptane (d) 4-*tert*-butyl-2-phenyloctane
(e) 2-phenyl-1-ethanol

5.9 CYCLOALKANES

A. Cyclopropane and Cyclobutane. Ring Strain

Cycloalkanes are hydrocarbons that have the general formula C_nH_{2n} and in which some of the carbon atoms form a ring. A cycloalkane is named by adding the prefix cyclo to the name of the alkane having the same number of carbon atoms as are in the ring. Alkanes are divided into two classes: the cycloalkanes, which contain a ring, and the **open-chain alkanes,** which are sometimes called **acyclic compounds** to distinguish them from the cyclic ones.

The three carbon atoms in cyclopropane, C_3H_6, the smallest cycloalkane, define a plane. Cyclopropane has internal carbon-carbon bond angles of 60° and external carbon-carbon-hydrogen and hydrogen-carbon-hydrogen bond angles of 116° and 118°, respectively. All of the bonds are eclipsed in cyclopropane.

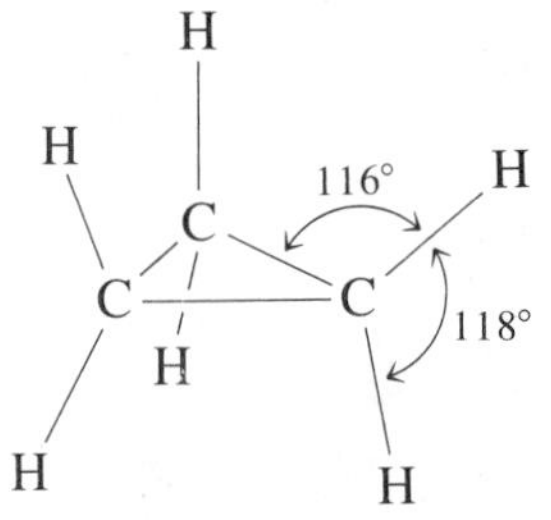
116°
118°

CH2
CH2—CH2
Cl

cyclopropane | cyclopropyl group | chlorocyclopropane cyclopropyl chloride

Cyclobutane, C_4H_8, is not planar. One of the atoms in the ring is bent out of the plane of the other three by about 20°. This causes the expected internal bond angles of 90° to be reduced to 88° and also minimizes the eclipsing of the hydrogen atoms on adjacent carbon atoms.

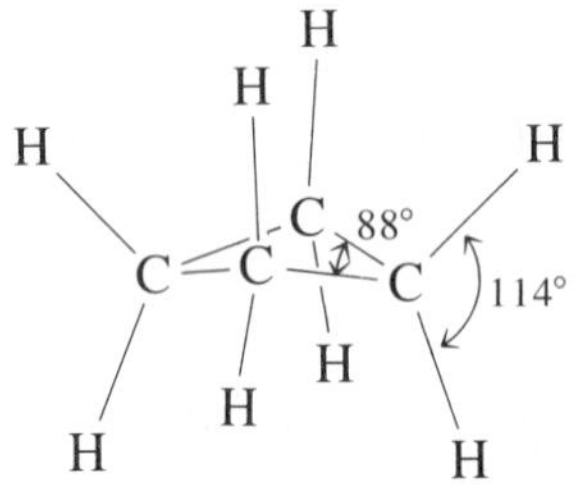
88°
114°

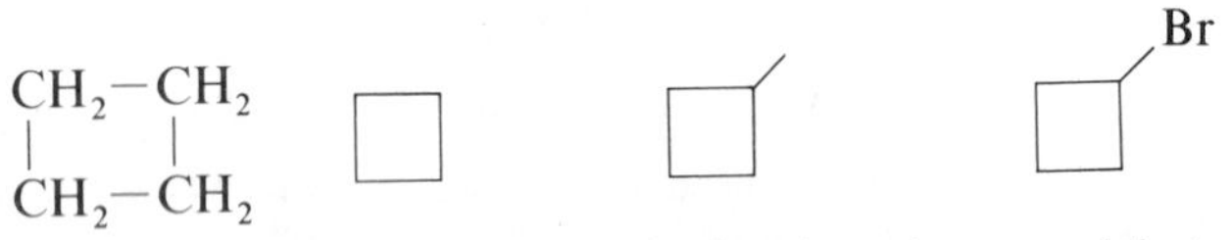
CH2—CH2
CH2—CH2
Br

cyclobutane | cyclobutyl group | bromocyclobutane cyclobutyl bromide

Cycloalkanes are usually symbolized in equations by regular polygons the corners of which correspond to the number of carbon atoms in the ring. Thus, cyclopropane is represented by a triangle, and cyclobutane by a square. It is understood that each corner of a polygon represents a carbon atom and two hydrogen atoms unless another substituent is shown bonded to that position. Cycloalkyl groups are derived from cycloalkanes, just as alkyl groups are derived from alkanes. Thus, cyclopropane gives a cyclopropyl group, and cyclobutane, a cyclobutyl group.

The bond angles in cyclopropane and cyclobutane are quite different from the normal tetrahedral bond angle of 109.5°. Cyclopropane and cyclobutane are unstable compared to the larger cycloalkanes such as cyclopentane and cyclohexane, which have tetrahedral bond angles. The two small cycloalkanes are said to have **ring strain,** to be destabilized by the deformation of their bond angles and by the nonbonding repulsions between the electrons in covalent bonds on adjacent atoms. This results in weaker carbon-carbon bonds. For example, the normal bond dissociation energy (p. 60) for a carbon-carbon single bond is about 88 kcal/mol. That for a carbon-carbon bond in cyclopropane is 65 kcal/mol. The strained state of the small ring compounds is demonstrated by their reactivity which is greater than is expected of alkanes. For example, compounds with double bonds, such as ethylene, react easily with hydrogen gas in the presence of metal catalysts (p. 304) to form alkanes.

$$\underset{\text{ethylene}}{CH_2{=}CH_2} \xrightarrow[\text{Ni}\ 20\ ^\circ\text{C}]{H_2} \underset{\text{ethane}}{CH_3CH_3}$$

Cyclopropane, even though it is not an alkene, also reacts with hydrogen. In the presence of a nickel catalyst, the ring opens and propane is formed.

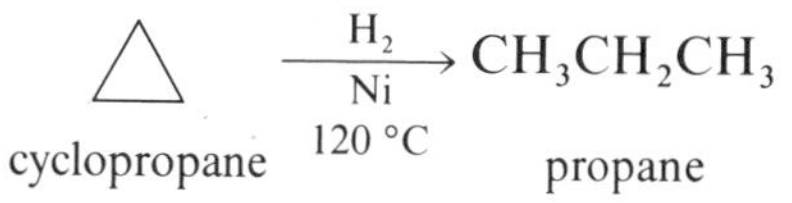
H2
Ni
120 °C
CH3CH2CH3
cyclopropane
propane

A higher temperature is required for this reaction than for the one with an alkene, and a still higher temperature is necessary to cleave a cyclobutane ring.

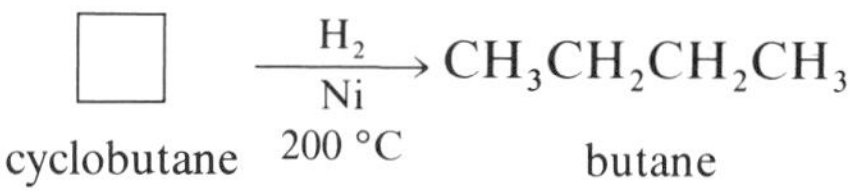
H2
Ni
200 °C
CH3CH2CH2CH3
cyclobutane
butane

The larger rings in cyclopentane and cyclohexane do not open when treated with hydrogen in the presence of a catalyst under these conditions

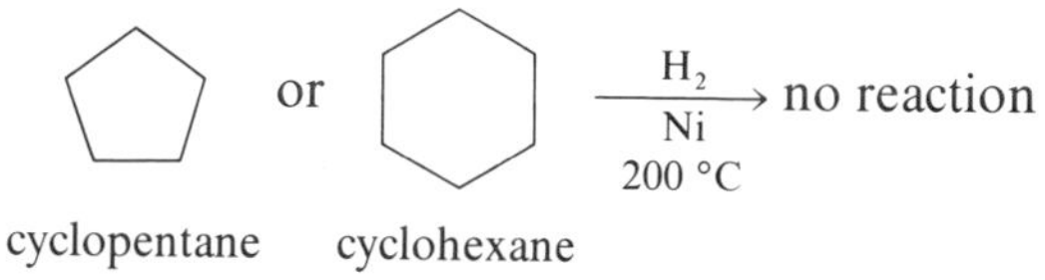
or
H2
Ni
200 °C
no reaction
cyclopentane
cyclohexane

The reactivity of compounds containing a three-membered ring raises questions about the exact nature of the bonding in such compounds. The H—C—H bond angles in cyclopropane, for example, appear to fit better with sp^2-hybridized (p. 48) than with sp^3-hybridized (p. 42) carbon atoms. The reactivity also suggests some unsaturated character to the ring. The small C—C—C bond angles in the molecule make overlap between orbitals on adjacent carbon atoms difficult, and certainly weaken the carbon-carbon bonds.

B. Cyclopentane

The bond angles in cyclopentane are very close to the tetrahedral angle and also close to the internal bond angle of 108° for a regular pentagon. Cyclopentane might be expected to be stable in a planar form except that the hydrogen atoms on adjacent carbon atoms are eclipsed in this form. Cyclopentane is most stable when one of the carbon atoms is out of the plane of the other four. This shape is called the **envelope**

form of cyclopentane. The carbon atom that is out of the plane moves around the ring in a phenomenon known as **pseudorotation,** so that the eclipsing of the hydrogen atoms at various points in the ring is relieved at least part of the time. Note, however, that the structure of the ring does not allow complete rotation around the carbon-carbon single bonds. The type of conformational isomerism seen for ethane or butane is not possible for a cyclic alkane.

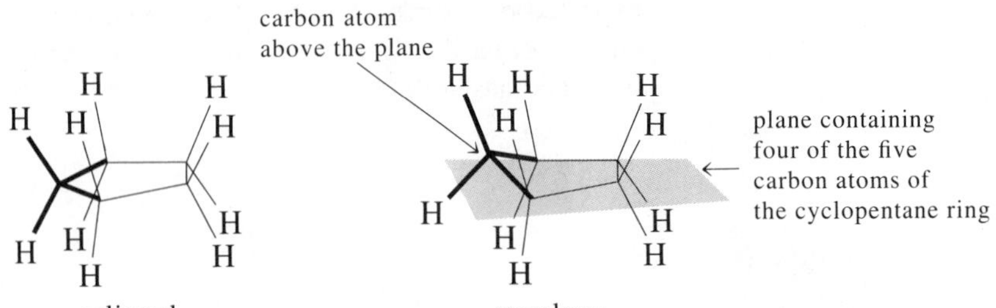
carbon atom
above the plane
plane containing
four of the five
carbon atoms of
the cyclopentane ring

eclipsed envelope

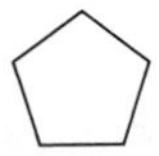

cyclopentane
planar representation

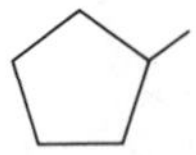

cyclopentyl
group

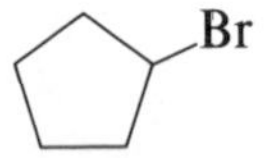
Br

bromocyclopentane
cyclopentyl bromide

Substitution of another group for a hydrogen atom on cyclopentane gives rise to cyclopentyl compounds.

PROBLEM 5.22

Name the following compounds.

(a) 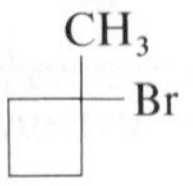
CH3
Br

(b) 
CH2CH3

(c) 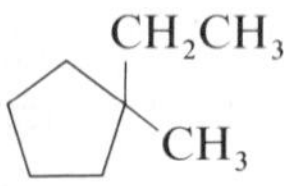
CH2CH3
CH3

(d) 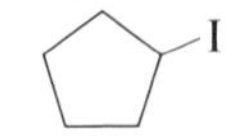
I

(e) 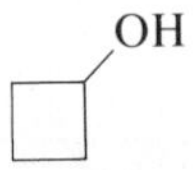
OH

(f) 
Br

PROBLEM 5.23

Draw structures for the following compounds.

(a) *tert*-butylcyclopentane (b) cyclopropanol (c) 1-ethyl-1-cyclopentanol
(d) isopropylcyclobutane (e) chlorocyclopentane (f) 2-cyclopentyloctane
(g) phenylcyclopentane (or cyclopentylbenzene)

C. Cyclohexane. Conformation in Cyclohexane and in Cyclohexanes with One Substituent

Cyclohexane, C_6H_{12}, was known to be an unstrained molecule in the late nineteenth century. This fact was puzzling because a planar regular hexagon with internal bond angles of 120°, which is larger than the tetrahedral angle, would be expected to exhibit some ring strain. In 1890, the German chemist Ulrich Sachse pointed out that cyclohexane need not be a planar molecule, that puckering of the ring (having one or more of the carbon atoms out of the plane) would allow for tetrahedral bond angles. He suggested that cyclohexane should exist in two forms, now called the chair form and the boat form. At that time, these forms could not be observed experimentally, so Sachse's ideas were not accepted until about 35 years later when experimental evidence was found for puckered six-membered rings in some more complex compounds. Then, in 1943, the Norwegian chemist Odd Hassel used electron diffraction to show that the chair form of cyclohexane was the predominant conformer in the gas phase. For this work Hassel shared the Nobel Prize in Chemistry in 1969 with Sir Derek Barton (Chapter 17).

In the **chair form** of cyclohexane, four of the carbon atoms in the ring are in a plane, a fifth carbon atom is above the plane, and a sixth is below it. The plane is shown in green in the structures below.

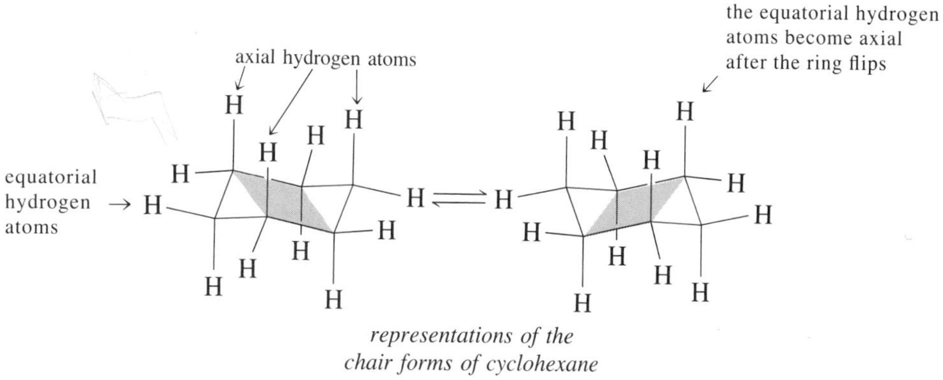
the equatorial hydrogen atoms become axial after the ring flips
axial hydrogen atoms
equatorial hydrogen atoms
representations of the chair forms of cyclohexane

Partial rotation around the carbon-carbon bonds in cyclohexane results in the conversion of one chair form to another. The energy barrier between the two forms is 10.8 kcal/mol, low enough that at room temperature the ring is constantly flipping (approximately 10^5 times a second) from one conformation to the other.

There are two types of hydrogen atoms in the chair form of cyclohexane. Six of the hydrogen atoms lie in a ring outside of the carbon skeleton and roughly in the plane of the molecule. These hydrogen atoms are called the **equatorial hydrogen atoms.** The bonds to the other six hydrogen atoms are parallel to an axis that goes through the center of the cyclohexane ring. Of these **axial hydrogen atoms,** three are above the plane of the carbon ring and three are below. The rapid flipping of the ring at room temperature converts the hydrogen atoms of one type into the other. You should use a molecular model to demonstrate to yourself how this works. The easiest way to convert a model of cyclohexane from one chair form to another is to move the carbon atom that is below the plane of the other four up, and the one that is above the plane down, as demonstrated in Figure 5.5.

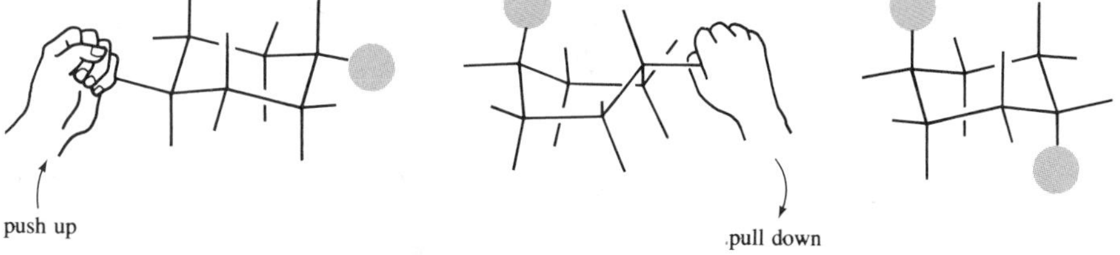
push up
pull down

FIGURE 5.5 The interconversion of axial and equatorial hydrogen atoms by flipping the ring in a model of cyclohexane.

All of the carbon-hydrogen bonds on adjacent carbon atoms in any chair form of cyclohexane are staggered, as shown by the Newman projection.

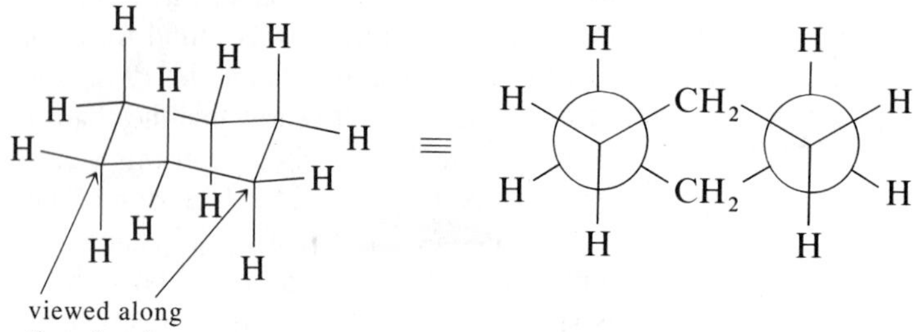
H
H
H
H
H
H
H
H
H
H
H
H
viewed along
these bonds
≡
H
H
H
H
CH_2
CH_2
H
H
H
H

Newman projection

Besides the chair forms, there are other conformations of cyclohexane that have higher energies (are less stable), such as the boat form and the twist form.

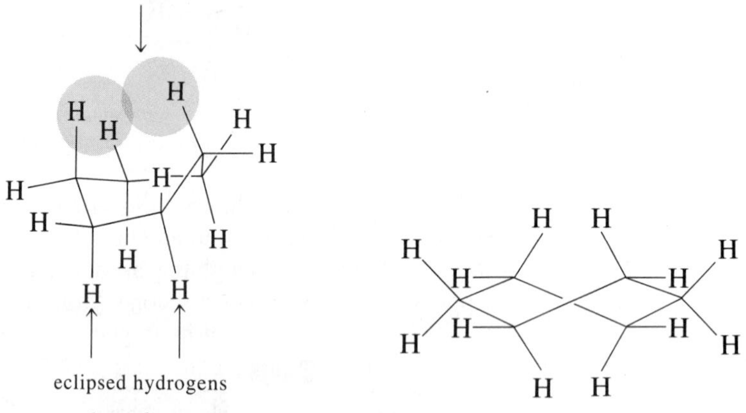
repulsion between
"flagpole" hydrogen
atoms
eclipsed hydrogens
boat form
twist form

high-energy conformations of cyclohexane

In the **boat form,** two of the carbon atoms lie above the plane defined by the other four carbon atoms. Two of the hydrogen atoms on these two carbon atoms (called the "flagpole" positions) are brought close enough that they repel each other. In addition, the carbon-hydrogen bonds on the carbon atoms in the plane are all eclipsed. The **twist form** is a little more stable than the boat form because some of these interactions are lessened. The eclipsed bonds move so they are out of alignment with one another, but the staggered bonds move closer. At room temperature, most cyclohexane molecules exist in the most stable conformation, the chair form. The twist form occurs as an intermediate stage in the conversion of one chair form to another.

If a substituent has replaced one of the hydrogen atoms of cyclohexane, the chair conformations are no longer equivalent. For example, molecules of methylcyclohexane are mostly in the form in which the methyl group is in an equatorial position (Figure 5.6). This form is more stable by approximately 1.7 kcal/mol than the chair form with the axial methyl group, which is created by flipping the ring. When the methyl group is in an axial position, it is close to the two axial hydrogen

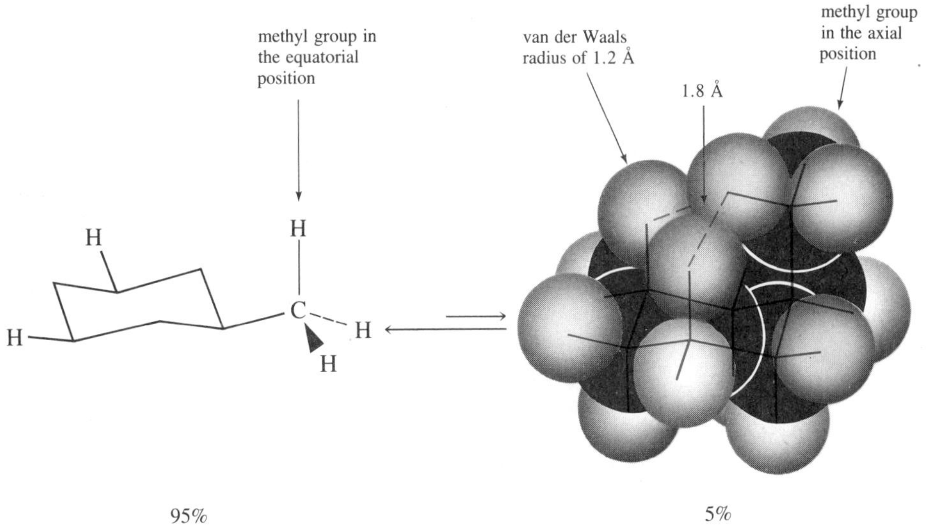
methyl group in
the equatorial
position
van der Waals
radius of 1.2 Å
methyl group
in the axial
position
1.8 Å
H
H
H
C
H
H
H
95%
5%

FIGURE 5.6 Chair conformations of methylcyclohexane.

atoms that are on the same side of the plane of the ring. The distance between a hydrogen atom of the methyl group and either one of these two axial hydrogen atoms is 1.8 Å, which is shorter than 2.4 Å, the sum of the van der Waals radii of two hydrogen atoms (p. 151). The steric interaction, the repulsion between the methyl group and each of these two hydrogen atoms, is called a **1,3-diaxial interaction.** The 1,3-diaxial interaction serves to make the conformer less stable than one in which the methyl group is equatorial.

A methyl group in the axial position on a cyclohexane ring is gauche (p. 151) to the largest substituent on each of the carbon atoms adjacent to the one to which it is bonded. These substituents happen to be other carbon atoms of the ring.

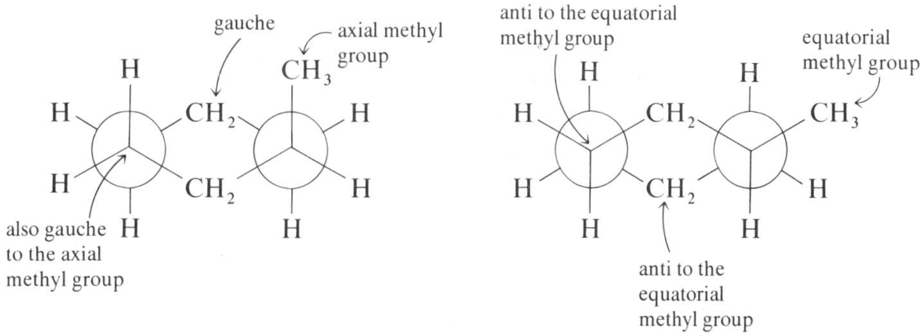
gauche
axial methyl
group
CH_3
H
H
CH_2
H
H
CH_2
H
H
also gauche
to the axial
methyl group
H
anti to the equatorial
methyl group
equatorial
methyl group
H
H
H
CH_2
CH_3
H
CH_2
H
H
H
anti to the
equatorial
methyl group

If a methyl group is equatorial on the ring, it is anti to the ring atoms that are the largest substituents on adjacent carbon atoms. This analysis also predicts a lower stability for the conformer of cyclohexane with a substituent in the axial position.

In general, the larger the substituent, the greater the repulsive interactions when it occupies an axial position. A *tert*-butyl group, for example, is so bulky that the conformation in which it is in the equatorial position is overwhelmingly favored (Figure 5.7). The difference in energy between the two chair forms of *tert*-butylcyclohexane is 5.6 kcal/mol.

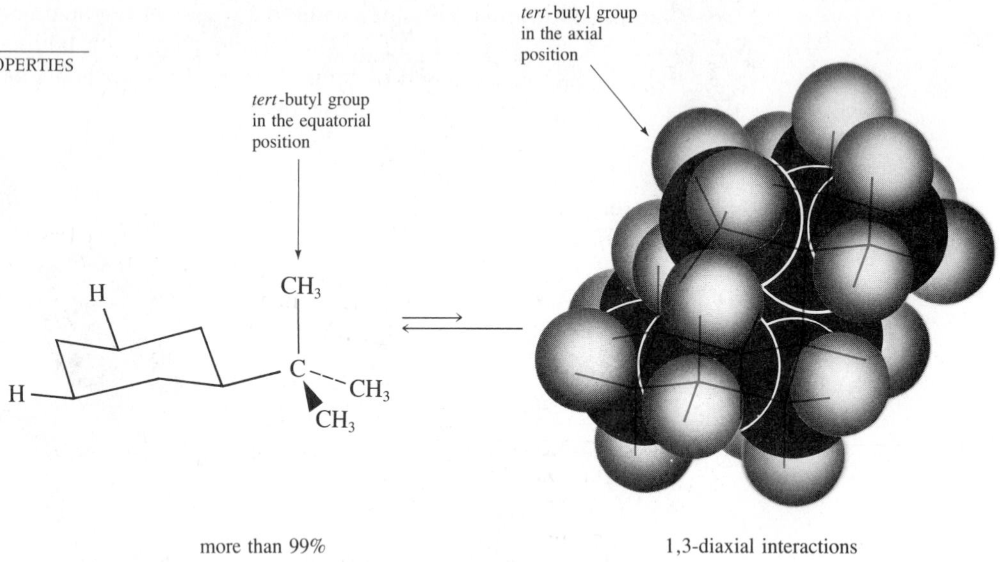
tert-butyl group in the equatorial position
tert-butyl group in the axial position
H
H
CH3
C
CH3
CH3
more than 99%
1,3-diaxial interactions

FIGURE 5.7 Chair conformations of *tert*-butylcyclohexane.

PROBLEM 5.24

Study Guide Concept Map 5.3

Draw structural formulas for chlorocyclohexane and cyclohexanol in the planar form (using a regular hexagon to represent the ring, p. 153) and in the two chair conformations.

5.10 THE CHEMICAL PROPERTIES OF ALKANES

Alkanes have structures in which all of the valence electrons of carbon are involved in σ bonds with carbon and hydrogen atoms. Saturated hydrocarbons are inert toward most reagents. Their reactions generally take place at high temperatures or in the presence of catalysts that promote the cleavage of single bonds.

Alkanes are important constituents of fuels, and their combustion in air provides much of the energy consumed in the modern world. Methane is the chief constituent of natural gas. Once ignited, it reacts with oxygen to evolve heat.

$$\underset{\text{methane}}{CH_4} + 2\,O_2 \longrightarrow CO_2 + 2\,H_2O \qquad \Delta H = -210.8 \text{ kcal/mol}$$

The equation represents the complete combustion of methane and assumes that sufficient oxygen is available to convert the methane to carbon dioxide and water. Often, however, incomplete combustion takes place, forming some carbon monoxide, which is highly toxic, and carbon, which gives the flame of a Bunsen burner its yellow color when the air supply is limited.

Petroleum is a mixture of a large variety of hydrocarbons, including alkanes and aromatic hydrocarbons (p. 63). The hydrocarbons are separated into fractions according to their boiling points. Propane and butane, for example, are gases that are liquefied under pressure and used as fuel that can be transported in small tanks such as in cigarette lighters. The hydrocarbons with boiling points up to 25 °C (Table 5.1) make up the fraction of petroleum known as gas and liquefied gas.

The next and largest fraction obtained from petroleum is gasoline, consisting of hydrocarbons with four to twelve carbon atoms and a boiling-point range of 20–200 °C. Hydrocarbons at the upper end of this range are converted into more useful mixtures of lower-molecular-weight alkanes by heating in the presence of hydrogen gas and catalysts in a process known as **hydrocracking.** For example, undecane, $C_{11}H_{24}$, has a boiling point of 196 °C. It is cracked to give a mixture of lower-weight alkanes in which butanes, pentanes, hexanes, and heptanes are the major constituents.

$$\underset{\substack{\text{undecane}\\ \text{bp 196 °C}}}{C_{11}H_{24}} \xrightarrow[\substack{\text{catalyst}\\ 275\ °\text{C}}]{H_2} \underset{\substack{\text{butanes}\\ \text{bp} \sim 0\ °\text{C}}}{C_4H_{10}} + \underset{\substack{\text{pentanes}\\ \text{bp} \sim 30\ °\text{C}}}{C_5H_{12}} + \underset{\substack{\text{hexanes}\\ \text{bp} \sim 68\ °\text{C}}}{C_6H_{14}} + \underset{\substack{\text{heptanes}\\ \text{bp} \sim 98\ °\text{C}}}{C_7H_{16}}$$

The boiling-point range of these compounds is approximately 0–100 °C; thus, they are more easily volatilized than undecane and can be used more efficiently in internal combustion engines.

Kerosene, consisting of hydrocarbons containing nine to sixteen carbon atoms and having a boiling range of 175–275 °C, is a higher-boiling fraction of petroleum. Gas oil and diesel oil come next, boiling at 200–400 °C and consisting of hydrocarbons with fifteen to twenty-five carbon atoms. Lubricating oil contains alkanes of even higher molecular weight, ones with twenty to seventy carbon atoms in the chain.

Compared to other functional group classes, alkanes have limited chemical reactivity. An important type of reaction of alkanes is their reaction with halogens at high temperatures or under the influence of light (light is symbolized in equations by $h\nu$ under the reaction arrow). Such reactions are complex and give mixtures of products depending on the ratios of reactants that are used. For example, methane reacts with an equal volume of chlorine in the presence of light to give a mixture of alkyl halides.

$$\underset{\text{methane}}{CH_4} + \underset{\text{chlorine}}{Cl_2} \xrightarrow[h\nu]{} \underset{\text{chloromethane}}{CH_3Cl} + \underset{\substack{\text{dichloromethane}\\ \\ \text{methylene}\\ \text{chloride}}}{CH_2Cl_2} + \underset{\substack{\text{trichloro-}\\ \text{methane}\\ \\ \text{chloroform}}}{CHCl_3} + \underset{\substack{\text{tetrachloro-}\\ \text{methane}\\ \\ \text{carbon}\\ \text{tetrachloride}}}{CCl_4} + \underset{\text{hexachloroethane}}{Cl_3CCCl_3}$$

1 volume	*1 volume*	*80%*	*10%*	*small amounts* (trichloromethane, tetrachloromethane, hexachloroethane)

If six times as much methane as chlorine is used, the product is almost completely chloromethane.

When a large excess of cyclopentane is heated with chlorine at 250 °C, chlorocyclopentane is formed, along with small amounts of dichlorocyclopentanes.

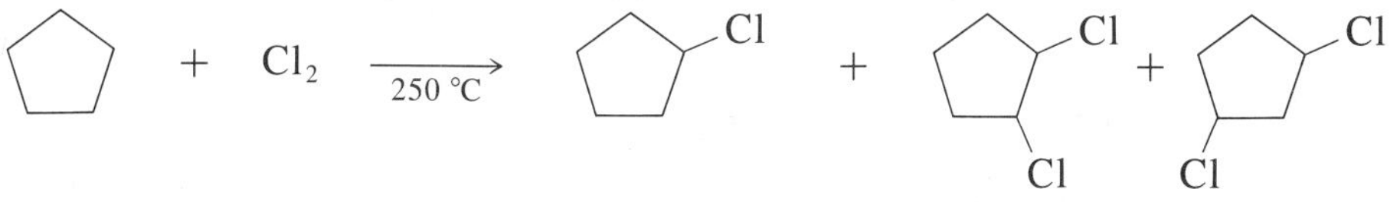
+
Cl2
250 °C
Cl
+
Cl
Cl
+
Cl
Cl

cyclopentane — chlorine — chlorocyclopentane *95%* — 1,2-dichloro-cyclopentane *4%* — 1,3-dichloro-cyclopentane *1%*

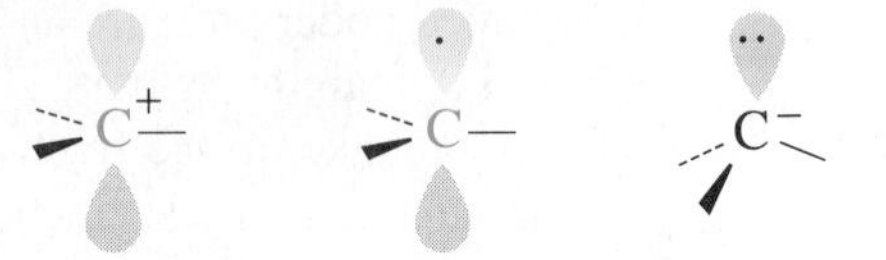

a carbocation   a radical   a carbanion

FIGURE 5.8 A carbocation, a radical, and a carbanion.

Halogenation reactions usually give mixtures of products. They are used to prepare alkyl halides in industry but not often in the laboratory.

The reactive intermediates in these halogenation reactions are radicals. A **radical** has an unpaired electron occupying a *p* orbital on an sp^2-hybridized carbon atom and no charge. It is shown in Figure 5.8 with two other reactive intermediates we have learned about, a carbocation and a carbanion. The chemistry of free radicals will be the subject of Chapter 20.

The alkyl portions of molecules that contain other functional groups usually remain unchanged in reactions that transform the functional groups. You should keep in mind this unreactivity of alkyl groups as you study the following chapters.

PROBLEM 5.25

The carbon atom in a radical has only seven electrons around it, which is not a complete octet. This atom is, therefore, electron-deficient. Using your knowledge of the relative stabilities of primary, secondary, and tertiary carbocations (p. 122), predict the relative stabilities of the following radicals.

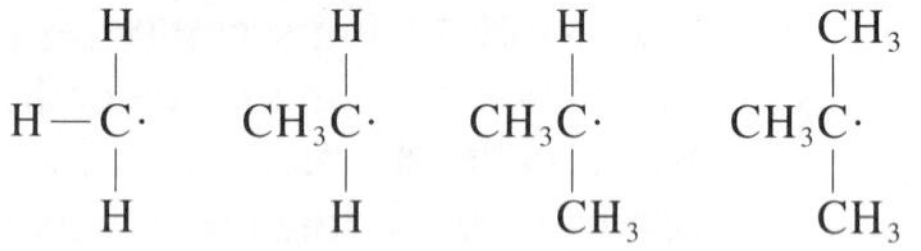

5.11 STRUCTURAL INFORMATION FROM MOLECULAR FORMULAS

A molecular formula tells how many atoms of each of the different elements are present in the molecules of a compound. It is not always easy, however, to go from a molecular formula to a structural formula. One of the pieces of information that helps in drawing a structural formula is knowing whether a compound is saturated, meaning that it has $2n + 2$ hydrogen atoms and n carbon atoms. If the molecular formula for a compound fits the formula C_nH_{2n+2}, the molecular structure cannot contain multiple bonds or rings. Every time a molecular formula has fewer hydrogen atoms per carbon atom than this formula requires, the compound is said to have units of unsaturation. The compound is assigned one **unit of unsaturation** for each two hydrogen atoms that are missing when the molecular formula of the compound is compared with the formula corresponding to C_nH_{2n+2}. A unit of unsaturation may be a double bond between two carbon atoms or between a carbon and an oxygen atom, or it may be a ring. A triple bond corresponds to two units of unsaturation. These concepts are easier to grasp from concrete examples.

Hexane has the molecular formula C_6H_{14}, so $n = 6$ and $2n + 2 = 14$. Hexane is a saturated compound, one containing no rings or double bonds. A compound having the molecular formula C_6H_{12} has one unit of unsaturation. The reasoning is as follows: A compound containing six carbon atoms is saturated when it has the molecular formula C_6H_{14} (C_nH_{2n+2} when $n = 6$). The compound C_6H_{12} is missing two hydrogen atoms, and therefore has *one* unit of unsaturation. The unit of unsaturation may be a ring, so C_6H_{12} may be cyclohexane or methylcyclopentane or ethylcyclobutane or a dimethylcyclobutane, among other possibilities. The unit of unsaturation may instead be a double bond, so C_6H_{12} may be any one of a number of open-chain alkenes such as 1-hexene, 3-hexene, 2-methyl-1-pentene, and so on.

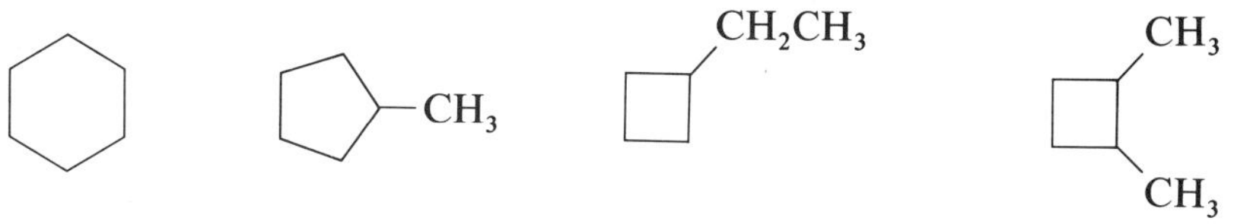
CH₂CH₃
CH₃
CH₃
CH₃

cyclohexane   methylcyclopentane   ethylcyclobutane   1,2-dimethylcyclobutane

some C_6H_{12} compounds in which the unit of unsaturation is a ring

$$\underset{\text{1-hexene}}{CH_3CH_2CH_2CH_2CH{=}CH_2} \qquad \underset{\text{2-hexene}}{CH_3CH_2CH_2CH{=}CHCH_3}$$

$$\underset{\text{3-hexene}}{CH_3CH_2CH{=}CHCH_2CH_3} \qquad \underset{\text{2-methyl-1-pentene}}{CH_3CH_2CH_2\overset{\overset{\displaystyle CH_3}{|}}{C}{=}CH_2}$$

some C_6H_{12} compounds in which the unit of unsaturation is a double bond

The molecular formula alone is not enough information to allow chemists to say whether a compound is an alkene or a cycloalkane, or which alkene or cycloalkane it might be. To make such a determination, chemists need further experimental evidence. For example, a chemist might find that the compound with the molecular formula C_6H_{12} rapidly decolorizes a solution of bromine in carbon tetrachloride (p. 63). These experimental results make it highly likely that the compound is an alkene, for example:

$$\underset{\text{1-hexene}}{CH_3CH_2CH_2CH_2CH{=}CH_2} + Br_2 \longrightarrow \underset{\text{1,2-dibromohexane}}{CH_3CH_2CH_2CH_2\underset{\underset{\displaystyle Br}{|}}{CH}{-}\underset{\underset{\displaystyle Br}{|}}{CH_2}}$$

To obtain further evidence, the chemist might investigate the hydrogenation of the compound (p. 165). If the compound reacts in the presence of a catalyst with one molar equivalent of hydrogen to give hexane, the chemist could state with some certainty that it is a hexene rather than a methylpentene.

$$\underset{\text{a hexene}}{\begin{matrix} CH_3CH_2CH{=}CHCH_2CH_3 \\ \text{or} \\ CH_3CH_2CH_2CH_2CH{=}CH_2 \\ \text{or} \\ CH_3CH_2CH_2CH{=}CHCH_3 \end{matrix}} \xrightarrow[\text{Pt}]{H_2} \underset{\text{hexane}}{CH_3CH_2CH_2CH_2CH_2CH_3}$$

Further physical, chemical, or spectroscopic evidence would be needed to make a final structural assignment.

The experiments just described for characterizing an alkene demonstrate the ways in which different kinds of unsaturation can be detected. Especially important is the hydrogenation reaction. Units of unsaturation that resist catalytic hydrogenation are identified as rings. For example, benzene has the molecular formula C_6H_6, and thus has four units of unsaturation (C_6H_6 has eight hydrogens less than C_6H_{14}). Benzene can be catalytically hydrogenated, usually at high temperatures or pressures or with special catalysts, to give cyclohexane, C_6H_{12}, which has one unit of unsaturation. Thus, three of the units of unsaturation in benzene are multiple bonds to which hydrogen will add; the fourth unit of unsaturation is a ring that is stable to hydrogenation.

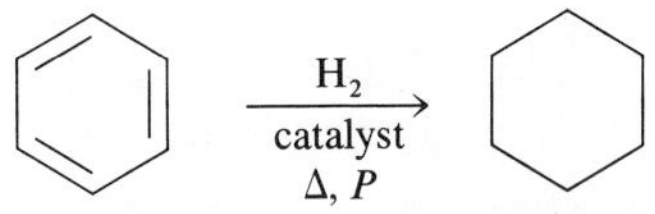
H_2
catalyst
Δ, P

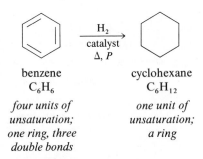

benzene
C_6H_6
four units of unsaturation; one ring, three double bonds

cyclohexane
C_6H_{12}
one unit of unsaturation; a ring

In general, compounds with very large numbers of units of unsaturation are likely to be aromatic hydrocarbons (p. 63). Cyclopropane rings may open on hydrogenation (p. 165). The problems in the rest of this chapter will assume, therefore, that rings have four or more carbon atoms in them.

No modification of the calculation of units of unsaturation is necessary if oxygen is present in a molecule. For example, $C_4H_{10}O$ is a saturated compound; its molecular formula corresponds to C_nH_{2n+2}. The oxygen atom can be disregarded in the calculation. A compound with the molecular formula $C_4H_{10}O$ cannot contain any rings or double bonds. It might be one of the butyl alcohols or an ether, such as diethyl ether.

A compound with the molecular formula C_4H_8O, on the other hand, has one unit of unsaturation. It might contain a carbon-carbon double bond, a carbon-oxygen double bond, or a ring containing only carbon atoms or including the oxygen atom. Some of the stable compounds representing these different possibilities are shown below.

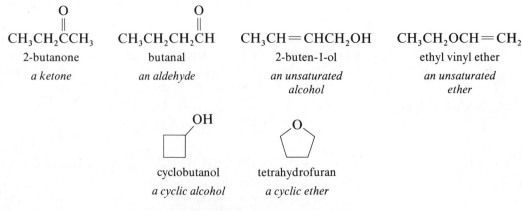

$CH_3CH_2C(=O)CH_3$
2-butanone
a ketone

$CH_3CH_2CH_2CH(=O)$
butanal
an aldehyde

$CH_3CH{=}CHCH_2OH$
2-buten-1-ol
an unsaturated alcohol

$CH_3CH_2OCH{=}CH_2$
ethyl vinyl ether
an unsaturated ether

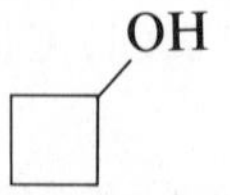
OH

cyclobutanol
a cyclic alcohol

O

tetrahydrofuran
a cyclic ether

some compounds having the molecular formula C_4H_8O

Other experimental evidence such as the origin of the compound or the results of qualitative tests would be necessary to assign it a definite structure.

PROBLEM 5.26

A compound with the molecular formula C_4H_8O decolorizes a solution of bromine in carbon tetrachloride and reacts readily with one equivalent of hydrogen in the presence of a catalyst. Which of the structures shown above are possible for it? Write equations showing the reactions of those compounds with bromine and with hydrogen.

The calculation of units of unsaturation can be done for an alkyl halide if hydrogen is first substituted for the halogen. For example, a compound with the molecular formula C_5H_9Br has one unit of unsaturation (C_5H_9Br corresponds to C_5H_{10}, which has two hydrogen atoms less than the saturated compound C_5H_{12}). The compound does not react with bromine in carbon tetrachloride; therefore, it does not have a double bond. The unit of unsaturation must be present as a ring.

PROBLEM 5.27

Write structural formulas for all possible compounds that have the molecular formula C_5H_9Br and contain a four-membered or larger ring.

A triple bond corresponds to two units of unsaturation. For example, a compound with the molecular formula C_7H_{12} has two units of unsaturation (C_7H_{12} would need four more hydrogen atoms to become C_7H_{16}). Such a compound may contain one triple bond, two double bonds, a ring and a double bond, or two rings.

PROBLEM 5.28

For each of the four categories of compounds given as possibilities for the molecular formula C_7H_{12} draw one structure.

If the compound with the molecular formula C_7H_{12} decolorizes bromine in carbon tetrachloride and adds hydrogen to give a new compound with the molecular formula C_7H_{16}, structures containing rings can be eliminated as a possibility.

PROBLEM 5.29

Adding hydrogen to the compound with the molecular formula C_7H_{12} gives a C_7H_{16} compound, which is identified as being 2-methylhexane. In light of this evidence, what structures are possible for the C_7H_{12} compound?

S U M M A R Y

Alkanes and cycloalkanes are organic compounds that have only carbon-hydrogen and carbon-carbon single bonds. The different ways in which the carbon atoms can be bonded together in alkanes gives rise to constitutional isomers, which are straight-chain or branched-chain compounds. The carbon atoms in alkanes are classified according to the number of other carbon atoms they are bonded to: primary (bonded to one other carbon atom), secondary (bonded to two other carbon atoms), tertiary (bonded to three other carbon atoms), and quaternary (bonded to four other carbon atoms). Hydrogen atoms are also classified as primary, secondary, and tertiary according to the type of carbon atom they are bonded to.

The carbon chains in alkanes assume different conformations through rotations around the carbon-carbon single bonds. Staggered conformations are more stable than eclipsed conformations because, in the latter, bonds on adjacent carbon atoms are aligned with each other and, therefore, repel each other. Anti conformations are staggered conformations in which the largest groups are as far apart as they can be. Anti conformations are more stable than other staggered conformations, such as the gauche conformation in which larger groups are close enough to repel each other.

The most stable conformation of cyclohexane is the chair form, which has six equatorial hydrogen atoms and six axial ones. A substituted cyclohexane is most stable when the substituent is in an equatorial position. A cyclohexane with a substituent in the axial position is destabilized by 1,3-diaxial interactions between the substituent and the two hydrogens also in axial positions on the same face of the ring.

Small ring compounds such as cyclopropane and cyclobutane are unstable in comparison to cyclopentane and cyclohexane because of ring strain. Such strain is due to bond angles that are smaller than the tetrahedral angle and to the eclipsing of bonds on adjacent atoms.

Alkanes are named according to the IUPAC rules of nomenclature. The names of alkanes are used as the basis for the names of other classes of organic compounds.

Alkanes have a low reactivity compared to that of compounds containing functional groups. Combustion and halogenation, both of which require the input of considerable energy, are the most important reactions of alkanes.

Alkanes are saturated hydrocarbons, meaning that they have $2n + 2$ hydrogen atoms for every n carbon atoms. Each pair of hydrogen atoms "missing" from a molecular formula means that the compound has a unit of unsaturation. A unit of unsaturation may be a ring or a double bond. Examining the molecular formula of a compound to detect units of unsaturation is one step toward determining what kind of structural formula is possible for the compound.

ADDITIONAL PROBLEMS

5.30 Name the following compounds.

(a) 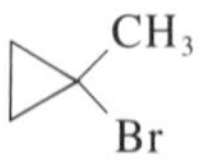
CH3
Br

(b) 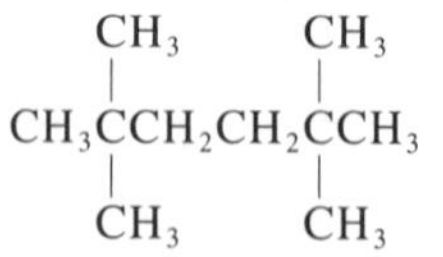
CH3 CH3
CH3CCH2CH2CCH3
CH3 CH3

(c) 
CH3
CH3

(d) 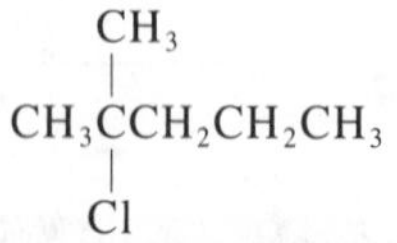
CH3
CH3CCH2CH2CH3
Cl

(e) 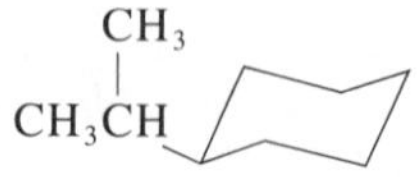
CH3
CH3CH

(f) 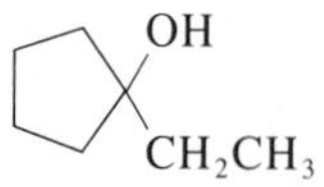
OH
CH2CH3

(g) 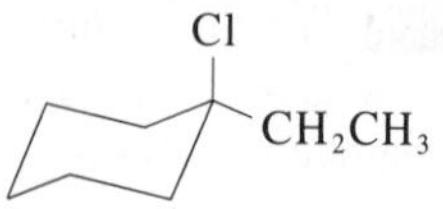
Cl
CH2CH3

(h) 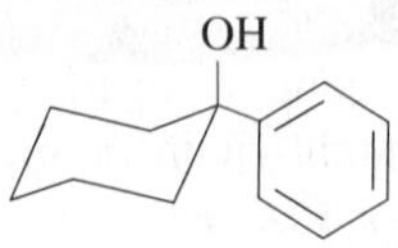
OH

(i) 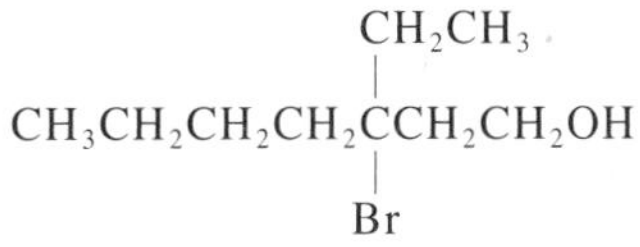

(j) 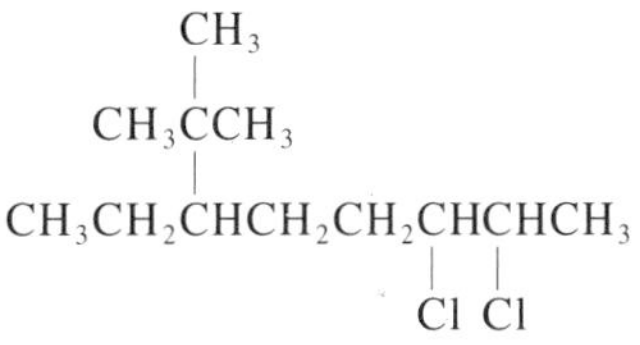

5.31 2-Bromo-2-chloro-1,1,1-trifluoroethane (bp 50.2 °C) is used as an inhalation anesthetic. What is the structure of this compound?

5.32 Draw a structural formula for each of the following compounds.

(a) 2-methyl-2-hexanol
(b) 1-methyl-1-cyclobutanol
(c) iodocyclopropane
(d) 4-*tert*-butylheptane
(e) 5-cyclopropylnonane
(f) isobutylcyclopentane
(g) 3-phenylheptane
(h) 1-bromo-1-methylcyclohexane
(i) 7-chloro-2,7-dimethyl-2-octanol

5.33 For each molecular formula, draw structural formulas of all possible constitutional isomers. Name each one.

(a) $C_5H_{12}O$ (Ethers may be named for the two alkyl groups they contain: for example, $CH_3OCH_2CH_3$ is methyl ethyl ether.)
(b) $C_3H_5Cl_3$

5.34 In each of the following sets of compounds, identify the one you expect will have the lowest boiling point. (Hint: You may want to review pp. 24–29 before answering.)

(a) 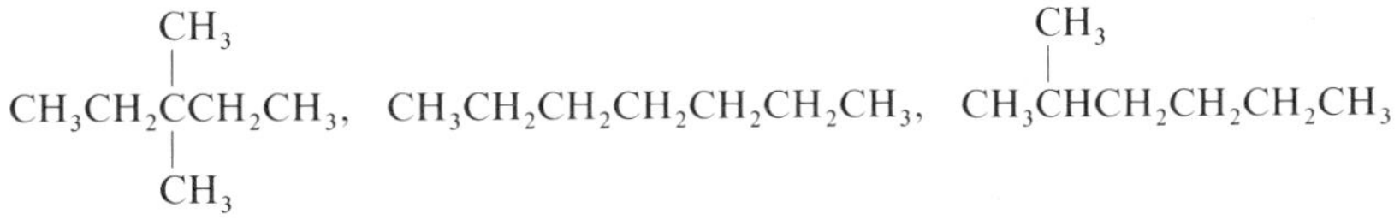

(b) 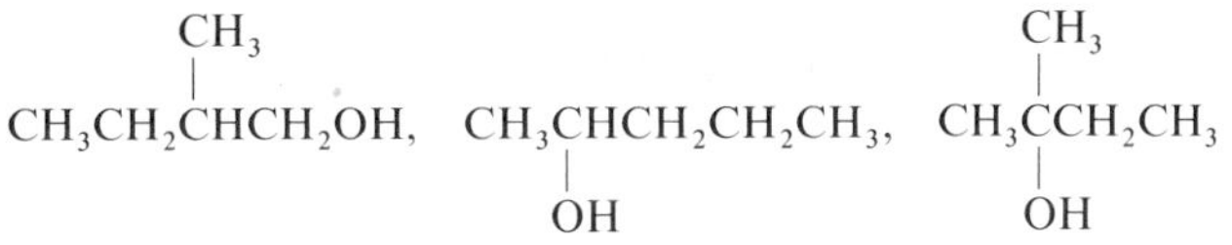

(c) 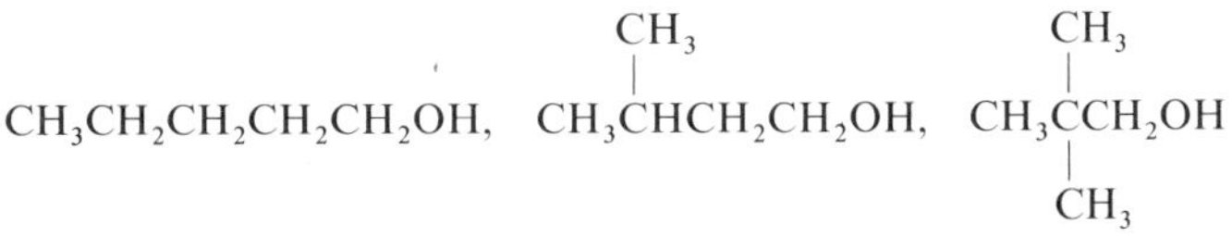

(d) 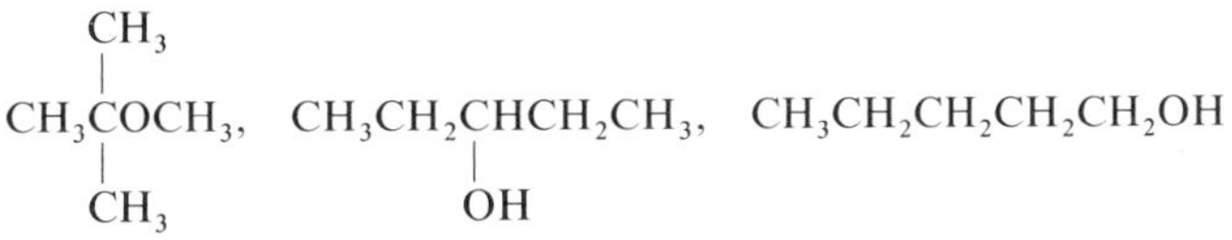

(e) 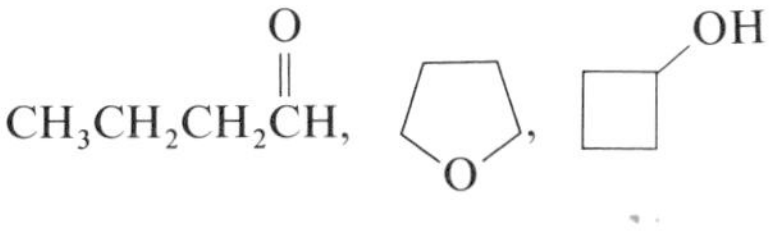

5.35 For each molecular formula, identify the number of units of unsaturation. For each molecular formula, draw structural formulas for three different constitutional isomers. Use examples that contain different structural features such as different functional groups, multiple bonds, and rings.

(a) C_5H_9Br
(b) $C_5H_{10}O$

(c) C_5H_8O

(d) An isomer of C_5H_9Br, does not react easily with bromine in carbon tetrachloride. What is a possible structure for it?

(e) An isomer of $C_5H_{10}O$ is treated with hydrogen in the presence of a catalyst and gives 3-methyl-1-butanol. What structures are possible for the original compound?

5.36 Show all the conformations possible for each of the following compounds and indicate which conformation would be the most stable one.

(a) isopropylcyclohexane

(b) bromocyclohexane

5.37 Predict the products of each of the following reactions.

(a) $CH_3CH_2NH_2 + HCl \longrightarrow$

(b) $CH_3CH_2C{\equiv}CH + NaNH_2 \longrightarrow$

(c) $(CH_3CH_2CH_2)_2NH + CH_3CH_2Br \longrightarrow$

(d) $CH_3CH_2CH{=}CHCH_2CH_3 + HBr \longrightarrow$

(e) $CH_3CH_2CH_2SH + NaOH \longrightarrow$

(f) 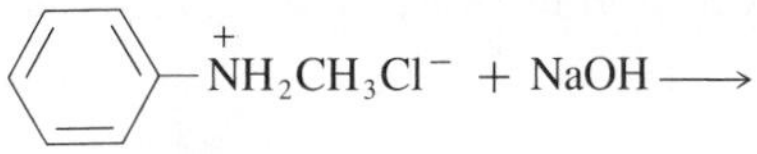
$\overset{+}{N}H_2CH_3Cl^- + NaOH \longrightarrow$

(g) $CH_3CH_2CH{=}CHCH_3 + Br_2 \longrightarrow$

(h) 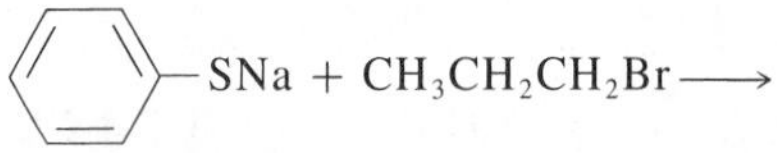
$SNa + CH_3CH_2CH_2Br \longrightarrow$

(i) 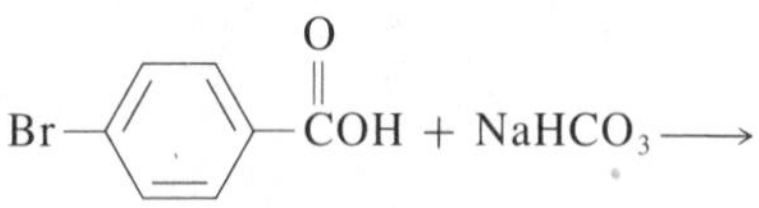
Br
O
COH + $NaHCO_3 \longrightarrow$

(j) $CH_3CH_2CH_2C{\equiv}CCH_3 + H_2 \text{ (2 equivalents)} \xrightarrow[PtO_2]{}$

(k) $\underset{\text{1 volume}}{CH_3CH_3} + \underset{\text{1 volume}}{Cl_2} \xrightarrow[h\nu]{}$

5.38 Tell what reagents could be used to make each of the following transformations. More than one step may be necessary in some cases.

(a) 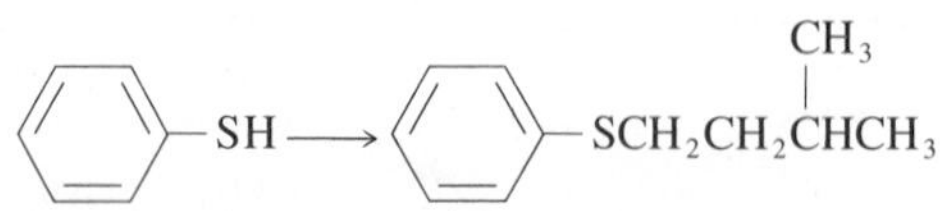
SH
SCH$_2$CH$_2$CHCH$_3$
CH$_3$

(b) $CH_3CH_2Br \longrightarrow CH_3CH_2\overset{+}{N}H_2CH_3\ Br^-$

(c) $CH_2{=}CH_2 \longrightarrow CH_3CH_2OCH_3$

(d) $CH_3CH_2CH_2CH_2NH_2 \longrightarrow CH_3CH_2CH_2CH_2\overset{+}{N}H_3\ Cl^-$

(e) $CH_3CH_2C{\equiv}CH \longrightarrow CH_3CH_2\overset{Br}{\underset{Br}{C}}-\overset{Br}{\underset{Br}{C}}H$

(f) $CH_3CH_2CH_2CH_2OH \longrightarrow CH_3CH_2CH_2CH_2OCH_3$

(g) $CH_3CH{=}CHCH_3 \longrightarrow CH_3\underset{\underset{\displaystyle CN}{|}}{C}HCH_2CH_3$

(h) $CH_3\underset{\underset{\displaystyle OH}{|}}{C}HCH_2CH{=}CH_2 \longrightarrow CH_3\underset{\underset{\displaystyle OH}{|}}{C}HCH_2CH_2CH_3$

(i) 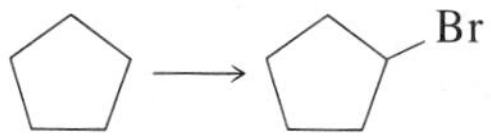

6

Stereochemistry

A • L O O K • A H E A D

Stereochemistry is chemistry in three-dimensional space. It is of immense importance, especially in the study of complex molecules that are biologically important, such as proteins (Chapter 26), carbohydrates (Chapter 25), and nucleic acids (Chapter 24). Much of the rest of the book will be concerned with the stereochemistry of chemical reactions, that is, how the reactions actually proceed in three dimensions. First, we must look more closely at the structures of organic molecules.

Chapter 5 explored the question of the equivalency of hydrogen atoms. Replacing different types of hydrogen atoms on a molecule with halogen atoms gives rise to constitutional isomers. Carrying this exercise a little further uncovers a more subtle kind of isomerism called stereoisomerism. Stereoisomers differ from each other only in the way their atoms are oriented in space. They cannot be distinguished by looking at their condensed structural formulas, which are identical.

Chapter 5 introduced one aspect of stereochemistry, the dynamic interconversions molecules undergo as the atoms within them move relative to one another by rotation around single bonds. Such motions give rise to one type of stereoisomers, called conformational isomers, or conformers (p. 150). This chapter explores another type of stereoisomerism called configurational isomerism. Configurational isomers differ from each other only in the arrangement of their atoms in space and cannot be converted from one into another by rotations about single bonds within the molecules. Configurational isomers may be enantiomers or diastereomers. Enantiomers are stereoisomers that are mirror images of one another. All stereoisomers that are not enantiomers are called diastereomers. Molecular models will help you to see this more subtle kind of isomerism, and you should use them as you read this chapter.

Study Guide Concept Map 6.1

6.1 ENANTIOMERS

The substitution of a chlorine atom for one of the secondary hydrogen atoms of butane has already been shown (on p. 146) to give rise to 2-chlorobutane. Butane exists as a single isomer, and at first glance the two hydrogen atoms on the secondary carbon atom may appear to be equivalent. But if first one and then the other of

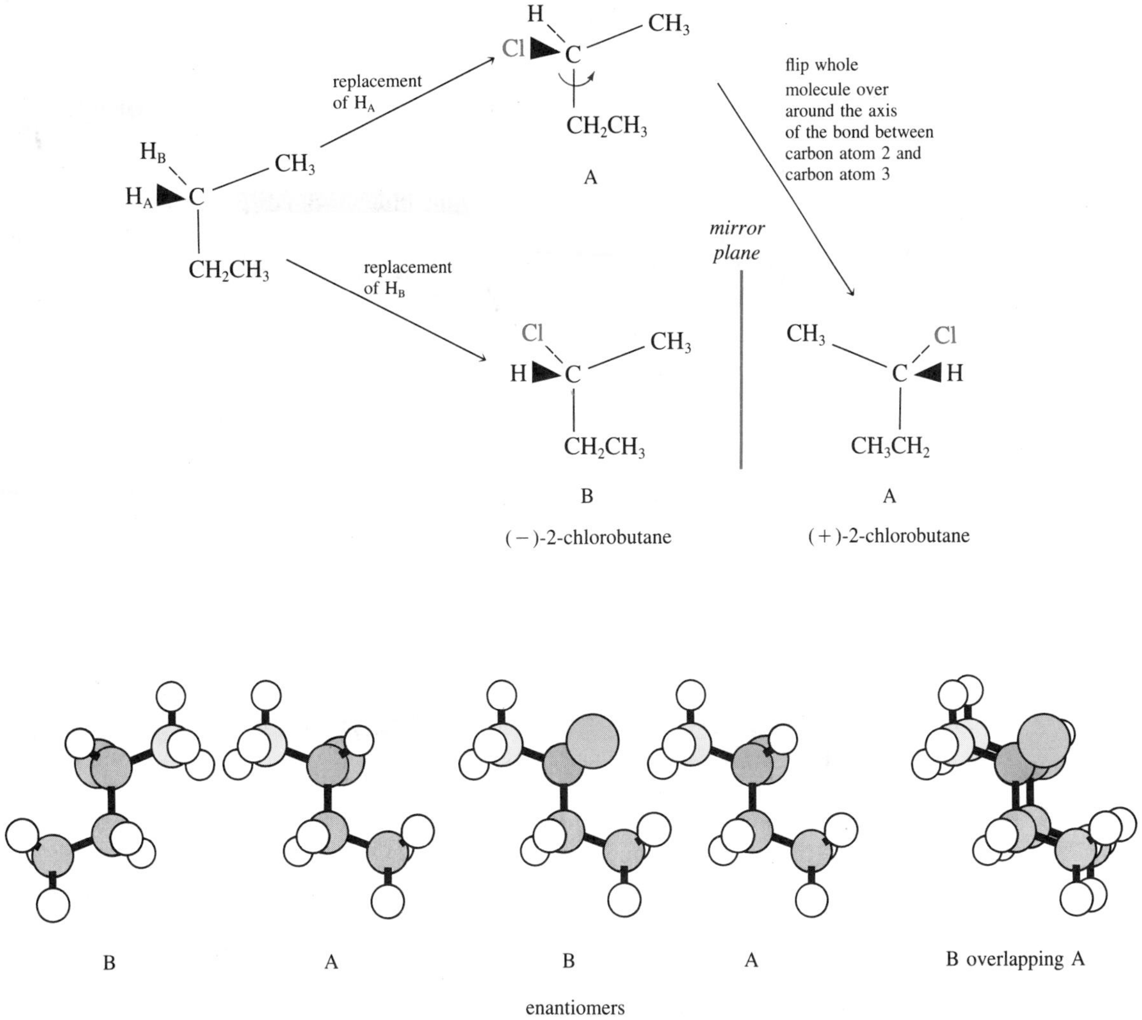
replacement of H_A
replacement of H_B
flip whole molecule over around the axis of the bond between carbon atom 2 and carbon atom 3
mirror plane
A
B
A
(−)-2-chlorobutane
(+)-2-chlorobutane
B
A
B
A
B overlapping A
enantiomers

FIGURE 6.1 Creation of nonsuperimposable mirror-image isomers by the replacement of one or the other of two secondary hydrogen atoms in butane by chlorine. Structures A and B are enantiomers.

these two hydrogen atoms is replaced by a chlorine atom (Figure 6.1), the two molecular species that are formed are not identical. It should be clear to you that structure A cannot be made to coincide perfectly with structure B. If you were to try to pick A up out of the page and place it over B, the methyl and ethyl groups and the central carbon atom would coincide, but the positions of the chlorine and hydrogen atoms would be reversed. Molecular models will help you see this. Thus, structure A and structure B represent two molecular species that differ from each other only in the orientation of the atoms in space. They are **stereoisomers** of 2-chlorobutane. If structure A is first flipped over by rotating its parts around the axis of the bond between carbon atom 2 and carbon atom 3 and then placed next to structure B, the relationship between the two isomers becomes more clear. Structures A and B are related as an object is related to its mirror image. If you imagine a mirror between the structures, you can see that one looks like the reflection of the other. Thus, the two molecular species represented by these structural formulas are mirror-image isomers of each other.

A mirror image may be drawn of any molecular structure. Isomerism exists only when mirror images are not superimposable on each other. Structure A in Figure 6.1 cannot be picked up and placed over structure B so that all points coincide, no matter how the two structures are rotated in space. Stereoisomers that are nonsuperimposable mirror images of each other are called **enantiomers.** Thus, the stereoisomers represented by structural formulas A and B in Figure 6.1 are enantiomers of 2-chlorobutane. Enantiomers have identical physical properties with one exception—they rotate plane-polarized light in opposite directions. Compounds that rotate the plane of polarized light are said to have **optical activity,** or to be **optically active** (p. 187). The two enantiomers of 2-chlorobutane rotate a plane of polarized light in opposite directions, as indicated by the plus and minus signs in their names in Figure 6.1. (−)-2-Chlorobutane rotates the plane of polarized light counterclockwise, and (+)-2-chlorobutane rotates it an equal amount clockwise. All other physical properties of enantiomers are identical. Therefore, they cannot be separated from each other by physical means such as distillation or crystallization. Enantiomers are one type of configurational isomer (p. 180).

6.2 CHIRALITY

A molecule or any other object that cannot be superimposed on its mirror image is said to be **chiral.** Thus, 2-chlorobutane (p. 181) is chiral. The concept of chirality is important in the study of the chemistry of biological systems. Many of the compounds that occur in living organisms, such as carbohydrates and proteins, are chiral. Living organisms, including human beings, are chiral. Our internal organs are arranged asymmetrically, and we have distinctive twists and whorls in the way our hair grows. Our left and right hands are nonsuperimposable mirror images of each other (Figure 6.2). In fact, the word chiral comes from the Greek word *cheir*, which means hand.

Many naturally occurring compounds exist as one of two possible enantiomers. Biological systems can distinguish between one enantiomer and its mirror image. For example, all amino acids that are present in proteins exhibit a certain spatial orientation of the groups around a central carbon atom. The natural enantiomer of the amino acid alanine and its mirror image are shown on the next page.

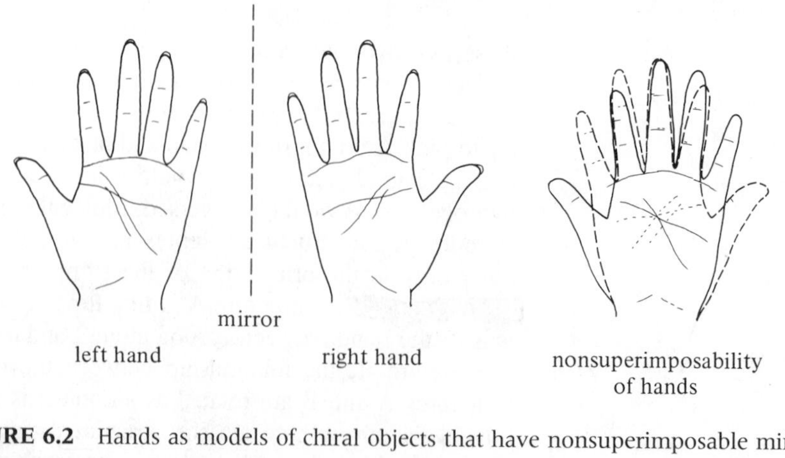
mirror
left hand
right hand
nonsuperimposability
of hands

FIGURE 6.2 Hands as models of chiral objects that have nonsuperimposable mirror images.

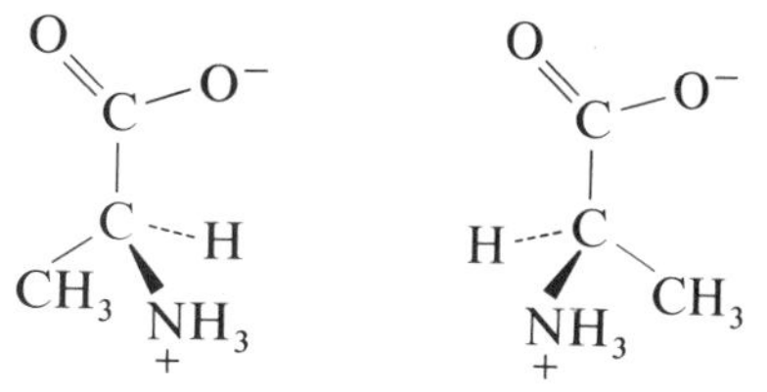

(+)-alanine
the natural enantiomer

(−)-alanine
the mirror image

The chirality of the human body, which is reflected at the most fundamental level in the stereochemistry of enzyme systems, requires that chemical reactions take place with a particular orientation in space. You will have some idea of what is involved if you try to put your left glove on your right hand or your left foot into your right shoe. Thus, one enantiomer of a compound may be a hormone or may be active as a medication, and its mirror image may be biologically inactive. The sensitivity of various biological systems to stereochemistry is one of the reasons why chemists are concerned both with the stereoisomerism of organic molecules and with the way their stereochemistry is determined and changed by chemical reactions.

An easy way to test for chirality is to look for a **plane of symmetry.** If an object or a molecule can be divided by a plane into two equal halves that are mirror images of each other, the plane is a plane of symmetry, and the object or molecule is not chiral. For example, butane in the anti conformation has a plane of symmetry that goes through all four carbon atoms and two primary hydrogen atoms. The plane bisects the angles between the pairs of secondary hydrogen atoms. Two planes of symmetry can be found in the eclipsed conformation of butane (Figure 6.3). 2-Chlorobutane does not have a plane of symmetry. No conformation of 2-chlorobutane can be divided by a plane into two halves that are mirror images of each other (see Figure 6.3).

As a consequence of its symmetry, a molecule of butane is superimposable upon its mirror image. In Figure 6.4, representation C is the mirror image of representation D, but these structural formulas do not represent different molecular species. If D is flipped over by a rotation around the axis of the bond between carbon atom 2 and carbon atom 3, it becomes identical with C; it could be lifted out of the page and fitted exactly over C. In other words, D is a superimposable mirror image of C. An object that is superimposable on its mirror image has a plane of symmetry and is **achiral.**

Some achiral molecules and common objects are shown in Figure 6.5. Bromochloromethane has a plane of symmetry that bisects the bromine atom, the chlorine atom, the carbon atom, and the angle between the two hydrogen atoms. (Remember that individual atoms are spherically symmetrical even though the symbols chemists use to represent them are not.) Acetone has a plane of symmetry that bisects the central carbon atom and the oxygen atom. The plane of symmetry of methylcyclobutane goes through the methyl group, the carbon atom to which it is bound, the hydrogen atom on that carbon atom, and the carbon atom and two hydrogen atoms on the opposite side of the ring. A pencil has many planes of symmetry; one is shown. The plane of symmetry for the mug bisects the handle of the mug. An object or molecule that has a plane of symmetry is achiral.

There are, however, complex molecules that do not have a plane of symmetry and yet are achiral because they possess other kinds of symmetry. For the molecules we are concerned with, the test as to whether or not a plane of symmetry is present is sufficient to distinguish between chiral and achiral systems.

Study Guide
Concept Maps 6.2 and 6.3

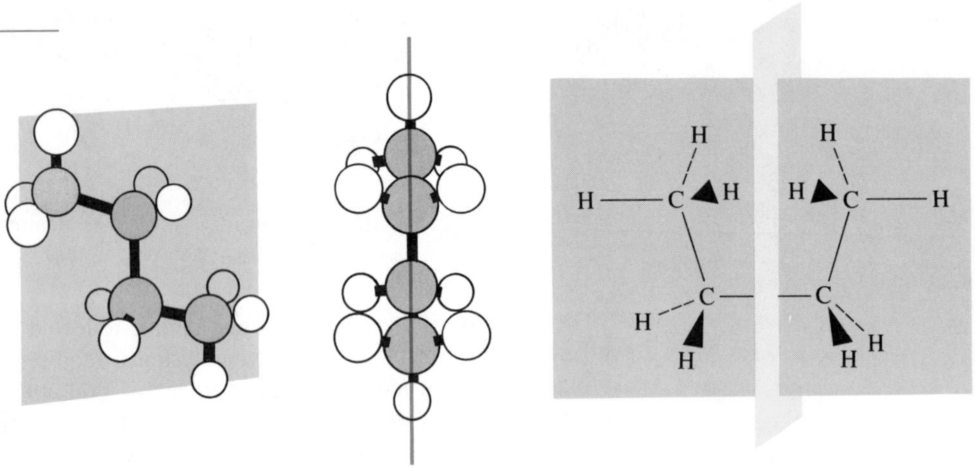

side view *front view*

one plane of symmetry

anti conformation

two planes of symmetry

eclipsed conformation

butane

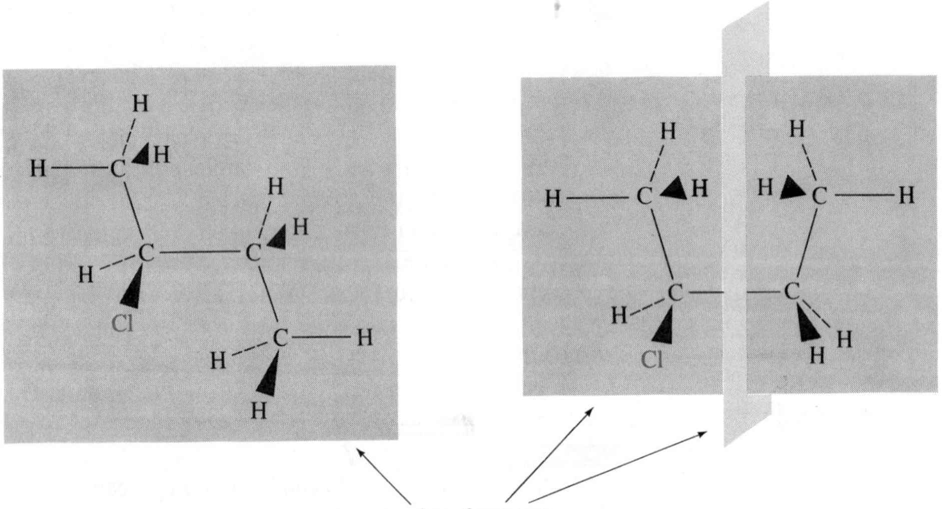

not a plane of symmetry

Cl on one side of each plane, H on the other side

anti conformation

eclipsed conformation

2-chlorobutane

FIGURE 6.3 Planes of symmetry in different conformations of butane; lack of symmetry in 2-chlorobutane.

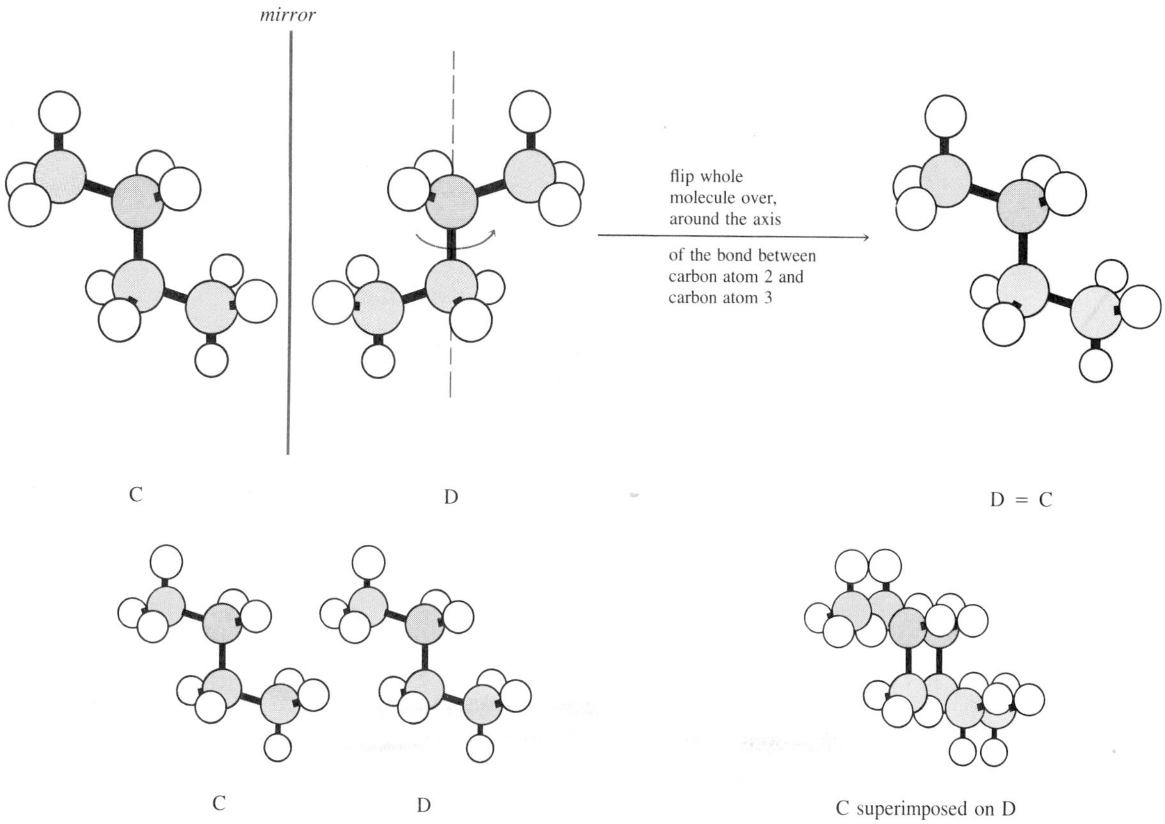
mirror
flip whole
molecule over,
around the axis
of the bond between
carbon atom 2 and
carbon atom 3
C
D
D = C
C
D
C superimposed on D

FIGURE 6.4 Two representations of butane; each is a superimposable mirror image of the other.

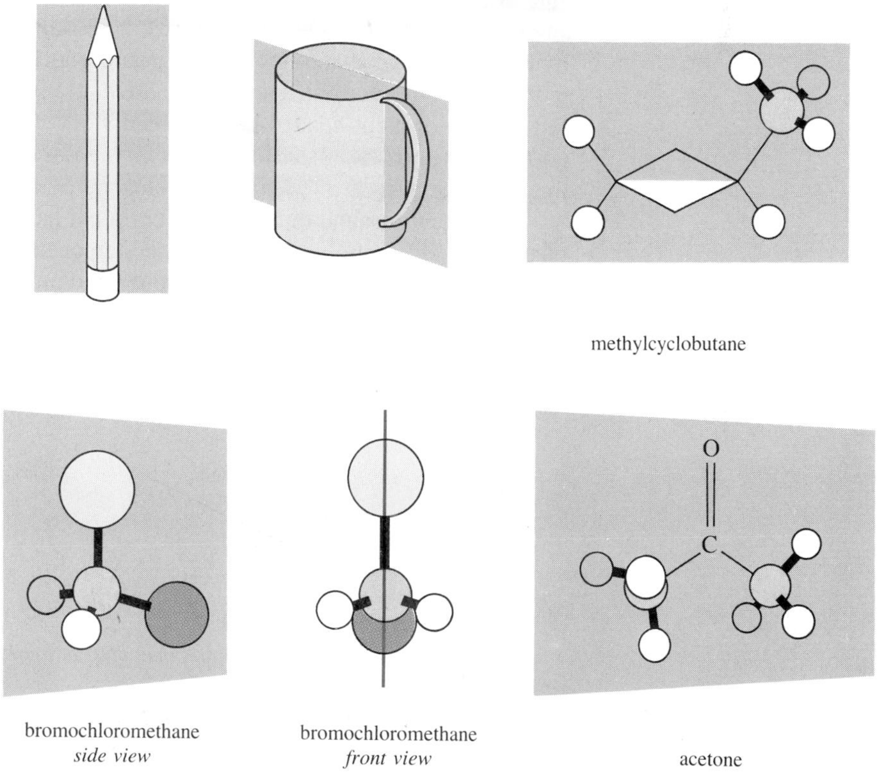
methylcyclobutane
O
C
bromochloromethane
side view
bromochloromethane
front view
acetone

FIGURE 6.5 Some achiral molecules and common objects, with planes of symmetry.

PROBLEM 6.1

(a) Suppose the mug in Figure 6.5 had the word MOTHER on one side. Would it be chiral or achiral? What if it had the word MOM?
(b) Would a spool with thread on it be chiral or achiral?
(c) What about an empty spool?

PROBLEM 6.2

Of the following common objects, which ones are chiral and which achiral?

(a) a screw (b) a nail (c) a hammer (d) a spade (e) socks
(f) mittens (g) a stocking cap (with no decoration on it) (h) a shirt
(i) a tee shirt with GO BLUE written on it

6.3 STEREOCENTERS

2-Chlorobutane (p. 181) and alanine (p. 183) are examples of chiral molecules, and they have one thing in common. One carbon atom in each of them has four different substituents. A carbon atom that is bonded to four different substituents constitutes a **stereocenter.** Such a center was also sometimes called an asymmetric carbon atom, but chemists no longer use this term.

In Section 6.8, we will see that molecules containing more than one stereocenter may or may not exhibit chirality depending on whether they have a plane of symmetry and are superimposable on their mirror images. In this and the next few sections, we are only concerned with molecules containing a single stereocenter. Such molecules do not have a plane of symmetry and are not superimposable on their mirror images. They are, therefore, chiral. Another way of looking at such molecules is to say that four different groups may be arranged around a carbon atom in two (and only two) different ways. Such arrangements may be considered to be left-handed or right-handed. You should work with models to convince yourself that any four groups around a carbon atom can only have two different arrangements and that these two different forms are nonsuperimposable mirror images.

The groups around the stereocenter do not have to be very different in order to qualify as different. For example, the stereocenter in 2-butanol has a hydrogen atom, a hydroxyl group, a methyl group, and an ethyl group as the four different substituents. Even though the methyl and ethyl groups are both alkyl groups, they are different.

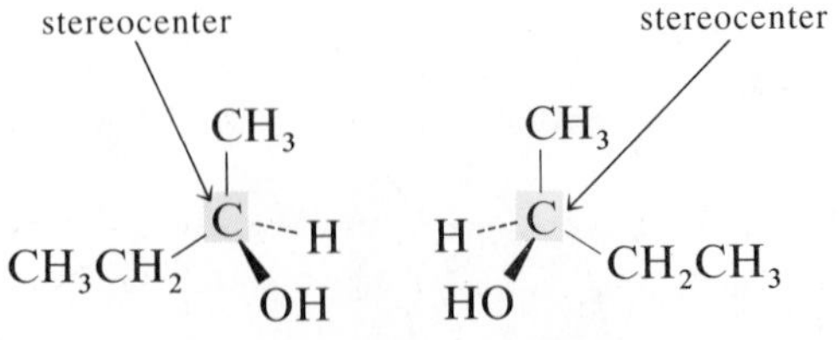
stereocenter
stereocenter
CH3
C
H
CH3CH2
OH
CH3
H
C
HO
CH2CH3

(−)-2-butanol (+)-2-butanol
enantiomers of 2-butanol

Isotopes of the same element are considered different, as are large alkyl groups that differ from each other in subtle ways at some distance from the stereocenter. To discover whether a carbon atom is a stereocenter involves searching carefully in all directions, moving away from the potential stereocenter until some difference is found or until it is established that at least two of the groups are the same.

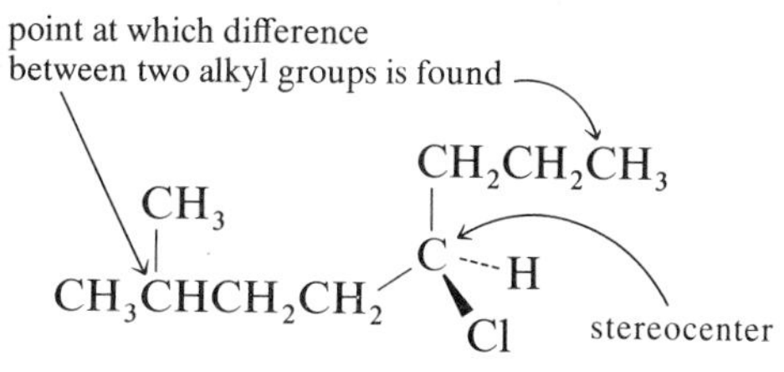
point at which difference
between two alkyl groups is found
stereocenter

one enantiomer of 5-chloro-2-methyloctane

PROBLEM 6.3

Determine whether or not each of the following compounds contains a stereocenter. If so, draw the two enantiomers of the compound. If not, draw the compound in such a way that you can identify the plane of symmetry.

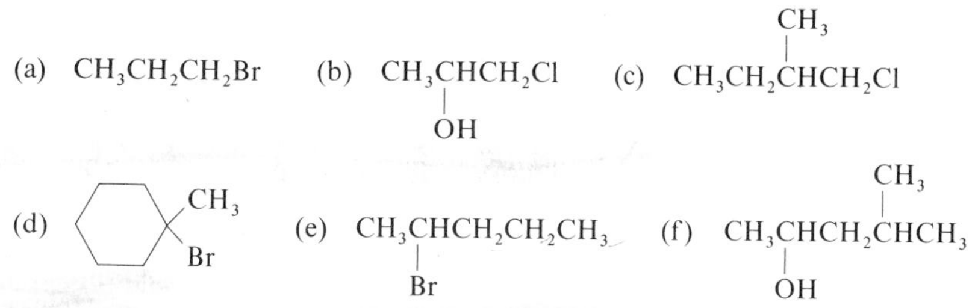

6.4 PLANE-POLARIZED LIGHT AND OPTICAL ACTIVITY

A. Plane-Polarized Light

Light is generally characterized as a wave with associated oscillating electric and magnetic fields. Thus, for a beam of light, one vector describes the electric field strength, and another the magnetic field strength. These vectors are in planes that are perpendicular to each other and to the direction of the propagation of the wave (Figure 6.6). In ordinary light, there are a great number of these electromagnetic waves; the planes of the electric and magnetic vectors of each individual wave are randomly oriented with respect to those of the other waves. Such light is said to be unpolarized.

Certain crystalline substances—calcite is a good example—only allow the passage of light waves with their electric vectors in one single plane. Prisms made of such crystals transmit light that is said to be **plane-polarized.**

B. The Experimental Determination of Optical Activity

When plane-polarized light interacts with molecules, the direction of the plane of polarization is changed. If the molecules are achiral, the interactions that take place with large numbers of molecules in many random orientations do not result in an

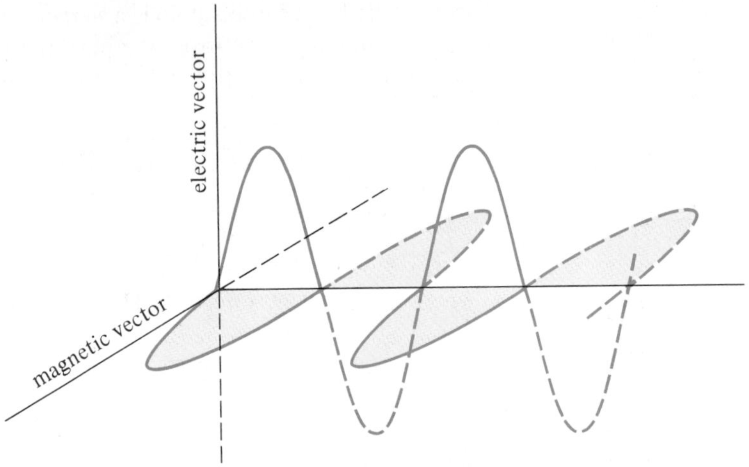
electric vector
magnetic vector

FIGURE 6.6 Electric and magnetic field strength vectors of a light wave.

overall change in the plane of polarization. A twist given to the plane of polarization by a molecule in one orientation is canceled out by an equal twist caused by a molecule in another orientation that looks like the mirror image of the first one. The result is that no rotation of the plane of polarization is observed. If the molecules are chiral, then no molecular orientation is the exact mirror image of any other. Therefore, the changes that take place in the plane of polarized light as it interacts with the molecules are not averaged out, but instead add up so that rotation of the plane of polarization is observed. The direction of this rotation may be counterclockwise, in which case the species is said to be **levorotatory.** If it is clockwise, the species is **dextrorotatory.** Whether an optically active compound rotates the plane of polarization clockwise or counterclockwise is indicated by the sign of rotation (+) or (−), at the beginning of the name of the compound.

The instrument that is used to measure optical activity is called a polarimeter. It has a source of light and two prisms—one is used to produce the plane-polarized light, and the other to detect any rotation in the plane of polarization. A tube containing a solution of the compound to be investigated is placed between the prisms. Finally, there is a viewing device with a scale on it, so that the angle by which the plane of polarized light has been rotated can be measured (Figure 6.7).

The light that goes through the polarizer prism emerges as plane-polarized light. If the axis of the analyzer prism is aligned with that of the polarizer prism and there is no optically active material in the sample tube, the light will go through to the scale unchanged. If the sample tube contains an optically active compound, however, the plane of polarization will be rotated. The analyzer prism is then

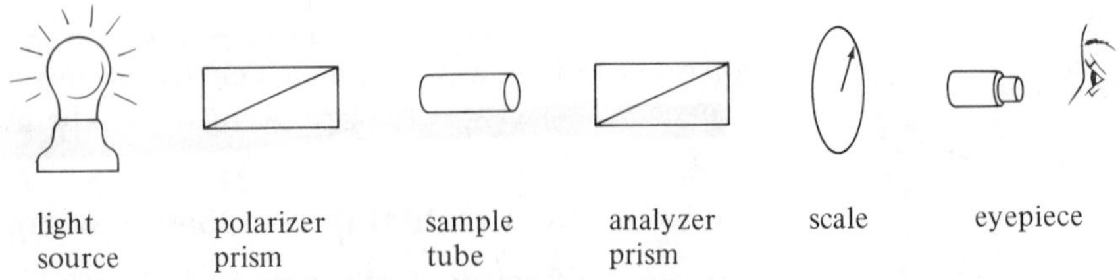
light source
polarizer prism
sample tube
analyzer prism
scale
eyepiece

FIGURE 6.7 Sketch of the main parts of a polarimeter.

rotated until its axis coincides with the new plane of polarization, and light is once more transmitted. The angle by which the analyzer prism has to be rotated in order to allow the transmission of light is the angle of rotation for the optically active compound (Figure 6.8).

How much the plane of polarization is rotated depends on several things. First, it depends on the particular optically active compound on which the measurement is being made. The sign of rotation for a compound is an experimental measurement. There are rules for predicting the direction and extent of the rotation for some compounds, but ultimately the optical rotation for a compound must be determined experimentally. Once the optical rotation for one enantiomer is known, however, it is certain that the other isomer will have the same degree of rotation but in the opposite direction.

Second, because the total rotation depends on the number of interactions between the light beam and the molecules, the concentration of the solution and the path length of the beam of polarized light through the solution must be considered. The particular conformations of the molecules being examined and any interactions between the molecules and the solvent are important, so temperature and solvent are also recorded.

Finally, optical rotation changes with the wavelength of the light used in the measurement, so a complete report must also include that information. For purposes of comparing different compounds, the specific rotation of a compound is calculated using this formula:

$$[\alpha]_{\mathrm{D}}^{T} = \frac{\alpha}{lc} \text{ (concentration; solvent)}$$

The symbol $[\alpha]_{\mathrm{D}}^{T}$ means the specific rotation, calculated from data measured at a temperature of T °C using the light from the yellow line (the so-called D line) in the visible spectrum of sodium. This wavelength of 589 nm corresponds to the yellow color seen when a sodium salt is heated in the flame of a Bunsen burner. In the nineteenth century, it was the only readily available source of light of a single wavelength. It is still used today so that contemporary results can be compared with

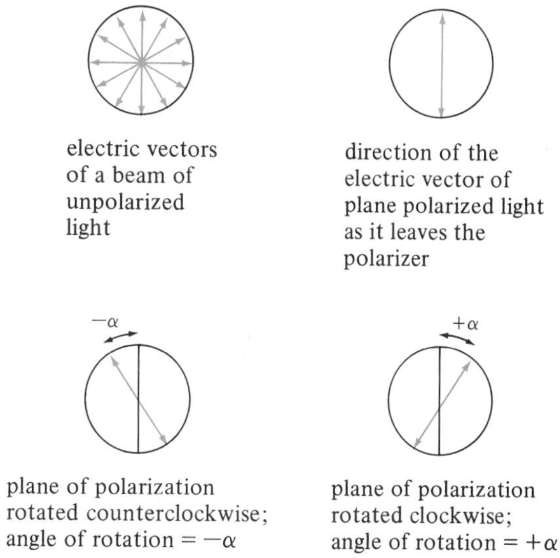
electric vectors of a beam of unpolarized light
direction of the electric vector of plane polarized light as it leaves the polarizer
−α
+α
plane of polarization rotated counterclockwise; angle of rotation = −α
plane of polarization rotated clockwise; angle of rotation = +α

FIGURE 6.8 Schematic representation of the rotation of the plane of polarized light.

those of earlier days. The measured angle of rotation is α. The length of the sample tube, l, is expressed in decimeters, and a standard tube is 10 cm or 1 dm long. The concentration of the sample, c, is given in g/mL of solution. In most cases, the concentration of the solution and the solvent used are reported in parentheses after the value for the specific rotation.

The optical rotation of a chiral compound is a specific physical property of the compound and is determined and reported just as other physical properties are, for example, boiling and melting points. Different chiral compounds have widely varying specific rotations. The specific rotations of some compounds isolated from natural sources illustrate this. In ethanol, cholesterol has $[\alpha]_D^{20}$ $-3.15°$, and for nicotine $[\alpha]_D^{20}$ is $+163.2°$. The specific rotation for cane sugar (sucrose) in water is $[\alpha]_D^{20}$ $+66.4°$.

Calculating specific rotation with the above formula is illustrated by the following problem.

Problem: A solution of 16.5 g of the levorotatory form of camphor in 100 mL of ethanol has an optical rotation of $-7.29°$ at 20 °C, using a 10-cm sample tube and a sodium lamp for the measurement. What is the specific rotation of this compound?

Solution:

$$[\alpha]_D^{20} = \frac{\alpha}{1\ \text{dm} \times 16.5\ \text{g}/100\ \text{mL}}$$

$$= \frac{\alpha \times 100}{1 \times 16.5} = \frac{-7.29° \times 100}{1 \times 16.5}$$

Thus, the specific rotation for camphor is

$$[\alpha]_D^{20} = -44.2° \quad (c\ 0.165;\ \text{ethanol})$$

Study Guide Concept Map 6.4

PROBLEM 6.4

The optical rotation of sugar is used in industry as a quick way to check on the concentration of sugar solutions. If 20 g of cane sugar dissolved in water to make up 100 mL of solution and placed in a tube 40 cm long rotates the plane of polarized light $+53.2°$ at 20 °C, what is the concentration of another solution of sugar, measured at the same temperature and in the same polarimeter tube, if the optical rotation is $+13.3°$?

PROBLEM 6.5

Menthol has an optical rotation of $+2.46°$ when the measurement is made with a sodium lamp at 20 °C on a solution containing 5 g of menthol in 100 mL of ethanol solution, using a sample tube 10 cm long. What is the specific rotation of menthol?

PROBLEM 6.6

When the analyzer prism is rotated 90° clockwise, it arrives at the same position as if it were rotated counterclockwise 270° ($-270°$). If the optical rotation were being determined for the first time, how would it be possible to establish whether the rotation should be reported as $+90°$ or $-270°$?

A. The Addition of Hydrogen Bromide to 1-Butene

Addition of hydrogen bromide to 1-butene produces 2-bromobutane.

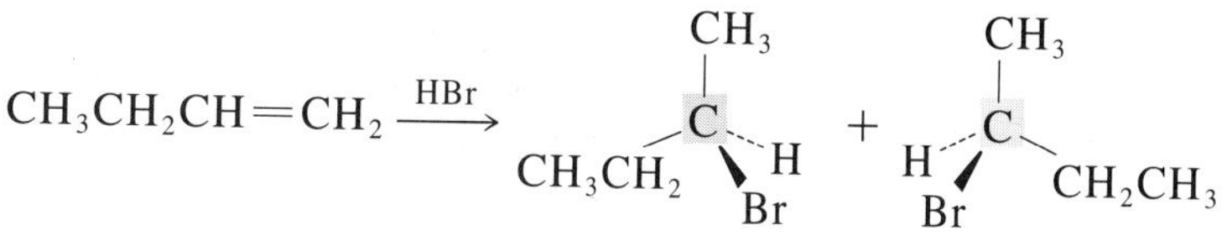
HBr

(−)-2-bromobutane (+)-2-bromobutane

An examination of the mechanism of this reaction shows that it should give equal amounts of the two stereoisomers. The reaction proceeds by way of a planar carbocation intermediate, a symmetrical achiral species. The bromide ion reacts with equal probability at either face of this cation.

VISUALIZING THE REACTION

Formation of enantiomers

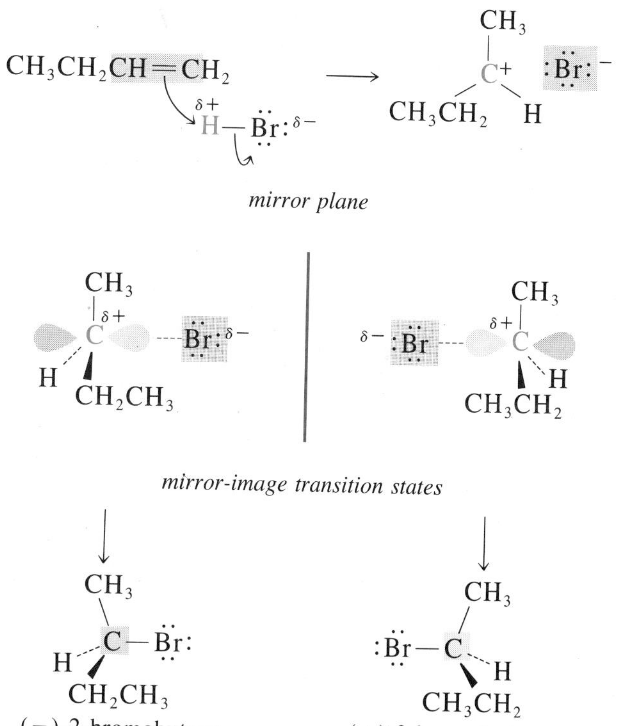
mirror plane
mirror-image transition states

(−)-2-bromobutane (+)-2-bromobutane

The transition state corresponding to the reaction of the cation with a bromide ion that approaches from the right is the enantiomer of the other transition state, which corresponds to the reaction of the cation with a bromide ion approaching from the left. The two transition states leading from the cation to the two different products are equal in energy. The two processes have the same energy of activation and, therefore, the same rate. (−)-2-Bromobutane and (+)-2-bromobutane are formed in equal amounts when hydrogen bromide adds to 1-butene. *Equal amounts of enantiomeric products are always formed when two achiral reagents react to give a chiral product.* The two enantiomers have the same physical properties, except for optical rotation, and cannot be separated from each other by ordinary physical methods, such as distillation.

The equations written in this section show the formation of both enantiomers of 2-bromobutane in order to increase your awareness of what is happening stereochemically in such a reaction. Normally, both enantiomers are not shown as products of a chemical reaction.

B. Racemic Mixtures and Enantiomeric Excess

The addition of hydrogen bromide to 1-butene results in the formation of a mixture consisting of equal amounts of (−)-2-bromobutane and (+)-2-bromobutane. Such a mixture is known as a racemic mixture. A **racemic mixture** contains equal numbers of molecules of two enantiomers and shows no optical rotation. Because a racemic mixture contains equal concentrations of molecules of opposite optical activity, plane-polarized light that is given a twist to the right by an encounter with a molecule of one enantiomer is twisted back by an encounter with a molecule of the mirror-image isomer. The plane of polarization remains unchanged as the light passes through the mixture. A racemic mixture is indicated by the use of the symbol (±) at the beginning of the name of the compound. For example, (−)-2-bromobutane has $[\alpha]_D^{22}$ −23.1° and (+)-2-bromobutane has $[\alpha]_D^{22}$ +23.1° when the determinations are done on the pure liquids. The mixture of equal quantities of the two enantiomers of 2-bromobutane obtained as the product of the addition of hydrogen bromide to 1-butene has $[\alpha]_D^{22}$ 0° and thus is not optically active. It is designated as (±)-2-bromobutane.

If one enantiomer of a pair is present to a greater extent, the mixture will show an optical rotation corresponding to the percentage of the species that is present in excess. The percentage of the enantiomer that is present in excess is known as the **enantiomeric excess.** It is calculated from a formula involving the rotation observed for a mixture and the known optical rotation of the pure enantiomer:

$$\frac{\text{measured specific rotation of mixture}}{\text{specific rotation for the pure enantiomer}} \times 100 = \%\ \text{enantiomeric excess}$$

Problem: A sample of 2-bromobutane has $[\alpha]_D^{22}$ +11.55°. The specific rotation of (+)-2-bromobutane at 22 °C is +23.1°. What is the enantiomeric excess of (+)-2-bromobutane in this sample?

Solution:

$$\frac{\text{measured specific rotation of mixture}}{\text{specific rotation of pure (+)-2-bromobutane}} \times 100 = \frac{+11.55}{+23.1} \times 100$$
$$= 50\%\ \text{enantiomeric excess of (+)-2-bromobutane}$$

The sample of 2-bromobutane described in the problem above is dextrorotatory because it has an excess of the dextrorotatory enantiomer. The exact composition of the mixture can be calculated from the enantiomeric excess. Leaving out the excess dextrorotatory molecules, the rest of the mixture has no optical rotation because it consists of equal numbers of dextrorotatory and levorotatory molecules. The optical rotations they cause cancel one another out.

Problem: If a mixture of 2-bromobutanes has an enantiomeric excess of 50% of (+)-2-bromobutane, what is the stereoisomeric composition of the mixture?

Solution: Of the total mixture, (100 − 50)% consists of equal numbers of dextrorotatory and levorotatory molecules. Thus, half of 50%, or 25%, of these molecules are dextrorotatory, and 25% of them are levorotatory. Therefore, the mixture is 25% (−)-2-bromobutane and 75% (50% enantiomeric excess + 25%) (+)-2-bromobutane.

A racemic mixture cannot be separated into its components by ordinary physical methods because the physical properties of the two components, except for optical activity, are identical. For example, (+)-2-bromobutane and (−)-2-bromobutane both boil at 91 °C. A separation of enantiomers must, therefore, always involve the use of chiral reagents, which will interact differently with molecules of differing chirality. Enzymes in biological systems are such reagents. One way of separating enantiomers is to use such living organisms to metabolize one form and leave the other form untouched.

A more general method for separating enantiomers, known as the resolution of a racemic mixture, involves the formation of compounds from the enantiomers by reaction of the mixture with a chiral reagent. Understanding this method requires that we first look at molecules containing more than one stereocenter. For that reason, we will postpone the discussion of this topic until Section 6.11.

Study Guide Concept Map 6.5

PROBLEM 6.7

(+)-2-Bromobutanoic acid,

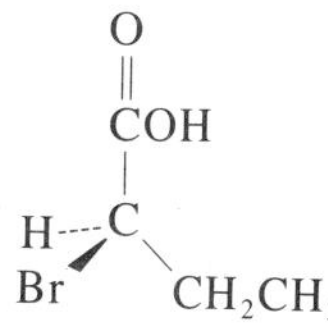
O
COH
H
C
Br
CH_2CH_3

has $[\alpha]_D^{25}$ +39.5° (*c* 9.1; ether). A sample of 2-bromobutanoic acid having $[\alpha]_D^{25}$ −14.70° was recovered by resolution of a racemic mixture of the acid.

(a) What is the enantiomeric excess of the sample of acid recovered from the racemic mixture?

(b) Draw the correct stereochemical formulas for (−)-2-bromobutanoic acid, and say what percentage of the mixture each enantiomer is.

6.6 THE DISCOVERY OF MOLECULAR DISSYMMETRY

Optical activity was known early in the nineteenth century to be a property of crystals, such as quartz, that are demonstrably dissymmetric (without symmetry) in appearance. In Paris in 1848, Louis Pasteur noticed that crystals of the sodium ammonium salt of (+)-tartaric acid, a by-product of winemaking, had dissymmetric crystals that could not be superimposed on their mirror images. He thought that this crystalline dissymmetry might indicate a similar lack of symmetry in the molecules of (+)-tartaric acid. Another form of tartaric acid, known as paratartaric acid, did not rotate the plane of polarized light. Pasteur studied the sodium ammonium salt of paratartaric acid, expecting to find that its crystals were symmetrical. Instead he saw that some of the crystals had a right-handed appearance and others a left-handed one. The two crystalline forms were mirror images of each other.

Pasteur separated the two crystalline forms of sodium ammonium paratartrate with tweezers as he looked through a microscope. A solution of the right-handed crystals in water rotated the plane of polarized light to the right, exactly the way the (+)-tartaric acid salt did. The left-handed crystals gave a solution with an optical rotation of equal magnitude but opposite sign. A mixture of equal weights of the two crystal forms, when dissolved in water, gave a solution with no optical rotation. With these experiments, Pasteur demonstrated that optical activity was the result of a molecular property that survived even when the crystal form was destroyed by dissolving it. He saw molecular dissymmetry as the cause of the phenomenon of optical activity, and he recognized that there were two dissimilar molecular forms of optically active tartaric acid.

In further experiments in 1854, Pasteur showed that the microorganism *Penicillum glaucum* consumed (+)-tartaric acid but not (−)-tartaric acid. His work with optically active compounds led Pasteur to say, "Life is dominated by dissymmetrical actions. I can even foresee that all living species are primordially in their structure, in their external forms, functions of cosmic dissymmetry."

Although the phenomenon of molecular dissymmetry was recognized in the 1840s, there was no clear picture of how it came about until 1874. Up to that time, chemists were still struggling with ways to represent molecular structures and had not yet made clear distinctions between structural isomers and stereoisomers. The idea that groups around a carbon atom are arranged in a tetrahedron was suggested in 1874 by the Dutch chemist J. H. van't Hoff, who was twenty-two years old at the time. He recognized that it was necessary to think of structures in three dimensions in order to solve the problems of isomerism that were being discovered in the laboratory. A carbon atom with four different substituents arranged tetrahedrally around it would account for the two isomers observed experimentally for compounds with a single stereocenter. The right-handed and left-handed arrangements that are possible for four groups around a tetrahedral carbon atom could be used to explain the phenomenon of optical activity.

The French chemist J. A. Le Bel started with the work of Pasteur and also arrived at the idea that a carbon atom with four different substituents around it is the basis for optical activity in organic molecules. He published his ideas in 1874, the same year as van't Hoff did. Le Bel emphasized the lack of symmetry, in particular, the lack of a plane of symmetry, at the molecular level as a necessary condition for optical activity. He hit upon the idea of a tetrahedral carbon atom by exploring the number of isomers that are formed as one, two, and then three different groups are substituted on a carbon that originally had four identical groups on it. The earlier discussion (Section 6.1) about creating isomers by substituting the hydrogen atoms on butane follows very closely his way of thinking about stereoisomerism. Le Bel's approach differed from van't Hoff's in that van't Hoff took a tetrahedral carbon atom as his starting point.

Le Bel's ideas remained more abstract than van't Hoff's. For example, van't Hoff created molecular models of tetrahedral carbon atoms and sent them to leading chemists of the time in an effort to gain acceptance of his ideas. He drew structural formulas in perspective to make his ideas clear. He made predictions about optical activity or the lack of it for compounds yet to be investigated, predictions that were found to be correct when the experimental work was done. In spite of this, his ideas were not generally accepted for a number of years. The opposition got some support from conflicting experimental results. In those days, many organic compounds were isolated from natural sources and contained small amounts of optically active compounds as impurities. Because of this contamination, confusing data were obtained that showed the presence of optical activity for compounds that did not have stereocenters. But van't Hoff himself supervised much of the experimental work that proved that pure compounds were not optically active unless molecular dissymmetry was present. In 1901, he received the first Nobel Prize in Chemistry for his work in this and other areas.

By the end of the nineteenth century, the tetrahedral carbon atom was accepted as the basis of structural organic chemistry. Much of the research being done by that time on sugars, which contain several stereocenters, would not have been possible without this basis. The experimental results confirmed the correctness of van't Hoff's ideas. His predictions about the number of isomers that are possible and the kinds of compounds that should have optical activity were demonstrated to be true.

6.7 CONFIGURATION. REPRESENTATION AND NOMENCLATURE OF STEREOISOMERS

A. Configuration of Stereoisomers

The actual orientation in space of the groups around a stereocenter is called the **absolute configuration** of the compound. The determination of this configuration for a particular compound is not a trivial matter. Later chapters will discuss reactions that take place with known stereochemical results. Experimental evidence from many such reactions has been accumulated over the years, so the stereochemical relationships among many series of compounds are known. The ultimate determination of exactly how these molecules look depended on the development of sophisticated x-ray diffraction techniques for determining the structures of crystals. The solution of this problem, known as the determination of the absolute configuration of stereoisomers, was completed in 1951 (see Section 6.7C, p. 200).

The configuration of a compound is unchanged unless at least one bond at the stereocenter is broken. Thus, configuration must not be confused with conformation, which changes continuously at room temperature as a result of rotation about single bonds and the flipping of rings in molecules. A stereoisomer has a single configuration but may exist in a number of conformations, depending on the solvent and the temperature. For example, four representations of (−)-2-chlorobutane are shown below; three of them represent different conformations of the molecule and one of them a rotation of the whole molecule in space. In all of these, the configuration of the molecule at the stereocenter remains unchanged.

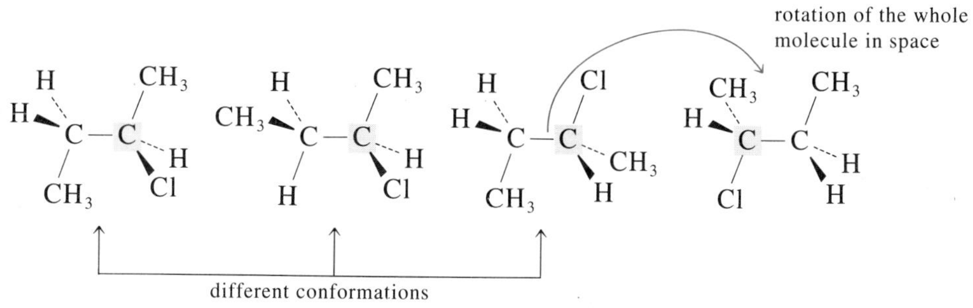

rotation of the whole molecule in space
different conformations

(−)-2-chlorobutane

Chemical reactions that involve breaking a bond at a stereocenter often result in a change in the configuration of a chiral compound. In the next chapter we will examine such reactions.

B. Nomenclature of Stereoisomers

The rules of nomenclature described in Section 5.8 are not adequate for naming stereoisomers. For example, unless you memorize the structures for (+)-2-chlorobutane and (−)-2-chlorobutane (p. 182), there is no way for you to draw a unique structure for each isomer from the names alone. There is no simple correlation between the sign of rotation and the structure. The sign of rotation by itself does not tell what the configuration of the compound is.

To solve this problem, another set of rules, the **Cahn-Ingold-Prelog rules,** were created. The configuration at a stereocenter is described as being ***R*****,** from the Latin *rectus* (or right-handed), or ***S*****,** from the Latin *sinister* (or left-handed), depending on the order in which the different substituents are arranged around the stereocenter. The rules that are applied to determine configuration are as follows:

1. Each group attached to the stereocenter is assigned a priority. The higher the atomic number of the atom bonded directly to the stereocenter, the higher the priority of the substituent; for example:

$$Cl > O > N > C > H$$

   Among isotopes, the one with the higher atomic weight takes priority. Thus, tritium, the isotope of hydrogen with an atomic weight of 3 has higher priority than deuterium, with an atomic weight of 2. Hydrogen, with an atomic weight of 1, has the lowest priority, not only of these three isotopes but in any case.

$$T > D > H$$

2. If two identical atoms are attached to the stereocenter, the next atoms in both chains are investigated, moving away from the stereocenter until some difference is found. A priority assignment is made at the first point at which atoms of different priorities are found.

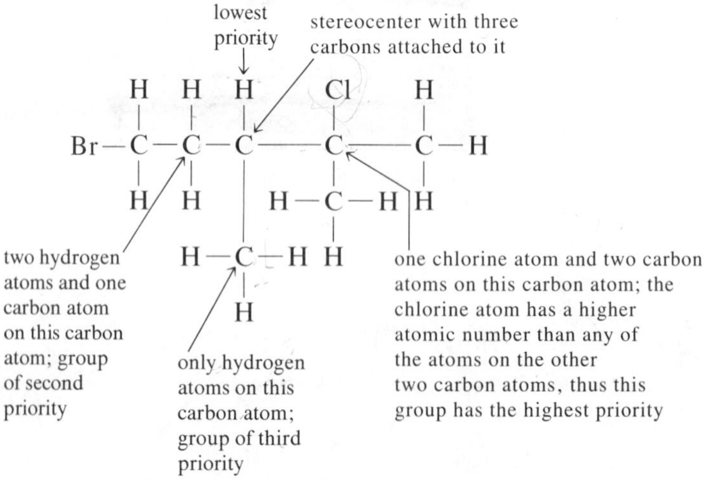
lowest priority
stereocenter with three carbons attached to it
two hydrogen atoms and one carbon atom on this carbon atom; group of second priority
only hydrogen atoms on this carbon atom; group of third priority
one chlorine atom and two carbon atoms on this carbon atom; the chlorine atom has a higher atomic number than any of the atoms on the other two carbon atoms, thus this group has the highest priority

Note that there is a bromine atom, an atom of higher atomic number than any other atom in the above molecule, at the end of the chain. This bromine atom does not influence the assignment of priorities because it is beyond the point of difference.

3. A double bond is counted as two single bonds for both of the atoms involved.

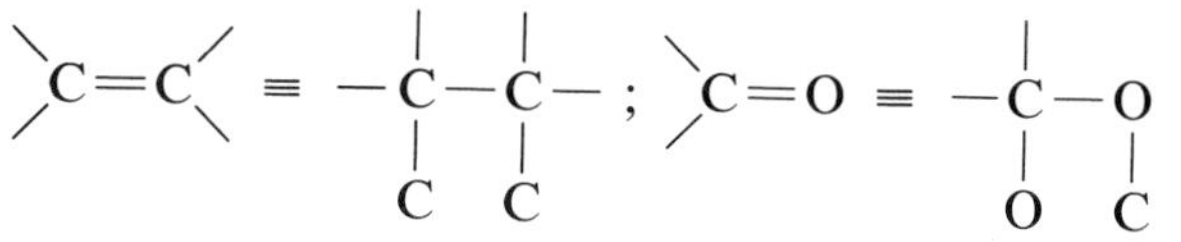

The same principle is extended to triple bonds.

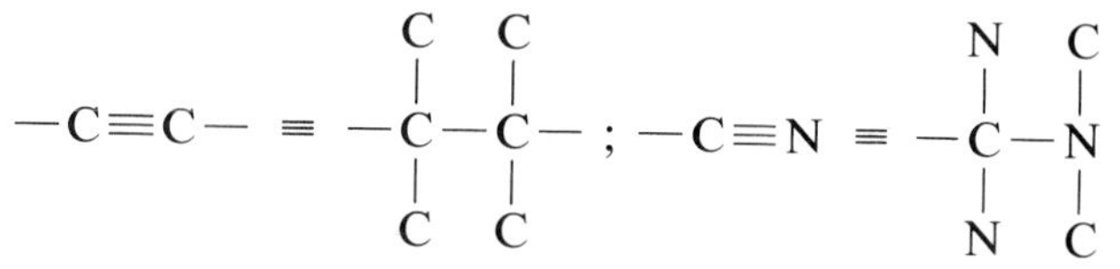

For example,

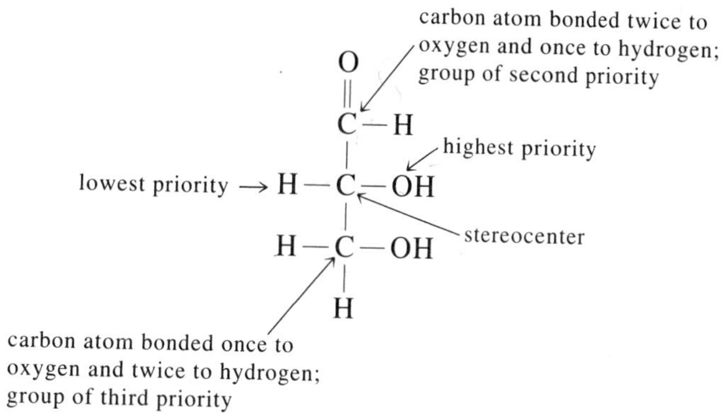
carbon atom bonded twice to oxygen and once to hydrogen; group of second priority
highest priority
lowest priority
stereocenter
carbon atom bonded once to oxygen and twice to hydrogen; group of third priority

4. After priorities have been assigned, the molecule is viewed with the substituent of lowest priority away from the viewer. If you can trace a clockwise path from the group of highest (or first) priority to the one of second priority and then to the one of third priority, the stereocenter is assigned the *R* configuration. If the arrangement of the groups in order of decreasing priority follows a counter-clockwise path, the configuration is *S*.

The Cahn-Ingold-Prelog rules can be clarified by applying them to assign configurations to stereoisomers of bromochloroiodomethane.

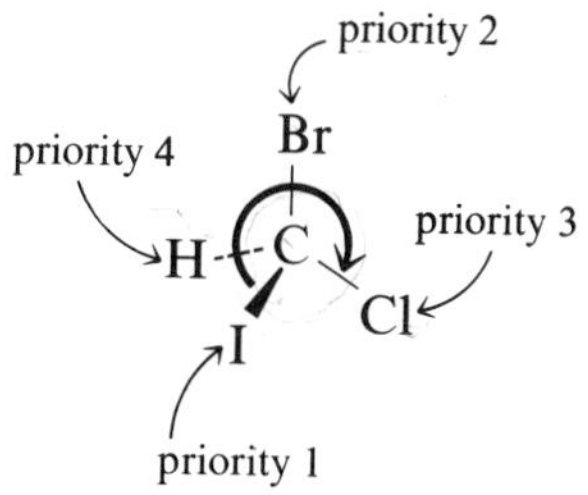
priority 2
priority 4
priority 3
priority 1

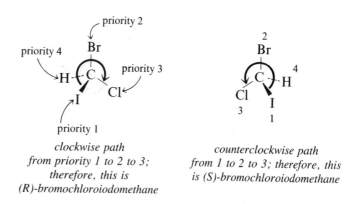

clockwise path from priority 1 to 2 to 3; therefore, this is (R)-bromochloroiodomethane

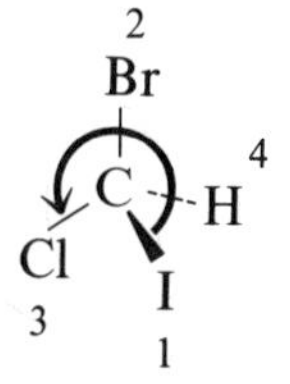

counterclockwise path from 1 to 2 to 3; therefore, this is (S)-bromochloroiodomethane

These rules can also be clarified by applying them to assign configurations to stereoisomers of 2-chlorobutane.

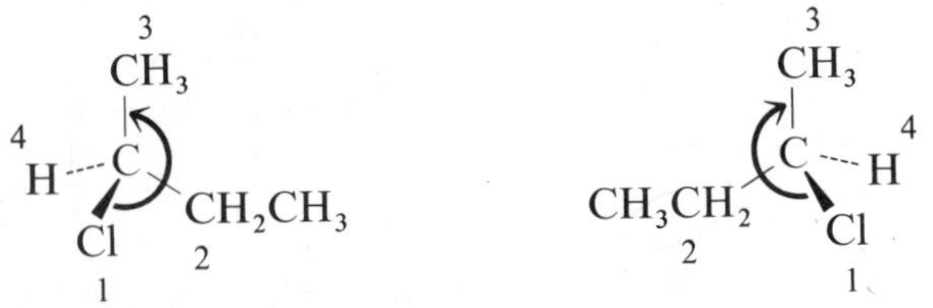

counterclockwise path from 1 to 2 to 3; therefore, this is (S)-(+)-2-chlorobutane

clockwise path from 1 to 2 to 3; therefore, this is (R)-(−)-2-chlorobutane

Note that the molecules are drawn so that the group of lowest priority is projecting away from the viewer. The assignment of priorities in bromochloroiodomethane is straightforward: the priorities are determined by looking up the atomic numbers for the atoms attached to the stereocenter. For 2-chlorobutane, two of the atoms bonded to the stereocenter are carbons. The ethyl group takes priority over the methyl group because the second carbon atom in the ethyl group has priority over any one of the three hydrogen atoms in the methyl group.

(+)-Alanine and (−)-lactic acid, two naturally occurring compounds, are assigned the configurations below.

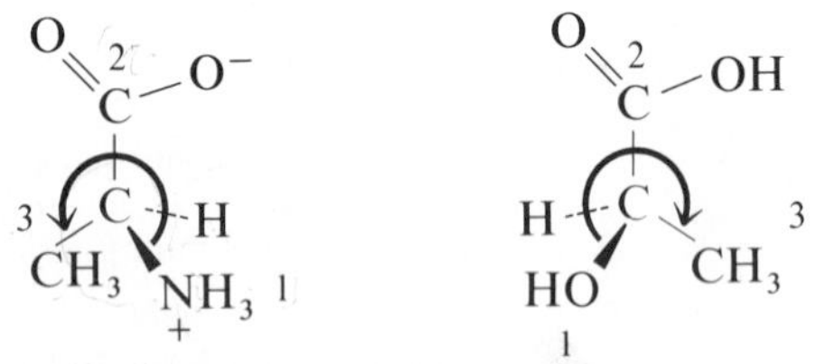

counterclockwise order of groups by priority
(*S*)-(+)-alanine

clockwise order of groups by priority
(*R*)-(−)-lactic acid

In alanine, the nitrogen atom takes priority over the methyl group and the carboxylic acid group. The carboxylic acid group, with oxygen bonded to its carbon atom, is of higher priority than the methyl group in which only hydrogens are bonded to the carbon atom. Thus, the enantiomer of alanine that has a positive sign of rotation is assigned the *S* configuration. Similar reasoning is used to assign the *R* configuration to the form of lactic acid that is levorotatory.

Note that the designation of a compound as *R* or *S* has nothing to do with the sign of rotation. The Cahn-Ingold-Prelog rules can be applied to any three-dimensional representation of a chiral compound to determine whether it is *R* or *S*. To find out whether a particular picture of a molecule corresponds to the dextrorotatory or levorotatory form requires much experimental work and cannot be determined just by looking at the structural formula. If the sign of rotation of one enantiomer is known, however, the other enantiomer will have the opposite sign of rotation. For example, knowing that (*S*)-(+)-alanine and (*R*)-(−)-lactic acid have the structures shown above means that the following must be the configurations of (*R*)-(−) alanine and (*S*)-(+)-lactic acid.

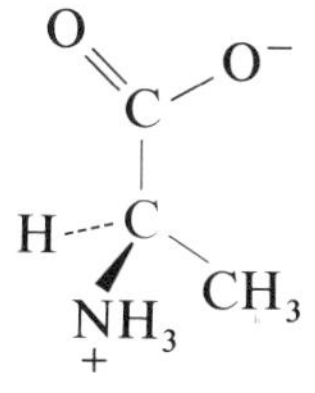

(*R*)-(−)-alanine
enantiomer of
(*S*)-(+)-alanine

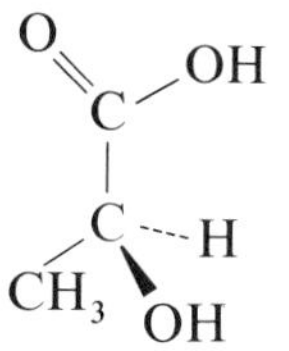

(*S*)-(+)-lactic acid
enantiomer of
(*R*)-(−)-lactic acid

The use of *R* or *S* in the name of a compound assigns a particular configuration, a specific orientation in space, to the atoms in the molecule. Thus, the Cahn-Ingold-Prelog rules enable chemists to describe the stereochemistry of a compound without having to draw a three-dimensional picture of the molecule. For example, the name (*R*)-2-bromo-1-propanol fully describes the orientation of the groups around the stereocenter in that compound.

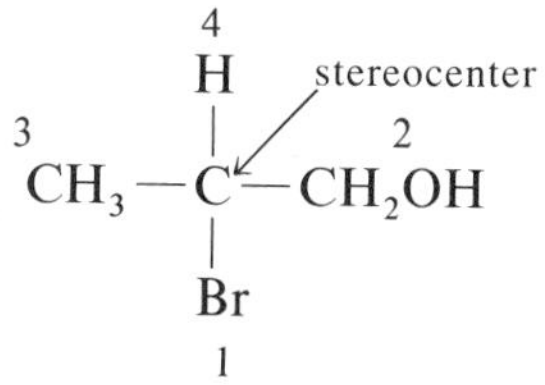
stereocenter

2-bromo-1-propanol
(no stereochemistry shown)
priorities assigned to the groups at the stereocenter

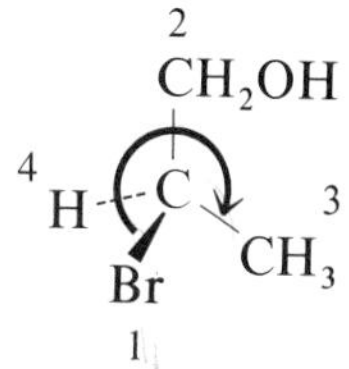

(*R*)-2-bromo-1-propanol
one possible three-dimensional picture showing stereochemistry

The Cahn-Ingold-Prelog rules fulfill the requirement for a good system of nomenclature: each name corresponds to a unique structure that can be reproduced from the name alone.

PROBLEM 6.8

Name each of the following compounds, including an assignment of configuration.

(a) 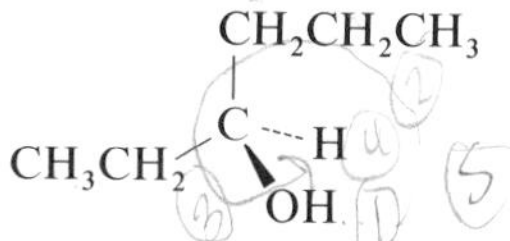

(b) 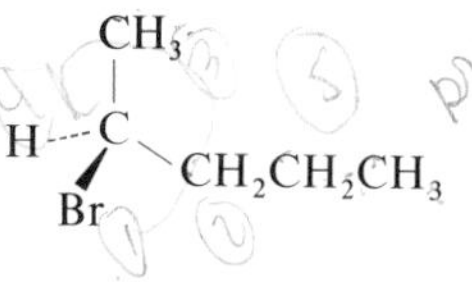

(c) 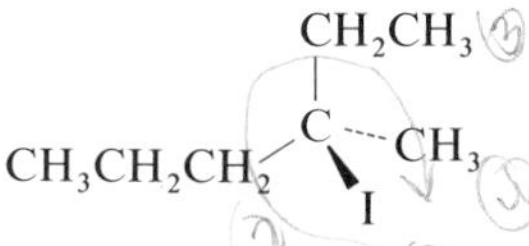

(d) 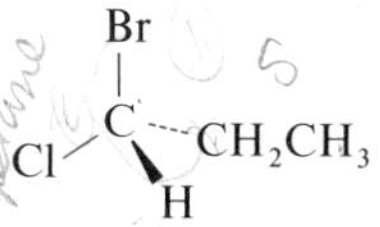

(e) 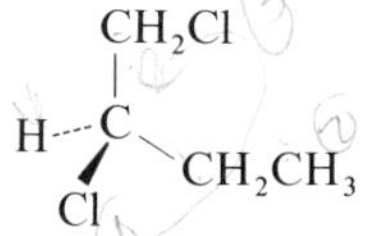

(f) 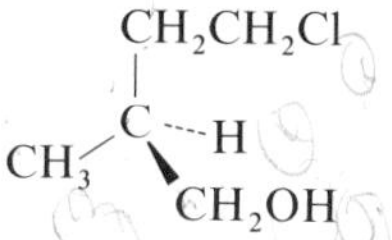

(g) 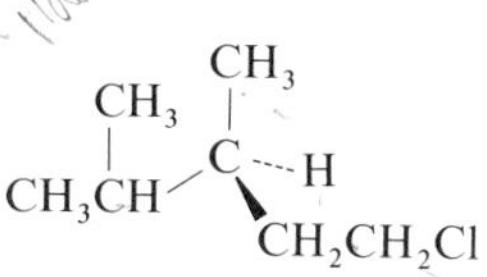

(h) 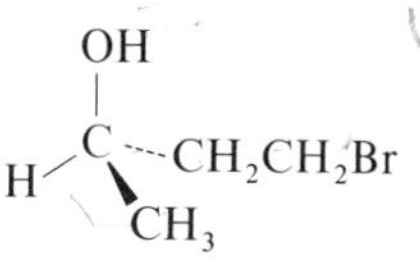

PROBLEM 6.9

Draw three-dimensional structural formulas for the following compounds.

(a) (*R*)-1-bromo-1-chloroethane (b) (*S*)-1-chloro-2-propanol
(c) (*S*)-2,3-dimethylpentane (d) (*R*)-2-butanol
(e) (*R*)-ethanol-1-*d* (CH_3CHDOH) (f) (*S*)-1-bromo-3-chloro-2-methylpropane

C. Relative and Absolute Configuration

The phenomenon of the optical activity of natural products, especially sugars, was known by chemists early in the nineteenth century. For over a century, however, the problem of determining absolute configuration (p. 195) stumped scientists who were working with optically active compounds. For example, the simplest optically active sugar is 2,3-dihydroxypropanal, usually called glyceraldehyde. It has a single stereocenter and exists as two enantiomers, (+)-glyceraldehyde and (−)-glyceraldehyde. The configurations of the two enantiomers are shown below.

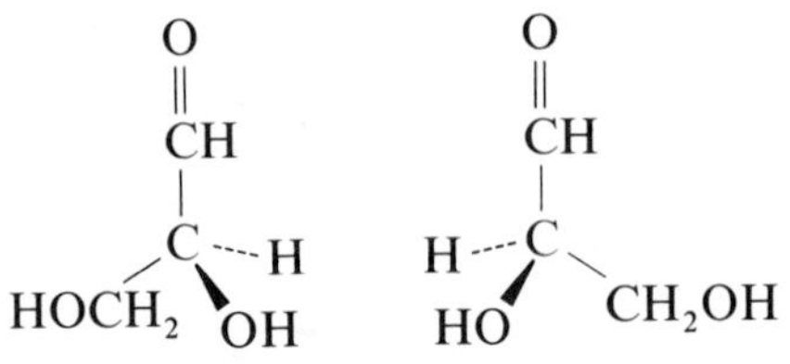
O
CH
C
H
$HOCH_2$
OH
O
CH
H
C
HO
CH_2OH

(*S*)-glyceraldehyde (*R*)-glyceraldehyde

There is no way of knowing just by looking at these structures whether the *R* or the *S* isomer corresponds to (+)-glyceraldehyde. In other words, the actual arrangement of the atoms around the stereocenter, the absolute configuration of a compound, cannot be determined from the sign of the optical rotation.

Chemists recognized that they could relate the configurations of various optically active compounds to each other by chemical interconversions, even when they did not know absolute configurations. Some of the ways in which this was done in the 1930s and 1940s are described in Chapter 7. Thus, chemists came to know the relative configurations of large numbers of compounds. In 1951, the absolute configuration of (+)-tartaric acid was determined by the Dutch chemist J. M. Bijvoet and his colleagues, using a sophisticated modification of x-ray diffraction. Their work was done in the laboratory named for van't Hoff, who had recognized that the existence of optically active enantiomers could be explained by postulating a tetrahedral carbon atom as the stereocenter (p. 186).

It was only after Bijvoet's determination that chemists were able to assign actual three-dimensional structures to the chiral compounds that they worked with. For example, it is now known that (+)-glyceraldehyde has the *R* configuration.

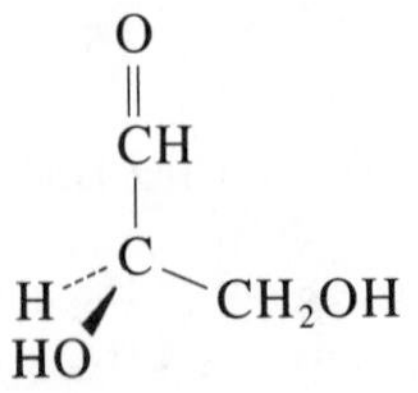
O
CH
C
H
CH_2OH
HO

(*R*)-(+)-glyceraldehyde

The story of how this came about is part of the history of the determination of the structure of glucose (Chapter 25).

(*R*)-(+)-Glyceraldehyde, in turn, has served as a reference compound for the assignment of configuration to many compounds. If the name of a compound includes both the sign of rotation and the designation *R* or *S*, then the actual three-dimensional arrangement of its atoms has been determined by careful experiments. In other words, the absolute configuration of that particular compound is known.

6.8 DIASTEREOMERS

A. Compounds with More Than One Stereocenter

Replacement of one of the secondary hydrogen atoms in butane with a chlorine atom creates one of the enantiomeric 2-chlorobutanes (Section 6.1, p. 181). Replacement of either of the hydrogen atoms on carbon atom 3 in each of the enantiomers of 2-chlorobutane with a bromine atom is shown below.

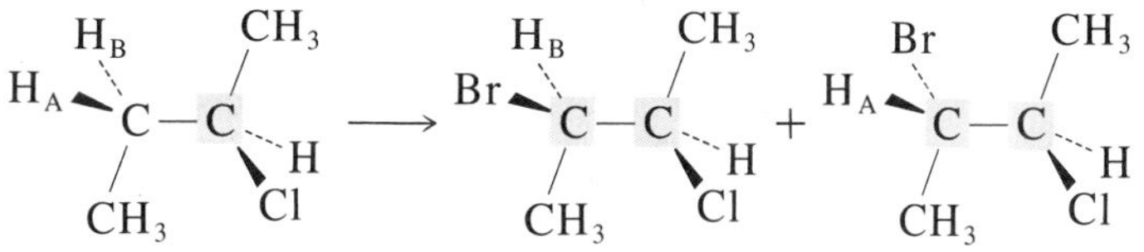

(*R*)-2-chlorobutane — (2*R*,3*R*)-2-bromo-3-chlorobutane *from replacement of* H_A + (2*S*,3*R*)-2-bromo-3-chlorobutane *from replacement of* H_B

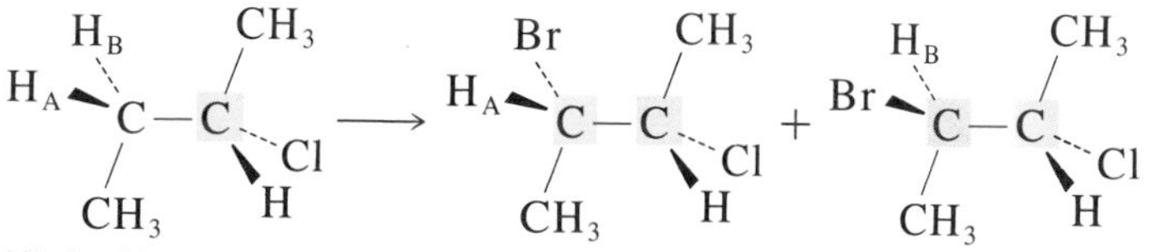

(*S*)-2-chlorobutane — (2*S*,3*S*)-2-bromo-3-chlorobutane enantiomer of (2*R*,3*R*)-2-bromo-3-chlorobutane + (2*R*,3*S*)-2-bromo-3-chlorobutane enantiomer of (2*S*,3*R*)-2-bromo-3-chlorobutane

Note that the original stereocenter is left undisturbed as one or the other of the two hydrogen atoms on carbon atom 3 is replaced by a bromine atom. Four compounds result. *The maximum number of stereoisomers for a compound with* n *stereocenters is* 2^n. A compound with two stereocenters will have 2^2, or 4, stereoisomers.

A close examination of the four 2-bromo-3-chlorobutanes reveals that they are two pairs of enantiomers. Stereoisomers that are not enantiomers are called **diastereomers.** Therefore, each of the four is an enantiomer of one of the other three compounds and a diastereomer of the other two.

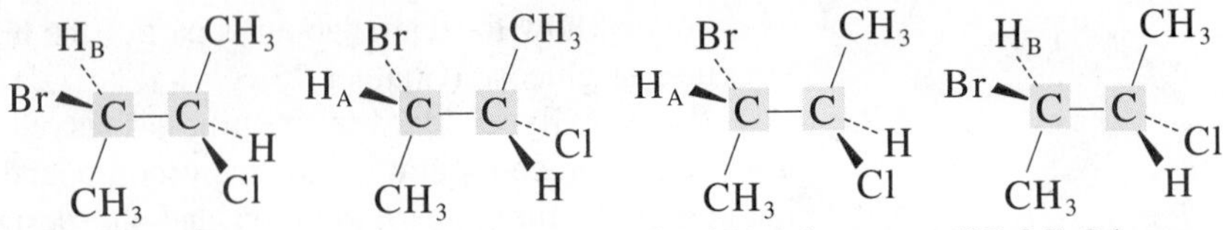

(2*R*,3*R*)-2-bromo-3-chlorobutane (2*S*,3*S*)-2-bromo-3-chlorobutane (2*S*,3*R*)-2-bromo-3-chlorobutane (2*R*,3*S*)-2-bromo-3-chlorobutane

enantiomers of each other
diastereomers of the compounds to the right

enantiomers of each other
diastereomers of the compounds to the left

The stereochemical relationships are apparent not only from the perspective formulas, but also from the names of the compounds. The enantiomers have the opposite configuration at each of the stereocenters. The diastereomers have the same configuration at one of the two centers and the opposite configuration at the other. You should examine these structures and their names until you convince yourself of these relationships.

Unlike enantiomers (p. 192), diastereomers have different physical properties, such as boiling points, melting points, dipole moments, and solubilities. Diastereomers can be separated from each other by the usual means of purification, such as fractional distillation. For example, a mixture of the four 2-bromo-3-chlorobutanes shown above could be separated into two racemic mixtures, each one consisting of a pair of enantiomers.

The examples in this section illustrate the stereochemical relationships that arise when two different stereocenters are present in a molecule. In the next section, we will explore the stereochemistry of a compound that contains two stereocenters that have the same substituents.

PROBLEM 6.10

Draw and name all the stereoisomers that are possible for each of the following compounds. Identify the enantiomers and diastereomers.

(a) $CH_3CH_2CHCH_2CH_2CH_3$
$\quad\quad\quad\quad|$
$\quad\quad\quad CH_3$

(b) $CH_3CH_2CHCH_2CHCH_3$
$\quad\quad\quad\quad|\quad\quad\quad|$
$\quad\quad\quad CH_3\quad Br$

(c) $CH_3CH_2CHCHCH_3$
$\quad\quad\quad\quad|\ \ |$
$\quad\quad\quad Br\ Br$

PROBLEM 6.11

Draw and name all the stereoisomers that are possible for 3-bromo-2-butanol and 2-chloro-3-methylheptane. Identify the enantiomers and diastereomers.

B. Compounds Containing Two Stereocenters with Identical Substituents. Meso Forms

The substitution of a bromine atom on the enantiomeric 2-chlorobutanes (described in Section 6.8A) can be done with a chlorine atom instead.

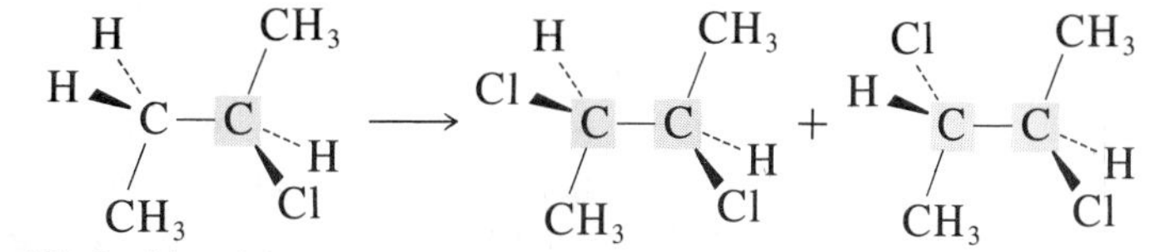

(*R*)-2-chlorobutane (2*R*,3*R*)-2,3-dichlorobutane (2*R*,3*S*)-2,3-dichlorobutane

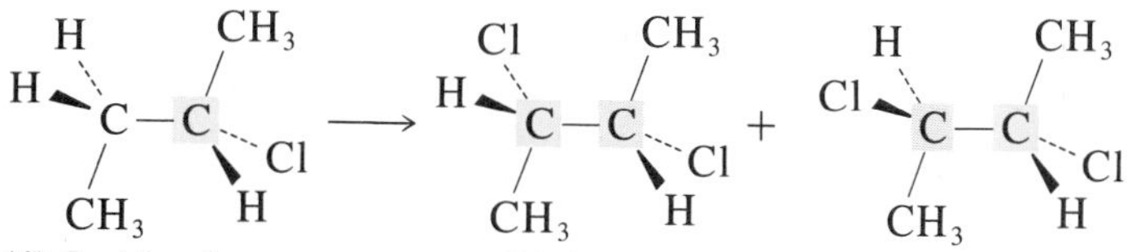

(*S*)-2-chlorobutane (2*S*,3*S*)-2,3-dichlorobutane (2*S*,3*R*)-2,3-dichlorobutane

enantiomer of (2R,3R)-2,3-dichlorobutane *apparent enantiomer of (2R,3S)-2,3-dichlorobutane*

It appears that four stereoisomers are formed; that is, two pairs of enantiomers. But a closer examination of the structures labeled (2*R*,3*S*)-2,3-dichlorobutane and (2*S*,3*R*)-2,3-dichlorobutane reveals that they are representations of the same molecule. These two structures are superimposable mirror images of each other. Two rotations of one of the structures in space makes it identical with the other one.

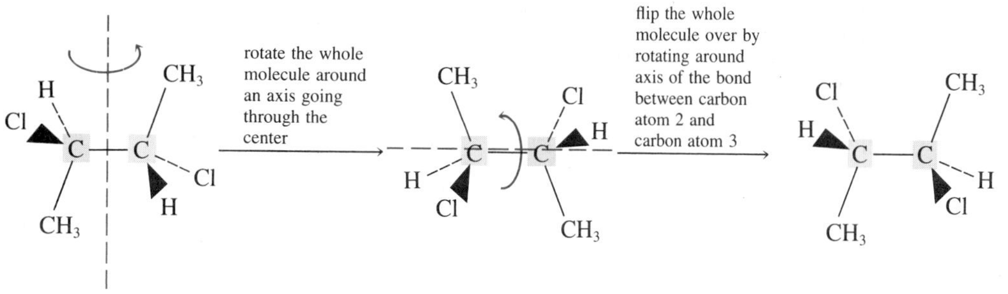

rotate the whole molecule around an axis going through the center
flip the whole molecule over by rotating around axis of the bond between carbon atom 2 and carbon atom 3

(2*S*,3*R*)-2,3-dichlorobutane (2*R*,3*S*)-2,3-dichlorobutane

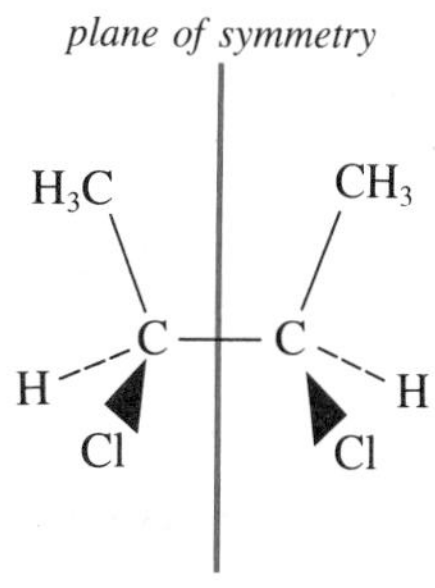

plane of symmetry

eclipsed conformation of (2*R*,3*S*)-2,3-dichlorobutane

The eclipsed conformation of (2*R*,3*S*)-2,3-dichlorobutane has a plane of symmetry because the two stereocenters have identical substituents. One half of the molecule is the mirror image of the other half. The molecule is not chiral and has no optical activity. It is known as a meso form. A **meso form** of a compound has stereocenters but no net chirality because of the existence of internal symmetry.

2,3-Dichlorobutane has three stereoisomers. They are the enantiomers (2*R*,3*R*)-2,3-dichlorobutane and (2*S*,3*S*)-2,3-dichlorobutane and the meso form *meso*-2,3-dichlorobutane, which can also be called either (2*R*,3*S*)- or (2*S*,3*R*)-2,3-dichlorobutane. The meso form is a diastereomer of either one of the enantiomers. In this case, because of symmetry in the compound, we find fewer than the 2^n or 4 stereoisomers expected of a compound having two stereocenters.

Study Guide Concept Map 6.6

PROBLEM 6.12

Draw perspective formulas for all stereoisomers possible for 2,3-butanediol, 2,3-pentanediol, and 2,4-pentanediol. Identify the enantiomers and diastereomers.

PROBLEM 6.13

Pasteur isolated the stereoisomers of tartaric acid, and the compound was used to establish absolute configuration for the first time (Chapter 25). Tartaric acid has the formula

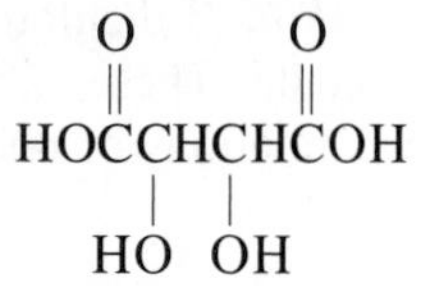
O
O
HOCCHCHCOH
HO
OH

Its IUPAC name is 2,3-dihydroxybutanedioic acid. (+)-Tartaric acid is (2*R*,3*R*)-2,3-dihydroxybutanedioic acid. Draw (+)-tartaric acid and any other possible stereoisomers for tartaric acid.

PROBLEM 6.14

How many stereoisomers are possible for a compound with three different stereocenters? 2,3,4-Tribromohexane is such a compound. Draw all possible stereoisomers for the compound, and identify the enantiomers and diastereomers. One stereoisomer is shown below to demonstrate how they can be drawn.

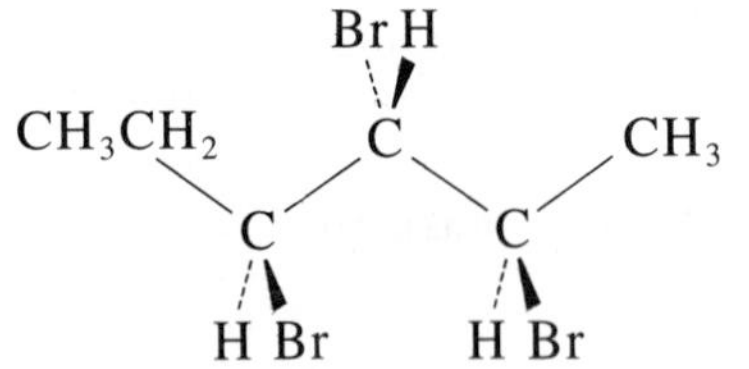
Br H
CH_3CH_2
C
CH_3
C
C
H Br
H Br

(2*R*,3*R*,4*S*)-2,3,4-tribromohexane

6.9 STEREOISOMERISM IN CYCLIC COMPOUNDS

A. Cis and Trans Compounds

The addition of bromine to the π bond in cyclopentene, a typical reaction of alkenes (Section 2.9), gives 1,2-dibromocyclopentanes in which the two bromine atoms are on opposite sides of the plane of the ring.

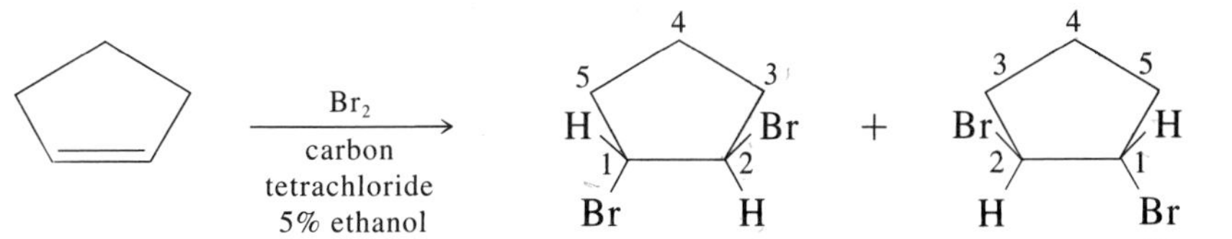

4
5
3
H
Br
+
Br
H
1
2
Br
H
Br_2
carbon
tetrachloride
5% ethanol
−8 °C

cyclopentene

(1*R*,2*R*)-1,2-dibromocyclopentane

(1*S*,2*S*)-1,2-dibromocyclopentane

trans-1,2-dibromocyclopentane

Two stereoisomers, enantiomers of each other, are formed in equal amounts in the reaction, resulting in a racemic mixture.

The rigidity of the ring and the lack of free rotation around the carbon-carbon bonds (p. 166) gives rise to stereoisomerism. Two kinds of 1,2-dibromocyclopentanes are possible. The two substituents may be on opposite sides of the plane of the ring, or **trans** to each other. Alternatively, they may be on the same side of the plane of the ring, or **cis** to each other (Figure 6.9).

Each of the *trans*-1,2-dibromocyclopentanes has two stereocenters; one isomer has the *R* configuration and the other has the *S* configuration at each stereocenter. Thus, the two isomers are mirror-image isomers, or enantiomers, of each other. The molecules of the isomers are not superimposable on each other and are, therefore, truly different species.

cis-1,2-Dibromocyclopentane is a stereoisomer of the two trans compounds, but is not a mirror image of either one. Having the *R* configuration at one stereocenter and the *S* configuration at the other one, it is a diastereomer (p. 201) of the trans compounds and a meso compound (p. 204). A plane of symmetry bisects carbon atom 4, the two hydrogen atoms on it, and the carbon-carbon bond between carbon atoms 1 and 2. *cis*-1,2-Dibromocyclopentane is not chiral and has no optical activity (Figure 6.10).

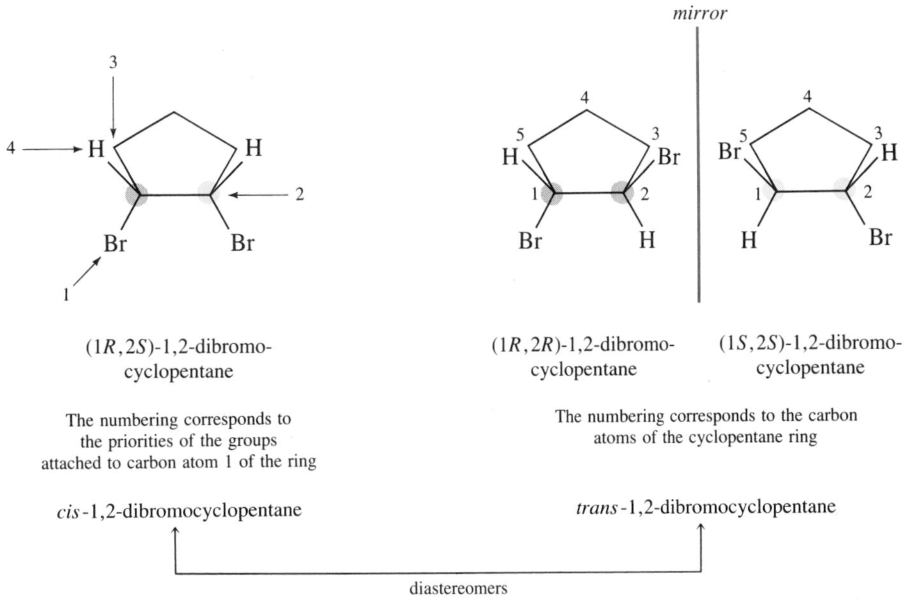

mirror
3
4
H
H
2
Br
Br
1
4
5
3
H
Br
1
2
Br
H
4
5
3
Br
H
1
2
H
Br
(1R,2S)-1,2-dibromo-
cyclopentane
(1R,2R)-1,2-dibromo-
cyclopentane
(1S,2S)-1,2-dibromo-
cyclopentane
The numbering corresponds to
the priorities of the groups
attached to carbon atom 1 of the ring
The numbering corresponds to the carbon
atoms of the cyclopentane ring
cis-1,2-dibromocyclopentane
trans-1,2-dibromocyclopentane
diastereomers

FIGURE 6.9 Different stereoisomers of 1,2-dibromocyclopentane.

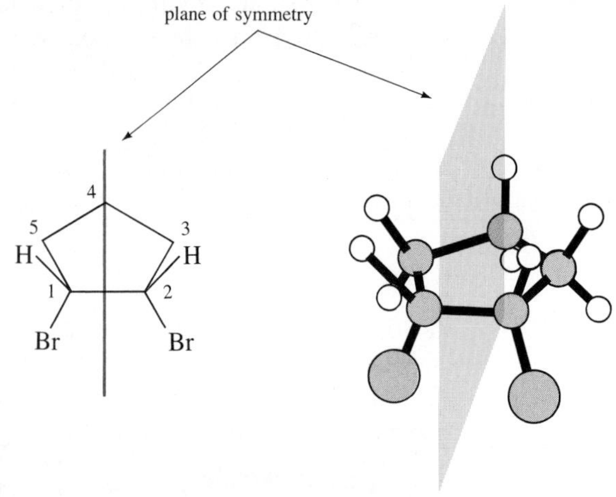
plane of symmetry
4
5
3
H
H
1
2
Br
Br

cis-1,2-dibromocyclopentane
achiral
a meso compound

FIGURE 6.10 The symmetry of *cis*-1,2-dibromocyclopentane.

PROBLEM 6.15

How many stereoisomers are possible for 1,3-dimethylcyclopentane? Draw the structural formulas, name the compounds, and discuss the stereochemical relationships among them.

PROBLEM 6.16

How many stereoisomers are possible for 1-bromo-2-methylcyclopentane? Draw structural formulas, and give them names that designate configurations at the stereocenters. You may want to use molecular models.

B. Configuration and Conformation of Disubstituted Cyclohexanes

For a cyclohexane with a single substituent on the ring, the preferred conformation is the chair form in which the substituent is in the equatorial position (p. 168). When there are two substituents on the ring, they may be either cis or trans to each other. For example, bromine adds to cyclohexene to give a mixture of enantiomeric *trans*-1,2-dibromocyclohexanes.

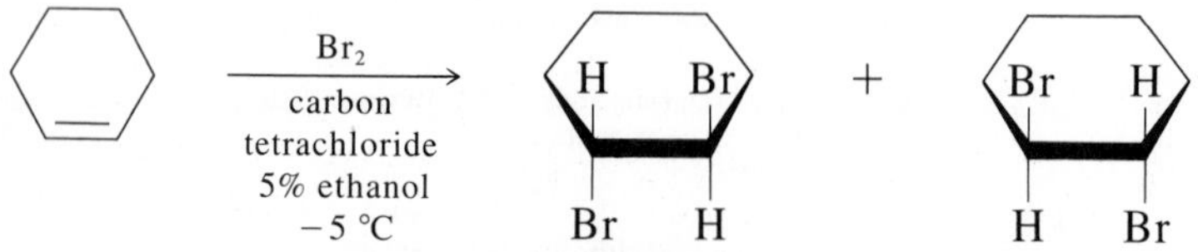
Br_2
carbon
tetrachloride
5% ethanol
−5 °C
H
Br
Br
H
+
Br
H
H
Br

cyclohexene

(1*R*,2*R*)-1,2-dibromocyclohexane

(1*S*,2*S*)-1,2-dibromocyclohexane

trans-1,2-dibromocyclohexane
95%

As in the *trans*-1,2-dibromocyclopentanes discussed in the preceding section, the two bromine atoms are on opposite sides of the ring. Because of the flexibility of the ring in cyclohexane and the interconversion of the ring between two chair forms, each enantiomer of 1,2-dibromocyclohexane exists mainly in two conformations. In one, the two bromine atoms occupy equatorial positions on the ring; in the other, the two bromine atoms are both in axial positions (Section 5.9C). The conformers of (1*S*,2*S*)-1,2-dibromocyclohexane are shown below.

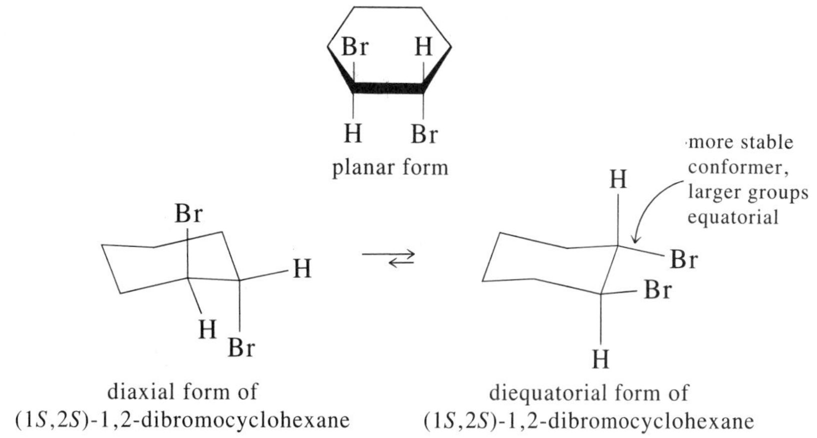

Br
H
H
Br
planar form
more stable conformer, larger groups equatorial
Br
H
H
Br
H
Br
Br
H
diaxial form of (1*S*,2*S*)-1,2-dibromocyclohexane
diequatorial form of (1*S*,2*S*)-1,2-dibromocyclohexane

The diequatorial form of *trans*-1,2-dibromocyclohexane predominates in the equilibrium mixture at room temperature.

In *cis*-1,2-dibromocyclohexane, one bromine atom is axial, and the other one is equatorial. Flipping of the ring puts the bromine atom that was originally equatorial into the axial position, and the one that was originally axial becomes equatorial.

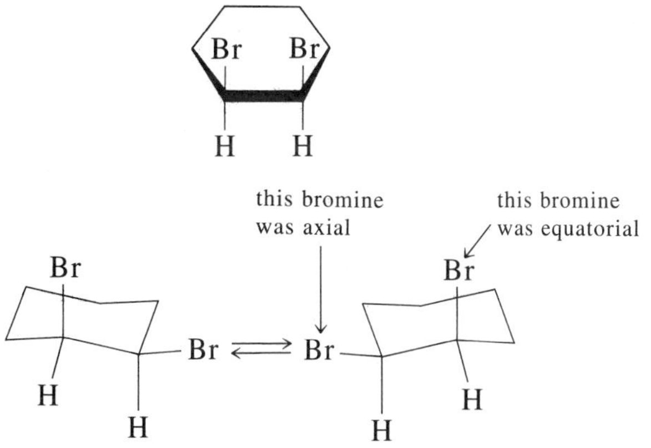

Br
Br
H
H
this bromine was axial
this bromine was equatorial
Br
Br
Br
Br
H
H
H
H

cis-1,2-dibromocyclohexane

These two conformers have equal energy and are present in equal amounts. Although each conformer is chiral, the rapid interconversion between the two at room temperature means that they cannot be separated. The compound is not optically active. *cis*-1,2-Dibromocyclohexane is a diastereomer of each of the *trans*-1,2-dibromocyclohexanes.

For 1,3-disubstituted cyclohexanes, the cis isomer is the more stable one. *cis*-1,3-Dimethylcyclohexane has two conformations, one in which both alkyl

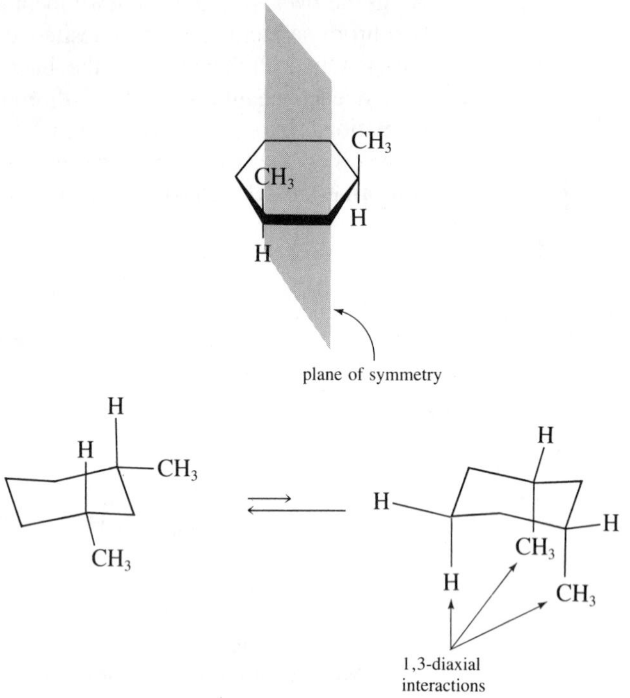
CH3
CH3
H
H
plane of symmetry
H
H
CH3
CH3
H
H
CH3
H
CH3
H
1,3-diaxial interactions

cis-1,3-dimethylcyclohexane

FIGURE 6.11 Conformers and symmetry of *cis*-1,3-dimethylcyclohexane.

groups are equatorial and one in which both are axial (Figure 6.11). The di-equatorial conformer of *cis*-1,3-dimethylcyclohexane is more stable than the diaxial one by about 5.4 kcal/mol, largely because of 1,3-diaxial interactions (p. 169) of the two methyl groups. *cis*-1,3-Dimethylcyclohexane has a plane of symmetry and is a meso compound.

trans-1,3-Dimethylcyclohexane exists as two enantiomers. Each enantiomer has one axial and one equatorial methyl group (Figure 6.12).

Neither *cis*- nor *trans*-1,4-dimethylcyclohexane has stereocenters, and both are optically inactive. Starting with either carbon atom 1 or carbon atom 4, an investigation of the substituents on that carbon atom turns up a hydrogen atom, a methyl group, and then two branches of the ring that are identical going in either direction. The molecules as a whole have planes of symmetry bisecting carbon atoms 1 and 4 and the methyl group and hydrogen atom on each one.

trans-1,4-Dimethylcyclohexane has two conformers, one in which the two methyl groups are equatorial and one in which they are axial (Figure 6.13). The form with two equatorial substituents is more stable than the other by about 3.6 kcal/mol.

In *cis*-1,4-dimethylcyclohexane, one methyl group is axial and the other equatorial (Figure 6.14). Flipping the ring gives a form that is identical with the starting structure.

In the above examples, the two substituents on the ring were identical, giving rise to symmetry in *cis*-1,2 and *cis*-1,3 compounds and reducing the number of

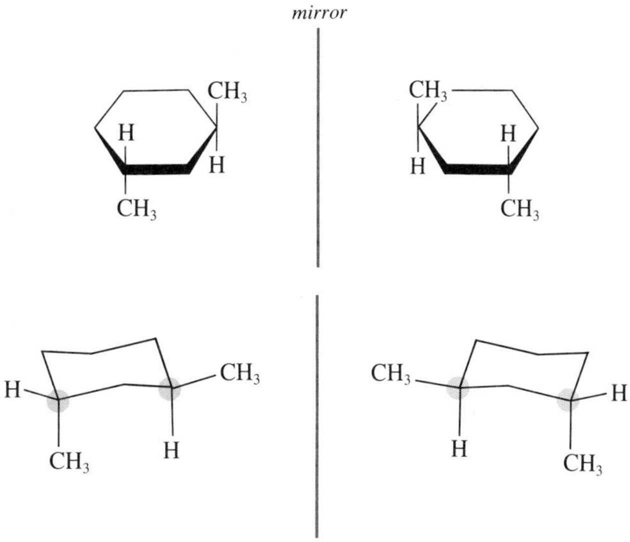
mirror
CH3
H
H
CH3
CH3
H
H
CH3
H
CH3
CH3
H
CH3
H
H
CH3

(1*R*,3*R*)-1,3-dimethylcyclohexane (1*S*,3*S*)-1,3-dimethylcyclohexane

FIGURE 6.12 The enantiomeric *trans*-1,3-dimethylcyclohexanes.

stereoisomers. If two different stereocenters are present, then there will be four stereoisomers. For example, 1-bromo-3-methylcyclohexane has four stereoisomers, two enantiomeric cis isomers and two enantiomeric trans isomers (Figure 6.15).

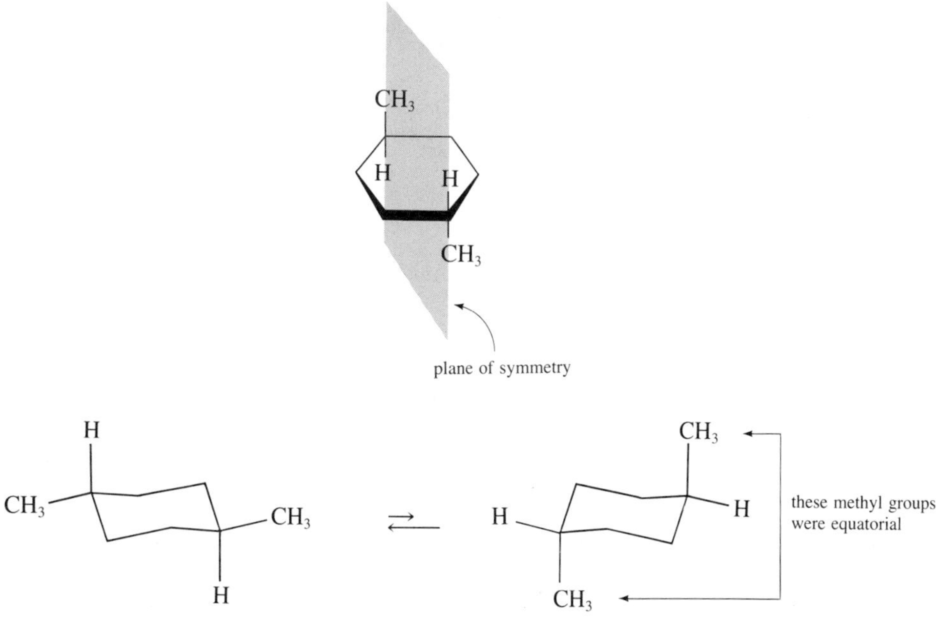
CH3
H
H
CH3
plane of symmetry
H
CH3
CH3
H
CH3
H
H
CH3
these methyl groups were equatorial

trans-1,4-dimethylcyclohexane

FIGURE 6.13 Conformers and symmetry of *trans*-1,4-dimethylcyclohexane.

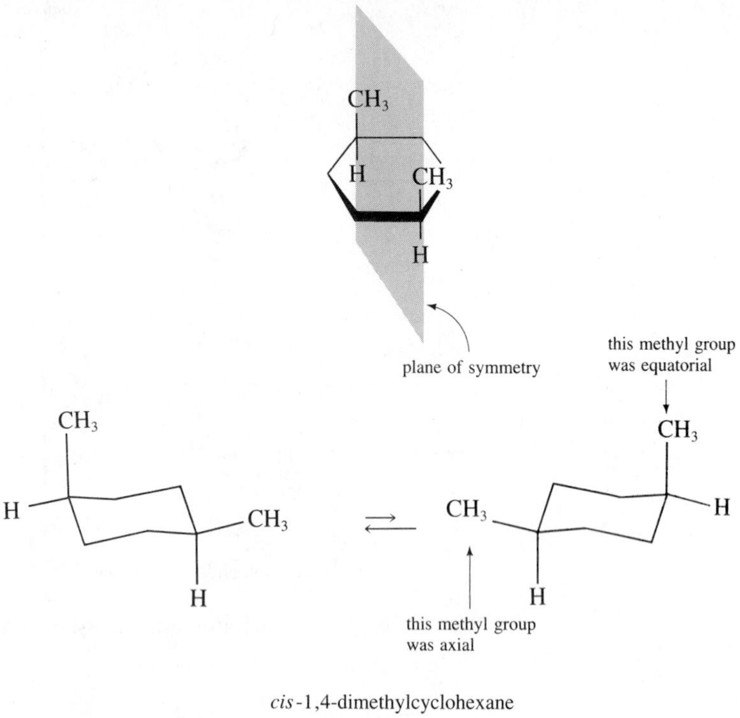
CH3
H
CH3
H
plane of symmetry
this methyl group
was equatorial
CH3
H
CH3
H
CH3
CH3
H
H
this methyl group
was axial
cis-1,4-dimethylcyclohexane

FIGURE 6.14 The symmetry of *cis*-1,4-dimethylcyclohexane.

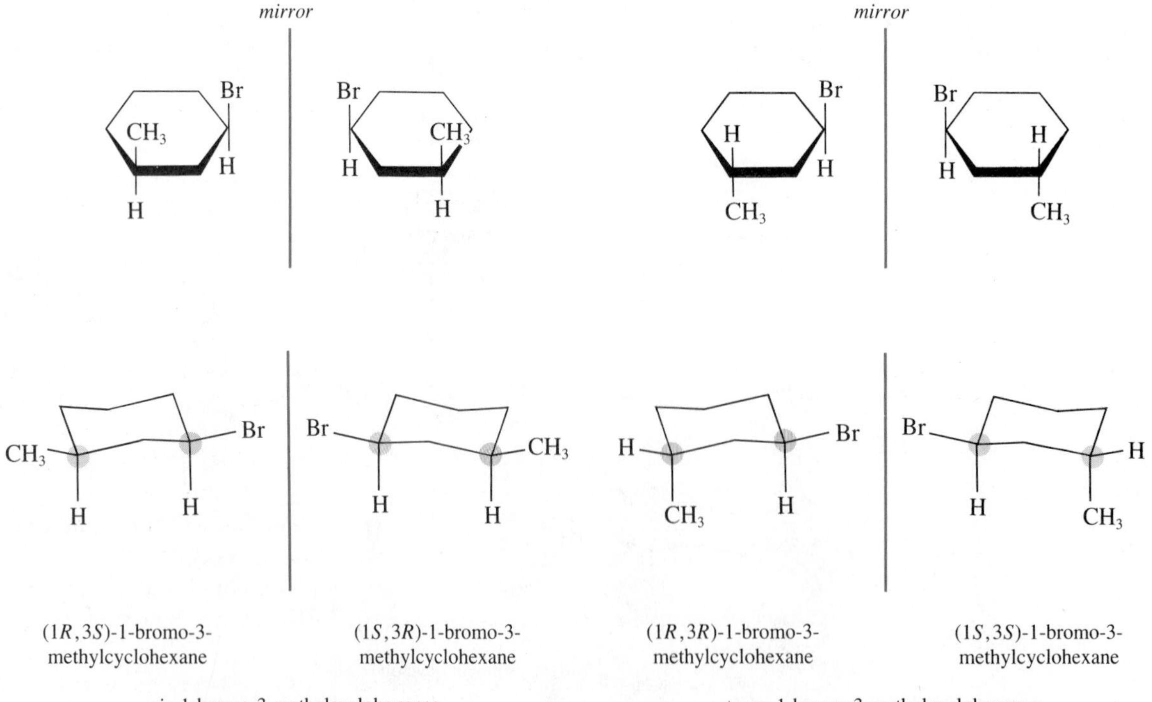
mirror
mirror
Br
CH3
H
H
Br
CH3
H
H
Br
H
H
CH3
Br
H
H
CH3
CH3
Br
H
H
Br
CH3
H
H
H
Br
CH3
H
Br
H
H
CH3
(1R,3S)-1-bromo-3-methylcyclohexane
(1S,3R)-1-bromo-3-methylcyclohexane
(1R,3R)-1-bromo-3-methylcyclohexane
(1S,3S)-1-bromo-3-methylcyclohexane
cis-1-bromo-3-methylcyclohexanes
trans-1-bromo-3-methylcyclohexanes

FIGURE 6.15 The stereoisomers of 1-bromo-3-methylcyclohexane.

PROBLEM 6.17

How many different configurational isomers are possible for 1-bromo-2-methylcyclohexane? Illustrate by drawing planar formulas and chair conformations. Do the same for 1-bromo-4-methylcyclohexane.

PROBLEM 6.18

(a) In a stereoisomer of 2-isopropyl-5-methylcyclohexanol, the methyl group is cis to the hydroxyl group, and the isopropyl group is trans. Draw the two chair forms for the compound, and decide which conformer would be more stable.

(b) Draw the chair conformations of a stereoisomer of the compound described in part a. There are several possible stereoisomers.

6.10 STEREOISOMERISM IN ALKENES

A. The Origin of Stereoisomerism in Alkenes

The bonding between two carbon atoms involved in a double bond has been described (p. 49) as consisting of a σ bond and a π bond, which is created by overlap of the p orbitals. The necessity that the p orbitals in the π bond must overlap imparts a certain rigidity to the double bond. Rotation of the carbons on the two ends of the double bond with respect to each other does not take place unless enough energy is supplied to break the π bond. A carbon-carbon single bond consisting of a σ bond with cyclindrical symmetry (p. 40), however, does allow free rotation (p. 148) of the bonded atoms with respect to each other.

The rigidity of the double bond gives rise to the possibility of stereoisomerism. For example, 2-butene can have two methyl groups on the same side of the molecule (or cis to each other) or on opposite sides (or trans to each other). Each isomer is converted to the other when enough energy is supplied, for example, by absorption of ultraviolet radiation or being heated to temperatures around 300 °C. The conversion takes place because the π bond breaks when energy is absorbed, and the two halves of the molecule can then rotate with respect to each other before the π bond forms again (Figure 6.16).

The interconversion of stereoisomers having a double bond requires much more energy (~65 kcal/mol) than does the interconversion of conformational isomers by rotations around single bonds (less than 10 kcal/mol; see p. 149). Conformational isomers cannot be separated and isolated as individual molecular species at ordinary temperatures (where ~20 kcal/mol of kinetic energy is available to molecules), but the stereoisomers created by different spatial arrangements of groups around double bonds can. Double-bond stereoisomers are also configurational isomers (p. 180). They are stereoisomers but not mirror-image isomers and, therefore, are diastereomers (p. 201). Alkenes, because of the planarity of the double bond, have a plane of symmetry and are not chiral. An alkene, therefore, does not show optical activity unless there is a stereocenter elsewhere in the molecule.

Even alkenes with relatively simple structures show several kinds of isomerism. For example, there are four different noncyclic structures that can be drawn for C_4H_8.

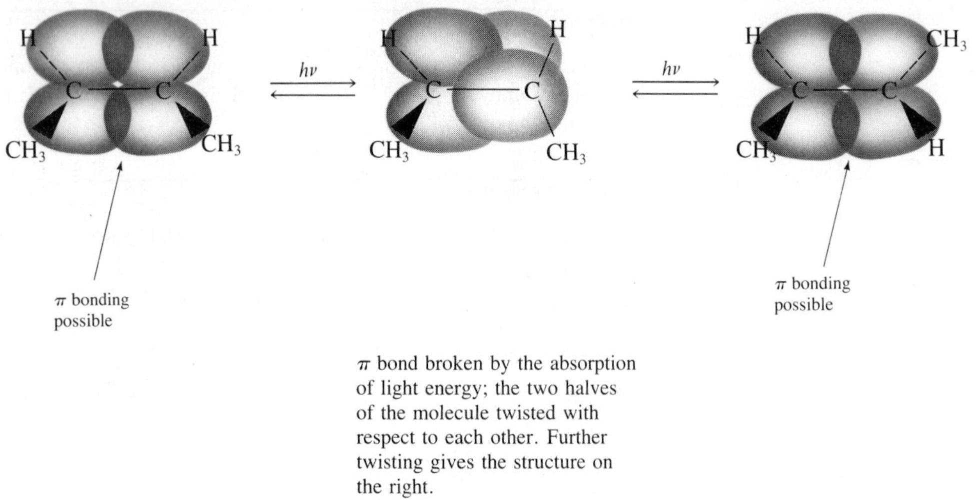
hν
hν
H
H
CH3
CH3
C
C
π bonding possible
π bonding possible
π bond broken by the absorption of light energy; the two halves of the molecule twisted with respect to each other. Further twisting gives the structure on the right.

FIGURE 6.16 The transformation of *cis*-2-butene into *trans*-2-butene.

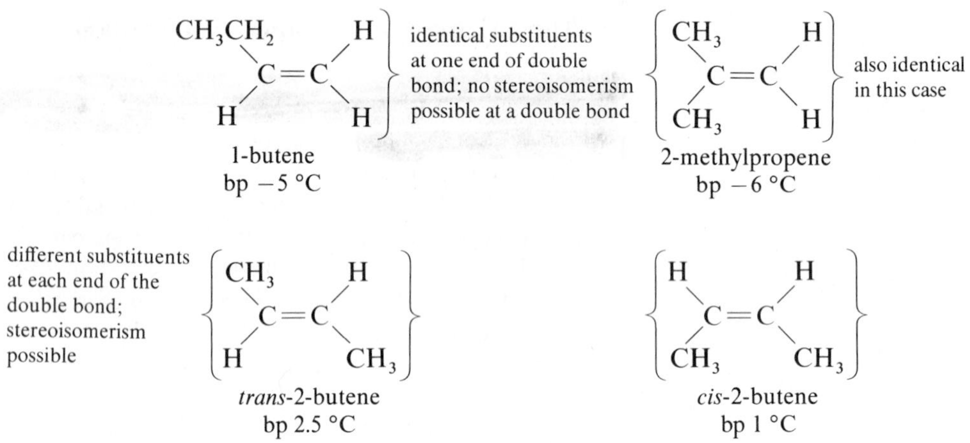
identical substituents at one end of double bond; no stereoisomerism possible at a double bond
also identical in this case
1-butene
bp −5 °C
2-methylpropene
bp −6 °C
different substituents at each end of the double bond; stereoisomerism possible
trans-2-butene
bp 2.5 °C
cis-2-butene
bp 1 °C

isomeric structures for alkenes with the molecular formula C_4H_8

1-Butene and 2-methylpropene are constitutional isomers of the 2-butenes. They have different arrangements of the carbon atoms and double bonds. They do not have stereoisomers. Stereoisomerism is not possible for an alkene unless two different substituents are present on the carbon at each end of the double bond. *cis*-2-Butene and *trans*-2-butene are stereoisomers of each other. They differ from each other only in the spatial arrangement of the methyl groups and the hydrogen atoms. This similarity is reflected in their names, which are identical except for the designation for spatial orientation. *cis*-2-Butene and *trans*-2-butene, being diastereomers (p. 211), have different physical properties, for example, their different boiling points.

B. Nomenclature of Stereoisomeric Alkenes

When an alkene has stereoisomers, the complete name of the compound should include a definition of the stereochemistry, if it is known. In simple alkenes such as the 2-butenes, the designations cis and trans are used with no ambiguity. In the cis isomer, similar groups are on the same side of the double bond; in the trans isomer, they are on opposite sides of the double bond. In some simple cases, such nomenclature is used with no confusion.

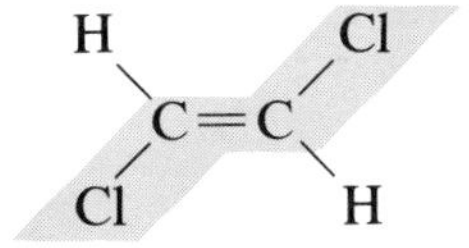

trans-1,2-dichloroethene

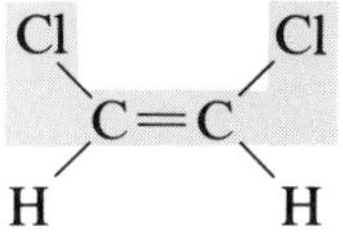

cis-1,2-dichloroethene

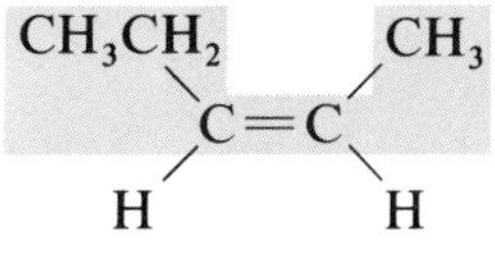

cis-2-pentene

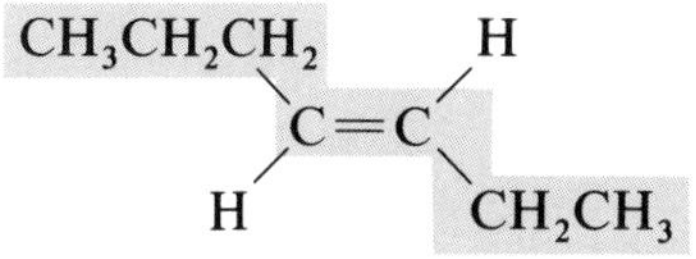

trans-3-heptene

The stereochemistry of more highly substituted alkenes is harder to define as cis or trans. The Cahn-Ingold-Prelog rules (p. 196) are used systematically to assign priorities to the substituents on a double bond. The double bond is assigned a ***Z*** (for *zusammen,* the German word for together) configuration if the two groups of higher priority at each end of the double bond are on the same side of the molecule. If the two groups of higher priority are on the opposite sides of the double bond, the configuration is denoted by an ***E*** (for *entgegen,* the German word for opposite).

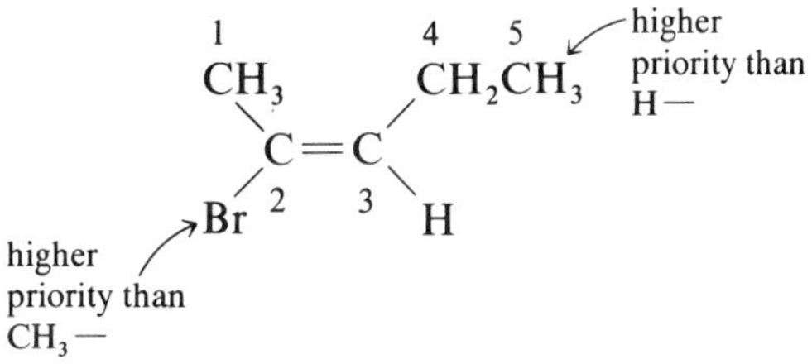
higher priority than H—
higher priority than CH_3—

groups of higher priority are on opposite sides of the double bond; therefore this is (E)-2-bromo-2-pentene

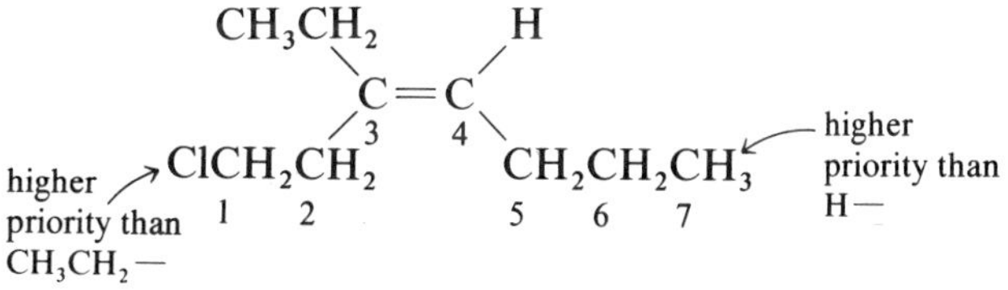
higher priority than CH_3CH_2—
higher priority than H—

groups of higher priority are on the same side of the double bond; therefore this is (Z)-1-chloro-3-ethyl-3-heptene

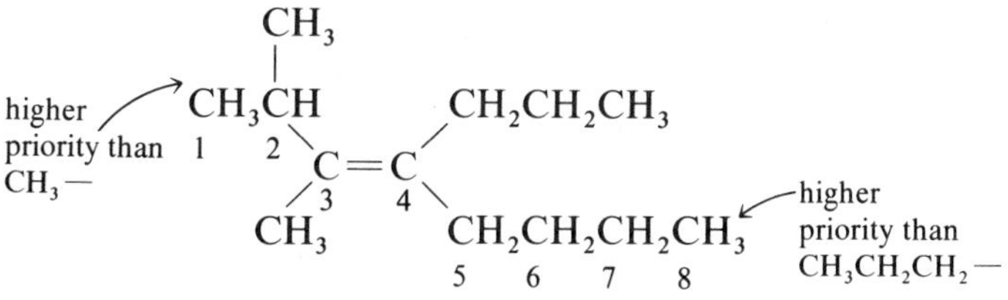
higher priority than CH_3—
higher priority than $CH_3CH_2CH_2$—

groups of higher priority are on opposite sides of the double bond; therefore this is (E)-2,3-dimethyl-4-propyl-3-octene

To avoid any possible confusion, this system of assigning configuration to stereoisomers of alkenes is used in this book. The terms cis and trans are reserved for descriptions of the spatial relationship between groups. For example, in (*E*)-2-bromo-2-pentene shown above, the methyl and the ethyl group on the carbon atoms of the double bond are cis to each other. The molecule, however, has the *E* configuration.

PROBLEM 6.19

Assign configuration to each of the following alkenes.

(a)

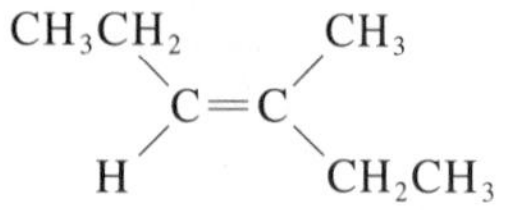

3-methyl-3-hexene

(b)

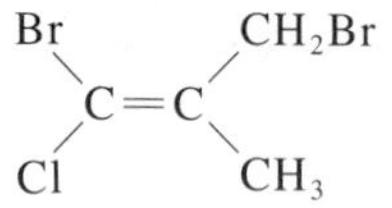

1,3-dibromo-1-chloro-2-methylpropene

Study Guide Concept Map 6.7

(c)

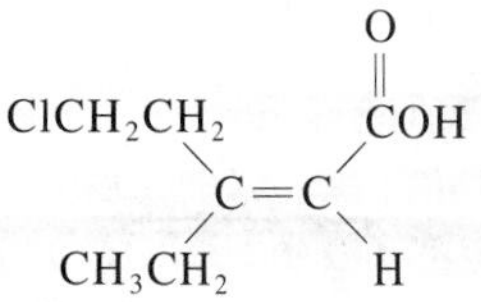

5-chloro-3-ethyl-2-pentenoic acid

(d)

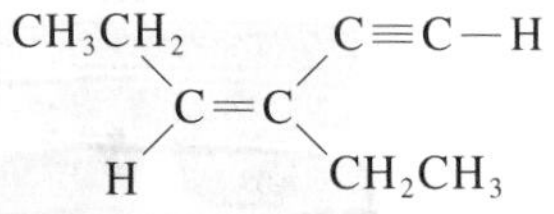

3-ethyl-3-hexen-1-yne

6.11 THE RESOLUTION OF A RACEMIC MIXTURE

A. Separation of Mixtures

Mixtures of organic compounds are separated by means of a variety of laboratory techniques. Compounds with different boiling points may be separated by distillation, for example. If compounds have different solubilities in solvents, they may be separable by extraction or recrystallization. An acidic compound can be removed from a mixture by washing the mixture with a dilute base, such as a solution of sodium bicarbonate. Chromatographic techniques will separate not only compounds belonging to different functional group classes, but members of the same functional group class with differing molecular structures. Diastereomers, such as cis and trans isomers of alkenes or cyclic compounds, can also be separated chromatographically.

What about compounds that differ from each other only in the configuration, *R* or *S*, of stereocenters? Compounds that have only one stereocenter exist as two mirror-image isomers, or enantiomers. Enantiomers have identical physical properties except for the direction in which they rotate the plane of polarized light (p. 187). Enantiomers are indistinguishable with respect to the polarity of their molecules and have identical solubilities in ordinary solvents, so they cannot be separated by recrystallization or chromatography in the usual way. They have identical boiling points, so distillation cannot be used to separate them.

When a chiral compound containing a single stereocenter is prepared in the laboratory, starting with achiral reagents, both enantiomers are formed in equal amounts. The product is a racemic mixture. An example of such a reaction is the addition of hydrogen bromide to 1-butene (p. 191). Once a racemic mixture is formed, the two enantiomers in the mixture are inseparable by the techniques normally used to separate constitutional isomers and diastereomers.

As discussed earlier (p. 201), compounds that have two different stereocenters have four stereoisomers, which are two sets of enantiomers. The members of one pair of enantiomers are diastereomers of the compounds in the other pair of enantiomers. The diastereomeric pairs have physical properties sufficiently different that they can be separated by usual techniques such as recrystallization, distillation, and chromatography.

B. Resolution of Racemic Mixtures

The separation of a mixture of enantiomers is called the **resolution of a racemic mixture.** Louis Pasteur performed a very unusual resolution using a pair of tweezers on the crystals of sodium ammonium tartrate to pick out left-handed and right-handed crystals under a microscope (p. 194). Living organisms have enzyme systems that are highly sensitive to the stereochemistry of the compounds with which they interact; these systems perform resolutions when they metabolize one enantiomer and reject another one. Yeast cells, for example, if allowed to ferment in a medium containing sugar and a racemic amino acid, consume the enantiomer of the amino acid that is found naturally in proteins and leave the other. Most vertebrates, including human beings, also separate enantiomers biologically. For example, if racemic alanine (p. 183) is fed to a human being, the *S* isomer is metabolized and the *R* isomer is excreted in the urine. Researchers have observed that in most biological resolutions, only one enantiomer, the one that cannot be used by the living organism, is recovered.

A common way to separate enantiomers is to convert them into compounds that are diastereomers, which have different physical properties and, as a result, are separable. Reactions of enantiomers with achiral reagents cause no change in their relationship as mirror-image isomers of each other. Only when the reagent itself is chiral is new stereochemistry introduced into the molecule and, with it, the possibility of a physical separation. The most widely used technique for the resolution of mixtures of enantiomers depends on the reaction of a chiral acid with a racemic base (or a chiral base with a racemic acid) to give a mixture of diastereomeric salts. The salts are separated by recrystallization, and the individual enantiomers, as well as the chiral reagent, are recovered by further acid-base reactions.

The next section contains a description of an actual resolution that will help to clarify these ideas.

C. Resolution of Amphetamine

2-Amino-1-phenylpropane is commonly known as amphetamine or benzedrine.

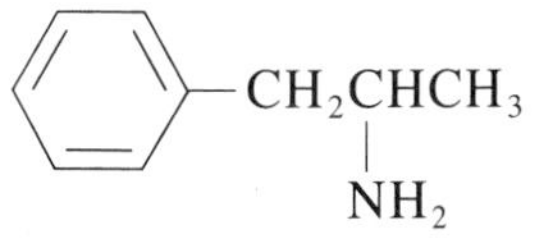
CH2CHCH3
NH2

2-amino-1-phenylpropane
or amphetamine
or benzedrine

Amphetamine has one stereocenter and, therefore, two stereoisomers. Each enantiomer has distinctive physiological properties. The dextrorotatory isomer has the *S* configuration and is the form that stimulates the central nervous system. It is sold as its sulfuric acid salt under names such as dextroamphetamine sulfate. The levorotatory isomer acts on the sympathetic nervous system.

Amphetamine is synthesized as a racemic mixture, which must be resolved to isolate the *S* isomer. (+)-Tartaric acid is used as the chiral reagent for the resolution. The mixture of the enantiomers of amphetamine and a molar equivalent of (+)-tartaric acid are dissolved in ethanol. A proton is transferred from one of the carboxylic acid groups of (+)-tartaric acid to the amine group of amphetamine. Two salts are formed, one from each enantiomer of the amphetamine.

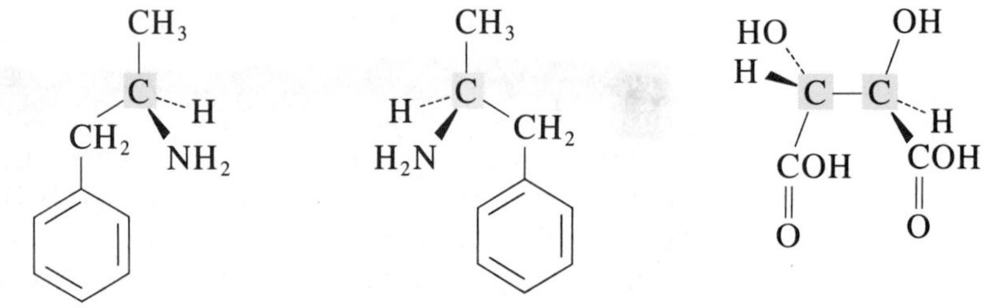
CH_3
CH_2
NH_2
H
H_2N
HO
OH
COH
O

(*R*)-(−)-amphetamine (*S*)-(+)-amphetamine
racemic mixture
$[\alpha]_D^{15}$ 0.0°

(2*R*,3*R*)-(+)-2,3-dihydroxybutanedioic acid
(+)- tartaric acid

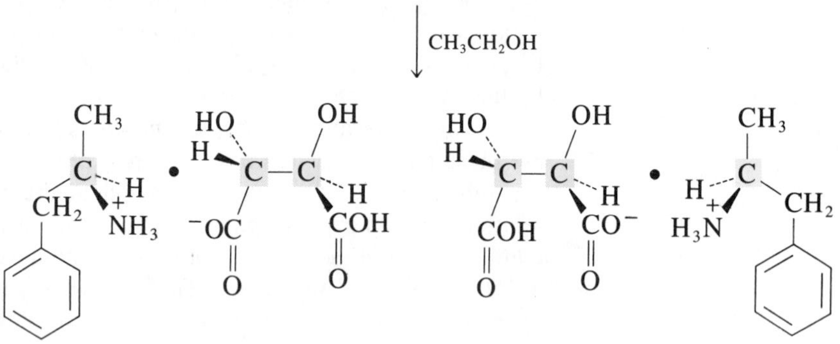
CH_3CH_2OH
CH_3
CH_2
NH_3
H_3N
HO
OH
^-OC
COH
CO^-
O

salt of (*R*)-(−)-amphetamine with (2*R*,3*R*)-(+)-tartaric acid
more soluble in the reaction mixture remains in solution

salt of (*S*)-(+)-amphetamine with (2*R*,3*R*)-(+)-tartaric acid
less soluble in the reaction mixture crystallizes out

diastereomers of each other

The products are diastereomers of each other. If you examine closely the structural formulas written for the salts, you will see that the two compounds are *not* mirror-image isomers of each other. The amine portions of the molecules remain enantiomeric, but attached to each one is the ion of an acid, and it is identical in both cases. The two salts have three stereocenters, but do not have the opposite configuration at each of these points. One salt could be called the *R*,*R*,*R* salt, and the other the *S*,*R*,*R* salt. The two compounds are thus stereoisomers but not enantiomers. By definition, all stereoisomers that are not enantiomers are diastereomers of each other (p. 201).

The diastereomeric salts have different physical properties. The salt derived from the *S* amine is less soluble in the reaction mixture than the other salt is and crystallizes out of solution. The free amine is recovered by treating this salt with potassium hydroxide.

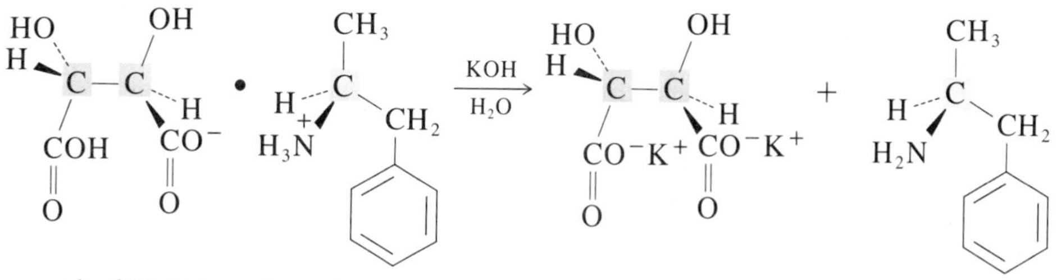
KOH
H_2O

salt of (*S*)-(+)-amphetamine with (2*R*,3*R*)-(+)-tartaric acid

soluble in water remains in solution

insoluble in water extracted out of solution and distilled
$[\alpha]_D^{15}$ +40.2°
(c. 8.7; benzene)

The strong base, hydroxide ion, removes a proton from the organic ammonium ion and regenerates the free water-insoluble amine, which is separated from the water-soluble, nonvolatile dipotassium salt of (+)-tartaric acid by extraction and distillation. The purified amine has an optical rotation of $[\alpha]_D^{15}$ +40.2° (*c* 8.7; benzene).

Crystals of the salt of the *R* amine with (+)-tartaric acid are also recovered from the solution. These are contaminated by the crystals of the salt of the *S* amine, so further purification is necessary. The *R* amine is recovered in the same way the *S* amine is, then treated with (−)-tartaric acid, which is also available from natural sources.

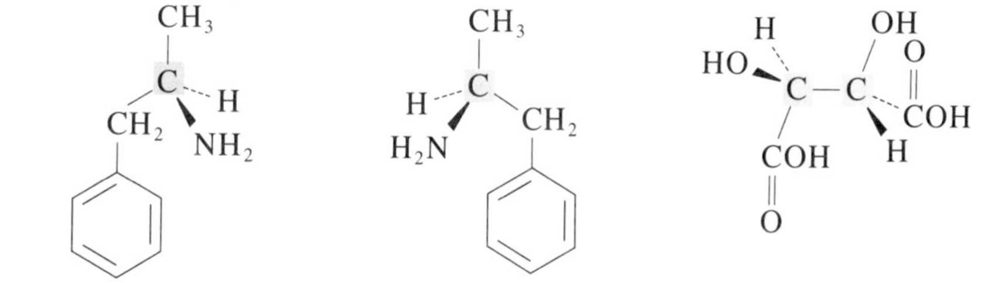

(*R*)-(−)-amphetamine
most of mixture

(*S*)-(+)-amphetamine
present as impurity

(2*S*,3*S*)-(−)-2,3-dihydroxybutanedioic acid
(−)-tartaric acid

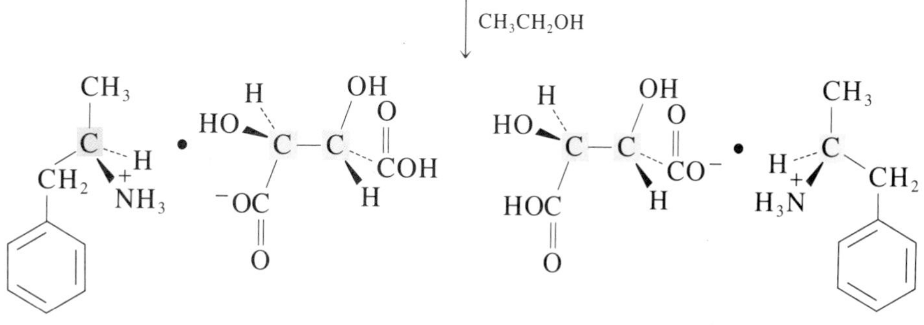
CH_3CH_2OH

salt of (*R*)-(−)-amphetamine with (2*S*,3*S*)-(−)-tartaric acid
less soluble in the reaction mixture crystallizes out

salt of (*S*)-(+)-amphetamine with (2*S*,3*S*)-(−)-tartaric acid
more soluble in the reaction mixture remains in solution

Here the salt of the *R* amine with (−)-tartaric acid is the less soluble one and crystallizes out of the solution. The *R* amine is recovered from the salt using exactly the same method used for recovering the *S* amine and, after being distilled, has $[\alpha]_D^{15}$ −40.1° (*c* 8.83; benzene). The two enantiomeric amines, having been separated from a racemic mixture, are seen to have optical rotations that are equal in magnitude (within experimental error) but opposite in sign.

The process described above is the resolution of a racemic mixture. The chiral reagent used, tartaric acid in this case, is called the **resolving agent.** A large number of resolving agents are available; many of them are acids and bases from natural sources or new ones created by modifications of naturally chiral compounds. Some synthetic compounds that are easily resolved are then used as resolving agents themselves.

Resolutions of racemic mixtures are not easy procedures. They require a great deal of patience and skill, and some good luck, too. In many cases, they depend on taking advantage of the properties of organic compounds as acids and bases. The most important step in any resolution, however, is the conversion of the mixture of enantiomers into a mixture of diastereomers that have physical properties that allow for their separation.

The separation of diastereomers need not be done so that isolable compounds end up in different flasks. For example, if a chromatography column were to contain a chiral adsorbent, one enantiomer would interact more strongly than the other with the material in the column and a separation would take place. Such methods are already being used and will certainly be used more and more in the future as new chiral adsorbents and complexing agents are developed.

Study Guide Concept Map 6.8

PROBLEM 6.20

Write an equation showing how (*R*)-(−)-amphetamine can be recovered from the salt it forms with (−)-tartaric acid.

PROBLEM 6.21

Write structural formulas showing the details for the resolution of a racemic mixture of 1-amino-1-phenylethane using (+)-tartaric acid. The separation takes place in methanol as the solvent. The salt of (*S*)-(−)-1-amino-1-phenylethane with (+)-tartaric acid is the one with the lower solubility.

S U M M A R Y

Stereochemistry deals with the spatial properties of compounds and chemical reactions. Compounds that differ from each other only in how their atoms are arranged in space are known as stereoisomers. Stereoisomers are classified into conformational isomers, which are interconvertible by rotations around single bonds at room temperature, and configurational isomers. Configurational isomers can be either enantiomers, which are related to each other as nonsuperimposable mirror images, or diastereomers, which include all other stereoisomers. An important structural feature of many stereoisomers is the stereocenter, a carbon atom bonded to four different substituents.

Enantiomers have identical physical properties except that they rotate a plane of polarized light in opposite directions. Compounds that rotate plane-polarized light are said to have optical activity. A mixture of equal amounts of two enantiomers has no optical activity and is called a racemic mixture. To separate the enantiomers in a racemic mixture, a resolution must be performed.

Diastereomers have different physical properties and may be separated from each other by ordinary physical means. Diastereomers may have optical activity if they contain stereocenters; however, if they have a plane of symmetry, they are meso compounds and are not optically active. Compounds that contain n stereocenters have 2^n stereoisomers unless symmetry reduces this number. Diastereomers such as cis and trans isomers of alkenes do not have optical activity.

Configurational isomers are named using the Cahn-Ingold-Prelog rules. Priorities are assigned to substituents on carbon atoms that are stereocenters or are involved in double bonds. Depending on the spatial relationships of the groups, stereocenters are assigned *R* or *S* configurations and double bonds, *E* or *Z* configurations. These configurations can be assigned by looking at structural formulas of compounds. Both the sign of optical rotation, (+) or (−), and the actual arrangement in space of the atoms in a compound that shows a particular rotation have to be determined experimentally. The spatial arrangement of the atoms is called the absolute configuration of the compound.

ADDITIONAL PROBLEMS

6.22 Name the following compounds.

(a) 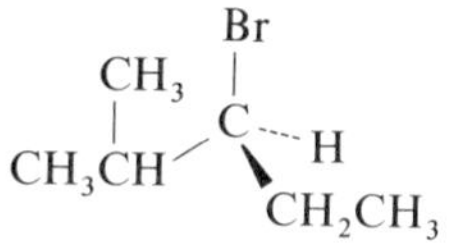

(b) 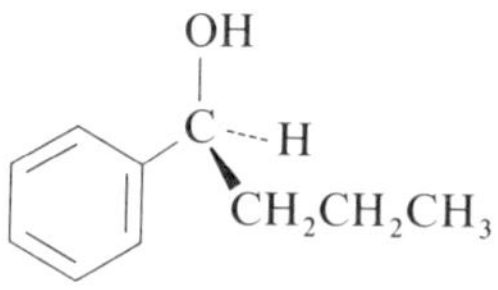

(c) 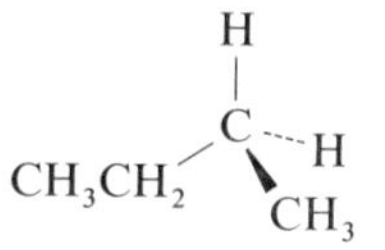

(d) 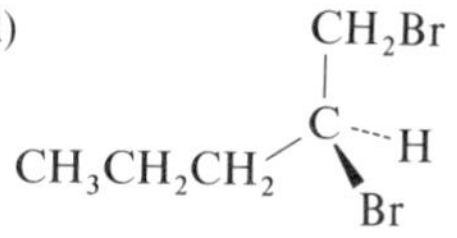

(e) 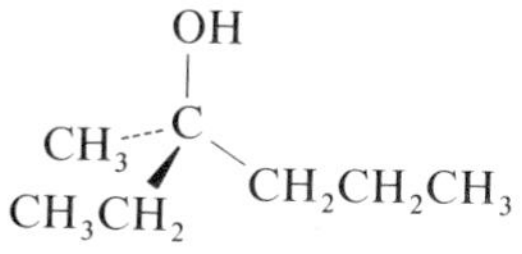

(f) 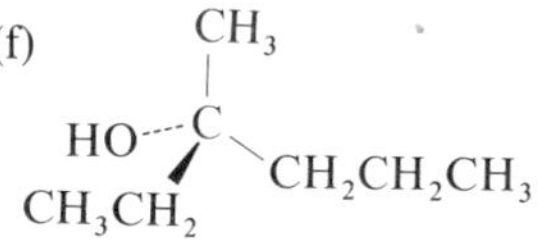

(g) 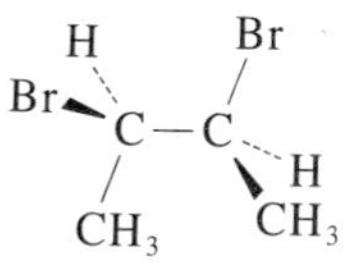

(h) 

(i) 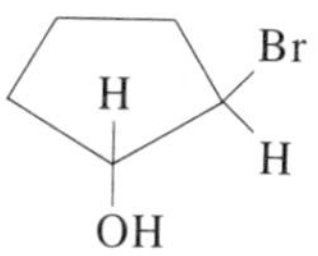

(j) 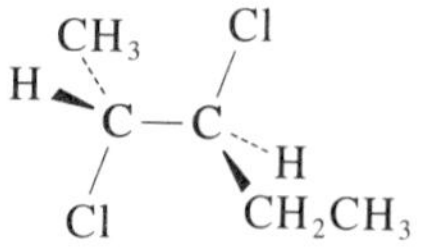

6.23 Which of the following compounds have stereocenters? Draw three-dimensional pictures of those that do, showing any enantiomers and diastereomers.

(a) 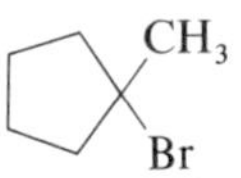

(b) $ClCH_2CH(Cl)CH_2CH_2Cl$

(c) 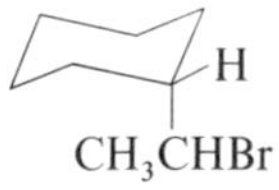

(d) 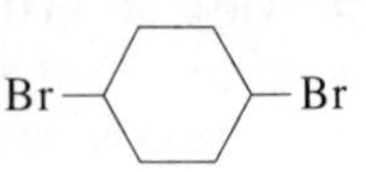

(e) 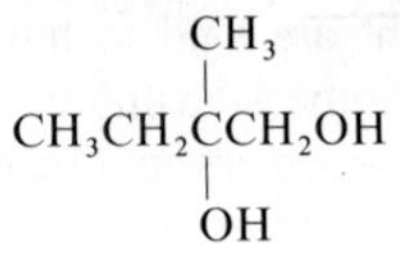

(f) 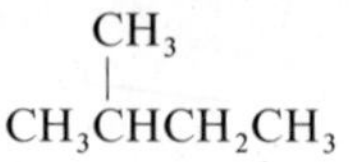

(g) 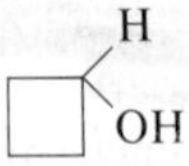

(h) 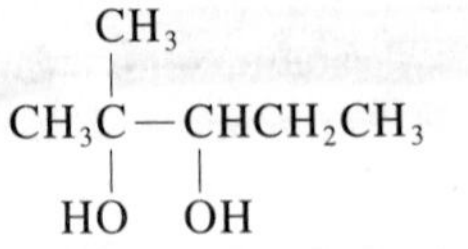

(i) 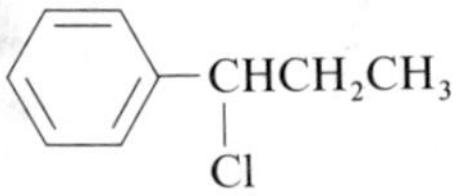

(j) 

(k) 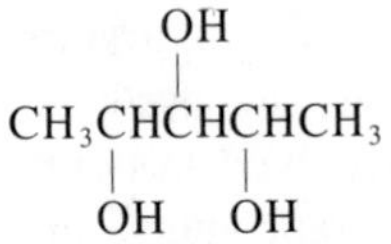

6.24 For each of the following pairs of structural formulas, tell whether the two represent identical molecular species, conformers of the same species, constitutional isomers, enantiomers, or diastereomers.

(a) 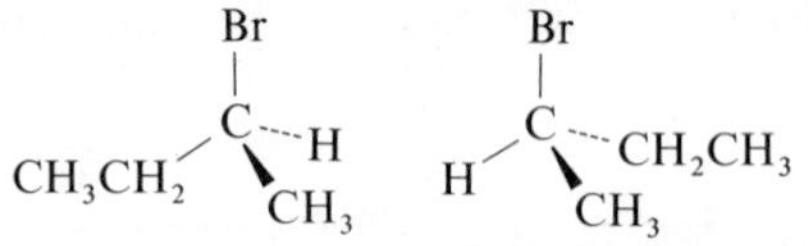

(b) 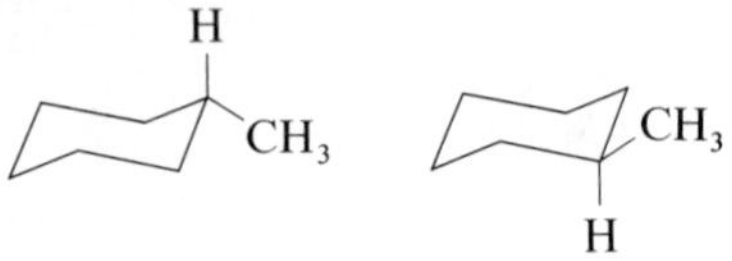

(c) 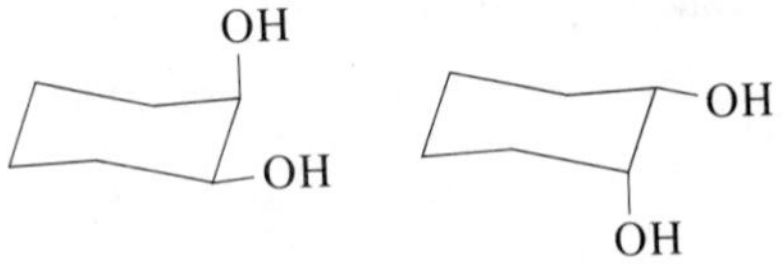

(d) 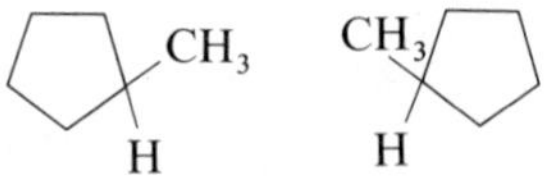

(e) 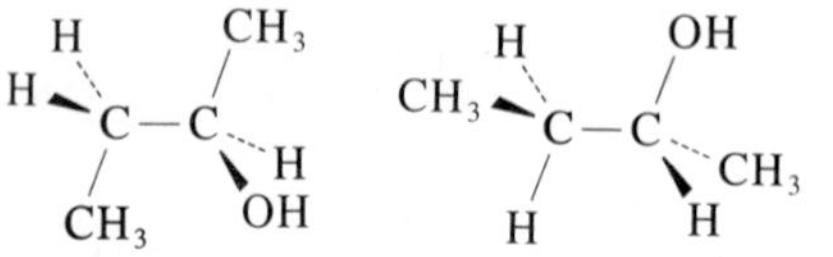

(f) 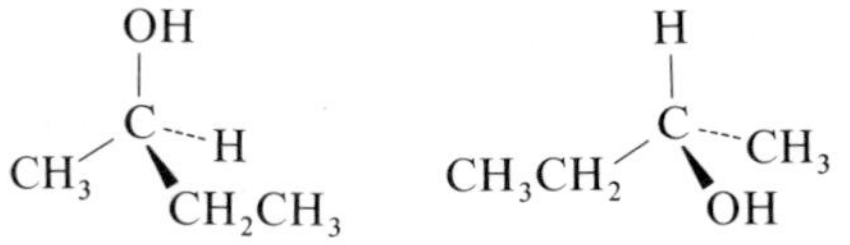

(g) 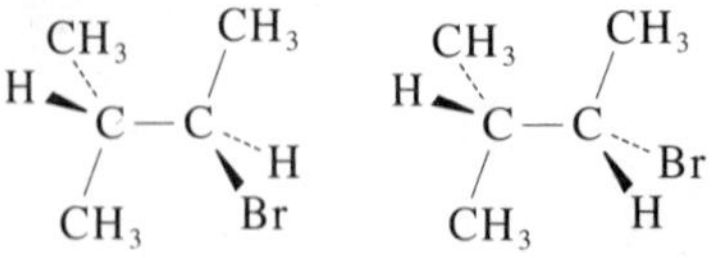

(h) 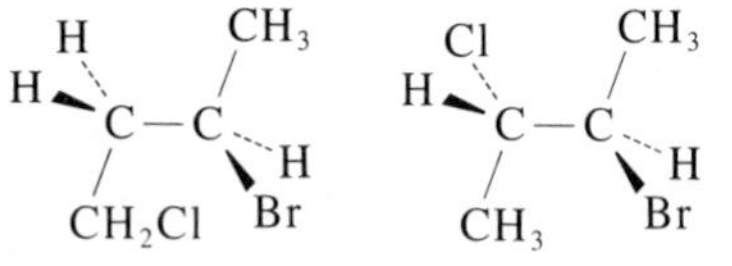

(i) 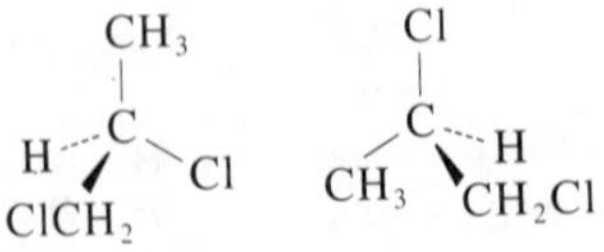

(j) 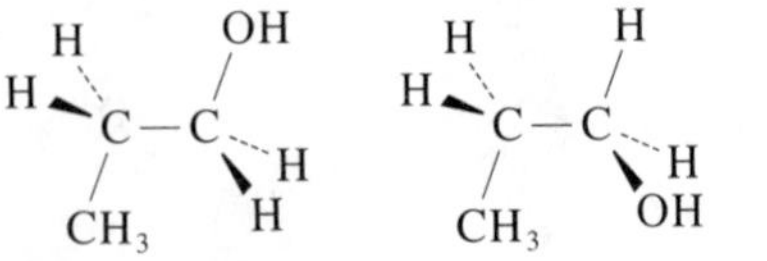

(k) 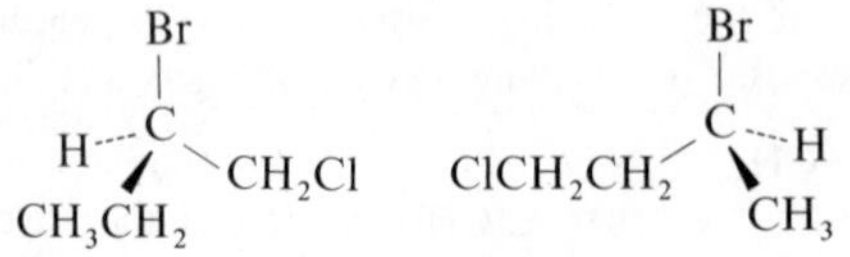

(l) 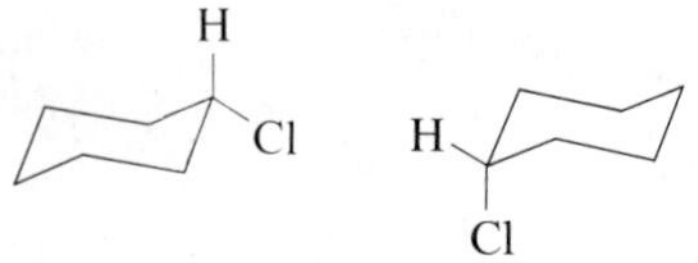

6.25 The acid shown below was synthesized and resolved using (+)-1-amino-1-phenylethane and was then used as a chiral intermediate in a synthesis of Vitamin B_{12}. Draw a three-dimensional representation of the *S* isomer of this compound.

$$\begin{array}{c} \quad\quad\quad CH_3 \quad O \\ \quad\quad\quad | \quad\quad \| \\ HC{\equiv}C{-}C{-\!-}COH \\ \quad\quad\quad | \\ \quad\quad\quad\quad CH{=}CH_2 \end{array}$$

6.26 Chiral acetic acid in which two of the hydrogen atoms on the methyl group have been replaced by deuterium, D, and tritium, T (p. 196), has been synthesized for studies of the stereochemistry of biological reactions. Write three-dimensional formulas for (*R*)- and (*S*)-DHTCCOH.

$$\begin{array}{c} DHTCCOH \\ \| \\ O \end{array}$$

6.27 Citronellol is a fragrant component of various plant oils, such as geranium oil and rose oil. A synthetic sample of (−)-citronellol with an enantiomeric excess of 88% has $[\alpha]_D^{20}$ −4.1°. What is the optical rotation of the pure enantiomer? (−)-Citronellol has the *S* configuration. The structure of citronellol is

$$\begin{array}{l} CH_3 \quad\quad\quad\quad\quad\quad\quad CH_3 \\ \quad \diagdown \quad\quad\quad\quad\quad\quad\quad\quad | \\ \quad\; C{=}CHCH_2CH_2CHCH_2CH_2OH \\ \quad / \\ CH_3 \end{array}$$

Draw a stereochemically correct structural formula for (−)-citronellol.

6.28 The enantiomers of 1-amino-2-propanol were separated and recovered as their hydrochloride salts:

$$\begin{array}{l} CH_3CHCH_2NH_3{}^+Cl^- \\ \quad\quad | \\ \quad\; OH \end{array}$$

The sample of (*R*)-(−)-1-amino-2-propanol hydrochloride had $[\alpha]_D^{25}$ −31.5° (*c* 0.01; methanol), and that of (*S*)-(+)-1-amino-2-propanol hydrochloride had $[\alpha]_D^{25}$ +35° (*c* 0.01; methanol).

(a) Write correct stereochemical formulas for the levorotatory and dextrorotatory isomers.
(b) Which enantiomer was recovered with the higher purity?
(c) Assuming that the optical rotation observed for the enantiomer of higher purity is the correct optical rotation for the compound, what is the enantiomeric excess in the sample of the other enantiomer?

6.29 Fruit sugar, or fructose, has $[\alpha]_D^{20}$ −92° (*c* 0.02; water). Calculate the rotation that would be observed for a solution made with 1 g of fructose in 100 mL of water and measured in a tube 5 cm long at 20 °C using a sodium lamp.

6.30 (+)-2-Butanol has $[\alpha]_D^{20}$ +13.90° when the measurement is made on the pure liquid. A sample of 2-butanol was found to have an optical rotation of −3.5°. What is the stereoisomeric composition of this mixture?

6.31 A widely prescribed analgesic (painkiller) is Darvon. The medically active form of Darvon is the propanoic ester of (2*S*,3*R*)-4-dimethylamino-1,2-diphenyl-3-methyl-2-butanol, which is sold as the hydrochloric acid salt of the amine. The structural formula of the amino-alcohol used in the preparation of Darvon is shown below. Identify the stereocenters, write structural formulas for all possible stereoisomers, and pick out the medicinally useful form of the compound.

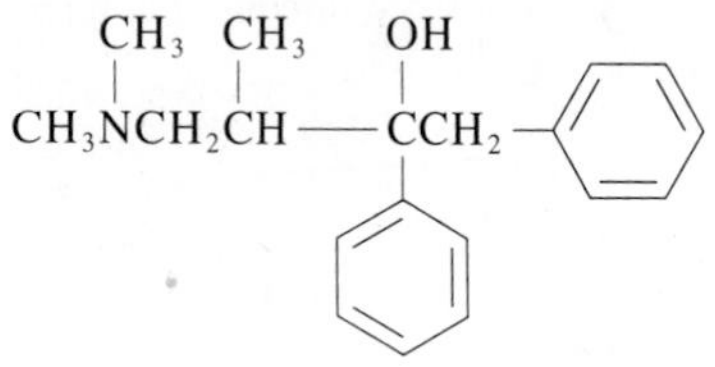

4-dimethylamino-1,2-diphenyl-3-methyl-2-butanol
Darvon alcohol

6.32 The compound decalin has two cyclohexane rings fused together. The hydrogen atoms at the ring junction are on the same side of the rings in *cis*-decalin and on opposite sides of the rings in *trans*-decalin, as shown below. Decide which stereoisomer of decalin is more stable. (Hint: Think of the one ring as substituents on the other ring. It may help you to construct the ring systems with models.)

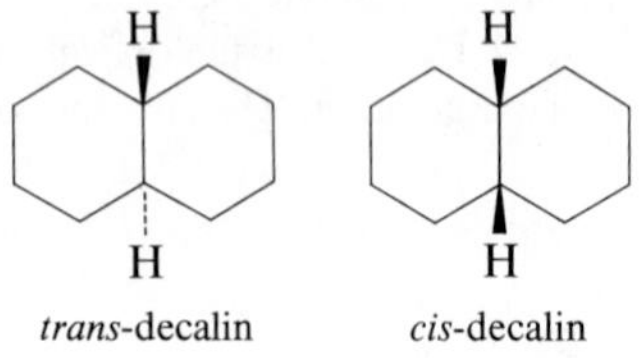

7

Nucleophilic Substitution and Elimination Reactions

A • L O O K • A H E A D

The displacement of a halide ion by a nucleophile is known as a nucleophilic substitution and is a reaction type we have already examined (Section 4.2). This important type of reaction allows chemists doing organic syntheses to transform alkyl halides into a large variety of other compounds.

An alkyl halide has two sites of reactivity.

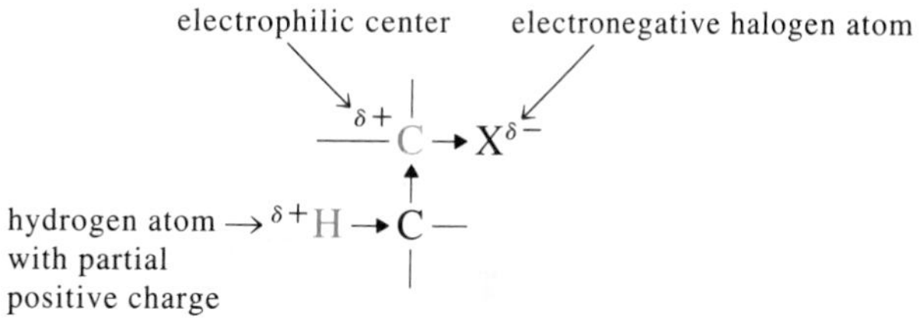
electrophilic center
electronegative halogen atom
hydrogen atom
with partial
positive charge

The electronegative halogen atom polarizes the carbon-halogen bond, giving the carbon atom and any hydrogen atoms close to it a partial positive character. The carbon atom bearing the partial positive charge is an electrophilic center. A nucleophile will react with an alkyl halide at that carbon atom, displacing the halogen atom in a nucleophilic substitution reaction (p. 112).

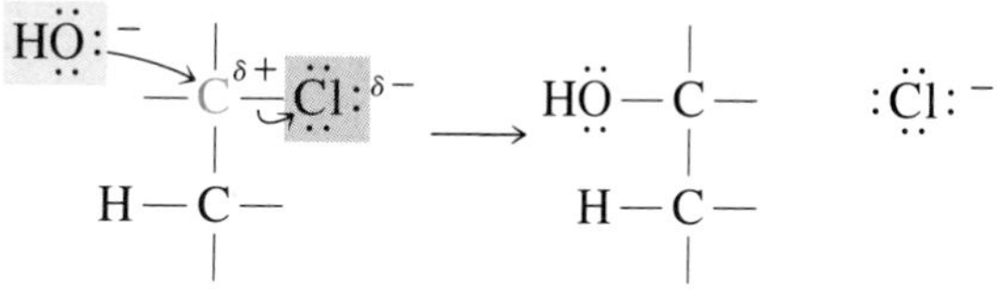

a nucleophilic substitution reaction

Nucleophiles that are strong bases also remove a proton from the carbon atom adjacent to the electrophilic center. The halide ion is lost and a double bond is formed. This reaction is called an elimination reaction, and is the reverse of the addition of a hydrogen halide to a double bond (p. 118).

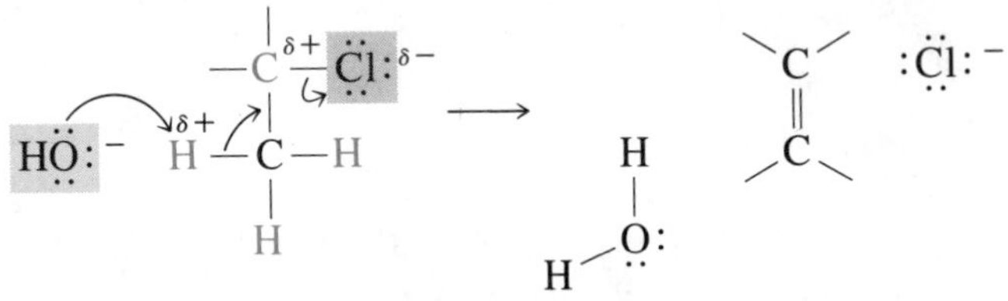

an elimination reaction

Elimination reactions usually accompany nucleophilic substitution reactions. In this chapter, we will study the mechanisms of these two kinds of reactions and the factors that influence them.

7.1 A COMPARISON OF SUBSTITUTION AND ELIMINATION REACTIONS

An alkyl halide such as isopropyl bromide undergoes two types of reactions. For example, iodide ion in a solution with acetone as the solvent, at room temperature, will substitute for bromide ion to give isopropyl iodide.

$$\underset{\text{isopropyl bromide}}{\underset{|\atop \text{Br}}{CH_3CHCH_3}} + Na^+ + I^- \xrightarrow[25\ ^\circ C]{\text{acetone}} \underset{\text{isopropyl iodide}}{\underset{|\atop \text{I}}{CH_3CHCH_3}} + \underset{\textit{insoluble in acetone}}{Na^+Br^-\downarrow}$$

The reaction is reversible. Treatment of isopropyl iodide with sodium bromide gives isopropyl bromide. However, sodium bromide, which is not as soluble in acetone as sodium iodide is, precipitates out of solution, drawing the forward reaction toward completion.

In the above reaction, iodide ion is the nucleophile, the species with electrons to donate. It reacts with isopropyl bromide at the carbon atom bearing the partial positive charge. That carbon atom is an **electrophilic center,** a site of electron deficiency.

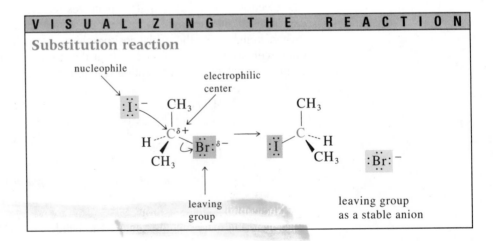

VISUALIZING THE REACTION

Substitution reaction

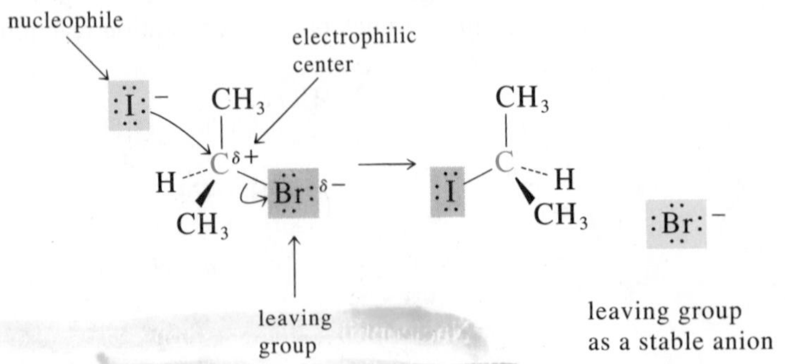
nucleophile
electrophilic center
leaving group
leaving group as a stable anion

The bromine atom of isopropyl bromide is lost as a bromide ion. A stable species that can be detached from a molecule during a reaction is called a **leaving group.** Bromide ion is the leaving group in the above reaction. The displacement of a leaving group by a nucleophile is called a **nucleophilic substitution reaction.**

When isopropyl bromide is heated with a strong base such as sodium ethoxide, in ethanol as the solvent, a mixture of products is formed.

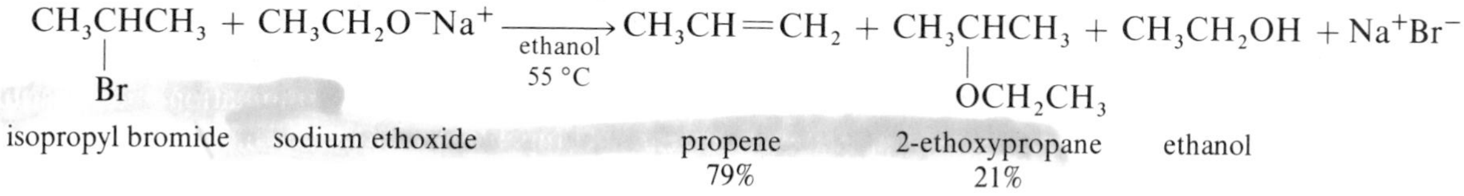
CH₃CHCH₃ + CH₃CH₂O⁻Na⁺ → CH₃CH=CH₂ + CH₃CHCH₃ + CH₃CH₂OH + Na⁺Br⁻
Br
ethanol
55 °C
OCH₂CH₃
isopropyl bromide
sodium ethoxide
propene
79%
2-ethoxypropane
21%
ethanol

The major product is propene, resulting from the removal of hydrogen bromide from isopropyl bromide. A reaction in which a stable species is lost and, usually, a multiple bond is also formed is called an **elimination reaction.** The strong base, ethoxide ion, removes a proton from isopropyl bromide and is thereby transformed into its conjugate acid, ethanol. A double bond is created, as bromide ion, a leaving group, breaks away. In this elimination reaction, a π bond is formed between two carbon atoms in the alkyl halide that is the starting material. The minor product of the reaction, 2-ethoxypropane, results from displacement of the bromide ion by ethoxide ion.

V I S U A L I Z I N G   T H E   R E A C T I O N

Elimination reaction

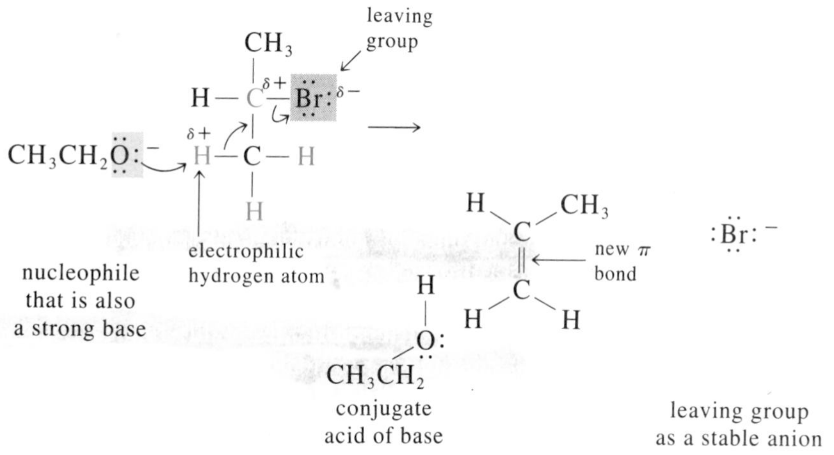
leaving
group
CH₃
H—C—Br:
δ+
δ−
CH₃CH₂O:⁻
H—C—H
H
electrophilic
hydrogen atom
nucleophile
that is also
a strong base
H
CH₃
C
new π
bond
C
H
H
:Br:⁻
H
O:
CH₃CH₂
conjugate
acid of base
leaving group
as a stable anion

PROBLEM 7.1

Isopropyl bromide is converted into mixtures of isopropyl alcohol,

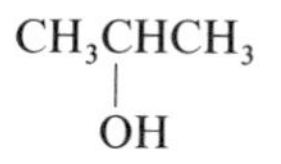
CH₃CHCH₃
OH

and propene by reaction with water or by treatment with sodium hydroxide dissolved in water. Do you expect the same ratio of isopropyl alcohol to propene to result under these two different conditions? (Hint: What is the nucleophile in each reaction? What are the relative basicities of the two nucleophiles?

Substitution and elimination reactions have tremendous potential for transforming organic molecules; as a result, they have been studied intensively. Early work on substitution reactions of alkyl halides was done by Sir Christopher Ingold and Edward Hughes and their collaborators at University College, London, in the 1930s and 1940s. Their ideas have influenced the thinking of organic chemists about the mechanisms of many different reactions and have inspired much experimental work and continuing vigorous debate.

The kinds of substituents that can be introduced into organic molecules depend on the availability of starting materials with suitable leaving groups and of nucleophiles that will give substitution reactions with them. Elimination and substitution reactions almost always appear in competition with each other. Factors such as the nature of the reagents and the solvents used play a part in determining which reaction is the major one in a given case. In the remainder of this chapter, we will explore the detailed mechanisms for these transformations so that you will understand the limitations on them and how to control them.

7.2 NUCLEOPHILICITY. SOLVENT EFFECTS

A. A Comparison of Basicity and Nucleophilicity

Ethoxide ion, a nucleophile that is also a strong base, gives chiefly the elimination product and only a small amount of the substitution product from isopropyl bromide (p. 225). Iodide ion, a good nucleophile but a weak base, gives the substitution product. Basicity and nucleophilicity describe different phenomena. A better understanding of the difference is necessary in order to further examine substitution and elimination reactions.

The basicity of a species as indicated by the pK_a of its conjugate acid is a measure of a thermodynamic property of the system (p. 93). The pK_a is derived from an equilibrium constant, the acidity constant, which is determined by the relative energy levels for the base and its protonated form. When a nucleophile is removing a proton from an alkyl halide (as on p. 225, for example), it is behaving as a Brønsted base. How effective a base will be at this task can be predicted by looking it up in the table of pK_a values.

The **nucleophilicities** of different bases are determined by measuring the rates of substitution reactions. For example, the relative nucleophilicities of various ions and neutral molecules have been determined using the rates of their substitution reactions with methyl bromide in water:

$SH^- \geqslant CN^- > I^- > OH^- > N_3^- > Br^- >$  $O^- > CH_3\overset{\overset{\displaystyle O}{\|}}{C}O^- \geqslant Cl^- > F^- > NO_3^- > H_2O$

In other words, the hydrosulfide anion, SH^-, is a better nucleophile than hydroxide ion, OH^-, in water because it reacts faster with the electrophilic center in methyl bromide. That is, the second of these two reactions is faster than the first:

$$H\ddot{\underset{\cdot\cdot}{O}}:^- + \overset{\delta+}{C}H_3\overset{\delta-}{Br} \xrightarrow[H_2O]{} H\ddot{\underset{\cdot\cdot}{O}}CH_3 + Br^-$$

$$H\ddot{\underset{\cdot\cdot}{S}}:^- + \overset{\delta+}{C}H_3\overset{\delta-}{Br} \xrightarrow[H_2O]{} H\ddot{\underset{\cdot\cdot}{S}}CH_3 + Br^-$$

Note that the electrophile in these reactions is a carbon atom; in contrast, in the reactions used to measure basicity, the electrophile is a proton.

An examination of the relative nucleophilicities given above shows several trends.

1. For nucleophiles of the same period in the periodic table, nucleophilicity decreases with increasing electronegativity and decreasing basicity; for example:

more basic / *more nucleophilic* — *less basic* / *less nucleophilic*

Oxygen is less electronegative than fluorine is, meaning that it holds the electrons around it less firmly than fluorine does. Thus, the oxygen atom in hydroxide ion is better able to donate its electrons than a fluoride ion is. Consequently, the hydroxide ion, in which there are nonbonding electrons on oxygen, is more basic and more nucleophilic than the fluoride ion. This reasoning applies to other elements in the same period as oxygen and fluorine.

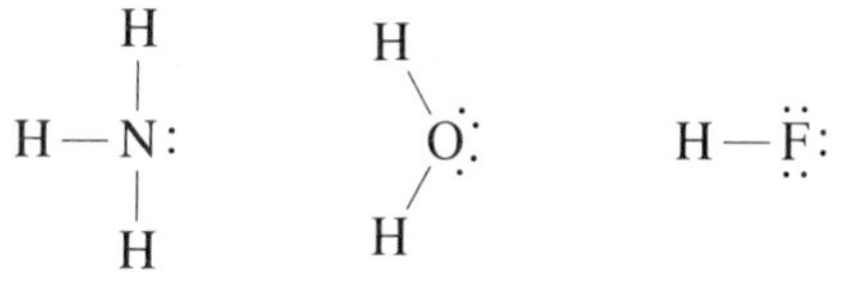

most basic / *most nucleophilic* — *least basic* / *least nucleophilic*

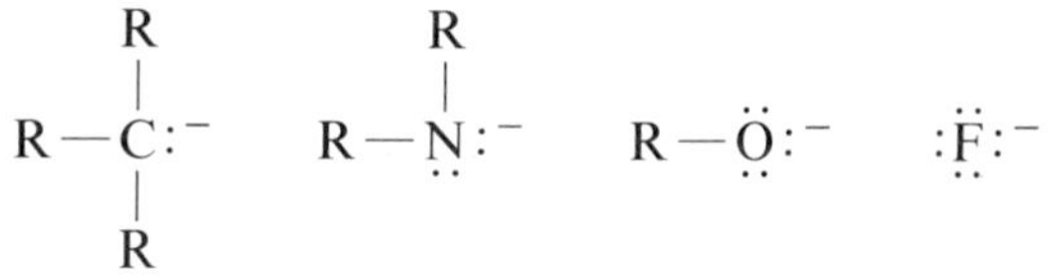

most basic / *most nucleophilic* — *least basic* / *least nucleophilic*

2. Anions are more powerful nucleophiles than their uncharged conjugate acids, for example:

more basic / *more nucleophilic* — *less basic* / *less nucleophilic*

The electrons on an atom bearing a negative charge are more loosely held and therefore more easily donated than are the electrons in a neutral molecule. Thus, alkoxide anions are better nucleophiles than alcohols, and amide anions are better nucleophiles than amines.

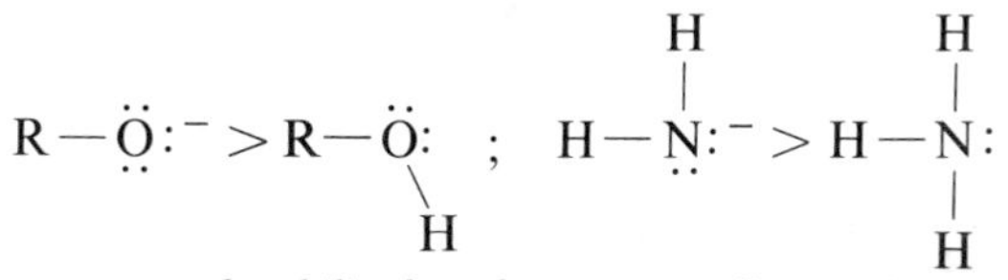

anions more nucleophilic than the corresponding conjugate acids

3. For anions from a given group in the periodic table, nucleophilicity increases going down the group, as seen in the structure below. This is the opposite of their order of basicity (p. 85).

$$HS:^- \quad > \quad HO:^-$$

less basic / *more nucleophilic* — *more basic* / *less nucleophilic*

$$:I:^- \quad :Br:^- \quad :Cl:^- \quad :F:^-$$

least basic / *most nucleophilic* — *most basic* / *least nucleophilic*

Two factors are responsible for these relative nucleophilicities. One is the effect of solvent. The experiments that led to these conclusions were carried out in water, a protic solvent that hydrogen bonds. Hydrogen bonding decreases the availability of electrons and, therefore, the nucleophilicity of an anion in solution. The effect of solvent will be discussed in greater detail in the next section.

The reactivity of an uncharged nucleophile is not much affected by changes in solvent. As with anions, the nucleophilicity of neutral molecules increases the further down in a given group of the periodic table is the atom that is the nucleophilic center. For example, if the reaction of an amine is compared to that of a phosphine, the phosphine reacts about 1000 times faster under the same conditions. That is, the second of these two reactions is 1000 times faster than the first:

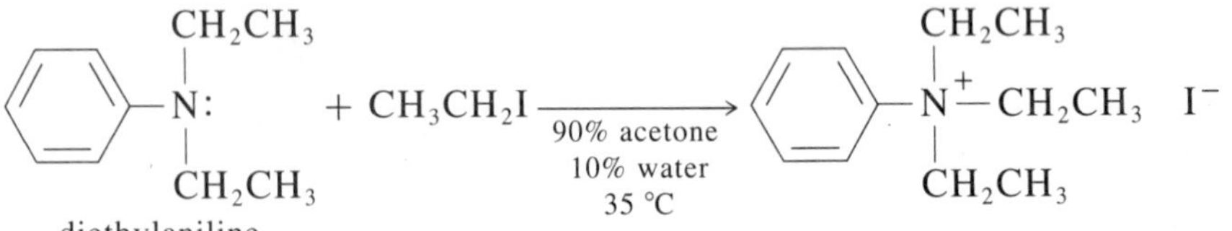
CH2CH3
N:
+ CH3CH2I
90% acetone
10% water
35 °C
CH2CH3
N+—CH2CH3 I−
CH2CH3

diethylaniline

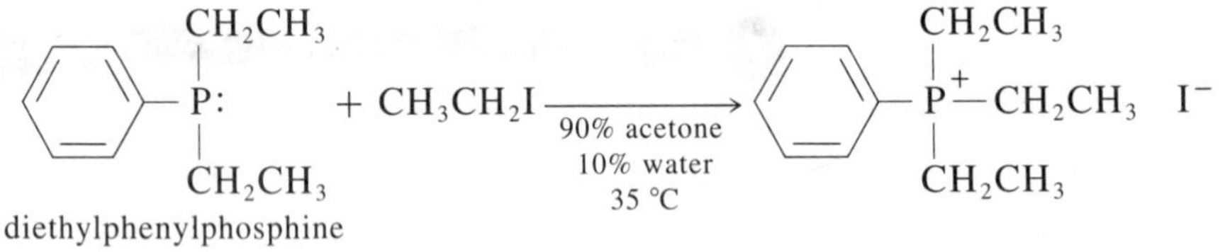
CH2CH3
P:
+ CH3CH2I
90% acetone
10% water
35 °C
CH2CH3
P+—CH2CH3 I−
CH2CH3

diethylphenylphosphine

The phosphorus atom is larger than the nitrogen atom. The outermost electrons on phosphorus are farther away from the nucleus and more loosely held. When phosphorus is close to a center of partial positive charge, its orbitals containing nonbonding electrons are easily distorted in the direction of the charge, allowing the bonding process to start more easily. The ease with which the electron cloud on an atom can be distorted is called the **polarizability** of the atom. In general, as atoms become larger going down a group in the periodic table, they also become more polarizable. They become less basic but more nucleophilic. In other words, polarizability is important in the process of bonding to a carbon atom that is an electrophilic center bearing a partial positive charge. However, polarizability is not so important in bonding to a proton, which is a center of highly concentrated positive charge.

PROBLEM 7.2

For each pair of reagents, decide which species is more nucleophilic.

(a) 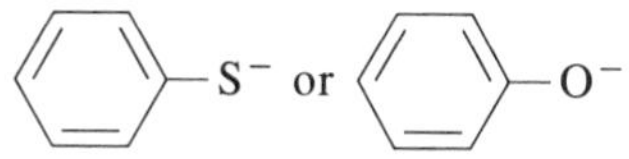 (b) OH^- or $CH_3\overset{O}{\overset{\|}{C}}O^-$

(c) OH^- or NO_3^- (d) $(CH_3)_2NH$ or NH_3 (e) OH^- or CH_3O^-

B. Solvent Effects

The basicity and nucleophilicity of a species are relative, not absolute; they depend on the other reagent in the reaction, on the other ions present in the solution, on the solvent, and on the degree of solvation of the ions. The importance of solvent in determining nucleophilicity has become much clearer in recent years as more reactions are studied in the gas phase, where relative nucleophilicities are quite different from those given on p. 226. For example, in the gas phase, fluoride ion is a much better nucleophile than bromide ion. The relative nucleophilicities given for the halide ions were determined in solvents such as water, ethanol, and methanol. These solvents have hydroxyl groups and participate in hydrogen bonding. The smaller the anion, the more concentrated is its negative charge, and the more strongly solvated it is by a solvent such as ethanol (p. 26). Thus, chloride ion will be more strongly hydrogen-bonded than bromide ion in such a solution. Hydrogen bonding diminishes the availability of the nonbonding electrons of the anion and decreases its nucleophilicity (Figure 7.1 on the next page). In hydroxylic solvents, chloride ion is a weaker nucleophile than bromide ion.

Some polar solvents lack a functional group that can serve as a proton donor in hydrogen bonding. Such polar but aprotic solvents are acetone, acetonitrile, dimethylformamide, dimethyl sulfoxide, and hexamethylphosphoric triamide.

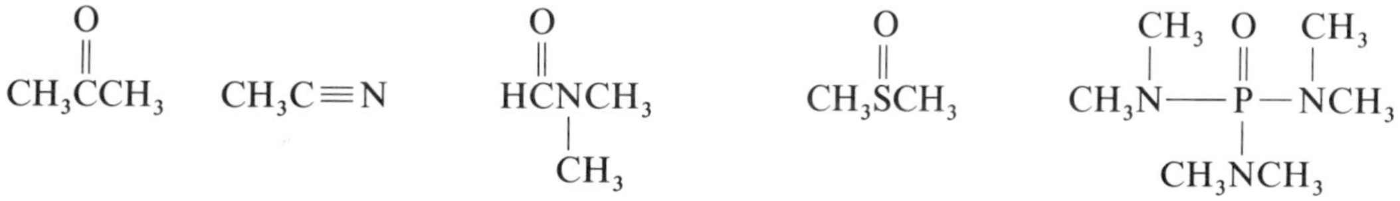

acetone	acetonitrile	dimethylformamide	dimethyl sulfoxide	hexamethylphosphoric triamide
bp 56.5 °C	bp 81.6 °C	bp 153 °C	bp 189 °C	bp 232 °C
μ 2.88	μ 3.92	μ 3.82	μ 3.96	μ 4.30
ε 20.7	ε 36.2	ε 36.7	ε 49	ε 30

All of these solvents have high dipole moments because of the presence of the nitrile group (the carbon-nitrogen triple bond) or of a polar group such as carbon, sulfur, or phosphorus doubly bonded to oxygen. These solvents also have relatively high dielectric constants (p. 27). Thus, they can be used as solvents for organic reactions involving ionic reagents or reaction intermediates. When ionic compounds are dissolved in these solvents, the anions are not solvated as strongly as they would be in hydroxylic solvents, so they are freer to participate in nucleophilic substitution reactions. Rates of nucleophilic substitution reactions in these solvents are higher than they are in alcohols.

In polar but aprotic solvents, nucleophilicity more closely approximates basicity. For example, bromide ion is a better nucleophile than chloride ion in methanol,

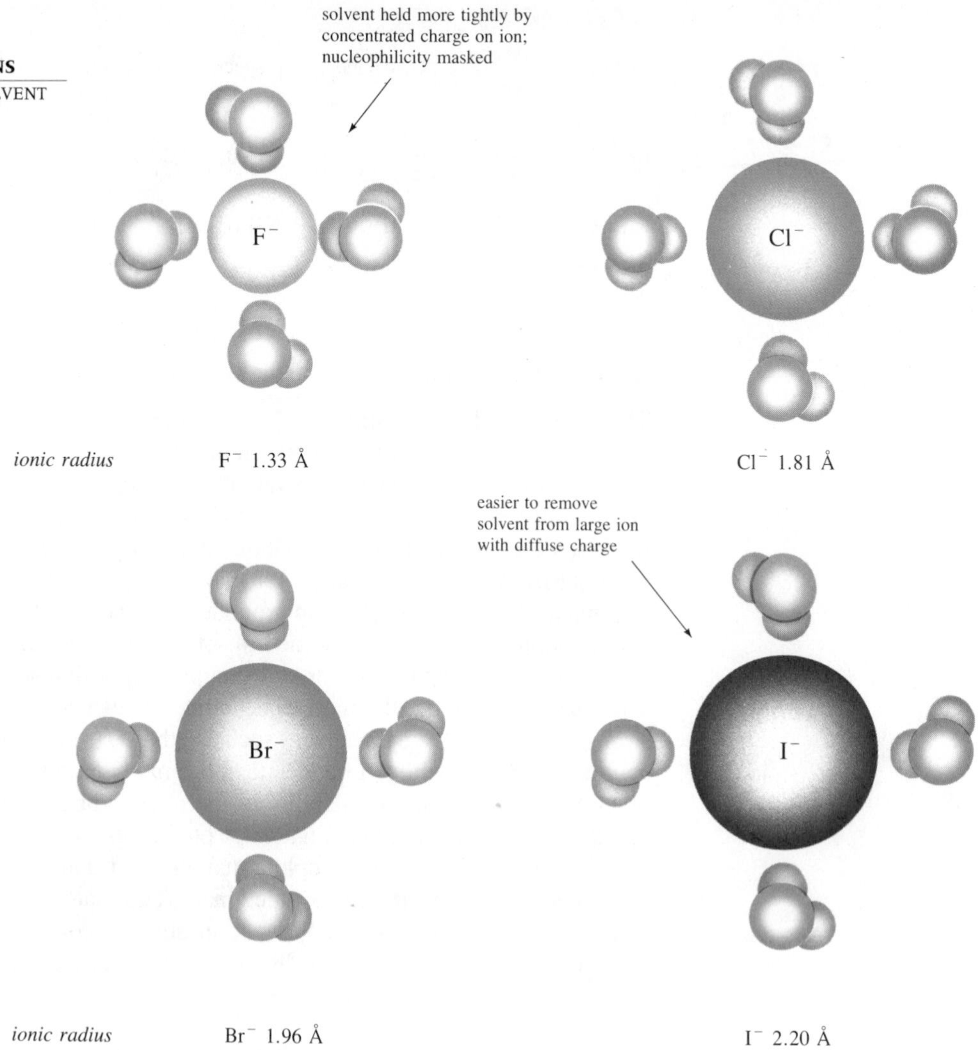
solvent held more tightly by concentrated charge on ion; nucleophilicity masked
F⁻
Cl⁻
ionic radius
F⁻ 1.33 Å
Cl⁻ 1.81 Å
easier to remove solvent from large ion with diffuse charge
Br⁻
I⁻
ionic radius
Br⁻ 1.96 Å
I⁻ 2.20 Å

FIGURE 7.1 Effect of the size of an anion on its interaction with water, a protic solvent.

but this order is reversed in dimethylformamide, where chloride ion is not involved in hydrogen bonding. The more basic chloride ion is, therefore, the better nucleophile in the aprotic solvent. Although it is not possible to assign absolute nucleophilicities to Lewis bases, the trends discussed in the last two sections provide useful generalizations.

Study Guide Concept Map 7.1

PROBLEM 7.3

The rate of the reaction of methyl iodide with chloride ion was measured in a series of solvents. The structures of the solvents and the relative rates are shown below the reaction equation. How would you explain these observations?

$$CH_3I + Cl^- \longrightarrow CH_3Cl + I^-$$

Solvent	CH_3OH	$HCNH_2$ (HC(=O)NH₂)	$HCN(CH_3)_2$ (HC(=O)N(CH₃)₂)	$CH_3CN(CH_3)_2$ (CH₃C(=O)N(CH₃)₂)
Relative rate	1	1.25×10^1	1.2×10^6	7.4×10^6

PROBLEM 7.4

(a) Hydroxylamine, H_2NOH, has two sites of potential basicity and nucleophilicity. Predict what the product will be for each of the following reactions.

1. $H_2NOH \xrightarrow{HCl}$ 2. $H_2NOH \xrightarrow{CH_3I\text{ (1 equivalent)}}$

(b) The conjugate acid of hydroxylamine has pK_a 5.97. What does this tell you about the structure of the conjugate acid? To which acid in the table of pK_a values is it comparable? How do you explain the difference in the acidities of the two species?

7.3 LEAVING GROUPS

A. Experimental Evidence for the Ionization of Alkyl Halides

When *tert*-butyl chloride is put into a solution of silver nitrate in ethanol at room temperature, a white precipitate of silver chloride forms instantly. This indicates that *tert*-butyl chloride readily gives up a chloride ion in solution. The ionization of *tert*-butyl chloride to give chloride ion suggests that a carbocation may also be formed at the same time.

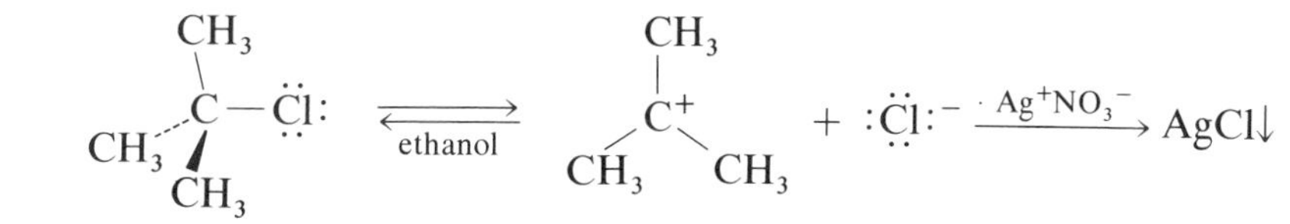

ethanol
$Ag^+NO_3^-$
AgCl↓

2-chloro-2-methylpropane *tert*-butyl chloride — *tert*-butyl cation — silver chloride

If *n*-butyl chloride is put into a solution of silver nitrate in ethanol, no precipitate forms.

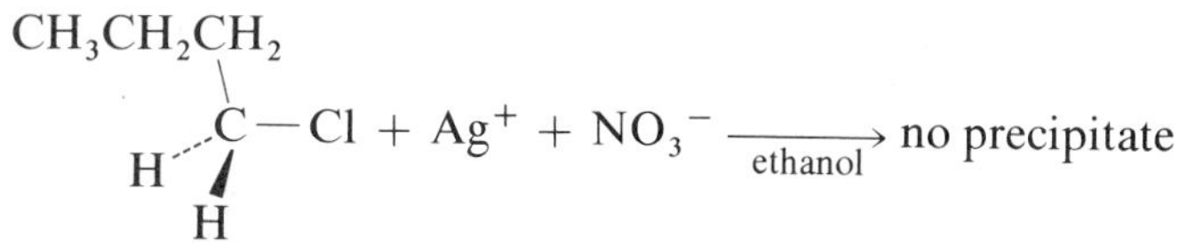

ethanol
no precipitate

n-butyl chloride

Chloride ions are not formed from *n*-butyl chloride, nor are *n*-butyl cations. Varying the structure of the alkyl halide from *tert*-butyl chloride, in which the chlorine atom is bonded to a tertiary carbon atom, to *n*-butyl chloride, in which the chlorine is bonded to a primary carbon atom, results in a dramatic difference in reactivity.

In the ionization of *tert*-butyl chloride in ethanol, the bond that is already polarized as a result of the difference in electronegativity between carbon and chlorine is broken to give *tert*-butyl cation and chloride anion. In this bond cleavage, both electrons that make up the covalent bond leave with the more electronegative atom, chlorine. This kind of bond cleavage does not take place easily in *n*-butyl chloride. The structure of *tert*-butyl chloride is such that the departing chloride ion leaves behind a tertiary carbocation, one that is stabilized by three methyl groups. In contrast, the corresponding ion from *n*-butyl chloride is a primary carbocation.

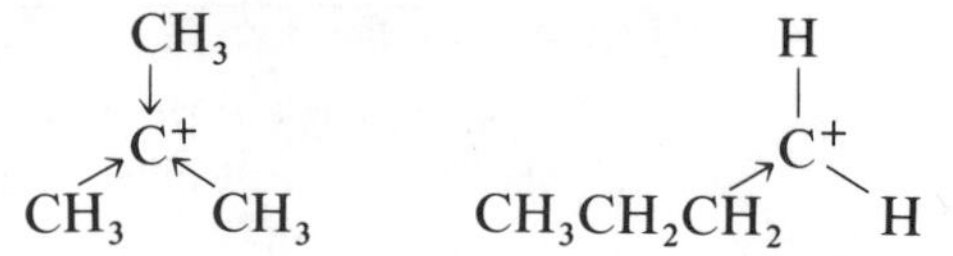

tert-butyl cation
a tertiary carbocation

n-butyl cation
a primary carbocation

The electron-releasing effect of the methyl groups sufficiently stabilizes the positive charge of the tertiary carbocation to make it a realistic intermediate for the reaction under these conditions. The primary carbocation is too unstable to be formed (p. 122). The chloride ion, as the conjugate base of the strong acid hydrochloric acid, is a stable species and a good leaving group in this ionization.

Another important factor in this reaction is the solvent. Ethanol has a dielectric constant of 24.3 and can interact with and further stabilize by solvation the ionic species that are formed (p. 229). In ethanol, energy is released by the electrostatic interactions of the negative ends of the dipoles of the oxygen-hydrogen bonds with the carbocation and the positive ends of those dipoles with the chloride ion.

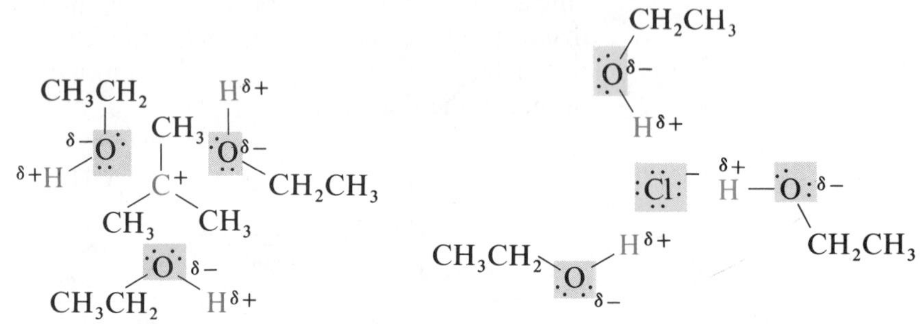

solvation of tert-*butyl cation and chloride ion by ethanol*

This energy from solvation helps to offset the energy that is necessary to break the carbon-chlorine bond. *tert*-Butyl chloride ionizes much less readily in another solvent, such as acetone. Ethanol, because of its polar oxygen-hydrogen bonds, stabilizes chloride anions, as well as carbocations, and is, thus, a better ionizing solvent than acetone. The importance of solvent in stabilizing ionic species has already been emphasized earlier in this book (pp. 27 and 93).

B. Leaving Groups and Basicity

The experimental facts presented in the last section indicate that *tert*-butyl chloride ionizes easily in an ionizing solvent such as ethanol and *n*-butyl chloride does not. This difference is attributed to the relative stabilities of the two carbocations that will be formed as intermediates. However, the stability of the potential carbocation is not the only factor that determines whether ionization takes place. For example, *tert*-butyl alcohol does not ionize in water to give *tert*-butyl cation and hydroxide ion. The difference between *tert*-butyl chloride and *tert*-butyl alcohol is in the leaving group, which is chloride ion in the one case and hydroxide ion in the other.

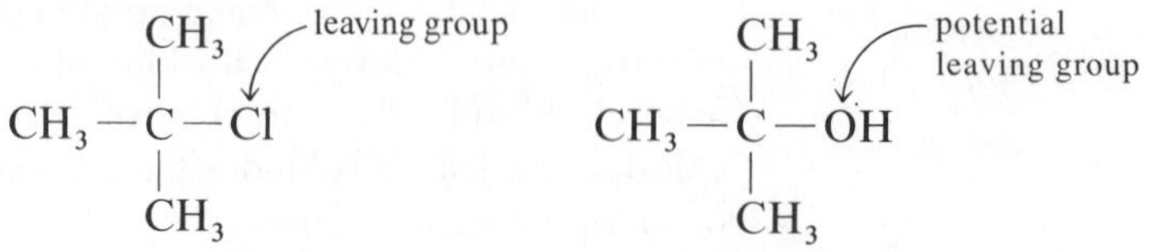
leaving group
potential
leaving group

tert-butyl chloride

tert-butyl alcohol

Hydroxide ion resembles chloride ion in being a stable species with an octet of electrons and a negative charge on an electronegative atom. It differs from chloride ion in being the conjugate base of water, a much weaker acid than hydrochloric acid, which is the conjugate acid of chloride ion.

In Chapter 3, acidity was discussed in terms of a competition between different conjugate bases for the same proton. The more successful a species was in this competition, the stronger a base it was and the weaker its conjugate acid. Strong bases with weak conjugate acids appear in the lower part of the table of pK_a values inside the front cover of the book. The table can be used to make predictions about how good an ion or molecule will be as a leaving group. The same factors that make a species a weak base also make it a good leaving group. Among the halogens, covalently bonded iodine (which becomes iodide ion, the weakest base) is the best leaving group; fluorine (which becomes fluoride ion, the strongest base) is the poorest.

$$-\mathrm{F} < -\mathrm{Cl} < -\mathrm{Br} < -\mathrm{I}$$

poorest leaving group ... *best leaving group*

For example, in aqueous ethanol, *tert*-butyl iodide reacts 2.5 times faster than *tert*-butyl bromide, which in turn reacts 44 times faster than *tert*-butyl chloride. That is, the following reaction is fastest when X = I and slowest when X = Cl.

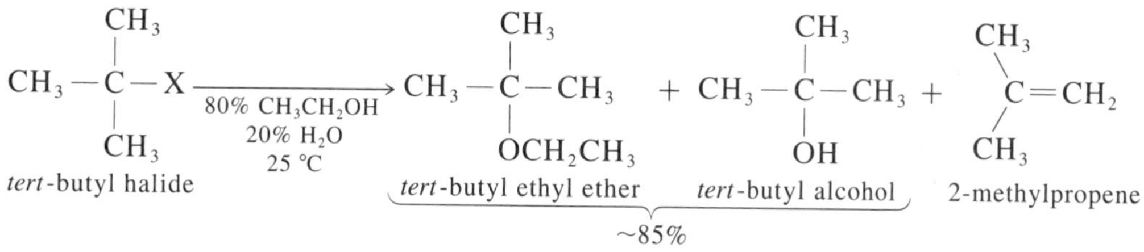
80% CH3CH2OH
20% H2O
25 °C
tert-butyl halide
tert-butyl ethyl ether
tert-butyl alcohol
~85%
2-methylpropene

Generally, the conjugate bases in the top half of the pK_a table, with pK_a values less than 8, are reasonably good leaving groups. More basic species generally are not very good leaving groups.

C. Making Leaving Groups Better

Certain kinds of species are good leaving groups, and reaction conditions can be tailored in order to make poor leaving groups into better ones. To bring about the transformation of *tert*-butyl alcohol into *tert*-butyl bromide it might appear that a hydroxyl group has to be replaced by a bromine atom.

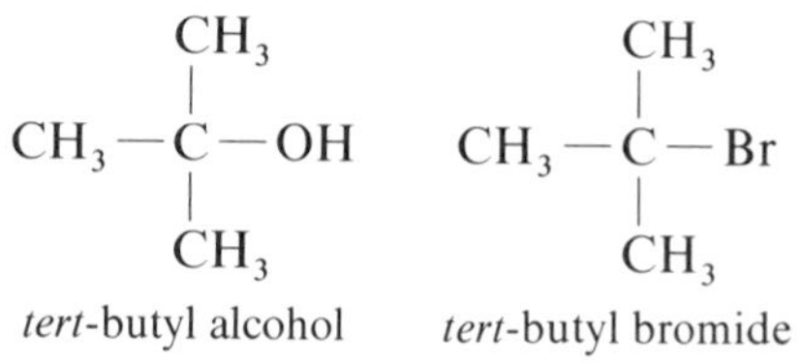
tert-butyl alcohol
tert-butyl bromide

The reaction called for seems to be a substitution achieved by addition of bromide ion to the reaction mixture to give displacement of hydroxide ion. In fact, nothing happens when sodium bromide is added to *tert*-butyl alcohol.

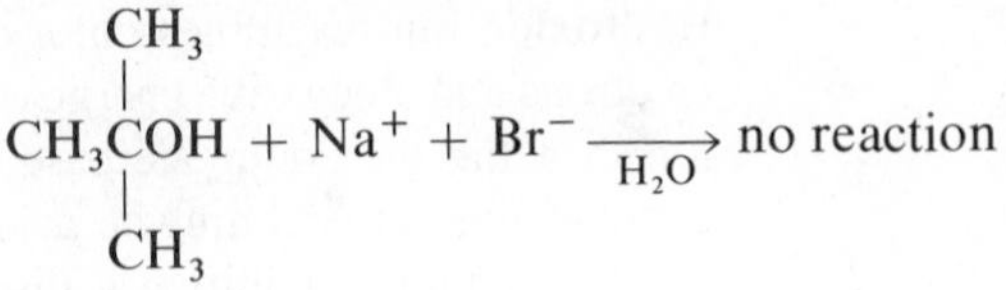
no reaction
H_2O

tert-butyl alcohol

Hydroxide ion is not a good leaving group, so the substitution reaction cannot take place.

If the reaction conditions are changed by using hydrobromic acid instead of sodium bromide, *tert*-butyl bromide is easily formed.

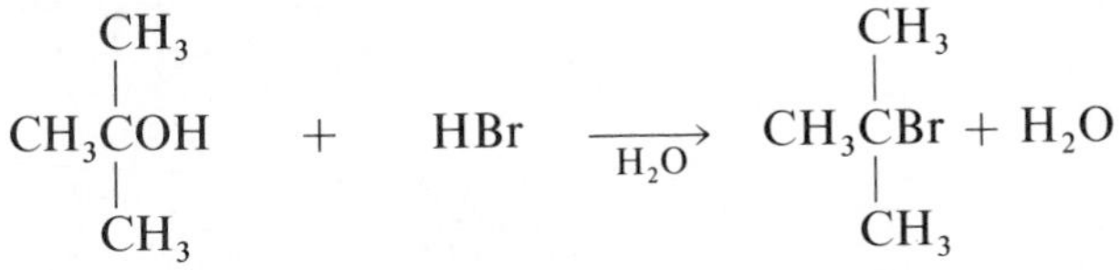

tert-butyl alcohol — hydrobromic acid — *tert*-butyl bromide

The acid protonates the unshared electrons on the oxygen atom of the alcohol to give a new leaving group, water, the conjugate base of the strong acid hydronium ion (pK_a −1.7). Water is a much better leaving group than hydroxide ion is, and ionization can take place. The *tert*-butyl cation can then combine with the nucleophile, bromide ion.

VISUALIZING THE REACTION

Converting a poor leaving group into a good one

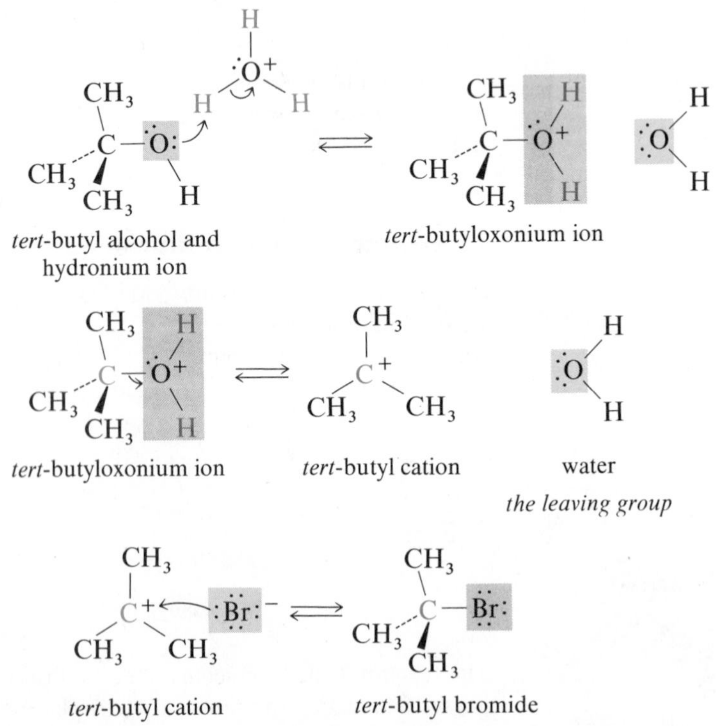
tert-butyl alcohol and hydronium ion
tert-butyloxonium ion
tert-butyloxonium ion
tert-butyl cation
water
the leaving group
tert-butyl cation
tert-butyl bromide

PROBLEM 7.5

Where did the hydronium ion in the first step of the mechanism on the previous page come from?

PROBLEM 7.6

What other nucleophiles are present in the reaction mixture created by combining *tert*-butyl alcohol, hydrobromic acid, and water? What other reactions are possible for the tertiary carbocation?

A hydroxyl group is frequently protonated in order to convert it into water, a good leaving group. But a protonated alcohol is an ionic intermediate that cannot be isolated and put in a bottle. Sometimes it is convenient to convert alcohols into stable, isolable compounds that have good leaving groups on them. A successful strategy is to convert the alcohol into the ester of the relatively strong acid *p*-toluenesulfonic acid, $pK_a \sim -0.6$. This acid is an organic derivative of sulfuric acid. *p*-Toluenesulfonyl chloride reacts easily with the alcohol.

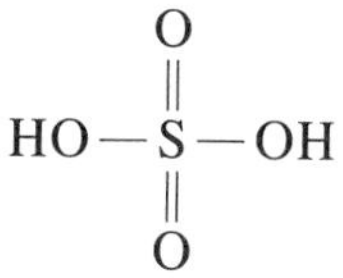

sulfuric acid

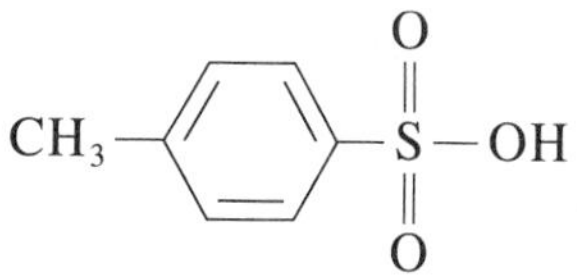

p-toluenesulfonic acid

TsOH

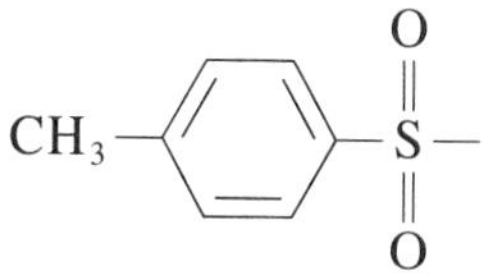

p-toluenesulfonyl group

Ts—

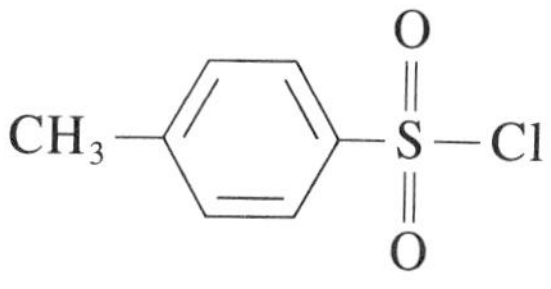

p-toluenesulfonyl chloride
tosyl chloride
TsCl

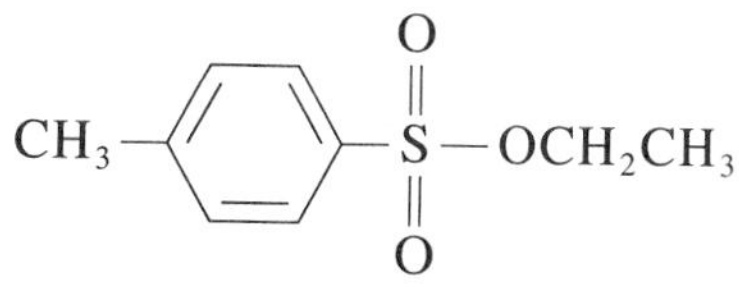

ethyl *p*-toluenesulfonate
ethyl tosylate
$TsOCH_2CH_3$

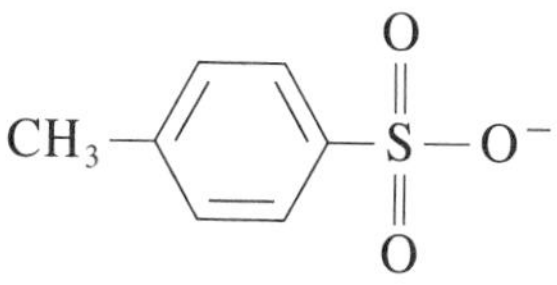

p-toluenesulfonate anion
tosylate
TsO^-

The *p*-toluenesulfonyl group has a large structure and a long name. Because it is used so often in organic reactions, it has been given an abbreviation for the sake of convenience. The name *p*-toluenesulfonyl has been condensed by taking only the underlined sections to form the short name **tosyl.** In structural formulas, the symbol Ts is used, as shown above.

An alcohol, cyclopentanol, reacts with tosyl chloride.

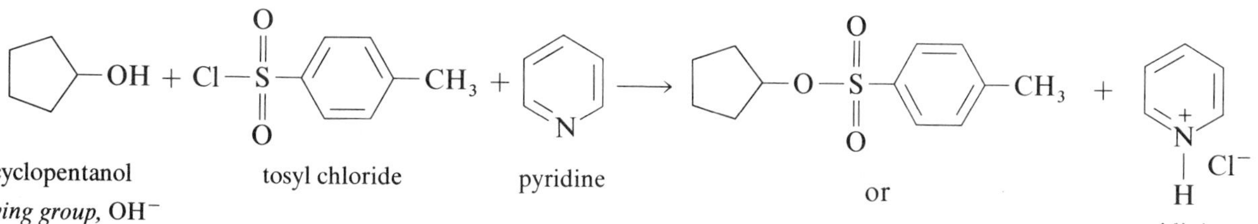

cyclopentanol
leaving group, OH^-

tosyl chloride

pyridine

or

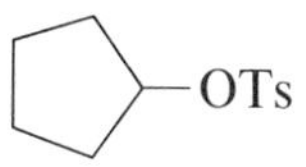

cyclopentyl tosylate
leaving group, TsO^-

pyridinium hydrochloride

A basic solvent, pyridine, is used to neutralize the hydrogen chloride that is formed. In the reaction, the poor leaving group, the hydroxide ion, is converted into a good leaving group, the tosylate anion.

VISUALIZING THE REACTION

Formation of a tosylate

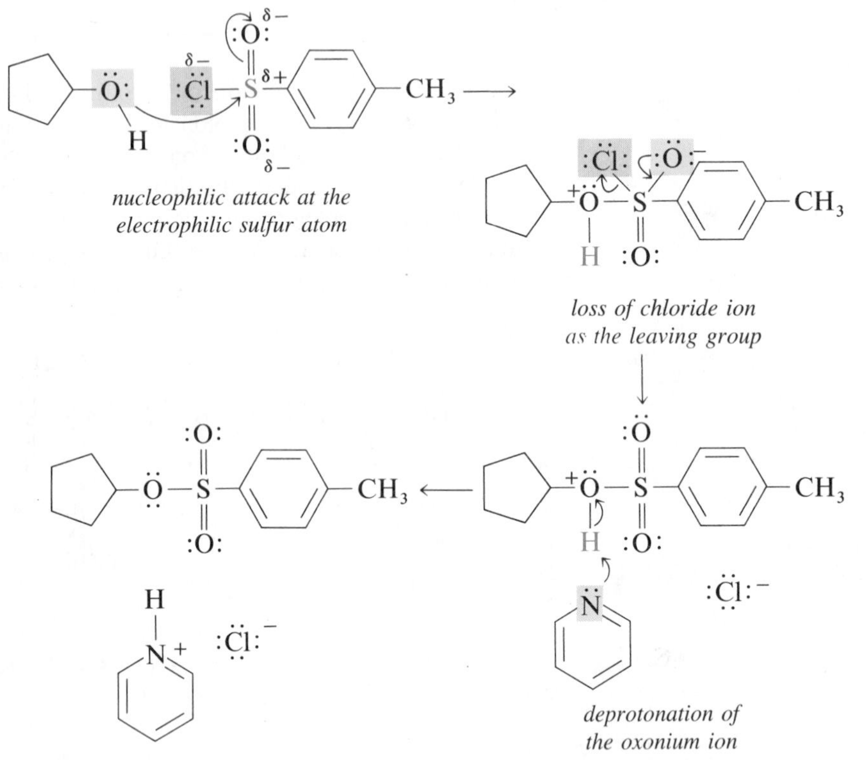
nucleophilic attack at the electrophilic sulfur atom
loss of chloride ion as the leaving group
deprotonation of the oxonium ion

Note that the alcohol is converted to a tosylate without breaking the carbon-oxygen bond in the alcohol. Later in this chapter, we will look closely at the stereochemistry of substitution reactions and will note when bonds to a stereocenter are broken. The tosylate of a chiral alcohol can be prepared without affecting stereochemistry because the bond to the stereocenter is not broken until the tosylate anion is displaced in a nucleophilic substitution reaction.

Cyclopentyl tosylate is converted mostly into ethoxycyclopentane in ethanol; *p*-toluenesulfonic acid is the other product of the reaction.

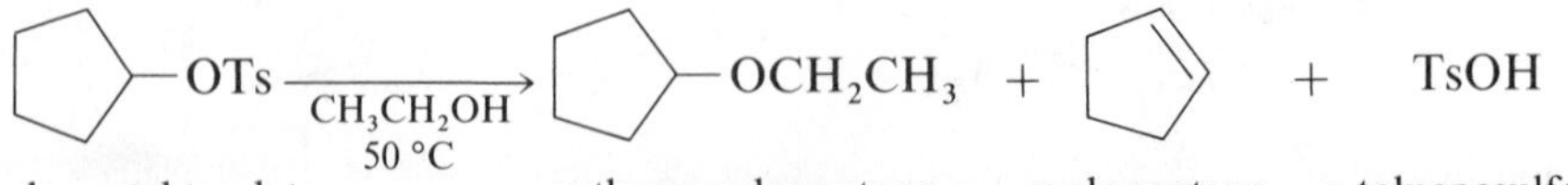
OTs
CH_3CH_2OH
50 °C
OCH_2CH_3
TsOH

cyclopentyl tosylate — ethoxycyclopentane 93% — cyclopentene 7% — *p*-toluenesulfonic acid

No ethoxycyclopentane is formed if cyclopentanol is put into ethanol. The substitution reaction takes place only when a poor leaving group has been converted into a good leaving group.

Study Guide Concept Map 7.2

A. Experimental Evidence for Substitution Without Carbocation Intermediates

When *n*-butyl bromide is put into a solution of sodium iodide in acetone, a white precipitate of sodium bromide appears. Sodium bromide precipitates because it is not very soluble in acetone.

$$\underset{n\text{-butyl bromide}}{CH_3CH_2CH_2CH_2Br} + Na^+ + I^- \xrightarrow[\text{25 °C}]{\text{acetone}} \underset{n\text{-butyl iodide}}{CH_3CH_2CH_2CH_2I} + Na^+Br^-\downarrow$$

fast reaction

Isopropyl bromide reacts with sodium iodide in acetone, but more slowly than *n*-butyl bromide does.

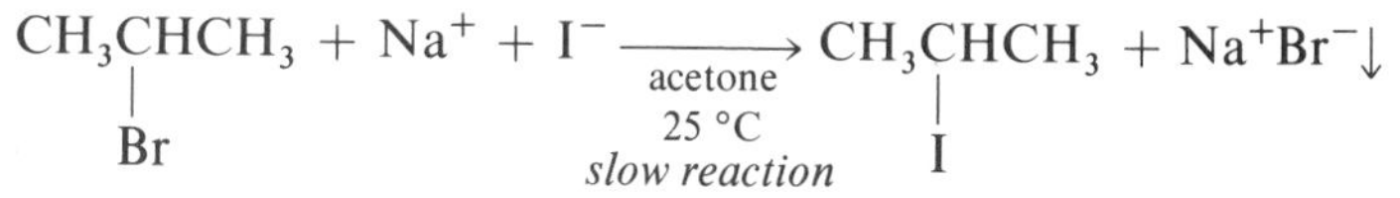

isopropyl bromide   isopropyl iodide

If *tert*-butyl bromide is placed in the same solution, no precipitate is observed and no reaction takes place.

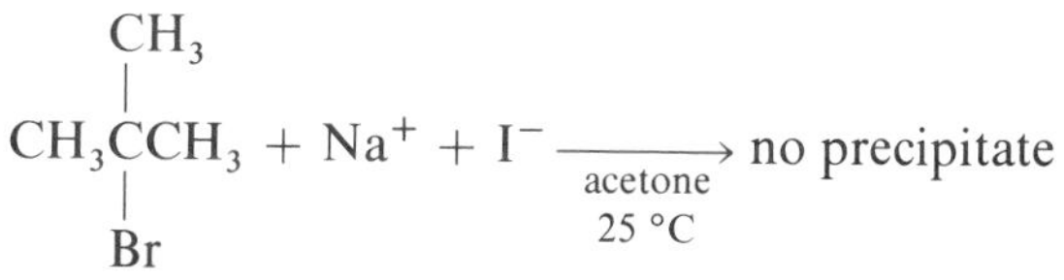

tert-butyl bromide

This order of reactivity does not reflect the order of stability of carbocations. *tert*-Butyl bromide, which would give the most stable cation in the series, does not react to a noticeable extent with this reagent. No sodium bromide precipitates, indicating that bromide ions are not formed from *tert*-butyl bromide under these conditions and no substitution reaction takes place. Acetone is not a good ionizing solvent, so any reaction proceeding in this solvent must go by a pathway that does not involve the formation of carbocations.

PROBLEM 7.7

The order of reactivity described above for the reactions of *n*-butyl bromide, isopropyl bromide, and *tert*-butyl bromide with sodium iodide in acetone comes from a laboratory experiment carried out by students in organic chemistry laboratory courses. Unfortunately, they very often do not see these results because they use *wet* test tubes for the experiment in spite of being given careful instructions that the test tubes must be *dry*. Why is the absence of water so important? What effect does water have?

B. The Kinetics of Substitution Reactions of Primary Alkyl Halides. The S_N2 Reaction

Chemists often study the mechanism of a reaction by investigating how the rate of the reaction depends on the concentrations of the reacting species (p. 113). Measurements of the rate of the reaction of *n*-butyl bromide with iodide ion in acetone show that it depends on the concentrations of both reactants. An increase in the concentration of either reactant causes a proportional increase in the rate of the reaction. The rate expression for this second-order reaction is

$$R = k[n\text{-BuBr}][\text{I}^-] \qquad \text{Bu} \equiv \text{butyl}$$

where R is the rate of the reaction and k is the second-order rate constant for this particular reaction at a given temperature and in a given solvent.

PROBLEM 7.8

The rate constant, k, for the reaction

$$CH_3CH_2Br + OH^- \xrightarrow[\text{55 °C}]{\text{80\% ethanol}} CH_3CH_2OH + Br^-$$

is 1.7×10^{-3} L/mol · s. What is the initial rate of the reaction when 0.05 M ethyl bromide is allowed to react with 0.07 M sodium hydroxide? What are the units in which the rate of the reaction, R, is expressed? What would the initial rate of the reaction be if 0.1 M ethyl bromide were used instead?

PROBLEM 7.9

Suppose 0.06 M *n*-butyl bromide is allowed to react with 0.02 M potassium iodide in acetone at 25 °C. The rate constant, k, for the reaction at that temperature is 1.09×10^{-1} L/mol · min. What is the overall rate at which *n*-butyl bromide disappears from the reaction mixture under these conditions?

Experimental evidence indicates that no reactive intermediate, such as a carbocation, is involved in the reaction of *n*-butyl bromide with iodide ion (p. 237). The energy diagram for the reaction would show the reactants being converted to products by way of one transition state, with no energy minimum representing a reactive intermediate (see Figure 4.1, p. 116).

For the reaction of *n*-butyl bromide and iodide ion, the transition state is thought to involve both species.

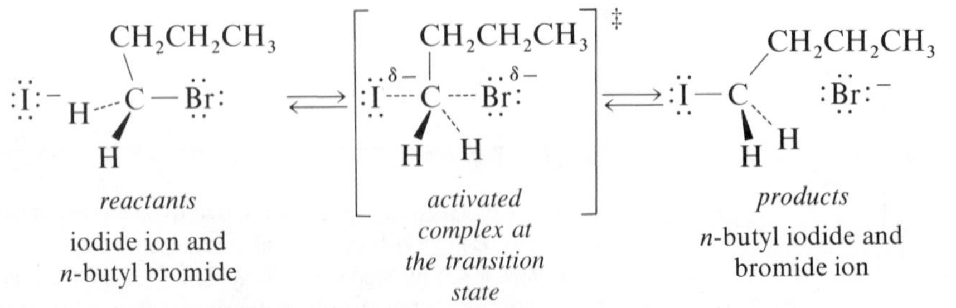
reactants
iodide ion and
n-butyl bromide
activated
complex at
the transition
state
products
n-butyl iodide and
bromide ion

Iodide ion, a nucleophile, approaches *n*-butyl bromide. The electrons on the iodide ion are drawn toward the electrophilic carbon atom, while the bond to the bromine atom loosens and lengthens. The lowest-energy way for the iodide ion to approach the alkyl halide molecule, and the least crowded, is from the side opposite the departing bromide ion. In this way, the negative charge on the iodide ion and the developing negative charge on the bromide ion can be kept as far apart as possible. At the transition state, there is partial bonding between carbon and iodine. The carbon atom is also still partially bonded to the bromine atom, which is beginning to become a bromide ion. In the activated complex at the transition state, there are five groups around the carbon atom; it is **pentacoordinated.** The activated complex can revert to starting materials or proceed to products.

PROBLEM 7.10

The energy of activation for the reaction of *n*-butyl chloride with iodide ion to give *n*-butyl iodide has been determined to be 22.2 kcal/mol. The frequency factor, *A* (p. 117), for the reaction is 2.24×10^{11} L/mol·s. What is the rate constant, *k*, for the reaction at 60 °C?

The reaction of *n*-butyl bromide with iodide ion belongs to a class of reactions known as **bimolecular nucleophilic substitution reactions.** Bimolecular refers to the fact that *two* species undergo bonding changes in the transition state of such a reaction. The reaction is initiated by the approach of a nucleophile and is therefore a nucleophilic substitution reaction. Chemists refer to these reactions as **S_N2 reactions,** where the letters and number stand for substitution, nucleophilic, and bimolecular, respectively. An S_N2 reaction takes place in one step. The reaction of hydroxide ion and chloromethane, discussed in Chapter 4, is another S_N2 reaction.

PROBLEM 7.11

When ethanol is heated with a small amount of sulfuric acid, diethyl ether is formed and can be distilled out of the reaction mixture. This is, in fact, how the ether once used for anesthesia was prepared.

$$2\,CH_3CH_2OH \xrightarrow[\Delta]{H_2SO_4} CH_3CH_2OCH_2CH_3 + H_2O$$

Write a detailed mechanism for this reaction, applying all the ideas developed so far about acids, bases, conversion of poor leaving groups into good leaving groups, and displacement of leaving groups by nucleophiles.

PROBLEM 7.12

When diethyl ether is heated with hydrogen iodide, iodoethane is the product.

$$CH_3CH_2OCH_2CH_3 + 2\,HI \xrightarrow{\Delta} 2\,CH_3CH_2I + H_2O$$

Write a detailed mechanism for this reaction.

C. Steric Effects in S_N2 Reactions

An experiment to explore the effect of structure on reactivity in S_N2 reactions used radioactive bromide ion to see how fast it was incorporated into different alkyl bromides in acetone at 25 °C. The relative rates of reaction for different alkyl groups are given in Table 7.1. Methyl bromide reacts the fastest. Even ethyl bromide, in which a methyl group has replaced one of the hydrogen atoms of methyl bromide, reacts much more slowly than methyl bromide. The reaction is

TABLE 7.1 Relative Rates for the Reaction of Alkyl Bromides, RBr, with Radioactive Bromide Ion in Acetone at 25 °C

$$Br^{*-} + R—Br \underset{25\ °C}{\overset{acetone}{\rightleftharpoons}} R—Br^{*} + Br^{-}$$

R	$CH_3—$	$CH_3CH_2—$	$CH_3CH_2CH_2—$	$CH_3CH(CH_3)—$	$CH_3C(CH_3)_2—$
	methyl	ethyl	*n*-propyl	isopropyl	*tert*-butyl
Relative rates	100	1.31	0.81	0.015	0.004
Nature of R		primary	primary	secondary	tertiary

very slow for the tertiary halide *tert*-butyl bromide, and what reaction does occur probably goes by way of the *tert*-butyl cation.

$CH_3—$ > primary > secondary > tertiary halides
fastest → *slowest*
decreasing rate of reaction in S_N2 reactions

Why is the bimolecular substitution reaction slower for a secondary halide than for a primary one, and why does the reaction hardly go at all with a tertiary halide? The transition states for these reactions become progressively more crowded as the number of alkyl groups on the carbon atom undergoing substitution increases. The transition state for the reaction of iodide ion with *tert*-butyl bromide is compared to that for its reaction with methyl bromide in Figure 7.2. A methyl group is considerably larger than a hydrogen atom in its effective radius. The three methyl groups on *tert*-butyl bromide hinder the approach of the iodide ion.

A transition state with five groups crowded around the central carbon atom is a high-energy transition state. Only a few molecules in a reaction system will have enough energy to react by that pathway. When the solvent does not help the process of ionization, there is no lower-energy pathway available for substitution at the tertiary carbon atom, and reaction is very slow. The structure of the alkyl halide, especially the number of alkyl groups bonded to the carbon atom undergoing substitution, determines how high the energy of the transition state for the reaction will be. Reactions that can go through transition states of relatively low energy will proceed faster than those that must reach high-energy states.

Whenever the structure of the alkyl part of the alkyl halide is branched in such a way as to hinder the approach of the nucleophile to the back side of the carbon atom that will undergo substitution, the rate of the reaction decreases. This effect is seen in the relative rates for two series of compounds reacting with the strong nucleophile ethoxide anion (Table 7.2). In the first series, the alkyl group becomes larger from left to right, from methyl to *n*-pentyl. The biggest drop in rate occurs between the methyl and ethyl groups. For the three larger alkyl groups, the rates do not differ much. In the second series, branching increases at the carbon next to the reaction site (the β-position), apparently hindering the bimolecular reaction severely. The rate of the S_N2 reaction falls off rapidly from the propyl group (with one methyl group at the β-carbon) to the isobutyl group (with two methyl groups at the β-carbon). In neopentyl bromide, with three methyl groups at the β-carbon,

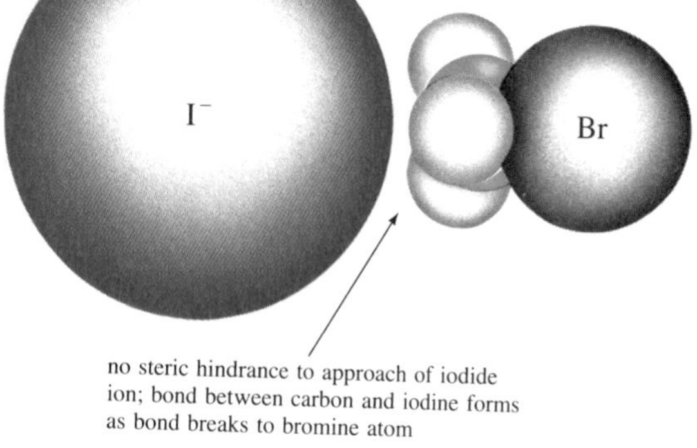
I⁻
Br
no steric hindrance to approach of iodide ion; bond between carbon and iodine forms as bond breaks to bromine atom

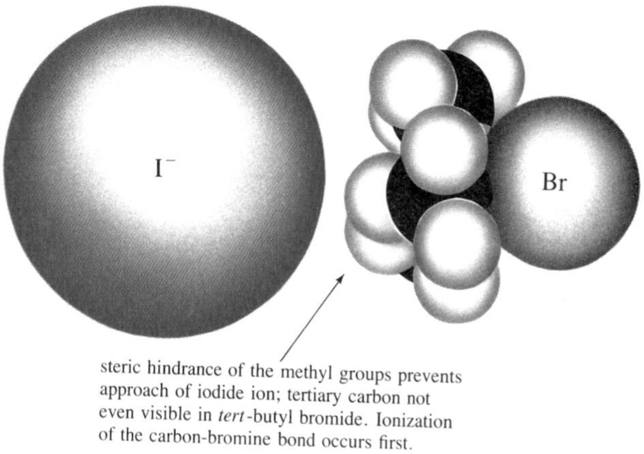
I⁻
Br
steric hindrance of the methyl groups prevents approach of iodide ion; tertiary carbon not even visible in tert-butyl bromide. Ionization of the carbon-bromine bond occurs first.

FIGURE 7.2 A comparison of the reactions of iodide ion with methyl bromide and with *tert*-butyl bromide.

TABLE 7.2 Relative Rates for the Reaction of Alkyl Bromides, RBr, with Ethoxide Anion in Ethanol at 55 °C

$$RBr + CH_3CH_2O^- \xrightarrow[55\ ^\circ C]{ethanol} ROCH_2CH_3 + Br^-$$

R	$CH_3—$ methyl	$CH_3CH_2—$ ethyl	$CH_3CH_2CH_2—$ *n*-propyl	$CH_3CH_2CH_2CH_2—$ *n*-butyl	$CH_3CH_2CH_2CH_2CH_2—$ *n*-pentyl
Relative rates	34.3	1.95	0.60	0.44	0.41
R	$CH_3—$ methyl	$\underset{\beta}{C}H_3\underset{\alpha}{C}H_2—$ ethyl	$CH_3\underset{\beta}{C}H_2\underset{\alpha}{C}H_2—$ *n*-propyl	$CH_3\underset{\beta}{C}H(CH_3)\underset{\alpha}{C}H_2—$ isobutyl	$CH_3\underset{\beta}{C}(CH_3)_2\underset{\alpha}{C}H_2—$ neopentyl
Relative rates	34.3	1.95	0.60	0.058	0.0000083

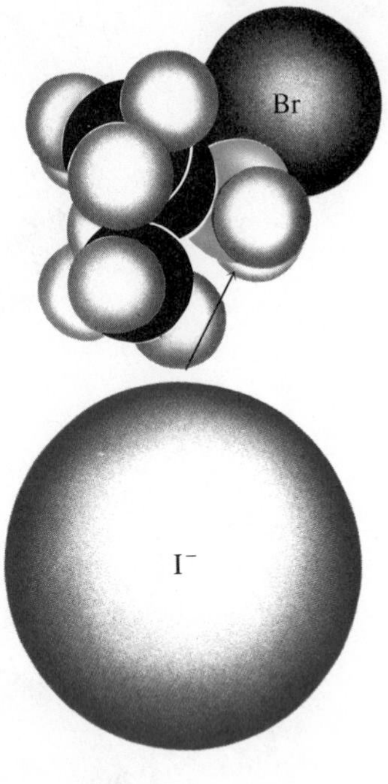
Br
I⁻

FIGURE 7.3 Methyl groups bonded to the β-carbon atom of neopentyl bromide hindering the approach of the nucleophile.

the bimolecular reaction is extremely slow (Figure 7.3). In fact, neopentyl halides react in substitution reactions only if they are placed in good ionizing solvents so that a carbocation intermediate can form.

D. Stereochemical Evidence for S_N2 Reactions

When (*S*)-(+)-2-bromobutane reacts with iodide ion in acetone, the 2-iodobutane that is formed has the *R* configuration. The iodine atom has become attached to the stereocenter on the side opposite to the position that was occupied by the bromine atom. This result is evidence in favor of the transition state for the S_N2 reaction that is an activated complex involving pentacoordinated carbon (p. 239).

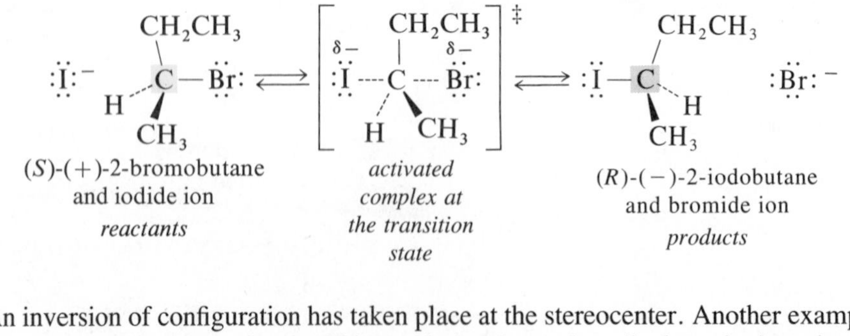
CH2CH3
:I:⁻
C
Br:
H
CH3
δ−
δ−
‡
(S)-(+)-2-bromobutane and iodide ion
reactants
activated complex at the transition state
(R)-(−)-2-iodobutane and bromide ion
products

An inversion of configuration has taken place at the stereocenter. Another example may make this a little clearer. If bromide ion is used to displace the bromine in (*S*)-(+)-2-bromobutane, the original chiral compound is converted into its enantiomer, (*R*)-(−)-2-bromobutane.

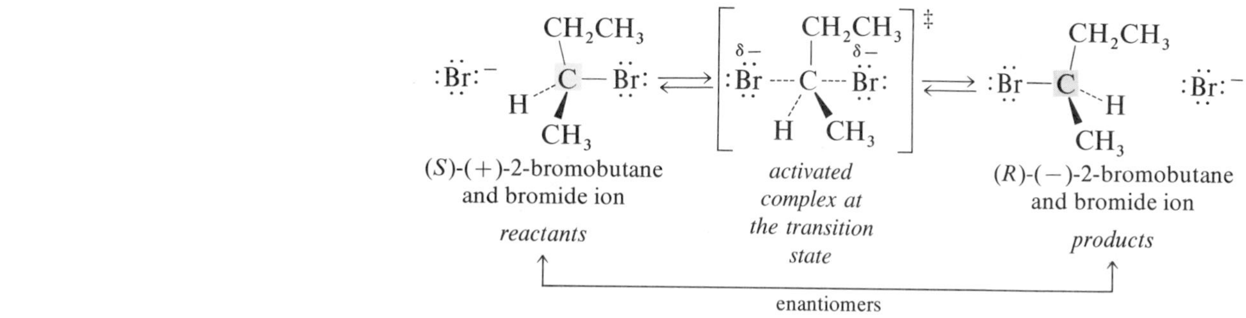
(S)-(+)-2-bromobutane
and bromide ion
reactants
activated
complex at
the transition
state
(R)-(−)-2-bromobutane
and bromide ion
products
enantiomers

Remember that the assignment of *S* and *R* designations can be made just by looking at the three-dimensional pictures. The use of (+) or (−) indicates whether the compound actually rotates the plane of polarized light to the right or left, an experimental determination that cannot be assigned by looking at the structure. Once the sign of rotation is determined for one enantiomer, the other must have optical rotation of the opposite sign (p. 198).

PROBLEM 7.13

What will happen to the optical rotation observed for pure (*S*)-(+)-2-bromobutane after it has been exposed to bromide ion in acetone over a period of time?

Inversion of configuration is the conversion of a stereocenter from one configuration to the opposite one. It was discovered in 1893 by Paul Walden at the University of Rostock. This stereochemical transformation is known as the Walden inversion. It can take place only if bonds are broken and reformed at the stereocenter. If the species that undergoes inversion has the same groups on it as before inversion, the product is the enantiomer of the starting compound. If the groups are different, an inversion has taken place if the point of attachment of the new group is on the opposite side of the stereocenter from that of the original leaving group. In the iodide displacement reaction, iodide ion displaces bromide ion with inversion of configuration because its point of attachment to the stereocenter is clearly on the opposite side of the molecule from where the bromine atom originally was.

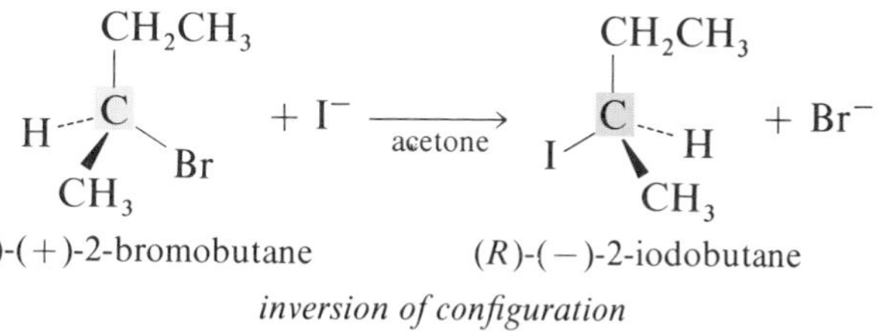
acetone
(S)-(+)-2-bromobutane
(R)-(−)-2-iodobutane

inversion of configuration

Not all substitution reactions proceed with inversion of configuration. If (*R*)-(−)-2-butanol is converted into its tosylate using tosyl chloride, the tosylate also has the *R* configuration because no bonds to the stereocenter are broken in this reaction (and nothing is done to any group that changes its priority in the rules for nomenclature of chiral compounds). The configuration of the molecule is untouched; the reaction takes place with **retention of configuration.** The tosylate anion is a good leaving group, and a nucleophilic substitution reaction with iodide ion gives 2-iodobutane with the *S* configuration. This reaction goes by an S_N2 mechanism with inversion of configuration at the chiral center.

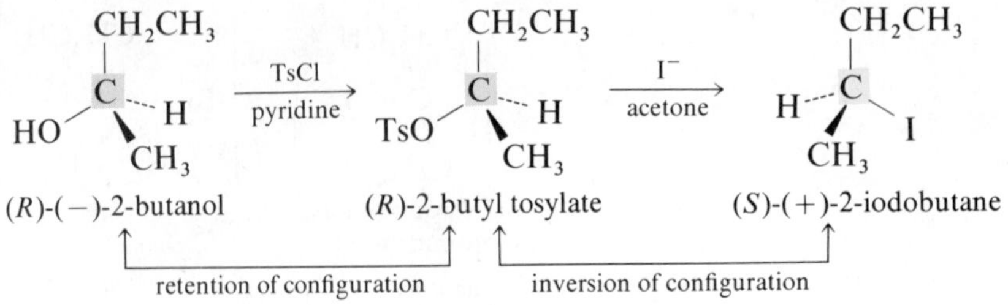
TsCl
pyridine
acetone
(R)-(−)-2-butanol
(R)-2-butyl tosylate
(S)-(+)-2-iodobutane
retention of configuration
inversion of configuration

Study Guide Concept Map 7.3

All experimental evidence shows that when a bimolecular nucleophilic substitution reaction takes place, there is a complete inversion of configuration at the carbon atom undergoing the substitution.

PROBLEM 7.14

You can obtain experimental evidence for inversion of configuration by carefully working out the configurational relationships of different optically active compounds. One such cycle is outlined below. Look at all of the reactions, decide when bonds are being broken at a stereocenter and when not, and predict for each reaction whether it goes with inversion or retention of configuration. (Hint: You already know the answers for the first and third reactions.)

1. 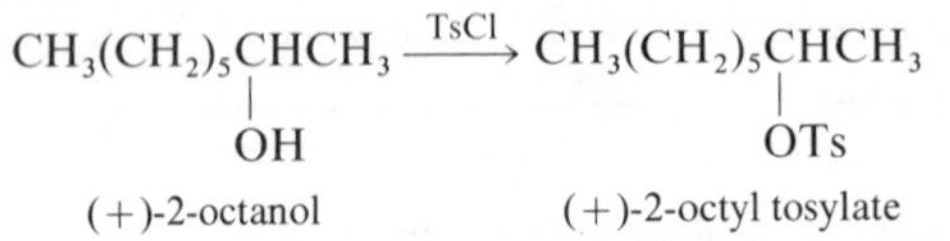
TsCl
(+)-2-octanol
(+)-2-octyl tosylate

2. 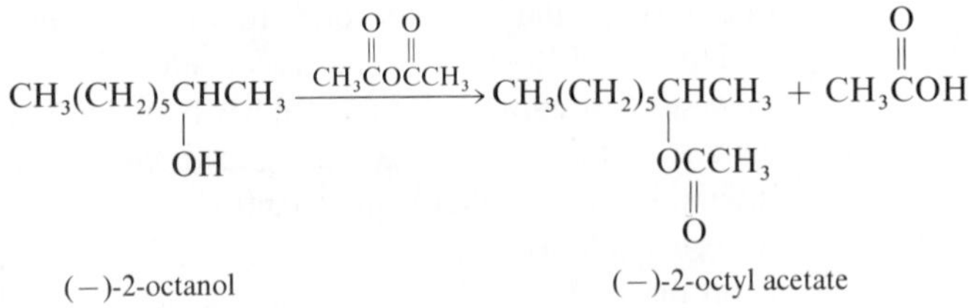
(−)-2-octanol
(−)-2-octyl acetate

3. 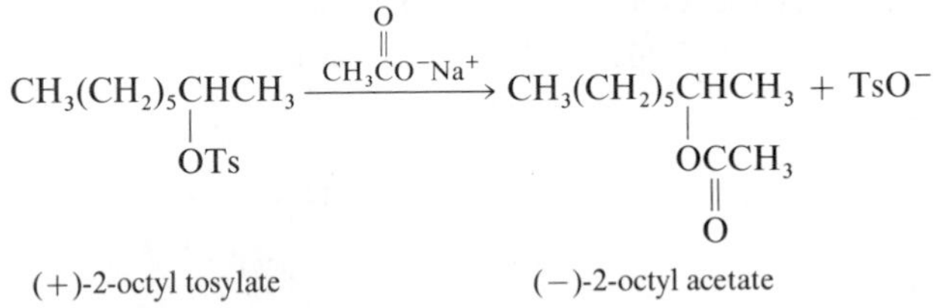
(+)-2-octyl tosylate
(−)-2-octyl acetate

PROBLEM 7.15

In one of the classic experiments used to explore the stereochemistry of S_N2 reactions, two different rates were measured for the reaction shown below. Radioactive iodide ion was used to measure the rate at which iodine was substituted in (*S*)-(+)-2-iodooctane. The rate at which the starting alkyl iodide lost its optical activity, the **rate of racemization,** was also measured. The rate of racemization is twice the rate of the substitution reaction. Explain these observations.

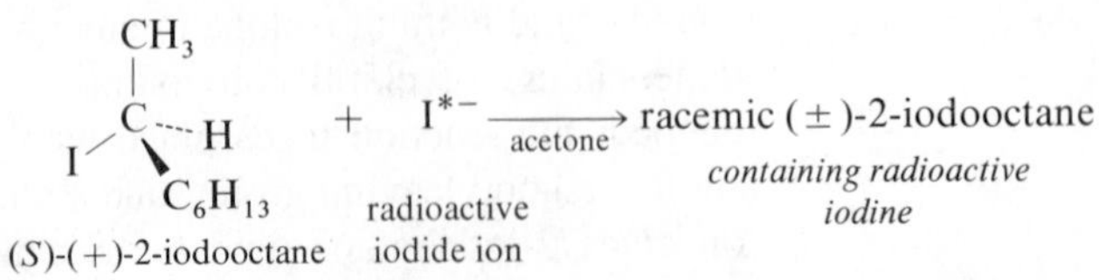
acetone
racemic (±)-2-iodooctane
containing radioactive iodine
radioactive iodide ion
(S)-(+)-2-iodooctane

7.5 KINETICS AND STEREOCHEMISTRY FOR SUBSTITUTION REACTIONS WITH CARBOCATIONS AS INTERMEDIATES

A. The Kinetics of Substitution Reactions of *tert*-Butyl Chloride. The S_N1 Reaction

When *tert*-butyl chloride reacts with potassium hydroxide in a solvent mixture of 80% ethanol and 20% water at 25 °C, it is observed that the rate at which the alkyl chloride is converted to product is independent of the concentration of the base. The reaction proceeds at the same rate even if the solution is allowed to become acidic.

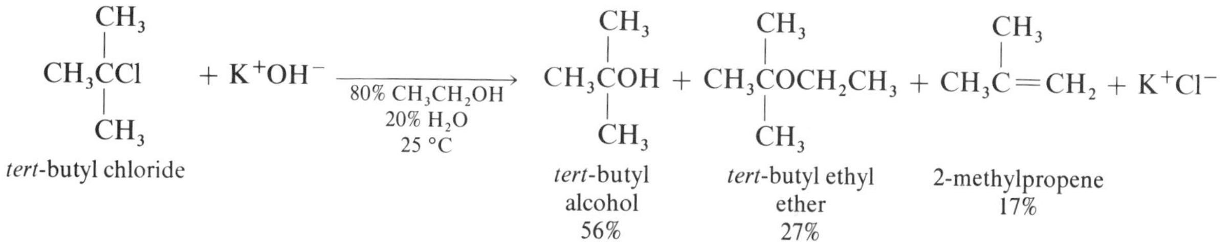

tert-butyl chloride
80% CH3CH2OH
20% H2O
25 °C
tert-butyl alcohol 56%
tert-butyl ethyl ether 27%
2-methylpropene 17%

The rate of the reaction is proportional only to the concentration of alkyl halide in the reaction mixture. The reaction is said to be first order in alkyl halide concentration.

$$R = k[tert\text{-BuCl}]$$

The rate-determining step (p. 126) is the ionization of the alkyl halide. Only one species is undergoing changes in bonding in the transition state for the rate-determining step, so this reaction is classified as a **unimolecular reaction.**

VISUALIZING THE REACTION

Slow step of the reaction of a tertiary alkyl halide

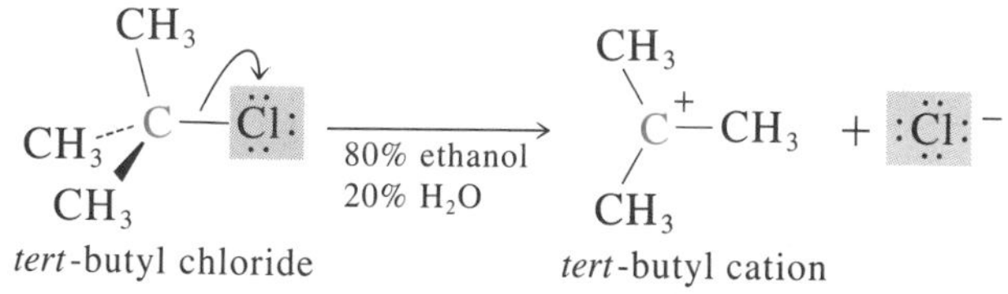

tert-butyl chloride
80% ethanol
20% H2O
tert-butyl cation

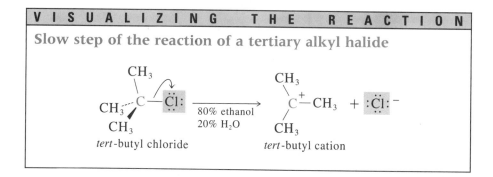

The high-energy step for the reaction is the breaking of the carbon-chlorine bond, assisted by the solvation of the ions that are formed. In an experiment to determine how solvent affects the rate of the unimolecular reaction, it was found that the rate of the reaction becomes 750 times faster if the solvent mixture is changed from 90% ethanol and 10% water to 40% ethanol and 60% water, indicating how important solvation is. Water, with a higher dielectric constant (ε 78.5) than ethanol (ε 24.3), increases the ionizing capacity of the solvent mixture and better solvates the ions that are formed. Because the transition state has some of the character of the product ions, the more polar solvent also helps to stabilize the transition state and lowers the energy of activation for the reaction. This increases the rate of the reaction.

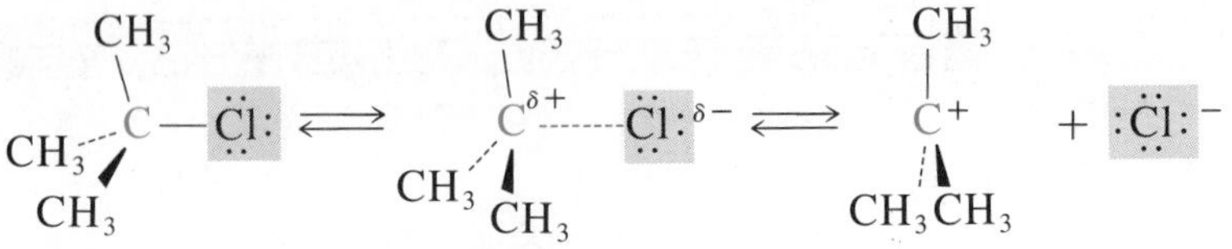
reactant
transition state
carbocation intermediate
stabilized by polar solvent
product of the slow step of the reaction

Once the carbocation is formed, it has two reaction pathways open to it: reaction with a nucleophile or loss of a proton. Both of these reactions are fast compared to the ionization that is the rate-determining step. The carbocation is a Lewis acid, which means it is an electrophile (p. 110) that lacks a pair of electrons and can accept them to attain a stable octet. When the *tert*-butyl cation is formed in ethanol, the cation and the alcohol, which is a nucleophile, react. The oxonium ion formed is a strong acid and loses a proton to the solvent.

VISUALIZING THE REACTION

Reaction of a carbocation with a nucleophilic solvent

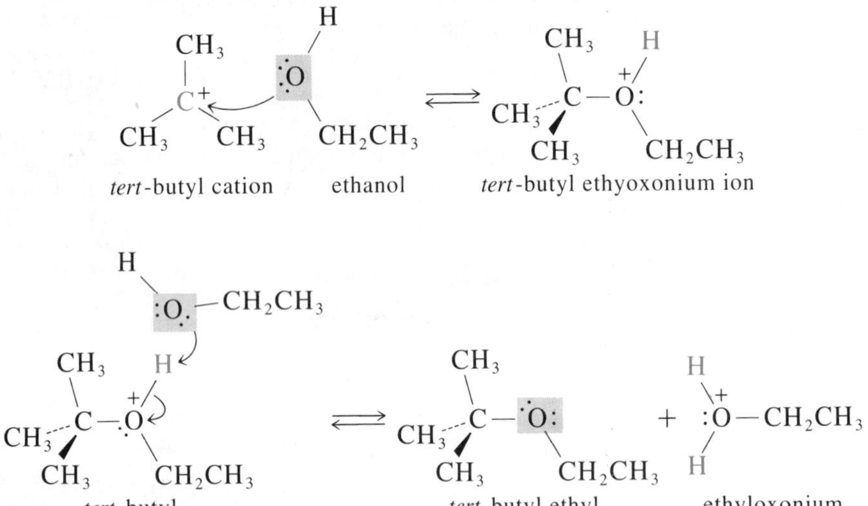
tert-butyl cation
ethanol
tert-butyl ethyoxonium ion
tert-butyl ethyloxonium ion
pKa ~ −3.6
tert-butyl ethyl ether
ethyloxonium ion
pKa ~ −2.4

In the above reaction, the substitution product is derived from the solvent. A substitution reaction in which the solvent acts as the nucleophile is called a **solvolysis reaction.**

When there are a number of potential bases in the reaction mixture, as there are in this case, the symbol :B is often used in equations to represent any one of them. The last step of the solvolysis reaction can be rewritten using this symbol, as shown on the following page.

VISUALIZING THE REACTION

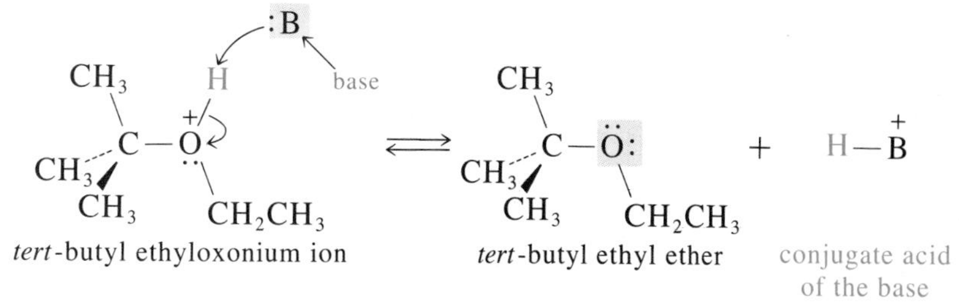

:B
base
CH_3
H
CH_3
CH_3
CH_2CH_3
tert-butyl ethyloxonium ion
tert-butyl ethyl ether
H—B
conjugate acid of the base

PROBLEM 7.16

Of all the species present in the reaction mixture for the first equation on p. 245, make a list of those that can act as bases.

The intermediate tertiary carbocation also reacts rapidly with the other nucleophiles (water and hydroxide ion).

VISUALIZING THE REACTION

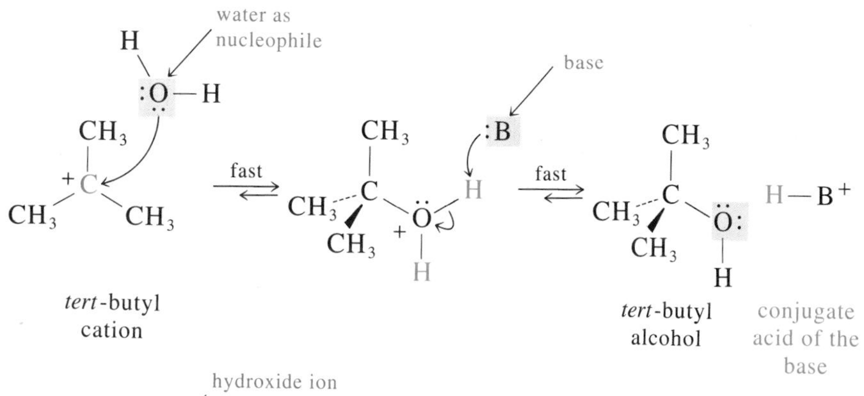

water as nucleophile
base
:B
fast
fast
H—B+
tert-butyl cation
tert-butyl alcohol
conjugate acid of the base

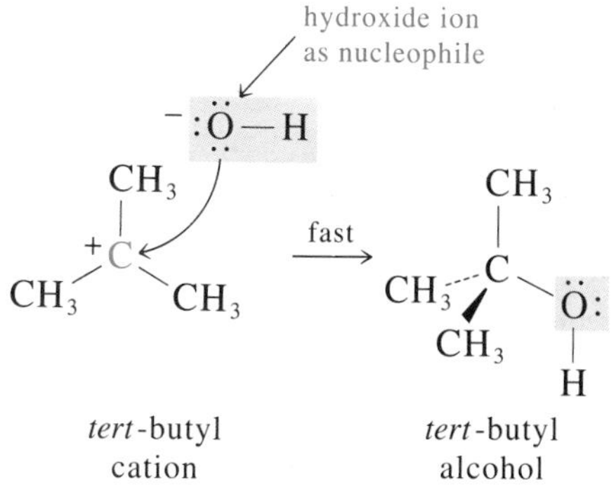

hydroxide ion as nucleophile
fast
tert-butyl cation
tert-butyl alcohol

The intermediate tertiary carbocation is also a Brønsted acid (p. 88). The cationic carbon atom gives a partial positive character to the hydrogen atoms on carbon atoms adjacent to it. These hydrogen atoms are acidic enough to be removed by any of the bases present in the reaction mixture to give the minor product, the product of the elimination reaction.

VISUALIZING THE REACTION

Deprotonation of a carbocation

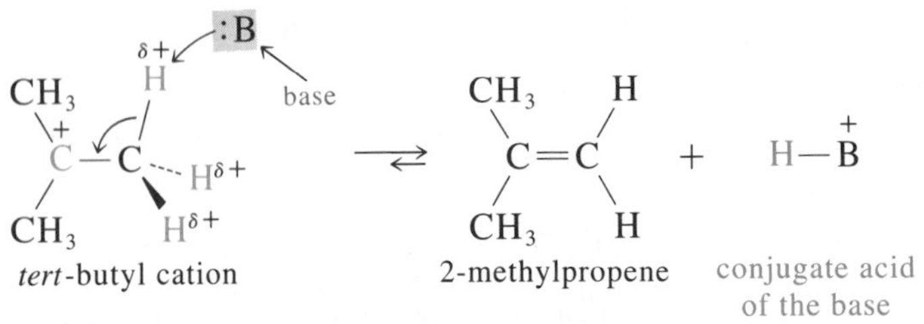
:B
base
tert-butyl cation
2-methylpropene
conjugate acid
of the base

When a carbocation is the intermediate for a substitution reaction, usually some alkene is one of the products because of the ease with which protons are lost from that intermediate.

A unimolecular nucleophilic substitution reaction is called an **S_N1 reaction,** where the sequence of letters and number stands for substitution, nucleophilic, and unimolecular.

PROBLEM 7.17

The rate constant, k, for the solvolysis of *tert*-butyl chloride in 70% aqueous ethanol at 25 °C has been determined to be 0.145/h. If the initial concentration of *tert*-butyl chloride is 0.0824 M, what is the inital rate of the reaction? What units are used to express this rate? Will this rate remain the same as the reaction progresses? What will it be when half of the *tert*-butyl chloride that was initially there has reacted?

B. Stereochemistry of the S_N1 Reaction

What kind of stereochemistry will characterize a reaction that goes through a carbocation intermediate? The simplest tertiary alkyl halide that is chiral is 3-bromo-3-methylhexane. The rate-determining step in the hydrolysis of (*S*)-3-bromo-3-methylhexane is expected to give a planar carbocation solvated on both sides by water molecules (Figure 7.4).

Two of the water molecules are in position to donate an electron pair to the empty *p* orbital. If reaction takes place with the water moelcule on the right, the alcohol (*S*)-3-methyl-3-hexanol is formed, with retention of configuration. If the other water molecule reacts, the alcohol with the inverted configuration, (*R*)-3-methyl-3-hexanol, is formed. When a symmetrical intermediate, such as the planar carbocation, is formed in a reaction at the stereocenter, the chirality of the starting material is lost. In this case, the products are enantiomers. If they are formed in equal amounts, there will be no optical activity detected in the product, which will be a racemic mixture of alcohols.

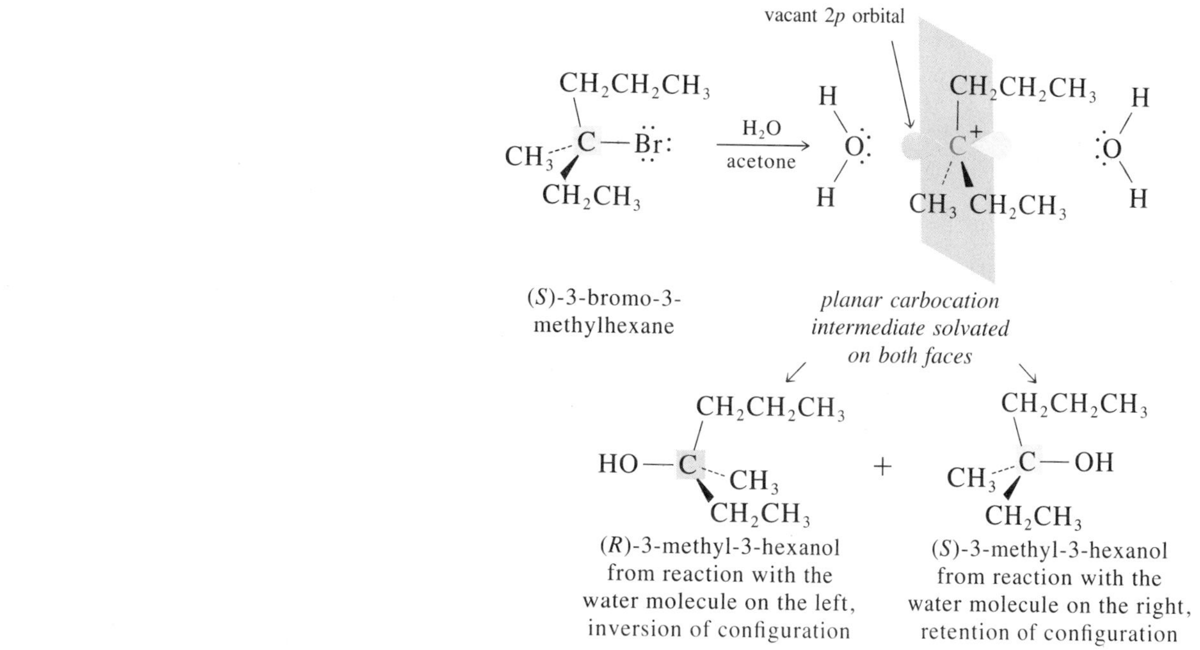
vacant 2p orbital
H2O
acetone
(S)-3-bromo-3-methylhexane
planar carbocation intermediate solvated on both faces
(R)-3-methyl-3-hexanol from reaction with the water molecule on the left, inversion of configuration
(S)-3-methyl-3-hexanol from reaction with the water molecule on the right, retention of configuration

FIGURE 7.4 The hydrolysis of *S*-3-bromo-3-methylhexane.

Experiments with chiral compounds that can undergo S_N1 reactions show that a great deal of racemization does take place. The reactions of (*S*)-(−)-1-chloro-1-phenylethane illustrate this best.

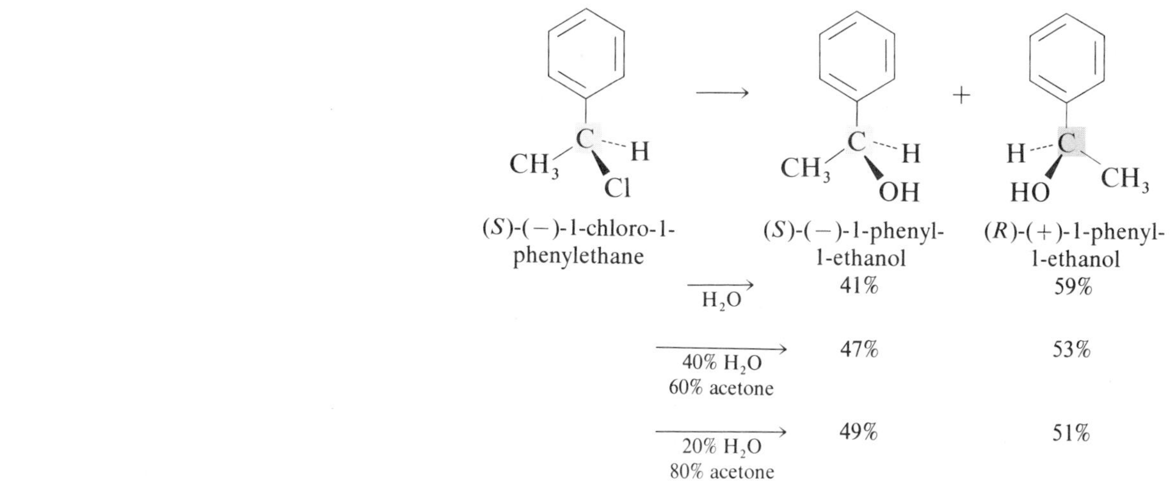
(S)-(−)-1-chloro-1-phenylethane
(S)-(−)-1-phenyl-1-ethanol
(R)-(+)-1-phenyl-1-ethanol
H2O
41%
59%
40% H2O
60% acetone
47%
53%
20% H2O
80% acetone
49%
51%

1-Chloro-1-phenylethane, though it is a secondary alkyl halide, ionizes easily because the carbocation that it forms has high stability. The halogen is bonded to an sp^3-hybridized carbon atom that is bonded to an aromatic ring. Such halides are called **benzylic halides** (see Problem 7.18, p. 251) and ionize to give **benzylic carbocations.** The stability of benzylic carbocations can be rationalized by writing resonance contributors in which the positive charge is delocalized to the aromatic ring.

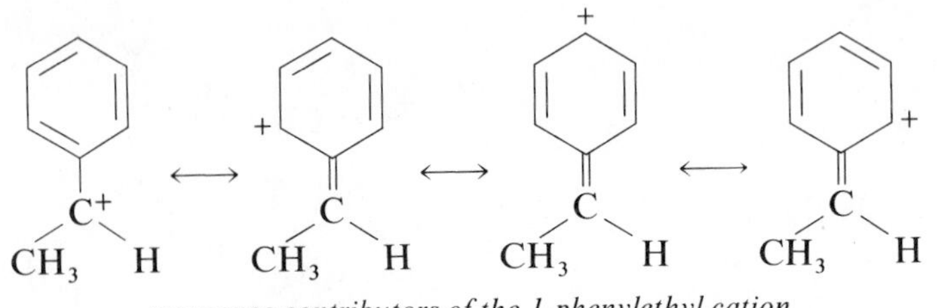

resonance contributors of the 1-phenylethyl cation

An examination of the experimental facts shows that 1-chloro-1-phenylethane undergoes the S_N1 reaction with extensive racemization, but with a slight excess of inversion. For example, when the reaction is carried out in water, there is approximately an 18% excess of the inverted product. As the percentage of water in the reaction mixture decreases to 40%, and then to 20%, and is replaced by an increasing amount of acetone, the observed product becomes more and more completely racemized.

We can interpret these facts to mean that the carbocation that is originally formed is not completely free to react with water on both sides. One side of the carbocation is shielded by the departing chloride ion for some time. In fact, the carbocation and the chloride ion form an **ion pair,** held together for a while by a **cage of solvent molecules.** At this stage, only the water molecules on the side of the cation opposite to the departing chloride ion can react with the cation, giving rise to inversion (Figure 7.5). If the cation is stable enough to survive this stage, the ions can diffuse farther apart, and water molecules can get close to the other side of the cation. Now the ion pair is described as being separated by the solvent molecules (Figure 7.6). Products with retention as well as inversion of configuration result.

If the solvent is pure water, the concentration of water molecules is high enough that some of the reaction takes place while the cation is still shielded by chloride ion in the front. As the concentration of water molecules in the solvent decreases from 100% to 40% to 20%, more and more of the cations can survive long enough without reacting with water that the ions can diffuse apart. Reaction with water can then take place on both sides of the cation.

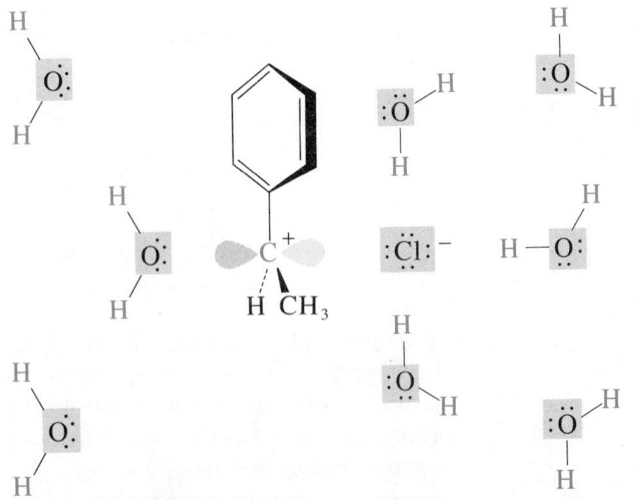

FIGURE 7.5 1-Phenylethyl cation and chloride ion as an ion pair in a solvent cage; reaction of the cation with water occurs with inversion of configuration.

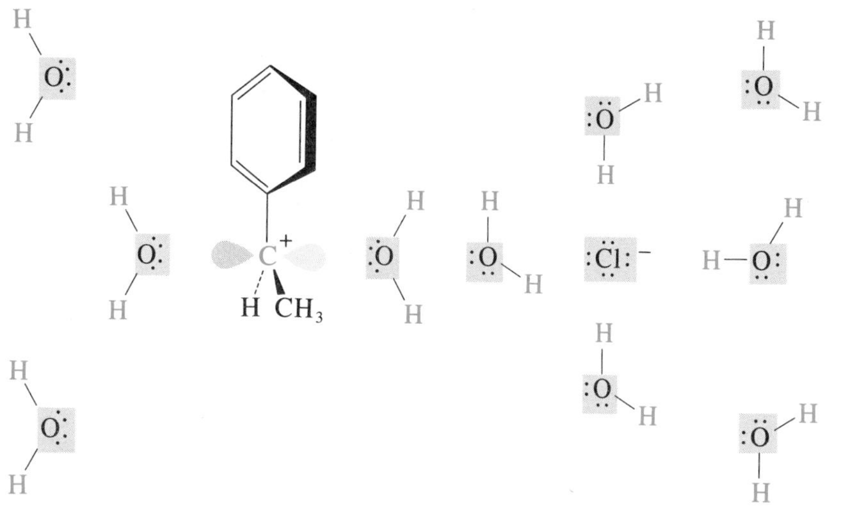

FIGURE 7.6 1-Phenylethyl cation and chloride ion separated by solvent molecules; reaction of the cation with water occurs with racemization.

Study Guide Concept Map 7.4

PROBLEM 7.18

Benzyl chloride has the structure shown below. Write an equation for the ionization of benzyl chloride, and draw resonance contributors showing the delocalization of charge in the benzylic cation.

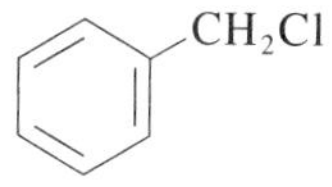

benzyl chloride

PROBLEM 7.19

Write detailed mechanisms showing the conversion of benzyl chloride to benzyl alcohol by both S_N1 and S_N2 pathways.

7.6 A UNIFIED VIEW OF NUCLEOPHILIC SUBSTITUTION REACTIONS

S_N1 and S_N2 reactions represent mechanistic extremes for nucleophilic substitution. Most substitution reactions of alkyl halides proceed by some intermediate mechanism involving different degrees of ionization of the carbon-halogen bond and different degrees of bonding with nucleophiles in the transition state. For example, *tert*-butyl chloride was shown on p. 245 undergoing ionization in a single step to give the *tert*-butyl cation and the chloride ion, which were represented as being completely separated from each other. Yet in the preceding section stereochemical evidence was presented that indicated that the ionization of an alkyl halide may actually occur in several stages. When ionization first occurs, the ions are held together by the surrounding solvent molecules in an ion pair. Then they are separated by one or more solvent molecules and become a **solvent-separated ion pair.** Finally they drift apart entirely and become free, totally solvated ions. These stages are shown schematically for an alkyl halide, RX, in an ionizing solvent, water.

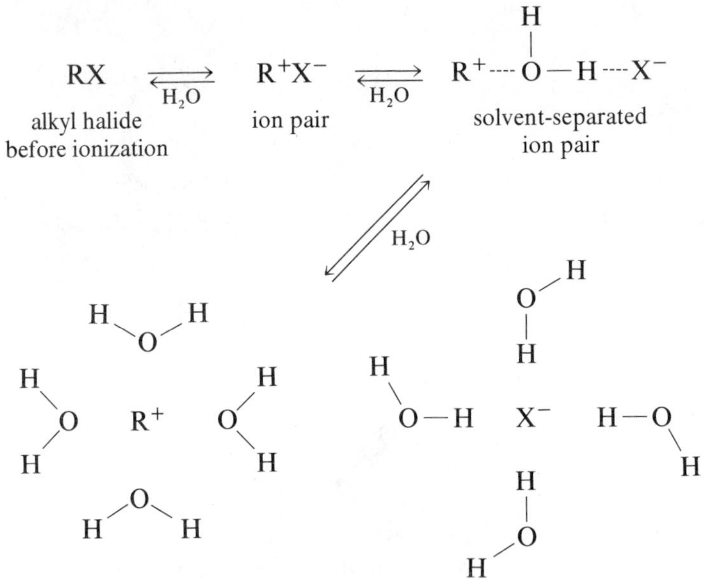
RX
H2O
alkyl halide
before ionization
R+X−
H2O
ion pair
R+----O—H----X−
solvent-separated
ion pair
H2O
R+
X−

totally ionized alkyl halide
ions are stabilized by solvation

The extent of ionization that occurs before reaction with a nucleophile depends on the stability of the potential carbocation, the nature of the leaving group, and the ionizing power of the solvent.

The alkyl halide may react with the nucleophile at any one of the stages shown above. For example, a primary alkyl halide, in most solvents, will interact with the nucleophile before there is any significant ionization. Such a reaction is the typical S_N2 reaction described earlier. Tertiary alkyl halides, or other compounds that give rise to stabilized carbocations, react with a nucleophile in a typical S_N1 reaction after the ions are separated by solvent molecules.

Note that reaction of a nucleophile with the alkyl halide before ionization or with an ion pair gives inversion of configuration. If the nucleophile is the solvent, reaction of the cation with a solvent molecule, either in a solvent-separated ion pair or as the free cation, gives extensive racemization. With careful experiments, chemists have shown that secondary alkyl compounds undergo nucleophilic substitution reactions with almost complete inversion of configuration, even in solvents that encourage ionization. Therefore, reaction of these secondary alkyl compounds must occur through nucleophilic attack either on the neutral alkyl halide or on the ion pair.

This unified way of looking at nucleophilic substitution reactions that we are describing here was proposed by Saul Winstein of the University of California at Los Angeles after much careful experimentation. Understanding the factors that govern these reactions allows chemists to manipulate the conditions such as the solvent and the type and concentration of nucleophile so as to obtain the desired products.

In summary, more nearly pure S_N2 reactions are favored with primary alkyl halides in solvents, such as acetone, that do not promote ionization and with good nucleophiles, such as iodide, bromide, hydroxide, and ethoxide ions. Under these conditions, methyl compounds react the fastest. The rate of the reaction is slower with bulky substituents and with secondary and tertiary halides. For S_N2 conditions, this is the order of reactivity for alkyl halides.

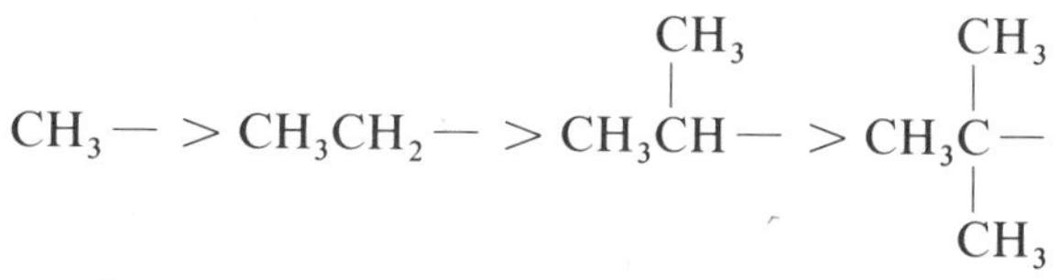

reacts fastest *reacts most slowly*

More nearly pure S_N1 reactions take place with tertiary alkyl halides, or other compounds that give relatively stable carbocations, in ionizing solvents such as water, aqueous alcohol, or aqueous acetone. In ionizing solvents, especially with no added nucleophiles, tertiary alkyl halides react the fastest. The rate of the reaction decreases with secondary and primary halides. For S_N1 conditions, this is the order of reactivity of alkyl halides:

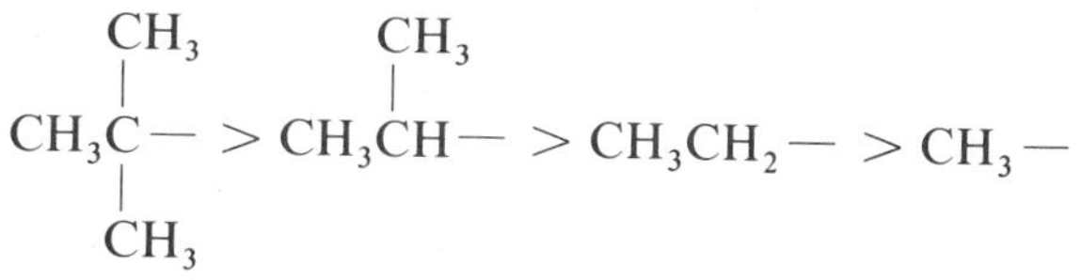

reacts fastest *reacts most slowly*

Some of the various reagents and solvents that are actually used in syntheses involving nucleophilic substitution reactions are shown in Section 7.8.

7.7 ELIMINATION REACTIONS

A. The E_1 Reaction in Competition with the S_N1 Reaction. Orientation in the E_1 Reaction

Elimination reactions accompany substitution reactions and occur to a lesser or greater degree. The intermediate carbocation that is formed in all S_N1 reactions loses a proton to a base in the reaction mixture to give an alkene in what is known as the **unimolecular elimination,** or **E_1 reaction.** The rate-determining step for this elimination reaction is the unimolecular ionization of the alkyl halide.

For the *tert*-butyl cation, only one alkene is possible as the product.

VISUALIZING THE REACTION

The E_1 reaction

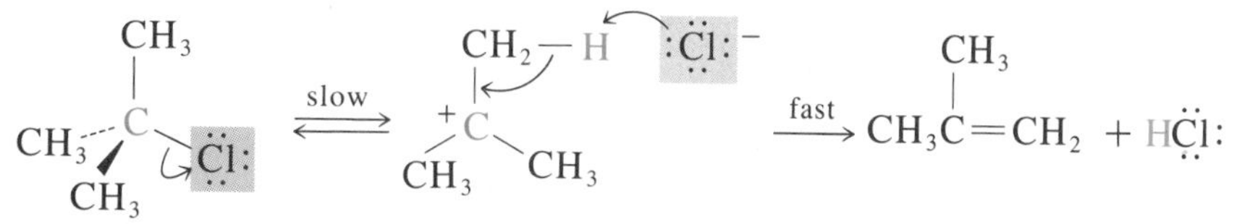
slow
fast

tert-butyl chloride *tert*-butyl cation 2-methylpropene

If the structure of the carbocation is such that different kinds of protons can be lost to give different alkenes, the more highly substituted alkene is formed in a larger amount.

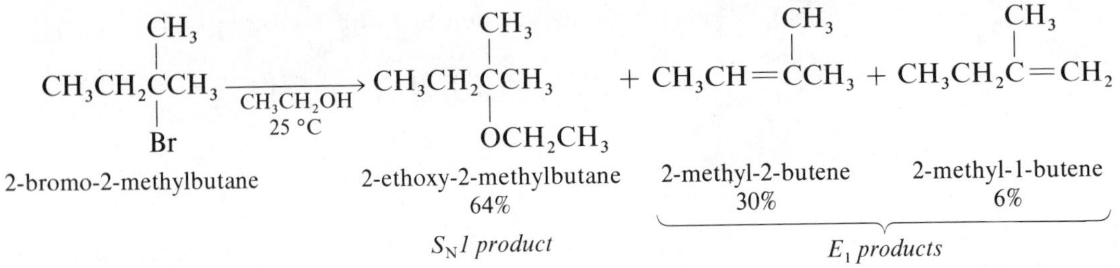
CH3CH2OH
25 °C
2-bromo-2-methylbutane
2-ethoxy-2-methylbutane
64%
2-methyl-2-butene
30%
2-methyl-1-butene
6%
SN1 product
E1 products

The more highly substituted alkene is defined as the one with fewer hydrogen atoms on the carbons involved in the double bond. In the above example, the major alkene product, 2-methyl-2-butene, has three alkyl groups and one hydrogen atom on the carbons of the double bond. It is, therefore, more highly substituted than 2-methyl-1-butene, which has two alkyl groups and two hydrogen atoms on the double bond. The more highly substituted an alkene is, the more stable it is. A further discussion of the relative stabilities of alkenes is given on p. 278.

The regioselectivity of an E_1 reaction is determined by the relative stabilities of the products. The molecular complex at the transition state for the formation of an alkene resembles the alkene in structure and is stabilized by the same factors that stabilize the alkene. The carbocation from the ionization of 2-bromo-2-methylbutane can follow two possible pathways for further reaction:

1. A proton can be removed from a primary carbon atom, giving 2-methyl-1-butene, an alkene with two substituents on the doubly bonded carbons.
2. A proton can be removed from the secondary carbon atom, giving 2-methyl-2-butene, an alkene with three substituents on the doubly bonded carbons.

The reaction goes chiefly by the pathway leading through the transition state of lower energy (p. 117) to give the more highly substituted alkene, which is the more stable one.

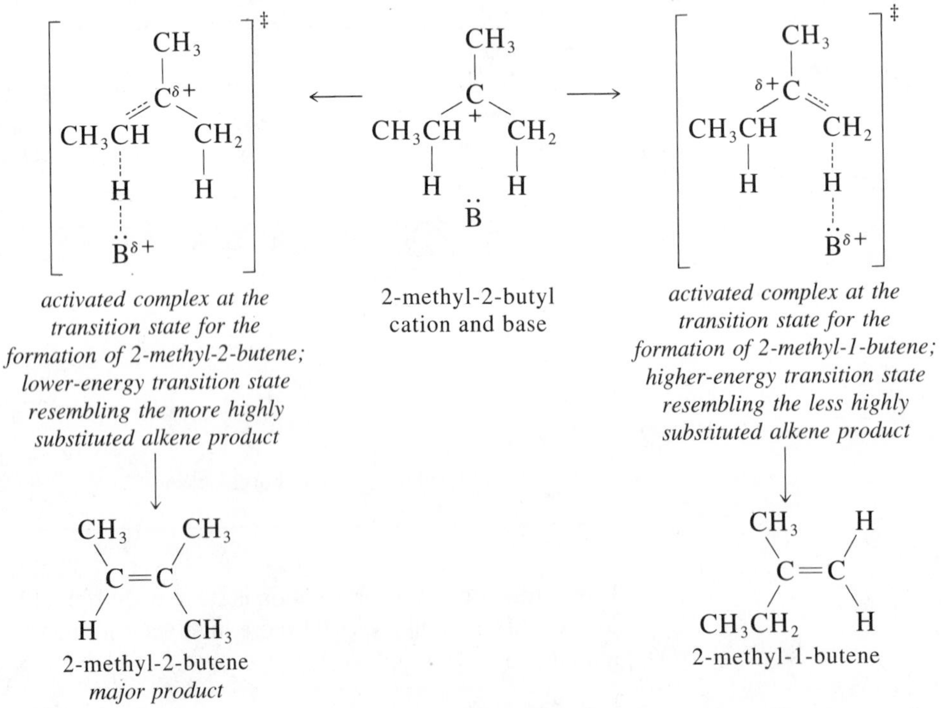
activated complex at the transition state for the formation of 2-methyl-2-butene; lower-energy transition state resembling the more highly substituted alkene product
2-methyl-2-butyl cation and base
activated complex at the transition state for the formation of 2-methyl-1-butene; higher-energy transition state resembling the less highly substituted alkene product
2-methyl-2-butene
major product
2-methyl-1-butene

B. Competition Between S_N2 and E_2 Reactions

When reactions of alkyl halides are carried out in the presence of a strong base, a **bimolecular elimination,** or **E_2 reaction,** accompanies the S_N2 reaction. How much elimination takes place depends on the structure of the alkyl halide. The three equations shown below demonstrate that it is highly impractical to try to carry out nucleophilic substitution reactions on secondary and tertiary alkyl halides with a strongly basic reagent because the chief product in such cases is the one due to elimination.

$$CH_3CH_2Br \xrightarrow[\text{ethanol, 55 °C}]{CH_3CH_2O^-Na^+} CH_3CH_2OCH_2CH_3 + CH_2{=}CH_2$$

ethyl bromide, *a primary alkyl halide* → diethyl ether, 99%, *S_N2 product*; ethylene, 1%, *E_2 product*

$$CH_3CH(Br)CH_3 \xrightarrow[\text{ethanol, 55 °C}]{CH_3CH_2O^-Na^+} CH_3CH(OCH_2CH_3)CH_3 + CH_3CH{=}CH_2$$

isopropyl bromide, *a secondary alkyl halide* → ethyl isopropyl ether, 21%, *S_N2 product*; propene, 79%, *E_2 product*

$$CH_3C(CH_3)(Br)CH_3 \xrightarrow[\text{ethanol, 55 °C}]{CH_3CH_2O^-Na^+} CH_3C(CH_3){=}CH_2$$

tert-butyl bromide, *a tertiary alkyl halide* → 2-methylpropene, 100%, *E_1 + E_2 product*

If the formation of more than one alkene is possible, the more highly substituted alkene is favored. Thus, *sec*-butyl bromide in base gives chiefly 2-butene.

$$CH_3CH_2CH(Br)CH_3 \xrightarrow[\text{ethanol, 25 °C}]{CH_3CH_2O^-Na^+(1\ M)} CH_3CH_2CH(OCH_2CH_3)CH_3 + CH_3CH{=}CHCH_3 + CH_3CH_2CH{=}CH_2$$

sec-butyl bromide → *sec*-butyl ethyl ether, 18%, *S_N2 product*; 2-butene, 66%; 1-butene, 16% (*E_2 products*)

The amount of this alkene that is formed is increased by increasing the concentration of the base and raising the temperature.

$$CH_3CH_2CH(Br)CH_3 \xrightarrow[\text{ethanol, 80 °C}]{K^+OH^-(4\ M)} CH_3CH_2CH(OH)CH_3 + CH_3CH{=}CHCH_3 + CH_3CH_2CH{=}CH_2$$

sec-butyl bromide → *sec*-butyl alcohol, 9%; 2-butene, 75%; 1-butene, 16%

The use of a bulky base, such as the *tert*-butoxide ion, also favors the elimination reaction over the substitution reaction. For example, when the primary alkyl halide 1-bromooctadecane is treated with sodium methoxide, methyl octadecyl ether is the major product.

$$\underset{\text{1-bromooctadecane}}{CH_3(CH_2)_{15}CH_2CH_2Br} \xrightarrow[\substack{\text{methanol} \\ 65\ ^\circ C \\ 12\ h}]{CH_3O^-Na^+\ (1\ M)} \underset{\substack{\text{methyl octadecyl ether} \\ 96\%}}{CH_3(CH_2)_{15}CH_2CH_2OCH_3} + \underset{\substack{\text{1-octadecene} \\ 1\%}}{CH_3(CH_2)_{15}CH{=}CH_2}$$

When potassium *tert*-butoxide is used as the base, the major product is the alkene.

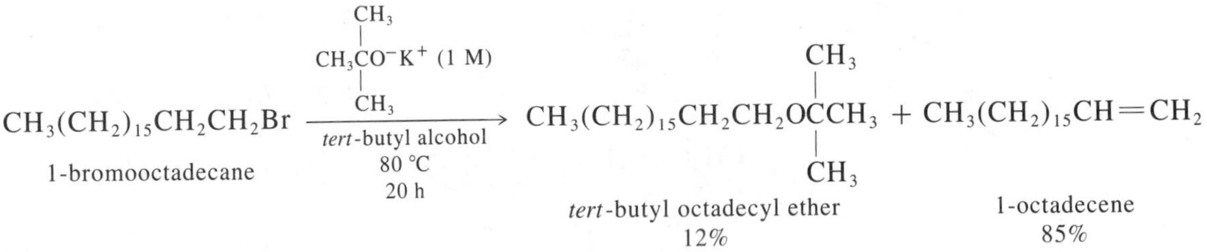
CH3(CH2)15CH2CH2Br
1-bromooctadecane
tert-butyl alcohol
80 °C
20 h
tert-butyl octadecyl ether
12%
1-octadecene
85%

This effect is primarily due to the higher basicity of the *tert*-butoxide ion (the pK_a of *tert*-butyl alcohol is 19; that of methanol is 15.5). The bulkiness of the base is also important. The rate of the elimination reaction, in which a proton is removed from the surface of the molecule, is increased relative to the rate of the substitution reaction, in which the bulky *tert*-butoxide ion has to form a bond to an electrophilic center surrounded by other groups.

In summary, elimination reactions of alkyl halides become prevalent when the reagent is a strong base, such as an alkoxide ion, in high concentration. Tertiary halides are more likely to undergo elimination reactions than are secondary halides, which, in turn, are more prone to elimination than are primary halides.

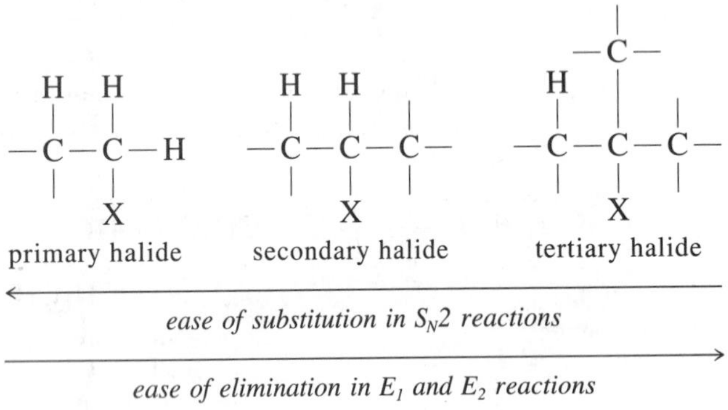
primary halide
secondary halide
tertiary halide
ease of substitution in SN2 reactions
ease of elimination in E1 and E2 reactions

PROBLEM 7.20

Write the structure(s) of the product(s) of the elimination reactions that you would expect in the following cases. If more than one product can be formed, predict which one will be the major product.

(a) $CH_3CH_2CH_2CH_2CH_2Br \xrightarrow[\substack{\text{ethanol} \\ \Delta}]{KOH}$

(b) 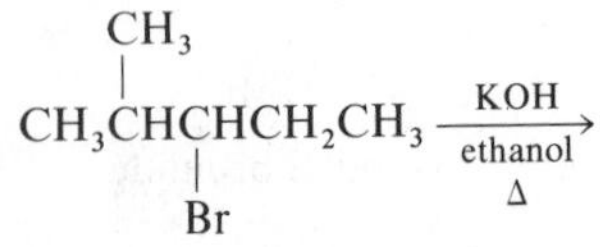
KOH
ethanol
Δ

(c) 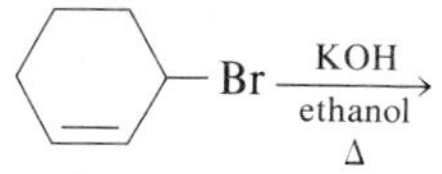

(d) 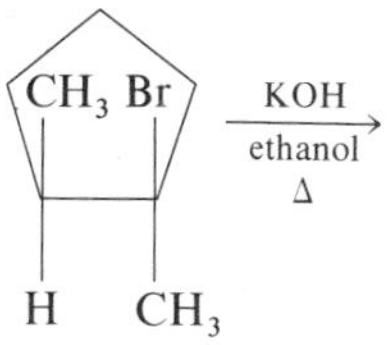

C. Mechanism and Stereochemistry for the E_2 Reaction

The rate of the E_2 reaction of *sec*-butyl bromide, like that of the S_N2 reaction, depends on the concentrations of both the alkyl halide and the base. Any mechanism for the reaction must involve both species in the rate-determining step.

$$R = k[sec\text{-BuBr}][OH^-]$$

VISUALIZING THE REACTION

The E_2 reaction

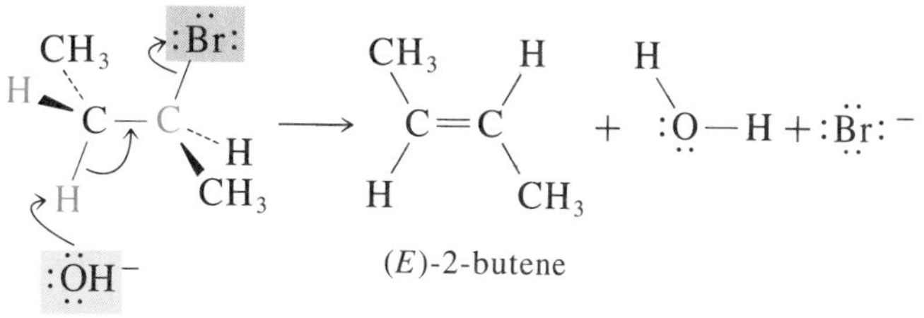
(E)-2-butene

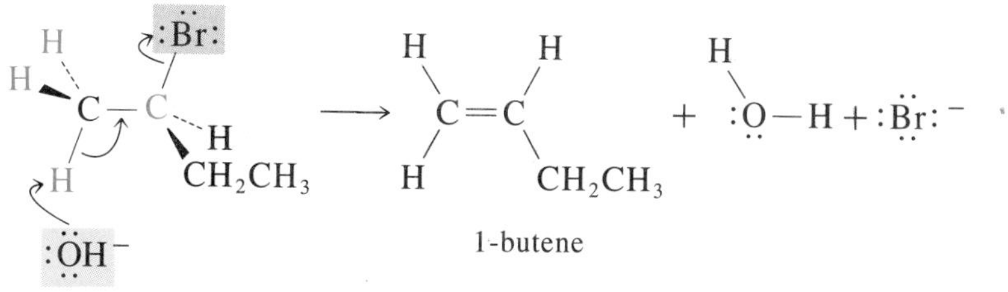
1-butene

The lower-energy transition state, and, therefore, the one that leads to most of the product, is the one giving rise to the more highly substituted alkene, 2-butene.

This generalization is usually true if the leaving group is a halide ion and the base is a hydroxide or ethoxide ion. Much experimental work has been done to show that the nature and size of the leaving group, the size of the base used to remove the proton, and, especially, the stereochemistry of the compound undergoing elimination can affect the exact proportions of the alkenes formed as products. How the orientation of the elimination reaction can change with the nature of the leaving group is the subject of Section 22.7A.

PROBLEM 7.21

Is another stereochemistry possible for the 2-butene that is formed in the elimination reaction of *sec*-butyl bromide? Draw any other conformations of the halide that will give 2-butene as the product.

A definite conformation for the species undergoing an E_2 reaction has been suggested. There is evidence that E_2 elimination proceeds best when the hydrogen to be removed and the leaving group are in an anti conformation with respect to each other. For example, (2*R*,3*R*)-3-phenyl-2-butyl tosylate gives exclusively (*E*)-2-phenyl-2-butene, and its diastereomer, (2*S*,3*R*)-3-phenyl-2-butyl tosylate gives (*Z*)-2-phenyl-2-butene.

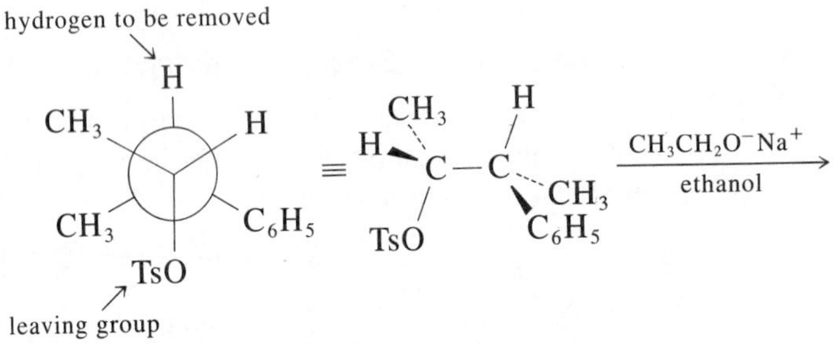
hydrogen to be removed
leaving group
$CH_3CH_2O^-Na^+$
ethanol

(2*R*,3*R*)-3-phenyl-2-butyl tosylate

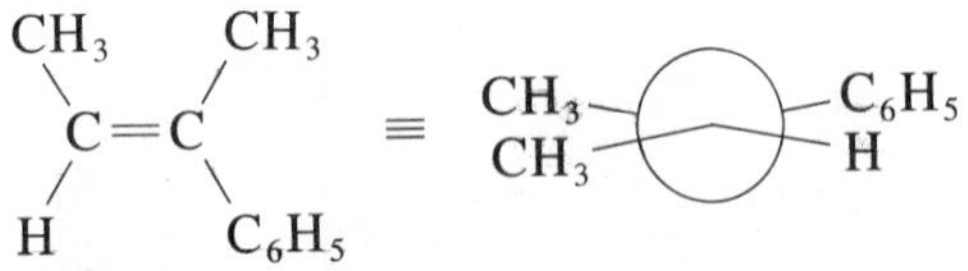

(*E*)-2-phenyl-2-butene

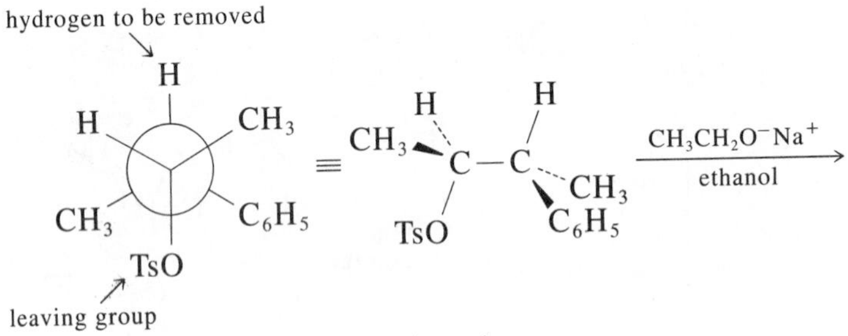
hydrogen to be removed
leaving group
$CH_3CH_2O^-Na^+$
ethanol

(2*S*,3*R*)-3-phenyl-2-butyl tosylate

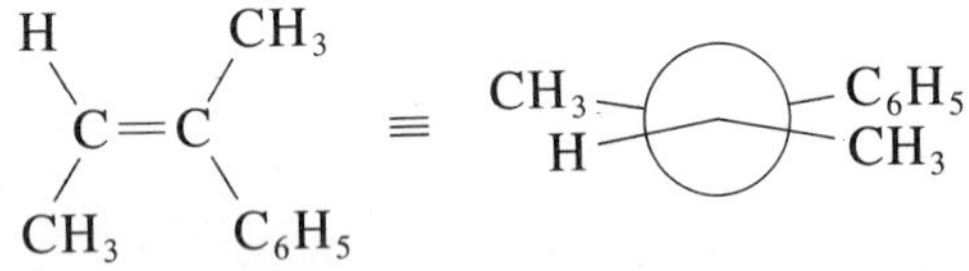

(*Z*)-2-phenyl-2-butene

In an E_2 reaction, the stereochemistry of the alkene that is formed is governed by the conformation that the starting material must adopt in the transition state so that the leaving group and the proton being removed by the base are anti to each other. There is no such stereochemical requirement for an E_1 reaction, because the leaving group is lost and a carbocation has been formed before the proton is removed.

PROBLEM 7.22

Two diastereomers of 1,2-dibromo-1,2-diphenylethane have been found to give two different bromoalkenes on undergoing E_2 reactions.

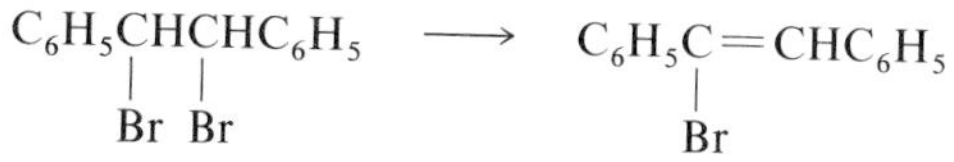

$C_6H_5CHCHC_6H_5$ ⟶ $C_6H_5C{=}CHC_6H_5$
Br Br
Br

1,2-dibromo-1,2-diphenylethane   1-bromo-1,2-diphenylethene

(1*R*,2*S*)-1,2-Dibromo-1,2-diphenylethane gives the *E* alkene. (1*R*,2*R*)-1,2-Dibromo-1,2-diphenylethane gives the *Z* alkene. Draw stereochemically correct representations of the starting materials and the products. Show the conformations that are necessary for the starting bromoalkanes for the E_2 reaction to occur.

An anti conformation for the leaving group and the proton that is lost is also necessary for elimination reactions of certain cyclic compounds. For example, *trans*-2-methylcyclohexyl tosylate gives only 3-methylcyclohexene when treated with potassium *tert*-butoxide.

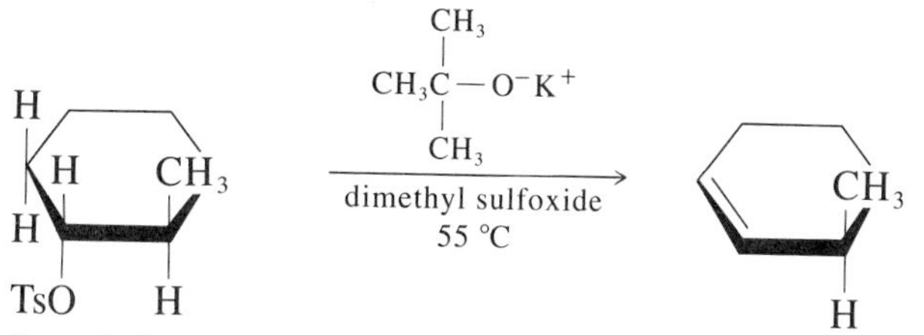

H
H
CH3
H
TsO
H
$CH_3C(CH_3)_2-O^-K^+$
dimethyl sulfoxide
55 °C
CH3
H

trans-2-methylcyclohexyl tosylate   3-methylcyclohexene

In the more stable diequatorial conformation of *trans*-2-methylcyclohexyl tosylate, no hydrogen atom is anti to the leaving group. If the cyclohexane ring flips, one hydrogen atom becomes anti to the tosylate group. The only observed product is the alkene that would be formed by the removal of that proton, suggesting that the molecule reacts in this conformation. There is always a small amount of the diaxial conformation in equilibrium with the more stable conformation at room temperature.

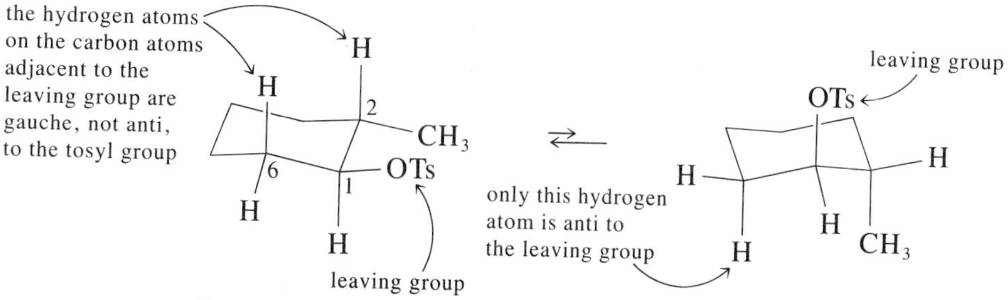

the hydrogen atoms on the carbon atoms adjacent to the leaving group are gauche, not anti, to the tosyl group
H
H
2
CH3
6
1
OTs
H
H
leaving group
only this hydrogen atom is anti to the leaving group
leaving group
OTs
H
H
H
H
CH3

diequatorial conformation   diaxial conformation

trans-2-methylcyclohexyl tosylate

Although the anti orientation between leaving group and hydrogen atom appears to be the most favorable one for elimination reactions, it is not the only one possible. In cyclopentyl compounds especially, the elimination of a hydrogen and a leaving group that are trans to each other is easier, but the elimination of groups that are cis to each other also takes place. *cis*-2-Phenylcyclopentyl tosylate and *trans*-2-phenylcyclopentyl tosylate both give 1-phenylcyclopentene as the product. The first reaction, in which the groups that must be eliminated to give the more stable alkene are trans to each other, is nine times faster than the second reaction, in which the groups eliminated are cis to each other. An elimination in which the groups that are lost are cis to each other is called a **syn elimination.** In an **anti elimination** reaction, the groups that are lost are trans to each other on a ring or anti to each other in one conformation of an acyclic compound.

Study Guide Concept Map 7.5

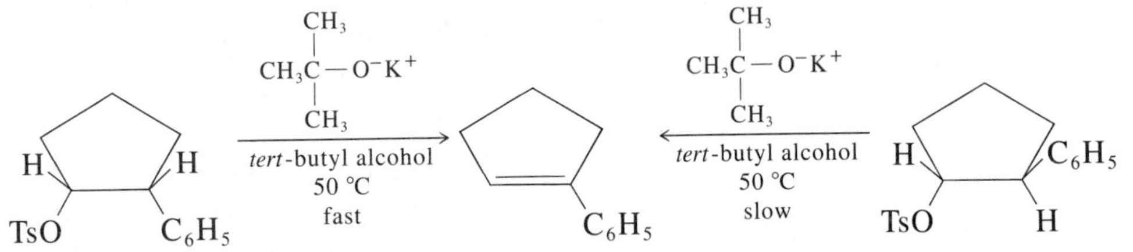
CH3
CH3C—O⁻K⁺
CH3
tert-butyl alcohol
50 °C
fast
H
H
TsO
C6H5
C6H5
CH3
CH3C—O⁻K⁺
CH3
tert-butyl alcohol
50 °C
slow
H
C6H5
TsO
H

anti elimination *syn elimination*

7.8 SYNTHESES. NUCLEOPHILIC SUBSTITUTION REACTIONS IN CHEMICAL TRANSFORMATIONS

New bonds are created whenever a nucleophile substitutes at an electrophilic center bearing a leaving group. Some of these transformations have already been shown in Section 4.4 (p. 127). Substitution reactions generally proceed with the fewest complications when primary alkyl compounds react with nucleophiles. Secondary alkyl halides or tosylates also undergo substitution reactions as long as a combination of a strongly basic nucleophile, such as alkoxide ions, and a high temperature is avoided. Tertiary alkyl halides are not good reagents for nucleophilic substitution reactions because they undergo elimination reactions instead.

Substitution reactions are useful for creating new carbon-oxygen, carbon-sulfur, carbon-nitrogen, and carbon-carbon bonds. Nucleophilic reagents that can be used to create bonds of each of these types are shown in Table 7.3. Most of these reagents are familiar from Section 4.4. The new reagents in this table are azide ion, N_3^-, and the two carbanions. Azide ion (Problem 1.28, p. 32) is a good nucleophile that is often used to introduce carbon-nitrogen bonds, for example:

TABLE 7.3 Common Nucleophilic Reagents for Creating New Bonds to Carbon

Carbon-Oxygen Bonds	Carbon-Sulfur Bonds	Carbon-Nitrogen Bonds	Carbon-Carbon Bonds
HO^-	HS^-	NH_3	CN^-
RO^-	RS^-	RNH_2	$RC{\equiv}C^-$
ArO^-	ArS^-	N_3^-	$(CH_3CH_2O\overset{\overset{\large O}{\|}}{C})_2CH^-$

$$CH_3CH_2CH_2{-}I + Na^+N_3^- \xrightarrow[H_2O]{carbitol} CH_3CH_2CH_2{-}N_3 + Na^+I^-$$

n-propyl iodide; sodium azide; *n*-propyl azide; sodium iodide

$$\text{carbitol} \equiv CH_3CH_2OCH_2CH_2OCH_2CH_2OH$$

an ether and an alcohol

Some of the most useful nucleophilic substitution reactions involve reagents in which carbon atoms are nucleophilic centers. Such reagents are used to create carbon-carbon bonds. A carbanion is generated when an alkyne with the triple bond at the end of the carbon chain is deprotonated by a strong base (p. 91).

$$CH_3CH_2CH_2C{\equiv}CH + Na^+NH_2^- \xrightarrow[NH_3(liq)]{} CH_3CH_2CH_2C{\equiv}C{:}^-Na^+ + {:}NH_3$$

1-pentyne, $pK_a \sim 26$; sodium amide; conjugate base of 1-pentyne; ammonia, pK_a 36

Such carbanions react with alkyl halides to give more highly substituted alkynes.

$$CH_3CH_2CH_2Br + CH_3CH_2CH_2C{\equiv}C{:}^-Na^+ \xrightarrow[NH_3(liq)]{} CH_3CH_2CH_2C{\equiv}CCH_2CH_2CH_3 + Na^+Br^-$$

n-propyl bromide; carbanion from 1-pentyne; 4-octyne

An important carbanion is prepared from the ester diethyl malonate. The closest analogy to diethyl malonate in the pK_a table (inside the cover of the book) has pK_a 11.

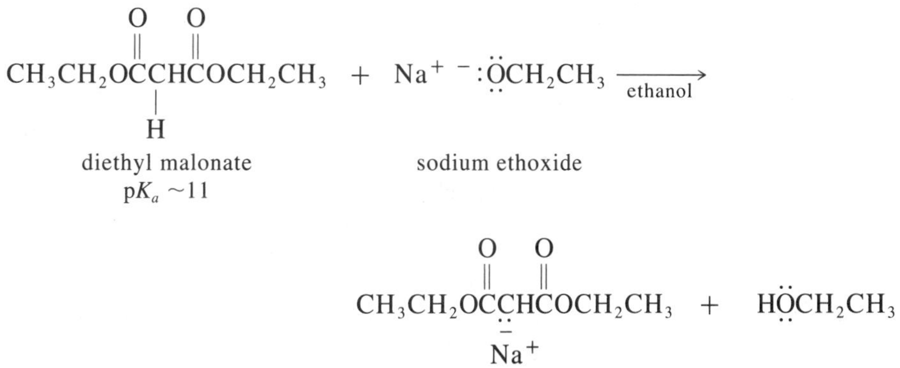

diethyl malonate, $pK_a \sim 11$; sodium ethoxide; conjugate base of diethyl malonate; ethanol, pK_a 17

The carbanion that is generated reacts with alkyl halides to give substituted esters.

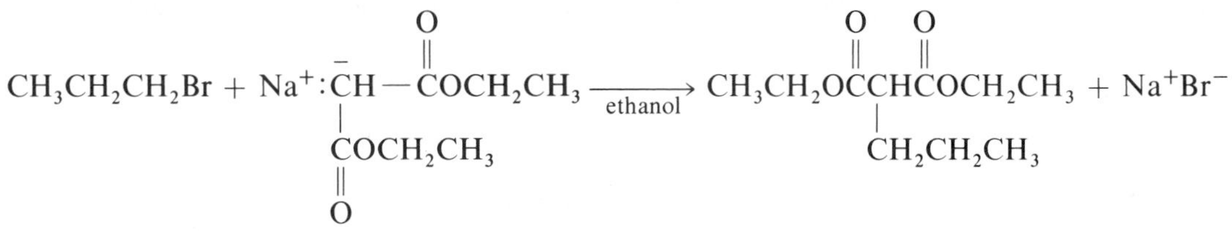

n-propyl bromide; carbanion from diethyl malonate; diethyl propylmalonate

Substituted diethyl malonates are important intermediates in the synthesis of barbituric acid derivatives, which are used in sleeping pills (p. 654). This is just one example of how substitution reactions are used to create compounds of medicinal and industrial usefulness.

PROBLEM 7.23

Use the curved-arrow notation to show how each of the following substitution reactions will take place; also, show what the products will be.

(a) 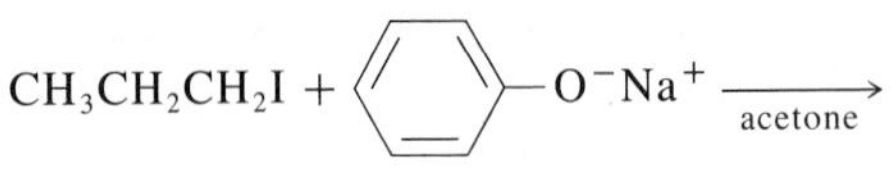

(b) 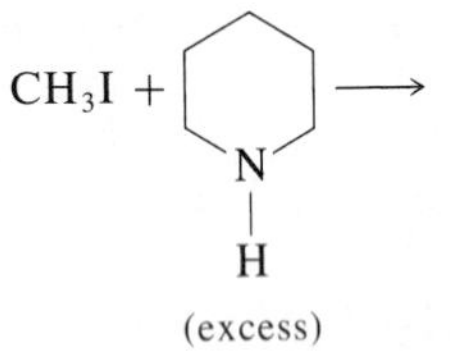

(c) 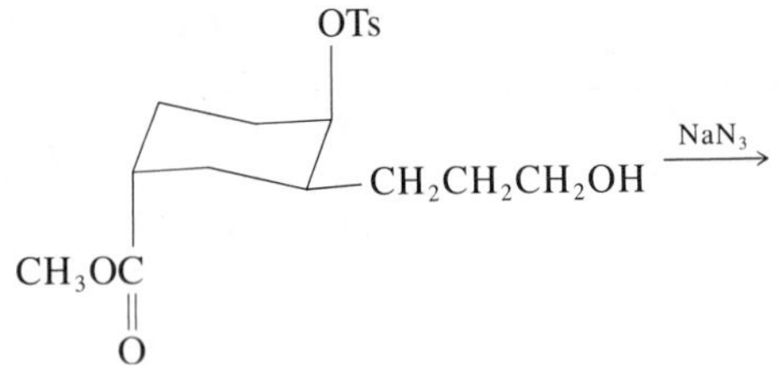

(d) $CH_3CH_2CH{=}CH_2 + HBr \longrightarrow$ 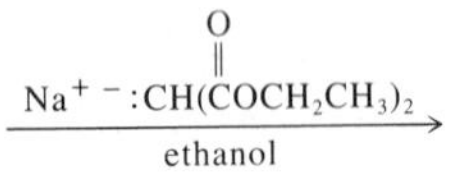

7.9 PROBLEM-SOLVING SKILLS

In Section 4.4B (p. 128), you learned some ways of analyzing the structures of reactants and products in order to decide how to transform one into another. This section offers some further guidelines on how to approach the solution of problems in organic chemistry.

Most of the problems in this book and on examinations fall into two categories: (1) predicting the products of a reaction, given the reagents; and (2) deciding what reagents are necessary to convert a given starting material into a desired product. You can solve such problems by systematically asking yourself a series of questions that will help you analyze the important features of a reaction. In this section, two problems are analyzed in detail in this way to illustrate the necessary thought processes.

Problem

Predict the product(s) of the following reaction, including the correct stereochemistry.

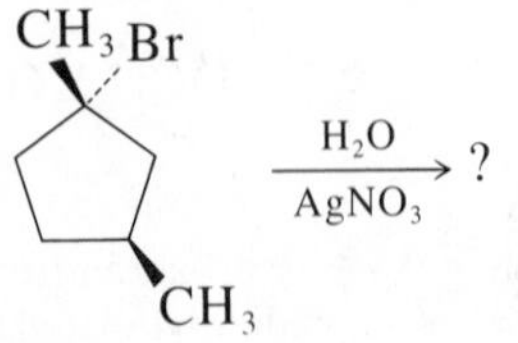

Solution

1. To what functional-group class does the reactant belong? What is the electronic character of the functional group?

   The reactant contains a halogen atom on a tertiary carbon atom; therefore, the compound is a tertiary alkyl halide.

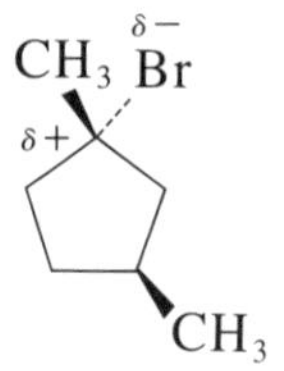

δ−
CH_3 Br
δ+
CH_3

2. Does the reactant have a good leaving group?

   Yes, the bromine atom will leave as the weakly basic bromide ion, Br^-.

3. What reagents are present? Are they good acids, bases, nucleophiles, or electrophiles? Is there an ionizing solvent present?

   Water (above the reaction arrow) is the chief reagent. It is weakly nucleophilic but is a good ionizing solvent. Silver nitrate (below the arrow) is present to catalyze the reaction. Silver ions react with halide ions to form silver halides that precipitate, so the silver ions help ionization.

4. What is the most likely first step for the reaction: protonation or deprotonation, ionization, attack by a nucleophile, or attack by an electrophile?

   A tertiary alkyl halide ionizes in an ionizing solvent.

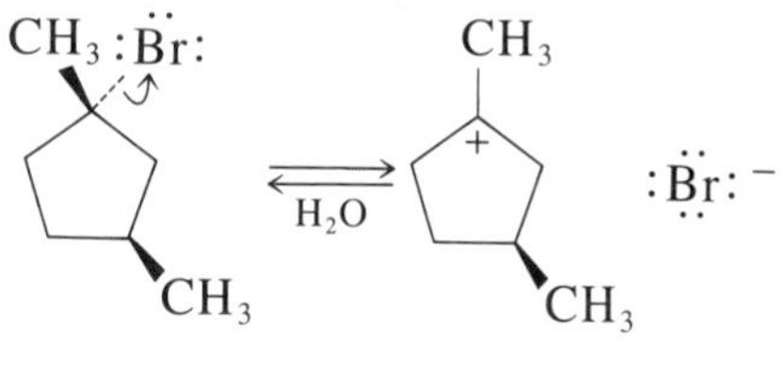

CH_3 :Br:
CH_3
H_2O
CH_3
+
:Br: −
CH_3

   This is the slow step of an S_N1 reaction.

5. What are the properties of the species present in the reaction mixture after the first step? What is likely to happen next?

   The carbocation is a powerful electrophile and will react with water in the second step of an S_N1 reaction, which is fast.

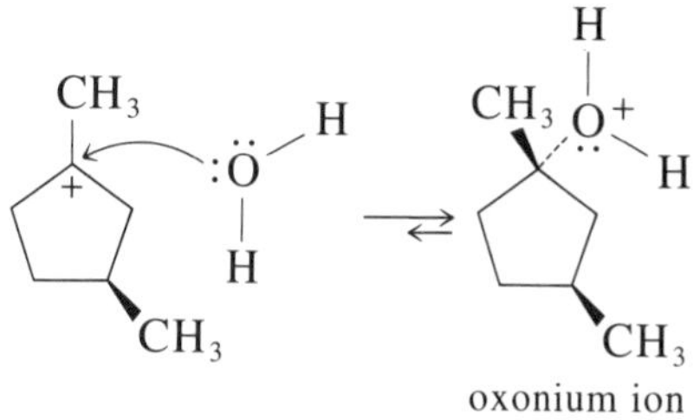

CH_3
+
:O
H
H
CH_3
H
CH_3 O+
H
CH_3
oxonium ion

The species created in this second step is an oxonium ion, a strong acid. A proton will be transferred to a solvent molecule.

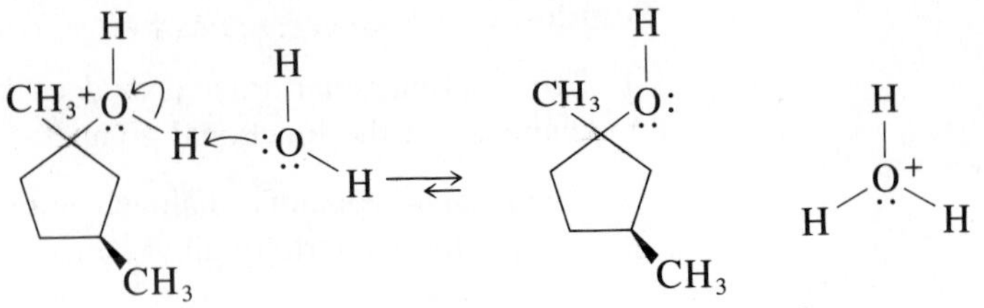

stable, neutral species; product

A side reaction involves the loss of a proton from the carbocation to give two possible alkenes.

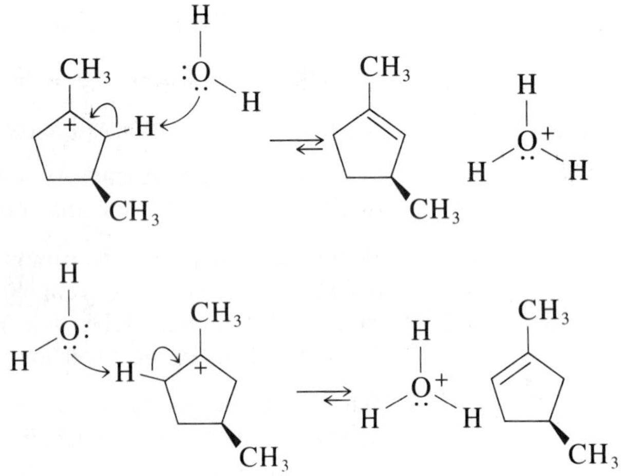

6. What is the stereochemistry of the reaction?

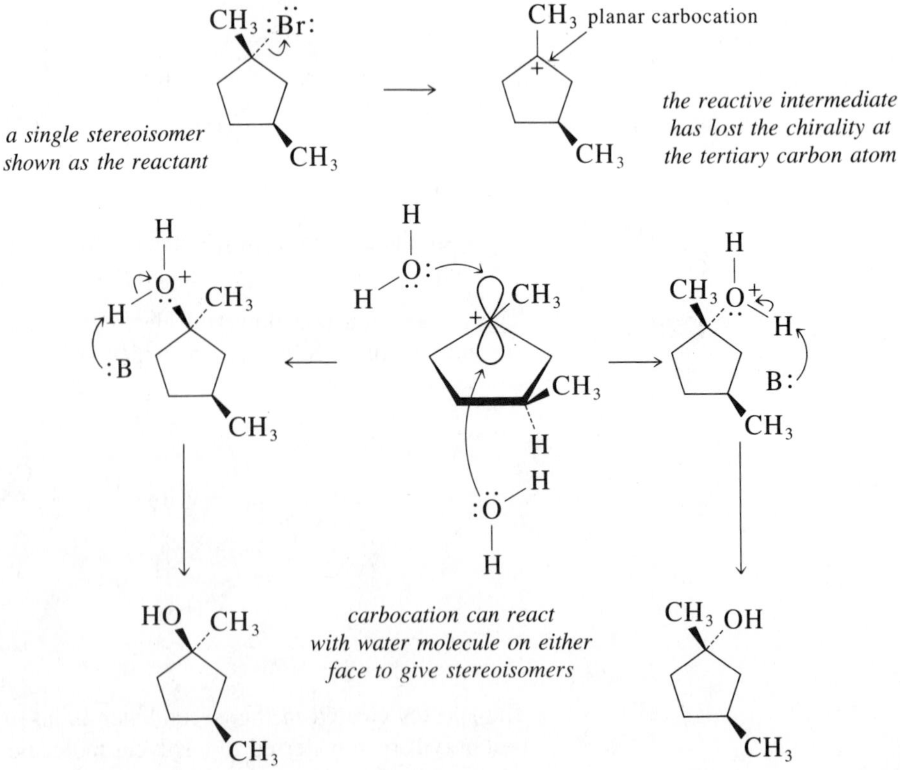
planar carbocation
a single stereoisomer shown as the reactant
the reactive intermediate has lost the chirality at the tertiary carbon atom
carbocation can react with water molecule on either face to give stereoisomers

The complete correct answer is as follows:

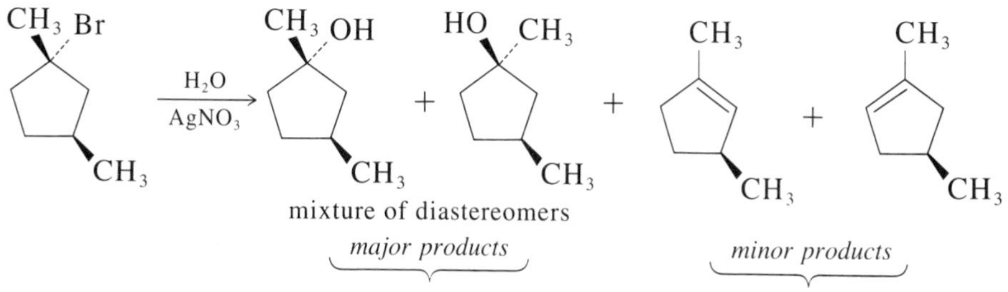

H₂O
AgNO₃
mixture of diastereomers
major products
minor products

The thought processes used in solving syntheses problems are similar but involve asking different questions.

Problem

The following compound labeled with radioactive carbon-14 at the cyano group was needed for a study of how long-chain acids are synthesized by marine organisms. How would you make it from 12-methyl-1-tridecanol? Potassium cyanide labeled with radioactive carbon is available.

$$\mathrm{CH_3\overset{\displaystyle CH_3}{\overset{|}{C}}H(CH_2)_{10}CH_2{}^{*}CN} \qquad \text{from} \qquad \mathrm{CH_3\overset{\displaystyle CH_3}{\overset{|}{C}}H(CH_2)_{10}CH_2OH}$$

Solution

1. What functional groups are present in the starting material and the product?

   The starting material is a primary alcohol; the product contains a cyano group.

2. How do the carbon skeletons of the two compounds compare? How many carbon atoms does each contain? Are there any rings? What are the positions of branches and functional groups on the carbon skeletons?

   The product has one more carbon atom than the starting material. That additional carbon atom is part of the cyano group. Otherwise, the carbon skeletons are identical.

3. How do the functional groups change in going from starting material to product? Does the starting material have a good leaving group?

   A hydroxyl group has been replaced by a cyano group. The starting material does not have a good leaving group, but the hydroxyl group can be converted into one.

4. Is it possible to dissect the structures of the starting material and product to see which bonds must be broken and which formed?

$$\underset{\textit{new bond formed}}{\mathrm{CH_3\overset{\displaystyle CH_3}{\overset{|}{C}}H(CH_2)_{10}CH_2 \wr CN}} \longleftarrow \underset{\textit{bond to be broken}}{\mathrm{CH_3\overset{\displaystyle CH_3}{\overset{|}{C}}H(CH_2)_{10}CH_2 \wr OH}}$$

5. Do we recognize any part of the product molecule as coming from a good nucleophile or an electrophilic addition?

   The cyano group, $-\mathrm{C{\equiv}N}$, is directly related to cyanide ion, $\mathrm{CN^-}$, a good nucleophile.

6. What type of compound would be a good precursor to the product?

   A good precursor would be a compound that has a good leaving group at the position the cyano group occupies in the product.

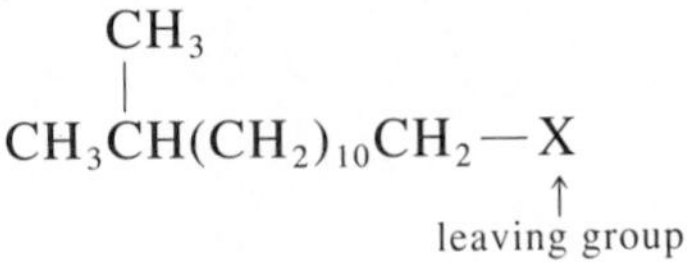

7. After this last step, do we see how to get from starting material to product? If not, we need to analyze the structure obtained in step 6 by applying questions 5 and 6 to it.

   The hydroxyl group of an alcohol can be converted into a good leaving group; so we are now able to solve the problem.

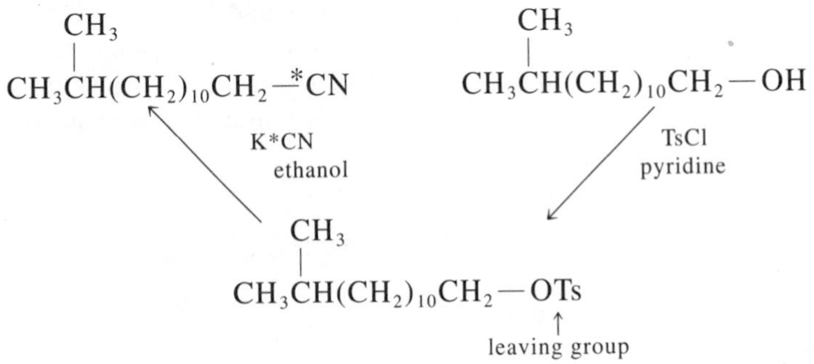

   (Protonation of the alcohol to make a leaving group will not work because cyanide ion is a stronger base than the alcohol and will deprotonate the oxonium ion to give HCN, a highly poisonous gas.)

These steps are a restatement of the way of thinking about problems that was introduced in Section 4.4B. It will be helpful to you to ask yourself these questions in a systematic way for each problem you work on, until this way of thinking becomes familiar and easy. Only consistent practice will develop the problem-solving skills you need to succeed in organic chemistry.

PROBLEM 7.24

Predict what the product(s) will be for each of the following reactions.

(a) 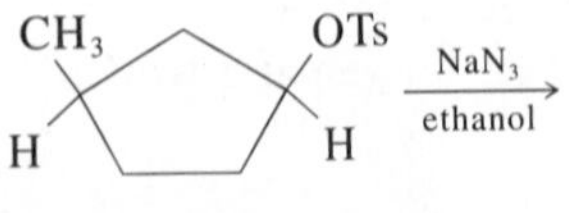

(b) 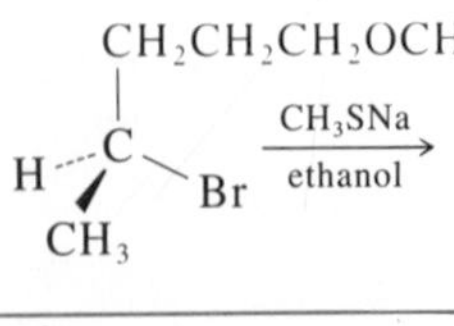

PROBLEM 7.25

How would you carry out the following transformation?

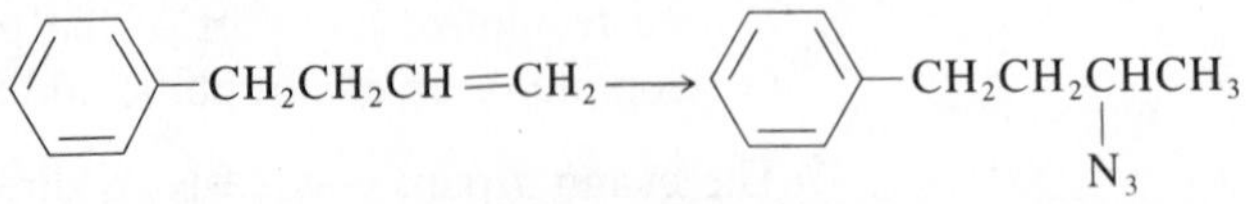

S U M M A R Y

Nucleophilic substitution reactions occur when nucleophiles displace leaving groups, resulting in the substitution of one group for another bonded to a carbon atom. Nucleophiles are species with nonbonding electrons that are available for donation to electrophilic centers. Nucleophilicity is affected by the structure of the nucleophile and by the solvent used for the reaction. Leaving groups are stable species that can be detached from a molecule during reaction.

S_N2 reactions are second-order reactions with rates that depend on the concentrations of both the alkyl halide and the nucleophile. An S_N2 reaction occurs in one step that involves a pentacoordinated carbon atom at the transition state and results in inversion of configuration. The order of reactivity of alkyl halides in S_N2 reactions is

$$CH_3 > \text{primary} > \text{secondary} > \text{tertiary}$$

S_N1 reactions are first order with respect to the concentration of the alkyl halide. The reaction proceeds in two steps: the rate-determining ionization of the alkyl halide gives a carbocation intermediate, which then reacts with nucleophiles. Racemization is common in S_N1 reactions. The order of reactivity of alkyl halides in S_N1 reactions is

$$\text{tertiary} > \text{secondary} > \text{primary} > CH_3$$

An elimination reaction may be either E_1, which involves a carbocation intermediate, or E_2, which is a second-order reaction. Elimination reactions occur in competition with substitution reactions and are promoted by strong bases and high temperatures. For elimination reactions that can give rise to more than one alkene, the more highly substituted alkene is the major product. E_2 reactions proceed most easily when the leaving group and the proton to be removed are in the anti conformation. The order of reactivity for alkyl halides in elimination reactions is

Study Guide Concept Map 7.6

$$\text{tertiary} > \text{secondary} > \text{primary}$$

ADDITIONAL PROBLEMS

7.26 Write equations for the reaction of *n*-butyl bromide as a typical primary alkyl halide, with the following reagents.

(a) KOH, in ethanol, heat (b) NaOH, 0.1 M, in 50% aqueous ethanol (c) NH_3

(d) NaN_3 (e) NaCN (f) CH_3CH_2SNa

(g) $CH_3\overset{\overset{O}{\|}}{C}ONa$ (h) $CH_3CH_2C\equiv CNa$ (i) $NaCH(\overset{\overset{O}{\|}}{C}OCH_2CH_3)_2$

7.27 The following equations represent some nucleophilic substitution reactions that are possible. Only primary alkyl groups are used, so the reactions will be S_N2, and competing elimination reactions will not be a problem. Complete the equations, showing the products that will form. In each case, it will be helpful to label the leaving group and the incoming nucleophile.

(a) 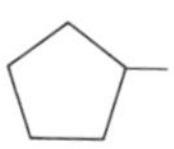 $—CH_2CH_2CH_2OTs + NaN_3 \longrightarrow$

(b) 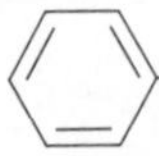 $-CH_2CH_2Cl + NaCN \longrightarrow$

(c) $CH_3CH_2CH_2CH_2Br + CH_3CH_2SNa \longrightarrow$

(d) $CH_3CH_2CH_2Cl + CH_3C{\equiv}CNa \longrightarrow$

(e) $CH_3CH(CH_3)CH_2CH_2Br + \underset{\text{(excess)}}{NH_3} \longrightarrow$

(f) $CH_3I + NaCH(\overset{O}{\overset{\|}{C}}OCH_2CH_3)_2 \longrightarrow$

(g) $CH_3CH_2I + (CH_3CH_2CH_2CH_2)_3P \longrightarrow$

(h) $HOCH_2CH_2Cl + CH_3SNa \longrightarrow$

7.28 Predict whether each of the following organic bromides will give a precipitate when treated with silver nitrate in ethanol or with sodium iodide in acetone. Some may react with both.

(a) $CH_3CH_2CH_2Br$ (b) 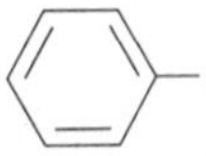 $-\underset{Br}{CH}CH_3$ (c) $CH_3CH_2\underset{Br}{CH}CH_3$

(d)  $-CH_2CH_2Br$ (e) $CH_3\overset{CH_3}{\underset{Br}{C}}CH_2CH_3$

7.29 The following equations represent elimination reactions that are possible. Complete the equations, showing the products from the elimination reactions. Wherever you can, indicate which product will be the major one and which the minor.

(a)  $-\underset{Br}{CH}CH_2CH_3 \xrightarrow[\text{ethanol, } \Delta]{\text{KOH}}$

(b) $CH_3CH_2\overset{CH_3}{\underset{Br}{C}}CH_2CH_2CH_3 \xrightarrow[\text{ethanol, } \Delta]{\text{KOH}}$

(c) 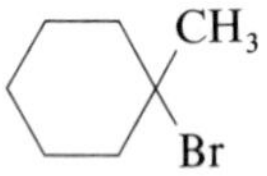
CH3
Br
$\xrightarrow[\text{ethanol, } \Delta]{\text{KOH}}$

(d) $CH_3CH_2CH_2\underset{Br}{CH}\overset{O}{\overset{\|}{C}}OH \xrightarrow[\text{ethanol, } \Delta]{\text{KOH}}$

(e) 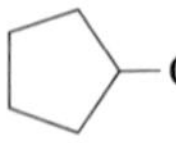 $-OTs \xrightarrow[\text{ethanol, } \Delta]{\text{KOH}}$

7.30 Complete the following reaction sequences, showing the major product you expect for each stage. When you know what the stereochemistry of a product will be, show it in your answer.

(a) $CH_3\underset{OH}{CH}CH_3 \xrightarrow[\text{pyridine}]{\text{TsCl}} A \xrightarrow{NaN_3} B$

(b) $CH_3\overset{CH_3}{CH}CH_2CH_2CH_2CH_2Br \xrightarrow[\text{ethanol}]{\text{NaCN}} C$

(c) 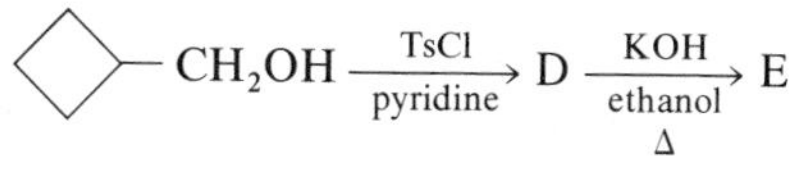

(d) 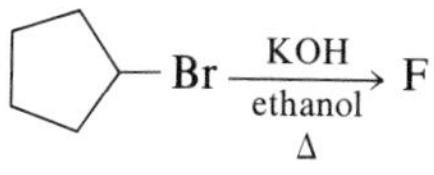

(e)
$$\mathrm{CH_3CH_2CH_2\overset{\overset{\displaystyle CH_3}{|}}{C}{=}CHCH_2CH_2OH \xrightarrow[\text{pyridine}]{\text{TsCl}} G \xrightarrow[\text{acetone}]{\text{NaI}} H}$$

(f) 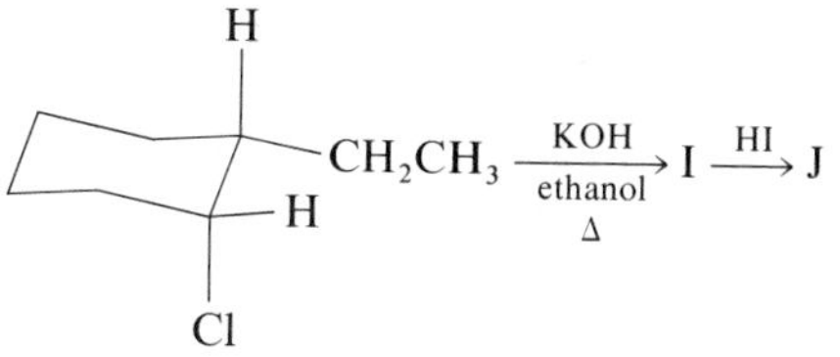

(g) 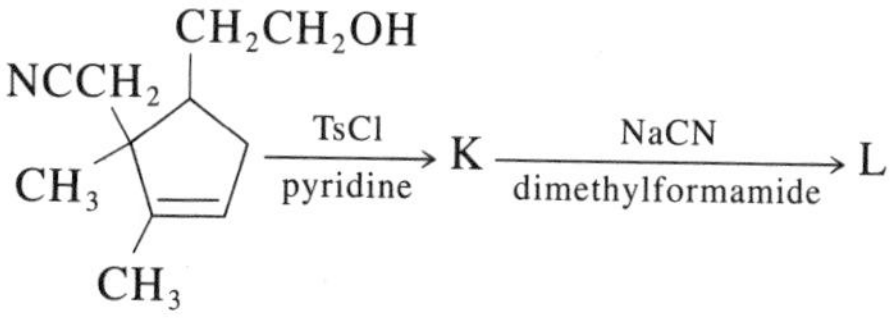

7.31 The following incomplete reaction sequences give only a starting material and a product that can be prepared from it. Show the reagents that would be necessary to carry out each transformation and any major products that would be formed in the intermediate stages. There may be more than one correct way to complete a sequence.

(a) 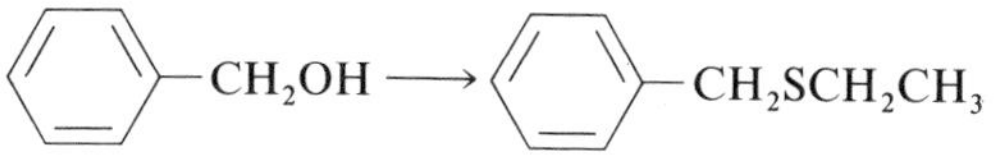

(b) $\mathrm{CH_3CH_2CH_2CH_2CH_2OH \longrightarrow CH_3CH_2CH_2CH_2CH_2I}$

(c) $\mathrm{CH_3CH_2CH_2CH_2Br \longrightarrow CH_3CH_2CH_2CH_2C{\equiv}CH}$

(d) 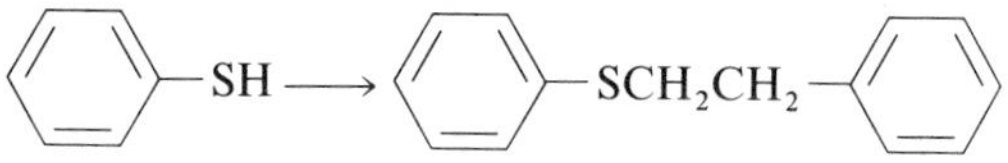

(e) 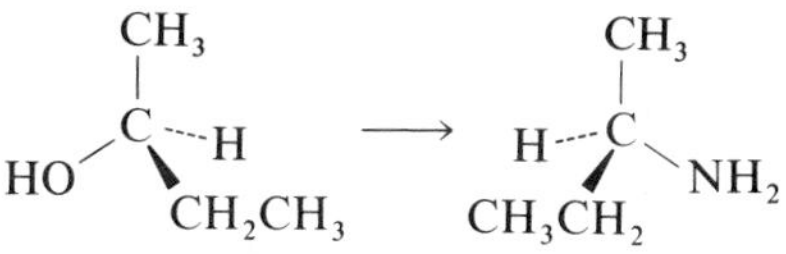

(f) $\mathrm{CH_3CH{=}CH_2 \longrightarrow CH_3\underset{\underset{\displaystyle SCH_2CH_3}{|}}{C}HCH_3}$

(g) $\mathrm{CH_3CH_2CH_2CH_2CH_2Br \longrightarrow CH_3CH_2CH_2CH_2CH_2CH(\overset{\overset{\displaystyle O}{\|}}{C}OCH_2CH_3)_2}$

7.32 Decide which species in each of the following pairs is more nucleophilic.

(a) SH^- or Cl^- (b) $(CH_3)_3B$ or $(CH_3)_3P$ (c) CH_3NH^- or CH_3NH_2

(d) CH_3CH_2OH or CH_3CH_2Cl (e) CH_3SCH_3 or CH_3OCH_3

7.33 Which of the reaction conditions listed below would give the most inversion of configuration for the following reaction if the starting material is optically active?

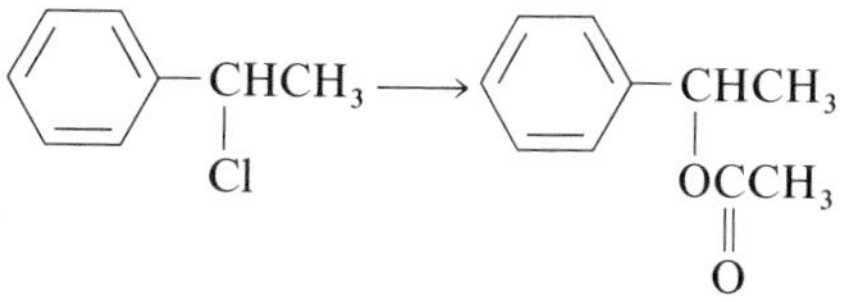

(a) $CH_3C(=O)O^-K^+$ in acetic acid, 50 °C (b) $(CH_3CH_2)_4N^+\ ^-OC(=O)CH_3$ in acetone, 50 °C

(c) $CH_3C(=O)O^-Na^+$ in 60% aqueous ethanol, 25 °C

7.34 Answer the questions below for this S_N2 reaction:

$$CH_3CH_2CH_2CH_2Br + NaOH \xrightarrow[\text{ethanol}]{} CH_3CH_2CH_2CH_2OH + NaBr$$

(a) What is the rate expression for the reaction?
(b) Draw an energy diagram for the reaction. Label all parts. You may assume that the products are lower in energy than the reactants.
(c) What will be the effect on the rate of the reaction of doubling the concentration of *n*-butyl bromide?
(d) What will be the effect on the rate of the reaction of halving the concentration of sodium hydroxide?
(e) Will the rate of the reaction change significantly if the solvent is changed to 80% ethanol, 20% water?

7.35 Answer the questions below for this S_N1 reaction:

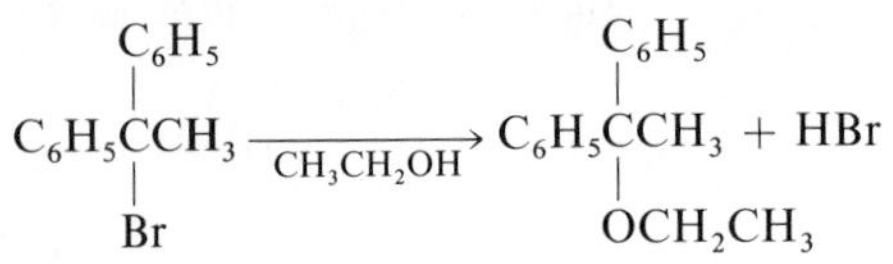
C6H5
C6H5CCH3
Br
CH3CH2OH
C6H5
C6H5CCH3 + HBr
OCH2CH3
1-bromo-1,1-diphenylethane

(a) What is the rate expression for the reaction?
(b) Draw an energy diagram for the reaction. Label all parts. You may assume that the products are lower in energy than the reactants.
(c) What will be the effect on the rate of the reaction of doubling the initial concentration of 1-bromo-1,1-diphenylethane?
(d) Will the rate of the reaction change significantly if some water is added to the solvent, which is ethanol?

7.36 When 1-chloro-2-butene reacts in 50% aqueous acetone at 47 °C, the product is a mixture of two alcohols.

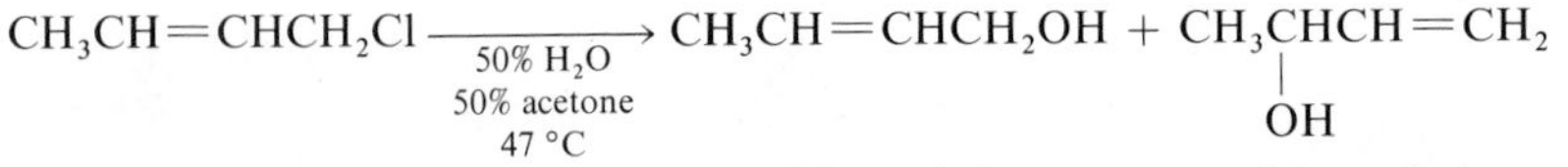
CH3CH=CHCH2Cl
50% H2O
50% acetone
47 °C
CH3CH=CHCH2OH + CH3CHCH=CH2
OH
1-chloro-2-butene   2-buten-1-ol 56%   3-buten-2-ol 44%

The reaction goes through a resonance-stabilized cation, called an **allylic cation.** What is the structure of the cation? Explain why it has high stability. Write a detailed mechanism that accounts for the observed experimental facts.

7.37 *cis*-2-Phenylcyclohexyl tosylate reacts with potassium *tert*-butoxide in *tert*-butyl alcohol at 50 °C to give exclusively 1-phenylcyclohexene. *trans*-2-Phenylcyclohexyl tosylate does not give any alkene under the same conditions. Draw structures for the two compounds, and explain these observations.

7.38 One of the classic experiments in which the stereochemistry of E_2 reactions was explored involved menthyl chloride and neomenthyl chloride. These two compounds differ from each other only in the stereochemistry at the carbon atom to which the chlorine atom is attached. The following equations show the reactions they undergo without revealing their stereochemistry.

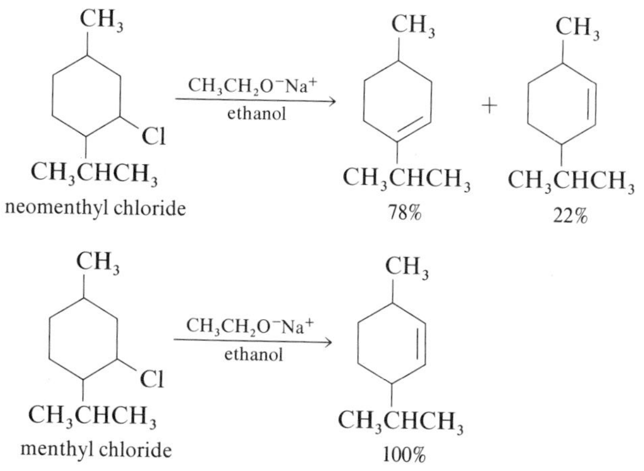
CH3
Cl
CH3CHCH3
neomenthyl chloride
CH3CH2O−Na+
ethanol
CH3
CH3CHCH3
78%
+
CH3
CH3CHCH3
22%
CH3
Cl
CH3CHCH3
menthyl chloride
CH3CH2O−Na+
ethanol
CH3
CH3CHCH3
100%

(a) Using the information given, assign relative stereochemistry to menthyl chloride and neomenthyl chloride. Menthyl chloride is derived from natural sources, and the substituents on the cyclohexane ring are in the most stable configuration.

(b) When the reaction with menthyl chloride is carried out in 80% aqueous ethanol with no added base, the following products from the elimination reaction are obtained. How do you explain these results?

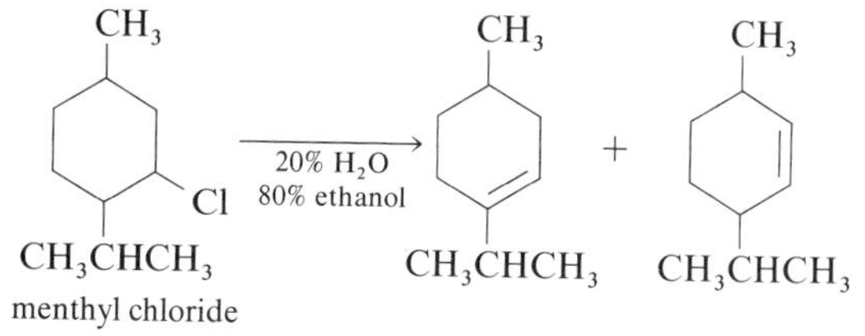
CH3
Cl
CH3CHCH3
menthyl chloride
20% H2O
80% ethanol
CH3
CH3CHCH3
+
CH3
CH3CHCH3

7.39 (2*S*,3*R*)-4-Dimethylamino-1,2-diphenyl-3-methyl-2-butanol is the alcohol from which the analgesic Darvon is made (Problem 6.31, p. 231). Several steps in the synthesis of this alcohol are shown below.

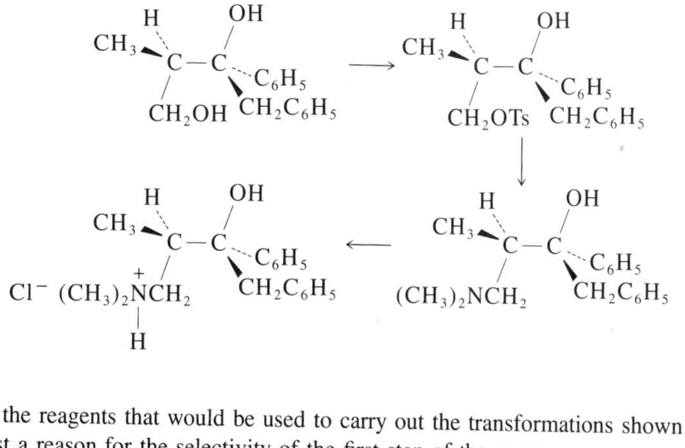
H
OH
CH3
C
C
C6H5
CH2OH
CH2C6H5
H
OH
CH3
C
C
C6H5
CH2OTs
CH2C6H5
H
OH
CH3
C
C
C6H5
(CH3)2NCH2
CH2C6H5
H
OH
CH3
C
C
C6H5
Cl− (CH3)2NCH2
+
H
CH2C6H5

(a) Fill in the reagents that would be used to carry out the transformations shown.

(b) Suggest a reason for the selectivity of the first step of the sequence.

7.40 The stereochemical requirements for the E_2 elimination reaction were explored using alkyl halides labeled with deuterium atoms. 2-Bromo-3-deuteriobutane was used in two isomeric forms. The products observed in each case are outlined on the following page.

1. (2*S*,3*R*)$CH_3CH(D)CH(Br)CH_3$ $\xrightarrow{CH_3CH_2O^-K^+,\ \text{ethanol},\ 70\ ^\circ C}$

(*E*)-2-butene with no deuterium (30%) + (*Z*)-2-butene with one deuterium (34%) + 1-butene with one deuterium (35%)

2. (2*S*,3*S*)$CH_3CH(D)CH(Br)CH_3$ $\xrightarrow{CH_3CH_2O^-K^+,\ \text{ethanol},\ 70\ ^\circ C}$

(*E*)-2-butene with one deuterium (58%) + (*Z*)-2-butene with no deuterium (13%) + 1-butene with one deuterium (29%)

(a) Draw stereochemically accurate representations of (2*S*,3*R*)- and (2*S*,3*S*)-2-bromo-3-deuteriobutane.

(b) Show the transition states for the reactions leading to (*E*)-2-butene and (*Z*)-2-butene in each case.

7.41 The following experimental observations were made. Even though the two reactions proceeded at very different rates, the relative amounts of the products obtained were practically identical.

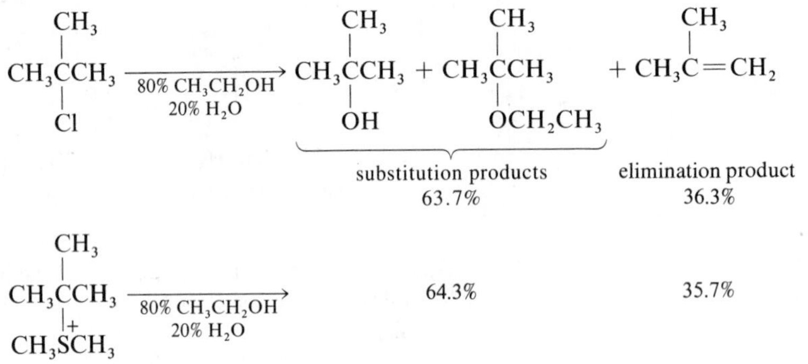
CH3CCH3 (CH3, Cl)
80% CH3CH2OH
20% H2O
CH3CCH3 (CH3, OH) + CH3CCH3 (CH3, OCH2CH3) + CH3C=CH2 (CH3)
substitution products
63.7%
elimination product
36.3%
CH3CCH3 (CH3, CH3SCH3 +)
80% CH3CH2OH
20% H2O
64.3%
35.7%

The first reaction is 7.5 times faster than the second reaction.

(a) What do these data suggest about the mechanism of the reaction?

(b) Why are the rates of the two reactions different?

7.42 When an alkyl halide reacts with a thiocyanate ion, SCN^-, reaction takes place at the sulfur atom. With the cyanate ion, OCN^-, reaction takes place at the nitrogen.

$$RX + \underset{\text{thiocyanate ion}}{SCN^-} \longrightarrow \underset{\text{alkyl thiocyanate}}{RSCN} + X^-$$

$$RX + \underset{\text{cyanate ion}}{OCN^-} \longrightarrow \underset{\text{alkyl isocyanate}}{RNCO} + X^-$$

Write structures for the thiocyanate and cyanate ions and their alkyl derivatives, and explain the difference in the reactivity of the ions. (Hint: You might find it useful to review Section 7.2.)

7.43 The following transformation was carried out in synthesizing an enantiomerically pure natural product. Show how you would do this transformation.

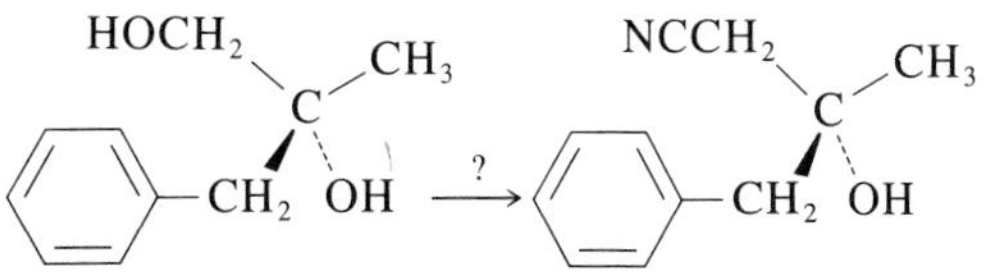

7.44 When the dibromo compound shown below is dissolved in methanol, one of the bromine atoms is replaced by a methoxy group. Predict the structure of the product of this reaction, and explain why this product is the one formed.

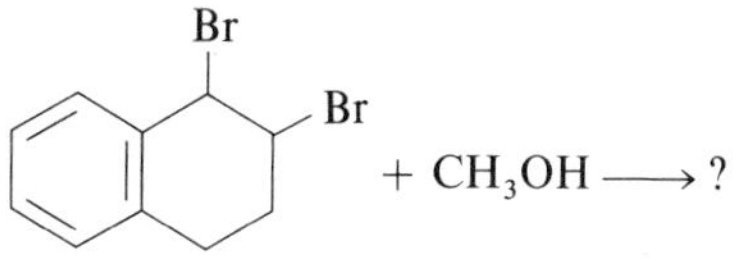

8

Alkenes

A • LOOK • AHEAD

The most important reaction of alkenes is electrophilic addition to the double bond. You are familiar with an example of this type of reaction from Chapter 4, the addition of hydrogen bromide to propene.

The functional group in alkenes is the double bond.

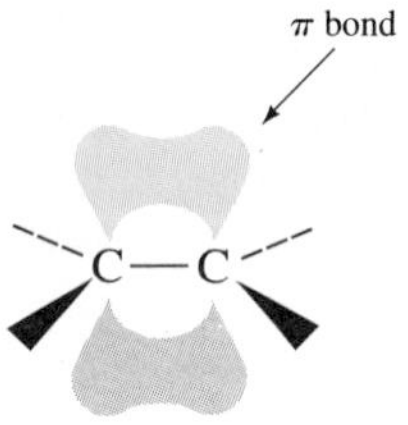
π bond
C—C

The electrons of the π bond are donated to electrophilic reagents, forming bonds to the carbon atoms that were involved in the double bond. An acid such as hydrogen bromide is an example of one type of electrophilic reagent that adds to the double bond.

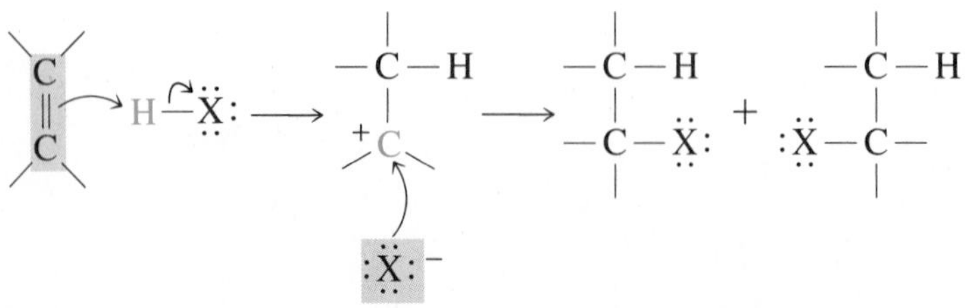

In this chapter, we will study the addition reactions of other electrophiles: Lewis acids, such as diborane, and oxidizing agents, such as the halogens, permanganate ion, and ozone. We will examine the mechanisms by which they react with alkenes and the stereochemistry of those reactions.

8.1 STRUCTURE AND ISOMERISM IN ALKENES

Alkenes are hydrocarbons that contain carbon-carbon double bonds. The carbon atoms joined by the double bond and the four atoms bonded to them lie in a plane, with bond angles of approximately 120° around the two carbon atoms. Such trigonal carbon atoms are described as being sp^2-hybridized (p. 48). The double bond in an alkene consists of a σ bond resulting from the overlap of sp^2 hybrid orbitals on two carbon atoms and a π bond resulting from the side-to-side overlap of the remaining p orbital on each carbon (Figure 8.1).

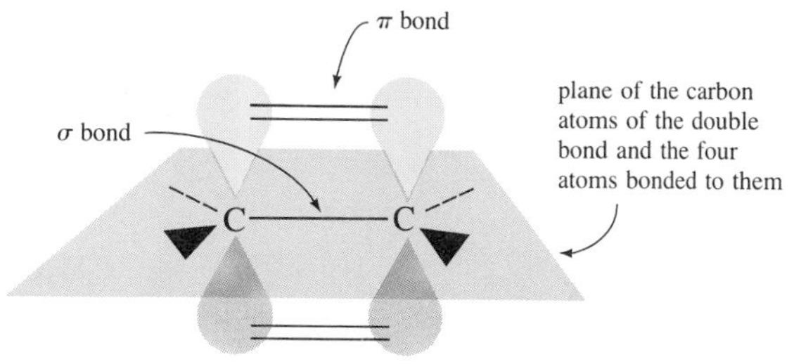

π bond
σ bond
C
C
plane of the carbon atoms of the double bond and the four atoms bonded to them

FIGURE 8.1 The orbitals and stereochemistry of a double bond.

Alkenes form a homologous series with the molecular formula C_nH_{2n} and are, thus, unsaturated hydrocarbons (p. 172). The smallest members of the series—ethene, usually called ethylene (C_2H_4), propene (C_3H_6), and butenes (C_4H_8)—are gases at room temperature. Larger alkenes up to $C_{20}H_{40}$ are liquids. As with alkanes, the boiling points and melting points of alkenes increase with molecular weight but show some variations that depend on the shape of the molecule. Alkenes with the same molecular formula may be isomers of one another if the position or the stereochemistry of the double bond differ. The different noncyclic isomers of C_4H_8—1-butene, 2-methylpropene, (E)-2-butene and (Z)-2-butene—were shown on p. 212.

PROBLEM 8.1

Write structural formulas for the six noncyclic isomers of C_5H_{10}. Show any stereoisomerism that exists for these compounds. (Hint: A review of Section 6.10 will be helpful.)

8.2 NOMENCLATURE OF ALKENES

A. IUPAC Names for Alkenes

The systematic name of an alkene is derived from the name of the alkane corresponding to the longest continuous chain of carbon atoms that contains the double bond. The **ane** ending of the name of the alkane is changed to **ene** for the alkene. Thus, the IUPAC name ethene comes from ethane. An alkene with five carbon atoms in a continuous chain is named as a pentene; one with six carbon atoms in the longest continuous chain is a hexene.

$CH_2{=}CH_2$ — ethene, ethylene

$\overset{5}{C}H_3\overset{4}{C}H_2\overset{3}{C}H_2\overset{2}{C}H{=}\overset{1}{C}H$ — 1-pentene

$\overset{6}{C}H_3\overset{5}{C}H_2\overset{4}{C}H{=}\overset{3}{C}H\overset{2}{C}H_2\overset{1}{C}H_3$ — 3-hexene

The chain is numbered so as to give the carbon atoms joined by the double bond the lowest possible numbers. The position of the double bond is indicated by the number of the *first* of the two carbon atoms. Thus, the examples shown above are 1-pentene, because the double bond is between carbon atoms 1 and 2 of a five-carbon chain, and 3-hexene, because the double bond is between carbon atoms 3 and 4 of a six-carbon chain. If other substitutents are present, they too are named and their positions are indicated by numbers.

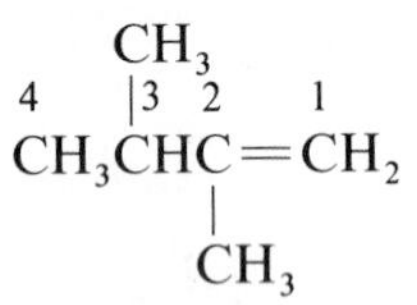
CH3
4
3
2
1
CH3CHC=CH2
CH3
2,3-dimethyl-1-butene

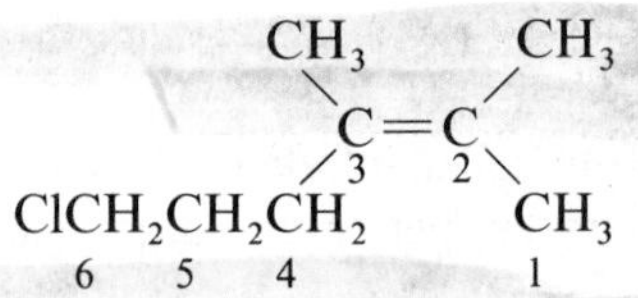
CH3
CH3
C=C
3
2
ClCH2CH2CH2
CH3
6
5
4
1
6-chloro-2,3-dimethyl-2-hexene

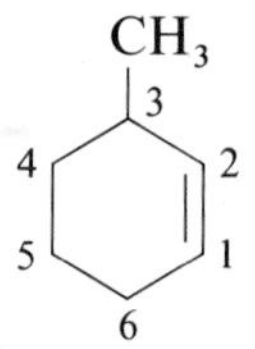
CH3
3
2
1
6
5
4
3-methylcyclohexene

the first carbon of the double bond is understood to be carbon 1 in the cyclic case

The nomenclature of stereoisomeric alkenes was discussed on pp. 212–214. You may want to review those pages before you do the next two problems.

PROBLEM 8.2

Name all of the isomers of C_5H_{10} for which you drew structural formulas in Problem 8.1.

PROBLEM 8.3

Draw a structural formula for each of the following compounds, showing the correct stereochemistry when it is designated.

(a) 1-phenylcyclohexene
(b) 3,3-dimethylcyclopentene
(c) (*Z*)-2-phenyl-2-butene
(d) (*E*)-1,3-dichloro-2-methyl-2-pentene

B. Vinyl and Allyl Groups

Two unsaturated groups have common names that are used as substituents or to indicate structural features in molecules. The **vinyl group** is formed when a hydrogen atom is removed from ethylene. C_2H_4

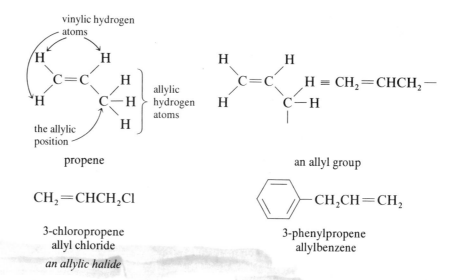

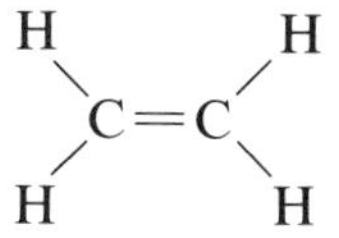

ethylene

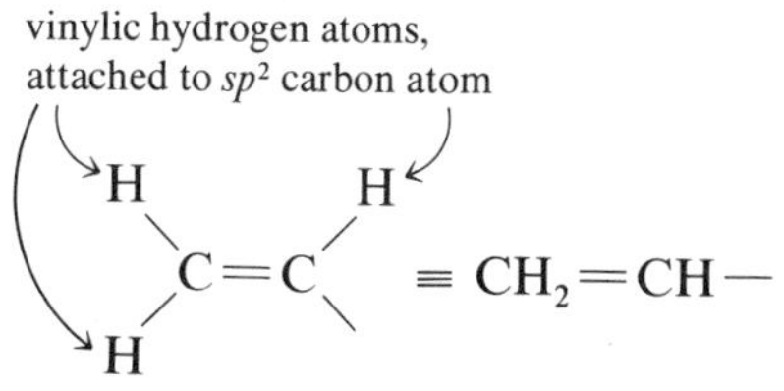
vinylic hydrogen atoms,
attached to sp² carbon atom
≡ CH₂=CH—

the vinyl group

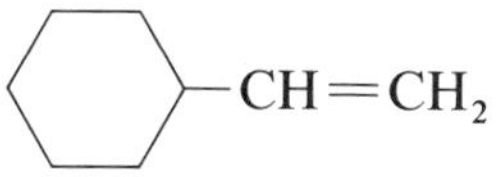

cyclohexylethene
vinylcyclohexane

$H_2C{=}CHCl$

chloroethene
vinyl chloride
a vinylic halide

Hydrogen atoms attached to the sp^2-hybridized carbon atoms of a double bond are described as **vinylic hydrogen atoms.** Similarly, an organic halide with a halogen atom bonded directly to an sp^2-hybridized carbon atom of a double bond is known as **vinylic halide.**

Removing a hydrogen atom from the sp^3-hybridized carbon atom of propene gives an **allyl group.**

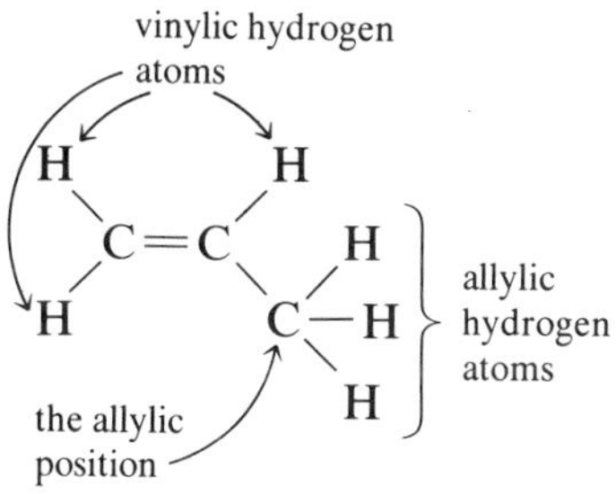
vinylic hydrogen
atoms
allylic
hydrogen
atoms
the allylic
position

propene

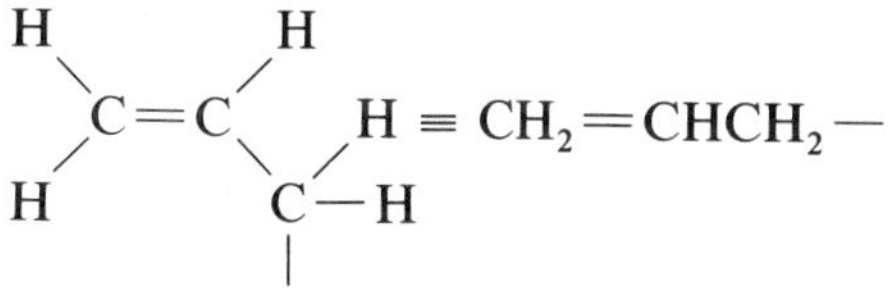
≡ CH₂=CHCH₂—

an allyl group

$CH_2{=}CHCH_2Cl$

3-chloropropene
allyl chloride
an allylic halide

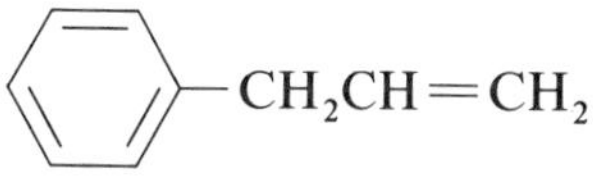

3-phenylpropene
allylbenzene

In general, sp^3-hybridized carbon atoms adjacent to a double bond are known as the **allylic positions** in the molecule. The hydrogen atoms on such a carbon atom are **allylic hydrogen atoms.** Replacement of one of the allylic hydrogen atoms by a halogen atom gives an **allylic halide.** Hydrogen atoms (or halogen atoms) in a vinylic position are much less reactive than hydrogen atoms (or halogen atoms) in an allylic position (pp. 829 and 689), so the kind of structural distinctions described here are important.

RELATIVE STABILITIES OF ALKENES

Alkenes react with hydrogen in the presence of a metal catalyst to give alkanes (pp. 304–310). This hydrogenation reaction is exothermic. The convention is to write the heat of an exothermic reaction as a negative quantity to emphasize that the total internal energy of the system decreases during the reaction. Some of the energy stored in the molecules on the left-hand side of the reaction equation has been released to the environment as heat. Experimental heats of hydrogenation of isomeric alkenes can be used to measure their relative stabilities.

1-Butene and (*Z*)- and (*E*)-2-butene each add one equivalent of hydrogen to become butane.

$$\underset{\text{1-butene}}{CH_3CH_2CH{=}CH_2} + H_2 \xrightarrow[\text{catalyst}]{} \underset{\text{butane}}{CH_3CH_2CH_2CH_3} \qquad \Delta H^\circ = -30.3\ \text{kcal/mol}$$

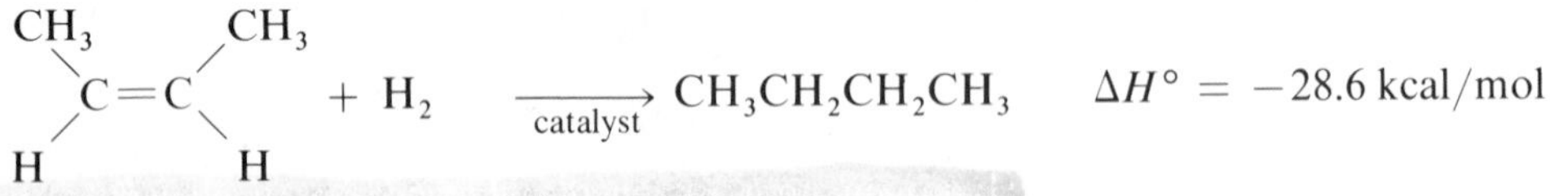

CH3
CH3
C=C
H
H
+ H2
catalyst
CH3CH2CH2CH3
ΔH° = −28.6 kcal/mol

(*Z*)-2-butene butane

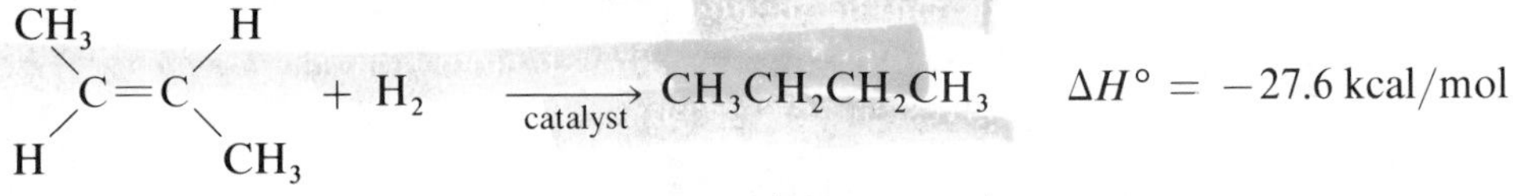

CH3
H
C=C
H
CH3
+ H2
catalyst
CH3CH2CH2CH3
ΔH° = −27.6 kcal/mol

(*E*)-2-butene butane

Each reaction has the same product, butane, and a single reagent, hydrogen. The differences in the heats of hydrogenation, the energy evolved as the alkene is converted to the alkane, must reflect the differences in energy in the alkenes. These differences are shown schematically in Figure 8.2. Less heat is evolved when (*E*)-2-butene is hydrogenated than when (*Z*)-2-butene undergoes the same reaction. Thus, the energy of the trans compound must be closer before the reaction to that of butane. (*E*)-2-Butene is at a lower energy level than (*Z*)-2-butene, or, in other words, it is more stable than (*Z*)-2-butene. The same argument demonstrates that 1-butene is less stable than either of the 2-butenes. More heat is evolved as 1-butene is transformed into butane than for either of the other two cases. 1-Butene starts at a higher energy level than either of the 2-butenes and has farther to drop to reach the energy level of butane.

A **terminal alkene,** in which the double bond is at the end of a chain, is less stable than an **internal alkene,** with the double bond somewhere in the middle of the chain. A trans alkene, in which the larger substituents on the double bond are farther apart from each other, is more stable than the corresponding cis alkene in which the larger substituents are closer together.

An important factor governing stability seems to be the number of substituents on the double bond. This is illustrated in the following reactions by the experimental heats of hydrogenation for three more alkenes, 3-methyl-1-butene, 2-methyl-1-butene, and 2-methyl-2-butene, all of which give the same alkane when hydrogenated.

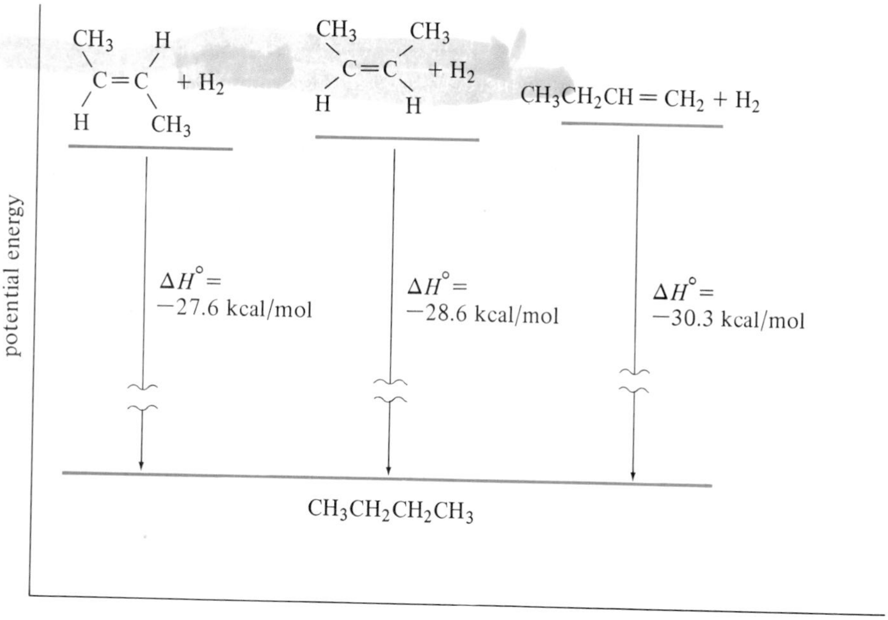
CH3 H C=C + H2 H CH3
CH3 CH3 C=C + H2 H H
CH3CH2CH=CH2 + H2
potential energy
ΔH° = −27.6 kcal/mol
ΔH° = −28.6 kcal/mol
ΔH° = −30.3 kcal/mol
CH3CH2CH2CH3

FIGURE 8.2 Difference in energy level of butane from that of hydrogen plus 1-butene, (*Z*)-2-butene, and (*E*)-2-butene, as reflected by heats of hydrogenation.

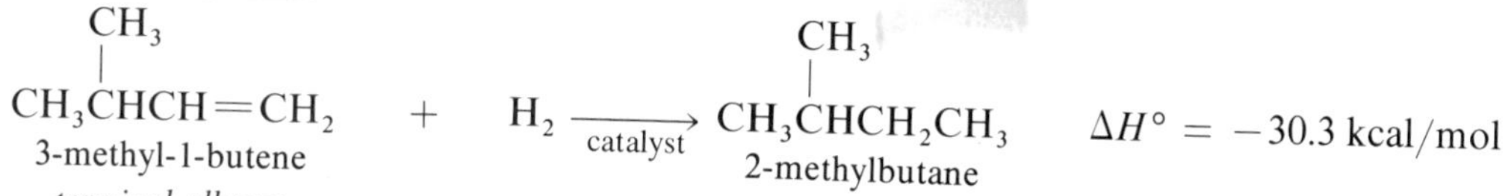
3-methyl-1-butene
catalyst
2-methylbutane
ΔH° = −30.3 kcal/mol

terminal alkene;
one substituent on double bond;
least stable

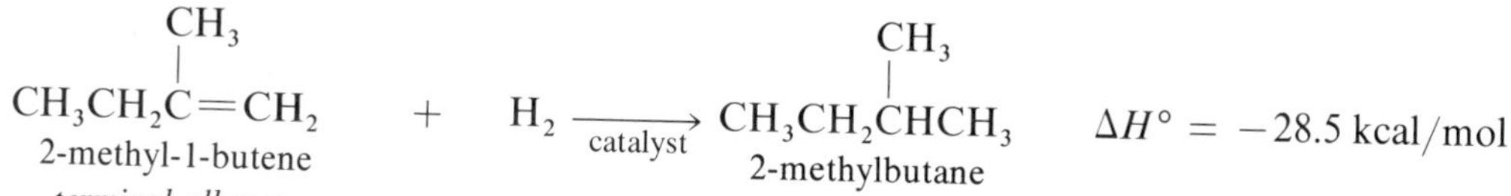
2-methyl-1-butene
catalyst
2-methylbutane
ΔH° = −28.5 kcal/mol

terminal alkene;
two substituents on double bond

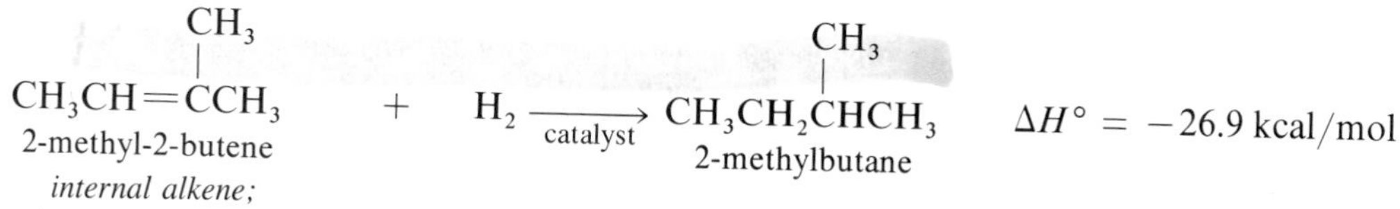
2-methyl-2-butene
catalyst
2-methylbutane
ΔH° = −26.9 kcal/mol

internal alkene;
three substituents on double bond;
most stable

The product in each case is the same compound, 2-methylbutane. One of the reactants, hydrogen, is the same for each reaction. The differences in the heats of hydrogenation must arise from the differences in energy of the other reactants, the three different alkenes.

In summary, the position of the double bond and the number of alkyl substituents on it seem to be more important in predicting the relative stability of alkenes than the nature of the alkyl groups on carbon atoms of the double bond. Internal alkenes are more stable than terminal alkenes. The stability of an alkene increases with the number of alkyl groups that are substituted on carbon atoms of the double bond. Alkenes in which bulky substituents are trans to each other are more stable than the corresponding cis alkenes.

PROBLEM 8.4

Are *Z* and *E* isomers possible for any of the methylbutenes shown above?

PROBLEM 8.5

Draw an energy diagram like Figure 8.2 to illustrate the relative stabilities of the methylbutenes in a schematic way.

8.4 ELECTROPHILIC ADDITION OF ACIDS TO ALKENES

A. Addition of Water to Alkenes

The addition of hydrogen bromide to propene was examined in detail in Chapter 4 as an example of an electrophilic addition to a double bond. Other acids, acting as electrophiles, also react with alkenes. For example, water adds to alkenes in the presence of acids to form alcohols. The addition of water is known as a **hydration reaction.** The hydration reactions of simple alkenes that are available from petroleum as by-products in the manufacture of gasoline are of commercial importance. They hydration of 2-methylpropene is one example.

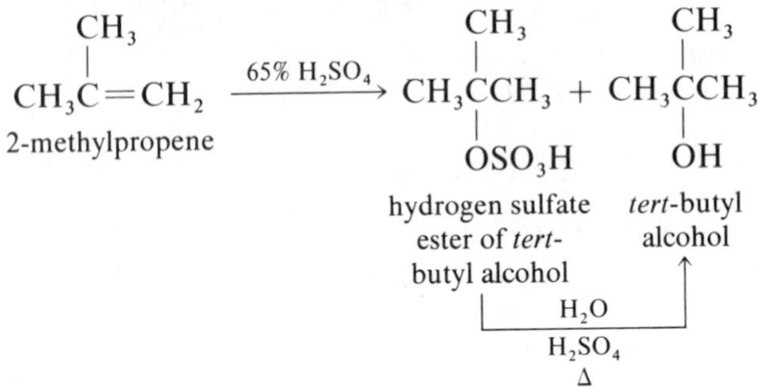
CH3
CH3C=CH2
2-methylpropene
65% H2SO4
CH3CCH3 + CH3CCH3
OSO3H
OH
hydrogen sulfate ester of tert-butyl alcohol
tert-butyl alcohol
H2O
H2SO4
Δ

The alkene reacts with the acid to give a tertiary carbocation. The nucleophiles present in the reaction mixture are water and hydrogen sulfate ion. Some of the carbocations combine with water to form *tert*-butyl alcohol; some combine with hydrogen sulfate ion to form an ester. In the industrial process, the sulfate ester is decomposed by heating with dilute aqueous acid, and the alcohol is isolated as the product.

VISUALIZING THE REACTION

A hydration reaction

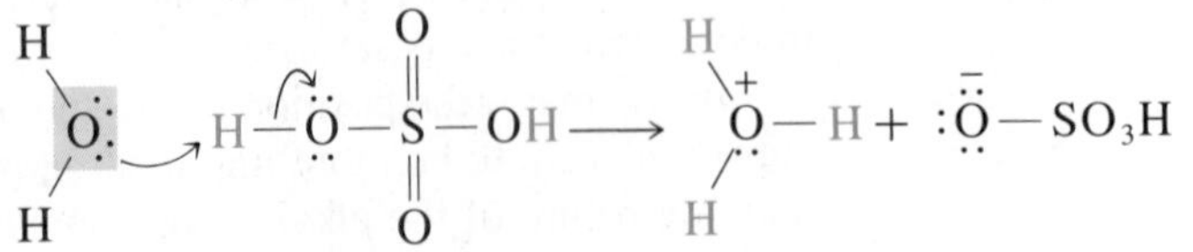
H
O
H
H—O—S—OH
O
O
H
O—H +
:O—SO3H
H

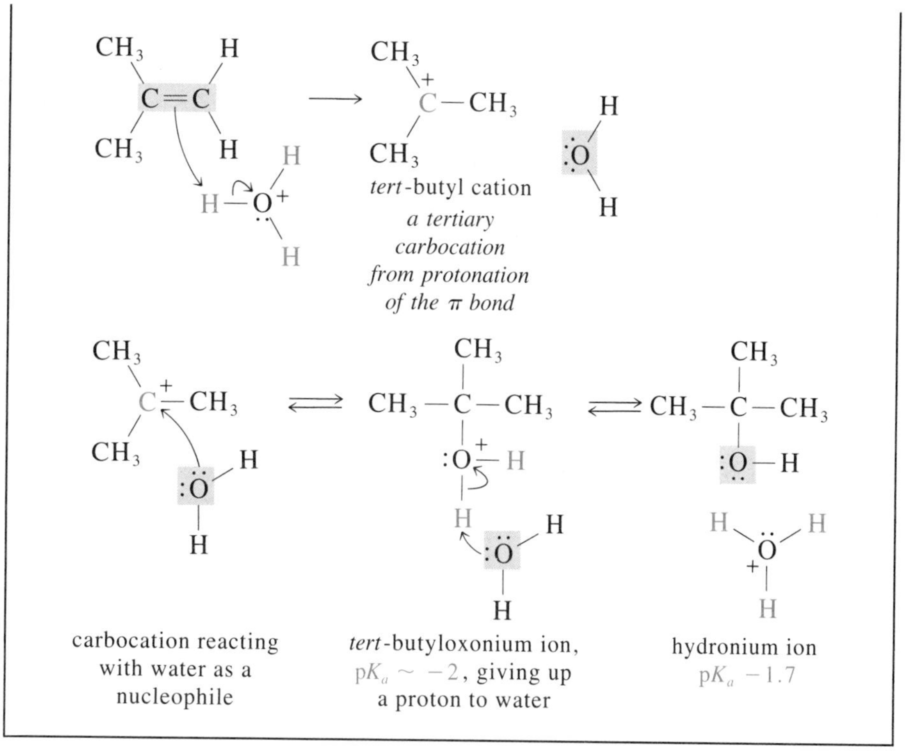
tert-butyl cation
a tertiary carbocation from protonation of the π bond
carbocation reacting with water as a nucleophile
tert-butyloxonium ion, pK_a ~ −2, giving up a proton to water
hydronium ion pK_a −1.7

The overall process results in Markovnikov addition of water to an alkene.

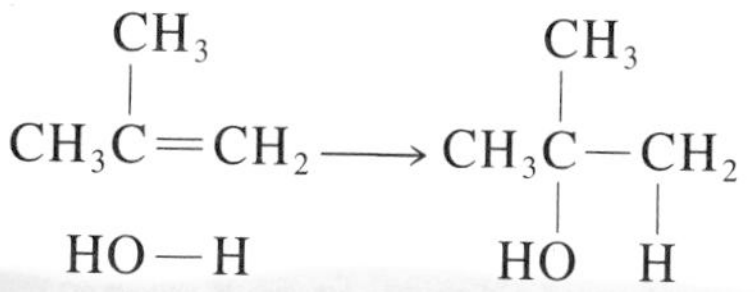

Markovnikov addition of H—OH to an alkene

The reactivity of an alkene toward acids is highly dependent on the stability of the carbocation that it forms on protonation. For example, for reactions in aqueous acid, 2,3-dimethyl-2-butene, 2-methyl-2-butene, and 2-methylpropene (which give tertiary carbocations on protonation) react about 10,000 times faster than (*E*)-2-butene and propene (which give secondary carbocations) do.

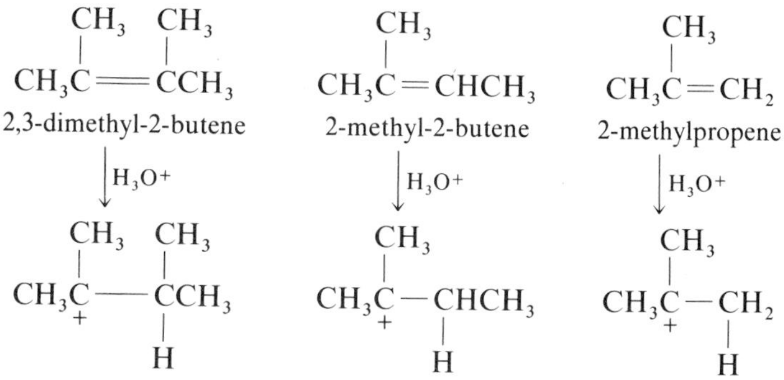
2,3-dimethyl-2-butene
2-methyl-2-butene
2-methylpropene

tertiary carbocations

That is, the above reactions are approximately 10,000 times faster than the following:

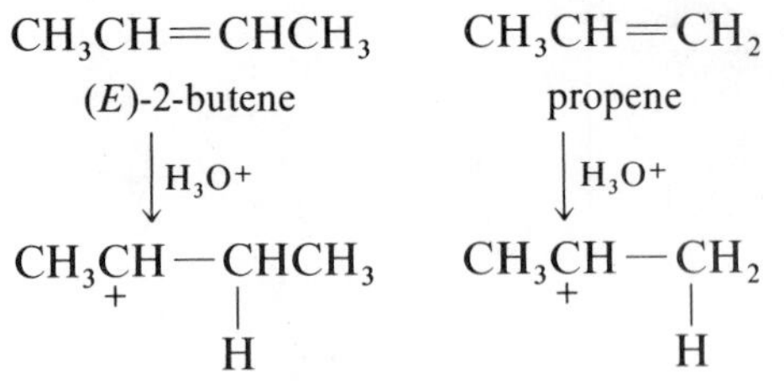
(E)-2-butene
propene
H_3O^+
H_3O^+

secondary carbocations

Such experimental evidence reinforces the idea that alkyl groups are electron-releasing when bonded to sp^2-hybridized carbon atoms and stabilize carbocation intermediates (p. 121).

The hydration reaction is a reversible one. Alcohols, when heated in acid, are converted to alkenes, and it is thought that the reaction goes through a series of steps that are the reverse of the ones shown for the hydration reaction. Cyclopentanol, for example, gives cyclopentene when heated with phosphoric acid.

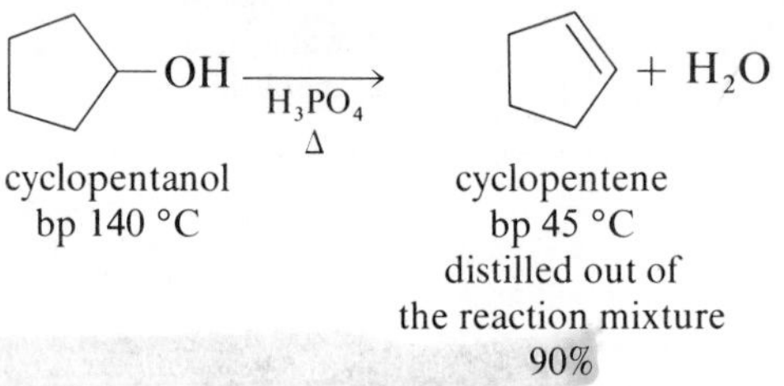
OH
H_3PO_4
Δ
+ H_2O
cyclopentanol
bp 140 °C
cyclopentene
bp 45 °C
distilled out of
the reaction mixture
90%

The alcohol is protonated by the acid, giving an oxonium ion. The oxonium ion loses a molecule of water to give a carbocation. The carbocation loses a proton to water, a base.

VISUALIZING THE REACTION

A dehydration reaction

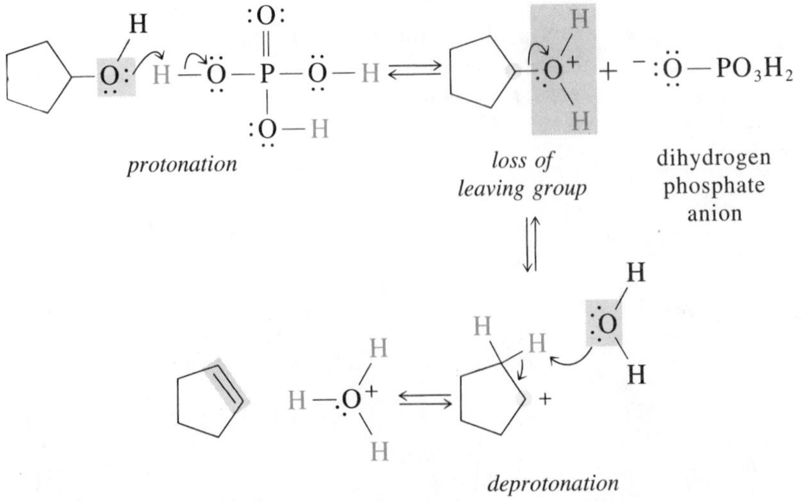
protonation
loss of
leaving group
dihydrogen
phosphate
anion
deprotonation

The hydration reaction, addition of water to an alkene, occurs when an alkene is treated with dilute acid. The reverse reaction, the **dehydration of an alcohol,** occurs when an alcohol is heated with concentrated acid. If the alkene that is formed is removed during the reaction by distillation, a good yield is obtained.

PROBLEM 8.6

Predict the products of the following reactions.

(a) 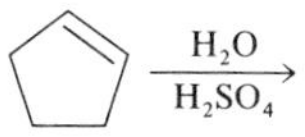

(b) 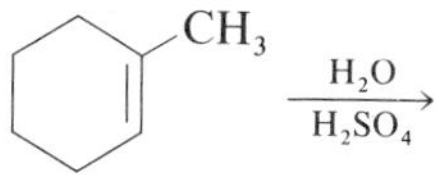

(c) $CH_3CH_2CH_2CH_2CH{=}CH_2 \xrightarrow[H_2SO_4]{H_2O}$

(d) $CH_3\overset{\overset{CH_3}{|}}{\underset{\underset{OH}{|}}{C}}CH_3 \xrightarrow[H_3PO_4,\ \Delta]{}$

(e) $CH_3\underset{\underset{OH}{|}}{C}HCH_3 \xrightarrow[H_3PO_4,\ \Delta]{}$

(f) $CH_2{=}CH_2 \xrightarrow[H_2SO_4]{H_2O}$

(g) $CH_3\overset{\overset{CH_3}{|}}{C}{=}CHCH_2CH_3 \xrightarrow[H_2SO_4]{H_2O}$

(h) 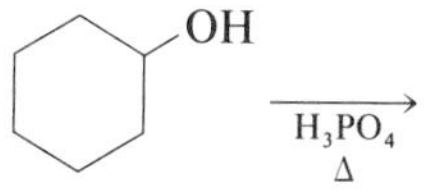

B. Reaction of Carbocations with Alkenes

Alkenes are electron-rich reagents; the electrons of the π bond are available for reaction with electrophiles. Whenever carbocations are formed in reaction mixtures containing alkenes, some reaction between these electron-deficient intermediates and the alkenes can be expected. Under some conditions, this reaction may be a major one; sometimes it is used in syntheses. An example is the dimerization of 2-methylpropene that takes place in sulfuric acid. The same product mixture is obtained whether the alkene or the corresponding alcohol, *tert*-butyl alcohol, is used as the starting material.

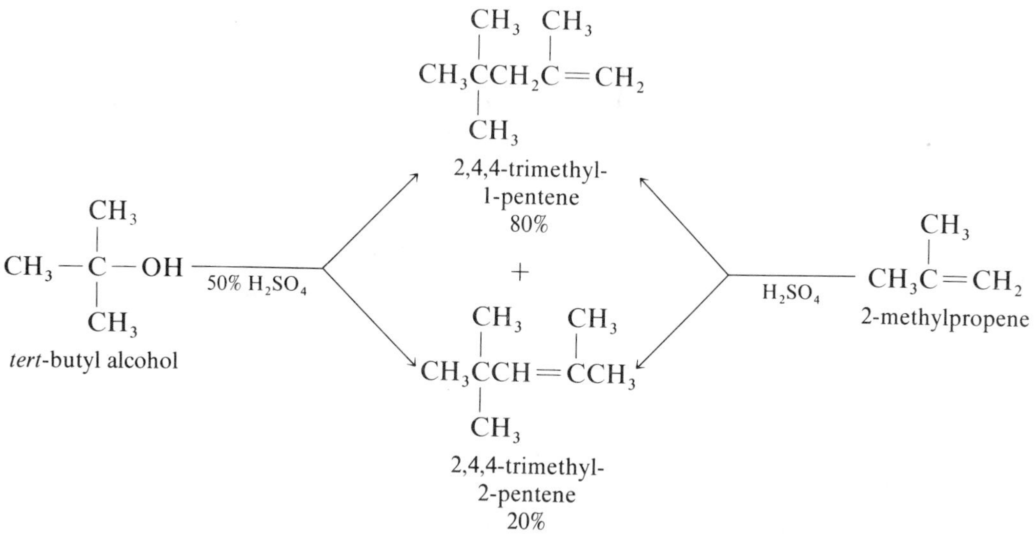
2,4,4-trimethyl-1-pentene
80%
50% H2SO4
tert-butyl alcohol
H2SO4
2-methylpropene
2,4,4-trimethyl-2-pentene
20%

This observed result indicates that the intermediate in both cases is the same. Each reaction goes through the *tert*-butyl cation, which adds to the double bond of the alkene to give a new carbocation. The two alkenes are formed by the loss of a proton in two different ways from the second carbocation.

VISUALIZING THE REACTION

Dimerization of an alkene or an alcohol

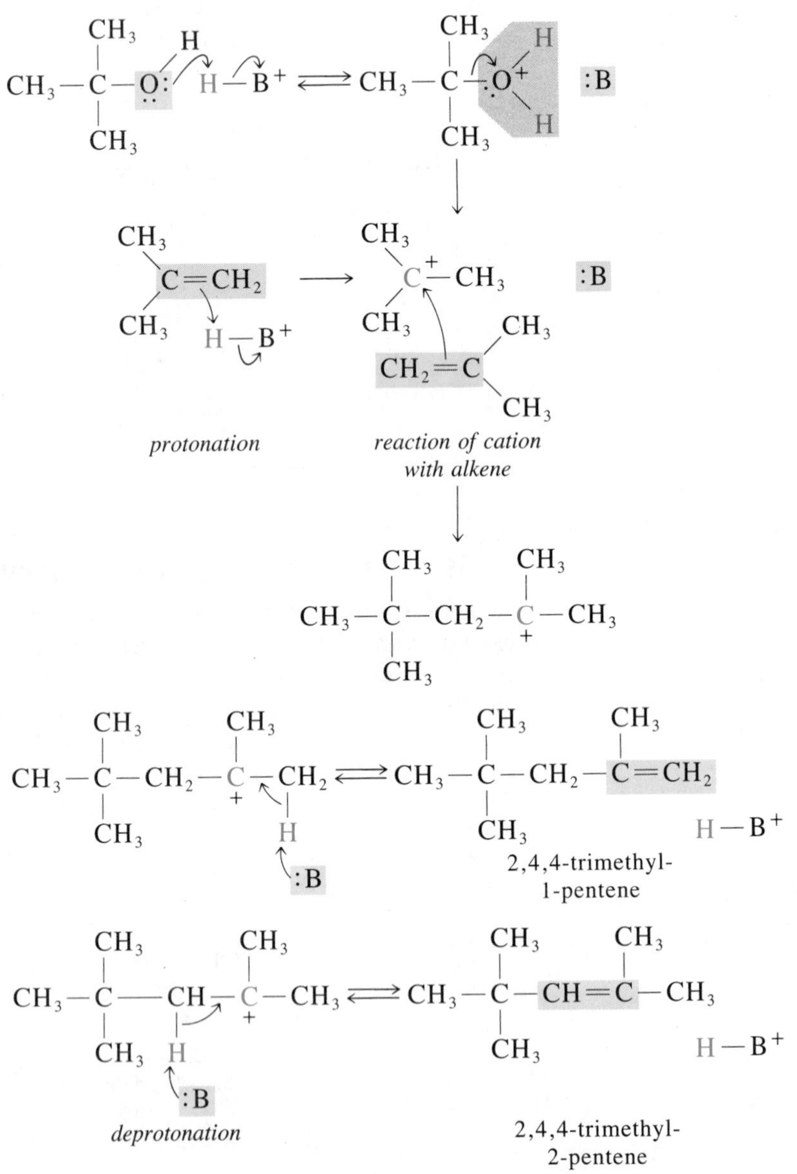
protonation
reaction of cation with alkene
2,4,4-trimethyl-1-pentene
deprotonation
2,4,4-trimethyl-2-pentene

The two products have the same carbon skeleton and can be converted, by the addition of hydrogen (p. 306), to the same alkane, 2,2,4-trimethylpentane, which is an important constituent of high-octane gasoline.

The reaction shown does not have to stop with the combination of two alkene units. Further additions of 2-methylpropene can occur. In fact, with longer heating, complex mixtures containing high-molecular-weight alkenes are formed.

VISUALIZING THE REACTION

Polymerization of an alkene

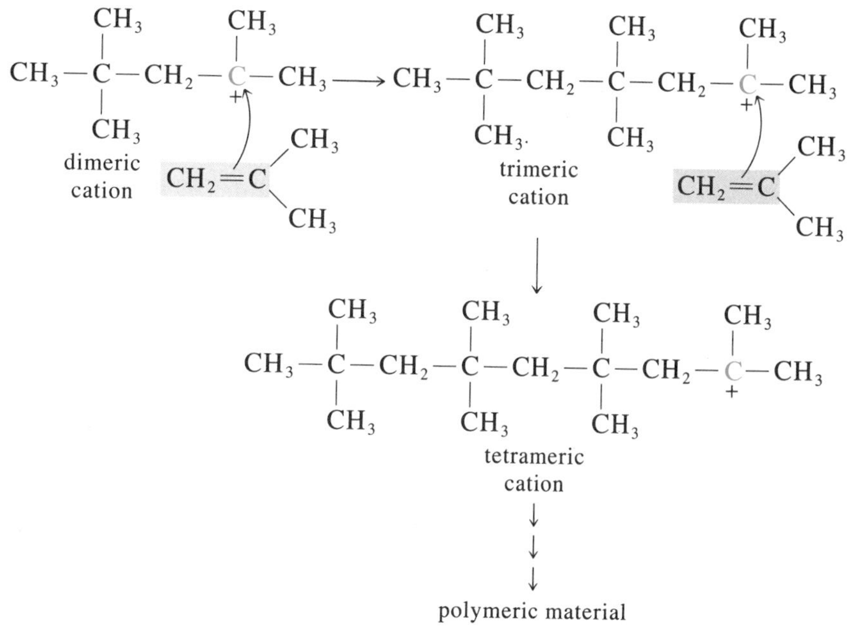

dimeric cation
trimeric cation
tetrameric cation
polymeric material

In these reactions, 2-methylpropene is the **monomer,** a low-molecular-weight unit that adds to itself in a repetitious fashion to give a molecule having a higher molecular weight. Two units of the monomer combine to give a **dimer;** three form a **trimer.** A large molecule consisting of many units of monomer bonded together is called a **polymer.** The synthesis of polymeric materials with useful properties is an important area of organic chemistry that we will consider in Chapter 27.

8.5 REARRANGEMENTS OF CARBOCATIONS

A. Shifts of Hydrogen Atoms

At the end of the nineteenth century, Georg Wagner, who was working at the University of Warsaw, investigated the reactions of some naturally occurring alkenes and alcohols with acidic reagents. He found that some of the products had carbon skeletons that differed from those of the starting materials. These reactions were further investigated by Hans Meerwein in Germany in the 1920s, and he confirmed that the bonding between carbon atoms was being rearranged.

Chemists have also observed unexpected products in some reactions of simple alkenes. For example, 3-methly-1-butene reacts with hydrogen chloride to give

2-chloro-2-methylbutene, in which the chlorine atom is bound to a tertiary carbon atom, as well as the expected product, 2-chloro-3-methylbutane.

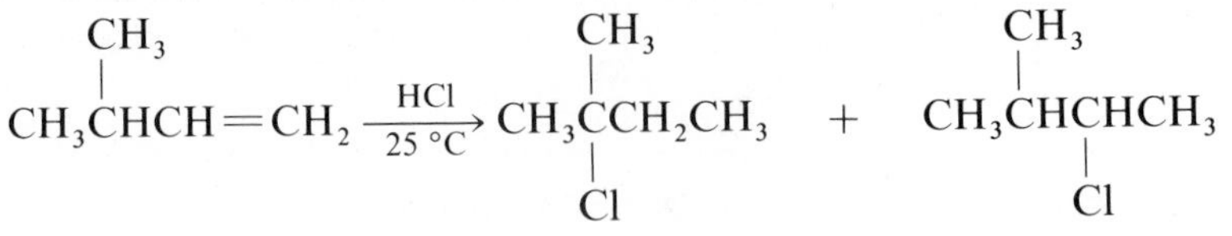

3-methyl-1-butene — 2-chloro-2-methylbutane ~50% — 2-chloro-3-methylbutane ~50%

Frank C. Whitmore, who did much early work on the chemistry of alkenes at Pennsylvania State University, suggested that the secondary carbocation that formed as an intermediate in this reaction rearranged to a more stable tertiary carbocation. The two products result from the combination of chloride ion with a secondary or tertiary carbocation intermediate.

VISUALIZING THE REACTION

1,2-Hydride shift

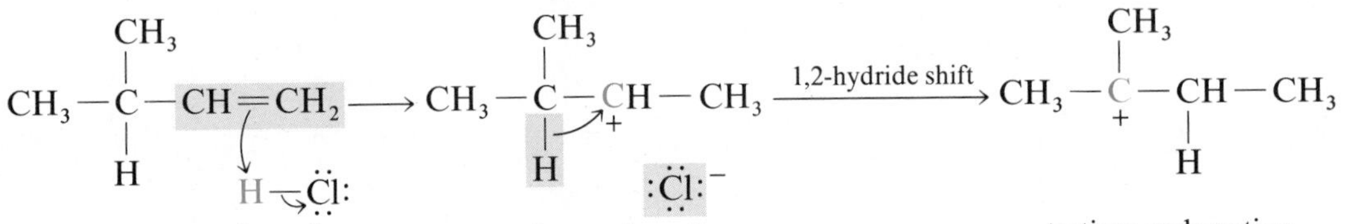

secondary carbocation

tertiary carbocation
more stable

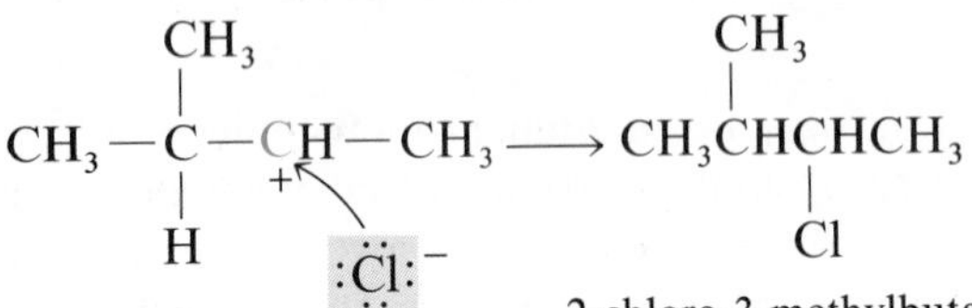

2-chloro-3-methylbutane
product from unrearranged secondary carbocation

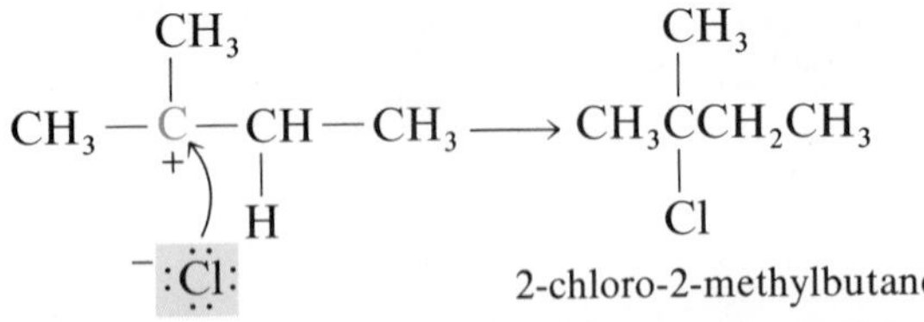

2-chloro-2-methylbutane
product from rearranged tertiary carbocation

The rearrangement shown above results when a hydrogen atom with the pair of electrons that bind it to a carbon atom moves to an adjacent carbon atom that has a deficiency of electrons. A hydrogen atom with a pair of electrons is a **hydride ion,** so the rearrangement is called a **hydride shift.** When it takes place between adjacent carbon atoms, it is a **1,2-hydride shift.** The loss of the hydrogen atom and its bonding electrons from one carbon atom leaves a new cationic center in the molecule. In the example shown above, a secondary carbocation is transformed into a more stable tertiary carbocation. Some of the secondary carbocations react with chloride ions without rearranging, so products derived from both secondary and tertiary carbocations are seen.

B. Shifts of Carbon Atoms

Carbon atoms can migrate within carbocations to give rearrangements of carbon skeletons. Another reaction studied by Frank C. Whitmore illustrates this.

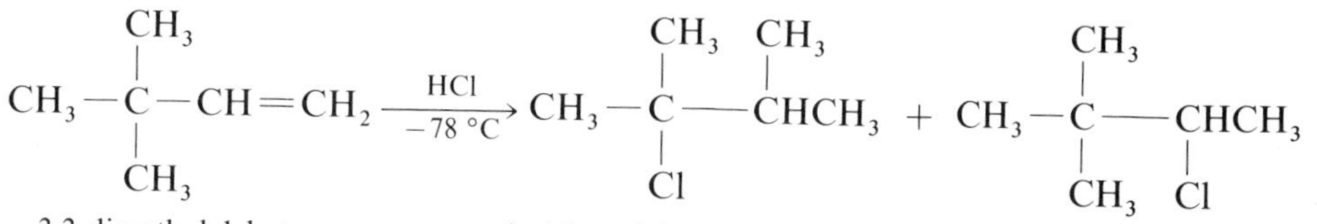

HCl
−78 °C

3,3-dimethyl-1-butene — 2-chloro-2,3-dimethylbutane 61% — 3-chloro-2,2-dimethylbutane 37%

3,3-Dimethyl-1-butene reacts with hydrogen chloride to give a secondary and a tertiary alkyl halide. The following mechanism has been proposed for the formation of these compounds.

V I S U A L I Z I N G   T H E   R E A C T I O N

1,2-Alkyl shift

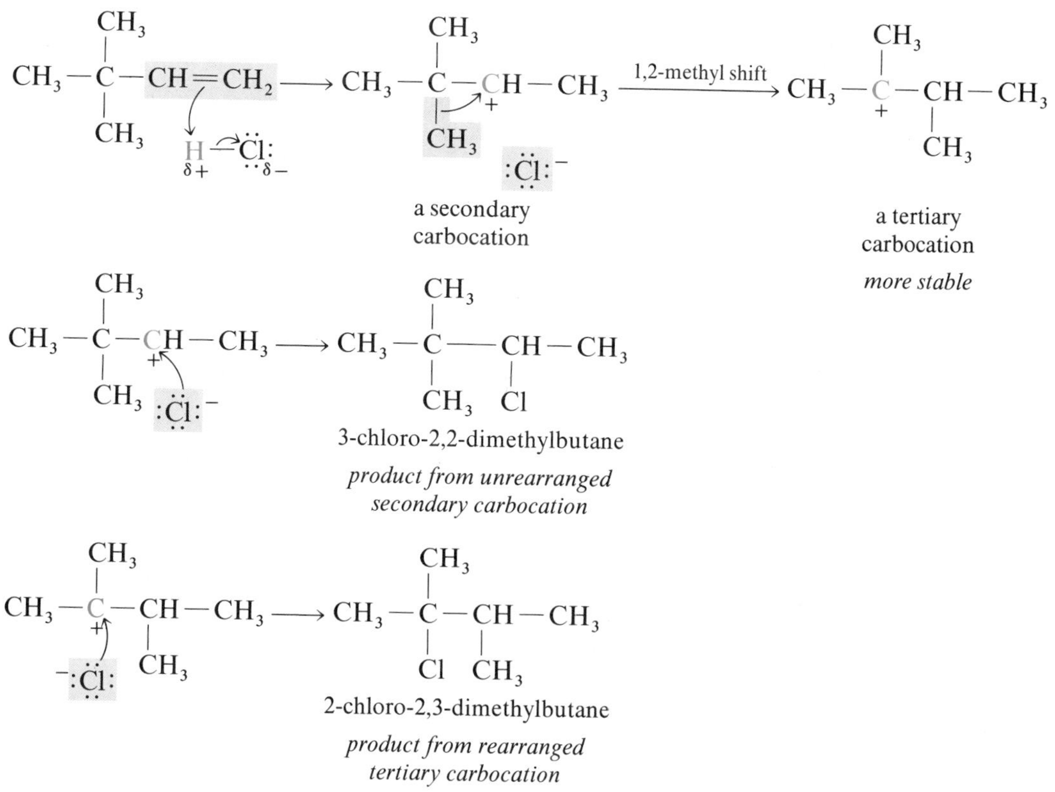

δ+
δ−
1,2-methyl shift
a secondary carbocation
a tertiary carbocation
more stable
3-chloro-2,2-dimethylbutane
product from unrearranged secondary carbocation
2-chloro-2,3-dimethylbutane
product from rearranged tertiary carbocation

A secondary carbocation forms when 3,3-dimethyl-1-butene reacts with hydrogen chloride. The shift of a methyl group, along with the pair of electrons that binds it, from one carbon atom to the adjacent cationic carbon atom creates a new tertiary carbocation. This rearrangement is called a **1,2-methyl shift.** More generally, alkyl or aryl groups of all sorts may participate in such rearrangements. Note that a 1,2-hydride shift does not change the carbon skeleton of the molecule, but a 1,2-alkyl shift does.

Carbocation intermediates may be formed by the loss of a water molecule from an alcohol in acid, as well as by the addition of a proton to an alkene. For example, the dehydration of cyclopentanol by phosphoric acid occurs by way of a carbocation (p. 282). Rearrangements occur in the conversion of some alcohols to alkenes, which is evidence for the existence of carbocation intermediates in these reactions. The dehydration of 3,3-dimethyl-2-butanol is an example. The unrearranged alkene, 3,3-dimethyl-1-butene, is formed in very small amounts. The major products, 2,3-dimethyl-2-butene and 2,3-dimethyl-1-butene, are formed by rearrangement.

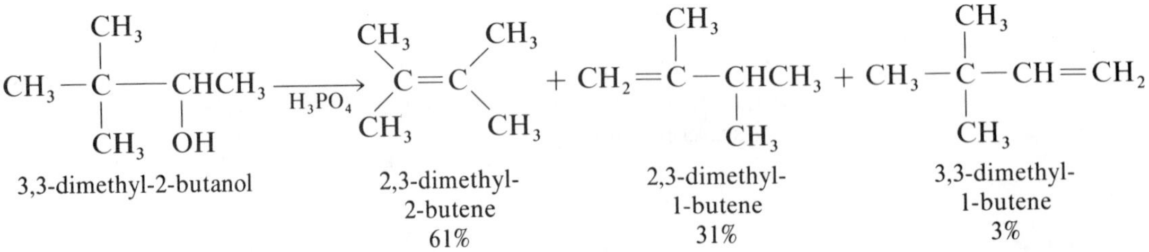
H_3PO_4
3,3-dimethyl-2-butanol
2,3-dimethyl-2-butene 61%
2,3-dimethyl-1-butene 31%
3,3-dimethyl-1-butene 3%

The above equation represents a fundamental reaction of carbocations. Carbocations are strong acids, and lose a proton easily, even to a weak base. Most reactions that proceed through a carbocation intermediate give rise to some alkene as a product (p. 253).

Interconversion of carbocations is a common phenomenon. The alkene mixture seen when 3,3-dimethyl-2-butanol is dehydrated is formed whenever any one of those three product alkenes isolated in the pure state is subjected to the acid conditions of the dehydration reaction. Such a product composition represents, therefore, an equilibrium mixture of secondary and tertiary carbocations.

Study Guide Concept Maps 8.1 and 8.2

In summary, two ways of forming alkyl cations are by the protonation of an alkene or by loss of water from a protonated alcohol. The major reactions of carbocations involve reacting as Lewis acids with electron-rich species or as Brønsted acids to lose a proton to a base. If their structure permits, they also rearrange to other carbocations in equilibrium reactions.

PROBLEM 8.7

The following products were obtained. Write a mechanism showing how these products arise.

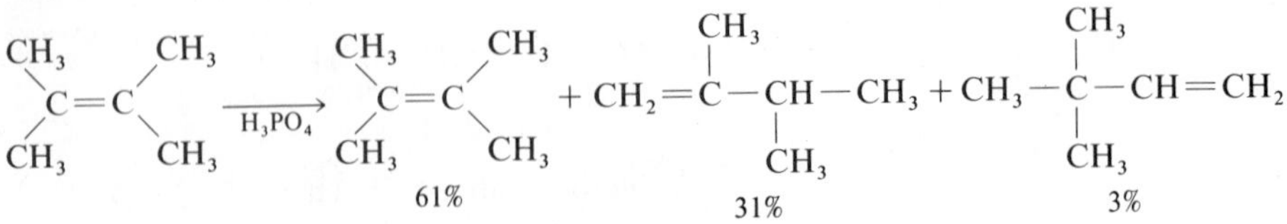
H_3PO_4
61%
31%
3%

PROBLEM 8.8

The following reaction was observed. Write a mechanism that accounts for the product that was obtained.

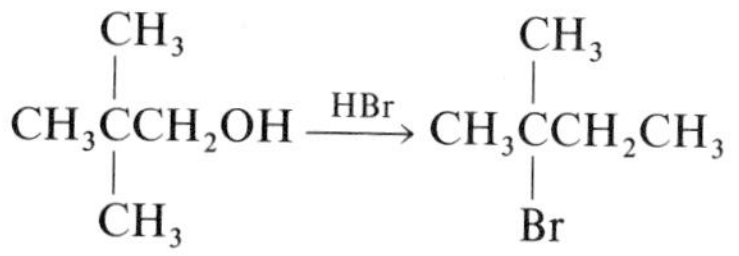
CH3
CH3CCH2OH
CH3
HBr
CH3
CH3CCH2CH3
Br

PROBLEM 8.9

A mixture of two primary alcohols obtained from natural sources, 3-methyl-1-butanol and 2-methyl-1-butanol, gives chiefly 2-methyl-2-butene when dehydrated with acid. 2-Methyl-2-butene is converted to 2-methyl-2-butanol with 90% yield by treatment with 50% aqueous sulfuric acid at 0 °C. Write equations showing the details of the conversion of the mixture of primary alcohols into a single tertiary alcohol.

8.6 ADDITION OF BROMINE TO ALKENES

A. Bromine as an Electrophile

Halogens are oxidizing agents that react by accepting electrons to form halide anions. Because they are electron acceptors, oxidizing agents are also classified as electrophiles. Bromine is therefore an electrophile and adds to carbon-carbon double bonds (pp. 63 and 204). A bromine molecule is normally symmetrical, but as it approaches the π electrons of an alkene the distribution of the electrons in the covalent bond changes. One of the bromine atoms becomes more positive, the other one more negative. This polarization of the bond in the bromine molecule enables bonding to take place between it and the alkene.

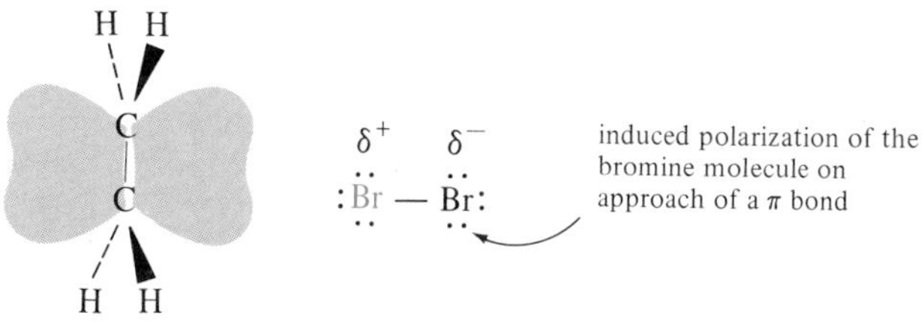
H H
C
C
H H
δ+ δ−
:Br — Br:
induced polarization of the bromine molecule on approach of a π bond

A solution of bromine in carbon tetrachloride has the reddish-brown color typical of elemental bromine. When such a solution is added to an alkene, the color of bromine rapidly disappears. This reaction serves as a test for the presence of carbon-carbon multiple bonds and distinguishes alkenes and alkynes, which have π bonds, from alkanes, which have none. Aromatic compounds such as benzene, in which electrons are present in especially stable orbitals, do not react with bromine under these conditions (p. 64).

The addition of bromine to ethylene appears to proceed in two steps. Evidence is provided by the following experiments. If bromination is carried out in the presence of species that can act as nucleophiles, such as negatively charged ions, or in water as the solvent, mixtures of products are obtained. When bromide ion or chloride ion is the nucleophile, the products are alkyl halides. If water is the nucleophile, the product has a hydroxyl group and a halogen atom on adjacent carbon atoms. Such compounds are known as **halohydrins.**

$$\underset{\text{ethylene}}{CH_2{=}CH_2} + \underset{\text{bromine}}{Br_2} \xrightarrow[H_2O]{\text{NaCl (saturated)}} \underset{\substack{\text{1,2-dibromoethane}\\54\%}}{BrCH_2CH_2Br} + \underset{\substack{\text{1-bromo-2-chloro-}\\\text{ethane}\\46\%}}{BrCH_2CH_2Cl}$$

$$\underset{\text{ethylene}}{CH_2{=}CH_2} + \underset{\text{bromine}}{Br_2} \xrightarrow[0\,°C]{H_2O} \underset{\substack{\text{2-bromoethanol}\\54\%}}{BrCH_2CH_2OH} + \underset{\substack{\text{1,2-dibromoethane}\\37\%}}{BrCH_2CH_2Br}$$

Note that, in the second reaction, water is both the solvent and the nucleophile and is present in a high concentration. Therefore, the halohydrin is the major product.

These facts support the proposal that reaction of bromine with the double bond gives an intermediate cation that then reacts with any nucleophilic species present.

B. The Bromonium Ion

If the addition of bromine to cyclopentene (p. 204) proceeds by nucleophilic attack of the double bond on bromine, with the resulting formation of a symmetrical planar carbocation as an intermediate, some *cis*-1,2-dibromopentane would be expected as a product because the second step, attack of bromide ion, can take place from either side of the planar carbocation.

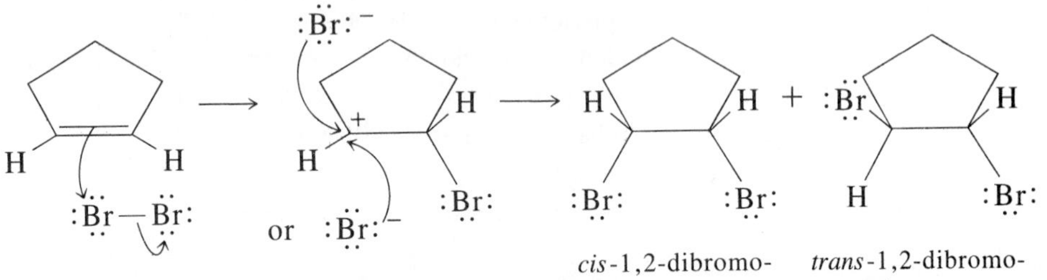

cis-1,2-dibromo-cyclopentane — *not observed*

trans-1,2-dibromo-cyclopentane, *racemic mixture* — *only product*

In fact, only the trans isomer is obtained from the reaction. This observation led to the idea that the intermediate cation is a cyclic **bromonium ion.** The bromine atom is so large in comparison to carbon, that the electron cloud of bromine overlaps with *p* orbitals of both of the doubly bonded carbon atoms (Figure 8.3).

As shown in Visualizing the Reaction on the next page, the opening of a bromonium ion is a nucleophilic substitution reaction. The nucleophile is the bromide ion, and the leaving group is the positively charged bromine atom of the bromonium ion. The reaction produces inversion of configuration at the carbon atom attacked by the bromide ion and retention of configuration at the carbon atom

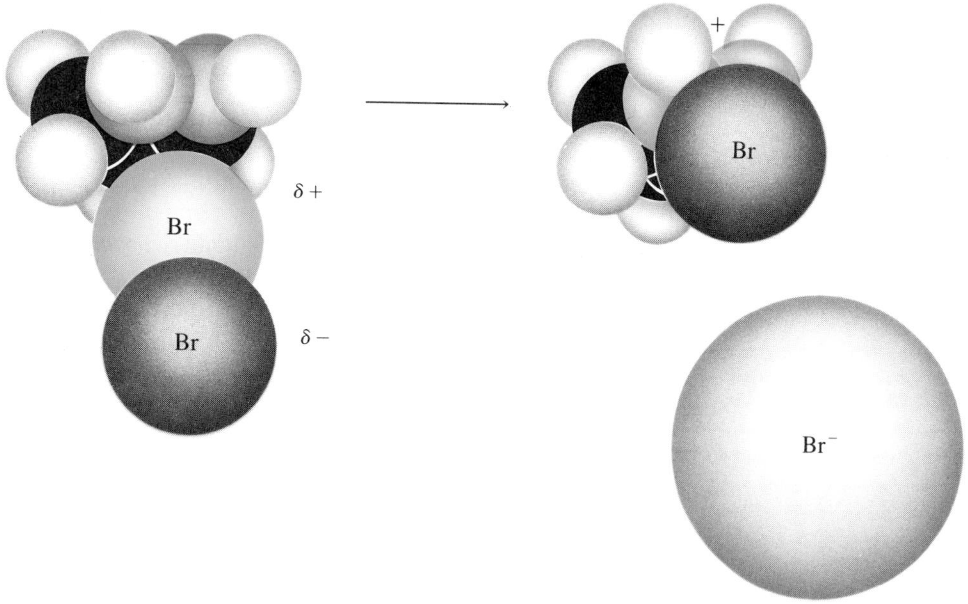
δ+
Br
Br
δ−
+
Br
Br⁻

FIGURE 8.3 Formation of the cyclopentylbromonium ion.

VISUALIZING THE REACTION

Addition of bromine to an alkene

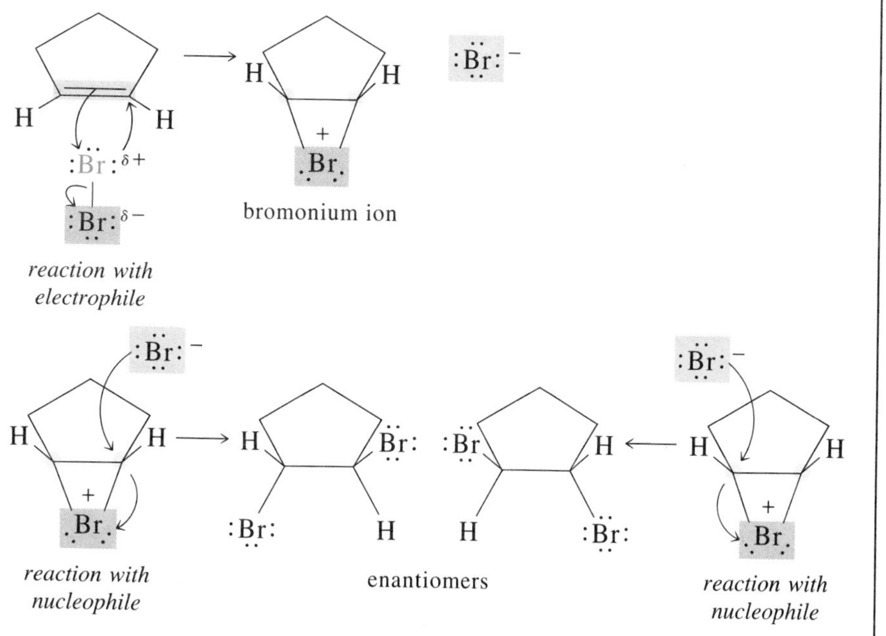
:Br:δ+
:Br:δ−
reaction with electrophile
bromonium ion
reaction with nucleophile
enantiomers
reaction with nucleophile

that winds up with the bromine atom of the bromonium ion. The overall addition to the double bond results in a racemic mixture with trans stereochemistry, as seen for 1,2-dibromocyclopentane.

C. Stereoselective Reactions. Stereochemistry of the Addition of Bromine to Alkenes

A reaction in which one of several possible stereochemical results predominates is called a **stereoselective reaction.** The addition of bromine to the double bond in cyclopentene is 100% stereoselective, leading only to *trans*-1,2-dibromocyclopentane as a product. The bromonium ion is symmetrical in the case of cyclopentene, so nucleophilic attack at either one of the carbon atoms is equally likely. A racemic mixture of the two enantiomeric *trans*-1,2-dibromocyclopentanes is formed.

An addition reaction in which the two components that add to the double bond end up trans to each other is said to proceed with anti stereochemistry, and is an **anti addition reaction.** The origin of this nomenclature is illustrated clearly by the addition of bromine to an acyclic alkene. The reaction of (*Z*)-2-pentene with bromine gives (2*S*,3*S*)-2,3-dibromopentane and (2*R*,3*R*)-2,3-dibromopentane. The products from the reaction of (*E*)-2-pentene with bromine are (2*S*,3*R*)-2,3-dibromopentane and (2*R*,3*S*)-2,3-dibromopentane.

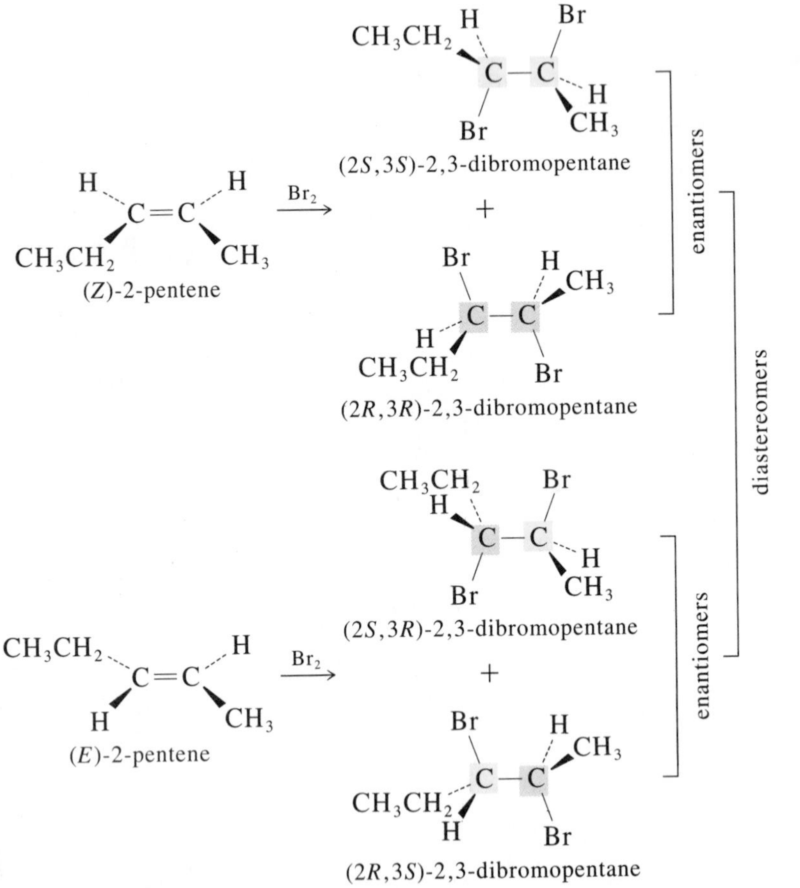
(2S,3S)-2,3-dibromopentane
(Z)-2-pentene
Br_2
(2R,3R)-2,3-dibromopentane
enantiomers
diastereomers
(2S,3R)-2,3-dibromopentane
(E)-2-pentene
Br_2
(2R,3S)-2,3-dibromopentane
enantiomers

Each of these stereoisomers is shown in a conformation that illustrates the anti relationship between the bromine atoms. (*Z*)-2-Pentene gives rise to one set of enantiomeric 2,3-dibromopentanes; (*E*)-2-pentene gives rise to another set. The addition of bromine to either of these two alkenes proceeds with high stereoselectivity.

There are four stereoisomers of 2,3-dibromopentane, which has two different stereocenters. They exist as the pairs of enantiomers shown above. (2*S*,3*S*)-2,3-Dibromopentane (or its enantiomer) is also a diastereomer of (2*S*,3*R*)-2,3-

dibromopentane (or its enantiomer). The presence of two stereocenters gives rise to 2^2, or 4, stereoisomers (p. 201), unless some symmetry is present to reduce the number observed. For example, 2^2, or 4, stereoisomers would be expected for the 1,2-dibromocyclopentanes, but the symmetry of *cis*-1,2-dibromocyclopentane reduces to three the number of isomers actually observed (p. 205).

The two pentenes and bromine are achiral. The bromonium ion that results from either pentene is chiral and is, therefore, formed as a racemic mixture (p. 192). The two forms arise from addition of bromine to each face of the planar double bond.

VISUALIZING THE REACTION

Formation of enantiomeric bromonium ions from (Z)-2-pentene

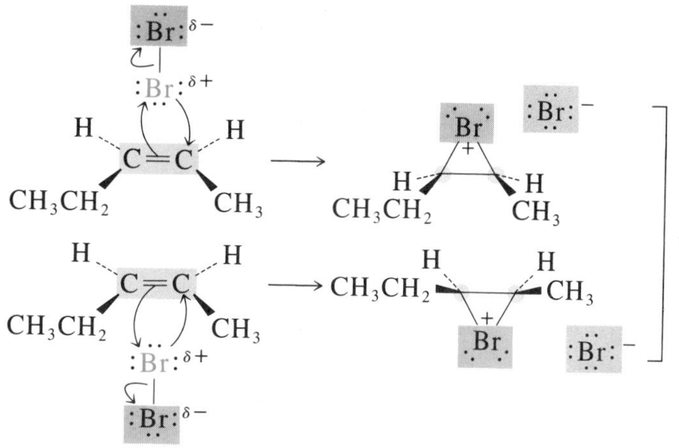
:Br:δ−
:Br:δ+
H
H
C=C
CH₃CH₂
CH₃
Br
+
:Br:−
CH₃
enantiomers

The enantiomeric bromonium ions are opened by attack of a bromide ion to give enantiomeric 2,3-dibromopentanes. The reaction is shown below with the bromide ion attacking carbon atom 3 of each bromonium ion.

VISUALIZING THE REACTION

Reaction of enantiomeric bromonium ions with bromide ion

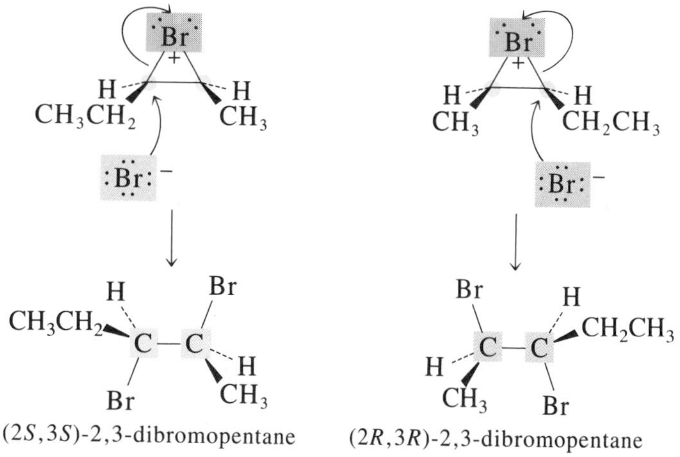
Br
+
H
CH₃CH₂
CH₃
CH₂CH₃
:Br:−
C—C
(2S,3S)-2,3-dibromopentane
(2R,3R)-2,3-dibromopentane

Attack at carbon atom 2 of each of the enantiomeric bromonium ions derived from (*Z*)-2-pentene will give the same mixture of enantiomeric 2,3-dibromopentanes. A similar sequence of reactions results in the enantiomers obtained from (*E*)-2-pentene.

PROBLEM 8.10

Prove to yourself that the two statements just above are true.

PROBLEM 8.11

Complete the following equations.

(a) (*E*)-$CH_3CH_2CH{=}CHCH_2CH_3$ $\xrightarrow[\text{carbon tetrachloride}]{Br_2}$

(b) (*Z*)-$CH_3CH_2CH{=}CHCH_2CH_3$ $\xrightarrow[\text{carbon tetrachloride}]{Br_2}$

(c) 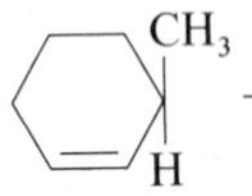 $\xrightarrow[\text{carbon tetrachloride}]{Br_2}$

(d) 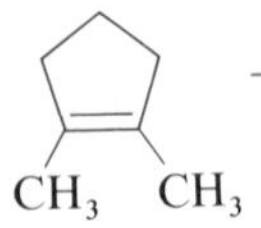 $\xrightarrow[\text{carbon tetrachloride}]{Br_2}$

D. The Bromonium Ion and Nucleophiles. Halohydrins

The addition of bromine to ethylene in the presence of high concentrations of chloride ions gives 1-bromo-2-chloroethane, as well as 1,2-dibromoethane (p. 290). Chloride ion does not add to the double bond unless bromine is also present, and no 1,2-dichloroethane is obtained. Similarly, the use of water instead of an unreactive solvent for the addition reaction results in the formation of 2-bromoethanol, a halohydrin (p. 290). These facts provide additional experimental evidence for a bromonium ion as the intermediate. The intermediate bromonium ion is attacked by a nucleophile—bromide ion, chloride ion, or water—to give the final products observed.

VISUALIZING THE REACTION

Reactions of bromonium ions with nucleophiles

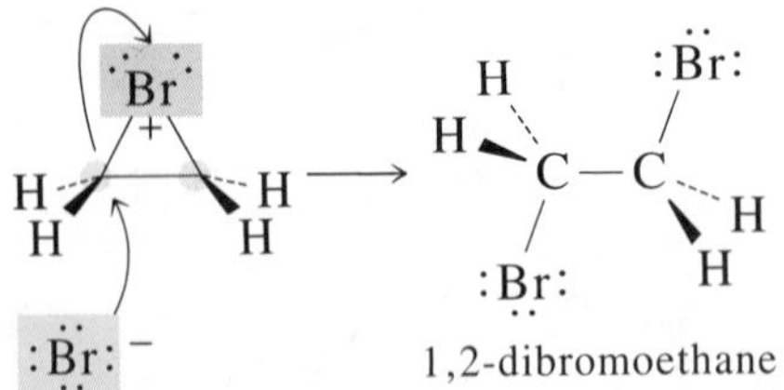

1,2-dibromoethane

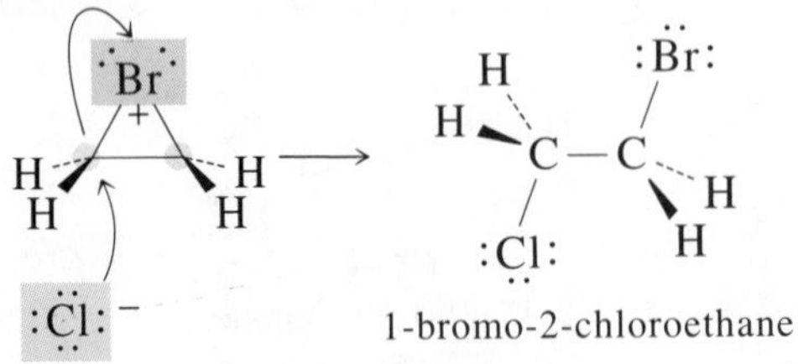

1-bromo-2-chloroethane

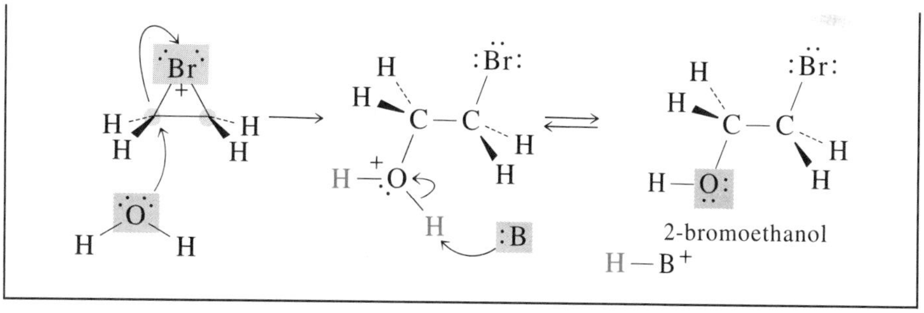
2-bromoethanol

Halohydrins are also formed by reaction of alkenes with hypochlorous acid, HOCl, or hypobromous acid, HOBr, prepared by the acidification of solutions of sodium or calcium hypohalites. Sodium hypochlorite is the chief constituent of laundry bleach and gives a solution of hypochlorous acid on treatment with cold dilute nitric acid.

$$\underset{\text{sodium hypochlorite}}{NaOCl} + HNO_3 \xrightarrow[\text{cold}]{H_2O} \underset{\text{hypochlorous acid}}{HOCl} + NaNO_3$$

Hypochlorous acid adds to an alkene, such as cyclohexene, to give a trans chlorohydrin.

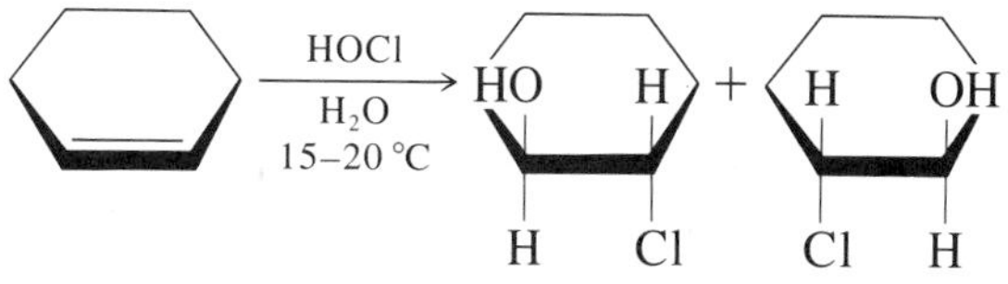

cyclohexene

trans-2-chlorocyclohexanol
chlorohydrin of cyclohexene
70%
racemic mixture

The reaction is believed to involve a chloronium ion intermediate that results from an attack of the double bond on chlorine, which is present in solutions of hypochlorous acid.

Ethylene and cyclohexene are symmetrical molecules; therefore, the same halohydrin is obtained no matter which carbon atom the halogen atom bonds to and which carbon the hydroxyl group bonds to. The reactions of unsymmetrical alkenes have also been studied. For example, 2-methylpropene reacts with hypochlorous acid to give 1-chloro-2-methyl-2-propanol.

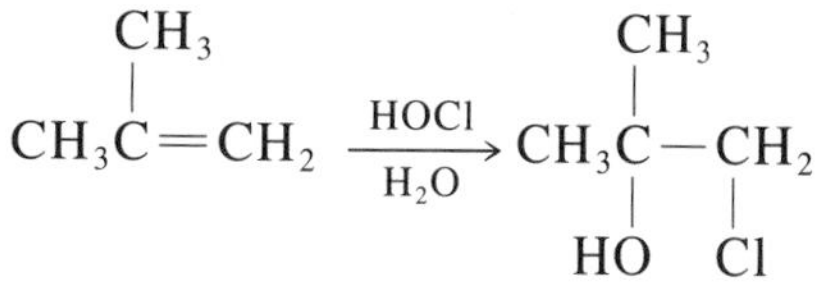

2-methylpropene | 1-chloro-2-methyl-2-propanol

Similarly, when styrene (phenylethene) is treated with chlorine water, 2-chloro-1-phenylethanol is observed.

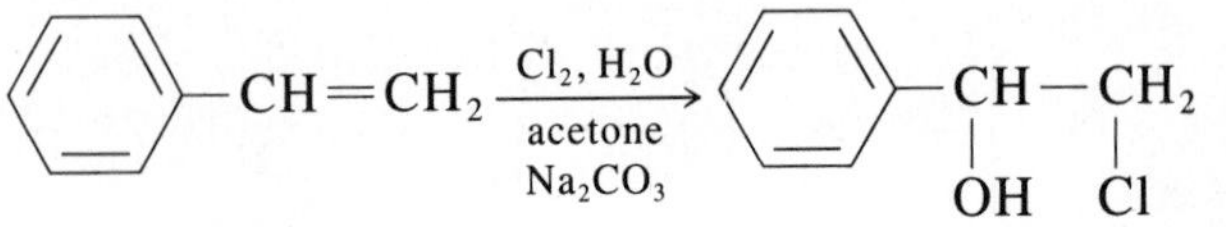
Cl2, H2O
acetone
Na2CO3

styrene 2-chloro-1-phenylethanol 72%

The products observed for these reactions can be explained by proposing that the intermediate chloronium ion is unsymmetrical, with considerable cationic character at the more highly substituted carbon atom.

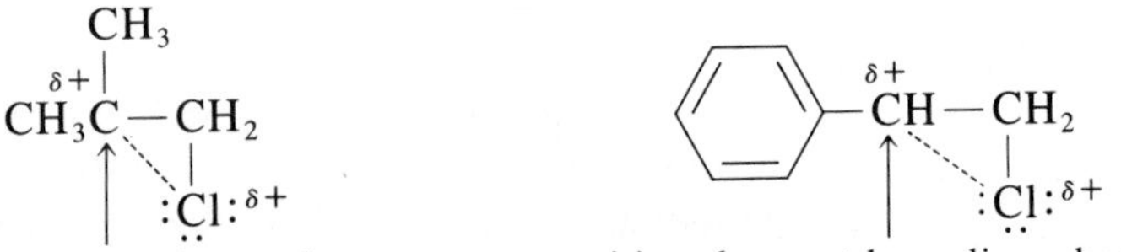

positive charge at tertiary carbon atom positive charge at benzylic carbon atom

unsymmetrical chloronium ion intermediates

The nucleophile, water in these examples, reacts with the relatively stable unsymmetrical intermediate, and the product obtained has the hydroxyl group on the carbon atom that had the partial positive charge in the intermediate.

Why couldn't the intermediate in such reactions be an ordinary carbocation with no bridging to the halogen atom? Evidence for a halonium ion intermediate comes from the stereochemistry that results from reactions in which halohydrins are formed. An example involving a bromonium ion intermediate is as follows. A brominated amide, *N*-bromosuccinimide, is often used as the source of electrophilic bromine.

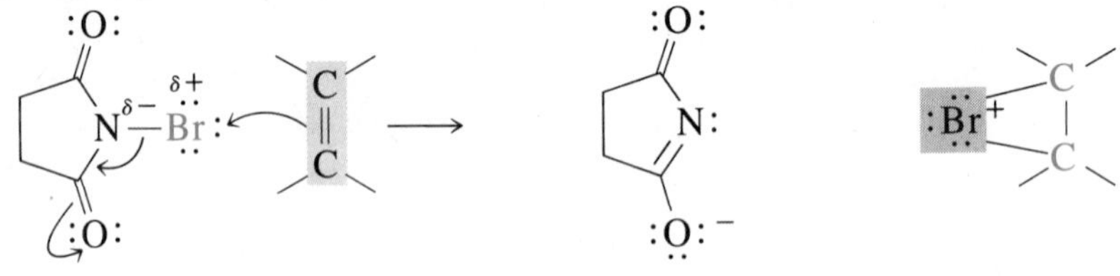

N-bromosuccinimide resonance-stabilized anion from *N*-bromosuccinimide bromonium ion

For example, when treated with *N*-bromosuccinimide in aqueous dimethyl sulfoxide, (*E*)-1-phenyl-1-propene gives a racemic mixture of (1*R*,2*S*)- and (1*S*,2*R*)-2-bromo-1-phenyl-1-propanol.

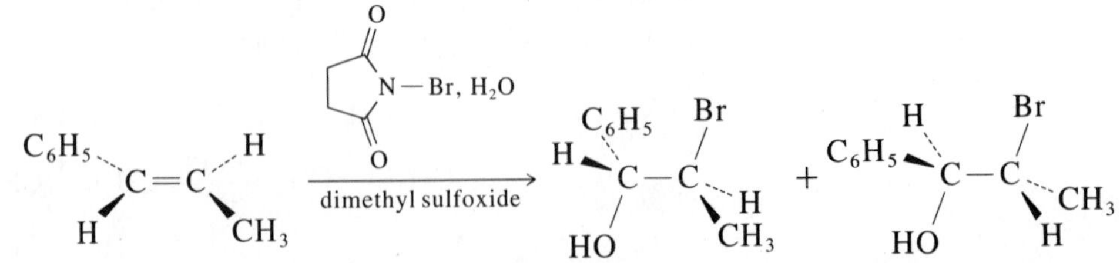
N—Br, H2O
dimethyl sulfoxide

(*E*)-1-phenyl-1-propene (1*R*,2*S*)-2-bromo-1-phenyl-1-propanol (1*S*,2*R*)-2-bromo-1-phenyl-1-propanol

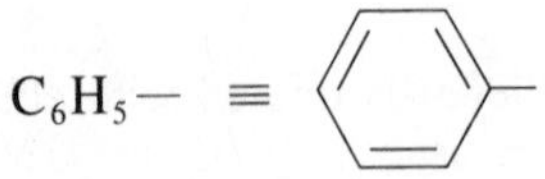

The stereochemistry of the products shows clearly that anti addition to the double bond has taken place. The intermediate, therefore, must be more like a bromonium ion than like a planar benzylic carbocation, which would be able to react with water on either side.

PROBLEM 8.12

Write a detailed mechanism showing an unsymmetrical bromonium ion and one showing a planar carbocation as the intermediate for the formation of the halohydrins from (*E*)-1-phenyl-1-propene. Prove to yourself that the stereochemistry that is observed experimentally would not be possible with a planar carbocation as the intermediate.

Study Guide Concept Map 8.3

A proton is small in size and has no nonbonding electrons, so it does not usually form bridged cations the way bromine does. Carbocations formed when acids add to alkenes therefore give more rearranged products (p. 285) and show less stereoselectivity than bromonium ions do (pp. 204 and 292).

PROBLEM 8.13

2-Methyl-1-phenyl-1-propene, when treated with *N*-bromosuccinimide and water in dimethyl sulfoxide, gives a mixture of two halohydrins in roughly equal amounts. What are the structures of the products? What accounts for the lack of regioselectivity in this case?

PROBLEM 8.14

trans-1-Bromo-2-chlorocyclopentane is synthesized in the laboratory by treating cyclopentene with bromine and dry hydrogen chloride in dichloromethane. Write a mechanism for its formation. Be sure to show the correct stereochemistry.

8.7 ADDITION OF DIBORANE TO ALKENES

A. Diborane as an Electrophile

Diborane is an interesting compound that has only six pairs of bonding electrons. The question of what holds the two boron atoms together has been controversial and has led to much lively discussion and research.

Diborane exists as B_2H_6, a dimer of BH_3, in the absence of Lewis bases. It dissociates easily in the presence of ethers, which have nonbonding electrons, to give ether-borane complexes.

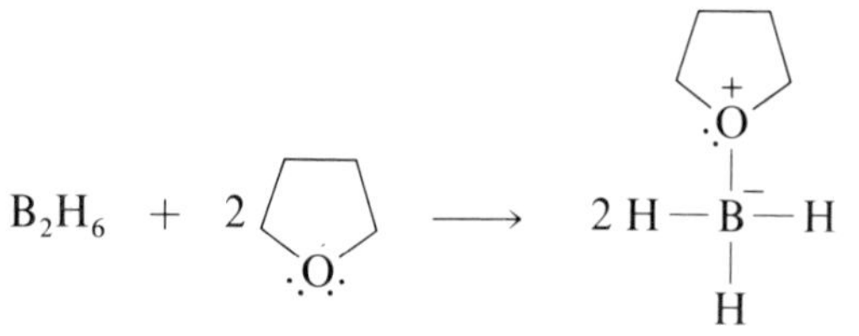

B₂H₆ + 2
2 H—B—H
H

diborane — tetrahydrofuran, *a cyclic ether* — borane-tetrahydrofuran complex

Ethers are used as solvents for reactions of diborane, so the reagent exists in the complexed form. Therefore, the formula BH_3 will be used in this book to represent the reactive species.

PROBLEM 8.15

Write the Lewis structure for BH_3. What shape do you expect the molecule to have? What kind of hybridization does the boron atom have in BH_3? Draw an orbital picture of the bonding in the molecule.

BH_3 is a Lewis acid and, therefore, an electrophile. It reacts with the π electrons of an alkene with an orientation that gives rise to a partial positive charge at the carbon atom that will lead to the more stable carbocation.

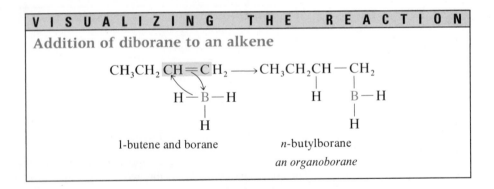

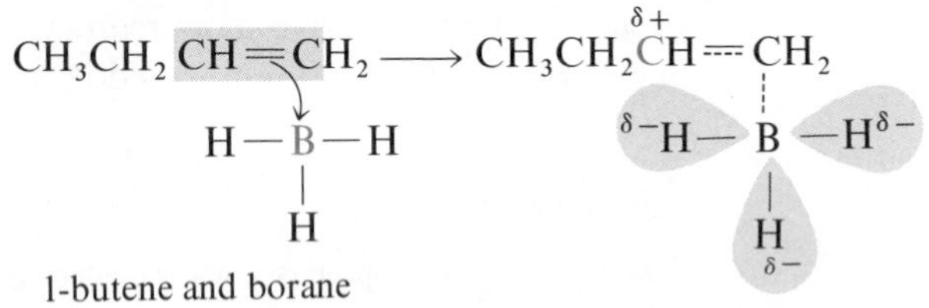

1-butene and borane

partial bonding between 1-butene and borane

At the same time, negative charge develops at the boron atom. The hydrogen atoms bonded to the boron acquire partial negative character, and one of them is transferred as a hydride ion to the carbon atom that has the partial positive charge. The reaction may be represented as a single step.

VISUALIZING THE REACTION

Addition of diborane to an alkene

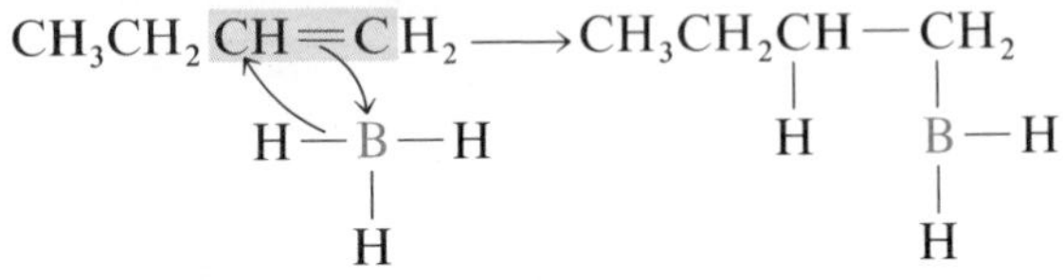

1-butene and borane

n-butylborane
an organoborane

The reaction is said to have a **four-center transition state,** with four atoms (two carbon, one boron, and one hydrogen) undergoing changes in bonding at the same time.

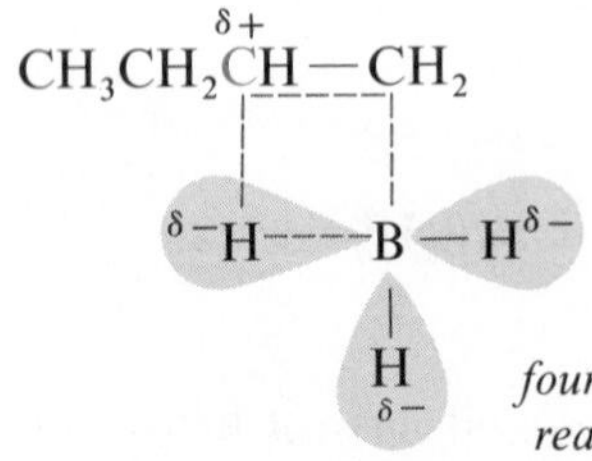

four-center transition state for the reaction of 1-butene with borane

The product from the addition of borane to an alkene is known as an organoborane. Such a compound, for example, *n*-butylborane, which is shown below, is still a Lewis acid in which the boron atom can accept another pair of electrons from an alkene and can transfer a hydride ion to it. The reaction proceeds until all three hydrogen atoms that were originally bonded to boron have been replaced by alkyl groups.

$$\underset{\text{1-butene}}{CH_3CH_2CH{=}CH_2} + \underset{n\text{-butylborane}}{H{-}\underset{\displaystyle H}{\overset{|}{B}}{-}CH_2CH_2CH_2CH_3} \longrightarrow \underset{\text{di}(n\text{-butyl})\text{borane}}{CH_3CH_2CH_2CH_2{-}\underset{\displaystyle H}{\overset{|}{B}}{-}CH_2CH_2CH_2CH_3}$$

$$\Big\downarrow CH_3CH_2CH{=}CH_2$$

$$\underset{\text{tri}(n\text{-butyl})\text{borane}}{CH_3CH_2CH_2CH_2{-}\underset{\displaystyle CH_2CH_2CH_2CH_3}{\overset{|}{B}}{-}CH_2CH_2CH_2CH_3}$$

PROBLEM 8.16

Write equations for the reactions of cyclopentene, 1-decene, and 3,3-dimethyl-1-butene with diborane.

B. The Hydroboration Reaction

The **hydroboration reaction,** which was discovered in 1956 by Herbert C. Brown at Purdue University, has made it possible to synthesize alcohols in high yields under mild conditions with high regioselectivity (p. 120) and stereoselectivity (p. 292). In this reaction, diborane adds to an alkene to give organoboranes. Oxidation of the carbon-boron bond in the organoboranes with hydrogen peroxide in basic solution gives alcohols, as illustrated by the overall reaction for 1-butene shown below.

$$\underset{\text{1-butene}}{CH_3CH_2CH{=}CH_2} \xrightarrow{BH_3} \underset{\text{tri}(n\text{-butyl})\text{borane}}{(CH_3CH_2CH_2CH_2)_3B} + \underset{\text{tri}(sec\text{-butyl})\text{borane}}{(CH_3CH_2\overset{\displaystyle CH_3}{\overset{|}{C}H}{-})_3B}$$

$$\Big\downarrow \begin{matrix} H_2O_2 \\ NaOH \\ H_2O \end{matrix}$$

$$\underset{\substack{n\text{-butyl alcohol} \\ 93\%}}{CH_3CH_2CH_2CH_2OH} + \underset{\substack{sec\text{-butyl alcohol} \\ 7\%}}{CH_3CH_2\underset{\displaystyle OH}{\underset{|}{C}H}CH_3}$$

Note that the chief product of the reaction sequence, *n*-butyl alcohol, arises from anti-Markovnikov addition of water to the double bond.

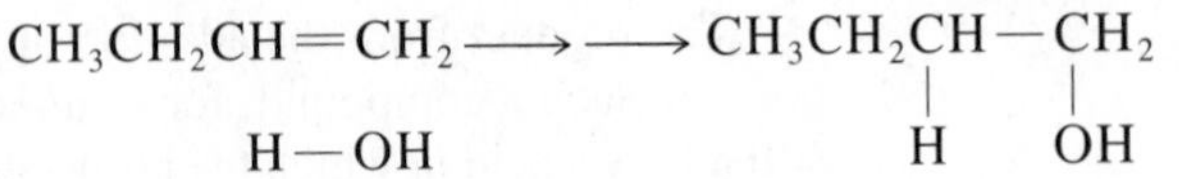

$$CH_3CH_2CH{=}CH_2 \longrightarrow \longrightarrow CH_3CH_2CH{-}CH_2$$

$$H{-}OH \qquad\qquad\qquad H \quad\; OH$$

anti-Markovnikov addition of H—OH to an alkene

In contrast, in the hydration reaction (p. 281), alcohols are produced by a Markov-nikov addition of water to the double bond.

C. Regioselectivity in Hydroboration Reactions

In the unsymmetrical terminal alkene 1-butene, the boron atom adds chiefly to the less highly substituted of the doubly bonded carbon atoms. In the case of internal alkenes in which one of the doubly bonded carbon atoms has two substituents and the other has one, the boron atom also attaches itself to the less substituted carbon atom. For example, 2-methyl-2-butene reacts with diborane to give chiefly the organoborane in which the boron atom is attached to the less highly substituted carbon atom.

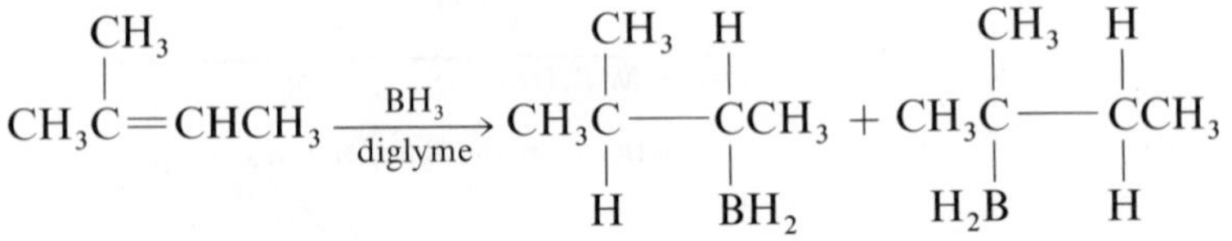

$$\underset{\text{2-methyl-2-butene}}{CH_3C{=}CHCH_3} \;\;\xrightarrow[\text{diglyme}]{BH_3}\;\; \underset{98\%}{CH_3C{-}CCH_3} \;+\; \underset{2\%}{CH_3C{-}CCH_3}$$

diglyme $\equiv$ $CH_3OCH_2CH_2OCH_2CH_2OCH_3$
an ether

For the more or less symmetrically disubstituted alkenes, however, two prod-ucts are possible. They are formed in roughly equal amounts even when the bulk-iness of one of the substituents would seem to favor addition of the boron atom to the other carbon atom, as illustrated by the reaction of (*E*)-4-methyl-2-pentene below.

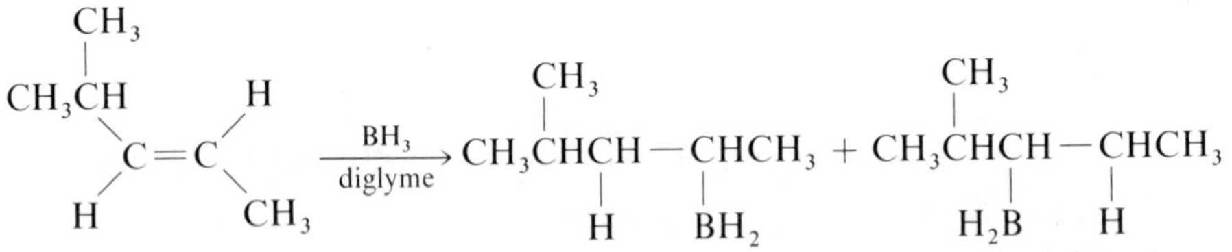

$$\underset{(E)\text{-4-methyl-2-pentene}}{C{=}C} \;\;\xrightarrow[\text{diglyme}]{BH_3}\;\; \underset{\substack{\text{1,3-dimethylbutylborane}\\57\%}}{CH_3CHCH{-}CHCH_3} \;+\; \underset{\substack{\text{1-isopropylpropylborane}\\43\%}}{CH_3CHCH{-}CHCH_3}$$

Herbert C. Brown reasoned that a bulkier reagent than borane might show greater selectivity in the way it added to a double bond, because steric interactions between substituents on the double bond and on the reagent would increase and become more important. One of the reagents he developed for greater selectivity is the organoborane 9-borabicyclo[3.3.1]nonane (9-BBN), which is formed when borane adds twice to the cyclic diene 1,5-cyclooctadiene, to give a bicyclic com-pound. In a bicyclic compound, two or more atoms are shared by two different rings (p. 694).

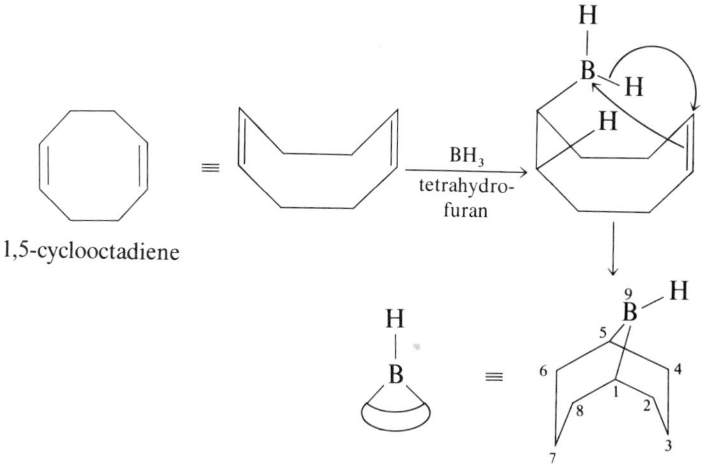
BH3
tetrahydro-
furan
1,5-cyclooctadiene
H
B
H
H
9
B
H
5
6
4
1
8
2
7
3

a symbol frequently used for 9-BBN

9-borabicyclo[3.3.1]nonane or 9-BBN

9-BBN is a reasonably stable compound that can be purchased in a bottle and handled with minor precautions in the air, in contrast to diborane, which is a gas and reacts violently with air. This organoborane is thus a more convenient reagent for hydroboration reactions. The bulkiness of substituents on the boron atom gives the reagent high selectivity in its reaction with substituted alkenes. With (*Z*)-4-methyl-2-pentene, for example, only the product in which the boron atom has bonded to the carbon atom that had the less bulky of the two substituents is formed.

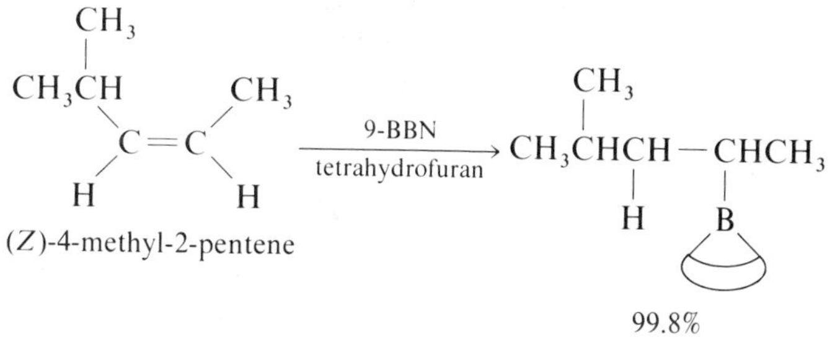
CH3
CH3CH
CH3
C=C
H
H
9-BBN
tetrahydrofuran
CH3
CH3CHCH—CHCH3
H
B
(Z)-4-methyl-2-pentene
99.8%

This contrasts with the equal distribution of products for the reaction of diborane with the *E* isomer, shown above.

PROBLEM 8.17

Write an equation showing the chief product from the reaction of 9-BBN with: (a) 2-methyl-1-pentene; (b) styrene (phenylethene); (c) (*Z*)-4,4-dimethyl-2-pentene; (d) 2,3-dimethyl-2-butene; and (e) (*Z*)-3-hexene.

D. Stereochemistry of Hydroboration Reactions

The transition state shown on p. 298 for the addition of borane to an alkene predicts that the boron atom and the hydrogen atom that is transferred from it to the adjacent carbon atom end up bonded to the same side of the molecule. An addition reaction in which the incoming groups are added to the same side of a molecule is called a

syn addition. A syn addition contrasts with an anti addition, an example of which is the addition of bromine to a π bond (p. 292). The words syn and anti refer to the mechanism of a reaction; the words cis and trans are used to describe the stereochemistry of the products of the reaction. Depending on the structure of the starting reagent, syn addition may give products with cis (see p. 315) or trans (see below) stereochemistry. In the hydroboration reaction, syn addition of boron and hydrogen to the double bond is thought to occur.

There is no easy way to determine the stereochemistry of organoboranes. The alcohols formed from the oxidation of the organoborane intermediates show overall syn addition of water to the double bond in the cases where stereochemistry can be determined. For example, 1-methylcyclohexene gives *trans*-2-methylcyclohexanol in the hydroboration reaction.

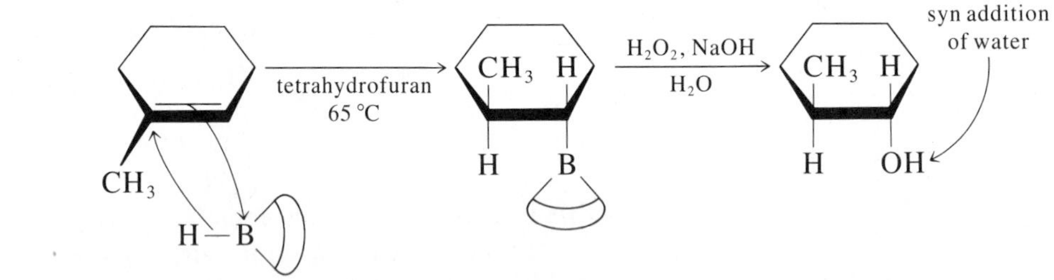
tetrahydrofuran
65 °C
CH3
H—B
CH3 H
H B
H2O2, NaOH
H2O
CH3 H
H OH
syn addition
of water

1-methylcyclohexene and 9-borabicyclo[3.3.1]nonane — organoborane from syn addition — *trans*-2-methylcyclohexanol 99.8%

Careful experiments with several different systems have led chemists to conclude that the overall syn addition of water is the result of syn addition of the boron hydride to the double bond followed by oxidation of the carbon-boron bond with retention of configuration. We will examine the details of the oxidation process in the next section.

E. Oxidation of Organoboranes

The oxidation of alkylboranes with hydrogen peroxide in basic solution gives alcohols. Both the carbon atom and the boron atom in the organoborane are oxidized, and hydrogen peroxide is reduced to water. The reaction requires the addition of 6 M sodium hydroxide and 30% hydrogen peroxide directly to the hydroboration reaction mixture, so that the conversion of an alkene to an alcohol takes place in one reaction vessel without the isolation of the organoborane.

The reaction is believed to start with the formation of the hydroperoxide anion, HOO^-, in base.

$$HOOH + Na^+OH^- \longrightarrow Na^+OOH^- + H_2O$$

hydrogen peroxide

more acidic than H_2O because of the inductive effect of the second oxygen atom

This nucleophilic anion then reacts with the boron atom. The important step is the migration of an alkyl group to oxygen, a rearrangement that takes place with retention of configuration.

VISUALIZING THE REACTION

Oxidation of an organoborane

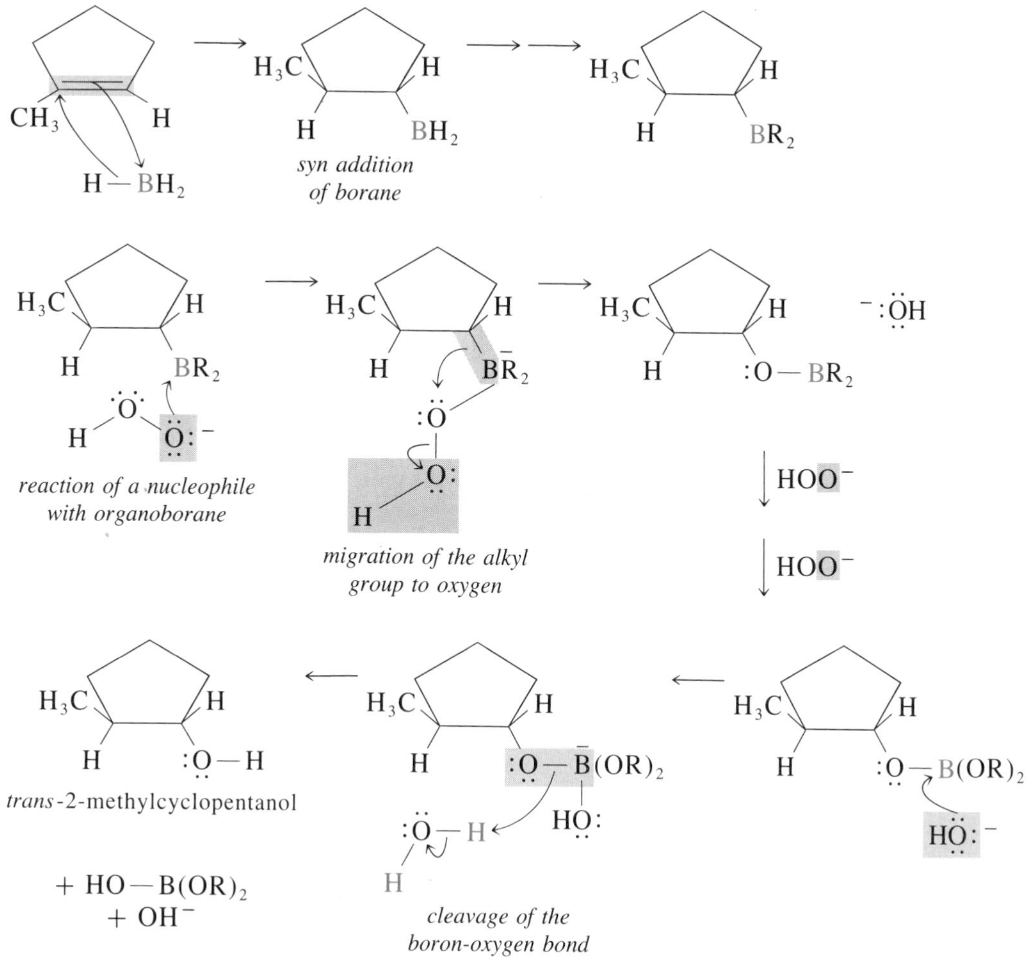

CH₃
H
H—BH₂
H₃C
H
H
BH₂
syn addition of borane
H₃C
H
H
BR₂
reaction of a nucleophile with organoborane
migration of the alkyl group to oxygen
⁻:ÖH
:O—BR₂
HOO⁻
HOO⁻
HO:⁻
:O—B(OR)₂
HO:
H
cleavage of the boron-oxygen bond
:O—H
trans-2-methylcyclopentanol
+ HO—B(OR)₂
+ OH⁻

The hydroxyl group in *trans*-2-methylcyclopentanol ends up on the side of the molecule where the boron atom was originally attached in the organoborane. Boron is oxidized to boric acid, present as the sodium salt, sodium borate.

Study Guide Concept Map 8.4

The hydroboration-oxidation sequence provides a very mild and convenient way to prepare alcohols from alkenes, with water adding to the double bond with high regioselectivity and stereoselectivity. The rearrangements and polymerization reactions typical of acid-catalyzed hydrations of alkenes do not take place under these conditions. The importance of this type of reaction was recognized in 1979 when Herbert C. Brown was honored with the Nobel Prize for his work with organoboranes.

PROBLEM 8.18

For the organoboranes formed in the reactions in Problems 8.16 and 8.17, show the products you would expect to get from peroxide oxidation in basic solution.

A. Heterogeneous Catalysis

Hydrogen adds to a double bond in the presence of a metallic catalyst in a **hydrogenation reaction.** (A hydrogenation reaction is one type of reduction reaction. Oxidation-reduction reactions are discussed more fully in Section 12.4A.) A hydrogenation reaction is usually carried out in some inert solvent. The alkene is stirred in the presence of the solid catalyst while hydrogen gas is introduced into the reaction mixture. The amount of hydrogen used up in the reaction can be monitored by watching the drop in pressure or the decrease in the volume of the gas in the system. Hydrogen gas does not add to double bonds unless a specially prepared metal surface is present.

The presence of the catalyst as a separate solid phase makes the reaction mixture a **heterogeneous** one, in contrast to reactions in which all reagents are present in a single **homogeneous** phase. Metals that are commonly used as **hydrogenation catalysts** are palladium, platinum, and nickel. Palladium and platinum catalysts are usually prepared by the reduction of a salt of the metal by hydrogen, very often in the presence of a larger amount of an inert material that serves to dilute and support the catalyst. An example is palladium chloride on carbon, which when reduced by hydrogen gives a solid that consists of palladium metal in a finely divided state supported on powdered carbon. Reduction of platinum oxide gives a platinum catalyst. A form of nickel called Raney nickel is made by using sodium hydroxide to dissolve out the aluminum in a nickel-aluminum alloy. The reaction of aluminum with sodium hydroxide gives hydrogen gas, which is adsorbed on finely divided nickel from the alloy.

Exactly how a catalyst functions in a hydrogenation reaction is a matter of debate and may vary with the particular catalyst, the relative amounts of alkene and hydrogen gas used, and the temperature. The important factor in catalysis is the nature and extent of the available metal surface. The alkene is believed to be adsorbed onto the surface of the catalyst, forming bonds to the metal atoms. The π bond is broken at this stage. This reaction may be viewed as a Lewis acid-base interaction between the metal atoms, which have room in their orbitals for electrons, and the π electrons of the multiple bond (Figure 8.4). In some cases, the hydrogen molecule may also be adsorbed to the surface of the metal near the organic molecule, with some loosening of the hydrogen-hydrogen bond. Reaction occurs so that a hydrogen atom is bonded to each of the carbon atoms.

The addition of hydrogen to an alkene to give an alkane is an exothermic process (p. 278). In the absence of a catalyst, however, the reaction has a high energy of activation and takes place at a negligible rate. The catalyst changes the nature of the transition state for the reaction and thereby lowers the energy of activation. The reaction can then proceed at a reasonable rate under practical temperature and pressure. Figure 8.5 is an energy diagram showing how the catalyzed reaction compares with the uncatalyzed reaction.

Because the uncatalyzed hydrogenation reaction has a very high energy of activation, no reaction takes place unless a catalyst is used. By showing the formation of an intermediate on the pathway from reactants to product, Figure 8.5 indicates that the mechanism of the catalyzed reaction is different from that of the uncatalyzed reaction. The interaction between the alkene and the surface of the catalyst, depicted in Figure 8.4, changes the nature of the bonding in the alkene and creates a species that shows greater reactivity toward hydrogen. The change in the energy of activation that accompanies catalysis occurs because the nature of the

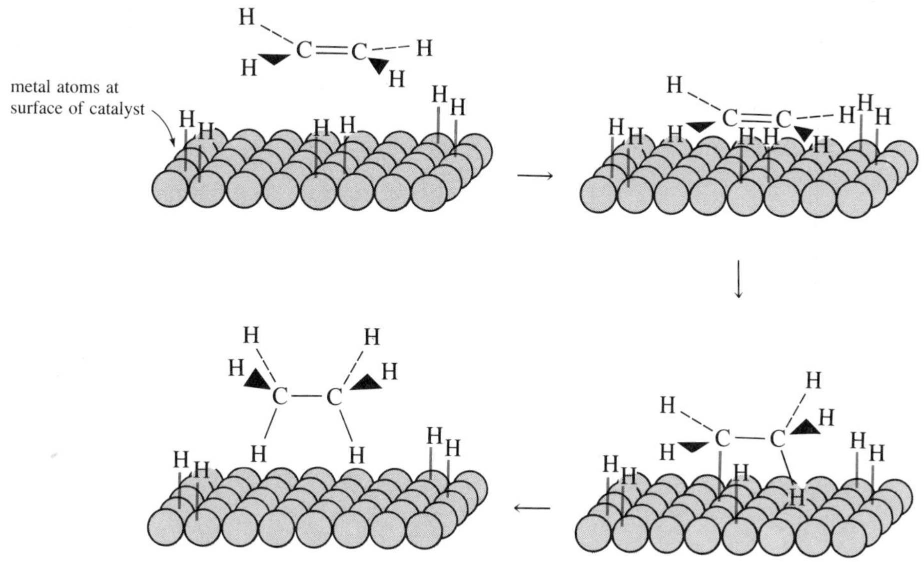
metal atoms at
surface of catalyst

FIGURE 8.4 A schematic representation of the interaction leading to the transfer of hydrogen atoms from the surface of a metallic catalyst to the carbon atoms of an alkene.

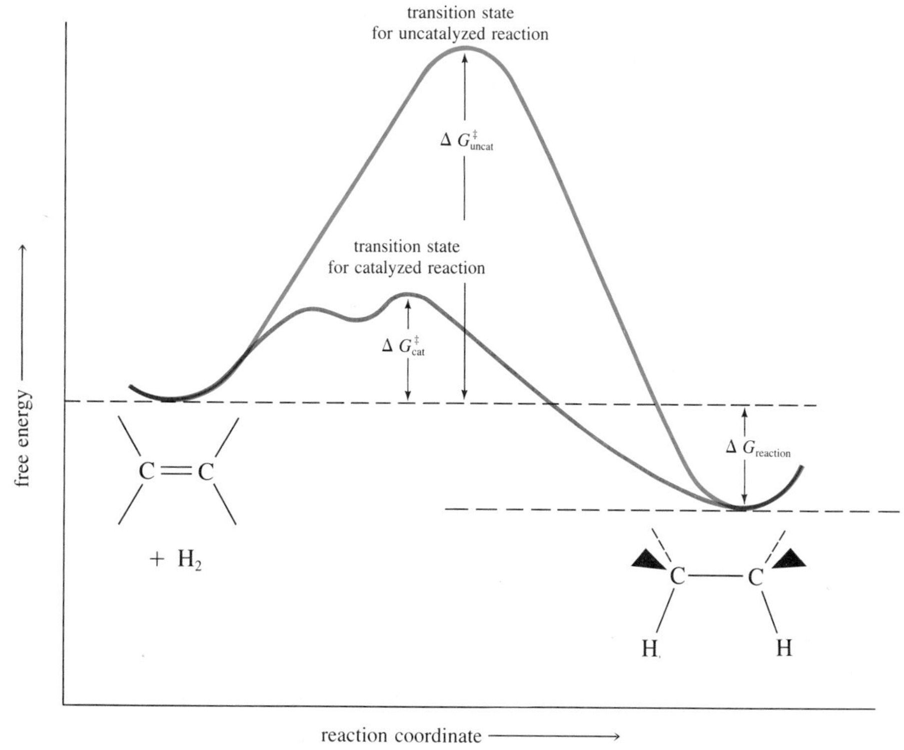
transition state
for uncatalyzed reaction
$\Delta G^{\ddagger}_{uncat}$
transition state
for catalyzed reaction
$\Delta G^{\ddagger}_{cat}$
$\Delta G_{reaction}$
+ H_2
free energy
reaction coordinate

FIGURE 8.5 Energy diagram comparing the free energy of activation for a catalyzed hydrogenation reaction with that for an uncatalyzed reaction.

transition state is changed by the presence of the catalyst. Note that the catalyst also lowers the energy of activation for the reverse reaction, the dehydrogenation of an alkane to an alkene. Such reactions do take place in the presence of hydrogenation catalysts if no external hydrogen gas is added to the reaction system.

B. Some Hydrogenation Reactions of Alkenes

Alkenes are converted to alkanes by catalytic hydrogenation. Some typical examples are the conversion of 2-methyl-2-pentene to 2-methylpentane and 3-methylcyclopentene to methylcyclopentane.

$$\underset{\text{2-methyl-2-pentene}}{CH_3\overset{\overset{CH_3}{|}}{C}{=}CHCH_2CH_3} \xrightarrow[\text{Pt, 25 °C}]{H_2} \underset{\text{2-methylpentane}}{CH_3\overset{\overset{CH_3}{|}}{C}HCH_2CH_2CH_3}$$

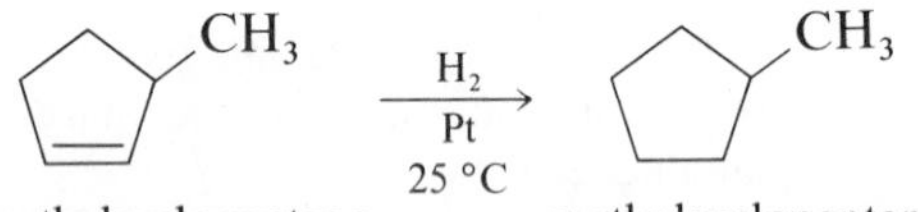

3-methylcyclopentene → methylcyclopentane (H_2, Pt, 25 °C)

Hydrogenation reactions are generally efficient reactions that give the products in high yield and are important in industrial processes. For example, hydrogenation of the mixture of trimethylpentenes obtained from the acid-catalyzed dimerization of 2-methylpropene (p. 283) gives 2,2,4-trimethylpentane, known commercially as isooctane and used as a component of high-octane gasoline.

$$\underset{\text{2,4,4-trimethyl-1-pentene}}{CH_3\overset{\overset{CH_3}{|}}{\underset{\underset{CH_3}{|}}{C}}CH_2\overset{\overset{CH_3}{|}}{C}{=}CH_2} + \underset{\text{2,4,4-trimethyl-2-pentene}}{CH_3\overset{\overset{CH_3}{|}}{\underset{\underset{CH_3}{|}}{C}}CH{=}\overset{\overset{CH_3}{|}}{C}CH_3} \xrightarrow[\text{Pt, acetic acid}]{H_2} \underset{\text{2,2,4-trimethylpentane (isooctane)}}{CH_3\overset{\overset{CH_3}{|}}{\underset{\underset{CH_3}{|}}{C}}CH_2\overset{\overset{CH_3}{|}}{C}HCH_3}$$

Another important industrial application of hydrogenation is the conversion of vegetable oils, which are mixtures of esters of unsaturated acids, into solid fats, which are esters of saturated acids. Oleic acid, the acid found in olive oil, is (Z)-9-octadecenoic acid. Hydrogenation of oleic acid converts it to stearic acid, also called octadecanoic acid, which is found in butter and beef fat.

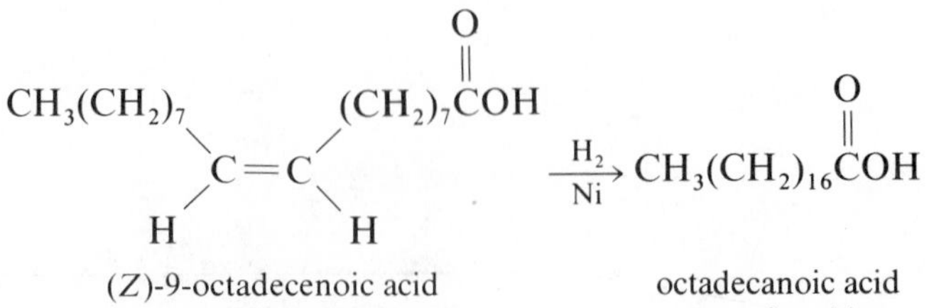

(Z)-9-octadecenoic acid
oleic acid
mp 14 °C
liquid at room temperature, a component of vegetable oil

octadecanoic acid
stearic acid
mp 69 °C
solid at room temperature, a component of fats

Oils and fats are similar in structure except for the presence of double bonds in the acid components of oils, which lowers their melting points and makes them liquids at room temperature. Because fats are more stable toward oxidation by air and more convenient to handle and store, hydrogenation is used to convert oils to fats such as margarine or vegetable shortening. The chemistry of fats and oils is further explored in Chapter 15.

PROBLEM 8.19

Complete the following equations.

(a) $\underset{\underset{\displaystyle CH_3}{|}}{\overset{\overset{\displaystyle CH_3}{|}}{CH_3CCH}}{=}CH_2 \xrightarrow{\underset{Ni}{H_2}} A$ (b) $CH_3\overset{\overset{\displaystyle CH_3}{|}}{C}{=}\overset{\overset{\displaystyle CH_3}{|}}{C}CH_3 \xrightarrow{\underset{Ni}{H_2}} B$

(c) $CH_3CH_2\underset{\underset{\displaystyle OH}{|}}{\overset{\overset{\displaystyle CH_3}{|}}{C}}CH_2CH_3 \xrightarrow[\underset{\Delta}{H_2SO_4}]{} C + D \xrightarrow{\underset{Pt}{H_2}} E$ (d) $CH_3\overset{\overset{\displaystyle CH_3}{|}}{CH}{-}\overset{\overset{\displaystyle CH_3}{|}}{C}{=}CH_2 \xrightarrow{\underset{Ni}{H_2}} F$

(e) $CH_3\underset{\underset{\displaystyle OH}{|}}{\overset{\overset{\displaystyle CH_3}{|}}{C}}CH_2CH_2CH_3 \xrightarrow[\underset{\Delta}{H_3PO_4}]{} G + H \xrightarrow{\underset{Pt}{H_2}} I$

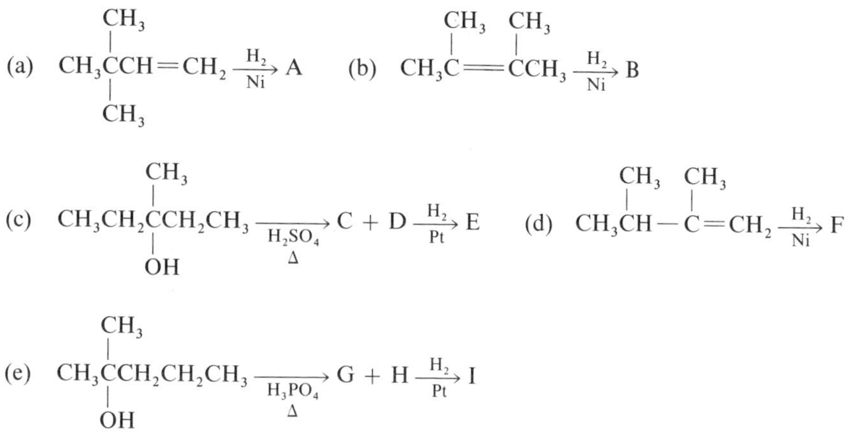

C. Homogeneous Catalysis. Organometallic Compounds as Hydrogenation Catalysts

Some hydrogenation catalysts contain transition metal atoms bonded to organic groups. Such compounds containing bonds between metal atoms and carbon atoms are known as **organometallic compounds.** The transition metals, appearing in the central portion of the periodic table, with up to twelve electrons in their s, p, and d orbitals, have room for a total of eighteen electrons in their valence shells. Consequently, transition metals function as Lewis acids and form covalent bonds with Lewis bases, such as anions or organic compounds with nonbonding electrons to donate. The Lewis bases that bind to the central metal atom are called **ligands.** The presence of the organic ligands around the metal makes the organometallic compound soluble in organic solvents.

Some soluble transition metal compounds containing organic ligands function as hydrogenation catalysts. Because the organometallic compound is soluble in the solvent in which the reaction is taking place, the process is known as **homogeneous catalysis** (p. 304). Catalysis occurs not at the surface of a metal layer, but throughout the solution. The catalytic properties of the transition metal compounds that serve as catalysts for hydrogenation reactions are due to the ease with which these metals accept and release electrons.

An organometallic compound that functions as a catalyst for hydrogenation reactions is chlorotris(triphenylphosphine)rhodium(I), $RhCl[P(C_6H_5)_3]_3$. A look at the periodic table reveals that rhodium has nine electrons in its valence shell and, as rhodium(I) (rhodium in the oxidation state of $+1$), retains eight of them. In the organometallic compound, the metal has accepted four pairs (a total of eight) electrons. One pair comes from a chloride ion and the three other pairs from three triphenylphosphine molecules. The rhodium metal thus has sixteen electrons around it in this organometallic compound.

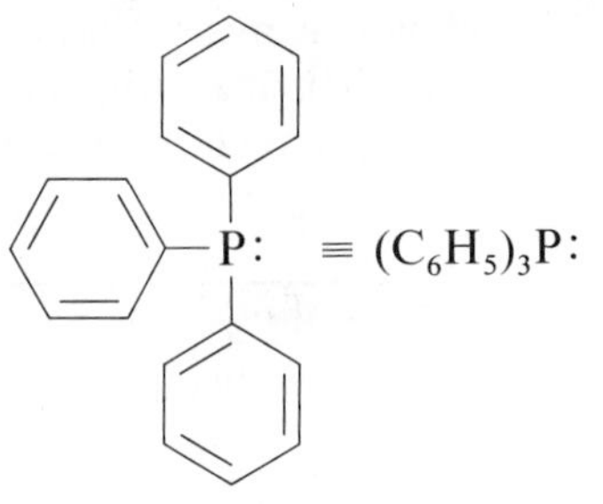
P:
≡ (C6H5)3P:

triphenylphosphine

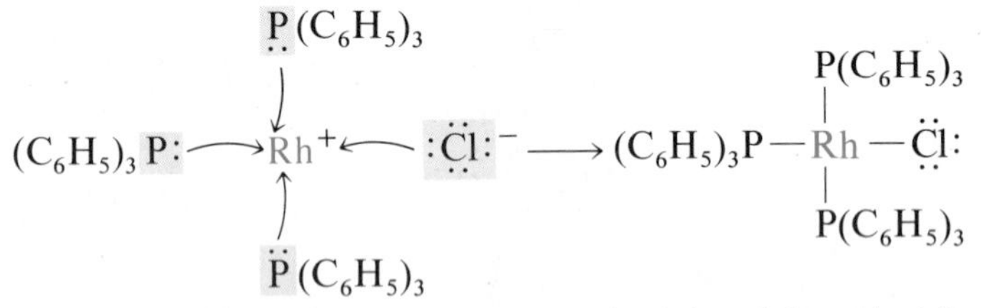
P(C6H5)3
(C6H5)3 P:
Rh+
:Cl:−
P(C6H5)3
(C6H5)3P—Rh—Cl:
P(C6H5)3
P(C6H5)3

rhodium ion with 8 electrons in its valence shell, gaining 8 electrons by coordination with 4 ligands, which each donate 2 electrons

chlorotris(triphenylphosphine)rhodium(I)

rhodium now has 16 electrons around it

Chlorotris(triphenylphosphine)rhodium(I) dissolves in benzene in the presence of hydrogen gas to give a new organometallic compound that contains metal-hydrogen bonds. The number of electrons around the central metal atom is then eighteen, the maximum number that its orbitals can hold. The geometry of the new complex is octahedral (Figure 8.6). When an alkene is added to the solution, it donates its π electrons to the rhodium, displacing one of the triphenylphosphine ligands. A reorganization of bonding occurs, with the hydrogen atoms being transferred from the metal to the carbon atoms of the double bond. For the sake of clarity, the process is shown on the next page in two steps, but it probably occurs in a single step.

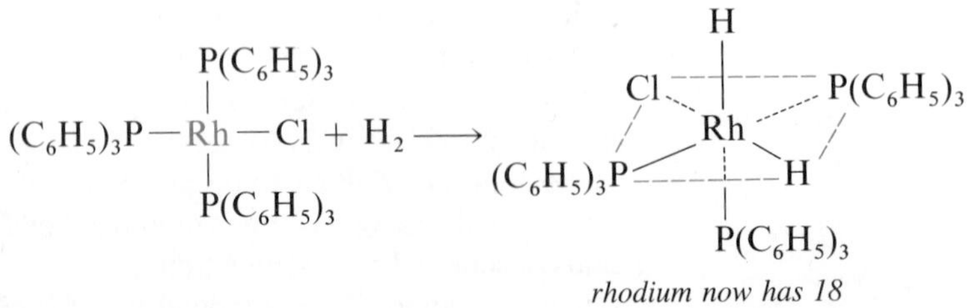
P(C6H5)3
(C6H5)3P—Rh—Cl + H2
P(C6H5)3
H
Cl
P(C6H5)3
Rh
(C6H5)3P
H
P(C6H5)3

rhodium now has 18 electrons around it; this is a gain of 2 electrons, one from each hydrogen atom

FIGURE 8.6 Octahedral geometry of the complex of chlorotris(triphenylphosphine)-rhodium(I) with hydrogen.

VISUALIZING THE REACTION

Catalysis of hydrogenation by a homogeneous catalyst

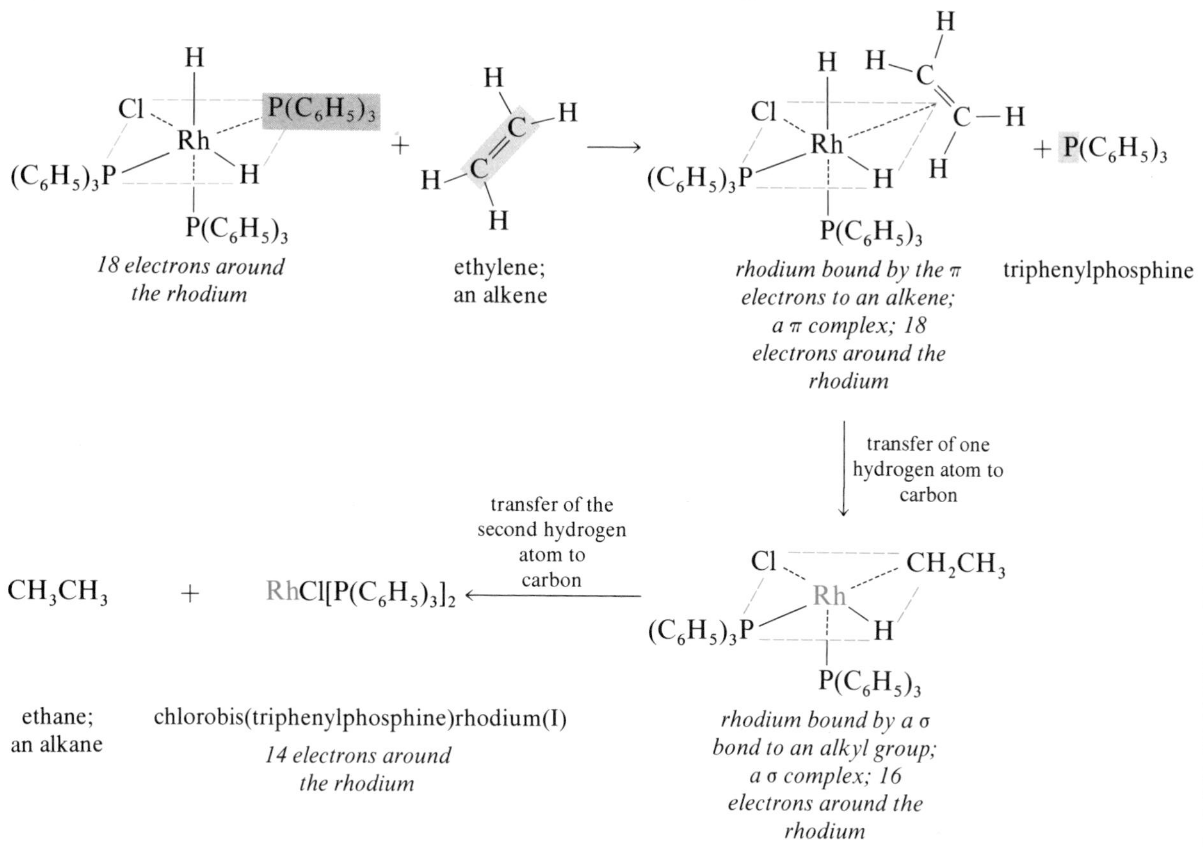

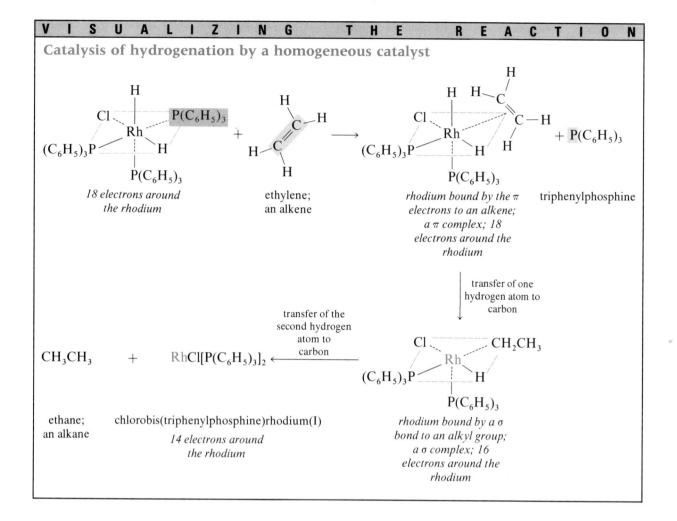

18 electrons around the rhodium
ethylene; an alkene
rhodium bound by the π electrons to an alkene; a π complex; 18 electrons around the rhodium
triphenylphosphine
transfer of one hydrogen atom to carbon
transfer of the second hydrogen atom to carbon
ethane; an alkane
chlorobis(triphenylphosphine)rhodium(I)
14 electrons around the rhodium
rhodium bound by a σ bond to an alkyl group; a σ complex; 16 electrons around the rhodium

The rhodium metal undergoes successive changes in the number of electrons in its valence shell. As the alkene is converted to an alkane in a reduction reaction, rhodium loses a ligand with its bonding electrons. The organometallic rhodium compound serves as a catalyst because it picks up triphenylphosphine and hydrogen from the solution to regenerate the eighteen-electron species that reacts with the alkene to start the cycle again.

Chlorotris(triphenylphosphine)rhodium(I) catalyzes the hydrogenation of alkenes to alkanes and usually does not affect other functional groups. A typical hydrogenation reaction with this catalyst is the conversion of vinylcyclopropane to ethylcyclopropane.

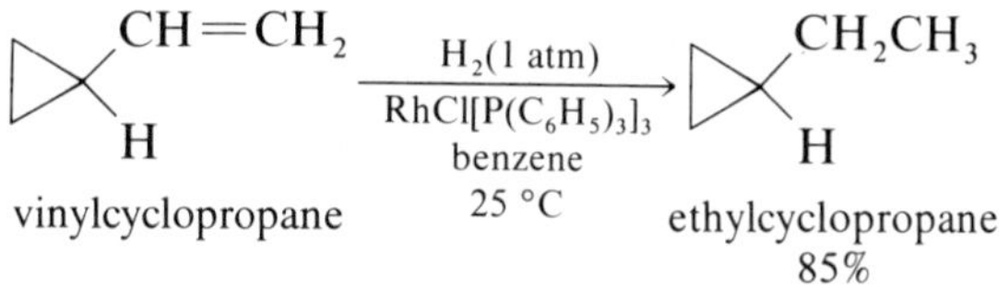

vinylcyclopropane
benzene
25 °C
ethylcyclopropane
85%

The rhodium catalyst differs from the heterogeneous catalysts discussed on pp. 304 and 306 in being more selective in its action. Terminal double bonds are hydrogenated much more rapidly by this bulky rhodium catalyst than are internal double bonds. In general, the higher the number of substituents on the doubly bonded carbon atoms, the slower the hydrogenation reaction is when this catalyst is used. In fact, it is possible to carry out selective reductions of one double bond in the presence of another one in the same molecule. The selectivity of this catalyst is illustrated by the reduction of 3,7-dimethyl-1,6-octadien-3-ol to 3,7-dimethyl-6-octen-3-ol.

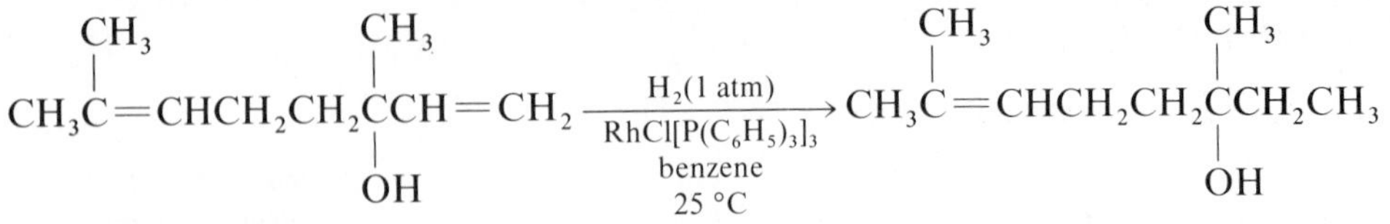

$$CH_3C\!=\!CHCH_2CH_2CCH\!=\!CH_2 \xrightarrow[\substack{RhCl[P(C_6H_5)_3]_3 \\ benzene \\ 25\,°C}]{H_2(1\ atm)} CH_3C\!=\!CHCH_2CH_2CCH_2CH_3$$

3,7-dimethyl-1,6-octadien-3-ol

3,7-dimethyl-6-octen-3-ol
80% yield

Study Guide
Concept Map 8.5

The rhodium compound catalyzes the hydrogenation of the terminal double bond, leaving the internal double bond untouched.

PROBLEM 8.20

Complete the following equations.

(a) $CH_3CH_2CH\!=\!CHCH_2CH_3 \xrightarrow{H_2}{Pt}$

(b) $CH_3CH_2CH_2\overset{\textstyle CH_3}{\underset{}{C}}\!=\!CH_2 \xrightarrow[benzene]{H_2}{RhCl[P(C_6H_5)_3]_3}$

(c) $\xrightarrow[Pd/C]{H_2}$

(d) $\xrightarrow[Pd/C]{H_2\ (excess)}$

(e) $CH_3\overset{\textstyle CH_3}{\underset{\textstyle CH_3}{C}}CH\!=\!CH_2 \xrightarrow[benzene]{H_2}{RhCl[P(C_6H_5)_3]_3}$

(f) $\xrightarrow[benzene]{H_2}{RhCl[P(C_6H_5)_3]_3}$

(g) $CH_3CH_2CH_2CH_2CH_2CH\!=\!CH_2 \xrightarrow[benzene]{H_2}{RhCl[P(C_6H_5)_3]_3}$

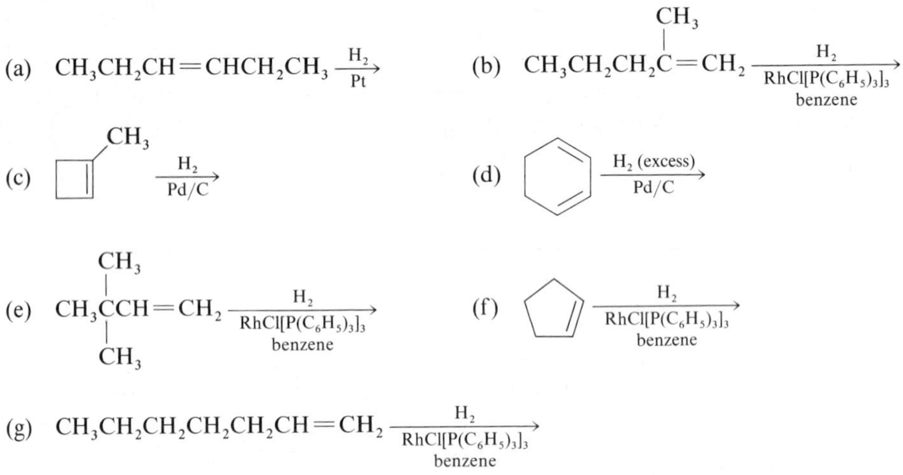

8.9
OXIDATION REACTIONS OF ALKENES

A. Reaction with Ozone. Ozonolysis

When **ozone, O_3,** reacts with an alkene the carbon-carbon double bond is broken and a carbon-oxygen double bond is formed on each fragment of the molecule. The overall reaction is called an **ozonolysis reaction,** meaning a cleavage (*lysis* means "to loosen") of bonds by ozone. For example, 2-methylpropene is converted to acetone and formaldehyde by ozone.

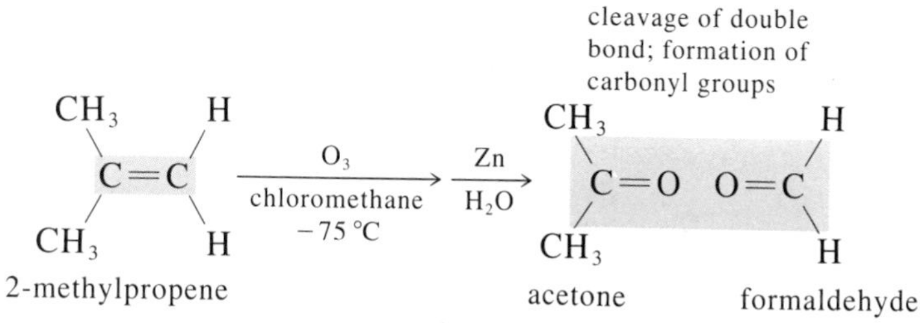
cleavage of double
bond; formation of
carbonyl groups
O3
chloromethane
−75 °C
Zn
H2O
2-methylpropene
acetone
formaldehyde
an ozonolysis reaction

Ozone is formed when an electric discharge is passed through oxygen gas. The ozone molecule, which cannot be symbolized satisfactorily by a single Lewis structure, is represented by a set of resonance contributors.

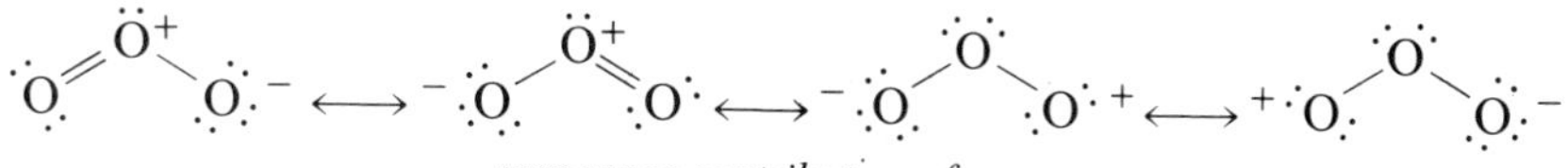
resonance contributors of ozone

Ozone is an electrophile that adds to an alkene to give an unstable cyclic compound called a **molozonide.** The molozonide decomposes to a charged intermediate and a carbonyl compound held close together in a solvent cage. If the carbonyl fragment is an aldehyde, the two pieces recombine to form an **ozonide.** These reactions are illustrated for 2-methylpropene.

VISUALIZING THE REACTION

Reaction of an alkene with ozone

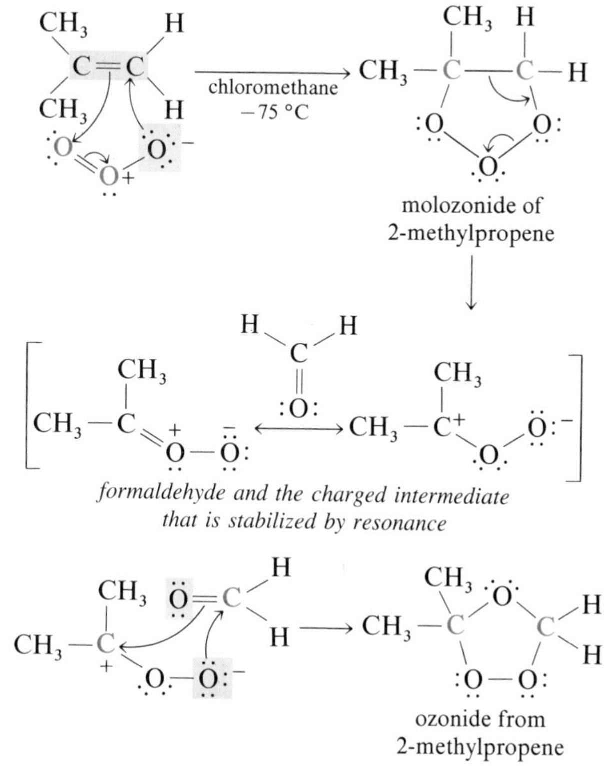
chloromethane
−75 °C
molozonide of
2-methylpropene
formaldehyde and the charged intermediate
that is stabilized by resonance
ozonide from
2-methylpropene

Ozonides and peroxides are unstable compounds that break up in water to give carbonyl compounds. The products obtained from an ozonolysis reaction depend on the conditions used. For example, 1-octene gives formic acid and heptanoic acid when an oxidizing agent such as hydrogen peroxide is present. Such conditions are called an **oxidative work-up** of the reaction mixture.

$$\underset{\text{1-octene}}{CH_3(CH_2)_5CH{=}CH_2} \xrightarrow[\text{water, 10 °C}]{O_3} \xrightarrow[NaOH]{H_2O_2} \xrightarrow{H_3O^+} \underset{\text{heptanoic acid}}{CH_3(CH_2)_5\overset{O}{\overset{\|}{C}}OH} + \underset{\text{formic acid}}{HO\overset{O}{\overset{\|}{C}}H}$$

oxidative work-up of ozonolysis mixture

If a reducing agent such as zinc or dimethyl sulfide is used, in what is called a **reductive work-up,** the products are aldehydes.

$$\underset{\text{1-octene}}{CH_3(CH_2)_5CH{=}CH_2} \xrightarrow[-60\ °C]{O_3,\ CH_3OH} \xrightarrow[-60\ °C]{(CH_3)_2S} \underset{\text{heptanal}}{CH_3(CH_2)_5\overset{O}{\overset{\|}{C}}H} + \underset{\text{formaldehyde}}{H\overset{O}{\overset{\|}{C}}H} + \underset{\text{dimethyl sulfoxide}}{CH_3\overset{O}{\overset{\|}{S}}CH_3}$$

oxidation product of dimethyl sulfide

reductive work-up of ozonolysis mixture

Dimethyl sulfide acts as a reducing agent because it reacts with oxygen to form a stable compound, dimethyl sulfoxide, and thereby prevents oxidation of the aldehyde products to carboxylic acids.

For the reaction of ozone with cyclohexene, different products are also obtained by changing the reaction conditions. Cyclohexene is a cyclic alkene. Thus, its reaction with ozone, when followed by treatment with dimethyl sulfide to decompose the ozonide, gives a compound in which two aldehyde groups are still attached to each other by a chain of carbon atoms that was the rest of the ring.

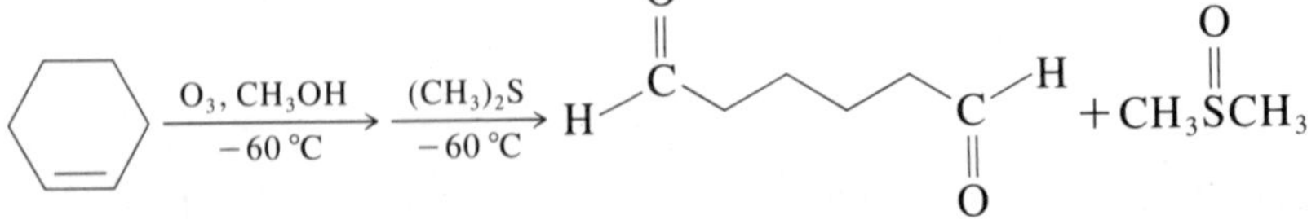
O₃, CH₃OH
−60 °C
(CH₃)₂S
−60 °C
H
C
O
C
H
O
O
+CH₃SCH₃

cyclohexene — *reductive work-up* — 1,6-hexanedial 62% *dialdehyde from cleavage of the double bond in cyclohexene by ozone* — dimethyl sulfoxide

If hydrogen peroxide is used after the addition of ozone, the corresponding dicarboxylic acid is formed.

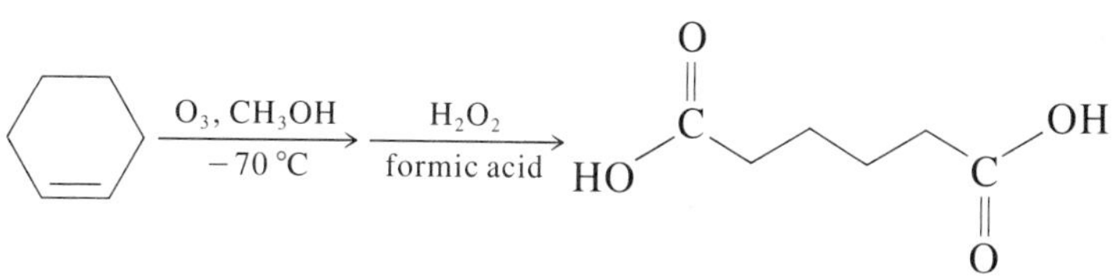

$$\text{cyclohexene} \xrightarrow[-70\,°C]{O_3,\ CH_3OH} \xrightarrow[\text{formic acid}]{H_2O_2} HO-\overset{\overset{\displaystyle O}{\|}}{C}-CH_2CH_2CH_2CH_2-\overset{\underset{\displaystyle O}{\|}}{C}-OH$$

cyclohexene   *oxidative*   1,6-hexanedioic acid
                     work-up   adipic acid
                                               85%

Thus, if a reducing agent is not used in the decomposition of an ozonide, the alde-hyde products are oxidized to carboxylic acids. In fact, ozone is used to synthesize carboxylic acids by cleavage of multiple bonds (p. 558).

If there are two organic groups as substituents on one of the doubly bonded carbon atoms in the alkene, a product of the reaction with ozone is a ketone. 2-Methylpropene, for example, gives acetone on treatment with ozone (p. 311). A ketone, in contrast to an aldehyde, cannot be easily oxidized. In order to add any more oxygen atoms to the carbonyl group, a carbon-carbon single bond would have to be broken.

Reactions that result in the cleavage of bonds are used to convert large mole-cules into smaller, more easily identifiable fragments. Such reactions are known as **degradation reactions.** Degradation reactions have been particularly important in determining the structures of complex molecules isolated from natural sources. Ozonolysis is a degradation reaction. The compounds formed as a result of ozonolysis—aldehydes, ketones, and carboxylic acids—have smaller molecules than the starting alkene as well as reactive functional groups. They are, therefore, more easily identified than the starting alkene. Knowing what the fragments are makes it possible to reconstruct the original molecule. For example, a chemist who identified heptanal and formaldehyde as ozonolysis products from an alkene of unknown structure would be able to conclude that the alkene was 1-octene.

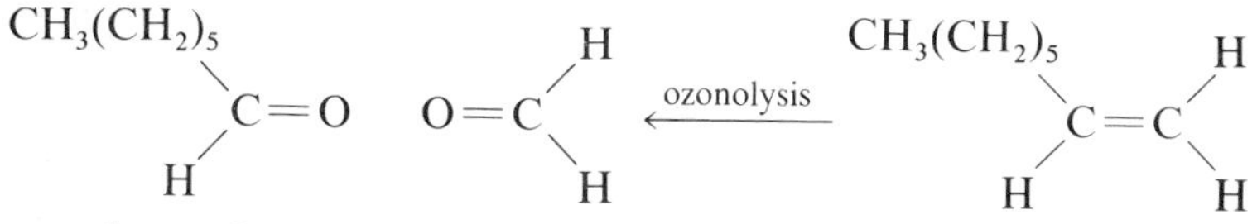

$$\underset{\text{heptanal}}{\overset{CH_3(CH_2)_5}{\underset{H}{>}}C=O} \qquad \underset{\text{formaldehyde}}{O=C\overset{H}{\underset{H}{<}}} \xleftarrow{\text{ozonolysis}} \underset{\text{1-octene}}{\overset{CH_3(CH_2)_5}{\underset{H}{>}}C=C\overset{H}{\underset{H}{<}}}$$

The structure of the alkene is easily arrived at by mentally removing the oxygen atoms and joining the carbon atoms of the two carbonyl groups with a double bond.

PROBLEM 8.21

Complete the following equations.

(a) $\underset{\underset{\displaystyle CH_3}{|}}{CH_3C}=CHCH_3 \xrightarrow{O_3} \xrightarrow[\text{water}]{Zn}$ 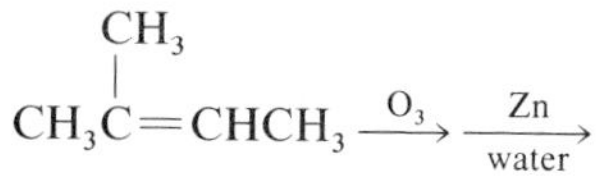

(b) $\underset{\underset{\displaystyle CH_3}{|}}{CH_3CHCH}=\underset{\underset{\displaystyle CH_2CH_3}{|}}{CCH_2CH_3} \xrightarrow{O_3,\ CH_3OH} \xrightarrow{(CH_3)_2S}$ 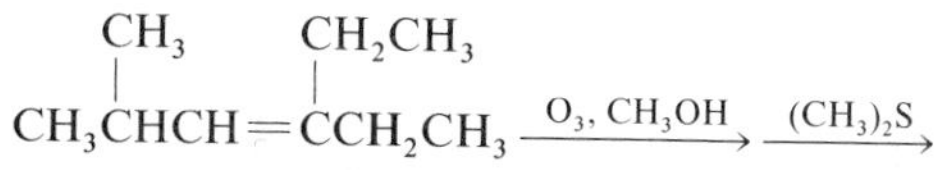

(c) $\underset{\underset{\displaystyle CH_3}{|}}{CH_3CHCH}=\underset{\underset{\displaystyle CH_2CH_3}{|}}{CCH_2CH_3} \xrightarrow[\text{water}]{O_3} \xrightarrow{H_2O_2,\ NaOH} \xrightarrow{H_3O^+}$ 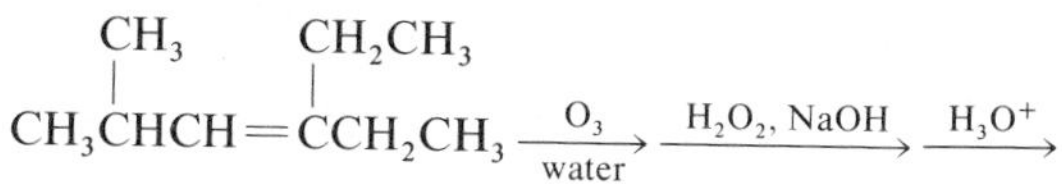

(d) 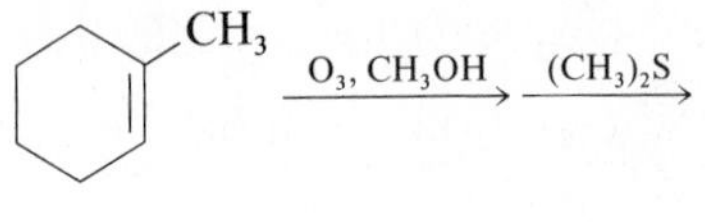

(e) 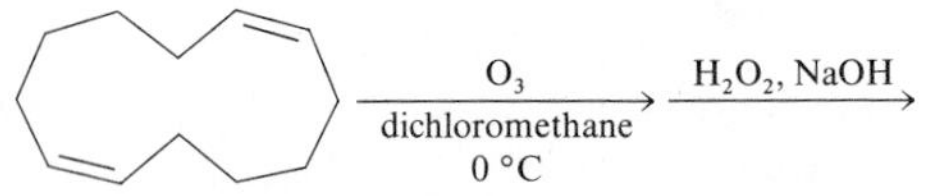

(f) $CH_3CH_2C(CH_3){=}CH_2 \xrightarrow{O_3} \xrightarrow[\text{water}]{Zn}$

(g) $CH_3(CH_2)_{13}CH{=}CH_2 \xrightarrow[\text{water}]{O_3} \xrightarrow{H_2O_2,\ NaOH} \xrightarrow{H_3O^+}$

PROBLEM 8.22

In early research into the structures of branched hydrocarbons, chemists determined the structures of a number of alkenes by ozonolysis. What products would you expect to get from the ozonolysis of a mixture of alkenes consisting of 90% 3-ethyl-5,5-dimethyl-2-hexene and 10% 4-ethyl-2,2-dimethyl-3-hexene? The reaction mixture was treated with zinc and water.

PROBLEM 8.23

A tertiary alcohol was dehydrated to give an alkene. Treatment of the alkene with ozone, then with water and zinc, gave equal amounts of 3-pentanone and acetaldehyde. What is the structure of the alcohol?

$CH_3CH_2C({=}O)CH_2CH_3$ 3-pentanone

$CH_3CH({=}O)$ acetaldehyde

B. Oxidation of Alkenes with Permanganate or Osmium Tetroxide

An aqueous solution of potassium permanganate oxidizes multiple bonds. In this reaction, the purple permanganate ion is reduced to manganese dioxide, a brown solid. The change in the color and appearance of the reaction mixture is the basis of a test for the presence of double and triple bonds known as the **Baeyer test for unsaturation.**

The first stage of the oxidation of an alkene leads to the formation of a compound with hydroxyl groups on the carbon atoms that were involved in the double bond, a 1,2-diol. For example, cyclopentene is oxidized to *cis*-1,2-cyclopentanediol by potassium permanganate.

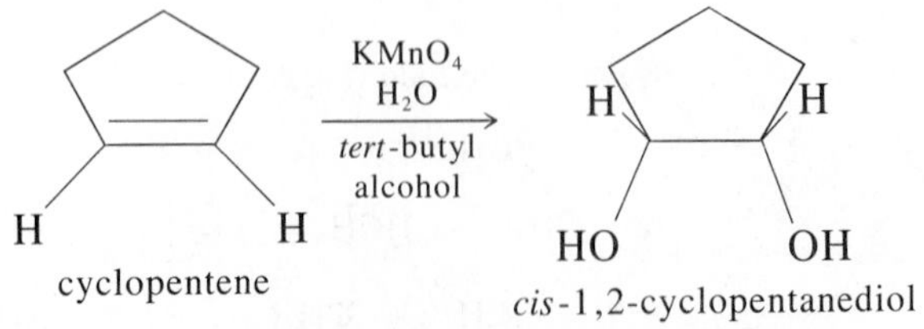
KMnO4
H2O
tert-butyl
alcohol
cyclopentene
cis-1,2-cyclopentanediol

Manganese(VII) in permanganate ion is ultimately reduced to manganese(IV) in manganese dioxide. The carbon atoms of the double bond are oxidized. Even if no base is added at first, the solution becomes progressively more basic as the reaction proceeds. Reactions that are being used to synthesize diols go best if some base is added initially.

In this oxidation reaction, the two hydroxyl groups become attached to the same face of the double bond. The permanganate ion is believed to add to the double bond to give a cyclic intermediate, a manganate ester. The first step of this reaction, the syn addition (p. 302) of permanganate ion to the double bond, is the important one to remember. This intermediate breaks down in water to give the *cis*-1,2-diol.

VISUALIZING THE REACTION

Oxidation of an alkene by permanganate ion

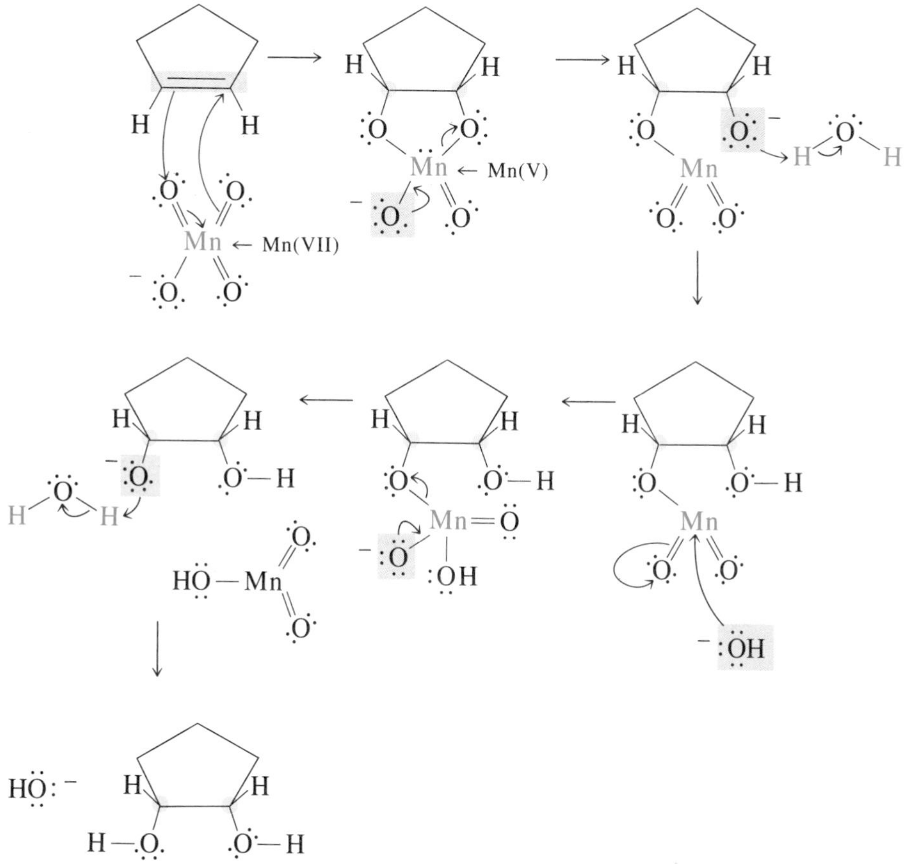

Mn(VII)
Mn(V)

The first stage of the reduction of permanganate ion gives manganate ion with Mn(V). Manganate ion is eventually converted to manganese dioxide, which precipitates. The oxygen atoms of the diol come from the permanganate ion, but water and hydroxide ion participate in breaking up the ester.

Another example of the oxidation of an alkene by potassium permanganate is the conversion of oleic acid to 9,10-dihydroxystearic acid.

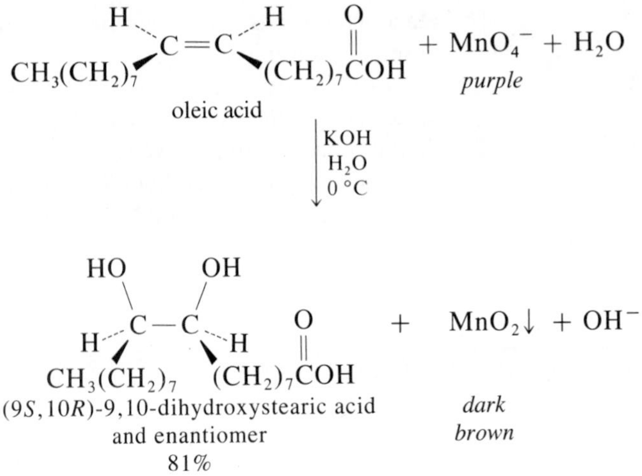
oleic acid
purple
KOH
H_2O
0 °C
(9*S*,10*R*)-9,10-dihydroxystearic acid
and enantiomer
81%
dark
brown

The product of the reaction is written in a conformation that emphasizes the syn addition of the hydroxyl groups to the double bond.

Another reagent that forms cis 1,2-diols by way of a similar intermediate is osmium tetroxide, an expensive and highly toxic reagent. This reagent is usually used when small amounts of precious alkenes have to be converted to cis diols in high yields. Although the cyclic manganate ester breaks down very rapidly in water and cannot be isolated, cyclic osmate esters, which can be prepared in organic solvents, have been isolated in some cases. The oxidation of 1,2-dimethylcyclopentene to *cis*-1,2-dimethyl-1,2-cyclopentanediol is an example of the use of osmium tetroxide.

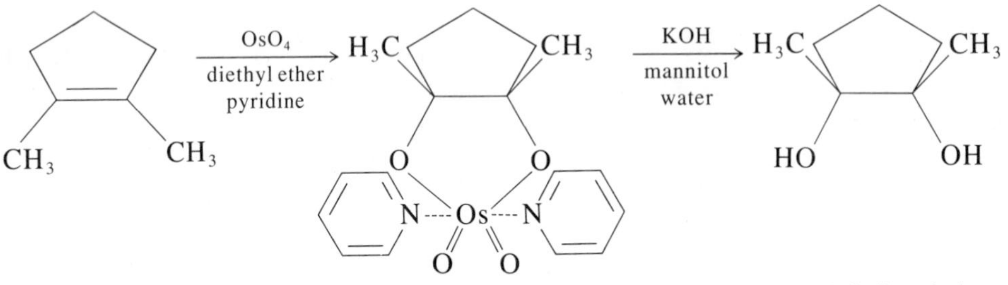
OsO_4
diethyl ether
pyridine
KOH
mannitol
water

cyclic osmate ester complexed with pyridine

cis-1,2-dimethyl-1,2-cylopentanediol

$$\text{mannitol} \equiv HOCH_2\left(\begin{matrix}OH\\|\\CH\end{matrix}\right)_4CH_2OH$$

The reaction is carried out in ether. Pyridine catalyzes the addition of osmium tetroxide to the double bond and complexes with the osmium in the ester. Various methods are used to decompose the esters. In this case, a base and mannitol, a polyhydroxy compound that complexes with osmium, are used. Because osmium tetroxide is such an expensive and toxic reagent, it is often used in catalytic amounts in the presence of another oxidizing agent that keeps recycling the osmium back to the tetroxide stage. For an example, see Problem 8.42 at the end of the chapter.

PROBLEM 8.24

Complete the following equations, predicting the products and showing stereochemistry wherever possible.

(a) 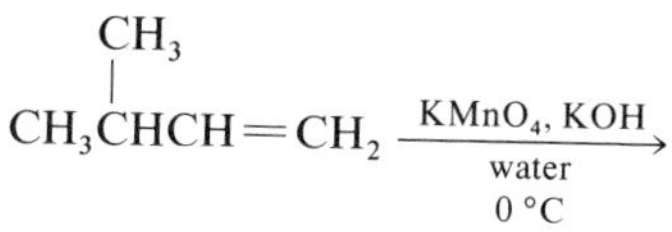

(b) 

(c) 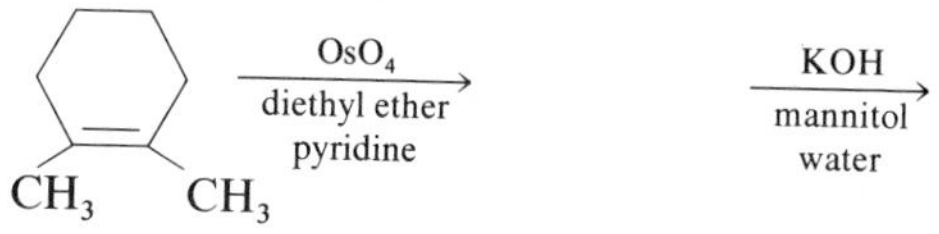

(d) 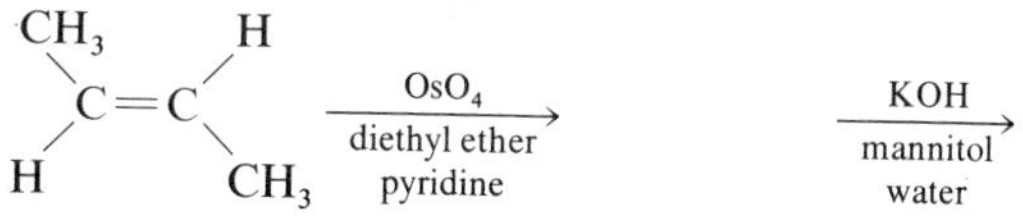

(e) 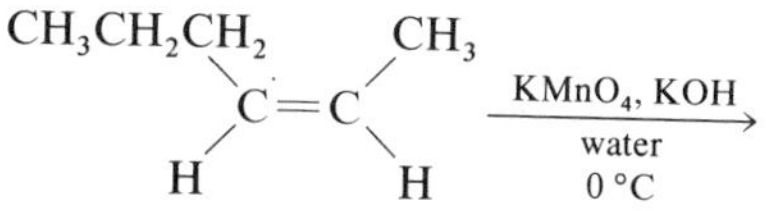

PROBLEM 8.25

Permanganate ion in which the oxygen atoms were labeled with ^{18}O was used to oxidize an alkene. Using

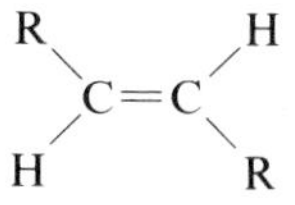

as the alkene, write out the mechanism for the oxidation reaction, showing where the labeled oxygen atoms will be found in the product and what the stereochemistry of the product will be.

C. Oxidation of Alkenes with Peroxyacids

Carboxylic acids have the general formula

and may be regarded as being derived from water, HOH, by the replacement of one of the hydrogen atoms by an acyl group,

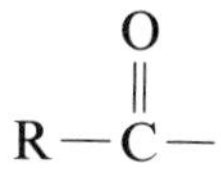

Carboxylic acids have an acidic hydrogen atom and act as Brønsted acids. They are *not* oxidizing agents.

Peroxycarboxylic acids have the general formula

$$R-\overset{\overset{\displaystyle O}{\|}}{C}-O-O-H$$

and may be viewed as being related to the oxidizing agent hydrogen peroxide, HOOH, in the same way that carboxylic acids are related to water. Replacement of one of the hydrogen atoms in hydrogen peroxide by an acyl group gives the formula for a peroxycarboxylic acid. Peroxyformic acid, for example, is an unstable compound that can be made by mixing formic acid with hydrogen peroxide.

$$\underset{\text{formic acid}}{H\overset{\overset{\displaystyle O}{\|}}{C}OH} + \underset{\text{hydrogen peroxide}}{HOOH} \rightleftharpoons \underset{\text{peroxyformic acid}}{H\overset{\overset{\displaystyle O}{\|}}{C}OOH} + \underset{\text{water}}{H_2O}$$

Peroxyacids are oxidizing agents and react with alkenes to give three-membered cyclic ethers known as **epoxides** or **oxiranes.** These small ring compounds are quite reactive; the ring opens easily, especially in acidic solutions. These reactions are considered in more detail in Section 12.6. This section examines the formation of oxiranes from alkenes using *m*-chloroperoxybenzoic acid as the reagent.

m-Chloroperoxybenzoic acid is sold commercially and is stable enough that it can be stored and used over a period of time. Its reactions are carried out in organic solvents, such as ethers or halogenated hydrocarbons, so the solubility of starting materials and products is not a problem. The oxiranes that are formed are stable in the reaction mixture and are isolated in reasonably good yields.

The reaction of *m*-chloroperoxybenzoic acid with alkenes is stereoselective. Stereoisomeric alkenes give the corresponding oxiranes with virtually no change in stereochemistry. For example, (*Z*)-2-butene is oxidized to *cis*-2,3-dimethyloxirane, and (*E*)-2-butene gives *trans*-2,3-dimethyloxirane with over 99% stereochemical purity.

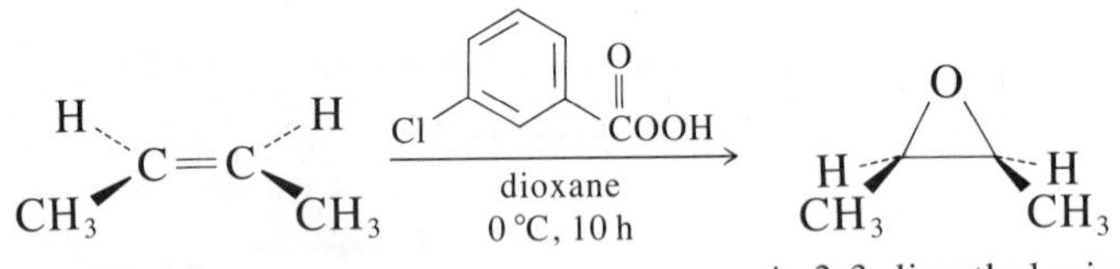
H
H
CH3
CH3
Cl
O
COOH
dioxane
0 °C, 10 h
O
H
H
CH3
CH3

(Z)-2-butene

cis-2,3-dimethyloxirane
60%
a meso form

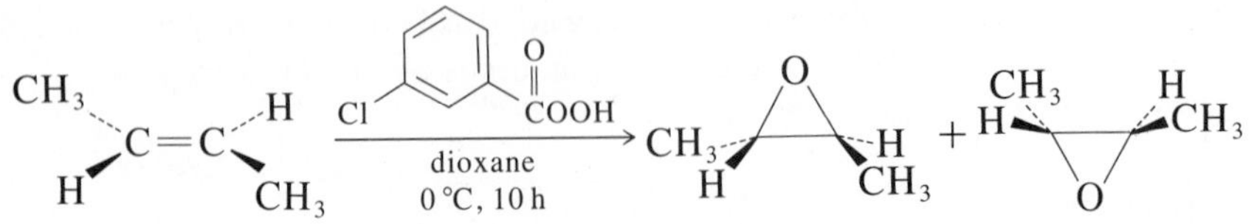
CH3
H
H
CH3
Cl
O
COOH
dioxane
0 °C, 10 h
O
CH3
H
H
CH3
+
CH3
H
H
CH3
O

(*E*)-2-butene

trans-2,3-dimethyloxirane
60%
racemic mixture

dioxane ≡ 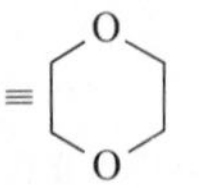
O
O

an ether solvent

cis-2,3-Dimethyloxirane is not optically active because it is a meso form; it has a plane of symmetry bisecting the oxygen atom and the carbon-carbon bond of the oxirane ring. *trans*-2,3-Dimethyloxirane is formed as a racemic mixture; the two enantiomers result because there is an equal probability of attack from above and below the plane of the double bond.

An oxygen atom is transferred from the peroxyacid to the double bond, giving the cyclic ether. The other product of the reaction is *m*-chlorobenzoic acid, no longer a peroxyacid. The mechanism of the reaction involves a single-step transfer of the peroxy oxygen atom to the double bond.

VISUALIZING THE REACTION

Reaction of a peroxyacid with an alkene

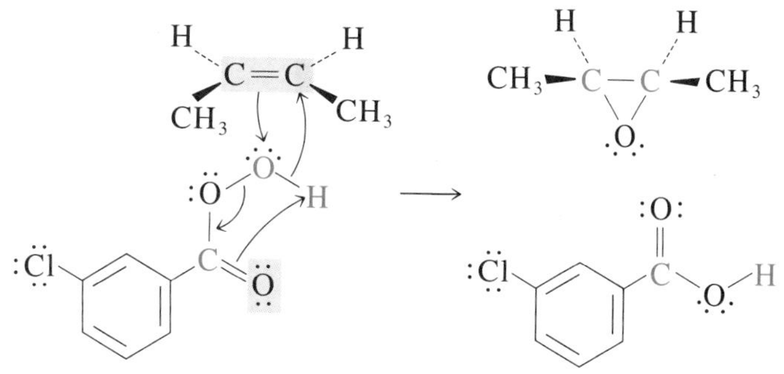

H
H
C=C
CH3
CH3
:O
O
H
:Cl
C
O
H
H
CH3
C
C
CH3
O
:O:
:Cl
C
O
H

The peroxyacid behaves as an electrophile in the reaction shown above. Its electrophilic character is revealed in the selectivity with which it reacts with double bonds having different degrees of substitution. The more alkyl substituents on a double bond, the more readily it is attacked by the peroxyacid. This fact suggests that higher electron density in a π bond favors the reaction with peroxyacids. 1,2-Dimethyl-1,4-cyclohexadiene, for example, reacts selectively with one molar equivalent of *m*-chloroperoxybenzoic acid at the more highly substituted of the two double bonds.

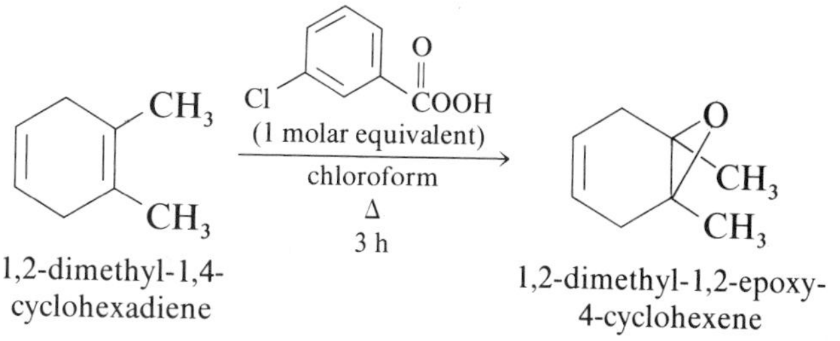

CH3
CH3
Cl
O
COOH
(1 molar equivalent)
chloroform
Δ
3 h
O
CH3
CH3
1,2-dimethyl-1,4-cyclohexadiene
1,2-dimethyl-1,2-epoxy-4-cyclohexene

Study Guide Concept Map 8.6

This observed result demonstrates the effect that alkyl groups have in increasing the electron density in the functionalities to which they are attached.

PROBLEM 8.26

Predict the major product of each of the following reactions. Write a structure showing the stereochemistry when it is pertinent.

(a) 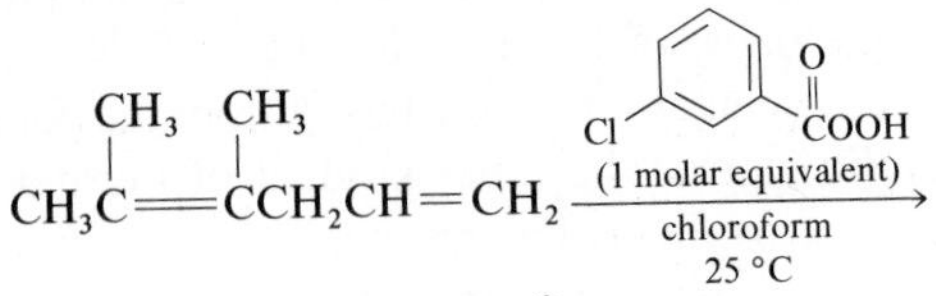

(b) 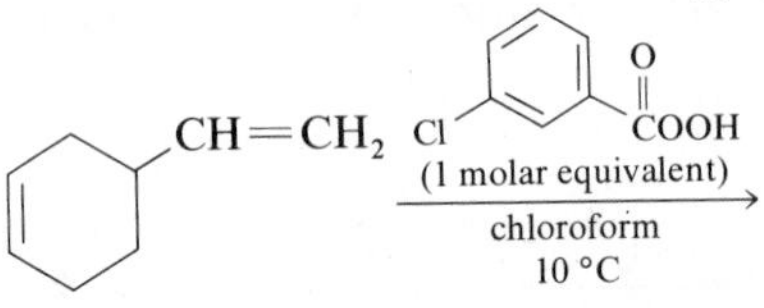

(c) 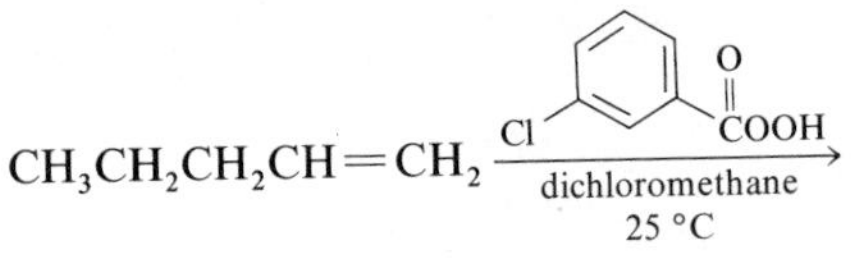

(d) 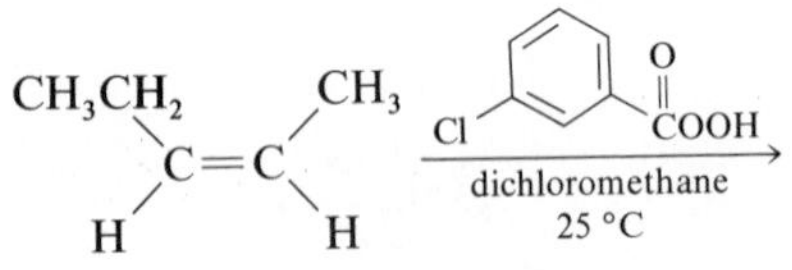

(e) 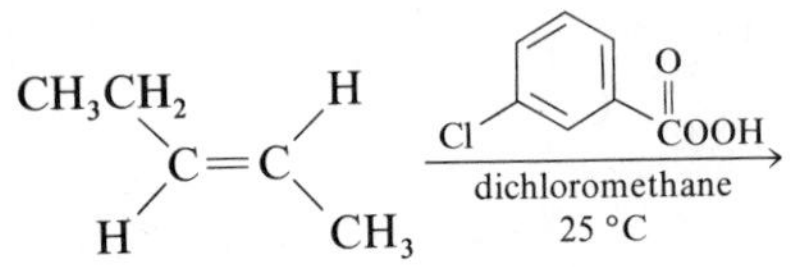

D. Problem-Solving Skills

Problem

meso-2,3-Butanediol is prepared from (*Z*)-2-butene. Suggest a reagent that can be used for this transformation.

Solution

1. What functional groups are present in the starting material and the product?

   The starting material is an alkene and thus contains a double bond. The product is a diol, with the two hydroxyl groups on adjacent carbon atoms. Specific stereochemistry is given for both the starting material and the product.

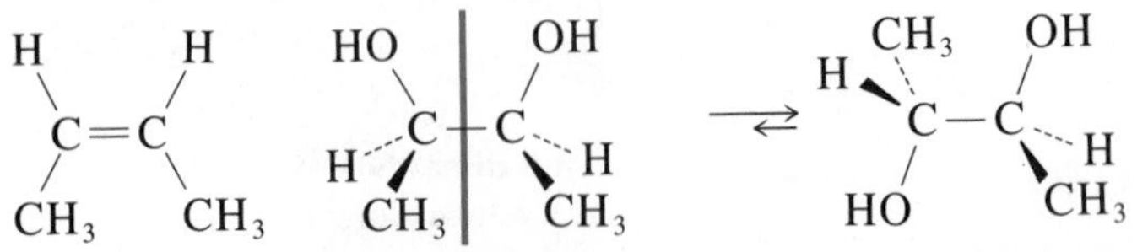

(*Z*)-2-butene; *meso*-2,3-butanediol drawn in the eclipsed conformation to show the plane of symmetry; more stable anti conformation of *meso*-2,3-butanediol

2. How do the carbon skeletons of the two compounds compare? How many carbon atoms does each contain? Are there any rings? What are the positions of branches and functional groups on the carbon skeletons?

   The carbon skeletons of the two compounds are identical. The two carbon atoms that are joined by a double bond in the starting material have hydroxyl groups on them in the product. The methyl groups in the starting alkene and in the eclipsed conformation of the resulting diol are on the same side of the two carbon atoms to which they are bonded.

3. How do the functional groups change in going from starting material to product? Does the starting material have a good leaving group?

   The double bond disappears, and the two carbon atoms of the double bond end up with hydroxyl groups on them. There is no leaving group present in the starting material.

4. Is it possible to dissect the structures of starting material and product to see which bonds must be broken and which formed?

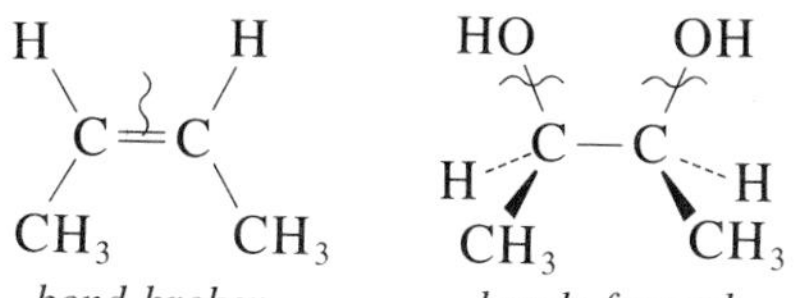

bond broken
bonds formed

5. Do we recognize any part of the product molecule as coming from a good nucleophile or an electrophilic addition?

   A diol results from reaction of an alkene with an oxidizing agent (an electrophile) such as permanganate ion or osmium tetroxide. The reaction of permanganate ion or osmium tetroxide is a syn addition and therefore will give the desired stereochemistry.

6. What type of compound would be a good precursor to the product?

   No other precursor is needed.

7. After this last step, do we see how to get from starting material to product? If not, we need to analyze the structure obtained in step 6 by applying questions 5 and 6 to it.

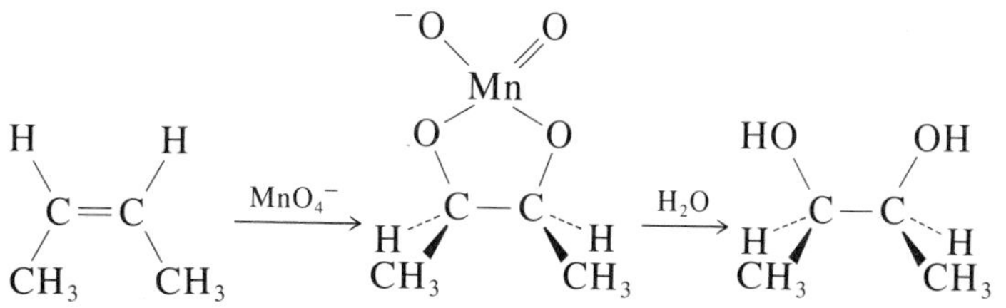

MnO_4^-
H_2O

PROBLEM 8.27

How would you synthesize racemic 2,3-butanediol from (*E*)-2-butene?

SUMMARY

Alkenes react with electrophiles in electrophilic addition reactions. In this type of reaction, the important step is the interaction between the electrons of the π bond of the alkene and an electron-seeking reagent. With some electrophiles, such as acids and halogens, a reactive intermediate, which is a carbocation or a halonium ion, is formed first. It then reacts with a nucleophile to give the final product.

When an alkene reacts with diborane, ozone, permanganate ion, or osmium tetroxide, the intermediate first formed is broken up in aqueous solution to give the products. In catalytic hydrogenations and peroxyacid oxidations, no discrete intermediates are observed. The reactions of alkenes covered in this chapter are outlined in Table 8.1 on the opposite page.

The addition of an unsymmetrical reagent, such as an acid or diborane, to an alkene with different substituents on the two carbons of the double bond shows high regioselectivity, which is determined by the formation of the most stable intermediate. A positive charge develops at the more highly substituted carbon atom. The more stable carbocation is formed in the addition of an acid; a hydride ion is transferred to the more highly substituted carbon atom in the addition of diborane. Experiments with alkenes of differing stereochemistry have shown that many addition reactions also have high stereoselectivity. Anti addition to the double bond occurs with halogens, for example. Syn addition results with hydrogen, permanganate ion, osmium tetroxide, diborane, and the oxygen from a peroxyacid.

ADDITIONAL PROBLEMS

8.28 Name each of the following compounds, assigning the correct configuration to any for which stereochemistry is shown.

(a) 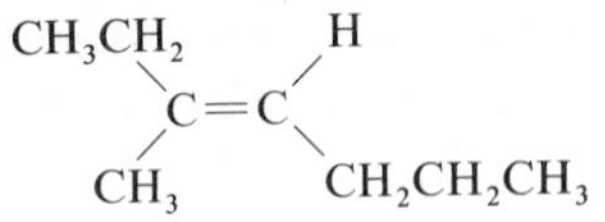

(b) 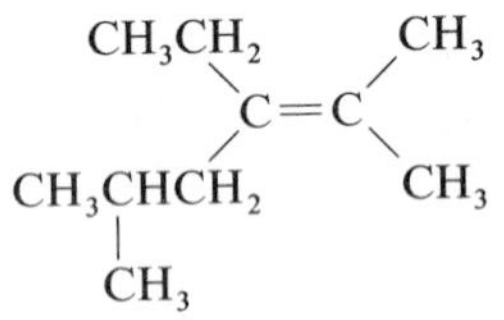

(c) 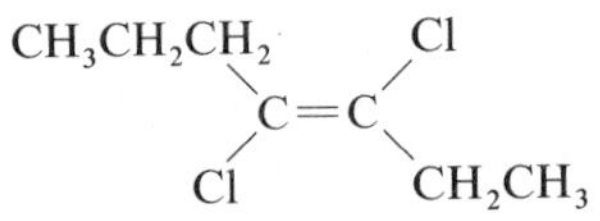

(d) 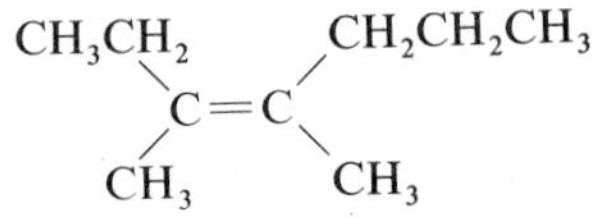

(e) 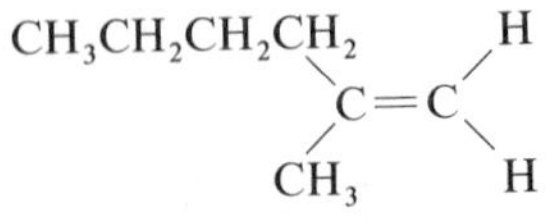

(f) 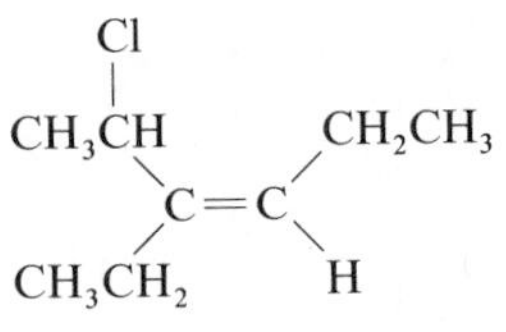

(g) 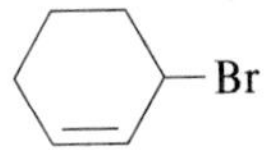

(h) 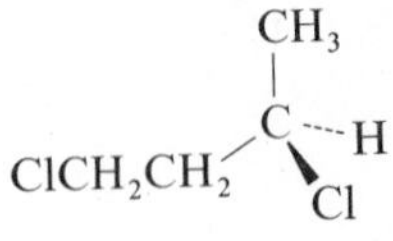

(i) 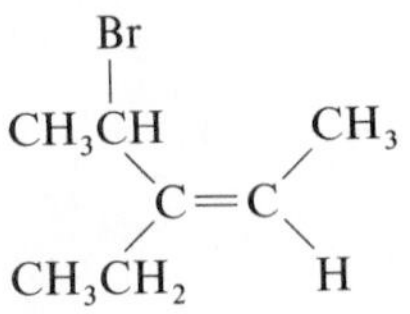

(j) 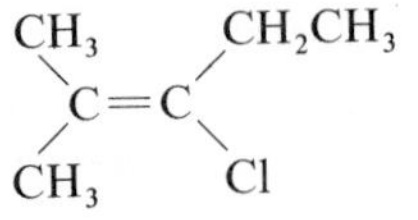

(k) 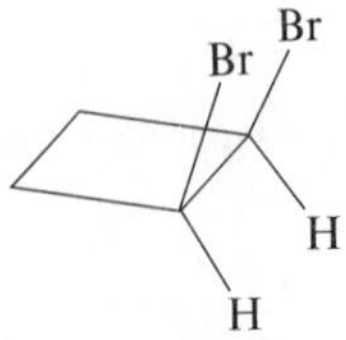

(l) 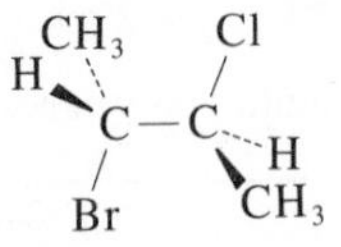

TABLE 8.1 Outline of the Reactions of Alkenes

Alkene	Electrophile	Intermediate	Reagent in Second Step	Product(s)	Stereochemistry
Electrophilic Addition Reactions					
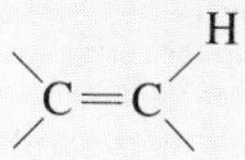	HX	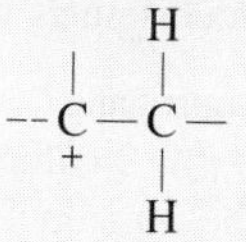	X^-	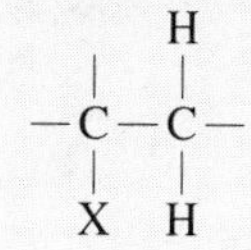	
	X_2	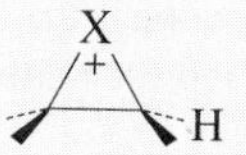	X^-, ROH	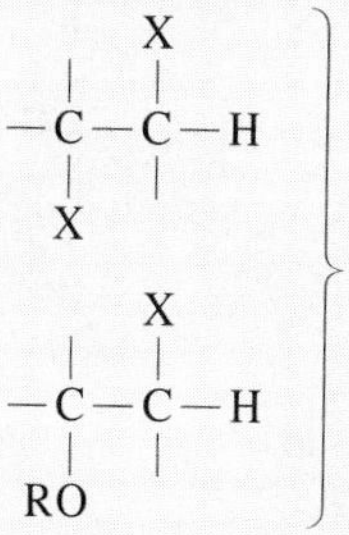	anti addition
	BH_3	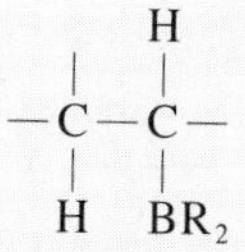	H_2O_2, OH^-	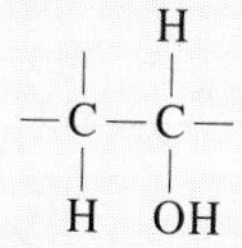	syn addition
Oxidation Reactions					
	O_3	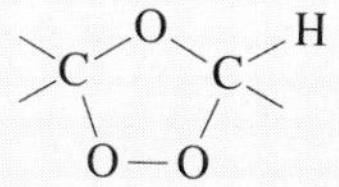	Zn, H_2O or CH_3SCH_3 (reductive work-up)	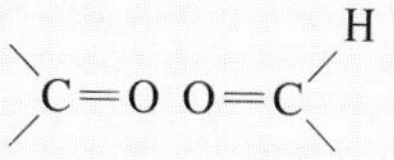	
			H_2O_2 (oxidative work-up)	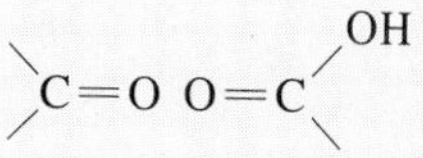	
	MnO_4^- (OsO_4)	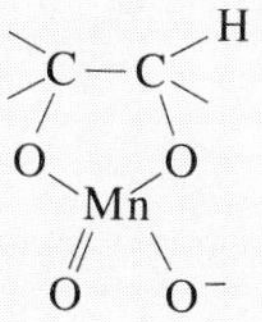	H_2O	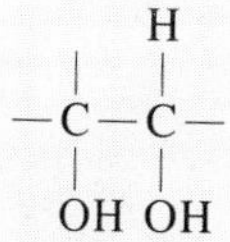	syn addition
	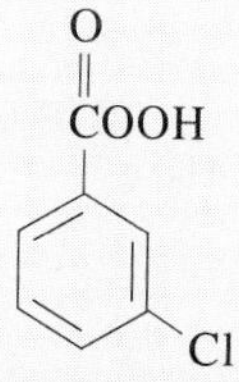	none observed	none	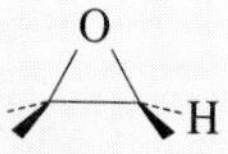	syn addition
Reduction Reaction					
	H_2, PtO_2 or $RhCl[P(C_6H_5)_3]_3$	none observed	none	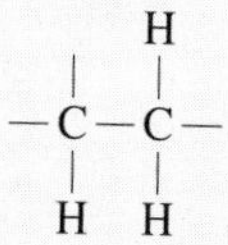	syn addition

8.29 Draw a structural formula for each of the following compounds. Be sure to show stereochemistry when it is indicated.

(a) (Z)-5-chloro-2-pentene (b) *trans*-1,2-dimethylcyclopropane
(c) *meso*-2,3-dibromobutane (d) 3-chloro-2-methyl-1-pentene
(e) *cis*-1,2-dichlorocyclopentane (f) 1,2-dimethylcyclohexene

8.30 Draw structural formulas for the stereoisomers of the following compounds.

(a) 3-hexene (b) 1,2-dimethylcyclopropane
(c) 1-bromo-2-methylcyclopentane (d) 2-chloropentane
(e) 3,4-dimethylhexane (f) 2-bromo-3-methylpentane
(g) 5-bromo-2-hexene

8.31 Write equations predicting the reaction of (*E*)-3-methyl-2-pentene with each of the following reagents. In cases where you know what the stereochemistry of the product(s) should be, illustrate it with a three-dimensional representation.

(a) H_2, Pt (b) H_2O, H_2SO_4 (c) HBr

(d) 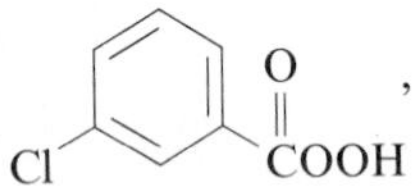
O
Cl
COOH
, chloroform (e) $KMnO_4$, OH^-, water

(f) (1) O_3, CH_3OH; (2) $(CH_3)_2S$ (g) (1) OsO_4, diethyl ether; (2) KOH, water

(h) (1) O_3, water; (2) H_2O_2, NaOH; (3) H_3O^+

(i) Br_2, carbon tetrachloride (j) Br_2, H_2O (k) H_2, $RhCl[P(C_6H_5)_3]_3$

8.32 Complete the following equations, showing the stereochemistry wherever it is known.

(a) 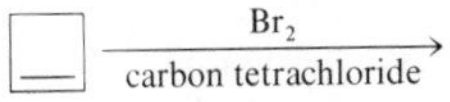
Br2
carbon tetrachloride

(b) 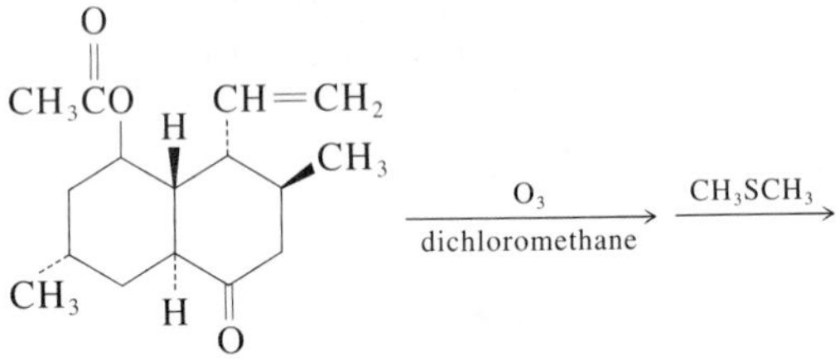
O
CH3CO
CH=CH2
H
CH3
CH3
H
O
O3
dichloromethane
CH3SCH3

(c) 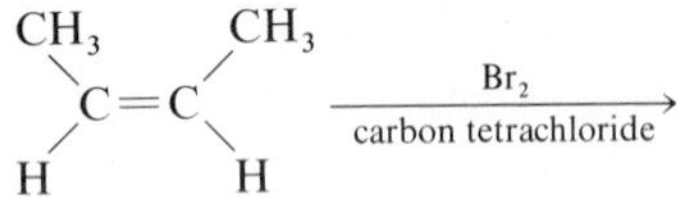
CH3
CH3
C=C
H
H
Br2
carbon tetrachloride

(d) 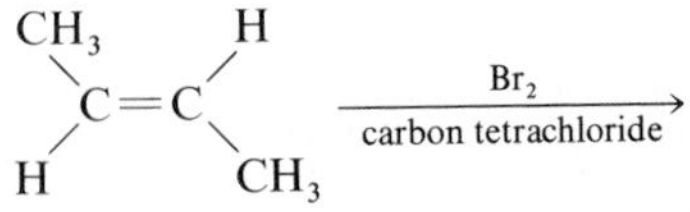
CH3
H
C=C
H
CH3
Br2
carbon tetrachloride

(e) 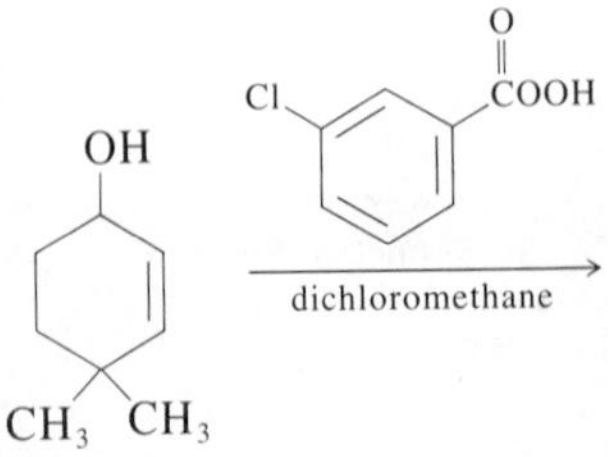
OH
CH3
CH3
Cl
O
COOH
dichloromethane

(f) 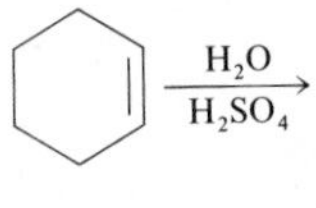
H2O
H2SO4

(g) 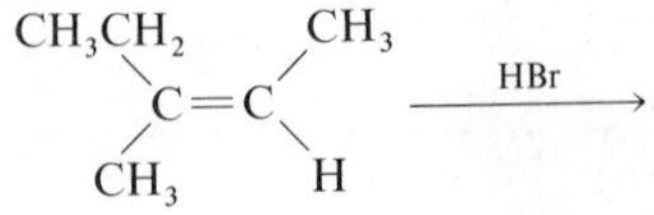
CH3CH2
CH3
C=C
CH3
H
HBr

(h) 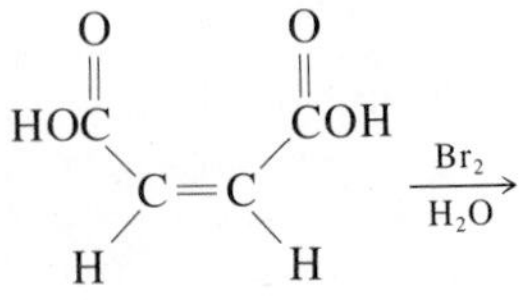
O
O
HOC
COH
C=C
H
H
Br2
H2O

(i) 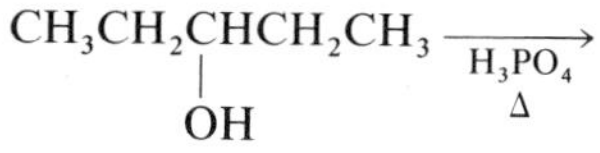
CH3CH2CHCH2CH3
OH
H3PO4
Δ

(j) 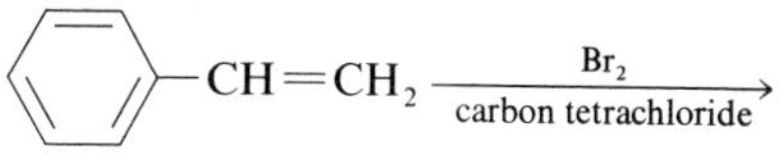
CH=CH2
Br2
carbon tetrachloride

(k) 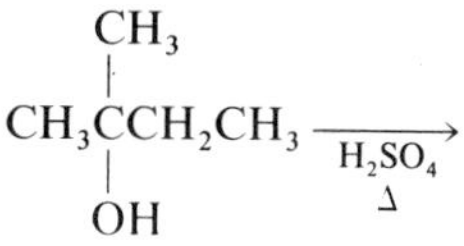
CH3
CH3CCH2CH3
OH
H2SO4
Δ

(l) 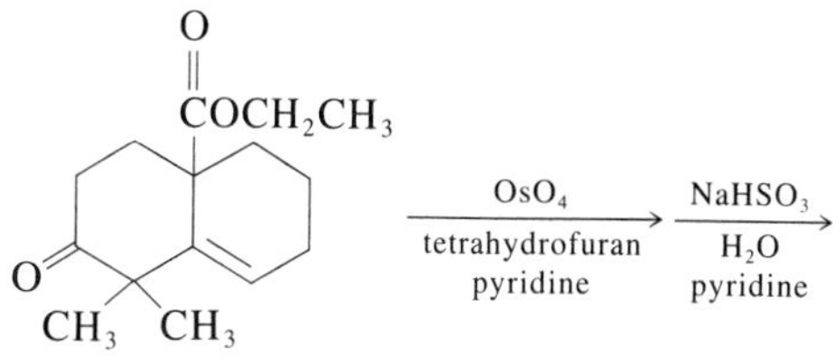
COCH2CH3
CH3 CH3
OsO4
tetrahydrofuran
pyridine
NaHSO3
H2O
pyridine

(m) 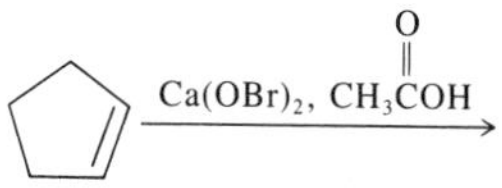
Ca(OBr)2, CH3COH

8.33 Supply structural formulas for the intermediates and products designated by capital letters. Show the stereochemistry of the product(s) when known.

(a) 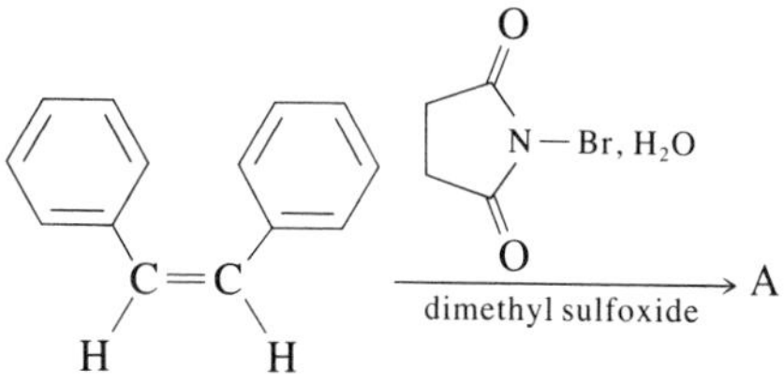
N—Br, H2O
dimethyl sulfoxide
A

(b) 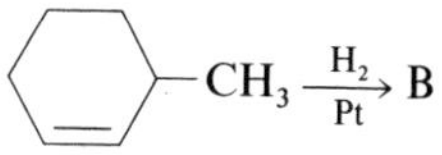
CH3
H2
Pt
B

(c) 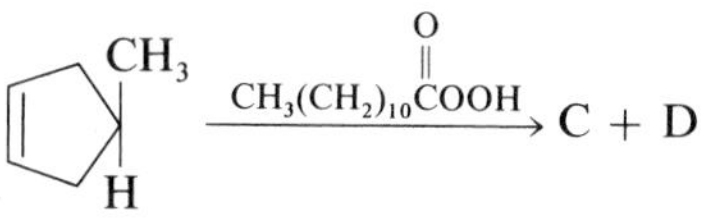
CH3
H
CH3(CH2)10COOH
C + D

(d) 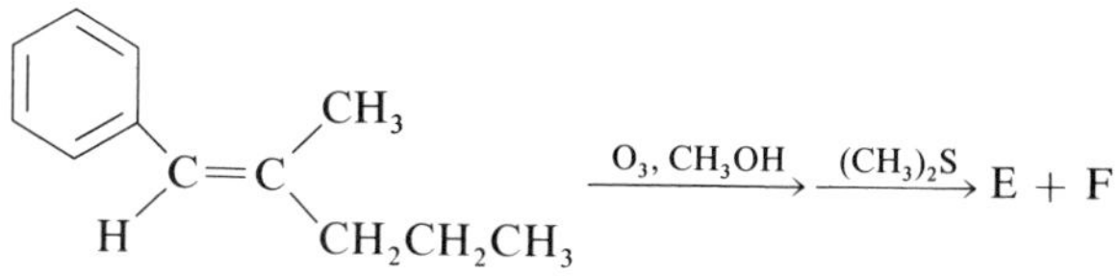
CH3
CH2CH2CH3
O3, CH3OH
(CH3)2S
E + F

(e) 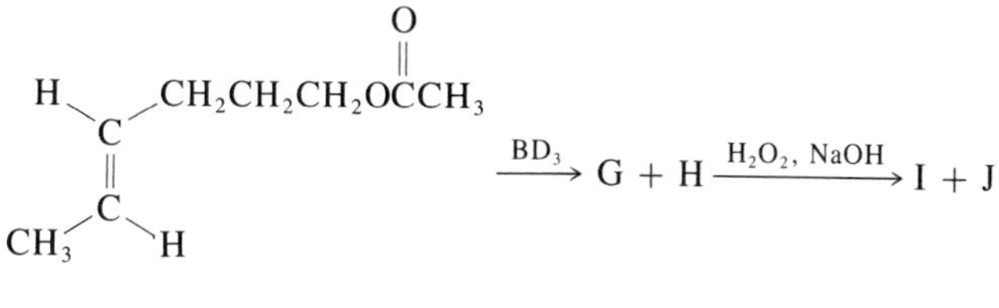
CH2CH2CH2OCCH3
BD3
G + H
H2O2, NaOH
I + J

(f) 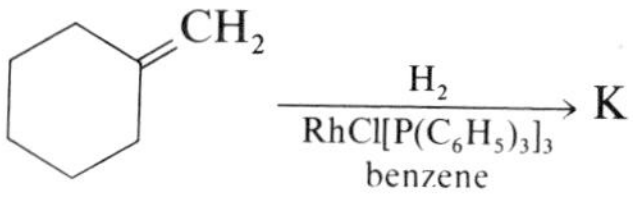
CH2
H2
RhCl[P(C6H5)3]3
benzene
K

(g) 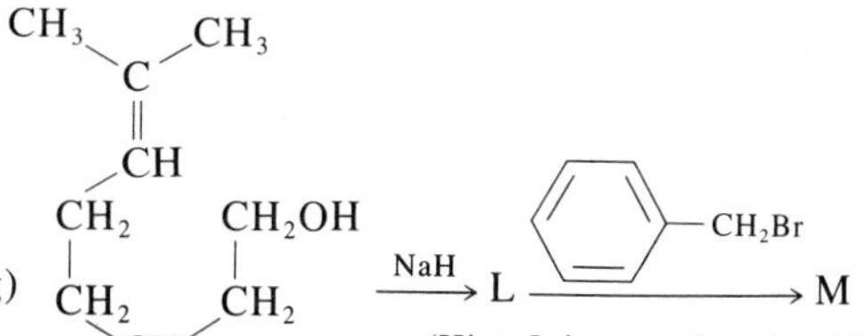
CH2OH
NaH
L
CH2Br
M

(Hint: L is a good nucleophile)

(h) 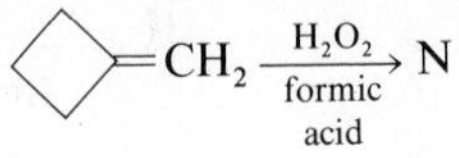

(i) 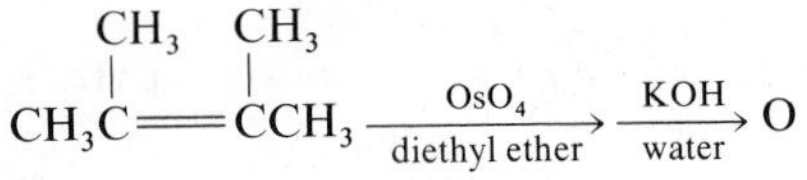

(j) 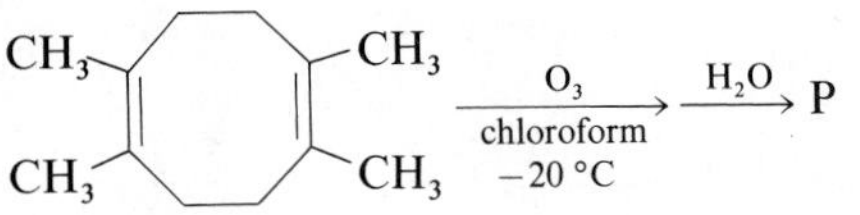

(k) 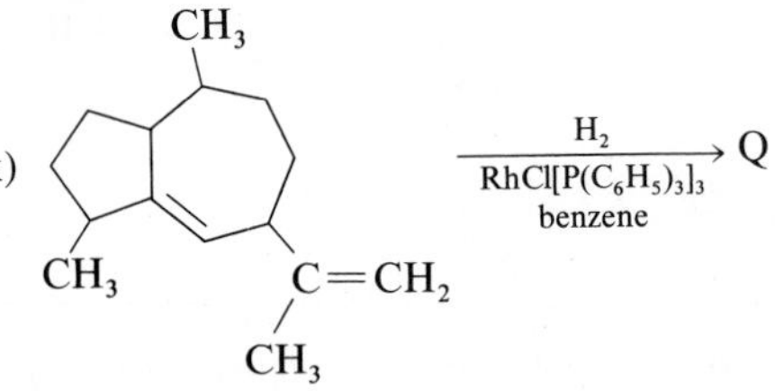

(l) 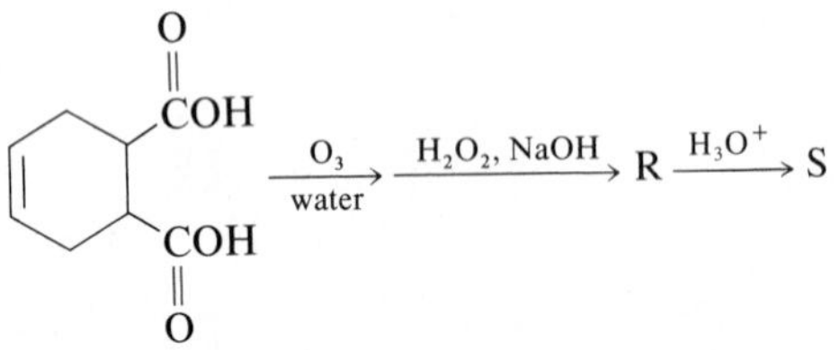

(m) 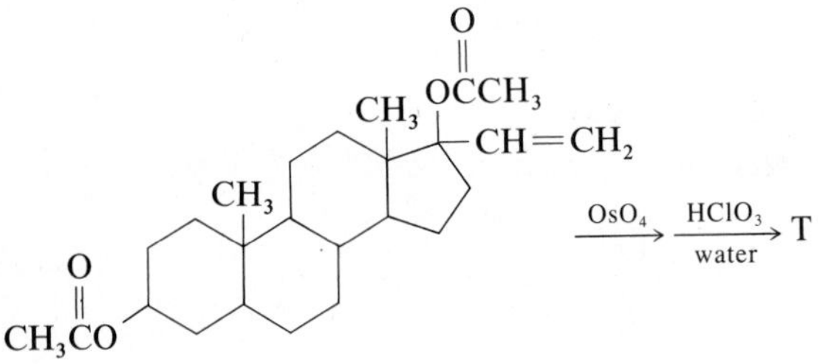

(n) 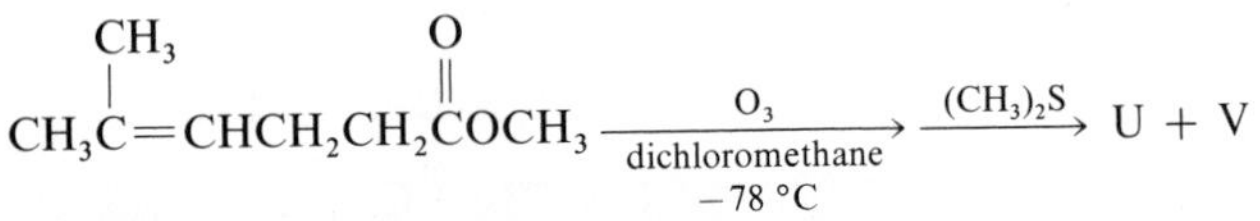

8.34 Fill in structural formulas for the starting material or reagents that would be necessary for the following transformations.

(a) 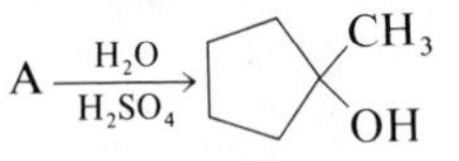

(b) 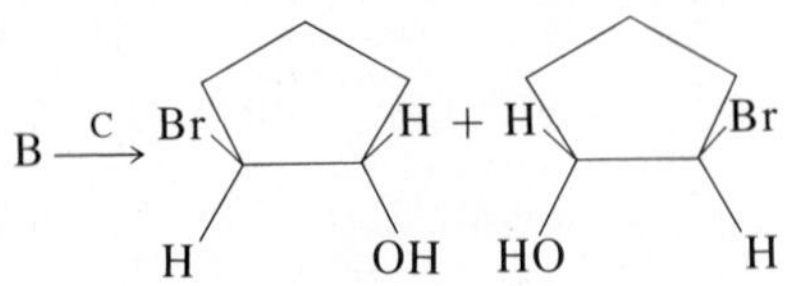

(c) 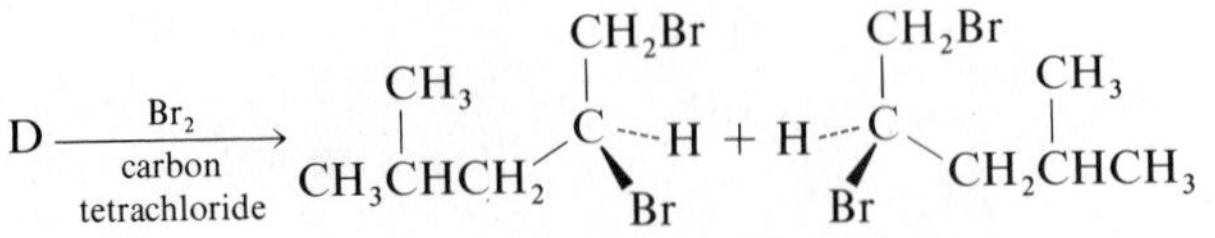

(d) 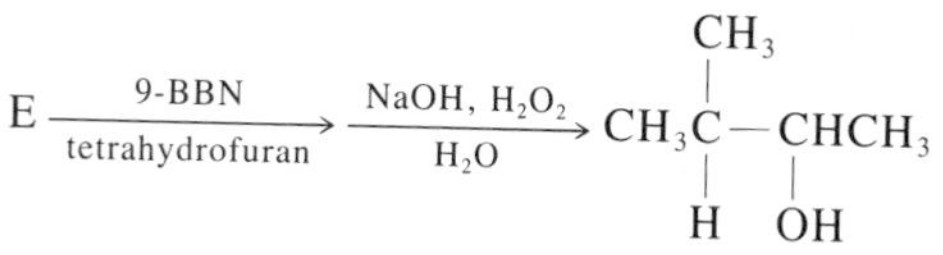
9-BBN
tetrahydrofuran
NaOH, H2O2
H2O

(e) 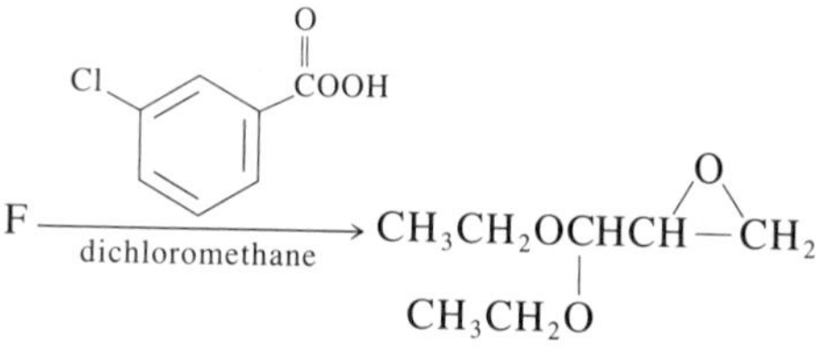
dichloromethane

(f) 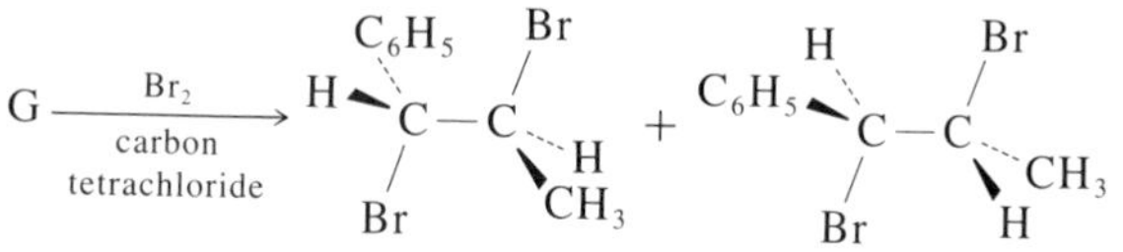
Br2
carbon
tetrachloride

(g) 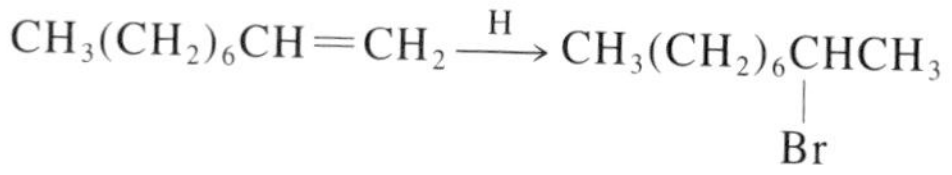

(h) 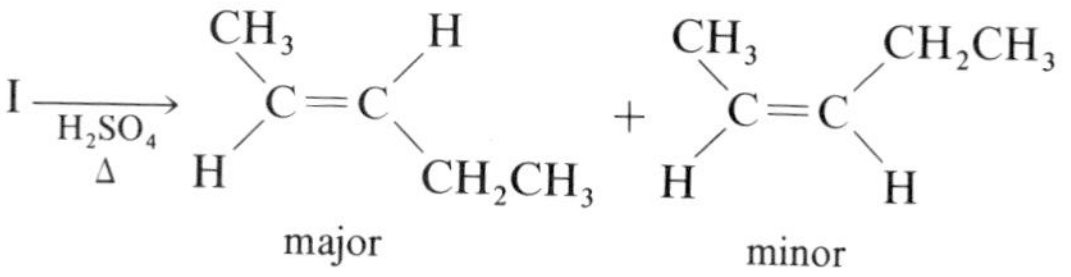
major
minor

(i) 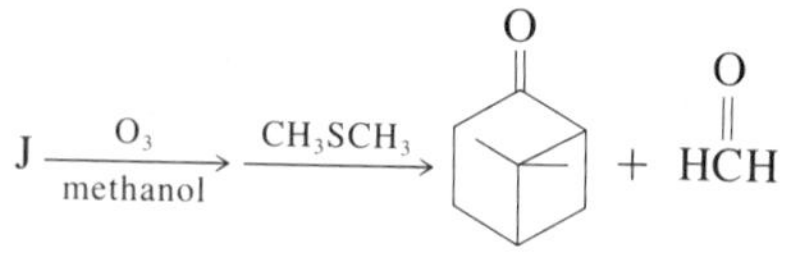
methanol

8.35 When 2,2-dimethylcyclohexanol is treated with acid, 1,2-dimethylcyclohexene and isopropylidenecyclopentane are the products obtained. Draw a detailed mechanism that explains this result.

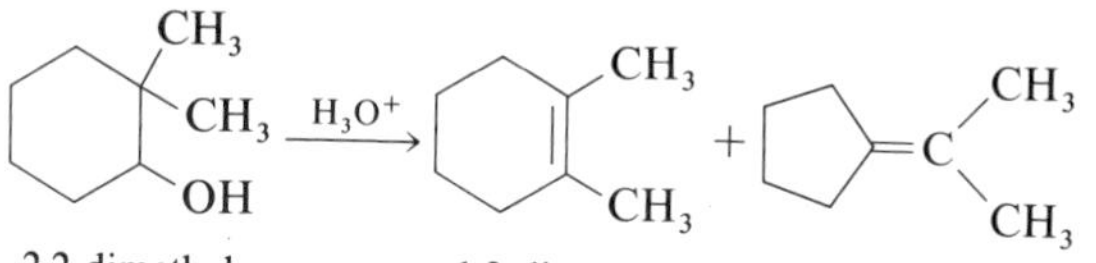

2,2-dimethylcyclohexanol — 1,2-dimethylcyclohexene — isopropylidenecyclopentane

8.36 When 2,2,4-trimethyl-3-pentanol was heated over alumina (which is an acidic reagent), it lost water to give a mixture of alkenes. The alkenes were identified as follows:

2,4,4-trimethyl-2-pentene	24%	2,3,4-trimethyl-2-pentene	18%
2,4,4-trimethyl-1-pentene	24%	2-isopropyl-3-methyl-1-butene	3%
2,3,4-trimethyl-1-pentene	29%	3,3,4-trimethyl-1-pentene	2%

(a) Write structural formulas for the alcohol and the product alkenes.
(b) Write a mechanism that accounts for the formation of the various products.

8.37 The use of lead in gasoline is being phased out because of its harmful environmental effects. Research is being done to find other additives that will improve the octane rating of the fuel. Methyl *tert*-butyl ether is a compound that has the properties being sought. It is synthesized from 2-methylpropene and methanol. Write the structural formula for methyl *tert*-butyl ether, and propose a synthesis for it. Show the mechanism for the reaction you propose.

8.38 Propose a mechanism for the following reaction that accounts for the regioselectivity that is observed.

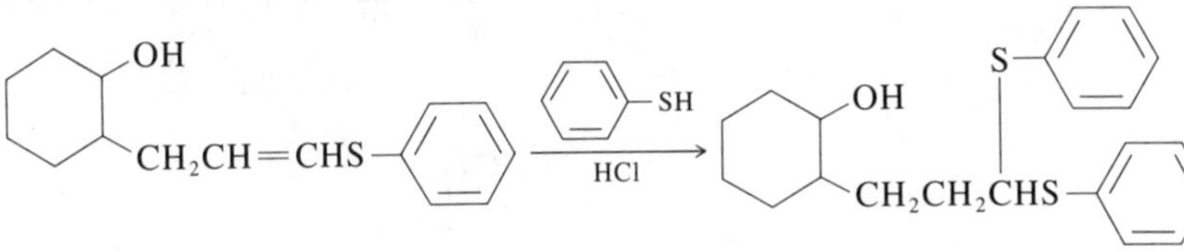
OH
CH₂CH=CHS
SH
HCl
OH
S
CH₂CH₂CHS

8.39 Neopentyl bromide reacts in 50% aqueous ethanol at a high temperature (125 °C) at a rate that is independent of the concentration of added sodium hydroxide. Under these conditions, the major products of the reaction are 2-ethoxy-2-methylbutane and 2-methyl-2-butene.

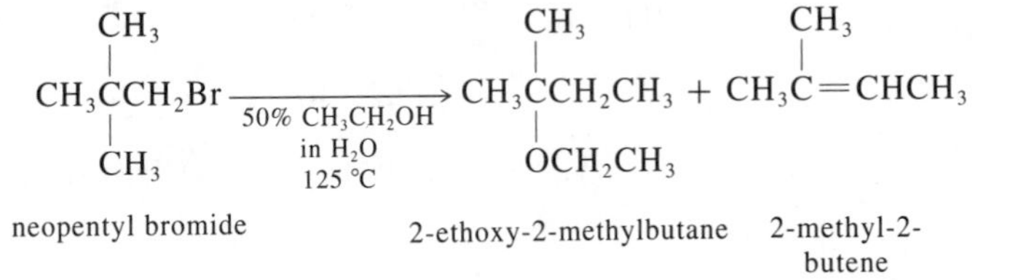
50% CH₃CH₂OH
in H₂O
125 °C
neopentyl bromide
2-ethoxy-2-methylbutane
2-methyl-2-butene

A carbocation has been suggested as the intermediate for this reaction. Write a mechanism for the reaction that explains the observed product.

8.40

(a) The structure of the following bicyclic compound was investigated by subjecting it to hydrogenation and ozonolysis. Give structural formulas for the products of the reactions shown.

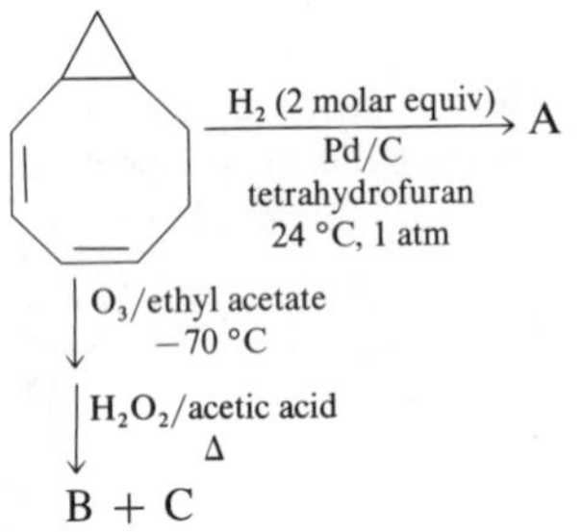
H₂ (2 molar equiv)
Pd/C
tetrahydrofuran
24 °C, 1 atm
A
O₃/ethyl acetate
−70 °C
H₂O₂/acetic acid
Δ
B + C

(b) Another bicyclic compound, Compound D, C_7H_{10}, also gives Compound B after ozonolysis under the same conditions. What is the structure of Compound D?

8.41 Reactions of *cis*-4,5-dimethylcyclohexene and 3-*tert*-butylcyclohexene with *m*-chloroperoxybenzoic acid give predominantly one product in each case. Show the conformation of each starting alkene, and predict, by showing the reaction with the peroxyacid, what the structure of the expected product should be. (Hint: Using molecular models will help.)

8.42 A number of different oxidizing agents have been used to recycle osmium tetroxide in reactions that convert alkenes to diols (p. 316). One of the most successful was developed by by Barry Sharpless at the Massachusetts Institute of Technology. In this reaction, *tert*-butyl hydroperoxide, $(CH_3)_3COOH$, in the presence of the base tetraethylammonium hydroxide, $(CH_3CH_2)_4N^+OH^-$, is used as the oxidizing agent. The osmate esters are cleaved by sodium bisulfite, $NaHSO_3$, and water. For each of the following alkenes, predict what the major products of these reactions will be: 2,3-dimethyl-2-butene, 1-decene, (*E*)-4-octene, and (*Z*)-4-octene. Include stereochemistry where pertinent.

8.43 Compound A, which is a degradation product of the antibiotic vermiculine (p. 608), has the following structure.

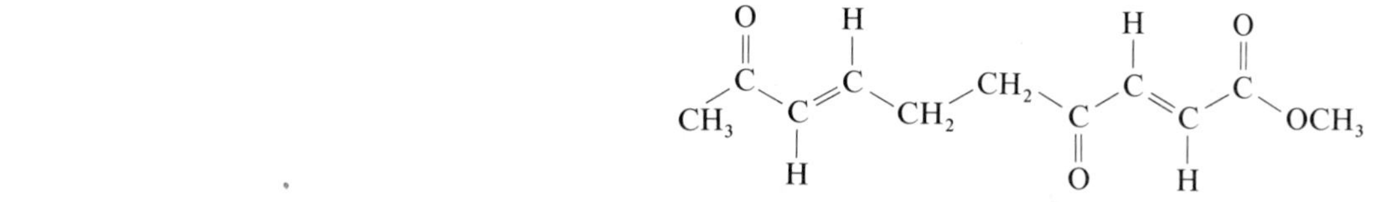

Compound A

The structure was confirmed by converting A to Compound B, $C_{11}H_{18}O_4$, which was also prepared by ozonolysis of Compound C, $C_{11}H_{18}O_2$.

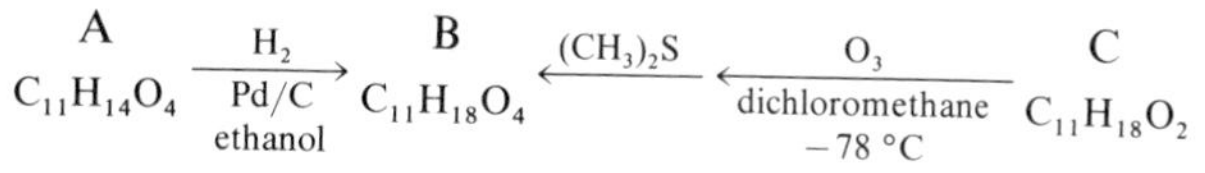
A
C11H14O4
H2
Pd/C
ethanol
B
C11H18O4
(CH3)2S
O3
dichloromethane
−78 °C
C
C11H18O2

Assign structures to Compounds B and C.

9

Alkynes

A • L O O K • A H E A D

Alkynes undergo addition reactions at the triple bond.

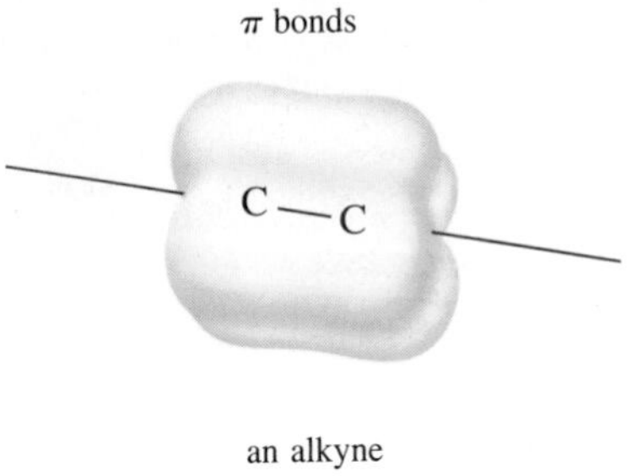

π bonds
C—C
an alkyne

The π bonds in alkynes react with the same electrophilic reagents that the π bonds in alkenes react with. Alkynes, however, are able to add two molar equivalents of a reagent such as hydrogen bromide, hydrogen, or bromine.

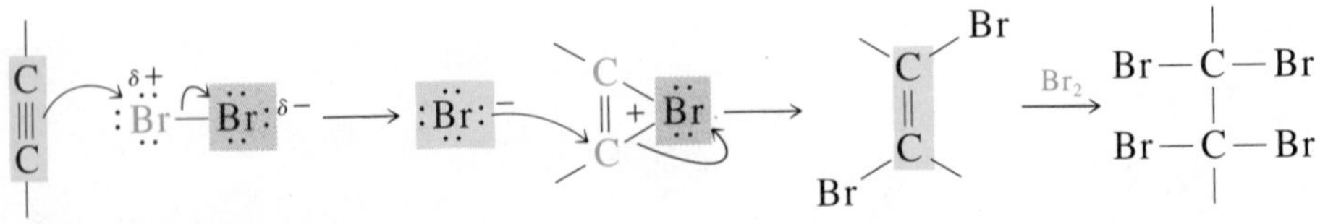

Alkynes differ from alkenes in that the hydrogen atom bonded to an *sp*-hybridized atom in an alkyne is acidic enough to be removed by a strong base, such as amide ion. The conjugate base of the alkyne is a good nucleophile and is useful in creating carbon-carbon bonds.

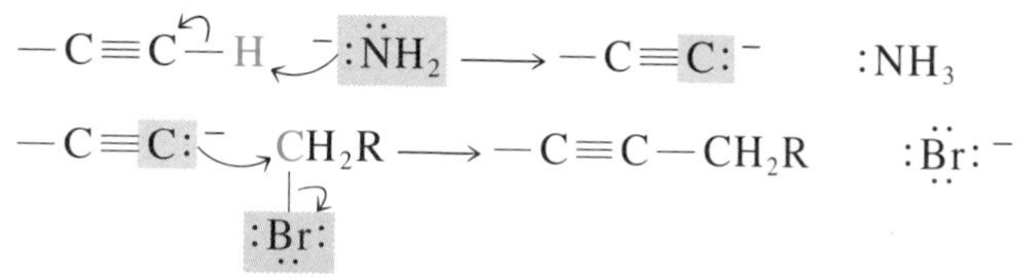

$$-C\equiv C-H \quad :\ddot{N}H_2 \longrightarrow -C\equiv C:^- \quad :NH_3$$

$$-C\equiv C:^- \quad CH_2R \longrightarrow -C\equiv C-CH_2R \quad :\ddot{B}r:^-$$
$$:\ddot{B}r:$$

In this chapter, you will study alkynes and their reactions, which in some ways are similar to and in other ways differ from the reactions of alkenes. You will also learn how to construct more complex molecules from simpler ones.

9.1
STRUCTURE AND ISOMERISM OF ALKYNES

Alkynes contain a carbon-carbon triple bond consisting of a σ bond and two π bonds between two *sp*-hybridized carbon atoms (p. 53). Because an alkyne group is linear, the only isomerism shown for those with the same formula is due to the position of the triple bond in the carbon chain.

$$CH_3CH_2CH_2C\equiv CH \qquad CH_3CH_2C\equiv CCH_3$$

1-pentyne	2-pentyne
bp 40 °C	bp 56 °C
a terminal alkyne	*an internal alkyne*

In a **terminal alkyne,** the triple bond is at the end of the carbon chain. A terminal alkyne has a hydrogen atom bonded to an *sp*-hybridized carbon atom, a structural feature that is important in its chemistry (p. 332). In an **internal alkyne,** the triple bond is at least one carbon atom away from the end of the carbon chain. There is no hydrogen atom bonded to either of the carbon atoms joined by the triple bond in an internal alkyne.

9.2
NOMENCLATURE OF ALKYNES

To name an alkyne, the **ane** ending of the name of the alkane corresponding to the longest continuous chain of carbon atoms that contains the triple bond is changed to **yne.** Thus, the IUPAC name for the simplest alkyne is ethyne (commonly known as acetylene). For more complex alkynes, the chain is numbered so that the carbon atoms involved in the triple bond have the lowest possible numbers. The position of the triple bond is indicated by the lower of the two numbers for the triply bonded carbon atoms. If substituents are present on the chain, they are named and their positions also indicated by numbers.

$$HC\equiv CH \qquad \overset{6}{C}H_3\overset{5}{C}H_2\overset{4}{C}\equiv \overset{3}{C}\overset{2}{C}H_2\overset{1}{C}H_3$$

ethyne
acetylene

3-hexyne

$$\begin{array}{c} CH_3 \\ | \\ \overset{5}{C}H_3\overset{4}{C}\overset{3}{C}\equiv \overset{2}{C}\overset{1}{C}H_3 \\ | \\ CH_3 \end{array}$$

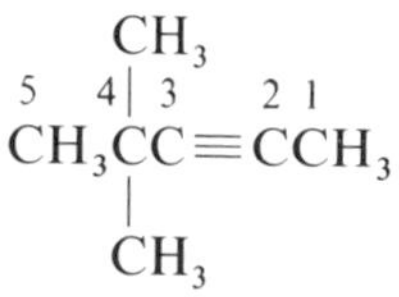

4,4-dimethyl-2-pentyne

PROBLEM 9.1

Name each of the following compounds.

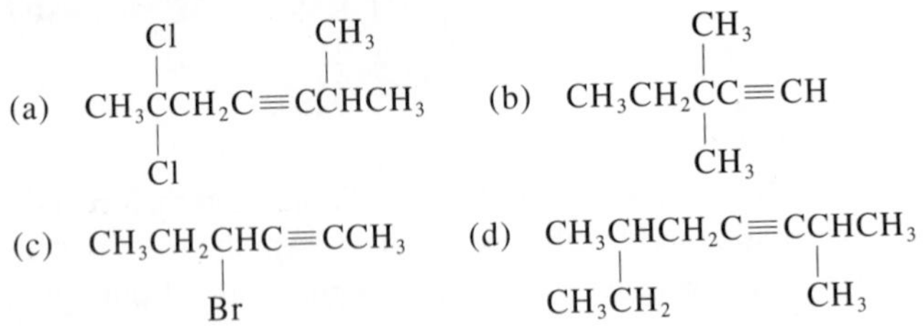

PROBLEM 9.2

Draw a structural formula for each of the following compounds.

(a) 2-chloro-2-methyl-3-heptyne (b) 5-phenyl-2-octyne (c) 2,2-dimethyl-3-hexyne

9.3 ALKYNES AS ACIDS

A. The Preparation of Sodium Acetylide, a Source of Nucleophilic Carbon

Anions derived from terminal alkynes are nucleophiles (p. 261). These anions are prepared by the deprotonation of terminal alkynes by strong bases. An inspection of the pK_a table shows that terminal alkynes have pK_a values of ~26. This means they can be deprotonated completely by bases such as amide anion, the conjugate base of ammonia, which has pK_a 36 (p. 91). Acetylene, for example, reacts with sodium amide, prepared by putting sodium metal in liquid ammonia, to form sodium acetylide.

$$\underset{\text{sodium metal}}{2\,Na} + \underset{\text{ammonia}}{2\,NH_3} \longrightarrow \underset{\text{sodium amide}}{2\,Na^+NH_2^-} + \underset{\text{hydrogen}}{H_2\uparrow}$$

$$\underset{\substack{\text{acetylene}\\ \text{acid}}}{HC\equiv CH} + \underset{\substack{\text{sodium}\\ \text{amide}\\ \text{base}}}{Na^+:\ddot{N}H_2^-} \xrightarrow[\text{−33 °C}]{NH_3(liq)} \underset{\substack{\text{sodium}\\ \text{acetylide}}}{HC\equiv CNa} + NH_3$$

The anion from a terminal alkyne is a very strong nucleophile and base and must be protected from water or other acids that will protonate it.

$$\underset{\text{base}}{HC\equiv C:^-\,Na^+} + \underset{\substack{\text{acid}\\ \text{p}K_a\ 15.7}}{H-\ddot{\underset{..}{O}}-H} \rightleftarrows \underset{\text{p}K_a\ 26}{HC\equiv CH} + Na^+\ {}^-:\ddot{\underset{..}{O}}-H$$

In other words, water is a much stronger acid than a terminal alkyne.

B. Reaction of Nucleophilic Acetylide Anions with Alkyl Halides

A nucleophile reacts at an electrophilic center that is a carbon atom bearing a good leaving group to create a new σ bond (pp. 260–262). Reagents that contain nucleophilic carbon atoms are used in nucleophilic substitution reactions to create new carbon-carbon bonds. Carbon-carbon bonds are formed by reactions between a carbon atom that is a nucleophilic center and a carbon atom that is an electrophilic center.

The reaction of nucleophilic acetylide anions with alkyl halides is a widely used way of synthesizing compounds that contain terminal triple bonds, for example:

$$\underset{\text{1-bromobutane}}{CH_3CH_2CH_2CH_2Br} + \underset{\text{sodium acetylide}}{HC{\equiv}C\!:^-\ Na^+} \xrightarrow[NH_3(liq)]{} \underset{\text{1-hexyne}}{CH_3CH_2CH_2CH_2C{\equiv}CH} + \underset{\text{sodium bromide}}{Na^+Br^-}$$

The acetylide anion is a strong base, so only primary alkyl halides can be used in this reaction. With secondary or tertiary alkyl halides, E_2 reactions (p. 255) lead to alkenes as the major products.

A terminal alkyne can be converted into its anion and used to make a larger internal alkyne. (Z)-9-Tricosene is a substance that enables the female housefly to attract the male housefly. (Chemical substances used by insects and animals to communicate with each other are known as pheromones; see p. 701). 9-Tricosyne, an intermediate in the synthesis of (Z)-9-tricosene, is made from 1-pentadecyne and 1-chlorooctane in two steps.

$$\underset{\text{1-pentadecyne}}{CH_3(CH_2)_{12}C{\equiv}CH} + \underset{\text{lithium amide}}{LiNH_2} \xrightarrow[\text{hexamethylphosphoric triamide}]{\text{tetrahydrofuran}} \underset{\text{conjugate base of 1-pentadecyne}}{CH_3(CH_2)_{12}C{\equiv}C\!:^-\ Li^+} + \underset{\text{ammonia}}{NH_3}$$

$$CH_3(CH_2)_{12}C{\equiv}C\!:^-\ Li^+ + \underset{\text{1-chlorooctane}}{CH_3(CH_2)_6CH_2Cl} \longrightarrow \underset{\substack{\text{9-tricosyne} \\ 54\%}}{CH_3(CH_2)_{12}C{\equiv}CCH_2(CH_2)_6CH_3}$$

Study Guide Concept Map 9.1

This synthesis is an example of the general method by which a terminal alkyne is converted into its anion, which is then used to make an internal alkyne. The conversion of 9-tricosyne to (Z)-9-tricosene is shown on p. 342.

PROBLEM 9.3

Complete the following equations by supplying structural formulas for the products and reagents symbolized by the letters A–G.

(a) $CH_3CH_2CH_2C{\equiv}CH \xrightarrow[NH_3(liq)]{NaNH_2} A \xrightarrow{CH_3CH_2CH_2Br} B$

(b) $C \xrightarrow[NH_3(liq)]{D} CH_3\overset{\overset{\large CH_3}{|}}{C}HCH_2CH_2C{\equiv}CH$

(c) $CH_3CH_2CH_2C{\equiv}CH \xrightarrow[NH_3(liq)]{E} F \xrightarrow{(CH_3)_2SO_4} G$

[Hint: What is the leaving group in $(CH_3)_2SO_4$?]

9.4 ELECTROPHILIC ADDITION REACTIONS OF ALKYNES

A. Addition of Hydrogen Halides

The ionic addition of hydrogen bromide to an alkyne is similar to its addition to an alkene, except that two equivalents of hydrogen bromide are added before a saturated compound is formed. The overall reaction is regioselective, and the product is the one that is predicted by Markovnikov's Rule. 1-Hexyne, for example, adds hydrogen bromide twice to give 2,2-dibromohexane.

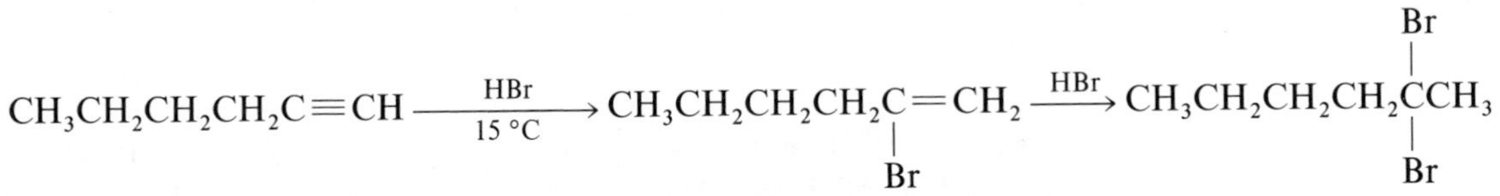

$CH_3CH_2CH_2CH_2C{\equiv}CH$
HBr
15 °C
$CH_3CH_2CH_2CH_2C{=}CH_2$
Br
HBr
$CH_3CH_2CH_2CH_2CCH_3$
Br
Br

1-hexyne 2-bromohexene 2,2-dibromohexane

The reaction goes by way of a vinyl cation formed by electrophilic attack on the triple bond.

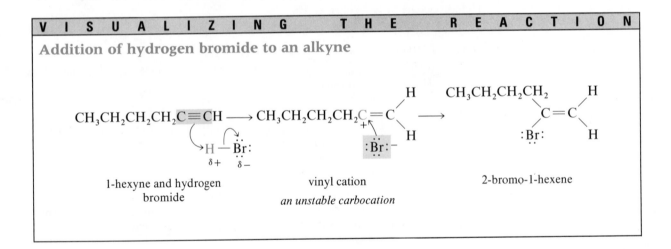

V I S U A L I Z I N G T H E R E A C T I O N

Addition of hydrogen bromide to an alkyne

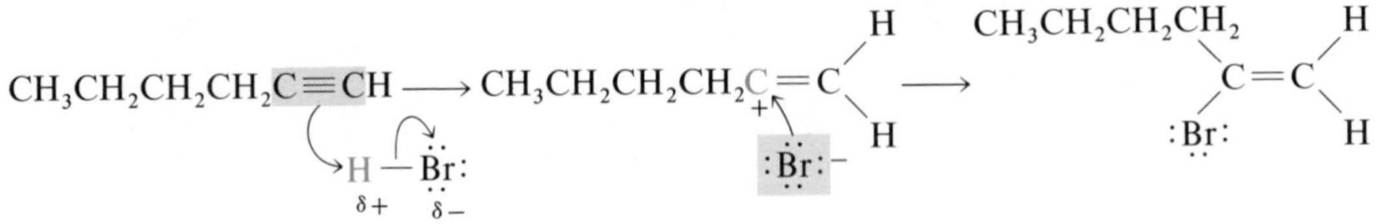

$CH_3CH_2CH_2CH_2C{\equiv}CH$
H—Br:
δ+ δ−
$CH_3CH_2CH_2CH_2C{=}C$
+
H
H
:Br:−
$CH_3CH_2CH_2CH_2$
H
C=C
:Br:
H

1-hexyne and hydrogen bromide

vinyl cation
an unstable carbocation

2-bromo-1-hexene

The electron-deficient carbon atom of the vinyl cation is bonded to two other atoms and may be considered to be *sp*-hybridized. It is more electronegative (p. 56) than the sp^2-hybridized carbon atom found in an alkyl cation, and thus less able to bear a positive charge. Vinyl cations are intermediate in energy between methyl and ethyl (or *n*-propyl) cations (p. 122) and less stable than secondary or tertiary cations. However, an alkyne is a higher-energy molecule than an alkene, so the energy of activation in going from an alkyne to a vinyl cation is not much larger than the energy of activation for the formation of an alkyl cation from an alkene. A π bond in an alkyne is, therefore, about as reactive toward acids as the π bond in an alkene is.

2-Bromo-1-hexene is less reactive toward electrophilic addition than a similar alkene without a halogen substituent would be. The halogen atom on the double bond is electron-withdrawing and decreases the availability of the π electrons for reaction with a second molecule of hydrogen bromide. The orientation of the second addition reaction is also interesting. Two intermediates are possible, as seen below.

VISUALIZING THE REACTION

Addition of hydrogen bromide to a vinyl halide

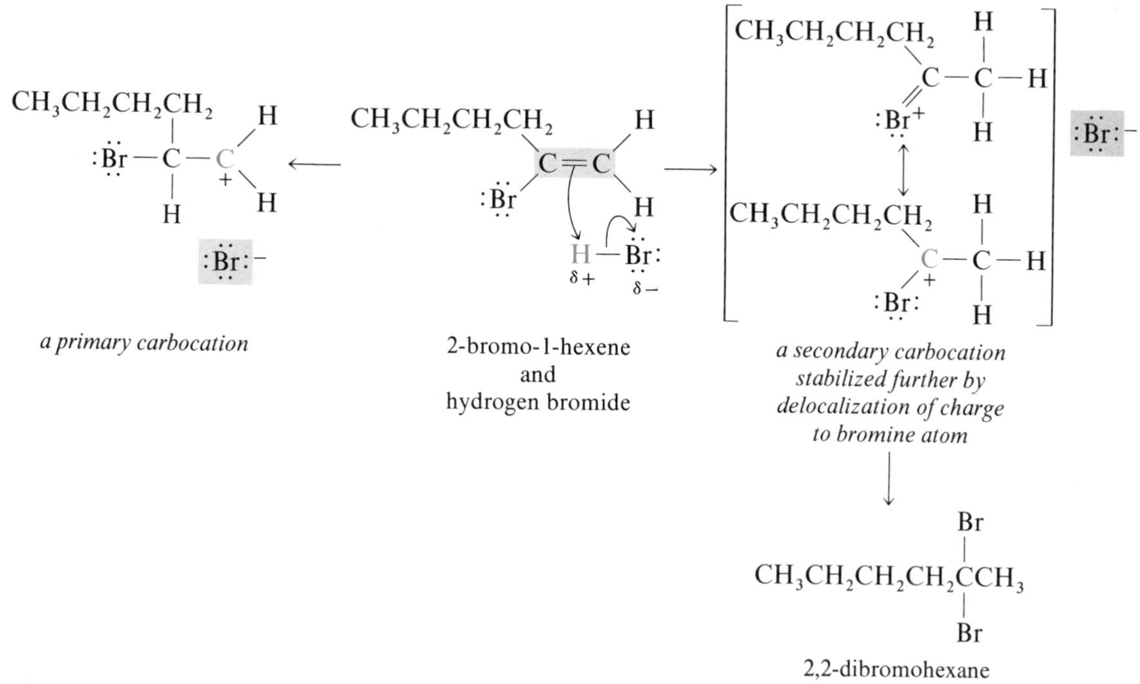

a primary carbocation
2-bromo-1-hexene
and
hydrogen bromide
a secondary carbocation
stabilized further by
delocalization of charge
to bromine atom
2,2-dibromohexane

One is a primary carbocation, and the other is a secondary carbocation with an electron-withdrawing bromine atom at the cationic center. The halogen, though electron-withdrawing in its inductive effect, does stabilize the carbocation by some delocalization of charge to the bromine atom by resonance. The product that is from the addition of hydrogen bromide to 2-bromo-1-hexene is the product derived from the reaction of bromide ion with the more stable of the two possible intermediates.

Resonance stabilization of a cation by a halogen atom is even more clearly present in the addition of hydrogen chloride to acetylene.

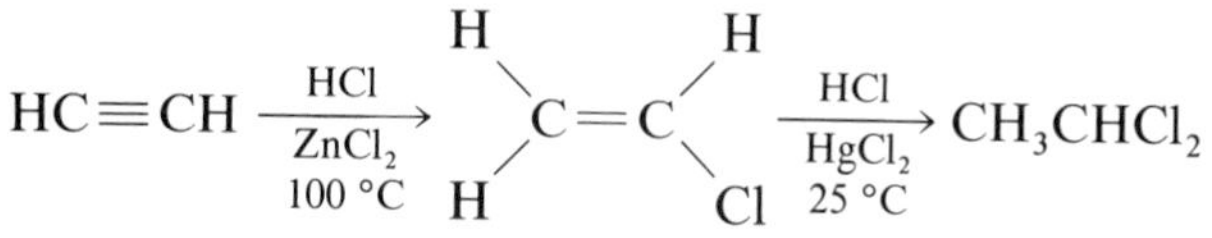

acetylene   vinyl chloride   1,1-dichloroethane

VISUALIZING THE REACTION

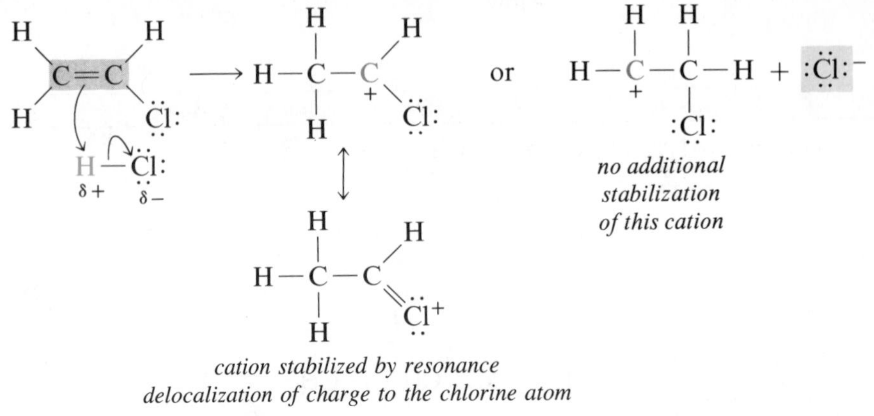
or
no additional stabilization of this cation
cation stabilized by resonance delocalization of charge to the chlorine atom

Both possible intermediates are primary cations, and in one case a chlorine atom with its electron-withdrawing inductive effect would be expected to further destabilize the cation. The observed product can be rationalized by predicting resonance stabilization of one of the cations by the chlorine atom. Metal salts such as zinc(II) chloride and mercury(II) chloride, which act as Lewis acids, are often used to catalyze the reactions of hydrogen chloride with alkynes and vinyl halides.

PROBLEM 9.4

Complete the following equations.

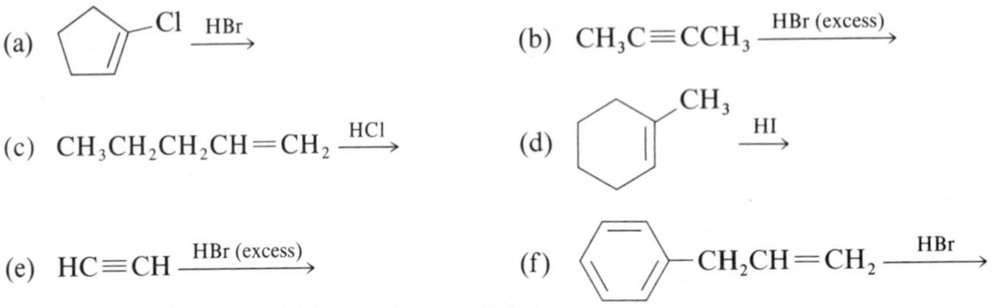
(a)
Cl
HBr
(b) $CH_3C{\equiv}CCH_3 \xrightarrow{\text{HBr (excess)}}$
(c) $CH_3CH_2CH_2CH{=}CH_2 \xrightarrow{\text{HCl}}$
(d)
CH_3
HI
(e) $HC{\equiv}CH \xrightarrow{\text{HBr (excess)}}$
(f)
$CH_2CH{=}CH_2$
HBr

B. Addition of Water to Alkynes

Water adds to acetylene in the presence of acid and mercury and iron salts to give acetaldehyde. This reaction is important industrially.

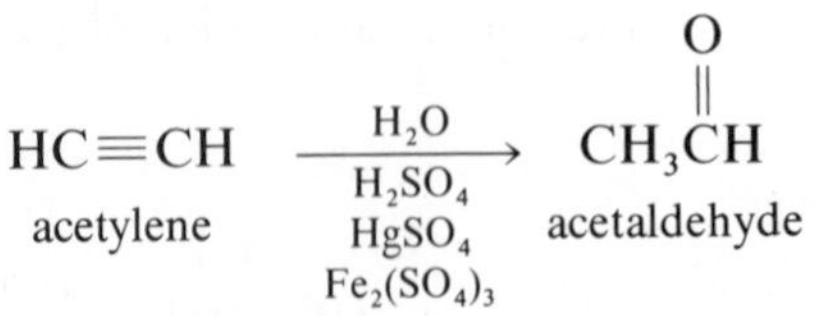
$HC{\equiv}CH$
acetylene
H_2O
H_2SO_4
$HgSO_4$
$Fe_2(SO_4)_3$
CH_3CH
O
acetaldehyde

All other alkynes give ketones. For example, 1-hexyne is converted into 2-hexanone.

$$\underset{\text{1-hexyne}}{CH_3CH_2CH_2CH_2C{\equiv}CH} \xrightarrow[\substack{H_2SO_4 \\ HgSO_4}]{H_2O} \underset{\substack{\text{2-hexanone} \\ 80\%}}{CH_3CH_2CH_2CH_2\overset{\overset{O}{\|}}{C}CH_3}$$

The hydration of an alkyne proceeds by electrophilic attack on the π electrons of the triple bond. Mercury(II) ion, Hg^{2+}, is a Lewis acid and serves as a catalyst in the hydration reaction. Alkynes that are substituted on both sides of the triple bond, such as 2-pentyne, are more reactive and can be hydrated without the use of a catalyst.

$$\underset{\text{2-pentyne}}{CH_3C{\equiv}CCH_2CH_3} \xrightarrow[\substack{H_2SO_4 \\ 0\ ^\circ C,\ 10\ \text{min}}]{H_2O} \underset{\substack{\text{3-pentanone} \\ \sim 50\%}}{CH_3CH_2\overset{\overset{O}{\|}}{C}CH_2CH_3} + \underset{\substack{\text{2-pentanone} \\ \sim 50\%}}{CH_3CH_2CH_2\overset{\overset{O}{\|}}{C}CH_3}$$

In summary, hydration of an alkyne gives rise to a ketone, except in the case of acetylene, where acetaldehyde is the product.

PROBLEM 9.5

9-Undecynoic acid, shown below, is hydrated in 80% sulfuric acid. Write an equation for the reaction.

$$CH_3C{\equiv}C(CH_2)_7\overset{\overset{O}{\|}}{C}OH$$

C. Problem-Solving Skills

Some of the mechanisms presented in this book, such as the ones for the S_N2 and S_N1 reactions, are supported by large amounts of experimental data, including careful studies of kinetics and stereochemistry. For many of the mechanisms presented in the features called "Visualizing the Reaction," however, all of the details are not rigorously supported by experimental data. These mechanisms represent attempts made by organic chemists to rationalize the transformations of the reactants into the products under the conditions of the reaction. Organic chemists do this by reasoning by analogy from other better-known and more thoroughly researched reactions. The kinds of questions chemists ask when proposing the mechanism for a reaction are a combination of those used in solving synthesis and transformation problems. The following problem illustrates how to figure out a mechanism.

Problem

Write a detailed mechanism for the addition of water to 2-butyne.

$$CH_3C{\equiv}CCH_3 \xrightarrow[H_2SO_4]{H_2O} CH_3\overset{\overset{O}{\|}}{C}CH_2CH_3$$

Solution

1. To what functional group class does the reactant belong? What is the electronic character of the functional group?

   The reactant is an alkyne. The triple bond has a cloud of π electrons, which are nucleophilic.

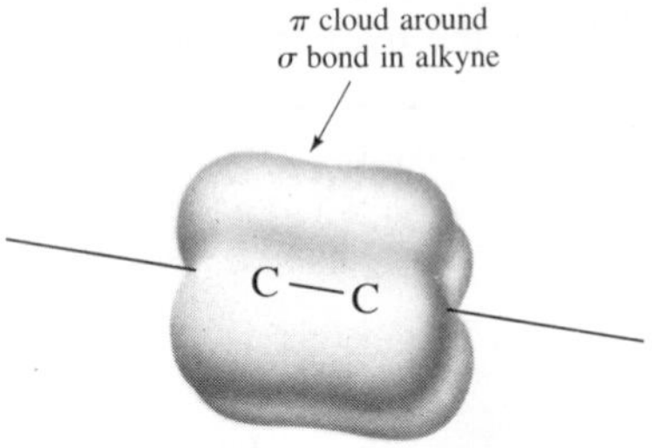
π cloud around
σ bond in alkyne
C—C

2. How do the structures of the reactant and the product compare? How many carbon atoms does each contain? What bonds must be broken and formed to transform reactant into product?

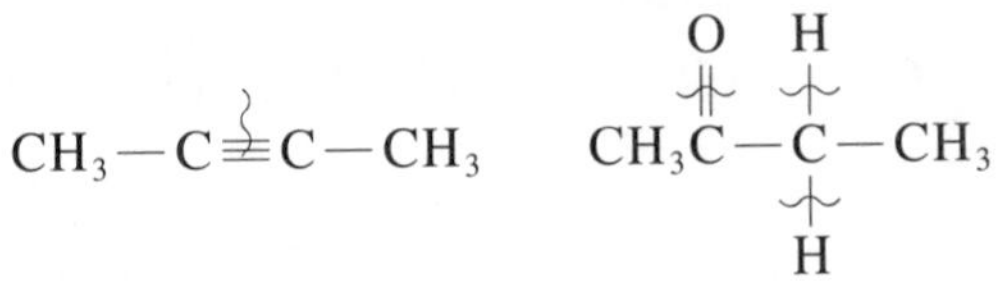

bonds broken *bonds formed*

   The reactant and the product have the same number of carbon atoms. The π bonds in the alkyne must be broken. A double bond between carbon and oxygen and two carbon-hydrogen bonds must be formed.

3. What reagents are present? Are they good acids, bases, nucleophiles, or electrophiles?

   H_2SO_4 is a strong acid. In the presence of a strong acid, H_2O behaves as a weak base. It is also a nucleophile. The mixture of water and sulfuric acid contains the hydronium ion, H_3O^+.

4. What is the most likely first step for the reaction: protonation or deprotonation, ionization, attack by a nucleophile, or attack by an electrophile?

   The π bond will be protonated.

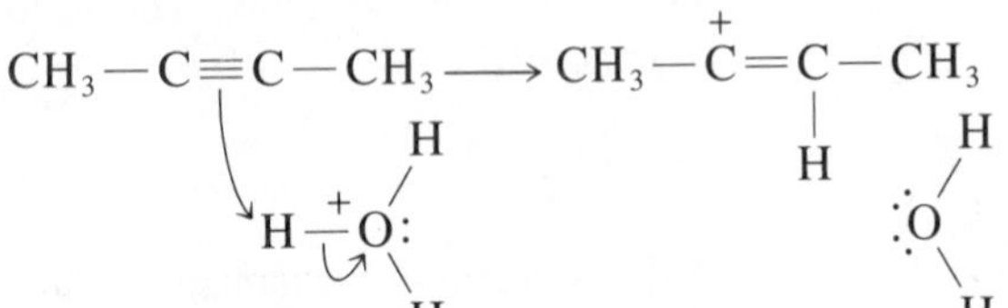

5. What are the properties of the species present in the reaction mixture after the first step? What is likely to happen next?

   The vinyl cation is an electrophile. It will react with the nucleophile, water.

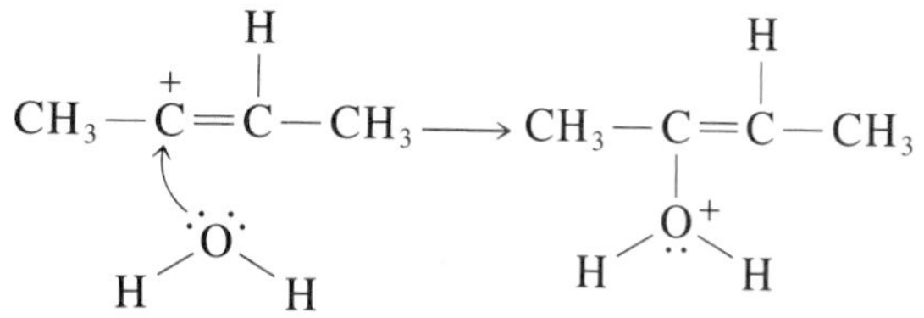

5. (repeated) What are the properties of the species present in the reaction mixture now? What is likely to happen next?

The species resulting from the preceding step is an oxonium ion, an acid. It will be deprotonated by the solvent.

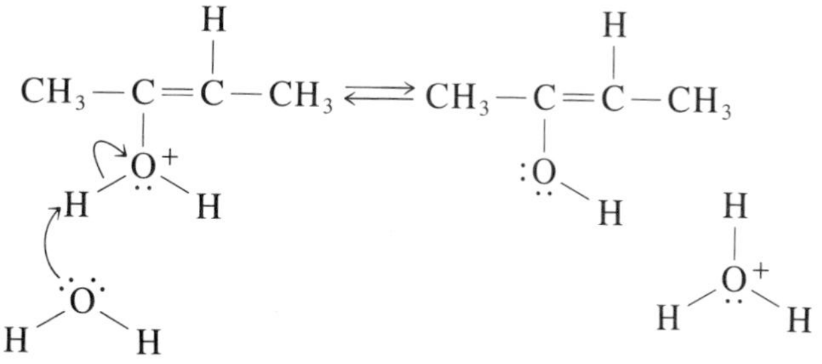

Comparing this species to the starting material and to the product, as shown below, reveals that it has a carbon-oxygen bond and a carbon-hydrogen bond at the right locations—on the carbon atoms that were originally part of the triple bond. This intermediate is called an **enol,** indicating that it contains a hydroxyl group (ol) on a double bond (ene).

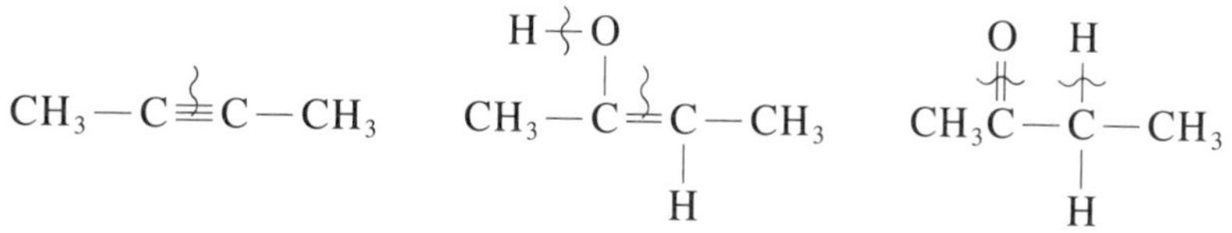

an enol

To get from the enol to the product, a hydrogen must be added at the double bond, a hydrogen must be removed from oxygen, and the double bond must be shifted.

5. (repeated) What are the properties of the species present in the reaction mixture now? What is likely to happen next?

The enol will be protonated in the acidic solution. Two sites of protonation are available.

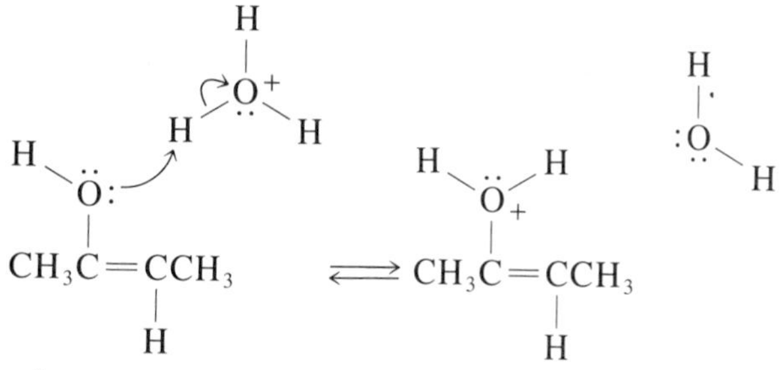

gain of a proton at oxygen

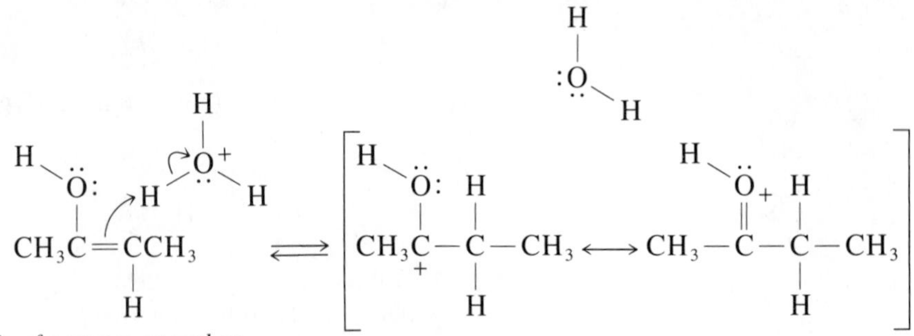

gain of a proton at carbon

Protonation at the oxygen is the reverse of the deprotonation reaction of the preceding step. It occurs, but does not further the overall reaction that gives the desired product. Protonation of the double bond in an enol is particularly easy because the positive charge in the resulting carbocation is adjacent to the oxygen atom and stabilized by resonance. In one of the resonance contributors for this cation, carbon and oxygen each have an octet of electrons, and the positive charge is delocalized to the oxygen atom. A close inspection of this resonance-stabilized cation reveals that it is a protonated ketone. Loss of the proton to a base gives 2-butanone.

The complete correct answer is as follows:

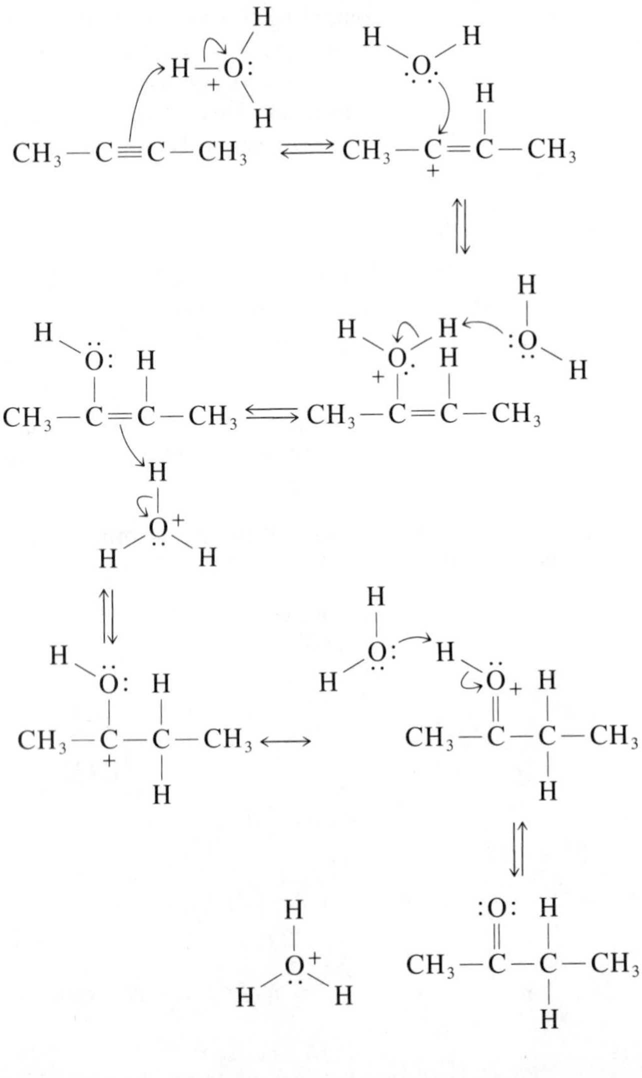

D. Tautomerism

The conversion of an enol to a ketone by protonation at the carbon atom of the double bond and deprotonation at the oxygen atom is known as **tautomerization.** The ketone and its enol form are examples of **tautomers,** readily interconvertible constitutional isomers that exist in equilibrium with each other. Isomers that differ from each other only in the location of a hydrogen atom and a double bond are proton tautomers. Proton tautomers are isomers in which a hydrogen atom and a double bond switch locations between a carbon atom and a **heteroatom,** which is an atom other than carbon, such as oxygen or nitrogen. Tautomers differ from each other in the locations of atoms as well as of electrons. Therefore, they are not resonance contributors, which are different representations of the same structure (p. 11).

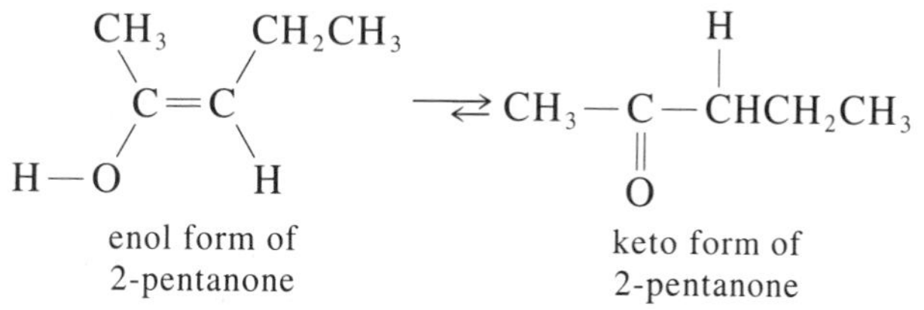
enol form of
2-pentanone
keto form of
2-pentanone

tautomers of each other

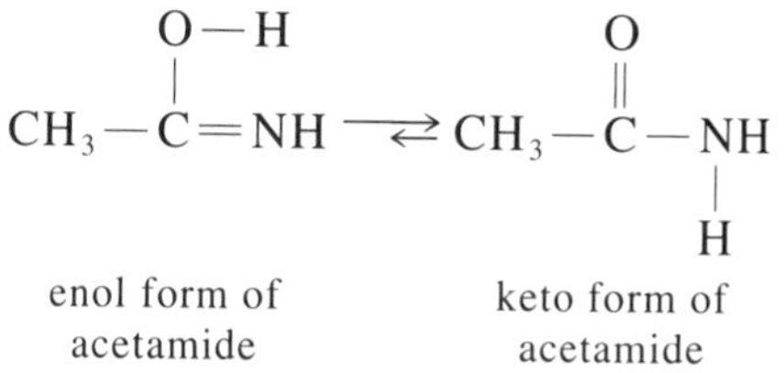
enol form of
acetamide
keto form of
acetamide

tautomers of each other

Study Guide Concept Map 9.2

When keto-enol tautomerization occurs, the keto form, the one in which a carbonyl group is present, is usually the more stable and predominates at equilibrium.

PROBLEM 9.6

Write a detailed mechanism for the conversion of 2-pentyne to 3-pentanone.

PROBLEM 9.7

Write a detailed mechanism for the formation of acetaldehyde from acetylene.

9.5 REDUCTION OF ALKYNES

A. Catalytic Hydrogenation of Alkynes

Addition of 1 molar equivalent of hydrogen to an alkyne gives an alkene. The catalysts that are used in such hydrogenation reactions are active toward alkynes but not alkenes, so the reaction stops when 1 equivalent of hydrogen has reacted with the triple bond. One such catalyst, known as a **poisoned catalyst,** is palladium on barium sulfate or calcium carbonate to which a small amount of lead or the organic base quinoline has been added to make the catalyst less reactive. For example, phenylethyne can be converted to phenylethene (styrene) by the use of such a catalyst.

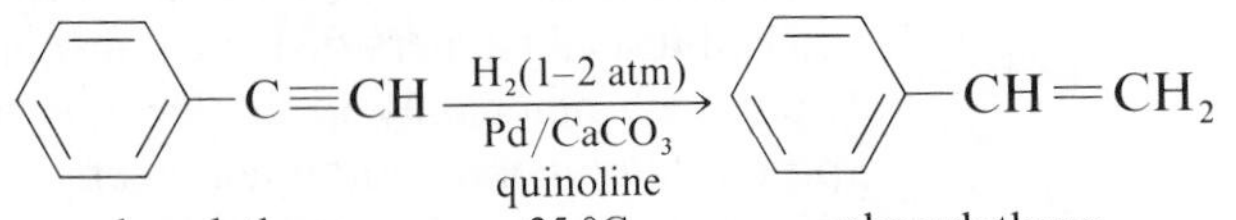
C≡CH
H2(1–2 atm)
Pd/CaCO3
quinoline
25 °C
CH=CH2

phenylethyne — phenylethene, styrene

quinoline ≡ 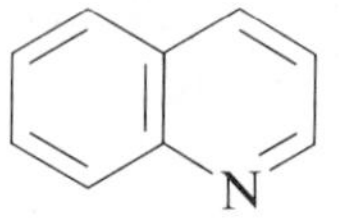
N

The reaction stops when 1 equivalent of hydrogen has been used up. The alkene is not reduced to an alkane on this catalyst, even though such a double bond would normally undergo that reaction (p. 306, for example). Note that the aromatic ring is not affected under these conditions. High temperatures (~100 °C) and high pressures (up to 100 atm) are usually required to add hydrogen to aromatic rings.

Various experiments with different metallic catalysts have shown that catalytic hydrogenation of internal alkynes gives predominantly cis alkenes. For example, 9-tricosyne (p. 333) is converted to (Z)-9-tricosene, the sex pheromone of the housefly, in this way.

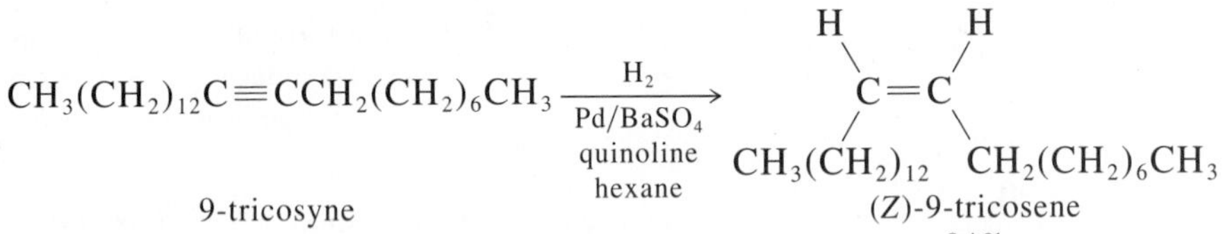
CH3(CH2)12C≡CCH2(CH2)6CH3
H2
Pd/BaSO4
quinoline
hexane
H
H
C=C
CH3(CH2)12
CH2(CH2)6CH3

9-tricosyne — (Z)-9-tricosene, 84%

The reaction stops cleanly when exactly 1 equivalent of hydrogen has reacted. The stereochemistry that results from catalytic reductions of alkynes indicates that syn addition occurs. Hydrogen atoms are added at the surface of the catalyst to the same side of the alkyne bond.

When the catalysts that are used to hydrogenate alkenes (pp. 304–310) are used, alkynes add 2 equivalents of hydrogen to give alkanes. For example, tricosane, another pheromone, is prepared by hydrogenating 10-tricosyne over platinum.

$$\underset{\text{10-tricosyne}}{CH_3(CH_2)_{11}C{\equiv}CCH_2(CH_2)_7CH_3} \xrightarrow[\text{hexane}]{\text{H}_2\text{ (excess)}\;/\;\text{PtO}_2} \underset{\text{tricosane}}{CH_3(CH_2)_{21}CH_3}$$

PROBLEM 9.8

10-Tricosyne has been prepared starting with 1-tetradecyne. How would you synthesize 10-tricosyne?

B. Reduction of Alkynes to Alkenes by Dissolving Metals

Alkynes are also reduced by sodium or lithium metal in liquid ammonia. In contrast to catalytic hydrogenation, this reaction produces trans alkenes. For example, 3-octyne is reduced to (*E*)-3-octene with sodium metal in ammonia.

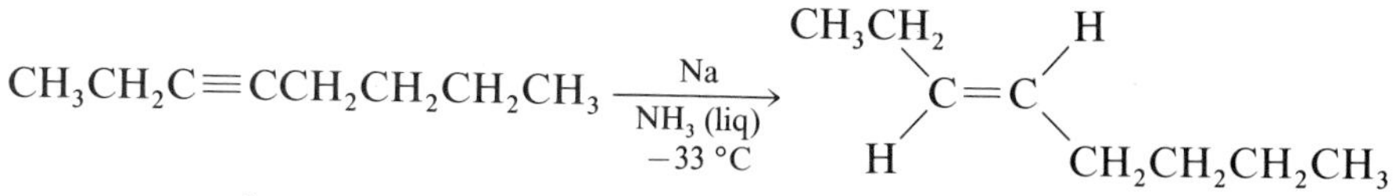

Na
NH3 (liq)
−33 °C

3-octyne (*E*)-3-octene 95%

This reaction occurs by a transfer of an electron from sodium metal to the alkyne. The intermediate formed is a **radical anion,** a species that bears a negative charge and has an unpaired electron. Such a strongly basic species is protonated by ammonia. The radical that is formed is reduced again by sodium metal and then protonated to give the most stable alkene (p. 278), having the *E* configuration.

VISUALIZING THE REACTION

Reduction of an alkyne by sodium in ammonia

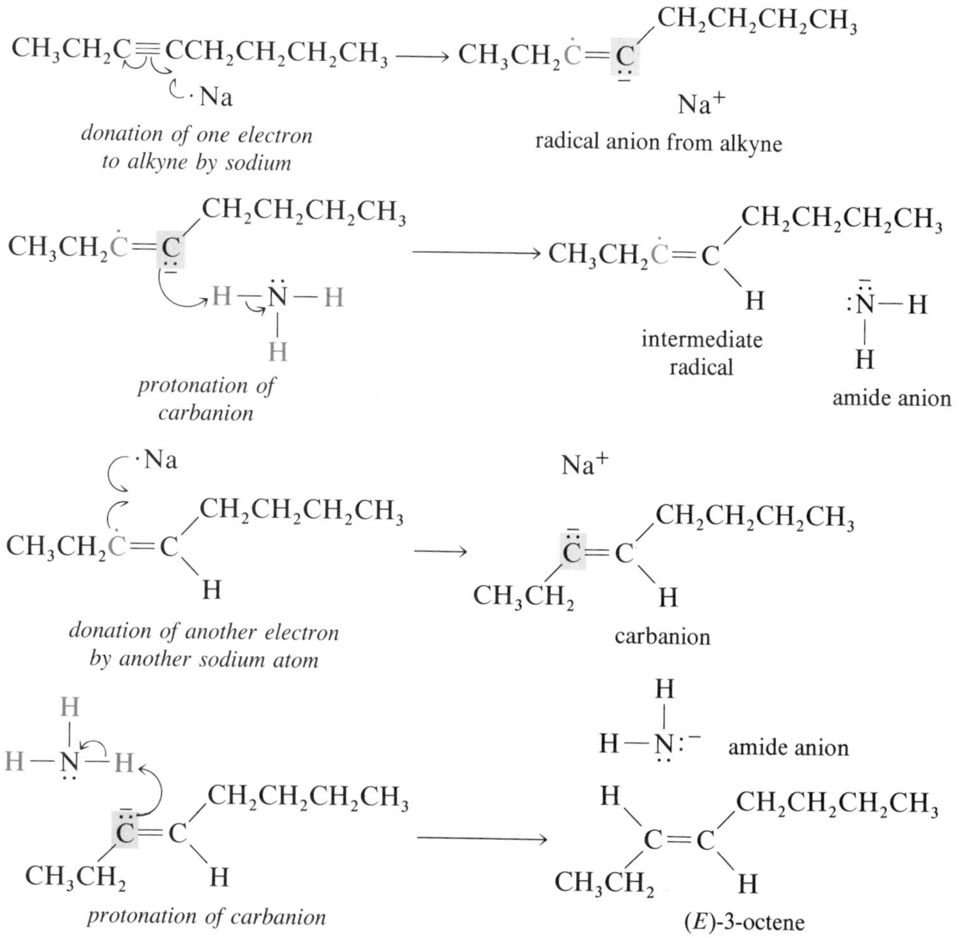

donation of one electron to alkyne by sodium
radical anion from alkyne
protonation of carbanion
intermediate radical
amide anion
donation of another electron by another sodium atom
carbanion
amide anion
protonation of carbanion
(E)-3-octene

In the above mechanism, a new type of arrow, called a **fishhook**, $\frown$, is used. A fishhook represents the motion of a single electron, such as the donation of one electron to the alkyne by sodium. By contrast, a regular curved arrow, $\frown$, is used to indicate the movements of pairs of electrons. The protonation of a carbanion, for example, is represented with regular arrows.

In contrast to the result of the preceding reaction, when 3-octyne is hydrogenated on a metal surface, the product is predominantly (Z)-3-octene.

$$CH_3CH_2C\equiv CCH_2CH_2CH_2CH_3 \xrightarrow[\text{Ni}]{H_2 \text{ (1 molar equivalent)}}$$

$$\begin{array}{cc} CH_3CH_2 & CH_2CH_2CH_2CH_3 \\ \diagdown \quad / \\ C = C \\ / \quad \diagdown \\ H & H \end{array}$$

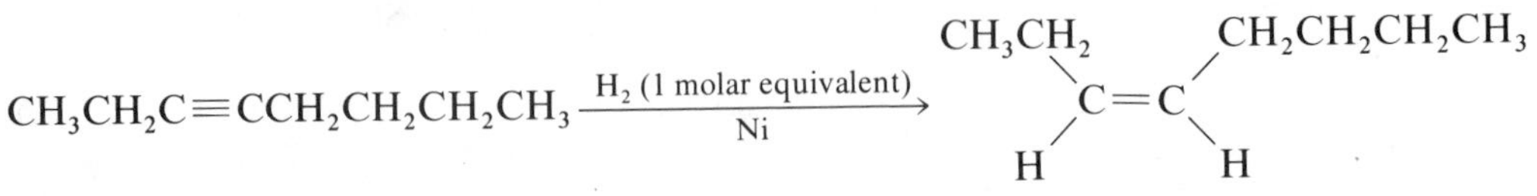

3-octyne

(Z)-3-octene
98% this isomer

Study Guide
Concept Map 3

The equations in this section demonstrate that alkynes can be converted with stereoselectivity (p. 292) to either cis or trans alkenes, depending on the reagents used.

PROBLEM 9.9

Complete the following equations. Be sure to show the stereochemistry of the product where it is pertinent.

(a) $CH_3CH_2CH_2CH_2C\equiv CH \xrightarrow[\substack{Pd/BaSO_4 \\ \text{quinoline}}]{H_2}$

(b) $CH_3CH_2CH_2CH_2C\equiv CCH_3 \xrightarrow[\substack{Pd/BaSO_4 \\ \text{quinoline}}]{H_2}$

(c) $CH_3CH_2CH_2CH_2C\equiv CCH_3 \xrightarrow[\substack{NH_3 \text{ (liq)}}]{Na}$

(d) $CH_3CH_2CH_2C\equiv CCH_3 \xrightarrow[RhCl[P(C_6H_5)_3]_3]{H_2}$

(e) $CH_3CH_2CH_2CH_2C\equiv CCH_3 \xrightarrow[\substack{Pt \\ \text{acetic acid}}]{H_2 \text{ (excess)}}$

PROBLEM 9.10

When 2-butyne in the gas phase is treated with deuterium gas (D_2), with palladium on an alumina support as the catalyst, one alkene is obtained in 99% yield. Predict the structure of the alkene.

9.6
OZONOLYSIS OF ALKYNES

Alkynes react with ozone to give carboxylic acids. Ozonolysis of 1-hexyne, for example, gives formic acid and pentanoic acid.

$$CH_3CH_2CH_2CH_2C\equiv CH \xrightarrow[\substack{\text{carbon} \\ \text{tetrachloride} \\ 0\,°C}]{O_3} \xrightarrow{H_2O} \overset{O}{\overset{\|}{CH_3CH_2CH_2CH_2COH}} + \overset{O}{\overset{\|}{HCOH}}$$

1-hexyne

pentanoic acid

formic acid

In another example, 9-octadecynedioic acid is converted to two molecules of nonanedioic acid by ozonolysis.

$$\mathrm{HO\overset{\displaystyle O}{\overset{\|}{C}}(CH_2)_7C{\equiv}C(CH_2)_7\overset{\displaystyle O}{\overset{\|}{C}}OH \xrightarrow[\substack{\text{acetic}\\ \text{acid}\\ 25\ ^\circ\text{C}}]{O_3} \xrightarrow{H_2O} 2\ HO\overset{\displaystyle O}{\overset{\|}{C}}(CH_2)_7\overset{\displaystyle O}{\overset{\|}{C}}OH}$$

9-octadecynedioic acid → nonanedioic acid 81%

Thus, ozonolysis breaks the triple bond in an alkyne to give fragments that have carboxylic acid groups at the carbon atoms that were originally joined by the multiple bond.

PROBLEM 9.11

Complete the following equations.

(a) $$\mathrm{CH_3\overset{\displaystyle CH_3}{\overset{|}{C}}HC{\equiv}CCH_2CH_3 \xrightarrow[\substack{\text{carbon}\\ \text{tetrachloride}}]{O_3} \xrightarrow{H_2O}}$$

(b) $$\mathrm{CH_3C{\equiv}C\overset{\displaystyle CH_3}{\overset{|}{C}}HCH_2CH_3 \xrightarrow[\substack{\text{carbon}\\ \text{tetrachloride}}]{O_3} \xrightarrow{H_2O}}$$

9.7 PLANNING SYNTHESES

Organic chemists are constantly developing new methods for synthesizing compounds having industrial, medicinal, or biological importance. For example, modifications of natural hormones and antibiotics are synthesized in attempts to understand how the natural substances function and to create more effective ones. The design of a synthesis of a complex molecule having several functional groups and a specific stereochemistry is an intellectual exercise that appeals to chemists. The actual synthesis of such a compound in the laboratory is also challenging and requires skill and care.

To design a synthesis, a chemist has to know what reactions can be used to give different types of functional groups. The reactions that create carbon-carbon bonds are among the most important. You know how to synthesize an alkyne with the triple bond in any position (pp. 332–333) and how to convert an alkyne to an alkene with either a cis or trans double bond (pp. 341–344). Other useful reactions you have learned include nucleophilic substitution reactions of halides and electrophilic addition reactions of alkenes. This is a good time to review these types of transformations in Table 7.3 (p. 260) and Table 8.1 (p. 323).

The thinking that goes into the design of syntheses (pp. 128, 265) is reviewed in the two problems that follow.

Problem

How would you synthesize (*E*)- and (*Z*)-3-heptene from acetylene and any other chemicals?

Solution

1. What functional groups are present in the starting material and the product?

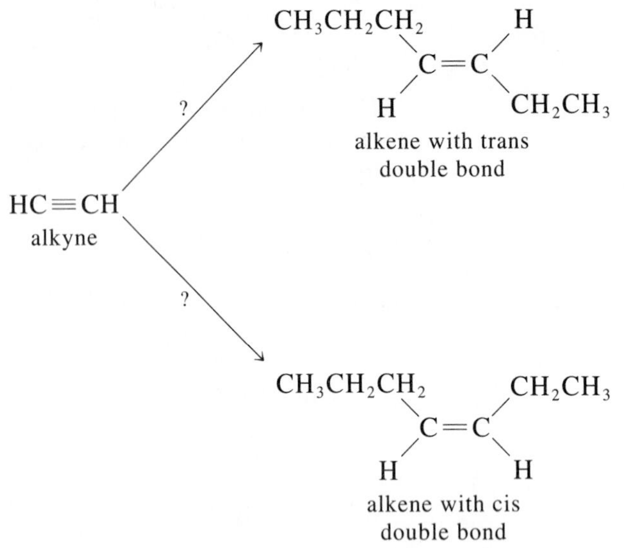
$CH_3CH_2CH_2$
H
C=C
H
CH_2CH_3
alkene with trans double bond
?
HC≡CH
alkyne
?
$CH_3CH_2CH_2$
CH_2CH_3
C=C
H
H
alkene with cis double bond

2. How do the carbon skeletons of the two compounds compare? How many carbon atoms does each contain? Are there any rings? What are the positions of branches and functional groups on the carbon skeletons?

   The starting material, acetylene, has only two carbon atoms. The products each have seven carbon atoms and a double bond in the middle of the chain.

3. How do the functional groups change in going from starting material to product? Does the starting material have a good leaving group?

   Alkyl groups have been added to both ends of the alkyne. The triple bond has been converted to a double bond.

4. Is it possible to dissect the structures of the starting material and product to see which bonds must be broken and which formed?

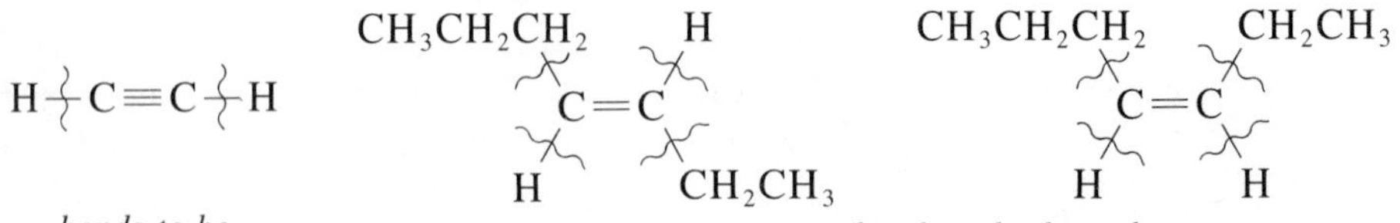
H—C≡C—H
$CH_3CH_2CH_2$
H
C=C
H
CH_2CH_3
$CH_3CH_2CH_2$
CH_2CH_3
C=C
H
H

bonds to be broken to give new C—C bonds

new bonds to be formed

5. Do we recognize any part of the product molecule as coming from a good nucleophile or an electrophilic addition?

   The carbon-carbon bonds must be the result of nucleophilic substitution reactions. Acetylene can be converted into a good nucleophile.

6. What type of compound would be a good precursor to the product?

   The alkyne $CH_3CH_2CH_2C{\equiv}CCH_2CH_3$ can be converted into either the cis or the trans alkene.

7. After this last step, do we see how to get from starting material to product? If not, we need to analyze the structure obtained in step 6 by applying questions 5 and 6 to it.

   We need to repeat the dissection.

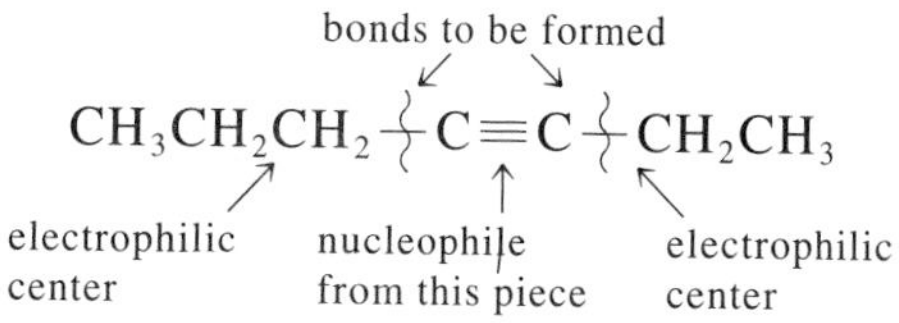
bonds to be formed
electrophilic center
nucleophile from this piece
electrophilic center

   This molecule can be put together from these reagents:

   $$CH_3CH_2CH_2Br \qquad HC{\equiv}CH \qquad BrCH_2CH_3$$

   The complete synthesis is as follows:

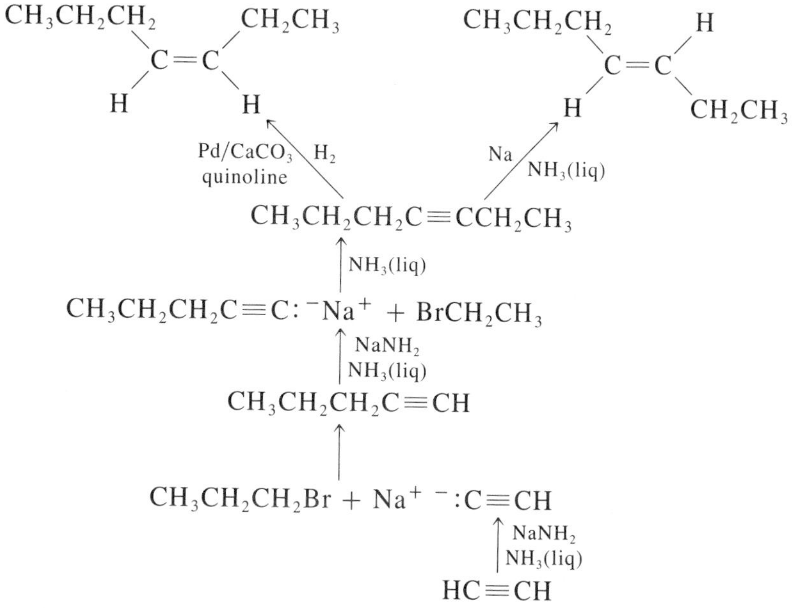
Pd/CaCO3 H2
quinoline
Na
NH3(liq)
NaNH2

Problem

Synthesize 5-methyl-1-hexanol from acetylene, using any other reagents you need.

Solution

1. What functional groups are present in the starting material and the product?

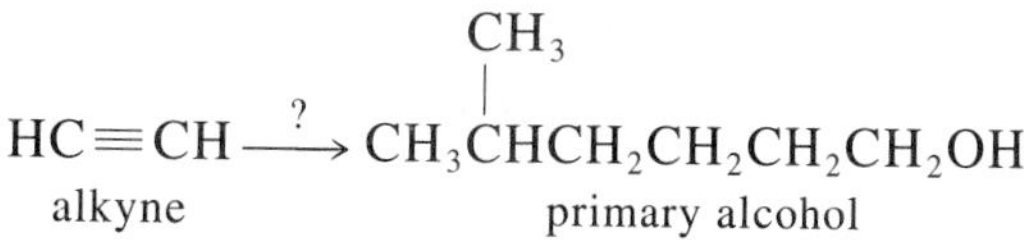
alkyne
primary alcohol

2. How do the carbon skeletons of the two compounds compare? How many carbon atoms does each contain? Are there any rings? What are the positions of branches and functional groups on the carbon skeletons?

   Acetylene has only two carbon atoms. The product has seven and a hydroxyl group at the end of the chain.

3. How do the functional groups change in going from starting material to product? Does the starting material have a good leaving group?

   The triple bond disappears. The rest of the chain must have been added to the alkyne, probably on one side.

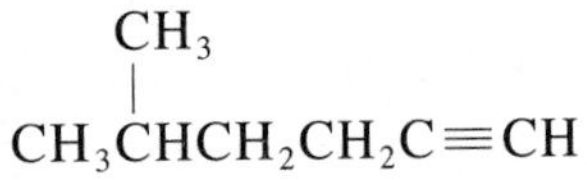

4. Is it possible to dissect the structures of starting material and product to see which bonds must be broken and which formed?

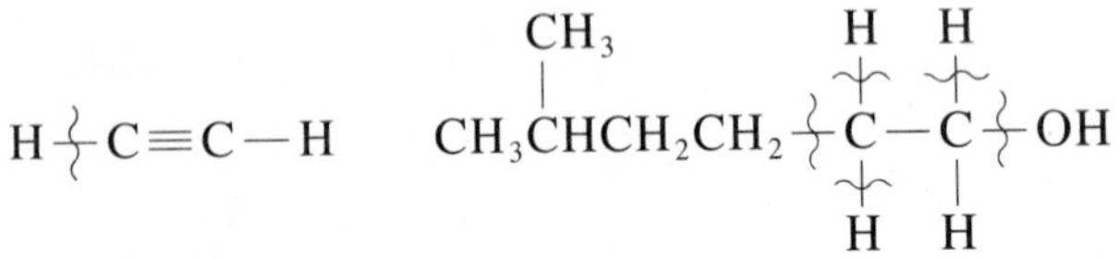

bond broken *new bonds formed*

5. Do we recognize any part of the product molecule as coming from a good nucleophile or an electrophilic addition?

   The carbon-carbon bond must have resulted from a nucleophilic substitution reaction. Acetylene can be converted into a good nucleophile. The hydroxyl group must have come from an electrophilic addition to a multiple bond.

6. What type of compound would be a good precursor to the product?

   Anti-Markovnikov addition of water to a double bond will give a primary alcohol. An alkene is the desired precursor.

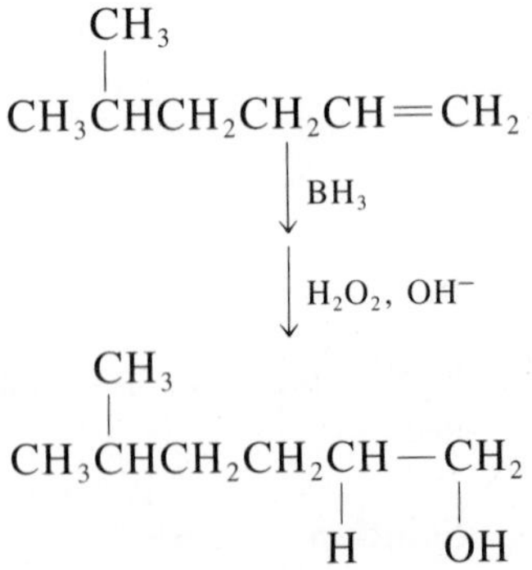

7. After this last step, do we see how to get from starting material to product? If not, we need to analyze the structure obtained in step 6 by applying questions 5 and 6 to it.

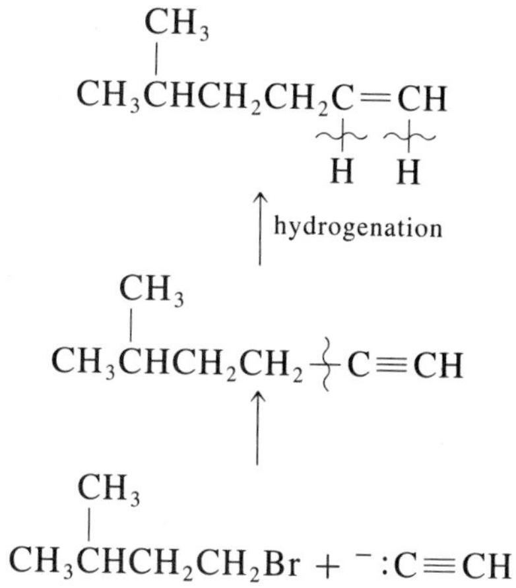

The complete synthesis is as follows:

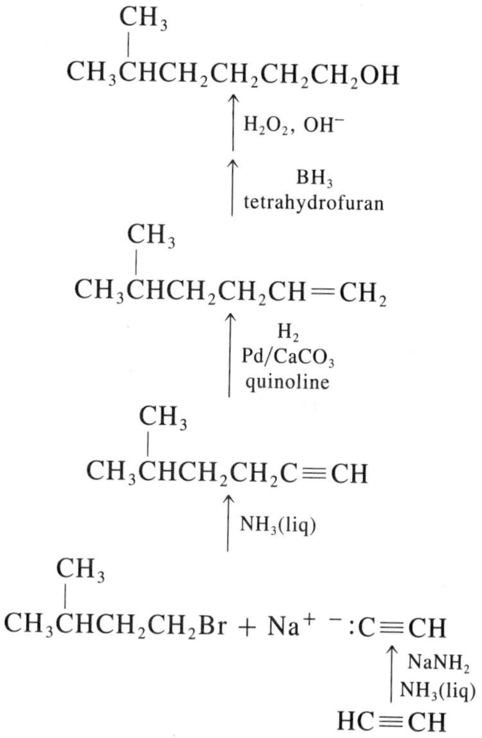

PROBLEM 9.12

Show how you would carry out each of the following transformations.

(a) 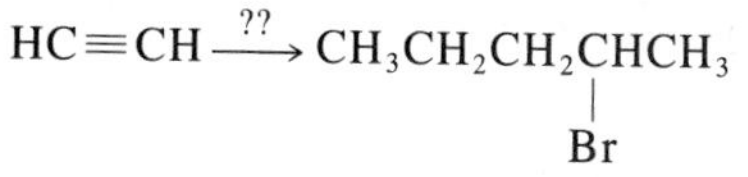

(b) 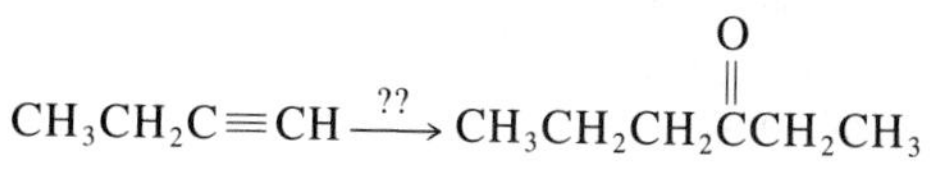

(c) $CH_3CH_2CH_2C{\equiv}CH \xrightarrow{??}$ 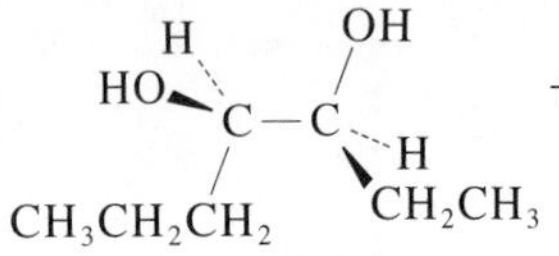 + enantiomer

S U M M A R Y

Alkynes undergo three important types of reactions. Like alkenes, all alkynes undergo electrophilic addition reactions with halogens, acids, and oxidizing reagents. When unsymmetrical reagents are used, the relative stabilities of the possible cationic intermediates determine the regioselectivity of the addition reaction. The addition of water to an alkyne gives an enol, which is a tautomer of the product, a ketone. Tautomers are easily interconvertible constitutional isomers that exist in equilibrium with each other.

Alkynes add 2 equivalents of hydrogen to give alkanes. Alkynes add 1 equivalent of hydrogen in the presence of a poisoned catalyst to give alkenes; the reduction of alkynes by sodium in ammonia also gives alkenes. When internal alkynes are reduced, catalytic hydrogenation produces the *Z* isomer, and reduction with sodium gives the *E* isomer.

Terminal alkynes are deprotonated by strong bases, giving nucleophilic carbanions that can react with alkyl halides to form new carbon-carbon bonds. This reaction can be used to produce more complex alkynes.

The reactions of alkynes are outlined in Table 9.1.

ADDITIONAL PROBLEMS

9.13 Name the following compounds, specifying the stereochemistry where indicated.

(a) 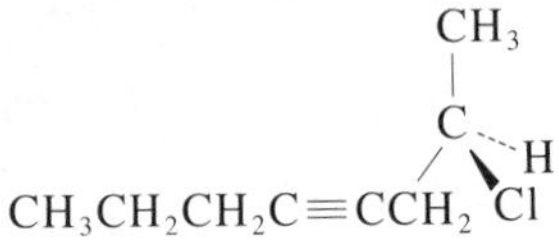

(b) 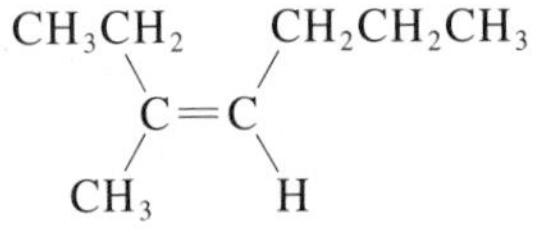

(c) 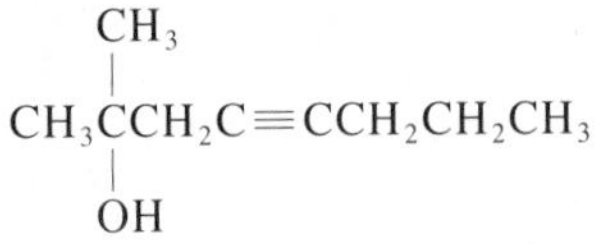

(d) 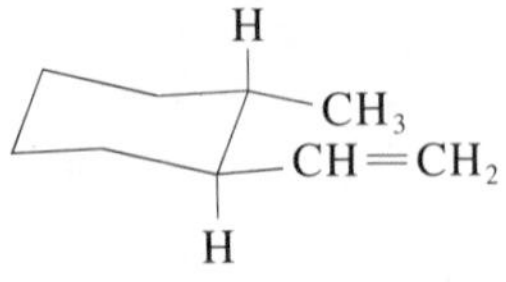 and enantiomer

9.14 Draw a structural formula for each of the following compounds.

(a) (*E*)-2-chloro-2-methyl-4-octene
(b) (*S*)-4-bromo-1-heptyne
(c) *cis*-3-methylcyclopentanol
(d) 3-hexyn-1-ol

9.15 Write equations predicting the reaction of 2-pentyne with each of the following reagents. In cases where you know what the stereochemistry of the product(s) should be, use appropriate conventions to illustrate it.

(a) H_2, $Pd/CaCO_3$, quinoline (b) H_2 (excess), Pt

TABLE 9.1 Reactions of Alkynes

Alkyne	Electrophile	Intermediates		Product
Electrophilic Addition Reactions				
$RC{\equiv}CR'$ R may be H	X_2	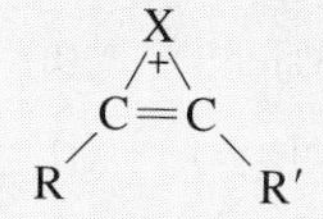	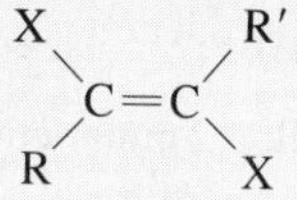	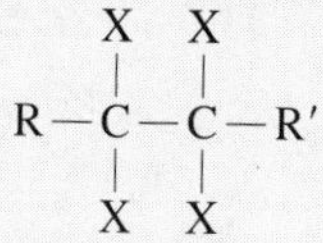
	HX	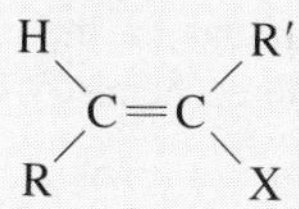	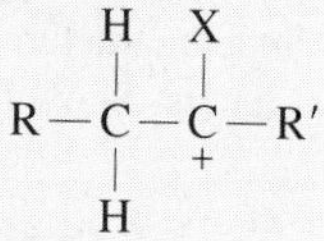	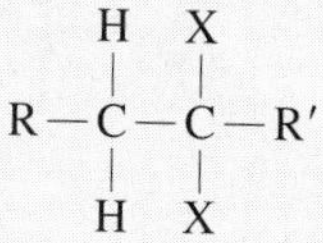
	H_3O^+	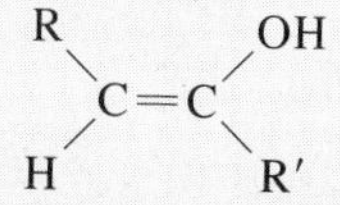		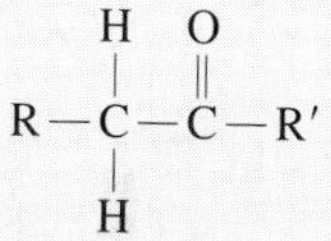
Oxidation Reaction				
	O_3			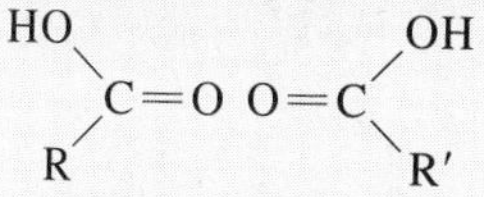
Reduction Reactions				
	H_2, PtO_2 or $RhCl[P(C_6H_5)_3]_3$			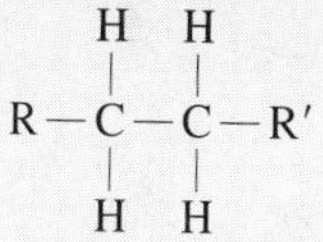
	H_2, Pd, $BaSO_4$ quinoline			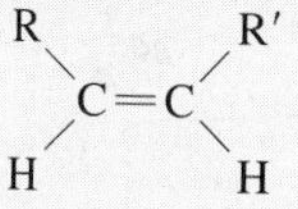 (syn addition)
	Na, NH_3	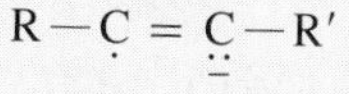	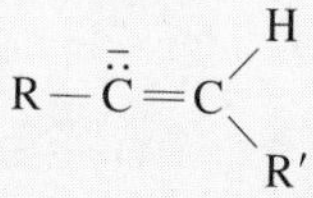	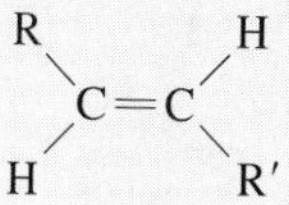 (anti addition)

Terminal alkyne	Reagent	Intermediate	Reagent	Product
$RC{\equiv}CH$	$NaNH_2$, NH_3(liq) strong base	$RC{\equiv}C{:}^-$ nucleophile	$R'CH_2X$ electrophile	$RC{\equiv}CCH_2R'$

(c) Na, NH_3(liq) (d) product of part a + 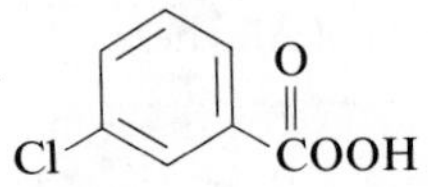

(e) product of part c + 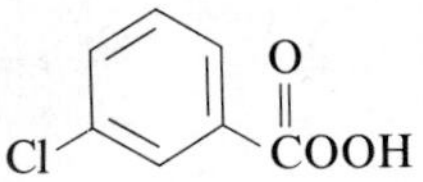 (f) HBr (1 molar equivalent)

(g) HBr (2 molar equivalents) (h) Br_2 (1 molar equivalent)

(i) Br_2 (2 molar equivalents) (j) (1) O_3, carbon tetrachloride; (2) H_2O

(k) H_2O, H_2SO_4, $HgSO_4$ (l) H_2, $RhCl[P(C_6H_5)_3]_3$

9.16 Supply structural formulas for the reagents, intermediates, and products designated by capital letters. Show the stereochemistry of the product(s) wherever it is known.

(a) 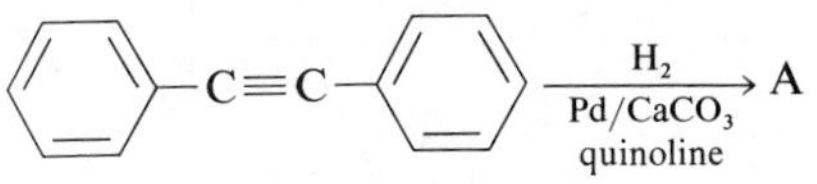

(b) $CH_2{=}CHBr \xrightarrow{HBr} B$

(c) $CH_2{=}CHCH_2Br \xrightarrow[\text{carbon tetrachloride, 0 °C}]{Br_2} C$

(d)  (e) 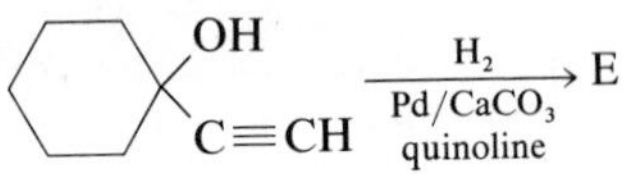

(f) $CH_3(CH_2)_7C{\equiv}C(CH_2)_7CO_2H \xrightarrow[\text{carbon tetrachloride}]{O_3} \xrightarrow{H_2O} F + G$

(g) 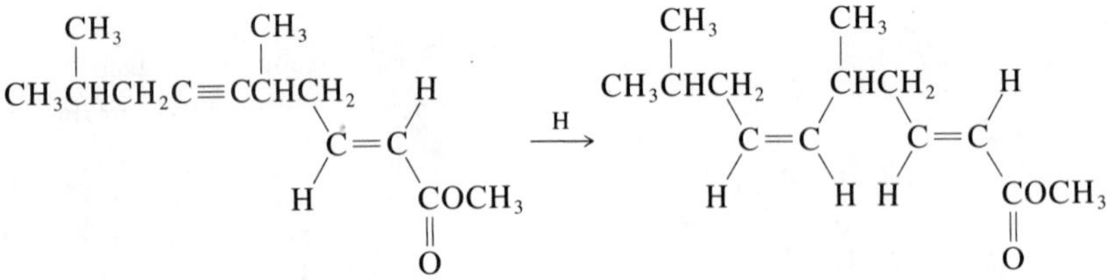

(h) 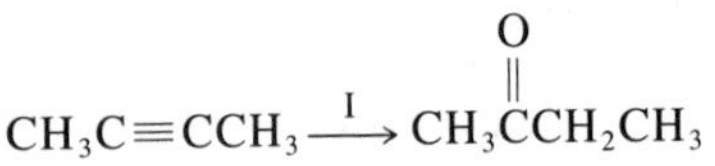

(i) 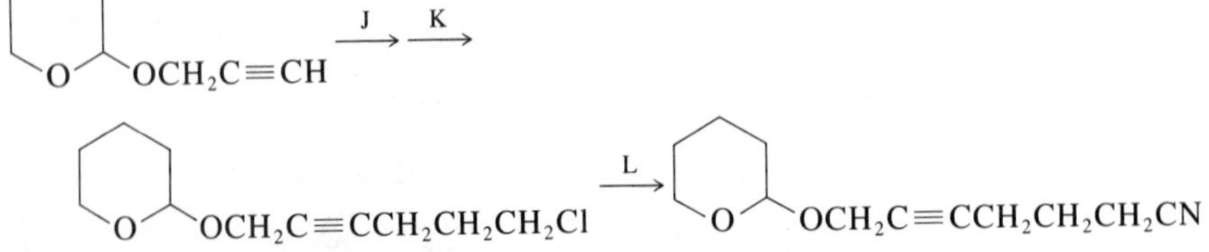

(j) $CH_2{=}CHCH_2CH_2Br \xrightarrow[NH_3(liq)]{HC{\equiv}C^-Na^+} M$

(k) 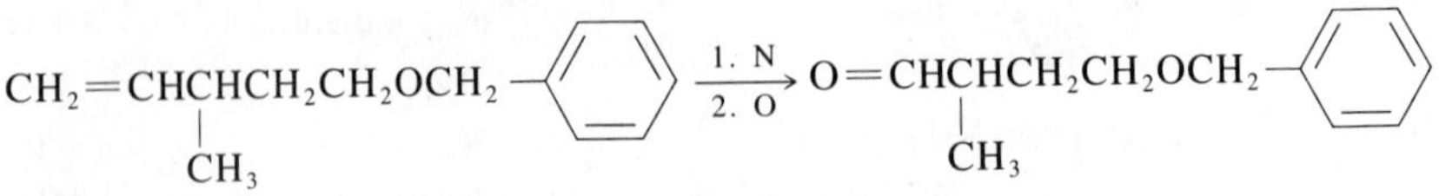

(l) 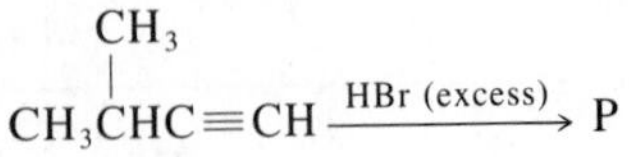

9.17 How would you carry out each of the following transformations? For some, more than one step may be necessary.

(a) 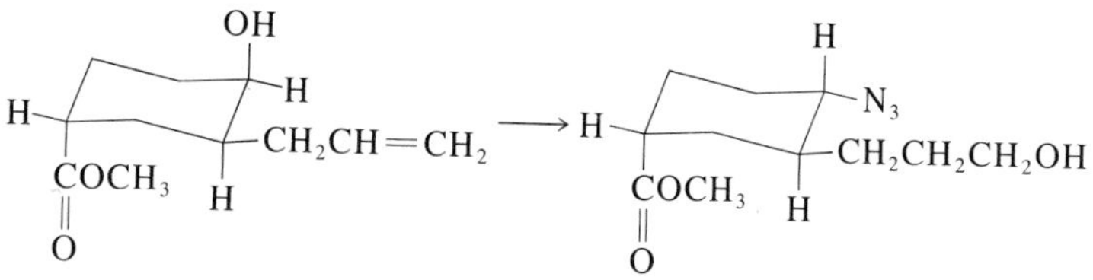

(b) 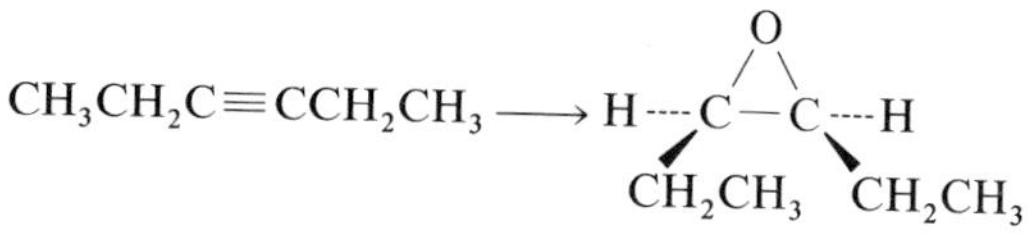

(c) 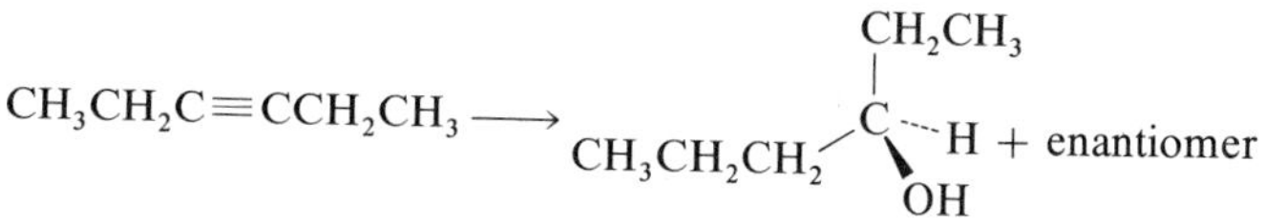

(d) 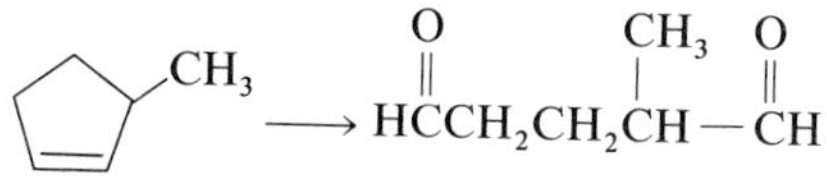

(e) 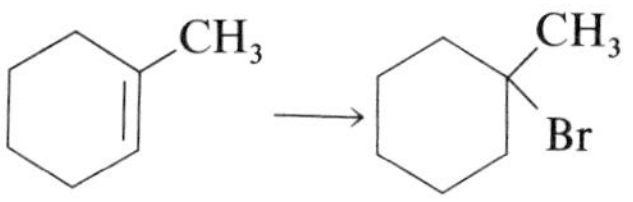

(f) 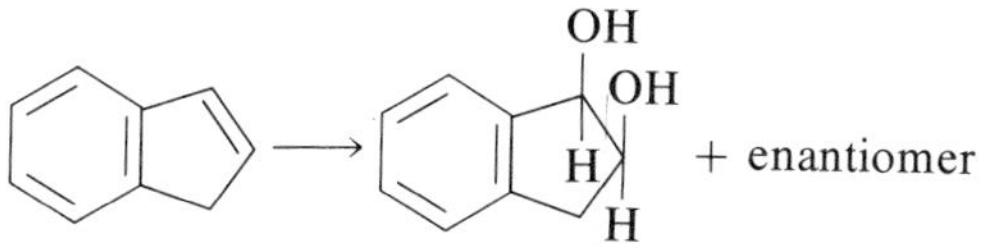

(g) 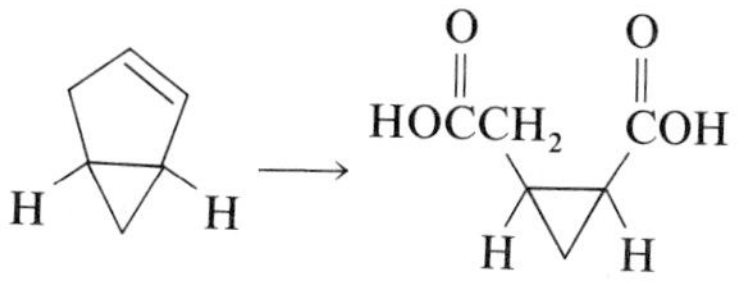

(h) 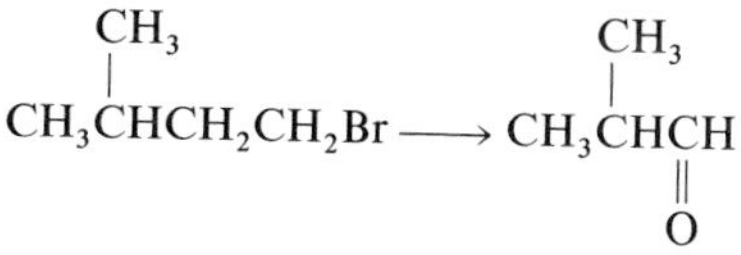

(i) 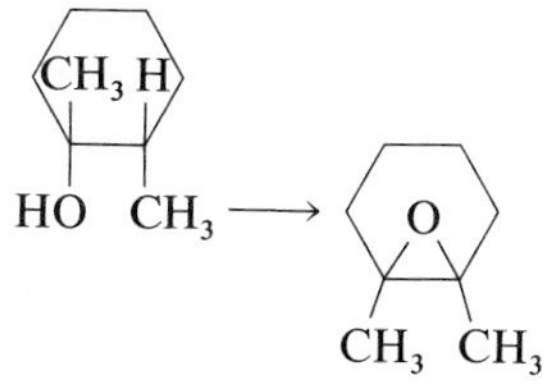

(j) 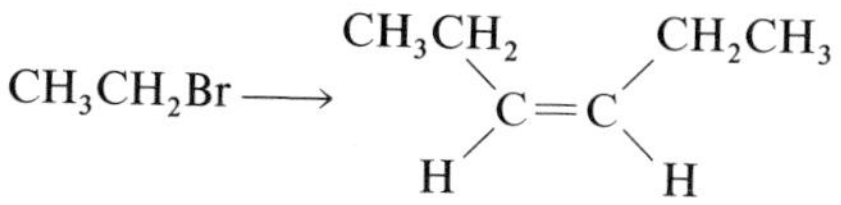

(k) 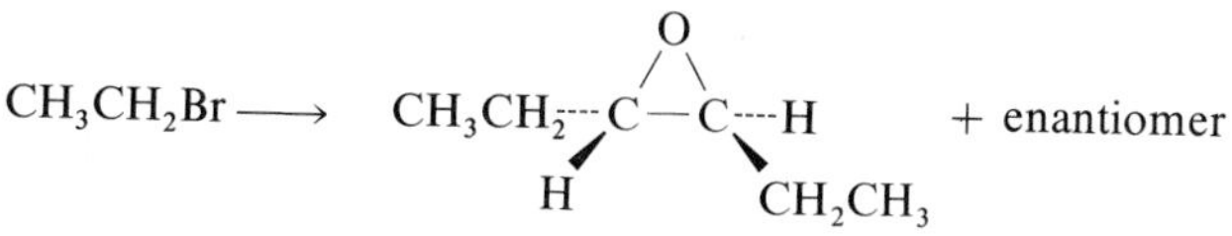

9.18 Compound A, $C_{13}H_{10}O$, isolated from the roots of *Carlina acaulis* (a thistle-like plant) and named Carlina-oxide, is hydrogenated. The product is Compound B, $C_{13}H_{14}O$, which has the following structural formula:

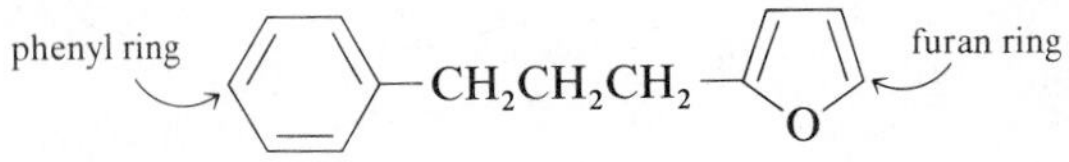
phenyl ring
CH2CH2CH2
furan ring
O

Compound B

Compound B has a three-carbon chain with a phenyl ring at one end and a furan ring at the other. The furan ring is similar to the phenyl ring in resistance to hydrogenation at low pressures and temperatures.

(a) How many units of unsaturation does Carlina-oxide contain? How many of them were saturated by reaction with hydrogen? What structures are possible for Carlina-oxide?
(b) Carlina-oxide was treated with ozone in acetic acid. After removal of the acetic acid, the ozonide was decomposed with water. No reducing agent was used. Phenylacetic acid was isolated. What must the structure of Carlina-oxide be?

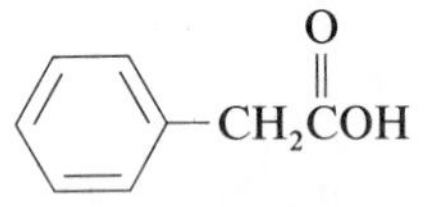
O
CH2COH

phenylacetic acid

9.19 2-Methylheptadecane is a sex pheromone for certain species of insects. It has been synthesized using 1-undecyne as the starting material. How would you carry out this synthesis?

9.20 Disparlure, the sex pheromone of the gypsy moth (p. 702), is the epoxide of (*Z*)-2-methyl-7-octadecene. The starting material for a synthesis of disparlure was 1-dodecyne.

(a) Outline a synthesis for disparlure.
(b) How would you modify the synthesis of part a to end up with the isomer of disparlure in which the alkyl groups are trans to each other?

9.21 Alkynes react with diborane just as alkenes do. The resulting vinylboranes are oxidized by hydrogen peroxide. 3-Hexyne is converted to 3-hexanone with a 68% yield by such a sequence of reactions. Write equations showing the intermediates in this conversion. A review of Section 8.7 may be helpful.

9.22 A vinyl ether reacts with an alcohol in the presence of a trace of acid as a catalyst. Predict what the product of the reaction will be by evaluating the relative stabilities of possible intermediates.

$$CH_3OCH{=}CH_2 + CH_3OH \xrightarrow{HCl}$$

9.23 The kinetics of the addition of hydrogen chloride in the gas phase to 2-methylpropene and to ethyl vinyl ether, $CH_2{=}CHOCH_2CH_3$, have been studied. The rate constants for the two reactions may be calculated using the Arrhenius equation (p. 117). The reaction of 2-methylpropene with hydrogen chloride has a frequency factor, A, of 1×10^{11} mL/mol·s and E_a of 28.8 kcal/mol. For the reaction of ethyl vinyl ether with hydrogen chloride, A is 5×10^{8} mL/mol·s and E_a is 14.7 kcal/mol. What is the second-order rate constant for each reaction at 25 °C? How do you account for the large difference in E_a values for the two reactions? Write equations and draw energy diagrams to illustrate your answer.

10

Infrared Spectroscopy

A • L O O K • A H E A D

Chemists use two general methods to obtain information about the structures of molecules. One is to carry out chemical transformations on molecules and observe the results. Qualitative analysis tests, such as the bromine (p. 173) and Baeyer tests for unsaturation (p. 314), give immediate information about the presence or absence of specific functional groups, in these cases, of double or triple bonds. Ozonolysis cleaves the molecules of alkenes (p. 311) or alkynes (p. 344) at double or triple bonds, giving aldehydes, ketones, or carboxylic acids, which can be identified more easily than the original hydrocarbons can. Combining such chemical information with knowledge about the molecular formula and the degree of unsaturation of a compound (p. 172) often leads to an assignment of structure.

Another way chemists gather information about the structures of molecules is by making physical measurements on compounds. As was discussed in Chapter 1, the dipole moment of a compound can be used to deduce the shape and the symmetry of its molecules. Electron diffraction is a method of obtaining information about bond lengths and bond angles. However, for organic chemists, spectroscopy is the most widely used physical method of investigation. A molecule interacts with electromagnetic radiation by absorbing energy that corresponds to transitions between fixed energy levels for the molecular species. The absorption of energy can be monitored by instruments called spectrophotometers, which record the changing absorption of energy as a function of the wavelength or the frequency of the radiation being used. The tracings obtained from such instruments are called spectra and are used to draw conclusions about the structures of the molecules that give rise to them.

As your knowledge of chemical structures grows, you will learn to analyze different types of spectra for specific structural information. In this chapter, you will learn about infrared spectroscopy. When chemists take the infrared spectrum of a compound, they are investigating the changes in the vibrational motions of the atoms in its molecules. The energy associated with each vibrational transition depends on the masses of the atoms that are bonded together and on the strength of the bond. The infrared spectrum of a compound is, therefore, a source of information about the types of bonds the compound contains and is most useful in identifying the presence of specific functional groups, such as hydroxyl and carbonyl groups.

THE ELECTROMAGNETIC SPECTRUM AND ABSORPTION SPECTROSCOPY

Molecules have several different kinds of energy levels and, therefore, absorb radiation in several regions of the electromagnetic spectrum. The idea of electronic energy levels, which are the orbitals of an atom, is a familiar one. When an atom absorbs light of a given frequency, an electron moves from a lower energy level to a higher one.

An atom that has absorbed energy is said to be in an **excited state,** which is at a higher energy than its normal **ground state.** It returns to the ground state by losing energy, usually by emitting heat or, less frequently, light. The difference in energy between two atomic energy levels is proportional to the frequency of the light absorbed:

$$E = h\nu$$

where h is Planck's constant, 6.624×10^{-27} erg·s, and ν is the frequency of the light in cycles/s (hertz, or Hz). The frequency of the light is related to its wavelength, λ:

$$\nu = \frac{c}{\lambda}$$

where c is the velocity of light in a vacuum, 2.998×10^{10} cm/s.

The electromagnetic spectrum has a tremendous range of energy. At one end of the spectrum are cosmic rays, which have energies of approximately 10^9 kcal/mol (1 erg/photon = 1.439×10^{13} kcal/mol of photons). They have very short wavelengths (10^{-12} cm) and high frequencies (10^{22} Hz). Medically useful x-rays have wavelengths of 10^{-8} to 10^{-6} cm and energies in the range of 10^3 kcal/mol. On the low-energy end of the electromagnetic spectrum are radiowaves, which have long wavelengths (10^3 to 10^6 cm) and low frequencies (10^7 to 10^4 Hz).

The regions of the electromagnetic spectrum that are used for investigating molecular structures are summarized in Table 10.1. Transitions between electronic energy levels occur when a molecule absorbs radiation in the **ultraviolet and visible regions** of the spectrum. A molecule goes from one vibrational energy level

TABLE 10.1 Regions of the Electromagnetic Spectrum Used in Spectroscopy

Region of Spectrum	Molecular Change	Energy, kcal/mol	Frequency, Hz	Wavelength, cm
ultraviolet	electronic transitions (Chapter 17)	100	10^{15}	2×10^{-5} to 4×10^{-5}
visible	electronic transitions	50	5×10^{14}	4×10^{-5} to 8×10^{-5}
infrared	molecular vibrations	5	10^{13} to 10^{14}	10^{-4} to 10^{-2}
microwave	molecular rotations	3×10^{-3}	3×10^{10}	1
radio-frequency	orientation of spin of nucleus in magnetic field (Chapter 11)	10^{-7}	10^6	5×10^5

to another upon absorbing **infrared radiation.** Absorption of **microwaves** causes changes in rotational energy levels. Finally, transitions between energy levels associated with the nuclei of certain elements occur upon the absorption of **radio-frequency waves.** Table 10.1 shows the approximate energy associated with each of these types of transitions and also indicates the chapters in which ultraviolet and nuclear magnetic resonance spectroscopy are discussed. The rest of this chapter focuses on the infrared region, to investigate how molecular interactions with radiation in that energy range give chemists information about molecular structure.

10.2 MOLECULAR VIBRATIONS AND ABSORPTION FREQUENCIES IN THE INFRARED REGION

The atoms within a molecule are constantly in motion, distorting the chemical bonds. Such motions are called **molecular vibrations.** One type of vibration possible for a molecule produces changes in bond length; such vibrations are called **stretching vibrations** (Figure 10.1). Other vibrations result in changes in bond angles and are called **bending vibrations** (Figure 10.2). Any given molecule has a number of energy levels corresponding to the different vibrational states possible for that molecule. The spacing between these energy levels corresponds to the energy of radiation in the infrared region of the electromagnetic spectrum. The differences in vibrational energy levels in the region of the infrared that is most useful to chemists correspond to radiation having frequencies, $\bar{\nu}$, wavenumber of 4000 to 666 cm^{-1} ($\bar{\nu} = \frac{1}{\lambda}$ with λ given in cm), or wavelengths of 2.5 to 15.0 μm (a micrometer, μm, is 10^{-4} cm, or 10^{-6} m).

When radiation having energy that is the same as the difference in energy between molecular vibrational energy levels impinges on a molecule, the radiation is absorbed and the amplitude of molecular vibration increases. The molecule moves from one vibrational energy level to a higher one. The energy that is absorbed is ultimately returned to the environment as heat. This energy is roughly 1 to 10 kcal/mol, not enough to break chemical bonds or cause chemical reactions. The fact that the radiation is absorbed is recorded as a tracing by an instrument called an **infrared spectrophotometer** (Figure 10.3).

The infrared spectra shown in this book were recorded on a type of instrument called a Fourier-Transform Infrared Spectrophotometer. Such spectra are called FT-IR spectra for short. With such a spectrophotometer, the spectrum is obtained

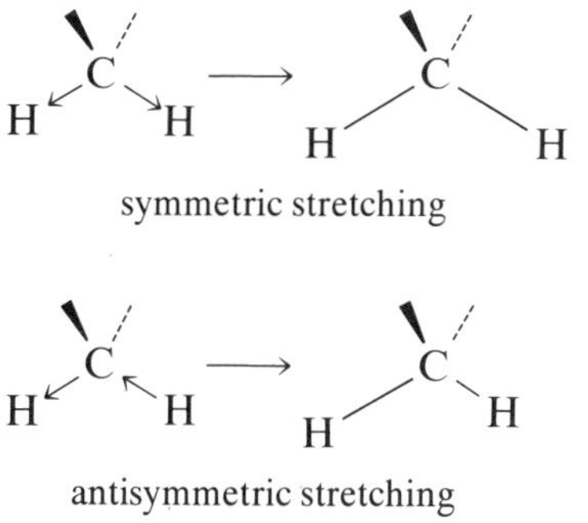
symmetric stretching
antisymmetric stretching

FIGURE 10.1 Typical stretching vibrations of a methylene group.

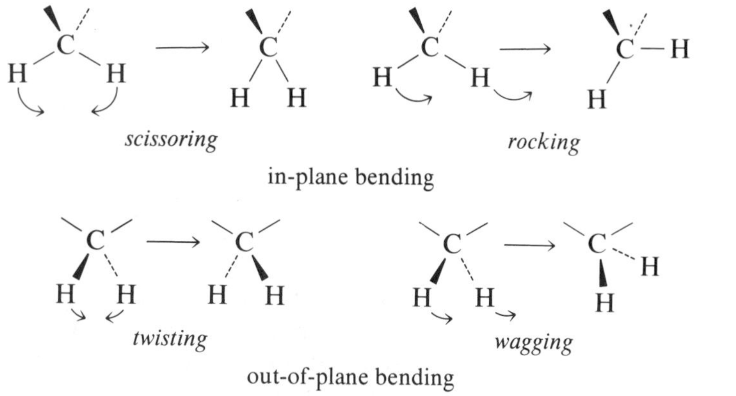
scissoring
rocking
in-plane bending
twisting
wagging
out-of-plane bending

FIGURE 10.2 Different kinds of bending vibrations for a methylene group.

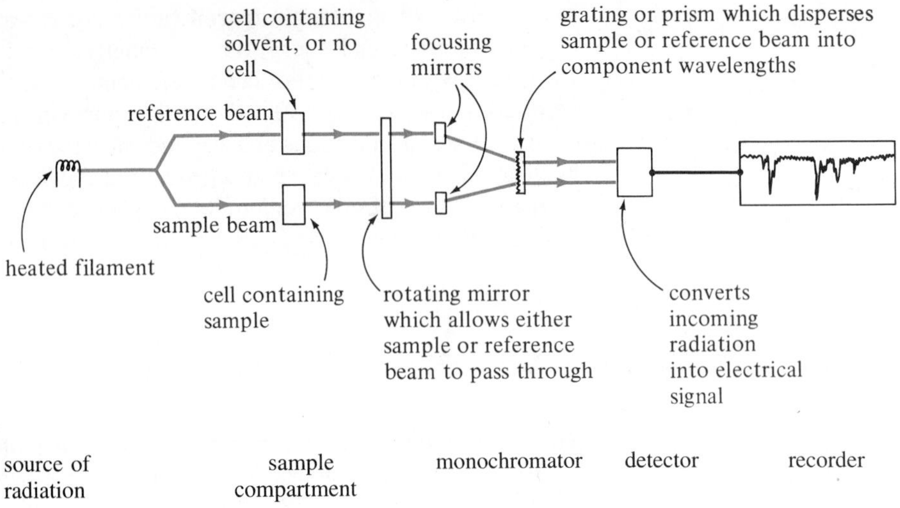
cell containing solvent, or no cell
focusing mirrors
grating or prism which disperses sample or reference beam into component wavelengths
reference beam
sample beam
heated filament
cell containing sample
rotating mirror which allows either sample or reference beam to pass through
converts incoming radiation into electrical signal
source of radiation
sample compartment
monochromator
detector
recorder

FIGURE 10.3 Schematic diagram of a typical infrared spectrophotometer.

in a second or less in the form of an interferogram, which is a plot of the sums of the cosine waves of all of the frequencies present in the source of infrared radiation as modified by passage through the sample. A computer stores these signals in its memory, carries out a mathematical operation known as a Fourier transformation on them, corrects for the frequencies generated by the source of the infrared radiation, and plots the FT-IR spectrum (Figure 10.4).

Even a simple molecule can have a large number of possible molecular vibrations, and an infrared spectrum usually has many absorption bands. Infrared spectra are useful to chemists because different functional groups absorb at different frequencies, corresponding to certain vibrations typical of that portion of a molecule. Tables that list the infrared absorption frequencies for different types of functional groups are available. Table 10.2 on page 360 is an abbreviated listing of such group frequencies. The frequencies at which functional groups absorb energy are related to the types of bonds present. Figure 10.5 makes this clearer by summarizing some trends that can be seen by carefully examining Table 10.2.

The **stretching frequency** of a bond is related to the masses of the two atoms involved in the bond and to the strength of the bond.

$$\bar{\nu} = \frac{1}{2\pi c}\sqrt{\frac{f(m_1 + m_2)}{m_1 m_2}}$$

Here $\bar{\nu}$ is the frequency in cm^{-1}, c is the velocity of light, m_1 and m_2 are the masses of the two atoms in grams, and f is the force constant in dyne/cm. The force constant for a single bond is approximately 5×10^5 dyne/cm. The force constant for a double bond is 10×10^5 dyne/cm, about twice that for a single bond; that for a triple bond is 15×10^5 dyne/cm. Table 10.2 and Figure 10.5 show, for example, that triple bonds absorb at higher frequencies (2260–2100 cm^{-1}) than do double bonds (1800–1390 cm^{-1}), which in turn absorb at higher frequencies than do single bonds (1360–1030 cm^{-1}). It takes more energy to stretch a stronger bond.

The stretching vibrations for bonds to hydrogen, the lightest of the elements, also occur at high frequencies (3650–2500 cm^{-1}). A bond between hydrogen and an sp-hybridized carbon is shorter and absorbs at a higher frequency (3300 cm^{-1}) than does one between hydrogen and an sp^2-hybridized carbon (3080–3020 cm^{-1}).

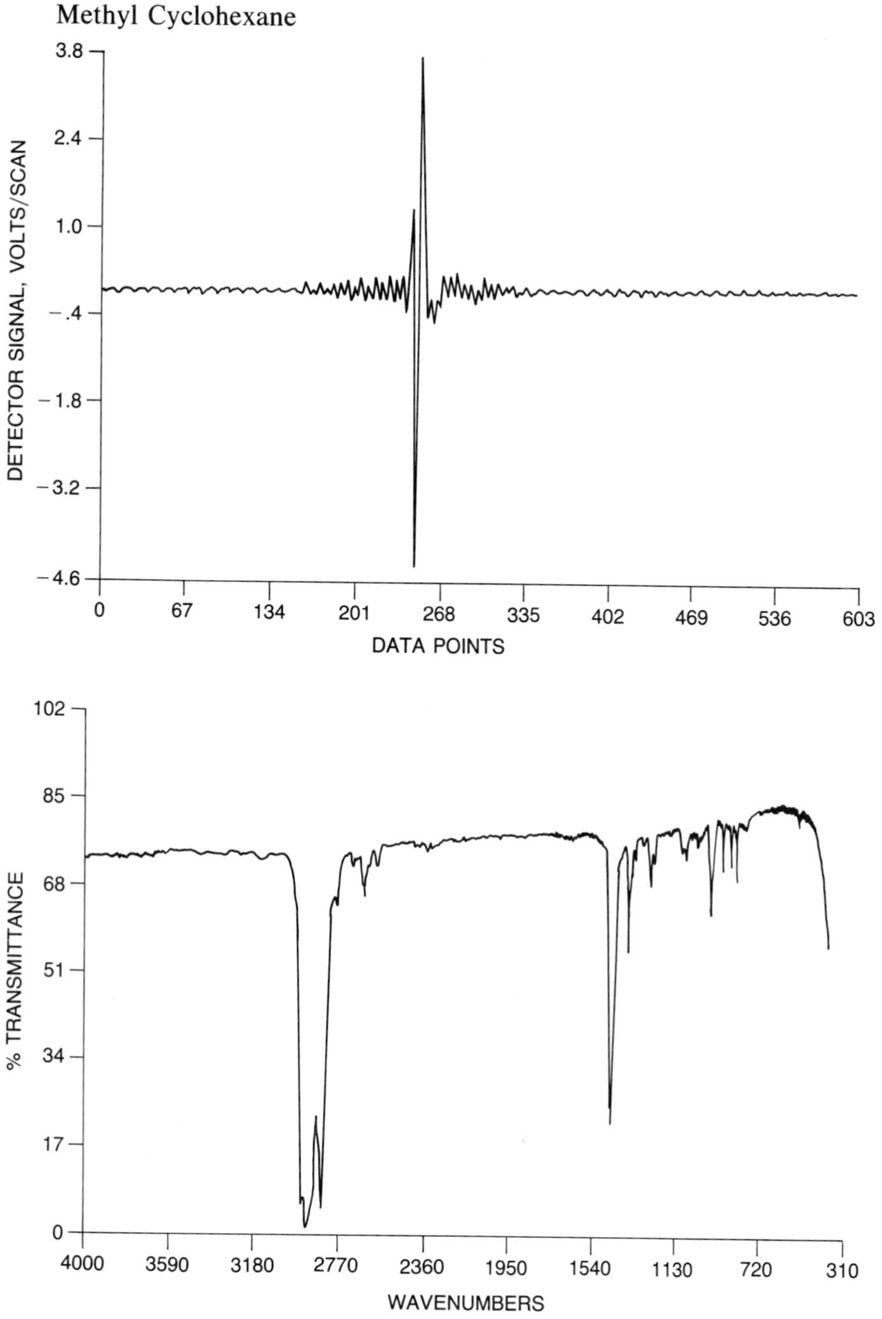
Methyl Cyclohexane
DETECTOR SIGNAL, VOLTS/SCAN
3.8
2.4
1.0
−.4
−1.8
−3.2
−4.6
0
67
134
201
268
335
402
469
536
603
DATA POINTS
% TRANSMITTANCE
102
85
68
51
34
17
0
4000
3590
3180
2770
2360
1950
1540
1130
720
310
WAVENUMBERS

FIGURE 10.4 Interferogram and spectrum for methylcyclohexane obtained on a Nicolet 60-SX Fourier-Transform Infrared Spectrophotometer.

The carbon-hydrogen bonds of alkanes absorb at the lowest frequencies (2960–2850 cm^{-1}).

The major frequencies that are typical of functional groups usually appear between 4000 and 1400 cm^{-1}. The portion of an infrared spectrum between 1400 and 200 cm^{-1} is called the **fingerprint region.** It is more difficult to make specific assignments for bands in that region as they are dependent on the structure of the molecule as a whole. This region is immensely useful in making a positive identification of a compound. If spectra of two species show the same bands, with

TABLE 10.2 Characteristic Infrared Absorption Frequencies

Bond Type	Stretching, cm^{-1}	Bending, cm^{-1}
C—H alkanes	2960–2850 (*s*)	1470–1350 (*s*)
C—H alkenes	3080–3020 (*m*)	1000–675 (*s*)
C—H aromatic	3100–3000 (*v*)	870–675 (*v*)
C—H aldehyde	2900, 2700 (*m*, 2 bands)	
C—H alkyne	3300 (*s*)	
C≡C alkyne	2260–2100 (*v*)	
C≡N nitrile	2260–2220 (*v*)	
C═C alkene	1680–1620 (*v*)	
C═C aromatic	1600–1450 (*v*)	
C═O ketone	1725–1705 (*s*)	
C═O aldehyde	1740–1720 (*s*)	
C═O α,β-unsaturated ketone	1685–1665 (*s*)	
C═O aryl ketone	1700–1680 (*s*)	
C═O ester	1750–1735 (*s*)	
C═O acid	1725–1700 (*s*)	
C═O amide	1690–1650 (*s*)	
O—H alcohols (not hydrogen bonded)	3650–3590 (*v*)	
O—H alcohols (hydrogen bonded)	3600–3200 (*s*, broad)	1620–1590 (*v*)
O—H acids	3000–2500 (*s*, broad)	1655–1510 (*s*)
N—H amines	3500–3300 (*m*)	
N—H amides	3500–3350 (*m*)	
C—O alcohols, ethers, esters	1300–1000 (*s*)	
C—N amines, alkyl	1220–1020 (*w*)	
C—N amines, aromatic	1360–1250 (*s*)	
NO_2 nitro	1560–1515 (*s*)	
	1385–1345 (*s*)	

s = strong absorption *w* = weak absorption
m = medium absorption *v* = variable absorption

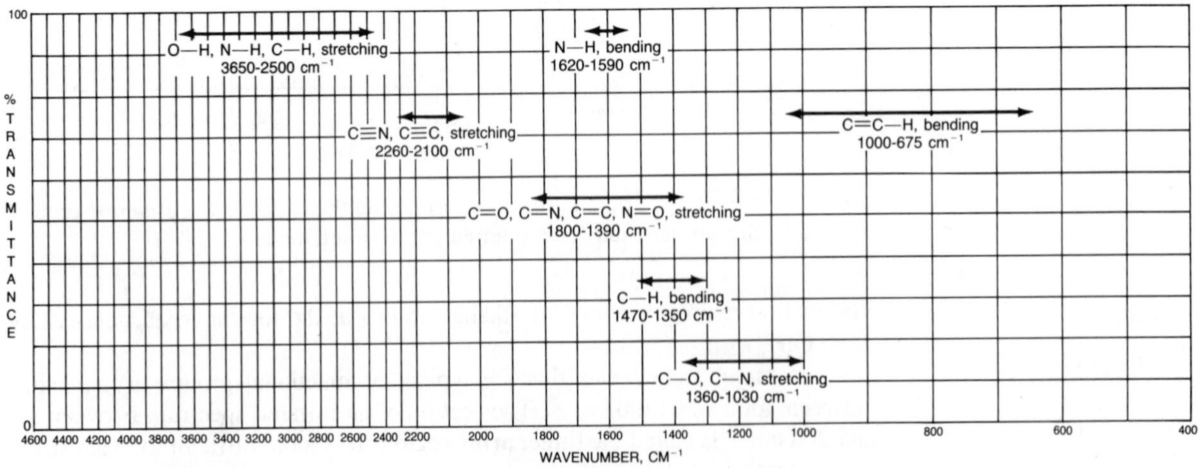
100
%
TRANSMITTANCE
0
O—H, N—H, C—H, stretching
3650-2500 cm⁻¹
N—H, bending
1620-1590 cm⁻¹
C≡N, C≡C, stretching
2260-2100 cm⁻¹
C═C—H, bending
1000-675 cm⁻¹
C═O, C═N, C═C, N═O, stretching
1800-1390 cm⁻¹
C—H, bending
1470-1350 cm⁻¹
C—O, C—N, stretching
1360-1030 cm⁻¹
4600 4400 4200 4000 3800 3600 3400 3200 3000 2800 2600 2400 2200 2000 1800 1600 1400 1200 1000 800 600 400
WAVENUMBER, CM⁻¹

FIGURE 10.5 Infrared regions for different types of bonds.

the same relative intensities, in this region as well as in the higher-frequency region, this is considered to be proof of the identity of the two species.

Not all bands in an infrared spectrum have the same intensity, and this too is useful in identifying functional groups. In general, during a vibration that corresponds to a change in the dipole of a molecule, the molecule absorbs radiation strongly; for a vibration in which there is a small or no change in the dipole of the molecule, an absorption band may not be seen. For example, stretching of a carbonyl group gives a strong absorption, and stretching of a carbon-carbon double bond, a weak one.

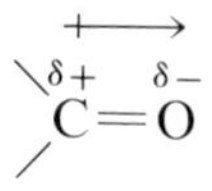

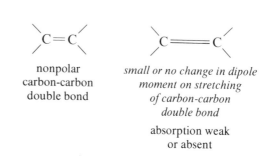

polar bond in carbonyl group

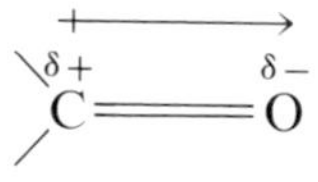

increase in dipole moment on stretching of bond in carbonyl group

intense absorption

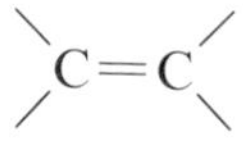

nonpolar carbon-carbon double bond

small or no change in dipole moment on stretching of carbon-carbon double bond

absorption weak or absent

Variations such as these are why the interpretation of infrared spectra is an art requiring careful observation of relative intensities and appearances of bands rather than a mere reading of numbers from tables. The examples of spectra of compounds with different functional groups presented in the next section will make the important points of this discussion clearer.

10.3 USING INFRARED SPECTROSCOPY TO STUDY CHEMICAL TRANSFORMATIONS. INFRARED SPECTRA OF ALKYNES, ALKENES, AND ALCOHOLS

The characteristic infrared absorption frequencies of major functional groups can be used to document the transformation of one chemical species into another. For example, the terminal alkyne 1-hexyne can be converted by hydrogenation (p. 341) to 1-hexene, which, in turn, gives 1-hexanol on hydroboration (p. 299). Infrared spectra for these three compounds are shown in Figure 10.6. The absorption bands typical of a terminal alkyne are the triple bond stretching frequency in the region of 2260–2100 cm^{-1} and the stretching frequency for the terminal carbon-hydrogen bond, the strong sharp band at 3300 cm^{-1}. These two features taken together identify a terminal alkyne. Both of these bands are missing in the spectrum of 1-hexene. The important band in the spectrum of the alkene comes from the stretching frequency of the carbon-carbon double bond at 1680–1640 cm^{-1}. In addition, the band for the carbon-hydrogen stretching frequency for the vinyl hydrogens appears near 3100 cm^{-1}. In the spectrum of 1-hexanol, the most important band arises from the hydroxyl (O—H) stretching frequency. Most of the time, this is a strong broad band at 3600–3200 cm^{-1}, indicating the presence of a hydrogen-bonded hydroxyl group. There is also a strong band around 1300–1080 cm^{-1}, which is attributable to carbon-oxygen single bond stretching vibrations.

In all three spectra in Figure 10.6, other bands appear; these are characteristic of carbon-hydrogen bonds involving sp^3-hybridized carbon. Among these bands

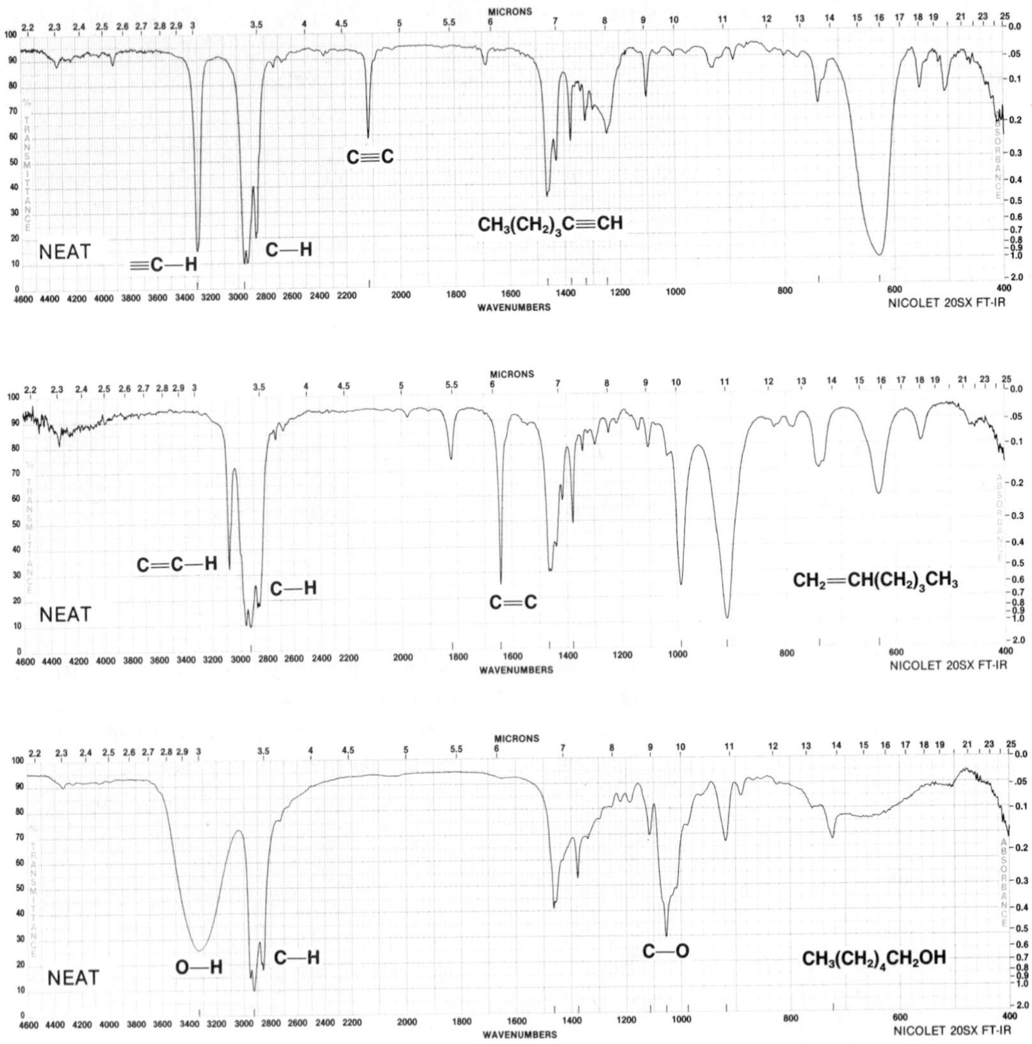
MICRONS
WAVENUMBERS
TRANSMITTANCE
ABSORBANCE
NEAT
≡C—H
C—H
C≡C
CH₃(CH₂)₃C≡CH
NICOLET 20SX FT-IR
NEAT
C=C—H
C—H
C=C
CH₂=CH(CH₂)₃CH₃
NICOLET 20SX FT-IR
NEAT
O—H
C—H
C—O
CH₃(CH₂)₄CH₂OH
NICOLET 20SX FT-IR

FIGURE 10.6 Infrared spectra of 1-hexyne, 1-hexene, and 1-hexanol. (From *The Aldrich Library of FT-IR Spectra*)

are those for carbon-hydrogen stretching frequencies around 2900 cm^{-1} and those for the bending vibrations for such bonds around 1400 cm^{-1}.

The bands that have been singled out in this description are the most important in these spectra. Many of the other bands cannot be assigned to any one simple vibration of the molecule. They are the result of combinations of vibrations or are overtones of other bands. These bands are, however, characteristic of each molecule and can be used for its identification.

The three infrared spectra shown in Figure 10.6 also illustrate the usefulness of such spectra in keeping track of chemical transformations. Once a chemist's eye is trained to recognize the characteristic absorption bands, he or she can use infrared spectra to "see" what is happening in a reaction mixture by noting the disappearance of bands that characterize the functional group in the starting material and the appearance of bands that characterize the functional group being created in the product. To get information from all kinds of spectra, you must learn to look beyond the tracings and numbers produced by spectrometers in order to see the structural features that cause them.

PROBLEM 10.1

The infrared spectra of Compounds A–D appear in Figure 10.7. The compounds may be alcohols, alkenes, or alkynes. Assign the correct functional group class to each compound. Point out the bands that you used in making each assignment.

Compound A

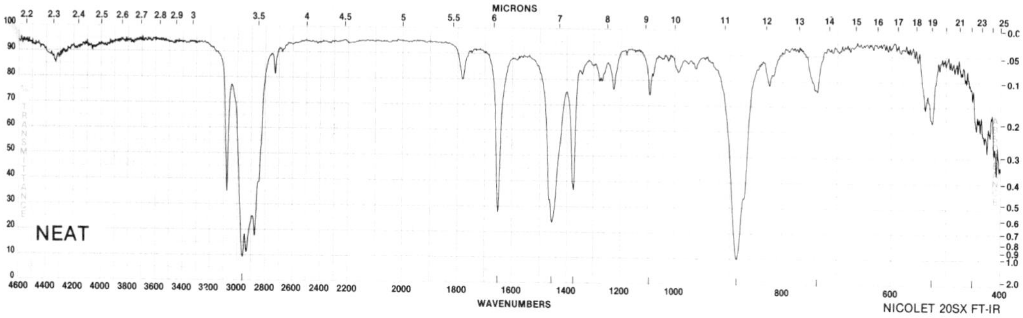
MICRONS
NEAT
WAVENUMBERS
NICOLET 20SX FT-IR

Compound B

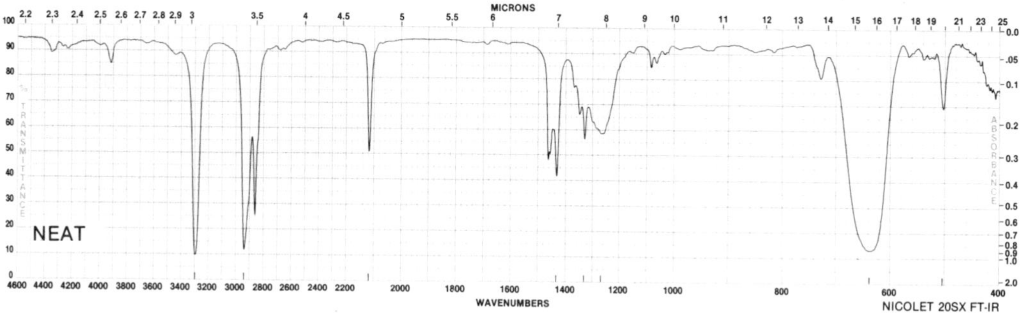
MICRONS
NEAT
WAVENUMBERS
NICOLET 20SX FT-IR

Compound C

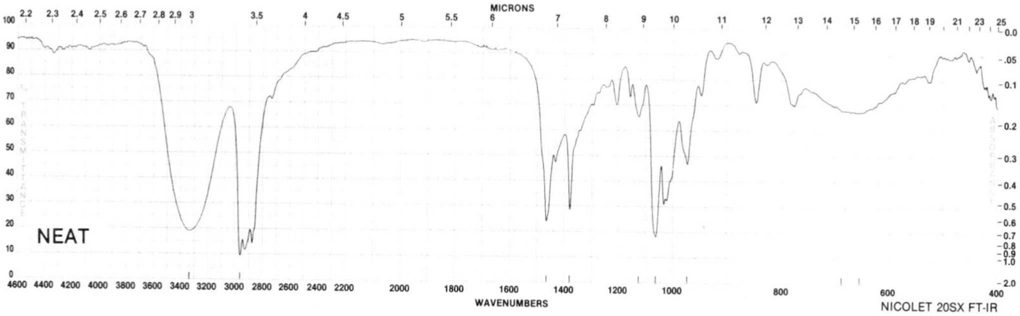
MICRONS
NEAT
WAVENUMBERS
NICOLET 20SX FT-IR

Compound D

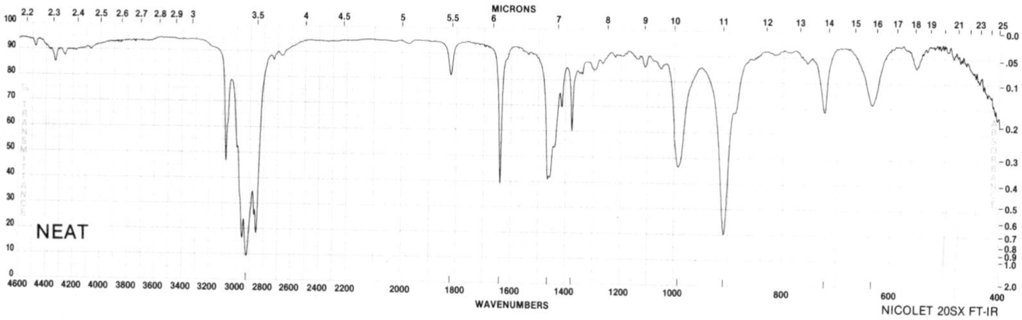
MICRONS
NEAT
WAVENUMBERS
NICOLET 20SX FT-IR

FIGURE 10.7

From *The Aldrich Library of FT-IR Spectra*

A. Infrared Spectra of Aldehydes and Ketones

The band for the stretching frequency of a carbonyl group is usually one of the strongest in the spectrum. The spectra of butanal and 2-butanone are given in Figure 10.8. Butanal has a strong absorption due to the carbonyl group at 1725 cm^{-1}. Another absorption that is typical of aldehydes is the carbon-hydrogen stretching frequency for the carbon-hydrogen bond on the carbonyl group. Two bands are actually present for the hydrogen on the carbonyl group; one is visible around 2700 cm^{-1} but the other is buried in the strong absorption band for alkane-type hydrogens around 2900 cm^{-1}. The spectrum of butanal from 1400 to 400 cm^{-1} is characterized by bands for the stretching and bending frequencies for the alkane portion of the molecule. The carbonyl band in 2-butanone appears at 1710 cm^{-1}. Note that the band for the carbon-hydrogen stretching frequency seen around 2700 cm^{-1} in the spectrum of the aldehyde is missing from the spectrum of the ketone.

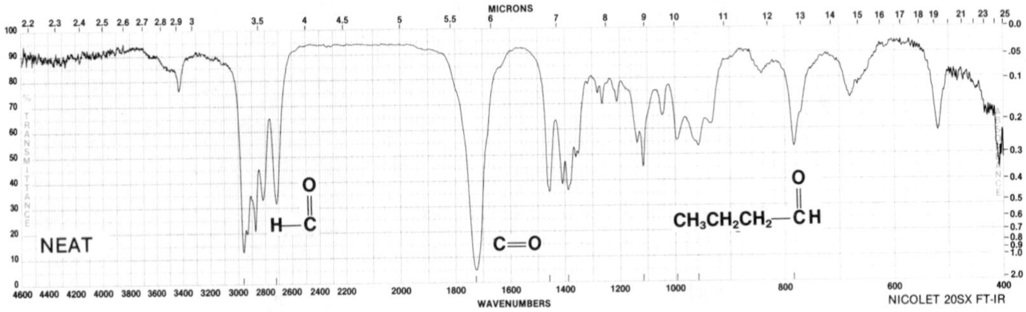
MICRONS
H—C
CH₃CH₂CH₂—CH
NEAT
C=O
WAVENUMBERS
NICOLET 20SX FT-IR

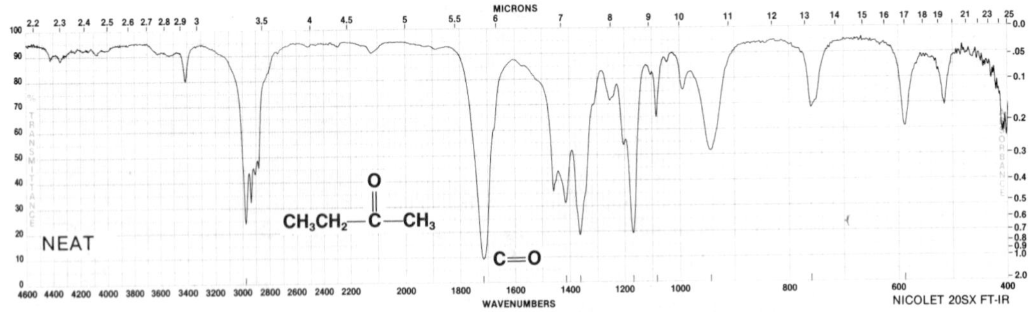
MICRONS
CH₃CH₂—C—CH₃
NEAT
C=O
WAVENUMBERS
NICOLET 20SX FT-IR

FIGURE 10.8 Infrared spectra of butanal and 2-butanone. (From *The Aldrich Library of FT-IR Spectra*)

PROBLEM 10.2

The infrared spectra and molecular formulas of Compounds E–H, all of which contain oxygen, are given in Figure 10.9. Decide whether each compound is an alcohol, an aldehyde, or a ketone. Propose a structure for each compound that is compatible with both the spectrum and the molecular formula.

Compound E, $C_4H_{10}O$

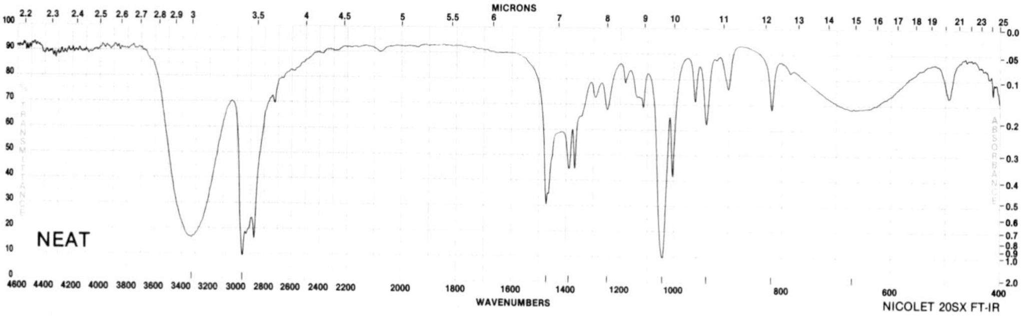
MICRONS
NEAT
WAVENUMBERS
NICOLET 20SX FT-IR

Compound F, $C_6H_{12}O$

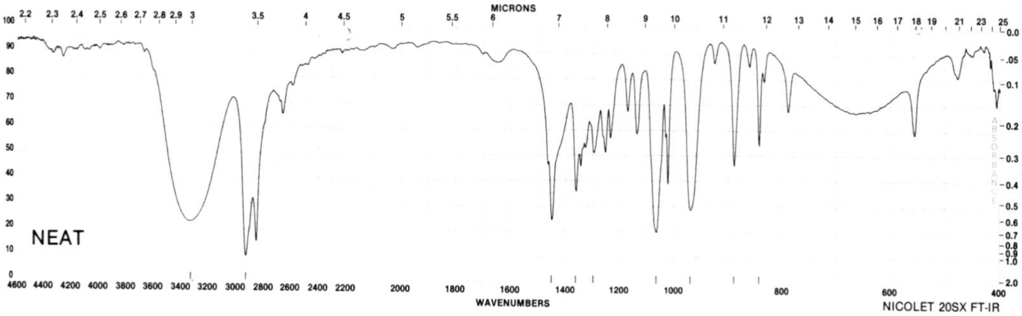
MICRONS
NEAT
WAVENUMBERS
NICOLET 20SX FT-IR

Compound G, $C_6H_{12}O$

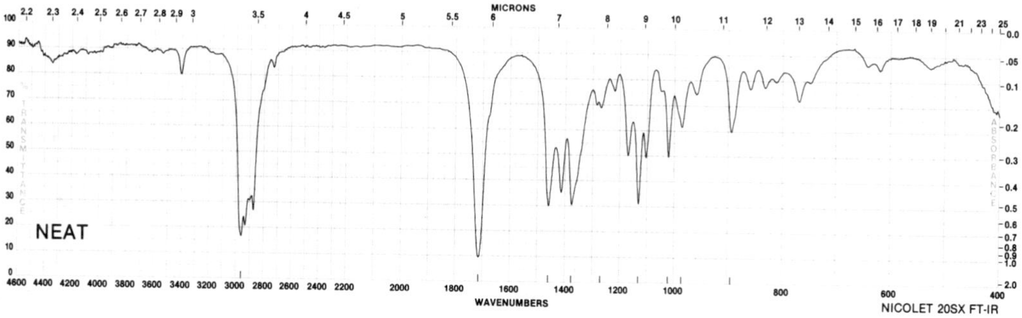
MICRONS
NEAT
WAVENUMBERS
NICOLET 20SX FT-IR

Compound H, C_4H_8O

FIGURE 10.9

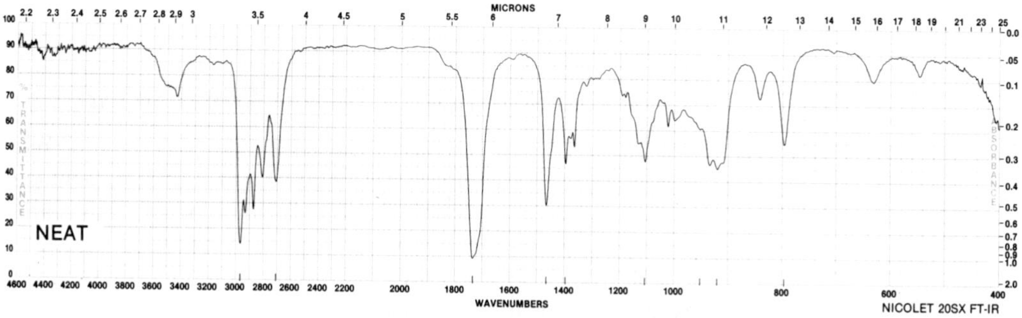
MICRONS
NEAT
WAVENUMBERS
NICOLET 20SX FT-IR

From *The Aldrich Library of FT-IR Spectra*

B. Infrared Spectra of Conjugated Carbonyl Compounds

When a carbonyl group in an aldehyde or ketone is separated from a carbon-carbon double bond (or an aromatic ring) by only a single bond, the carbonyl group is said to be **conjugated** with the double bond (or the ring).

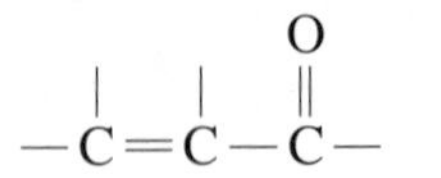

carbonyl group conjugated with a double bond

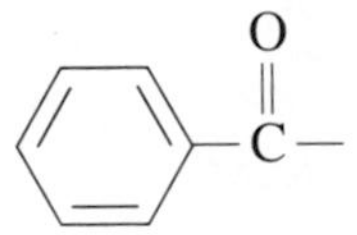

carbonyl group conjugated with an aromatic ring

A ketone or aldehyde in which the carbonyl group is conjugated has a lower carbonyl stretching frequency than does a ketone or aldehyde in which there is no such conjugation. This fact is rationalized by writing resonance contributors, of which one shows single-bond character for the carbon-oxygen bond.

Single bonds are easier to stretch than double bonds and, therefore, absorb at lower frequencies (p. 358). For example, comparing the infrared spectrum of 5-hexen-2-one, in which the carbonyl group is not conjugated with the double bond, with that of 2-cyclohexen-1-one, in which it is, shows clearly the shift of the carbonyl band to a lower frequency (Figure 10.10). In 5-hexen-2-one, the carbonyl stretching frequency is 1710 cm^{-1}; in 2-cyclohexen-1-one, it is 1675 cm^{-1}. Note, too, in the spectrum of 5-hexen-2-one the band for the carbon-carbon double bond stretching frequency at 1630 cm^{-1} and the band for the stretching frequency at 3080 cm^{-1} for the hydrogen atoms on the double bond (p. 361).

A similar comparison may be made between the infrared spectrum of 1-phenyl-2-butanone and that of 1-phenyl-1-butanone (Figure 10.11). The carbonyl group in 1-phenyl-2-butanone is separated from the aromatic ring by an sp^3-hybridized carbon atom. Its carbonyl absorption band appears at 1715 cm^{-1}, a typical stretching frequency for an alkyl ketone. The carbonyl stretching frequency for 1-phenyl-1-butanone is lower, at 1695 cm^{-1}, indicating that in this compound the carbonyl group is conjugated with an unsaturated system. Note the strong, sharp bands between 1600 and 1450 cm^{-1} and between 850 and 700 cm^{-1} in the spectra in Figure 10.11. These bands indicate the presence of aromatic rings.

C. Infrared Spectra of Carboxylic Acids and Esters

The infrared spectra of carboxylic acids display two important features, which are illustrated in the spectra of hexanoic acid and benzoic acid (Figure 10.12). First, because of the very strong hydrogen bonding between the carboxyl groups of acid molecules, a strong and broad absorption band appears from 3300 cm^{-1} to as low as 2500 cm^{-1} in the region of the spectrum for stretching frequencies for the oxygen-hydrogen single bond. The stretching frequencies for the carbon-hydrogen bonds of carboxylic acids are usually buried within this band. Second, the stretching frequency for the carbonyl group of a carboxylic acid appears in one of two regions. Acids in which the carbonyl group is bonded to a tetrahedral carbon atom

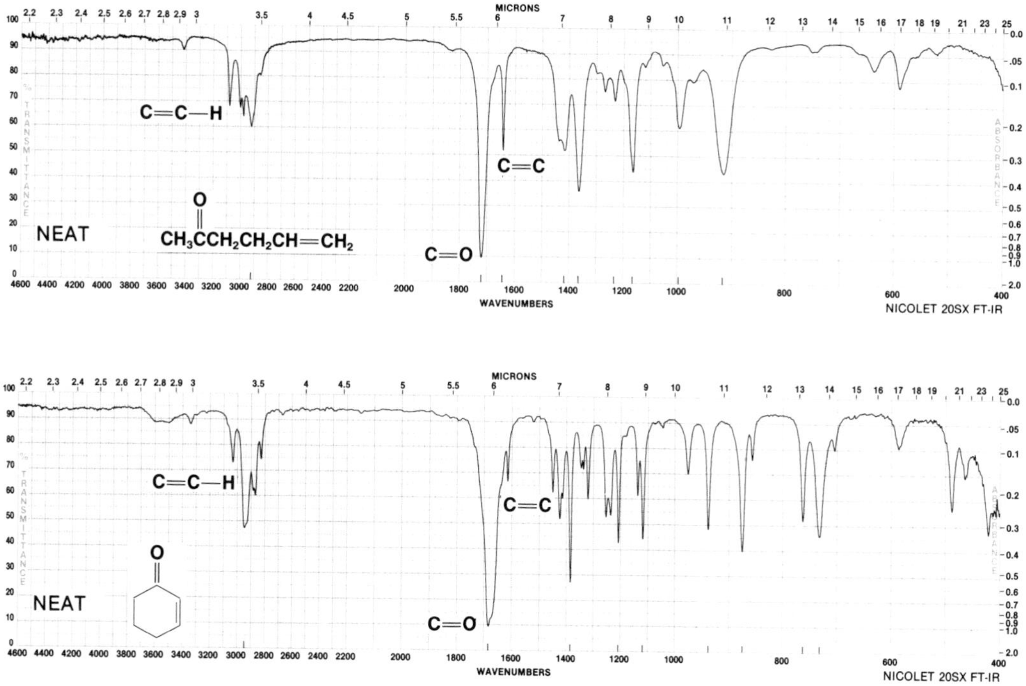
MICRONS
WAVENUMBERS
TRANSMITTANCE
ABSORBANCE
C=C—H
C=C
NEAT
$CH_3\overset{O}{\overset{\|}{C}}CH_2CH_2CH{=}CH_2$
C=O
NICOLET 20SX FT-IR
C=C—H
C=C
NEAT
C=O
NICOLET 20SX FT-IR

FIGURE 10.10 The infrared spectra of 5-hexen-2-one and 2-cyclohexen-1-one. (From *The Aldrich Library of FT-IR Spectra*)

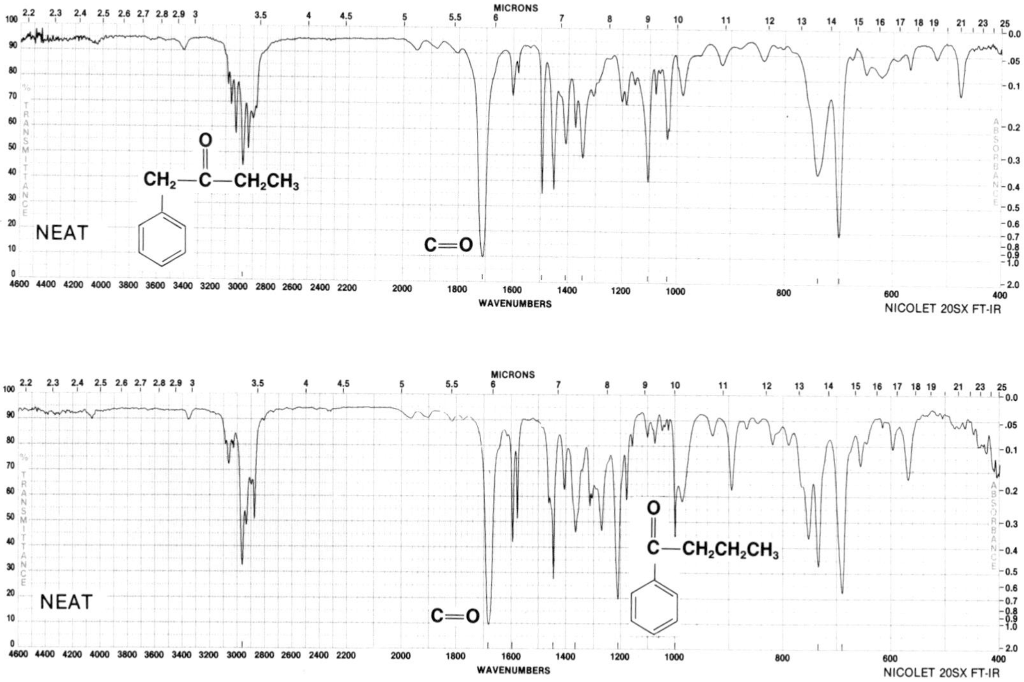
MICRONS
WAVENUMBERS
TRANSMITTANCE
ABSORBANCE
$CH_2—\overset{O}{\overset{\|}{C}}—CH_2CH_3$
NEAT
C=O
NICOLET 20SX FT-IR
$\overset{O}{\overset{\|}{C}}—CH_2CH_2CH_3$
NEAT
C=O
NICOLET 20SX FT-IR

FIGURE 10.11 The infrared spectra of 1-phenyl-2-butanone and 1-phenyl-1-butanone. (From *The Aldrich Library of FT-IR Spectra*)

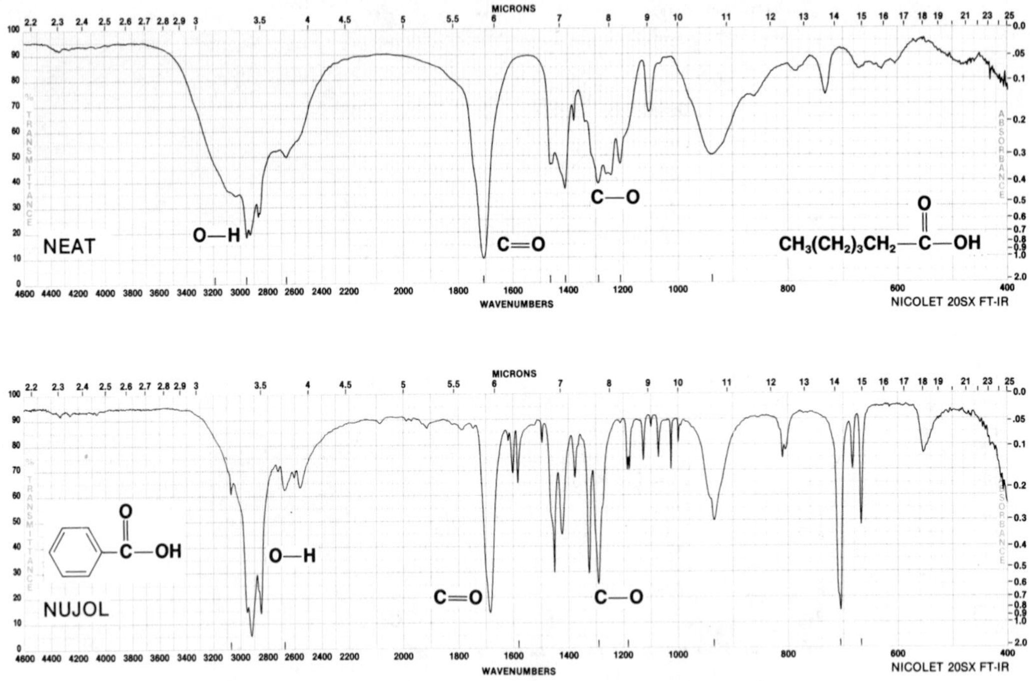
MICRONS
WAVENUMBERS
TRANSMITTANCE
ABSORBANCE
NEAT
O—H
C=O
C—O
$CH_3(CH_2)_3CH_2—\overset{O}{\overset{\|}{C}}—OH$
NICOLET 20SX FT-IR
NUJOL
C—OH
O—H
C=O
C—O
NICOLET 20SX FT-IR

FIGURE 10.12 Infrared spectra of hexanoic and benzoic acids. (From *The Aldrich Library of FT-IR Spectra*)

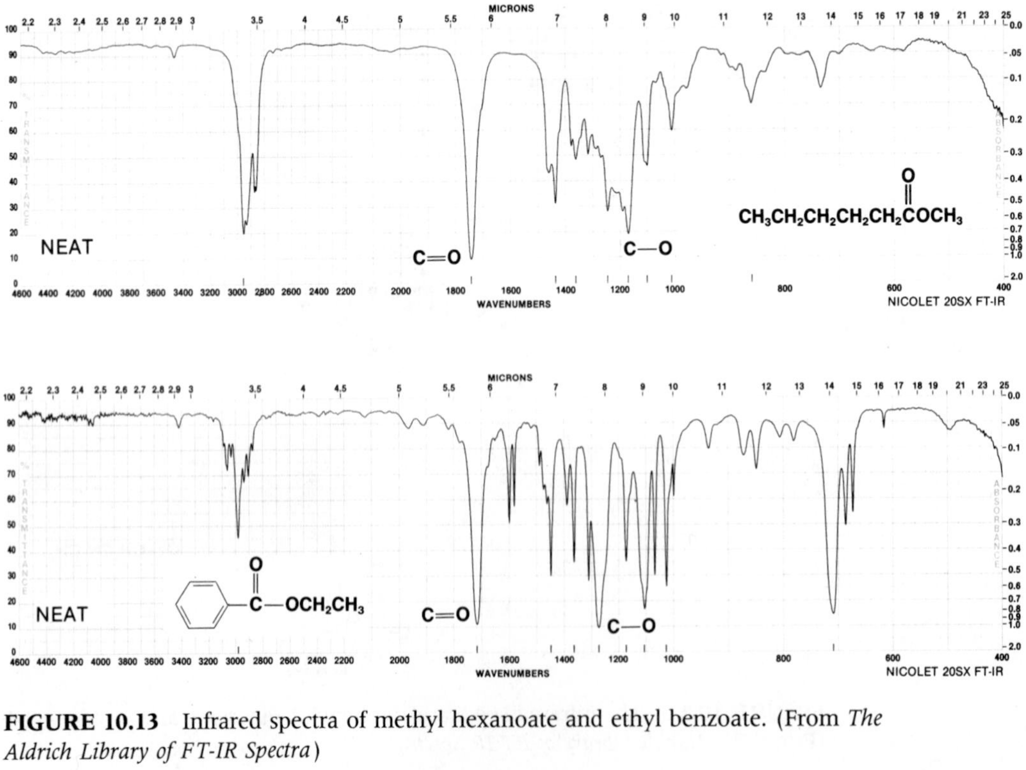
MICRONS
WAVENUMBERS
NEAT
C=O
C—O
$CH_3CH_2CH_2CH_2CH_2\overset{O}{\overset{\|}{C}}OCH_3$
NICOLET 20SX FT-IR
NEAT
C—OCH₂CH₃
C=O
C—O
NICOLET 20SX FT-IR

FIGURE 10.13 Infrared spectra of methyl hexanoate and ethyl benzoate. (From *The Aldrich Library of FT-IR Spectra*)

absorb around 1720–1706 cm^{-1}. For acids in which the carbonyl group is conjugated with a double bond or with an aromatic ring, of which benzoic acid is an example, the carbonyl stretching frequency is around 1710–1680 cm^{-1}.

The important absorption bands for esters lie in the carbonyl stretching region and the carbon-oxygen single bond stretching region of the spectrum. The carbonyl stretching frequency for alkyl esters is around 1750–1735 cm^{-1}. If the carbonyl group of the ester is conjugated, absorption occurs at 1730–1715 cm^{-1}. The carbon-oxygen single bond stretching frequency varies with the nature of the group attached to the oxygen atom. In all esters, two bands are seen in the region between 1300 and 1000 cm^{-1}, the same region in which a similar absorption is seen for alcohols (p. 361). Typical spectra of esters are those of methyl hexanoate and ethyl benzoate (Figure 10.13).

Note that the infrared spectra of esters do not show a band for the stretching frequency for the hydroxyl group. Also, note the greater complexity of the spectra of the aromatic acid and ester compared to those of the alkyl acid and ester. The several sharp bands from 1600 to 1450 cm^{-1} and the strong bands between 870 and 675 cm^{-1} are typical of compounds that contain aromatic rings (p. 366).

PROBLEM 10.3

Spectra numbered 1–6 are given in Figure 10.14. Match each spectrum with one of the following compounds.

(a) $CH_3CH_2CH_2CH_2CH_2CH_2CH_2CH_2OH$ (1-octanol)

(b) 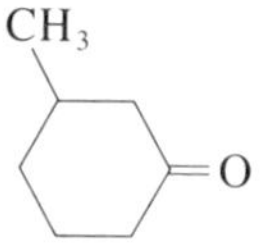 (3-methylcyclohexanone)

(c) $CH_3CH_2CH_2CH_2CH_2CH_2CH_2\overset{O}{\overset{\|}{C}}OH$ (octanoic acid)

(d) $CH_3CH_2CH_2CH_2\overset{O}{\overset{\|}{C}}OCH_3$ (methyl pentanoate)

(e) $CH_2{=}CHCH_2\overset{O}{\overset{\|}{C}}OH$ (3-butenoic acid)

(f) $CH_2{=}CHCH_2CH_2CH_2CH_2CH_2CH_2CH_2CH_2\overset{O}{\overset{\|}{C}}H$ (10-undecenal)

Spectrum 1

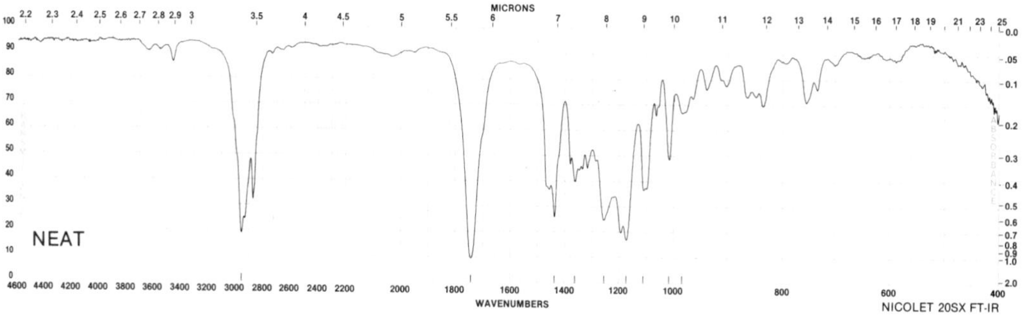
MICRONS
WAVENUMBERS
NEAT
NICOLET 20SX FT-IR

FIGURE 10.14

(Continued)

FIGURE 10.14
(Continued)

Spectrum 2

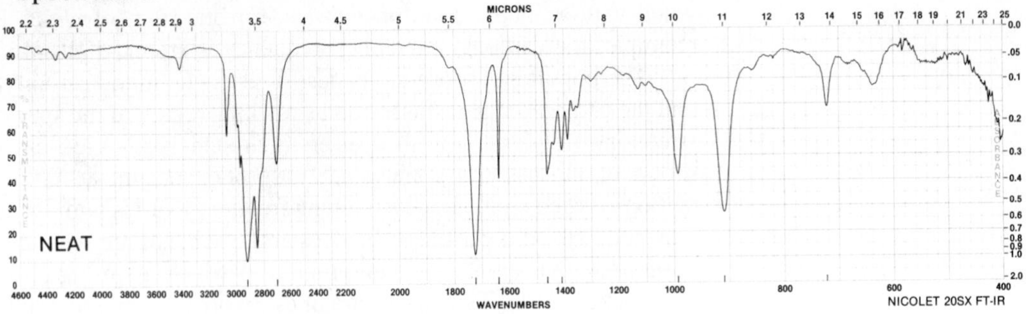
MICRONS
% TRANSMITTANCE
ABSORBANCE
NEAT
WAVENUMBERS
NICOLET 20SX FT-IR

Spectrum 3

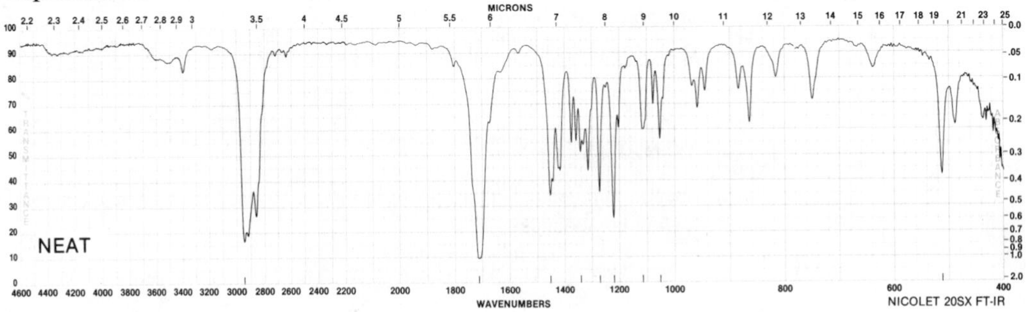
MICRONS
% TRANSMITTANCE
ABSORBANCE
NEAT
WAVENUMBERS
NICOLET 20SX FT-IR

Spectrum 4

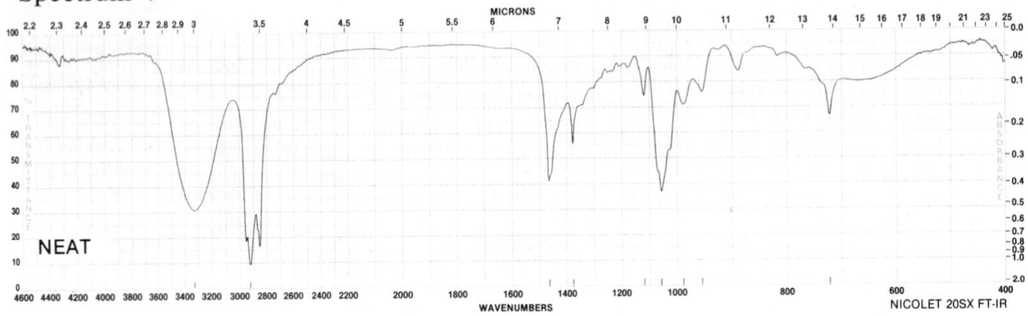
MICRONS
% TRANSMITTANCE
ABSORBANCE
NEAT
WAVENUMBERS
NICOLET 20SX FT-IR

Spectrum 5

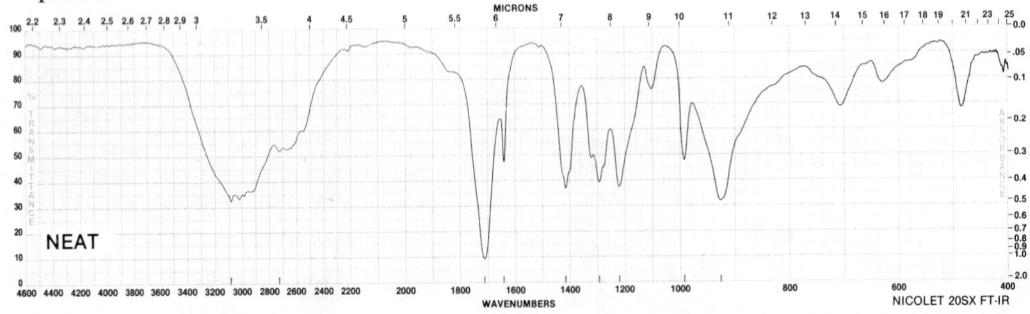
MICRONS
% TRANSMITTANCE
ABSORBANCE
NEAT
WAVENUMBERS
NICOLET 20SX FT-IR

Spectrum 6

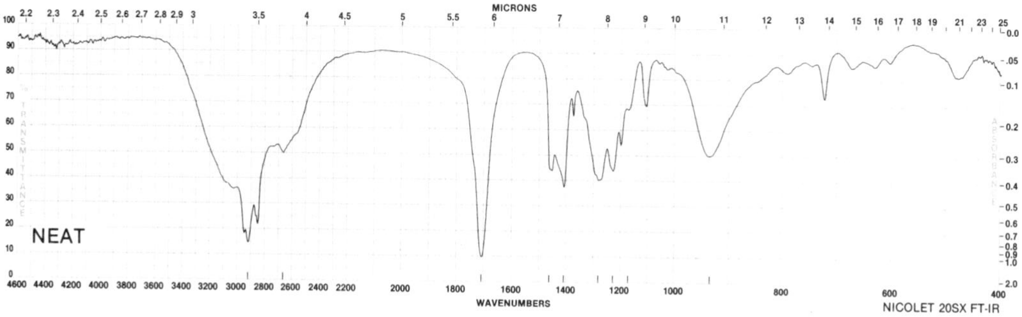

MICRONS
NEAT
WAVENUMBERS
NICOLET 20SX FT-IR

FIGURE 10.14
(Continued)

From *The Aldrich Library of FT-IR Spectra*

SUMMARY

A molecule absorbs electromagnetic radiation that corresponds to transitions between fixed energy levels for that species. Transitions occur between vibrational energy levels in a molecule upon absorption of infrared radiation, usually at a frequency from 4000 to 600 cm^{-1}. Chemical bonds undergo various stretching and bending vibrations. Vibrations that give rise to a change in the dipole moment result in absorption of radiation that can be monitored by a spectrophotometer. A spectrum is a record of the change in the absorption of energy by the compound plotted against the wavelength or the frequency of the radiation being used.

The frequencies at which a molecule absorbs depend on the types of bonds present. The stretching frequency of a bond is related to the masses of the two atoms involved in the bond and to the strength of the bond. For example, it is more difficult to stretch a triple bond than it is to stretch a double bond; therefore, a triple bond absorbs at a higher frequency than a double bond does.

An infrared spectrum is generated by the molecule as a whole, but certain absorption bands are typical of particular groups of atoms, such as doubly bonded carbons, a carbonyl group, or a hydroxyl group. Typical group frequencies are compiled in tables (Table 10.2, p. 360, for example) that can be consulted in the interpretation of spectra. The group frequency is sensitive to the environment around the functional group. For example, the absorption band for a carbonyl group is at a different frequency for a ketone, an ester, and a conjugated carbonyl compound. Such variations in absorption allow chemists to use infrared spectra to assign structures to organic compounds. The frequencies that are useful in detecting the presence of functional groups are between 4000 and 1400 cm^{-1}. The portion of an infrared spectrum from 1400 to 600 cm^{-1} is called the fingerprint region. Here it is not possible to assign all of the bands, but the number of bands and their relative intensities are typical of the compound and can be used to identify it.

The interpretation of infrared spectra requires both training and practice. Only by looking at a number of spectra until you learn to see beyond the tracing on the page, to visualize the structural features that give rise to the spectrum, can you achieve proficiency.

ADDITIONAL PROBLEMS

10.4 Figure 10.15 presents infrared spectra of Compounds I–L, which may be alkenes, alkynes, alcohols, aldehydes, ketones, acids, or esters. Assign each compound to a functional group class.

Compound I

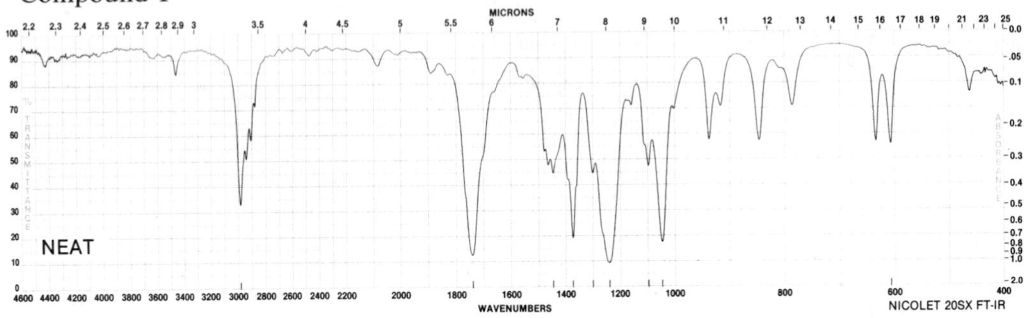
MICRONS
NEAT
WAVENUMBERS
NICOLET 20SX FT-IR

Compound J

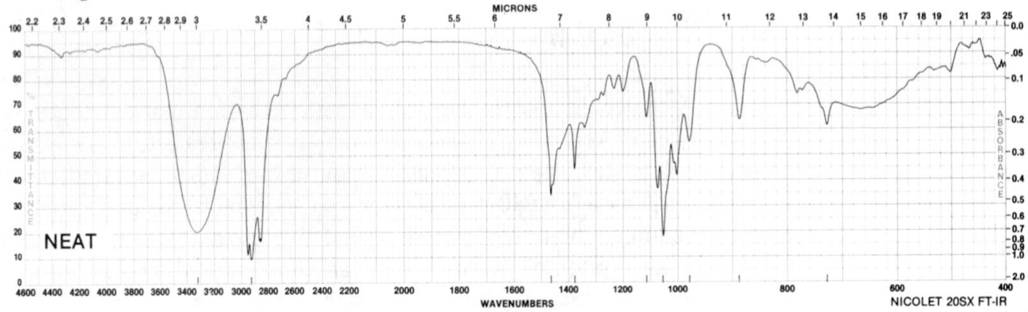
MICRONS
NEAT
WAVENUMBERS
NICOLET 20SX FT-IR

Compound K

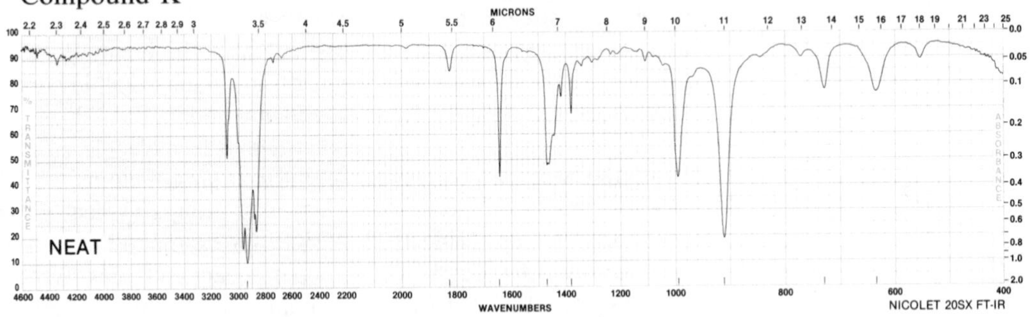
MICRONS
NEAT
WAVENUMBERS
NICOLET 20SX FT-IR

Compound L

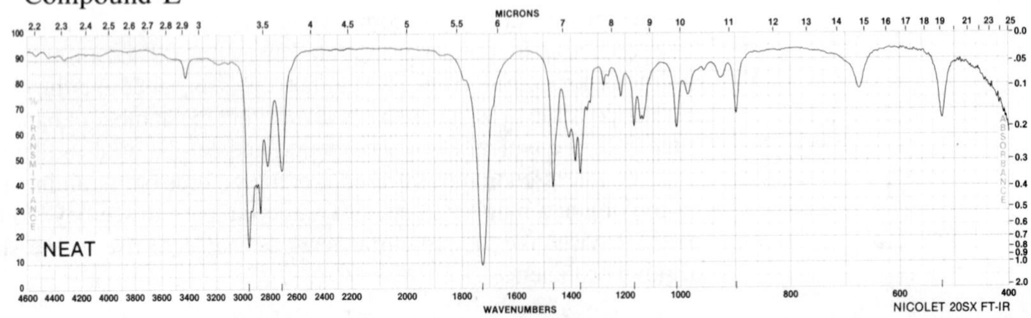
MICRONS
NEAT
WAVENUMBERS
NICOLET 20SX FT-IR

FIGURE 10.15

From *The Aldrich Library of FT-IR Spectra*

10.5 Infrared spectra and molecular formulas are given in Figure 10.16 for Compounds M–P. Assign structures that are compatible with the data. (Hint: Calculating the units of unsaturation is a good way to start to solve these problems.)

Compound M, C_3H_3Cl

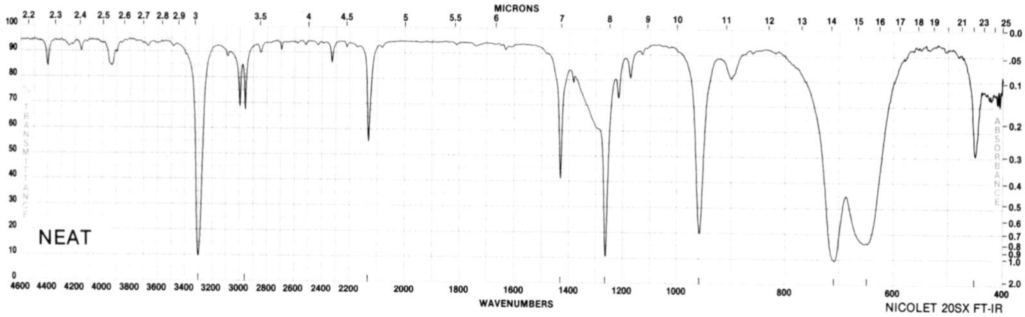
MICRONS
NEAT
WAVENUMBERS
NICOLET 20SX FT-IR

Compound N, $C_5H_{10}O$

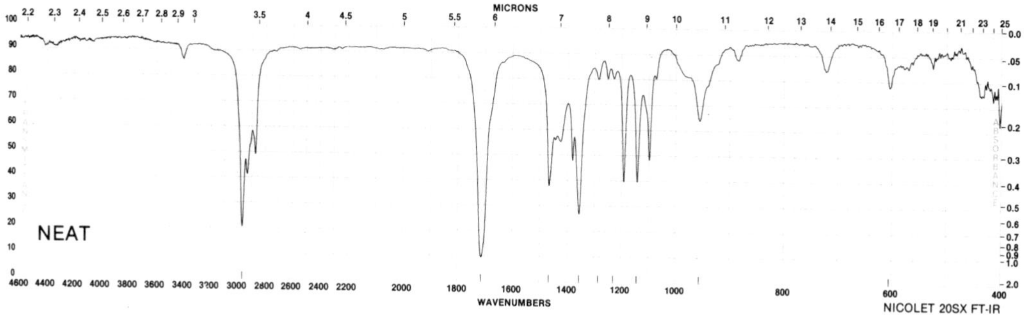
MICRONS
NEAT
WAVENUMBERS
NICOLET 20SX FT-IR

Compound O, $C_5H_{10}O_2$

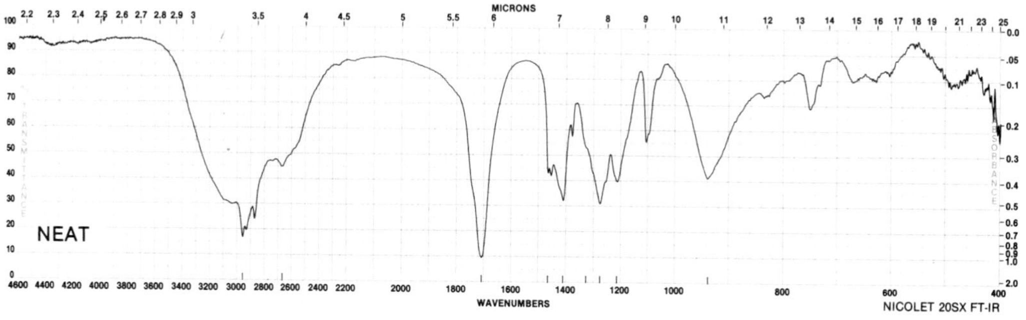
MICRONS
NEAT
WAVENUMBERS
NICOLET 20SX FT-IR

Compound P, $C_8H_8O_2$

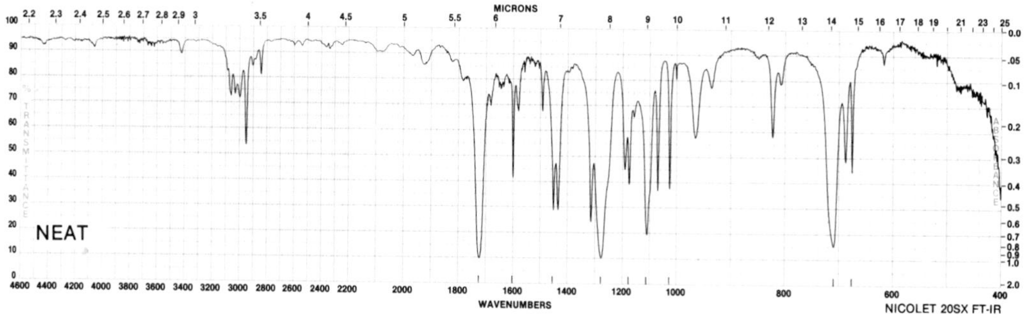
MICRONS
NEAT
WAVENUMBERS
NICOLET 20SX FT-IR

FIGURE 10.16

From *The Aldrich Library of FT-IR Spectra*

10.6 The infrared spectra in Figure 10.17 are of Compounds Q–T, each of which has two major functional groups. Use the spectra to identify the functional groups that are present.

Compound Q

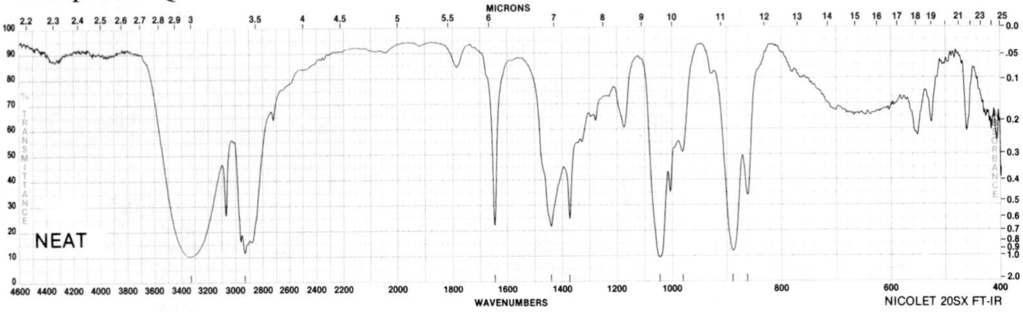
MICRONS
% TRANSMITTANCE
ABSORBANCE
NEAT
WAVENUMBERS
NICOLET 20SX FT-IR

Compound R

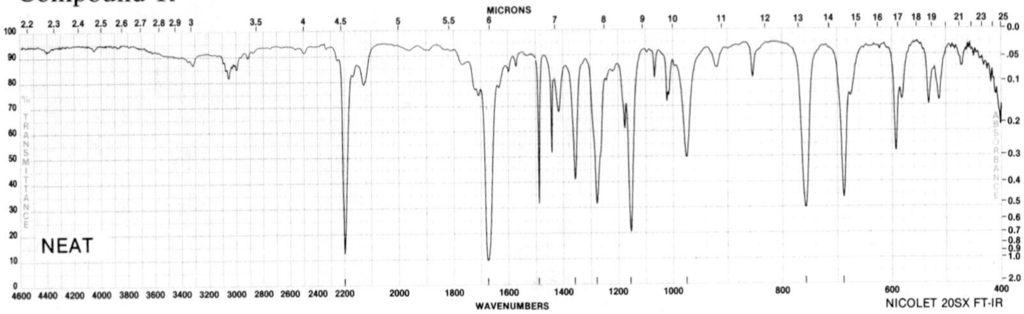
MICRONS
% TRANSMITTANCE
ABSORBANCE
NEAT
WAVENUMBERS
NICOLET 20SX FT-IR

Compound S

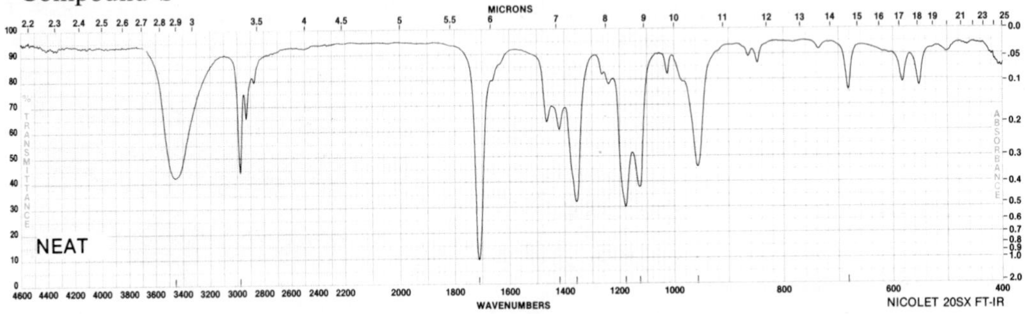
MICRONS
% TRANSMITTANCE
ABSORBANCE
NEAT
WAVENUMBERS
NICOLET 20SX FT-IR

Compound T

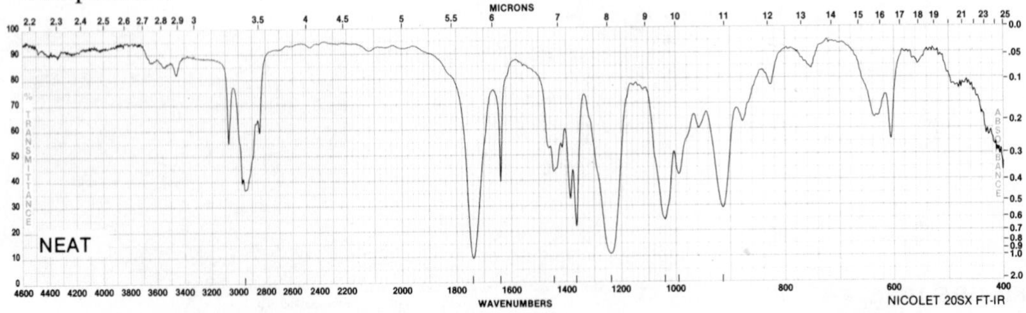
MICRONS
% TRANSMITTANCE
ABSORBANCE
NEAT
WAVENUMBERS
NICOLET 20SX FT-IR

FIGURE 10.17

From *The Aldrich Library of FT-IR Spectra*

10.7 (+)-17-Methyl testosterone is a steroid related to the male sex hormones. Its infrared spectrum and structural formula appear in Figure 10.18. Assign as many of the absorption bands as you can to specific vibrational transitions in the molecule.

FIGURE 10.18

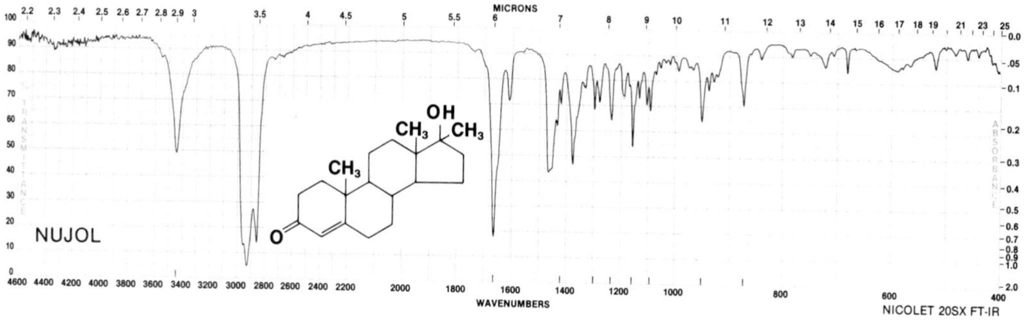

MICRONS
WAVENUMBERS
NUJOL
OH
CH3
CH3
CH3
O
NICOLET 20SX FT-IR

From *The Aldrich Library of FT-IR Spectra*

11

Nuclear Magnetic Resonance Spectroscopy

A • L O O K • A H E A D

The nuclei of atoms such as hydrogen and carbon-13 behave like small magnets. If a sample of a compound containing these elements is placed in a strong magnetic field, slightly more than half of the nuclei align themselves with the field. Nuclei in this state absorb radiation in the radiofrequency range of the electromagnetic spectrum (p. 356) and are raised to a higher-energy state, in which they are aligned against the external magnetic field. The record of their transitions to a higher-energy state is a nuclear magnetic resonance spectrum.

The exact amount of energy necessary to cause a nucleus to undergo a transition from a lower energy state to a higher one depends on the strength of the magnetic field used and on the environment of the atom in the molecule. For example, a hydrogen atom on a methyl group bonded to an electronegative oxygen atom experiences the external magnetic field in a slightly different way than does a hydrogen atom on a methyl group bonded to a carbon atom, and, therefore, absorbs a different amount of energy in undergoing a nuclear transition. A nuclear magnetic resonance spectrum of a compound thus has a number of peaks corresponding to the different types of hydrogen atoms in the molecule. This chapter will describe the different kinds of information that chemists obtain from nuclear magnetic resonance spectra and how to use such spectra in determining the structures of organic compounds.

11.1 THE EXPERIMENTAL OBSERVATIONS

The nuclei of hydrogen atoms absorb radiofrequency radiation when placed in a magnetic field. The spectra that record these absorptions of energy are known as **proton magnetic resonance** spectra. Two examples of such spectra are shown in Figure 11.1. Both of these spectra have a signal at the far right, which is the

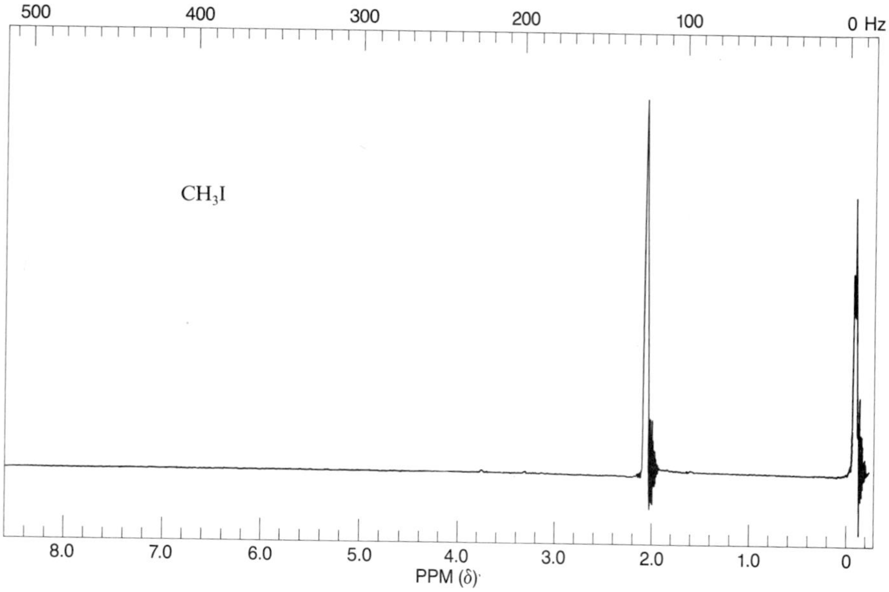
500
400
300
200
100
0 Hz
CH3I
8.0
7.0
6.0
5.0
4.0
3.0
2.0
1.0
0
PPM (δ)

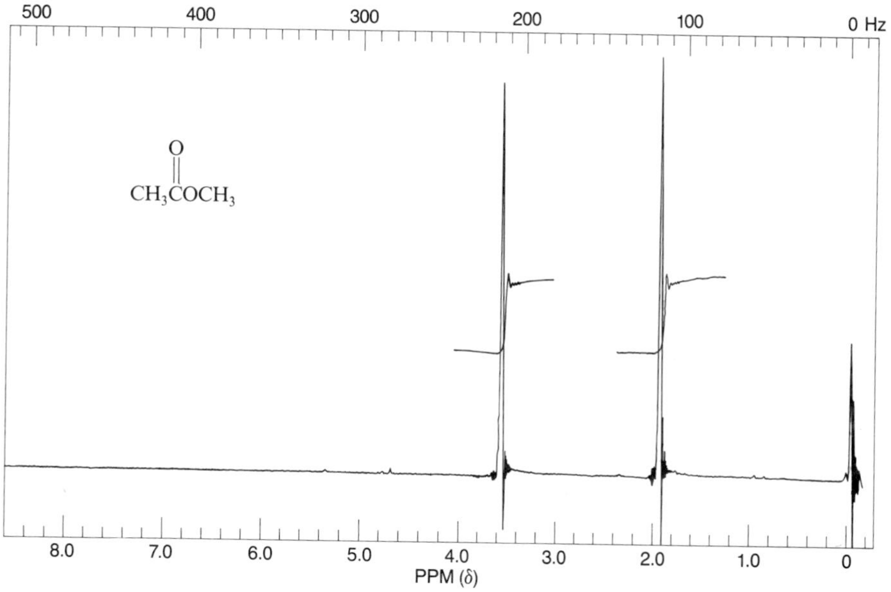
500
400
300
200
100
0 Hz
CH3COCH3
8.0
7.0
6.0
5.0
4.0
3.0
2.0
1.0
0
PPM (δ)

FIGURE 11.1 Proton magnetic resonance spectra of iodomethane and methyl acetate.

absorption band for the hydrogen atoms in tetramethylsilane, $(CH_3)_4Si$. Tetramethylsilane, usually abbreviated TMS, is added to the solution of a compound for which a proton magnetic spectrum is being run to give a reference point. The divisions on the scale at the bottom of the spectra are in units called δ (delta) and represent chemical shift values (p. 384). The signal for the hydrogen atoms in TMS appears at δ 0. The spectrum of iodomethane has one other signal, at δ 2.10, and that of methyl acetate has two other signals, at δ 1.95 and δ 3.60.

The three hydrogen atoms in iodomethane are said to be **chemical-shift equivalent.** In methyl acetate, the hydrogen atoms of the methyl group adjacent to the carbonyl group are different from those of the methyl group adjacent to the oxygen atom. Whether hydrogen atoms are chemical-shift equivalent can be determined by mentally substituting another species for each one of them in turn and seeing if the same or a different compound results. For example, with iodomethane, substitution of a bromine atom for any of the three hydrogen atoms gives rise to the same compound.

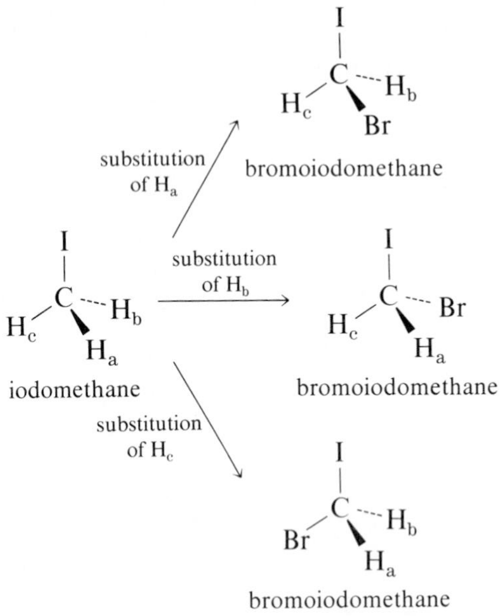
substitution of H_a
bromoiodomethane
substitution of H_b
iodomethane
bromoiodomethane
substitution of H_c
bromoiodomethane

The three hydrogen atoms are, therefore, chemical-shift equivalent and give rise in the proton magnetic resonance spectrum to a single sharp peak, called a **singlet.**

The three hydrogen atoms of the acetyl group in methyl acetate are chemical-shift equivalent but are different from the three hydrogen atoms of the methoxyl group. For example, substitution of a methyl group for a hydrogen atom on each of these parts of the molecule leads to two entirely different compounds.

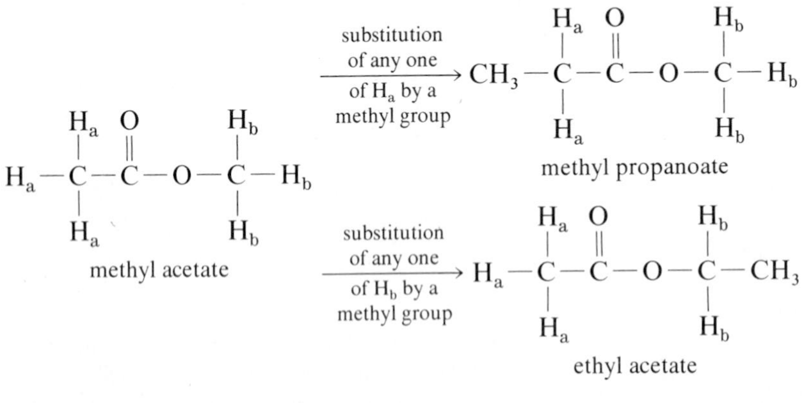
substitution of any one of H_a by a methyl group
methyl propanoate
methyl acetate
substitution of any one of H_b by a methyl group
ethyl acetate

The hydrogen atoms of the acetyl group and of the methoxyl group thus give rise to different absorption bands, each one a singlet, in the proton magnetic resonance spectrum.

The spectrum of methyl acetate in Figure 11.1 shows another feature. The steplike tracing over the two bands of the spectrum is known as the **integration.** The instrument measures the area under each absorption band and records it as the height of a step in the tracing. An examination of the spectrum of methyl acetate shows that the two steps are of the same height. In other words, the areas under the

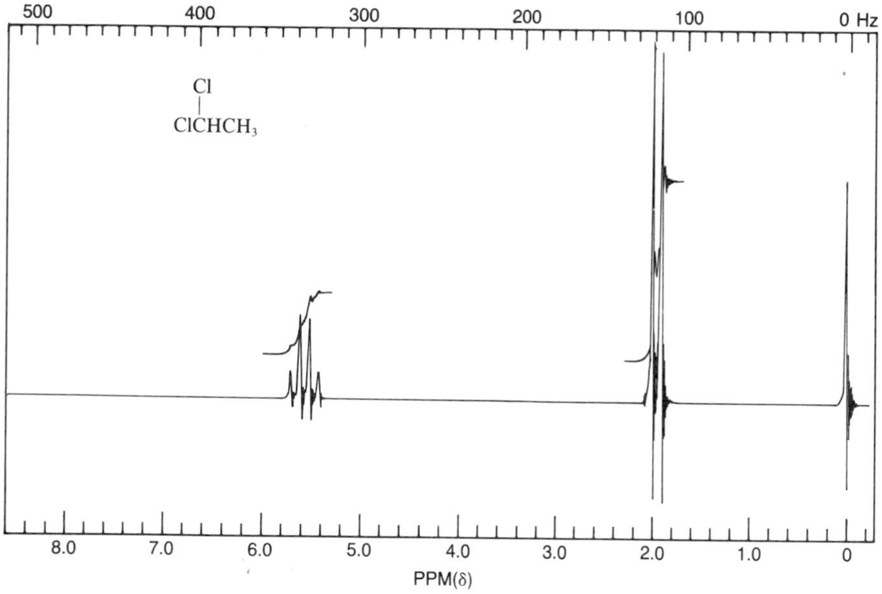
500
400
300
200
100
0 Hz
Cl
ClCHCH$_3$
8.0
7.0
6.0
5.0
4.0
3.0
2.0
1.0
0
PPM(δ)

FIGURE 11.2 Proton magnetic resonance spectrum for 1,1-dichloroethane.

two peaks are equal. In a proton magnetic resonance spectrum, the area under an absorption band is proportional to the relative number of hydrogen atoms giving rise to that signal. Thus, the integration of this spectrum reveals that there are equal numbers of protons of two types in methyl acetate. Note that the integration does *not* give the absolute number of protons of each kind, but only their relative abundance.

Not all proton magnetic resonance spectra are as simple as the ones shown in Figure 11.1. In those compounds, even though there are a number of hydrogen atoms, they are either chemical-shift equivalent or isolated from one another by five σ bonds. Figure 11.2 shows the proton magnetic resonance spectrum for 1,1-dichloroethane. In this compound, adjacent carbon atoms bear different types of hydrogen atoms.

The spectrum of 1,1-dichloroethane shows a number of peaks. Note, however, that the peaks appear in two groups, one centered at δ 1.95 and the other at δ 5.60. The relative intensities of these bands (for all the small peaks in one group counted together) are about 3:1. The band at δ 1.95 appears, therefore, to be the absorption band for the hydrogen atoms of the methyl group. The one at δ 5.60 belongs to the hydrogen atom on the carbon atom that is also bonded to two electronegative chlorine atoms.

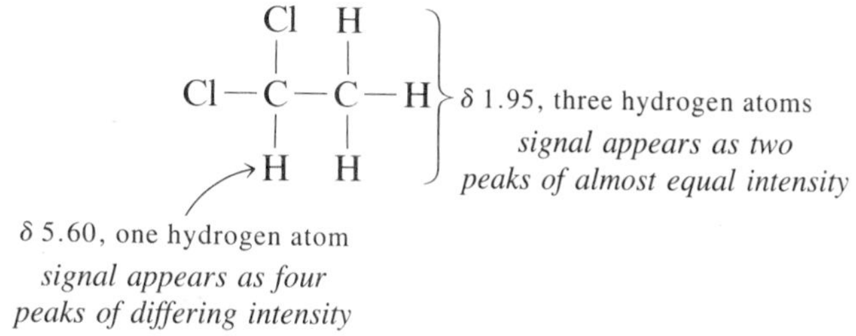
δ 1.95, three hydrogen atoms
signal appears as two
peaks of almost equal intensity
δ 5.60, one hydrogen atom
signal appears as four
peaks of differing intensity

The bands in the spectrum in Figure 11.2 are said to be split. This splitting arises from the interaction of the nuclei of the hydrogen atoms on neighboring carbon atoms and is known as **spin-spin coupling.** Section 11.4A examines how such interactions create the observed patterns.

Four essential pieces of information about a compound are obtained from its proton magnetic resonance spectrum.

1. The number of different groupings of peaks tells how many different kinds of hydrogen atoms are present in the compound.
2. The δ values (chemical shifts) observed for the hydrogen atoms give information about the environment of the hydrogen atoms.
3. The integration of the spectrum reveals the relative numbers of hydrogen atoms of each type that are present in the molecule.
4. The splitting patterns that appear in the spectrum tell which hydrogen nuclei are interacting with one another.

PROBLEM 11.1

For each of the following compounds, pick out the hydrogen atoms that are chemical-shift equivalent to each other.

(a) $CH_3CH_2OCH_2CH_3$ (b) $CH_3CH_2CH_2Br$ (c) 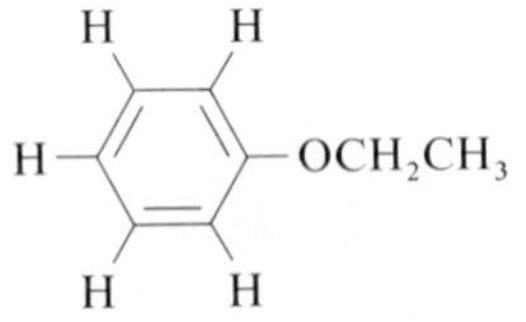

(d) $CH_3CH(CH_3)CH_2Cl$ (e) $CH_3CH_2C(=O)CH_2CH_3$ (f) $CH_3CH_2CH_2CH_2CH_3$

PROBLEM 11.2

The three spectra shown in Figure 11.3 are of the following compounds. Assign each spectrum to the correct compound, and discuss how you made each decision. You can determine the structures by comparing the spectra with those already given in this section and reasoning by analogy.

(a) acetone (b) 1,2-dichloroethane (c) 1,1,2-trichloroethane

Compound A

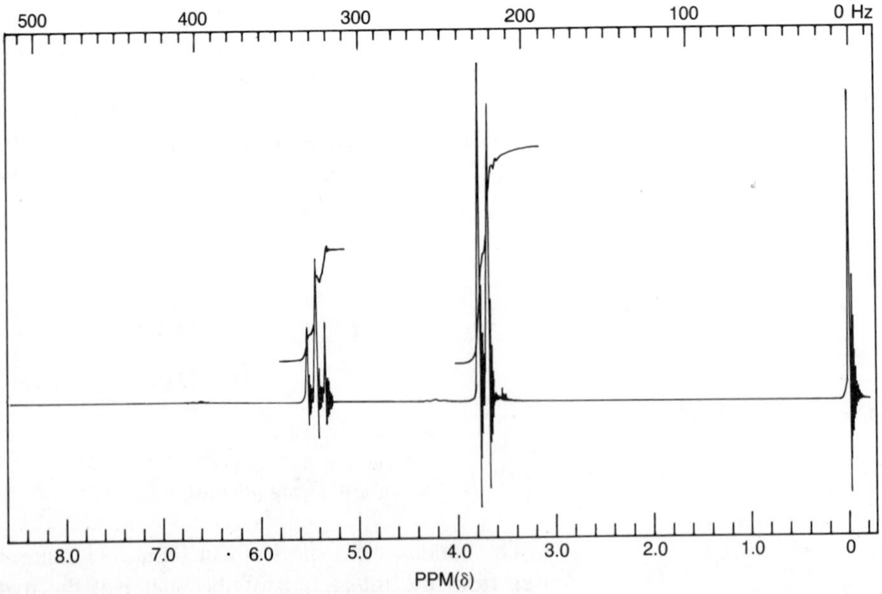
500
400
300
200
100
0 Hz
8.0
7.0
6.0
5.0
4.0
3.0
2.0
1.0
0
PPM(δ)

FIGURE 11.3

FIGURE 11.3 *(Continued)*

Compound B

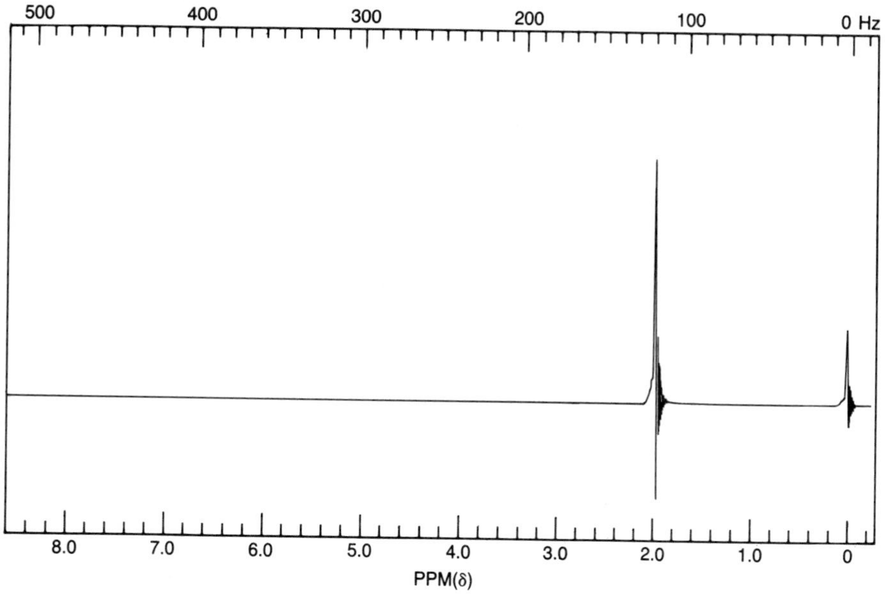

500
400
300
200
100
0 Hz
8.0
7.0
6.0
5.0
4.0
3.0
2.0
1.0
0
PPM(δ)

Compound C

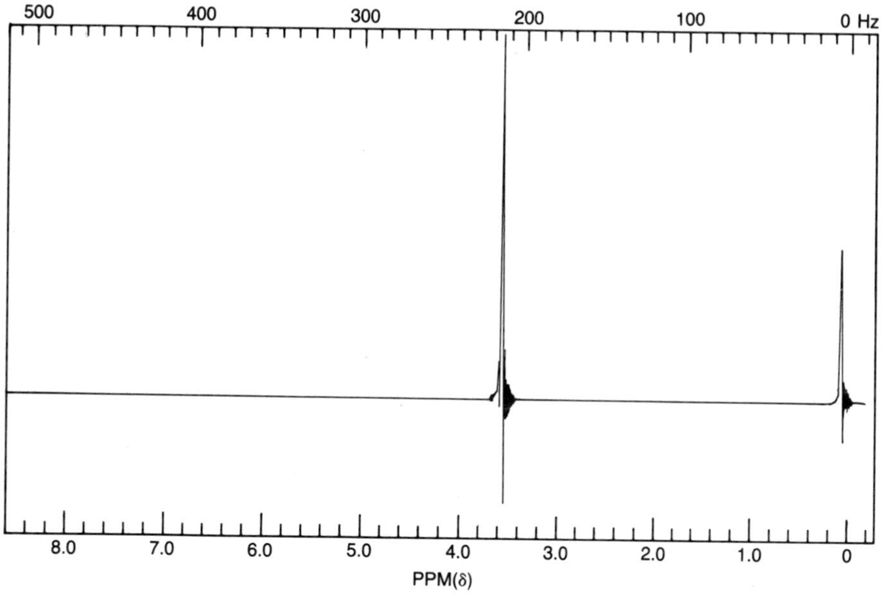

500
400
300
200
100
0 Hz
8.0
7.0
6.0
5.0
4.0
3.0
2.0
1.0
0
PPM(δ)

11.2 THE ORIGIN OF NUCLEAR MAGNETIC RESONANCE SPECTRA

Just as electrons in an atom are assigned spin quantum numbers, nuclei of atoms with an odd mass number have **nuclear spin** and are given a **spin number, *I*,** of $\frac{1}{2}$. Thus, the nuclei of the common isotope of hydrogen, ^{1}H, and the stable isotope of carbon, ^{13}C, have a spin number, I, of $\frac{1}{2}$. Nuclei of atoms that have an even mass number but an odd atomic number also possess nuclear spin. Deuterium, ^{2}H, and the common isotope of nitrogen, ^{14}N, belong in this category, and their nuclei have a spin number, I, of 1. Nuclei of atoms that have an even mass number and an even atomic number have no spin. The nuclei of the most abundant isotopes of carbon, ^{12}C, and oxygen, ^{16}O, fall in this category; their spin number, I, is 0.

Only atoms with nuclear spin give rise to nuclear magnetic resonance. The nucleus is a charged particle, so a spinning nucleus creates a magnetic field. Such a nucleus behaves like a small magnet; it has a **magnetic moment** and is affected by an external magnetic field. When atoms containing such nuclei are placed in a strong magnetic field, $(2I + 1)$ energy levels, where I is the spin number of the nuclei, become available to the nuclei. For example, when a sample of a compound containing hydrogen is put in an external magnetic field, the hydrogen nuclei exist in two energy states. Slightly more than half of the nuclei line up so that their magnetic moments are aligned with the external magnetic field. This situation is the lower-energy state for the nuclei. The difference between this lower-energy level and the higher-energy one corresponds to the energy of radiation in the radio-frequency range of the electromagnetic spectrum (p. 356). If an external magnetic field of 14,092 gauss is used, the sample absorbs radiation having a wavelength of approximately 5 meters or a frequency of 60,000,000 cycles/s, or hertz (Hz). Nuclei are raised to the higher-energy level, in which their magnetic moments are opposed to the external magnetic field. The exact amount of energy required for the transition depends on the strength of the magnetic field experienced by the nuclei (Figure 11.4).

Once a nucleus is in the higher-energy level, it returns to the lower one by various processes that involve losing energy to its surroundings. These processes

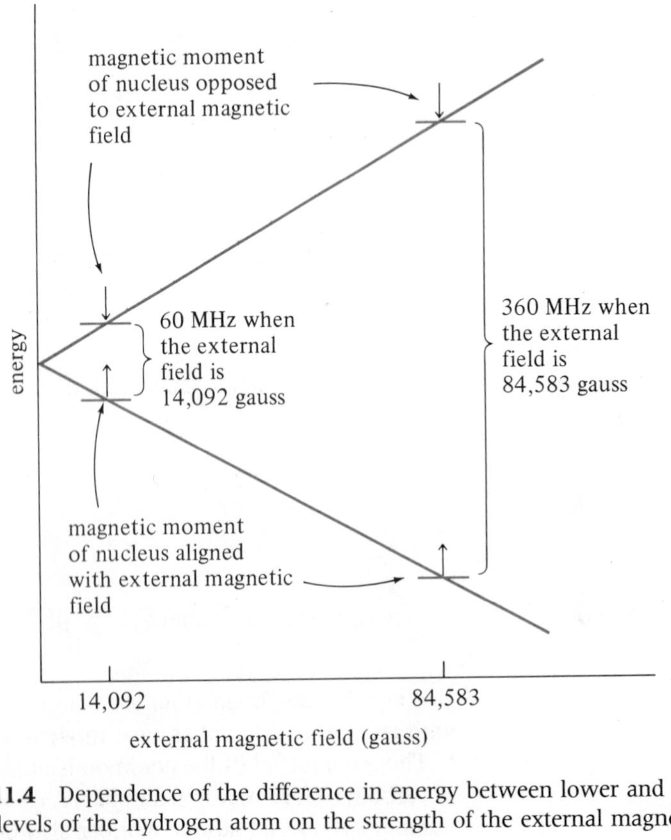
magnetic moment of nucleus opposed to external magnetic field
60 MHz when the external field is 14,092 gauss
360 MHz when the external field is 84,583 gauss
energy
magnetic moment of nucleus aligned with external magnetic field
14,092
84,583
external magnetic field (gauss)

FIGURE 11.4 Dependence of the difference in energy between lower and higher nuclear spin levels of the hydrogen atom on the strength of the external magnetic field.

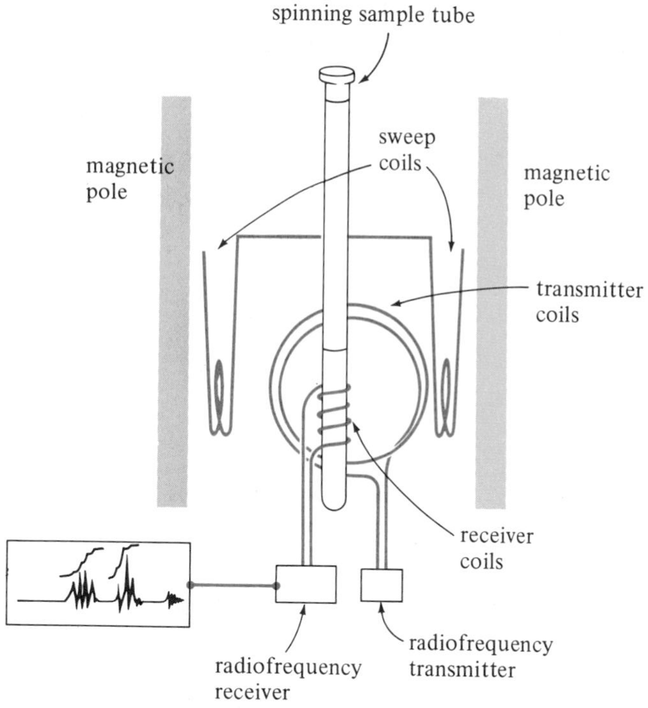
spinning sample tube
sweep
coils
magnetic
pole
magnetic
pole
transmitter
coils
receiver
coils
radiofrequency
receiver
radiofrequency
transmitter

FIGURE 11.5 Schematic diagram of a typical nuclear magnetic resonance spectrometer.

are known as **relaxation,** and are especially important to the biological and medical applications of nuclear magnetic resonance spectroscopy (p. 416).

Not all of the hydrogen nuclei in a molecule of an organic compound experience the same external magnetic field. The electrons around a nucleus **shield** it from the effects of the magnetic field. Thus, hydrogen atoms in different electronic environments in a molecule experience an external magnetic field to different degrees. Slightly different amounts of energy are, therefore, needed to promote the nuclei of different types of hydrogen atoms from the lower-energy level to the higher-energy level.

With most nuclear magnetic resonance spectrometers, the sample is irradiated at a constant radiofrequency, for example, at 60,000,000 Hz (or 60 MHz), while the magnetic field is varied slightly. Whenever the magnetic field actually felt at a nucleus reaches the value at which a transition between energy levels becomes possible with the absorption of energy at 60 MHz, absorption takes place. The instrument (Figure 11.5) records this series of events as a spectrum.

The magnetic field in nuclear magnetic resonance spectrometers must be highly homogeneous if small differences in energy are to be observed. One way to average out the field experienced by all of the molecules in a sample is to spin the tube containing the sample inside a cavity in the magnet. In practice, then, nuclear magnetic resonance spectra are usually taken of liquid samples or of solids in solution. The ideal solvent for proton magnetic resonance spectroscopy is one such as carbon tetrachloride that does not contain any hydrogen atoms. But carbon tetrachloride is not polar enough to dissolve many organic compounds, so other solvents in which hydrogen atoms have been replaced by deuterium, which does not absorb in the same region of the spectrum as hydrogen does, are used. Among these, deuteriochloroform, $CDCl_3$, is the most generally useful. Because it is difficult to measure the absolute value of an applied magnetic field, all nuclear magnetic resonance spectra are measured with reference to a standard compound, usually tetramethylsilane (p. 377).

PROBLEM 11.3

Predict whether or not each of the following isotopes has a nuclear magnetic moment.

(a) ^{28}Si (b) ^{15}N (c) ^{19}F (d) ^{31}P (e) ^{11}B (f) ^{32}S

11.3 CHEMICAL SHIFT

A. The Origin of the Chemical Shift. Chemical Shift Values for Hydrogen Atoms on Tetrahedral Carbon Atoms

The absorption of radiation of a given radiofrequency by the nucleus of a hydrogen atom depends on the effective magnetic field that it experiences. This, in turn, depends on the environment of the hydrogen atom in the molecule as well as on the strength of the external magnetic field. The hydrogen atoms in tetramethylsilane, the commonly used reference compound, have a higher electron density around them and are said to be more shielded than most hydrogen atoms in organic compounds. This is because hydrogen and carbon are more electronegative (2.1 and 2.5) than silicon (1.8), and electron density flows from silicon to the methyl groups. Hydrogen atoms bonded to carbon atoms that are also bonded to electronegative elements such as oxygen or the halogens have less electron density around them. They are said to be **deshielded** relative to the hydrogen atoms on the methyl groups in tetramethylsilane (TMS). The effect is cumulative; a hydrogen atom on a carbon atom bearing two halogen atoms is more deshielded than one on a carbon bearing only a single halogen. For example, the hydrogen atoms in iodomethane absorb at δ 2.10, and those in diiodomethane at δ 3.88.

The effect of electronegativity can also be seen in a series of compounds that differ only in the type of halogen substituent. The hydrogen atoms on the carbon atom bearing the halogen atom become progressively more deshielded in iodomethane (δ 2.10), bromomethane (δ 2.70), chloromethane (δ 3.05), and fluoromethane (δ 4.30), as the electronegativity of the halogen increases. The deshielding of the hydrogen atoms is reflected in larger δ values.

CH_3F	CH_3Cl	CH_3Br	CH_3I
δ 4.30	δ 3.05	δ 2.70	δ 2.10
fluoromethane	chloromethane	bromomethane	iodomethane

←

increasing deshielding of the hydrogen atoms with increasing electronegativity of the halogen

The spectrum of 2,2-dimethoxypropane (Figure 11.6) has two major peaks of equal intensity at δ 1.20 and 2.93. The signal at δ 1.20 is assigned to the hydrogen atoms of the methyl groups bonded to carbon, and that at δ 2.93 to the hydrogen atoms of the methyl groups bonded to oxygen. The hydrogen atoms of these methoxyl groups are more highly deshielded and absorb at lower field than do the

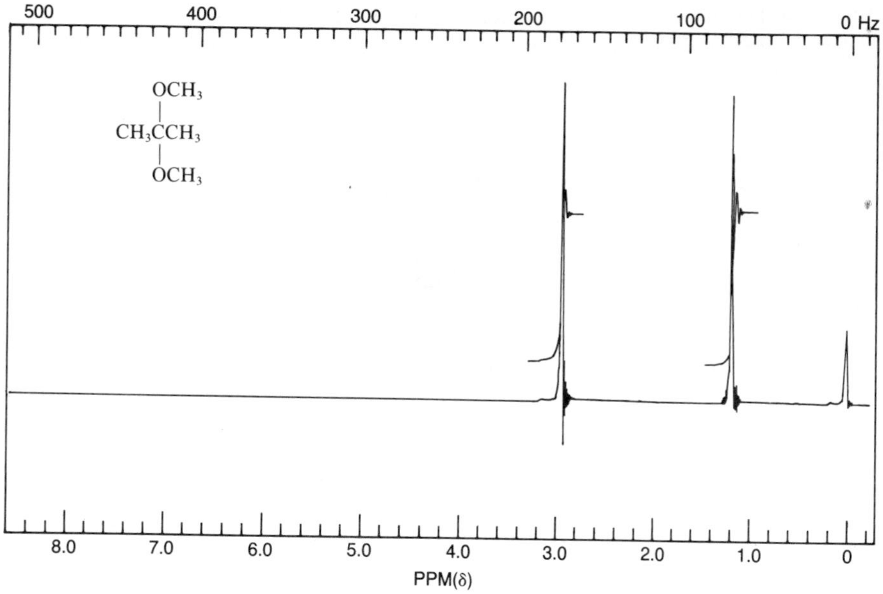
500
400
300
200
100
0 Hz
OCH3
CH3CCH3
OCH3
8.0
7.0
6.0
5.0
4.0
3.0
2.0
1.0
0
PPM(δ)

FIGURE 11.6 Proton magnetic resonance spectrum of 2,2-dimethoxypropane.

other hydrogen atoms. The other hydrogen atoms, those giving rise to the peak at δ 1.20, are said to be more shielded and to absorb at higher field. Both these groups of hydrogen atoms, however, are more deshielded and absorb energy at lower field than do the hydrogen atoms in TMS.

The difference between the position of the absorption band for a given type of hydrogen atom and the position of the peak for TMS is called the **chemical shift** for that hydrogen. The difference in energy between the peak for TMS and those for the hydrogen atoms in 2,2-dimethoxypropane can be read off the spectrum in either hertz or delta units. The spectrum is recorded so that the space between the peak for TMS and the left-hand edge of the chart represents 515 Hz. Each small division from right to left, therefore, corresponds to 10 Hz. The positions of the absorption bands for the hydrogen atoms may be read directly off the chart. For example, the first absorption band occurs at a frequency, ν, of 70 Hz and the second one at 176 Hz, **downfield** from the signal for TMS (to the left of TMS). These chemical shift values depend on the strength of the magnetic field in the instrument and, therefore, on the frequency, ν, of the radiation used to make the measurements. In most cases, chemical shifts measured in Hz are converted to δ, which is independent of instrumentation and has units of ppm:

$$\delta = \frac{\nu \text{ sample} - \nu \text{ TMS}}{\nu \text{ applied of the instrument}} \times 10^6 \text{ ppm}$$

For a 60-MHz instrument, for example,

$$\delta = \frac{\nu \text{ sample} - \nu \text{ TMS}}{60 \times 10^6} \times 10^6 \text{ ppm}$$

Thus, for a band appearing 176 Hz downfield from TMS,

$$\delta = \frac{176 - 0}{60 \times 10^6} \times 10^6 \text{ ppm}$$
$$= 2.93 \text{ ppm}$$

Note that the differences in chemical shift that are being measured in a proton magnetic resonance spectrum are very small, on the order of parts per million in the frequency of the radiation being used to give rise to the nuclear transition. The entire range in chemical shift values for hydrogen atoms is approximately 20 ppm. Note also that δ values increase from right to left, which is opposite to what is usually seen on graphs.

B. Special Effects Seen in Proton Magnetic Resonance Spectra of Compounds with π Bonds

The hydrogen atoms that are on sp^2-hybridized carbon atoms involved in double bonds and aromatic rings and the hydrogen atom on the carbonyl group of an aldehyde are much more deshielded than would be expected just from the effect of the electronegativity of the atoms to which they are bonded. For example, the hydrogen atoms on the carbons of the double bond in cyclohexene absorb at δ 5.30; those on the aromatic ring in toluene, at δ 6.75 (Figure 11.7). In the spectrum for benzaldehyde, the signal for the hydrogen atom on the carbonyl group appears at δ 9.80 (Figure 11.8). This peak appears far downfield from the range normally recorded for proton magnetic spectra, so the spectrum in Figure 11.8 has an expanded scale, with the left-hand edge being 1030 Hz downfield from TMS rather than 515 Hz.

The hydrogen atom bonded to the sp-hybridized carbon atom of a terminal alkyne, on the other hand, absorbs at δ 2.5, far upfield from a vinylic hydrogen atom. The hydrogen atom bonded to an sp-hybridized carbon atom should, in fact, be more deshielded than a hydrogen atom on an sp^2-hybridized carbon atom if electronegativity (p. 56) is the only factor that is operative.

To explain the effects described here, it has been proposed that the motion of electrons in a multiple bond sets up an induced magnetic field around the molecule containing such a bond. For certain orientations of the molecule, the induced magnetic field may either reinforce the external magnetic field or oppose it. Any

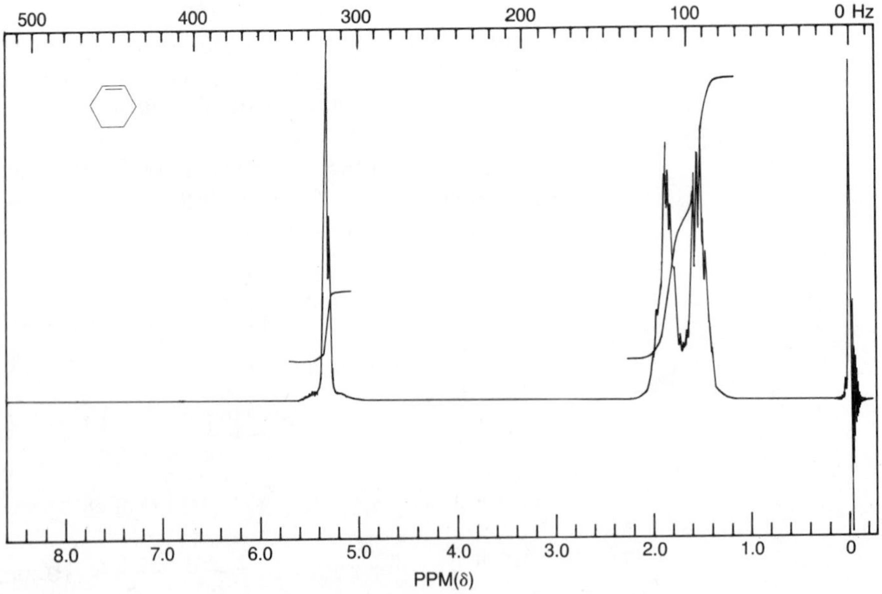
500
400
300
200
100
0 Hz
8.0
7.0
6.0
5.0
4.0
3.0
2.0
1.0
0
PPM(δ)

FIGURE 11.7 Proton magnetic resonance spectra for cyclohexene and toluene.

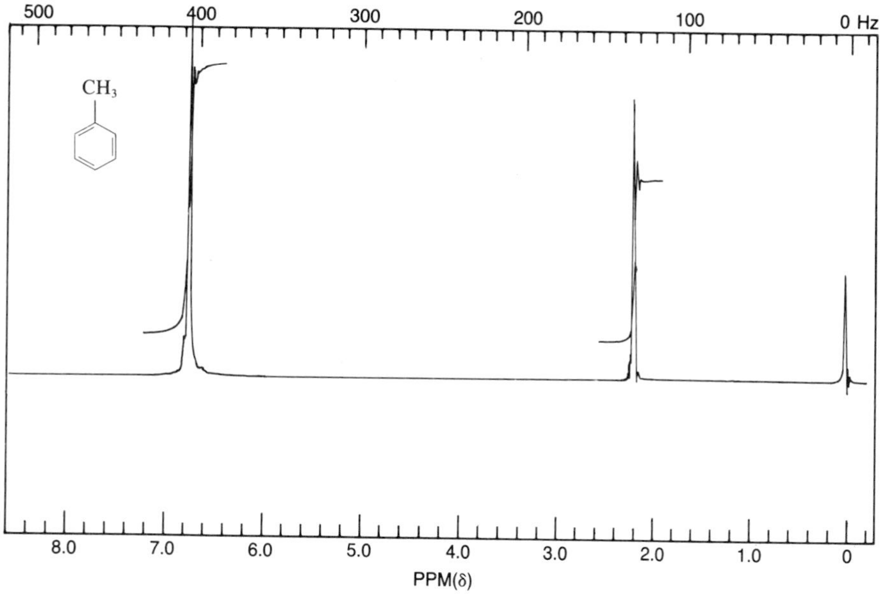

500
400
300
200
100
0 Hz
CH3
8.0
7.0
6.0
5.0
4.0
3.0
2.0
1.0
0
PPM(δ)

FIGURE 11.7 (*Continued*)

hydrogen atoms that fall in the region of space around the molecule where the induced magnetic lines of force reinforce the external field appear to be deshielded in the spectrum. In other words, a lower external field strength is necessary for those nuclei to absorb energy and be promoted to the higher-energy level. Hydrogen atoms that fall in the region of space in which the induced magnetic field opposes the external field are shielded. The external magnetic field must be increased before absorption of energy takes place. Thus, molecules with multiple bonds have deshielding and shielding regions about them. These regions are most easily seen for benzene. In Figure 11.9, the flow of electrons around the aromatic ring, called the **ring current,** the induced magnetic lines of force that result from

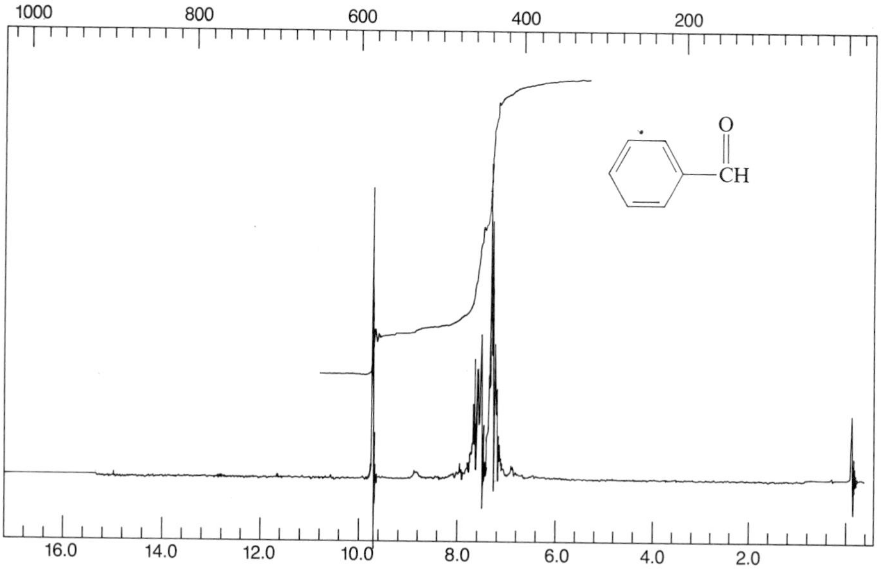

1000
800
600
400
200
O
CH
16.0
14.0
12.0
10.0
8.0
6.0
4.0
2.0

FIGURE 11.8 The proton magnetic resonance spectrum of benzaldehyde.

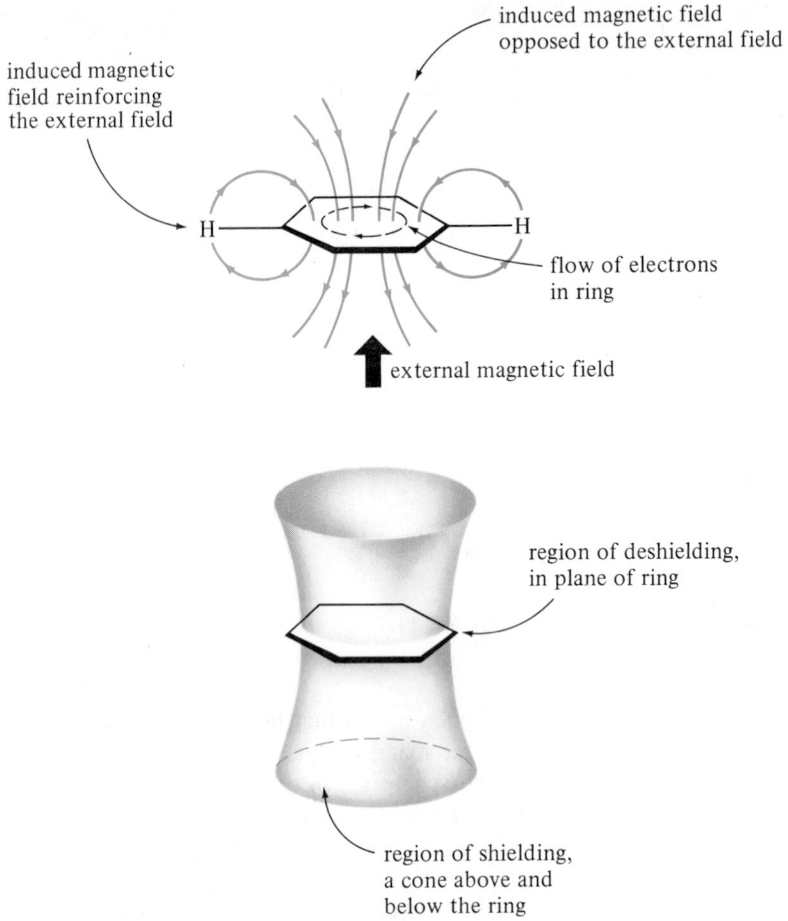
induced magnetic field
opposed to the external field
induced magnetic
field reinforcing
the external field
H
H
flow of electrons
in ring
external magnetic field
region of deshielding,
in plane of ring
region of shielding,
a cone above and
below the ring

FIGURE 11.9 Simplified representations of the ring current and induced magnetic lines of force and the shielding and deshielding regions around the aromatic ring of benzene.

this flow of charge, and the shielding and deshielding regions around the benzene molecule are shown.

A difference in magnetic properties at different points in space is called **magnetic anisotropy.** Molecules that have π bonds are magnetically anisotropic. Thus, benzene is a molecule that has magnetic anisotropy.

Functional groups that contain multiple bonds also have shielding and deshielding regions around them. Basically, these regions arise in the same way as do those around an aromatic ring, but it is a little more difficult to give a simple picture of the circulation of electrons in the bonds and of the induced magnetic fields that result. The conical surfaces in Figure 11.10 show shielding and deshielding regions around a carbonyl group, a double bond, and a triple bond. The hydrogen atom on a carbonyl group of an aldehyde or the hydrogen atoms on the carbons of a double bond lie in the plane of the respective molecules and in the deshielding regions. In an alkyne, the absence of a nodal plane and the circular symmetry of the π electrons around the linear axis of the molecule (p. 53) means that the major electronic current is around the axis. In this case, the shielding regions lie along the axis and the deshielding regions around it. A hydrogen atom on a triply bonded carbon atom is therefore more shielded than one on a doubly bonded carbon but deshielded relative to one on a tetrahedral carbon.

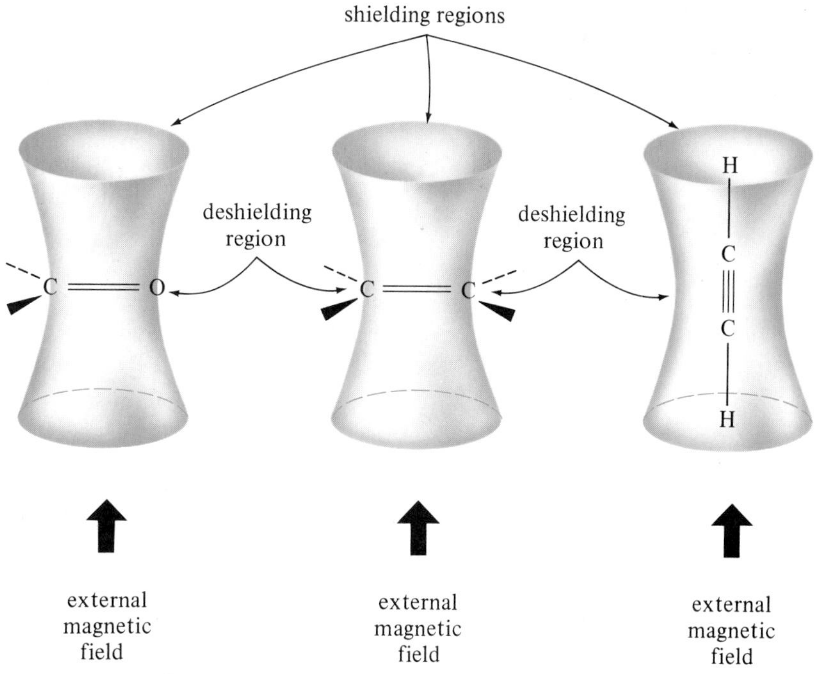
shielding regions
deshielding
region
deshielding
region
C
O
C
C
H
C
C
H
external
magnetic
field
external
magnetic
field
external
magnetic
field

FIGURE 11.10 Shielding and deshielding regions around functional groups containing multiple bonds.

C. Chemical Shifts of Hydrogen Atoms of Hydroxyl Groups

In general, the chemical shifts of hydrogen atoms bonded to oxygen are quite variable because they depend on the extent to which hydrogen bonding is taking place, and this, in turn, depends on the concentration of the solution. The proton magnetic resonance spectrum of methanol with no solvent (Figure 11.11 below and on the next page) shows the typical chemical shift for the hydrogen atom on a

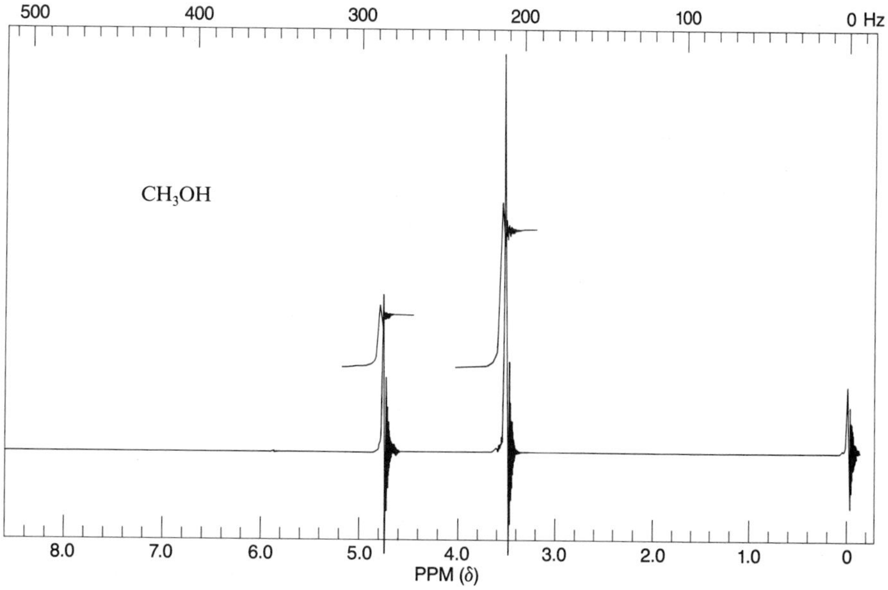
500
400
300
200
100
0 Hz
CH_3OH
8.0
7.0
6.0
5.0
4.0
3.0
2.0
1.0
0
PPM (δ)

FIGURE 11.11 Proton magnetic resonance spectra for methanol and phenol.

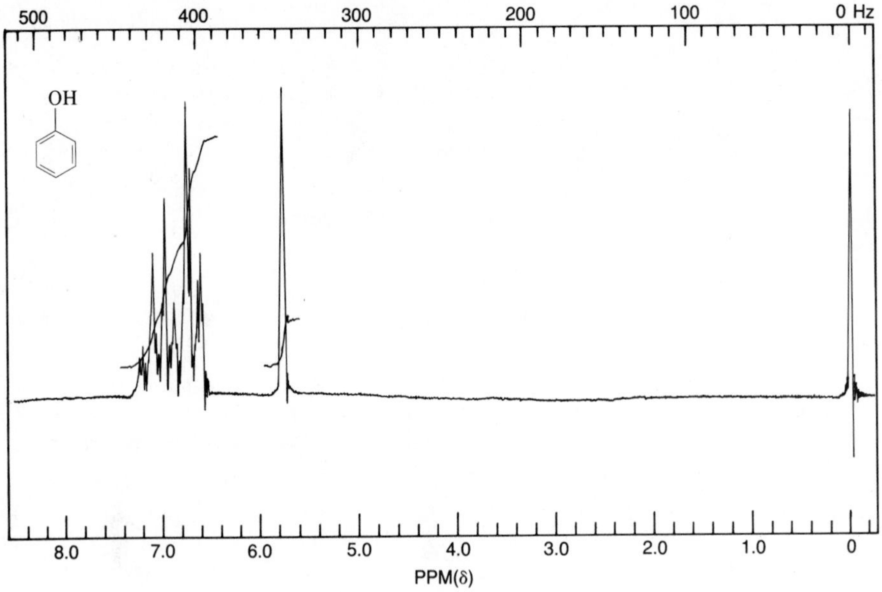
500
400
300
200
100
0 Hz
OH
8.0
7.0
6.0
5.0
4.0
3.0
2.0
1.0
0
PPM(δ)

FIGURE 11.11 *(Continued)*

hydroxyl group. The spectrum shows two bands, at δ 3.40 for the three hydrogen atoms of the methyl group and at δ 4.50 for the hydrogen atom of the hydroxyl group. The hydrogen atom of the hydroxyl group in a phenol, in which the hydroxyl group is on an aromatic ring, is usually more deshielded than that of an alcohol. In Figure 11.11, the band for the hydrogen atom of the hydroxyl group in phenol appears at δ 5.76, and those for the aromatic hydrogen atoms appear at δ 6.5–7.3.

The proton on the carboxyl group of an acid is the most deshielded of all hydrogen atoms bonded to oxygen. The spectrum of acetic acid, which, like that for benzaldehyde, has an expanded scale, provides an example (Figure 11.12). The spectrum has two singlets, at δ 2.00 for the hydrogen atoms of the methyl group and at δ 11.57 for the acidic hydrogen atom.

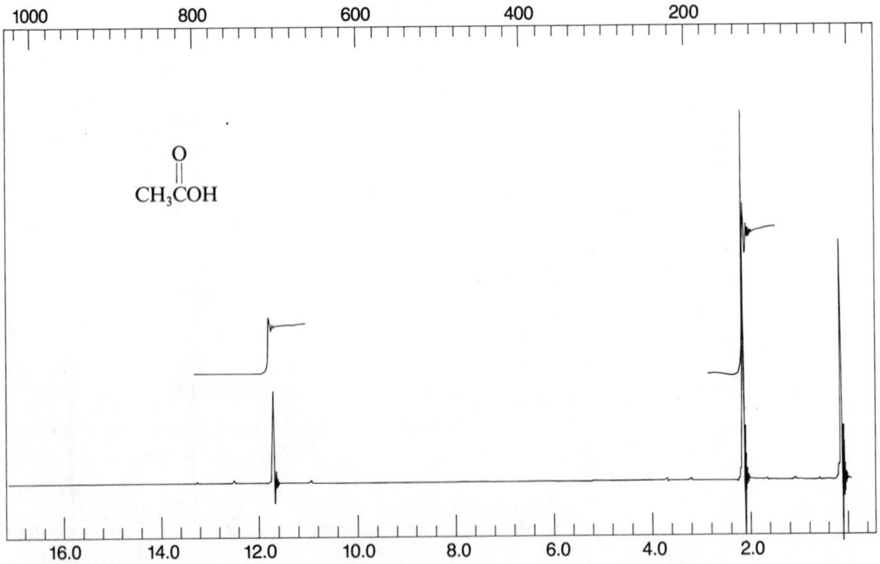
1000
800
600
400
200
O
CH3COH
16.0
14.0
12.0
10.0
8.0
6.0
4.0
2.0

FIGURE 11.12 Proton magnetic resonance spectrum of acetic acid.

PROBLEM 11.4

If an alcohol is diluted with $CDCl_3$, one of the bands in the proton magnetic resonance spectrum moves upfield as dilution increases. Explain what is happening. Why does hydrogen bonding further deshield a hydrogen atom?

D. Typical Chemical Shifts in Proton Magnetic Resonance Spectra

A regularity in the chemical shifts shown by hydrogen atoms in different environments is apparent if the spectra given so far in this chapter are examined carefully. A summary of the chemical shifts usually observed for these different types of hydrogen atoms is given in Table 11.1.

Note that the hydrogen atom of a **methine group,** a tertiary carbon atom with a hydrogen, is more deshielded than those of a methylene group, which in turn are more deshielded than the hydrogen atoms of a methyl group. The effects of substituents are cumulative, so the observed chemical shift of a hydrogen atom may be different from the value given in the table if several factors influence it at once. Just as with infrared spectra, it is important to pay attention to all the information available when interpreting proton magnetic resonance spectra, including the molecular formula of the compound, the integration of the spectrum, and the appearance of spin-spin coupling, as well as the chemical shift values.

TABLE 11.1 Typical Chemical Shifts for Types of Hydrogen Atoms, Seen in Proton Magnetic Resonance Spectra

Type of Hydrogen Atom	δ*	Type of Hydrogen Atom	δ*
$\mathrm{R\mathbf{CH_3}}$	0.9	$\mathrm{R_2C{=}\mathbf{CH_2}}$	5.0
$\mathrm{R\mathbf{CH_2}R}$ acyclic	1.3	$\mathrm{R\mathbf{CH}{=}CR_2}$	5.3
cyclic	1.5	$\mathrm{Ar\mathbf{H}}$	7.3
$\mathrm{R_3C\mathbf{H}}$	1.5–2.0	$\mathrm{R\overset{O}{\overset{\|}{C}}\mathbf{H}}$	9.7
$\mathrm{R_2C{=}\underset{\underset{R'}{\|}}{C}\mathbf{CH_3}}$	1.8	$\mathrm{R\mathbf{NH_2}}$	1–3
$\mathrm{R\overset{O}{\overset{\|}{C}}\mathbf{CH_3}}$	2.0–2.3	$\mathrm{Ar\mathbf{NH_2}}$	3–5
$\mathrm{Ar\mathbf{CH_3}}$	2.3	$\mathrm{R\overset{O}{\overset{\|}{C}}\mathbf{NH}R}$	5–9
$\mathrm{RC{\equiv}\mathbf{CH}}$	2.5	$\mathrm{RO\mathbf{H}}$	1–5
$\mathrm{RNH\mathbf{CH_3}}$	2–3	$\mathrm{ArO\mathbf{H}}$	4–7
$\mathrm{R\mathbf{CH_2}X(X{=}Cl, Br, I)}$	3.5	$\mathrm{R\overset{O}{\overset{\|}{C}}O\mathbf{H}}$	10–13
$\mathrm{RO\mathbf{CH_3}}$, $\mathrm{R\overset{O}{\overset{\|}{C}}O\mathbf{CH_3}}$	3.8		

*The chemical shift values are given in ppm relative to tetramethylsilane at δ 0.00 and are for the hydrogen atoms shown in boldface in the formulas. The values for hydrogen atoms on oxygen and nitrogen are highly dependent on solvent, concentration, and temperature.

PROBLEM 11.5

Using the chemical shift values given in Table 11.1, assign the absorption bands between δ 1.0 and δ 3.0 in the spectrum of cyclohexene (Figure 11.7). Remember that values given in such a table are approximate and that you have to reason by analogy in making assignments for the spectrum of a particular compound.

PROBLEM 11.6

Molecular formulas and proton magnetic resonance spectra for Compounds D–G are given in Figure 11.13. Assign structures to the compounds. (Hint: Calculating the units of unsaturation for a compound is a good way to start solving this kind of problem.)

FIGURE 11.13

Compound D, $C_8H_{10}O_2$

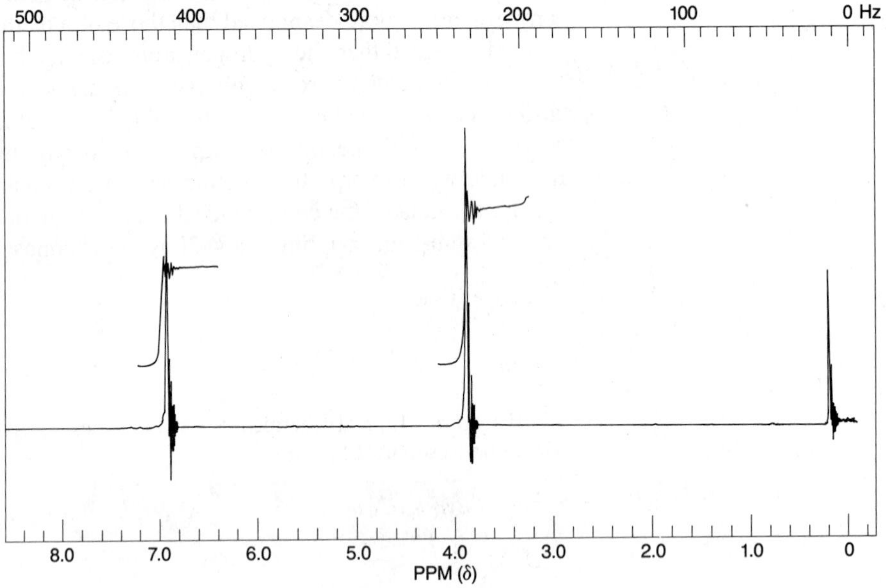
500
400
300
200
100
0 Hz
8.0
7.0
6.0
5.0
4.0
3.0
2.0
1.0
0
PPM (δ)

Compound E, C_8H_{10}

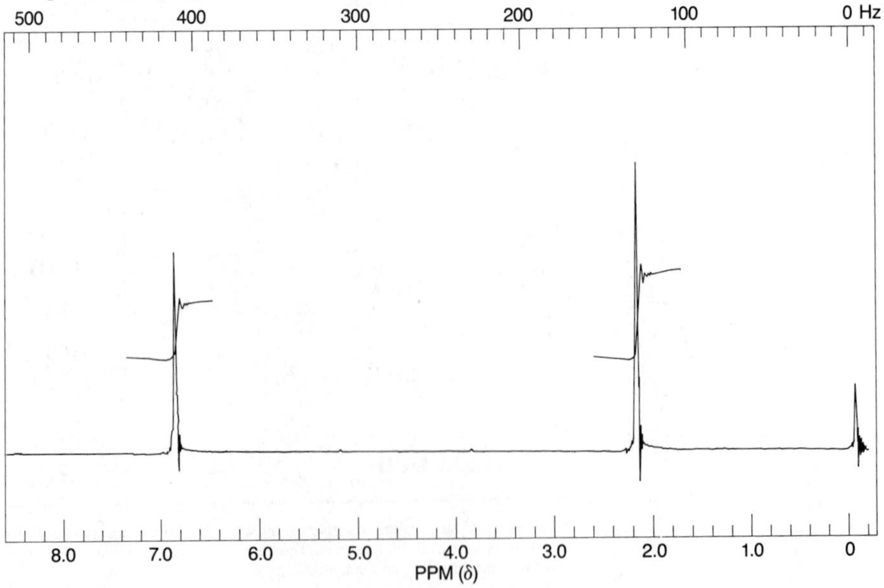
500
400
300
200
100
0 Hz
8.0
7.0
6.0
5.0
4.0
3.0
2.0
1.0
0
PPM (δ)

FIGURE 11.13 *(Continued)* Compound F, C_7H_7Cl

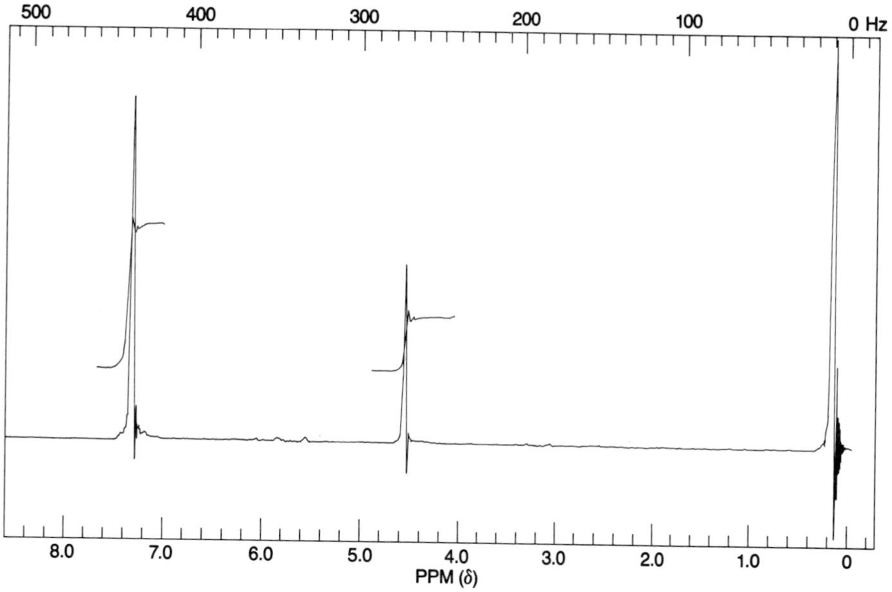
500
400
300
200
100
0 Hz
8.0
7.0
6.0
5.0
4.0
3.0
2.0
1.0
0
PPM (δ)

Compound G, C_8H_8O

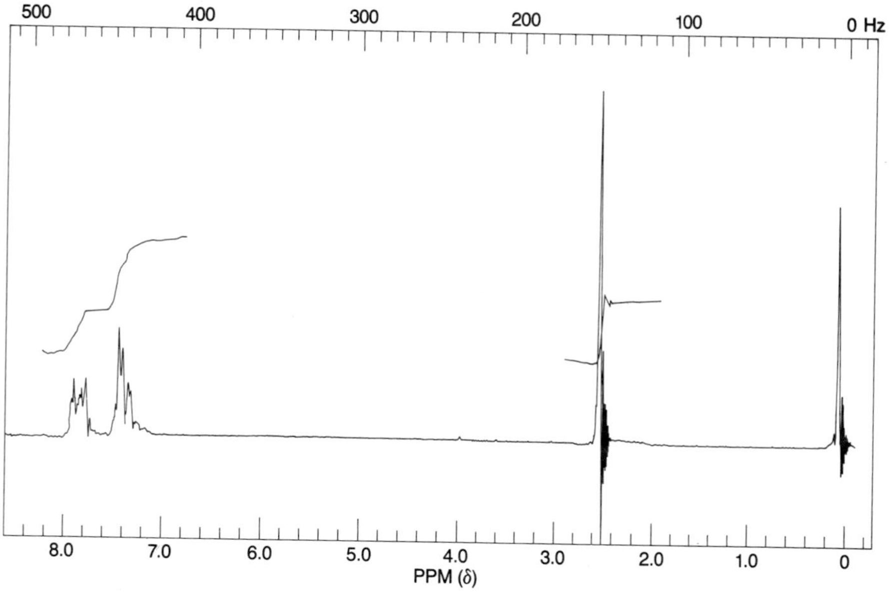
500
400
300
200
100
0 Hz
8.0
7.0
6.0
5.0
4.0
3.0
2.0
1.0
0
PPM (δ)

11.4 SPIN-SPIN COUPLING

A. The Origin of Spin-Spin Coupling

Many of the spectra shown in the first three sections of this chapter have bands that are split into smaller peaks. These patterns arise from interactions among the nuclei of hydrogen atoms that are close enough that their nuclear spins affect each other. This effect is transmitted through the electrons of the covalent bonds and so is dependent on the number and types of bonds separating the hydrogen atoms.

In the spectrum of 1,1-dichloroethane (Figure 11.2, p. 379), the hydrogen atoms of the methyl group appear as two peaks, called a **doublet,** separated by 7 Hz. All of the methyl hydrogen atoms are, of course, chemical-shift equivalent. The value for their chemical shift is read off the spectrum as the central point of the doublet, δ 1.95. Hydrogen atoms that have the same chemical shift do not split each other. The signal for the hydrogen atoms of the methyl group of 1,1-dichloroethane appears as a doublet because of the influence of the single hydrogen atom on the other carbon atom in the molecule. The magnetic field experienced by the hydrogen atoms of the methyl group is influenced by the spin of the hydrogen on the neighboring carbon. At any given time, about half of the hydrogen atoms in a sample have their spins aligned with the external field and half against it. As a result, the net magnetic field experienced by the hydrogen atoms of the methyl group of 1,1-dichloroethane is slightly stronger or slightly weaker than the external field supplied by the instrument (Figure 11.14). Therefore, absorption of energy occurs at two slightly different frequency values. Two absorption bands of approximately equal intensity appear in the spectrum. If the hydrogen atoms on the methyl group in 1,1-dichloroethane are influenced by the hydrogen atom on the other carbon atom, the signal for the single hydrogen atom on the other carbon of 1,1-dichloroethane, at δ 5.60, must show the effect of the three hydrogen atoms of the methyl group. The presence of three hydrogen atoms on the adjacent carbon splits this downfield absorption band into four peaks, having relative intensities of 1:3:3:1. Such a pattern is called a **quartet.**

Three nuclei have four different ways in which they can be oriented with respect to an external magnetic field. Because individual spin orientations are indistinguishable, all the cases in which one nucleus is oriented against the field and two with the field are equal in energy, as are those in which two nuclei are against the field and one with the field. The low-field hydrogen atom of 1,1-dichloroethane may thus experience one of four slightly different magnetic fields. The probability that any one of these situations will occur is 1:3:3:1 (Figure 11.14). Thus, four peaks, a quartet, having those relative intensities arise. The distance between these peaks is the same as that between the peaks of the methyl doublet.

The interaction between the nuclei of different hydrogen atoms is called **spin-spin coupling.** The interaction occurs through the intervening bonds. The strength

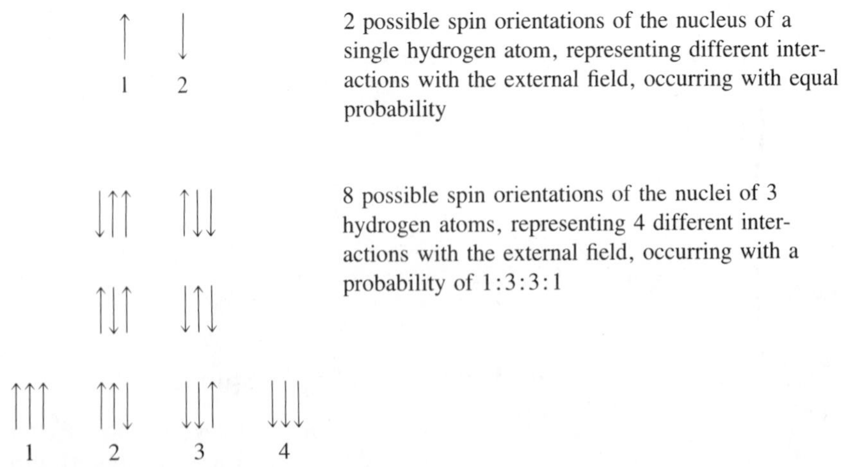
1 2
2 possible spin orientations of the nucleus of a single hydrogen atom, representing different interactions with the external field, occurring with equal probability
8 possible spin orientations of the nuclei of 3 hydrogen atoms, representing 4 different interactions with the external field, occurring with a probability of 1:3:3:1
1 2 3 4

FIGURE 11.14 The interactions possible between the magnetic moments of the nuclei of the hydrogen atoms in 1,1-dichloroethane and an external magnetic field.

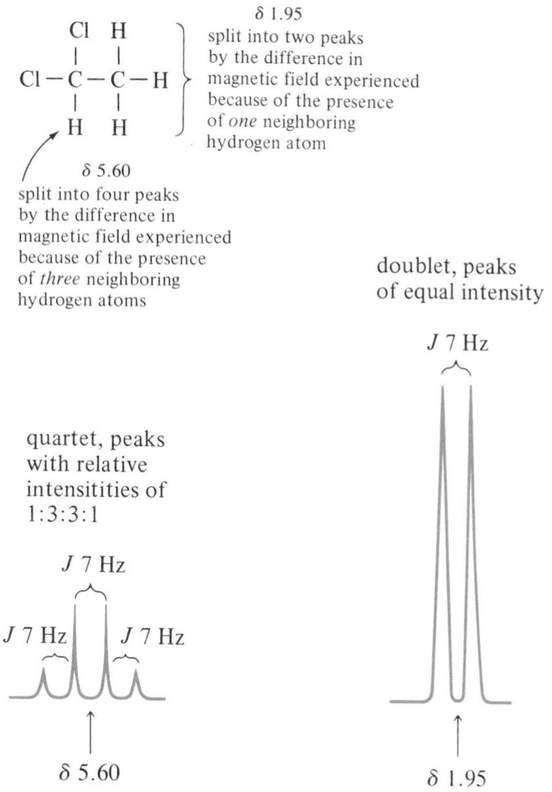
δ 1.95
split into two peaks by the difference in magnetic field experienced because of the presence of *one* neighboring hydrogen atom
δ 5.60
split into four peaks by the difference in magnetic field experienced because of the presence of *three* neighboring hydrogen atoms
doublet, peaks of equal intensity
J 7 Hz
quartet, peaks with relative intensitities of 1:3:3:1
J 7 Hz
J 7 Hz
J 7 Hz
δ 5.60
δ 1.95

FIGURE 11.15 A representation of the coupling between hydrogen atoms on neighboring carbon atoms in 1,1-dichloroethane.

of the interaction thus depends on factors such as the number of bonds between the interacting nuclei, the types of those bonds, and the stereochemical relationship between the atoms. The measure of the interaction is the **coupling constant, *J*,** given in Hz. The value seen in the spectrum of 1,1-dichloroethane, approximately 7 Hz, is typical of coupling constants for alkyl hydrogen atoms on adjacent carbon atoms (Figure 11.15). Note that the bands arising from hydrogen atoms that are coupled to each other can be identified because the distances between the small peaks are always the same. The coupling between the different hydrogen atoms gives rise to one coupling constant, *J*.

For hydrogen atoms that have differing chemical shifts and are on adjacent carbon atoms, the splitting patterns seen are determined by the number of hydrogen atoms on the neighboring carbon atoms. The number of peaks seen is $N + 1$, where N equals the number of neighboring hydrogen atoms that are chemical-shift equivalent or that have equivalent coupling constants. The patterns that are to be expected, and the relative intensities of the peaks within each pattern are given in Table 11.2 on the next page.

PROBLEM 11.7

The spectrum of bromoethane is given in Figure 11.16. Assign chemical shifts to the different hydrogen atoms in bromoethane and analyze the spin-spin coupling. What is the coupling constant, J? Make sure that you understand why the methyl hydrogen atoms in bromoethane appear as a triplet by demonstrating to yourself the number of different orientations possible for the nuclei of the hydrogen atoms of the methylene group. What gives rise to the relative intensities seen for the three peaks in the triplet?

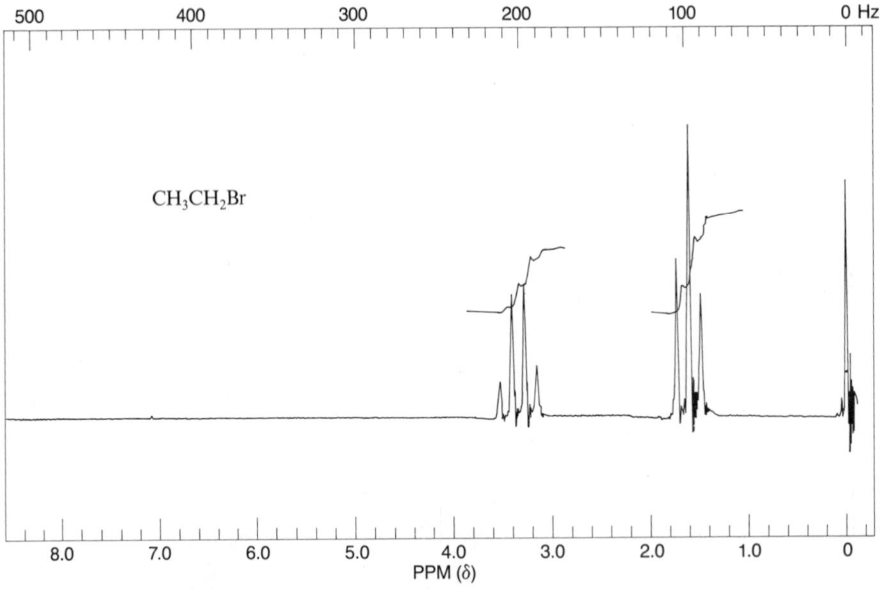
500
400
300
200
100
0 Hz
CH₃CH₂Br
8.0
7.0
6.0
5.0
4.0
3.0
2.0
1.0
0
PPM (δ)

FIGURE 11.16

Looking at other spectra that show the effects of spin-spin coupling will help to clarify them. Figure 11.17 shows the spectra of 2-bromopropane and 1-bromopropane. The spectrum of 2-bromopropane indicates the presence of two types of hydrogen atoms, in a ratio of 1:6. Those on the two methyl groups absorb at δ 1.60, and the one on the carbon atom bearing the bromine atom absorbs at δ 4.00. The signal for the methyl hydrogen atoms, which are all equivalent to each other, appears as a doublet because of the presence of one hydrogen atom on the adjacent carbon atom. It is very important to understand that the methyl signal in this spectrum is a doublet *not* because there are two methyl groups but because of spin-spin coupling. The influence of the six equivalent hydrogen atoms of the

TABLE 11.2 Splitting Patterns That Result from the Presence of N Neighboring Hydrogen Atoms

N	Multiplet (N + 1)	Relative Intensities of Peaks
0	singlet (1)	1
1	doublet (2)	1:1
2	triplet (3)	1:2:1
3	quartet (4)	1:3:3:1
4	quintet (5)	1:4:6:4:1
5	sextet (6)	1:5:10:10:5:1
6	septet (7)	1:6:15:20:15:6:1

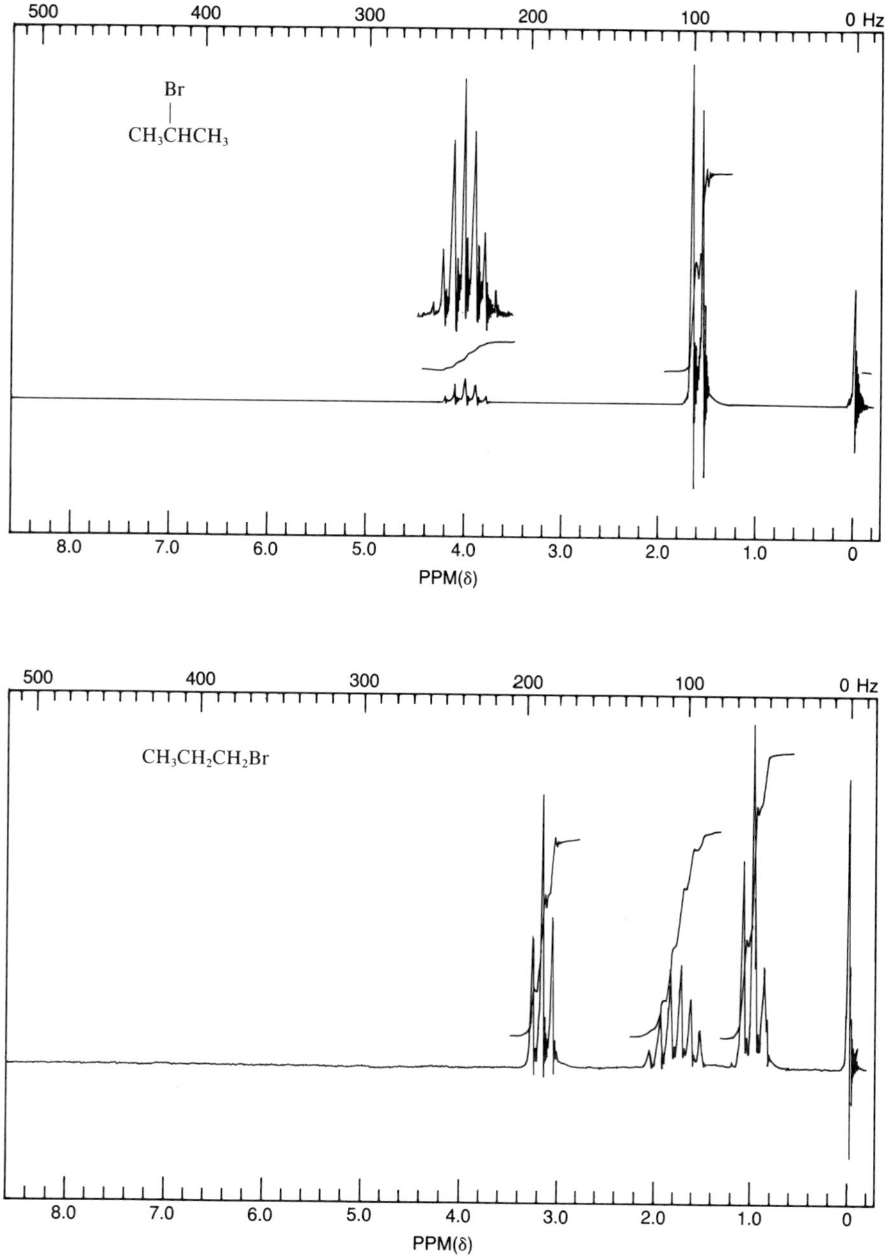
500
400
300
200
100
0 Hz
Br
CH₃CHCH₃
8.0
7.0
6.0
5.0
4.0
3.0
2.0
1.0
0
PPM(δ)
500
400
300
200
100
0 Hz
CH₃CH₂CH₂Br
8.0
7.0
6.0
5.0
4.0
3.0
2.0
1.0
0
PPM(δ)

FIGURE 11.17 Proton magnetic resonance spectra of 2-bromopropane and 1-bromopropane.

methyl groups, in turn, splits the absorption band for the hydrogen atom on the secondary carbon atom into 6 + 1, or 7, peaks. The intensities of the outer peaks of the septet are very weak; they do not appear unless that portion of the spectrum is enhanced. They can only be seen in the insert in the spectrum, which was recorded at a higher sensitivity. The distance between adjacent peaks in the septet is equal to the distance between the two peaks of the doublet. The value, read off the spectrum, is about 7 Hz, the usual coupling constant for hydrogen atoms on neighboring sp^3-hybridized carbon atoms.

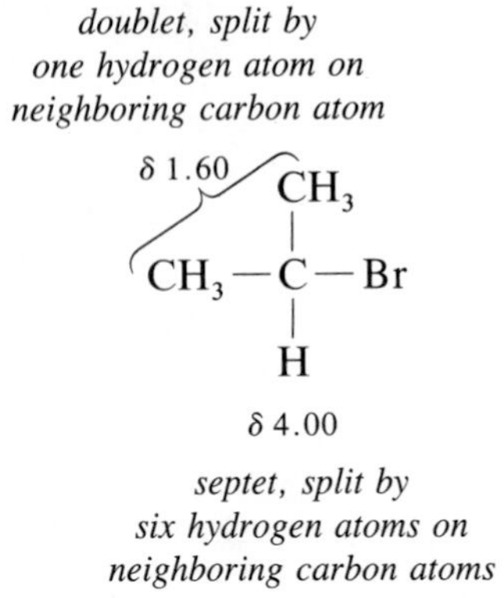
doublet, split by
one hydrogen atom on
neighboring carbon atom
δ 1.60
CH3
CH3—C—Br
H
δ 4.00
septet, split by
six hydrogen atoms on
neighboring carbon atoms

2-bromopropane

The spectrum of 1-bromopropane has three bands, a triplet centered at δ 1.00 (*J* 7 Hz), a sextet at δ 1.80 (*J* 7 Hz), and another triplet at δ 3.20 (*J* 7 Hz). These bands have relative intensities of 3:2:2. The band at δ 1.00 is assigned to the methyl group, and the one at δ 3.20 to the hydrogen atoms on the carbon atom bearing the bromine atom. The hydrogen atoms on carbon atom 2 fall in between these. The interactions that give rise to the splitting patterns are analyzed as follows:

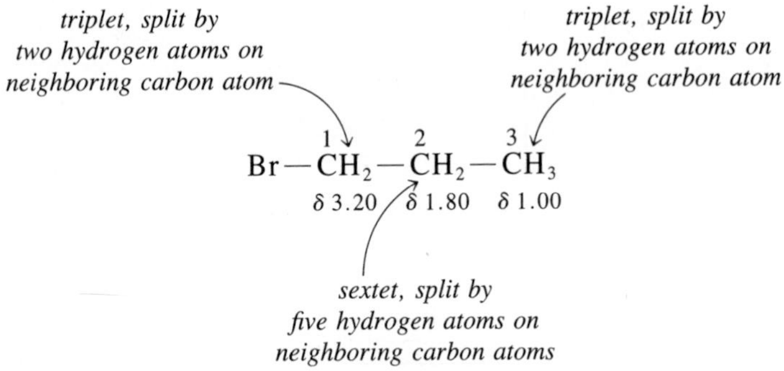
triplet, split by
two hydrogen atoms on
neighboring carbon atom
triplet, split by
two hydrogen atoms on
neighboring carbon atom
1
2
3
Br—CH2—CH2—CH3
δ 3.20
δ 1.80
δ 1.00
sextet, split by
five hydrogen atoms on
neighboring carbon atoms

1-bromopropane

The band for the hydrogen atoms on carbon 2 appears as a sextet because the coupling constants for the interactions of those hydrogen atoms with those on carbons 1 and 3 happen to be of the same magnitude. Thus, the hydrogen atoms on carbon 2 appear to be coupled with five equivalent hydrogen atoms. If the hydrogen atoms on carbon 2 and those on carbon 1 and carbon 3 had different coupling constants, a more complex pattern would have resulted.

PROBLEM 11.8

Compound H, C_9H_{12}, gives the proton magnetic resonance spectrum shown in Figure 11.18. Assign a structure to the compound.

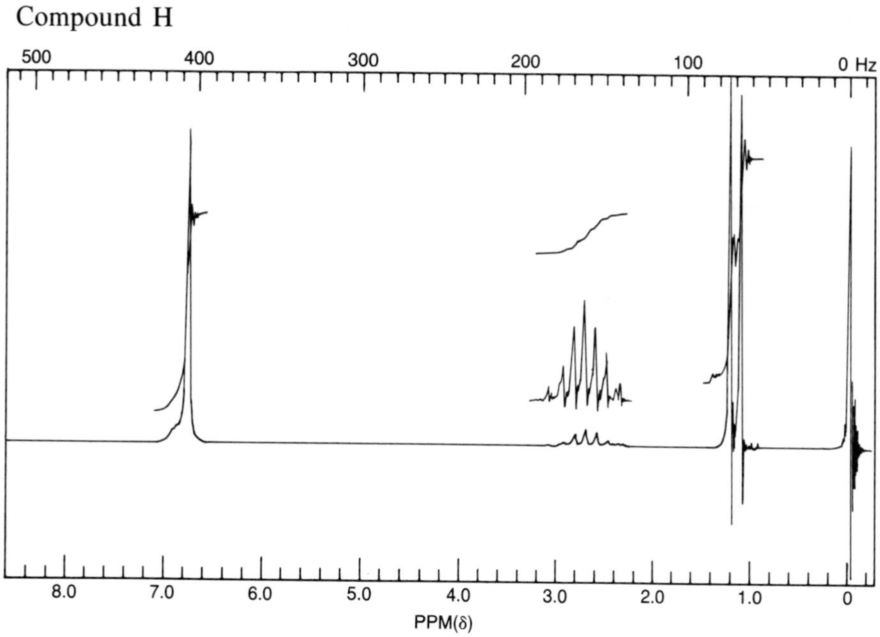

Compound H
500
400
300
200
100
0 Hz
8.0
7.0
6.0
5.0
4.0
3.0
2.0
1.0
0
PPM(δ)

FIGURE 11.18

B. Spin-Spin Coupling in Compounds with Multiple Bonds

Not all hydrogen atoms have a coupling constant of 7 Hz. It is easiest to see different types of coupling in the spectra of alkenes. The spectrum of vinyl acetate is shown in Figure 11.19. Besides the singlet for the hydrogen atoms of the acetyl group, the spectrum of vinyl acetate has complex bands for three other hydrogen atoms. These bands are centered at δ 4.25, 4.78, and 6.90. Each band is split into

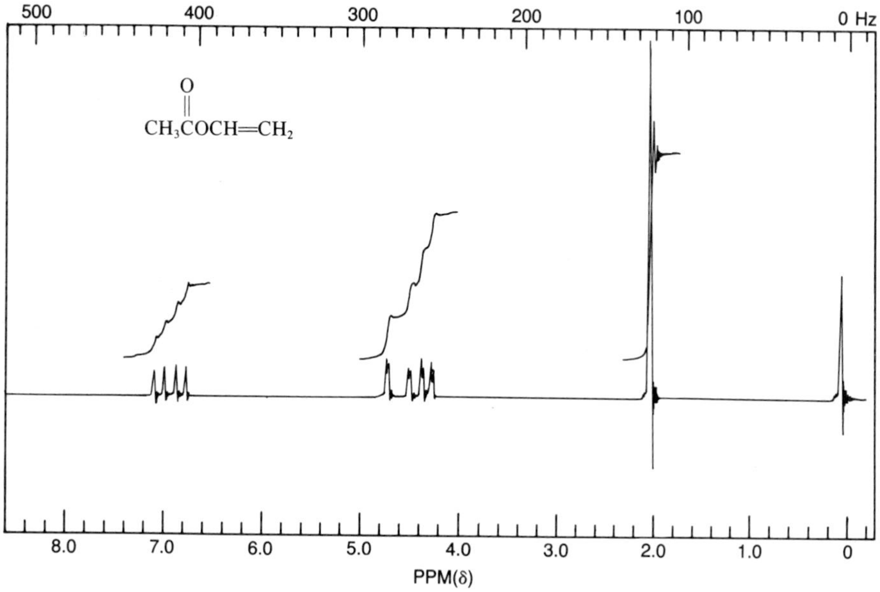

500
400
300
200
100
0 Hz
CH3COCH=CH2
8.0
7.0
6.0
5.0
4.0
3.0
2.0
1.0
0
PPM(δ)

FIGURE 11.19 Proton magnetic resonance spectrum of vinyl acetate.

a doublet, and the peaks of each doublet are split again into another doublet. This pattern is called a **doublet of doublets.** None of these groupings of four peaks is a quartet because the four peaks are not evenly spaced and do not have relative intensities of 1:3:3:1.

The hydrogen atoms on the terminal carbon atom of the double bond in vinyl acetate are distinct from each other. One (δ 4.25) is trans to the oxygen atom and cis to the hydrogen atom on the other carbon atom of the double bond, and the other (δ 4.78) is cis to the oxygen and trans to the hydrogen. Three coupling constants are seen in the spectrum. The hydrogen atoms that are trans to each other across the double bond have the strongest interaction, J 14 Hz. The coupling constant for the hydrogen atoms that are cis to each other is 6.4 Hz. The two hydrogen atoms on the same carbon atom have a small coupling constant, J 1.4 Hz. J_{trans} values are always larger than J_{cis} values for hydrogen atoms on doubly bonded carbons.

The splitting patterns seen in the spectrum of vinyl acetate can be duplicated by constructing a diagram. The diagram shows the lines that will result if the band for a hydrogen atom is split by interaction with two other hydrogen atoms with two different coupling constants. If a ruler or graph paper is used so that the distances in the diagram are really proportional to the coupling constants, the patterns seen in the spectrum emerge (Figure 11.20).

Several different coupling constants are also seen in the spectrum of an aromatic compound, represented here by 2-bromo-4-nitrotoluene (Figure 11.21). There are four different kinds of hydrogen atoms in 2-bromo-4-nitrotoluene. The methyl group at δ 2.40 is clearly separated from the aromatic hydrogen atoms, which absorb at δ 7.00, 7.60, and 7.85. In general, very little coupling is seen between hydrogen atoms on the side chains of aromatic rings and the hydrogen

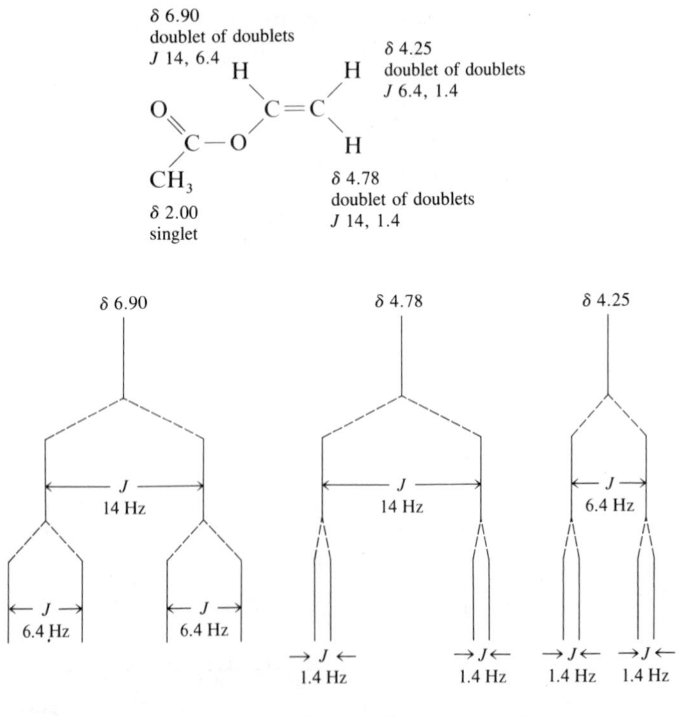
δ 6.90
doublet of doublets
J 14, 6.4
δ 4.25
doublet of doublets
J 6.4, 1.4
δ 4.78
doublet of doublets
J 14, 1.4
δ 2.00
singlet
δ 6.90
δ 4.78
δ 4.25
J
14 Hz
J
14 Hz
J
6.4 Hz
J
6.4 Hz
J
6.4 Hz
J
1.4 Hz
J
1.4 Hz
J
1.4 Hz
J
1.4 Hz

FIGURE 11.20 Analysis of the spin-spin coupling seen in the proton magnetic resonance spectrum of vinyl acetate.

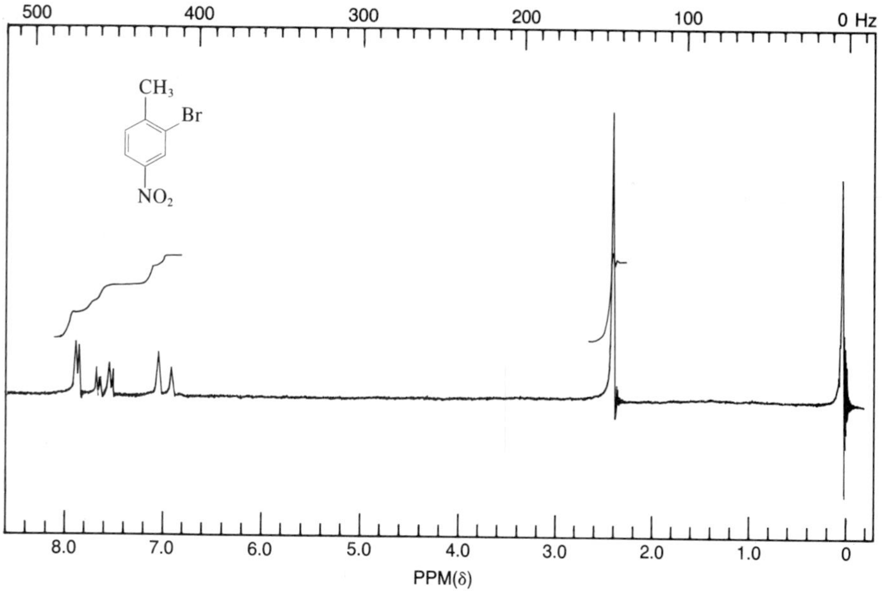
500
400
300
200
100
0 Hz
CH₃
Br
NO₂
8.0
7.0
6.0
5.0
4.0
3.0
2.0
1.0
0
PPM(δ)

FIGURE 11.21 Proton magnetic resonance spectrum of 2-bromo-4-nitrotoluene.

atoms on the ring. The signal farthest downfield is assigned to the hydrogen atom that is between the nitro group and the bromine atom on the ring, labeled H_C in the structural formula. It is split into a doublet by H_B, with a very small coupling constant, J 2 Hz. The signal at highest field is assigned to the hydrogen atom next to the methyl group, H_A. It is split into a doublet by H_B, with J 8 Hz. Note that coupling between hydrogen atoms next to each other is stronger than coupling between the more distant hydrogens. Very little coupling is seen between hydrogen atoms H_A and H_C, with J about 0.7 Hz. The hydrogen atom labeled H_B gives rise to a doublet of doublets. This hydrogen atom interacts with H_A with one coupling constant, J 8 Hz, and with H_C with another coupling constant, J 2 Hz.

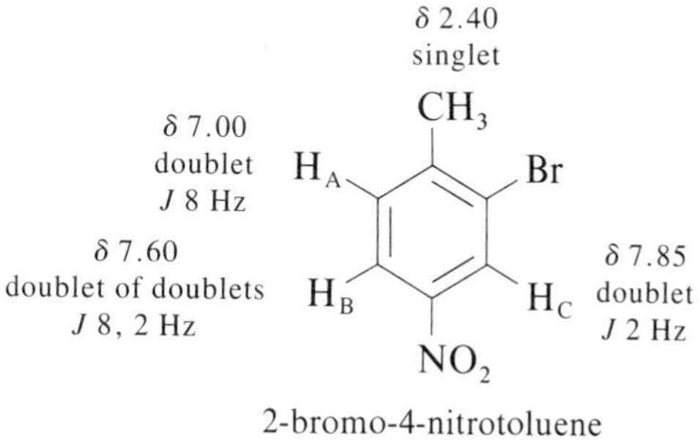
δ 2.40
singlet
CH₃
δ 7.00
doublet
J 8 Hz
H_A
Br
δ 7.60
doublet of doublets
J 8, 2 Hz
H_B
H_C
δ 7.85
doublet
J 2 Hz
NO₂

2-bromo-4-nitrotoluene

Note that the proton magnetic resonance spectra of compounds that contain multiple bonds show coupling between hydrogen atoms that have as many as four or five bonds between them. In general, coupling constants for these interactions are small but do give rise to complexity in the spectra.

A list of coupling constants for hydrogen atoms in various environments is given in Table 11.3.

TABLE 11.3 Coupling Constants Seen in Proton Magnetic Resonance Spectra

Type of Hydrogen Atoms	J, Hz	Type of Hydrogen Atoms	J, Hz
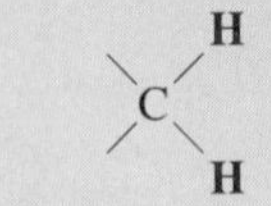	12–15	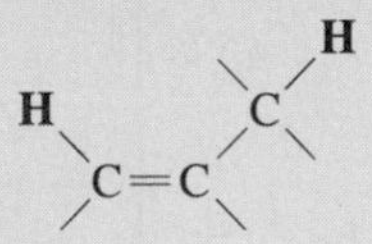	0–3
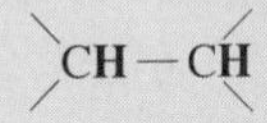	6–8	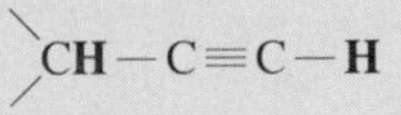	2–3
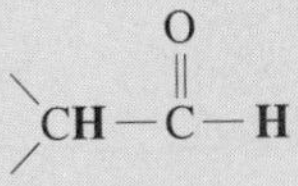	2–3	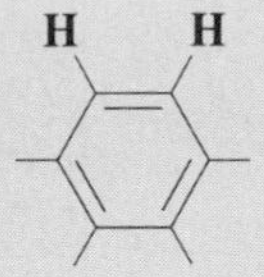	6–10
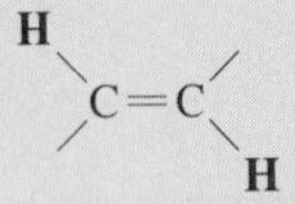	12–18	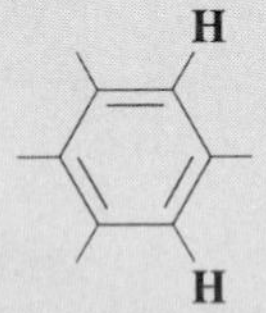	1–3
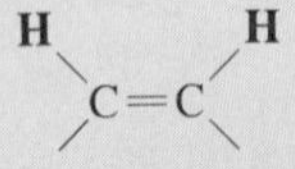	6–12	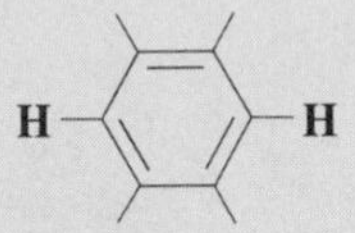	0–1
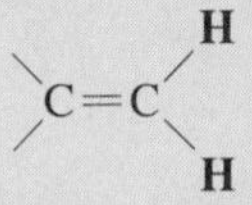	0–2		
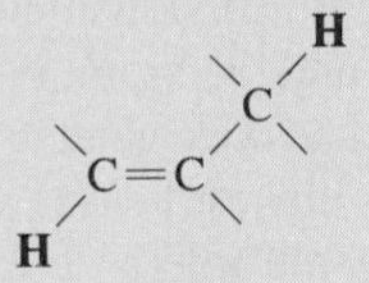	0–3		

PROBLEM 11.9

Use the coupling constants given for 2-bromo-4-nitrotoluene (p. 401) to construct a diagram like the one in Figure 11.20, showing the pattern of lines you predict for the aryl hydrogen atoms.

PROBLEM 11.10

The signal for the hydrogen atoms on carbon atom 2 of 1-bromopropane (p. 397) may be regarded as being split into a quartet by the three hydrogen atoms on carbon 3, with each one of those four peaks further split into a triplet by the hydrogen atoms on carbon 1. Assume that the coupling constant for both types of interactions is 7 Hz and construct a diagram like the one in Figure 11.20 to determine the number of lines you expect to see in the pattern that results.

C. More Complex Splitting Patterns

The spin-spin splitting patterns such as are seen in the spectra used as examples so far arise only when the difference in chemical shifts between the hydrogen atoms that are interacting is large compared with their coupling constant, *J*. When hydro-

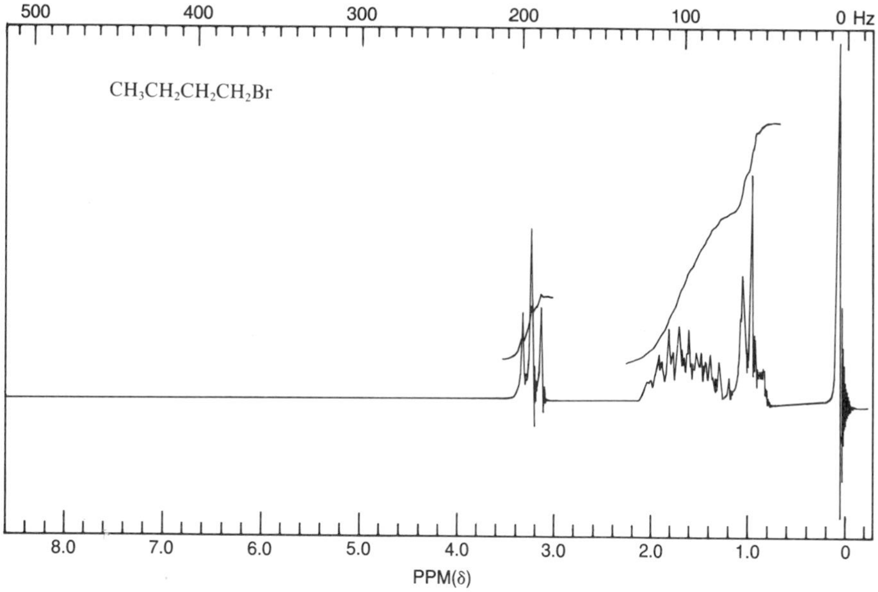
500
400
300
200
100
0 Hz
$CH_3CH_2CH_2CH_2Br$
8.0
7.0
6.0
5.0
4.0
3.0
2.0
1.0
0
PPM(δ)

FIGURE 11.22 Proton magnetic resonance spectrum of 1-bromobutane.

gen atoms of similar chemical shift interact, the patterns become much more complex. The spectrum for 1-bromobutane in Figure 11.22 above demonstrates this. In the spectrum of 1-bromobutane, only the absorption band for the hydrogen atoms on carbon atom 1 (δ 3.20) is a recognizable triplet. These hydrogen atoms are deshielded by the bromine atom and have a chemical shift quite different from those for the rest of the hydrogen atoms in the molecule. The hydrogen atoms on carbon atoms 2 and 3 are similar in chemical shift, but not identical. They couple with each other and also with the hydrogen atoms on carbon 1 or those of the methyl group. Thus, their signal appears as a complex pattern called a **multiplet.** The signal for the methyl group, however, which would be expected to appear as a triplet, is so distorted that it is hard to detect the pattern. This distortion of the triplet arises because the chemical shift for the hydrogen atoms of the methyl group is too close to that of the methylene hydrogen atoms. The appearance of the spectrum of 1-bromobutane is typical of that for substituted alkanes in which most of the hydrogen atoms have similar chemical shifts. The closer the chemical shifts of the interacting hydrogen atoms, the more distorted are the patterns in the spectrum.

1,4-Substituted, or *para*-substituted (p. 479) aromatic compounds in which the two substituents differ from each other in electronic effects typically show symmetrical patterns in their proton magnetic resonance spectra (Figure 11.23). For 1-bromo-4-methoxybenzene, because the chemical shifts of the hydrogen atoms next to the methoxy group (δ 6.38) and of those next to the bromine atom (δ 6.95) are quite different, the pattern is spread apart. Note the resemblance of the splitting patterns seen for the aromatic hydrogen atoms in the spectra of 1-(4-chlorophenyl)-1-propanone and 1-bromo-4-methoxybenzene. Both spectra show a complex pattern that is roughly a symmetrical doublet of doublets. The pattern typical of an ethyl group bonded to a deshielding atom such as a halogen (Problem 11.7, p. 395) or a carbonyl group is also shown clearly in the spectrum of 1-(4-chlorophenyl)-1-propanone.

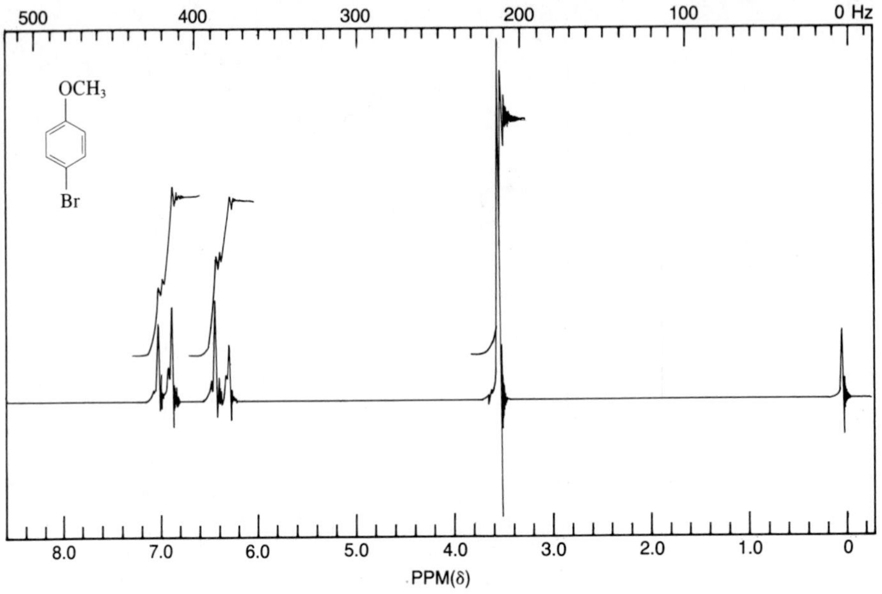
500
400
300
200
100
0 Hz
OCH₃
Br
8.0
7.0
6.0
5.0
4.0
3.0
2.0
1.0
0
PPM(δ)

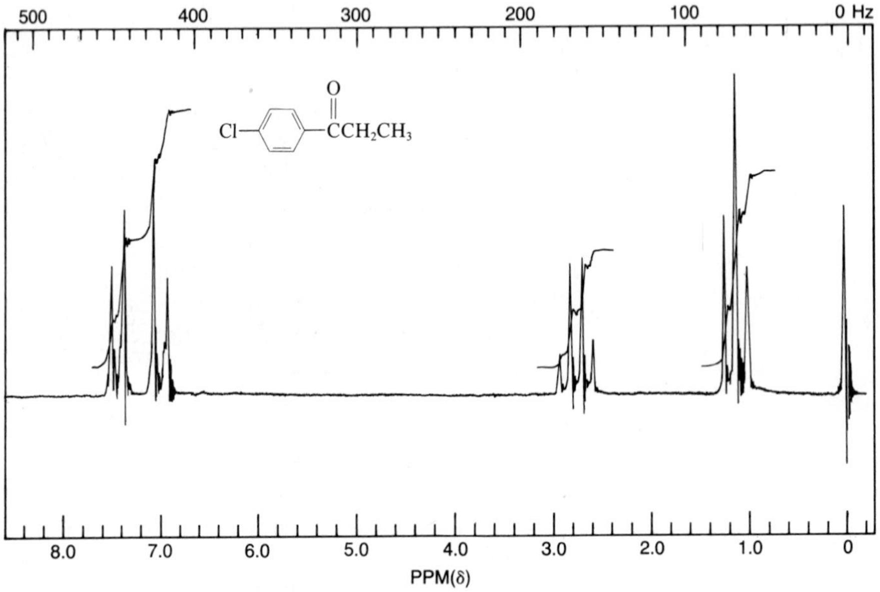
500
400
300
200
100
0 Hz
O
Cl
CCH₂CH₃
8.0
7.0
6.0
5.0
4.0
3.0
2.0
1.0
0
PPM(δ)

FIGURE 11.23 Proton magnetic resonance spectra of 1-bromo-4-methoxybenzene and 1-(4-chlorophenyl)-1-propanone.

PROBLEM 11.11

Proton magnetic resonance spectra (Figure 11.24) and structural formulas for 3-methylbenzonitrile and 4-methylbenzonitrile are given below. Assign each spectrum to the correct compound.

(a) 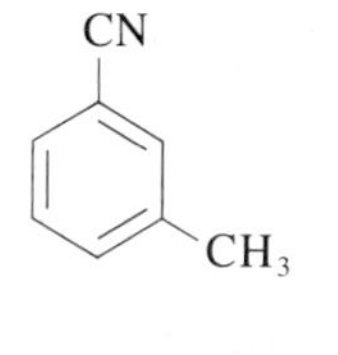
CN
CH3

3-methylbenzonitrile

(b) 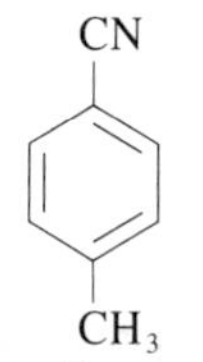
CN
CH3

4-methylbenzonitrile

Spectrum 1

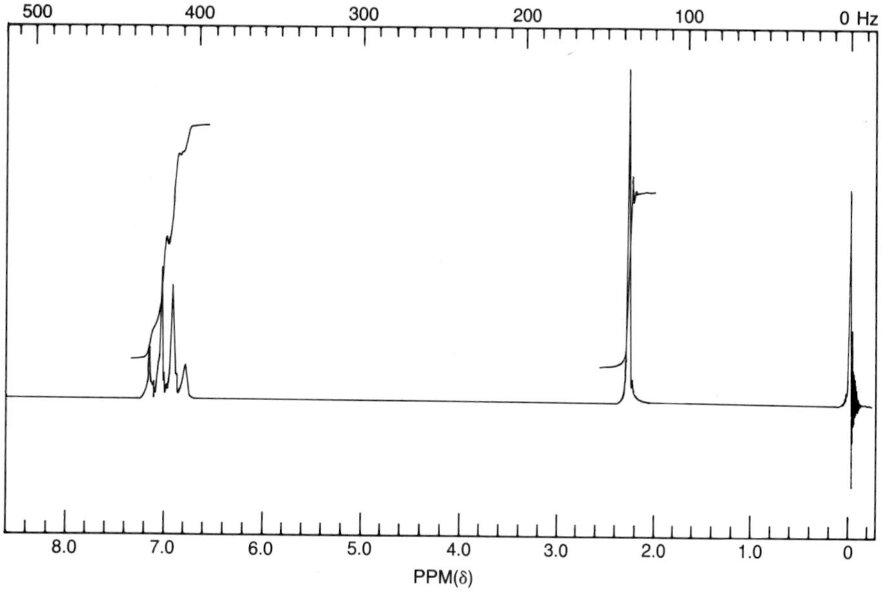
500
400
300
200
100
0 Hz
8.0
7.0
6.0
5.0
4.0
3.0
2.0
1.0
0
PPM(δ)

Spectrum 2

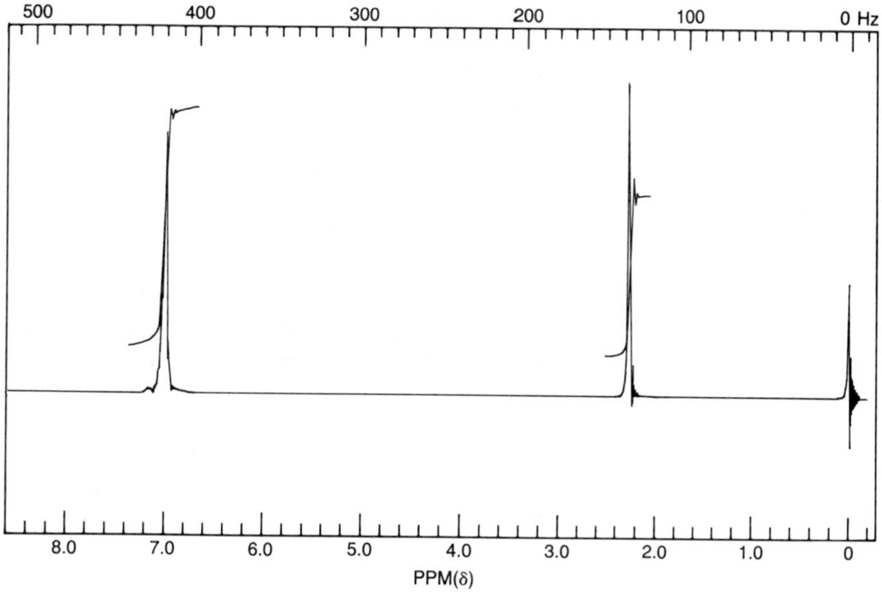
500
400
300
200
100
0 Hz
8.0
7.0
6.0
5.0
4.0
3.0
2.0
1.0
0
PPM(δ)

FIGURE 11.24

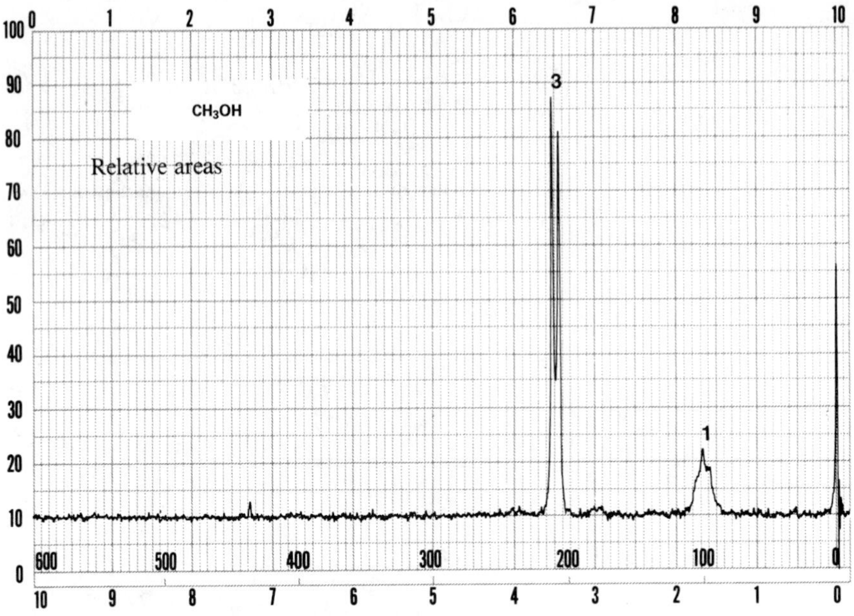
CH3OH
Relative areas
3
1

FIGURE 11.25 Proton magnetic resonance spectrum of highly purified methanol. (From *The Aldrich Library of NMR Spectra*)

D. The Effect of Chemical Exchange on Spin-Spin Coupling

The spectrum of methanol, shown in Figure 11.11 (p. 389), has two singlets, one for the hydrogen atoms of the methyl group and one for the hydrogen atom bonded to oxygen. These two different types of hydrogen atoms are on adjacent atoms but do not appear to be coupled. This phenomenon is quite general for hydrogen atoms on oxygen atoms and on the nitrogen atoms of amino groups. Such hydrogen atoms hydrogen bond strongly and are transferred rapidly from one molecule to another. As a result, no hydrogen atom spends much time on a given molecule and the effect of the nuclear spin of the hydrogen atom is averaged out. Thus, the hydrogen atoms on an adjacent carbon atom experience only an average change of the external magentic field, and no spin-spin coupling appears.

If a sample of an alcohol is carefully purified so that no trace of acid, which catalyzes the exchange reaction, remains, coupling between the hydrogen atom of the hydroxyl group and hydrogen atoms on an adjacent carbon atom can be observed. The spectrum of a highly purified sample of methanol appears above in Figure 11.25. In this spectrum, the band for the hydrogen atom of the hydroxyl group is a multiplet, and that for the methyl group is clearly a doublet. The coupling constant for the interaction between the two types of hydrogen atoms is approximately 5 Hz.

PROBLEM 11.12

Assign structures to Compounds I–K, for which spectra and molecular formulas are given in Figure 11.26. Calculation of units of unsaturation and careful examination of chemical shifts, integration, and splitting patterns will be helpful.

FIGURE 11.26

Compound I, $C_5H_{10}O$

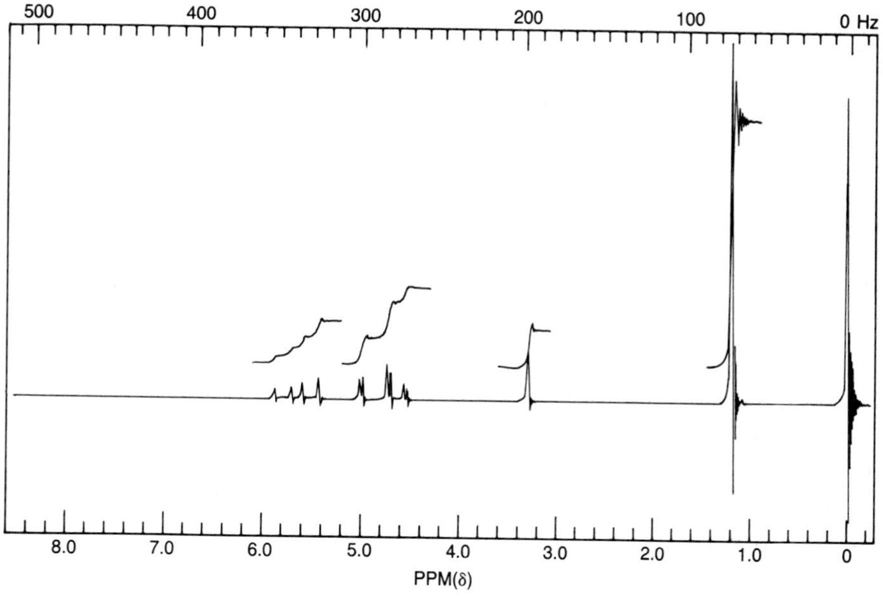
500
400
300
200
100
0 Hz
8.0
7.0
6.0
5.0
4.0
3.0
2.0
1.0
0
PPM(δ)

Compound J, C_4H_8O

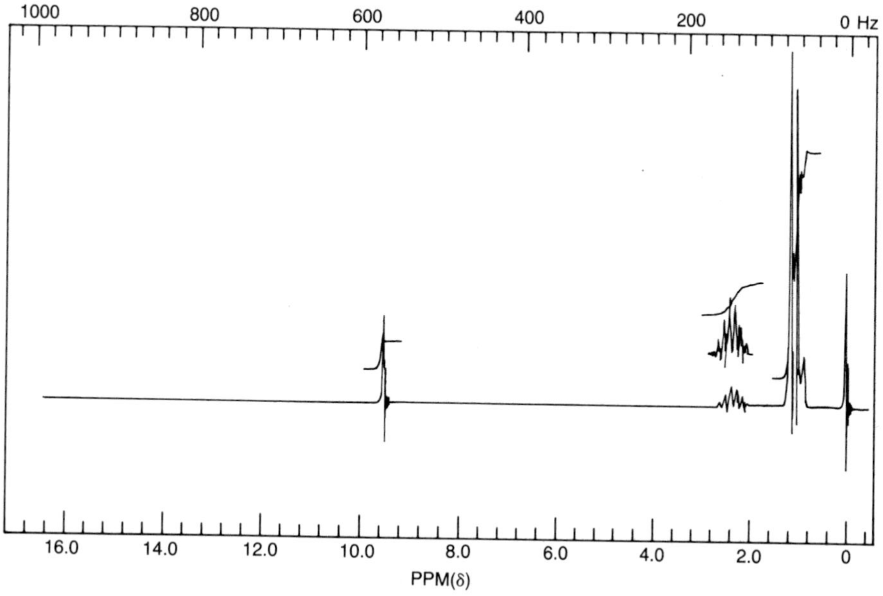
1000
800
600
400
200
0 Hz
16.0
14.0
12.0
10.0
8.0
6.0
4.0
2.0
0
PPM(δ)

(Continued)

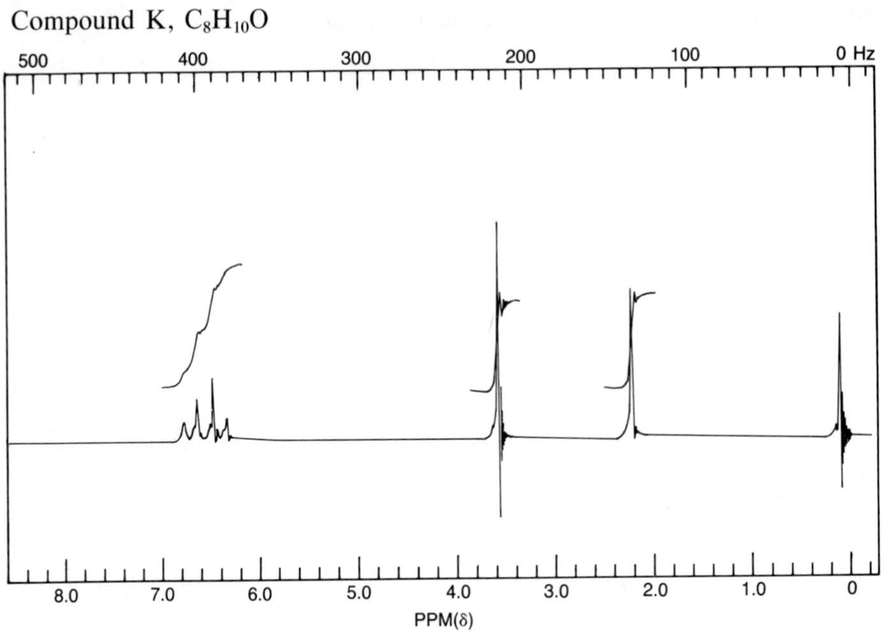
Compound K, $C_8H_{10}O$
500
400
300
200
100
0 Hz
8.0
7.0
6.0
5.0
4.0
3.0
2.0
1.0
0
PPM(δ)

FIGURE 11.26 *(Continued)*

PROBLEM 11.13

Compound L has the molecular formula $C_5H_{12}O$. Its proton magnetic resonance spectrum is shown in Figure 11.27. The important bands in its infrared spectrum are at 3350, 3000, 1460, and 1120 cm^{-1}. Assign a structure to Compound L.

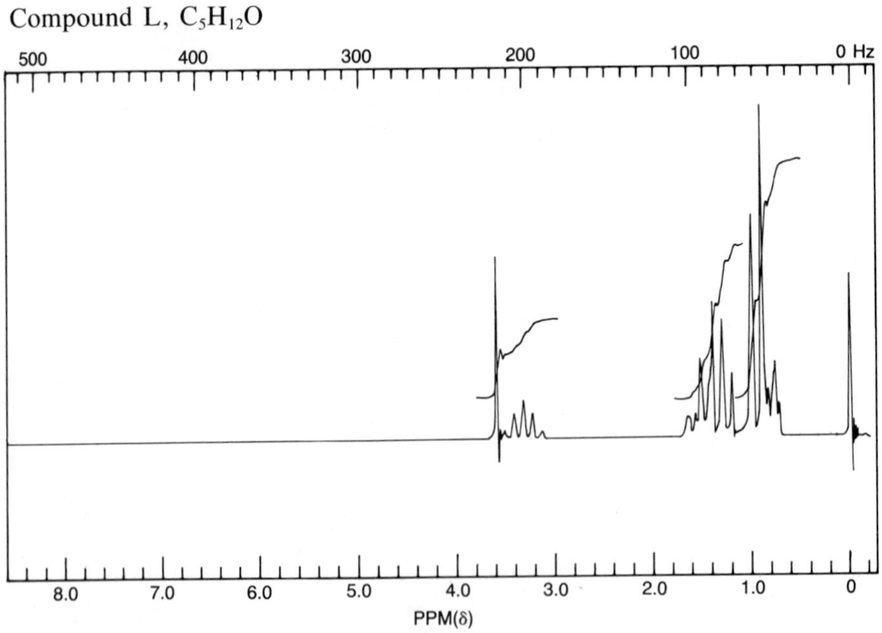
Compound L, $C_5H_{12}O$
500
400
300
200
100
0 Hz
8.0
7.0
6.0
5.0
4.0
3.0
2.0
1.0
0
PPM(δ)

FIGURE 11.27

PROBLEM 11.14

The infrared spectrum of Compound M, $C_5H_8O_2$, has bands at 3100, 1760, 1675, and 1200 cm^{-1}. The proton magnetic resonance spectrum of the compound is shown in Figure 11.28. Assign a structure to Compound M.

Compound M, $C_5H_8O_2$

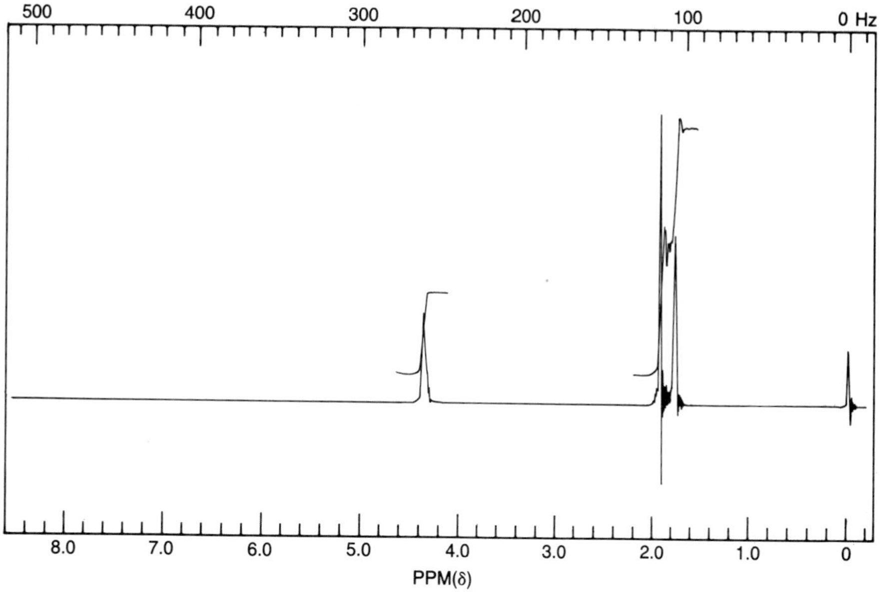
500
400
300
200
100
0 Hz
8.0
7.0
6.0
5.0
4.0
3.0
2.0
1.0
0
PPM(δ)

FIGURE 11.28

11.5 CARBON-13 NUCLEAR MAGNETIC RESONANCE SPECTROSCOPY

The nucleus of an atom of carbon-13, the isotope of carbon with an atomic mass of 13, has a spin number, I, of $\frac{1}{2}$, just as the nucleus of a hydrogen atom does. This isotope of carbon has a natural abundance of about 1%; that is, approximately 1% of all the carbon atoms in any sample of an organic compound are carbon-13 atoms. It is possible to record the nuclear magnetic resonance of these carbon atoms; because they occur randomly throughout molecules of a given compound, the spectrum is representative of the compound. The principles of carbon-13 nuclear magnetic resonance spectroscopy are the same as those described for proton magnetic resonance spectroscopy. When placed in an external magnetic field, carbon nuclei absorb energy in the radiofrequency region of the electromagnetic spectrum in going from one energy level to a higher one. These carbon nuclei may be shielded or deshielded depending on their environment, just as the nuclei of hydrogen atoms are. Tetramethylsilane is used as a reference compound and the positions of the absorption bands for carbon atoms are reported in parts per million from the TMS signal, just as in proton nuclear magnetic resonance.

Modern Fourier-transform spectrometers use short pulses of radiofrequency radiation to excite all of the carbon nuclei at once. The data are collected as an exponentially decaying sine wave (shown in Figure 11.29) containing information about the frequencies at which the carbon atoms in the sample absorb energy. A computer stores the data and converts them into a spectrum. Figure 11.29 contains the original data for 4-methyl-2-pentanone and two spectra derived from them. One of the spectra has five peaks in it. (The three closely spaced peaks at δ 78 come from the solvent $CDCl_3$.) These peaks are the absorption bands for the five different kinds of carbon atoms in 4-methyl-2-pentanone.

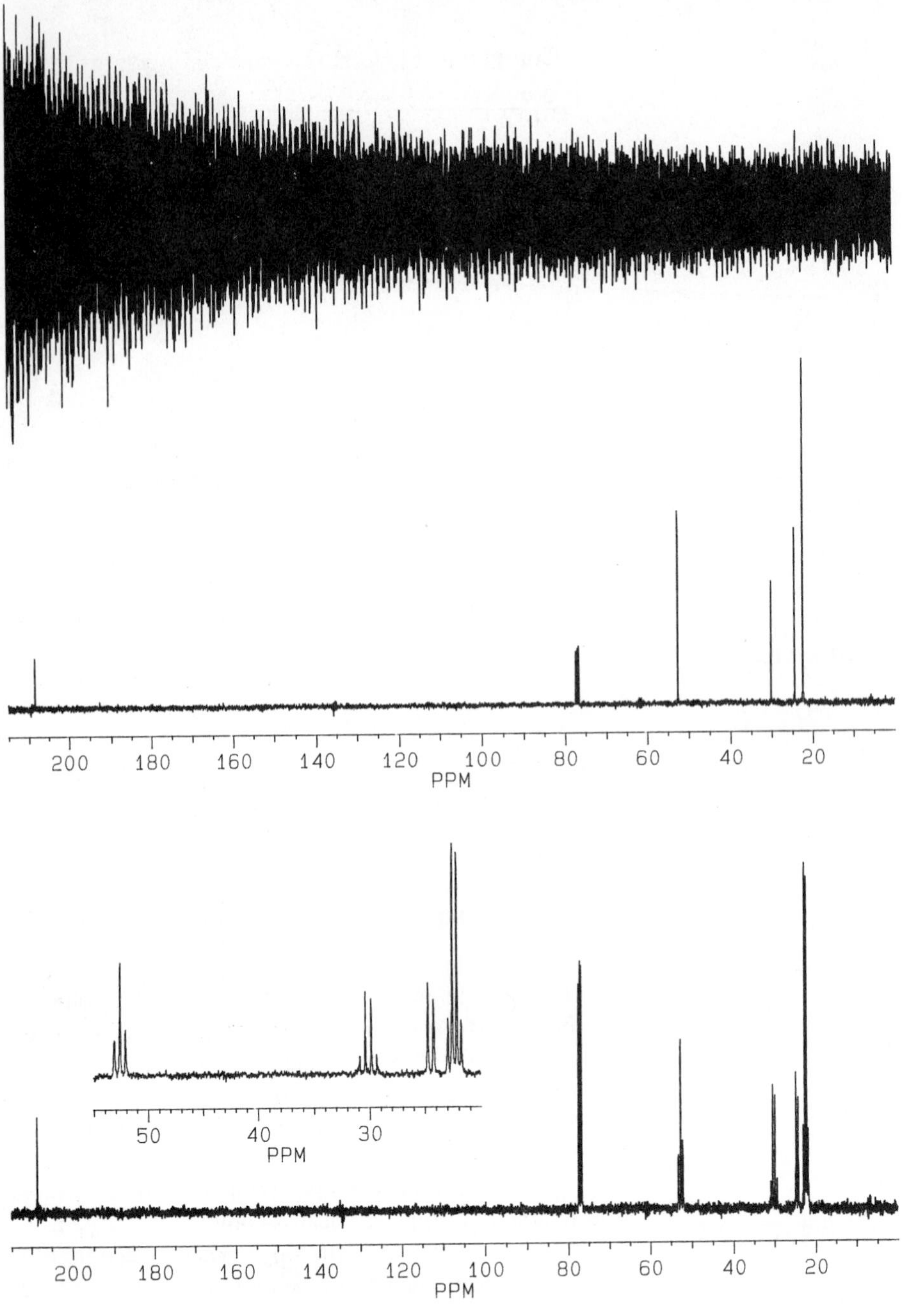
200
180
160
140
120
100
80
60
40
20
PPM
50
40
30
PPM
200
180
160
140
120
100
80
60
40
20
PPM

FIGURE 11.29 Free-induction decay data and proton-decoupled and off-resonance proton-decoupled carbon-13 nuclear magnetic resonance spectra for 4-methyl-2-pentanone, recorded on a Bruker WM-360 nuclear magnetic resonance spectrometer.

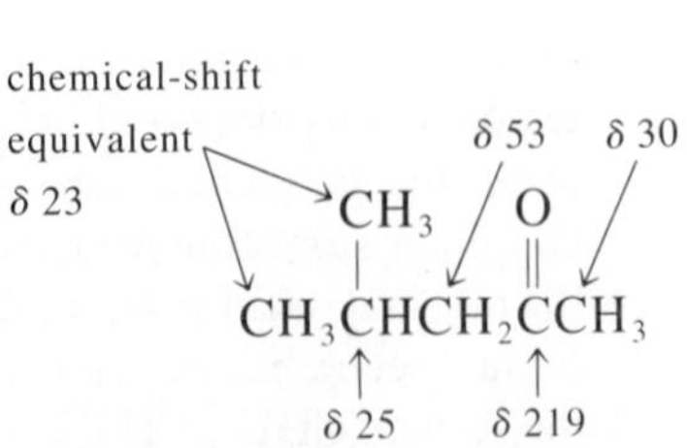
chemical-shift
equivalent
δ 23
δ 53
δ 30
CH3
O
CH3CHCH2CCH3
δ 25
δ 219

4-methyl-2-pentanone

The carbon atom that is deshielded the most, with a signal at 219 ppm downfield from TMS (δ 0), is the carbon atom of the carbonyl group. The carbon atoms that are the most highly shielded (δ 23) are the ones in the methyl groups that are farthest away from the carbonyl group. The carbon atom of the other methyl group (δ 30) is deshielded by the carbonyl group. The signal for the methylene carbon is at δ 53, and that for the methine carbon is at δ 25.

The nuclei of carbon atoms couple with the nuclei of hydrogen atoms that are bonded to them in exactly the same way that the nuclei of hydrogen atoms on neighboring carbon atoms couple with each other. For example, the signal for a carbon atom with one hydrogen atom on it will be a doublet. The coupling constants for carbon atoms and hydrogens bonded to them are so large (J 100–320 Hz), however, that the spectra that result are often of great complexity. To simplify the spectra, strong radiation of a second radiofrequency range is used to excite all of the hydrogen atoms in the molecule. Because hydrogen nuclei are constantly flipping to the higher-energy spin state and back again, they do not spend enough time at either energy level to affect the magnetic field felt by the neighboring carbon nuclei. The neighboring nuclei are **decoupled.** The spectrum in Figure 11.29 in which five singlets are seen for the carbon atoms of 4-methyl-2-pentanone is a **proton-decoupled spectrum.** The second spectrum in Figure 11.29 shows the coupling between the carbon-13 and hydrogen nuclei and was taken using a technique called **off-resonance proton decoupling,** which greatly decreases the value of the coupling constant for the carbon and hydrogen nuclei bonded to each other and also eliminates any longer-range coupling between a carbon nucleus and hydrogen nuclei farther away. In this spectrum, the signals for methyl carbons appear as quartets, that for the methylene carbon as a triplet, and that for the methine carbon as a doublet. The peak for the carbon atom of the carbonyl group is a singlet, because that carbon has no hydrogen atoms on it.

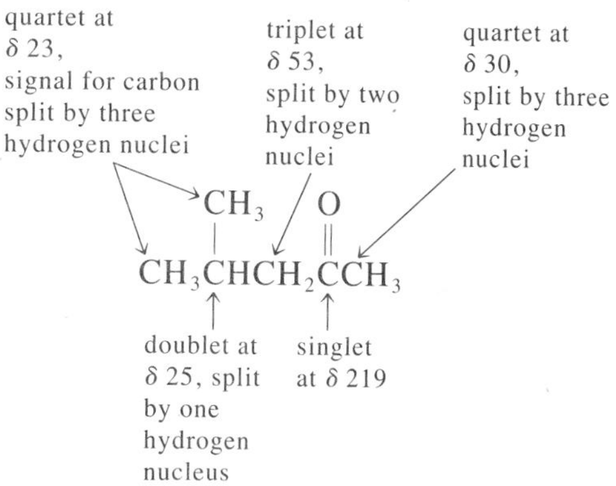
quartet at δ 23, signal for carbon split by three hydrogen nuclei
triplet at δ 53, split by two hydrogen nuclei
quartet at δ 30, split by three hydrogen nuclei
CH3
O
CH3CHCH2CCH3
doublet at δ 25, split by one hydrogen nucleus
singlet at δ 219

4-methyl-2-pentanone
off-resonance proton-decoupled spectrum

The spectrum does not show any spin-spin coupling between carbon nuclei because the natural abundance of carbon-13 is so low that there is a very low probability that the same molecule will contain two carbon-13 atoms close enough to each other to allow interaction between their nuclei.

Two other features of carbon-13 nuclear magnetic resonance spectra are worth mentioning. The range of chemical shifts for carbon atoms (~220 ppm) is much larger than that for hydrogen atoms (~20 ppm). This means that it is much easier to identify individual carbon atoms from a spectrum, even for compounds with complex molecules. Carbon-13 nuclear magnetic resonance spectroscopy has become, therefore, a powerful tool for structure determination in organic chemistry. Carbon-13 nuclear magnetic resonance spectra also differ from proton magnetic resonance spectra in not having integration curves. The intensity of a band, in this case the height of one of the very narrow peaks, is not directly related to the number of carbon atoms undergoing that energy transition. In general, carbon atoms that have hydrogen atoms bonded to them give rise to more intense bands than those that have no hydrogen atoms on them. In the proton-decoupled spectrum of 4-methyl-2-pentanone, for example, the peak for the carbon atom of the carbonyl group is much smaller than the peak for the carbon atom of the methine group, which has a hydrogen atom on it. A carbon-13 nuclear magnetic resonance spectrum, therefore, gives information about the different types of carbon atoms but *not* about their relative numbers.

The spectrum of methyl benzoate is shown in Figure 11.30. A comparison of this spectrum with that of 4-methyl-2-pentanone (Figure 11.29) demonstrates that the carbon atom of the carbonyl group of an ester is more shielded than the carbon atom of the carbonyl group of a ketone. The signal for the carbonyl carbon of the ester appears at 166.8 ppm downfield from TMS; that for the carbonyl carbon of the ketone appears 211.8 ppm downfield from TMS. This observed difference in the chemical shifts of the two carbons is reflected in the chemistry of ketones and esters (pp. 473 and 562). The carbon atom of the carbonyl group of a ketone is more electrophilic than the carbon atom of the carbonyl group of an ester.

The carbon atom of the methyl group bonded to the oxygen atom in methyl benzoate is more deshielded than the carbon atoms of the methyl groups in 4-methyl-2-pentanone and absorbs at 51.8 ppm. The four different types of carbon atoms of the aromatic ring in the ester are also highly deshielded, and their signals cluster from 128.5 to 132.9 ppm. Again, large differences are seen in the intensities of the bands. The two bands of the lowest intensity, at 130.3 and 166.8 ppm, belong to carbon atom 1 in the aromatic ring and the carbon atom of the carbonyl group, respectively. Neither of them bears a hydrogen atom.

The spectrum of 2-octanol is given in Figure 11.31. The eight different carbon atoms of 2-octanol all have different absorption bands in this spectrum. The most highly shielded carbon atom (14.1 ppm) is in the methyl group farthest away from the hydroxyl group. The most highly deshielded carbon atom (67.8 ppm) is, of course, the one bonded to the oxygen atom. The others fall somewhere in between.

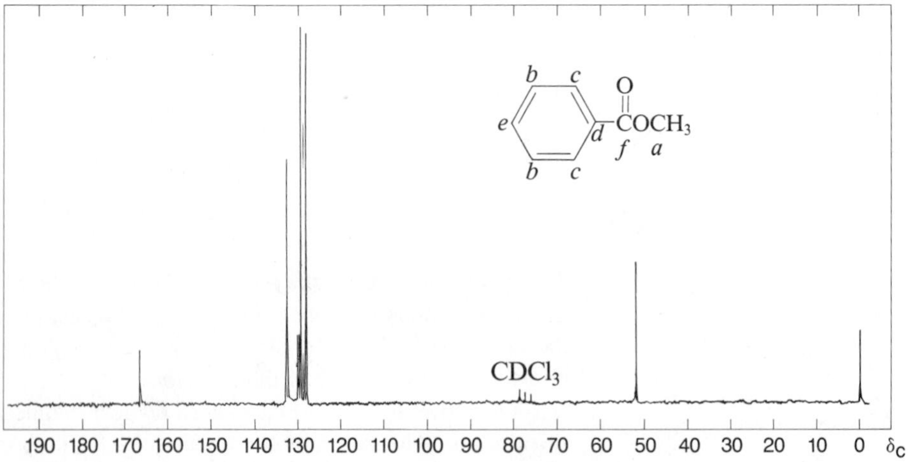
b c
O
e
COCH3
d
f a
b c
CDCl3
190 180 170 160 150 140 130 120 110 100 90 80 70 60 50 40 30 20 10 0 δC

FIGURE 11.30 Carbon-13 nuclear magnetic resonance spectrum of methyl benzoate. (Adapted from Johnson and Jankowski. Reprinted by permission.)

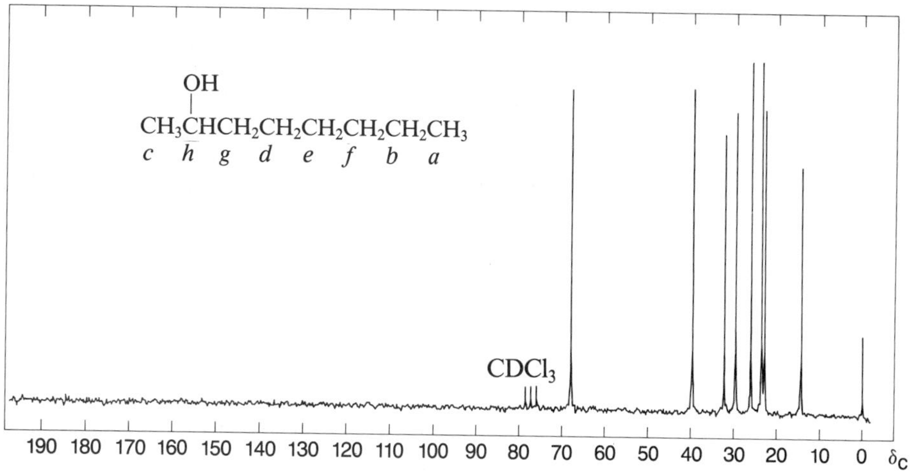

OH
$CH_3CHCH_2CH_2CH_2CH_2CH_2CH_3$
c h g d e f b a
$CDCl_3$
190 180 170 160 150 140 130 120 110 100 90 80 70 60 50 40 30 20 10 0 δ_C

FIGURE 11.31 Carbon-13 nuclear magnetic resonance spectrum of 2-octanol. (Adapted from Johnson and Jankowski. Reprinted by permission.)

The spectrum illustrates the power of carbon-13 nuclear magnetic resonance in giving structural information about an organic compound.

Typical chemical shifts observed for different types of carbon atoms are given in Table 11.4. As the first three entries in the table show, a more highly substituted carbon atom usually has a higher chemical shift than a less substituted one with a

TABLE 11.4 Chemical Shifts for Carbon Atoms in Carbon-13 Nuclear Magnetic Resonance Spectra

Type of Carbon Atom	δ*	Type of Carbon Atom	δ*
$RCH_2\mathbf{CH_3}$	13–16	$RCH{=}\mathbf{CH_2}$	115–120
$R\mathbf{CH_2}CH_3$	16–25	$R\mathbf{CH}{=}CH_2$	125–140
$R_3\mathbf{CH}$	25–38	$R\mathbf{C}{\equiv}N$	117–125
$\mathbf{CH_3}CR$ (with C=O)	~30	Ar**H**	125–150
$\mathbf{CH_3}COR$ (with C=O)	~20	$R\mathbf{C}OR'$ (with C=O)	170–175
$R\mathbf{CH_2}Cl$	40–45	$R\mathbf{C}OH$ (with C=O)	177–185
$R\mathbf{CH_2}Br$	28–35	$R\mathbf{C}H$ (with C=O)	190–200
$R\mathbf{CH_2}NH_2$	37–45	$R\mathbf{C}R'$ (with C=O)	205–220
$R\mathbf{CH_2}OH$	50–64		
$RC{\equiv}\mathbf{CH}$	67–70		
$R\mathbf{C}{\equiv}CH$	74–85		

*The chemical shift values are given in ppm relative to tetramethylsilane at δ 0.00 and are for the carbon atoms shown in boldface in the formulas.

similar environment. The numbers in the table are, as usual, to be applied with caution and only after taking into account all of the available information.

PROBLEM 11.15

2-Methylpentane, 3-methylpentane, and 2,3-dimethylbutane are isomeric hydrocarbons. Carbon-13 nuclear magnetic resonance spectral data for the compounds are given below.

Isomer 1: 19.5, 34.0 ppm
Isomer 2: 11.4, 18.8, 29.3, 36.4 ppm
Isomer 3: 14.3, 20.6, 22.6, 27.9, 41.6 ppm

Draw a structural formula for each compound, and decide which set of data corresponds to which compound.

PROBLEM 11.16

Five carbon-13 nuclear magnetic resonance spectra are given in Figure 11.32. Decide which spectrum belongs to which of the following compounds.

(a) 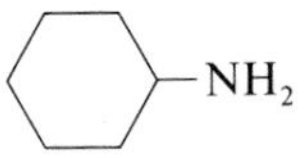
NH2
(b) 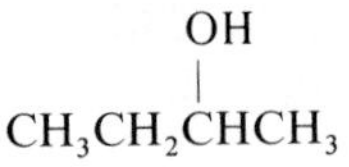
OH
CH3CH2CHCH3
(c) 
O O
CH3CH2COCCH2CH3

(d) 
O
(e) 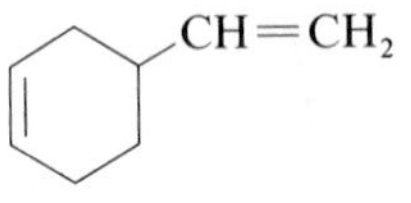
CH=CH2

FIGURE 11.32

Spectrum 1

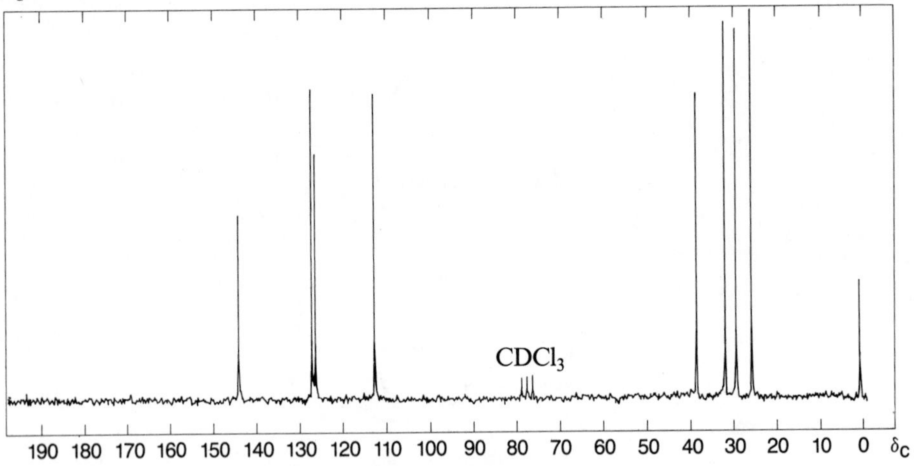
CDCl3
190 180 170 160 150 140 130 120 110 100 90 80 70 60 50 40 30 20 10 0 δC

Spectrum 2

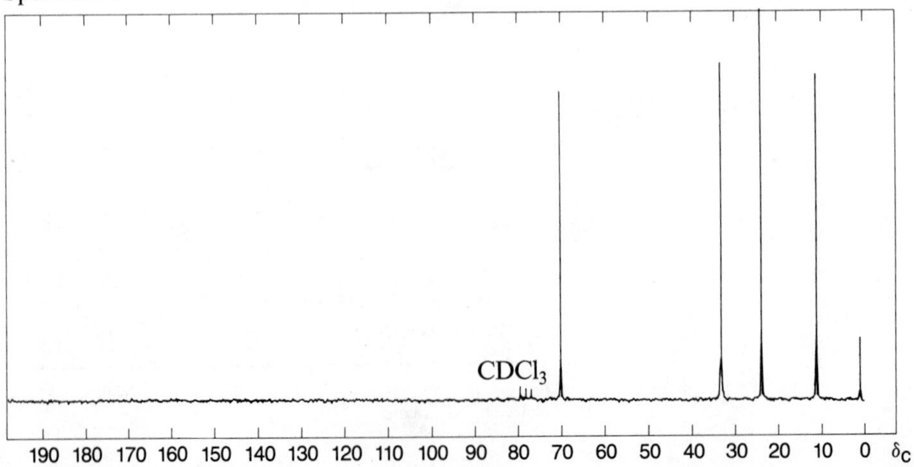
CDCl3
190 180 170 160 150 140 130 120 110 100 90 80 70 60 50 40 30 20 10 0 δC

FIGURE 11.32 (*Continued*) Spectrum 3

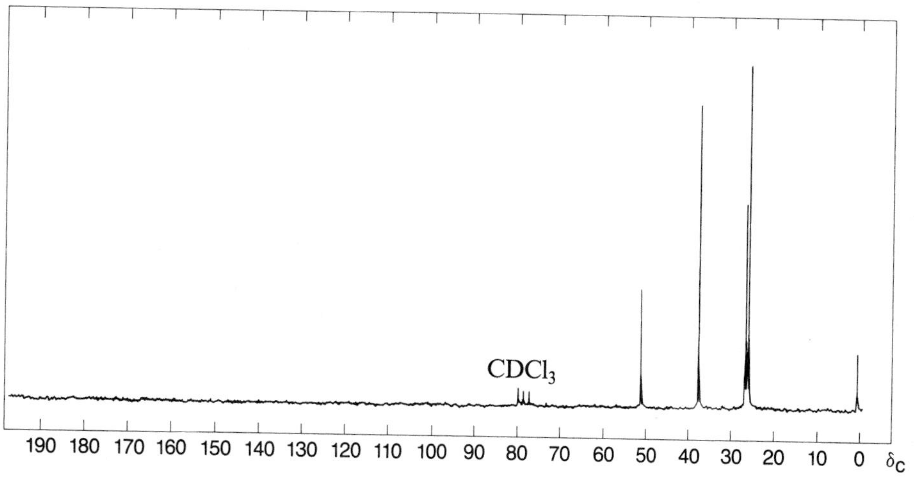
CDCl3
190 180 170 160 150 140 130 120 110 100 90 80 70 60 50 40 30 20 10 0 δC

Spectrum 4

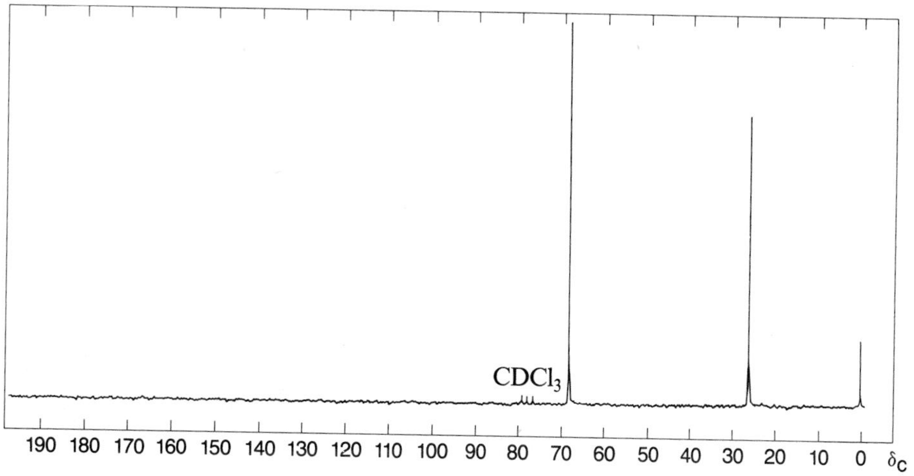
CDCl3
190 180 170 160 150 140 130 120 110 100 90 80 70 60 50 40 30 20 10 0 δC

Spectrum 5

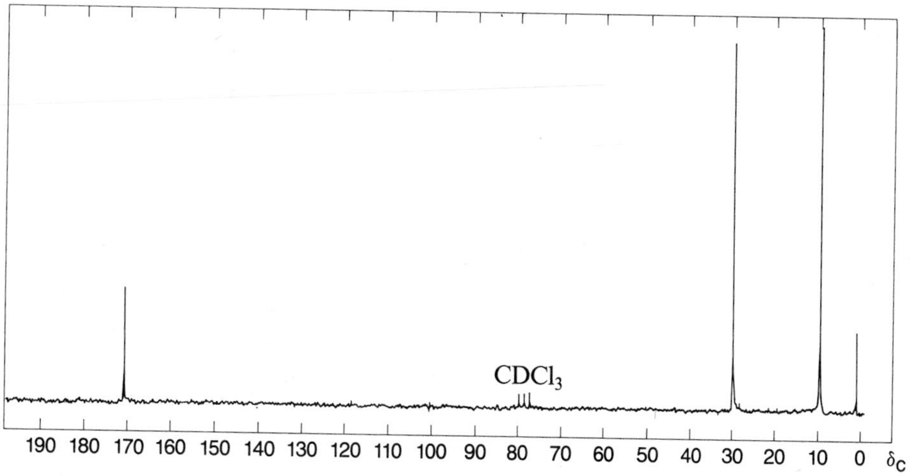
CDCl3
190 180 170 160 150 140 130 120 110 100 90 80 70 60 50 40 30 20 10 0 δC

Adapted from Johnson and Jankowski. Reprinted by permission.

PROBLEM 11.17

Compound N, C_5H_8O, has bands in the carbon-13 nuclear magnetic resonance spectrum at δ 23.4, 38.2, and 219.6. Assign a structure to this compound.

11.6 MEDICAL APPLICATIONS OF NUCLEAR MAGNETIC RESONANCE

Proton nuclear magnetic resonance spectroscopy has been used since the early 1980s to obtain images from living beings. Figure 11.33 shows an image of the brain of a patient with multiple sclerosis obtained by a technique called magnetic resonance imaging (MRI). The white regions near the center of the brain correspond to the lesions produced by the disease. Signals are collected from many different locations in the brain in the form of a free-induction decay signal (Figure

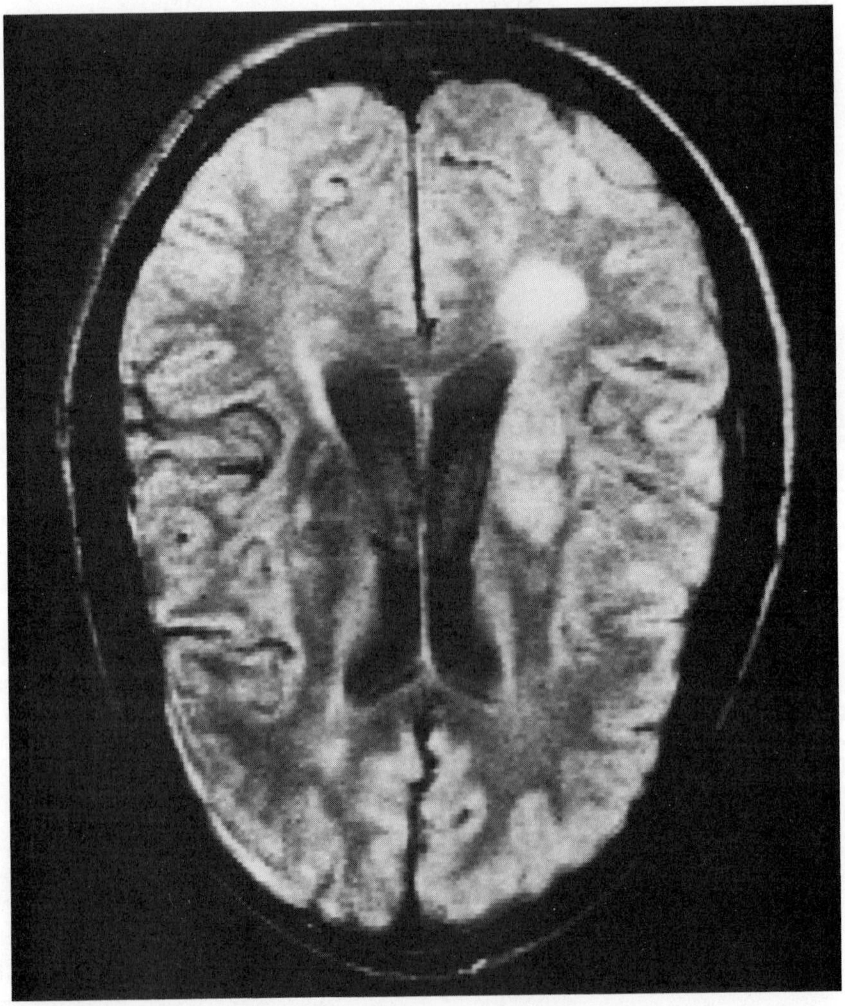

FIGURE 11.33 Image of the brain of a patient with multiple sclerosis, obtained by the technique of magnetic resonance imaging. (From Dr. Alex Aisen, Department of Radiology, the University of Michigan)

11.29, p. 410) containing many frequencies. A computer then performs a Fourier transformation and translates these frequencies into an image. Such an image is essentially a map showing the density of the protons in water molecules and lipid molecules (p. 609) contained in the tissues of the brain.

Two factors contribute to forming this image. One, of course, is the number of protons in the substances in the head. For example, the bones of the skull do not show up in the image because they do not contain much water or other substances containing hydrogen. The most important factor for distinguishing between healthy and diseased tissues are the **relaxation times.** Relaxation is the process by which a nucleus that has been raised to the higher-energy level returns to the lower-energy level. Two types of relaxation occur. In one, the energy of a nucleus at the upper level is dissipated to the atoms of adjacent molecules. This is called spin-lattice relaxation, and T_1 is the time it takes for a nucleus to return to the lower-energy level by this route. Another type of relaxation occurs when a nucleus at the upper level transfers its energy to the nucleus of a nearby atom. This type of relaxation is called spin-spin relaxation. The time it takes for a nucleus to lose energy by spin-spin relaxation is called T_2. T_1 and T_2 for protons in pure water are both about 3 seconds; these times are much shorter (from a few hundred milliseconds to 2 seconds) for water in biological systems.

The principal factor in determining T_1 and T_2 is the rate at which molecules tumble in solution. Small molecules can change their orientation relative to each other easily and thus have inefficient relaxation processes and relatively large values for T_1 and T_2. Large molecules, such as cholesterol or other lipids, tumble more slowly because of their greater inertia and the greater friction between themselves and their neighbors. They undergo relaxation more efficiently and have smaller values of T_1 and T_2. The ordering of water molecules in normal cells is different from that in tumor cells, for example, which also contributes to a difference in T_1 and T_2 values for normal and diseased tissues.

The timing of the radiofrequency signal used to excite nuclei to the higher-energy level is used to differentiate between tissues with components that have different values of T_1 and T_2. If, for example, the interval between radiofrequency pulses is short compared to the T_1 of a sample, the pulse arrives before the nuclei have had a chance to return to the lower-energy level. No absorption of energy can take place; therefore, no signal is detected from that portion of the tissue. The lesion from multiple sclerosis produces edema, an abnormal accumulation of fluid with a high water content. That region has a large T_2 compared to adjoining tissues. A technique known as T_2-weighting combines different timings and orientations of the radiofrequency pulse to make areas with large T_2 values look bright. Magnetic resonance imaging, therefore, generates detailed images based on the chemical composition and the ordering of molecules in different types of tissues. Physicians use this technique to get images of the spinal cord, the brain, and the liver without the use of ionizing radiation (such as x-rays) and without the injection of contrasting agents, which sometimes cause allergic reactions or paralysis when used in the central nervous system.

The nuclear magnetic resonance of phosphorus, ^{31}P, is used in another area of biomedical research. Compounds that are esters of phosphoric acid play a central role in cellular metabolism. Adenosine triphosphate (ATP, p. 1079) and adenosine diphosphate (ADP, p. 1079) are particularly important to the supply and use of chemical energy in cells. The role of phosphate esters of glucose in the metabolism of glucose is described on p. 661. Phosphate esters are present in high enough concentration (~0.5 mM) that they can be detected in intact tissues by ^{31}P nuclear magnetic resonance. In addition, phosphorus-31 nuclear magnetic resonance is used to study DNA (p. 1013), which also contains phosphate groups.

Of particular interest is the use of phosphorus-31 nuclear magnetic resonance to follow cellular metabolism during exercise, as a way of identifying normal metabolic pathways so that this knowledge can be applied to diagnose muscle diseases. Knowledge of what went on in muscle used to be obtained only by doing muscle biopsies, removal of small amounts of muscle for testing. Athletes were understandably reluctant to provide many samples. Nuclear magnetic resonance provides a noninvasive way to follow cellular metabolism. For example, during strenuous exercise, lactic acid accumulates in muscle tissue (p. 661), producing a decrease in pH. This process can be followed using phosphorus-31 nuclear magnetic resonance because the chemical shift of the phosphorus atom in phosphate ion depends on the degree of protonation of the ion.

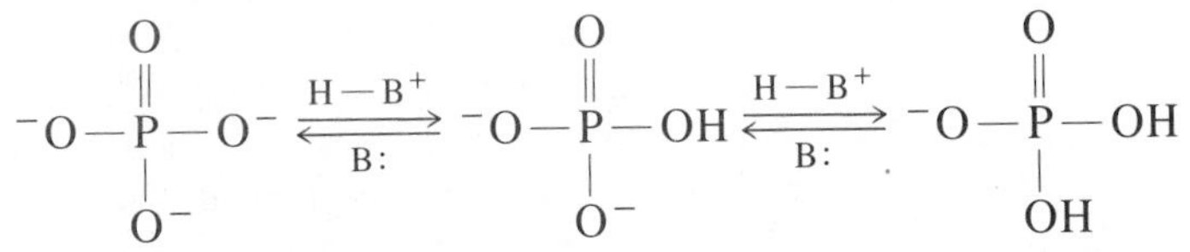
H—B+
B:
H—B+
B:

high pH
appears at lower field in phosphorus-31 nuclear magnetic resonance spectrum

low pH
appears at higher field in phosphorus-31 nuclear magnetic resonance spectrum

In addition, different chemical shifts are seen for ^{31}P depending on the nature of the organic group bound to the phosphate. Therefore, it is possible to follow the decrease of compounds being metabolized and the build-up of the products of metabolism in muscle tissue. Such studies are beginning to lead to a better understanding of the enzymatic processes that are important in metabolism. Faulty metabolic pathways that lead to disease and to the failure of muscles can thus be identified.

SUMMARY

When a sample of an organic compound is placed in a strong magnetic field, slightly more than half of the nuclei of atoms such as hydrogen and carbon-13 align themselves with the external field. In this state, the nuclei are in the lower of the two energy levels available to them; they absorb energy in the radiofrequency range in going to a higher-energy level in which they are aligned against the external magnetic field. The exact amount of energy necessary for this transition depends on the strength of the external magnetic field and on the environment of the hydrogen and carbon atoms in the compound being investigated. A record of these transitions is a nuclear magnetic resonance spectrum, which can be a proton or a carbon-13 nuclear magnetic resonance spectrum.

A magnetic resonance spectrum gives four different kinds of information. The chemical shifts (δ values) tell how many different kinds of hydrogen (Table 11.1, p. 391) or carbon atoms (Table 11.4, p. 413) are present in the compound and something about their environments. Proton magnetic resonance spectra are integrated, which reveals the relative numbers of the different hydrogen atoms in the compound. The splitting patterns (Table 11.2, p. 396) tell which atoms are close enough that their nuclei can interact. The magnitudes of the coupling constants (Table 11.3, p. 402) give information about the relative positions of the hydrogen atoms.

In carbon-13 nuclear magnetic resonance spectra, coupling between the nucleus of a carbon atom and the nuclei of the hydrogen atoms on it gives a splitting pattern that can be used to identify methyl, methylene, and methine carbon atoms, as well as those that have no hydrogen atoms on them.

Nuclear magnetic resonance is also important in biological and medical fields. Proton magnetic resonance can be used to create images of tissues that distinguish diseased from normal tissues. Phosphorus-31 nuclear magnetic resonance is used to follow cellular metabolism.

ADDITIONAL PROBLEMS

11.18 Molecular formulas and proton magnetic resonance spectra for Compounds O–T are given in Figure 11.34. Assign structures to the compounds.

FIGURE 11.34

Compound O, $C_8H_8O_2$

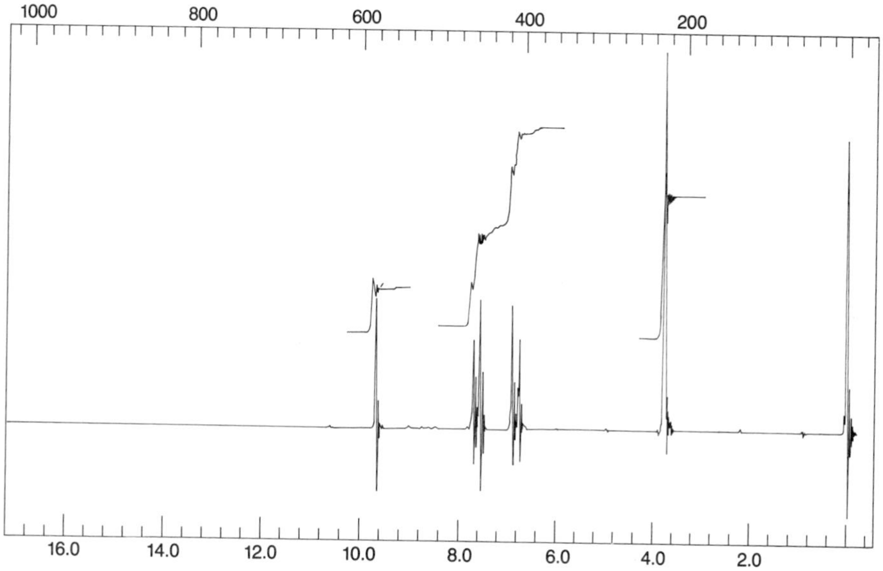
1000
800
600
400
200
16.0
14.0
12.0
10.0
8.0
6.0
4.0
2.0

Compound P, $C_4H_8O_2$

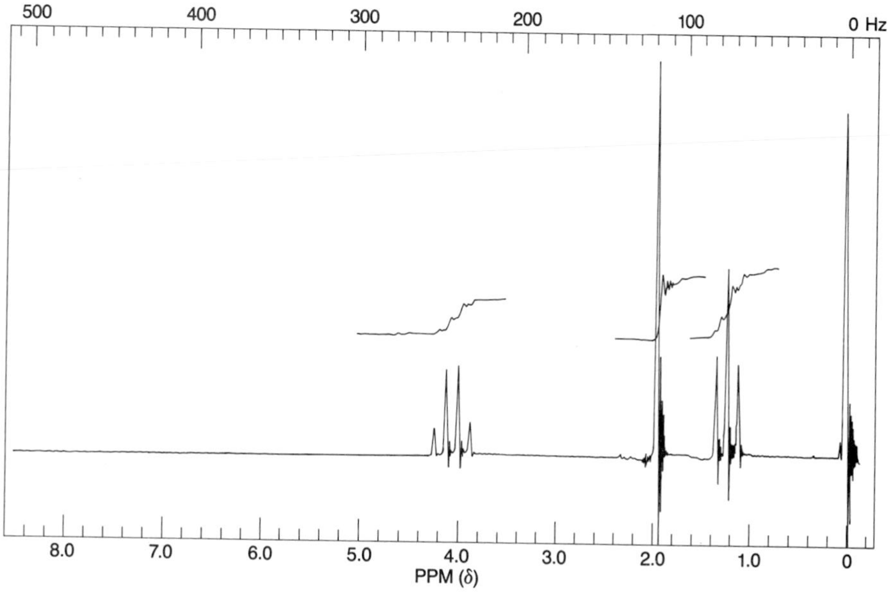
500
400
300
200
100
0 Hz
8.0
7.0
6.0
5.0
4.0
3.0
2.0
1.0
0
PPM (δ)

Compound Q, $C_5H_8O_2$

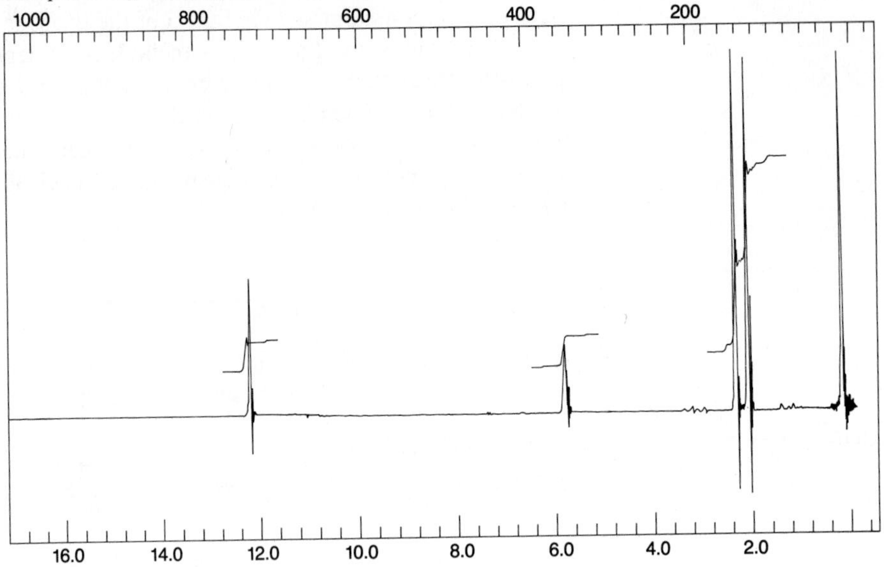
1000
800
600
400
200
16.0
14.0
12.0
10.0
8.0
6.0
4.0
2.0

Compound R, C_7H_9N

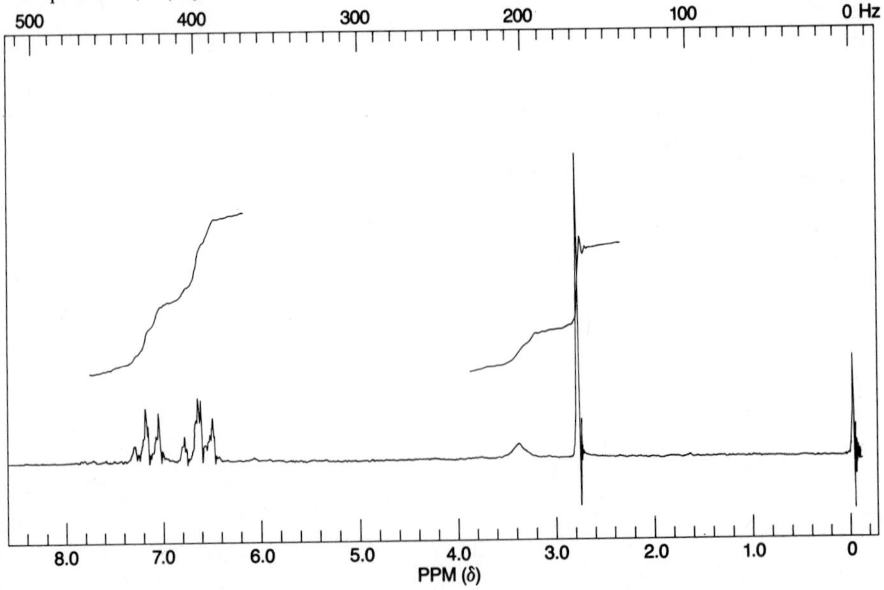
500
400
300
200
100
0 Hz
8.0
7.0
6.0
5.0
4.0
3.0
2.0
1.0
0
PPM (δ)

FIGURE 11.34 *(Continued)*

FIGURE 11.34 (*Continued*) Compound S, $C_7H_{12}O_3$

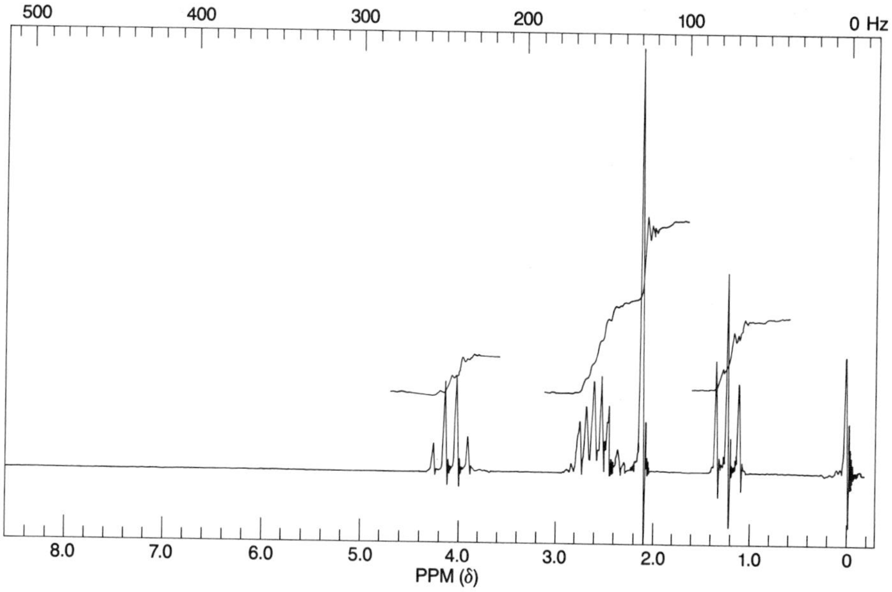

500
400
300
200
100
0 Hz
8.0
7.0
6.0
5.0
4.0
3.0
2.0
1.0
0
PPM (δ)

Compound T, $C_6H_{15}N$

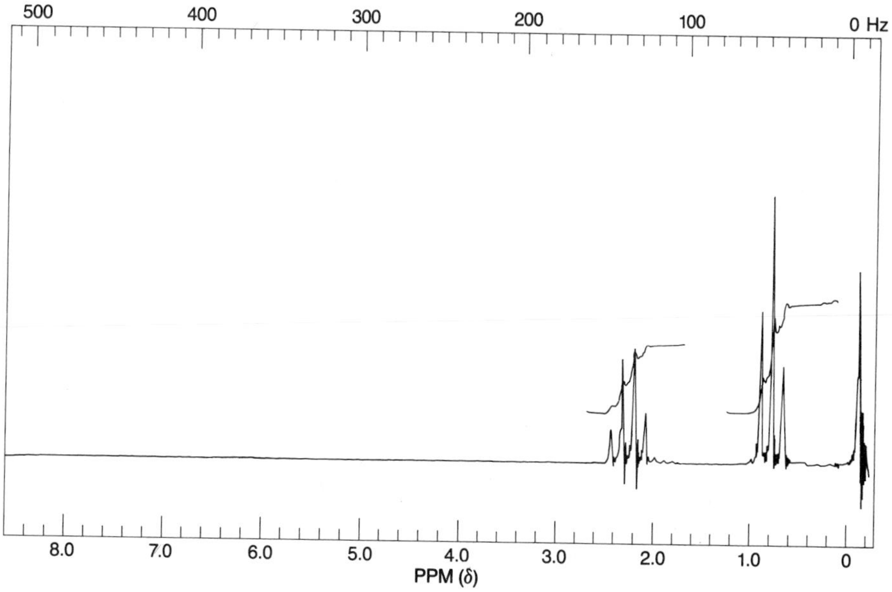

500
400
300
200
100
0 Hz
8.0
7.0
6.0
5.0
4.0
3.0
2.0
1.0
0
PPM (δ)

11.19 The proton magnetic resonance spectrum of Compound U, $C_{10}H_{12}O$, is shown in Figure 11.35. The infrared spectrum of Compound U has a strong band at 1717 cm^{-1}. Assign a structure to the compound.

Compound U, $C_{10}H_{12}O$

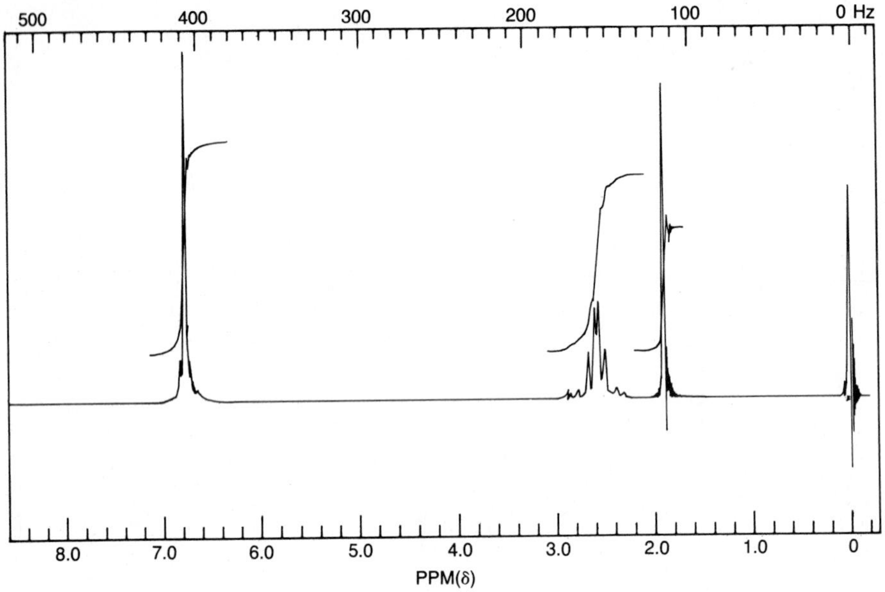
500
400
300
200
100
0 Hz
8.0
7.0
6.0
5.0
4.0
3.0
2.0
1.0
0
PPM(δ)

FIGURE 11.35

11.20 Compound V, $C_5H_{12}O$, has the proton magnetic resonance spectrum shown in Figure 11.36. Important bands appear at 3320 and 1060 cm^{-1} in its infrared spectrum. What is the structure of Compound V?

Compound V, $C_5H_{12}O$

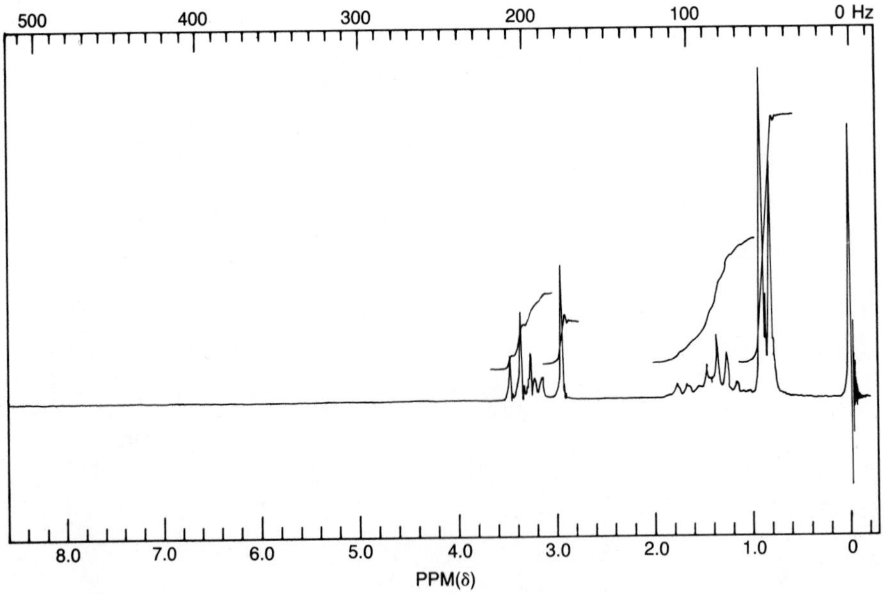
500
400
300
200
100
0 Hz
8.0
7.0
6.0
5.0
4.0
3.0
2.0
1.0
0
PPM(δ)

FIGURE 11.36

11.21 The proton magnetic resonance spectrum of *sec*-butylbenzene is given in Figure 11.37. Assign all of the bands in the spectrum and analyze the splitting patterns.

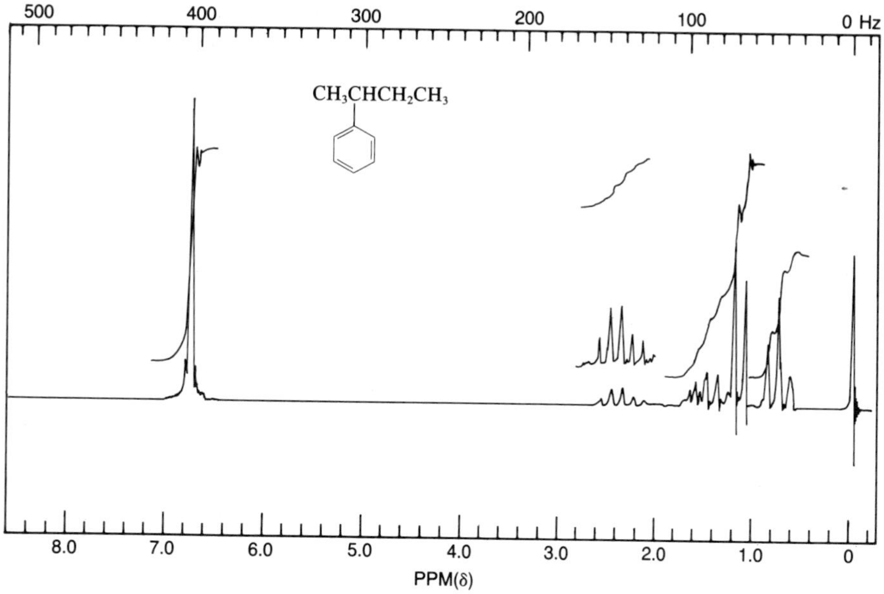

500
400
300
200
100
0 Hz
$CH_3CHCH_2CH_3$
8.0
7.0
6.0
5.0
4.0
3.0
2.0
1.0
0
PPM(δ)

FIGURE 11.37

11.22 Compound W has a strong band in its infrared spectrum at 1685 cm^{-1}. Its proton magnetic resonance spectrum is shown in Figure 11.38. Assign a structure to Compound W.

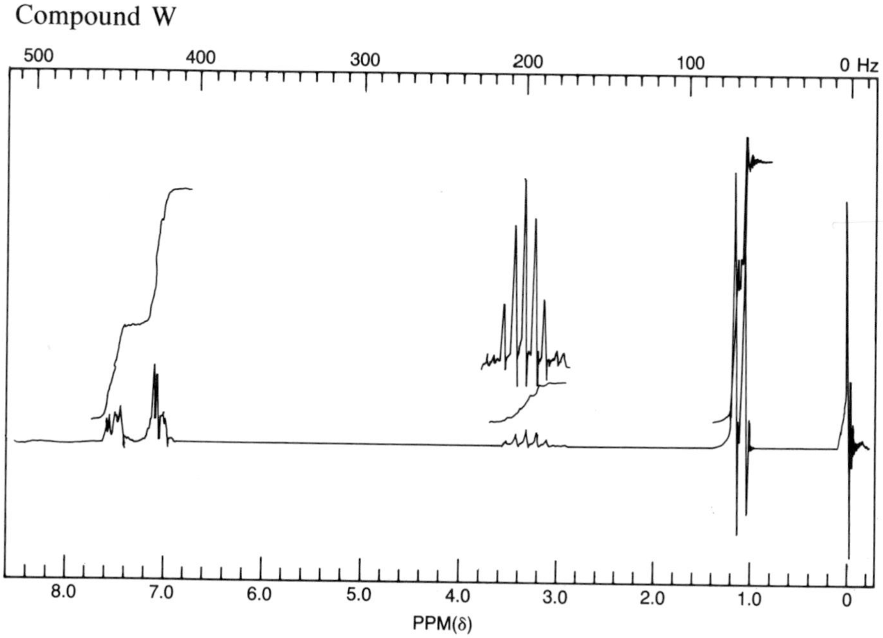

Compound W
500
400
300
200
100
0 Hz
8.0
7.0
6.0
5.0
4.0
3.0
2.0
1.0
0
PPM(δ)

FIGURE 11.38

11.23 The proton magnetic resonance spectrum of 2,4-dibromoaniline appears in Figure 11.39. Assign all the bands in the spectrum and analyze the splitting patterns.

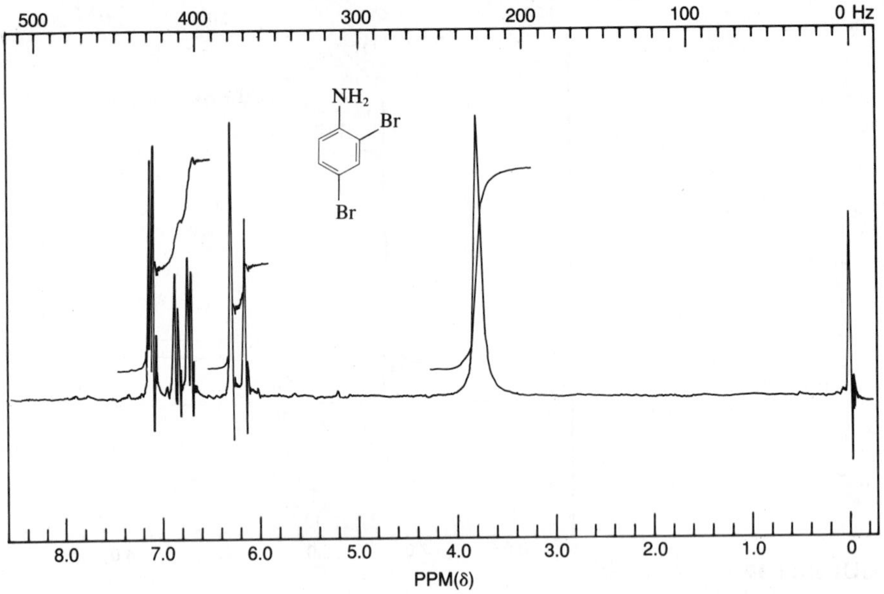
500
400
300
200
100
0 Hz
NH₂
Br
Br
8.0
7.0
6.0
5.0
4.0
3.0
2.0
1.0
0
PPM(δ)

FIGURE 11.39

11.24 The proton magnetic resonance spectrum of *N,N*-dimethylformamide (Figure 11.40) shows two different chemical shifts for the hydrogen atoms on the methyl groups. The phenomenon observed is not due to spin-spin coupling. How do you know this? What does this observation reveal about the nature of the bond between the nitrogen atom and the carbonyl group? Draw structural formulas that account for the observation. (Hint: Writing resonance contributors for the compound may be helpful.)

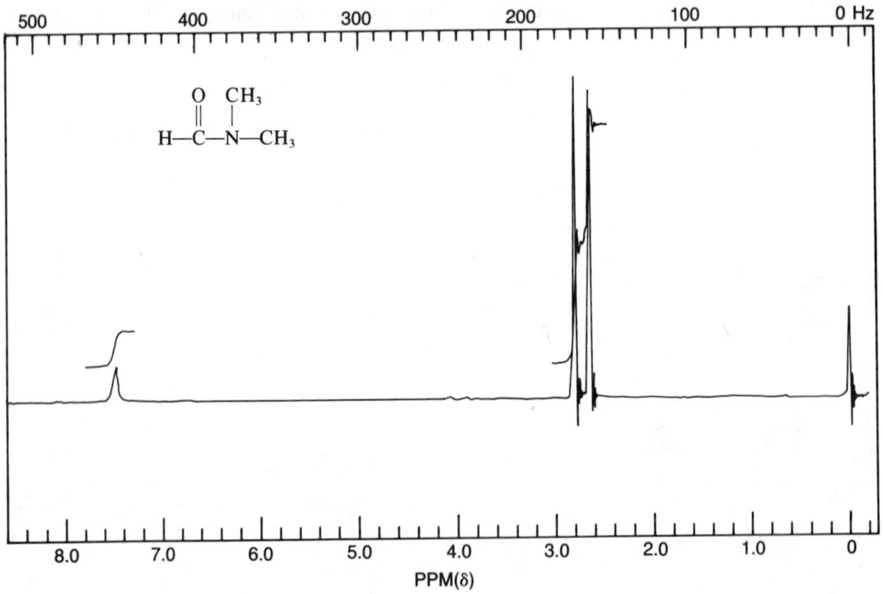
500
400
300
200
100
0 Hz
O CH₃
H—C—N—CH₃
8.0
7.0
6.0
5.0
4.0
3.0
2.0
1.0
0
PPM(δ)

FIGURE 11.40

11.25 Compound X has the molecular formula $C_4H_8Cl_2$ and shows carbon-13 nuclear magnetic resonance absorption bands at 30.0 and 44.1 ppm. Assign a structure to Compound X.

11.26 Compound Y has only two singlets in its proton magnetic resonance spectrum, at δ 1.42 and 1.96, with relative intensities of 3:1. Its carbon-13 nuclear magnetic resonance spectrum has bands at δ 22.3, 28.1, 79.9, and 170.2. The important bands in its infrared spectrum are at 1738, 1256, and 1173 cm^{-1}. Assign a structure to this compound and tell how the various spectra support your assignment.

11.27 Compound Z has three bands in its proton magnetic resonance spectrum: δ 2.5 (1 H, t, J 3 Hz), 3.1 (1 H, s), and 4.3 (2 H, d, J 3 Hz). (Chemists use this shorthand way of describing nuclear magnetic resonance spectra. For each chemical shift value, the number of hydrogen atoms, the multiplicity of the band, and the coupling constant are given in parentheses: s = singlet, d = doublet, m = multiplet, q = quartet, and t = triplet). Three bands at δ 50 (t), 73.8 (d), and 83.0 (s) appear in the carbon-13 magnetic resonance spectrum. The infrared spectrum of Z is shown in Figure 11.41. Assign a structure to Z, showing all of your reasoning.

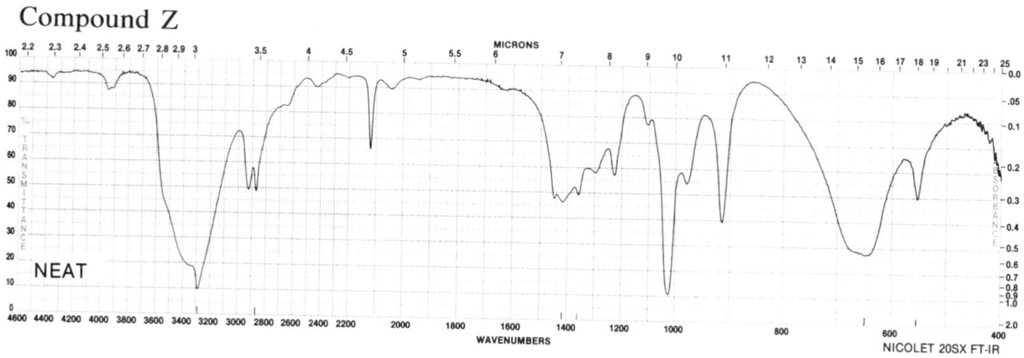

Compound Z
MICRONS
% TRANSMITTANCE
ABSORBANCE
NEAT
WAVENUMBERS
NICOLET 20SX FT-IR

FIGURE 11.41

12

Alcohols, Diols, and Ethers

A • L O O K • A H E A D

The functional group in alcohols is the hydroxyl group.

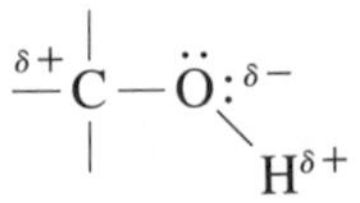

Alcohols resemble water in their acidity and basicity and in their ability to hydrogen bond. Alcohols are easily converted by substitution reactions to alkyl halides when the hydroxyl group is first converted into a good leaving group. Primary alcohols are oxidized to aldehydes, and secondary alcohols to ketones. Deprotonation of alcohols gives alkoxide ions, which are both good nucleophiles and strong bases. Alkoxide ions react with primary and secondary alkyl halides to give ethers.

Ethers have two hydrocarbon groups bonded to oxygen.

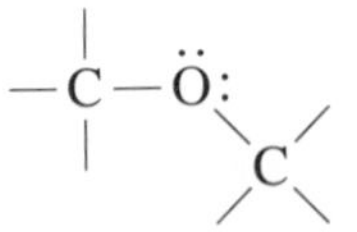

Ethers serve as good solvents. The carbon-oxygen bond can be broken by some strong acids, but ethers are otherwise not very reactive. Oxiranes, in which the oxygen atom is part of a small strained ring, are an exception. They are easily opened in acids and by nucleophiles.

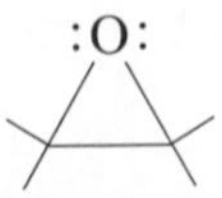

In this chapter, you will study these reactions of alcohols and ethers.

A. Some Typical Alcohols and Ethers and Their Properties

Alcohols are compounds that contain a hydroxyl group attached to a tetrahedral carbon atom. The presence of this hydroxyl group greatly influences the physical and chemical properties of alcohols. The polarity of the oxygen-hydrogen bond leads to hydrogen bonding between alcohol molecules, which results in the high boiling points characteristic of alcohols. The water solubility of low-molecular-weight alcohols and of larger compounds with several hydroxyl groups can also be attributed to hydrogen bonding (p. 26). Their properties make alcohols good solvents for a variety of reactions (pp. 232 and 255).

The simplest alcohol is methanol, or methyl alcohol, a compound that is highly toxic and can cause blindness and death. Methanol can be generated by heating wood; and is thus also called **wood alcohol.** Ethanol, or ethyl alcohol, is the familiar **grain alcohol.** Some alcohols have more than one hydroxyl group. One of these is 1,2-ethanediol, also known as ethylene glycol, which is used as automobile antifreeze. Other alcohols are large complicated molecules of which the hydroxyl group is a small but important part. The major constituent of most gallstones, cholesterol, which has also been implicated as a possible risk factor in arterial and heart disease, is such an alcohol.

CH_3OH	CH_3CH_2OH	$HOCH_2CH_2OH$
methanol	ethanol	1,2-ethanediol
methyl alcohol	ethyl alcohol	ethylene glycol
bp 64.7 °C	bp 78.5 °C	bp 197.2 °C

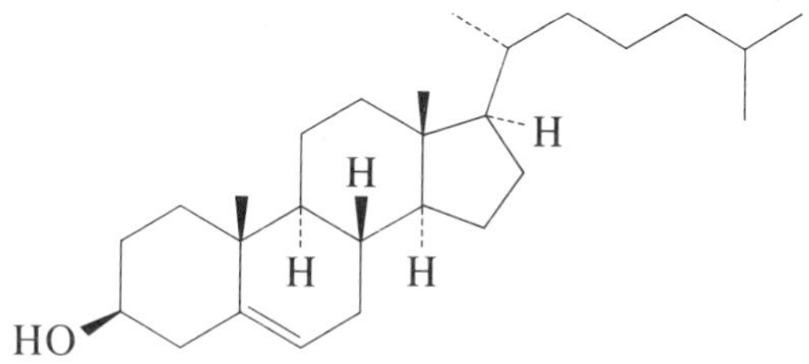
H
H
H
H
HO

cholesterol
mp 148.5 °C

PROBLEM 12.1

The boiling point of 1,2-ethanediol is 197.2 °C. Why is it a good compound to use as an antifreeze?

PROBLEM 12.2

Would you expect cholesterol to be particularly soluble in water? Explain your answer.

PROBLEM 12.3

Cholesterol has more than one functional group. To what other class of compounds does it belong? Write equations using cholesterol to illustrate three reactions that are typical of that other functional group class.

Ethers act only as hydrogen-bond acceptors (p. 27) and, therefore, have much lower boiling points than do alcohols of comparable molecular weight. Because ethers can donate their nonbonding electrons, they complex with Lewis acids such as diborane (p. 297) and serve as good solvents for them. Some common ethers are diethyl ether, tetrahydrofuran, and diglyme. Diethyl ether was formerly used as an anesthetic. The name diglyme is a contraction of the more descriptive name diethylene glycol dimethyl ether.

$CH_3CH_2OCH_2CH_3$

diethyl ether
bp 34.5 °C

tetrahydrofuran
bp 67 °C

$CH_3OCH_2CH_2OCH_2CH_2OCH_3$
diethylene glycol dimethyl ether
diglyme
bp 162 °C

B. Nomenclature of Alcohols

Simple alcohols can be named by using the name of the alkyl group they contain with the word *alcohol*. Methyl alcohol, ethyl alcohol, isopropyl alcohol, *tert*-butyl alcohol, allyl alcohol are common names that are created this way. But this system becomes cumbersome when the alkyl groups are large and complicated or when other substituents are also present in the molecule.

Alcohols are named systematically according to the IUPAC rules by substituting the ending **ol** for the final **e** of the systematic name of the hydrocarbon portion of the molecule. The location of the hydroxyl group on the hydrocarbon chain is indicated by a number, usually placed at the beginning of the name. Below are some examples of IUPAC nomenclature for simple alcohols. Common names are also shown for some compounds.

CH_3OH
methanol
methyl alcohol

CH_3CH_2OH
ethanol
ethyl alcohol
a primary alcohol

$CH_3CH_2CH_2OH$
1-propanol
n-propyl alcohol
a primary alcohol

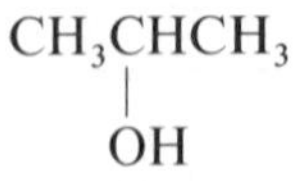

2-propanol
isopropyl alcohol
a secondary alcohol

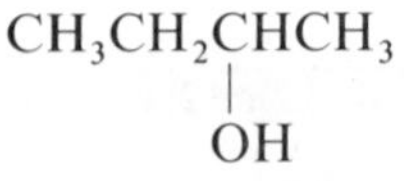

2-butanol
sec-butyl alcohol
a secondary alcohol

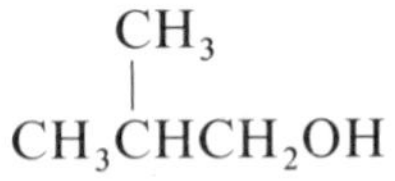

2-methyl-1-propanol
isobutyl alcohol
a primary alcohol

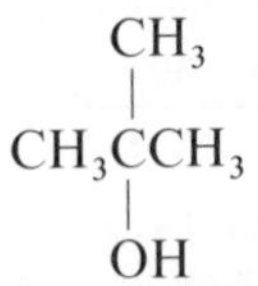

2-methyl-2-propanol
tert-butyl alcohol
a tertiary alcohol

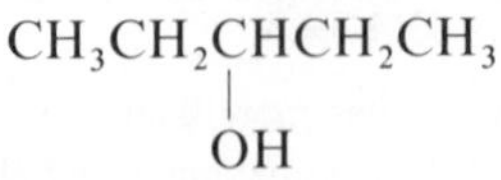

3-pentanol
a secondary alcohol

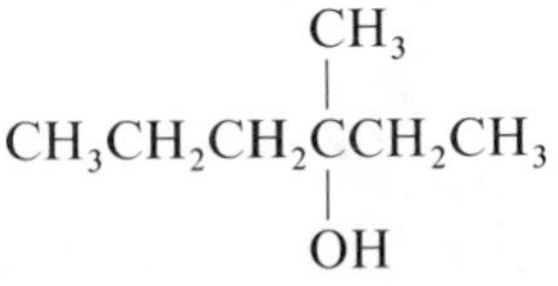

3-methyl-3-hexanol
a tertiary alcohol

The reactivity of alcohols, like that of alkyl halides, varies according to whether the functional group is on a primary, secondary, or tertiary carbon atom (pp. 237 and 252). Therefore, alcohols are designated as primary, secondary, or tertiary alcohols, as shown above.

The alcohol functional group is considered to be more important than double or triple bonds or halogen substituents and is consequently given the lower number in naming compounds with several functional groups.

$$\begin{array}{c} CH_3 \\ | \\ ClCH_2CH_2CHCH_2OH \end{array}$$

4-chloro-2-methyl-1-butanol

a primary alcohol

$$\begin{array}{c} \quad CH_3 \\ \quad | \\ CH_3CHCHCH_2CH_2CH_2OH \\ | \quad\quad\quad \\ OH \quad\quad\quad \end{array}$$

4-methyl-1,5-hexanediol

a primary and secondary alcohol

$$HOCH_2C{\equiv}CCH_2OH$$

2-butyne-1,4-diol

presence of a triple bond indicated by changing the name of the backbone from butane to butyne

a primary alcohol

$$CH_2{=}CHCH_2OH$$

2-propen-1-ol
allyl alcohol

presence of a double bond indicated by changing the name of the backbone from propane to propene

a primary alcohol, also an allylic alcohol

note location of numbers for the hydroxyl groups in these two names; necessary for clarity

C. Nomenclature of Ethers

Ethers have two hydrocarbon groups bonded to an oxygen atom. There are two ways of naming ethers. For simple ethers, the name is obtained by naming each of the organic groups on the oxygen atom and adding the word ether, for example:

$$\begin{array}{c} CH_3\ CH_3 \\ |\quad\ \ | \\ CH_3CHOCHCH_3 \end{array}$$

diisopropyl ether

$$CH_3OCH_2CH_2CH_2CH_3$$

methyl butyl ether

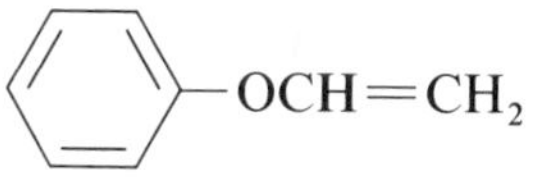

phenyl vinyl ether

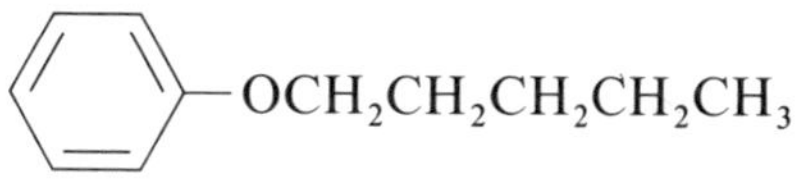

pentyl phenyl ether

For more complex molecules, the simplest organic group along with the oxygen atom is named as an **alkoxy or aryloxy group** and considered to be a substituent on a more complex chain. Important alkoxy or aryloxy groups are the following:

CH_3O- methoxy

CH_3CH_2O- ethoxy

$CH_3CH_2CH_2O-$ propoxy

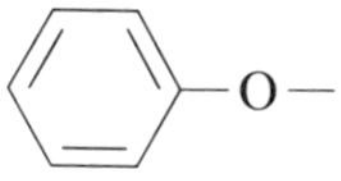

phenoxy

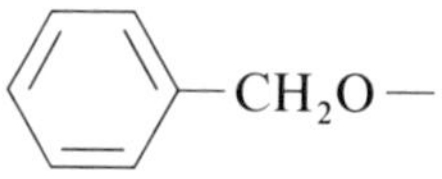

benzyloxy

Some examples of this nomenclature are as follows:

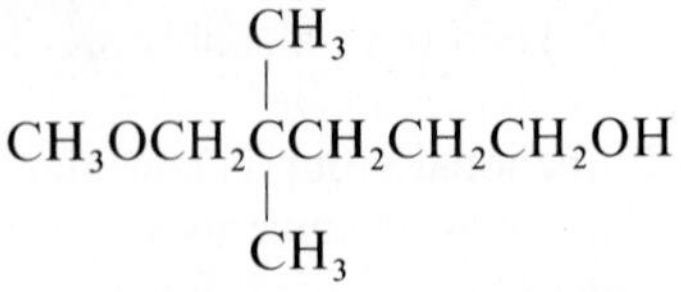
5-methoxy-4,4-dimethyl-1-pentanol

$CH_3OCH_2CH_2OCH_2CH_2CH_3$
1-methoxy-2-propoxyethane

1,4-diethoxybenzene

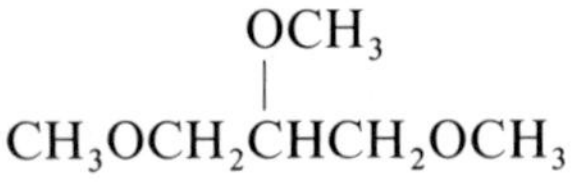
1,2,3-trimethoxypropane

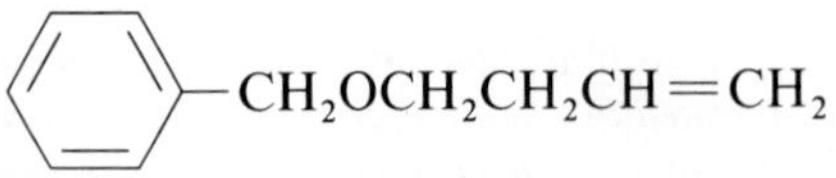
4-benzyloxy-1-butene

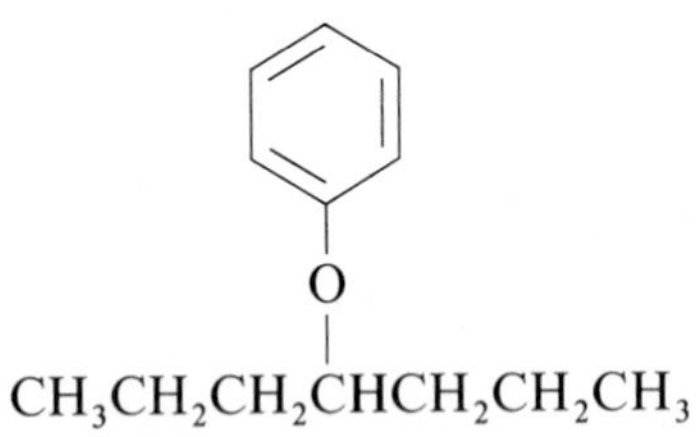
4-phenoxyheptane

PROBLEM 12.4

Name the following compounds, including stereochemistry if shown, according to IUPAC rules.

(a) 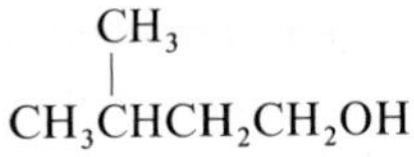

(b) 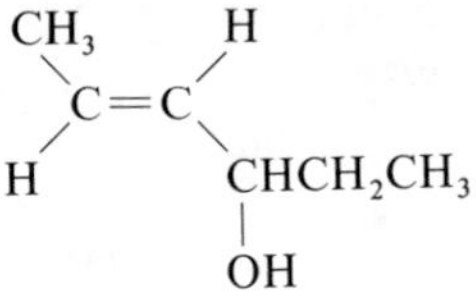

(c) $CH_3CH_2C{\equiv}CCH_2OH$

(d) 

(e) $CH_3CH_2OCH_2CH_2CH_2Cl$

(f) 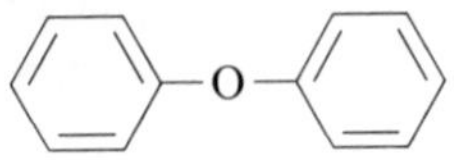

(g) 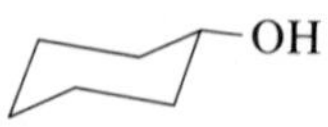

(h) 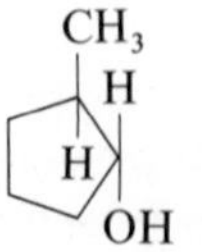

(i) 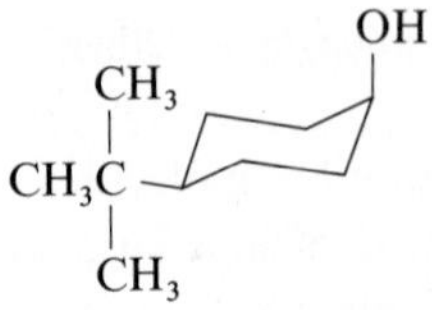

(j) 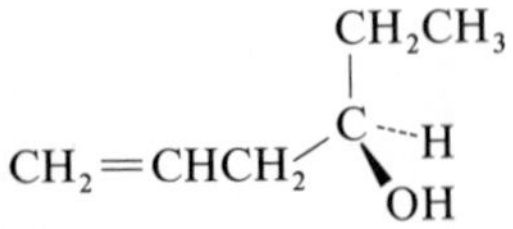

(k) $CH_3OCH_2CH_2CH_2OCH_3$

(l) 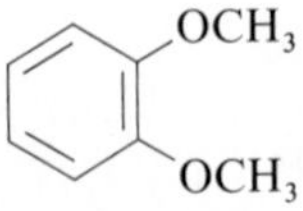

An alcohol is formed when water adds to the double bond in an alkene. For example, propene is converted to isopropyl alcohol in dilute acid (p. 280).

$$\underset{\text{propene}}{CH_3CH{=}CH_2} + H_2O \xrightarrow[H_2SO_4]{} \underset{\text{isopropyl alcohol}}{CH_3CH(OH)CH_3}$$

The alcohol that results is from Markovnikov addition of water to the double bond. The reaction is initiated by the reaction of the electrophile, hydronium ion, with the double bond.

The above reaction is the basis of an important industrial process for the production of simple alcohols but is not useful for the preparation of more complex alcohols. The acidic reagent can attack other functional groups in the molecule and cause unwanted side reactions. Alkenes also isomerize in acids via carbocation intermediates (p. 285). Attempts to add water to such alkenes with aqueous acid give rise to alcohols with rearranged carbon skeletons. For example, the Markovnikov addition product is not obtained from the reaction of 3-methyl-1-butene with 60% aqueous sulfuric acid. The only alcohol recovered is 2-methyl-2-butanol.

$$\underset{\text{3-methyl-1-butene}}{CH_3CH(CH_3)CH{=}CH_2} \xrightarrow[60\%\ H_2SO_4]{H_2O} \underset{\text{2-methyl-2-butanol}}{CH_3C(CH_3)(OH)CH_2CH_3}$$

PROBLEM 12.5

(a) What would be the structure of the Markovnikov addition product for the reaction of 3-methyl-1-butene with dilute acid?

(b) Write a detailed mechanism that accounts for the observed product.

PROBLEM 12.6

3,3-Dimethyl-1-butene is isomerized by acid (p. 287). Predict, by writing a mechanism, what the structures of hydration products would be if a reaction of this alkene with aqueous acid were attempted.

Hydroboration followed by oxidation (p. 299) converts alkenes to alcohols, with an overall anti-Markovnikov addition of water to the double bond. The reaction does not involve carbocation intermediates, so no rearrangements are observed. 4-Penten-1-ol, for example, is converted into 1,5-pentanediol in high yield by hydroboration-oxidation.

$$\underset{\text{4-penten-1-ol}}{H_2C{=}CHCH_2CH_2CH_2OH} \xrightarrow[\text{tetrahydrofuran}]{\text{9-BBN}} \xrightarrow[H_2O]{H_2O_2,\ NaOH} \underset{\substack{\text{1,5-pentanediol} \\ 98\%}}{HOCH_2(CH_2)_3CH_2OH}$$

PROBLEM 12.7

If 4-penten-1-ol were treated with aqueous acid, what products would be obtained?

Alcohols are also formed when a leaving group is displaced from an alkyl halide or alkyl tosylate by hydroxide ion or by water in nucleophilic substitution reactions. Elimination reactions often accompany these reactions, so they are not useful for the preparation of high-purity alcohols in good yield (p. 245). Also, because alkyl halides or tosylates are often prepared from the corresponding alcohol in the first place, they are not appropriate starting materials in an independent synthesis of an alcohol.

Two important methods for preparing alcohols are reduction of aldehydes or ketones and the Grignard reaction. Each of these involves the addition of a nucleophile to a carbonyl group. These reactions are discussed on pp. 484 and 491 as part of the chemistry of aldehydes and ketones.

Study Guide
Concept Map 12.1

12.3 CONVERTING ALCOHOLS TO ALKYL HALIDES

A. Reactions of Alcohols with Hydrogen Halides

Alkyl halides, which participate in many different substitution reactions (p. 260), may be prepared from alcohols in different ways. The hydroxyl group in an alcohol can be converted into a good leaving group (p. 234) so that it can be displaced by an incoming halide ion. For example, dissolving gaseous hydrogen bromide in 1-heptanol converts the alcohol to 1-bromoheptane.

$$\underset{\text{1-heptanol}}{CH_3(CH_2)_5CH_2OH} + \underset{\text{hydrogen bromide}}{HBr} \longrightarrow \underset{\text{1-bromoheptane, 90\%}}{CH_3(CH_2)_5CH_2Br} + \underset{\text{water}}{H_2O}$$

Alcohols have basic properties because of the presence of nonbonding electrons on the oxygen atom; these electrons can be donated to a proton. The protonation of the hydroxyl group creates a new leaving group, a water molecule. If this reaction is carried out on a primary alcohol, the water is displaced by halide ion in an S_N2 reaction, giving the alkyl halide.

VISUALIZING THE REACTION

Conversion of an alcohol to an alkyl halide by hydrogen bromide

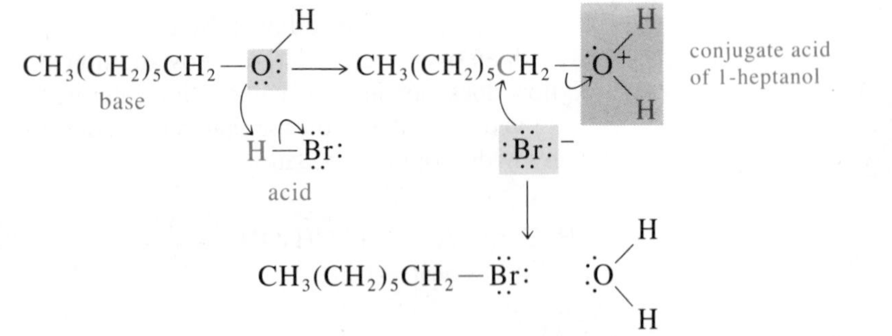
CH3(CH2)5CH2—O:
base
H—Br:
acid
CH3(CH2)5CH2—O+
conjugate acid of 1-heptanol
:Br:−
CH3(CH2)5CH2—Br:
H
O
H

In secondary alcohols or other alcohols in which S_N2 reactions are hindered, the substitution must proceed by way of a carbocation intermediate, which can rearrange. For example, when 3-pentanol is treated with hydrogen bromide, 3-bromopentane and 2-bromopentane are obtained in a mixture of varying composition depending on whether aqueous or gaseous hydrogen bromide is used as the reagent.

$$CH_3CH_2CHCH_2CH_3 \xrightarrow[\Delta]{HBr} CH_3CH_2CHCH_2CH_3 + CH_3CH_2CH_2CHCH_3$$

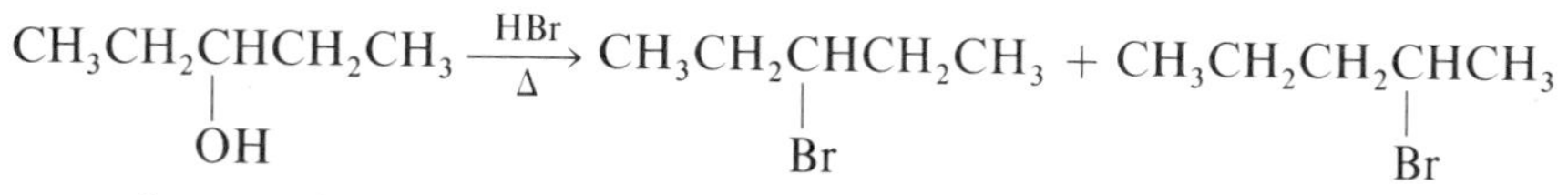

OH	Br	Br
3-pentanol	3-bromopentane	2-bromopentane

PROBLEM 12.8

Write the full mechanism for the conversion of 3-pentanol to the two bromopentanes in 48% aqueous hydrobromic acid.

PROBLEM 12.9

Whenever either 2-chloropentane or 3-chloropentane is allowed to stand in concentrated hydrochloric acid in which zinc chloride has been dissolved, a mixture of the two compounds consisting of 66% 2-chloropentane and 34% 3-chloropentane is obtained. Write a mechanism for this interconversion, starting with 3-chloropentane. What is the role of zinc chloride? What does the fact that the equilibrium mixture is 66% 2-chloropentane tell you about the relative stabilities of 2-pentyl and 3-pentyl cations? (Hint: What are the relative numbers of 2-pentyl and 3-pentyl cations that can form?)

Another way to convert the hydroxyl group in an alcohol into a good leaving group is to make the tosylate ester, from which the tosylate group can be displaced by halide ion in a nucleophilic substitution reaction (p. 235). In this way, 3-pentanol can be converted to 3-bromopentane free of 2-bromopentane; this result suggests that a carbocation intermediate is not formed under these conditions.

$$CH_3CH_2CHCH_2CH_3 + TsCl \xrightarrow[\text{pyridine}]{} CH_3CH_2CHCH_2CH_3 \xrightarrow[\text{dimethyl sulfoxide}]{Na^+Br^-} CH_3CH_2CHCH_2CH_3$$

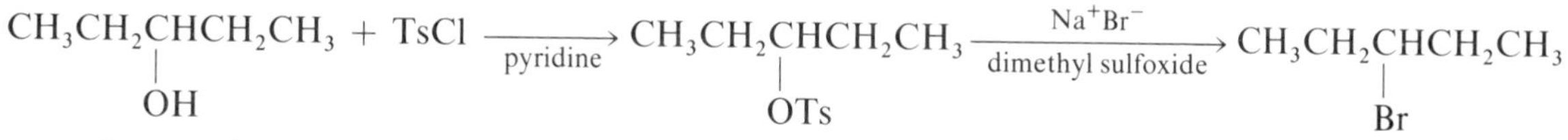

OH		OTs	Br
3-pentanol	*p*-toluenesulfonyl chloride	3-pentyl tosylate	3-bromopentane 85%

Tertiary alcohols react readily with hydrogen halides to give alkyl halides. For example, *tert*-butyl alcohol is converted to *tert*-butyl chloride by shaking it with cold concentrated hydrochloric acid for a few minutes.

$$CH_3-\overset{\overset{\displaystyle CH_3}{|}}{\underset{\underset{\displaystyle CH_3}{|}}{C}}-OH \xrightarrow[25\ °C]{HCl} CH_3-\overset{\overset{\displaystyle CH_3}{|}}{\underset{\underset{\displaystyle CH_3}{|}}{C}}-Cl$$

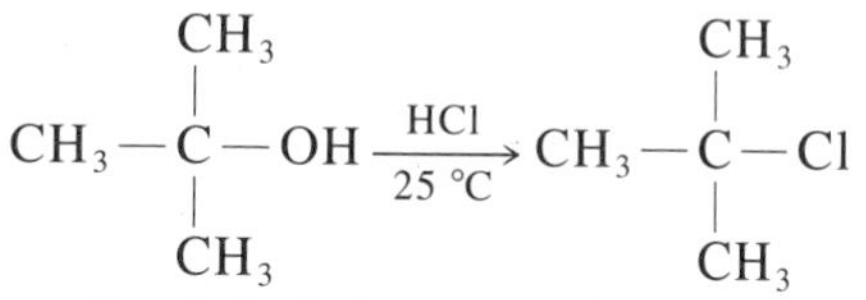

tert-butyl alcohol   *tert*-butyl chloride

This is an S_N1 reaction. As long as the temperature is kept low, not much of the product from the competing E_1 reaction is obtained. Even if 2-methylpropene forms under these conditions, it also reacts with hydrochloric acid to give *tert*-butyl chloride.

PROBLEM 12.10

Write a mechanism for the conversion of *tert*-butyl alcohol to *tert*-butyl chloride using hydrochloric acid. How would 2-methylpropene be formed? How would that alkene react with hydrochloric acid?

B. Reactions with Thionyl Chloride and Phosphorus Halides

Other reagents can also be used to avoid carbocation intermediates and minimize rearrangements in the conversion of alcohols to alkyl halides. One of these is thionyl chloride, $SOCl_2$. 3-Pentanol will react with thionyl chloride in pyridine to give 3-chloropentane, and no 2-chloropentane.

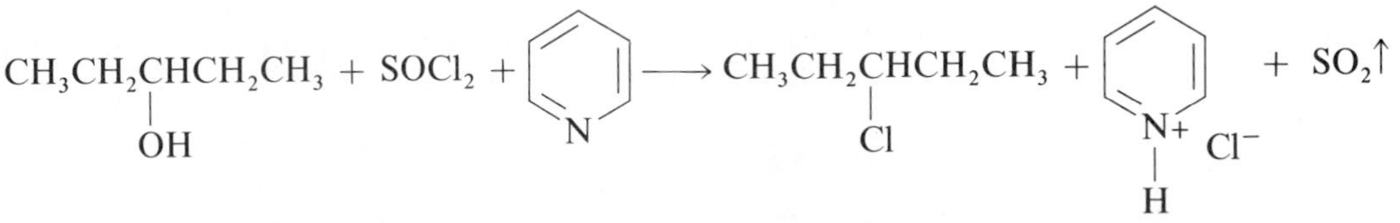
CH₃CH₂CHCH₂CH₃ + SOCl₂ +
OH
N
CH₃CH₂CHCH₂CH₃ +
Cl
N+
H
Cl⁻
+ SO₂↑

3-pentanol | thionyl chloride | pyridine | 3-chloropentane | pyridine hydrochloride | sulfur dioxide

However, this reaction is often accompanied by elimination reactions and does not give high yields.

This reaction also starts with the conversion of the alcohol to a compound with a good leaving group, an intermediate chlorosulfite ester, which decomposes to give the alkyl halide. In the presence of organic bases such as pyridine, inversion of configuration is seen for chiral compounds.

VISUALIZING THE REACTION

Conversion of an alcohol to an alkyl halide by thionyl chloride

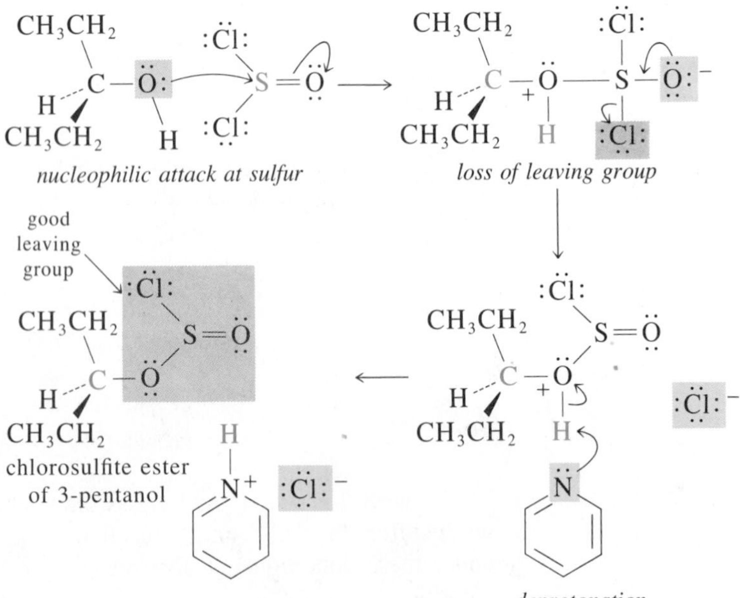
nucleophilic attack at sulfur
loss of leaving group
good leaving group
chlorosulfite ester of 3-pentanol
deprotonation

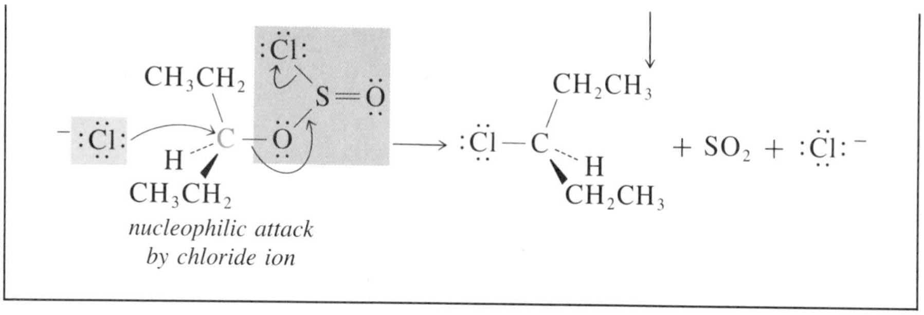

CH₃CH₂
:Cl:
S=O
:Cl:
H
CH₃CH₂
nucleophilic attack
by chloride ion
CH₂CH₃
:Cl—C
H
CH₂CH₃
+ SO₂ + :Cl:⁻

PROBLEM 12.11

When 2-ethyl-1-butanol is treated with zinc chloride in concentrated hydrochloric acid, a mixture of chloroalkanes forms, including chiefly 2-ethyl-1-chlorobutane, 3-chlorohexane, 2-chlorohexane, and 3-chloro-3-methylpentane. When 2-ethyl-1-butanol is treated with thionyl chloride in pyridine, only 1-chloro-2-ethylbutane is formed. Write detailed mechanisms that account for these observations. What is the function of zinc chloride?

Another reagent that is used to substitute a halogen for a hydroxyl group with a minimum of rearrangement is phosphorus tribromide. For example, it is used to convert the primary alcohol isobutyl alcohol into isobutyl bromide.

$$3\ CH_3\underset{\displaystyle CH_3}{\overset{|}{C}}HCH_2OH + PBr_3 \xrightarrow[0\ ^\circ C]{} 3\ CH_3\underset{\displaystyle CH_3}{\overset{|}{C}}HCH_2Br + P(OH)_3 \equiv H_3PO_3$$

isobutyl alcohol — phosphorus tribromide — isobutyl bromide 60% — phosphorous acid

The mechanism suggested for the conversion of an alcohol to an alkyl halide using phosphorus tribromide resembles the one proposed for the conversion using thionyl chloride. Its first step is the formation of a phosphorus-oxygen bond between the alcohol and phosphorus tribromide, giving a phosphorus ester that has a good leaving group.

Study Guide Concept Map 12.2

PROBLEM 12.12

When optically active 2-octanol is treated with phosphorus tribromide, 2-bromooctane is formed with almost complete inversion of configuration. The other product of the reaction is phosphorous acid, $P(OH)_3$. Write a detailed mechanism for the reaction that accounts for these facts.

PROBLEM 12.13

Complete the following equations.

(a) 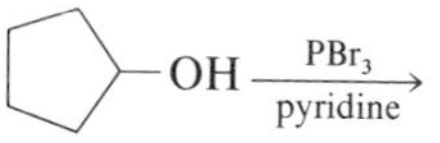

OH
PBr₃
pyridine

(b) 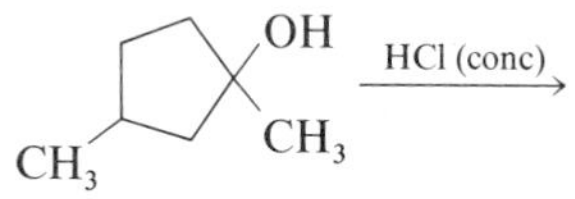

OH
CH₃
CH₃
HCl (conc)

(c) $CH_3(CH_2)_9CH_2OH \xrightarrow[\Delta]{SOCl_2}$

(d) $CH_3(CH_2)_8CH_2OH \xrightarrow[\Delta]{HBr}$

(e) $CH_3SCH_2CH_2OH \xrightarrow[\text{chloroform}]{SOCl_2}$

(f) $CH_3CH_2OCH_2CH_2OH \xrightarrow[\text{pyridine}]{PBr_3}$

(g) 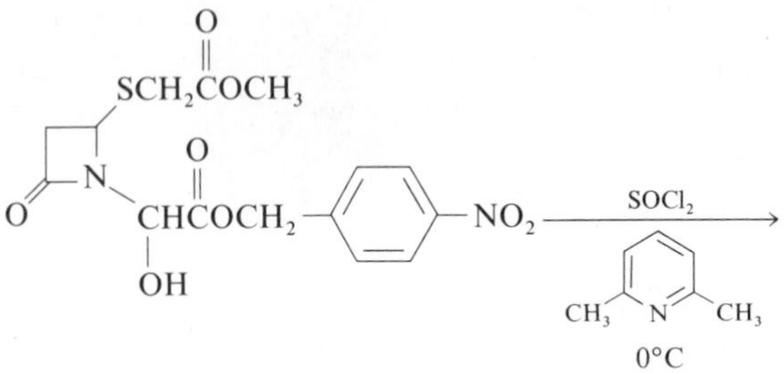
SOCl2
0°C

12.4 OXIDATION REACTIONS OF ALCOHOLS

A. Oxidation-Reduction of Organic Compounds

In Chapter 3, we explored the properties of molecules from the two opposite poles of acidity and basicity. Oxidation and reduction reactions are another pair of polar opposites. For many metals and their ions, the processes of oxidation and reduction can be easily seen. When copper metal is put into a solution of silver nitrate, crystals of silver metal grow on the surface of the copper and the solution turns the blue color that is typical of water solutions of copper (II) ion. Copper metal loses electrons to silver ions and is oxidized, and silver ions are reduced to silver metal by the gain of electrons.

$$\underset{\textit{shiny red metal}}{Cu} + \underset{\textit{colorless}}{2\,Ag^+} \xrightarrow[\text{water}]{} \underset{\textit{blue}}{Cu^{2+}} + \underset{\textit{silver-grey crystals}}{2\,Ag}$$

For organic compounds, however, changes in oxidation states are generally not observable in this way. Yet, it is of great importance in organizing reaction types to be able to decide whether a given compound has been oxidized or reduced. You are already familiar with several oxidation (pp. 310 and 314, for example) and reduction (pp. 306 and 341) reactions. In being oxidized, carbon "loses" electrons by forming bonds with elements that are more electronegative than it is, elements such as oxygen, nitrogen, or the halogens. In being reduced, carbon "gains" electrons by giving up bonds to more electronegative elements and forming bonds with hydrogen atoms instead.

In deciding whether an organic compound is being oxidized or reduced, the first thing to look for is changes in the number of bonds to hydrogen or to oxygen (or other electronegative elements) at the carbon atoms undergoing reaction. Loss of bonds to a hydrogen atom or gain of bonds to an oxygen atom at a carbon atom is an oxidation, for example:

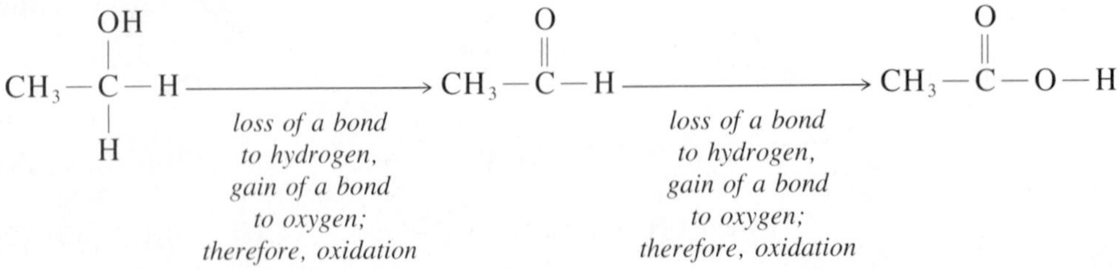
loss of a bond to hydrogen, gain of a bond to oxygen; therefore, oxidation
loss of a bond to hydrogen, gain of a bond to oxygen; therefore, oxidation

The changes that ethanol must undergo to be converted first to acetaldehyde and then to acetic acid are oxidations; these reactions require the use of an oxidizing agent. The reverse reactions must be reduction reactions, carried out by some reducing agent. Formally, they involve the loss of bonds to oxygen or the gain of bonds to hydrogen.

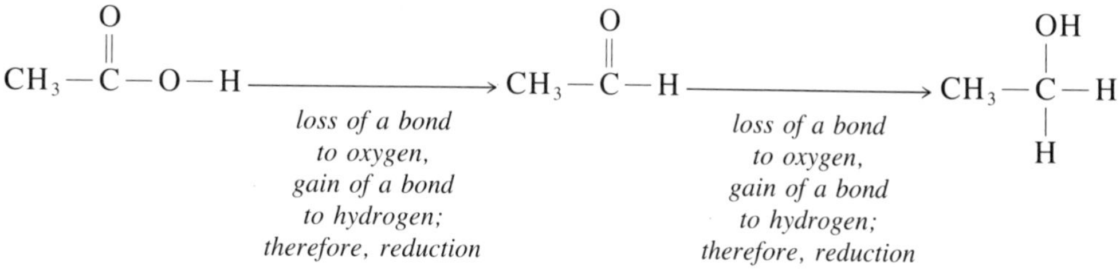
loss of a bond
to oxygen,
gain of a bond
to hydrogen;
therefore, reduction
loss of a bond
to oxygen,
gain of a bond
to hydrogen;
therefore, reduction

Note that only one of the carbon atoms in this series of compounds undergoes oxidation or reduction. The methyl group is unchanged in these reactions.

A similar reasoning process allows us to conclude that the carbon atoms in an alkyne are more highly oxidized than the carbon atoms in an alkene, which in turn are more oxidized than those in an alkane.

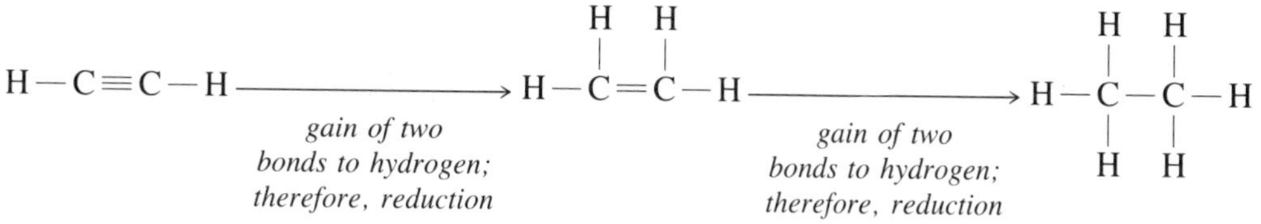
gain of two
bonds to hydrogen;
therefore, reduction
gain of two
bonds to hydrogen;
therefore, reduction

A reducing agent is necessary to convert an alkyne to an alkene and then to an alkane. This reduction would be carried out by adding hydrogen to the multiple bonds (pp. 306, 341). However, it is not necessary to know the reagents that are used for the transformations that are shown in this section. Your goal here is to learn how to recognize which reactions involve oxidation and which reduction. The ability to distinguish between these two types of reactions is important.

Giving oxidation numbers to various substituents allows us to be quantitative about the oxidation states of carbon. A carbon-carbon bond contributes nothing to the oxidation state of a carbon atom because the bonding electrons are shared equally by the two carbon atoms. Atoms that are more electronegative than carbon pull bonding electrons toward themselves and are given negative oxidation numbers. A halogen or an oxygen atom in a hydroxyl group is in a −1 oxidation state, and a doubly bonded oxygen atom is in a −2 oxidation state. Each hydrogen atom is in a +1 oxidation state. Using these numbers, we can assign oxidation states to the carbon atoms of the compounds shown earlier in this section, for example:

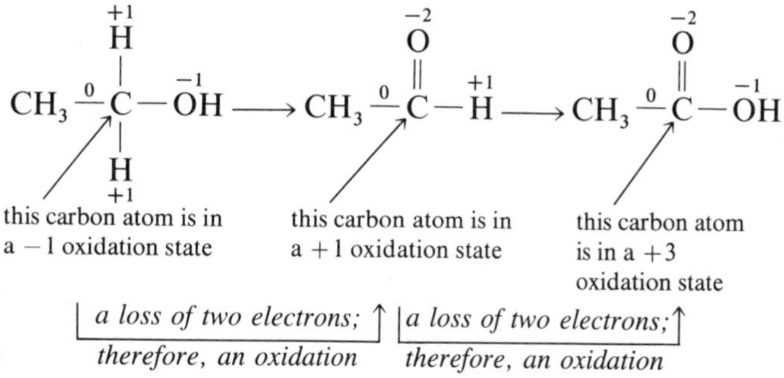
this carbon atom is in
a −1 oxidation state
this carbon atom is in
a +1 oxidation state
this carbon atom
is in a +3
oxidation state
a loss of two electrons;
therefore, an oxidation
a loss of two electrons;
therefore, an oxidation

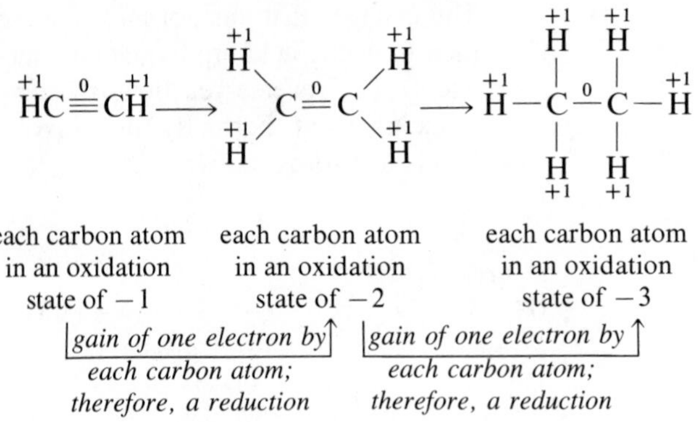
each carbon atom in an oxidation state of −1
each carbon atom in an oxidation state of −2
each carbon atom in an oxidation state of −3
gain of one electron by each carbon atom; therefore, a reduction
gain of one electron by each carbon atom; therefore, a reduction

Both the qualitative approach presented earlier and this quantitative one lead to the same conclusions about oxidation and reduction.

You should work to develop skill in deciding when processes involve oxidation-reduction. If you are able to recognize the different oxidation states of carbon, you will be able to make predictions about the kinds of reagents needed for the transformations you wish to achieve.

Study Guide Concept Map 12.3

PROBLEM 12.14

For each of the following transformations, decide whether the starting material has undergone oxidation or reduction. Specify whether an oxidizing or reducing agent would be needed to carry out the change that is shown.

(a) 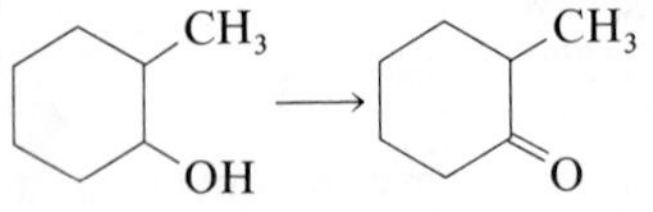

(b) 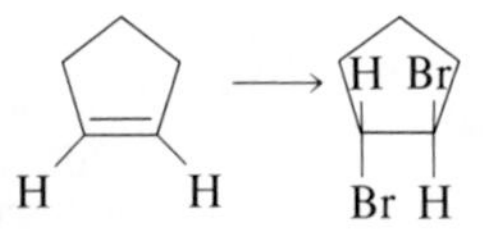

(c) $CH_3CH_2Br \longrightarrow CH_3CH_2NH_2$

(d) 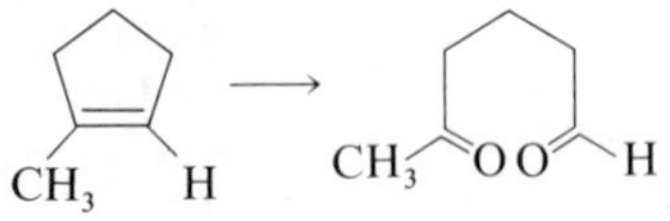

(e) 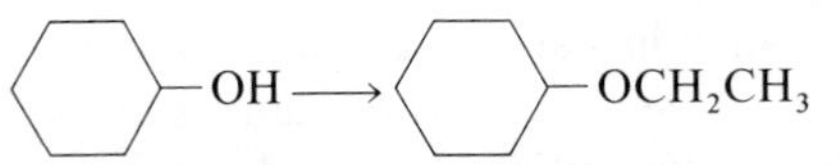

(f) $CH_3CH_2\overset{O}{\overset{\|}{C}}H \longrightarrow CH_3CH_2CH_2OH$

(g) $CH_3CH{=}CHCH_3 \longrightarrow CH_3\underset{OH}{\underset{|}{C}}H{-}\underset{OH}{\underset{|}{C}}HCH_3$

(h) $CH_3C{\equiv}CCH_2CH_3 \longrightarrow CH_3CH{=}CHCH_2CH_3$

(i) $CH_3C{\equiv}CCH_3 \longrightarrow CH_3\overset{O}{\overset{\|}{C}}{-}\overset{O}{\overset{\|}{C}}CH_3$

(j) $CH_3CH_2CH_2OH \longrightarrow CH_3CH_2\overset{O}{\overset{\|}{C}}OH$

(k) $CH_3\overset{O}{\overset{\|}{C}}H \longrightarrow CH_3\overset{OCH_3}{\overset{|}{C}}HOCH_3$

PROBLEM 12.15

If you have not already done so in answering Problem 12.14, assign oxidation states to the carbon atoms that undergo reactions. Do your quantitative assignments support the conclusions you reached qualitatively?

PROBLEM 12.16

Methane is converted to carbon tetrachloride by successive replacement of hydrogen atoms by chlorine atoms. Methane represents the most highly reduced state of carbon, and carbon tetrachloride, the most highly oxidized. Calculate oxidation numbers for the carbon atom in methane, in carbon tetrachloride, and in the intermediate alkyl halides.

B. Oxidation with Sodium Dichromate

Primary and secondary alcohols are easily oxidized by reagents containing chromium(VI). For low-molecular-weight alcohols that are water-soluble, water solutions of sodium dichromate in the presence of acid are used. A primary alcohol such as *n*-butyl alcohol is oxidized first to an aldehyde, which is oxidized further by the same reagent to a carboxylic acid.

$$3\ CH_3CH_2CH_2CH_2OH + Cr_2O_7^{2-} + 4\ H_2SO_4 \longrightarrow 3\ CH_3CH_2CH_2\overset{\overset{O}{\|}}{C}H + 7\ H_2O + 2\ Cr^{3+} + 4\ SO_4^{2-}$$

n-butyl alcohol, bp 118 °C, *a primary alcohol*; dichromate ion, *orange*; butanal, bp 75 °C, *distilled out of the reaction mixture*; Cr^{3+} *blue-green*

$$3\ CH_3CH_2CH_2\overset{\overset{O}{\|}}{C}H + Cr_2O_7^{2-} + 4\ H_2SO_4 \longrightarrow 3\ CH_3CH_2CH_2\overset{\overset{O}{\|}}{C}OH + 2\ Cr^{3+} + 4\ SO_4^{2-} + 4\ H_2O$$

butanal; dichromate ion, *orange*; butanoic acid; Cr^{3+} *blue-green*

This process is not of practical use for preparing aldehydes from primary alcohols except for a few cases where the aldehyde that is formed has a boiling point low enough that it can be distilled out of the reaction mixture before further reaction takes place. Butanal is such an aldehyde.

A secondary alcohol, such as *sec*-butyl alcohol, is oxidized to a ketone, which usually does not undergo further oxidation. For a ketone to be oxidized, carbon-carbon bonds would have to be broken because there are no more carbon-hydrogen bonds available on the carbonyl carbon atom. Ketones, therefore, survive under conditions in which aldehydes are readily oxidized to carboxylic acids.

$$3\ CH_3CH_2\underset{\underset{OH}{|}}{C}HCH_3 + Cr_2O_7^{2-} + 4\ H_2SO_4 \longrightarrow 3\ CH_3CH_2\overset{\overset{O}{\|}}{C}CH_3 + 2\ Cr^{3+} + 4\ SO_4^{2-} + 7\ H_2O$$

sec-butyl alcohol, *a secondary alcohol*; dichromate ion, *orange*; 2-butanone; Cr^{3+} *blue-green*

The tertiary alcohol *tert*-butyl alcohol is not oxidized by acidic sodium dichromate solution. The carbon atom bearing the hydroxyl group in this alcohol does not have a hydrogen atom on it. Here, too, carbon-carbon bonds would have to be broken to allow oxidation to take place.

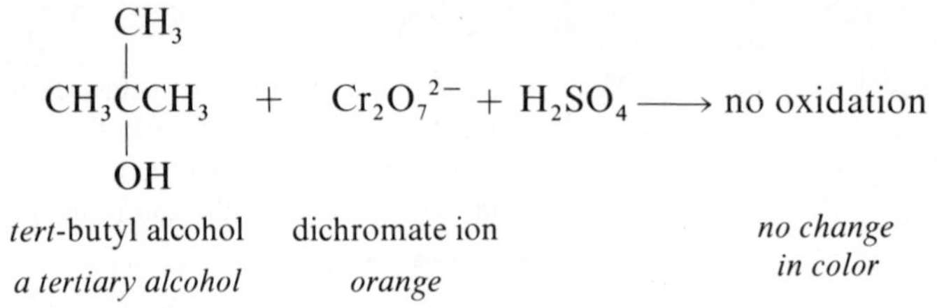
CH₃
CH₃CCH₃
OH
+ Cr₂O₇²⁻ + H₂SO₄ ⟶ no oxidation
tert-butyl alcohol
a tertiary alcohol
dichromate ion
orange
no change in color

The oxidizing solution in the above examples, an acidic solution of sodium dichromate, contains a variety of species in equilibrium. Most important of these is the acid chromate ion, $HCrO_4^-$.

$$H_2O + Cr_2O_7^{2-} \rightleftarrows 2\,HCrO_4^-$$

dichromate ion — acid chromate ion

Chromium(VI), present in both the dichromate and the acid chromate ions, is the oxidizing agent. It is reduced to chromium(III) in several steps, with chromium(IV) and chromium(V) present at some stages of the reaction. It has been suggested that the reaction involves, first, the formation of a complex between the alcohol and the chromium atom followed by a series of deprotonation and protonation steps leading to the loss of water from the complex and the formation of the chromate ester of the alcohol. The mechanism for this part of the reaction is shown below.

VISUALIZING THE REACTION

Formation of a chromate ester

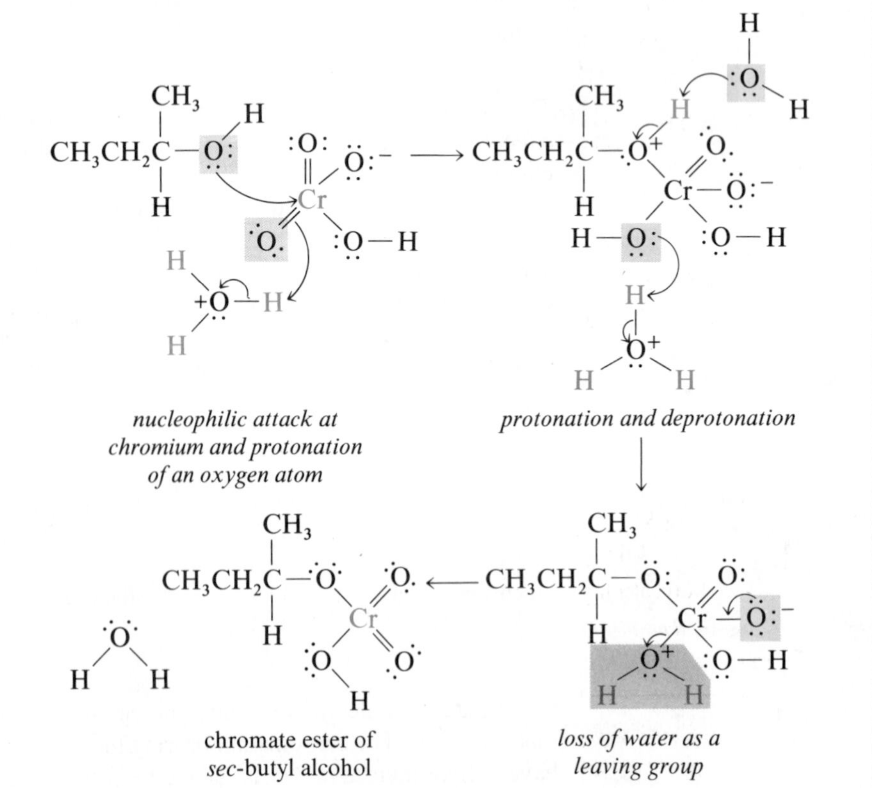
nucleophilic attack at chromium and protonation of an oxygen atom
protonation and deprotonation
chromate ester of *sec*-butyl alcohol
loss of water as a leaving group

If you examine the flow of electrons indicated by the curved arrows, you will recognize the individual processes as acid-base reactions, even though the reaction looks complicated at first glance. A hydroxyl group (on the chromium atom) is converted to a good leaving group, and the chromate ester of the alcohol results. The oxidation states of carbon and chromium do not change during the formation of the chromate ester. The removal of the hydrogen atom from the carbon atom that originally bore the hydroxyl group is the oxidation-reduction step of the reaction. Chromium(IV) is formed in this step and is ultimately reduced to chromium(III). The exact details of all the processes involved are quite complex. It is sufficient if you understand that a hydroxyl group and a hydrogen atom on the same carbon atom are necessary for the oxidation reaction that proceeds through the formation of the chromate ester of the alcohol.

VISUALIZING THE REACTION

The oxidation reaction

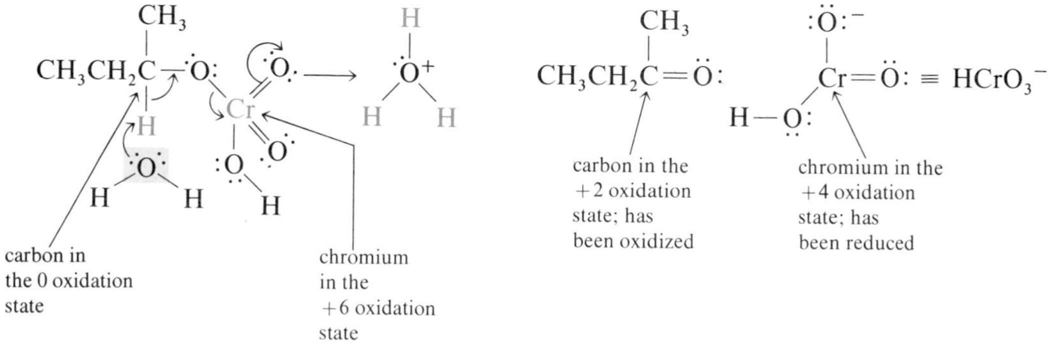

carbon in the 0 oxidation state
chromium in the +6 oxidation state
carbon in the +2 oxidation state; has been oxidized
chromium in the +4 oxidation state; has been reduced
$\equiv HCrO_3^-$

PROBLEM 12.17

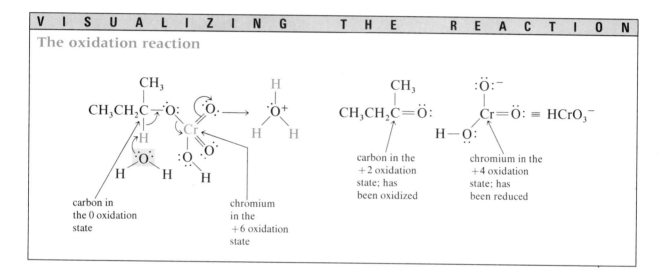

The orange color of chromium(VI) and the blue-green color of chromium(III) have been used as the basis for a qualitative test for primary and secondary alcohols. The alcohol is dissolved in acetone, and a solution of chromium trioxide, CrO_3, dissolved in water and sulfuric acid is added drop by drop. Primary and secondary alcohols give an instant blue-green cloudiness. Write the structures of the following alcohols, and predict whether each one will give a positive test with chromium trioxide reagent. For each alcohol you expect to give a positive test, write an equation for the reaction.

(a) 4-methyl-2-pentanol (b) 2-*tert*-butylcyclohexanol (c) cholesterol (p. 427)
(d) 1-ethylcyclopentanol (e) 1-pentanol (f) 2,3,3-trimethyl-2-butanol

---

C. Selective Oxidations. Pyridinium Chlorochromate

Aqueous acidic dichromate solutions are not always suitable for oxidizing alcohols to carbonyl compounds. Some alcohols are not soluble in water, so other solvents have to be used. With complex molecules having multiple functional groups, use of strong acids may lead to undesirable side reactions. For such systems, milder and more selective reagents have been developed.

A highly selective reagent is made by dissolving chromium trioxide in hydrochloric acid and adding the basic solvent pyridine to the solution.

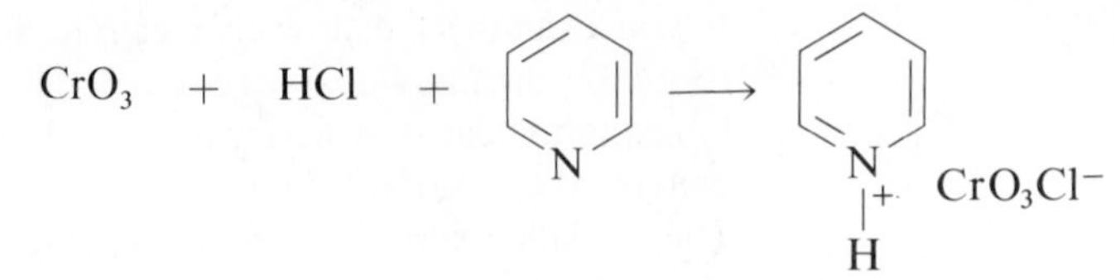

chromium trioxide — hydrochloric acid — pyridine — pyridinium chlorochromate 84%

yellow-orange crystals

This reagent is called pyridinium chlorochromate and is used in organic solvents such as dichloromethane to oxidize primary and secondary alcohols. For example, 1-decanol is oxidized to decanal with a high yield.

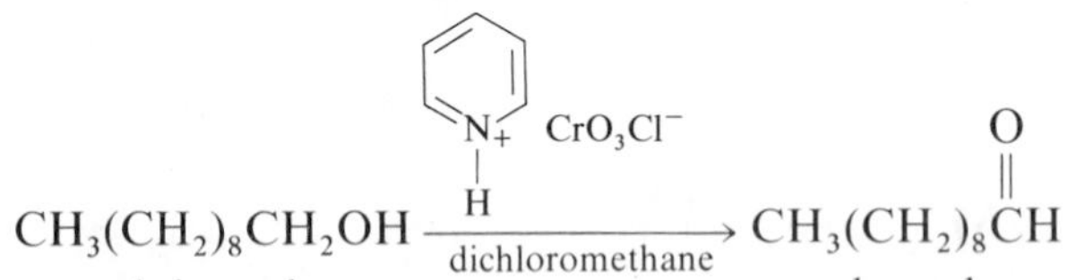

1-decanol — decanal 92%

The oxidation stops at the aldehyde stage because no water is present to form the hydrate of the aldehyde, which is the necessary intermediate for further oxidation to the carboxylic acid by chromium(VI).

$$\mathrm{R\overset{O}{\overset{\|}{C}}H} \underset{}{\overset{H_2O}{\rightleftharpoons}} \mathrm{R-\underset{OH}{\overset{OH}{C}}-H} \xrightarrow{\text{oxidation}} \mathrm{R\overset{O}{\overset{\|}{C}}OH}$$

aldehyde — hydrate of the aldehyde — carboxylic acid

Similarly, 4-*tert*-butylcyclohexanol is oxidized to 4-*tert*-butylcyclohexanone.

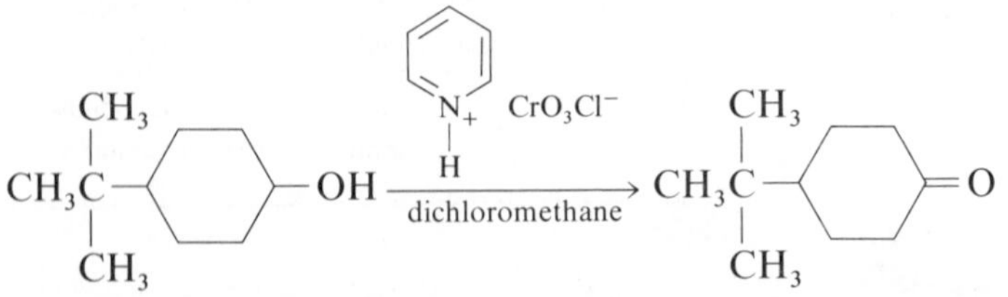

4-*tert*-butylcyclohexanol — 4-*tert*-butylcyclohexanone

Pyridinium chlorochromate is also useful for oxidizing alcohols containing other functional groups that may react with aqueous acid, such as double bonds. For example, the unsaturated alcohol citronellol, which is one of the fragrant compounds in rose oil and geranium oil, is oxidized by this reagent to the corresponding aldehyde in good yield.

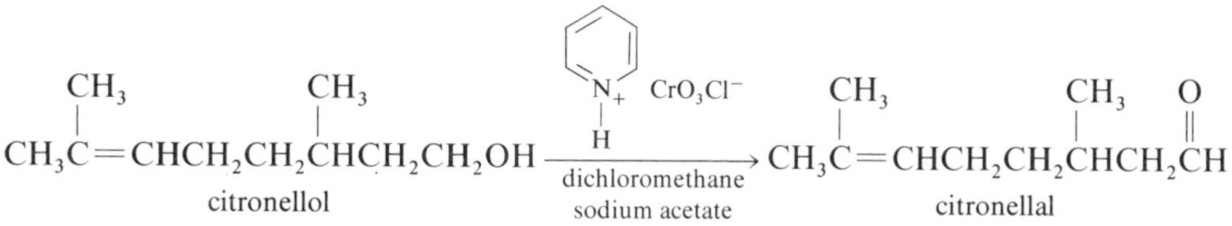

citronellol
dichloromethane
sodium acetate
citronellal

Study Guide Concept Map 12.4

The double bond is unaffected by this reaction.

PROBLEM 12.18

Write an equation showing the product you would expect if the oxidation of citronellol were carried out with chromic acid in the presence of water. Can you foresee any complications from having a strong acid such as sulfuric acid in the reaction mixture?

PROBLEM 12.19

Complete the following equations.

(a) 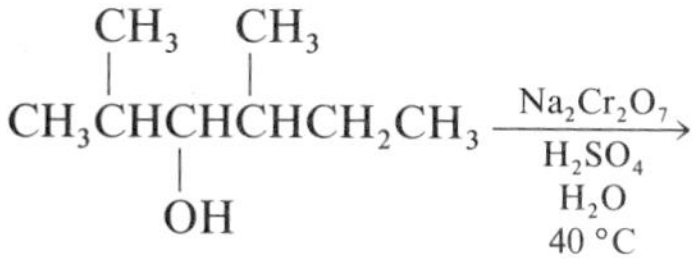

(b) 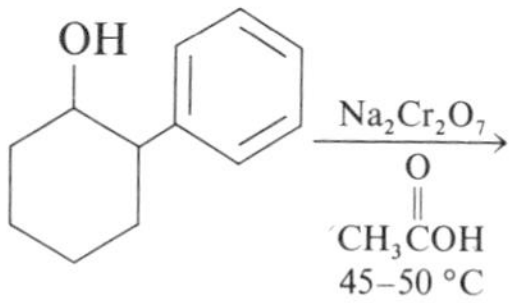

(c) 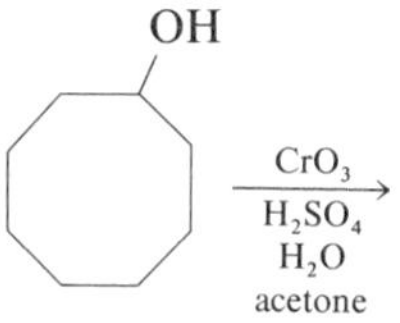

(d) 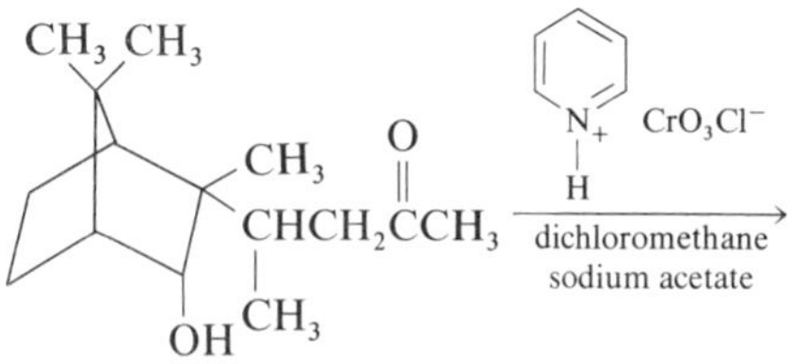

dichloromethane
sodium acetate

(e) 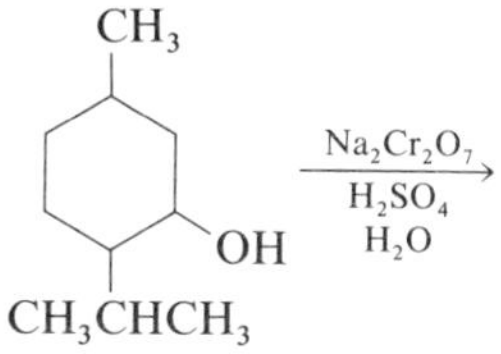

D. Problem-Solving Skills

Problem

How could we carry out the following transformation?

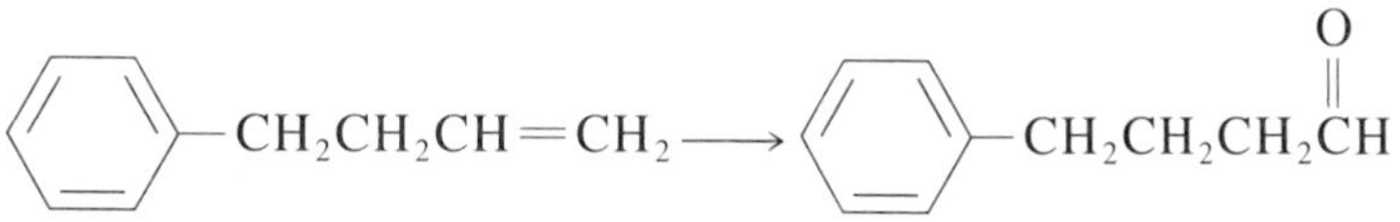

Solution

1. What functional groups are present in the starting material and the product?

   The starting material is an alkene, and the product is an aldehyde.

2. How do the carbon skeletons of the two compounds compare? How many carbon atoms does each contain? Are there any rings? What are the positions of branches and functional groups on the carbon skeletons?

   Both compounds have a straight chain of four carbon atoms with an aryl group at one end.

3. How do the functional groups change in going from starting material to product? Does the starting material have a good leaving group?

   An aldehyde functional group is formed at the first carbon atom of the double bond. There is no good leaving group in the starting material.

4. Is it possible to dissect the structures of the starting material and product to see which bonds must be broken and which formed?

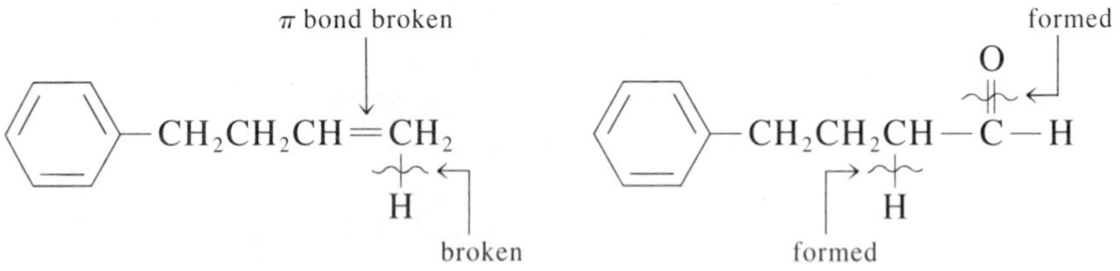
π bond broken
broken
formed
formed

   A π bond and a carbon-hydrogen bond must be broken. Carbon-hydrogen and carbon-oxygen bonds are formed.

5. Do we recognize any part of the product molecule as coming from a good nucleophile or an electrophilic addition?

   The formation of bonds to a hydrogen and an oxygen on adjacent carbon atoms looks like an addition of H—OH to the double bond, in an anti-Markovnikov fashion.

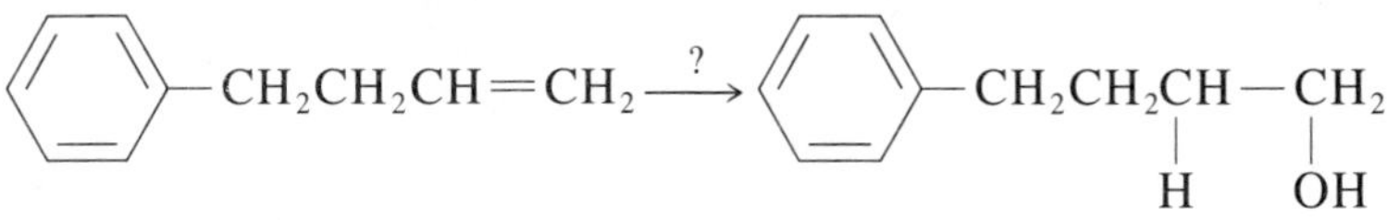

6. What type of compound would be a good precursor to the product?

   An aldehyde is formed by the oxidation of a primary alcohol.

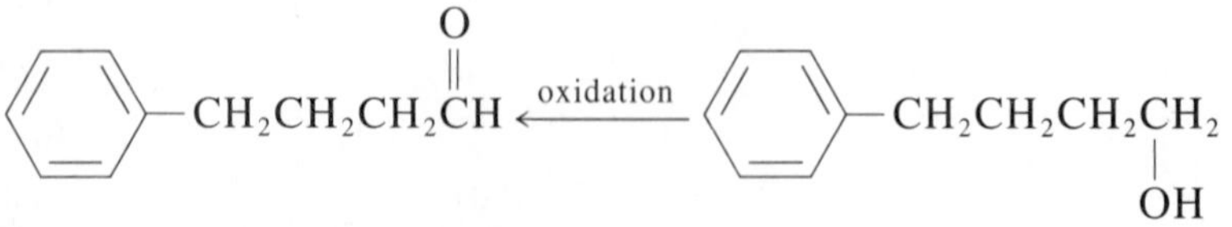
oxidation

7. After this step, do we see how to get from starting material to product? If not, we need to analyze the structure obtained in step 6 by applying questions 5 and 6 to it.

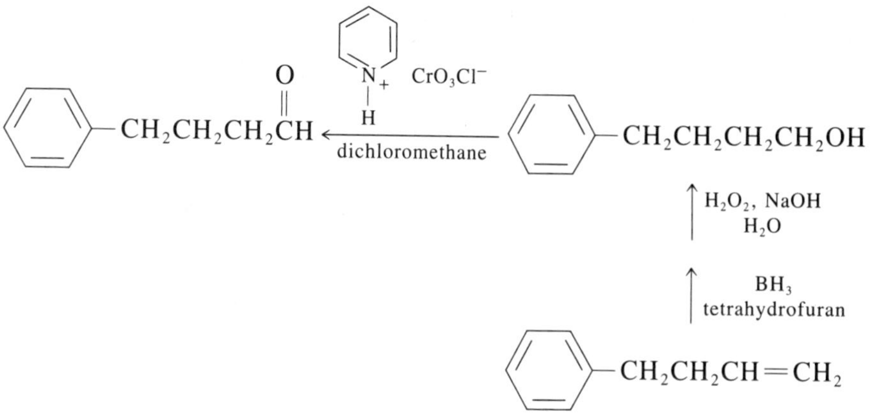

CrO3Cl−
dichloromethane
H2O2, NaOH
H2O
BH3
tetrahydrofuran
CH2CH2CH2CH
CH2CH2CH2CH2OH
CH2CH2CH=CH2

PROBLEM 12.20

How would you carry out each of the following transformations?

(a) 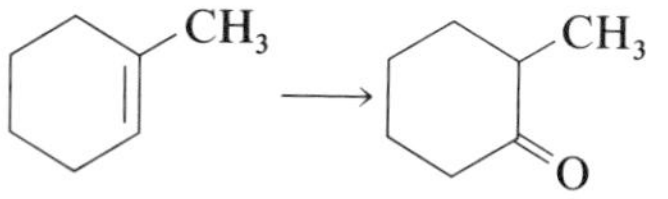

(b) $CH_3CH_2CH{=}CH_2 \longrightarrow CH_3CH_2\underset{\displaystyle NH_2}{\underset{|}{C}}HCH_3$

(c) 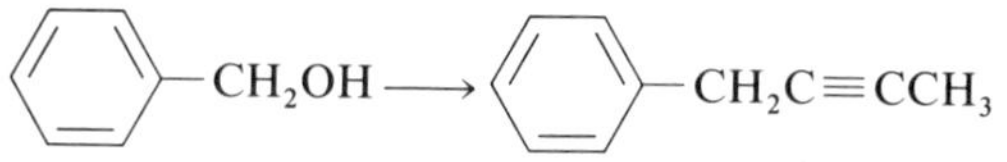

12.5 REACTIONS OF ALKOXIDE ANIONS

A. Preparation of Alkoxide Anions

Alcohols have acidic properties because of the presence of the hydrogen atom bonded to the oxygen atom in the hydroxyl group. This hydrogen atom is approximately as acidic as the hydrogen atom in water (p. 84). Thus, the concentrations of hydroxide ion and methoxide anion are roughly equal when sodium hydroxide is dissolved in methanol.

CH_3OH	+	OH^-	$\rightleftharpoons$	CH_3O^-	+	H_2O
methanol		hydroxide ion		methoxide anion		water
acid pK_a ~15.5		*base*		*conjugate base of methanol*		*conjugate acid of hydroxide ion* pK_a ~15.7

The comparable acidities of water and alcohols make it impossible to achieve high concentrations of alkoxide anions from alcohols by using hydroxide ions. When a reaction requires high concentrations of alkoxide anions (as reagents in nucleophilic substitution reactions, for example), they are prepared by the reaction of an alkali metal with an excess of dry alcohol. For example, potassium metal in *tert*-butyl alcohol gives potassium *tert*-butoxide.

$$2\,CH_3\overset{\overset{\large CH_3}{|}}{\underset{\underset{\large CH_3}{|}}{C}}OH + 2\,K \longrightarrow 2\,CH_3\overset{\overset{\large CH_3}{|}}{\underset{\underset{\large CH_3}{|}}{C}}O^-K^+ + H_2\uparrow$$

tert-butyl alcohol — potassium metal — potassium *tert*-butoxide — hydrogen

This reaction for the preparation of potassium *tert*-butoxide is an oxidation-reduction reaction, with potassium metal being oxidized and the hydrogen atom of the hydroxyl group being reduced to elemental hydrogen. The reaction is not reversible; therefore, alkoxide anions can be generated in good yields. In a similar way, metallic sodium reacts with methanol to give sodium methoxide and hydrogen gas.

$$2\,CH_3OH + 2\,Na \longrightarrow 2\,CH_3O^-Na^+ + H_2\uparrow$$

methanol — sodium metal — sodium methoxide — hydrogen

Potassium is more reactive than sodium in this type of reaction, and primary alcohols are more reactive than secondary or tertiary alcohols. The reaction of potassium with methanol occurs so rapidly and is so exothermic that the hydrogen is ignited and an explosion occurs. Therefore, sodium is used with methanol. On the other hand, sodium reacts only very slowly with *tert*-butyl alcohol, so potassium is used with that alcohol.

Sodium hydride, which is a source of the strongly basic hydride ion, is also used frequently to deprotonate alcohols to make alkoxide ions. Hydrogen gas is evolved, so the reaction goes to completion. For example, when 2-methyl-1,3-propanediol is treated with 1 equivalent of sodium hydride, the monoalkoxide ion is formed.

$$HOCH_2\overset{\overset{\large CH_3}{|}}{C}HCH_2OH + NaH \longrightarrow HOCH_2\overset{\overset{\large CH_3}{|}}{C}HCH_2O^-\,Na^+ + H_2\uparrow$$

2-methyl-1,3-propanediol (*acid*) — sodium hydride 1 equivalent (*base*) — *conjugate base of 2-methyl-1,3-propanediol*

B. Reactions of Alkoxide Anions with Alkyl Halides. The Williamson Synthesis of Ethers

Alkoxide ions, as good nucleophiles, react at the electrophilic center of primary alkyl halides in an S_N2 reaction. The substitution reaction of *n*-butyl iodide, for example, with isopropoxide anion, gives an unsymmetrical ether, an ether with two different organic groups bonded to the oxygen atom.

VISUALIZING THE REACTION

The Williamson synthesis

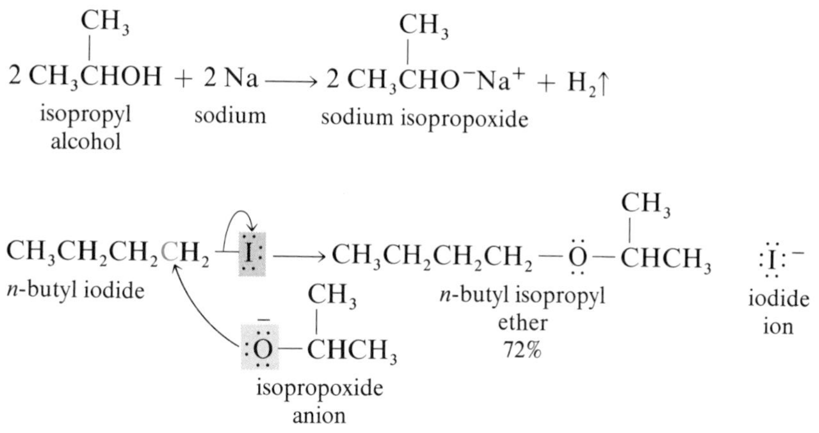

2 CH3CHOH (CH3) + 2 Na ⟶ 2 CH3CHO⁻Na⁺ (CH3) + H2↑
isopropyl alcohol
sodium
sodium isopropoxide
CH3CH2CH2CH2—I: ⟶ CH3CH2CH2CH2—O—CHCH3 (CH3) + :I:⁻
n-butyl iodide
:O⁻—CHCH3 (CH3)
isopropoxide anion
n-butyl isopropyl ether
72%
iodide ion

This way of making ethers is known as the **Williamson synthesis.** For these reactions, the alkoxide ion and the alkyl halide must be carefully chosen. In the example above, a primary alkyl halide is used so that the accompanying E_2 reaction will be minimized. In another example, the alkoxide ion from 2-methyl-1,3-propanediol (prepared on p. 446) reacts with benzyl bromide to give a benzyl ether.

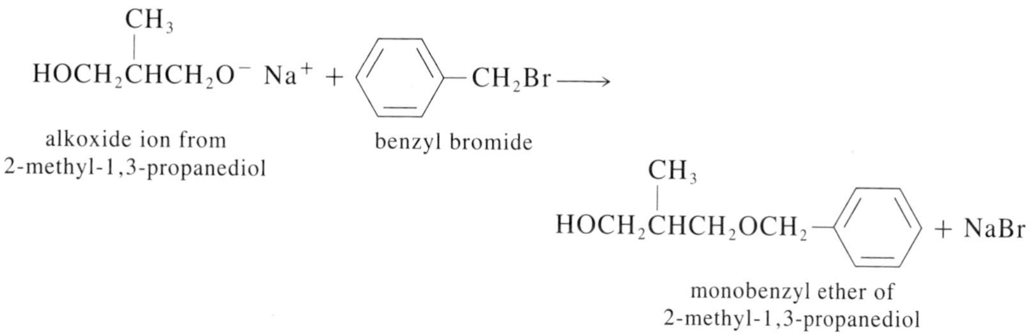

HOCH2CHCH2O⁻ Na⁺ (CH3) + C6H5—CH2Br ⟶
alkoxide ion from
2-methyl-1,3-propanediol
benzyl bromide
HOCH2CHCH2OCH2—C6H5 (CH3) + NaBr
monobenzyl ether of
2-methyl-1,3-propanediol

PROBLEM 12.21

Synthesize each of the following ethers by the Williamson method, choosing the alkoxide anion and alkyl halide that will give the best yield. Show the preparation of the alkoxide anions.

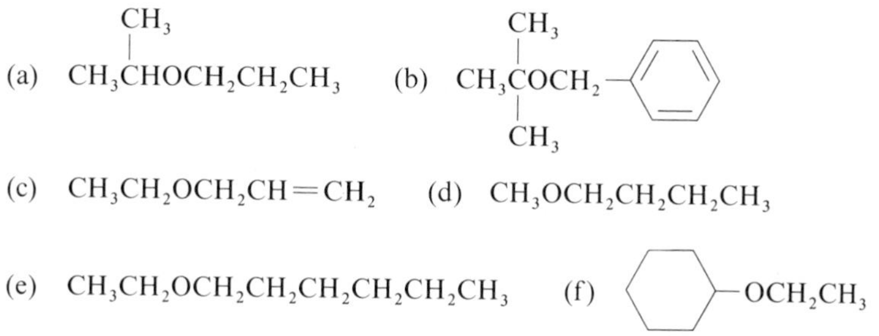

(a) CH3CHOCH2CH2CH3 (CH3)
(b) CH3COCH2—C6H5 (CH3, CH3)
(c) CH3CH2OCH2CH=CH2
(d) CH3OCH2CH2CH2CH3
(e) CH3CH2OCH2CH2CH2CH2CH2CH3
(f) C6H11—OCH2CH3

C. Intramolecular Reactions of Alkoxide Anions. The Preparation of Cyclic Ethers

If 4-chloro-1-butanol is treated with sodium hydroxide, a cyclic ether called tetrahydrofuran is formed in high yield.

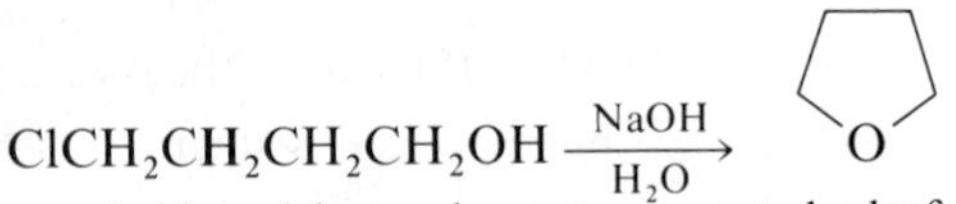
$ClCH_2CH_2CH_2CH_2OH$
NaOH
H_2O
O

4-chloro-1-butanol — tetrahydrofuran 95%

The formation of this cyclic ether is an **intramolecular reaction,** a reaction that occurs between two functional groups in the same molecule. Note that the intramolecular reaction is favored over an **intermolecular reaction,** a reaction between two different molecules. No 1,4-butanediol from the displacement of chloride ion by hydroxide ion is obtained.

The first and fastest step of the reaction is the deprotonation of the alcohol by the base. The resulting alkoxide ion then displaces chloride ion. The reacting centers in 4-chloro-1-butanol are held in close proximity by the intervening chain of four carbon atoms so that collision and reaction between them is more probable than reaction between the same reactive groups on separate molecules. Note that hydroxide ion can be used as the base for this reaction because each alkoxide ion that is formed reacts very rapidly with the electrophilic center at the other end of the molecule, and the overall reaction goes to completion. Hydroxide ion cannot be used to make alkoxide ions for intermolecular Williamson syntheses (p. 446).

VISUALIZING THE REACTION

An intramolecular S_N2 reaction

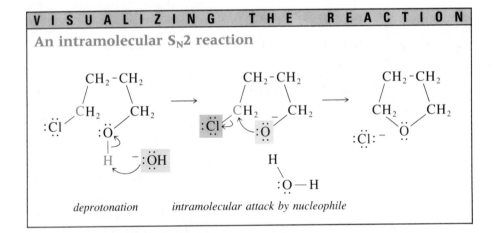

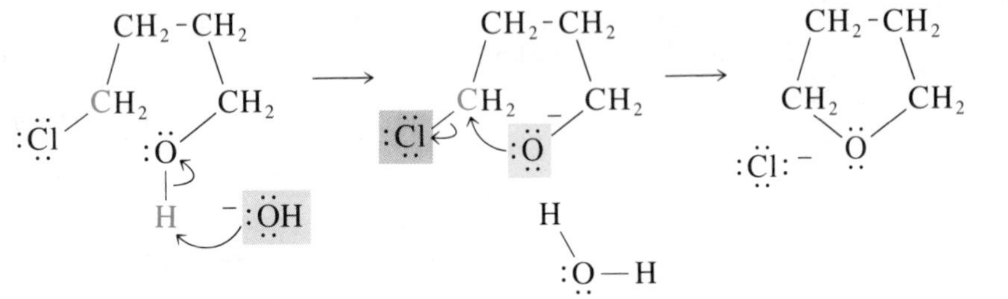

deprotonation *intramolecular attack by nucleophile*

It is particularly easy to form strainless rings of five and six carbon atoms (p. 166). A three-membered ring also forms easily in an intramolecular reaction. For example, when the trans halohydrin from cyclohexene (p. 295) is treated with base, an oxirane is the product.

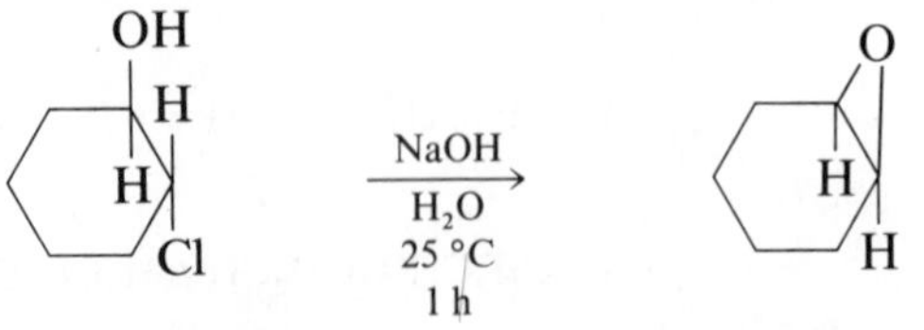
OH
H
H
Cl
NaOH
H_2O
25 °C
1 h
O
H
H

trans-2-chlorocyclohexanol — cyclohexene oxide 70%

The reaction occurs easily because in one conformation of the molecule the nucleophilic alkoxide ion is in position to displace halide ion in an intramolecular S_N2 reaction.

VISUALIZING THE REACTION

Stereochemistry of the formation of an oxirane by an intramolecular S_N2 reaction

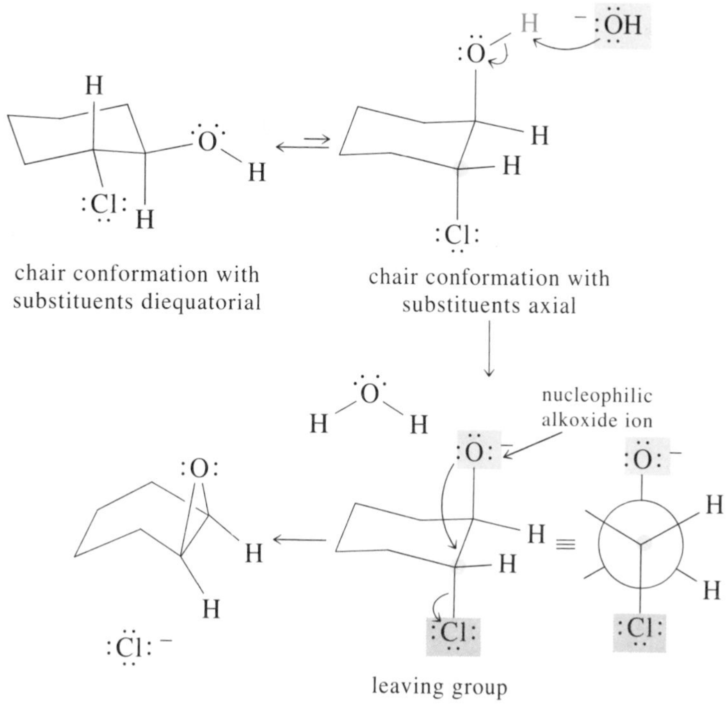
chair conformation with
substituents diequatorial
chair conformation with
substituents axial
nucleophilic
alkoxide ion
leaving group

Study Guide Concept Map 12.5

In this conformation the alkoxide ion and the leaving group are anti to each other as shown in the Newman projection for that portion of the cyclohexane ring. Halohydrins that cannot achieve an anti orientation between the nucleophile and the leaving group do not give oxiranes.

PROBLEM 12.22

Write a detailed mechanism for the following reaction. Be sure to show the conformation of the molecule in the transition state.

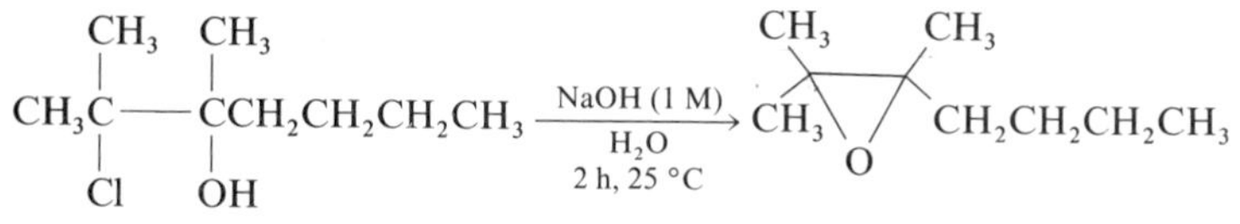
NaOH (1 M)
H_2O
2 h, 25 °C

2-chloro-2,3-dimethyl-3-heptanol

2,2,3-trimethyl-3-butyloxirane

PROBLEM 12.23

Give structures for the products of the following reactions. Show stereochemistry if it is known.

(a) 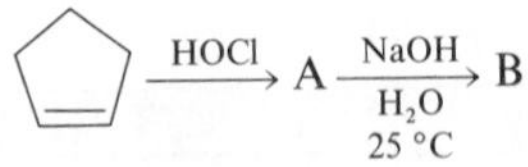
HOCl
A
NaOH
H_2O
25 °C
B

(b) $\underset{\text{Cl}\ \ \text{OH}}{CH_3CH_2CH_2CHCHCH_2CH_2CH_3} \xrightarrow[H_2O,\ 25\ °C]{NaOH} C$

(c) $ClCH_2C(CH_3)(OH)CH_2CH_2CH_2CH_3 \xrightarrow[H_2O,\ 25\ °C]{NaOH} D$

(d) $BrCH_2CH_2CH_2CH_2CH_2OH \xrightarrow[H_2O,\ 25\ °C]{NaOH} E$

(e) 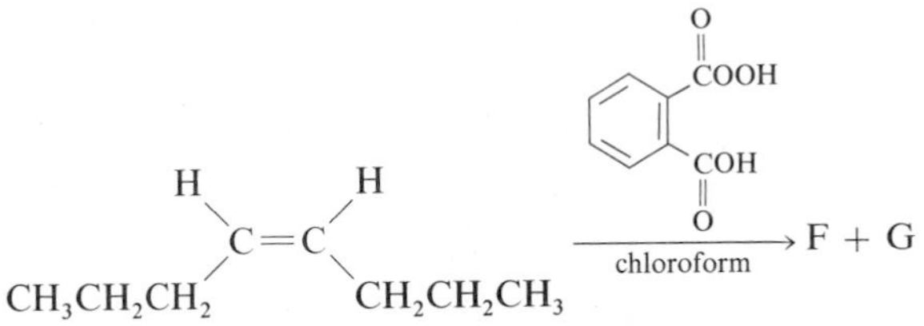
H
H
C=C
$CH_3CH_2CH_2$
$CH_2CH_2CH_3$
O
COOH
COH
O
chloroform
F + G

12.6 CLEAVAGE REACTIONS OF ETHERS

A. Reaction with Hydriodic Acid

An ether can be cleaved by heating it with a strong acid (p. 239). Hydriodic acid is often used for this purpose. It is a strong enough acid to protonate the oxygen atom in the ether, creating a good leaving group. Iodide ion is a good nucleophile, so an S_N2 reaction of the protonated ether occurs. For example, anisole (methyl phenyl ether) is cleaved in this way to give phenol and methyl iodide.

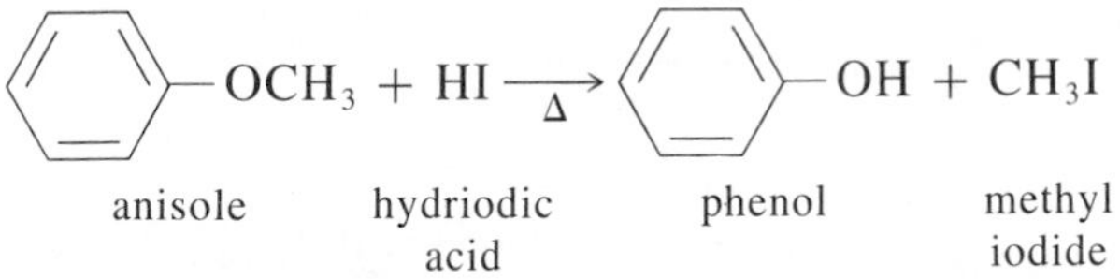
OCH_3 + HI
Δ
OH + CH_3I
anisole
hydriodic acid
phenol
methyl iodide

VISUALIZING THE REACTION

Cleavage of an ether

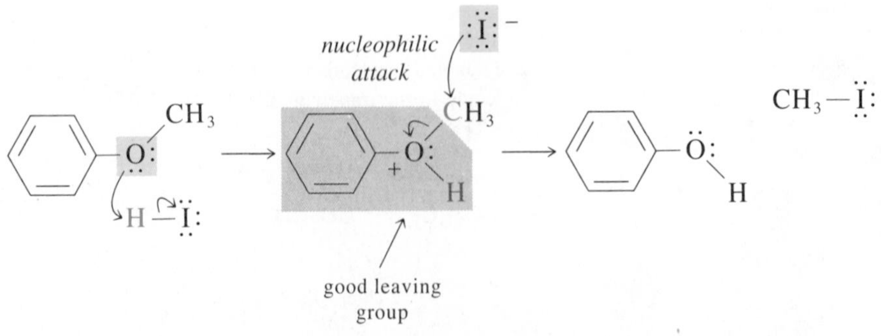
nucleophilic attack
:I:⁻
CH_3
O:
H—I:
CH_3
O
+
H
O:
H
CH_3—I:
good leaving group

In this reaction, the methyl group and not the phenyl group (p. 162) undergoes nucleophilic attack.

If an ether has two alkyl groups bonded to its oxygen atom, the smaller group, such as a methyl or an ethyl group, is more likely to be converted to the alkyl iodide in the first step of the reaction. If there is an excess of hydriodic acid, however, both alkyl groups will eventually become alkyl iodides, for example:

$$CH_3CH_2CH_2CH_2OCH_2CH_3 \xrightarrow[\Delta]{HI} [CH_3CH_2CH_2CH_2OH + CH_3CH_2I]$$

$$\downarrow \; HI, \Delta$$

$$CH_3CH_2CH_2CH_2I + CH_3CH_2I$$

Phenols, compounds in which the hydroxyl group is bonded directly to a carbon of the aromatic benzene ring, are not converted to aryl halides by hydrogen halides.

PROBLEM 12.24

Tetrahydrofuran is a stable ether, but the ring will open when the molecule is heated with acid.

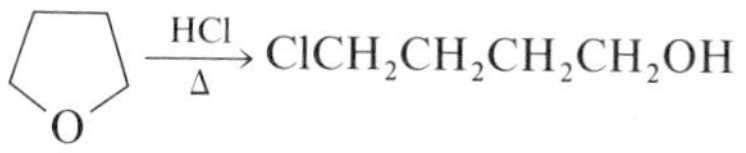

HCl
Δ
ClCH₂CH₂CH₂CH₂OH
O

Propose a mechanism for this reaction.

B. Acid-Catalyzed Ring-Opening Reactions of Oxiranes

The ring of an oxirane opens readily when acidic reagents are used. For example, *cis*- and *trans*-2,3-dimethyloxirane (p. 318) both give a pair of enantiomeric 3-bromo-2-butanols with 48% aqueous hydrobromic acid. The reaction of *cis*-2,3-dimethyloxirane gives one racemic mixture.

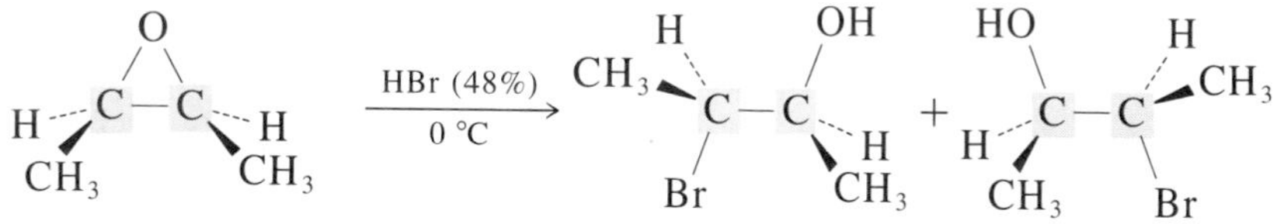

HBr (48%)
0 °C

cis-2,3-dimethyloxirane

(2*S*,3*S*)-3-bromo-2-butanol

(2*R*,3*R*)-3-bromo-2-butanol

racemic mixture
85%

trans-2,3-Dimethyloxirane, which exists as a racemic mixture, gives another pair of enantiomeric 3-bromo-2-butanols.

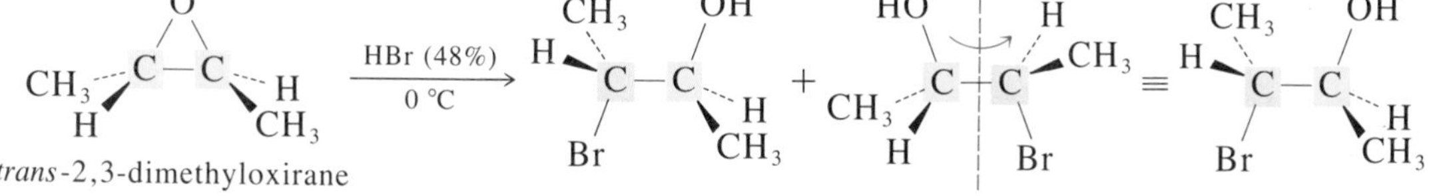

HBr (48%)
0 °C

trans-2,3-dimethyloxirane
one enantiomer

(2*S*,3*R*)-3-bromo-2-butanol

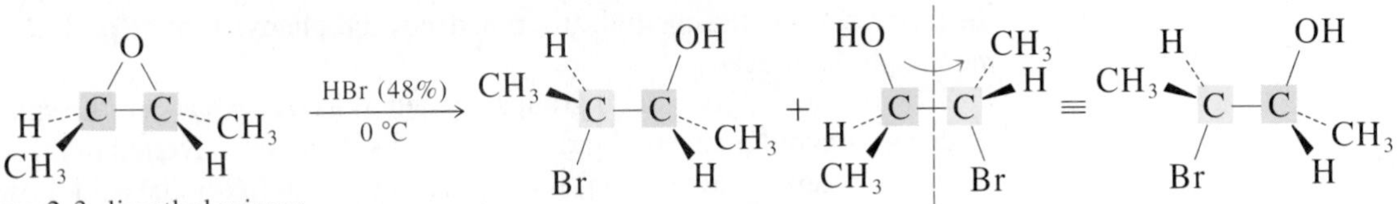

HBr (48%)
0 °C

trans-2,3-dimethyloxirane
the other enantiomer

(2*R*,3*S*)-3-bromo-2-butanol

The 3-bromo-2-butanols obtained from *cis*-2,3-dimethyloxirane are diastereomers of those obtained from the trans oxirane.

This reaction starts with the protonation of the oxygen atom in the oxirane ring. The protonated oxirane is an oxonium ion and is electronically similar to the bromonium ion suggested as an intermediate in the addition of bromine to a double bond (p. 290). The oxonium ion intermediate is attacked at either carbon atom by the nucleophilic halide ion, as illustrated below for *cis*-2,3-dimethyl oxirane.

VISUALIZING THE REACTION

Acid-catalyzed ring-opening reaction of an oxirane

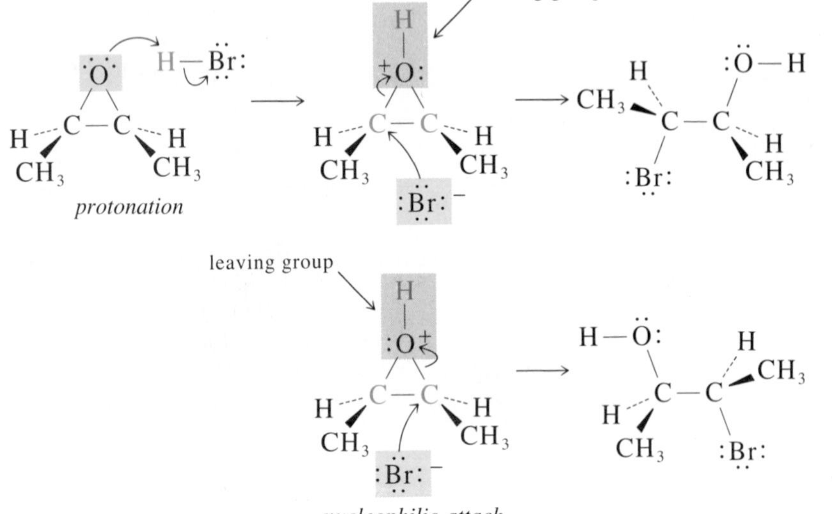

leaving group
protonation
leaving group
nucleophilic attack

The reactions of the 2,3-dimethyloxiranes are examples of stereoselectivity in chemical transformations. Two starting materials with differing stereochemistry are converted into products that are also stereochemically different from each other.

PROBLEM 12.25

When *cis*-2,3-dimethyloxirane is treated with water containing a trace of perchloric acid, $HClO_4$, a racemic mixture of 2,3-butanediols is formed. *trans*-2,3-Dimethyloxirane gives *meso*-2,3-butanediol under the same conditions. Write mechanisms for these reactions, showing all the stereochemistry.

PROBLEM 12.26

When one of the two enantiomeric *trans*-2,3-dimethyloxiranes, (2*R*,3*R*)-(+)-2,3-dimethyloxirane, was treated with methanol containing a trace of sulfuric acid, there was obtained a 57% yield of a single, optically active 3-methoxy-2-butanol.

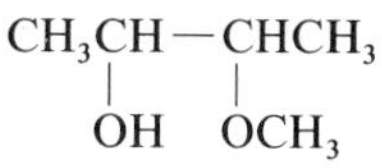

Draw the correct structure for the oxirane and write a stereochemically correct mechanism showing how the 3-methoxy-2-butanol is formed. Assign configurations to the stereocenters in the product and give the correct name for the compound.

The 2,3-dimethyloxiranes have symmetrical structures. Regioselectivity is possible only in ring-opening reactions of unsymmetrical oxiranes. 2,2,3-Trimethyloxirane reacts with methanol in the presence of sulfuric acid to give 2-methoxy-2-methyl-3-butanol as the major product.

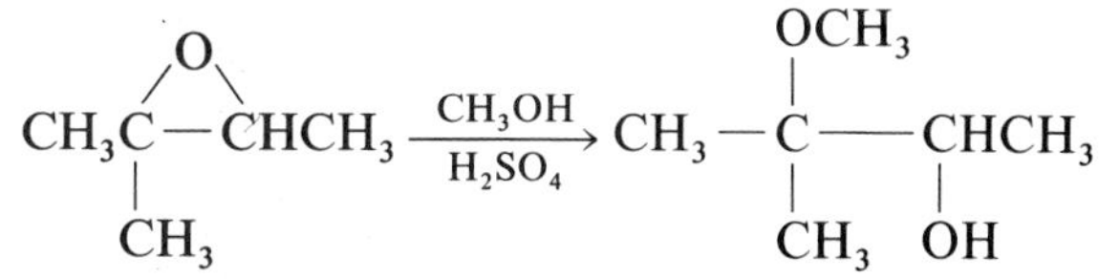

2,2,3-trimethyloxirane

2-methoxy-2-methyl-3-butanol
76%

The protonated oxygen atom of the oxirane ring is a good leaving group, and the carbon-oxygen bonds in the strained ring may start to cleave before the approach of the nucleophile. In an unsymmetrically substituted oxirane, one carbocationic intermediate will be favored over the other. For 2,2,3-trimethyloxirane, the tertiary carbocation rather than the secondary one is the favored intermediate. The nucleophile in this reaction, methanol, attacks at the tertiary carbon atom.

VISUALIZING THE REACTION

Acid-catalyzed ring-opening reaction of an unsymmetrical oxirane

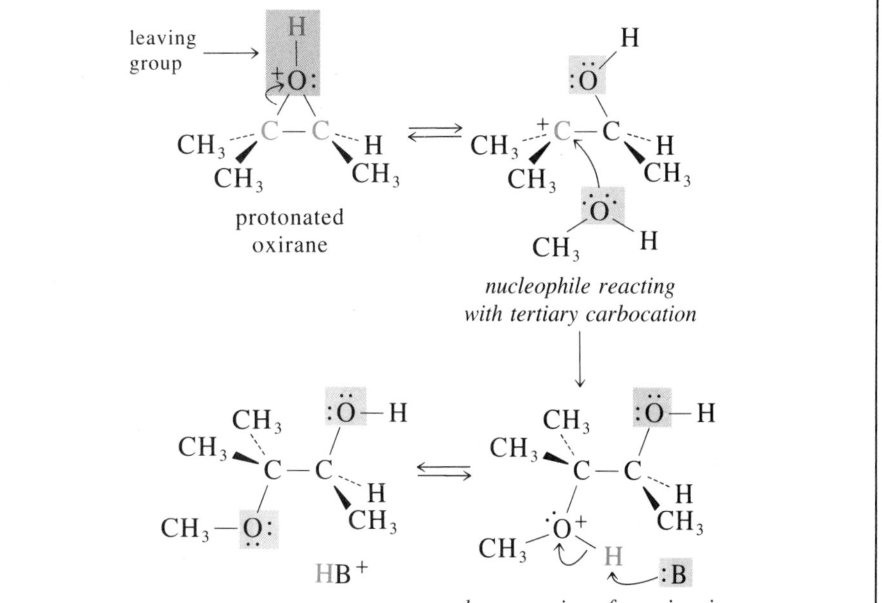
leaving group
protonated oxirane
nucleophile reacting with tertiary carbocation
deprotonation of oxonium ion
HB+
:B

C. Ring-Opening Reactions of Oxiranes with Nucleophiles

The strained ring of an oxirane is opened easily by nucleophilic reagents even without protonation of the oxygen atom. These reactions are important in syntheses, especially when the attacking reagent is a nucleophilic carbon atom, as will be seen in Section 13.6C. However, other nucleophiles, such as alkoxide ions or amines, also react with oxiranes. For example, 2,2,3-trimethyloxirane reacts with sodium methoxide to give 3-methoxy-2-methyl-2-butanol.

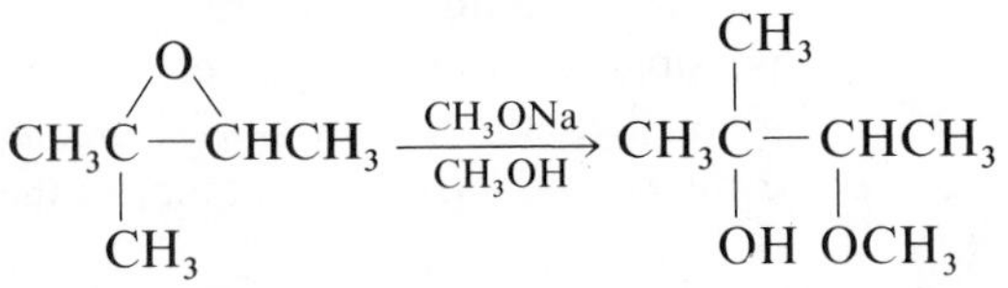
CH3ONa
CH3OH

2,2,3-trimethyloxirane — 3-methoxy-2-methyl-2-butanol 53%

In the absence of a strong acid, no cationic intermediate for the reaction can form. An S_N2 reaction occurs at the less highly substituted carbon atom. The new alkoxide ion that forms is protonated by the solvent.

VISUALIZING THE REACTION

Opening of an oxirane by a nucleophile

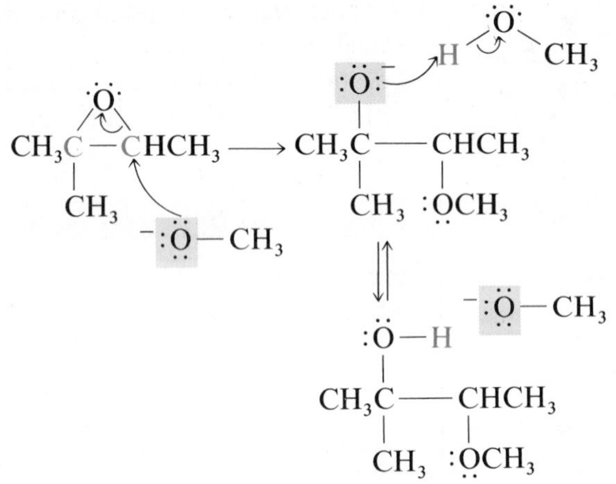

PROBLEM 12.27

Suggest reagents for carrying out the following transformations.

(a) 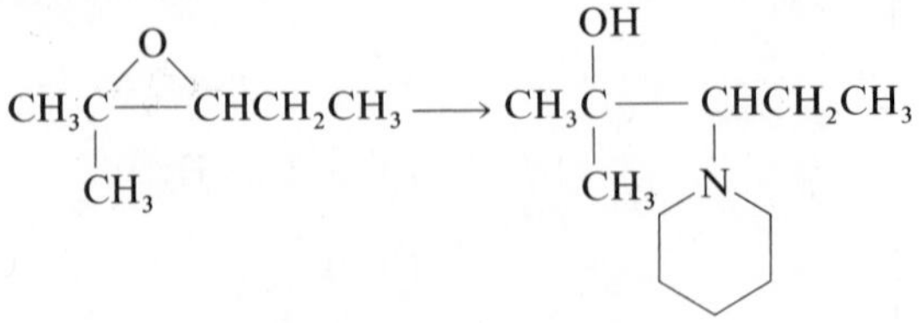

(b) 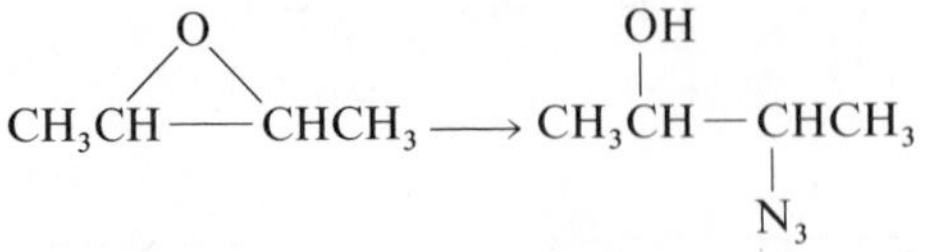

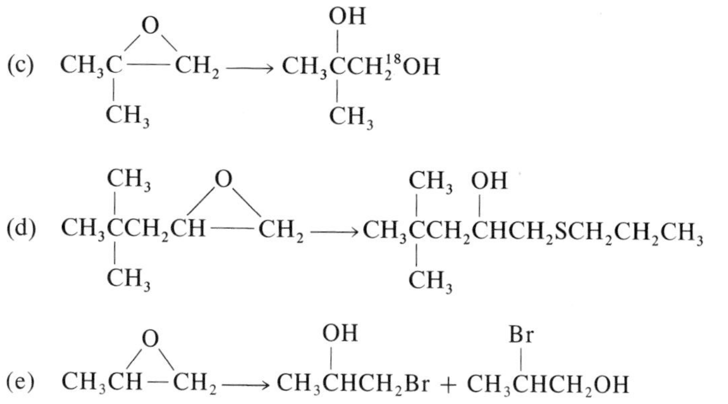
(c)
O
OH
CH3C—CH2 ⟶ CH3CCH2 18OH
CH3
CH3
(d)
CH3
O
CH3 OH
CH3CCH2CH—CH2 ⟶ CH3CCH2CHCH2SCH2CH2CH3
CH3
CH3
(e)
O
OH
Br
CH3CH—CH2 ⟶ CH3CHCH2Br + CH3CHCH2OH

D. Stereoselective Formation of *trans*-1,2-Diols from Oxiranes

The oxirane from cyclopentene gives a racemic mixture of *trans*-1,2-cyclopentanediols when treated with acid in water.

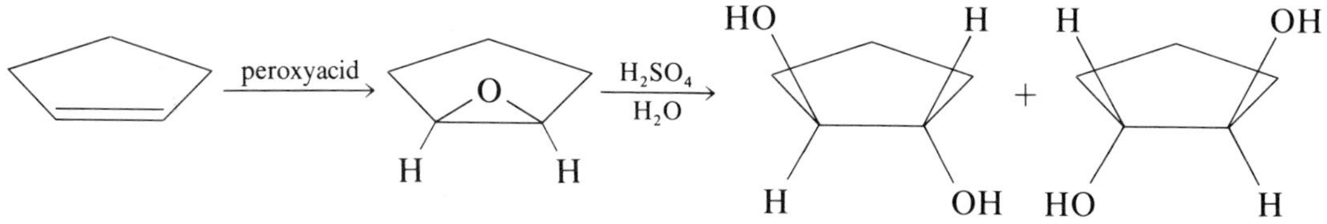
peroxyacid
O
H
H
H_2SO_4
H_2O
HO
H
H
OH
+
H
OH
HO
H

cyclopentene

cyclopentene oxide
1,2-epoxycyclopentane

trans-1,2-cyclopentanediol
a racemic mixture

Here, again, the ring-opening reaction starts with protonation of the oxygen atom of the oxirane to give a reactive oxonium ion. Nucleophilic attack by a water molecule on either one of the carbon atoms of the oxirane ring opens the ring to give the trans orientation of the two hydroxyl groups. The reaction is stereoselective, giving only the trans diol.

You can think of this above sequence of reactions as being complementary to the oxidation of cyclopentenes with potassium permanganate or osmium tetroxide, which gives *cis*-1,2-cyclopentanediols (pp. 314 and 316). Cyclohexene, like cyclopentene, reacts with aqueous potassium permanganate or osmium tetroxide to give a cis-1,2-diol.

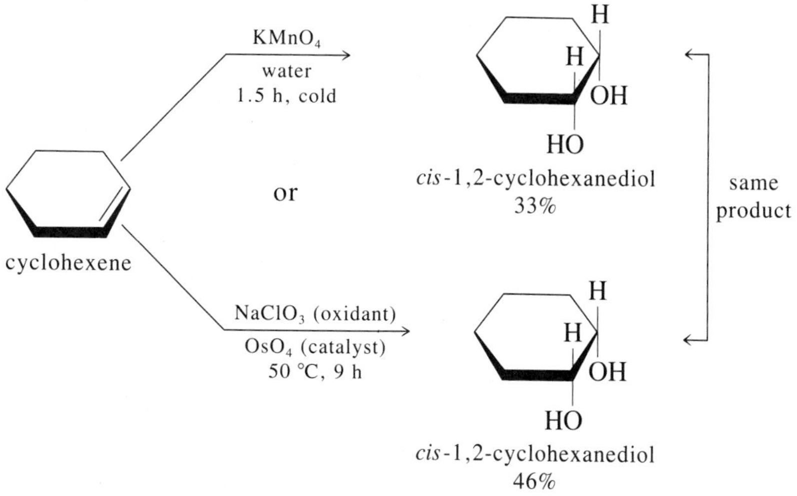
$KMnO_4$
water
1.5 h, cold
H
H
OH
HO
cis-1,2-cyclohexanediol
33%
or
same
product
cyclohexene
$NaClO_3$ (oxidant)
OsO_4 (catalyst)
50 °C, 9 h
H
H
OH
HO
cis-1,2-cyclohexanediol
46%

The stereochemistry of the cis diol is determined by the cyclic intermediate formed when either potassium permanganate or osmium tetroxide reacts with the alkene.

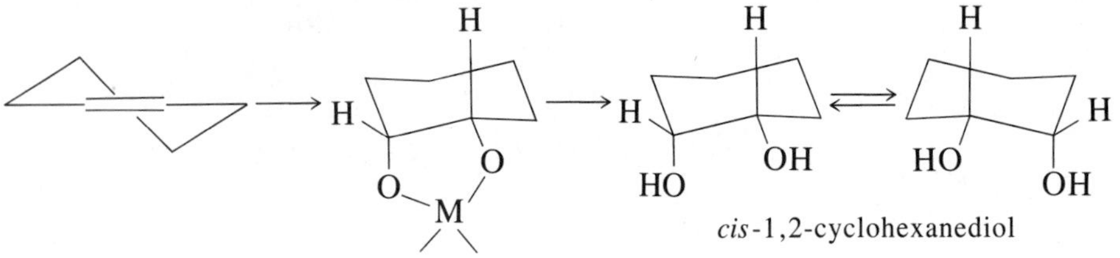
H
H
O
O
M
H
H
HO
OH
H
H
HO
OH

cis-1,2-cyclohexanediol

If, on the other hand, cyclohexene is oxidized by a peroxyacid, and the resulting epoxide opened in aqueous acid, *trans*-1,2-cyclohexanediol is formed as a racemic mixture.

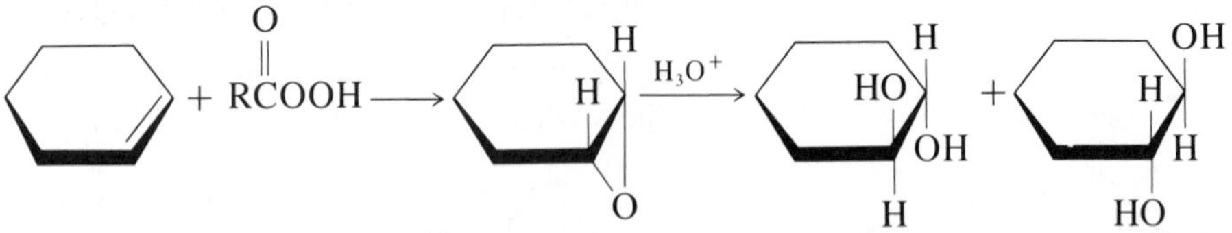
O
+ RCOOH
H
H
O
H_3O^+
H
HO
OH
H
+
OH
H
H
HO

trans-1,2-cyclohexanediol
racemic mixture

VISUALIZING THE REACTION

Stereochemistry of the opening of the oxirane ring in cyclohexene oxide

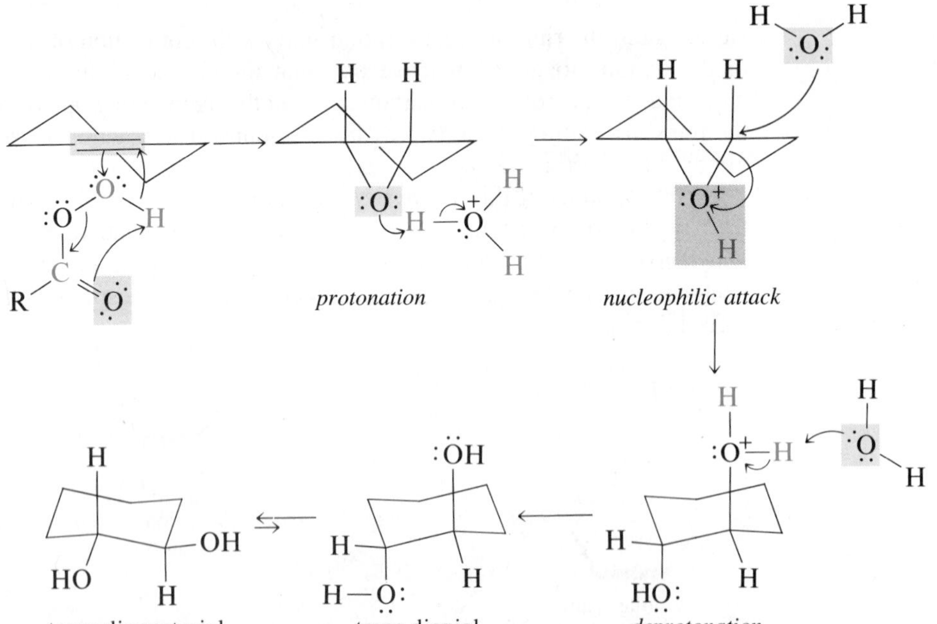
R
C
O
O
O
H
H
H
O
H
O
H
H
protonation
H
O
H
H
H
O
H
nucleophilic attack
H
O
H
H
O
H
H
H
HO
deprotonation
OH
H
H—O
H
trans diaxial
H
OH
HO
H
trans diequatorial

As the above conformational formulas demonstrate, the opening of the three-membered oxirane ring first gives the two hydroxyl groups in a trans diaxial

Study Guide Concept Map 12.6

orientation. The more stable conformation for the diol is the diequatorial one, which predominates at equilibrium. Attack by the nucleophile, water, is equally probable at either carbon atom of the oxirane ring, so equal amounts of the two enantiomeric *trans*-1,2-cyclohexanediols are formed.

PROBLEM 12.28

Because of the bulk of the *tert*-butyl group, *tert*-butylcyclohexane exists mainly in one conformation (p. 169). *trans*-4-*tert*-Butylcyclohexene oxide, when treated with methanethiol, CH_3SH, in ethanol in the presence of sodium ethoxide, gives *trans*-2-methylthio-*trans*-5-*tert*-butylcyclohexanol. Write a mechanism for this reaction showing the conformation you expect the product to have.

E. Problem-Solving Skills

Problem

How could we carry out the following transformation?

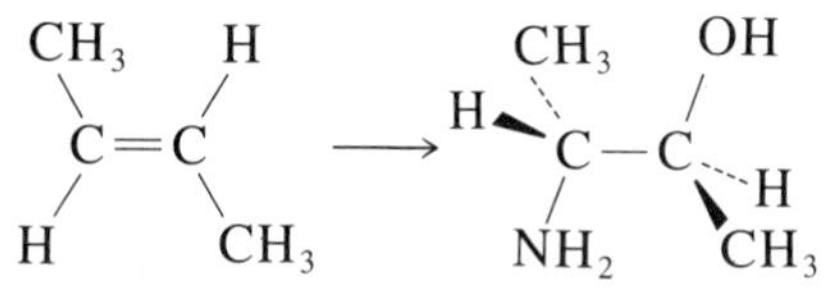

racemic mixture

Solution

1. What functional groups are present in the starting material and the product?

   The starting material is an alkene; the product contains an amino group and a hydroxyl group on adjacent carbon atoms.

2. How do the carbon skeletons of the two compounds compare? How many carbon atoms does each contain? Are there any rings? What are the positions of branches and functional groups on the carbon skeletons?

   The carbon skeletons are the same. In the alkene, two methyl groups are trans to each other across the double bond. In the product, the methyl groups are on opposite sides of the plane of the page, which bisects the hydroxyl group, carbon atoms 2 and 3, and the amino group.

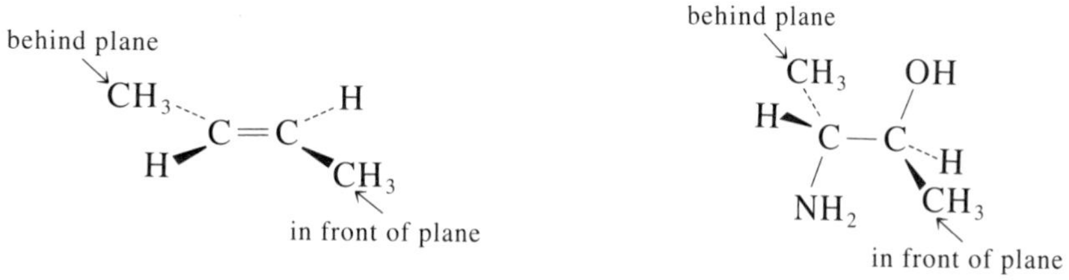
behind plane
in front of plane
behind plane
in front of plane

   The amino group and the hydroxyl group are anti to each other.

3. How do the functional groups change in going from starting material to product? Does the starting material have a good leaving group?

   The double bond of the alkene disappears. A hydroxyl group and an amino group are attached to the carbon atoms that were joined by the double bond. The alkene does not contain a leaving group.

4. Is it possible to dissect the structures of the starting material and product to see which bonds must be broken and which formed?

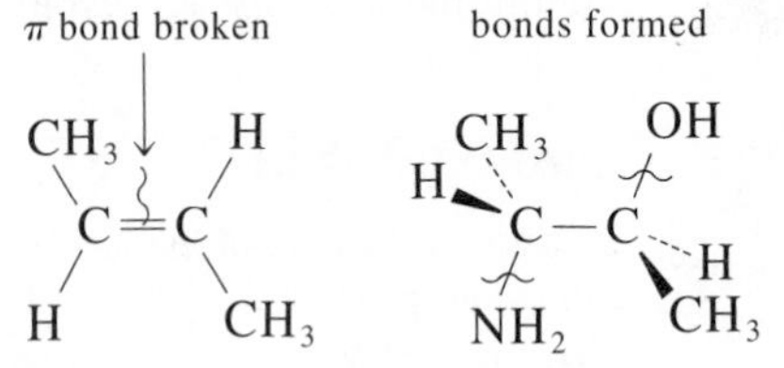
π bond broken
bonds formed
CH3
H
C=C
H
CH3
CH3
OH
H
C—C
H
NH2
CH3

A π bond must be broken; carbon-oxygen and carbon-nitrogen bonds must be formed anti to each other. In the product, carbon is bonded to elements that are more electronegative than it is, which suggests an oxidation reaction (see p. 436).

5. Do we recognize any part of the product molecule as coming from a good nucleophile or an electrophilic addition?

The amino group, $—NH_2$, is related to ammonia, NH_3, a good nucleophile. An alkene can be converted to a compound containing oxygen by oxidizing it with the electrophilic reagent *m*-chloroperoxybenzoic acid.

6. What type of compound would be a good precursor to the product?

An oxirane could be converted to the amino alcohol.

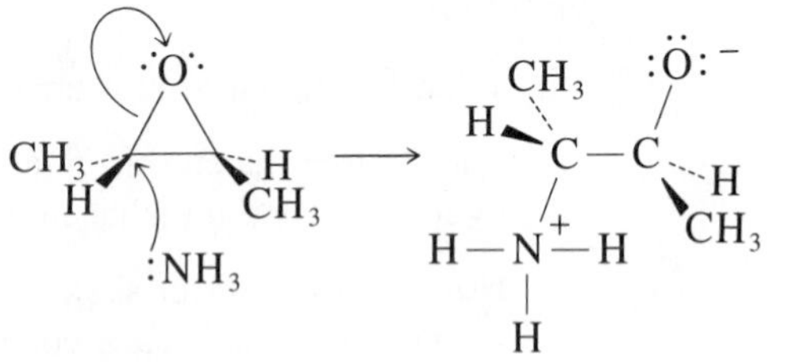
:O:
CH3
H
H
CH3
:NH3
CH3
:O:−
H
C—C
H
CH3
H—N—H
H

7. After this last step, do we see how to get from starting material to product? If not, we need to analyze the structure obtained in step 6 by applying questions 5 and 6 to it.

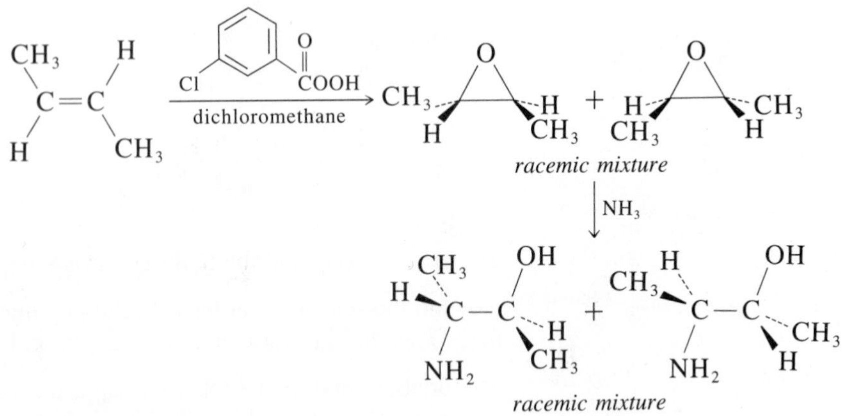
CH3
H
C=C
H
CH3
Cl
O
COOH
dichloromethane
O
CH3
H
H
CH3
+
O
H
CH3
CH3
H
racemic mixture
NH3
CH3
OH
H
C—C
H
NH2
CH3
+
H
OH
CH3
C—C
CH3
NH2
H
racemic mixture

PROBLEM 12.29

(a) Predict the products of the following reactions, including the stereochemistry.

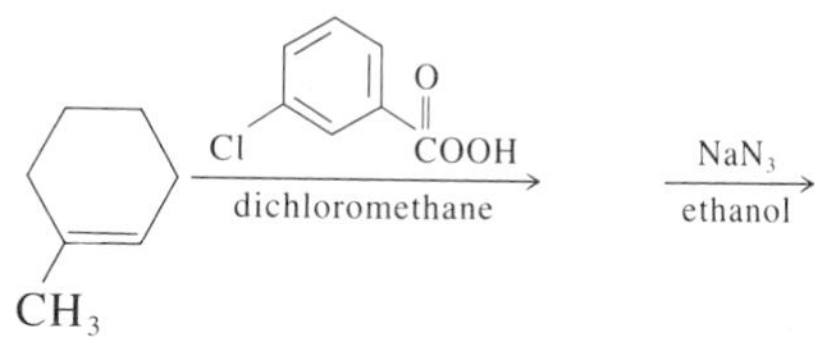

(b) How would you carry out the following transformation?

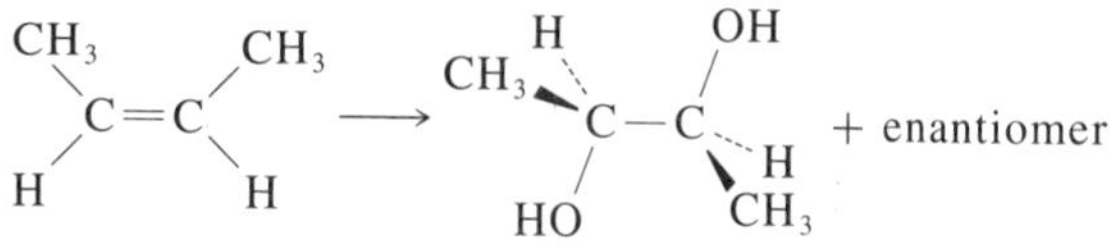

12.7 CLEAVAGE OF CARBON-CARBON BONDS IN DIOLS WITH PERIODIC ACID

Diols are cleaved by the oxidizing agent periodic acid, usually given the formula HIO_4. This reagent is used in the determination of the structures of carbohydrates, which are polyhydroxy compounds (p. 1085). The reaction of a diol with periodic acid is carried out at or below room temperature. For example, 2,3-butanediol is converted quantitatively into acetaldehyde.

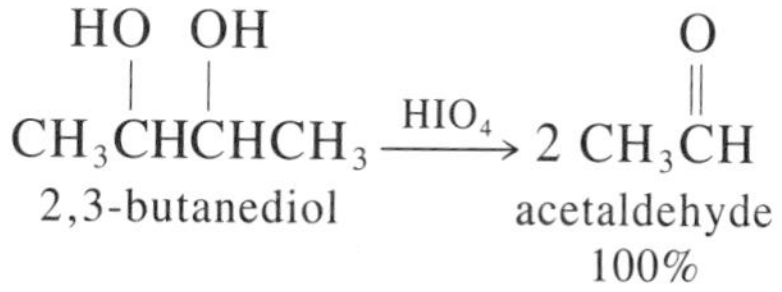

This reaction proceeds by way of a cyclic intermediate.

VISUALIZING THE REACTION

Cleavage of carbon-carbon bonds with periodic acid

1. Formation of a cyclic periodate ester

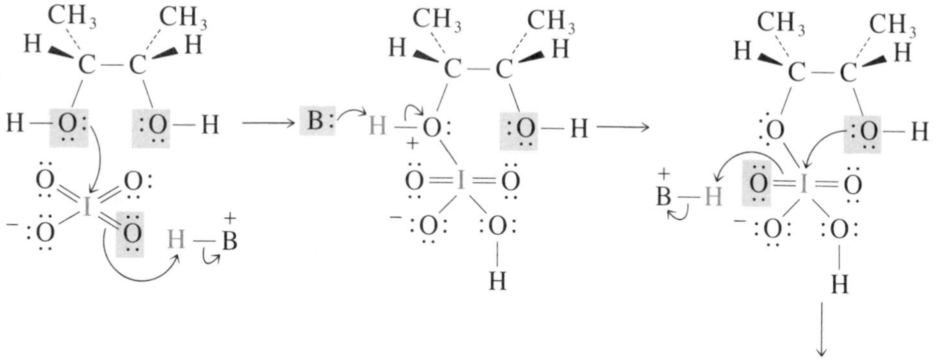

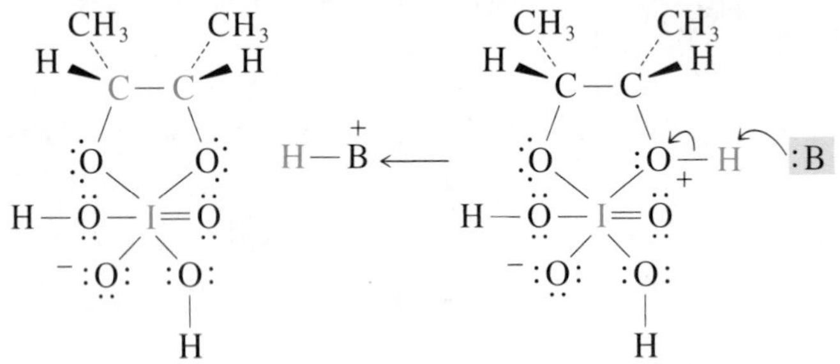

2. Cleavage of a carbon-carbon bond

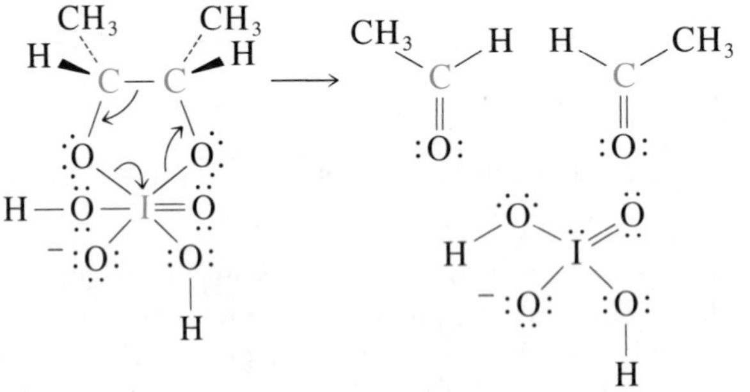

The secondary alcohol groups in 2,3-butanediol are converted to aldehyde functions. Overall, hydroxyl groups are converted to carbonyl groups as a result of cleavage of a carbon-carbon bond.

In a related reaction, a periodate, in combination with osmium tetroxide, is used to cleave double bonds. A catalytic amount of osmium tetroxide reacts with a double bond to give an osmate ester, which then decomposes in the presence of sodium or potassium periodate to regenerate osmium tetroxide and the cleavage product of the diol. 1-Dodecene, for example, is converted to undecanal by this method.

$$\underset{\text{1-dodecene}}{CH_3(CH_2)_9CH{=}CH_2} \xrightarrow[\substack{H_2O \\ \text{dioxane} \\ 25\ ^\circ C}]{OsO_4,\ NaIO_4} \underset{\text{undecanal}}{CH_3(CH_2)_9\overset{\overset{\large O}{\|}}{C}H} + \underset{\text{formaldehyde}}{H\overset{\overset{\large O}{\|}}{C}H}$$

Formaldehyde, the other product of this reaction, is not isolated. Note that the overall result of this reaction is the same as that from ozonolysis with a reductive work-up (p. 312).

PROBLEM 12.30

Predict the products of the following reactions.

(a) 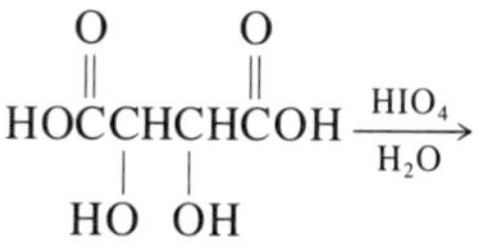

(b) 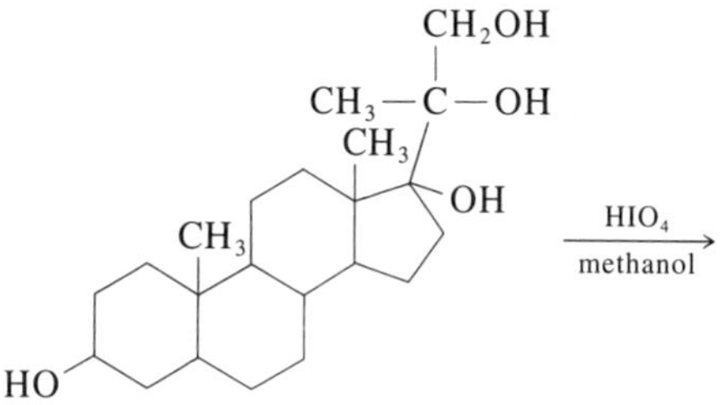

(c) 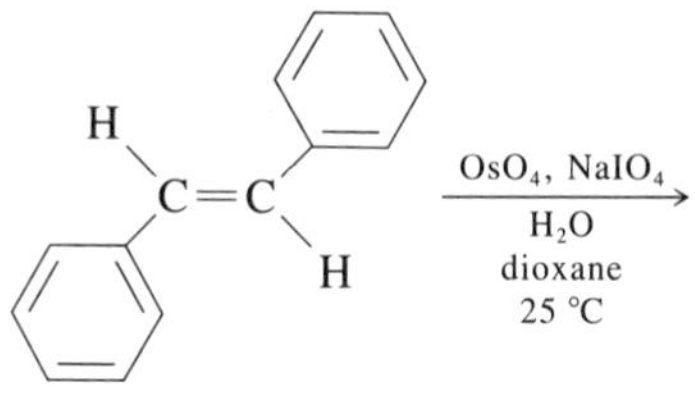

S U M M A R Y

Alcohols contain a hydroxyl group, —OH. Alcohols are converted to alkyl halides by reagents such as hydrogen halides, thionyl chloride, and phosphorus tribromide. Oxidation of primary alcohols with aqueous dichromate solutions first produces aldehydes but these are further oxidized to carboxylic acids. Aldehydes are obtained from primary alcohols when pyridinium chlorochromate is used. Secondary alcohols are oxidized by either of these oxidizing agents to ketones. Tertiary alcohols are not oxidized under these conditions.

Alcohols are converted to alkoxide ions by strong bases such as hydride ion or by alkali metals such as sodium or potassium. Alkoxide ions are good nucleophiles and react with primary or secondary alkyl halides to give ethers. If the halogen is part of the same molecule as the alkoxide ion and a three-, five-, or six-membered ring can form, a fast intramolecular reaction occurs to form a cyclic ether. The reactions of alcohols are outlined in Table 12.1.

Acyclic ethers (and cyclic ethers with rings larger than three-membered) are cleaved by strong acid. Three-membered cyclic ethers, oxiranes, readily undergo

TABLE 12.1 Reactions of Alcohols

Alcohol	Reagent	Intermediate	Product
	Conversion to Alkyl Halides		
ROH	HX	$ROH_2^+\ X^-$	RX
	$SOCl_2$	$R{-}O{-}S({=}O){-}Cl\quad Cl^-$	RCl
	PBr_3	$R{-}O{-}PBr_2\quad Br^-$	RBr

(*Continued*)

TABLE 12.1 *(Continued)*

Alcohol	Reagent	Intermediate	Product
	Oxidation Reactions		
RCH_2OH	$Cr_2O_7^{2-}$, H_2O, H_3O^+	$RCH(H)-O-Cr(=O)_2-OH$	$RC(=O)H$, $RC(=O)OH$
	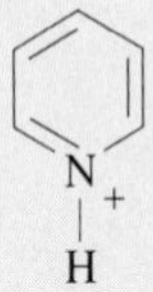 CrO_3Cl^-	$RCH(H)-O-Cr(=O)_2-OH$	$RC(=O)H$
$RC(R)(H)-OH$	$Cr_2O_7^{2-}$, H_2O, H_3O^+ or 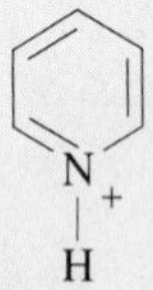 CrO_3Cl^-	$RC(R)(H)-O-Cr(=O)_2-OH$	$RC(=O)R$
$RC(R)(R)-OH$	$Cr_2O_7^{2-}$, H_2O, H_3O^+ or 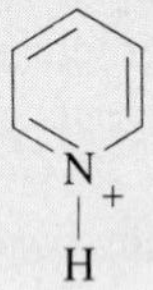 CrO_3Cl^-	not oxidized	
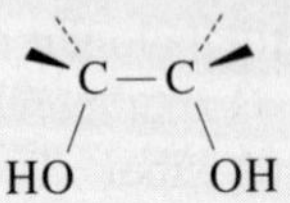	HIO_4	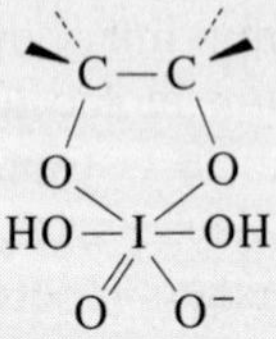	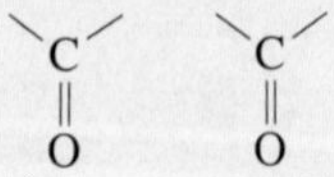
	Conversion to Ethers		
ROH	Na or NaH	RO^- + $R'X$ (primary or secondary)	ROR′
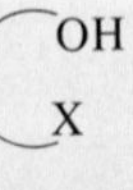	NaOH	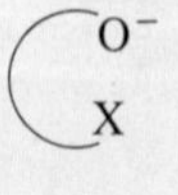	 (five- or six-membered ring)
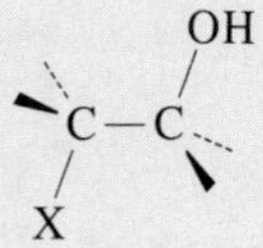	NaOH	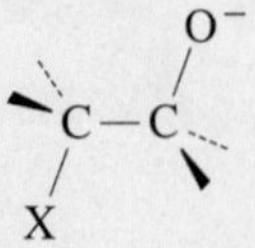	

ring-opening reactions with dilute acids to give 1,2-diols and with hydrogen halides to give halohydrins. Oxiranes also react with nucleophiles to give alcohols in which a group derived from the nucleophile is substituted at the carbon next to the one carrying the hydroxyl group. These ring-opening reactions occur stereoselectively, so the hydroxyl group derived from the oxygen atom of the oxirane and the incoming group are anti to each other. The reactions of ethers are outlined in Table 12.2.

TABLE 12.2 Cleavage Reactions of Ethers

Ether	Reagent	Intermediate	Product
	Acids		
ROR′	HI	$R\overset{+}{O}R'\ I^-$ (with H on O)	ROH + R′I RI + R′OH
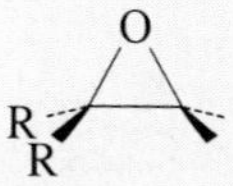	H_2O, H_3O^+	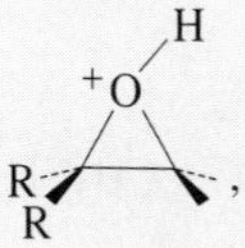	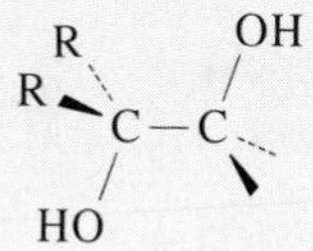
	HX	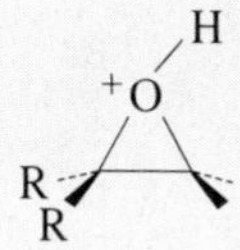	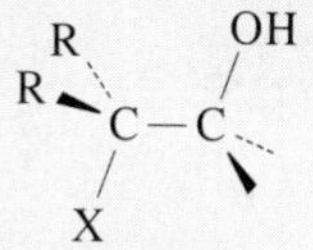 major product 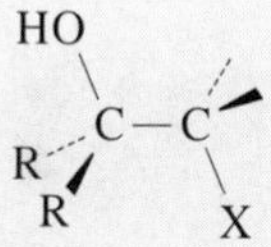 minor product
	Nucleophiles		
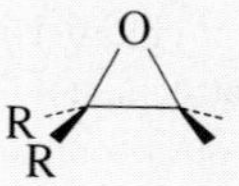	RO^-, ROH	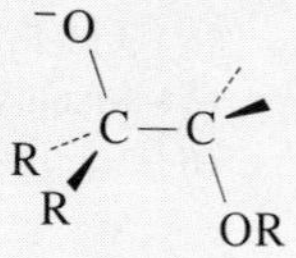 ROH	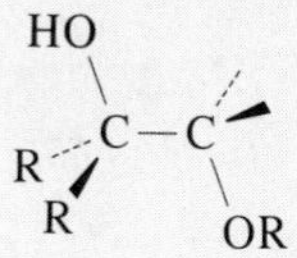
	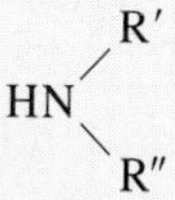	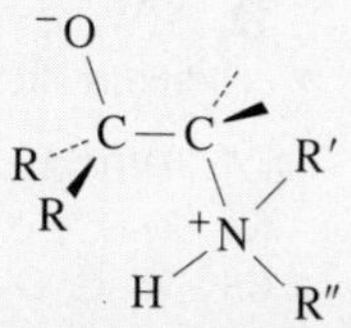	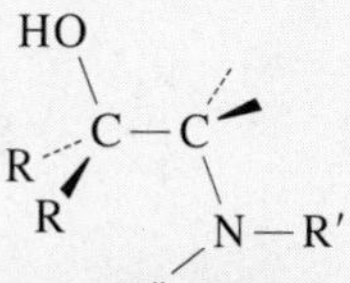
	HS—R′	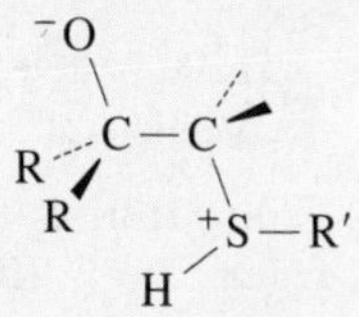	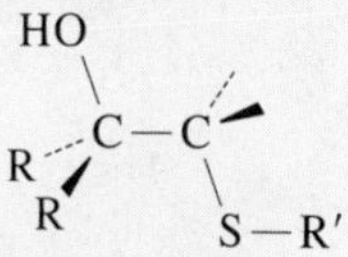

Study Guide Concept Map 12.7

1,2-Diols are prepared by ring-opening reactions of oxiranes or by permanganate or osmium tetroxide oxidation of alkenes. These diols are oxidized by periodic acid with cleavage of the carbon-carbon bond between the two carbon atoms bearing the hydroxyl groups.

ADDITIONAL PROBLEMS

12.31 Name each of the following compounds according to the IUPAC rules, including stereochemistry when shown.

(a) $HOCH_2CH_2CH_2OH$ (b) $CH_3OCH_2CH_2OCH_3$

(c) 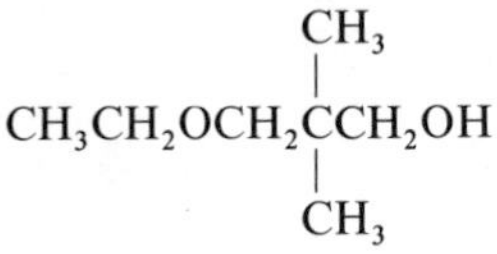 (d) $CH_3OCH_2CH_2Br$ (e) 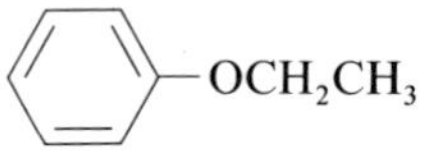

(f) $CH_3CH_2OCH_2CH_2OH$ (g) 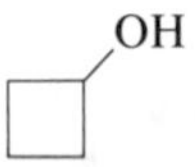 (h) 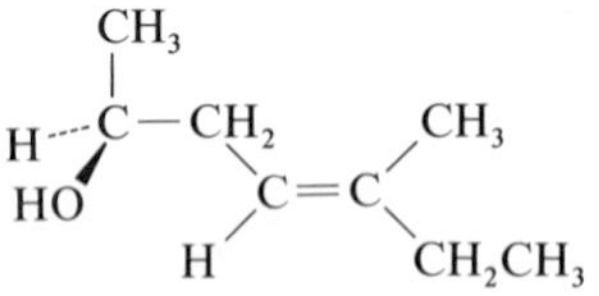

(i) 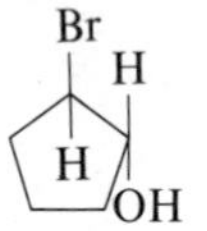 (j) 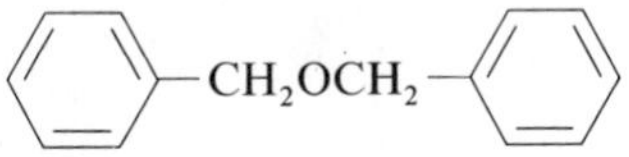

(k)  (l) 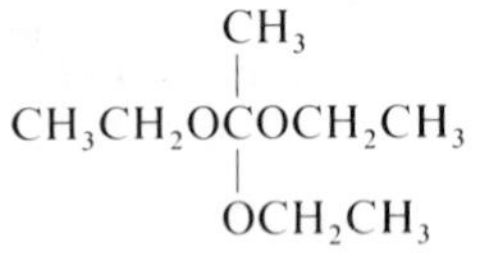

12.32 Draw structures for the following compounds. When necessary, show stereochemistry by the use of appropriate conventions.

(a) 1-chloro-1-ethoxyethene (b) (*E*)-1-methoxy-2-propoxyethene
(c) 1,3-dichloro-2-propanol (d) cyclobutylmethanol
(e) (*R*)-3-methyl-5-hexen-3-ol (f) (*S*)-2-chloro-1-propanol
(g) (*S*)-3-methyl-1-pentyn-3-ol (h) 2-nitroethanol
(i) (*R*)-5,5-dimethyl-3-heptanol (j) (3*S*,4*R*)-4-methyl-3-hexanol

12.33 Write equations for the reactions of 1-propanol with the following reagents under the conditions shown.

(a) HBr, Δ (b) PBr_3 (c) $SOCl_2$, pyridine (d) NaH
(e) H_3O^+, cold (f) H_2SO_4, Δ (g) Na metal, then $CH_3CH_2CH_2CH_2Br$
(h) $Na_2Cr_2O_7$, H_2SO_4, Δ (i) pyridinium chlorochromate, dichloromethane

12.34 Write equations for the reactions of 2-hexanol with the following reagents under the conditions shown.

(a) HBr, Δ (b) PBr_3 (c) $SOCl_2$, pyridine (d) NaH
(e) $ZnCl_2$, HCl (f) H_2SO_4, Δ (g) $Na_2Cr_2O_7$, H_2SO_4, Δ
(h) pyridinium chlorochromate, dichloromethane
(i) CrO_3, H_2O, H_2SO_4, acetone as solvent (see Problem 12.17, p. 441)

12.35 The two columns on the left contain structural formulas for some organic compounds. The right-hand column is a list of reagents that may react with some of them. For each compound on the left, list all the reagents on the right that would produce a reaction detectable by the human senses when mixed with the compound in a test tube. (The human senses, for example, can detect a change in color, the formation of a solution, the evolution of a gas, the evolution of heat, or a change in odor.)

(a) 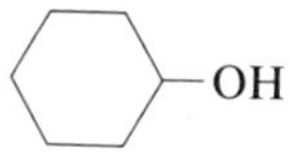

(b) 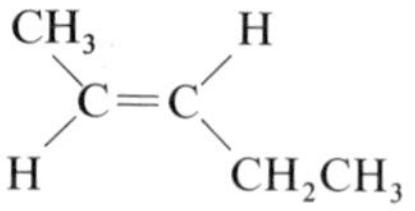

(c) 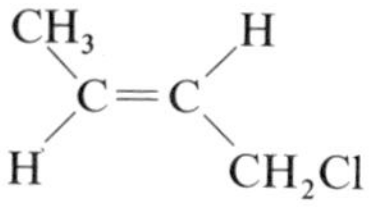

(d) 

(e) 

(f) $CH_3CH_2CH_2CH_2CH_2OH$

(g) $CH_3C{\equiv}CCH_2CH_3$

(h) 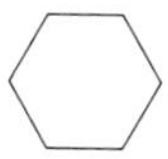

(i) $HOCH_2CH_2OH$

(j) $CH_3CH_2C(CH_3)(Br)CH_3$

(k) 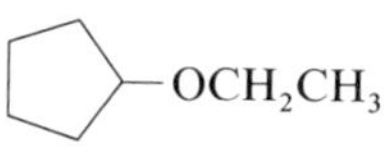

(l) $CH_3CH_2CH_2CH_2CH_2CH_2Cl$

1. H_2O, cold (solubility)
2. H_2SO_4, cold, conc (protonation)
3. Br_2, carbon tetrachloride
4. CrO_3, H_2O, H_2SO_4
5. $AgNO_3$, CH_3CH_2OH
6. NaI, acetone

12.36 For each set of experimental facts presented below, give an explanation for the observed trend, based on the relationships between structure and physical properties.

(a)

$CH_3CH_2CH_2CH_2OH$	$CH_3CH_2CH(OH)CH_3$	$CH_3C(CH_3)(OH)CH_3$
solubility 7.9 g in 100 mL H_2O	solubility 12.5 g in 100 mL H_2O	infinitely soluble in H_2O

(b)

$CH_3CH_2CH(OH)CH_3$	$CH_3CH_2CH(Cl)CH_3$	$CH_3CH_2CH(Br)CH_3$
solubility 12.5 g in 100 mL H_2O	very slightly soluble in H_2O	insoluble in H_2O

(c)

$CH_3CH_2CH(CH_2CH_3)CH_2OH$	$CH_3CH_2CH(CH_2CH_3)OC(=O)CH_3$	$CH_3CH_2CH(NH_2)CH_2OH$
solubility 0.63 g in 100 mL H_2O	solubility 0.06 g in 100 mL H_2O	infinitely soluble in H_2O

(d)

CH_3CH_2OH	$CH_3C(CH_3)(OH)CH_3$	CF_3CH_2OH
pK_a 15.9	$pK_a \sim 18$	pK_a 12.4

(e)

$CH_3CH_2CH_2CH_2OH$	$CH_3CH_2CH_2CH_2CH_2CH_2OH$	$CH_3CH(OH)CH(OH)CH_2CH_2CH_3$
bp 118 °C solubility 7.9 g in 100 mL H_2O	bp 157 °C solubility 0.59 g in 100 mL H_2O	bp 207°C infinitely soluble in H_2O

(f)

$CH_3CH_2CH_2CH_2OH$	$CH_3CH_2CH_2CH_2SH$
solubility 7.9 g in 100 mL H_2O	slightly soluble in H_2O

12.37 Complete the following equations.

(a) $CH_3CH(CH_3)CH_2OH \xrightarrow[\text{pyridine}]{SOCl_2}$

(b) $CH_3C(CH_3)(OH)CH_2CH_3 \xrightarrow{\text{HCl (conc)}}$

(c) $CH_3CH_2CH(OH)CH_3 \xrightarrow{PBr_3}$

(d) 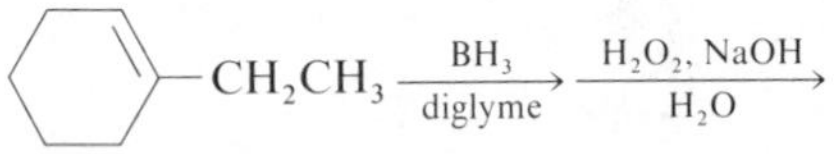

(e) $HOCH_2(CH_2)_4CH_2OH \xrightarrow{\text{HBr(g), excess}}$

(f) 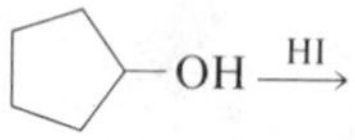

(g) 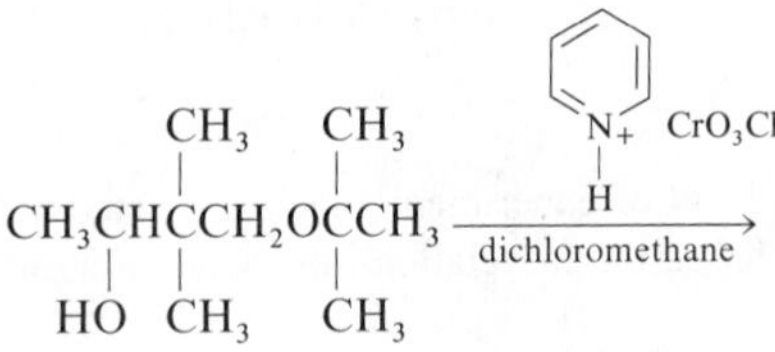

(h) 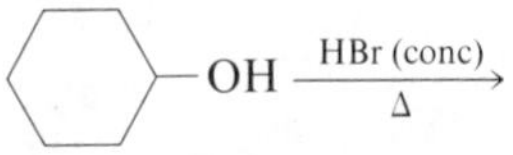

(i) 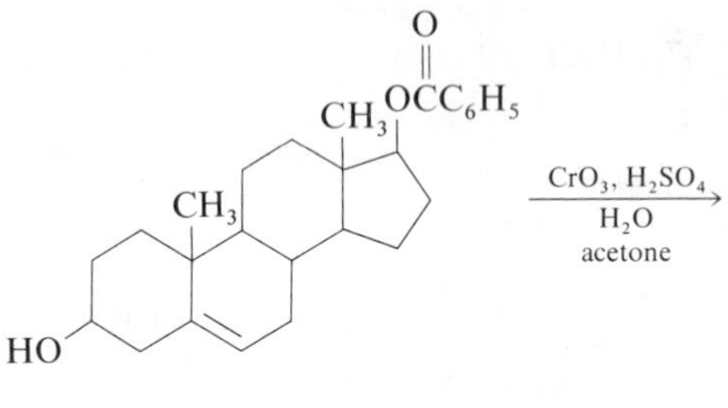

(j) 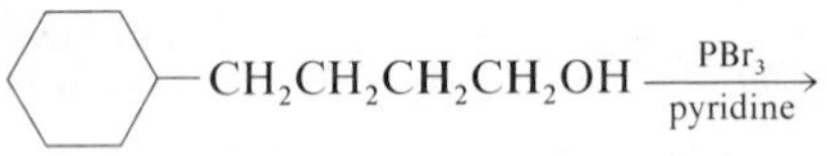

(k) 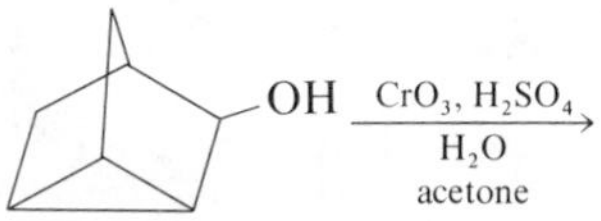

(l) 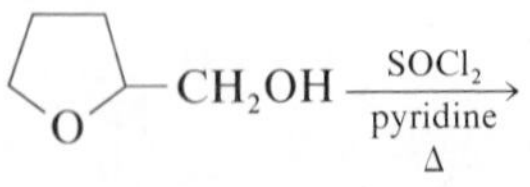

(m) 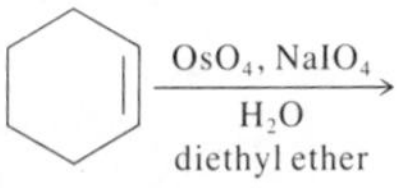

(n) 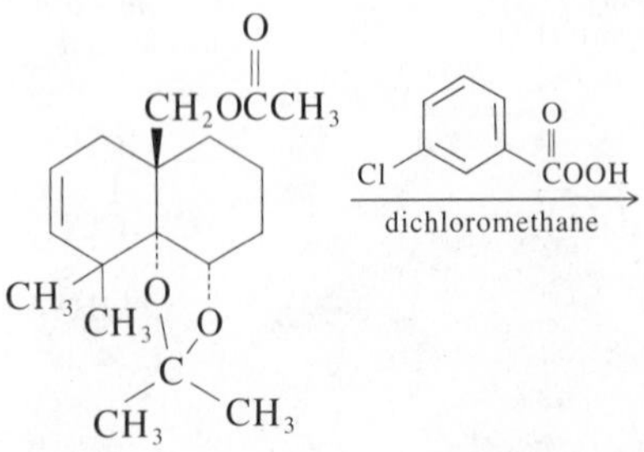

(o) 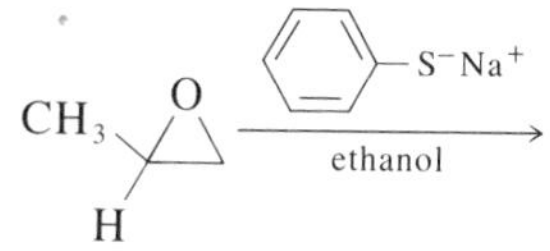
CH3
O
H
S⁻Na⁺
ethanol

(p) 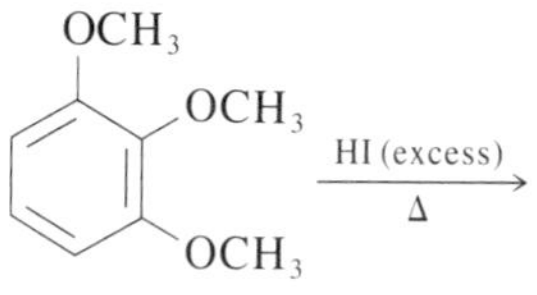
OCH3
OCH3
OCH3
HI (excess)
Δ

12.38 Predict the products corresponding to each of the capital letters. Show stereochemistry by appropriate conventions. When a racemic mixture forms, show the structure for one enantiomer and write "and enantiomer" below it.

(a) 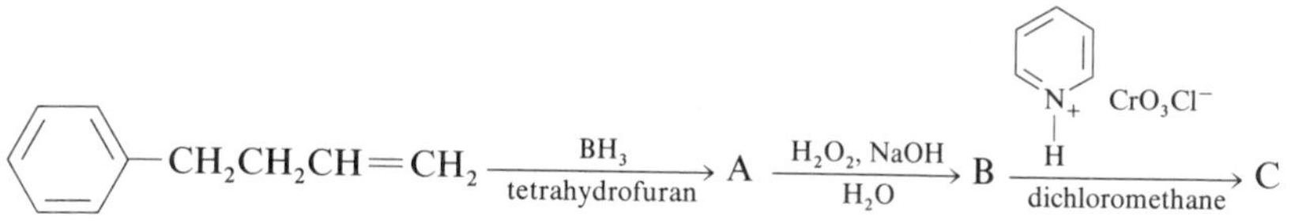
CH2CH2CH=CH2
BH3
tetrahydrofuran
A
H2O2, NaOH
H2O
B
N+ CrO3Cl⁻
H
dichloromethane
C

(b) 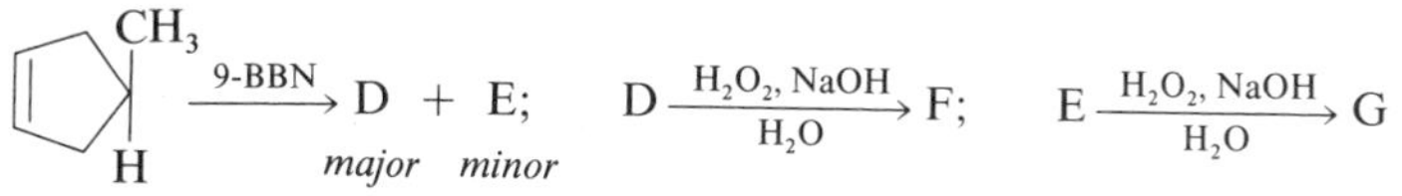
CH3
H
9-BBN
D + E;
major minor
D
H2O2, NaOH
H2O
F;
E
H2O2, NaOH
H2O
G

(c) 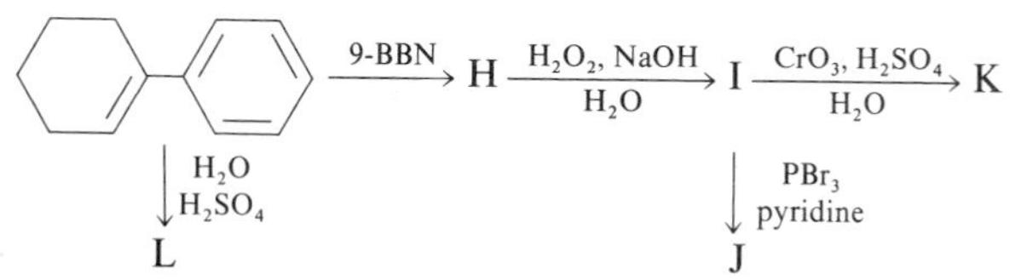
9-BBN
H
H2O2, NaOH
H2O
I
CrO3, H2SO4
H2O
K
H2O
H2SO4
L
PBr3
pyridine
J

(d) 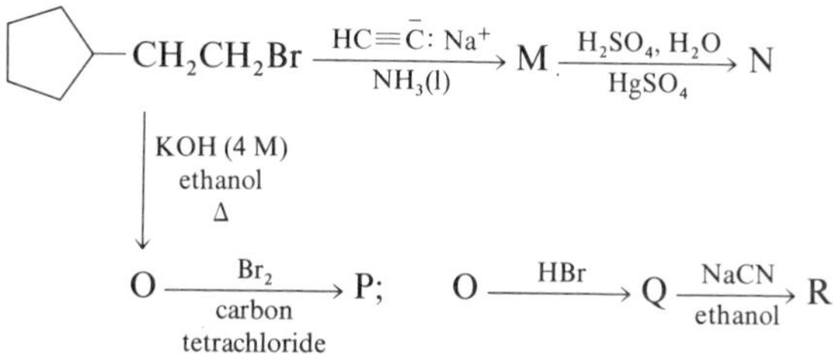
CH2CH2Br
HC≡C: Na⁺
NH3(l)
M
H2SO4, H2O
HgSO4
N
KOH (4 M)
ethanol
Δ
O
Br2
carbon
tetrachloride
P;
O
HBr
Q
NaCN
ethanol
R

(e) 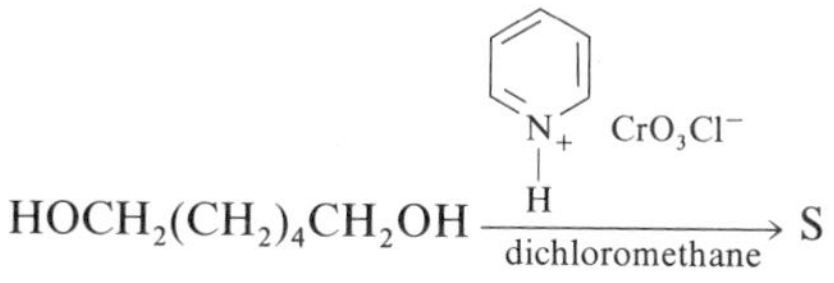
HOCH2(CH2)4CH2OH
N+ CrO3Cl⁻
H
dichloromethane
S

(f) 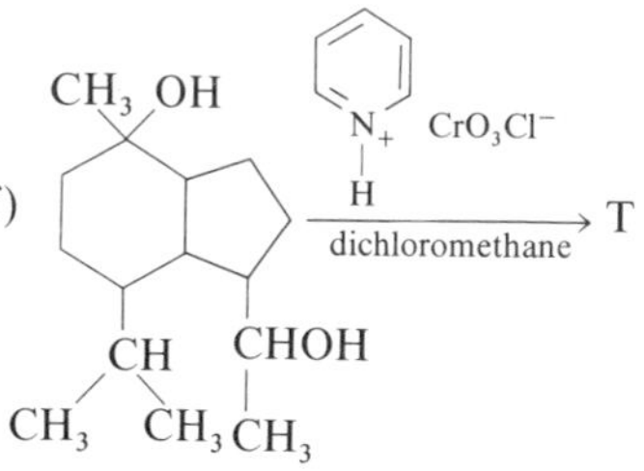
CH3 OH
CH
CHOH
CH3 CH3 CH3
N+ CrO3Cl⁻
H
dichloromethane
T

(g) 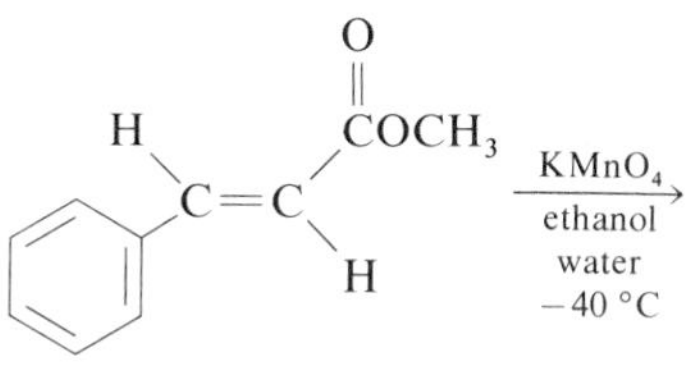
O
H
COCH3
C=C
H
KMnO4
ethanol
water
−40 °C
U

(h) 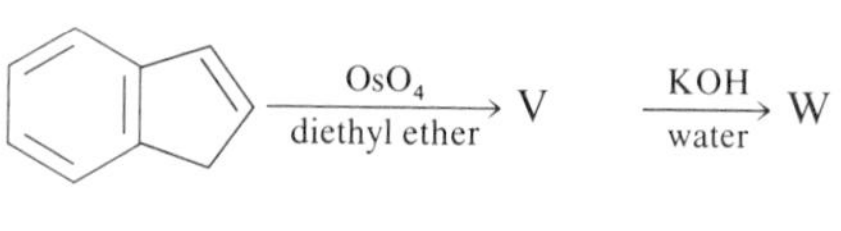
OsO4
diethyl ether
V
KOH
water
W

(i) 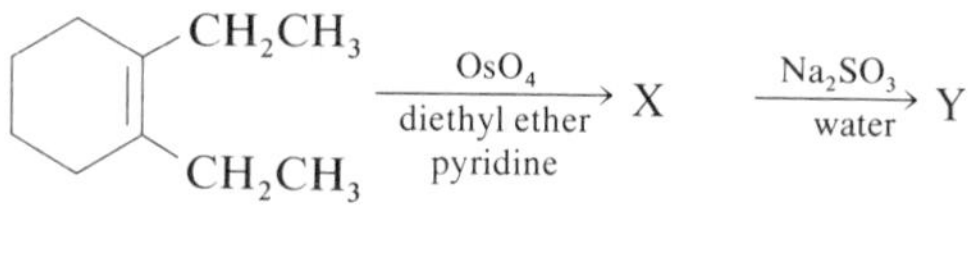
CH2CH3
CH2CH3
OsO4
diethyl ether
pyridine
X
Na2SO3
water
Y

(j) 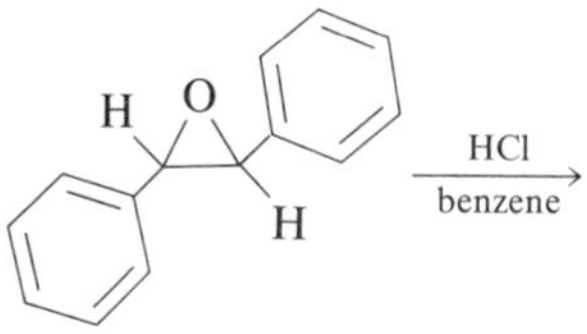
H
O
H
HCl
benzene
Z

12.39 Supply the reagents that are necessary and the intermediate compounds that will form in the following transformations. There may be more than one good way to carry out each synthesis.

(a) $CH_3CH(CH_3)CH_2CH{=}CH_2 \longrightarrow CH_3CH(CH_3)CH_2C(=O)CH_3$

(b) $CH_3(CH_2)_4CH{=}CH_2 \longrightarrow CH_3(CH_2)_4CH_2CH{=}O$

(c) 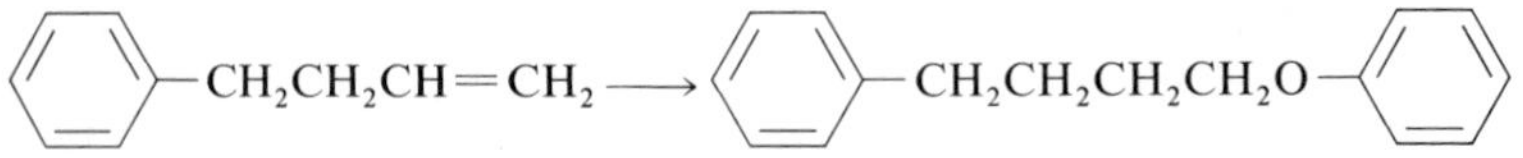

(d) $CH_3CH(CH_3)CH_2CH{=}CH_2 \longrightarrow (CH_3CH(CH_3)CH_2CH_2CH_2)_2O$

(e) 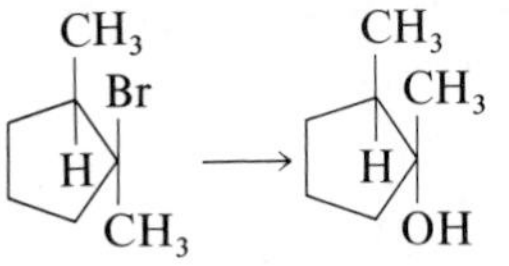

(f) 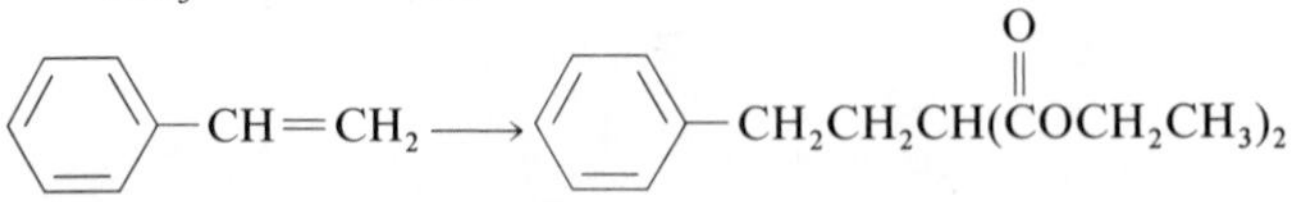

(g) $CH_3(CH_2)_3CH{=}CH_2 \longrightarrow CH_3(CH_2)_4CH_2OCH_3$

(h) 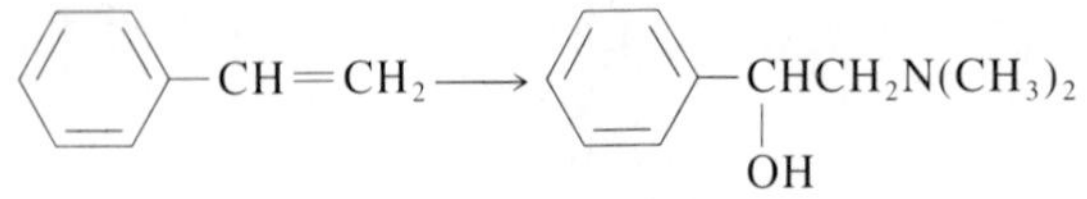

12.40 The following sequence of reactions is a good way to convert a 1-alkylcyclohexene to a 3-alkylcyclohexene. Draw structural formulas showing the stereochemistry for each step of the reaction sequence, and show why it gives the desired product.

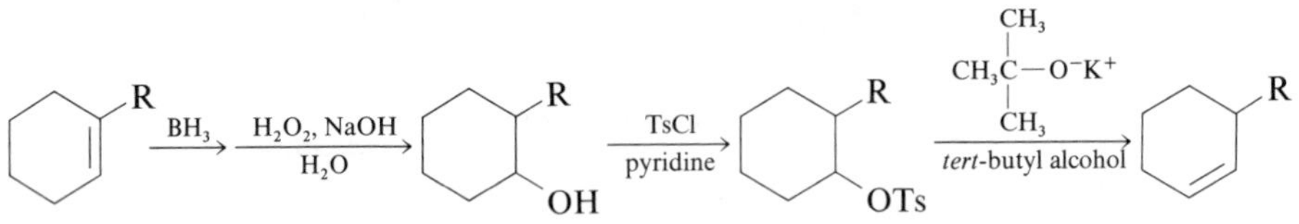
R
BH_3
H_2O_2, NaOH
H_2O
R
OH
TsCl
pyridine
R
OTs
CH_3
$CH_3C{-}O^-K^+$
CH_3
tert-butyl alcohol
R

12.41 An important component of the visual pigment in the eye is retinal, the aldehyde structurally related to Vitamin A. Retinal, as it exists in the retina before absorption of light, has the *Z* configuration at the double bond at carbon 11. Absorption of light converts it to the *E* configuration at that double bond.

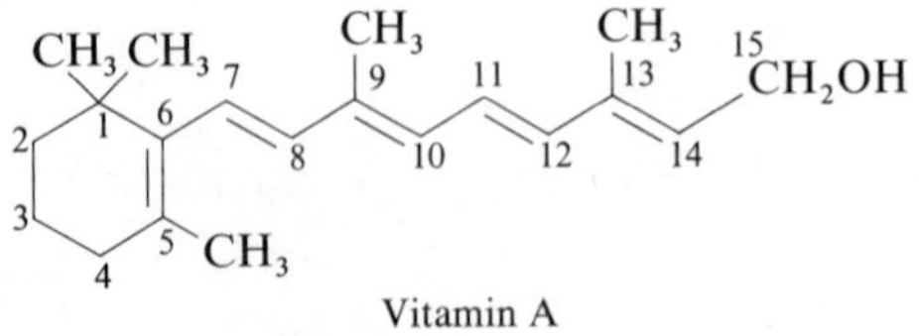
CH_3 CH_3
CH_3
CH_3
CH_2OH
CH_3
1 2 3 4 5 6 7 8 9 10 11 12 13 14 15

Vitamin A

(a) Draw the structure of (11*E*)-retinal.
(b) Draw the structure of (11*Z*)-retinal.
(c) What reagent would you use to oxidize Vitamin A to retinal?

12.42 Cholesterol gives primarily one product when it is treated first with diborane and then with alkaline peroxide. Predict the structure of the product, assuming that the side of the molecule away from the angular methyl groups (carbons 18 and 19), is less hindered.

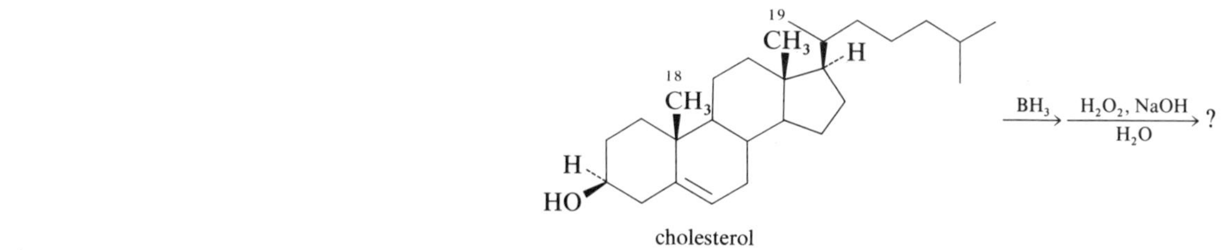

19
CH3
H
18
CH3
H
HO
BH3
H2O2, NaOH
H2O
?

cholesterol

12.43 The following transformation has been observed. Assign a structure to Compound A and suggest a mechanism for its conversion of A to the final product.

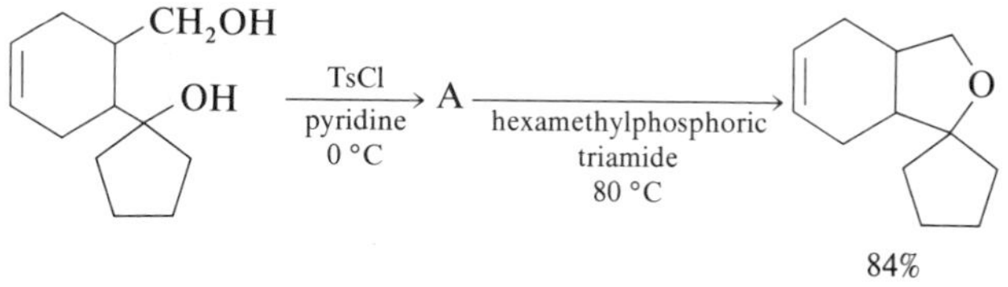

CH2OH
OH
TsCl
pyridine
0 °C
A
hexamethylphosphoric
triamide
80 °C
O
84%

12.44 An intermediate in the synthesis of cerulein, an antibiotic, was prepared in the following way. Fill in structural formulas for the missing compounds.

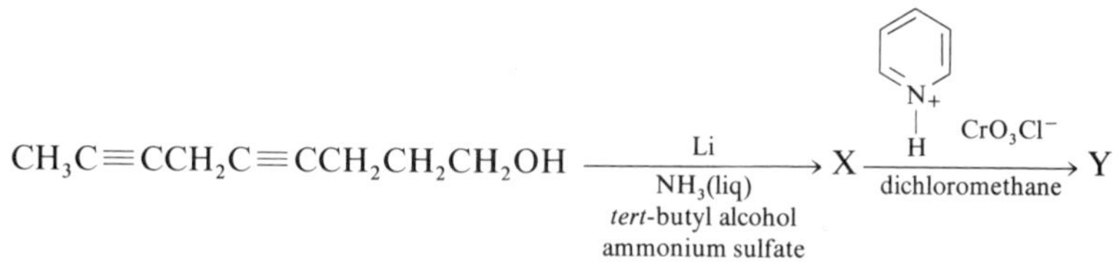

CH3C≡CCH2C≡CCH2CH2CH2OH
Li
NH3(liq)
tert-butyl alcohol
ammonium sulfate
X
N+
H
CrO3Cl−
dichloromethane
Y

12.45 2-Phenyloxirane is treated with sulfuric acid in methanol and, in a separate reaction, with sodium methoxide in methanol. Predict what the major product in each reaction will be and write detailed mechanisms to rationalize your predictions.

12.46 Cholesterol is converted into a saturated ketone called cholestanone in several steps. Supply the reagents necessary to make this transformation.

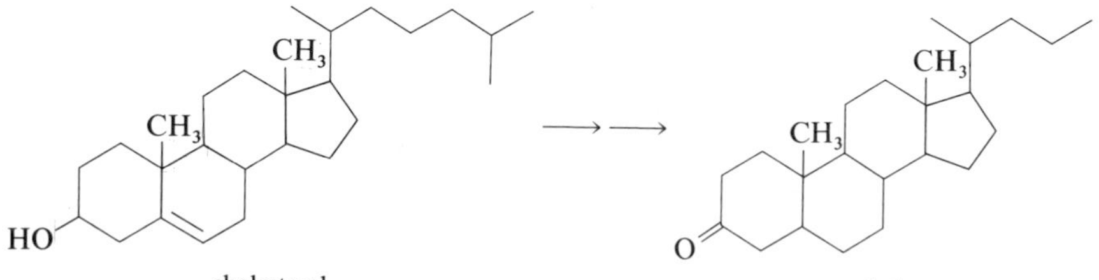

CH3
CH3
HO
cholesterol
CH3
CH3
O
cholestanone

12.47 Ethers can be prepared by the Markovnikov addition of alcohols to alkenes in the presence of strong acids with nonnucleophilic conjugate bases. *tert*-Butyl ethers are often prepared in this way. Write a complete mechanism for the addition of methanol to 2-methylpropene in the presence of fluoboric acid, HBF_4. Why must an acid with a non-nucleophilic conjugate base be used as the catalyst for this reaction?

12.48 Microorganisms reduce carbonyl compounds to alcohols stereoselectively. 5-Chloro-2-pentanone is reduced to (*S*)-5-chloro-2-pentanol in this way. Distillation of this alcohol over sodium hydride gives a chiral cyclic ether. Write stereochemically correct structural formulas for the alcohol and the cyclic ether.

12.49 2-Octyn-1-ol is needed as an intermediate in the synthesis of certain polyunsaturated acids that are related to prostaglandins (p. 844), which are implicated in the clotting of blood, inflammation, and allergic responses. 2-Octyn-1-ol is prepared by treating 2-propyn-1-ol with 2 equivalents of lithium amide and then treating the resulting intermediate with 1 equivalent of 1-bromopentane. The reaction mixture is then treated with dilute acid to convert organic ions into neutral species.

(a) Provide structures for the intermediate, A, and for 2-octyn-1-ol.

$$HC{\equiv}CCH_2OH + 2\,Li^+NH_2^- \xrightarrow[NH_3\,(liq)]{} A \xrightarrow{CH_3(CH_2)_3CH_2Br} \xrightarrow[H_2O]{H_3O^+} \text{2-octyn-1-ol}$$

(b) Another product could have been formed from the reaction of the intermediate with 1-bromopentane. What is its structure?
(c) Explain briefly why 2-octyn-1-ol and not the alternative product is the major one.
(d) What would have been the structure of the intermediate formed if *only* 1 equivalent of lithium amide had been used in part a?

12.50 The following hydroboration reaction takes place with almost total regioselectivity. Predict what the product of the reaction is and explain the observed regioselectivity.

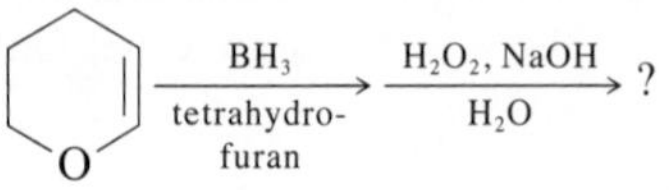
O
BH3
tetrahydro-
furan
H2O2, NaOH
H2O
?

12.51 Compounds having the following general structure are known as aryloxypropanolamines and belong to a class of medications known as β-blockers, which are used to treat hypertension and heart disease.

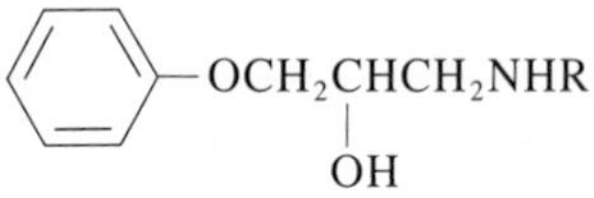
OCH2CHCH2NHR
OH

Using the following reactions, chemists synthesized model compounds in which the conformational relationships between the functional groups of the medications were known.

(a) Give structural formulas for the compounds indicated by letters, being sure to show correct stereochemistry.

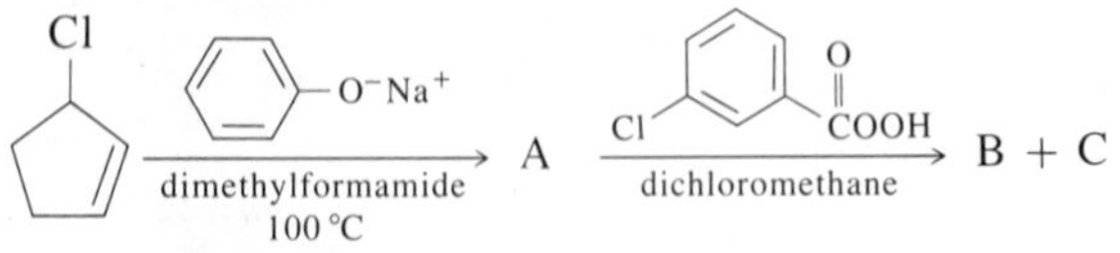
Cl
O⁻Na⁺
dimethylformamide
100 °C
A
Cl
O
COOH
dichloromethane
B + C

$$B \xrightarrow[\text{ethanol}]{CH_3CH(CH_3)NH_2} D \qquad C \xrightarrow[\text{ethanol}]{CH_3CH(CH_3)NH_2} E$$

major product   major product

(b) Compounds D and E are stereoisomers of each other. What is the stereochemical relationship between them? How many stereoisomers does D have altogether? What are they?

12.52 Consider the kinetics of the formation of oxirane from 2-chloroethanol and the formation of tetrahydrofuran from 4-chlorobutanol. These reactions were found to have the following heats and entropies of activation at 30 °C. Which reaction do you expect to be faster? Why? (Hint: $\Delta G^{\ddagger} = \Delta H^{\ddagger} - T\Delta S^{\ddagger}$)

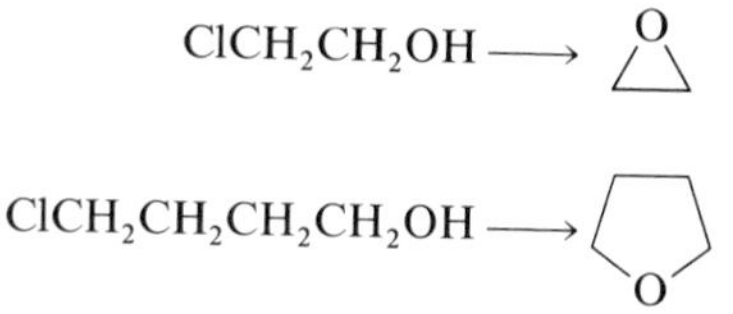

ClCH2CH2OH
O
ClCH2CH2CH2CH2OH
O

at 30 °C

$\Delta H^{\ddagger} = 23.2$ kcal/mol
$\Delta S^{\ddagger} = 9.9 \times 10^{-3}$ kcal/mol · deg

$\Delta H^{\ddagger} = 19.8$ kcal/mol
$\Delta S^{\ddagger} = -5.0 \times 10^{-3}$ kcal/mol · deg

12.53 Compound A, $C_5H_{10}O$, is oxidized to Compound B, C_5H_8O, with chromic acid. Their infrared spectra are given in Figure 12.1. Neither A nor B reacts with bromine in carbon tetrachloride or with dilute aqueous potassium permanganate solution. Assign possible structures to the compounds.

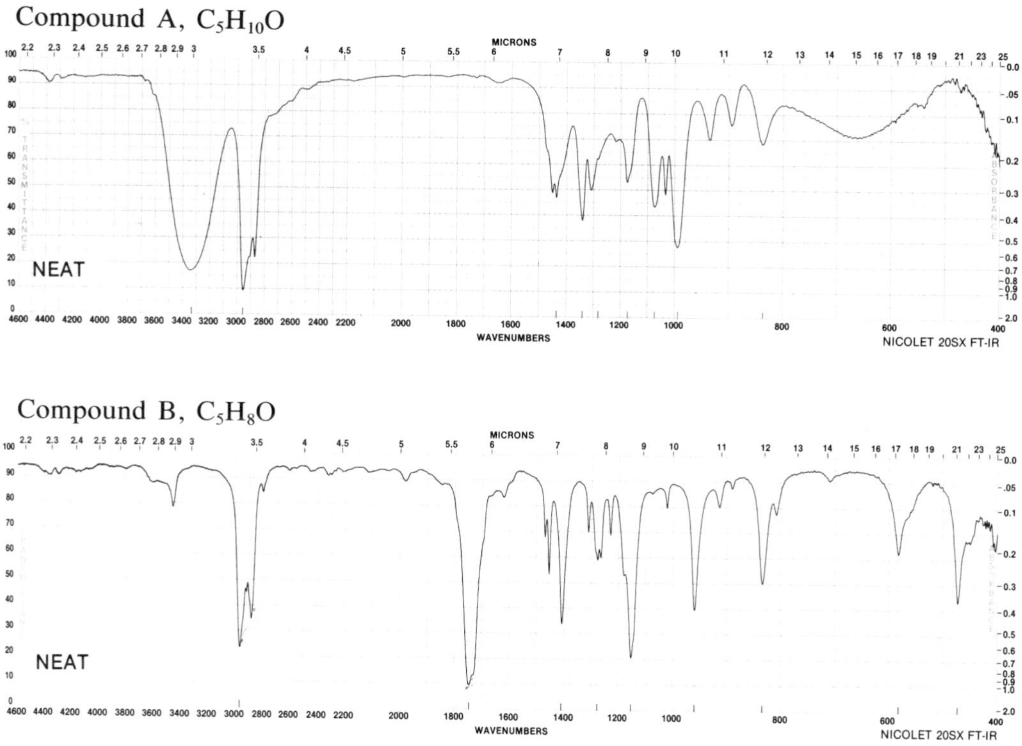

Compound A, C5H10O
MICRONS
TRANSMITTANCE
ABSORBANCE
NEAT
WAVENUMBERS
NICOLET 20SX FT-IR
Compound B, C5H8O
MICRONS
NEAT
WAVENUMBERS
NICOLET 20SX FT-IR

FIGURE 12.1

12.54 Compound C has the molecular formula $C_7H_{16}O$. Oxidizing it with pyridinium chlorochromate in dichloromethane gives Compound D, $C_7H_{14}O$. The infrared spectra of Compounds C and D are given in Figure 12.2. The proton magnetic resonance spectrum of C has bands at δ 0.9 (3 H, t), 1.3 (10 H, m), 2.2 (1 H, s), and 3.6 (2 H, t). Its carbon-13 nuclear magnetic resonance spectrum has absorption bands at 14.2, 23.1, 26.4, 29.7, 32.4, 33.2, and 62.2 ppm. The proton magnetic resonance spectrum of D has bands at δ 0.9 (3 H, t), 1.3 (8 H, m), 2.4 (2 H, m), and 9.8 (1 H, t). The most distinctive band in its carbon-13 spectrum is at 202.2 ppm. Assign structures to the compounds that are compatible with their spectra. What structures for Compound C are ruled out by the nuclear magnetic resonance data?

Compound C, $C_7H_{16}O$

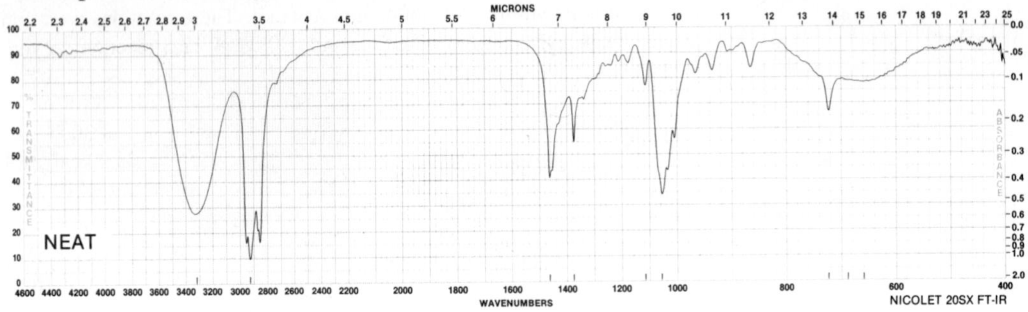
MICRONS
NEAT
WAVENUMBERS
NICOLET 20SX FT-IR

Compound D, $C_7H_{14}O$

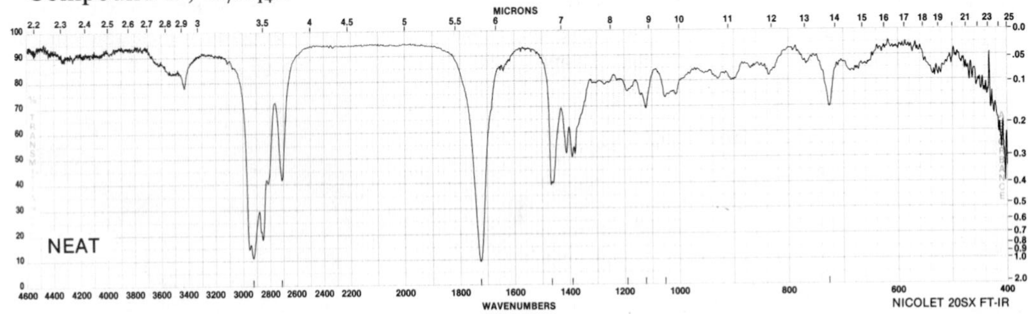
MICRONS
NEAT
WAVENUMBERS
NICOLET 20SX FT-IR

FIGURE 12.2

12.55 Compound E, $C_5H_{12}O$, has a strong band at 3400 cm^{-1} in its infrared spectrum. When it is dissolved in acetone and a solution of chromium trioxide in sulfuric acid and water is added, the solution remains orange. On the other hand, Compound F, also $C_5H_{12}O$, with a band at 3400 cm^{-1} in its infrared spectrum, turns the solution of chromium trioxide green under the same conditions. The proton magnetic resonance spectrum of E has a sharp, strong singlet at δ 1.2 and two multiplets at δ 0.9 and 1.5. Its carbon-13 nuclear magnetic resonance spectrum has bands at 8.8, 28.9, 36.8, and 70.6 ppm. The proton magnetic resonance spectrum of F has a complex multiplet at δ 0.9–1.4, a doublet at δ 1.2, a multiplet at δ 3.7. Bands appear in its carbon-13 nuclear magnetic resonance spectrum at 14.3, 19.4, 23.6, 41.9, and 67.3 ppm. Assign structures to the two compounds that are compatible with all the information given.

13

Aldehydes and Ketones. Reactions at Electrophilic Carbon Atoms

A • LOOK • AHEAD

Aldehydes and ketones contain the carbonyl group, in which a carbon atom is doubly bonded to an oxygen atom. The carbonyl group is highly polarized, with a very electrophilic carbon atom.

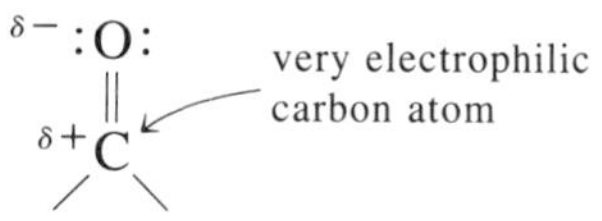
very electrophilic
carbon atom

Many reactions of aldehydes and ketones start with a nucleophilic attack at the carbon atom of the carbonyl group, by a nucleophile such as a cyanide ion:

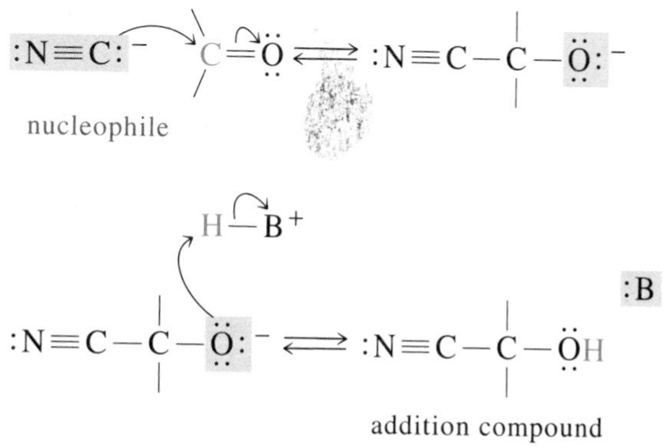
nucleophile
addition compound

As a result of the nucleophilic attack at the carbonyl group, the electrons of the carbon-oxygen double bond become localized on the oxygen atom. The alkoxide ion that forms is protonated in a subsequent step.

Many nucleophiles will react with the carbonyl group, and, depending on the reagents, the reaction conditions, and the nature of the intermediate formed, further reactions are possible before the final product is obtained. But the most important step in all of these reactions is the bonding between a nucleophile and the carbon atom of the carbonyl group.

13.1 INTRODUCTION TO CARBONYL COMPOUNDS

A. The Carbonyl Group

The carbonyl group, a carbon atom doubly bonded to an oxygen atom, determines most of the physical and chemical properties of aldehydes and ketones. In aldehydes, the carbonyl group is bonded to an organic group on one side and a hydrogen atom on the other. In ketones, it is bonded to two organic groups.

The carbon atom of the carbonyl group is sp^2-hybridized; therefore, the carbonyl group and the two atoms bonded to it lie in a plane (Figure 13.1). Oxygen is more electronegative than carbon, so the carbonyl group has a permanent dipole moment with the negative end of the dipole at the oxygen atom. For example, acetone, a typical ketone, has a dipole moment of 2.88 D. The high polarity of the carbonyl group is rationalized by pointing to a resonance contributor in which the carbon atom bears a positive formal charge and the oxygen atom a negative one.

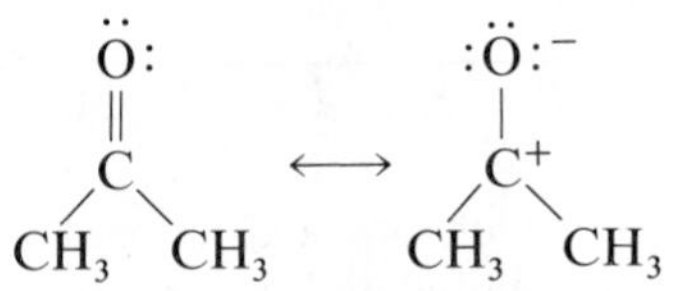

resonance contributors for acetone

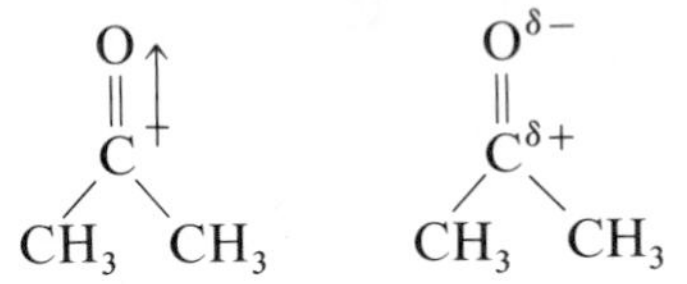

$\mu = 2.88$ D

different representations of the polarity of the carbonyl group

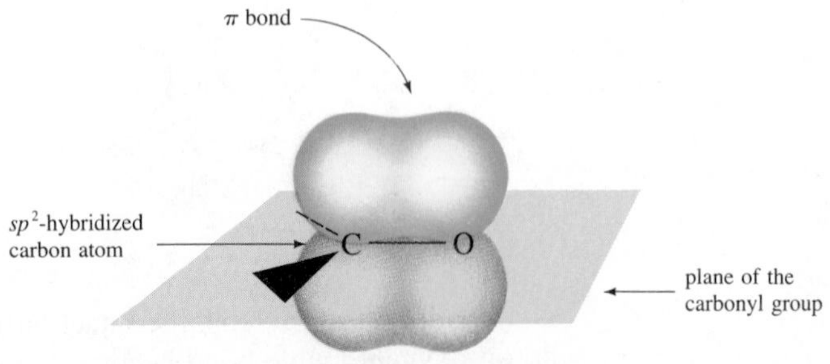
π bond
sp²-hybridized carbon atom
C
O
plane of the carbonyl group

FIGURE 13.1 The bonding and geometry of the carbonyl group.

The polarity of the carbonyl group makes low-molecular-weight aldehydes and ketones miscible with water. The oxygen atom acts as an acceptor of hydrogen bonds, and the carbonyl compound dissolves in water as long as the hydrocarbon portion of the molecule does not contain more than four or five carbon atoms. The kinds of interactions that occur between water molecules and acetone, an example of a low-molecular-weight carbonyl compound, are shown below.

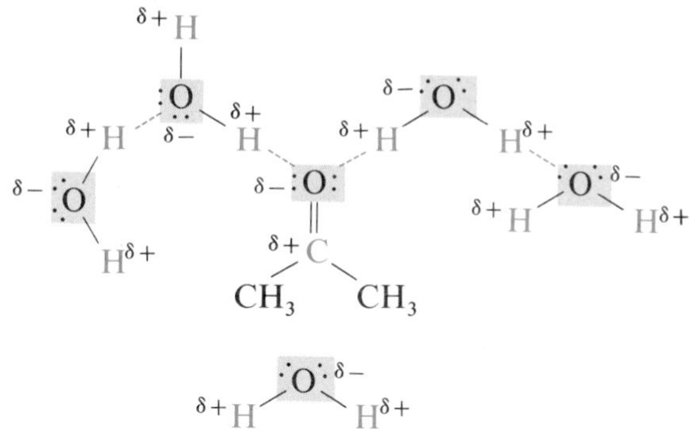

B. Carbonyl Compounds as Acids and Bases

Chapter 3 explored the basicity of the carbonyl group and showed how the non-bonding electrons on the oxygen atom can accept a proton from a protic acid or be shared with a Lewis acid. Protonation of the carbonyl group creates a positive charge at the oxygen atom and thereby enhances the electrophilicity of the carbonyl group and increases its reactivity.

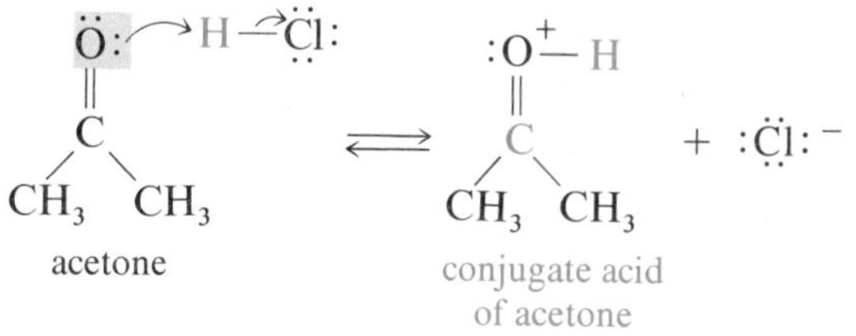
acetone
conjugate acid of acetone

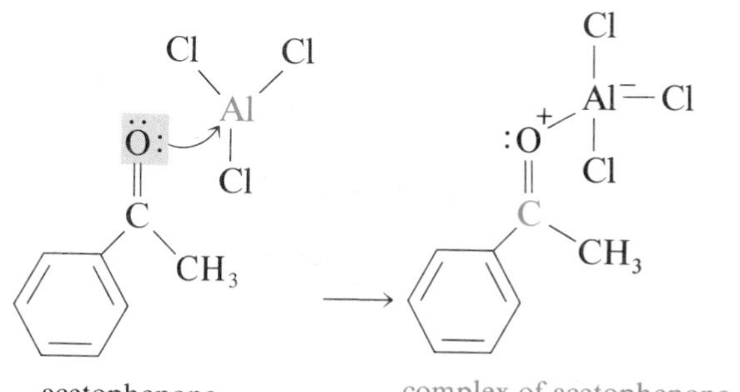
acetophenone
complex of acetophenone with a Lewis acid, aluminum chloride

Aldehydes and ketones are also weak acids. Acetone, for example, has a pK_a of 19. A hydrogen atom can be removed from the carbon atom adjacent to the carbonyl group by a base to give a carbanion. The carbanion is stabilized by the delocalization of electrons to the oxygen atom of the carbonyl group. The other resonance contributor for the carbanion is called an **enolate anion.**

VISUALIZING THE REACTION

Enolization of acetone

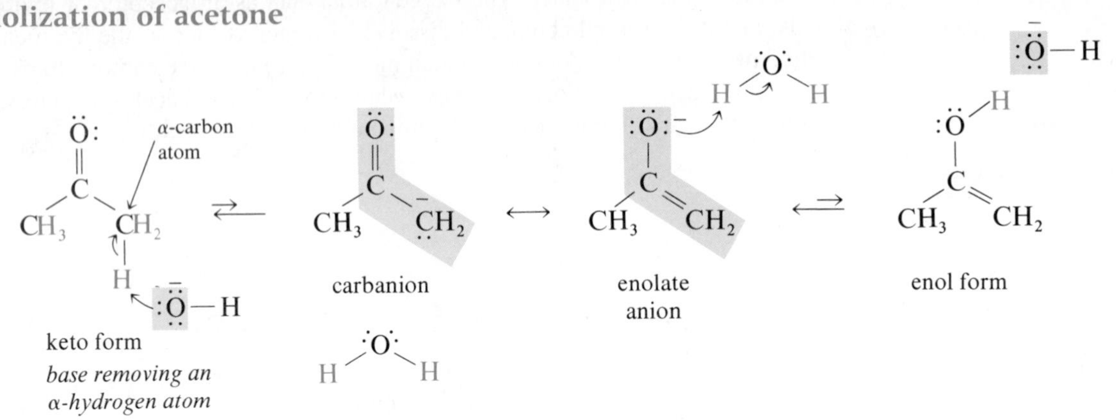
α-carbon atom
keto form
base removing an α-hydrogen atom
carbanion
enolate anion
enol form

The enolate anion, in which the negative charge is on the oxygen atom, is protonated to give an enol. The enol and keto forms of a carbonyl compound are tautomers (p. 341) existing in equilibrium with each other. For acetone, the form in which the hydrogen atom is on the carbon atom is the more stable one and predominates in the equilibrium mixture. For some compounds in which a hydrogen atom may be removed from a carbon atom that is between two carbonyl groups, the enol form becomes important. The reactions that carbonyl compounds undergo because of the acidity of the hydrogen atoms bonded to the carbon adjacent to the carbonyl group are explored extensively in Chapters 16 and 18.

PROBLEM 13.1

Decide whether each set of structural formulas shown below represents resonance contributors or tautomers.

(a) 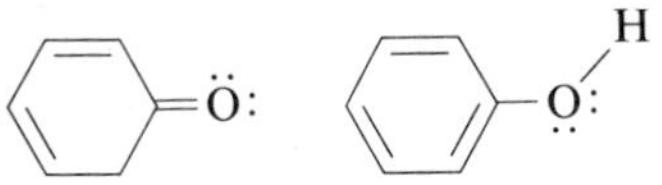

(b) 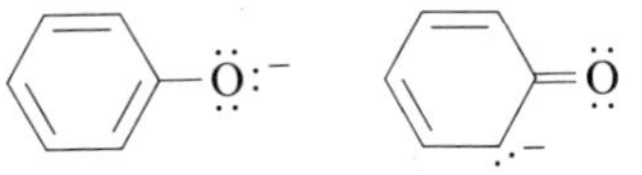

(c) 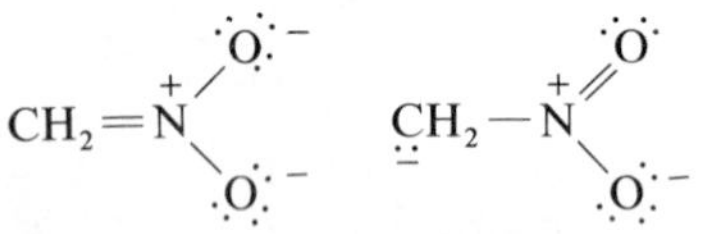

(d) 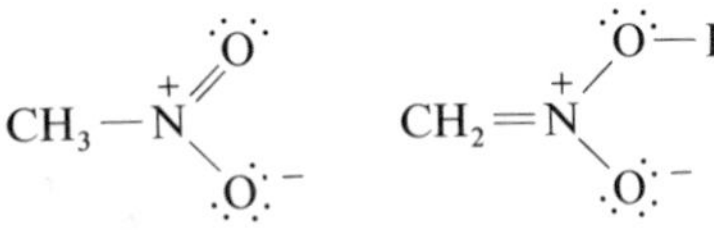

PROBLEM 13.2

For each of the following compounds, write equations showing the reactions you expect with (1) concentrated sulfuric acid and (2) sodium ethoxide. Show resonance contributors that explain the stabilization of any ions formed.

(a) $CH_3CH_2CH_2\overset{\overset{O}{\|}}{C}H$

(b) 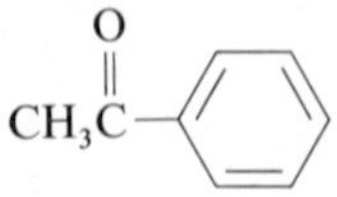

(c) $CH_3\overset{\overset{CH_3}{|}}{C}{=}CHCH_2CH_2\overset{\overset{O}{\|}}{C}CH_3$

(d) 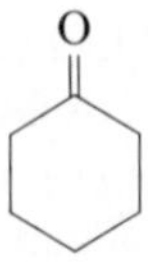

C. The Occurrence of Aldehydes and Ketones in Nature

Many compounds of biological interest are aldehydes or ketones. The most abundant natural aldehyde is glucose, the carbohydrate that is metabolized in the human body to produce energy. Many of the steroid hormones are ketones, including testosterone, the hormone that controls the development of male sex characteristics; progesterone, the hormone secreted at the time of ovulation in females; and cortisone, a hormone of the adrenal glands that is used medicinally to relieve inflammation. Another hormone from the adrenal glands, aldosterone, is an aldehyde as well as a ketone. Aldosterone is important in regulating the concentration of sodium ions in the body and thus water retention. Prednisone is an example of a synthetic drug; it is designed to be a substitute for cortisone in relieving the symptoms of rheumatoid arthritis with fewer undesirable side effects.

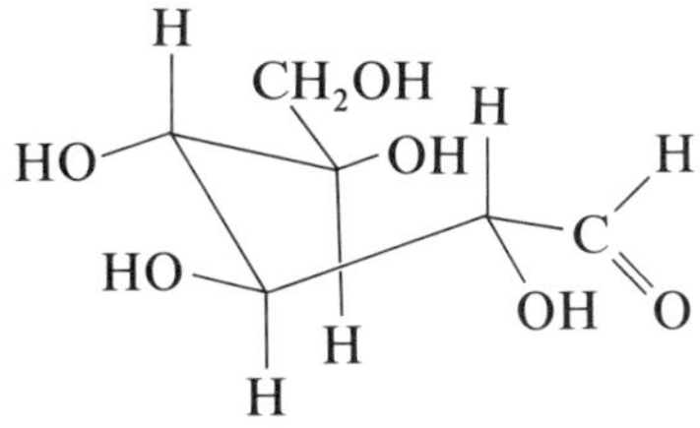

glucose

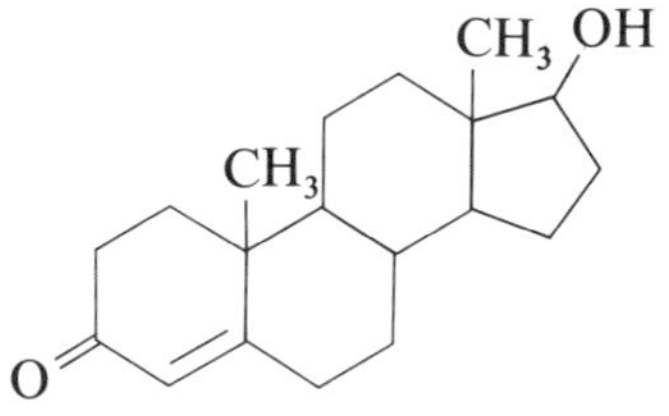

testosterone

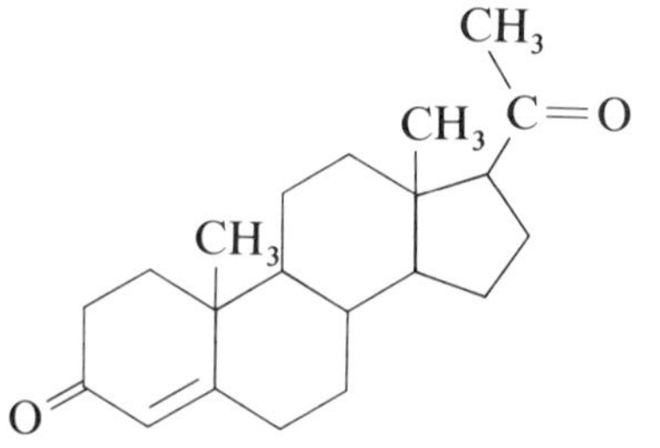

progesterone

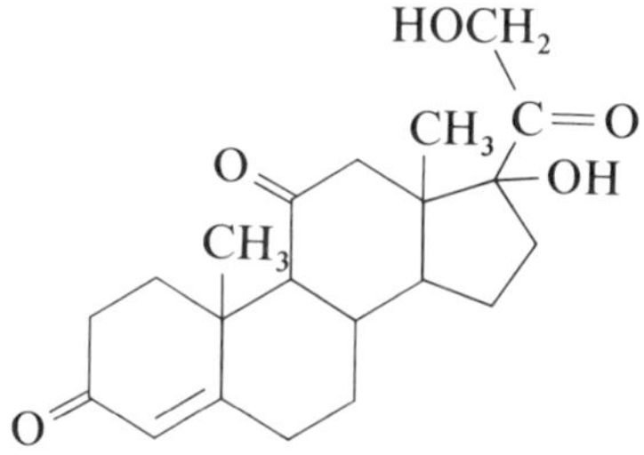

cortisone

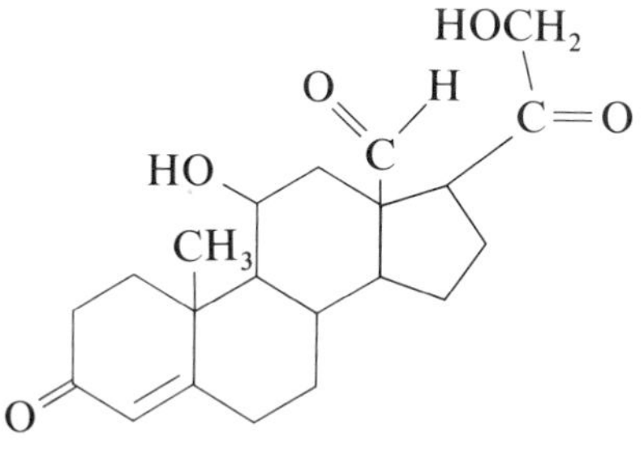

aldosterone

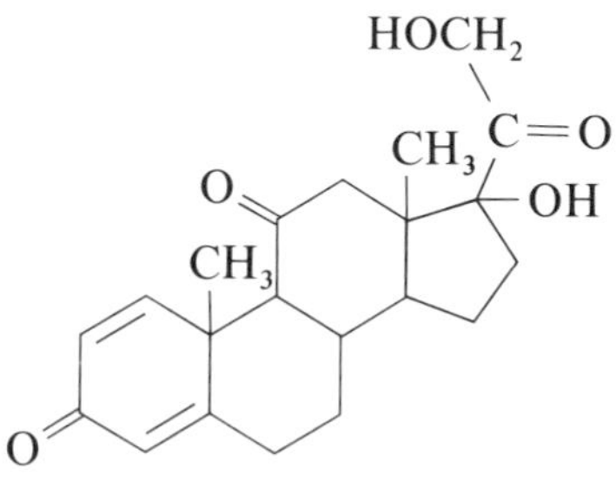

prednisone

Careful inspection of the structural formulas of the steroids shown above may lead you to wonder how molecules that are so similar to each other in structure can perform such different functions in the body. What makes testosterone a male hormone and progesterone a female one, for example, and the function of cortisone something else entirely? Contemplating such questions will give you an appreciation of the finely discriminating systems that exist in living organisms. Scientists have been unable to explain fully how the structures of hormones enable them to perform their physiological functions. Organic chemists and biochemists constantly refine ways of looking at such compounds in order to understand their interactions with large molecules, such as the proteins known as hormone receptors that occur at the surface of cells or inside them. The binding of the hormone to the receptor triggers a series of reactions that ultimately result in the physiological changes associated with that hormone. Interactions between hormones and receptors are described using the same concepts that chemists use in describing the reactivity of simpler compounds. Among the ideas chemists use as they try to understand the multiple reactions in the human body are polarity of bonds, hydrogen bonding, ionic interactions, van der Waals forces between nonpolar portions of molecules, transfer of protons from one basic site to another one, and nucleophilic centers reacting with electrophilic sites.

13.2 NOMENCLATURE OF CARBONYL COMPOUNDS

A. Nomenclature of Aldehydes and Ketones

Simple aldehydes are named systematically by changing the **e** at the end of the name of the hydrocarbon having the same number of carbon atoms to **al.** The aldehyde function is at the first carbon atom of the chain. Other substituents are named using prefixes and numbered to indicate their positions relative to the aldehyde group. The common names of the two simplest aldehydes, formaldehyde and acetaldehyde, are almost always used rather than the systematic names.

methanal
formaldehyde

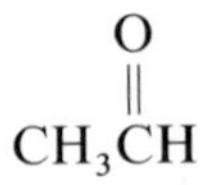

ethanal
acetaldehyde

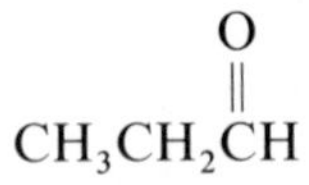

propanal

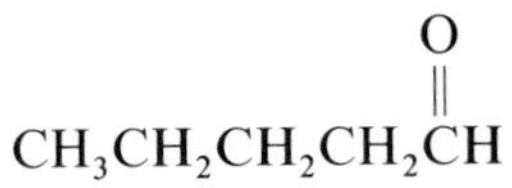

pentanal

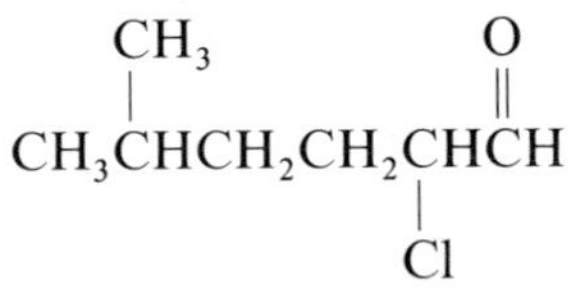

2-chloro-5-methylhexanal

If the aldehyde group is a substituent on a ring, the suffix **-carbaldehyde** is used in the name.

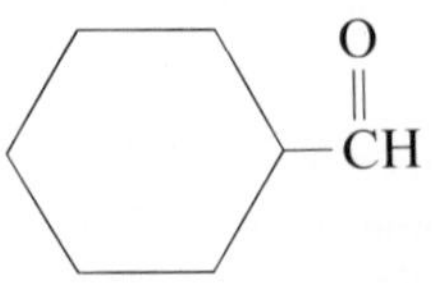

cyclohexanecarbaldehyde

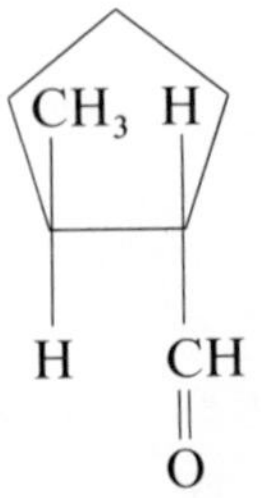

trans-2-methylcyclopentanecarbaldehyde

Note that in these cases the aldehyde group is considered to be a substituent. The carbon atom of the carbonyl group is not counted in assigning a name to the hydrocarbon portion of the molecule.

Benzene on which an aldehyde group is substituted is usually called benzaldehyde. A second substituent on the aromatic ring of benzaldehyde can have one of three positions in relation to the aldehyde group. A second substituent at carbon atom 2 of the ring is designated as **ortho** (abbreviated *o*- in the name of the compound to the carbonyl group). One at carbon atom 3 is **meta** (*m*-), and one at carbon atom 4 is **para** (*p*-).

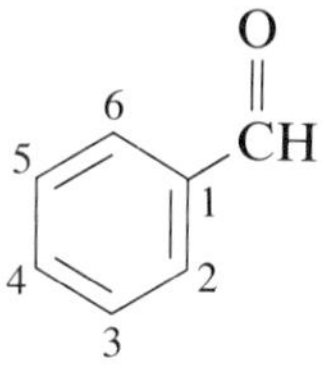

benzaldehyde

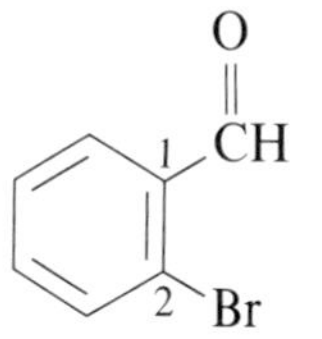

o-bromobenzaldehyde

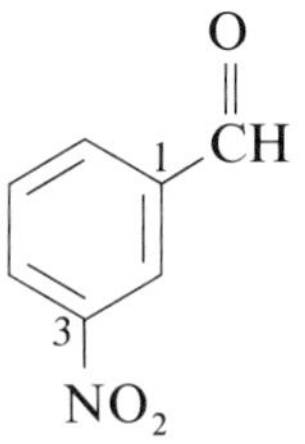

m-nitrobenzaldehyde

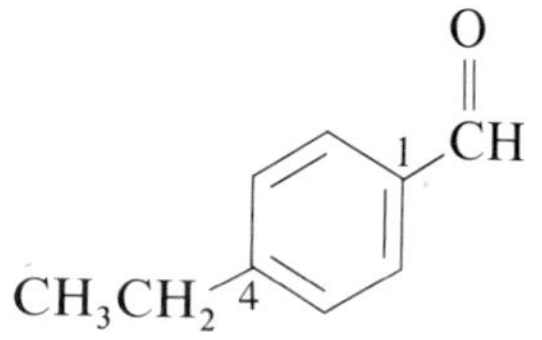

p-ethylbenzaldehyde

Ketones are named according to the IUPAC rules by substituting the ending **one** for the final **e** of the name of the hydrocarbon having the same number of carbon atoms and using a number to indicate the position of the carbonyl group. Other substituents are named using prefixes, and their positions are indicated by numbers. In a few cases, it is useful to name the organic groups that are present as substituents on the carbonyl group, which is then designated by the word **ketone.** A few ketones, such as acetone, are generally known by their common names.

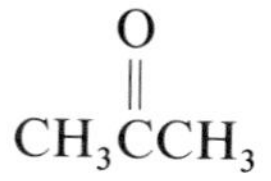

propanone
acetone

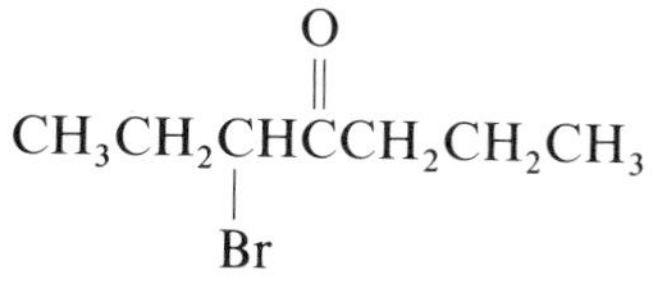

3-bromo-4-heptanone

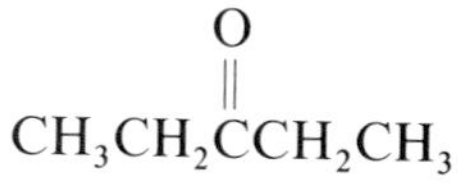

3-pentanone

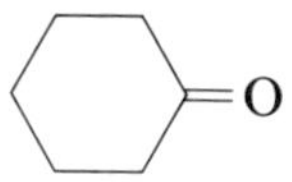

cyclohexanone

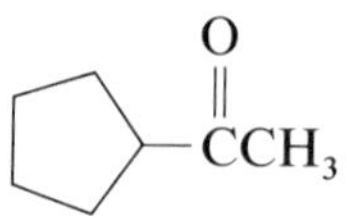

cyclopentyl methyl ketone

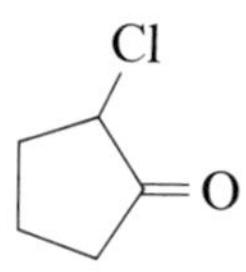

2-chlorocyclopentanone

Ketones having a phenyl group adjacent to the carbonyl group have common names ending in **phenone.** Two of these common names that are widely used by chemists are acetophenone and benzophenone. Other ketones containing aromatic rings are generally named systematically or as substituted ketones.

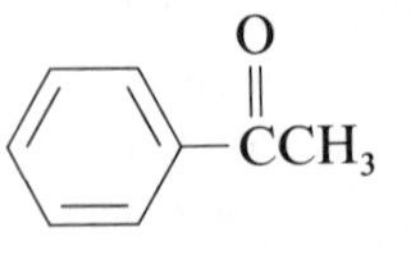

acetophenone

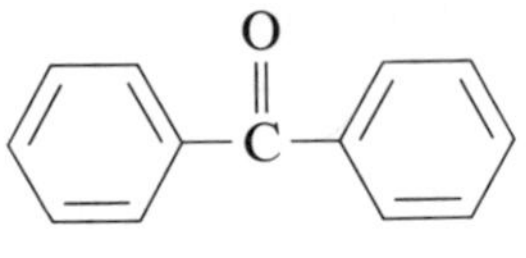

benzophenone

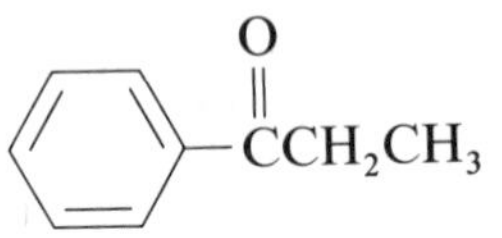

1-phenyl-1-propanone
ethyl phenyl ketone
propiophenone

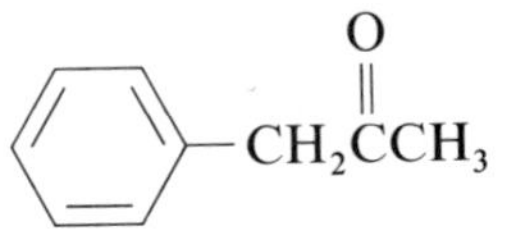

1-phenyl-2-propanone
benzyl methyl ketone

B. Nomenclature of Polyfunctional Compounds

When a compound contains more than one major functional group, the suffix for only one of them can be used as the ending of the name. The IUPAC rules establish priorities that specify which suffix is used. The priorities among the common functional groups are given in Table 13.1. The presence of double or triple bonds in a molecule is indicated by naming the chain (or ring) as an alkene or alkyne. Halogen substituents and alkoxyl groups are always named using prefixes. Thus, an aldehyde or a ketone function takes precedence over an alkene or an alcohol function as the suffix used in naming the compound. These examples illustrate these priorities of nomenclature:

TABLE 13.1 Decreasing Order of Precedence of Principal Functional Groups Used as Suffixes in Nomenclature

Chemical Structure	Functional Group	Chemical Structure	Functional Group
$-\overset{\overset{\displaystyle O}{\|}}{C}OH$	carboxylic acid	$-\overset{\overset{\displaystyle O}{\|}}{C}H$	aldehyde
$-\underset{\underset{\displaystyle O}{\|}}{\overset{\overset{\displaystyle O}{\|}}{S}}OH$	sulfonic acid	$-\overset{\overset{\displaystyle O}{\|}}{C}-$	ketone
$-\overset{\overset{\displaystyle O}{\|}}{C}OR$	ester	$-OH$	alcohol
$-\overset{\overset{\displaystyle O}{\|}}{C}NH_2$	amide	$-NH_2$	amine
$-C\equiv N$	nitrile	$-OR$	ether

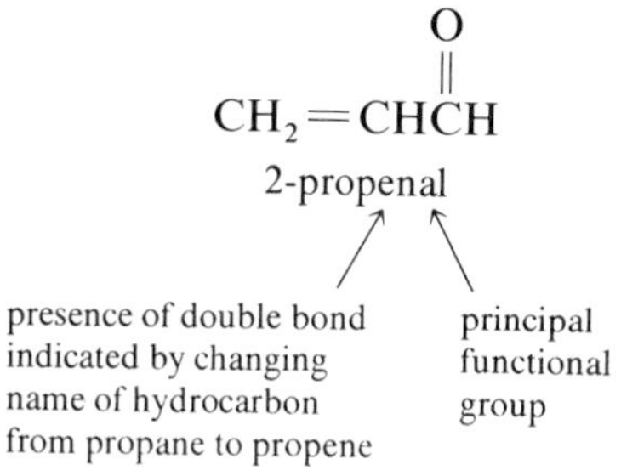
CH2=CHCH with C=O
2-propenal
presence of double bond indicated by changing name of hydrocarbon from propane to propene
principal functional group

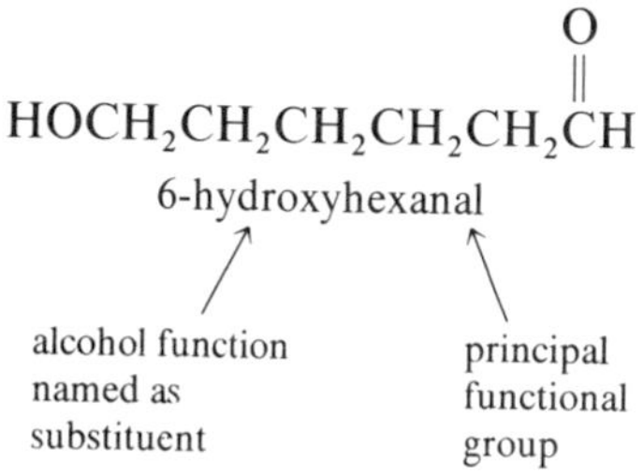
HOCH2CH2CH2CH2CH2CH with C=O
6-hydroxyhexanal
alcohol function named as substituent
principal functional group

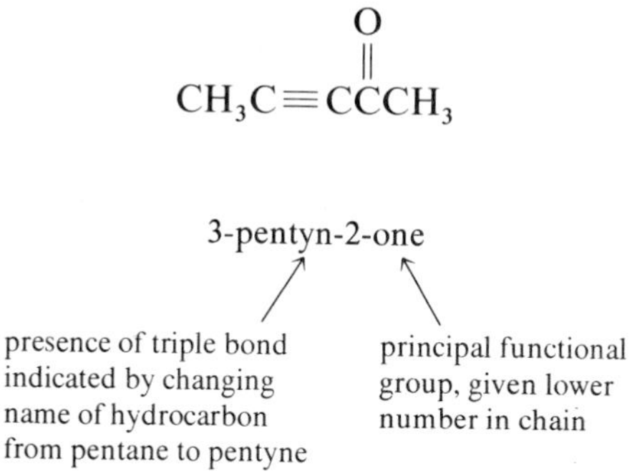
CH3C≡CCCH3 with C=O
3-pentyn-2-one
presence of triple bond indicated by changing name of hydrocarbon from pentane to pentyne
principal functional group, given lower number in chain

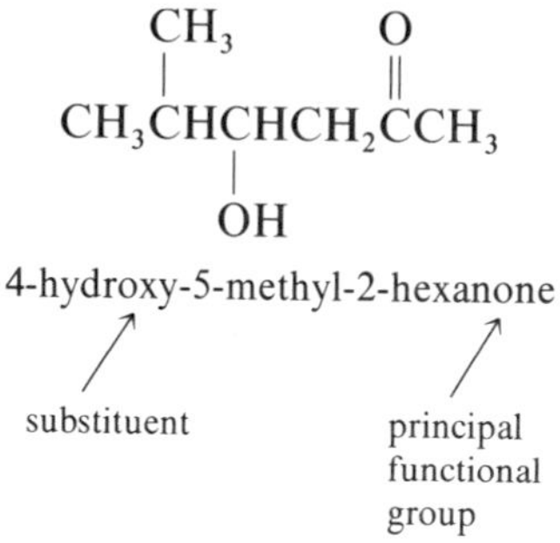
CH3
O
CH3CHCHCH2CCH3
OH
4-hydroxy-5-methyl-2-hexanone
substituent
principal functional group

When it is necessary to name the carbonyl group as a substituent, the prefix **oxo** with a number to indicate the position is used. The prefix oxo is used to indicate either an aldehyde or a ketone. This type of nomenclature is needed when a functional group of higher priority than an aldehyde or ketone, such as a carboxylic acid or ester, is present.

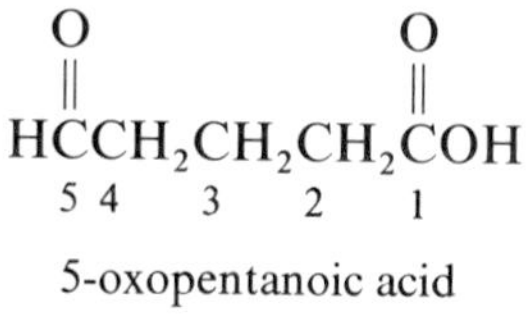
O O
HCCH2CH2CH2COH
5 4 3 2 1
5-oxopentanoic acid

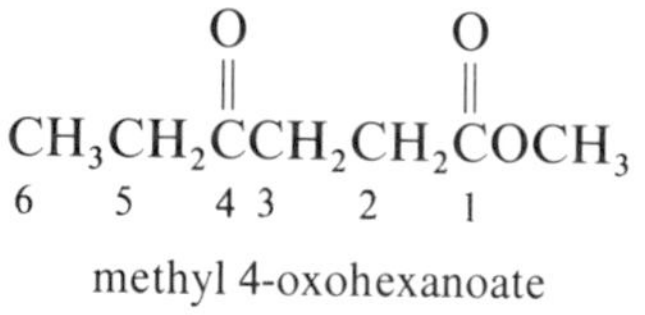
O O
CH3CH2CCH2CH2COCH3
6 5 4 3 2 1
methyl 4-oxohexanoate

Greek letters are used to designate positions on a carbon chain in relation to a carbonyl group for describing positions of special reactivity or relationships between functional groups. For example, the carbon atoms adjacent to the carbonyl group are called the α-carbon atoms and the hydrogen atoms on them, the α-hydrogen atoms (p. 475).

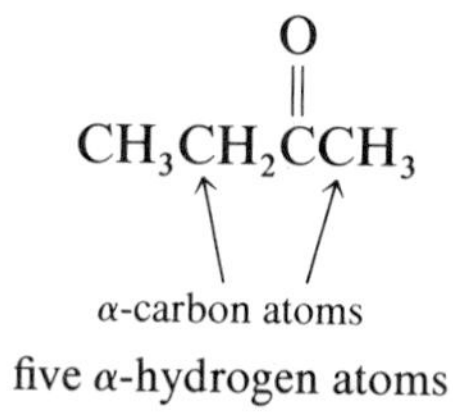
O
CH3CH2CCH3
α-carbon atoms
five α-hydrogen atoms

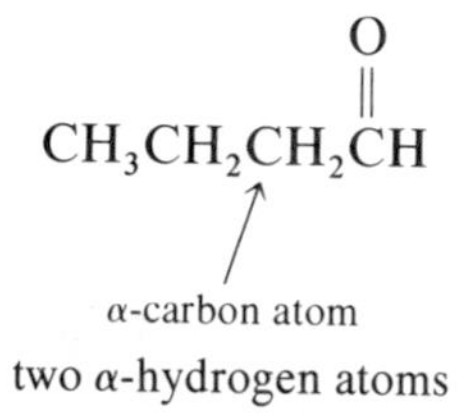
O
CH3CH2CH2CH
α-carbon atom
two α-hydrogen atoms

The positions alpha to a carbonyl group are significant because of the high reactivity of the α-hydrogen atoms (p. 475). The four positions farther from the carbonyl group than the α-carbon are designated, in order, by the Greek letters β (beta), γ (gamma), δ (delta), and ϵ (epsilon). This nomenclature is also used with substituted carboxylic acids.

$$-\underset{\epsilon}{CH_2}\underset{\delta}{CH_2}\underset{\gamma}{CH_2}\underset{\beta}{CH_2}\underset{\alpha}{CH_2}-\overset{\overset{\displaystyle O}{\|}}{C}-$$

$$\underset{\gamma}{CH_3}\overset{\overset{\displaystyle OH}{|}}{\underset{\beta}{C}}H\underset{\alpha}{CH_2}\overset{\overset{\displaystyle O}{\|}}{C}H \qquad \underset{\alpha}{CH_3}\overset{\overset{\displaystyle O}{\|}}{C}\underset{\alpha}{CH_2}Br \qquad \underset{\delta}{CH_3}\overset{\overset{\displaystyle O}{\|}}{C}\underset{\gamma}{CH_2}\underset{\beta}{CH_2}\overset{\overset{\displaystyle O}{\|}}{\underset{\alpha}{C}}OH$$

3-hydroxybutanal, *a β-hydroxyaldehyde* — 1-bromo-2-propanone, *an α-bromoketone* — 4-oxopentanoic acid, *a γ-ketoacid*

PROBLEM 13.3

Name the following compounds according to the IUPAC rules, including stereochemistry wherever it is shown.

(a) $CH_3CH_2\underset{\underset{\displaystyle OH}{|}}{C}H\overset{\overset{\displaystyle CH_3}{|}}{C}H-\overset{\overset{\displaystyle O}{\|}}{C}H$ (b) $CH_3\overset{\overset{\displaystyle O}{\|}}{C}CH_2CH_2CH_3$

(c) 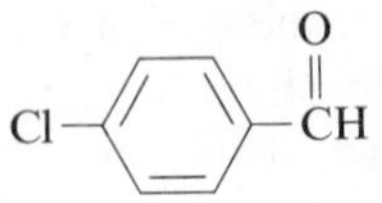 (d) $CH_3CH_2C\equiv C\overset{\overset{\displaystyle O}{\|}}{C}CH_2CH_3$

(e) 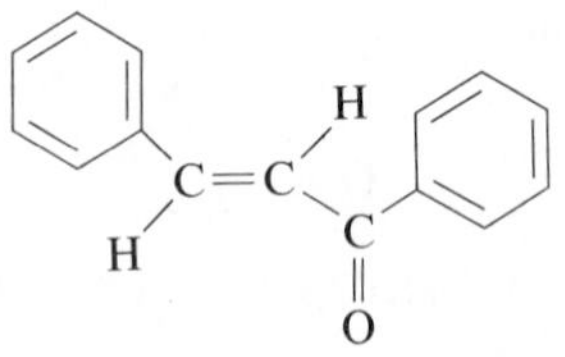 (f) 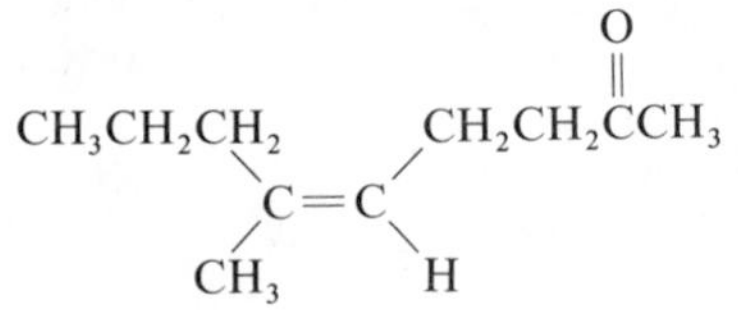

(g) 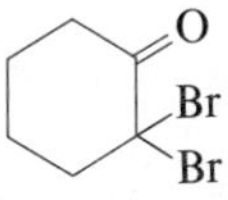 (h) $Cl_3C\overset{\overset{\displaystyle O}{\|}}{C}H$ (i) 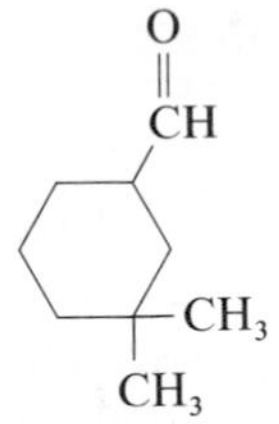

(j) $CH_3\overset{\overset{\displaystyle O}{\|}}{C}CH_2\overset{\overset{\displaystyle O}{\|}}{C}CH_3$ (k) 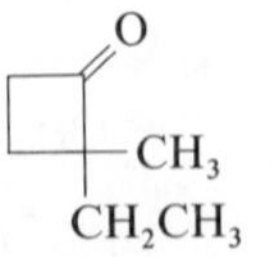 (l) 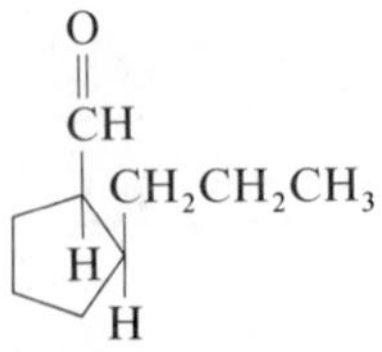

PROBLEM 13.4

Draw structural formulas for the following compounds:

(a) 3-methylcyclobutanone (b) 4,4-dimethyl-2,5-cyclohexadienone
(c) phenyl cyclopentyl ketone (d) (*E*)-4-heptenal (e) *p*-nitrobenzaldehyde
(f) 1-hydroxy-3-pentanone (g) hexanedial (h) 2-hydroxy-1,2-diphenylethanone
(i) 1,4-cyclohexanedione

Aldehydes and ketones are often prepared by the oxidation of alcohols. Earlier (pp. 439 and 441) we saw how the conditions for the reaction of alcohols with compounds of chromium(VI) could be controlled to allow different types of alcohols to be oxidized to carbonyl compounds. For example, cyclohexanol is a secondary alcohol, there are no other functional groups in the molecule, and it is low enough in molecular weight to be soluble in water. Cyclohexanol is oxidized to cyclohexanone by an aqueous, acidic solution of sodium dichromate.

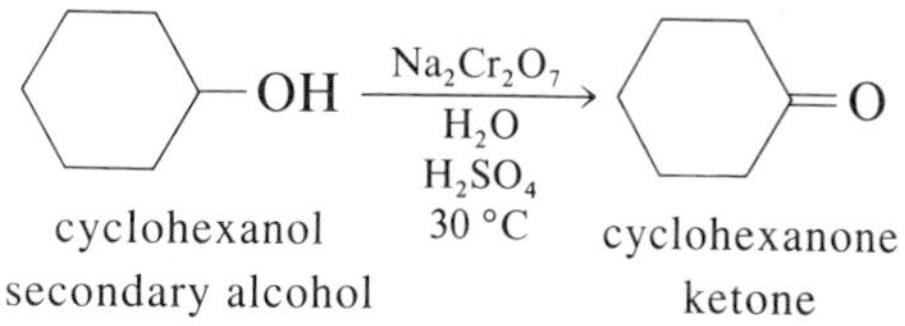
OH
$Na_2Cr_2O_7$
H_2O
H_2SO_4
30 °C
O
cyclohexanol
secondary alcohol
cyclohexanone
ketone

5-Methyl-4-hexen-1-ol is a primary alcohol. Oxidation of a primary alcohol in the presence of water leads to the formation of the corresponding carboxylic acid (p. 439). Pyridinium chlorochromate is used to oxidize this alcohol so that the reaction stops at the aldehyde stage (p. 442).

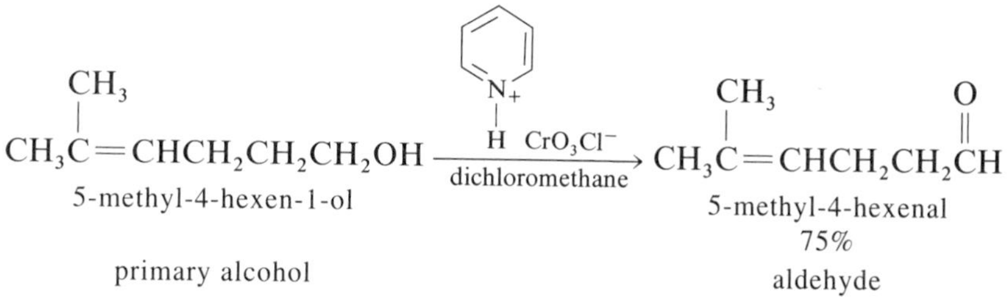
CH_3
$CH_3C{=}CHCH_2CH_2CH_2OH$
N+
H CrO_3Cl^-
dichloromethane
CH_3
O
$CH_3C{=}CHCH_2CH_2CH$
5-methyl-4-hexen-1-ol
primary alcohol
5-methyl-4-hexenal
75%
aldehyde

PROBLEM 13.5

Write equations showing the side reactions that would take place if 5-methyl-4-hexen-1-ol were oxidized with sodium dichromate in aqueous acid.

Ozonolysis is also used to prepare aldehydes and ketones. Ozonolysis is most useful as a method of preparing an aldehyde or a ketone when the other fragment from the ozonolysis is a low-molecular-weight compound that is easily separated from the desired product of the reaction. The cleavage of 1-octene into heptanal and formaldehyde (p. 312) is a case where the desired aldehyde is easily separated from the other product. Cyclic alkenes are frequently cleaved by ozonolysis to make dicarbonyl compounds; the preparation of 1,6-hexanedial from cyclohexene (p. 312) is an example.

Aldehydes can be prepared by reduction of esters or nitriles. These reactions will be studied in Chapter 15. The reverse of the oxidation of an alcohol to an aldehyde or a ketone is a reduction reaction (p. 436), in which an aldehyde or a ketone is converted to an alcohol.

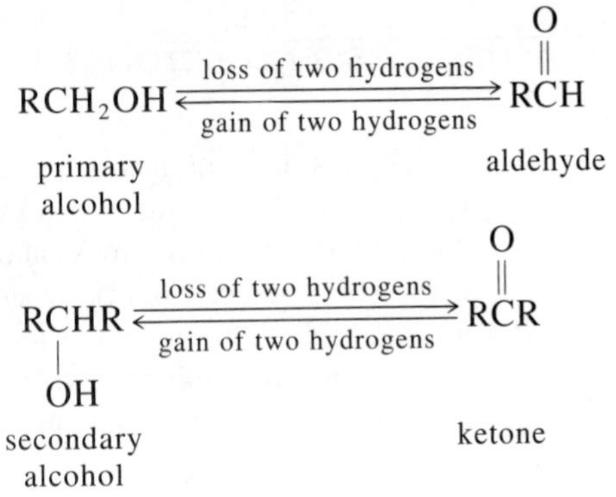
loss of two hydrogens
gain of two hydrogens
primary alcohol
aldehyde
loss of two hydrogens
gain of two hydrogens
secondary alcohol
ketone

Chapter 12 showed how alkyl halides are made from alcohols. Alkyl halides undergo a variety of transformations through nucleophilic substitution reactions. Alkyl halides are also converted into organometallic reagents that react with aldehydes and ketones to give alcohols (p. 491). Being able to go back and forth between carbonyl compounds and alcohols gives chemists great flexibility in planning syntheses of organic compounds.

Study Guide Concept Map 13.1

13.4 ADDITION OF THE NUCLEOPHILE HYDRIDE ION. REDUCTION OF ALDEHYDES AND KETONES TO ALCOHOLS

Nucleophilic addition of the hydride ion to the electrophilic carbon atom of the carbonyl group results in the reduction of aldehydes to primary alcohols and ketones to secondary alcohols. These reductions are carried out using **sodium borohydride** or **lithium aluminum hydride.** In both of these reagents, the anion is a source of hydride ion, which is very basic and a powerful nucleophile. Lithium aluminum hydride reacts violently with water or alcohols to generate hydrogen gas.

$$\underset{\text{lithium aluminum hydride}}{LiAlH_4} + \underset{\text{water}}{4\,H_2O} \xrightarrow[\text{fast}]{} Li^+\ Al(OH)_4^- + 4\,H_2\uparrow$$

Sodium borohydride reacts in the same way as lithium aluminum hydride does, though much more slowly.

$$\underset{\text{sodium borohydride}}{NaBH_4} + \underset{\text{methanol}}{4\,CH_3OH} \xrightarrow[\text{slow}]{} Na^+\ B(OCH_3)_4^- + 4\,H_2\uparrow$$

Because sodium borohydride reacts slowly with water, it is easier to handle than lithium aluminum hydride is and may be used with water or an alcohol as the solvent. The reactions of this reagent with carbonyl compounds are fast enough to compete with its decomposition reaction. The reduction of butanal to *n*-butyl alcohol and of 2-butanone to 2-butanol by sodium borohydride in water are typical reactions of this reagent.

$$4\ CH_3CH_2CH_2\overset{\overset{\displaystyle O}{\|}}{C}H + NaBH_4 \xrightarrow[H_2O]{} 4\ CH_3CH_2CH_2CH_2OH + Na^+ + BO_3^{3-}$$

butanal → *n*-butyl alcohol, 85%

$$4\ CH_3CH_2\overset{\overset{\displaystyle O}{\|}}{C}CH_3 + NaBH_4 \xrightarrow[H_2O]{} 4\ CH_3\underset{\underset{\displaystyle OH}{|}}{C}HCH_2CH_3 + Na^+ + BO_3^{3-}$$

2-butanone → 2-butanol, 87%

The mechanism for the reduction of 2-butanone with sodium borohydride is shown below. A hydride ion, a nucleophile, is transferred to the electrophilic carbon atom of the carbonyl group. The alkoxide ion that forms is protonated by the solvent. The carbonyl group, having a trigonal carbon atom, is planar and thus symmetrical. It is equally probable that a hydride ion will be donated to the carbonyl group from either side of the molecule, giving rise to equal numbers of molecules of the two enantiomers of 2-butanol. The alcohol recovered as the product of the reaction is a racemic mixture and not optically active.

VISUALIZING THE REACTION

Reduction of a carbonyl group

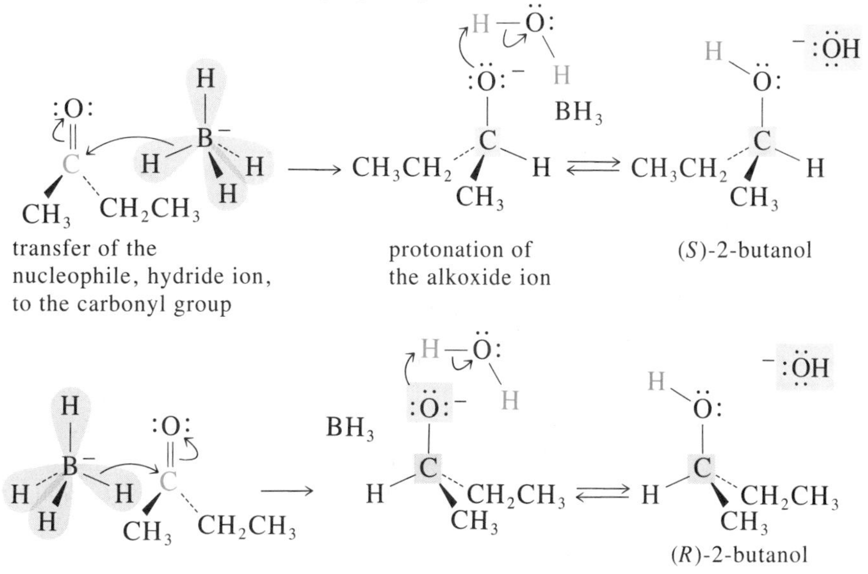
transfer of the nucleophile, hydride ion, to the carbonyl group
protonation of the alkoxide ion
(S)-2-butanol
(R)-2-butanol

For the sake of simplicity, the transfer of only one of the four hydride ions possible from sodium borohydride is shown in the above mechanism. Actually, each molecule of sodium borohydride or lithium aluminum hydride can reduce four carbonyl groups. The reduction of an aldehyde to a primary alcohol by lithium aluminum hydride in diethyl ether proceeds in two steps. The reduction step gives an alkoxide anion, which complexes with the metal ions present. To recover the alcohol, the complex is treated with water and dilute acid. For example, heptanal is converted to 1-heptanol.

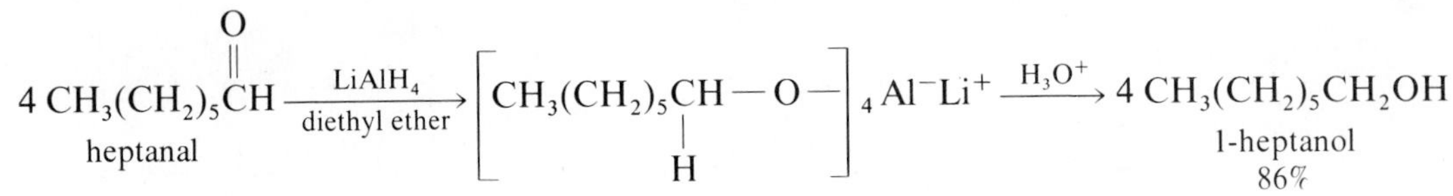

$$4\ CH_3(CH_2)_5\overset{\overset{\displaystyle O}{\|}}{CH} \xrightarrow[\text{diethyl ether}]{LiAlH_4} \left[CH_3(CH_2)_5\underset{\underset{\displaystyle H}{|}}{CH}-O- \right]_4 Al^-Li^+ \xrightarrow{H_3O^+} 4\ CH_3(CH_2)_5CH_2OH$$

heptanal — 1-heptanol 86%

A similar sequence of reactions gives a secondary alcohol from 4-methyl-2-pentanone.

$$4\ CH_3\overset{\overset{\displaystyle CH_3}{|}}{CH}CH_2\overset{\overset{\displaystyle O}{\|}}{C}CH_3 \xrightarrow[\text{diethyl ether}]{LiAlH_4} \left[CH_3\overset{\overset{\displaystyle CH_3}{|}}{CH}CH_2\overset{\overset{\displaystyle CH_3}{|}}{\underset{\underset{\displaystyle H}{|}}{C}}-O- \right]_4 Al^-Li^+ \xrightarrow{H_3O^+} CH_3\overset{\overset{\displaystyle CH_3}{|}}{CH}CH_2\underset{\underset{\displaystyle OH}{|}}{CH}CH_3$$

4-methyl-2-pentanone — 4-methyl-2-pentanol 85%

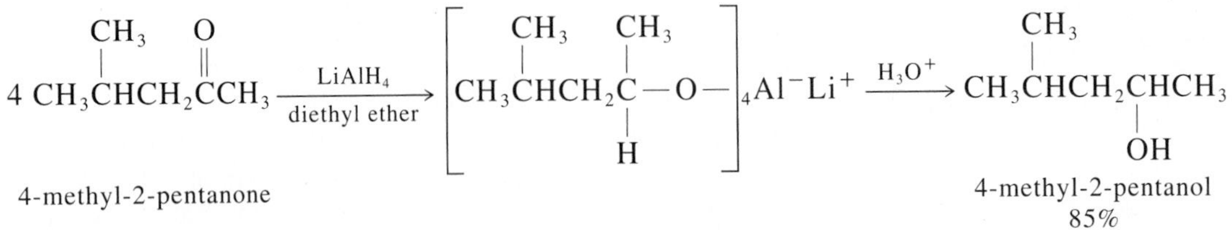

Lithium aluminum hydride is used in dry ether solvents because of its reactivity with water and alcohols. It is a much more reactive donor of hydride ions than sodium borohydride is, and it reacts with carboxylic acids and esters (p. 596) as well as with aldehydes and ketones.

Sodium borohydride and lithium aluminum hydride do not reduce isolated carbon-carbon double bonds. Thus, it is possible to selectively reduce a carbonyl group in a molecule containing a double bond. An example of such a selective reduction is the synthesis of 6-methyl-5-hepten-2-ol, a pheromone (pp. 333 and 701) of the ambrosia beetle.

$$CH_3\overset{\overset{\displaystyle CH_3}{|}}{C}=CHCH_2CH_2\overset{\overset{\displaystyle O}{\|}}{C}CH_3 \xrightarrow[\text{ethanol}]{NaBH_4} CH_3\overset{\overset{\displaystyle CH_3}{|}}{C}=CHCH_2CH_2\overset{\overset{\displaystyle OH}{|}}{C}HCH_3$$

6-methyl-5-hepten-2-one — 6-methyl-5-hepten-2-ol 95%

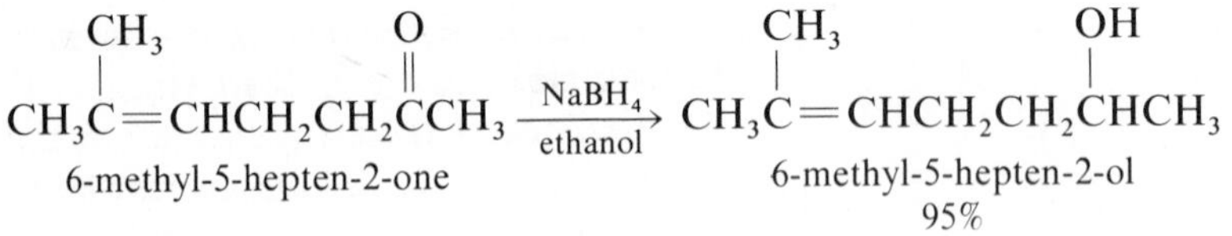

Reductions of double bonds adjacent to carbonyl groups do sometimes occur with metal hydride reagents, but are dependent on conditions such as solvent, temperature, and the presence of excess hydride. Compounds in which a double bond is adjacent to the carbonyl group can be reduced selectively at the carbonyl group if mild reaction conditions are used. Thus, 2-butenal is reduced to 2-buten-1-ol by either sodium borohydride or lithium aluminum hydride.

$$CH_3CH=CH\overset{\overset{\displaystyle O}{\|}}{CH}$$
2-butenal

NaBH_4 / H_2O

LiAlH_4 / H_3O^+ diethyl ether

$$CH_3CH=CHCH_2OH$$
2-buten-1-ol

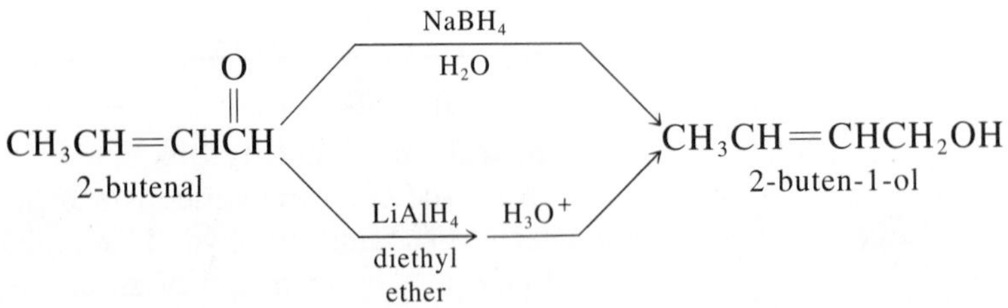

PROBLEM 13.6

Complete the following equations.

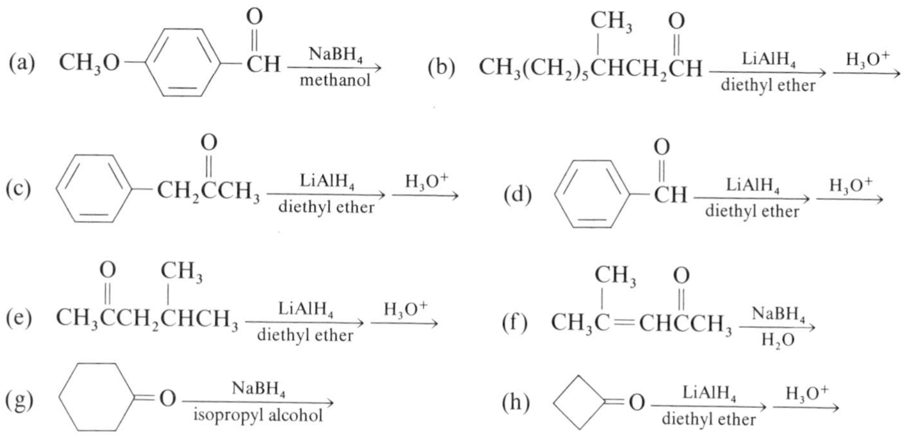
(a)
NaBH4
methanol
(b)
LiAlH4
diethyl ether
H3O+
(c)
LiAlH4
diethyl ether
H3O+
(d)
LiAlH4
diethyl ether
H3O+
(e)
LiAlH4
diethyl ether
H3O+
(f)
NaBH4
H2O
(g)
NaBH4
isopropyl alcohol
(h)
LiAlH4
diethyl ether
H3O+

13.5 ADDITION OF THE NUCLEOPHILE CYANIDE ION TO THE CARBONYL GROUP. CYANOHYDRIN FORMATION

Another nucleophilic addition reaction of aldehydes and ketones is the addition of the elements of hydrogen cyanide. The reagents generally used are sodium or potassium cyanide followed by acid. This type of reaction is illustrated below for acetone.

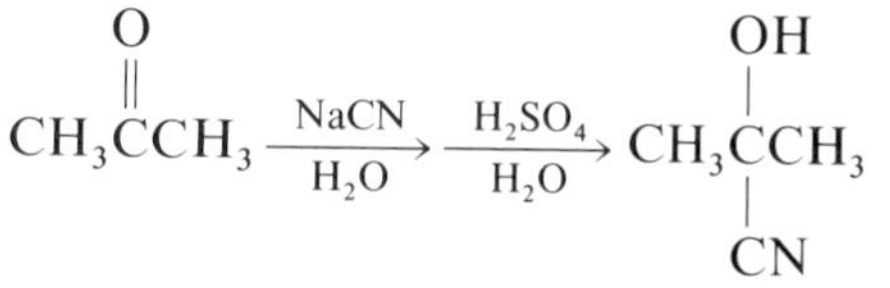
NaCN
H2O
H2SO4
H2O

acetone — acetone cyanohydrin

The products of such a reaction are called **cyanohydrins** because they include a hydroxyl and a cyano group in the same molecule. The reaction is initiated when the cyanide ion attacks the electrophilic carbon of the carbonyl group. Then, the alkoxide ion that is generated is protonated.

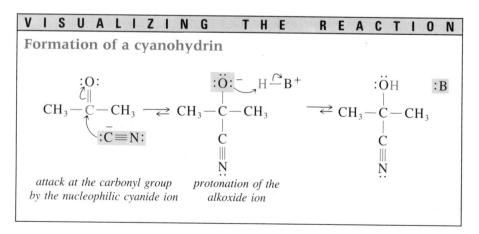

VISUALIZING THE REACTION

Formation of a cyanohydrin

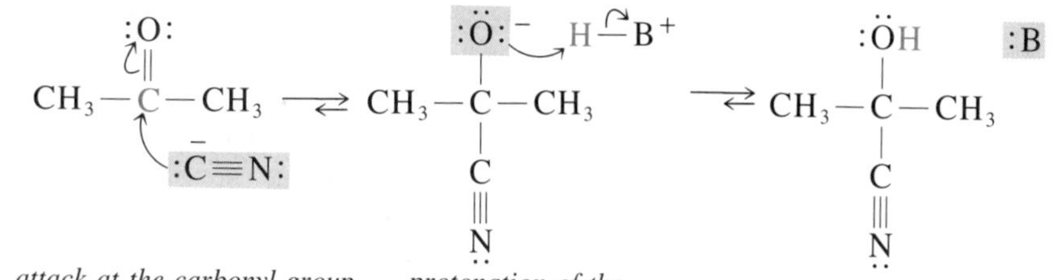
:B

attack at the carbonyl group by the nucleophilic cyanide ion

protonation of the alkoxide ion

The addition of hydrogen cyanide to an aldehyde or ketone increases the number of carbon atoms in the molecule by one and introduces two functional groups that can undergo further reactions. An important reaction of the cyano group is the addition of water to the carbon-nitrogen triple bond (p. 564). An example of the formation and further transformation of a cyanohydrin is the conversion of acetaldehyde to lactic acid, which is one of the products of the metabolism of glucose in muscle tissue and also gives sour milk its acid taste.

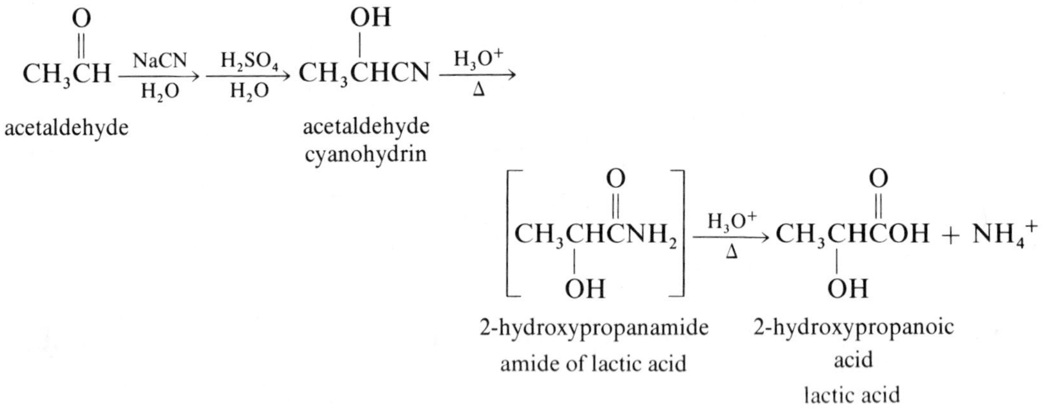
NaCN
H₂O
H₂SO₄
H₂O
H₃O⁺
Δ
acetaldehyde
acetaldehyde cyanohydrin
2-hydroxypropanamide
amide of lactic acid
2-hydroxypropanoic acid
lactic acid

PROBLEM 13.7

The lactic acid present in muscle tissue is (*R*)-(−)-lactic acid, and that found in sour milk is (*S*)-+)-lactic acid. What is the stereochemistry of the addition of hydrogen cyanide to acetaldehyde? What is the stereochemistry of the lactic acid formed by the hydrolysis of the cyano group?

13.6 ADDITION OF OTHER CARBON NUCLEOPHILES TO THE CARBONYL GROUP

A. Preparation of Organometallic Reagents

Reactions in which carbon-carbon bonds are formed are important in synthesizing new organic compounds. A carbon-carbon bond forms when a reagent containing a carbon that is a nucleophilic center reacts with one containing a carbon that is an electrophilic center. For example, the reactions of alkyl halides with carbanions derived from terminal alkynes are used to create new carbon-carbon bonds (p. 333). In this section, we will look at other ways to create carbon nucleophiles and, in the next sections, will examine their reactions with compounds containing an electrophilic carbon center.

One way of creating nucleophiles is to deprotonate an organic compound using a strong base. Acetylide anions (p. 332) and the anion of diethyl malonate (p. 261) are examples of nucleophiles prepared in this way. Another way of preparing carbon nucleophiles is by the reaction of organic halides with metals such as lithium, sodium, potassium, or magnesium to give **organometallic reagents.** Alkyl, aryl, and vinyl halides all react with metals. The organometallic reagents prepared in this way with magnesium are named **Grignard reagents,** after Victor Grignard, a French chemist who developed them at the beginning of this century.

The reaction of methyl iodide with magnesium metal in ether gives methylmagnesium iodide, a methyl Grignard reagent.

$$CH_3I + Mg \xrightarrow{\text{diethyl ether}} CH_3MgI$$

methyl iodide — magnesium metal — methylmagnesium iodide

a methyl Grignard reagent

Grignard reagents are also formed from aryl and vinyl halides. Phenylmagnesium bromide, for example, is a useful Grignard reagent.

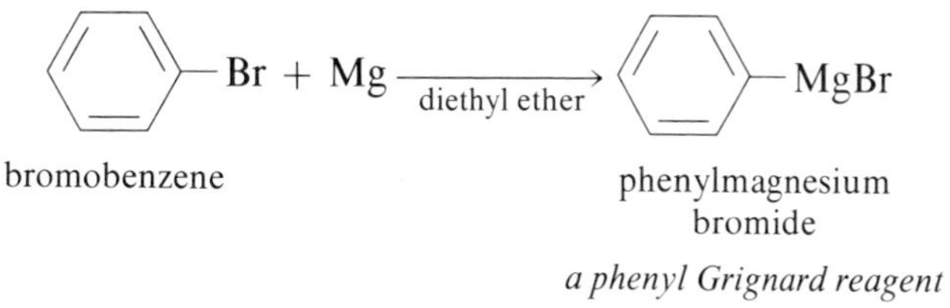

Br + Mg
diethyl ether
MgBr
bromobenzene
phenylmagnesium bromide
a phenyl Grignard reagent

The organometallic compound is formed by a transfer of electrons from the magnesium atom to the more electronegative carbon atom. As a result of this electron transfer, the carbon atom that is an electrophilic center in the alkyl halide becomes a nucleophilic center in the Grignard reagent.

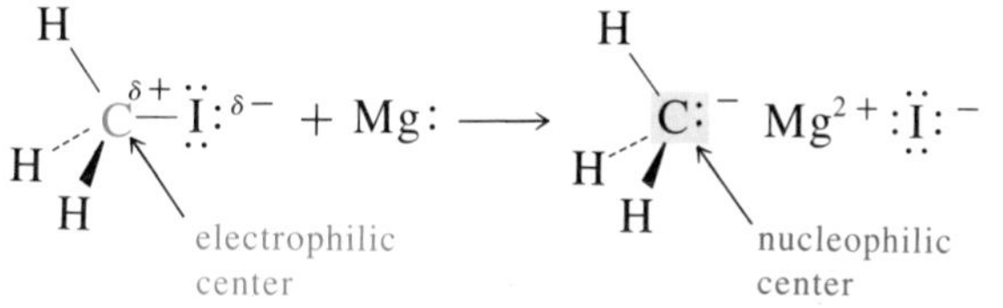

+ Mg:
electrophilic center
nucleophilic center

In all organometallic compounds, the carbon atom bonded to the metal reacts as a Lewis base and is a strong nucleophile. All organometallic reagents must be prepared in the absence of water or alcohols, which are acidic enough to protonate carbanions.

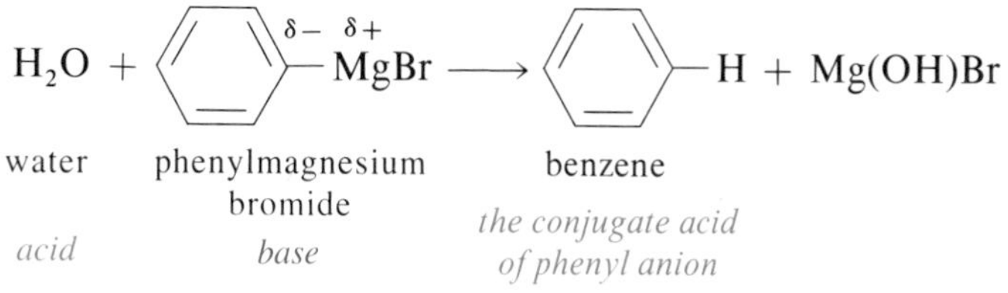

H₂O +
MgBr
H + Mg(OH)Br
water
acid
phenylmagnesium bromide
base
benzene
the conjugate acid of phenyl anion

Therefore, *Grignard reagents cannot be prepared from compounds that contain acidic hydrogen atoms.* An organic halide that has a hydroxyl, carboxyl, thiol, or amino group elsewhere in the molecule cannot be used to make a Grignard reagent.

The basicity of the carbon atom bound to magnesium in the Grignard reagent can be utilized to make Grignard reagents from terminal alkynes. The relatively acidic hydrogen atom of a terminal alkyne is removed by an alkyl Grignard reagent.

$$CH_3(CH_2)_3C{\equiv}CH + CH_3CH_2MgBr \xrightarrow[\text{ether}]{\text{diethyl}} CH_3(CH_2)_3C{\equiv}CMgBr + CH_3CH_3\uparrow$$

1-hexyne; *acid*, *pK_a ~26*

ethylmagnesium bromide; *base*

1-hexynylmagnesium bromide; *from the conjugate base of 1-hexyne*

ethane; *the conjugate acid of ethyl anion*, *pK_a ~49*

In a competition between the two bases, the proton is transferred from the alkyne to the ethyl group. Ethane gas evolves from the reaction mixture, pulling the reaction to completion.

When lithium metal reacts with an organic halide, an **organolithium reagent** is formed. The reaction is illustrated below for isopropyl bromide and bromobenzene.

$$\underset{\text{isopropyl bromide}}{CH_3\overset{\overset{\displaystyle CH_3}{|}}{C}HBr} + \underset{\text{lithium metal}}{2\,Li} \xrightarrow{\text{pentane}} \underset{\text{isopropyllithium}}{CH_3\overset{\overset{\displaystyle CH_3}{|}}{C}HLi} + Li^+Br^-$$

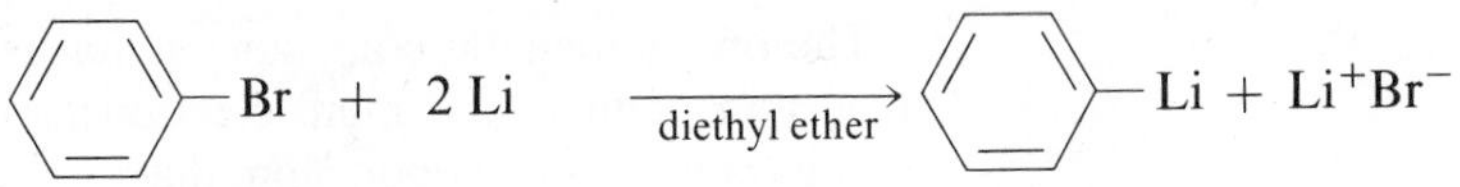

bromobenzene lithium metal phenyllithium

Organolithium reagents are also used as a source of basic and nucleophilic carbon atoms.

PROBLEM 13.8

Complete the following equations.

(a) $CH_3CH_2CH_2CH_2Br \xrightarrow[\text{tetrahydrofuran}]{\text{Li}}$

(b) 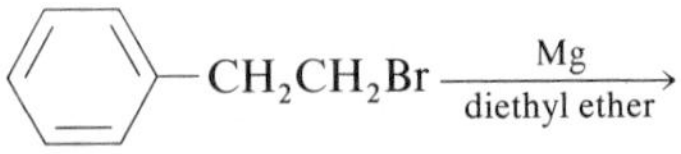

(c) $CH_2{=}CHCl \xrightarrow[\text{tetrahydrofuran, } \Delta]{\text{Mg}}$

(d) 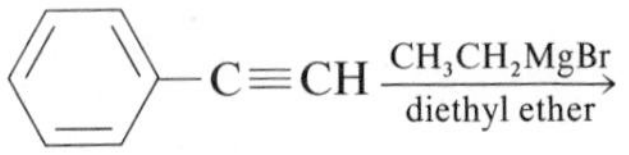

(e) 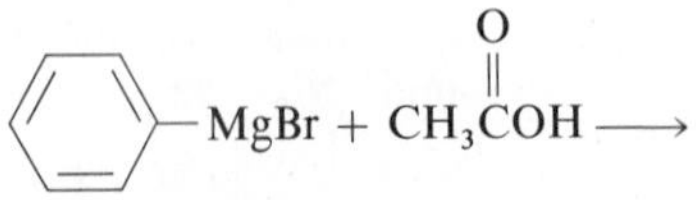

(f) $CH_3CH_2C{\equiv}CNa + H_2O \longrightarrow$

(g) 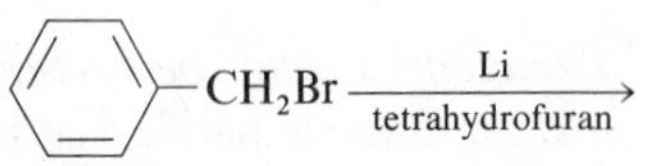

(h) $CH_3CH_2CH_2Li + NH_3 \longrightarrow$

(i) $CH_3CH_2MgI + CH_3OH \longrightarrow$

(j) 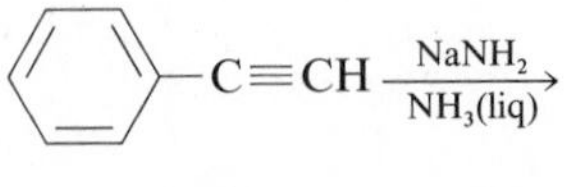

B. Reactions of Organometallic Reagents with Aldehydes or Ketones as Electrophiles

All of the organometallic reagents prepared in Section 13.6A will react with aldehydes or ketones to produce alcohols. The reactions take place in two steps. The first involves attack by the nucleophilic carbon atom of the organometallic reagent on the electrophilic carbon atom of the carbonyl group and gives an alkoxide anion complexed with a metal ion. In the second step, this complex is broken up with water, and usually some dilute acid, to produce the alcohol.

VISUALIZING THE REACTION

Reaction of an organometallic reagent with a carbonyl compound

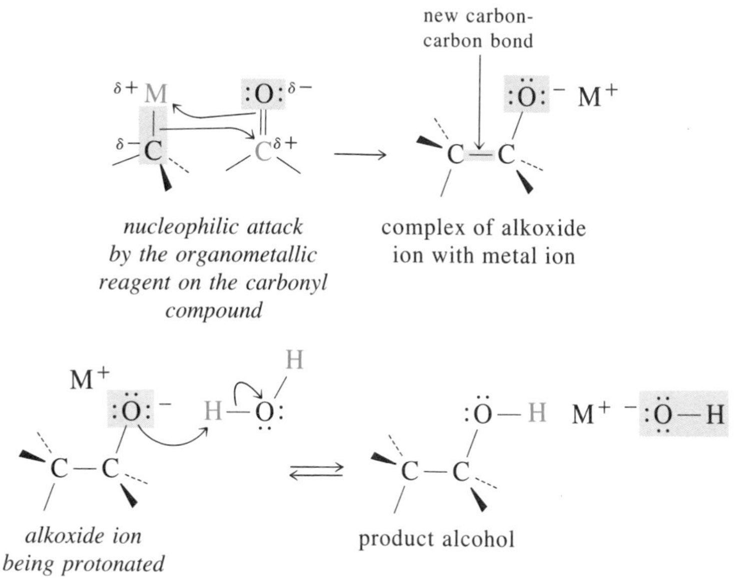
new carbon-carbon bond
nucleophilic attack by the organometallic reagent on the carbonyl compound
complex of alkoxide ion with metal ion
alkoxide ion being protonated
product alcohol

Because it is possible to vary the structure of the aldehyde or ketone as well as that of the organometallic reagent, many different alcohols may be prepared by such reactions. Historically, Grignard reagents have been used extensively to synthesize alcohols by means of the **Grignard reaction.** In each of the three reactions shown below, the Grignard reagent is prepared, an aldehyde or a ketone is added to it to give an alkoxide ion complexed with magnesium ion, and the alcohol is formed by protonation of the alkoxide ion. In the last equation, a very weak acid, ammonium chloride, is used because a stronger acid would cause dehydration of the tertiary alcohol.

$$\underset{\text{1-hexyne}}{CH_3(CH_2)_3C{\equiv}CH} + CH_3CH_2MgBr \xrightarrow[\text{diethyl ether}]{} \underset{\text{1-hexynylmagnesium bromide}}{CH_3(CH_2)_3C{\equiv}CMgBr} + CH_3CH_3\uparrow \xrightarrow{\overset{O}{\|}{HCH}\ \text{(formaldehyde)}}$$

$$CH_3(CH_2)_3C{\equiv}CCH_2OMgBr \xrightarrow{H_3O^+} \underset{\substack{\text{2-heptyn-1-ol} \\ 82\%}}{CH_3(CH_2)_3C{\equiv}CCH_2OH}$$

$$\underset{\text{isopropyl bromide}}{CH_3\overset{\overset{\displaystyle CH_3}{|}}{C}HBr} + Mg \xrightarrow[\text{diethyl ether}]{} \underset{\text{isopropylmagnesium bromide}}{CH_3\overset{\overset{\displaystyle CH_3}{|}}{C}HMgBr} \xrightarrow{CH_3\overset{\overset{\displaystyle O}{\|}}{C}H \text{ (acetaldehyde)}}$$

$$CH_3\overset{\overset{\displaystyle CH_3}{|}}{C}H\underset{\underset{\displaystyle OMgBr}{|}}{C}HCH_3 \xrightarrow{H_3O^+} \underset{\substack{\text{3-methyl-2-butanol} \\ 50\%}}{CH_3\overset{\overset{\displaystyle CH_3}{|}}{C}H\underset{\underset{\displaystyle OH}{|}}{C}HCH_3}$$

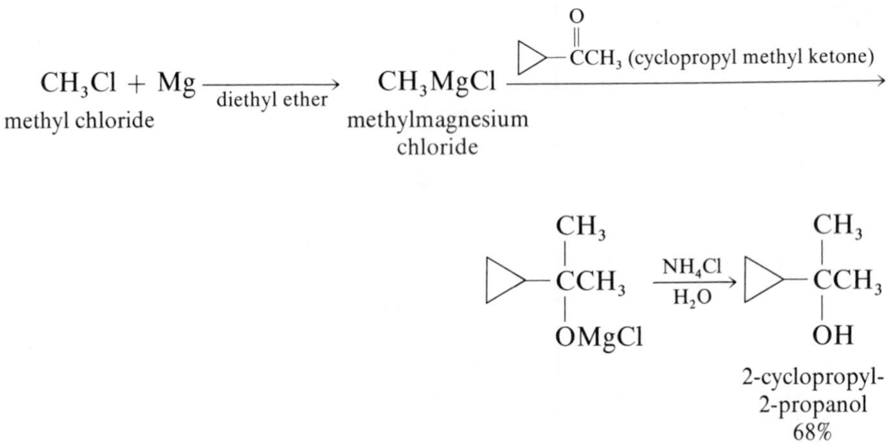
CH3Cl + Mg
diethyl ether
CH3MgCl
CCH3 (cyclopropyl methyl ketone)
methyl chloride
methylmagnesium chloride
CH3
CCH3
OMgCl
NH4Cl
H2O
CH3
CCH3
OH
2-cyclopropyl-2-propanol
68%

The type of alcohol formed depends on the type of carbonyl compound that is used. Formaldehyde produces a primary alcohol that has one more carbon atom in it than the alkyl halide used to make the Grignard reagent. Other aldehydes such as acetaldehyde produce secondary alcohols. A ketone, such as cyclopropyl methyl ketone, produces a tertiary alcohol.

A secondary or tertiary alcohol formed in a Grignard reaction may contain stereocenters, depending on the nature of the organic groups bonded to the carbon attached to the hydroxyl group. As in the reduction of a carbonyl group by sodium borohydride (p. 485), the organometallic reagent attacks with equal probability either face of the carbonyl group. A racemic mixture of product alcohol results, unless, of course, chirality was already present in the Grignard reagent or existed elsewhere in the carbonyl compound; in that case, a mixture of diastereomers is formed.

Organolithium reagents also add to aldehydes or ketones to form alcohols. The reactions parallel those of Grignard reagents. In many cases, organolithium reagents give better yields than Grignard reagents do. In fact, organolithium reagents form alcohols with some sterically hindered ketones that do not react with Grignard reagents. Some representative reactions of organolithium reagents are shown on the following page.

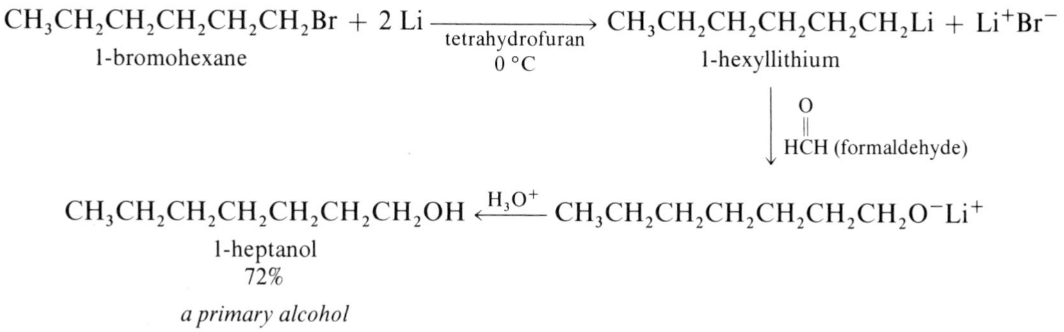

$$CH_3CH_2CH_2CH_2CH_2CH_2Br + 2\,Li \xrightarrow[\text{0 °C}]{\text{tetrahydrofuran}} CH_3CH_2CH_2CH_2CH_2CH_2Li + Li^+Br^-$$

1-bromohexane   1-hexyllithium

$$\downarrow \overset{O}{\overset{\|}{HCH}} \text{ (formaldehyde)}$$

$$CH_3CH_2CH_2CH_2CH_2CH_2CH_2OH \xleftarrow{H_3O^+} CH_3CH_2CH_2CH_2CH_2CH_2CH_2O^-Li^+$$

1-heptanol
72%

a primary alcohol

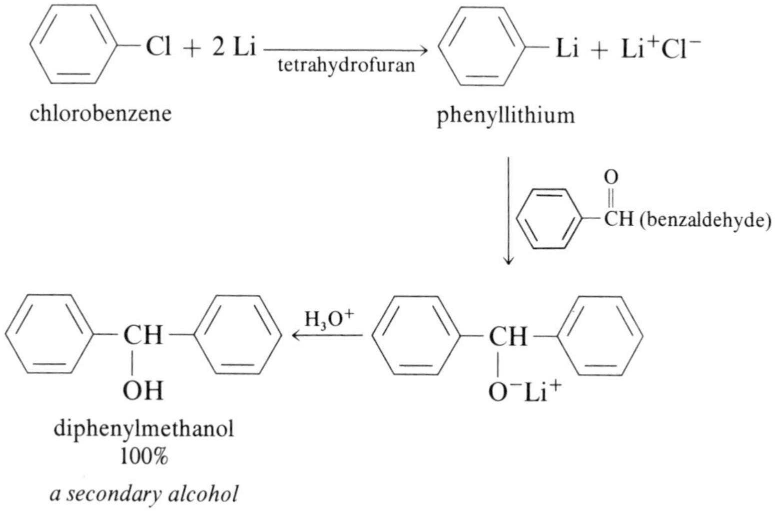

$$\langle\bigcirc\rangle\!-\!Cl + 2\,Li \xrightarrow{\text{tetrahydrofuran}} \langle\bigcirc\rangle\!-\!Li + Li^+Cl^-$$

chlorobenzene   phenyllithium

$$\downarrow \langle\bigcirc\rangle\!-\!\overset{O}{\overset{\|}{CH}} \text{ (benzaldehyde)}$$

$$\langle\bigcirc\rangle\!-\!\underset{OH}{\overset{|}{CH}}\!-\!\langle\bigcirc\rangle \xleftarrow{H_3O^+} \langle\bigcirc\rangle\!-\!\underset{O^-Li^+}{\overset{|}{CH}}\!-\!\langle\bigcirc\rangle$$

diphenylmethanol
100%

a secondary alcohol

$$CH_3CH_2CH_2CH_2Br + 2\,Li \xrightarrow[\text{0 °C}]{\text{tetrahydrofuran}} CH_3CH_2CH_2CH_2Li + Li^+Br^-$$

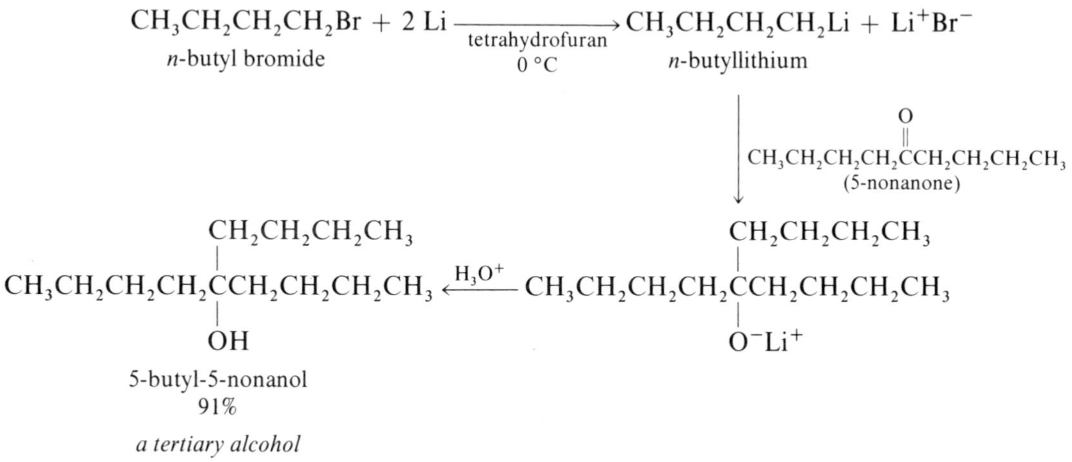

n-butyl bromide   *n*-butyllithium

$$\downarrow \underset{\text{(5-nonanone)}}{CH_3CH_2CH_2CH_2\overset{O}{\overset{\|}{C}}CH_2CH_2CH_2CH_3}$$

$$CH_3CH_2CH_2CH_2\underset{OH}{\overset{\overset{\displaystyle CH_2CH_2CH_2CH_3}{|}}{C}}CH_2CH_2CH_2CH_3 \xleftarrow{H_3O^+} CH_3CH_2CH_2CH_2\underset{O^-Li^+}{\overset{\overset{\displaystyle CH_2CH_2CH_2CH_3}{|}}{C}}CH_2CH_2CH_2CH_3$$

5-butyl-5-nonanol
91%

a tertiary alcohol

In each case, the organometallic reagent adds to the carbonyl group, and the resulting alkoxide anion is protonated by dilute acid to recover the alcohol. Formaldehyde yields a primary alcohol, other aldehydes give secondary alcohols, and ketones give tertiary alcohols.

Sodium acetylide, prepared from sodium amide and acetylene in liquid ammonia, also adds to carbonyl groups. It reacts with benzaldehyde to give an acetylenic secondary alcohol and with cyclohexanone to form a tertiary alcohol.

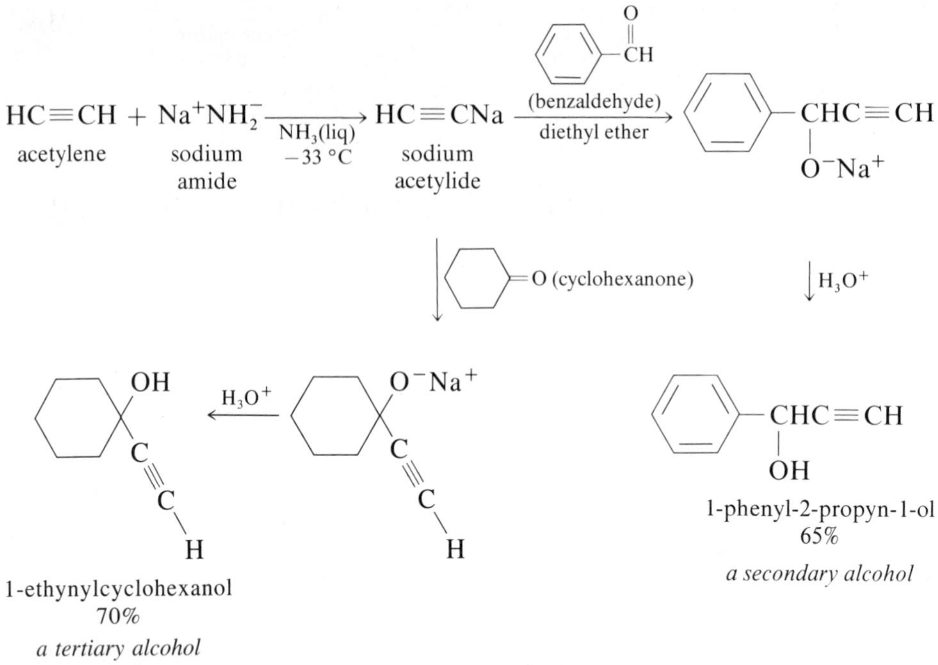
HC≡CH + Na⁺NH₂⁻
NH₃(liq) −33 °C
HC≡CNa
(benzaldehyde)
diethyl ether
CHC≡CH
O⁻Na⁺
acetylene
sodium amide
sodium acetylide
O (cyclohexanone)
H₃O⁺
OH
O⁻Na⁺
C
C
H
CHC≡CH
OH
1-phenyl-2-propyn-1-ol
65%
a secondary alcohol
1-ethynylcyclohexanol
70%
a tertiary alcohol

C. Reaction of Organometallic Reagents with Oxirane as an Electrophile

The three-membered ring of an oxirane is strained enough that it will open under nucleophilic attack (p. 454) by an organometallic reagent. In this respect, the oxirane behaves as though it were a carbonyl group: the nucleophilic carbon atom of the organometallic reagent bonds to one of the carbon atoms of the oxirane ring, and the metal ion complexes with the alkoxide ion that forms when the ring opens. Because of this similarity, the reaction of oxirane is discussed here where it can be compared with the reactions of aldehydes and ketones with carbon nucleophiles. *n*-Butylmagnesium bromide, for example, reacts with oxirane to give 1-hexanol, a primary alcohol; the carbon chain in the product is two carbon atoms longer than the chain in the starting organometallic reagent.

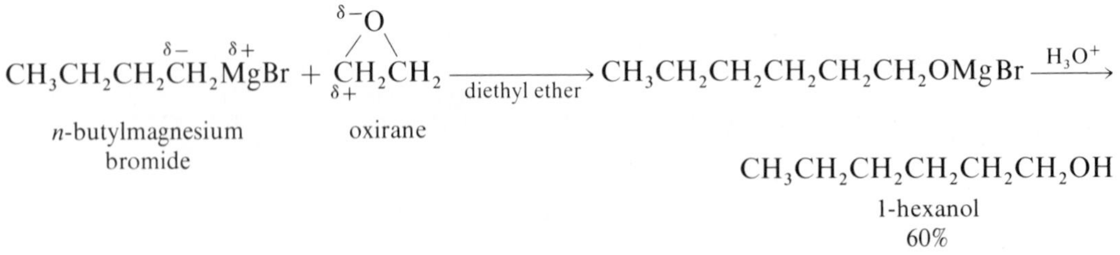
CH₃CH₂CH₂CH₂MgBr
δ− δ+
δ−O
CH₂CH₂
δ+
diethyl ether
CH₃CH₂CH₂CH₂CH₂CH₂OMgBr
H₃O⁺
n-butylmagnesium bromide
oxirane
CH₃CH₂CH₂CH₂CH₂CH₂OH
1-hexanol
60%

Study Guide
Concept Maps 13.3 and 13.4

The structural changes that occur when organometallic reagents react with formaldehyde, other aldehydes, ketones, and oxiranes are summarized in Table 13.2.

D. Problem-Solving Skills

Problem

Predict the products of the following reaction, including their stereochemistry.

TABLE 13.2 Alcohols from Reactions of Organometallic Reagents with Aldehydes, Ketones, and Oxirane

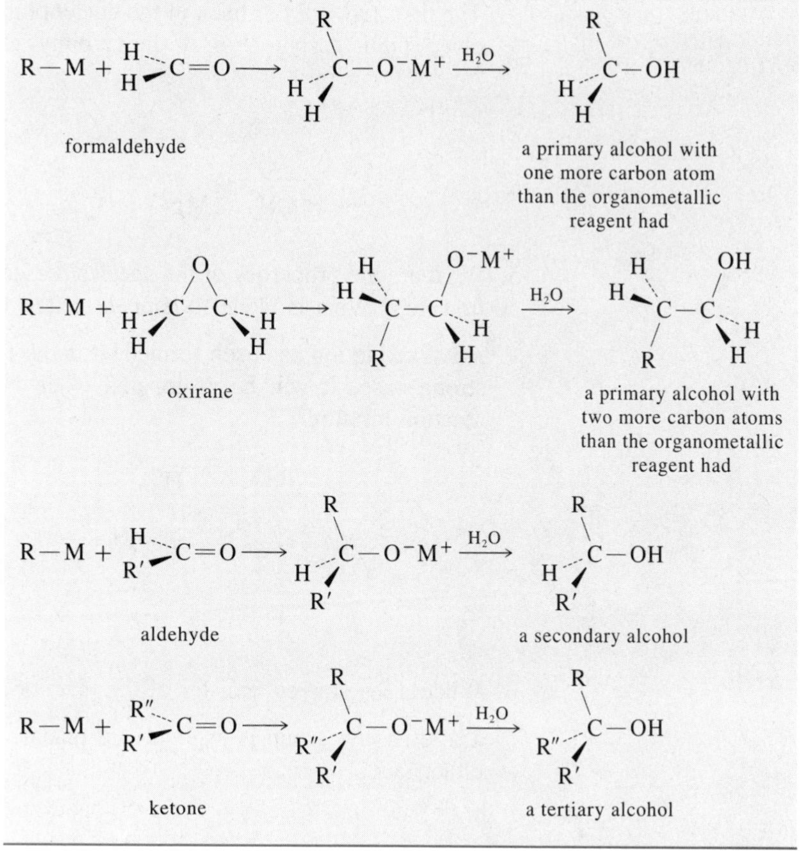
formaldehyde
a primary alcohol with one more carbon atom than the organometallic reagent had
oxirane
a primary alcohol with two more carbon atoms than the organometallic reagent had
aldehyde
a secondary alcohol
ketone
a tertiary alcohol

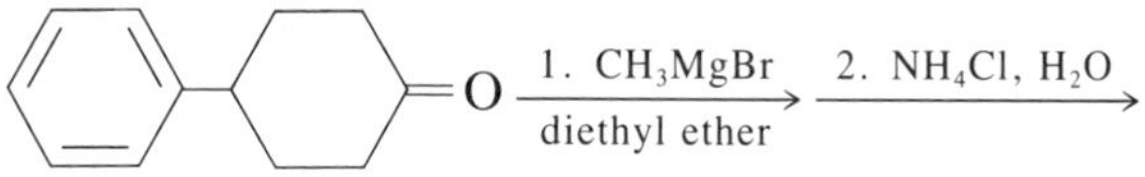
1. CH3MgBr
diethyl ether
2. NH4Cl, H2O

Solution

1. To what functional group class does the reactant belong? What is the electronic character of the functional group?

   The reactant contains a carbonyl group bonded to two carbon atoms. It is a ketone.

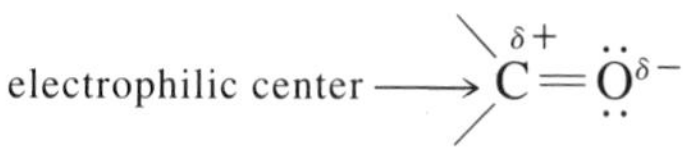
electrophilic center

2. Does the reactant have a good leaving group?

   No.

3. What reagents are present? Are they good acids, bases, nucleophiles, or electrophiles? Is there an ionizing solvent present?

   CH_3MgBr, a Grignard reagent, contains a strongly nucleophilic carbon atom.

4. What is the most likely first step for the reaction: protonation or deprotonation, ionization, attack by a nucleophile, or attack by an electrophile?

   The first step will be attack of the nucleophilic carbon atom of CH_3MgBr at the electrophilic carbon atom of the carbonyl group.

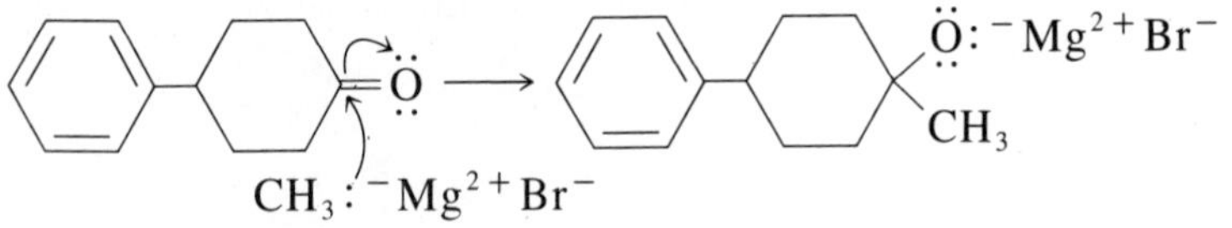

5. What are the properties of the species present in the reaction mixture after the first step? What is likely to happen next?

   An alkoxide ion has been formed. It is the conjugate base of an alcohol and a strong base. It will be protonated when NH_4Cl and H_2O are added to the reaction mixture.

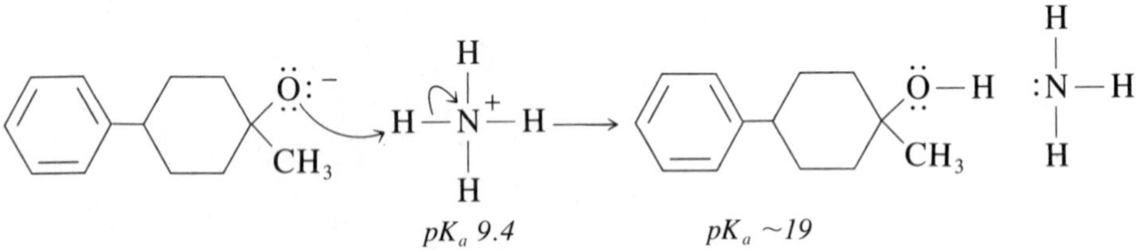
pKa 9.4
pKa ~19

6. What is the stereochemistry of the reaction?

   The carbonyl group is trigonal and planar. The Grignard reagent can add to either face.

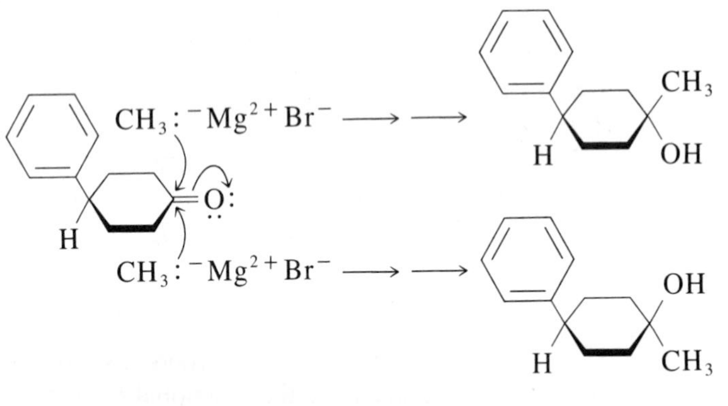

The product is formed as two isomers, one with the methyl group cis to the phenyl group and the other with the methyl group trans to the phenyl group.

The complete correct solution is as follows:

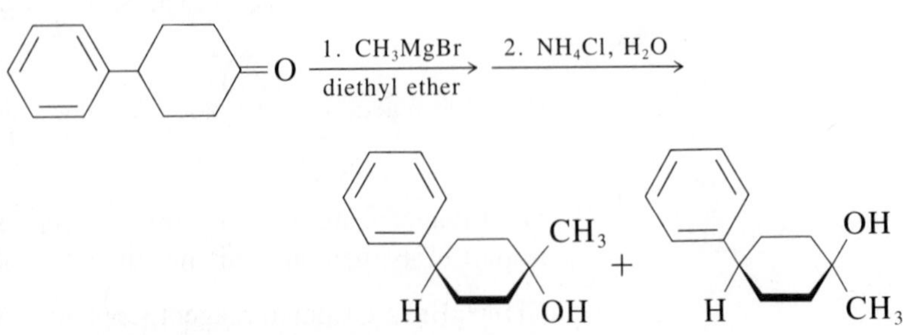
1. CH3MgBr
diethyl ether
2. NH4Cl, H2O

PROBLEM 13.9

Complete the following equations, showing the structures of all the intermediates and final products in the reactions.

(a) 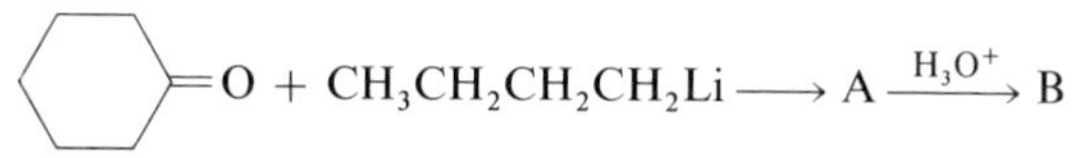

(b) 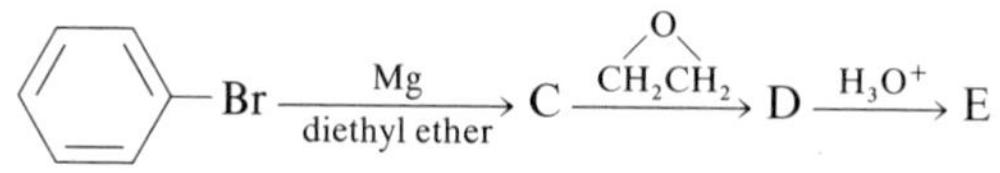

(c) 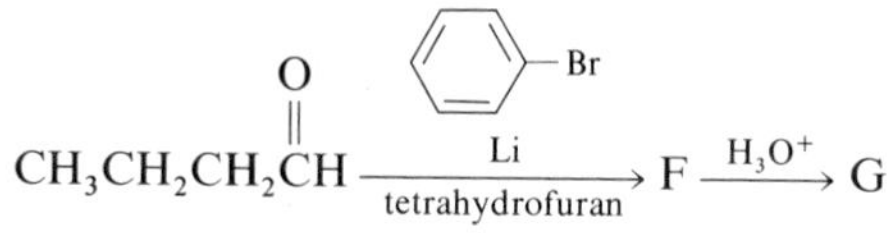

(d) 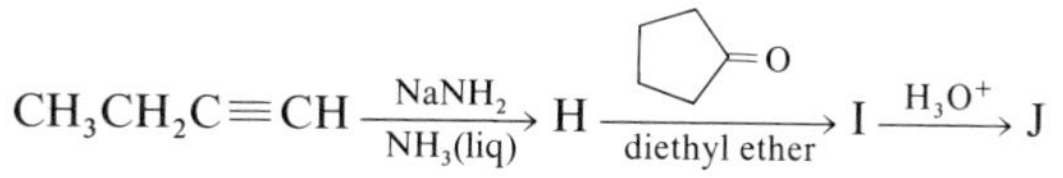

(e) 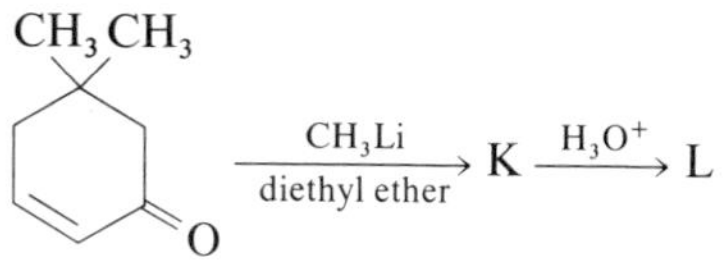

(f) $\underset{\displaystyle CH_3\overset{\displaystyle |}{C}=CH\overset{\displaystyle \|}{C}CH_3}{\overset{\displaystyle CH\quad\;\; O}{}} \xrightarrow[\text{diethyl ether}]{CH_3CH_2CH_2CH_2Li} M \xrightarrow{H_3O^+} N$

(g) $\underset{\displaystyle CH_3\overset{\displaystyle |}{C}=CHCH_2CH_2\overset{\displaystyle |}{C}=CHCH_2CH_2\overset{\displaystyle \|}{C}H}{\overset{\displaystyle CH_3 \qquad\qquad\quad CH_3 \qquad\qquad\quad O}{}} \xrightarrow[\text{dichloromethane}]{HC\equiv CMgBr} O \xrightarrow{H_3O^+} P$

13.7 REACTIONS OF ALDEHYDES AND KETONES WITH THE NUCLEOPHILES WATER AND ALCOHOLS

A. Hydrates

As nucleophiles, water and alcohols react with carbonyl compounds. The oxidation of an aldehyde such as butanal, for example, to a carboxylic acid by aqueous chromic acid is thought to involve an intermediate, the **hydrate** (p. 442) of the aldehyde.

$$\underset{\text{butanal}}{CH_3CH_2CH_2\overset{\displaystyle O}{\overset{\|}{C}}H} \underset{}{\overset{H_2O}{\rightleftarrows}} \underset{\text{butanal hydrate}}{CH_3CH_2CH_2\underset{\displaystyle \overset{|}{OH}}{\overset{\displaystyle OH}{\overset{|}{C}}}H} \xrightarrow[H_3O^+]{Cr_2O_7^{2-}} \underset{\text{butanoic acid}}{CH_3CH_2CH_2\overset{\displaystyle O}{\overset{\|}{C}}OH}$$

The interaction of water with the carbonyl group is reversible, and the free carbonyl compound predominates at equilibrium in most cases. Using water labeled with an

13 ALDEHYDES AND KETONES.
REACTIONS AT ELECTROPHILIC
CARBON ATOMS

13.7 REACTIONS OF ALDEHYDES
AND KETONES WITH THE
NUCLEOPHILES WATER AND
ALCOHOLS

isotope of oxygen, ^{18}O, shows that the hydrate must be present in equilibrium with the carbonyl compound, even if only in very low concentration. When acetone is put in such labeled water, some of the oxygen atoms of the carbonyl group become ^{18}O atoms.

$$\overset{16}{O} \\ \underset{}{\overset{\|}{CH_3CCH_3}} \ + \ H_2\overset{18}{O} \rightleftharpoons \ CH_3\overset{\overset{16}{OH}}{\underset{\underset{18}{OH}}{C}}CH_3 \ \rightleftharpoons \ CH_3CCH_3 \ + \ H_2\overset{16}{O}$$

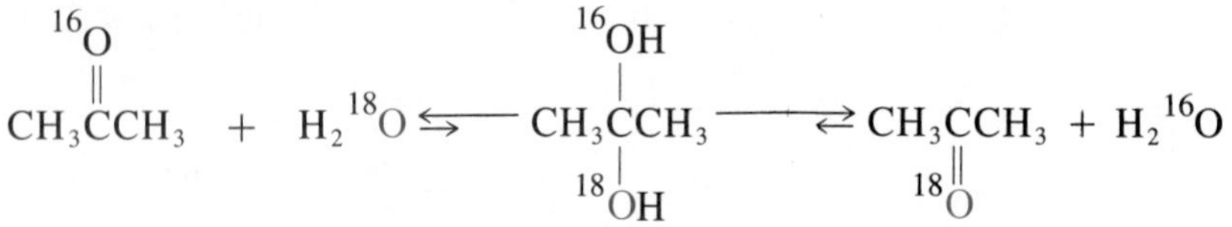

acetone
shown with the most
abundant isotope of oxygen

hydrate of acetone
a symmetrical compound

acetone
labeled with ^{18}O

VISUALIZING THE REACTION

Formation and decomposition of a hydrate

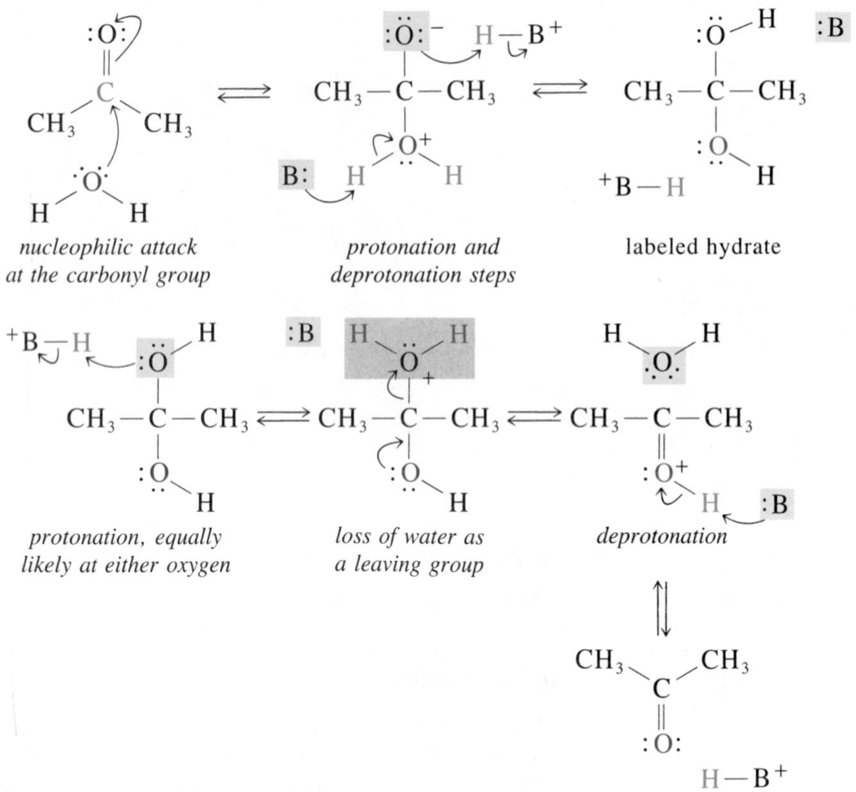

nucleophilic attack
at the carbonyl group

protonation and
deprotonation steps

labeled hydrate

protonation, equally
likely at either oxygen

loss of water as
a leaving group

deprotonation

labeled
carbonyl
group

Nucleophilic attack by water at the carbonyl group forms a new carbon-oxygen bond. Protonation of the alkoxide ion and deprotonation of the oxonium ion complete the formation of the hydrate. A reversal of each of these steps gives back the carbonyl compound. Because the hydrate is a symmetrical compound, its decomposition is equally likely to give a labeled ketone or an unlabeled one.

B. Acetals and Ketals

Alcohols will interact with aldehydes in an equilibrium process, just as water does, to yield compounds known as **hemiacetals.** In a hemiacetal, a hydroxyl group and an alkoxyl group are bonded to the same carbon atom. Hemiacetals, after losing a molecule of water, react with yet another molecule of alcohol to give **acetals.** In an acetal, two alkoxyl groups are bonded to the same carbon atom. When acetaldehyde is placed in an excess of methanol, the aldehyde, its hemiacetal, and its acetal are all present at equilibrium.

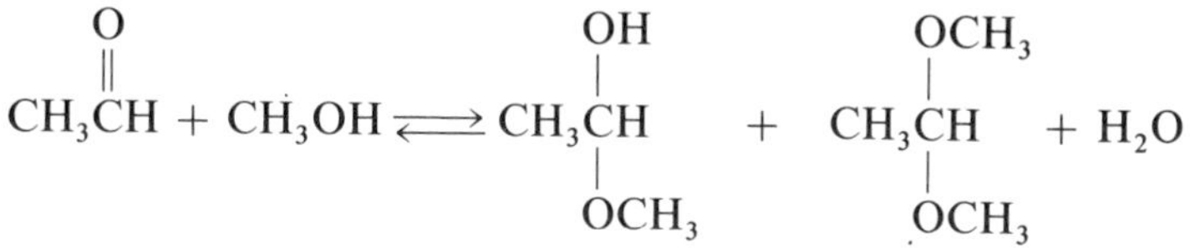

acetaldehyde methanol 1-methoxy-1-ethanol *a hemiacetal* 1,1-dimethoxyethane *an acetal*

Ketones give **hemiketals** and **ketals** with alcohols. Hemiketals and ketals do not form as spontaneously as hemiacetals and acetals do. The amount of the hemiketal in equilibrium with acetone dissolved in methanol, for example, is very small.

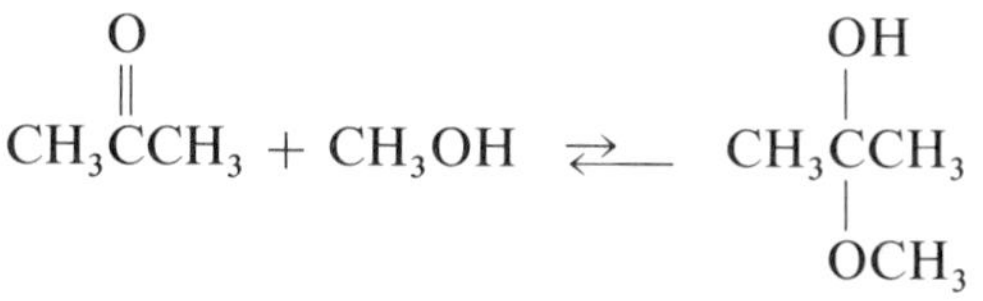

acetone methanol 2-methoxy-2-propanol 0.3% at equilibrium *a hemiketal*

Hemiacetals and hemiketals are generally unstable compounds unless the alcohol and carbonyl groups are part of the same molecule, in which case stable cyclic hemiacetals form. The most important examples of these structures are found in carbohydrates. For example, glucose exists as a cyclic hemiacetal with a six-membered ring.

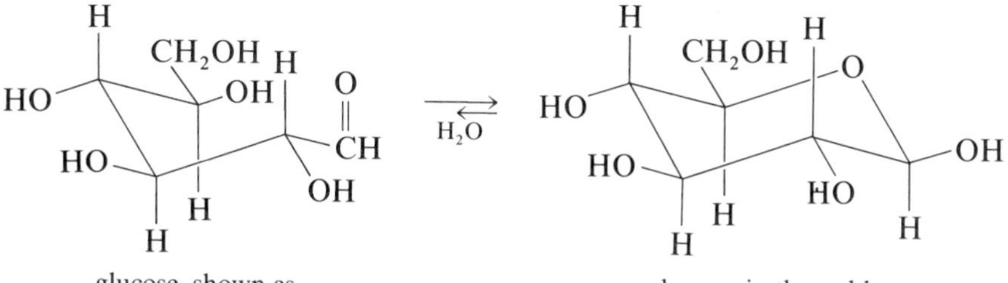

glucose, shown as a hydroxyaldehyde

glucose, in the stable hemiacetal form

Other simpler hydroxyaldehydes also form stable cyclic hemiacetals. 4-Hydroxybutanal, for example, forms a hemiacetal that is a five-membered ring.

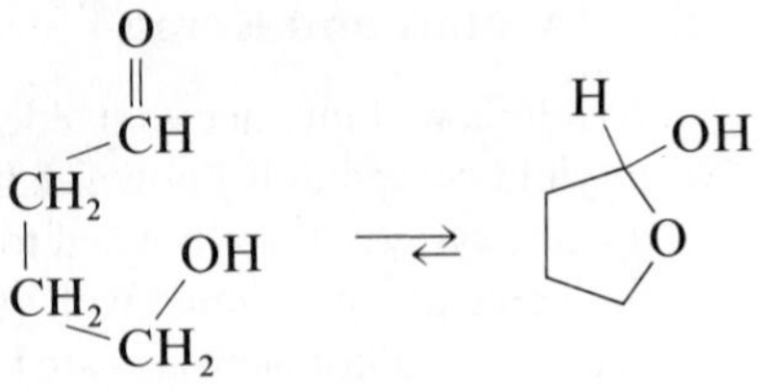

4-hydroxybutanal — 11% at equilibrium; 2-hydroxytetrahydrofuran — 89% at equilibrium

PROBLEM 13.10

5-Hydroxypentanal gives a six-membered cyclic hemiacetal. Write an equation for the formation of the hemiacetal, showing the six-membered ring in the chair form. How many stereoisomers can the hemiacetal have?

If an aldehyde or ketone is placed in an excess of an alcohol in the presence of an acid, an acetal or ketal is formed. Acetone and methanol, in a reaction that is catalyzed by an arylsulfonic acid, are converted into the dimethyl ketal of acetone, 2,2-dimethoxypropane.

$$\underset{\text{acetone}}{CH_3\overset{\overset{\displaystyle O}{\|}}{C}CH_3} + \underset{\text{methanol}}{2\,CH_3OH} \xrightarrow[ArSO_3H]{} \underset{\substack{\text{2,2-dimethoxypropane}\\ \textit{a ketal}}}{CH_3\overset{\overset{\displaystyle OCH_3}{|}}{\underset{\underset{\displaystyle OCH_3}{|}}{C}}CH_3} + H_2O$$

An acetal or ketal structurally resembles a diether and as such is stable to bases and nucleophilic reagents. Acetals and ketals, however, are sensitive to acid and are easily converted to the original carbonyl compound and alcohol, as shown for 2,2-dimethoxypropane.

$$\underset{\text{2,2-dimethoxypropane}}{CH_3\overset{\overset{\displaystyle OCH_3}{|}}{\underset{\underset{\displaystyle OCH_3}{|}}{C}}CH_3} + H_2O \xrightarrow[HCl]{} \underset{\text{acetone}}{CH_3\overset{\overset{\displaystyle O}{\|}}{C}CH_3} + \underset{\text{methanol}}{2\,CH_3OH}$$

In order to get good yields of acetals and ketals, the water formed in the equilibrium reaction must be removed.

The conversion of an acetal or ketal to the original carbonyl compound and alcohol is an example of a hydrolysis reaction. A **hydrolysis reaction** is one in which a σ bond is cleaved by the addition of the elements of water to the fragments formed in the cleavage. Note that **hydrolysis** in which *lysis* (meaning *to loosen*) takes place contrasts with **hydration,** in which water is added to a multiple bond, but no fragmentation of the molecule occurs (p. 280).

C. Mechanism of Acetal or Ketal Formation and Hydrolysis

Any mechanism proposed for the formation of an acetal or ketal must account for the ease with which such a compound is hydrolyzed by aqueous acid. The first stage of the reaction of acetone with methanol is similar to the formation of the hydrate of acetone (p. 498) but gives the hemiketal.

PROBLEM 13.11

(a) Using the mechanism for the formation of the hydrate of acetone (p. 498) as a guide, write a mechanism for the formation of the hemiketal of acetone and methanol.
(b) How would the mechanism change if the reaction were carried out in the presence of a strong acid?

The hemiketal of acetone is converted to the ketal in the presence of excess methanol and a strong acid.

VISUALIZING THE REACTION

Conversion of a hemiketal to a ketal

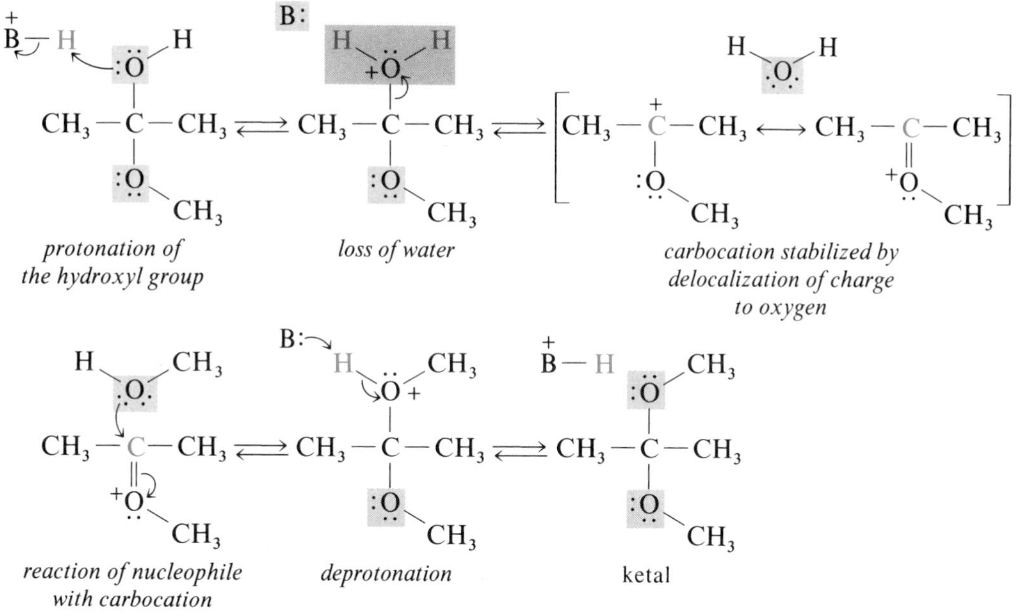
protonation of the hydroxyl group
loss of water
carbocation stabilized by delocalization of charge to oxygen
reaction of nucleophile with carbocation
deprotonation
ketal

Protonation of the hydroxyl group in the hemiketal and loss of water forms a carbocation that is particularly stable because the positive charge can be delocalized to an oxygen atom in a resonance contributor in which carbon and oxygen atoms both have octets of electrons. A pair of nonbonding electrons from the oxygen atom of methanol is donated to the carbocation, and deprotonation of the new intermediate gives the dimethyl ketal of acetone. Continuous removal from the reaction mixture of the water that is formed draws the equilibrium toward the product. In practice, these reactions are often carried out in the presence of solvents such as benzene or toluene that codistill with water, carrying it out of the reaction mixture.

The hydrolysis of the ketal is a reversal of all of the steps in its formation. If the ketal is put in excess of water with a trace of acid, an ether oxygen atom is protonated and methanol is lost. Water adds to the carbocation intermediate, the hemiketal is obtained, and the reverse process continues until ketone and alcohol are regenerated.

PROBLEM 13.12

When acetals or ketals are hydrolyzed in water containing ^{18}O, the resulting alcohols contain very little ^{18}O. Write a mechanism for the hydrolysis of 2,2-dimethoxypropane in $H_2{}^{18}O$ with a trace of hydrochloric acid added, and show where the ^{18}O will turn up in the products.

PROBLEM 13.13

Acetals, though they resemble ethers in structure, are cleaved by acids much more easily than ethers are (p. 450). Vinyl ethers, also called **enol ethers,** are hydrolyzed easily with dilute acid, just as acetals are, to give aldehydes or ketones as products. Propose a mechanism for the following reaction.

$$CH_2{=}CH{-}OCH_2CH_3 \xrightarrow[HClO_4\ (0.1\ M)]{H_2O} CH_3\overset{O}{\overset{\|}{C}}H + CH_3CH_2OH$$

(A review of Problem 9.22 may be helpful.) Why does this hydrolysis reaction occur so easily?

D. Cyclic Acetals or Ketals

Because acetals or ketals can be hydrolyzed under very mild conditions to regenerate the original carbonyl compound, they are often prepared in order to protect carbonyl groups in multifunctional compounds while reactions are carried out elsewhere in the molecule. The acetal function is stable to bases, nucleophiles, and reducing agents, so it can be used with a wide variety of reagents as long as acidic conditions are avoided. The acetals or ketals of simple aldehydes and ketones with low-molecular-weight alcohols are relatively easy to form, but the reaction becomes difficult if the ketone or alcohol is large and sterically hindered.

Ethylene glycol is used to make cyclic acetals or ketals in which both alcohol groups that react with the carbonyl group are on the same molecule. The reaction is shown below for cyclohexanone.

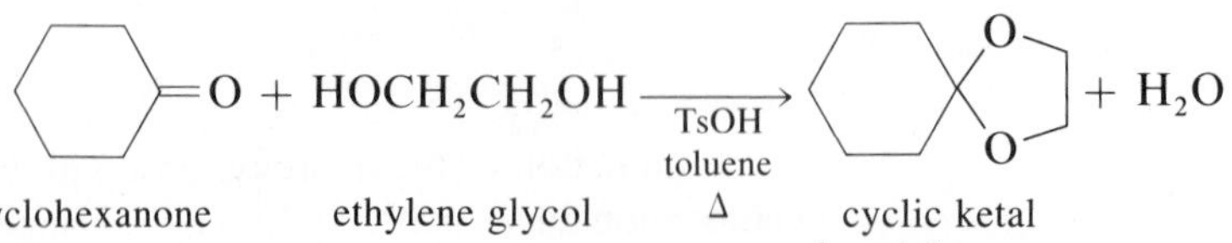
=O + HOCH₂CH₂OH
TsOH
toluene
Δ
O
O
+ H₂O

cyclohexanone ethylene glycol cyclic ketal of cyclohexanone

Study Guide Concept Map 13.5

The second alcohol function is held in close proximity to the intermediate carbocation by a two-carbon chain, and a ketal is produced easily because of the ease with which five-membered rings form.

PROBLEM 13.14

Predict the products of the following reactions.

(a) 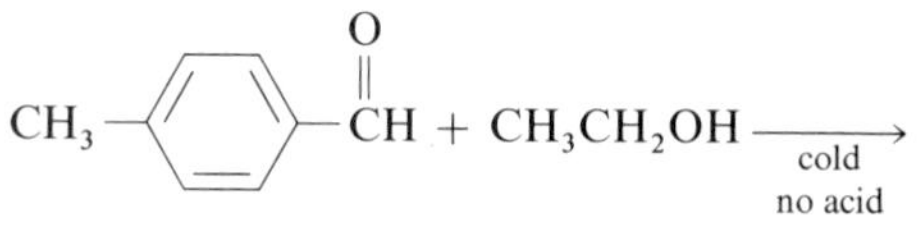

(b) 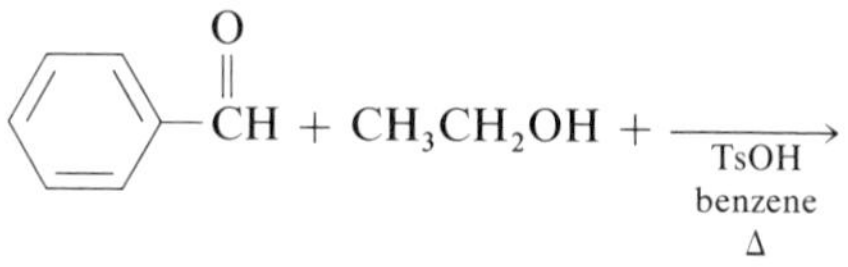

(c) $CH_3CH_2CH_2\overset{O}{\overset{\|}{C}}H + CH_3OH \xrightarrow[\text{benzene, }\Delta]{\text{TsOH}}$

(d) 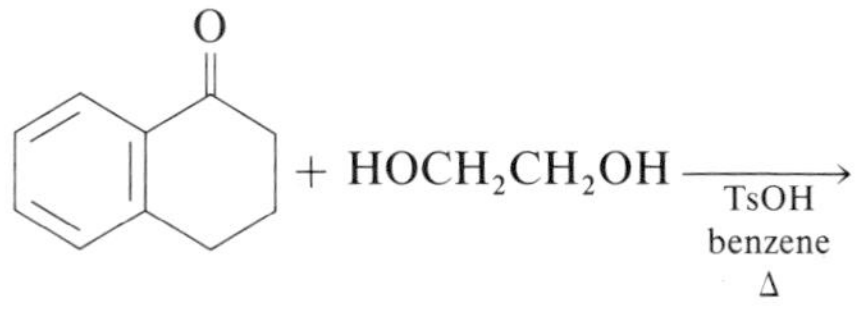

PROBLEM 13.15

Acetone reacts with glycerol (1,2,3-propanetriol) in the presence of *p*-toluenesulfonic acid to give a product that has the molecular formula $C_6H_{12}O_3$. Propose a structure for this compound.

13.8 ADDITION REACTIONS OF NUCLEOPHILES RELATED TO AMMONIA

A. Compounds Related to Ammonia and Their Use in the Characterization of Aldehydes and Ketones

Amines and other compounds related to ammonia make up an important class of reagents that behave as nucleophiles toward the carbonyl group. Some typical compounds with this kind of reactivity are shown below.

RNH_2

alkylamine

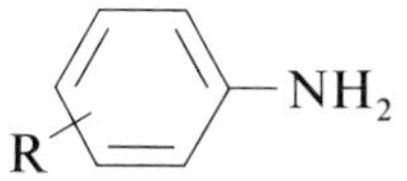

arylamine

may have other substituents on the ring

$HONH_2$

hydroxylamine

H_2NNH_2

hydrazine

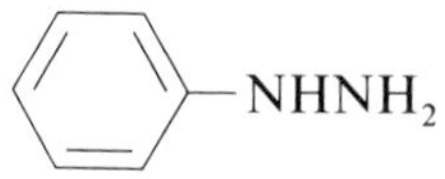

phenylhydrazine

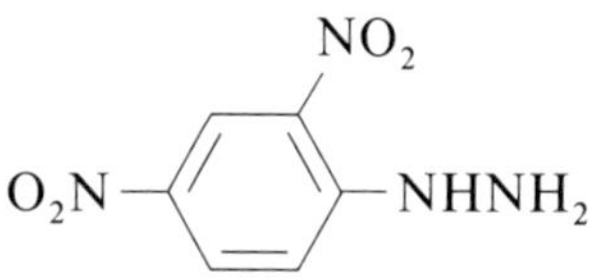

2,4-dinitrophenylhydrazine

Some of these reagents convert low-molecular-weight, liquid aldehydes and ketones into solid compounds. A compound prepared from another compound by some standard reaction typical of a functional group is called a **derivative.** A derivative is used to identify a compound by comparing the physical properties of the derivative, such as the melting point, with those reported for derivatives of similar known compounds until a match in properties is found. Solid derivatives are easier to handle in small quantities than liquids are, and it is easier to determine an accurate melting point on a small amount of material than it is to determine an accurate boiling point. Solid derivatives are, therefore, useful in characterizing compounds.

Cyclohexanone, for example, is a liquid that reacts with hydroxylamine (available in the laboratory as its hydrochloride salt) to form a solid derivative called an **oxime.**

$$\underset{\text{hydroxylamine hydrochloride}}{HONH_3^+\ Cl^-} + \underset{\text{sodium acetate}}{CH_3\overset{O}{\overset{\|}{C}}O^-Na^+} \longrightarrow \underset{\text{hydroxylamine}}{HONH_2} + CH_3\overset{O}{\overset{\|}{C}}OH + Na^+Cl^-$$

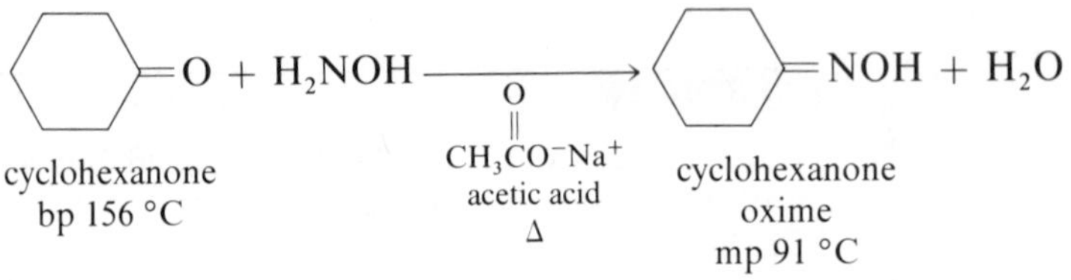
O + H2NOH
NOH + H2O
CH3CO⁻Na⁺
acetic acid
Δ
cyclohexanone
bp 156 °C
cyclohexanone
oxime
mp 91 °C

The reaction shown above is typical of the reactions of amines and similar compounds with aldehydes and ketones. A nitrogen atom is attached to the carbon atom of the carbonyl group by a double bond, and water is eliminated. The reaction usually goes best with mild acid catalysis at a pH of about 5. A mechanism for the reaction is shown in the next section.

B. A Mechanism for the Reaction of Compounds Related to Ammonia with Aldehydes and Ketones

The reaction of cyclohexanone with hydroxylamine (see above) can be used as a model for the general reaction of aldehydes and ketones with such nitrogen compounds. Sodium acetate is used as a base to remove a proton from the nitrogen atom in hydroxylamine hydrochloride to form the weak acid acetic acid. Sodium acetate and acetic acid together form a buffer system, so the acidity of the reaction mixture remains close to pH 5. It is important that the solution not be too acidic because the unprotonated nitrogen atom in hydroxylamine is the nucleophile that reacts with the electrophilic carbon atom of the carbonyl group. On the other hand, acid is necessary in order to assist in the loss of water in the final stages of the reaction, so the solution must be somewhat acidic. The rate of the reaction is greatest at pH ~5.

The reaction of hydroxylamine with cyclohexanone takes place in two steps. In the first step, illustrated in the box below, the nucleophile attacks the carbonyl group.

VISUALIZING THE REACTION

Addition of the nucleophile to the carbonyl group

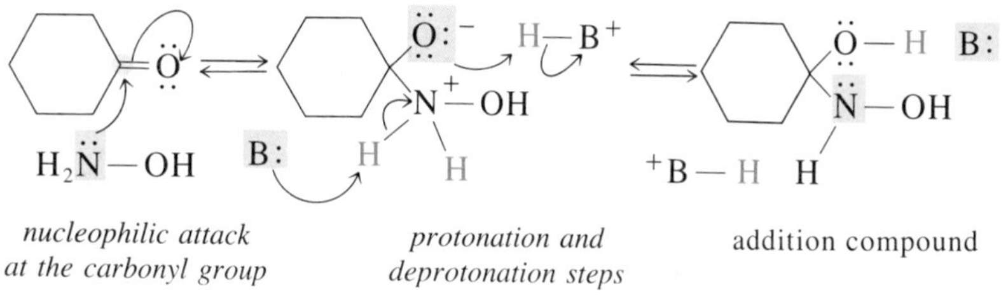
H₂N̈—OH
B:
H
H
H—B⁺
+B—H
Ö—H
B:
N̈—OH
N—OH
H
nucleophilic attack
at the carbonyl group
protonation and
deprotonation steps
addition compound

The intermediate that forms loses a proton from the positively charged nitrogen atom and is protonated at the oxygen atom. The addition compound with a hydroxyl group and an amino group on the same carbon atom loses water easily, and a carbon-nitrogen double bond forms. The loss of water is catalyzed by acid and is the rate-determining step for the reaction at moderate acidity.

VISUALIZING THE REACTION

Loss of water from the addition compound

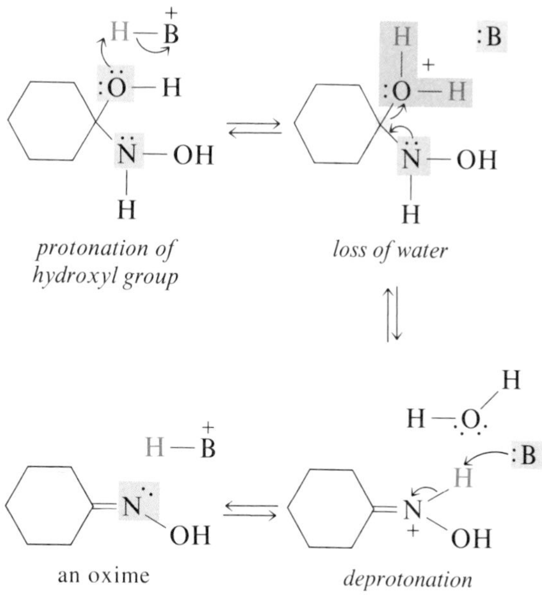
H—B
:Ö—H
N̈—OH
H
:B
N—OH
protonation of
hydroxyl group
loss of water
H—B
OH
:B
an oxime
deprotonation

The hydroxyl group is converted into a good leaving group by protonation, and the water molecule is displaced by the nonbonding electrons on the nitrogen atom. Removal of a proton from the nitrogen atom gives the oxime.

All of the processes shown above for the formation of cyclohexanone oxime may be applied to reactions of other compounds that are related to ammonia with aldehydes and ketones.

C. Reactions of Carbonyl Compounds with Amines

Aldehydes react readily with both alkylamines and arylamines to give compounds known as **imines** or **Schiff bases.** Usually heating the aldehyde and the amine together while distilling off the water that forms is all that is necessary, as illustrated below by the formation of *N*-ethylpropanalimine and *N*-methylbenzaldimine.

$$\underset{\text{propanal}}{CH_3CH_2\overset{O}{\overset{\|}{C}}H} + \underset{\text{ethylamine}}{H_2NCH_2CH_3} \xrightarrow[\Delta]{} \underset{\substack{N\text{-ethylpropanalimine} \\ 81\%}}{CH_3CH_2CH{=}NCH_2CH_3} + H_2O\uparrow$$

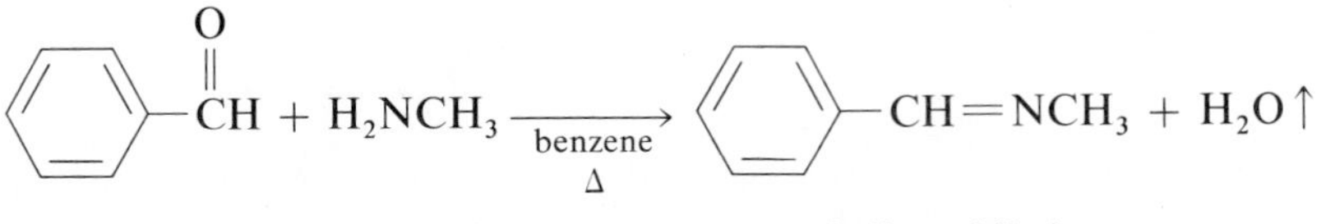

benzaldehyde methylamine *N*-methylbenzaldimine 90%

The reactions are reversible; the imines are converted back to aldehydes and amines by reaction with water. The imines derived from aromatic aldehydes and amines are more stable than the ones derived from alkyl components.

Ketones form imines, too, but generally not as readily as aldehydes do. A typical reaction is that of acetone with propylamine. The reaction is catalyzed by hydrochloric acid, and the acid is then neutralized by sodium hydroxide.

$$\underset{\text{acetone}}{CH_3\overset{O}{\overset{\|}{C}}CH_3} + \underset{\text{propylamine}}{CH_3CH_2CH_2NH_2} \xrightarrow[HCl]{} \xrightarrow{NaOH} \underset{\substack{\text{imine of acetone and propylamine} \\ 67\%}}{CH_3\overset{CH_3}{\overset{|}{C}}{=}NCH_2CH_2CH_3} + H_2O$$

PROBLEM 13.16

Suggest a mechanism for the formation of *N*-methylbenzaldimine.

PROBLEM 13.17

The steps shown for the mechanism for the formation of cyclohexanone oxime (p. 505) are all reversible. Using that mechanism as a guide, propose a mechanism for the reaction of *N*-methylbenzaldimine with water to give benzaldehyde and methylamine.

PROBLEM 13.18

Suggest a mechanism for the formation of the imine of acetone and propylamine. What role does the catalytic amount of hydrochloric acid play? Why is sodium hydroxide used in the second step of the reaction?

D. Reactions of Carbonyl Compounds with Hydrazines

Phenylhydrazine and 2,4-dinitrophenylhydrazine are used to prepare derivatives of aldehydes and ketones known as **phenylhydrazones.** Not all phenylhydrazones are solids. For example, acetone phenylhydrazone has a low melting point and is often isolated as a liquid.

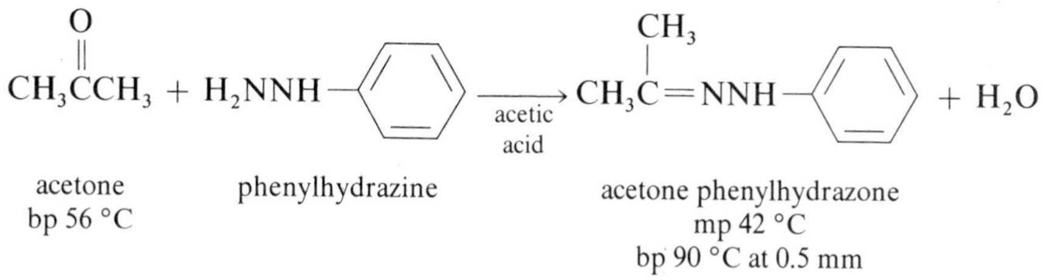
acetic acid
acetone
bp 56 °C
phenylhydrazine
acetone phenylhydrazone
mp 42 °C
bp 90 °C at 0.5 mm

For this reason, 2,4-dinitrophenylhydrazine is more often used in the preparation of derivatives. The two nitro groups increase both the molecular weight and the polarity of the reagent so that the derivatives it forms are more crystalline and have higher melting points than the corresponding unsubstituted phenylhydrazones. The difference is illustrated below for 2-methylpropanal.

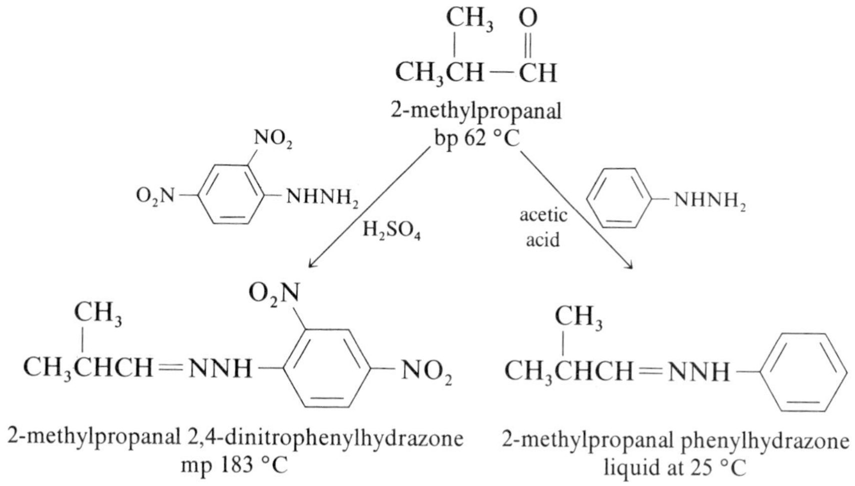
2-methylpropanal
bp 62 °C
H2SO4
acetic acid
2-methylpropanal 2,4-dinitrophenylhydrazone
mp 183 °C
2-methylpropanal phenylhydrazone
liquid at 25 °C

Hydrazines are less basic but more nucleophilic than amines. The nitrogen atom that is not bonded directly to the aromatic ring is the more nucleophilic one in phenylhydrazine, and it reacts with the carbonyl group. Note that the reaction is catalyzed by acetic acid. For the reaction with 2,4-dinitrophenylhydrazine, a more powerful acid catalyst, sulfuric acid, is used. 2,4-Dinitrophenylhydrazine is less nucleophilic than phenylhydrazine, and the conjugate acid of the carbonyl compound must be present for a reaction to take place.

The differing nucleophilicity of the two nitrogen atoms in phenylhydrazine and the weaker nucleophilicity of the nitrogen atoms in 2,4-dinitrophenylhydrazine can both be rationalized by drawing resonance contributors for the molecules.

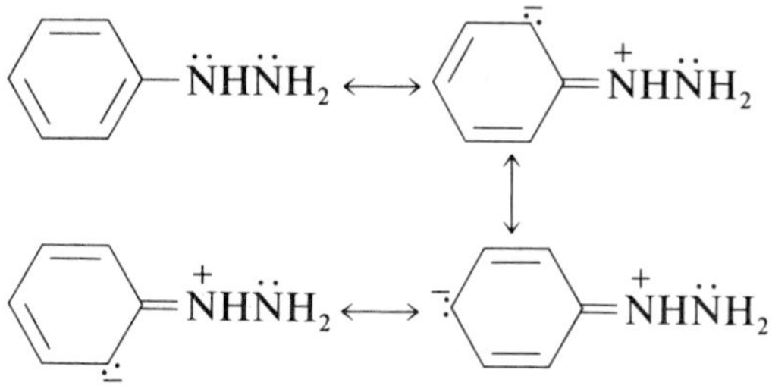

resonance contributors for phenylhydrazine

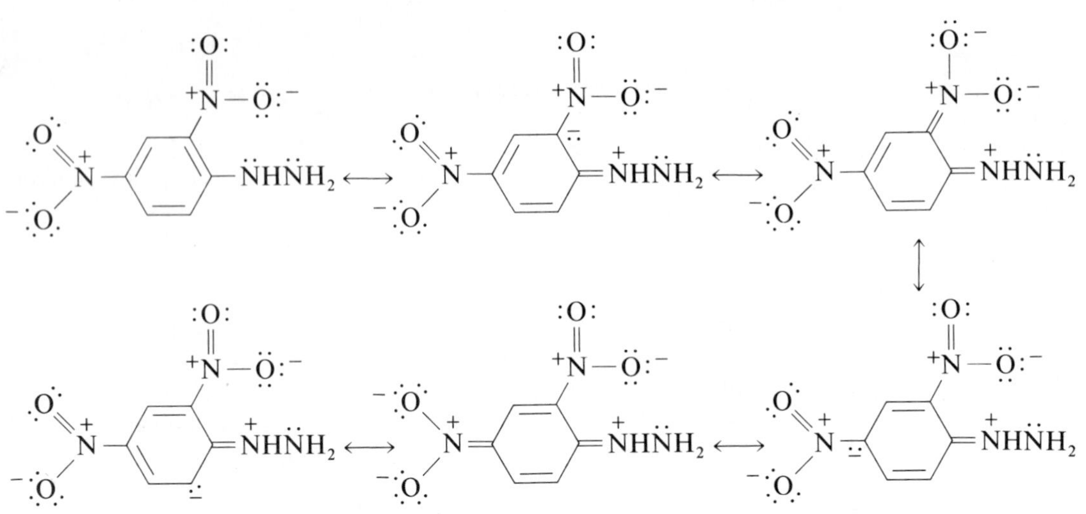

resonance contributors for 2,4-dinitrophenylhydrazine

The nonbonding electrons on the nitrogen atom adjacent to the aromatic ring in phenylhydrazine can be delocalized to the ortho and para positions of the ring by resonance. They are, therefore, less available for reaction. The delocalization possible for the nonbonding electrons from the analogous nitrogen atom in 2,4-dinitrophenylhydrazine is even more extensive. Resonance contributors can be written delocalizing the electrons to oxygen atoms of the nitro groups. The electron density at this nitrogen atom adjacent to the aromatic ring is less in 2,4-dinitrophenylhydrazine than it is in phenylhydrazine. The partial positive character at the nitrogen atom next to the ring in 2,4-dinitrophenylhydrazine also reduces electron density to some extent at the nitrogen atom next to it. For this reason, reactions with 2,4-dinitrophenylhydrazine are carried out under strongly acidic conditions so that the carbonyl group, by protonation, is made more reactive toward this weaker nucleophile.

Study Guide
Concept Map 13.6

PROBLEM 13.19

Complete the following equations by writing structures for the expected products.

(a) 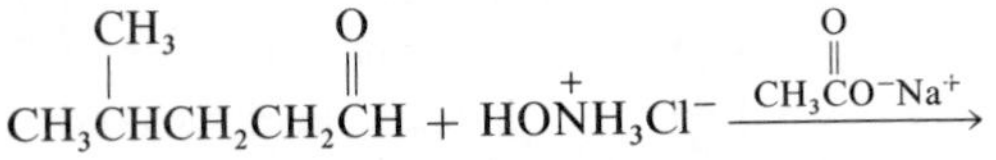

(b) 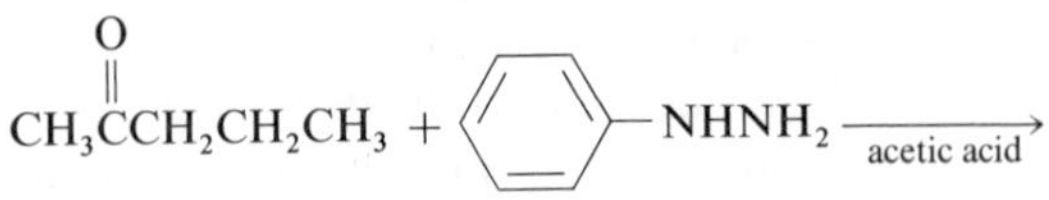

(c) 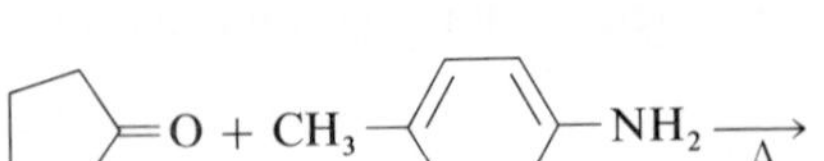

(d) 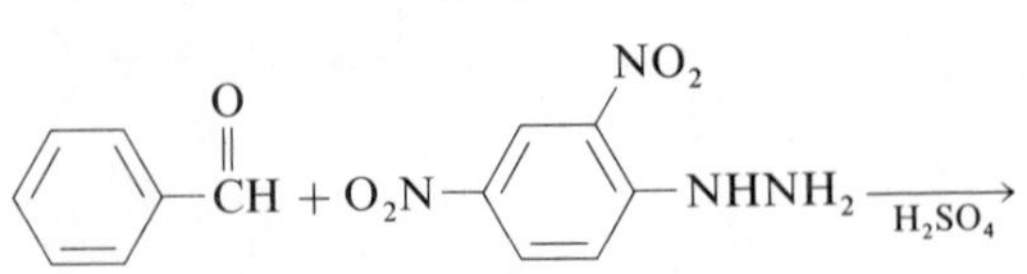

(e) 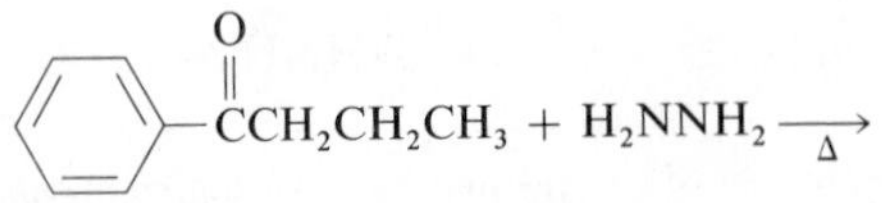

(f) $CH_3\overset{O}{\overset{\|}{C}}CH_2CH_3 + CH_3\overset{CH_3}{\overset{|}{C}}HCH_2NH_2 \xrightarrow[HCl]{} \xrightarrow{NaOH}$

(g) 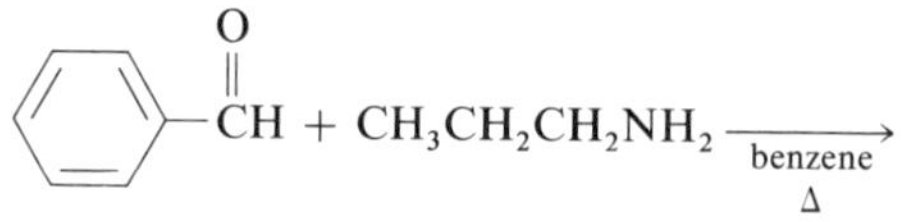

13.9 REDUCTION OF CARBONYL GROUPS TO METHYLENE GROUPS

A. The Clemmensen Reduction

A general reaction of aldehydes and ketones is the reduction of the carbonyl group all the way to a methylene, $—CH_2—$, group. For example, heptanal is reduced to the hydrocarbon heptane.

$$\underset{\text{heptanal}}{CH_3CH_2CH_2CH_2CH_2CH_2\overset{O}{\overset{\|}{C}}H} \xrightarrow[Zn(Hg)]{HCl} \underset{\substack{\text{heptane} \\ 72\%}}{CH_3CH_2CH_2CH_2CH_2CH_2CH_3}$$

loss of two bonds to oxygen, gain of two bonds to hydrogen; therefore, a reduction at carbon 1

This reaction, known as the **Clemmensen reduction,** is carried out with zinc amalgam (zinc dissolved in mercury) and concentrated hydrochloric acid. The reaction is selective for a carbonyl group of an aldehyde (or ketone) and does not affect the carbonyl group of a carboxylic acid. Therefore, 4-oxo-4-phenylbutanoic acid is reduced to 4-phenylbutanoic acid.

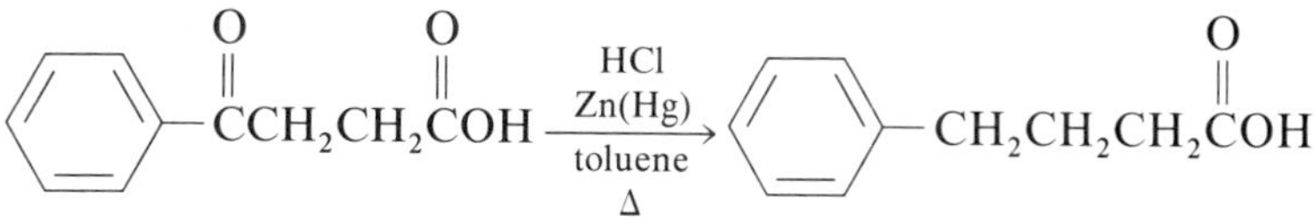
HCl
Zn(Hg)
toluene
Δ

4-oxo-4-phenylbutanoic acid — 24–30 h — 4-phenylbutanoic acid 90%

The Clemmensen reduction involves a transfer of electrons from the metal and protonation of intermediate species by the strongly acidic reaction mixture. An alcohol is formed as one intermediate. The mechanism is complex and not completely understood. It is sufficient that you recognize that the conversion of a carbonyl group to a methylene group is a reduction reaction and requires a reducing agent, zinc metal in an acidic solution.

B. The Wolff-Kishner Reduction

The conditions for the Clemmensen reduction are strongly acidic and may give rise to side reactions. An alternative way of reducing carbonyl groups to methylene groups is the **Wolff-Kishner reduction,** which uses a basic reagent. In the Wolff-

Kishner reduction, the hydrazone of an aldehyde or ketone is prepared and decomposed under basic conditions. An example is the reduction of 2-octanone to octane.

$$\underset{\text{2-octanone}}{CH_3CH_2CH_2CH_2CH_2CH_2\overset{O}{\overset{\|}{C}}CH_3} \xrightarrow[\text{diethylene glycol}\ \Delta]{H_2NNH_2,\ NaOH} \underset{\substack{\text{octane}\\ 75\%}}{CH_3CH_2CH_2CH_2CH_2CH_2CH_2CH_3}$$

The reaction starts with the formation of a hydrazone. Under the strongly basic conditions, the hydrazone is deprotonated at the nitrogen atom.

VISUALIZING THE REACTION

The Wolff-Kishner reduction

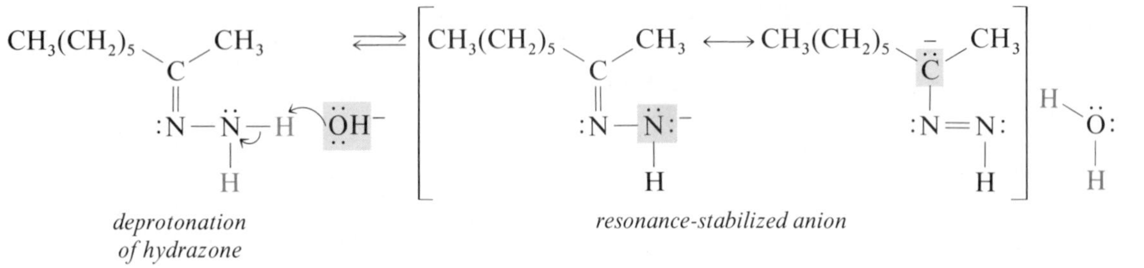
deprotonation of hydrazone
resonance-stabilized anion

Such a deprotonation is possible because there is delocalization of the charge to give anionic character to the carbon atom that was part of the original carbonyl group. A negatively charged carbon atom is more basic than a negatively charged nitrogen atom because carbon is less electronegative than nitrogen (p. 86). The anion accepts a proton at the carbon atom and loses another one from the nitrogen atom.

VISUALIZING THE REACTION

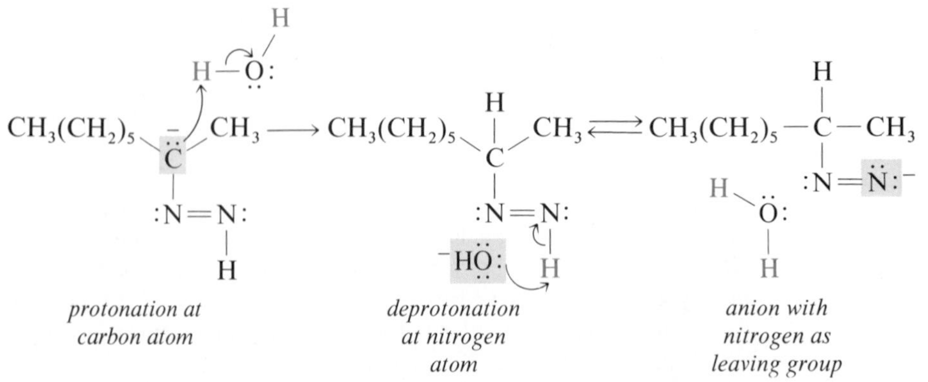
protonation at carbon atom
deprotonation at nitrogen atom
anion with nitrogen as leaving group

For this new anion, the very stable nitrogen molecule is a good leaving group. Usually the loss of a stable molecule as a leaving group creates a cation. In this case, the species as a whole is negatively charged, so the loss of the neutral molecule, nitrogen, leaves behind a carbanion. Protonation of this carbanion completes the reduction.

V I S U A L I Z I N G   T H E   R E A C T I O N

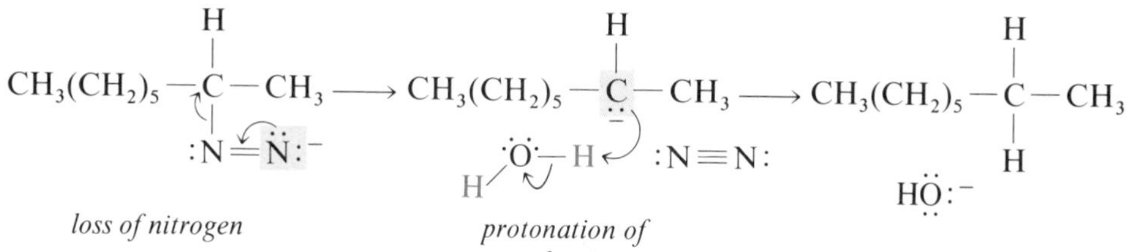
loss of nitrogen
protonation of carbanion

An example of a class of compounds for which reduction with zinc amalgam fails but Wolff-Kishner reduction succeeds is the sapogenins. The sapogenins are toxic compounds found in a variety of plants, especially desert plants. Their name derives from their ability to form soapy solutions in water. Chemists are interested in sapogenins because they contain a functionalized steroid structure that can be chemically modified to give medicinal steroids with hormonal activity. The sapogenin hecogenin, found in a number of agave plants, has the structure shown below. The carbonyl group on the ring designated by the letter C (p. 709) is reduced to a methylene group by the Wolff-Kishner reaction to give another naturally occurring sapogenin, tigogenin.

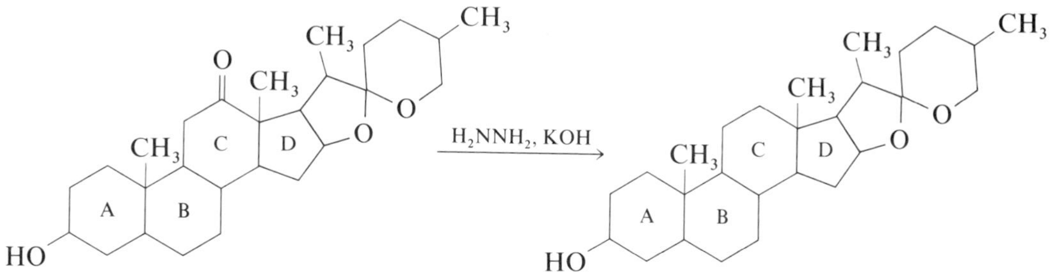
H_2NNH_2, KOH
hecogenin
tigogenin

PROBLEM 13.20

What side reaction(s) might occur if hecogenin were put in hydrochloric acid?

C. Raney Nickel Reduction of Thioacetals and Thioketals

A way of converting a carbonyl group to a methylene group without the use of strong acid or base is to make the thioacetal or thioketal of the carbonyl compound and then to cleave the carbon-sulfur bonds with hydrogen in the presence of Raney nickel (p. 304). Because thiols are more nucleophilic than alcohols (p. 226), 1,2-ethanethiol also can be used to make cyclic acetals or ketals. The reaction is catalyzed by a solution of the Lewis acid boron trifluoride in diethyl ether. The steroid diketone cholestan-3,6-dione is reduced to cholestane by this method.

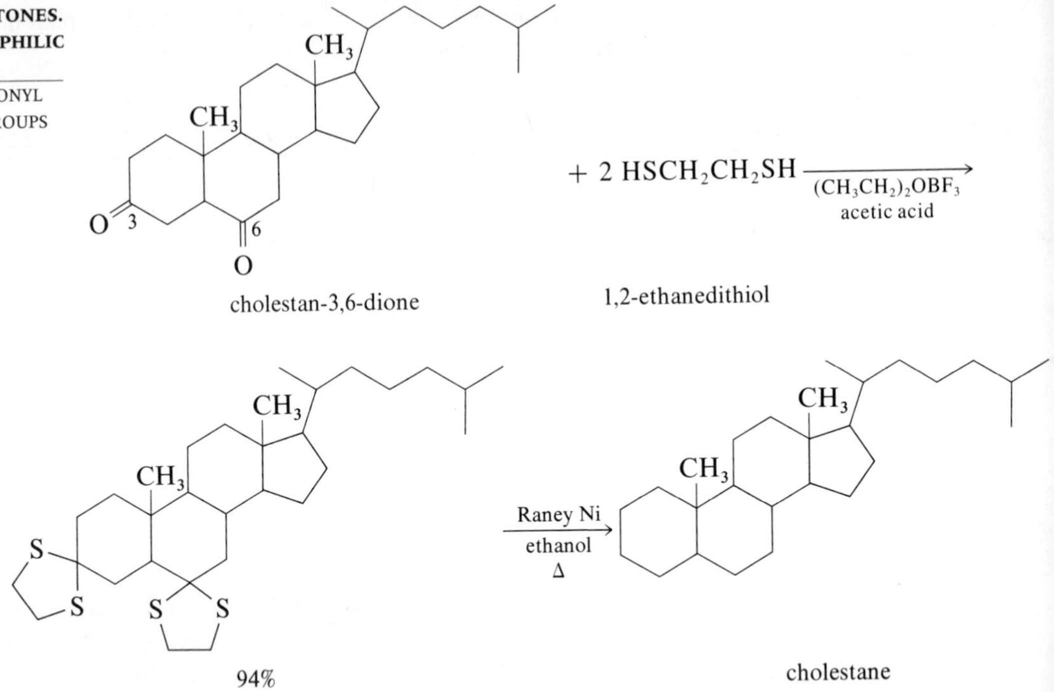
CH3
CH3
O
3
6
O
+ 2 HSCH2CH2SH
(CH3CH2)2OBF3
acetic acid
cholestan-3,6-dione
1,2-ethanedithiol
CH3
CH3
S
S
S
S
94%
Raney Ni
ethanol
Δ
CH3
CH3
cholestane

As seen in the structure above, cholestan-3,6-dione is converted into cyclic dithioketals at both ketone groups, and the sulfur atoms are replaced by hydrogen atoms to yield the hydrocarbon cholestane. This way of simplifying the structure of a steroid can be used to prove that a series of reactions did not change the carbon skeleton of the molecule or to establish the stereochemistry at various ring junctions.

Study Guide Concept Map 13.7

PROBLEM 13.21

Predict what the products of the following reactions will be.

(a) 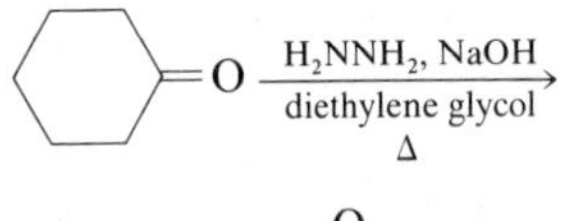
O
H2NNH2, NaOH
diethylene glycol
Δ

(b) 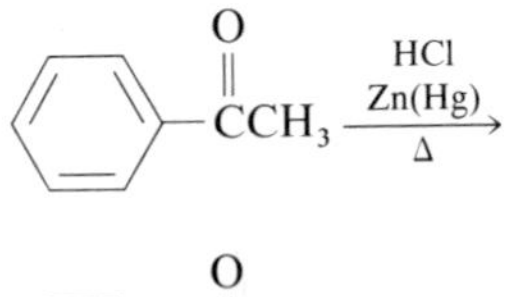
O
CCH3
HCl
Zn(Hg)
Δ

(c) 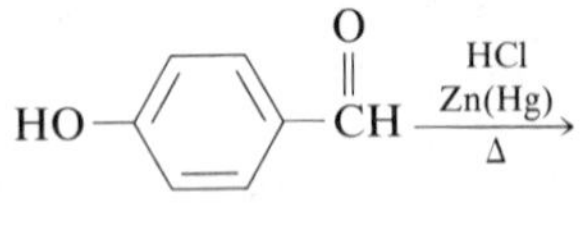
HO
O
CH
HCl
Zn(Hg)
Δ

(d) 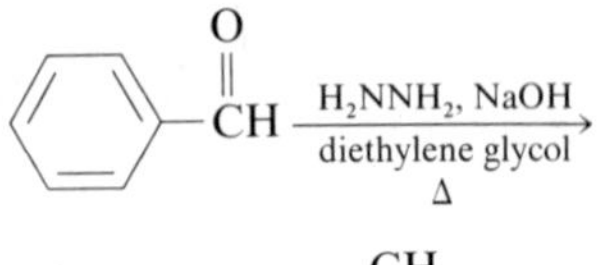
O
CH
H2NNH2, NaOH
diethylene glycol
Δ

(e) 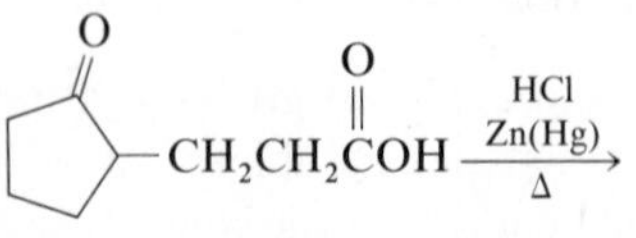
O
O
CH2CH2COH
HCl
Zn(Hg)
Δ

(f) 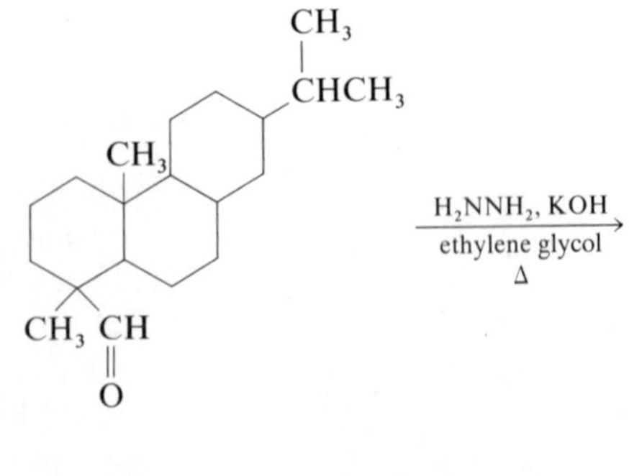
CH3
CHCH3
CH3
CH3 CH
O
H2NNH2, KOH
ethylene glycol
Δ

(g) 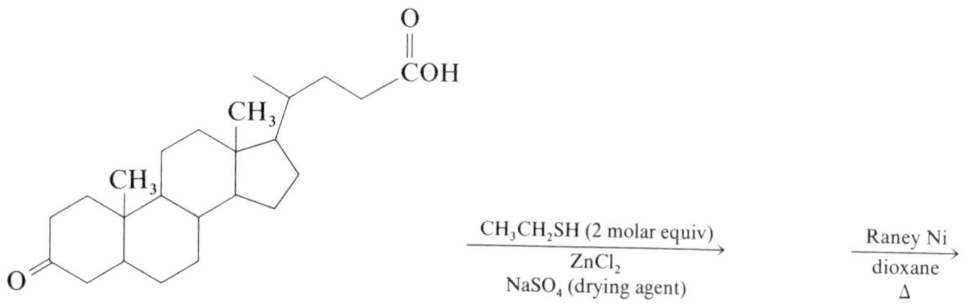

(h) $CH_3\overset{\overset{\displaystyle O}{\|}}{C}CH_2CH_2CH_3 + HSCH_2CH_2SH \xrightarrow[(CH_3CH_2)_2OBF_3]{}$

(i) 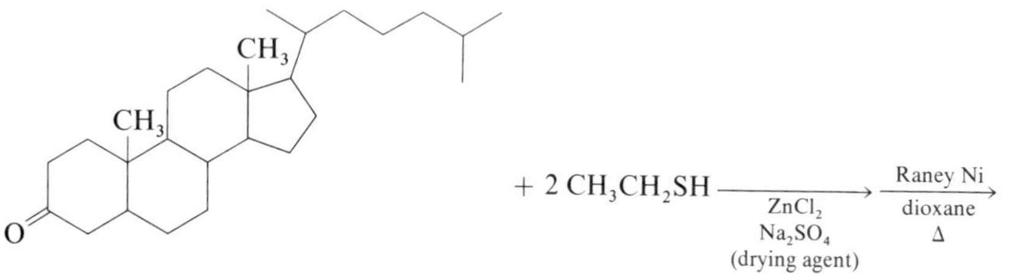

13.10 PROTECTING GROUPS IN SYNTHESIS

A. Acetals and Ketals

Most naturally occurring compounds with interesting biological properties have a variety of functional groups in them. One functional group may be adversely affected by or interfere with a reaction that a chemist wishes to carry out at another one. For example, it is not possible to prepare a Grignard reagent from an alkyl halide containing an alcohol function. And yet it may be useful to make such an organometallic reagent in a complex synthesis (p. 514). To do this, chemists use protecting groups. A **protecting group** converts a reactive functional group into a different group that is inert to the conditions of some reaction (or reactions) that is to be carried out as part of a synthetic pathway. For example, a hydroxyl group is too acidic to be present while a Grignard reagent is formed. Conversion of an alcohol to an ether would prevent it from reacting with the Grignard reagent. But not just any ether will do, because the protecting group must be easily removable once its job is done so that the original functional group can be restored. Ordinary alkyl ethers require strongly acidic conditions to be cleaved and are usually converted to alkyl iodides in the process (p. 450).

Acetals and ketals are very useful as protecting groups because of the ease with which they are formed and removed. The acetal function is stable to bases (no acidic protons), nucleophiles (no good leaving groups), and reducing agents (no multiple bonds to which hydrogen can add). Acetals can, therefore, be used with a variety of reagents as long as acidic conditions are avoided. The preparation of cyclic ketals from carbonyl compounds was shown in Section 13.7D. Diols in which the two hydroxyl groups are on adjacent carbon atoms (1,2-diols) or in which they are on carbon atoms separated by a single carbon (1,3-diols) can also be protected as cyclic ketals. Acetone is usually used to make the ketal. An example of the protection of a 1,3-diol in order to be able to make a Grignard reagent is shown below.

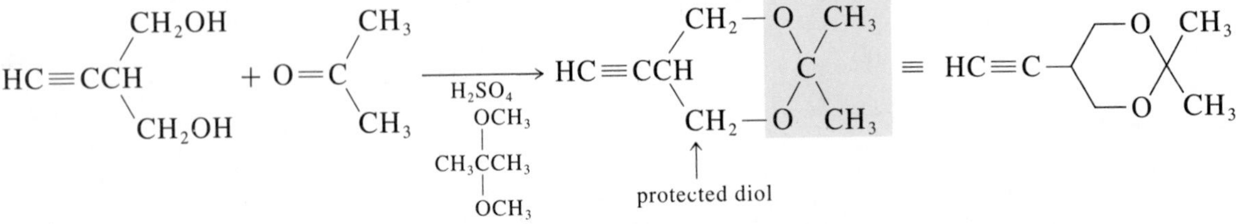
H_2SO_4
protected diol

This reaction is carried out in the presence of 2,2-dimethoxypropane, the dimethyl ketal of acetone. This ketal serves to remove the water formed as the diol is converted to the cyclic ketal and generates more acetone in the process.

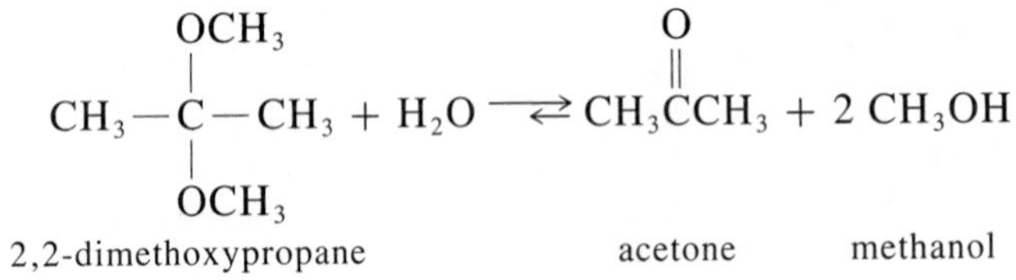
2,2-dimethoxypropane
acetone
methanol

The protected diol is converted into a Grignard reagent, which then reacts with an aldehyde that also contains a protected ketone function.

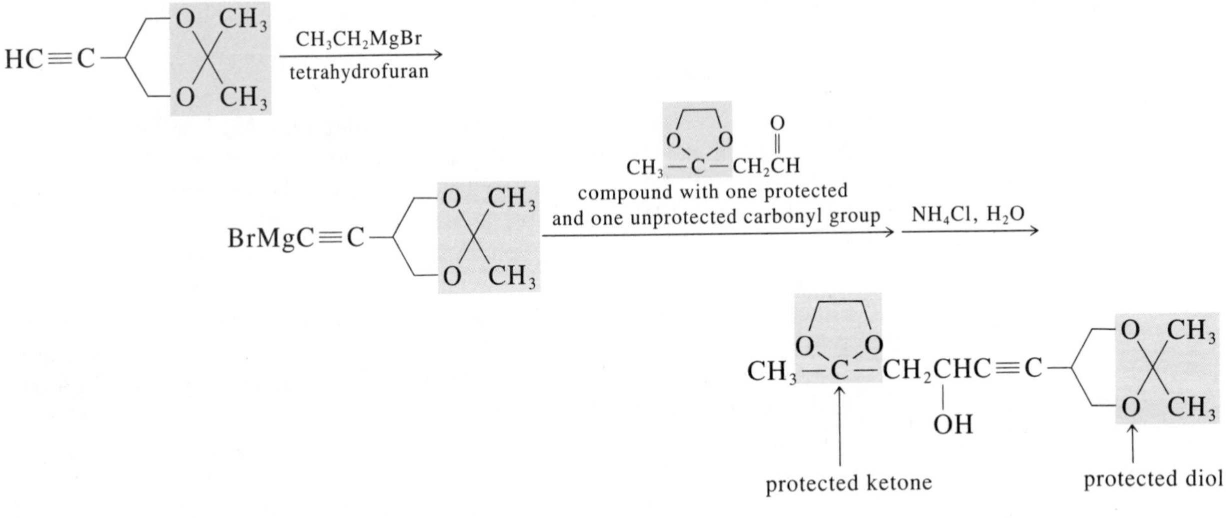
CH_3CH_2MgBr
tetrahydrofuran
compound with one protected
and one unprotected carbonyl group
NH_4Cl, H_2O
protected ketone
protected diol

PROBLEM 13.22

The compound synthesized above is converted into the following compound as part of the synthesis of a pseudomonic acid, an antibiotic. How would you carry out this conversion?

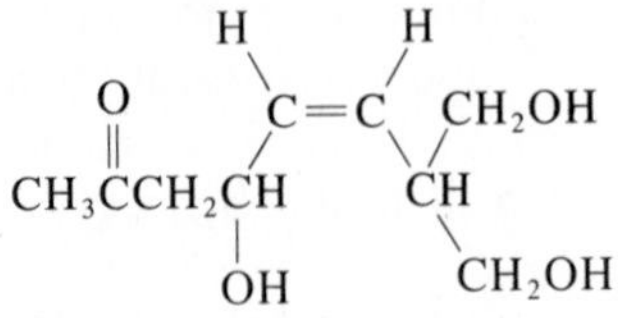

(Hint: It will be helpful if you first describe in words the changes that you see between the two structures. The problem on p. 265 will remind you of the systematic steps you need to take in solving such a problem.)

The cyclic ketals used as protecting groups in the above synthesis serve three functions.

1. One ketal protects two hydroxyl groups so that a Grignard reagent can be made in another part of the molecule.
2. The second ketal protects one carbonyl group in a molecule so that it can be differentiated from a second one.
3. Both ketals can be removed easily so that the two alcohol functions and the ketone function can be restored later in the synthesis (Problem 13.22).

Single alcohol functions can also be protected as acetals. The most common way this is done is by forming what is known as the tetrahydropyranyl ether of the alcohol. For example, the Grignard reagent from 5-chloro-1-pentanol is needed in the synthesis of a constituent of civet, which is a substance isolated from the civet cat and used in making perfumes. The hydroxyl group must be protected before the Grignard reagent can be made. This is done by treating the alcohol with dihydropyran in the presence of p-toluenesulfonic acid.

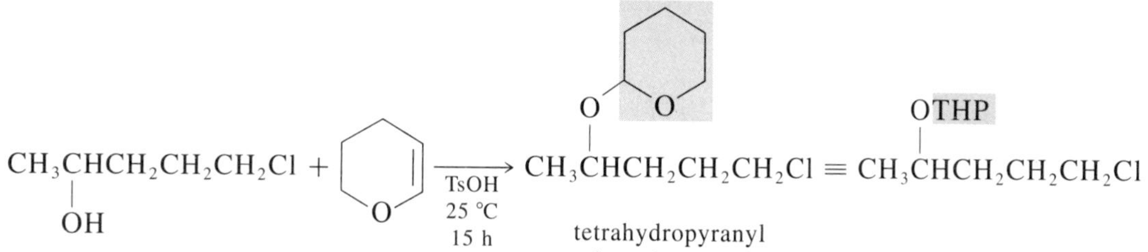
CH3CHCH2CH2CH2Cl +
OH
TsOH
25 °C
15 h
O
O
CH3CHCH2CH2CH2Cl ≡ CH3CHCH2CH2CH2Cl
OTHP

5-chloro-2-pentanol　　dihydropyran　　tetrahydropyranyl ether of 5-chloro-2-pentanol

The abbreviation THP is often used to represent the tetrahydropyranyl group in structural formulas, just as Ts is used to represent the p-toluenesulfonyl group (p. 235).

How does the tetrahydropyranyl ether form under the conditions of the reaction? Dihydropyran is a cyclic vinyl ether. It is easily protonated to give a stabilized carbocation in which the positive charge can be delocalized to the adjacent oxygen. This cation then reacts with the alcohol, which behaves as a nucleophile.

VISUALIZING THE REACTION

Addition of an alcohol to dihydropyran

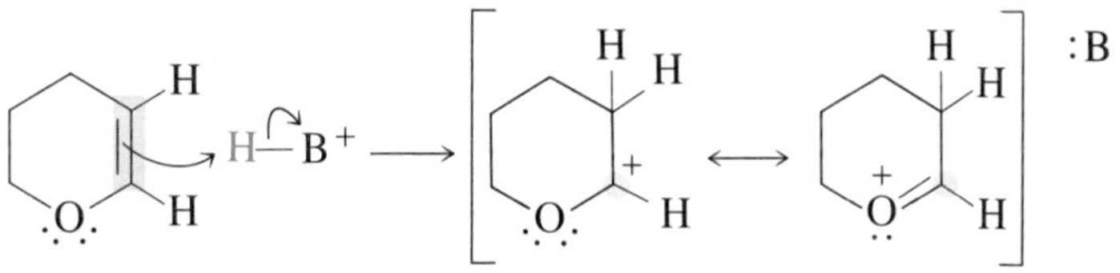
H
H
H—B+
H
H
H
H
H
H
:B

protonation of double bond　　resonance-stabilized carbocation

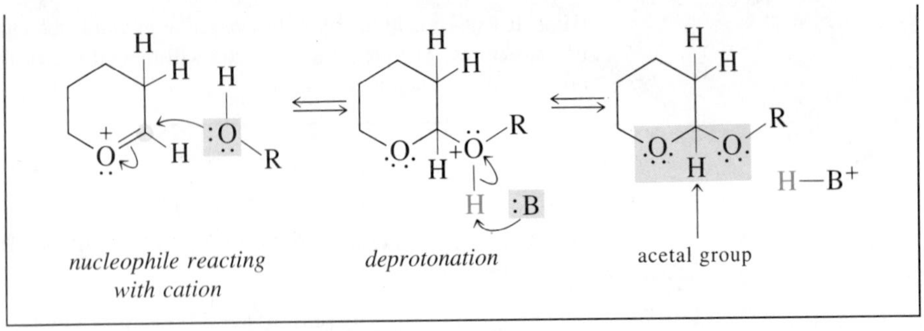
nucleophile reacting with cation
deprotonation
acetal group
H—B+

The product, though usually called an ether, is really an acetal, with two ether linkages to the same carbon atom. It is easily hydrolyzed with dilute acid, to get back the original alcohol when the need for protection is over. (Such a deprotection step is used later in the synthesis started above.)

The protected 5-chloro-2-pentanol can be converted to a Grignard reagent, which then reacts with cinnamaldehyde (3-phenyl-2-propenal).

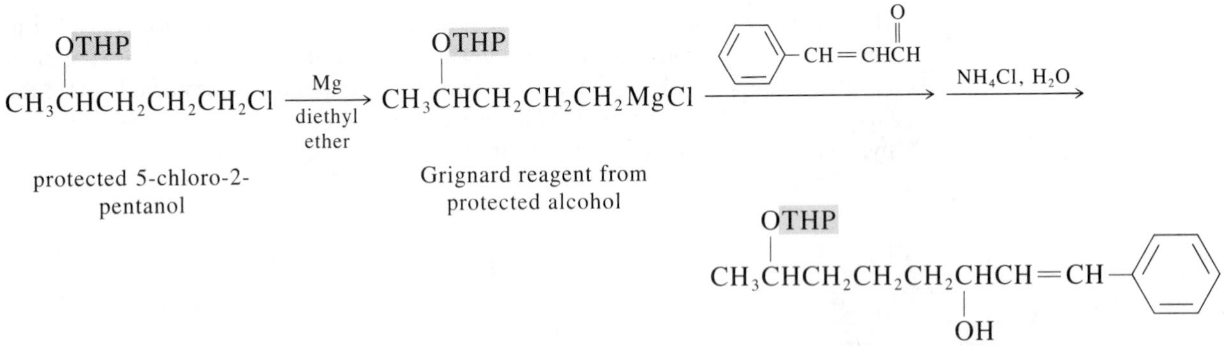
Mg
diethyl ether
NH4Cl, H2O
protected 5-chloro-2-pentanol
Grignard reagent from protected alcohol

This product has two hydroxyl groups, one protected and the other one not. The unprotected hydroxyl group is converted into the acetic acid ester (p. 576), and then the other hydroxyl group is deprotected.

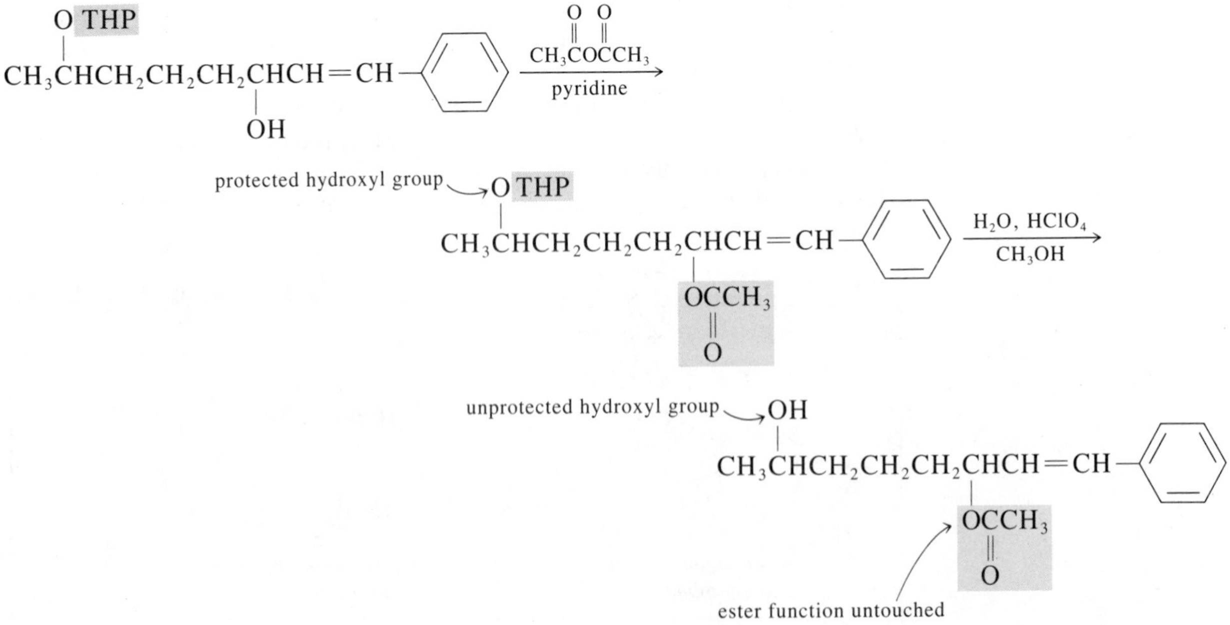
pyridine
protected hydroxyl group
H2O, HClO4
CH3OH
unprotected hydroxyl group
ester function untouched

The acetal group is hydrolyzed under such mild conditions that the ester group is untouched, even though esters generally undergo hydrolysis reactions (p. 563). The acetoxy ester group serves as a leaving group in a later stage of the synthesis.

The protection of the hydroxyl group, therefore, serves three functions in the above synthesis.

1. It allows a Grignard reagent to be made from an alkyl halide containing a hydroxyl group.
2. It differentiates that hydroxyl group from a new hydroxyl group created in the Grignard reaction.
3. It allows the original hydroxyl group to be restored at a later stage of the synthesis.

B. Ethers

Even some ethers that are not acetals or ketals are easily cleaved. Benzyl ethers are often used as protecting groups for alcohols. The benzylic ether bond is cleaved by hydrogenation reactions. In Section 13.11, an example of the use of this protecting group will be presented.

Silyl ethers, in which there is an oxygen-silicon bond, are usually cleaved by fluoride ion. The high bond energy of the silicon-fluorine bond (Table 2.5, p. 61) serves as a thermodynamic driving force for this cleavage. Silyl ethers are prepared by the reaction of an alcohol and a silyl halide in the presence of a base to aid in the deprotonation of the alcohol. For example, the *tert*-butyldimethylsilyl group (abbreviated TBDMS) is used to protect the hydroxyl group of 3-butyn-1-ol so that the terminal alkyne can be selectively deprotonated. The carbanion resulting from that deprotonation adds to acetaldehyde.

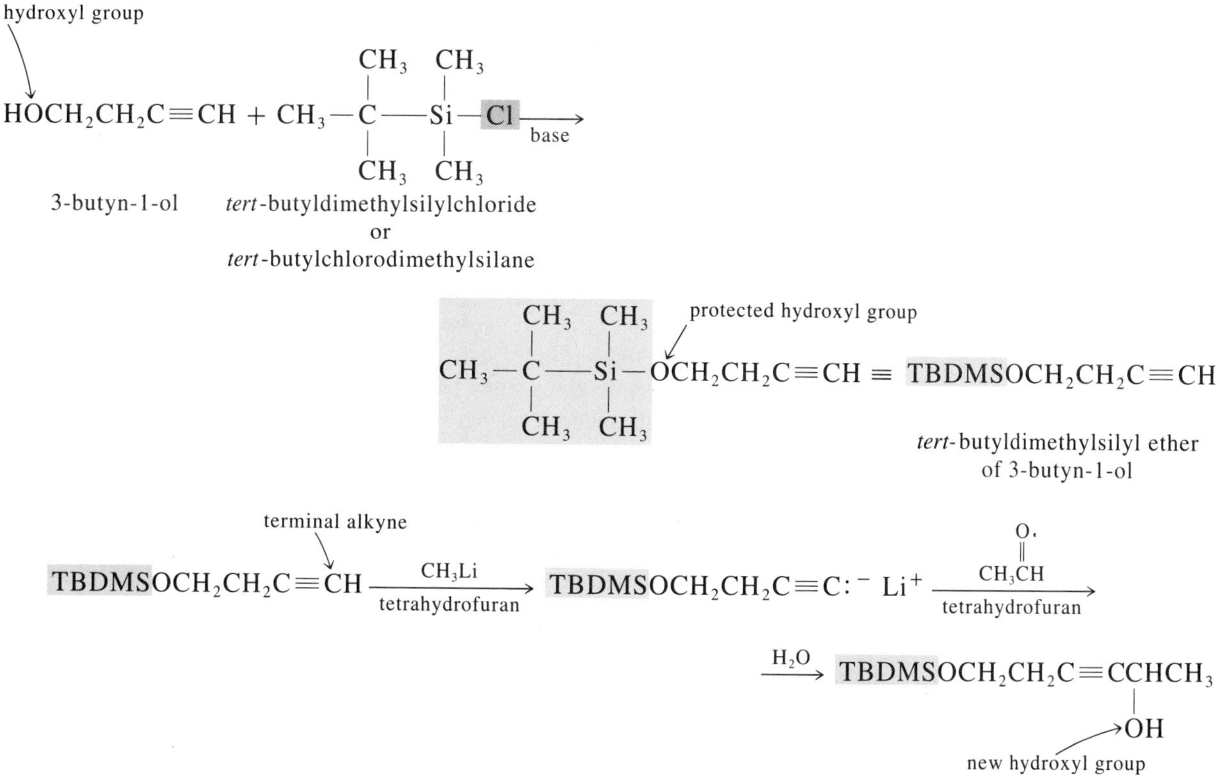
hydroxyl group
HOCH2CH2C≡CH + CH3—C(CH3)2—Si(CH3)2—Cl
base
3-butyn-1-ol
tert-butyldimethylsilylchloride
or
tert-butylchlorodimethylsilane
protected hydroxyl group
CH3—C(CH3)2—Si(CH3)2—OCH2CH2C≡CH ≡ TBDMSOCH2CH2C≡CH
tert-butyldimethylsilyl ether
of 3-butyn-1-ol
terminal alkyne
TBDMSOCH2CH2C≡CH
CH3Li
tetrahydrofuran
TBDMSOCH2CH2C≡C:⁻ Li⁺
CH3CH
tetrahydrofuran
H2O
TBDMSOCH2CH2C≡CCHCH3
OH
new hydroxyl group

This synthesis is continued, as shown in the reactions below, by hydrogenating the alkyne to a cis alkene, then protecting the secondary alcohol as a tetrahydropyranyl ether, and finally using fluoride ion to deprotect the primary alcohol for further reactions.

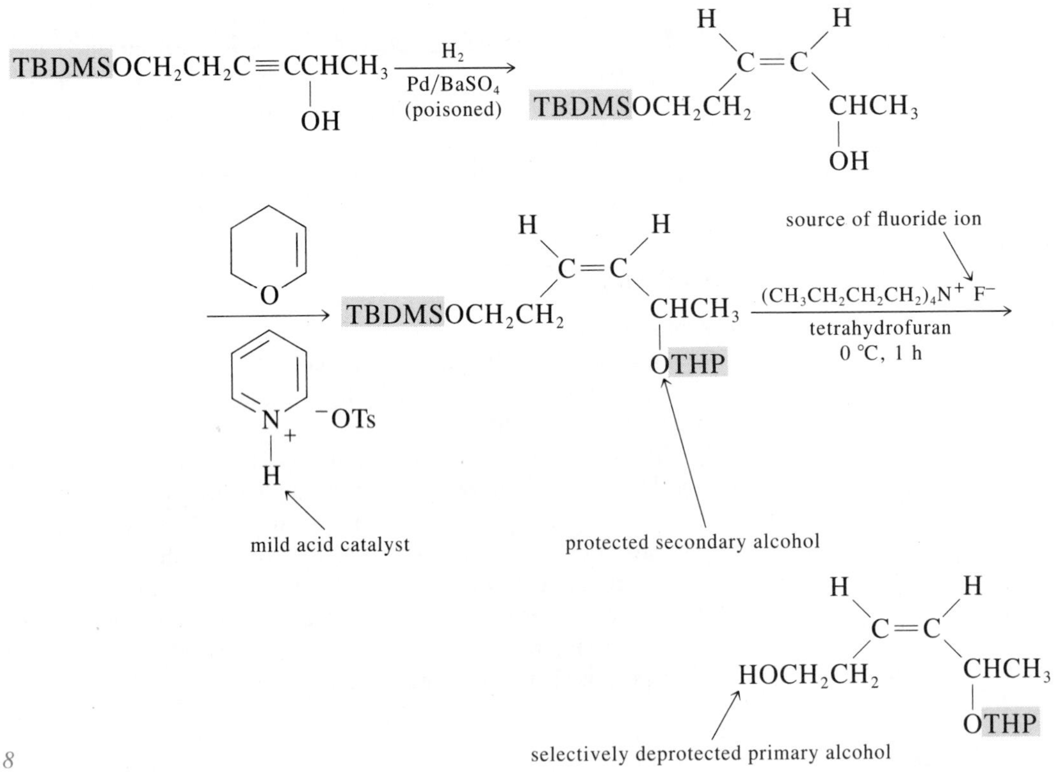
TBDMSOCH2CH2C≡CCHCH3
OH
H2
Pd/BaSO4
(poisoned)
TBDMSOCH2CH2
CHCH3
OH
mild acid catalyst
OTs
TBDMSOCH2CH2
OTHP
protected secondary alcohol
source of fluoride ion
(CH3CH2CH2CH2)4N+ F−
tetrahydrofuran
0 °C, 1 h
HOCH2CH2
selectively deprotected primary alcohol

Study Guide
Concept Map 13.8

13.11 DESIGNING SYNTHESES

The addition of organometallic compounds to carbonyl groups is a versatile reaction that chemists can employ to create carbon-carbon bonds using a wide variety of reagents. The alcohols formed in these reactions are easily converted into alkenes, alkyl halides, and, in some cases, new carbonyl compounds. Each of these functional group classes will undergo further reactions, so a wide range of compounds can be synthesized.

The type of reasoning that goes into designing a transformation of one compound into another has already been demonstrated in Sections 4.4B and 9.7 and in problems on pp. 265 and 457. The synthesis of a chiral ketone from a naturally occurring chiral alcohol will serve as another example. The chiral alcohol (*R*)-(+)-citronellol is used as the starting material for the synthesis of (*R*)-(+)-6-benzyloxy-4-methyl-1-phenyl-1-hexanone. How is this transformation accomplished? An analysis of the problem starts with an examination of the two structures for functional groups and for any changes in the carbon skeleton.

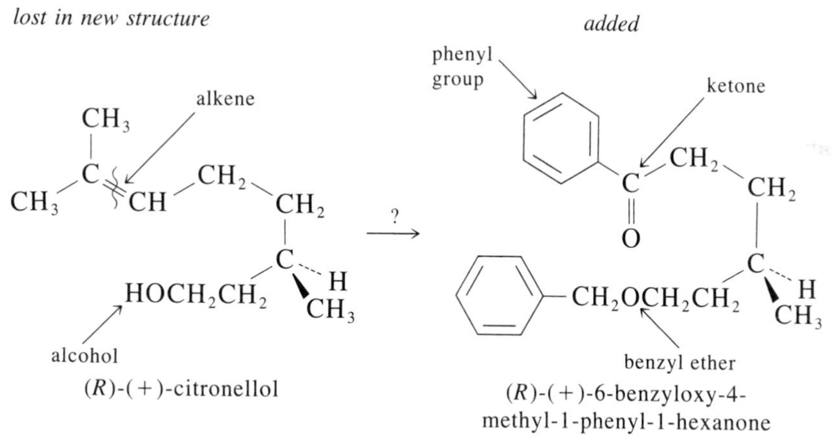
lost in new structure
added
alkene
phenyl group
ketone
?
alcohol
benzyl ether
(R)-(+)-citronellol
(R)-(+)-6-benzyloxy-4-methyl-1-phenyl-1-hexanone

The alcohol function of citronellol has been converted to a benzyl ether in the product. The alkene function in citronellol disappears, and one of the carbon atoms involved in the double bond is part of a ketone function in the product. A new phenyl group is bonded to this carbonyl group.

Dissecting the structures of starting material and product shows which bonds have been broken and which formed in the transformation.

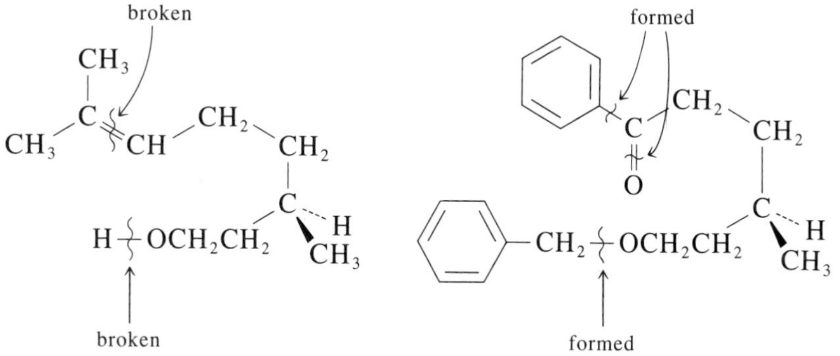
broken
broken
formed
formed

One of the changes is easy to recognize, the conversion of an alcohol function to an ether. The alcohol can be converted into a good nucleophile using a strong base. This nucleophile will react with the electrophilic carbon atom of a benzyl halide.

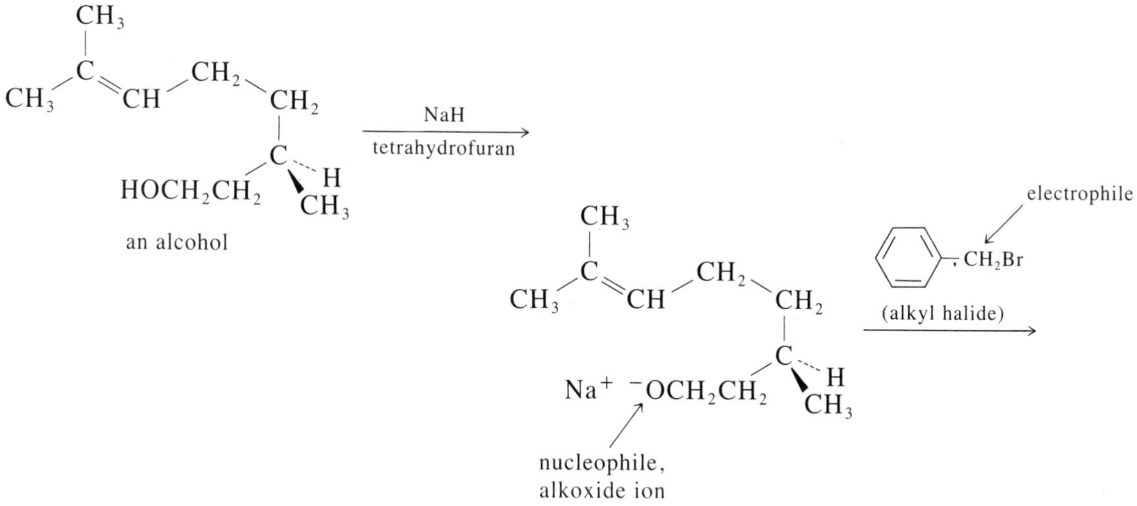
NaH
tetrahydrofuran
an alcohol
electrophile
(alkyl halide)
nucleophile,
alkoxide ion

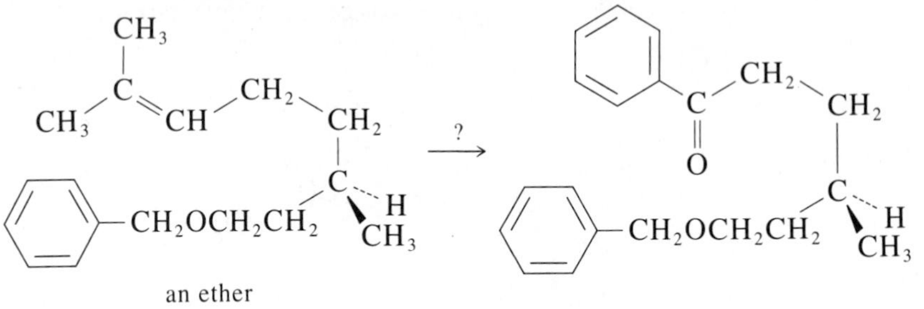

The easiest way to solve the rest of the problem is to work backwards in steps—for each compound isolated, asking what would be a good precursor to it and then how to get to that precursor from the starting material. For example, the ketone function in the final product can be prepared by oxidation of the corresponding secondary alcohol.

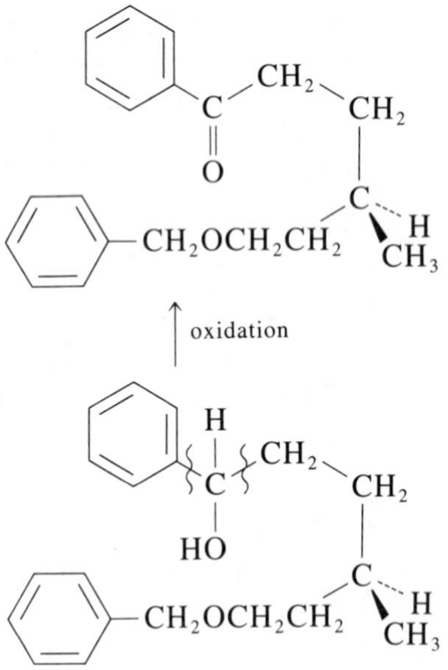

A secondary alcohol can be prepared by the reaction of an organometallic reagent with an aldehyde (p. 492). Dissection of the alcohol at either side of carbon atom to which the hydroxyl group is bonded gives two sets of possible reagents. Each set must contain a nucleophilic center at a carbon atom and an electrophilic one.

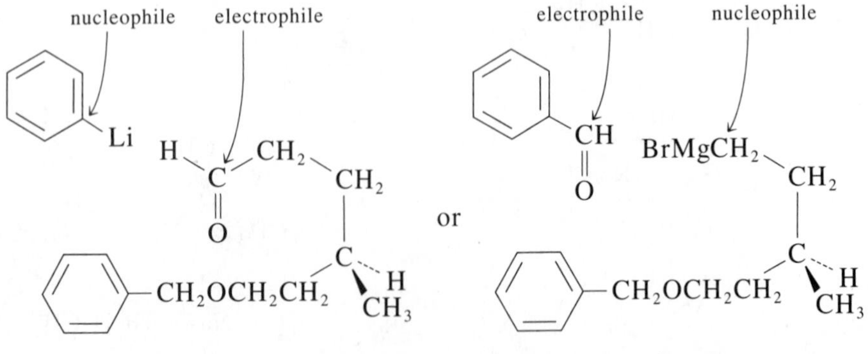

The set of reagents on the left is the better choice because the given starting material for this synthesis has a double bond where the carbonyl group of the aldehyde is and there are ways to cleave double bonds to make aldehydes. For example, ozonolysis with a reductive work-up will convert the benzyl ether of citronellol into the aldehyde shown on the left above.

The complete synthesis, working backwards, is outlined below.

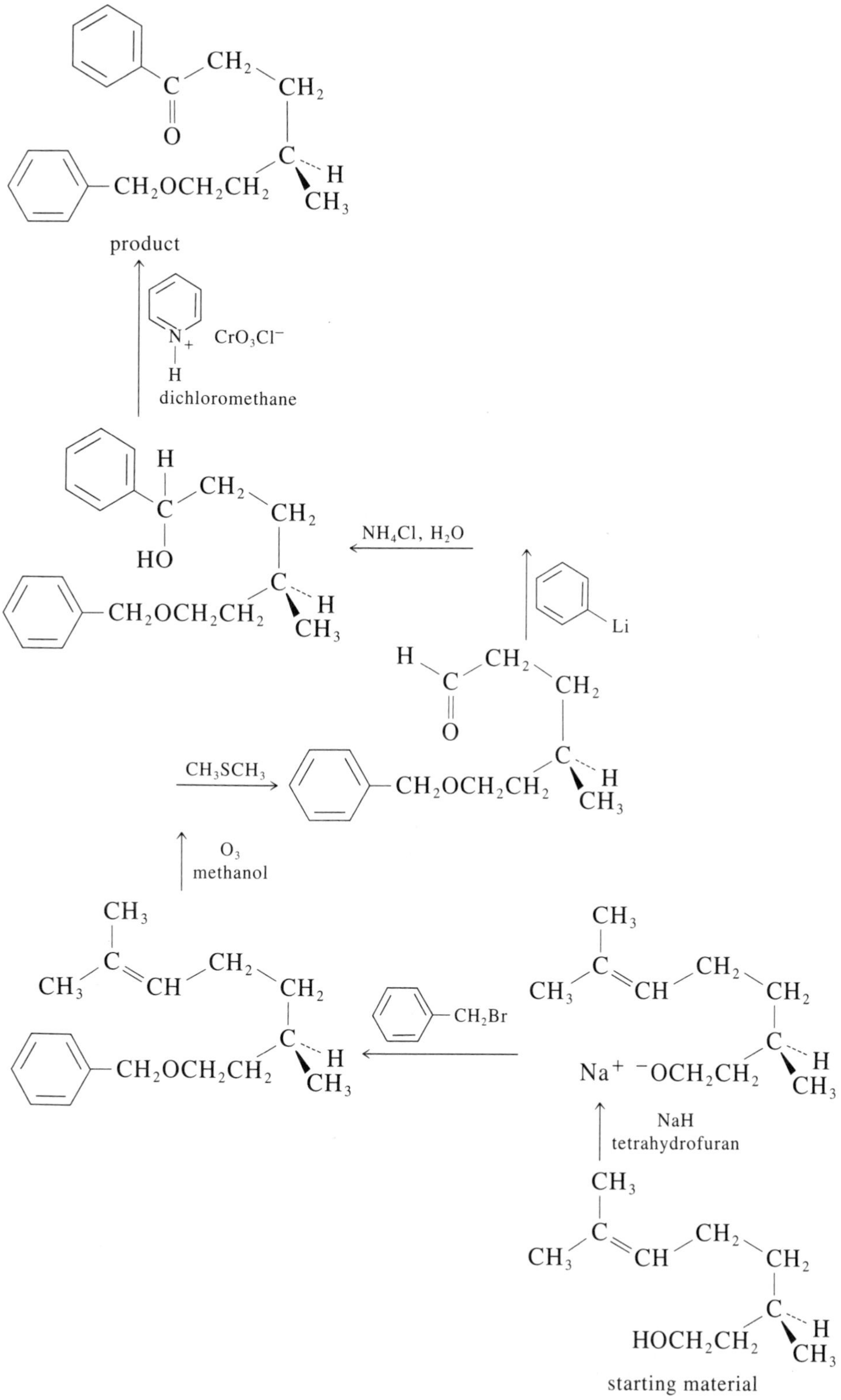
product
CrO3Cl−
dichloromethane
NH4Cl, H2O
Li
CH3SCH3
O3
methanol
CH2Br
NaH
tetrahydrofuran
starting material

Note that the benzyl group protects the alcohol function during the reaction with phenyllithium and during the oxidation reaction in which the secondary alcohol is converted to a ketone, while the primary alcohol remains protected as the ether. In the next steps of this synthesis, the benzyl ether remains in place for two more reactions, one involving strong bases and the other a Grignard reagent, and is finally removed by catalytic hydrogenation to give the ultimate product, (3*R*)-3,5-dimethyl-6-phenylheptane-1,6-diol.

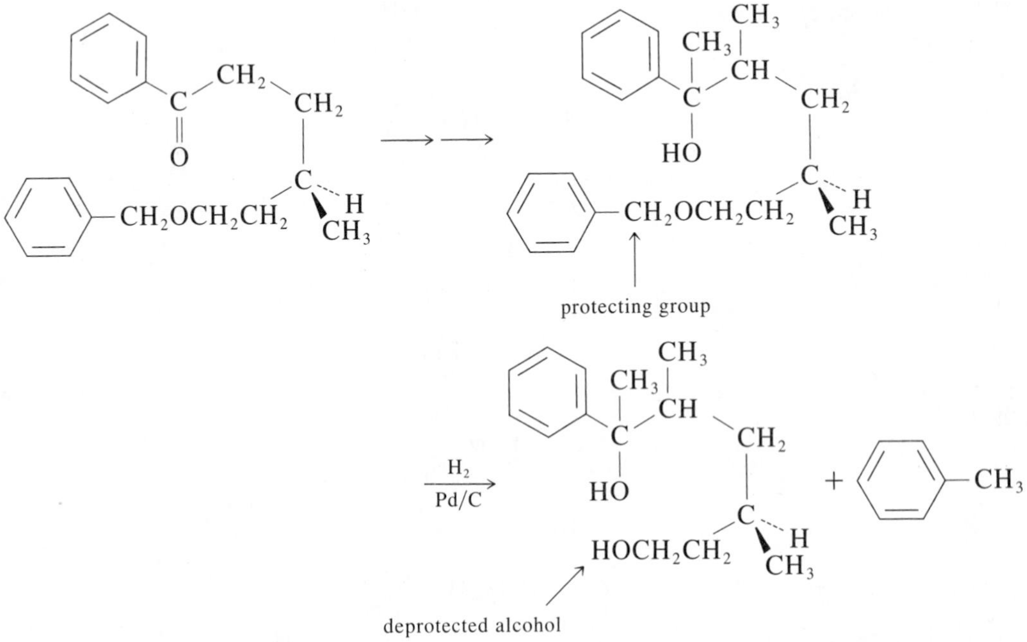
protecting group
deprotected alcohol

The following problems will provide practice in applying this type of reasoning to some other situations.

PROBLEM 13.23

The following transformation is part of a synthesis of a compound believed to be an intermediate in the biosynthesis of monensin, an antibiotic widely used in chicken feed. Show how the starting material can be converted into the product.

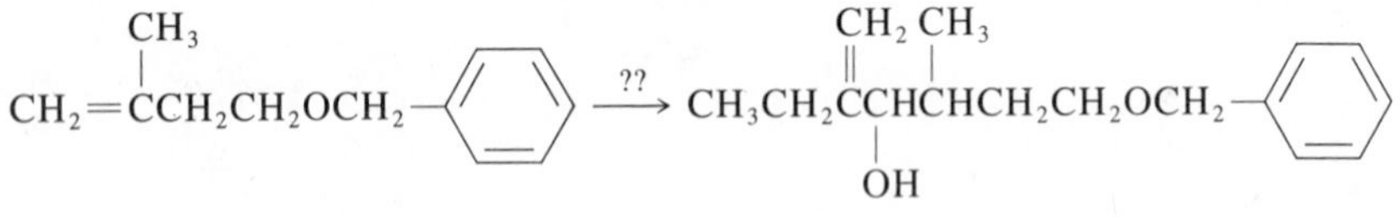

PROBLEM 13.24

The pheromone of the ambrosia beetle, 6-methyl-5-hepten-2-ol, can be prepared by the sodium borohydride reduction of 6-methyl-5-hepten-2-one (p. 486), but has also been synthesized by another route:

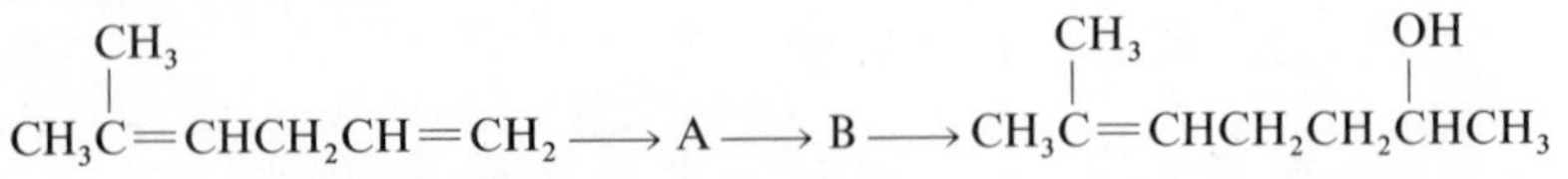

Supply the structures of Compounds A and B and the reagents necessary to transform 5-methyl-1,4-hexadiene into 6-methyl-5-hepten-2-ol.

PROBLEM 13.25

A highly mutagenic compound containing five double bonds has been isolated from the feces of people living in industrialized countries. These types of compounds, called fecapentaenes, are suspected of causing colon cancer. Chemists are currently attempting to synthesize fecapentaenes in large enough amounts to test them for carcinogenicity. A portion of a synthesis is shown below. How would you carry out this transformation?

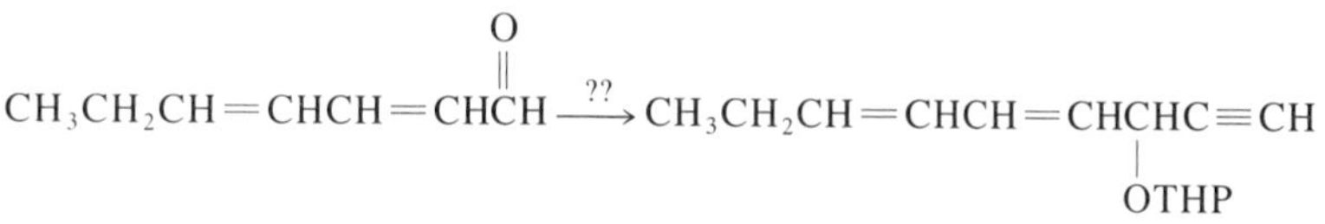

O
$CH_3CH_2CH{=}CHCH{=}CHCH$
??
$CH_3CH_2CH{=}CHCH{=}CHCHC{\equiv}CH$
OTHP

PROBLEM 13.26

A synthesis of a talaromycin, a toxic metabolite of a fungus that grows on chicken litter, involved the transformation shown below. How would you carry it out?

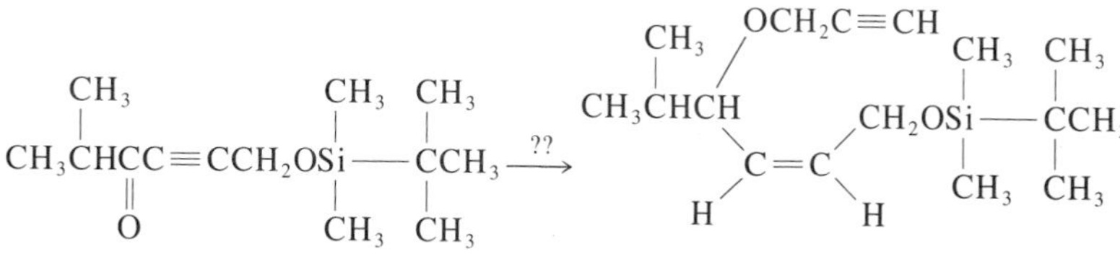

CH_3
$CH_3CHCC{\equiv}CCH_2OSi$
O
CH_3 CH_3
CCH_3
CH_3 CH_3
??
$OCH_2C{\equiv}CH$
CH_3
CH_3CHCH
C=C
H
H
CH_2OSi
CCH_3
CH_3 CH_3
CH_3 CH_3

S U M M A R Y

Aldehydes and ketones are usually prepared by oxidizing alcohols. Oxidation of a primary alcohol with pyridinium chlorochromate gives an aldehyde. A secondary alcohol gives a ketone. Aldehydes and ketones are also products of the ozonolysis of alkenes (Table 13.3).

The carbonyl group, the functional group in aldehydes and ketones, is reactive toward nucleophiles. Some reagents, such as the nucleophilic carbon of organometallic reagents, hydride ion, and cyanide ion add to the carbonyl group to give, after a subsequent protonation step, products containing the hydroxyl group. Reactions in which a carbon-carbon bond is formed between a nucleophilic center at a carbon atom and the electrophilic center of a carbonyl group are used to synthesize alcohols. A careful analysis of the structure of the desired alcohol allows a chemist to pick the carbonyl compound and the organometallic reagent that will give that product. The reactions of organometallic reagents that produce alcohols are summarized in Table 13.2 (p. 495).

Addition of hydride ion to a carbonyl group results in reduction of an aldehyde to a primary alcohol or of a ketone to a secondary alcohol. Addition of cyanide ion to a carbonyl group gives cyanohydrins (Table 13.4).

Weak nucleophiles such as water and alcohols also add reversibly to carbonyl compounds to give hydrates, hemiacetals, and hemiketals. Only if a strong acid is present and water is removed from the equilibrium mixture will an aldehyde be converted to a stable acetal and a ketone to a stable ketal. Acetals and ketals do not react with bases or nucleophilic reagents and are, therefore, useful as protecting groups.

A protecting group converts a reactive functional group into a different functional group that is unaffected by the conditions of a reaction or reactions to be

TABLE 13.3 Some Ways to Prepare Aldehydes and Ketones

Starting Material	Reagent	Intermediate	Reagent in Second Step	Product
		Oxidation of Alcohols		
RCH_2OH	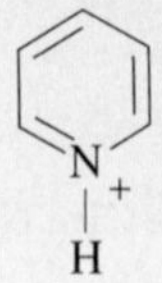 CrO_3Cl^-	$RCH_2OCr(=O)_2O^-$		$RC(=O)H$
$RR'CHOH$	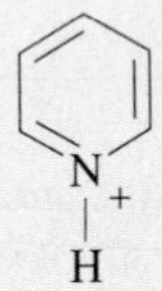 CrO_3Cl^- or $Cr_2O_7^{2-}$, H_3O^+	$RR'CHOCr(=O)_2O^-$		$RC(=O)R'$
		Ozonolysis of Alkenes		
$RR'C{=}CHR''$	O_3	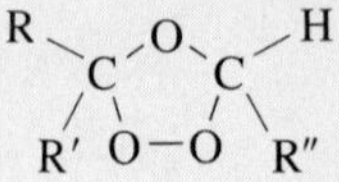	Zn, H_2O or CH_3SCH_3	$RR'C{=}O$ $O{=}CHR''$
			H_2O_2	$RR'C{=}O$ $O{=}C(OH)R''$

TABLE 13.4 An Outline of Reactions of Aldehydes and Ketones

Starting Material	Reagent	Intermediate	Reagent in Second Step	Product
		Nucleophilic Addition Reactions		
$RC(=O)R''$ R″ may be H	$NaBH_4$ $LiAlH_4$	$R-C(O^-)(H)-R''$	H_2O	$R-C(OH)(H)-R''$
	NaCN	$R-C(O^-)(R'')-C\equiv N$	H_3O^+	$R-C(OH)(R'')-C\equiv N$
	R′MgX or R′Li	$R-C(O^-)(R'')-R'$	H_3O^+	$R-C(OH)(R'')-R'$
[	R′MgX or R′Li	$R'CH_2CH_2O^-$	H_3O^+	$R'CH_2CH_2OH$]

(Continued)

TABLE 13.4 (*Continued*)

Starting Material	Reagent	Intermediate	Reagent in Second Step	Product
		Nucleophilic Addition Reactions		
$R-C(=O)-R''$	H_2O			$R-C(OH)(OH)-R''$
	$R'OH$	$R-C(O^-)(\overset{+}{H}OR')-R''$		$R-C(OH)(OR')-R''$
	$R'OH$, HB^+	$R-C(OH)(OR')-R''$	$R'OH$, HB^+	$R-C(OR')(OR')-R''$
		Nucleophilic Addition Followed by Elimination of Water		
$R-C(=O)-R''$ R″ may be H	H_2NOH	$R-C(OH)(HN-OH)-R''$		$RR''C=N-OH$
	$H_2NNH-C_6H_4-R'$ HB^+	$R-C(OH)(HNNH-C_6H_4-R')-R''$		$RR''C=NNH-C_6H_4-R'$
	H_2NR'	$R-C(OH)(HNR')-R''$		$RR''C=NR'$
		Reduction to Methylene Group		
$R-C(=O)-R''$ R″ may be H	Zn(Hg), HCl, Δ			$R-C(H)(H)-R''$
	H_2NNH_2, OH^-	$RR''C=NNH_2$	OH^-	$R-C(H)(H)-R''$
	$HSCH_2CH_2SH$, BF_3	$RR''C(SCH_2CH_2S)$ (cyclic dithioacetal)	H_2, Raney Ni	$R-C(H)(H)-R''$

TABLE 13.5 Protecting Groups

Functional Group Protected	Protecting Group	Reagents Used to Make Protecting Group	Reagents Used to Remove Protecting Group
Alcohol ROH	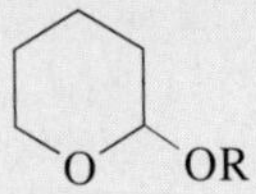 tetrahydropyranyl ether	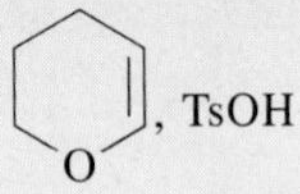, TsOH	H_3O^+
	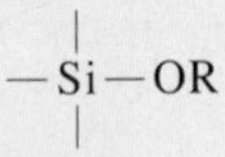 silyl ether	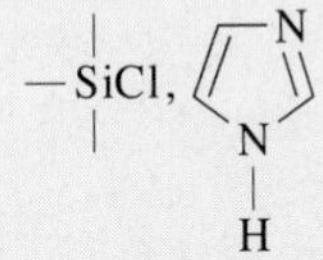	F^-
	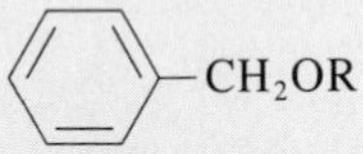 benzyl ether	NaH, then 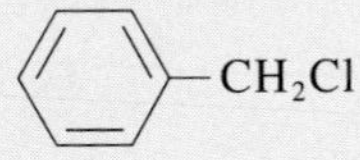	H_2 with Pd catalyst
1,2-Diol 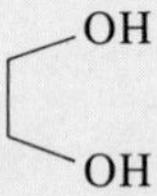	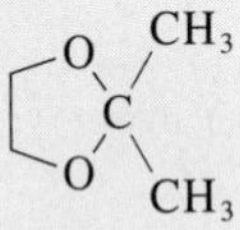 cyclic ketal	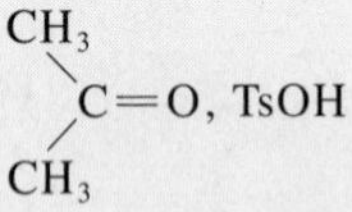	H_3O^+
1,3-Diol 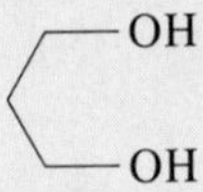	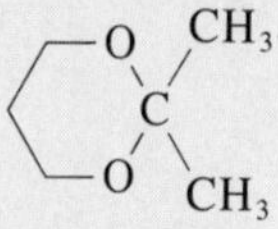 cyclic ketal	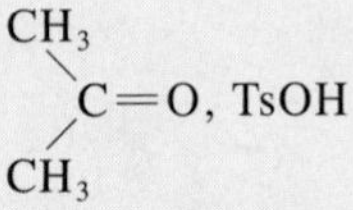	H_3O^+
Aldehyde 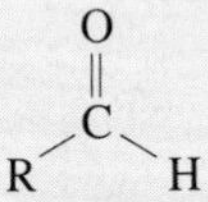	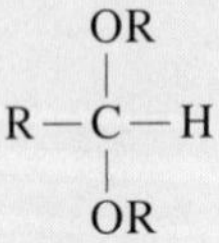 acetal	ROH, TsOH	H_3O^+
	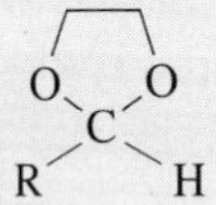 cyclic acetal	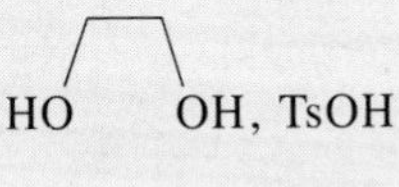	H_3O^+
Ketone 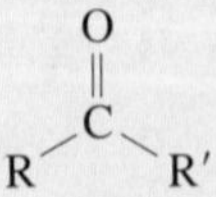	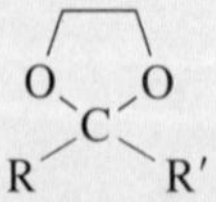 cyclic ketal	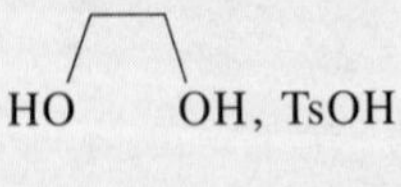	H_3O^+

carried out. A protecting group must be easily removed so that the original functional group can be recovered. Besides acetals and ketals, tetrahydropyranyl ethers (which are also acetals), silyl ethers, and benzyl ethers are useful protecting groups (Table 13.5).

Carbonyl compounds also react with nucleophiles that are compounds related to ammonia. Addition of the nucleophile to the carbonyl group followed by elimination of water gives products containing a carbon-nitrogen double bond. Carbonyl groups may be replaced by hydrogen in reduction reactions under acidic conditions (the Clemmensen reduction), basic conditions (Wolff-Kishner reduction), or neutral conditions (Raney nickel reduction of a thioacetal). All of these reactions are also summarized in Table 13.4.

ADDITIONAL PROBLEMS

13.27 Name the following compounds according to the IUPAC rules, specifying stereochemistry if it is indicated in the structure.

(a) 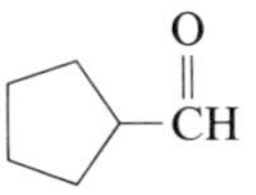
(b) 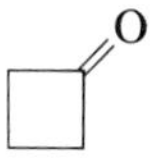
(c) 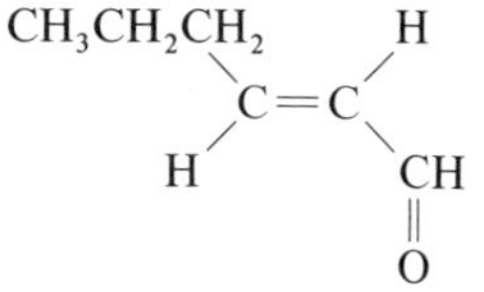

(d) 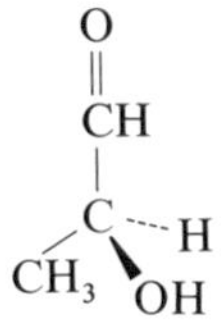
(e) 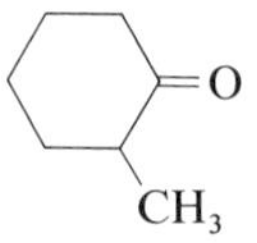
(f) 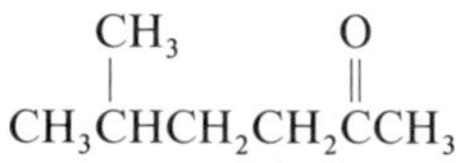

(g) 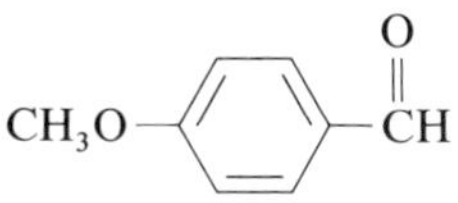
(h) 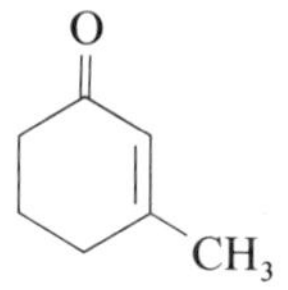
(i) 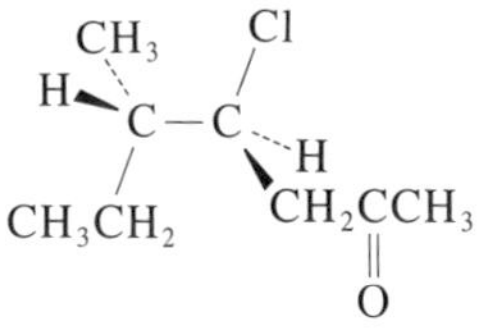

(j) 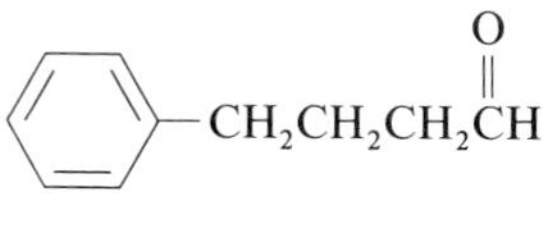

13.28 Draw structural formulas for the following compounds.

(a) 3,3-dimethylcyclopentanone (b) 1-bromo-2-hexanone
(c) *m*-chlorobenzaldehyde (d) *p*-methylacetophenone (e) 3,5-hexadien-2-one
(f) 2-methylpentanal (g) (*R*)-2-chlorocyclobutanone (h) (*S*)-3-bromobutanal

13.29 Using butanal as a typical aldehyde, write equations showing the reactions that you predict will occur if the following reagents are used. Write "no reaction" if you expect none.

(a) H_2SO_4, cold, conc (b) $NaBH_4$, H_2O

(c) 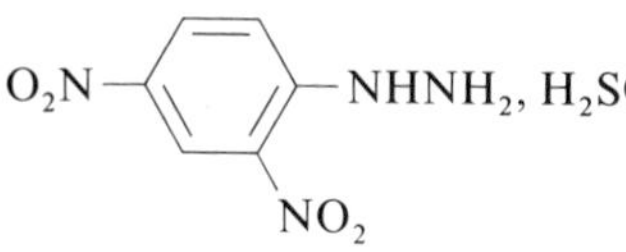, H_2SO_4 (d) 1. $CH_3CH_2CH_2CH_2Li$; 2. H_3O^+

(e) 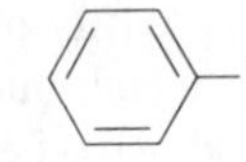—NH_2, Δ (f) NaCN, H_2SO_4(aq) (g) $Na_2Cr_2O_7$, H_2O, H_3O^+

(h) NaI, acetone (i) $HONH_3{}^+Cl^-$, $CH_3\overset{O}{\overset{\|}{C}}O^-Na^+$

(j) 1. CH_3MgI, diethyl ether; 2. H_3O^+ (k) 5% $NaHCO_3$, H_2O

(l) 1. $LiAlH_4$, diethyl ether; 2. H_3O^+

(m) H_2NNH_2, KOH, diethylene glycol, Δ (n) Zn(Hg), HCl, Δ

13.30 Using 2-pentanone as a typical ketone, write equations showing the reactions that you predict will occur if the following reagents are used. Write ''no reaction'' if you expect none.

(a) H_2SO_4, cold, conc (b) $NaBH_4$, H_2O

(c) O_2N—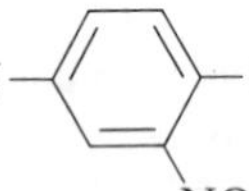—$NHNH_2$, H_2SO_4 (with NO_2 on the ring) (d) 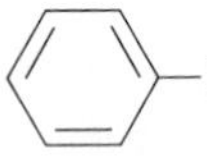—NH_2, Δ

(e) 1. 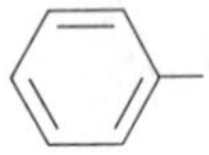—Li, tetrahydrofuran; 2. NH_4Cl, H_2O (f) NaCN, H_2SO_4(aq)

(g) $Na_2Cr_2O_7$, H_2O, H_3O^+ (h) $AgNO_3$, ethanol

(i) $HONH_3{}^+Cl^-$, $CH_3\overset{O}{\overset{\|}{C}}O^-Na^+$

(j) 1. CH_3CH_2MgBr, diethyl ether; 2. H_3O^+ (k) 5% $NaHCO_3$, H_2O

(l) 1. $CH_3C{\equiv}CNa$; 2. NH_4Cl, H_2O

(m) H_2NNH_2, KOH, diethylene glycol, Δ (n) Zn(Hg), HCl, Δ

13.31 Complete each of the following equations, showing the structures of the products that will be formed.

(a) 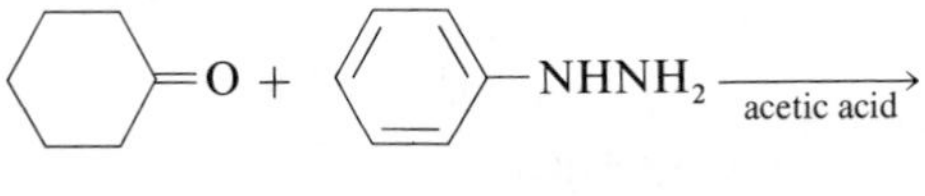
=O +
NHNH2
acetic acid

(b) 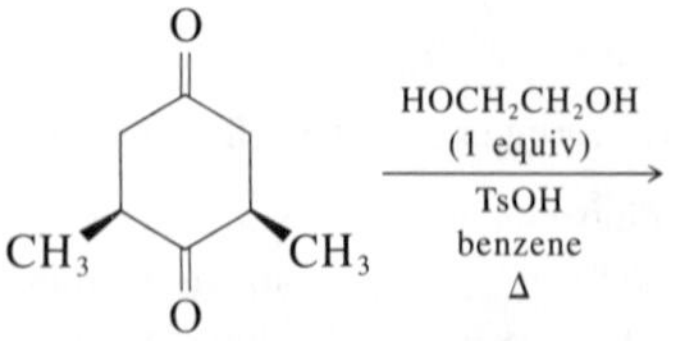
O
CH3
CH3
O
HOCH2CH2OH
(1 equiv)
TsOH
benzene
Δ

(c) 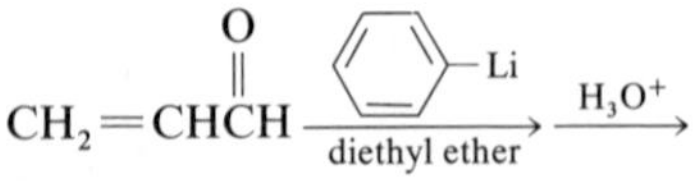
O
CH2=CHCH
Li
diethyl ether
H3O+

(d) 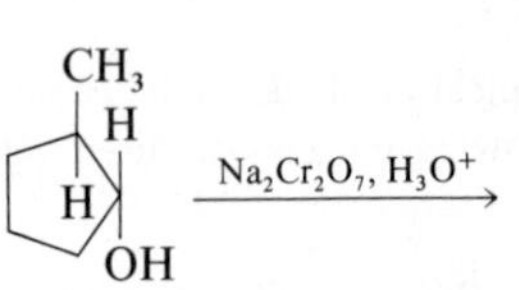
CH3
H
H
OH
Na2Cr2O7, H3O+

(e) 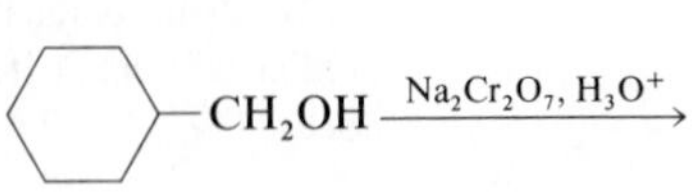
CH2OH
Na2Cr2O7, H3O+

(f) 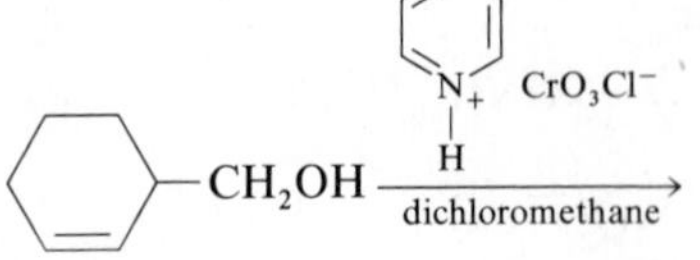
CH2OH
N+ CrO3Cl−
H
dichloromethane

(g) 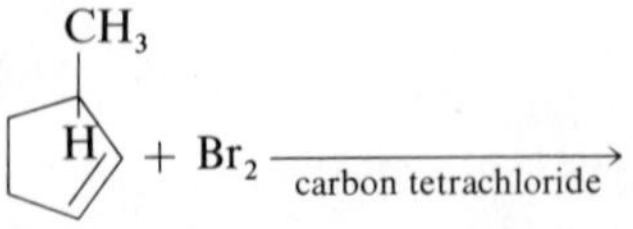
CH3
H
+ Br2
carbon tetrachloride

(h) 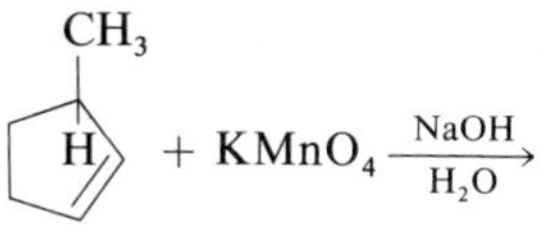

(i) 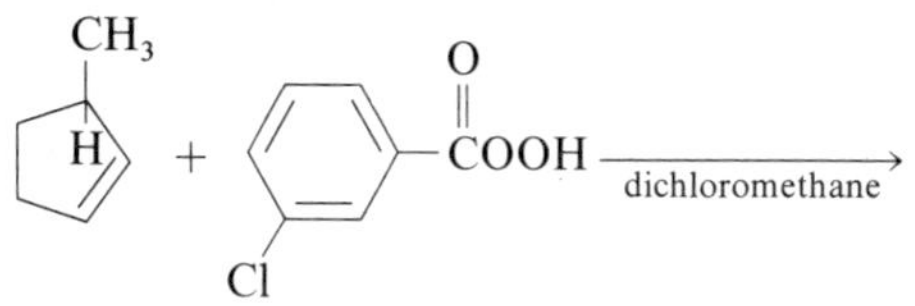

(j) 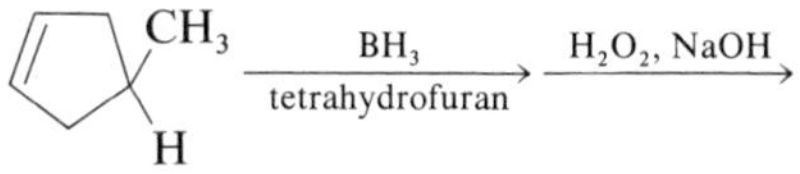

(k) 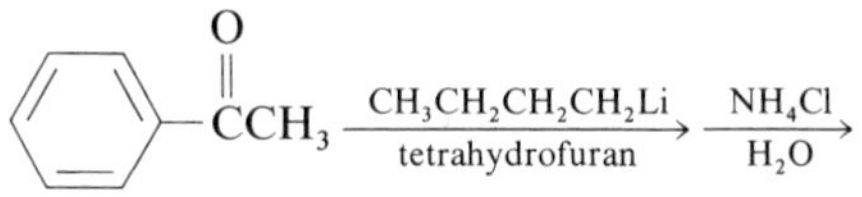

(l) 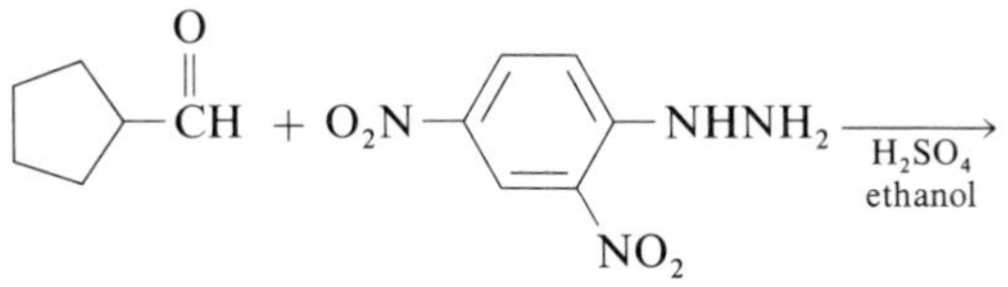

(m) 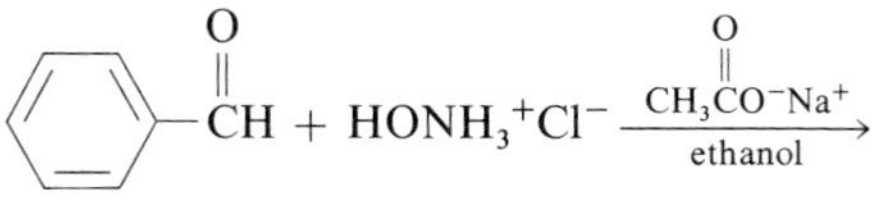

(n) 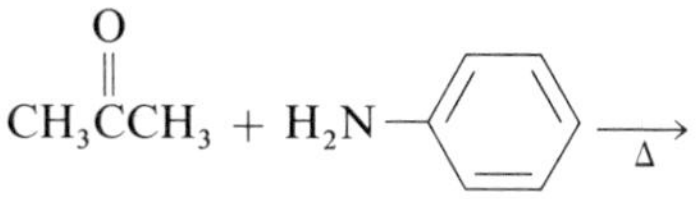

(o) $CH_3CH_2CH_2C{\equiv}CH + NaNH_2 \xrightarrow[NH_3(liq)]{}$

(p) $CH_3CH_2CH_2C{\equiv}CH + CH_3CH_2MgBr \xrightarrow[\text{diethyl ether}]{}$

(q) 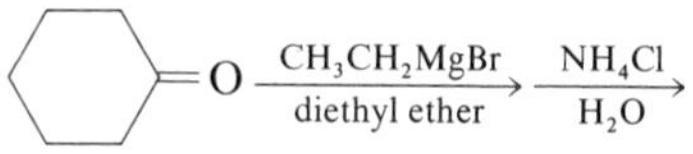

(r) 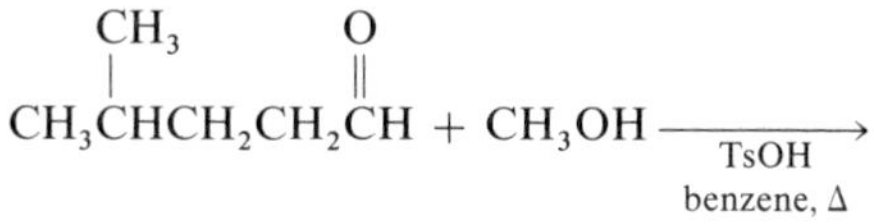

(s) 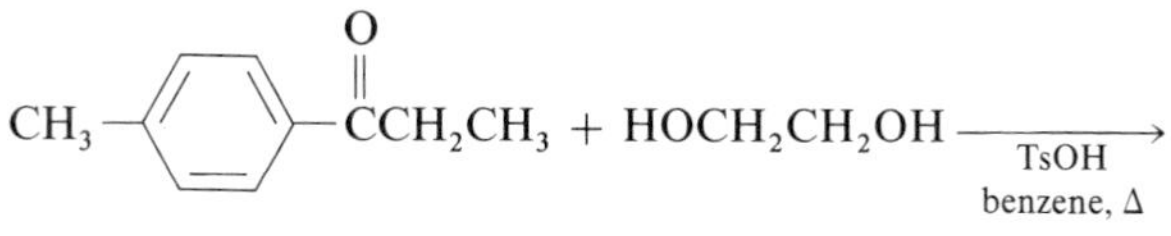

(t) 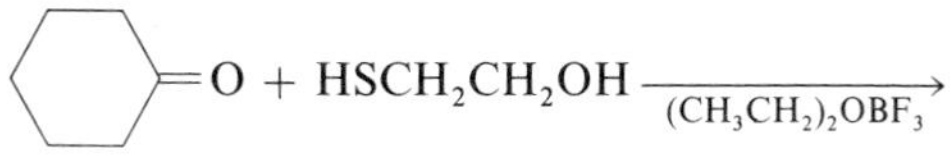

(u) 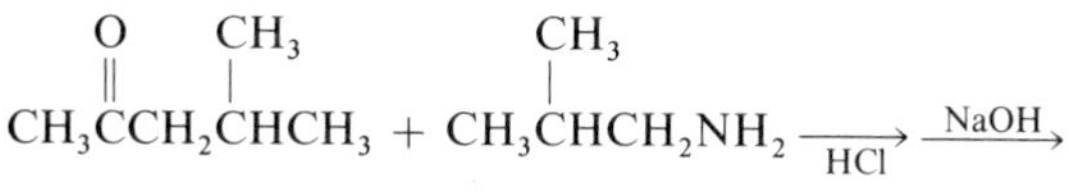

(v) 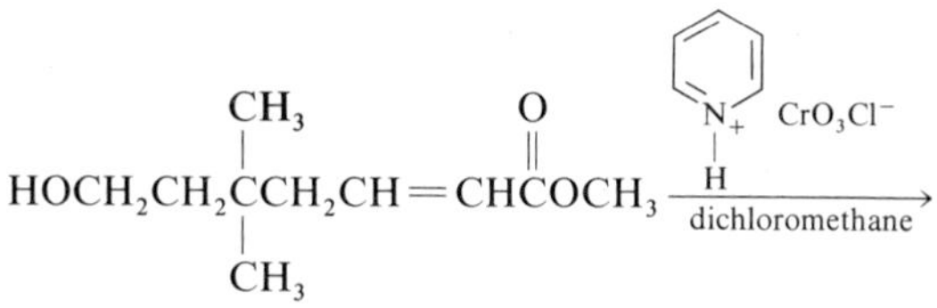

(w) 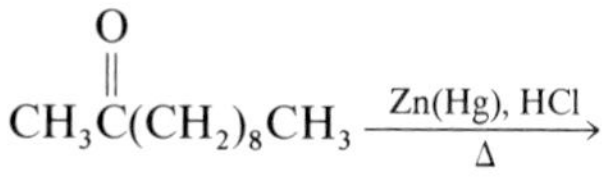

(x) 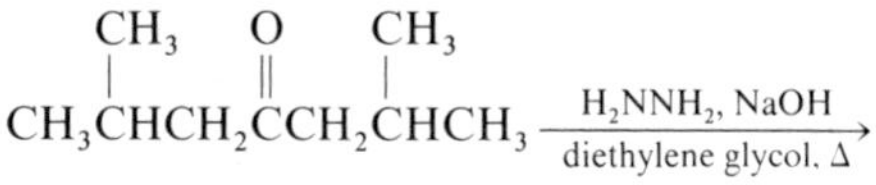

13.32 Predict the structures of the products or intermediates designated by letters in the following equations. Show stereochemistry when it is known.

(a) 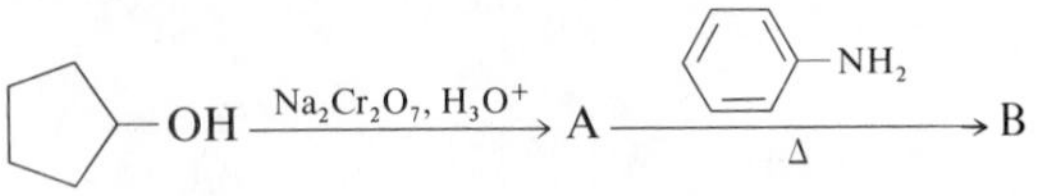

(b) 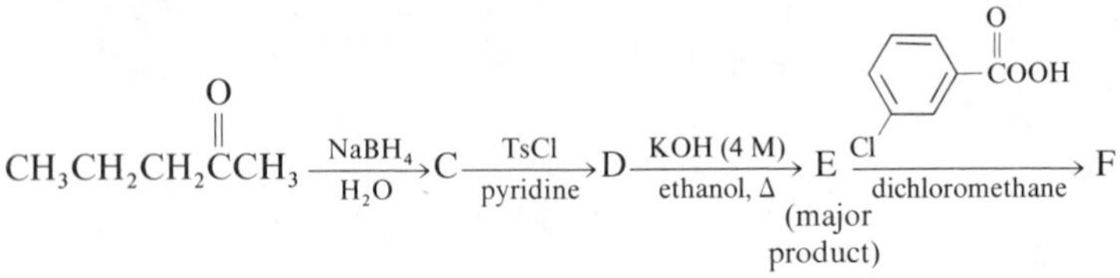

(c) 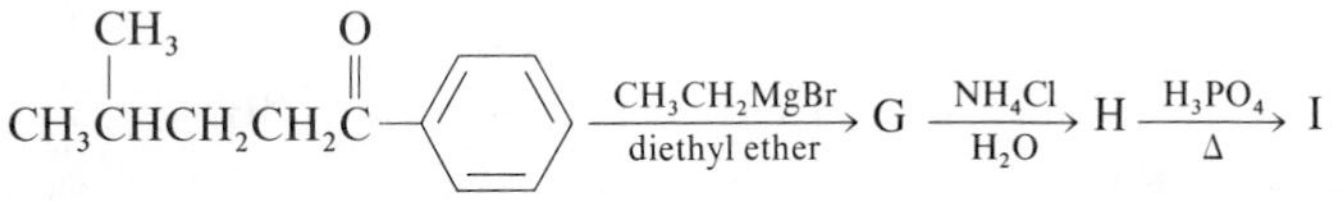

(d) 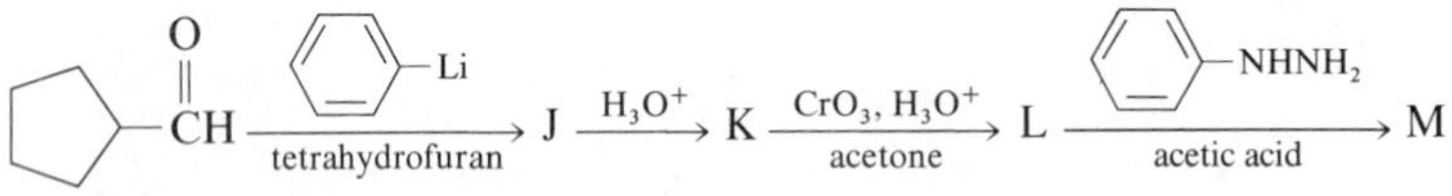

(e) $CH_3\overset{CH_3}{\overset{|}{C}}{=}CHCH_3 \xrightarrow[\text{tetrahydrofuran}]{BH_3} N \xrightarrow{H_2O_2,\ NaOH} O \xrightarrow[H_2O]{Na_2Cr_2O_7,\ H_2SO_4} P \xrightarrow{NaCN} Q \xrightarrow{H_3O^+} R$

(f) 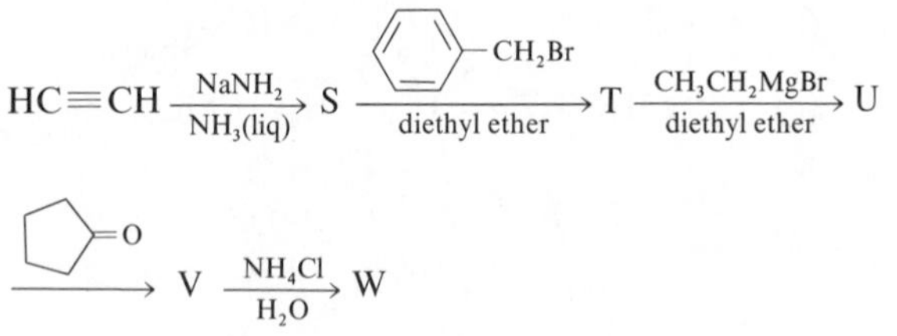

(g) 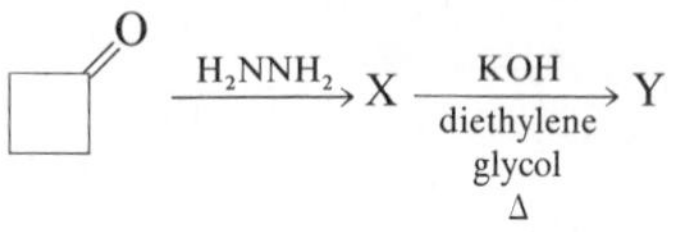

13.33 For each product designated by a letter or letters in the following reactions, give the structure.

(a) 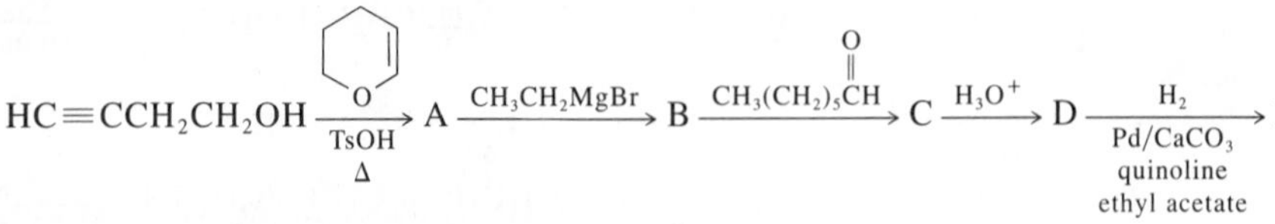

(b) 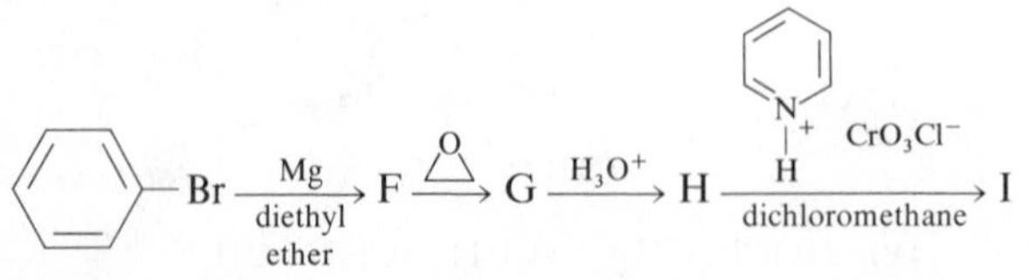

(c) $CH_3\overset{CH_3}{\overset{|}{C}}{=}CHCH_2CH_2\overset{O}{\overset{\|}{C}}CH_3 \xrightarrow[\text{TsOH, benzene, }\Delta]{HOCH_2CH_2OH} J \xrightarrow[\text{tetrahydrofuran}]{BH_3} K \xrightarrow{H_2O_2,\ NaOH} L \xrightarrow{H_3O^+} M$

(d) 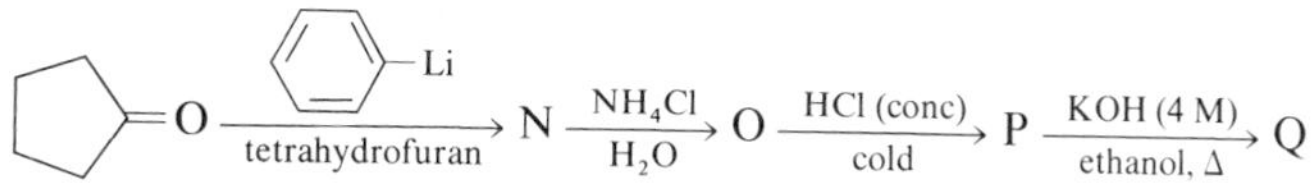

(e) 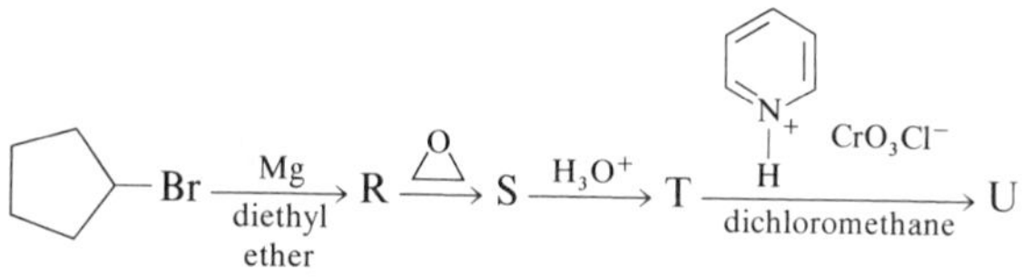

(f) 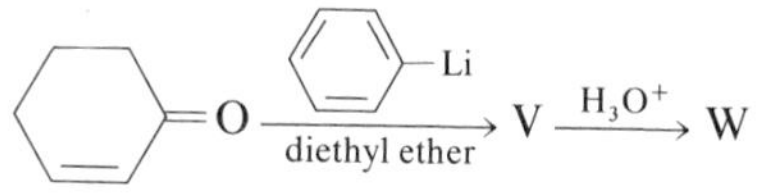

(g) 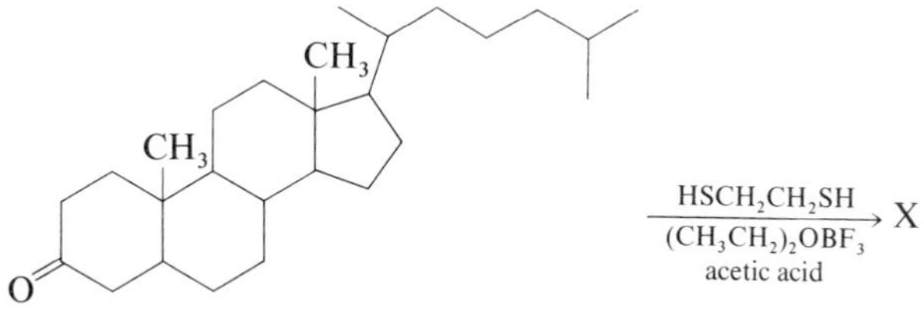

(h) 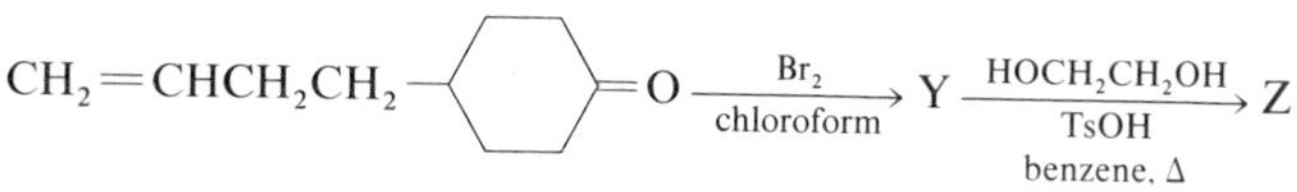

(i) $CH_3CH(CH_3)CH_2CH_2\overset{O}{\overset{\|}{C}}CH_2CH_2\overset{O}{\overset{\|}{C}}OH \xrightarrow[\text{diethylene glycol, } \Delta\Delta]{H_2NNH_2,\ NaOH} AA \xrightarrow{H_3O^+} BB$

(j) 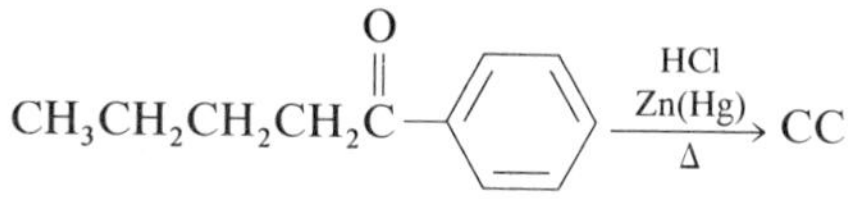

(k) 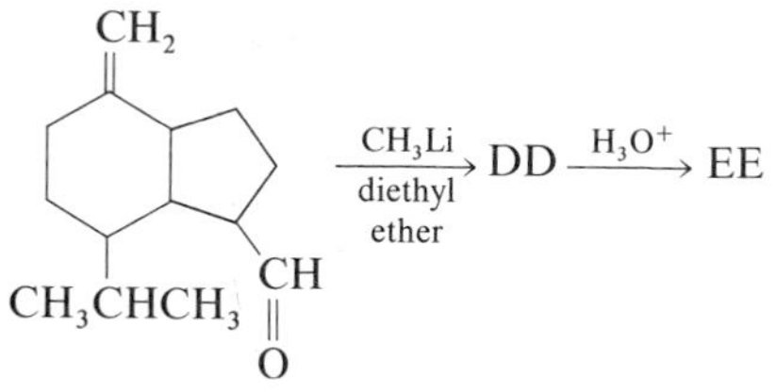

13.34 Plan a synthesis for each of the following compounds. You have bromobenzene, any organic reagents containing three or fewer carbon atoms, and any inorganic reagents you need. There may be more than one acceptable route for a given product. Try to find the shortest route.

(a) 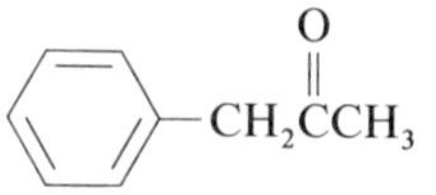
(b) 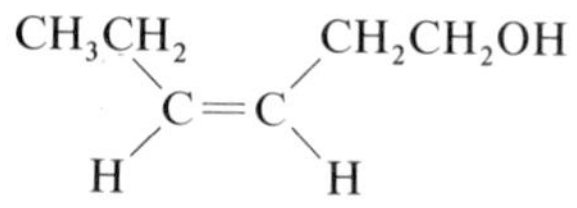

(c) $CH_3CH_2\overset{O}{\overset{\|}{C}}CH_2CH_2CH_3$
(d) $CH_3\overset{CH_3}{\overset{|}{C}}{=}CHCH_2CH_2CH_3$

(e) 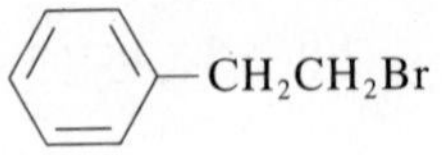 (f) 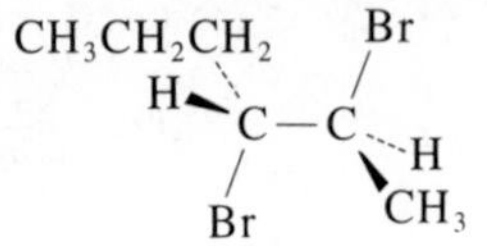

(g) $(CH_3(CH_2)_8\overset{\overset{\displaystyle O}{\|}}{C}CH{=}CH_2$ (h) 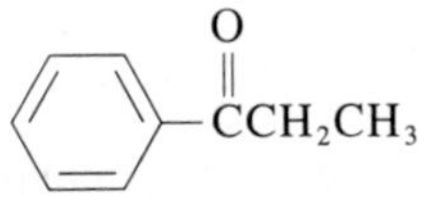

(i) 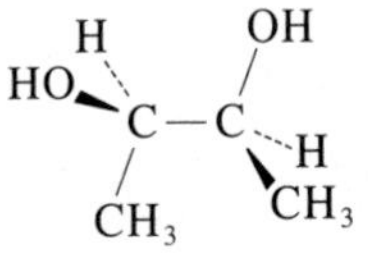 (j) 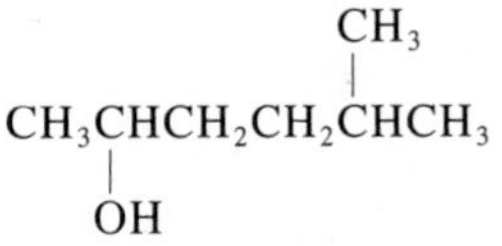

(k) 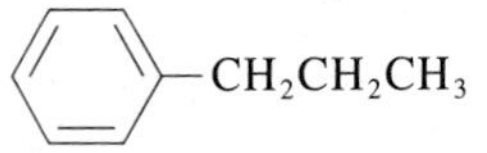 (l) 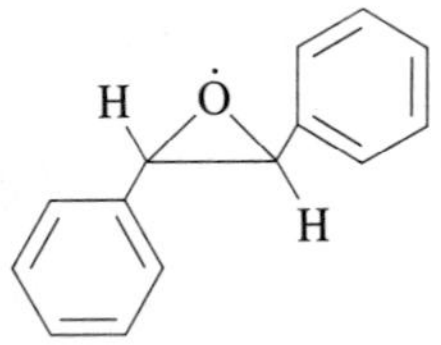 + enantiomer

13.35 Specify the reagents that are necessary to carry out the following transformations.

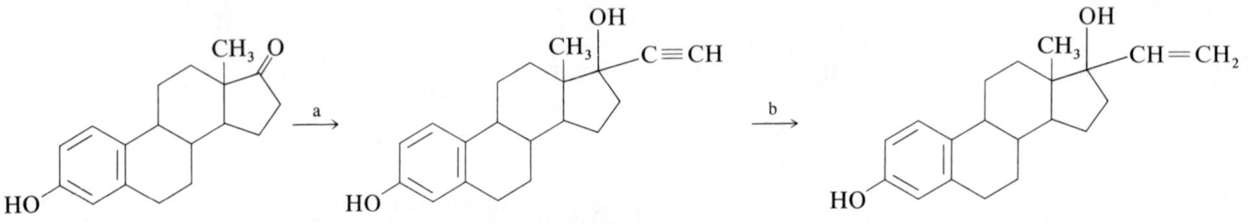

estrone
a female sex hormone

a synthetic modification of estrone that is as active as the natural hormone

a synthetic hormone that is half as active as estrone

13.36 Specify the reagents that are necessary to carry out the following transformations among members of the steroid family.

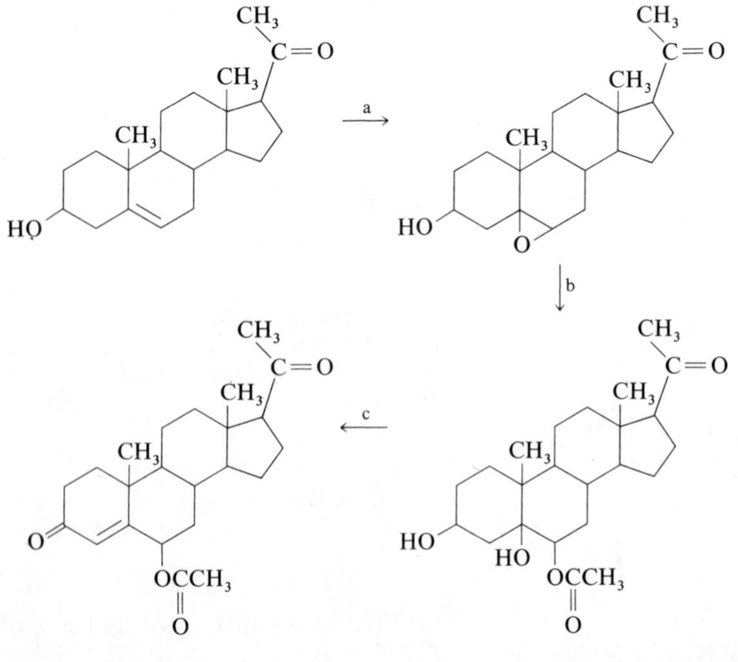

13.37 A laboratory synthesis of chrysomelidial, a compound secreted by the larvae of some beetles to defend themselves from attack, starts in the following way.

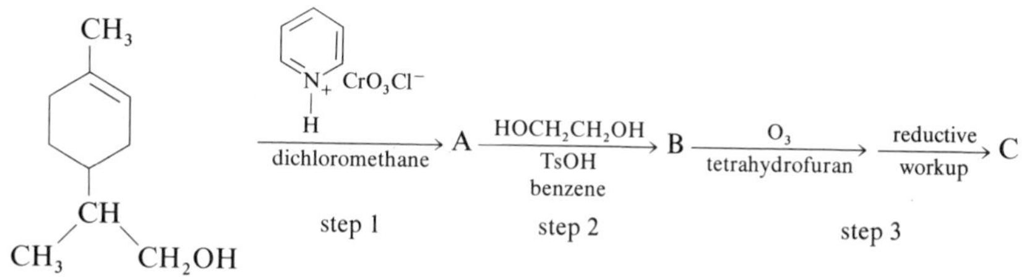

Fill in the structures of Compounds A, B, and C. Why is the second step of this synthesis necessary?

13.38 2,2-Dimethoxypropane is converted into 2,2-dibutoxypropane when it is heated with 1-butanol and a trace of acid. Some experimental facts are summarized in the following equation.

$$\underset{\text{bp } 83\ ^\circ\text{C}/760 \text{ mm}}{CH_3C(OCH_3)_2CH_3} + \underset{\text{bp } 118\ ^\circ\text{C}/760 \text{ mm}}{2\,CH_3CH_2CH_2CH_2OH} \xrightarrow[\text{benzene, bp } 80\ ^\circ\text{C}/760 \text{ mm}]{\text{TsOH}} \underset{\substack{\text{bp } 90\ ^\circ\text{C}/20 \text{ mm} \\ 75\%}}{CH_3C(OCH_2CH_2CH_2CH_3)_2CH_3} + \underset{\text{bp } 65\ ^\circ\text{C}/760 \text{ mm}}{2\,CH_3OH}$$

Write a mechanism for the reaction and suggest some practical measures that could be taken to ensure a good yield of 2,2-dibutoxypropane.

13.39 The right-hand column below contains a list of some common laboratory reagents. For each compound in the left-hand columns, list all the reagents with which that compound will show a change detectable by the human senses when it is mixed with the reagent in a test tube.

Compounds

(a) $CH_3CH_2CH_2OH$

(b) 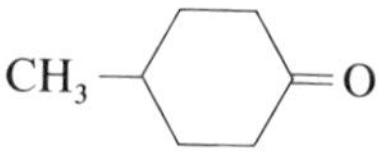

(c) 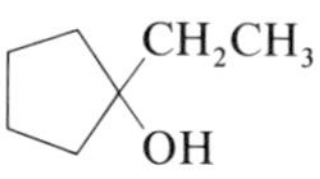

(d) 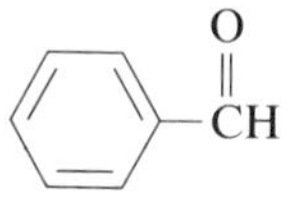

(e) 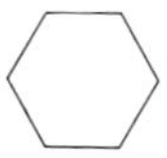

(f) $CH_3CH_2CH_2C(CH_3)(Br)CH_3$

(g) $HOCH_2CH_2OH$

(h) $CH_3C{\equiv}CCH_2CH_2OH$

(i) 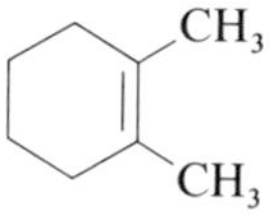

(j) 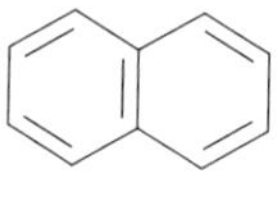

(k) 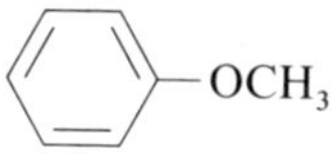

(l) $CH_2{=}CHCH(OH)CH_3$

(m) $CH_3CH_2CH_2Br$

(n) $CH_3CH_2C(=O)CH_3$

Reagents

(1) H_2SO_4, cold, concentrated (protonation)

(2) $KMnO_4$, H_2O, cold

(3) CrO_3, H_2SO_4, acetone

(4) Br_2, carbon tetrachloride

(5) $AgNO_3$, ethanol

(6) H_2O, cold (solubility)

(7) NaI, acetone

(8) 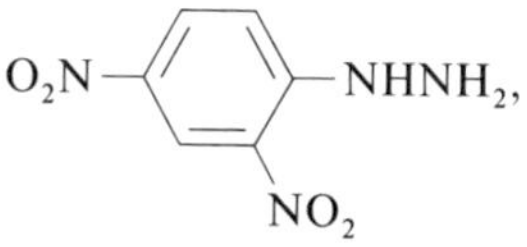
H_2SO_4, ethanol

13.40 When 4-*tert*-butylcyclohexanone is reduced with sodium borohydride, the product is 86% *trans*-4-*tert*-butylcyclohexanol and 14% *cis*-4-*tert*-butylcyclohexanol. When 3,3,5-trimethylcyclohexanone is reduced with sodium borohydride, the product is a mixture, with 48% in which the hydroxyl group is cis to the methyl group at carbon 5 and 52% in which the hydroxyl group is trans to that methyl group. Draw structures showing the correct conformations for the starting materials and the products of these reactions. Offer an explanation for the facts observed experimentally.

13.41 An intermediate in the synthesis of a natural product that has antitumor activity is prepared by the following sequence of steps.

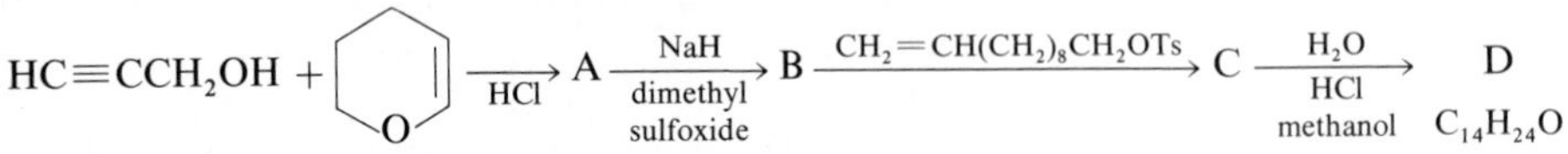
HC≡CCH₂OH +
HCl
A
NaH
dimethyl sulfoxide
B
CH₂=CH(CH₂)₈CH₂OTs
C
H₂O
HCl
methanol
D
C₁₄H₂₄O

Assign structures to Compounds A, B, C, and D.

13.42 During research into methods for synthesizing natural products, the following sequence of reactions was carried out.

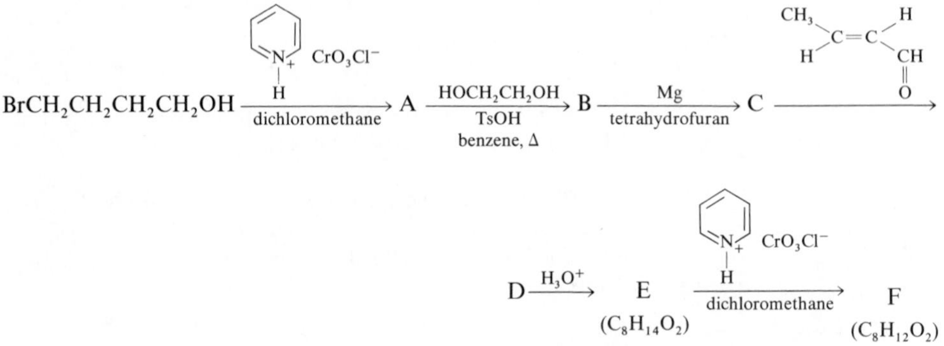
BrCH₂CH₂CH₂CH₂OH
CrO₃Cl⁻
dichloromethane
A
HOCH₂CH₂OH
TsOH
benzene, Δ
B
Mg
tetrahydrofuran
C
CH₃
H
CH
O
D
H₃O⁺
E
(C₈H₁₄O₂)
F
(C₈H₁₂O₂)

Assign structures to Compounds A, B, C, D, E, and F. (Hint: A review of Section 13.7B may be helpful in assigning a structure to E.) How many units of unsaturation does F have? How does the structure you assigned account for all of them?

13.43 Supply reagents for each of the following transformations. More than one step may be necessary for some.

(a) $CH_3CH(OH)CH_2CH_2CH_2CH_3 \longrightarrow CH_3CH_2CH_2CH_2CH_2CH_3$

(b) 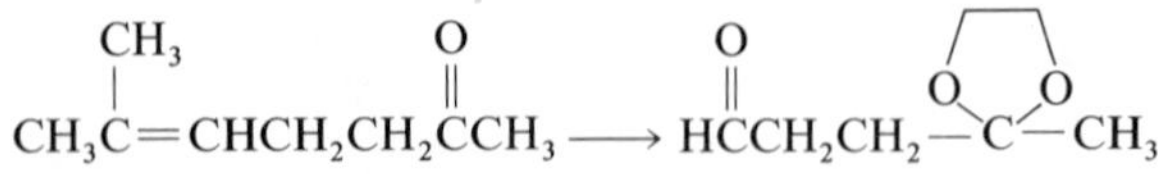

(c) $CH_2{=}CHCH_2CH_2C(=O)CH{=}CHC(=O)OCH_3 \longrightarrow CH_2{=}CHCH_2CH_2C(OCH_3)_2CH{=}CHC(=O)OCH_3$

(d) 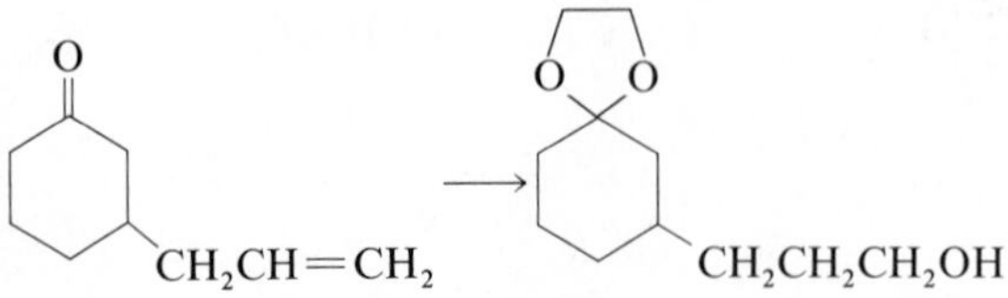
CH₂CH=CH₂
CH₂CH₂CH₂OH

(e)
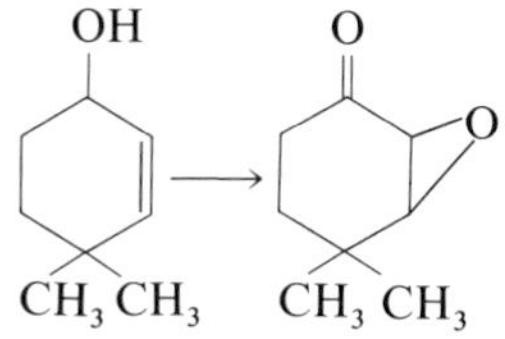

(f)
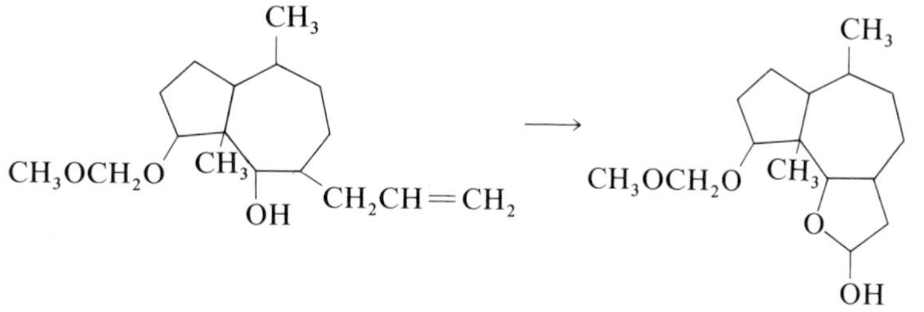

(g)
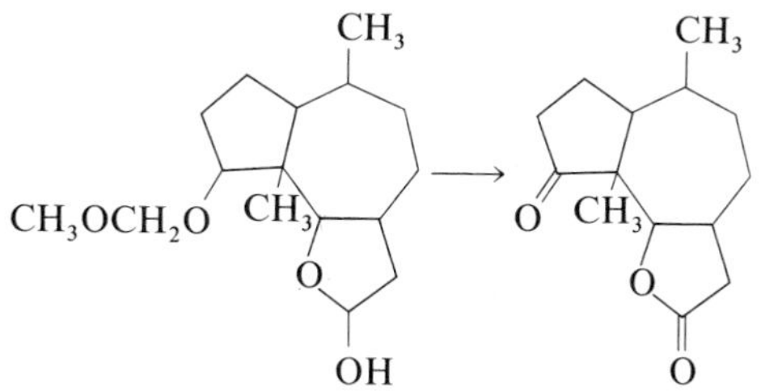

(h)
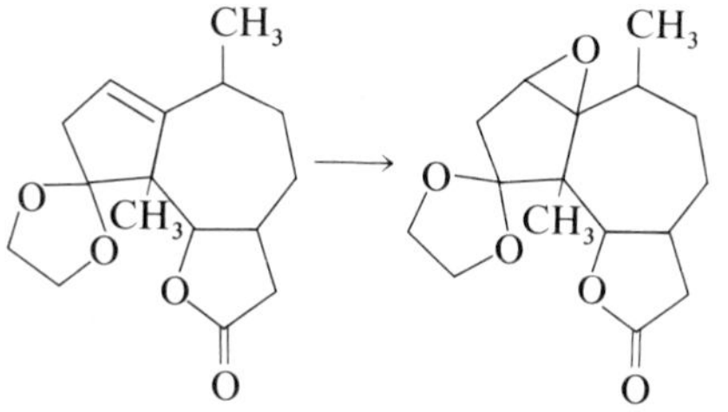

(i)
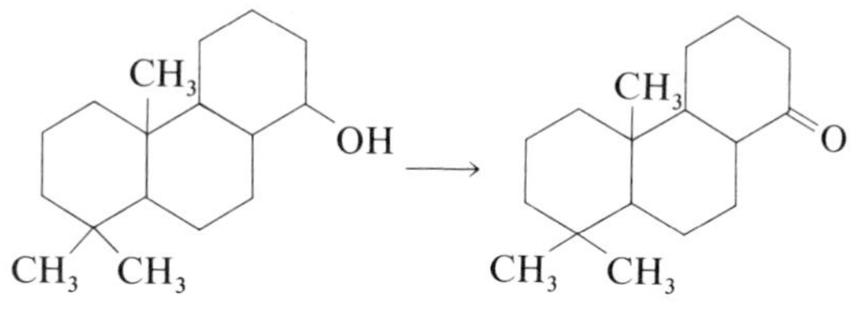

(j)
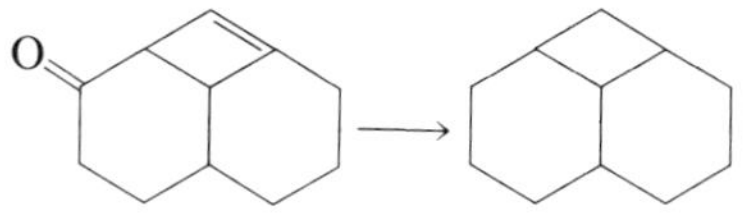

13.44 The following sequence of reactions has been carried out in a synthesis of pentalenene, a natural product related to an antibacterial and antifungal agent. What reagents would you use to accomplish the transformations shown?

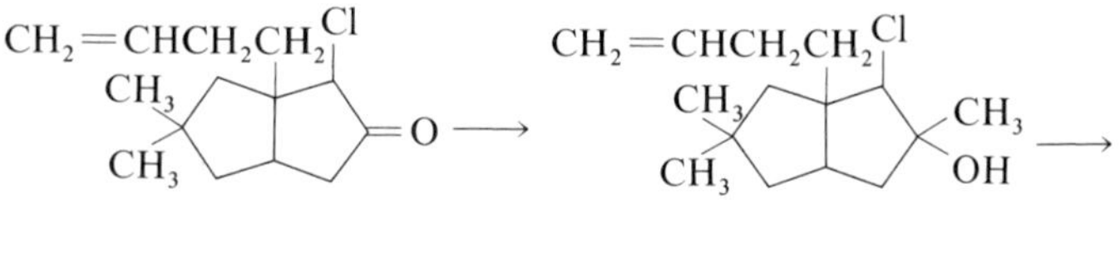

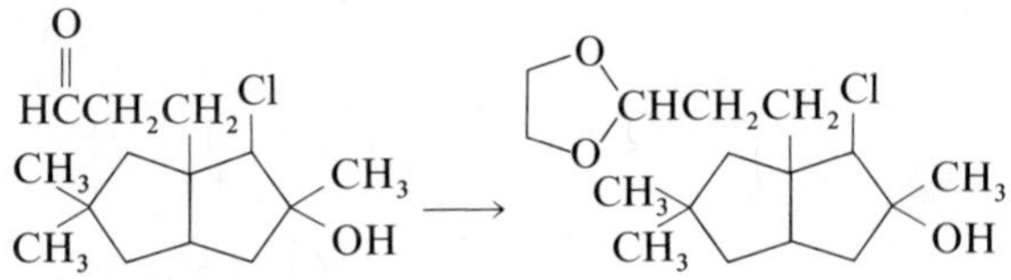

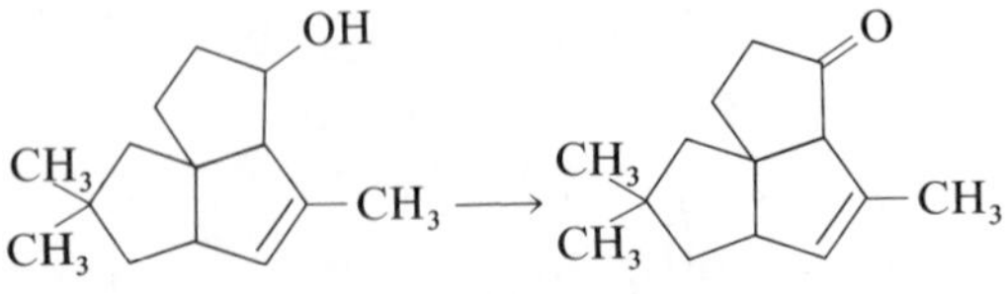

13.45 Write detailed mechanisms that account for the products observed for the following transformations.

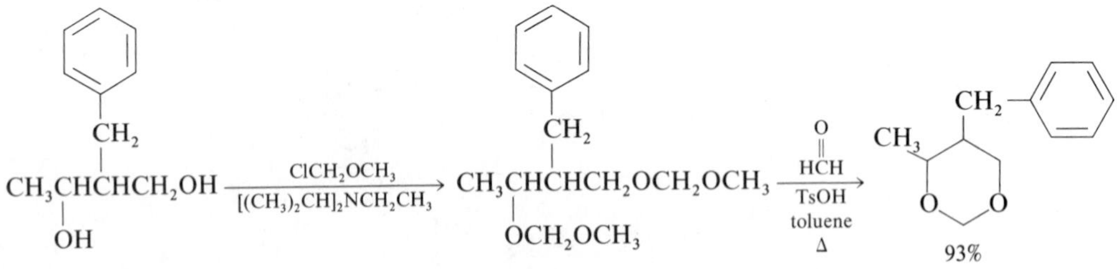

13.46 Alkyl halides are converted to alkanes by being treated with lithium aluminum hydride in tetrahydrofuran. The following equation illustrates the reaction.

$$\underset{\text{1-bromooctane}}{CH_3(CH_2)_6CH_2Br} \xrightarrow[\text{tetrahydrofuran, } 25\,^\circ\text{C, 30 min}]{LiAlH_4} \underset{\text{octane, 96\%}}{CH_3(CH_2)_6CH_3}$$

Rate studies on this reaction have established the following orders of reactivity.

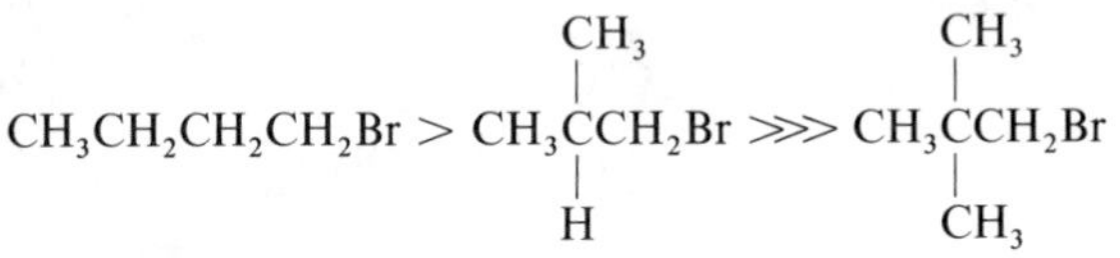

$$CH_3CH_2CH_2CH_2I > CH_3CH_2CH_2CH_2Br > CH_3CH_2CH_2CH_2Cl$$

Propose a mechanism for the reaction, and discuss how the rate data support your mechanism.

13.47 A synthesis of natural products that have antileukemic and cytotoxic properties involves the following transformation. How would you carry it out?

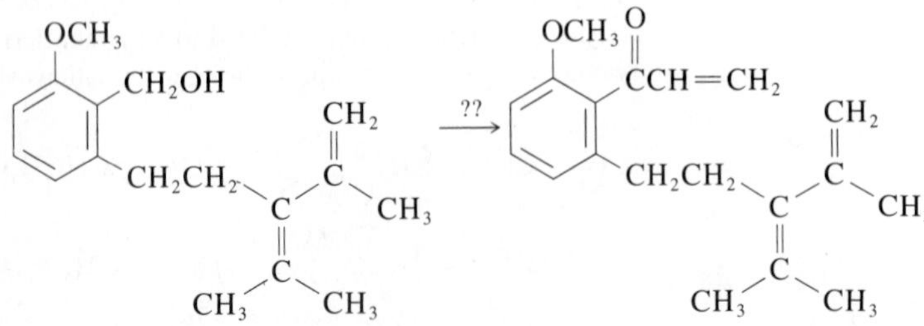

13.48 Further steps in the synthesis of a fecapentaene (Problem 13.25, p. 523) are shown below. Give structures for all compounds represented by letters.

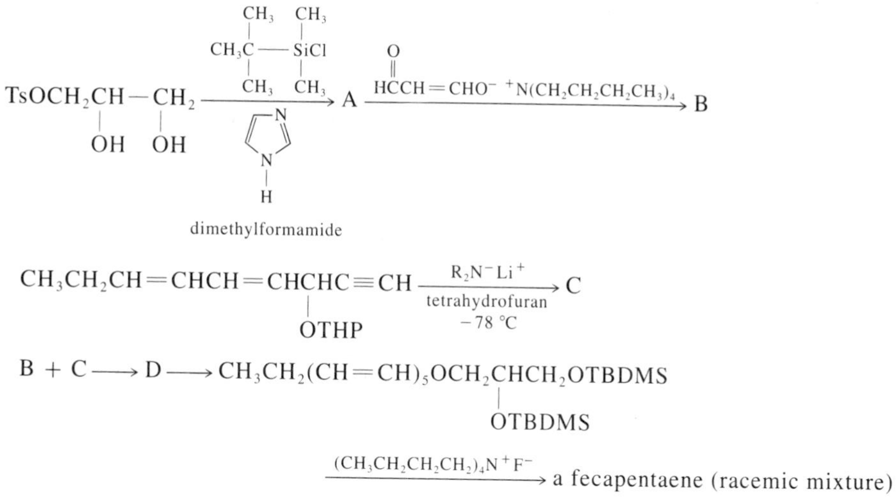
TsOCH2CH—CH2
OH OH
CH3 CH3
CH3C—SiCl
CH3 CH3
A
HCCH=CHO− +N(CH2CH2CH2CH3)4
B
N
H
dimethylformamide
CH3CH2CH=CHCH=CHCHC≡CH
OTHP
R2N−Li+
tetrahydrofuran
−78 °C
C
B + C → D → CH3CH2(CH=CH)5OCH2CHCH2OTBDMS
OTBDMS
(CH3CH2CH2CH2)4N+F−
a fecapentaene (racemic mixture)

13.49 The oxirane derived from 7-methyl-6-octen-3-ol is used as an intermediate in the synthesis of the pheromone of the square-necked grain beetle, an insect that causes great damage to corn.

(a) What is the structure of 7-methyl-6-octen-3-ol, and how would you prepare it from 5-bromo-2-methyl-2-pentene?

(b) Before the double bond is oxidized, the hydroxyl group of 7-methyl-6-octen-3-ol is protected as the *tert*-butyldimethylsilyl ether. How would you do this?

(c) How would you complete the synthesis of the oxirane? Why was the hydroxyl group protected before the oxirane was made? What type of reaction is possible between an unprotected alcohol and an oxirane ring?

13.50 The following tosylhydrazone is required for a study of reaction mechanisms.

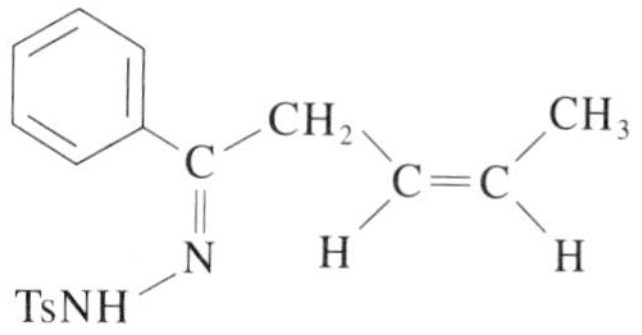
CH2
CH3
C
C=C
N
H
H
TsNH

Show how you would synthesize the compound, assuming you have the following starting materials.

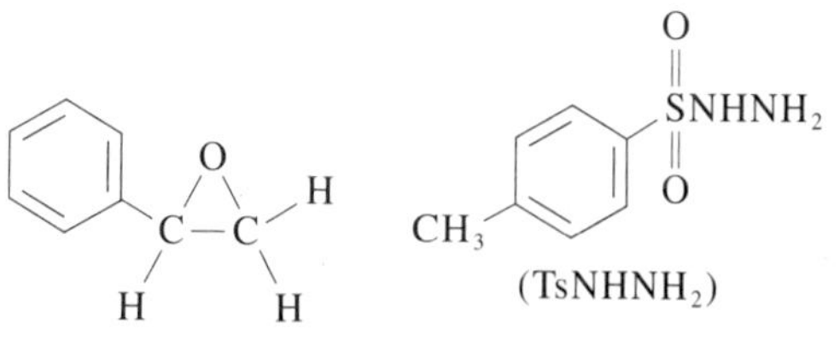
O
C—C
H
H
H
O
SNHNH2
O
CH3
(TsNHNH2)

styrene oxide

tosylhydrazine

Your stockroom also contains any other organic compounds containing three or fewer carbon atoms and a good supply of solvents and inorganic reagents.

13.51 Supply a structural formula for Compound A, and write equations showing how it is converted into the final product.

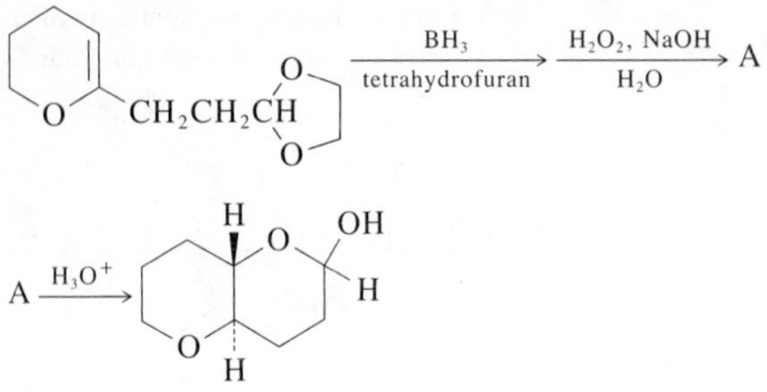

13.52 The following synthesis was carried out during a study of the mechanism by which Vitamin B_{12} functions in the human body.

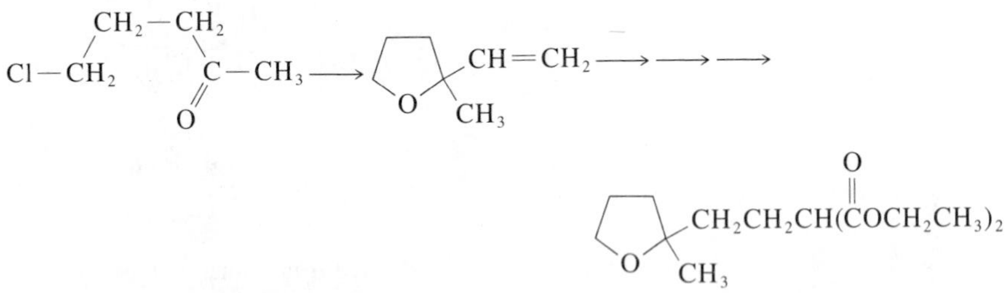

Provide a reagent and a mechanism for the first step. How would you carry out the rest of the transformation? (Hint: A review of p. 261 may be helpful with respect to the last step of the synthesis.)

13.53 Compound A, $C_8H_{12}O$, is prepared by the reaction of Compound B, $C_6H_{10}O$, with an organometallic reagent. Assign structures to Compounds A and B using the infrared spectra given in Figure 13.2, and write an equation for the conversion of the one to the other.

Compound A, $C_8H_{12}O$

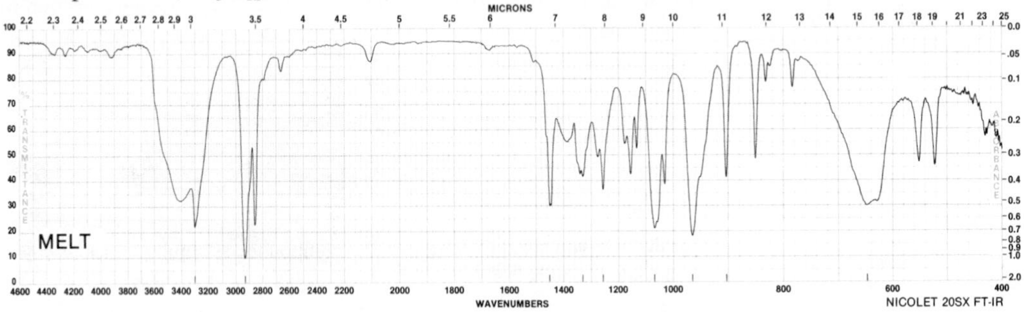
MICRONS
MELT
WAVENUMBERS
NICOLET 20SX FT-IR

Compound B, $C_6H_{10}O$

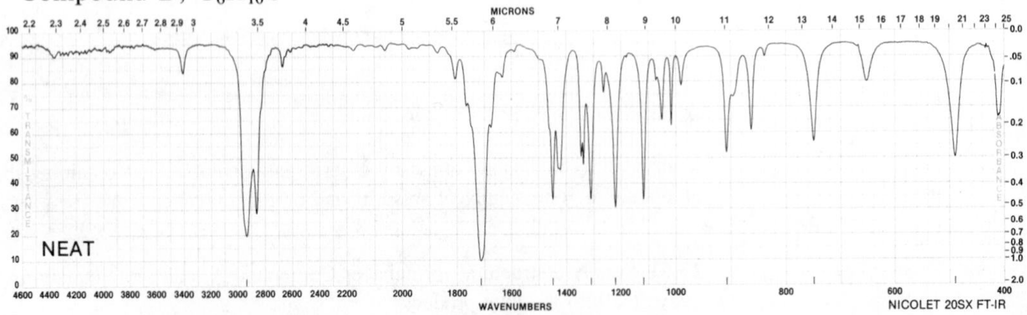
MICRONS
NEAT
WAVENUMBERS
NICOLET 20SX FT-IR

FIGURE 13.2

13.54 Compounds C, D, and E all have the molecular formula $C_5H_{10}O$. They all show absorption in their infrared spectra at 1717 cm^{-1}. Proton magnetic resonance spectra for the compounds are given in Figure 13.3. Carbon-13 nuclear magnetic resonance data for the compounds are summarized below.

Compound C: 13.5, 17.5, 29.3, 45.2, and 206.6 ppm
Compound D: 7.3, 35.3, and 209.3 ppm
Compound E: 18.1, 27.3, 41.5, and 211.8 ppm

Assign structures to Compounds C, D, and E, showing specifically how each piece of spectral data supports your assignments.

Compound C

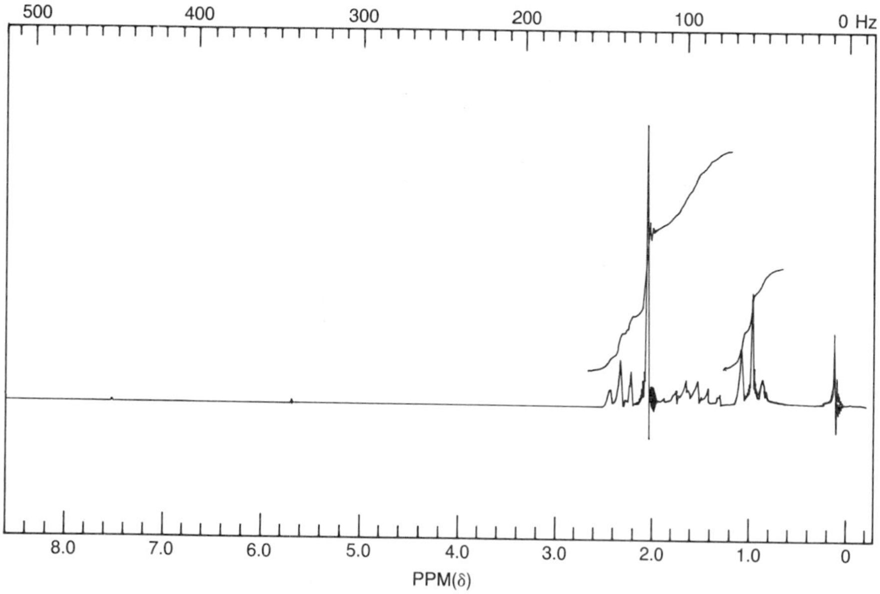
500
400
300
200
100
0 Hz
8.0
7.0
6.0
5.0
4.0
3.0
2.0
1.0
0
PPM(δ)

Compound D

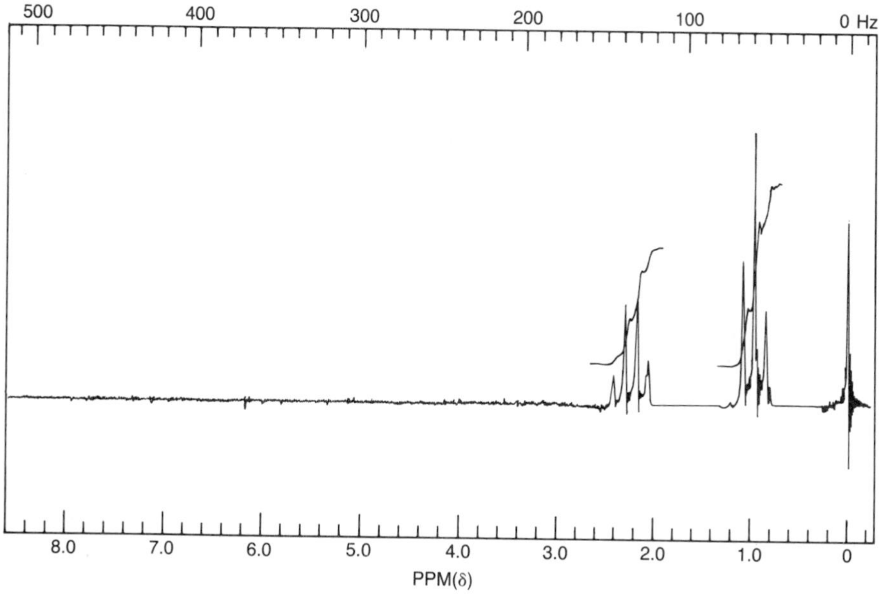
500
400
300
200
100
0 Hz
8.0
7.0
6.0
5.0
4.0
3.0
2.0
1.0
0
PPM(δ)

FIGURE 13.3

(Continued)

Compound E

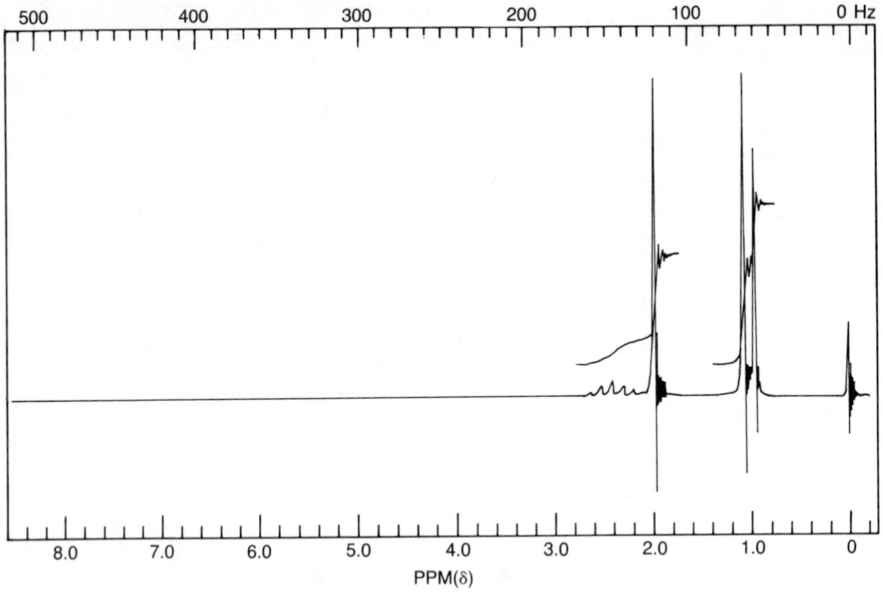
500
400
300
200
100
0 Hz
8.0
7.0
6.0
5.0
4.0
3.0
2.0
1.0
0
PPM(δ)

FIGURE 13.3

14

Carboxylic Acids and Their Derivatives I. Nucleophilic Substitution Reactions at the Carbonyl Group

A • L O O K • A H E A D

Carboxylic acids and their derivatives are compounds in which a carbonyl group is bonded to an atom that has at least one pair of nonbonding electrons on it. Acetic acid and its derivatives are examples.

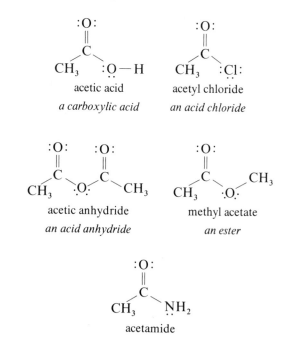

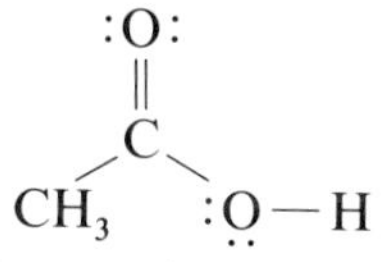

acetic acid

a carboxylic acid

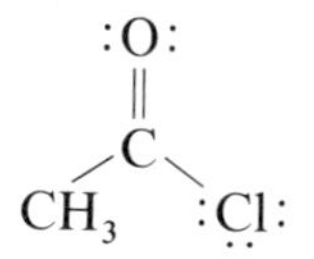

acetyl chloride

an acid chloride

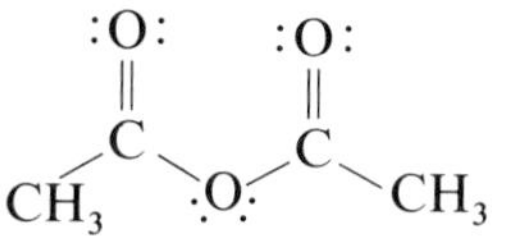

acetic anhydride

an acid anhydride

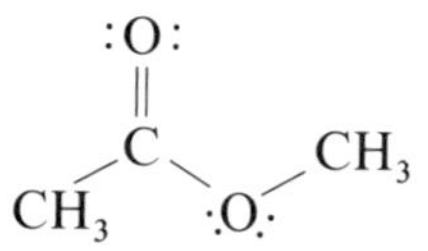

methyl acetate

an ester

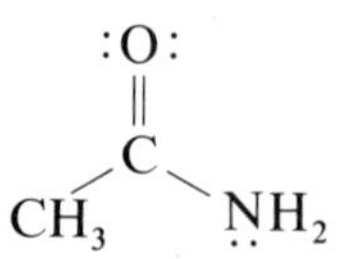

acetamide

an amide

Carboxylic acids are strong organic acids. Also, the carbon atom of the carbonyl group is electrophilic and reacts with nucleophiles.

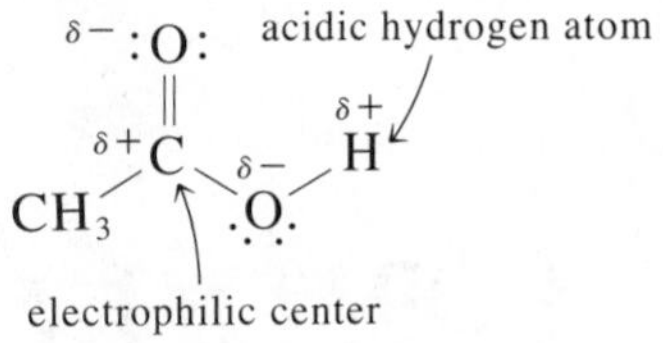
acidic hydrogen atom
electrophilic center

An acid derivative may be thought of as having been created from a carboxylic acid by replacement of the hydroxyl group of the carboxyl group by another atom or group. This group either is a good leaving group or may be converted to a good leaving group by protonation. Acids and acid derivatives, therefore, undergo nucleophilic substitution reactions, an example of which is the reaction of acetyl chloride with ammonia.

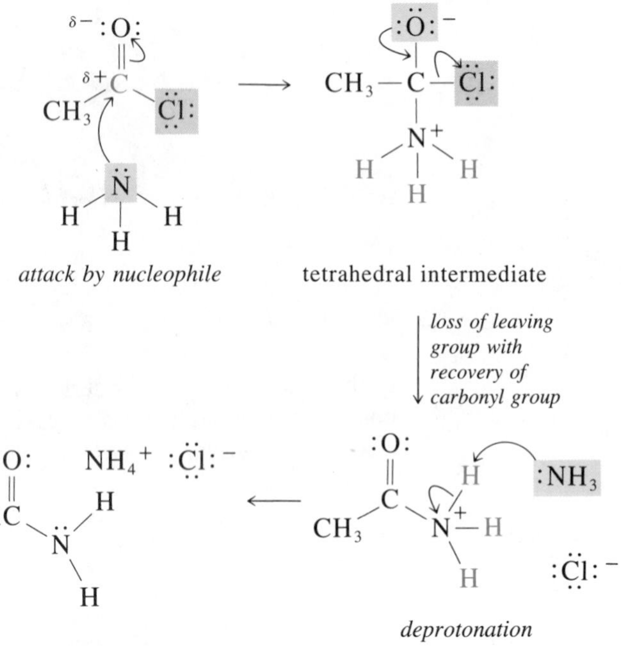
attack by nucleophile
tetrahedral intermediate
loss of leaving group with recovery of carbonyl group
deprotonation

Most nucleophilic substitution reactions of acids and acid derivatives have two steps.

1. Nucleophilic attack on the carbon atom of the carbonyl group, with formation of a tetrahedral intermediate.
2. Loss of a leaving group, with the recovery of the carbonyl group.

In the reaction shown above, the acetyl group, an acyl group, is transferred from a chlorine atom to a nitrogen atom.

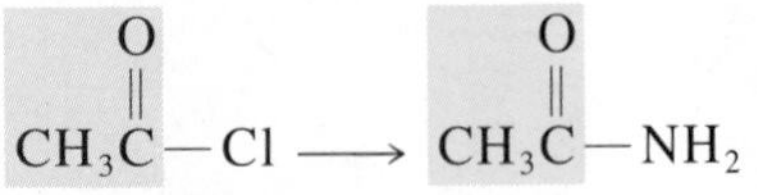

the acetyl group in
acetyl chloride
an acyl group

These important reactions of acid derivatives are called acylation, or acyl-transfer, reactions.

The reactions of acid derivatives differ from those of aldehydes and ketones, which do not have good leaving groups bonded to the carbonyl group. The first step of the reaction with nucleophiles is the same for acid derivatives as it is for aldehydes and ketones (p. 504, for example). Unlike aldehydes and ketones, however, acid derivatives undergo nucleophilic substitution rather than nucleophilic addition.

This chapter will emphasize the interconversions of the different acid derivatives through nucleophilic substitution reactions.

14.1 PROPERTIES OF THE FUNCTIONAL GROUPS IN CARBOXYLIC ACIDS AND THEIR DERIVATIVES

A. The Functional Groups in Carboxylic Acids and Their Derivatives

Carboxylic acids are organic compounds that contain the carboxyl group, a functional group in which a hydroxyl group is directly bonded to the carbon atom of a carbonyl group. Interaction between the carbonyl group and the hydroxyl group affects the properties of both. For example, the carbonyl group in acids is not as electrophilic as the carbonyl group in aldehydes and ketones. A comparison of the resonance contributors possible for a carbonyl group and for a carboxyl group shows why this is so.

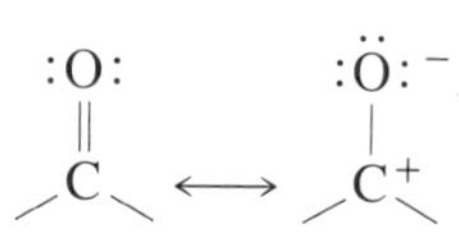

resonance contributors for a carbonyl group

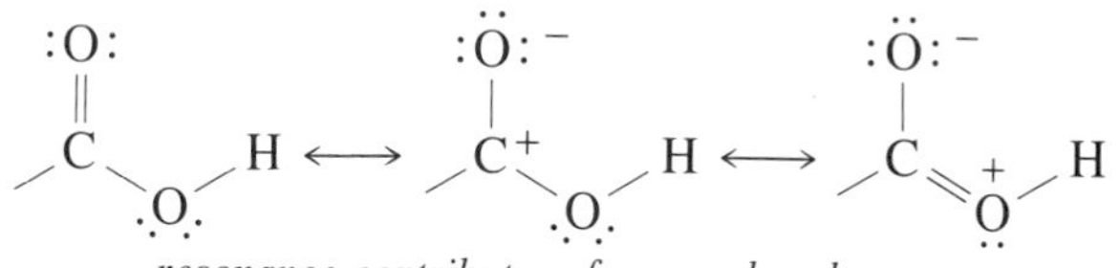

resonance contributors for a carboxyl group

The carbon atom of the carbonyl group in an aldehyde or ketone is an electrophilic center that reacts with a variety of nucleophiles, such as alcohols (p. 499), amine derivatives (p. 503), and organometallic reagents (p. 491). In carboxylic acids, the electrophilicity of the carbonyl group is modified by the presence of nonbonding electrons on the oxygen atom of the hydroxyl group. Donation of these electrons to the carbonyl group transfers some of the positive character of the carbonyl carbon atom to that oxygen atom. For that reason, many reagents that react easily with the carbonyl group of aldehydes or ketones react more slowly or only in the presence of powerful catalysts when attacking the carbonyl group of a carboxylic acid or an acid derivative.

The hydroxyl group of a carboxylic acid is unlike the hydroxyl group of an alcohol. The drain of electrons away from the hydroxyl group by the carbonyl group increases the positive character of the hydrogen atom and stabilizes the carboxylate anion (p. 95). The hydrogen atom of the hydroxyl group of a carboxylic acid is much more easily lost as a proton than is the hydrogen atom of the hydroxyl group of an alcohol. The acidity of carboxylic acids is discussed further in Section 14.3.

In an **acid chloride,** the hydroxyl group of a carboxylic acid has been replaced by a chlorine atom. In an **acid anhydride,** the anion corresponding to a carboxylic acid has taken the place of the original hydroxyl group. In an **ester,** an alkoxyl group replaces the hydroxyl group. In an **amide,** an amino group is the replacement.

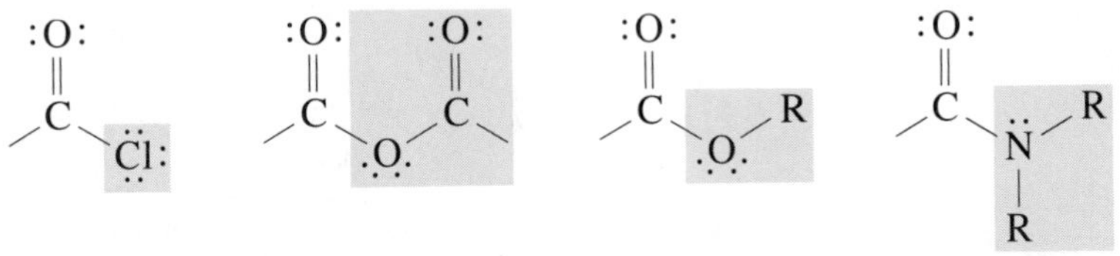

acid chloride   acid anhydride   ester   amide

In each acid derivative, the atom bonded directly to the carbonyl group has at least one pair of nonbonding electrons on it and can therefore interact with the carbonyl group in the same way the hydroxyl group does in carboxylic acids. Also, each of the groups shaded above either is a good leaving group or may be converted into a good leaving group by protonation. These structural features are important in the chemistry of acids and acid derivatives.

PROBLEM 14.1

(a) Write structural formulas for propanoic acid, $CH_3CH_2CO_2H$, and its acid chloride, acid anhydride, ethyl ester, and amide.
(b) Write equations for the reactions that you would expect between propanoic acid and concentrated sulfuric acid. Repeat the process for ethyl propanoate and propanamide. (Reviewing Section 3.2 may be helpful.)
(c) Encircle any good leaving groups that you see in the structural formulas you have written in parts a and b.

PROBLEM 14.2

Write resonance contributors for propanoic acid and its acid chloride, acid anhydride, ethyl ester, and amide, showing in each case how the polarity of the carbonyl group is affected by the presence of an adjacent atom having a pair of nonbonding electrons.

PROBLEM 14.3

Propanamide is much less basic than propylamine.

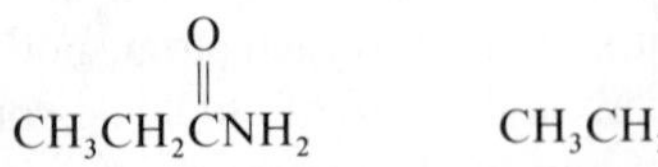

How would you explain this fact in light of the resonance contributors that you wrote for the amide in Problem 14.2? (You may want to review the factors affecting basicity on pp. 102–104).

B. Physical Properties of Low-Molecular-Weight Acids and Acid Derivatives

Carboxylic acids of low molecular weight have boiling points that are relatively high, and they are very soluble in water. Molecular weight determinations indicate that carboxylic acids exist as dimers even in the vapor state. All of these data suggest that the carboxyl group participates both as a donor and an acceptor in extensive hydrogen bonding, as illustrated below for acetic acid in the vapor state, in the liquid state, and in solution in water.

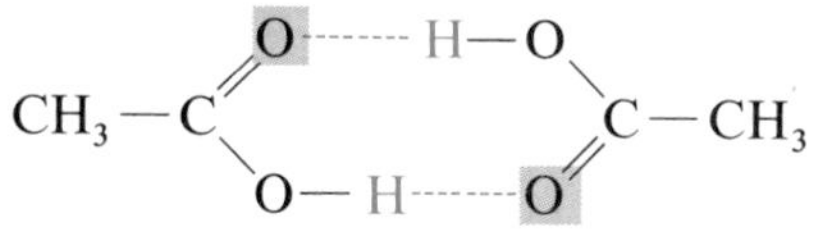

dimer of acetic acid held together by hydrogen bonding in the vapor state

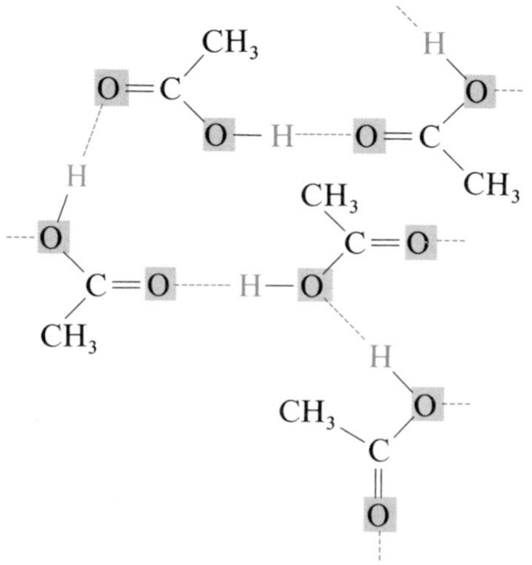

network of hydrogen bonding between molecules of acetic acid in the liquid state

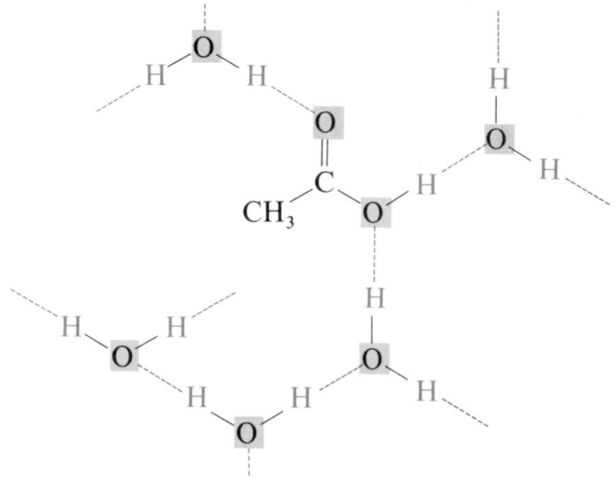

acetic acid, hydrogen bonded to water molecules in aqueous solution

Carboxylic acids with no other functional group and fewer than ten carbon atoms in the chain are liquids at room temperature. Acetic acid has a particularly high melting point, 16.7 °C, for a compound with such a low molecular weight and is known as **glacial acetic acid** in its pure state. It is a liquid at room temperature but freezes easily in an ice bath, a phenomenon that has practical importance in the laboratory. Oxalic acid and the larger dicarboxylic acids, as well as the aromatic carboxylic acids, are all solids at room temperature.

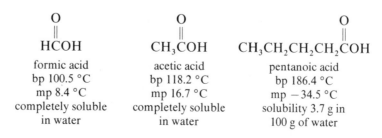

$HC(=O)OH$	$CH_3C(=O)OH$	$CH_3CH_2CH_2CH_2C(=O)OH$
formic acid	acetic acid	pentanoic acid
bp 100.5 °C	bp 118.2 °C	bp 186.4 °C
mp 8.4 °C	mp 16.7 °C	mp −34.5 °C
completely soluble in water	completely soluble in water	solubility 3.7 g in 100 g of water

$CH_3(CH_2)_8\overset{O}{\overset{\|}{C}}OH$
decanoic acid
bp 270.0 °C
mp 31.3 °C
solubility 0.015 g
in 100 g of water

$HO\overset{O}{\overset{\|}{C}}-\overset{O}{\overset{\|}{C}}OH$
oxalic acid
mp 187 °C
solubility 9.0 g
in 100 g of water

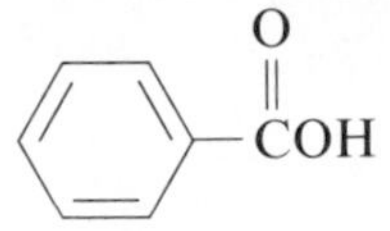
benzoic acid
mp 122 °C
solubility 0.29 g
in 100 g of water

Monocarboxylic acids and dicarboxylic acids of low molecular weight are soluble in water. When the hydrocarbon portion of the molecule has more than about five carbon atoms for each carboxyl group, solubility decreases. The high-molecular-weight carboxylic acids are almost insoluble in water.

Carboxylic acids that have low solubility in water, such as benzoic acid, are converted to water-soluble salts by reaction with aqueous base (p. 95). Protonation of the carboxylate anion by a strong acid regenerates the water-insoluble acid. These properties of carboxylic acids are useful in separating them from reaction mixtures containing neutral and basic compounds.

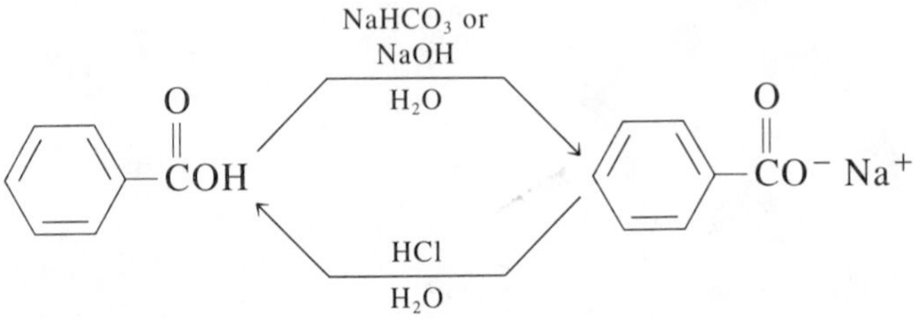
NaHCO3 or
NaOH
H2O
HCl
H2O

benzoic acid
covalent,
insoluble in water

sodium benzoate
ionic,
soluble in water

The importance of hydrogen bonding to the physical properties and solubility in water of carboxylic acids is demonstrated by comparing acetic acid with two of its derivatives, an ester and an amide.

$CH_3\overset{O}{\overset{\|}{C}}OH$
acetic acid
bp 118 °C
completely soluble
in water

$CH_3\overset{O}{\overset{\|}{C}}OCH_2CH_3$
ethyl acetate
bp 77 °C
solubility 8.6 g
in 100 g of water

$CH_3\overset{O}{\overset{\|}{C}}NH_2$
acetamide
mp 82 °C
solubility 97.5 g
in 100 g of water

Acetic acid boils at 118 °C and is fully miscible with water, but its ethyl ester has a boiling point of 77 °C and a solubility of 8.6 g in 100 g of water. Ethyl acetate cannot hydrogen bond to itself in the liquid state. In water, it can serve only as a hydrogen-bond acceptor at its oxygen atoms. Therefore, it has a low boiling point and relatively low solubility in water. Acetamide, on the other hand, is a solid (mp 82 °C) with a very high solubility in water. The hydrogen atoms on the nitrogen atom of an amide participate strongly in hydrogen bonding, a fact of crucial importance to the structure of proteins, which are polyamides (p. 1150).

PROBLEM 14.4

Predict which compound in each of the following series will have the highest solubility in water and which will have the lowest.

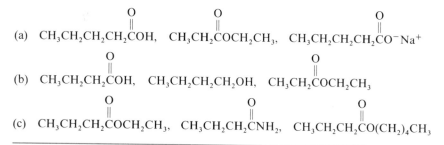

(a) $CH_3CH_2CH_2CH_2\overset{O}{\overset{\|}{C}}OH$, $CH_3CH_2\overset{O}{\overset{\|}{C}}OCH_2CH_3$, $CH_3CH_2CH_2CH_2\overset{O}{\overset{\|}{C}}O^-Na^+$

(b) $CH_3CH_2CH_2\overset{O}{\overset{\|}{C}}OH$, $CH_3CH_2CH_2CH_2OH$, $CH_3CH_2\overset{O}{\overset{\|}{C}}OCH_2CH_3$

(c) $CH_3CH_2CH_2\overset{O}{\overset{\|}{C}}OCH_2CH_3$, $CH_3CH_2CH_2\overset{O}{\overset{\|}{C}}NH_2$, $CH_3CH_2CH_2\overset{O}{\overset{\|}{C}}O(CH_2)_4CH_3$

14.2 NOMENCLATURE OF CARBOXYLIC ACIDS AND THEIR DERIVATIVES

A. Naming Carboxylic Acids

The systematic name of an alkyl carboxylic acid is derived by replacing the **e** at the end of the name of the hydrocarbon having the same number of carbon atoms in the chain with **-oic acid.** The carboxyl function is always assumed to be the first carbon atom of the chain. The presence of other substituents is indicated by assigning a name and a position number to each one. The two smallest carboxylic acids, formic acid (from *formica,* Latin for ant) and acetic acid (from *acetum,* Latin for vinegar), are usually known by their common names.

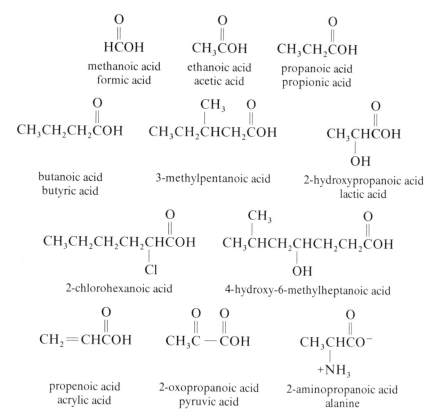

$H\overset{O}{\overset{\|}{C}}OH$
methanoic acid
formic acid

$CH_3\overset{O}{\overset{\|}{C}}OH$
ethanoic acid
acetic acid

$CH_3CH_2\overset{O}{\overset{\|}{C}}OH$
propanoic acid
propionic acid

$CH_3CH_2CH_2\overset{O}{\overset{\|}{C}}OH$
butanoic acid
butyric acid

$CH_3CH_2\overset{CH_3}{\overset{|}{C}}HCH_2\overset{O}{\overset{\|}{C}}OH$
3-methylpentanoic acid

$CH_3\underset{OH}{\underset{|}{C}}H\overset{O}{\overset{\|}{C}}OH$
2-hydroxypropanoic acid
lactic acid

$CH_3CH_2CH_2CH_2\underset{Cl}{\underset{|}{C}}H\overset{O}{\overset{\|}{C}}OH$
2-chlorohexanoic acid

$CH_3\overset{CH_3}{\overset{|}{C}}HCH_2\underset{OH}{\underset{|}{C}}HCH_2CH_2\overset{O}{\overset{\|}{C}}OH$
4-hydroxy-6-methylheptanoic acid

$CH_2{=}CH\overset{O}{\overset{\|}{C}}OH$
propenoic acid
acrylic acid

$CH_3\overset{O}{\overset{\|}{C}}-\overset{O}{\overset{\|}{C}}OH$
2-oxopropanoic acid
pyruvic acid

$CH_3\underset{^+NH_3}{\underset{|}{C}}H\overset{O}{\overset{\|}{C}}O^-$
2-aminopropanoic acid
alanine

Biologically important carboxylic acids are also usually known by their common names; the hydroxyacid lactic acid, the α-ketoacid pyruvic acid, and the amino acid alanine are shown above.

The presence of two carboxyl groups is indicated in a systematic name by using **-dioic acid** after the full name (including the final **e**) of the hydrocarbon having the same number of carbon atoms in the chain. The dicarboxylic acids shown below are generally known by their common names, however.

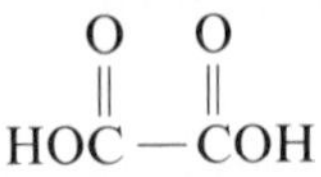

ethanedioic acid
oxalic acid

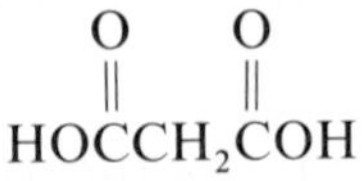

propanedioic acid
malonic acid

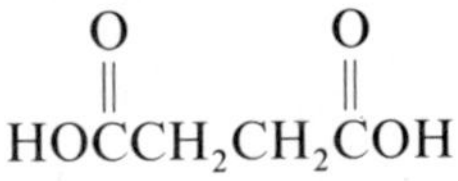

butanedioic acid
succinic acid

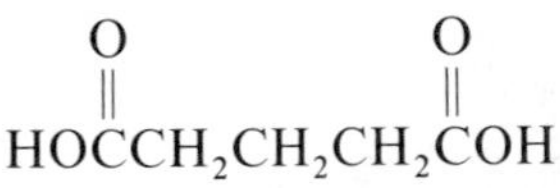

pentanedioic acid
glutaric acid

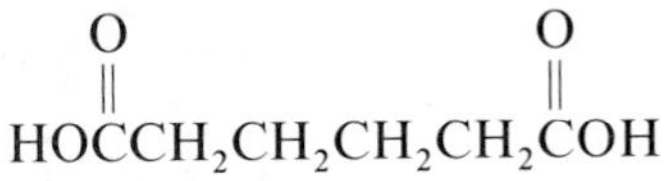

hexanedioic acid
adipic acid

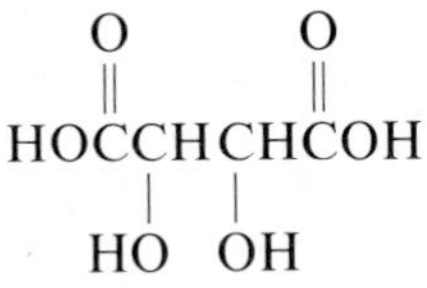

2,3-dihydroxybutanedioic acid
tartaric acid

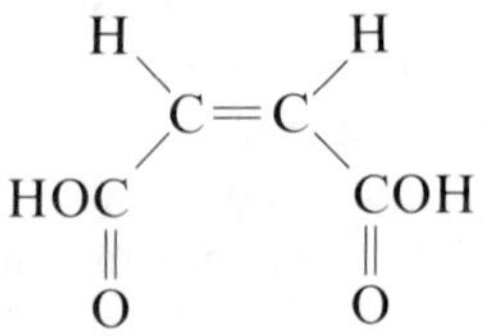

(*Z*)-butenedioic acid
maleic acid

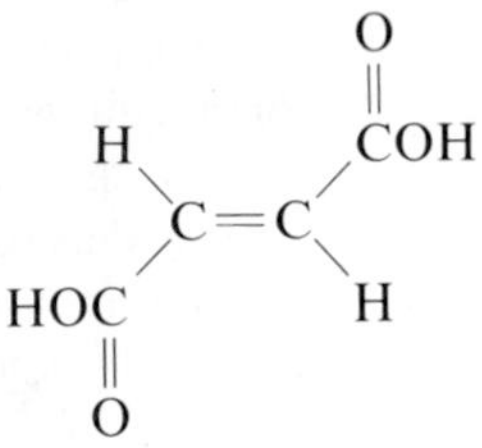

(*E*)-butenedioic acid
fumaric acid

In aromatic carboxylic acids, the carboxyl group is attached to an aromatic ring. Benzoic acid, which is the simplest unsubstituted aromatic acid, and some other examples are shown below.

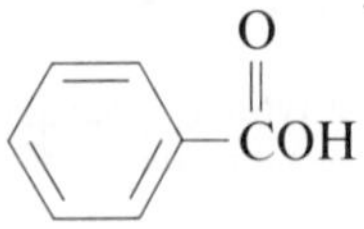

benzoic acid

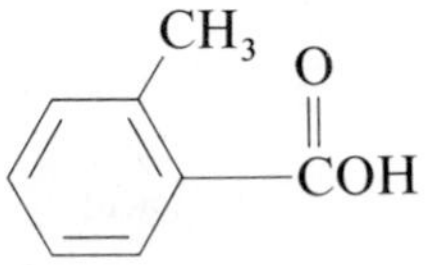

o-methylbenzoic acid
o-toluic acid

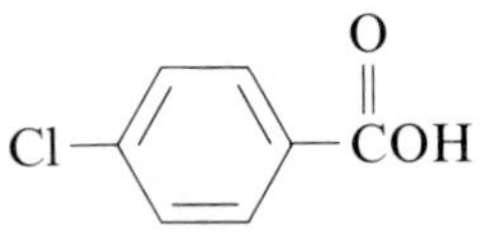

p-chlorobenzoic acid

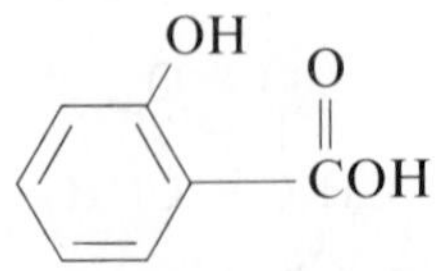

o-hydroxybenzoic acid
salicylic acid

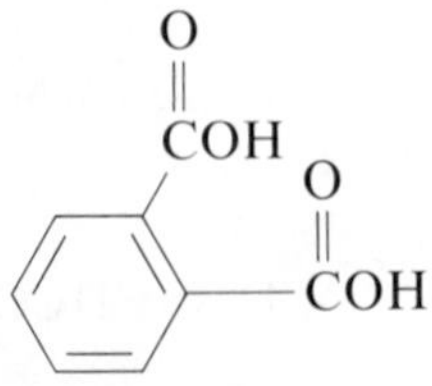

phthalic acid

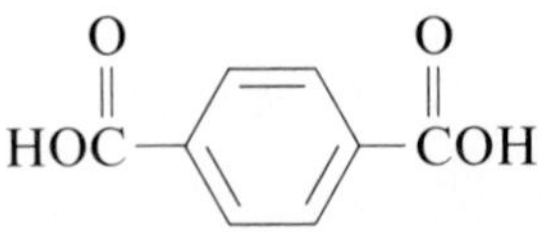

terephthalic acid

When the carboxyl group is attached to a cycloalkane, **-carboxylic acid** is added to the name of the hydrocarbon constituting the rest of the molecule.

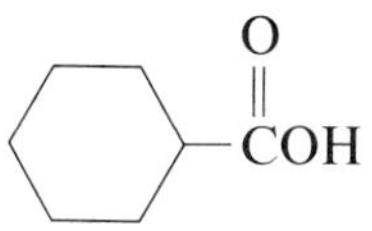

cyclohexanecarboxylic acid

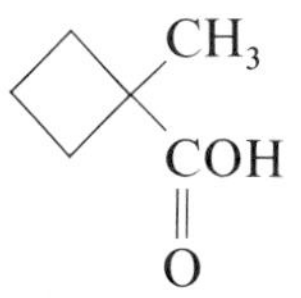

1-methylcyclobutanecarboxylic acid

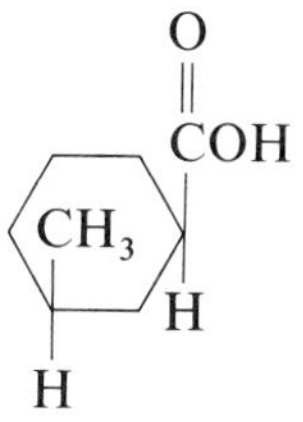

cis-3-methylcyclohexanecarboxylic acid

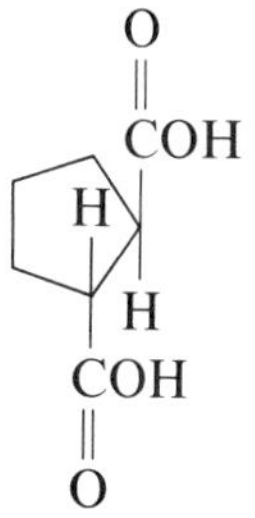

trans-1,2-cyclopentanedicarboxylic acid

PROBLEM 14.5

Some of the carboxylic acids shown in this section have more than one stereoisomer. Identify the compounds for which this is true and draw all possible stereoisomers, naming each one correctly.

PROBLEM 14.6

Name the following compounds, including an indication of the stereochemistry where appropriate.

(a) 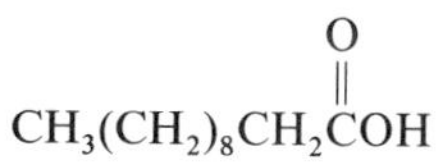

(b) 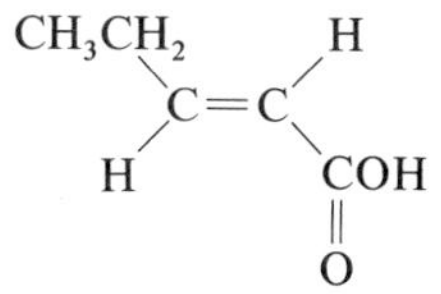

(c) 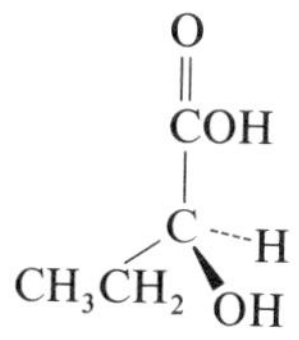

(d) 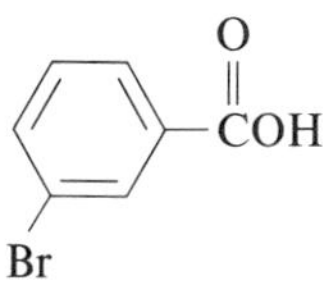

(e) 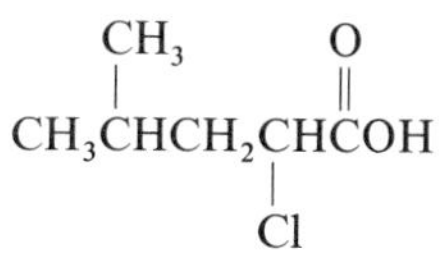

(f) 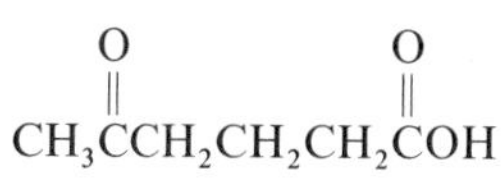

(g) 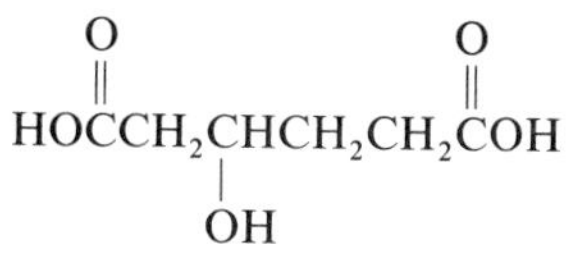

(h) 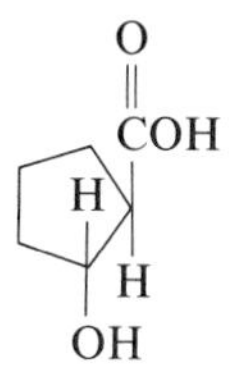

B. Naming Acyl Groups, Acid Chlorides, and Anhydrides

The group obtained from a carboxylic acid by the removal of the hydroxyl portion is known as an **acyl group.** The name of an acyl group is created by changing the **ic** at the end of the name of the carboxylic acid to **yl.** This applies to common as well as systematic names of acids. When **-carboxylic acid** is used in the name of a compound, this ending is changed to **-carbonyl** for the corresponding acyl group. Some important acyl groups are shown below with their related acids.

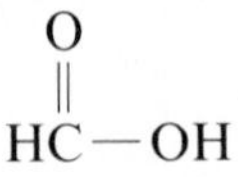
formic acid

formyl group

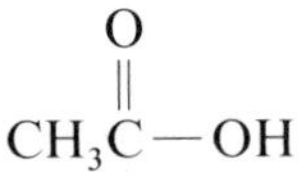
acetic acid

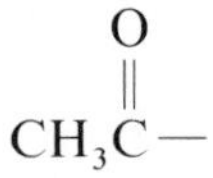
acetyl group

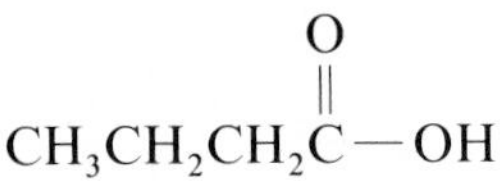
butanoic acid

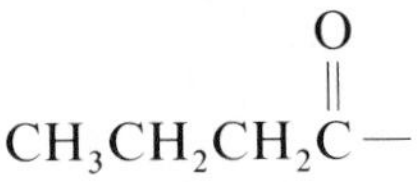
butanoyl group

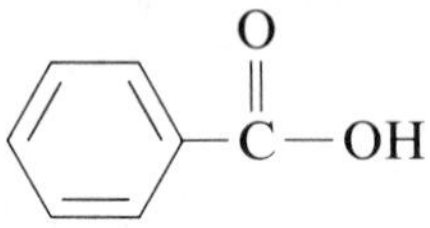
benzoic acid

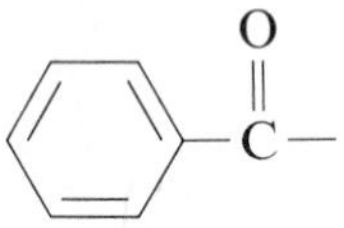
benzoyl group

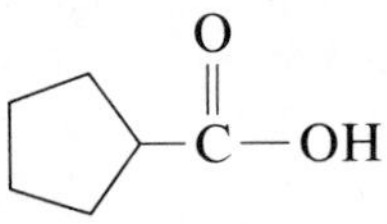
cyclopentanecarboxylic acid

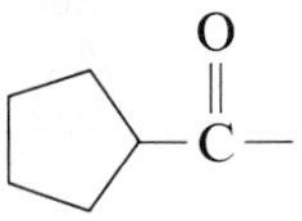
cyclopentanecarbonyl group

Acid chlorides are named systematically as acyl chlorides.

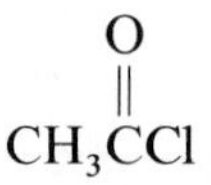
acetyl chloride

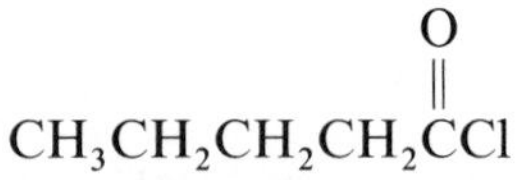
pentanoyl chloride

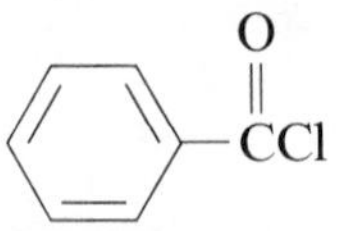
benzoyl chloride

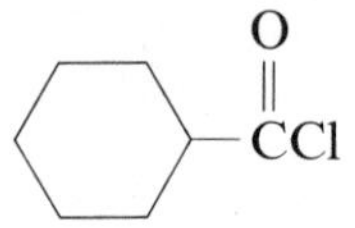
cyclohexanecarbonyl chloride

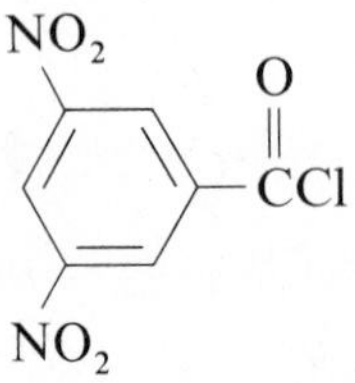
3,5-dinitrobenzoyl chloride

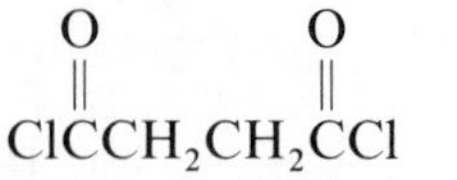
butanedioyl dichloride
succinyl dichloride

Important acid anhydrides are the anhydride of acetic acid and some cyclic anhydrides formed from dicarboxylic acids. An acid anhydride is named by substituting **anhydride** for **acid** in the name of the acid from which it is derived. Cyclic anhydrides containing five- or six-membered rings are stable and easily formed. Of the aromatic dicarboxylic acids, only the ones with the carboxylic groups in adjacent positions on the aromatic ring form cyclic anhydrides.

acetic anhydride

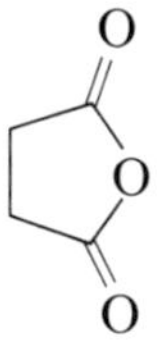
succinic anhydride

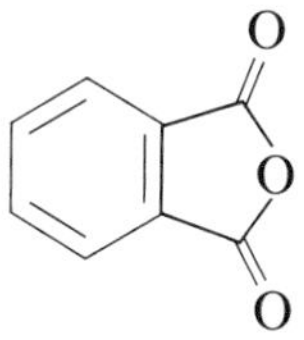
phthalic anhydride

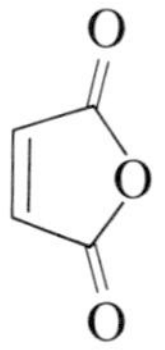
maleic anhydride

C. Naming Salts and Esters

Salts and esters of carboxylic acids are named in the same way. The name of the cation (in the case of a salt) or the name of the organic group attached to the oxygen of the carboxyl group (in the case of an ester) precedes the name of the acid. The **-ic acid** part of the name of the acid is converted to **-ate.**

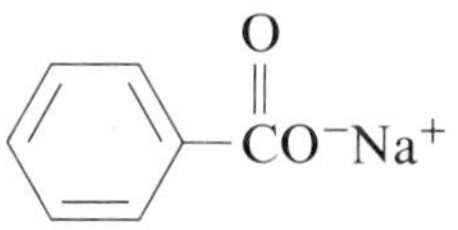
sodium benzoate

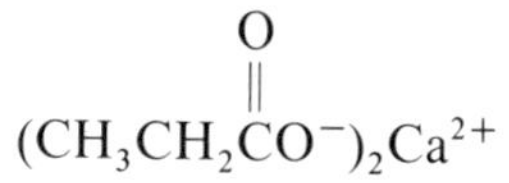
calcium propanoate
calcium propionate

both salts used to retard spoilage in foods

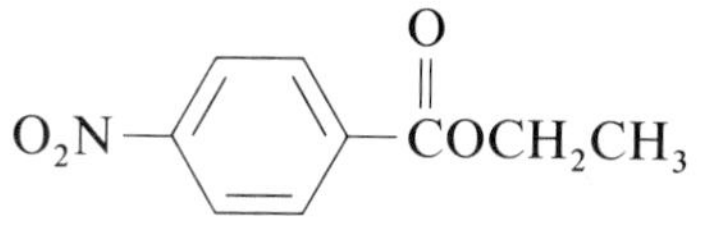
ethyl *p*-nitrobenzoate

dimethyl succinate

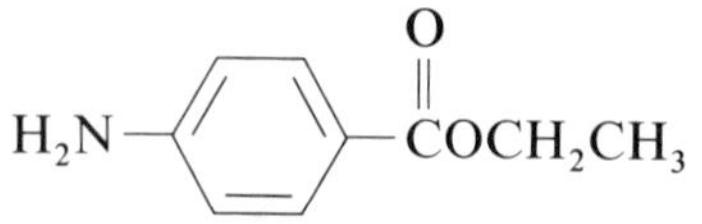
ethyl *p*-aminobenzoate
a local anesthetic

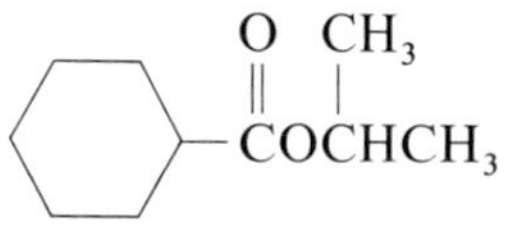
isopropyl cyclohexanecarboxylate

PROBLEM 14.7

Name the following compounds, including an indication of the stereochemistry where appropriate.

(a) $CH_3CH_2CH_2CH_2CH_2\overset{O}{\overset{\|}{C}}Cl$ (b) $CH_3CH_2CH_2\overset{O}{\overset{\|}{C}}O\overset{O}{\overset{\|}{C}}CH_2CH_2CH_3$

(c) 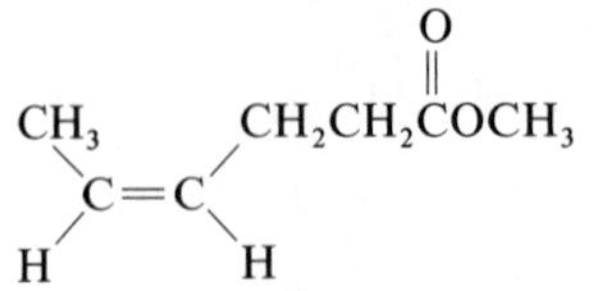 (d) 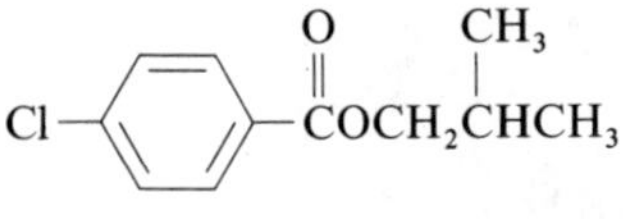

(e) 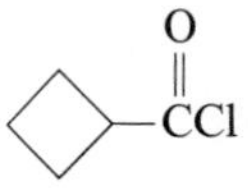 (f) 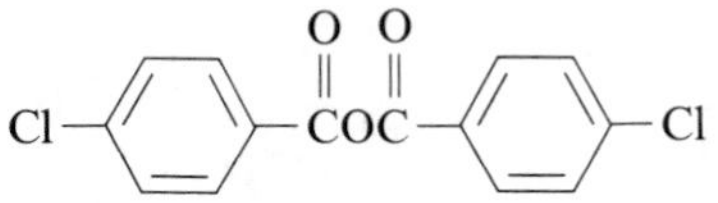

D. Naming Amides, Imides, and Nitriles

The names of amides are formed by replacing **-oic acid** (or **-ic acid** for common names) by **-amide** or **-carboxylic acid** by **-carboxamide.**

$H\overset{O}{\overset{\|}{C}}NH_2$	$CH_3\overset{O}{\overset{\|}{C}}NH_2$	$CH_3CH_2CH_2CH_2\overset{O}{\overset{\|}{C}}NH_2$
methanamide formamide	ethanamide acetamide	pentanamide

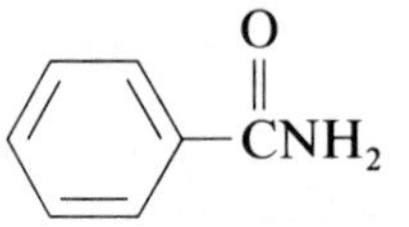

benzamide

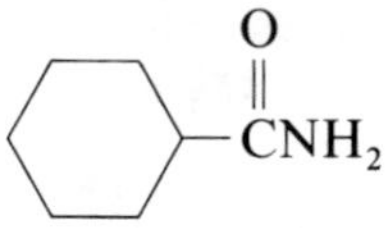

cyclohexanecarboxamide

If the nitrogen atom of the amide has any alkyl groups as substituents, the name of the amide is prefixed by the capital letter *N*-, to indicate substitution on nitrogen, followed by the name(s) of the alkyl group(s).

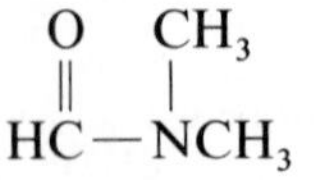

N,N-dimethylformamide
abbreviated DMF

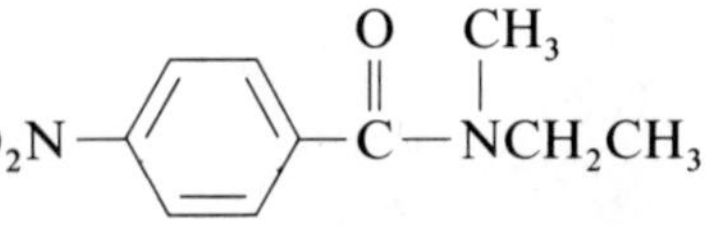

N-ethyl-*N*-methyl-*p*-nitrobenzamide

If the substituent on the nitrogen atom of an amide is a phenyl group, the ending for the name of the carboxylic acid is changed to **-anilide.**

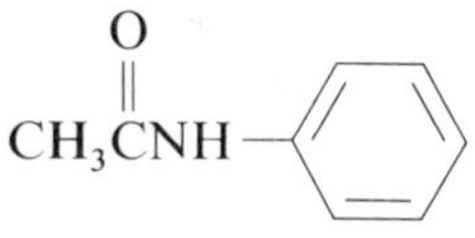

acetanilide

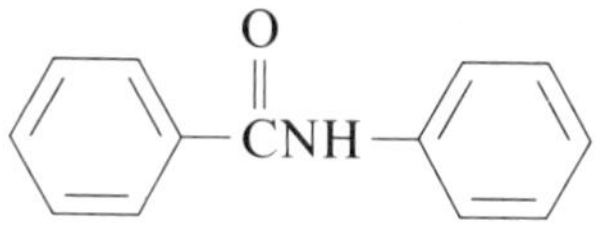

benzanilide

Some dicarboxylic acids form cyclic amides in which two acyl groups are bonded to the nitrogen atom. The suffix **-imide** is given to such compounds, easily produced when five- or six-membered rings may form.

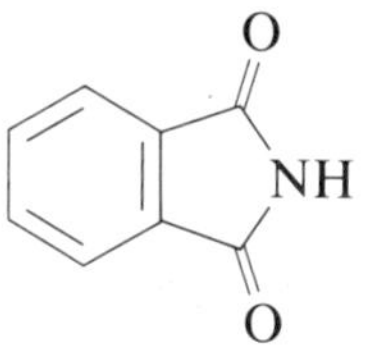

phthalimide

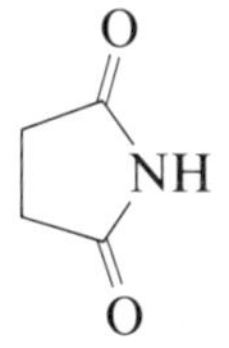

succinimide

N-bromosuccinimide
abbreviated NBS

Nitriles, compounds containing the cyano group, —C≡N, are considered to be acid derivatives because they can be hydrolyzed to form amides and carboxylic acids (p. 563). In the systematic nomenclature of these compounds, the suffix **-nitrile** is added to the name of the hydrocarbon containing the same number of carbon atoms, counting the carbon atom of the cyano group.

$$\overset{5}{C}H_3\overset{4}{C}H_2\overset{3}{C}H_2\overset{2}{C}H_2\overset{1}{C}\equiv N$$

pentanenitrile

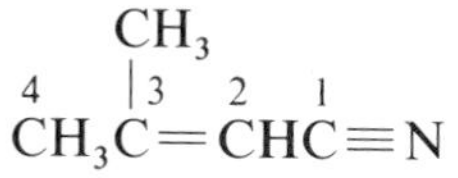

3-methyl-2-butenenitrile

The nitriles related to acetic acid and benzoic acid are called acetonitrile and benzonitrile.

$$CH_3C\equiv N$$

acetonitrile

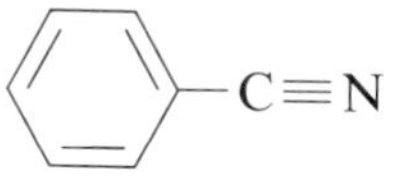

benzonitrile

When the cyano group is on a cycloalkane, the suffix **-carbonitrile** is used with the name of the hydrocarbon.

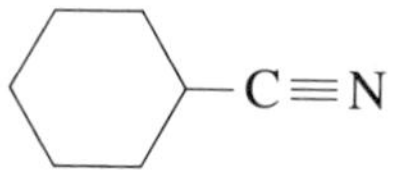

cyclohexanecarbonitrile

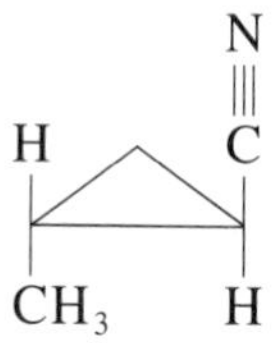

trans-2-methylcyclo-
propanecarbonitrile

If other functional groups are present, the cyano group is treated as a substituent, and its presence is indicated by the prefix **cyano-**. Note that when the suffix

-nitrile is used, the carbon atom bonded to the nitrogen is included in the count of carbon atoms that determines the name. When -cyano (or -carbonitrile) is used, that carbon atom is part of a substituent and is not included in the numbering of the rest of the chain (or ring).

4-cyanobutanoic acid

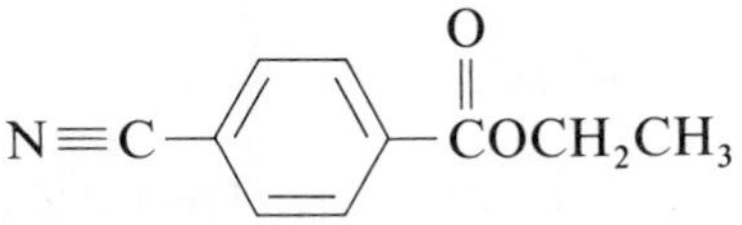

ethyl *p*-cyanobenzoate

PROBLEM 14.8

Write a structural formula for each of the following compounds.

(a) (*Z*)-4-heptenoic acid (b) *trans*-2-methylcyclobutanecarboxylic acid
(c) (*R*)-2-bromopentanoic acid (d) octanedioic acid
(e) dimethyl propanedioate (dimethyl malonate) (f) benzoic anhydride
(g) *N,N*-dimethylbenzamide (h) pentanedioyl dichloride (glutaryl dichloride)
(i) disodium ethanedioate (sodium oxalate) (j) methyl 3-nitrobenzoate

PROBLEM 14.9

Name the following compounds, including an indication of the stereochemistry where appropriate.

(a) 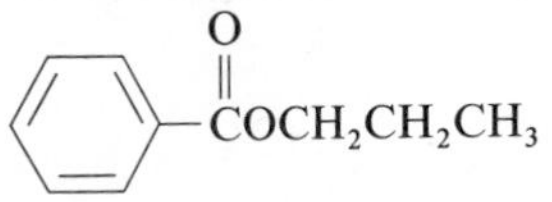 (b) 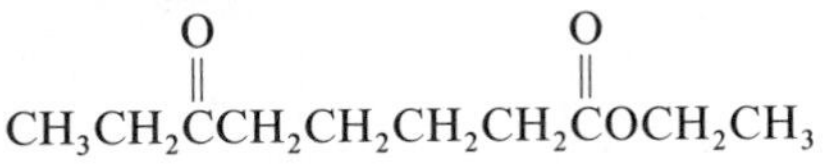

(c)  (d) 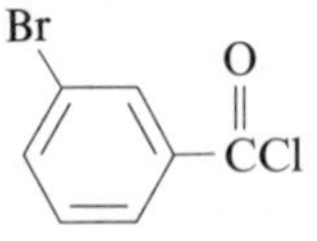

(e) 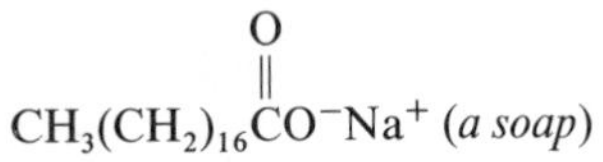 (f) 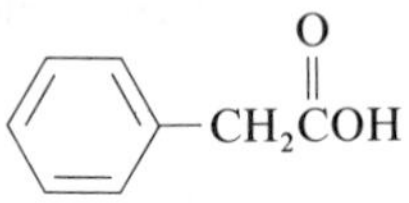

(g) 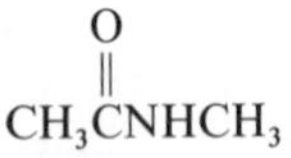 (h) 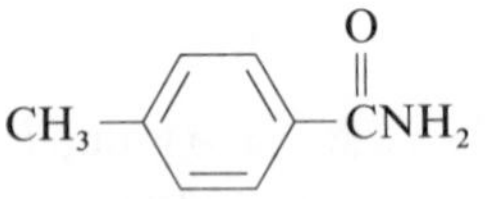 (i) 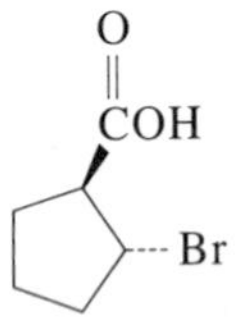

(j) 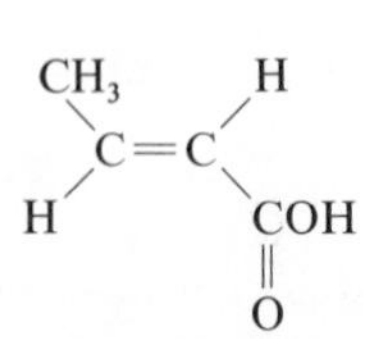 (k) 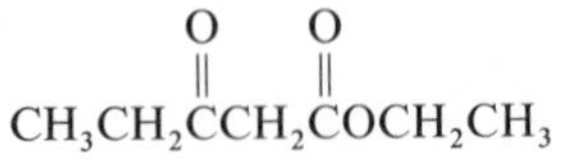 (l) 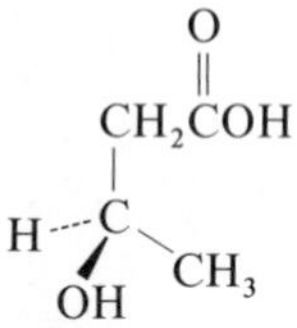

(m) $CH_3CH_2CH_2CH_2CH_2CH_2C\equiv N$ (n) 

(o) 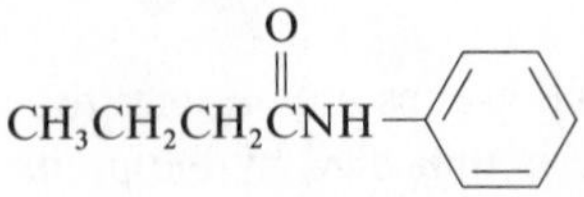

Carboxylic acids (pK_a ~5) are much weaker acids than are the strong inorganic acids, such as hydrogen chloride (pK_a −7) and sulfuric acid (pK_a −9). They are, however, strong for organic acids. The strength of a carboxylic acid depends on the other groups that are substituted on the hydrocarbon portion of the molecule. The effect of substitution on the acidity of alkyl carboxylic acids is discussed extensively in Chapter 3 and is summarized in Tables 3.1 (p. 96), 3.2 (p. 97), and 3.3 (p. 99). Inductive effects are used to explain the observed differences in acidity for substituted acids (p. 98).

Aromatic carboxylic acids are stronger acids than alkyl carboxylic acids. Benzoic acid, for example, has a pK_a of 4.2 and is a slightly stronger acid than acetic acid, pK_a 4.8. In benzoic acid, the carboxyl group is bonded to an sp^2-hybridized carbon atom, which is more electronegative than an sp^3-hybridized carbon atom (p. 56).

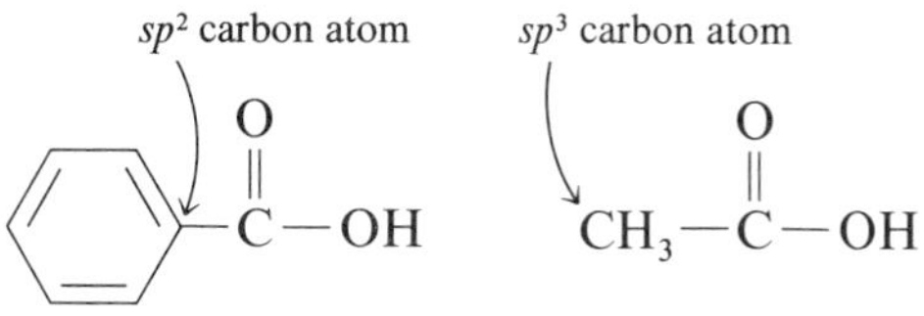
sp^2 carbon atom
sp^3 carbon atom

benzoic acid
pK_a 4.2

stronger acid than acetic acid, carboxyl group bonded to a more electronegative carbon atom

acetic acid
pK_a 4.8

The greater electronegativity of an sp^2-hybridized carbon atom is used to explain the apparent electron-withdrawing inductive effect that the phenyl group demonstrates in stabilizing the carboxylate anion.

Just as substituents on acetic acid (Table 3.1, p. 96) change its acidity, so do substituents on the aromatic ring in benzoic acid. For example, all the fluorobenzoic acids are more acidic than benzoic acid itself.

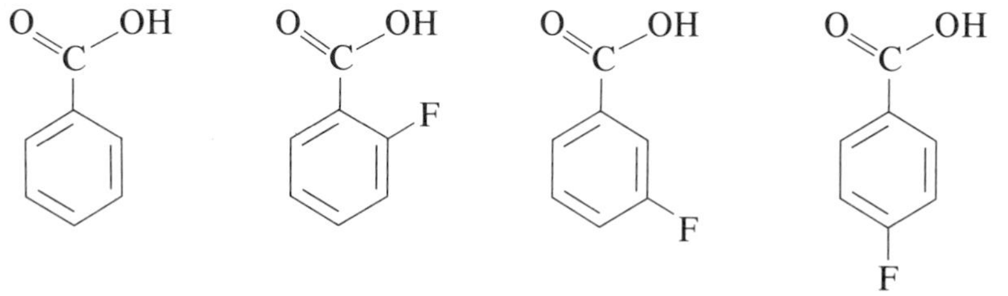

benzoic acid	*o*-fluorobenzoic acid	*m*-fluorobenzoic acid	*p*-fluorobenzoic acid
pK_a 4.2	pK_a 3.3	pK_a 3.9	pK_a 4.1

o-Fluorobenzoic acid, in which the fluorine atom is closest to the carboxyl group, is the strongest acid of the four; *p*-fluorobenzoic acid, in which the fluorine atom is farthest from the carboxyl group, is only slightly more acidic than benzoic acid. The electronegative fluorine atom exerts its effect through bonds and through space, withdrawing electronic density from the vicinity of the carboxyl group and thus stabilizing the conjugate base.

For some substituents, the inductive effect is not sufficient to explain the experimental observations. *p*-Nitrobenzoic acid is a stronger acid than *m*-nitrobenzoic acid. This observation contrasts with the observed acidities for the fluorobenzoic acids, for which acid strength decreases as the electron-withdrawing group is moved farther away from the carboxyl group.

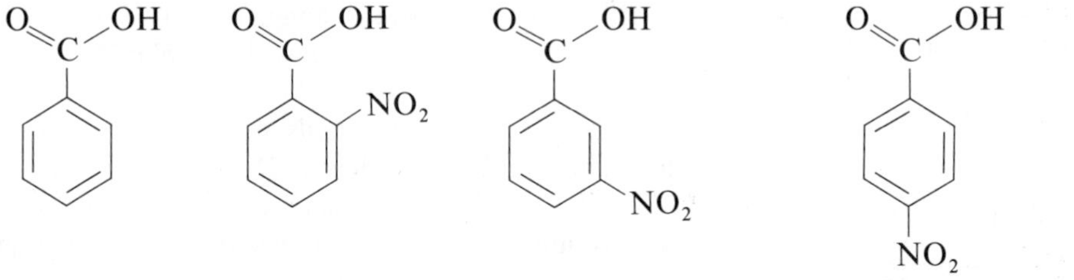
O
OH
C
NO2
NO2
NO2

benzoic acid	*o*-nitrobenzoic acid	*m*-nitrobenzoic acid	*p*-nitrobenzoic acid
pK_a 4.2	pK_a 2.2	pK_a 3.5	pK_a 3.4
	strong inductive effect and resonance effect	*inductive effect*	*weak inductive effect and strong resonance effect*

Resonance contributors having a positive charge at the carbon atom bearing the carboxylate anion can be written for the ortho and para isomers of nitrobenzoic acid, whereas the nitro group in the meta position exerts primarily an inductive effect. The carboxylate anion in each case is stabilized by combinations of electron-withdrawing effects. All of the nitrobenzoic acids are stronger than benzoic acid. In *o*-nitrobenzoic acid, the strongest acid of the three isomers, the inductive effect is strong and a resonance effect is also present. In *p*-nitrobenzoic acid, a resonance effect operates, but the inductive effect is weak.

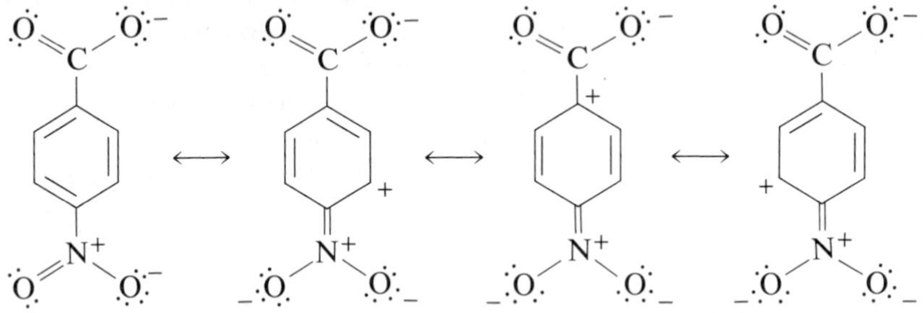
O
O−
C
N+

this resonance contributor puts a positive charge on the carbon atom to which the carboxylate group is bound

The fact that *m*-nitrobenzoic acid is a weaker acid than *p*-nitrobenzoic acid reconfirms that when the resonance effect operates, it is more important than the inductive effect.

PROBLEM 14.10

Draw resonance contributors for the anions of *o*- and *m*-nitrobenzoic acid. Point out the contributor that is important in explaining the acidity of the ortho compound, and prove to yourself that no such effect is present in the meta isomer.

A. Carboxylic Acids as Products of Oxidation Reactions

Carboxylic acids are formed as products of certain oxidation reactions. You have already studied many of these reactions in earlier chapters.

Primary alcohols and aldehydes are oxidized to carboxylic acids containing the same number of carbon atoms (p. 439). Oxidations of alcohols with acidic chromic acid solutions often give esters, the result of the reaction of the starting material alcohol with the product carboxylic acid (p. 577). For this reason, base-catalyzed oxidation with potassium permanganate followed by acidification is preferred when the acid is the desired product. 2-Ethyl-1-hexanol and the corresponding aldehyde have both been converted to 2-ethylhexanoic acid by this method.

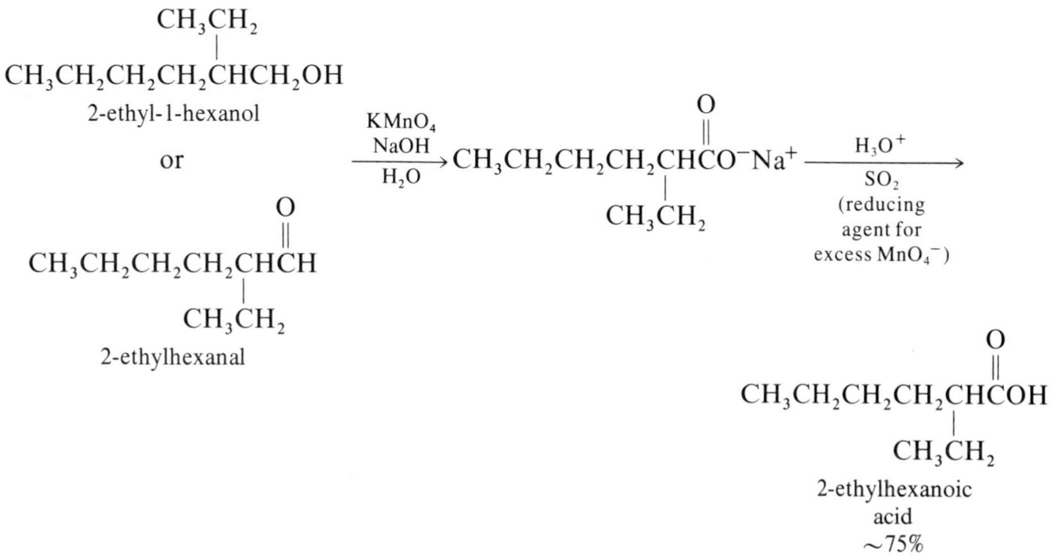

$$
\begin{array}{c}
\text{CH}_3\text{CH}_2 \\
|\\
\text{CH}_3\text{CH}_2\text{CH}_2\text{CH}_2\text{CHCH}_2\text{OH} \\
\text{2-ethyl-1-hexanol}
\end{array}
$$

or

$$
\begin{array}{c}
\overset{\displaystyle O}{\overset{\|}{}} \\
\text{CH}_3\text{CH}_2\text{CH}_2\text{CH}_2\text{CHCH} \\
|\\
\text{CH}_3\text{CH}_2 \\
\text{2-ethylhexanal}
\end{array}
$$

$$\xrightarrow[\text{H}_2\text{O}]{\substack{\text{KMnO}_4 \\ \text{NaOH}}}$$

$$
\begin{array}{c}
\overset{\displaystyle O}{\overset{\|}{}} \\
\text{CH}_3\text{CH}_2\text{CH}_2\text{CH}_2\text{CHCO}^-\text{Na}^+ \\
|\\
\text{CH}_3\text{CH}_2
\end{array}
$$

$$\xrightarrow[\substack{\text{SO}_2 \\ \text{(reducing} \\ \text{agent for} \\ \text{excess MnO}_4^-)}]{\text{H}_3\text{O}^+}$$

$$
\begin{array}{c}
\overset{\displaystyle O}{\overset{\|}{}} \\
\text{CH}_3\text{CH}_2\text{CH}_2\text{CH}_2\text{CHCOH} \\
|\\
\text{CH}_3\text{CH}_2 \\
\text{2-ethylhexanoic} \\
\text{acid} \\
\sim 75\%
\end{array}
$$

In basic solution, the acid is formed as its salt and therefore is not reactive toward unused alcohol in the reaction mixture. The free carboxylic acid is generated by adding sulfuric acid after the oxidation is completed (p. 546).

A very mild oxidizing agent for aldehydes is moist silver oxide. Heptanal is oxidized to heptanoic acid in very high yield with this reagent.

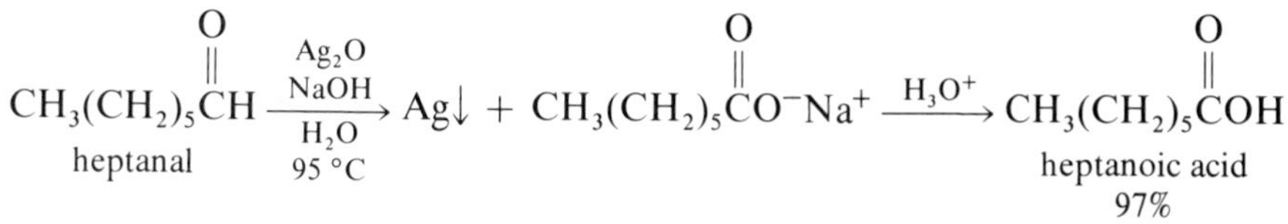

$$
\overset{\displaystyle O}{\overset{\|}{\text{CH}_3(\text{CH}_2)_5\text{CH}}} \xrightarrow[\substack{\text{H}_2\text{O} \\ 95\,^\circ\text{C}}]{\substack{\text{Ag}_2\text{O} \\ \text{NaOH}}} \text{Ag}\downarrow + \overset{\displaystyle O}{\overset{\|}{\text{CH}_3(\text{CH}_2)_5\text{CO}^-\text{Na}^+}} \xrightarrow{\text{H}_3\text{O}^+} \overset{\displaystyle O}{\overset{\|}{\text{CH}_3(\text{CH}_2)_5\text{COH}}}
$$

$$\text{heptanal} \qquad\qquad\qquad\qquad\qquad\qquad\qquad\qquad\qquad \text{heptanoic acid} \\ 97\%$$

In this reaction, silver(I) is reduced to metallic silver. If the reaction is carried out in a clean test tube or flask, a silver mirror deposits on the glass. The reaction is, therefore, also used as a test to distinguish aldehydes, by the ease with which they are oxidized, from ketones. The reagent is known as **Tollens reagent** and the test as the **Tollens test** or the **silver mirror test.**

In compounds containing other functional groups that would be sensitive to stronger oxidizing agents, such as potassium permanganate (p. 314), an aldehyde function can be successfully oxidized to a carboxyl group using silver oxide. For example, the unsaturated aldehyde, 9,12-octadecadiynal is converted to the corresponding acid.

$$\underset{\text{9,12-octadecadiynal}}{CH_3(CH_2)_4C{\equiv}CCH_2C{\equiv}C(CH_2)_7\overset{O}{\overset{\|}{C}}H} \xrightarrow[\text{ethanol}\ N_2\ \text{atmosphere}]{Ag_2O\ NaOH} CH_3(CH_2)_4C{\equiv}CCH_2C{\equiv}C(CH_2)_7\overset{O}{\overset{\|}{C}}O^-Na^+$$

$$\xrightarrow{H_3O^+} \underset{\substack{\text{9,12-octadecadiynoic acid} \\ 78\%}}{CH_3(CH_2)_4C{\equiv}CCH_2C{\equiv}C(CH_2)_7\overset{O}{\overset{\|}{C}}OH}$$

The reaction shown above is carried out in an atmosphere of nitrogen gas, because these unsaturated compounds are sensitive to oxidation, even by the oxygen in air. (The reasons for the great sensitivity of such polyunsaturated compounds to oxygen are discussed on p. 844.) The oxidation reaction with silver oxide is highly selective. The aldehyde is converted to a carboxyl group, and the triple bonds are untouched.

Ozonolysis is one of the most effective ways to cleave a carbon-carbon double bond in order to produce acids by way of aldehyde intermediates (p. 312). The conversion of 1-tridecene to dodecanoic acid is an example. In this reaction, silver oxide oxidizes the aldehydes that are the products of the breakdown of 1-tridecene ozonide.

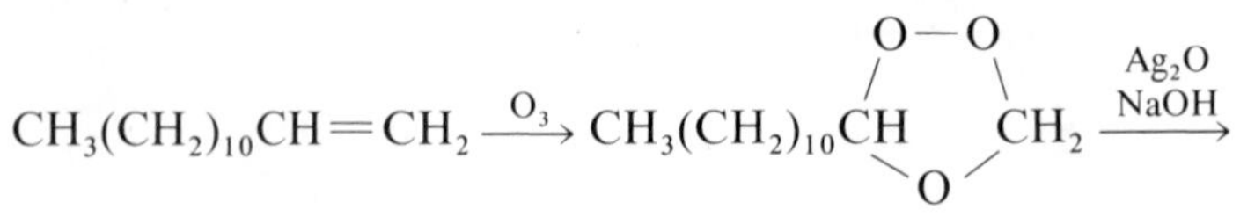

1-tridecene 1-tridecene ozonide

$$\underset{\substack{\text{sodium}\\ \text{formate}}}{H\overset{O}{\overset{\|}{C}}O^-Na^+} + \underset{\substack{\text{sodium}\\ \text{dodecanoate}}}{CH_3(CH_2)_{10}\overset{O}{\overset{\|}{C}}O^-Na^+} \xrightarrow{H_3O^+} \underset{\substack{\text{formic}\\ \text{acid}}}{H\overset{O}{\overset{\|}{C}}OH} + \underset{\substack{\text{dodecanoic acid}\\ 94\%}}{CH_3(CH_2)_{10}\overset{O}{\overset{\|}{C}}OH}$$

Oxidative cleavage reactions give rise to carboxylic acids having fewer carbon atoms than the starting alkene does, unless, of course, a cyclic alkene is the starting material.

B. Reactions of Organometallic Reagents with Carbon Dioxide

Organometallic reagents react with carbon dioxide to give salts of carboxylic acids. The salt is treated with a strong mineral acid to recover the carboxylic acid. The Grignard reaction is used to prepare acids that have one more carbon atom than the alkyl or aryl halide used to make the organomagnesium reagent. The conversion of 2-chlorobutane to 2-methylbutanoic acid illustrates this synthesis.

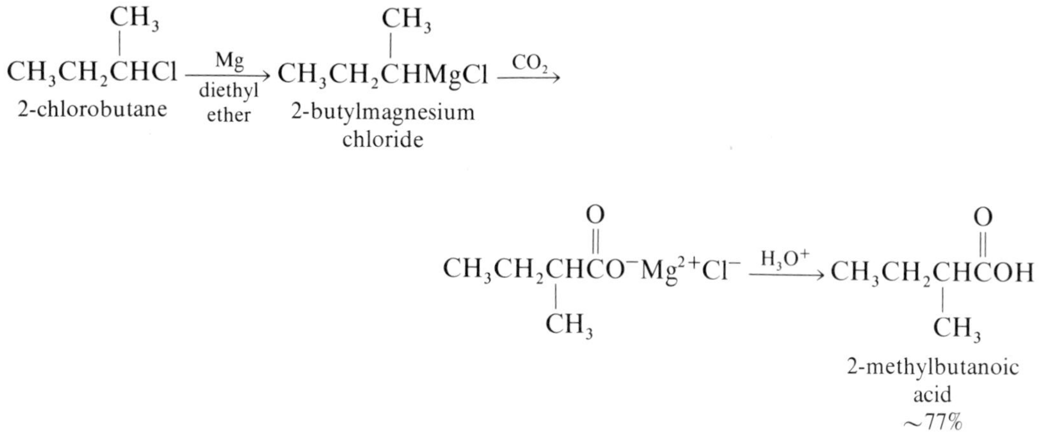

2-chlorobutane
Mg
diethyl ether
2-butylmagnesium chloride
CO_2
H_3O^+
2-methylbutanoic acid
~77%

Aromatic as well as alkyl Grignard reagents undergo these reactions. For example, 1-bromonaphthalene is converted to 1-naphthoic acid.

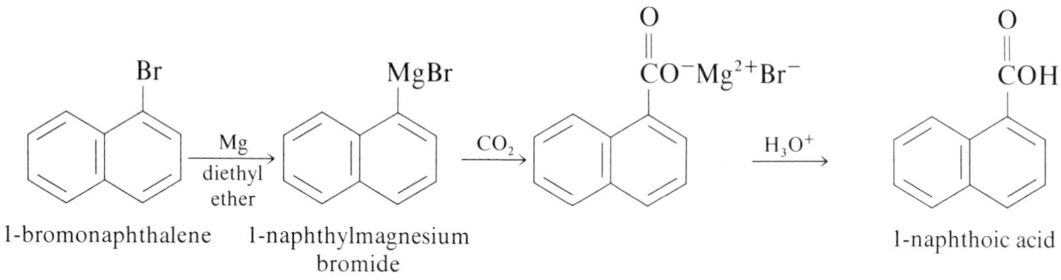

Br
MgBr
Mg
diethyl ether
CO_2
H_3O^+
1-bromonaphthalene
1-naphthylmagnesium bromide
1-naphthoic acid

Study Guide Concept Map 14.1

PROBLEM 14.11

Assign structures to all compounds symbolized by a letter in the following equations.

(a) 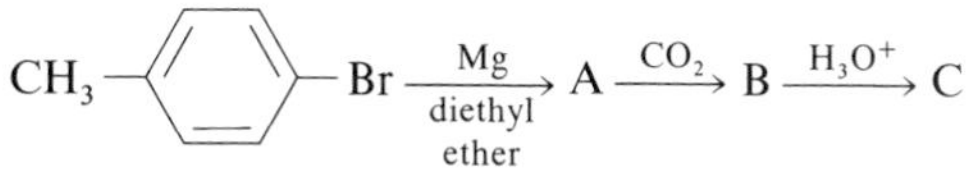

(b) $CH_3CH_2CH_2CH(CH_3)CH_2CH_2OH \xrightarrow[H_2O,\ \Delta]{KMnO_4,\ NaOH} D \xrightarrow{H_3O^+} E$

(c) 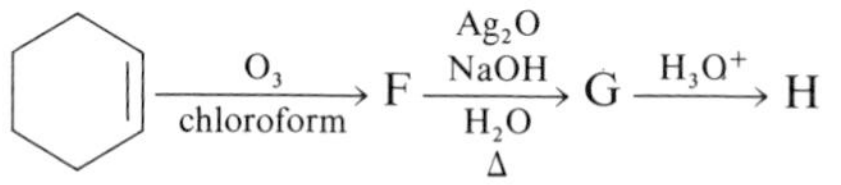

(d) 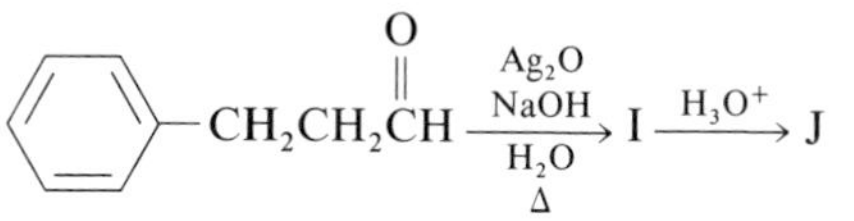

(e) 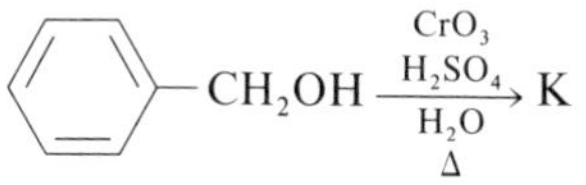

(f) $CH_3(CH_2)_5CHO \xrightarrow[H_2O]{CrO_3,\ H_2SO_4} L$

(g) $CH_3(CH_2)_9CH{=}CH_2 \xrightarrow[chloroform]{O_3} \xrightarrow{H_2O_2,\ NaOH} \xrightarrow{H_3O^+} M + N$

A. Preparation of Acid Chlorides

Phosphorus halides and thionyl chloride, the same reagents that convert alcohols into alkyl halides (p. 434), change carboxylic acids into acid halides. For example, when glacial acetic acid is heated with phosphorus trichloride, it is converted to acetyl chloride.

$$3\,CH_3\overset{\overset{\displaystyle O}{\|}}{C}OH + PCl_3 \xrightarrow[\Delta]{} 3\,CH_3\overset{\overset{\displaystyle O}{\|}}{C}Cl + P(OH)_3$$

acetic acid; phosphorus trichloride; acetyl chloride 67% *distilled out of reaction mixture*; phosphorous acid

Thionyl chloride is particularly useful for the preparation of acid halides because it has a low boiling point, 79 °C, and the other products of the reaction, sulfur dioxide and hydrogen chloride, can be easily removed from the reaction mixture because they are gases. Acid chlorides of higher molecular weight can be purified sufficiently for most uses by heating to expel those gases and then distilling out the excess thionyl chloride. Benzoyl chloride is prepared in this way.

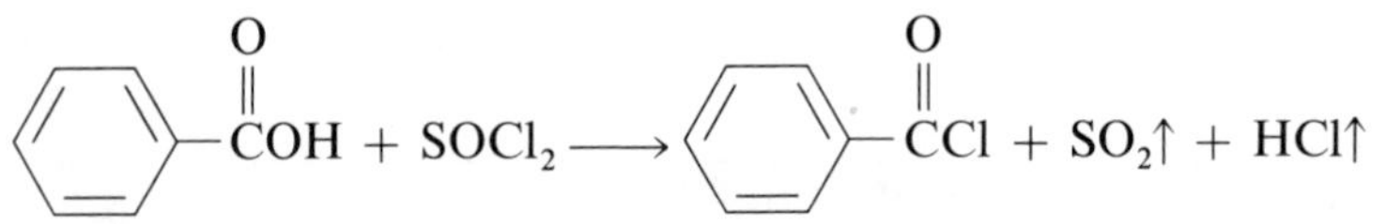
O
COH
+ SOCl₂
O
CCl
+ SO₂↑ + HCl↑

benzoic acid, mp 122 °C; thionyl chloride, bp 79 °C; benzoyl chloride, bp 198 °C, 91%

Acid chlorides are attacked by water or other nucleophiles (pp. 562 and 583), so in most cases they are prepared just before they are used in reactions with alcohols or amines. Note, therefore, that the water solutions of hydrohalic acids, such as HCl and HBr, used to convert alcohols into alkyl halides cannot be used to make acid halides.

B. Preparation of Acid Anhydrides

As the name indicates, acid anhydrides are compounds that are derived from carboxylic acids by the loss of water. Many of them, especially cyclic anhydrides formed from dicarboxylic acids, can actually be prepared by heating the acid and driving off the water. Acetic anhydride, a compound of great commercial importance, is prepared industrially by heating acetic acid to 800 °C.

$$2\,CH_3\overset{\overset{\displaystyle O}{\|}}{C}OH \xrightarrow[\text{quartz tube, porcelain chips, }\Delta,\ 800\ ^\circ C]{} CH_3\overset{\overset{\displaystyle O}{\|}}{C}O\overset{\overset{\displaystyle O}{\|}}{C}CH_3 + H_2O\uparrow$$

acetic acid; acetic anhydride

Acetic anhydride is used as a dehydrating agent. Other acid anhydrides that have boiling points higher than that of acetic acid can be prepared by heating the required acid with acetic anhydride and distilling out acetic acid as it forms. For example, dodecanoic anhydride is prepared by heating dodecanoic acid with an excess of acetic anhydride.

$$2\ CH_3(CH_2)_{10}\overset{\overset{\displaystyle O}{\|}}{C}OH\ +\ CH_3\overset{\overset{\displaystyle O}{\|}}{C}O\overset{\overset{\displaystyle O}{\|}}{C}CH_3\ \underset{\substack{6-8\ h}}{\overset{\Delta}{\longrightarrow}}\ \begin{matrix}CH_3(CH_2)_{10}\overset{\overset{\displaystyle O}{\|}}{C}\\ \diagdown\\ \diagup\\ CH_3(CH_2)_{10}\overset{\underset{\displaystyle O}{\|}}{C}\end{matrix}O\ +\ 2\ CH_3\overset{\overset{\displaystyle O}{\|}}{C}OH$$

dodecanoic acid	acetic anhydride	dodecanoic	acetic acid
bp 131 °C	bp 138 °C	anhydride	bp 118 °C
mp 44 °C		mp 42 °C	

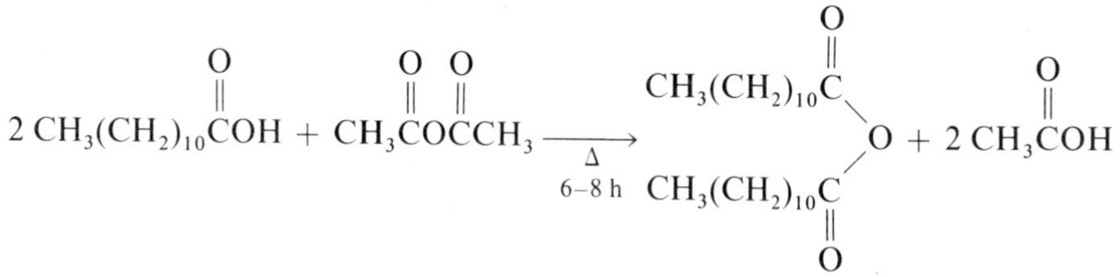

As mentioned earlier, cyclic anhydrides, especially those having a five- or six-membered anhydride ring, form easily. Consider, for example, the preparation of maleic anhydride from maleic acid.

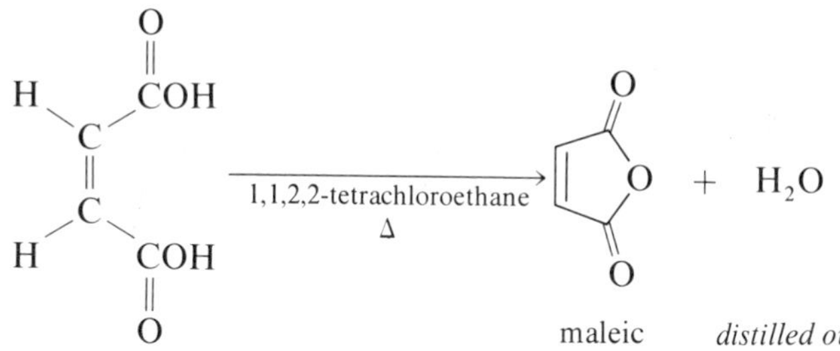

maleic acid → (1,1,2,2-tetrachloroethane, Δ) → maleic anhydride 89% + H₂O

maleic
anhydride   *distilled out
89%   of the reaction
             mixture with
             the solvent*

With all of the above preparations, some method to remove water from the reaction mixture must be used in order to obtain a high yield of the acid anhydride. Otherwise, the reverse reaction, the hydrolysis of an anhydride by water (p. 562), will take place. In the preparation of acetic anhydride, the water is vaporized because of the high temperature and distills out of the reaction mixture. The water formed when dodecanoic acid is converted into its anhydride reacts with the readily available acetic anhydride that is present in excess and is thus prevented from hydrolyzing the dodecanoic anhydride. In the preparation of maleic anhydride, 1,1,2,2-tetrachloroethane, a solvent that co-distills with water, is used to remove the water from the reaction mixture at relatively low temperatures.

Another way of making an acid anhydride is to carry out a nucleophilic substitution reaction on an acid chloride using the salt of the same acid. This reaction is illustrated by the preparation of 2-methylpropanoic anhydride.

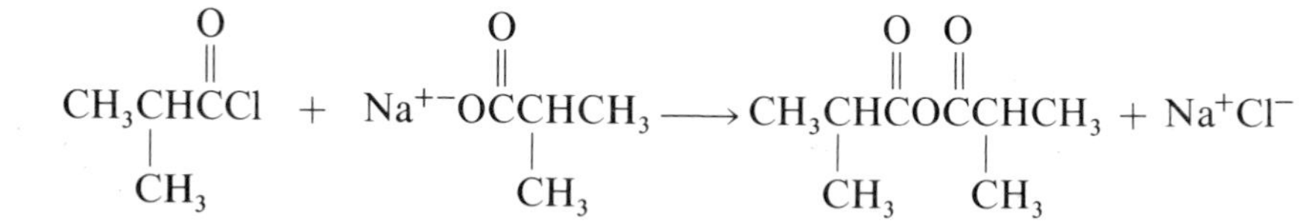

$$\underset{\substack{CH_3}}{CH_3\overset{\overset{\displaystyle O}{\|}}{C}HCCl}\ +\ Na^{+-}O\overset{\overset{\displaystyle O}{\|}}{C}\underset{\substack{CH_3}}{CHCH_3}\ \longrightarrow\ CH_3\underset{\substack{CH_3}}{\overset{\overset{\displaystyle O}{\|}}{C}H}C\overset{\overset{\displaystyle O}{\|}}{O}C\underset{\substack{CH_3}}{CHCH_3}\ +\ Na^+Cl^-$$

2-methylpropanoyl	sodium 2-methyl-	2-methylpropanoic
chloride	propanoate	anhydride

PROBLEM 14.12

Complete the following equations.

(a) 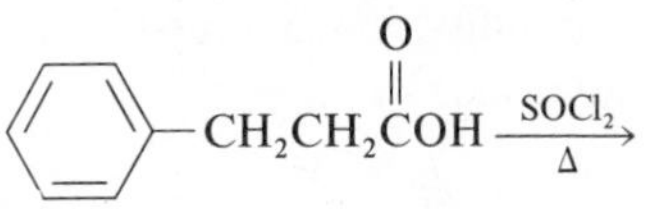

(b) $CH_3CH_2CH_2\overset{\overset{O}{\|}}{C}OH \xrightarrow{PCl_3}$

(c) $CH_3CH_2CH_2\overset{\overset{O}{\|}}{C}Cl + Na^{+-}O\overset{\overset{O}{\|}}{C}CH_2CH_2CH_3 \longrightarrow$

(d) $CH_3(CH_2)_{10}\overset{\overset{O}{\|}}{C}OH \xrightarrow[\Delta]{SOCl_2}$

(e) $HO\overset{\overset{O}{\|}}{C}CH_2CH_2\overset{\overset{O}{\|}}{C}OH \xrightarrow{CH_3\overset{\overset{O}{\|}}{C}O\overset{\overset{O}{\|}}{C}CH_3}$

(f) $CH_3CH_2\overset{\overset{O}{\|}}{C}OH \xrightarrow[\text{clay}, \Delta, 650\ ^\circ C]{}$

(g) $HO\overset{\overset{O}{\|}}{C}(CH_2)_5\overset{\overset{O}{\|}}{C}OH \xrightarrow[\Delta]{\text{excess } SOCl_2}$

14.6 REACTIONS OF CARBOXYLIC ACID DERIVATIVES WITH WATER AS NUCLEOPHILE

A. Hydrolysis Reactions

All acid derivatives react with water to give carboxylic acids. Low-molecular-weight acid chlorides and acid anhydrides, such as acetyl chloride and acetic anhydride, are soluble in water and react very rapidly with it.

$$\underset{\text{acetyl chloride}}{CH_3\overset{\overset{O}{\|}}{C}Cl} + H_2O \longrightarrow \underset{\text{acetic acid}}{CH_3\overset{\overset{O}{\|}}{C}OH} + HCl$$

$$\underset{\text{acetic anhydride}}{CH_3\overset{\overset{O}{\|}}{C}O\overset{\overset{O}{\|}}{C}CH_3} + H_2O \longrightarrow 2\ \underset{\text{acetic acid}}{CH_3\overset{\overset{O}{\|}}{C}OH}$$

Each of these compounds has a good leaving group, chloride ion for the acid chloride and acetate ion for the anhydride. Water is the nucleophile in these reactions.

PROBLEM 14.13

Using the mechanism on p. 542 as a guide, write a detailed mechanism for the reaction of acetyl chloride with water.

Esters are much less reactive toward nucleophilic substitution than either acid chlorides or acid anhydrides. The carbon atom of the carbonyl group is less electrophilic in an ester than it is in an acid chloride or anhydride.

PROBLEM 14.14

Why is the carbonyl carbon in an ester less electrophilic than the one in an acid chloride and the one in an acid anhydride? (Drawings of resonance contributors will be useful.)

An ester has to be heated with water and an acid or base as a catalyst to be hydrolyzed to a carboxylic acid. For example, when ethyl acetate is heated with water in the presence of acid, it is converted to acetic acid and ethanol.

$$\underset{\text{ethyl acetate}}{CH_3\overset{\overset{O}{\|}}{C}OCH_2CH_3} + H_2O \underset{\Delta}{\overset{}{\underset{H_2SO_4}{\rightleftarrows}}} \underset{\text{acetic acid}}{CH_3\overset{\overset{O}{\|}}{C}OH} + \underset{\text{ethanol}}{CH_3CH_2OH}$$

This type of reaction is reversible. Acids react with alcohols to give esters (p. 577). Using a large excess of water pushes the hydrolysis reaction toward completion.

If a hydrolysis reaction is carried out in a basic solution, it is called a **saponification reaction** because such a reaction is used to make soap. Pioneer women, for example, made soap by heating animal fat saved from their cooking together with wood ashes, a source of potassium hydroxide. Animal fats are mixtures of esters of long-chain acids and glycerol (p. 611). Heating the esters in the presence of a base and water gives glycerol and the salts of the long-chain acids, which form micelles and thus have the property of emulsifying and solubilizing grease (p. 612). The conversion of a typical fat to glycerol and soap is shown below.

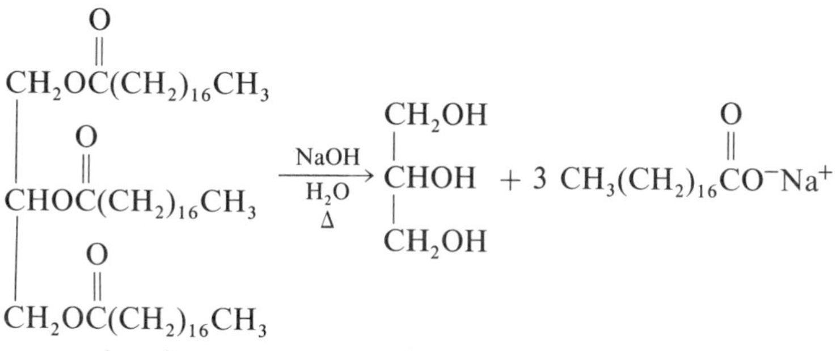
CH2OC(CH2)16CH3
CHOC(CH2)16CH3
CH2OC(CH2)16CH3
NaOH
H2O
Δ
CH2OH
CHOH
CH2OH
+ 3 CH3(CH2)16CO⁻Na⁺

tristearin *from beef fat* | glycerol | sodium stearate *soap*

Hydrolysis of an ester using a base gives the salt of a carboxylic acid. The acid itself is not formed until the reaction mixture is acidified with a stronger acid such as hydrochloric or sulfuric acid. Saponification goes essentially to completion because one of the reactants necessary for the reverse reaction, the carboxylic acid, is removed from the reaction mixture as its salt.

Nitriles can also be hydrolyzed to carboxylic acids and can be prepared by nucleophilic substitution reactions of alkyl halides with cyanide ion. Thus, alkyl halides can be converted to carboxylic acids by these two steps. The conversion of benzyl chloride to phenylacetonitrile and then to phenylacetic acid, shown below, is an example. Amides are formed as intermediates in the hydrolysis reaction and may be isolated under certain conditions. For example, phenylacetonitrile is converted to phenylacetamide if treated with hydrochloric acid for 1 hour at 40 °C but to phenylacetic acid if heated to boiling with aqueous sulfuric acid for 3 hours.

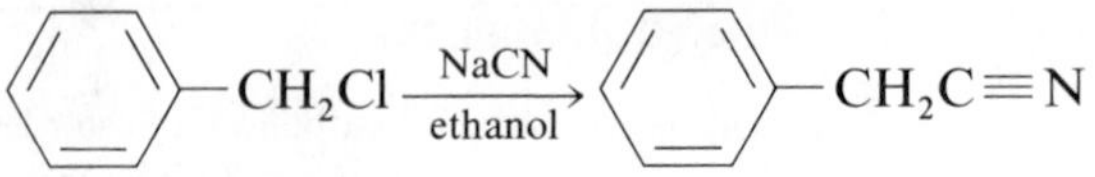

benzyl chloride — phenylacetonitrile
contains one more carbon atom than starting material

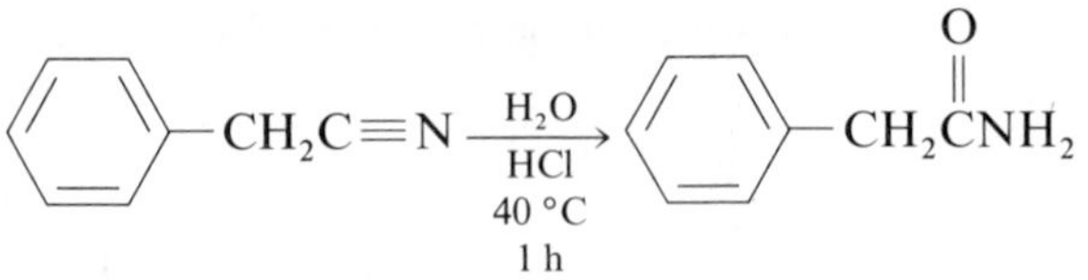

phenylacetonitrile — phenylacetamide 80%

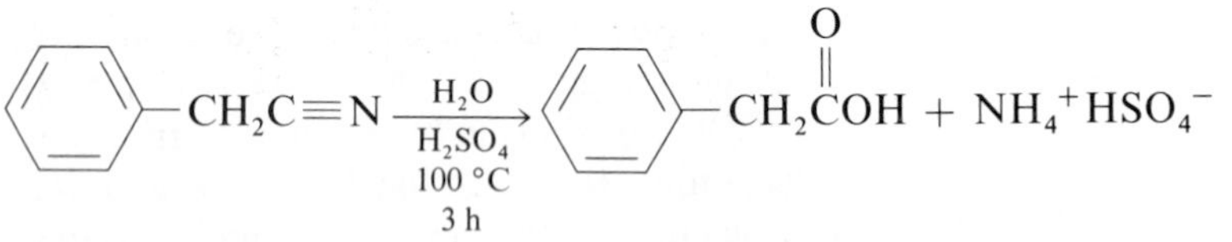

phenylacetonitrile — phenylacetic acid — ammonium hydrogen sulfate

The hydrolysis of amides to amines and carboxylic acids is one of the most important types of chemical reactions. Proteins are large molecules held together chiefly by amide groups known as peptide linkages. A large part of Chapter 26 is devoted to a study of the peptide bond and the structure of proteins. Digestion breaks down proteins into smaller units by the hydrolytic cleavage of amide bonds. The smallest units resulting from such hydrolysis of proteins are the amino acids; the amino group of one amino acid forms an amide bond with the carboxylic acid of another one. Molecules made up of a few amino acids held together by amide bonds are called **peptides.** Glycylglycylglycine is a simple peptide made up of three units of the amino acid glycine. Hydrolysis of the peptide with aqueous acid gives the free amino acid units.

called a peptide linkage when in proteins

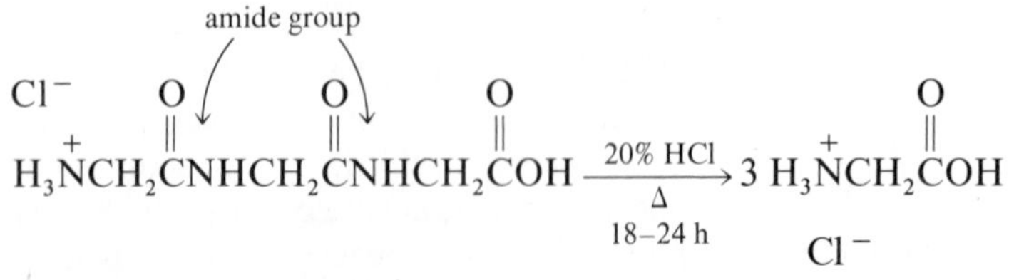
amide group

glycylglycylglycine — glycine hydrochloride
a peptide made up of three units of glycine

In the body, the cleavage of the peptide linkages occurs under very mild conditions and is catalyzed by enzymes (pp. 1132 and 1154).

The hydrolysis of nitriles or amides is also catalyzed by base. For example, chloroacetic acid in the form of its sodium salt is converted to sodium cyanoacetate.

The nitrile function is then hydrolyzed in base, and malonic acid is recovered by careful acidification of the solution containing its sodium salt.

$$\underset{\text{chloroacetic acid}}{\text{ClCH}_2\overset{\displaystyle O}{\overset{\|}{\text{C}}}\text{OH}} \xrightarrow[\text{H}_2\text{O}]{\text{Na}_2\text{CO}_3} \underset{\substack{\text{sodium} \\ \text{chloroacetate}}}{\text{ClCH}_2\overset{\displaystyle O}{\overset{\|}{\text{C}}}\text{O}^-\text{Na}^+} \xrightarrow[\substack{\text{H}_2\text{O} \\ \Delta}]{\text{NaCN}} \underset{\substack{\text{sodium} \\ \text{cyanoacetate}}}{\text{N}\equiv\text{CCH}_2\overset{\displaystyle O}{\overset{\|}{\text{C}}}\text{O}^-\text{Na}^+} \xrightarrow[\substack{\text{H}_2\text{O} \\ \Delta}]{\text{NaOH}}$$

$$\underset{\substack{\text{disodium} \\ \text{malonate}}}{\text{Na}^+{}^-\text{O}\overset{\displaystyle O}{\overset{\|}{\text{C}}}\text{CH}_2\overset{\displaystyle O}{\overset{\|}{\text{C}}}\text{O}^-\text{Na}^+} \xrightarrow[\text{cold}]{\text{H}_3\text{O}^+} \underset{\text{malonic acid}}{\text{HO}\overset{\displaystyle O}{\overset{\|}{\text{C}}}\text{CH}_2\overset{\displaystyle O}{\overset{\|}{\text{C}}}\text{OH}}$$

Study Guide
Concept Map 14.2

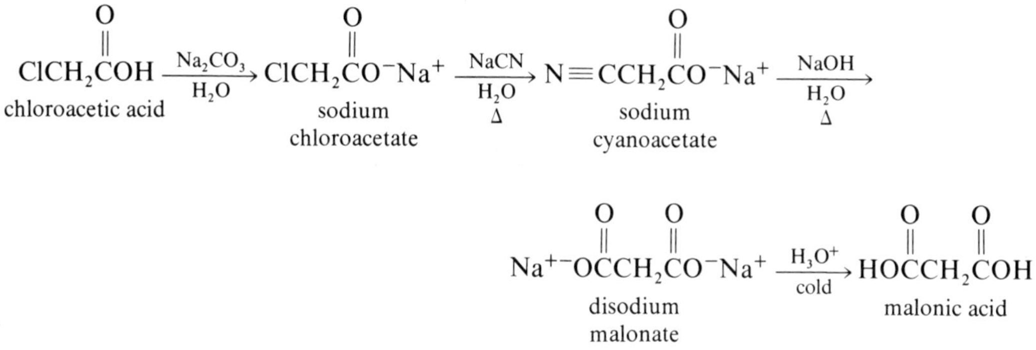

PROBLEM 14.15

In the above sequence of reactions, why is chloroacetic acid converted to its salt before sodium cyanide is added to the reaction mixture? (Hint: The table of pK_a values inside the cover of the book may be helpful.)

B. Problem-Solving Skills

Problem

How would you carry out the following transformation?

$$\underset{}{\overset{\overset{\displaystyle \text{CH}_3}{|}}{\text{CH}_3\text{CHCH}_2\text{CH}_2\text{OH}}} \longrightarrow \overset{\overset{\displaystyle \text{CH}_3}{|}}{\text{CH}_3\text{CHCH}_2\text{CH}_2}\overset{\displaystyle O}{\overset{\|}{\text{C}}}\text{OH}$$

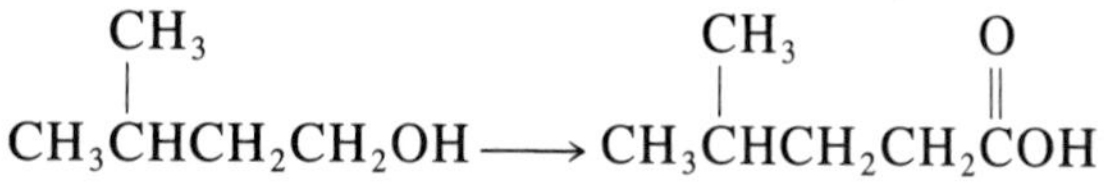

Solution

1. What functional groups are present in the starting material and the product?

   The starting material is a primary alcohol; the product is an acid.

2. How do the carbon skeletons of the two compounds compare? How many carbon atoms does each contain? Are there any rings? What are the positions of branches and functional groups on the carbon skeletons?

   The main chain of the carbon skeleton of the starting material is four carbons long with a methyl group on the third carbon. The main chain of the product is five carbons long with a branch at the fourth carbon atom. The chain has been lengthened by one carbon atom.

3. How do the functional groups change in going from starting material to product? Does the starting material have a good leaving group?

   A hydroxyl group has been converted into a carboxylic acid group.

$$-\text{OH} \longrightarrow -\overset{\displaystyle O}{\overset{\|}{\text{C}}}\text{OH}$$

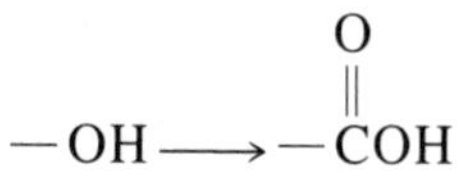

14 CARBOXYLIC ACIDS AND
THEIR DERIVATIVES I.
NUCLEOPHILIC SUBSTITUTION
REACTIONS AT THE CARBONYL
GROUP

14.6 REACTIONS OF CARBOXYLIC
ACID DERIVATIVES WITH WATER
AS NUCLEOPHILE

The starting material does not have a good leaving group, but the hydroxyl group can be converted into one.

4. Is it possible to dissect the structures of the starting material and product to see which bonds must be broken and which formed?

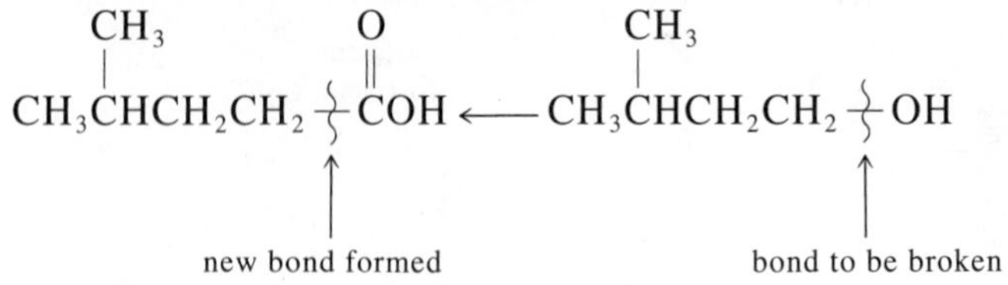

$$\underset{\text{new bond formed}}{\underset{\uparrow}{\overset{\overset{\displaystyle CH_3}{|}}{CH_3CHCH_2CH_2}\!\!\{\overset{\overset{\displaystyle O}{\|}}{C}OH}}} \longleftarrow \underset{\text{bond to be broken}}{\underset{\uparrow}{\overset{\overset{\displaystyle CH_3}{|}}{CH_3CHCH_2CH_2}\!\!\{OH}}}$$

5. Do we recognize any part of the product molecule as coming from a good nucleophile or an electrophilic addition?

A carboxylic acid can be prepared from a nitrile.

$$-\overset{\overset{\displaystyle O}{\|}}{C}OH \longleftarrow -C\equiv N$$

The cyano group comes from cyanide ion, a good nucleophile.

6. What type of compound would be a good precursor to the product?

The corresponding cyano compound would be a good precursor.

$$\overset{\overset{\displaystyle CH_3}{|}}{CH_3CHCH_2CH_2}-\overset{\overset{\displaystyle O}{\|}}{C}OH \longleftarrow \overset{\overset{\displaystyle CH_3}{|}}{CH_3CHCH_2CH_2}-C\equiv N$$

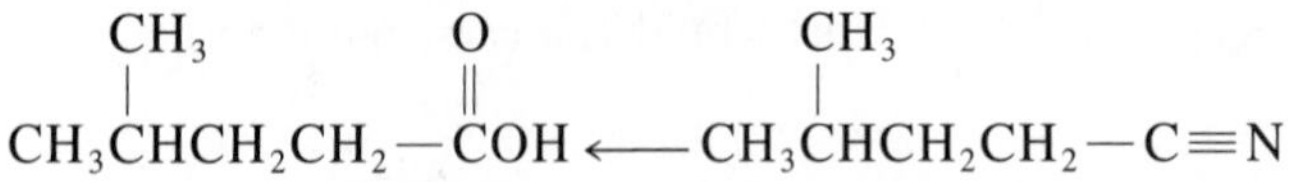

The best precursor for the cyano compound would have a good leaving group where the $-C\equiv N$ group is.

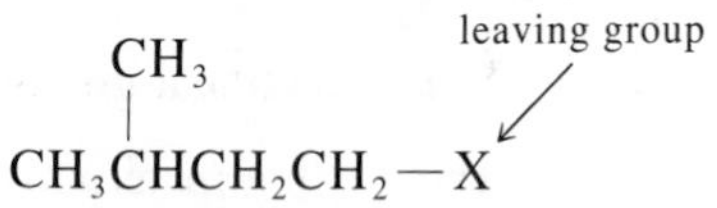

$$\overset{\overset{\displaystyle CH_3}{|}}{CH_3CHCH_2CH_2}-X\overset{\text{leaving group}}{\nwarrow}$$

7. After this step, do we see how to get from starting material to product? If not, we need to analyze the structure obtained in step 6 by applying questions 5 and 6 to it.

The precursor with a good leaving group can be obtained directly from the starting primary alcohol, so a complete synthetic pathway is as follows:

$$\overset{\overset{\displaystyle CH_3}{|}}{CH_3CHCH_2CH_2}\overset{\overset{\displaystyle O}{\|}}{C}OH \xleftarrow[\Delta]{H_3O^+} \overset{\overset{\displaystyle CH_3}{|}}{CH_3CHCH_2CH_2}C\equiv N$$

$$\uparrow NaC\equiv N$$

$$\overset{\overset{\displaystyle CH_3}{|}}{CH_3CHCH_2CH_2}OH \xrightarrow[\text{pyridine}]{TsCl} \overset{\overset{\displaystyle CH_3}{|}}{CH_3CHCH_2CH_2}OTs$$

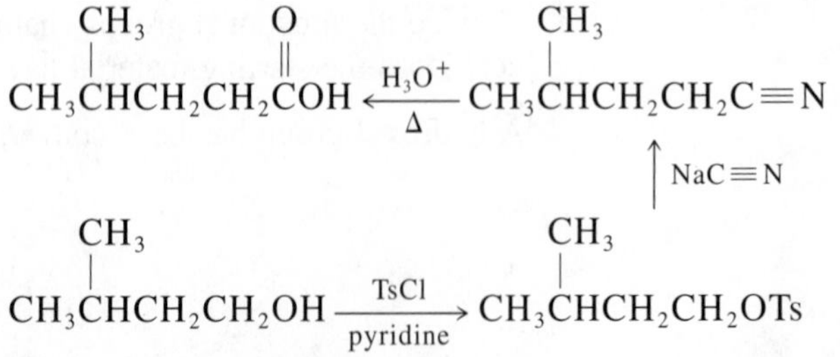

There is at least one other set of answers to questions 5 through 7.

5. Do we recognize any part of the product molecule as coming from a good nucleophile or an electrophilic addition?

A carboxylic acid having one more carbon atom than the starting material can also be prepared by the reaction of an organometallic reagent with carbon dioxide.

$$\underset{\substack{\parallel \\ O}}{RCOH} \longleftarrow \underset{\substack{\parallel \\ O}}{RCOMgX} \longleftarrow RMgX + CO_2$$

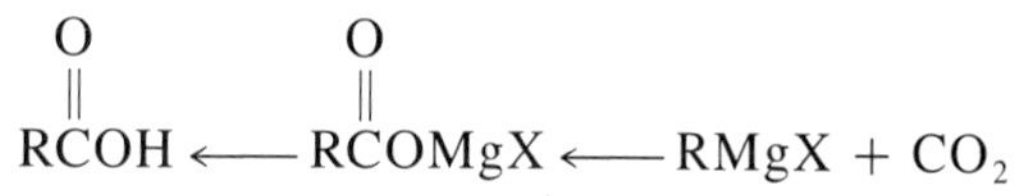

The Grignard reagent is a good nucleophile, and the carbon atom of carbon dioxide is electrophilic.

6. What type of compound would be a good precursor to the product?

A good precursor would be an alkyl halide, from which a Grignard reagent can be prepared.

$$\underset{\overset{|}{CH_3}}{CH_3CHCH_2CH_2Br}$$

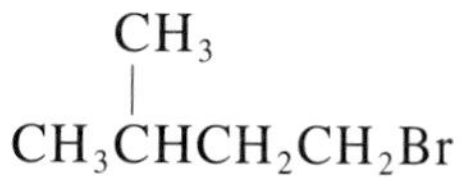

7. After this last step, do we see how to get from starting material to product? If not, we need to analyze the structure obtained in step 6 by applying questions 5 and 6 to it.

Another complete synthesis is as follows:

$$\underset{\overset{|}{CH_3}}{CH_3CHCH_2CH_2\underset{\substack{\parallel \\ O}}{C}OH} \xleftarrow{\;H_3O^+\;} \underset{\overset{|}{CH_3}}{CH_3CHCH_2CH_2\underset{\substack{\parallel \\ O}}{C}OMgBr}$$

$$\uparrow CO_2$$

$$\underset{\overset{|}{CH_3}}{CH_3CHCH_2CH_2OH} \xrightarrow{PBr_3} \underset{\overset{|}{CH_3}}{CH_3CHCH_2CH_2Br} \xrightarrow[\substack{diethyl \\ ether}]{Mg} \underset{\overset{|}{CH_3}}{CH_3CHCH_2CH_2MgBr}$$

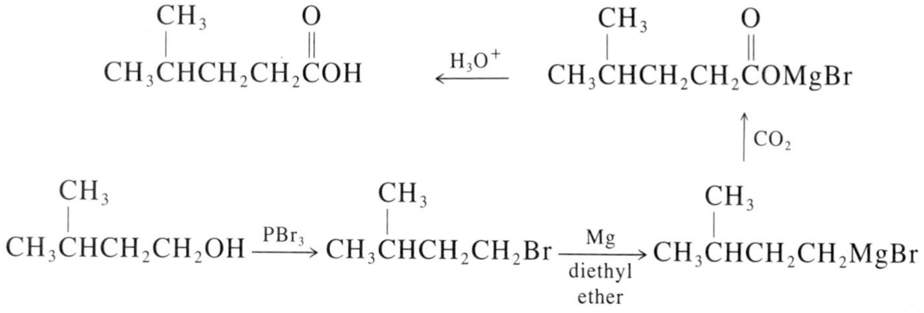

PROBLEM 14.16

How would you carry out the following transformations?

(a) $\longrightarrow$ $Na^+\; {}^-O\underset{\substack{\parallel \\ O}}{C}CH_2CH_2\underset{\substack{\parallel \\ O}}{C}O^-\; {}^+Na$ 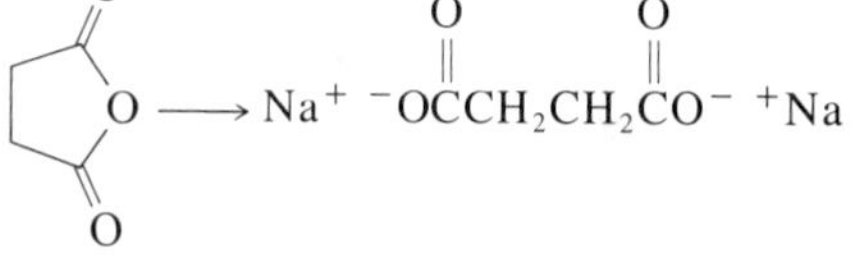

(b) $CH_3CH_2CH_2\underset{\substack{\parallel \\ O}}{C}H \longrightarrow CH_3CH_2CH_2\underset{\overset{|}{CH_3}}{C}H\underset{\substack{\parallel \\ O}}{C}OH$ 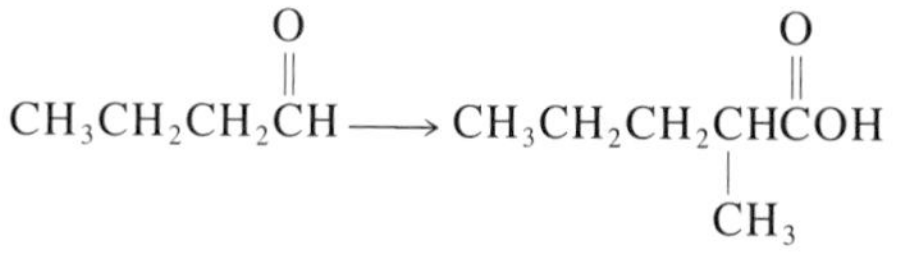

C. Mechanisms of Hydrolysis Reactions

The nucleophilic substitution reaction involving acetyl chloride was shown on p. 542 as a two-step reaction with the formation of a tetrahedral intermediate instead of as a simple S_N2 reaction. Evidence for the formation of such an intermediate comes from studies of the hydrolysis of isotopically labeled esters. Ethyl benzoate labeled with oxygen-18 in the carbonyl group is allowed to react with water in the presence of acid or base. The reaction is stopped before all of the ester has been hydrolyzed, and it is found that some of the unreacted ester molecules no longer contain oxygen-18.

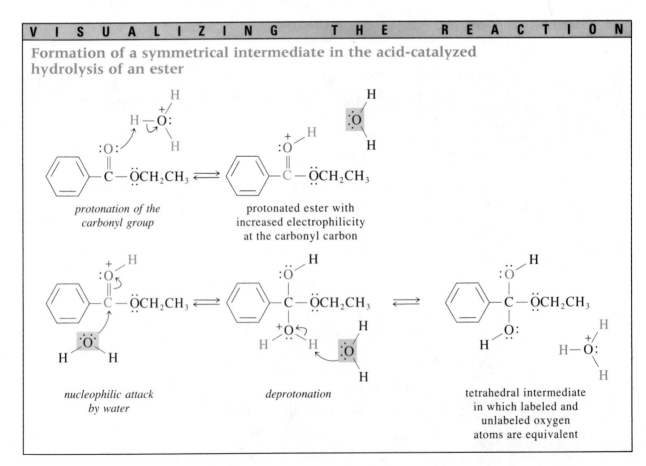

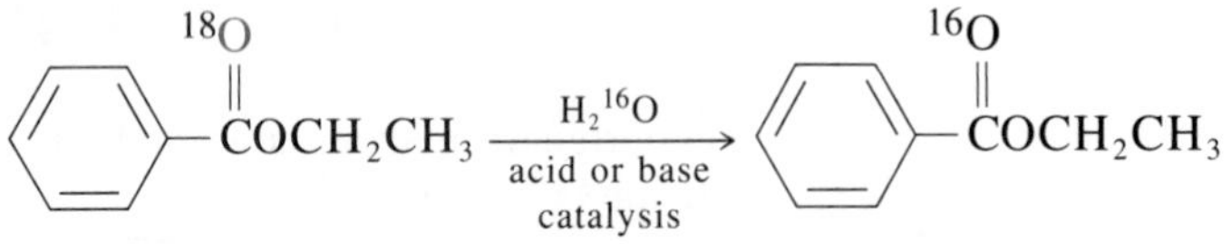
^{18}O
COCH$_2$CH$_3$
H$_2$^{16}O
acid or base
catalysis
^{16}O
COCH$_2$CH$_3$

ethyl benzoate *labeled with ^{18}O*

ethyl benzoate *recovered from the reaction mixture; has lost its label*

This observation can be explained only by assuming the reversible formation of a symmetrical intermediate in which ^{18}O and ^{16}O are both bonded to the carbon atom of the carbonyl group. The formation of such an intermediate is shown below for the acid-catalyzed hydrolysis reaction.

VISUALIZING THE REACTION

Formation of a symmetrical intermediate in the acid-catalyzed hydrolysis of an ester

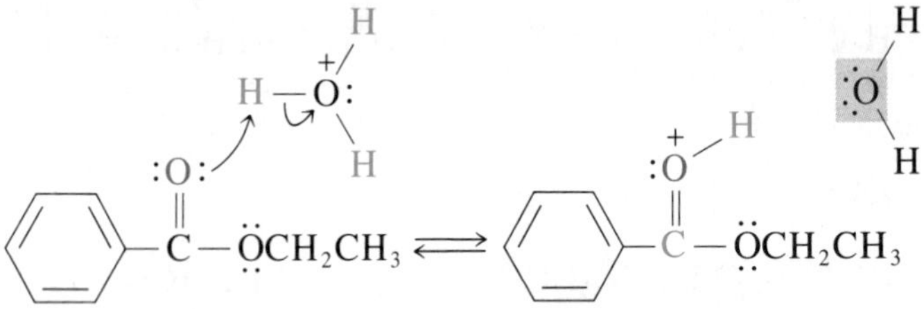

protonation of the carbonyl group

protonated ester with increased electrophilicity at the carbonyl carbon

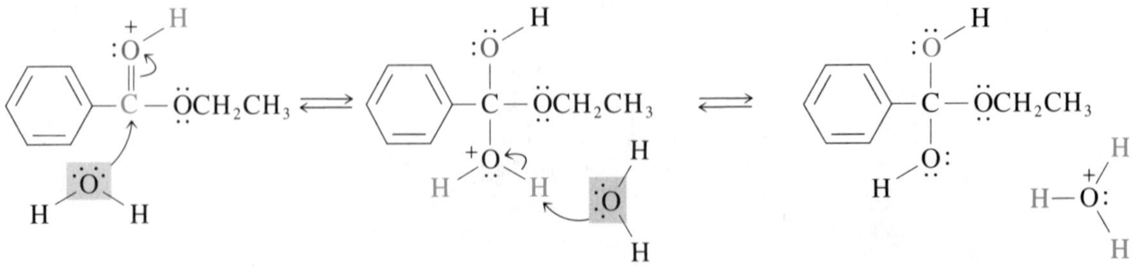

nucleophilic attack by water

deprotonation

tetrahedral intermediate in which labeled and unlabeled oxygen atoms are equivalent

In aqueous acid, the most abundant nucleophile is the water molecule itself, which is a weak nucleophile. Therefore, the uncatalyzed hydrolysis of an ester in pure water is slow. The reversible protonation of the carbonyl group is the first step in the hydrolysis of an ester in acidic solution. The electrophilicity of the carbonyl group is increased by protonation, and it is attacked by water. Loss of a proton from the resulting intermediate gives a symmetrical tetrahedral intermediate. All of these steps are reversible, so some of the tetrahedral intermediate is converted to the ester having the unlabeled oxygen atom.

PROBLEM 14.17

Write a mechanism that shows how the symmetrical tetrahedral intermediate shown above is converted into ethyl benzoate that has no label in the carbonyl group.

In the hydrolysis reaction, the tetrahedral intermediate is cleaved to give a carboxylic acid and an alcohol.

VISUALIZING THE REACTION

Acid-catalyzed hydrolysis of an ester

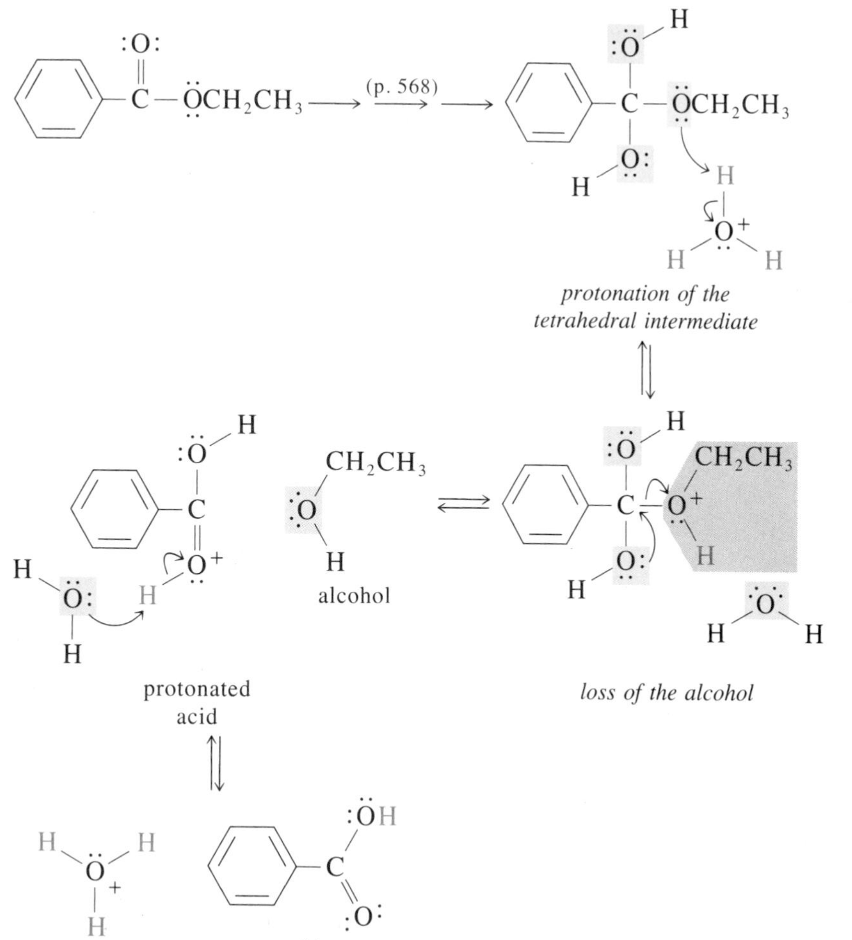

(p. 568)
protonation of the tetrahedral intermediate
loss of the alcohol
alcohol
protonated acid
acid

The tetrahedral intermediate has three sites at which it can be protonated. Protonation at either of the hydroxyl groups is simply a reversal of the preceding step and could lead, by loss of water, back to the protonated ester. Only when the oxygen atom of the alkoxyl group is protonated is a further reaction possible. The protonated alkoxyl group is a good leaving group, and the bond between the alkoxyl group and the carbonyl group is cleaved.

Further evidence for the mechanism shown above comes from other studies using isotopic tracers. When pentyl acetate is allowed to stand in water labeled with oxygen-18 in the presence of sodium hydroxide, none of the isotopic oxygen shows up in the alcohol recovered from the reaction mixture.

$$2\,H_2{}^{18}O + 2\,Na \longrightarrow 2\,Na^+\ {}^{18}OH^- + H_2\uparrow$$

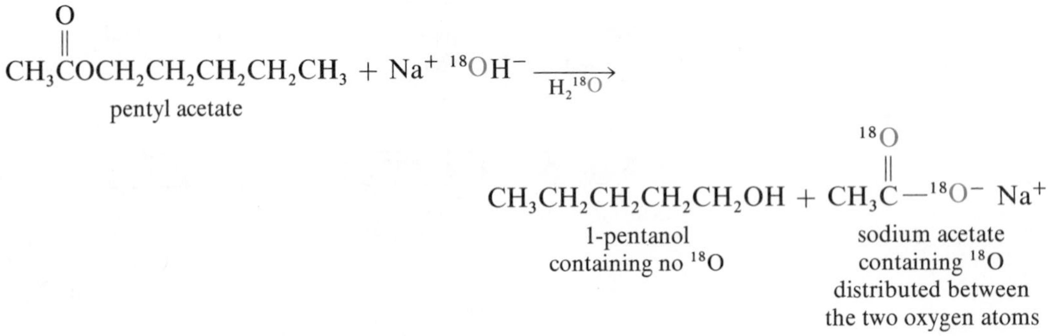
pentyl acetate
$H_2{}^{18}O$
1-pentanol
containing no ^{18}O
sodium acetate
containing ^{18}O
distributed between
the two oxygen atoms

This experimental evidence excludes a mechanism that involves an S_N2 reaction between hydroxide ion and the first carbon of the pentyl group.

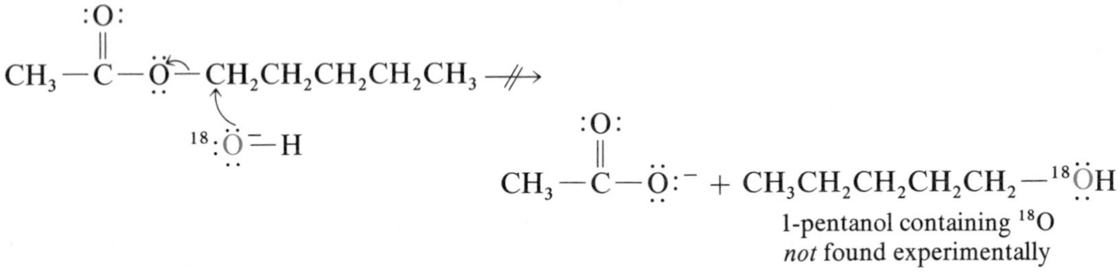
1-pentanol containing ^{18}O
not found experimentally

The observed results are consistent with an attack by hydroxide ion containing oxygen-18 at the carbonyl group, followed by cleavage of the bond between the carbonyl group and the alkoxyl group.

PROBLEM 14.18

Oxygen-18 is shown above as being present as either one of the two oxygen atoms in acetate ion after the hydrolysis of pentyl acetate in $H_2{}^{18}O$. Why is this so?

In aqueous base, the best nucleophile present is the hydroxide ion. It attacks the carbonyl group directly to give a tetrahedral intermediate, which breaks down into an acid and an alkoxide ion.

VISUALIZING THE REACTION

Base-catalyzed hydrolysis of an ester

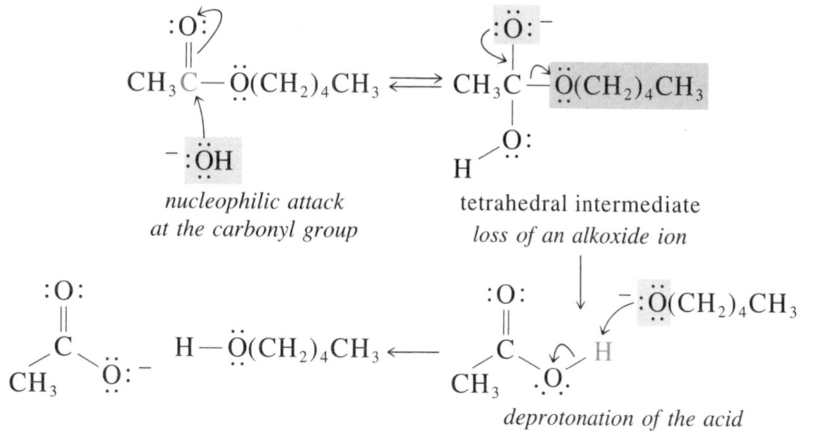

nucleophilic attack
at the carbonyl group
tetrahedral intermediate
loss of an alkoxide ion
deprotonation of the acid

An alkoxide ion is not a good leaving group for an S_N2 reaction. The bond energy of a carbonyl group is so high (~179 kcal/mol), however, that there is a powerful driving force for the expulsion of the alkoxide ion and the formation of the carbonyl group in this type of reaction. The last step of the reaction goes essentially to completion because alkoxide ion (or hydroxide ion) is a much stronger base than acetate ion is.

Another way chemists gather evidence for the mechanism of a reaction is by studying the stereochemistry of the reaction. For example, for an S_N2 reaction at an alkyl group, using a chiral alkyl group makes it possible to discover whether the reaction involves inversion at the stereocenter. Racemization of a chiral alkyl group, on the other hand, would suggest that a carbocation intermediate had formed. In an actual experiment, optically active (*S*)-(−)-2-hydroxybutanedioic acid (malic acid), isolated from natural sources, was converted into its acetic acid ester by a reaction that did not break the carbon-oxygen bond at the stereocenter. The ester was then hydrolyzed with aqueous potassium hydroxide.

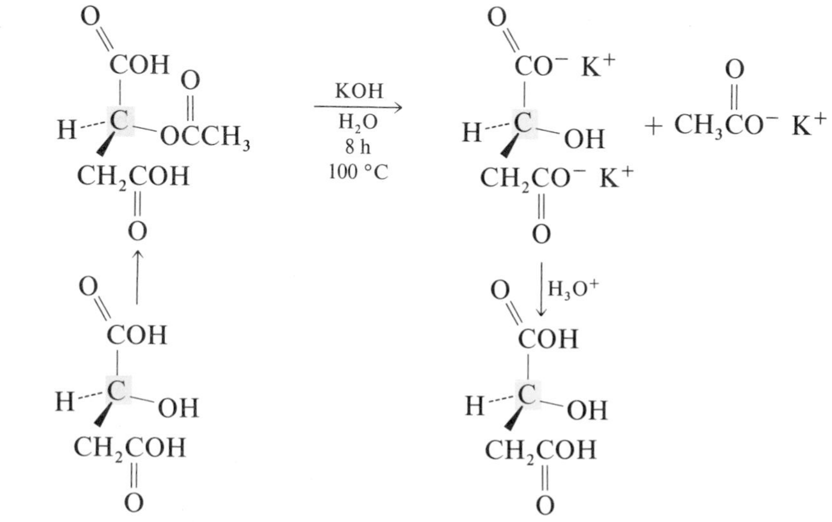

KOH
H2O
8 h
100 °C
H3O+

(*S*)-(−)-2-hydroxybutanedioic acid, used as starting material

(*S*)-(−)-2-hydroxybutanedioic acid, recovered

The optical rotation of the 2-hydroxybutanedioic acid recovered after this sequence of reactions was found to be essentially identical to that of the 2-hydroxybutanedioic acid used as the starting material. This experiment confirms that nucleophilic attack on an ester takes place at the carbon atom of the carbonyl group and not at the alkyl carbon atom. Cleavage of the ester takes place at the carbon-oxygen bond between the acyl group and the alkoxyl group.

Study Guide Concept Map 14.3

PROBLEM 14.19

Write detailed mechanisms for the base-catalyzed hydrolysis of the acetate ester of (*S*)-(−)-2-hydroxybutanedioic acid, showing first the stereochemistry that would result from attack at the carbonyl carbon atom and then that from attack at the stereocenter. Prove to yourself that the experimental result obtained above does indeed show that nucleophilic attack takes place at the carbonyl carbon.

The steps shown for the acid-catalyzed and base-catalyzed hydrolysis of esters apply to the nucleophilic substitution reactions of all acid derivatives.

D. Problem-Solving Skills

The following problem provides practice in figuring out the mechanism of a reaction.

Problem

Write a detailed mechanism for the acid-catalyzed hydrolysis of a nitrile to an amide.

$$CH_3C{\equiv}N \xrightarrow[H_3O^+]{H_2O} CH_3\overset{\overset{\displaystyle O}{\|}}{C}NH_2$$

Solution

1. To what functional group class does the reactant belong? What is the electronic character of the functional group?

   The reactant is a nitrile. The cyano group is polarized, with partial positive character at the carbon atom and partial negative character at the nitrogen atom. The carbon atom is electrophilic, and the nitrogen atom has nonbonding electrons on it.

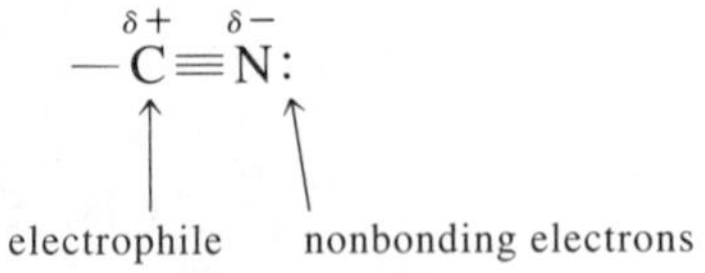
δ+ δ−
—C≡N:
electrophile
nonbonding electrons

2. How do the structures of the reactant and the desired product compare? How many carbon atoms does each contain? What bonds must be broken and formed to transform reactant into product?

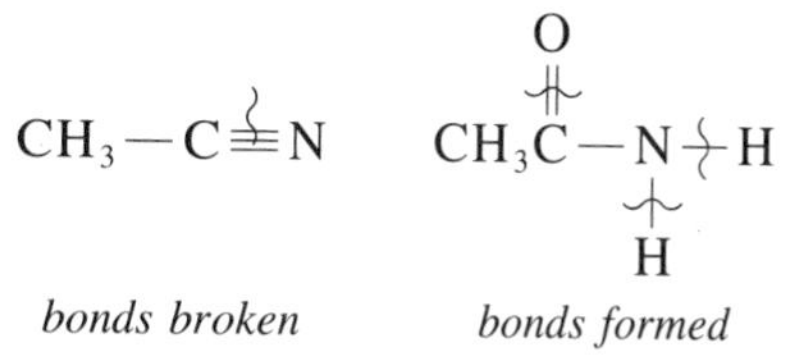

The reactant and product have the same number of carbon atoms. Two of the bonds between carbon and nitrogen have been broken. A carbon-oxygen double bond and two nitrogen-hydrogen single bonds have been formed.

3. What reagents are present? Are they good acids, bases, nucleophiles, or electrophiles?

   H_3O^+ is an acid; H_2O is a weak nucleophile.

4. What is the most likely first step for the reaction: protonation or deprotonation, ionization, attack by a nucleophile, or attack by an electrophile?

   An acid is present; therefore, protonation is the most likely first step. Water is a weak nucleophile and is unlikely to attack the unprotonated nitrile.

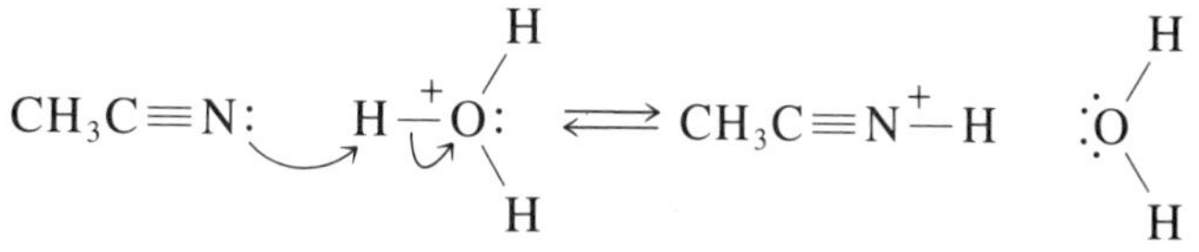

5. What are the properties of the species present in the reaction mixture after the first step? What is likely to happen next?

   The electrophilicity of the carbon bonded to nitrogen is greatly increased by the protonation of the nitrile; this carbon will now react with water.

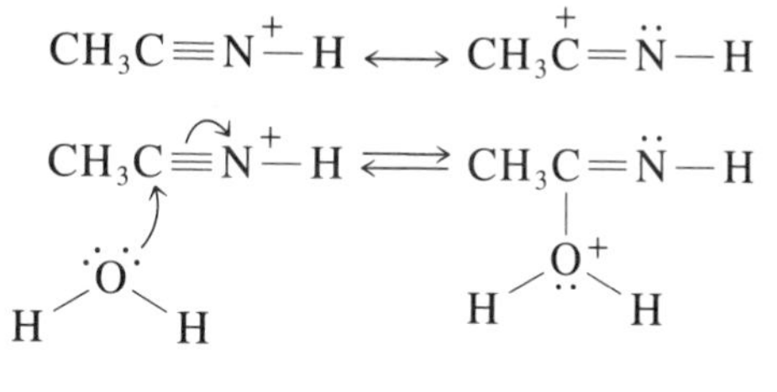

5. (repeated) What are the properties of the species present in the reaction mixture after this step? What is likely to happen next?

   The oxonium ion resulting from the attack by water will be deprotonated by the solvent.

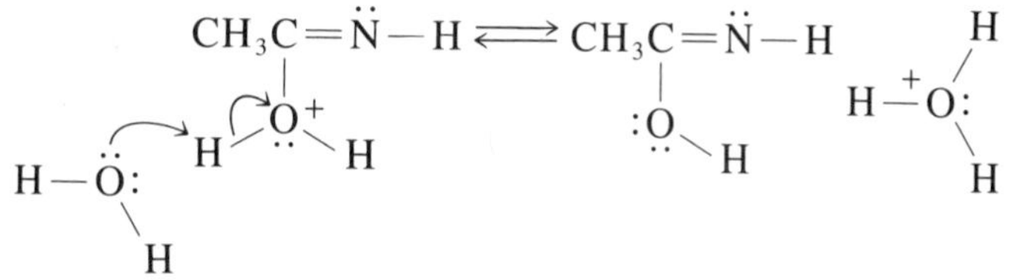

Comparing this new species to the desired product shows that it has a carbon-oxygen bond and a nitrogen-hydrogen bond in the right locations. To get to the product requires protonation at nitrogen, deprotonation at oxygen, and a shifting of the double bond.

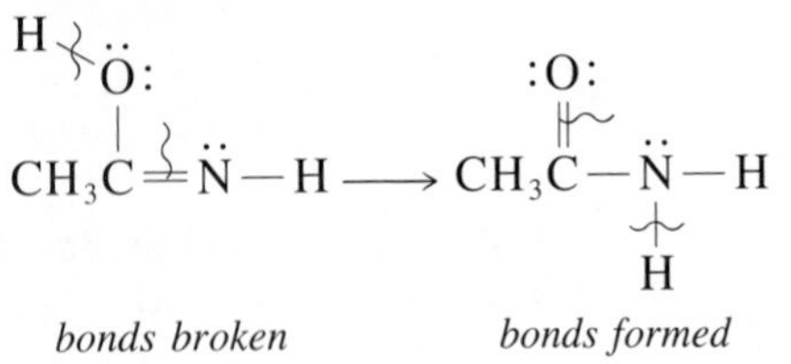
bonds broken
bonds formed

This is a tautomerization (p. 341).

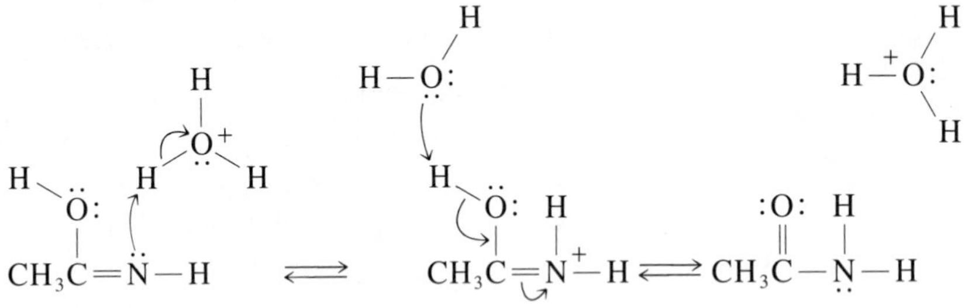

The complete mechanism is as follows:

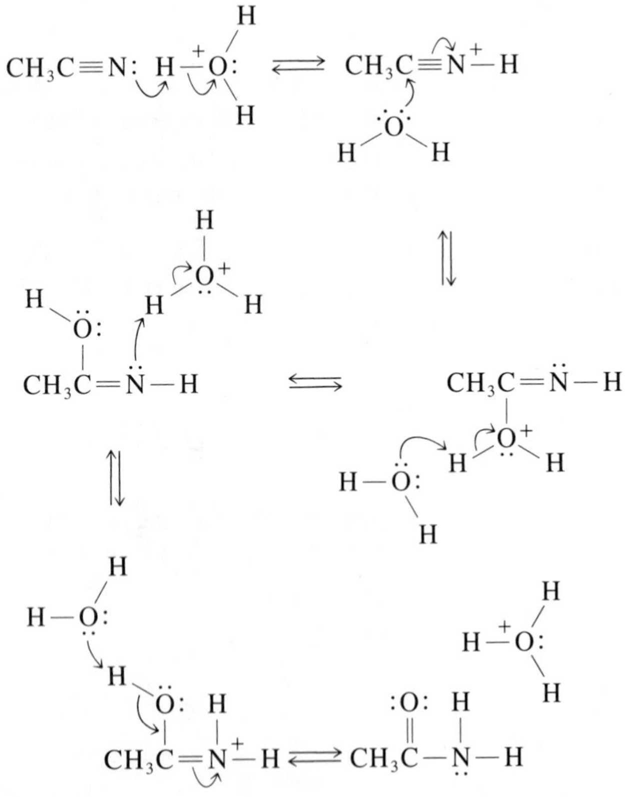

PROBLEM 14.20

Using acetonitrile, CH_3CN, write a mechanism for the base-catalyzed hydrolysis of a nitrile to an amide.

PROBLEM 14.21

Write a detailed mechanism for the acid-catalyzed hydrolysis of acetamide to acetic acid and ammonium ion. (Hint: A close look at the mechanism for the acid-catalyzed hydrolysis of an ester on p. 569 will be helpful.)

PROBLEM 14.22

In sodium hydroxide solution, acetamide is hydrolyzed to sodium acetate and ammonia. Write a detailed mechanism for this reaction.

E. Relative Reactivities of Acid Derivatives

Acid derivatives vary with respect to reactivity in nucleophilic substitution reactions. Acid chlorides, for example, react fast with cold water (p. 562), but esters and amides have to be heated with aqueous acid or base to hydrolyze them (pp. 563 and 564). The reactivity of acid derivatives toward nucleophilic substitution reactions at the carbonyl group follows this order:

$$\mathrm{R\overset{\overset{O}{\|}}{C}Cl > R\overset{\overset{O}{\|}}{C}O\overset{\overset{O}{\|}}{C}R > R\overset{\overset{O}{\|}}{C}OR' \geq R\overset{\overset{O}{\|}}{C}OH > R\overset{\overset{O}{\|}}{C}NH_2 > R\overset{\overset{O}{\|}}{C}O^-}$$

Reactivity increases as the electrophilicity of the carbon atom of the carbonyl group increases and decreases as the effect of resonance stabilization grows stronger. The carbonyl group is most electrophilic in an acid chloride, where it is bonded to the electronegative chlorine atom. It is least electrophilic in the carboxylate anion, which bears a negative charge. The species that is least reactive toward nucleophilic substitution reactions is, therefore, the carboxylate anion. Of all the acid derivatives, this anion is the most stabilized by resonance and is formed whenever any of the other compounds shown above react with water in the presence of base. An acid chloride is the derivative least stabilized by resonance.

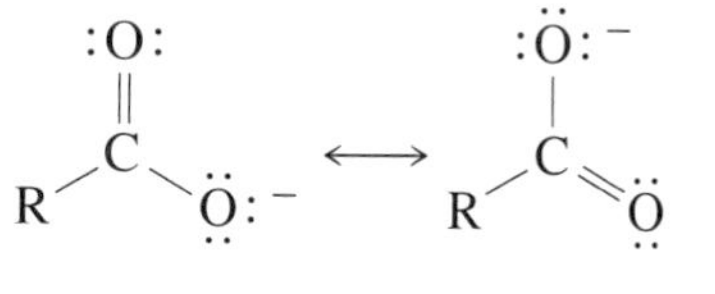

both structures contribute equally to resonance hybrid;
greater resonance stabilization

resonance contributors of a carboxylate anion

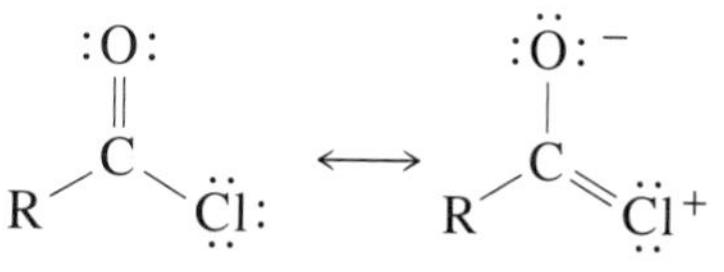

minor contributor;
positive charge on electronegative atom

resonance contributors of an acid chloride

Resonance stabilization is lost when the carbonyl group is converted to a tetrahedral intermediate. The energy of activation (p. 116) for the conversion of an acid chloride, which has little resonance stabilization, to a tetrahedral intermediate

Study Guide Concept Map 14.4

is lower than that for the conversion of the more stable amide, for example, to a similar intermediate. The ease with which a given acid derivative can be converted to the other acid derivatives by nucleophilic substitution reactions decreases going from left to right in the reactivity series shown above. That is, an acid chloride can be rather easily converted to any of the other compounds in the series; an amide, on the other hand, can be hydrolyzed to a carboxylic acid or a carboxylate anion but is not easily transformed into an ester, an acid anhydride, or an acid chloride. These relationships are summarized in Table 14.2 at the end of this chapter.

14.7 REACTION OF CARBOXYLIC ACIDS AND ACID DERIVATIVES WITH ALCOHOLS AS NUCLEOPHILES

A. Preparation of Esters

Nucleophilic substitution reactions of acid chlorides and acid anhydrides with alcohols give esters. For example, cyclohexanol is converted to a 3,5-dinitrobenzoic ester by reaction with 3,5-dinitrobenzoyl chloride.

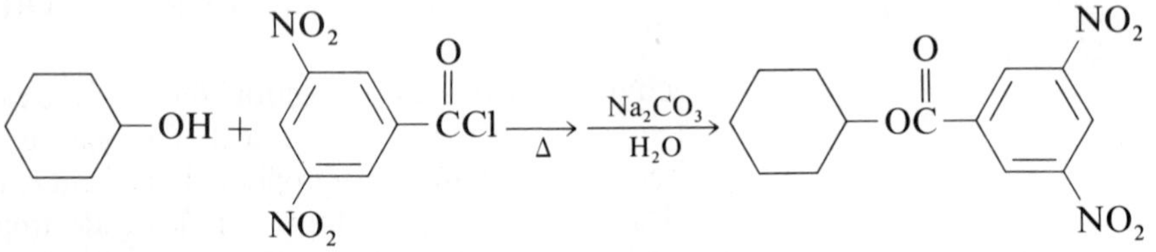
NO2
O
OH +
CCl
Δ
Na2CO3
H2O
NO2
O
OC
NO2
NO2

cyclohexanol bp 160 °C; 3,5-dinitrobenzoyl chloride; cyclohexyl 3,5-dinitrobenzoate mp 112 °C

PROBLEM 14.23

The Schotten-Baumann reaction is another way of converting an alcohol into its ester with benzoyl chloride. The alcohol and the acid chloride are mixed in 10% aqueous sodium hydroxide. Write an equation for this reaction using *n*-butyl alcohol. Show any other products that will be formed under these conditions.

Acid anhydrides are generally less reactive than acid chlorides. The carbonyl group in an acid anhydride is not as electrophilic as is one in an acid chloride. Anhydrides will, however, react readily with alcohols. Acetic anhydride, for example, is often used to make derivatives of natural products such as cholesterol.

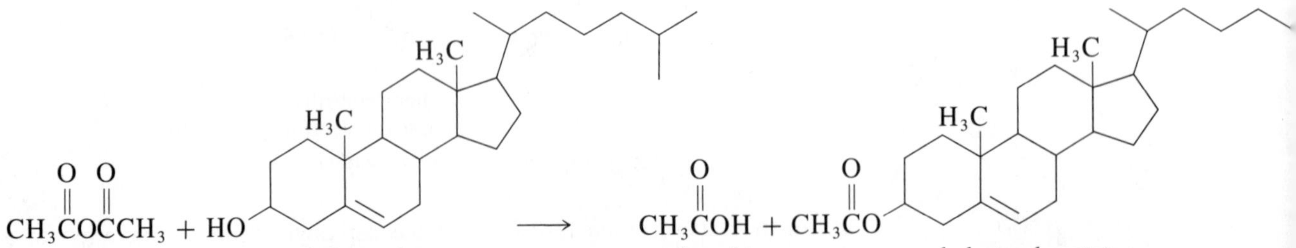
H3C
H3C
O O
CH3COCCH3 + HO
O
O
CH3COH + CH3CO
H3C
H3C

acetic anhydride; cholesterol *an alcohol*; acetic acid; cholesteryl acetate *an ester*

Acetic anhydride converts polyhydroxy alcohols into the corresponding polyacetates. Such reactions are especially important in the chemistry of carbohydrates, which are polyhydroxy aldehydes or ketones and usually have high melting points because of very strong hydrogen bonding between molecules. Carbohydrates are notorious for not crystallizing easily; again this is because they hydrogen bond so strongly with water that they form syrups instead. The formation of esters of such compounds with acetic anhydride often lowers the melting point and makes them easier to crystallize and thus to purify. The application of these reactions to the chemistry of carbohydrates is discussed in Chapter 25.

A simple example of the reaction of acetic anhydride with a polyhydroxy compound involves 2,2-bis(hydroxymethyl)-1,3-propanediol (pentaerythritol).

$$\mathrm{HOCH_2C(CH_2OH)_2CH_2OH} + 4\,\mathrm{CH_3\overset{O}{\overset{\|}{C}}O\overset{O}{\overset{\|}{C}}CH_3} \xrightarray{} \xrightarrow[\Delta]{\mathrm{CH_3\overset{O}{\overset{\|}{C}}O^-Na^+}} \mathrm{C(CH_2O\overset{O}{\overset{\|}{C}}CH_3)_4} + 4\,\mathrm{CH_3\overset{O}{\overset{\|}{C}}OH}$$

2,2-bis(hydroxymethyl)-1,3-propanediol, pentaerythritol, mp 253 °C — acetic anhydride — tetraacetate of 2,2-bis(hydroxymethyl)-1,3-propanediol, mp 84 °C — acetic acid

2,2-Bis(hydroxymethyl)-1,3-propanediol has a relatively low molecular weight but a high melting point. It is a symmetrical molecule that packs tightly into a crystal lattice and can participate extensively in intermolecular hydrogen bonding. The tetraacetate has a much lower melting point (84 °C) than the parent tetraol (253 °C). The loss of hydrogen bonding is clearly reflected in the change in this physical property.

PROBLEM 14.24

Complete the following equations.

(a) $\mathrm{CH_3C(CH_3)_2OH + CH_3\overset{O}{\overset{\|}{C}}O\overset{O}{\overset{\|}{C}}CH_3 \longrightarrow}$

(b) $\mathrm{CH_3CH_2CH_2CH_2OH + CH_3CH_2CH_2\overset{O}{\overset{\|}{C}}Cl \longrightarrow}$

(c) 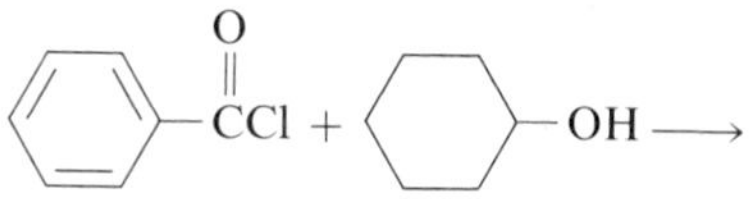

(d) 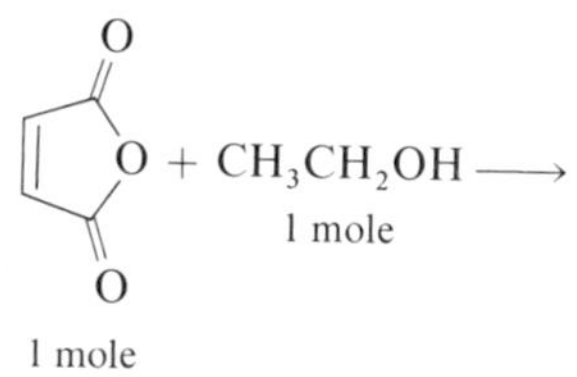
1 mole
1 mole

Carboxylic acids react with alcohols in the presence of a strong acid, such as dry hydrogen chloride, concentrated sulfuric acid, or *p*-toluenesulfonic acid, to give esters. The reaction reaches equilibrium. For example, when 1 mole of acetic acid and 1 mole of ethanol are mixed, 0.667 mole of ester and 0.667 mole of water are produced, and 0.333 mole of acid and 0.333 mole of alcohol remain at equilibrium.

$$CH_3\overset{\overset{O}{\|}}{C}OH + CH_3CH_2OH \rightleftharpoons CH_3\overset{\overset{O}{\|}}{C}OCH_2CH_3 + H_2O$$

	CH_3COOH	CH_3CH_2OH	$CH_3COOCH_2CH_3$	H_2O
At start	1 mole	1 mole	0 mole	0 mole
At equilibrium	0.333 mole	0.333 mole	0.667 mole	0.667 mole

The yield of ester can be increased either by removing one of the products of the reaction as it is formed or by increasing the concentration of one of the reactants. In practice, either water is distilled out of the reaction mixture or a greatly increased amount of alcohol, which is relatively inexpensive if it is methanol or ethanol, is used. In some particularly tough cases, both measures are taken. For example, adipic acid is converted into its diester by heating it with ethanol in the presence of sulfuric acid and toluene. Toluene, ethanol, and water form a constant-boiling mixture that allows for the removal of water from the reaction mixture as it is formed.

$$\underset{\substack{\text{adipic acid} \\ \text{1 mole}}}{HO\overset{\overset{O}{\|}}{C}(CH_2)_4\overset{\overset{O}{\|}}{C}OH} + \underset{\substack{\text{ethanol} \\ \text{3 moles}}}{2\,CH_3CH_2OH} \xrightarrow[\substack{\text{toluene} \\ \Delta}]{H_2SO_4} \underset{\substack{\text{diethyl adipate} \\ 96\%}}{CH_3CH_2O\overset{\overset{O}{\|}}{C}(CH_2)_4\overset{\overset{O}{\|}}{C}OCH_2CH_3} + 2\,H_2O$$

PROBLEM 14.25

Assign a structure to each compound indicated by a letter in the following equations.

(a) $$HO\overset{\overset{O}{\|}}{C}(CH_2)_9\overset{\overset{O}{\|}}{C}OH \xrightarrow[\substack{H_2SO_4 \\ \Delta}]{CH_3OH\ (\text{excess})} A$$

(b) 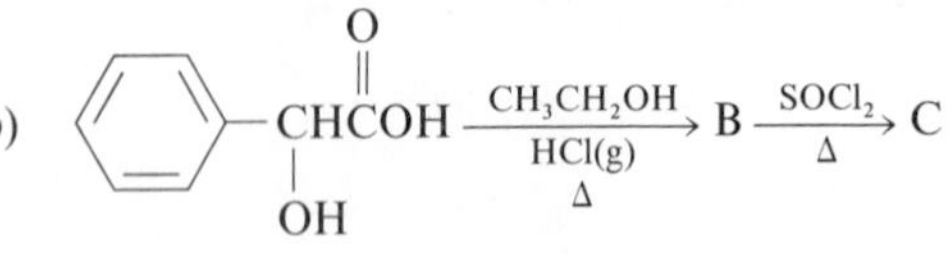
CHCOH
O
OH
CH3CH2OH
HCl(g)
Δ
B
SOCl2
Δ
C

(c) 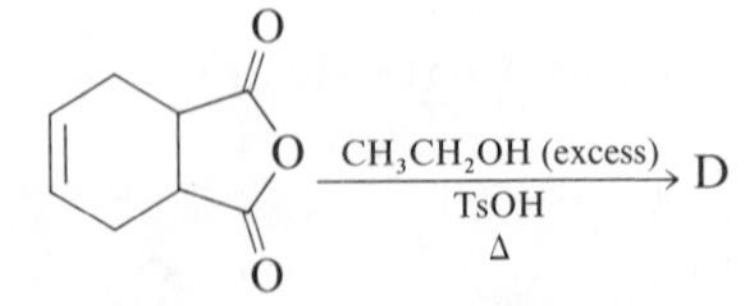
O
O
O
CH3CH2OH (excess)
TsOH
Δ
D

(d) $$HO\overset{\overset{O}{\|}}{C}C{\equiv}C\overset{\overset{O}{\|}}{C}OH \xrightarrow[\substack{H_2SO_4 \\ \Delta}]{CH_3OH\ (\text{excess})} E$$

(e) $$BrCH_2\overset{\overset{O}{\|}}{C}OH \xrightarrow[\substack{H_2SO_4 \\ \Delta}]{CH_3CH_2OH} F$$

(f) $$CH_3CH{=}CH\overset{\overset{O}{\|}}{C}OH \xrightarrow[\substack{H_2SO_4 \\ \text{benzene} \\ \Delta}]{CH_3CH_2\underset{OH}{\underset{|}{C}}HCH_3} G$$

PROBLEM 14.26

Methyl benzoate is prepared by heating 10 g of benzoic acid with 25 mL of methanol and 3 mL of concentrated sulfuric acid. (a) Write an equation for the equilibrium reaction. (b) How would you remove excess methanol, sulfuric acid, and unreacted benzoic acid from the reaction mixture in order to isolate pure methyl benzoate? (Hint: Think about the physical and chemical properties of each reagent.)

Transesterification is the conversion of an ester into another ester by heating it with an excess of either an alcohol or a carboxylic acid in the presence of an acidic or a basic catalyst. In an equilibrium reaction, either the alcohol or the acid portion of the original ester is freed. Such a reaction is carried out when there are practical reasons why an ordinary hydrolysis reaction would not work. For example, 2-chloro-2-phenylacetic acid is freed from its ethyl ester by a transesterification reaction.

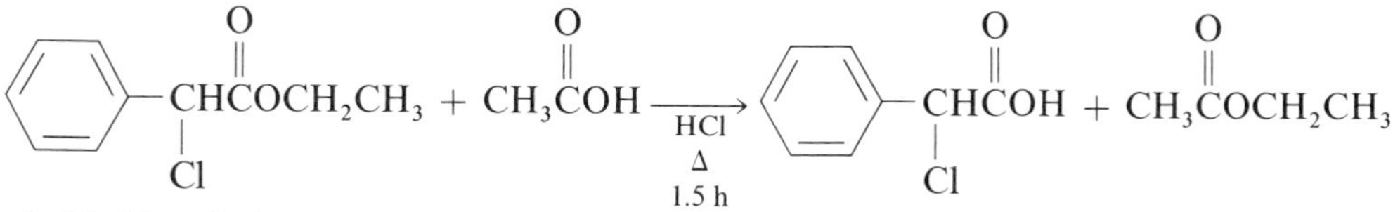

ethyl 2-chloro-2-phenylacetate	acetic acid	2-chloro-2-phenylacetic acid	ethyl acetate
1 mole	7 moles		
	bp 118 °C	mp 78 °C	bp 77 °C

The presence of the chlorine substituent on the carboxylic acid makes prolonged heating in aqueous base unwise; a nucleophilic substitution at that carbon atom as well as the saponification reaction may take place. Even long heating in dilute acid may lead to some substitution at the benzylic position. Concentrated hydrochloric acid is used to catalyze the transesterification reaction, in which the ethoxyl group is transferred from 2-chloro-2-phenylacetic acid to acetic acid. Ethyl acetate and the excess of acetic acid are easily separated from the solid 2-chloro-2-phenylacetic acid by distillation.

In another example, methyl acrylate is converted into *n*-butyl acrylate by heating it with *n*-butyl alcohol in the presence of *p*-toluenesulfonic acid as the catalyst.

$$CH_2{=}CH\overset{\overset{\displaystyle O}{\|}}{C}OCH_3 + CH_3CH_2CH_2CH_2OH \xrightarrow[\Delta]{TsOH} CH_2{=}CH\overset{\overset{\displaystyle O}{\|}}{C}OCH_2CH_2CH_2CH_3 + CH_3OH$$

methyl acrylate	*n*-butyl alcohol	*n*-butyl acrylate	methanol
bp 81 °C	bp 117 °C	bp 145 °C	bp 65 °C

Study Guide Concept Map 14.5

The difference in the boiling points of the alcohols allows the equilibrium to be shifted toward the higher-molecular-weight ester by distillating the methanol out of the reaction mixture.

PROBLEM 14.27

Coconut oil consists mainly of triglycerides (p. 611) of octanoic, dodecanoic, and tetradecanoic acids and is used to synthesize the ethyl esters of those acids. Write an equation showing how this could be done.

B. The Mechanism of the Esterification Reaction

The steps in the mechanism for the formation of an ester from an acid and an alcohol are the reverse of the steps for the acid-catalyzed hydrolysis of an ester (p. 569). As the equation for the equilibrium reaction (p. 578) indicates, the reaction can go in either direction depending on the conditions used. A carboxylic acid does not react with an alcohol unless a strong acid is used as a catalyst. Protonation makes the carbonyl group more electrophilic and enables it to react with the alcohol, which is a weak nucleophile.

VISUALIZING THE REACTION

Esterification

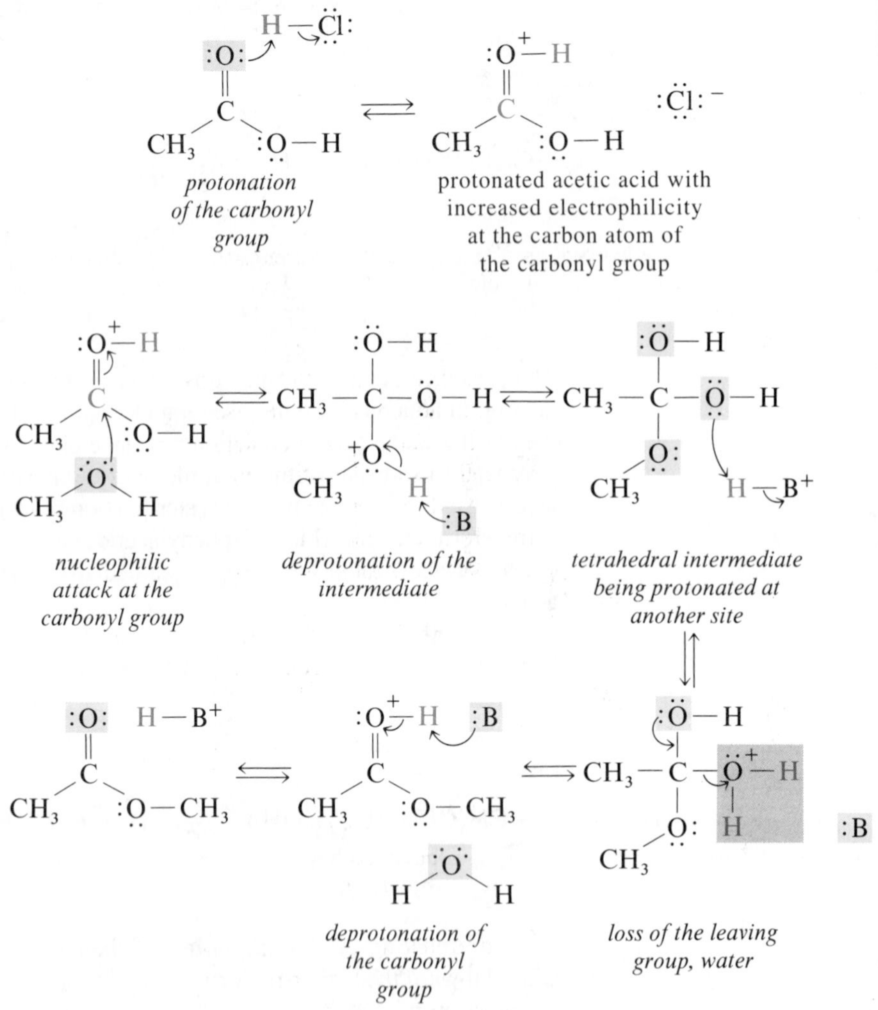
protonation of the carbonyl group
protonated acetic acid with increased electrophilicity at the carbon atom of the carbonyl group
nucleophilic attack at the carbonyl group
deprotonation of the intermediate
tetrahedral intermediate being protonated at another site
deprotonation of the carbonyl group
loss of the leaving group, water

Note that the above reaction has the same two steps described earlier (p. 542). The carbon atom of the carbonyl group is attacked by a nucleophile with formation of a tetrahedral intermediate, and the carbonyl group is regenerated by the loss of a leaving group. Because this reaction is carried out in strongly acidic solution, protonation and deprotonation steps start and finish it.

PROBLEM 14.28

Write a detailed mechanism for the following transesterification reaction.

$$CH_3\overset{\overset{O}{\|}}{C}OCH_2CH_3 + CH_3OH \underset{H_2SO_4}{\rightleftharpoons} CH_3\overset{\overset{O}{\|}}{C}OCH_3 + CH_3CH_2OH$$

C. Biological Transesterification Reactions

The process of transesterification goes on all the time in the human body, usually involving thioesters. **Thioesters** are compounds in which the oxygen atom of the alkoxyl group of an ester has been replaced by a sulfur atom. An important thioester that participates in many physiological processes is acetyl coenzyme A.

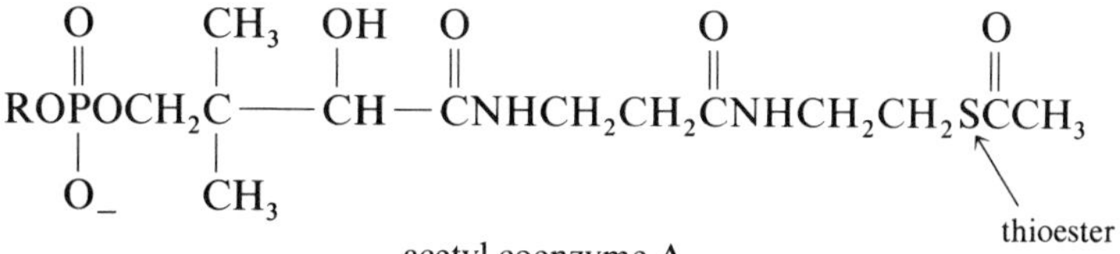
thioester

acetyl coenzyme A

usually abbreviated $CH_3\overset{\overset{O}{\|}}{C}S{-}CoA$ *or* acetyl CoA

A **coenzyme** is a relatively small molecule, usually incorporating a vitamin, that is essential for the activity of an enzyme. The vitamin present in coenzyme A is pantothenic acid, one of the B vitamins. It is linked by an amide bond at one end to 2-aminoethanethiol and by a carbon-oxygen bond at the other end to a phosphate group of a nucleotide, adenosine 3′,5′-diphosphate. (The structures of nucleotides are discussed on p. 1013.)

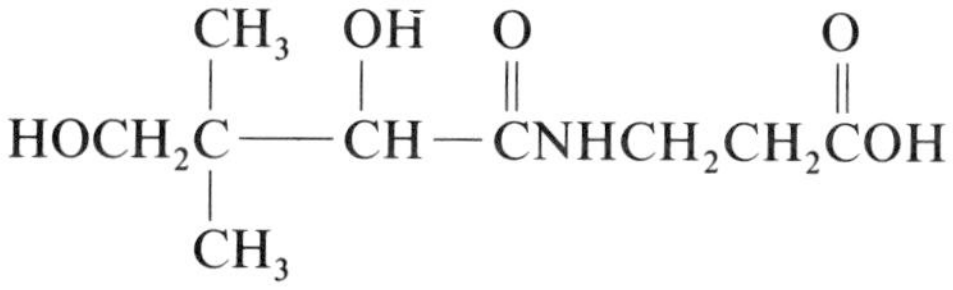

pantothenic acid

a B vitamin

$H_2NCH_2CH_2SH$

2-aminoethanethiol

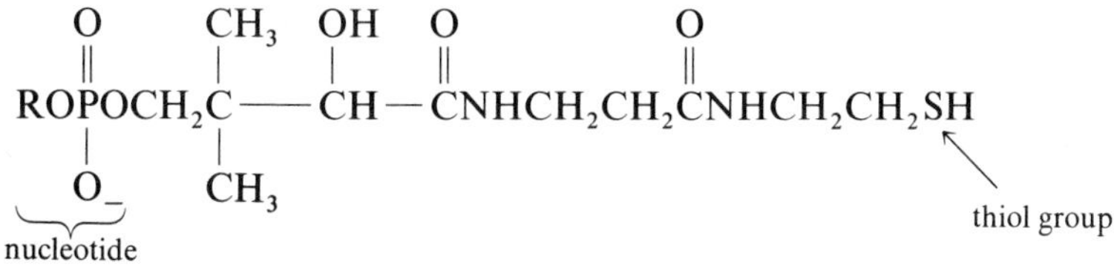
nucleotide
thiol group

coenzyme A

abbreviated CoA—SH

A thioester is more reactive than the corresponding oxyester. There is less resonance interaction between the larger sulfur atom and the adjacent carbonyl group, so the acyl group is more vulnerable to nucleophilic attack. Thus, the acyl group of a thioester is easily transferred to another sulfur, oxygen, or nitrogen atom. Compounds that lose their acyl group easily in such reactions are called **acyl-transfer agents.**

Acetyl coenzyme A is a good acyl-transfer agent. For example, a reaction that is important to the smooth functioning of the nervous system is the synthesis of acetylcholine at the nerve synapses as it is needed (p. 926). Acetyl coenzyme A participates in what is essentially a transesterification reaction.

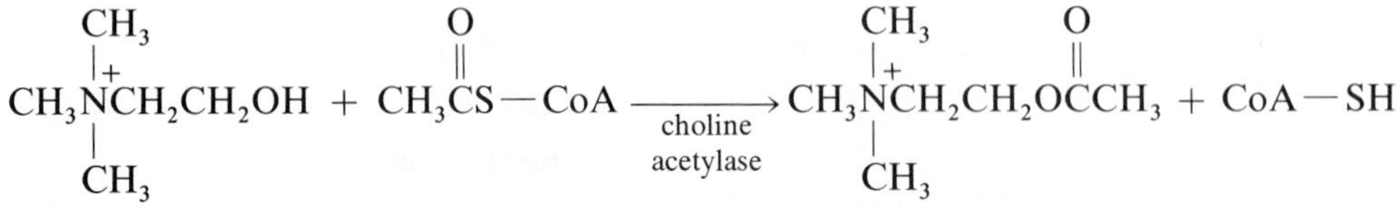
choline
acetylase

2-(trimethylammonium)ethanol — choline; acetyl CoA; acetylcholine — *essential in the transmission of nerve impulses*; coenzyme A

The thioester, acetyl coenzyme A, is converted into an oxyester by the alcohol function of choline. Coenzyme A is the leaving group. In the body, the process is catalyzed by an enzyme called, appropriately, choline acetylase.

PROBLEM 14.29

Write a complete mechanism for the acyl-transfer reaction shown above. You may assume that the enzyme is a good source of protons, HB^+, and of bases to remove protons, B:.

PROBLEM 14.30

The vitamin pantothenic acid is sold over the counter as calcium pantothenate. Write the structural formula for calcium pantothenate.

14.8 REACTIONS OF CARBOXYLIC ACID DERIVATIVES WITH AMMONIA OR AMINES AS NUCLEOPHILES. PREPARATION OF AMIDES

Acid chlorides and acid anhydrides react with ammonia or amines to give amides. Most amides are solids and are therefore useful derivatives of amines. For example, *p*-toluidine is converted to *N*-acet-*p*-toluidide by a reaction with acetic anhydride.

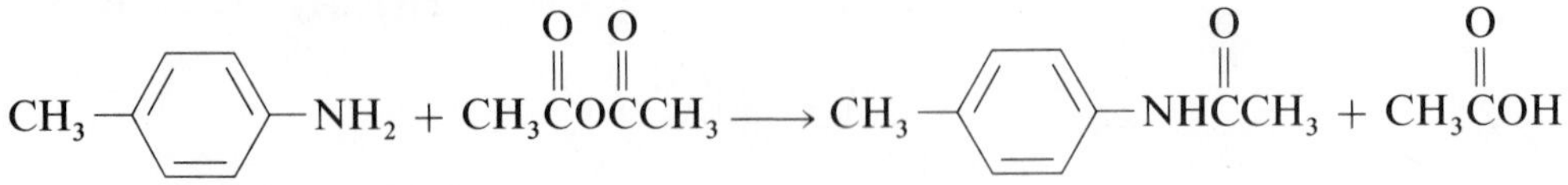

p-toluidine; acetic anhydride; *N*-acet-*p*-toluidide; acetic acid

N,N-Disubstituted amides are formed when acid chlorides or anhydrides react with amines having two substituents on the nitrogen atom. An example is the conversion of cyclohexanecarboxylic acid to *N,N*-dimethylcyclohexanecarboxamide; as is usual, the acid is converted into the acid chloride, which then reacts with the amine.

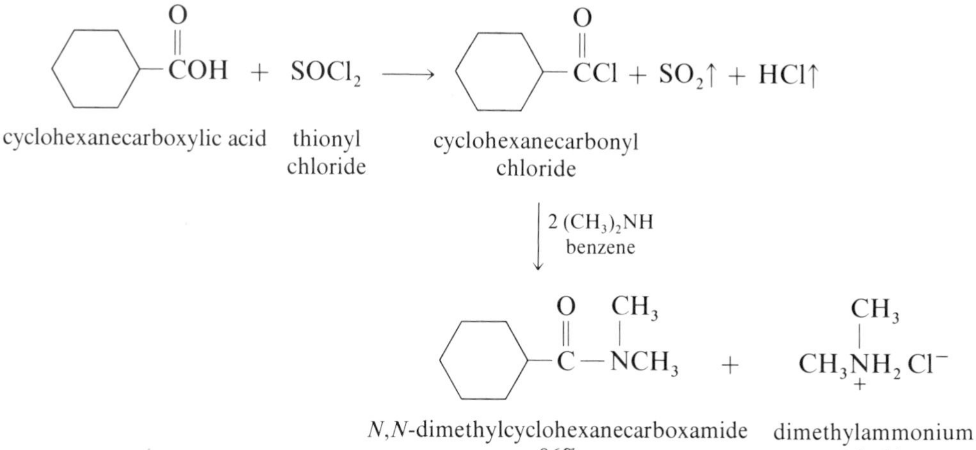

cyclohexanecarboxylic acid, thionyl chloride, cyclohexanecarbonyl chloride

N,N-dimethylcyclohexanecarboxamide 86%, dimethylammonium chloride

If an amine has no hydrogen atom on the nitrogen atom, an isolable amide is not formed when the amine reacts with an acid chloride or an acid anhydride. Such amines do, however, form unstable reactive intermediates with acid chlorides, as illustrated for the reaction of benzoyl chloride and pyridine.

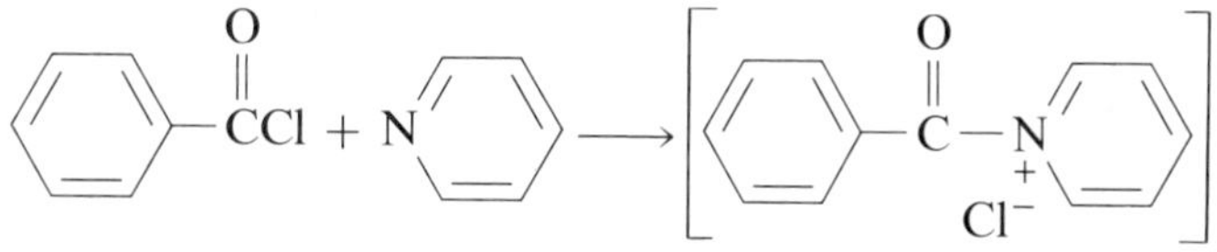

benzoyl chloride, pyridine, unstable reactive intermediate

Esters react with ammonia or amines to give amides and alcohols. For example, ethyl lactate is converted into lactamide and ethanol by treatment with liquid ammonia.

$$CH_3CH(OH)COCH_2CH_3\ (C{=}O) + NH_3(liq) \xrightarrow[-70\,^{\circ}C]{} CH_3CH(OH)CNH_2\ (C{=}O) + CH_3CH_2OH$$

ethyl lactate, ammonia, lactamide 70%, ethanol

Another example demonstrates clearly the differing reactivity toward nucleophilic substitution reactions of a carbonyl group and a carbon-halogen bond. Ethyl chloroacetate, when treated with aqueous ammonium hydroxide at low temperatures and for short periods of time, undergoes substitution quite selectively at the carbonyl group.

$$\underset{\text{ethyl chloroacetate}}{\mathrm{ClCH_2\overset{\overset{\Large O}{\|}}{C}OCH_2CH_3}} + \underset{\text{ammonia}}{\mathrm{NH_3}} \xrightarrow[\substack{H_2O \\ -10\,°C \\ 1\,h}]{} \underset{\text{chloroacetamide}}{\mathrm{ClCH_2\overset{\overset{\Large O}{\|}}{C}NH_2}} + \underset{\text{ethanol}}{\mathrm{CH_3CH_2OH}}$$

PROBLEM 14.31

Study Guide Concept Map 14.6

Suggest a mechanism for the reaction of ethyl acetate with ammonia. What is the nucleophile, and what is the leaving group?

PROBLEM 14.32

Assign structures to the compounds designated by letters in the following equations.

(a) $\mathrm{CH_3CH_2O\overset{\overset{O}{\|}}{C}CH{=}CH\overset{\overset{O}{\|}}{C}OCH_2CH_3} \xrightarrow[\substack{NH_4Cl \\ H_2O}]{NH_3\ (excess)} \mathrm{A}$

(b) 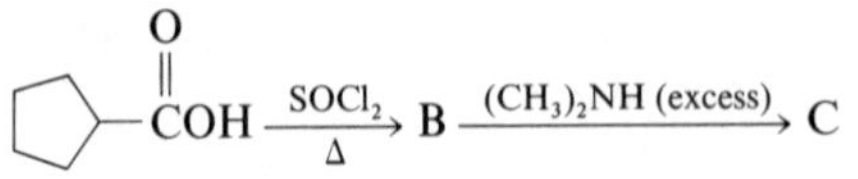

(c) $\mathrm{CH_3CH_2O\overset{\overset{O}{\|}}{C}CH_2CH_2CH_2CH_2\overset{\overset{O}{\|}}{C}OCH_2CH_3} \xrightarrow{H_2NNH_2\ (excess)} \mathrm{D}$

(d) $\mathrm{CH_3(CH_2)_4\underset{\underset{CH_2CH_3}{|}}{C}H\overset{\overset{O}{\|}}{C}OH} \xrightarrow[\Delta]{SOCl_2} \mathrm{E} \xrightarrow[H_2O]{NH_3} \mathrm{F}$

(e) 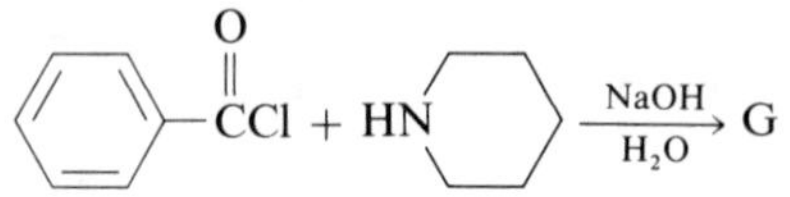

(f) $\mathrm{N{\equiv}CCH_2\overset{\overset{O}{\|}}{C}OCH_2CH_3} \xrightarrow[\substack{H_2O \\ 0\,°C}]{NH_3} \mathrm{H}$

S U M M A R Y

Carboxylic acids are strong organic acids. The functional group in carboxylic acids is the carboxyl group, in which a hydroxyl group is bonded to the carbon of a carbonyl group. Carboxylic acids are prepared by oxidation of primary alcohols or aldehydes, by ozonolysis of alkenes followed by oxidative work-ups and by the Grignard reaction (Table 14.1).

Carboxylic acids are converted to acid chlorides by treatment with thionyl chloride or phosphorus trichloride. Dicarboxylic acids will lose water to give cyclic acid anhydrides, especially readily when the result is a five- or six-membered ring. Anhydrides are also prepared from acid chlorides and acid salts (Table 14.2).

Acid derivatives undergo nucleophilic substitution reactions at the carbonyl group. The first step is attack by the nucleophile at the carbon atom of the carbonyl

TABLE 14.1 Reactions Used to Prepare Acids

Starting Material	Reagent	Intermediate	Reagent in Second Step	Product
RCH_2OH	$Cr_2O_7^{2-}$, H_3O^+	$R\overset{O}{\overset{\|}{C}}H$	$Cr_2O_7^{2-}$, H_3O^+	$R\overset{O}{\overset{\|}{C}}OH$, with $R\overset{O}{\overset{\|}{C}}OCH_2R$ as a side product
	MnO_4^-, OH^-	$R\overset{O}{\overset{\|}{C}}O^-$	H_3O^+	$R\overset{O}{\overset{\|}{C}}OH$
$R\overset{O}{\overset{\|}{C}}H$	MnO_4^-, OH^-	$R\overset{O}{\overset{\|}{C}}O^-$	H_3O^+	$R\overset{O}{\overset{\|}{C}}OH$
	Ag_2O, H_2O	$R\overset{O}{\overset{\|}{C}}O^-$	H_3O^+	$R\overset{O}{\overset{\|}{C}}OH$
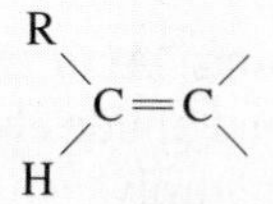	O_3	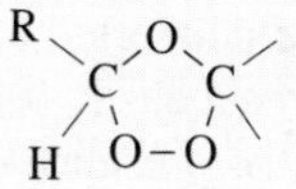	H_2O_2	$R\overset{O}{\overset{\|}{C}}OH$
RMgX (from RX)	CO_2	$R\overset{O}{\overset{\|}{C}}O^-$	H_3O^+	$R\overset{O}{\overset{\|}{C}}OH$

TABLE 14.2 Reagents for Interconverting Acids and Acid Derivatives

From ↓ / To make →	$R\overset{O}{\overset{\|}{C}}Cl$	$R\overset{O}{\overset{\|}{C}}O\overset{O}{\overset{\|}{C}}R$	$R\overset{O}{\overset{\|}{C}}OR'$	$R\overset{O}{\overset{\|}{C}}OH$	$R\overset{O}{\overset{\|}{C}}NR_2$	$R\overset{O}{\overset{\|}{C}}O^-$
$R\overset{O}{\overset{\|}{C}}Cl$	—	$R\overset{O}{\overset{\|}{C}}O^-$	R′OH	H_2O	R_2NH	H_2O, OH^-
$R\overset{O}{\overset{\|}{C}}O\overset{O}{\overset{\|}{C}}R$	—	—	R′OH	H_2O	R_2NH	H_2O, OH^-
$R\overset{O}{\overset{\|}{C}}OR''$	—	—	R′OH, Δ HB^+ or B:	H_2O, H_3O^+ Δ	R_2NH	H_2O, OH^- Δ
$R\overset{O}{\overset{\|}{C}}OH$	$SOCl_2$ or PCl_3	Δ, $-H_2O$ (cyclic anhydride)	R′OH HB^+, Δ	—	R_2NH, Δ	OH^-
$R\overset{O}{\overset{\|}{C}}NR_2$	—	—	—	H_2O, H_3O^+ Δ	—	H_2O, OH^- Δ
$R\overset{O}{\overset{\|}{C}}O^-$	—	$R\overset{O}{\overset{\|}{C}}Cl$	—	H_3O^+	—	—

group, which leads to localization of the electrons of the π bond on the oxygen atom in a tetrahedral intermediate. The carbonyl group forms again when a leaving group is lost. The relative reactivities of the acid derivatives toward nucleophilic substitution are related to the electrophilicity of the carbon atom of the carbonyl group and the resonance stabilization of the species.

$$R\overset{O}{\overset{\|}{C}}Cl > R\overset{O}{\overset{\|}{C}}O\overset{O}{\overset{\|}{C}}R > R\overset{O}{\overset{\|}{C}}OR' \geq R\overset{O}{\overset{\|}{C}}OH > R\overset{O}{\overset{\|}{C}}NH_2 > R\overset{O}{\overset{\|}{C}}O^-$$

most electrophilic carbon, least resonance stabilization, most reactive — *least electrophilic carbon, most resonance stabilization, least reactive*

Acid chlorides and acid anhydrides are easily converted into esters by reaction with alcohols and into amides by reaction with ammonia and amines. Acids can be converted to esters by heating with an excess of an alcohol in the presence of a strong acid; such esterification reactions are equilibrium processes. Esters are converted to other esters in transesterification reactions when heated with an alcohol and either an acid or a base as catalyst (Table 14.2).

All acid derivatives are converted to acids by hydrolysis reactions. Acid chlorides and acid anhydrides react easily with water. Esters, nitriles, and amides have to be heated with water in the presence of acid or base. The basic hydrolysis of an ester is known as saponification. This type of reaction goes to completion because the salt of the carboxylic acid is formed, thereby removing the acid from the reaction mixture and driving the equilibrium toward the products. Nitriles and amides also give carboxylate salts when hydrolyzed in base but give the free acids when hydrolyzed in acid (Table 14.2).

ADDITIONAL PROBLEMS

14.33 Name the following compounds, specifying the stereochemistry where appropriate.

(a) 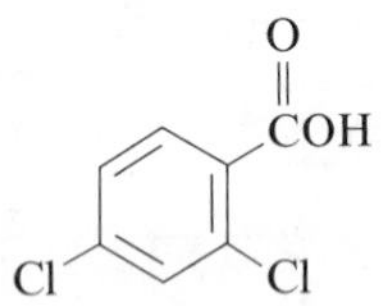

(b) $CH_3CH_2\underset{Br}{\underset{|}{C}}HCH_2CH_2\overset{O}{\overset{\|}{C}}OH$

(c) 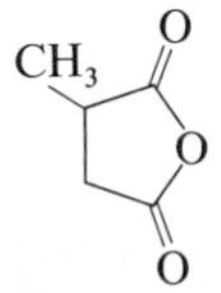

(d) $CH_3CH_2CH_2\overset{CH_3}{\overset{|}{C}}HC{\equiv}N$

(e) 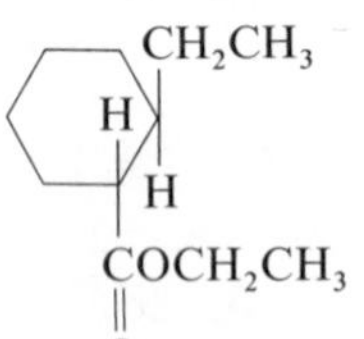

(f) $CH_3CH_2CH_2CH_2\overset{O}{\overset{\|}{C}}\underset{CH_3}{\underset{|}{N}}CH_3$

(g) 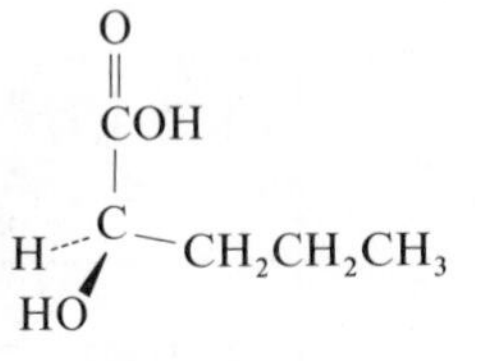

(h) $CH_3CH_2CH_2\underset{Cl}{\underset{|}{C}}H\overset{O}{\overset{\|}{C}}Cl$

(i) 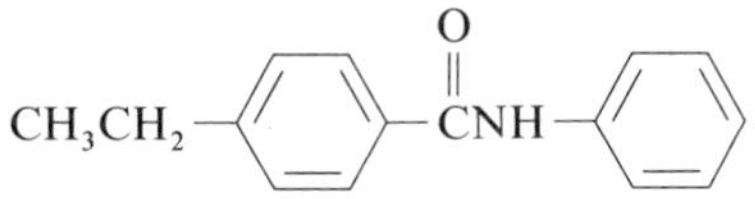

(j) 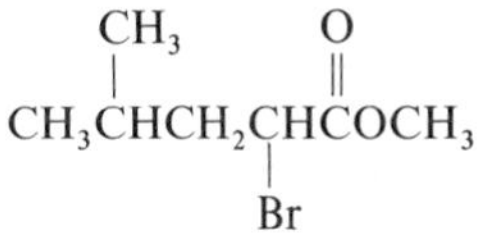

(k) 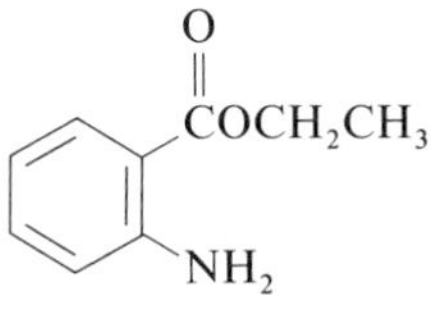

(l) 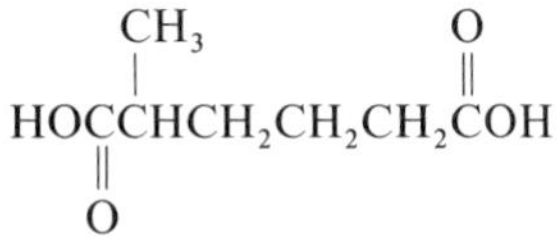

(m) 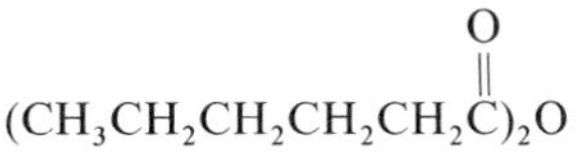

(n) 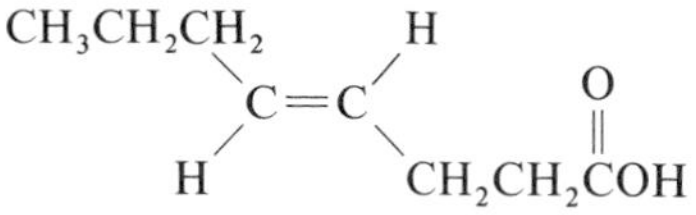

(o) 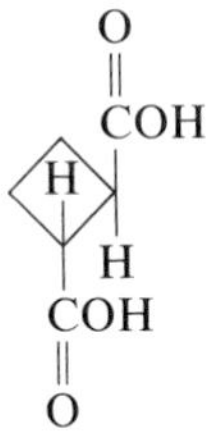

14.34 Draw structural formulas for the following compounds.

(a) methyl (*S*)-3-bromopentanoate (b) *cis*-3-hydroxycyclopentanecarboxylic acid
(c) 2,4-dibromobenzamide (d) ethyl (*Z*)-3-hexenoate (e) heptanenitrile
(f) *N*,*N*-diethylhexanamide (g) diethyl 2,2-dimethylpentanedioate
(h) methyl *p*-nitrobenzoate (i) 3-hydroxy-5,5-dimethyldecanoic acid
(j) *p*-bromobenzoic anhydride (k) *p*-methoxybenzoyl chloride

14.35 Give structural formulas for all of the organic intermediates and products indicated by letters in the following equations.

(a) 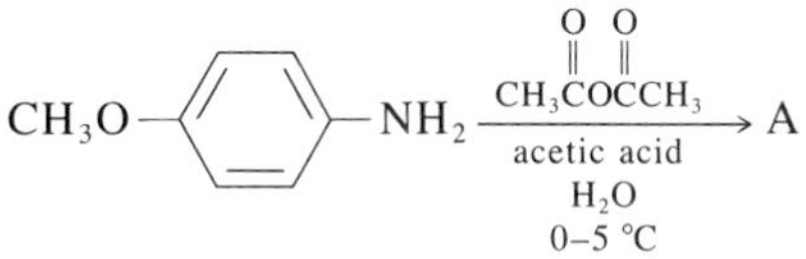

(b) 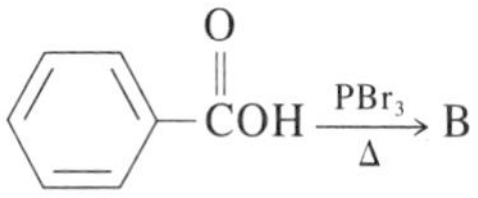

(c) 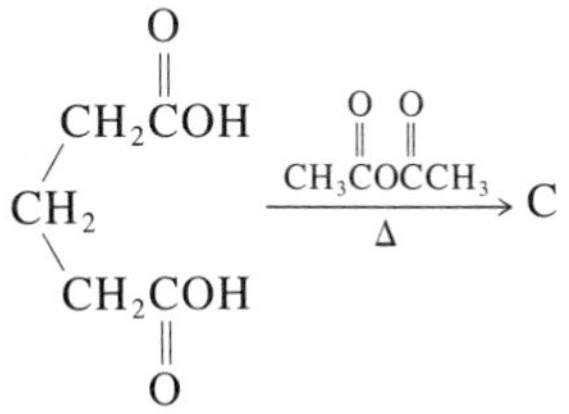

(d) 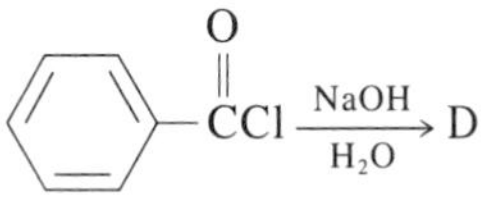

(e) 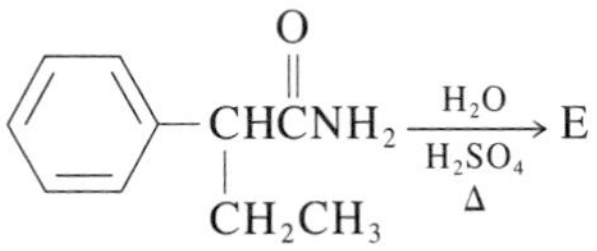

(f) 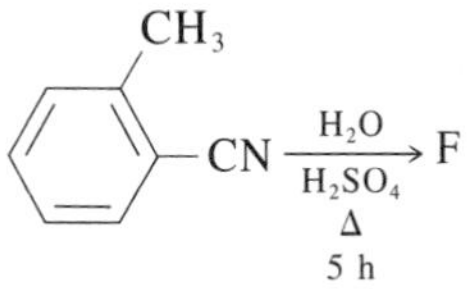

(g) 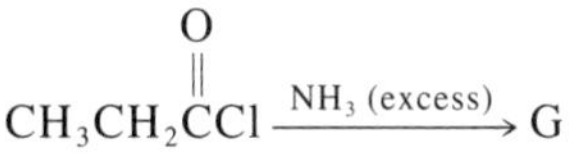

(h) 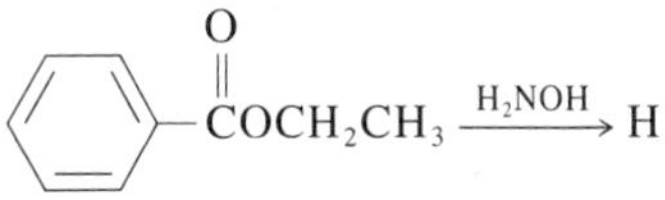

(i) 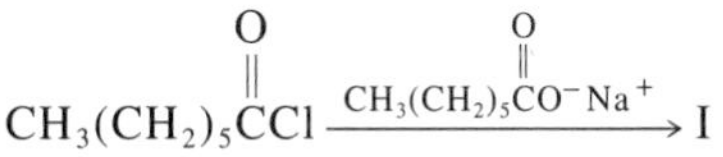

(j) 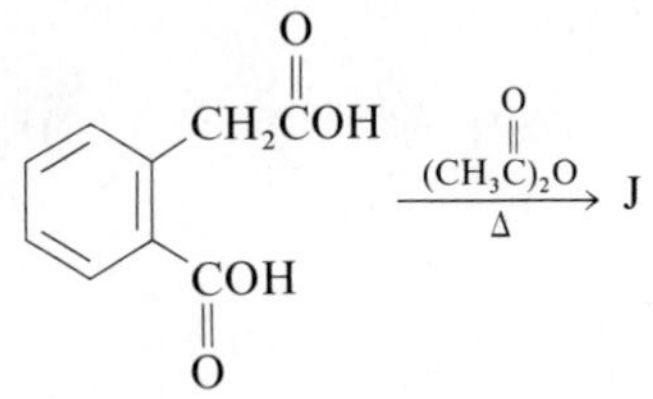

(k) $CH_3(CH_2)_{10}\overset{O}{\overset{\|}{C}}OCH_2(CH_2)_{10}CH_3 + CH_3OH \xrightarrow[\Delta]{H_2SO_4} K + L$

(l) $HO\overset{O}{\overset{\|}{C}}(CH_2)_4\overset{O}{\overset{\|}{C}}OH \xrightarrow{SOCl_2\ (excess)} M$

(m) 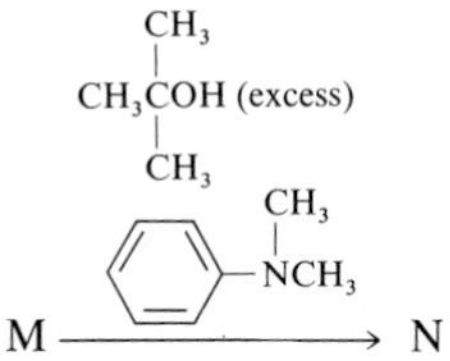

(n) 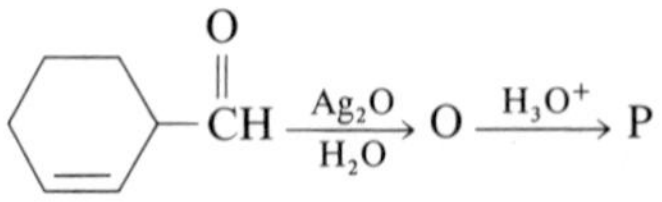

(o) $CCl_3CH_2CH_2CH_2CH_2OH \xrightarrow[H_2O,\ \Delta]{KMnO_4} Q \xrightarrow{H_3O^+} R$

14.36 Assign structures to all the organic intermediates and products designated by letters in the following equations.

(a) 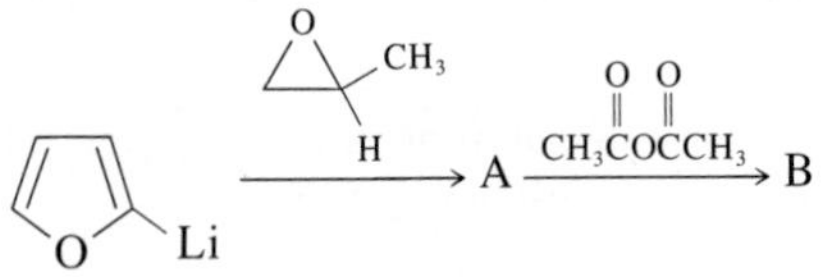

(Hint: What regioselectivity do you expect for the reaction of an unsymmetrical oxirane with a nucleophile?)

(b) $CH_3\overset{CH_3}{\overset{|}{C}}H(CH_2)_{10}CH_2OH \xrightarrow[(CH_3CH_2)_3N,\ dichloromethane]{CH_3SO_2Cl} C \xrightarrow[tetrahydrofuran,\ dimethyl\ sulfoxide]{K^{14}CN} D$

$D \xrightarrow[ethanol,\ \Delta]{KOH,\ H_2O} E \xrightarrow{H_3O^+} F$

(c) 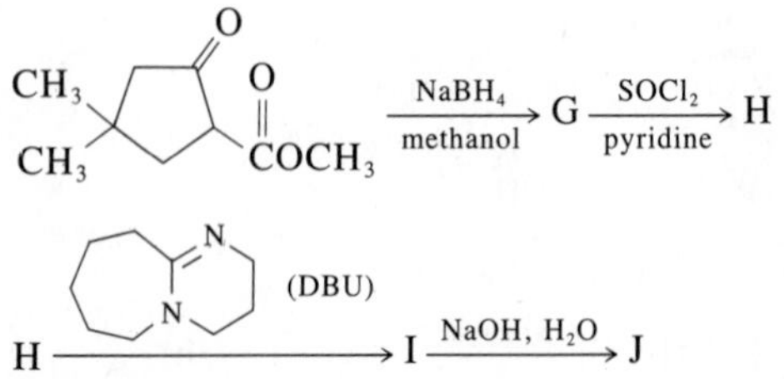

(Hint: DBU, 1,8-diazabicyclo[5.4.0]undec-7-ene, is a strong base but not a good nucleophile.)

14.37

(a) When pyridine is used as a solvent for the reactions of acid chlorides, a reactive intermediate is formed (p. 583). This intermediate is more reactive than benzoyl chloride is toward nucleophilic substitution at the carbon atom of the carbonyl group. Explain this fact.

(b) If water were added to the system resulting from mixing equimolar amounts of benzoyl chloride and pyridine, what would happen? Illustrate by writing a mechanism.

(c) When 1 mole of heptanoic acid, 0.5 mole of thionyl chloride, and 1 mole of pyridine are mixed at − 10 °C in ether solution, a 97% yield of heptanoic anhydride is formed. Propose a reaction scheme that accounts for this observation.

14.38 The following steps were used in the synthesis of prostaglandins, biologically important fatty acids (p. 844). Assign structures to all intermediates indicated by letters.

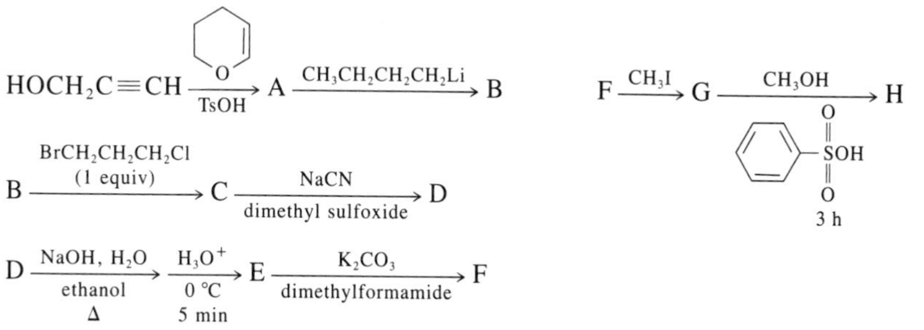

Compound H has a strong absorption bond at 3450 cm^{-1} and another at 1720 cm^{-1} in its infrared spectrum. There are bands at δ 1.6–2.0 (2H, m), 2.2–2.6 (5H, m), 3.68 (3H, s) and 4.2–4.3 (2H, m) in its proton magnetic resonance spectrum.

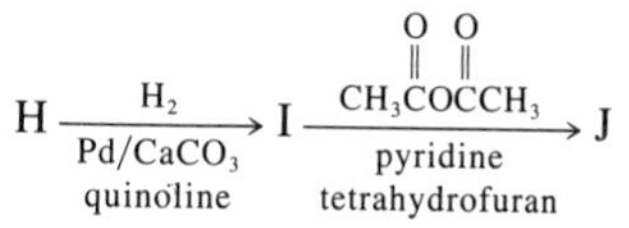

14.39 The following transformation was observed. Propose a mechanism for it.

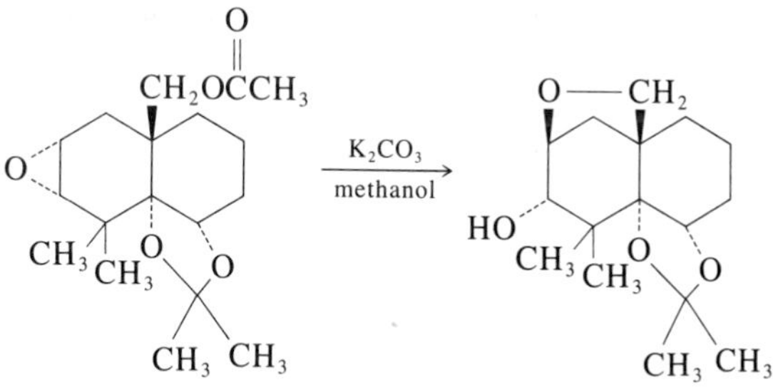

14.40 Dimethyl (2*R*,3*S*)-2-acetoxy-3-bromobutanedioate is converted into a chiral alcohol on treatment with methanol in the presence of a trace of hydrogen chloride. The chiral alcohol, in turn, reacts with potassium carbonate in acetone to give a chiral oxirane. Provide stereochemically correct structures for the ester, alcohol, and oxirane, and show how the transformations described occur.

14.41 Compound A, $C_9H_{10}O_2$, has sharp singlets in its proton magnetic resonance spectrum at δ 2.2 (3H), 5.2 (2H) and 7.4 (5H). The important bands in its infrared spectrum are at 1743, 1229, and 1027 cm^{-1}. Its carbon-13 nuclear magnetic resonance spectrum has bands at 20.7, 66.1, 128.1, 128.4, 136.2, and 170.5 ppm. Assign a structure to Compound A, showing how the spectroscopic data support it.

15

Carboxylic Acids and Their Derivatives II. Synthetic Transformations and Compounds of Biological Interest

A • L O O K • A H E A D

The carbonyl group in carboxylic acid derivatives, like the carbonyl group in aldehydes and ketones, reacts with nucleophiles, such as metal hydrides and organometallic reagents. Acid derivatives differ from aldehydes and ketones in that a leaving group is present on the carbonyl carbon of an acid derivative. Therefore, further reactions are possible. For example, when a Grignard reagent reacts with an ester, loss of the alkoxide group and reaction with a second molecule of the Grignard reagent gives an alcohol as the end product.

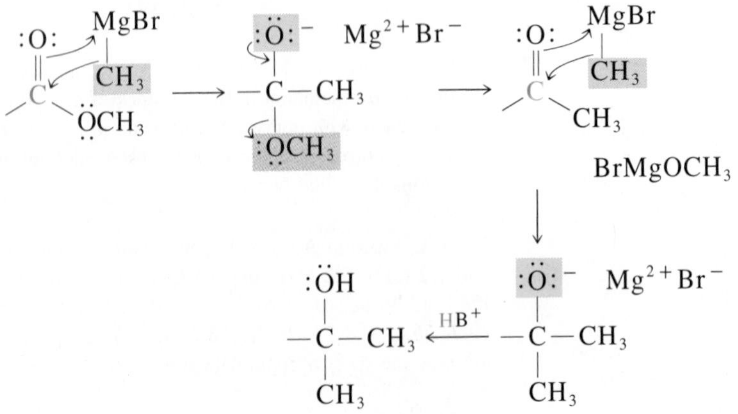

This chapter will explore the reactions of organometallic compounds and metal hydrides with carboxylic acids and their derivatives. Also, some acid derivatives are of biological and practical interest, and the structure and properties of certain antibiotics, polyunsaturated fats, and soaps will be considered.

15.1 REACTIONS OF ORGANOMETALLIC REAGENTS WITH CARBOXYLIC ACIDS AND THEIR DERIVATIVES

A. Grignard Reagents

Esters are the only acid derivatives that react with Grignard reagents in a synthetically useful way. A Grignard reagent (p. 489) is a strong enough base that it reacts with carboxylic acids (pK_a ~5) and unsubstituted amides (pK_a ~15, and thus as acidic as water) to deprotonate them. The anions that result are stabilized by resonance, which produces greatly reduced electrophilicity at the carbon atom of the carbonyl group. A Grignard reagent is not sufficiently nucleophilic to react with these ions; no reaction occurs even when an excess of the Grignard reagent is present.

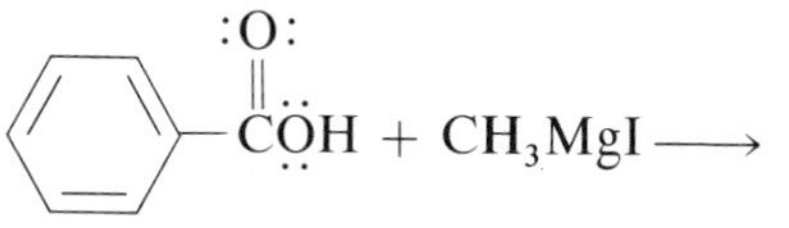

benzoic acid   methylmagnesium iodide

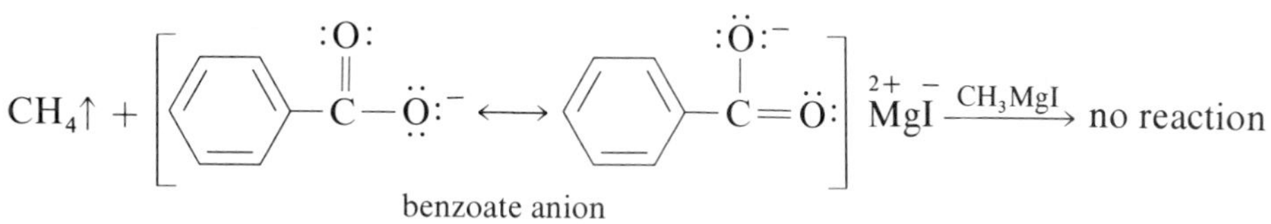

benzoate anion

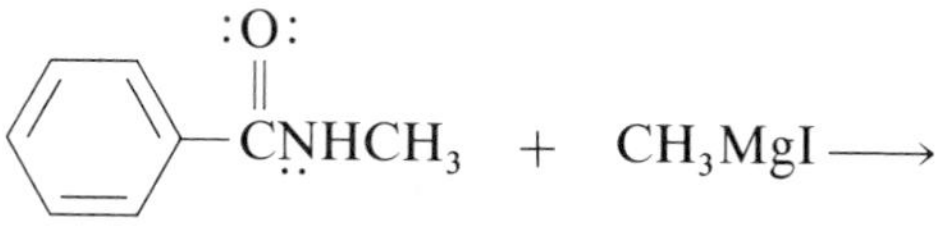

N-methylbenzamide   methylmagnesium iodide

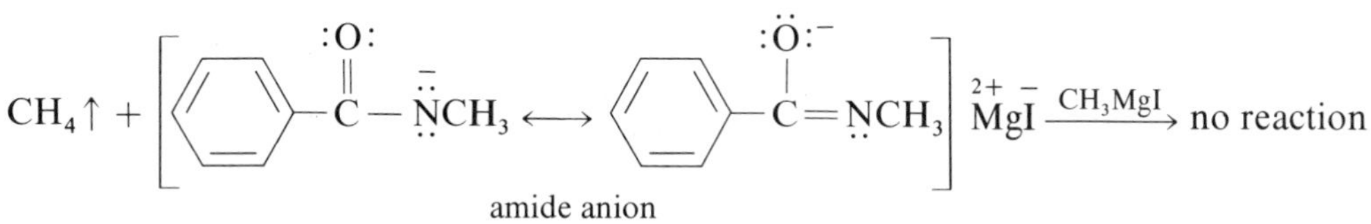

amide anion

An ester, however, has no acidic proton. Two equivalents of Grignard reagent react with one equivalent of ester to give an alcohol. A classic application of this reaction is the preparation of triphenylmethanol from bromobenzene and methyl benzoate.

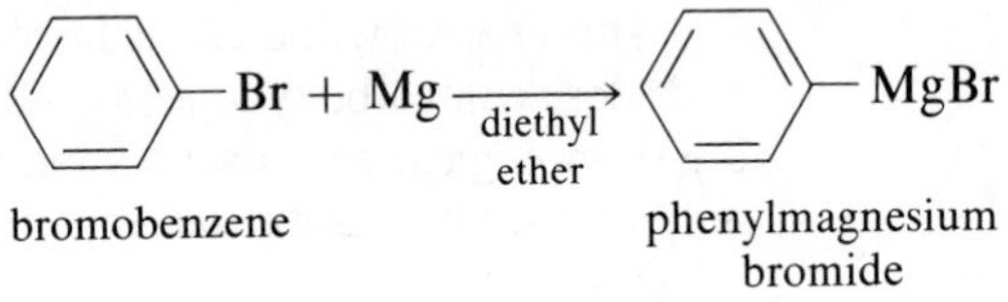
Br + Mg
diethyl ether
MgBr
bromobenzene
phenylmagnesium bromide

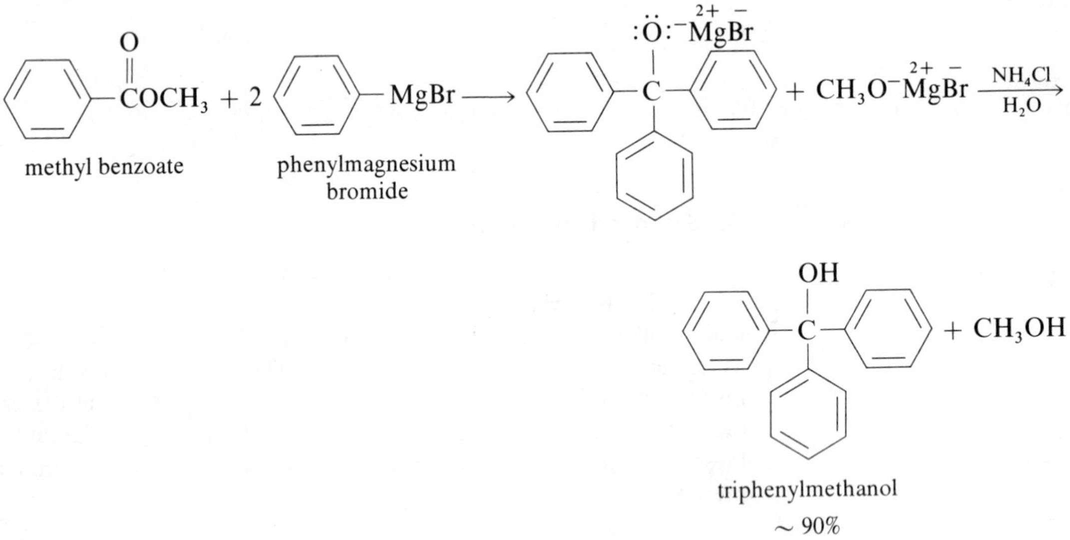
COCH3 + 2
MgBr
methyl benzoate
phenylmagnesium bromide
+ CH3O⁻ MgBr
NH4Cl
H2O
OH
+ CH3OH
triphenylmethanol
~ 90%

The first addition of the Grignard reagent to the carbonyl group of the ester leads to an intermediate that loses methoxide ion, thus regenerating a carbonyl group, which reacts again with the Grignard reagent.

VISUALIZING THE REACTION

Reaction of a Grignard reagent with an ester

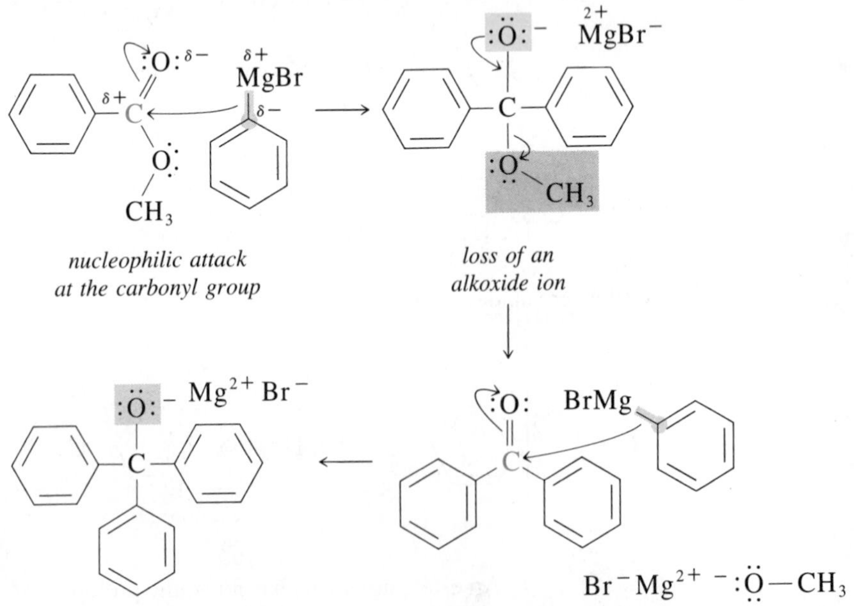
MgBr
CH3
MgBr⁻
CH3
BrMg
Mg2+ Br⁻
Br⁻ Mg2+ ⁻:Ö—CH3

nucleophilic attack at the carbonyl group

loss of an alkoxide ion

alkoxide ion complexed with metal

nucleophilic attack at the new carbonyl group

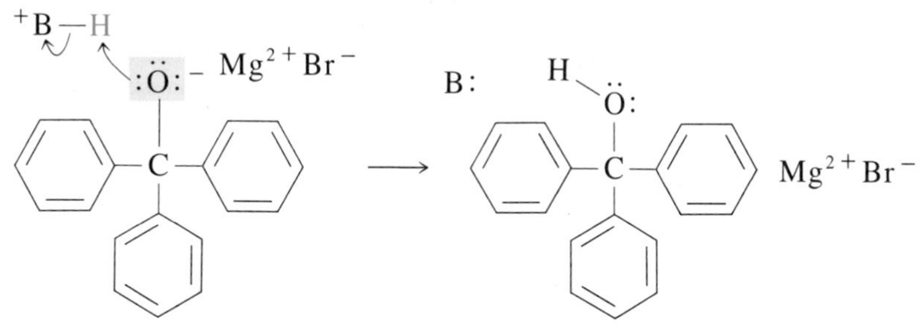

$^{+}$B—H
:Ö:$^{-}$ Mg^{2+}Br^{-}
C
B:
H
Ö:
C
Mg^{2+}Br^{-}

protonation of the alkoxide ion by acid added in a second step

The resulting complex contains an alkoxide ion, which is protonated when aqueous acid is added to give the free alcohol. You should be familiar with all of these steps from the reactions of aldehydes and ketones with Grignard reagents (p. 491). The reactions of esters and those of aldehydes and ketones differ because of the presence in an ester of the alkoxyl group, which can act as a leaving group.

Reactions of Grignard reagents with most esters give tertiary alcohols in which at least two identical groups are bonded to the tertiary carbon atom. In triphenylmethanol, all three groups are the same because an ester with a phenyl group was used as a starting material. If a formate ester is used, a secondary alcohol with two identical groups bonded to the secondary carbon atom results, as in the synthesis of 5-nonanol.

$$\underset{n\text{-butyl bromide}}{CH_3CH_2CH_2CH_2Br} + Mg \xrightarrow[\text{diethyl ether}]{} \underset{n\text{-butylmagnesium bromide}}{CH_3CH_2CH_2CH_2MgBr}$$

$$\underset{n\text{-butylmagnesium bromide}}{CH_3CH_2CH_2CH_2MgBr} + \underset{\text{ethyl formate}}{H\overset{O}{\overset{\|}{C}}OCH_2CH_3} \xrightarrow[\text{diethyl ether}]{} CH_3CH_2CH_2CH_2\overset{O^{-}\overset{2+}{Mg}Br^{-}}{\overset{|}{C}}HCH_2CH_2CH_2CH_3 + CH_3CH_2O^{-}\overset{2+}{Mg}Br^{-}$$

$$\downarrow H_3O^{+}$$

$$CH_3CH_2CH_2CH_2\underset{\underset{OH}{|}}{C}HCH_2CH_2CH_2CH_3$$

5-nonanol
84%

PROBLEM 15.1

Write a mechanism for the formation of 5-nonanol from butylmagnesium bromide and ethyl formate.

PROBLEM 15.2

Suppose you want to synthesize 3-ethyl-3-pentanol with a radioactive carbon-14 atom as the tertiary carbon atom. Carbon dioxide labeled with carbon-14, $^{14}CO_2$, is readily available. Devise a synthesis for 3-ethyl-3-pentanol-3-^{14}C.

PROBLEM 15.3

It is possible to use a Grignard reagent to make triphenylmethanol by a different route from the one shown on p. 592. Work out another synthesis.

B. Organolithium Reagents

Organolithium reagents are more powerfully nucleophilic than Grignard reagents are and therefore react with carboxylate anions as well as with other carbonyl compounds (p. 492). This reaction is a useful way to synthesize ketones. In a typical example, shown below, hexanoic acid is converted to 2-heptanone by methyllithium.

$$\underset{\text{hexanoic acid}}{CH_3CH_2CH_2CH_2CH_2\overset{\overset{\displaystyle O}{\|}}{C}OH} \xrightarrow[\text{diethyl ether}]{CH_3Li\ (2\ \text{molar equivalents})} \xrightarrow{H_3O^+} \underset{\substack{\text{2-heptanone} \\ 83\%}}{CH_3CH_2CH_2CH_2CH_2\overset{\overset{\displaystyle O}{\|}}{C}CH_3}$$

Methyllithium is, of course, a strong base as well as a good nucleophile. The first stage of the reaction is the conversion of the carboxylic acid to its anion, which then reacts a second time with the strongly nucleophilic methyllithium, as seen below.

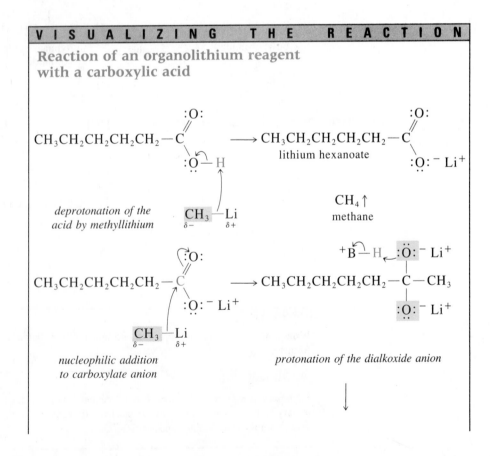

VISUALIZING THE REACTION

Reaction of an organolithium reagent with a carboxylic acid

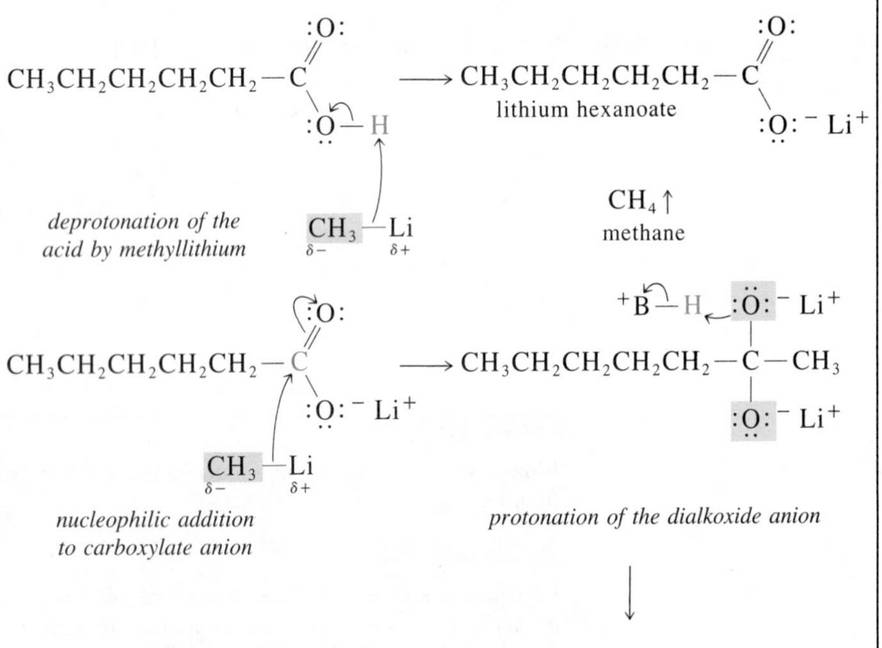
:O:
CH₃CH₂CH₂CH₂CH₂—C
:O—H
CH₃CH₂CH₂CH₂CH₂—C
lithium hexanoate
:O:⁻ Li⁺
deprotonation of the acid by methyllithium
CH₃—Li
δ−
δ+
CH₄↑
methane
CH₃CH₂CH₂CH₂CH₂—C—CH₃
⁺B—H
nucleophilic addition to carboxylate anion
protonation of the dialkoxide anion

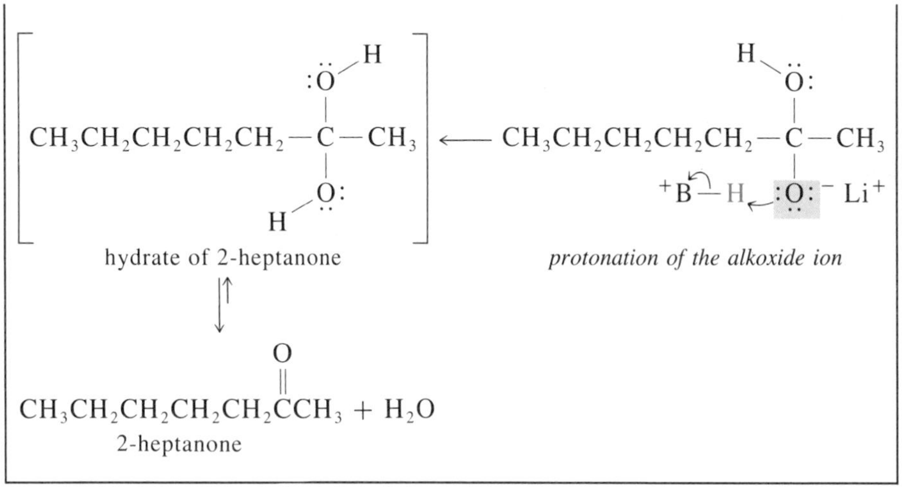
CH₃CH₂CH₂CH₂CH₂—C—CH₃
hydrate of 2-heptanone
protonation of the alkoxide ion
CH₃CH₂CH₂CH₂CH₂CCH₃ + H₂O
2-heptanone

The addition of methyllithium to the carbonyl group of the acid salt produces a dialkoxide ion. Protonation of the oxygen atoms of this intermediate gives the hydrate of the ketone, which is in equilibrium with the ketone itself (p. 498).

The organolithium reagent used does not have to be methyllithium. For example, phenyllithium reacts with lithium butanoate, prepared from butanoic acid.

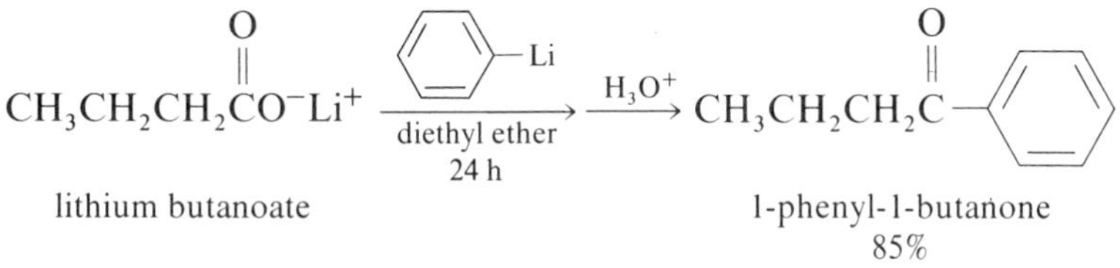
diethyl ether
24 h
lithium butanoate
1-phenyl-1-butanone
85%

The lithium intermediates in this type of reaction have properties that differ from those formed in a Grignard reaction. For example, lithium carboxylate salts are less highly ionic and, therefore, more soluble in ether than their magnesium counterparts (p. 591). Because such a salt is less ionic, the carbonyl group retains more of its electrophilic character. Therefore, a second organolithium reagent can add to it, whereas magnesium carboxylate salts do not react with Grignard reagents. A dialkoxide anion is more stable with lithium as its counter-ion; the lithium-oxyanion, LiO^-, seems to be a poorer leaving group than the corresponding magnesium ion. The dialkoxide anion survives in the reaction mixture without reacting further with the organolithium reagent; then it is protonated when water is added. Therefore, ketones, not tertiary alcohols, are produced in this reaction.

Study Guide Concept Map 15.1

PROBLEM 15.4

Give structural formulas for the intermediates and products indicated by letters in the following equations.

(a) 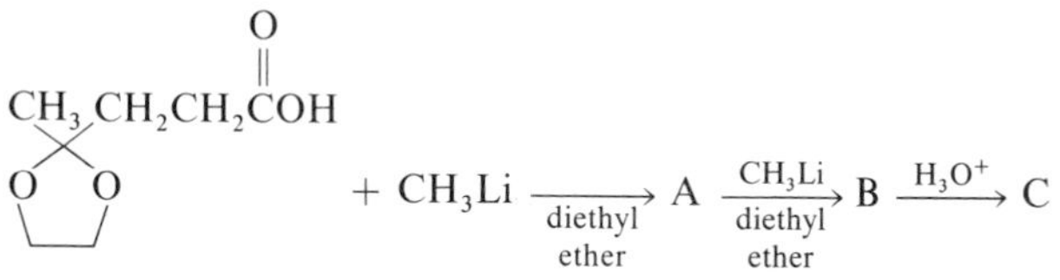
diethyl ether
diethyl ether

(b) $CF_3\overset{O}{\overset{\|}{C}}O^-Li^+ + CH_3CH_2CH_2CH_2Li \xrightarrow[\text{ether}]{\text{diethyl}} D \xrightarrow{H_3O^+} E$

(c) 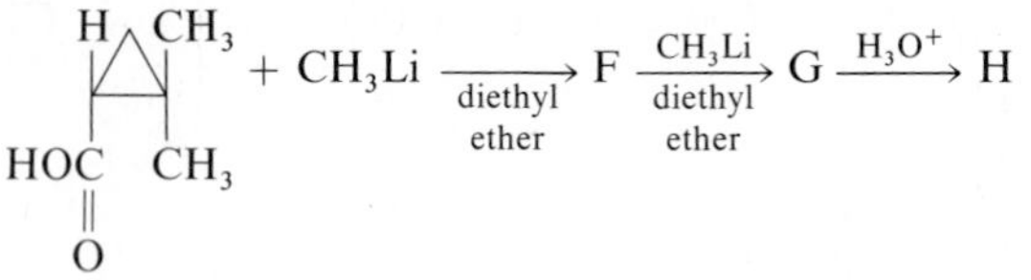

(d) 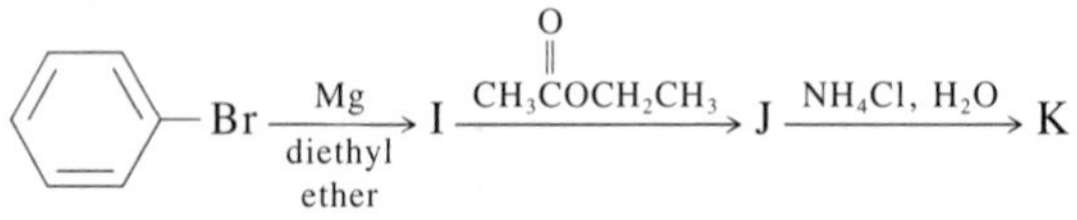

15.2 REDUCTION OF CARBOXYLIC ACIDS AND THEIR DERIVATIVES

A. Lithium Aluminum Hydride

Carboxylic acids and their derivatives are reduced by metal hydrides just as aldehydes and ketones are. Sodium borohydride reduces aldehydes and ketones with ease (p. 485) but usually reacts with acid derivatives only with difficulty. Thus, it is possible to reduce aldehydes or ketones selectively with sodium borohydride in the presence of an acid or ester if reaction times are short and temperatures are kept low. Such a selective reduction of a cyclic ketone in the presence of an ester group is shown below.

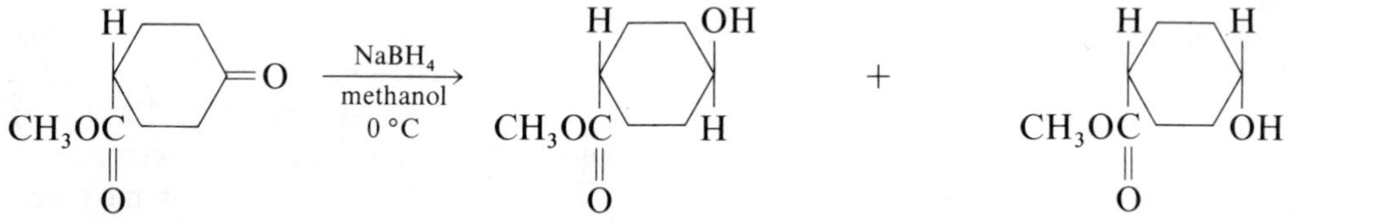

methyl 4-oxocyclohexanecarboxylate — methyl *trans*-4-hydroxycyclohexanecarboxylate *major product* — methyl *cis*-4-hydroxycyclohexanecarboxylate *minor product*

Lithium aluminum hydride is generally useful in reducing acids and acid derivatives. It reduces acids, esters, amides, and nitriles.

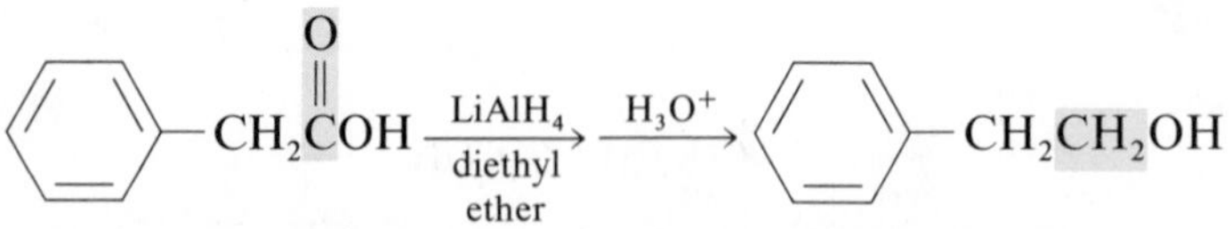

phenylacetic acid — 2-phenyl-1-ethanol 92%

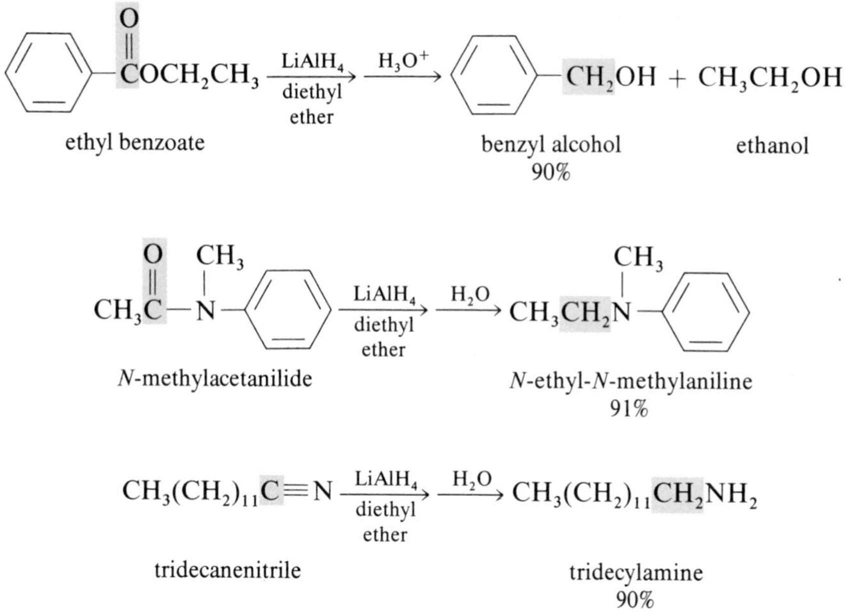

ethyl benzoate → benzyl alcohol 90% + ethanol

$$\text{C}_6\text{H}_5-\overset{\overset{\textstyle O}{\|}}{\text{C}}\text{OCH}_2\text{CH}_3 \xrightarrow[\text{diethyl ether}]{\text{LiAlH}_4} \xrightarrow{\text{H}_3\text{O}^+} \text{C}_6\text{H}_5-\text{CH}_2\text{OH} + \text{CH}_3\text{CH}_2\text{OH}$$

N-methylacetanilide → N-ethyl-N-methylaniline 91%

$$\text{CH}_3\overset{\overset{\textstyle O}{\|}}{\text{C}}-\overset{\overset{\textstyle CH_3}{|}}{\text{N}}-\text{C}_6\text{H}_5 \xrightarrow[\text{diethyl ether}]{\text{LiAlH}_4} \xrightarrow{\text{H}_2\text{O}} \text{CH}_3\text{CH}_2\overset{\overset{\textstyle CH_3}{|}}{\text{N}}-\text{C}_6\text{H}_5$$

tridecanenitrile → tridecylamine 90%

$$\text{CH}_3(\text{CH}_2)_{11}\text{C} \equiv \text{N} \xrightarrow[\text{diethyl ether}]{\text{LiAlH}_4} \xrightarrow{\text{H}_2\text{O}} \text{CH}_3(\text{CH}_2)_{11}\text{CH}_2\text{NH}_2$$

A close examination of these chemical transformations reveals that in each case the carbonyl group (or the carbon atom of the nitrile) has been reduced to a methylene group, $-\text{CH}_2-$. A carboxylic acid is reduced to a primary alcohol. In the case of the ethyl ester, clearly, ethanol is formed by protonation of the alkoxyl leaving group. An ester is reduced to two alcohols: a primary alcohol corresponding to the carboxylic acid portion of the molecule and an alcohol corresponding to the alkoxyl group in the ester. Amides and nitriles are reduced to amines.

In the reaction of acid derivatives with lithium aluminum hydride, the first attack is on the electrophilic carbon atom of the carbonyl group by the nucleophilic metal hydride ion in a step analogous to that seen for aldehydes and ketones (p. 485).

VISUALIZING THE REACTION

Reduction of an ester by lithium aluminum hydride

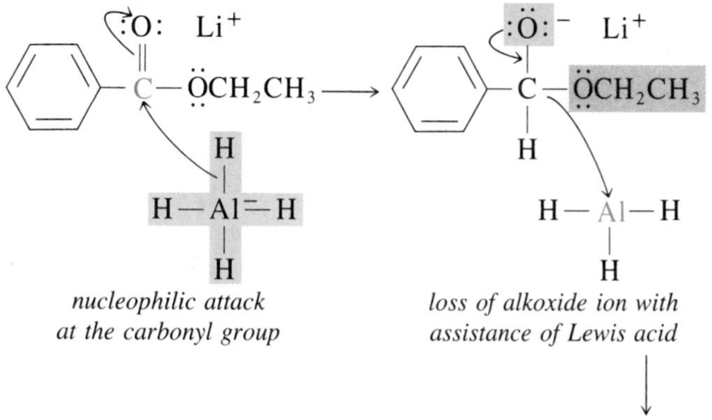

nucleophilic attack
at the carbonyl group

loss of alkoxide ion with
assistance of Lewis acid

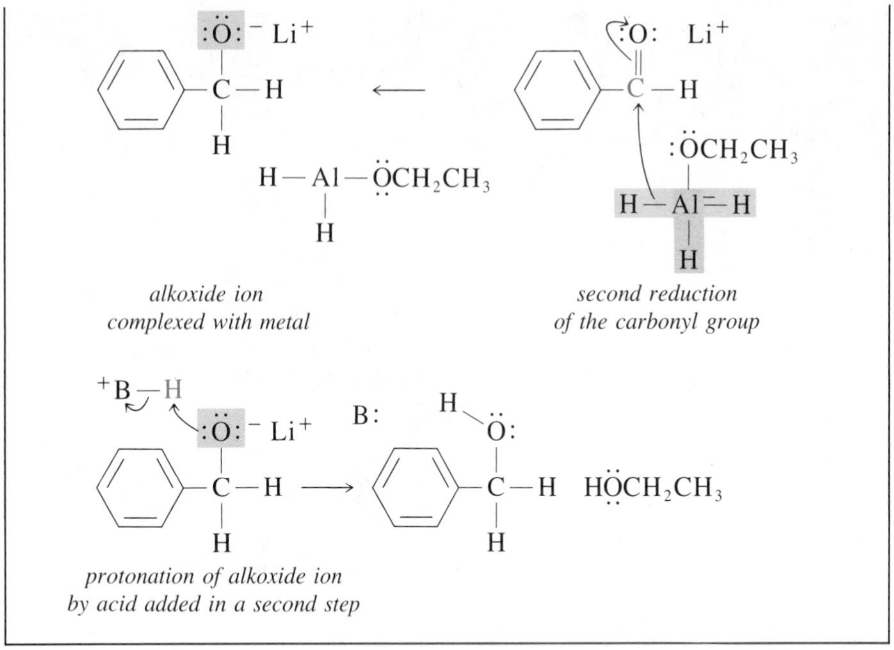
alkoxide ion
complexed with metal
second reduction
of the carbonyl group
protonation of alkoxide ion
by acid added in a second step

The addition of hydride ion to the carbonyl group of the ester gives a tetrahedral intermediate that resembles an acetal in structure and is therefore at the oxidation state of an aldehyde. It loses ethoxide ion with the assistance of the Lewis acid aluminum hydride. The carbon atom of the carbonyl group accepts another hydride ion and is reduced to a primary alcohol. The complexes involving aluminum hydride and alkoxide ions from the ester are sources of hydride ion and will continue to reduce carbonyl groups until all the hydride ions are used up. Cautious addition of a dilute acid protonates the alkoxide ions.

The reductions of carboxylic acids and unsubstituted amides with lithium aluminum hydride do not proceed as smoothly as does the reduction of esters. The hydride ion is a powerful base as well as being a good nucleophile, and it deprotonates the acid or the amide. The resulting salts are often insoluble in the reaction mixture and, therefore, react slowly.

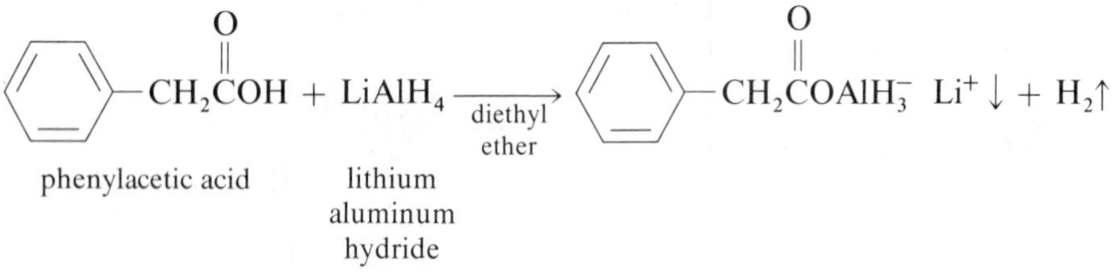
phenylacetic acid
lithium
aluminum
hydride
diethyl
ether

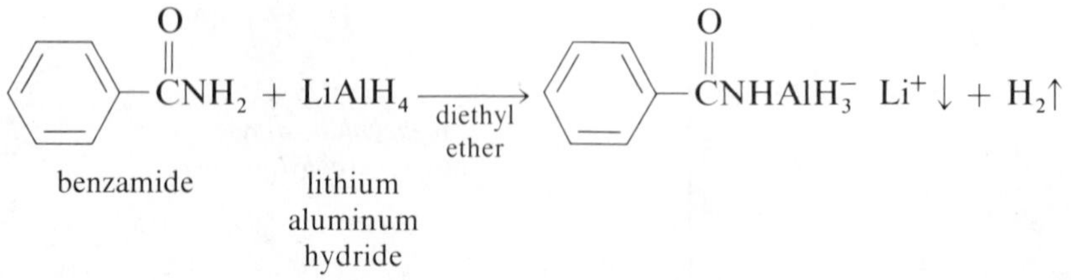
benzamide
lithium
aluminum
hydride
diethyl
ether

The reduction of an amide by lithium aluminum hydride probably proceeds through an iminium ion intermediate.

VISUALIZING THE REACTION

Reduction of an amide by lithium aluminum hydride

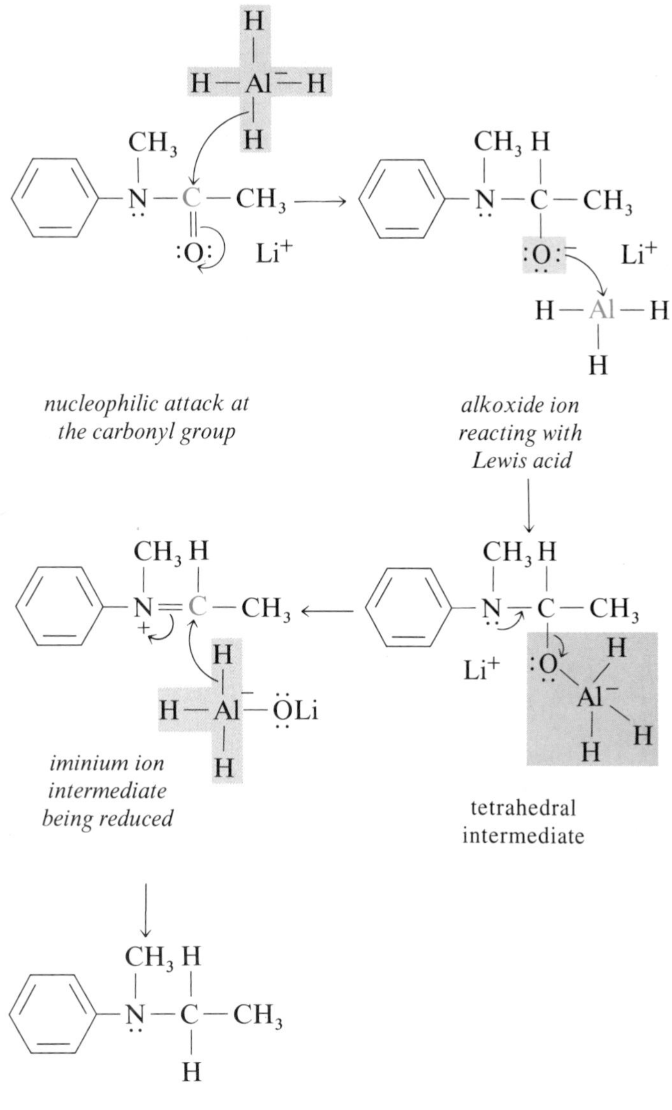
nucleophilic attack at the carbonyl group
alkoxide ion reacting with Lewis acid
tetrahedral intermediate
iminium ion intermediate being reduced

The loss of the oxygen atom of the carbonyl group is thought to take place by the donation of nonbonding electrons from the nitrogen atom of the amide accompanied by the formation of an iminium ion, which is then reduced in another step. The reduction of a nitrile by lithium aluminum follows a mechanism that is similar.

The bond between the carbonyl carbon atom and the nitrogen atom in an amide, which is cleaved by hydrolysis reactions (p. 564), is unaffected by lithium aluminum hydride. Lithium aluminum hydride does cleave the carbon-oxygen bond between the carbonyl carbon and the alkoxyl group in an ester. This reaction frees the alcohol portion of the ester and reduces the carbonyl group to a primary alcohol function. Lithium aluminum hydride is sometimes used to remove acyl groups from valuable alcohols that would be harmed by hydrolysis reactions (Problem 15.7 on p. 600, for example).

PROBLEM 15.5

Write a complete mechanism for the reduction of benzonitrile with lithium aluminum hydride.

PROBLEM 15.6

Complete the following equations.

(a) 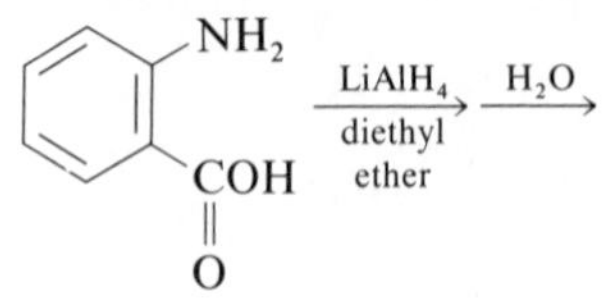
NH2
COH
O
LiAlH4
diethyl
ether
H2O

(b) 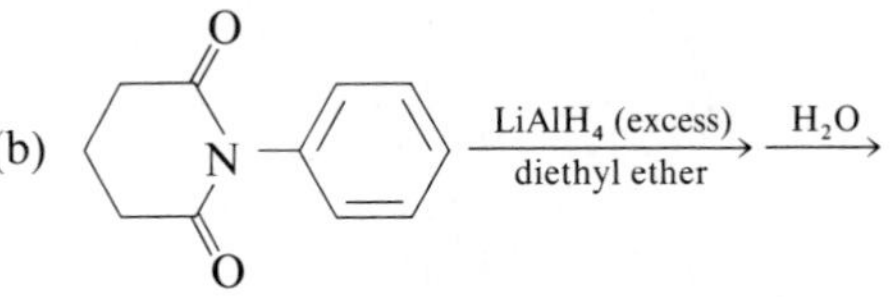
O
N
O
LiAlH4 (excess)
diethyl ether
H2O

(c) $CH_3CH{=}CHCH_2CH_2\overset{O}{\overset{\|}{C}}OCH_3 \xrightarrow[\text{diethyl ether}]{LiAlH_4} \xrightarrow{H_3O^+}$

(d) $CH_3CH_2O\overset{O}{\overset{\|}{C}}(CH_2)_4\overset{O}{\overset{\|}{C}}OCH_2CH_3 \xrightarrow[\text{diethyl ether}]{LiAlH_4} \xrightarrow{H_3O^+}$

(e) $CH_3CH_2\overset{O}{\overset{\|}{C}}NHCH_2CH_3 \xrightarrow[\text{diethyl ether}]{LiAlH_4} \xrightarrow{H_2O}$

(f) 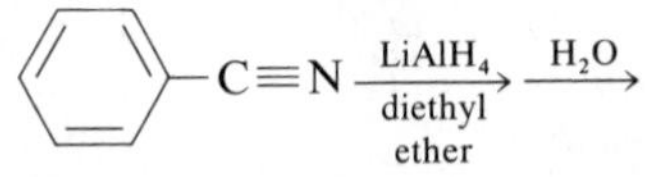
C≡N
LiAlH4
diethyl
ether
H2O

PROBLEM 15.7

The following sequence of reactions is part of a synthesis of compounds that are similar to Vitamin D (p. 1218). What are the structures of Compounds A and B?

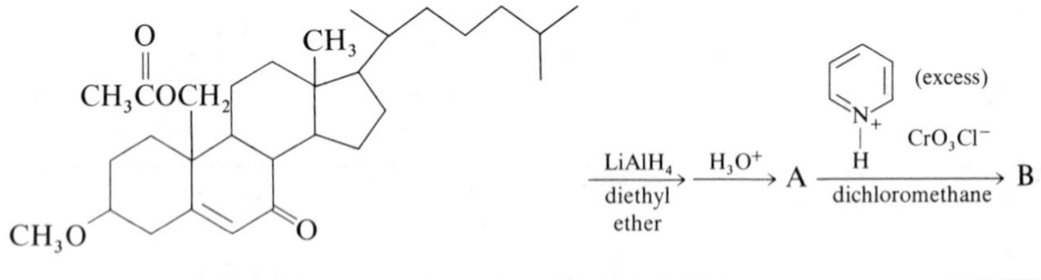
O
CH3
CH3COCH2
CH3O
O
LiAlH4
diethyl
ether
H3O+
A
(excess)
N
H
CrO3Cl−
dichloromethane
B

B. Diisobutylaluminum Hydride

Acids, esters, and nitriles can also be reduced with diisobutylaluminum hydride. Using an excess of this reagent leads to the same products as would be obtained from reduction by lithium aluminum hydride. For example, butyl hexanoate is reduced to a mixture of 1-hexanol and 1-butanol by 4 equivalents of diisobutyl-aluminum hydride.

$$\underset{\text{butyl hexanoate}}{CH_3(CH_2)_4\overset{\overset{\displaystyle O}{\|}}{C}OCH_2CH_2CH_2CH_3} + \underset{\substack{\text{diisobutylaluminum}\\ \text{hydride}\\ \text{4 equivalents}}}{[(CH_3)_2CHCH_2]_2AlH} \xrightarrow[\substack{\text{benzene}\\ 45\ ^\circ C,\ 8\ h}]{N_2}$$

$$\xrightarrow{CH_3OH,\ H_2O} \underset{\substack{\text{1-hexanol}\\ 82\%}}{CH_3(CH_2)_4CH_2OH} + \underset{\substack{\text{1-butanol}\\ 58\%}}{CH_3CH_2CH_2CH_2OH}$$

If, however, only 1 equivalent of this reducing agent is used and the temperature is kept around $-70\ ^\circ C$, an ester will be reduced to an aldehyde in good yield.

$$\underset{\substack{\text{ethyl laurate}\\ \text{ethyl dodecanoate}}}{CH_3(CH_2)_{10}\overset{\overset{\displaystyle O}{\|}}{C}OCH_2CH_3} + \underset{\substack{\text{diisobutylaluminum}\\ \text{hydride}\\ \text{1 equivalent}}}{[(CH_3)_2CHCH_2]_2AlH} \xrightarrow[-70\ ^\circ C]{\text{hexane}}$$

$$\underset{\substack{\text{complex from reaction of}\\ \text{ester with metal hydride}}}{CH_3(CH_2)_{10}\underset{\underset{\displaystyle H}{|}}{\overset{\overset{\displaystyle [(CH_3)_2CHCH_2]_2AlO}{|}}{C}}OCH_2CH_3} \xrightarrow{H_2O}$$

$$\underset{\substack{\text{dodecanal}\\ 88\%}}{CH_3(CH_2)_{10}\overset{\overset{\displaystyle O}{\|}}{C}H} + \underset{\text{isobutane}}{CH_3\overset{\overset{\displaystyle CH_3}{|}}{C}HCH_3} + \underset{\text{ethanol}}{CH_3CH_2OH} + Al(OH)_3\downarrow$$

Nitriles are also reduced to aldehydes with high yields when only 1 equivalent of diisobutylaluminum hydride is used.

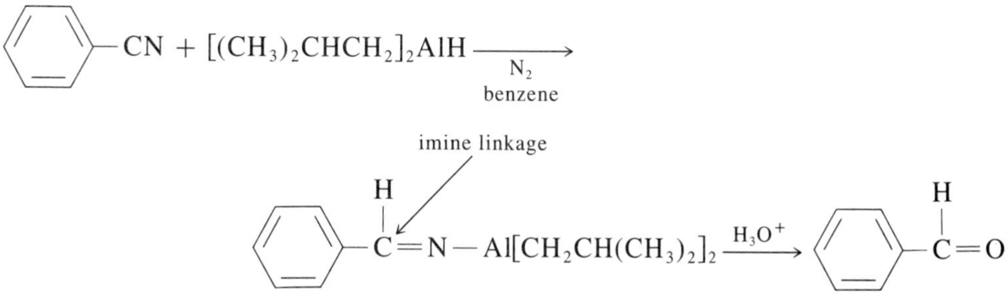
CN + [(CH3)2CHCH2]2AlH
N2
benzene
imine linkage
H
C=N—Al[CH2CH(CH3)2]2
H3O+
H
C=O

complex from reaction of nitrile with metal hydride

benzaldehyde 90%

The product from the reaction of the nitrile with the metal hydride is an imine (p. 506) complexed with aluminum. When dilute acid is added to the reaction mixture, this complex is hydrolyzed, and a carbonyl group is generated.

The reduction of nitriles to aldehydes is a particularly useful reaction because nitriles can be prepared by S_N2 reactions of cyanide ion with alkyl halides or alkyl tosylates. Therefore, an alcohol can be converted by this sequence of reactions into an aldehyde that has one more carbon atom.

Study Guide Concept Map 15.2

PROBLEM 15.8

Pick an alcohol and convert it into an aldehyde containing one more carbon atom, using the sequence of reactions described above.

PROBLEM 15.9

Complete the following equations.

(a) $$CH_3CH_2OCH_2(CH_2)_5\overset{O}{\overset{\|}{C}}OCH_2CH_3 \xrightarrow[\text{hexane}, -70\ °C]{[(CH_3)_2CHCH_2]_2AlH\ (1\ \text{equiv})} \xrightarrow{H_2O}$$

(b) 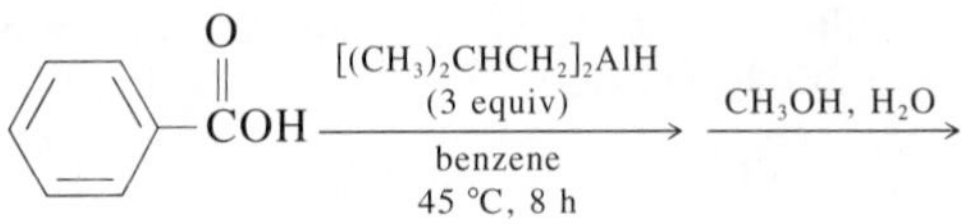

(c) 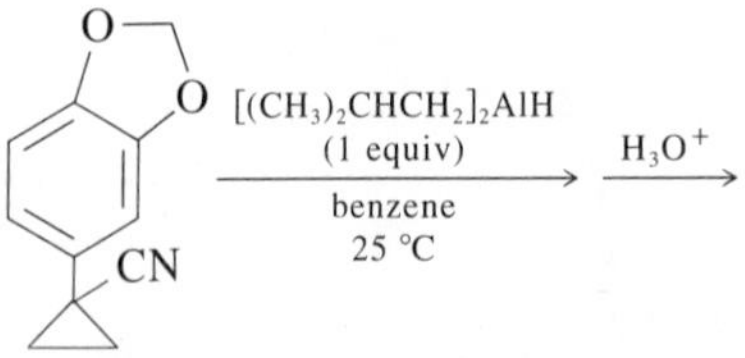

C. Problem-Solving Skills

Problem

The following transformation was used in the synthesis of compounds needed in a study of reaction mechanisms. How would you carry out this transformation?

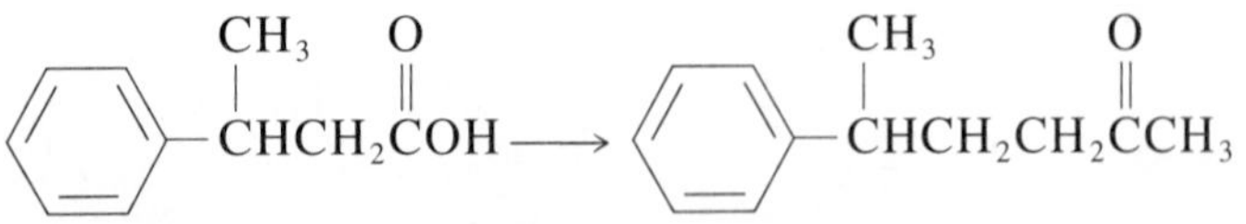

Solution

1. What functional groups are present in the starting material and the product?

   The starting material is an acid. The product is a ketone.

2. How do the carbon skeletons of the two compounds compare? How many carbon atoms does each contain? Are there any rings? What are the positions of branches and functional groups on the carbon skeletons?

The starting material has a four carbon chain with a phenyl group on the third carbon atom and a carboxyl group for the first one. The product has a six carbon chain with a phenyl group on the fifth carbon atom. The chain has been lengthened by two carbon atoms and contains a carbonyl group as the second carbon atom.

3. How do the functional groups change in going from starting material to product? Does the starting material have a good leaving group?

The carbon atom of the carboxylic acid has been changed into a methylene group in the product.

$$-\overset{\overset{\textstyle O}{\|}}{C}OH \longrightarrow -CH_2\overset{\overset{\textstyle O}{\|}}{C}CH_3$$

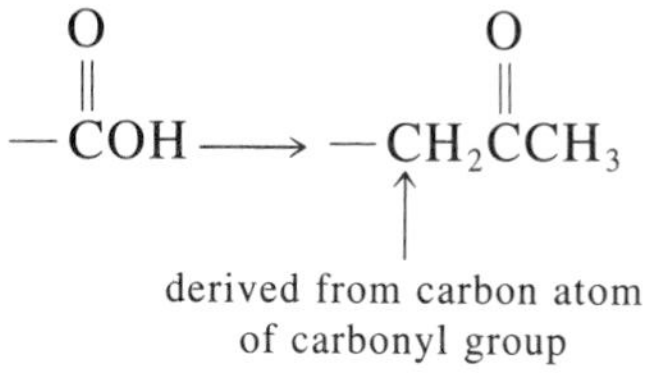

derived from carbon atom
of carbonyl group

The hydroxyl group of the acid can be converted into a leaving group.

4. Is it possible to dissect the structures of the starting material and product to see which bonds must be broken and which formed?

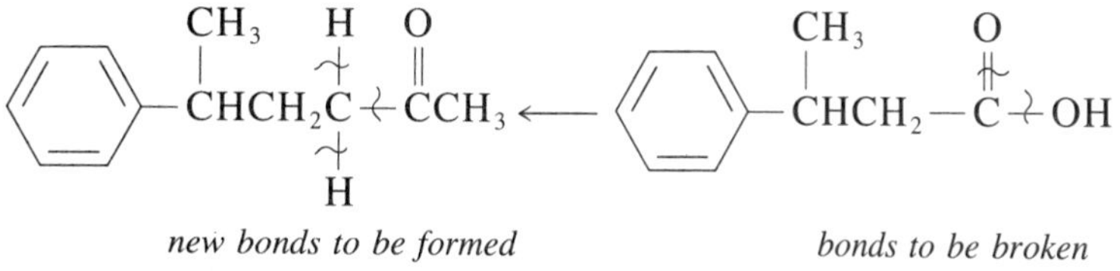

new bonds to be formed   bonds to be broken

5. Do we recognize any part of the product molecule as coming from a good nucleophile or an electrophilic addition?

No, but the replacement of carbon-oxygen bonds in the carboxylic acid with carbon-hydrogen bonds in the product requires a reduction.

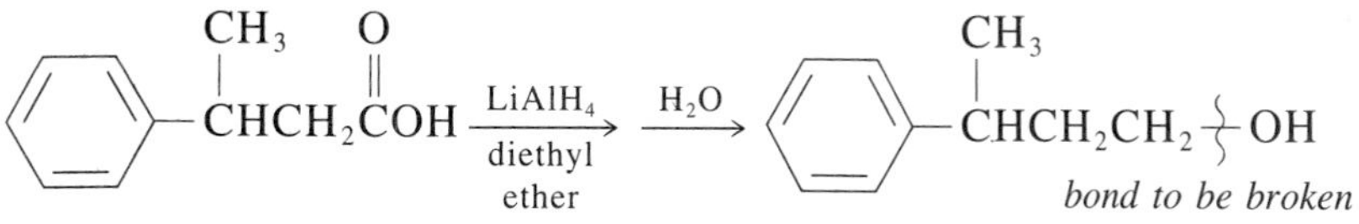

$$\text{Ph}-\overset{\overset{\textstyle CH_3}{|}}{C}HCH_2\overset{\overset{\textstyle O}{\|}}{C}OH \xrightarrow[\text{diethyl}]{\text{LiAlH}_4} \xrightarrow{\text{H}_2\text{O}} \text{Ph}-\overset{\overset{\textstyle CH_3}{|}}{C}HCH_2CH_2 \}OH$$

bond to be broken

6. What type of compound would be a good precursor to the product?

A ketone is obtained by oxidizing a secondary alcohol.

$$\text{Ph}-\overset{\overset{\textstyle CH_3}{|}}{C}HCH_2CH_2 \} \overset{\overset{\textstyle O}{\|}}{C}CH_3 \xleftarrow{oxidation} \text{Ph}-\overset{\overset{\textstyle CH_3}{|}}{C}HCH_2CH_2 \} \overset{\overset{\textstyle OH}{|}}{C}HCH_3$$

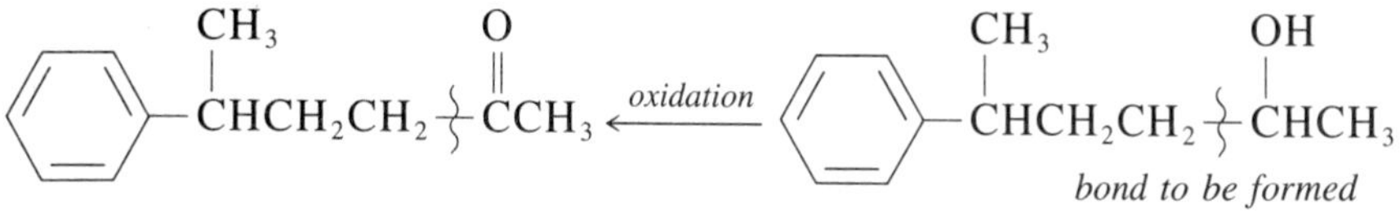

bond to be formed

7. After this last step, do we see how to get from starting material to product? If not, we need to analyze the structure obtained in step 6 by applying questions 5 and 6 to it.

The main bond to be formed in the secondary alcohol can come from a reaction of an organometallic reagent with an aldehyde. This completes the synthesis.

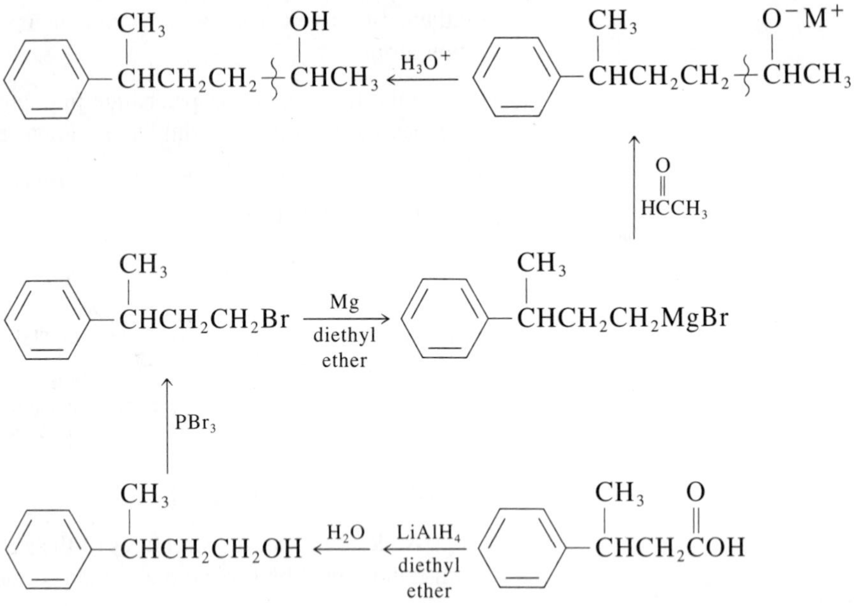
CH3 OH
CHCH2CH2CHCH3
H3O+
CH3 O−M+
CHCH2CH2CHCH3
O
HCCH3
CH3
CHCH2CH2Br
Mg
diethyl ether
CH3
CHCH2CH2MgBr
PBr3
CH3
CHCH2CH2OH
H2O
LiAlH4
diethyl ether
CH3 O
CHCH2COH

PROBLEM 15.10

How would you carry out the following transformations?

(a)

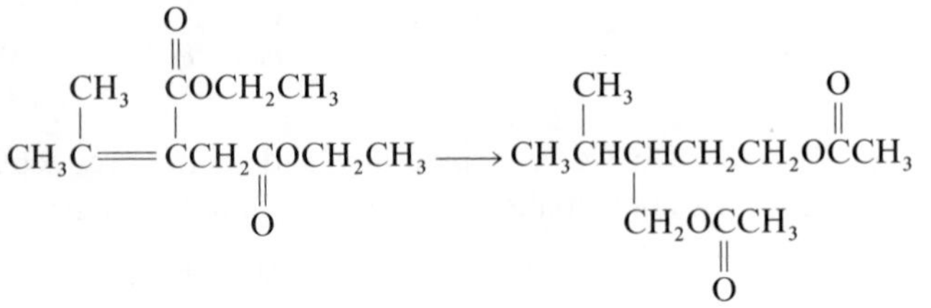

(b)

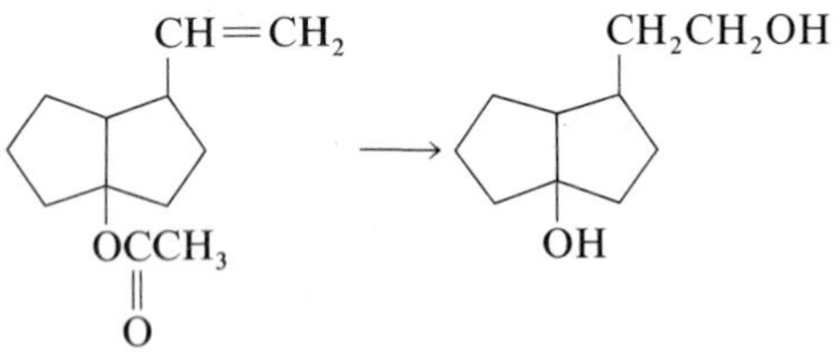

15.3 LACTONES

Lactones are cyclic esters formed when a carboxyl and a hydroxyl group are present within the same molecule. In fact, the formation of a five- or six-membered ring is so highly favored that reactions leading to a compound in which a carboxyl group and a hydroxyl group are separated by three or four carbon atoms usually give a lactone directly as the product. An example of such a reaction is the reduction of

4-oxopentanoic acid with sodium borohydride. Sodium borohydride reduces the ketone function (p. 485) but not the carboxylic acid. The product of the reaction after acidification is the lactone of 4-hydroxypentanoic acid.

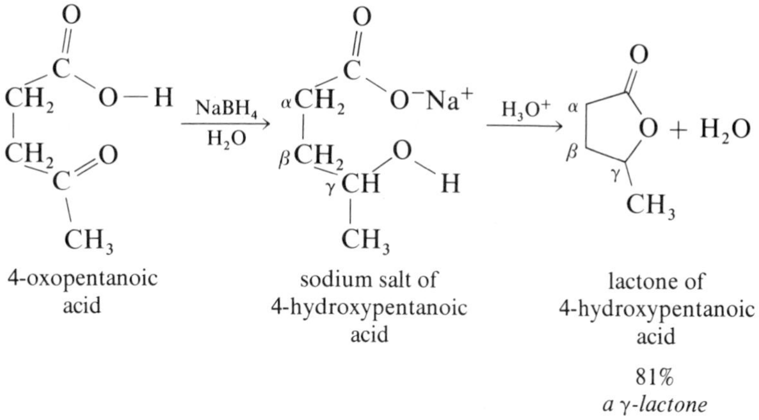
NaBH$_4$
H$_2$O
H$_3$O$^+$
4-oxopentanoic acid
sodium salt of 4-hydroxypentanoic acid
lactone of 4-hydroxypentanoic acid
81%
a γ-lactone

A lactone having a five-membered ring is called a γ-lactone because it is formed when an intramolecular reaction takes place between a carboxyl group and the hydroxyl group on the γ-carbon atom of the acid (p. 481). A six-membered or δ-lactone is formed when ethyl 5-acetoxypentanoate, for example, synthesized by a nucleophilic substitution reaction of ethyl 5-chloropentanoate, is hydrolyzed.

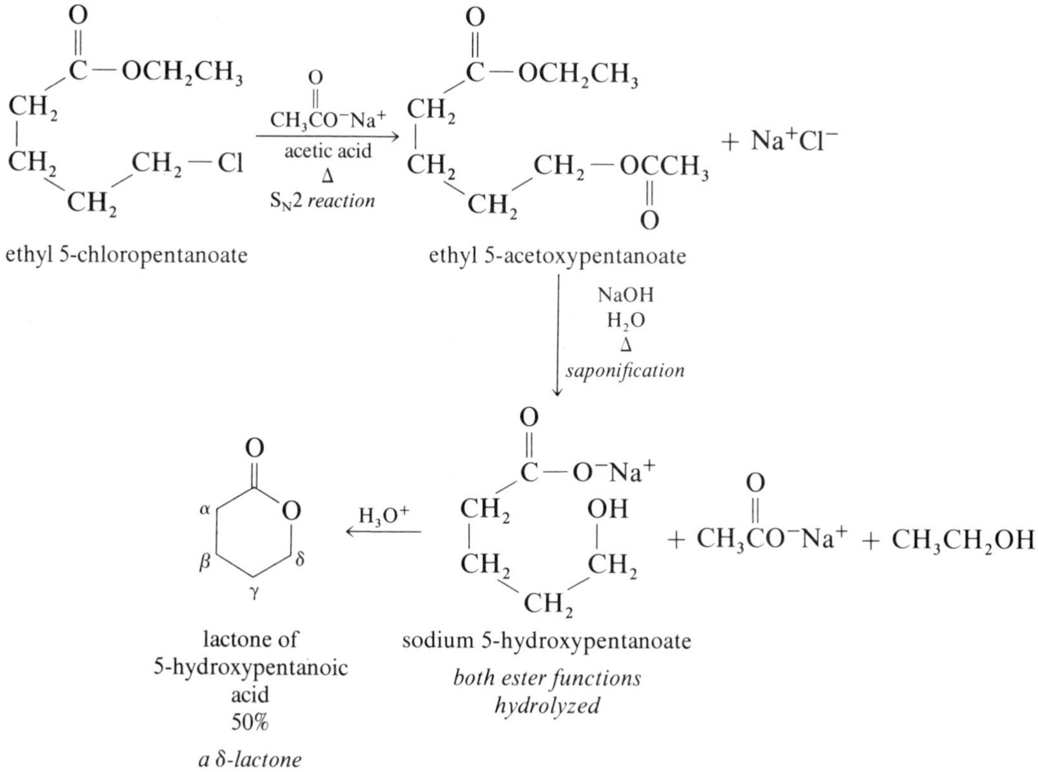
acetic acid
Δ
S_N2 *reaction*
+ Na$^+$Cl$^-$
ethyl 5-chloropentanoate
ethyl 5-acetoxypentanoate
NaOH
H$_2$O
Δ
saponification
H$_3$O$^+$
+ CH$_3$CO$^-$Na$^+$ + CH$_3$CH$_2$OH
lactone of 5-hydroxypentanoic acid
50%
a δ-lactone
sodium 5-hydroxypentanoate
both ester functions hydrolyzed

Note that a lactone does not form as long as the hydroxyacid exists as the acid salt. Only on acidification of the solution does ring closure take place.

Lactones are also produced when cyclic ketones are oxidized with peroxyacids. For example, cyclohexanone is converted to the lactone of 6-hydroxyhexanoic acid by oxidation with peroxybenzoic acid.

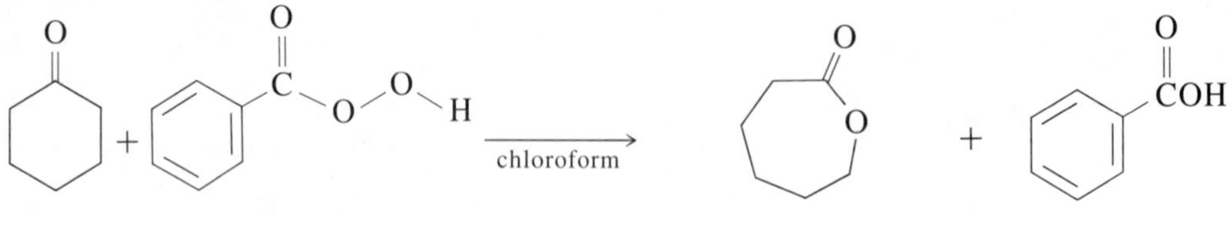
O
O
C
O
O
H
chloroform
O
O
O
COH
+
+

cyclohexanone peroxybenzoic acid lactone of 6-hydroxyhexanoic acid 71% benzoic acid

This reaction, in which a ketone is converted into an ester, is called the **Baeyer-Villiger reaction** after the German chemists who discovered it in 1899. Those chemists' original experiments showed that if the ketone being oxidized has two different alkyl groups on the two sides of the carbonyl group, the more highly substituted carbon atom ends up bonded to the new oxygen atom. For example, if 5-isopropyl-2-methylcyclohexanone is oxidized with monoperoxysulfuric acid, the more highly substituted of the carbon atoms adjacent to the carbonyl group ends up bonded to the oxygen atom that has been inserted into the ring.

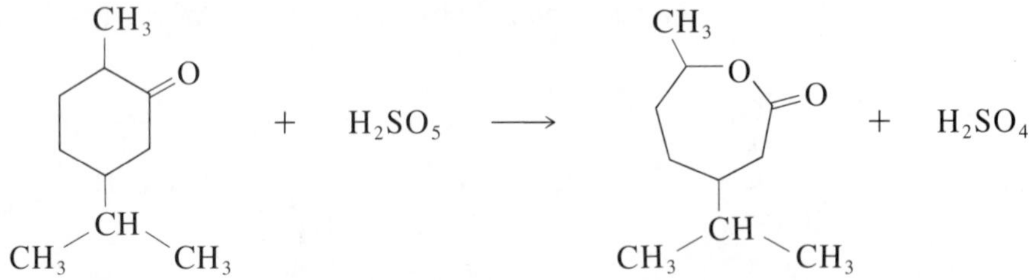
CH3
O
CH
CH3
CH3
+
H2SO5
CH3
O
O
CH
CH3
CH3
+
H2SO4

5-isopropyl-2-methyl-cyclohexanone monoperoxy-sulfuric acid lactone of 6-hydroxy-3-isopropyl-heptanoic acid sulfuric acid

The Baeyer-Villiger reaction involves a rearrangement to oxygen. The reaction is acid-catalyzed and is pictured as starting by protonation of the carbonyl group followed by nucleophilic attack by the peroxyacid.

VISUALIZING THE REACTION

The Baeyer-Villiger reaction

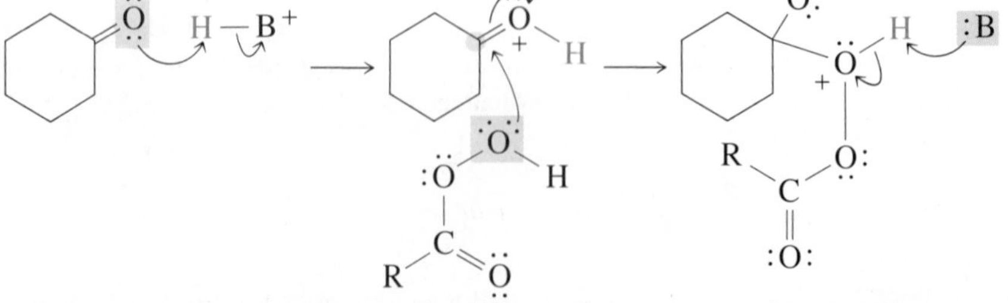
O
H—B+
B:
O
H
O
H
O
C
R
O
H
O
O
H
:B
R
C
O
O
protonation
nucleophilic attack
deprotonation

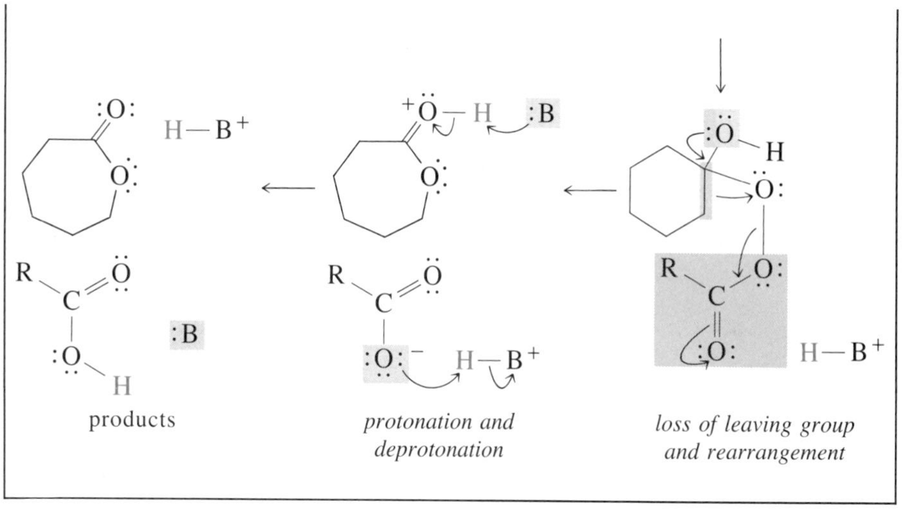

H—B+
:B
H—B+
:B
H—B+
products
protonation and deprotonation
loss of leaving group and rearrangement

The intermediate that results from the addition of the peroxyacid to the carbonyl group has a good leaving group, a carboxylate anion, on an oxygen atom. An oxygen-oxygen bond is weak (~47 kcal/mol) and thus not very difficult to break. As the leaving group departs, partial positive character develops at the oxygen atom, and a 1,2-alkyl shift from a carbon atom to the adjacent oxygen atom takes place. The recovery of the bond energy of the carbonyl group is also a factor in driving the reaction forward. The reaction is similar to a carbocation rearrangement (p. 285), except that in this case the electron-deficient atom is oxygen rather than carbon. The more highly substituted carbon atom which is best able to stabilize a partial positive charge in the transition state, migrates in the rearrangement.

The Baeyer-Villiger reaction is particularly useful for the preparation of lactones having large rings, which are hard to make from intramolecular esterifications of long-chain hydroxyacids. Although 4- and 5-hydroxycarboxylic acids readily form lactones (p. 605), it is difficult to make lactones from hydroxyacids with longer chains. A long-chain hydroxyacid molecule has many conformations in which the hydroxyl and the carboxyl groups are far apart from each other.

Large-ring lactones are found in nature. Some of them as well as the large-ring ketones muscone, isolated from the scent glands of the musk deer, and civetone, from the civet cat, are highly valued by perfume manufacturers. In concentrated form, all of these compounds have rather unpleasant odors. The odors become attractive when the compounds are highly diluted. In perfumes, these substances enhance and fix scents derived from other sources. The structures of four of these large-ring compounds, two ketones and two lactones, are shown below.

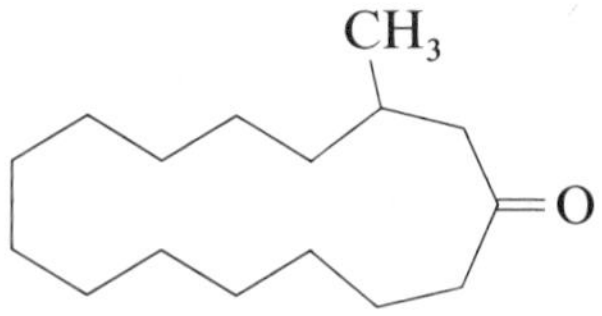

muscone
3-methylcyclopentadecanone
from the musk deer
15-membered ring

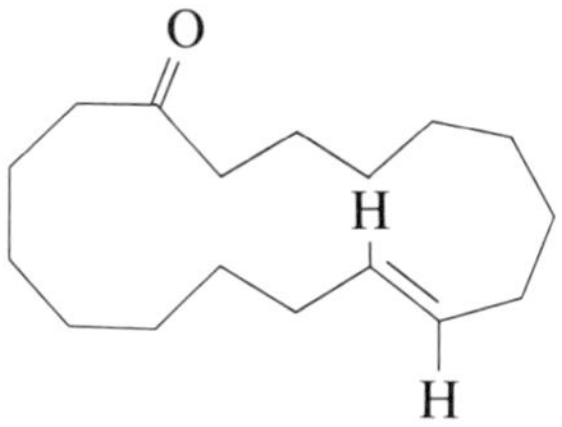

civetone
(*E*)-9-cycloheptadecenone
from the civet cat
17-membered ring

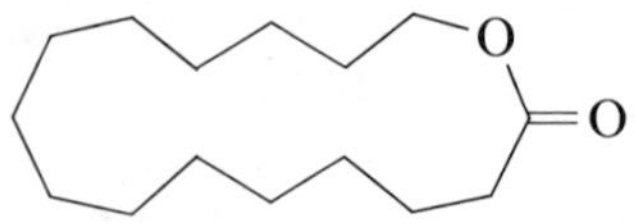

lactone of 15-hydroxypentadecanoic acid from angelica oil, from the roots of *Angelica archangelica*

16-membered ring

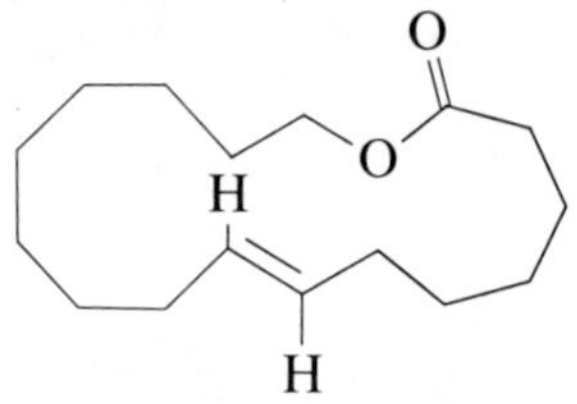

lactone of (*E*)-16-hydroxy-7-hexadecenoic acid musk ambrette from the oil from the seeds of *Hibiscus abelmoschus*

17-membered ring

The lactone from angelica oil has been synthesized from cyclopentadecanone by the Baeyer-Villiger reaction.

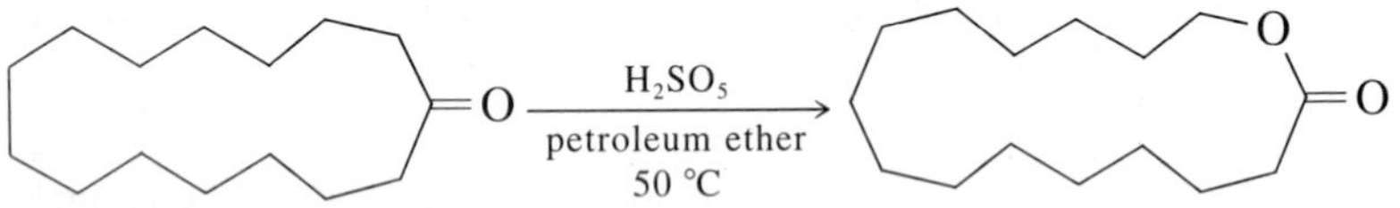

The rings in these lactones and ketones contain 15 to 17 atoms. The odor of cyclic ketones has been found to correlate with ring size; the musky odors desirable for perfumes are characteristic of compounds having 14 to 17 members in the ring. Compounds with larger rings have much less odor—at least, to the human nose.

Many other large-ring lactones that have been isolated from natural sources are interesting because they have antibiotic or antitumor activity. These lactones are known as **macrolides.** Two macrolides are shown below.

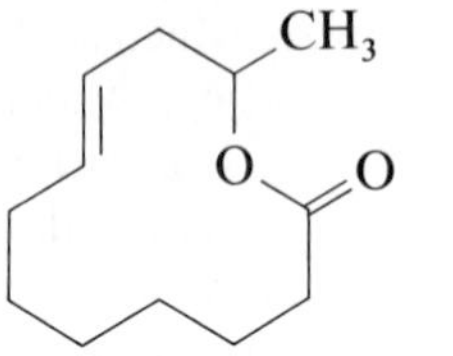

recifeiolide
from the fungus
Cephalosporium recifei

a 12-membered ring lactone natural product

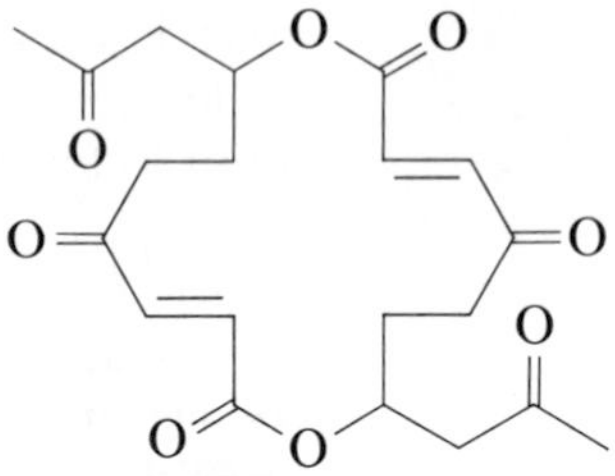

vermiculine
from the microorganism
Penicillium vermiculatum

a 16-membered ring dilactone

PROBLEM 15.11

Complete the following equations.

(a) 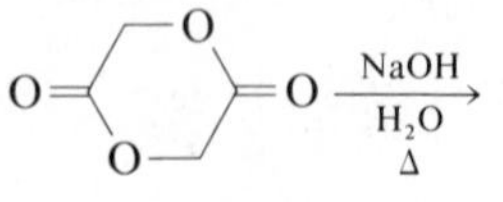

(b) $CH_3\overset{O}{\overset{\|}{C}}CH_2CH_2CH_2\overset{O}{\overset{\|}{C}}OH \xrightarrow[H_2O]{NaBH_4} \xrightarrow{H_3O^+}$

(c) 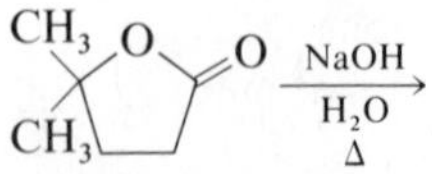

(d) 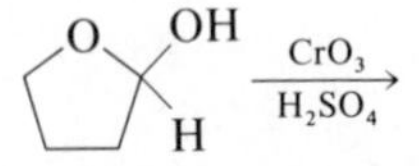

(e) 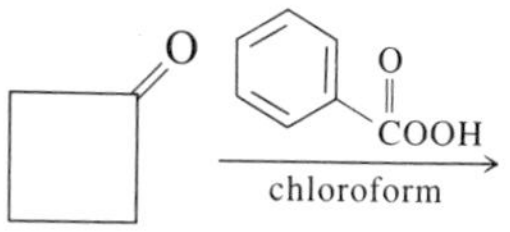

PROBLEM 15.12

An intermediate in the synthesis of the vitamin biotin is prepared by the following sequence of reactions.

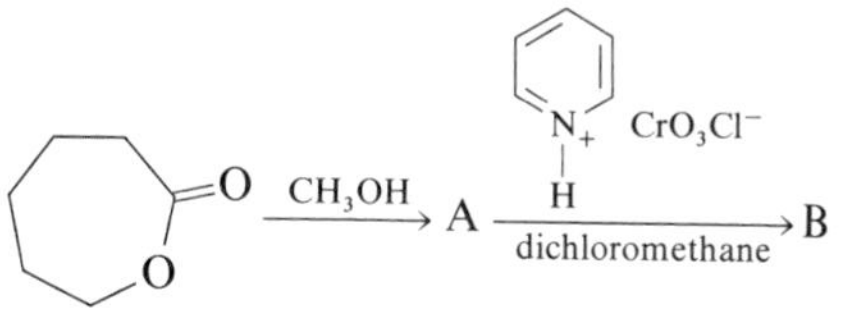
O
O
CH3OH
A
N+
CrO3Cl−
H
dichloromethane
B

Assign structures to Compounds A and B. Biotin has the following structure.

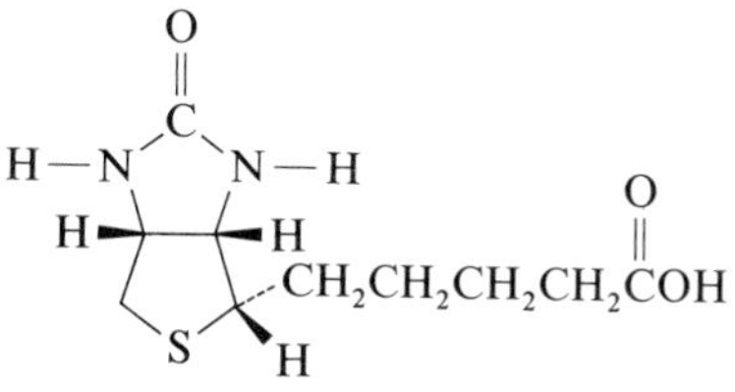
O
C
H—N
N—H
H
H
CH2CH2CH2CH2COH
O
S
H

PROBLEM 15.13

How would you carry out the following transformation, which requires several steps?

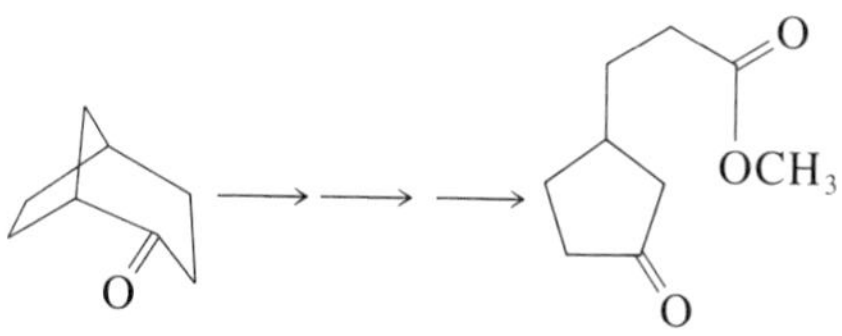
O
O
OCH3
O

(Hint: The reactions shown in Problem 15.12 will be helpful.)

15.4 LIPIDS, FATS, OILS, AND WAXES

Lipids are substances that are found in living organisms and are insoluble in water but soluble in organic solvents. A lipid contains a long hydrocarbon chain as part of its structure but can have a variety of functional groups. The fat-soluble vitamins A (p. 712), D (p. 1218), E (p. 854), and K (p. 854) are classified as lipids, as are cholesterol (p. 427), its esters, and other similar steroids (p. 709). Other lipids, however, are fats in body tissues and the bloodstream, oils isolated from plant sources, such as sunflower seed oil, safflower seed oil, peanut oil, and olive oil, or waxes, which can have either animal or plant sources. All of these compounds are esters.

Fats and **oils** are esters of glycerol (1,2,3-propanetriol) and long-chain carboxylic acids, usually having twelve or more carbon atoms (p. 611). Carboxylic acids became known as fatty acids because they were originally isolated from fats. The acids found in fats are predominantly saturated; those in oils tend to be unsaturated, with one or more double bonds in the chain. The following are the important fatty acids.

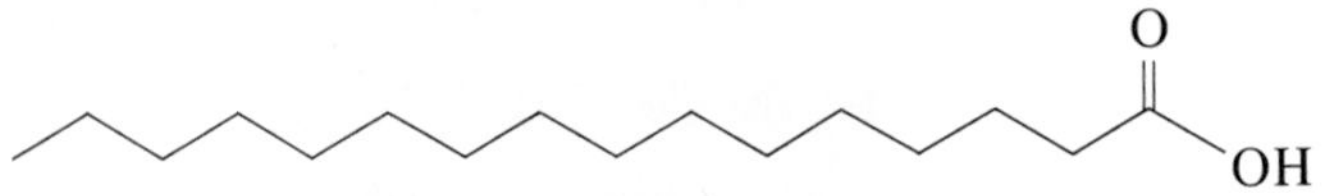
O
OH

palmitic acid
hexadecanoic acid
mp 62.9 °C

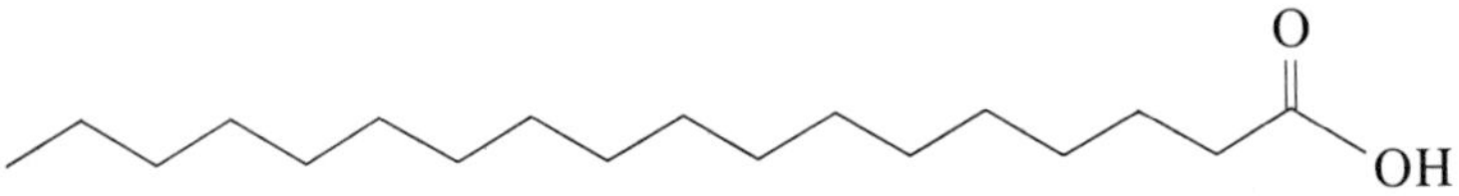
O
OH

stearic acid
octadecanoic acid
mp 69.6 °C

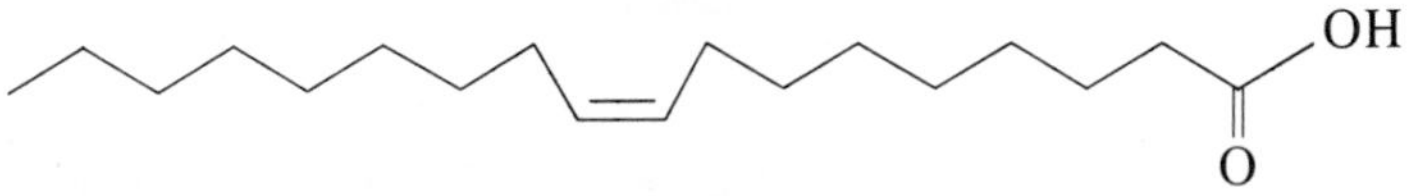
OH
O

oleic acid
(Z)-9-octadecenoic acid
mp 13 °C

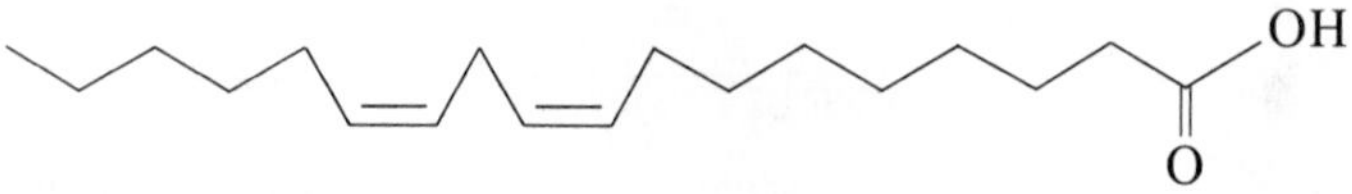
OH
O

linoleic acid
(9Z,12Z)-9,12-octadecadienoic acid
mp −5 °C

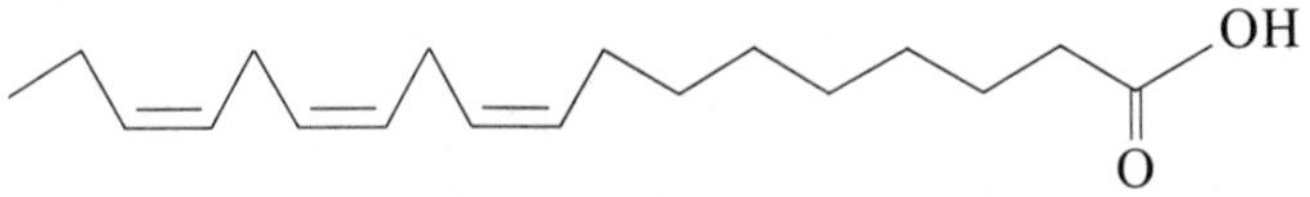
OH
O

linolenic acid
(9Z,12Z,15Z)-9,12,15-octadecatrienoic acid
mp −16 °C

OH
O

arachidonic acid
(5Z,8Z,11Z,14Z)-5,8,11,14-icosatetraenoic acid

The long-chain acids, palmitic acid and stearic acid, occur extensively in nature as the **saturated fatty acids** in solid fats. Note that they are solids at room temperature and at body temperature (37 °C). The melting points of long-chain fatty acids decrease with the introduction of double bonds into the chain. Oleic acid melts below body temperature, but will solidify in a refrigerator. It is the chief fatty acid in olive oil and also in human body fat. As unsaturation increases in linoleic acid, linolenic acid, and arachidonic acid, these **polyunsaturated fatty acids** have lower and lower melting points. The melting point of arachidonic acid is so low that it has not yet been determined accurately. Polyunsaturated fatty acids are found as constituents of oils from the seeds of plants, such as safflower seed oil, sesame oil, and peanut oil. Hydrogenation of the double bonds in oils (p. 306) converts the liquid oils into fats.

The properties of fatty acids are reflected in the properties of their esters. Medical researchers are interested in the esters formed from cholesterol and fatty acids. Esters of cholesterol with saturated fatty acids are solids and have been implicated in the formation of solid deposits on the walls of blood vessels, causing diseases of the heart and arteries. Physicians have recommended the substitution of vegetable oils containing polyunsaturated fatty acids for animal fats and margarines, which are rich in saturated fatty acids.

The glycerol esters of fatty acids are known as **triglycerides.** Two triglycerides, one containing three units of a saturated fatty acid and the other an unsaturated one, are shown below.

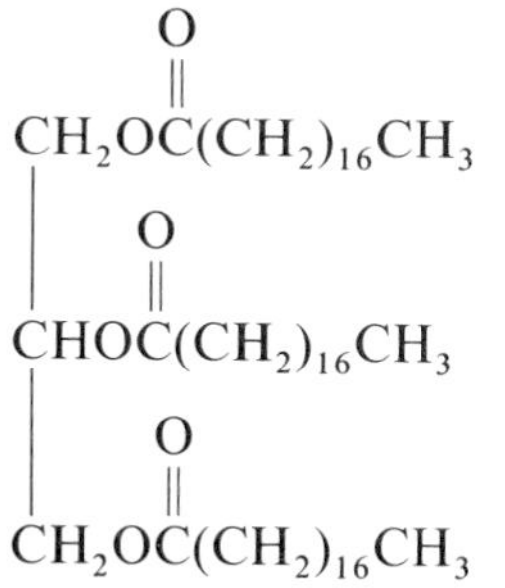

the ester of glycerol with
three molecules of stearic acid
tristearin

a saturated triglyceride
a component of beef fat

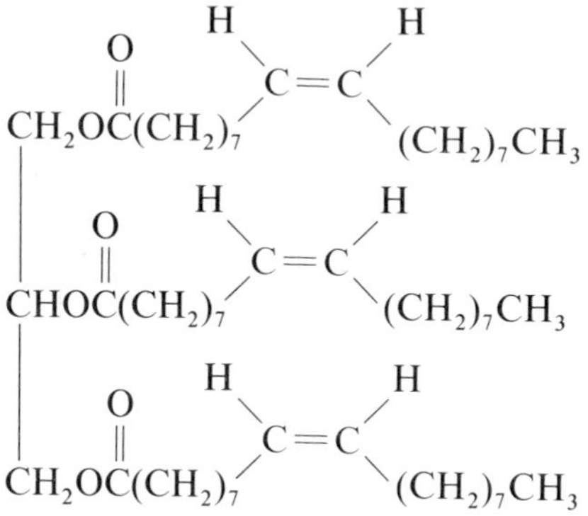

the ester of glycerol with three
molecules of oleic acid
triolein

an unsaturated triglyceride
a component of olive oil

Because they were chosen for the sake of simplicity, the triglycerides shown have the same fatty acid esterifying each hydroxyl group in glycerol. In nature, two or three different fatty acids are often found in a single triglyceride.

As a result of a growing consciousness of the connections between nutrition and heart disease, many people monitor the levels of triglycerides and cholesterol in their blood. High levels of these two components of the blood have been correlated with higher risks of heart and artery disease. Researchers have discovered, however, that some of the cholesterol and blood fat can be tied up with blood proteins in **high-density lipoproteins,** or **HDL.** High-density lipoproteins seem to have some protective value against heart disease, as opposed to **low-density lipoproteins,** or **LDL,** which do not have such properties. Regular exercise is especially helpful in raising the ratio of high-density lipoproteins to low-density lipoproteins in the blood.

Waxes are esters formed from long-chain fatty acids, usually containing 24 to 28 carbon atoms, and long-chain (16 to 36 carbon atoms) primary alcohols or alcohols of the steroid series. Waxes are solids used for lubricating and protecting surfaces. In plants, for example, leaves and fruits are coated with a waterproof layer of such waxes.

$$CH_3(CH_2)_{15}OC(CH_2)_{16}CH_3$$

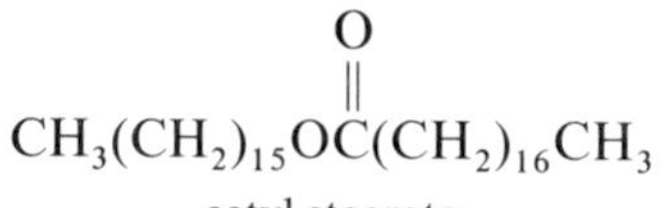

cetyl stearate

common in bacteria

$$CH_3(CH_2)_{15}OC(CH_2)_{14}CH_3$$

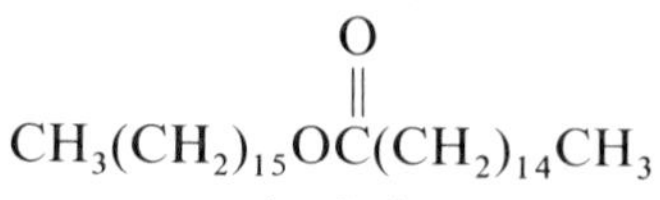

cetyl palmitate

chief wax ester of spermaceti
from the sperm whale

$$CH_3(CH_2)_{29}OC(CH_2)_{14}CH_3$$

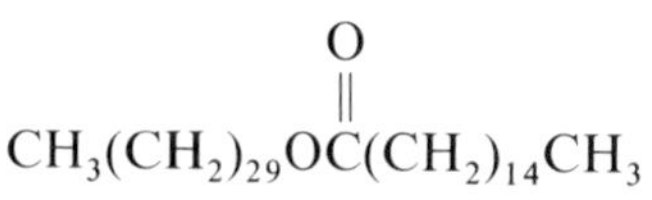

triacontyl palmitate

a wax ester typical of those
found in beeswax

PROBLEM 15.14

9,12-Octadecadiynoic acid (p. 558) is a synthetic precursor to one of the polyunsaturated fatty acids. Which one? How can 9,12-octadecadiynoic acid be transformed into the natural product?

15.5 SURFACE-ACTIVE COMPOUNDS. SOAPS

The salt of a carboxylic acid is usually more soluble in water than the parent acid itself, but an interesting phenomenon arises when the hydrocarbon portion of the acid is very large in comparison with the carboxyl group. The ionic portion of the molecule interacts well with water and dissolves in it, but the rest of the long molecule will not go into solution. The hydrocarbon portions of adjacent molecules are more attracted to each other by van der Waals forces than they are to the polar water molecules. They are in fact hydrophobic, or water-repelling, in their behavior. A long-chain acid salt thus has two domains: a **hydrophilic** head, the ionic carboxylate group that is soluble in water, and a **hydrophobic** tail, the long hydrocarbon portion that is repelled by water molecules and attracted to other hydrocarbon residues.

The structure of such a compound allows for a particular orientation of its molecules at the surface of water; the heads are in water and the hydrocarbon tails stick up into the air. The concentration of molecules at the surface of the water has the effect of lowering the surface tension of water. Compounds that behave in this way are said to be **surface-active compounds** or **surfactants.** Soaps are one category of surface-active compound. All good surfactants have structures with a hydrophilic head and a long hydrophobic tail. The property of surfactancy is exhibited by acids with twelve or more carbon atoms in the hydrocarbon portion of the molecule.

At a certain critical concentration of the surfactant, the surface layer is broken up into smaller units, clusters of ions called micelles. **Micelles** are particles in which the long hydrocarbon tails, repelled by the water molecules and attracted to each other, make up the interior and the negatively charged heads coat the surface and interact with the surrounding water molecules and positive ions. Figure 15.1 on the next page depicts a spherical micelle. Such micelles are formed when sodium octadecanoate (sodium stearate), which is a common ingredient of soap, is put into water.

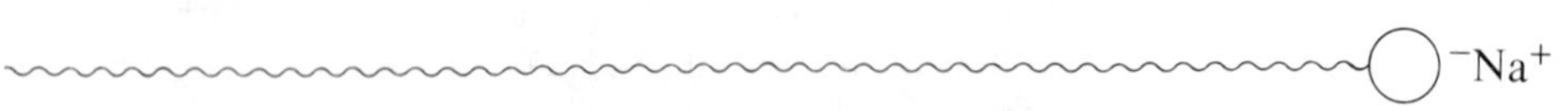
⁻Na⁺

$$CH_3CH_2CH_2CH_2CH_2CH_2CH_2CH_2CH_2CH_2CH_2CH_2CH_2CH_2CH_2CH_2CH_2\overset{O}{\overset{\|}{C}}O^-Na^+$$

← hydrophobic tail → ionic head

sodium octadecanoate
sodium stearate

a soap

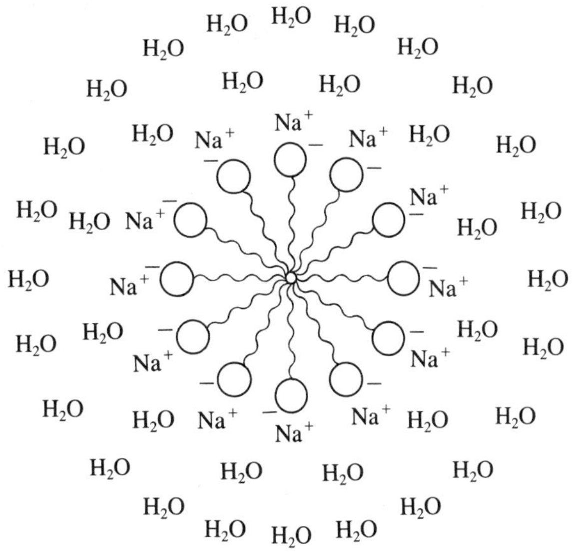
H_2O
Na^+

FIGURE 15.1 Cross section of a spherical micelle.

The opalescent appearance of a soap solution, which is colloidal, is evidence that the particles present are larger than individual molecules. The particles in a colloidal solution are large enough that light is scattered instead of being transmitted as is the case with a true solution, which appears clear to the eye.

Grease, in its chemical composition, resembles the hydrocarbon tails in the micelle. Rubbing a greasy spot with a soap solution causes the grease to break up into small enough particles to be trapped inside the micelles. The particles are held in solution by the hydrocarbon portion of the soap but are kept in colloidal suspension by the interaction of the ionic surface of the micelles with the surrounding water. The grease is said to be **emulsified,** held in suspension in a medium in which it would not normally be soluble.

Soaps are usually the sodium salts of long-chain carboxylic acids. Ordinary soaps have disadvantages in hard water. Hard water has dissolved calcium and magnesium ions in it, so when soap is used in such water, the calcium and magnesium salts of the carboxylic acids in the soap precipitate out. This precipitate is the scum seen in hard water and causes the ring around the bathtub. A great variety of surfactants with more soluble calcium and magnesium salts have been developed. An example of a simple one prepared from nut oils is sodium dodecanyl sulfate (sodium lauryl sulfate), the sodium salt of the ester of 1-dodecanol with sulfuric acid.

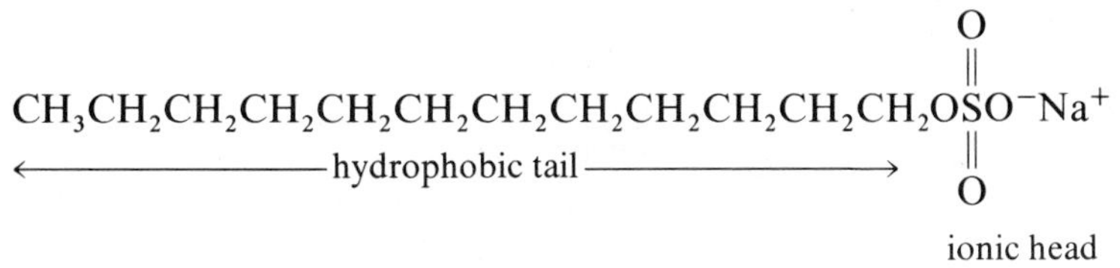
$CH_3CH_2CH_2CH_2CH_2CH_2CH_2CH_2CH_2CH_2CH_2CH_2OSO^-Na^+$
O
O
hydrophobic tail
ionic head

sodium dodecanyl sulfate
sodium lauryl sulfate

a surfactant used in commercial detergents

In sodium dodecanyl sulfate, sulfuric acid instead of a carboxylic acid group provides the ionic portion of the molecule.

Molecules that have hydrophilic heads and long-chain hydrophobic tails made up of the hydrocarbon portions of saturated and unsaturated fatty acids also play an important part in the structure of cell membranes, which provide the boundaries between the outer and inner worlds of the cell. The functioning of these molecules in membranes will be described in Section 22.7B.

PROBLEM 15.15

Which of the following compounds are surfactants?

(a) 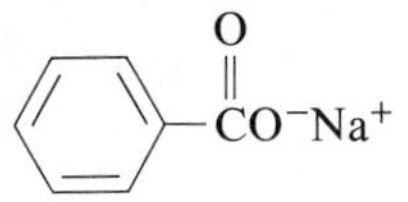

(b) $CH_3(CH_2)_{12}CH_2O\overset{O}{\underset{O}{S}}O^-Na^+$

(c) $CH_3(CH_2)_{10}CH_2\overset{+}{N}(CH_3)_3\ Cl^-$

(d) 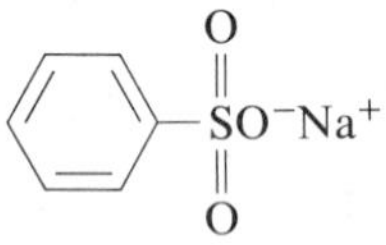

(e) 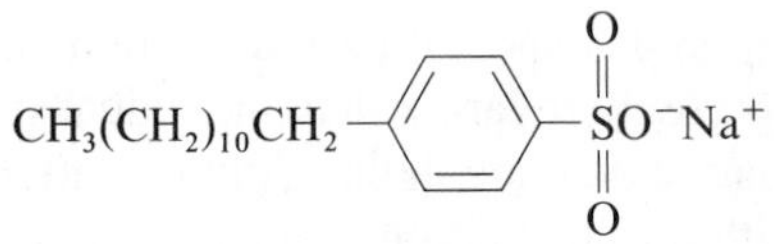

(f) $CH_3(CH_2)_{15}CH_2\overset{O}{\overset{\|}{C}}O^-\ H\overset{+}{N}(CH_2CH_2OH)_3$

(g) $Na^+\ {}^-O\overset{O}{\overset{\|}{C}}(CH_2)_{12}\overset{O}{\overset{\|}{C}}O^-\ Na^+$

SUMMARY

Derivatives of carboxylic acids resemble aldehydes and ketones in their reactions with nucleophiles such as metal hydrides and organometallic reagents. Acids and esters are reduced by lithium aluminum hydride to primary alcohols. Nitriles and amides are reduced by that reagent to amines. Esters and nitriles treated with 1 equivalent of diisobutylaluminum hydride give intermediates that hydrolyze to form aldehydes (Table 15.1).

Grignard reagents are protonated by carboxylic acids or by amides with acidic hydrogen atoms on the nitrogen. The organomagnesium compound is not nucleophilic enough to react with the resulting anions. Two equivalents of a Grignard reagent react with an ester to give an alcohol, usually a tertiary one.

Organolithium reagents also deprotonate carboxylic acids. The acid salt that results is less ionic than the magnesium salt, and an organolithium reagent is more nucleophilic than a Grignard reagent. Therefore, addition of the organolithium reagent to the carbonyl group takes place, giving a dialkoxide ion, which is converted by protonation to a ketone (Table 15.1).

Many acid derivatives are found in living organisms and have important biological properties. Cyclic esters, called lactones, are among these. Reactions that lead to a hydroxyl group γ or δ to a carboxylic acid group in the same molecule are used

TABLE 15.1 More Reactions of Acids and Acid Derivatives*

Starting Material	Reagent	Intermediate	Reagent in Second Step	Product
		Organometallic Reagents		
RC(=O)OR′	R″MgX	$R{-}C(O^-Mg^{2+}X^-)(R'')_{}{-}OR'$ $R{-}C(O^-Mg^{2+}X^-)(R''){-}R''$	H_3O^+	R—C(OH)(R″)—R″
RC(=O)OH	R′Li	$RC(=O)O^-Li^+$ $R{-}C(O^-Li^+)(O^-Li^+){-}R'$	H_2O	R—C(OH)(OH)—R′ ⇄ RC(=O)R′
		Metal Hydrides		
RC(=O)OH	$LiAlH_4$	R—C(O—M)(H)—H	H_3O^+	RCH_2OH
RC(=O)NR′$_2$	$LiAlH_4$	$RC(H){=}\overset{+}{N}R'_2$	$LiAlH_4$	$RCH_2NR'_2$
RC(=O)OR′	$LiAlH_4$	$R{-}C(O{-}M)(H){-}H + R'O^-$	H_3O^+	$RCH_2OH + R'OH$
RC(=O)OR′	$[(CH_3)_2CHCH_2]_2AlH$ excess	$R{-}C(O{-}M)(H){-}H + R'O^-$	H_3O^+	$RCH_2OH + R'OH$
	$[(CH_3)_2CHCH_2]_2AlH$ 1 equivalent	R—C(O—M)(H)—OR′	H_2O	RC(=O)H + R′OH
RC≡N	$LiAlH_4$	R—C(H)(H)—N(M)—M	H_2O	RCH_2NH_2
	$[(CH_3)_2CHCH_2]_2AlH$ 1 equivalent	RC(H)=N—M	H_3O^+	RC(H)=O

*See also Table 14.2, p. 585.

TABLE 15.2 Some Reactions Used in the Preparation of Lactones

Starting Material	Reagent	Intermediate	Reagent in Second Step	Product
Through Hydroxyacid Intermediates				
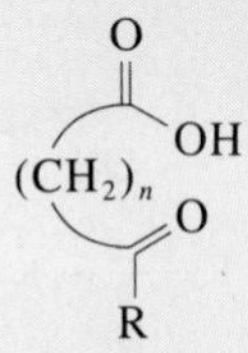 n = 2 or 3	$NaBH_4$, H_2O	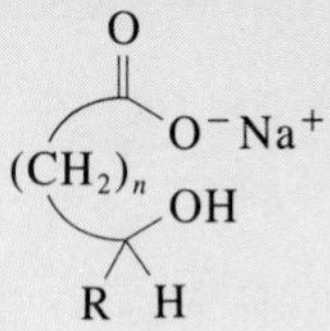	H_3O^+	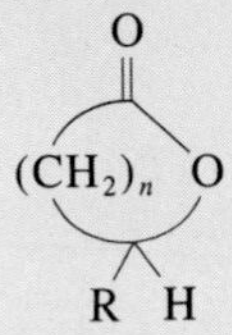
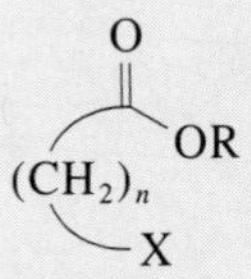 n = 3 or 4	$R'\overset{O}{\overset{\|}{C}}O^-$	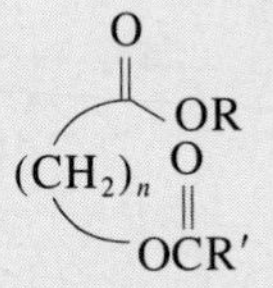	OH^-, H_2O, Δ then H_3O^+	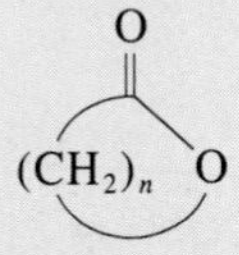
Peroxyacid Oxidation of Cyclic Ketones: Baeyer-Villiger Reaction				
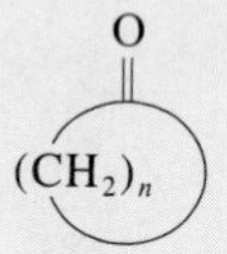 n can be a large number	$R\overset{O}{\overset{\|}{C}}OOH$ or H_2SO_5	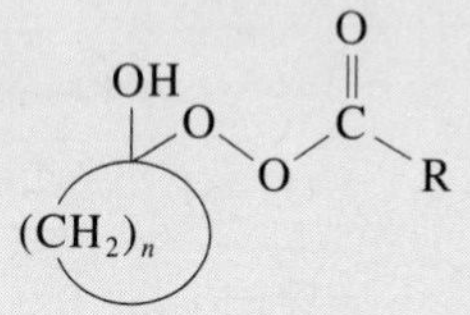		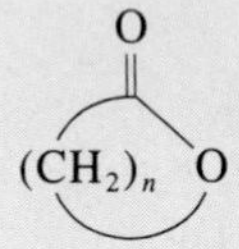

to prepare five- or six-membered ring lactones. Lactones are also formed in the Baeyer-Villiger reaction by peroxyacid oxidation of cyclic ketones (Table 15.2). Some large-ring lactones, called macrolides, are antibiotics.

Lipids are water-insoluble but soluble in organic solvents. They are isolated from living organisms. The esters of long-chain saturated acids, called fatty acids, are constituents of solid fats. Polyunsaturated fatty acids are found in oils. Waxes are the esters formed from long-chain acids and large alcohols.

The salts of long-chain fatty acids are surfactants. They form colloidal solutions consisting of micelles, which have the property of emulsifying and solubilizing grease.

ADDITIONAL PROBLEMS

15.16 Give structural formulas for all the organic intermediates and products indicated by letters in the following equations.

(a) 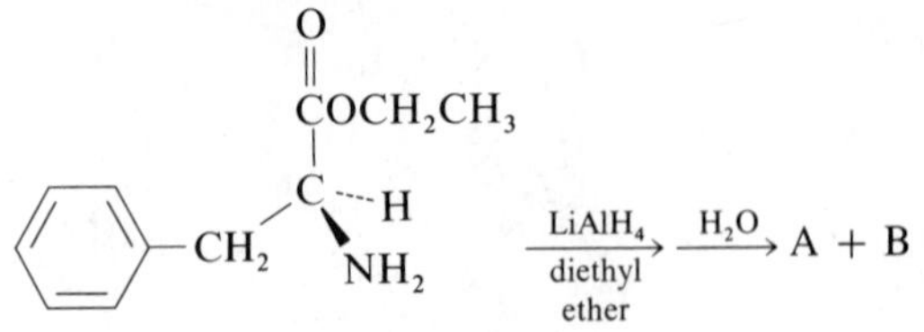
COCH2CH3
H
CH2
NH2
LiAlH4
diethyl ether
H2O
A + B

(b) 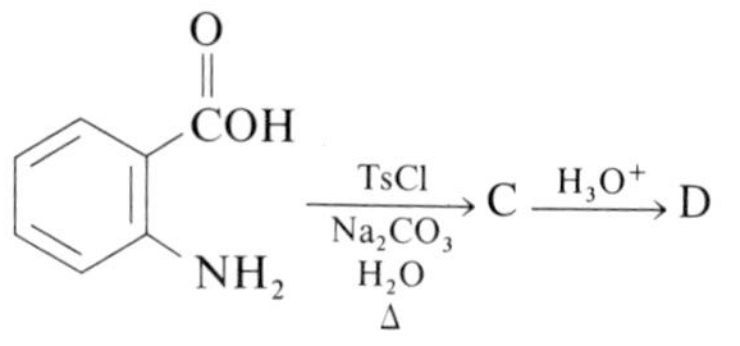

(c) 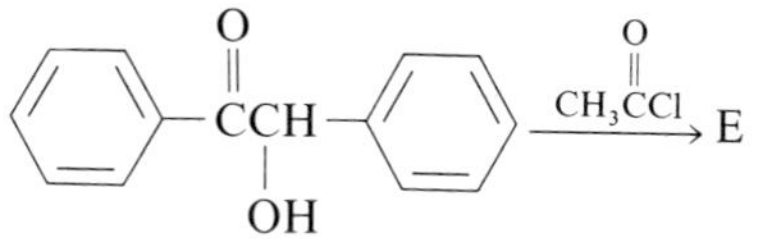

(d) 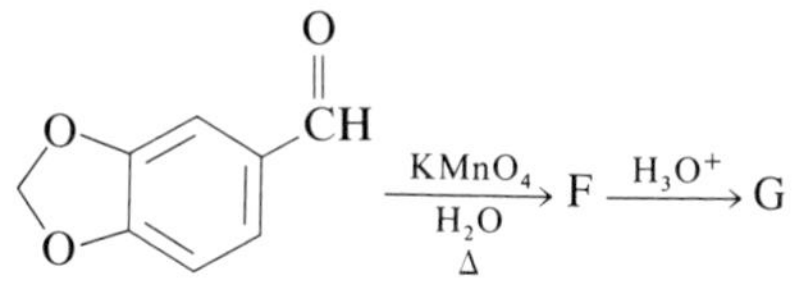

(e) $CH_3(CH_2)_{14}CH_2OH \xrightarrow[\text{pyridine}]{\text{TsCl}} H$

(f) 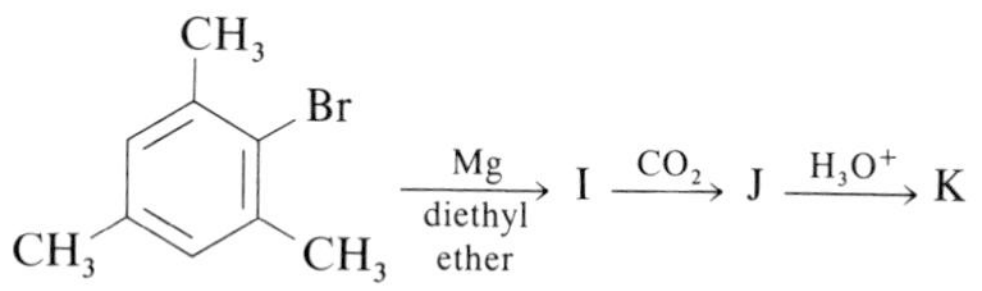

(g) $BrCH_2(CH_2)_8CH_2OH \xrightarrow[\substack{H_2O \\ \text{acetic acid}}]{CrO_3} L$

(h) 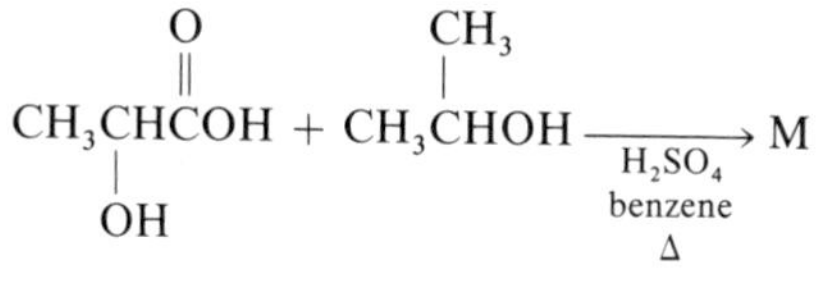

(i) 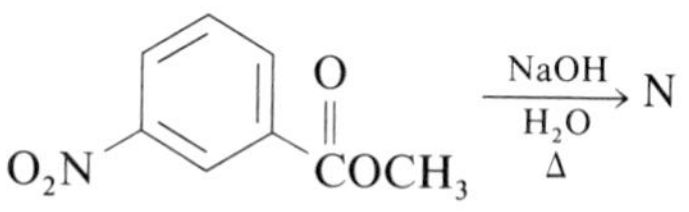

(j) 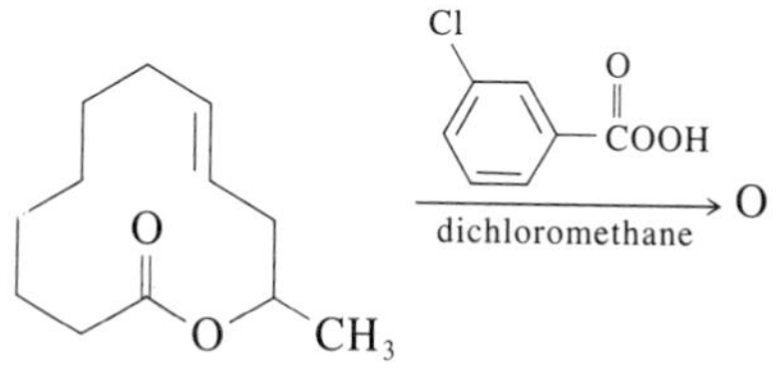

(k) 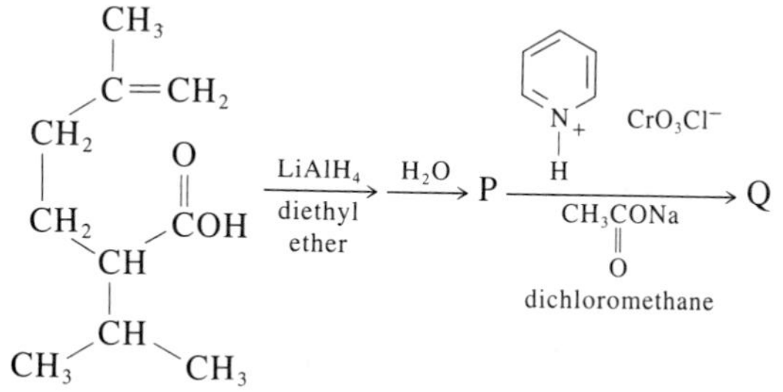

(l) 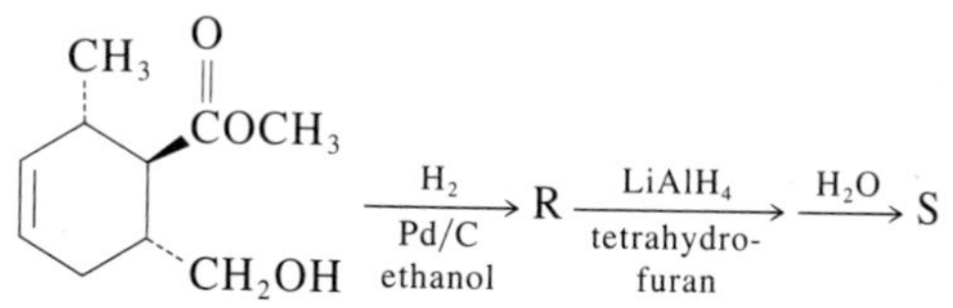

(m) 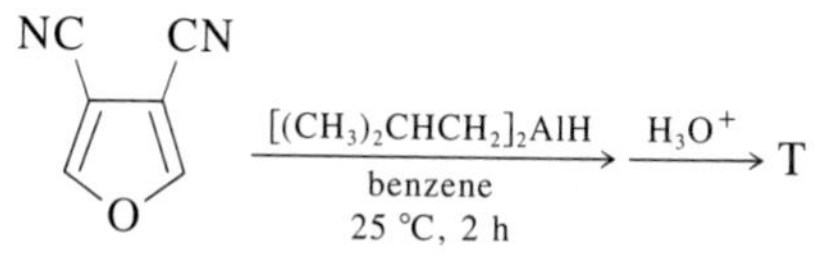

(n) 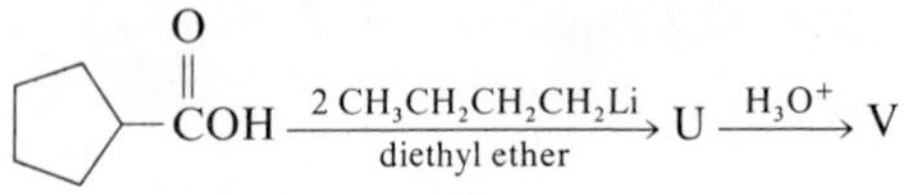

(o) 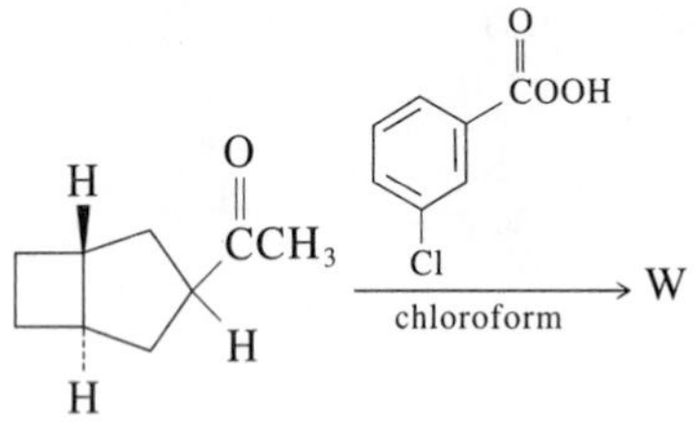

15.17 Give structural formulas for the intermediates and products indicated by letters in the following equations.

(a) 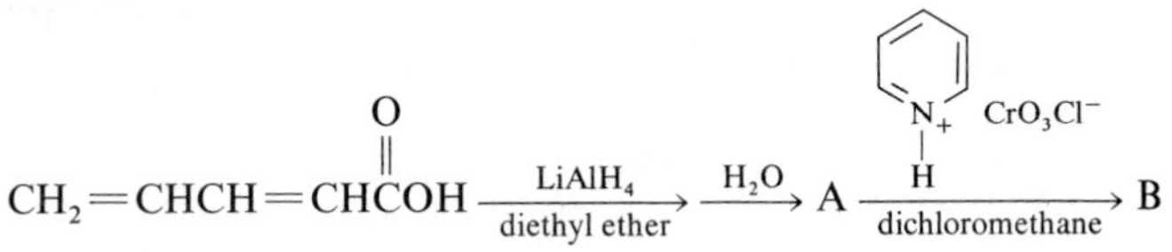

(b) $CH_3CH_2CH_2CH_2\overset{O}{\overset{\|}{C}}OCH_2CH_3 \xrightarrow[\text{diethyl ether}]{2\ CH_3CH_2MgBr} C \xrightarrow[H_2O]{NH_4Cl} D$

(c) $CH_3CH_2\overset{O}{\overset{\|}{C}}NHCH_2CH_3 \xrightarrow[\text{diethyl ether}]{LiAlH_4} \xrightarrow{H_2O} E$

(d) $CH_3CH{=}CHCH_2C{\equiv}CCH_2CH_2\overset{O}{\overset{\|}{C}}OH \xrightarrow[\substack{NH_3(liq)\\ \text{ethanol}}]{Li} \xrightarrow{H_3O^+} F$

(e) 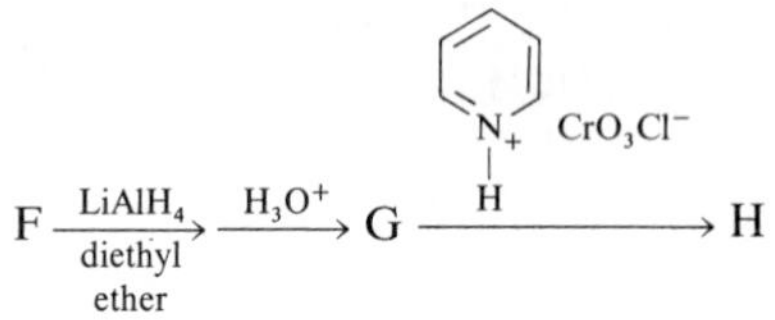

(f) 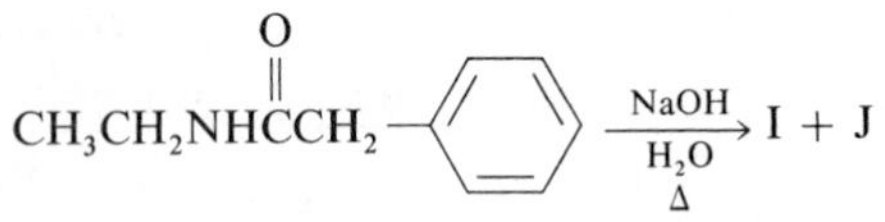

(g) $CH_3\overset{O}{\overset{\|}{C}}CH_2O\overset{O}{\overset{\|}{C}}H + CH_3OH \xrightarrow[\Delta]{KOH} K + L$

(h) $CH_3\underset{}{\overset{CH_3}{\overset{|}{C}H}}CH_2\overset{O}{\overset{\|}{C}}O^-Na^+ + CH_3CH_2CH_2\overset{O}{\overset{\|}{C}}Cl \longrightarrow M$

(i) 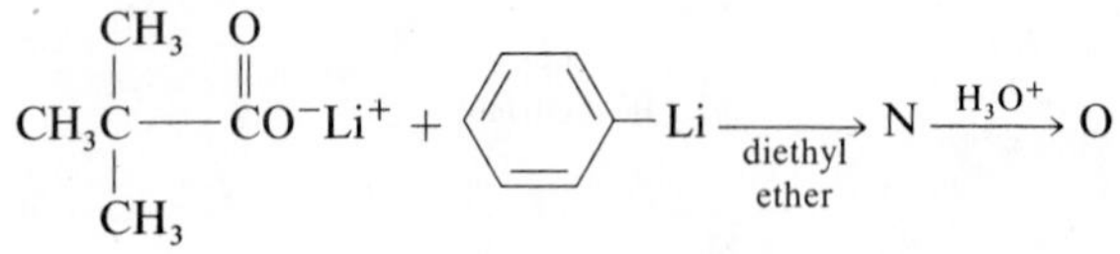

(j) 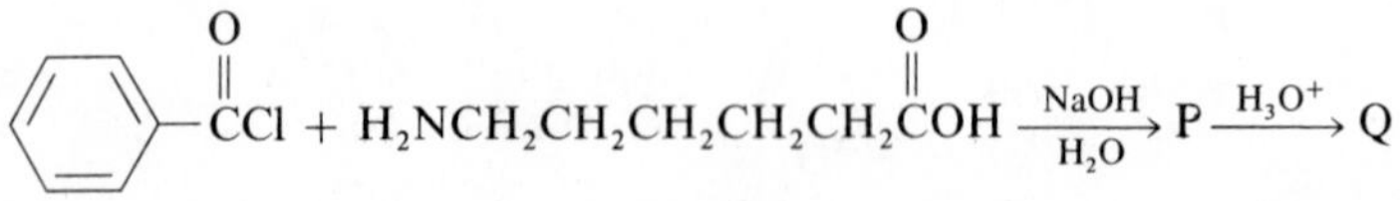

15.18 Polycyclic hydrocarbon derivatives are used as medication for viral infections. A compound that inactivates viruses was synthesized by the following sequence of reactions. What is its structure?

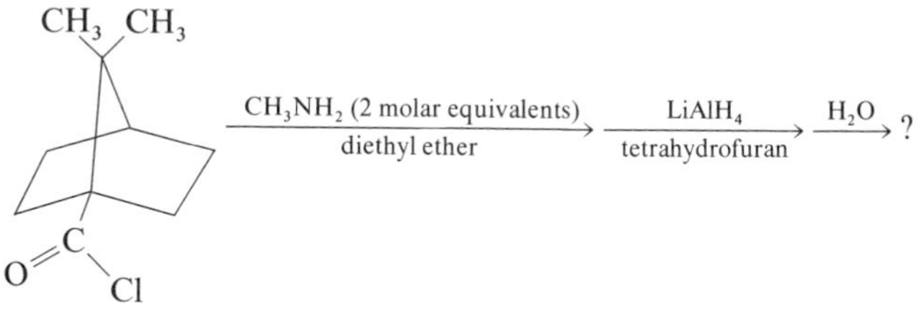
CH3 CH3
O
C
Cl
CH3NH2 (2 molar equivalents)
diethyl ether
LiAlH4
tetrahydrofuran
H2O
?

15.19 Suggest synthetic routes for the following transformations, showing all reagents and intermediate products.

(a)

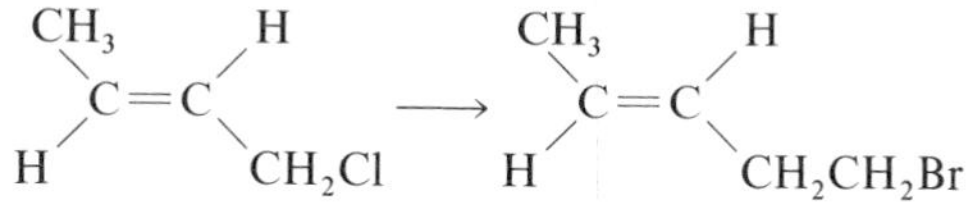

(b)

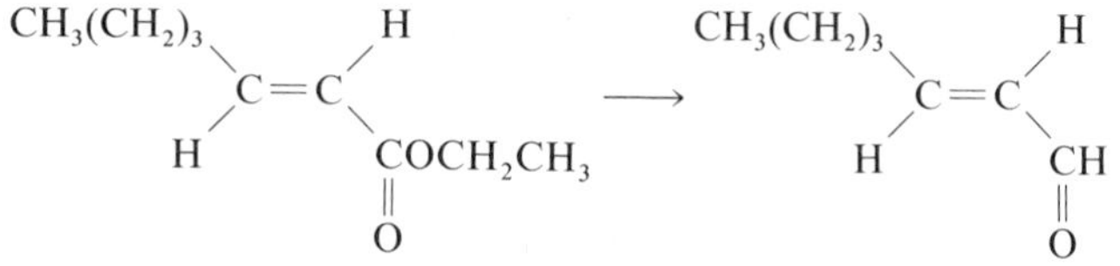

(c)

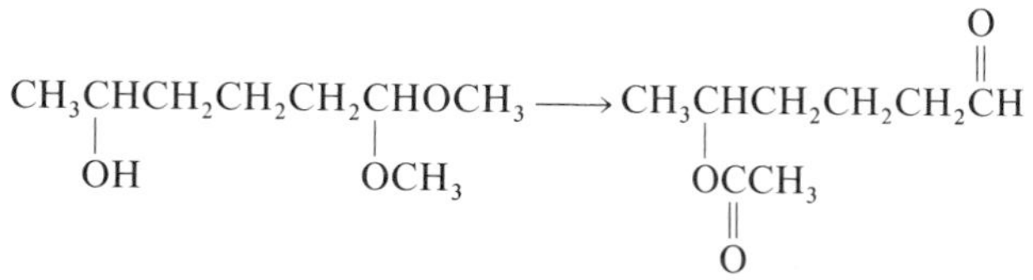

(d)

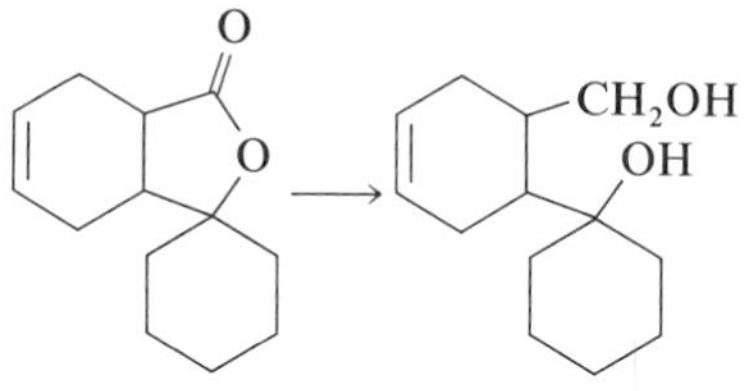

(e)

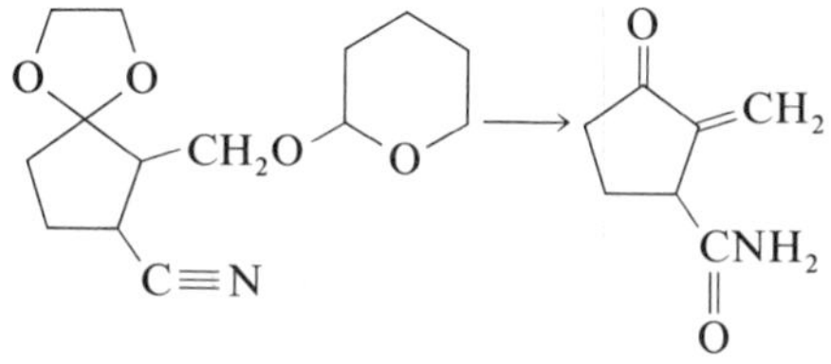

(f)

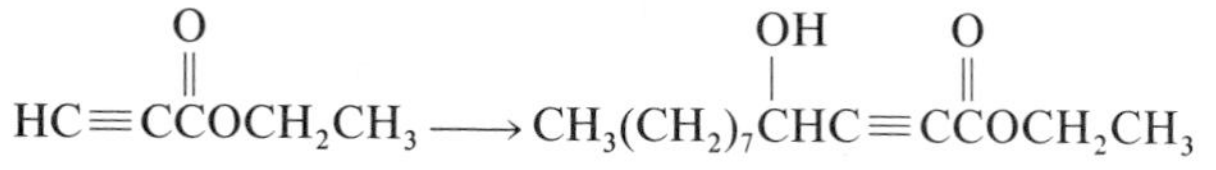

(g)

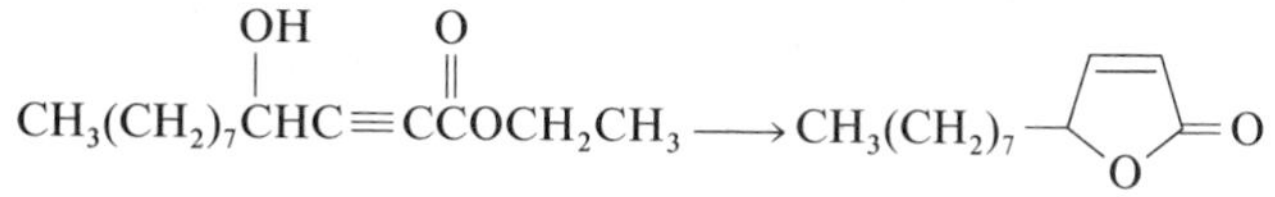

(h)

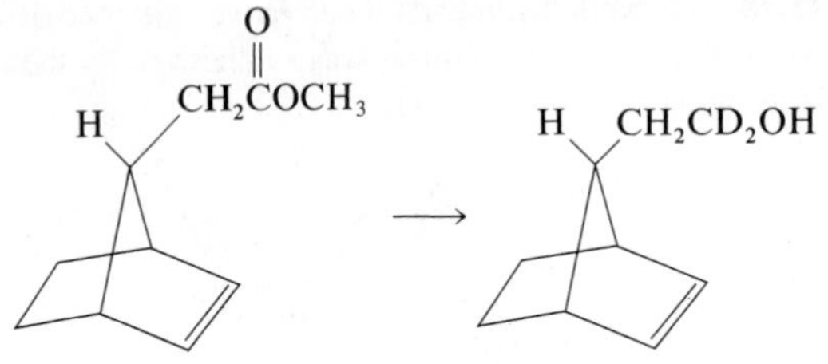

(i)

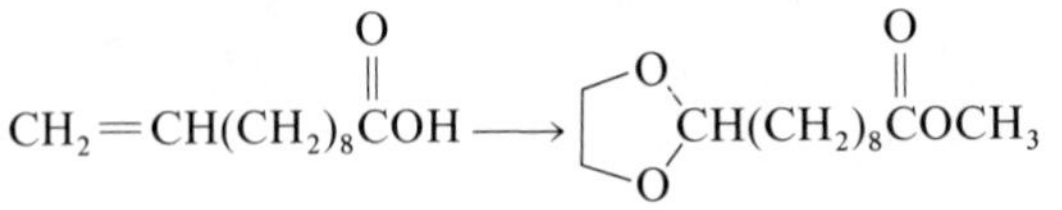

(j)

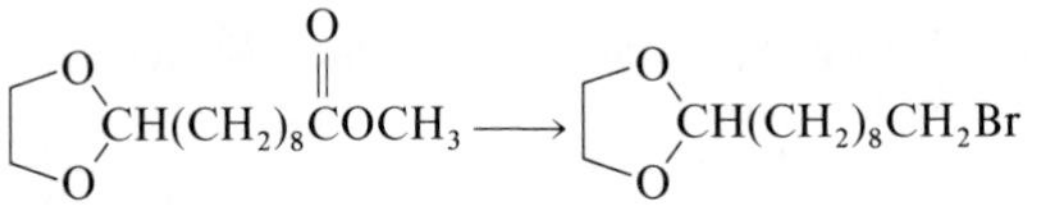

(k)

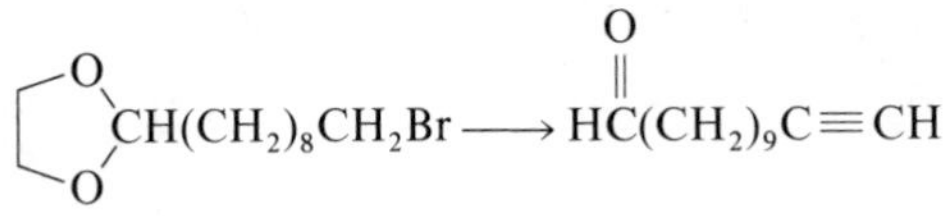

(l)

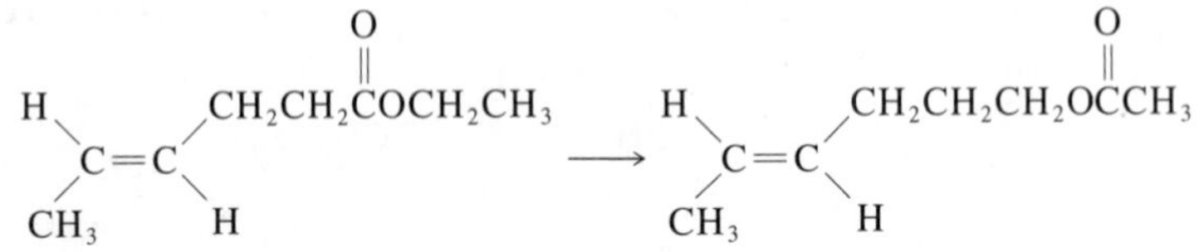

15.20 Suggest a mechanism for the following sequence of reactions.

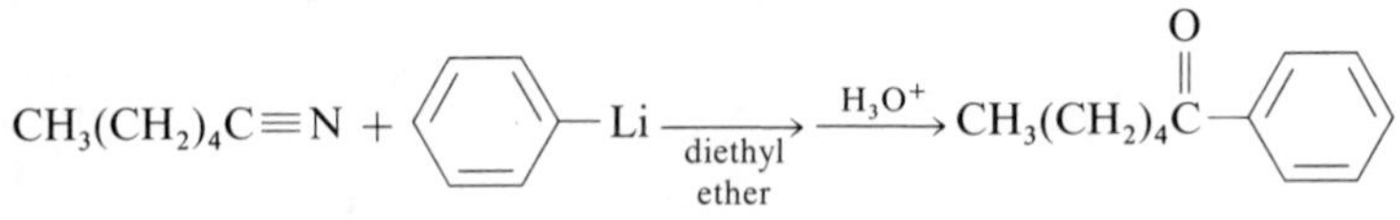

15.21 The following reaction is observed. Offer an explanation for the selectivity of the reduction.

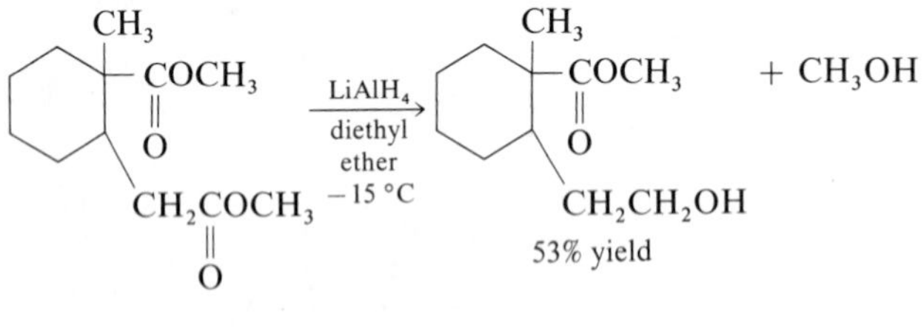

15.22 A general reaction (the **Reformatsky reaction**) for the preparation of 3-hydroxycarboxylic acid esters involves the reaction of an aldehyde or ketone with a 2-bromoester in the presence of zinc. A typical example is shown below.

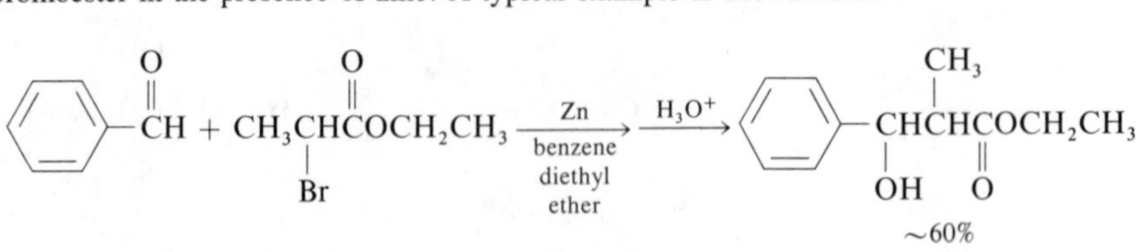

Suggest a mechanism for this reaction.

15.23 The reactions used in the determination of the structure of musk ambrette (p. 608) serve as a review of the chemistry of alkenes, alcohols, aldehydes, and acid derivatives. Some of the reactions are given in the form of the incomplete equations below. Complete the equations, giving structures for all compounds indicated by Roman numerals.

(a) $\text{musk ambrette} \xrightarrow[\text{H}_2\text{O},\ \Delta]{\text{NaOH}} \xrightarrow{\text{H}_3\text{O}^+} \text{I}$ (b) $\text{I} \xrightarrow[\text{Pt, ethanol}]{\text{H}_2} \text{II} \xrightarrow{(\text{CH}_3\overset{\text{O}}{\overset{\|}{\text{C}}})_2\text{O}} \text{III } (\text{C}_{18}\text{H}_{34}\text{O}_4)$

(c) $\text{II} \xrightarrow[\text{H}_2\text{SO}_4,\ \text{acetic acid}]{\text{CrO}_3} \text{IV } (\text{C}_{16}\text{H}_{30}\text{O}_4)$ (d) $\text{I} \xrightarrow{\text{O}_3} \xrightarrow{\text{reductive work up}} \text{V} + \text{VI}$

(e) $\text{VI} \xrightarrow[\text{H}_2\text{O}]{\text{KMnO}_4} \xrightarrow{\text{H}_3\text{O}^+} \text{VII } (\text{C}_9\text{H}_{18}\text{O}_3)$ (f) $\text{VII} \xrightarrow[\text{H}_2\text{SO}_4]{\text{CrO}_3} \text{VIII } (\text{C}_9\text{H}_{16}\text{O}_4)$

(g) $\text{V} \xrightarrow[\text{Na}_2\text{CO}_3]{\text{HONH}_3{}^+\text{Cl}^-} \text{IX } (\text{C}_7\text{H}_{13}\text{O}_3\text{N})$ (h) $\text{IX} \xrightarrow[\Delta]{(\text{CH}_3\overset{\text{O}}{\overset{\|}{\text{C}}})_2\text{O}} \text{X } (\text{C}_7\text{H}_{11}\text{O}_2\text{N})$

(i) $\text{X} \xrightarrow[\text{H}_2\text{O},\ \Delta]{\text{NaOH}} \xrightarrow{\text{H}_3\text{O}^+} \text{heptanedioic acid}$

15.24 The following series of reactions was carried out during research into methods for synthesizing macrolides, such as recifeiolide (p. 608).

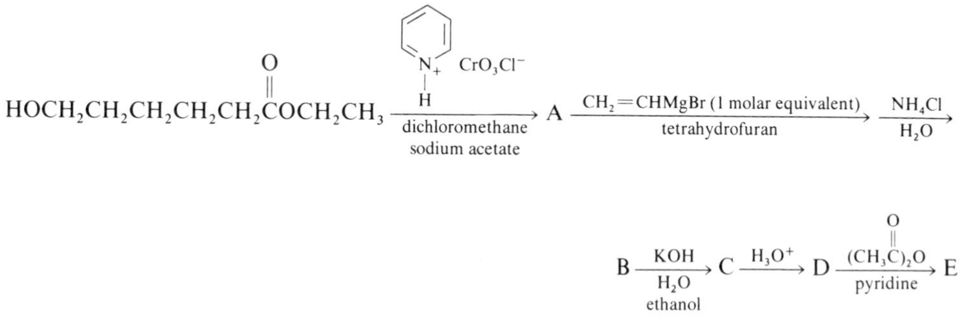

What are the structures of Compounds A, B, C, D, and E? How do you account for the selectivity of the Grignard reaction?

15.25 The stereochemistry of the Baeyer-Villiger reaction has been studied using two types of reactions.

(a) (S)-(+)-3-Phenyl-2-butanone was oxidized with peroxybenzoic acid. The resulting ester was hydrolyzed to (S)-(−)-1-phenylethanol. Write equations for these reactions, being careful to show the correct stereochemistry all the way through. What does this result reveal about the stereochemistry of the Baeyer-Villager reaction?

(b) Both the cis and trans isomers of 1-acetyl-2-methylcyclopentane were oxidized with peroxybenzoic acid in chloroform. Draw structural formulas for the starting materials and predict the product of each reaction, basing your answer on the stereochemical results obtained in part a.

15.26 The following sequence of reactions was used to synthesize an optically active cyclic ether, Compound E. Give the structural formula, showing stereochemistry, for each product designated by a letter in these transformations.

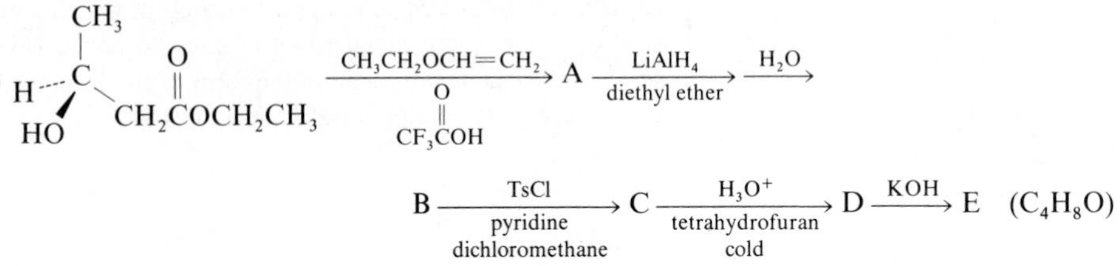

15.27 Coriolic acid, isolated from the hearts of cows, is of interest because of its ability to transport calcium ions through membranes. It was synthesized by the following sequence of steps. Supply structural formulas for the compounds designated by letters.

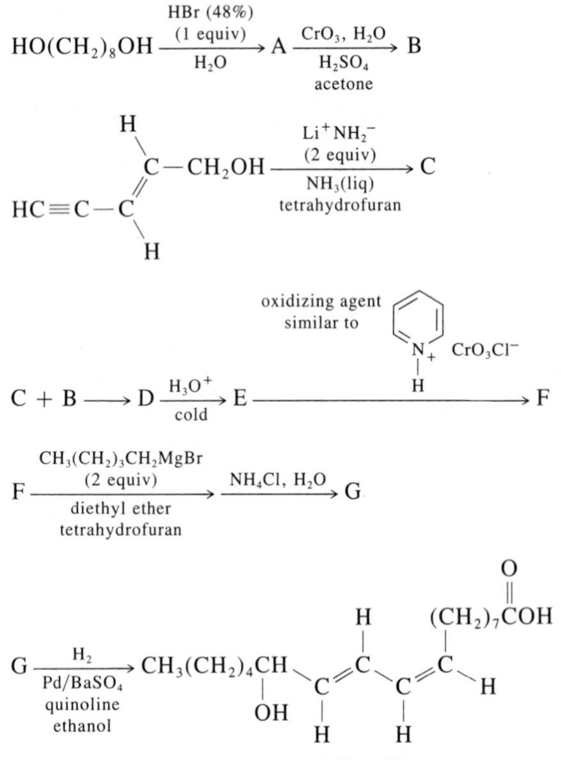

coriolic acid

15.28 A furancarboxylic acid was converted to two esters as shown below. How would you carry out these transformations?

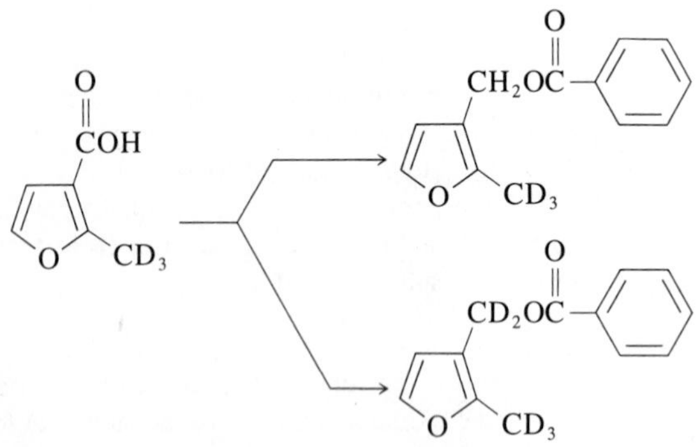

15.29 A synthesis of certain antibiotics called the pseudomonic acids included the following transformations. Suggest reagents for these reactions.

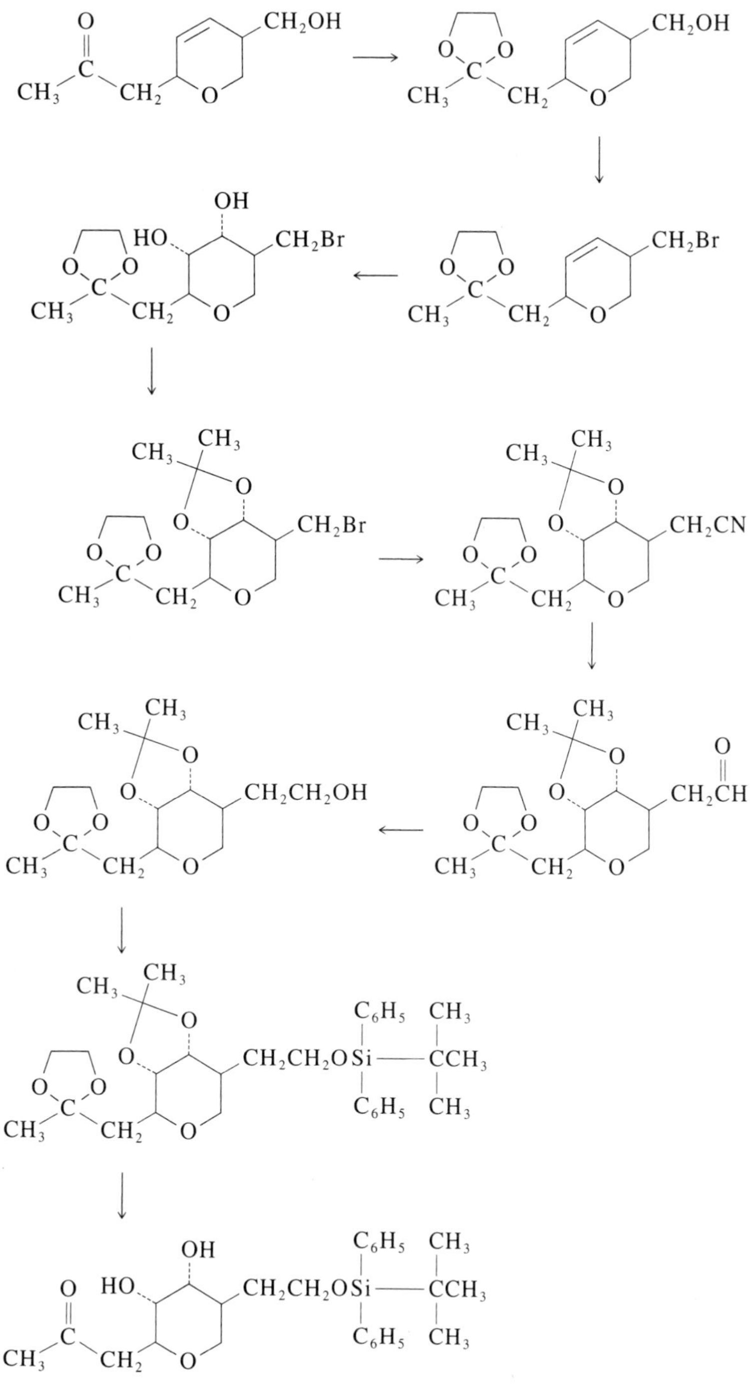

15.30 The proton magnetic resonance spectra of Compounds A and B, two isomeric esters with the molecular formula $C_6H_{12}O_2$, are shown in Figure 15.2. Assign structures to the compounds.

Compound A

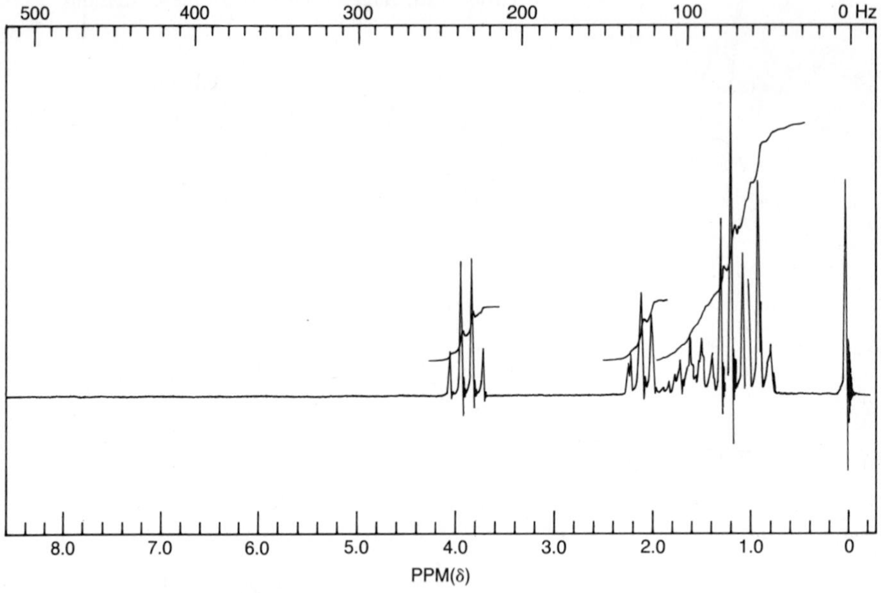
500
400
300
200
100
0 Hz
8.0
7.0
6.0
5.0
4.0
3.0
2.0
1.0
0
PPM(δ)

Compound B

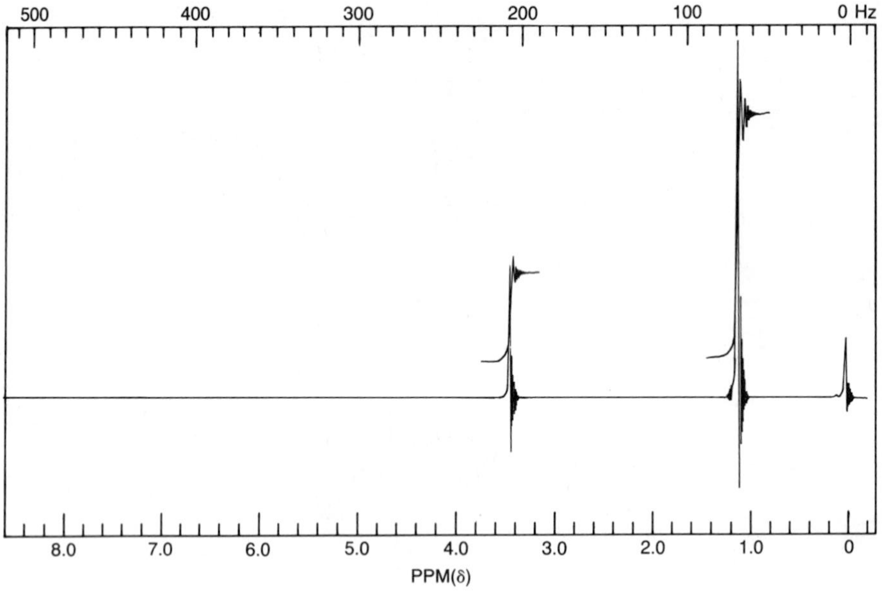
500
400
300
200
100
0 Hz
8.0
7.0
6.0
5.0
4.0
3.0
2.0
1.0
0
PPM(δ)

FIGURE 15.2

15.31 The proton magnetic resonance spectrum of Compound C, $C_9H_{10}O_3$, is shown in Figure 15.3. Important bands in this compound's infrared spectrum are at 3300–2600 and 1719 cm^{-1}. Assign a structure to Compound C and show how the spectral data support it.

Compound C

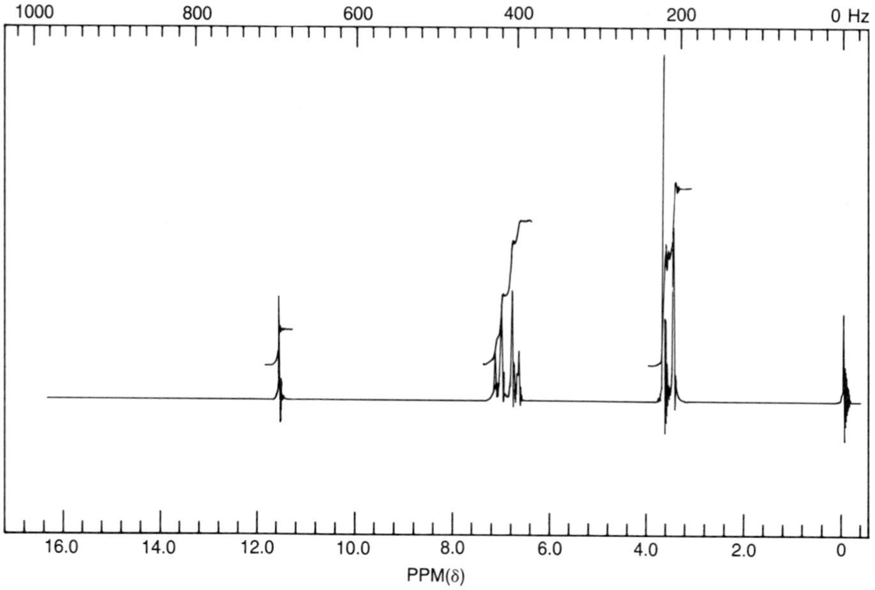
1000
800
600
400
200
0 Hz
16.0
14.0
12.0
10.0
8.0
6.0
4.0
2.0
0
PPM(δ)

FIGURE 15.3

15.32 The proton magnetic resonance spectra for two esters, Compounds D and E, are given in Figure 15.4. Assign a structure to each compound that is compatible with its spectrum.

Compound D

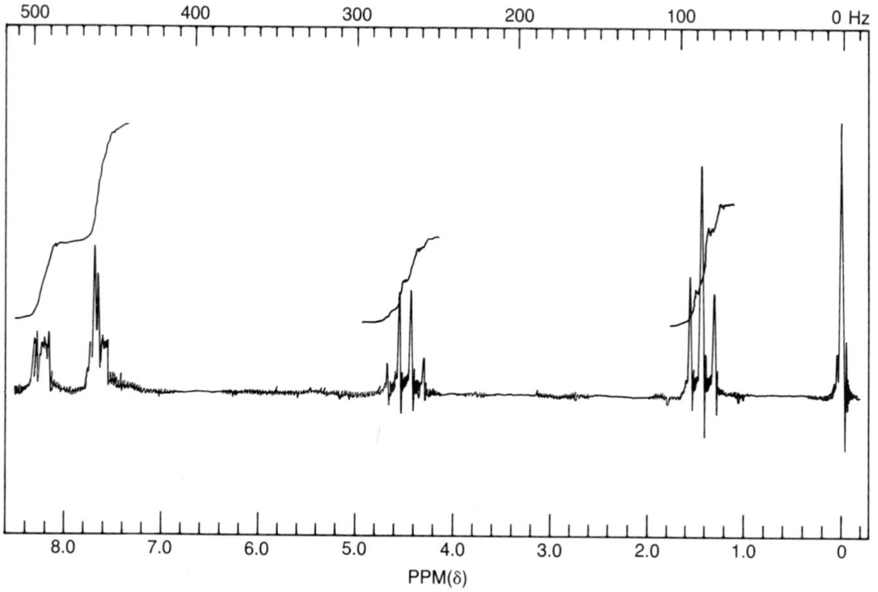
500
400
300
200
100
0 Hz
8.0
7.0
6.0
5.0
4.0
3.0
2.0
1.0
0
PPM(δ)

FIGURE 15.4

(Continued)

Compound E

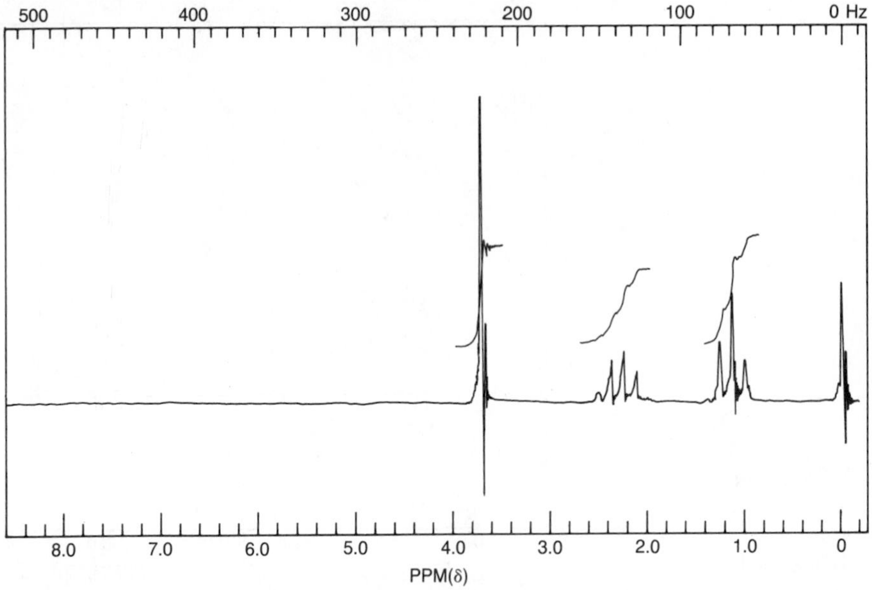
500
400
300
200
100
0 Hz
8.0
7.0
6.0
5.0
4.0
3.0
2.0
1.0
0
PPM(δ)

FIGURE 15.4 *(Continued)*

15.33 Compound F, $C_5H_8O_2$, has industrial importance, being widely used in polymerization reactions. The proton magnetic resonance spectrum of F is shown in Figure 15.5. In its infrared spectrum, absorption bands appear at 1731, 1635, 1279, 1207, 1069, and 989 cm^{-1}. What is the structure of F? Analyze all the spectral data, being as precise as possible in assigning the bands appearing in the spectra.

Compound F

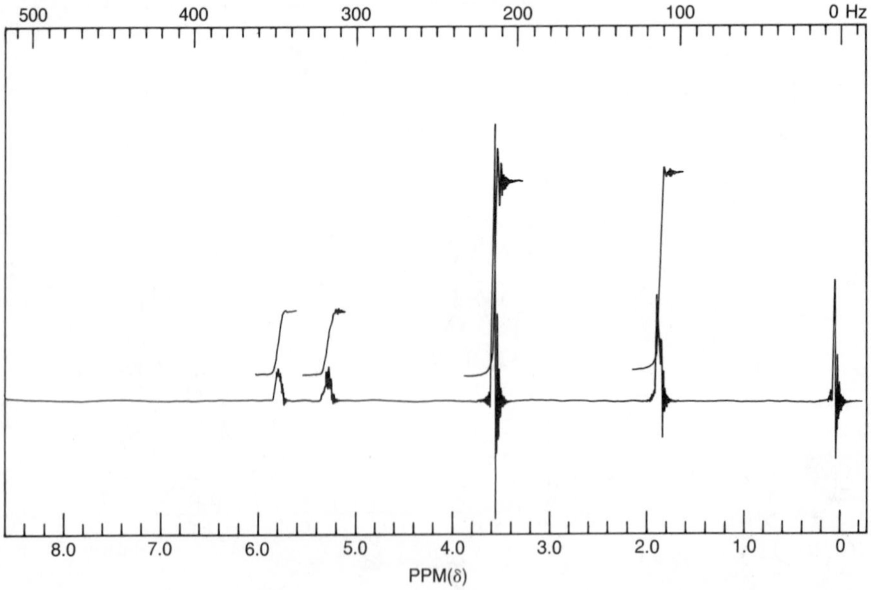
500
400
300
200
100
0 Hz
8.0
7.0
6.0
5.0
4.0
3.0
2.0
1.0
0
PPM(δ)

FIGURE 15.5

15.34 Compound G, $C_8H_{14}O_4$, has absorption bands in its infrared spectrum at 1736, 1160, and 1032 cm^{-1}. Its proton magnetic resonance spectrum is shown in Figure 15.6. What is the structure of Compound G?

Compound G

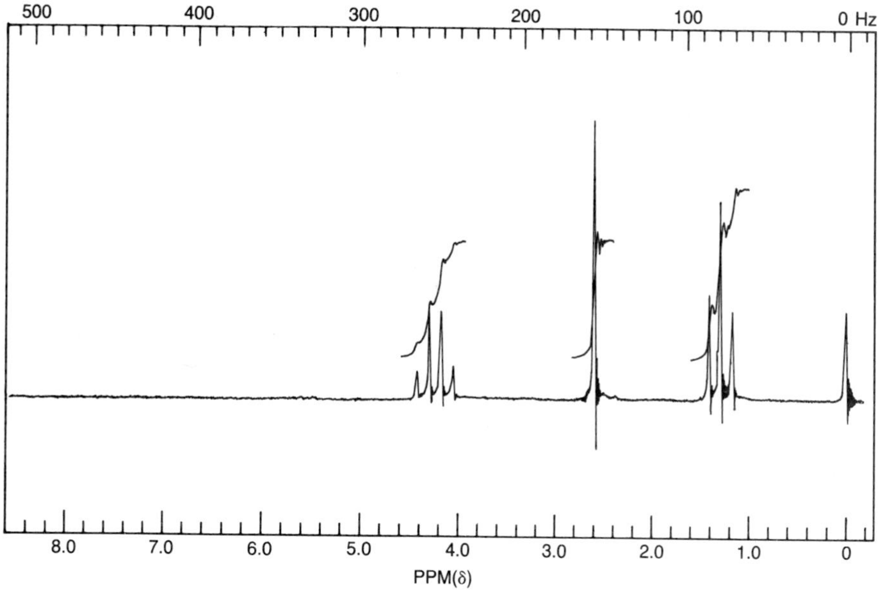

500
400
300
200
100
0 Hz
8.0
7.0
6.0
5.0
4.0
3.0
2.0
1.0
0
PPM(δ)

FIGURE 15.6

16

Enols and Enolate Anions as Nucleophiles I. Halogenation, Alkylation, and Condensation Reactions

A • L O O K • A H E A D

The hydrogen atoms bonded to the carbon atom adjacent to a carbonyl group in an aldehyde, a ketone, or an ester are acidic. The carbanion resulting from deprotonation at such a carbon atom is stabilized by delocalization of the negative charge to the oxygen atom of the carbonyl group.

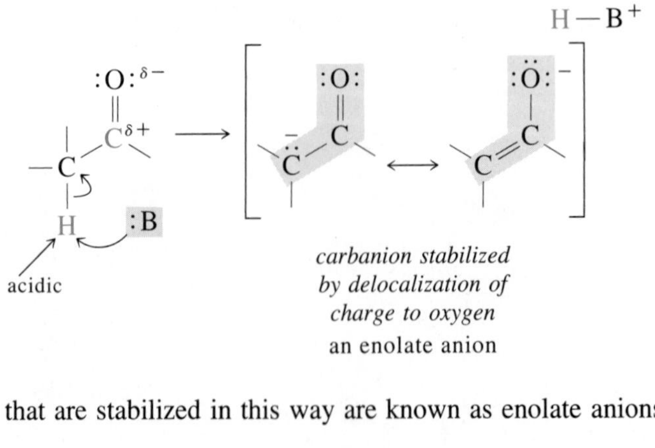
H—B⁺
acidic
:B
carbanion stabilized
by delocalization of
charge to oxygen
an enolate anion

Carbanions that are stabilized in this way are known as enolate anions.

Enols are the conjugate acids of enolate anions. Both enols and enolate anions are nucleophiles and react with electrophiles. For example, they react with halogens, alkyl halides, and carbonyl groups.

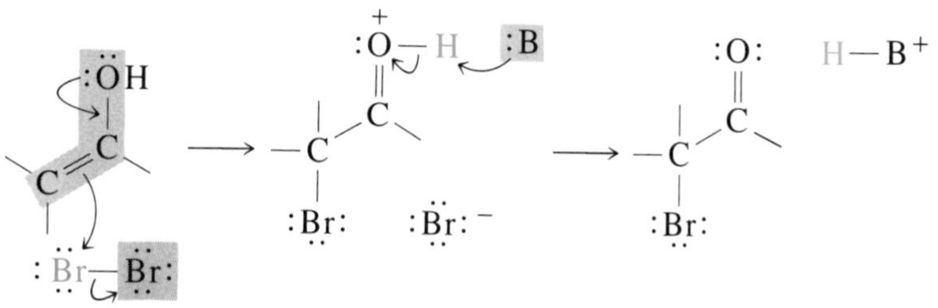

enol reacting with halogen

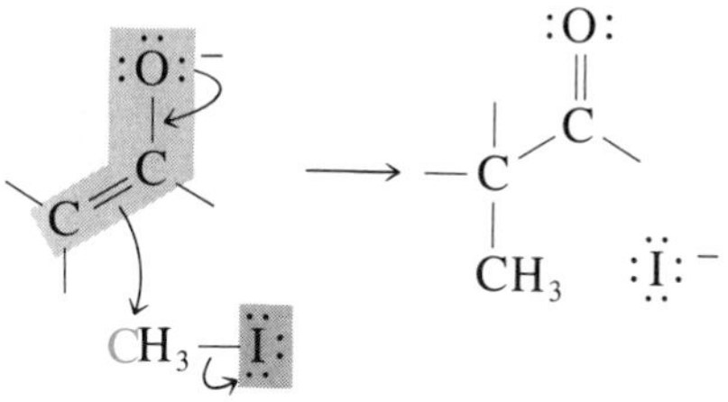

enolate reacting with alkyl halide

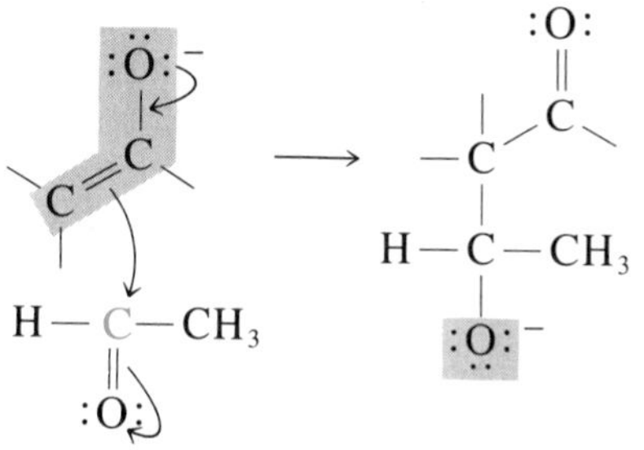

enolate reacting with carbonyl compound

When enolate anions react as nucleophiles with electrophilic carbon atoms, carbon-carbon bonds are formed. Such reactions are, therefore, important in syntheses.

16.1 ENOLS

A. Carbanions as Reactive Intermediates

Carbanions are reactive intermediates in which a carbon atom has a pair of nonbonding electrons and bears a negative charge. Carbanions are strong bases and nucleophiles (pp. 100 and 261).

Carbanions are generated in two ways. One method is to remove a proton from an organic compound using a base. Most carbon-hydrogen bonds are not acidic enough to be broken easily in this way unless there are electron-withdrawing groups on the carbon atom to help stabilize the negative charge on the carbanion. The most effective way to stabilize the negative charge on a carbanion is by delocalization of the charge to adjacent functional groups. For example, the anions derived from the deprotonation of nitromethane (pK_a 10.2, p. 101) and the deprotonation of acetone (pK_a 19, p. 475) are stabilized by resonance. If a strong enough base is used, even the terminal hydrogen atom of a terminal alkyne can be removed to give a carbanionic intermediate. Such synthetically useful intermediates are formed in the preparation of organosodium or organomagnesium derivatives of alkynes (pp. 332, 489).

The other common method of generating carbanions is by the reaction of organic halides with metals to give organometallic reagents. Prepared in this way, Grignard reagents and organolithium reagents react as sources of groups containing strongly basic and nucleophilic carbon atoms. Their reactions both as bases and as nucleophiles were explored in Section 13.6. Such reagents have great versatility because they are capable of forming new carbon-carbon bonds.

Carbanions generated by the deprotonation of carbonyl compounds also react as nucleophiles with the electrophilic carbon atoms of carbonyl groups or alkyl halides to give new carbon-carbon bonds. The remainder of this chapter shows the synthetic importance of these reactions.

B. Enols and Enolate Anions

All carbonyl compounds are in equilibrium with their enol forms. The percentage of enol present depends on the structure of the carbonyl compound as well as other factors, such as the solvent. For a compound with a single carbonyl group, such as acetone, the equilibrium lies far on the side of the ketone. For 1,3-dicarbonyl compounds, such as 2,4-pentanedione, the enol form is a significant part of the equilibrium mixture.

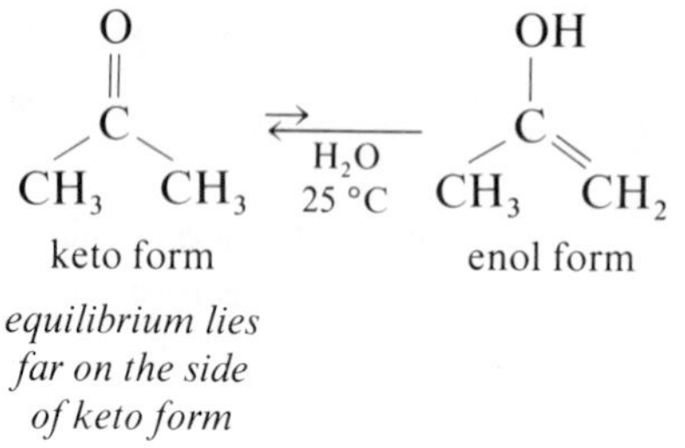
O
C
CH3 CH3
H2O
25 °C
OH
C
CH3 CH2
keto form
enol form
equilibrium lies far on the side of keto form

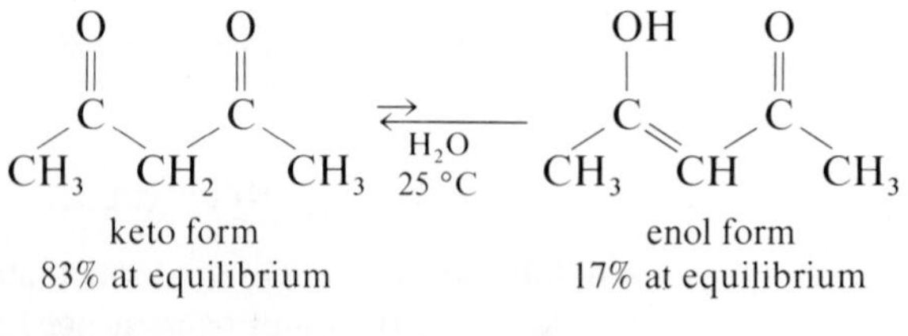
O O
C C
CH3 CH2 CH3
H2O
25 °C
OH O
C C
CH3 CH CH3
keto form
83% at equilibrium
enol form
17% at equilibrium

The process of converting a carbonyl compound into its enol is called enolization. The extent of enolization that occurs for a carbonyl compound depends on the structure of the compound, which determines the acidity of the α-hydrogens. The following data for 2,4-pentanedione and acetone, for example, are as given in Visualizing the Reaction on the next page (and in the table of pK_a values inside the cover of the book).

VISUALIZING THE REACTION

Base-catalyzed enolization reactions

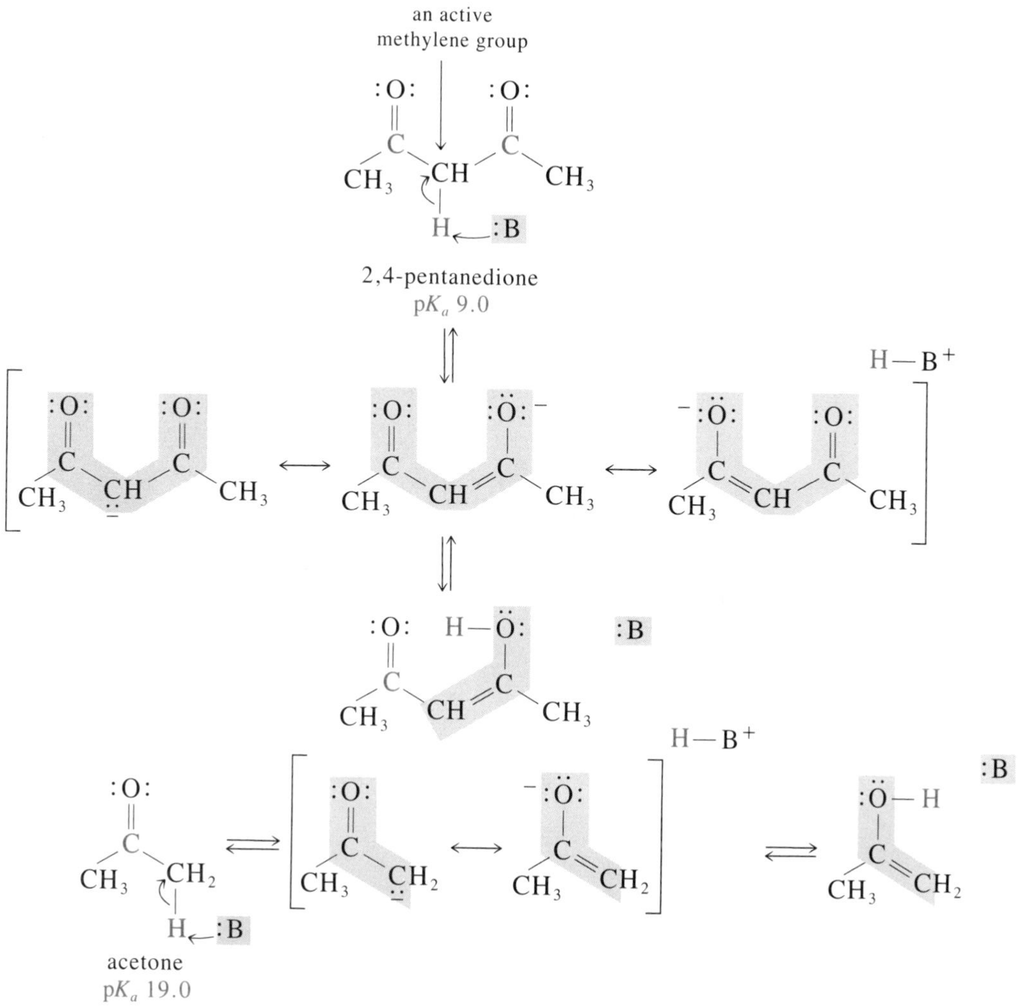

an active methylene group
2,4-pentanedione
pK_a 9.0
acetone
pK_a 19.0

For both of these compounds, removal of a proton from an α-carbon atom results in the formation of an anion that is stabilized by delocalization of the negative charge to the electronegative oxygen atom of the carbonyl group. Such an anion is called an **enolate anion.** Protonation of the enolate anion at the oxygen atom gives rise to an enol. The most acidic hydrogen atom in 2,4-pentanedione (pK_a 9.0) is one of those bonded to the carbon atom that lies between the two carbonyl groups. Resonance contributors can be drawn for the enolate anion of 2,4-pentanedione that show the delocalization of charge to two different oxygen atoms. A methylene group that is alpha to two carbonyl groups is said to be an **active methylene group.** The hydrogen atoms of an active methylene group are easily removed by bases such as alkoxide anions because the resulting conjugate base, the enolate anion, is highly stabilized by resonance. Acetone (pK_a 19) is much less acidic than 2,4-pentanedione. Removal of a proton from acetone also creates an enolate anion, but in this case the ion is stabilized by the delocalization of charge to only one oxygen atom. The conjugate base of acetone is not stabilized relative to its acid as much as the conjugate base of 2,4-pentanedione is.

Esters also enolize. Ethyl 3-oxobutanoate (ethyl acetoacetate, pK_a 11.0) has an active methylene group and is, therefore, more acidic than ethyl acetate (pK_a 23), for example.

VISUALIZING THE REACTION

Base-catalyzed enolization reactions of esters

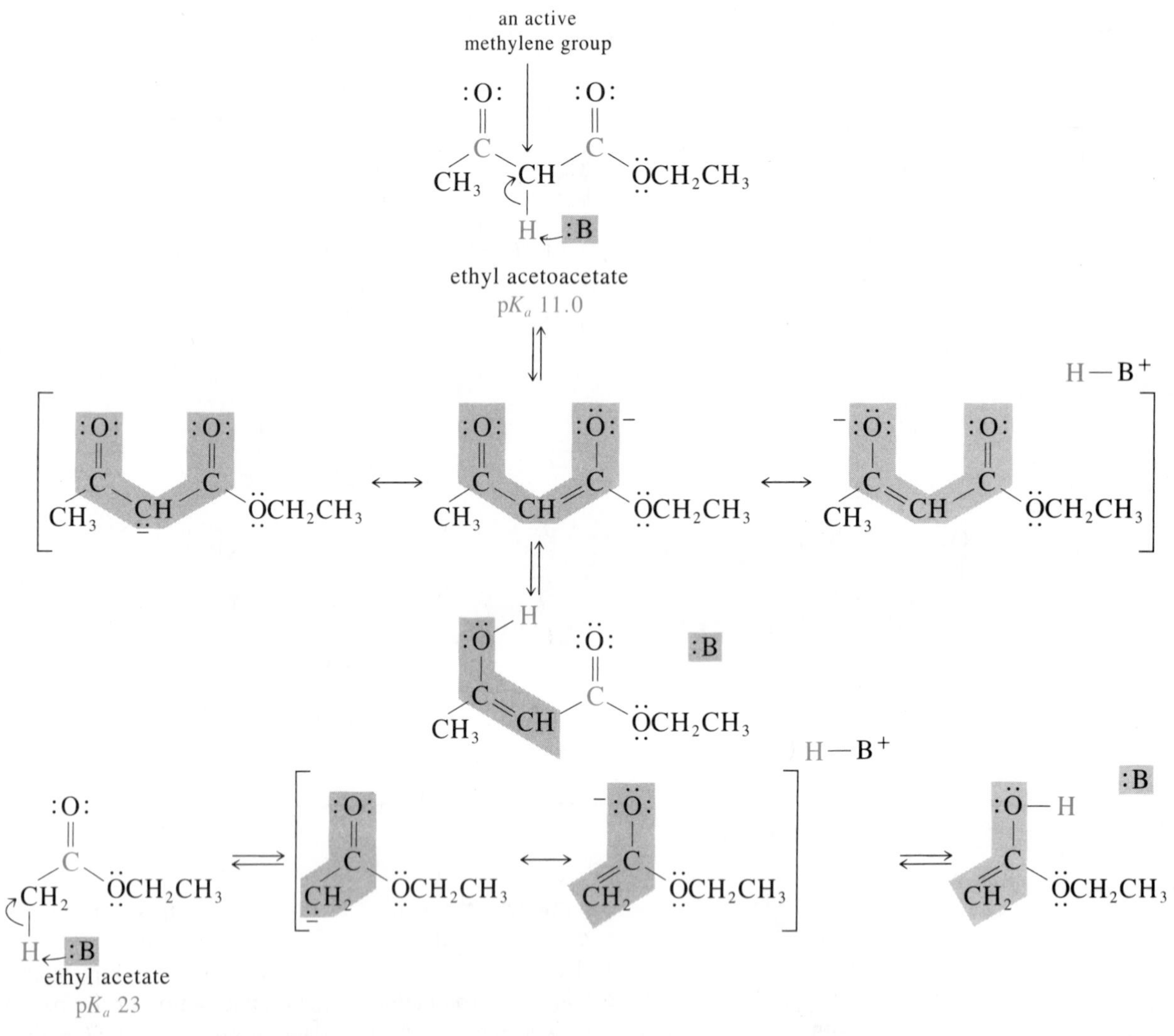
an active
methylene group
ethyl acetoacetate
pKa 11.0
ethyl acetate
pKa 23

The carbonyl group of an ester does not contribute as much to the stabilization of an enolate anion as the carbonyl group of a ketone does. Ethyl acetoacetate is a weaker acid than 2,4-pentanedione, and ethyl acetate is a weaker acid than acetone. The differences in polarity and electrophilicity between carbonyl groups in ketones (and aldehydes) and those in acid derivatives have been thoroughly discussed in several places, especially in connection with the reactivity of acid derivatives (p. 543). The presence of unshared pairs of electrons on the oxygen atom of the alkoxyl group of an ester diminishes the effectiveness of the carbonyl group of the ester in delocalizing charge.

Other groups besides carbonyl groups increase the acidity of hydrogen atoms bonded to the carbon atoms adjacent to them. Two of the most important are the nitro group, which is even more effective than a carbonyl group in delocalizing charge (p. 101), and the cyano group, which is not quite as effective as a ketone in stabilizing charge but is better than an ester group.

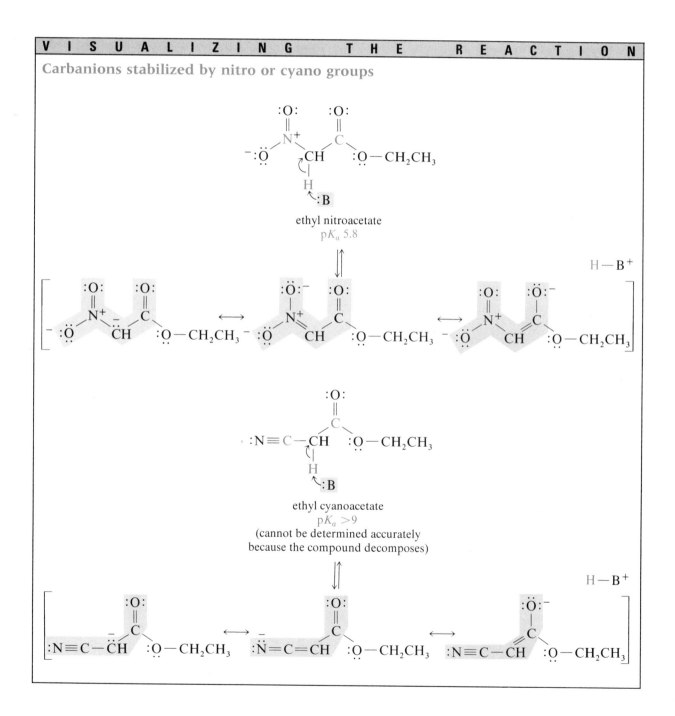

VISUALIZING THE REACTION

Carbanions stabilized by nitro or cyano groups

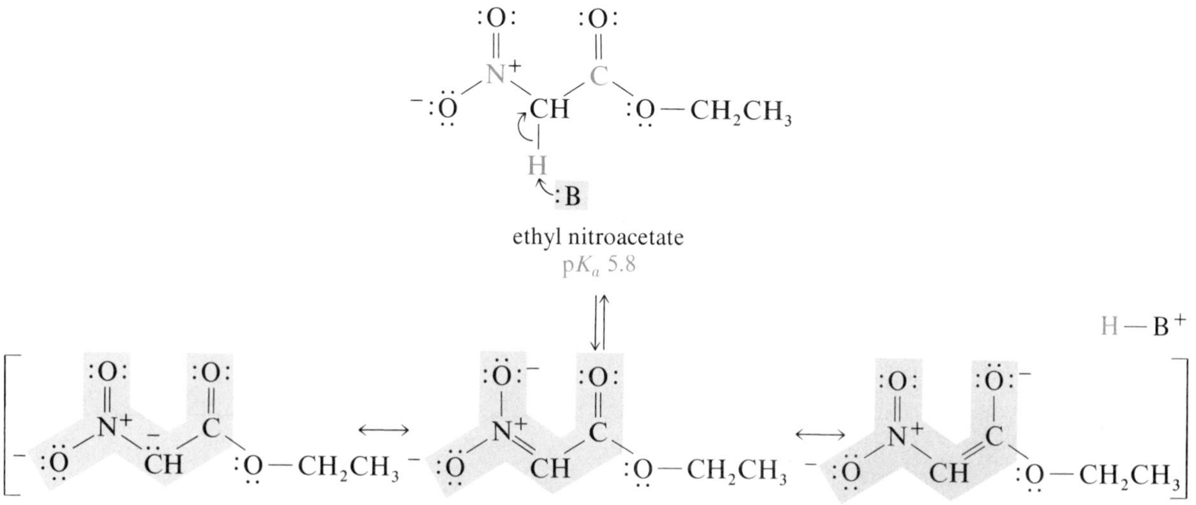

ethyl nitroacetate
pK_a 5.8
H—B⁺

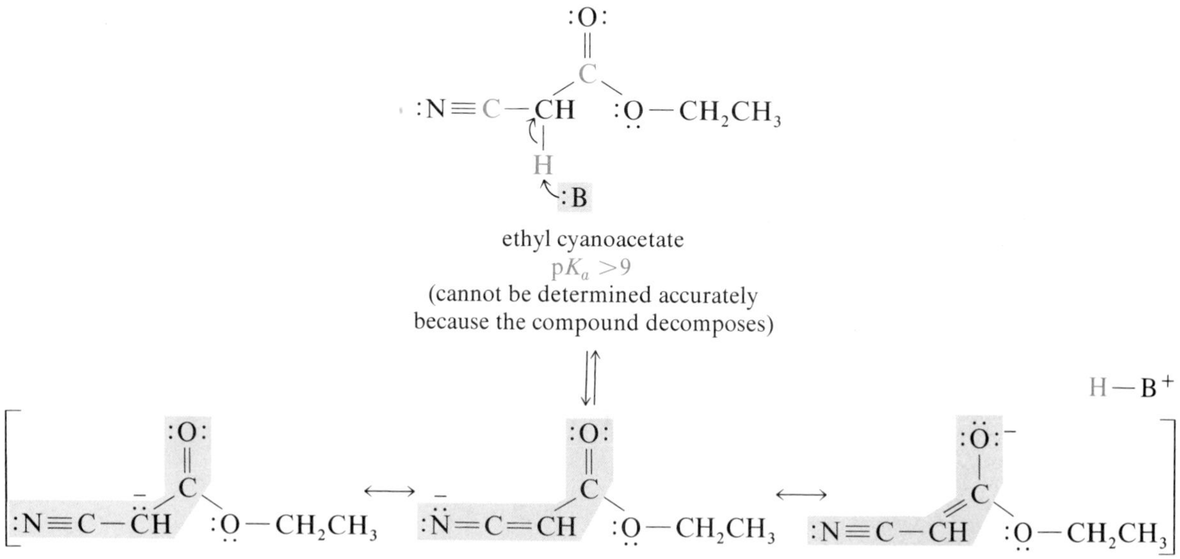

ethyl cyanoacetate
pK_a >9
(cannot be determined accurately
because the compound decomposes)
H—B⁺

The enolization reactions shown above are all base-catalyzed and are all shown starting with a deprotonation step. Acids also promote enolization, as demonstrated by the acid-catalyzed enolization of acetone.

VISUALIZING THE REACTION

Acid-catalyzed enolization of acetone

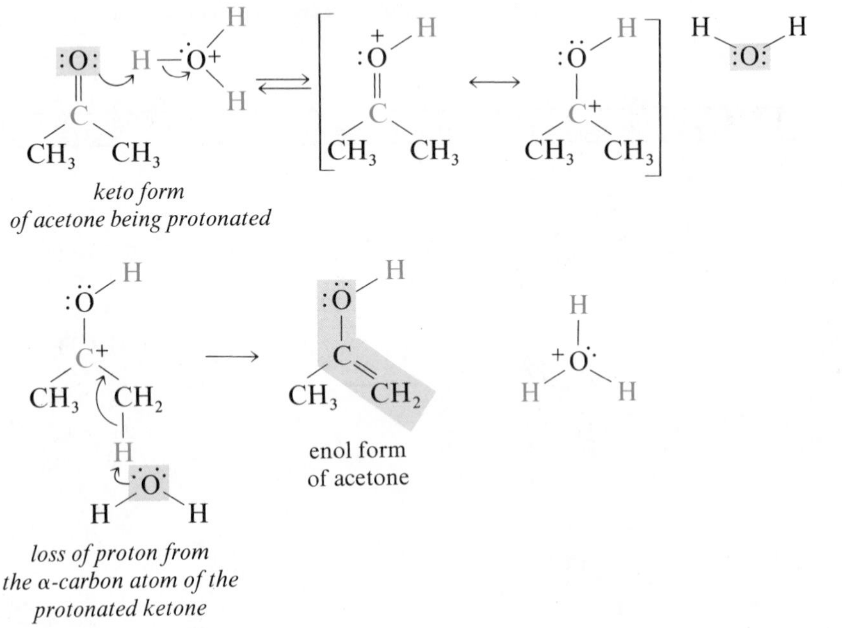
keto form
of acetone being protonated
enol form
of acetone
loss of proton from
the α-carbon atom of the
protonated ketone

Study Guide Concept Map 16.1

The ketone, protonated at the carbonyl group in acidic solution, loses a proton from the α-carbon atom to give an enol.

PROBLEM 16.1

For each of the following pairs of compounds, predict which one will be more extensively enolized.

(a) 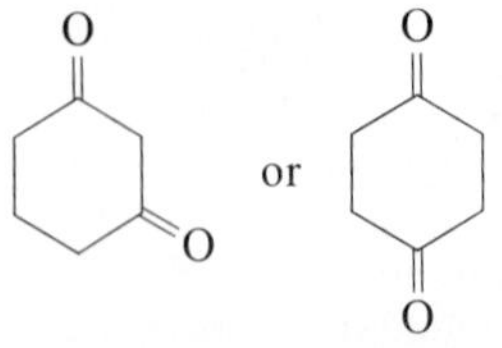

(b) 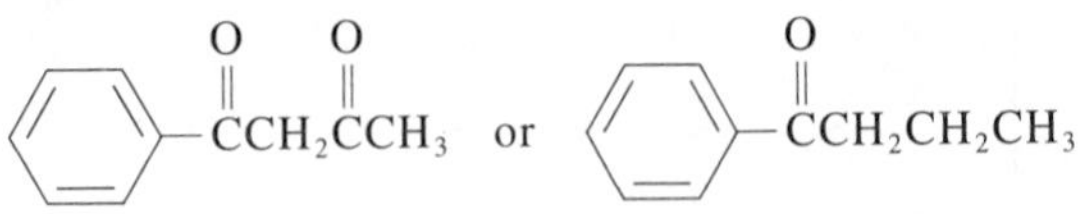

(c) $CH_3C(=O)N(CH_3)CH_3$ or $CH_3C(=O)OCH_2CH_3$

(d) $CH_3OC(=O)CH_2CH_2C(=O)OCH_3$ or $CH_3OC(=O)CH_2C(=O)OCH_3$

C. Exchanging α-Hydrogen Atoms for Deuterium Atoms

One of the easiest ways to test for the presence of enolizable α-hydrogen atoms is to see whether any of the hydrogen atoms in a molecule can be exchanged for deuterium atoms that come from deuterium oxide. The hydrogen atom of an active methylene group exchanges so readily that the reaction will take place in water with no added catalyst.

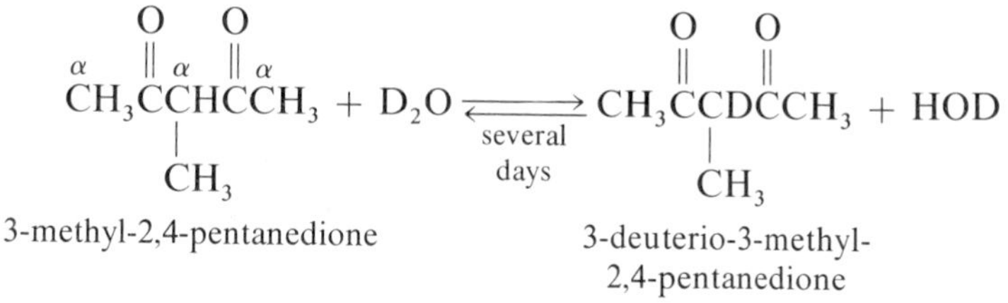
several days
3-methyl-2,4-pentanedione
3-deuterio-3-methyl-2,4-pentanedione

If 3-methyl-2,4-pentanedione is allowed to stand in deuterium oxide at room temperature for several days, one of its hydrogen atoms is exchanged for a deuterium atom. The reaction is an equilibrium reaction, and fresh deuterium oxide must be added to the reaction mixture from time to time for every molecule of the diketone to exchange the hydrogen atom. Note that the α-hydrogen atom alpha to two carbonyl groups exchanges under these conditions.

The α-hydrogen atoms of ordinary ketones can also be exchanged for deuterium atoms, but acid or base catalysis is usually necessary to make such a reaction proceed at a reasonable rate. Acetone exchanges its hydrogens, which are all α-hydrogens, for deuterium atoms when it is placed in deuterium oxide in the presence of a trace of sulfuric acid.

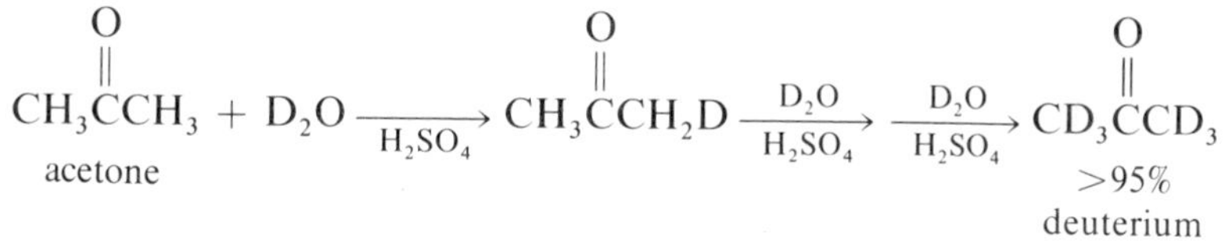
acetone
>95% deuterium

If the reaction time is extended and fresh deuterium oxide is periodically added to the reaction mixture, eventually all of the α-hydrogen atoms in acetone are replaced by deuterium atoms.

PROBLEM 16.2

Complete the following equations.

(a) 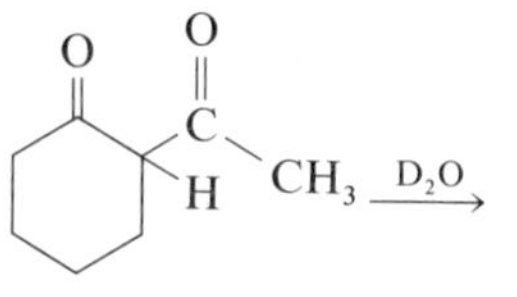

(b) 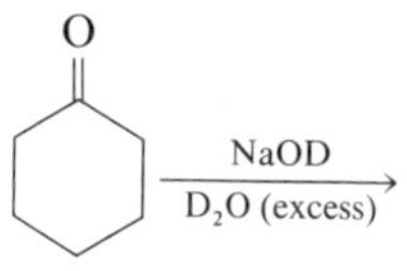

(c) 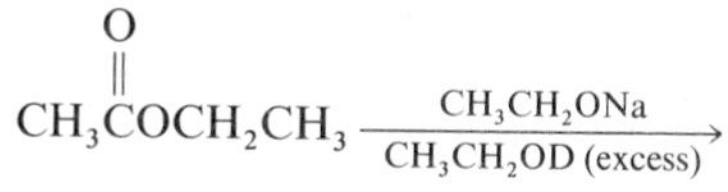

(d) $CH_2(\overset{O}{\overset{\|}{C}}OCH_2CH_3)_2 \xrightarrow[D_2O\ (excess)]{NaOD}$

D. Ambident Nucleophiles

An examination of the structure of an enolate anion reveals that it has electron density and negative charge at two sites. An enolate anion is basic and nucleophilic at the α-carbon atom and at the oxygen atom of the carbonyl group; it can be protonated or can react with other electrophiles at either of those sites. An anion that has the capacity to react at either of two positions is called an **ambident anion** or an **ambident nucleophile.** Whether such an anion reacts at the carbon atom or at the oxygen atom depends on the reaction conditions, such as the nature of the positive ion associated with the anion, the solvent, and the nature of the electrophile that is present.

The enolate anion derived from diethyl malonate was used as an example of a typical nucleophile in nucleophilic substitution reactions in Section 7.8. Enolate anions react with alkyl halides in nucleophilic substitution reactions known as **alkylation reactions.** Alkylation reactions are important in syntheses because carbon-carbon bonds are formed; these reactions will be examined further starting on p. 645. The enolate anion from ethyl acetoacetate also reacts with alkyl halides in nucleophilic substitution reactions. When such a reaction is carried out in ethanol, which hydrogen-bonds to the negatively charged oxygen atom of the enolate anion, the anion reacts as a carbon nucleophile to form a new bond at the α-carbon atom of the ester.

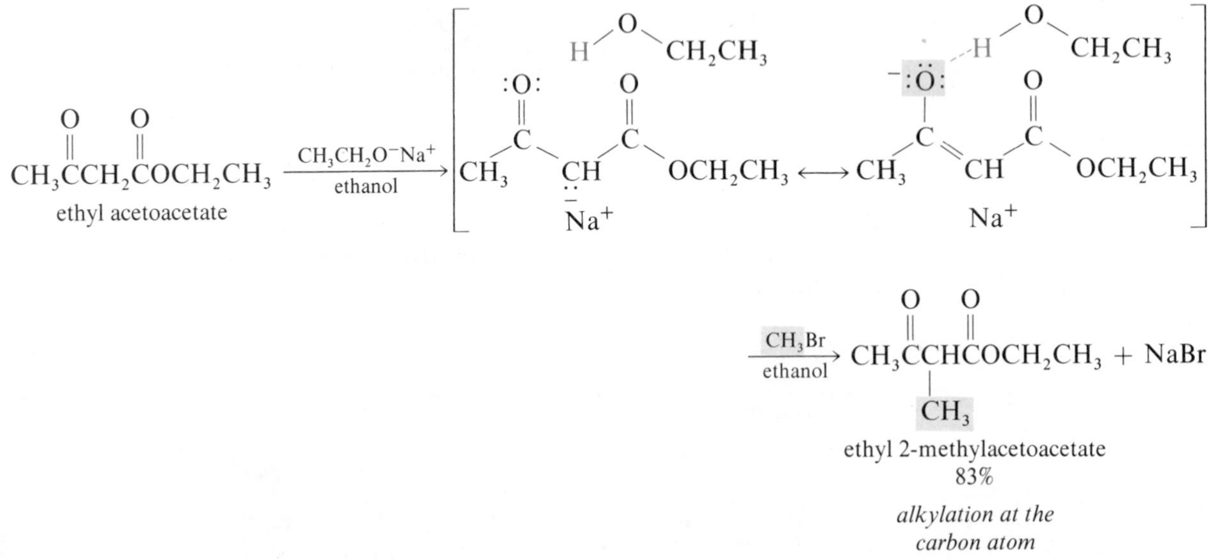
ethyl acetoacetate
CH3CH2O−Na+
ethanol
CH3Br
ethanol
+ NaBr
ethyl 2-methylacetoacetate
83%
alkylation at the carbon atom

The ambident nucleophile undergoes alkylation at the α-carbon atom. If the enolate anion is generated in dimethylformamide, a polar solvent (p. 229) that does not hydrogen-bond, reaction with n-butyl iodide gives 99% alkylation at the α-carbon atom. If the alkyl halide used is n-butyl bromide instead, 67% alkylation at the carbon atom and 33% alkylation at the oxygen atom result.

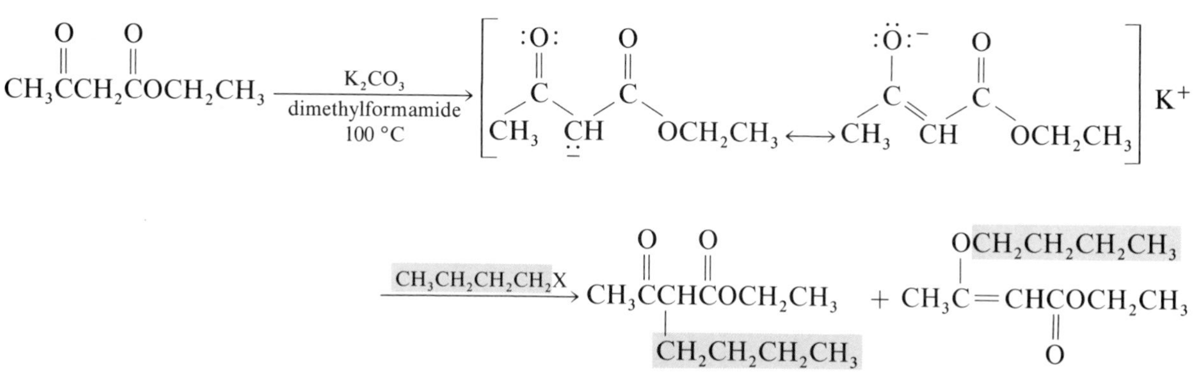

ethyl 2-butylacetoacetate

product from alkylation at the carbon atom
X = I 99%
X = Br 67%

an enol ether

product from alkylation at the oxygen atom
1%
33%

These experimental data indicate that the nature of the leaving group on the alkylating agent and the nature of the solvent are important in determining whether the enolate anion will react as a carbon nucleophile or as an oxygen nucleophile. Protic solvents that hydrogen-bond with the oxygen anion increase the percentage of alkylation at the carbon atom. Reagents with large polarizable leaving groups, such as the iodide ion, react chiefly at the carbon atom. Other experiments have shown that the nature of the metal ion associated with the enolate anion and whether or not it coordinates tightly with the oxygen atom also influence the extent of carbon alkylation versus oxygen alkylation.

The reaction conditions for syntheses using enolate anions are usually chosen so as to give mostly alkylation at the carbon atom.

E. Relative Stabilities of Enolate Anions

An unsymmetrical carbonyl compound enolizes to give two different enolate anions. For example, when 2-methylcyclohexanone is heated with the base triethylamine in dimethylformamide in the presence of trimethylchlorosilane, a mixture of two trimethylsilyl enol ethers is formed.

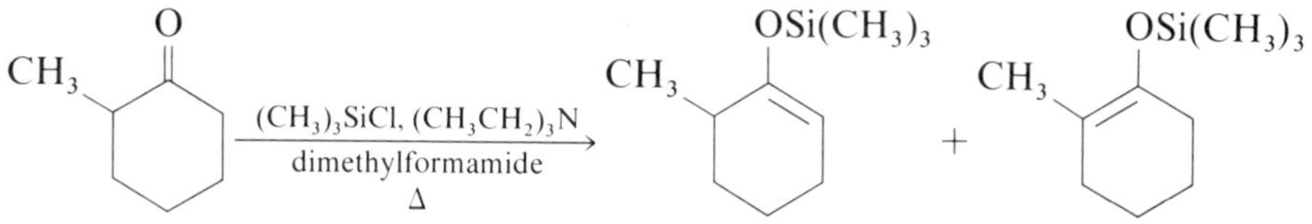

2-methylcyclohexanone 22% 78%

Trimethylchlorosilane is a reagent that reacts exclusively with the enolate anion at the oxygen atom because a strong silicon-oxygen bond is formed. Therefore, the reagent is used to trap enolate anions as stable trimethylsilyl enol ethers. The composition of the mixture of enol ethers reflects the relative stabilities of the two enolate anions derived from 2-methylcyclohexanone.

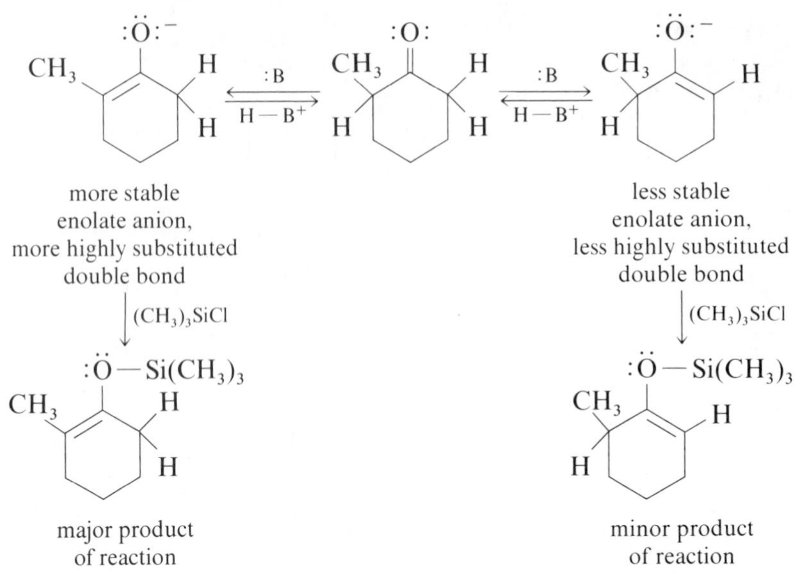
:B
H—B+
:B
H—B+
more stable enolate anion, more highly substituted double bond
less stable enolate anion, less highly substituted double bond
(CH3)3SiCl
(CH3)3SiCl
major product of reaction
minor product of reaction

The enolate anion with the more highly substituted double bond is the more stable anion and predominates under conditions that allow equilibrium to be established.

The conditions described for the reaction above allow an equilibrium to exist between the two enolate anions. The reaction is run at a high temperature. Triethylamine is not a strong base; therefore, the conjugate acid of triethylamine and unreacted ketone molecules serve as acids to protonate the enolate anions and interconvert them by way of the ketone. All of these factors ensure that the mixture that is observed reflects the relative thermodynamic stabilities of the anions. The more highly substituted enolate anion is said to be the **thermodynamic enolate** of the ketone. Under conditions that allow for equilibrium, free-energy considerations, that is, thermodynamics (p. 112), control what the major product of a reaction will be.

Enolate anions are versatile intermediates in reactions that result in new carbon-carbon bonds being formed. Therefore, it is important that chemists know how to control the compositions of mixtures obtained from unsymmetrical carbonyl compounds. Gilbert Stork at Columbia University and Herbert O. House at the Georgia Institute of Technology have made major contributions by devising methods for regioselective generation of enolate anions. For example, 2-methylcyclohexanone is added slowly to a solution of the very strong base lithium diisopropylamide in 1,2-dimethoxyethane at 0 °C. The resulting mixture is then treated with trimethylchlorosilane in the presence of triethylamine, and the trimethylsilyl ether of the less highly substituted enol is obtained almost exclusively.

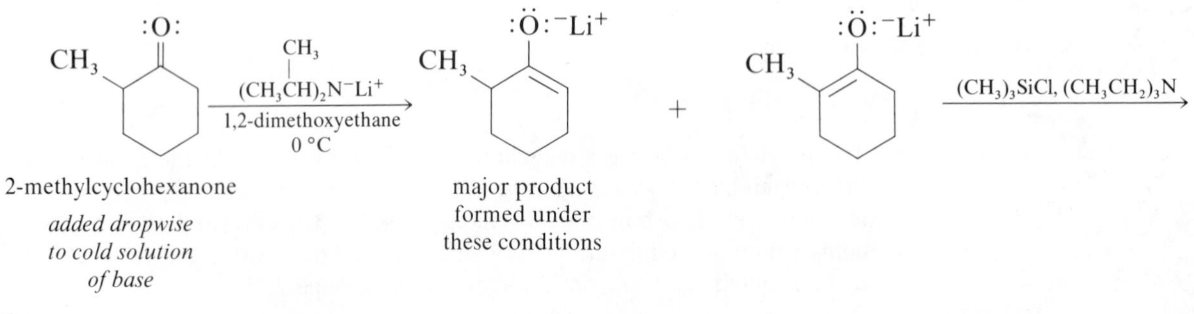
(CH3CH)2N−Li+
1,2-dimethoxyethane
0 °C
2-methylcyclohexanone
added dropwise to cold solution of base
major product formed under these conditions
(CH3)3SiCl, (CH3CH2)3N

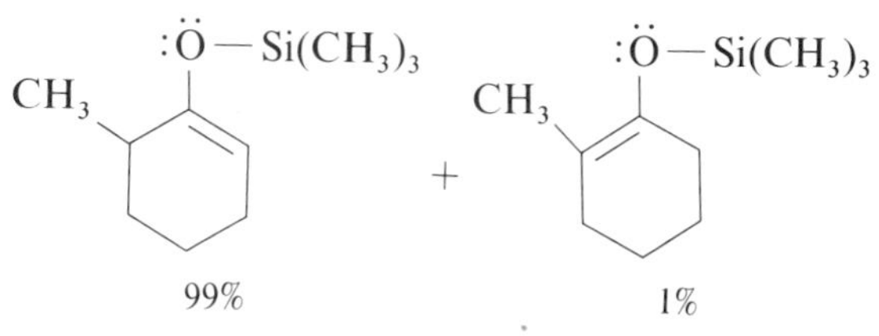
:Ö—Si(CH3)3
CH3
99%
+
:Ö—Si(CH3)3
CH3
1%

The reaction conditions just described do not allow equilibration between the enolate anions. A very strong base, the diisopropylamide anion, is used to deprotonate the ketone. The conjugate acid of this anion, diisopropylamine (pK_a ~36), is not a strong enough acid to protonate the enolate anion that is formed. The other acid in the system, the ketone 2-methylcyclohexanone (pK_a ~16), is never effectively present at the same time as the enolate anion because it is added slowly to the solution of the base. An excess of base is always present, and the ketone is deprotonated completely. The temperature is also kept low. Under these conditions, the major product is the **kinetic enolate,** the enolate that is formed the fastest rather than the one with the greatest thermodynamic stability. Kinetic enolates tend to be the less highly substituted ones. The rate-determining step in an enolization reaction is the breaking of a carbon-hydrogen bond at the α-carbon atom of the carbonyl compound. The hydrogen atoms on the less highly substituted α-carbon atom are less sterically hindered and react fastest with a strong base. Under conditions that do not allow equilibrium to be attained, rate considerations, that is, kinetics (p. 125), control what the major product of the reaction will be.

The reasons for choosing lithium diisopropylamide as the strong base in the above reaction are interesting. This reagent is easily prepared by treating diisopropylamine in an ether or hydrocarbon solvent with an alkyllithium, usually *n*-butyllithium. The *n*-butyl anion is a powerful base and deprotonates diisopropylamine completely; the result is a solution of lithium diisopropylamide.

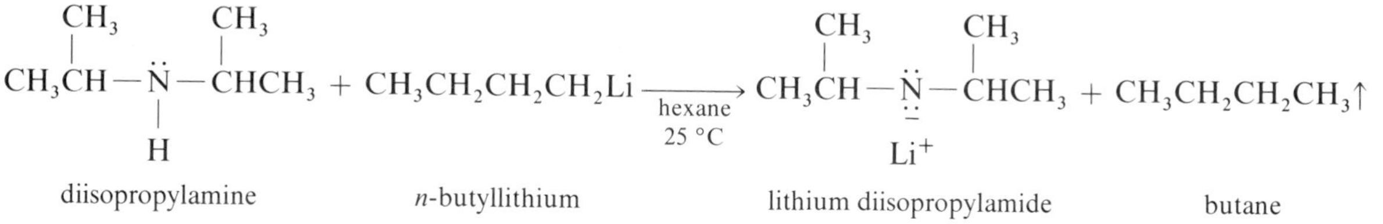
CH3 CH3
CH3CH—N̈—CHCH3
H
+ CH3CH2CH2CH2Li
hexane
25 °C
CH3 CH3
CH3CH—N—CHCH3
Li+
+ CH3CH2CH2CH3↑
diisopropylamine
n-butyllithium
lithium diisopropylamide
butane

Solutions of lithium diisopropylamide in hydrocarbons are stable, but those in ether solvents are not. The base is strong enough to deprotonate ethers, so its ether solutions must be kept cold and used as soon as they are prepared.

The diisopropylamide anion is a strong base but is hindered enough that it is not a good nucleophile. Thus, it does not react with alkyl halides or other reagents that react with the enolate anions in subsequent steps. The conjugate acid of the diisopropylamide anion, diisopropylamine, boils at 86 °C and is easily separated from the other products of these reactions. The small lithium ion, with its high concentration of positive charge, coordinates strongly with the oxygen atom and enhances the regioselectivity of the enolization.

Study Guide Concept Map 16.2

PROBLEM 16.3

Predict the major product of each of the following reactions.

(a) $$CH_3CH_2\overset{\overset{\large O}{\|}}{C}CH_3 \xrightarrow[\text{dimethylformamide}\ \Delta]{(CH_3)_3SiCl,\ (CH_3CH_2)_3N}$$

(b) $$\underset{\text{(added to base)}}{CH_3CH_2\overset{\overset{\large O}{\|}}{C}CH_3} \xrightarrow[\text{1,2-dimethoxyethane}]{(CH_3\overset{\overset{CH_3}{|}}{C}H)_2N^-Li^+} \xrightarrow{(CH_3)_3SiCl,\ (CH_3CH_2)_3N}$$

16.2 REACTIONS OF ENOLS AND ENOLATE ANIONS WITH HALOGENS AS ELECTROPHILES

A. Halogenation Reactions in Acidic Media

Ketones react with halogens to give products in which a halogen atom is substituted for one of the α-hydrogen atoms. For example, acetone reacts with bromine to give chiefly 1-bromo-2-propanone.

$$\underset{\text{acetone}}{CH_3\overset{\overset{\large O}{\|}}{C}CH_3} + \underset{\text{bromine}}{Br_2} \xrightarrow[65\ ^\circ C]{H_2O} \underset{\substack{\text{1-bromo-2-}\\ \text{propanone}\\ \sim 50\%}}{CH_3\overset{\overset{\large O}{\|}}{C}CH_2Br} + \underbrace{\underset{\substack{\text{1,1-dibromo-}\\ \text{2-propanone}}}{CH_3\overset{\overset{\large O}{\|}}{C}CHBr_2} + \underset{\substack{\text{1,3-dibromo-}\\ \text{2-propanone}}}{BrCH_2\overset{\overset{\large O}{\|}}{C}CH_2Br}}_{\text{small amounts}} + \underset{\substack{\text{hydrobromic}\\ \text{acid}}}{HBr}$$

A solution of bromine in water is acidic.

$$Br_2 + H_2O \rightleftharpoons HBr + HOBr$$

Thus, the reaction of acetone with bromine proceeds via an acid-catalyzed enolization of acetone (p. 634).

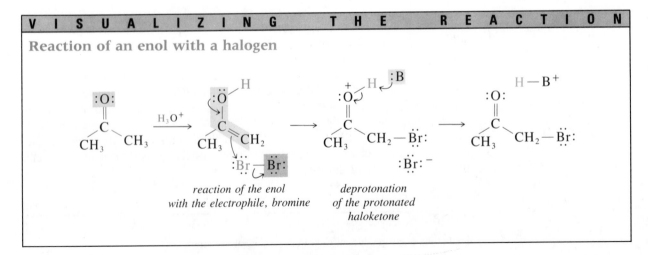

VISUALIZING THE REACTION

Reaction of an enol with a halogen

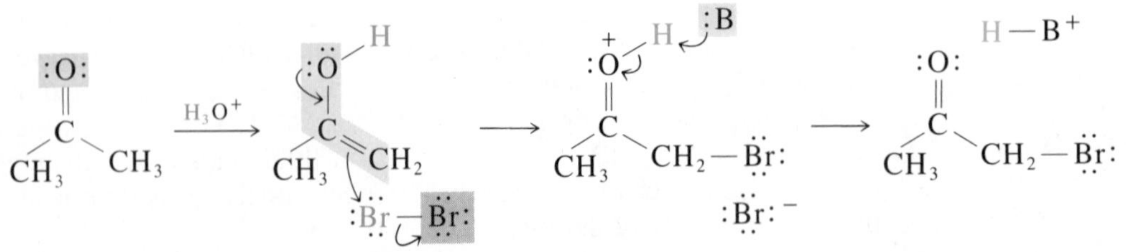

reaction of the enol with the electrophile, bromine

deprotonation of the protonated haloketone

As illustrated, the nucleophilic enol reacts with the electrophilic halogen to give the protonated haloketone. Deprotonation of this intermediate gives the final product.

Chemists have studied the mechanisms of this type of halogenation reaction in great detail. The rate of halogenation depends only on the concentration of the ketone. Also, the rate of halogenation is equal to the rate at which deuterium is introduced into acetone (p. 635). These facts suggest that the formation of the enol from the ketone is the rate-determining step for both of these reactions and that once the enol is formed, it reacts quickly with any electrophile that may be present.

This type of halogenation reaction is applicable to a wide range of compounds; another example is the chlorination of cyclohexanone.

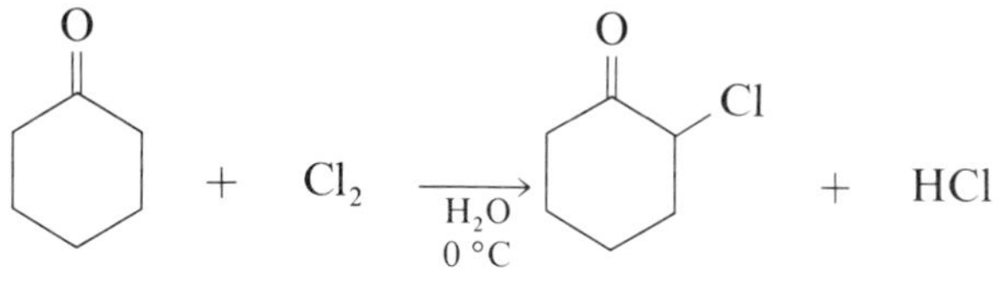
O
Cl₂
H₂O
0 °C
O
Cl
HCl

cyclohexanone chlorine 2-chlorocyclohexanone ~60% hydrochloric acid

In this case, too, one of the α-hydrogen atoms on the ketone is replaced by a halogen atom.

PROBLEM 16.4

(a) 3-Methyl-2,4-pentanedione reacts with bromine in water to give a product containing only one bromine atom. Write an equation for this reaction.
(b) If 3-methyl-2,4-pentanedione is dissolved in water and allowed to stand for a day, the initial rate at which this solution reacts with bromine is very fast. After awhile the bromine color disappears much more slowly but steadily. Explain these experimental observations.

PROBLEM 16.5

An α-hydrogen atom in a carboxylic acid can also be replaced by a halogen atom to give an α-haloacid. This reaction, which is known as the **Hell-Volhard-Zelinsky reaction,** works best if the acid is first converted in catalytic amounts into its acid halide. Write a mechanism for the reaction and explain why the halogenation works better with the acid halide than with the acid itself.

B. Halogenation Reactions in Basic Solution. The Haloform Reaction

When halogenation occurs in acidic solution, only one of the α-hydrogen atoms in a ketone is replaced by a halogen atom in the major product. The reaction does give small amounts of dihalogenated products, as seen for acetone (p. 640), and the addition of more halogen and longer reaction times lead to further substitution. It is not difficult, however, to stop the halogenation when only one halogen atom has been introduced into the ketone.

If a carbonyl compound is treated with halogen in the presence of base, however, the situation is quite different, as illustrated by the reaction of acetone with iodine in sodium hydroxide solution.

$$\underset{\text{acetone}}{CH_3\overset{\overset{\displaystyle O}{\|}}{C}CH_3} + \underset{\text{iodine}}{I_2 \text{ (excess)}} + \underset{\substack{\text{sodium}\\ \text{hydroxide}}}{NaOH} \xrightarrow[H_2O]{} \underset{\substack{\text{triiodomethane}\\ \text{iodoform}}}{CHI_3\downarrow} + \underset{\substack{\text{sodium}\\ \text{acetate}}}{CH_3\overset{\overset{\displaystyle O}{\|}}{C}O^-Na^+}$$

A basic solution of acetone gives a yellow precipitate almost as soon as iodine is added to the reaction mixture. The precipitate is triiodomethane, also known as iodoform. Many hospitals smell of iodoform, which is used as an antiseptic.

The reaction of a methyl ketone with a halogen in base is known as the **haloform reaction.** With chlorine, the product is trichloromethane (chloroform), and with bromine, it is tribromomethane (bromoform). The haloform reaction is typical of most methyl ketones as well as of compounds that can be oxidized to methyl ketones by the basic solution of halogen. For example, secondary alcohols such as 2-pentanol, in which the hydroxyl group is one carbon away from the end of the chain, also give a precipitate of iodoform when treated with iodine and sodium hydroxide.

$$\underset{\text{2-pentanol}}{CH_3CH_2CH_2\underset{\underset{\displaystyle OH}{|}}{C}HCH_3} \xrightarrow[\substack{H_2O\\ \textit{oxidizing}\\ \textit{agent}}]{I_2,\ NaOH} \underset{\text{2-pentanone}}{CH_3CH_2CH_2\overset{\overset{\displaystyle O}{\|}}{C}CH_3} \xrightarrow[H_2O]{I_2,\ NaOH}$$

$$\underset{\substack{\text{iodoform}\\ \textit{yellow}}}{CHI_3\downarrow} + \underset{\text{sodium butanoate}}{CH_3CH_2CH_2\overset{\overset{\displaystyle O}{\|}}{C}O^-Na^+} \xrightarrow{H_3O^+} \underset{\substack{\text{butanoic acid}\\ \textit{carboxylic acid with one}\\ \textit{fewer carbon atom than}\\ \textit{the starting alcohol or}\\ \textit{ketone}}}{CH_3CH_2CH_2\overset{\overset{\displaystyle O}{\|}}{C}OH}$$

Removing the methyl group from a methyl ketone as iodoform leaves the remainder of the molecule in the form of a carboxylate anion, which gives a carboxylic acid on protonation. Thus, the haloform reaction may be used to synthesize a carboxylic acid having one fewer carbon atom than the starting ketone.

Iodoform is very insoluble in water; even small amounts of it precipitate and can be detected. This reaction, called the **iodoform reaction,** is useful in qualitative analysis as a test for methyl ketones and compounds that can be oxidized to methyl ketones. Acetaldehyde is the only aldehyde that gives a positive test, and ethanol the only primary alcohol.

$$\underset{\text{ethanol}}{CH_3CH_2OH} \xrightarrow[\substack{NaOH\\ H_2O}]{I_2} \underset{\text{acetaldehyde}}{CH_3\overset{\overset{\displaystyle O}{\|}}{C}H} \xrightarrow[\substack{NaOH\\ H_2O}]{I_2} CHI_3\downarrow + \underset{\substack{\text{sodium}\\ \text{formate}}}{H\overset{\overset{\displaystyle O}{\|}}{C}O^-Na^+}$$

A haloform reaction proceeds by rapid replacement of the α-hydrogen atoms via the enolate ion present in the basic solution.

VISUALIZING THE REACTION

The haloform reaction

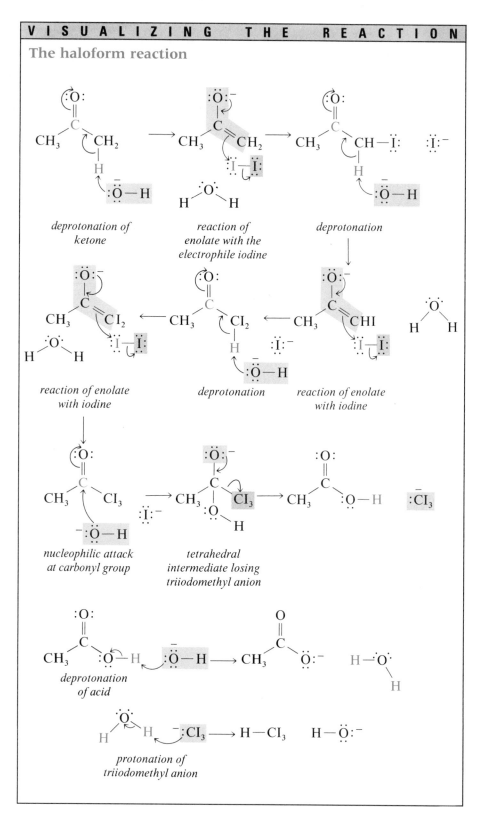

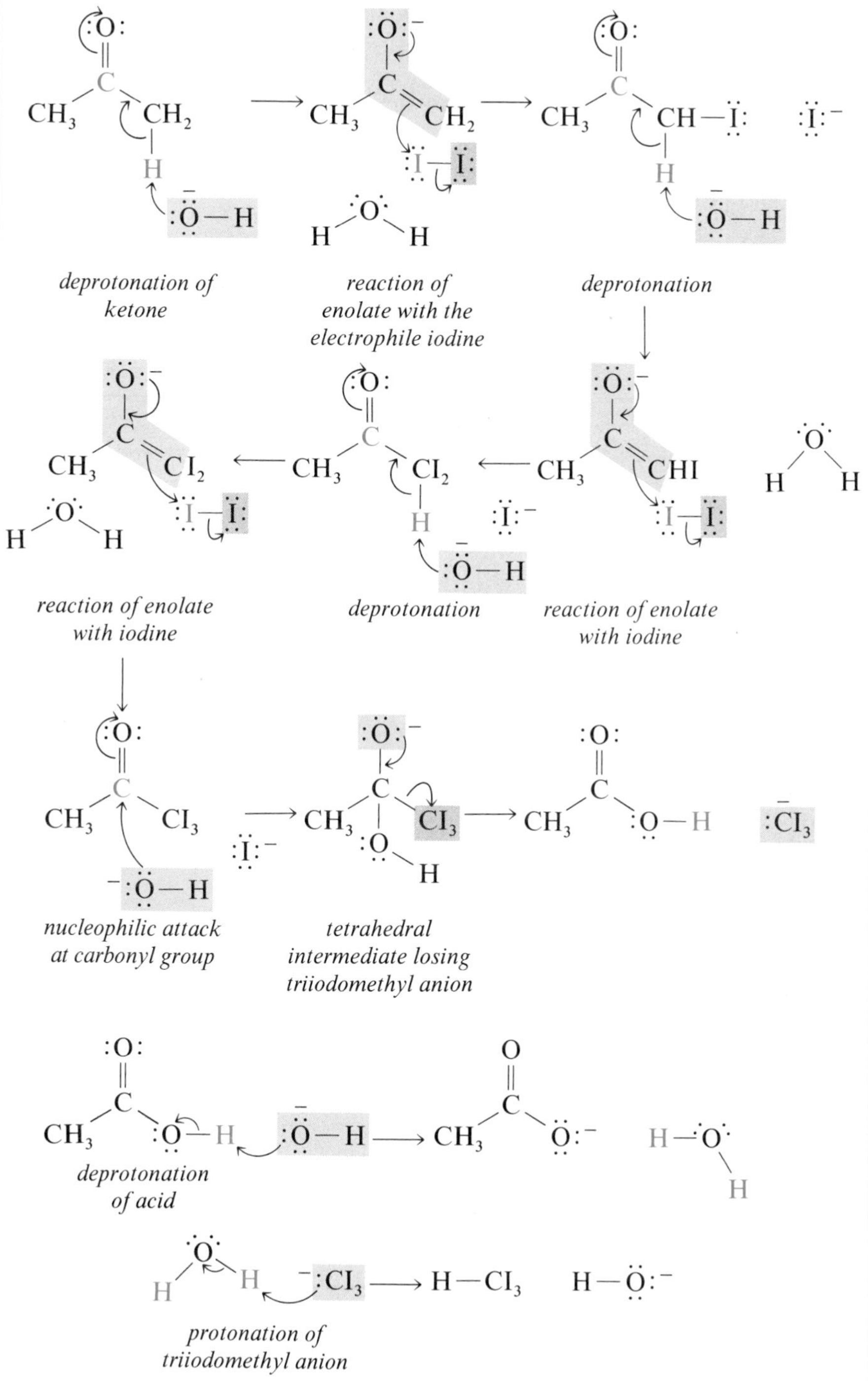
deprotonation of ketone
reaction of enolate with the electrophile iodine
deprotonation
reaction of enolate with iodine
deprotonation
reaction of enolate with iodine
nucleophilic attack at carbonyl group
tetrahedral intermediate losing triiodomethyl anion
deprotonation of acid
protonation of triiodomethyl anion

Once one of the α-hydrogen atoms is replaced by a halogen atom, the remaining hydrogen atoms on the same carbon atom become more acidic because of the electron-withdrawing effect of the halogen. Each successive hydrogen atom is more easily substituted, and the trihalogenated compound is the product. In some cases, such a product can be isolated, but most of the time it is so unstable that it falls apart when the carbonyl group is attacked by base. The methyl group adjacent to the carbonyl group is converted into a reasonably good leaving group by substitution of halogen atoms for hydrogen atoms. In the case of iodine, especially, the bulkiness of the iodine atoms results in a very crowded molecule. The electron-withdrawing effect of the three halogen atoms also increases the electrophilicity of the carbon atom of the carbonyl group. All these factors combine to make it possible to break a carbon-carbon bond in the ketone by nucleophilic attack of a hydroxyl ion on the carbonyl group. The triiodomethyl anion is a stabilized carbanion and picks up a proton from the solvent to give iodoform.

Study Guide Concept Map 16.3

PROBLEM 16.6

Complete the following equations.

(a) $CH_3CH_2OH + NaOH + Br_2 \xrightarrow[H_2O]{}$

(b) 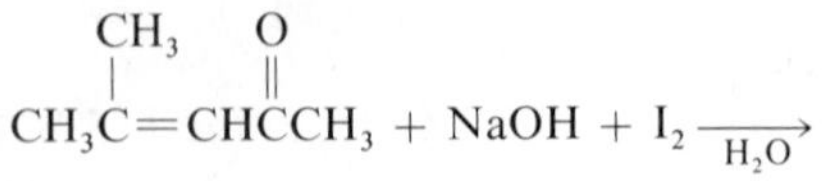

(c) 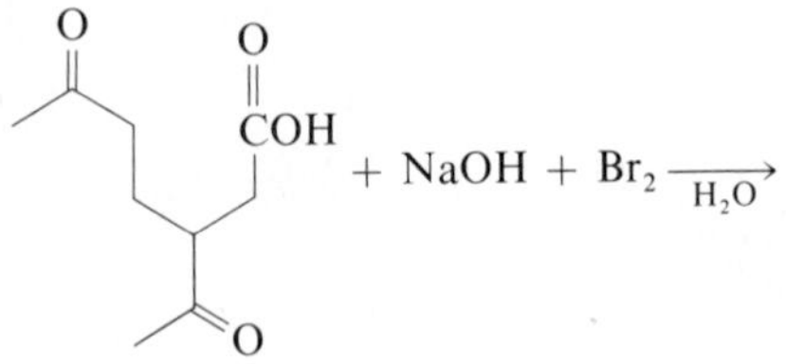

(d) 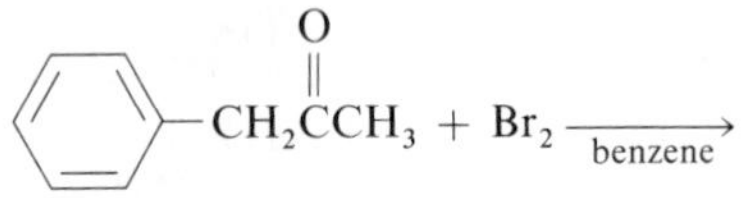

(Hint: Which will be the more stable enol form for 1-phenyl-2-propanone?)

(e) 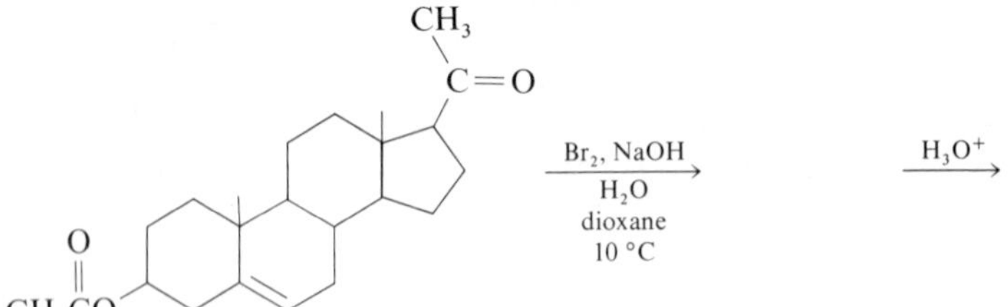

(f) 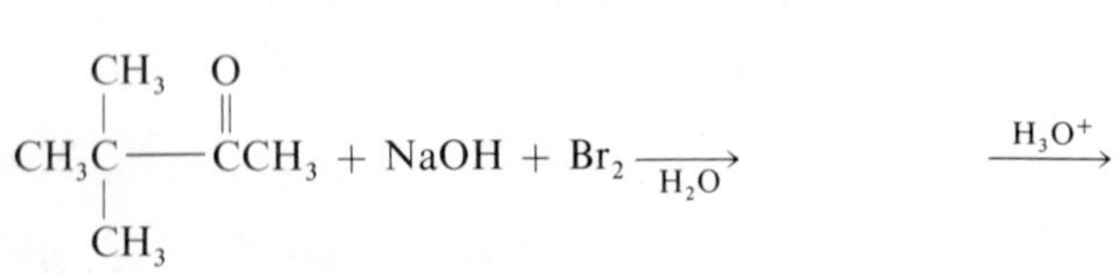

(g) $CH_3\overset{\overset{O}{\|}}{C}CH_3 + NaOH + Cl_2 \xrightarrow[H_2O]{}$

A. Alkylation of Ketones

Alkylation of an enolate anion gives a compound with a new substituent at the α-carbon atom (p. 636). For example, the enolate anion generated when 2-benzylcyclohexanone is deprotonated by lithium diisopropylamide reacts in a nucleophilic substitution reaction with methyl iodide.

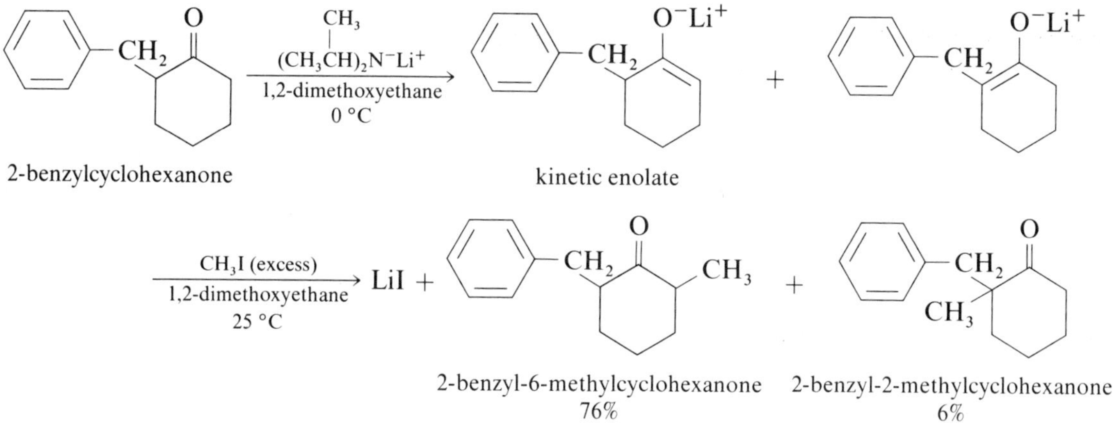
$(CH_3CH)_2N^-Li^+$
1,2-dimethoxyethane
0 °C
2-benzylcyclohexanone
kinetic enolate
CH_3I (excess)
1,2-dimethoxyethane
25 °C
LiI +
2-benzyl-6-methylcyclohexanone
76%
2-benzyl-2-methylcyclohexanone
6%

The major product is the one derived from the kinetic enolate (p. 639).

Intramolecular alkylation reactions of enolate anions have been used to make fused rings. For example, as shown below, cyclization of a disubstituted cyclopentane gives two different fused-ring compounds depending on whether the reaction conditions used favor the kinetic enolate or the thermodynamic enolate.

VISUALIZING THE REACTION

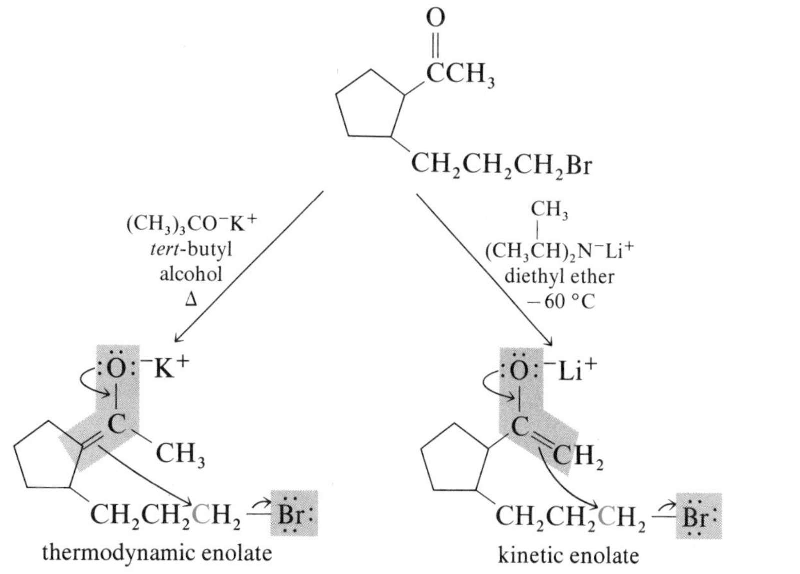
$CH_2CH_2CH_2Br$
$(CH_3)_3CO^-K^+$
tert-butyl
alcohol
Δ
$(CH_3CH)_2N^-Li^+$
diethyl ether
−60 °C
thermodynamic enolate
kinetic enolate

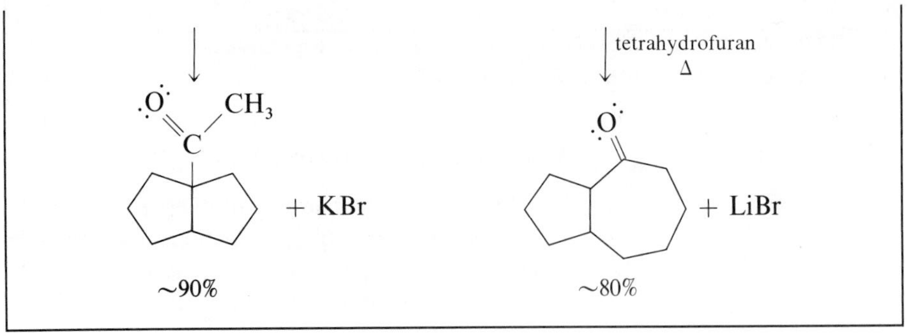
tetrahydrofuran
Δ
+ KBr
~90%
+ LiBr
~80%

B. Alkylation of Esters

Enolate anions of esters can also be prepared using lithium diisopropylamide. Methyl butanoate is converted into methyl-2-ethylbutanoate by alkylation of its enolate.

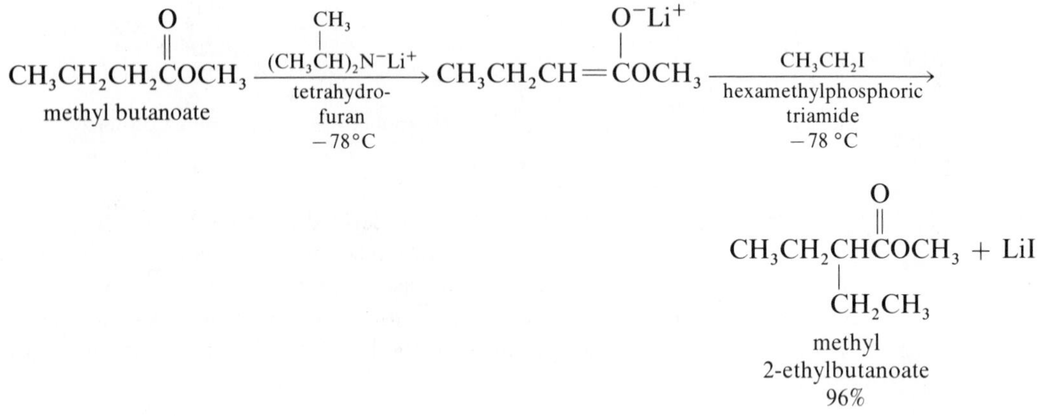
methyl butanoate
tetrahydro-
furan
−78°C
hexamethylphosphoric
triamide
−78 °C
+ LiI
methyl
2-ethylbutanoate
96%

Lactones undergo similar reactions. For example, the lactone from 4-hydroxybutanoic acid is substituted at carbon atom 2 of the ring by an allyl group via deprotonation and alkylation.

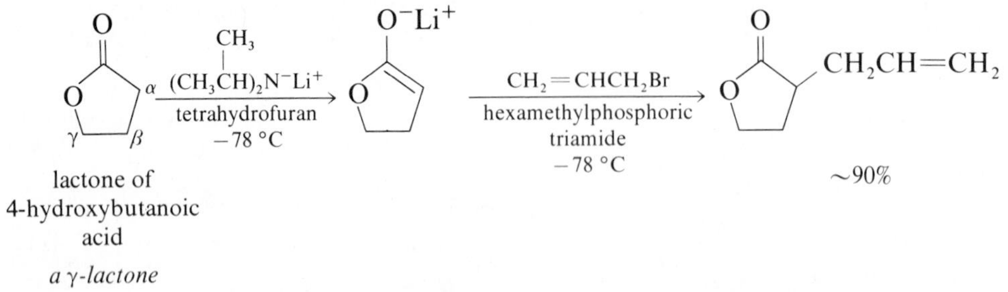
tetrahydrofuran
−78 °C
hexamethylphosphoric
triamide
−78 °C
~90%
lactone of
4-hydroxybutanoic
acid
a γ-lactone

The fact that lithium diisopropylamide reacts with esters and lactones to give products of deprotonation reactions rather than those of nucleophilic substitution reactions at the carbonyl group (p. 582) is further evidence of how low the nucleophilicity of the diisopropylamide anion is.

C. Problem-Solving Skills

Problem

Predict the product(s) of the following sequence of reactions, including the stereochemistry.

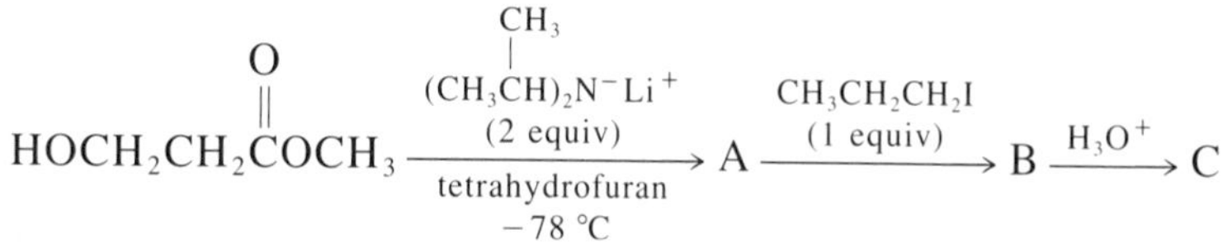

HOCH2CH2COCH3
CH3
(CH3CH)2N− Li+
(2 equiv)
tetrahydrofuran
−78 °C
A
CH3CH2CH2I
(1 equiv)
B
H3O+
C

Solution

1. To what functional group class does the reactant belong? What is the electronic character of the functional group?

   The reactant has both alcohol and ester functional groups.

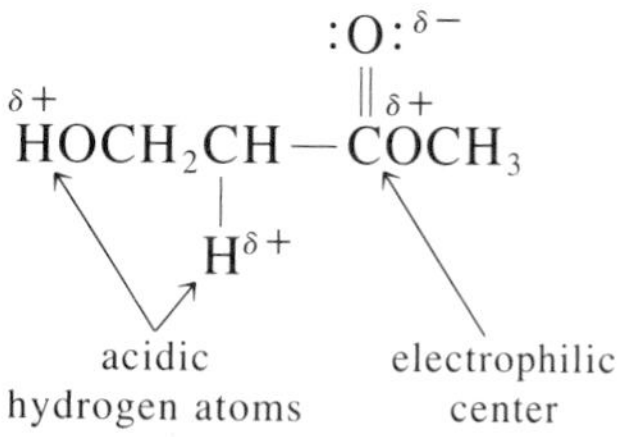

acidic
hydrogen atoms
electrophilic
center

   The hydrogen of the hydroxyl group is more acidic (pK_a ~17) than the hydrogens alpha to the carbonyl group (pK_a ~23).

2. Does the reactant have a good leaving group?

   No, there is no good leaving group present.

3. What reagents are present? Are they good acids, bases, nucleophiles, or electrophiles? Is there an ionizing solvent present?

   Lithium diisopropylamide is a strong base. Two equivalents are used.

4. What is the most likely first step for the reaction: protonation or deprotonation, ionization, attack by a nucleophile, or attack by an electrophile?

   The first step will be deprotonation at the hydroxyl group, which has the most acidic proton, followed by deprotonation at the α-carbon.

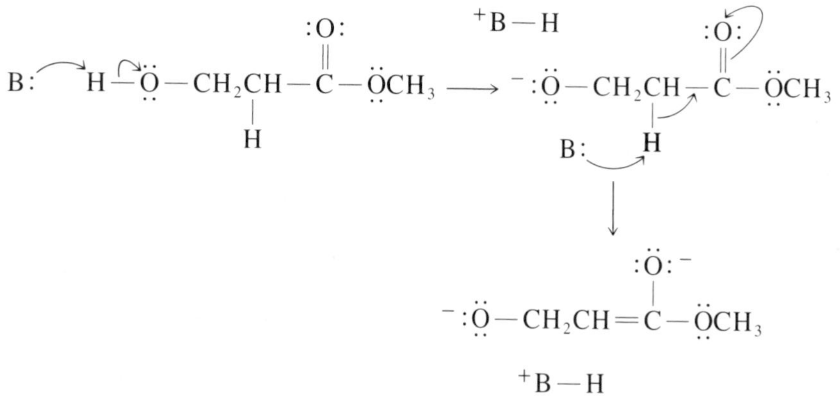

5. What are the properties of the species present in the reaction mixture after the first step? What is likely to happen next?

   A dianion that is an alkoxide ion and an enolate anion is formed. It is strongly basic and nucleophilic. It will react with the electrophilic carbon atom of the alkyl halide, the reagent for the second step.

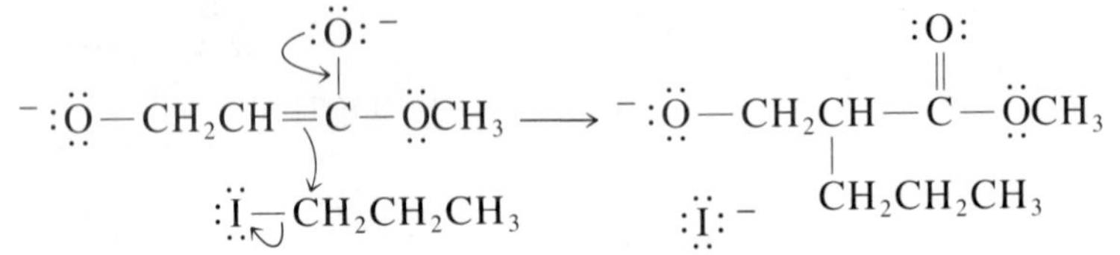

   Reaction occurs at the carbon atom; a carbanion is more nucleophilic than an oxygen anion (p. 227). The product of this reaction is an alkoxide ion, which is protonated by aqueous acid, the reagent in the third step of the reaction.

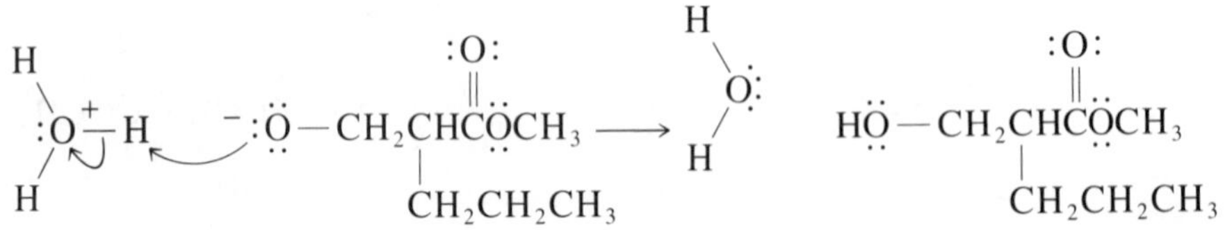

6. What is the stereochemistry of the reaction?

   The starting compound is achiral. Substitution at the α-carbon atom creates a stereocenter. The product is a racemic mixture.

The complete correct answer is as follows:

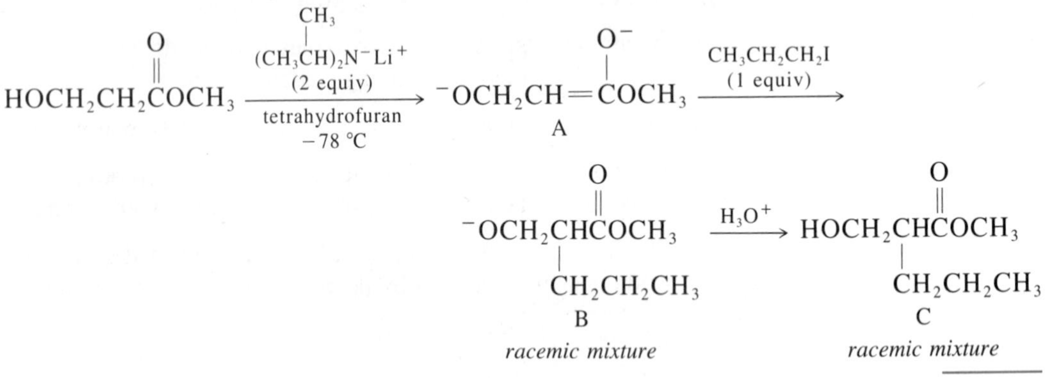

PROBLEM 16.7

Supply structural formulas for the reagents, intermediates, and products indicated by letters.

(a) 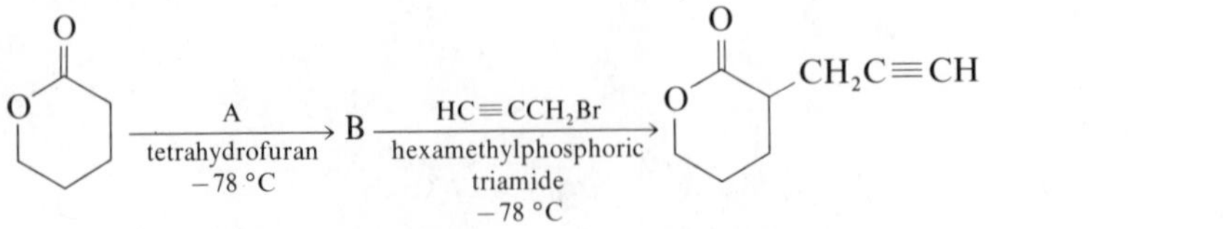

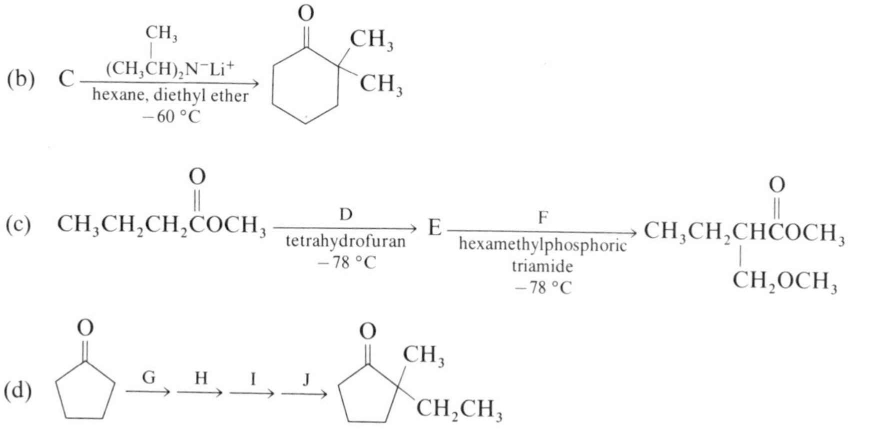
(b)
C
$(CH_3CH)_2N^-Li^+$
hexane, diethyl ether
−60 °C
(c)
$CH_3CH_2CH_2COCH_3$
D
tetrahydrofuran
−78 °C
E
F
hexamethylphosphoric triamide
−78 °C
$CH_3CH_2CHCOCH_3$
CH_2OCH_3
(d)
G H I J
CH_2CH_3

16.4 REACTION OF STABILIZED ENOLATE ANIONS WITH ALKYL HALIDES AS ELECTROPHILES

A. Alkylation of Ethyl Acetoacetate. Decarboxylation of β-Ketoacids

The enolate anions of compounds containing active methylene groups are formed easily and with high regioselectivity. For example, on p. 635 we saw that only one of the seven α-hydrogen atoms in 3-methyl-2,4-pentanedione is exchanged with deuterium oxide in an uncatalyzed reaction.

Ethyl acetoacetate enolizes regioselectively to one of the two possible enolate anions, the one stabilized by delocalization of charge to two oxygen atoms (p. 632). The regioselectivity of this enolization allows ethyl acetoacetate to be alkylated selectively at the active methylene group. The product of the alkylation reaction is a β-ketoester. Saponification (p. 563) of the ester group with dilute base followed by careful acidification with cold dilute acid causes an unstable β-ketocarboxylic acid to be formed. This acid loses carbon dioxide on heating. A ketone is the product. The overall two-step sequence starting from a β-ketoester is a synthesis for ketones. Ethyl acetoacetate, for example, is converted to 2-heptanone by this sequence of reactions. The alkylation of ethyl acetoacetate with n-butyl bromide (p. 637) gives ethyl 2-butylacetoacetate. Hydrolysis of this ester and decarboxylation of the β-ketoacid give 2-heptanone.

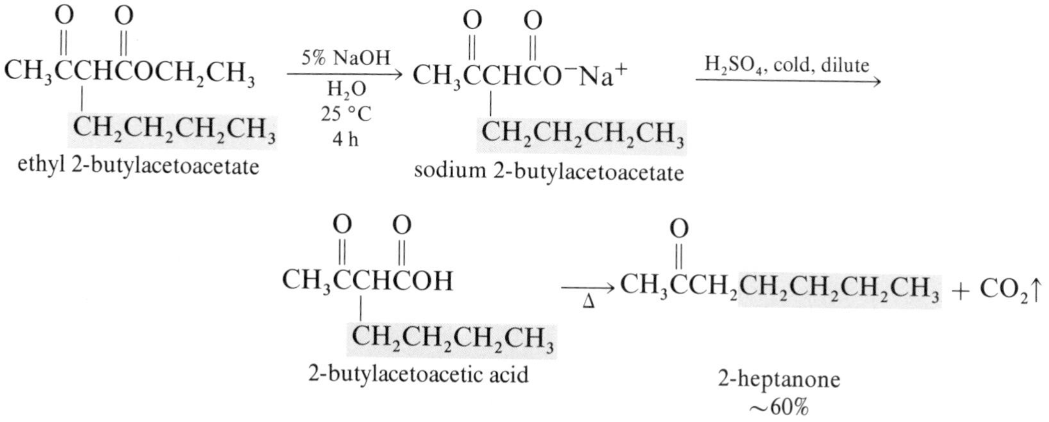
$CH_3CCHCOCH_2CH_3$
$CH_2CH_2CH_2CH_3$
ethyl 2-butylacetoacetate
5% NaOH
H_2O
25 °C
4 h
$CH_3CCHCO^-Na^+$
$CH_2CH_2CH_2CH_3$
sodium 2-butylacetoacetate
H_2SO_4, cold, dilute
$CH_3CCHCOH$
$CH_2CH_2CH_2CH_3$
2-butylacetoacetic acid
Δ
$CH_3CCH_2CH_2CH_2CH_2CH_3 + CO_2\uparrow$
2-heptanone
~60%

Thus, the ester group in ethyl acetoacetate serves to activate the α-hydrogen atoms, making their regioselective substitution possible. The ester group is removed easily once it has performed this function.

Carboxylic acids with carbonyl groups beta to the carboxyl function are unstable because they can react by means of a cyclic transition state. Transfer of a proton, loss of carbon dioxide, and the formation of an enol all occur simultaneously. For example, when carbon dioxide is lost from 2-butylacetoacetic acid, the species formed is the enol form of 2-heptanone. Deprotonation at the oxygen atom and protonation of the α-carbon atom converts it to the ketone.

VISUALIZING THE REACTION

Decarboxylation of a β-ketoacid

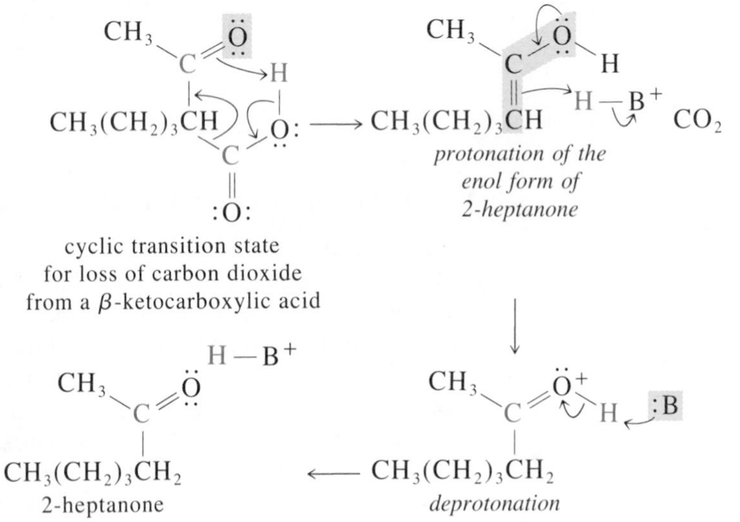
cyclic transition state
for loss of carbon dioxide
from a β-ketocarboxylic acid
protonation of the
enol form of
2-heptanone
deprotonation
2-heptanone

Methyl ketones are formed when ethyl acetoacetate is the starting material, but other β-ketoesters can also be used. For example, 2-ethylcyclopentanone is synthesized from ethyl 2-oxocyclopentanecarboxylate using similar reactions. The synthesis of the β-ketoesters that are the starting materials for these reactions is described on p. 662.

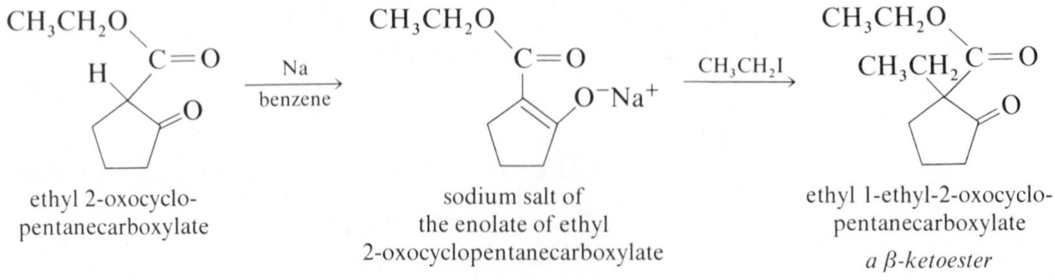
Na
benzene
CH_3CH_2I
ethyl 2-oxocyclo-
pentanecarboxylate
sodium salt of
the enolate of ethyl
2-oxocyclopentanecarboxylate
ethyl 1-ethyl-2-oxocyclo-
pentanecarboxylate
a β-ketoester

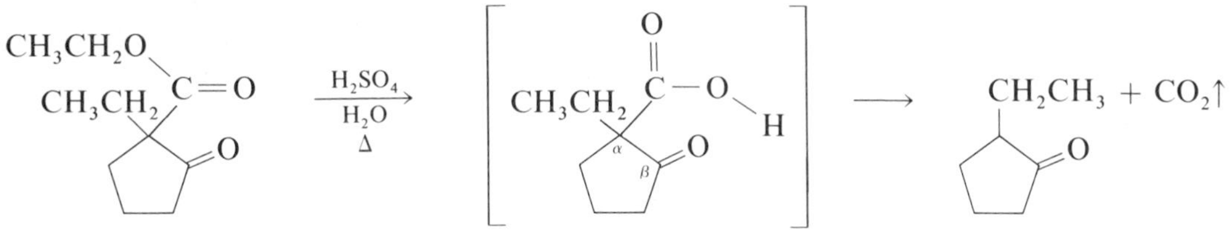

ethyl 1-ethyl-2-oxocyclopentanecarboxylate — 1-ethyl-2-oxocyclopentanecarboxylic acid, *a β-ketocarboxylic acid* — 2-ethylcyclopentanone

---

PROBLEM 16.8

Write a mechanism for the last step of the above reaction sequence, the formation of 2-ethylcyclopentanone from 1-ethyl-2-oxocyclopentanecarboxylic acid.

PROBLEM 16.9

Give structural formulas for all compounds symbolized by letters in the following equations.

(a) $$CH_3\overset{O}{\overset{\|}{C}}CH_2\overset{O}{\overset{\|}{C}}OCH_2CH_3 \xrightarrow[\text{ethanol}]{CH_3CH_2ONa} A \xrightarrow[\Delta]{ClCH_2\overset{O}{\overset{\|}{C}}OCH_2CH_3} B \xrightarrow[\Delta]{H_3O^+} C$$

(b) 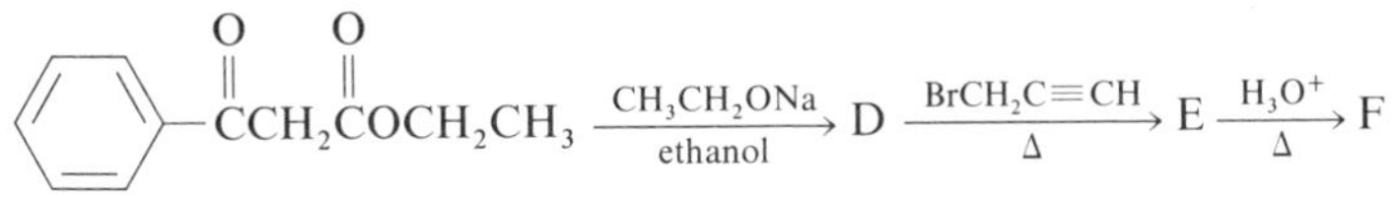

(c) $$CH_3\overset{O}{\overset{\|}{C}}CH_2\overset{O}{\overset{\|}{C}}OCH_2CH_3 \xrightarrow[\text{ethanol}]{CH_3CH_2ONa} G \xrightarrow{BrCH_2CH_2\overset{O}{\overset{\|}{C}}OCH_2CH_3} H$$

(d) $$CH_3\overset{O}{\overset{\|}{C}}CH_2\overset{O}{\overset{\|}{C}}OCH_2CH_3 \xrightarrow[\text{ethanol}]{CH_3CH_2ONa\text{ (1 molar equivalent)}} I \xrightarrow{ClCH_2CH_2CH_2Br} J$$

(e) $$CH_3\overset{O}{\overset{\|}{C}}CH_2\overset{O}{\overset{\|}{C}}OCH_2CH_3 \xrightarrow[\text{ethanol}]{CH_3CH_2ONa} K \xrightarrow{CH_2{=}CHCH_2CH_2Br} L \xrightarrow[\text{toluene}]{M} N \xrightarrow{CH_3\overset{I}{\overset{|}{C}}HCH_3} O$$

B. Alkylation of Diethyl Malonate

Diethyl malonate also has an active methylene group and is a useful starting material for a number of syntheses. The enolate anion displaces bromide ion from *sec*-butyl bromide to give the substituted ester.

$$\underset{\text{diethyl malonate}}{CH_2(\overset{O}{\overset{\|}{C}}OCH_2CH_3)_2} + \underset{\text{sodium ethoxide}}{CH_3CH_2O^-Na^+} \longrightarrow \underset{\substack{\text{sodium salt of}\\ \text{the enolate of}\\ \text{diethyl malonate,}\\ \text{shown as carbanion}}}{Na^+ :\overset{-}{C}H(\overset{O}{\overset{\|}{C}}OCH_2CH_3)_2} + CH_3CH_2OH$$

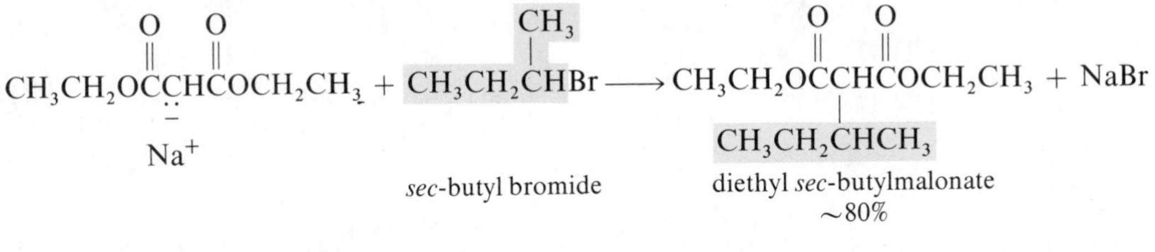

$$\underset{\underset{\underset{\text{Na}^+}{|}}{\overset{\overset{\overset{\displaystyle O}{\parallel}}{}}{\text{CH}_3\text{CH}_2\text{OCCHCOCH}_2\text{CH}_3}} + \underset{\textit{sec}\text{-butyl bromide}}{\overset{\overset{\displaystyle CH_3}{|}}{\text{CH}_3\text{CH}_2\text{CHBr}}} \longrightarrow \underset{\underset{\underset{\sim 80\%}{\text{diethyl \textit{sec}-butylmalonate}}}{\text{CH}_3\text{CH}_2\text{CHCH}_3}}{\overset{\overset{\displaystyle O\ \ O}{\parallel\ \parallel}}{\text{CH}_3\text{CH}_2\text{OCCHCOCH}_2\text{CH}_3}} + \text{NaBr}$$

The substituted malonic ester is interesting for several reasons. When the diester is hydrolyzed, the resulting dicarboxylic acid is unstable and loses carbon dioxide on heating. This reaction is another illustration of the instability of molecules in which a carboxyl group has a carbonyl group beta to it (p. 650). Thus, diethyl *sec*-butylmalonate can be saponified and then heated with acid to remove one of the carboxyl groups.

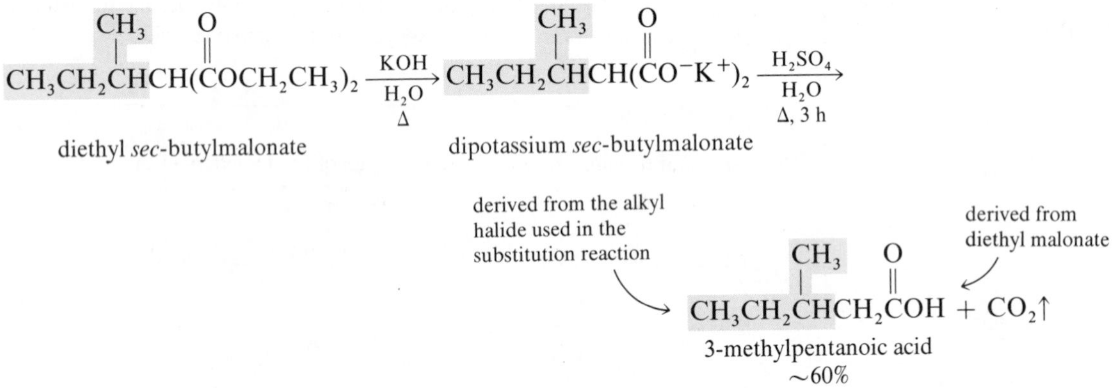

$$\underset{\text{diethyl \textit{sec}-butylmalonate}}{\overset{\overset{\displaystyle CH_3\quad O}{|\qquad\parallel}}{\text{CH}_3\text{CH}_2\text{CHCH(COCH}_2\text{CH}_3)_2}} \xrightarrow[\substack{\text{H}_2\text{O} \\ \Delta}]{\text{KOH}} \underset{\text{dipotassium \textit{sec}-butylmalonate}}{\overset{\overset{\displaystyle CH_3\quad O}{|\qquad\parallel}}{\text{CH}_3\text{CH}_2\text{CHCH(CO}^-\text{K}^+)_2}} \xrightarrow[\substack{\text{H}_2\text{O} \\ \Delta,\ 3\ \text{h}}]{\text{H}_2\text{SO}_4}$$

derived from the alkyl halide used in the substitution reaction

derived from diethyl malonate

$$\underset{\underset{\sim 60\%}{\text{3-methylpentanoic acid}}}{\overset{\overset{\displaystyle CH_3\quad O}{|\qquad\parallel}}{\text{CH}_3\text{CH}_2\text{CHCH}_2\text{COH}}} + \text{CO}_2\uparrow$$

The loss of carbon dioxide does not start until the reaction mixture is acidified and heated; then the evolution of a gas is observed.

The reaction sequence that consists of alkylation of diethyl malonate, hydrolysis of the new diester, and then decarboxylation of the diacid results in the conversion of *sec*-butyl bromide to a carboxylic acid that has two more carbon atoms than the halide. This sequence is a general synthetic method for carboxylic acids. These acids may be regarded as substituted acetic acids. The carboxyl group and the α-carbon atom are derived from diethyl malonate, and the rest of the molecule from the alkyl halide used in the substitution reaction.

A monosubstituted malonic ester still has a hydrogen atom on the carbon atom between the two ester groups, so another enolization and alkylation are possible. The presence of one alkyl group on that carbon makes the remaining hydrogen less acidic, and sometimes a stronger base must be used to remove it. For example, sodium *tert*-butoxide, which is a stronger base than sodium ethoxide, is used to form the enolate from diethyl isopropylmalonate in order to introduce a second alkyl group.

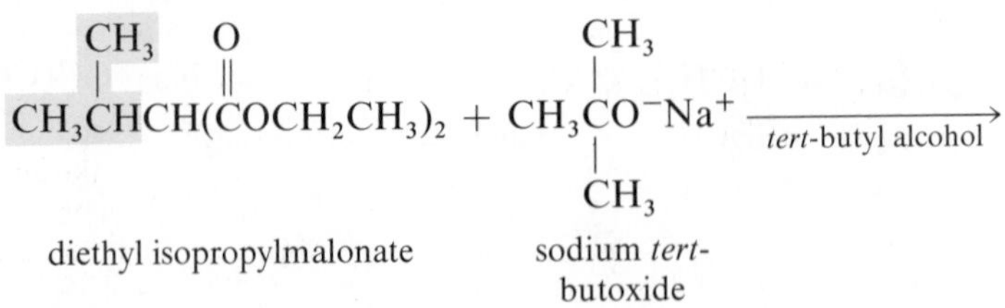

$$\underset{\text{diethyl isopropylmalonate}}{\overset{\overset{\displaystyle CH_3\quad O}{|\qquad\parallel}}{\text{CH}_3\text{CHCH(COCH}_2\text{CH}_3)_2}} + \underset{\substack{\text{sodium \textit{tert-}}\\\text{butoxide}}}{\overset{\overset{\displaystyle CH_3}{|}}{\underset{\underset{\displaystyle CH_3}{|}}{\text{CH}_3\text{CO}^-\text{Na}^+}}} \xrightarrow[\textit{tert}\text{-butyl alcohol}]{}$$

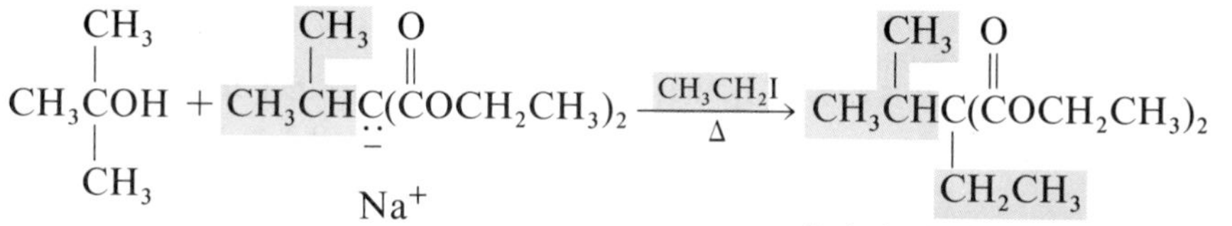

diethyl ethylisopropylmalonate
65%

Hydrolysis and decarboxylation of such a disubstituted malonic ester produces a carboxylic acid branched at the α-carbon atom.

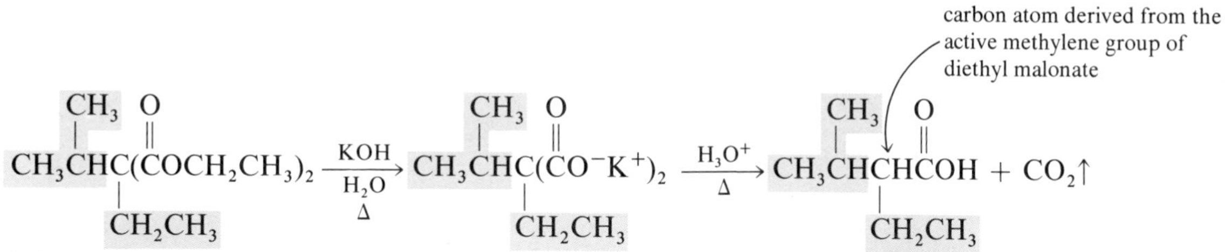
carbon atom derived from the active methylene group of diethyl malonate

diethyl ethylisopropylmalonate

2-ethyl-3-methylbutanoic acid

Study Guide Concept Map 16.4

PROBLEM 16.10

Write a mechanism for the decarboxylation of malonic acid.

PROBLEM 16.11

Supply structural formulas for products indicated by letters in the following equations.

(a) 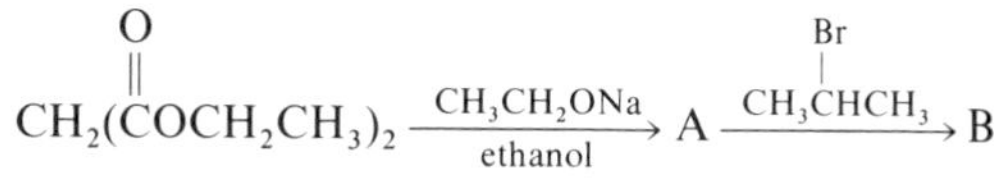

(b) 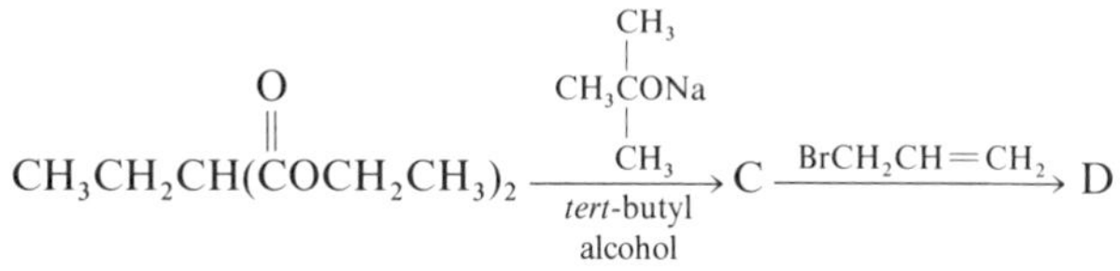

(c) 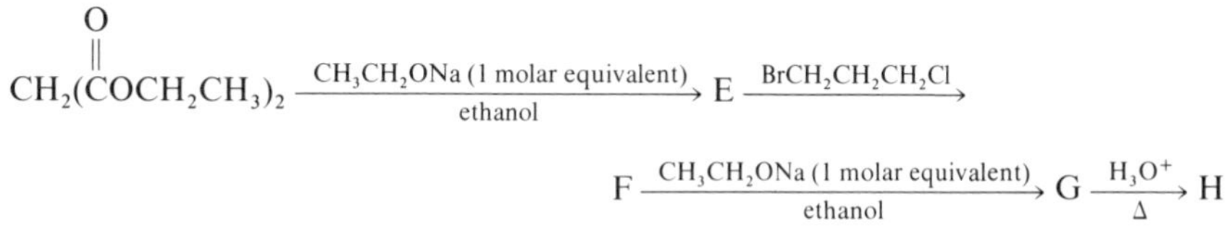

(d) 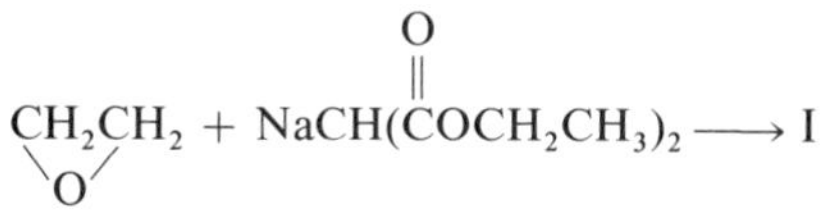

(Hint: A review of p. 454 may be helpful.)

(e) 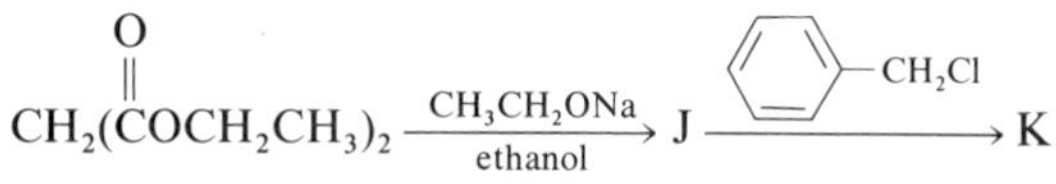

C. Barbituric Acid Derivatives

Another reason why reactions that introduce substituents on the carbon of an active methylene group are interesting is because substituted malonic esters react with urea to give compounds of great medical interest, the **barbituric acid derivatives,** or barbiturates.

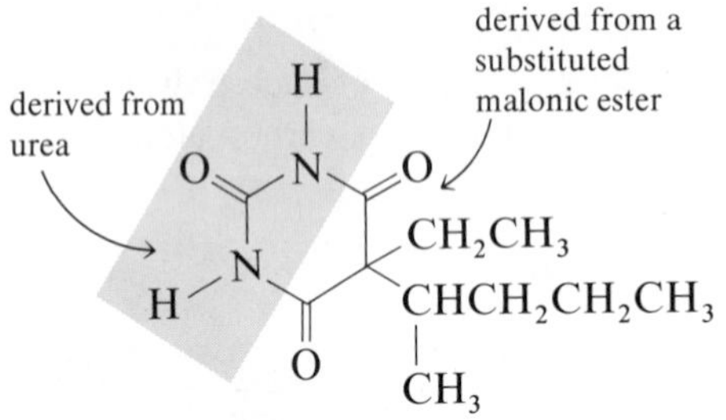
derived from a substituted malonic ester
derived from urea

pentobarbital
Nembutal

intermediate duration of action

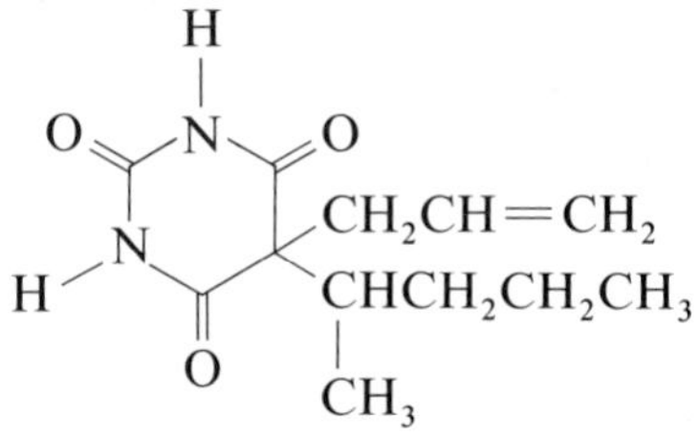

secobarbital
Seconal

short duration of action

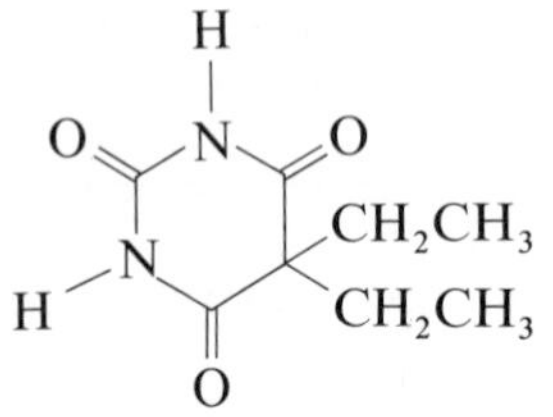

barbital
Veronal

long duration of action

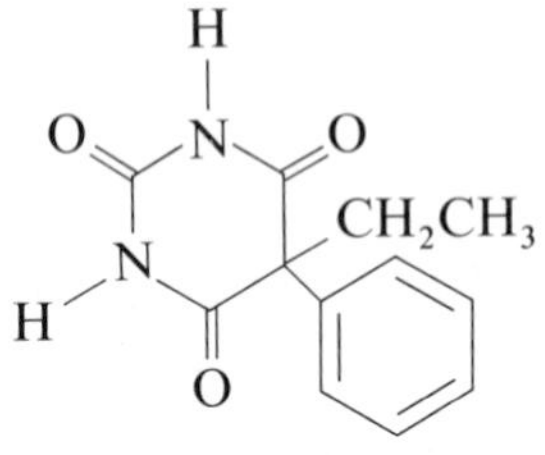

phenobarbital
Luminal

long duration of action

The physiological properties of the barbituric acid derivatives vary, depending on the substituents that are present on the carbon atom that was part of the active methylene group of diethyl malonate. Barbiturates act as sedatives and hypnotics, that is, sleeping pills. Some of them act rapidly to induce sleep but do not have long-lasting effects. Others last longer, but are slower to establish their effect. Some, such as phenobarbital, also prevent convulsions. Barbiturates act by depressing the activity of the central nervous system and, in high doses, also depress the respiratory system, which accounts for their toxicity. They are all addictive, meaning that a person who takes them with any regularity will suffer withdrawal symptoms when their use is discontinued.

The barbiturate Veronal is synthesized by the reaction of the diester of diethylmalonic acid with urea in the presence of sodium methoxide.

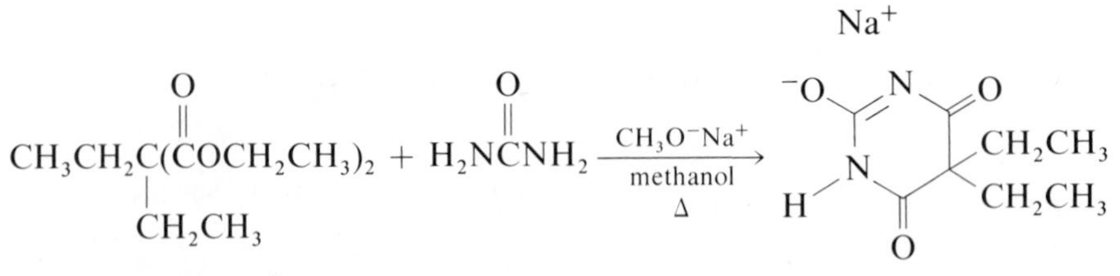
$CH_3O^-Na^+$
methanol
Δ

diethyl diethylmalonate — urea — sodium salt of 5,5-diethylbarbituric acid Veronal

A six-membered ring containing two nitrogen atoms separated by one carbon atom is called a **pyrimidine ring.** This ring system is found in barbituric acid derivatives. Thymine and uracil, bases found in DNA and RNA, are also pyrimidines and are discussed in greater detail in Section 24.6A.

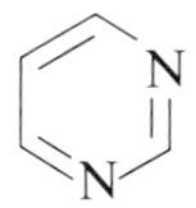

pyrimidine

The pyrimidines formed when malonic esters react with urea are acidic, which gives rise to the name barbituric acid derivatives. If the carbon atom between the two carbonyl groups still has an enolizable hydrogen atom, an enol form having aromatic character can be written for the molecule. The relationship to pyrimidine shows up most clearly in such a case. Even barbituric acid derivatives that have two substituents on that carbon atom ionize readily because the conjugate base formed by the loss of a proton from a nitrogen atom is stabilized by delocalization of charge to two carbonyl groups.

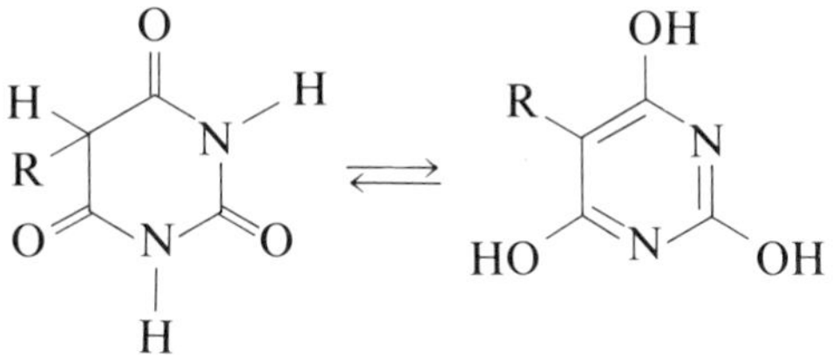

keto form of a monosubstituted barbituric acid

enol form of a monosubstituted barbituric acid

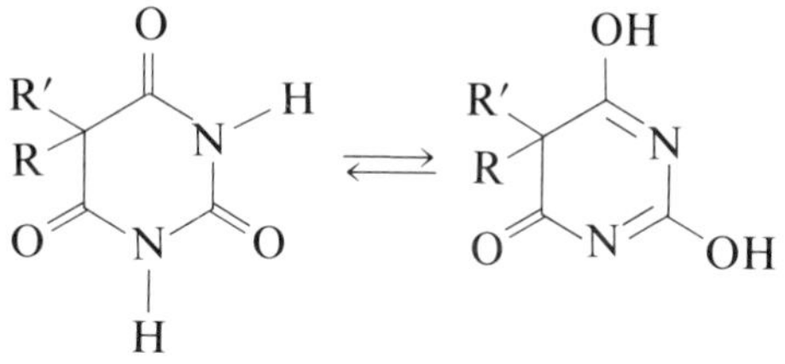

keto form of a disubstituted barbituric acid

enol form of a disubstituted barbituric acid

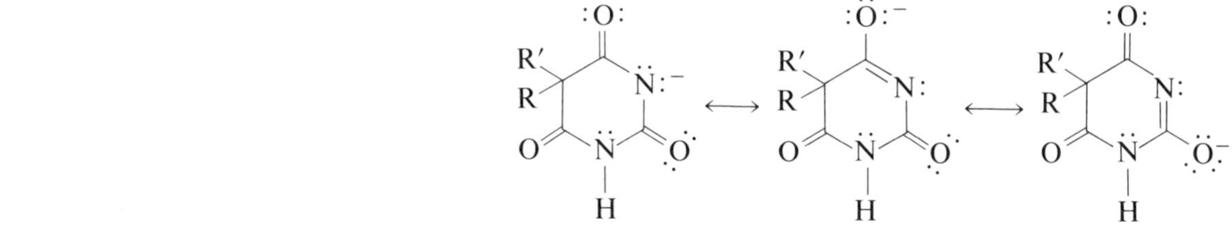

resonance contributors of the conjugate base of a disubstituted barbituric acid

Most barbituric acid derivatives are prepared and sold as their sodium salts. For example, Veronal is isolated as a salt (p. 654).

PROBLEM 16.12

The barbiturate Amytal has the structural formula shown below. Write equations for its synthesis, starting from urea, diethyl malonate, and any other compounds containing not more than three carbon atoms.

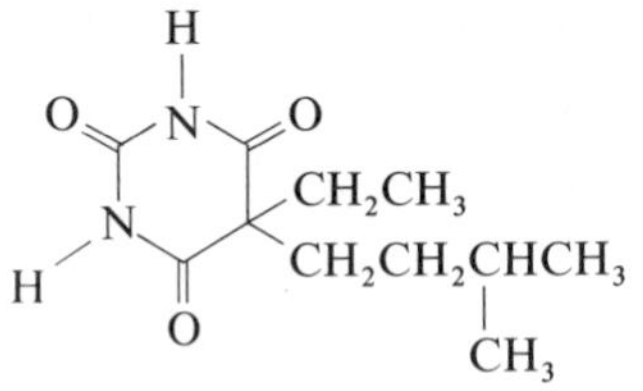

amobarbital
Amytal

16.5 REACTIONS OF ENOLATE ANIONS WITH CARBONYL COMPOUNDS

A. The Aldol Condensation

The reaction of nucleophilic carbon atoms with electrophilic carbon atoms to form carbon-carbon bonds is quite general. Reactions of this type form the basis of many synthetic transformations, including the ones described in the last two sections. Reactions of carbanions with carbonyl groups are particularly important in the creation of complex structures. Such reactions have already been discussed in connection with the addition of organometallic reagents to the carbonyl group of aldehydes or ketones (p. 491), of acids (p. 594), and of acid derivatives (p. 591).

Enolate ions also react with carbonyl groups to form new carbon-carbon bonds. For example, acetaldehyde reacts with itself in acid or base to give the β-hydroxyaldehyde that is called **aldol** because it is both an **ald**ehyde and an alcoh**ol.**

$$2\,CH_3\overset{\overset{O}{\|}}{C}H \xrightarrow[\substack{H_2O \\ 5\,°C \\ 1\,h}]{NaOH} \xrightarrow{H_3O^+} CH_3\overset{\overset{OH}{|}}{C}HCH_2\overset{\overset{O}{\|}}{C}H$$

acetaldehyde → 3-hydroxybutanal, aldol, 50%

The reaction involves the enolization of a molecule of acetaldehyde and then attack by that enolate ion on the carbonyl group of another molecule of acetaldehyde.

VISUALIZING THE REACTION

The aldol condensation

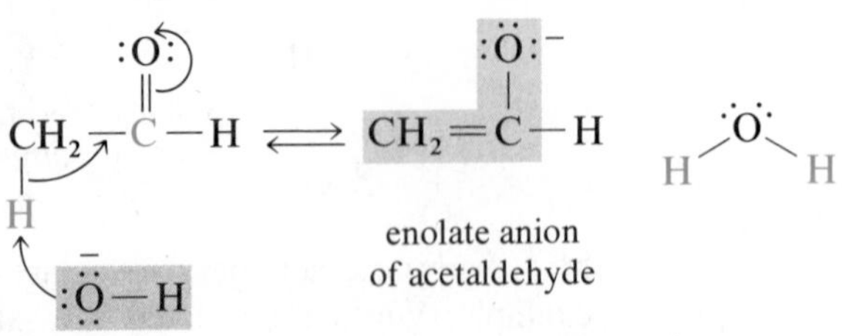
enolate anion of acetaldehyde

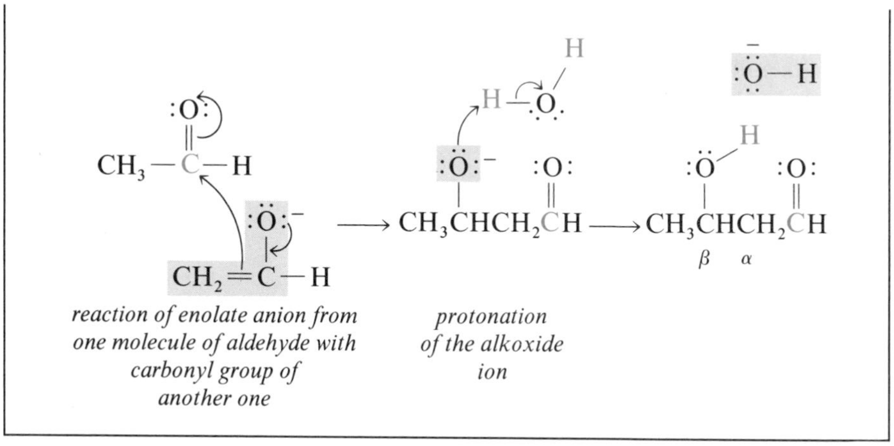
reaction of enolate anion from one molecule of aldehyde with carbonyl group of another one
protonation of the alkoxide ion

The resulting alkoxide ion is protonated by the solvent, water, which regenerates the catalyst, hydroxide ion. The product of the reaction shown here is a β-hydroxyaldehyde.

A reaction in which the enolate ion of one carbonyl compound reacts with the carbonyl group of another one is called an **aldol condensation.** This type of reaction is called a condensation because one larger molecule is created from the union of two smaller ones.

β-Hydroxyaldehydes are rather unstable and are easily dehydrated to compounds in which the double bond is in conjugation with the carbonyl group. If an aldol condensation is carried out at higher temperatures, the unsaturated compound is the product.

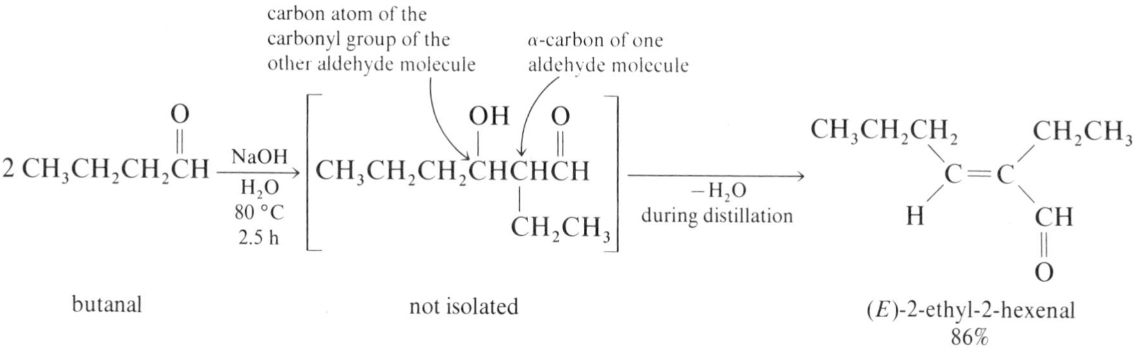
carbon atom of the carbonyl group of the other aldehyde molecule
α-carbon of one aldehyde molecule
NaOH
H2O
80 °C
2.5 h
−H2O
during distillation
butanal
not isolated
(E)-2-ethyl-2-hexenal
86%

Note that it is the α-carbon atom of one molecule of aldehyde that reacts with the carbonyl group of another molecule of aldehyde. Because acetaldehyde contains no more carbon atoms beyond the α-carbon, it gives a straight-chain product. Other aldehydes give a product that is branched at the α-carbon atom.

Ketones do not condense with themselves in aldol reactions as readily as aldehydes do. The carbonyl group of a ketone is more hindered and less electrophilic than the carbonyl group of an aldehyde. In the equilibrium between a ketone and its aldol condensation product, the ketone is favored unless the product is continually removed so that it cannot undergo the base-catalyzed retroaldol reaction (p. 660).

The simplest ketone, acetone, has been condensed with itself; the product is an α,β-unsaturated ketone.

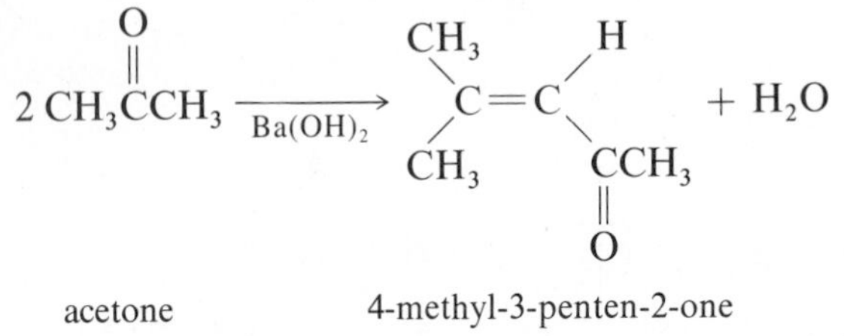

16 ENOLS AND ENOLATE
ANIONS AS NUCLEOPHILES I.
HALOGENATION, ALKYLATION,
AND CONDENSATION
REACTIONS

16.5 REACTIONS OF ENOLATE
ANIONS WITH CARBONYL
COMPOUNDS

$$2 \ CH_3\overset{\overset{\displaystyle O}{\|}}{C}CH_3 \xrightarrow{Ba(OH)_2} \ \underset{CH_3}{\overset{CH_3}{\diagdown}}C=C\underset{\underset{\overset{\|}{O}}{CCH_3}}{\overset{H}{\diagup}} + H_2O$$

acetone   4-methyl-3-penten-2-one

The two carbonyl compounds that participate in an aldol condensation do not have to be the same. Aldol reactions between two different compounds are possible. One of the compounds must be a source of enolate anions, and the other must have a carbonyl group for them to attack. For example, 2-pentanone is converted into the kinetic enolate by lithium diisopropylamide at low temperatures. Under these conditions, self-condensation of the ketone, which is an unfavorable reaction anyway, does not occur. The enolate anion reacts cleanly with the carbonyl group of butanal in the second step of the reaction.

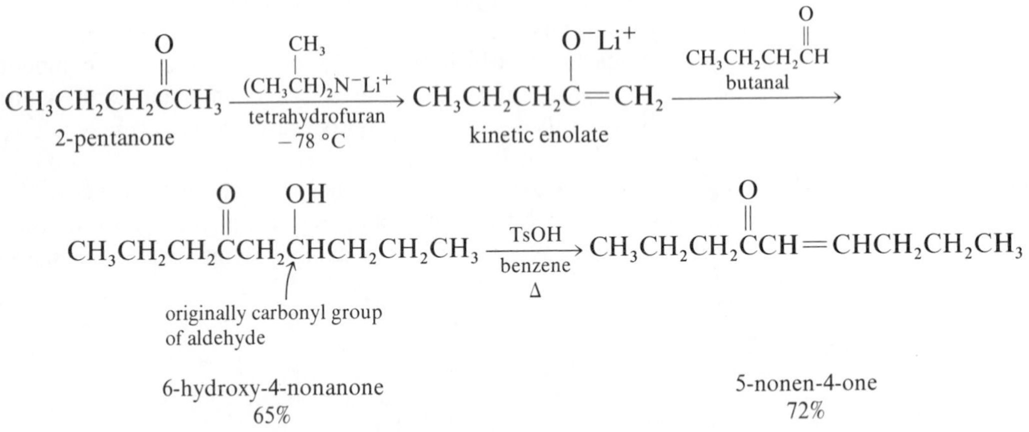

$$CH_3CH_2CH_2\overset{\overset{\displaystyle O}{\|}}{C}CH_3 \xrightarrow[\substack{\text{tetrahydrofuran} \\ -78\,°C}]{\overset{CH_3}{\overset{|}{(CH_3CH)_2N^-Li^+}}} CH_3CH_2CH_2\overset{\overset{\displaystyle O^-Li^+}{|}}{C}=CH_2 \xrightarrow[\text{butanal}]{CH_3CH_2CH_2\overset{\overset{\displaystyle O}{\|}}{CH}}$$

2-pentanone   kinetic enolate

$$CH_3CH_2CH_2\overset{\overset{\displaystyle O}{\|}}{C}CH_2\overset{\overset{\displaystyle OH}{|}}{C}HCH_2CH_2CH_3 \xrightarrow[\substack{\text{benzene} \\ \Delta}]{TsOH} CH_3CH_2CH_2\overset{\overset{\displaystyle O}{\|}}{C}CH=CHCH_2CH_2CH_3$$

originally carbonyl group
of aldehyde

6-hydroxy-4-nonanone   5-nonen-4-one
65%   72%

When one of the participants in an aldol condensation does not have any α-hydrogen atoms, the possibility for competing reactions is reduced. One reactant supplies enolate ions, and the other one provides the carbonyl group. Benzaldehyde is an aldehyde that reacts with enolate ions but cannot form one itself. It reacts with the enolate of acetone, for example, to give an α,β-unsaturated ketone.

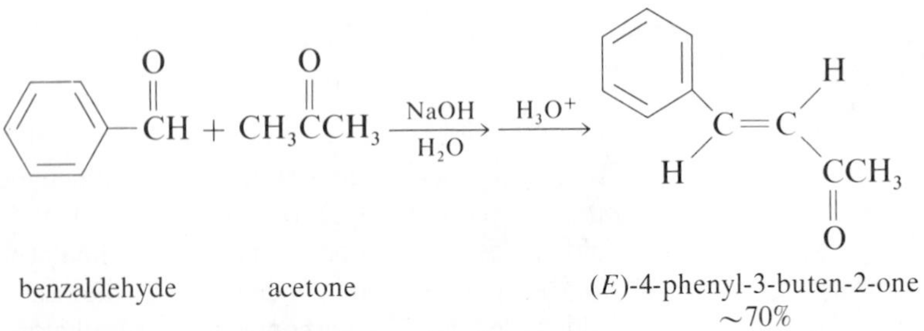

$$\text{C}_6\text{H}_5\overset{\overset{\displaystyle O}{\|}}{\text{CH}} + CH_3\overset{\overset{\displaystyle O}{\|}}{C}CH_3 \xrightarrow[H_2O]{NaOH} \xrightarrow{H_3O^+} \underset{H}{\overset{C_6H_5}{\diagdown}}C=C\underset{\underset{\overset{\|}{O}}{CCH_3}}{\overset{H}{\diagup}}$$

benzaldehyde   acetone   (E)-4-phenyl-3-buten-2-one
                                                            ~70%

In condensation reactions involving aryl aldehydes, the intermediate loses water with great ease because the product of that dehydration has a double bond in conjugation not only with the carbonyl group but with the aromatic ring as well. In most cases, the unsaturated compounds obtained from aldol condensations have the E configuration.

Any compound with an α,β-unsaturated carbonyl function in it is likely to have been synthesized by an aldol condensation. Each product shown in the four preceding equations can be dissected to show the carbonyl compounds from which it was created.

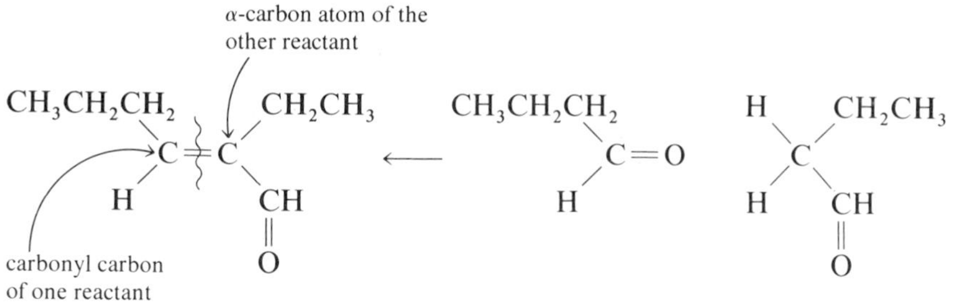
α-carbon atom of the other reactant
carbonyl carbon of one reactant

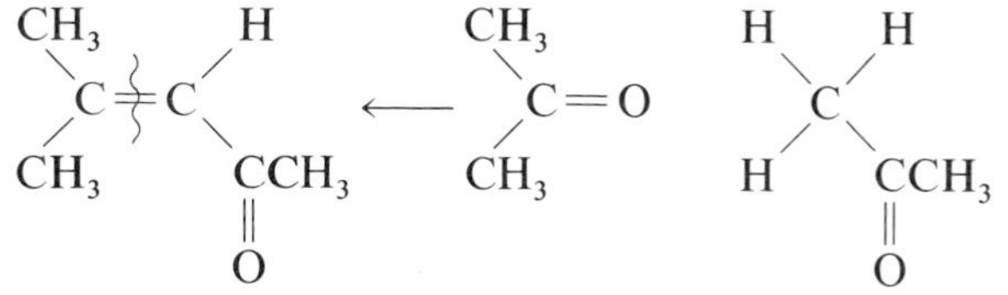

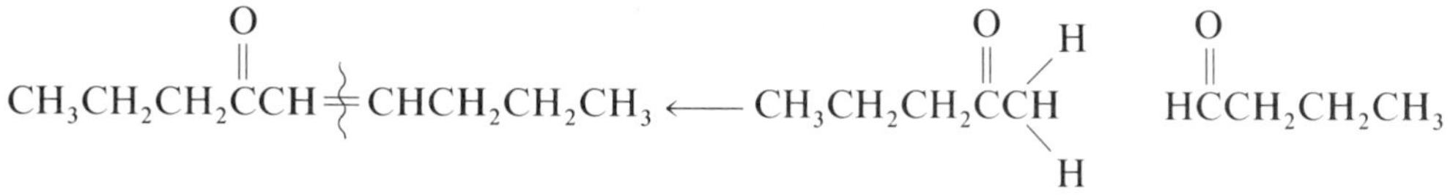

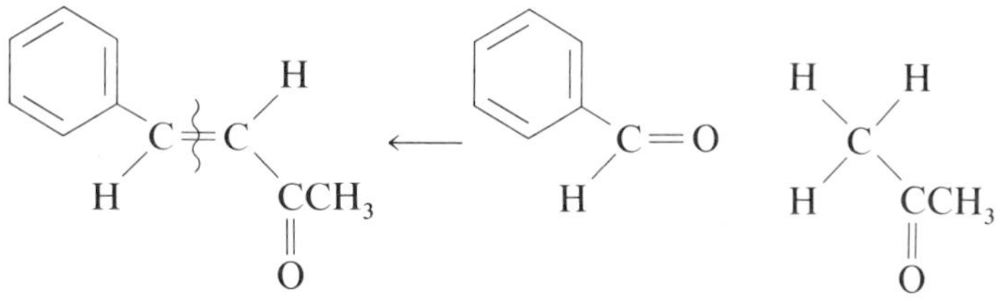

Study Guide Concept Map 16.5

Note that the carbonyl group of one reactant is reconstructed by replacing a double bond to carbon with a double bond to oxygen. Two hydrogen atoms are also put back on the α-carbon atom of the other reactant.

PROBLEM 16.13

Complete the following equations.

(a) 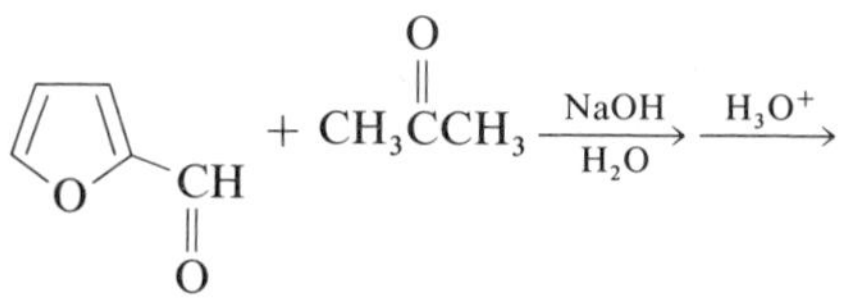

(b) 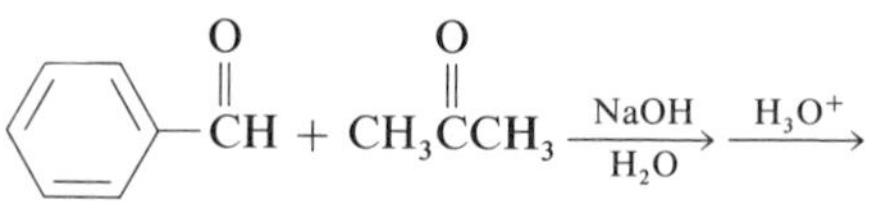

2 equivalents 1 equivalent

(c) 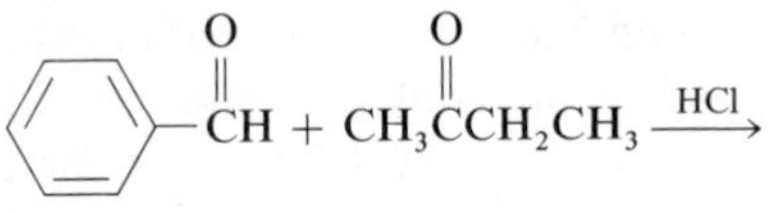

(Hint: Which is the more stable enol?)

(d) 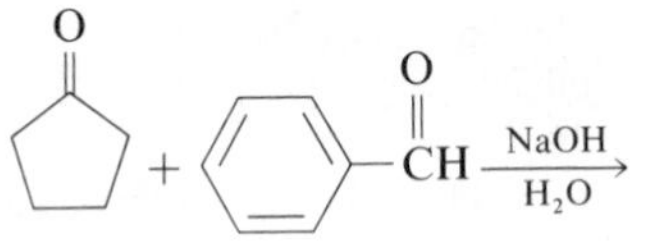

(e) 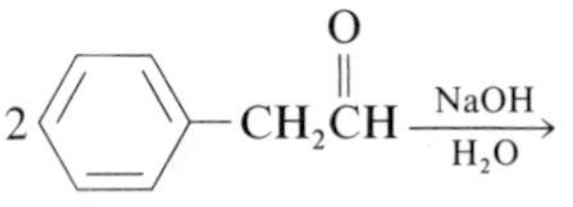

PROBLEM 16.14

Give the structures of the starting materials that would be required to synthesize the following compounds.

(a) 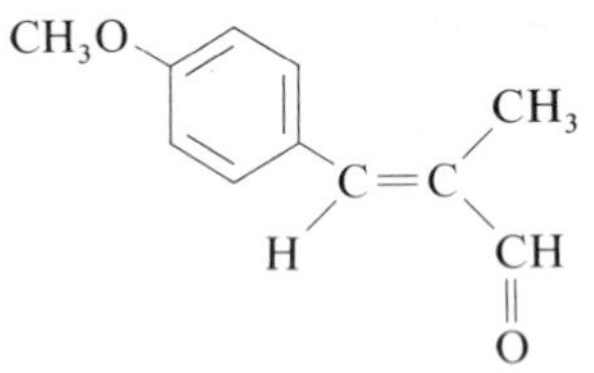

(b) 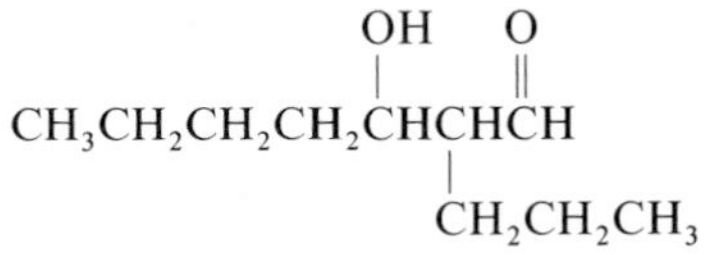

(c) 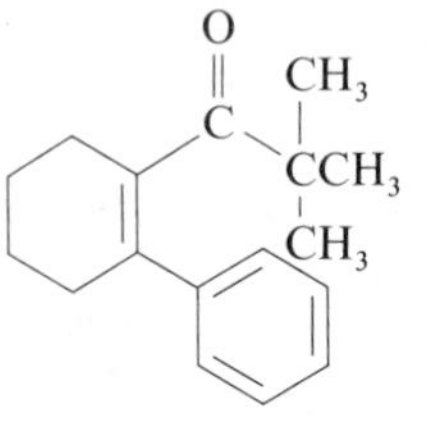

(d) 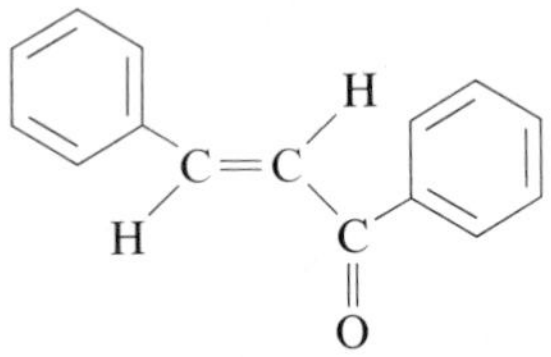

B. Biological Significance of the Retroaldol Reaction

Aldol condensations are reversible. The reverse reaction, illustrated below for 3-hydroxybutanal, is called the **retroaldol reaction.**

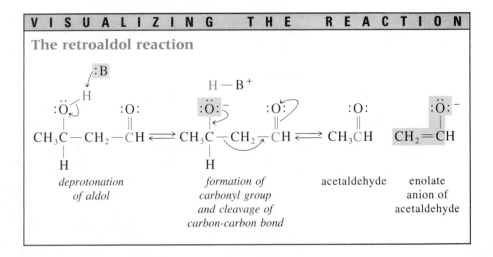

VISUALIZING THE REACTION

The retroaldol reaction

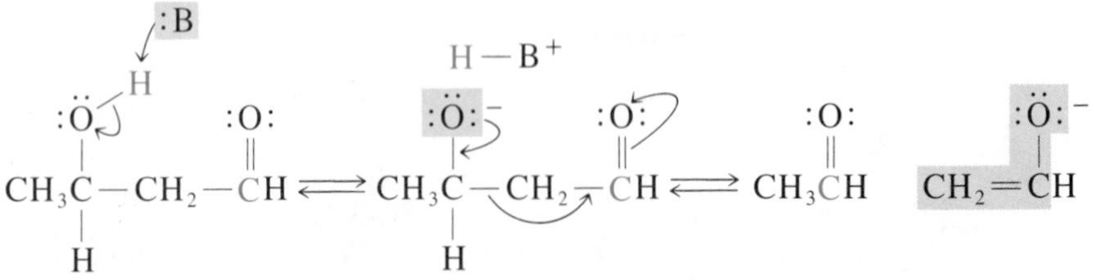

deprotonation of aldol

formation of carbonyl group and cleavage of carbon-carbon bond

acetaldehyde

enolate anion of acetaldehyde

In basic solution, the hydroxyl group in the product of an aldol reaction loses a proton to give an alkoxide ion. The formation of a carbonyl group provides the driving force for cleaving this alkoxide ion into the two fragments from which the aldol was originally formed.

The reversibility of the aldol condensation is crucial to the metabolism of glucose. In the body, glucose is broken down into lactic acid by means of a process known as **glycolysis.** Fructose 1,6-diphosphate, which is formed in the body from glucose (pp. 1074, 1079), is the intermediate that undergoes cleavage in a retro-aldol reaction. The cyclic form of fructose is a ketal and is in equilibrium with the noncyclic form, which is a β-hydroxyketone. An enzyme-catalyzed removal of a proton from the hydroxyl group starts the reverse aldol reaction, which cleaves the molecule into two smaller carbohydrates, each containing one of the phosphate groups.

VISUALIZING THE REACTION

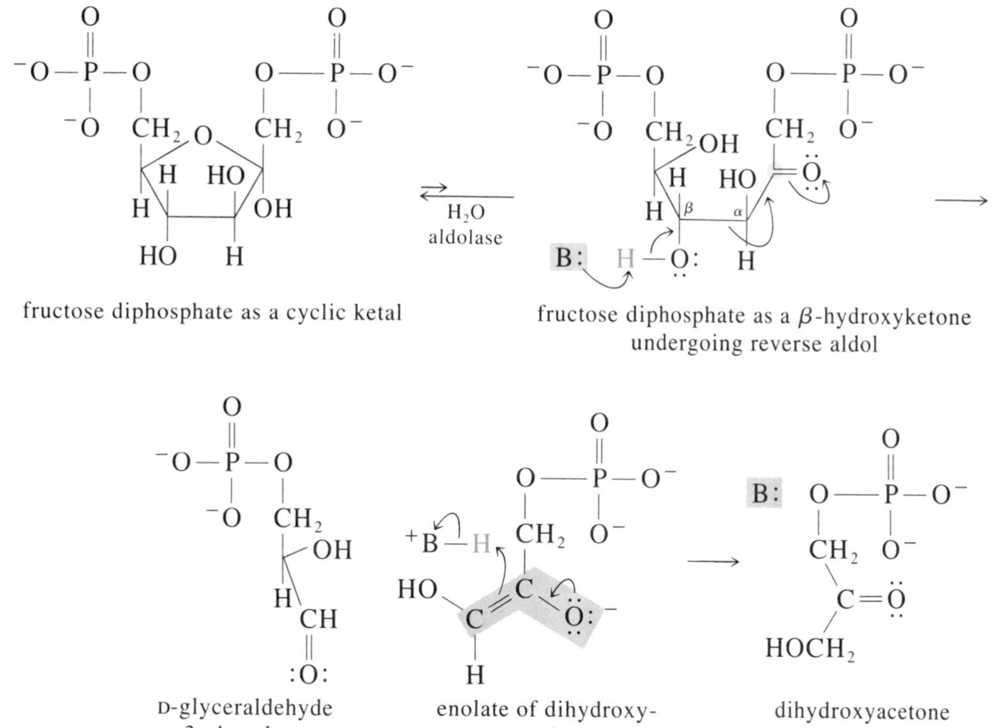
H2O
aldolase
B:

fructose diphosphate as a cyclic ketal

fructose diphosphate as a β-hydroxyketone undergoing reverse aldol

D-glyceraldehyde 3-phosphate

enolate of dihydroxyacetone phosphate

dihydroxyacetone phosphate

One fragment is glyceraldehyde phosphate; the other is dihydroxyacetone phosphate. The later stages of glycolysis convert dihydroxyacetone phosphate into glyceraldehyde phosphate and glyceraldehyde phosphate into the anion of lactic acid. The muscle soreness experienced during extended vigorous exercise is the result of the accumulation of lactic acid.

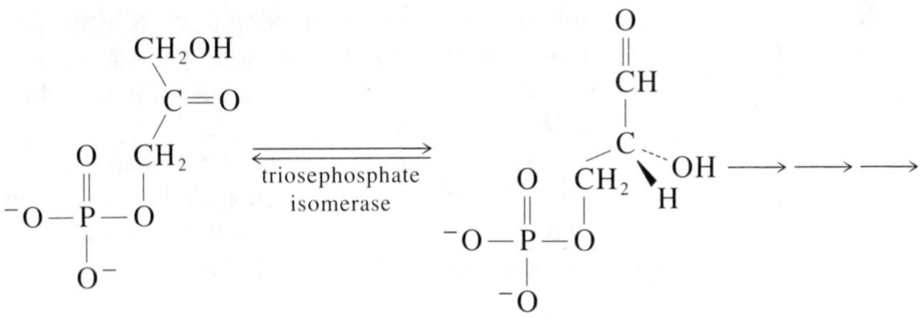
triosephosphate
isomerase

dihydroxyacetone phosphate

D-glyceraldehyde 3-phosphate

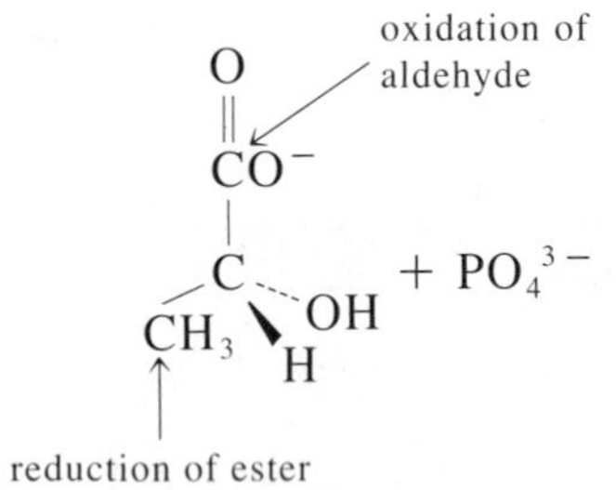
oxidation of aldehyde
reduction of ester

lactate ion

phosphate ion

PROBLEM 16.15

Write a detailed mechanism for the conversion of dihydroxyacetone phosphate into D-glyceraldehyde 3-phosphate. (Hint: Try applying the questions given in the problem-solving skills section on p. 572.)

C. The Formation of β-Ketoesters. The Claisen Condensation

The aldol condensation involves the reaction of an enolate ion with the carbonyl group of an aldehyde or ketone. In a similar reaction, enolate ions from esters react with the carbonyl groups of acid derivatives to form β-ketoesters. Such reactions are **acylation reactions of enolate anions.** In the reaction, the α-carbon atom of the compound giving rise to the enolate ion ends up bonded to an acyl group. For example, the enolate anion of *tert*-butyl 2-methylpropanoate reacts with benzoyl chloride to give *tert*-butyl 2,2-dimethyl-3-oxo-3-phenylpropanoate.

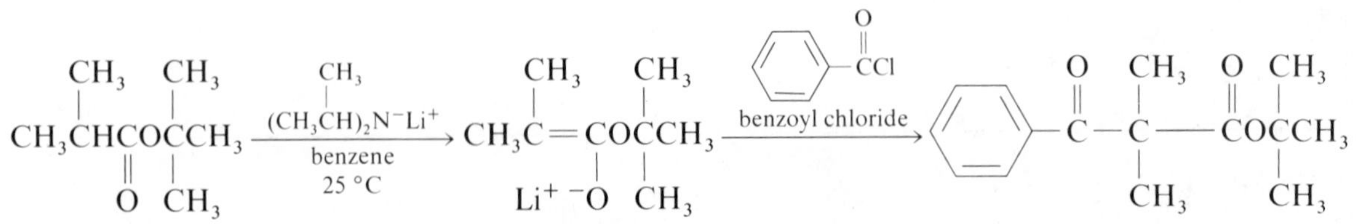
$(CH_3CH)_2N^-Li^+$
benzene
25 °C
benzoyl chloride

tert-butyl 2-methylpropanoate

tert-butyl 2,2-dimethyl-3-oxo-3-phenylpropanoate
78%

An ester is acylated at the carbon alpha to its carbonyl group by this sequence of reactions. The product of an acylation reaction of an enolate anion is always a dicarbonyl compound in which the carbonyl groups are separated by one carbon atom. Such compounds are called **1,3 dicarbonyl compounds.**

When the enolate anion from an ester reacts with the carbonyl group of another ester, the reaction is known as the **Claisen condensation.** The classic example of this type of reaction is the condensation of two molecules of ethyl acetate to give the enolate of ethyl acetoacetate. In a second step, acid is added to the reaction mixture to form the β-ketoester.

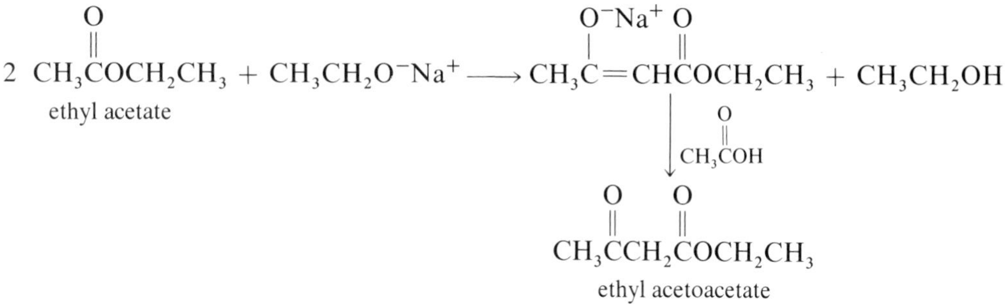

$$2 \ \underset{\text{ethyl acetate}}{CH_3\overset{\displaystyle O}{\overset{\|}{C}}OCH_2CH_3} + CH_3CH_2O^-Na^+ \longrightarrow CH_3\overset{\displaystyle O^-Na^+}{\overset{|}{C}}{=}CH\overset{\displaystyle O}{\overset{\|}{C}}OCH_2CH_3 + CH_3CH_2OH$$

$$\downarrow CH_3\overset{\displaystyle O}{\overset{\|}{C}}OH$$

$$\underset{\text{ethyl acetoacetate}}{CH_3\overset{\displaystyle O}{\overset{\|}{C}}CH_2\overset{\displaystyle O}{\overset{\|}{C}}OCH_2CH_3}$$

The alkoxyl group on the ester function serves as a leaving group from the tetrahedral intermediate formed when the carbonyl group is attacked by an enolate ion. The formation of an enolate ion by deprotonation at the active methylene group of the product provides the driving force for the completion of the reaction.

VISUALIZING THE REACTION

The Claisen condensation

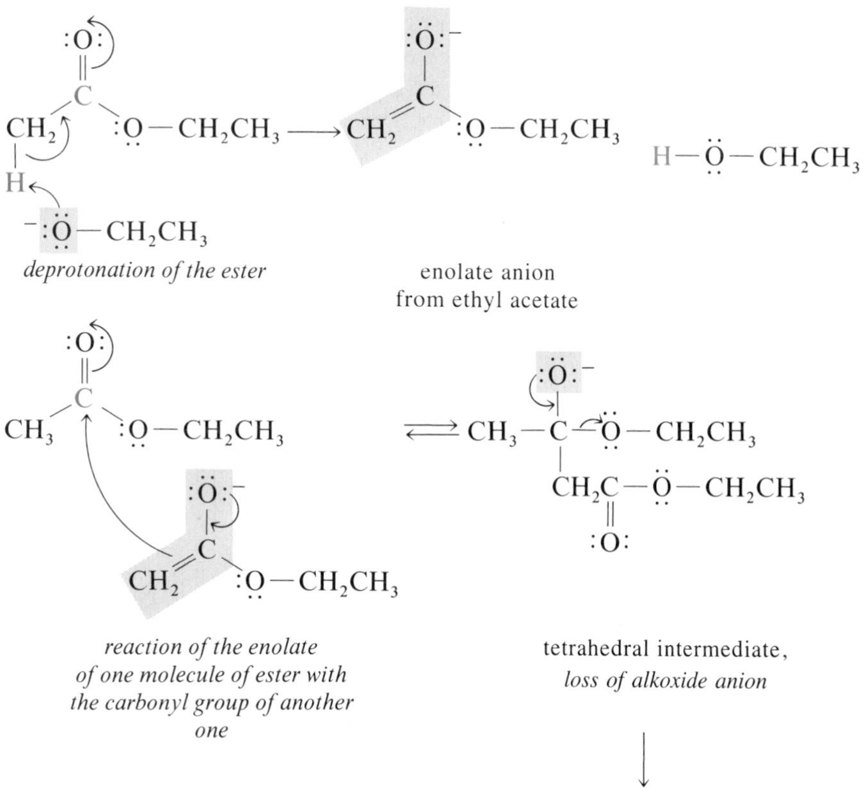

deprotonation of the ester

enolate anion
from ethyl acetate

*reaction of the enolate
of one molecule of ester with
the carbonyl group of another
one*

tetrahedral intermediate,
loss of alkoxide anion

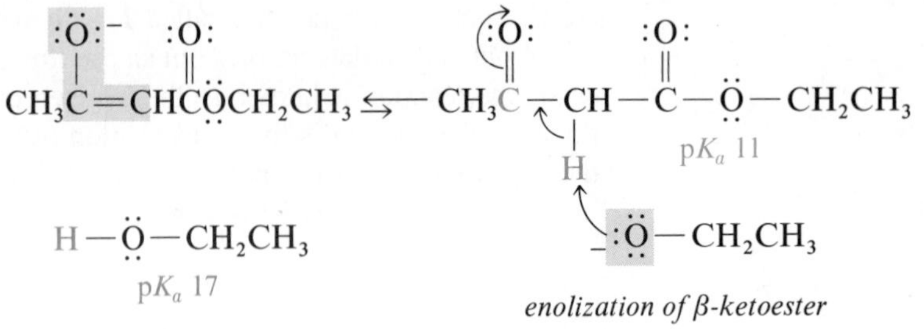
pK_a 11
pK_a 17
enolization of β-ketoester

Note that the reaction of an enolate anion with an ester is a reversible reaction. For this reason, the Claisen condensation does not give a good yield of the β-ketoester unless that product is converted to its enolate in the basic reaction mixture. If the β-ketoester is not converted completely to its enolate, it is attacked at the ketone functional group by an alkoxide ion to give the tetrahedral intermediate shown above. The intermediate decomposes to give the enolate anion and the ester. The last stage of the reaction, the reversible protonation and deprotonation of the active methylene group in the β-ketoester, is written as an equilibrium. At equilibrium, the enolate anion of the β-ketoester predominates because the compound with the active methylene group is much more acidic (pK_a 11) than ethanol (pK_a 17). The enolate anion of the ketoester is converted to the desired product by adding an organic acid, such as acetic acid.

Just as aldol reactions are possible between different aldehydes or between an aldehyde and a ketone, Claisen condensations between different esters are possible. If one of the esters does not have enolizable α-hydrogen atoms, the reaction works quite well. For example, a mixed condensation between ethyl benzoate and ethyl acetate is used in industry to synthesize ethyl 3-phenyl-3-oxopropanoate, shown in the reaction below. Ethyl acetate and sodium are added slowly to ethyl benzoate. Under these conditions, the enolate ion formed from a molecule of ethyl acetate is more likely to encounter a molecule of ethyl benzoate than another molecule of ethyl acetate.

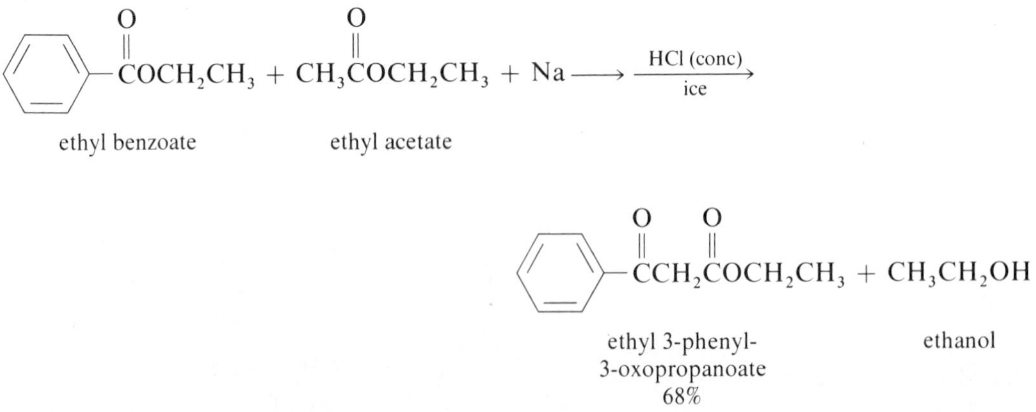
HCl (conc)
ice
ethyl benzoate
ethyl acetate
ethyl 3-phenyl-3-oxopropanoate 68%
ethanol

Sodium metal is used to generate the enolate ion in order to avoid the problem of the reversibility of the reaction that is seen when alkoxide ions are used as the catalysts. Some ethyl acetoacetate is formed as a side product.

D. The Dieckmann Condensation

If two ester functions are in the same molecule and separated by four or five carbon atoms, the carbonyl group of one is in an ideal position to accept the enolate anion created alpha to the other. A cyclic β-ketoester is the product because of the ease with which five- and six-membered rings form (p. 448). This variation of the Claisen condensation is called the **Dieckmann condensation** and is best illustrated by the formation of ethyl 2-oxocyclopentanecarboxylate.

VISUALIZING THE REACTION

The Dieckmann condensation

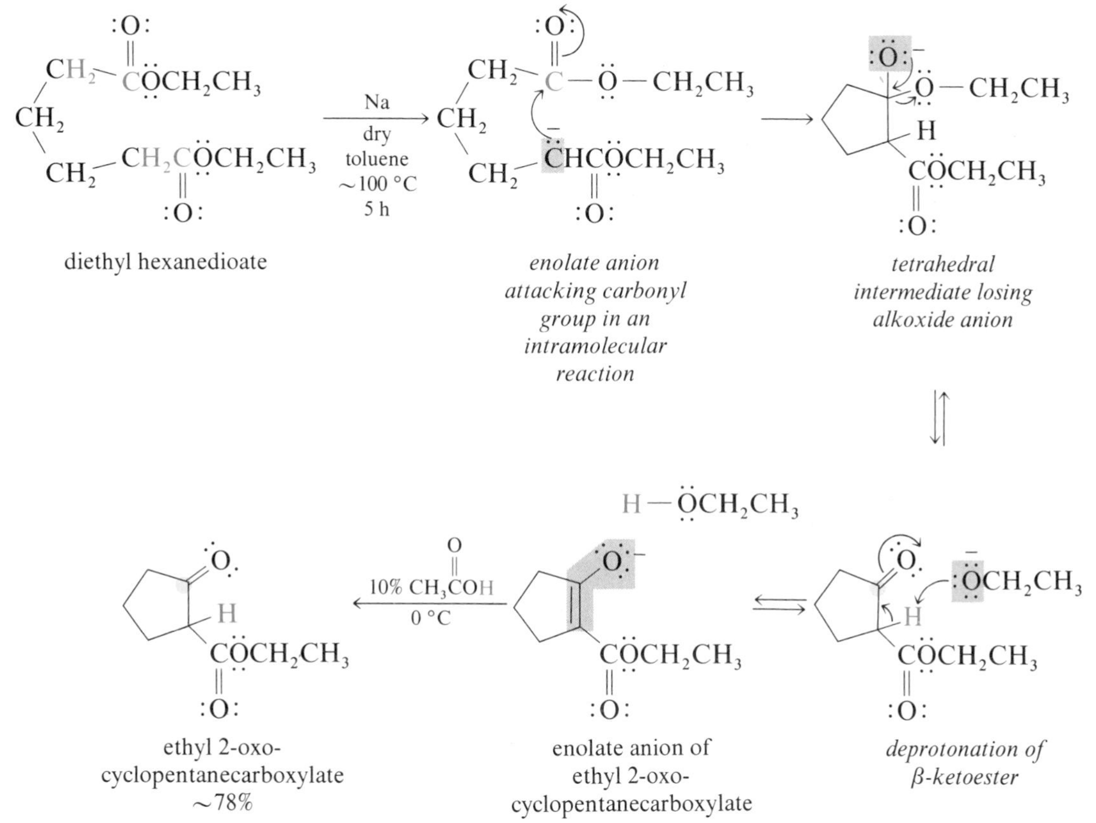

Na
dry
toluene
~100 °C
5 h
diethyl hexanedioate
enolate anion attacking carbonyl group in an intramolecular reaction
tetrahedral intermediate losing alkoxide anion
H—ÖCH2CH3
10% CH3COH
0 °C
ethyl 2-oxo-cyclopentanecarboxylate ~78%
enolate anion of ethyl 2-oxo-cyclopentanecarboxylate
deprotonation of β-ketoester

All the factors that affect the Claisen condensation also affect the Dieckmann condensation. The reaction is reversible, and the β-ketoester is stable in basic solution only as its enolate anion.

E. A Unified Look at Condensation Reactions

Anions from esters can also be used in condensations with aldehydes or ketones. For example, ethyl acetate will condense with benzaldehyde to give an unsaturated carboxylic acid derivative.

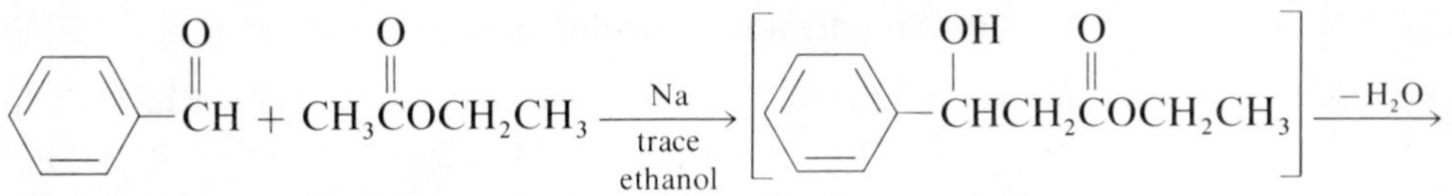

benzaldehyde ethyl acetate β-hydroxyester intermediate

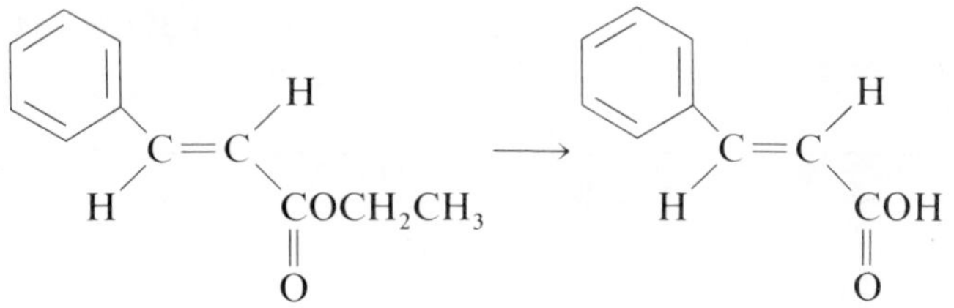

ethyl (*E*)-3-phenyl-2-propenoate
α, β-unsaturated ester
~70%

(*E*)-3-phenyl-2-propenoic acid
trans-cinnamic acid
α, β-unsaturated acid

In another example, the active methylene group of diethyl malonate condenses with formaldehyde under mild conditions.

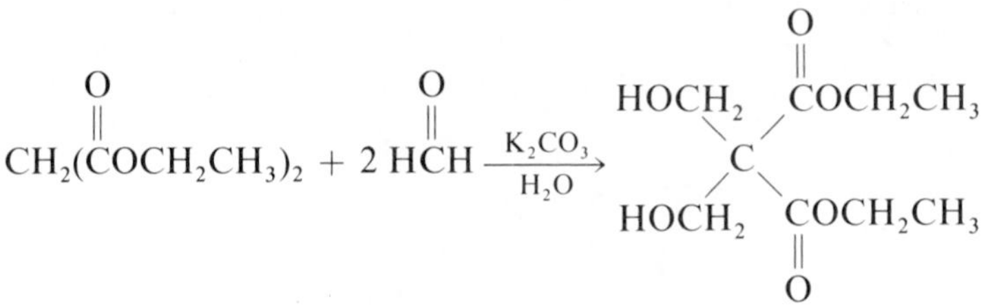

diethyl malonate formaldehyde diethyl bis(hydroxymethyl)-propanedioate

For each of the reactions above, only one of the reactants has an enolizable hydrogen atom, so the reaction goes one way in high yield. Also, note that in both cases the enolate ion reacts much more readily with the carbonyl group of the aldehyde that is present than with the carbonyl group of the ester. In the reaction with benzaldehyde, the intermediate hydroxy compound dehydrates easily to give a conjugated alkene. With formaldehyde, the product retains the β-hydroxy groups.

Clearly, a large variety of reactions are possible between enolate ions and compounds containing carbonyl groups. Many of them are known by the names of their discoverers or developers. But such an array of different names and structures tends to hide the underlying unity and simplicity of this type of reaction. In deciding what kinds of condensation reactions are possible, you have to ask four questions:

1. Is there an enolizable hydrogen atom in one of the reactants? Hydrogens alpha to nitro and cyano groups also count as acidic hydrogen atoms.
2. Is there a carbonyl group that can be attacked by the enolate anion? The carbonyl group of an aldehyde is more reactive (because it is less hindered) than the carbonyl group of a ketone. The carbonyl group of an ester is the least reactive because its carbonyl carbon atom is less electrophilic than that of either of the other two (p. 543).
3. Is the carbonyl group in the same molecule as the enolizable hydrogen? If a five- or six-membered ring can form, it usually will.

4. Is the carbonyl carbon that is being attacked also bonded to a leaving group (usually the alkoxyl group of an ester)? If there is such a group present, it is lost in going to the product. These types of reactions may be classified as Claisen condensations. If there is no good leaving group, the product is a β-hydroxy-carbonyl compound or its dehydration product. These reactions are aldol-type reactions.

Study Guide Concept Map 16.6

All you need to bring order to what appears to be a bewildering array of reactions is a sharp eye for these features and a willingness to practice making predictions on a wide variety of examples. The next problem provides such practice.

PROBLEM 16.16

Give structures for the intermediates and products designated by the letters.

(a) 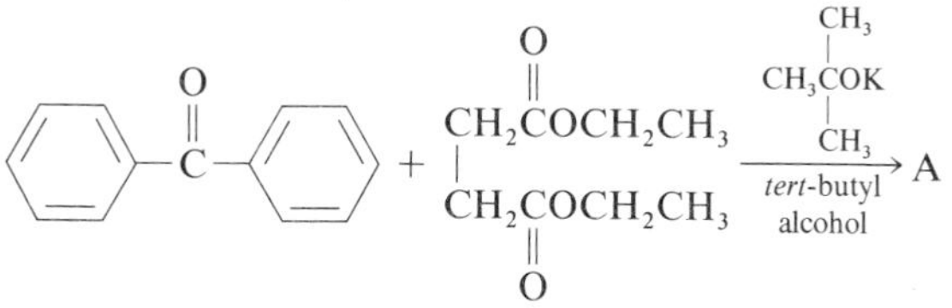

(b) 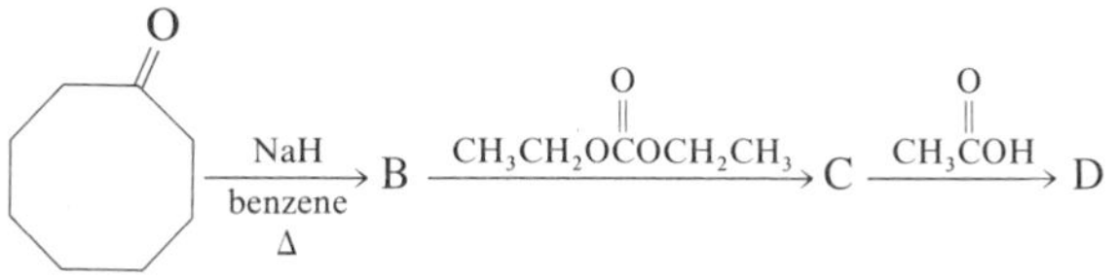

(c) 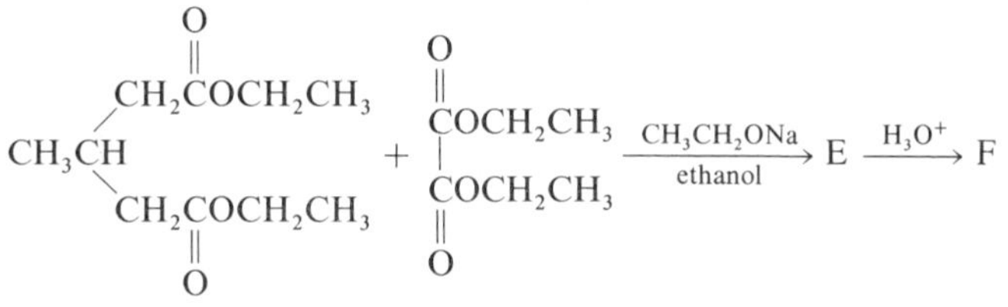

(d) 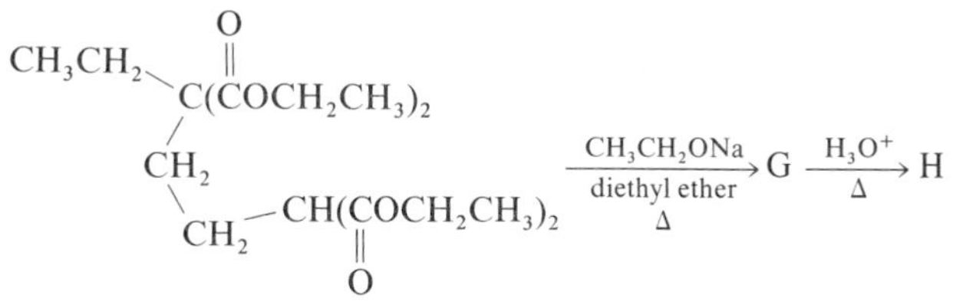

PROBLEM 16.17

Condensation reactions between a compound containing an active methylene group and an aldehyde or ketone are called **Knoevenagel reactions** when they are catalyzed by amines or amine salts. Predict the products of the examples shown below.

(a) 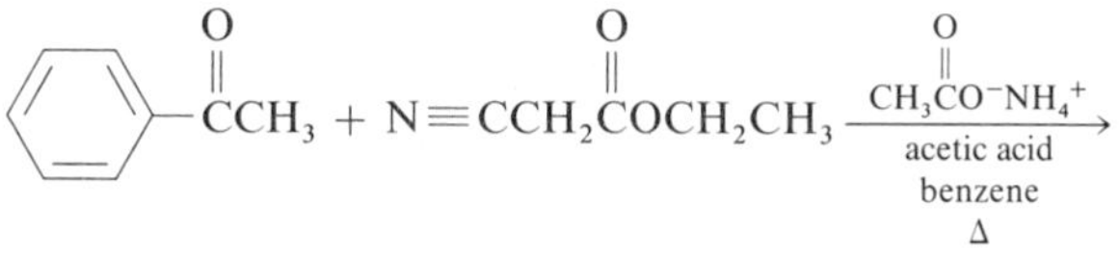

(b) 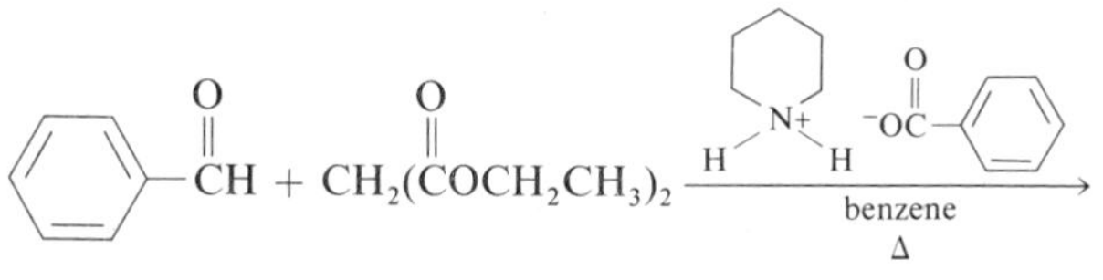

F. Problem-Solving Skills

Problem

The synthesis of a natural product involved the following transformation. How would you carry it out?

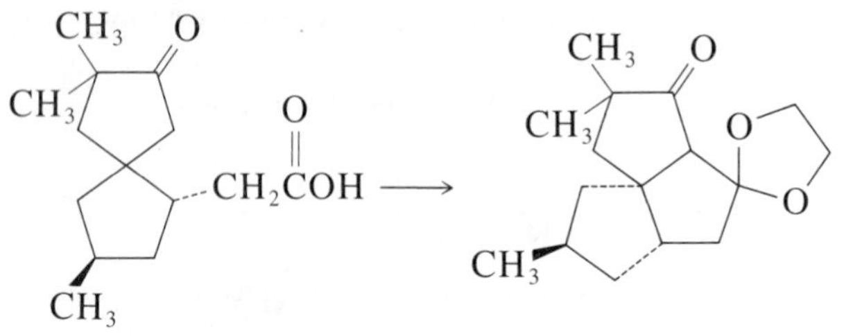
CH3
O
CH3
O
CH2COH
CH3
CH3
O
CH3
O
O
CH3

Solution

1. What functional groups are present in the starting material and the product?

   The starting material has both a ketone function and a carboxyl group. The product retains the ketone function but also has a cyclic ketal function.

2. How do the carbon skeletons of the two compounds compare? How many carbon atoms does each contain? Are there any rings? What are the positions of branches and functional groups on the carbon skeletons?

   The starting material has two five-membered rings that share a carbon atom. The product has two more carbon atoms than the starting material, the two carbon atoms of the cyclic ketal. The two-carbon side chain of the starting material is part of a third ring in the product. The carbon atom of the carbonyl group of the acid is the carbon atom of a protected ketone function in the product.

3. How do the functional groups change in going from starting material to product? Does the starting material have a good leaving group?

   The carboxyl group of the starting material has been converted to a cyclic ketone, protected as a ketal. The hydroxyl group of the carboxylic acid can be converted into a leaving group. It is lost in the transformation.

4. Is it possible to dissect the structures of the starting material and product to see which bonds must be broken and which formed?

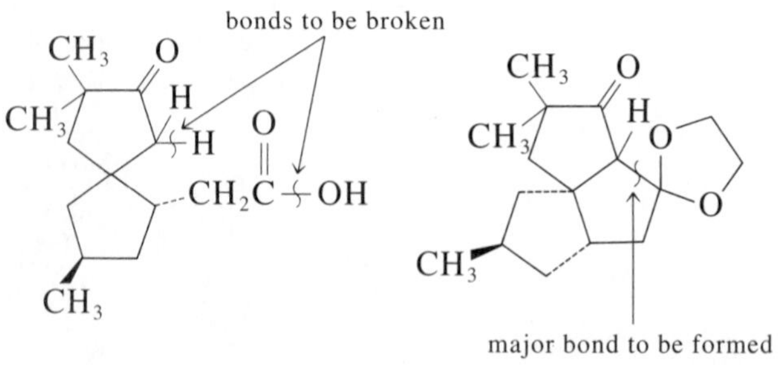
bonds to be broken
CH3
O
H
CH3
H
O
CH2C
OH
CH3
CH3
O
H
CH3
O
O
CH3
major bond to be formed

The major transformation is the formation of a bond between the carbon atom alpha to the ketone function and the carbon atom of the carboxyl group.

5. Do we recognize any part of the product molecule as coming from a good nucleophile or an electrophilic addition?

   The enolate anion from the ketone is a good nucleophile and will react with the electrophilic carbon atom of a carbonyl group. For this reaction to take place, the acid must be converted to an ester. Any base strong enough to deprotonate the ketone will deprotonate the acid, and the carboxylate anion is not reactive enough toward substitution at the carbonyl group (p. 575).

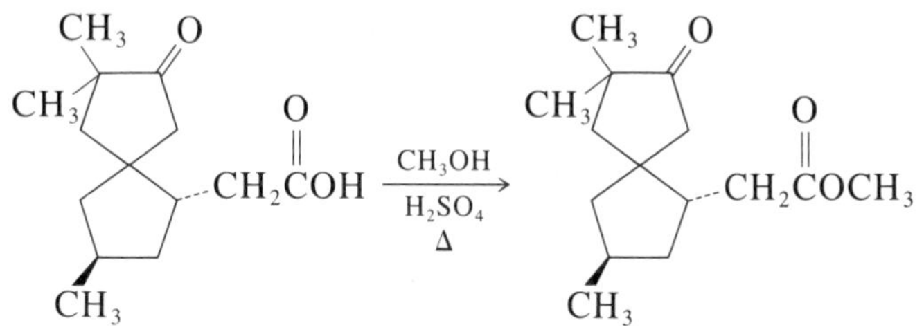

CH3
O
CH3
O
CH2COH
CH3OH
H2SO4
Δ
CH3
CH3
O
CH3
O
CH2COCH3
CH3

6. What type of compound would be a good precursor to the product?

   The ketone from which the ketal is prepared is a good precursor.

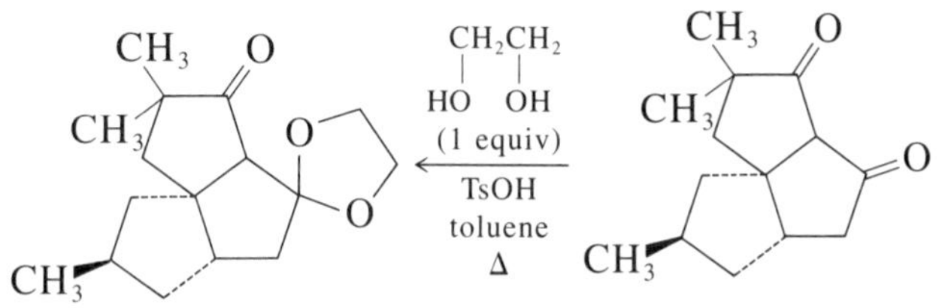

CH3
O
CH3
O
O
CH3
CH2CH2
HO OH
(1 equiv)
TsOH
toluene
Δ
CH3
O
CH3
O
CH3

7. After this last step, do we now see how to get from starting material to product? If not, we need to analyze the structure obtained in step 6 by applying questions 5 and 6 to it.

   The enolate anion from the ketone will react intramolecularly with the carbonyl group of the ester to give the diketone needed to complete the synthesis.

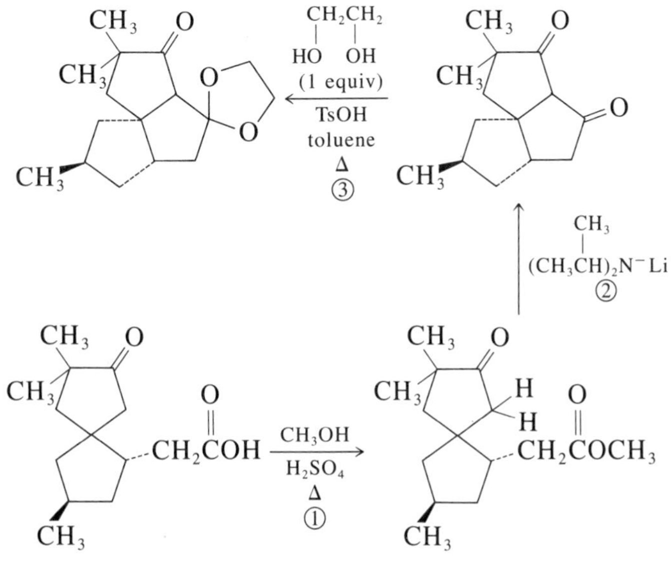

CH3
O
CH3
O
O
CH3
CH2CH2
HO OH
(1 equiv)
TsOH
toluene
Δ
③
CH3
O
CH3
O
CH3
CH3
(CH3CH)2N⁻ Li⁺
②
CH3
O
CH3
O
CH2COH
CH3OH
H2SO4
Δ
①
CH3
CH3
O
CH3
H
H
O
CH2COCH3
CH3

Two steps of the above transformation show interesting regioselectivity. After step 1, there are hydrogen atoms alpha to the ketone function and others alpha to the ester function. In step 2, the more acidic hydrogens (pK_a ~20) adjacent to the ketone are removed rather than those adjacent to the ester (pK_a ~23). The ring formed in this condensation is also more stable than the one that would have formed if the enolate ion of the ester had attacked the ketone. In step 3, it is possible to selectively protect one of the ketone carbonyl groups. The one that reacts is less sterically hindered; the other one has two methyl groups adjacent to it, making its reaction slower.

PROBLEM 16.18

The following transformation is carried out during the synthesis of a natural product. How would you accomplish this transformation?

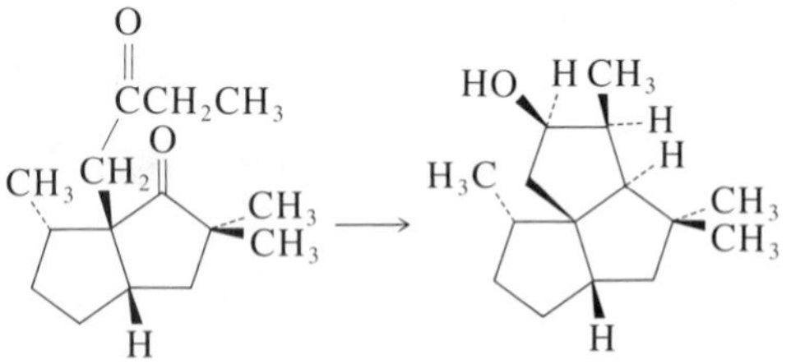

PROBLEM 16.19

(a) Write a mechanism for the following reaction. (Hint: Reviewing the questions in the problem-solving skills section on pp. 572–574 may be helpful.)

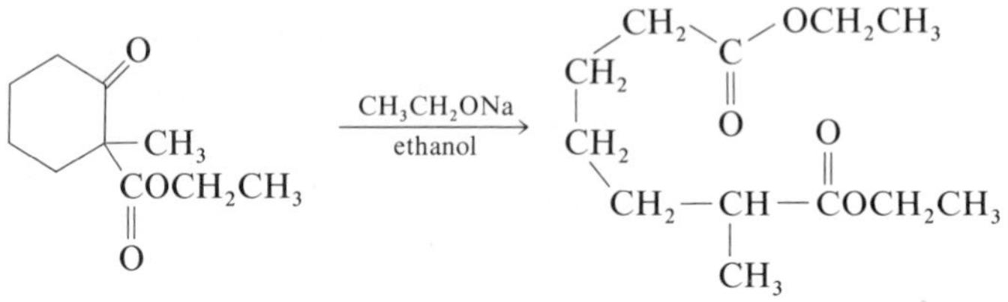

(b) When the diester produced in the above reaction is treated with 1 equivalent of sodium hydride, carefully excluding any alcohol, ethyl 3-methyl-2-oxocyclohexanecarboxylate is formed in 90% yield. Explain why that compound, and not the original ethyl 1-methyl-2-oxocyclohexanecarboxylate, is formed under these conditions.

PROBLEM 16.20

The disubstituted malonic ester required for the synthesis of phenobarbital (p. 654) cannot be prepared by the usual route involving the enolate anion from diethyl malonate in a nucleophilic substitution reaction, because ordinary aryl halides do not undergo such substitution reactions. Work out a synthesis for diethyl ethylphenylmalonate starting from phenylacetonitrile using condensation reactions and alkylation reactions of compounds containing active methylene groups.

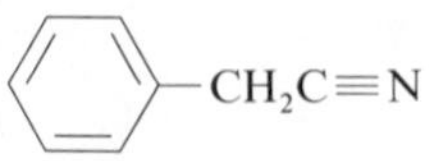

phenylacetonitrile

SUMMARY

Carbonyl compounds exist in equilibrium with their tautomers, which are enols. The concentration of the enol found in equilibrium with a carbonyl compound is minute unless the compound has an active methylene group, meaning that it has hydrogen atoms on a carbon atom that is alpha to two carbonyl groups or to nitro or cyano groups, which also stabilize anions. The conversion of a carbonyl compound to its enol is catalyzed by either acid or base.

When unsymmetrical carbonyl compounds are converted to their enols under conditions that allow equilibrium to be established (high temperatures and weak bases or acids as catalysts) the enolate anion with the more highly substituted double bond is formed. This ion is the more stable of the two possible enolates and is called the thermodynamic enolate. If the carbonyl compound is added to an excess of a strong base (usually lithium diisopropylamide) at low temperatures, the least hindered hydrogen atom is removed. The ion with the less substituted double bond forms; it is the less stable of the two possible enolates but is the one that forms faster. For this reason, it is called the kinetic enolate.

Enols and enolate anions are nucleophiles. They react with electrophiles, such as halogens, alkyl halides, carbonyl compounds, and acid derivatives. Enols react with halogens to give α-halocarbonyl compounds. If such a reaction is run under basic conditions, multiple substitutions by the halogen and carbon-carbon bond cleavage occur to give a haloform and a carboxylate ion.

Reactions of enolates with alkyl halides, carbonyl compounds, or acid derivatives are important because a carbon-carbon bond is formed. The enolates from β-ketoesters and 1,3-diesters are alkylated regioselectively at the active methylene group. Hydrolysis of β-ketoesters gives β-ketocarboxylic acids, which lose carbon dioxide. Ketones are synthesized from β-ketoesters, such as ethyl acetoacetate. Acids containing two more carbon atoms than the starting alkyl halide are synthesized from 1,3-diesters such as diethyl malonate.

The enolate ion formed from one molecule of an aldehyde adds to the carbonyl group of another molecule of the aldehyde to give a β-hydroxyaldehyde in the aldol condensation. Similar reactions take place with ketones, but more slowly. β-Hydroxyaldehydes and ketones dehydrate easily to α,β-unsaturated carbonyl compounds, which are the usual products of aldol condensations. Enolate anions also react with carbonyl groups in acid derivatives, such as acid chlorides and esters. The reaction of an enolate with an ester function is called the Claisen condensation when it is intermoelcular and the Dieckmann condensation when it is intramolecular. The products of the reactions of enolates with acid derivatives are 1,3-dicarbonyl compounds.

The reactions leading to the formation of enols and enolate anions, and the subsequent reactions of these species with electrophiles are summarized in Tables 16.1 through 16.4.

TABLE 16.1 Reactions of Enols or Enolate Anions with Halogens

Carbonyl Compound	Conditions for Enolization	Structure of Reactive Intermediate	Reagents for Second Step	Product
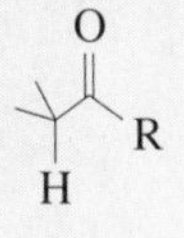	H_3O^+	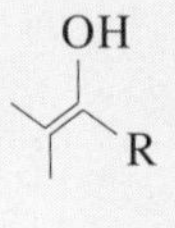	X_2, H_2O	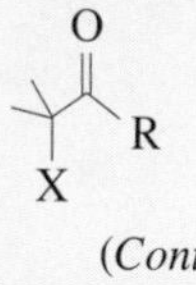

(*Continued*)

TABLE 16.1 *(Continued)*

Carbonyl Compound	Conditions for Enolization	Structure of Reactive Intermediate	Reagents for Second Step	Product
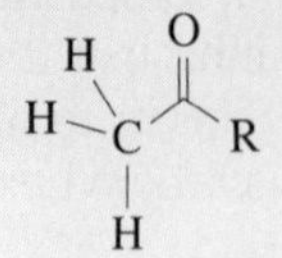	NaOH, H_2O	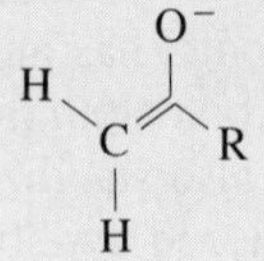	X_2, NaOH, H_2O	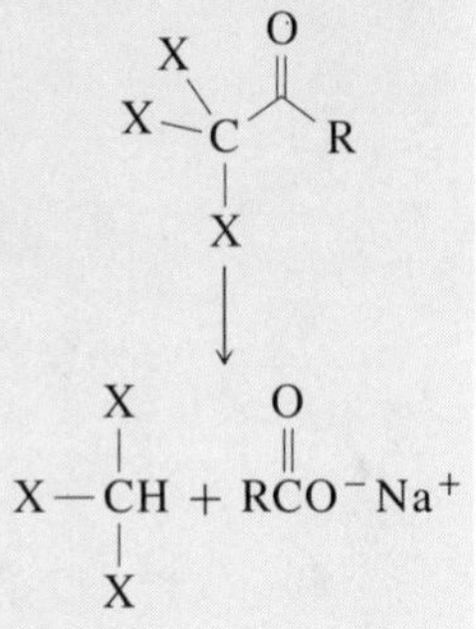

TABLE 16.2 Reactions of Enolate Anions with Alkyl Halides

Carbonyl Compound	Conditions for Enolization	Structure of Reactive Intermediate	Reagents for Second Step	Product
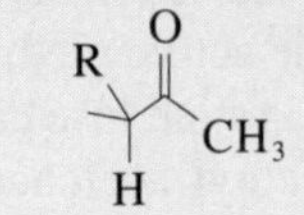	$(CH_3CH_2)_3N$, Δ or $R''O^-$	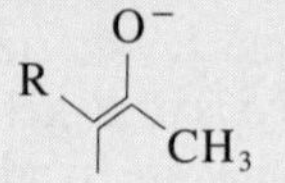	R′X (R′ is primary or secondary)	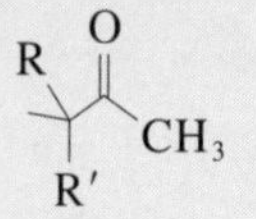
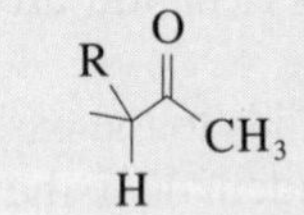	$(CH_3CH(CH_3))_2N^- Li^+$ −78 °C	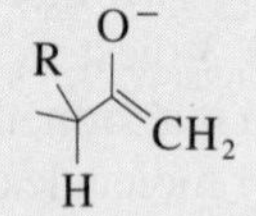	R′X	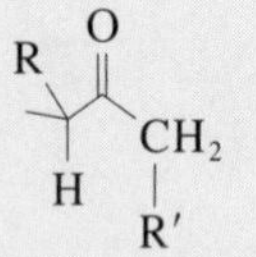
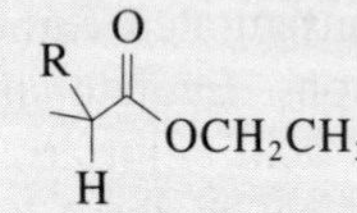	$(CH_3CH(CH_3))_2N^- Li^+$ −78 °C	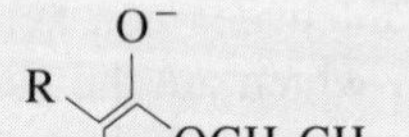	R′X	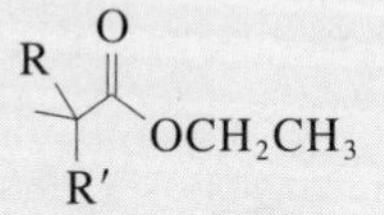
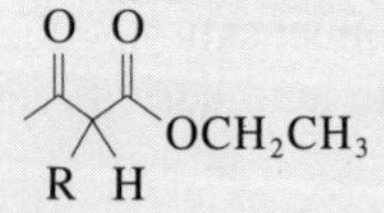	$CH_3CH_2O^-$	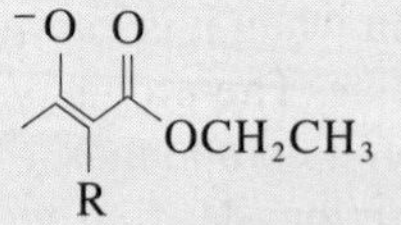	R′X	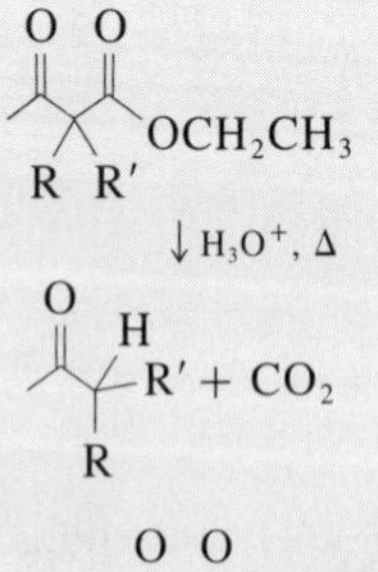
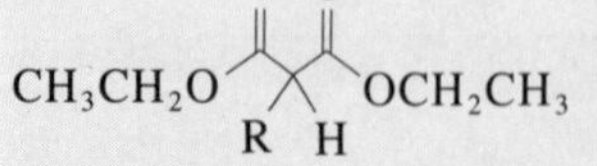	$CH_3CH_2O^-$	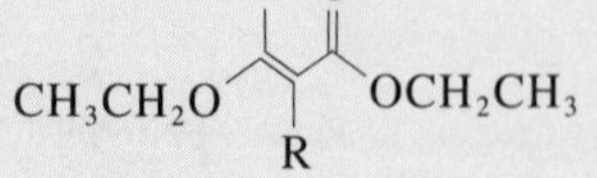	R′X	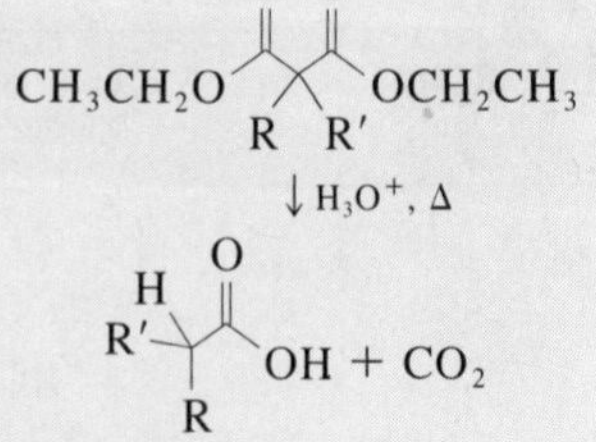

TABLE 16.3 The Aldol Condensation

Carbonyl Compound	Conditions for Enolization	Structure of Reactive Intermediate	Reagents for Second Step	Product
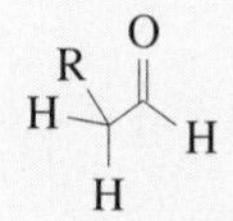	NaOH	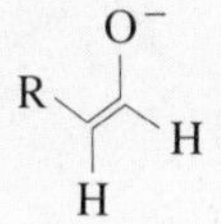	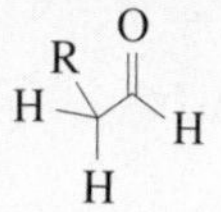	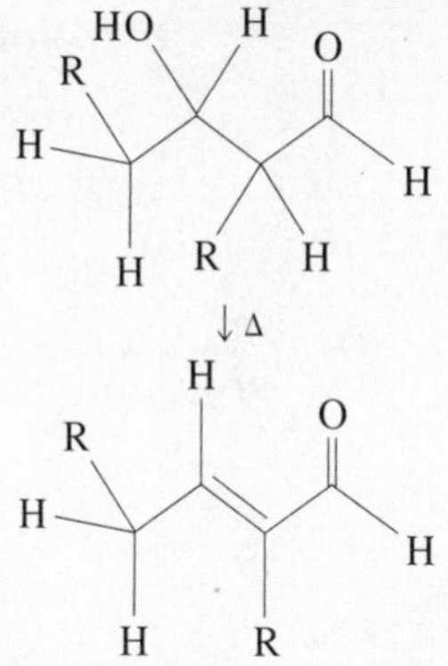
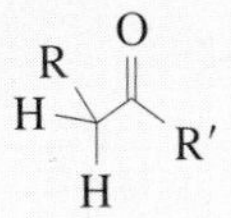	NaOH	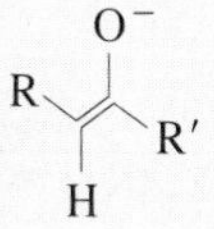	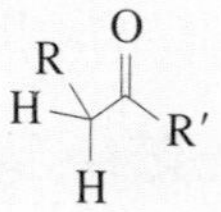	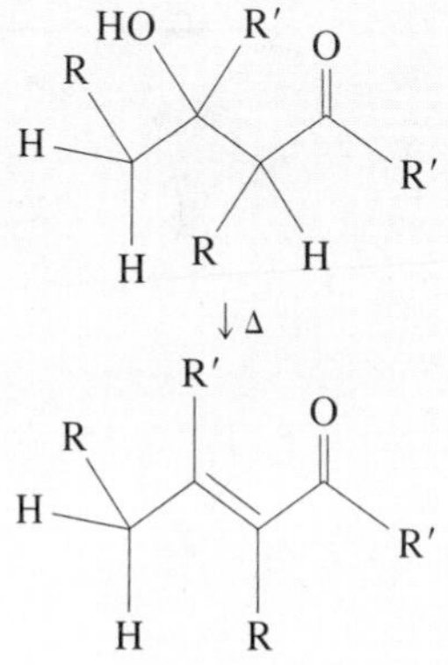
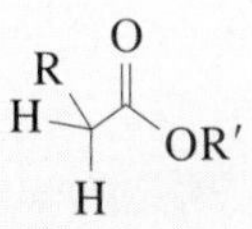	Na or NaH or CH_3 $(CH_3CH)_2N^-$ Li^+	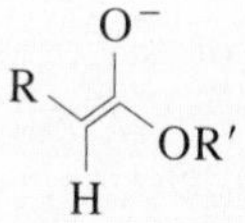	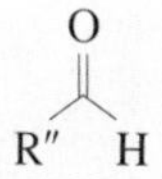	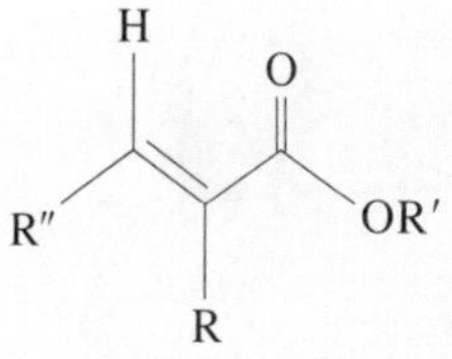

TABLE 16.4 Reactions of Enolate Anions with Acid Derivatives

Carbonyl Compound	Conditions for Enolization	Structure of Reactive Intermediate	Reagents for Second Step	Product
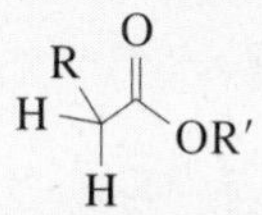	CH_3 $(CH_3CH)_2N^-$ Li^+	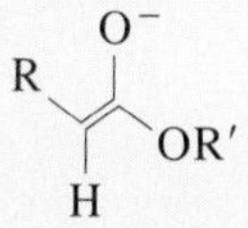	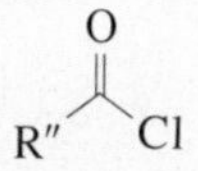	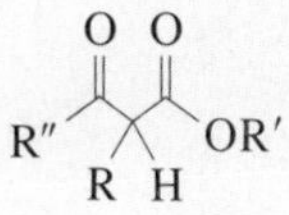
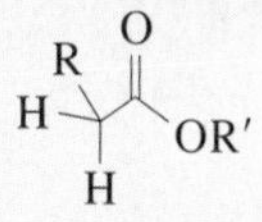	$R'O^-$	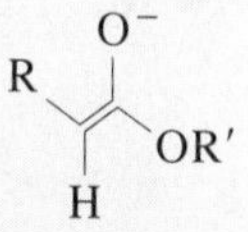 (Claisen condensation)	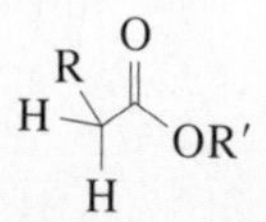	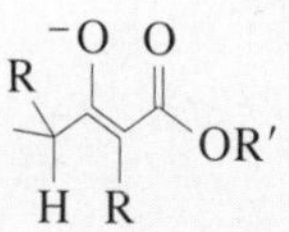 ↓ HB^+ *(Continued)*

TABLE 16.4 *(Continued)*

Carbonyl Compound	Conditions for Enolization	Structure of Reactive Intermediate	Reagents for Second Step	Product
				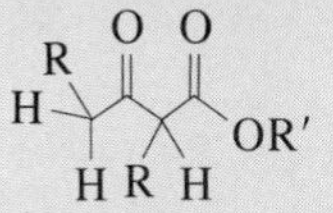
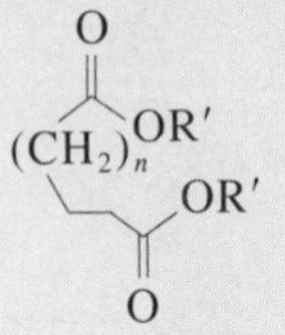 $n = 2$ or 3	Na or NaH or $R'O^-$	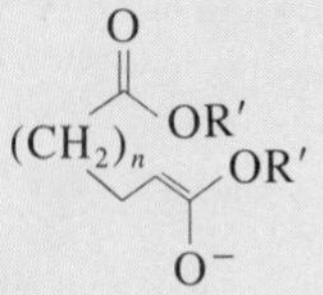 (Dieckmann condensation)		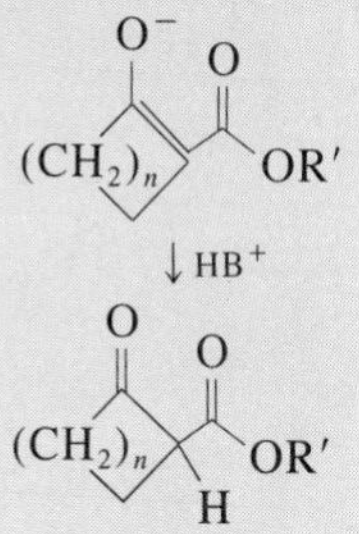

ADDITIONAL PROBLEMS

16.21 For each of the following sets of compounds, decide which one is the most acidic.

(a) $CH_3\overset{O}{\overset{\|}{C}}CH_2\overset{O}{\overset{\|}{C}}CH_3$ or $CH_3\overset{O}{\overset{\|}{C}}CH_2\overset{O}{\overset{\|}{C}}CF_3$

(b) 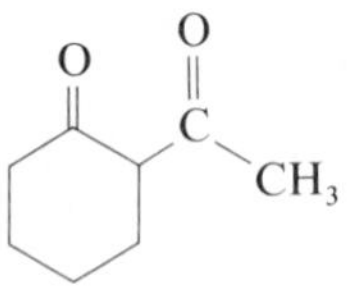 or 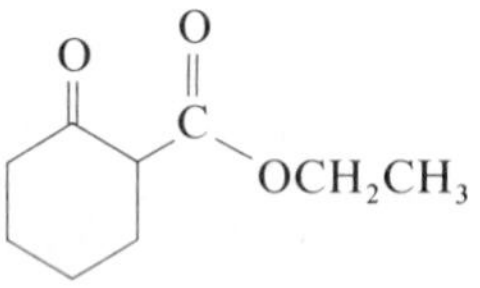

(c) $CH_3\overset{O}{\overset{\|}{C}}CH_2\overset{O}{\overset{\|}{C}}OCH_2CH_3$ or $CH_3\overset{O}{\overset{\|}{C}}\underset{CH_2CH_3}{\underset{|}{CH}}\overset{O}{\overset{\|}{C}}OCH_2CH_3$

(d) $O_2NCH_2\overset{O}{\overset{\|}{C}}OCH_2CH_3$, $O_2NCH_2NO_2$, or $O_2NCH_2\overset{O}{\overset{\|}{C}}CH_3$

(e) $H\overset{O}{\overset{\|}{C}}CH_2\overset{O}{\overset{\|}{C}}H$, $CH_3CH_2O\overset{O}{\overset{\|}{C}}CH_2\overset{O}{\overset{\|}{C}}OCH_2CH_3$, or $CH_3CH_2O\overset{O}{\overset{\|}{C}}\underset{CH_2CH_3}{\underset{|}{CH}}\overset{O}{\overset{\|}{C}}OCH_2CH_3$

16.22 Give structural formulas for all intermediates and products designated by letters in the following equations.

(a) 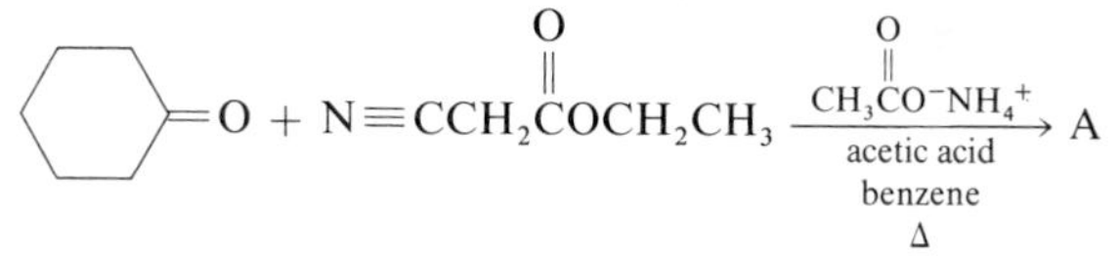

(b) $$CH_2(\overset{\overset{\large O}{\|}}{C}OCH_2CH_3)_2 \xrightarrow[\text{ethanol}]{CH_3CH_2ONa} B \xrightarrow{CH_3\overset{\overset{\large CH_3}{|}}{C}HCH_2Br} C \xrightarrow[\Delta]{H_3O^+} D$$

(c) 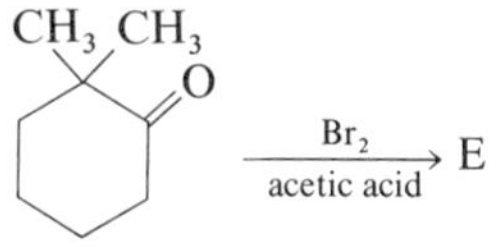

(d) 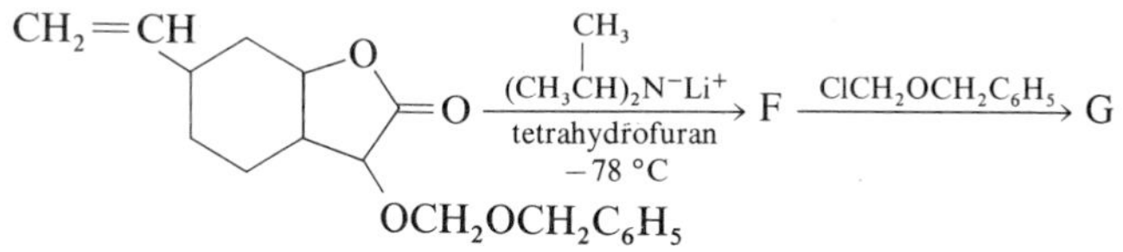

(e) 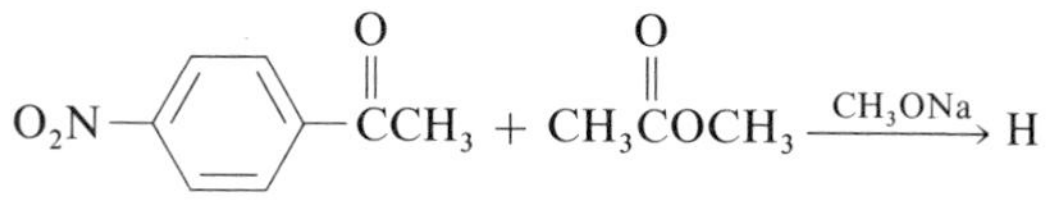

(f) 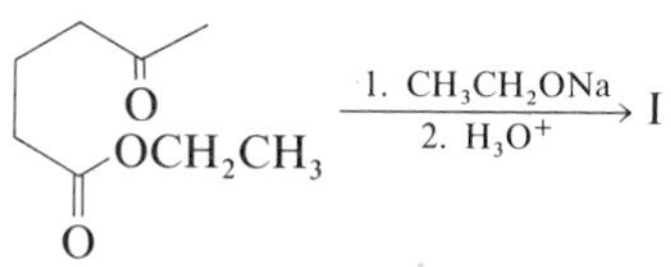

(g) 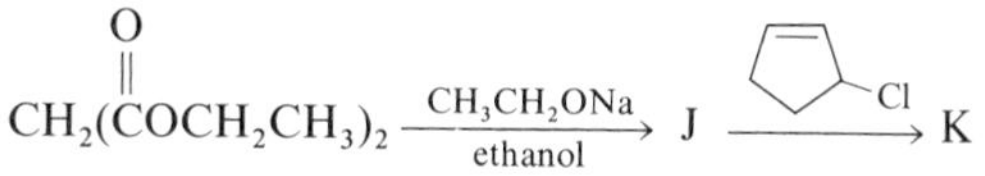

(h) $$CH_3\overset{\overset{\large O}{\|}}{C}H + CH_2(\overset{\overset{\large O}{\|}}{C}OCH_2CH_3)_2 \xrightarrow[\Delta]{(CH_3\overset{\overset{O}{\|}}{C})_2O} L$$

(i) 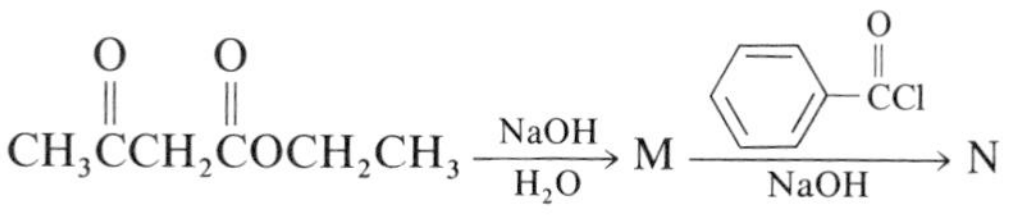

(j) 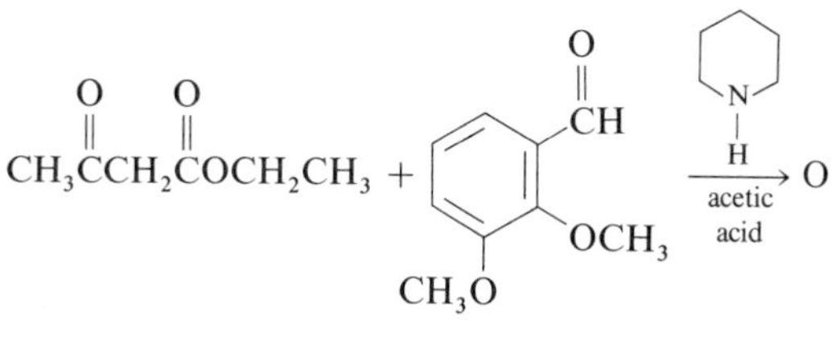

(k) 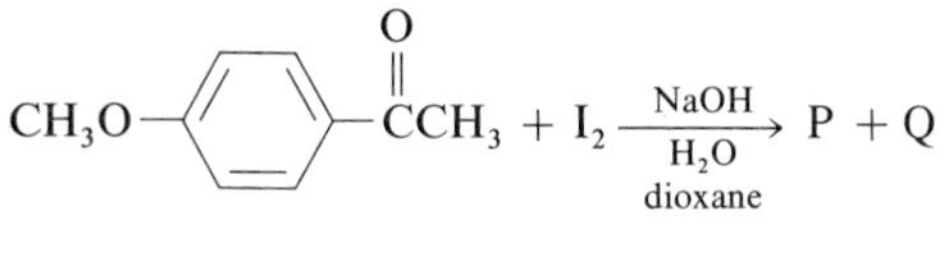

(l) 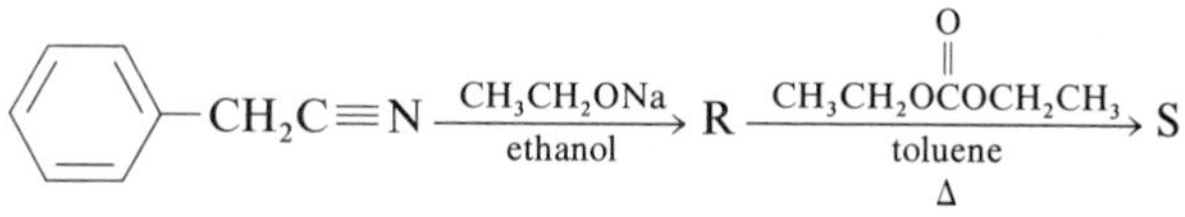

(m) 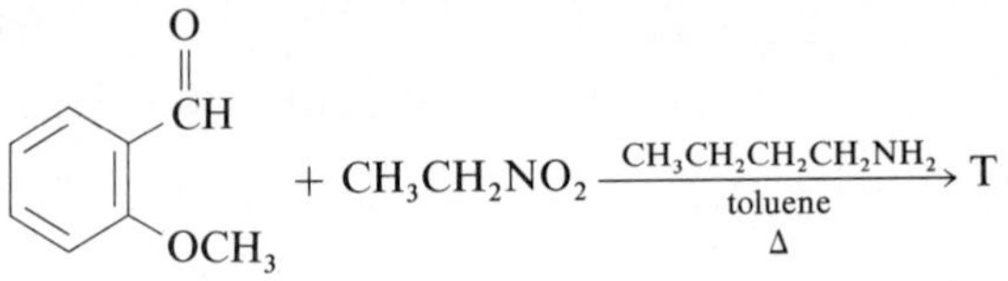
O
CH
OCH3
+ CH3CH2NO2
CH3CH2CH2CH2NH2
toluene
Δ
T

(n) 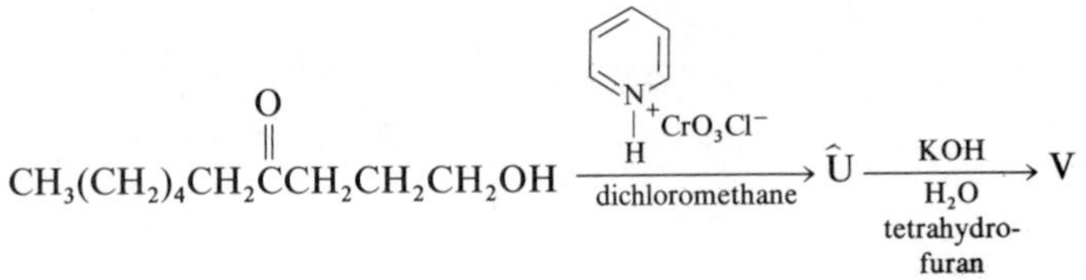
O
CH3(CH2)4CH2CCH2CH2CH2OH
N+
CrO3Cl−
H
dichloromethane
U
KOH
H2O
tetrahydro-
furan
V

16.23 Gilbert Stork of Columbia University discovered that **enamines,** nitrogen-containing compounds that are analogous to enolate anions, can be used as sources of nucleophilic carbon atoms in syntheses. The following equations show the preparation and reactions of an enamine. Provide a mechanism for each step.

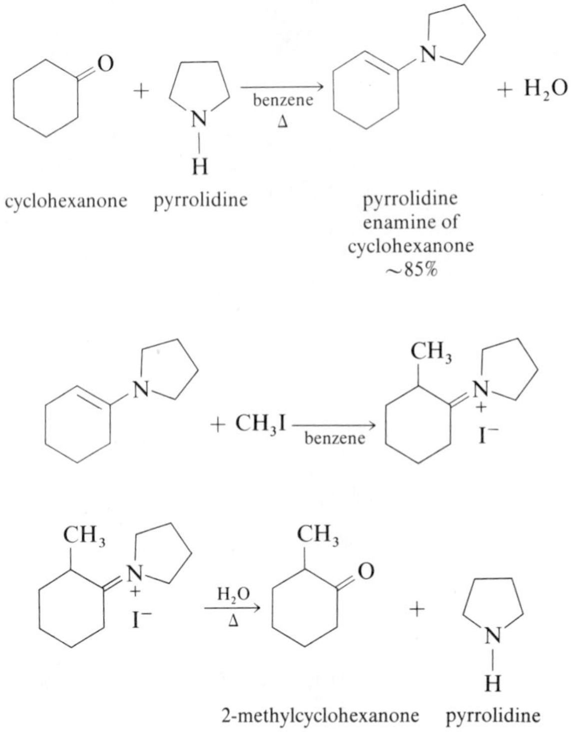
benzene
Δ
+ H2O
cyclohexanone
pyrrolidine
pyrrolidine
enamine of
cyclohexanone
~85%
+ CH3I
benzene
CH3
I−
H2O
Δ
2-methylcyclohexanone
pyrrolidine

16.24 Fill in structural formulas for the intermediates and products indicated by letters in the following sequence of reactions.

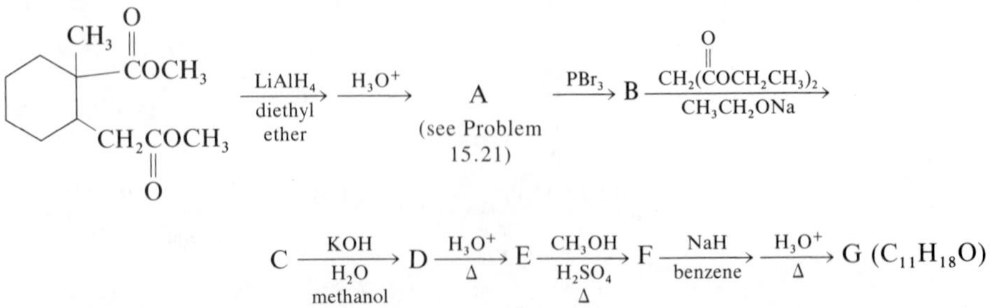
CH3
COCH3
CH2COCH3
LiAlH4
diethyl
ether
H3O+
A
(see Problem
15.21)
PBr3
B
CH2(COCH2CH3)2,
CH3CH2ONa
C
KOH
H2O
methanol
D
H3O+
Δ
E
CH3OH
H2SO4
Δ
F
NaH
benzene
H3O+
Δ
G (C11H18O)

16.25 Suggest a mechanism that accounts for the product observed for the following reaction.

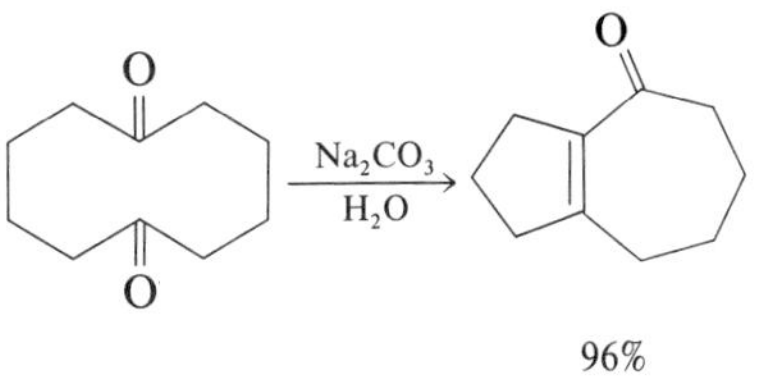
Na2CO3
H2O
96%

16.26 Tell what reagents would be necessary to carry out each step of the following transformation.

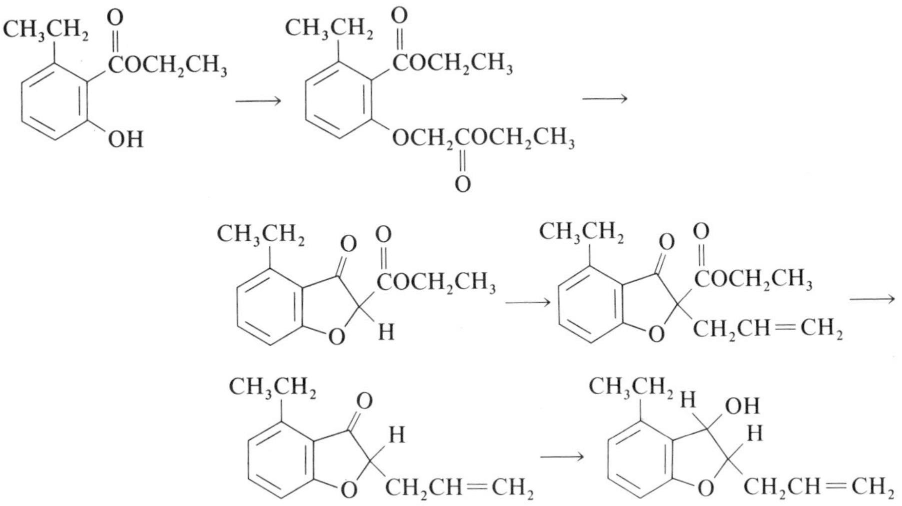

16.27

(a) The following sequence of reactions was used to convert the isobutyl enol ether of 1,3-cyclohexadione to 4-propylcyclohexanone. Give structural formulas for the intermediates, products, and reagent indicated by the letters.

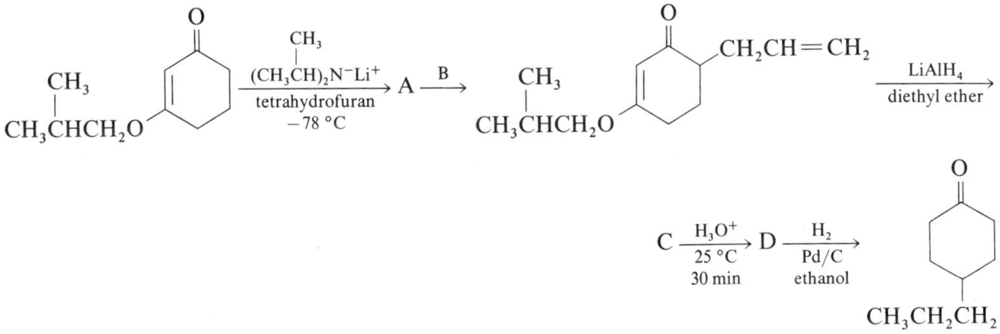
tetrahydrofuran
−78 °C
LiAlH4
diethyl ether
25 °C
30 min
Pd/C
ethanol

(b) The isobutyl enol ether of 1,3-cyclohexadione is prepared in 90% yield by heating the diketone with isobutyl alcohol and *p*-toluenesulfonic acid in benzene. Write a mechanism for the reaction that shows why it is so easy to make this enol ether.

16.28 1,3-Diketones decompose when heated with base, for example:

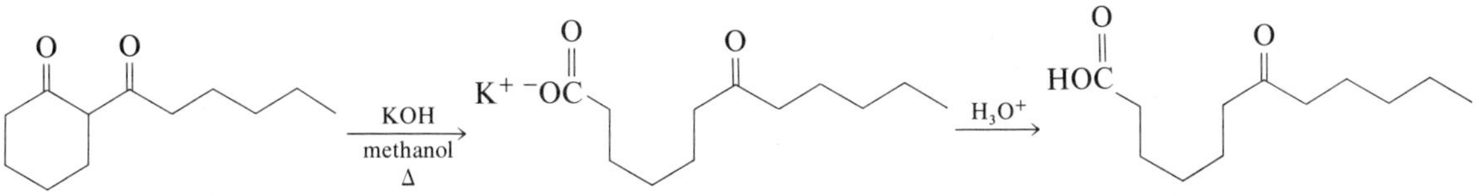
KOH
methanol
Δ

Propose a mechanism for this reaction.

16.29 Propose a mechanism that accounts for the product obtained in the following cyclization reaction.

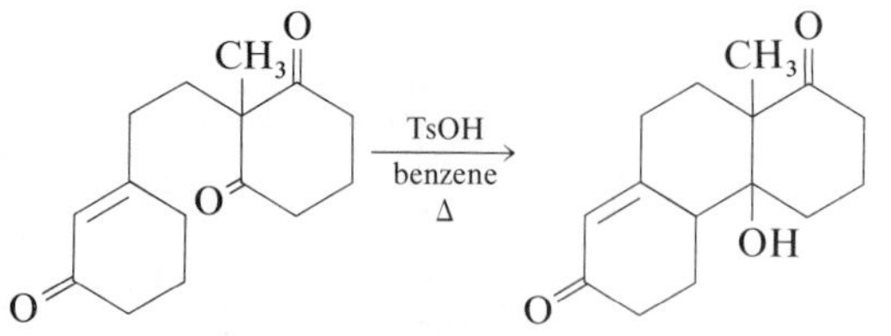
CH3
O
TsOH
benzene
Δ
OH

16.30 Write mechanisms that rationalize the following experimental observations.

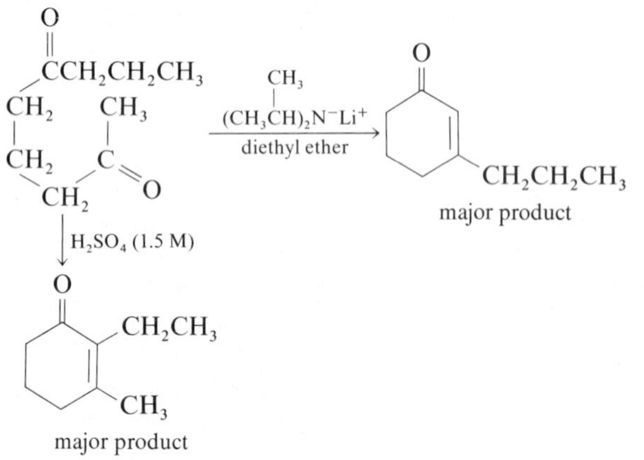
CCH2CH2CH3
CH2
CH3
CH2
C
CH2
O
CH3
(CH3CH)2N−Li+
diethyl ether
CH2CH2CH3
major product
H2SO4 (1.5 M)
CH2CH3
CH3
major product

16.31 Organic chemists have tried to discover whether the metal ion in an enolate salt is primarily associated with a negatively charged oxygen atom or with a negatively charged carbon atom in the ambident nucleophile (p. 636). One of the research tools chemists have used in attempting to answer this question is nuclear magnetic resonance spectroscopy. The proton magnetic resonance spectra of the lithium salt of the enolate ion from 1-phenyl-2-methyl-1-propanone in a variety of solvents (such as benzene and tetrahydrofuran) have two singlets in the region δ 1.0–2.0. Compare structural formulas for the two resonance contributors for the enolate ion from 1-phenyl-2-methyl-1-propanone and decide what conclusion can be drawn from the proton magnetic resonance data.

16.32 Propose a detailed mechanism for the following reaction.

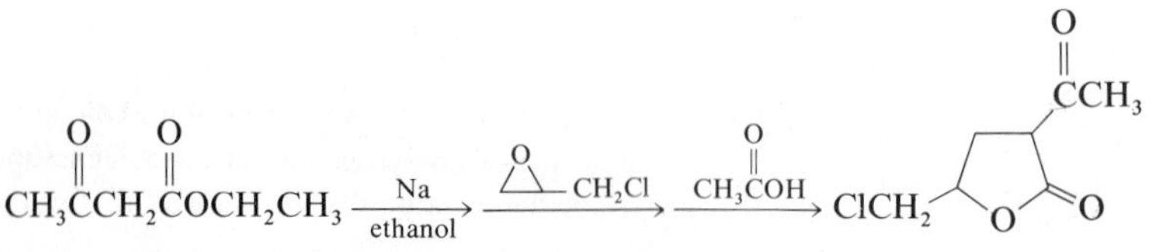
CH3CCH2COCH2CH3
Na
ethanol
CH2Cl
CH3COH
ClCH2
CCH3
O

16.33 When the ketolactone shown below is heated with concentrated hydrochloric acid, the product is 5-chloro-2-pentanone. Treatment of 5-chloro-2-pentanone with aqueous sodium hydroxide gives cyclopropyl methyl ketone in about 80% yield. How would you rationalize these experimental observations?

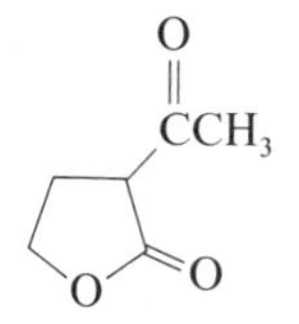

ketolactone used as
starting material

16.34 It has been discovered that it is possible to prepare dianions of carbonyl compounds if very strong bases are used. One such dianion derived from a substituted acetoacetic ester was used in the synthesis of dihydrojasmone, a compound used in making perfumes. The synthesis is outlined below, and some hints are given. Supply structural formulas for the intermediates and products designated by letters.

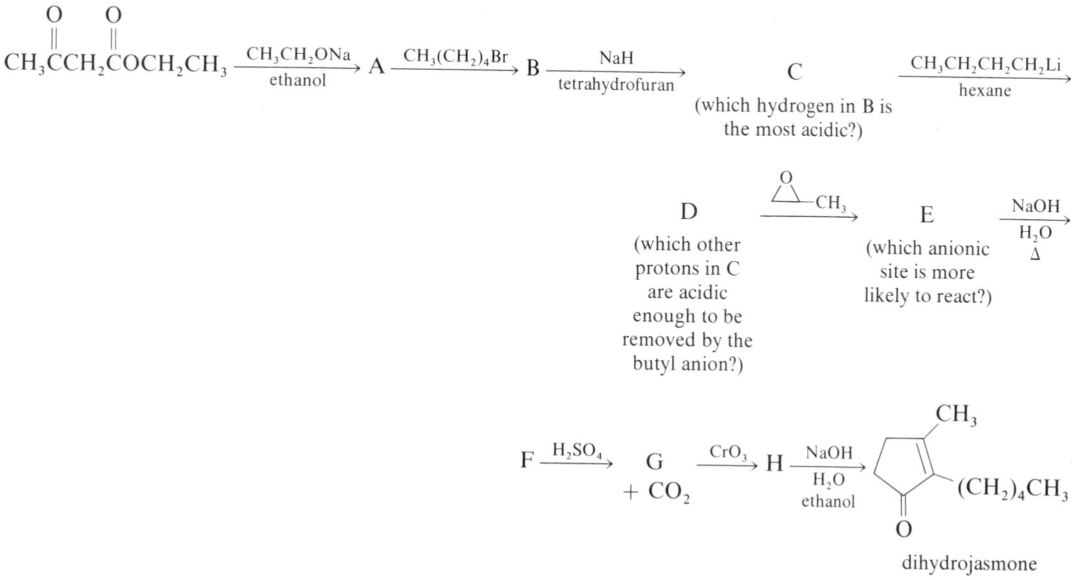

CH3CCH2COCH2CH3
CH3CH2ONa
ethanol
A
CH3(CH2)4Br
B
NaH
tetrahydrofuran
C
(which hydrogen in B is the most acidic?)
CH3CH2CH2CH2Li
hexane
D
(which other protons in C are acidic enough to be removed by the butyl anion?)
CH3
E
(which anionic site is more likely to react?)
NaOH
H2O
Δ
F
H2SO4
G
+ CO2
CrO3
H
NaOH
H2O
ethanol
CH3
(CH2)4CH3
dihydrojasmone

What other product is possible in the last step of the synthesis? How do you explain the regioselectivity of the reaction?

16.35 Compound A, having the formula $C_{12}H_{20}O_2$, is isolated from a fungus. The compound has bands at 2978(s), 2935(s), 2850(s), 1735(s), 1450(m), 1365(m), 1225(s), and 975(s) cm^{-1} in its infrared spectrum. Its proton magnetic resonance spectrum shows a doublet corresponding to three hydrogens at δ 1.2 and a band corresponding to two hydrogens at δ 5.3 ppm.

Compound A was treated with ozone in ethanol and ethyl acetate; the reaction mixture was then heated with formic acid and hydrogen peroxide and finally heated under reflux with potassium hydroxide in alcohol. Acidification of the reaction mixture gave octanedioic acid and 3-hydroxybutanoic acid.

Compound A reacts with hydrogen in the presence of platinum to give Compound B, $C_{12}H_{22}O_2$. When Compound B is heated with aqueous base, then acidified, Compound C, $C_{12}H_{24}O_3$, is obtained. Treatment of Compound C with chromium trioxide in acetic acid gives Compound D, $C_{12}H_{22}O_3$. Compound D reacts with iodine in base to give iodoform and undecanedioic acid.

Assign structures to Compounds A, B, C, and D that are compatible with these facts.

16.36 How would you carry out the following transformation?

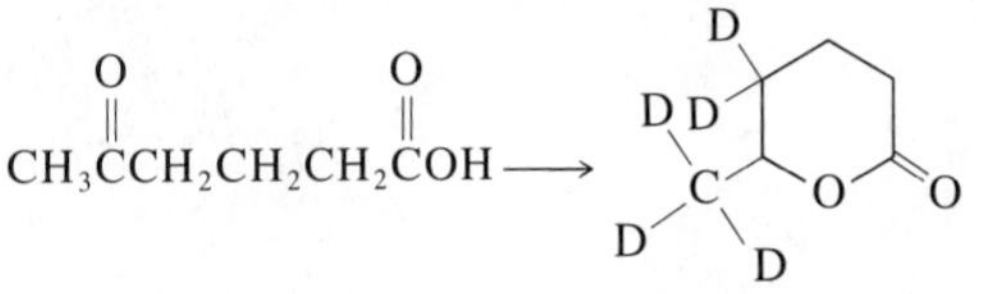

16.37 Write a complete mechanism for the following reaction.

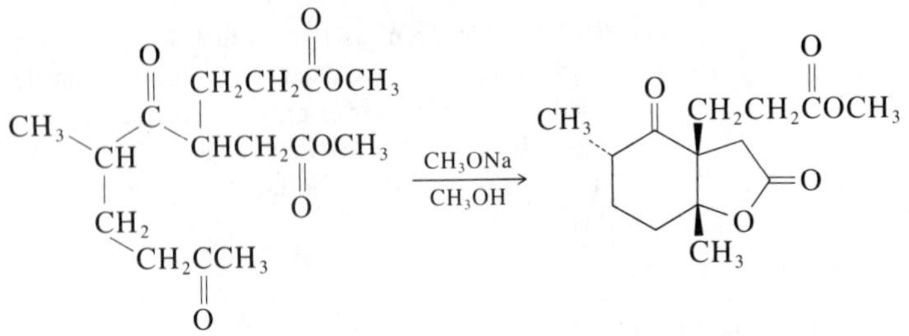

16.38 What reagents would you use to carry out each step of the following transformation?

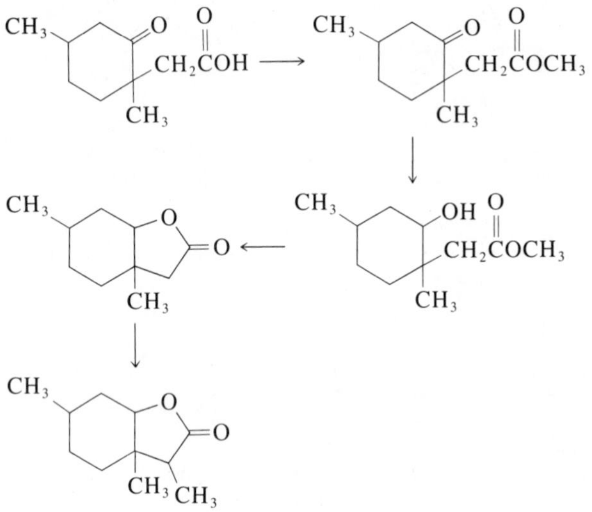

16.39 Tell what reagents would be required to carry out the following sequence of transformations.

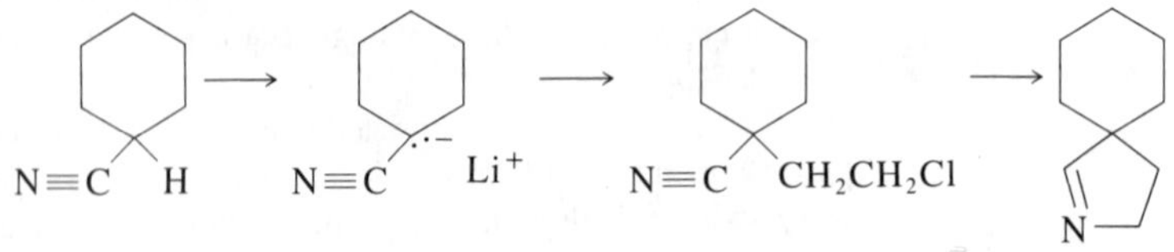

17

Polyenes

A • L O O K • A H E A D

Multiple bonds that are separated from each other by a single bond are said to be conjugated.

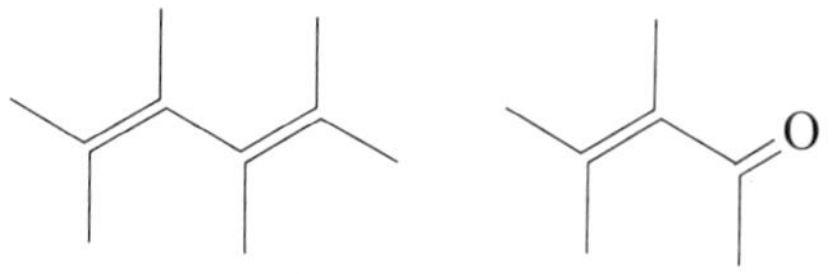

conjugated systems

Chapter 18 will cover the reactions of compounds in which a carbonyl group is conjugated with a carbon-carbon double bond. This chapter will look at the reactions of conjugated alkenes.

Compounds having conjugated multiple bonds undergo the reactions that are typical of the individual functional groups. They also have special reactivity resulting from the interaction of the multiple bonds. The most important consequence of this is the ability to undergo reactions at the ends of the conjugated system.

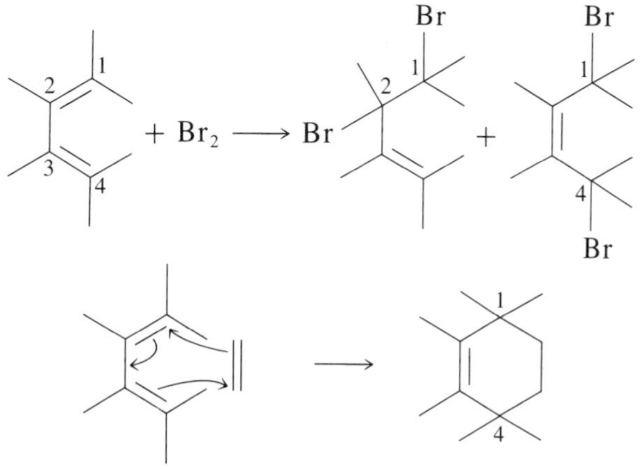

Br
Br
1
2
3
4
+ Br2
Br
2
1
+
1
4
Br
1
4

Many biologically important compounds have multiple double bonds or are derived from such compounds. In this chapter, you will see how such complex compounds are built up from simpler units in living organisms.

17.1 ISOLATED, SKIPPED, CONJUGATED, AND CUMULATIVE MULTIPLE BONDS

The chemistry of carbon-carbon multiple bonds is familiar by this point. Compounds containing double or triple bonds add halogens and acids. They are reduced by hydrogen and oxidized by reagents such as ozone, potassium permanganate, and peroxyacids. They react with Lewis acids such as diborane to give intermediates that are converted to alcohols. This chemistry is also characteristic of compounds that contain more than one multiple bond, as many examples in preceding chapters have demonstrated. In all of those cases, however, the multiple bonds were separated from each other by one or more tetrahedral carbon atoms.

Multiple bonds that are separated from each other by two or more tetrahedral carbon atoms are said to be **isolated multiple bonds.** Some examples of compounds containing isolated multiple bonds are given below with their names.

$CH_2{=}CHCH_2CH_2CH{=}CH_2$
1,5-hexadiene

$HC{\equiv}CCH_2CH_2CH_2C{\equiv}CH$
1,6-heptadiyne

$CH_2{=}CHCH_2CH_2CH{=}CHCH_2CH_2CH{=}CH_2$
1,5,9-decatriene

$HC{\equiv}CCH_2CH_2CH{=}CH_2$
1-hexen-5-yne

Such compounds are named by indicating the number of multiple bonds in the chain by the suffixes **-adiene** and **-atriene** or **-adiyne** and **-atriyne,** and also indicating the position of the multiple bond by giving the number in the chain of the first carbon atom of each multiple bond. When a compound contains both a double bond and a triple bond, the triple bond is named as the suffix.

Multiple bonds that are separated from each other by only one tetrahedral carbon atom are called **skipped multiple bonds.** The polyunsaturated fatty acids (p. 610), for example, contain skipped double bonds. This structural feature makes such compounds especially reactive in ways that are explored in Section 20.6B.

If the multiple bonds are separated from each other by one single bond, the *p* orbitals on adjacent carbon atoms can interact. The prime example of this kind of interaction occurs in benzene. Sideways overlap of a *p* orbital from each of the six carbon atoms on the six-membered ring results in delocalization of six electrons over the entire benzene ring (p. 65). Multiple bonds that are separated from each other by one single bond are said to be **conjugated.** The concept of conjugation was first introduced in connection with a carbonyl group separated by a single bond from a double bond or an aromatic ring (p. 366). Some other examples of conjugated systems are shown below.

$CH_2{=}CH{-}CH{=}CH_2$
1,3-butadiene

$CH_2{=}CH{-}CH{=}CH{-}CH{=}CH_2$
1,3,5-hexatriene

$HC{\equiv}C{-}C{\equiv}CCH_2CH_3$
1,3-hexadiyne

$HC{\equiv}C{-}CH{=}CHCH_3$
2-penten-4-yne

$$\begin{array}{c} \qquad\qquad\quad CH_3 \\ \qquad\qquad\quad | \\ CH_2{=}CH{-}C{=}O \end{array}$$
3-buten-2-one

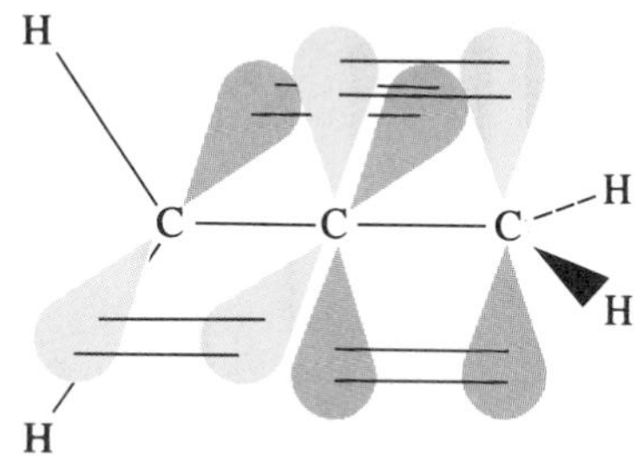
H
C
C
C
H
H
H

FIGURE 17.1 Bonding in allene.

When at least three adjacent carbon atoms in a molecule are joined by double bonds, the compound is called a **cumulene** or is said to have **cumulative double bonds.** The simplest cumulene is 1,2-propadiene, usually called allene.

$$\overset{3}{C}H_2=\overset{2}{C}=\overset{1}{C}H_2$$

1,2-propadiene
allene

Study Guide Concept Map 17.1

The central carbon atom is *sp*-hybridized and is bonded to the other two carbon atoms of the system by π bonds that are at right angles to each other (Figure 17.1).

PROBLEM 17.1

Name the following compounds.

(a) $CH_3C{\equiv}CCH_2C{\equiv}CH$ (b) $CH_3CH{=}CHC{\equiv}CCH_3$ (c) 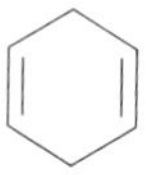

(d) 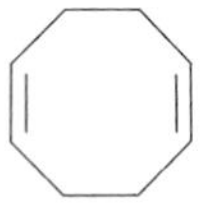 (e) 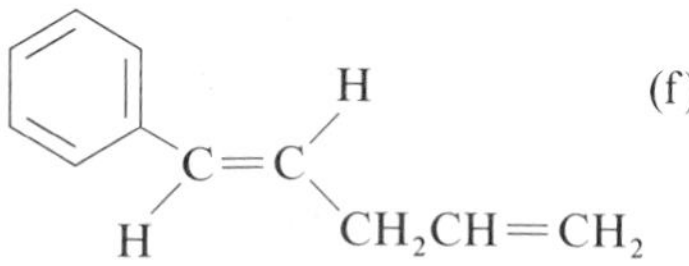
H
C=C
H
CH2CH=CH2

(f) $ClCH_2C{\equiv}CC{\equiv}CCH_2CH_3$

PROBLEM 17.2

Write structural formulas for the following compounds.

(a) 5-hexen-2-one (b) 1,5-hexadiyne (d) 1,4-pentadiene (d) 2,5-heptadiyne
(e) (*E*)-4,4-dimethyl-1,6-octadiene (f) 1,3-hexadien-5-yne

17.2 1,3-BUTADIENE

1,3-Butadiene is the simplest compound that contains two double bonds separated by a single bond. Electron diffraction studies have shown 1,3-butadiene to be a planar molecule. The double bonds in 1,3-butadiene are 1.34 Å long, just about the length of the double bond in ethylene; the length of the single bond is 1.48 Å, slightly shorter than a single bond (1.49 Å) adjacent to the double bond in propene and significantly shorter than the single bond in ethane (1.54 Å).

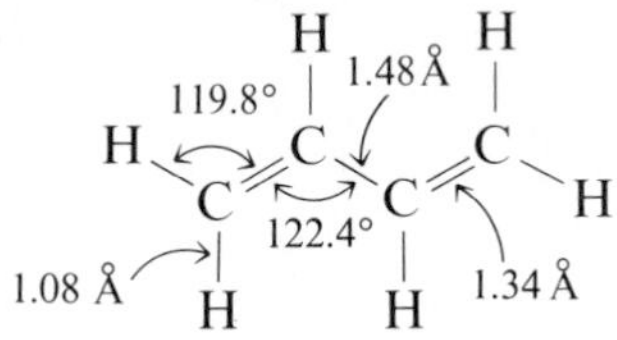
H
H
1.48 Å
119.8°
H
H
122.4°
1.08 Å
1.34 Å
H
H
structure of 1,3-butadiene

The heat of hydrogenation (p. 278) for 1,3-butadiene (−57.1 kcal/mol) is 4 kcal/mol less than expected for a compound containing two isolated double bonds (~61 kcal/mol). 1,3-Butadiene is thus more stable than expected by that amount. The difference between the stability determined experimentally for a compound (in this case 1,3-butadiene) and that predicted for a hypothetical molecule with isolated double bonds is called the **empirical resonance energy** for the system.

1,3-Butadiene has a skeleton of four carbon atoms in a chain bonded to six hydrogen atoms by σ bonds. The carbon atoms are sp^2 hybridized, and a p orbital containing one electron is available to each carbon atom in the chain (p. 48). The double bonds in 1,3-butadiene can be pictured as arising from the overlap of pairs of p orbitals on adjacent carbon atoms (Figure 17.2). But this model for 1,3-butadiene, with double bonds localized between carbon atoms 1 and 2 and carbon atoms 3 and 4, does not explain the planarity of the molecule, the shorter than usual single bond, and the empirical resonance energy determined for the compound.

A better model for the bonding in 1,3-butadiene has been developed using molecular orbital theory (p. 49). In this model, the four atomic p orbitals on the carbon atoms are combined to give four π molecular orbitals, designated as ψ_1, ψ_2, ψ_3, and ψ_4 (Figure 17.3).

The molecular orbitals ψ_1 and ψ_2 are bonding molecular orbitals, and each has two electrons in it. The other two molecular orbitals are antibonding and do not contain any electrons. According to this picture of 1,3-butadiene, the molecule is planar to allow interaction between the p orbitals on carbon atoms 2 and 3. There is some double bond character between those carbon atoms due to the contribution of the lowest-energy bonding molecular orbital, ψ_1, and, therefore, a shortening of the carbon-carbon single bond. 1,3-Butadiene is more stable by 4 kcal/mol than a molecule with two isolated double bonds because of the delocalization of electrons over the whole chain, represented by molecular orbital ψ_1.

1,3-Butadiene is shown in Figure 17.2 in an extended conformation in which the two double bonds are trans to each other across the central single bond. This conformation is known as the ***s*-trans conformation,** the *s* indicating that the stereochemistry refers to the single bond. A higher-energy conformation that is important in the reactions of butadiene is the ***s*-cis conformation.** Both of these conformations exist in equilibrium with each other. In both of them, all four of the carbon atoms and all the hydrogen atoms lie in the same plane.

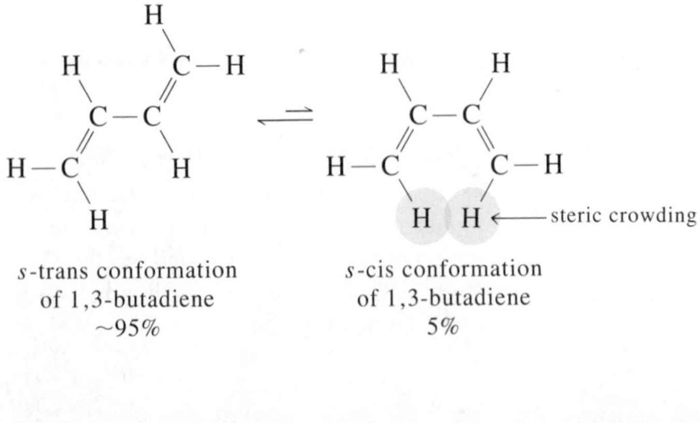
steric crowding
s-trans conformation
of 1,3-butadiene
~95%

s-cis conformation
of 1,3-butadiene
5%

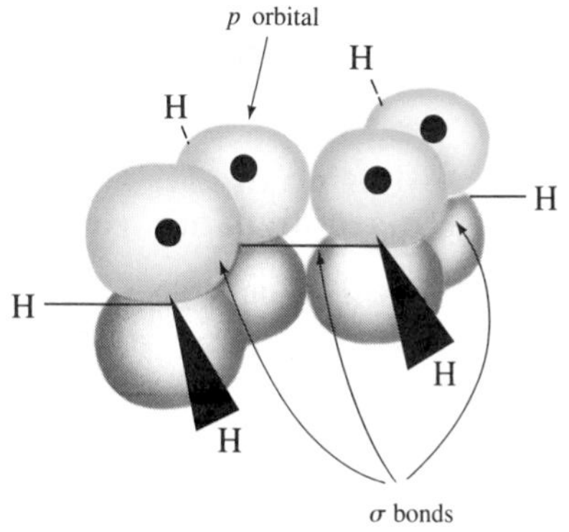
p orbital
H
H
H
H
H
H
σ bonds

1,3-butadiene showing σ bonds and p orbitals, each orbital containing one electron

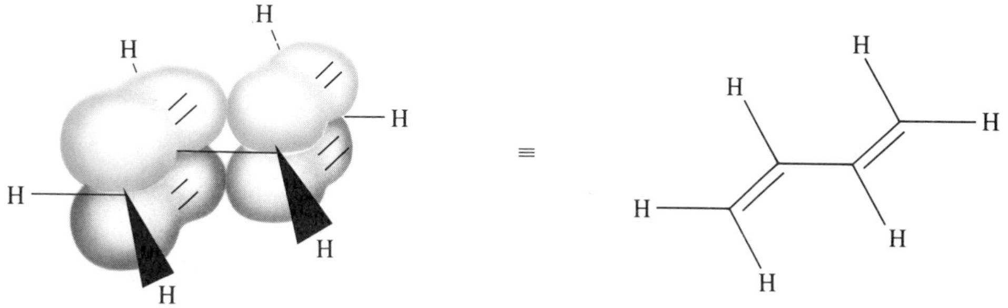
H
H
H
H
H
H
≡
H
H
H
H
H
H

1,3-butadiene shown with localized double bonds

FIGURE 17.2 An orbital picture of the structure of 1,3-butadiene.

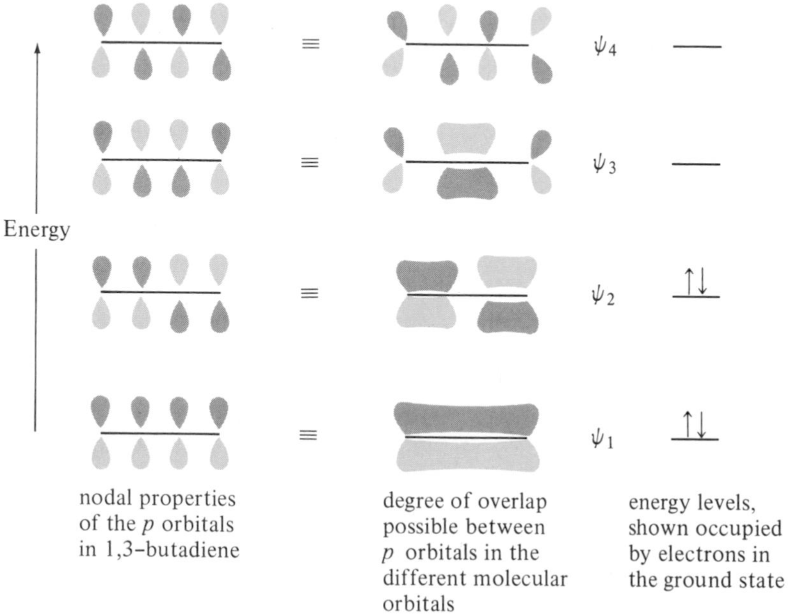
Energy
≡
≡
≡
≡
ψ_4
ψ_3
ψ_2
ψ_1
nodal properties of the p orbitals in 1,3–butadiene
degree of overlap possible between p orbitals in the different molecular orbitals
energy levels, shown occupied by electrons in the ground state

Study Guide Concept Map 17.2

FIGURE 17.3 The four molecular orbitals of 1,3-butadiene, created from four p atomic orbitals.

A. Addition of Electrophiles. 1,2- and 1,4-Addition. Thermodynamic versus Kinetic Control of a Reaction

1,3-Butadiene undergoes reactions in which electrophilic reagents such as chlorine and hydrogen chloride add to either or both of the double bonds. If 1 equivalent of the electrophilic reagent is used, a mixture of two major products is obtained. The experimental observations about the addition of chlorine to butadiene are given below.

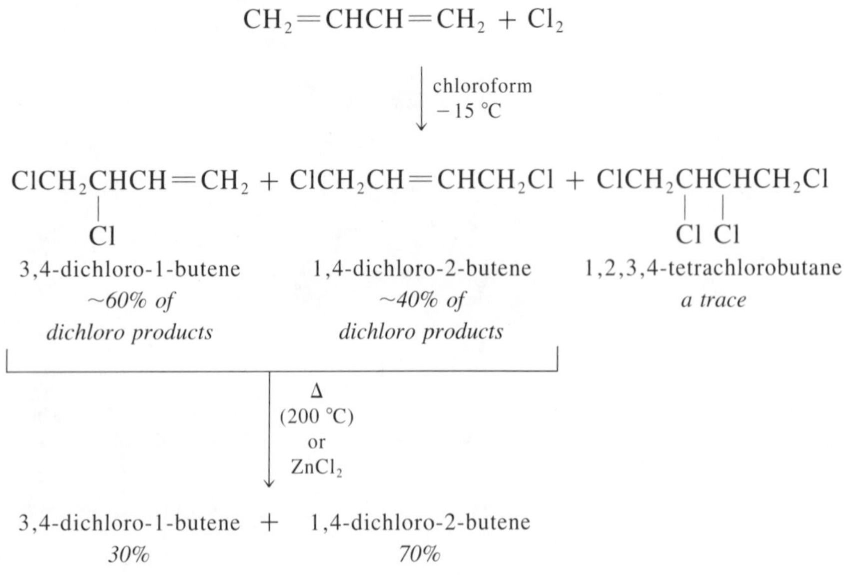
CH2=CHCH=CH2 + Cl2
chloroform
−15 °C
ClCH2CHCH=CH2 + ClCH2CH=CHCH2Cl + ClCH2CHCHCH2Cl
Cl
Cl Cl
3,4-dichloro-1-butene
~60% of dichloro products
1,4-dichloro-2-butene
~40% of dichloro products
1,2,3,4-tetrachlorobutane
a trace
Δ
(200 °C)
or
ZnCl2
3,4-dichloro-1-butene + 1,4-dichloro-2-butene
30%
70%

3,4-Dichloro-1-butene is the product from the addition of chlorine to one of the double bonds in 1,3-butadiene. It is called the **1,2-addition product.** In 1,4-dichloro-2-butene, the chlorine atoms have been attached to the first and fourth carbon atoms of the conjugated system, and the double bond that remains has shifted position in the molecule. This compound is called the **1,4-addition product.**

Experimentally, at low temperatures, 3,4-dichloro-1-butene is formed in larger amounts than is 1,4-dichloro-2-butene. The mixture formed at low temperatures can be equilibrated at high temperatures or in the presence of a Lewis acid, $ZnCl_2$, which aids in the removal of chloride ion. At equilibrium, the mixture contains 70% 1,4-dichloro-2-butene. 1,4-Dichloro-2-butene is thermodynamically more stable than 3,4-dichloro-1-butene, as indicated by its predominance under equilibrium conditions. Yet, at low temperatures, under reaction conditions that do not allow for equilibrium to be established, 3,4-dichloro-1-butene is formed to a larger extent. In other words, 3,4-dichloro-1-butene is the kinetic product of the reaction, the one that is formed faster at low temperatures (p. 639).

A conjugated diene system is always attacked at one end by an electrophile, because this gives the most highly stabilized cationic intermediate, as shown in the following.

VISUALIZING THE REACTION

Electrophilic addition to a diene. First step

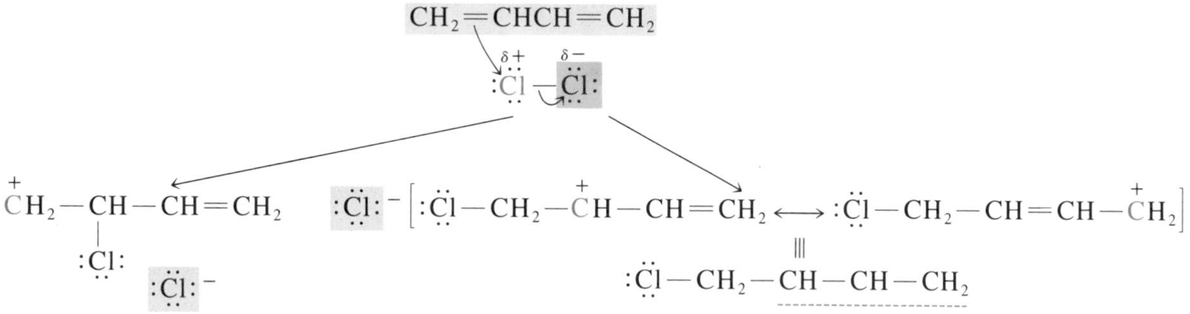

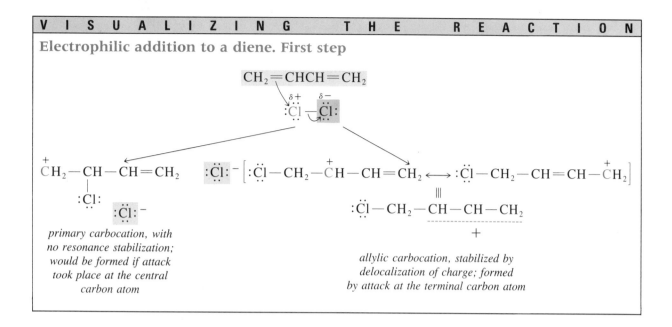

primary carbocation, with no resonance stabilization; would be formed if attack took place at the central carbon atom

allylic carbocation, stabilized by delocalization of charge; formed by attack at the terminal carbon atom

The intermediate formed by addition of a chlorine atom to the end of the conjugated system is an allylic carbocation. An allylic cation is stabilized by delocalization of the positive charge. Addition of a chlorine atom to one of the central carbon atoms in 1,3-butadiene would result in an unstable primary carbocation that had no possibility of resonance stabilization.

An examination of the two contributing resonance structures to the allylic cation shows why 1,2- and 1,4-addition products are formed. The allylic cation has positive charge delocalization to carbon atoms 2 and 4 in the chain. The negatively charged chloride ion can react with the cation at either one of these sites.

VISUALIZING THE REACTION

Electrophilic addition to a diene. Second step

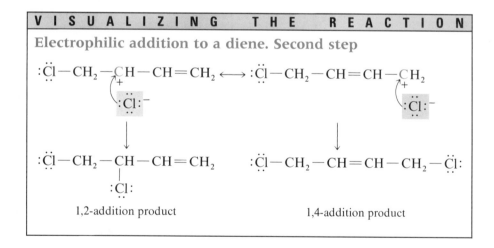

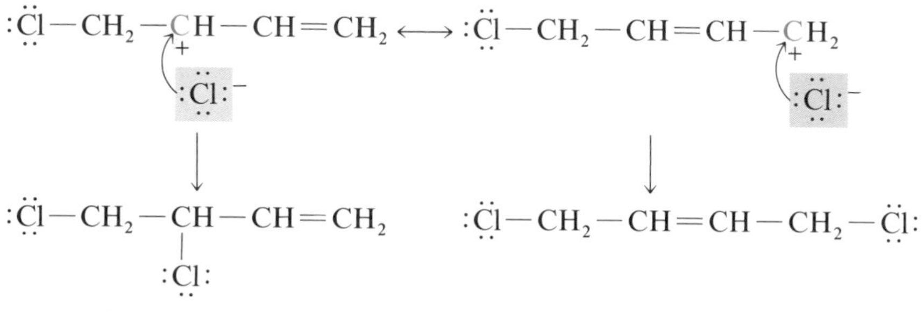

1,2-addition product

1,4-addition product

The 1,4-addition product, with the internal double bond, is thermodynamically more stable than the 1,2-addition product, with the terminal double bond (p. 278). It is not so obvious why the 1,2-addition product should be the one favored kinetically. In other words, the free energy of activation (p. 115) for the formation of

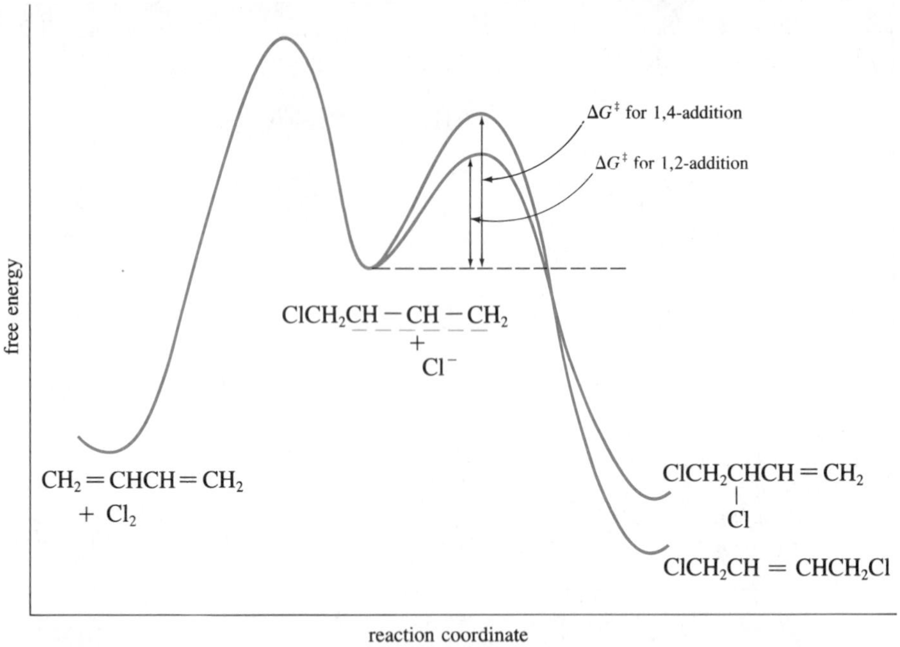
ΔG‡ for 1,4-addition
ΔG‡ for 1,2-addition
free energy
ClCH₂CH — CH — CH₂
+
Cl⁻
CH₂=CHCH=CH₂
+ Cl₂
ClCH₂CHCH=CH₂
Cl
ClCH₂CH = CHCH₂Cl
reaction coordinate

FIGURE 17.4 Energy diagram illustrating the relative energies of activation for 1,2- and 1,4-addition of chlorine to 1,3-butadiene.

the 1,2-addition product from the intermediate is lower than the free energy of activation for the 1,4-addition product from the same intermediate. These energy relationships are shown in Figure 17.4.

The energy diagram shows 1,4-dichloro-2-butene to be more stable and of lower energy than 3,4-dichloro-1-butene. However, the free energy of activation, the hill that must be climbed to get to the transition state from the common cationic intermediate, is higher for the 1,4-addition product. The highest free energy of activation of all is, of course, the one leading to the formation of the reactive intermediate, the cation, from the starting reagents butadiene and chlorine. That is the rate-determining step (p. 126) for the overall reaction.

Why should the free energy of activation for 1,4-addition be higher than it is for 1,2-addition? Chemists have offered two explanations. One is that the intermediate has some cyclic chloronium ion character (p. 290), which means that attack at carbon atom 1 to open the three-membered ring is favored.

VISUALIZING THE REACTION

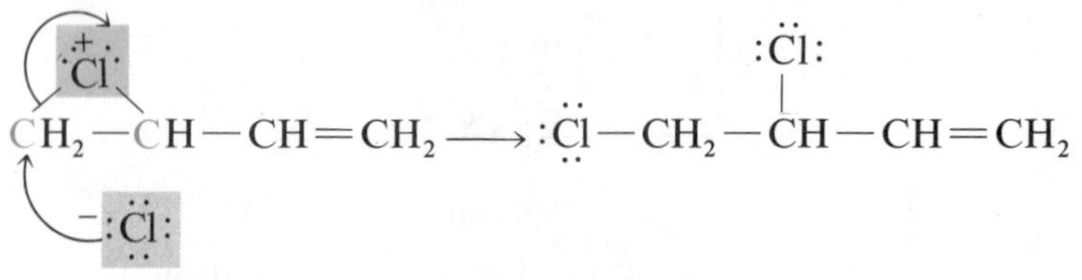
Cl
CH₂—CH—CH=CH₂ ⟶ :Cl—CH₂—CH—CH=CH₂
:Cl:
⁻:Cl:

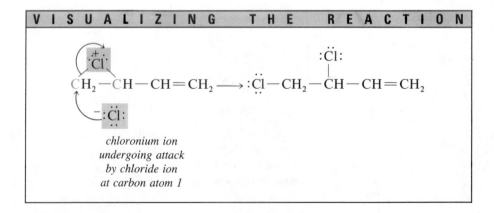

chloronium ion undergoing attack by chloride ion at carbon atom 1

The other explanation is that the two resonance contributors for the allylic cation are not equivalent; the resonance contributor with the charge localized at the secondary carbon atom is the major one. Positive charge density is greater at this secondary carbon, and the chloride ion reacts faster at that site.

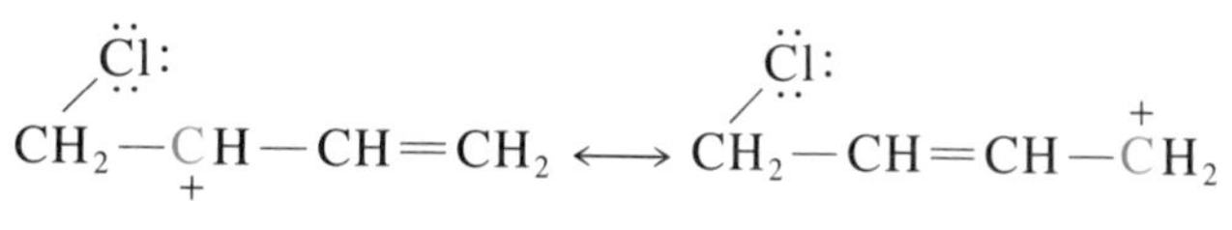

major resonance contributor *minor resonance contributor*

Because an allylic halide ionizes easily, either one of the products can lose a halide ion and again become an allylic cation, allowing an equilibrium to be set up between the two products.

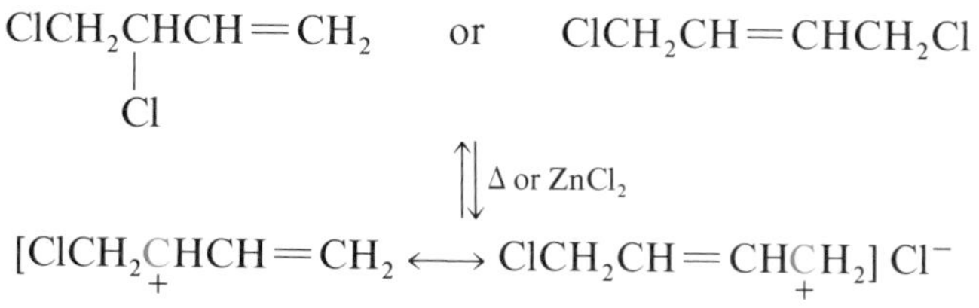
ClCH2CHCH=CH2
Cl
or
ClCH2CH=CHCH2Cl
Δ or ZnCl2
[ClCH2CHCH=CH2 ⟷ ClCH2CH=CHCH2] Cl−

The competition between 1,2- and 1,4-addition just described for 1,3-butadiene applies to additions of halogens or hydrogen halides to other conjugated dienes. In general, the 1,4-addition products are favored under equilibrium conditions, but the 1,2-addition products are the major ones at low temperatures.

B. Addition of Hydrogen to Conjugated Dienes

Hydrogen adds to dienes in the presence of a catalyst (p. 304). Catalytic hydrogenation reactions that are stopped before 2 equivalents of hydrogen have added to the diene give mixtures of alkenes with very small amounts of the corresponding alkane, indicating that the diene reacts more readily with hydrogen than isolated double bonds do. For example, when butadiene is hydrogenated to the point where none of it is left, it has absorbed 54% of the hydrogen that would be required to hydrogenate it completely and only 6% of the product mixture is butane. The rest is a mixture of 1-butene and (*E*)- and (*Z*)-2-butene.

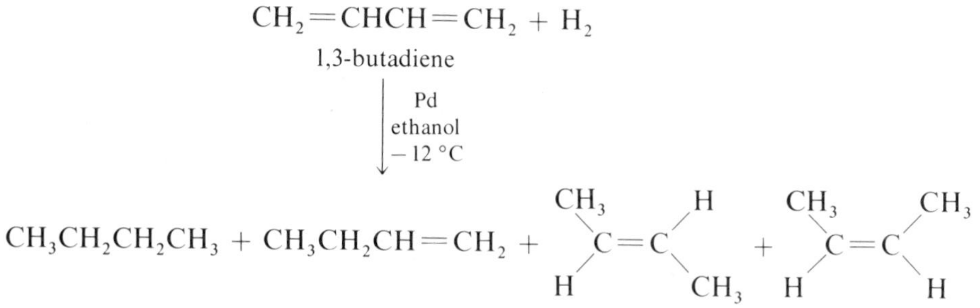
CH2=CHCH=CH2 + H2
1,3-butadiene
Pd
ethanol
−12 °C
CH3CH2CH2CH3 + CH3CH2CH=CH2 +
CH3
H
C=C
H
CH3
+
CH3
CH3
C=C
H
H

Study Guide Concept Map 17.3

butane 6% | 1-butene 45% | (*E*)-2-butene 38% | (*Z*)-2-butene 10%

PROBLEM 17.3

Complete the following equations.

(a)  $\xrightarrow[\text{carbon tetrachloride, 0 °C}]{Br_2\text{ (1 molar equivalent)}}$

(b) $CH_2{=}CHCH{=}CH_2 \xrightarrow{\text{HCl (1 molar equivalent)}}$

(c) $CH_2{=}C(CH_3){-}C(CH_3){=}CH_2 \xrightarrow[\text{carbon tetrachloride, 0 °C}]{Br_2\text{ (1 molar equivalent)}}$

(d) 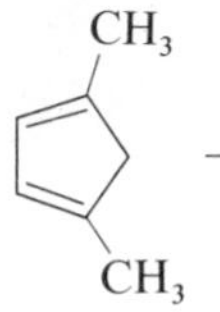 $\xrightarrow[Pd]{H_2\text{ (1 molar equivalent)}}$

17.4 ELIMINATION REACTIONS IN WHICH DIENES ARE FORMED

Dehydrohalogenation of alkyl halides by base (p. 255) and dehydration of alcohols under acidic conditions (p. 282) lead to the formation of alkenes. These reactions are used to synthesize dienes from suitably substituted starting materials. If the product dienes are conjugated, they tend to be highly reactive and their isolation is not always easy, so low yields are often obtained.

3-Bromocyclohexene, for example, is converted to 1,3-cyclohexadiene by elimination of hydrogen bromide using the organic base quinoline.

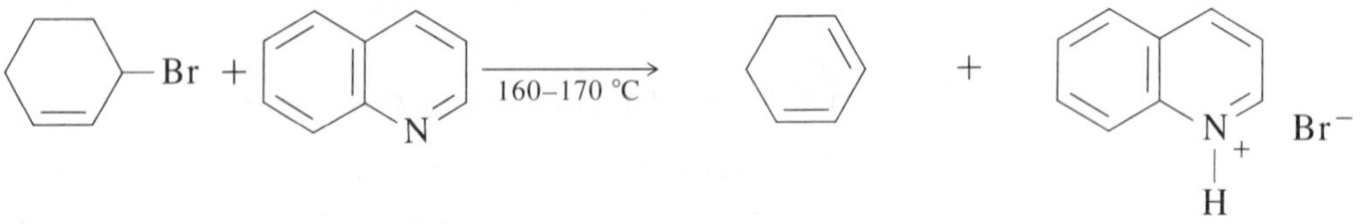

3-bromocyclohexene — quinoline — 1,3-cyclohexadiene — quinoline hydrobromide

Another preparation of 1,3-cyclohexadiene is carried out by dehydrohalogenating 1,2-dibromocyclohexane. The base used is the anion of isopropyl alcohol, prepared from the alcohol with sodium hydride (p. 446) in a high-boiling ether solvent.

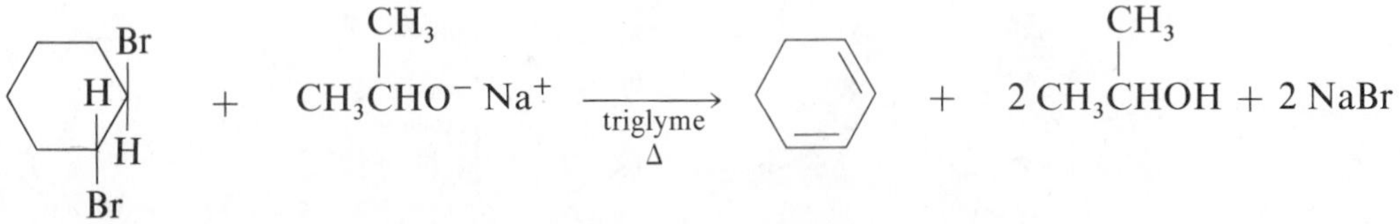

1,2-dibromocyclohexane, 1 equivalent — sodium isopropoxide, 2.2 equivalents — 1,3-cyclohexadiene, bp 81 °C, 70% — isopropyl alcohol

triglyme ≡ $CH_3OCH_2CH_2OCH_2CH_2OCH_2CH_2OCH_3$
bp 222 °C

1,3-Cyclohexadiene is distilled out of the reaction mixture as it is formed so that it is not subjected to prolonged heating and contact with the other reagents.

Because dienes are particularly sensitive to electrophiles, acidic conditions are rarely used for their synthesis. 2,3-Dimethyl-1,3-butadiene is prepared, along with an interesting side product, by the double dehydration of 2,3-dimethyl-2,3-butanediol.

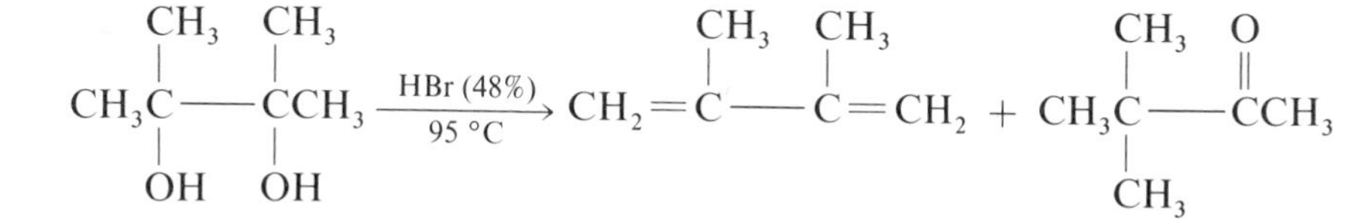
HBr (48%)
95 °C

2,3-dimethyl-2,3-butanediol pinacol — 2,3-dimethyl-1,3-butadiene 57% — 3,3-dimethyl-2-butanone pinacolone ~25%

PROBLEM 17.4

2,3-Dimethyl-2,3-butanediol has the common name pinacol. The side product of its dehydration reaction, 3,3-dimethyl-2-butanone, commonly known as pinacolone, comes from the loss of water and a molecular rearrangement, known as the **pinacol-pinacolone rearrangement.** Write a detailed mechanism for the dehydration of pinacol in acid to 2,3-dimethyl-1,3-butadiene. Look closely at the structure of pinacolone and decide what type of rearrangement has occurred. Why does it occur? Write a mechanism for the formation of pinacolone. (Hint: You may find it helpful to review Section 8.5B.)

17.5 THE DIELS-ALDER REACTION

A. Introduction

One of the most important reactions of conjugated dienes is the 1,4-addition of another multiple bond to the conjugated system to give a six-membered ring. A classic example of this reaction is the formation of a substituted cyclohexene from 1,3-butadiene and maleic anhydride.

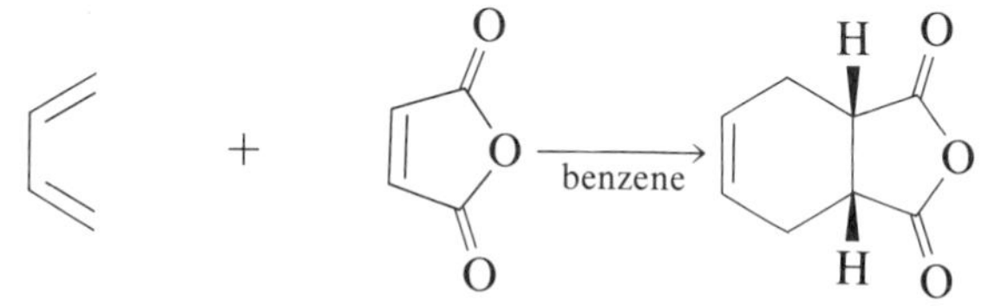
benzene

s-cis conformation of 1,3-butadiene *diene* — maleic anhydride *dienophile* — tetrahydrophthalic anhydride ~95%

The reaction of 1,3-butadiene with maleic anhydride is an example of the **Diels-Alder reaction,** named after the two German chemists, Otto Diels and Kurt Alder, who recognized the generality of that reaction and jointly received the Nobel Prize in 1950 for their work. This type of addition reaction always has two reactants. One is a conjugated diene, which may have many different types of substituents on it. The other reactant always has a double or a triple bond in it and is known as the **dienophile,** a compound that is attracted to and reacts with the diene. The dienophile may be a simple alkene or part of a diene system. The most reactive dienophiles usually have a carbonyl group or another electron-withdrawing group such as a cyano or nitro group conjugated with a carbon-carbon double bond.

B. Stereochemistry of the Diels-Alder Reaction

The Diels-Alder reaction involves a redistribution of electrons and bonds. It occurs in one step; no reactive intermediate is formed. Two double bonds disappear, two new single bonds are formed, and a double bond appears between two atoms that formerly shared a single bond.

VISUALIZING THE REACTION

The Diels-Alder reaction of 1,3-butadiene and maleic anhydride

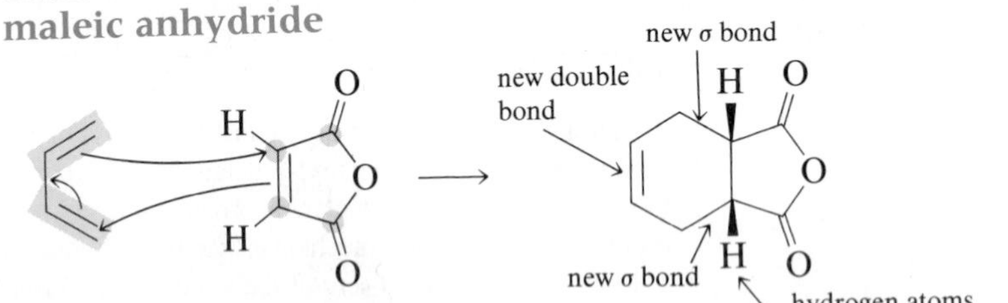
new σ bond
new double bond
new σ bond
hydrogen atoms from maleic anhydride retain cis stereochemistry

diene in *s*-cis conformation

dienophile, a multiple bond conjugated with electron-withdrawing groups

The essential aspects of a Diels-Alder reaction are illustrated by the above reaction. A conjugated diene reacts with a dienophile to give a six-membered ring with a double bond in it. The reaction is thought to proceed via a cyclic transition state. The reaction is highly stereoselective. Substituents on the dienophile as well as those on the diene retain the stereochemistry that they had relative to each other before the reaction. The diene, in order to react, must be in the *s*-cis conformation.

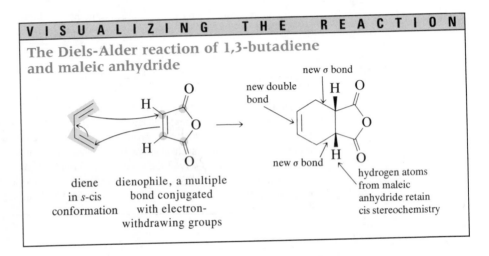

The dienophile in a Diels-Alder reaction may contain a triple bond instead of a double bond. For example, esters of acetylenedicarboxylic acid are highly reactive dienophiles, which is demonstrated by the reaction of diethyl acetylenedicarboxylate with (1*E*,3*E*)-1,4-diphenyl-1,3-butadiene.

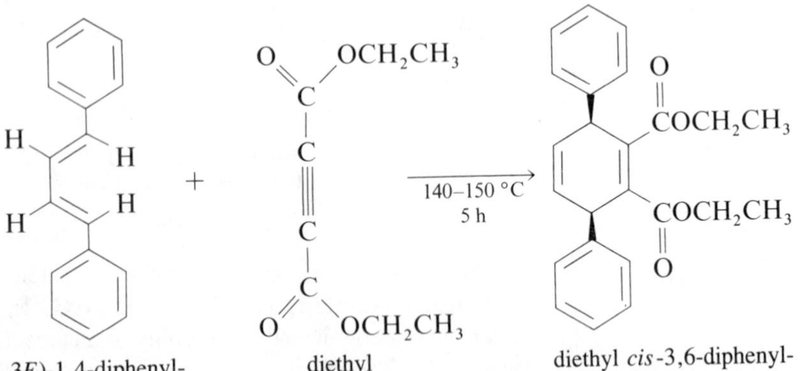
140–150 °C
5 h

(1*E*,3*E*)-1,4-diphenyl-1,3-butadiene

diethyl acetylenedicarboxylate

diethyl *cis*-3,6-diphenyl-1,4-cyclohexadiene-1,2-dicarboxylate
90%

Note that the phenyl groups in the product of the above Diels-Alder reaction have a specific stereochemical relationship that reflects the stereochemistry of the diene used as a starting material.

C. Bicyclic Compounds from Diels-Alder Reactions. Endo and Exo Stereochemistry

Cyclic dienes are particularly reactive in Diels-Alder reactions because the two double bonds are held in an *s*-cis conformation in five- or six-membered rings. (The *s*-trans conformation is favored for open-chain dienes; see p. 684.) Cyclopentadiene is so reactive that it forms a Diels-Alder adduct with itself in a reaction that is reversed at high temperatures.

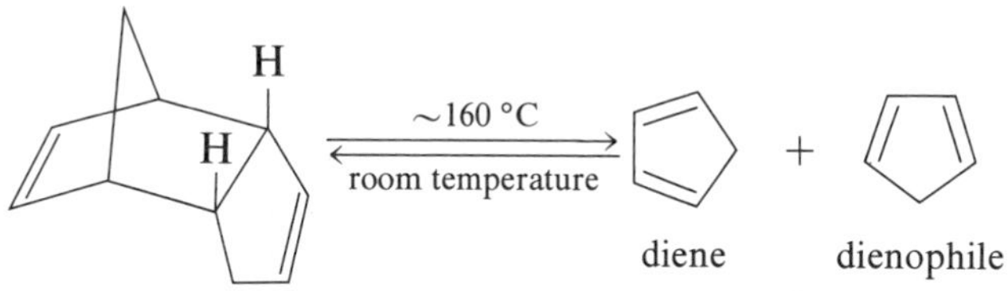
H
H
~160 °C
room temperature

cyclopentadiene dimer

diene | dienophile

bp 42 °C

kept in ice

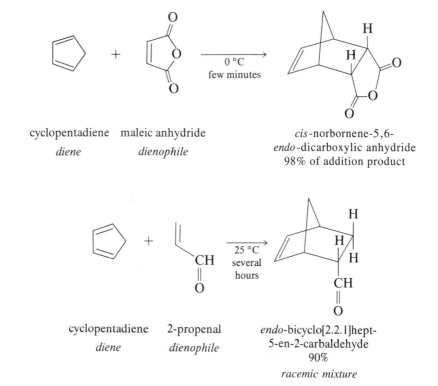

In this reaction, one cyclopentadiene molecule acts as the diene, and one of the two double bonds in another molecule is the dienophile. Cyclopentadiene is stored in the form of its dimer. The dimer is heated to approximately 160 °C and the cyclopentadiene that distills out is collected in an ice-cooled container and used immediately for reactions. On standing, cyclopentadiene reverts to its dimeric form. Freshly distilled cyclopentadiene reacts with maleic anhydride or propenal to give Diels-Alder adducts.

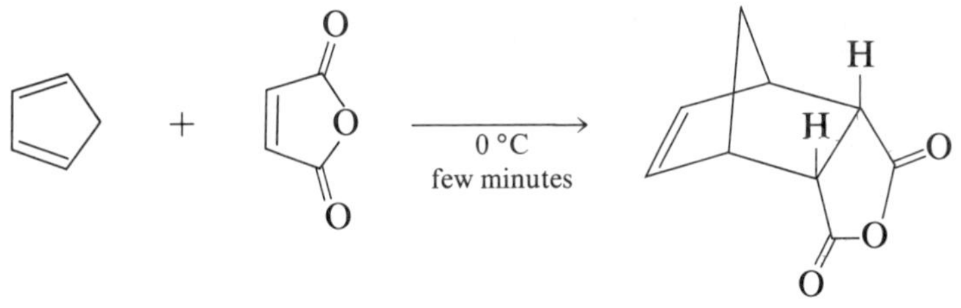
O
O
O
0 °C
few minutes
H
H
O
O
O

cyclopentadiene *diene*

maleic anhydride *dienophile*

cis-norbornene-5,6-*endo*-dicarboxylic anhydride
98% of addition product

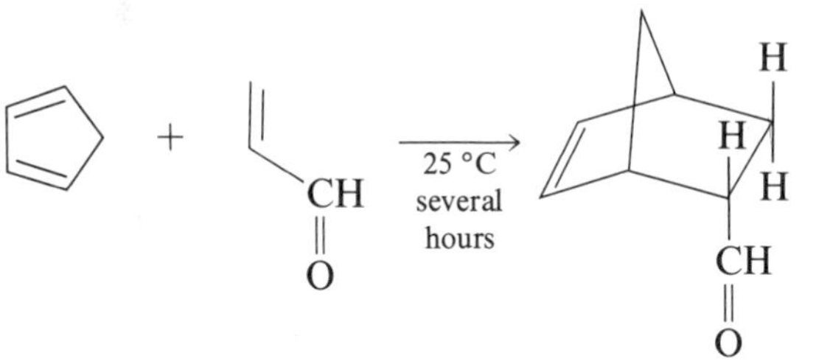
CH
O
25 °C
several
hours
H
H
H
CH
O

cyclopentadiene *diene*

2-propenal *dienophile*

endo-bicyclo[2.2.1]hept-5-en-2-carbaldehyde
90%
racemic mixture

Reactions of cyclopentadiene with dienophiles give rise to compounds containing a cyclohexene ring bridged by a single carbon atom. The common name of the saturated hydrocarbon having such a carbon skeleton is norbornane, and the systematic name is bicyclo[2.2.1]heptane.

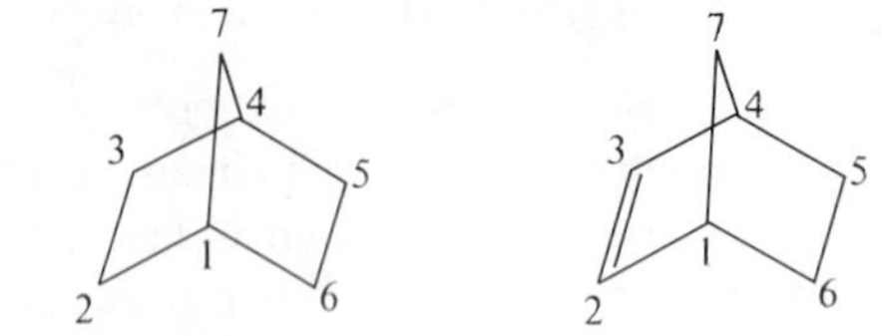
7
4
3
5
1
2
6
7
4
3
5
1
2
6
bicyclo[2.2.1]heptane
norbornane

bicyclo[2.2.1]hept-2-ene
norbornene

The systematic name is derived from the way the seven atoms are held together in a structure with two rings, a **bicyclic structure.** The two rings have two carbon atoms in common, numbered 1 and 4 in the structural formula. These positions are called the **bridgehead positions.** Carbon atoms 1 and 4 are tied together by three bridges, two of which have two carbon atoms and one of which has a single carbon atom. The numbers in the brackets, [2.2.1], represent the numbers of atoms in the bridges. The presence of a substituent or a double bond in a bicyclic compound is indicated by using prefixes or suffixes just as for other types of compounds. Thus, the compound that results when ethylene adds to cyclopentadiene has the common name norbornene and the systematic name bicyclo[2.2.1]hept-2-ene.

The products of the reactions of cyclopentadiene with dienophiles bring up an interesting question of stereochemistry. The substituents on the bicyclic ring created by the Diels-Alder reaction are oriented away from the carbon bridge on the other face of the six-membered ring. Such an orientation is said to be **endo,** meaning that the substituent projects *into* the cavity on the concave side of the bicyclic ring. Another orientation in which the substituent extends *out of* the cavity is called the **exo** orientation.

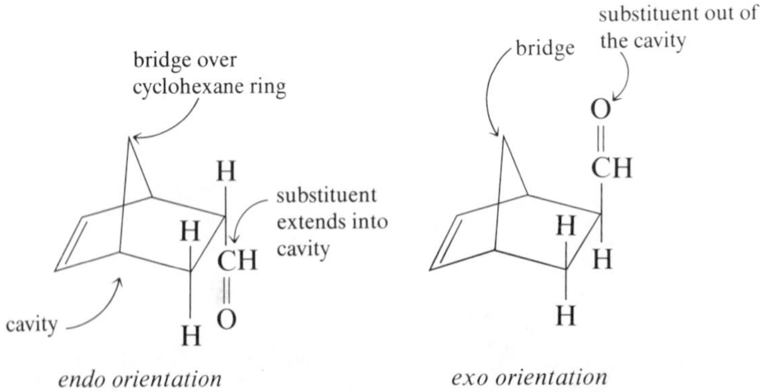
bridge over
cyclohexane ring
H
H
CH
O
H
substituent
extends into
cavity
cavity
substituent out of
the cavity
bridge
O
CH
H
H
H
endo orientation of aldehyde group on bicyclic ring

exo orientation of aldehyde group on bicyclic ring

The reactions of cyclopentadiene with maleic anhydride, with propenal, and with itself give predominantly the endo orientation in the products, a stereochemistry that is quite general for many dienes and dienophiles. This phenomenon is discussed in much greater detail in Section 28.2B, which examines the nature of the interaction between the π bonds of the diene and that of the dienophile.

The electronic character of the diene and dienophile have significance with regard to the ease with which a Diels-Alder reaction takes place. Comparing the reaction of cyclopentadiene and maleic anhydride (p. 694) with the reaction of cyclopentadiene and ethylene makes this evident. A mixture of cyclopentadiene and maleic anhydride must be cooled in ice to prevent the reaction from becoming so vigorous that the low-boiling cyclopentadiene (bp 42 °C) is lost. The reaction of cyclopentadiene with ethylene, on the other hand, requires the use of a steel reaction vessel called a bomb, in which the mixture of reagents can be maintained at 800–900 pounds per square inch (psi) at approximately 200 °C for seven hours.

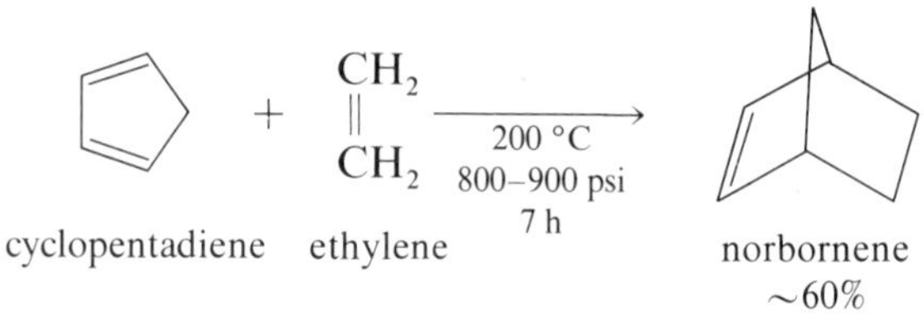
CH2
CH2
200 °C
800–900 psi
7 h
cyclopentadiene
ethylene
norbornene
~60%

Clearly, the electron-withdrawing effect of the conjugated carbonyl groups in maleic anhydride makes that molecule a much better dienophile than ethylene, which has only a double bond.

Electron-rich dienes act as nucleophiles. Dienophiles containing multiple bonds conjugated with electron-withdrawing groups are electrophiles. Classic Diels-Alder reactions occur when these two types of compounds interact. This type of reaction is tremendously useful as a way to synthesize functionalized six-membered rings with high stereoselectivity.

PROBLEM 17.5

All of the following pairs of reactants undergo Diels-Alder reactions. Complete the equations, showing the stereochemistry of the product when it can be predicted.

(a) 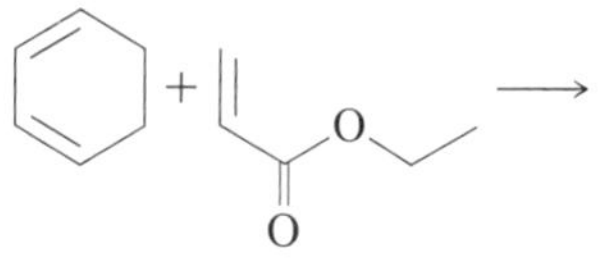
O
O

(b) 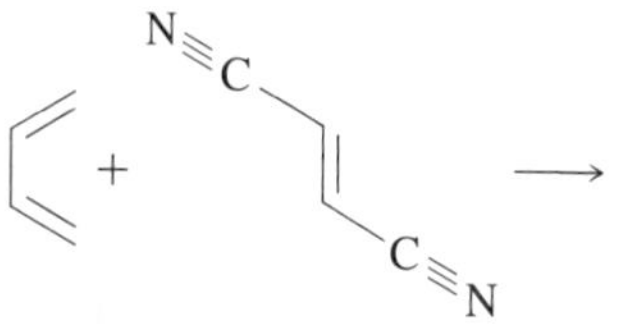
N
C
C
N

(c) 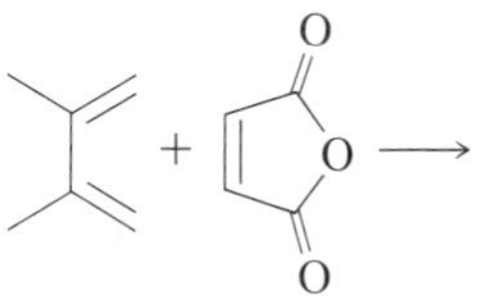
O
O
O

(d) 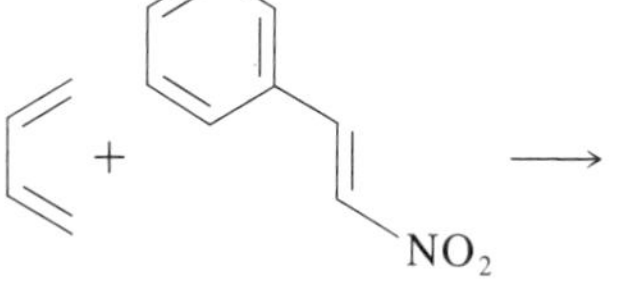
NO2

(e) 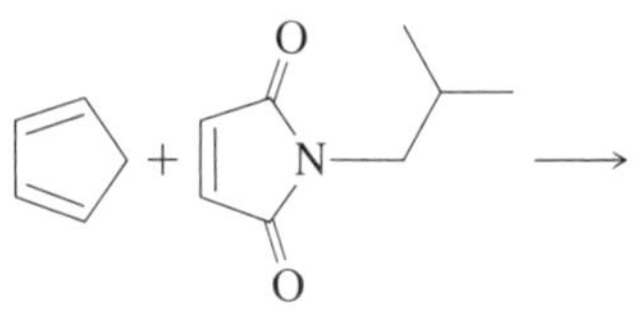
O
N
O

(f) 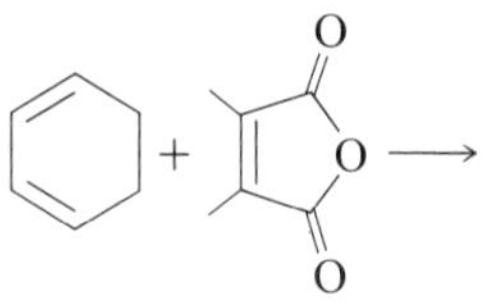
O
O
O

(g) 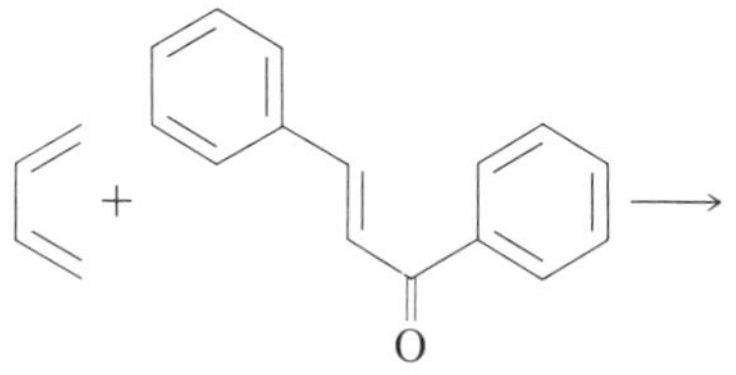
O

(h) 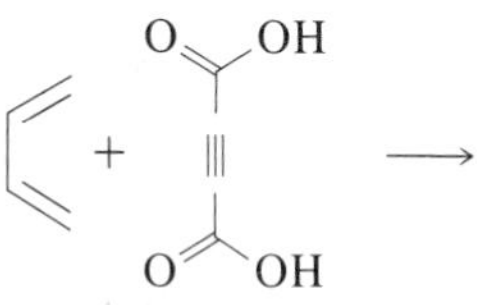
O
OH
O
OH

PROBLEM 17.6

Diels-Alder reactions are reversible. Cyclohexene, for example, is broken down into a diene and a dienophile when it is exposed to a red-hot wire in the absence of air. Write an equation showing the products of this reaction.

D. Diels-Alder Reactions of Unsymmetrical Dienes and Dienophiles

The examples of Diels-Alder reactions discussed so far have involved symmetrical dienes or dienophiles, so the question of regioselectivity has not come up. However, a Diels-Alder reaction does have high regioselectivity as well as high stereoselectivity. Some examples of reactions between unsymmetrically substituted dienes and dienophiles are given below.

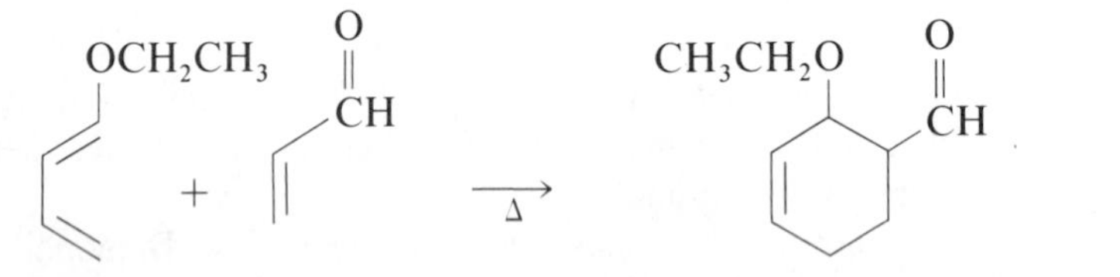
OCH₂CH₃
O
CH
+
Δ
CH₃CH₂O
O
CH

1-ethoxy-1,3-butadiene — propenal — 2-ethoxy-3-cyclohexenecarbaldehyde 58%

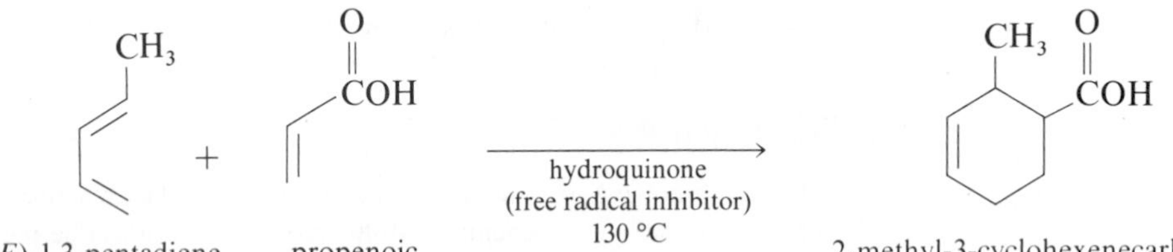
CH₃
O
COH
+
hydroquinone
(free radical inhibitor)
130 °C
CH₃
O
COH

(*E*)-1,3-pentadiene — propenoic acid — 2-methyl-3-cyclohexenecarboxylic acid ~58%

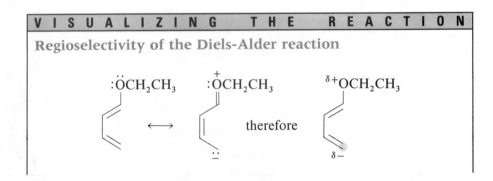

The above experimental observations indicate that when a diene substituted on the first carbon atom reacts with an unsymmetrical dienophile, the major product usually has the two substituents on adjacent carbons.

The regioselectivity observed in Diels-Alder reactions can be rationalized for the examples shown above by writing resonance contributors for the diene and dienophile and comparing the transition states. For example, the course of the reaction of 1-ethoxy-1,3-butadiene with propenal is quite easy to explain in this way.

VISUALIZING THE REACTION

Regioselectivity of the Diels-Alder reaction

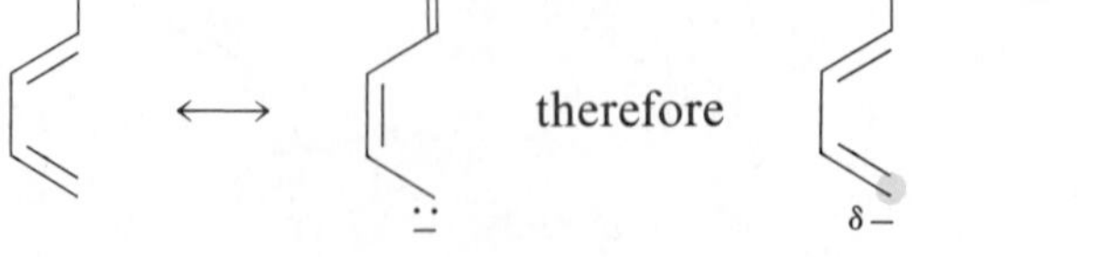
:ÖCH₂CH₃
+
:ÖCH₂CH₃
therefore
δ+OCH₂CH₃
δ−

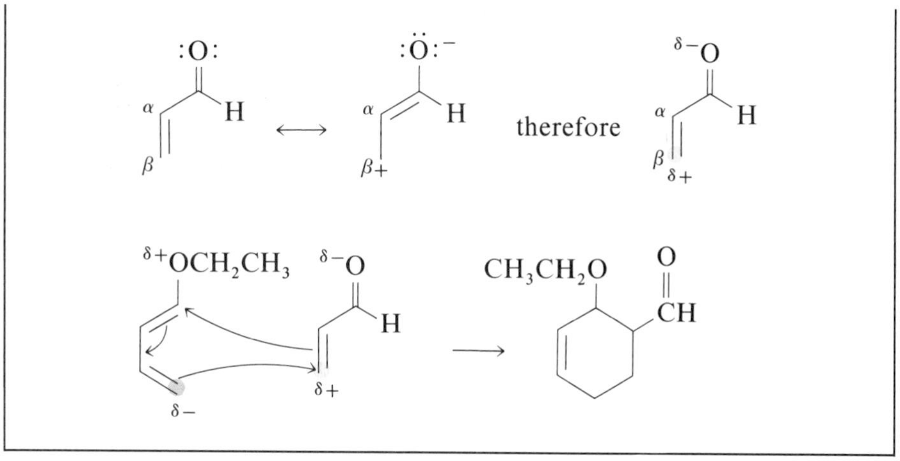

Polarization of 1-ethoxy-1,3-butadiene is most likely to increase electron density at carbon 4 of the diene. The β-carbon atom of the α,β-unsaturated carbonyl compound, meanwhile, is deficient in electrons. In the transition state, the molecules line up so that the partial charges interact with each other, and the product in which the ethoxyl and aldehyde groups are next to each other is the result.

If the substituent is on the second carbon atom of the diene, most frequently another type of product is obtained.

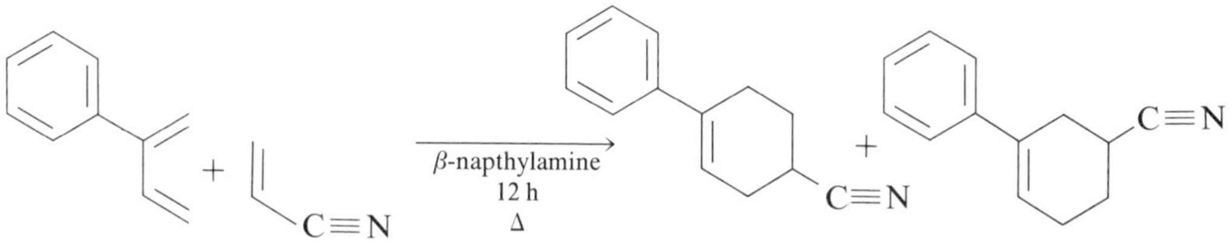

2-phenyl-1,3-butadiene + propenenitrile → 4-cyano-1-phenylcyclohexene + 4-cyano-2-phenylcyclohexene

4 : 1

32.8%

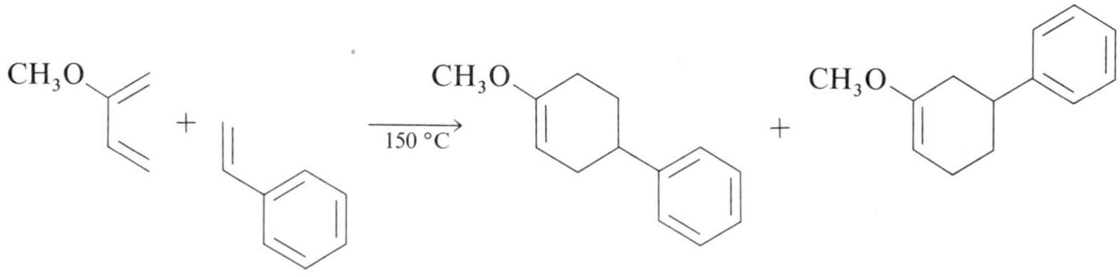

2-methoxy-1,3-butadiene + phenylethene → 1-methoxy-4-phenylcyclohexene + 2-methoxy-4-phenylcyclohexene

12 : 1

59%

In these cases, the cyclohexene ring that is formed usually has the substituents located at positions 1 and 4 on the ring. These products are more difficult to rationalize simply. Nevertheless, theoretical considerations outlined in Section

28.2B do make remarkably good predictions that the 1,4-orientation of the substituents should be favored. It is sufficient for you to know that this is the experimental result in most cases.

The addition of 2-methoxy-1,3-butadiene to alkenes is an excellent way to synthesize substituted cyclohexanones. The Diels-Alder products in these cases are enol ethers and are easily hydrolyzed to the corresponding ketones (Problem 13.13, p. 502).

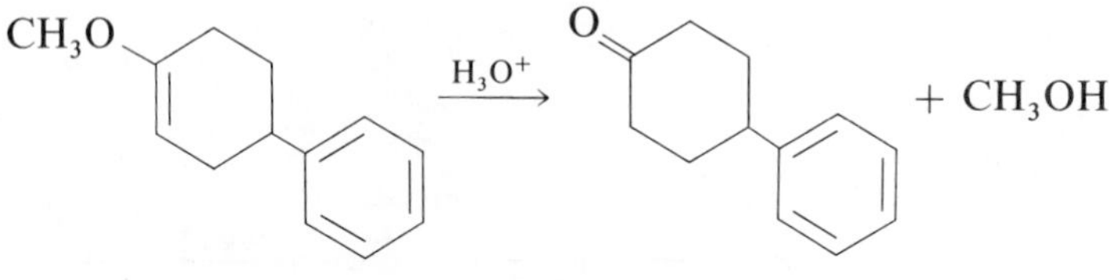
CH₃O
H₃O⁺
O
+ CH₃OH

1-methoxy-4-phenylcyclohexene   4-phenylcyclohexanone 88%

Study Guide
Concept Map 17.4

PROBLEM 17.7

Predict what the product of the following reaction will be.

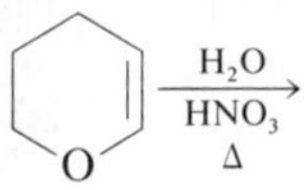
H₂O
HNO₃
Δ
O

E. Problem-Solving Skills

Problem

How would you carry out the following transformation?

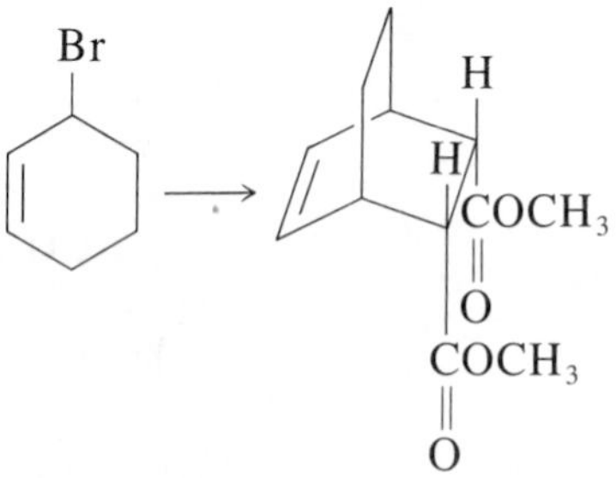
Br
H
H
COCH₃
O
COCH₃
O

Solution

1. What functional groups are present in the starting material and the product?

   The starting material is a cyclic alkene with a bromine at the allylic position; the product is an alkene with two ester groups also present.

2. How do the carbon skeletons of the two compounds compare? How many carbon atoms does each contain? Are there any rings? What are the positions of branches and functional groups on the carbon skeletons?

The starting material is a six-membered ring. The product is a bicyclic compound in which two carbons have added to the opposite corners (at positions 1 and 4) of the six-membered ring. Each carbon atom that has been added also has an ester group on it.

3. How do the functional groups change in going from starting material to product? Does the starting material have a good leaving group?

   The bromine atom is a good leaving group, and it is lost during the transformation. Ester groups are added in the reaction.

4. Is it possible to dissect the structures of the starting material and product to see which bonds must be broken and which formed?

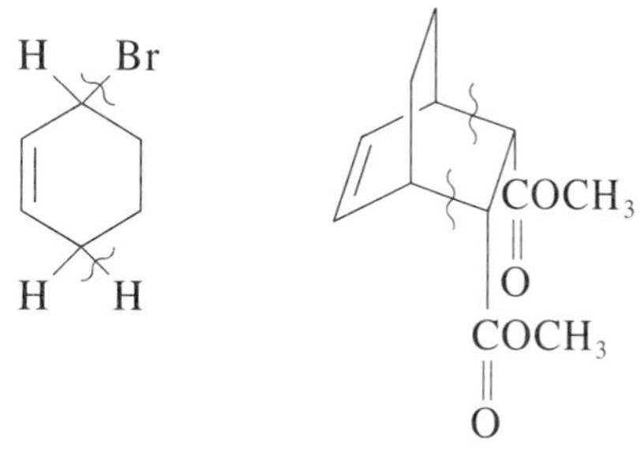

H
Br
H
H
COCH3
O
COCH3
O

bonds to be broken *bonds to be formed*

5. Do we recognize any part of the product molecule as coming from a good nucleophile or an electrophilic addition?

   No, the product has no nucleophilic center or any parts resulting from an electrophilic addition.

6. What type of compound would be a good precursor to the product?

   The structure of the product, with 1,4-orientation of substituents on a cyclohexene ring, suggests a Diels-Alder reaction. Recognizing that a diene and a dienophile are required for this reaction is the most important step of this analysis.

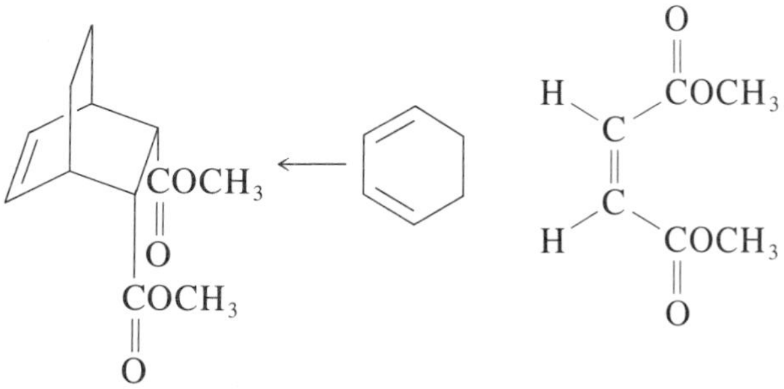

COCH3
O
COCH3
O
O
H
COCH3
C
C
H
COCH3
O

7. After this last step, do we see how to get from starting material to product? If not, we need to analyze the structure obtained in step 6 by applying questions 5 and 6 to it.

   Repeating steps 5 and 6 is necessary to figure out how to make cyclohexadiene from 3-bromocyclohexene.

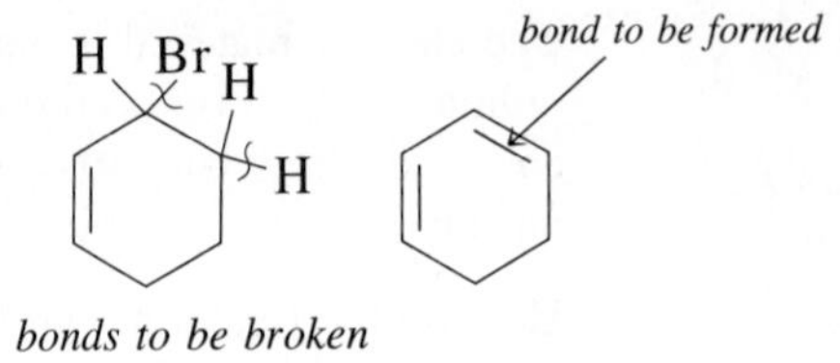
H
Br
H
H
bond to be formed
bonds to be broken

An elimination reaction to make a π bond completes the synthesis.

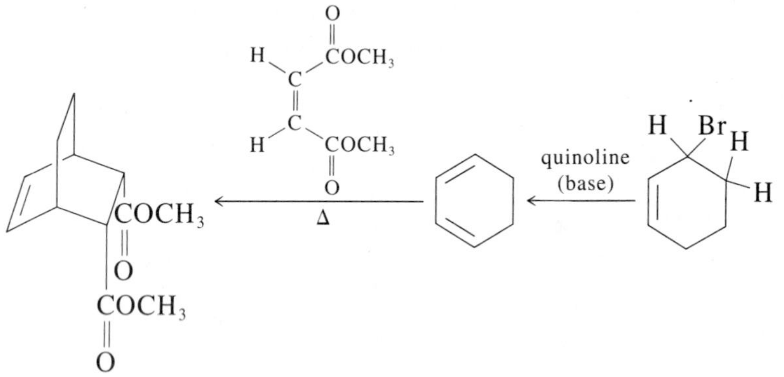
O
COCH3
H
C
C
H
COCH3
O
Δ
COCH3
O
COCH3
O
quinoline
(base)
H
Br
H
H

PROBLEM 17.8

In the synthesis of an antibiotic, an intermediate was prepared from cyclohexanone. How would you carry out this transformation?

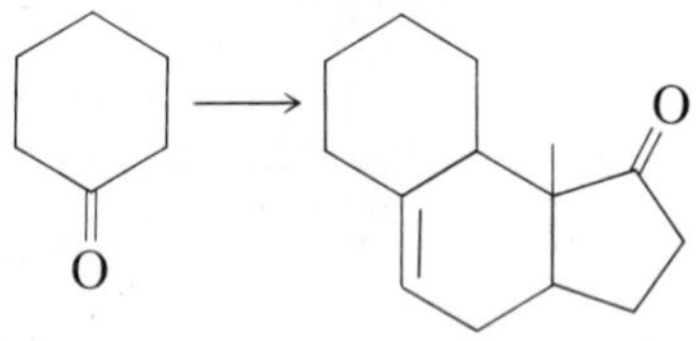

PROBLEM 17.9

Assign structures to the major products designated by letters in the following reactions.

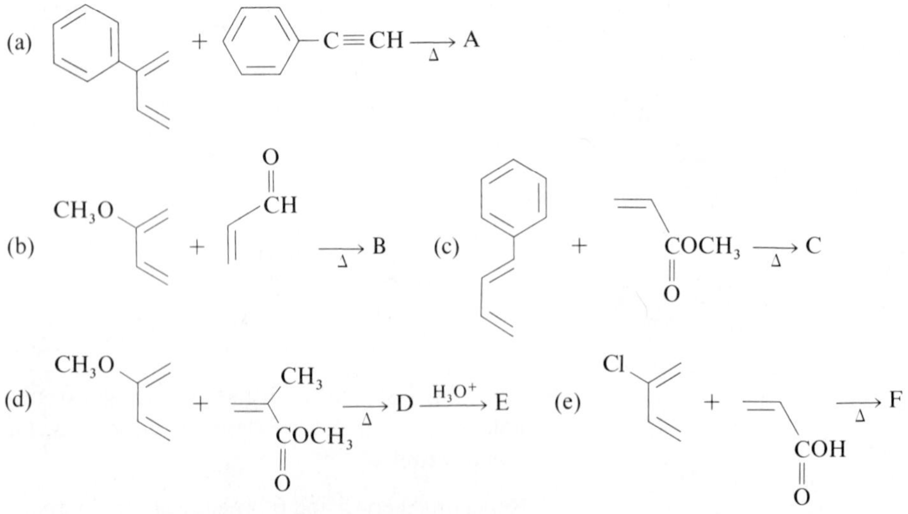
(a)
+
C≡CH
Δ
A
(b)
CH3O
+
O
CH
Δ
B
(c)
+
COCH3
O
Δ
C
(d)
CH3O
+
CH3
COCH3
O
Δ
D
H3O+
E
(e)
Cl
+
COH
O
Δ
F

PROBLEM 17.10

What reagents would be required to obtain each Diels-Alder adduct shown below?

(a)

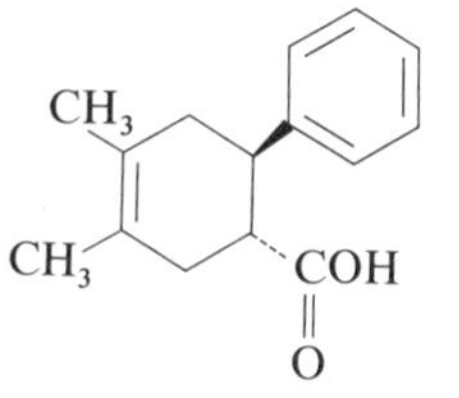
CH3
CH3
COH
O

(b)

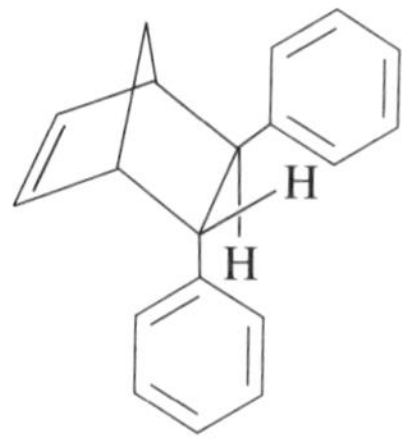
H
H

(c)

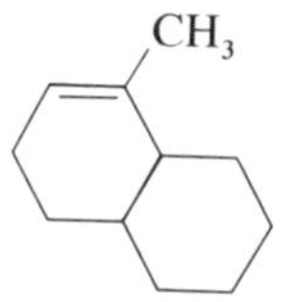
CH3

(d)

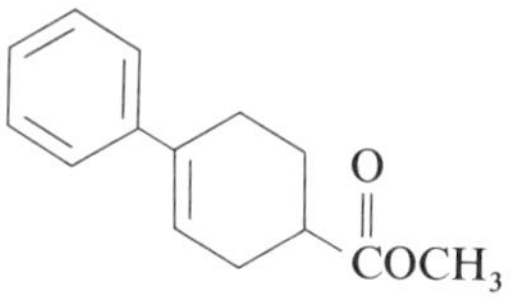
O
COCH3

PROBLEM 17.11

The stereochemistry of the product of a Diels-Alder reaction is not always easy to determine. In one case, the decision about the stereochemistry of the reaction was based on the following set of reactions.

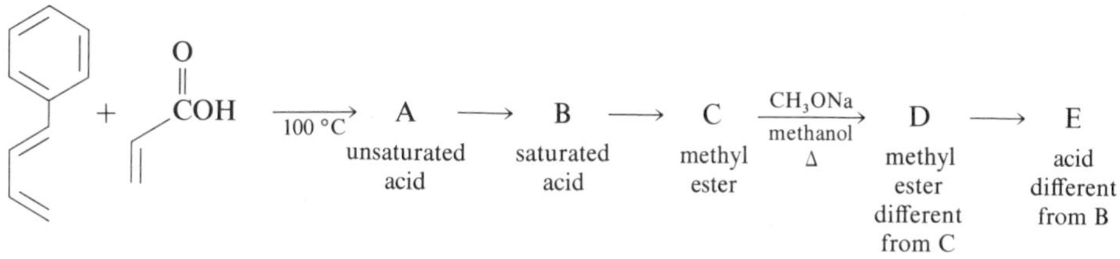
O
COH
100 °C
A
unsaturated acid
B
saturated acid
C
methyl ester
CH3ONa
methanol
Δ
D
methyl ester different from C
E
acid different from B

Supply reagents to carry out the transformations where they are missing, and write full structures, including stereochemistry, for Compounds A through E. What do these reactions prove about the stereochemistry of the Diels-Alder addition product, Compound A?

17.6 BIOLOGICALLY INTERESTING ALKENES AND POLYENES

A. Pheromones

Insects communicate with each other, and with their environment, by means of organic chemicals secreted in minute amounts. These substances, known as **pheromones,** have a wide range of structures and include a variety of functional groups.

Because chemists must work with such small amounts of material, the chemistry associated with the isolation of these compounds and the determination of their structures is particularly challenging. Many of the structural determinations were done on only a few milligrams of isolated material. Methods of separation and isolation, as well as spectroscopic techniques, were refined in order to deal with these very small quantities. For example, extraction of the abdominal tips of 500,000 virgin gypsy moth females yielded only 75 mg of the pheromone that attracts males. This quantity of active material had to be separated from approximately 250 g of fatty acids and their esters and 70 g of steroids, mostly cholesterol. This process of purification, which is like looking for the proverbial needle in a haystack, was aided by biological assay methods in which male gypsy moths were exposed to the fractions obtained in various separation steps. The fractions that did not attract the male moths were discarded and those that did were purified further until finally a single component that is highly attractive to the male was isolated.

The sex pheromone for the gypsy moth, which is a devastating pest in hardwood forests, is a chiral oxirane. The dextrorotatory compound, known as disparlure, is a potent attractant for male gypsy moths, but the enantiomer, the levorotatory compound, has no such activity.

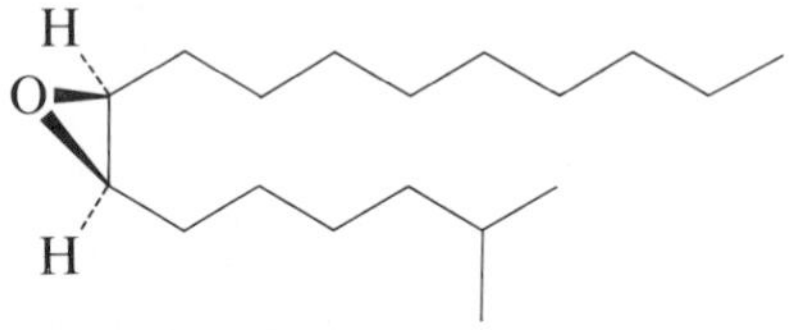
H
O
H

(+)-disparlure

the sex pheromone of the gypsy moth

A sex pheromone is usually specific for one species of insect. Understanding how pheromones work might allow scientists to devise ways of controlling certain insect populations without harming the environment. At present, sex pheromones are used as lures in traps that allow foresters to estimate the population of a given insect in a region. For example, in 1980, a single Mediterranean fruit fly, caught in such a trap in a citrus grove in California, served as a warning that a new infestation of the destructive pest was on the way and alerted the agricultural experts in that state to be prepared. The infestation became a disastrous reality in the summer of 1981.

The sex pheromones of many insects are long-chain alcohols or esters, with one or more double bonds in them. In some cases, the double bonds are conjugated; in others, they are not. They may have cis or trans stereochemistry. The sex pheromones of the pink bollworm, the silkworm, and the red-banded leaf roller are among those that have been isolated and their structures determined.

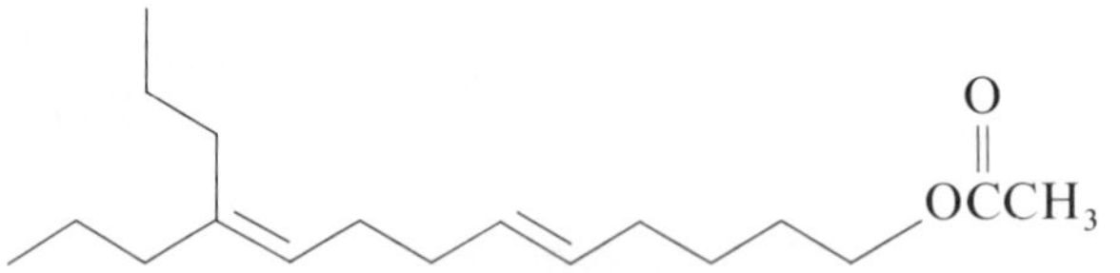
O
OCCH3

(*E*)-10-propyl-5,9-tridecadienyl acetate

from the pink bollworm

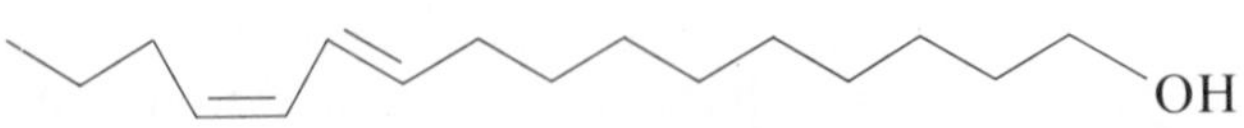
OH

(10*E*,12*Z*)-10,12-hexadecadien-1-ol

from the silkworm

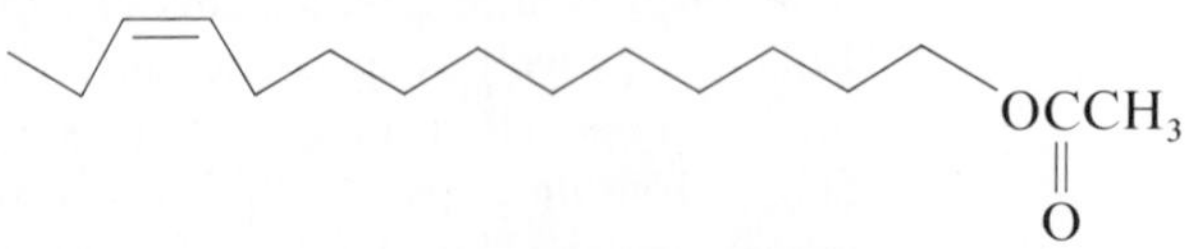
OCCH3
O

(*Z*)-11-tetradecenyl acetate

from the red-banded leaf roller

The insect pheromone released in the greatest concentrations is the alarm pheromone, a chemical substance used to signal a disturbance in an insect colony. The simple ester 3-methylbutyl acetate is the alarm pheromone for the honey bee. Many other insect species use terpenes (p. 704) such as citronellal and limonene to signal alarm.

Even the fact of death is communicated by chemical signals in the insect world. Ants, for example, do not recognize that a freshly frozen ant is dead and treat it quite differently at first than they do after a while. Apparently, bacterial decomposition in the dead insect releases a number of fatty acids, such as oleic and linoleic acids. If mixtures of acids such as these are placed on a small piece of paper and put into the colony, the ants rapidly remove the paper and deposit it on a refuse pile. They do the same thing to a fellow ant once it has been dead for a while.

Organic compounds are used to regulate the relationship between an insect and its environment in many other ways. Plants synthesize compounds that prevent insects from eating them, and insects produce defensive secretions that keep birds and mammals away. Mammals, too, have chemical methods of communication that are just beginning to be recognized as important. Perfume makers, of course, have always understood the importance of scent as a means of communication.

PROBLEM 17.12

A chemical that protects a plant from being eaten by insects has been isolated and its structure determined in Japan. The compound, called shiromodiol diacetate, has the structure shown below.

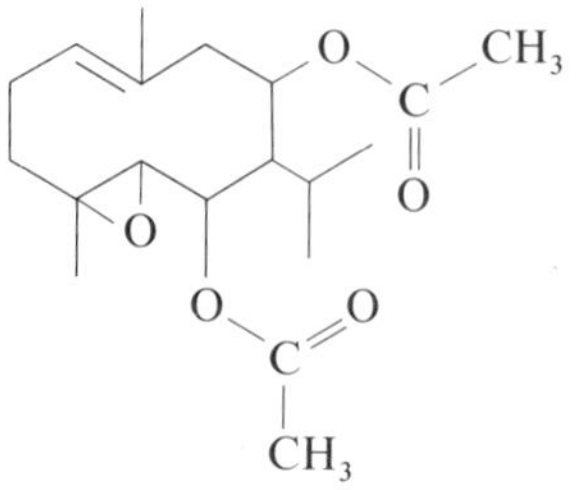
O
CH3
C
O
O
O
O
C
CH3

Shiromodiol diacetate has significant bands in its infrared spectrum at 1735 and 1240 cm^{-1}. Among the reactions used in the proof of the structure of the compound was hydrolysis with dilute aqueous base to give Compound A, with absorption in the infrared at 3400 cm^{-1} but no absorption between 1800 and 1700 cm^{-1}. When Compound A was treated with chromium trioxide in pyridine, a reagent that behaves like pyridinium chlorochromate (p. 441), Compound B, having absorption at 1720 and 1695 but not at 3400 cm^{-1}, was formed. Write equations predicting what the structures of A and B are. Account for the changes observed in the infrared spectra for all three compounds.

PROBLEM 17.13

The sex pheromone of the cabbage looper has the structure shown below.

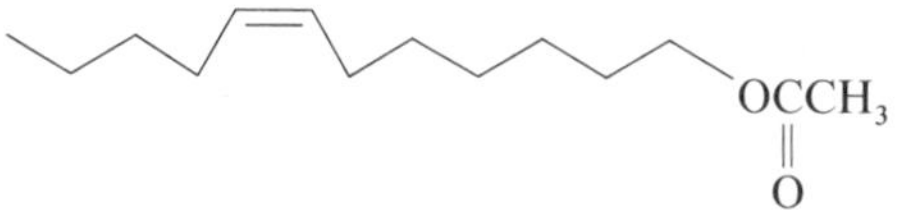
OCCH3
O

(Z)-7-dodecenyl acetate

(a) Devise a synthesis for it. Organic reagents that are available to you are ethylene glycol, acetylene, butyl bromide, and 6-hydroxyhexanal. You may assume that the laboratory is well stocked with solvents, acids, bases, and catalysts. (Hint: You may find it helpful to review pp. 342 and 513.)

(b) The E- isomer of the pheromone has also been synthesized. How would you modify your synthesis to obtain that isomer?

B. Terpenoids. Isoprene and the Isoprene Rule

For centuries, human beings have known that volatile oils with a variety of fragrances and flavors could be isolated from certain plants. These compounds occur in all parts of the plants and are called **essential oils.** They are of great commercial importance in the perfume and flavoring industries, and many have also been used medicinally. The chemical constituents of essential oils have a wide variety of structures; many of them contain rings or one or more double bonds. Some of them are alcohols or ethers; others are ketones or aldehydes. After many years of carrying out structural determinations on these interesting and challenging compounds, chemists began to detect some patterns. For example, most of these compounds are composed of multiples of five carbon atoms. Whole families of compounds were discovered and classified according to their molecular formulas. Compounds with 10 carbon atoms, the **monoterpenes,** those with 15 carbon atoms, the **sesquiterpenes,** and those with 20 carbon atoms, the **diterpenes,** are the chief constituents of essential oils. Steroids are **triterpenes,** compounds with 30 carbon atoms, or are derived from them. Materials that give plants colors, such as carotene (from carrots), are **tetraterpenes,** containing 40 carbon atoms. Rubber is a polymeric terpenoid (p. 1204).

The structures of a few essential oils as well as some other plant constituents are shown below.

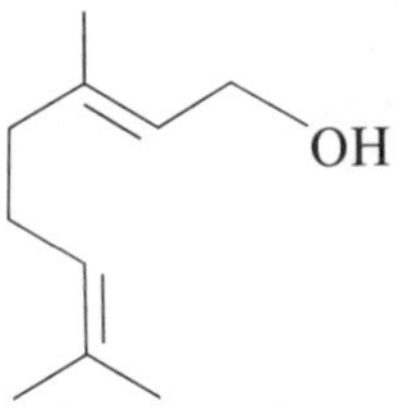
OH

geraniol

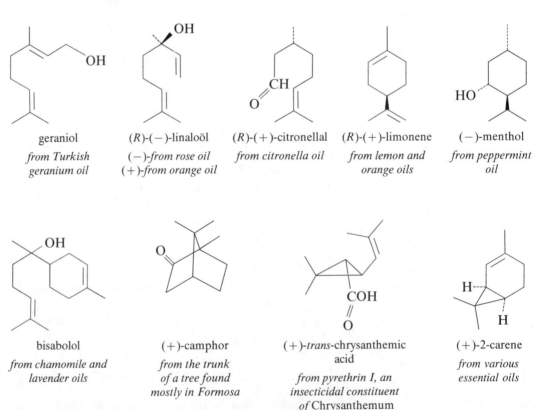

from Turkish geranium oil

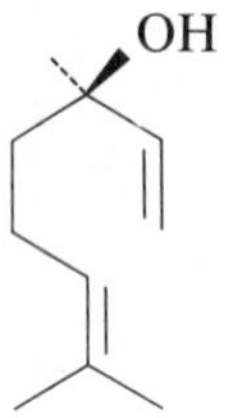
OH

(*R*)-(−)-linaloöl

(−)-*from rose oil*
(+)-*from orange oil*

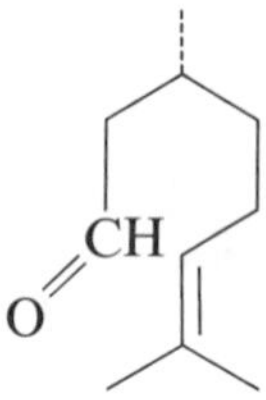
CH
O

(*R*)-(+)-citronellal

from citronella oil

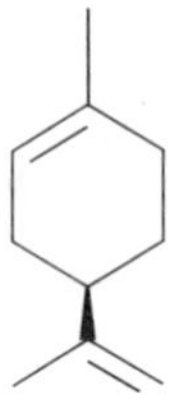

(*R*)-(+)-limonene

from lemon and orange oils

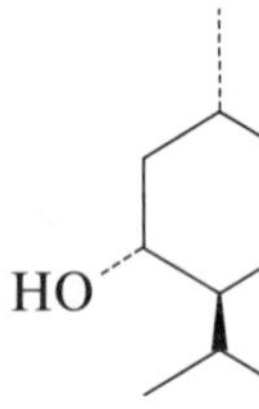
HO

(−)-menthol

from peppermint oil

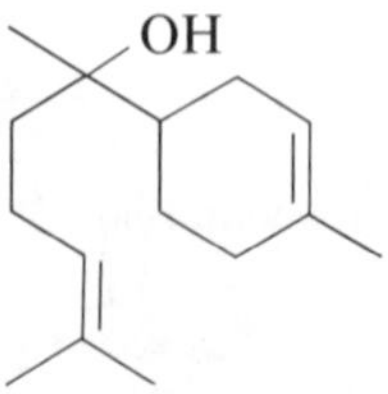
OH

bisabolol

from chamomile and lavender oils

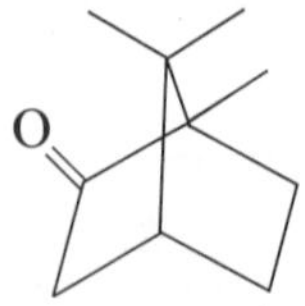
O

(+)-camphor

from the trunk of a tree found mostly in Formosa

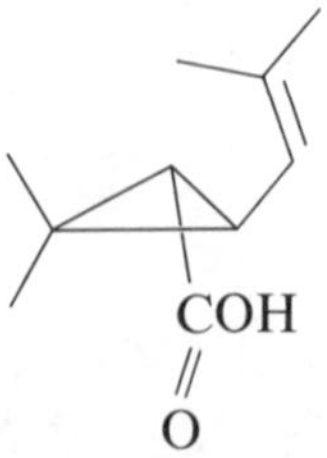
COH
O

(+)-*trans*-chrysanthemic acid

from pyrethrin I, an insecticidal constituent of Chrysanthemum cinerariifolium

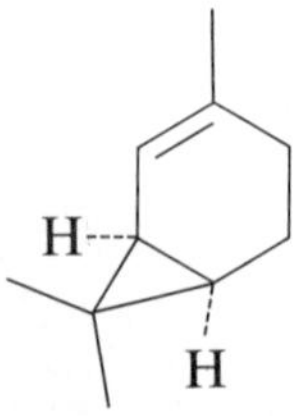
H
H

(+)-2-carene

from various essential oils

Chrysanthemic acid and 2-carene have historical significance. In 1920, the Yugoslav chemist Leopold Ruzicka, who did his scientific work in Switzerland and later won the Nobel Prize for discoveries in terpene chemistry, finished a proof of the structure of chrysanthemic acid. This compound is a constituent of a pyrethrin, a component of an insecticidal extract from a type of chrysanthemum that grows primarily in East Africa. (Pyrethrins have become important commercially as highly selective and biodegradable insecticides.) In the same year, the structure of 2-carene was determined by the British chemist John L. Simonsen. Ruzicka was struck by the structural resemblance between the two compounds and could see in both of them the elements of a five-carbon structural unit known as **isoprene.**

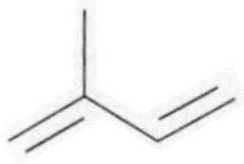

2-methyl-1,3-butadiene
isoprene

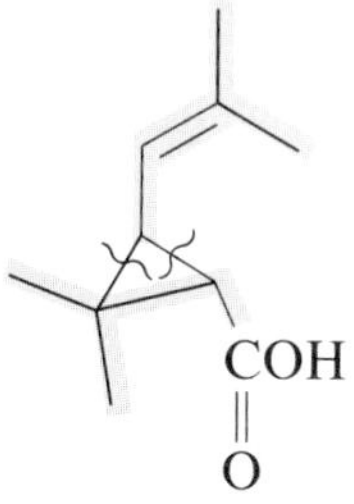
COH
O

chrysanthemic
acid

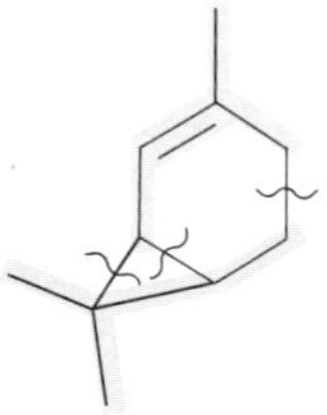

2-carene

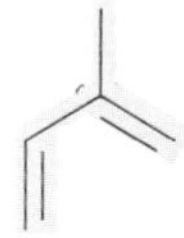

isoprene in the
s-cis conformation

In other naturally occurring compounds that were known at that time, especially certain ones that did not contain a ring, another regularity could be seen. Referring to the branched end of the isoprene molecule as its head and the other end as its tail, molecules that contained two or more isoprene units had the head of one isoprene unit attached to the tail of the next one in the chain.

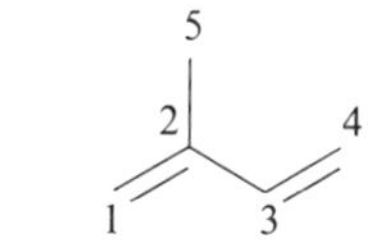
5
2
4
1
3

head of isoprene *tail of isoprene*

four-carbon chain with
methyl branch on second carbon

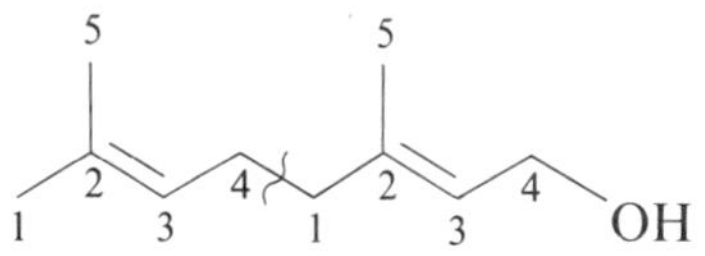
5
5
2
4
2
4
1
3
1
3
OH

geraniol, a terpene made up of
two isoprene units attached
head-to-tail

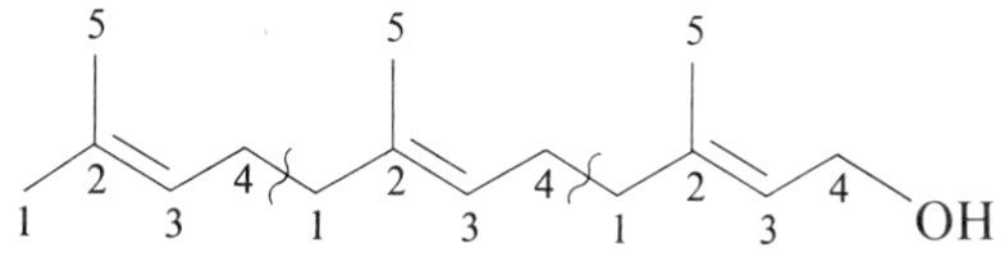
5
5
5
2
4
2
4
2
4
1
3
1
3
1
3
OH

farnesol, scent of lily of the valley,
a sesquiterpene made up of three
isoprene units attached head-to-tail

Ruzicka hypothesized that these plant constituents were synthesized in nature by a head-to-tail connection of isoprene units. He examined a large number of terpenoid compounds over the years and found that the structures of most of them showed this regularity. Thus, he formulated the isoprene rule.

The **isoprene rule** means that given a choice of possible structures for a natural product that contains a multiple of five carbon atoms, chemists will favor one that appears to be made of isoprene units connected in a head-to-tail fashion. The isoprene rule says nothing about whether the compound contains double bonds or functional groups containing oxygen. A wide variety of patterns of rings, unsaturation, and alcohol, carbonyl, and carboxyl groups are present in these natural products. The isoprene rule applies only to the carbon skeleton of the molecule. When rings are present, they indicate that additional points of attachment between various isoprene units have been formed. In those cases, it is still possible to trace the main outline of the chain of isoprene units around the molecule. Some cyclic terpenoid compounds are dissected into isoprene units below.

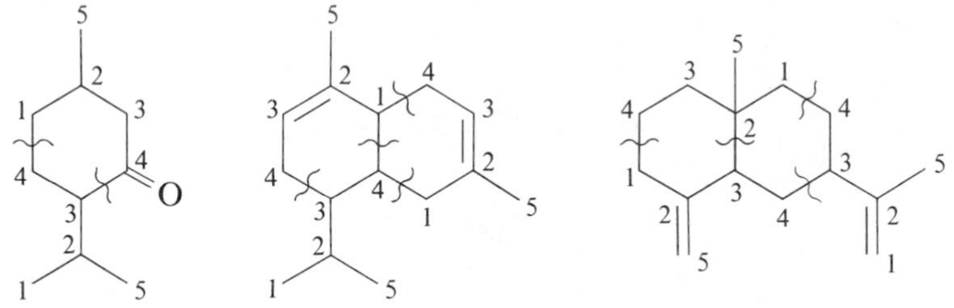

(−)-menthone *peppermint oils* | α-cadinene *oil of citronella* | β-selinene *oil of celery*

In some structures, it is possible to find more than one way to mark off the isoprene units.

As molecules become larger, the final structure is often achieved only after some molecular rearrangements during the biosynthesis. The isoprene rule does not apply to all parts of such structures, though it is still possible to find portions of the molecule that follow the rule.

PROBLEM 17.14

Mark off the isoprene units in the following compounds.

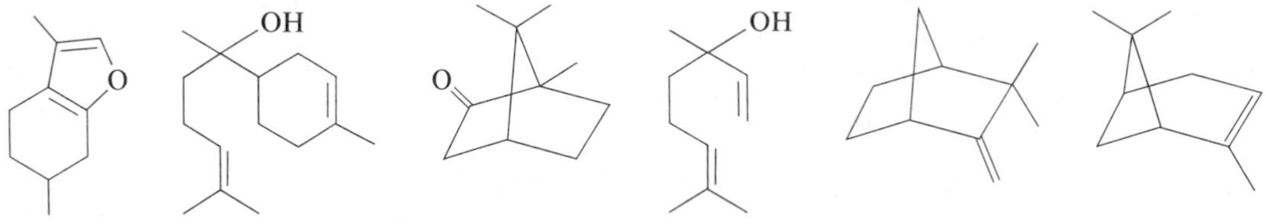

menthofuran | bisabolol | camphor | linaloöl | camphene | α-pinene

C. Biosynthesis of Terpenes

The pyrophosphate ester of an unsaturated five-carbon alcohol, 3-methyl-3-buten-1-ol, is the structural building block for naturally occurring terpenes. Pyrophosphoric acid is an anhydride of phosphoric acid and seems to be nature's tool for creating good leaving groups. 3-Methyl-3-buten-1-yl pyrophosphate, known in the biochemical literature as isopentenyl pyrophosphate, is isomerized enzymatically to 3-methyl-2-buten-1-yl (dimethylallyl) pyrophosphate in a reaction that may be regarded as a protonation at one sp^2-hybridized carbon atom and a deprotonation of the incipient carbocation at another site to give the more highly substituted alkene. The participation of an enzyme, a highly specific biological catalyst (see p. 1154 for an example of how an enzyme functions), ensures that no high-energy intermediate is formed at any point of the reaction.

VISUALIZING THE REACTION

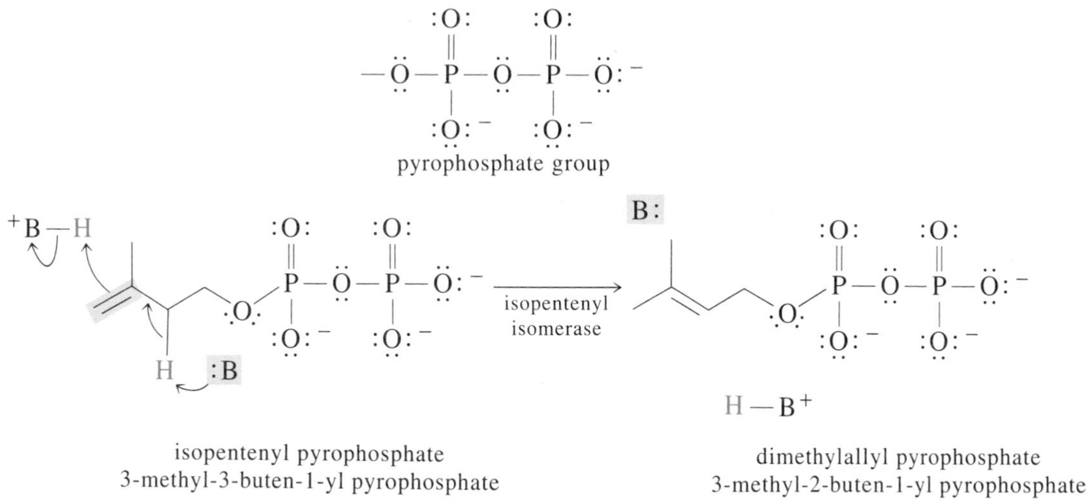
pyrophosphate group
isopentenyl
isomerase
isopentenyl pyrophosphate
3-methyl-3-buten-1-yl pyrophosphate
dimethylallyl pyrophosphate
3-methyl-2-buten-1-yl pyrophosphate

The π electrons in isopentenyl pyrophosphate serve as a nucleophile to displace pyrophosphate anion from dimethylallyl pyrophosphate, with the simultaneous loss of a proton to create a double bond.

VISUALIZING THE REACTION

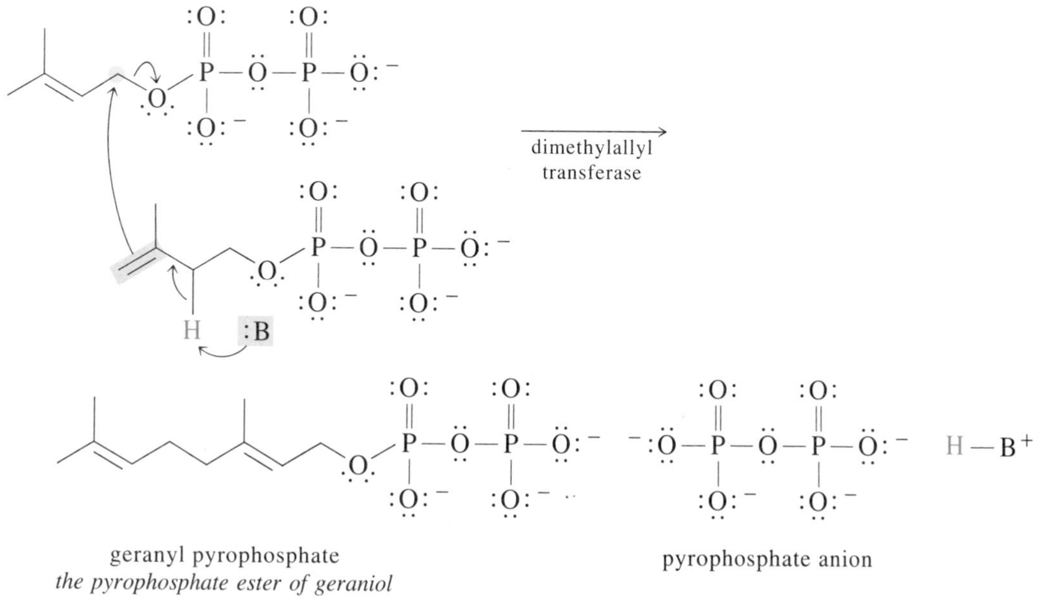
dimethylallyl
transferase
geranyl pyrophosphate
the pyrophosphate ester of geraniol
pyrophosphate anion

The product is a ten-carbon terpene, geraniol.

Note that the mechanism proposed for terpene synthesis leads in each case to the head-to-tail connection of isoprene units. Five carbon atoms are added, and the end of the chain is left functionalized as a pyrophosphate ester that can serve as a leaving group for further reactions. Further protonations and deprotonations would lead to isomerizations of the positions and of the stereochemistry of double bonds. The oxygen-containing functional groups are modified by oxidation and reduction reactions, and new hydroxyl groups are created by the addition of water to double bonds or by enzymatic oxidation reactions at unactivated sites. Conversely, dehydration of alcohol functions introduces unsaturation into the systems. Biological reduction reactions convert alkenes into alkanes. Carbocations created by protonation of a double bond or loss of water from a protonated alcohol can attack a double bond in another part of the molecule to create rings. The next problem reviews this familiar chemistry in the context of terpene reactions.

PROBLEM 17.15

In all parts of this problem, you may use HB^+ as an acid and B: as a base as necessary. Water is always present as a nucleophile.

(a) Geraniol (p. 704) has a diasteromer called nerol. Nerol is also a natural product but occurs somewhat less abundantly than geraniol. Write a structural formula for nerol, and propose a mechanism for the isomerization of geraniol to nerol.

(b) Treatment of geraniol or nerol with aqueous acid gives rise to α-terpineol and terpin, shown below. Nerol is converted into these compounds faster than geraniol is. Write mechanisms for the formation of α-terpineol and terpin from nerol.

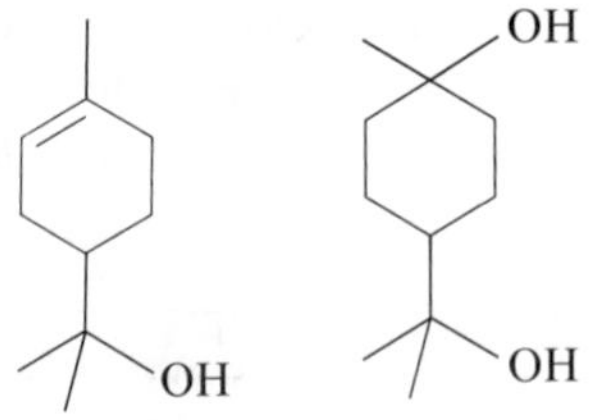
OH
OH
OH

α-terpineol terpin

(c) Limonene, a flavor constituent of citrus fruits (p. 704), is formed from α-terpineol as well as from terpin. Write equations showing how this would take place.

(d) Linaloöl (p. 704) may be considered to arise from either geraniol or nerol. Propose a mechanism for its formation.

PROBLEM 17.16

Isoprene undergoes a Diels-Alder reaction with itself to give rise to an optically inactive compound known for some time as dipentene. Dipentene was finally identified as the racemic form of a naturally occurring terpene. Write an equation for the reaction and identify the terpene, which appears on p. 704.

PROBLEM 17.17

Show how the sesquiterpene farnesol (p. 705) could be synthesized from geranyl pyrophosphate (p. 707).

PROBLEM 17.18

Diterpenes, compounds having 20 carbon atoms, are derived from a pyrophosphate that has the following structure.

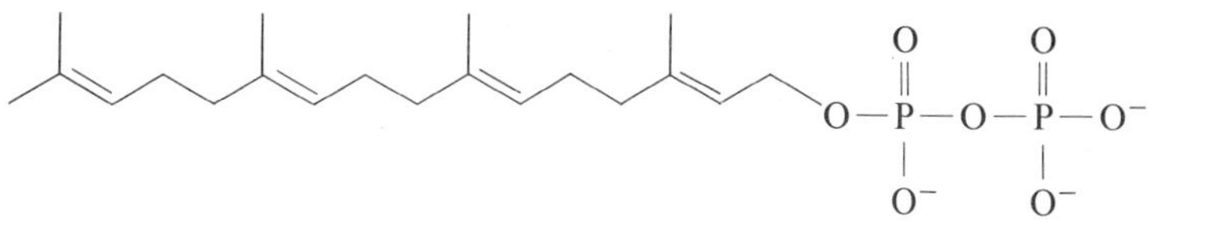

Propose a biosynthesis of this compound, sometimes called geranylgeranyl pyrophosphate.

D. Steroids

Steroids are triterpenoids, a group of compounds produced in plants or animals. They are widespread in nature and have important biological functions. Some hormones and cholesterol (pp. 477 and 711) are steroids.

Steroids are characterized by a tetracyclic structure. A typical steroid ring skeleton consists of three six-membered rings and a five-membered ring fused together. Depending on their source and biological function, steroids may also have a variety of functional groups substituted on the skeleton (p. 477).

The rings in a steroid skeleton are designated as the A, B, C, and D rings.

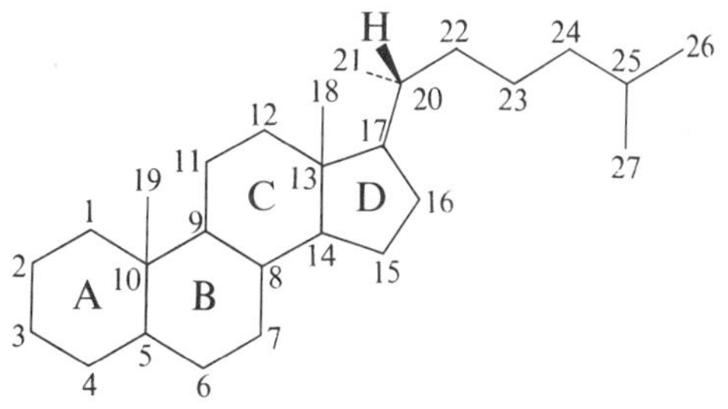

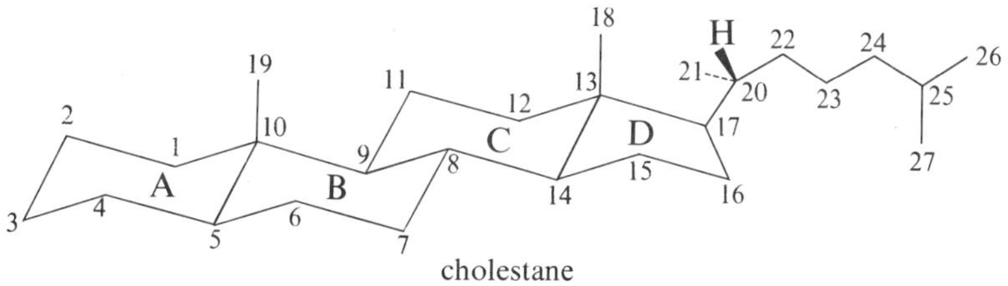

cholestane

The fusion of the four rings gives rigidity to a steroid molecule. The conformational changes (p. 166) that are easy for individual cyclohexane rings are not possible at room temperature for fused rings. Steroids have, therefore, been used for research on the influence of conformation on the course of organic reactions. Sir Derek H. R. Barton of Great Britain received the Nobel Prize in 1969 for recognizing that functional groups could vary in reactivity depending on whether they occupied an axial or an equatorial position on a ring. This way of thinking about stereochemistry, called **conformational analysis,** has greatly influenced the way chemists analyze reactivity.

Squalene, $C_{30}H_{50}$, was first isolated from shark liver oil in 1916. Since that time, research has established its role in the formation of the steroid ring skeleton in living matter. Squalene is a triterpene, but one in which the isoprene rule is violated in one place. Rather than a head-to-tail arrangement of six units of isoprene, there appear to be two farnesyl units that have been connected tail-to-tail (see the next page).

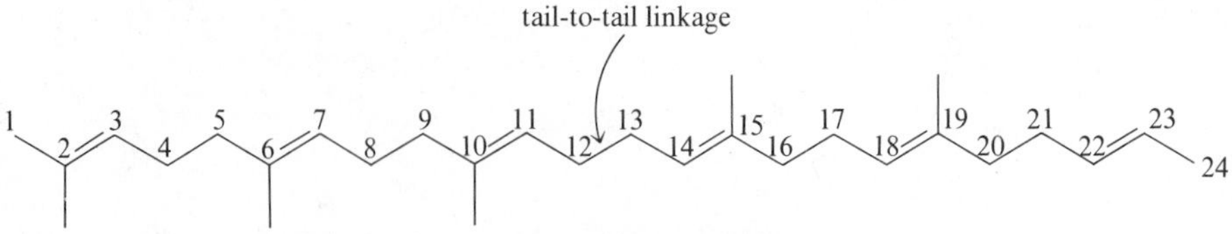
tail-to-tail linkage
1
2
3
4
5
6
7
8
9
10
11
12
13
14
15
16
17
18
19
20
21
22
23
24

squalene
(6*E*,10*E*,14*E*,18*E*)-2,6,10,15,19,23-hexamethyl-2,6,10,14,18,22-tetraicosahexaene
a triterpene

A great deal of research, using selective labeling of carbon atoms and hydrogen atoms, has confirmed the tail-to-tail connection. Other evidence indicates that the formation and then the opening of a cyclopropane ring in the middle of the chain occurs as two farnesyl pyrophosphate units interact.

The critical intermediate in the conversion of squalene to the steroid skeleton is an oxirane, squalene-2,3-oxide, which is transformed by enzymes into lanosterol, a steroid alcohol found in wool fat. Similar cyclization reactions take place in the laboratory with Lewis or Brønsted acids as catalysts. The squalene molecule is believed to be folded in a conformation that allows one double bond after another to react as a nucleophile with a cationic center that develops nearby in the molecule. The initial leaving group that starts the process is the oxygen atom of the oxirane ring protonated or complexed in a way that makes it a better leaving group. Oxiranes, as strained small ring compounds, are reactive toward nucleophiles, especially in systems where the oxygen atom develops a positive charge by protonation (p. 451).

VISUALIZING THE REACTION

Cyclization of squalene oxide

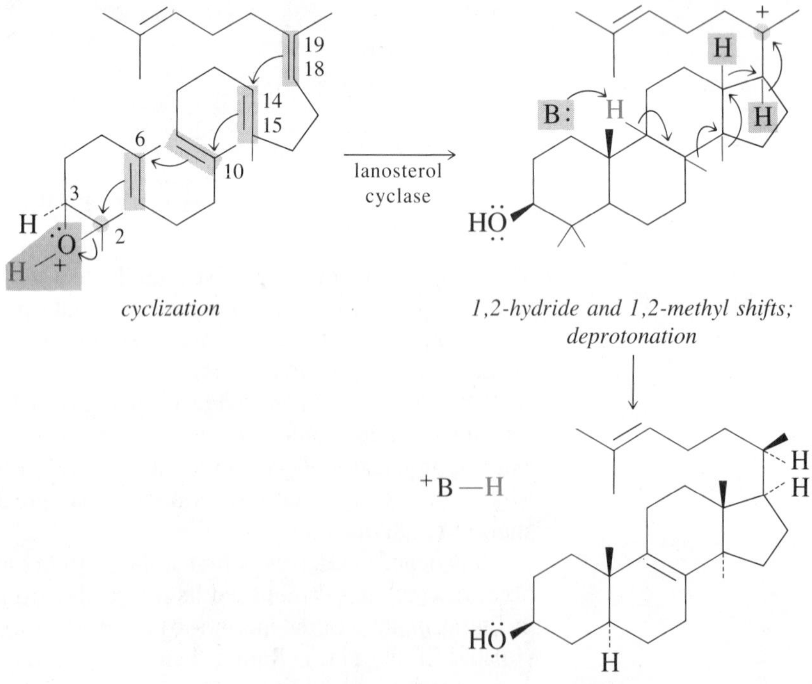
19
18
14
15
6
10
3
H
O
2
H
+
lanosterol
cyclase
+
H
B:
H
H
HO
⁺B—H
H
H
HO
H

cyclization

1,2-hydride and 1,2-methyl shifts; deprotonation

lanosterol

The whole process, catalyzed by an enzyme, is highly stereoselective. The individual steps of the transformation are familiar as typical carbocation reactions (p. 283). A carbocationic center, as it is generated, behaves as a Lewis acid toward the π electrons of an adjacent double bond, which in turn creates a new cationic center, so the cyclization progresses. The initial cyclization product is a cation that differs from lanosterol in the placement of two methyl groups. The double bond is created in the B ring by a series of 1,2-shifts that result in a tertiary cation that undergoes the final deprotonation.

Squalene is the biological precursor of many triterpenoids. Important among these are the steroids, one of which is cholesterol. During the biological conversion of lanosterol to cholesterol, three methyl groups are lost, the hydrocarbon chain is reduced to a saturated one, and the position of the double bond in ring B of the steroid skeleton changes. The exact details of how these transformations are carried out are not known. More than one pathway may be involved. That lanosterol is converted to cholesterol by enzymes from the liver has been proved by a number of experiments. For example, when lanosterol, labeled with radioactive carbon-14 and carefully purified so that it contains no cholesterol, is incubated with cells from the liver of a rat, cholesterol containing radioactivity is recovered.

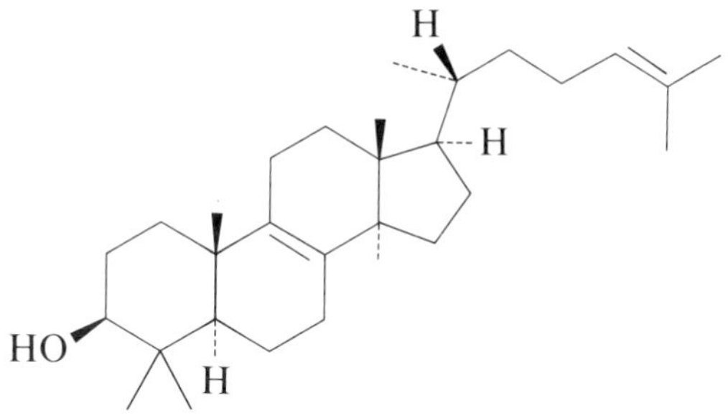
H
H
HO
H

lanosterol

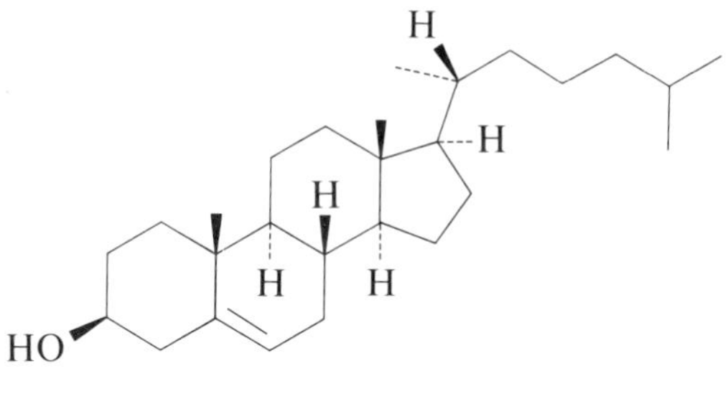
H
H
H
H
H
HO

cholesterol

E. Carotenoids. Vitamin A

The **carotenoids** are compounds made up of eight isoprene units and usually containing some rings and a number of conjugated double bonds. The backbone of the chain appears to be constructed by a tail-to-tail union of two geranylgeranyl pyrophosphate units (Problem 17.18), similar to the way squalene is derived from a combination of farnesyl pyrophosphate units. Carotenoids are insoluble in water but soluble in fat and hydrocarbon solvents. All the carotenoids, which are generally yellow to red in color, are widely distributed in nature, in both plants and animals.

Lycopene is the parent carotenoid, related to all other known carotenoids by changes such as reduction of some of the double bonds, cyclization, isomerization of the position of a double bond, and introduction of functional groups containing oxygen.

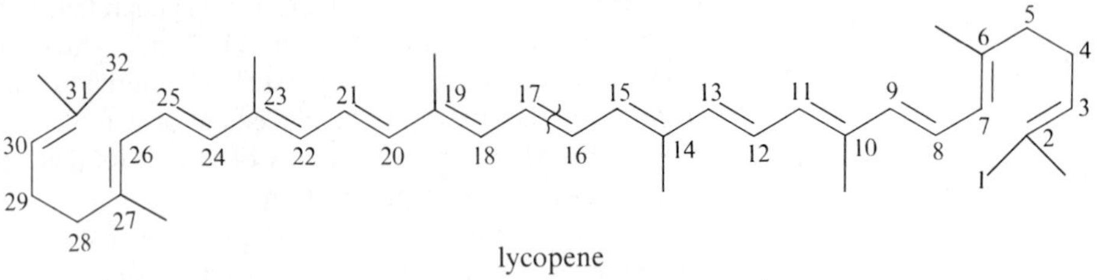

lycopene

dissected at the point of tail-to-tail connection of two 20-carbon units

Note that the double bonds in lycopene are all trans in configuration and that, except for the ones nearest the ends of the molecule, they are all in conjugation. Section 17.7A will show how this extended system of conjugation is responsible for the color of these compounds, which are the pigments in many plants. Lycopene is abundant in tomatoes, for example.

Carotene, the pigment of carrots, is especially interesting because it is closely related to Vitamin A, a diterpenoid essential to growth, to the health of membranes, and to vision. Two isomers of carotene, differing from each other in the position of one of the double bonds, are abundant in carrots. Both of them show strong Vitamin A activity in biological tests.

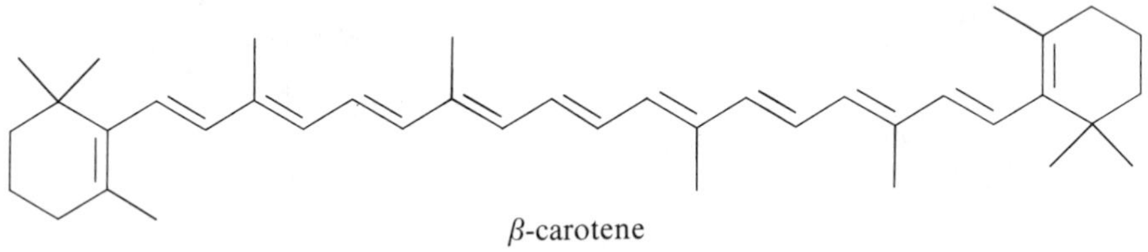

β-carotene

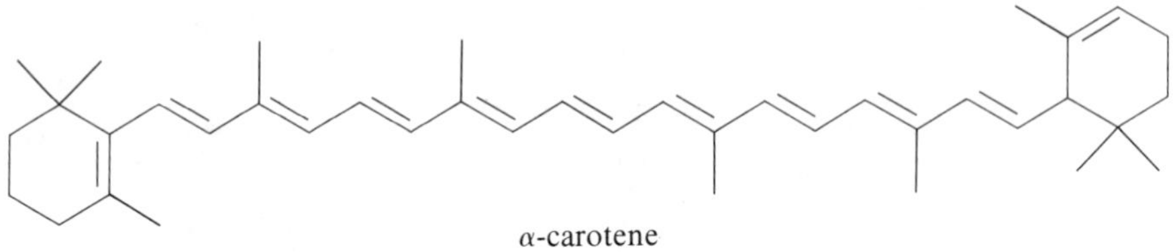

α-carotene

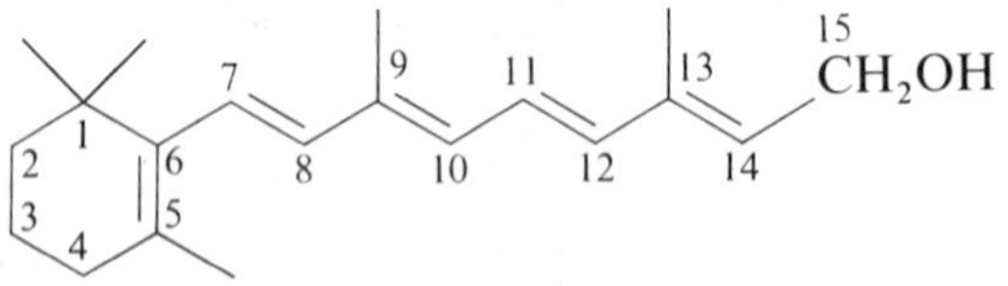
CH_2OH

vitamin A
retinol

How the carotenes are derived from lycopene by cyclization of the two ends of the molecule is evident. β-Carotene, in which all of the double bonds are in conjugation, is more abundant than α-carotene, in which one of the double bonds (the one in the right-hand ring above) is not conjugated with the rest. A molecule of Vitamin A corresponds to half a molecule of carotene and is an alcohol. It is pale yellow, in comparison with the deep red-orange color of crystalline carotene. Animal feeding experiments show that carotenes are transformed into Vitamin A in the bodies of most mammals, but to differing extents.

Vitamin A, also called **retinol,** is involved in the process of vision as its aldehyde with the cis configuration at the double bond at the 11 position, numbered as is traditional for these systems. This compound, (11*Z*)-retinal, is bonded with an imine linkage (p. 506) to a protein, opsin, in the visual pigment in the retina of the human eye. The complex molecule that results, called **rhodopsin,** absorbs light energy in the visible region of the spectrum. When light is absorbed, (11*Z*)-retinal undergoes isomerization to the all-trans configuration of retinal. In this form it can no longer stay bound to the protein and dissociates into (11*E*)-retinal and opsin, neither of which absorbs light in the visible region of the spectrum. This conversion is known as the bleaching of the visual pigment. (11*E*)-Retinal is reconverted to the cis form by an enzyme, binds once more to opsin, and the visual cycle begins again. Somehow the change in configuration of the double bond and the resulting dissociation of the visual pigment is translated by the retina into a message that travels through the optic nerve to the brain, to be interpreted there as sight and color vision.

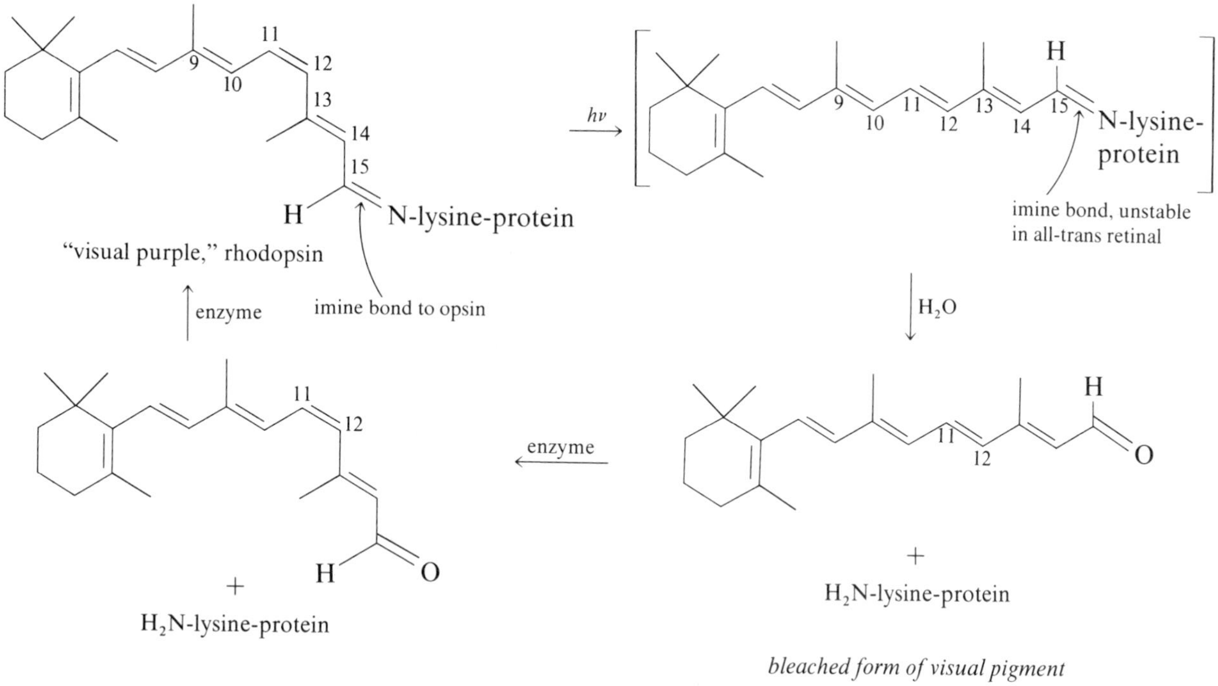

"visual purple," rhodopsin
N-lysine-protein
imine bond to opsin
enzyme
hν
N-lysine-protein
imine bond, unstable in all-trans retinal
H_2O
enzyme
+ H_2N-lysine-protein
+ H_2N-lysine-protein
bleached form of visual pigment

All the details of the changes that take place between the moment when light falls on the retina and the creation of a visual image are not yet understood. We can only wonder at the way nature uses simple chemical reactions in systems of such precise form that one stereochemical change in a small part of a large molecule sets in motion events of such consequence.

The absorption of light by conjugated systems such as Vitamin A and how this can lead to isomerization of a double bond are discussed on p. 715.

PROBLEM 17.19

(a) Vitamin A, as the free alcohol, is sensitive to air. Its esters are more stable. It is found in fish oils as its hexadecanoate (palmitate) ester. What is the structure of this ester?

(b) Vitamin A is often sold as its acetate ester. Write an equation showing how the vitamin could be converted into its acetate ester in the laboratory.

PROBLEM 17.20

β-Carotene is converted by monoperoxyphthalic acid into a mixture of a monoepoxide and a diepoxide. Which double bonds in β-carotene would be most vulnerable to oxidation by a peroxyacid? (Hint: See p. 319.) Write structural formulas for the mono- and diepoxides of β-carotene.

17.7 ULTRAVIOLET SPECTROSCOPY

A. Transitions Between Electronic Energy Levels

Absorption of energy corresponding to the ultraviolet and visible regions of the electromagnetic spectrum results in transitions between electronic energy levels in molecules. A simple alkene such as ethylene, for example, absorbs radiation at a wavelength of 171 nm. A nanometer, abbreviated nm, is 10^{-7} cm (10^{-9} m or 10 Å). The energy of the transition can be calculated from the relationship $\Delta E = h\nu$, or

$$\Delta E = \frac{hc}{\lambda}$$

where h is Planck's constant, 6.624×10^{-27} erg·s, and c is the velocity of light, 2.998×10^{10} cm/s. Radiation having a wavelength of 171 nm corresponds to 1.16×10^{-11} erg/molecule, 6.99×10^{12} erg/mol, or 166 kcal/mol. Therefore, the energy absorbed in the ultraviolet and visible region of the spectrum is of the same order as covalent bond energies (Table 2.5, p. 61) and often causes chemical changes. The branch of chemistry that deals with the transformations that organic compounds undergo when they absorb ultraviolet or visible radiation is known as **photochemistry.** One well known photochemical reaction is the primary event in vision, the isomerization of (11*Z*)-retinal to (11*E*)-retinal upon the absorption of light, described in Section 17.6E.

Molecular orbital theory is used to picture what happens when light is absorbed by a molecule. The π bond of an alkene, for example, is made up of a combination of two $2p$ orbitals on carbon atoms, each containing one electron. The two atomic orbitals combine to give two molecular orbitals: a bonding π orbital that is lower in energy than the atomic orbitals, and an antibonding orbital that is higher in energy and is designated as π^* (p. 49). The two electrons from the carbon atoms are usually found in the bonding π orbital, the state of lowest energy. If radiation corresponding to the difference in energy between the bonding orbital and the antibonding orbital is absorbed by a molecule, one of the electrons moves to the higher-energy level. The molecule is said to undergo a $\pi \rightarrow \pi^*$ transition and to be raised from the **ground state** to an **excited state,** which is an unstable situation of high energy (Figure 17.5).

A bonding π molecular orbital has two lobes, which are above and below the plane defined by the two carbon atoms of the alkene and the four atoms bonded to them. There is electron density between the two carbon atoms, and the fact that parallel overlap of the p orbitals must occur if the π bond is to be maintained

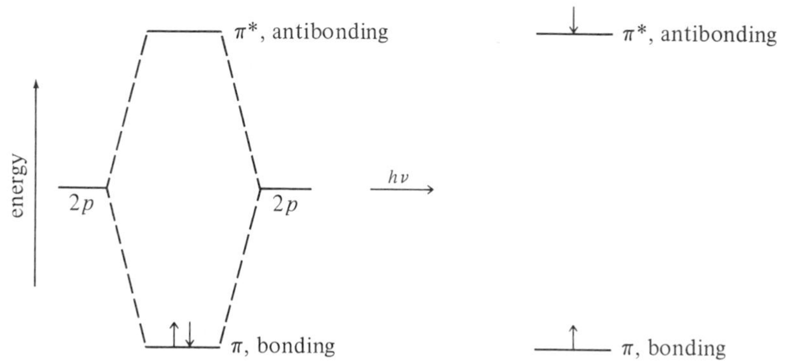

π^*, antibonding
energy
2p
2p
$h\nu$
π, bonding
π^*, antibonding
π, bonding

FIGURE 17.5 Schematic representation of the change in the electronic configuration of an alkene molecule on absorbing electromagnetic radiation. A $\pi \rightarrow \pi^*$ transition.

prevents rotation of the two carbon atoms in relation to each other. These phenomena are responsible for the existence of cis and trans isomers of alkenes (p. 211).

An antibonding π molecular orbital not only has a node in the plane of the alkene, but also has a nodal plane bisecting the bond between the two carbon atoms (Figure 2.16, p. 50). If one of the electrons of a π orbital is promoted to a π^* orbital, the electron density between the two carbon atoms decreases. The bond between them becomes more like a single bond, and rotation of one carbon atom in relation to the other becomes easier. Thus, one of the reactions seen for alkenes when radiation is absorbed is cis-trans isomerization about the double bond.

A carbonyl group has a π bond and also undergoes $\pi \rightarrow \pi^*$ transitions. In addition, the carbonyl group has two pairs of nonbonding electrons (represented by n) on the oxygen atom. The nonbonding electrons occupy orbitals designated as n orbitals, which are higher in energy than the bonding π orbital but lower in energy than the antibonding π^* orbital. The energy levels for the π, n, and π^* orbitals for a carbonyl compound, formaldehyde, are shown in Figure 17.6. Just as nonbonding electrons are more readily available for protonation than are π electrons, so are they more easily promoted to the π^* orbital of the carbonyl group. Thus, the carbonyl

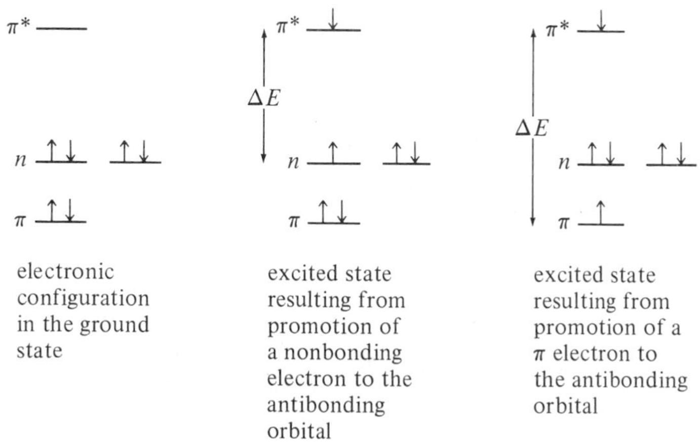

π^*
n
π
electronic configuration in the ground state
π^*
ΔE
n
π
excited state resulting from promotion of a nonbonding electron to the antibonding orbital
π^*
ΔE
n
π
excited state resulting from promotion of a π electron to the antibonding orbital

FIGURE 17.6 Electronic configurations and relative energy levels of the π and nonbonding (n) electrons in formaldehyde for the ground state and for the $n \rightarrow \pi^*$ and $\pi \rightarrow \pi^*$ excited states.

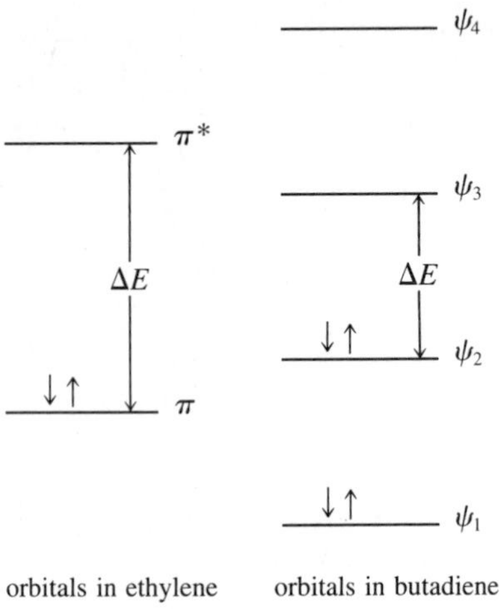
π*
ΔE
π
ψ4
ψ3
ΔE
ψ2
ψ1
orbitals in ethylene
orbitals in butadiene

FIGURE 17.7 A comparison of the electronic configuration and energy levels for a simple alkene to those for a conjugated alkene.

group absorbs ultraviolet radiation in two regions. A lower-energy transition corresponds to the promotion of one of the nonbonding electrons to the antibonding π orbital, that is, an $n \rightarrow \pi^*$ transition. The transition corresponding to the promotion of an electron from a bonding π orbital to an antibonding π orbital is the $\pi \rightarrow \pi^*$ transition, which requires more energy than the $n \rightarrow \pi^*$ transition.

Typical ultraviolet and visible spectrophotometers do not record at wavelengths below 200 nm. The ultraviolet spectra of most organic compounds become interesting when there are two or more multiple bonds in conjugation. The extended interaction between the conjugated π orbitals decreases the gap in energy between bonding and antibonding molecular orbitals (Figure 17.7). This lowers the amount of energy necessary for a $\pi \rightarrow \pi^*$ transition and increases the intensity of the absorption of radiation. If the conjugated system is extended enough, absorption takes place in the visible region of the spectrum. The carotene in vegetables like squash, carrots, and red peppers, for example, appears red-orange to us because it is absorbing the violet-blue portion of the visible spectrum, allowing most of the red-orange portion to reach our eyes.

B. Chromophores

The infrared spectrum of a compound is sensitive to exact structure and can be used to distinguish similar compounds on the basis of differences in the fingerprint region (p. 359), but the ultraviolet spectrum of a compound shows only the presence or absence of certain portions of the molecule that undergo $\pi \rightarrow \pi^*$ or $n \rightarrow \pi^*$ transitions. These distinctive groupings that absorb ultraviolet or visible radiation are usually conjugated double bonds, double bonds conjugated with carbonyl groups, or aromatic rings and are known as **chromophores.**

Ultraviolet spectroscopy reveals the presence of chromophores in complicated molecules but tells nothing about the large differences in structure that may exist. For example, cholesta-4-en-3-one and 4-methyl-3-penten-2-one have essentially the same ultraviolet spectrum, with a $\pi \rightarrow \pi^*$ transition around 240 nm and an $n \rightarrow \pi^*$ transition around 310 nm.

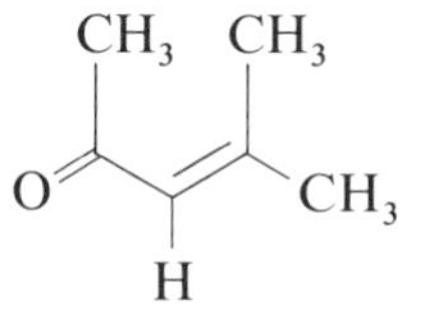
CH₃ CH₃
O
CH₃
H
4-methyl-3-penten-2-one

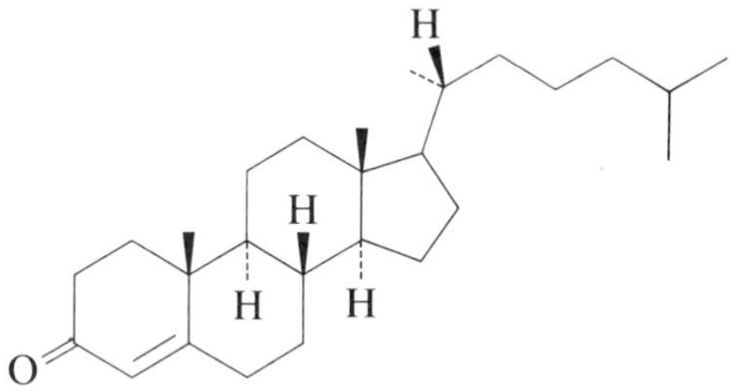
H
H
H H
O
cholesta-4-en-3-one

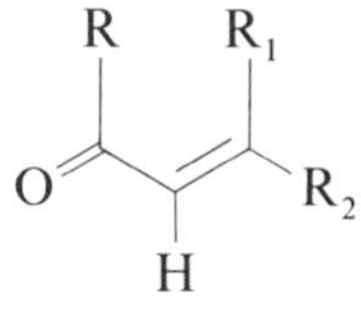
R R₁
O
R₂
H
the chromophore common to both compounds

Both compounds have a carbonyl group conjugated with a double bond. The carbonyl group is part of an alkyl ketone, and two alkyl groups are substituted on the carbon atom at the end of the double bond that is farther from the carbonyl group. These things taken together contribute to the distinctive wavelength and intensity of the absorption of radiation, which is quite similar for both compounds. Only the chromophore in the large steroid molecule interacts with ultraviolet radiation in a way that results in absorption of energy in the region of the spectrum that an ultraviolet spectrophotometer scans. The rest of the molecule is invisible as far as this spectroscopic technique is concerned. This simplifying quality of ultraviolet spectroscopy has been extraordinarily useful in identifying compounds that have similar chromophores, regardless of the complexities and differences in the rest of their molecules. The same quality makes it useful as an analytical tool when the compound sought has an intense and distinctive absorption spectrum.

C. The Absorption Spectrum

An ultraviolet or visible spectrum of a compound is obtained by comparing the radiation absorbed by a solution of the compound with the radiation absorbed by a similar thickness of pure solvent. A sketch of an ultraviolet spectrophotometer is shown in Figure 17.8.

The two critical aspects of an ultraviolet spectrum are the wavelength at which absorption takes place and the intensity of the absorption. The ultraviolet absorption spectrum of 1,3-pentadiene shown in Figure 17.9 is typical of a conjugated diene. First, note the simplicity of this spectrum in comparison with the usual infrared spectrum. One absorption appears as a broad band. Unless the radiation supplied to the compound is all of a single wavelength, molecular transitions will take place not only between one electronic energy level and a higher one, but also between the closely spaced vibrational energy levels belonging to each electronic level. The infrared spectrum of a compound is a record of the transitions between those

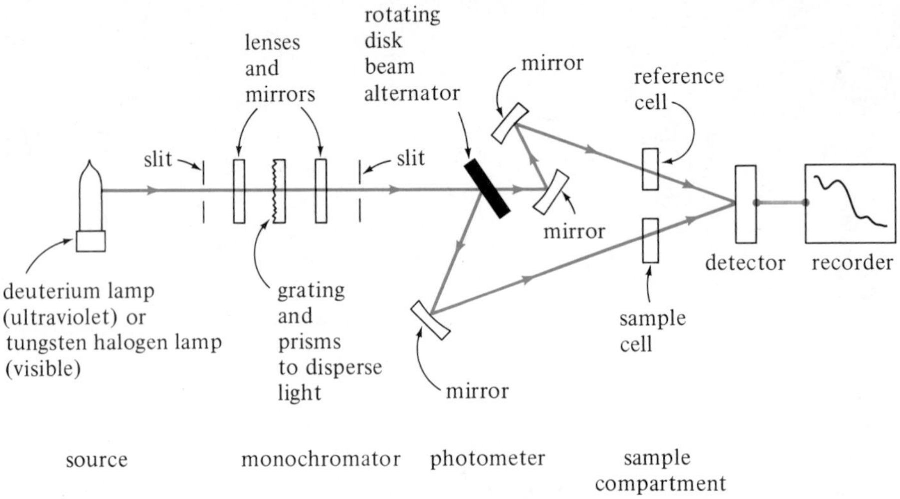
rotating disk beam alternator
lenses and mirrors
mirror
reference cell
slit
slit
mirror
detector
recorder
deuterium lamp (ultraviolet) or tungsten halogen lamp (visible)
grating and prisms to disperse light
sample cell
mirror
source
monochromator
photometer
sample compartment

FIGURE 17.8 Schematic diagram of a typical ultraviolet-visible spectrophotometer.

vibrational levels (p. 357). The ultraviolet spectrum is really the envelope for a series of closely spaced transitions clustered around a major electronic transition.

The wavelength corresponding to the energy of the electronic transition is read off the spectrum, which plots **molar absorptivity, ϵ,** a measure of the intensity of the absorption of radiation characteristic of the compound, against wavelength (or wavenumber). The position of maximum absorption is recorded as λ_{max}, 224 nm for 1,3-pentadiene. Because the solvent can affect the spectrum (an alcohol interacts with the nonbonding electrons on a carbonyl group, for example), the solvent is always specified, too. The complete designation of the wavelength of maximum

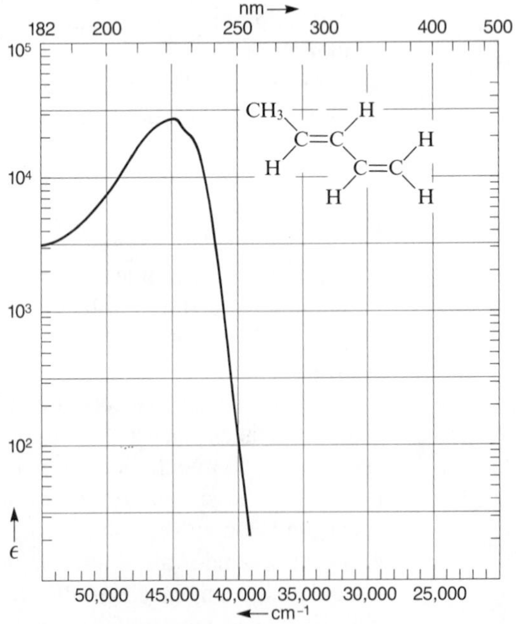
nm
182
200
250
300
400
500
10⁵
10⁴
10³
10²
ϵ
CH₃
H
C=C
H
C=C
H
H
H
50,000
45,000
40,000
35,000
30,000
25,000
cm⁻¹

FIGURE 17.9 Ultraviolet spectrum of 1,3-pentadiene in heptane. (From the *UV Atlas of Organic Compounds*)

absorption for the sample in Figure 17.9 is

$$\lambda_{max}^{heptane}\ 224\ \text{nm}$$

The molar absorptivity, ϵ, for 1,3-pentadiene is determined from the absorbance measured for the solution of the compound in heptane. The absorbance is related to the number of molecules of 1,3-pentadiene that were in the path of the light. This depends on the concentration of the solution and on the thickness of the cell through which the radiation passes. The thicker the cell, the longer the path length for the beam, and the more molecules that will be encountered by the radiation on its way through, which means a greater potential for absorption of energy. The molar absorptivity is related to the absorbance, A, by this equation:

$$\epsilon = \frac{A}{bc}$$

Here A is the experimentally determined absorbance, b is the path length in centimeters, and c is the concentration in moles per liter. For 1,3-pentadiene, ϵ is 26,000 L/mol·cm.

Information about the ultraviolet spectrum of a compound is thus given in two parts: the wavelength or wavelengths of maximum absorption, and the molar absorptivity for the compound at those wavelengths. For the ultraviolet spectrum of 1,3-pentadiene, the complete information is $\lambda_{max}^{heptane}$ 224 nm (ϵ 26,000). Note that the units of ϵ are usually not given.

D. The Relationship Between Structure and the Wavelength of Maximum Absorption

The wavelength of maximum absorption, λ_{max}, is dependent on the exact structure of the chromophore. For example, 2,5-dimethyl-2,4-hexadiene has the ultraviolet spectrum shown in Figure 17.10. Note the general similarity of this spectrum to that

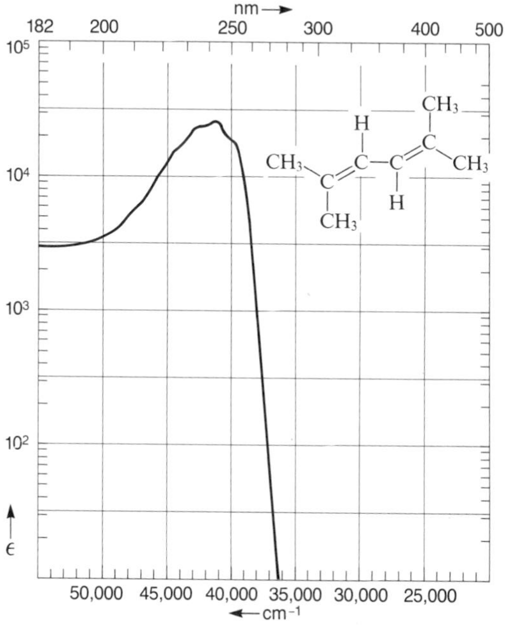

nm
182
200
250
300
400
500
10^5
10^4
10^3
10^2
ϵ
50,000 45,000 40,000 35,000 30,000 25,000
cm^{-1}
CH₃
H
C
CH₃
C—C
CH₃
C
H
CH₃

FIGURE 17.10 Ultraviolet spectrum of 2,5-dimethyl-2,4-hexadiene. (From the *UV Atlas of Organic Compounds*)

of 1,3-pentadiene (Figure 17.9). The difference consists of the shift of the position of maximum absorption to a longer wavelength, $\lambda_{max}^{heptane}$ 243 mn (ϵ 24,500). It takes less energy for the $\pi \rightarrow \pi^*$ transition to take place in this more highly substituted diene. This phenomenon is quite general; the position of λ_{max} is dependent on the extent of conjugation and the degree of substitution on the conjugated system.

For ketones, the $\pi \rightarrow \pi^*$ transition gives rise to a more intense absorption band than the $n \rightarrow \pi^*$ transition does. Acetone has two absorption bands: λ_{max}^{hexane} 189 nm (ϵ 900) and 279 nm (ϵ 15). The shorter-wavelength, higher-energy absorption is the $\pi \rightarrow \pi^*$ transition; the absorption at 279 nm corresponds to the $n \rightarrow \pi^*$ transition. Neither band is intense in a simple alkyl ketone.

The effect of conjugation on the spectrum of a ketone is shown in the spectra of 3-penten-2-one and 4-methyl-3-penten-2-one (Figure 17.11). For 3-penten-2-one, two bands are seen at $\lambda_{max}^{ethanol}$ 220 (ϵ 13,000) and 311 nm (ϵ 35). The chromophore in this compound is different from that in acetone and this $\pi \rightarrow \pi^*$ transition occurs over the entire conjugated system and not just in any one π bond. The absorption of this conjugated system at 220 nm has a high intensity. When substitution is increased on the carbon-carbon double bond of the conjugated system, the wavelength of the major absorption increases. For 4-methyl-3-penten-2-one, the absorption bands have $\lambda_{max}^{ethanol}$ 236 (ϵ 12,600) and 314 nm (ϵ 58).

When a system of really extended conjugation is present, such as in Vitamin A or carotene, λ_{max} moves into the region of the spectrum from approximately 380 nm to 780 nm, classified as visible. The intensity of the absorption also increases. The major absorption bands in β-carotene, for example, occur at λ_{max}^{hexane} 425 (ϵ 103,000), 450 (ϵ 145,000), and 477 nm (ϵ 130,000). This absorption is in the blue-violet region of the visible spectrum.

You should be aware of the following important points:

1. Ultraviolet spectra reveal the presence of distinctive groupings of atoms, usually involving conjugated systems, called chromophores.

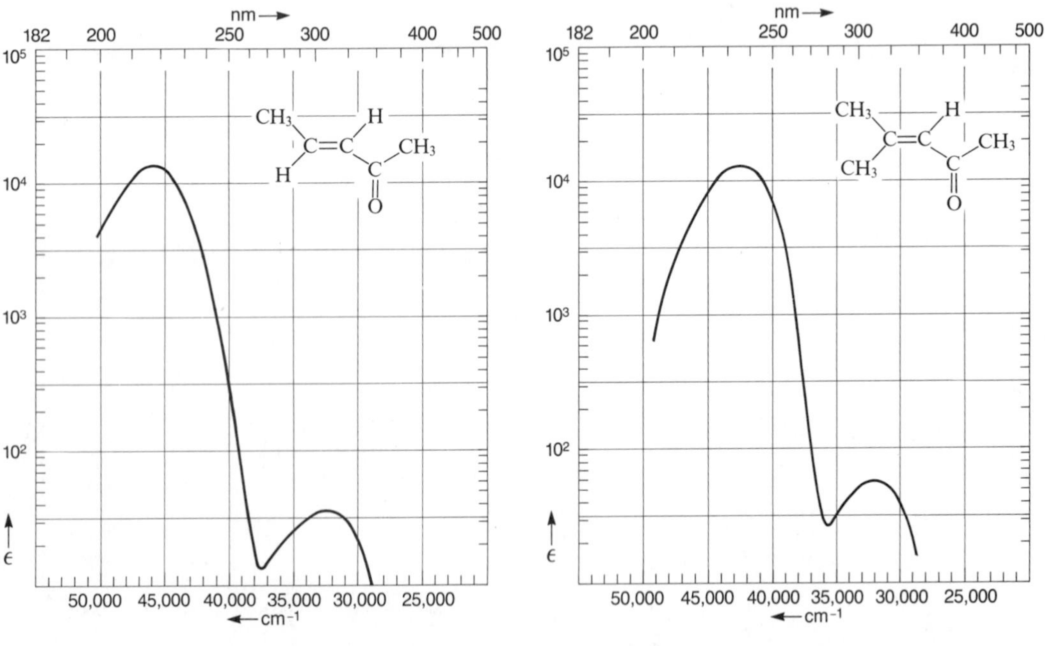
nm→
182
200
250
300
400
500
10⁵
10⁴
10³
10²
ε
50,000 45,000 40,000 35,000 30,000 25,000
←cm⁻¹
CH₃
H
C=C
CH₃
H
C
O
CH₃
H
C=C
CH₃
CH₃
C
O

FIGURE 17.11 Ultraviolet spectra of 3-penten-2-one and 4-methyl-3-penten-2-one in ethanol. (From the *UV Atlas of Organic Compounds*)

2. In general, the more extended the conjugation, the longer the wavelength at which absorption takes place and the greater the intensity of the absorption.
3. Added substitution, even of alkyl groups, on the conjugated system also increases the wavelength of the absorption.

At a more advanced level, it is possible to be much more precise about how conjugation and substitution affect absorption of energy in the ultraviolet region of the electromagnetic spectrum. However, it is sufficient for you to appreciate and be able to apply the principles listed above.

Study Guide Concept Map 17.5

Other systems that have useful features to their ultraviolet spectra are discussed in Section 23.7.

PROBLEM 17.21

Which of the following compounds will absorb at the longest wavelength?

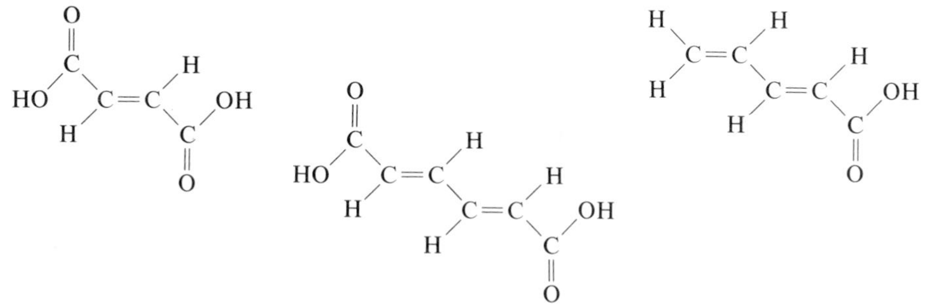

HO
OH

PROBLEM 17.22

The ultraviolet spectrum of 1-acetyl-2-methyl-1-cyclohexene in ethanol is shown in Figure 17.12. Identify the electronic transitions that are responsible for the absorption bands in the spectrum. Determine λ_{max} and ϵ for the compound and present the information about the ultraviolet absorption of 1-acetyl-2-methyl-1-cyclohexene in the standard way.

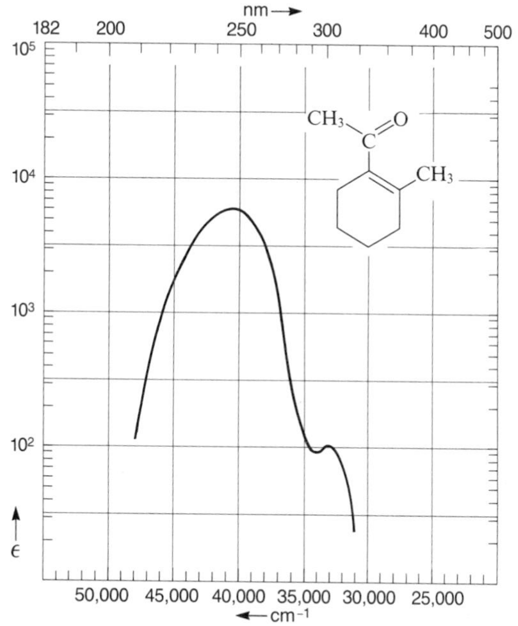

nm
182
200
250
300
400
500
10^5
10^4
10^3
10^2
ϵ
CH_3
CH_3
50,000
45,000
40,000
35,000
30,000
25,000
cm^{-1}

FIGURE 17.12 (From the *UV Atlas of Organic Compounds*)

TABLE 17.1 Reactions of Dienes

Diene	Reagent	Intermediate	Product(s)
Additions to Dienes			
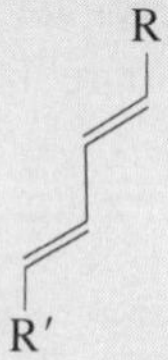 R R′	X_2	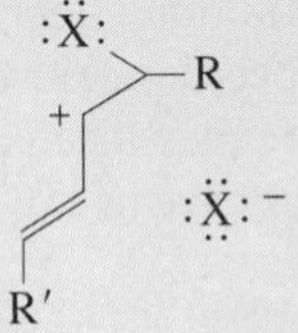 :X: R + :X:⁻ R′	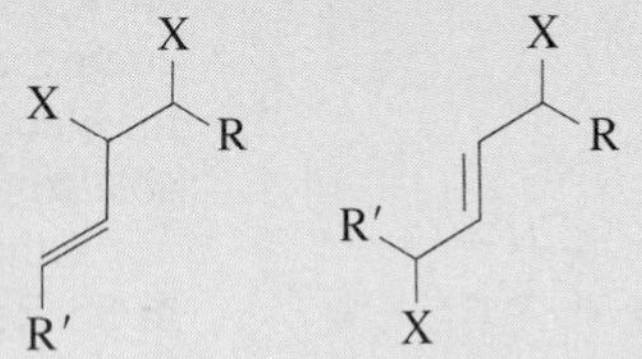 X X R R′ X R R′ X
	H_2, catalyst		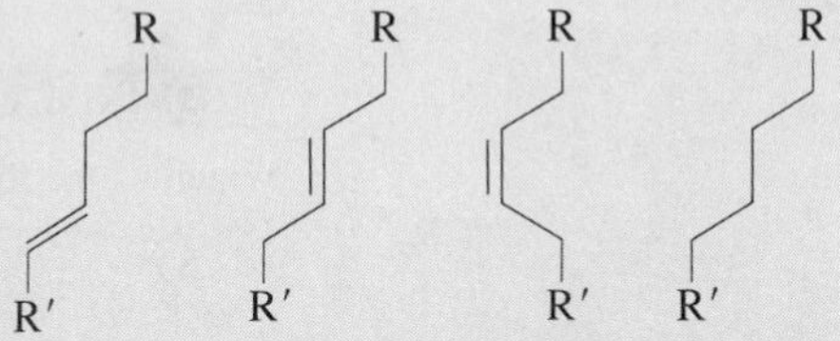 R R R R R′ R′ R′ R′
Diels-Alder Reaction			
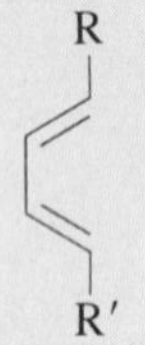 R R′	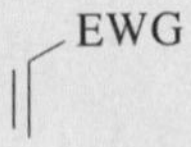 EWG EWG = electron-withdrawing group		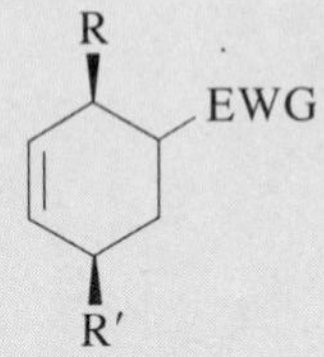 R EWG R′
	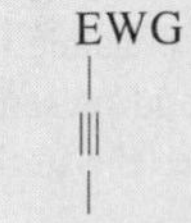 EWG		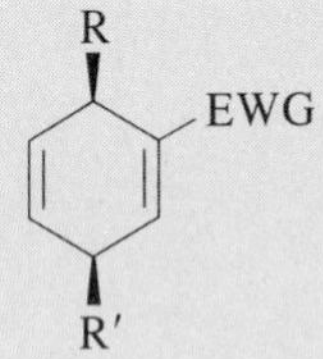 R EWG R′
 Z Z = electron-donating group	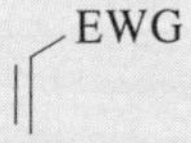 EWG		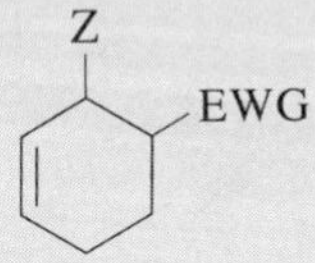 Z EWG
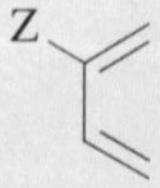 Z	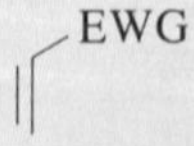 EWG		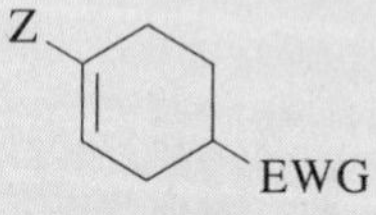 Z EWG
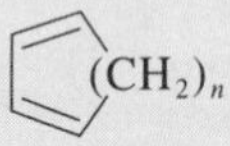 (CH₂)ₙ	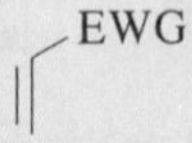 EWG		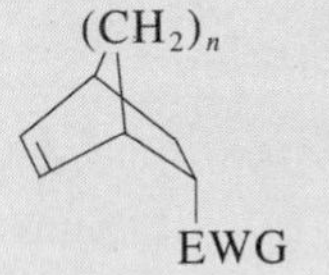 (CH₂)ₙ EWG

S U M M A R Y

Conjugated systems contain multiple bonds alternating with single bonds. The interaction of the p orbitals that form two or more π bonds results in the delocalization of the π electrons over the entire conjugated system. The multiple bonds in a conjugated system may react as isolated multiple bonds do or in new ways that are due to the interaction of the bonds in conjugation. For example, halogens add to conjugated dienes to give 1,2- and 1,4-addition products. In general, the 1,2-addition product is formed faster and predominates at low temperatures. It is called the kinetic product of the reaction. The 1,4-addition product is more stable. The 1,2-addition product is converted to the 1,4-product when equilibration can occur. Therefore, the 1,4-addition product is the thermodynamic product of the reaction.

The Diels-Alder reaction is an important 1,4-addition to 1,3-dienes. A diene in its *s*-cis conformation reacts with a dienophile, usually an alkene with an electron-withdrawing group on it. The product is a substituted cyclohexene. The reaction occurs with retention of the stereochemistry of the dienophile and of the substituents on the diene. If the diene is cyclic, a bicyclic compound is formed, and the substituent on the dienophile usually ends up in the endo orientation. The reactions of dienes are outlined in Table 17.1.

Many important classes of natural products are related to polyenes. Among the most important are the terpenoids, which have five-carbon units known as isoprene units. The isoprene rule states that, given a choice of possible structures for a natural product that contains five carbon atoms, chemists assume that the one that appears to be made of isoprene units connected in a head-to-tail fashion is most likely to be correct. The carotenoids, to which Vitamin A is related, and the steroids are among the compounds that have been shown to be built up in nature from isoprene units. The carotenoids and the steroids often have one tail-to-tail connection.

Ultraviolet spectroscopy is used to detect and identify compounds having conjugated systems. The distinctive grouping of conjugated bonds in a molecule that absorbs ultraviolet or visible radiation is known as a chromophore. Each chromophore absorbs radiation at a distinctive wavelength and intensity. The wavelength at which a chromophore absorbs energy is related to the energy gap between the bonding and antibonding molecular orbitals for the conjugated system. An increase in conjugation or of alkyl substitution on the multiple bonds brings the bonding and antibonding molecular orbitals closer in energy, which decreases the energy necessary to promote an electron from a bonding to an antibonding orbital. This, in turn, increases λ_{max} and, usually, the intensity (given as ϵ, the molar absorptivity) of the absorption band.

ADDITIONAL PROBLEMS

17.23 Give structural formulas for all products symbolized by letters in the following equations. If the stereochemistry of a reaction is known, be sure to show it in your answer.

(a) 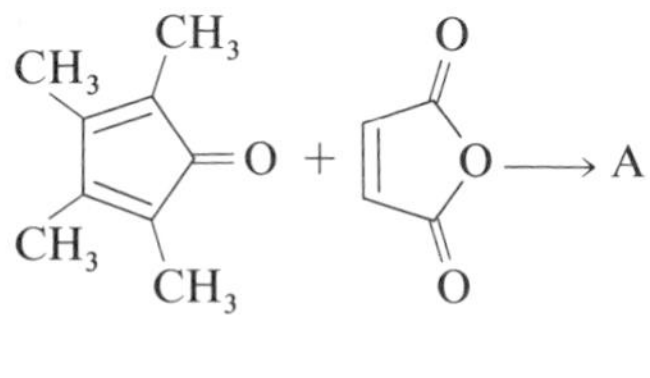
CH3
CH3
O
=O +
O
A
CH3
CH3
O

(b) 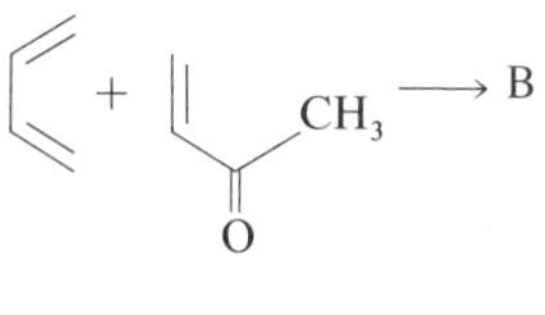
+
CH3
B
O

(c) 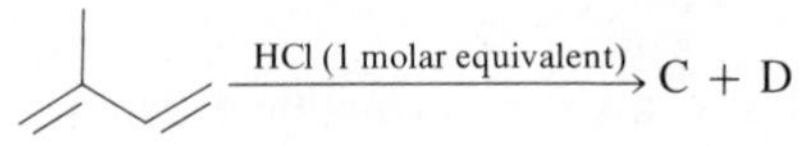
HCl (1 molar equivalent)
C + D

(d) 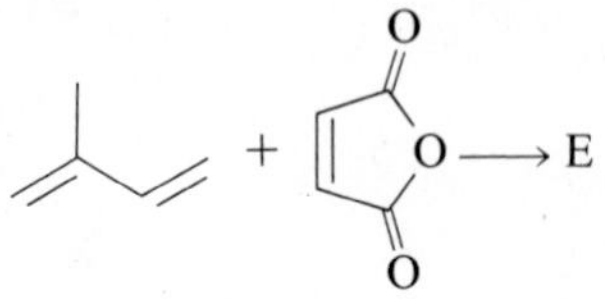
O
O
O
E

(e) 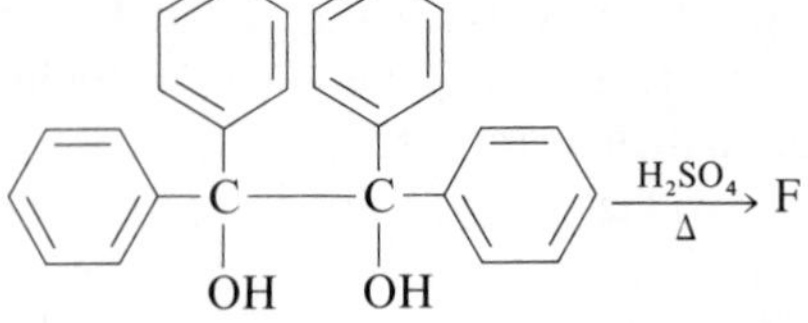
C
C
OH
OH
H₂SO₄
Δ
F

(f) 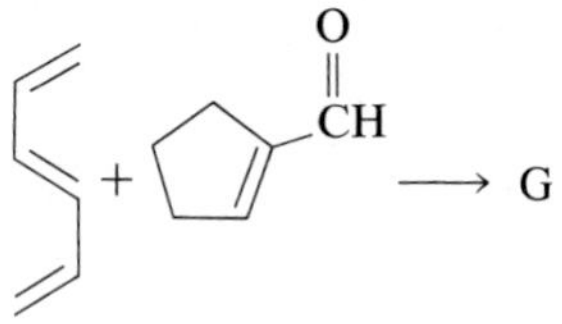
O
CH
G

(g) 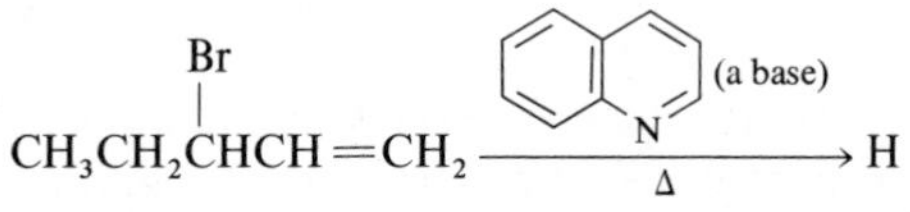
Br
CH₃CH₂CHCH=CH₂
N
(a base)
Δ
H

(h) $CH_3C{\equiv}CCH_2C{\equiv}CCH_2CH_3 \xrightarrow[\text{Pd/BaSO}_4,\ \text{quinoline}]{H_2}$ I

(i) 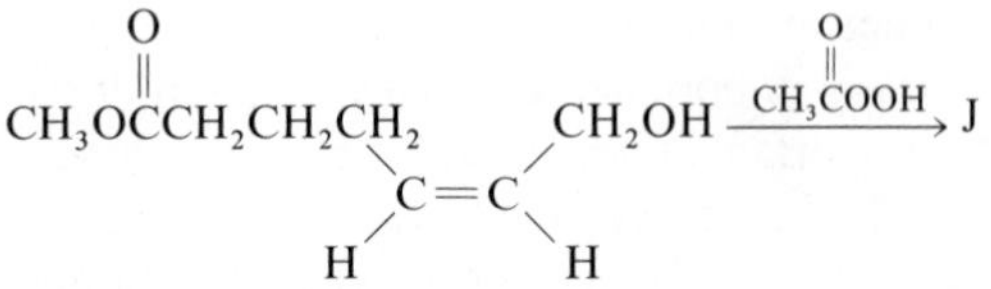
O
CH₃OCCH₂CH₂CH₂
CH₂OH
C=C
H
H
O
CH₃COOH
J

(j) 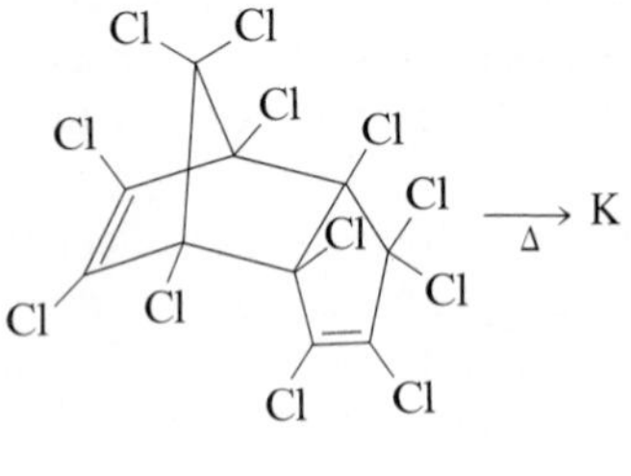
Cl
Cl
Cl
Cl
Cl
Cl
Cl
Cl
Cl
Cl
Cl
Cl
Δ
K

(k) 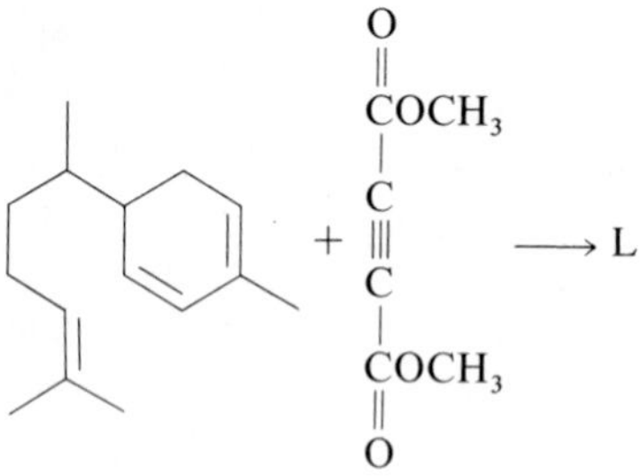
O
COCH₃
C
C
COCH₃
O
L

(l) 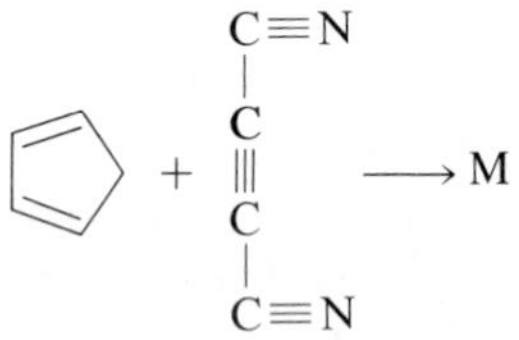
C≡N
C
C
C≡N
M

(m) 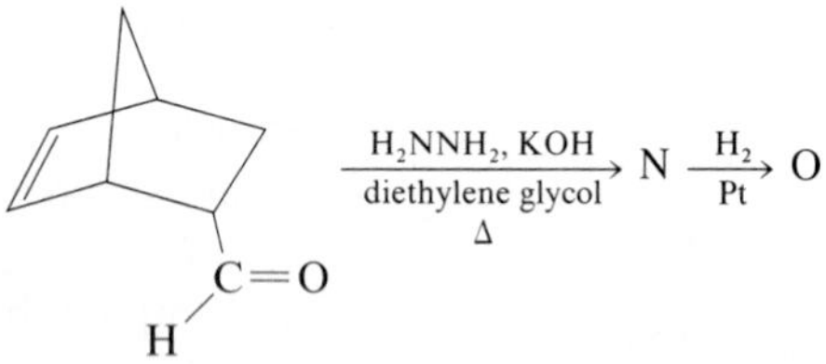
C=O
H
H₂NNH₂, KOH
diethylene glycol
Δ
N
H₂
Pt
O

(n) 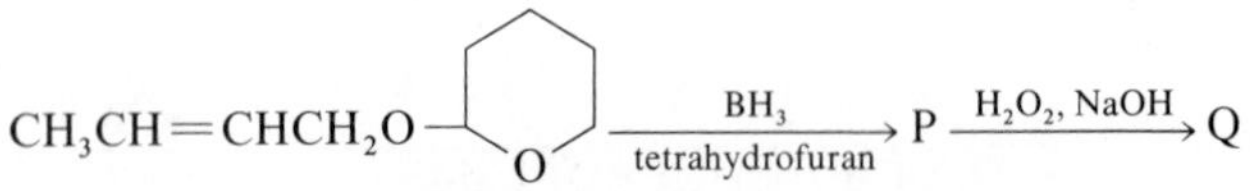
CH₃CH=CHCH₂O
O
BH₃
tetrahydrofuran
P
H₂O₂, NaOH
Q

(o) 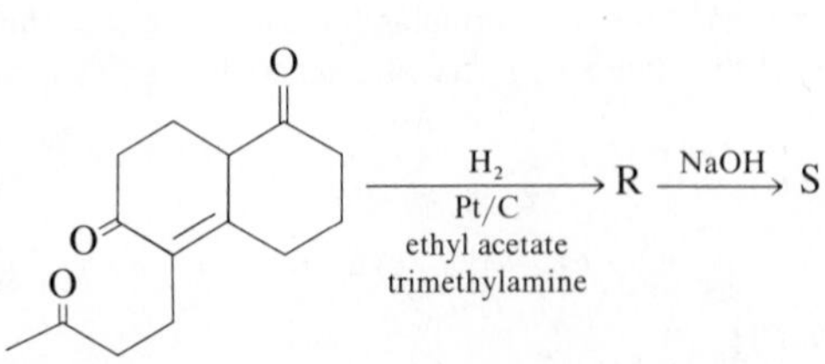
O
O
O
H₂
Pt/C
ethyl acetate
trimethylamine
R
NaOH
S

(p)
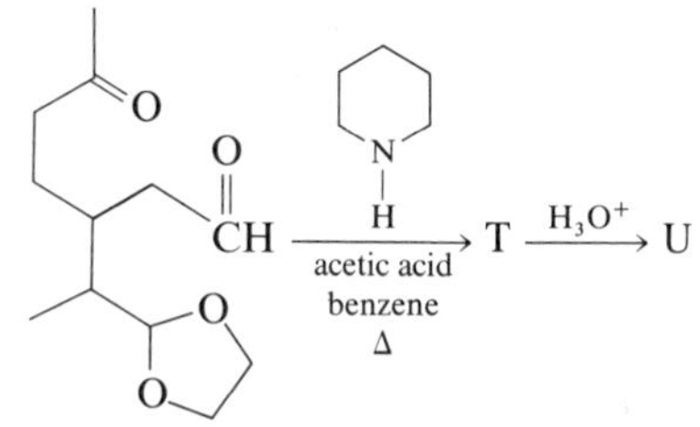
acetic acid
benzene
Δ
T
U

(q)
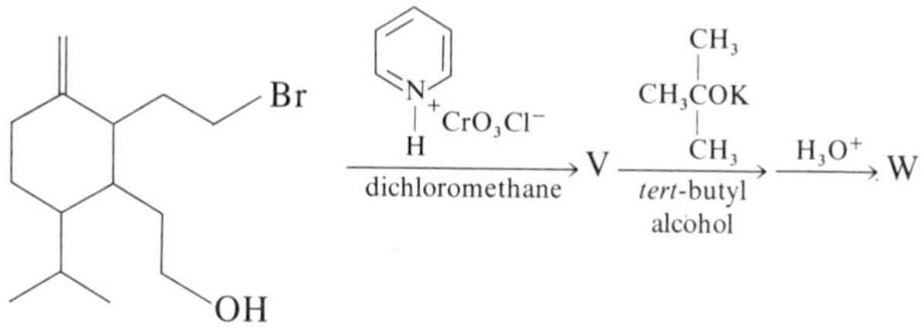
dichloromethane
V
tert-butyl
alcohol
W

17.24 Tell what reagents are necessary to carry out the following transformations. More than one step may be required in some cases.

(a)
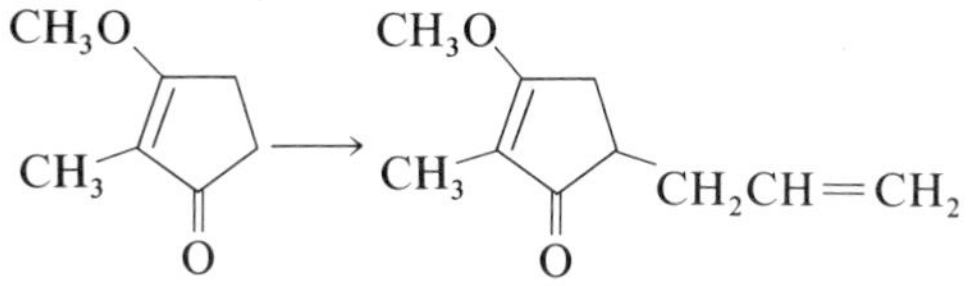

(b)
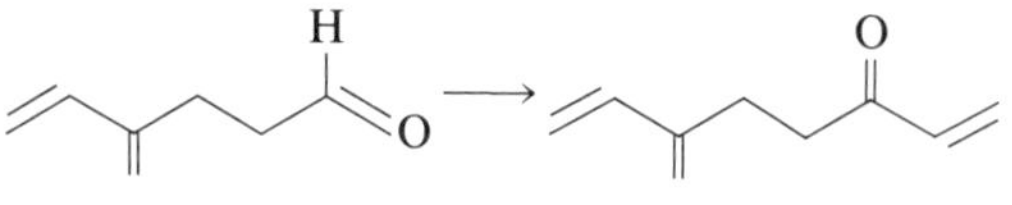

(c)
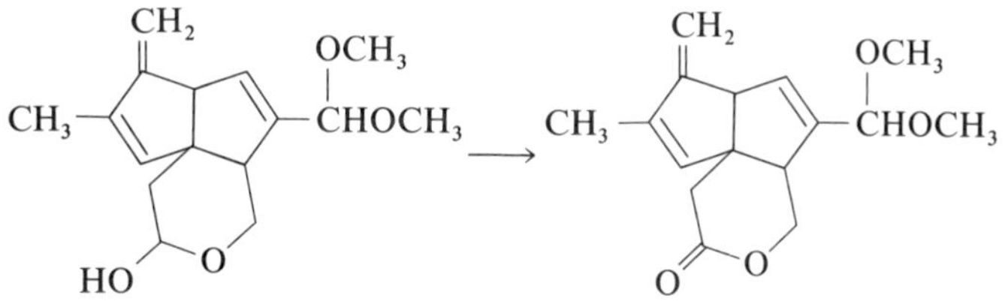

(d)
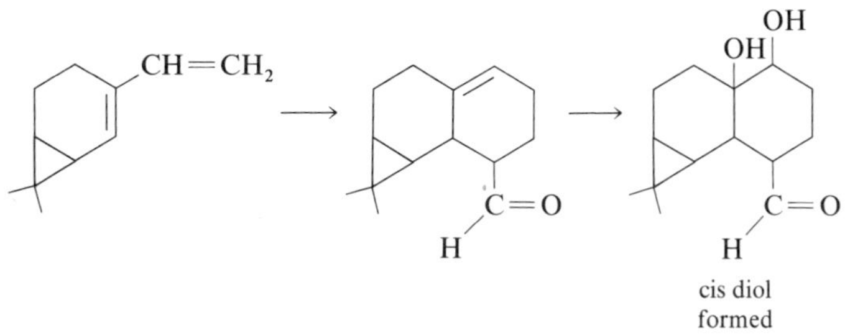
cis diol
formed

(e)
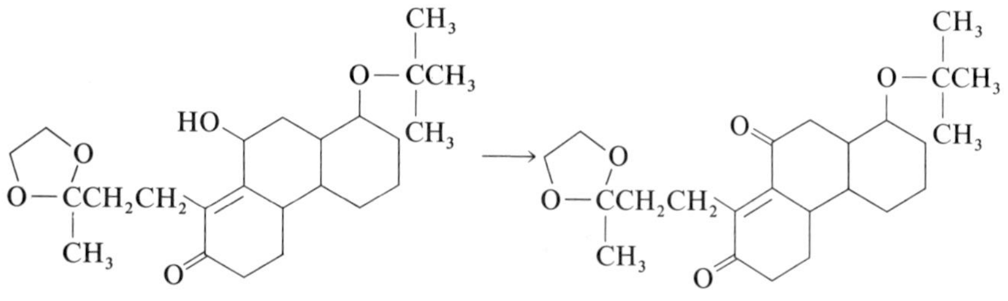

17.25 The following compounds are synthesized by Diels-Alder reactions. Write equations for their syntheses, showing the diene and dienophile that would have to be used in each case.

(a)
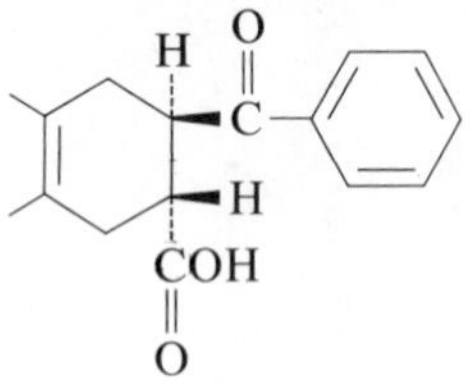

(b)
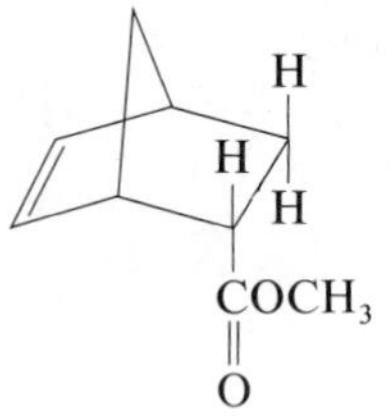

(c)
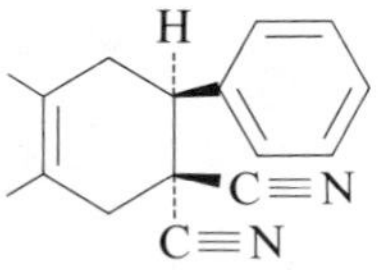

(d)
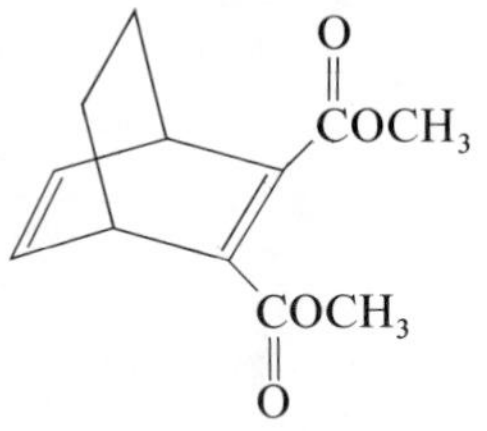

(e)
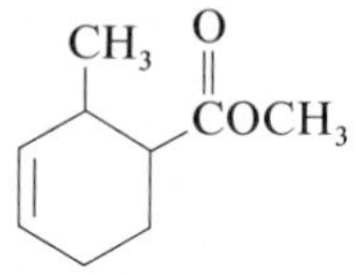

(f)
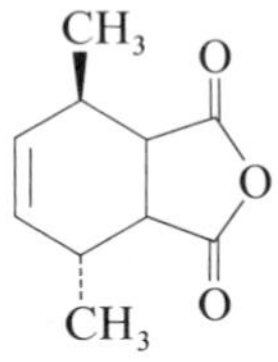

(g)
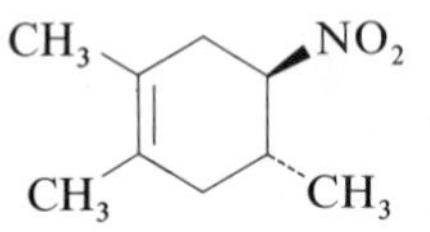

(h)
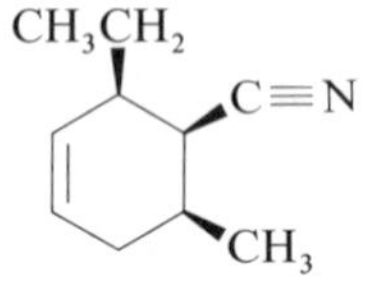

17.26 Write structural formulas for the products and intermediates designated by letters in the following sequence of reactions.

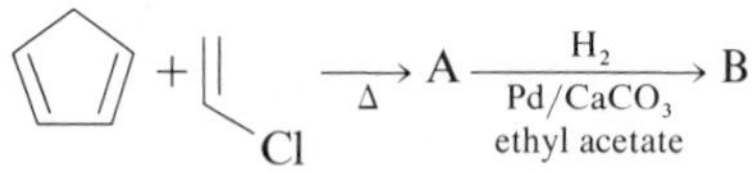

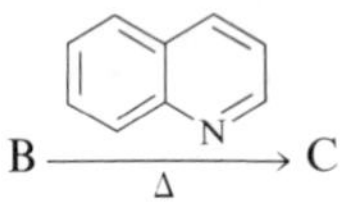

$$\mathrm{B} \xrightarrow[\text{diethyl ether}]{\text{Mg}} \mathrm{D} \xrightarrow{CO_2} \mathrm{E} \xrightarrow{H_3O^+} \mathrm{F}$$

$$\mathrm{C} \xrightarrow[\text{chloroform}]{Br_2} \mathrm{G}$$

$$\mathrm{C} \xrightarrow[H_2O]{KMnO_4,\ NaOH} \mathrm{H}$$

$$\mathrm{C} \xrightarrow{O_3} \xrightarrow{H_2O_2,\ NaOH} \mathrm{I}$$

17.27 1,3-Butadiene adds to maleic anhydride at a reasonable rate at 100 °C and slowly at room temperature. 2,3-Di-*tert*-butyl-1,3-butadiene does not react with maleic anhydride at all under these conditions. Offer an explanation for these experimental observations.

17.28 1,3,5-Hexatriene reacts with 1 equivalent of bromine to give a mixture of 1,2- and 1,6-addition products. No 1,4-addition product is obtained. Explain this phenomenon by writing equations and showing the structures of the relevant intermediates.

17.29 Give the structural formulas, including the correct stereochemistry, for the major products represented by letters in the following synthetic sequence.

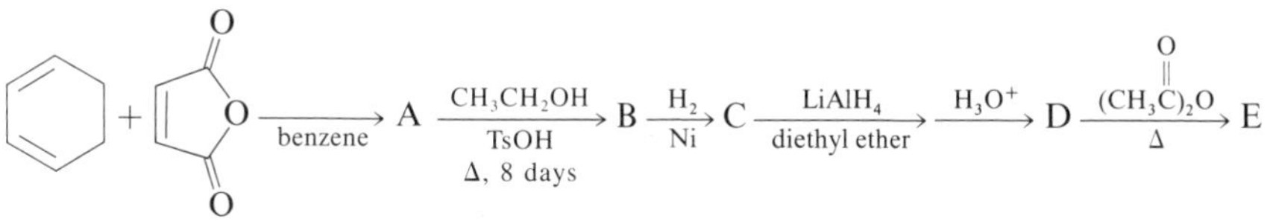
benzene
A
CH3CH2OH
TsOH
Δ, 8 days
B
H2
Ni
C
LiAlH4
diethyl ether
H3O+
D
(CH3C)2O
Δ
E

17.30 Predict the products of the following reactions of the terpene chrysanthemic acid and its esters by completing the equations.

(a) 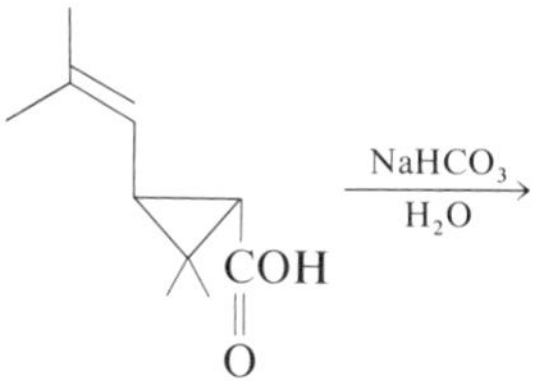
COH
O
NaHCO3
H2O

(b) 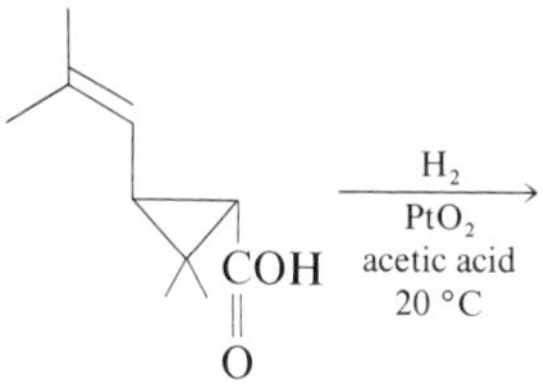
COH
O
H2
PtO2
acetic acid
20 °C

(c) 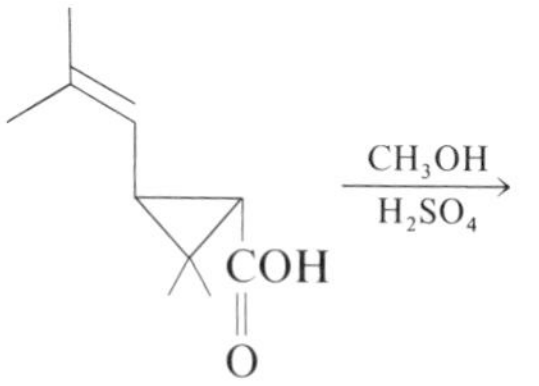
COH
O
CH3OH
H2SO4

(d) Product of part c $\xrightarrow[H_2O \text{, dioxane}]{OsO_4}$

(e) Product of part c $\xrightarrow{O_3} \xrightarrow[H_2O]{Zn}$

(f) 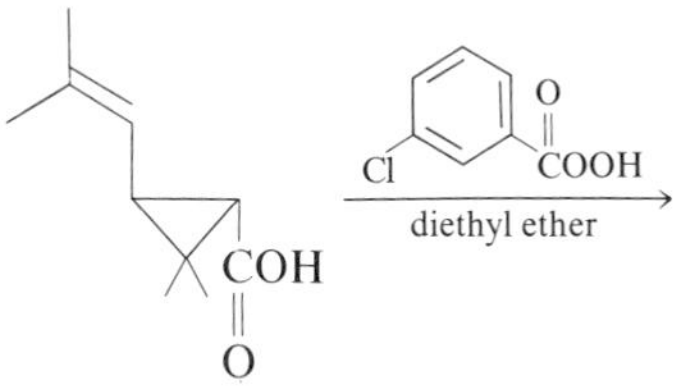
COH
O
Cl
COOH
diethyl ether

(g) Product of part f $\xrightarrow[H_2O, \Delta]{CH_3NHCH_3}$

17.31

(a) Bisabolene, $C_{15}H_{24}$, is found widely distributed in nature, especially in myrrh and oil of bergamot. Hydrogenation over platinum in acetic acid gives Compound X, $C_{15}H_{30}$. How many units of unsaturation does bisabolene have? How many of these are rings?

(b) Bisabolene undergoes incomplete hydrogenation in cyclohexane to give Compound Y, $C_{15}H_{28}$. Ozonolysis of Y gives 6-methyl-2-heptanone and 4-methylcyclohexanone. What is the structure of Compound Y?

(c) Ozonolysis of bisabolene gives, among other products, acetone and 4-oxopentanoic acid. Assign a partial structure on the basis of these data. What uncertainty is there about the structure of bisabolene at this point?

(d) Bisabolene and an alcohol derived from bisabolene are among the products obtained from the acid-catalyzed cyclization of nerolidol, obtained from the flowers of bitter orange. The structural formula for nerolidol is shown below.

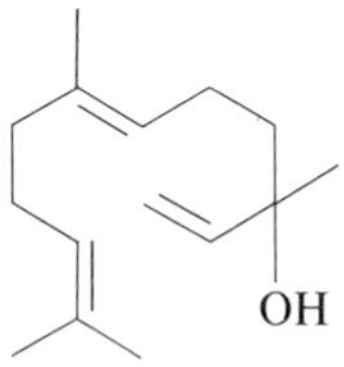
OH

Propose a mechanism for an acid-catalyzed cyclization of nerolidol to bisabolene. What does this, in combination with the data given above, suggest about the structure of bisabolene?

(e) What is the structure of the bisabolol also formed in the acid-catalyzed cyclization of nerolidol?

(f) Bisabolene reacts easily with hydrogen chloride to give Compound Z, $C_{15}H_{27}Cl_3$. Compound Z is also formed when the mixture of bisabolene and bisabolol from nerolidol is treated with hydrogen chloride. What is the structure of Compound Z? Write equations showing its formation from bisabolene and bisabolol.

17.32 Disparlure, the sex pheromone of the gypsy moth (p. 702), was synthesized as a racemic mixture by the following sequence of reactions. Give structures for the products or reagents designed by letters.

$$HC{\equiv}CH \xrightarrow[\text{tetrahydrofuran}]{\text{Na}} A \xrightarrow{B} CH_3(CH_2)_9C{\equiv}CH \xrightarrow[\text{diglyme}]{CH_3CH_2CH_2CH_2Li}$$

$$C \xrightarrow{D} CH_3(CH_2)_9C{\equiv}C(CH_2)_4\overset{\overset{\large CH_3}{|}}{C}HCH_3 \xrightarrow{E} F \xrightarrow{G} \text{racemic disparlure}$$

$$CH_3\overset{\overset{\large CH_3}{|}}{C}HCH_2CH_2CH_2CH_2OH \xrightarrow{H} D$$

17.33 Assign structures to the compounds designated by letters. Infrared spectral data are given for the compounds. Assign as many of the bands as you can.

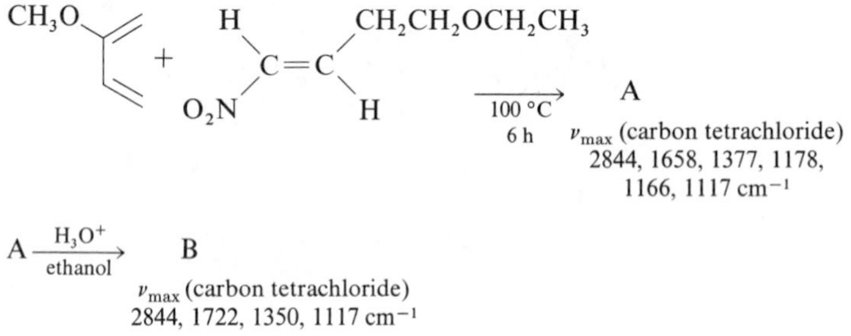
CH3O
H
CH2CH2OCH2CH3
+
C=C
O2N
H
100 °C
6 h
A
νmax (carbon tetrachloride)
2844, 1658, 1377, 1178,
1166, 1117 cm−1
A
H3O+
ethanol
B
νmax (carbon tetrachloride)
2844, 1722, 1350, 1117 cm−1

17.34 The following reactions were carried out in the synthesis of the terpene neointermedeol. What are the structures of neointermederol and of the intermediates in its synthesis?

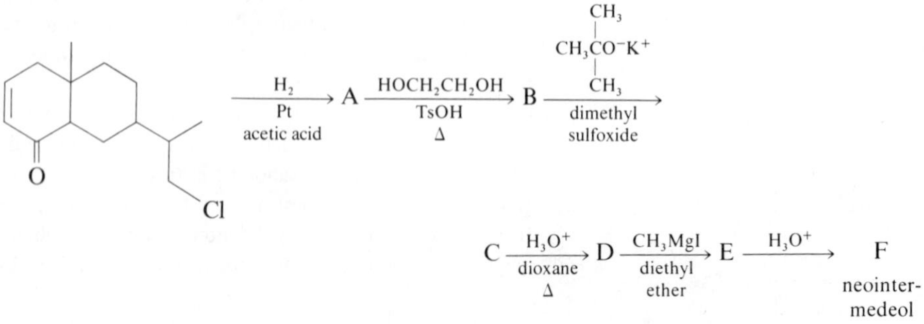
O
Cl
H2
Pt
acetic acid
A
HOCH2CH2OH
TsOH
Δ
B
CH3
CH3CO−K+
CH3
dimethyl
sulfoxide
C
H3O+
dioxane
Δ
D
CH3MgI
diethyl
ether
E
H3O+
F
neointer-
medeol

17.35 The monoepoxide of β-carotene (Problem 17.20, p. 714) rearranges in dilute hydrochloric acid to a compound having the structure shown below. Propose a mechanism for its formation from β-carotene monoepoxide.

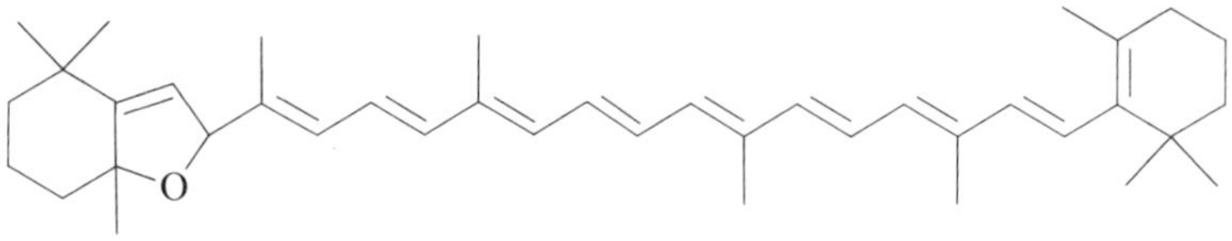

17.36 Steroids, because of the rigidity of their fused six-membered rings, have been used to study the stereochemistry of many reactions.

(a) Predict which bromonium ion (p. 290) is the intermediate in the addition of bromine to cholesterol (p. 711). What is the stereochemistry of the dibromide that forms? (Hint: It may help you to review Section 12.6D.)

(b) Two stereoisomeric bromohydrins derived from a steroid react with base at different rates to give oxiranes (p. 448). Partial structures, A and B, for the two compounds are given below. Predict which compound produces a high yield of oxirane in 30 seconds and which one requires 24 hours to give the same yield. How do you account for the difference in reactivity between the two isomers?

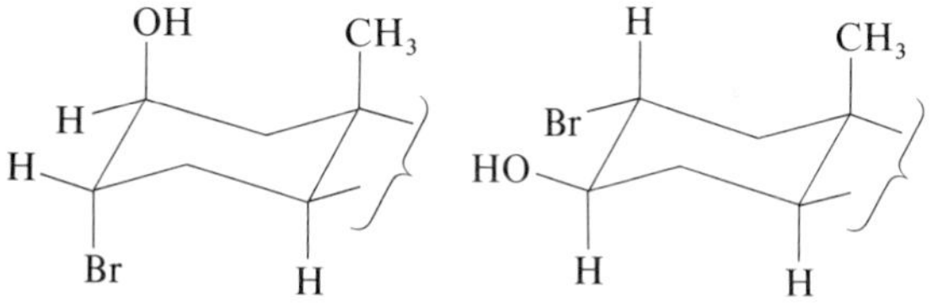
OH
CH$_3$
H
H
Br
H
H
CH$_3$
Br
HO
H
H
A B

17.37 Helenalin, a natural product isolated from *Helenium autumnale,* sneezeweed, is of interest because it is cytotoxic. Some steps in a recent synthesis of the compound are shown below. What reagents could be used for these transformations?

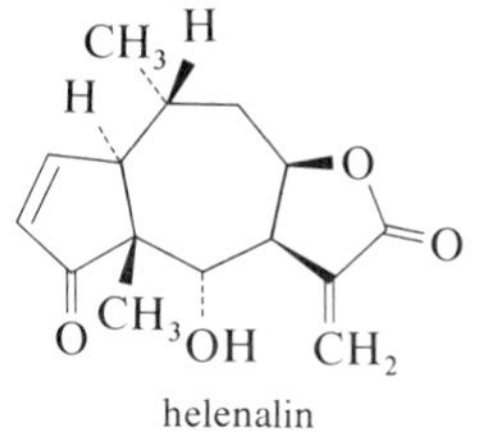
CH$_3$ H
H
O
O
O
CH$_3$
OH
CH$_2$
helenalin

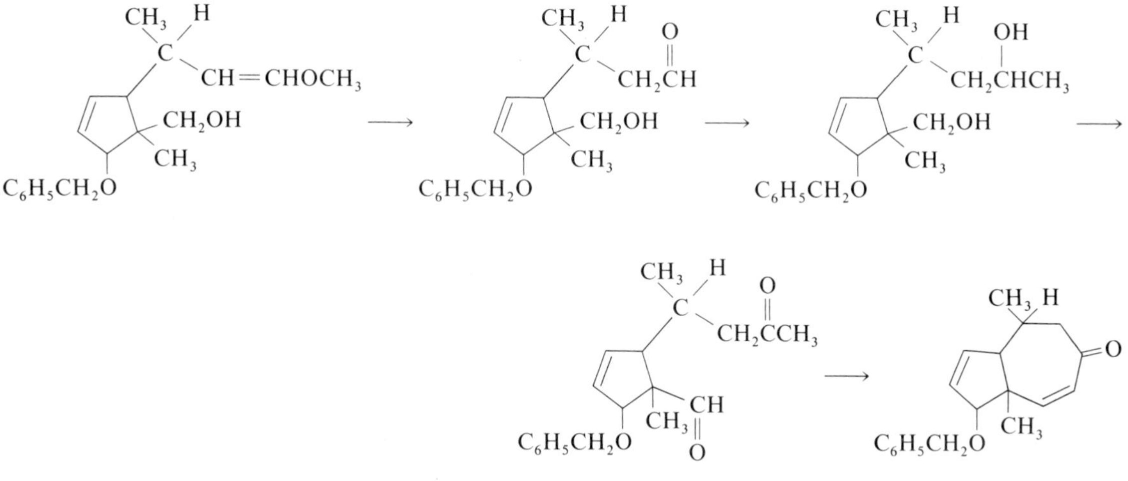
CH$_3$ H
C
CH=CHOCH$_3$
CH$_2$OH
CH$_3$
C$_6$H$_5$CH$_2$O
CH$_3$ H
C
O
CH$_2$CH
CH$_2$OH
CH$_3$
C$_6$H$_5$CH$_2$O
CH$_3$ H
C
OH
CH$_2$CHCH$_3$
CH$_2$OH
CH$_3$
C$_6$H$_5$CH$_2$O
CH$_3$ H
C
O
CH$_2$CCH$_3$
CH
CH$_3$
O
C$_6$H$_5$CH$_2$O
CH$_3$ H
O
CH$_3$
C$_6$H$_5$CH$_2$O

17.38 An intermediate in the synthesis of quadrone, a sesquiterpene with possible antitumor activity, was prepared by the following reactions. Write structural formulas for the compounds represented by letters.

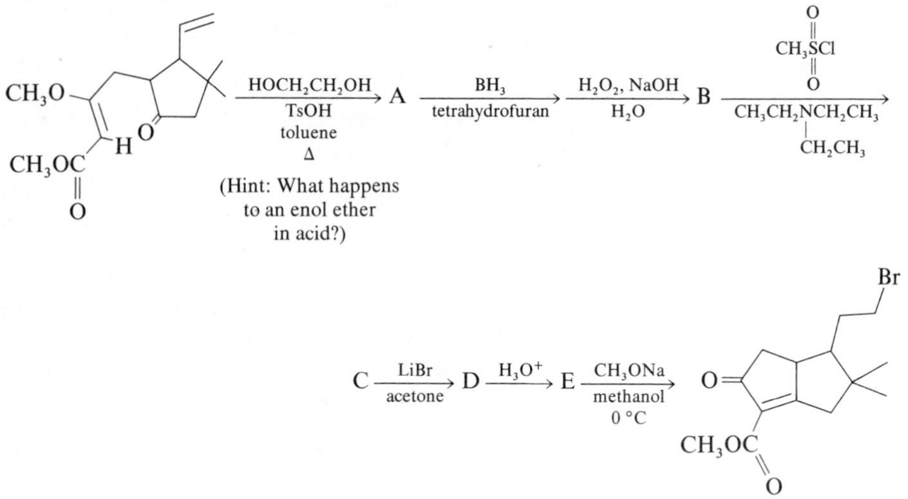
CH3O
CH3OC
O
H
O
HOCH2CH2OH
TsOH
toluene
Δ
A
(Hint: What happens to an enol ether in acid?)
BH3
tetrahydrofuran
H2O2, NaOH
H2O
B
CH3SCl
CH3CH2NCH2CH3
CH2CH3
C
LiBr
acetone
D
H3O+
E
CH3ONa
methanol
0 °C
Br

17.39 The following steps are part of a synthesis of dihydroxycalciferols, which are the hormonally active forms of Vitamin D. Supply reagents that will accomplish the transformations. TBDMS is the *tert*-butyldimethylsilyl, and THP, the tetrahydropyranyl protecting group (Section 13.10).

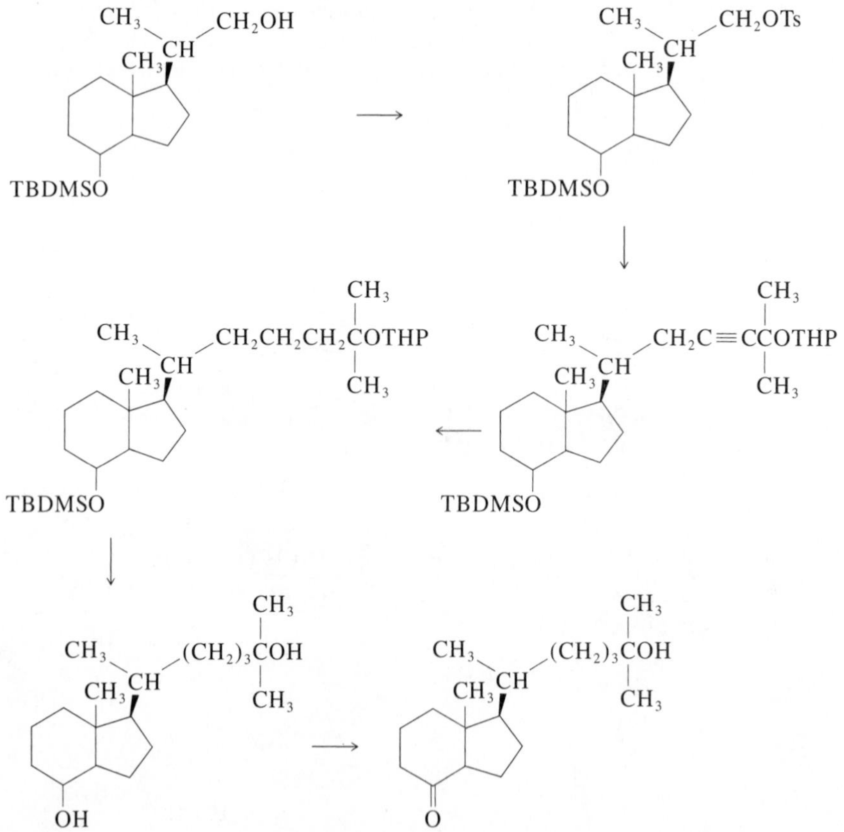
CH3
CH2OH
CH
TBDMSO
CH2OTs
CH2C≡CCOTHP
CH2CH2CH2COTHP
(CH2)3COH
OH
O

17.40 Three samples, labeled Compounds A, B, and C, are known to be unsaturated carboxylic acids. Ultraviolet spectra were taken of the three samples with the following results: Compound A, λ_{max}^{hexane} 302 nm (ϵ 36,500); Compound B, 208 nm (ϵ 12,000); Compound C, 261 nm (ϵ 25,000). Which structural formula given below belongs to Compound A, B, and C?

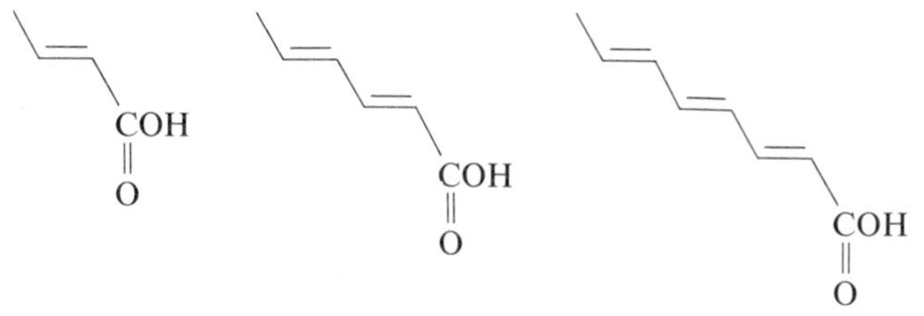

17.41 A compound with the molecular formula C_4H_6O has $\lambda_{max}^{ethanol}$ 219 nm (ϵ 16,600) and 318 nm (ϵ 30).

(a) How many units of unsaturation does the compound have?
(b) Draw some structures compatible with its molecular formula and its ultraviolet spectrum.
(c) What are some structures that would not be compatible with the ultraviolet spectrum, even though they fit the molecular formula?

17.42 Of the following three isomeric compounds, which one will absorb at the longest wavelength?

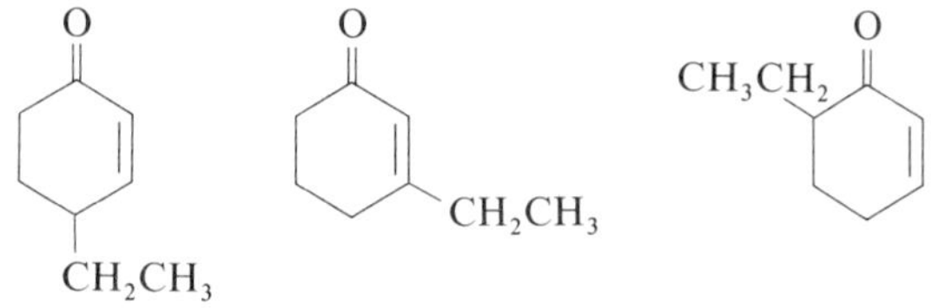

17.43 Irone, responsible for the fragrance of violets, is a mixture of isomers for which structural formulas are given below. One of these compounds, β-irone, absorbs at $\lambda_{max}^{ethanol}$ 294.5 nm. The other two, α-irone and γ-irone, have absorption bands at $\lambda_{max}^{ethanol}$ 229 nm and 226.5 nm. Which of the structural formulas below belongs to β-irone?

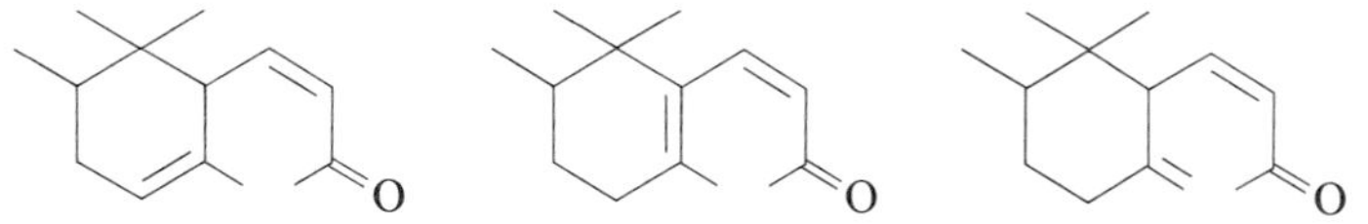

17.44 Of the two isomeric hexadienols, 5-methyl-2,4-hexadien-1-ol and 2-methyl-3,5-hexadien-2-ol, one has $\lambda_{max}^{ethanol}$ 223 nm and the other $\lambda_{max}^{ethanol}$ 236 nm. Draw structural formulas for the two compounds and assign the ultraviolet absorption bands to the correct compounds.

18

Enols and Enolate Anions as Nucleophiles II. Conjugate Addition Reactions; Ylides

A • L O O K • A H E A D

When a carbon-carbon double bond is conjugated with the carbon-oxygen double bond of a carbonyl group, the electrophilic character of the carbonyl carbon is transmitted to the β-carbon atom of the α,β-unsaturated carbonyl system. A compound in which a carbon-carbon double bond is conjugated with an electron-withdrawing group is called an electrophilic alkene.

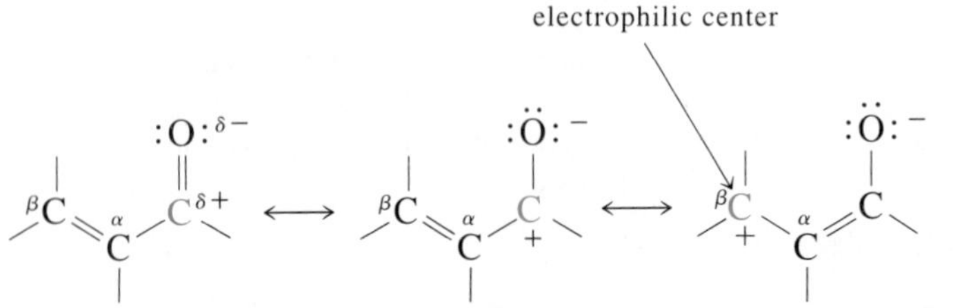
electrophilic center

resonance contributors for an α,β-unsaturated carbonyl system

Nucleophiles add to the double bond of an electrophilic alkene by attacking the β-carbon atom. An enolate anion, for example, may serve as the nucleophile in a reaction known as the Michael addition.

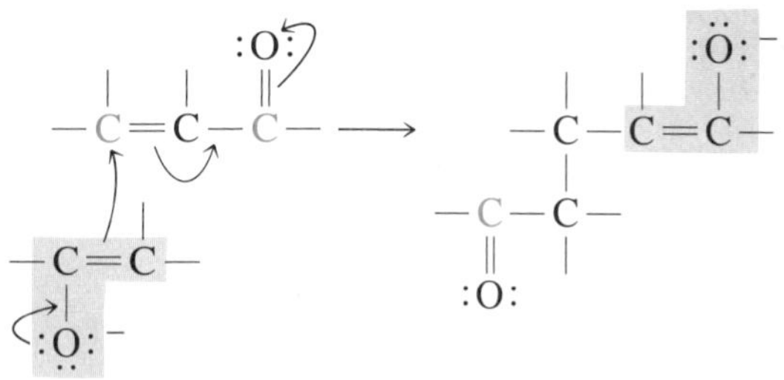

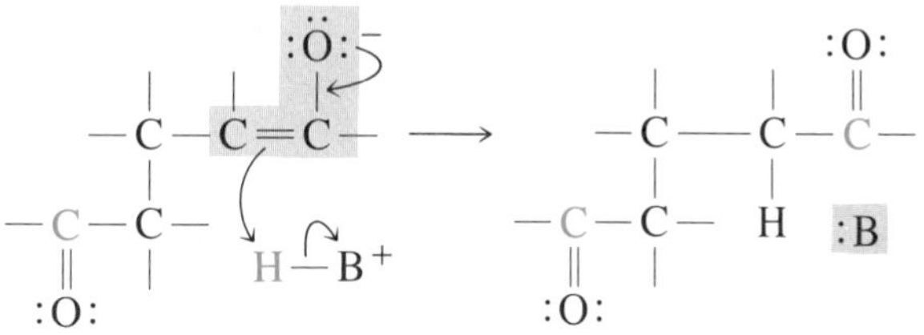

The product of a Michael addition always has two carbonyl groups that are separated by three tetrahedral carbon atoms. These products may undergo intramolecular aldol reactions to give cyclic compounds in an important synthetic process.

This chapter will also cover the reactions of ylides, carbanions stabilized by phosphorus or sulfur.

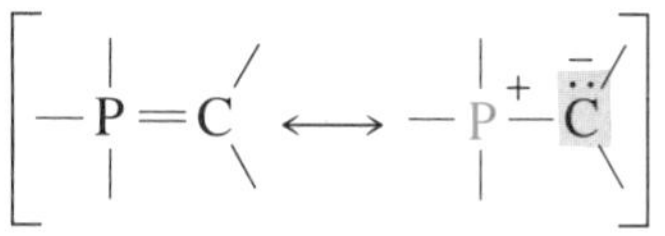

a phosphorus ylide

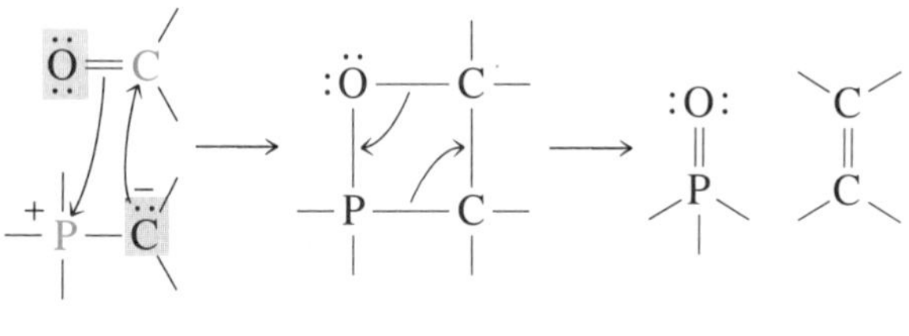

A phosphorus ylide is useful for replacing the carbonyl group of aldehydes and ketones with a double bond.

18.1 REACTIONS OF NUCLEOPHILES WITH α,β-UNSATURATED CARBONYL COMPOUNDS AS ELECTROPHILES

A. Reactions of Nitrogen, Oxygen, and Sulfur Nucleophiles with α,β-Unsaturated Carbonyl Compounds

α,β-Unsaturated carbonyl compounds, which contain a carbon-carbon double bond that is conjugated with a carbonyl group, react with nucleophiles at the double bond. The behavior of compounds with this type of structure contrasts with that of ordinary alkenes, which usually react with electrophiles but not with nucleophiles. The reactions of alkenes with electrophiles such as acids, halogens, and oxidizing

18 ENOLS AND ENOLATE
ANIONS AS NUCLEOPHILES II.
CONJUGATE ADDITION
REACTIONS; YLIDES

18.1 REACTIONS OF
NUCLEOPHILES WITH
α,β-UNSATURATED CARBONYL
COMPOUNDS AS ELECTROPHILES

agents were presented in Chapter 8, but there were no reactions that began with the attack of a nucleophile on a carbon-carbon double bond.

However, diethylamine does add to the double bond that is in conjugation with a carbonyl group in methyl vinyl ketone.

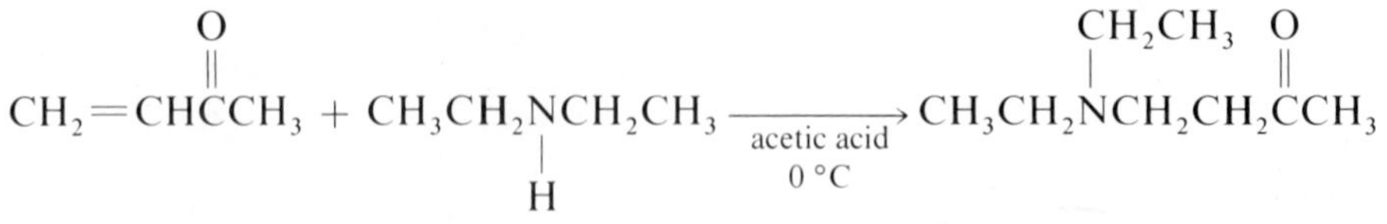

$$CH_2=CHCCH_3 + CH_3CH_2NCH_2CH_3 \xrightarrow[\substack{acetic\ acid \\ 0\ °C}]{} CH_3CH_2NCH_2CH_2CCH_3$$

methyl vinyl ketone   diethylamine   4-(N,N-diethylamino)-2-butanone
68%

The reaction requires mild acid catalysis, and the nucleophilic nitrogen atom binds to the carbon atom beta to the carbonyl group.

V I S U A L I Z I N G   T H E   R E A C T I O N

Addition of a nucleophile to an α,β-unsaturated carbonyl compound

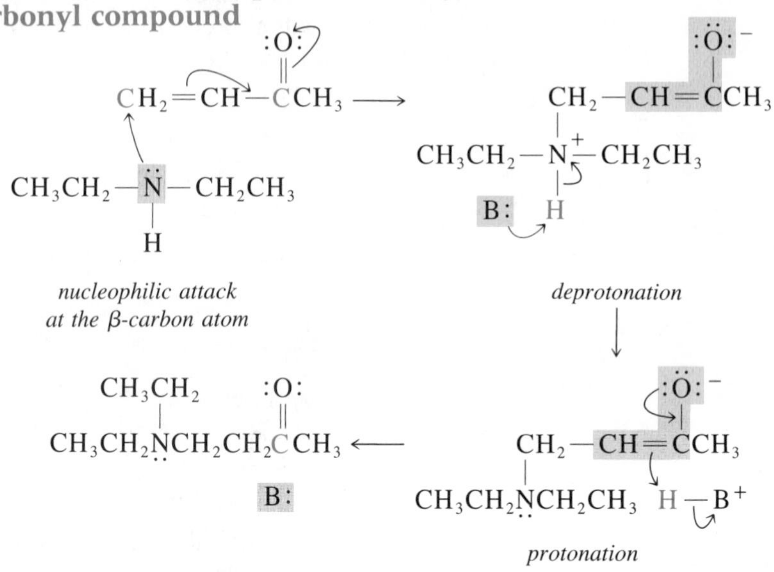

nucleophilic attack
at the β-carbon atom

deprotonation

protonation

The reaction starts with the bonding between the nucleophilic nitrogen atom of the amine and the β-carbon atom of methyl vinyl ketone to give an enolate. Deprotonation at the nitrogen atom and protonation at the α-carbon atom complete the reaction.

An examination of the polarity of an α,β-unsaturated carbonyl compound shows why this reaction takes place.

$$CH_2=CH-CCH_3 \longleftrightarrow CH_2=CH-\overset{+}{C}CH_3 \longleftrightarrow \overset{+}{C}H_2-CH=CCH_3$$

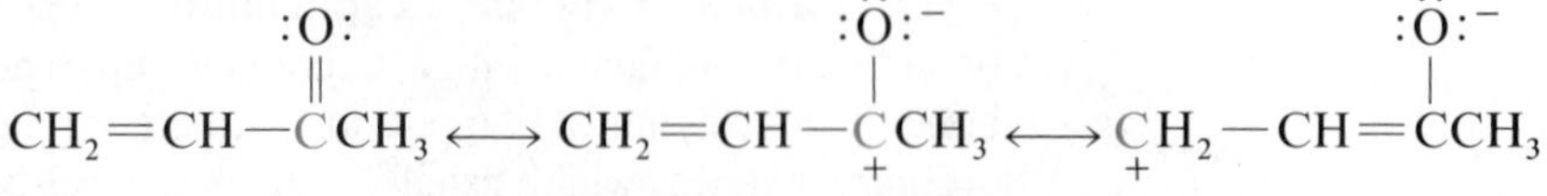

resonance contributors for methyl vinyl ketone

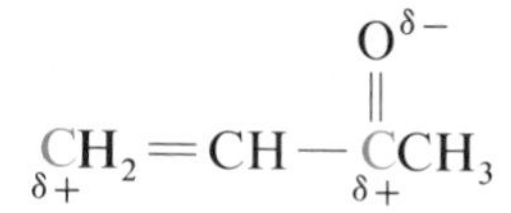

the polarization of an
α,β-unsaturated carbonyl compound

Resonance contributors can be written for methyl vinyl ketone that show that the presence of the carbonyl group in conjugation with the carbon-carbon double bond leads to a polarization such that the β-carbon atom has a partial positive charge. The electrophilicity of the carbon atom of the carbonyl group is transferred to the β-carbon atom by conjugation. Methyl vinyl ketone is the simplest possible α,β-unsaturated ketone, but the electronic properties described for it can be extended to all other α,β-unsaturated carbonyl compounds. The same electronic properties are also found in alkenes in which the double bond is next to other functional groups analogous in their polarization to the carbonyl group, such as the nitro and cyano groups. In all these compounds, the β-carbon atom will be vulnerable to attack by nucleophiles. Such alkenes are often called **electrophilic alkenes.**

Study Guide Concept Map 18.1

The carbon-carbon double bond in an α,β-unsaturated carbonyl compound is oxidized slowly by peroxyacids (p. 317), which are electrophiles. It undergoes rapid oxidation when a nucleophilic oxidizing agent, a basic solution of hydrogen peroxide, is used. Propenal is oxidized to an oxirane in this type of reaction.

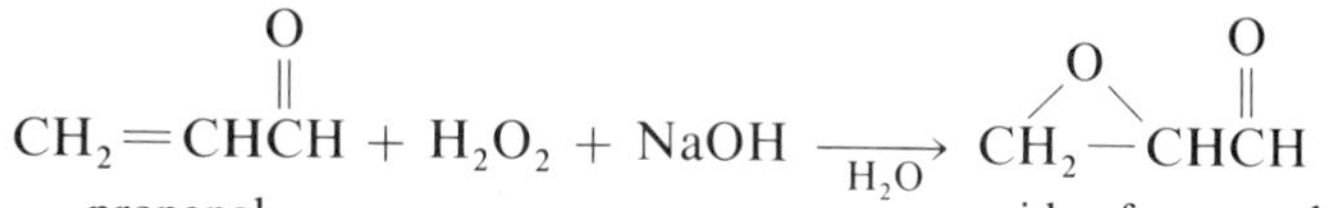

propenal

oxide of propenal
~80%
a substituted oxirane

This reaction is another example of the vulnerability to nucleophilic attack of a double bond conjugated with a carbonyl group. Hydrogen peroxide is converted into its anion in basic solution. The anion reacts at the β-carbon atom of the unsaturated system to give an enolate ion as an intermediate.

VISUALIZING THE REACTION

Epoxidation of an α,β-unsaturated carbonyl compound

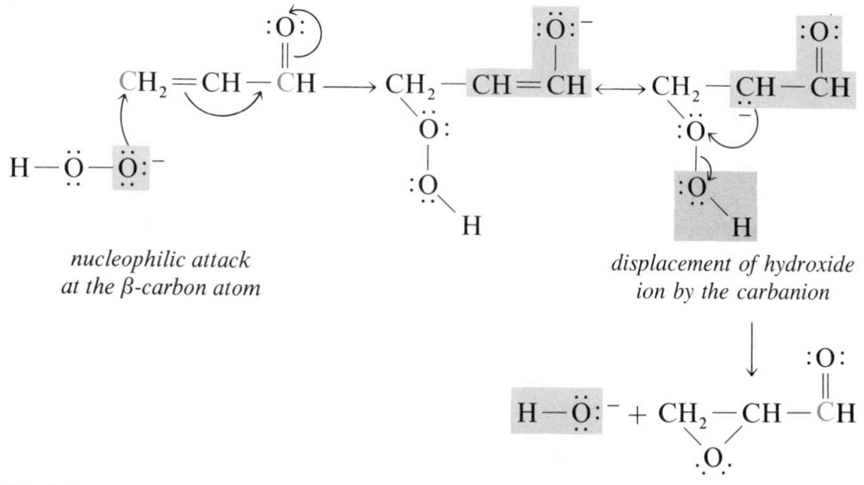
nucleophilic attack
at the β-carbon atom
displacement of hydroxide
ion by the carbanion

The carbanionic α-carbon atom of the enolate ion displaces hydroxide ion, breaking the weak oxygen-oxygen bond in the hydroperoxy group to give the oxirane.

Sulfur compounds also act as nucleophiles toward α,β-unsaturated carbonyl compounds. For example, hydrogen sulfide reacts with two molecules of methyl propenoate to give an organic sulfide.

$$\underset{\text{methyl propenoate}}{2\,CH_2{=}CH\overset{\overset{O}{\|}}{C}OCH_3} + H_2S \xrightarrow[\text{ethanol}\ \Delta]{CH_3\overset{\overset{O}{\|}}{C}O^-Na^+} \underset{\substack{\text{bis(2-carbomethoxyethyl)}\\ \text{sulfide}\\ \sim 75\%}}{CH_3O\overset{\overset{O}{\|}}{C}CH_2CH_2SCH_2CH_2\overset{\overset{O}{\|}}{C}OCH_3}$$

The first step of the reaction gives a thiol, which then adds to another molecule of methyl propenoate to give the final product.

Why does a nucleophile react with the β-carbon atom of an α,β-unsaturated carbonyl system and not with the electrophilic carbon atom of the carbonyl group? In some cases, of course, it does react at the carbonyl group. For example, in many instances sodium borohydride or lithium aluminum hydride will reduce the carbonyl group in an α,β-unsaturated carbonyl compound without touching the carbon-carbon double bond, especially if the reaction is carried out quickly at low temperatures (p. 486). Many α,β-unsaturated carbonyl compounds react with Grignard reagents at the carbonyl group or with reagents such as hydroxylamine to give carbonyl derivatives.

Whether a reaction takes place at the carbonyl group or at the β-carbon atom depends on the nature of the nucleophile and on the structure of the carbonyl compound. For example, the hydride ion, a small species with a high concentration of charge, reacts primarily at the carbonyl group. A sulfur atom, large and polarizable, is more likely to react at the β-carbon atom. The next section explores how the change in the nature of an organometallic reagent changes the site at which it reacts with an α,β-unsaturated carbonyl compound.

PROBLEM 18.1

Show the details of the mechanism for the reaction of hydrogen sulfide with methyl propenoate.

PROBLEM 18.2

Give structural formulas for the reagents and products represented by letters in the following equations.

(a) 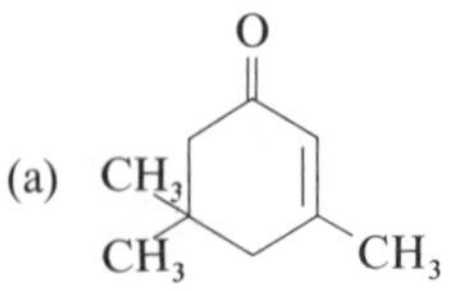 $+ H_2O_2 \xrightarrow[\text{methanol}]{NaOH} A$

(b) $B + C \longrightarrow CH_3\overset{\overset{CH_3}{|}}{N}{-}\overset{\overset{CH_3}{|}}{\underset{\underset{CH_3}{|}}{C}}{-}CH_2\overset{\overset{O}{\|}}{C}CH_3$

(c) $CH_2{=}CHC{\equiv}N$ (2 molar equiv) $+ NH_3 \xrightarrow[H_2O]{} D$

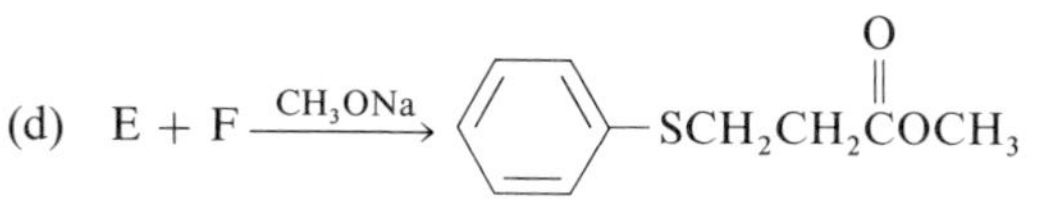

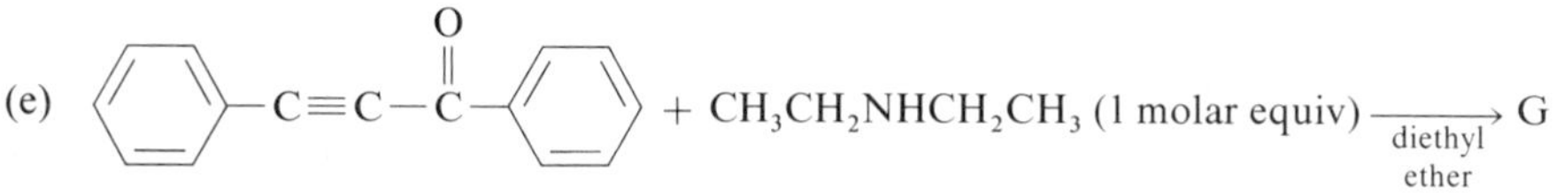

B. Organocuprate Reagents

Grignard reagents and organolithium compounds are reactive toward carbonyl groups (p. 491) but do not react readily with organic halides to give new carbon-carbon bonds. Copper, however, promotes reactions at electrophilic carbon atoms other than carbonyl carbons. Organocuprate reagents that are particularly useful are prepared by treating an organolithium compound with a copper(I) halide, usually copper(I) iodide.

$$\underset{\text{methyllithium}}{2\,CH_3Li} + \underset{\text{copper(I) iodide}}{CuI} \xrightarrow[0\,°C]{\text{diethyl ether}} \underset{\text{lithium dimethylcuprate}}{(CH_3)_2CuLi} + \underset{\text{lithium iodide}}{LiI}$$

Just as with the Grignard reagent, the exact structure of the organocuprate reagent is not known; the formula given above represents the stoichiometry observed for the reaction. Two organic groups appear to be associated with the copper atom in a negatively charged species that is a source of nucleophilic carbon atoms. The name lithium dimethylcuprate indicates that copper is associated with the anion in the compound. Primary alkyl halides give reasonably stable organocuprates. Vinyl and aryl organocuprates can also be prepared.

Organocuprates react with organic halides to give compounds with longer carbon chains. For example, lithium dibutylcuprate reacts with 1-bromopentane to give nonane.

$$\underset{\text{1-bromopentane}}{CH_3CH_2CH_2CH_2CH_2Br} \xrightarrow[\text{tetrahydrofuran, 25 °C, 1 h}]{(CH_3CH_2CH_2CH_2)_2CuLi} \underset{\substack{\text{nonane} \\ 98\%}}{CH_3CH_2CH_2CH_2CH_2CH_2CH_2CH_2CH_3}$$

Organocuprates may be used to synthesize alkenes or polyenes. The unsaturation in the product may be derived from the organocuprate, from the halide with which it reacts, or from both. 3-Bromo-1-methylcyclohexene, for example, reacts with lithium diisopropenylcuprate to give 3-isopropenyl-1-methylcyclohexene.

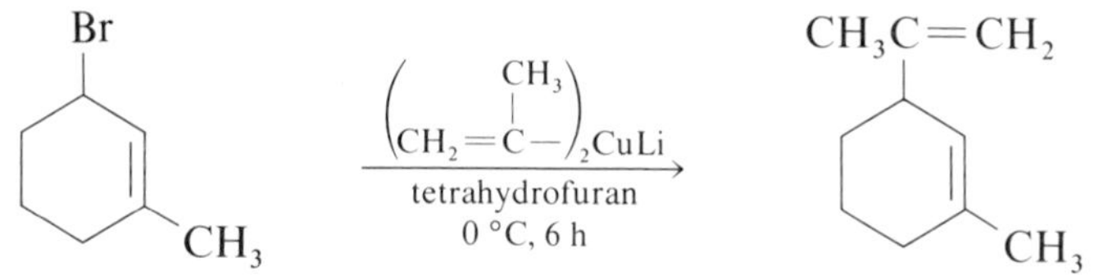

3-bromo-1-methylcyclohexene → 3-isopropenyl-1-methylcyclohexene 75%

The differing reactivities of lithium organocuprates and organolithium reagents are illustrated by the reaction of 4-bromocyclohexanone with lithium diisopropenylcuprate.

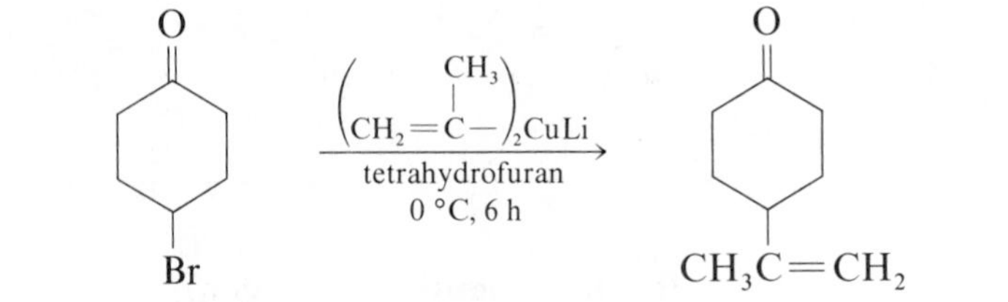
O
CH3
CH2=C—)2CuLi
tetrahydrofuran
0 °C, 6 h
Br
CH3C=CH2

4-bromocyclohexanone

4-isopropenylcyclohexanone
65%

An organolithium reagent would react with the carbonyl group (p. 492) rather than with the alkyl halide function in 4-bromocyclohexanone.

PROBLEM 18.3

Complete the following equations.

(a) $CH_3CH_2CH_2CH_2Li + CuI \xrightarrow[0\ °C]{\text{diethyl ether}}$

(b) $CH_3CH_2CH_2CH_2CH_2I \xrightarrow[\text{diethyl ether},\ 25\ °C]{(CH_3)_2CuLi}$

(c) 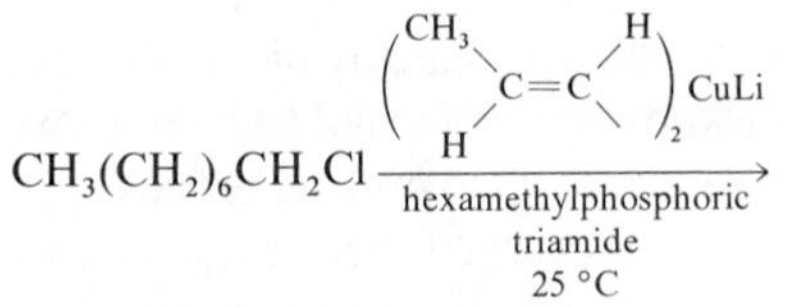
CH3(CH2)6CH2Cl
CH3
H
C=C
H
CuLi
2
hexamethylphosphoric
triamide
25 °C

(d) 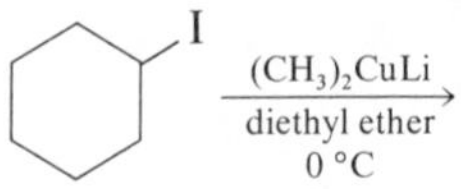
I
(CH3)2CuLi
diethyl ether
0 °C

(e) 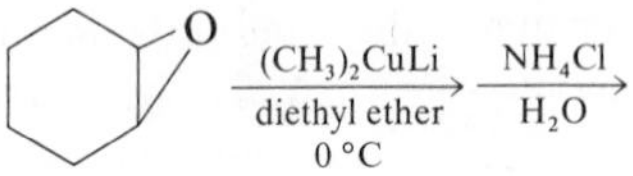
O
(CH3)2CuLi
diethyl ether
0 °C
NH4Cl
H2O

PROBLEM 18.4

Juvenile hormones are substances that block the maturation of the pupae of insects. The juvenile hormone of the giant silkworm moth has the following structure.

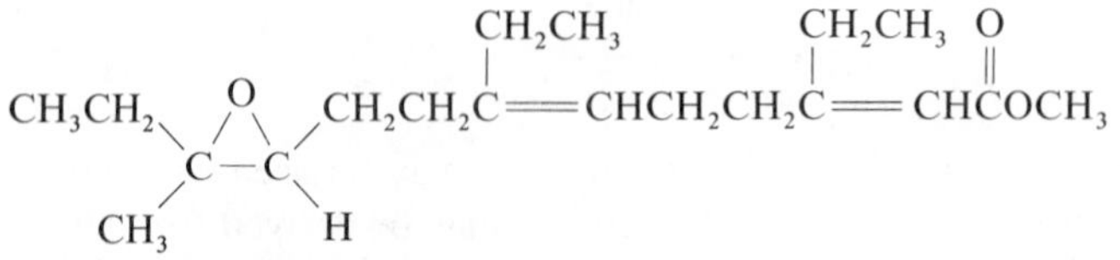
CH2CH3
CH2CH3
O
CH3CH2
O
CH2CH2C=CHCH2CH2C=CHCOCH3
C—C
CH3
H

An important step in the synthesis of the compound involves the following transformation. Suggest a reagent for the reaction.

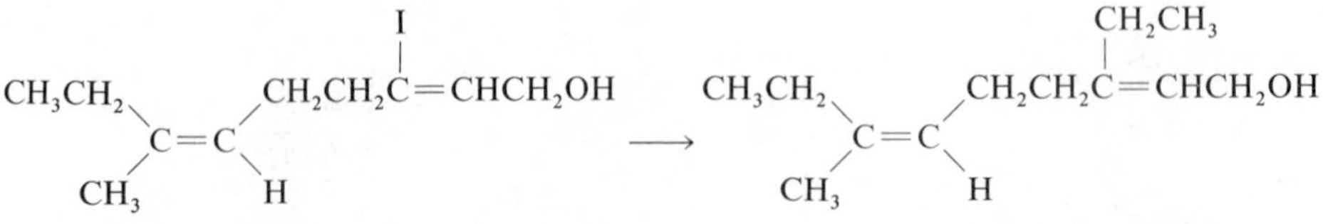
I
CH3CH2
CH2CH2C=CHCH2OH
C=C
CH3
H
CH2CH3
CH3CH2
CH2CH2C=CHCH2OH
C=C
CH3
H

C. Reaction of Organocuprate Reagents with α,β-Unsaturated Carbonyl Compounds

An α,β-unsaturated carbonyl compound can react with a nucleophile at the carbonyl group to give a product in which the carbonyl function has been modified. This reaction is called a 1,2-addition to the conjugated system, by analogy with the reactions of conjugated dienes (p. 686). Alternatively, the nucleophile may attack the β-carbon atom, giving a product in which the carbonyl group is regenerated from an enolate intermediate. The case in which the nucleophile becomes bonded to the β-carbon atom is known as 1,4-addition or **conjugate addition.**

Grignard reagents react with α,β-unsaturated ketones mainly by 1,2-addition.

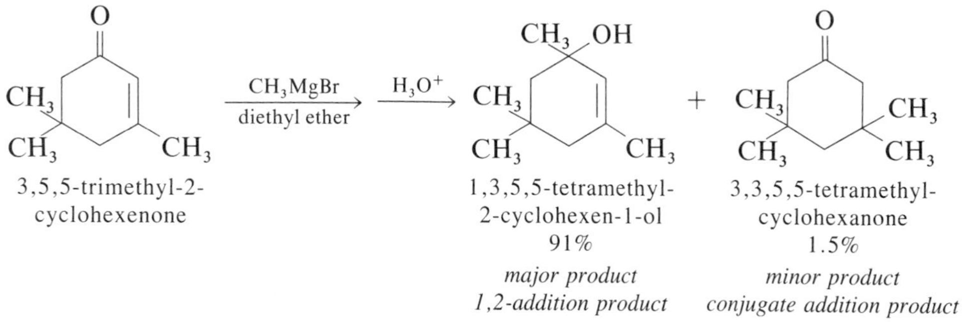
O
CH₃
CH₃
CH₃
CH₃MgBr
diethyl ether
H₃O⁺
CH₃ OH
+
3,5,5-trimethyl-2-cyclohexenone
1,3,5,5-tetramethyl-2-cyclohexen-1-ol
91%
major product
1,2-addition product
3,3,5,5-tetramethylcyclohexanone
1.5%
minor product
conjugate addition product

Organocuprates react with α,β-unsaturated carbonyl compounds at the β-carbon atom to give high yields of conjugate addition products. For example, 2-cyclohexenone is converted almost quantitatively to 3-methylcyclohexanone by lithium dimethylcuprate; the reaction is run at low temperatures to improve regioselectivity.

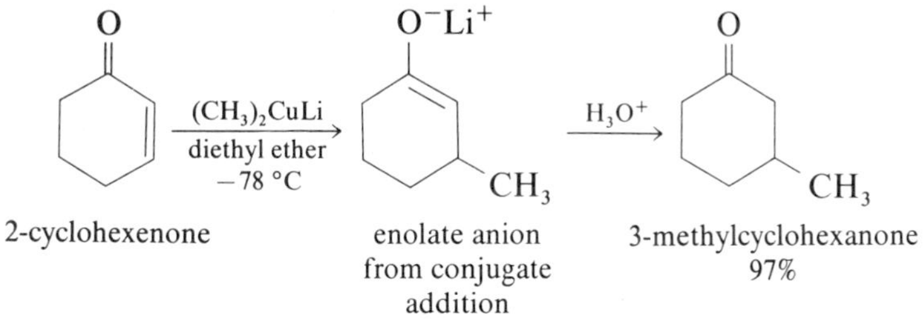
O
(CH₃)₂CuLi
diethyl ether
−78 °C
O⁻Li⁺
CH₃
H₃O⁺
2-cyclohexenone
enolate anion from conjugate addition
3-methylcyclohexanone
97%

The exact mechanism of the reaction is not known. An excess of the organocuprate is required, and chemists have proposed mechanisms that involve formation of a complex between the organometallic reagent and the carbonyl compound. An enolate anion is formed as an intermediate and can be used in further reactions that form carbon-carbon bonds, for example:

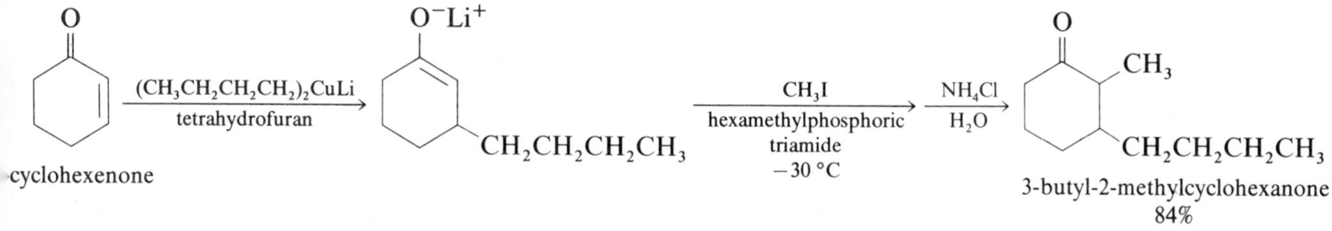
O
(CH₃CH₂CH₂CH₂)₂CuLi
tetrahydrofuran
O⁻Li⁺
CH₂CH₂CH₂CH₃
CH₃I
hexamethylphosphoric triamide
−30 °C
NH₄Cl
H₂O
CH₃
cyclohexenone
3-butyl-2-methylcyclohexanone
84%

The product that is obtained can be rationalized by postulating nucleophilic attack at the β-carbon atom giving an enolate anion, which is then alkylated by methyl iodide.

Conjugate addition of organocuprate reagents is widely used for creating new carbon skeletons in organic syntheses.

PROBLEM 18.5

Give structural formulas for the reagents, intermediates, and products symbolized by letters in the following equations.

(a) 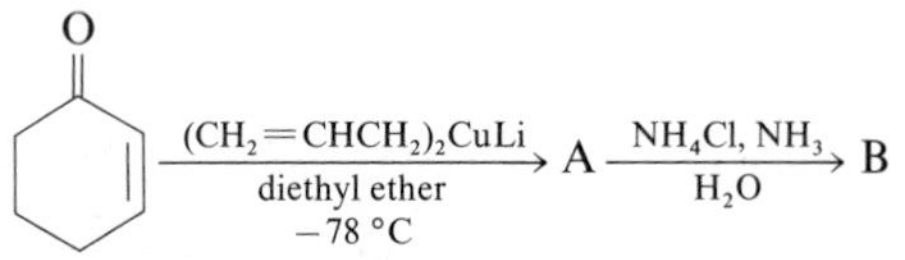
O
$(CH_2{=}CHCH_2)_2CuLi$
diethyl ether
−78 °C
A
NH_4Cl, NH_3
H_2O
B

(b) 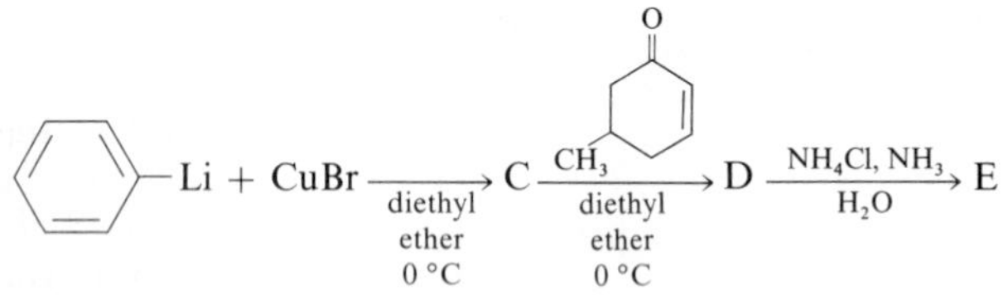
Li + CuBr
diethyl
ether
0 °C
C
O
CH_3
diethyl
ether
0 °C
D
NH_4Cl, NH_3
H_2O
E

(c) 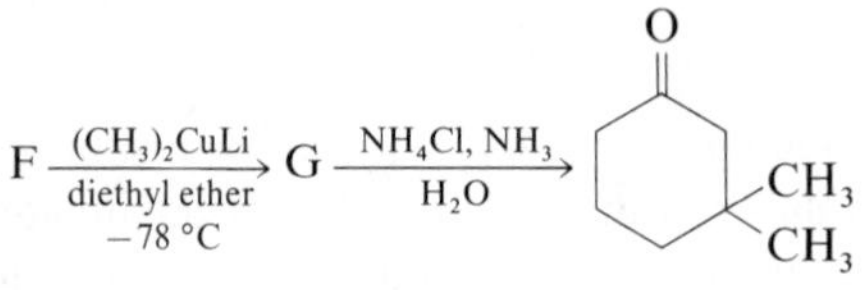
F
$(CH_3)_2CuLi$
diethyl ether
−78 °C
G
NH_4Cl, NH_3
H_2O
O
CH_3
CH_3

(d) 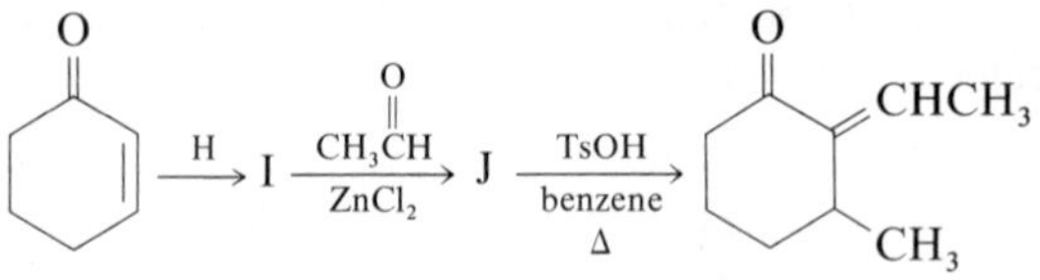
O
H
I
O
CH_3CH
$ZnCl_2$
J
TsOH
benzene
Δ
O
$CHCH_3$
CH_3

PROBLEM 18.6

Pentalenene is a hydrocarbon related to pentalenolactone, an antibiotic that inhibits the synthesis of nucleic acids in bacterial cells. Part of a synthesis of pentalenene involves the following reaction. Write structural formulas for the compounds indicated by letters.

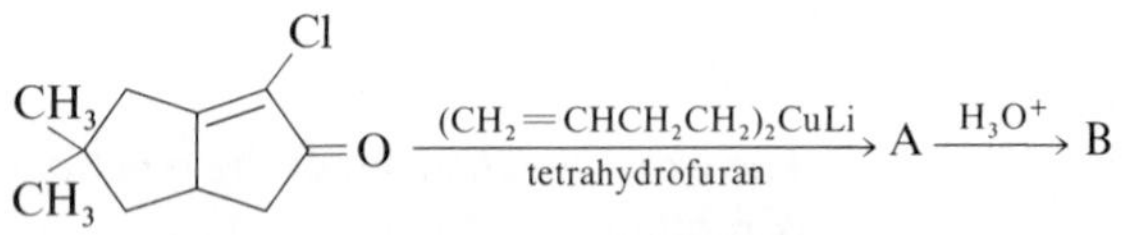
Cl
CH_3
CH_3
O
$(CH_2{=}CHCH_2CH_2)_2CuLi$
tetrahydrofuran
A
H_3O^+
B

18.2 REACTIONS OF ENOLATE ANIONS WITH α,β-UNSATURATED CARBONYL COMPOUNDS

A. The Michael Reaction

Enolate anions react with α,β-unsaturated carbonyl compounds at the β-carbon atom in conjugate additions. For example, the anion generated by treating diethyl malonate with sodium metal adds to methyl propenoate.

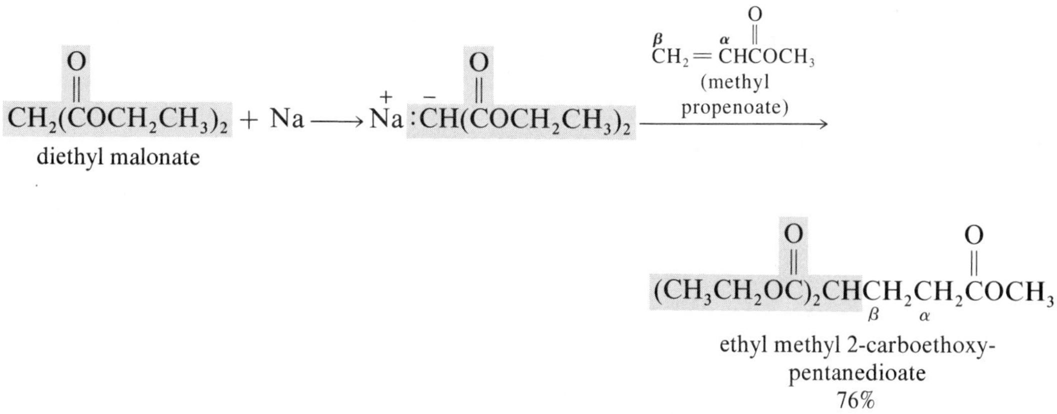

CH₂(COCH₂CH₃)₂ + Na ⟶ Na :CH(COCH₂CH₃)₂
diethyl malonate
(methyl
propenoate)
(CH₃CH₂OC)₂CHCH₂CH₂COCH₃
ethyl methyl 2-carboethoxy-
pentanedioate
76%

The product of the addition of an enolate anion to an α,β-unsaturated carbonyl compound is a **1,5-dicarbonyl compound.**

In a similar way, both the hydrogen atoms from the active methylene group in ethyl acetoacetate are replaced by alkyl groups through the addition of anions to methyl vinyl ketone.

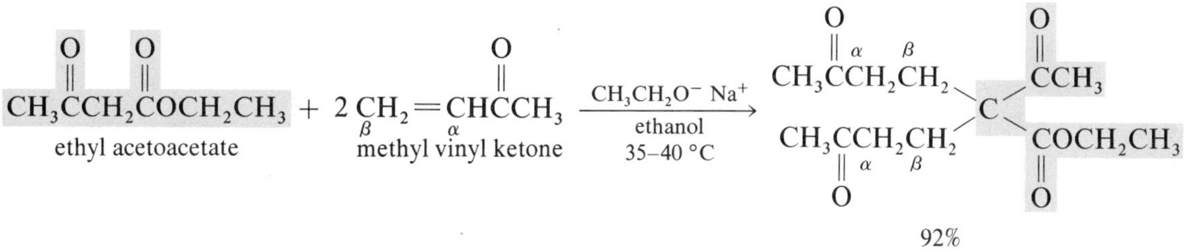

ethyl acetoacetate
methyl vinyl ketone
ethanol
35–40 °C
92%

The carbanion does not have to be derived from a carbonyl compound. Nitro groups also stabilize carbanions (p. 101). For example, 2-nitropropane will add to methyl vinyl ketone in the presence of sodium methoxide.

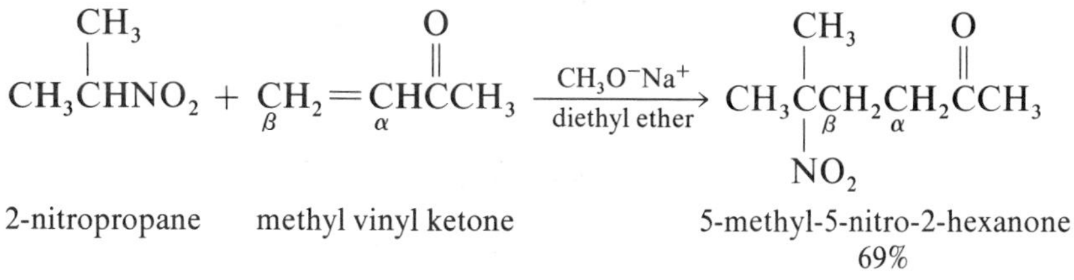

diethyl ether
2-nitropropane
methyl vinyl ketone
5-methyl-5-nitro-2-hexanone
69%

A nitro group can also polarize a double bond and make it susceptible to attack by an anion. The enolate anion from dimethyl malonate adds to 1-nitro-2-phenylethene with a high yield.

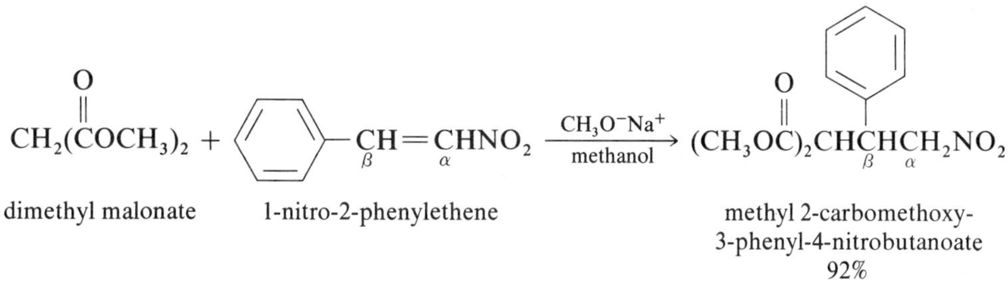

methanol
dimethyl malonate
1-nitro-2-phenylethene
methyl 2-carbomethoxy-
3-phenyl-4-nitrobutanoate
92%

In all these examples, a stabilized carbanion adds to the β-carbon atom of a double bond that is made electrophilic by an electron-withdrawing substituent.

The reactions described above are known as **Michael reactions,** after Arthur Michael, of Tufts University. The compound containing the nucleophilic carbon atom is called the **Michael donor,** and the compound containing the polarized double bond is called the **Michael acceptor.** The reactions of Michael acceptors with oxygen, nitrogen, and sulfur nucleophiles used as examples in Section 18.1 are not strictly Michael reactions, but they are frequently called Michael-type reactions. Such additions of nucleophilic atoms other than carbon to Michael acceptors are biologically important. Proteins and components of RNA and DNA contain nucleophilic amino, thiol, or hydroxyl groups. Michael acceptors are toxic to living organisms because of their ability to react with these groups. For example, one of the defensive secretions of the East African termite is dodeca-1-en-3-one. The termite uses it to kill other insects. The α,β-unsaturated ketone acts as a solvent that penetrates the cuticle of the victim. Chemists believe that the substance is toxic because it reacts with biological nucleophiles. It might, for example, deactivate an enzyme by reacting with an amino or thiol group in a critical part of the molecule.

Michael acceptors are one example of a class of toxic compounds known as **alkylating agents** because they introduce alkyl groups onto nucleophilic sites in biologically important molecules. Other compounds such as methyl iodide and dimethyl sulfate have considerable toxicity for the same reason.

PROBLEM 18.7

Write an equation showing how a thiol, RSH, might react with dodeca-1-en-3-one.

PROBLEM 18.8

Show the structure and stabilization of the carbanion derived from 2-nitropropane.

PROBLEM 18.9

Write structural formulas that rationalize the polarity of the double bond in 1-nitro-2-phenylethene.

PROBLEM 18.10

Complete the following equations.

(a) $$CH_2(C{\equiv}N)_2 + 2\,CH_2{=}CH\overset{O}{\overset{\|}{C}}CH_3 \xrightarrow[\text{benzene}]{\text{Na}}$$

(b) $$CH_3\underset{\underset{CH_3}{|}}{CH}\overset{O}{\overset{\|}{C}}CH_2\overset{O}{\overset{\|}{C}}OCH_2CH_3 + CH_3\overset{O}{\overset{\|}{C}}{-}\overset{CH_3}{\overset{|}{C}}{=}CH_2 \xrightarrow[\text{ethanol}]{\text{KOH}}$$

(c) 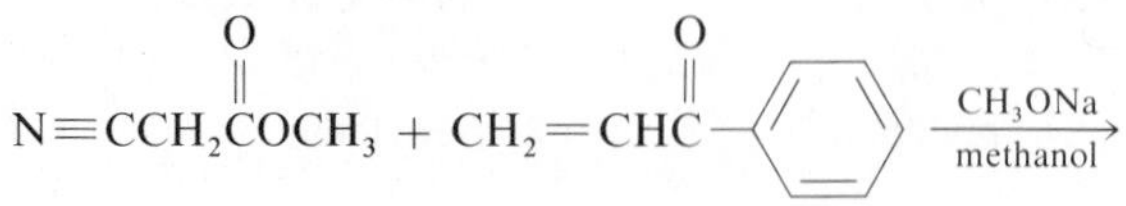

(d) 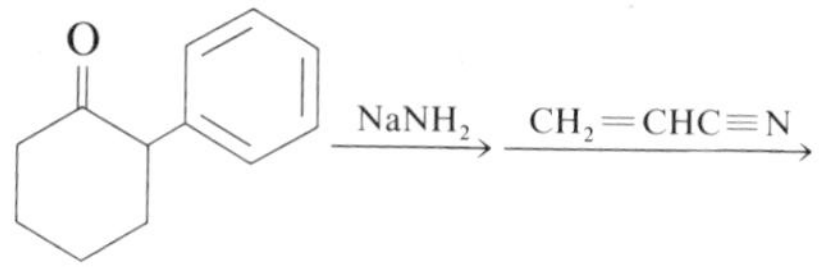

(Hint: Which enolate anion will be more stable?)

(e) 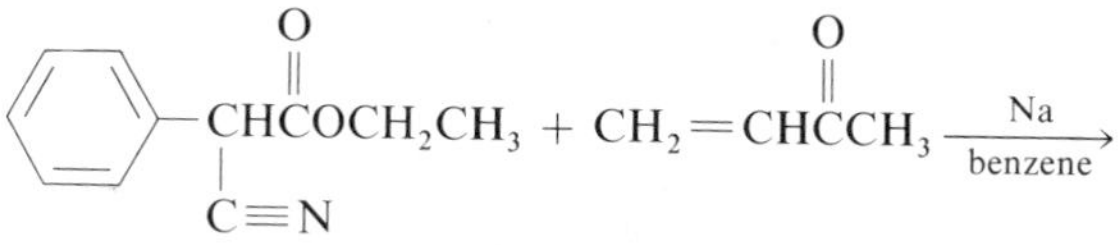

(f) $CH_3NO_2 + CH_3\overset{CH_3}{\overset{|}{C}}{=}CH\overset{O}{\overset{\|}{C}}CH_3 \xrightarrow{CH_3CH_2NHCH_2CH_3}$

(g) 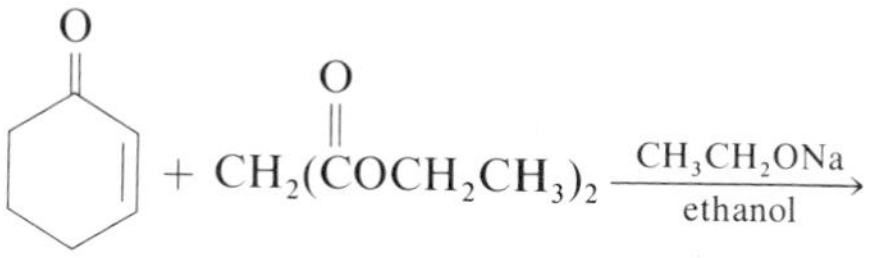

PROBLEM 18.11

The following sequence of reactions was used in a synthesis of pentalenolactone (Problem 18.6, p. 740). Supply structural formulas for reagents, intermediates, and products designated by letters.

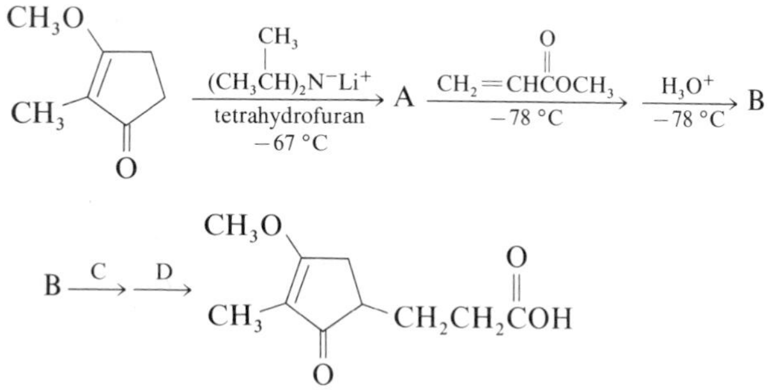

PROBLEM 18.12

The ketoacid that is the product of the reaction sequence in Problem 18.11 is converted into another ketoacid when it is treated successively with methyllithium and dilute hydrochloric acid. Write equations that show how this conversion occurs.

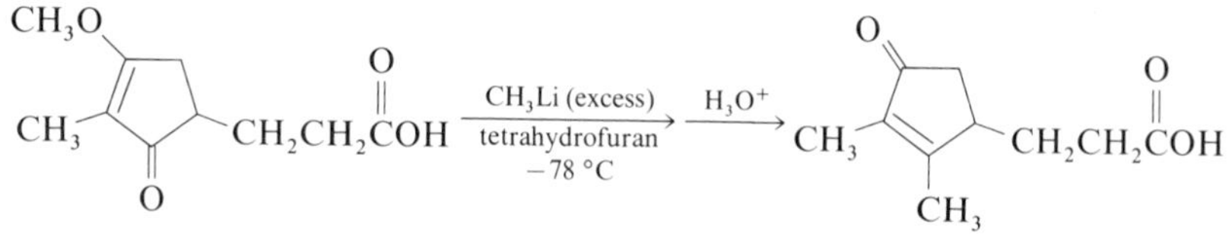

How do you explain the regioselectivity of the reaction with methyllithium?

PROBLEM 18.13

The synthesis of pentalenolactone is continued by the conversion of the product ketoacid from Problem 18.12 to a bicyclic diketone. Assign structural formulas to the reagents used in the two steps of this conversion, Compounds A and B in the following equation.

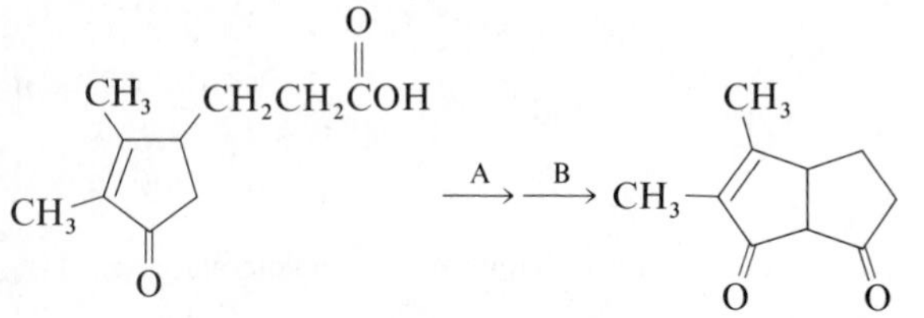

B. The Robinson Annulation: A Cyclization Reaction of Products of Michael Reactions

The product of a Michael reaction is often a carbonyl compound that will enolize further under the conditions of the reaction or in the presence of added base. If the structure permits, the new enolate ion reacts with other functional groups within the same molecule to form rings. For example, 2-methyl-1,3-cyclohexadione adds to methyl vinyl ketone to give a methyl ketone that is in equilibrium with several enolate ions.

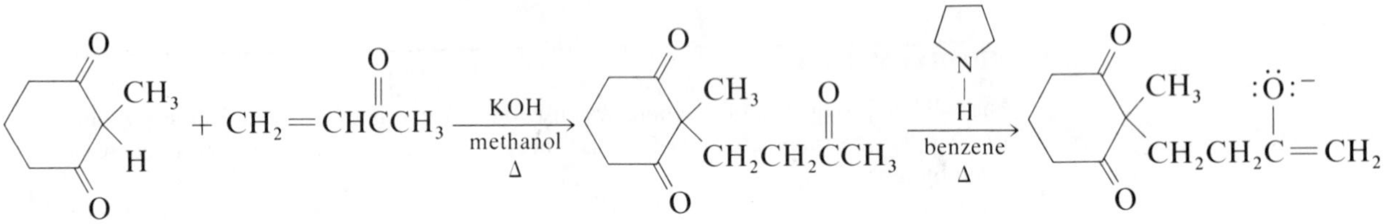

2-methyl-1,3-cyclohexadione

methyl vinyl ketone

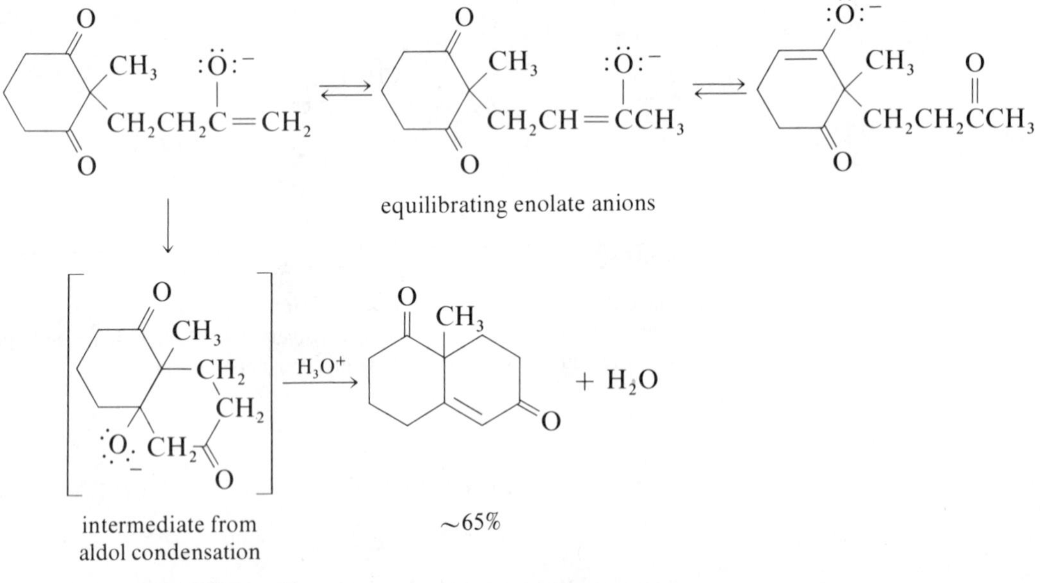

equilibrating enolate anions

intermediate from aldol condensation

~65%

In one of the enolate anions, the carbanionic center is in a position to form a six-membered ring by an aldol condensation with one of the two equivalent carbonyl groups of the cyclohexadione. Protonation and loss of a molecule of water gives a cyclic α,β-unsaturated ketone as the final product of the reaction.

This method of attaching new rings to functionalized cyclopentane or cyclohexane rings is known as the **Robinson annulation,** after Sir Robert Robinson who developed it and used it extensively in his research. In recognition of his many contributions to organic chemistry, Robinson was awarded a Nobel Prize in 1947.

In the Robinson annulation, a single six-membered ring is converted into a fused pair of six-membered rings in essentially one step. The bicyclic product contains an α,β-unsaturated ketone function and may undergo further transformations. Also, the starting material may contain other functional groups that are retained in the product. Such is the case with 2-methyl-1,3-cyclohexadione; thus, the product of its Robinson annulation is reactive at several points. For example, reaction of sodium acetylide (p. 493) with the ketone group in that product gives an alcohol with an alkyne function that is converted in acid to an α,β-unsaturated methyl ketone.

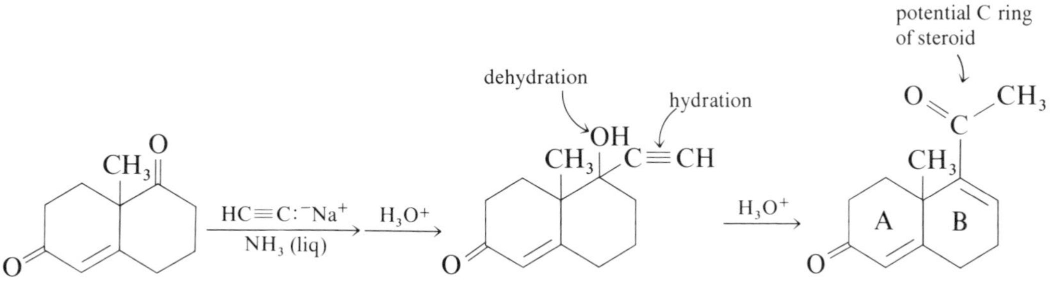

potential C ring of steroid
dehydration
hydration
$HC\equiv C:^-Na^+$
NH_3 (liq)
H_3O^+
A
B

The final stage of the transformation may be visualized as an acid-catalyzed hydration of the alkyne to a methyl ketone (p. 336) and dehydration of the tertiary alcohol. The resulting α,β-unsaturated methyl ketone can be used as a Michael acceptor, so it is possible to build still another ring onto the system.

Even the small portion of the synthesis outlined above shows how structural elements that are important in the synthesis of steroids can be introduced. For example, the final product, as it is written, shows the A and B rings of a steroid system with the beginning of the construction of the C ring. The methyl group in the angle between rings A and B and the α,β-unsaturated ketone function of ring A are features that are found in naturally occurring steroids (p. 477).

In many Robinson annulations, a compound that will give rise to methyl vinyl ketone in basic solution is used instead of the α,β-unsaturated ketone itself, which is unstable. The Michael acceptor is generated easily from compounds that have good leaving groups on the β-carbon atom. In one example, shown below, methyl vinyl ketone is generated in the reaction mixture from 4-(*N,N*-diethylamino)-2-butanone. The Michael donor is the enolate ion from ethyl 2-oxocyclohexanecarboxylate.

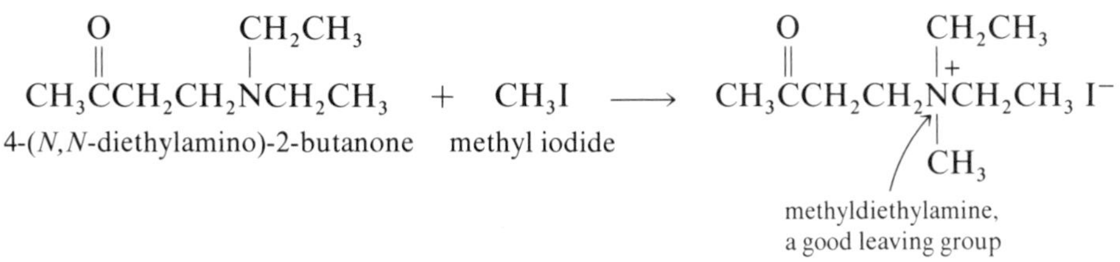

4-(N,N-diethylamino)-2-butanone
methyl iodide
methyldiethylamine, a good leaving group

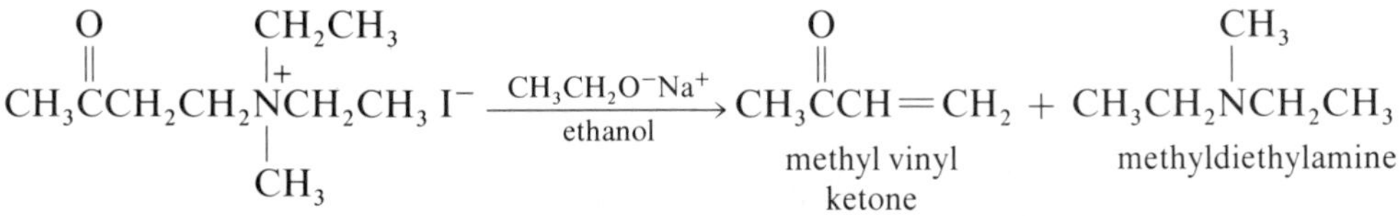

$CH_3CH_2O^-Na^+$
ethanol
methyl vinyl ketone
methyldiethylamine

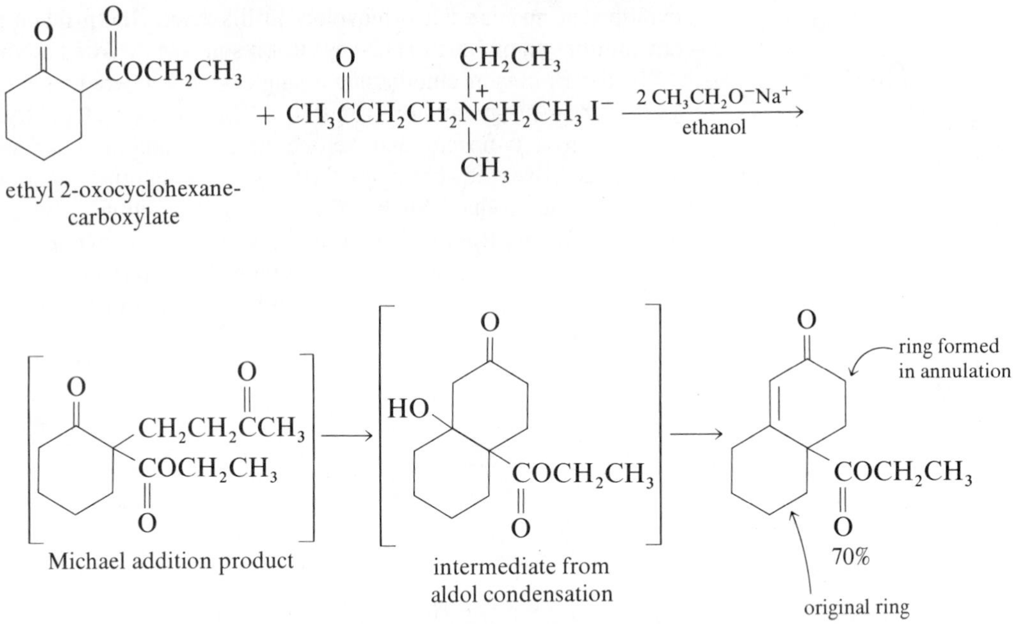
2 $CH_3CH_2O^-Na^+$
ethanol
ethyl 2-oxocyclohexane-carboxylate
Michael addition product
intermediate from aldol condensation
ring formed in annulation
70%
original ring

4-(*N*,*N*-Diethylamino)-2-butanone reacts with methyl iodide to give a compound with a positively charged nitrogen atom bonded to four organic groups. Such compounds, called quaternary ammonium compounds, contain a good leaving group, an amine, and thus readily undergo elimination reactions. These reactions are discussed in detail in Section 22.7A. The anion from ethyl 2-oxocyclohexanecarboxylate adds to the α,β-unsaturated ketone that is generated in the reaction mixture, and the intermediate ionizes under the reaction conditions, so the final product is the result of the aldol cyclization reaction.

Study Guide Concept Map 18.2

The Robinson annulation builds a six-membered ring containing an α,β-unsaturated carbonyl function by combining a Michael reaction of methyl vinyl ketone (or an equivalent compound) with an intramolecular condensation reaction of the product of the first step. The products of Robinson annulations are fused-ring systems containing carbonyl functions that allow further synthetic transformations.

C. Problem-Solving Skills

Problem

How would you carry out the following transformation?

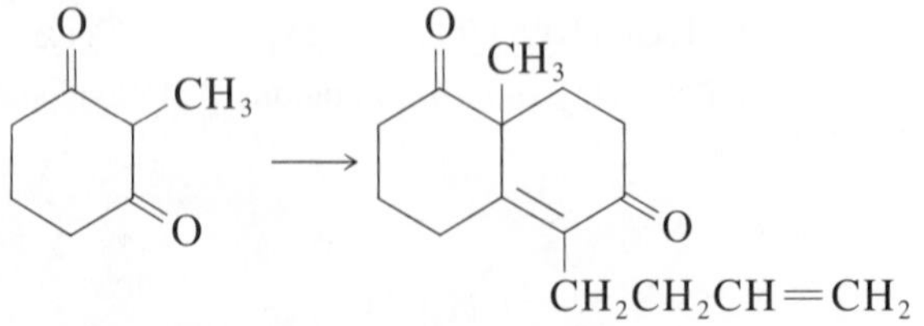

Solution

1. What functional groups are present in the starting material and the product?

The starting material is a cyclic 1,3-diketone. The product is a bicyclic compound containing an α,β-unsaturated carbonyl, with a side chain that has a carbon-carbon double bond.

2. How do the carbon skeletons of the two compounds compare? How many carbon atoms does each contain? Are there any rings? What are the positions of branches and functional groups on the carbon skeletons?

   A second six-membered ring with a four-carbon chain as a substituent has been built onto the starting ring.

3. How do the functional groups change in going from starting material to product? Does the starting material have a good leaving group?

   The double bond of the α,β-unsaturated system in the product is located where one of the carbonyl groups of the starting diketone is. No leaving group is present in the starting material.

4. Is it possible to dissect the structures of the starting material and product to see which bonds must be broken and which formed?

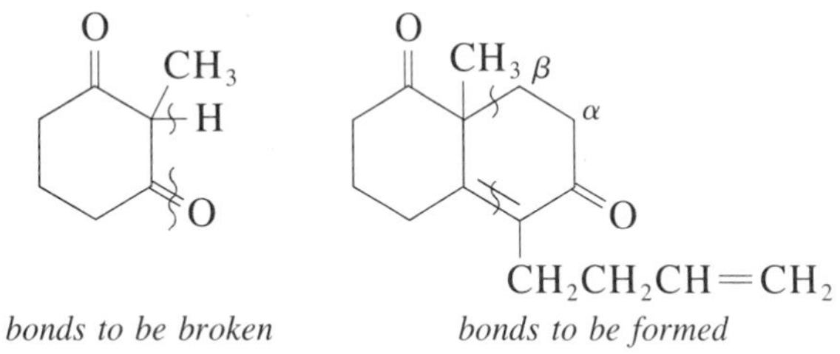
O
CH3
H
O
bonds to be broken
O
CH3 β
α
O
CH2CH2CH=CH2
bonds to be formed

5. Do we recognize any part of the product molecule as coming from a good nucleophile or an electrophilic addition?

   The starting material can be deprotonated to give an enolate anion, a good nucleophile. This nucleophile must then react with an electrophilic carbon beta to the carbonyl group in the second ring.

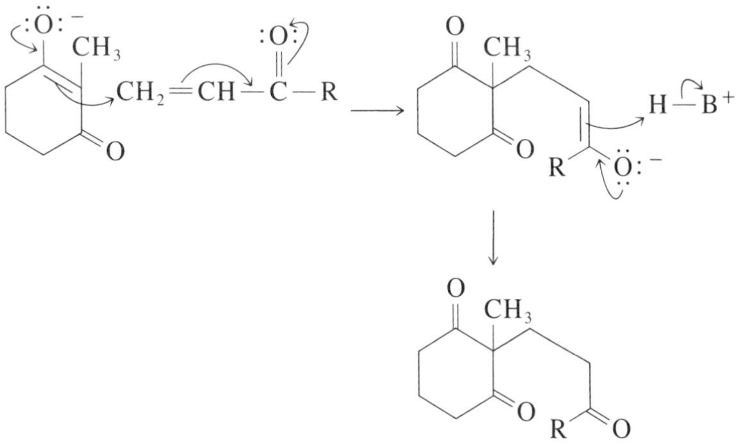
:Ö:⁻
CH3
O
:O:
CH2=CH—C—R
O
CH3
O
R
:Ö:⁻
H—B+
O
CH3
O
R
O

6. What type of compound would be a good precursor to the product?

   The final product is an α,β-unsaturated carbonyl compound. Such compounds come from aldol condensation reactions. Dissection of the aldol product gives the precursor, which is also the product of step 5.

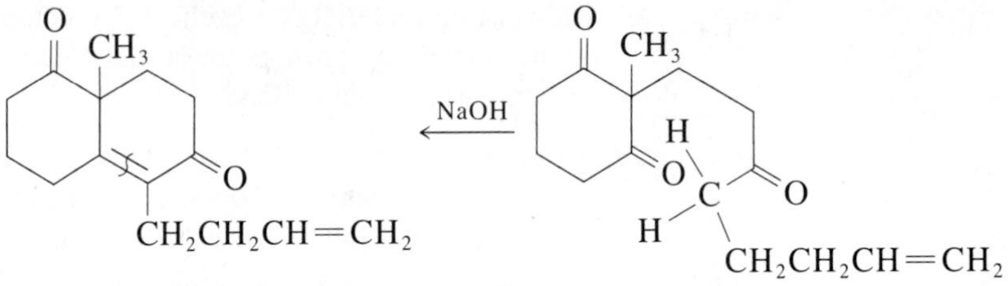

7. After the last step, do we see how to get from starting material to product? If not, we need to analyze the structure obtained in step 6 by applying questions 5 and 6 to it.

The complete synthesis is as follows:

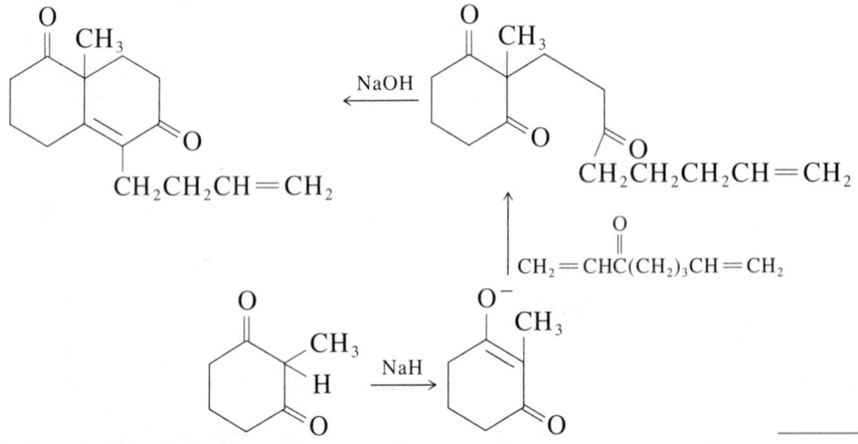

PROBLEM 18.14

Complete the following equations, showing first the product of the Michael reaction and then the Robinson annulation product.

(a) 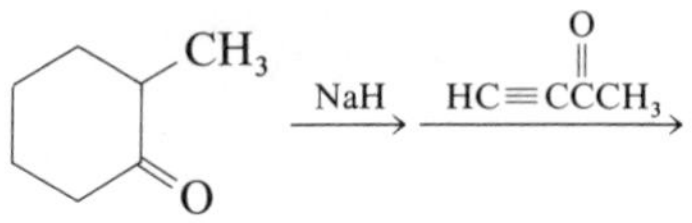

(b) 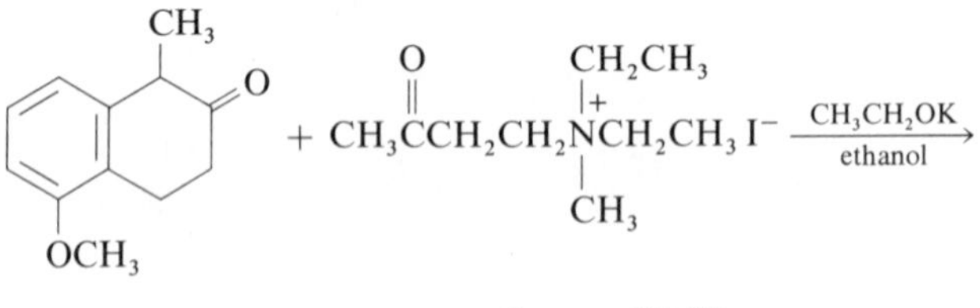

(c) 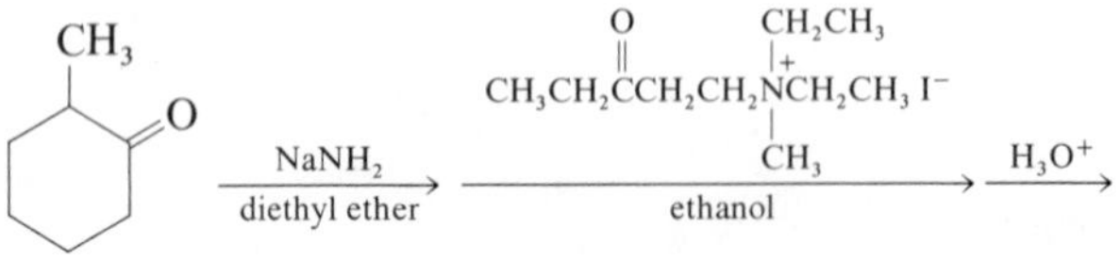

(d) 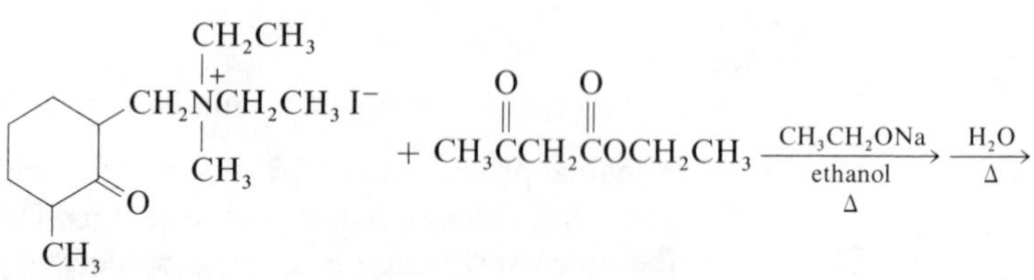

PROBLEM 18.15

How would you carry out the following transformation?

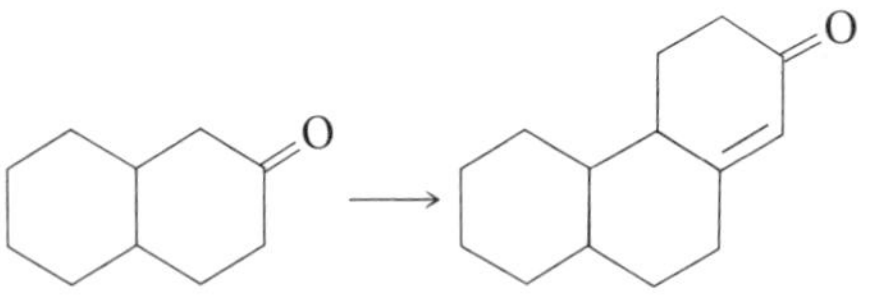

O
O

18.3 CARBANIONS STABILIZED BY PHOSPHORUS

A. Phosphonium Ylides

Chemists continue to search for new ways of stabilizing carbanions so that they can be generated under conditions that allow them to be used in different types of syntheses. Researchers have found that elements from the third row of the periodic table, such as sulfur and phosphorus, stabilize negatively charged carbon atoms that are adjacent to them in a variety of structures.

A positively charged phosphorus atom is formed when a phosphine, the phosphorus analog of an amine, reacts with an alkyl halide. The phosphonium compound can be deprotonated at the carbon atom adjacent to the phosphorus atom to give a carbanion stabilized by the positively charged phosphorus atom. This sequence of reactions is shown for triphenylphosphine.

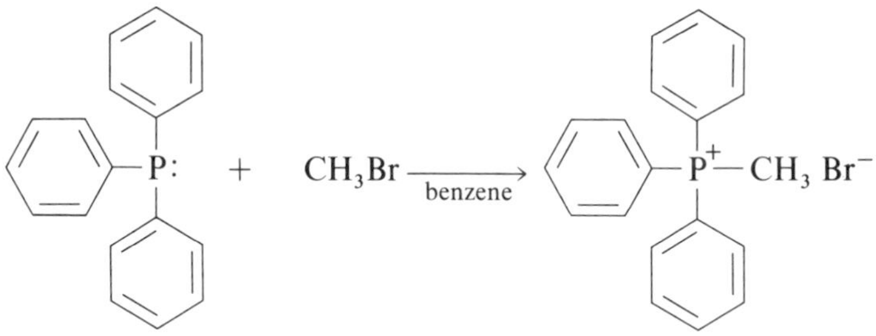

P:
+
CH3Br
benzene
P+—CH3 Br−

triphenylphosphine   methyl bromide   methyltriphenylphosphonium bromide

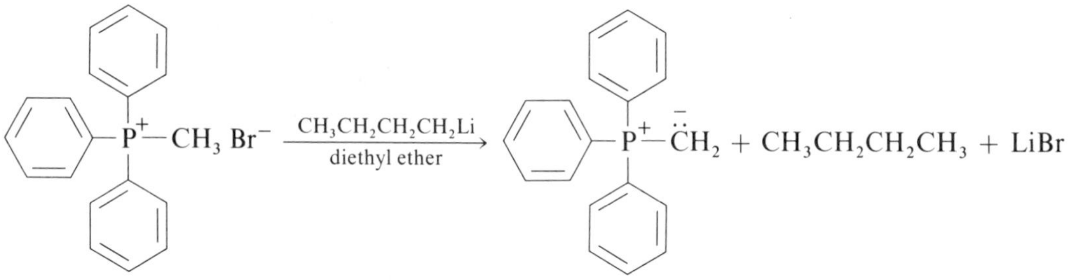

P+—CH3 Br−
CH3CH2CH2CH2Li
diethyl ether
P+—CH2 + CH3CH2CH2CH3 + LiBr

methyltriphenylphosphonium bromide   a phosphonium ylide   butane

Such a stabilized carbanion is known as a **phosphonium ylide.** An **ylide** (pronounced "ill•id") is a reactive species having a positively charged atom next to one that is negatively charged. The phosphonium ylide can be written with a double bond symbolizing a donation of electrons from the 2*p* orbital on the carbon atom to the empty 3*d* orbitals of the phosphorus atom.

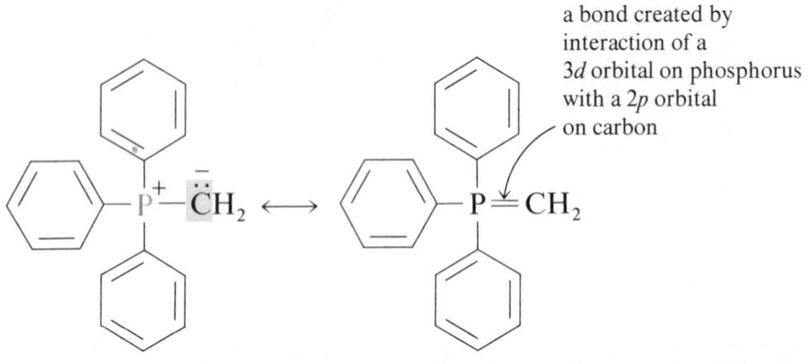
a bond created by interaction of a 3d orbital on phosphorus with a 2p orbital on carbon

resonance contributors for a triphenylphosphonium ylide

In the resonance contributor that has a double bond between phosphorus and carbon, neither atom has a formal charge. There are ten electrons around phosphorus in this resonance contributor, but this is permissible for an element in the third row.

Phosphorus ylides with a wide range of structures can be prepared. A particularly interesting class of ylides comes from α-haloesters such as ethyl bromoacetate.

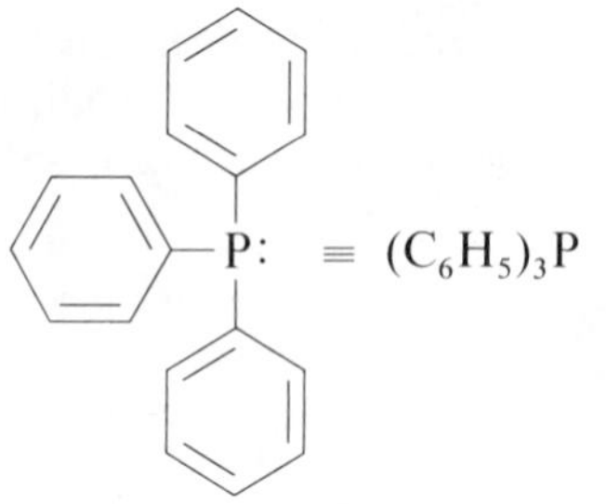

$$\underset{\text{triphenylphosphine}}{(C_6H_5)_3P} + \underset{\text{ethyl bromoacetate}}{BrCH_2\overset{O}{\overset{\|}{C}}OCH_2CH_3} \longrightarrow \underset{\text{phosphonium salt}}{(C_6H_5)_3\overset{+}{P}CH_2\overset{O}{\overset{\|}{C}}OCH_2CH_3 \; Br^-} \xrightarrow[H_2O]{NaOH} \underset{\text{phosphonium ylide}}{(C_6H_5)_3\overset{+}{P}-\overset{-}{\ddot{C}}H\overset{O}{\overset{\|}{C}}OCH_2CH_3}$$

A comparison of the above reaction with the one on p. 749 indicates that to remove a proton from the phosphonium salt derived from ethyl bromoacetate is much easier than to remove one from methyltriphenylphosphonium bromide. The base used to deprotonate methyltriphenylphosphonium bromide is butyllithium, a strong base. The reaction is carried out in a nonprotic solvent, dry ether. The hydrogens of the methylene group between the positively charged phosphorus atom and the carbonyl group of the ester, are much more acidic than the hydrogens of the methyl group in methyltriphenylphosphonium bromide. Aqueous sodium hydroxide is a strong enough base to remove a proton from the phosphonium salt of the ester, and the resulting ylide is stable in water, a medium to be avoided with most other types of ylides. Phosphorus ylides with carbonyl groups on the carbanionic center are generally more stable and less reactive than other types of phosphorus ylides.

Study Guide Concept Map 18.3

B. The Wittig Reaction

Phosphorus ylides react with aldehydes or ketones to introduce a carbon-carbon double bond selectively in place of the carbonyl group. For example, the ylide from methyltriphenylphosphonium bromide reacts with cyclohexanone to give methylenecyclohexane.

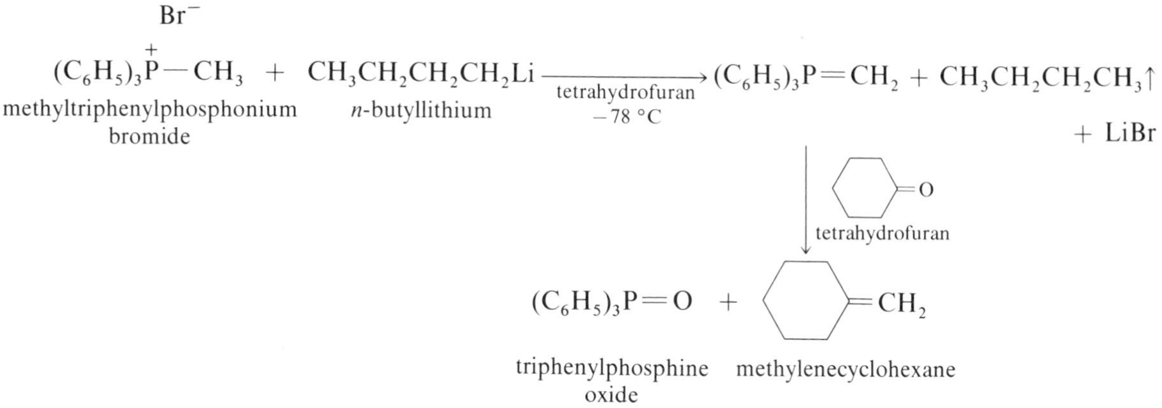

methyltriphenylphosphonium bromide
n-butyllithium
tetrahydrofuran
−78 °C
+ LiBr
tetrahydrofuran
triphenylphosphine oxide
methylenecyclohexane

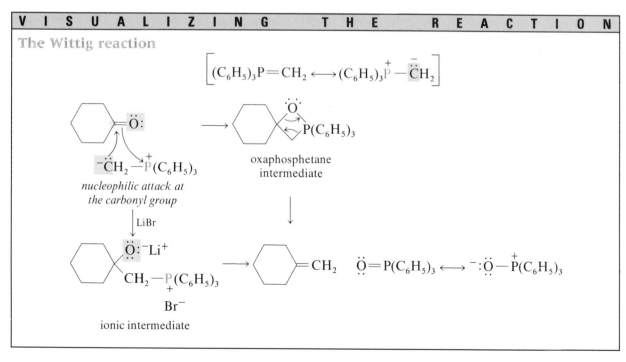

The base used to deprotonate the phosphonium salt in this reaction is the anion from n-butyllithium. The ylide reacts quickly with the ketone to give an alkene and triphenylphosphine oxide. The reaction is believed to start by nucleophilic attack of the ylide on the carbonyl group. The nature of the intermediate in such reactions depends on the conditions used. At low temperatures, an unstable intermediate having a four-membered ring, an **oxaphosphetane,** has been observed using nuclear magnetic resonance spectroscopy. In the presence of lithium bromide, an ionic intermediate stabilized by interaction with lithium and bromide ions also forms. Bond reorganization in either intermediate leads to the products.

VISUALIZING THE REACTION

The Wittig reaction

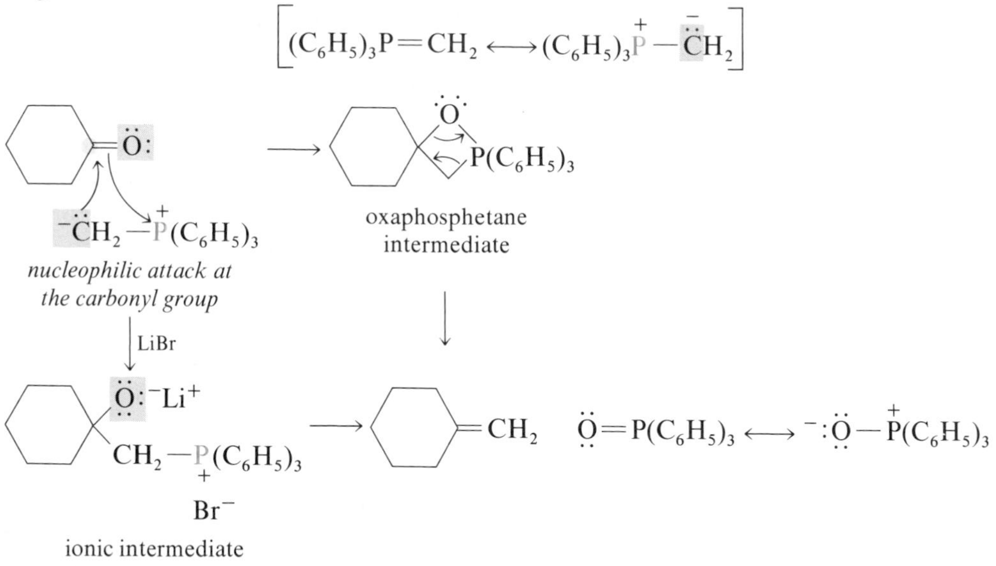

nucleophilic attack at the carbonyl group
oxaphosphetane intermediate
LiBr
ionic intermediate

Triphenylphosphine oxide is a stable compound and the formation of the phosphorus-oxygen bond is part of the driving force for the reaction. In the product the group that can be written as being doubly bonded to the phosphorus atom in the ylide is always doubly bonded to the carbon atom of the carbonyl group in the aldehyde or ketone that is used in the reaction. For example, benzaldehyde is converted into an α,β-unsaturated ester by reaction with the phosphonium ylide from ethyl bromoacetate.

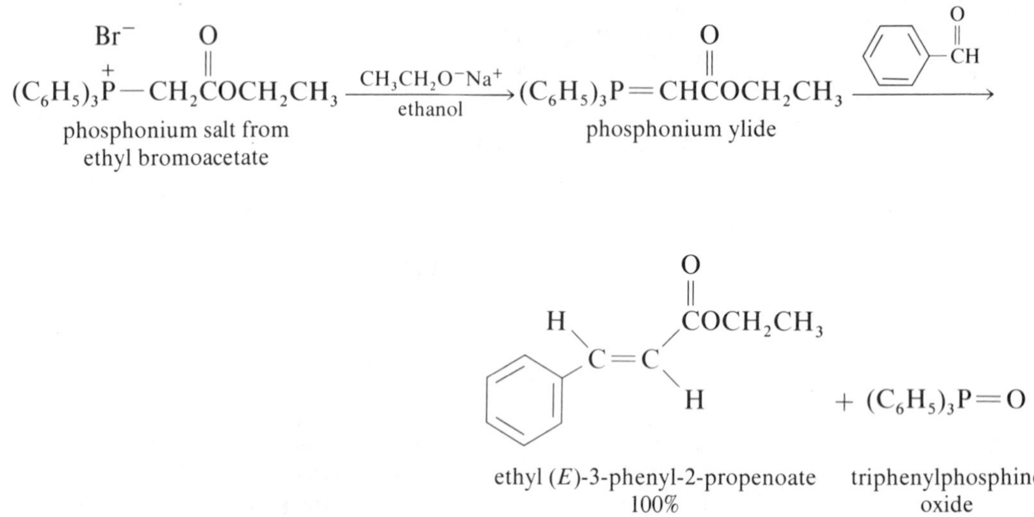
Br⁻
$(C_6H_5)_3\overset{+}{P}-CH_2\overset{O}{\overset{\|}{C}}OCH_2CH_3$
$CH_3CH_2O^-Na^+$
ethanol
phosphonium salt from ethyl bromoacetate
$(C_6H_5)_3P{=}CH\overset{O}{\overset{\|}{C}}OCH_2CH_3$
phosphonium ylide
CH
H
COCH₂CH₃
C=C
H
+ $(C_6H_5)_3P{=}O$
ethyl (E)-3-phenyl-2-propenoate
100%
triphenylphosphine oxide

The product of the above reaction is formed exclusively with the trans configuration around the double bond, but this is not always so. In most cases where stereoisomerism is possible, some of each isomer forms. The composition of the mixture depends on factors such as the solvent used, the presence of inorganic salts in the reaction mixture, and the temperature at which the reaction is carried out.

In both reactions shown above, the carbonyl group has been replaced by a double bond to a carbon atom that was the anionic center of a phosphonium ylide. This method of forming carbon-carbon double bonds quite selectively is called the **Wittig reaction** because it was the German chemist Georg Wittig who developed it. Phosphonium ylides are called **Wittig reagents.** The reaction is useful because it introduces double bonds into molecules at specific locations. The Wittig reaction is especially important in preparing compounds such as methylenecyclohexane (p. 751), which has an **exocyclic double bond,** a carbon-carbon double bond that is outside a ring, but which shares one carbon atom with the ring. Such a compound is thermodynamically less stable than its isomer in which the double bond is inside the ring and, therefore, difficult to prepare; methods such as the dehydration of alcohols (p. 282) or elimination reactions (p. 253) often give mixtures of products. Because Wittig reactions usually take place at low temperatures and under basic conditions, rearrangements of complex molecules do not generally occur. Note that the Wittig reaction resembles the aldol condensation (p. 656) in its essentials.

Study Guide Concept Map 18.4

PROBLEM 18.16

Mentally review the difficulties involved in preparing methylenecyclohexane selectively by other methods starting with cyclohexanone. How would you introduce first a carbon atom, then a double bond?

PROBLEM 18.17

Complete the following equations.

(a) 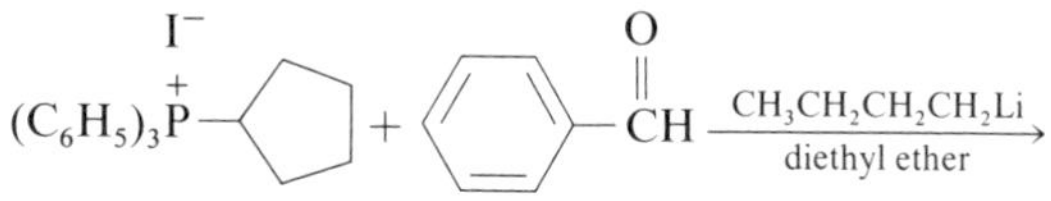

(b) $(C_6H_5)_3\overset{+}{P}CH_2\overset{O}{\overset{\|}{C}}OCH_2CH_3\ Br^- + CH_3CH{=}CH\overset{O}{\overset{\|}{C}}H \xrightarrow[\text{ethanol}]{CH_3CH_2ONa}$

(c) 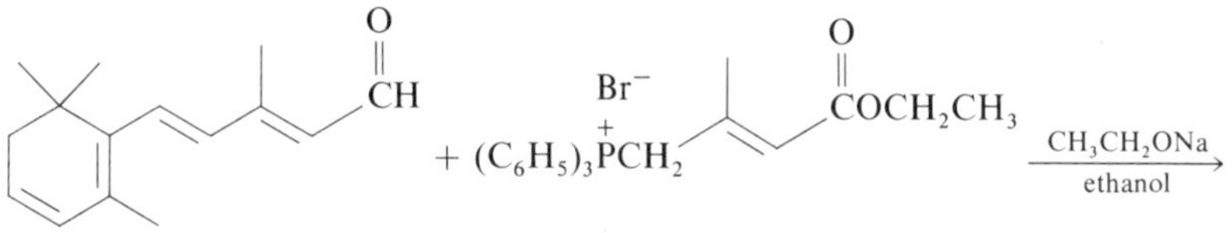

(d) 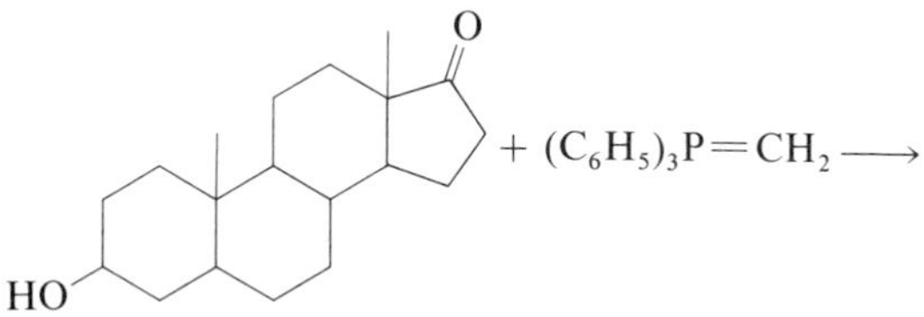

C. Problem-Solving Skills

Problem

5-Hexen-1-ol can be converted to methyl 2,7-octadienoate in two steps. How would you carry out this conversion?

$$\overset{1}{C}H_2{=}\overset{2}{C}H\overset{3}{C}H_2\overset{4}{C}H_2\overset{5}{C}H_2\overset{6}{C}H_2OH \longrightarrow \overset{1}{C}H_2{=}\overset{2}{C}H\overset{3}{C}H_2\overset{4}{C}H_2\overset{5}{C}H_2\overset{6}{C}H{=}CH\overset{O}{\overset{\|}{C}}OCH_3$$

Solution

1. What functional groups are present in the starting material and the product?

   The starting material is an alkene and an alcohol. The product is an alkene and an α,β-unsaturated ester.

2. How do the carbon skeletons of the two compounds compare? How many carbon atoms does each contain? Are there any rings? What are the positions of branches and functional groups on the carbon skeletons?

   The product has two more carbon atoms in the main chain than the starting material does. The original double bond remains at the end of the chain.

3. How do the functional groups change in going from starting material to product? Does the starting material have a good leaving group?

   The carbon atom that is bonded to the hydroxyl group in the starting material is involved in a carbon-carbon double bond in the product.

4. Is it possible to dissect the structures of the starting material and product to see which bonds must be broken and which formed?

$$CH_2{=}CHCH_2CH_2CH_2\overset{H}{\underset{H}{C}}{\text{-}}OH \qquad CH_2{=}CHCH_2CH_2CH_2CH{=}CH\overset{O}{\overset{\|}{C}}OCH_3$$

bonds to be broken *bonds to be formed*

5. Do we recognize any part of the product molecule as coming from a good nucleophile or an electrophilic addition?

   The new carbon-carbon double bond can be formed by a nucleophilic attack of a phosphorus ylide on a carbonyl group.

$$-CH{=}O + {>}P{=}CH\overset{O}{\overset{\|}{C}}OCH_3 \longrightarrow -CH{=}CH\overset{O}{\overset{\|}{C}}OCH_3$$

6. What type of compound would be a good precursor to the product?

   An aldehyde will react with an ylide to give the desired product.

$$CH_2{=}CHCH_2CH_2CH_2CH{=}O + (C_6H_5)_3P{=}CH\overset{O}{\overset{\|}{C}}OCH_3$$
$$\downarrow$$
$$CH_2{=}CHCH_2CH_2CH_2CH{=}CH\overset{O}{\overset{\|}{C}}OCH_3$$

7. After this last step, do we see how to get from starting material to product? If not, we need to analyze the structure obtained in step 6 and apply questions 5 and 6 to it.

   The aldehyde can be obtained from oxidation of the primary alcohol, which completes the synthesis.

$$CH_2{=}CHCH_2CH_2CH_2CH{=}CH\overset{O}{\overset{\|}{C}}OCH_3$$
$$\uparrow \; (C_6H_5)_3P{=}CH\overset{O}{\overset{\|}{C}}OCH_3$$

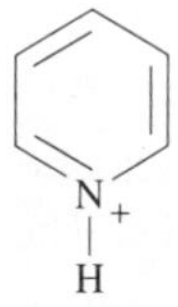
N
+
H

$$CH_2{=}CHCH_2CH_2CH_2CH_2OH \xrightarrow[\text{dichloromethane}]{\text{pyridinium } CrO_3Cl^-} CH_2{=}CHCH_2CH_2CH_2\overset{O}{\overset{\|}{C}}H$$

PROBLEM 18.18

How would you carry out the following transformation?

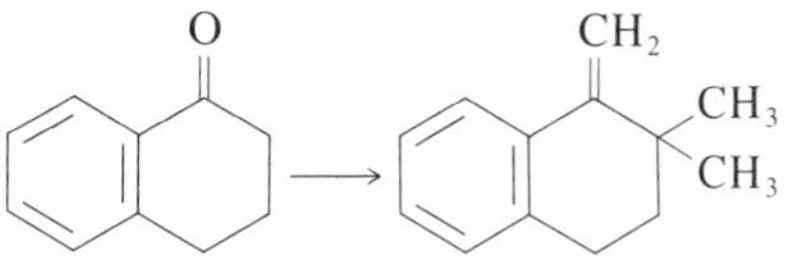

18.4 CARBANIONS STABILIZED BY SULFUR

A. Sulfonium Ylides

A sulfur atom, like a phosphorus atom, can stabilize a negative charge on an adjacent carbon atom by accepting electrons into its 3*d* orbitals. Several types of carbanions containing sulfur have been developed as synthetic intermediates by E. J. Corey at Harvard University. Among these, **sulfonium ylides** most closely resemble the Wittig reagent. For example, a sulfonium ylide forms on removal of a proton from trimethylsulfonium iodide (prepared from dimethyl sulfide and methyl iodide) by a strong base such as sodium hydride.

$$\underset{\text{dimethyl sulfide}}{CH_3SCH_3} + \underset{\text{methyl iodide}}{CH_3I} \longrightarrow \underset{\text{trimethylsulfonium iodide}}{CH_3\overset{+}{S}(CH_3)CH_3\ I^-}$$

$$CH_3\overset{+}{\ddot{S}}(CH_3)-CH_3\ I^- + NaH \xrightarrow[\text{sulfoxide}]{\text{dimethyl}} CH_3\overset{+}{\ddot{S}}(CH_3)-\overset{-}{\ddot{C}}H_2 + H_2\uparrow + NaI$$

In a sulfonium ylide, the sulfur atom is positively charged and the carbon atom is negatively charged.

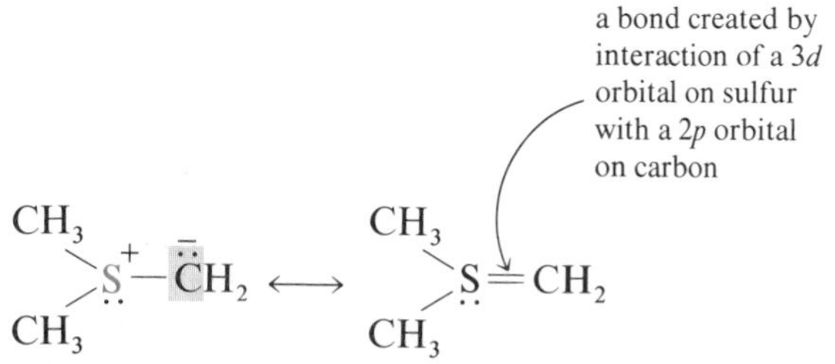
a bond created by interaction of a 3d orbital on sulfur with a 2p orbital on carbon

resonance contributors for the sulfonium ylide generated from trimethylsulfonium iodide

Sulfonium ylides react with carbonyl compounds to give oxiranes. For example, 2-phenyloxirane is formed when the sulfonium ylide from trimethylsulfonium iodide reacts with benzaldehyde.

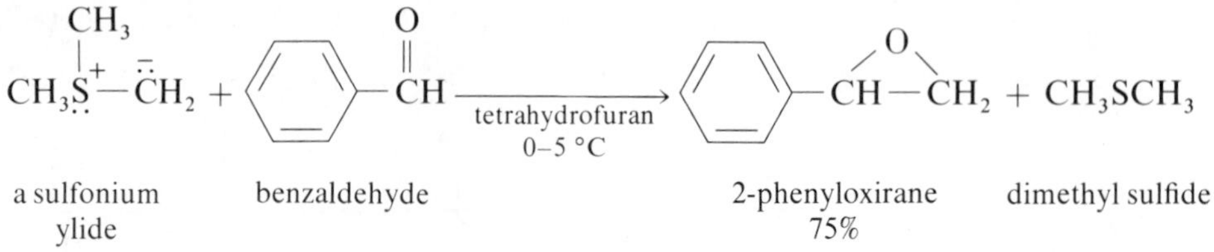
tetrahydrofuran
0–5 °C
a sulfonium ylide
benzaldehyde
2-phenyloxirane
75%
dimethyl sulfide

The products obtained from this reaction can be rationalized by recognizing that sulfonium ylides are nucleophilic at the carbanionic carbon atom and also contain a good leaving group, dimethyl sulfide in this case. The carbanion attacks the electrophilic carbon atom of the carbonyl compound to give an alkoxide ion intermediate.

VISUALIZING THE REACTION

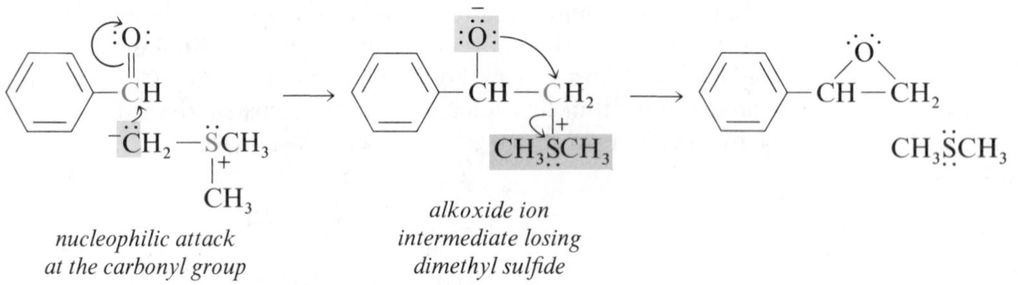
nucleophilic attack at the carbonyl group
alkoxide ion intermediate losing dimethyl sulfide

An intramolecular nucleophilic substitution reaction by the alkoxide ion on the carbon atom bearing the good leaving group completes the reaction. The overall reaction can be described as an addition of a methylene group to the carbon-oxygen double bond of the aldehyde.

It is interesting to compare the reactions of phosphonium ylides and carbonyl compounds with those of sulfonium ylides and carbonyl compounds. A phosphonium ylide replaces the carbon-oxygen double bond with a carbon-carbon double bond. A sulfonium ylide gives an oxirane. The difference in the reactions of these two types of ylides is due to the relative energies of the phosphorus-oxygen and sulfur-oxygen bonds. The formation of a phosphine oxide with a strong phosphorus-oxygen bond favors creation of the carbon-carbon double bond in the Wittig reaction.

B. Dithiane Anions

In searching for the synthetic equivalent of a carbonyl group to serve as a nucleophile in a nucleophilic substitution reaction, E. J. Corey developed another type of carbanion stabilized by sulfur, the **1,3-dithiane anion.** The carbon atom of a carbonyl group is an electrophilic center and is attacked by nucleophiles. Corey thought it would be useful to have a synthetic method to introduce a carbonyl group by a nucleophilic substitution reaction initiated by the potential carbonyl group. He chose the cyclic thioacetal of aldehydes, the 1,3-dithiane system, as the source of a carbanionic equivalent of a carbonyl group.

The anion derived from an aldehyde, called an **acyl anion,** usually cannot be prepared directly by removing the hydrogen bonded to the carbonyl carbon. For example, the base butyllithium would add to the carbonyl group of formaldehyde (p. 492) instead of deprotonating it. Deprotonation of most aldehydes occurs at the α-carbon atom (p. 656), not at the carbonyl group. Conversion of the aldehyde to a 1,3-dithiane protects the carbonyl group and increases the acidity of the hydrogen atom to the point that it is removed preferentially by a strong base.

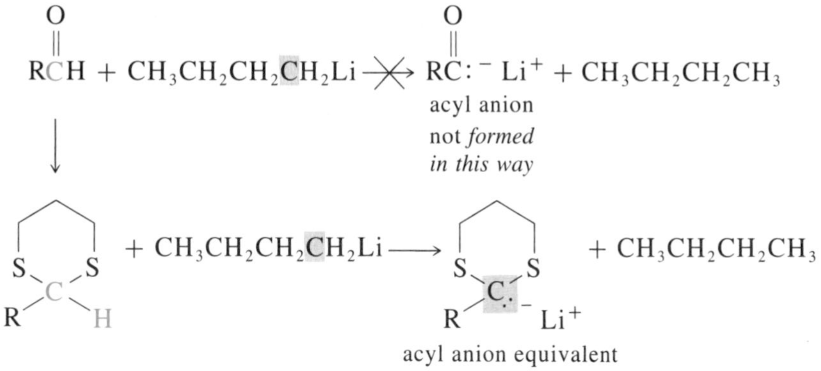
acyl anion
not formed
in this way
acyl anion equivalent

The electrophilic carbon atom of the carbonyl group is thus converted into a nucleophile.

The conversion of formaldehyde to 6-undecanone illustrates the chief features of Corey's synthetic method.

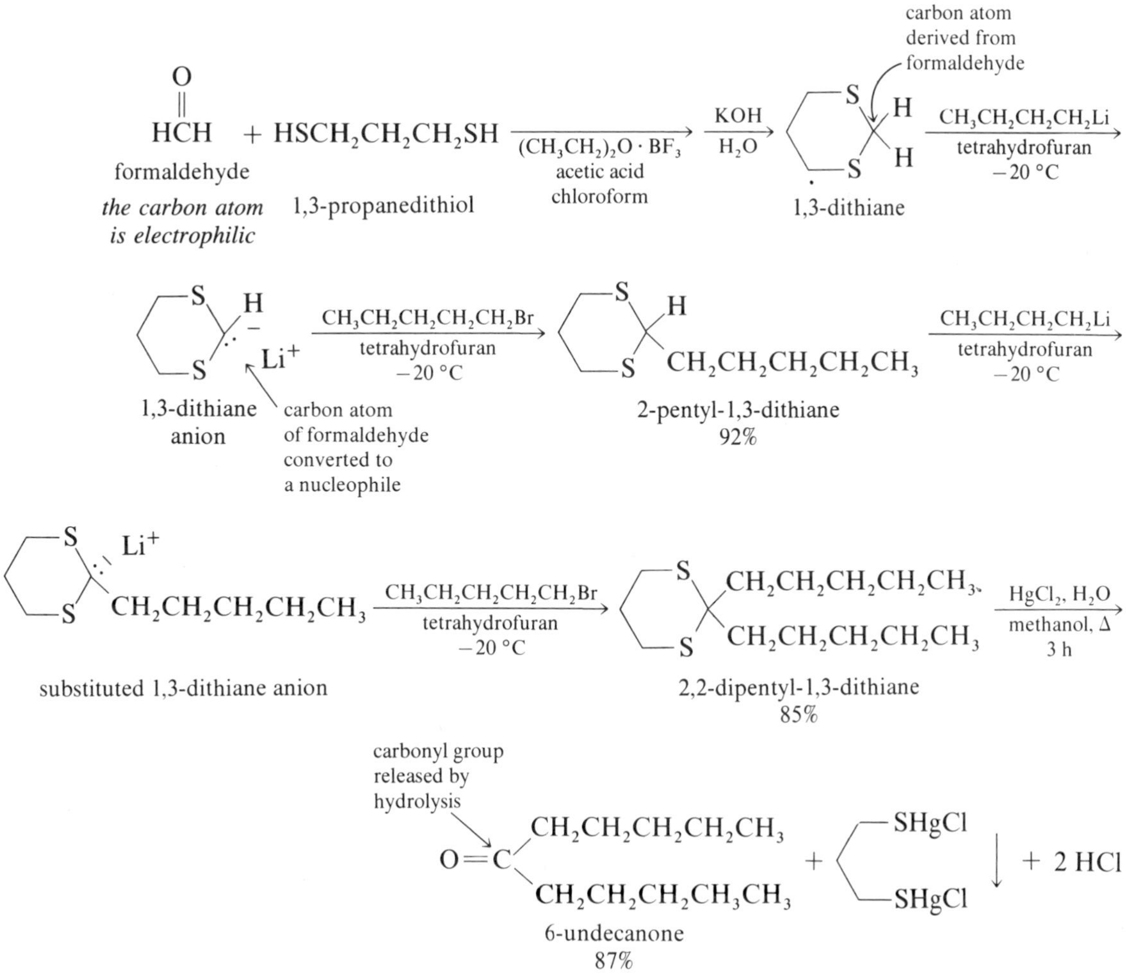
carbon atom
derived from
formaldehyde
formaldehyde
the carbon atom
is electrophilic
1,3-propanedithiol
acetic acid
chloroform
1,3-dithiane
tetrahydrofuran
1,3-dithiane
anion
carbon atom
of formaldehyde
converted to
a nucleophile
2-pentyl-1,3-dithiane
92%
substituted 1,3-dithiane anion
2,2-dipentyl-1,3-dithiane
85%
methanol, Δ
3 h
carbonyl group
released by
hydrolysis
6-undecanone
87%

In this sequence of reactions, polarity of the carbonyl group is reversed, resulting in great versatility in the design of syntheses.

The thioacetal (p. 511) is prepared from the aldehyde by treating it with 1,3-propanedithiol in the presence of an acid catalyst. A 1,3-dithiane is deprotonated at the carbon atom between the two sulfur atoms by a strong base, such as *n*-butyllithium, to give a carbanion stabilized by delocalization of charge to the sulfur atoms. The carbanion is a strong nucleophile and reacts with alkyl halides in nucleophilic substitution reactions. In the synthesis shown above, the 1,3-dithiane is deprotonated and alkylated twice. The product, 2,2-dipentyl-1,3-dithiane, contains a carbonyl group hidden in a thioacetal function. The carbonyl group is released by hydrolysis in the presence of mercury(II) salts, which serve as Lewis acids in coordinating with the sulfur atoms and help cleave the sulfur-carbon bonds.

1,3-Dithiane anions react with other types of electrophilic carbon atoms, such as those in carbonyl groups. The addition of such a carbanion to a carbonyl compound must be carried out at very low temperatures if the carbonyl compound has enolizable hydrogen atoms. The reaction of the anion from 1,3-dithiane with benzaldehyde, which has no α-hydrogen atoms, goes well.

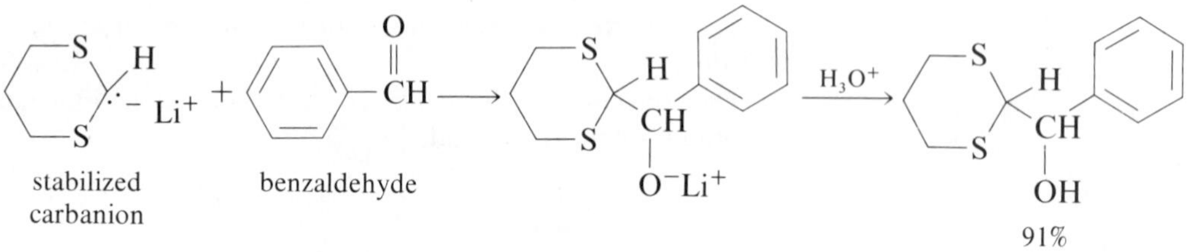
stabilized carbanion
benzaldehyde
H_3O^+
91%

That carbanion also reacts with oxiranes, attacking the less highly substituted carbon atom of unsymmetrical ones (p. 454).

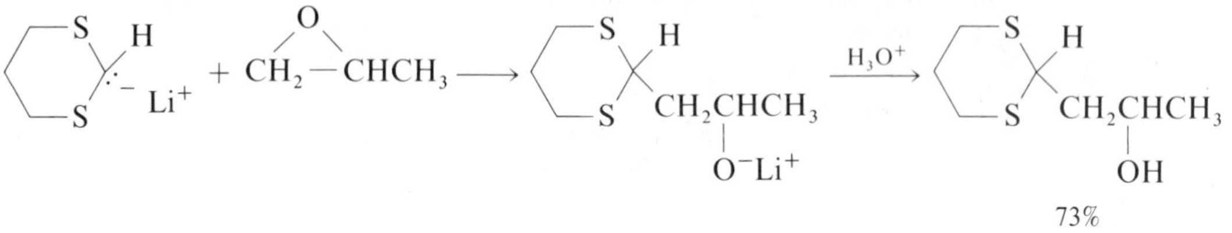
H_3O^+
73%

Study Guide Concept Map 18.5

The reactions of dithiane anions with carbonyl compounds or oxiranes are routes to potential hydroxycarbonyl compounds.

PROBLEM 18.19

Show how each of the following conversions can be carried out.

(a) 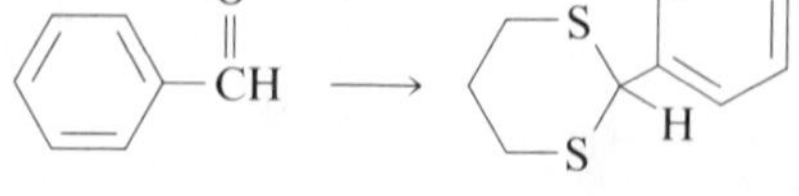

(b) 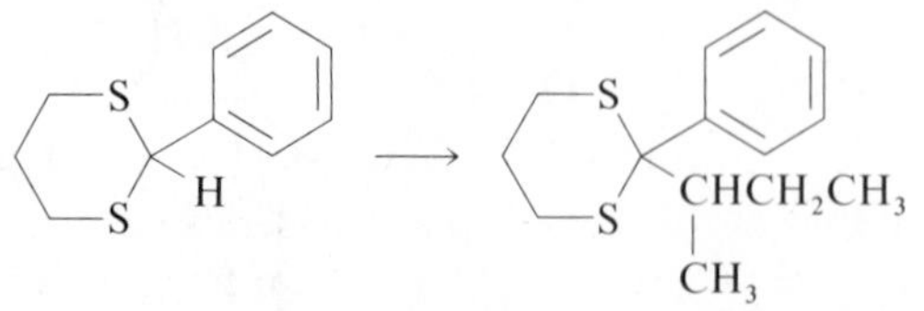

(c) 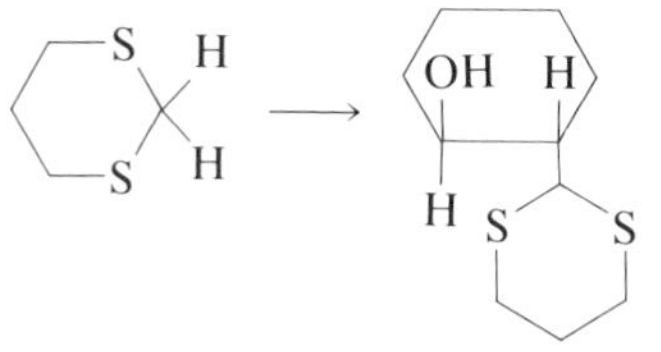

(d) 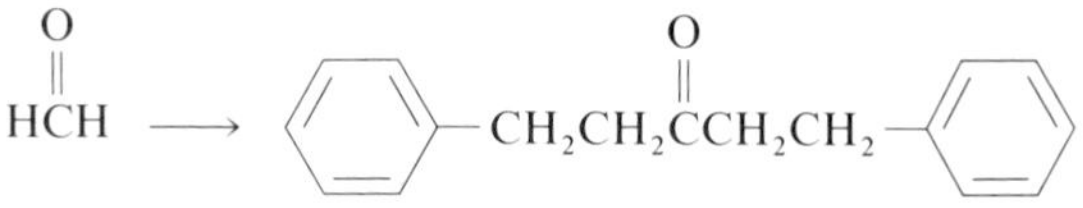

(e) 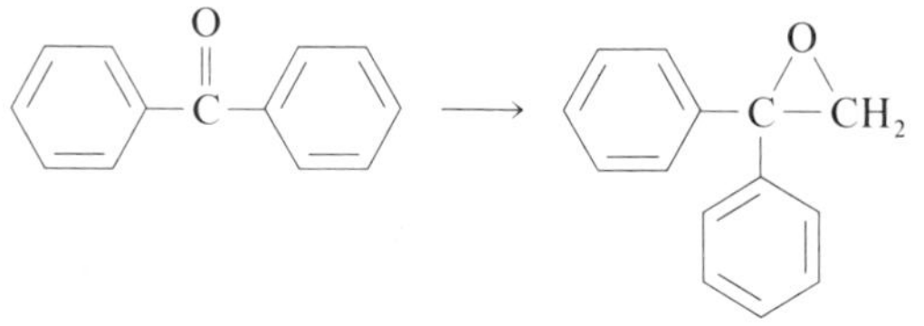

PROBLEM 18.20

Dimethyl sulfoxide, CH_3SOCH_3, is deprotonated by sodium hydride to give the methylsufinyl carbanion. Write an equation for the formation of this anion. How do you account for the acidity of the hydrogen atoms in dimethyl sulfoxide? What other species does the anion resemble? Predict how it will react with benzophenone and with ethyl cyclohexanecarboxylate.

S U M M A R Y

A compound in which a carbon-carbon double bond is conjugated with an electron-withdrawing group, such as a carbonyl, nitro, or cyano group, is known as an electrophilic alkene. Electrophilic alkenes are attacked by nucleophiles at the carbon atom beta to the electron-withdrawing group; the resulting enolate anion is protonated to give a β-substituted compound. Depending on the nature of the nucleophile, further reactions may occur. For example, when the nucleophile is the peroxide ion, the enolate ion displaces hydroxide ion, and the final product is an oxirane.

Organocuprate reagents (but not Grignard reagents or organolithium reagents) add to the β-carbon atom of α,β-unsaturated carbonyl compounds to give substituted enolates that may then react with alkyl halides or carbonyl compounds. Enolate anions, especially those from compounds having active methylene groups, are also used as nucleophiles in reactions with α,β-unsaturated carbonyl compounds. The products of these reactions, known as Michael reactions, are 1,5-dicarbonyl compounds. Under basic reaction conditions, Michael products undergo intramolecular aldol condensations to give six-membered rings, also containing α,β-unsaturated carbonyl functions. This cyclization reaction is called the Robinson annulation.

The reactions of electrophilic alkenes are summarized in Table 18.1.

Carbanions stabilized by the presence of phosphorus or sulfur atoms are useful synthetically. Triphenylphosphine reacts with alkyl halides to give a phosphonium salt that can be deprotonated to give an ylide. An ylide is a reactive intermediate that has a positive and a negative charge on adjacent atoms. Phosphorus ylides react with aldehydes and ketones, and the oxygen atom of the carbonyl group is replaced

TABLE 18.1 Reactions of Electrophilic Alkenes

Alkene	Nucleophile	Intermediate	Reagents for Second Step	Product
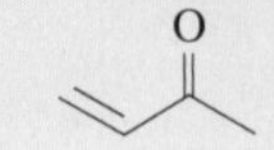	R_2NH	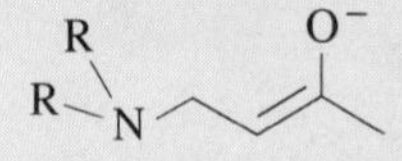	HB^+	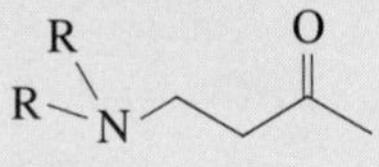
	RSH	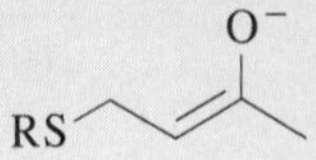	HB^+	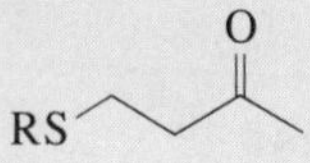
	HOOH, OH^-	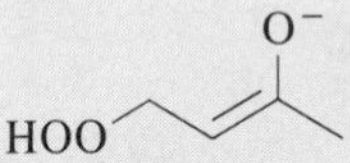		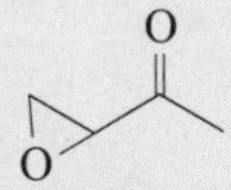
	R_2CuLi	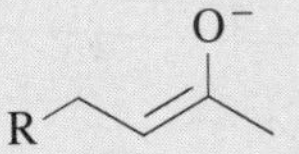	HB^+	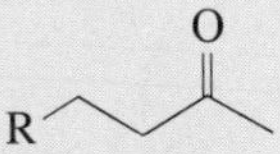
			R′X	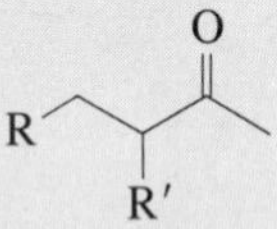
			$R'\overset{\overset{O}{\|}}{C}R''$ then HB^+	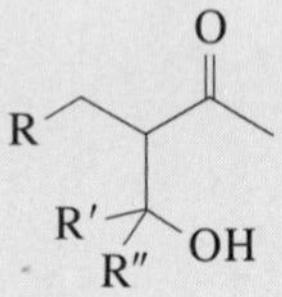
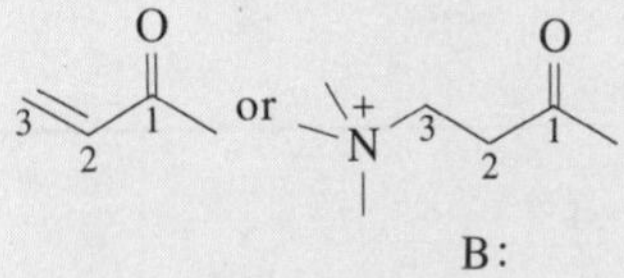	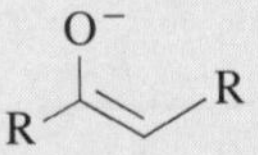	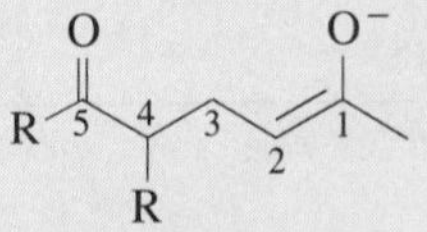	HB^+	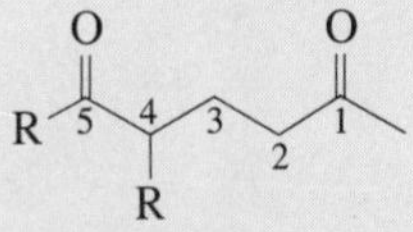 Michael product
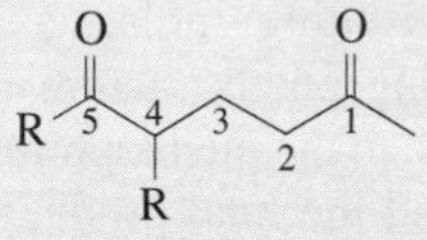 Michael product	B:	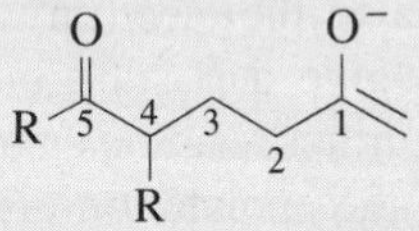	HB^+	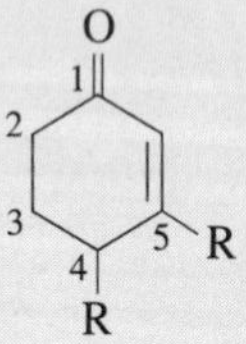 Robinson annulation product

by a double bond to carbon. This reaction, known as the Wittig reaction, is especially useful for making alkenes that are thermodynamically less stable than their isomers and, therefore, hard to make by other methods. Sulfonium ylides react with carbonyl compounds to give oxiranes as products.

The carbon atom of a carbonyl group is an electrophilic center. Converting an aldehyde to a 1,3-dithiane by reacting with a dithiol gives a compound that can be deprotonated. This creates a nucleophilic center at the carbon atom that was originally part of the carbonyl group. Dithiane anions react with electrophiles to give substituted dithianes, which can be hydrolyzed to recover the carbonyl group.

The reactions of phosphorus- and sulfur-stabilized carbanions are summarized in Table 18.2.

TABLE 18.2 Preparation and Reactions of Heteroatom-Stabilized Carbanions

Wittig Reaction, Phosphonium Ylides			
Reactive Intermediate	**Reagents That Give Reactive Intermediate**	**Reagent That Reacts with Ylide**	**Product**
$(C_6H_5)_3P{=}C(R)(H)$	$(C_6H_5)_3P$ RCH_2X BuLi	$R'\overset{O}{\overset{\|}{C}}H$ $R'\overset{O}{\overset{\|}{C}}R''$	$R'CH{=}CHR$ $(R')(R'')C{=}CHR$
$(C_6H_5)_3P{=}C(H)\overset{O}{\overset{\|}{C}}OCH_2CH_3$	$(C_6H_5)_3P$ $XCH_2\overset{O}{\overset{\|}{C}}OCH_2CH_3$ NaOH	$R'\overset{O}{\overset{\|}{C}}H$ $R'\overset{O}{\overset{\|}{C}}R''$	$R'CH{=}CH\overset{O}{\overset{\|}{C}}OCH_2CH_3$ $(R')(R'')C{=}CH\overset{O}{\overset{\|}{C}}OCH_2CH_3$

Sulfonium Ylides			
$R_2S{=}C(R')(H)$	R_2S $R'CH_2X$ NaH	$R''\overset{O}{\overset{\|}{C}}H$ $R'''\overset{O}{\overset{\|}{C}}R''$	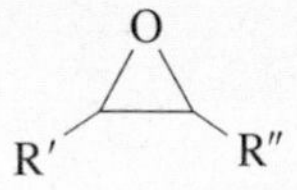 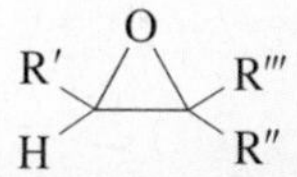

Dithiane Anions					
Reactive Intermediate	**Reagents That Give Reactive Intermediate**	**Reagent for First Step**	**Product**	**Reagent for Second Step**	**Product**
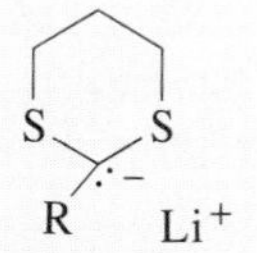	$R\overset{O}{\overset{\|}{C}}H$  BF_3 then BuLi	$R'X$	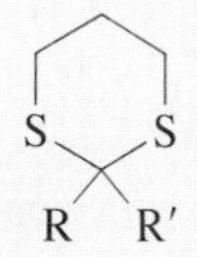	H_2O $HgCl_2$ Δ	$R\overset{O}{\overset{\|}{C}}R'$
		$R''\overset{O}{\overset{\|}{C}}H$	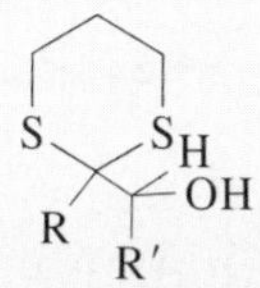	H_2O $HgCl_2$ Δ	$R\overset{O}{\overset{\|}{C}}\underset{OH}{CHR'}$
		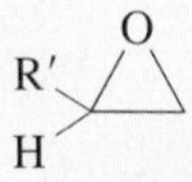	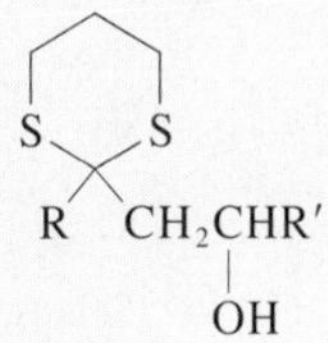	H_2O $HgCl_2$ Δ	$R\overset{O}{\overset{\|}{C}}CH_2\underset{OH}{CHR'}$

ADDITIONAL PROBLEMS

18.21 Give structural formulas for all intermediates and products designated by letters in the following equations.

(a) 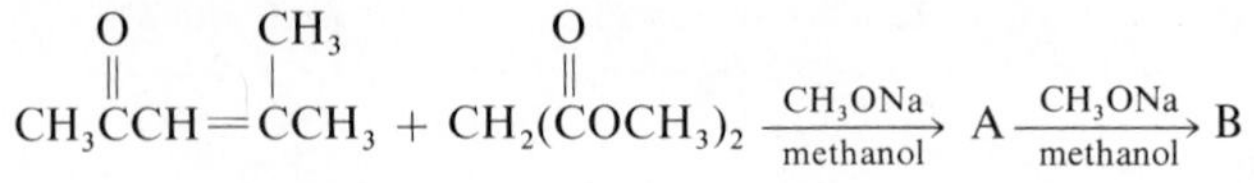

(b) 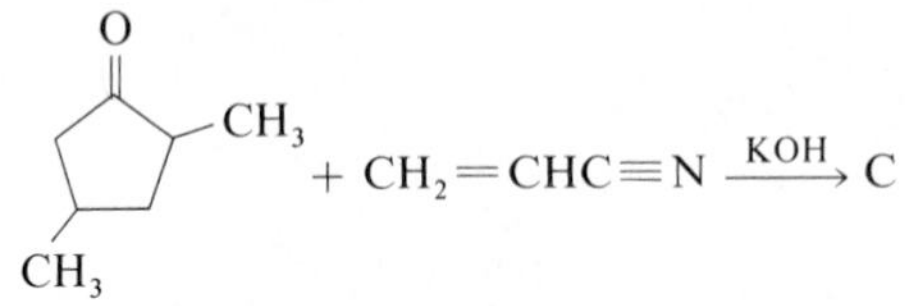

(c) 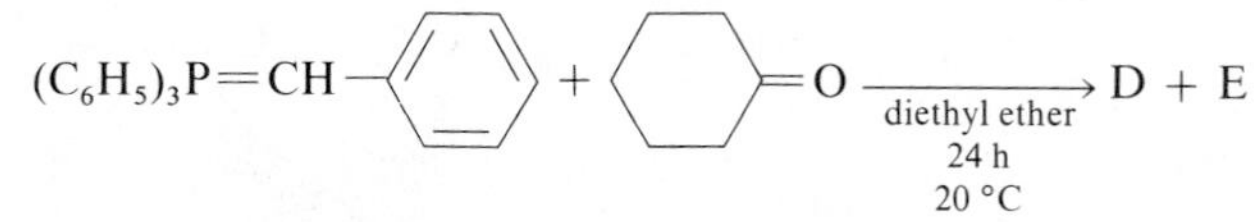

(d) 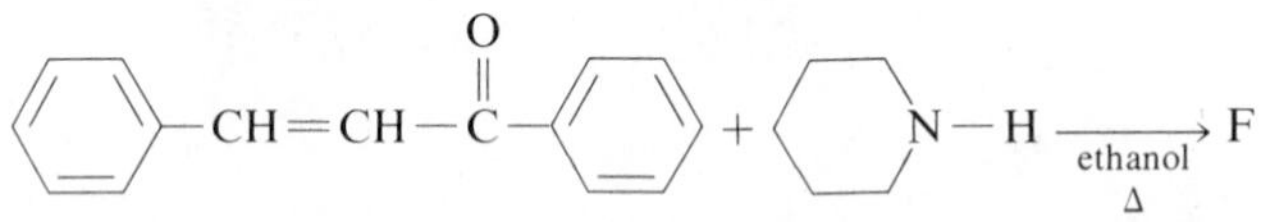

(e) 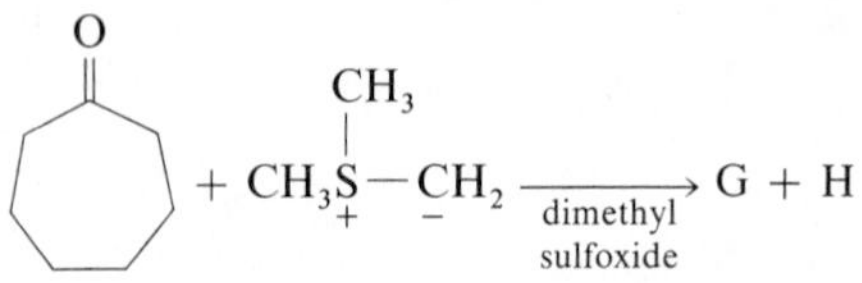

(f) 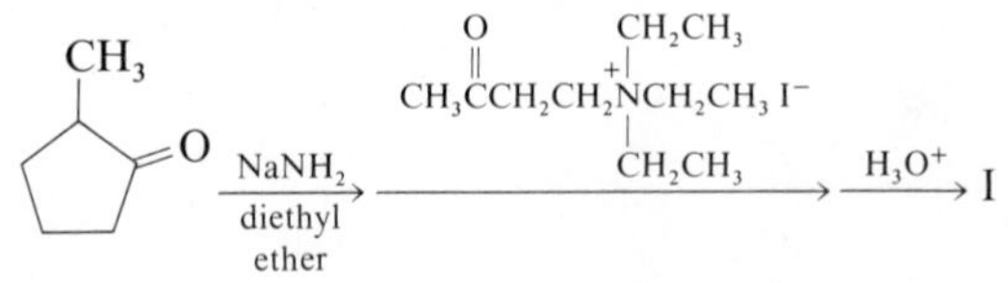

(g) 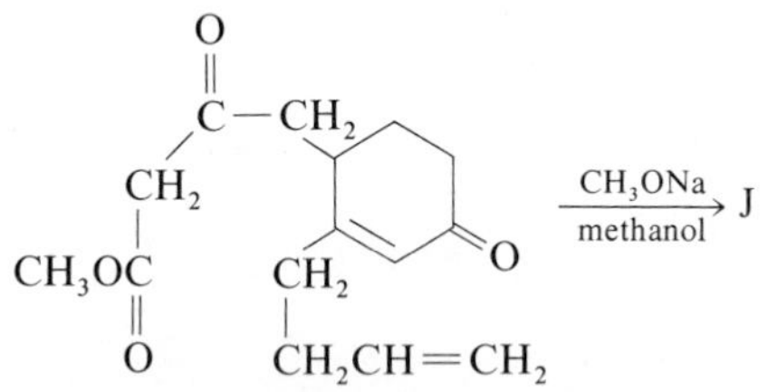

(h) 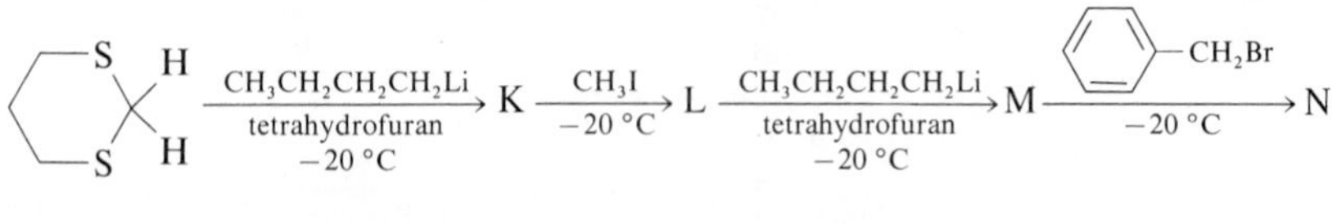

(i) 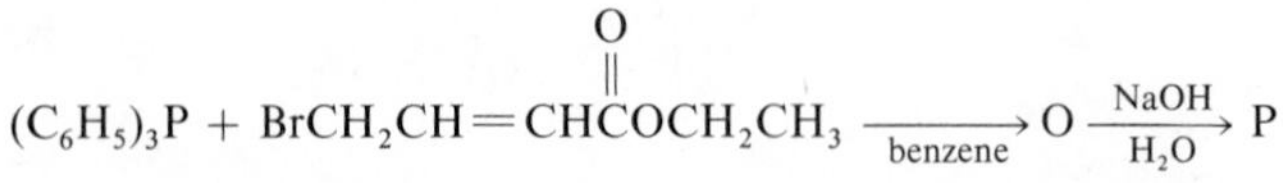

(j) 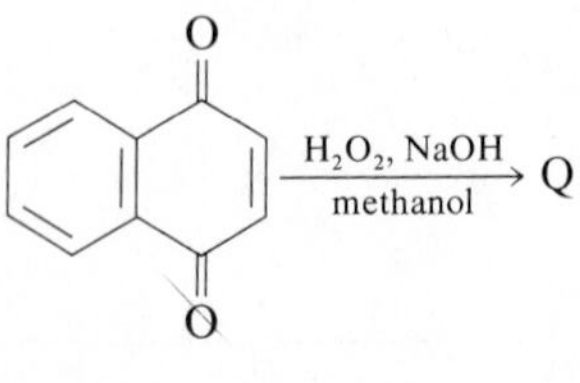

(k) 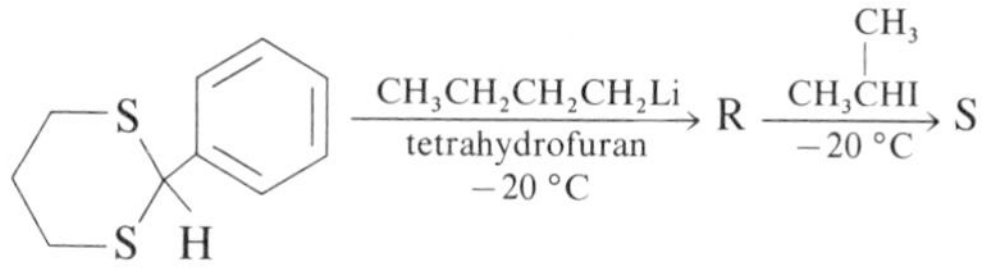

(l) 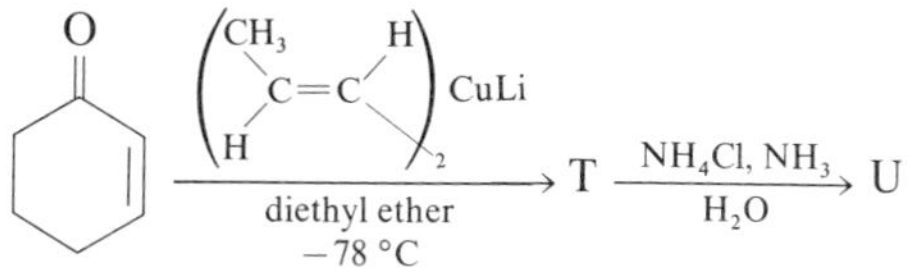

18.22 For further practice in recognizing reactions, assign structures to all compounds designated by letters in the following equations.

(a) 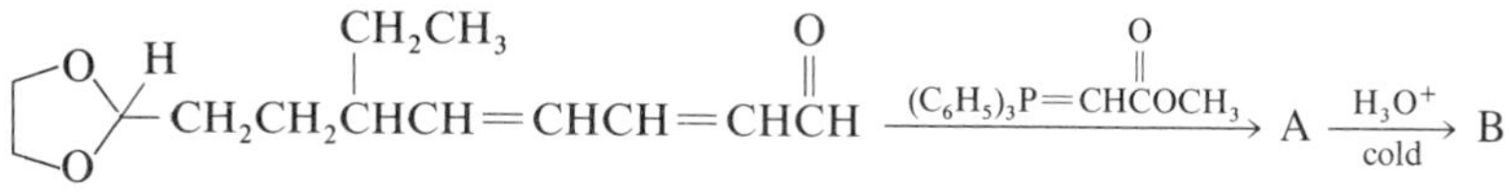

(b) $CH_3CH_2CH(CH_3)CHO \xrightarrow[\text{benzene, } \Delta]{(C_6H_5)_3P{=}CHCOCH_3} C \xrightarrow[\text{dichloromethane; 2. } H_3O^+]{\text{1. } [(CH_3)_2CHCH_2]_2AlH \text{ (excess)}} D$

(c) 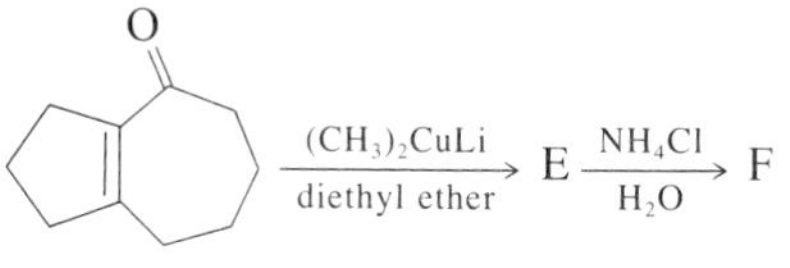

(d) 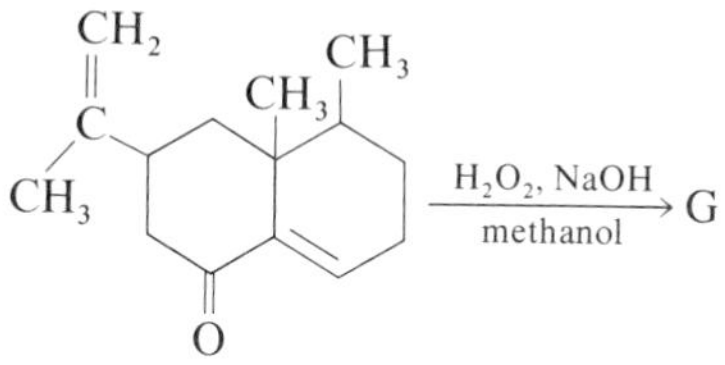

(e) 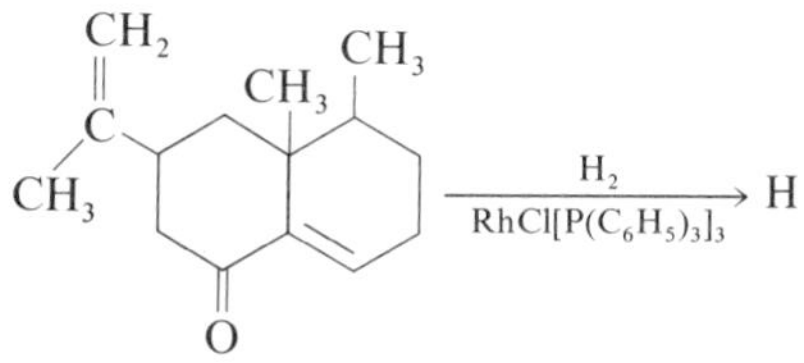

(f) 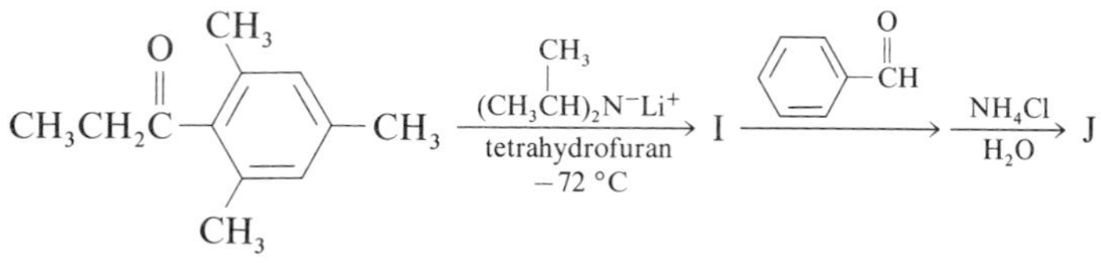

(g) 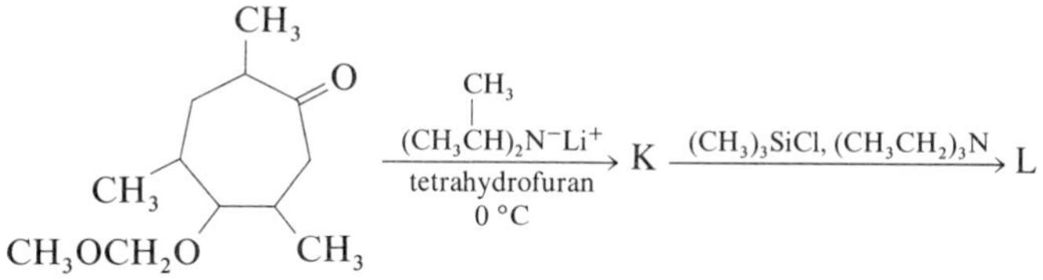

(h) 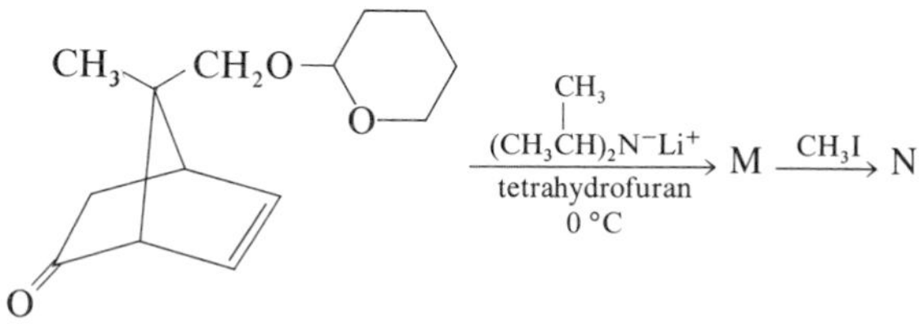

(i) 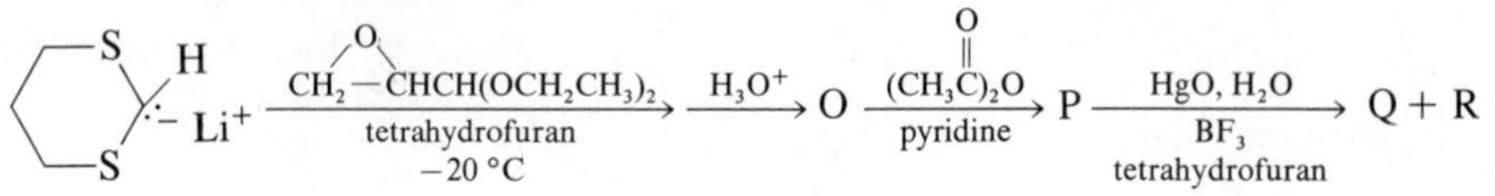

(j) 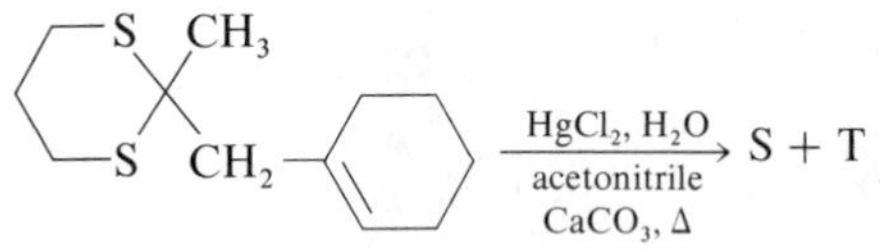

(k) 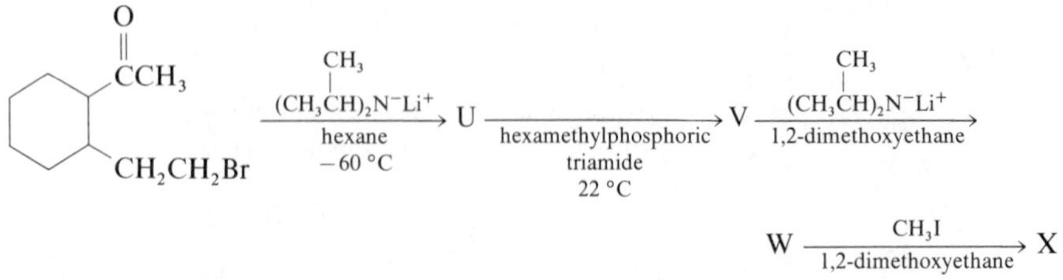

(l) $CH_3\overset{O}{\overset{\|}{C}}H \xrightarrow[\text{HCl (g)}\ \text{chloroform}]{HSCH_2CH_2CH_2SH} Y \xrightarrow[\text{tetrahydrofuran}\ -20\ °C]{CH_3CH_2CH_2CH_2Li} Z \xrightarrow{CH_3\overset{CH_3}{\overset{|}{C}}HBr} AA \xrightarrow[\text{methanol}]{HgCl_2,\ H_2O} BB + CC$

(m) 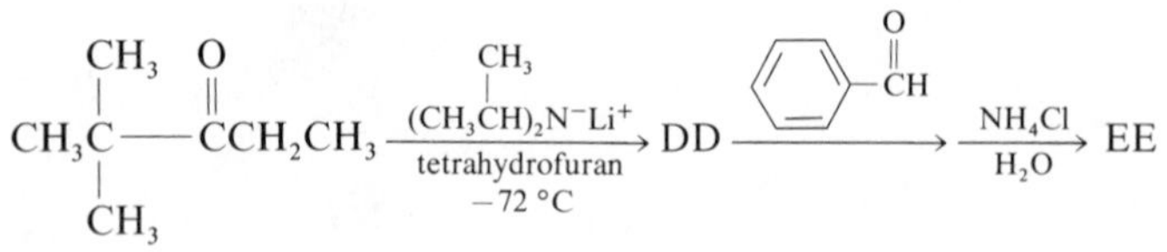

-72\,°C}]{(\text{CH}_3\text{CH})_2\text{N}^-\text{Li}^+}$ DD $\xrightarrow{\text{C}_6\text{H}_5\text{CH (O)}}$ $\xrightarrow[\text{H}_2\text{O}]{\text{NH}_4\text{Cl}}$ EE

18.23 Tell what reagents could be used to carry out the following transformations. More than one step may be required.

(a) 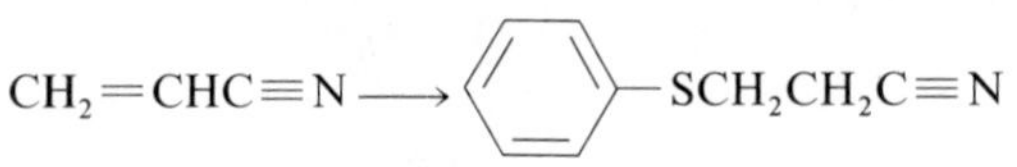

(b) 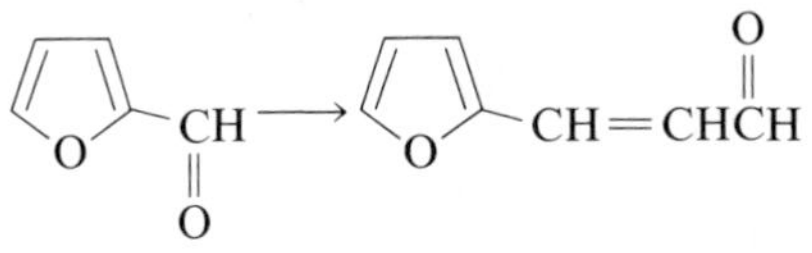

(c) $CH_3CH_2CH_2CH_2CH_2I \longrightarrow CH_3CH_2CH_2CH_2CH_2CH_2CH{=}CH_2$

(d) 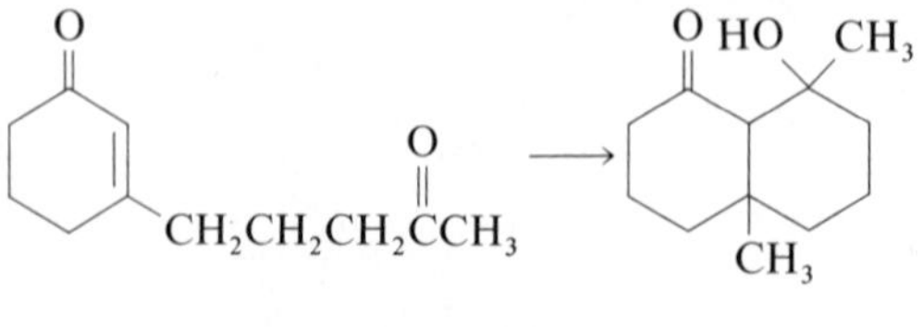

(e) 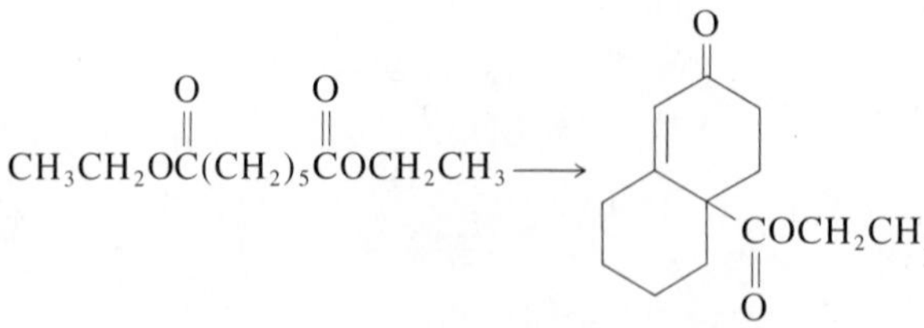

(f) 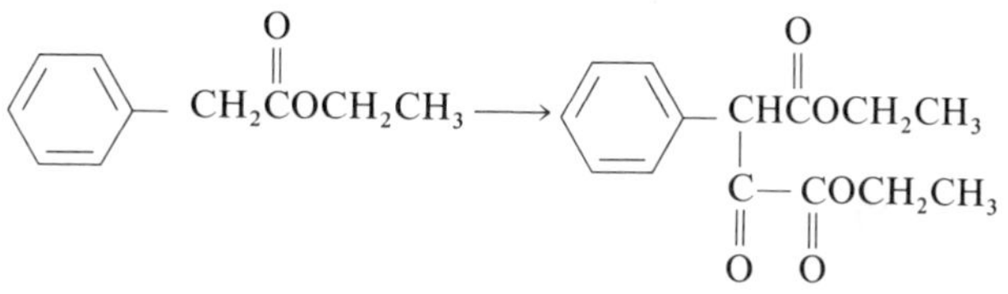

(g) 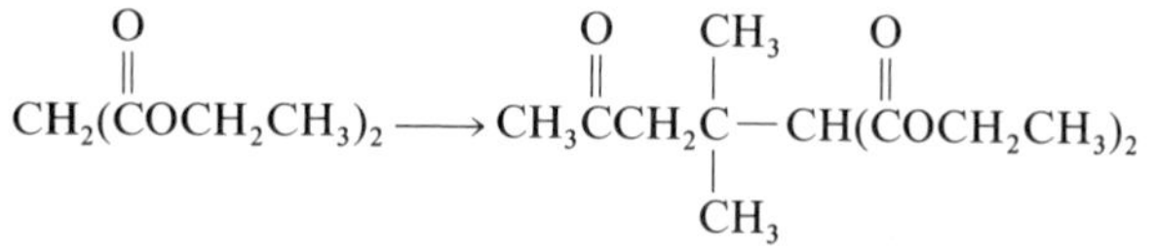

(h) 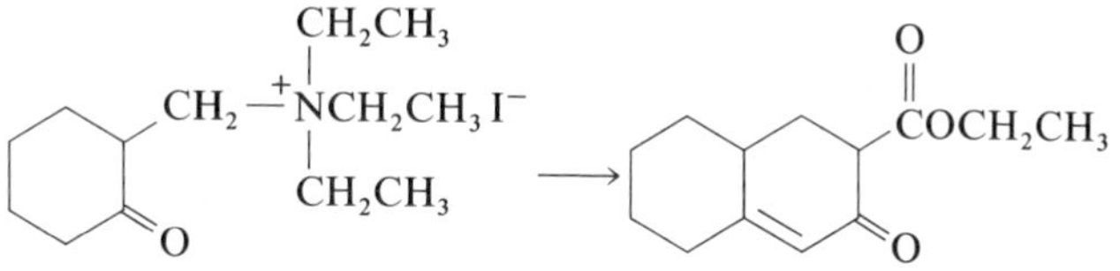

(i) 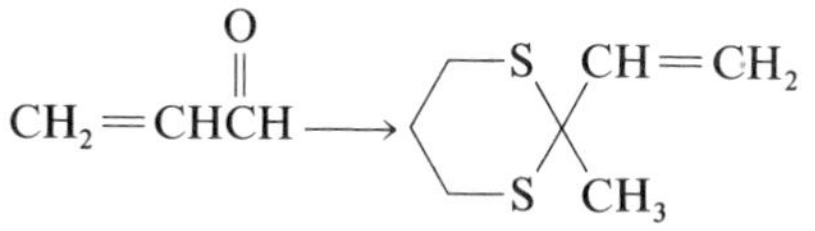

(j) 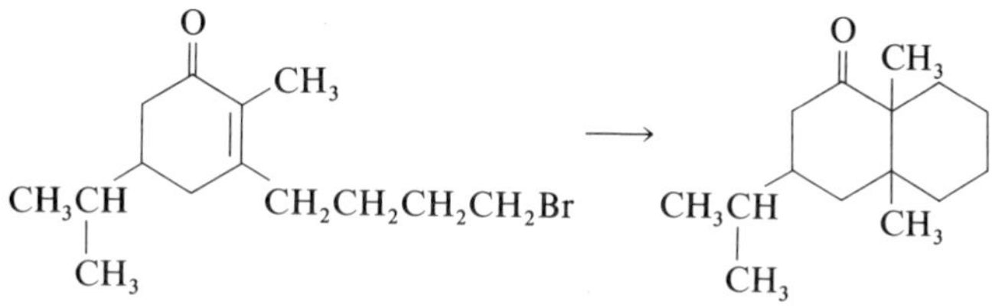

(k) 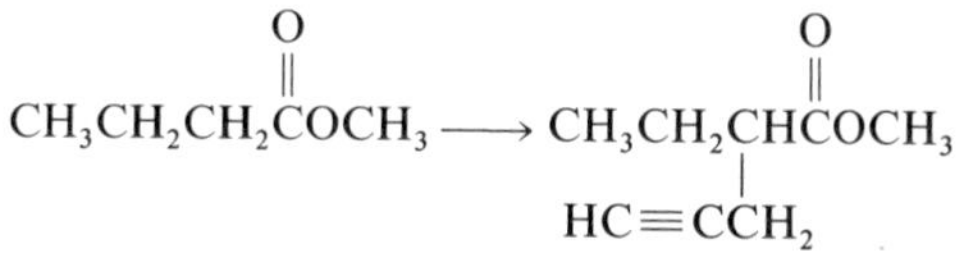

(l) 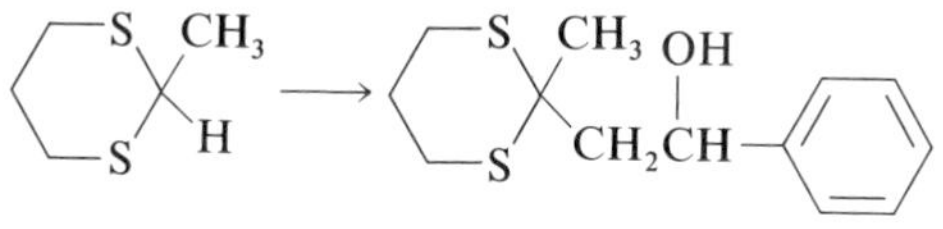

18.24 Write structural formulas for the intermediates and products indicated by the letters in the following reaction sequence.

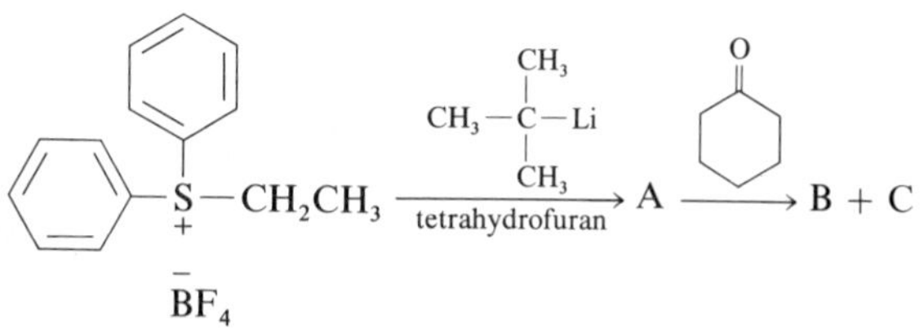

tetrahydrofuran
A
B + C

18.25 The preparation of intermediates in the synthesis of coriolin, a natural product with antibacterial and antitumor activity, involves the epoxidation of double bonds. Predict what the products of the two steps of the synthesis, shown below, will be.

(a)

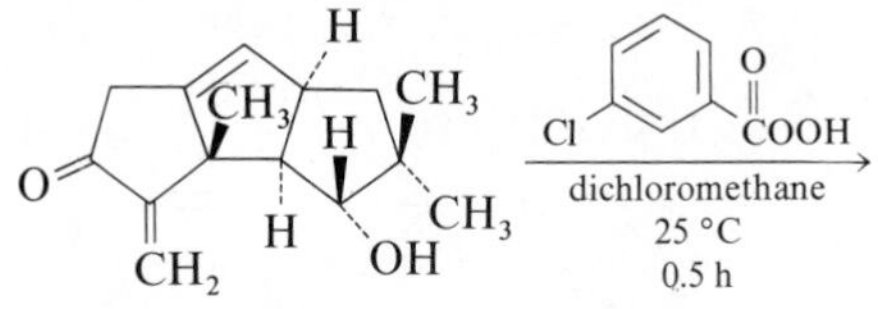
H
CH3
CH3
H
O
CH3
H
OH
CH2
Cl
COOH
dichloromethane
25 °C
0.5 h

(b)

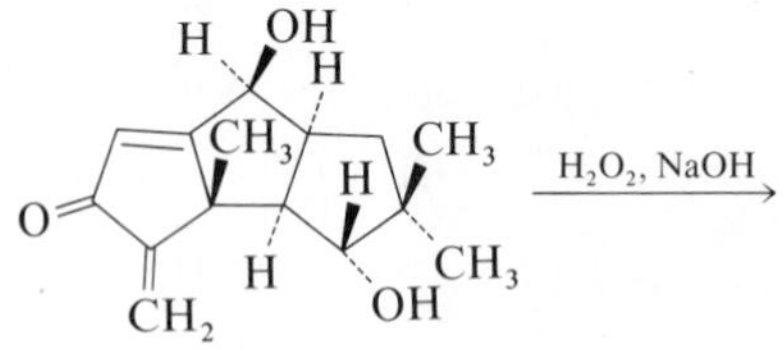
H
OH
H
CH3
CH3
H
O
H
CH3
OH
CH2
H2O2, NaOH

18.26 A carbanion in which the negatively charged carbon atom is adjacent to a sulfone, which has a sulfur atom bonded to two oxygen atoms, is a weaker base than a carbanion in which the negatively charged carbon is adjacent to a sulfoxide, which has only one oxygen atom bonded to sulfur. How would you rationalize this fact?

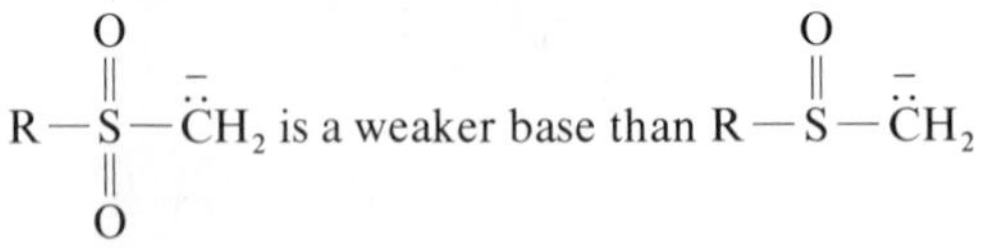
R—S—CH2 is a weaker base than R—S—CH2

18.27 Write a mechanism for the following reaction that accounts for the product obtained.

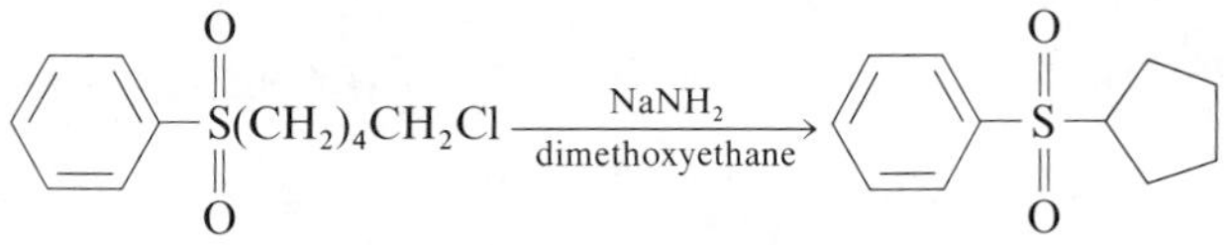
S(CH2)4CH2Cl
NaNH2
dimethoxyethane

18.28 Brevetoxins, isolated from "red tide" algae, are potent nerve and heart poisons. Some intermediates in a synthesis of a brevetoxin are shown below. Show how you would carry out the various steps.

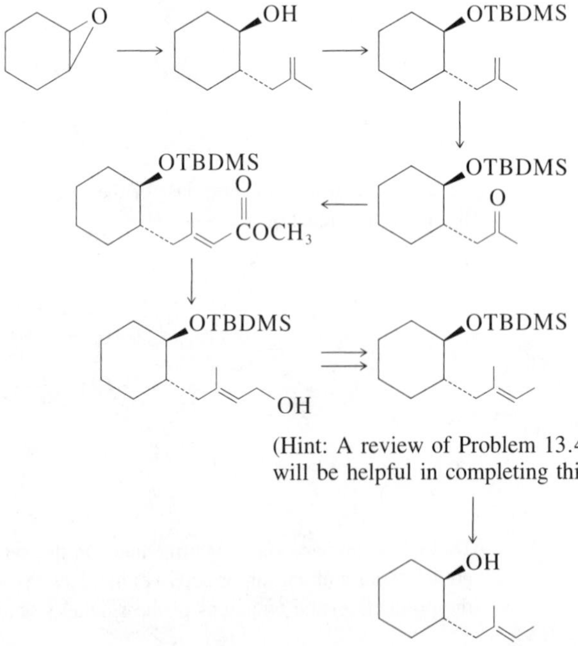
OH
OTBDMS
OTBDMS
COCH3
OTBDMS
OTBDMS
OH
OTBDMS
OH

(Hint: A review of Problem 13.46, p. 536 will be helpful in completing this step.)

18.29 A study of how long-chain branched fatty acids are synthesized by marine organisms required acids labeled with radioactive carbon-14 at the carbon atom of the carbonyl group. A synthesis of 12-methyltetradecanoic acid was carried out, with 9-bromo-1-nonanol and 2-methylbutanal as the starting materials and $K^{14}CN$ as the source of radioactive carbon. Outline a synthesis that employs the Wittig reaction. The tetrahydropyranyl ether also plays an important role as a protecting group in this synthesis.

18.30 A number of plants produce chemical substances that prevent insects from feeding on them. Some steps in the synthesis of such a compound, called an **antifeedant,** are shown below. Supply reagents for the transformations. (The wavy lines indicate either that stereochemistry is not known or that a mixture of stereoisomers may exist.)

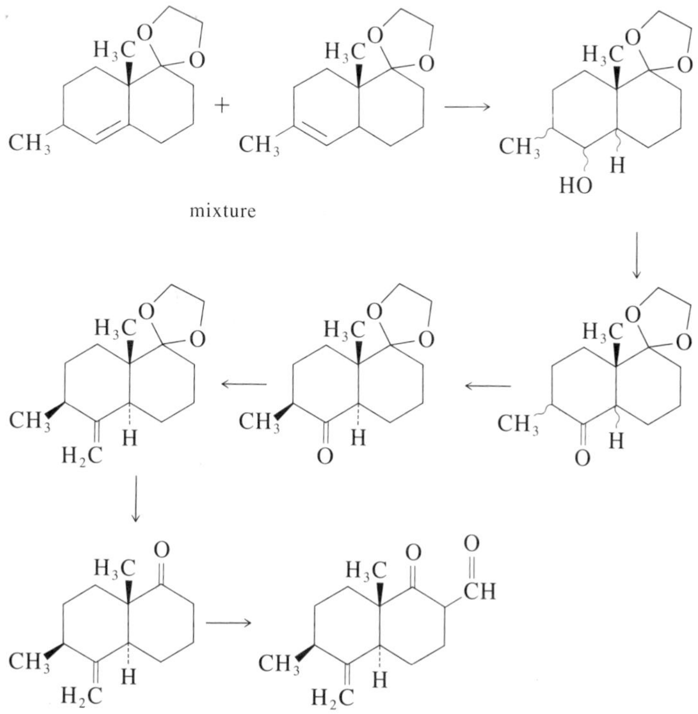

mixture

18.31 Macrocyclic ethers and lactones (p. 608) have interesting properties as antibiotics and probably play a role in the transport of ions through membranes. Some steps in the synthesis of such a compound are shown below. How would you carry out the reactions shown?

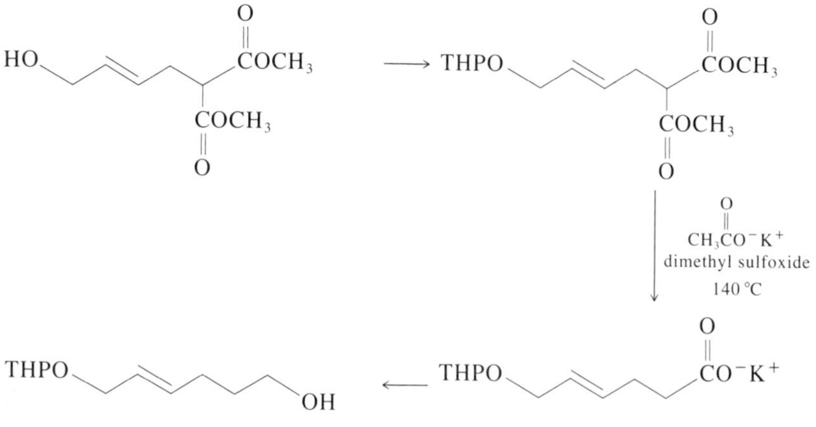

dimethyl sulfoxide
140 °C

(Propose a mechanism for this step of the reaction.)

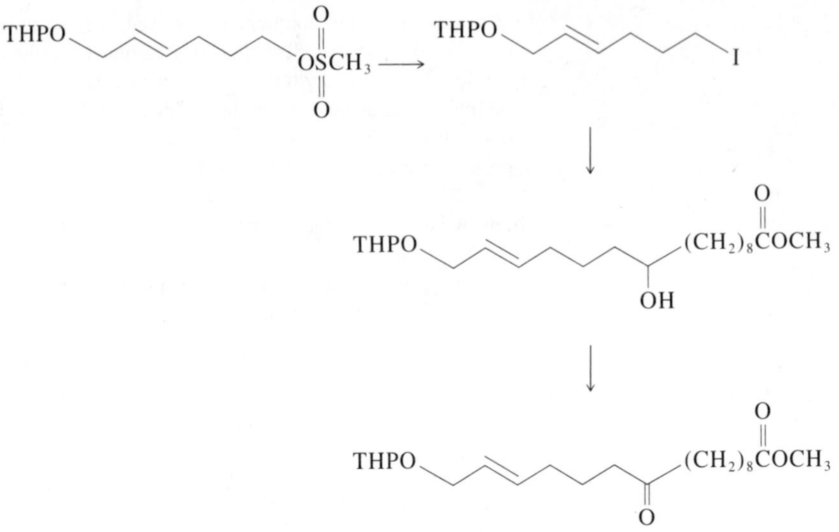
THPO
OSCH3
O
O
THPO
I
THPO
(CH2)8COCH3
O
OH
THPO
(CH2)8COCH3
O
O

18.32 Much current research is focused on the chemistry of vision. Compounds similar to rhodopsin (p. 713) have been synthesized for study. One such synthesis is outlined below. Supply the necessary reagents for the conversions shown.

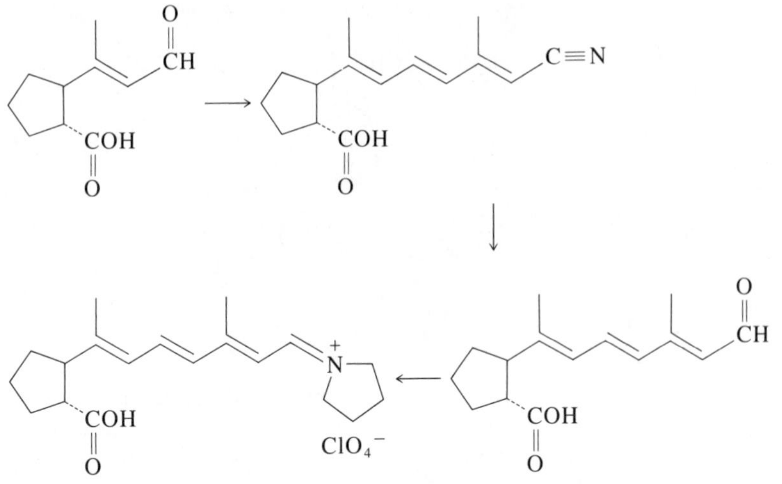
O
CH
COH
O
C≡N
COH
O
+
N
ClO4−
COH
O
O
CH
COH
O

18.33 A synthesis of β-carotene (p. 712) uses the following two reagents.

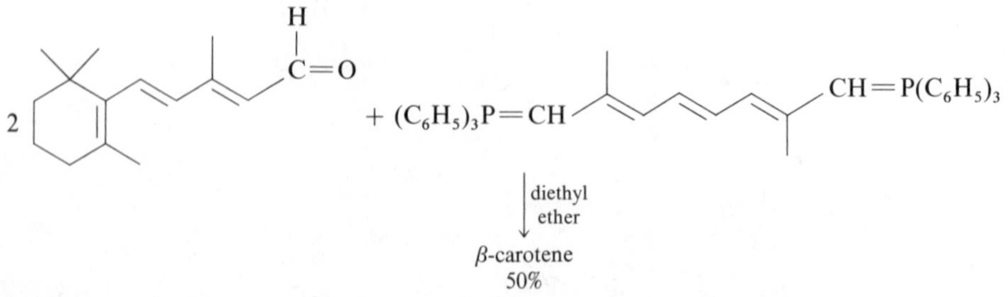
H
C=O
2
+ (C6H5)3P=CH
CH=P(C6H5)3
diethyl
ether
β-carotene
50%

The double Wittig reagent is synthesized by the following sequence of reactions. Give structural formulas for the compounds and intermediates indicated by letters.

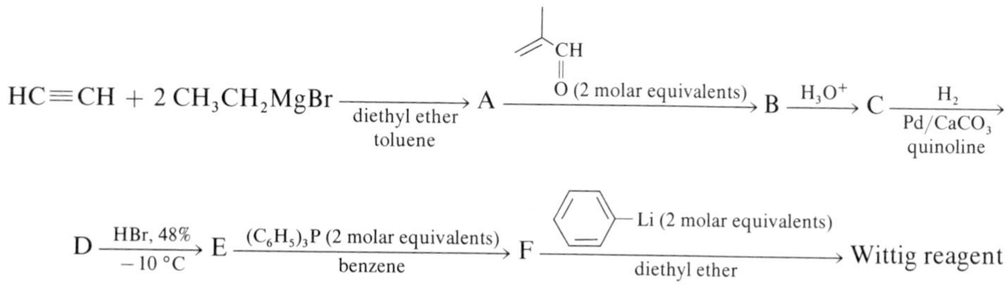
HC≡CH + 2 CH3CH2MgBr
diethyl ether
toluene
A
CH
O (2 molar equivalents)
B
H3O+
C
H2
Pd/CaCO3
quinoline
D
HBr, 48%
−10 °C
E
(C6H5)3P (2 molar equivalents)
benzene
F
Li (2 molar equivalents)
diethyl ether
Wittig reagent

18.34 A synthesis of disparlure, the sex pheromone of the gypsy moth (p. 702), was carried out in the following way. Provide structural formulas for all compounds or intermediates that are designated by letters.

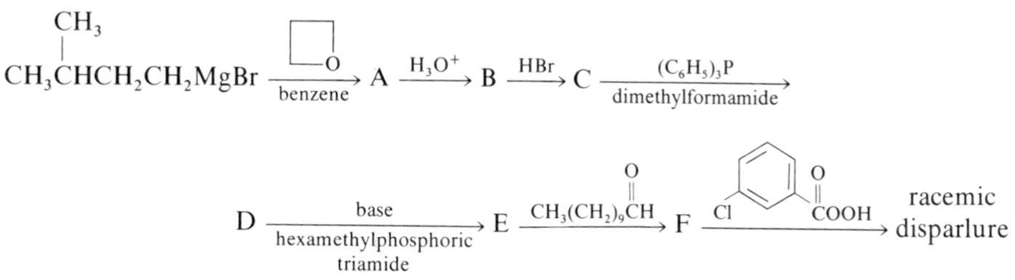
CH3
CH3CHCH2CH2MgBr
O
benzene
A
H3O+
B
HBr
C
(C6H5)3P
dimethylformamide
D
base
hexamethylphosphoric
triamide
E
O
CH3(CH2)9CH
F
Cl
O
COOH
racemic
disparlure

The Wittig reaction in hexamethylphosphoric triamide (p. 229) gives mostly the *Z* alkene.

18.35 The sex attractant of a female beetle that bores into the ponderosa pine has an unusual bicyclic structure, an internal cyclic ketal between a methyl ketone and a diol. The compound, called brevicomin, exists in its bicyclic form as the exo and endo isomers, each of which is a racemic mixture. Both diastereomeric forms are found in nature but only the exo isomer has biological activity.

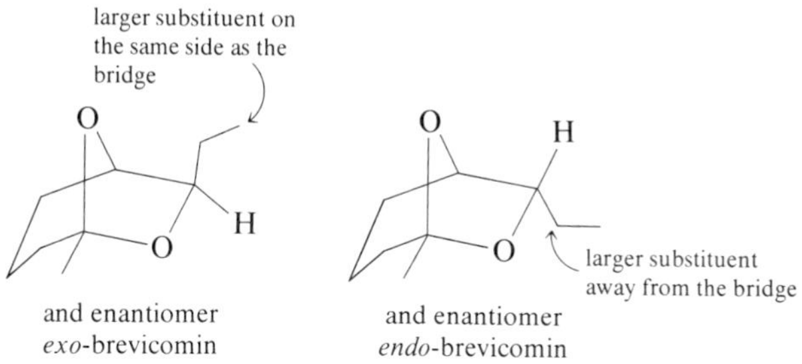
larger substituent on
the same side as the
bridge
O
O
H
O
H
O
larger substituent
away from the bridge
and enantiomer
exo-brevicomin
and enantiomer
endo-brevicomin

Both diastereomers of brevicomin were synthesized by the following sequence of reactions. Give structures for the intermediates and products indicated by letters. Be as specific as you can about stereochemistry at each stage.

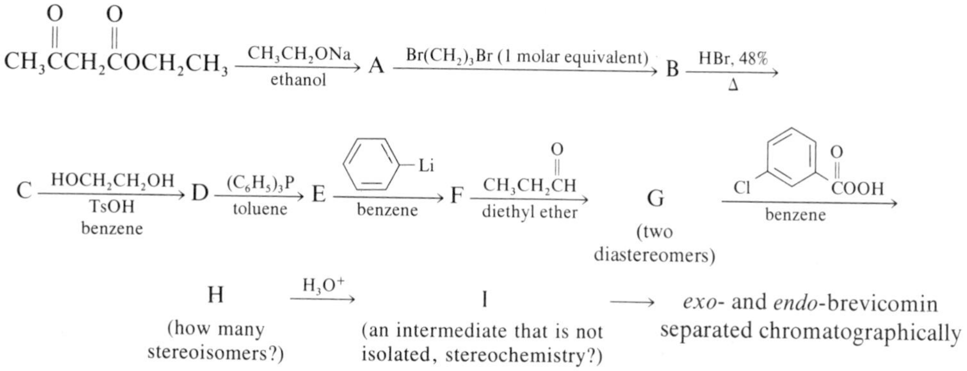
O O
CH3CCH2COCH2CH3
CH3CH2ONa
ethanol
A
Br(CH2)3Br (1 molar equivalent)
B
HBr, 48%
Δ
C
HOCH2CH2OH
TsOH
benzene
D
(C6H5)3P
toluene
E
Li
benzene
F
O
CH3CH2CH
diethyl ether
G
(two
diastereomers)
Cl
O
COOH
benzene
H
(how many
stereoisomers?)
H3O+
I
(an intermediate that is not
isolated, stereochemistry?)
exo- and endo-brevicomin
separated chromatographically

Another synthesis of *exo*-brevicomin is given below. Complete the steps for this synthesis also.

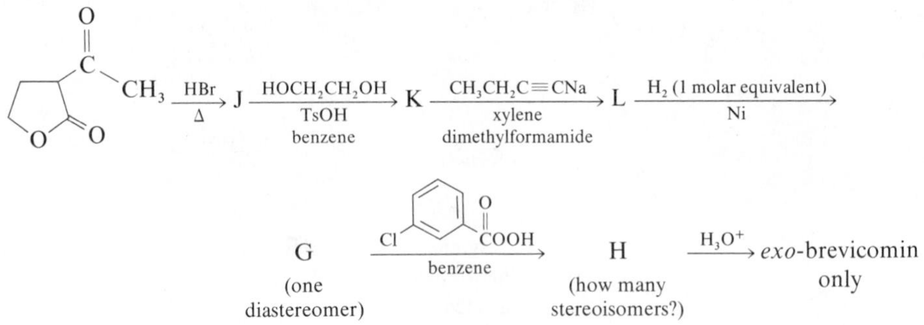
CH3
HBr
Δ
J
HOCH2CH2OH
TsOH
benzene
K
CH3CH2C≡CNa
xylene
dimethylformamide
L
H2 (1 molar equivalent)
Ni
G
(one
diastereomer)
Cl
COOH
benzene
H
(how many
stereoisomers?)
H3O+
exo-brevicomin
only

18.36 The proton magnetic resonance spectrum of ethyl (*E*)-3-phenylpropenoate is shown in Figure 18.1. Assign the bands in the spectrum, and analyze any splitting that is observed. How do you account for the large difference in chemical shift for the two vinylic hydrogen atoms?

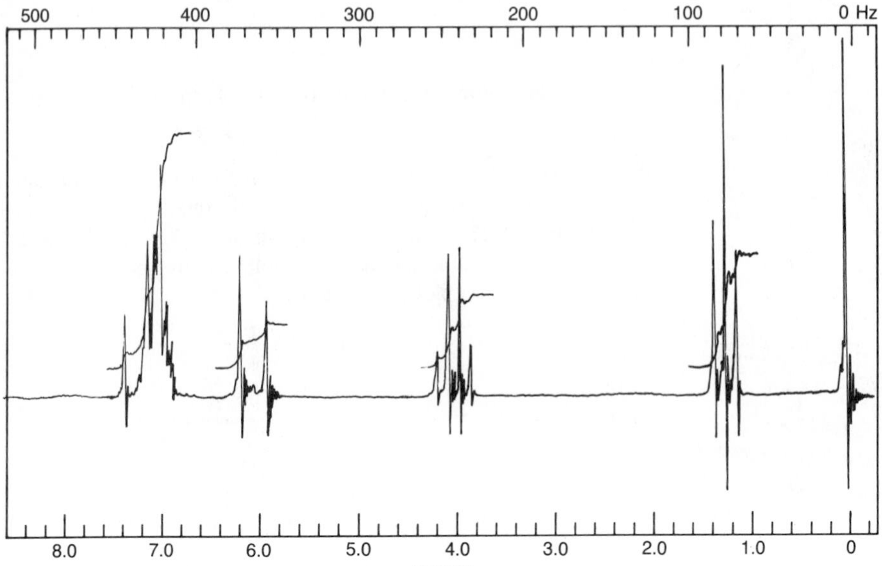
500
400
300
200
100
0 Hz
8.0
7.0
6.0
5.0
4.0
3.0
2.0
1.0
0
PPM(δ)

FIGURE 18.1

18.37 The carbon-13 nuclear magnetic resonance spectrum of (*E*)-2-butenoic acid has bands at 18.0, 122.6, 147.5, and 172.3 ppm downfield from TMS. Assign these bands to the carbon atoms in (*E*)-2-butenoic acid, and explain your reasoning in making these assignments. Table 11.4 (p. 413) may be useful.

19

The Chemistry of Aromatic Compounds I. Electrophilic Aromatic Substitution

A • L O O K • A H E A D

Aromatic hydrocarbons have multiple π bonds and yet are stable toward the electrophilic addition reactions that alkenes undergo. Aromatic hydrocarbons undergo electrophilic substitution reactions instead, an example being the reaction of benzene with bromine.

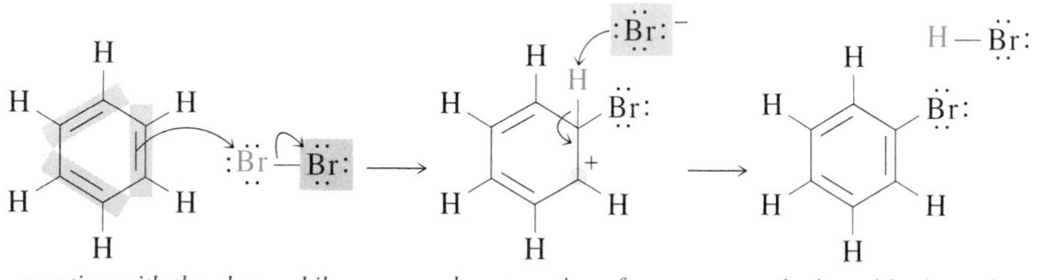

reaction with the electrophile
deprotonation of cationic intermediate
substituted hydrocarbon

Many electrophiles react with aromatic hydrocarbons, so electrophilic aromatic substitution reactions are among the most important reactions for transforming organic compounds. The structure of the compound undergoing the substitution determines its reactivity and the regioselectivity of the reaction. This chapter will look at the different electrophiles that react with aromatic compounds and show how to predict the major product of an electrophilic aromatic substitution reaction.

Chemists have long debated the question of what makes a compound aromatic. This chapter will also examine the current model for aromaticity.

A. Introduction. Resonance Stabilization of Benzene

Vanilla and oil of wintergreen are representatives of a class of flavoring agents isolated from plants and containing chemical substances called aromatic because of their characteristic fragrances.

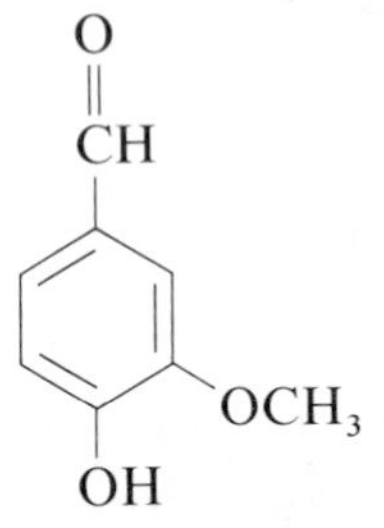

vanillin

constituent of vanilla

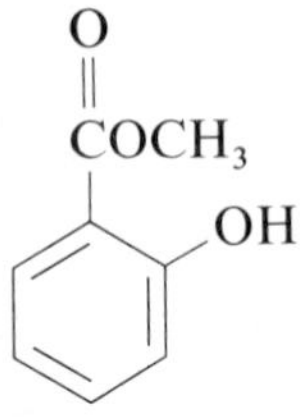

methyl salicylate

constituent of oil of wintergreen

When the structures of a number of these compounds were determined, they were all found to contain substituted benzene rings. In time, benzene and its derivatives were called **aromatic hydrocarbons** to distinguish them from saturated acyclic and cyclic hydrocarbons and simple unsaturated compounds. Still later, the term was redefined to include compounds that do not have a benzene ring but have chemical properties resembling those of benzene. Such compounds are said to have **aromaticity.**

Chemists are still arguing about what aromaticity means. At first the term was used to indicate that a compound had special stability, usually defined as resonance energy, and that it resisted the addition reactions expected of a molecule with that degree of unsaturation (p. 174). Benzene and substituted benzenes do not react with a large number of reagents that do react with other functional groups. This unreactivity has been demonstrated many times in earlier chapters in reactions that transform substituents on aromatic rings but leave the rings untouched.

The experimental heat of hydrogenation (p. 278) of benzene is −49.8 kcal/mol. This is 36 kcal/mol less than the heat of hydrogenation expected for ''1,3,5-cyclohexatriene,'' a hypothetical six-membered ring containing three localized double bonds (shown below as containing alternating short double bonds and long single bonds). This value is calculated as three times the value of the heat of hydrogenation of cyclohexene.

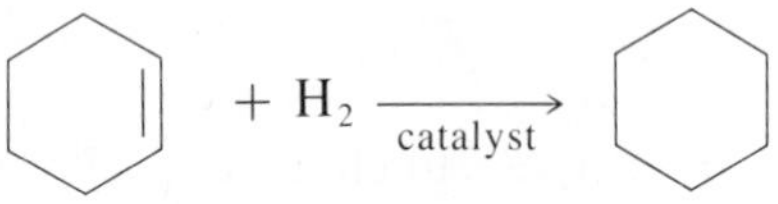

$\Delta H^\circ = -28.6$ kcal/mol

cyclohexene — cyclohexane — experimental

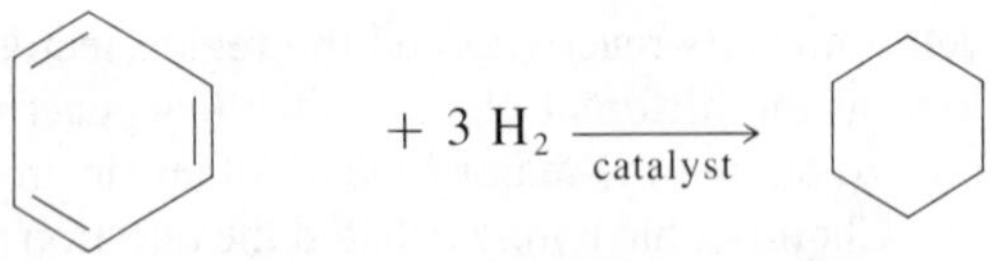

$\Delta H^\circ = -85.8$ kcal/mol

''1,3,5-cyclohexatriene'' — cyclohexane — calculated

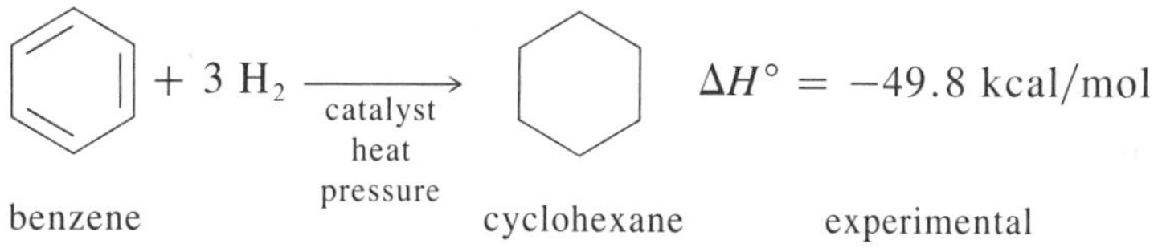
+ 3 H_2
catalyst
heat
pressure
$\Delta H° = -49.8$ kcal/mol
benzene
cyclohexane
experimental

Benzene thus is more stable by 36 kcal/mol than expected for a compound containing three double bonds in a six-membered ring. The value 36 kcal/mol is called the empirical resonance energy (p. 684) of benzene. It reflects the degree of stabilization of the aromatic ring that is attributed to the delocalization of π electrons. This resonance stabilization of benzene and other aromatic compounds is responsible for the lower reactivity of their π bonds in comparison to that of the π bonds of alkenes and alkynes.

B. The Aromatic Sextet

In 1926, Sir Robert Robinson, a British chemist who was among the first to suggest and use the ideas that form the basis for the mechanisms chemists write for organic reactions, recognized that the compounds that had been classified as aromatic always had six electrons, in either π or nonbonding orbitals, in a planar ring. Benzene is such a compound. Robinson suggested that there was a special stability associated with what he called an **aromatic sextet.** Some other compounds that exhibit the experimentally observed properties that have been assigned to aromaticity, such as resistance to addition reactions, are shown below.

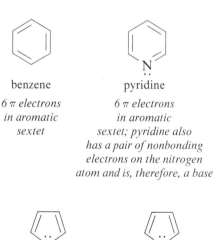

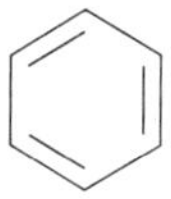

benzene

6 π electrons in aromatic sextet

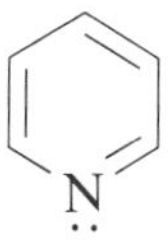
N

pyridine

6 π electrons in aromatic sextet; pyridine also has a pair of nonbonding electrons on the nitrogen atom and is, therefore, a base

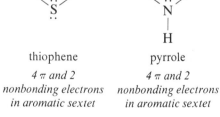

S

thiophene

4 π and 2 nonbonding electrons in aromatic sextet

N
H

pyrrole

4 π and 2 nonbonding electrons in aromatic sextet

As the theory of molecular bonding developed further, it became clear that a sextet was indeed a significant number, corresponding to the number of electrons that fill the three bonding molecular orbitals of benzene (p. 794). Fewer electrons would leave unpaired electrons in the bonding orbitals and create a species with radical character. Any more electrons would have to go in antibonding molecular orbitals and would create an unstable, high-energy species. Benzene has a filled shell of molecular orbitals, just as an element such as neon has a filled shell of atomic orbitals (Figure 19.1).

In 1931, German chemist Erich Hückel pointed out that there were other numbers of electrons that corresponded to filled shells for smaller or larger rings. He proposed what has since become known as **Hückel's Rule:** any conjugated monocyclic polyene that is planar and has $(4n + 2)$ π electrons, with $n = 0, 1, 2$, etc. will exhibit the special stability associated with aromaticity. Since then, researchers have tried to synthesize compounds that will test Hückel's prediction.

Benzene is the ideal aromatic compound. It has $4n + 2$, or 6, π electrons with $n = 1$. The six carbon atoms in benzene are sp^2-hybridized, so their regular bond angles of 120° create a perfect hexagon. The molecule is planar and totally symmetrical. Because all the ring atoms are carbon, no differences in polarity are introduced by the presence of other elements in the ring. The electrons are, therefore, delocalized evenly over the entire system.

Cyclobutadiene, a conjugated cyclic polyene with four carbon atoms in the ring, is unstable, and cyclooctatetraene, with an eight-membered ring, behaves as a polyene.

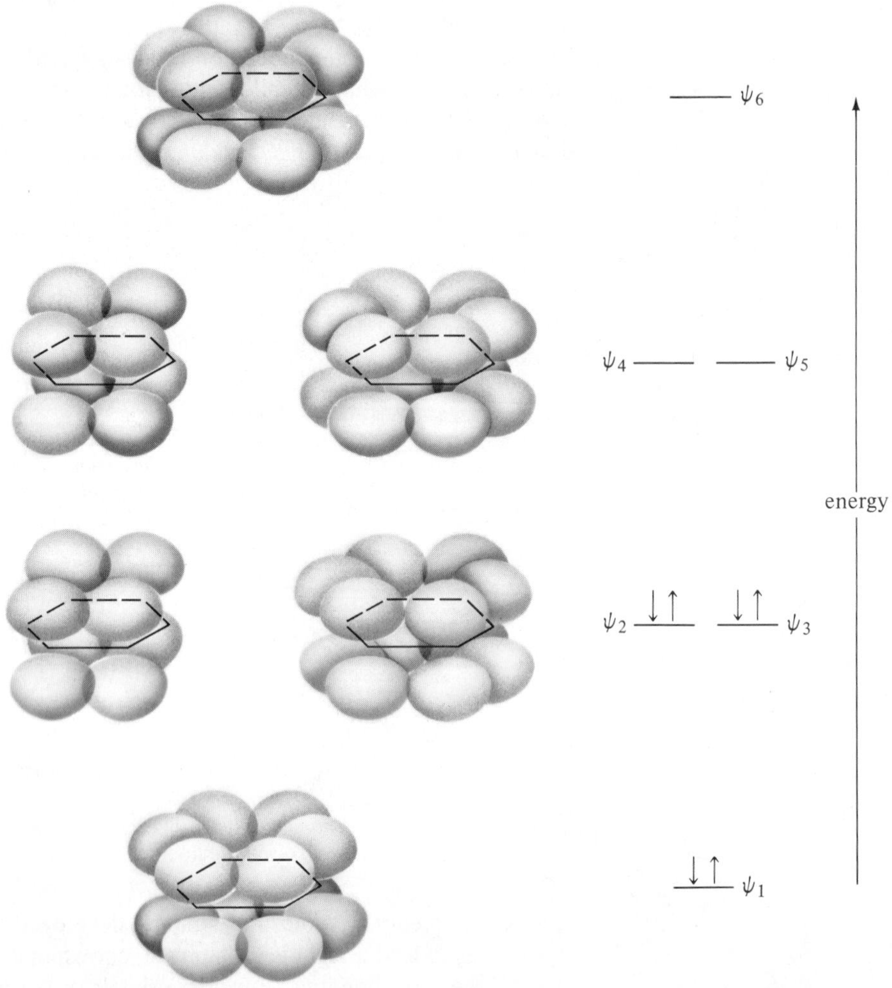
ψ_6
ψ_4
ψ_5
energy
ψ_2
ψ_3
ψ_1

FIGURE 19.1 Combinations of the six $2p$ atomic orbitals of benzene that form the six molecular π orbitals, with the relative energies of the three bonding and three antibonding orbitals.

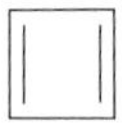

cyclobutadiene

4 π electrons, unstable

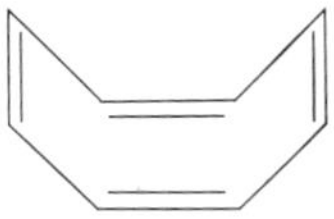

cyclooctatetraene

8 π electrons, reacts as a polyene

two monocyclic conjugated polyenes that do not have (4n + 2) π electrons and are not aromatic

Cyclobutadiene has 4 π electrons and thus does not meet the requirement of Hückel's Rule that an aromatic compound have $(4n + 2)$ π electrons. The compound has never been isolated but has been detected by infrared spectroscopy when it is generated at temperatures such as 4 K. Cyclooctatetraene is an ordinary polyene in its reactivity. It is not planar (a regular octagon would have to have bond angles of 135°), so there cannot be continuous conjugation of the double bonds around the ring. It, too, lacks the proper number of electrons to be aromatic, according to Hückel's Rule.

Larger rings with conjugated cis double bonds cannot be planar because the bond angles required for such geometrical figures are even larger than the 135° of a regular octagon. However, trans double bonds are possible for large rings. Such a configuration puts hydrogen atoms inside the ring where they interfere with each other and prevent planarity. Only for rings consisting of eighteen or more carbon atoms is this steric hindrance removed. None of these larger ring compounds, called **annulenes,** has the stability of benzene, although some of them have some of the properties associated with aromaticity, such as high resonance energy. [18]Annulene is shown below.

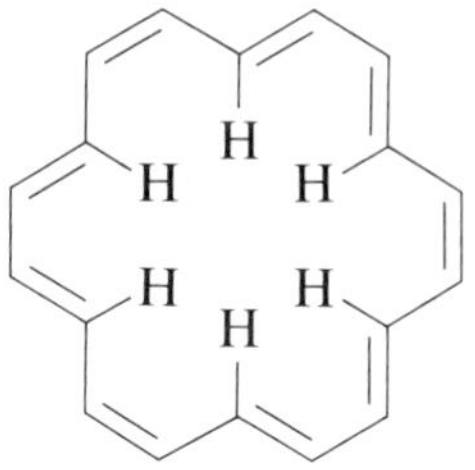
H
H
H
H
H
H

[18]annulene

planar

[18]annulene has (4n + 2) π electrons with n = 4, but lacks the stability of benzene

PROBLEM 19.1

Decide whether or not each of the following species is aromatic.

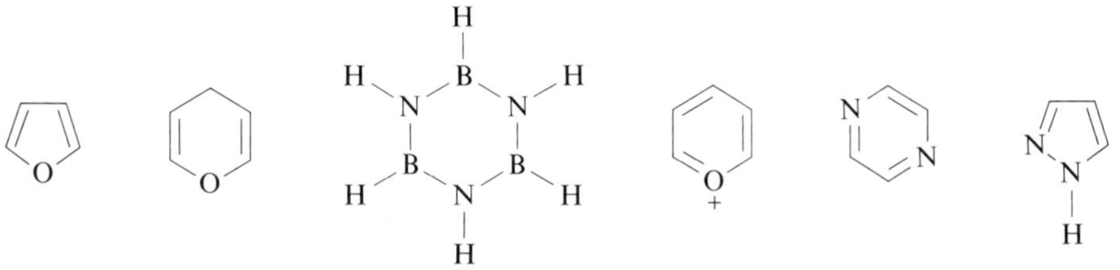
O
O
H
H
B
H
N
N
B
B
H
N
H
H
O
+
N
N
N
N
H

C. Detection of Aromaticity by Nuclear Magnetic Resonance Spectroscopy

A modern criterion of aromaticity for an unsaturated cyclic compound is the detection of a ring current (p. 387) by nuclear magnetic resonance spectroscopy. A comparison of the spectra of benzene and 1,3,5,7-cyclooctatetraene (Figure 19.2) shows quite clearly the extra deshielding of the hydrogen atoms on the benzene

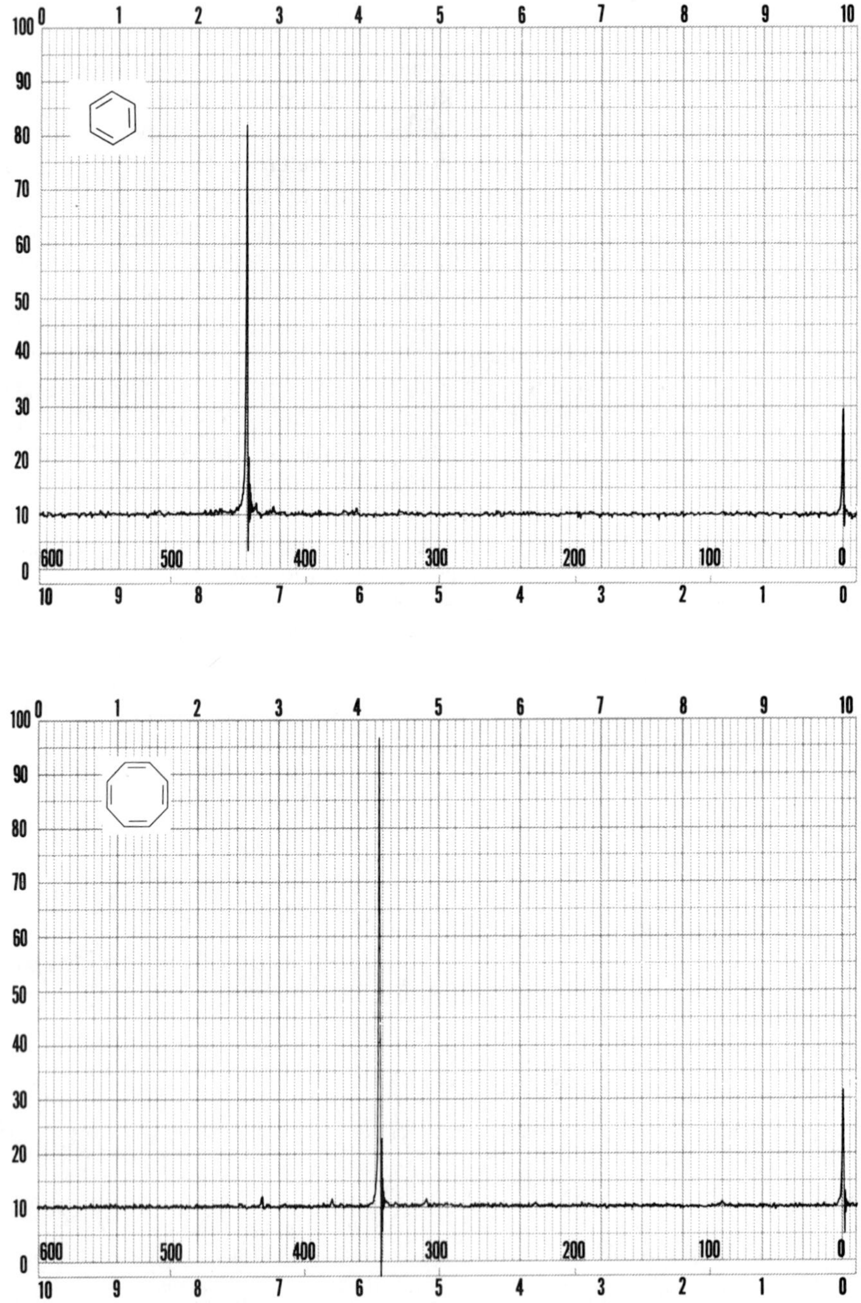

FIGURE 19.2 Proton magnetic resonance spectra for benzene and 1,3,5,7-cyclooctatetraene. (From *The Aldrich Library of NMR Spectra*)

ring. The chemical shift for the hydrogen atoms in 1,3,5,7-cyclooctatetraene is similar to that seen for the vinylic hydrogen atoms in cyclohexene (Figure 11.7, p. 386). The spectral evidence for cyclooctatetraene supports the earlier conclusion, which is based on chemical evidence and also on theory, that the double bonds in this cyclic polyene are localized and the compound does not have aromatic character.

One of the most dramatic demonstrations of the ring-current effect in a conjugated cyclic polyene can be seen in the proton magnetic resonance spectrum of [18]annulene (p. 775). The spectrum of this compound has two bands, one at δ 8.9 and the other at δ −1.8 (1.8 ppm to the right of, or upfield) from TMS. These bands have relative intensities of 2:1. Therefore, the low-field signal is assigned to the hydrogen atoms around the outside of the ring, which lie in the deshielding region around the molecule. The high-field signal is assigned to the hydrogen atoms inside the ring. They lie in the shielding region created by the induced magnetic field. [18]Annulene satisfies the nuclear magnetic resonance criterion for aromaticity.

PROBLEM 19.2

The proton magnetic resonance spectrum of thiophene (p. 773) is shown in Figure 19.3. In comparison, the hydrogen atoms on the sp^2-hybridized carbon atoms of methyl vinyl sulfide, $CH_3SCH{=}CH_2$, absorb at δ 4.95, 5.18, and 6.43. What conclusions can you draw from these data?

FIGURE 19.3

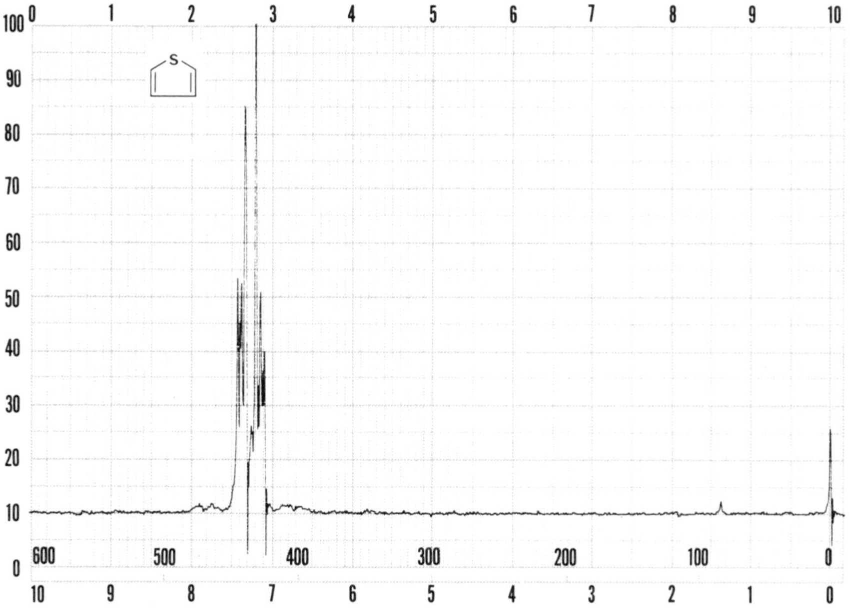
S
0 1 2 3 4 5 6 7 8 9 10
100 90 80 70 60 50 40 30 20 10 0
600 500 400 300 200 100 0
10 9 8 7 6 5 4 3 2 1 0

(From *The Aldrich Library of NMR Spectra*)

D. Aromaticity of Hydrocarbon Anions and Cations

Ions, as well as neutral molecules, can be said to have aromaticity. For example, cyclopentadiene (pK_a 15) is a strong acid compared to other types of alkenes and can be deprotonated by *tert*-butoxide anion.

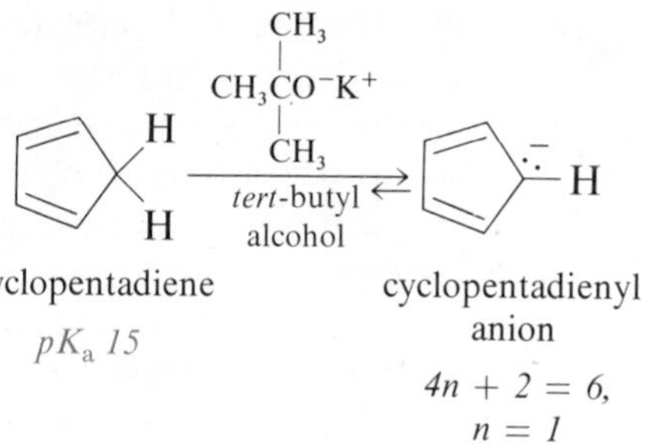
CH_3
$CH_3CO^-K^+$
CH_3
tert-butyl alcohol
H
H
H
cyclopentadiene
pK_a 15
cyclopentadienyl anion
$4n + 2 = 6,$
$n = 1$

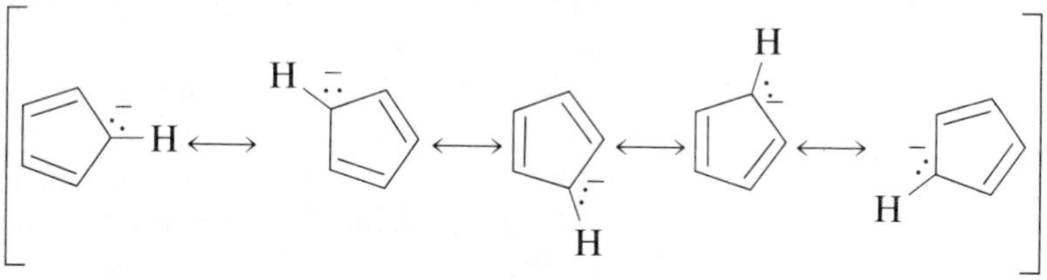
H
H
H
H
H

resonance contributors for the cyclopentadienyl anion

The ease with which cyclopentadiene loses a proton must mean that the resulting anion is particularly stable. The cyclopentadienyl anion is stabilized by delocalization of charge. However, if this were the full explanation for the special stability of the cyclopentadienyl anion, the corresponding cation, for which just as many resonance contributors can be written, should also be stable. But attempts to prepare such a cation have failed. The cyclopentadienyl anion has six electrons in a planar ring with the possibility of continuous delocalization of the electrons. It meets the criteria of Hückel's Rule. The cyclopentadienyl cation, on the other hand, is a system with 4 π electrons and is expected to be unstable.

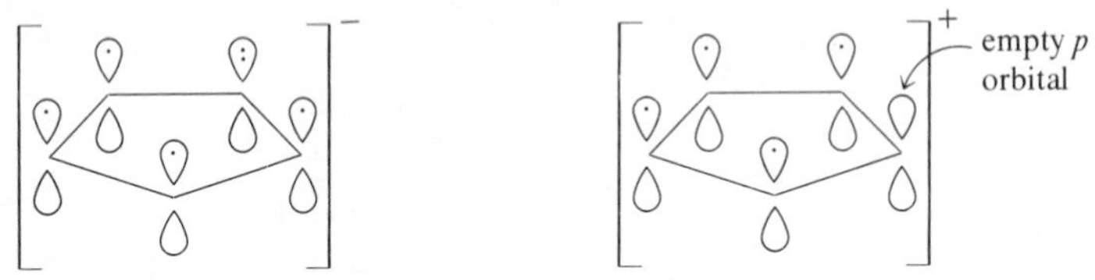
empty *p* orbital

p orbitals of cyclopentadienyl anion showing the presence of 6 π electrons ($4n + 2, n = 1$)

p orbitals of cyclopentadienyl cation showing the presence of only 4 π electrons

Cycloheptatriene is a compound with 6 π electrons in a ring. It is not aromatic because there cannot be continuous delocalization of the π electrons around the ring as long as one of the carbon atoms is an sp^3-hybridized carbon atom with no unhybridized *p* orbital to overlap with the *p* orbitals on the other carbon atoms. Cycloheptatriene adds bromine then easily loses hydrogen bromide to give a crystalline compound. This compound is insoluble in nonpolar organic solvents such as ether but dissolves readily in cold water and gives an instant precipitate of silver bromide when silver nitrate is added to its solution. The compound melts at 203 °C with decomposition. All of these properties suggest that the monobromo compound ionizes as shown.

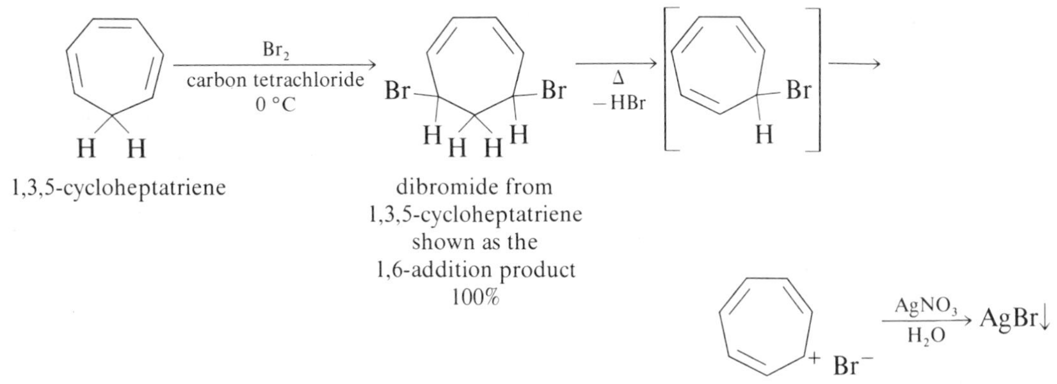
Br2
carbon tetrachloride
0 °C
Br
Br
H H
Δ
−HBr
Br
H
AgNO3
H2O
AgBr↓
Br−

1,3,5-cycloheptatriene

dibromide from 1,3,5-cycloheptatriene shown as the 1,6-addition product 100%

cycloheptatrienyl bromide
tropylium bromide
mp 203 °C with decomposition, insoluble in nonpolar organic solvents, soluble in water

Study Guide Concept Map 19.1

In other words, the compound is the salt of a stable carbocation, known as the tropylium ion, with bromide ion. The aromaticity possible for the cation stabilizes it so much that the normal tendency for covalent bonds to form between carbon and bromine is overcome.

PROBLEM 19.3

Write a mechanism showing how tropylium bromide is formed from the dibromide of 1,3,5-cycloheptatriene.

PROBLEM 19.4

What other dibromides of 1,3,5-cycloheptatriene besides the 1,6-addition product would be possible?

PROBLEM 19.5

Draw a representation of the tropylium ion showing the skeleton of σ bonds for the molecule and the p orbitals used for delocalization of charge. Include all of the available π electrons.

19.2 KEKULE STRUCTURES AND NOMENCLATURE FOR AROMATIC COMPOUNDS

Michael Faraday isolated benzene in 1825 from the residues of gaseous fractions of coal. The structural formula that chemists now use for benzene was first written in 1865 by the German chemist August Kekulé after a dream in which he saw rows of atoms twisting with a snakelike motion until finally, as he said: "One of the snakes had seized hold of its own tail, and the form whirled mockingly before my eyes. As if by a flash of lightning, I awoke, and this time also I spent the rest of the night in working out the consequences of the hypothesis." Kekulé recognized that the chemical properties of benzene were not compatible with the alternating double and single bonds indicated by his structural formula, and by 1872 he arrived at the idea that the double bonds were not localized. He wrote two structures, the ones now known as resonance contributors for benzene (p. 66), to show the changing nature of the double bonds, a remarkable feat of chemical imagination for his time, when electrons had not yet been discovered.

Along with benzene, other hydrocarbons that are unsaturated in their molecular formulas but stable toward reagents that add to double bonds have been isolated from various sources, chiefly from fractions of coal tar. Some of these compounds are shown below with their names.

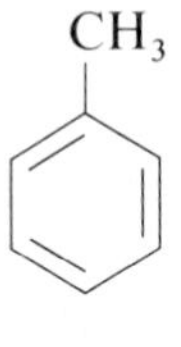
CH3

methylbenzene
toluene

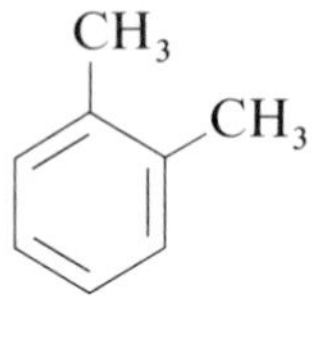
CH3
CH3

1,2-dimethylbenzene
o-xylene

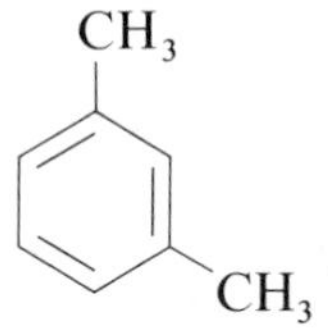
CH3
CH3

1,3-dimethylbenzene
m-xylene

CH3
CH3

1,4-dimethylbenzene
p-xylene

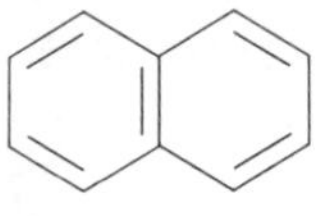

naphthalene

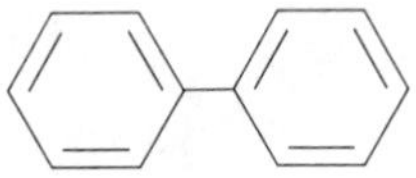

biphenyl

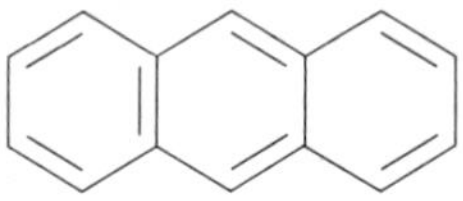

anthracene

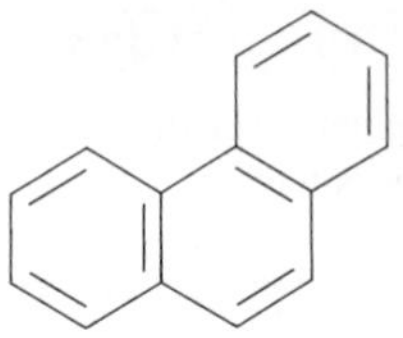

phenanthrene

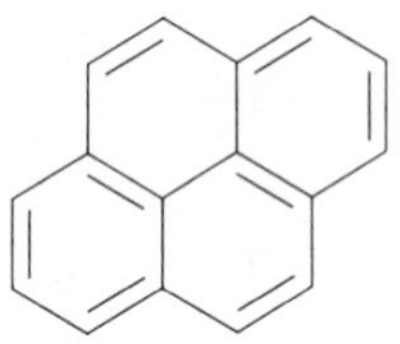

pyrene

Benzene, other aromatic hydrocarbons, and their alkyl derivatives as a class are known as **arenes.** Besides hydrocarbons, coal tar also contains aromatic compounds containing nitrogen, sulfur, or oxygen. Among the following, those having elements other than carbon in a ring—pyridine, quinoline, indole, and thiophene—are also known as **heterocyclic compounds.**

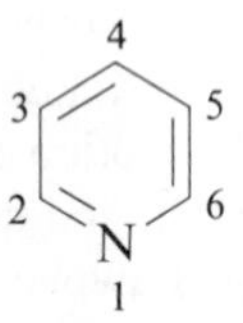
4
3
5
2
6
N
1

pyridine

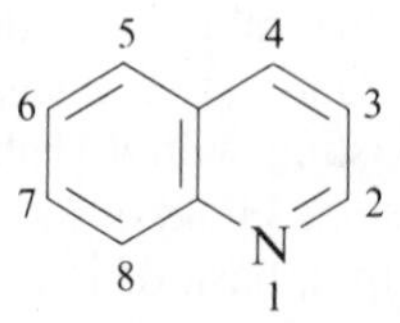
5
4
6
3
7
2
8
N
1

quinoline

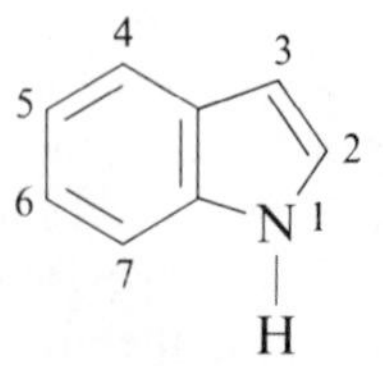
4
3
5
2
6
N
1
7
H

indole

S

thiophene

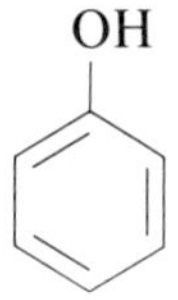
OH

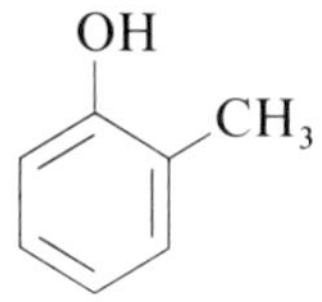
OH
CH_3

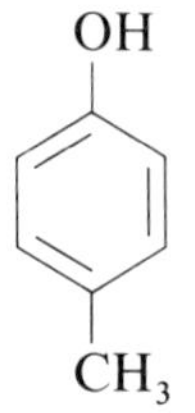
OH
CH_3

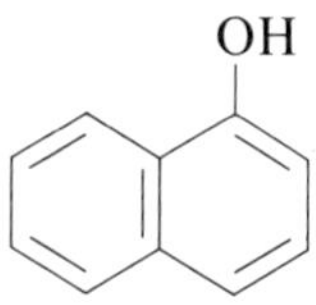
OH

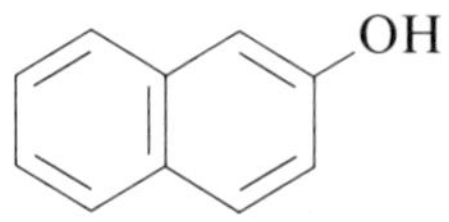
OH

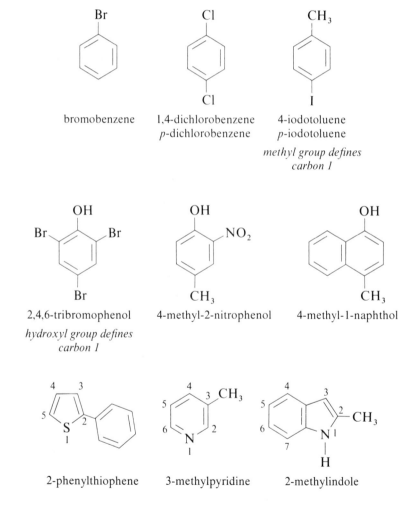

phenol | *o*-cresol | *p*-cresol | 1-naphthol α-naphthol | 2-naphthol β-naphthol

Other aromatic compounds are named as substitution products of benzene and the compounds shown above. The name of the parent compound is given a prefix with the name of each substituent and numbers indicating relative positions. For benzene derivatives, ortho, meta, and para are also used to designate 1,2, 1,3, and 1,4 relationships between two substituents (p. 479). The following examples illustrate these rules.

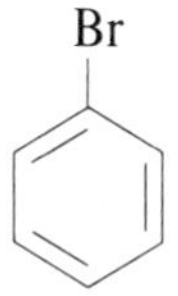
Br

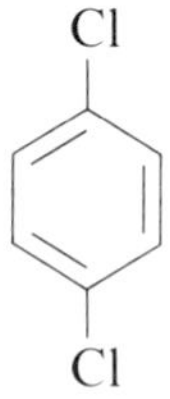
Cl
Cl

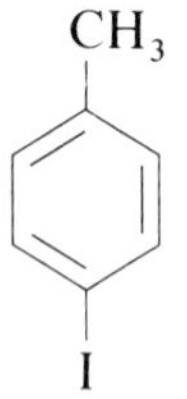
CH_3
I

bromobenzene | 1,4-dichlorobenzene *p*-dichlorobenzene | 4-iodotoluene *p*-iodotoluene *methyl group defines carbon 1*

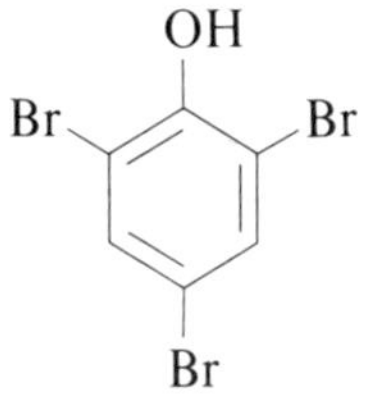
OH
Br
Br
Br

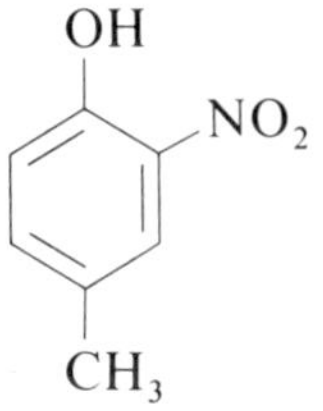
OH
NO_2
CH_3

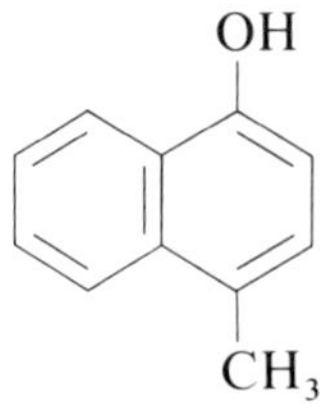
OH
CH_3

2,4,6-tribromophenol *hydroxyl group defines carbon 1* | 4-methyl-2-nitrophenol | 4-methyl-1-naphthol

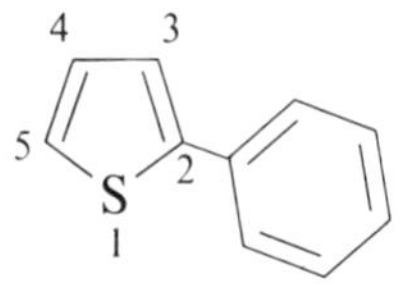
4 3
5 2
S
1

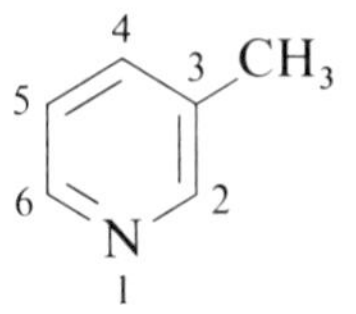
4
3 CH_3
5
6 2
N
1

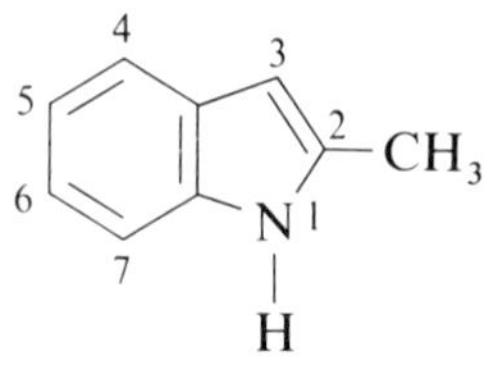
4
3
5
2 CH_3
6
N 1
7
H

2-phenylthiophene | 3-methylpyridine | 2-methylindole

Some of the compounds that are shown above contain more than one aromatic ring. In biphenyl (p. 780), for example, the two rings are joined by a single bond and are conjugated with each other. In other cases, such as naphthalene, anthracene, phenanthrene, and pyrene, the aromatic rings share at least one side and are said to contain **fused-ring systems** or to be **polycyclic aromatic hydrocarbons.**

The fused-ring aromatic hydrocarbons do not have completely identical resonance contributors, as benzene does. For example, the Kekulé structures, the structural formulas with localized double and single bonds, that can be written as resonance contributors for naphthalene are not equivalent.

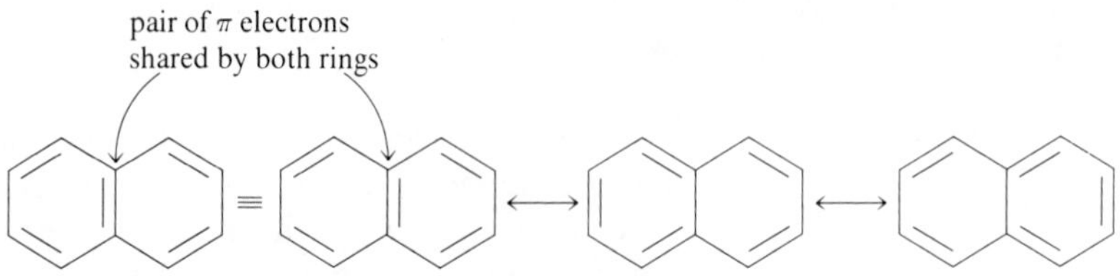
pair of π electrons
shared by both rings

resonance contributors for naphthalene

This nonequivalence of the resonance contributors is reflected in the chemical properties of polycyclic aromatic hydrocarbons. The difference in reactivity between rings in such compounds is clearly illustrated by anthracene and phenanthrene. The center ring in anthracene is highly reactive. For example, anthracene reacts with maleic anhydride in a Diels-Alder reaction as though its central ring were a diene (p. 691).

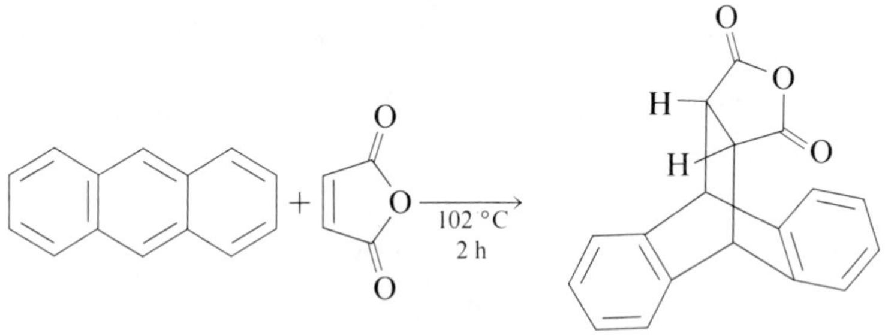
O
O
O
H
H
O
O
O
102 °C
2 h

anthracene maleic anhydride Diels-Alder adduct of anthracene and maleic anhydride 67%

Note that when maleic anhydride adds to the central ring, the rings on either side remain aromatic. Similarly, when bromine is added to anthracene, it adds to the central ring in a reaction that resembles 1,4-addition to a diene (p. 686).

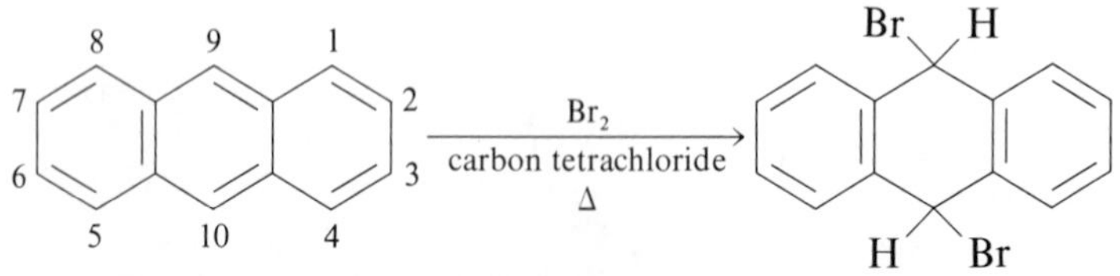
8
9
1
7
2
6
3
5
10
4
Br₂
carbon tetrachloride
Δ
Br
H
H
Br

anthracene 9,10-dibromo-9,10-dihydroanthracene

Phenanthrene also has a central ring that is more vulnerable to chemical reagents than the other two rings are. Most of the resonance contributors that can be written for phenanthrene have a double bond localized between the two carbon atoms usually numbered 9 and 10 in this system. Phenanthrene adds bromine at these positions as though it were an alkene (p. 290).

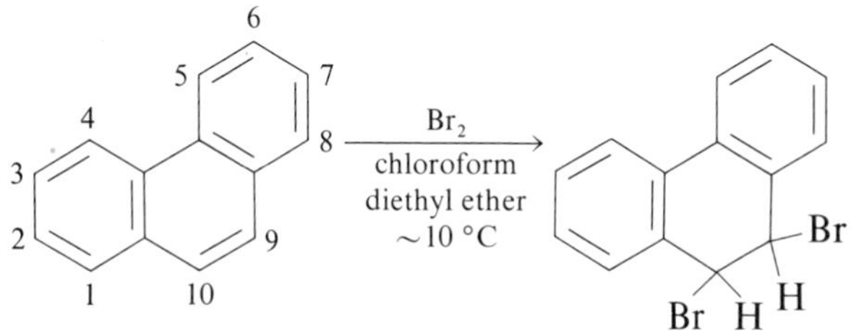
6
5
7
4
8
3
2
9
1
10
Br_2
chloroform
diethyl ether
~10 °C
Br
Br H
H

phenanthrene

9,10-dibromo-9,10-dihydrophenanthrene

Thus, although it is possible to talk in theoretical terms of molecular energy levels and the delocalization of electrons over entire molecules, Kekulé structures, with their alternating double and single bond character, are reminders of the experimental facts of the chemistry of aromatic compounds. Indeed, chemists have been unable to synthesize compounds for which electron delocalization over large molecules is possible but for which Kekulé structures cannot be written. For these reasons, Kekulé structures are used for benzene in this book, not the convention of a circle in a hexagon that is sometimes used to symbolize benzene.

PROBLEM 19.6

Name the following compounds.

(a) 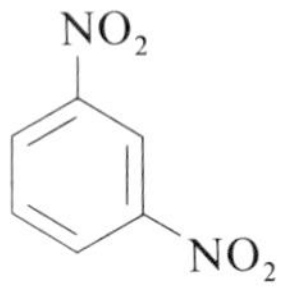
NO_2
NO_2

(b) 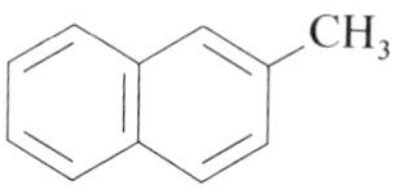
CH_3

(c) 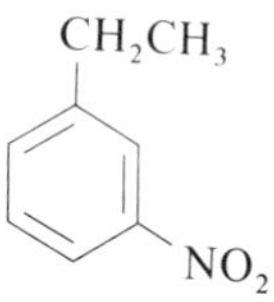
CH_2CH_3
NO_2

(d) 
OH
Cl

(e) 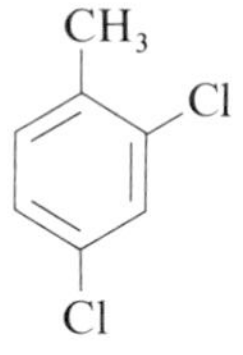
CH_3
Cl
Cl

(f) 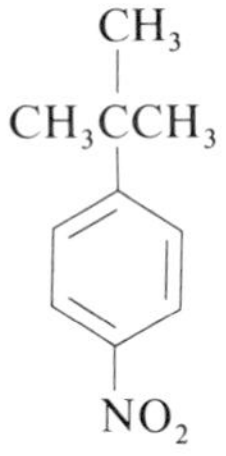
CH_3
CH_3CCH_3
NO_2

PROBLEM 19.7

Draw a structural formula for each of the following compounds.

(a) 2,4-dinitrophenol (b) 2,4,6-trimethylpyridine (c) 8-hydroxyquinoline
(d) 2,4,6-trinitrotoluene (TNT) (e) 4-bromo-2-methylnaphthalene
(f) 4,4′-dibromobiphenyl (g) 8-methyl-1-naphthol (h) 3-ethylthiophene
(i) 2,4-dibromo-1-pentylbenzene

PROBLEM 19.8

Draw resonance contributors for phenanthrene. Prove to yourself that the 9,10 bond in phenanthrene would be expected to behave more like a double bond than any of the other bonds in the molecule.

A. Experimental Observations from Halogenation and Nitration Reactions

Benzene is rich in electrons yet inert toward electrophilic addition reactions. Addition reactions would break up the aromatic sextet that gives benzene its high stability. Benzene and other aromatic hydrocarbons will react, however, with electrophilic reagents in reactions in which an incoming group substitutes for one of the hydrogen atoms on the ring. Such reactions are called **electrophilic aromatic substitution reactions.**

For example, benzene reacts with bromine in the presence of a catalyst, such as iron, or a Lewis acid, such as aluminum chloride, to give chiefly bromobenzene. Small amounts of *o*-dibromobenzene and *p*-dibromobenzene are also formed; the other product of the reaction is hydrogen bromide.

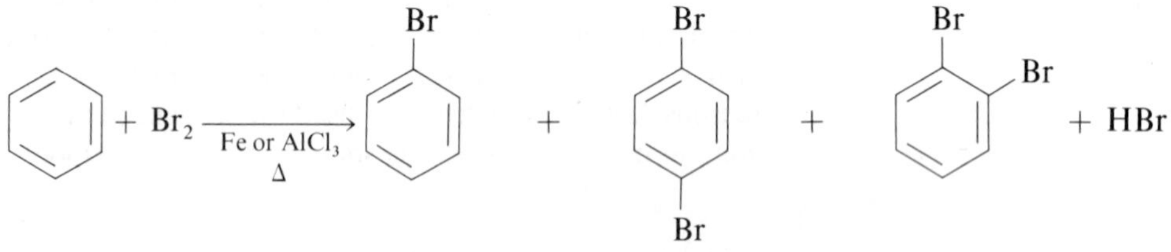
Br
Br
Br
Br
Br
+ Br_2
Fe or $AlCl_3$
Δ
+
+
+ HBr

benzene — bromobenzene *major* — *p*-dibromobenzene *minor* — *o*-dibromobenzene *trace*

In a similar fashion, treating benzene with nitric acid in concentrated sulfuric acid converts it to nitrobenzene.

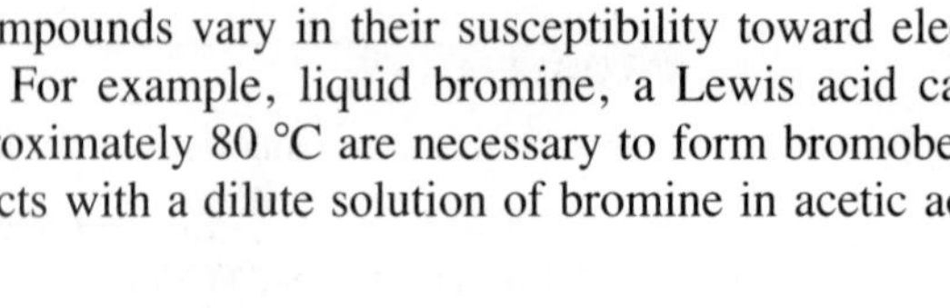
NO_2
+ $HONO_2$
H_2SO_4
~60 °C
+ H_2O

benzene — nitric acid — nitrobenzene

Aromatic compounds vary in their susceptibility toward electrophilic substitution reactions. For example, liquid bromine, a Lewis acid catalyst, and temperatures of approximately 80 °C are necessary to form bromobenzene from benzene. Phenol reacts with a dilute solution of bromine in acetic acid at 30 °C.

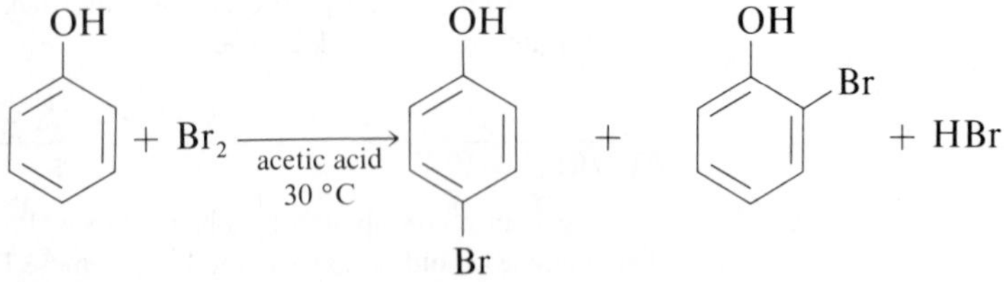
OH
OH
OH
Br
Br
+ Br_2
acetic acid
30 °C
+
+ HBr

phenol — *p*-bromophenol 88% — *o*-bromophenol 12%

p-Bromophenol is the major product of the reaction, along with some o-bromophenol. If phenol is treated with bromine in water, 2,4,6-tribromophenol is formed instantly.

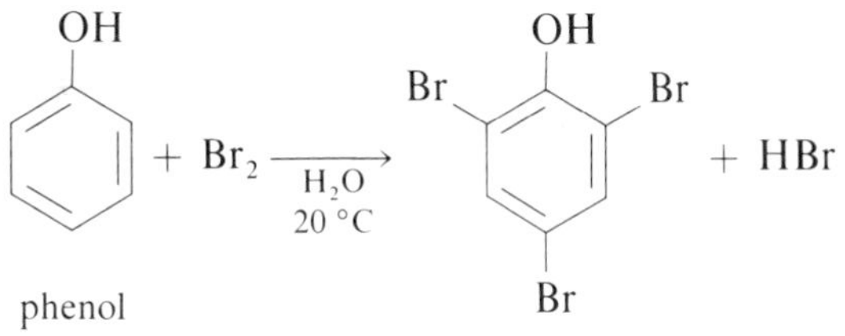

OH
Br
+ Br_2
H_2O
20 °C
+ HBr

phenol 2,4,6-tribromophenol

2,4,6-Tribromophenol is much less soluble in water than phenol is and precipitates from the solution. The reaction is so rapid that it is sometimes used as a test for phenols in qualitative analysis.

Some substituents make it harder to introduce a second group onto an aromatic ring. For example, to introduce a second nitro group onto nitrobenzene, fuming nitric acid, which is a more powerful reagent than ordinary concentrated nitric acid, and a temperature of 100 °C must be used.

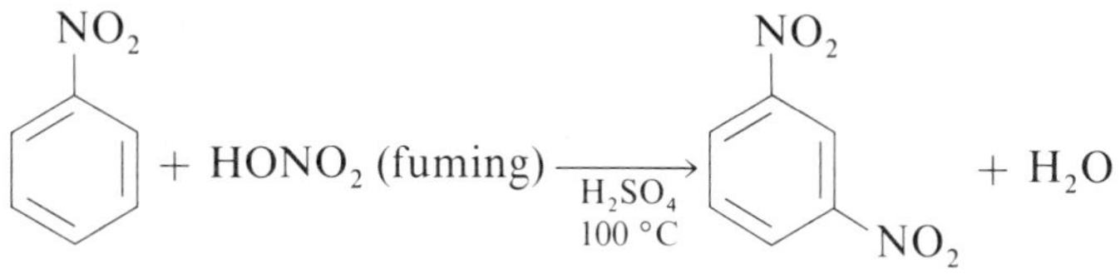

NO_2
+ $HONO_2$ (fuming)
H_2SO_4
100 °C
+ H_2O

nitrobenzene nitric acid m-dinitrobenzene 88%

These conditions are more severe than those used to substitute the first nitro group on the benzene ring (above). The second nitro group takes a position meta to the first one.

On the other hand, phenol can be nitrated with dilute nitric acid at room temperature. The products are o-nitrophenol and p-nitrophenol.

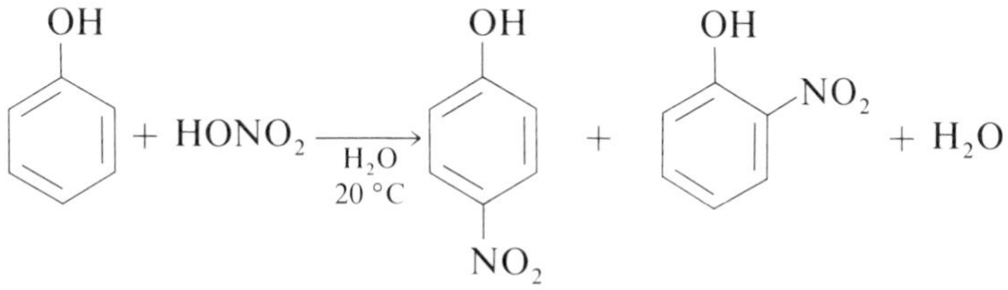

OH
+ $HONO_2$
H_2O
20 °C
NO_2
+
+ H_2O

phenol nitric acid p-nitrophenol 60% o-nitrophenol 40%

The experiments summarized in the equations above lead to two important observations. The first is that a group already present on the aromatic ring may make it easier or harder to introduce a second substituent onto the ring. A group that makes it easier to introduce new substituents is said to activate the ring toward electrophilic substitution or to be **ring-activating.** The hydroxyl group in phenol, for example, is a ring-activating substituent. Groups that make the introduction of a second substituent harder are said to deactivate the ring toward electrophilic substitution or to be **ring-deactivating.** The nitro group is such a ring-deactivating substituent.

The second observation is that the position that a second substituent takes on the ring is influenced by the group that is already there. When phenol undergoes electrophilic substitution, the new group takes up a position ortho or para to the hydroxyl group. The hydroxyl group is said to direct the incoming substituent to those positions or to be **ortho,para-directing.** The nitro group, on the other hand, directs the new substituent to the meta position and is said to be **meta-directing.**

Substituents on the aromatic ring generally fall into one or the other of these categories. Some are ortho,para-directing; others are meta-directing. For example, an alkyl group on the aromatic ring is predominantly ortho,para-directing, as shown by the nitration of ethylbenzene.

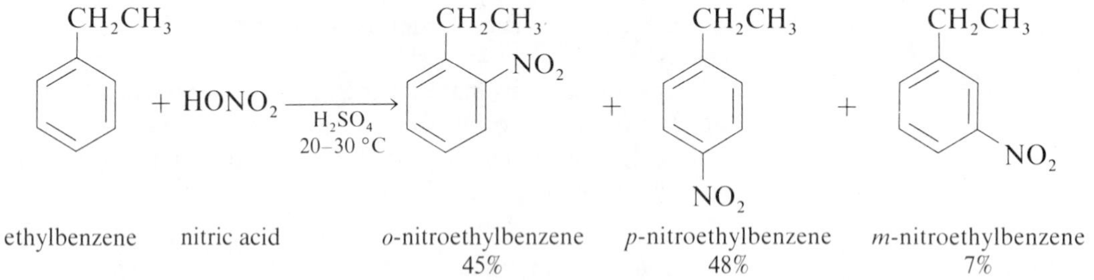
CH₂CH₃
+ HONO₂
H₂SO₄
20–30 °C
CH₂CH₃
NO₂
+
CH₂CH₃
NO₂
+
CH₂CH₃
NO₂
ethylbenzene
nitric acid
o-nitroethylbenzene
45%
p-nitroethylbenzene
48%
m-nitroethylbenzene
7%

The reaction gives a mixture, the chief components of which are *o*-nitroethylbenzene and *p*-nitroethylbenzene in approximately equal amounts. Very little meta-substituted product is formed. When methyl benzoate is nitrated, on the other hand, methyl *m*-nitrobenzoate is formed as the major product.

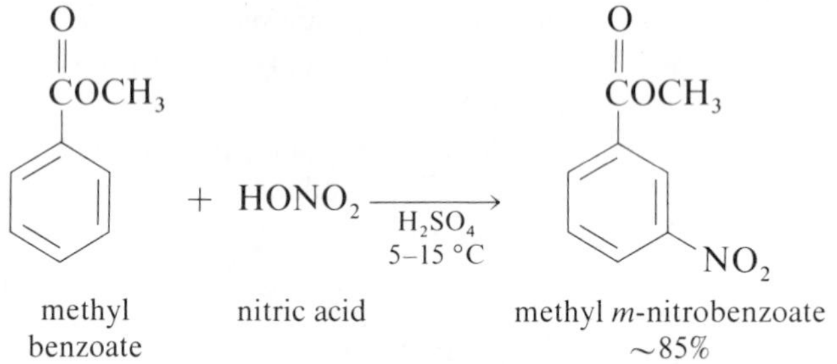
O
COCH₃
+ HONO₂
H₂SO₄
5–15 °C
O
COCH₃
NO₂
methyl benzoate
nitric acid
methyl m-nitrobenzoate
~85%

Thus, the ester function on the aromatic ring is meta-directing. This reaction, however, does not require the harsh conditions that are necessary to introduce a second nitro group into nitrobenzene.

The results of many experiments are summarized in Table 19.1, in which different substituents on the aromatic ring are classified as ortho,para-directing or meta-directing. Most of the ortho,para-directing groups, except for the halogens, activate the ring toward further substitution. This effect is strong only for the hydroxyl group and the unprotonated amino group. An examination of the structures of the groups that are ortho-para-directing shows that, except for alkyl groups, they have at least one pair of nonbonding electrons available on the atom directly bonded to the aromatic ring. All the meta-directing groups either bear a positive charge on the atom directly bonded to the aromatic ring or have a structure that can be polarized to put partial positive charge there. The meta-directing groups at the

TABLE 19.1 Substituents on Aromatic Rings Classified as Ortho, Para- and Meta-Directing Groups

Ortho,Para-Directing		Meta-Directing	
$-N(CH_3)_2$	very strongly activating	$-\overset{+}{N}(CH_3)_3$	very strongly deactivating
$-NH_2$		$-NO_2$	
$-OH$		$-C{\equiv}N$	
$-OCH_3$		$-SO_3H$	
$-NH\overset{O}{\overset{\Vert}{C}}CH_3$		$-\overset{O}{\overset{\Vert}{C}}H$	
$-O\overset{O}{\overset{\Vert}{C}}CH_3$		$-\overset{O}{\overset{\Vert}{C}}CH_3$	
$-R$		$-\overset{O}{\overset{\Vert}{C}}OH$	
$-Cl, Br, I$	mildly deactivating	$-\overset{O}{\overset{\Vert}{C}}OCH_3$	
		$-\overset{O}{\overset{\Vert}{C}}NH_2$	
		$-NH_3^+$	

top of the list are strongly ring-deactivating. Experimentally, the others, even though they are meta-directing, allow substitution reactions under relatively mild conditions. In other words, they do not deactivate the ring strongly.

Study Guide Concept Map 19.2

The next section examines the mechanism proposed to explain the trends just summarized.

PROBLEM 19.9

Complete the following equations.

(a) 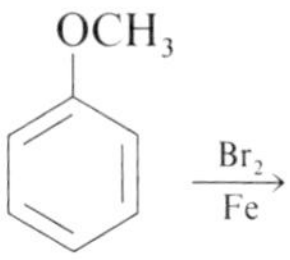
OCH3
Br2
Fe

(b) 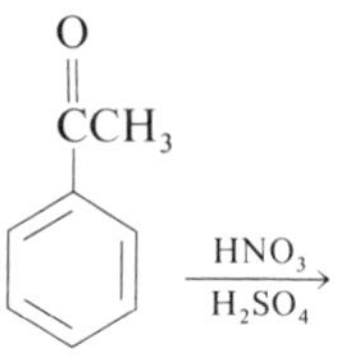
O
CCH3
HNO3
H2SO4

(c) 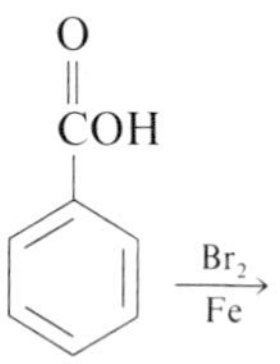
O
COH
Br2
Fe

(d) 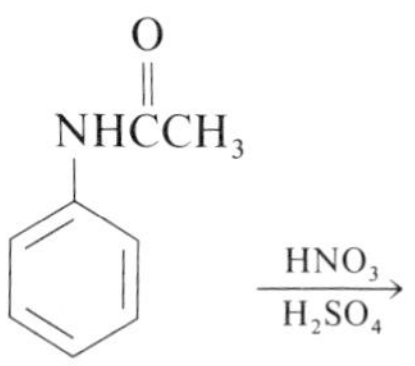
O
NHCCH3
HNO3
H2SO4

(e) 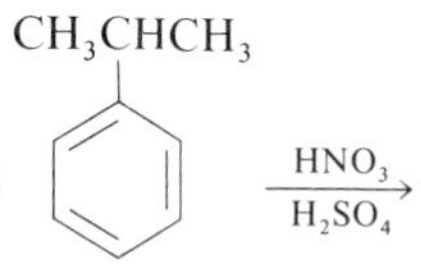
CH3CHCH3
HNO3
H2SO4

B. The Mechanism of Electrophilic Aromatic Substitution

Bromine and strong acids are examples of reagents that introduce substituents on the aromatic ring; they are electrophilic, electron-seeking reagents. Unless the ring is strongly activated, a Lewis acid must also be used with bromine. For example, a piece of iron is added to the reaction vessel in the bromination of benzene (p. 784). Iron reacts with bromine to give ferric bromide, which is a Lewis acid. The Lewis acid coordinates with the nonbonding electrons of the bromine molecule to polarize the bromine-bromine bond, making one of the bromine atoms more electrophilic and the other one a better leaving group.

$$3\,Br_2 + 2\,Fe \longrightarrow 2\,FeBr_3$$

bromine iron ferric bromide *a Lewis acid*

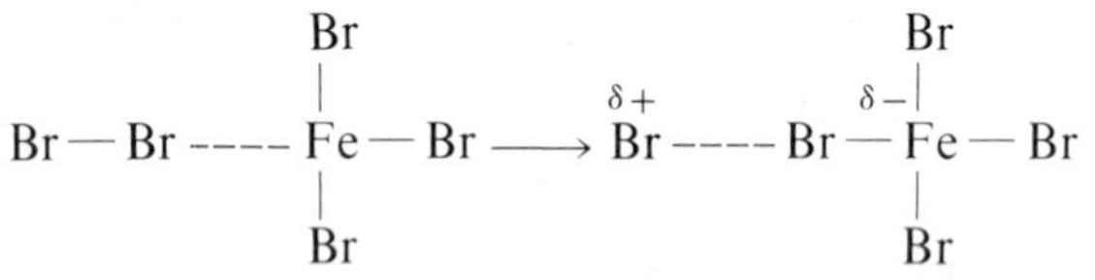
Br
Br — Br ---- Fe — Br
Br
δ+
Br ---- Br — Fe — Br
δ−
Br
Br

a complex of bromine with a Lewis acid

The π electrons of benzene are attracted to the bromine atom at which positive charge is developing. A covalent bond forms between a carbon atom in benzene and that bromine atom, as the other bromine atom becomes part of a complex with the Lewis acid.

VISUALIZING THE REACTION

First step of an electrophilic aromatic substitution reaction

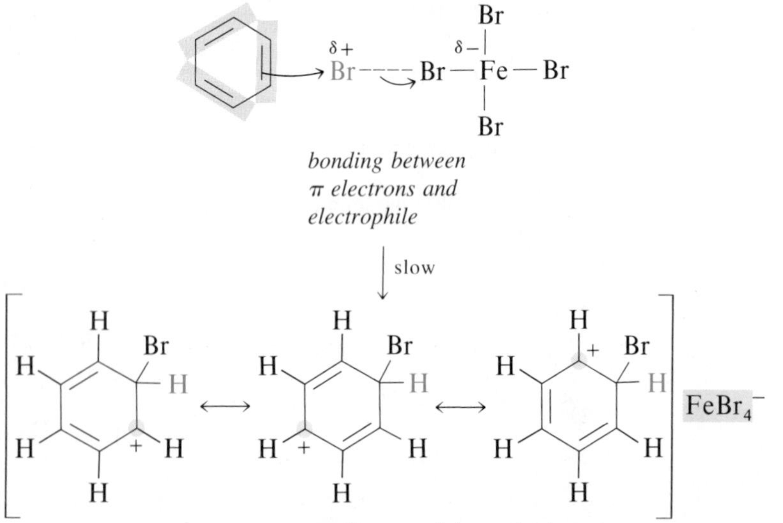
δ+
Br
δ−
Br — Fe — Br
Br
Br
bonding between π electrons and electrophile
slow
H
Br
H
FeBr$_4^-$

resonance contributors of the cationic intermediate from the reaction of benzene with an electrophilic bromine atom

The resulting species is a cation that loses a proton readily to restore the aromatic ring. Bromide ion (or $FeBr_4^-$) is a base strong enough to remove a proton from such an intermediate.

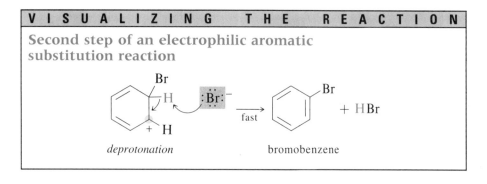

VISUALIZING THE REACTION

Second step of an electrophilic aromatic substitution reaction

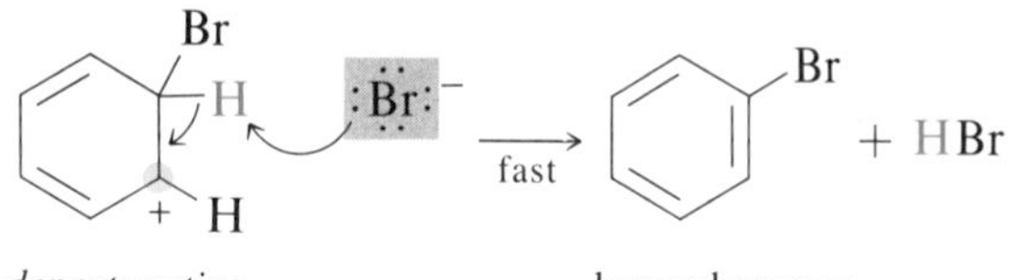

Br
H
:Br:⁻
H
fast
Br
+ HBr
deprotonation
bromobenzene

The equations above represent the essential steps in all electrophilic substitution reactions.

1. The substituting reagent is polarized or ionized in such a way as to create an electron-deficient species.
2. A cationic intermediate forms by bonding between the electrons of the aromatic ring acting as the nucleophile and the electrophilic reagent.
3. A proton is lost from the carbocation to restore the aromatic ring, on which a substituent has replaced a hydrogen atom.

The rate-determining step in most electrophilic aromatic substitution reactions is the formation of the carbocation intermediate. Going from benzene and bromine to the transition state involves not only breaking bonds but disrupting the aromaticity in the benzene ring as well. The energy barrier for this reaction is high (Figure 19.4).

The loss of a proton from the intermediate is a fast reaction; this is because the aromatic sextet is regenerated. It is possible to test whether the loss of the proton is the rate-determining step by substituting deuterium atoms for the hydrogen atoms on the aromatic ring. A deuterium atom is chemically similar to a hydrogen atom because it, too, has one proton and one electron. A deuterium atom, however, has twice the mass of a hydrogen atom. The breaking of a covalent bond results from an increase in the vibrational energy of the bond to a level represented by the bond dissociation energy (p. 59). The vibrational energy, in turn, depends on the masses of the atoms involved in the bond (p. 358). The vibrational energy of a bond is lower when the atoms joined by the bond have greater mass. Therefore, a C—D bond has a lower ground-state vibrational energy than does a C—H bond. For this reason, a reaction involving the breaking of a bond between carbon and deuterium has a higher energy of activation and is slower than a reaction involving the breaking of a carbon-hydrogen bond. If the removal of the proton from the cationic intermediate were the rate-determining step, a decrease in the rate of the reaction should be seen if deuteriobenzene, C_6D_6, is used in an electrophilic substitution reaction. The experiment has been tried, and no difference in the rate of nitration has been observed for C_6H_6 and C_6D_6 with a variety of reaction conditions. These experimental results confirm that the rate-determining step does not involve the breaking of the carbon-hydrogen bond.

In the nitration reaction, the electrophile is the nitronium ion, NO_2^+, which is created by protonation of nitric acid by the strong acid sulfuric acid, followed by the loss of water.

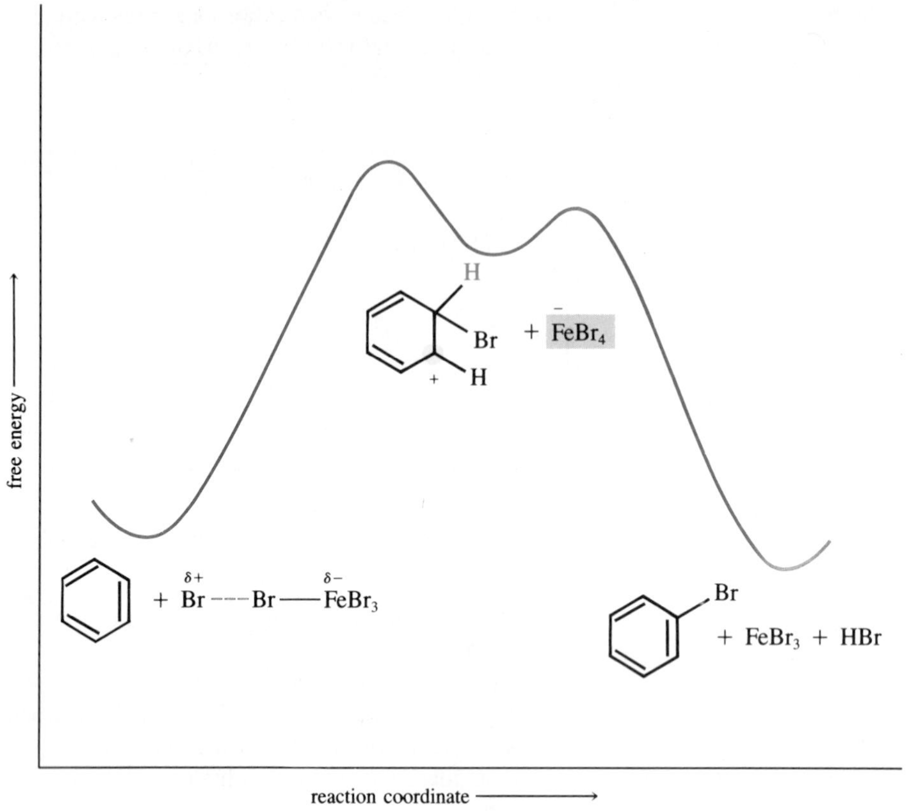
H
Br
+ $\overset{-}{FeBr_4}$
H
free energy
δ+
δ−
+ Br---Br—FeBr$_3$
Br
+ FeBr$_3$ + HBr
reaction coordinate

FIGURE 19.4 An energy diagram for the bromination of benzene.

VISUALIZING THE REACTION

Formation of the nitronium ion

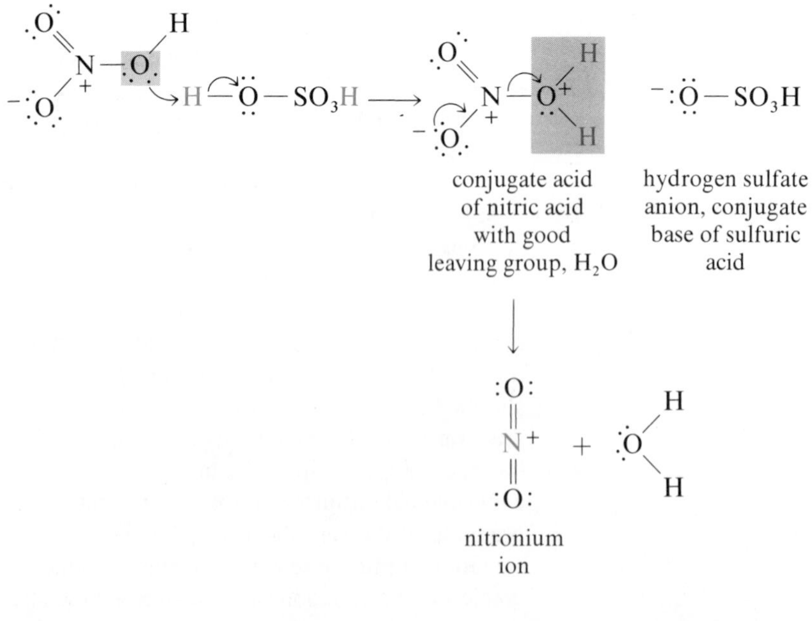
H
N
O
SO$_3$H
conjugate acid of nitric acid with good leaving group, H$_2$O
hydrogen sulfate anion, conjugate base of sulfuric acid
nitronium ion
H

Nitronium ion reacts with the π electrons of the aromatic ring to give a cationic intermediate, which loses a proton to recreate the aromatic sextet.

VISUALIZING THE REACTION

The nitration reaction

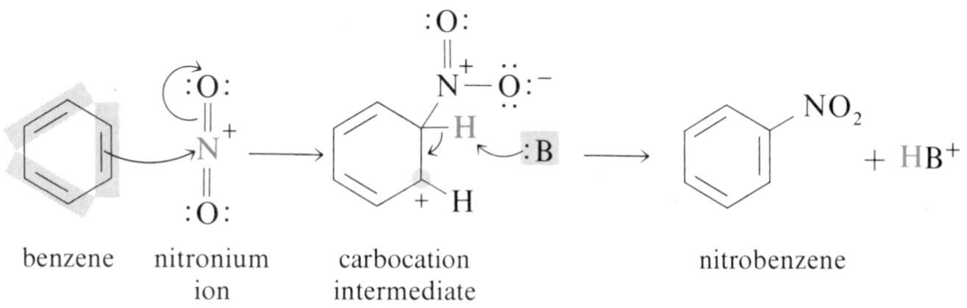

:O:
N
:O:
:O:
N—O:
H
:B
H
NO2
+ HB+
benzene
nitronium ion
carbocation intermediate
nitrobenzene

The reactivity of the aromatic ring toward acids is demonstrated by putting benzene in which deuterium has been substituted for one of the hydrogen atoms into aqueous sulfuric acid at room temperature.

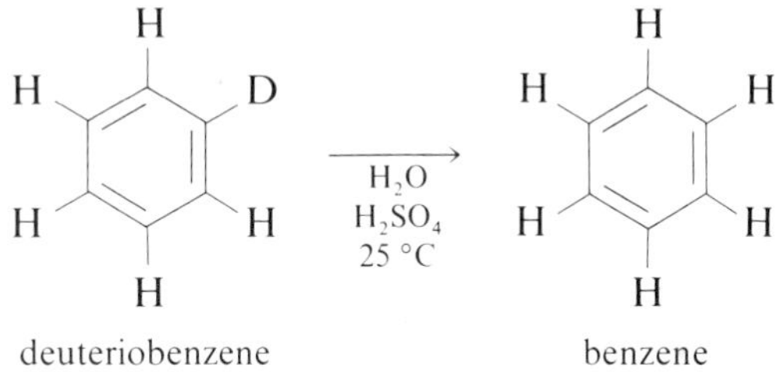

H2O
H2SO4
25 °C
deuteriobenzene
benzene

The deuterium atom is exchanged for a hydrogen atom.

VISUALIZING THE REACTION

Acid-catalyzed exchange of hydrogen atoms

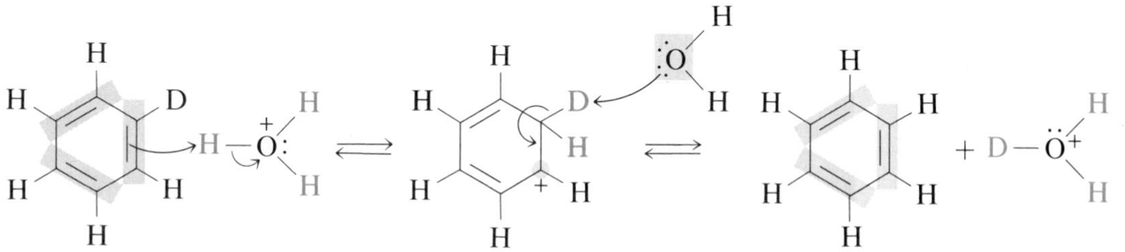

This observation means that in an acid medium the aromatic ring reacts constantly with protons. The protonation and deprotonation of the ring are usually not detected because the starting material and the product of the reaction are the same. The reaction can be detected, however, if the ring is labeled with isotopes of hydrogen. The reactive intermediate for this exchange reaction is a cation, arising from the reaction of a proton with the π electrons of the aromatic ring.

Study Guide Concept Map 19.3

C. Orientation in Electrophilic Aromatic Substitution on Compounds Having a Ring-Activating Substituent

How does the mechanism proposed in the preceding section account for the different orientations observed for substituents when the aromatic ring already has a group on it? This is a question regarding the regioselectivity of a reaction, a question that has come up again and again in many different contexts, for example, the addition of hydrogen bromide to propene (p. 120), the formation of alkenes in elimination reactions (p. 254), and the addition of diborane to double bonds (p. 300). In each case, the answer came from a consideration of the transition states that would lead to the different products and a selection of the transition state of lowest energy, or greatest stability, as the one that would give rise to the major product. The relative energies of different transition states have been assumed to be parallel to the relative energies of the possible reactive intermediates or of the products.

The rationalization of the regioselectivity seen for electrophilic aromatic substitution is similar. A consideration of the relative energies of transition states that would lead to the different products expected from a given substitution reaction should reveal which ones are the most stable and, therefore, most likely to yield products. In this discussion, the intermediates are assumed to resemble the transition states in structure and so are used as models.

When phenol reacts with bromine, the bromine atom can become bonded to a ring carbon at one of three possible positions: ortho, meta, and para to the hydroxyl group. The cationic intermediates that would result from bonding of the bromine atom to these positions, with their resonance contributors, are shown below.

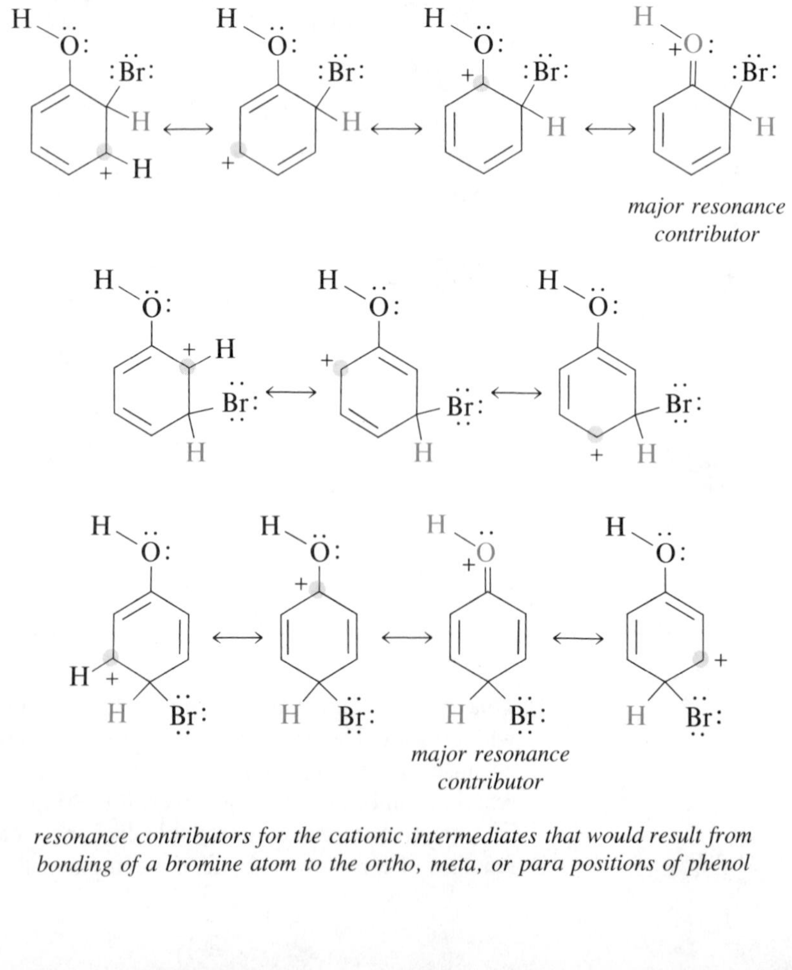
major resonance contributor
major resonance contributor

resonance contributors for the cationic intermediates that would result from bonding of a bromine atom to the ortho, meta, or para positions of phenol

The intermediates that result from the bonding of the bromine atom to the ortho and para positions of the ring have more resonance contributors than does the intermediate that would give rise to *m*-bromophenol. Also, for the intermediates arising from attack at the ortho and para positions, one of the resonance contributors has eight electrons around each atom, which is a particularly stable situation. The reasons just mentioned lead to the conclusion that the transition states for ortho and para substitution on a phenol ring are of lower energy than the one for meta substitution. The energies of activation for the formation of *o*-bromophenol and *p*-bromophenol are lower than that for the formation of *m*-bromophenol. Most of the reactant molecules will be channeled through the intermediates leading to those products.

The conclusions arrived at for phenol will be true for any substituent in which there is a pair of nonbonding electrons on the atom bonded to the aromatic ring. For an alkyl group, the electron-donating effect of the group stabilizes the intermediates resulting from ortho and para substitution more than the intermediate from meta substitution.

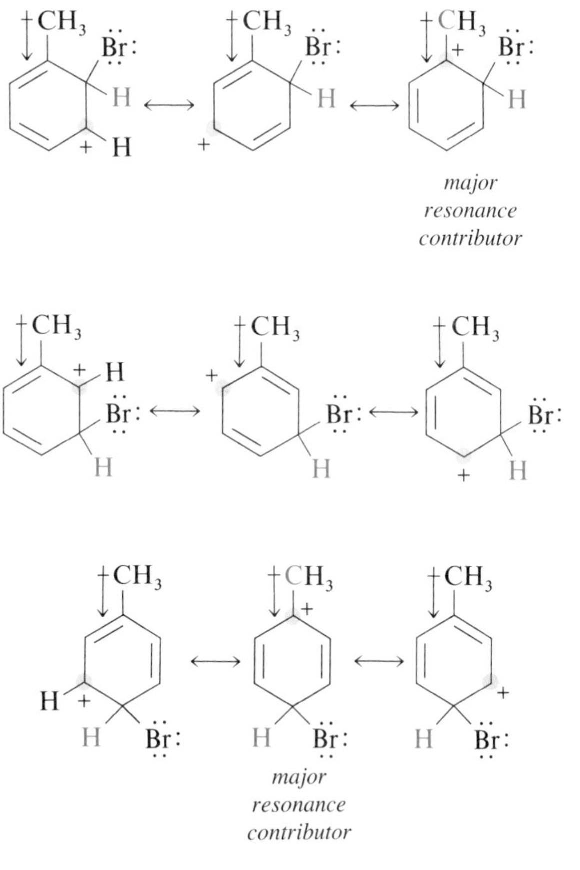

major resonance contributor
major resonance contributor

resonance contributors for the cationic intermediates that would result from bonding of a bromine atom to the ortho, meta, or para positions of toluene

For a second substituent on an aromatic ring, the orientation that gives rise to the most stable reactive intermediate is favored.

D. Orientation in Electrophilic Aromatic Substitution on Compounds Having a Ring-Deactivating Substituent

When a strongly electron-withdrawing substituent is on the aromatic ring, it is harder to form the carbocation intermediate. The cation is destabilized by the charge distribution in the polar electron-withdrawing group. The intermediate formed by reaction at the meta position is the most stable one because it is least destabilized. This conclusion is illustrated by the nitration of nitrobenzene.

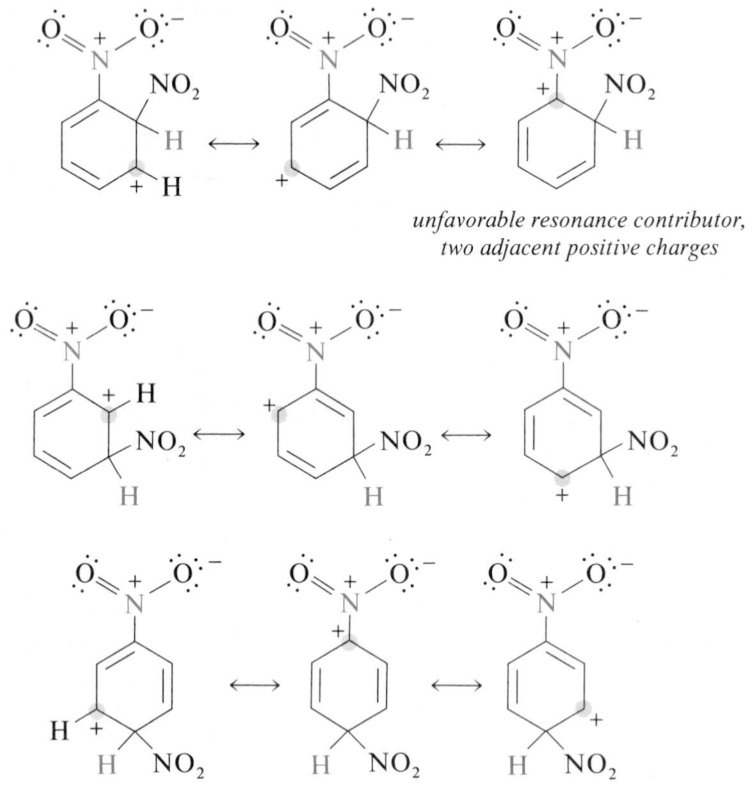
NO2
H
unfavorable resonance contributor, two adjacent positive charges
unfavorable resonance contributor, two adjacent positive charges

resonance contributors for the cationic intermediates that would result from the bonding of a nitronium ion to the ortho, meta, and para positions of nitrobenzene

None of the intermediates shown above has any special stabilization other than that due to delocalization of charge in general. Among the resonance contributors for the intermediates resulting from bonding of a nitronium ion at the ortho and para positions of nitrobenzene, however, one is a particularly unfavorable one, having a juxtaposition of two positive charges. As a result, the transition state arising from an initial approach of the nitronium ion to the meta position is the most stable among the three possible intermediates, and substitution on nitrobenzene takes place mostly at the meta position.

The substitution of a second group on a halobenzene is an interesting case. The halogens are ring-deactivating; their presence makes it harder for electrophilic aromatic substitution to take place. This would be expected from their electronegativity and thus their negative inductive effect. Yet the halogens are ortho, para-directing. As a positive charge develops on the aromatic ring during substitution, the nonbonding electrons of a halogen are drawn toward the charge and help to stabilize it. Thus, just as a halogen atom stabilizes a carbocation intermediate in the addition of hydrogen halides to an alkyne (p. 334), it is possible for halogen atoms to stabilize the carbocation intermediate in an electrophilic substitution reaction if the incoming substituent takes an ortho or para position.

For example, the intermediate leading to *p*-dibromobenzene is favored over the one leading to *m*-dibromobenzene.

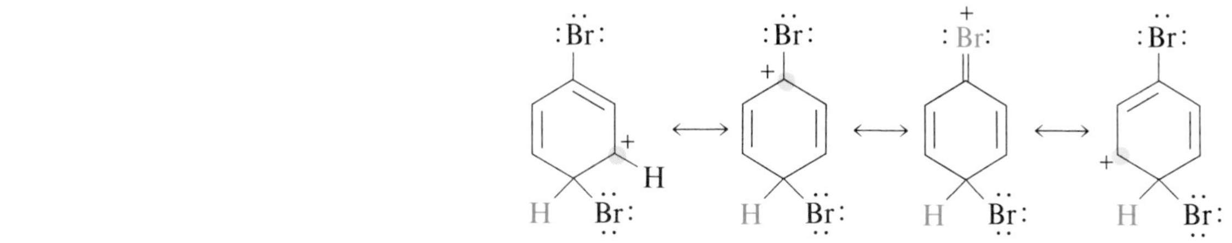
delocalization of
charge to bromine
atom

resonance contributors for the intermediate for p-dibromobenzene

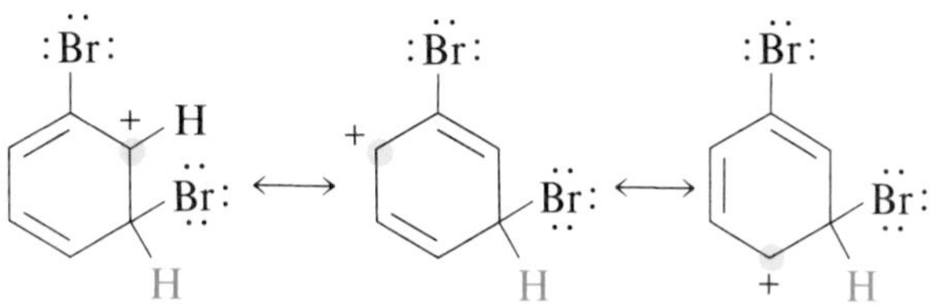
no resonance contributor in which charge
can be delocalized to bromine atom

resonance contributors for the intermediate for m-dibromobenzene

The positive charge in the carbocation intermediate for *p*-dibromobenzene can be delocalized to a bromine atom. Such a resonance contributor has eight electrons around each atom and is thus a major resonance contributor. No such delocalization is possible for the intermediate that would give rise to *m*-dibromobenzene. Only a very small amount of the meta isomer (1.8%) is formed in the reaction of bromobenzene with bromine.

E. Steric Effects in Electrophilic Aromatic Substitution

The arguments presented on p. 792 to rationalize the fact that the hydroxyl group in phenol or the methyl group in toluene is ortho,para-directing did not indicate any distinction between the ortho and para position on the basis of the energies of the transition state. If these two were really indistinguishable, twice as much ortho as para isomer should be formed because there are two ortho positions on the ring and only one para position. Statistically, it would be twice as likely that collision would occur with the reagents correctly oriented for reaction at the ortho position. The experimental observations prove otherwise (p. 784). Actually, the ratio of ortho to para product can vary greatly, but substitution at the para position is generally favored over what is predicted statistically. The group already on the aromatic ring has some bulk and hinders the approach of the incoming reagent to the positions closest to it, the ortho positions. This idea has been tested by systematically increasing the bulk of a substituent and measuring the change in the composition of the product mixture formed. Toluene, ethylbenzene, isopropylbenzene, and *tert*-butylbenzene are shown below with numbers that indicate the percentage of substitution that takes place at each position on the ring in a nitration reaction.

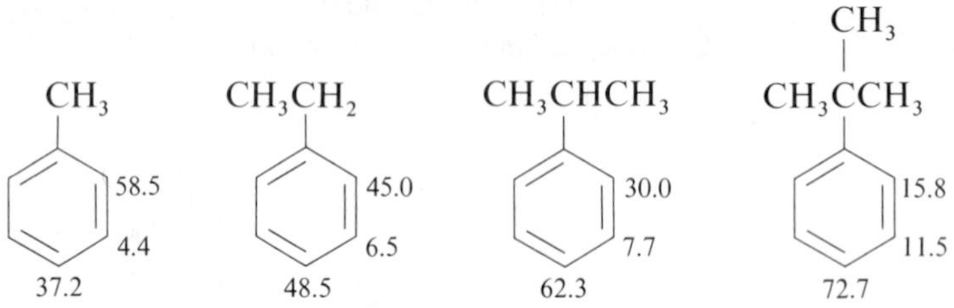
CH3
CH3CH2
CH3CHCH3
CH3
CH3CCH3
58.5
4.4
37.2
45.0
6.5
48.5
30.0
7.7
62.3
15.8
11.5
72.7

toluene ethylbenzene isopropylbenzene *tert*-butylbenzene

percentage of ortho, meta, and para isomers in the nitration of a series of alkylbenzenes

From these experimental facts, it is quite clear that the proportion of ortho isomer in the product mixture becomes smaller as the alkyl group on the aromatic ring becomes larger.

There is also some evidence that the size of the incoming substituent affects the ratio of ortho to para isomers formed as products. For example, when bromobenzene reacts with chlorine, 42% of the product is *o*-bromochlorobenzene and 53% is the para isomer. If bromine, a larger atom, is the incoming group, only 13% *o*-dibromobenzene is formed, and 85% of the product mixture is *p*-dibromobenzene.

Study Guide Concept Map 19.4

These data are a reminder that electronic factors are not the only ones to consider in thinking about a reaction. Steric interactions of the reagents are also important in determining relative energies of transition states and the resulting reaction pathways.

PROBLEM 19.10

The trimethylammonium ion, $(CH_3)_3\overset{+}{N}—$, is a strongly deactivating group that is also meta-directing. Explain this fact. (Hint: A picture is worth a thousand words!)

PROBLEM 19.11

The following groups are not in Table 19.1. Predict whether they will be ortho,para- or meta-directing.

(a) $CF_3—$ (b) $(CH_3)_3\overset{+}{P}—$ (c) $O_2NCH_2—$ (d) $(CH_3)_2\overset{+}{S}—$

(e) 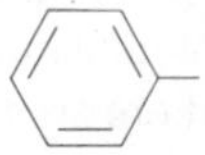 (f) $CH_3CH{=}CH—$

PROBLEM 19.12

How would you rationalize the experimental result given for the following reaction? (Hint: What are the reacting species when *N,N*-dimethylaniline is dissolved in a mixture of concentrated nitric and sulfuric acids?)

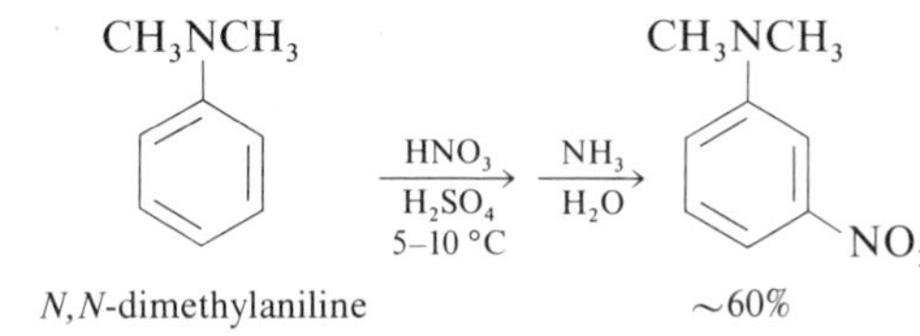
CH3NCH3
HNO3
H2SO4
5–10 °C
NH3
H2O
CH3NCH3
NO2
N,N-dimethylaniline
~60%

PROBLEM 19.13

p-Nitrophenol is shaken with 1 equivalent of D_2O in the presence of the strong acid perchloric acid, $HClO_4$, at 100 °C for 100 hours. The product has two deuterium atoms in it. One of these is lost instantly when the product is treated with ordinary water, H_2O. Propose a structure for the original product and explain the difference in the ease with which the two deuterium atoms can be removed from the molecule.

F. Problem-Solving Skills

Problem

How would you carry out the following conversion?

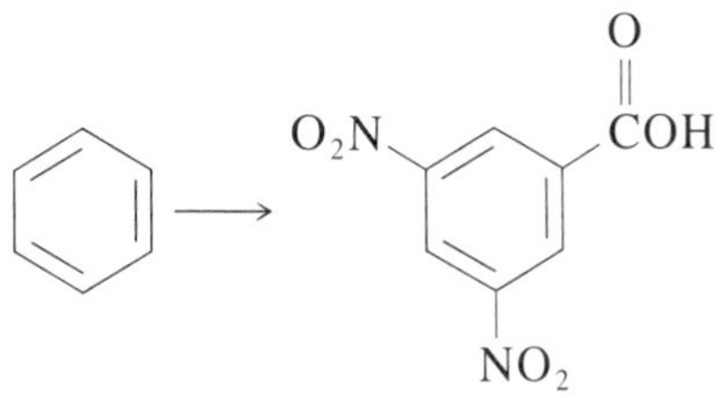
O
O2N
COH
NO2

Solution

1. What functional groups are present in the starting material and the product?

   The starting material is benzene, an aromatic hydrocarbon. The product is an aromatic carboxylic acid with two nitro groups meta to the carboxyl group.

2. How do the carbon skeletons of the two compounds compare? How many carbon atoms does each contain? Are there any rings? What are the positions of branches and functional groups on the carbon skeletons?

   One carbon atom, that of the carboxyl group, has been added to benzene. Two nitro groups have been substituted meta to the carboxylic acid function.

3. How do the functional groups change in going from starting material to product? Does the starting material have a good leaving group?

   The aromatic hydrocarbon has undergone substitution at three positions.

4. Is it possible to dissect the structures of the starting material and product to see which bonds must be broken and which formed?

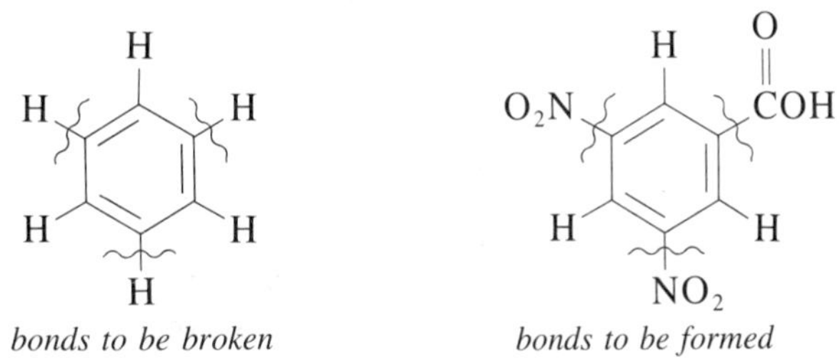
H
H
H
H
H
H
bonds to be broken
H
O
O2N
COH
H
H
NO2
bonds to be formed

5. Do we recognize any part of the product molecule as coming from a good nucleophile or an electrophilic addition?

The nitro groups can be introduced by an electrophilic aromatic substitution reaction on the carboxylic acid, which is meta-directing. The carbon-carbon bond to the carboxylic acid group would be most easily formed by the reaction of a nucleophilic carbon atom with the electrophilic carbon atom of carbon dioxide. An organometallic reagent such as a Grignard reagent has a nucleophilic carbon atom.

6. What type of compound would be a good precursor to the product?

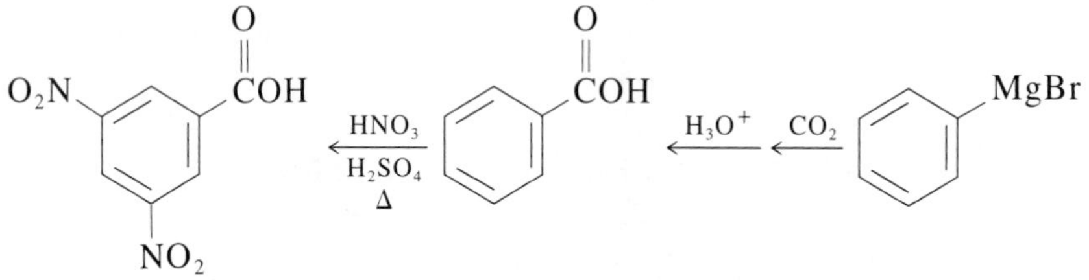
O
O₂N
COH
NO₂
HNO₃
H₂SO₄
Δ
O
COH
H₃O⁺
CO₂
MgBr

7. After this last step, do we see how to get from starting material to product? If not, we need to analyze the structure obtained in step 6 by applying questions 5 and 6 to it.

An electrophilic substitution reaction on benzene will give bromobenzene, which can be used to make the organometallic reagent necessary for the synthesis.

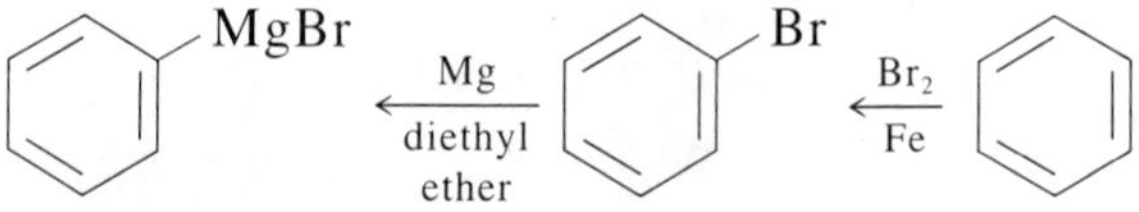
MgBr
Mg
diethyl
ether
Br
Br₂
Fe

A potential alternate pathway that might seem feasible is preparation of 3,5-dinitrobromobenzene and then formation of the acid.

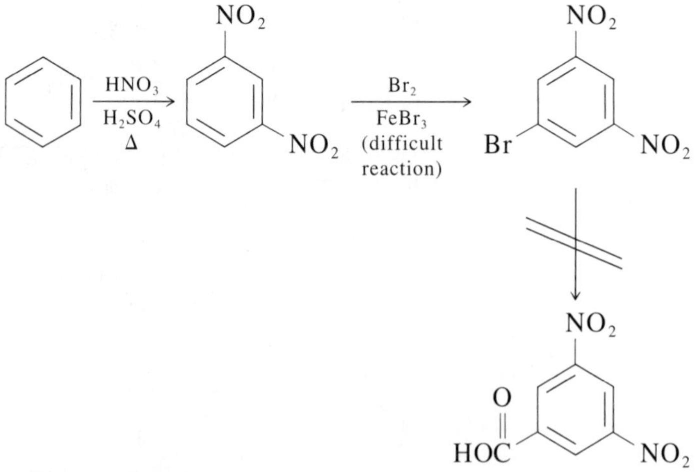
HNO₃
H₂SO₄
Δ
NO₂
NO₂
Br₂
FeBr₃
(difficult
reaction)
NO₂
Br
NO₂
NO₂
O
HOC
NO₂

This path must be rejected for two reasons: (1) two nitro groups deactivate the ring strongly, making bromination difficult; and (2) organometallic reagents cannot be prepared from compounds containing nitro groups.

PROBLEM 19.14

How would you synthesize the following compounds from benzene or toluene and any other organic reagents containing not more than three carbon atoms? You may assume that ortho and para isomers are easily separable in the laboratory.

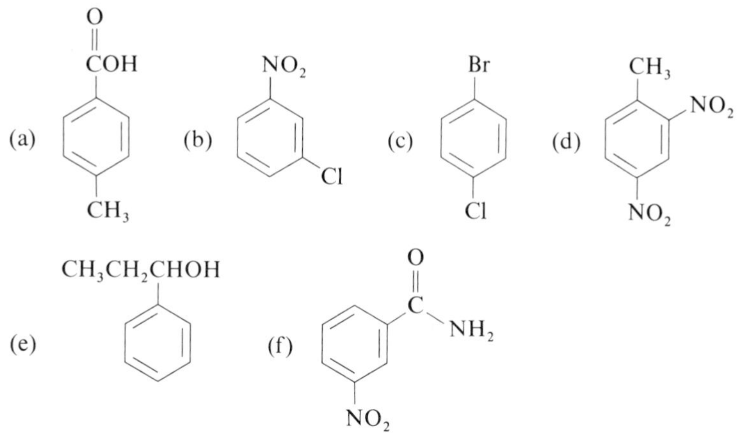
(a)
O
COH
CH3
(b)
NO2
Cl
(c)
Br
Cl
(d)
CH3
NO2
NO2
(e)
CH3CH2CHOH
(f)
O
C
NH2
NO2

19.4 AROMATIC SUBSTITUTION REACTIONS WITH CARBOCATIONS AS ELECTROPHILES

A. Friedel-Crafts Alkylation Reactions

In the reactions described in Section 19.3A, the electrophilic reagent was bromine or a nitronium ion. When carbon is cationic, it is also an electrophile; therefore, reactions that give rise to carbocations when carried out in the presence of an aromatic ring, create carbon-carbon bonds. Three types of reactions give rise to carbocations: the protonation of an alkene, the protonation of an alcohol with the subsequent loss of water, and the ionization of an alkyl halide. All of these are used to substitute carbon side chains on aromatic rings in the **Friedel-Crafts reactions,** named for Charles Friedel, a French chemist, and James Crafts, an American chemist, who developed these methods.

For example, sulfuric acid catalyzes the reaction between the alkene cyclohexene and benzene, which gives mostly cyclohexylbenzene and some *p*-dicyclohexylbenzene, even when an excess of benzene is used.

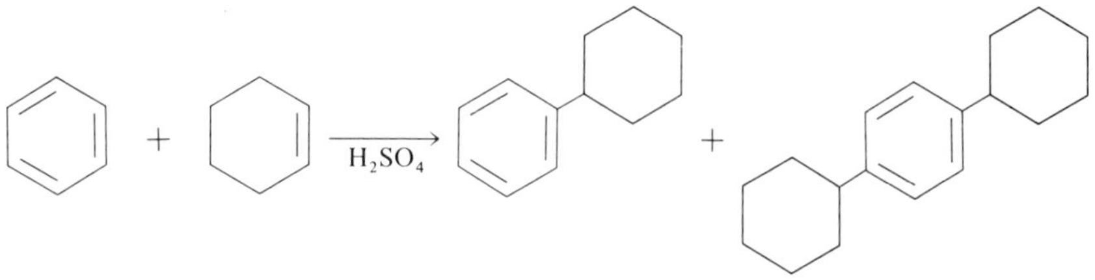
H_2SO_4

benzene *(used in excess)* cyclohexene cyclohexylbenzene *major* *p*-dicyclohexylbenzene *minor*

This reaction, an example of a **Friedel-Crafts alkylation,** illustrates one of the difficulties of such reactions. The product of the reaction, an alkylbenzene, is more reactive toward electrophilic aromatic substitution than is benzene itself. The alkyl group is ring-activating and ortho,para-directing. As a result, as soon as any quantity of mono-alkylated benzene accumulates in the reaction mixture, it starts to undergo further alkylation to give di- and even trisubstituted products.

The mechanism for the alkylation reaction of benzene with cyclohexene is straightforward and follows the steps outlined on p. 789. First, a good electrophile is created by protonation of cyclohexene by sulfuric acid, the catalyst in the reaction. A carbon-carbon bond forms when the electrophile reacts with the π electrons of the aromatic ring. A proton is lost from the cationic intermediate that is formed.

VISUALIZING THE REACTION

A Friedel-Crafts alkylation

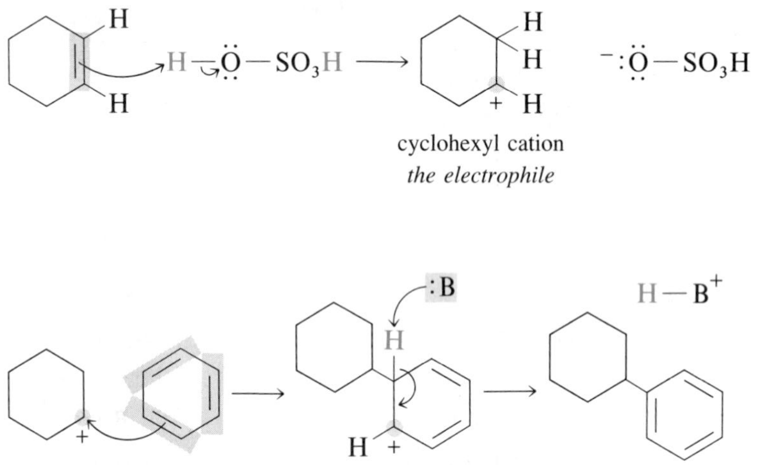
H
H
H—Ö—SO₃H
H
H
+ H
⁻:Ö—SO₃H
cyclohexyl cation
the electrophile
:B
H—B⁺
H
H +

reaction of the electrophile with the π electrons

the cationic intermediate losing a proton

Because carbocations are the electrophiles in the Friedel-Crafts alkylation, the types of compounds that can be synthesized by this method are limited. For example, it is difficult to put a primary alkyl chain on an aromatic ring. When benzene reacts with *n*-propyl chloride with aluminum chloride as a catalyst, a mixture of two alkylbenzenes is formed. The composition of the mixture depends on the temperature at which the reaction is run.

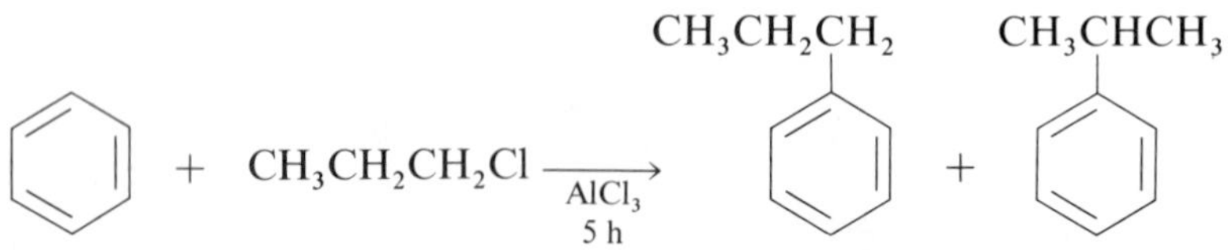
CH₃CH₂CH₂
CH₃CHCH₃
+
CH₃CH₂CH₂Cl
AlCl₃
5 h
+

benzene	*n*-propyl chloride	propylbenzene	isopropylbenzene
	at −6 °C	60%	40%
	at 35 °C	40%	60%

Aluminum chloride, which is a Lewis acid, complexes with nonbonding electrons of the chlorine atom in *n*-propyl chloride, polarizing the carbon-chlorine bond. The intermediate formed must have enough cationic character at the primary carbon atom that some rearrangement to the more stable secondary carbocation takes place by a 1,2-hydride shift (p. 285). The carbocation intermediate is seen as being tightly complexed with the aluminum halide in an ion pair.

VISUALIZING THE REACTION

Formation of the electrophile in a Friedel-Crafts alkylation

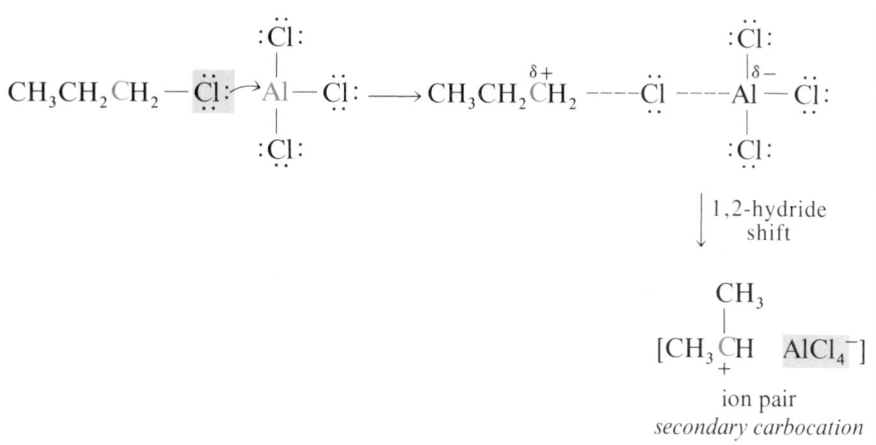
1,2-hydride shift
ion pair
secondary carbocation

The ratios of the two products from benzene and *n*-propyl chloride are evidence that free primary carbocations are not easily formed, even with a powerful Lewis acid such as aluminum chloride to complex with the chloride ion. At the lower temperature, the reaction is kinetically controlled (p. 686), and most of the product comes from reaction at the first carbon of the propyl chain. Only at the higher temperature, where there is a better chance of breaking the carbon-chlorine bond, does the rearranged product derived from the thermodynamically more stable cation become the major one. Clearly, the reaction is not a good one for the synthesis of either propylbenzene or isopropylbenzene.

There is no problem obtaining pure products if the alkyl halide used is one that ionizes to form the most stable carbocation directly. Thus, *tert*-butyl chloride reacts with ethylbenzene to give *p-tert*-butylethylbenzene quantitatively.

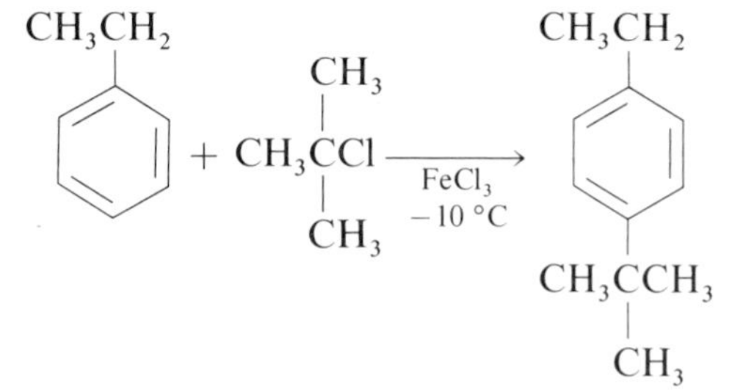
$FeCl_3$
−10 °C

ethylbenzene | *tert*-butyl chloride | *p-tert*-butylethylbenzene

Carbocations are created by the reaction of alcohols with protic acids or with Lewis acids. Alcohols, therefore, are used as reagents in Friedel-Crafts alkylations.

In these reactions, the stability of the carbocation formed as the electrophile is again important. Isopropyl alcohol reacts with benzene to give isopropylbenzene.

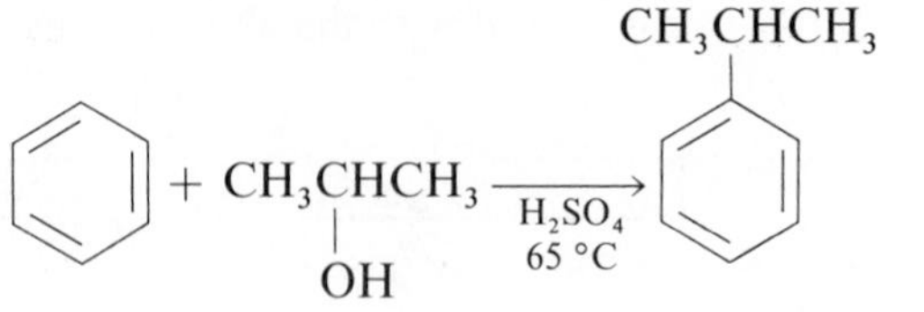
CH3CHCH3
+ CH3CHCH3
OH
H2SO4
65 °C

benzene — isopropyl alcohol — isopropylbenzene cumene 65%

PROBLEM 19.15

Complete the following equations.

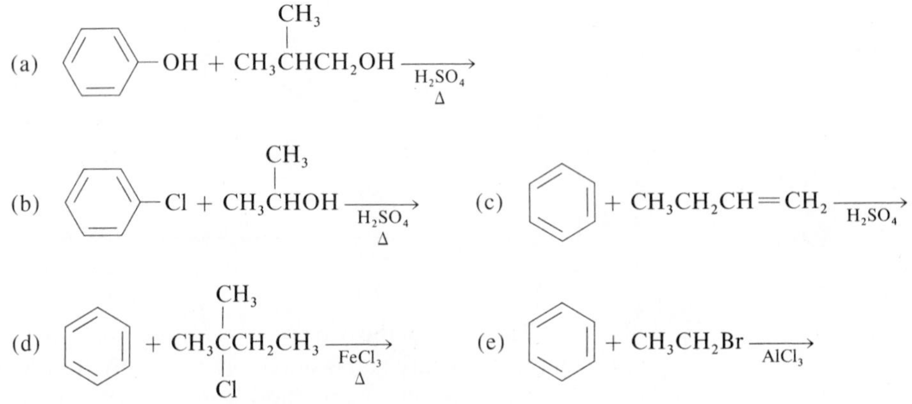
(a)
OH + CH3CHCH2OH
CH3
H2SO4
Δ
(b)
Cl + CH3CHOH
CH3
H2SO4
Δ
(c)
+ CH3CH2CH=CH2
H2SO4
(d)
+ CH3CCH2CH3
CH3
Cl
FeCl3
Δ
(e)
+ CH3CH2Br
AlCl3

B. Friedel-Crafts Acylation Reactions

Besides alkenes, alkyl halides, and alcohols, acid chlorides and acid anhydrides also serve as sources of electrophiles for Friedel-Crafts reactions. For example, benzene reacts with acetic anhydride in the presence of aluminum chloride to give acetophenone.

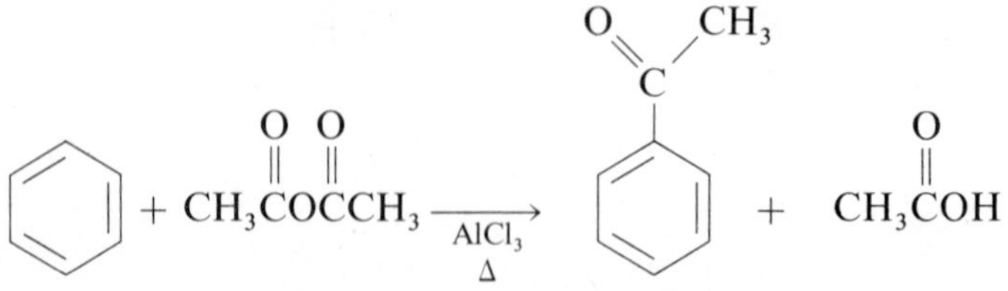
+ CH3COCCH3
O O
AlCl3
Δ
O
CH3
C
+
CH3COH
O

benzene — acetic anhydride — acetophenone ~80% — acetic acid

The product is a ketone. In effect, an acyl group has been substituted on the aromatic ring, so the reaction is called a **Friedel-Crafts acylation.**

An example in which an acid chloride is used is the acylation of bromobenzene with acetyl chloride.

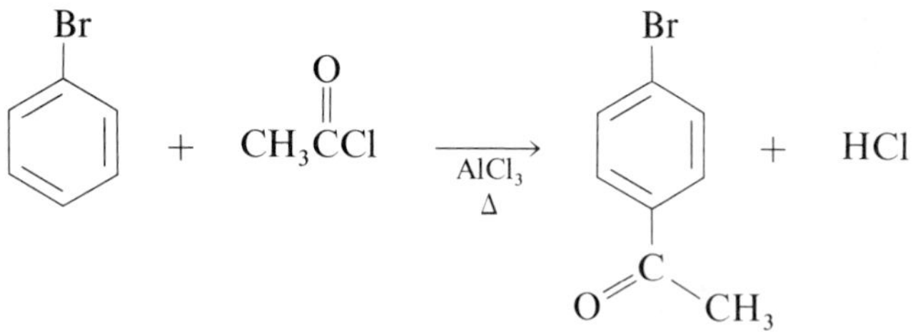

bromobenzene   acetyl chloride   *p*-bromoaceto-
phenone
70%

Both acid chlorides and acid anhydrides have a good leaving group (p. 562). Aluminum chloride can complex with either of these types of compounds to help remove the leaving group, creating an acylium cation.

V I S U A L I Z I N G   T H E   R E A C T I O N

Formation of the electrophile in a Friedel-Crafts acylation

$$CH_3\overset{\displaystyle :O:}{\overset{\|}{C}}-\overset{..}{\underset{..}{Cl}}: \rightarrow AlCl_3 \longrightarrow CH_3\overset{\displaystyle :O:}{\overset{\|}{C}}\overset{\curvearrowright}{\overset{..+}{Cl}}-\overset{-}{Al}Cl_3 \longrightarrow [CH_3\overset{+}{C}=\overset{..}{\underset{..}{O}} \quad \overset{..}{\underset{..}{Cl}}-\overset{-}{Al}Cl_3]$$

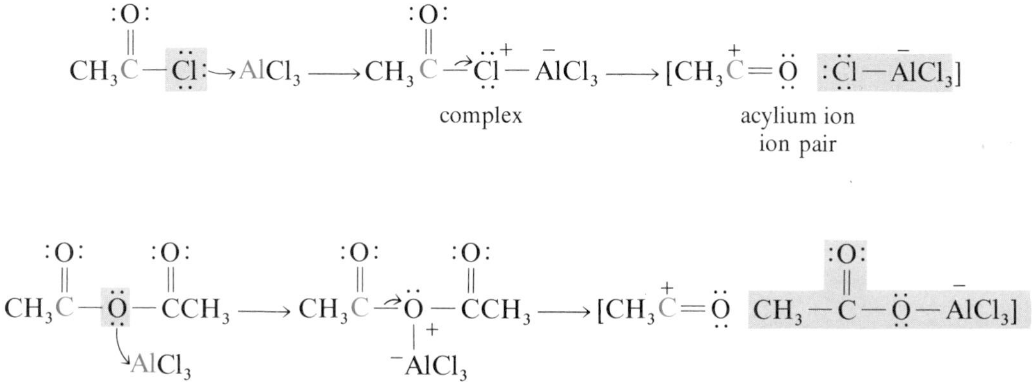

complex   acylium ion
ion pair

$$CH_3\overset{\displaystyle :O:}{\overset{\|}{C}}-\overset{\displaystyle :O:}{\underset{\displaystyle \underset{AlCl_3}{\searrow}}{\overset{..}{\underset{..}{O}}}}-\overset{\displaystyle :O:}{\overset{\|}{C}}CH_3 \longrightarrow CH_3\overset{\displaystyle :O:}{\overset{\|}{C}}\overset{\curvearrowright}{\underset{\underset{-AlCl_3}{\overset{+}{|}}}{\overset{..}{O}}}-\overset{\displaystyle :O:}{\overset{\|}{C}}CH_3 \longrightarrow [CH_3\overset{+}{C}=\overset{..}{\underset{..}{O}} \quad CH_3-\overset{\displaystyle :O:}{\overset{\|}{C}}-\overset{..}{\underset{..}{O}}-\overset{-}{Al}Cl_3]$$

complex   acylium ion
ion pair

The acylium ion is particularly stable because the positive charge can be delocalized to the oxygen atom.

$$\left[CH_3\overset{+}{C}=\overset{..}{\underset{..}{O}} \longleftrightarrow CH_3C\equiv\overset{+}{O}:\right]$$

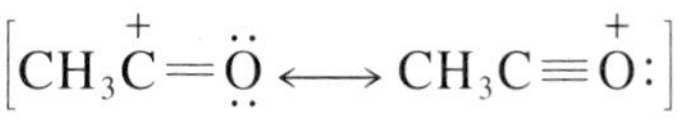

particularly good
resonance contributor,
8 electrons around
each atom

resonance contributors of the acylium ion

Acylium ions do not undergo rearrangement. Thus, the acid chloride from a long straight-chain carboxylic acid can be used to put a straight carbon chain on an aromatic ring as an acyl group. The acid chloride of propanoic acid has been used to put a straight chain of three carbon atoms on the benzene ring.

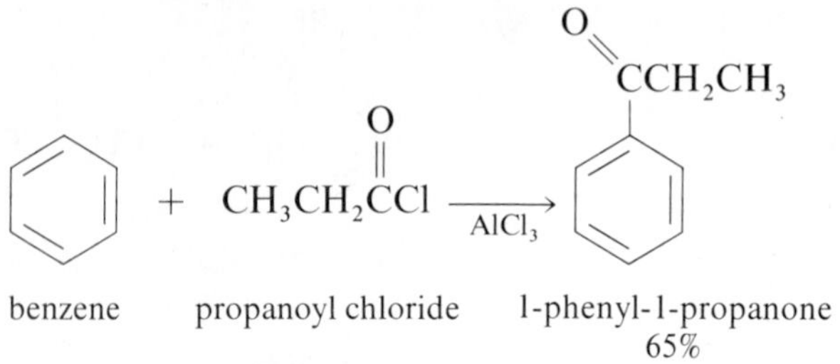
CCH2CH3
CH3CH2CCl
AlCl3
benzene
propanoyl chloride
1-phenyl-1-propanone
65%

A Friedel-Crafts acylation followed by reduction of the carbonyl group to a methylene group is the best way to introduce unbranched alkyl groups onto an aromatic ring. Thus, Wolff-Kishner reduction (p. 509) of 1-phenyl-1-propanone gives propylbenzene.

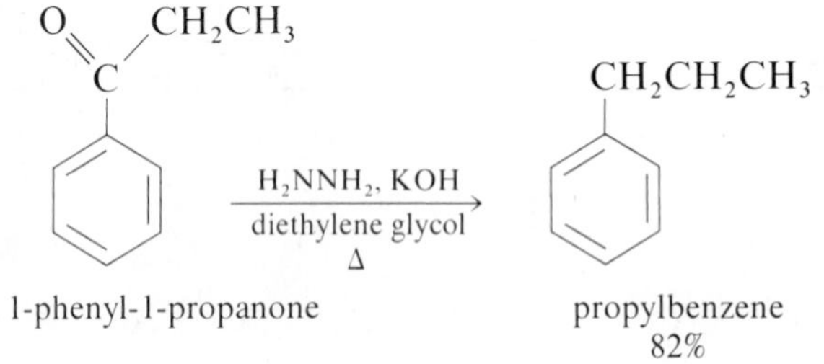
CH2CH3
CH2CH2CH3
H2NNH2, KOH
diethylene glycol
Δ
1-phenyl-1-propanone
propylbenzene
82%

Another advantage of Friedel-Crafts acylation is that the carbonyl group attached to the ring deactivates it toward further substitution. Side products resulting from multiple substitution reactions are not observed with Friedel-Crafts acylations.

Cyclic acid anhydrides are particularly useful as the source of the electrophile for a Friedel-Crafts reaction, because the leaving group remains bonded to the rest of the molecule and is present as a functional group that can undergo further reactions. Succinic anhydride reacts with benzene to give a ketone containing a carboxylic acid function.

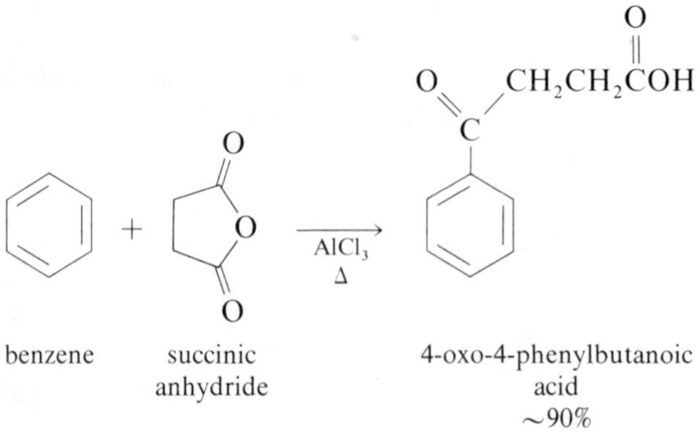
CH2CH2COH
AlCl3
Δ
benzene
succinic anhydride
4-oxo-4-phenylbutanoic acid
~90%

The electrophile that participates in this reaction has two functional groups on it. At one end is an acylium ion and at the other a carboxylate anion complexed with aluminum chloride.

VISUALIZING THE REACTION

A Friedel-Crafts acylation

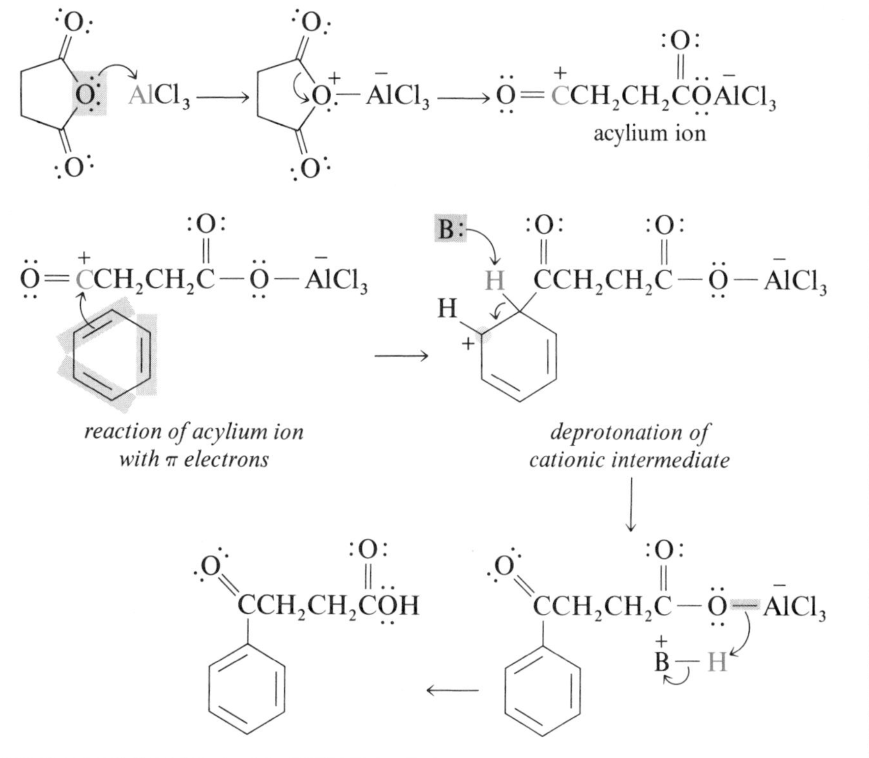

acylium ion
reaction of acylium ion with π electrons
deprotonation of cationic intermediate

Such a bifunctional molecule can undergo another Friedel-Crafts reaction between the carboxylic acid group and the ring. Clemmensen reduction (p. 509) of the ketone function to a methylene group gives 4-phenylbutanoic acid. This acid is converted into its acid chloride, which, on treatment with aluminum chloride, acylates the aromatic ring to give a cyclic ketone.

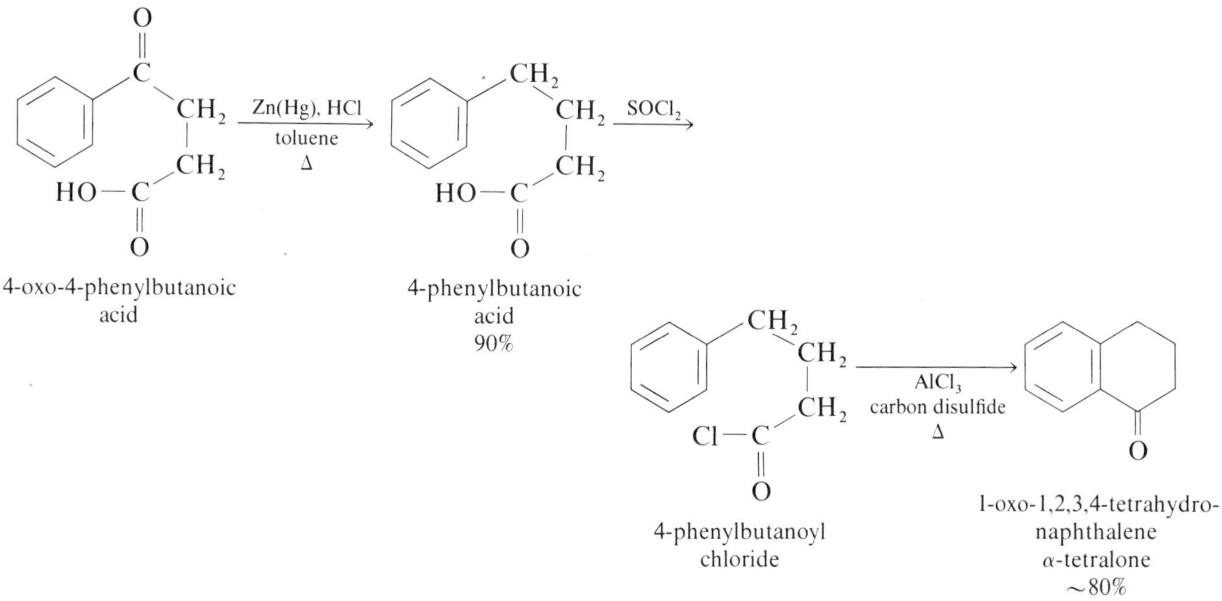

Zn(Hg), HCl
toluene
Δ
SOCl2
AlCl3
carbon disulfide
Δ
4-oxo-4-phenylbutanoic acid
4-phenylbutanoic acid
90%
4-phenylbutanoyl chloride
1-oxo-1,2,3,4-tetrahydro-naphthalene
α-tetralone
~80%

Steric factors ensure that the new bond to the aromatic ring is formed at the position ortho to the original point of attachment of the chain, and a stable six-membered ring is formed. This sequence of reactions is a standard way of making fused-ring compounds. The presence of a reactive carbonyl group in the final product makes it possible to extend the synthesis.

PROBLEM 19.16

Predict the products, designated by letters, that will be formed when 1-oxo-1,2,3,4-tetrahydronaphthalene (above) reacts with the following reagents.

(a) $NaBH_4$, $CH_3CH_2OH \longrightarrow A$

(b) CH_3CH_2MgBr, diethyl ether $\longrightarrow B \xrightarrow[H_2O]{NH_4Cl} C$

(c) $NaC{\equiv}CCH_3$, NH_3(liq) $\longrightarrow D \xrightarrow[H_2O]{NH_4Cl} E$

(d) 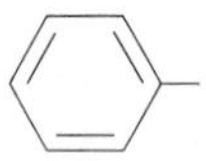 $-NHNH_2$, acetic acid $\longrightarrow F$

(e) $HOCH_2CH_2OH$, TsOH, toluene $\xrightarrow[\Delta]{} G$

Write equations showing the following further transformations.

(f) $A \xrightarrow[H_3PO_4,\ \Delta]{} H \xrightarrow[\text{tetrahydrofuran}]{BH_3} \xrightarrow{H_2O_2,\ OH^-} I$ (g) $C \xrightarrow[H_3PO_4,\ \Delta]{} J$

(h) $E \xrightarrow[\text{Pd/BaSO}_4,\ \text{quinoline}]{H_2} K \xrightarrow{\text{9-BBN}} L \xrightarrow{H_2O_2,\ OH^-} M$

PROBLEM 19.17

Complete the following equations.

(a) 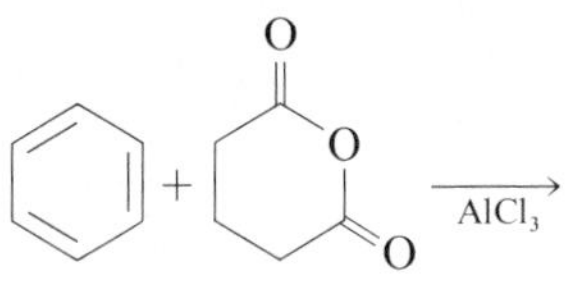
O
O
O
+
AlCl3

(b) 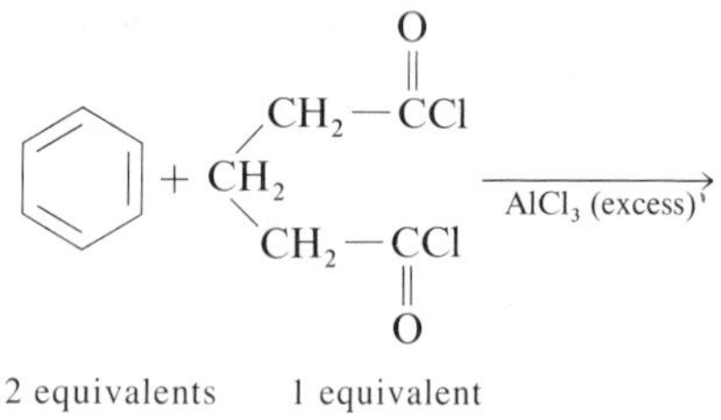
+ CH2
CH2—CCl
CH2—CCl
O
O
AlCl3 (excess)
2 equivalents
1 equivalent

(c) 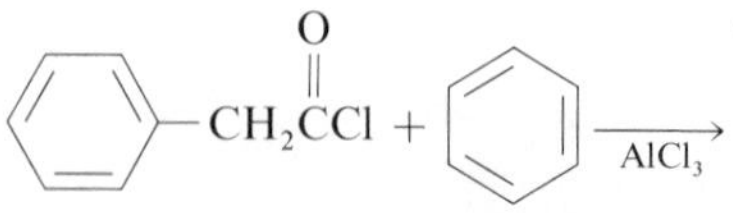
CH2CCl +
O
AlCl3

(d) 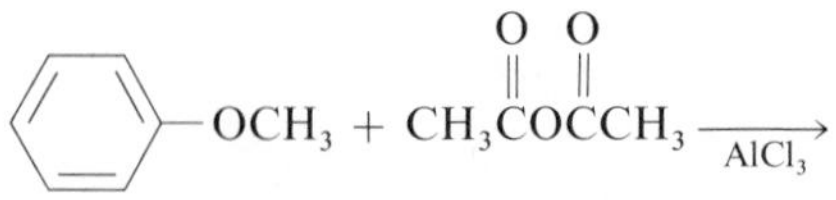
OCH3 + CH3COCCH3
O O
AlCl3

PROBLEM 19.18

Complete the following equations.

(a) 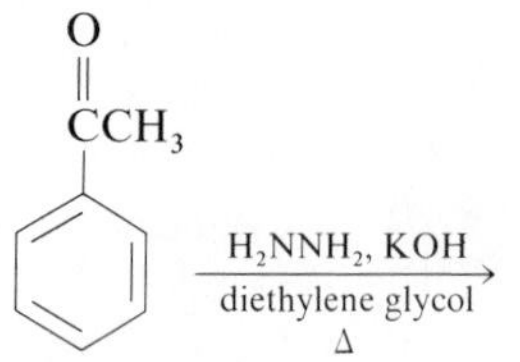
CCH3
O
H2NNH2, KOH
diethylene glycol
Δ

(b) 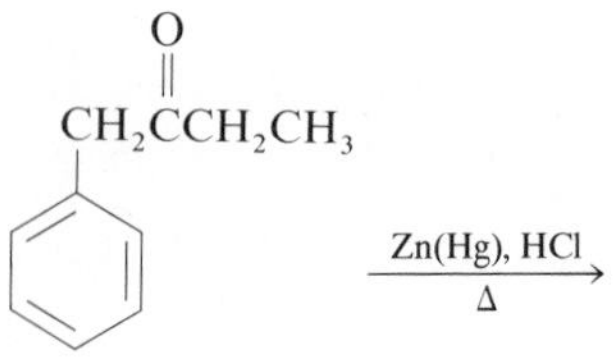
CH2CCH2CH3
O
Zn(Hg), HCl
Δ

(c) 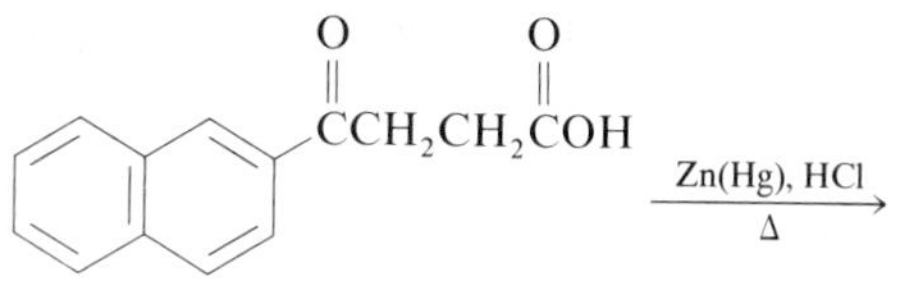

(d) 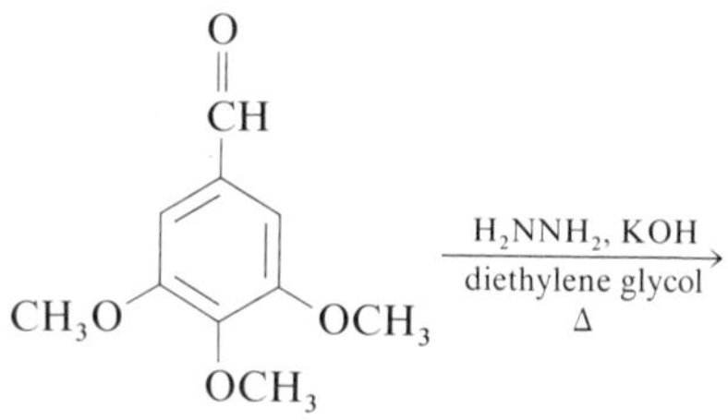

(e) 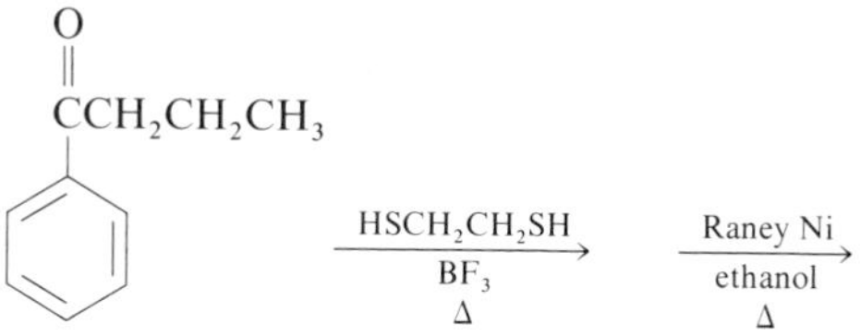

19.5 SULFONATION REACTIONS OF AROMATIC COMPOUNDS

A. Sulfonation. Sulfur Trioxide as the Electrophile

Benzene reacts with sulfuric acid to give benzenesulfonic acid. The reaction is reversible, but high yields of the sulfonic acid are obtained when water is removed from the reaction mixture by distillation with benzene or when sulfur trioxide is used as the electrophile.

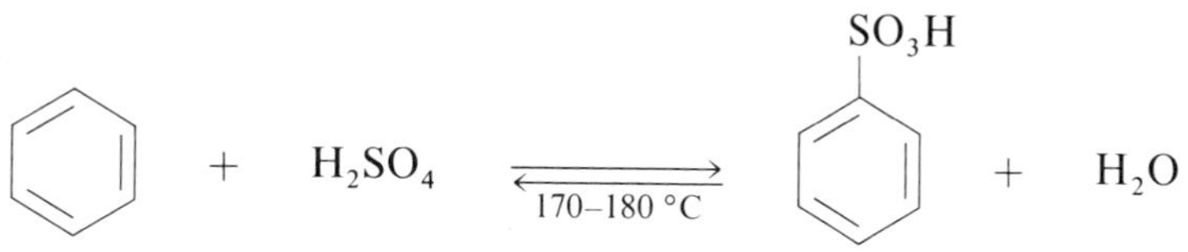

benzene — sulfuric acid — benzenesulfonic acid 95% — water *removed from the reaction mixture by distillation with benzene*

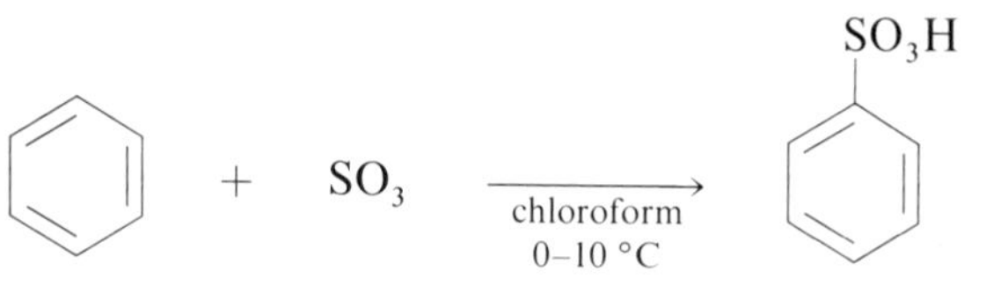

benzene — sulfur trioxide — benzenesulfonic acid 90%

Even when sulfuric acid is used as the reagent, the electrophile is believed to be sulfur trioxide, which is the anhydride of sulfuric acid. The fact that the rate of the reaction increases with increasing sulfur trioxide concentration is evidence for this conclusion.

VISUALIZING THE REACTION

The sulfonation reaction

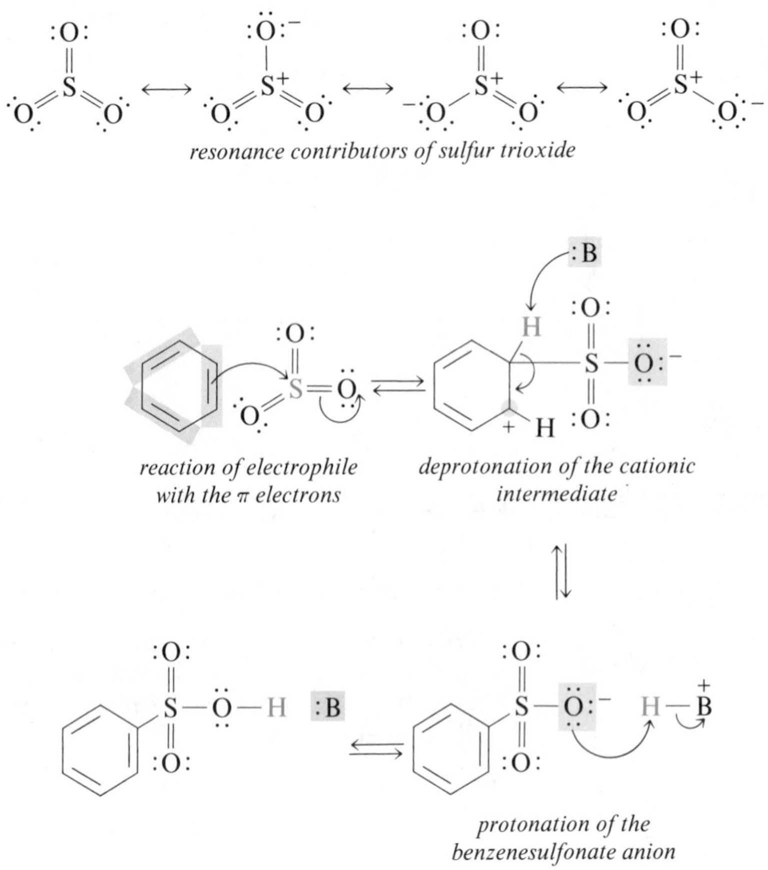
resonance contributors of sulfur trioxide
reaction of electrophile with the π electrons
deprotonation of the cationic intermediate
protonation of the benzenesulfonate anion

The reverse reaction occurs when a sulfonic acid is heated in water.

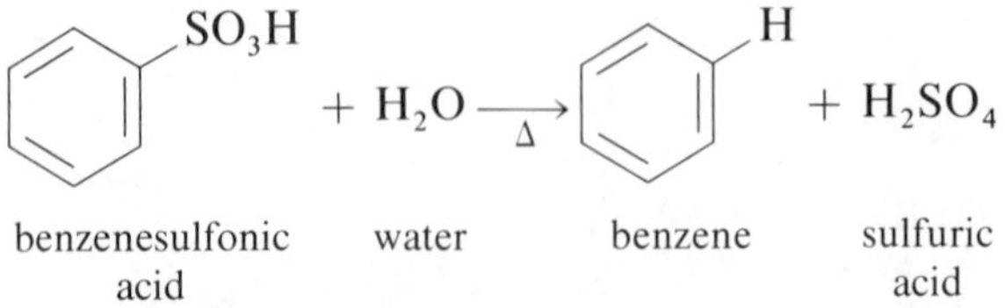
benzenesulfonic acid
water
benzene
sulfuric acid

The reaction is pictured as taking place by the loss of sulfur trioxide from the sulfonate anion and protonation of the developing anion.

VISUALIZING THE REACTION

Acid-catalyzed desulfonation reaction

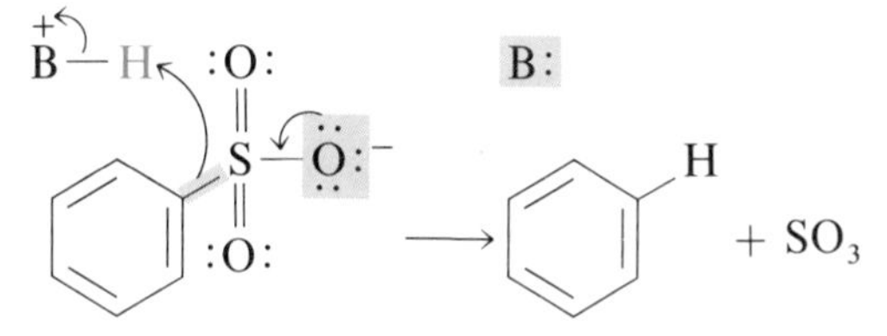

B—H
:O:
S
O:⁻
:O:
B:
H
+ SO₃

$$SO_3 + H_2O \rightleftarrows H_2SO_4$$

Study Guide Concept Map 19.5

PROBLEM 19.19

Complete the following equations.

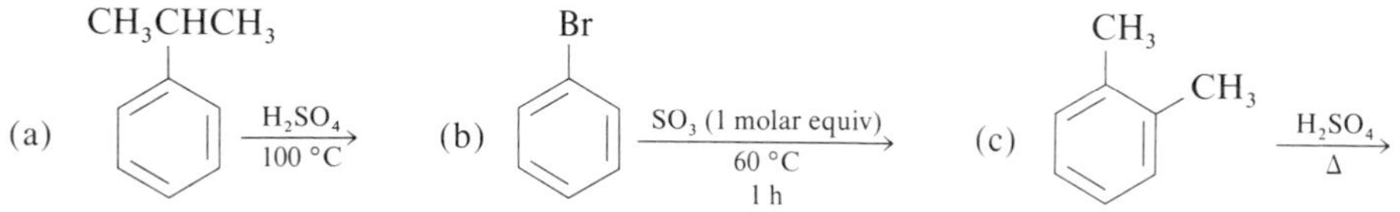

CH₃CHCH₃
(a)
H₂SO₄
100 °C
(b)
Br
SO₃ (1 molar equiv)
60 °C
1 h
(c)
CH₃
CH₃
H₂SO₄
Δ

PROBLEM 19.20

When toluene is sulfonated with concentrated sulfuric acid at 0 °C, the product mixture consists of 53% *p*-toluenesulfonic acid, 43% *o*-toluenesulfonic acid, and 4% *m*-toluenesulfonic acid. If the sulfonation is carried out at 100 °C, *p*-toluenesulfonic acid constitutes 79% of the product mixture. How do you explain the correlation of the change in the composition of the product mixture with the temperature of the reaction? (Hint: You may find it helpful to review p. 686.)

B. The Chemistry of Sulfonic Acids and Their Derivatives

The products of sulfonation reactions are sulfonic acids, organic acids related to sulfuric acid. Sulfonic acids, because of the resonance stabilization possible for the anion resulting from the loss of the proton on the hydroxyl group, are strong acids with pK_a values of approximately −0.5. The organic portion of the molecule ensures its solubility in organic solvents. Thus, a sulfonic acid, *p*-toluenesulfonic acid, for example, is often used when a strong acid that is soluble in an organic reaction mixture is required.

Sulfonic acids and carboxylic acids are closely related in their chemistry. For example, the acid chlorides of sulfonic acids are prepared by heating the sulfonic acids with thionyl chloride or phosphorus pentachloride (p. 560). The conversions of methanesulfonic acid and sodium benzenesulfonate to the corresponding sulfonyl chlorides are examples of these reactions.

$$CH_3S(=O)_2OH + SOCl_2 \xrightarrow[3.5\ h]{90\ °C} CH_3S(=O)_2Cl + SO_2\uparrow + HCl\uparrow$$

methanesulfonic acid — thionyl chloride — methanesulfonyl chloride

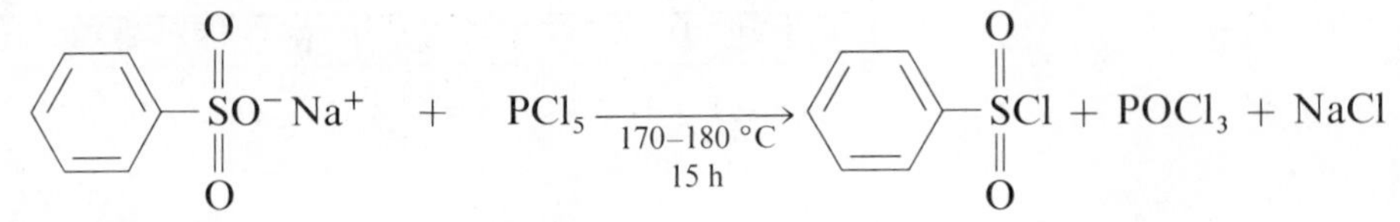

sodium benzenesulfonate — phosphorus pentachloride — benzenesulfonyl chloride

Sulfonyl chlorides react with nucleophiles in the way acyl chlorides do. Thus, reactions with alcohols give sulfonic acid esters and reactions with ammonia or amines give amides of sulfonic acids, called **sulfonamides.** All sulfonamides are solids with reasonably high melting points. Reactions of sulfonyl chlorides with alcohols or with ammonia or amines are carried out in the presence of a base to neutralize the hydrochloric acid that is the other product. In the reaction of *trans*-2-vinylcyclohexanol with *p*-toluenesulfonyl chloride, pyridine serves as both solvent and base.

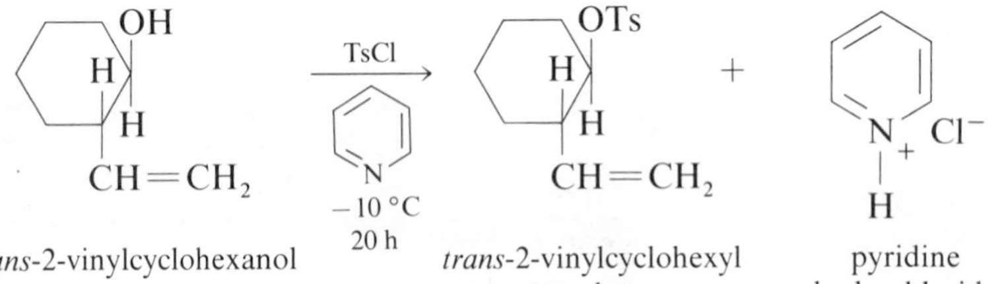

trans-2-vinylcyclohexanol — *trans*-2-vinylcyclohexyl tosylate 88% — pyridine hydrochloride

As the above equation shows, the formation of the tosylate occurs with retention of configuration at the carbon atom to which the hydroxyl group is bonded. Because tosylates are often used in syntheses and in studies of the mechanisms of organic reactions, the stereochemistry of their formation (p. 243) is important.

p-Toluenesulfonyl chloride heated with concentrated ammonium hydroxide solution gives *p*-toluenesulfonamide quantitatively.

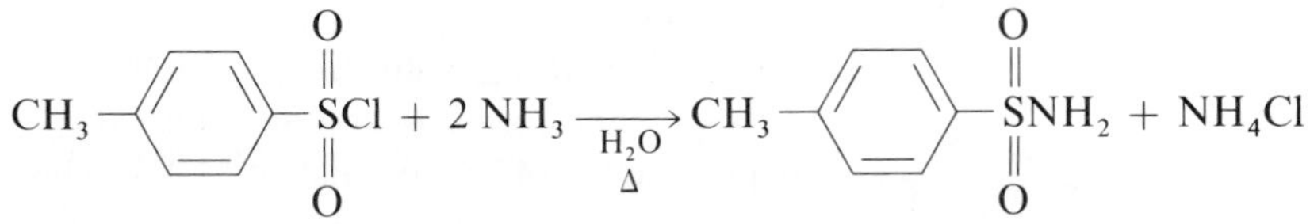

p-toluenesulfonyl chloride — *p*-toluenesulfonamide 100%

The excess ammonia serves as the base for the reaction. When an amine is used instead of ammonia, another base is used to absorb the hydrochloric acid. For example, benzenesulfonyl chloride reacts with amines in the presence of aqueous sodium hydroxide.

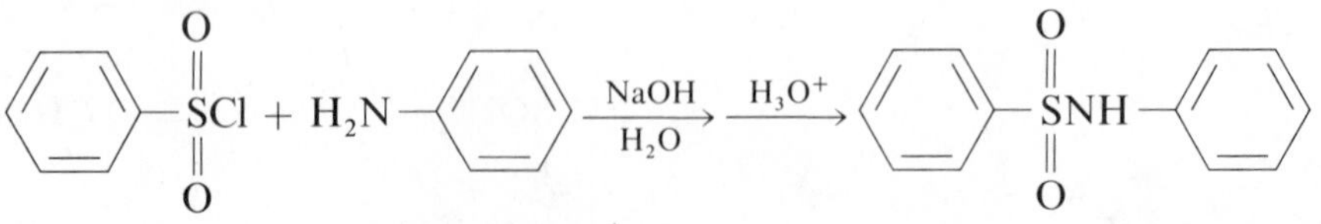

benzenesulfonyl chloride — aniline — *N*-phenylbenzenesulfonamide

There is one important difference between a sulfonamide and the amide of a carboxylic acid. A sulfonamide is acidic if it has at least one hydrogen atom on the nitrogen atom of the amide group. Such a sulfonamide is a reasonably strong acid. If the hydrocarbon portion of the amine is not large, the sulfonamide will dissolve in dilute sodium hydroxide. Thus, the product from the reaction of benzenesulfonyl chloride with aniline is soluble in aqueous sodium hydroxide and will precipitate when the solution is acidified. The acidity of a sulfonamide can be rationalized by pointing to the delocalization of electrons to the two oxygen atoms of the sulfonamide function.

VISUALIZING THE REACTION

The acidity of a sulfonamide

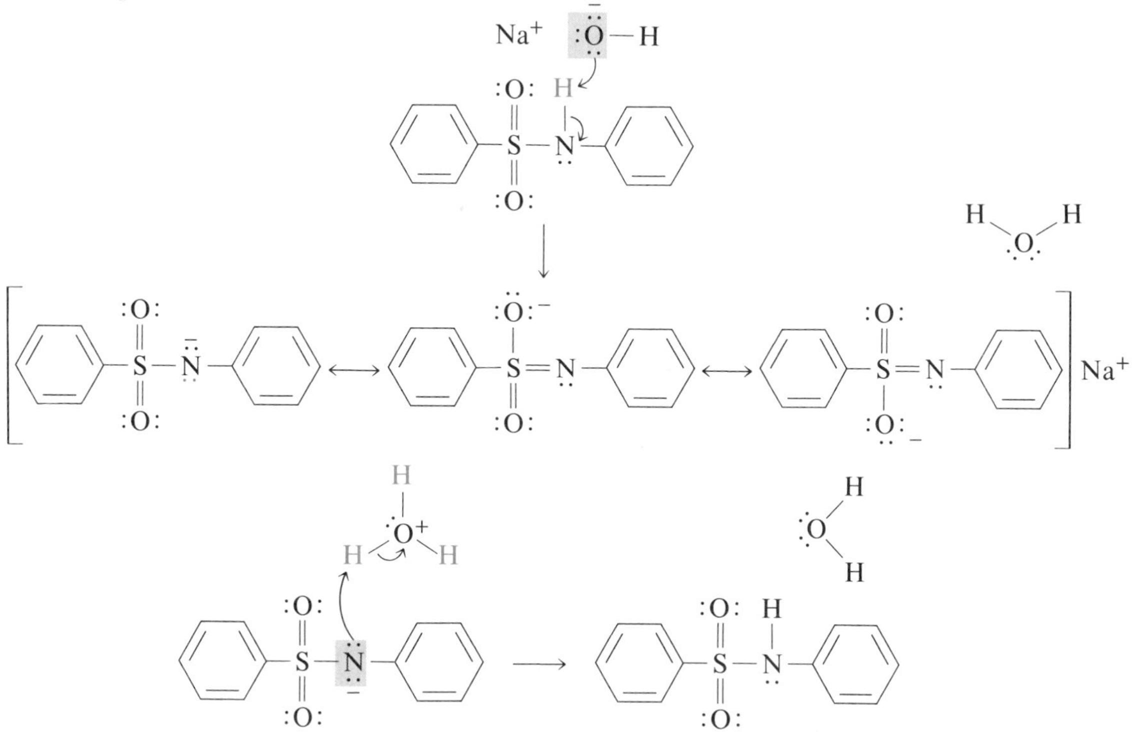

sodium salt of *N*-phenylbenzenesulfonamide
ionic compound, soluble in water

N-phenylbenzenesulfonamide
covalent compound, insoluble in water

Because *N*-methyl-*N*-phenylbenzenesulfonamide does not have a hydrogen atom on the nitrogen atom of the amide group, it is insoluble in aqueous sodium hydroxide. It forms as a precipitate in the solution.

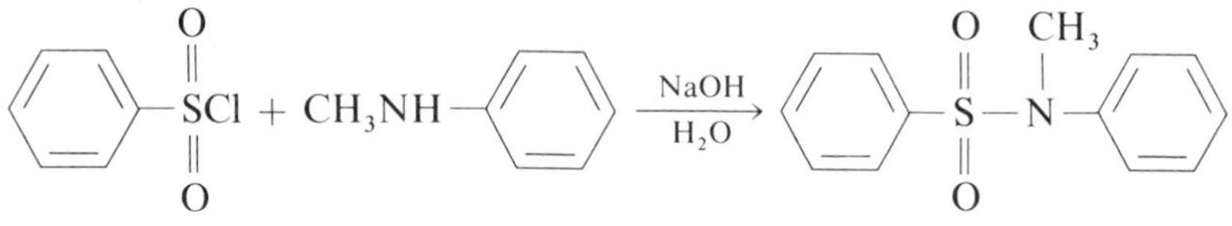

benzenesulfonyl chloride

N-methylaniline

N-methyl-*N*-phenylbenzenesulfonamide
not an acid, insoluble in base

Sulfonyl chlorides, like acid chlorides and acid anhydrides, do not form stable compounds in reactions with amines that have no hydrogen atoms on the nitrogen atom.

PROBLEM 19.21

Write a detailed mechanism for the reaction of ammonia with *p*-toluenesulfonyl chloride. Compare the steps of your mechanism with those for the reaction of an acyl chloride (p. 542).

PROBLEM 19.22

Complete the following equations.

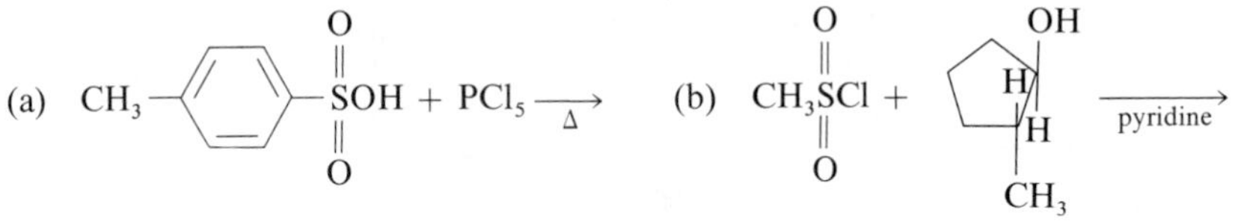
(a) CH3 SOH + PCl5 Δ
(b) CH3SCl + OH H H CH3 pyridine

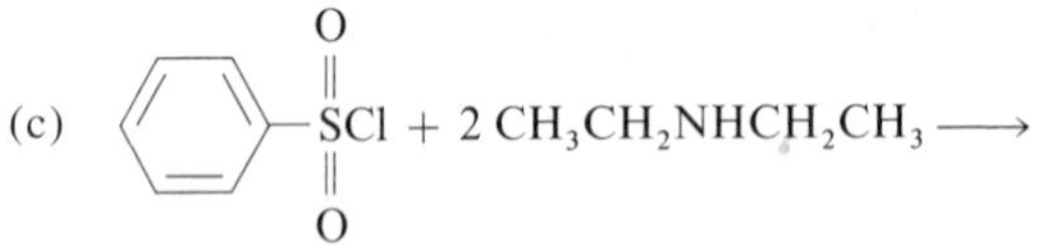
(c) SCl + 2 CH3CH2NHCH2CH3

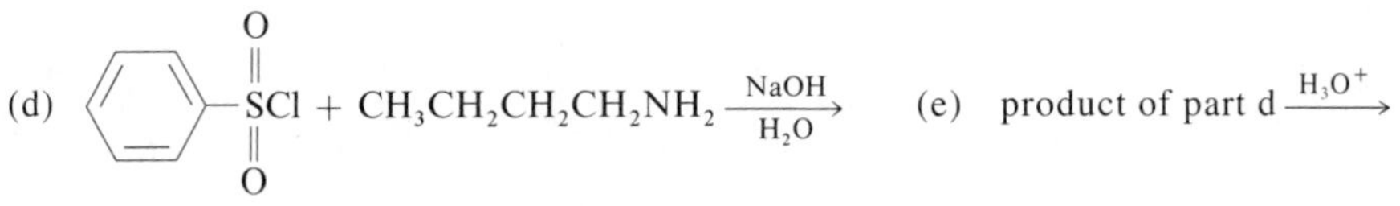
(d) SCl + CH3CH2CH2CH2NH2 NaOH H2O
(e) product of part d H3O+

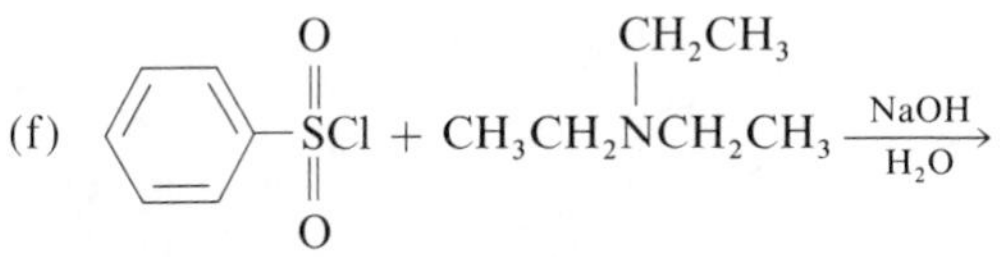
(f) SCl + CH3CH2NCH2CH3 CH2CH3 NaOH H2O

PROBLEM 19.23

Sections 19.3, 19.4, and 19.5 give detailed discussions of several types of electrophilic aromatic substitution reactions. The examples used do not represent all the reagents that can act as electrophiles in aromatic substitution. In each of the following cases, predict the structure of the electrophile and of the product of the reaction.

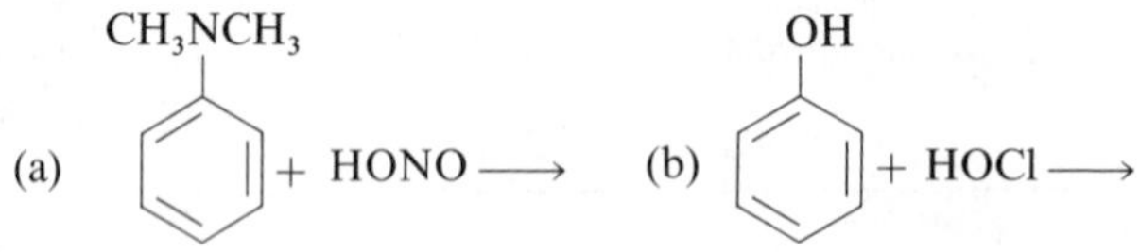
CH3NCH3
(a) + HONO
OH
(b) + HOCl

19.6 ELECTROPHILIC SUBSTITUTION REACTIONS OF MULTIPLY SUBSTITUTED AROMATIC COMPOUNDS. REACTIVITY AND ORIENTATION

When there are two substituents already on an aromatic ring and both of them act to direct an incoming electrophile to the same position, then predicting where the new substituent will end up is relatively easy. For example, *p*-nitrotoluene is brominated ortho to the methyl group.

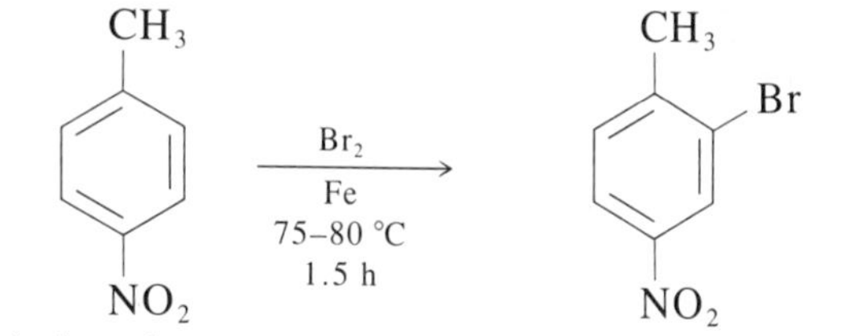

CH3
Br2
Fe
75–80 °C
1.5 h
Br
NO2

4-nitrotoluene → 2-bromo-4-nitrotoluene 90%

The methyl group is ortho,para-directing, and the nitro group is meta-directing. The bromine substitutes ortho to the methyl group and meta to the nitro group.

When the directing influences of the substituents already on the ring are in conflict, the one that is strongly activating and ortho,para-directing determines the orientation of the new substituent. For example, 2-fluoromethoxybenzene is nitrated ortho and para to the methoxy group.

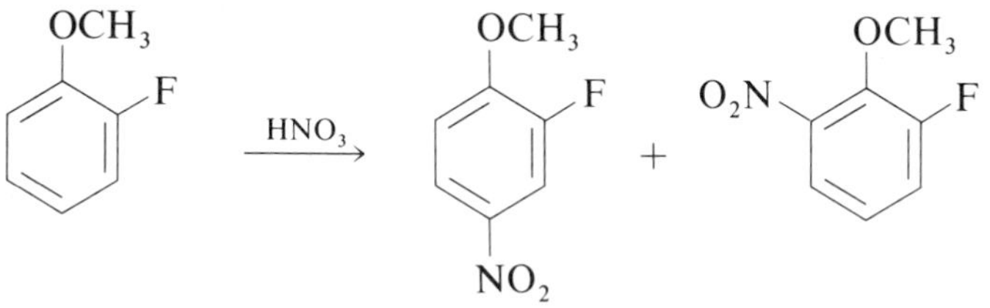

OCH3
F
HNO3
O2N
NO2

2-fluoromethoxybenzene → 2-fluoro-4-nitro-methoxybenzene + 2-methoxy-3-nitro-fluorobenzene

The following gives the relative effectiveness of substituents in directing an incoming electrophile to the ortho or para position.

$$-OH \geq -OCH_3 > -NH_2 > -NH\overset{\overset{\displaystyle O}{\|}}{C}CH_3 > -I > -Br > -Cl > -CH_3$$

For example, when 2-chlorotoluene is nitrated, the mixture that is formed shows that more substitution has occurred ortho and para to the chlorine than to the methyl group.

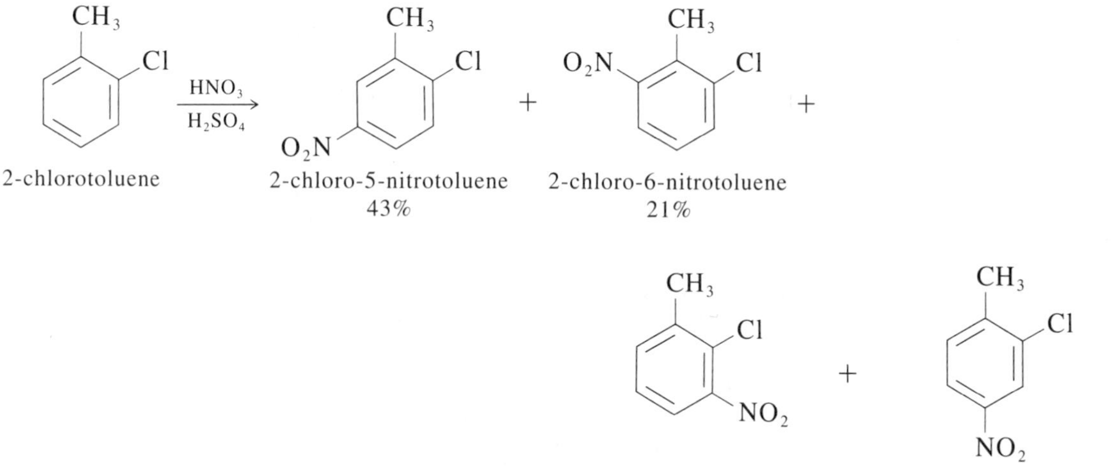

CH3
Cl
HNO3
H2SO4
O2N
NO2

2-chlorotoluene → 2-chloro-5-nitrotoluene 43% + 2-chloro-6-nitrotoluene 21% + 2-chloro-3-nitrotoluene 19% + 2-chloro-4-nitrotoluene 17%

This example also demonstrates the complexity of the mixtures that result when the directing effects of the groups on the ring are more or less equal.

PROBLEM 19.24

Predict the major product(s) of the following reactions.

(a) 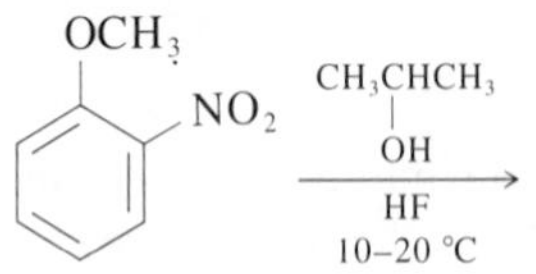
OCH_3
NO_2
CH_3CHCH_3
OH
HF
10–20 °C

(b) 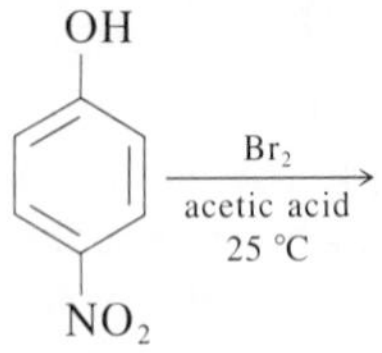
OH
Br_2
acetic acid
25 °C
NO_2

(c) 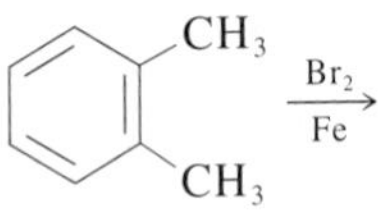
CH_3
CH_3
Br_2
Fe

PROBLEM 19.25

Synthesize the following compounds from benzene and any other organic reagents containing not more than three carbon atoms. There may be more than one way to carry out each synthesis; aim for the fewest steps.

(a) 
$CH_2CH_2CH_2CH_3$

(b) 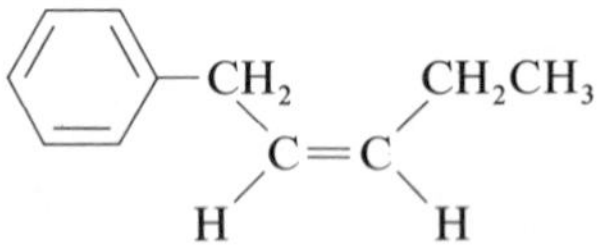
CH_2
CH_2CH_3
C=C
H
H

(c) 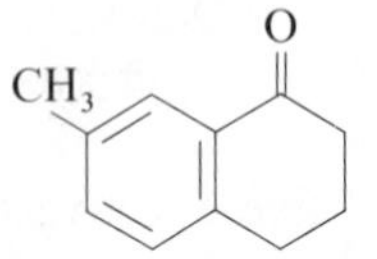
CH_3
O

(d) 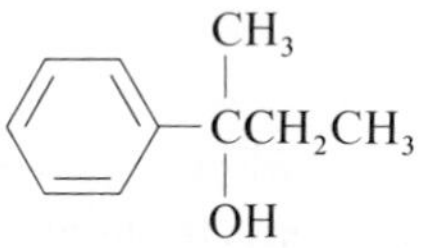
CH_3
CCH_2CH_3
OH

SUMMARY

Benzene and compounds related to it are stable toward reagents that react with alkenes. This special property of cyclic polyenes is called aromaticity. Aromaticity is observed for cyclic, planar polyenes that have $(4n + 2)$ π electrons in the ring, where n is 0, 1, 2, 3, and so on. Ions as well as uncharged compounds may be aromatic.

Aromatic compounds undergo substitution reactions with electrophiles rather than addition reactions. Electrophiles that react with aromatic rings include halogens (usually in the presence of Lewis acids), nitronium ions generated in a mixture of nitric and sulfuric acids, sulfur trioxide or concentrated sulfuric acid, carbocations from protonation of alkenes or alcohols, and acylium cations from the reaction of acid chlorides or anhydrides with Lewis acids. Electrophilic aromatic substitution reactions are summarized in Tables 19.2 and 19.3.

When aromatic compounds other than benzene undergo electrophilic substitution reactions, questions regarding reactivity and regioselectivity arise. A substituent already on an aromatic ring may make further substitution easier; if so, it is ring-activating. If the first substituent on the ring makes further substitution more difficult, it is ring-deactivating. Most ring-activating groups direct an incoming electrophile to the ortho and para positions of the ring. Ring-deactivating substituent are meta-directing. The halogens are an exception; they deactivate the ring but

TABLE 19.2 Electrophilic Aromatic Substitution Reactions

Aromatic Compound	Electrophile*	Reactive Intermediate(s)	Product(s)
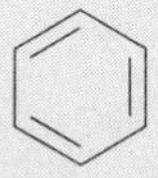	E^+	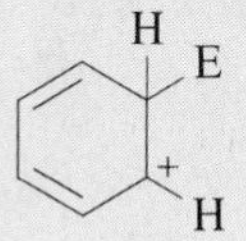	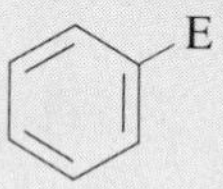
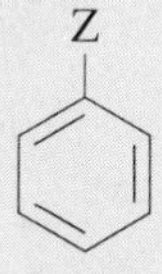 Z = —OH, —OR, —NR_2, R—, —X, etc.	E^+	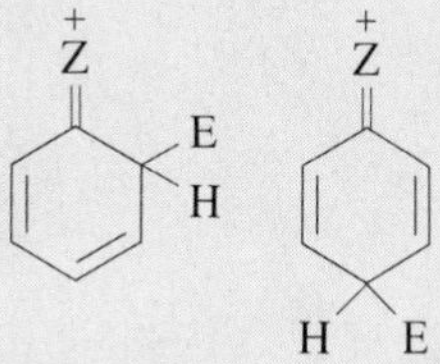	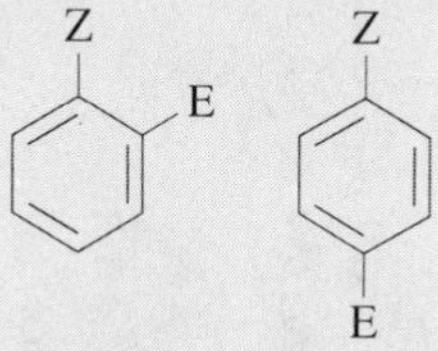 *relative amounts depend on sizes of Z and E*
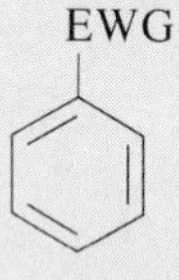 EWG = —NO_2, —$\overset{+}{N}R_3$, —CN, —SO_3H, —$\overset{O}{\overset{\|\|}{C}}R$, etc.	E^+	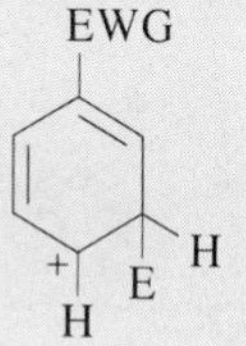	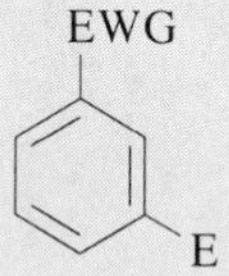

*See Table 19.3 for different electrophiles.

TABLE 19.3 Electrophiles for Electrophilic Aromatic Substitution Reactions

Reaction	Reagents	Electrophile	Substitution Product of Benzene
Halogenation	Cl_2 or Br_2 FeX_3	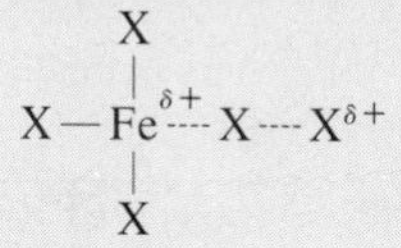	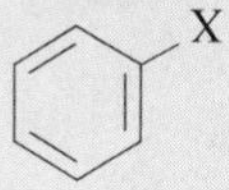
Nitration	HNO_3 H_2SO_4	$\overset{+}{N}O_2$	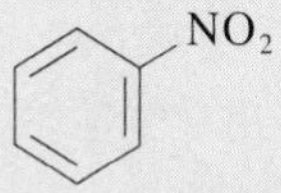
Sulfonation	SO_3 H_2SO_4	SO_3	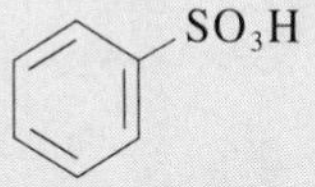
Friedel-Crafts alkylation	ROH HB^+ or RX $AlCl_3$	R^+	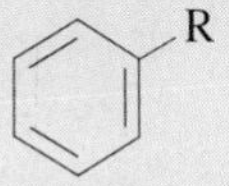
	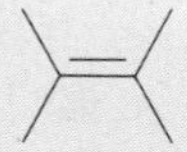 HB^+	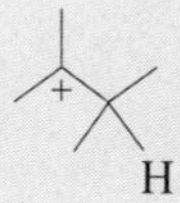	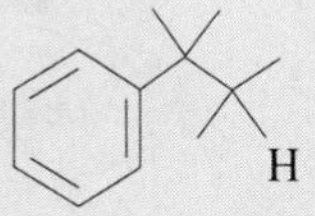

(*Continued*)

TABLE 19.3 *(Continued)*

Reaction	Reagents	Electrophile	Substitution Product of Benzene
Friedel-Crafts acylation	$R\overset{\overset{O}{\|}}{C}Cl$ $AlCl_3$	$RC\equiv O^+$	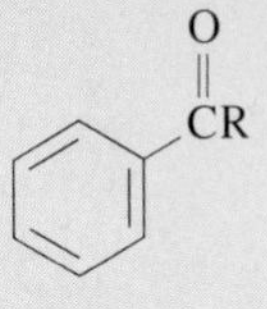
	$R\overset{\overset{O}{\|}}{C}O\overset{\overset{O}{\|}}{C}R$ $AlCl_3$	$RC\equiv O^+$	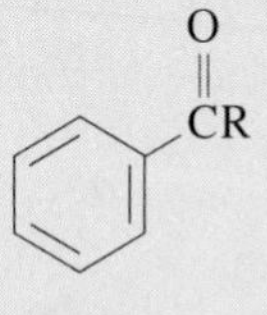
	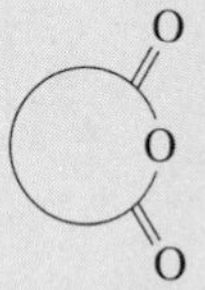 $AlCl_3$	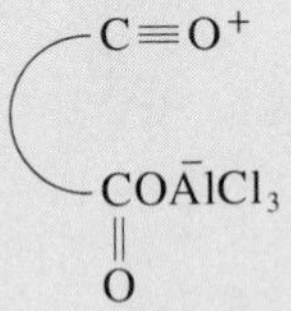	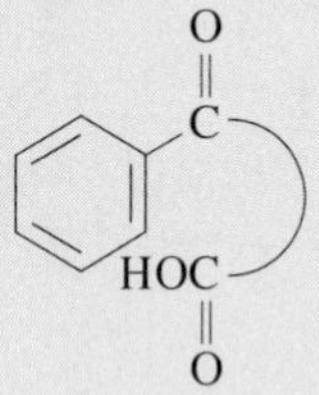

are ortho,para-directing. Table 19.1 (p. 787) summarizes the directing effects of common groups on aromatic compounds. These directing effects are rationalized by looking at the resonance stabilization of the carbocation intermediate formed in the rate-determining step of the substitution reaction. Ortho,para-directing groups stabilize the intermediate formed by attack of the electrophile at those positions. Meta-directing groups do not.

When more than one substituent is already present on an aromatic ring undergoing further substitution, the nature of the substituents and their orientation to each other will determine what happens. Generally, in these cases, ring-activating substituents are more important than ring-deactivating substituents in directing the incoming electrophile.

ADDITIONAL PROBLEMS

19.26 Name the following compounds.

(a) 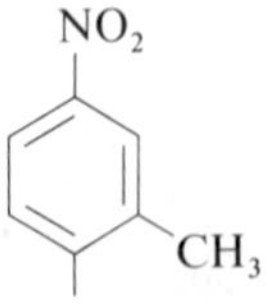

(b) 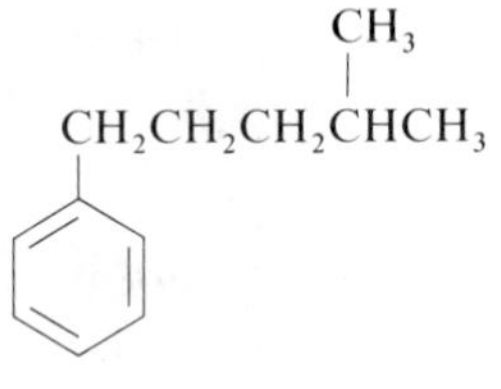

(c) 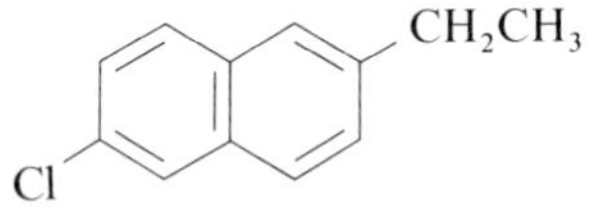

(d) 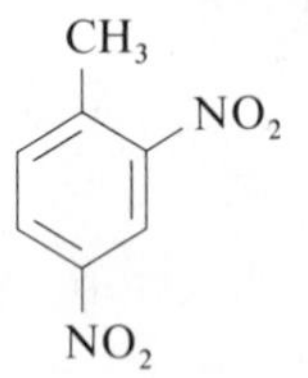

(e) 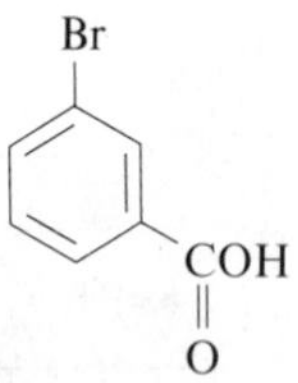

(f) 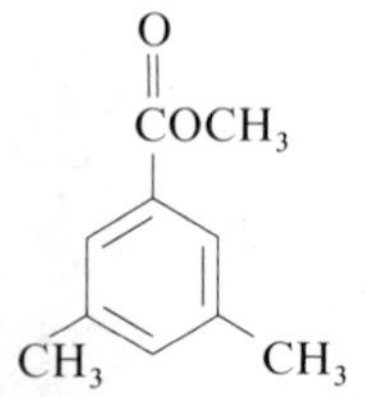

(g) 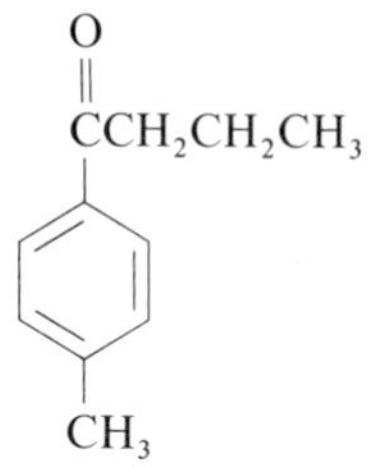

(h) 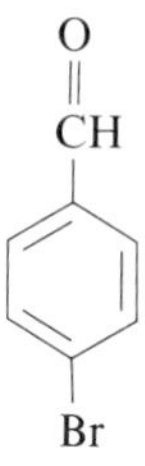

19.27 Write structural formulas for the following compounds.

(a) ethyl *m*-chlorobenzoate (b) 1,4-dimethylnaphthalene
(c) 1-phenyl-3-methyl-1-hexanone (d) 1-phenyl-1-chloropropane
(e) *m*-bromonitrobenzene (f) *p*-chlorobromobenzene

19.28 Give structures for compounds designated by letters in the following equations.

(a) 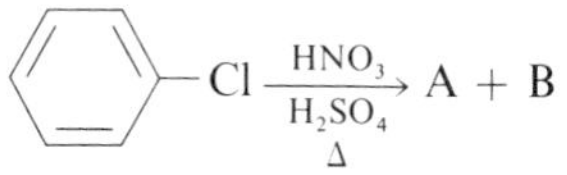

(b) 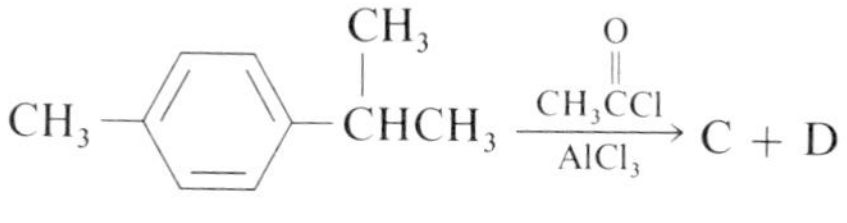

(c) 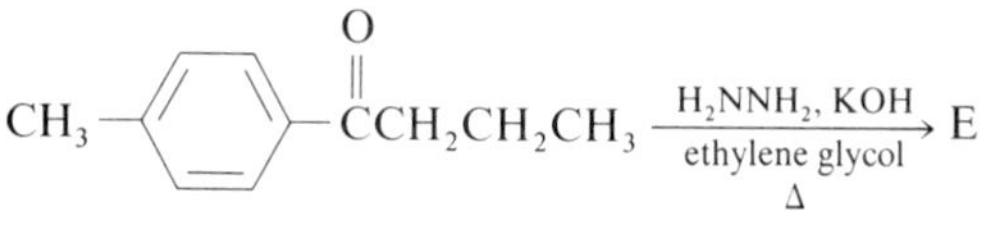

(d) 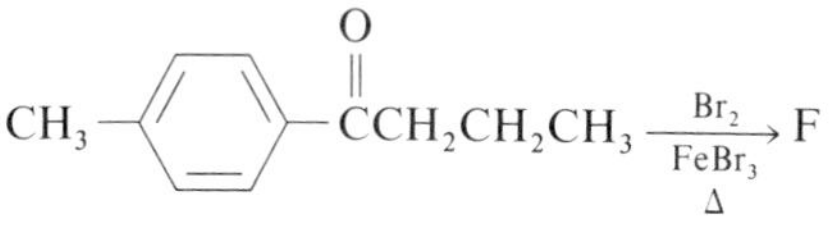

(e) 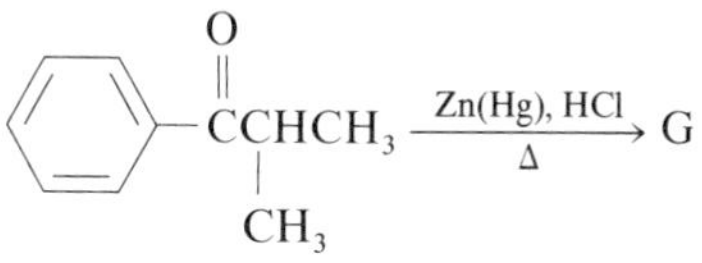

(f) 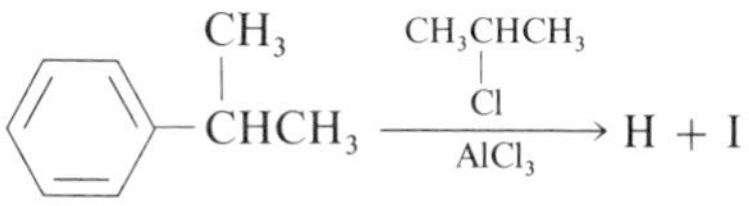

(g) 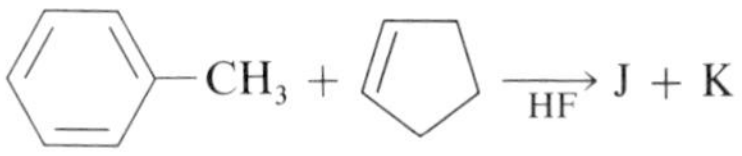

(h) 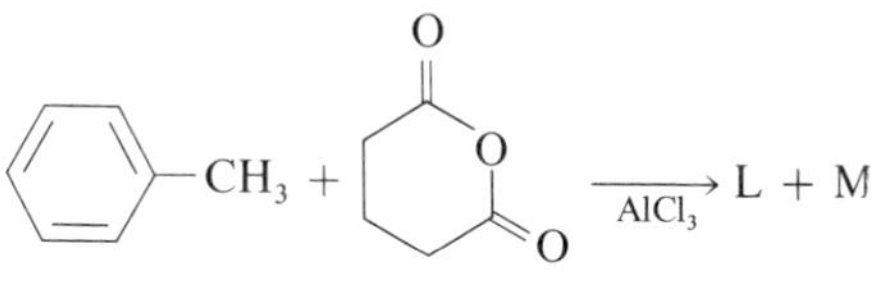

(i) 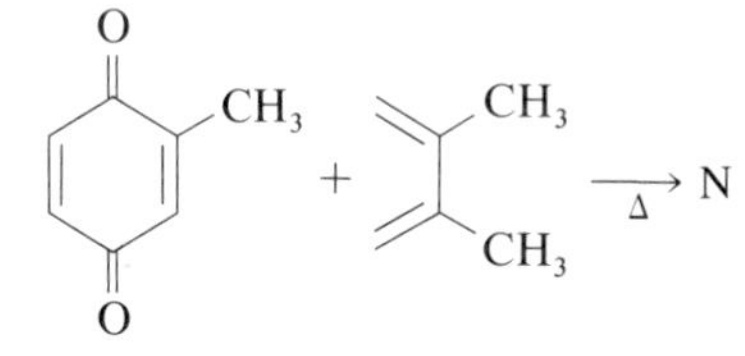

(j) 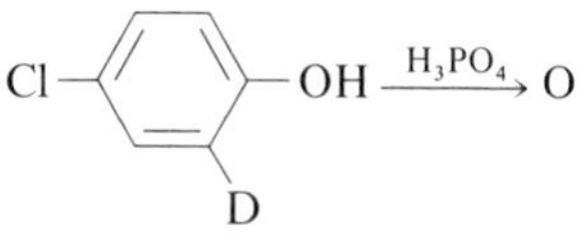

(k) 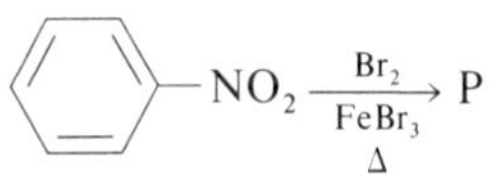

(l) 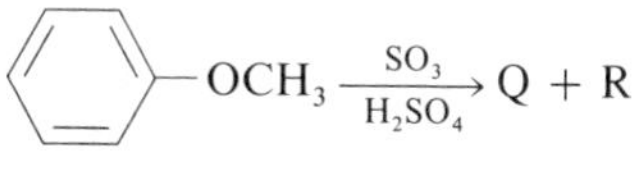

(m) 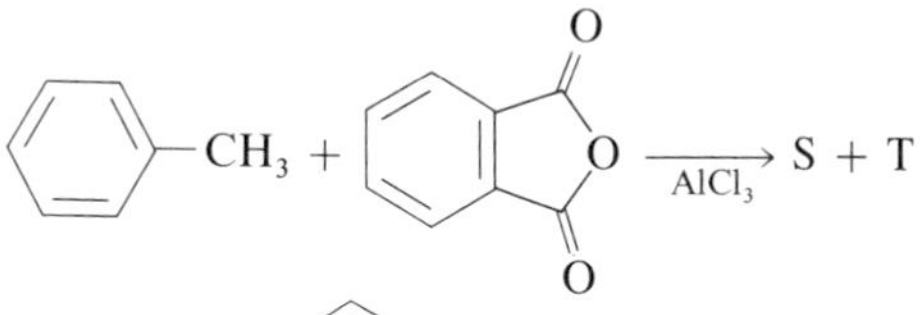

(n) 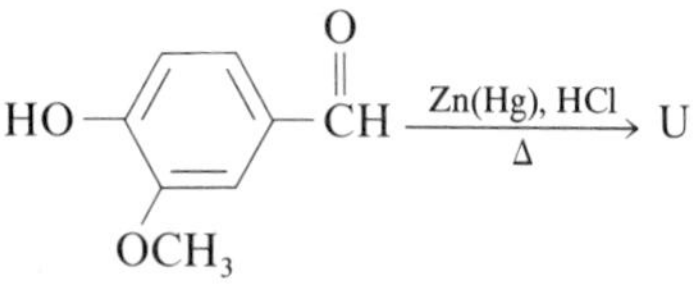

(o) 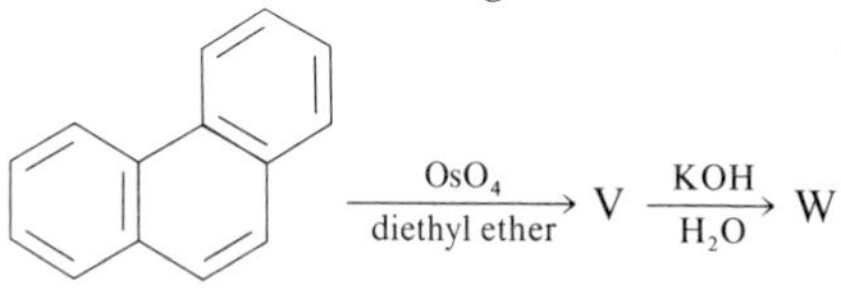

(p) 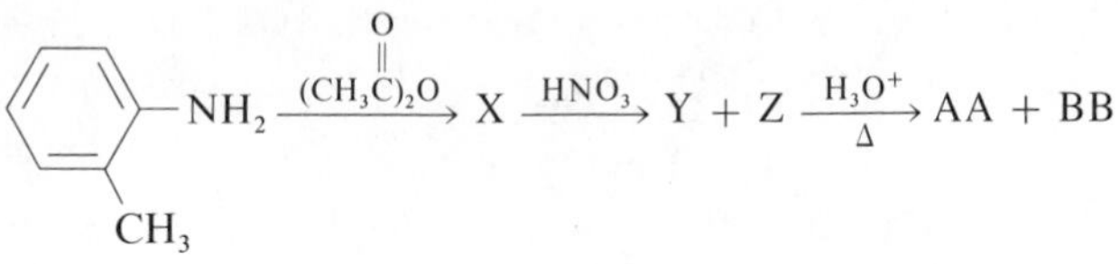

(q) 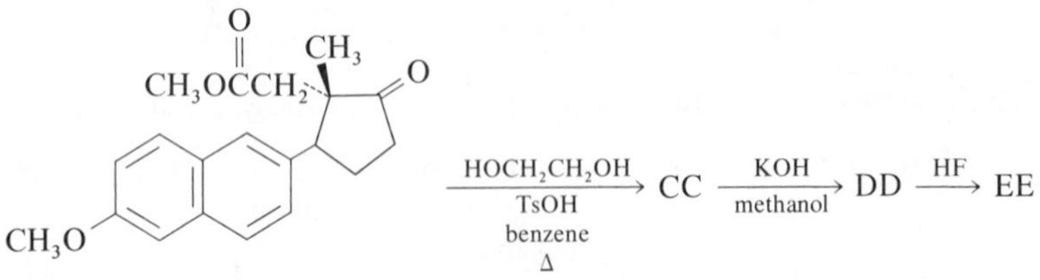

(r) 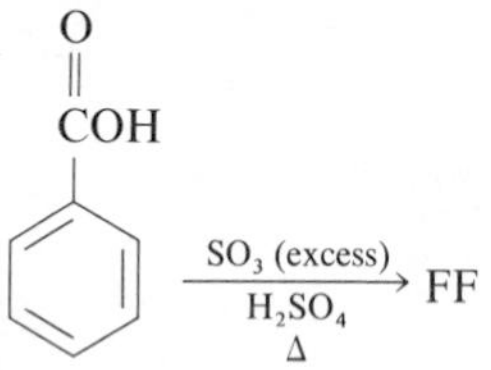

19.29 1,3,5-Trimethylbenzene is converted into 1,3,5-trimethylphenylacetic acid by the following sequence of reactions. Tell what the reagents designated A and B are, and propose a mechanism for the first step.

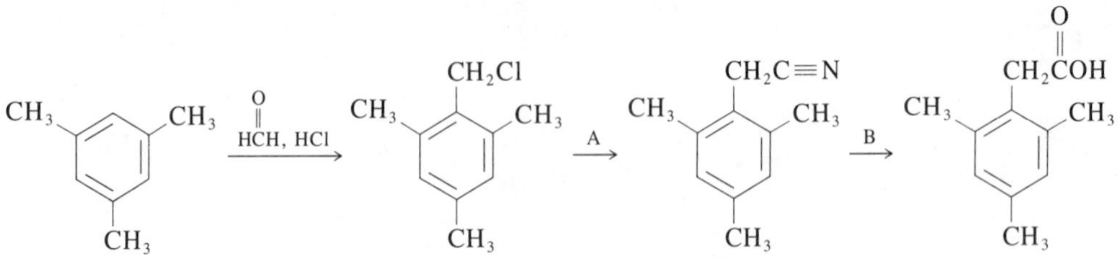

19.30

(a) Dewar benzene (Problem 2.25, p. 67) has actually been synthesized and shown to have reactivity that differs from that of benzene. The following reactions have been carried out with Dewar benzene; predict what the products, represented by letters, will be.

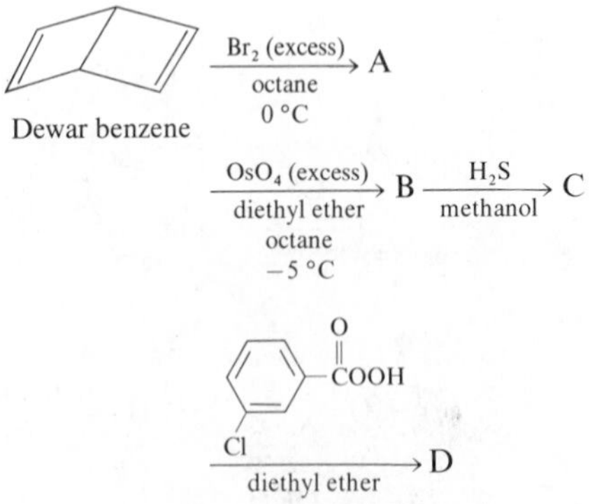

(b) When Dewar benzene is treated with either a Lewis acid, $AlCl_3$, or a protic acid, H_2SO_4, benzene is formed. Suggest a mechanism for the conversion of Dewar benzene to benzene in sulfuric acid.

19.31 Propose syntheses for the following compounds starting with benzene, toluene, and any other organic compound having no more than three carbon atoms. There may be more than one way to synthesize each compound; try to find the shortest route.

(a)
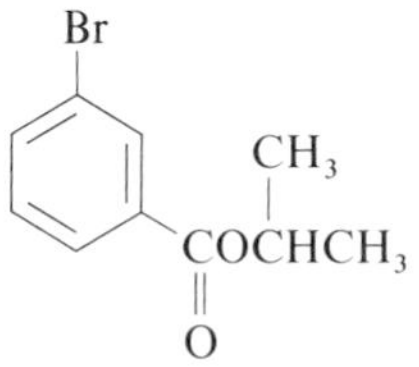
Br
CH3
COCHCH3
O

(b)
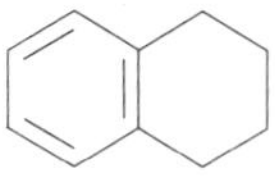

(c)
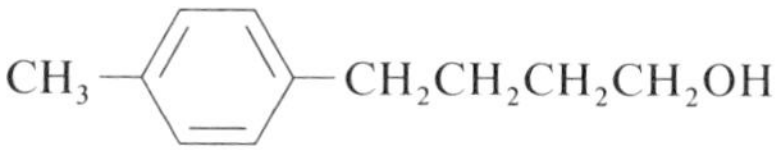
CH3
CH2CH2CH2CH2OH

(d)

CH2CH2CH2CH3

(e)
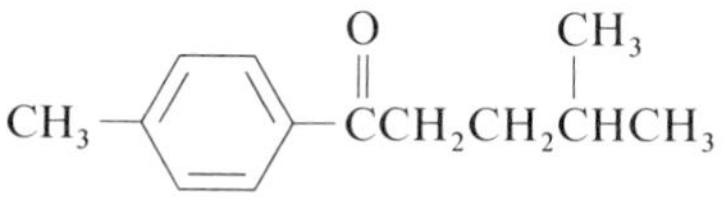
O
CH3
CH3
CCH2CH2CHCH3

(f)
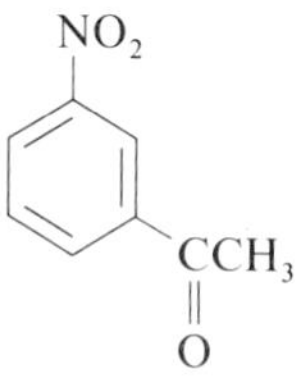
NO2
CCH3
O

(g)
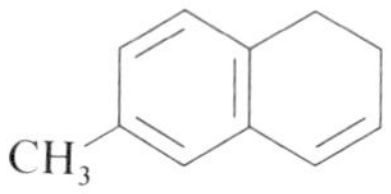
CH3

(h)
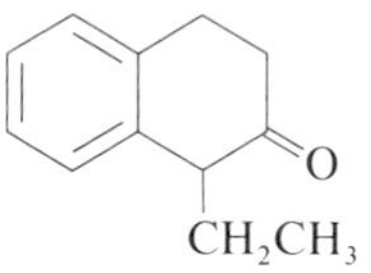
O
CH2CH3

(i)
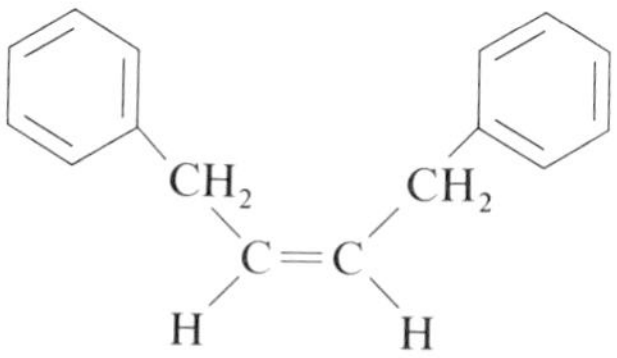
CH2
CH2
C=C
H
H

19.32 1,3,5-Trimethylbenzene undergoes electrophilic aromatic substitution with iodine monochloride, ICl. Write an equation for this reaction showing the product you expect. What is your reasoning in deciding what the product will be?

19.33 When allyl alcohol is treated with hydrogen fluoride in the presence of benzene, two products are formed: 3-phenyl-1-propene and 1,2-diphenylpropane. Write equations showing the mechanisms for the formation of these products.

19.34 When 2-hydroxybenzoic acid is heated with isobutyl alcohol in the presence of sulfuric acid, Compound A is formed. The same product is obtained if *tert*-butyl alcohol and sulfuric acid are used. What is the structure of Compound A? Write a mechanism that accounts for these experimental facts.

19.35 An intensely blue hydrocarbon, called azulene, has the structure shown below. Predict whether it has aromaticity.

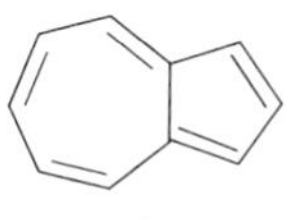

azulene

19.36 The heterocyclic compounds pyrrole and indole, shown below, are not very basic. Pyrrole has pK_a ~15 and thus has considerable acidity. Explain why pyrrole is so much less basic (so much more acidic) than ammonia (pK_a ~36).

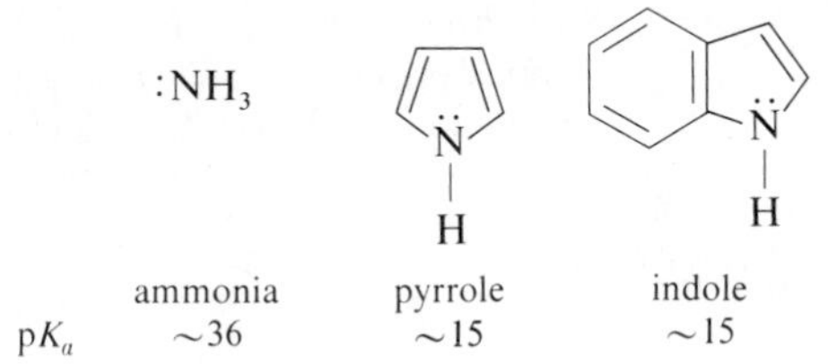
:NH₃
N
H
N
H
ammonia
pyrrole
indole
pKa
~36
~15
~15

19.37 Experimentally, it has been found that the cyclononatetraenyl anion and the cyclooctatetraenyl dianon can be prepared and are reasonably stable species. The reaction of cyclononatetraene with the carbanion generated when dimethyl sulfoxide, CH_3SOCH_3, is treated with sodium hydride gives cyclononatetraenyl anion. Cyclooctatetraenyl dianion is formed when cyclooctatetraene reacts with 2 equivalents of potassium metal in tetrahydrofuran. Write equations for the reactions described here, and explain the source of the stability of these hydrocarbon anions.

19.38 Heptafulvenes are compounds in which a carbon atom of the cycloheptatriene ring is doubly bonded to a carbon outside the ring. The parent compound, shown below, is highly unstable, but substitution of two cyano groups for the hydrogen atoms on the double bond outside the ring gives a stable compound. How would you rationalize these experimental observations? (Hint: Writing resonance contributors for both compounds will be helpful.)

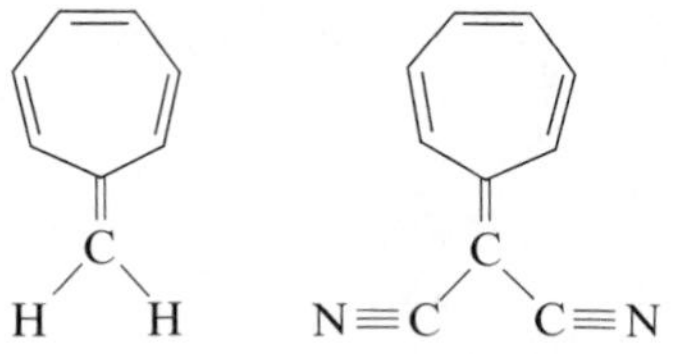
C
H
H
C
N≡C
C≡N

heptafulvene
unstable

a dicyanoheptafulvene
stable

19.39 Propose a mechanism for the following reaction.

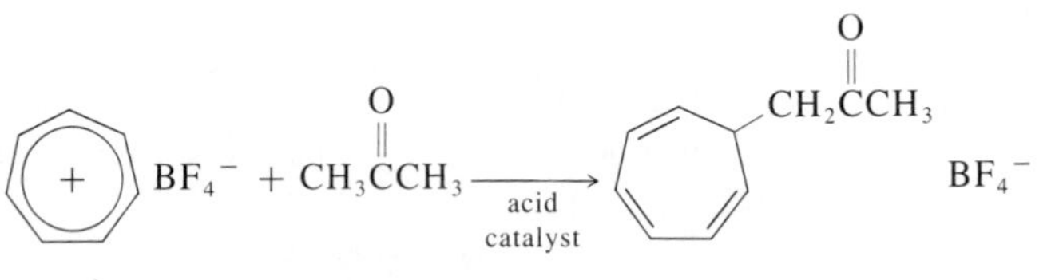
+
BF₄⁻ + CH₃CCH₃
O
acid
catalyst
CH₂CCH₃
O
BF₄⁻

20

Free Radicals

A • L O O K • A H E A D

A free radical is a reactive intermediate with an unpaired electron. Free radicals are important in many reactions. Chapter 5 introduced them briefly as the reactive intermediates involved in the halogenation reactions of alkanes (p. 172). The structure of a carbon radical was compared to those of a carbocation and of a carbanion. A radical has an incomplete octet and no charge. Halogen atoms, with seven electrons in the valence shell, may be considered to be radicals.

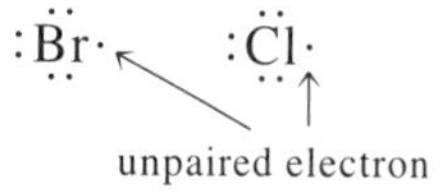

:Br·
:Cl·
unpaired electron

The oxygen molecule, O_2, is a diradical in its ground state. Many important consequences follow from this fact.

A radical behaves as an electrophile that is seeking only a single electron. This electron is often obtained from a sigma bond, usually a carbon-hydrogen bond, in a hydrogen-abstraction reaction.

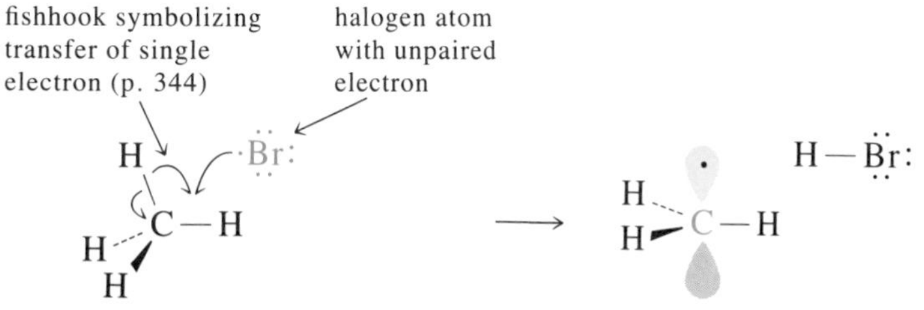

fishhook symbolizing transfer of single electron (p. 344)
halogen atom with unpaired electron
H—Br:

bromine atom abstracting a hydrogen atom

methyl radical

The product of a hydrogen-abstraction reaction is a new radical, which can itself abstract an atom. For example, the methyl radical may abstract a bromine atom from a molecule of bromine, forming bromomethane and a bromine atom.

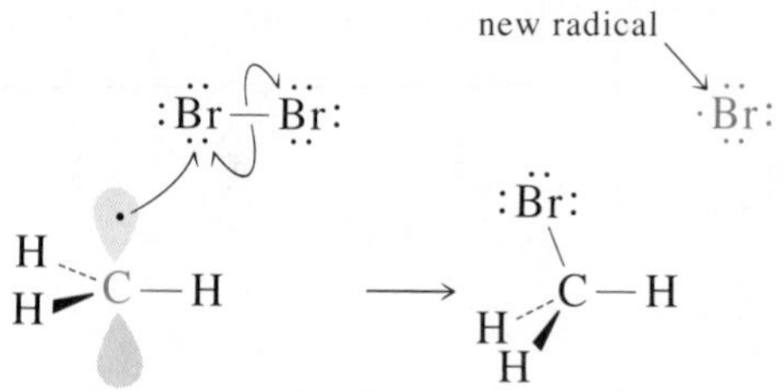
new radical

Free-radical reactions are often chain reactions. A product of one step of the reaction is a reactant in the next step. For example, the bromine atom formed in the second reaction above is a reactant in the first reaction shown. It will react with another molecule of methane to keep the sequence going.

This chapter will discuss the reactions of radicals and will examine some of the practical consequences of their reactivity.

20.1 FREE-RADICAL REACTIONS OF ALKANES

A. Reactions of Chlorine with Alkanes

Chlorine or bromine will react with alkanes in the presence of light to give alkyl halides. For example, methane reacts with chlorine to give a mixture of chlorinated methanes.

$$\underset{\text{methane}}{CH_4} + \underset{\text{chlorine}}{Cl_2} \xrightarrow[h\nu]{} \underset{\text{chloromethane}}{CH_3Cl} + \underset{\text{dichloromethane}}{CH_2Cl_2} + \underset{\text{trichloromethane}}{CHCl_3} + \underset{\text{tetrachloromethane}}{CCl_4}$$

The composition of the product mixture depends on the ratio of starting materials and the temperature, but even when large excesses of the alkane are used, mixtures are formed.

When an alkane with different types of hydrogen atoms is chlorinated, any of the hydrogens will be substituted. For example, when 2-methylbutane is chlorinated at 300 °C, the following mixture is observed.

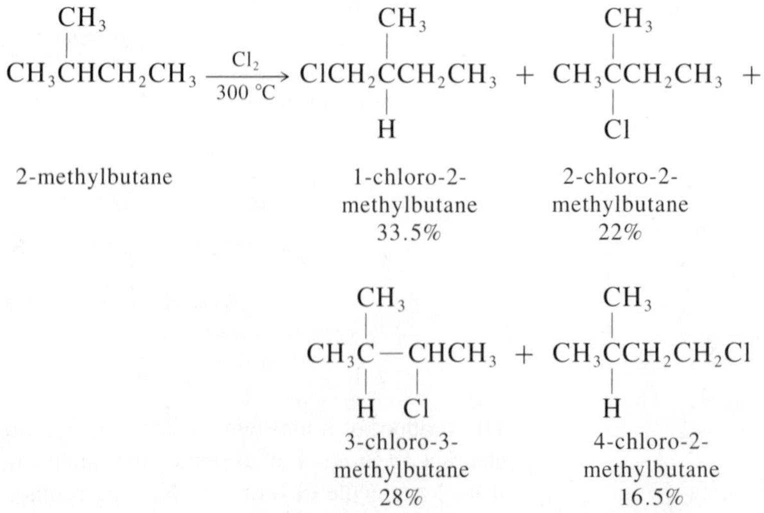
Cl2
300 °C
2-methylbutane
1-chloro-2-methylbutane 33.5%
2-chloro-2-methylbutane 22%
3-chloro-3-methylbutane 28%
4-chloro-2-methylbutane 16.5%

If the relative numbers of different types of hydrogen atoms are taken into account, these experimental results reveal that a tertiary hydrogen atom is about 4 times more likely to be replaced than a primary one, and a secondary hydrogen is about 2.5 times as reactive as a primary one.

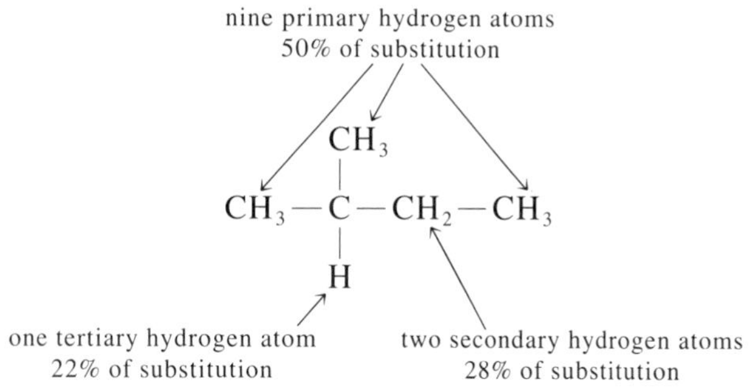
nine primary hydrogen atoms
50% of substitution
CH3
CH3—C—CH2—CH3
H
one tertiary hydrogen atom
22% of substitution
two secondary hydrogen atoms
28% of substitution

$$\frac{\text{tertiary}}{\text{primary}} = \frac{22/1}{50/9} = \frac{4}{1}$$

$$\frac{\text{secondary}}{\text{primary}} = \frac{28/2}{50/9} = \frac{2.5}{1}$$

B. Free-Radical Chain Reactions

Research has shown that the reactions of halogens with alkanes proceed via free-radical intermediates. On absorbing light energy, the halogen molecule dissociates into two halogen atoms.

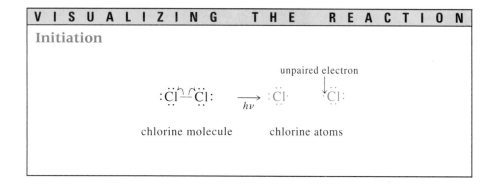

VISUALIZING THE REACTION

Initiation

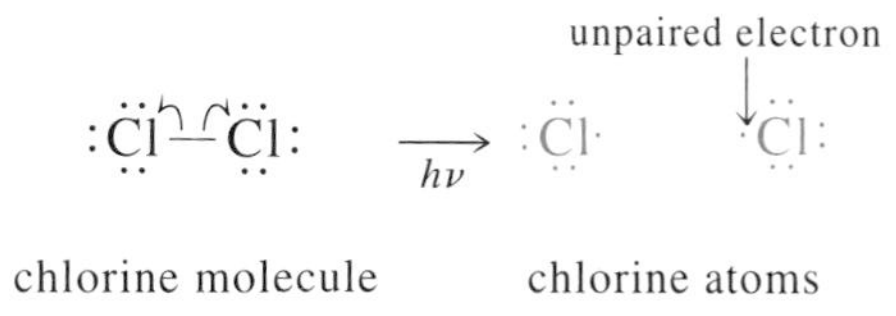
unpaired electron
:Cl—Cl:
hν
:Cl·
·Cl:
chlorine molecule
chlorine atoms

This bond cleavage is a homolytic cleavage (p. 59) in which one electron of the covalent bond goes to each atom. This cleavage is symbolized using fishhooks (p. 344).

A chlorine atom is highly reactive because of the presence of an unpaired electron in its outermost shell. It is electrophilic, seeking a single electron to complete the octet. It acquires this electron by abstracting a hydrogen atom from the alkane, methane, for example.

VISUALIZING THE REACTION

Propagation

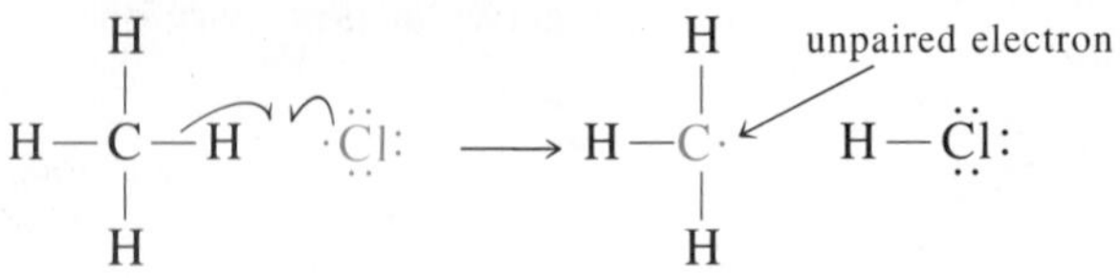
unpaired electron
methane
chlorine atom
methyl radical
hydrogen chloride

This reaction gives rise to a new electrophilic species, the methyl radical, which has an unpaired electron. The methyl radical abstracts a chlorine atom from a chlorine molecule.

VISUALIZING THE REACTION

Propagation

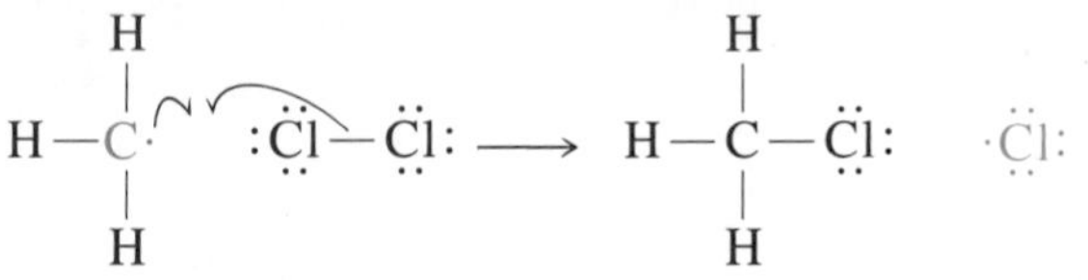
methyl radical
chlorine
chloromethane
chlorine atom

The products of this reaction are chloromethane and a chlorine atom, which abstracts a hydrogen atom from another molecule of methane and keeps the reaction going. This type of reaction, in which a product of one step is a reactant in the next step is known as a **chain reaction.** The halogenation of an alkane is a **free-radical chain reaction** because the reactive intermediates formed in different steps of the reaction are free radicals, species having an unpaired electron. Free radicals are highly reactive intermediates that are involved in oxidation reactions, combustion reactions, and many biological reactions, some of which will be investigated later in this chapter (pp. 842–846).

A chain reaction has several steps:

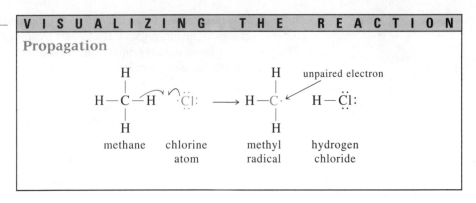

1. **Initiation** is a step in which the first reactive intermediate is formed. For the free-radical halogenation reaction, formation of halogen atoms by the dissociation of a halogen molecule is the initiation step.
2. **Propagation** consists of steps that are repeated many times to carry the reaction forward. The abstraction of a hydrogen atom by the halogen atom and the abstraction of a halogen atom by the methyl radical are the propagation steps of the free-radical halogenation reaction. For the chlorination of an alkane, about 10,000 propagation steps occur for each initiation step.
3. **Termination** consists of steps that occur when two free radicals happen to collide with each other and form a covalent bond, stopping the chain of reactions. Possible termination steps for the halogenation of an alkane are shown

below. In each of these termination reactions, two radicals are destroyed, which stops two chains. However, the concentration of radicals in the reaction mixture is low compared to the concentrations of the other reagents, so the odds greatly favor collision between a radical and a molecule over collision between two radicals. Propagation steps are therefore much more common than termination steps.

VISUALIZING THE REACTION

Termination

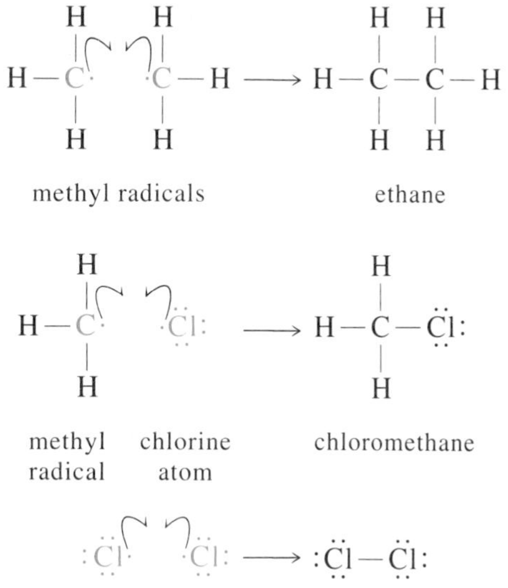
methyl radicals
ethane
methyl radical
chlorine atom
chloromethane
chlorine atoms
chlorine

Radicals are formed in initiation steps. The number of radicals stays constant in propagation steps. Radicals are destroyed in termination steps.

Reactions that occur by way of free-radical intermediates are different in many ways from those that go by way of ionic intermediates. Because free-radical intermediates do not have positive or negative charges, their reactions are not seriously affected by solvent polarity. However, the rates of the reactions are affected by substances, called **inhibitors,** that react with free radicals and thus act to terminate chains. The chlorination reaction, for example, is sensitive to the presence of oxygen, which is a diradical (p. 842). No rearrangements are observed for free-radical chain reactions, as they are for reactions that proceed by carbocation intermediates (p. 285). All of these facts make it clear that the intermediates involved in free-radical chain reactions are different from those involved in other substitution reactions.

Study Guide Concept Map 20.1

C. The Selectivity of Chlorination Reactions

The selectivity observed for the chlorination of 2-methylbutane (p. 822) can be explained by an examination of the relative stabilities of the different alkyl radicals that can be intermediates. The relative stabilities of radicals parallel the stabilities of the corresponding carbocations (pp. 122 and 172). A tertiary alkyl radical is more stable than a secondary one, which in turn is more stable than a primary one.

For radicals formed from 2-methylbutane, for example, the relative stabilities are as follows:

$$
\overset{\overset{\displaystyle CH_3}{|}}{CH_3\overset{\cdot}{C}CH_2CH_3} > \overset{\overset{\displaystyle CH_3}{|}}{CH_3CH\overset{\cdot}{C}HCH_3} > \overset{\overset{\displaystyle CH_3}{|}}{\cdot CH_2CHCH_2CH_3}, \overset{\overset{\displaystyle CH_3}{|}}{CH_3CHCH_2\overset{\cdot}{C}H_2}
$$

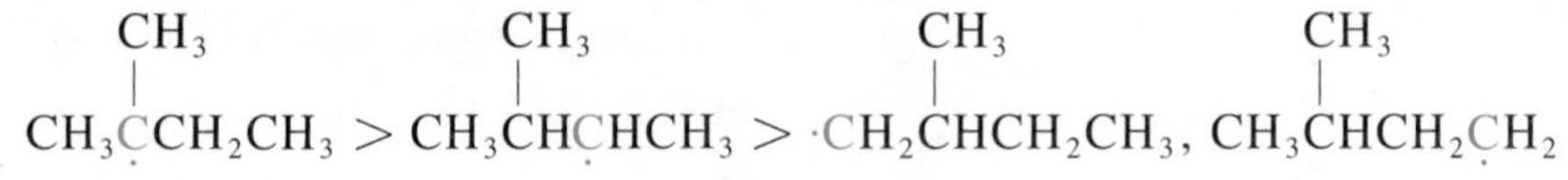

tertiary radical   secondary radical   primary radicals

The ease with which the different types of hydrogen atoms are abstracted reflects the bond dissociation energies for the different carbon-hydrogen bonds (p. 60). An examination of the energy changes that occur during the abstraction of a hydrogen atom by a chlorine atom illustrates the differences.

$$
\overset{\overset{\displaystyle CH_3}{|}}{CH_3CHCH_2CH_2-H} + \cdot \overset{\cdot\cdot}{\underset{\cdot\cdot}{Cl}}: \longrightarrow \overset{\overset{\displaystyle CH_3}{|}}{CH_3CHCH_2\overset{\cdot}{C}H_2} + HCl
$$

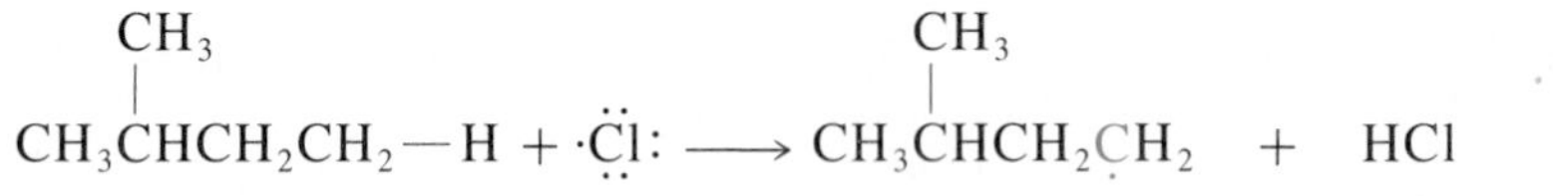

$DH°$ 98 kcal/mol   primary radical   103 kcal/mol

$$\Delta H_r = -5 \text{ kcal/mol}$$

$$
\overset{\overset{\displaystyle CH_3}{|}}{\underset{\underset{\displaystyle H}{|}}{CH_3CHCHCH_3}} + \cdot \overset{\cdot\cdot}{\underset{\cdot\cdot}{Cl}}: \longrightarrow \overset{\overset{\displaystyle CH_3}{|}}{CH_3CH\overset{\cdot}{C}HCH_3} + HCl
$$

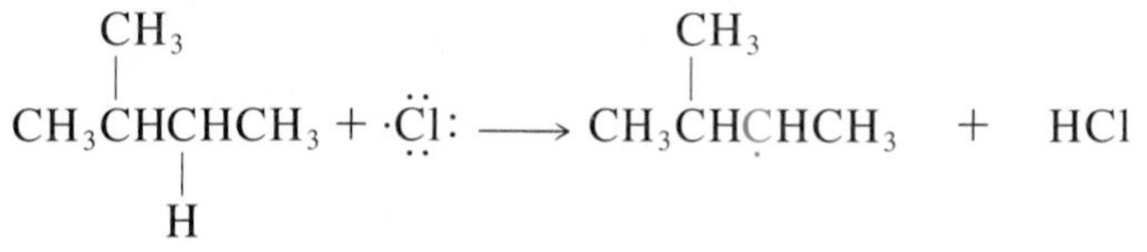

$DH°$ 95 kcal/mol   secondary radical   103 kcal/mol

$$\Delta H_r = -8 \text{ kcal/mol}$$

$$
\overset{\overset{\displaystyle CH_3}{|}}{\underset{\underset{\displaystyle H}{|}}{CH_3CCH_2CH_3}} + \cdot \overset{\cdot\cdot}{\underset{\cdot\cdot}{Cl}}: \longrightarrow \overset{\overset{\displaystyle CH_3}{|}}{CH_3\overset{\cdot}{C}CH_2CH_3} + HCl
$$

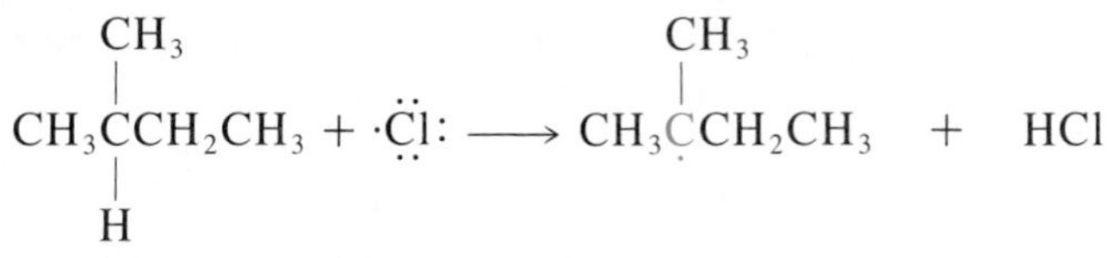

$DH°$ 91 kcal/mol   tertiary radical   103 kcal/mol

$$\Delta H_r = -12 \text{ kcal/mol}$$

In each case, enough energy must be supplied to break a carbon-hydrogen bond, and energy is recovered from the formation of a hydrogen-chlorine bond. The difference between these two energies is the enthalpy of each reaction. The three reactions involving a chlorine atom shown above are all exothermic, but the amount of energy given off as heat increases going from the abstraction of a primary hydrogen atom to that of a tertiary one. The hydrogen-abstraction reactions also have a small energy of activation, ranging from 3.8 kcal/mol for methane to about 1 kcal/mol for the abstraction of primary hydrogens and about 0.7–0.9 kcal/mol for the abstraction of secondary and tertiary hydrogens. The difference between the energy of activation for the formation of the primary radical and that for the formation of the tertiary radical is also small. Therefore, the regioselectivity of the hydrogen-abstraction reaction is low.

PROBLEM 20.1

Assume that the relative reactivities determined for the reaction of 2-methylbutane with chlorine at 300 °C hold for other alkanes. What kind of product mixture would result from the chlorination of propane under the same conditions? How about the chlorination of 2-methylpropane?

PROBLEM 20.2

The regioselectivity of the chlorination reaction is dependent on temperature. At 600 °C, the relative reactivities for primary, secondary, and tertiary hydrogens are 1 : 2.1 : 2.6, instead of 1 : 2.5 : 4 as observed at 300 °C. How would you explain this experimental result?

D. Reaction of Bromine with Alkanes

Bromination reactions are much more selective than chlorination reactions. For example, when 2-methylpropane is treated with bromine in the presence of light at 127 °C, the product is almost exclusively 2-bromo-2-methylpropane.

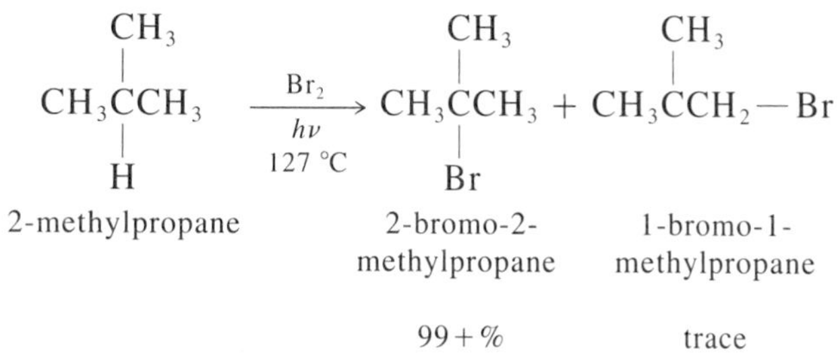
Br_2
$h\nu$
127 °C
2-methylpropane
2-bromo-2-methylpropane
1-bromo-1-methylpropane
99+%
trace

When the above bromination reaction is carried out in a reaction vessel in which butane is also present as a reactant, the relative reactivities for primary, secondary, and tertiary hydrogen atoms are 1 : 82 : 1640.

The bond dissociation energies and enthalpies for the abstractions of primary and tertiary hydrogens by bromine are shown below.

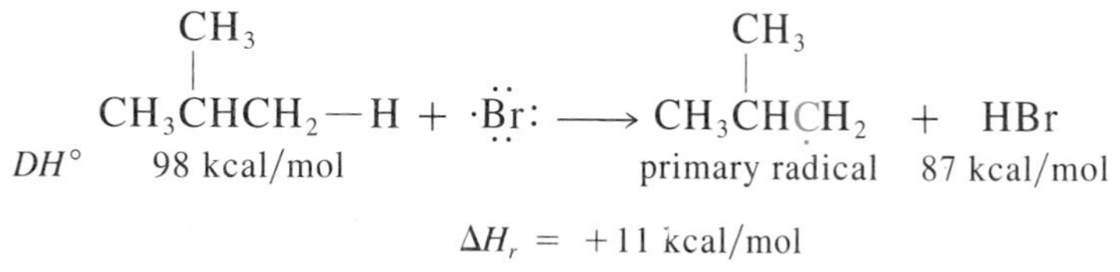
$DH°$
98 kcal/mol
primary radical
87 kcal/mol
$\Delta H_r = +11$ kcal/mol

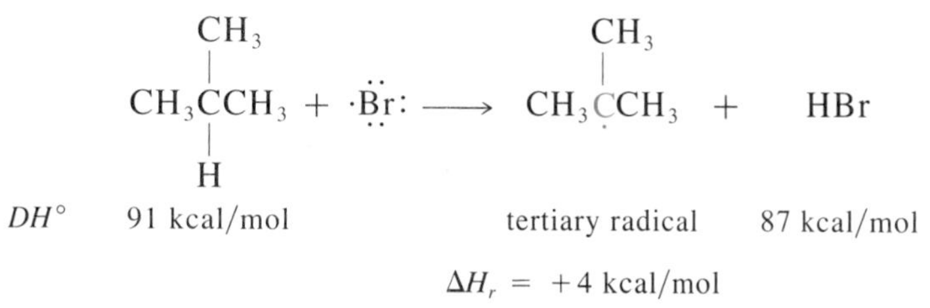
$DH°$
91 kcal/mol
tertiary radical
87 kcal/mol
$\Delta H_r = +4$ kcal/mol

Both reactions are endothermic. The first reaction has an energy of activation of at least 11 kcal/mol, and that for the formation of the tertiary radical is at least 4 kcal/mol. More importantly, the difference between the two is at least 7 kcal/mol, leading to a significant difference in the rates of abstraction of primary and tertiary hydrogen atoms. The high regioselectivity of the bromination of alkanes is derived from the differing endothermicity of the hydrogen-abstraction reactions with bromine atoms.

PROBLEM 20.3

Study Guide Concept Map 20.2

Neopentane (2,2-dimethylpropane) reacts very slowly with bromine in the presence of light even at relatively high temperatures but reacts easily with chlorine under the same conditions. How would you explain this difference in reactivity?

20.2 FREE-RADICAL SUBSTITUTION REACTIONS OF ALKENES

N-Bromosuccinimide in a polar solvent such as a mixture of dimethyl sulfoxide with water is used as a source of electrophilic bromine, which reacts with alkenes to give halohydrins by way of a bromonium ion intermediate (p. 296). In a nonpolar solvent such as carbon tetrachloride, however, *N*-bromosuccinimide reacts with alkenes that have allylic hydrogen atoms to substitute bromine for one of those hydrogens. For example, 2-heptene reacts with *N*-bromosuccinimide in boiling carbon tetrachloride in the presence of benzoyl peroxide (an initiator) to give 4-bromo-2-heptene.

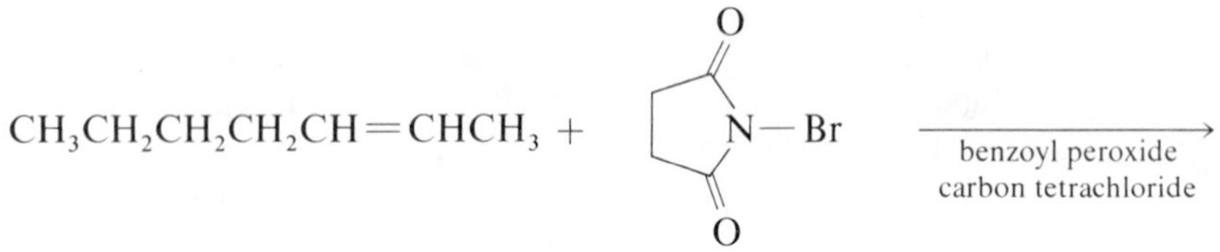
CH3CH2CH2CH2CH=CHCH3 +
O
N—Br
O
benzoyl peroxide
carbon tetrachloride

2-heptene — *N*-bromosuccinimide NBS

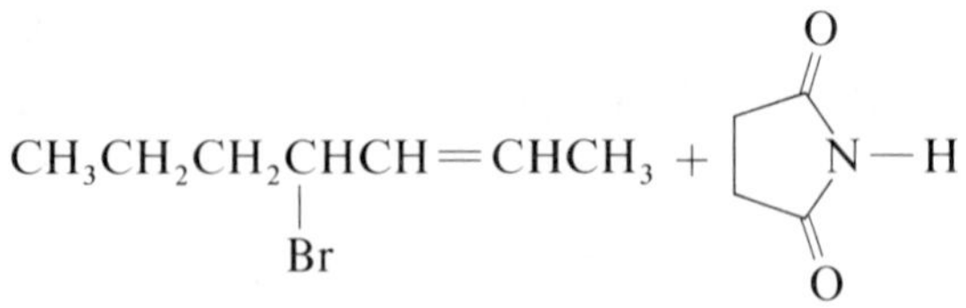
CH3CH2CH2CHCH=CHCH3 +
Br
O
N—H
O

4-bromo-2-heptene ~60% — succinimide

This reaction is quite general. *N*-Bromosuccinimide (often abbreviated in equations as NBS) reacts with a variety of alkenes to substitute bromine for an allylic hydrogen atom. The reaction is catalyzed by peroxides, heat, or light. It is believed that the reaction involves the formation of a bromine atom, which abstracts an allylic hydrogen atom (p. 277) from the alkene.

VISUALIZING THE REACTION

Abstraction of an allylic hydrogen

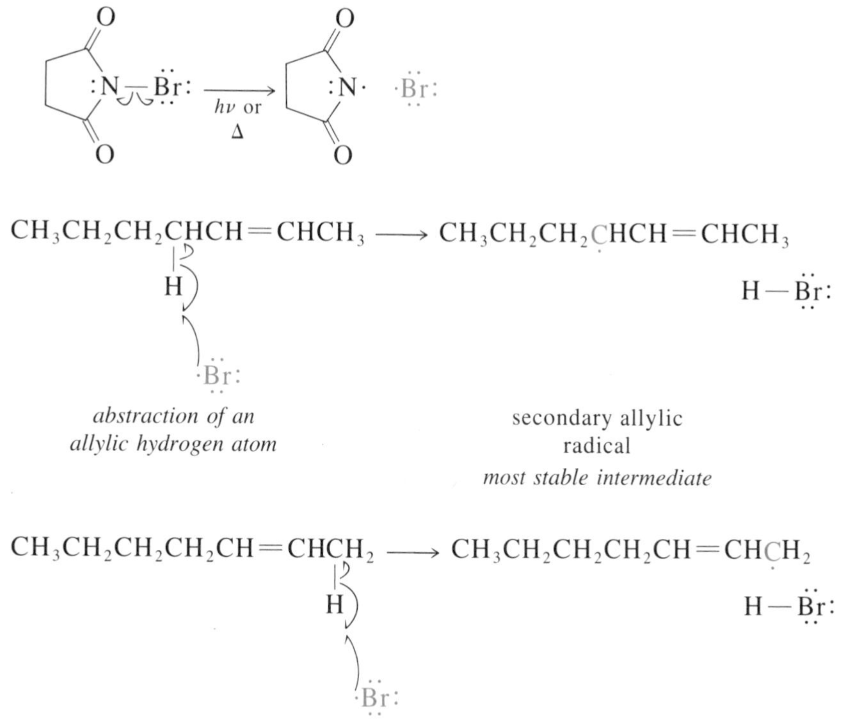

:N—Br:
hν or
Δ
:N·
·Br:
CH₃CH₂CH₂CHCH=CHCH₃
CH₃CH₂CH₂CHCH=CHCH₃
H
H—Br:
·Br:
CH₃CH₂CH₂CH₂CH=CHCH₂
CH₃CH₂CH₂CH₂CH=CHCH₂
H
H—Br:
·Br:

abstraction of an allylic hydrogen atom

secondary allylic radical
most stable intermediate

abstraction of an allylic hydrogen atom

primary allylic radical

The stabilities of various alkyl radicals, which are electron-deficient species, parallel those of carbocations (p. 825). A tertiary alkyl radical is more stable than a secondary one, which is in turn more stable than a primary one. Free radicals stabilized by resonance delocalization of the radical character are the most stable of all and, therefore, the most easily formed. A radical having the electron-deficient carbon adjacent to a double bond, at an allylic position, is stabilized by resonance.

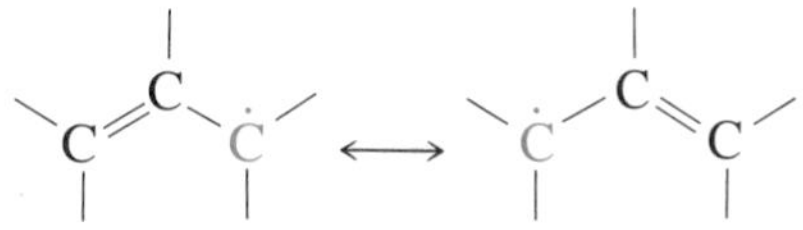

resonance contributors for an allylic radical

The most stable intermediate for the reaction of a bromine atom with 2-heptene is the radical that is secondary and allylic. The major product of the bromination reaction is derived from that intermediate, formed by the abstraction of an allylic hydrogen atom by a bromine atom. The other product of the hydrogen-abstraction reaction is hydrogen bromide, which reacts with *N*-bromosuccinimide to give bromine and succinimide.

VISUALIZING THE REACTION

Reaction of hydrogen bromide with *N*-bromosuccinimide

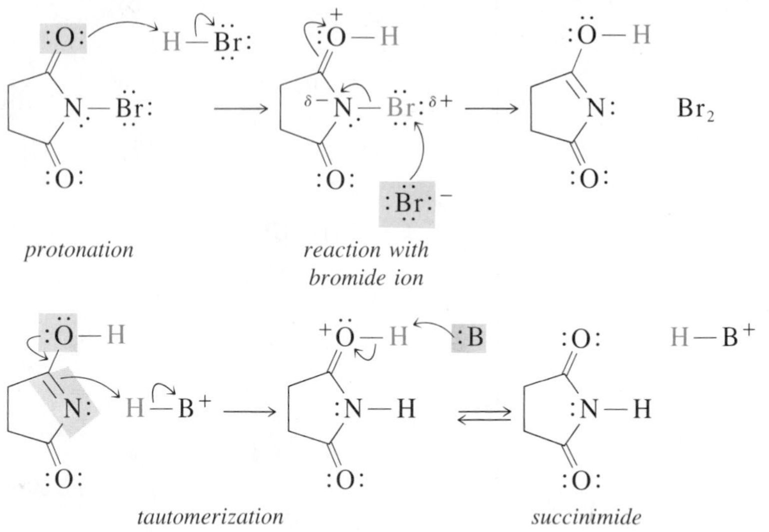
protonation
reaction with bromide ion
tautomerization
succinimide

As a result of this sequence of reactions, hydrogen bromide, which is a product of the hydrogen-abstraction reaction, is converted to bromine. The amount of bromine formed in the reaction mixture is limited by the amount of hydrogen bromide that has been formed in the earlier step of the reaction. The allylic radical formed in the first step of the reaction abstracts a bromine atom from a molecule of bromine, to give 4-bromo-2-heptene and a bromine atom, which starts the cycle all over again.

VISUALIZING THE REACTION

Reaction of an allylic radical with bromine

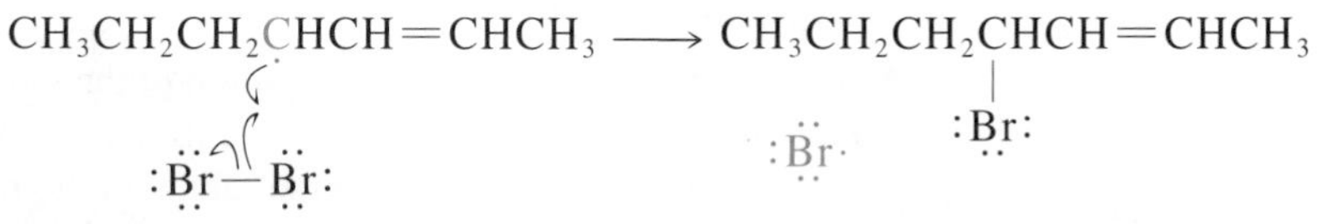

When molecular bromine is used as a reagent with alkenes, it adds to the double bond. With *N*-bromosuccinimide, addition reactions are not observed. Why does the bromine formed from the reaction of hydrogen bromide with *N*-bromosuccinimide not add to the double bond in heptene? One explanation is that a very low concentration of bromine is present in the reaction mixture at any one time. The initial addition of bromine to a double bond involves only one of the two bromine atoms in a cyclic bromonium ion intermediate (p. 290). The formation of this intermediate is reversible. If there is not a bromide ion nearby, the formation of the dibromide cannot be completed. *N*-Bromosuccinimide competes with the bromonium ion for bromide ions. Under these conditions, the allylic substitution reaction is favored.

The mechanism of halogenation reactions that employ *N*-bromosuccinimide has been intensively studied but is not yet completely understood. The ideas presented above are one picture of what happens under one set of experimental conditions.

The reactions illustrated for 2-heptene have been widely applied to other systems. The selectivity of the secondary over the primary allylic position shown for 2-heptene is quite general. For example, 1,1-diphenyl-1-propene, which has only primary allylic hydrogen atoms, reacts with *N*-bromosuccinimide much more slowly than does 1,1-diphenyl-1-butene, in which secondary allylic hydrogen atoms are available.

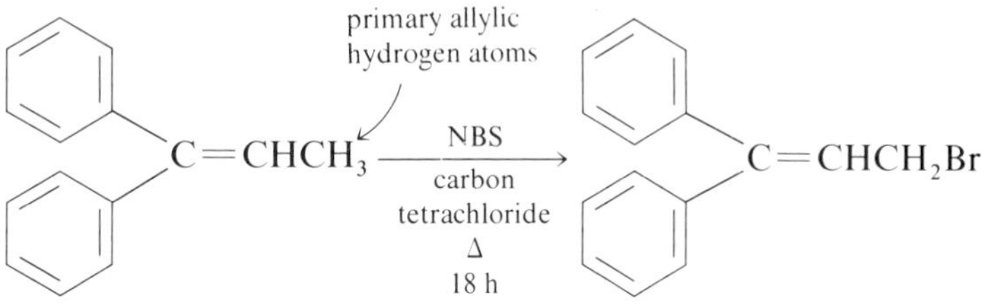

primary allylic
hydrogen atoms
C=CHCH3
NBS
carbon
tetrachloride
Δ
18 h
C=CHCH2Br

1,1-diphenyl-1-propene

3-bromo-1,1-diphenyl-1-propene 86%

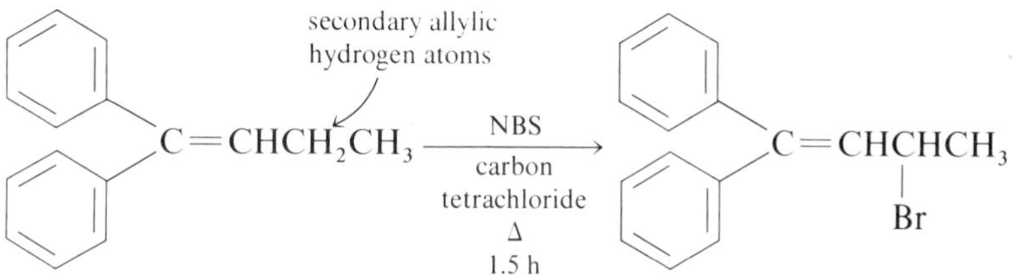

secondary allylic
hydrogen atoms
C=CHCH2CH3
NBS
carbon
tetrachloride
Δ
1.5 h
C=CHCHCH3
Br

1,1-diphenyl-1-butene

3-bromo-1,1-diphenyl-1-butene yield not known

PROBLEM 20.4

Predict the product(s) of the following reactions.

(a) 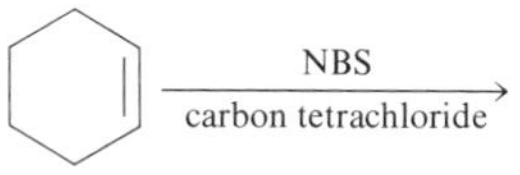

NBS
carbon tetrachloride

(b) 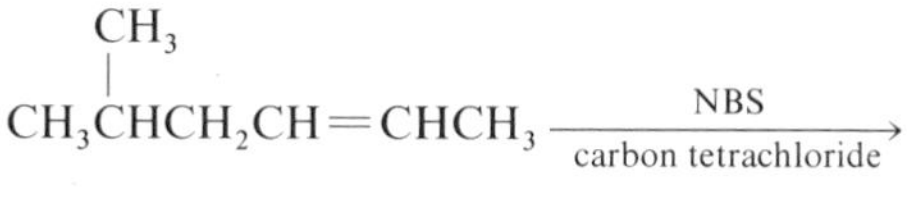

CH3
CH3CHCH2CH=CHCH3
NBS
carbon tetrachloride

(c) 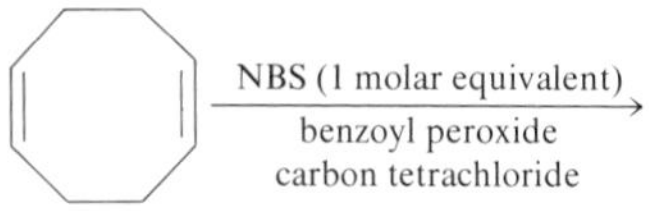

NBS (1 molar equivalent)
benzoyl peroxide
carbon tetrachloride

(d) 

O
OCCH3
NBS
carbon tetrachloride

PROBLEM 20.5

When 3-phenyl-1-propene is treated with *N*-bromosuccinimide in the presence of benzoyl peroxide, 3-bromo-1-phenyl-1-propene (50%) and 3-bromo-3-phenyl-1-propene (10%) are formed. Write a mechanism for the reaction, and explain why the major product is 3-bromo-1-phenyl-1-propene.

PROBLEM 20.6

When 2-heptene reacts with *N*-bromosuccinimide (p. 828), 4-bromo-2-heptene is obtained in only 60% yield. What are other possible substitution products of this reaction? Write equations supporting your answer.

20.3 FREE-RADICAL HALOGENATION REACTIONS AT THE BENZYLIC POSITION

When toluene reacts with bromine in the presence of a Lewis acid, two products are formed: *o*-bromotoluene and *p*-bromotoluene (p. 793).

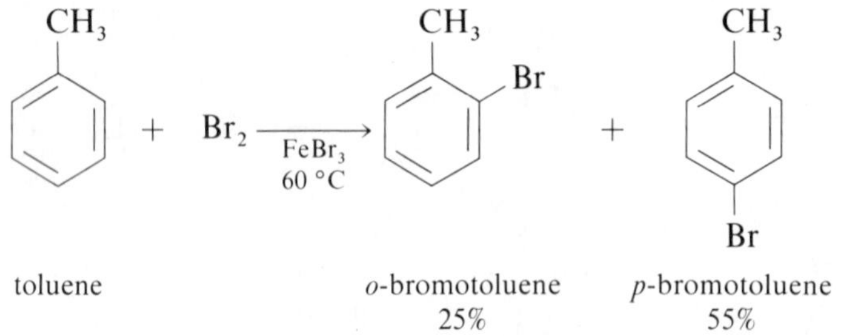
CH3
+ Br2
FeBr3
60 °C
CH3
Br
+
CH3
Br
toluene
o-bromotoluene
25%
p-bromotoluene
55%

Besides the hydrogen atoms bonded to carbons of the aromatic ring, however, toluene has hydrogen atoms in the methyl group. If bromine and toluene are brought together in the absence of a catalyst, a third substitution product, in which one of the hydrogen atoms of the methyl group has been replaced, is also formed.

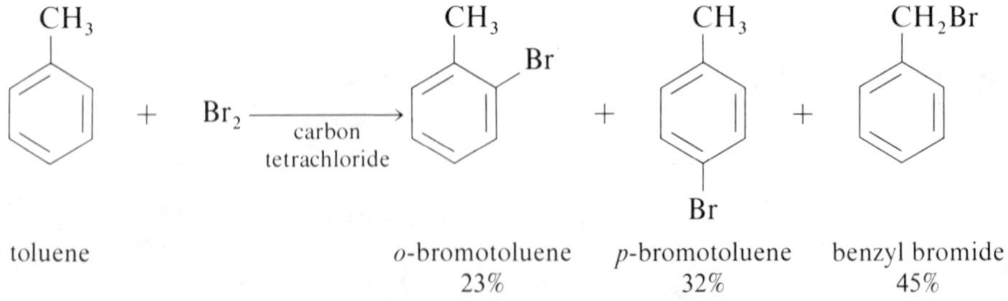
CH3
+ Br2
carbon
tetrachloride
CH3
Br
+
CH3
Br
+
CH2Br
toluene
o-bromotoluene
23%
p-bromotoluene
32%
benzyl bromide
45%

The hydrogen atoms of the methyl group of toluene are bonded to a benzylic carbon atom, one adjacent to an aromatic ring. Benzylic hydrogen atoms resemble allylic hydrogen atoms in reactivity (p. 829). They are easily substituted in free-radical reactions because the radical that is formed when a benzylic hydrogen atom is abstracted is stabilized by resonance.

VISUALIZING THE REACTION

Substitution of a benzylic hydrogen by halogen

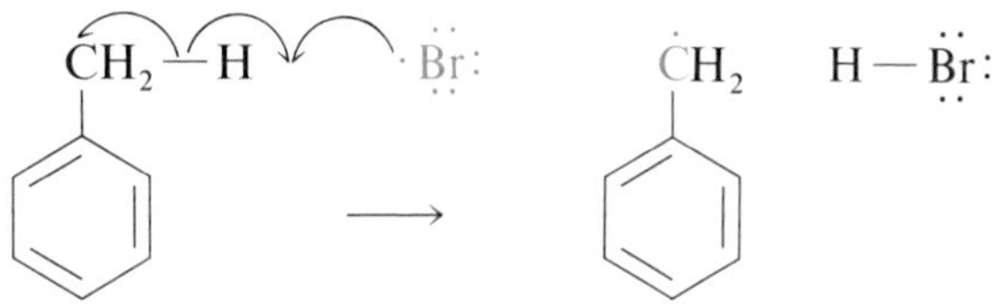

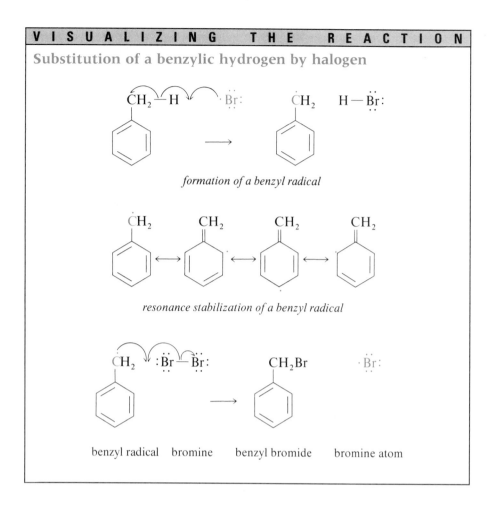

formation of a benzyl radical

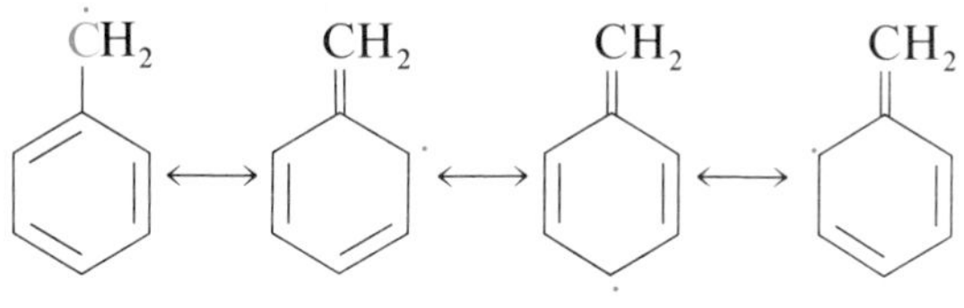

resonance stabilization of a benzyl radical

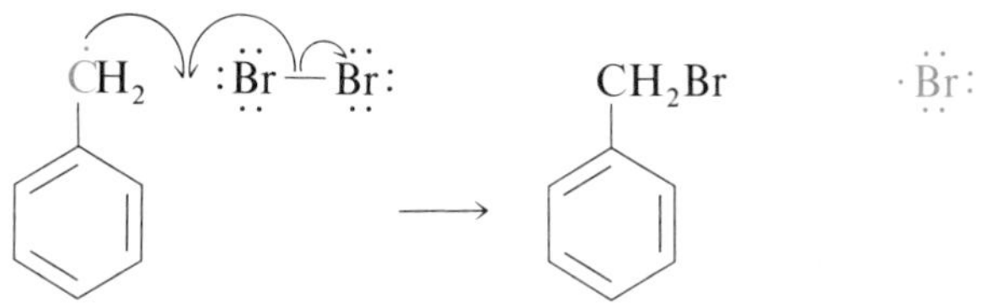

benzyl radical   bromine   benzyl bromide   bromine atom

The deficiency of an electron is delocalized over the aromatic ring, giving rise to a relatively stable species. The radical abstracts a bromine atom from a bromine molecule, forming benzyl bromide and a new bromine atom that can continue the reaction.

Bromine can act both as an electrophile toward the aromatic ring in toluene and as a source of bromine atoms, which give rise to free-radical reactions at the side chain. Free-radical reactions with bromine are more frequent when heat or light is used to cause the dissociation of molecular bromine into two bromine atoms.

N-Bromosuccinimide, which is used to substitute bromine for an allylic hydrogen atom (p. 828), has the same kind of reactivity toward benzylic hydrogen atoms. It can be used to substitute selectively at that position, without any competing substitution on the aromatic ring. For example, a benzylic hydrogen atom bonded to the carbon atom between two phenyl rings in diphenylmethane is selectively substituted by *N*-bromosuccinimide.

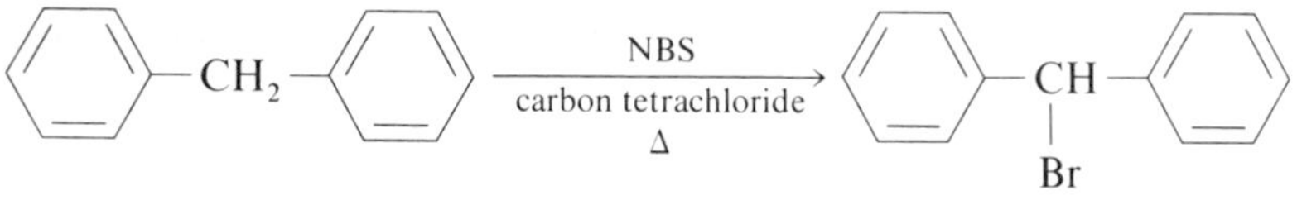
NBS
carbon tetrachloride
Δ

diphenylmethane   bromodiphenylmethane 81%

The hydrogen atoms of a methyl group on naphthalene are also benzylic in nature and undergo this substitution reaction.

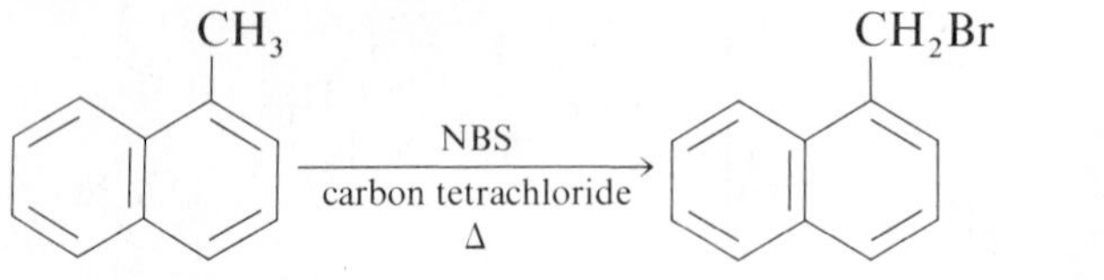
CH3
NBS
carbon tetrachloride
Δ
CH2Br

1-methylnaphthalene → 1-(bromomethyl)naphthalene 90%

Note that aromatic hydrogen atoms are not affected by this substitution reaction.

Benzylic radicals, like allylic radicals, are much more stable than alkyl radicals. Transition states leading to the formation of benzylic radicals are of lower energy than the transition states for competing reactions, so substitution reactions are channeled through such intermediates when possible.

Study Guide Concept Map 20.3

PROBLEM 20.7

Complete the following equations.

(a) 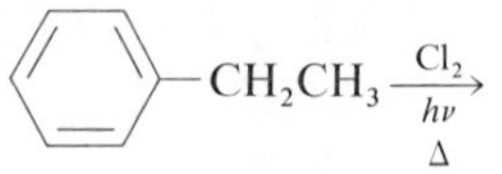
CH2CH3
Cl2
hν
Δ

(b) 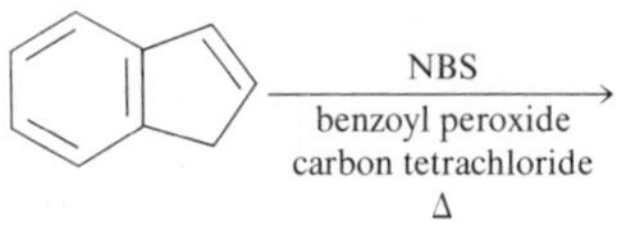
NBS
benzoyl peroxide
carbon tetrachloride
Δ

(c) 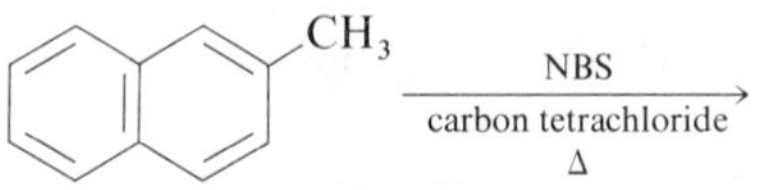
CH3
NBS
carbon tetrachloride
Δ

(d) 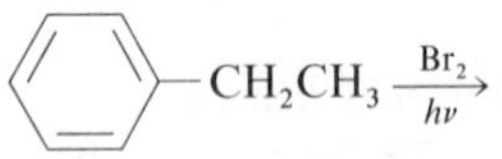
CH2CH3
Br2
hν

20.4 THE TRIPHENYLMETHYL RADICAL, A STABLE FREE RADICAL

A free radical that is stable enough to be observed as a discrete species was first reported in 1900 by Moses Gomberg of the University of Michigan. He was trying to synthesize hexaphenylethane by the reaction of bromotriphenylmethane with silver. When he ran the reaction under an atmosphere of carbon dioxide, he obtained a white solid that was highly reactive toward oxygen and halogens, including iodine. A solution of the solid in benzene prepared in the absence of air was yellow. The yellow color disappeared when a small amount of air was admitted into the reaction flask, but then developed again. Gomberg postulated that the yellow color was due to the formation of the triphenylmethyl radical in equilibrium with hexaphenylethane and that the radical reacted with molecular oxygen, which is a diradical (p. 842), to give a peroxide.

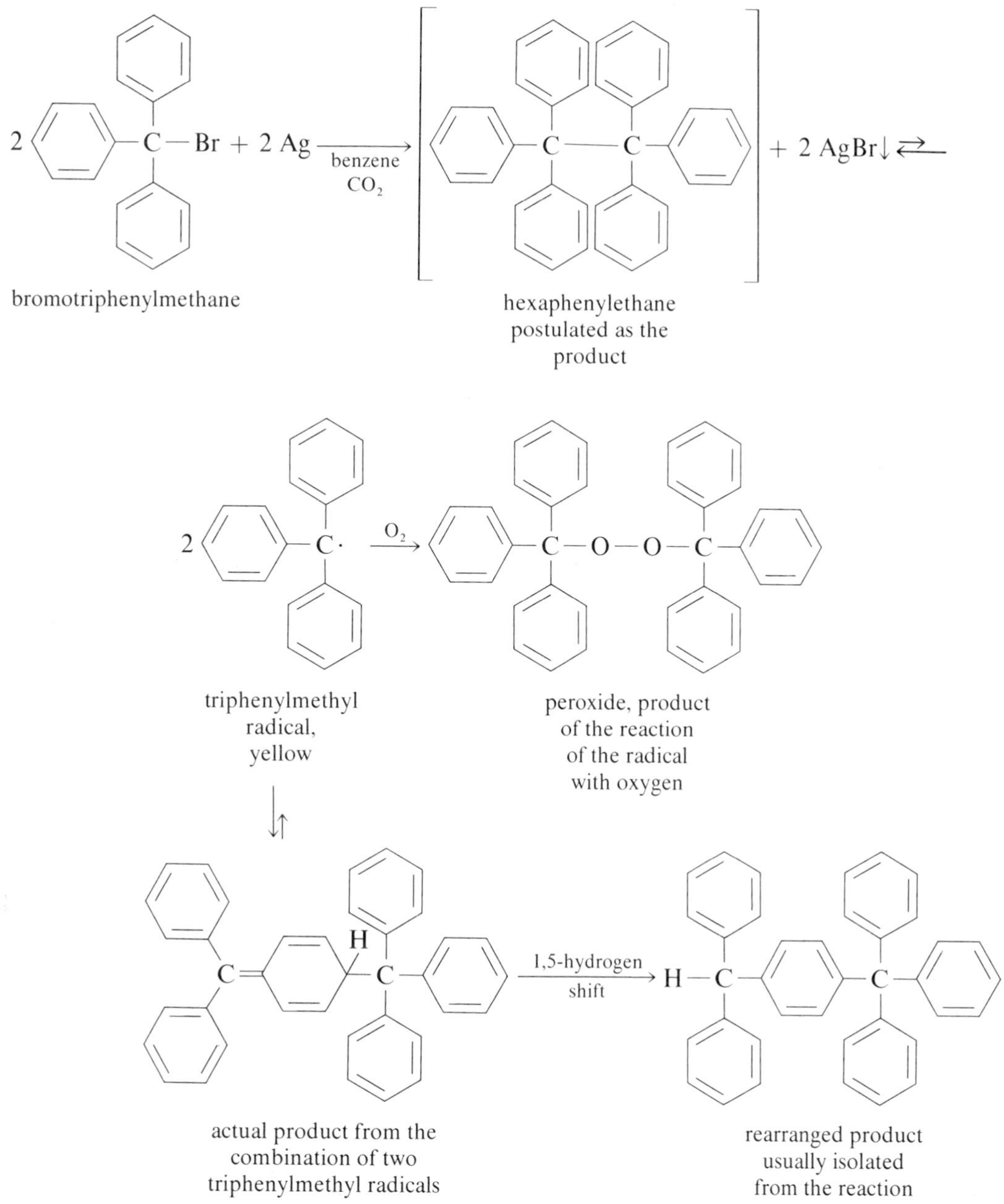

2
C—Br + 2 Ag
benzene
CO2
+ 2 AgBr↓
bromotriphenylmethane
hexaphenylethane postulated as the product
C·
O2
C—O—O—C
triphenylmethyl radical, yellow
peroxide, product of the reaction of the radical with oxygen
H
1,5-hydrogen shift
H—C
actual product from the combination of two triphenylmethyl radicals
rearranged product usually isolated from the reaction

Initially, many chemists rejected Gomberg's idea that a free radical could be stable enough to be observed as an intermediate in a reaction, but by 1911 a number of similar radicals had been prepared and characterized. Many years later it was shown that a triphenylmethyl radical adds to the para position of a phenyl ring in a second triphenylmethyl radical to give the product observed and that hexaphenylethane, which would have a highly strained carbon-carbon single bond, does not form.

Radicals that are not only stable, but positively inert, have been synthesized since Gomberg's experiments. One of these is the perchlorotriphenylmethyl radical.

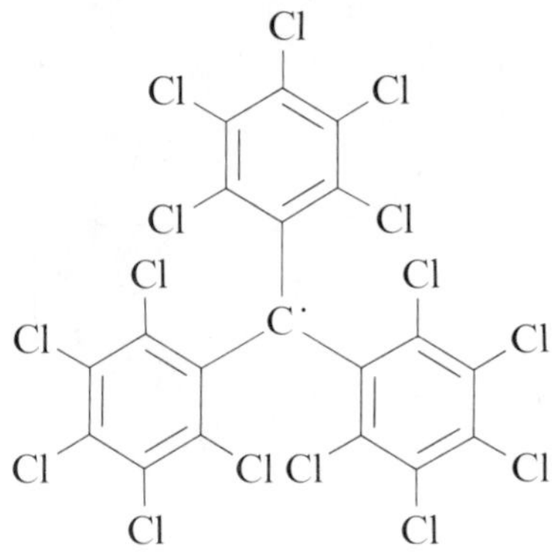

perchlorotriphenylmethyl radical
an inert free radical

This radical does not react with oxygen and is stable at temperatures up to 300 °C. The stability of the species is attributed to the presence of the large chlorine atoms that shield the electron-deficient carbon atom in the center from contact with reagents.

PROBLEM 20.8

Explain the stability of the triphenylmethyl radical.

20.5 FREE-RADICAL ADDITION REACTIONS OF ALKENES

A. Anti-Markovnikov Addition of Hydrogen Bromide

Specific reaction conditions for the addition of hydrogen bromide to propene were described when electrophilic addition reactions were first discussed (p. 118). Changing the conditions results in a different product for this reaction. When hydrogen bromide and propene are mixed at −78 °C in the presence of benzoyl peroxide and air, a very rapid reaction takes place; the product is a mixture of 96% *n*-propyl bromide and 4% isopropyl bromide.

$$CH_3CH{=}CH_2 + HBr \xrightarrow[\substack{\text{benzoyl} \\ \text{peroxide} \\ -78\ ^\circ C}]{} CH_3CH_2CH_2Br + CH_3\underset{\substack{\mid \\ Br}}{C}HCH_3$$

propene	hydrogen bromide	*n*-propyl bromide 96%	isopropyl bromide 4%

This reaction involves free-radical intermediates. The reaction starts with the homolytic cleavage of the relatively weak oxygen-oxygen single bond in benzoyl peroxide.

VISUALIZING THE REACTION

Homolytic cleavage of a covalent bond

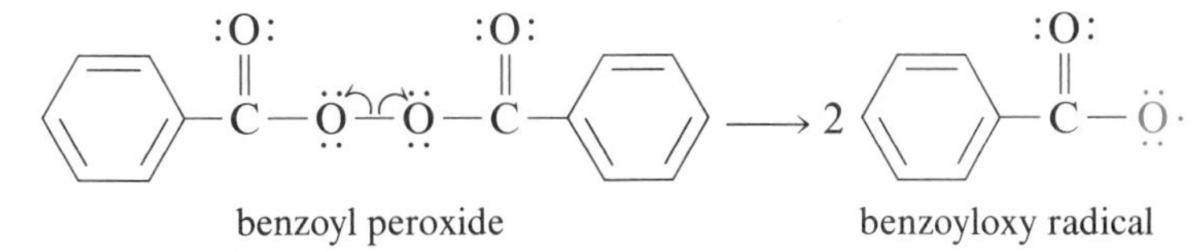

The benzoyloxy radicals from benzoyl peroxide abstract hydrogen atoms from hydrogen bromide, giving rise to bromine atoms in a second initiation step.

VISUALIZING THE REACTION

Initiation

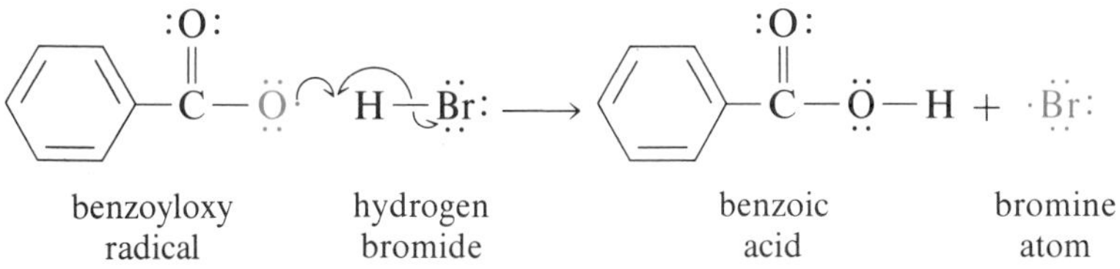

The bromine atom is an electrophile and reacts with the π electrons of the double bond. Carbon radicals form.

VISUALIZING THE REACTION

Free radical addition to a double bond. Propagation

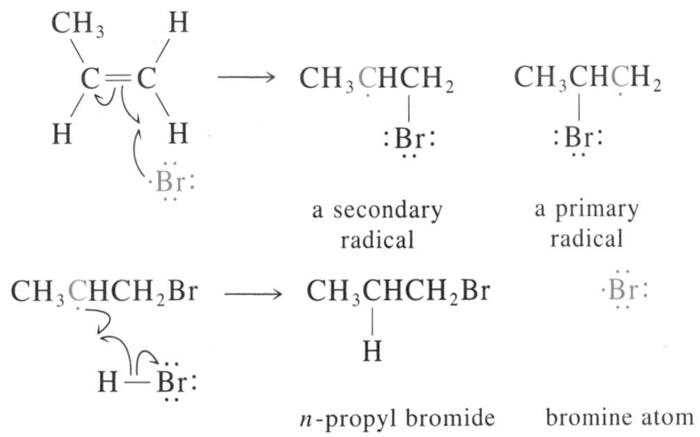

Whether a primary or secondary radical forms depends on which carbon atom of the double bond becomes bonded to the bromine atom. The carbon radicals are new electrophiles in the reaction mixture. The major product, *n*-propyl bromide, is formed when the secondary carbon radical abstracts a hydrogen atom from hydrogen bromide. A bromine atom is also formed to carry on the chain reaction.

The major result of this above sequence of steps is that hydrogen bromide adds to the double bond of propene with an orientation opposite to that seen when air and peroxides are carefully excluded from the reaction mixture. This orientation is said to result from the anti-Markovnikov addition (p. 299) of hydrogen bromide to the alkene. The change of orientation is due to the change in the mechanism of the reaction. This reversal of orientation is known as the **peroxide effect** and is seen only with hydrogen bromide and not with other halogen acids. (Problem 20.28 at the end of this chapter will give you a chance to explore why this is so.)

The order of the relative stabilities of alkyl radicals parallels that for carbocations (p. 826), but the differences in stability between radicals are much smaller. The *tert*-butyl radical, for example, is only 7 kcal/mol more stable than the isobutyl radical.

PROBLEM 20.9

The difference between the stability of the *tert*-butyl radical and that of the isobutyl radical is obtained by using bond dissociation energies (Table 2.4, p. 60). Show how the difference is derived from those data. What is the difference in stability between the isopropyl and propyl radicals?

Experiments have shown that the regioselectivity of free-radical addition to alkenes is mostly due to steric factors rather than to the relative stabilities of the radicals formed. The less highly substituted side of the double bond or the side with the smaller substituents is attacked preferentially by the incoming radical.

For a long time, the addition of hydrogen bromide to alkenes caused much confusion among chemists because alkenes form peroxides in the presence of oxygen (air). Reactions carried out with impure alkenes and solvents gave widely varying mixtures of products. This generated a lively controversy in the literature of organic chemistry until Morris S. Kharasch and his coworkers at the University of Chicago, working with carefully purified reagents under conditions that excluded air, discovered the source of the problem in the 1930s.

PROBLEM 20.10

Do the facts that the mechanism for the addition reaction changes when air and peroxides are excluded and that *n*-propyl bromide becomes the major product affect the relative energy levels for propene and hydrogen bromide and *n*-propyl bromide and isopropyl bromide (Figure 4.5, p. 125)? Explain your answer.

B. Other Free-Radical Additions to Alkenes

Many other reagents add to alkenes under conditions that generate free radicals. High temperatures, ultraviolet radiation, gamma-radiation from radioactive sources, peroxides, and oxygen all act as free-radical initiators. Three examples of this type of addition reaction are given below.

$$\mathrm{Cl_3CBr} + \mathrm{CH_2{=}CH(CH_2)_5CH_3} \xrightarrow[h\nu]{\mathrm{CH_3\overset{\overset{O}{\|}}{C}OO\overset{\overset{O}{\|}}{C}CH_3}} \mathrm{Cl_3CCH_2\underset{\underset{Br}{|}}{C}H(CH_2)_5CH_3}$$

bromotrichloromethane   1-octene   3-bromo-1,1,1-trichlorononane 88%

$$CH_3CH_2SH + CH_2{=}\underset{\displaystyle}{\overset{\displaystyle CH_3 \atop |}{C}}CH_3 \xrightarrow{\Delta} CH_3CH_2SCH_2\overset{\displaystyle CH_3 \atop |}{C}HCH_3$$

ethanethiol — 2-methylpropene — ethyl isobutyl sulfide 94%

$$Cl_3SiH + CH_2{=}CH(CH_2)_5CH_3 \xrightarrow[\Delta]{CH_3\overset{O}{\overset{\|}{C}}OO\overset{O}{\overset{\|}{C}}CH_3} Cl_3SiCH_2CH_2(CH_2)_5CH_3$$

trichlorosilane — 1-octene — octyltrichlorosilane 99%

In each case, one of the reactants has a bond that is weak enough to break homolytically under the reaction conditions. Products resulting from addition of a carbon, sulfur, or silicon radical to a double bond are obtained in high yields.

PROBLEM 20.11

Write a mechanism for each of the three reactions shown above.

PROBLEM 20.12

Predict the product(s) of each of the following reactions.

(a) 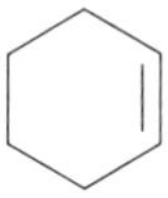 $+ CH_3SH \xrightarrow[h\nu]{\text{acetone}}$

(b) $CH_3(CH_2)_5CH{=}CH_2 \xrightarrow[\text{peroxides}]{HBr}$

(c) $Cl_3CBr + CH_2{=}\overset{\displaystyle CH_2CH_3 \atop |}{C}CH_2CH_3 \xrightarrow{\Delta}$

(d) $CH_3(CH_2)_4SiH_3 + CH_2{=}CH(CH_2)_5CH_3 \xrightarrow[\text{peroxides}]{}$

Industrially, the most important free-radical addition reactions to alkenes are polymerization reactions, which are used to make large quantities of polymeric materials such as plastics and synthetic fibers. The free-radical polymerization of vinyl chloride has the following steps.

$$R\cdot + CH_2{=}CHCl \longrightarrow RCH_2\dot{C}HCl \xrightarrow{CH_2=CHCl} RCH_2\underset{\displaystyle | \atop Cl}{C}HCH_2\dot{C}HCl$$

initiator — vinyl chloride *monomer* — *dimer*

$$\longrightarrow \longrightarrow R\left[CH_2\underset{\displaystyle | \atop Cl}{C}H\right]_n R'$$

poly[vinyl chloride] *polymer*

In these reactions, the carbon radical that is the product of one step adds to another molecule of alkene giving a new carbon radical to carry the chain. Polymerization reactions are favored when a small amount of initiator and a large amount of alkene are mixed together; then all the propagation steps involve additions of radicals to alkenes. Polymerization reactions are discussed in greater detail in Chapter 27.

C. Intramolecular Free-Radical Addition Reactions

Intramolecular addition of a free radical to a double bond produces a ring. Five-membered rings are usually formed in preference to six-membered rings. For example, when 6-bromo-1-hexene reacts with a tributyltin radical, methylcyclopentane is formed as the major product.

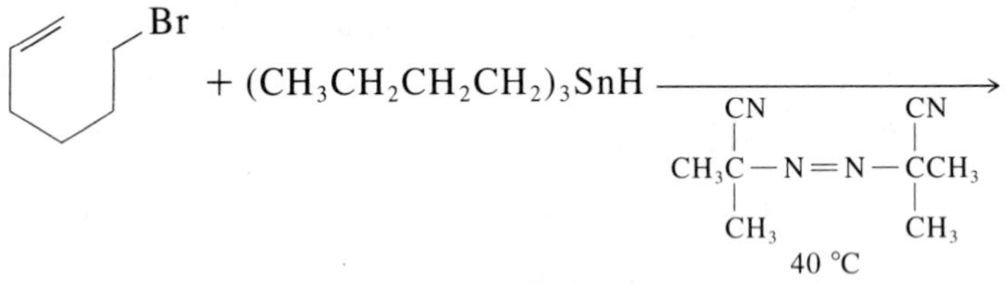
Br
+ (CH₃CH₂CH₂CH₂)₃SnH
CN
CN
CH₃C—N=N—CCH₃
CH₃
CH₃
40 °C

6-bromo-1-hexene   tributylstannane   azobis(isobutyronitrile)

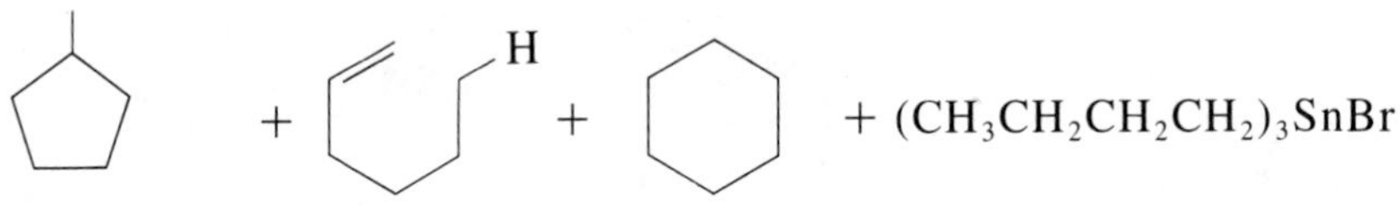
H
+
+
+ (CH₃CH₂CH₂CH₂)₃SnBr

methylcyclopentane 90%   1-hexene 10%   cyclohexane trace   tributyltin bromide

The initiator for the reaction is azobis(isobutyronitrile), widely used because it readily decomposes on heating to give molecular nitrogen and carbon radicals that are stabilized by a substituent cyano group. These carbon radicals abstract hydrogen from tributylstannane (tributyltin hydride).

VISUALIZING THE REACTION

Initiation

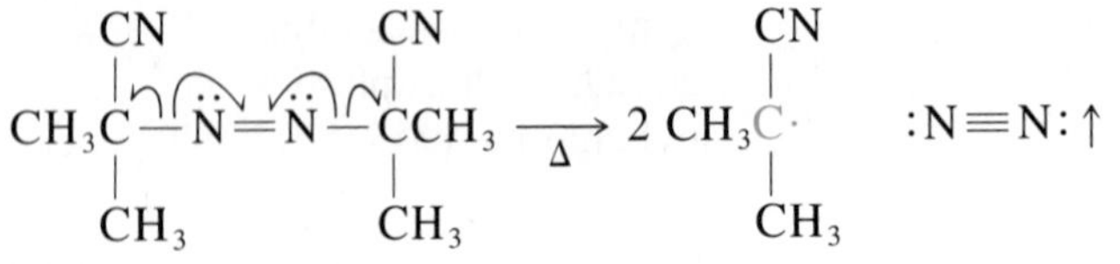
CN
CN
CH₃C—N=N—CCH₃
CH₃
CH₃
Δ
2 CH₃C·
CN
CH₃
:N≡N:↑

decomposition of azobis(isobutyronitrile)   stabilized radicals

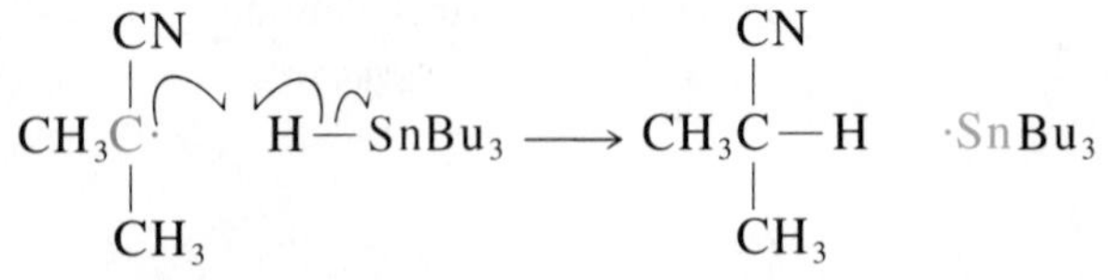
CN
CH₃C·
CH₃
H—SnBu₃
CH₃C—H
CN
CH₃
·SnBu₃

abstraction of hydrogen   tributyltin radical

The tin radical then abstracts a halogen atom from the haloalkane, and the resulting carbon radical reacts to give the observed products.

VISUALIZING THE REACTION

Free-radical cyclization of a haloalkene

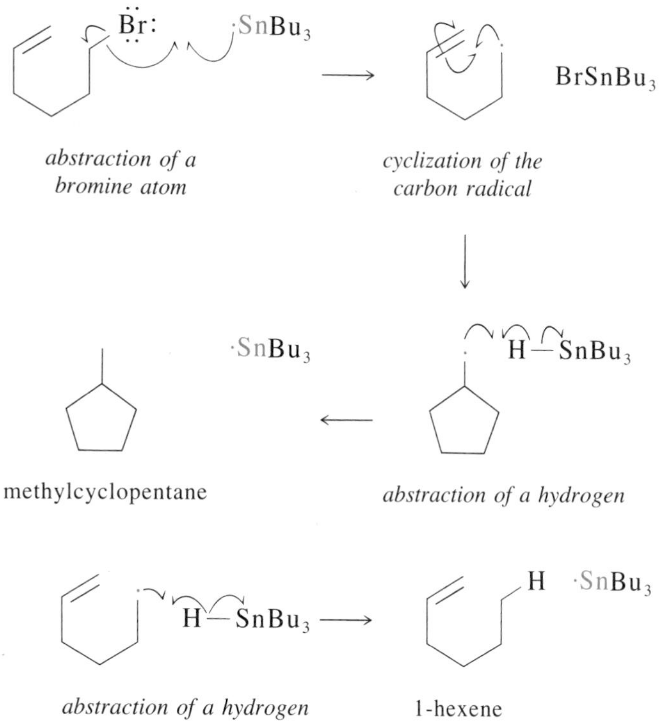

Br:
SnBu$_3$
BrSnBu$_3$
abstraction of a bromine atom
cyclization of the carbon radical
·SnBu$_3$
H—SnBu$_3$
methylcyclopentane
abstraction of a hydrogen
H—SnBu$_3$
H ·SnBu$_3$
abstraction of a hydrogen
1-hexene

Study Guide Concept Map 20.4

Intramolecular free-radical addition reactions are used in organic syntheses. The following problem concerns an example of such an application.

PROBLEM 20.13

Silphiperfol-6-ene, a natural product, was synthesized from the compound shown below. How would you carry out this transformation?

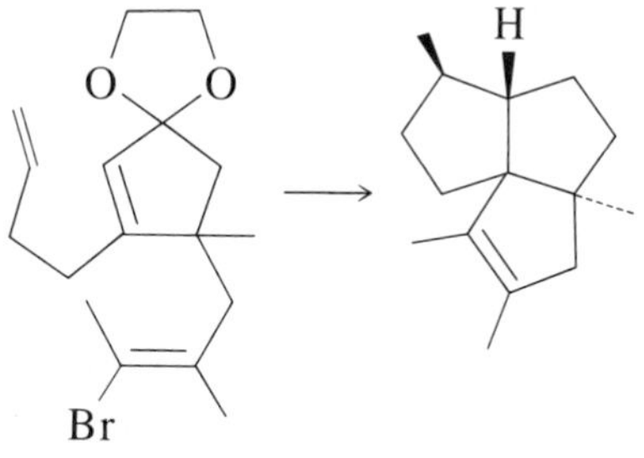

O
O
Br
H

silphiperfol-6-ene

A. Autooxidation

When organic compounds are exposed to air, they react slowly with oxygen to give hydroperoxides. This slow oxidation reaction is called **autooxidation.** This reaction is responsible for the slow deterioration in air of foods (below and p. 854), rubber (p. 1205), and paints. The reaction is catalyzed by light, which is why organic reagents are stored in cans or dark-colored bottles. The ease with which a compound undergoes autooxidation is directly related to the ease with which it forms free radicals. For that reason, autooxidation takes place especially easily at allylic or benzylic positions. Cumene, for example, is easily converted to a hydroperoxide.

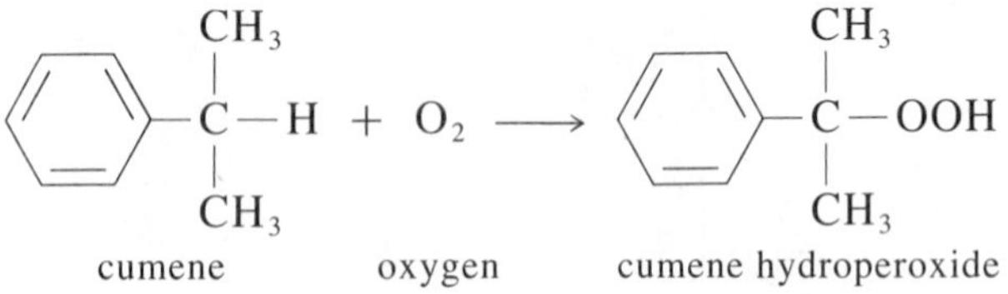
CH3
C—H + O2
CH3
CH3
C—OOH
CH3
cumene
oxygen
cumene hydroperoxide

The hydrogen atoms bonded to carbons that are adjacent to the oxygen atom in ethers and alcohols are also easily replaced by hydroperoxide groups, shown in the reaction below.

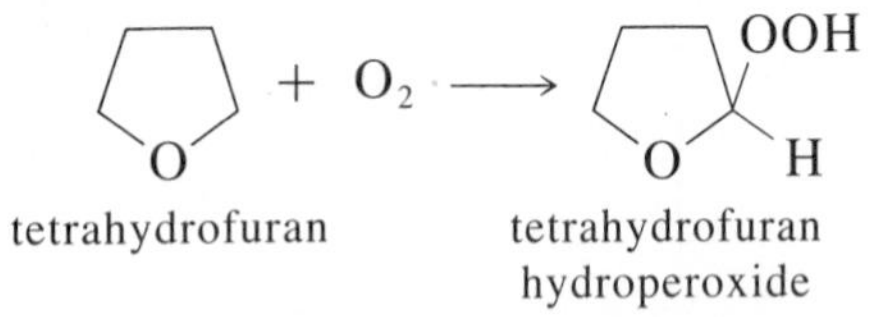
+ O2
O
OOH
O
H
tetrahydrofuran
tetrahydrofuran hydroperoxide

If such solvents have been stored for some time, they almost certainly contain some hydroperoxides. Hydroperoxides are unstable compounds and may decompose violently when heated. For safety, ethers are tested in the laboratory for peroxides and, if necessary, are purified before being used.

In the next section, the details of the formation of hydroperoxides are illustrated by the reaction of oxygen with polyunsaturated fatty acids.

B. Oxidation of Polyunsaturated Fatty Acids

Polyunsaturated fatty acids (p. 610) are converted by reactions with oxygen to compounds containing oxygen and conjugated double bonds. The formation of such compounds is responsible for the development of rancidity in oils and, therefore, for the spoilage of any food that contains an oil. Polyunsaturated fatty acids, mostly as their esters, are components of the human body. The role of free-radical chemistry in the body and especially its relation to aging are subjects of vigorous debate. Molecular oxygen is a **diradical,** having two unpaired electrons, so the reaction of an unsaturated acid with oxygen involves free-radical intermediates. Oxidation of linoleic acid, catalyzed by an enzyme from soybeans, gives (9*Z*,11*E*)-13-hydroperoxy-9,11-octadecadienoic acid

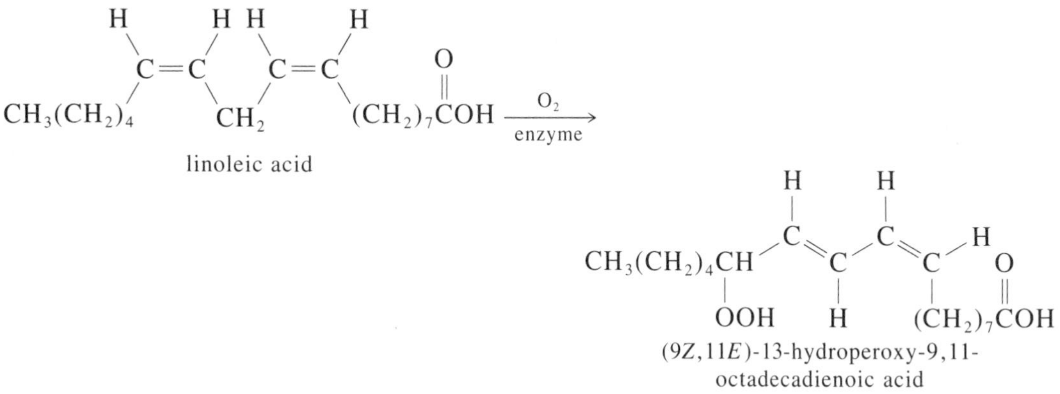

Experiments involving isotopic labeling have shown that the oxygen atoms in the product come from oxygen molecules, not from water molecules.

VISUALIZING THE REACTION

Formation of a hydroperoxide

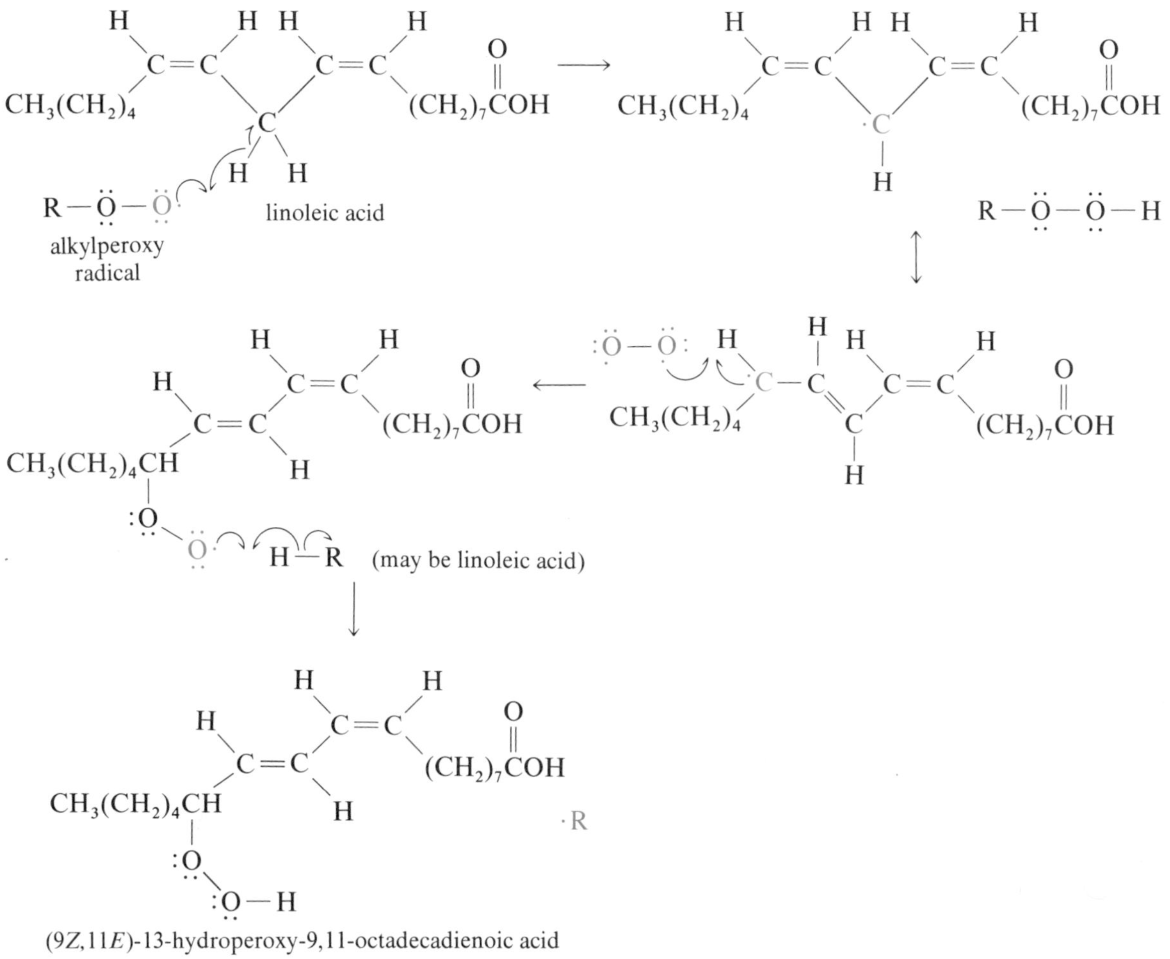

The exact nature of the radical that abstracts the first hydrogen from linoleic acid is not known. It is thought to be some peroxy species. The hydrogen atoms that are allylic to two double bonds are particularly vulnerable to attack by a radical initiator. The carbon radical that forms is allylic and can be stabilized by delocalization of the radical character, giving rise to a conjugated double bond system. The new double bond formed between carbon atoms 11 and 12 is always trans. Combination of the carbon radical with molecular oxygen creates an alkylperoxy radical, which starts the chain reaction going again by abstracting an allylic hydrogen atom. The more double bonds a fatty acid contains, the more sites there are for hydrogen-abstraction reactions, and the greater the complexity of the products. It is also possible for the alkylperoxy radicals to attack double bonds in other molecules of the unsaturated fatty acid, giving rise to large, complex molecules in which many acid units are held together by oxygen bridges. The characteristic drying of linseed oil on exposure to air, which gives a tough protective coating to paint, is believed to be due to formation of just such a linkage of linolenic acid molecules by oxygen.

C. Biosynthesis of Prostaglandins

Prostaglandins and related polyunsaturated acids containing twenty carbon atoms constitute a class of compounds that have a wide range of significant biological functions. The name prostaglandins comes from the fact that these compounds were first isolated from seminal fluid. Prostaglandins have since been found to be widely distributed in all kinds of body tissues and play roles in reproduction, the nervous system, the intestinal system, blood clotting, and the production of allergic and inflammatory reactions.

Arachidonic acid (p. 610) is the biological precursor of prostaglandin E_2. The structures of arachidonic acid, prostaglandin E_2, and a key intermediate in the transformation of the former into the latter are shown below.

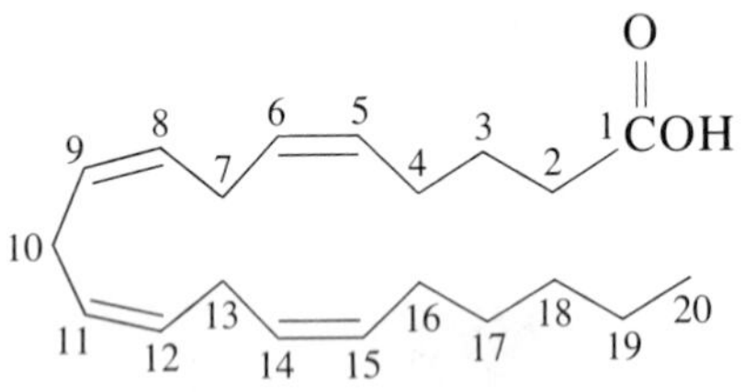
O
COH

arachidonic acid

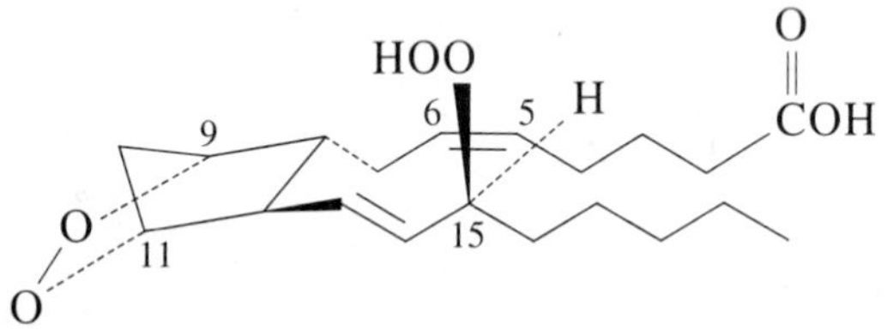
HOO
H
O
COH

endoperoxide

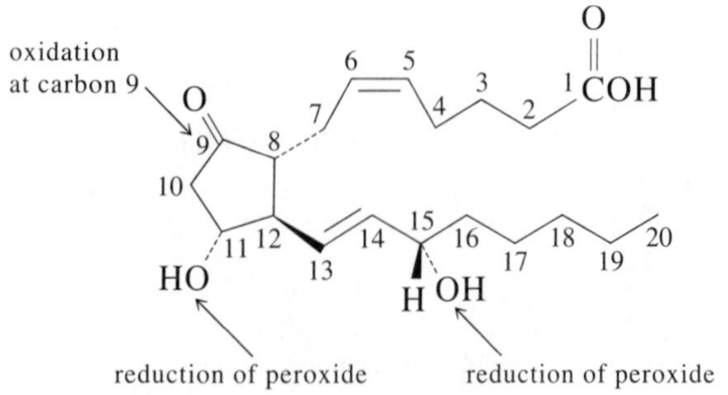
oxidation at carbon 9
O
COH
HO
H OH
reduction of peroxide
reduction of peroxide

prostaglandin E_2

It has been established that two molecules of oxygen are used in the biological transformation of arachidonic acid to prostaglandin E_2 and that the oxygen atoms at carbon atoms 9 and 11 of the prostaglandin are derived from the same molecule of oxygen. The enzymatic cyclization of arachidonic acid to prostaglandin involves a number of intermediates and starts with the abstraction of a hydrogen atom at carbon 13 to give a free radical. This radical sets off a chain of free-radical reactions, including the formation of a five-membered ring (p. 840) and the addition of oxygen across the ring at carbons 9 and 11 to give the endoperoxide. The endoperoxide is converted to prostaglandin E_2 by a series of oxidation and reduction steps.

The various prostaglandin structures shown below are also derived from the endoperoxide intermediate by familiar reactions, such as reduction of the double bond between carbon atoms 5 and 6, loss of water from the cyclopentane ring, and reduction of ketone functions to alcohols. These structures are not all derived biologically from arachidonic acid, however.

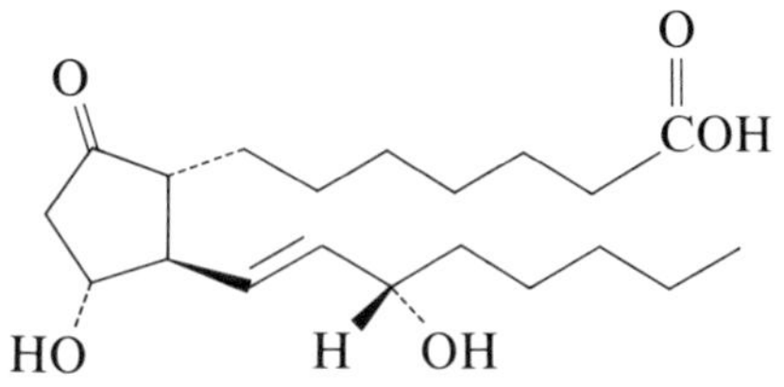
O
O
COH
HO
H
OH
prostaglandin E_1

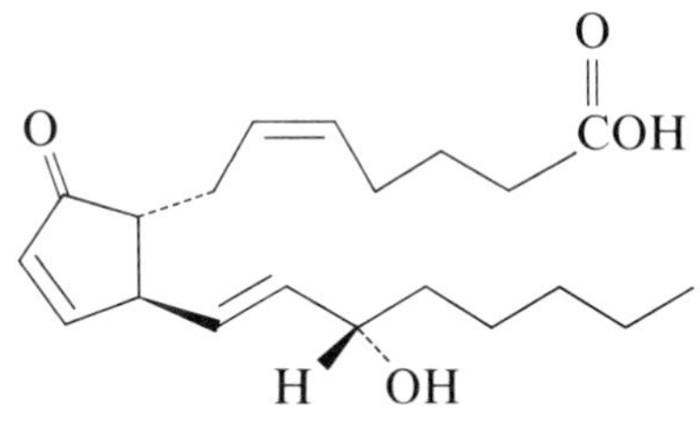
O
O
COH
H
OH
prostaglandin A_2

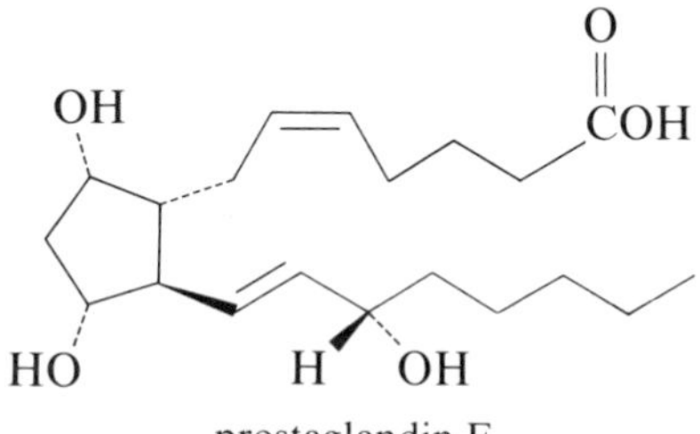
O
OH
COH
HO
H
OH
prostaglandin $F_{2\alpha}$

Another series of compounds that have high biological activity and are also polyunsaturated twenty-carbon acids are the leukotrienes, which despite their name have four double bonds. Of these, leukotriene C is known to be the slow-reacting substance involved in anaphylactic shock, the body's response to a foreign substance that provokes a severe allergic reaction, as in the case of asthma or insect stings. Exactly how leukotriene C is involved is not known. Apparently, cells are stimulated to release it by the arrival of antibodies formed in response to the presence of allergens, substances causing allergies.

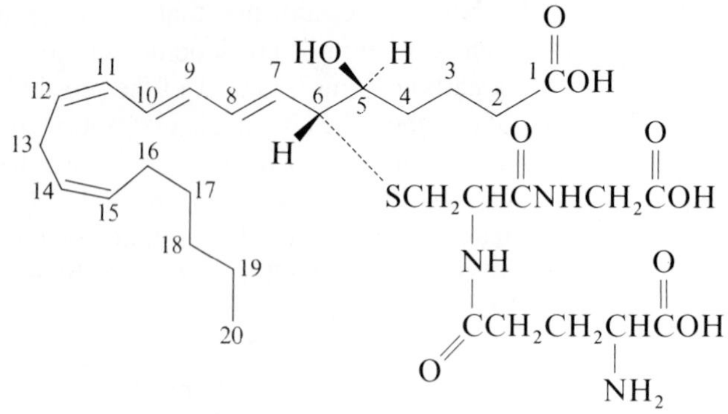

leukotriene C

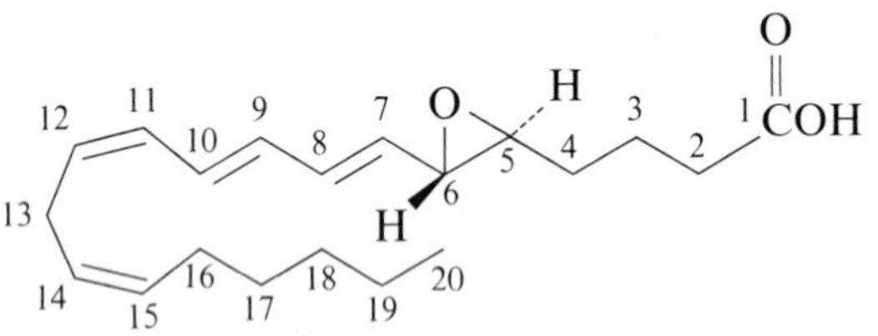

leukotriene A

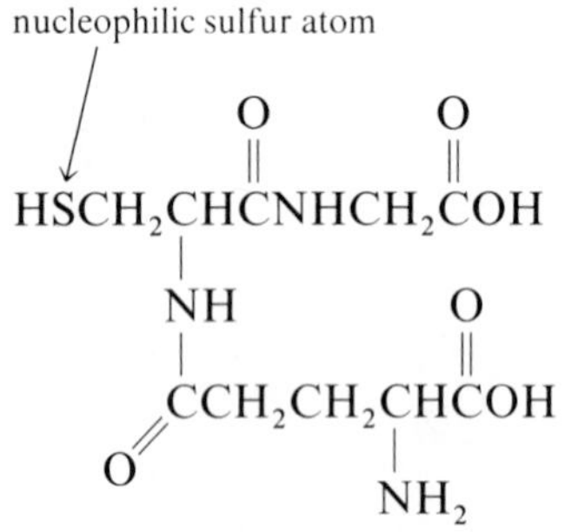
nucleophilic sulfur atom

glutathione

a peptide containing three amino acids and two peptide linkages

Both leukotriene A and leukotriene C have an oxygen bonded to carbon atom 5 of the chain. Leukotriene A is an intermediate in the biosynthesis of leukotriene C. The transformation of A to C involves addition of glutathione (a peptide) to the oxirane. Glutathione includes the amino acid cysteine, which has a thiol group. The opening of the oxirane ring in leukotriene A results from attack by the nucleophilic sulfur atom of glutathione (p. 454).

PROBLEM 20.14

(8*Z*,11*Z*,14*Z*)-8,11,14-Icosatrienoic acid is converted by an enzyme called soybean lipoxidase and molecular oxygen to (8*Z*,11*Z*,13*E*)-15-hydroperoxy-8,11,13-isosatrienoic acid. Write structural formulas for starting material and product, and propose a mechanism for the conversion.

PROBLEM 20.15

Write a mechanism for the reaction of the oxirane ring in leukotriene A with the peptide glutathione, paying close attention to the stereochemistry. The abbreviated structures shown below represent the reactants. In an enzyme system, acids (HB^+) and bases (B:) are readily available.

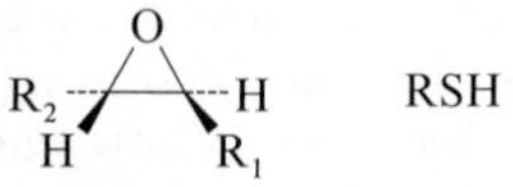

leukotriene A glutathione

A. Quinones

The readiness of phenols to undergo electrophilic aromatic substitution reactions (p. 784) indicates that the electrons of the aromatic ring must be available to react with electron-deficient species. Oxidizing agents, which, like electrophiles, can be regarded as electron acceptors, react with phenols as well as with aromatic amines. Two types of compounds that are oxidized with particular ease are aromatic compounds that have hydroxyl and/or amino groups ortho or para to each other. These compounds are oxidized to brightly colored carbonyl compounds known as **quinones.**

An important example of this reactivity is the oxidation of 1,4-dihydroxybenzene, commonly known as hydroquinone, by silver ion. The reaction is used in developing photographic film, which contains finely divided grains of silver bromide in a gelatin layer. Light activates particles of silver bromide, so that they become especially susceptible to reduction to free silver. The development process converts these silver bromide particles to silver metal, which creates a darkening of the film at those points. A basic aqueous solution of hydroquinone is used as the reducing agent. The essential chemistry of the process is shown below.

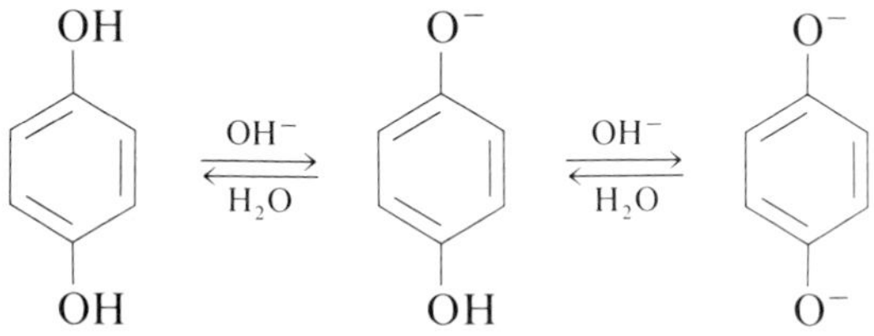

hydroquinone — anions of hydroquinone in equilibrium with it in basic solution

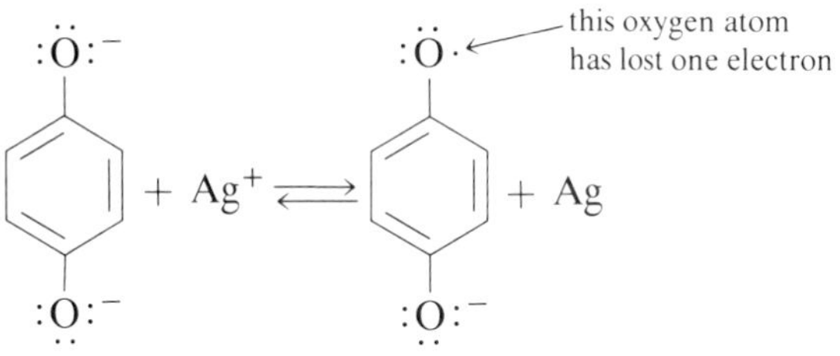
this oxygen atom
has lost one electron

dianion of hydroquinone — semiquinone radical anion

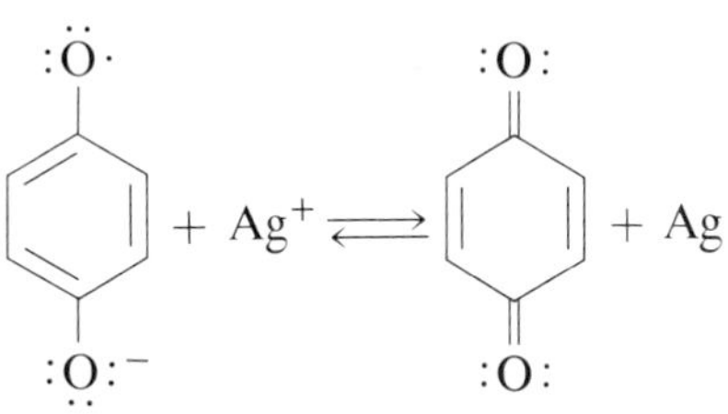

semiquinone radical anion — *p*-benzoquinone *yellow*

Hydroquinone is converted into its anion by base (p. 965). The anion is oxidized in two steps by silver ion, losing an electron at each stage. Two silver ions gain one electron apiece to become metallic silver. The intermediate in the oxidation is a radical anion known as semiquinone. One of its oxygen atoms has an unpaired electron and thus has radical character; the other bears a negative charge. Both the radical character and the negative charge are delocalized to the ring carbon atoms and to the other oxygen atom.

Hydroquinone and *p*-benzoquinone are interconvertible by oxidation and reduction reactions. The metal ion is similarly related to the free metal. The reactions are summarized below as two half reactions.

reduction half reaction: $Ag^{+} + e^{-} \rightleftharpoons Ag$

silver ion — metallic silver

oxidation half reaction:

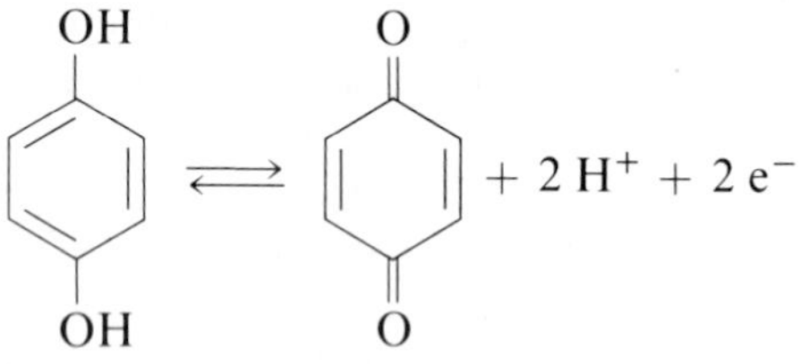
OH
OH
O
O
+ 2 H⁺ + 2 e⁻

hydroquinone — *p*-benzoquinone

It is obvious from an examination of the two half reactions that it will take two silver ions to oxidize one molecule of hydroquinone to quinone.

Hydroquinone, which is colorless, forms a deeply colored complex with quinone. This complex, called quinhydrone, is much less soluble than either of its components. A molecule of hydroquinone and a molecule of quinone are held together by the attraction of the electron-deficient quinone ring for the electrons of the electron-rich hydroquinone ring. Such a complex is called a **charge-transfer complex,** indicating that some actual transfer of electronic charge takes place between the rings, so the bond between them has some ionic character.

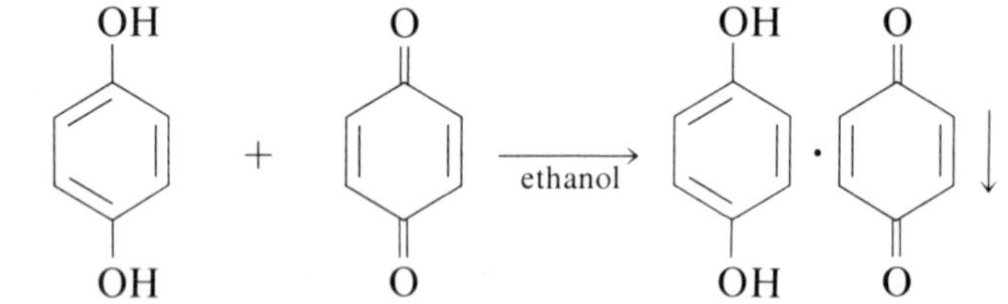
OH
OH
O
O
ethanol
OH
OH
O
O

hydroquinone *colorless* — *p*-benzoquinone *yellow* — quinhydrone *deep green*

Quinones may have the two carbonyl groups para to each other, as in *p*-benzoquinone, or ortho to each other, as in *o*-benzoquinone.

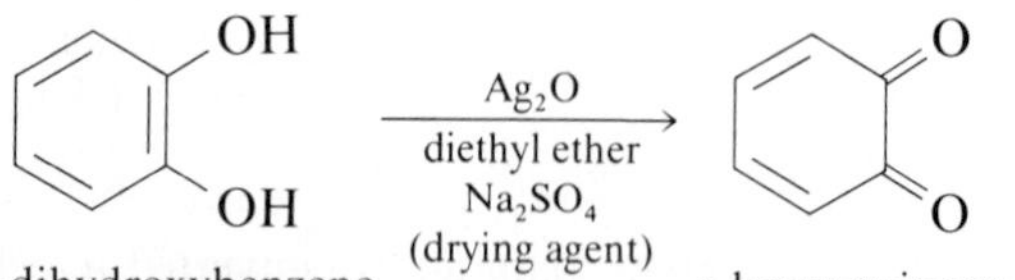
OH
OH
Ag₂O
diethyl ether
Na₂SO₄
(drying agent)
O
O

1,2-dihydroxybenzene
catechol — *o*-benzoquinone

o-Benzoquinone is rather unstable, especially in the presence of moisture, and is prepared with very dry reagents. Some quinones are prepared by the direct oxidation of hydrocarbons. The vulnerability of the para positions in the central ring of anthracene, demonstrated by the addition of bromine and the Diels-Alder reaction in that ring (p. 782), allows the easy oxidation of anthracene to anthraquinone, a para quinone.

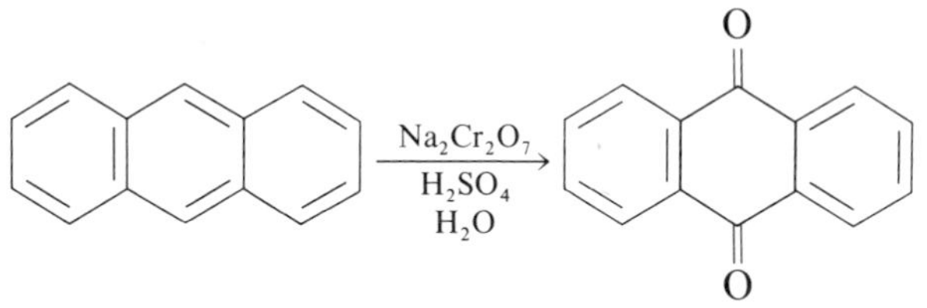

O
$Na_2Cr_2O_7$
H_2SO_4
H_2O
O

anthracene

anthraquinone
~90%
yellow

Similarly, the central ring in phenanthrene undergoes oxidation to give an ortho quinone.

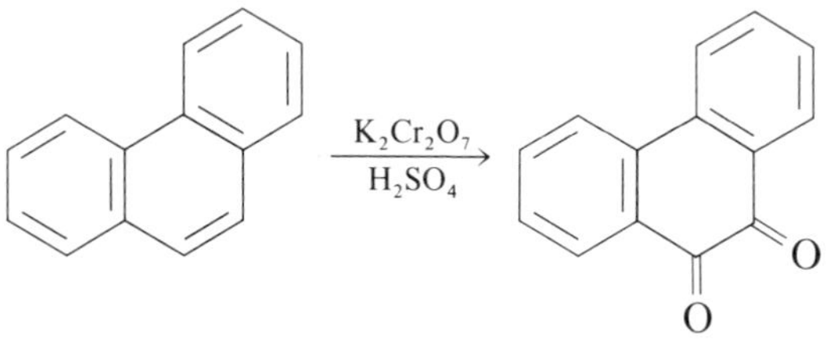

$K_2Cr_2O_7$
H_2SO_4
O
O

phenanthrene

phenanthraquinone
~50%
orange

Note that there is no aromatic character in a quinone ring. It has a carbonyl group conjugated with a double bond, and many of its reactions are those of such compounds. Quinones can serve as dienophiles in Diels-Alder reactions (p. 691). *p*-Naphthoquinone, for example, adds to 2,3-dimethyl-1,3-butadiene to give a reduced anthraquinone.

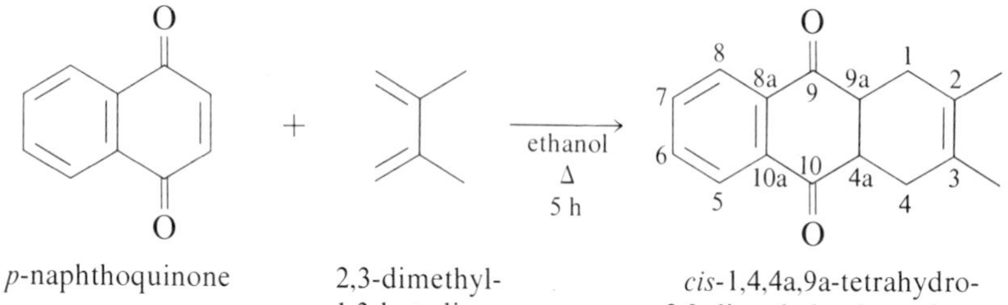

O
O
+
ethanol
Δ
5 h
O
8
8a
9
9a
1
2
7
6
10a
10
4a
3
5
4
O

p-naphthoquinone
dienophile

2,3-dimethyl-
1,3-butadiene
diene

cis-1,4,4a,9a-tetrahydro-
2,3-dimethylanthraquinone
90%

All quinones are intensely colored, but ortho quinones are more highly colored than para quinones. The chromophore responsible for the color is the arrangement of conjugated double bonds inside and outside the six-membered ring. The arrangements found in *p*-benzoquinone and *o*-benzoquinone are given the name **quinoid structures** regardless of whether there are oxygen atoms at the ends of the system.

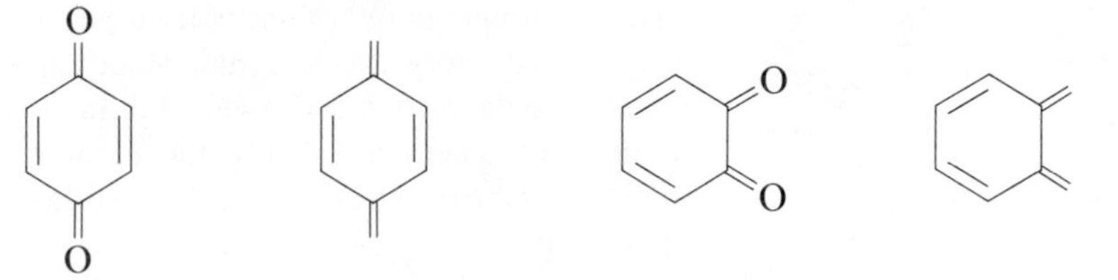

p-benzoquinone | *p*-quinoid arrangement of double bonds | *o*-benzoquinone | *o*-quinoid arrangement of double bonds

For example, the resonance contributors of polycyclic aromatic hydrocarbons can be analyzed in terms of whether the rings are benzenoid or quinoid in character. Resonance contributors that contain more benzenoid rings than quinoid ones are the major ones for polycyclic aromatic hydrocarbons.

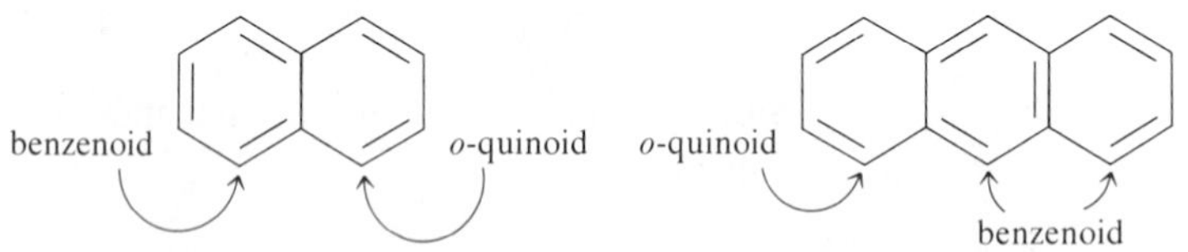
benzenoid
o-quinoid
o-quinoid
benzenoid

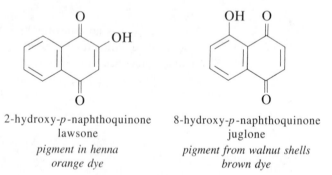

a resonance contributor of naphthalene with one benzenoid and one *o*-quinoid ring

a resonance contributor of anthracene with one *o*-quinoid and two benzenoid rings

Many important natural products are phenols or quinones. A number of the more complex structures, such as Vitamin E and Vitamin K, are discussed in Section 20.6D. Two simple plant products that are *p*-naphthoquinones are 2-hydroxy-*p*-naphthoquinone and 8-hydroxy-*p*-naphthoquinone.

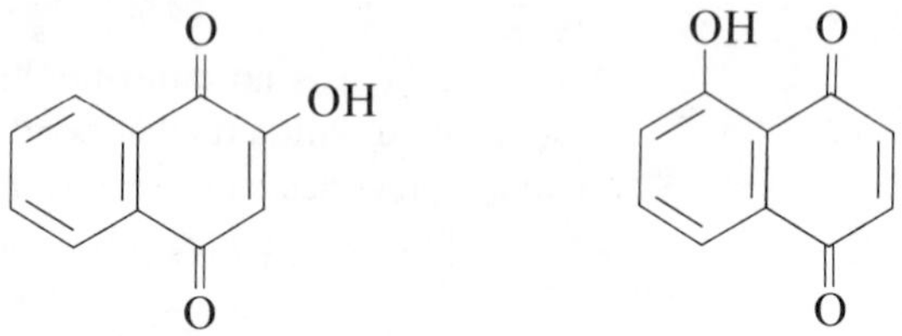

2-hydroxy-*p*-naphthoquinone
lawsone
pigment in henna
orange dye

8-hydroxy-*p*-naphthoquinone
juglone
pigment from walnut shells
brown dye

Lawsone is extracted from the leaves of the henna plant and is used by women in the Middle East to dye their hair, their palms, and their fingertips a reddish color. Juglone is the brown stain found in the outer soft shells of walnuts.

Quinones are also important biologically because of the ease with which they are reduced to phenols. Such pairs of compounds serve as mediators of oxidation and reduction reactions in living organisms (p. 855).

B. Oxidative Coupling Reactions of Phenols

Phenols that have only one hydroxyl group are oxidized by a wide variety of oxidizing agents to give compounds in which carbon-carbon bonds have been formed between aromatic rings. Oxidation reactions that result in the formation of carbon-carbon bonds are called **oxidative coupling reactions.** Unless the phenol is carefully chosen, a variety of products is obtained. 1-Naphthol, for example, gives three products when oxidized with ferric chloride.

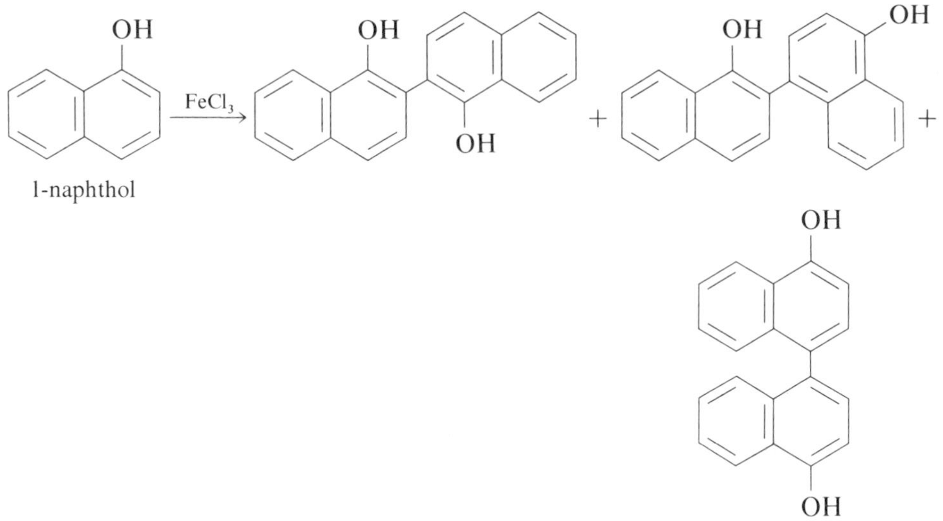

OH
FeCl3
1-naphthol
OH
OH
+
OH
OH
+
OH
OH

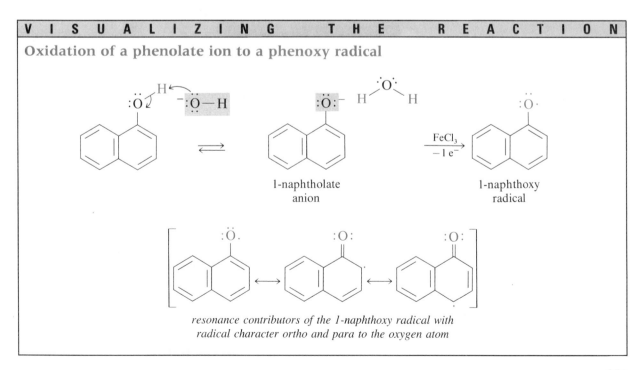

The new carbon-carbon bond between aromatic rings is formed ortho or para to their hydroxyl groups.

The reaction starts with the equivalent of the removal of a hydrogen atom from the phenol to give a radical. Usually this happens by removal of a proton from the phenol by the basic reaction medium and then the removal of an electron from the phenolate ion by the oxidizing agent. The phenoxy radical is stabilized by delocalization of radical character to the ortho and para positions of the ring.

VISUALIZING THE REACTION

Oxidation of a phenolate ion to a phenoxy radical

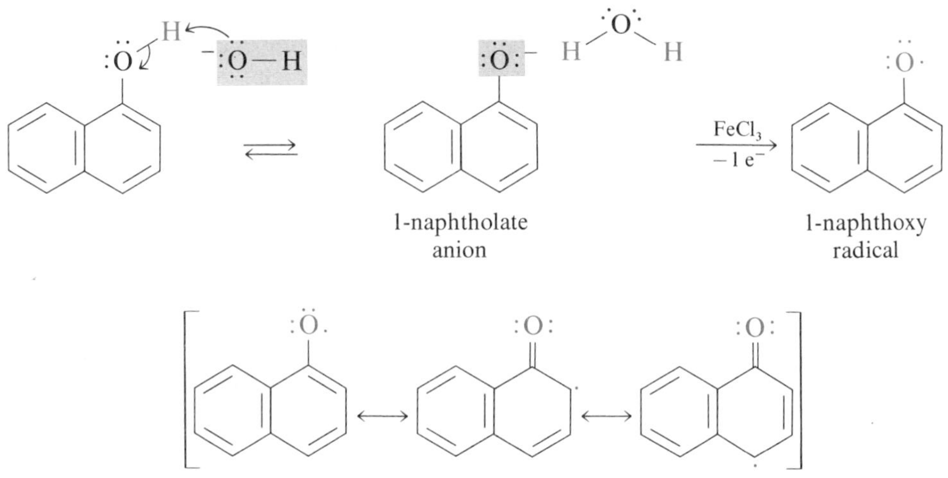

H
O–H
FeCl3
–1 e⁻
1-naphtholate anion
1-naphthoxy radical

resonance contributors of the 1-naphthoxy radical with radical character ortho and para to the oxygen atom

Two radicals then combine to give an unstable intermediate, which tautomerizes to give the aromatic diphenol product.

VISUALIZING THE REACTION

Coupling of phenoxy radicals

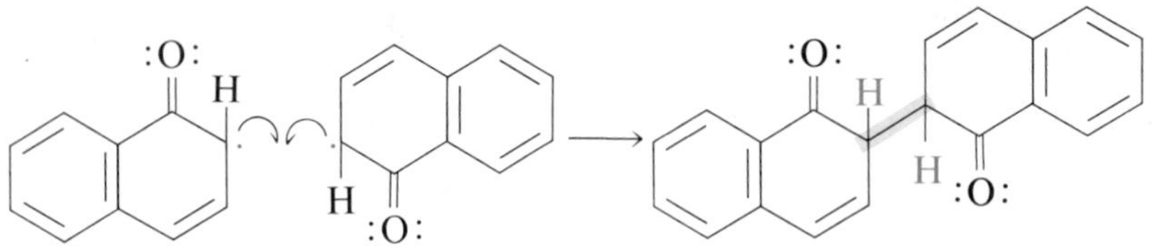

unstable intermediate from
the coupling of 2
radicals at the ortho position

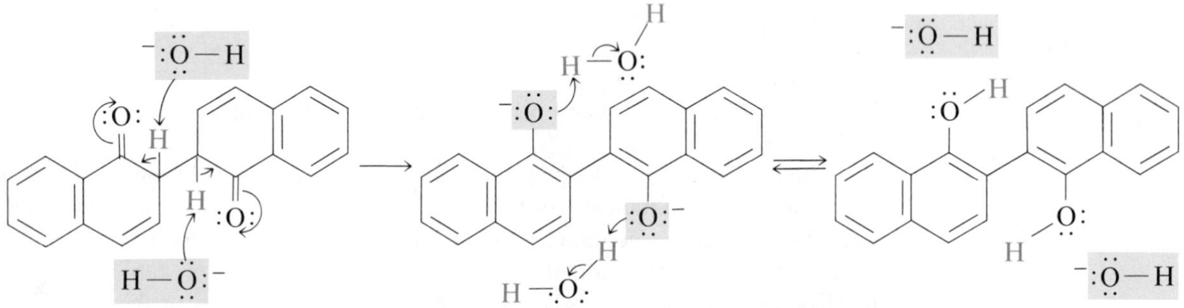

keto form of a phenol

enol form of a phenol

The oxidative coupling reactions of phenols are synthetically useful only when some of the positions on the ring of the starting compound are blocked by substituents so that a single product is formed in high yield. A good example of such a reaction is the oxidation of 2,6-di-*tert*-butylphenol by oxygen in a basic solution.

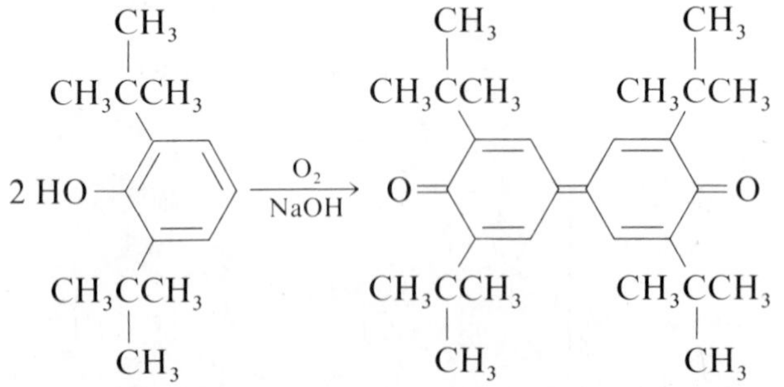
2 HO
O_2
NaOH
O
O
CH_3
CH_3CCH_3

2,6-di-*tert*-butylphenol

a diphenyl quinone
98%

red-brown

Oxygen functions as an oxidizing agent at two stages: first, to create the phenoxy radical, and, second, to remove hydrogen atoms from the intermediate formed in the coupling reaction and oxidize it to the quinone.

PROBLEM 20.16

Complete the following equations by first deciding in each case whether the reagent over the arrow is an oxidizing agent or a reducing agent.

(a) 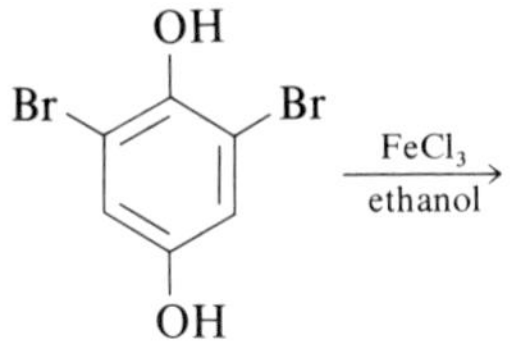
OH
Br
Br
OH
FeCl3
ethanol

(b) 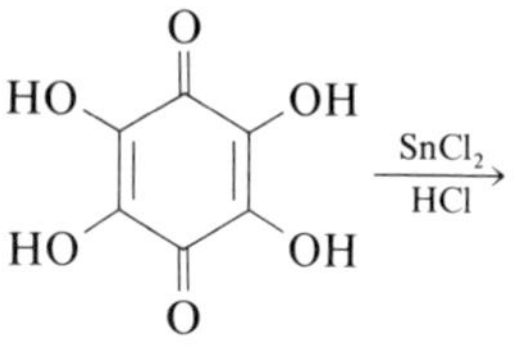
O
HO
OH
HO
OH
O
SnCl2
HCl

(c) 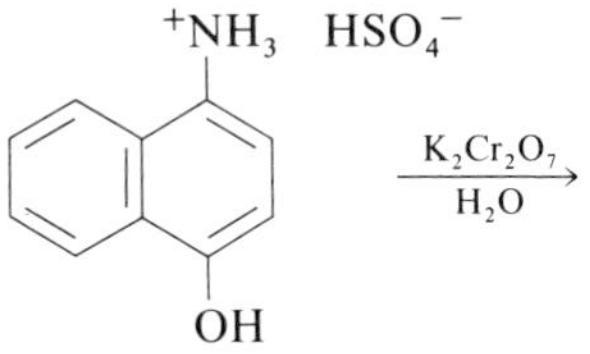
+NH3 HSO4−
OH
K2Cr2O7
H2O

(d) 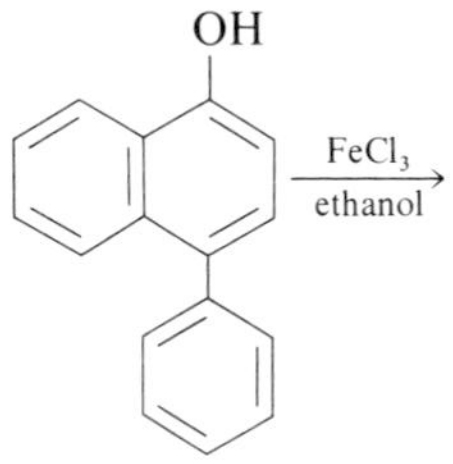
OH
FeCl3
ethanol

(e) 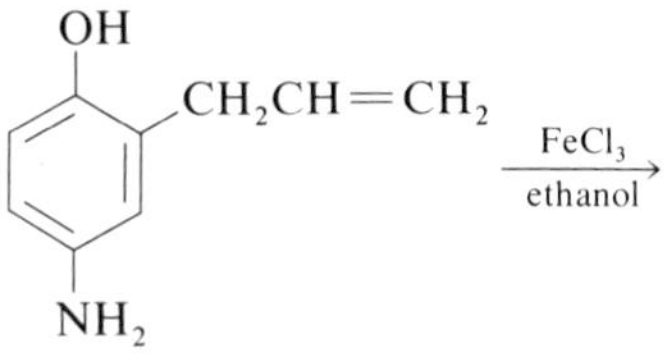
OH
CH2CH=CH2
NH2
FeCl3
ethanol

C. Phenols as Antioxidants

Substituted phenols are used as **antioxidants** in processed foods. On labels on cereals, cookies, rice products, and many other grocery items, BHA or BHT is frequently listed as an ingredient. The label will sometimes say that the substance was added to retard spoilage or rancidity. BHA stands for *b*utylated *h*ydroxy*a*nisole and is a mixture of *tert*-butylmethoxyphenols. BHT is *b*utylated *h*ydroxy*t*oluene, or 2,6-di-*tert*-butyl-4-methylphenol.

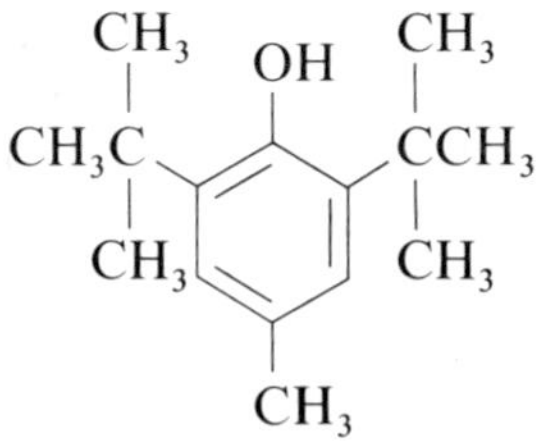
CH3
OH
CH3
CH3C
CCH3
CH3
CH3
CH3

2,6-di-*tert*-butyl-4-methylphenol

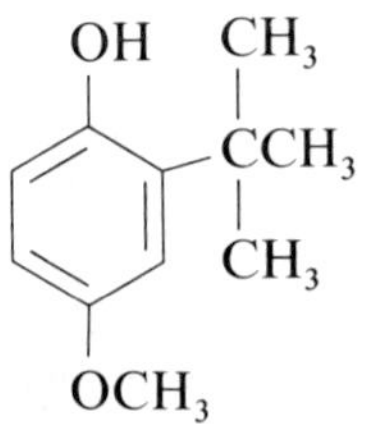
OH
CH3
CCH3
CH3
OCH3

2-*tert*-butyl-4-methoxyphenol

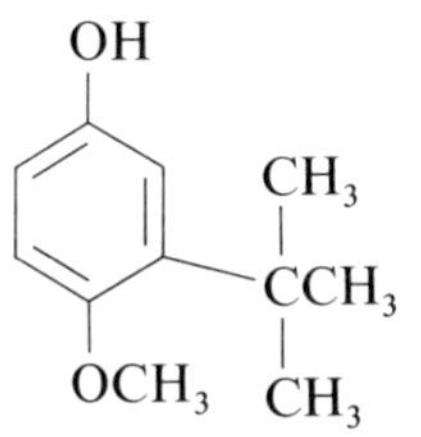
OH
CH3
CCH3
OCH3
CH3

3-*tert*-butyl-4-methoxyphenol

*b*utylated *h*ydroxy*t*oluene
BHT

*b*utylated *h*ydroxy*a*nisole
BHA

two commercially important antioxidants

Oxygen from the air reacts with compounds containing double bonds by free-radical chain reactions (p. 842). If these reactions are not stopped, the oils in food products stored at room temperature in warehouses and on supermarket shelves are gradually oxidized and become rancid. Not only do rancid oils taste bad, they are also toxic. As the food-processing industry became more and more centralized, it became necessary for foods to have a long shelf life. Therefore, chemists developed additives that break up the free-radical chain reactions by reacting themselves with any radicals that form. Phenols are reactive toward radicals (p. 850). BHA and BHT give hindered phenoxy radicals that are unreactive and thus terminate free-radical chain reactions. As a result, they are used to retard free-radical oxidation reactions in food products, among other things. A natural antioxidant found in foods that have a high content of polyunsaturated fatty acids, such as vegetable oils, is Vitamin E, discussed in the next section.

PROBLEM 20.17

(a) BHA is prepared commercially from *p*-methoxyphenol and 2-methylpropene. Suggest how this reaction could be carried out.
(b) A similar reaction is used to make BHT from *p*-methylphenol and 2-methylpropene. Why does the BHA sold commercially consist of a mixture of isomers, whereas the reaction to give BHT has greater regioselectivity?

D. Vitamin E and Vitamin K

Vitamin E, also called α-tocopherol, has phenolic and isoprenoid components. Its structure can be dissected into a trimethylhydroquinone and a chain containing four isoprene units, a phytyl group.

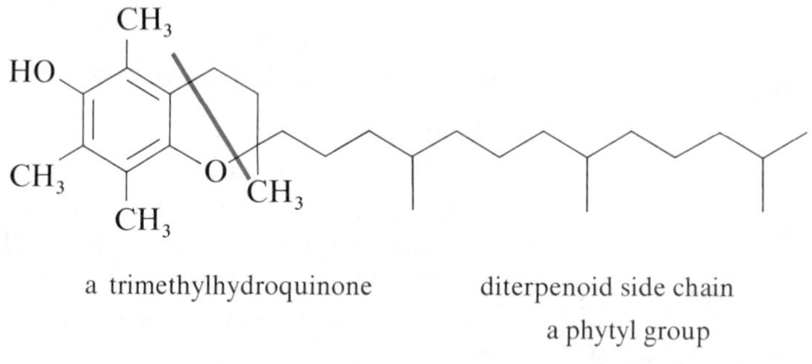
CH3
HO
CH3
O
CH3
CH3
a trimethylhydroquinone
diterpenoid side chain
a phytyl group

α-tocopherol, or Vitamin E

Vitamin E has been recognized as having an important biological role as a potent antioxidant. It is especially important in preventing the formation of hydroperoxides from polyunsaturated fatty acids (p. 610) and is found most abundantly in seeds that are rich in such fatty acids.

In 1930, researchers discovered that chicks developed hemorrhages when fed a highly artificial diet from which all fatty components had been removed. This observation led to the discovery that a fat-soluble vitamin, named Vitamin K, was important in the clotting of blood. Later, it was determined that a family of compounds has the activity shown by Vitamin K. These compounds have a 2-methyl-1,4-naphthoquinone structure with different isoprenoid side chains attached at carbon 3.

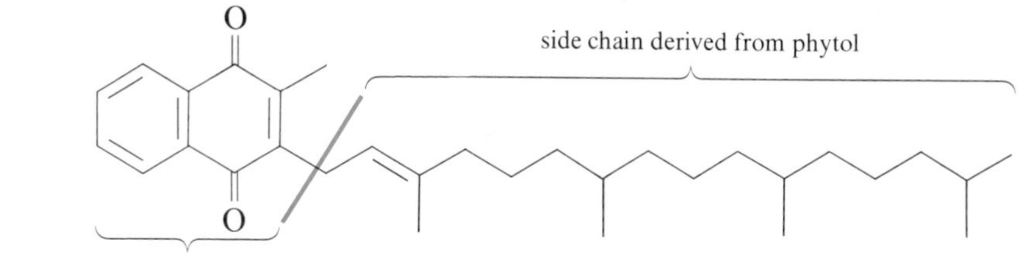
side chain derived from phytol
2-methyl-1,4-naphthoquinone

Vitamin K_1
phylloquinone

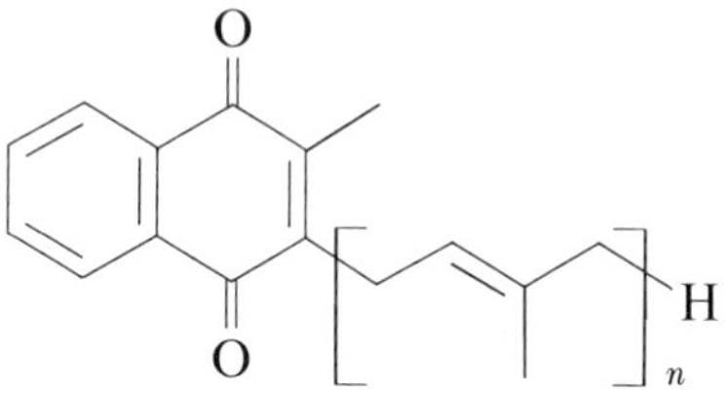

Vitamin K_2
menaquinones
n = 6, 7, 8, or 9

Vitamin K_1, also known as phylloquinone, has a side chain derived from the diterpene alcohol phytol; the side chain in Vitamin E is also derived from phytol. Vitamin K_2 is actually a group of compounds, the menaquinones, which have different numbers of isoprene units in the side chain. Vitamin K_1 is involved in **electron transport,** the transfer of electrons from one site to another, which is essential to the occurrence of oxidation and reduction reactions in cells.

The K vitamins are found widely in the leaves of green plants. Vitamin K is also synthesized by bacteria in the lower intestines. Because the compounds are so widespread, it is unlikely that a person would develop a deficiency unless he or she refused to eat any green vegetables at all.

A number of compounds are antagonists of Vitamin K; they increase the clotting time of blood. Such compounds are known as **anticoagulants.** For example, salicylic acid, a product of the hydrolysis of aspirin, has this effect. Two more powerful anticoagulants, dicoumarol and Warfarin, are also shown below.

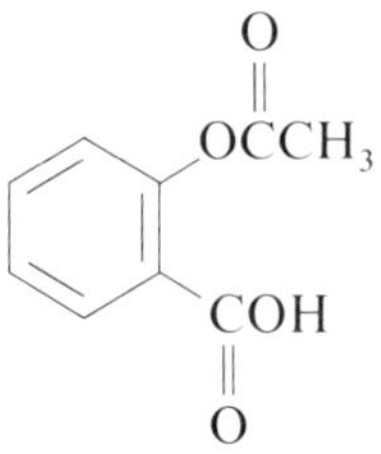

acetylsalicylic acid
aspirin

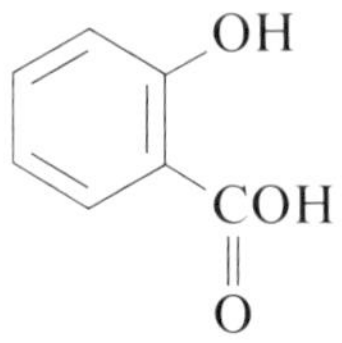

o-hydroxybenzoic acid
salicylic acid

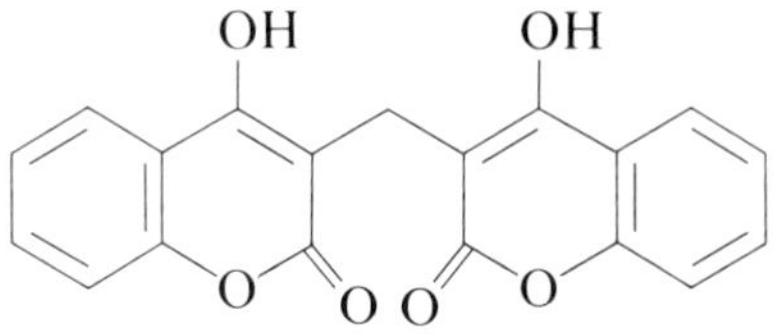

dicoumarol

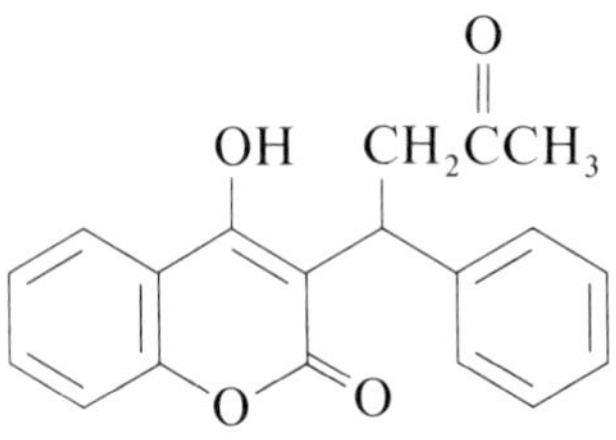

Warfarin

some compounds that increase blood clotting times
anticoagulants

Dicoumarol is a component of spoiled clover hay that was discovered because animals that ate it started to bleed abnormally. It is used medically to prevent blood clots. Warfarin is a synthetic compound, deliberately designed to be an anticoagulant and used as a rat poison. Note that all the anticoagulants shown have a phenolic or enolic hydroxyl group. Dicoumarol and Warfarin are also lactones of phenols. The phenol lactone system found in these two compounds is also found in coumarin, a fragrant constituent of clover.

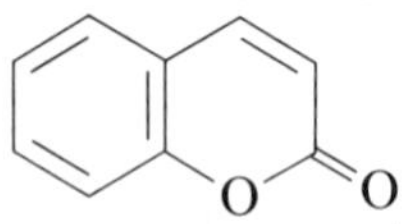
O
O

coumarin
a phenol lactone

Study Guide Concept Map 20.5

Coumarin itself does not interfere with the clotting of blood and has been used as a flavoring agent.

PROBLEM 20.18

The anticoagulant Warfarin is used in the form of its sodium salt. What is the structure of this compound?

PROBLEM 20.19

2-Methyl-1,4-naphthoquinone is a Vitamin K analog. A sample of the compound labeled with radioactive carbon-14 at the methyl group was needed for tracer studies on the metabolism of Vitamin K. The synthesis used 2-bromonaphthalene as the starting material; $^{14}CO_2$ is also readily available. The last two steps of the synthesis are shown below. Show, with equations, how you would synthesize the isotopically labeled halide.

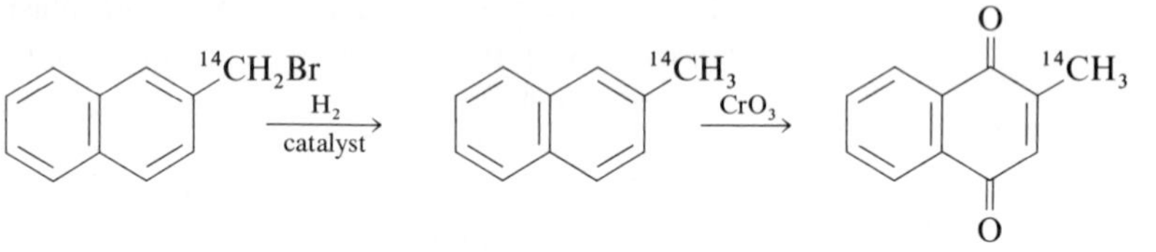
$^{14}CH_2Br$
H_2
catalyst
$^{14}CH_3$
CrO_3
O
$^{14}CH_3$
O

PROBLEM 20.20

The carbon skeleton of 2-methyl-1,4-naphthoquinone has been synthesized by a Diels-Alder reaction between a diene and a quinone. What diene and quinone would you use? What further steps would be necessary to convert the Diels-Alder product to the naphthaquinone?

S U M M A R Y

Free radicals are important reactive intermediates. A free radical is a species with an unpaired electron that reacts as an electrophile, seeking a single electron to complete its octet.

Free radicals are formed in initiation reactions under conditions that cause the homolytic cleavage of bonds. Free radicals abstract hydrogen or halogen atoms to create new radicals in propagation steps. If two radicals encounter each other, a relatively rare event, they combine in termination steps. Initiation, propagation, and termination steps are typical of chain reactions.

Alkanes are converted to alkyl halides by free-radical halogenation reactions. Whether such a reaction has synthetic usefulness depends on the structure of the alkane, whether chlorine or bromine is used as the halogen, and reaction conditions. Bromine is more selective in its reactions than chlorine is. The ease with which different types of hydrogen atoms are abstracted reflects the relative stabilities of the radicals formed.

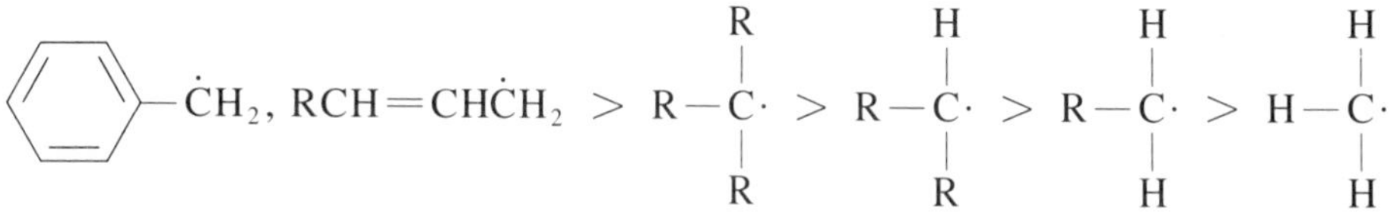

Synthetically useful free-radical halogenations are carried out at allylic and benzylic positions using *N*-bromosuccinimide.

Free radicals, as electrophiles, add to alkenes, attacking the least substituted side of the double bond. The carbon radical that forms abstracts a hydrogen or halogen atom from a molecule of another reagent. If the reagent is hydrogen bromide, this results in an anti-Markovnikov orientation for the addition reaction. Free-radical addition to alkenes gives high yields of products with new carbon-carbon, carbon-sulfur, or carbon-silicon bonds. If the carbon radical generated in the initiation step is in the same molecule as a double bond, an intramolecular free-radical addition gives a cyclic compound, usually a five-membered ring. Chain reactions propagated by the addition of carbon radicals to alkenes result in polymerization, an industrially important process.

Oxygen is a diradical. All organic compounds are slowly oxidized by oxygen in the presence of traces of free-radical initiators or light to give hydroperoxides. These autooxidation reactions are responsible for the deterioration of certain substances, such as rubber and oils, when exposed to air. Oxidation reactions often involve radical intermediates; examples are the oxidation of phenols to quinones and the oxidative coupling reactions of phenols. Hindered phenols such as BHT and BHA are inhibitors of chain reactions and thus serve as antioxidants.

Chemical transformations that involve free-radical intermediates are summarized in Tables 20.1, 20.2, and 20.3.

TABLE 20.1 Halogenation Reactions

Compound Reacting	Reagent	Initiation Conditions	Radical Intermediate(s)	Product
—C—H tertiary > secondary > primary > methane *fastest* *slowest*	Cl_2 or Br_2	$h\nu$ or Δ	:Cl· or :Br· —C·	—C—Cl or —C—Br
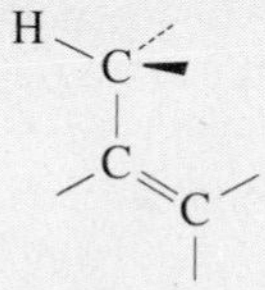	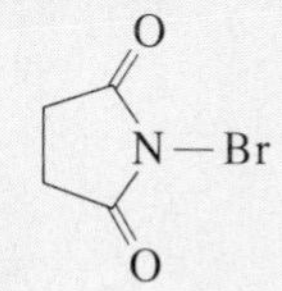	$h\nu$, Δ peroxides	:Br· 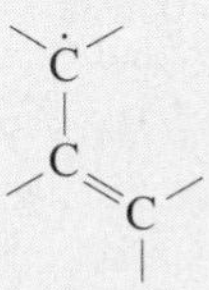	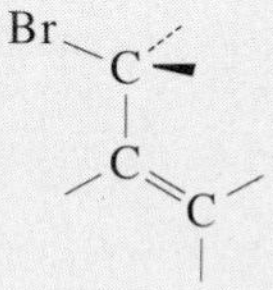

(*Continued*)

TABLE 20.1 *(Continued)*

Compound Reacting	Reagent	Initiation Conditions	Radical Intermediate(s)	Product
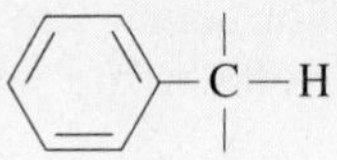	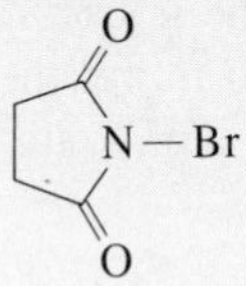	$h\nu$, Δ peroxides	:Br· 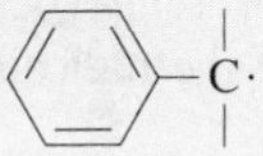	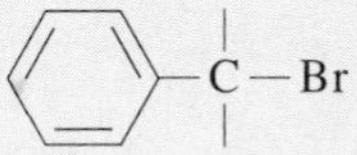
	Cl_2 or Br_2	$h\nu$ or Δ	:Cl· or :Br· 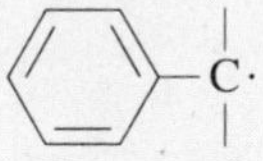	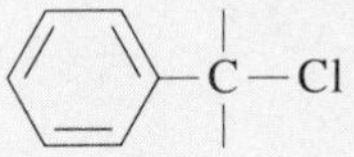 or 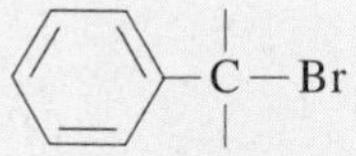

TABLE 20.2 Addition Reactions to Alkenes

Alkene	Reagent	Initiation Conditions	Radical Intermediate(s)	Product(s)
$R-C(R')=CH_2$	HBr	peroxides	:Br· and $R-\dot{C}(R')-CH_2Br$	$R-C(R')(H)-CH_2Br$
	Cl_3CBr or CCl_4	peroxides, $h\nu$	$Cl_3C\cdot$ and $R-\dot{C}(R')-CH_2CCl_3$	$R-C(R')(X)-CH_2CCl_3$ X = Br or Cl
	$R''SH$	Δ	$R''\ddot{S}\cdot$ and $R-\dot{C}(R')-CH_2SR''$	$R-C(R')(H)-CH_2SR''$
	R''_3SiH	peroxides, Δ	$R''_3Si\cdot$ and $R-\dot{C}(R')-CH_2SiR''_3$	$R-C(R')(H)-CH_2SiR''_3$
$R-C(R')=CH_2$	peroxides or $CH_3C(CN)(CH_3)N=NC(CN)(CH_3)CH_3$		$R''CH_2C(R')(R)\cdot$ and $R''CH_2C(R')(R)CH_2C(R')(R)\cdot$	$R''-[CH_2C(R')(R)]_n-R'''$
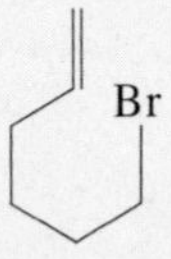	(intramolecular reaction)	$CH_3C(CN)(CH_3)N=NC(CN)(CH_3)CH_3$ and $(CH_3CH_2CH_2CH_2)_3SnH$		$CH_3C(CN)(CH_3)\cdot$ and $(CH_3CH_2CH_2)_3Sn\cdot$ and 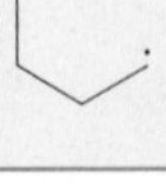 and 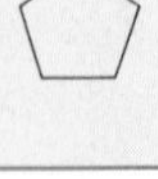	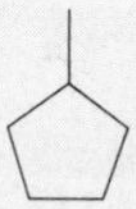 and 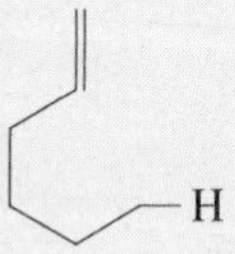

TABLE 20.3 Oxidation Reactions

Compound Reacting	Reagent(s)	Radical Intermediate(s)	Product
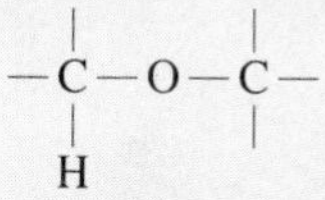	O_2, other R·, metal ion	ROO· and 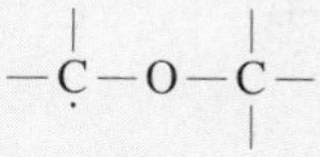	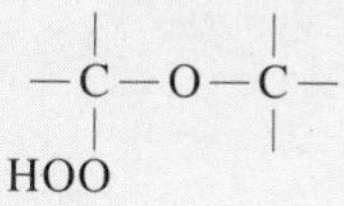
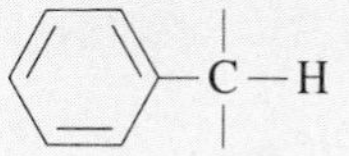	O_2, other R·, metal ion	ROO· and 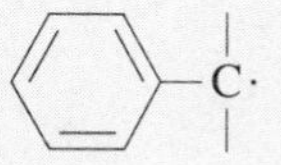	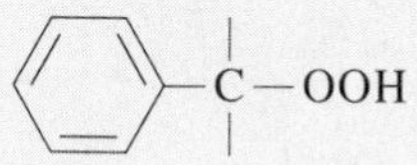
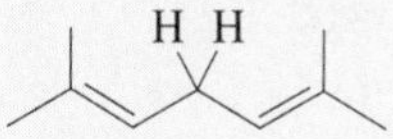	O_2, other R·	ROO· and 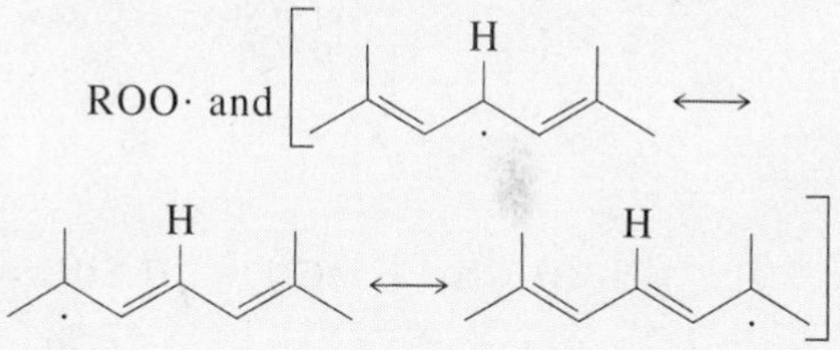	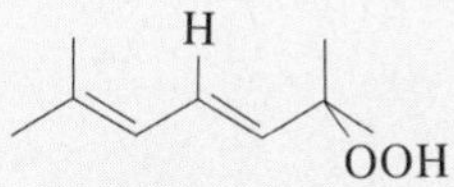
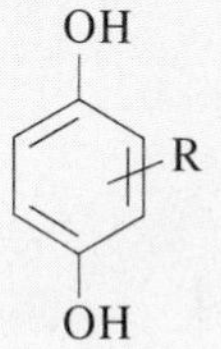	Ag^+, OH^-	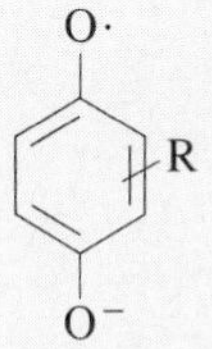	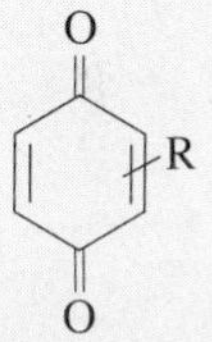
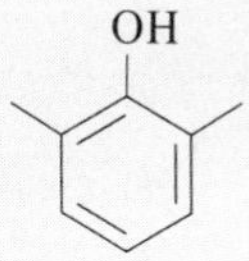	$FeCl_3$	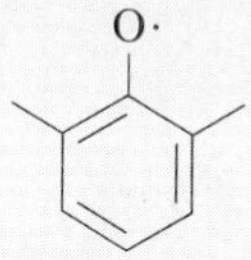	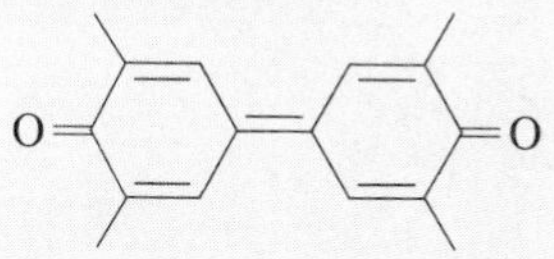

ADDITIONAL PROBLEMS

20.21 Complete the following equations, showing the product(s) you expect. If the stereochemistry is known for a reaction, make sure your answer shows it.

(a) $CH_3CH{=}CHCH{=}CHCH_3 \xrightarrow[\text{carbon tetrachloride}]{Br_2\text{ (1 molar equivalent)}}$

(b) 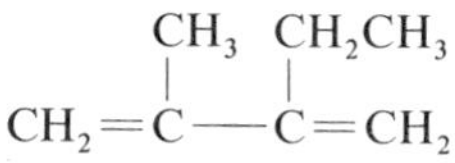 $+$ NBS $\xrightarrow[\text{carbon tetrachloride}]{}$

(c) 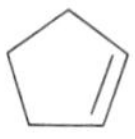 $\xrightarrow[\text{carbon tetrachloride}]{\text{NBS}}$

(d) 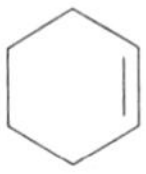 $\xrightarrow{O_2}$

(e) $CH_3CH_2CH_3 \xrightarrow[h\nu]{Cl_2}$

(f) 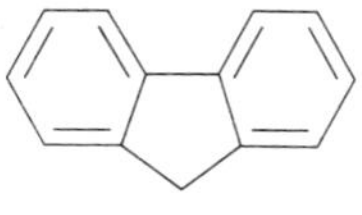 $\xrightarrow[\text{carbon tetrachloride}, \Delta]{\text{NBS}}$

(g) 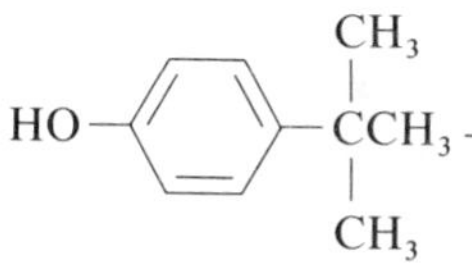 $\xrightarrow[\Delta]{O_2}$

(h) 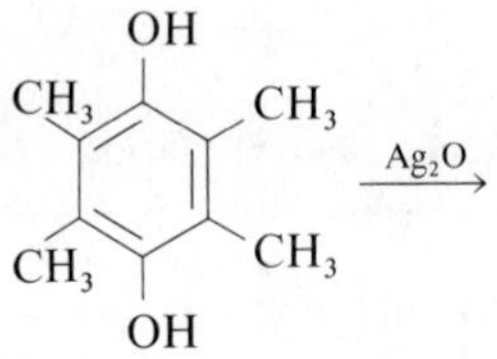

(i) 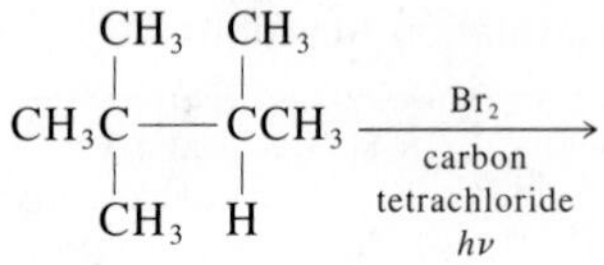

(j) 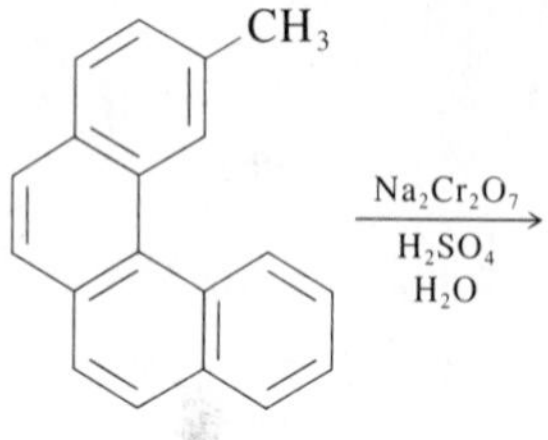

(k) 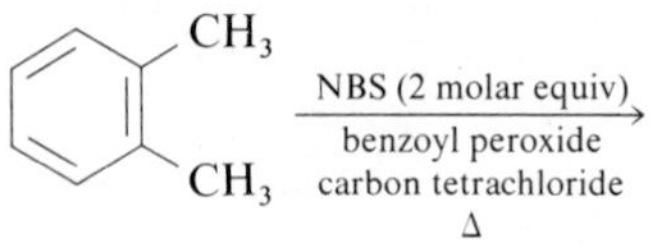

(l) $CH_3CH_2CH_2CH{=}CH_2 + Cl_3SiH \xrightarrow[280\ °C]{}$

(m) 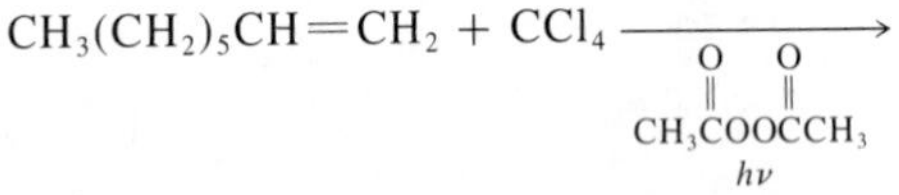

(n) 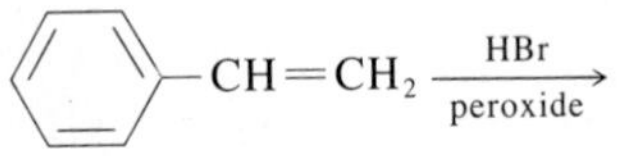

(o) 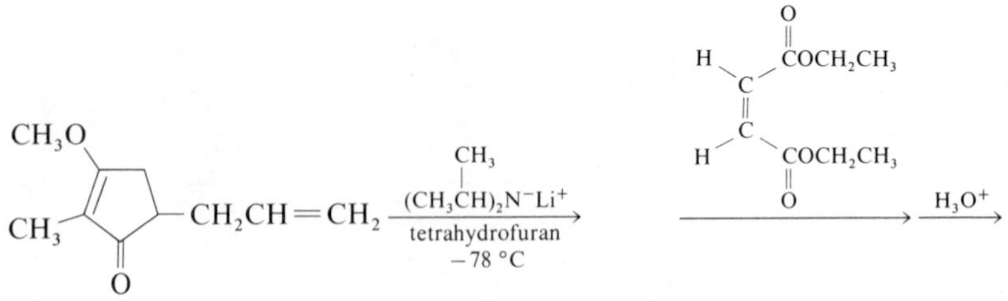

(p) $CH_3CH_2CH_2CH_3 \xrightarrow[h\nu]{Cl_2}$

(q) 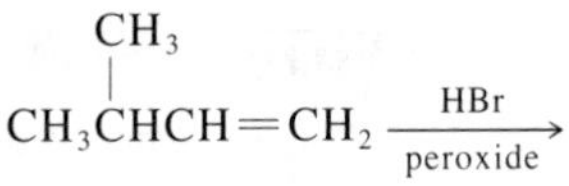

(r) $CH_3CH_2SH + CH_2{=}CHOCH_2CH_3 \xrightarrow[\Delta]{}$

(s) $BrCH_2CH{=}CH_2 \xrightarrow[O_2]{HBr}$

(t) 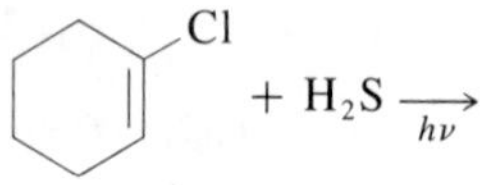

(u) 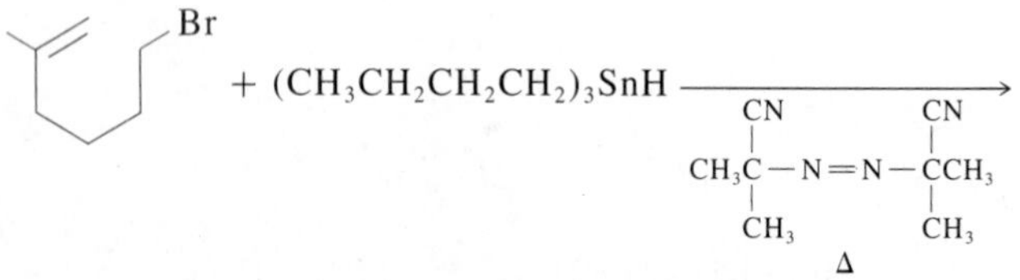

(v) 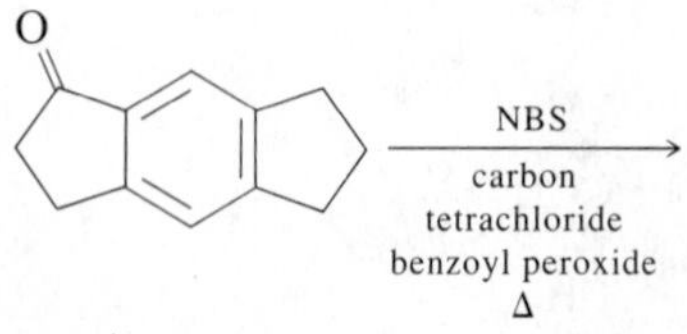

20.22 The yellow pigment of cottonseed, gossypol, has the structure shown below. There have been reports from the People's Republic of China that this compound acts as a male contraceptive. Suggest how 5-isopropyl-6,7-dimethoxy-3-methyl-1-naphthol could be used to create the backbone of gossypol.

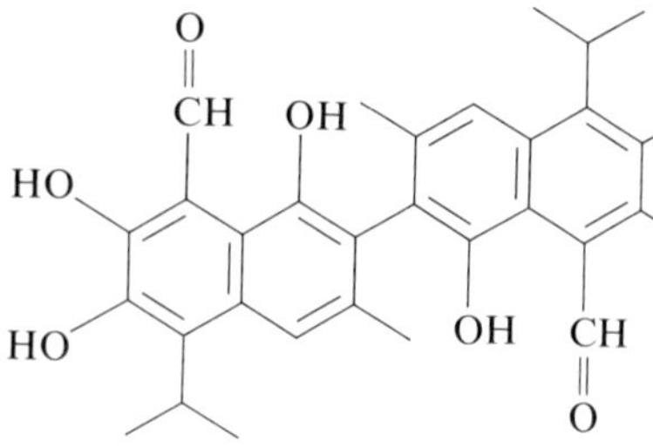

O
CH
OH
HO
HO
OH
OH
OH
OH
CH
O

gossypol

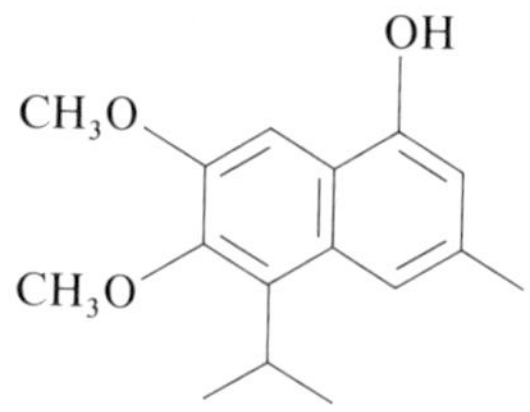

OH
CH3O
CH3O

5-isopropyl-6,7-dimethoxy-3-methyl-1-naphthol

20.23 Propose a mechanism that accounts for the formation of both of the products of the following reaction.

$$CH_3SiCl_2{-}H + CH_3CH{=}CH_2 \xrightarrow{\Delta} \underset{\text{major product}}{CH_3SiCl_2CH_2CH_2CH_3} + \underset{\text{minor product}}{CH_3SiCl_2CH_2CH_2CH_2CH_2CH_2CH_3}$$

20.24 The following reactions take place.

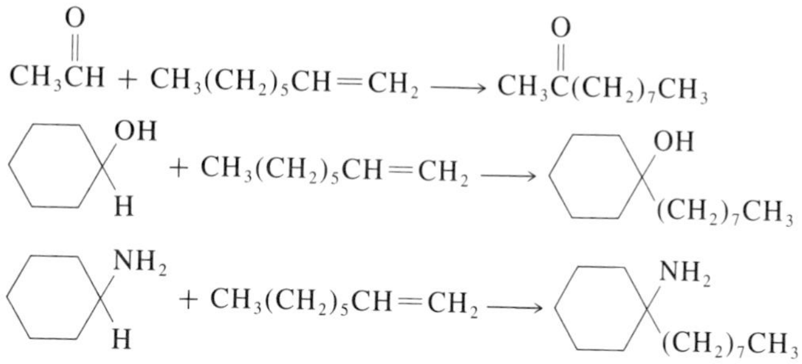

O
CH3CH + CH3(CH2)5CH=CH2
O
CH3C(CH2)7CH3
OH
H
+ CH3(CH2)5CH=CH2
OH
(CH2)7CH3
NH2
H
+ CH3(CH2)5CH=CH2
NH2
(CH2)7CH3

What conditions do you think are necessary for these reactions? Predict whether *tert*-butyl alcohol will react with 1-octene under these conditions.

20.25 How would you carry out each of the following transformations?

(a)

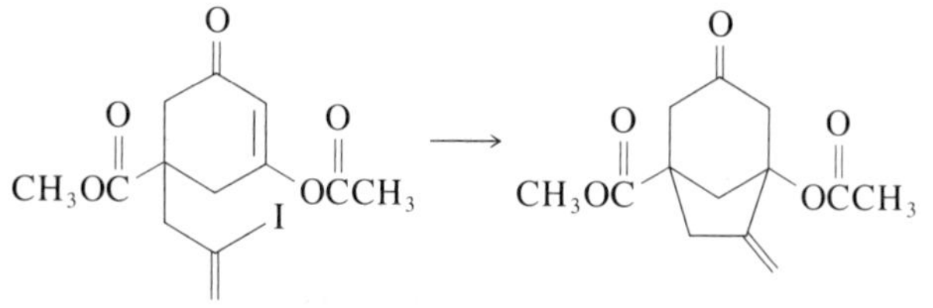

O
O
CH3OC
O
OCCH3
I
O
O
CH3OC
O
OCCH3

(b)

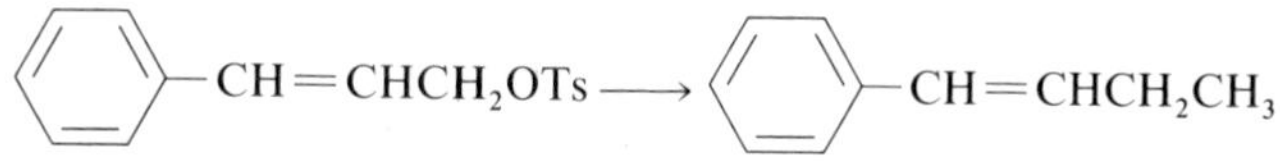

CH=CHCH2OTs
CH=CHCH2CH3

(c) 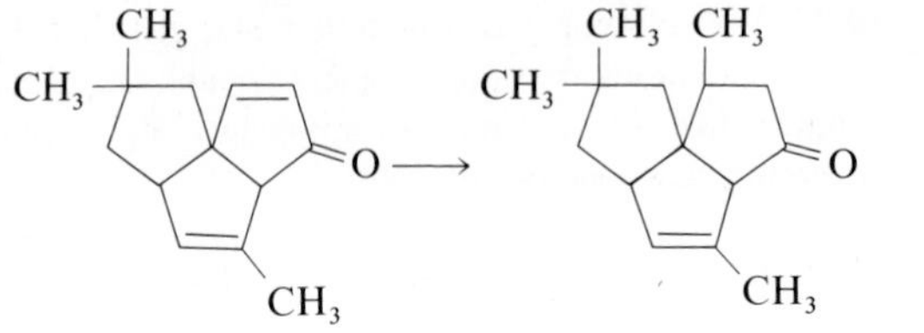

(d) 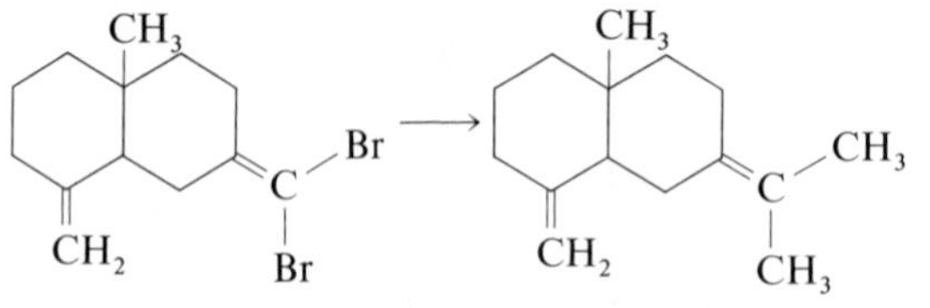

(e) 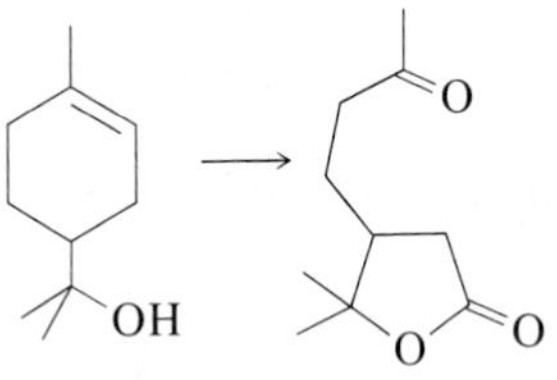

20.26 The following reactions were used in the synthesis of a prostaglandin. Supply the reagents that are necessary for each transformation.

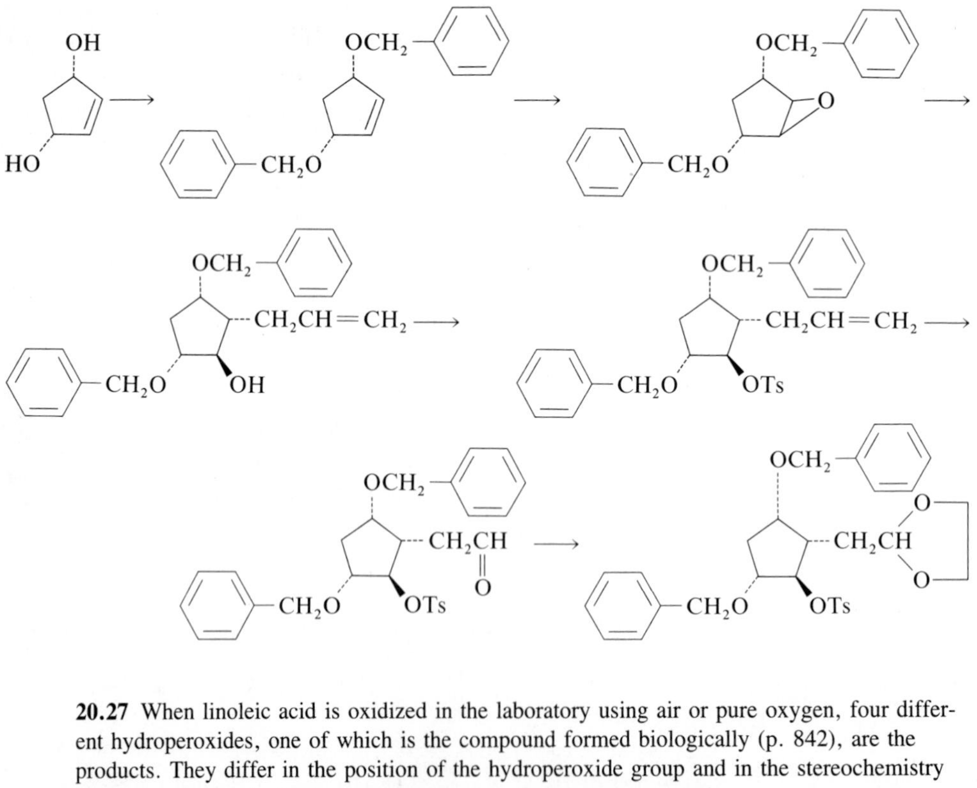

20.27 When linoleic acid is oxidized in the laboratory using air or pure oxygen, four different hydroperoxides, one of which is the compound formed biologically (p. 842), are the products. They differ in the position of the hydroperoxide group and in the stereochemistry of the double bonds. Write structures for the other three compounds.

20.28 Use the bond energies shown in Tables 2.4 and 2.5 (pp. 60, 61) to calculate the enthalpy of reaction, ΔH_r, for each of the following steps of the general mechanism proposed for the free-radical addition of hydrogen halide to propene, when X is Cl, Br, and I. It takes 59 kcal/mol to break just the π bond in an alkene.

(1) $CH_3CH{=}CH_2 + :\ddot{\underset{\cdot\cdot}{X}}\cdot \longrightarrow CH_3\dot{C}HCH_2{-}\ddot{\underset{\cdot\cdot}{X}}:$

(2) $CH_3\dot{C}HCH_2\ddot{\underset{\cdot\cdot}{X}}: + H{-}\ddot{\underset{\cdot\cdot}{X}}: \longrightarrow CH_3\underset{\displaystyle H}{\underset{|}{C}}HCH_2X + :\ddot{\underset{\cdot\cdot}{X}}\cdot$

How do your calculations explain why the peroxide effect occurs only with hydrogen bromide?

20.29 The following compounds were used in studies of the cyclization reactions of radicals.

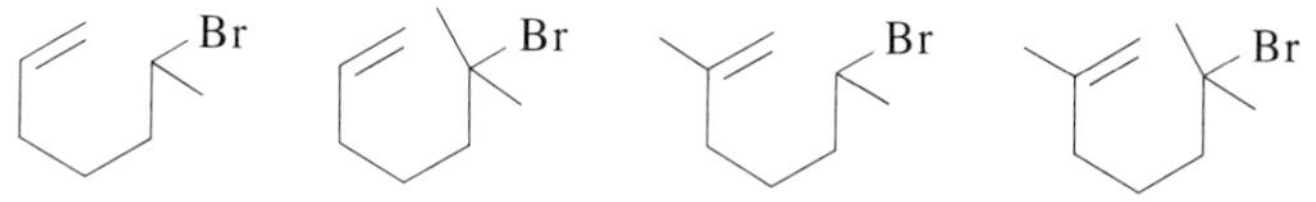

Br
Br
Br
Br

Assuming that 5-bromo-1-pentene and 6-methyl-6-hepten-2-one are available as starting materials, how would you synthesize the bromoalkenes shown above?

20.30 The starting material used in the synthesis of silphiperfol-6-ene (Problem 20.13, p. 841) was synthesized in the following steps. Supply reagents for each of the transformations.

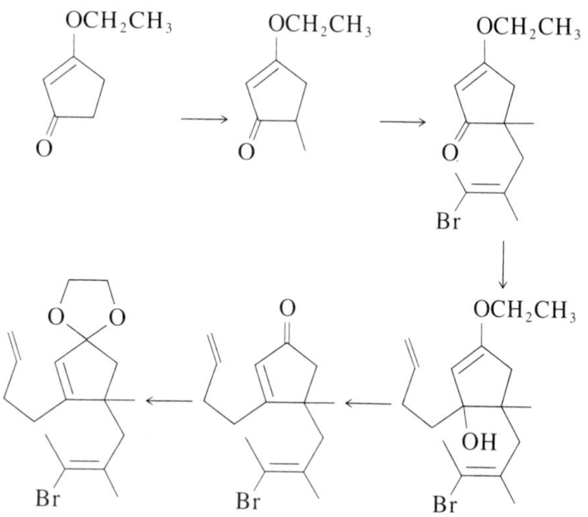

OCH2CH3
O
OCH2CH3
O
OCH2CH3
O
Br
OCH2CH3
OH
Br
O
Br
O
O
Br

20.31 Cholesteryl acetate is converted in the laboratory into the acetic ester of 7-dehydrocholesterol, a compound that is converted by sunlight to Vitamin D_3 in the skin (p. 1218). 7-Dehydrocholesterol has a conjugated system in the B ring of the steroid nucleus, with double bonds between carbon atoms 5 and 6 and carbon atoms 7 and 8 (p. 709). How would you convert cholesterol to 7-dehydrocholesteryl acetate?

21

Mass Spectrometry

A • L O O K • A H E A D

Mass spectrometry differs from the other types of spectroscopy discussed so far in that a mass spectrum is not a record of the energy absorbed by a molecule in going from one energy level to another. A mass spectrum is a record of the exact masses of a series of ions that are formed by fragmentation of a molecular species that is created by the collision of a molecule with a high-energy particle, usually an electron. The collision knocks an electron out of the molecule, giving rise to a radical-cation (a species with a positive charge and an unpaired electron) called the molecular ion.

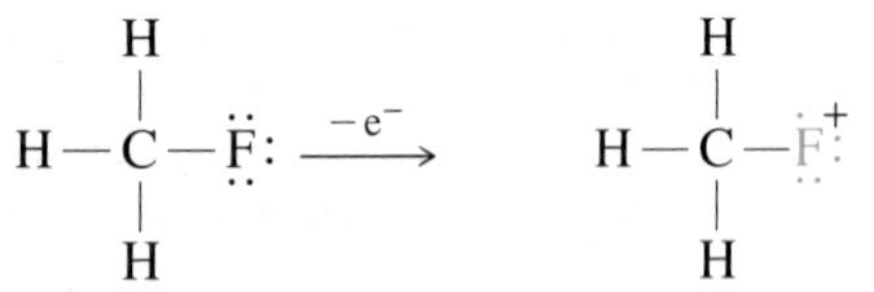

fluoromethane

radical-cation formed by loss of electron; the molecular ion of fluoromethane m/z 34

The molecular ion is unstable and breaks up to give a number of other species, such as smaller radical-cations, carbocations, radicals, and uncharged molecules. The positively charged species are separated according to their mass-to-charge ratios, m/z, and their masses are recorded to give the spectrum. The major peaks in the spectrum of fluoromethane are at m/z 34 (the molecular ion), 33, and 15. The peak at m/z 33 corresponds to the loss of a hydrogen atom from the molecular ion.

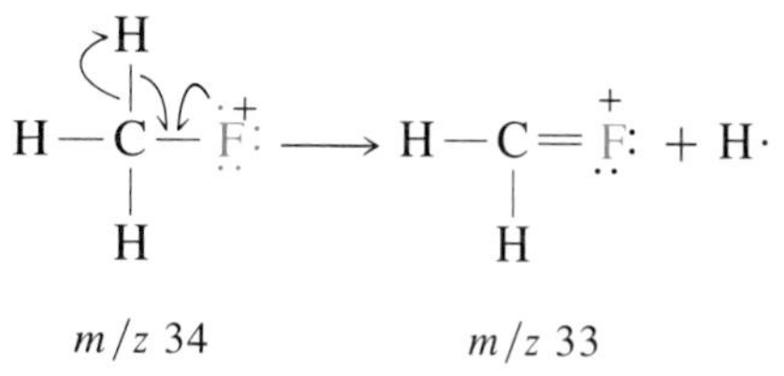

Loss of a fluorine atom from the molecular ion gives the methyl cation, which has m/z 15.

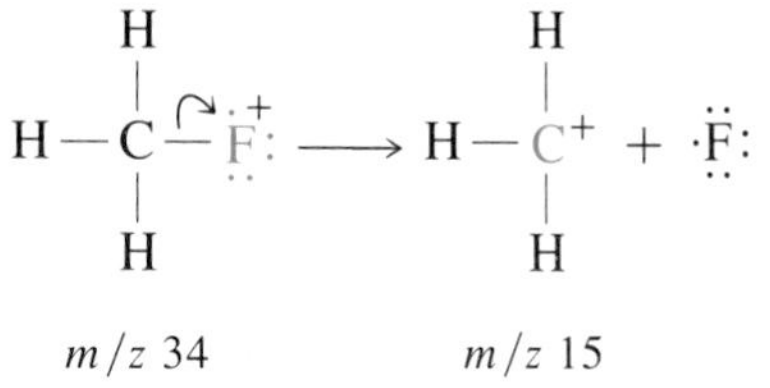

Only the positively charged species are recorded in the spectrum.

This chapter will show how mass spectrometry is used to determine the structure of organic compounds.

21.1 THE MASS SPECTRUM

A **mass spectrum** is obtained by injecting a very small sample (a millionth of a gram, 10^{-6} g, will do) of a compound into a mass spectrometer (Figure 21.1), where it is subjected to bombardment by a high-energy electron beam. The first reaction that occurs in the mass spectrometer is the formation of a **molecular ion**

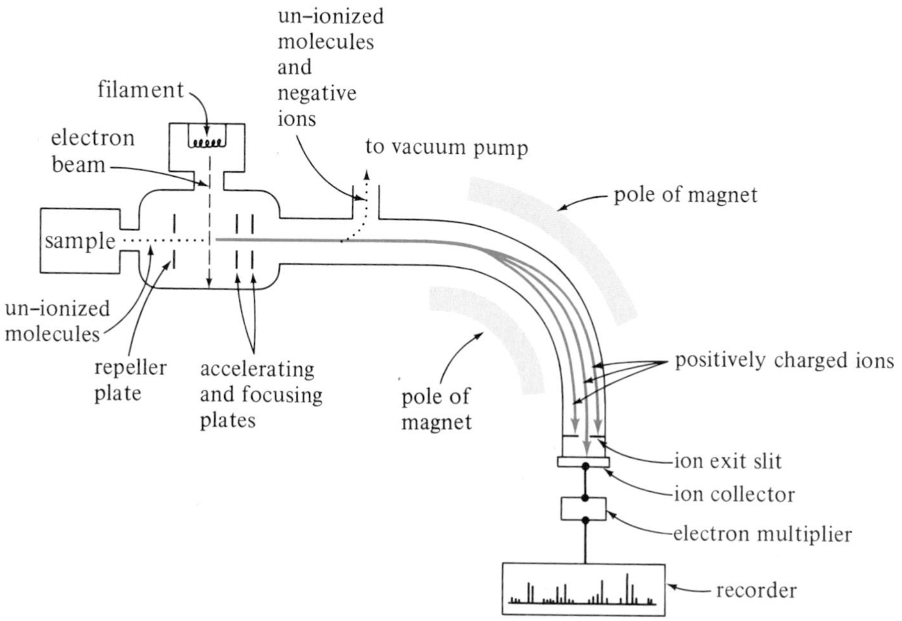
un-ionized molecules and negative ions
filament
electron beam
to vacuum pump
pole of magnet
sample
un-ionized molecules
repeller plate
accelerating and focusing plates
pole of magnet
positively charged ions
ion exit slit
ion collector
electron multiplier
recorder

FIGURE 21.1 Schematic diagram of a typical mass spectrometer.

by loss of an electron from the molecule. This electron may come from any bond in the molecule. A nonbonding electron is in a higher-energy molecular orbital than an electron in a π bond, which in turn is in a higher-energy molecular orbital than an electron in a σ bond (Figure 17.6, p. 715). Therefore, a nonbonding electron is removed more easily than an electron of a π bond, which is lost more easily than an electron of a σ bond. For example, one of the nonbonding electrons of the fluorine atom is removed in forming the molecular ion of fluoromethane (p. 864).

Fragmentation of the molecular ion gives other radical-cations, carbocations, and neutral molecules. The positively charged species are separated according to their **mass-to-charge ratios, m/z,** by a variety of techniques. Most ions created in a mass spectrometer have a single positive charge; therefore, the separation is essentially done according to mass. One way this is done is that the ions of different masses are accelerated in an electric field and are then deflected by different amounts as they travel through a magnetic field, as shown in Figure 21.1. Thus, ions of different masses impinge separately upon an electron multiplier and are recorded on a chart. The calibration of the instrument allows not only the masses of the ions to be recorded but the relative number of each kind.

The mass spectra of acetone and acetaldehyde, plotted from the original recordings as bar graphs, are shown in Figure 21.2. Vertical lines appear in the spectra above numbers on the horizontal axis that correspond to the masses of the ions that are formed when molecules of these compounds are bombarded with a high-energy

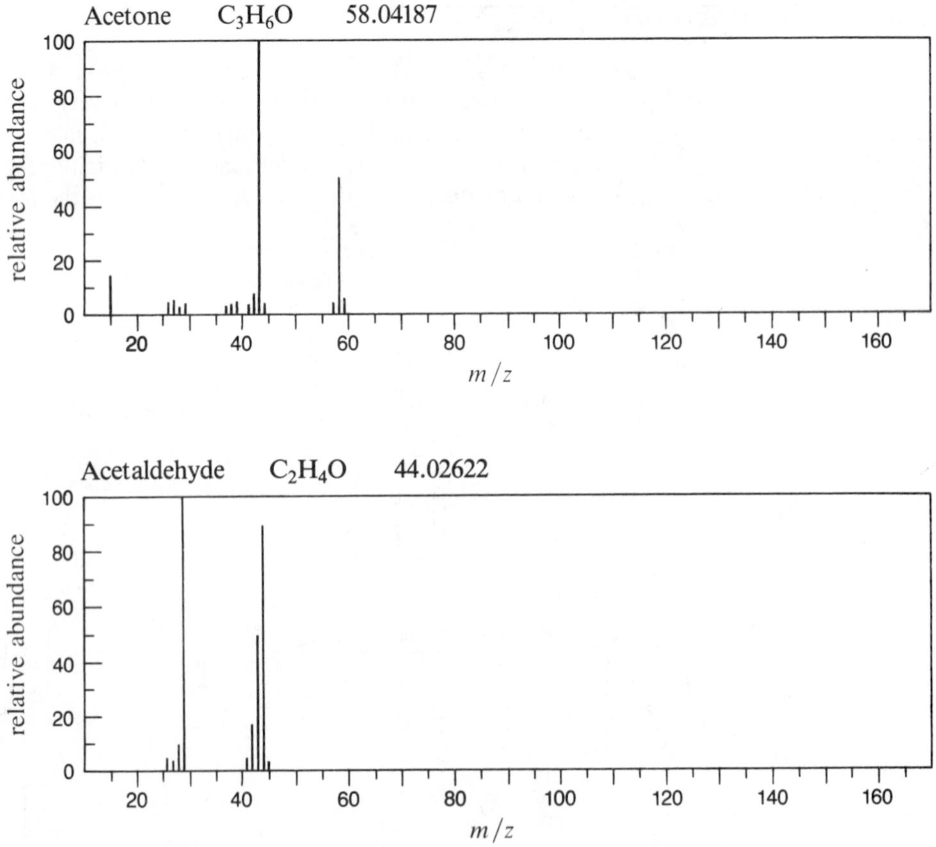
Acetone C3H6O 58.04187
relative abundance
0
20
40
60
80
100
20
40
60
80
100
120
140
160
m/z
Acetaldehyde C2H4O 44.02622
relative abundance
0
20
40
60
80
100
20
40
60
80
100
120
140
160
m/z

FIGURE 21.2 Mass spectra of acetone and acetaldehyde. (Adapted from Stenhagen et al. Reprinted by permission.)

beam of electrons. The first reaction for both acetone and acetaldehyde is the loss of a nonbonding electron from the oxygen atom, to give the molecular ions, which are radical-cations.

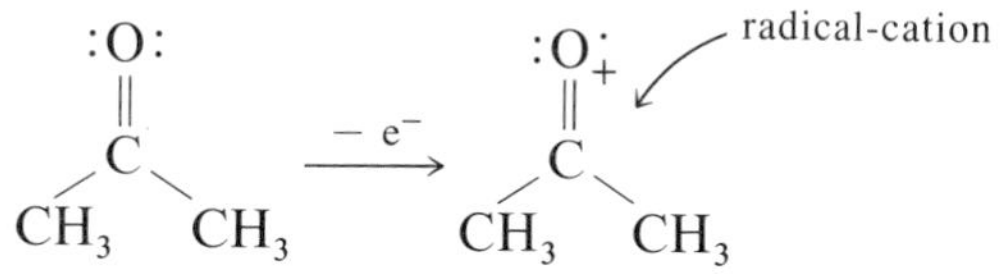

$$
\underset{\substack{\text{:O:} \\ \| \\ \text{C} \\ \mathrm{CH_3} \quad \mathrm{CH_3}}}{} \xrightarrow{-\,e^-} \underset{\substack{\text{:O:}^{+} \\ \| \\ \text{C} \\ \mathrm{CH_3} \quad \mathrm{CH_3}}}{\text{radical-cation}}
$$

molecular ion of acetone
formed by loss of a
nonbonding electron
from oxygen
$C_3H_6O^{+\cdot}$, m/z 58

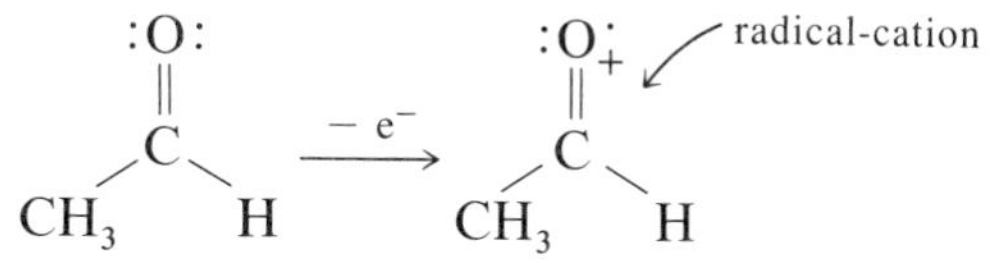

$$
\underset{\substack{\text{:O:} \\ \| \\ \text{C} \\ \mathrm{CH_3} \quad \mathrm{H}}}{} \xrightarrow{-\,e^-} \underset{\substack{\text{:O:}^{+} \\ \| \\ \text{C} \\ \mathrm{CH_3} \quad \mathrm{H}}}{\text{radical-cation}}
$$

molecular ion of
acetaldehyde
$C_2H_4O^{+\cdot}$, m/z 44

The radical-cations are created in the gas phase with a large amount of excess energy. They are unstable species and fragment in a series of unimolecular reactions. An inspection of the mass spectrum of acetone indicates that the most abundant ion has m/z 43. The most abundant ion in the mass spectrum of a compound is called the **base peak.** The bar graphs are designed so that the base peak always has an intensity of 100%. The abundance of the other ions is shown relative to that of the base peak. Thus, the molecular ion of acetone, m/z 58, has an abundance of approximately 40% of the ion with m/z 43.

The base peak in the spectrum of acetone corresponds to the loss of a fragment with a mass of 15 units ($58 - 43 = 15$). A methyl group, CH_3, has a mass of 15. Thus, it appears that the molecular ion of acetone fragments primarily by the loss of a methyl group.

V I S U A L I Z I N G　　T H E　　R E A C T I O N

Homolytic cleavage of a bond in a radical-cation

$$
\underset{\substack{\text{:O:}^{+} \\ \| \\ \text{C} \\ \mathrm{CH_3} \quad \mathrm{CH_3}}}{} \longrightarrow \mathrm{CH_3-C\equiv\ddot{O}^{+}} \;+\; \cdot\mathrm{CH_3}
$$

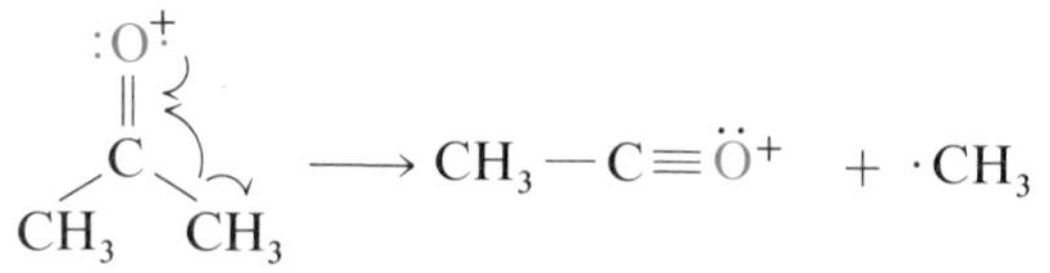

| molecular ion of acetone m/z 58 | $C_2H_3O^{+}$ m/z 43 | methyl radical *not charged, does not appear in the mass spectrum* |

The reaction is shown as a homolytic cleavage (represented by fishhooks) of one of the bonds between the carbonyl group and a methyl group. As a consequence, an acylium ion, an intermediate in Friedel-Crafts reactions of acid chlorides and anhydrides (p. 803), and a methyl radical are formed. The methyl radical is not a charged particle and thus does not appear in the mass spectrum. There is, however, a small peak at m/z 15, which corresponds to the formation of a small amount of methyl cations by an alternative cleavage.

VISUALIZING THE REACTION

Heterolytic cleavage of a bond in a radical-cation

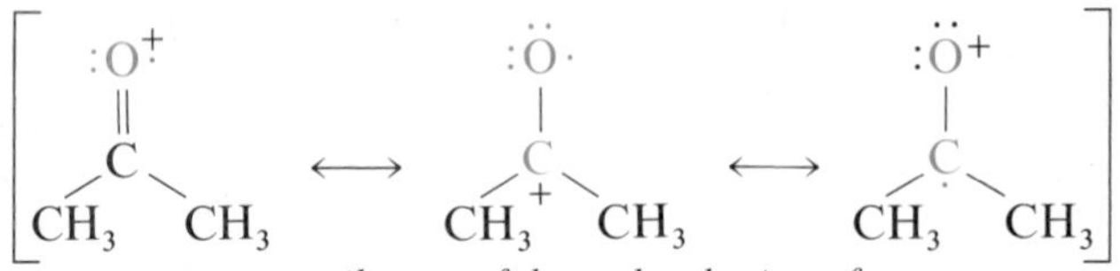

resonance contributors of the molecular ion of acetone

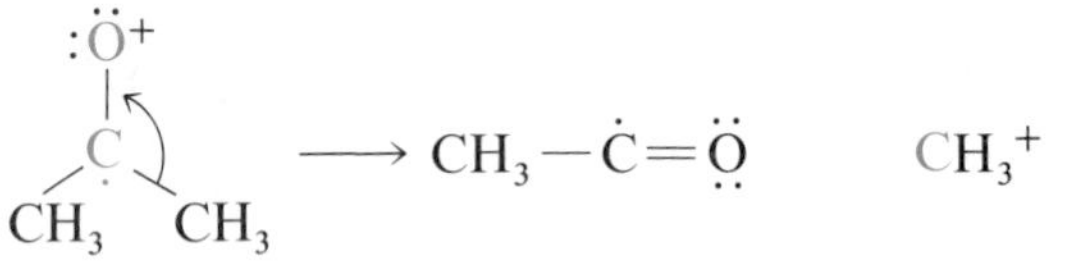

m/z 58 — an acyl radical — methyl cation m/z 15

The relative abundances of the acylium cation (m/z 43, 100%) and methyl cation (m/z 15, 15%) reflect the relative stabilities of the two species. In any case, it is important that the reactions are taking place in the gas phase in species that have much excess energy. Note, too, that in any fragmentation the numbers of unpaired electrons and charges present in the original species must be preserved.

Besides the molecular ion, m/z 44, the mass spectrum of acetaldehyde shows important peaks at m/z 43 and m/z 29 (base peak). These peaks may be assigned to the ions that form on the loss of a hydrogen atom (44 − 43 = 1) and a methyl radical (44 − 29 = 15) from the molecular ion.

VISUALIZING THE REACTION

Homolytic cleavages of bonds in a radical-cation

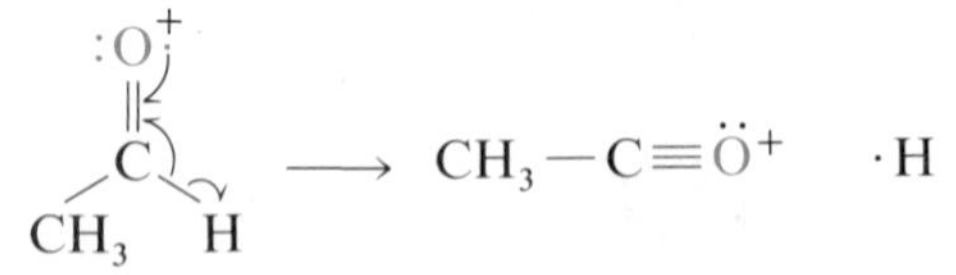

molecular ion of acetaldehyde m/z 44 — acylium ion m/z 43 — hydrogen atom

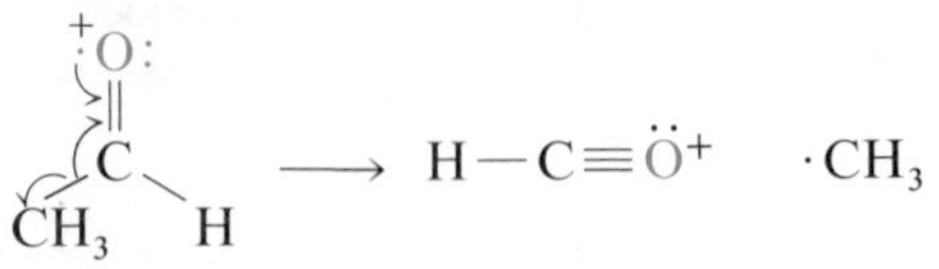

molecular ion of acetaldehyde m/z 44 — acylium ion m/z 29 — methyl radical

The mass spectrum is a record of the cationic species that are formed when a beam of electrons is used to create radical-cations called molecular ions, which then fragment into smaller ions. The masses of these ions may be used in making judgments about the structure of the compound that gives rise to them. The mass spectrum of a compound of any complexity is unique in the pattern of peaks and the relative intensities of those peaks. As such, it can serve as a fingerprint for the compound. Because a very small sample of a compound is sufficient to obtain a mass spectrum, mass spectrometry is a powerful tool for identifying minute quantities of compounds. Mass spectrometers are often connected to gas-phase chromatographs, which separate mixtures efficiently, and computers, which automatically identify the components of mixtures by comparing mass spectral information with stored patterns for known compounds. Such techniques make it possible to obtain information about compounds present at trace levels in complex mixtures, such as physiological fluids or discharges from industrial plants.

PROBLEM 21.1

The mass spectrum in Figure 21.3 is that of an aldehyde. Assign a structure to the aldehyde, and write equations showing the formation of the molecular ion and the ion giving rise to the base peak.

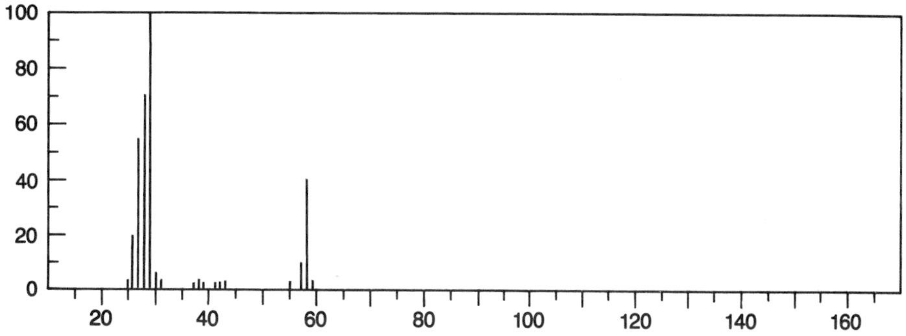

100
80
60
40
20
0
20
40
60
80
100
120
140
160

FIGURE 21.3

(Adapted from Stenhagen et al. Reprinted by permission.)

PROBLEM 21.2

The mass spectrum of a carboxylic acid is shown in Figure 21.4. Assign a structure to the acid, and write equations showing the formation of any ions having an abundance greater than 40%.

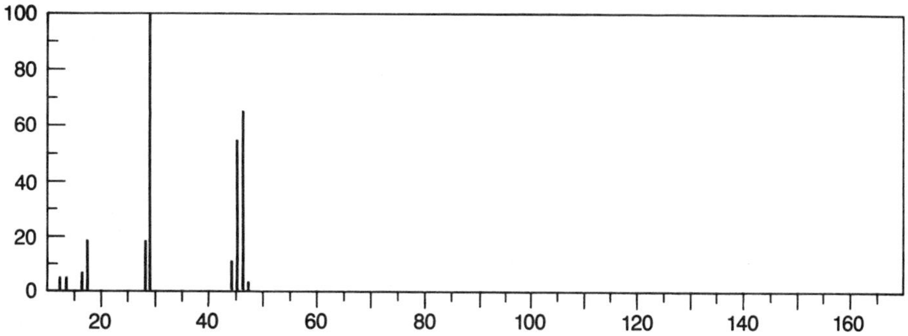

100
80
60
40
20
0
20
40
60
80
100
120
140
160

FIGURE 21.4

(Adapted from Stenhagen et al. Reprinted by permission.)

PROBLEM 21.3

Two mass spectra are shown in Figure 21.5. One is that of 2-pentanone; the other is of 3-pentanone. Which is which? Write equations showing how you made your decision.

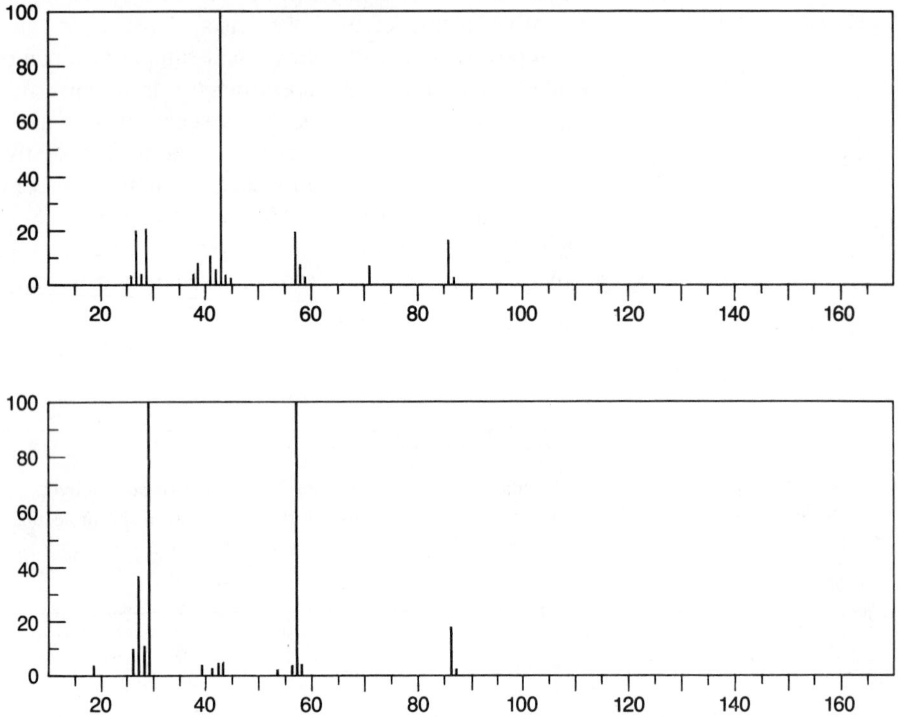
100
80
60
40
20
0
20
40
60
80
100
120
140
160

FIGURE 21.5

(Adapted from Stenhagen et al. Reprinted by permission.)

PROBLEM 21.4

The mass spectrum of 2,2-dimethylpropanal is given in Figure 21.6. Assign a structure to the base peak in the spectrum, and write an equation for the formation of that species from the molecular ion. What would be the comparable species from the fragmentation of the molecular ion of acetaldehyde? Is such a species evident in the mass spectrum of acetaldehyde (Figure 21.2, p. 866)? How do you explain the difference in the reactions of the molecular ions for these two compounds?

2,2-Dimethylpropanal $C_5H_{10}O$ 86.07317

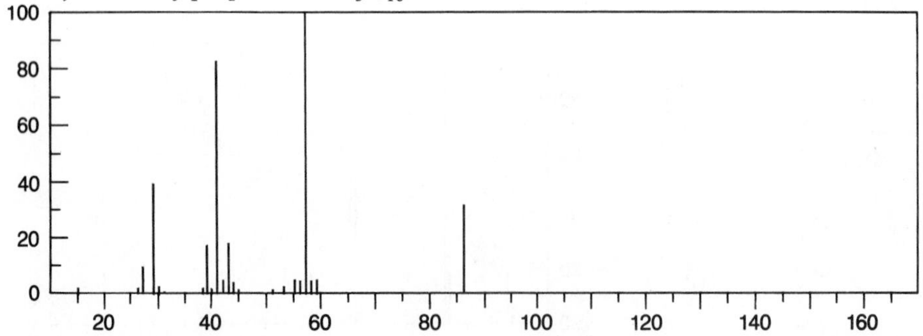
100
80
60
40
20
0
20
40
60
80
100
120
140
160

FIGURE 21.6

(Adapted from Stenhagen et al. Reprinted by permission.)

In the mass spectra of acetone and acetaldehyde (Figure 21.2, p. 866), there are small peaks next to the peaks assigned to the molecular ions, at m/z 59 and m/z 45. These peaks arise from the presence of isotopes of carbon, hydrogen, and oxygen in these compounds. The common isotopes of elements that are of importance in organic chemistry, their atomic weights rounded off to the nearest mass unit, and their abundance relative to that of the most abundant isotope of that element (shown as 100%) are given in Table 21.1. Thus, any compound that contains a number of carbon, hydrogen, and oxygen atoms will have some peaks in its mass spectrum at mass values one or two (or more) units above the mass calculated from its molecular formula using the atomic weights of the most common isotopes. For example, a compound containing ten carbon atoms is expected to have an ion having an intensity of about 11% (10 $\times$ 1.08%) of that of the peak for its molecular ion (M) at a mass of (M + 1). This peak would correspond to all of the species in which one of the carbon atoms in the molecule is ^{13}C instead of ^{12}C. The intensity of this peak, relative to the intensity of the molecular ion, would increase with increasing numbers of hydrogen and nitrogen atoms. Large numbers of oxygen atoms would increase the intensity of the (M + 2) peak, arising from molecular species in which ^{18}O instead of ^{16}O is found. The (M + 2) peak would also arise from species in which both ^{13}C and ^{2}H are found. The probability that there are such molecules increases with the total number of carbon and hydrogen atoms in the compound.

The most distinctive (M + 2) peaks in mass spectra arise when the halogen bromine or chlorine is present in a molecule. Bromine has two isotopes, ^{79}Br and ^{81}Br, of nearly equal abundance. Thus, in a sample of a compound containing a bromine atom, half of the molecules contain ^{79}Br and half ^{81}Br. Every species containing bromine will appear as a pair of peaks separated by two mass units. The spectrum of methyl bromide is shown in Figure 21.7 on the following page. The molecular ion for $CH_3{}^{79}Br$ appears at m/z 94 and that for $CH_3{}^{81}Br$ at m/z 96. These two peaks have roughly the same intensity, reflecting the almost equal abundance of the two isotopes of bromine in nature. Peaks corresponding to $^{79}Br^+$ and $^{81}Br^+$ are seen at m/z 79 and m/z 81. The base peak in the spectrum is at m/z 15, corresponding to the methyl cation.

TABLE 21.1 Natural Isotopic Abundance of Some Elements

	Most Common Isotopes			
Element	**Mass**	**%**	**Mass**	**%**
H	1	100	2	0.016
C	12	100	13	1.08
N	14	100	15	0.36
O	16	100	18	0.20
F	19	100	—	—
Cl	35	100	37	32.5
Br	79	100	81	98.0
I	127	100	—	—

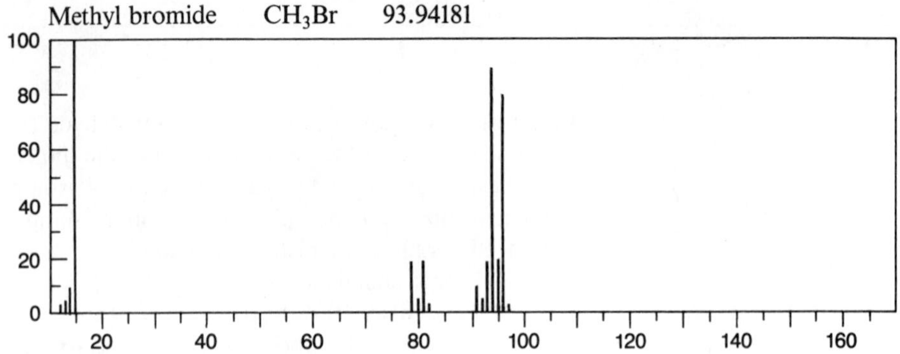
Methyl bromide CH3Br 93.94181
100
80
60
40
20
0
20
40
60
80
100
120
140
160

FIGURE 21.7 Mass spectrum of methyl bromide. (Adapted from Stenhagen et al. Reprinted by permission.)

VISUALIZING THE REACTION

Isotopic ions from methyl bromide

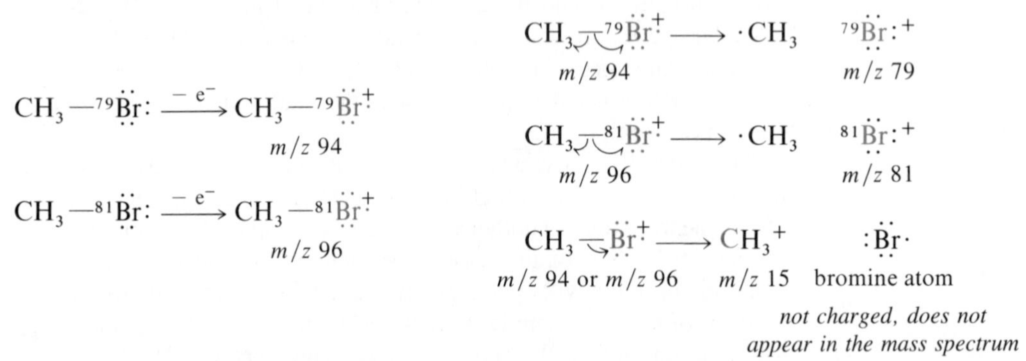
m/z 94
m/z 96
m/z 94
m/z 79
m/z 96
m/z 81
m/z 94 or m/z 96
m/z 15
bromine atom
not charged, does not appear in the mass spectrum

It is important to understand that the mass spectrum shows ions having different isotopic composition as individual peaks. The mass spectrum of a compound contains peaks corresponding to individual species containing all possible combinations of isotopes. The mass of a species as determined by mass spectrometry is *not* the same as the molecular weight calculated by using the average atomic weights found in the periodic table.

Compounds containing a chlorine atom show an (M + 2) peak having an intensity about a third of that of the peak for the molecular ion. The spectrum of methyl chloride shows this feature (Figure 21.8). The molecular ion of methyl chloride with the composition $CH_3{}^{35}Cl$ has its peak at m/z 50. A peak having an intensity of about a third of the one at m/z 50 appears at m/z 52 and corresponds to the molecular ion for the composition $CH_3{}^{37}Cl$. The chief fragmentation seen for these molecular ions is the loss of a chlorine atom giving rise to a methyl cation, m/z 15.

PROBLEM 21.5

Write equations showing the formation of the molecular ions for methyl chloride and their fragmentation.

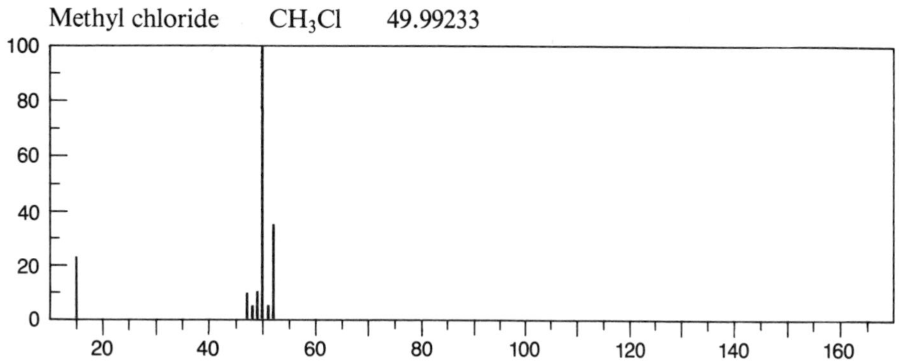
Methyl chloride CH3Cl 49.99233
100
80
60
40
20
0
20
40
60
80
100
120
140
160

FIGURE 21.8 Mass spectrum of methyl chloride. (Adapted from Stenhagen et al. Reprinted by permission.)

Many of the spectra used as examples in this chapter give the molecular weight of the compound to five decimal places. This is the accuracy with which masses can be determined in modern high-resolution mass spectrometers. At that accuracy the mass of the molecular ion can be used to determine the molecular formula of the compound. For example, acetone, C_3H_6O ($M^{+\cdot}$58.04187), can be distinguished from butane, C_4H_{10} ($M^{+\cdot}$58.07825), even though both have a mass of 58 amu, rounded off to the nearest mass unit. In fact, the elemental composition of each ion in a mass spectrum is also determined in the same way.

Not all compounds give molecular ions that are easily detectable in a mass spectrometer. The molecular ions formed by some compounds are so unstable that they fragment before they reach the detector. Alcohols and amines are among the compounds with very weak molecular ion peaks. The molecular ion from butylamine, for example, is so unstable that only a very small number of these ions survive to be recorded by the detector; careful inspection of the spectrum (Figure 21.11, p. 874) is necessary to see the peak (m/z 73). The molecular ion for *tert*-butyl alcohol does not appear at all on its spectrum (Figure 21.11).

PROBLEM 21.6

The mass spectrum of a haloalkane is given in Figure 21.9. Assign a structure to the compound.

FIGURE 21.9

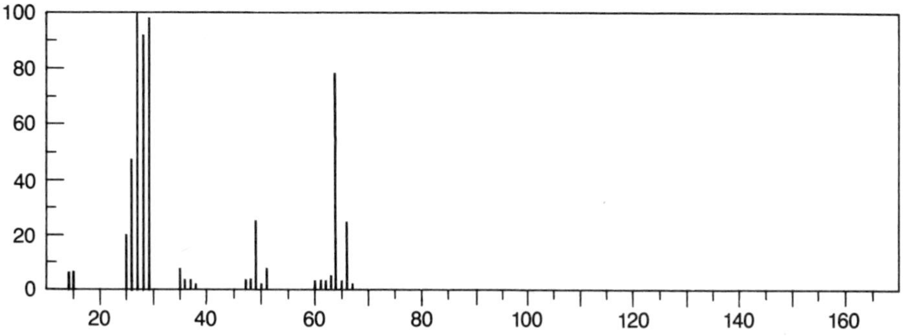
100
80
60
40
20
0
20
40
60
80
100
120
140
160

(Adapted from Stenhagen et al. Reprinted by permission.)

PROBLEM 21.7

The compound for which a mass spectrum appears in Figure 21.10 contains only one carbon atom and two types of halogen. After consulting Table 21.1 (p. 871), assign a structure to the compound, and account for the peaks that appear in the spectrum.

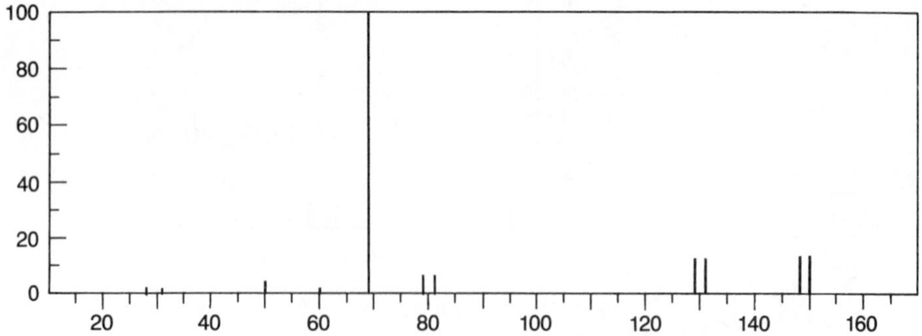
100
80
60
40
20
0
20
40
60
80
100
120
140
160

FIGURE 21.10

(Adapted from Stenhagen et al. Reprinted by permission.)

21.3 IMPORTANT FRAGMENTATION PATHWAYS

Molecular ions may fragment unimolecularly by either heterolytic or homolytic cleavage of bonds adjacent to the site of the radical-cation formed by the loss of an electron. In the first two sections of this chapter, examples were shown of how the loss of a nonbonding electron from an oxygen atom or a halogen atom creates a radical-cation. The molecular ion then loses different radical species and is converted to the fragment ions. Other types of reactions are also possible, but this section will concentrate on the simpler fragmentation pathways.

The most abundant ions recorded in a mass spectrum are those formed either because the ion itself is very stable (such as the acylium ion, m/z 43, in the mass spectrum of acetone, p. 866) or because the uncharged fragment that is lost is stable. For example, the methyl cation (m/z 15) in the spectrum of fluoromethane (p. 865), though not very stable itself, is formed by the loss of a stable fluorine atom from the molecular ion.

The fragmentation of an amine or an alcohol is determined by the groups adjacent to the nitrogen or the oxygen atom. For example, in the mass spectra of butylamine and *tert*-butyl alcohol (Figure 21.11), the base peak in each case can

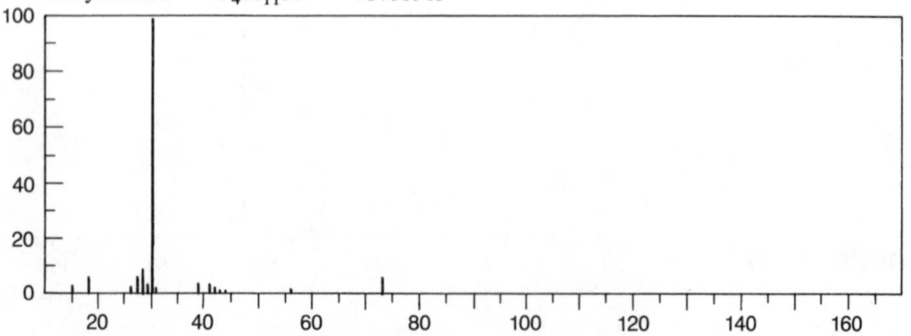
Butylamine $C_4H_{11}N$ 73.08915
100
80
60
40
20
0
20
40
60
80
100
120
140
160

FIGURE 21.11 Mass spectra of butylamine and *tert*-butyl alcohol (Adapted from Stenhagen et al. Reprinted by permission.)

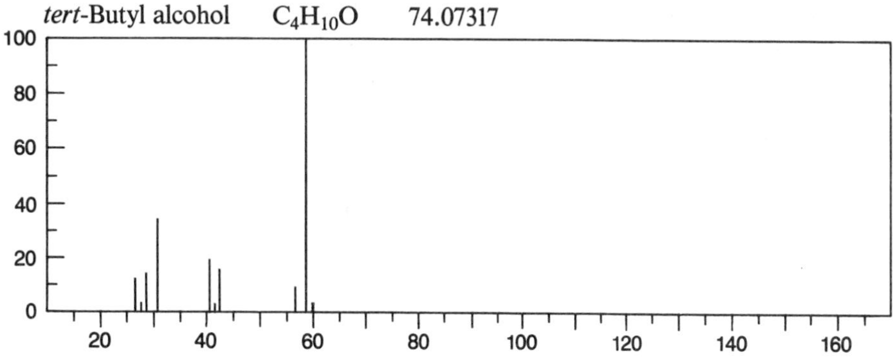

tert-Butyl alcohol C4H10O 74.07317
100
80
60
40
20
0
20
40
60
80
100
120
140
160

FIGURE 21.11 *(Continued)*

be rationalized as arising from the homolytic cleavage of a bond one removed (α-cleavage) from the radical-cation centered on the oxygen or the nitrogen atom.

VISUALIZING THE REACTION

The α-cleavage reaction of radical-cations

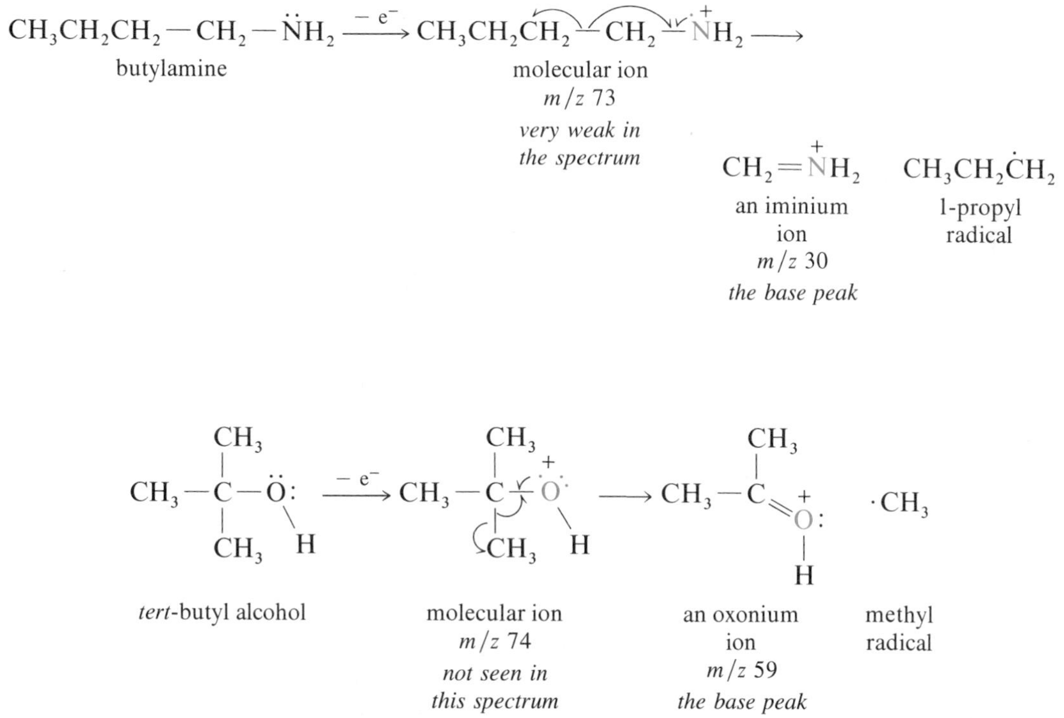

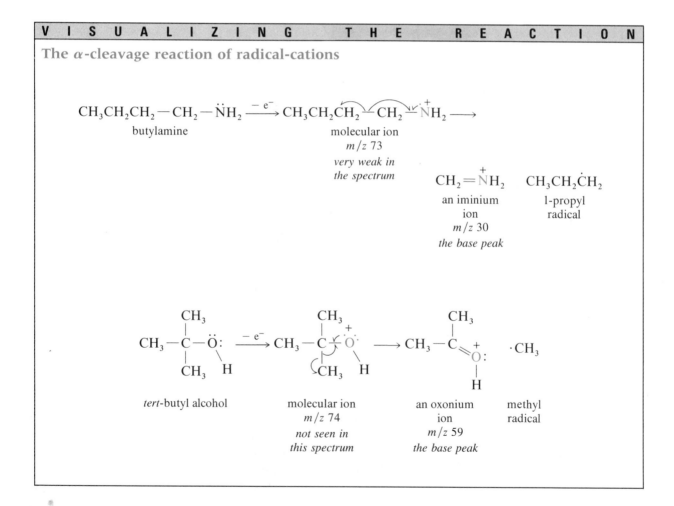

butylamine
molecular ion
m/z 73
very weak in the spectrum
an iminium ion
m/z 30
the base peak
1-propyl radical
tert-butyl alcohol
molecular ion
m/z 74
not seen in this spectrum
an oxonium ion
m/z 59
the base peak
methyl radical

Stable cations, oxonium or iminium ions, are formed in these fragmentations. The molecular ion peak is usually weak or nonexistent in the spectra of alcohols and amines because of the ease with which the fragmentations occur.

PROBLEM 21.8

The mass spectra of three isomeric amines, *sec*-butylamine, isobutylamine, and *tert*-butylamine, are given in Figure 21.12. Decide which spectrum belongs to which amine, and write equations justifying your conclusions.

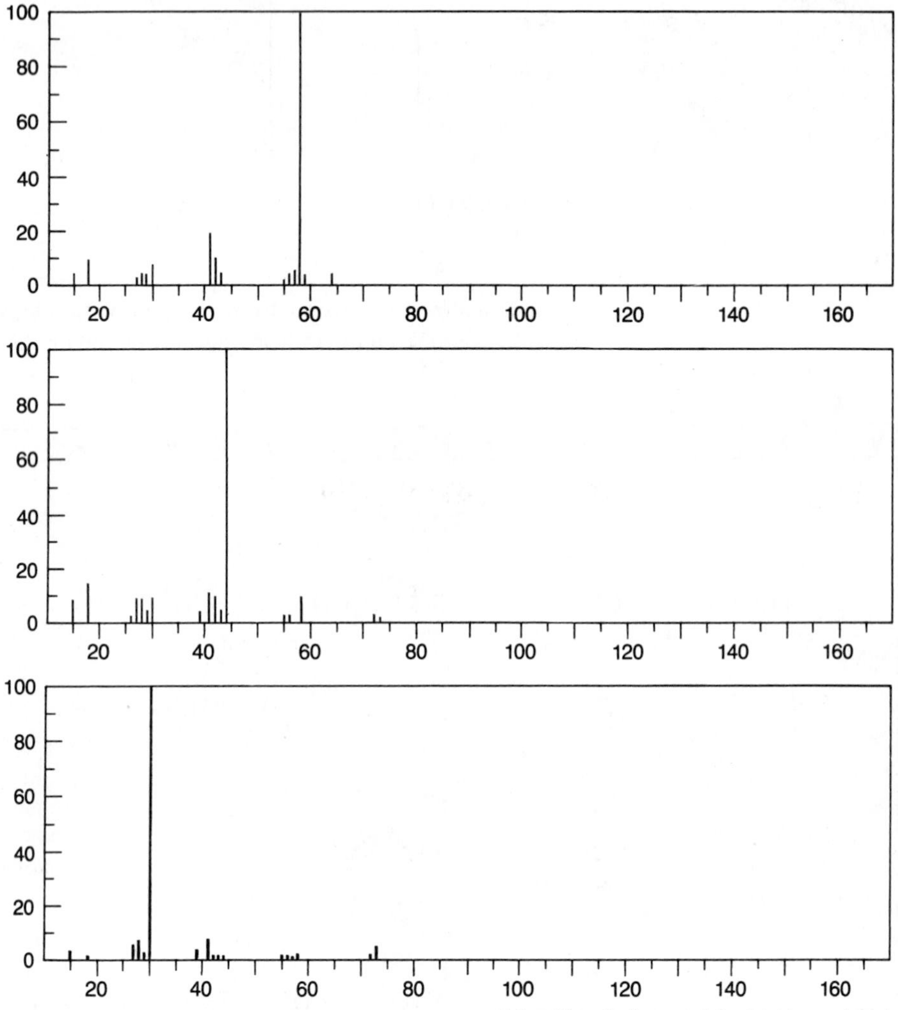

FIGURE 21.12

(Adapted from Stenhagen et al. Reprinted by permission.)

The cleavage of bonds adjacent to carbonyl groups and the loss of halogens as atoms or ions have already been discussed with examples on pp. 867 and 871.

The mass spectra of alkanes and alkenes are complicated because rearrangements of intermediate cations and radicals occur when the molecule is of any size. The molecular ion of alkenes arises from the loss of an electron from the π bond. Then an allylic bond in the molecular ion often cleaves to give an alkyl radical and an allylic cation. The ion arising from such processes is seen in the spectrum of 2-methyl-2-pentene (Figure 21.13). The formation of the molecular ion and the cleavage of the allylic bond in 2-methyl-2-pentene are shown on the next page.

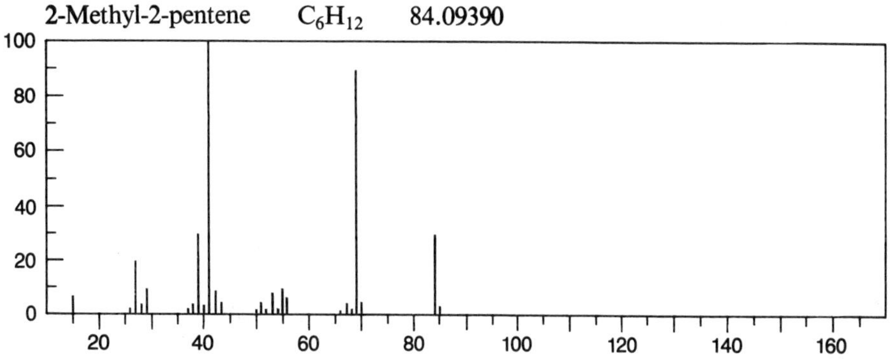
2-Methyl-2-pentene C6H12 84.09390
100
80
60
40
20
0
20
40
60
80
100
120
140
160

FIGURE 21.13 Mass spectrum of 2-methyl-2-pentene. (Adapted from Stenhagen et al. Reprinted by permission.)

VISUALIZING THE REACTION

α-Cleavage reaction of the radical-cation from an alkene

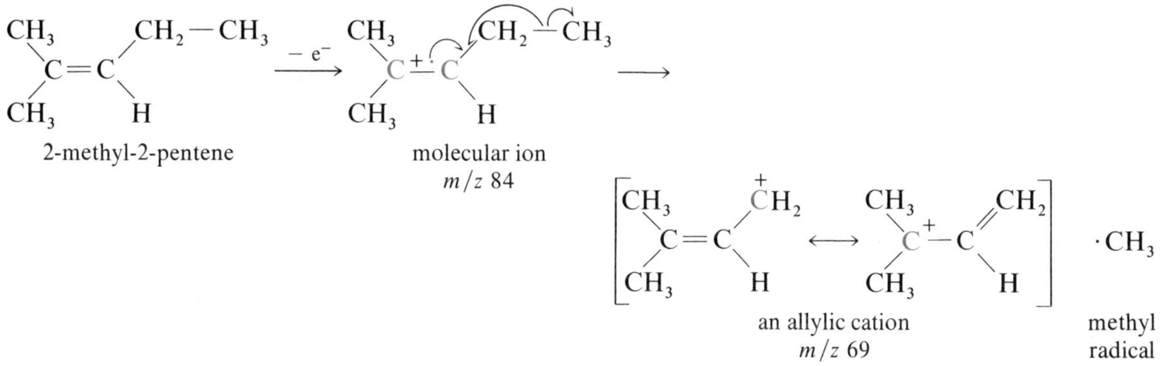
2-methyl-2-pentene
− e⁻
molecular ion
m/z 84
an allylic cation
m/z 69
methyl
radical

PROBLEM 21.9

The mass spectrum of 1-hexene is shown in Figure 21.14. Write equations showing how the ion giving rise to the base peak is formed.

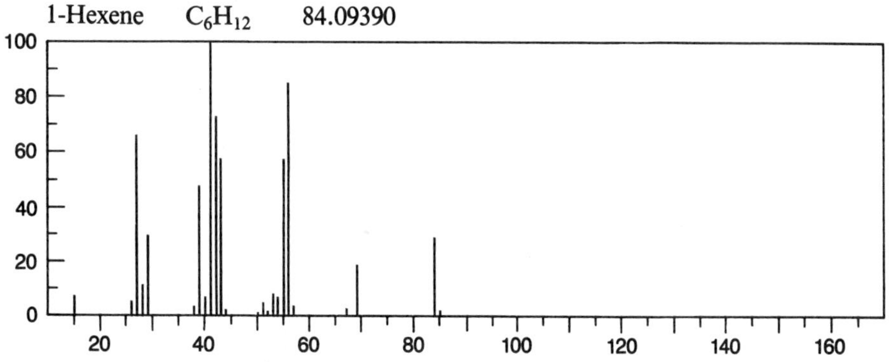
1-Hexene C6H12 84.09390
100
80
60
40
20
0
20
40
60
80
100
120
140
160

FIGURE 21.14

(Adapted from Stenhagen et al. Reprinted by permission.)

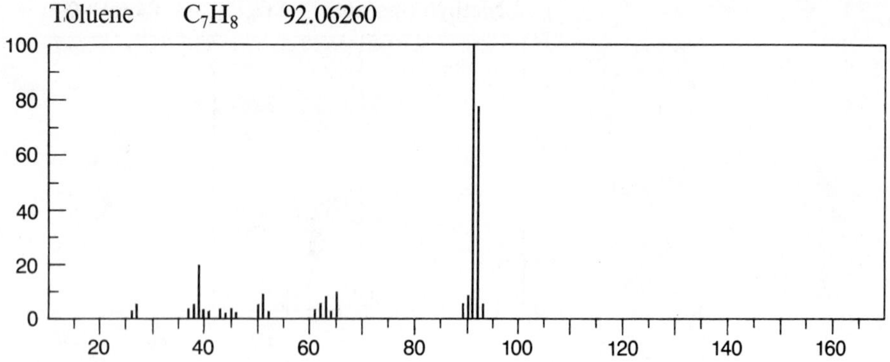
Toluene
C7H8
92.06260
100
80
60
40
20
0
20
40
60
80
100
120
140
160

FIGURE 21.15 Mass spectrum of toluene. (Adapted from Stenhagen et al. Reprinted by permission.)

Another category of fragmentation reaction that is important is cleavage at benzylic bonds. The mass spectrum of toluene illustrates such fragmentation (Figure 21.15). The mass spectrum of toluene shows the peak for the molecular ion at m/z 92 and the base peak at m/z 91, which arises from the loss of a hydrogen atom from the molecular ion.

VISUALIZING THE REACTION

α-Cleavage reaction of the radical-cation from an arene

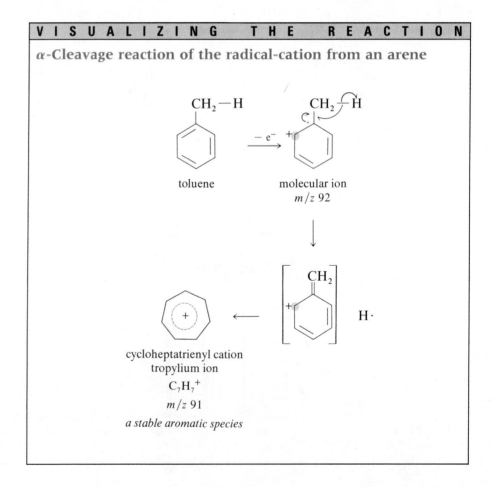

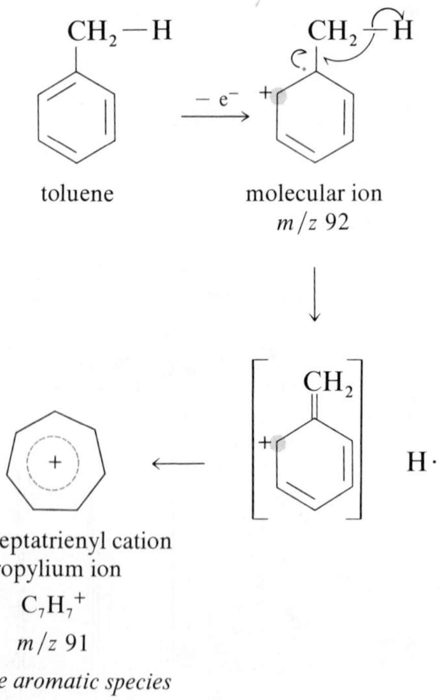
CH2—H
toluene
− e−
CH2—H
molecular ion
m/z 92
CH2
H·
cycloheptatrienyl cation
tropylium ion
C7H7+
m/z 91
a stable aromatic species

The cation formed on the loss of the hydrogen atom is believed to have the stable ring-expanded structure of the cycloheptatrienyl cation (p. 779) rather than that of a typical benzyl cation.

PROBLEM 21.10

Mass spectra for the isomeric compounds 2,2-dimethyl-1-phenylpropane and 2-methyl-3-phenylbutane are shown in Figure 21.16. Decide which spectrum belongs to which compound, and write equations showing the formation of the ions giving rise to the base peaks.

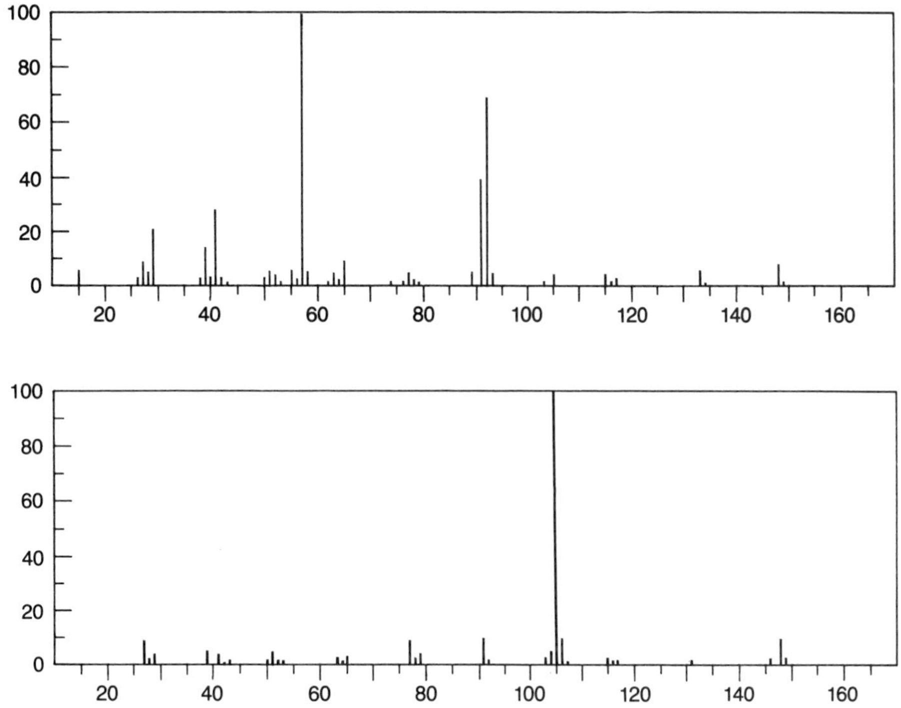

100
80
60
40
20
0
20
40
60
80
100
120
140
160
100
80
60
40
20
0
20
40
60
80
100
120
140
160

FIGURE 21.16

(Adapted from Stenhagen et al. Reprinted by permission.)

21.4 REARRANGEMENTS OF MOLECULAR IONS

One of the most common and most important rearrangements observed in the mass spectrometer is the transfer of a hydrogen atom from one part of a chain to a radical site at another part of the chain, usually via a six-membered cyclic transition state. The reaction is particularly easy to see with carbonyl compounds but is also one of the mechanisms by which rearrangements of alkenes occur. Such rearrangements are postulated to account for radical-cations that give rise to peaks in the mass spectra of 2-methyl-1-pentene and 2-hexanone (Figure 21.17).

The base peak in the spectrum of 2-methyl-1-pentene is at m/z 56 and results from the loss of a fragment having a mass of 28 units ($84 - 56 = 28$). Such a fragment could be ethylene, C_2H_4.

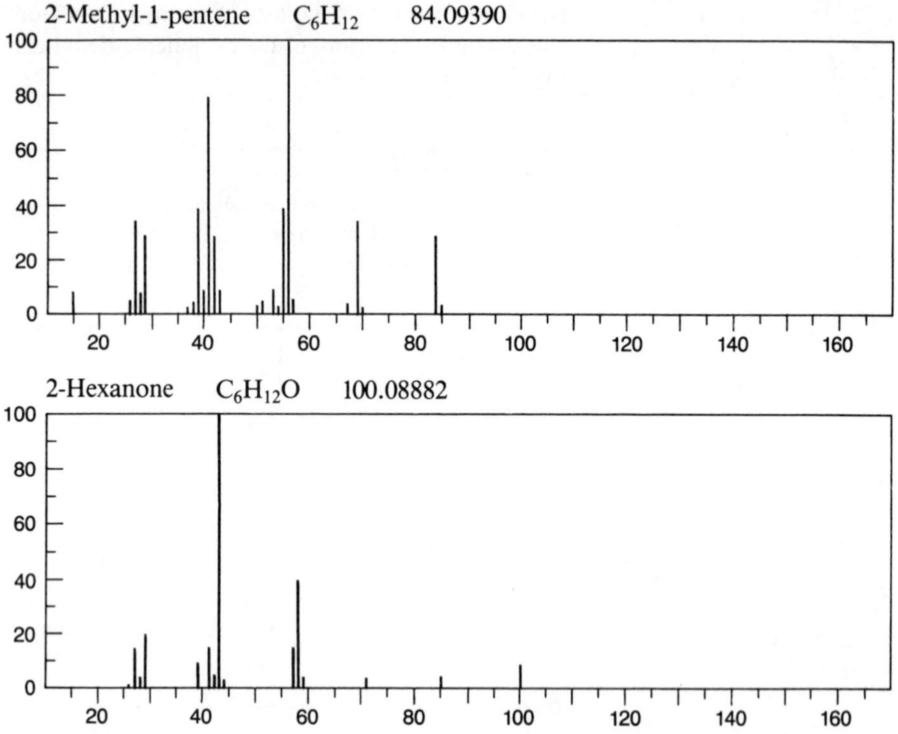
2-Methyl-1-pentene C6H12 84.09390
100
80
60
40
20
0
20
40
60
80
100
120
140
160
2-Hexanone C6H12O 100.08882
100
80
60
40
20
0
20
40
60
80
100
120
140
160

FIGURE 21.17 Mass spectra of 2-methyl-1-pentene and 2-hexanone. (Adapted from Stenhagen et al. Reprinted by permission.)

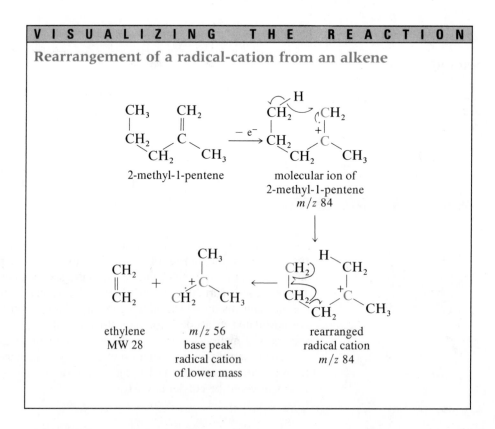

VISUALIZING THE REACTION

Rearrangement of a radical-cation from an alkene

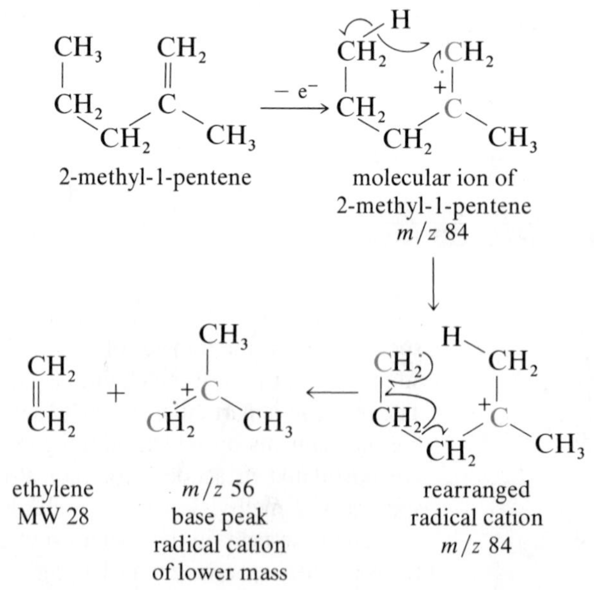
− e⁻
2-methyl-1-pentene
molecular ion of
2-methyl-1-pentene
m/z 84
ethylene
MW 28
m/z 56
base peak
radical cation
of lower mass
rearranged
radical cation
m/z 84

A hydrogen atom is transferred from one carbon atom to another carbon atom, a radical site at another part of the chain, by means of a six-membered ring transition state. The new radical cation fragments to give a stable alkene and another radical cation of lower mass.

The base peak in the spectrum of 2-hexanone, m/z 43, arises from cleavage of a bond next to the carbonyl group. A significant peak appearing at m/z 58 arises from a migration of a hydrogen atom from the γ-carbon atom to the carbonyl group, as seen below.

VISUALIZING THE REACTION

The McLafferty rearrangement

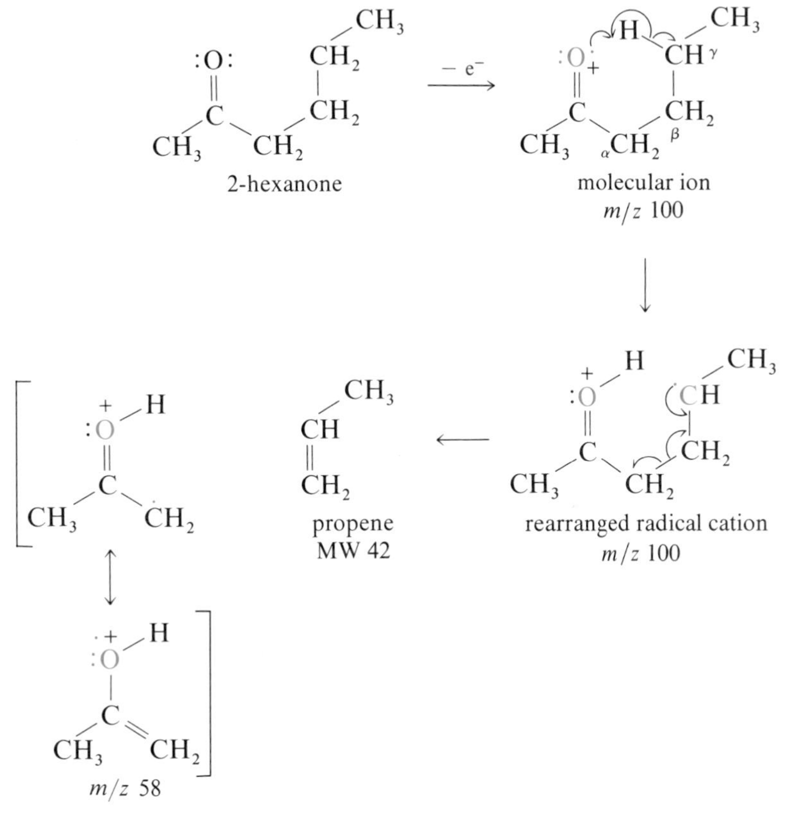

2-hexanone
− e⁻
molecular ion
m/z 100
rearranged radical cation
m/z 100
propene
MW 42
m/z 58

The radical cation with m/z 58 may be seen as being a tautomer of the radical cation of acetone and will undergo further fragmentations typical of acetone. This type of rearrangement is seen for a wide variety of compounds containing carbon-oxygen double bonds and is known as the **McLafferty rearrangement** for Fred W. McLafferty of Cornell University, who first discovered how general it is.

PROBLEM 21.11

The mass spectrum of 3-methyl-2-pentanone is shown in Figure 21.18. Write equations accounting for formation of the ions with m/z 29, 43, 56, and 72. (Hint: A review of the equations in Section 21.1 may be helpful.)

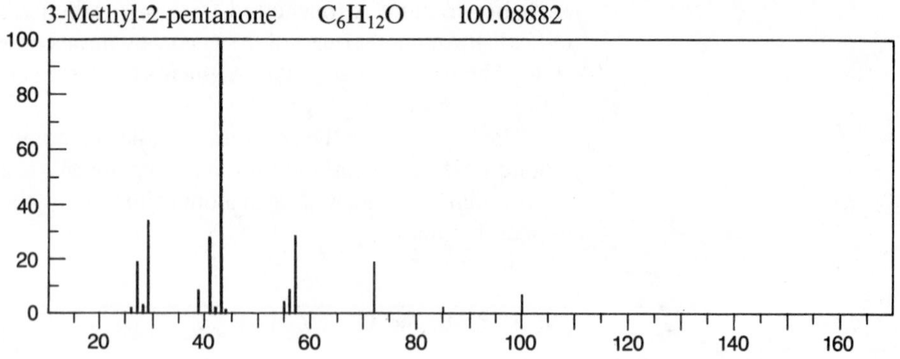
3-Methyl-2-pentanone $C_6H_{12}O$ 100.08882
100
80
60
40
20
0
20
40
60
80
100
120
140
160

FIGURE 21.18

(Adapted from Stenhagen et al. Reprinted by permission.)

PROBLEM 21.12

Two of the important peaks in the mass spectrum of 1-hexene (Figure 21.14, p. 877), at m/z 56 and 42 can be explained by writing mechanisms of the kind shown on p. 880 for 2-methyl-1-pentene. See whether you can use rearrangement and fragmentation reactions to arrive at structures for these radical-cations. (Hint: Either of the carbon atoms originally involved in the double bond may have radical character or cation character.)

S U M M A R Y

A mass spectrum is a record of the relative abundances of a series of ions of different masses that form when a sample of a compound is bombarded by high-energy electrons in a mass spectrometer. This bombardment causes an electron to be lost from a molecule of the compound, resulting in the formation of a radical-cation known as the molecular ion. The molecular ion fragments by the loss of radicals or neutral molecules to give positively charged ions. The mass spectrometer separates these cations according to their mass-to-charge ratios, m/z, and records their masses and their relative abundances. The most abundant ion in the spectrum is recorded as the base peak with an abundance of 100%.

The masses recorded by a mass spectrometer are exact masses, therefore, molecules of different isotopic composition give different molecular ions. For example, methyl bromide has two molecular ions of about equal abundance, one arising from $CH_3{}^{79}Br$ and the other from $CH_3{}^{81}Br$.

The fragmentation patterns seen for different types of compounds are typical of their structures. Fragmentations occur so as to produce stable cations or by the loss of stable neutral fragments. Rearrangements also occur, such as the McLafferty rearrangement, in which a hydrogen atom from the γ-carbon atom is transferred to the oxygen atom in a radical-cation derived from a carbonyl compound.

Mass spectra can be obtained from very small samples. Computer techniques that allow comparison of fragmentation patterns with those of known compounds make mass spectrometry a powerful tool for identifying organic compounds.

ADDITIONAL PROBLEMS

21.13 The proton magnetic resonance and mass spectra of Compound A, $C_9H_{12}O$, are shown in Figure 21.19. The compound's infrared spectrum has strong bands at 3338 and 1031 cm^{-1}. Peaks appear at 32.1 (t), 34.2 (t), 61.6 (t), 125.8 (d), 128.4 (d), and 141.9 (s) ppm in its carbon-13 nuclear magnetic resonance spectrum. Assign a structure to Compound A. Analyze the splitting patterns that appear in its nuclear magnetic resonance spectra in terms of your structure. Write an equation showing the formation of the ion that gives rise to the base peak in its mass spectrum.

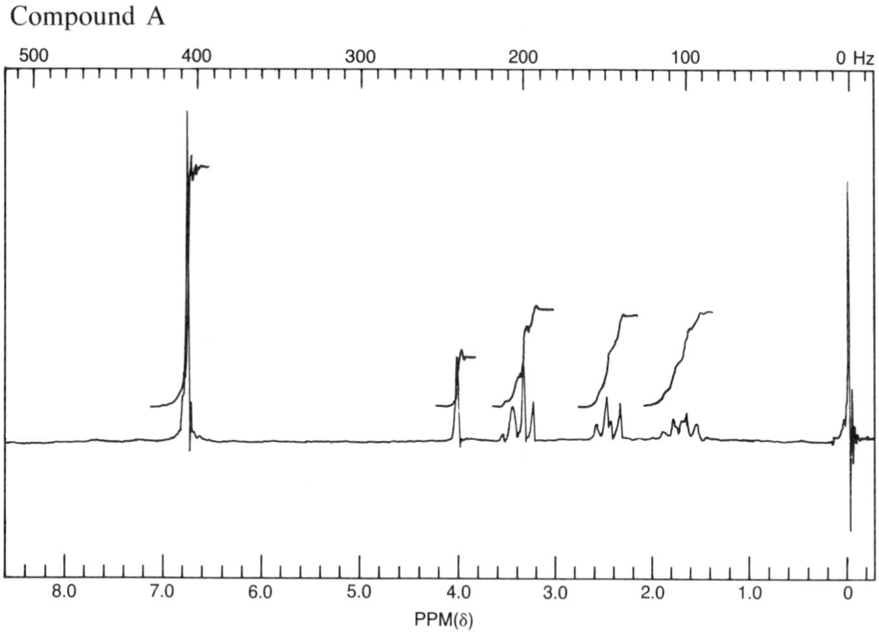

Compound A
500
400
300
200
100
0 Hz
8.0
7.0
6.0
5.0
4.0
3.0
2.0
1.0
0
PPM(δ)

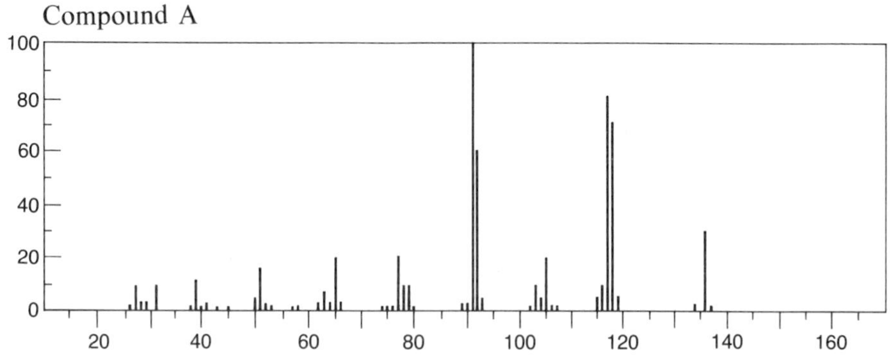

Compound A
100
80
60
40
20
0
20
40
60
80
100
120
140
160

FIGURE 21.19

21.14 Compound B contains only carbon, hydrogen, and a halogen. The mass spectrum is shown in Figure 21.20. Assign a structure to the compound.

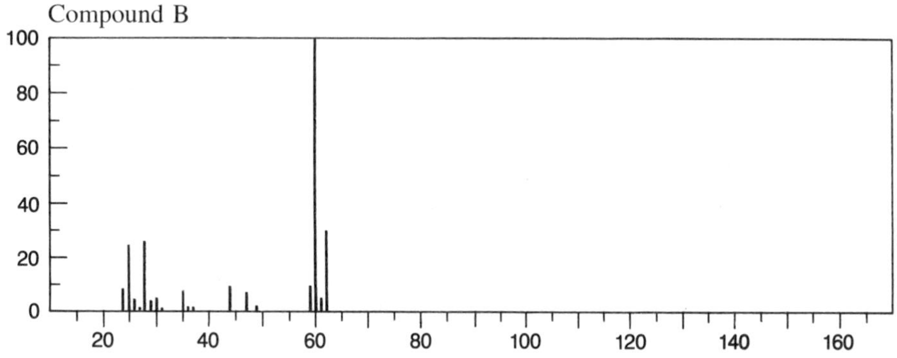

Compound B
100
80
60
40
20
0
20
40
60
80
100
120
140
160

FIGURE 21.20

(Adapted from Stenhagen et al. Reprinted by permission.)

21.15 Compound C is an ester. It has a singlet in its proton magnetic resonance spectrum at δ 3.69. Identify the molecular ion from its mass spectrum, shown in Figure 21.21. Assign a structure to that ion and to the ions having an intensity greater than 50%.

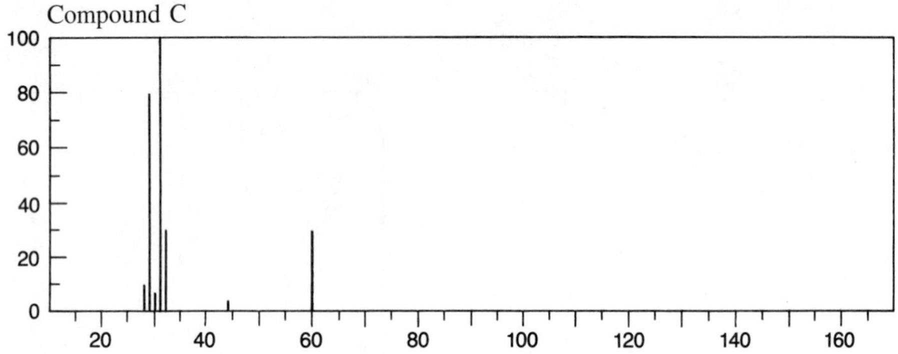
Compound C
100
80
60
40
20
0
20
40
60
80
100
120
140
160

FIGURE 21.21

(Adapted from Stenhagen et al. Reprinted by permission.)

21.16 The molecular ion for Compound D does *not* appear in its mass spectrum. The peaks for the important fragments from the molecular ion, however, are clearly seen in Figure 21.22. Assign a structure to Compound D.

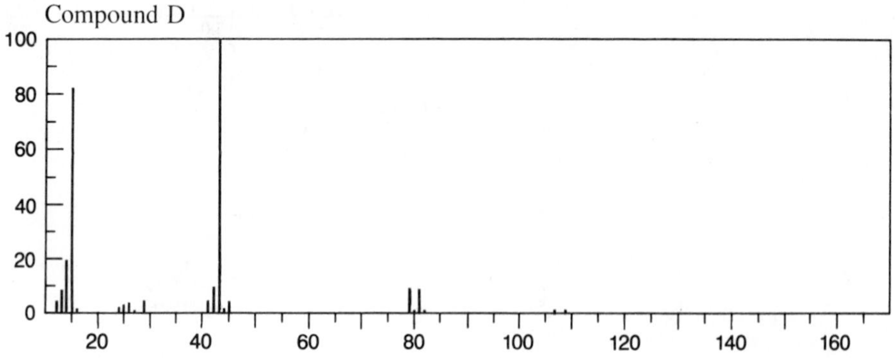
Compound D
100
80
60
40
20
0
20
40
60
80
100
120
140
160

FIGURE 21.22

(Adapted from Stenhagen et al. Reprinted by permission.)

21.17 The mass spectra of two isomeric alcohols, Compounds E and F, are given in Figure 21.23. The molecular ion shows up in the spectrum of one but not the other. Assign structures to the two alcohols, and write equations accounting for the formation of the ion responsible for the base peak in each case.

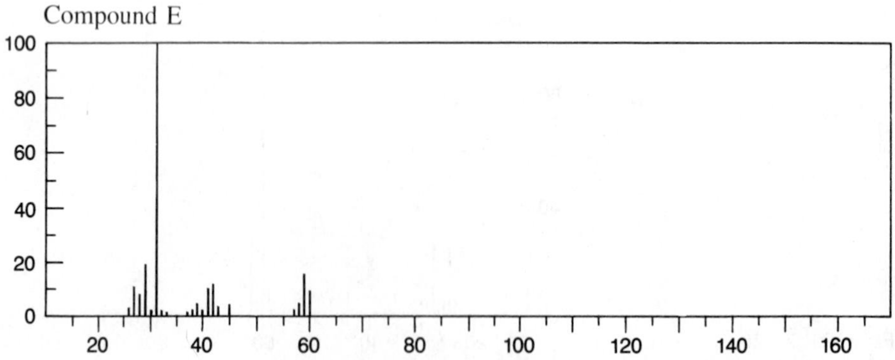
Compound E
100
80
60
40
20
0
20
40
60
80
100
120
140
160

FIGURE 21.23

(Adapted from Stenhagen et al. Reprinted by permission.)

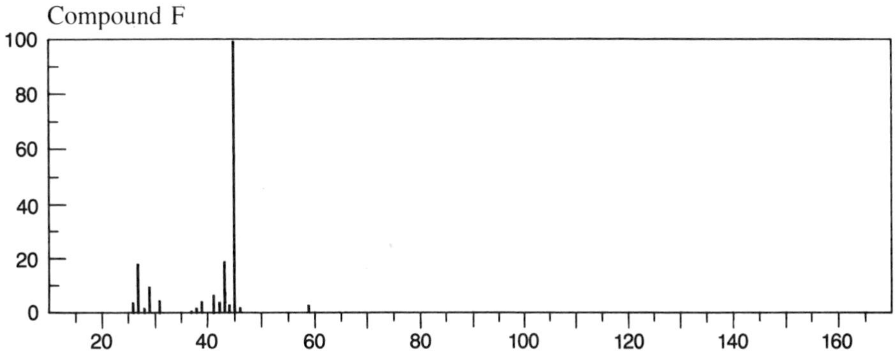
Compound F
100
80
60
40
20
0
20
40
60
80
100
120
140
160

FIGURE 21.23 *(Continued)*

(Adapted from Stenhagen et al. Reprinted by permission.)

21.18 The base peak in the mass spectrum of pentanal, shown in Figure 21.24, is the result of a rearrangement. Write an equation that shows how the ion represented by the base peak comes into being.

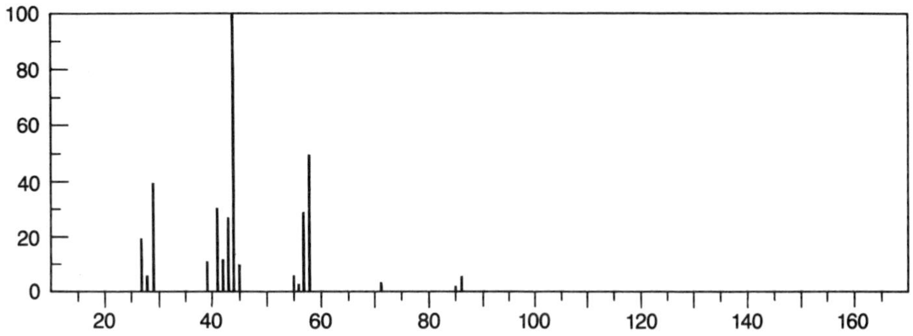
100
80
60
40
20
0
20
40
60
80
100
120
140
160

FIGURE 21.24

(Adapted from Stenhagen et al. Reprinted by permission.)

21.19 Compound G has the proton magnetic resonance and mass spectra shown in Figure 21.25. There is a peak for the molecular ion in the mass spectrum. The carbon-13 nuclear magnetic resonance spectrum of the compound has bands at 33, 120, and 135 ppm. Tell which spectra bands belong to each carbon and each hydrogen atom in the compound. Write an equation for the formation of the ion responsible for the base peak in the mass spectrum.

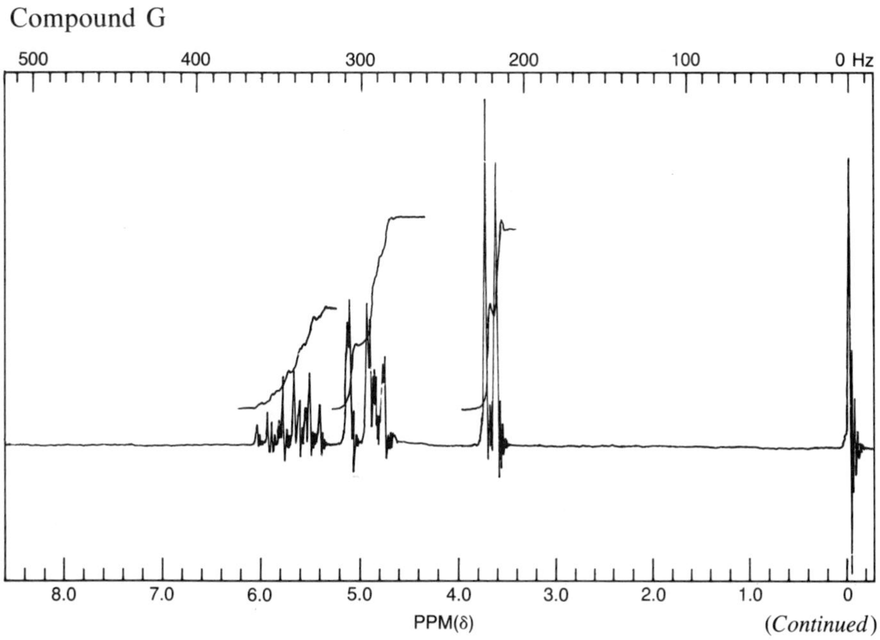
Compound G
500
400
300
200
100
0 Hz
8.0
7.0
6.0
5.0
4.0
3.0
2.0
1.0
0
PPM(δ)

FIGURE 21.25

(Continued)

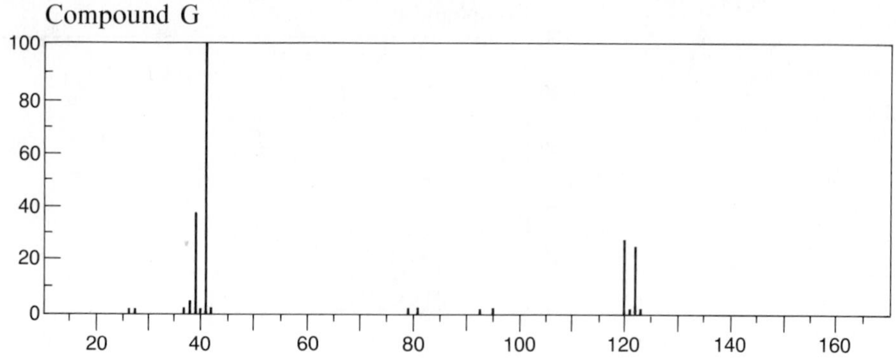
Compound G
100
80
60
40
20
0
20
40
60
80
100
120
140
160

FIGURE 21.25 *(Continued)*

21.20 The proton magnetic resonance spectrum of Compound H is shown in Figure 21.26. The compound absorbs strongly in the infrared at 1676 cm^{-1}. In its mass spectrum, the peak due to the molecular ion is at m/z 150, and the base peak is at m/z 135. Assign a structure to Compound H, showing how you used each piece of spectral information.

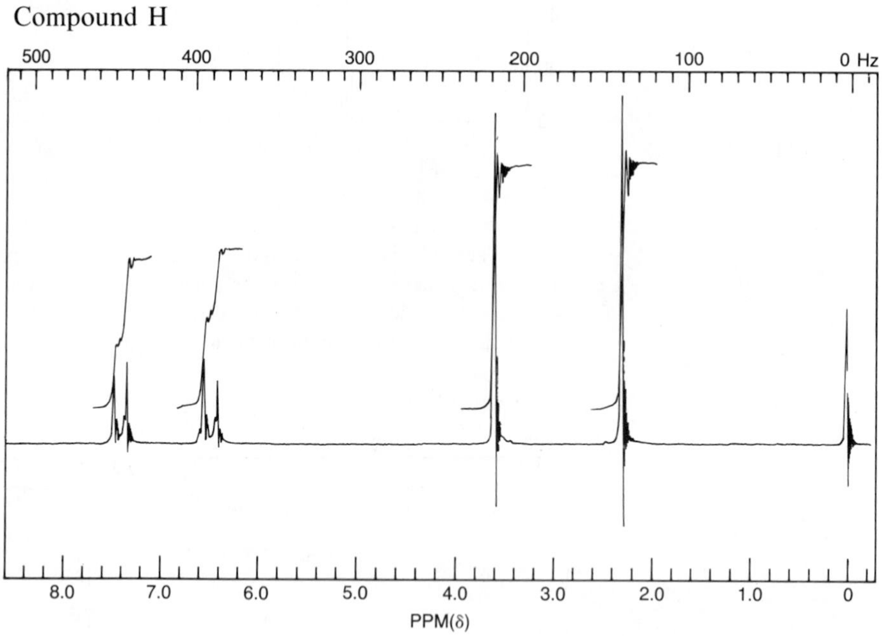
Compound H
500
400
300
200
100
0 Hz
8.0
7.0
6.0
5.0
4.0
3.0
2.0
1.0
0
PPM(δ)

FIGURE 21.26

21.21 Compound I absorbs at 1703 and 2730 cm^{-1} in the infrared. Its molecular ion has m/z 120, and there are two base peaks, at m/z 119 and 91, in its mass spectrum. Its carbon-13 nuclear magnetic resonance spectrum has bands at 21.6 (q), 129.6 (d), 134.4 (s), 145.3 (s), and 191.4 (d) ppm. The proton magnetic resonance spectrum of Compound I is shown in Figure 21.27. Assign a structure to the compound. Analyze all the spectral data, and write equations that account for the presence of the two base peaks in the mass spectrum.

Compound I

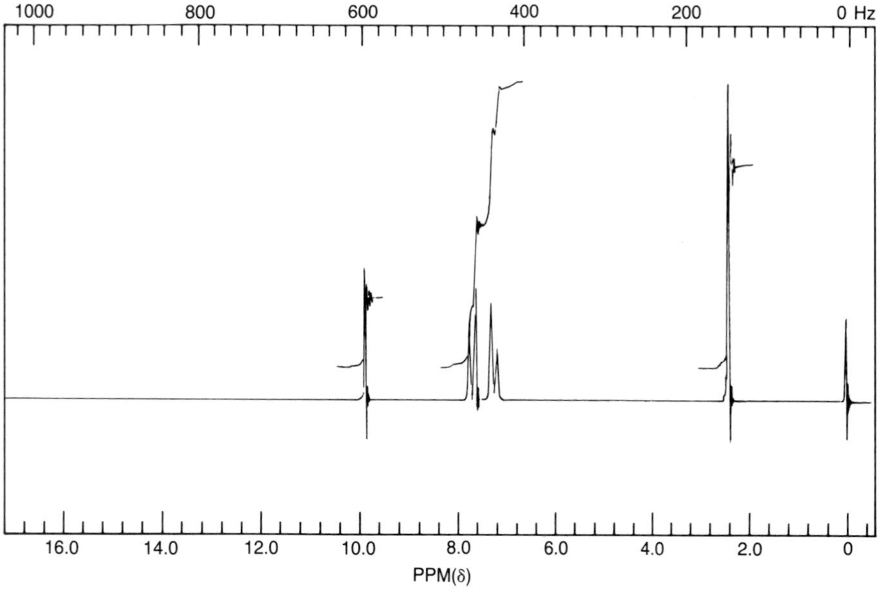
1000
800
600
400
200
0 Hz
16.0
14.0
12.0
10.0
8.0
6.0
4.0
2.0
0
PPM(δ)

FIGURE 21.27

22

The Chemistry of Amines

A • L O O K • A H E A D

Earlier chapters presented methods for the preparation of many compounds containing nitrogen. Among these were azides (p. 260), oximes (p. 504), imines (p. 506), amides (p. 582), and nitro compounds (p. 784). This chapter will describe how to convert these types of compounds to amines through reduction reactions.

$$\underset{\text{azide}}{R-\overset{-}{N}-\overset{+}{N}\equiv N} \xrightarrow{\textit{reduction}} R-NH_2 + N_2$$

$$\underset{\text{imine or oxime}}{R-\overset{\displaystyle R}{\overset{|}{C}}=N-} \xrightarrow{\textit{reduction}} R-\underset{\displaystyle H}{\underset{|}{\overset{\displaystyle R}{\overset{|}{C}}}}-NH-$$

$$\underset{\text{amide}}{R'-\overset{\displaystyle O}{\overset{\|}{C}}-NH-R} \xrightarrow{\textit{reduction}} R'-CH_2-NH-R$$

$$\underset{\text{nitro compound}}{Ar-NO_2} \xrightarrow{\textit{reduction}} Ar-NH_2$$

Amines may also be prepared by nucleophilic substitution reactions of ammonia or protected ammonia derivatives.

Because of the pair of nonbonding electrons on the nitrogen atom, amines are important organic bases (pp. 102–104) and nucleophiles. Their reactions as nucleophiles with electrophilic centers in acid derivatives are familiar (p. 582). Amines also react with the electrophile nitrous acid. Amines with only one organic group bonded to the nitrogen atom can be converted to diazonium ions, reactive species that lose N_2 as a leaving group.

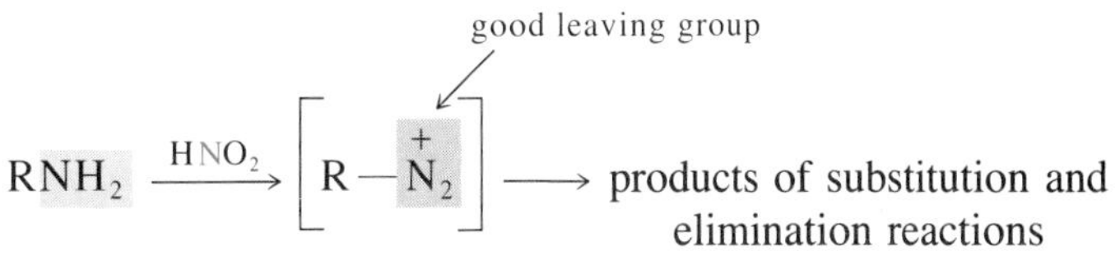
good leaving group
RNH_2
HNO_2
$R-\overset{+}{N_2}$
products of substitution and
elimination reactions

Molecular rearrangements, analogous to rearrangements of carbocations (p. 287), occur in compounds containing nitrogen whenever a reaction gives rise to an electron-deficient nitrogen atom. This chapter will look at some examples of such rearrangements to nitrogen.

22.1 STRUCTURE AND NATURAL OCCURRENCE OF AMINES

Amines, compounds in which one or more of the hydrogen atoms of an ammonia molecule have been replaced by an organic group, are found widely in nature. Many plants synthesize complex amines called **alkaloids,** some of which have medicinal or poisonous properties. Morphine, one of the most effective painkillers known; quinine, an important medication for malaria; and nicotine, the toxic component in tobacco, are among some of the alkaloids that have been isolated and used by humans.

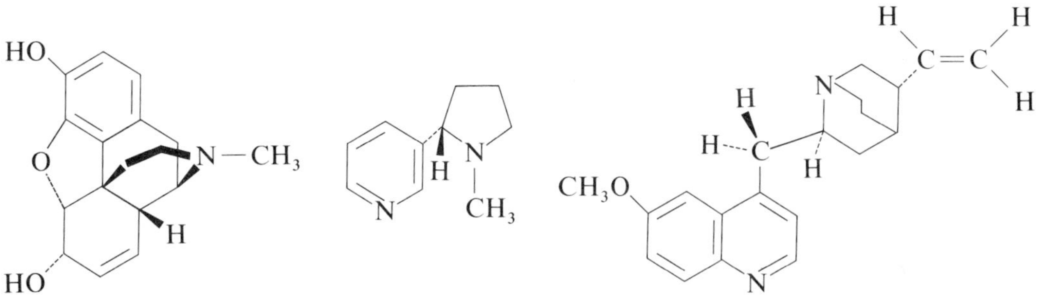
HO
O
HO
N—CH_3
H
H
N
N
CH_3
CH_3O
N
N
H
H
H
C
H
C=C
H
H

(−)-morphine (−)-nicotine (−)-quinine

some naturally occurring amines present in plants

Proteins, the complex constituents of animal bodies and of the enzyme systems that catalyze chemical processes in the body, are made up of **amino acids.** In amino acids, an amino group is substituted on the second carbon atom of a carboxylic acid. Proteins are formed when amino acids are bonded together by amide bonds, called **peptide linkages** (p. 564), between the amino group of one amino acid and the carboxylic acid group of another.

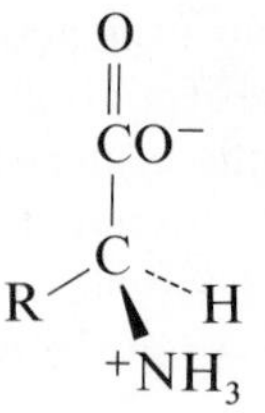

an amino acid

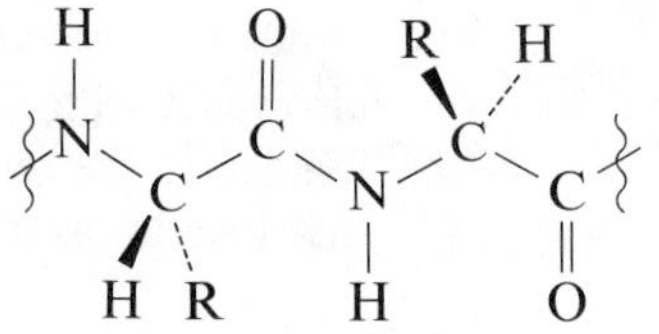

fragment of a protein molecule showing two amino acid units

amino groups, in a free amino acid and in an amide bond in part of a protein chain

The synthetic methods for preparing amino acids resemble those used for simple alkylamines and are covered in this chapter. The structural features of amino acids that are significant in the chemistry of peptides and proteins are discussed in Chapter 26.

The decomposition of amino acids and proteins in decaying animal matter produces many simple amines such as methylamine, 1,4-butanediamine, and 1,5-pentanediamine. The common names for 1,4-butanediamine and 1,5-pentanediamine are putrescine and cadaverine, respectively. The odor of fish comes from low-molecular-weight amines, such as methylamine.

CH_3NH_2 — methylamine

$NH_2CH_2CH_2CH_2CH_2NH_2$ — 1,4-butanediamine, putrescine

$NH_2CH_2CH_2CH_2CH_2CH_2NH_2$ — 1,5-pentanediamine, cadaverine

some amines found in decaying animal tissues

As shown in the structures below, amines are classified as **primary, secondary,** or **tertiary** depending on the number of organic groups bonded to the nitrogen atom. Amines are also divided into **aryl amines,** in which at least one of the organic substituents is an aryl group, and **alkyl amines,** in which all of the substituents are alkyl groups.

CH_3NH_2

methylamine

a primary alkyl amine

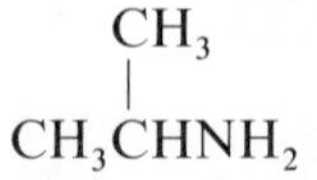

isopropylamine

a primary alkyl amine

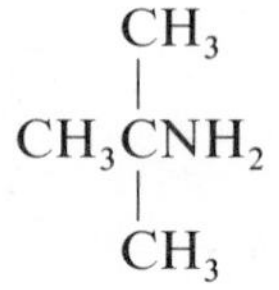

tert-butylamine

a primary alkyl amine

$CH_3CH_2NHCH_2CH_3$

diethylamine

a secondary alkyl amine

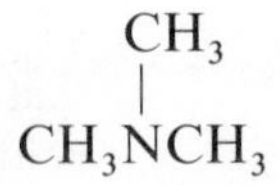

trimethylamine

a tertiary alkyl amine

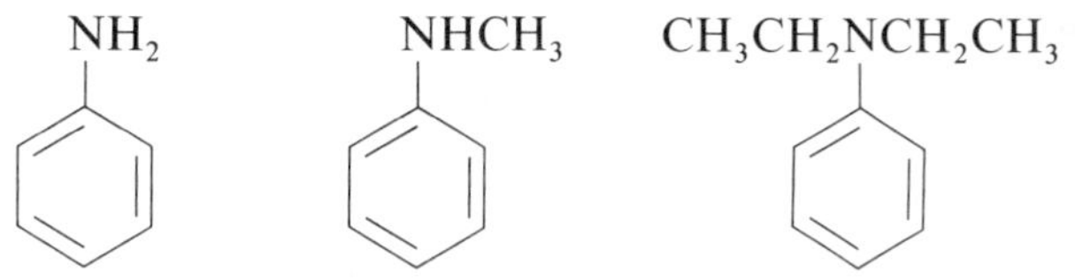

aniline	N-methylaniline	N,N-diethylaniline
a primary aryl amine	*a secondary aryl amine*	*a tertiary aryl amine*

Note that the designations primary, secondary, and tertiary refer to the degree of substitution on the nitrogen atom and not to the carbon atom to which the nitrogen is attached, as in the case of alcohols and alkyl halides.

Primary and secondary amines participate in hydrogen bonding, as both donors and acceptors (p. 27). Nitrogen is not as electronegative as oxygen; therefore, the nitrogen-hydrogen bond is less polar than the oxygen-hydrogen bond, and hydrogen bonding in amines is weaker than it is in alcohols. The boiling points of amines are lower than those of alcohols of similar molecular weight, but higher than the boiling points of comparable hydrocarbons and other compounds for which no hydrogen bonding is possible. The trend in boiling points is illustrated with pentane, butylamine, and *n*-butyl alcohol.

$CH_3CH_2CH_2CH_2CH_3$	$CH_3CH_2CH_2CH_2NH_2$	$CH_3CH_2CH_2CH_2OH$
pentane	butylamine	*n*-butyl alcohol
MW 72	MW 73	MW 74
bp 36 °C	bp 78 °C	bp 118 °C
no hydrogen bonding		strong hydrogen bonding

the effect of hydrogen bonding on the boiling points of compounds of comparable molecular weight

Primary, secondary, and tertiary amines in which the organic groups are not too large (p. 26) are soluble in water. The nonbonding electrons of a nitrogen atom are more available to the hydrogen atom of water than are the nonbonding electrons of an oxygen atom. The basicity of amines is related to this phenomenon (pp. 102–104).

22.2 NOMENCLATURE OF AMINES

Simple amines are named by combining the name of the alkyl group that is present with the suffix **-amine.** If there are several alkyl groups bonded to the nitrogen atom, they are named in order of increasing complexity. The prefixes **di-** and **tri-** are used to indicate the presence of two or three alkyl groups of the same kind. In IUPAC nomenclature, the ending **amine** is substituted for the final **e** in the name of the alkane and a number is used to indicate the position of the amino group on the chain or ring.

For more complex amines, the prefix **amino-** is used with a number to indicate the position of the amino group on the hydrocarbon chain. Other substituents that are attached to the nitrogen atom are indicated by an *N* before the name of the substituent.

$CH_3CH_2NH_2$
ethylamine

CH_3NHCH_3
dimethylamine

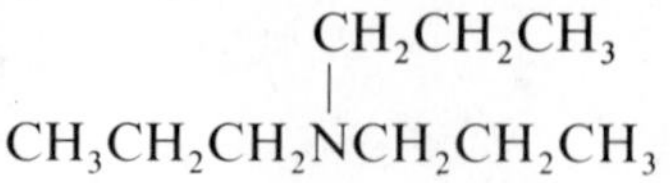
tripropylamine

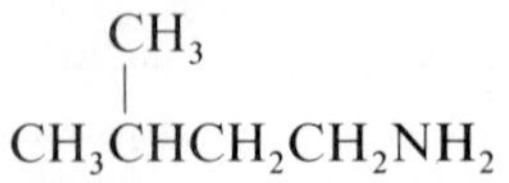
3-methyl-1-butylamine

$HOCH_2CH_2CH_2CH_2NH_2$
4-amino-1-butanol

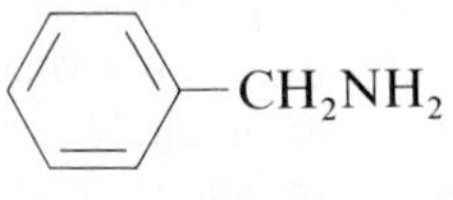
benzylamine

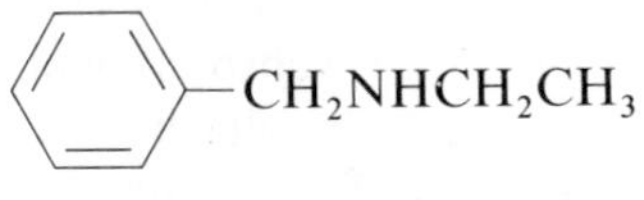
N-ethylbenzylamine

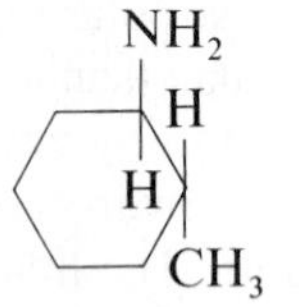
trans-2-methyl-1-cyclohexanamine
trans-(2-methylcyclohexyl)amine

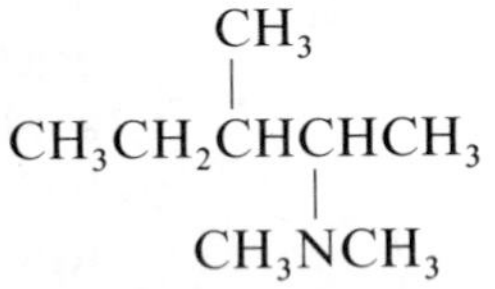
N,*N*-dimethyl-3-methyl-2-pentanamine
2-(*N*,*N*-dimethylamino)-3-methylpentane

The simplest aryl amine is **aniline,** and many aryl amines are named as substituted anilines. Exceptions are the amino derivatives of toluene, which are usually called **toluidines.** When there are a large number of substituents on the aromatic ring, the prefix amino- is used to locate the amino group. In some amines, nitrogen is part of a ring; such compounds are called heterocycles, meaning that they have atoms other than carbon in a ring (p. 987). Many of them have common names, and some that you should recognize are shown below.

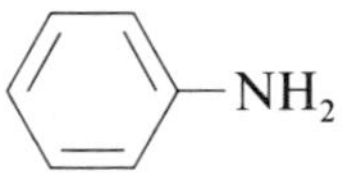
aniline

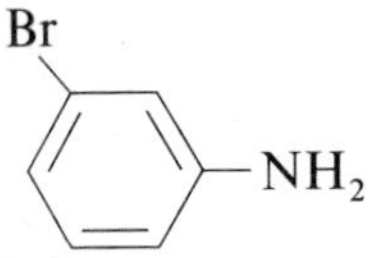
m-bromoaniline

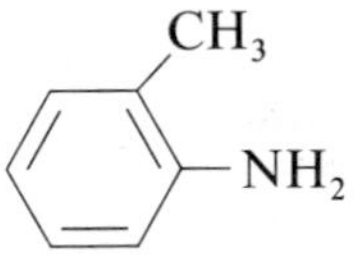
o-toluidine

p-toluidine

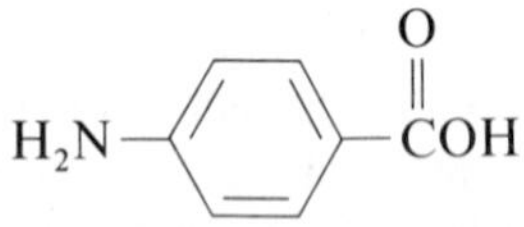
p-aminobenzoic acid

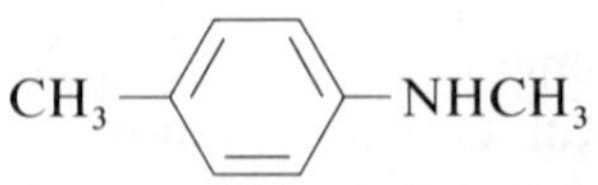
N-methyl-*p*-toluidine

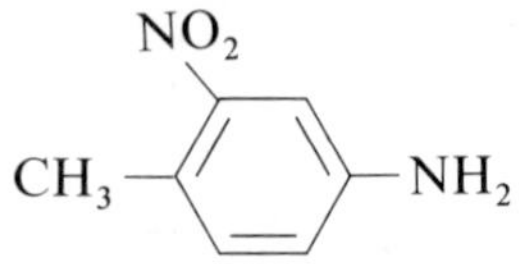
4-methyl-3-nitroaniline

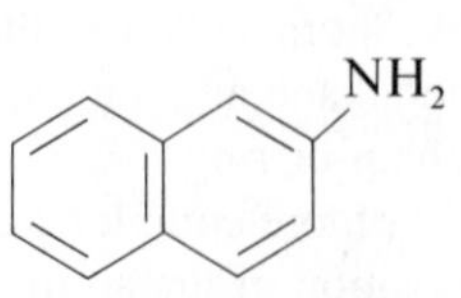
2-naphthylamine
2-aminonaphthalene
β-naphthylamine

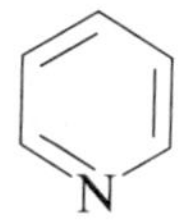

pyridine

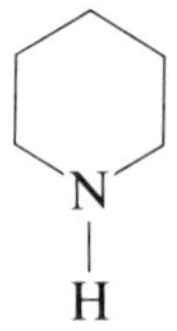

piperidine

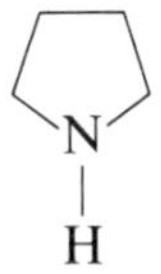

pyrrolidine

The chemistry of heterocyclic amines is discussed in Chapter 24.

PROBLEM 22.1

Name the following compounds.

(a) $\underset{\displaystyle CH_3CHCH_2NH_2}{\overset{\displaystyle CH_3}{|}}$ (b) 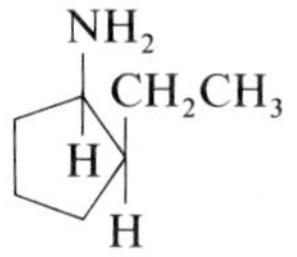

(c) $CH_3CH_2\underset{|}{\overset{CH_2CH_3}{N}}CH_2CH_2CH_2CH_3$ (d) 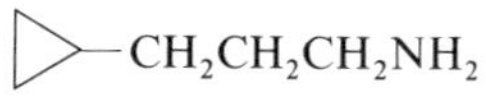

(e) 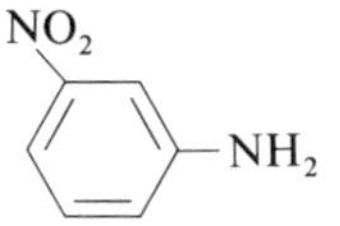 (f) 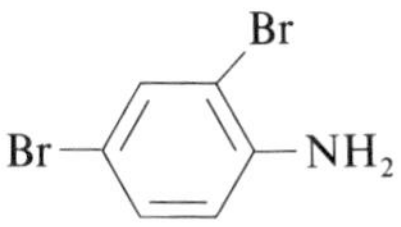

22.3 BASICITY OF AMINES

Alkyl amines are somewhat more basic than ammonia (p. 102). Their basicity depends on the number of alkyl groups bonded to the nitrogen atom; tertiary amines are less basic than primary and secondary amines. The presence of three organic groups around the nitrogen atom gives rise to steric hindrance that makes both protonation of the amine and solvation of the cation that results more difficult. These effects lead to a decrease in the basicity of a tertiary amine in comparison with that of secondary and primary amines.

PROBLEM 22.2

For the reaction

$$R_3\overset{+}{N}:\overset{-}{B}R'_3 \rightleftharpoons R_3N: + BR'_3$$

the following dissociation constants were measured:

$H_3\overset{+}{N}:\overset{-}{B}(CH_3)_3$	$CH_3\overset{+}{N}H_2:\overset{-}{B}(CH_3)_3$	$(CH_3)_2\overset{+}{N}H:\overset{-}{B}(CH_3)_3$	$(CH_3)_3\overset{+}{N}:\overset{-}{B}(CH_3)_3$
K_{diss} 4.6	K_{diss} 0.0350	K_{diss} 0.0214	K_{diss} 0.477

(a) Write the expression for the dissociation constant for the reaction.
(b) Give a brief rationalization of the observed experimental facts.

PROBLEM 22.3

The following pK_a values were determined for the conjugate acids of the amines shown. How would you rationalize the trend observed?

$CH_3(CH_2)_3NH_2$	$CH_3O(CH_2)_3NH_2$	$CH_3OCH(CH_3)CH_2NH_2$	$N{\equiv}CCH_2CH_2NH_2$	$N{\equiv}CCH_2NH_2$
pK_a 10.60	pK_a 9.92	pK_a 8.54	pK_a 7.8	pK_a 5.34

Aryl amines are much weaker bases than ammonia (p. 103). The nonbonding electrons of the nitrogen atom interact with the π electrons of the aromatic ring, creating electron density at the ring carbons that are ortho and para to the amino group. For this reason, the amino group is a powerful ring activator in electrophilic aromatic substitution reactions (p. 787). The basicity of aromatic amines is affected by substituents on the aromatic ring. For example, the conjugate acid of *p*-nitroaniline has a pK_a value of 1.0, indicating that that amine is a very weak base. The low basicity of an aryl amine with a nitro group substituted on the ring is rationalized by showing resonance contributors in which there is further delocalization of the nonbonding electrons of the amino group.

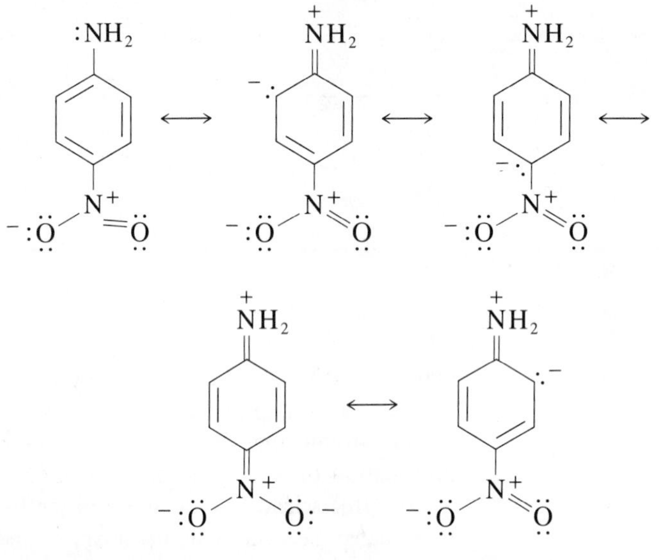

However, the conjugate acid of *p*-anisidine, also called 4-methoxyaniline, has a pK_a value of 5.34. Although an oxygen substituent on an aromatic ring has an electron-withdrawing inductive effect, it has an electron-donating resonance effect. The methoxy group is, therefore, expected to increase electron density near the amino group and to stabilize the conjugate acid of the amine.

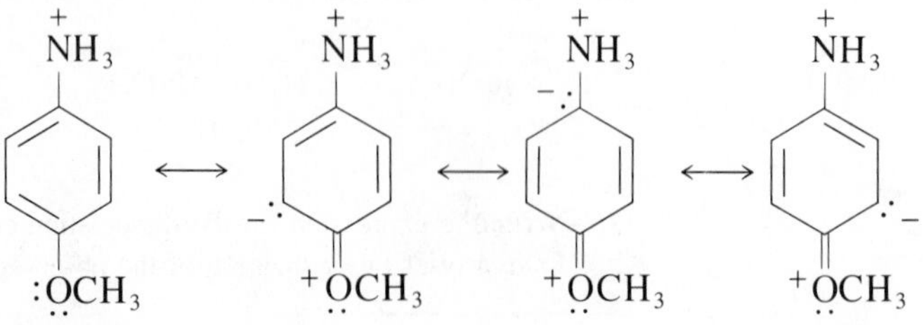

PROBLEM 22.4

The following pK_a values were measured for the conjugate acids of aniline and substituted anilines. Rationalize the trends that are apparent in the data.

(a)

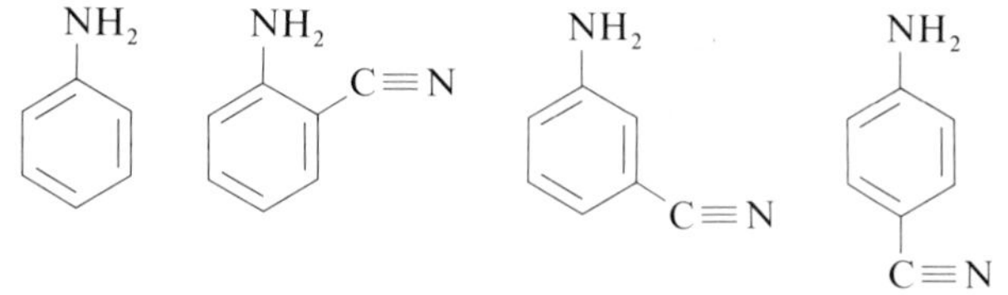

NH$_2$
NH$_2$
C≡N
NH$_2$
C≡N
NH$_2$
C≡N

pK_a	4.60	0.95	2.75	1.74

(b)

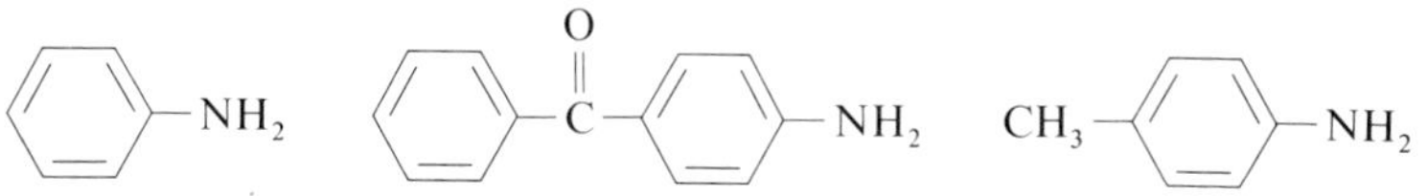

NH$_2$
O
C
NH$_2$
CH$_3$
NH$_2$

pK_a	4.60	2.17	5.10

22.4 PREPARATION OF AMINES

A. Reactions of Ammonia and Amines with Alkyl Halides

Ammonia reacts as a nucleophile with alkyl halides to form amines. The reaction of methyl iodide with ammonia illustrates the course of this nucleophilic substitution reaction.

$$\underset{\text{methyl iodide}}{CH_3I} + \underset{\text{ammonia}}{NH_3} \longrightarrow \underset{\text{methylammonium iodide}}{CH_3\overset{+}{N}H_3\ I^-}$$

$$\underset{\text{methylammonium iodide}}{CH_3\overset{+}{N}H_3\ I^-} + \underset{\text{ammonia}}{NH_3} \rightleftharpoons \underset{\text{methylamine}}{CH_3NH_2} + \underset{\text{ammonium iodide}}{\overset{+}{N}H_4\ I^-}$$

Ammonia first displaces iodide ion from a molecule of the alkyl halide to give methylammonium iodide. The reaction stops here if there is no excess ammonia present. Another molecule of ammonia deprotonates the methylammonium ion and frees methylamine, which is also a nucleophile and reacts further with methyl iodide. Thus, dimethylamine and trimethylamine are also products of this reaction.

$$\underset{\text{methylamine}}{CH_3NH_2} \xrightarrow[NH_3]{CH_3I} \underset{\text{dimethylamine}}{CH_3NHCH_3} \xrightarrow[NH_3]{CH_3I} \underset{\text{trimethylamine}}{CH_3\overset{\overset{\displaystyle CH_3}{|}}{N}CH_3}$$

It is difficult to prepare pure primary amines using this method.

When this type of reaction is used for synthetic purposes, a large excess of ammonia is used, as illustrated in the preparation of the amino acid glycine from chloroacetic acid.

$$\mathrm{ClCH_2\overset{O}{\overset{\|}{C}}OH + 2\,NH_3 \xrightarrow[25\,^\circ C]{H_2O} H_3\overset{+}{N}CH_2\overset{O}{\overset{\|}{C}}O^- + \overset{+}{N}H_4\ Cl^-}$$

chloroacetic acid, 1 equivalent; ammonia, 60 equivalents; glycine, 65%; ammonium chloride

The same precaution must be used when preparing a secondary amine so as to avoid getting a tertiary amine. For example, an excess of aniline is used in the reaction with benzyl chloride to prevent the formation of *N*,*N*-dibenzylaniline.

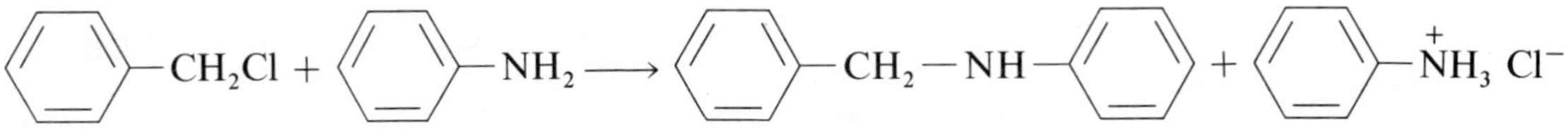
CH_2Cl +
NH_2
CH_2-NH
+
$\overset{+}{N}H_3\ Cl^-$

benzyl chloride, 1 equivalent; aniline, 4 equivalents; *N*-benzylaniline, 85%; aniline hydrochloride

PROBLEM 22.5

Tell what starting materials would be necessary to synthesize the following compounds. Assume that aniline, benzene, and any other organic compound containing not more than three carbon atoms are available.

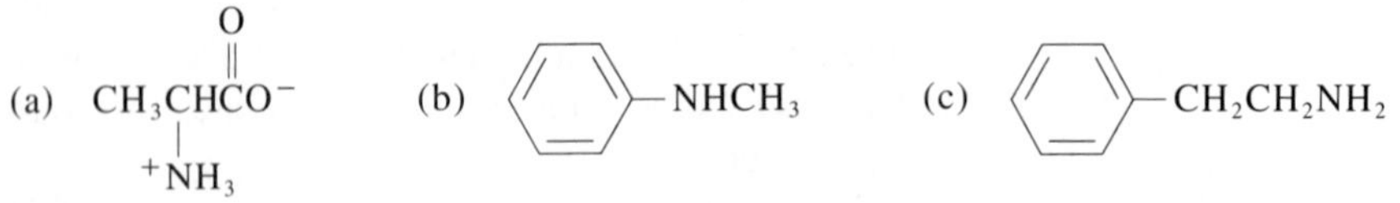
(a) CH_3CHCO^- (O, $^+NH_3$)
(b) $NHCH_3$
(c) $CH_2CH_2NH_2$

B. The Gabriel Phthalimide Synthesis

Nucleophilic substitution reactions of ammonia with alkyl halides result in mixtures of primary, secondary, and tertiary amines unless a large excess of ammonia is used. Pure primary amines can be prepared more conveniently if the nitrogen atom is protected so that alkylation can take place only once. Such a protected nitrogen atom is present in phthalimide, which is acidic enough (pK_a 7.4) to be deprotonated rather easily to give a nitrogen anion in a salt.

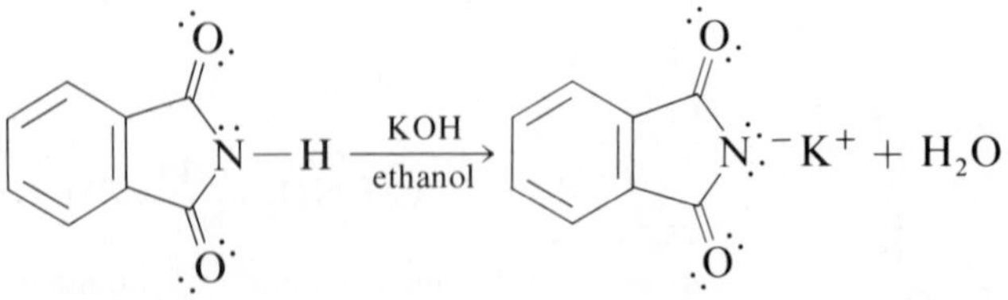
N—H
KOH
ethanol
K^+ + H_2O

phthalimide, pK_a 7.4; potassium phthalimidate, 85%

The acidity of phthalimide can be rationalized by the possibilities for delocalization of the negative charge in its conjugate base.

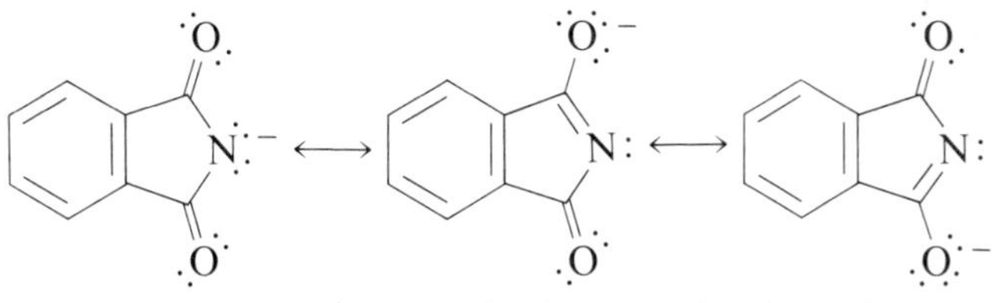

resonance contributors for the anion of phthalimide

The phthalimidate anion is a good nucleophile that participates in a wide variety of nucleophilic substitution reactions. It can be used to prepare primary amines that are difficult to obtain in high yields from the simple alkylation of ammonia. Benzyl chloride reacts with potassium phthalimidate to give *N*-benzylphthalimide, which is hydrolyzed in acid to the primary amine and phthalic acid.

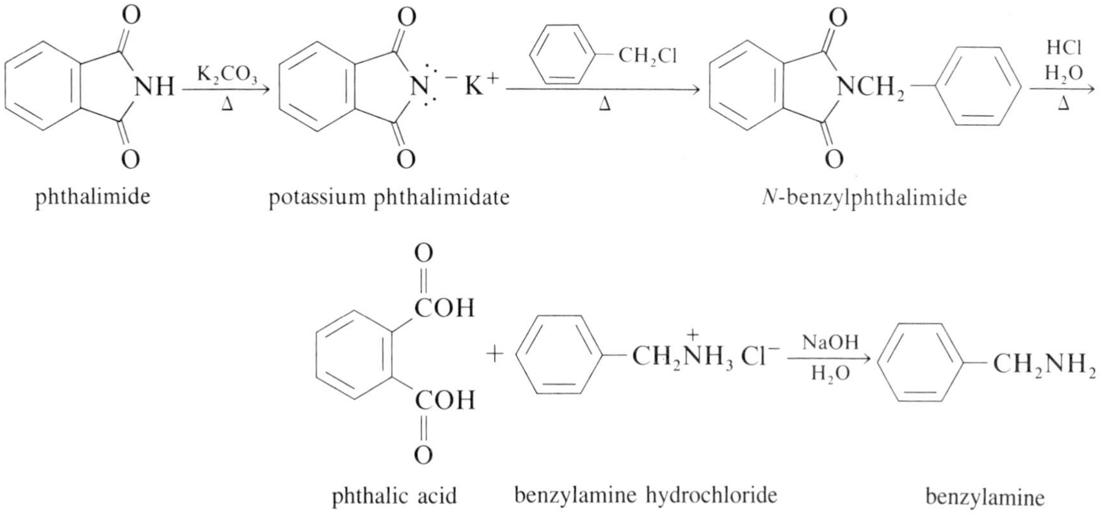

phthalimide
K_2CO_3
Δ
potassium phthalimidate
CH_2Cl
N-benzylphthalimide
HCl
H_2O
phthalic acid
benzylamine hydrochloride
NaOH
benzylamine

Such a sequence of reactions is called the **Gabriel phthalimide synthesis** of primary amines.

The phthalimide synthesis is most useful for making amines that also contain other functional groups. 4-Aminobutanoic acid, which functions as an agent in the transmission of nerve impulses and is also known as GABA (for *g*amma-*a*mino*b*utyric *a*cid, its common name), is synthesized by the alkylation of potassium phthalimidate with 4-chlorobutanenitrile.

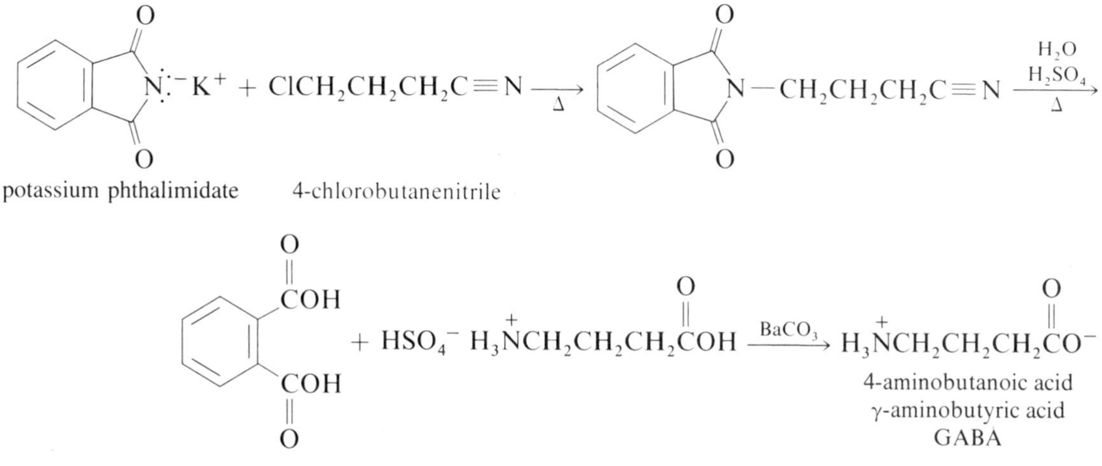

potassium phthalimidate
4-chlorobutanenitrile
H_2O
H_2SO_4
$BaCO_3$
4-aminobutanoic acid
γ-aminobutyric acid
GABA

The carboxylic acid group of the amino acid is protected as the nitrile, which is hydrolyzed in the second step of the synthesis.

PROBLEM 22.6

4-Chlorobutanoic acid was not used as the reagent to alkylate potassium phthalimidate in the synthesis shown above. Why not?

PROBLEM 22.7

4-Chlorobutanenitrile is synthesized from 1,3-propanediol. How would you carry out the synthesis?

C. Reduction of Azides

The azide ion, N_3^-, is a good nucleophile and is used to create carbon-nitrogen bonds (p. 260) in S_N2 reactions. Reduction of azides gives primary amines. Hydrogen in the presence of a metallic catalyst or lithium aluminum hydride is the reducing agent used most frequently. For example, (*S*)-2-octanol is converted into (*R*)-2-octylamine by the following sequence of reactions.

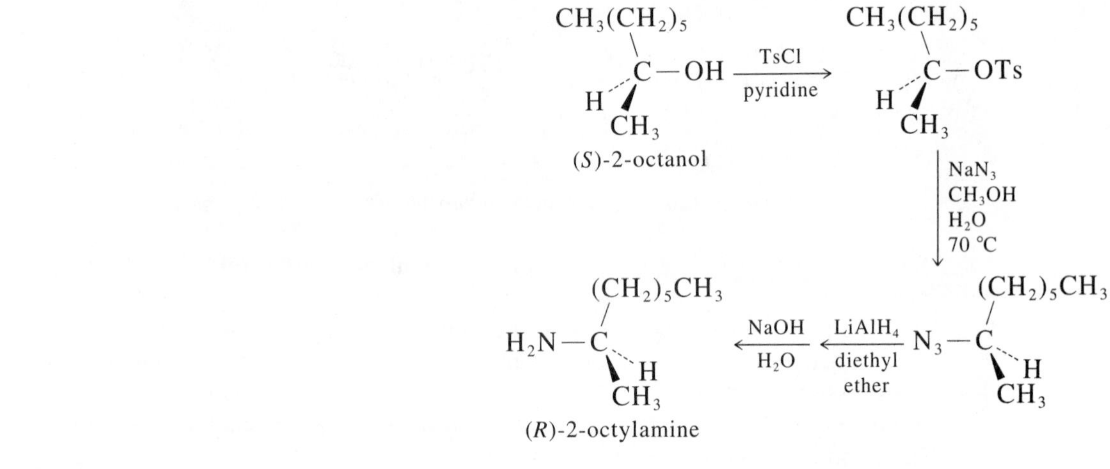
CH3(CH2)5
C—OH
H
CH3
(S)-2-octanol
TsCl
pyridine
CH3(CH2)5
C—OTs
H
CH3
NaN3
CH3OH
H2O
70 °C
(CH2)5CH3
N3—C
H
CH3
LiAlH4
diethyl
ether
NaOH
H2O
(CH2)5CH3
H2N—C
H
CH3
(R)-2-octylamine

The displacement of tosylate ion by azide ion is accompanied by inversion of configuration. The azide is reduced to the amine with retention of configuration. The other two nitrogen atoms of the azide group are lost as N_2, nitrogen gas.

Azides are also formed when oxiranes undergo a ring-opening reaction with sodium azide. For example, cyclohexene oxide is converted to *trans*-2-aminocyclohexanol in the following way.

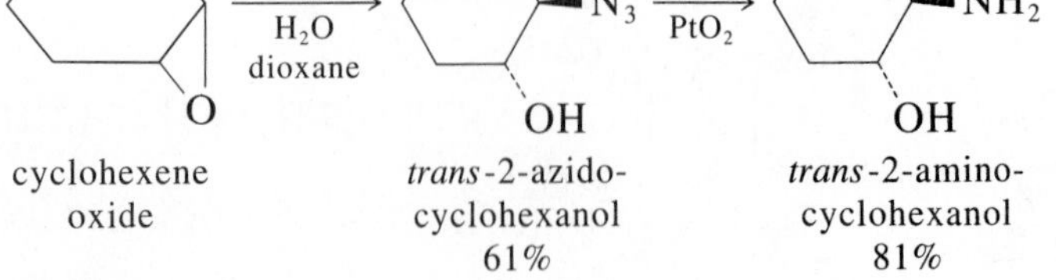
O
cyclohexene
oxide
NaN3
H2O
dioxane
N3
OH
trans-2-azido-
cyclohexanol
61%
H2
PtO2
NH2
OH
trans-2-amino-
cyclohexanol
81%

Note the stereochemistry that results from the opening of the oxirane ring by a nucleophile (p. 454).

Poisoned catalysts may be used in reducing an azide if the product amine is to retain other functional groups that would be affected by the reducing agents shown above, such as carbon-carbon double bonds or carbonyl groups. (*E*)-3-Bromo-1-phenyl-1-propene is converted to (*E*)-3-amino-1-phenyl-1-propene, for example.

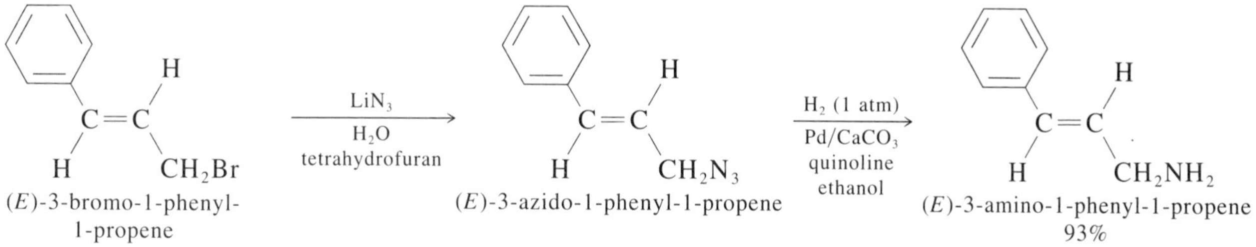
LiN3
H2O
tetrahydrofuran
H2 (1 atm)
Pd/CaCO3
quinoline
ethanol
(E)-3-bromo-1-phenyl-1-propene
(E)-3-azido-1-phenyl-1-propene
(E)-3-amino-1-phenyl-1-propene
93%

PROBLEM 22.8

Predict the products of the following reactions.

(a) 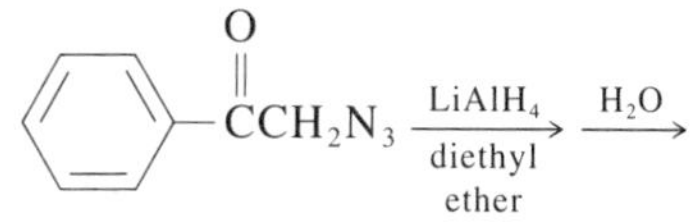
LiAlH4
diethyl ether
H2O

(b) $CH_3(CH_2)_4C{\equiv}C(CH_2)_8N_3 \xrightarrow[\text{Pd/CaCO}_3\text{, quinoline}]{H_2}$

D. Reduction of Imines or Acid Derivatives

Imines and oximes, compounds containing carbon-nitrogen double bonds, are formed when amines or hydroxylamines react with aldehydes or ketones (pp. 506 and 504). Nitriles, compounds containing carbon-nitrogen triple bonds, are made easily in nucleophilic substitution reactions with cyanide ion. These compounds can all be reduced to amines by metal hydrides or by hydrogen and a catalyst.

The phenylimine of benzaldehyde, which is formed when benzaldehyde and aniline condense, is reduced by sodium borohydride to a secondary amine.

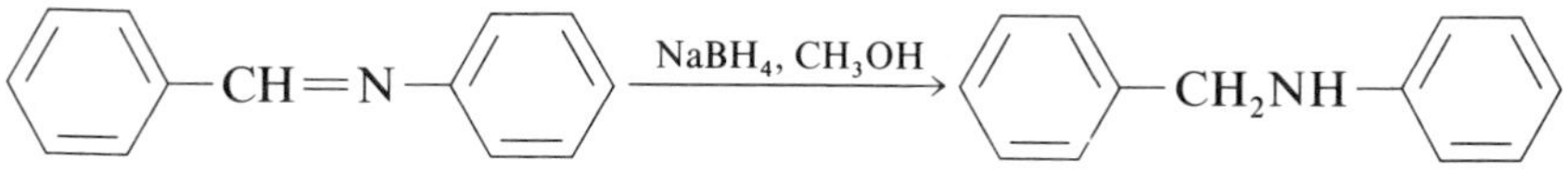
CH=N
NaBH4, CH3OH
CH2NH

phenylimine of benzaldehyde

N-benzylaniline
97%

The imine function is like a carbonyl group in its polarization and undergoes hydride reduction in the same way.

VISUALIZING THE REACTION

Reduction of an imine

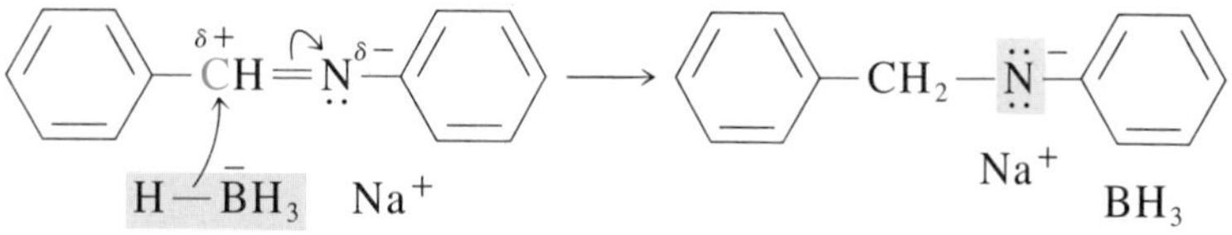
δ+
δ−
CH=N
H—BH3 Na+
CH2—N
Na+
BH3

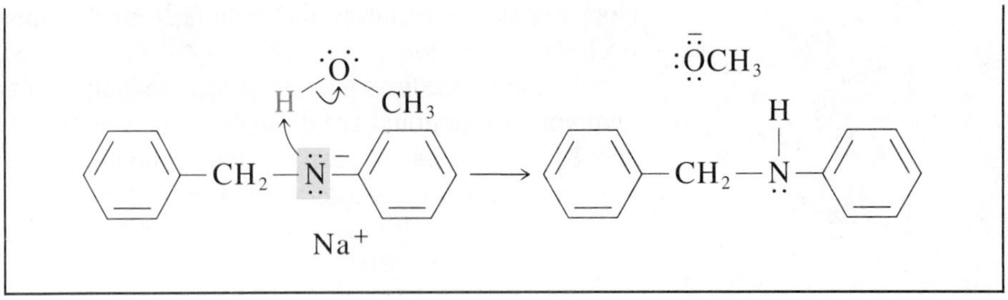

The phenylimine of benzaldehyde can also be reduced catalytically.

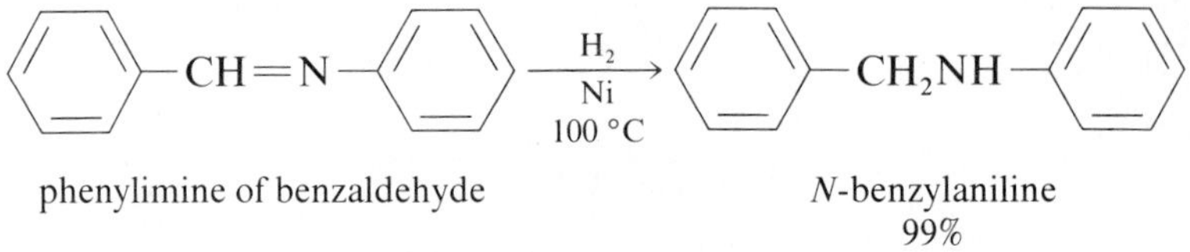

It is not usually necessary to isolate an imine in order to prepare the corresponding amine. The imines generated from aldehydes and ketones with ammonia are highly unstable. In a process known as **reductive amination,** an imine is formed and reduced in the same step by mixing a carbonyl compound and ammonia in the presence of hydrogen gas and a catalyst. Thus, acetophenone is converted into α-phenylethylamine.

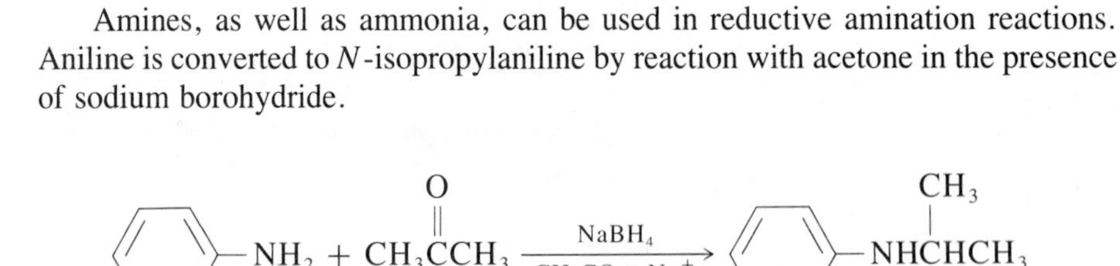

Amines, as well as ammonia, can be used in reductive amination reactions. Aniline is converted to *N*-isopropylaniline by reaction with acetone in the presence of sodium borohydride.

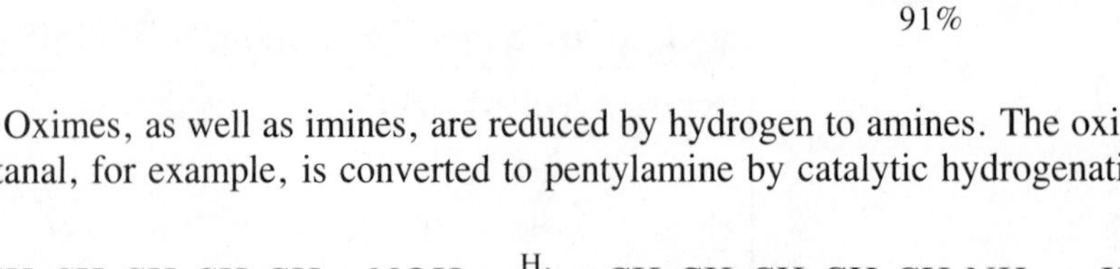

Oximes, as well as imines, are reduced by hydrogen to amines. The oxime of pentanal, for example, is converted to pentylamine by catalytic hydrogenation.

$$CH_3CH_2CH_2CH_2CH{=}NOH \xrightarrow[\text{Ni, 100 °C}]{H_2} CH_3CH_2CH_2CH_2CH_2NH_2 + H_2O$$

oxime of pentanal — pentylamine 62%

Remember that nitriles are reduced by 1 equivalent of diisobutylaluminum hydride to imines, which are hydrolyzed to aldehydes (p. 601). Nitriles can be reduced to amines by the more powerful hydride reagent lithium aluminum hydride (p. 597). Octanenitrile, for example, is reduced to octylamine by this reagent.

$$\underset{\text{octanenitrile}}{CH_3(CH_2)_6C{\equiv}N} \xrightarrow[\text{diethyl ether}]{LiAlH_4} \xrightarrow{H_2O} \underset{\substack{\text{octylamine}\\ 90\%}}{CH_3(CH_2)_6CH_2NH_2}$$

Catalytic reduction of nitriles is also possible. Thus, butyl bromide is converted to a primary amine containing one more carbon atom by a two-step sequence of nucleophilic substitution and catalytic reduction reactions.

$$\underset{\text{butyl bromide}}{CH_3CH_2CH_2CH_2Br} + K^+C{\equiv}N^- \longrightarrow \underset{\text{pentanenitrile}}{CH_3CH_2CH_2CH_2C{\equiv}N} + K^+Br^- \xrightarrow[Ni]{H_2}$$

$$\underset{\substack{\text{pentylamine}\\ 90\%}}{CH_3CH_2CH_2CH_2CH_2NH_2}$$

Amides, formed from the reaction of ammonia or amines with acid derivatives are reduced to amines by lithium aluminum hydride. The kind of amine that results depends on the structure of the amide undergoing reduction. The details of these reactions were explored on p. 599, so the reduction of α-phenoxyacetamide serves as a reminder of this route to amines.

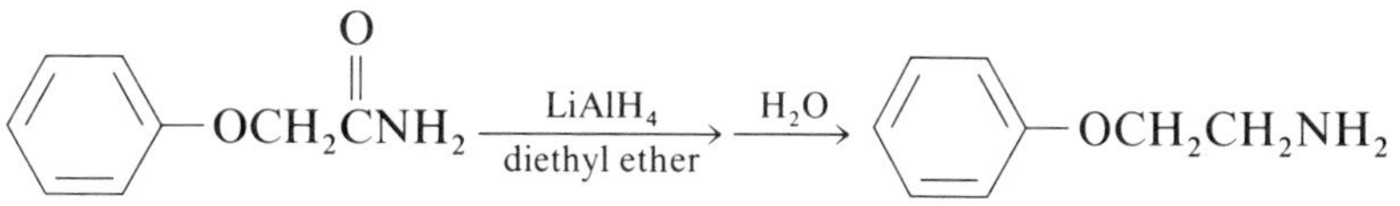

α-phenoxyacetamide $\qquad$ β-phenoxyethylamine 80%

PROBLEM 22.9

How could each of the following transformations be carried out?

(a) 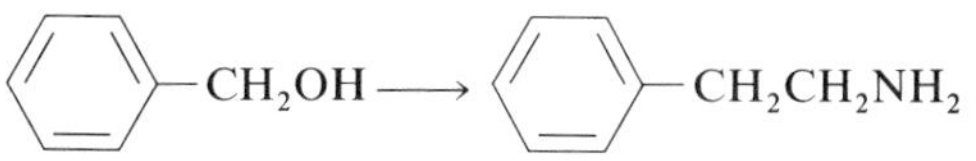

(b) $CH_3CH_2CH_2CH_2OH \longrightarrow CH_3CH_2CH_2CH(NH_2)CH_2CH_3$

(c) 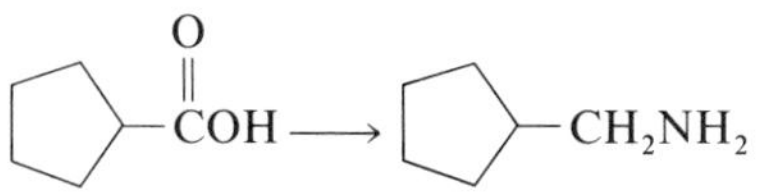

(d) 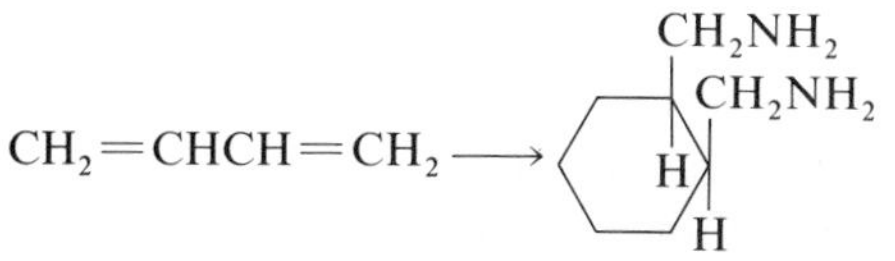

E. Reduction of Nitro Compounds

Aromatic amines are synthesized most conveniently by nitration of an aromatic ring and reduction of the nitro group to an amino group. The reduction may be carried out using tin or iron and hydrochloric acid, in which case the amine is obtained as its hydrochloric acid salt. Catalytic reduction of a nitro group with hydrogen gas and a catalyst gives the free amine directly. The reduction of nitrobenzene to aniline is a classic example of the reduction of a nitro group with a metal and an acid.

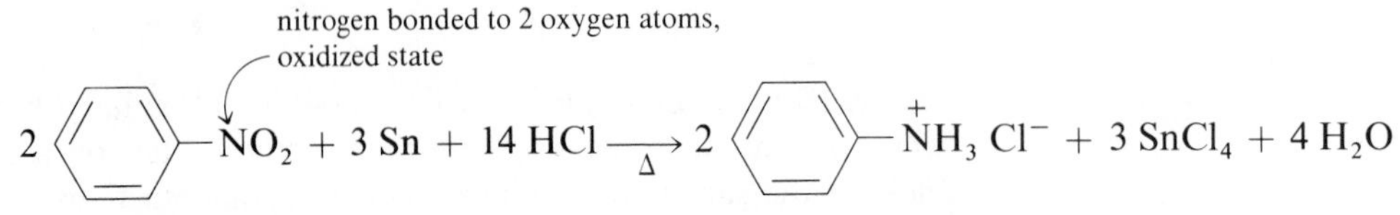
nitrogen bonded to 2 oxygen atoms, oxidized state
2 —NO2 + 3 Sn + 14 HCl → 2 —NH3 Cl⁻ + 3 SnCl4 + 4 H2O
Δ

nitrobenzene — aniline hydrochloride

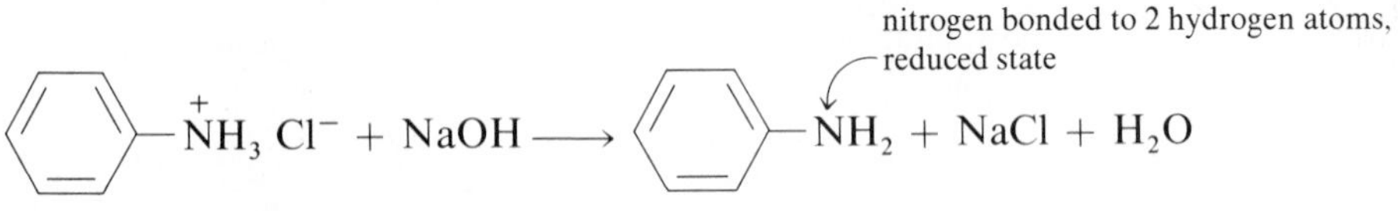
nitrogen bonded to 2 hydrogen atoms, reduced state
—NH3 Cl⁻ + NaOH → —NH2 + NaCl + H2O

aniline hydrochloride — aniline

Tin(0) is oxidized to tin(IV), and the nitrogen atom is reduced. The amine is in the form of its hydrochloride salt at the end of the reduction and must be treated with a strong base, such as sodium hydroxide, to be set free as aniline.

Catalytic reduction is part of a synthesis of *p*-aminobenzoic acid from *p*-nitrobenzoic acid.

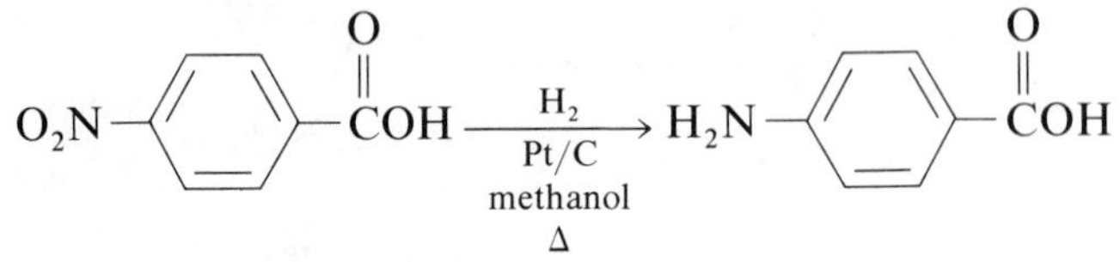
O2N— —COH
H2
Pt/C
methanol
Δ
H2N— —COH

p-nitrobenzoic acid — *p*-aminobenzoic acid

p-Aminobenzoic acid is a compound with many interesting physiological properties. It is incorporated by bacteria into folic acid, a vitamin that is essential to bacterial growth.

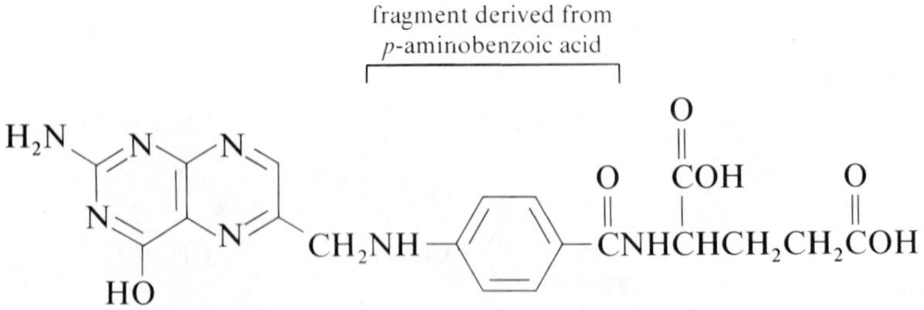
fragment derived from p-aminobenzoic acid
H2N
HO
CH2NH—
—CNHCHCH2CH2COH
COH

folic acid

The ability of sulfa drugs to inhibit bacterial growth results from the structural resemblance of these compounds, for example, sulfanilamide, to *p*-aminobenzoic acid.

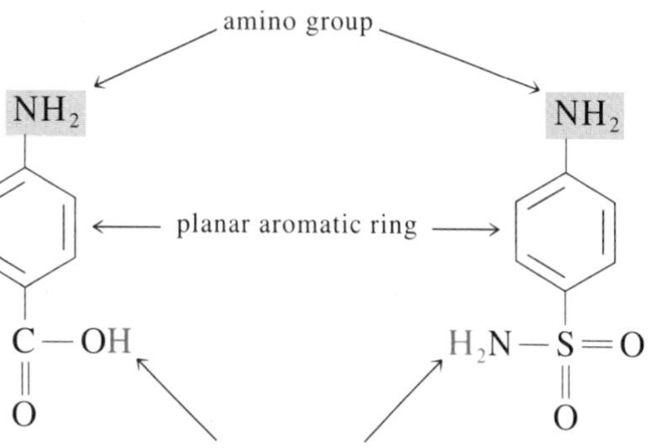
amino group
NH2
NH2
planar aromatic ring
C—OH
O
H2N—S=O
O
acidic hydrogen atom

Both *p*-aminobenzoic acid and sulfanilamide have the weakly basic amino group on an aromatic ring para to a functional group in which a central atom is doubly bonded to oxygen and an acidic hydrogen atom is bonded to an electronegative atom (oxygen or nitrogen). The sites of basicity and acidity and the spatial arrangement of the functional groups are similar enough in the two molecules that the bacterial enzyme that synthesizes folic acid mistakes the sulfa drug for *p*-aminobenzoic acid. The sulfa drug takes the place of *p*-aminobenzoic acid at the catalytic surface of the enzyme, preventing the synthesis of the essential vitamin and disrupting metabolic processes in the bacteria. Because human beings do not synthesize folic acid, sulfa drugs do not disrupt human metabolism.

Esters of *p*-aminobenzoic acid are local anesthetics. Ethyl *p*-aminobenzoate is commonly known as Benzocaine, and 2-(*N*,*N*-dimethylamino)ethyl *p*-aminobenzoate is the local anesthetic procaine, widely used in dentistry as its hydrochloride salt, Novocain.

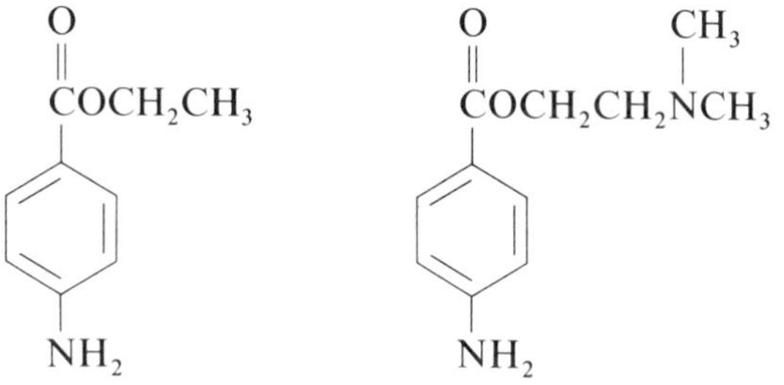
O
COCH2CH3
NH2
O
CH3
COCH2CH2NCH3
NH2

ethyl *p*-aminobenzoate
Benzocaine

2-(*N*,*N*-dimethylamino)ethyl
p-aminobenzoate
procaine

Other derivatives of *p*-aminobenzoic acid are used widely in suntan lotions to absorb ultraviolet radiation and to screen the skin from the more harmful wavelengths of sunlight.

PROBLEM 22.10

(a) A laboratory synthesis of *p*-aminobenzoic acid starts with *p*-toluidine and involves oxidation of the methyl group to a carboxylic acid (p. 963). The first step of the synthesis is protection of the amino group as the acetamide. Why is protection necessary?

(b) At the end of the synthesis the reaction mixture is basic. A careful adjustment of the acidity of the reaction mixture is necessary to isolate *p*-aminobenzoic acid. Why is the pH so important?

PROBLEM 22.11

Write the structural formula for the local anesthetic Novocain, which is procaine hydrochloride and has the molecular formula $C_{11}H_{16}N_2O_2 \cdot HCl$. (Hint: Which is the most basic site in procaine, the structure of which is shown above?)

Dinitration of benzene results in *m*-dinitrobenzene, an aromatic ring with two nitrogen substituents meta to each other (p. 785). It is possible to reduce one of the nitro groups selectively to an amine using ammonium sulfide, $(NH_4)_2S$.

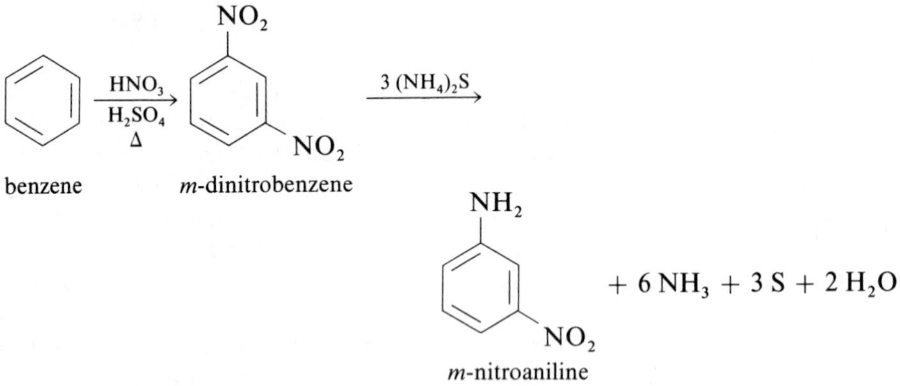
NO2
HNO3
H2SO4
Δ
NO2
3 (NH4)2S
benzene
m-dinitrobenzene
NH2
+ 6 NH3 + 3 S + 2 H2O
NO2
m-nitroaniline

This reaction is useful because it creates one amino group that can be transformed in a variety of ways while retaining the option of reducing the nitro group to create a new amino group at a later stage of synthesis. Section 23.1A will present some examples of how such manipulations are employed.

The regioselectivity of this reaction is dependent on the nature of substituents adjacent to the nitro groups. When there are alkyl substituents, for example, the least hindered nitro group is reduced. Such a selective reduction is used in the synthesis of 2-methyl-6-trifluoromethylaniline. The synthesis starts with the nitration of *m*-trifluoromethyltoluene.

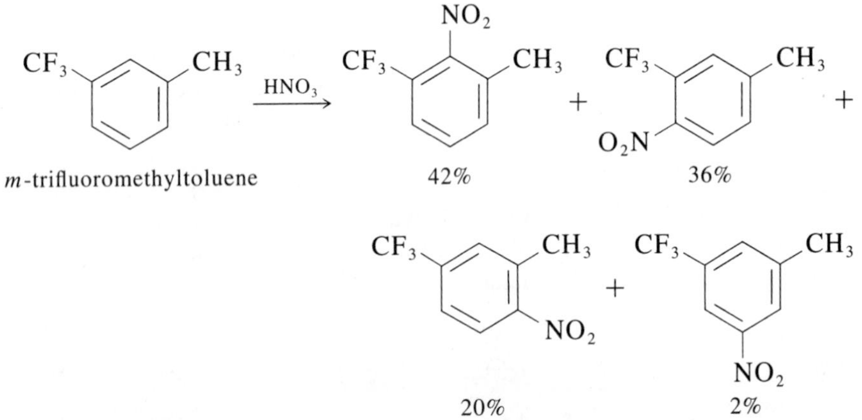
CF3
CH3
HNO3
m-trifluoromethyltoluene
NO2
42%
O2N
36%
NO2
20%
NO2
2%

The nitration gives a mixture of isomers, which is treated with sodium polysulfide in the presence of ammonium chloride in a strongly basic solution. The compound with the nitro group ortho to both alkyl substituents is not reduced in this reaction and can, therefore, be separated from the other isomers, which are converted to amines.

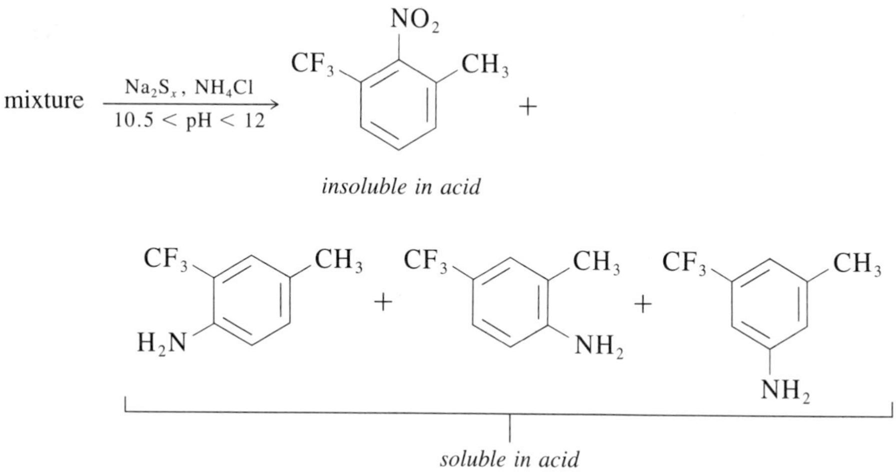
mixture
Na2Sx, NH4Cl
10.5 < pH < 12
NO2
CF3
CH3
insoluble in acid
CF3
CH3
H2N
CF3
CH3
NH2
CF3
CH3
NH2
soluble in acid

The separated nitro compound is then reduced with hydrogen, yielding the desired amine.

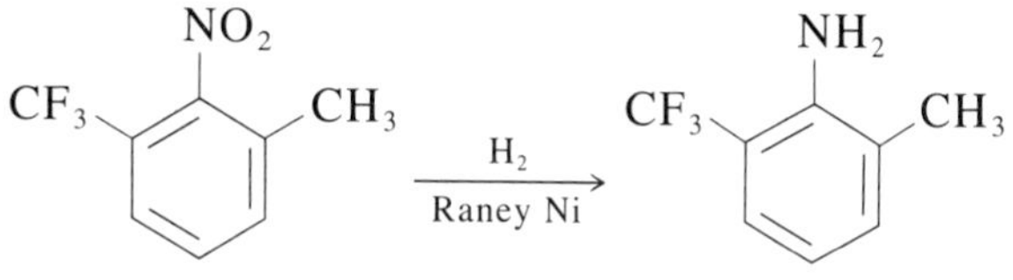
NO2
CF3
CH3
H2
Raney Ni
NH2
CF3
CH3

2-methyl-6-trifluoromethylaniline

An amino or hydroxyl group, on the other hand, causes a nitro group adjacent to it to be the one reduced predominantly. The selective reduction of 2,3-dinitrophenol to 2-amino-4-nitrophenol illustrates this.

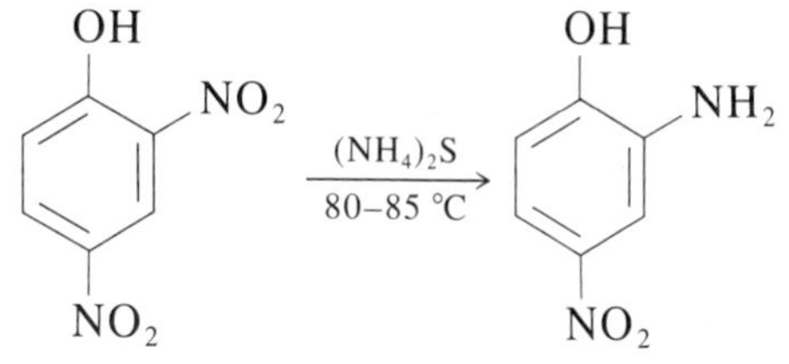
OH
NO2
NO2
(NH4)2S
80–85 °C
OH
NH2
NO2

2,4-dinitrophenol 2-amino-4-nitrophenol ~65%

The nitrite anion can be used as a nucleophile to make alkyl nitro compounds from primary and secondary alkyl halides. Bromocyclopentane is converted to nitrocyclopentane in this way.

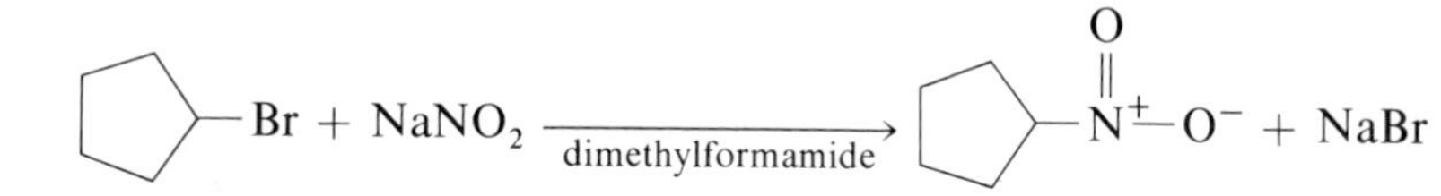
Br + NaNO2
dimethylformamide
O
N+ O− + NaBr

bromocyclopentane sodium nitrite nitrocyclopentane 57% sodium bromide

Alkyl nitro compounds can also be reduced to alkyl amines in high yield with metal and acid or by catalytic hydrogenation. Butylamine is prepared by either of these two methods.

$$CH_3CH_2CH_2CH_2NO_2 + 2\,Fe + 7\,HCl \longrightarrow CH_3CH_2CH_2CH_2\overset{+}{N}H_3\ Cl^- + 2\,FeCl_3 + 2\,H_2O$$

1-nitrobutane — butylamine hydrochloride

$$CH_3CH_2CH_2CH_2NO_2 \xrightarrow[\text{Ni, high pressure}]{H_2} CH_3CH_2CH_2CH_2NH_2 + 2\,H_2O$$

1-nitrobutane — butylamine

The reduction reactions are entirely analogous to those seen for aromatic nitro compounds.

F. Biologically Important Amines from Nitro Compounds

Nitroalkenes are easily prepared from the condensation of aldehydes with nitromethane. This reaction is important in the synthesis of **β-phenylethylamines,** a class of amines that play important roles in biological systems. Adrenalin, the hormone secreted by the adrenal glands to mobilize the body to energetic action when either danger or pleasure is anticipated, is a β-phenylethylamine, as is norepinephrine, another amine involved in the transmission of nerve impulses, and mescaline, the hallucinogenic alkaloid of the peyote cactus. These structures appear below.

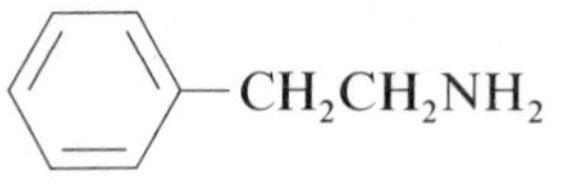

$CH_2CH_2NH_2$

β-phenylethylamine

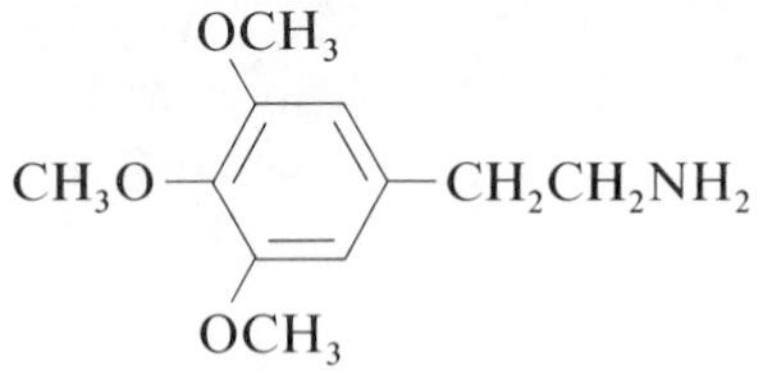

OCH_3
CH_3O
$CH_2CH_2NH_2$
OCH_3

mescaline

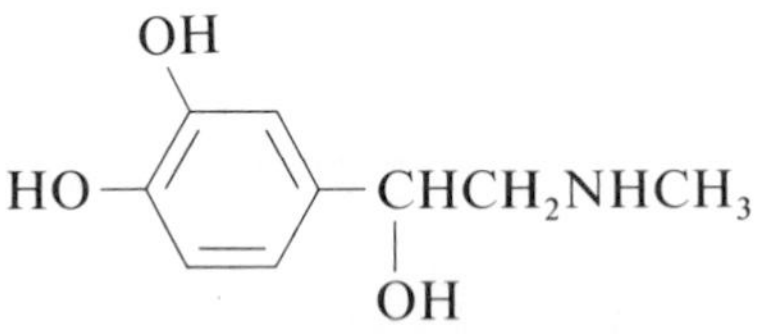

OH
HO
$CHCH_2NHCH_3$
OH

adrenalin
epinephrine

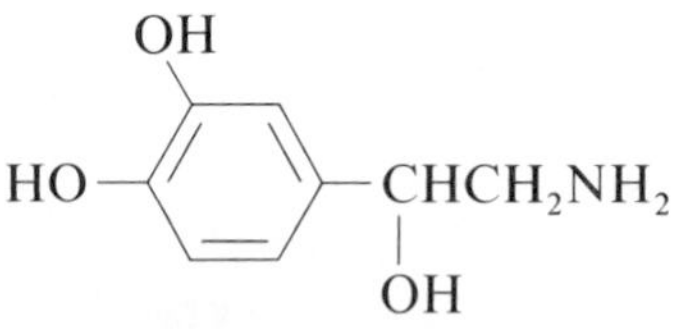

OH
HO
$CHCH_2NH_2$
OH

norepinephrine

biologically important β-phenylethylamines

The ways in which compounds such as adrenalin are generated in the body and how they are deactivated and removed from the body when the momentary need is over have been studied intensively. A large number of compounds that are structurally similar to these hormones have been synthesized so that the relationships between structure and biological activity can be studied. A synthesis of one biologically active β-phenylethylamine starts with the condensation of nitromethane with 3,4-methylenedioxybenzaldehyde.

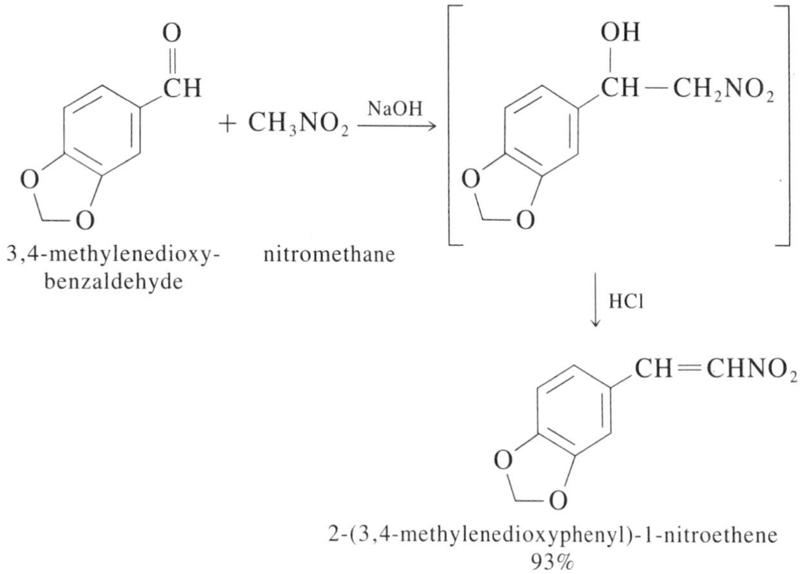
3,4-methylenedioxy-
benzaldehyde
nitromethane
NaOH
HCl
2-(3,4-methylenedioxyphenyl)-1-nitroethene
93%

The reaction is catalyzed by base and involves the addition of the carbanion from nitromethane to the carbonyl group of the aldehyde. The intermediate alcohol loses water easily to give a double bond conjugated with both the aromatic ring and the nitro group.

VISUALIZING THE REACTION

The condensation of nitromethane with an aldehyde

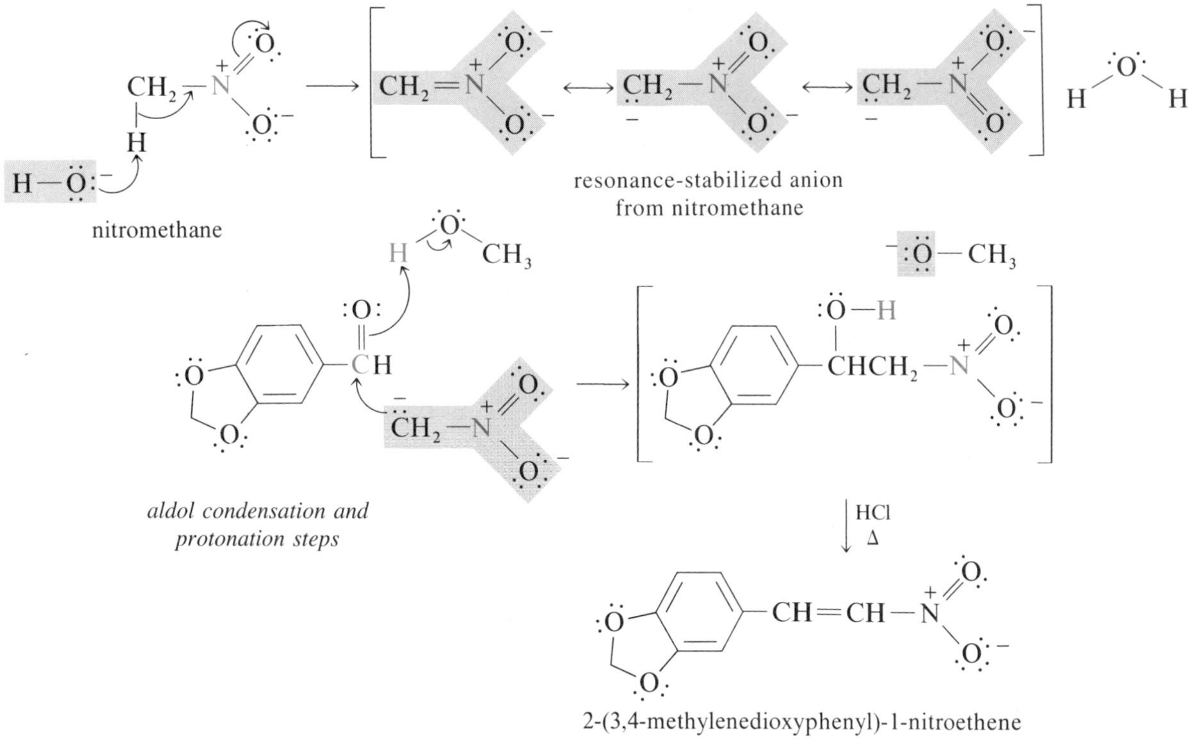
nitromethane
resonance-stabilized anion
from nitromethane
aldol condensation and
protonation steps
HCl
Δ
2-(3,4-methylenedioxyphenyl)-1-nitroethene

Study Guide Concept Map 22.1

The nitroalkene is converted to the β-phenylethylamine by catalytic hydrogenation.

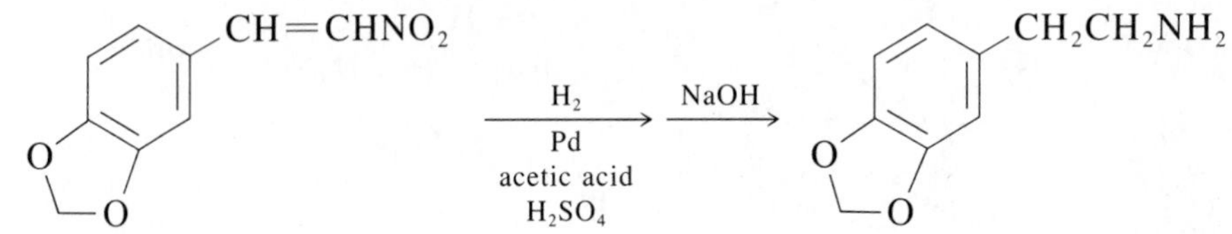

2-(3,4-methylenedioxyphenyl)-1-nitroethene → β-(3,4-methylenedioxyphenyl)-ethylamine

PROBLEM 22.12

Complete the following equations.

(a) 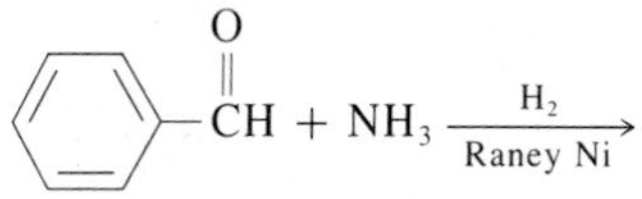

(b) 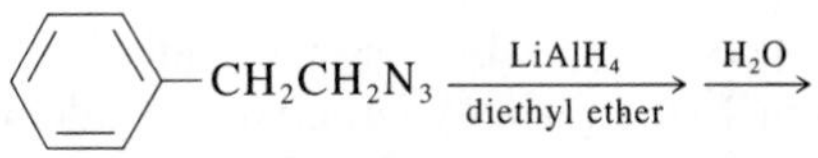

(c) 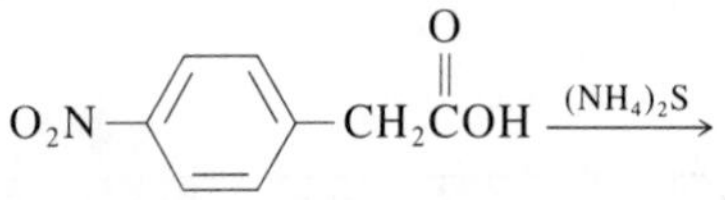

(d) 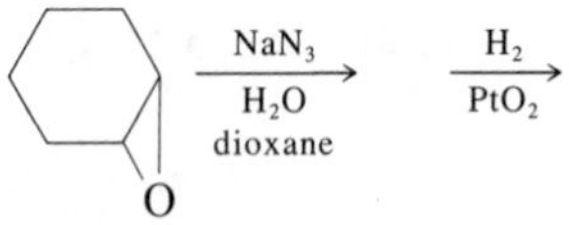

(e) 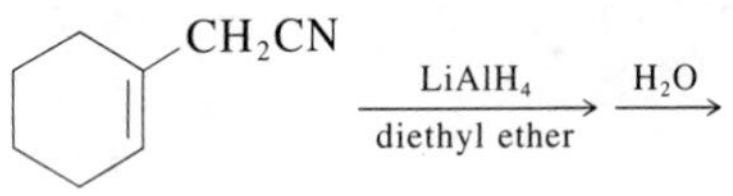

PROBLEM 22.13

Show how each of the following transformations could be carried out. Several steps may be required.

(a) 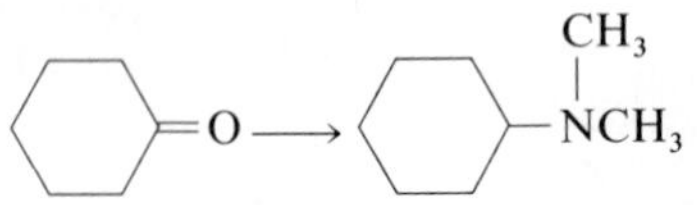

(b) 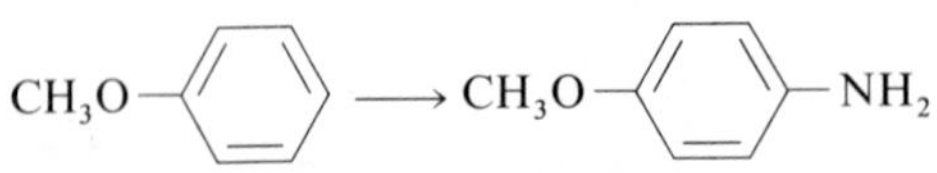

(c) 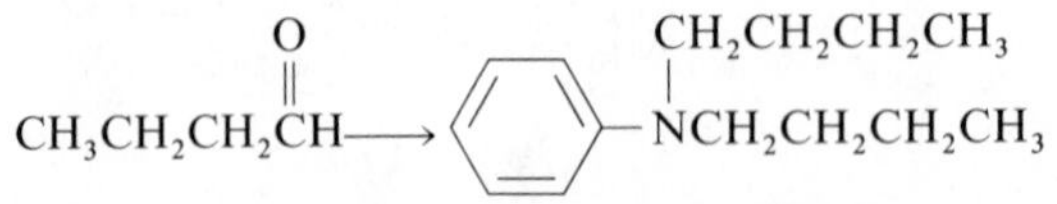

(d) $CH_3CH_2CH_2CH_2Br \longrightarrow CH_3(CH_2)_6CH_2NH_2$ (three different syntheses required)

PROBLEM 22.14

Starting with benzene, toluene, and any other organic compound containing no more than three carbon atoms, synthesize each of the following amines. Use the reduction of a nitro compound at some stage of each synthesis.

(a) 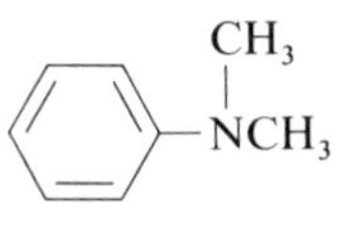
CH3
NCH3

(b) 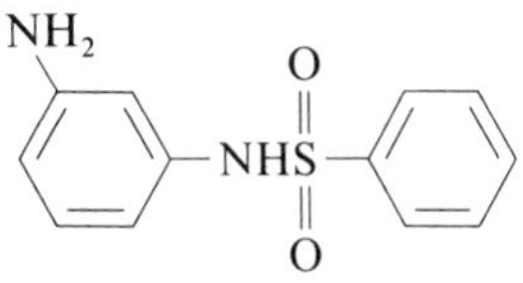
NH2
O
NHS
O

(c) 
O2N
NH2

(d) $CH_3CH(CH_3)CH_2CH_2NH_2$

22.5 NITROSATION REACTIONS OF AMINES

A. Nitrous Acid

Nitrous acid, HNO_2, is an unstable species that exists only as its salts or in solution in equilibrium with a number of other species, depending on the acidity of the solution and the other ions present. In the laboratory, it is generated as needed by treating sodium nitrite with a strong mineral acid, usually hydrochloric acid, at 0–5 °C.

$$Na^+ + NO_2^- + H_3O^+ + Cl^- \xrightarrow[H_2O,\ 0\,^\circ C]{} \underset{\substack{\text{nitrous acid} \\ pK_a\ 3.23}}{HO{-}N{=}O} + Na^+ + Cl^-$$

It is difficult to pin down the exact nature of the reacting species in a solution of nitrous acid. The acid is in equilibrium with its anhydride, dinitrogen trioxide, which, in turn, is in equilibrium with nitric oxide and nitrogen dioxide.

$$2\,\underset{\text{nitrous acid}}{HO{-}N{=}O} \rightleftharpoons H_2O + \underset{\text{dinitrogen trioxide}}{O{=}N{-}O{-}N{=}O} \rightleftharpoons \underset{\substack{\text{nitric} \\ \text{oxide}}}{N{=}O} + \underset{\substack{\text{nitrogen} \\ \text{dioxide} \\ \textit{a brown gas}}}{NO_2}$$

In strongly acidic solutions, nitrous acid is protonated; its conjugate acid then loses water to give the nitrosonium ion. In the presence of halide ions, nitrosyl halides also form.

$$\underset{\text{nitrous acid}}{HO{-}N{=}O} + H_3O^+ \rightleftharpoons \underset{\substack{\text{conjugate acid} \\ \text{of nitrous acid}}}{H\overset{+}{O}(H){-}N{=}O} \longrightarrow H_2O + \underset{\text{nitrosonium ion}}{\overset{+}{N}{=}O}$$

$$\underset{\substack{\text{nitrosonium} \\ \text{ion}}}{\overset{+}{N}{=}O} + Cl^- \rightleftharpoons \underset{\substack{\text{nitrosyl} \\ \text{chloride}}}{Cl{-}N{=}O}$$

All of these species are potentially sources of nitrosonium ion.

The nitrosonium ion is an electrophilic species, like the nitronium ion postulated as the reacting species in electrophilic aromatic substitution reactions (p. 791).

$$:N{\equiv}\overset{+}{O}: \longleftrightarrow :\overset{+}{N}{=}\ddot{O}$$

resonance contributors of the nitrosonium ion

Nitrosonium ion is a weaker electrophile with respect to aromatic rings than nitronium ion is but reacts readily with the nonbonding electrons on nitrogen atoms to initiate reactions that vary with the structure of the amine. Reactions of amines with nitrosonium ions are known as **nitrosation reactions.**

B. Nitrosation of Alkyl Amines. *N*-Nitrosamines

The first step of the reaction of nitrous acid with an amine is the formation of an *N*-nitrosamine. In the case of secondary alkylamines, the resulting *N*-nitrosamines are stable and are of great biological interest because they are known to be mutagens and carcinogens. One of the most potent is *N*-nitrosodimethylamine, formed when dimethylamine reacts with nitrous acid.

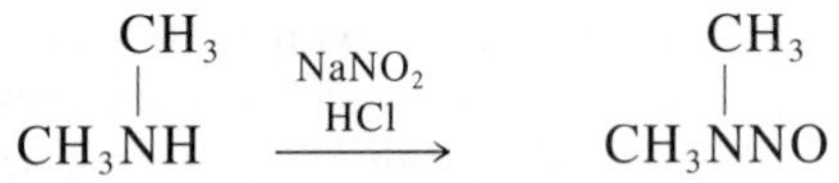
CH3
CH3NH
NaNO2
HCl
CH3
CH3NNO

dimethylamine → *N*-nitrosodimethylamine

The reaction is initiated by attack of the nonbonding electrons of the nitrogen atom of the amine on the electrophilic nitrosonium ion.

VISUALIZING THE REACTION

Nitrosation of a secondary amine

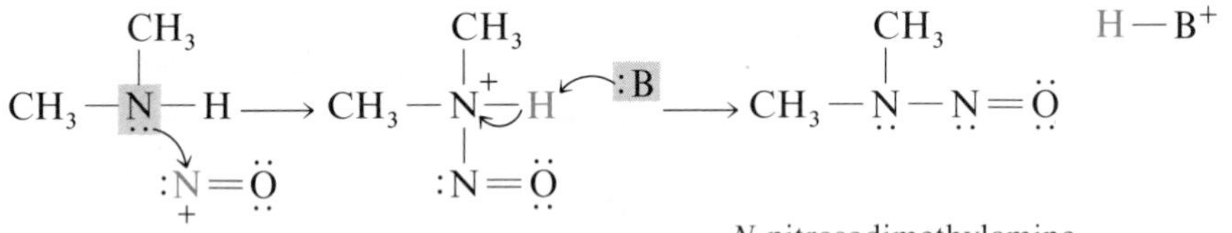
CH3
CH3—N—H
:N=O
CH3—N—H
:B
CH3—N—N=O
H—B+

N-nitrosodimethylamine

Since the discovery of the carcinogenicity of nitrosamines, extensive research has been conducted on the presence in biological systems of secondary amines capable of forming nitrosamines and on sources of nitrites in food. Hydrochloric acid in gastric juice generates nitrous acid from nitrites that are eaten. The two types of precursors to nitrosamines are found with disturbing frequency. Dimethylamine is found in a number of fish and meat products. Dimethylamine, methylethylamine, and the cyclic secondary amine pyrrolidine (p. 893) are found in tobacco smoke. Sodium nitrite is used as a preservative in meats such as bacon, cold cuts, and frankfurters. Nitrates, which are used widely as fertilizers, are also reduced to nitrites by the body and by some plants, so residues of these nitrates may also contribute to nitrite intake. Concern about the cancer-causing effects of nitrosamines has led to attempts to find ways of preserving foods without using nitrites.

The reaction of a primary alkyl amine, represented below by butylamine, with nitrous acid, generated from sodium nitrite and hydrochloric acid, yields a mixture of products including alcohols, alkenes, and alkyl halides.

$$\underset{\text{butylamine}}{CH_3CH_2CH_2CH_2NH_2} \xrightarrow[0\,°C]{NaNO_2,\ HCl} N_2\uparrow + \underset{\substack{n\text{-butyl alcohol}\\25\%}}{CH_3CH_2CH_2CH_2OH} + \underset{\substack{sec\text{-butyl alcohol}\\13\%}}{CH_3CH_2CH(OH)CH_3} +$$

$$\underset{\substack{n\text{-butyl chloride}\\5\%}}{CH_3CH_2CH_2CH_2Cl} + \underset{\substack{sec\text{-butyl chloride}\\3\%}}{CH_3CH_2CH(Cl)CH_3} + \underset{\substack{\text{1-butene}\\26\%}}{CH_3CH_2CH{=}CH_2} + \underset{\substack{\text{2-butene}\\10\%}}{CH_3CH{=}CHCH_3}$$

The reaction starts in the same way the nitrosation of a secondary amine does. Deprotonation and protonation then lead to the formation of an **alkyl diazonium ion,** which is an unstable species containing a very good leaving group, a nitrogen molecule.

V I S U A L I Z I N G   T H E   R E A C T I O N

Formation of a diazonium ion

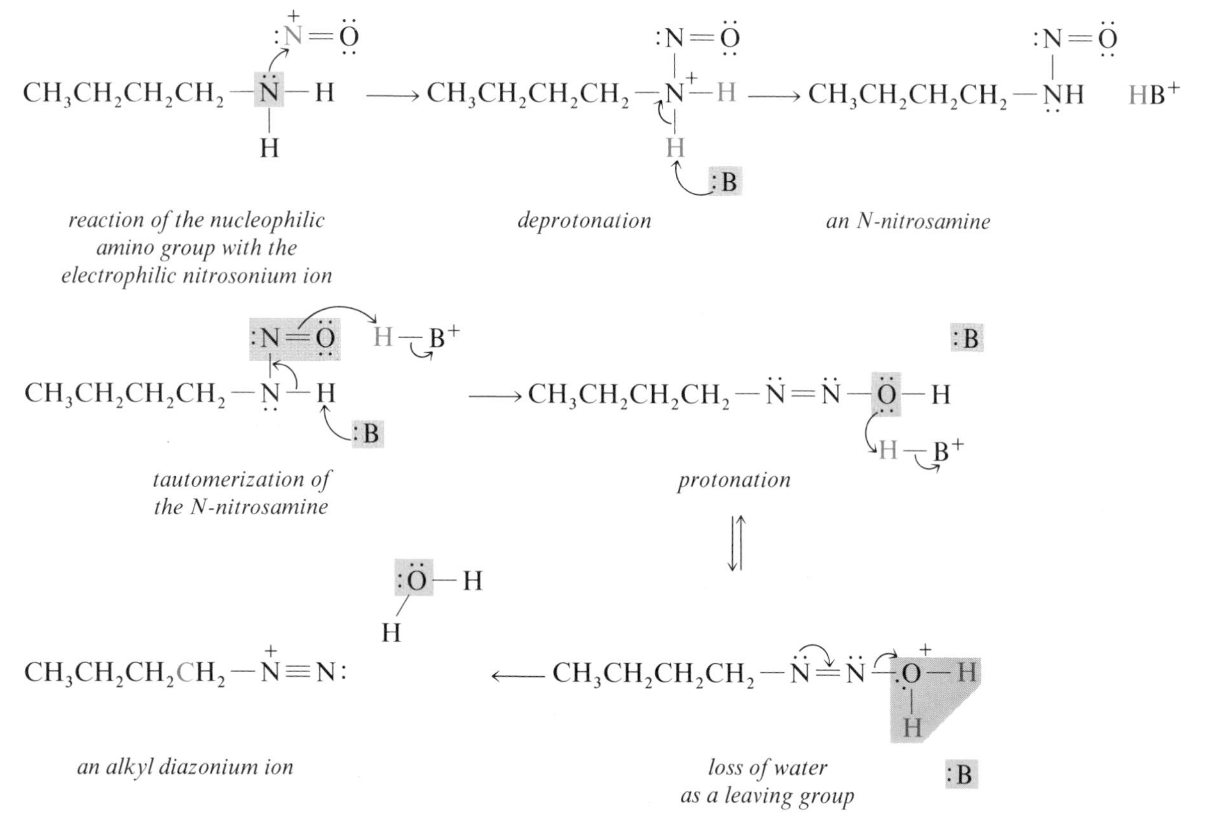
reaction of the nucleophilic amino group with the electrophilic nitrosonium ion
deprotonation
an N-nitrosamine
tautomerization of the N-nitrosamine
protonation
an alkyl diazonium ion
loss of water as a leaving group

Alkyl diazonium ions lose nitrogen spontaneously even at low temperatures. The nitrogen molecule may be displaced in an S_N2 reaction by a nucleophile present in the reaction mixture or lost as the leaving group in an E_2 elimination reaction.

VISUALIZING THE REACTION

S_N2 and E_2 reactions of diazonium ions

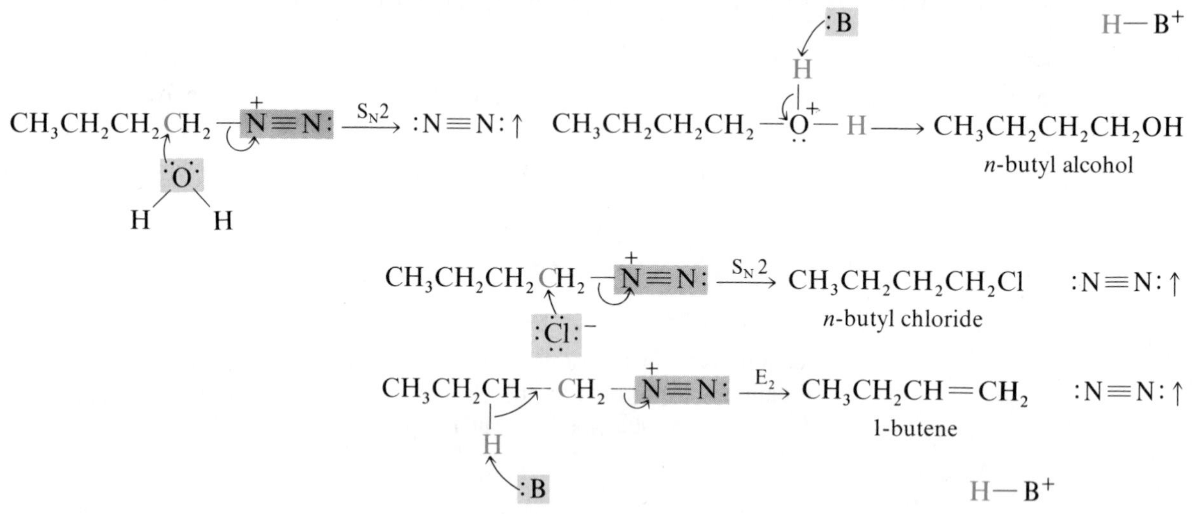
n-butyl alcohol
n-butyl chloride
1-butene

Alternatively, the alkyl part of the diazonium ion may rearrange to a more stable carbocation as nitrogen is lost. This cation, in turn, can react with a nucleophile in an S_N1 reaction or lose a proton in an E_1 reaction.

VISUALIZING THE REACTION

S_N1 and E_1 reactions of diazonium ions

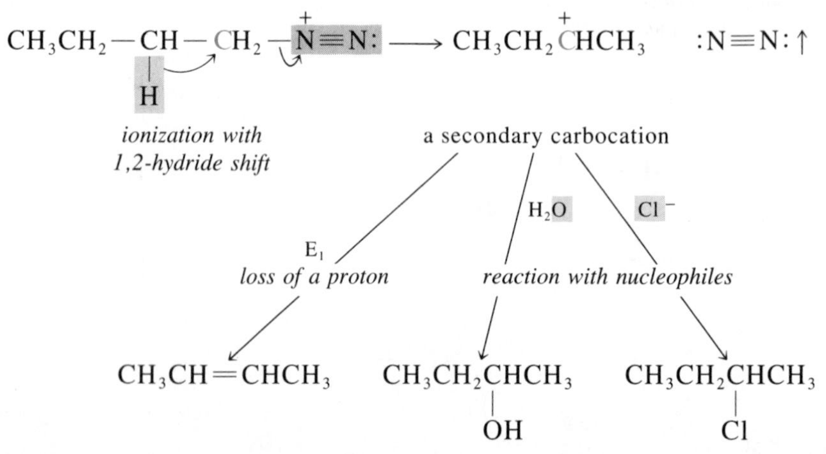
ionization with
1,2-hydride shift
a secondary carbocation
loss of a proton
reaction with nucleophiles

In the butyldiazonium ion, the leaving group is bonded to a primary carbon atom, and most of the products of the reaction are those formed by the S_N2 and E_2 reactions.

Tertiary amines also react with nitrous acid at low pH and low temperatures, though not as readily as primary and secondary amines do. At higher pH and higher temperatures, a tertiary amine is cleaved to give an *N*-nitroso secondary amine. The third organic group of the amine is converted into a carbonyl compound if there is a hydrogen atom on the carbon atom bonded to the nitrogen. The reaction of tribenzylamine with nitrous acid is typical.

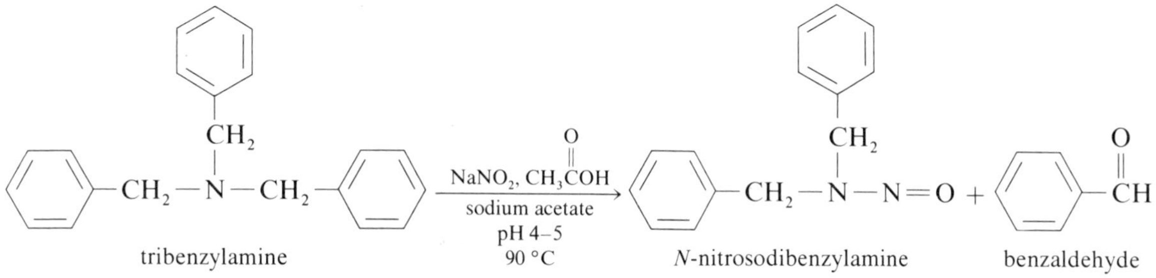
NaNO2, CH3COH
sodium acetate
pH 4–5
90 °C
tribenzylamine
N-nitrosodibenzylamine
benzaldehyde

PROBLEM 22.15

Complete the following equations, showing what product(s) you expect in each case.

(a) $\mathrm{CH_3CH_2CH_2CH(NH_2)CH_3 \xrightarrow[\text{H}_2\text{O, 0 °C}]{\text{NaNO}_2\text{, HCl}}}$

(b) 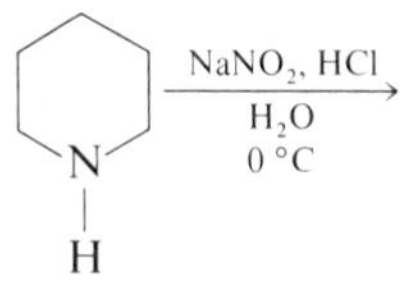
NaNO2, HCl
H2O
0 °C

(c) $\mathrm{CH_3CH_2CH_2N(CH_3)CH_3 \xrightarrow[\text{CH}_3\text{CO}^-\text{Na}^+ (\text{O}),\ \Delta]{\text{NaNO}_2\text{, CH}_3\text{COH (O)}}}$

(d) $\mathrm{CH_3NHC(=O)NHCH_3 \xrightarrow[\text{H}_2\text{O, 0 °C}]{\text{NaNO}_2\text{, HCl}}}$

(e) $\mathrm{CH_3C(CH_3)_2NH_2 \xrightarrow[\text{H}_2\text{O, 0 °C}]{\text{NaNO}_2\text{, HCl}}}$

PROBLEM 22.16

A secondary amine is soluble in the acidic solution used for a nitrosation reaction. The *N*-nitroso secondary amine that is the product of the reaction usually separates out of the solution as an insoluble oil or precipitate. Why?

C. Nitrosation of Aryl Amines. Aryl Diazonium Ions

Primary, secondary, and tertiary aryl amines undergo the same kinds of reactions with nitrous acid as alkyl amines do. However, in contrast to alkyl diazonium ions, **aryl diazonium ions,** which are formed from primary aromatic amines, are stable enough in solution at low temperatures that they can serve as reagents in reactions in which the nitrogen molecule is replaced in a controlled way. These reactions are of major synthetic importance and are discussed in detail in Section 23.1.

The mechanism described on p. 911 for the formation of an alkyl diazonium ion also applies to the reaction of an aryl amine with nitrous acid. Aniline, for example, is converted to benzenediazonium chloride when treated with sodium nitrite and hydrochloric acid at 0 °C. This is called a **diazotization reaction.**

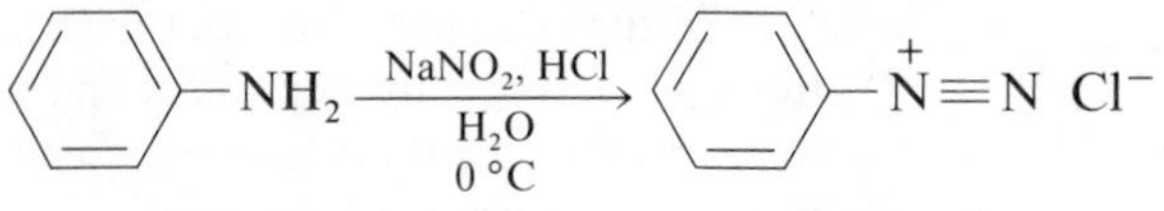

aniline — benzenediazonium chloride

soluble in water
stable at 0 °C

N-Methylaniline, on the other hand, is a secondary aryl amine and gives an *N*-nitrosamine under the same conditions.

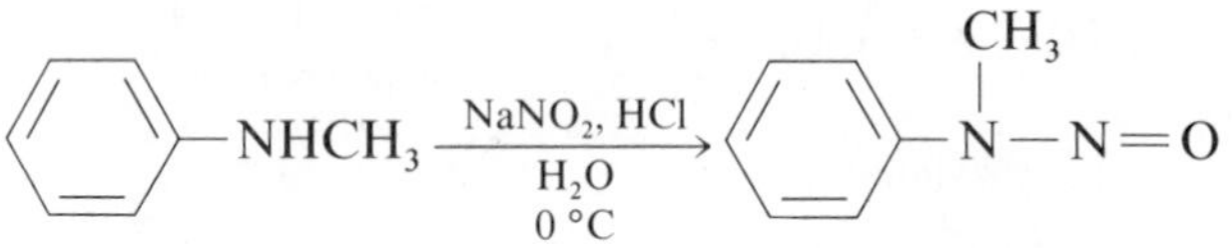

N-methylaniline — *N*-methyl-*N*-nitrosoaniline

insoluble in water

Tertiary aryl amines react in two ways. If the carbon at the position para to the amino group on the aromatic ring is unsubstituted, an electrophilic substitution reaction takes place with the nitrosonium ion acting as the electrophile.

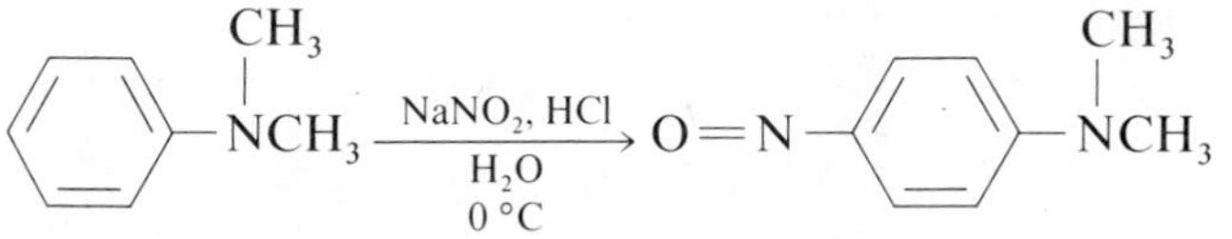

N,*N*-dimethylaniline — *N*,*N*-dimethyl-*p*-nitrosoaniline

If there is a substituent at that para position, reaction at the amino group (p. 910) takes place. For example, *N*,*N*-dimethyl-*p*-nitroaniline gives formaldehyde and *N*-methyl-*N*-nitroso-*p*-nitroaniline.

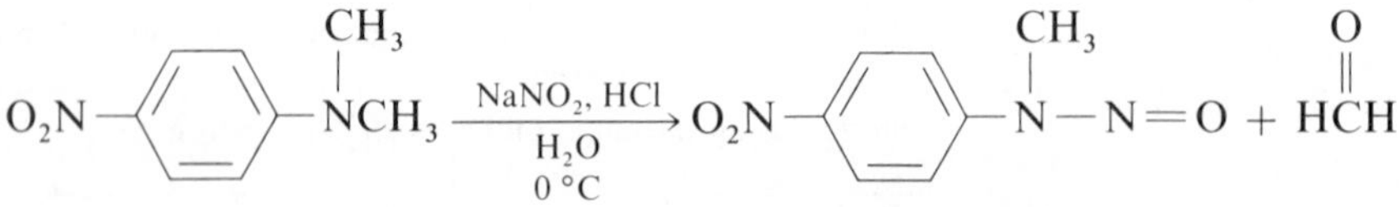

N,*N*,-dimethyl-*p*-nitroaniline — *N*-methyl-*N*-nitroso-*p*-nitroaniline 90% — formaldehyde

Study Guide Concept Map 22.2

PROBLEM 22.17

Complete the following equations.

(a) 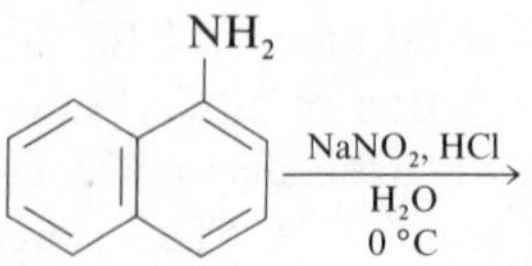

(b) 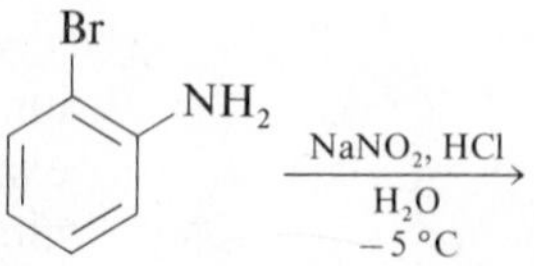

(c) 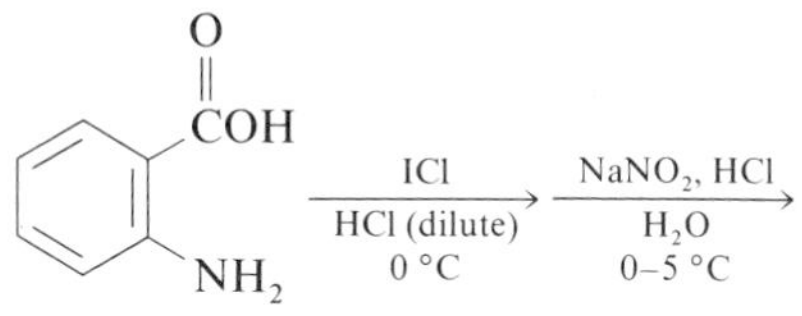

(d) 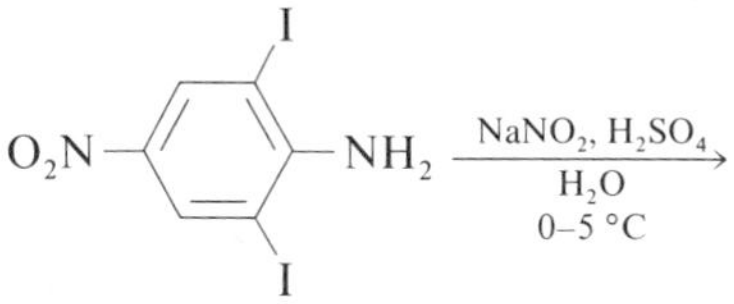

22.6 REARRANGEMENTS TO NITROGEN ATOMS

A. The Beckmann Rearrangement

When oximes are treated with a strong protic acid or a Lewis acid, they are converted to amides in a reaction known as the **Beckmann rearrangement.** For example, acetanilide is formed when the oxime of acetophenone is treated with trifluoroacetic acid.

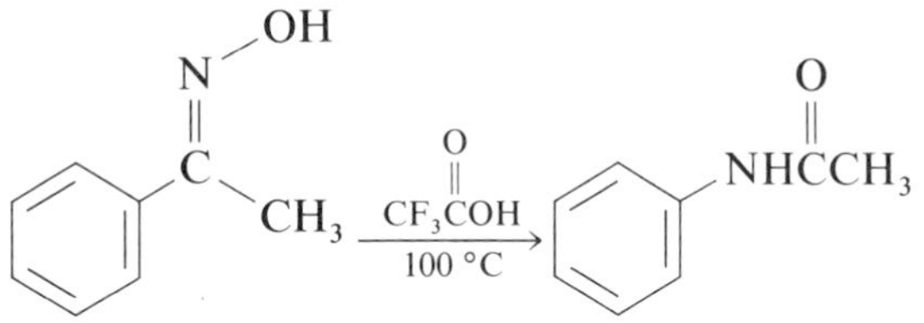

acetophenone oxime — acetanilide 91%

An inspection of the structural formulas shown for the oxime and the amide suggests that the phenyl group migrates from a carbon atom to a nitrogen atom during this reaction. Such a migration is somewhat similar to the 1,2-shift of an alkyl group to an electron-deficient carbon atom in rearrangements of carbocations (p. 287).

VISUALIZING THE REACTION

The Beckmann rearrangement

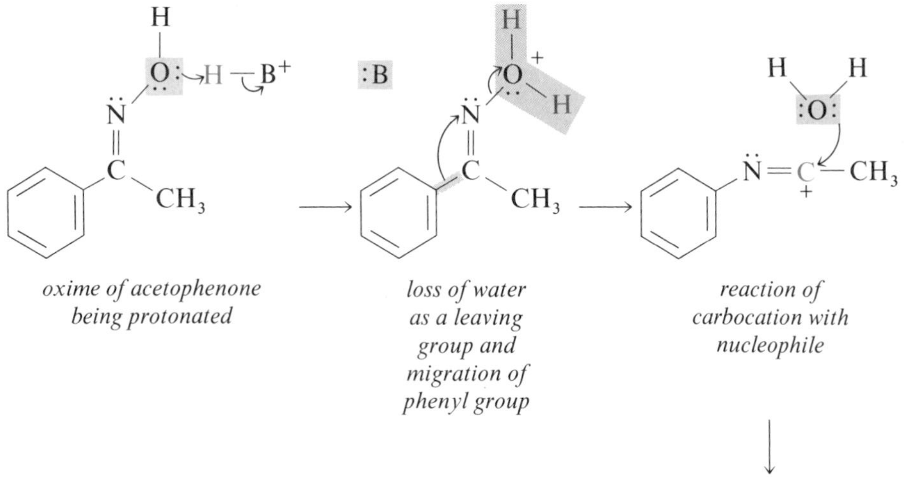

oxime of acetophenone being protonated

loss of water as a leaving group and migration of phenyl group

reaction of carbocation with nucleophile

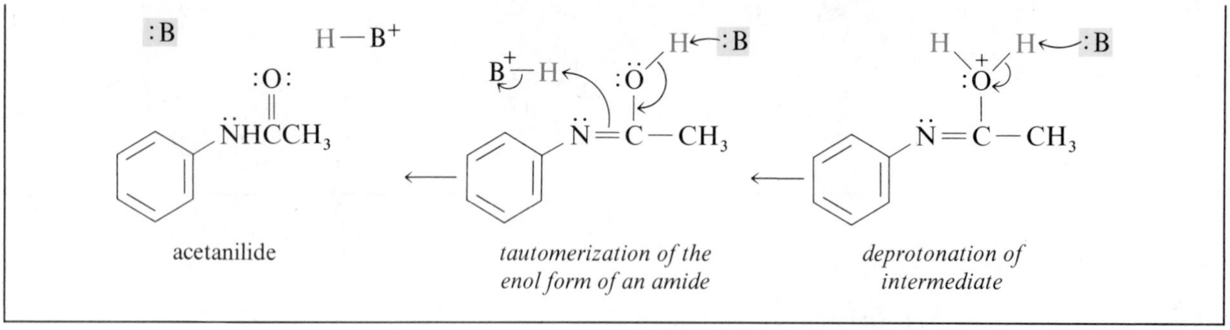

The first step of the reaction is a familiar one. The hydroxyl group of the oxime is converted into a good leaving group. As the leaving group moves away, the substituent that can approach the nitrogen atom from the rear and assist in the departure of the leaving group migrates to the nitrogen atom, creating an electron-deficient carbon atom. The carbocation reacts with water to give an amide. Hydrolysis of the amide will give an amine derived from one portion of the oxime molecule and a carboxylic acid derived from the other portion.

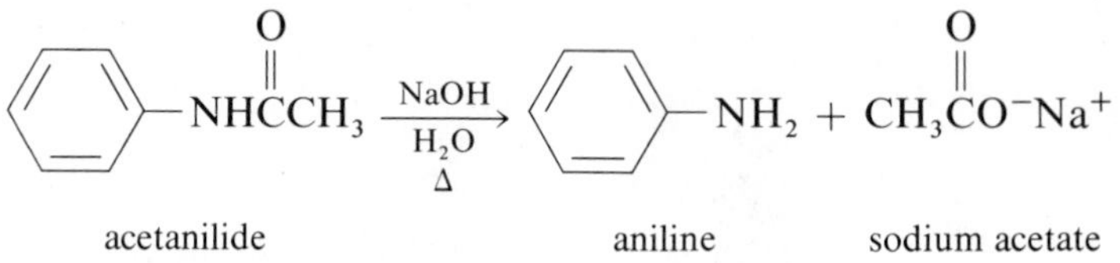

In one reaction of great commercial importance, the amide that is formed is of interest in its own right. ϵ-Caprolactam, the starting material for producing the polymer nylon 6, which is used extensively in tire cords, is made from cyclohexanone oxime.

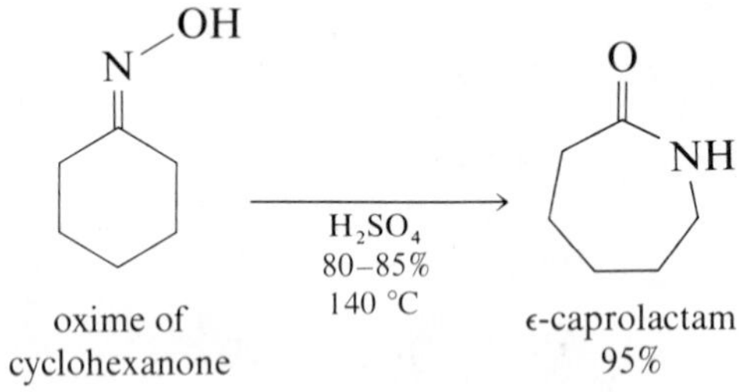

The Beckmann rearrangement of the oxime of a cyclic ketone gives rise to a **lactam,** a cyclic amide, which has one more atom in its ring than did the original ketone. ϵ-Caprolactam is polymerized (p. 1185) to give a linear polyamide.

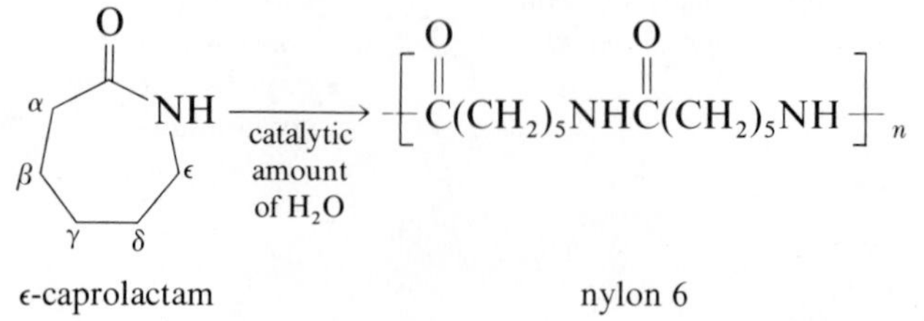

B. The Hofmann Rearrangement

In the **Hofmann rearrangement,** a primary amide is transformed into a primary amine containing one less carbon atom. The preparation of 3,4-dimethoxyaniline from 3,4-dimethoxybenzamide illustrates this reaction.

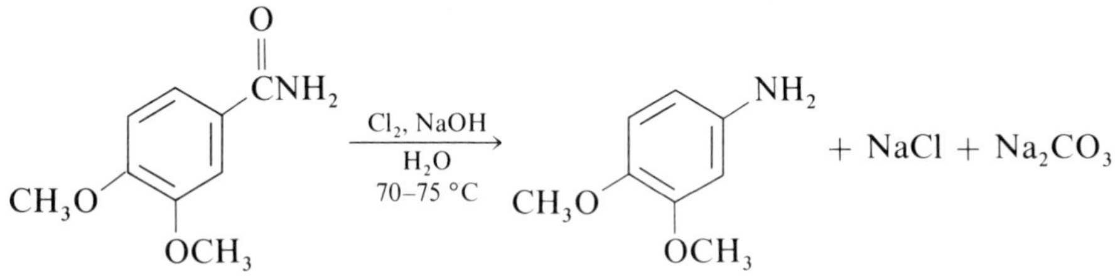

Cl_2, NaOH
H_2O
70–75 °C
+ NaCl + Na_2CO_3

3,4-dimethoxybenzamide

3,4-dimethoxyaniline 80%

Like the Beckmann rearrangement, the Hofmann rearrangement involves the migration of an organic group from a carbon atom to a nitrogen atom. The leaving group during the migration is a halide ion. The mechanism is shown below.

VISUALIZING THE REACTION

The Hofmann rearrangement

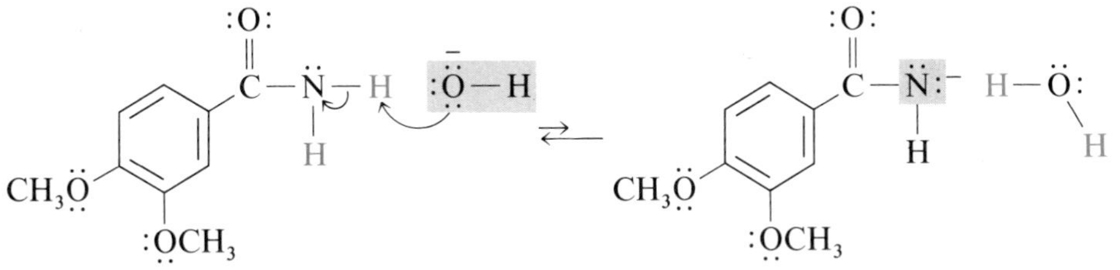

deprotonation of the amide

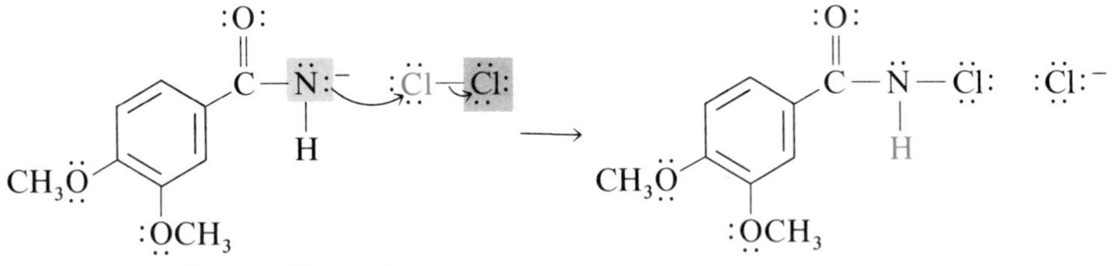

reaction of anion of the amide with the electrophilic halogen

an *N*-chloroamide

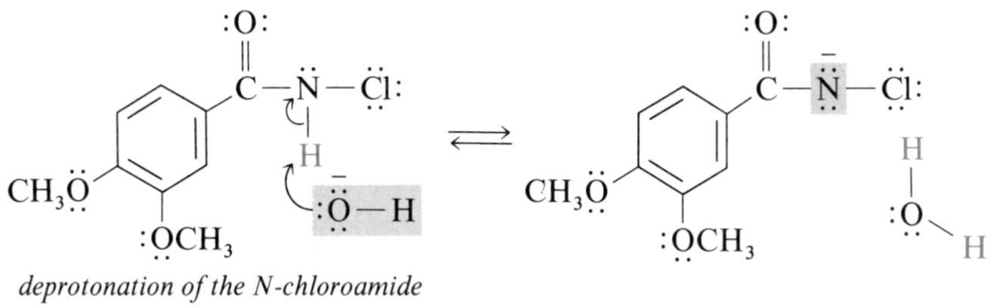

deprotonation of the N-chloroamide

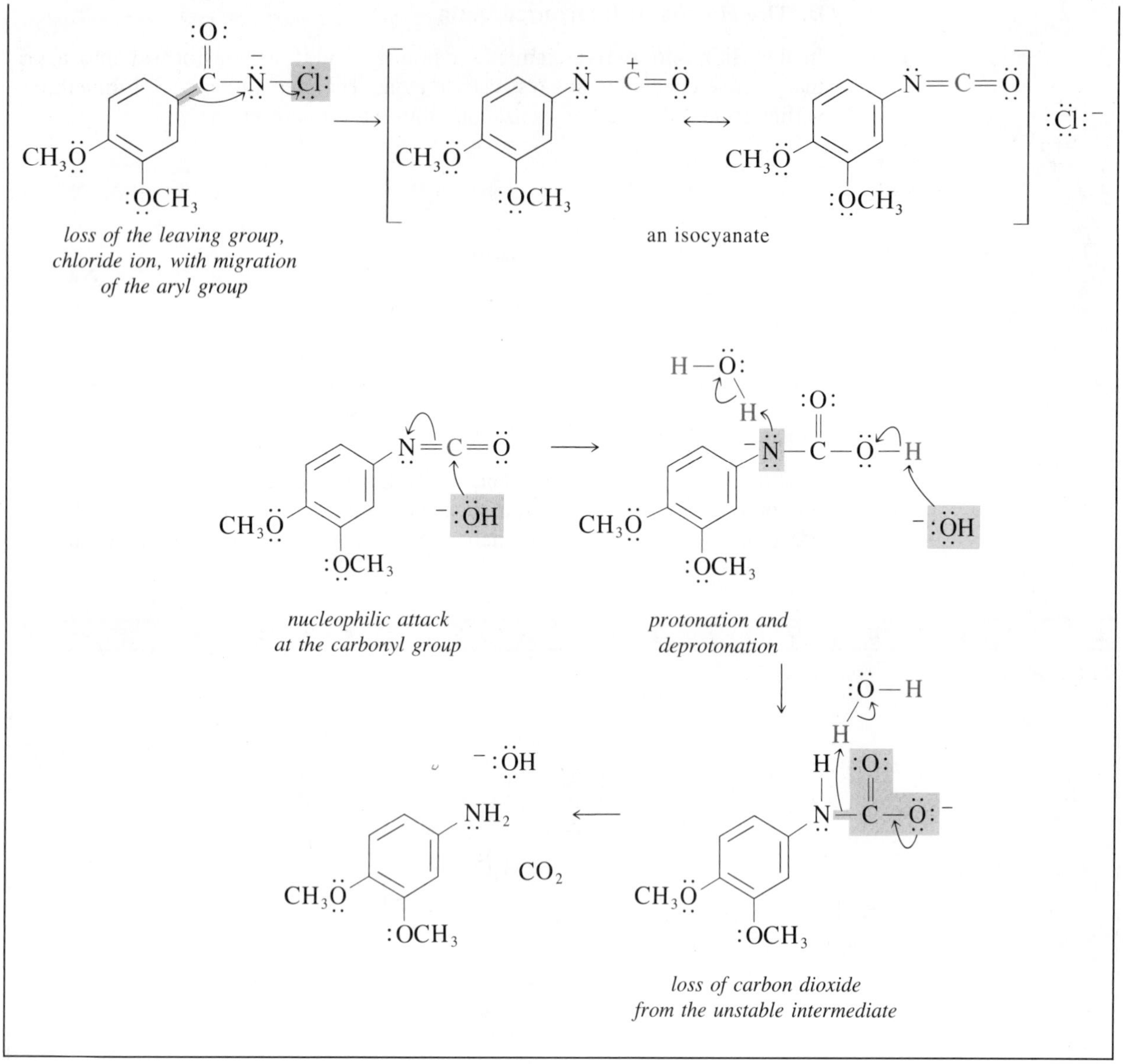
loss of the leaving group, chloride ion, with migration of the aryl group
an isocyanate
nucleophilic attack at the carbonyl group
protonation and deprotonation
loss of carbon dioxide from the unstable intermediate

The hydrogen atoms on the nitrogen atom of an amide are weakly acidic and can be removed by a base to give a nitrogen anion, which will react with chlorine. Because the *N*-chloroamide is much more acidic than the original amide, the loss of the second proton from the nitrogen atom takes place more easily than the loss of the first one. The anion of the *N*-chloroamide is unstable and loses chloride ion, with the simultaneous migration of the aryl group from the carbon atom to the nitrogen atom. The intermediate formed at this stage is an organic isocyanate, a compound that reacts readily with base and water to give the free amine and carbon dioxide. The carbon atom in the carbon dioxide was the carbonyl carbon of the amide. This reaction works with other halogens, but the amide must be unsubstituted on the nitrogen atom.

PROBLEM 22.18

Why is the *N*-chloroamide more acidic than the original unsubstituted amide?

Amines that are difficult to prepare by other methods can be made by the Hofmann rearrangement. An example is neopentylamine. Direct nucleophilic substitution of neopentyl bromide with ammonia is not possible because of steric hindrance to the S_N2 reaction by the methyl groups on the β-carbon atom (p. 242). *tert*-Butyl cyanide, which could be a precursor to the amine via a reductive pathway, cannot be easily prepared either because the tertiary halide, *tert*-butyl bromide, undergoes elimination rather than substitution reactions with cyanide ion (p. 255). Treatment of 3,3-dimethylbutanamide with bromine and sodium hydroxide converts it to neopentylamine.

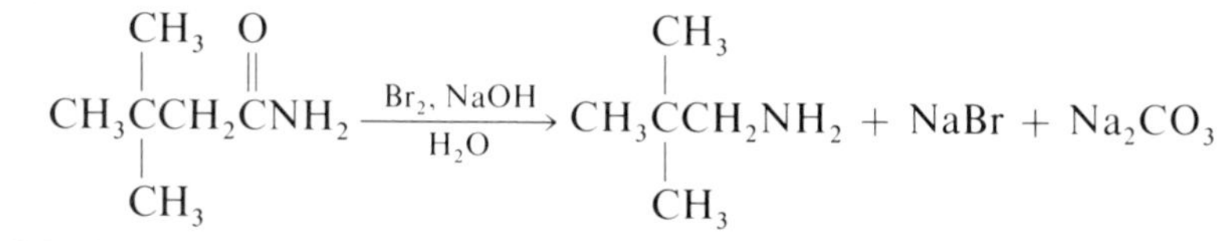

Br2, NaOH
H2O
+ NaBr + Na2CO3

3,3-dimethylbutanamide | neopentylamine 94%

Molecular rearrangements are intriguing to organic chemists, and much research is focused on the exact details of such reactions. The timing of the migration of a group from one atom to another and the nature of the bonding during the migration process are especially interesting. An important question is whether the migrating group is ever totally detached from the molecule during the rearrangement. Optically active compounds in which a carbon atom that is a stereocenter is the point of attachment of the migrating group have been used to investigate this question for both the Beckmann and the Hofmann rearrangements. Some results of such experiments are given below.

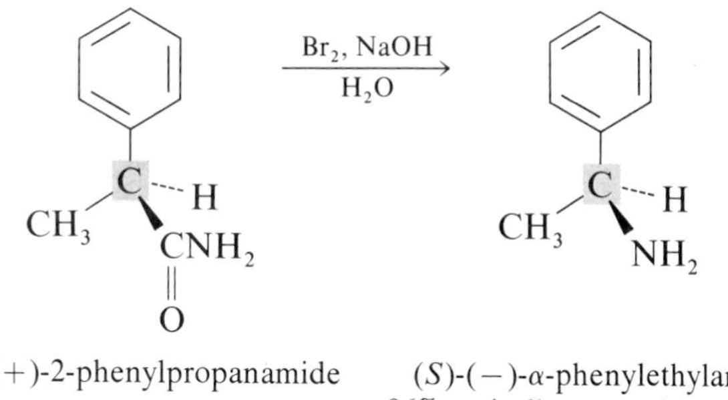

Br2, NaOH
H2O

(*S*)-(+)-2-phenylpropanamide | (*S*)-(−)-α-phenylethylamine 96% optically pure; characterized as its acetamide

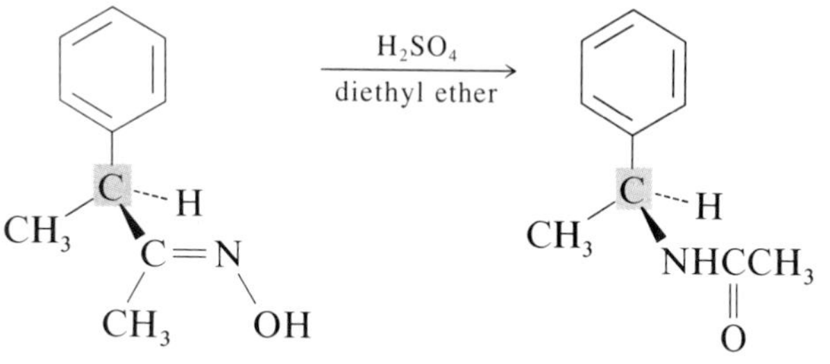

H2SO4
diethyl ether

(*S*)-(+)-3-phenyl-2-butanone oxime | (*S*)-(−)-*N*-acetyl-α-phenylethylamine 99% optically pure

In each reaction above, an α-phenylethyl group with a pair of bonding electrons migrates from a carbon atom to a nitrogen atom, essentially with complete retention of configuration. The alkyl group must be partially bonded first to the carbon atom and then to the nitrogen atom at the same face of the stereocenter at all times during the rearrangement. A detached intermediate would quickly lose its chirality. That migration takes place without a loss of the original point of attachment is also indicated by the formation of 3,4-dimethoxyaniline from 3,4-dimethoxybenzamide (p. 917).

Study Guide Concept Map 22.3

PROBLEM 22.19

In the Beckmann rearrangement, a water molecule is the leaving group attached to the nitrogen atom. A water molecule also reacts with the carbocation intermediate that forms to give the final amide product (p. 915). A mechanistic question that chemists have frequently asked is whether the water is ever free of the molecule during the migration. The following experimental observation was made when water containing oxygen-18 was added to the reaction mixture.

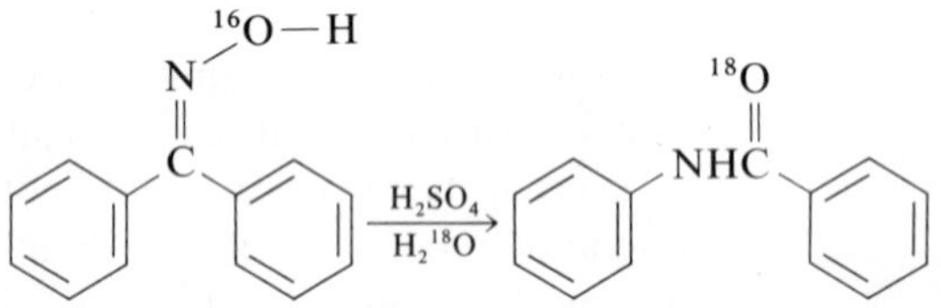
$^{16}O-H$
N
C
H_2SO_4
$H_2{}^{18}O$
^{18}O
NHC

Write a mechanism that accounts for this observation, and explain how it clarifies whether water reacts intramolecularly or intermolecularly in the rearrangement.

C. Nitrogen Derivatives of Carbonic Acid

The aryl isocyanate that is the reaction intermediate shown on p. 918 for the Hofmann rearrangement is related to compounds that are nitrogen-containing derivatives of carbonic acid.

$HOC(=O)OH$	$ClC(=O)Cl$	$H_2NC(=O)NH_2$
carbonic acid	phosgene	urea
	acid chloride of carbonic acid	*amide of carbonic acid*
$CH_3CH_2OC(=O)OCH_2CH_3$	$H_2NC(=O)OH$	$(CH_3)_3C-OC(=O)Cl$
diethyl carbonate	carbamic acid	*tert*-butyl chlorocarbonate
ester of carbonic acid	*half amide of carbonic acid*	*half ester, half acid chloride of carbonic acid*

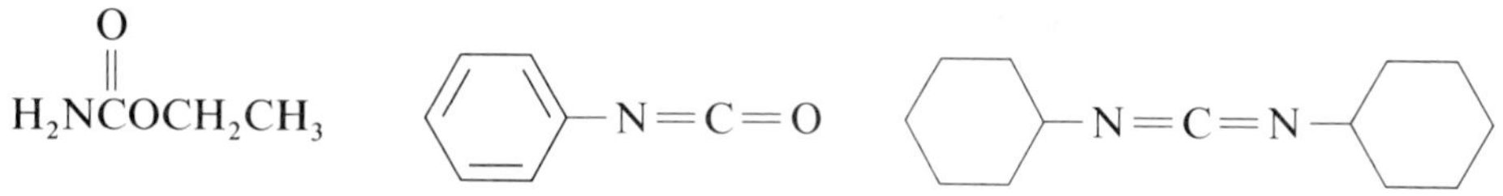

ethyl carbamate
urethane
half amide, half ester of carbonic acid

phenyl isocyanate
an isocyanate

dicyclohexylcarbodiimide

some derivatives of carbonic acid

Carbonic acid itself is unstable and decomposes into carbon dioxide and water. Carbamic acid is also unstable, giving carbon dioxide and ammonia. In fact, derivatives of carbonic acid in which only one of the two acid groups has been substituted all lose carbon dioxide with ease. The ones in which both sides are substituted, such as phosgene, urea, and ethyl carbamate, are stable. Phosgene was used in World War I as a poison gas. It reacts with water to give carbon dioxide and hydrogen chloride, which causes fluid to accumulate rapidly in the lungs. Urea is tremendously important in the metabolism of mammals. It is the chief form in which materials from the breakdown of proteins are excreted from the human body; an average man excretes about 30 grams of urea a day.

The unsymmetrical derivatives of carbonic acid retain the reactivity of each type of functional group present. For example, *tert*-butyl chlorocarbonate is an ester and an acid chloride. It reacts readily as an acid chloride with amines to give carbamates, which in turn are easily hydrolyzed to regenerate the amino group. This final step is another demonstration of the instability of the derivatives of carbonic acid from which carbon dioxide can be lost.

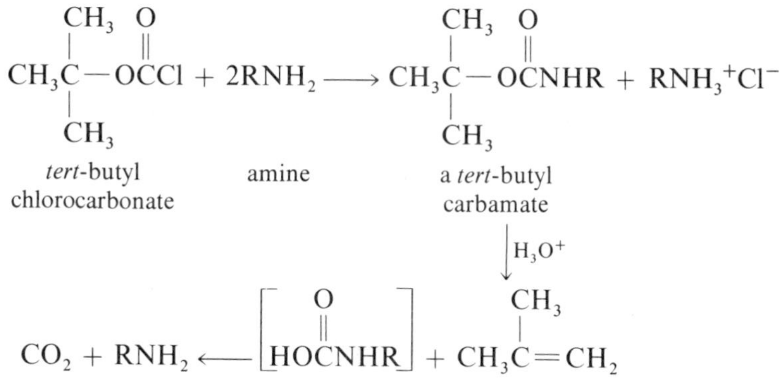

tert-butyl chlorocarbonate — amine — a *tert*-butyl carbamate

amine regenerated — an *N*-substituted carbamic acid — methylpropene from *tert*-butyl cation

This sequence of reactions is important in syntheses involving amino acids, in which it is necessary to protect the amino group while reactions are taking place at the carboxylic acid group (p. 1136). The protecting group must be removable under mild conditions that do not destroy amide bonds. Compounds related to *tert*-butyl chlorocarbonate serve to introduce a protecting group in such cases.

Isocyanates and carbodiimides are reactive compounds to which nucleophiles, such as water, alcohols, and amines, add with ease. 1-Naphthylisocyanate is used, for example, to convert liquid alcohols into solid derivatives that are useful for identifying the alcohols.

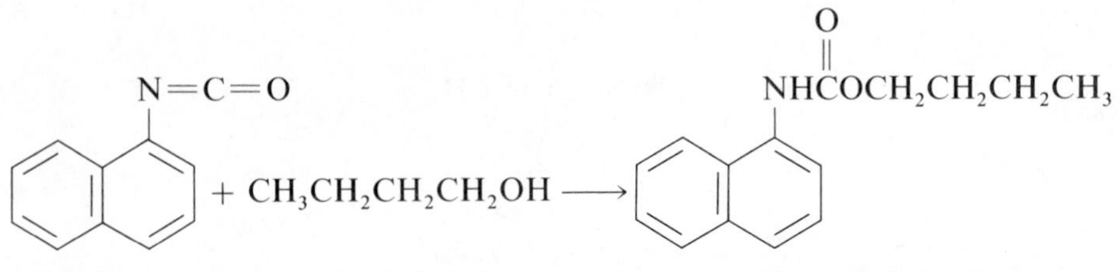

1-naphthylisocyanate   *n*-butyl alcohol   butyl *N*-(1-naphthyl)carbamate
                   bp 118 °C   butyl naphthylurethane
                                       mp 71 °C

The reaction of polyfunctional isocyanates with polyfunctional alcohols is of considerable industrial importance. Such reactions give polyurethane polymers, which are used to make foam cushions, fibers with elastic qualities, tire treads, and coatings for floors, among many other things (p. 1196).

Dicyclohexylcarbodiimide is useful in promoting the formation of amide bonds. It reacts with carboxylic acids to give intermediates that then react with amines to give amides. The reaction is used in the synthesis of peptides (p. 1139).

PROBLEM 22.20

Complete the following equations.

(a) $\overset{\overset{\displaystyle O}{\|}}{Cl\,C\,Cl}$ + NH$_3$ (excess) $\longrightarrow$ (b) $\overset{\overset{\displaystyle O}{\|}}{Cl\,C\,Cl}$ + ⟨ ⟩—CH$_2$OH (1 molar equiv) $\longrightarrow$ 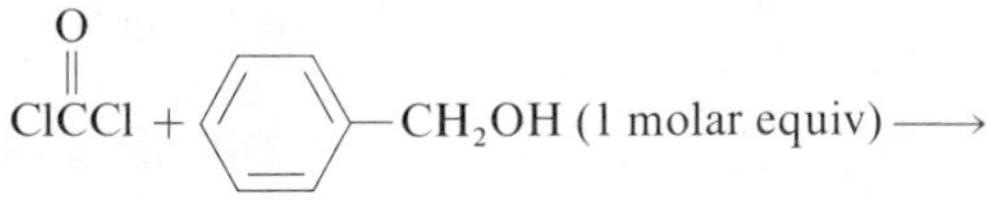

(c)   ⟨ ⟩—N=C=O + CH$_3$CH$_2$CH$_2$NH$_2$ $\longrightarrow$ 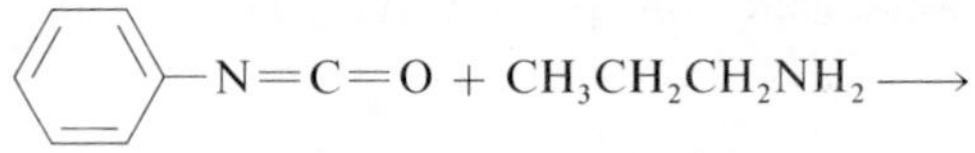

(d)   CH$_3$CH$_2$O$\overset{\overset{\displaystyle O}{\|}}{C}NH_2$ + H$_2$O $\xrightarrow[\Delta]{H_3O^+}$

(e)   ⟨ ⟩—N=C=S + ⟨ ⟩—NH$_2$ $\longrightarrow$ 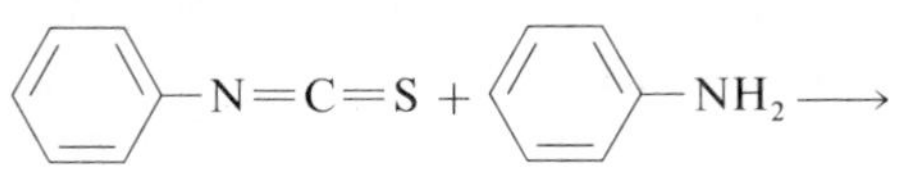
     phenyl isothiocyanate

PROBLEM 22.21

One of the pesticides developed after the discovery that chlorinated hydrocarbons such as DDT accumulate in the environment is carbaryl. It is a carbamate with the structure below.

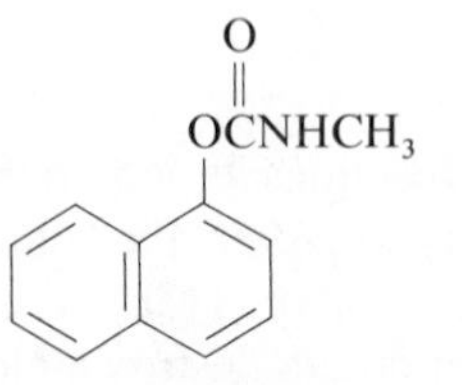

        carbaryl   1,1,1-trichloro-2,2-bis(*p*-chlorophenyl)ethane
     an insecticide that is   DDT
     biodegradable   *an insecticide that is not biodegradable*

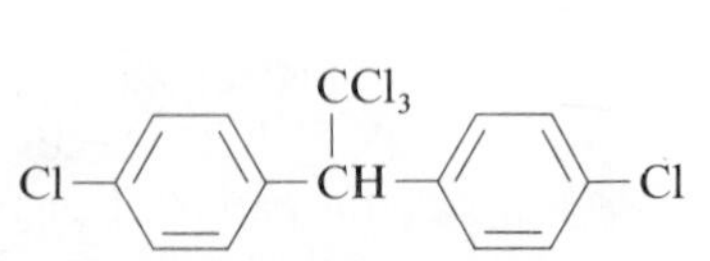

Write equations showing a mechanism for the degradation of carbaryl by water in the environment. You may assume that the pH of the soil is either below or above 7.

A. The Hofmann Elimination

Ammonia can react with an alkyl halide such as methyl iodide to form a mixture of primary, secondary, and tertiary amines (p. 895). If an excess of methyl iodide is used, a tertiary amine will react still further with the methyl iodide to give an ammonium salt with four organic groups bonded to the nitrogen atom. Such compounds are known as **quaternary ammonium salts.**

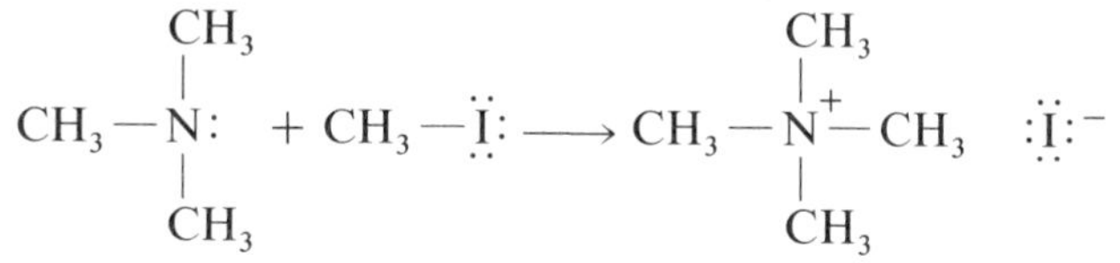

trimethylamine — methyl iodide — tetramethylammonium iodide

a quaternary ammonium salt

Quaternary ammonium salts are interesting for two reasons. First, the tertiary amine that is part of their structure is a good leaving group, so these salts undergo substitution and elimination reactions (p. 924, for example). Second, a quaternary ammonium ion is often incorporated into systems responsible for the transmission of nerve impulses in the human body, and many poisonous compounds and drugs that affect the nervous system also contain such a functional group. Some of these compounds of biological interest are examined in the next section.

Tetramethylammonium iodide is converted into the corresponding hydroxide when it is treated with moist silver oxide.

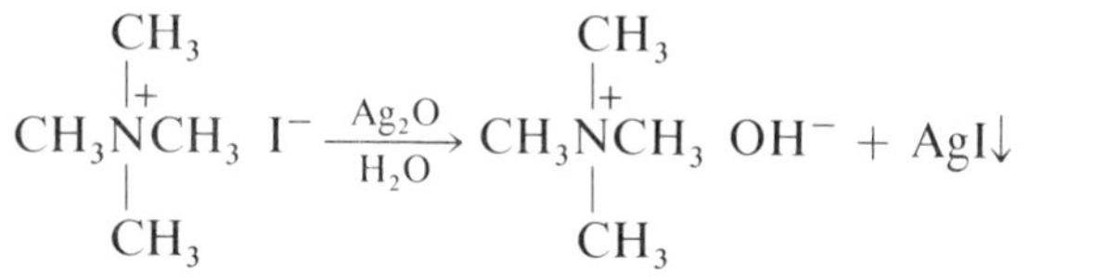

tetramethylammonium iodide — tetramethylammonium hydroxide

When tetramethylammonium hydroxide is heated, methanol and trimethylamine are formed in a typical S_N2 reaction.

VISUALIZING THE REACTION

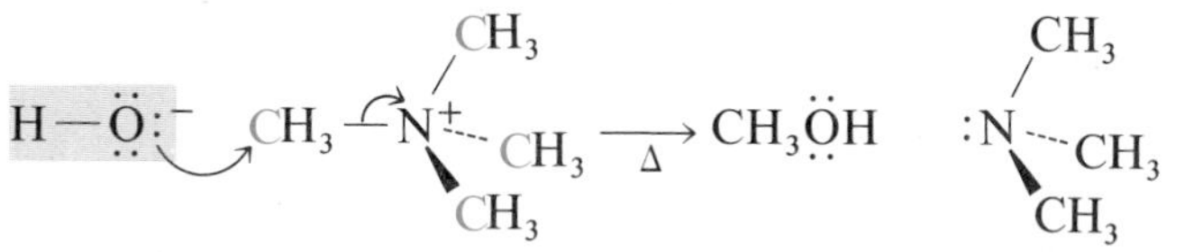

methanol — trimethylamine

A quaternary ammonium hydroxide derived from an amine having alkyl groups that are larger than the methyl group will also undergo some S_N2 reaction, but the chief reaction for such a compound will be an elimination reaction. *sec*-Butylamine, for example, can be converted into *N*,*N*,*N*-trimethyl-*sec*-butylammonium hydroxide and heated to give mostly 1-butene, with a very small amount of 2-butene in the product mixture.

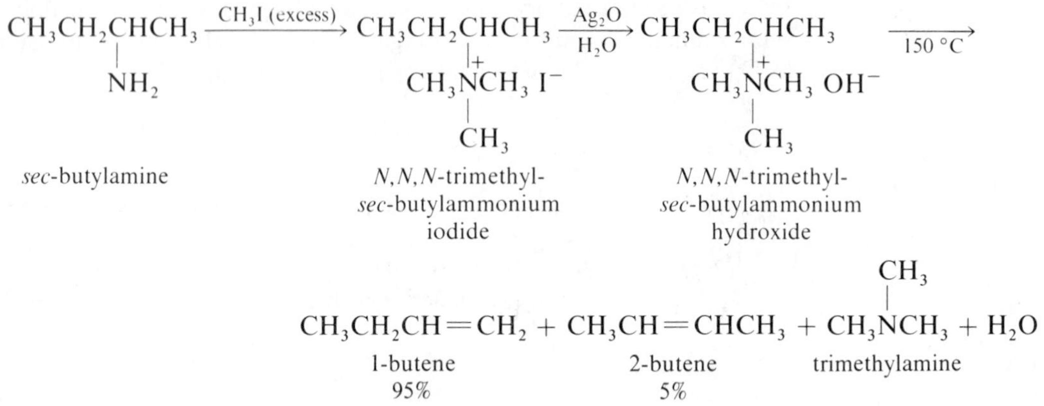
CH₃I (excess)
Ag₂O
H₂O
150 °C
sec-butylamine
N,N,N-trimethyl-sec-butylammonium iodide
N,N,N-trimethyl-sec-butylammonium hydroxide
1-butene
95%
2-butene
5%
trimethylamine

This type of elimination reaction is known as the **Hofmann elimination.** Experimental evidence, including the kind of stereochemical evidence presented in Section 7.7C for other elimination reactions, suggests that the reaction follows mainly the E_2 mechanism. Yet, in this elimination reaction, the formation of the less substituted alkene, instead of the thermodynamically more stable internal alkene, is favored. The reason for this orientation has been highly disputed, with steric arguments and electronic factors being used to rationalize the observations. The leaving group is a trialkylamine, rather than a halide ion as in the dehydrohalogenation reactions covered in Chapter 7. An important factor seems to be the degree to which the transition state for the elimination reaction resembles the product alkene. A relatively poor leaving group, such as the amine, allows a transition state to form in which the carbon-hydrogen bond has been broken more than the bond to the leaving group.

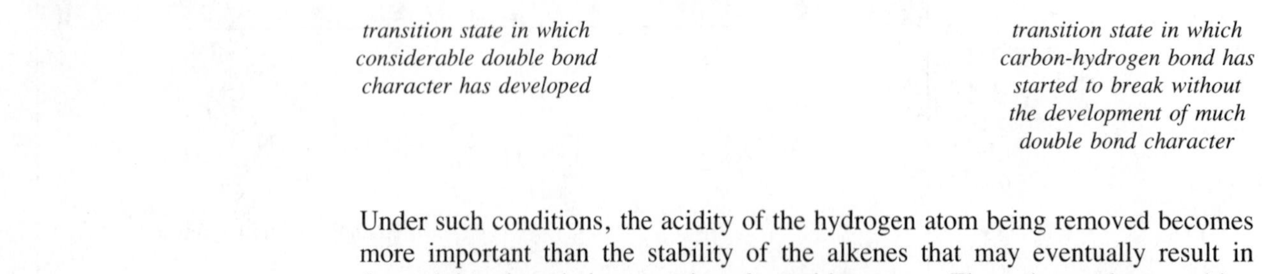

transition state in which considerable double bond character has developed

transition state in which carbon-hydrogen bond has started to break without the development of much double bond character

Under such conditions, the acidity of the hydrogen atom being removed becomes more important than the stability of the alkenes that may eventually result in determining the relative energies of transition states. Thus, the product resulting from the removal of a hydrogen atom from the less highly substituted β-carbon atom is the major one.

The Hofmann elimination is useful because it allows less highly substituted alkenes to be prepared. For example, we expect treatment of 1-bromo-1-methylcyclopentane with alcoholic potassium hydroxide (p. 255) to give chiefly 1-methylcyclopentene.

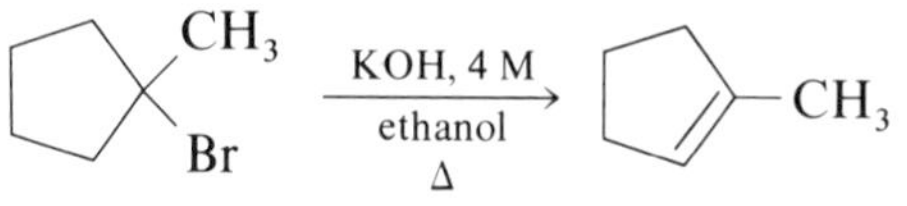

CH3
Br
KOH, 4 M
ethanol
Δ
CH3

1-bromo-1-methylcyclopentane 1-methylcyclopentene

When 1-amino-1-methylcyclopentane is converted into a quaternary ammonium hydroxide, which is then heated, methylenecyclopentane is the major product.

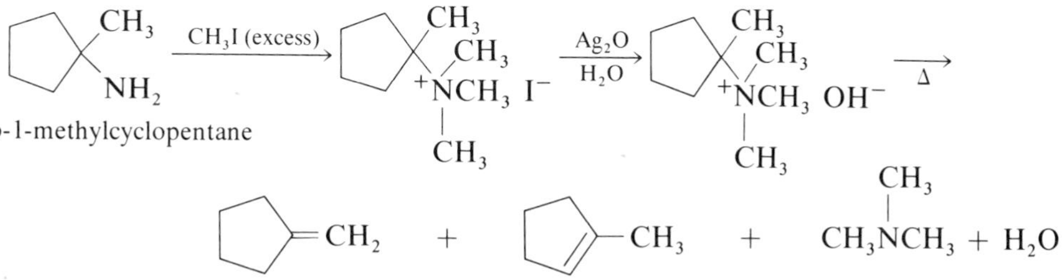

CH3
NH2
CH3I (excess)
+NCH3 I−
Ag2O
H2O
+NCH3 OH−
Δ
CH2
CH3
CH3NCH3 + H2O

1-amino-1-methylcyclopentane

methylenecyclopentane 91% 1-methylcyclopentene 9%

Historically, the Hofmann elimination has been useful in determining the structure of alkaloids because it gives information about the position of the nitrogen atom in cyclic compounds. The structure of piperidine, a cyclic amine found as part of many alkaloids, was proved in 1881 by August Wilhelm von Hofmann, for whom the reaction is named. Piperidine was converted to a quaternary ammonium hydroxide by treatment with excess methyl iodide and then with moist silver oxide. Heating the hydroxide resulted in elimination.

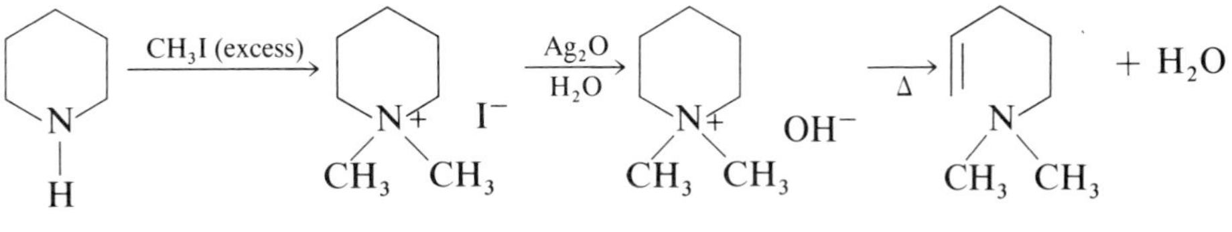

N
H
CH3I (excess)
N+ I−
CH3 CH3
Ag2O
H2O
N+ OH−
Δ
+ H2O

After the first elimination reaction, the trialkylamine leaving group is still attached to the product molecule. A second methylation and decomposition of the quaternary ammonium hydroxide were necessary to remove the nitrogen atom from the molecule as a tertiary amine.

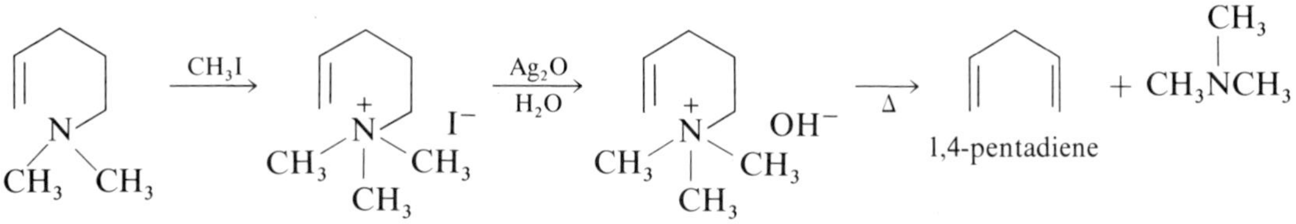

CH3I
Ag2O
H2O
I−
OH−
Δ
1,4-pentadiene
+ CH3NCH3

The structures of many alkaloids were investigated using these reactions. The number of steps necessary to eliminate the nitrogen atom from the molecule provided information about its position in the molecule and especially about the number of rings in which it was a member. The location of the nitrogen atom was also narrowed down by the determination of the structure of the final alkene.

Study Guide Concept Map 22.4

PROBLEM 22.22

Predict what the chief products will be when *N*-ethyl-*N*-methylpiperidinium iodide is treated with moist silver oxide and then heated.

PROBLEM 22.23

The quinuclidine ring system is part of the structure of quinine (p. 889). How many times would the Hofmann elimination have to be repeated to free the nitrogen atom from the compound? What would be the structure of the alkene that results from the entire sequence of reactions? Write equations illustrating your answer.

N

quinuclidine

PROBLEM 22.24

Hofmann applied his elimination reaction to an investigation of the structure of coniine, the poisonous alkaloid in hemlock. The compound causes a paralysis of the motor nerves; death results from the victim's inability to breathe. Hemlock was the potion used to execute the Greek philosopher Socrates.

(a) Coniine is 2-propylpiperidine. Write out equations for its degradation by the Hofmann method. Coniine is known to give a mixture of alkenes by this method. What structures are possible for these alkenes?

(b) Coniine occurs in nature in both the levorotatory and dextrorotatory forms. Levorotatory coniine has the *R* configuration. Draw a correct three-dimensional structural formula for (*R*)-coniine.

B. Biologically Active Quaternary Nitrogen Compounds

Choline, the systematic name of which is 2-(trimethylammonium)ethanol, is an important participant in the transmission of nerve impulses. It is also a component of lecithin, a major constituent of cell membranes. Choline is converted in nerve endings to its acetic ester, acetylcholine, by an enzyme with the participation of acetyl coenzyme A (p. 581).

$$\underset{\text{choline}}{(CH_3)_3\overset{+}{N}CH_2CH_2OH} + \underset{\text{acetyl coenzyme A}}{CoA{-}S\overset{O}{\overset{\|}{C}}CH_3} \xrightarrow[\text{choline acetylase}]{} \underset{\text{acetylcholine}}{(CH_3)_3\overset{+}{N}CH_2CH_2O\overset{O}{\overset{\|}{C}}CH_3} + \underset{\text{coenzyme A}}{CoA{-}SH}$$

At the exact instant of the transmission of the nerve impulse, acetylcholine moves across the synapse, the point where a nervous impulse passes from one nerve cell to the other. In motor nerves, this triggers the contraction of the muscle. An enzyme at the receptor site called acetylcholine esterase catalyzes the hydrolysis of acetylcholine back to choline and acetic acid; the nerve impulse stops and the muscle relaxes.

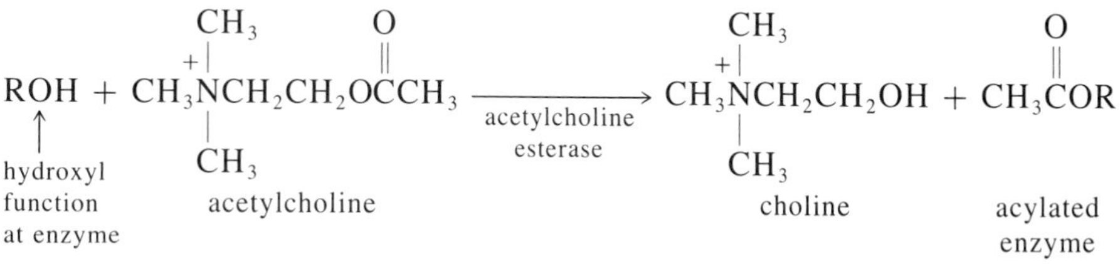
ROH + $CH_3\overset{+}{N}(CH_3)CH_2CH_2OCCH_3$ (O above C)
acetylcholine
esterase
$CH_3\overset{+}{N}(CH_3)CH_2CH_2OH$ + CH_3COR (O above C)
hydroxyl
function
at enzyme
acetylcholine
choline
acylated
enzyme

Anything that interferes with this alternation of the synthesis and destruction of acetylcholine at the nerve junction creates paralysis. The enzyme acetylcholine esterase is particularly vulnerable. Although the bond that is hydrolyzed at the enzyme active site is the ester linkage, the positively charged quaternary ammonium center is necessary to position acetylcholine so that hydrolysis can take place. The structure of proteins will not be fully explored until Chapter 26, but the exact locations of acidic, basic, charged, or even hydrophobic portions of molecules are crucial to interactions between enzymes and the molecules on which they act.

The importance of the quaternary ammonium group to the action of acetylcholine esterase is made apparent by the fact that a number of compounds that are quaternary ammonium salts have a paralyzing effect. (+)-Tubocurarine, the active component of the curare poison used by natives of South America on the tips of arrows to paralyze their prey, is an example of such a naturally occurring quaternary ammonium compound. A synthetic drug, decamethonium bromide, mimics tubocurarine in the presence of two quaternary ammonium salt centers and the distance between them.

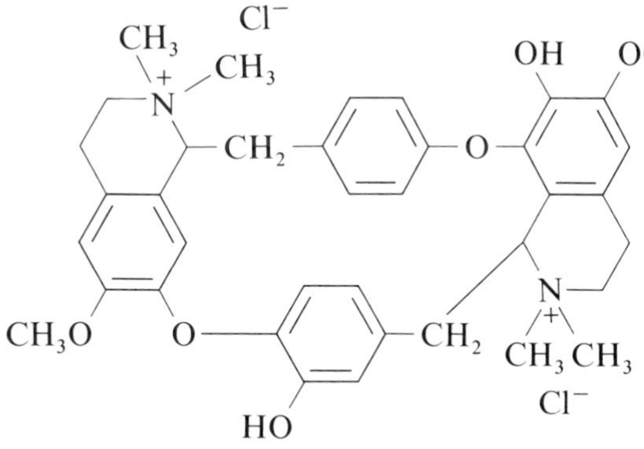
Cl^-
CH_3
CH_3
N
CH_2
O
OH
OCH_3
CH_3O
O
CH_2
N
CH_3 CH_3
Cl^-
HO

tubocurarine

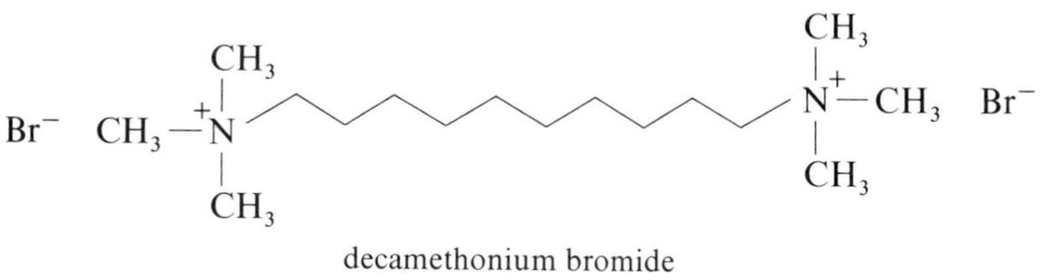
Br^-
CH_3
N
CH_3
CH_3
N
CH_3
CH_3
CH_3
Br^-

decamethonium bromide

Both tubocurarine and decamethonium bromide are used in surgery as muscle relaxants. They act as inhibitors of acetylcholine esterase, which means that they occupy the active site of the enzyme and interact with it because of the presence of the quaternary ammonium groups. They prevent the enzyme from carrying out its function of destroying acetylcholine and thus interfere with the proper transmission of nerve impulses. An excess of tubocurarine kills by paralyzing the respiratory muscles.

Another potent poison is found in the fungus *Amanita muscaria,* commonly called the fly agaric because it was used as a fly poison at one time. The alkaloid muscarine is found in this fungus, along with choline and acetylcholine.

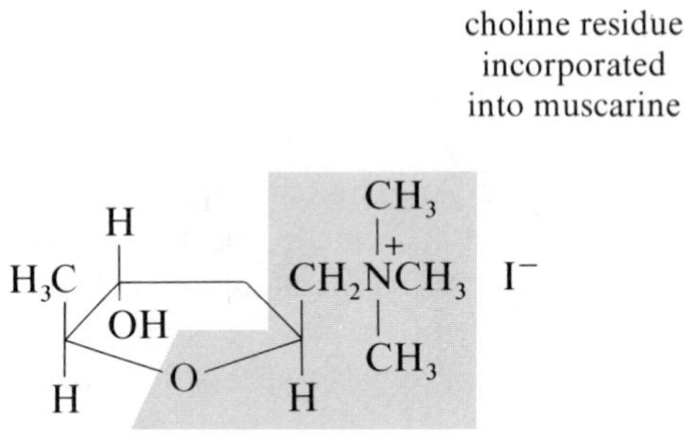
choline residue incorporated into muscarine
H
H_3C
OH
O
H
H
CH_3
$CH_2\overset{+}{N}CH_3$
CH_3
I^-
(+)-muscarine

Choline is incorporated into the structure of muscarine as indicated by the shaded portion of the formula. (+)-Muscarine inhibits acetylcholine esterase, but (−)-muscarine has very little toxicity, a good example of the high stereoselectivity of enzymatic reactions.

The function of choline as a component of lecithin is quite different from its function in the transmission of nerve impulses. Lecithin is a triglyceride (p. 611) in which one of the primary alcohol groups of glycerol has been esterified by phosphoric acid. The phosphoric acid also forms an ester link to choline. In lecithin, the other two alcohol functions of glycerol have been esterified by fatty acids, usually stearic acid and oleic acid.

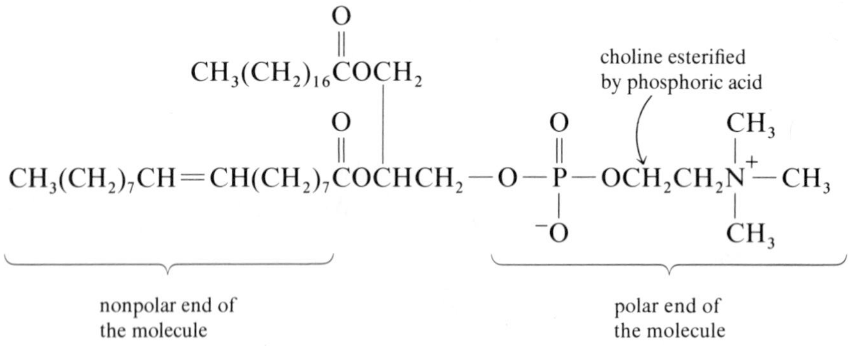
$CH_3(CH_2)_{16}COCH_2$
$CH_3(CH_2)_7CH{=}CH(CH_2)_7COCHCH_2{-}O{-}P{-}OCH_2CH_2\overset{+}{N}{-}CH_3$
choline esterified by phosphoric acid
nonpolar end of the molecule
polar end of the molecule
choline phosphoglyceride
lecithin

Lecithin is one of a series of compounds known as **phospholipids.** At the pH prevalent in the human body, the phosphoric acid group is negatively charged, and the quaternary ammonium group in the choline residue is positively charged. The molecule of lecithin as a whole is neutral, but quite highly charged at one end and hydrocarbon-like and nonpolar at the other. The hydrocarbon portions of phospholipid molecules are believed to interact with each other to form cell membranes consisting of two layers of phospholipids with the charged ends oriented toward

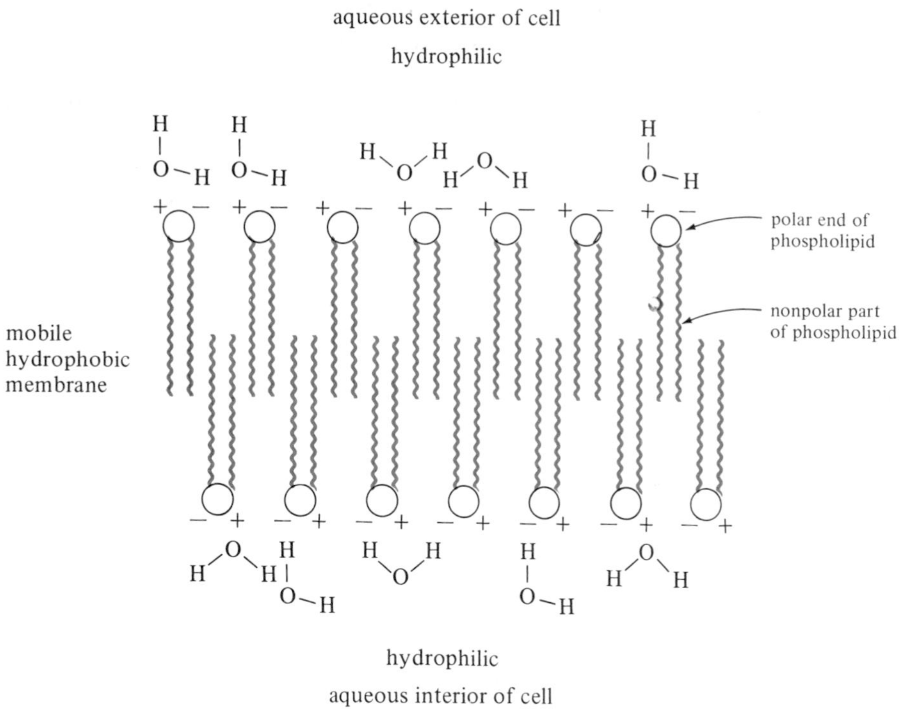
aqueous exterior of cell
hydrophilic
polar end of phospholipid
nonpolar part of phospholipid
mobile hydrophobic membrane
hydrophilic
aqueous interior of cell

FIGURE 22.1 Model of a cell membrane composed of a phospholipid bilayer.

both the exterior and interior of the cell (Figure 22.1). The layers have a high electrical resistance and are not easily penetrated by highly polar molecules. The polar parts of the membrane are placed where they interact with water in the cell and in surrounding fluids. The two layers of lipid molecules can move sideways past each other, giving the membranes flexibility.

S U M M A R Y

Amines, the organic derivatives of ammonia, are strong organic bases. The basicity of alkyl amines resembles that of ammonia. Aryl amines, in which the nitrogen atom is bonded to a carbon of the aromatic ring, are much weaker bases. The basicity of aryl amines is strongly affected by other substituents on the aromatic ring; electron-withdrawing groups, such as the nitro group, further weaken the base.

Amines can be prepared by nucleophilic substitution reactions of alkyl halides. Reaction of an alkyl halide with ammonia may seem to be the most direct way to make an amine, but side reactions detract from the usefulness of this method. The reaction of phthalimidate anion with an alkyl halide, followed by hydrolysis, gives a primary amine. Cyanide ion, azide ion, and nitrite ion react with alkyl halides to give nitriles, azides, and nitro compounds, respectively, which can be reduced to amines. Carbonyl compounds are transformed into amines by conversion to oximes or imines, which also give amines on reduction. Imines need not be isolated before reduction; mixtures of a carbonyl compound with ammonia or a primary amine in the presence of a reducing agent give amines directly in a process that is known as reductive amination. Amides and aromatic nitro compounds are also converted to amines by reduction. The reactions used to prepare amines are summarized in Table 22.1.

Amines, as nucleophiles, react with nitrous acid, a source of the electrophilic nitrosonium ion. Primary alkyl amines are converted by nitrous acid to unstable alkyl diazonium ions, which lose nitrogen to give mixtures of products from both substitution and elimination reactions. The aryl diazonium ions derived from primary aromatic amines are more stable and serve as useful intermediates in syntheses (see Section 23.1). Secondary amines react with nitrous acid to give *N*-nitrosamines. Tertiary aromatic amines having no substituent para to the amino group give *p*-nitrosoamines from electrophilic substitution reactions.

Reactions that give rise to electron-deficient nitrogen atoms often lead to a molecular rearrangement involving the migration of a group from carbon to nitrogen. In the Beckmann rearrangement, treatment of an oxime with acid creates a good leaving group on the nitrogen atom. Loss of the leaving group is accompanied by migration of a group from a nearby carbon to the nitrogen, which creates a carbocation. Reaction of the cation with water gives an amide as the final product. In the Hofmann rearrangement, a primary amide is converted to an amine that has one fewer carbon atom than the starting amide. This reaction is carried out by treating the amide with a halogen in the presence of hydroxide ion and involves the formation of an *N*-haloamide and an isocyanate as intermediates. Molecular rearrangements to carbon, oxygen, and nitrogen are summarized in Table 22.2 on page 932.

TABLE 22.1 Preparation of Amines

From Alkyl Halides				
Starting Material	**Nucleophile**	**Product of Substitution Reaction**	**Reagents for Second Step**	**Final Product**
RX R = primary or secondary alkyl group	NH_3 (1 equivalent)	$R\overset{+}{N}H_3X^-$		$R\overset{+}{N}H_3X^-$
	NH_3 (excess)	RNH_2	RX	RNH_2 R_2NH R_3N
	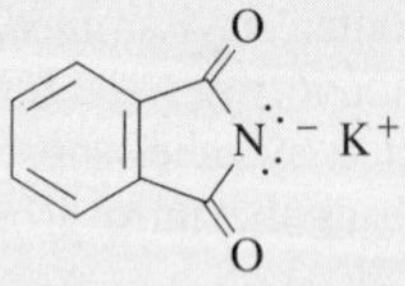	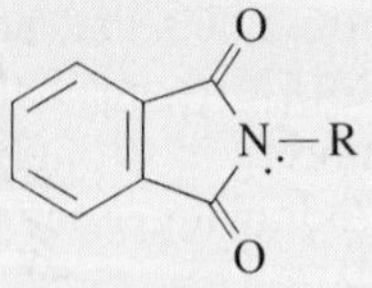	H_2O, H_3O^+, Δ then ^-OH	RNH_2
	NaCN	RCN	H_2/catalyst or $LiAlH_4$	RCH_2NH_2
	NaN_3	RN_3	H_2/catalyst or $LiAlH_4$	RNH_2
	$NaNO_2$	RNO_2	Fe/HCl, then OH^- or H_2/catalyst	RNH_2

(Continued)

TABLE 22.1 *(Continued)*

From Carbonyl Compounds and Acid Derivatives				
Starting Material	**Nucleophile**	**Product of Substitution Reaction**	**Reagents for Second Step**	**Final Product**
$R'C(=O)R''$	RNH_2	$R'(R'')C{=}NR$	$NaBH_4$, CH_3OH or H_2/catalyst *(called reductive amination if imine is not isolated and reduction is carried out on mixture of amine and carbonyl compound)*	$R''CH(R')NHR$
	H_2NOH	$R'(R'')C{=}NOH$	H_2/catalyst	$R''CH(R')NH_2$
$R'C(=O)Cl$ or $R'C(=O)OC(=O)R'$	R_2NH	$R'C(=O)NR_2$	$LiAlH_4$	$R'CH_2NR_2$
From Aromatic Compounds				
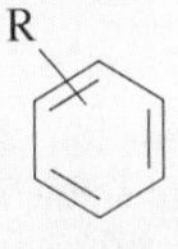 R	HNO_3, H_2SO_4	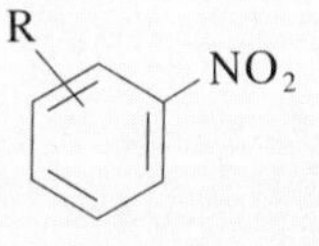 R NO2	Sn/HCl, then ^-OH or Fe/HCl, then ^-OH or H_2/catalyst	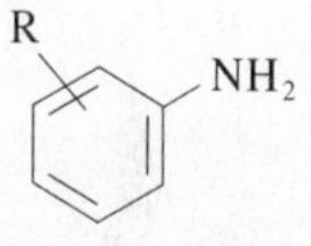 R NH2
	HNO_3, H_2SO_4, Δ	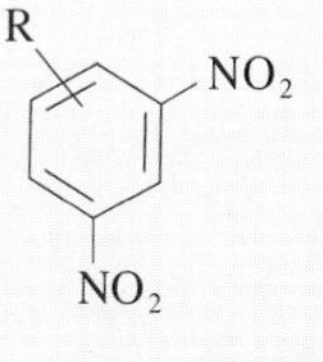 R NO2 NO2	$(NH_4)_2S$	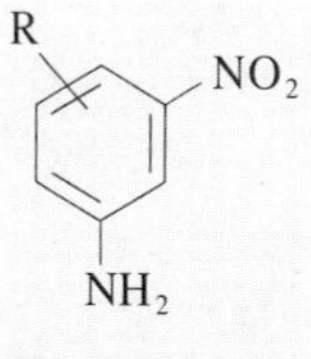 R NO2 NH2 *(selectivity depends on the nature of R)*

The reaction of tertiary amines with alkyl halides produces quaternary ammonium salts. In the presence of a base, a tertiary amine is lost as a leaving group from a quaternary ammonium ion. Elimination reactions of quaternary ammonium hydroxides, known as the Hofmann elimination, give the less highly substituted alkene, a regioselectivity opposite to that for elimination reactions of alkyl halides. Quaternary ammonium compounds are important biologically; for example, choline plays a role in the transmission of nerve impulses and is a component of phospholipids. Reactions of amines are summarized in Table 22.3, and reactions of quaternary ammonium salts are summarized in Table 22.4.

TABLE 22.2 Rearrangements Involving Electron-Deficient Species

Generation of Electron-Deficient Species	Rearranging Species	Product(s)	Driving Force
	Rearrangement to Electron-Deficient Carbon Atoms		
	A. 1,2-Hydride Shifts		
		Alkenes, alcohols, and/or halides derived from:	
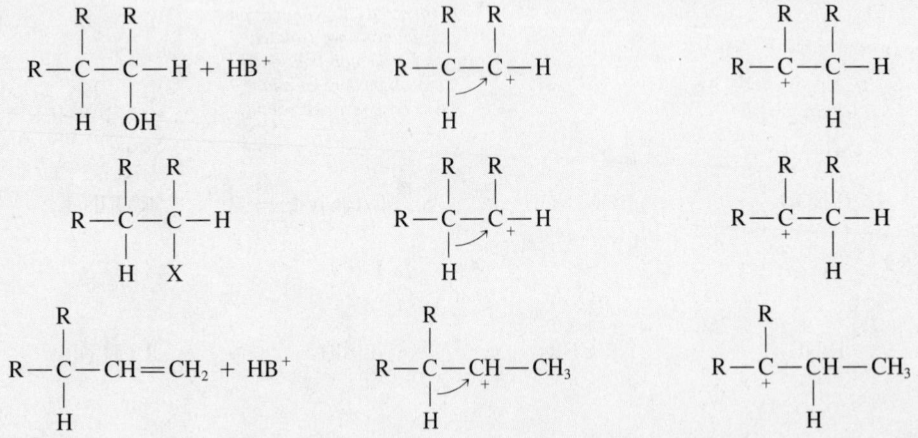			Formation of the more stable cation
	B. 1,2 Alkyl Shifts		
		Alkenes, alcohols, and/or halides, derived from:	
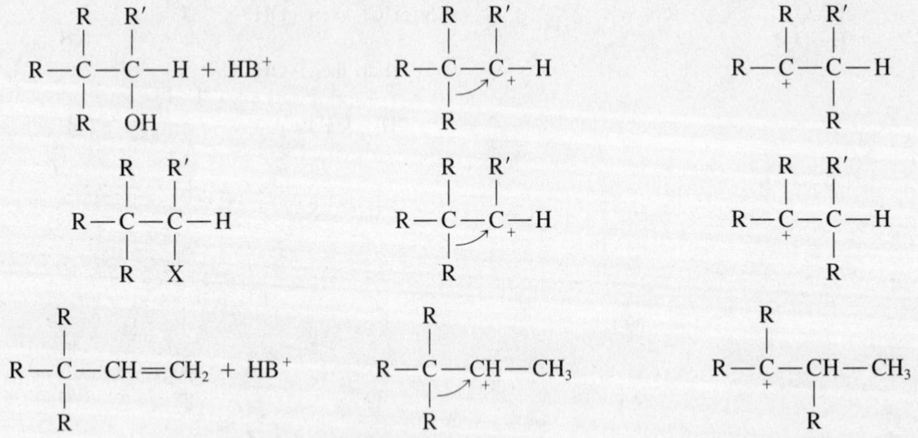			Formation of the more stable cation
	Rearrangement to Electron-Deficient Oxygen Atoms		
	Baeyer-Villiger Reaction		
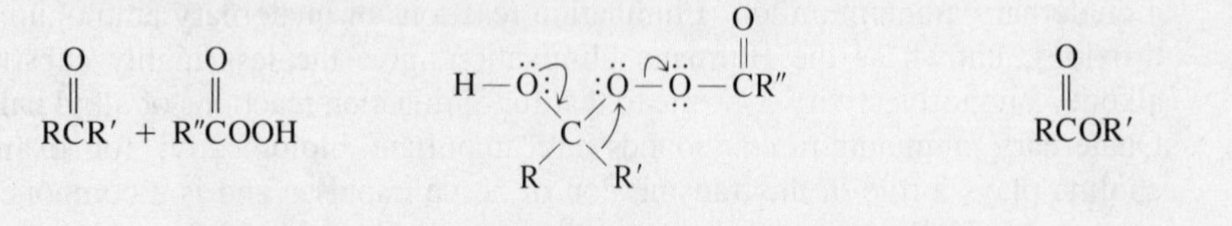			Formation of the more stable cation; gaining the bond energy of the carbonyl group

(*Continued*)

TABLE 22.2 (*Continued*)

Generation of Electron-Deficient Species	Rearranging Species	Product(s)	Driving Force
	Rearrangement to Electron-Deficient Nitrogen Atoms		
	A. Beckmann Rearrangement		
$RC(=NOH)R' + HB^+$	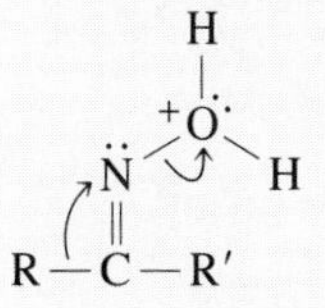	$R-N=\overset{+}{C}-R' \xrightarrow{H_2O} RNHC(=O)R'$	Formation of the more stable cation
	B. Hofmann Rearrangement		
$RC(=O)NH_2 + X_2 + NaOH$	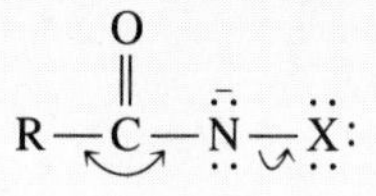	$RN=C=O \xrightarrow{H_2O} RNH_2 + CO_2$	Achieving octets around all atoms

TABLE 22.3 Reactions of Amines as Nucleophiles

Amine	Electrophile	Intermediate	Reagents for Second Step	Product
RNH_2	$R'X$	$R'\overset{+}{N}H_2R\ X^-$	RNH_2	$RNHR'$
R_2NH	$R'X$	$R_2\overset{+}{N}HR'\ X^-$	R_2NH	R_2NR'
R_3N	$R'X$	—	—	$R_3\overset{+}{N}R'\ X^-$
RNH_2	HONO (NO^+) from $NaNO_2$, HCl	$R-\overset{+}{N_2}$	Cl^-, H_2O	RCl, ROH
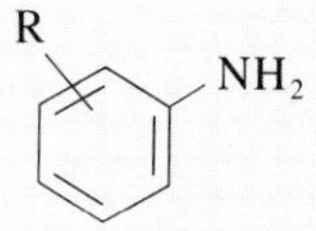	NO^+	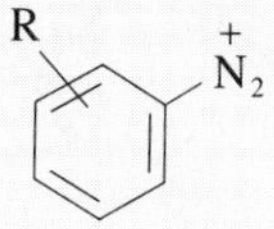		(see Section 23.1)
R_2NH	NO^+	—	—	R_2N-NO
RCH_2NR_2'	NO^+	—	—	$RC(=O)H + R_2'N-NO$
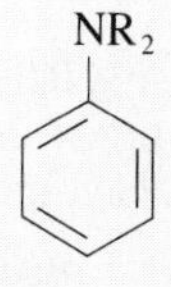	NO^+	—	—	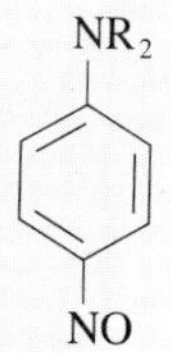

TABLE 22.4 Reactions of Quaternary Ammonium Compounds

Ammonium Salt	Reagent	Product of First Reaction	Reaction Conditions	Products
$(CH_3)_4N^+\ I^-$	Ag_2O, H_2O	$(CH_3)_4N^+\ OH^-$	Δ	$(CH_3)_3N + CH_3OH$
$RCH_2\underset{^-I\ \overset{+}{N}(CH_3)_3}{CH}{-}CH_3$	Ag_2O, H_2O	$RCH_2\underset{HO^-\ \overset{+}{N}(CH_3)_3}{CH}{-}CH_3$	Δ	$RCH_2CH{=}CH_2$ $+ (CH_3)_3N$
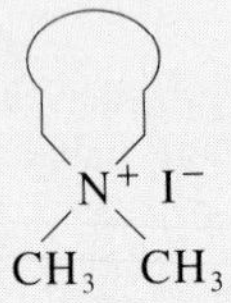	Ag_2O, H_2O	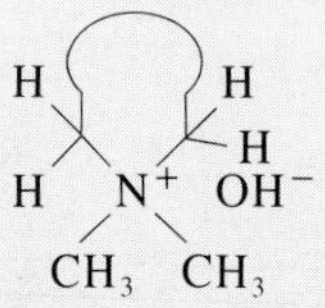	Δ	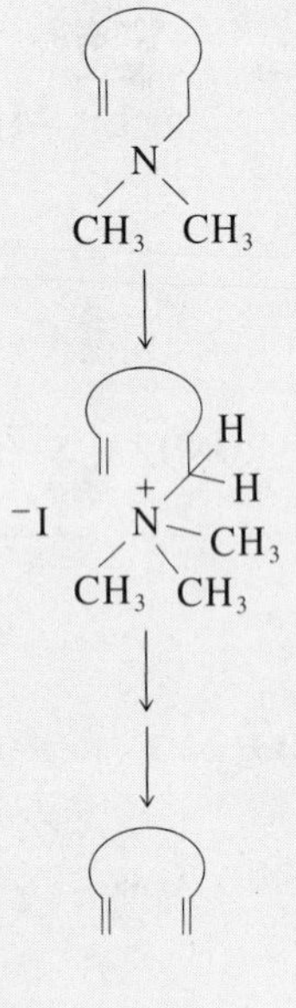 $+ (CH_3)_3N$

ADDITIONAL PROBLEMS

22.25 Name the following compounds.

(a) 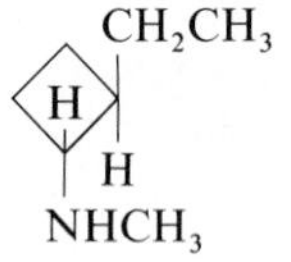
(b) 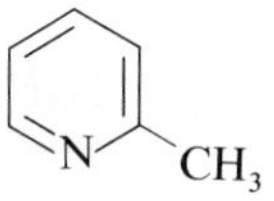
(c) $CH_3CH_2\underset{}{\overset{CH_2CH_3}{N}}CH_2CH_3$

(d) 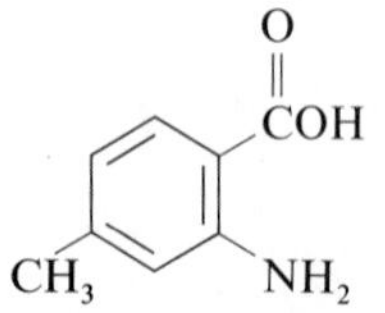
(e) $H_3\overset{+}{N}CH_2CH_2CH_2\overset{\overset{O}{\|}}{C}O^-$
(f) 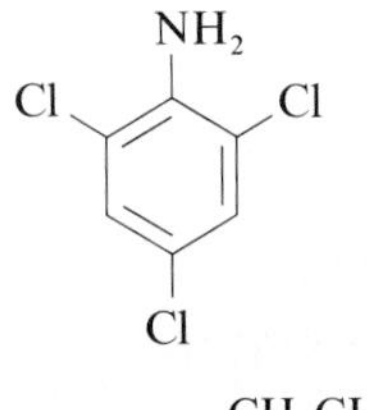

(g) $CH_3\overset{CH_3}{CH}CH_2CH_2\underset{NH_2}{CH}CH_3$
(h) 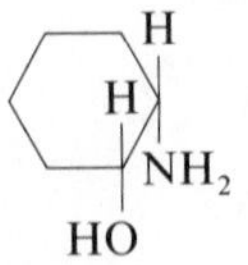
(i) 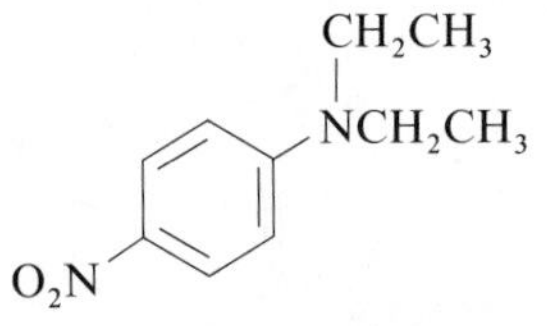

22.26 Write structural formulas for the following compounds.

(a) *trans*-2-aminocyclopentanol (b) *N*-methyl-*N*-propylcyclohexylamine
(c) 3,5-dinitroaniline (d) *N,N*-dimethyl-*p*-methoxyaniline (e) *m*-toluidine
(f) 2,5-diaminooctane (g) 3-hexanamine
(h) 4-amino-2,2-dimethylpentanoic acid

22.27 Give structural formulas for all intermediates and products designated by letters in the following equations.

(a) 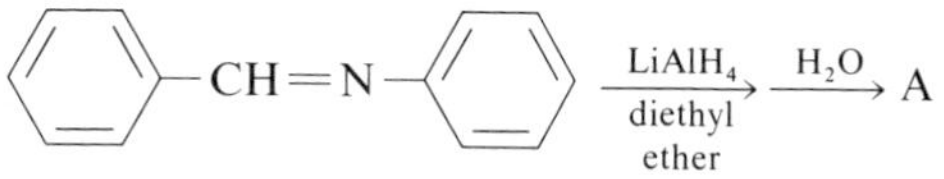

(b) 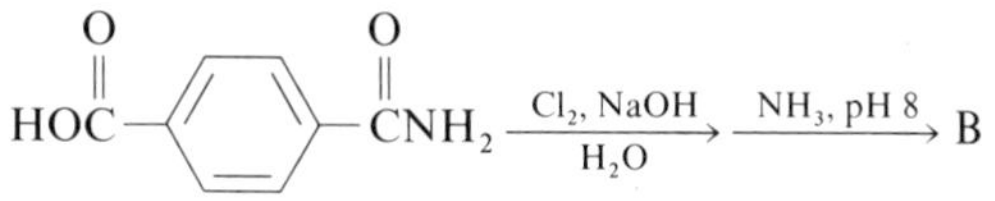

(c) $CH_3\overset{O}{\overset{\|}{C}}CH{=}\overset{CH_3}{\overset{|}{C}}CH_3 \xrightarrow[H_2O]{NH_3} C$

(d) $CH_3\overset{CH_3}{\underset{NO_2}{C}}CH_2CH_2\overset{O}{\overset{\|}{C}}OCH_3 \xrightarrow[Ni,\ \Delta]{H_2} D \quad (C_6H_{11}NO)$

(e) 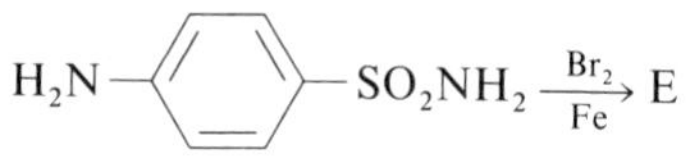

(f) $CH_3CH_2\overset{O}{\overset{\|}{C}}CH_2CH_2CH_3 \xrightarrow[Na_2CO_3]{HO\overset{+}{N}H_3\ HSO_4^-} F + G \xrightarrow{H_2SO_4} H + I$

(g) $CH_3\overset{O}{\overset{\|}{C}}CH_2CH_2CH_2CH_2CH_3 \xrightarrow[Ni,\ \Delta]{NH_3,\ H_2} J$

(h) $CH_3CH_2CH_2NO_2 \xrightarrow[NaOH]{H\overset{O}{\overset{\|}{C}}H} K \xrightarrow{H_2SO_4,\ Fe} L \xrightarrow{Ca(OH)_2} M$

(i) 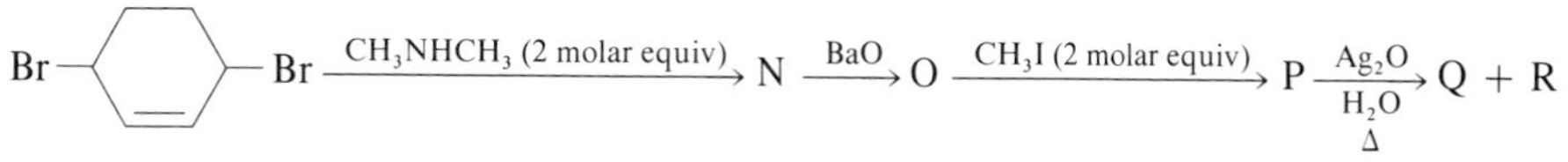

(j) 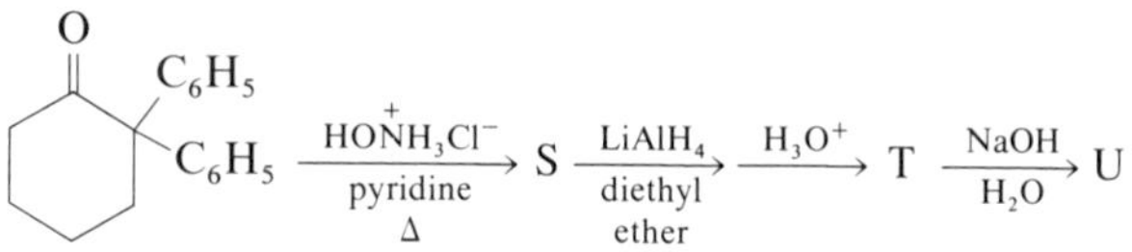

(k) 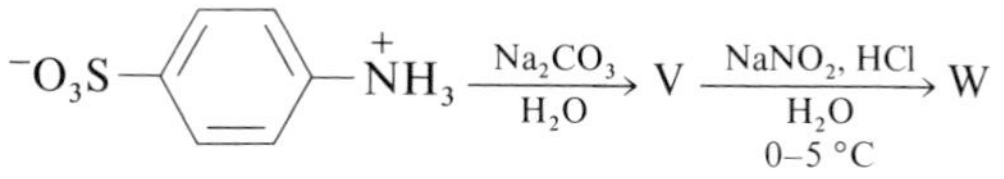

(l) 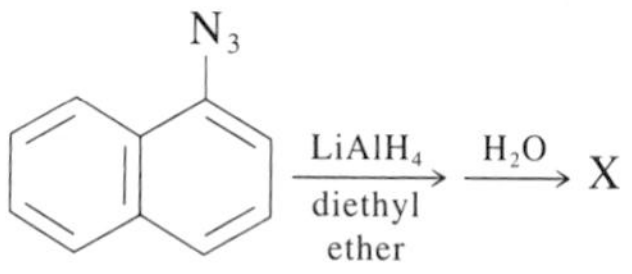

22.28 For more practice in recognizing reactions, write structural formulas for all compounds symbolized by letters in the following equations.

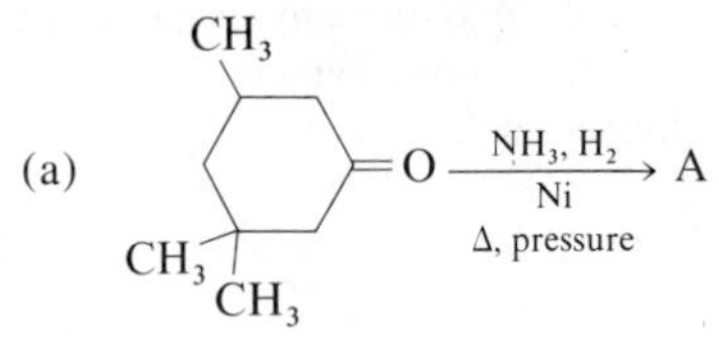

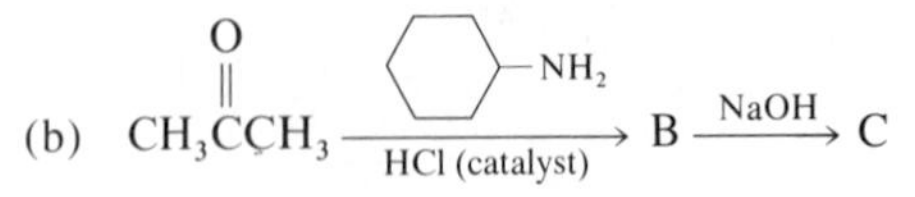

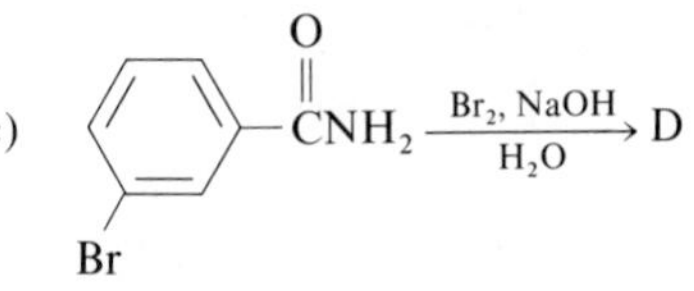

(d) $CH_3CH(Cl)COCH_2CH_3$ (ester: $CH_3CHClC(=O)OCH_2CH_3$) $\xrightarrow[\text{dimethyl sulfoxide}]{NaNO_2}$ E

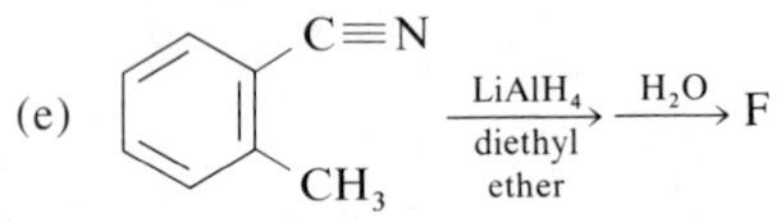

(f) $CH_3C(=O)NH_2 \xrightarrow[H_2O]{Br_2,\ KOH}$ G

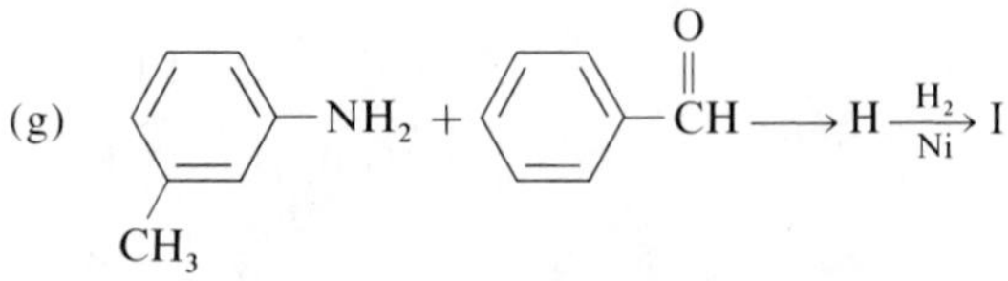

(h) $CH_3CH_2CH(OH)CH_2CH_2CH_3 \xrightarrow[\text{pyridine}]{TsCl}$ J $\xrightarrow{CH_3NHCH_3}\ \xrightarrow{NaOH}$ K $\xrightarrow[\text{acetonitrile}]{CH_3I}$ L

(i) $Br(CH_2)_{10}Br$ (1 equivalent) + $CH_3N(CH_3)CH_3$ (2 equivalents) $\xrightarrow[\text{methanol, 25 °C, 2 weeks}]{}$ M $\xrightarrow[H_2O]{Ag_2O}$ N $\xrightarrow[100\text{–}130\ °C]{}$ O

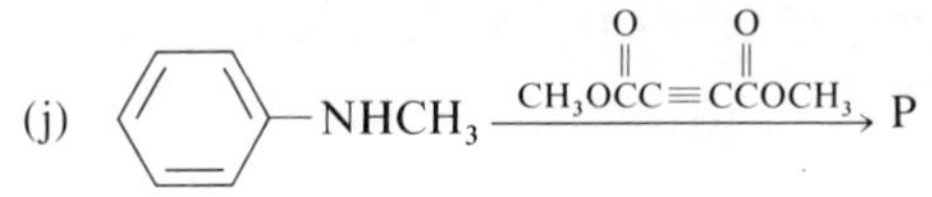

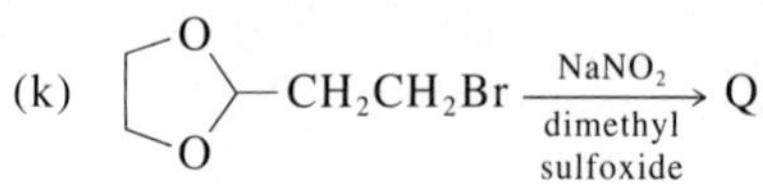

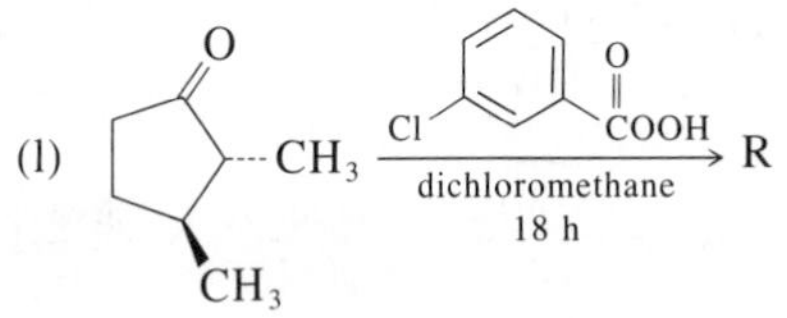

22.29 Show how you would carry out each of the following transformations. Some of them may require more than one step.

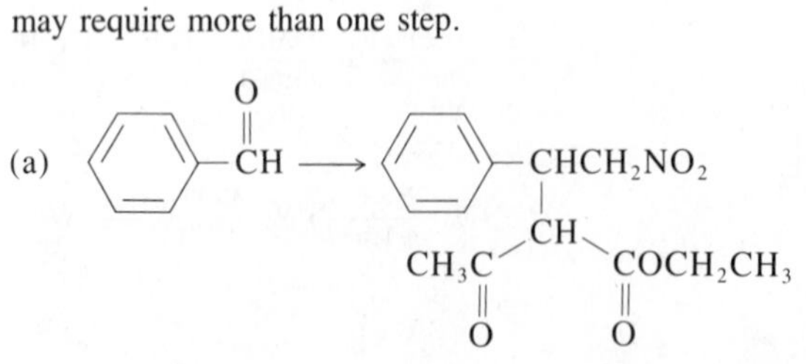

(b) $HOC(=O)CH_2CH(CH_3)CH_2CH_2C(=O)OH \longrightarrow H_2NCH_2CH(CH_3)CH_2CH_2NH_2$

(c) $CH_3CH_2C(=O)CH_2CH_3 \longrightarrow CH_3CH_2C(=O)NHCH_2CH_3$

(d) 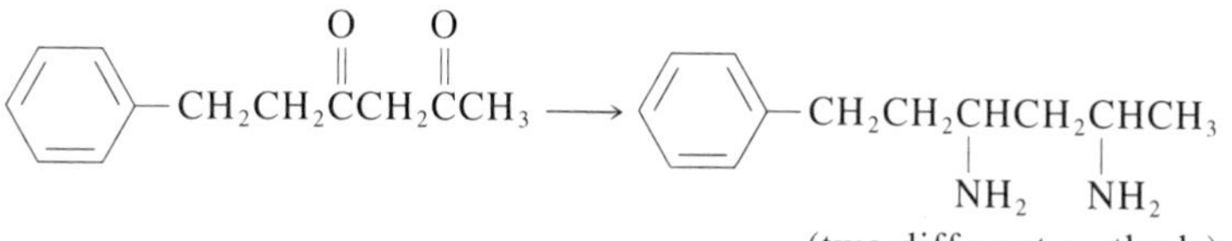

(two different methods)

(e) 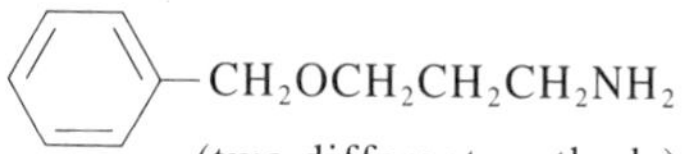

(two different methods)

22.30 Phthalimide is used to introduce an α-amino group in many syntheses of amino acids. These syntheses start with the reaction of diethyl bromomalonate with phthalimidate anion. They continue with alkylation (p. 651) of the resulting substituted diethyl malonate with a suitable substituent and end with hydrolysis and decarboxylation steps. The amino acids methionine, phenylalanine, and aspartic acid were synthesized in this way. Show how these syntheses could be carried out.

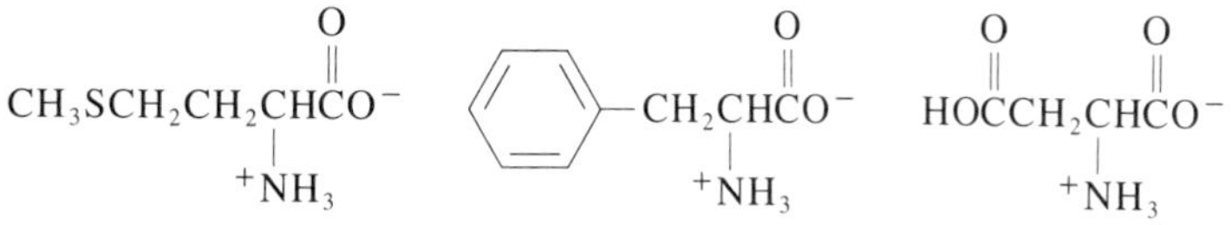

methionine phenylalanine aspartic acid

22.31 The following sequence of reactions was used to synthesize an amino acid. Write structural formulas for intermediates or products represented by letters.

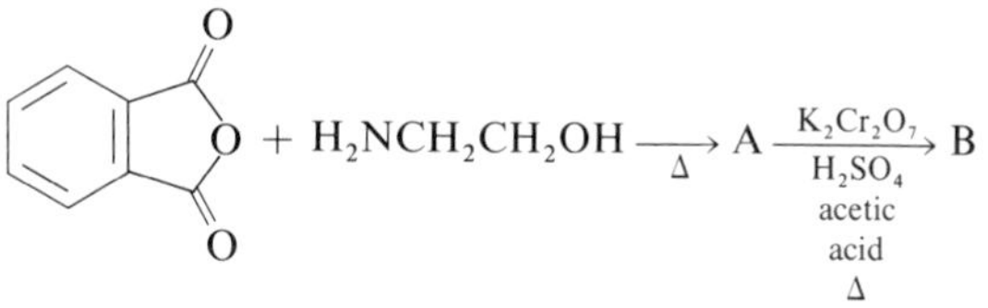

$$B \xrightarrow[HCl,\ \Delta]{H_2O} C + D$$

$$D \xrightarrow[pyridine]{} \text{an amino acid}$$

22.32 Alanine, which can also be called 2-aminopropanoic acid, has an isomer, 3-aminopropanoic acid, also called β-alanine. β-Alanine is prepared by a Gabriel phthalimide synthesis in which the phthalimidate anion is added to propenenitrile in a Michael-type reaction (p. 740). Write equations showing the mechanism of this reaction. What other reactions must be carried out to get 3-aminopropanoic acid by this route?

$$CH_2{=}CHC{\equiv}N$$

propenenitrile

22.33 When (aminomethyl)cyclopentane was treated with sodium nitrite in water with sodium dihydrogen phosphate as the acid, the following alcohol products were obtained.

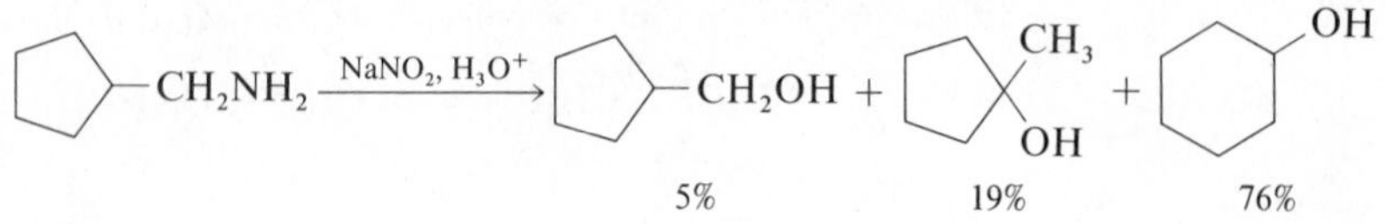
CH2NH2
NaNO2, H3O+
CH2OH +
CH3
OH
+
OH
5%
19%
76%

Write a detailed mechanism for the reaction showing how each of the products could have been formed.

22.34 The five-membered D ring of the steroid estrone is converted into a six-membered ring by the following sequence of reactions. Give structures for all intermediates designated by letters, and propose a mechanism for the step in which expansion of the D ring occurs.

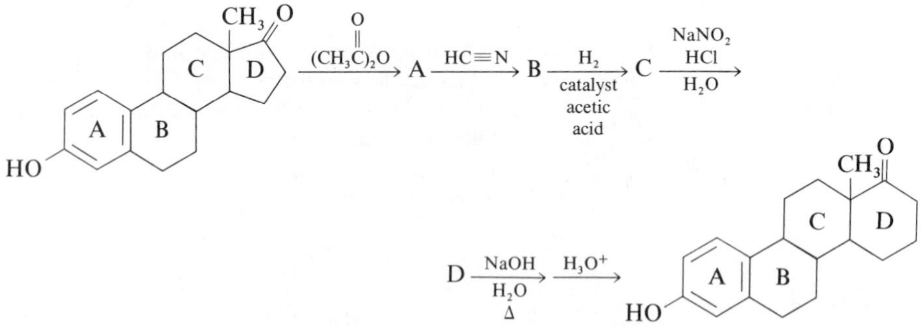
CH3 O
C
D
A
B
HO
(CH3C)2O
A
HC≡N
B
H2
catalyst
acetic
acid
C
NaNO2
HCl
H2O
D
NaOH
H2O
Δ
H3O+
CH3
O

22.35 In Section 19.5B, the reactions of benzenesulfonyl chloride with aniline and *N*-methylaniline and its lack of reactivity with a tertiary amine were described. The acid-base properties of the sulfonamides were emphasized. The reaction of benzenesulfonyl chloride with amines in the presence of aqueous sodium hydroxide is the basis of a qualitative analysis test known as the **Hinsberg test,** which distinguishes among low-molecular-weight primary, secondary, and tertiary amines. Using hexylamine, dipropylamine, and triethylamine to represent those classes of amines, write equations showing what you expect will happen with each one under the following conditions. At each point, describe what you expect to see in the test tube. All of the above amines are liquids; all sulfonamides are solids.

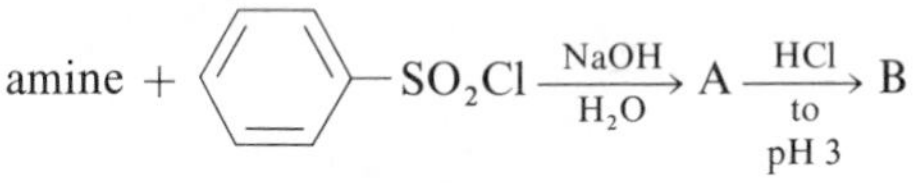
amine +
SO2Cl
NaOH
H2O
A
HCl
to
pH 3
B

How could these reactions be used to distinguish among primary, secondary, and tertiary amines?

22.36 A bicyclic amine has been synthesized by the following reactions. Provide structural formulas for the compounds designated by letters.

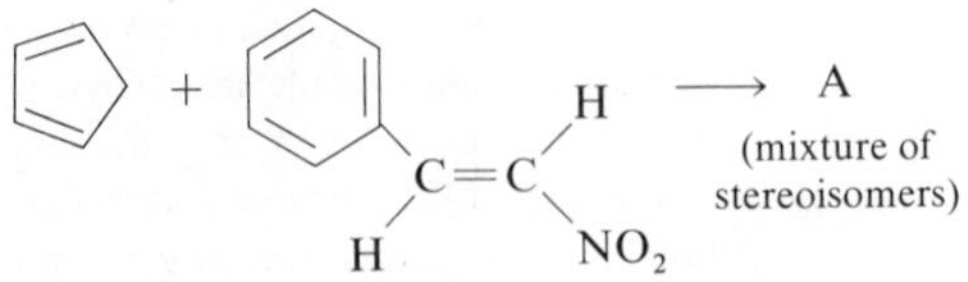
+
H
C=C
H
NO2
A
(mixture of
stereoisomers)

$$\text{A (with endo nitro group)} \xrightarrow[\text{H}_2\text{O}]{\text{Fe, HCl}} \text{B}$$

$$\text{A} \xrightarrow[\text{Pd/C, 4 atm, ethanol}]{\text{H}_2} \text{C}$$

The structure of Compound C was confirmed by the reactions that follow. Give structural formulas for Compounds D to H.

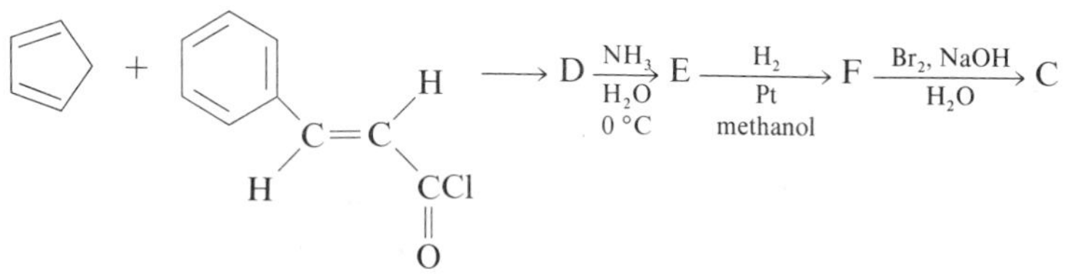

+
H
C=C
H
CCl
O
D
NH_3
H_2O
0 °C
E
H_2
Pt
methanol
F
Br_2, NaOH
H_2O
C

$$\mathrm{D} \xrightarrow[\mathrm{H_2O}]{\mathrm{NaOH}} \mathrm{G} \xrightarrow{\mathrm{H_3O^+}} \mathrm{H}$$

22.37 Trimethylenecyclopropane, C_6H_6, is a constitutional isomer of benzene. A sample of the compound was synthesized in order to study its physical and chemical properties.

(a) The synthesis is outlined below. Give structural formulas for the compounds designated by letters.

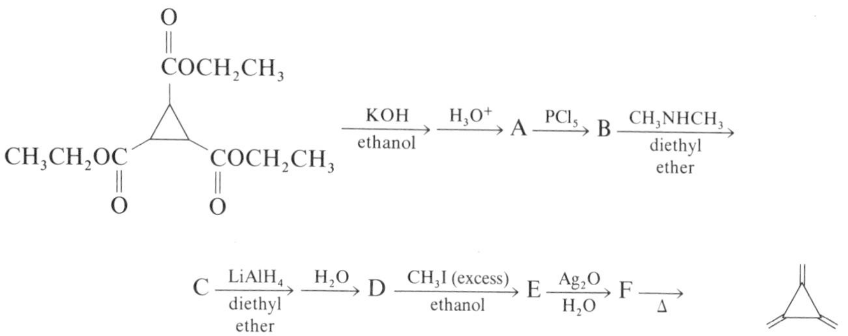

$COCH_2CH_3$
CH_3CH_2OC
$COCH_2CH_3$
KOH
ethanol
H_3O^+
A
PCl_5
B
CH_3NHCH_3
diethyl
ether
C
$LiAlH_4$
diethyl
ether
H_2O
D
CH_3I (excess)
ethanol
E
Ag_2O
H_2O
F
Δ

trimethylenecyclopropane

(b) Compound E was also prepared by the following route.

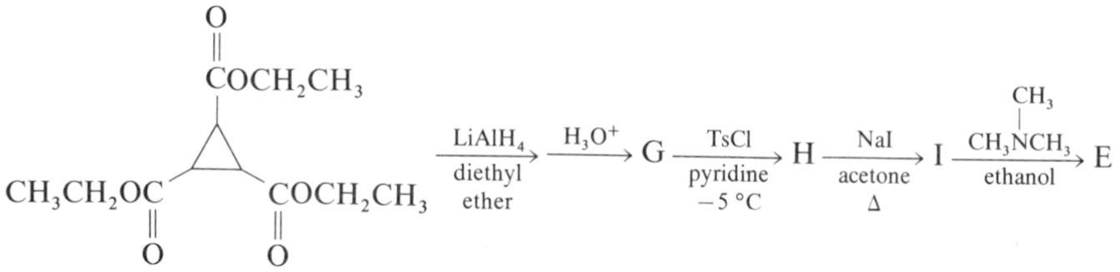

$COCH_2CH_3$
CH_3CH_2OC
$COCH_2CH_3$
$LiAlH_4$
diethyl
ether
H_3O^+
G
TsCl
pyridine
−5 °C
H
NaI
acetone
Δ
I
CH_3
CH_3NCH_3
ethanol
E

(c) The conversion of Compound G to Compound I could in theory also be carried out with hydriodic acid. That acid was not used because it was thought that molecular rearrangement might result. Write equations showing what rearrangement is possible, and explain why special care was necessary in this synthesis.

(d) What stereoisomerism is possible for the triethyl cyclopropanetricarboxylate used as the starting material for these syntheses? Draw structural formulas that clearly illustrate your answer.

(e) Trimethylenecyclopropane can also be prepared by another route from Compound I. Write equation(s) showing how you would carry out this transformation.

22.38 In research on chair conformations, compounds labeled at certain positions with carbon-13 are to be studied by nuclear magnetic resonance spectroscopy. A compound to be synthesized and the available labeled starting material are shown below. How would you carry out this transformation? (Hint: The Diels-Alder reaction is useful in synthesizing six-membered rings.)

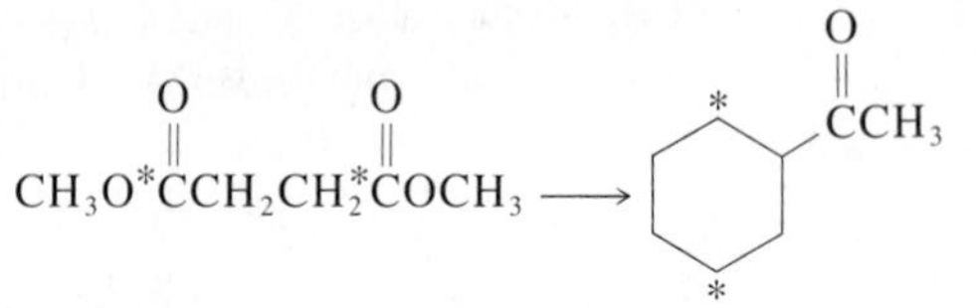

22.39 The following compounds were prepared as intermediates in the synthesis of an alkaloid-like compound. Supply reagents for all transformations shown.

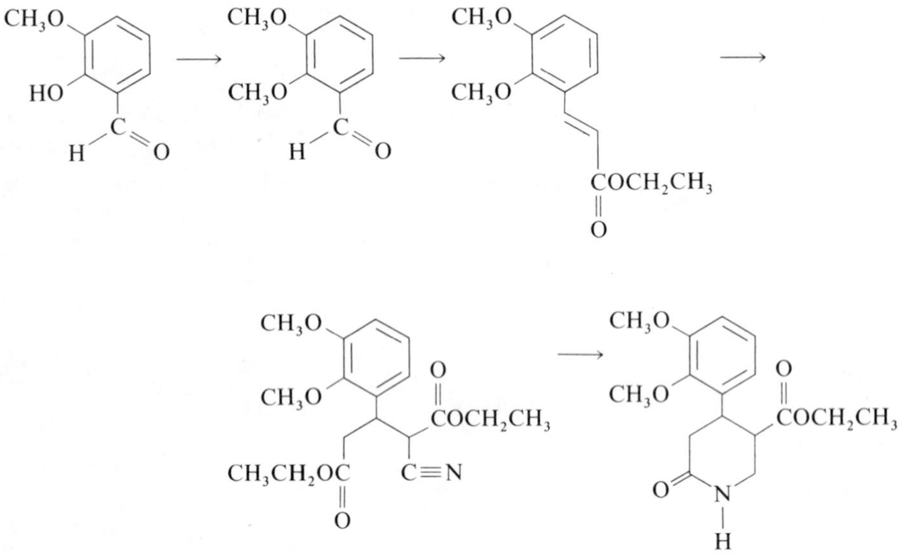

22.40 The following transformations were carried out during a study involving the quantitative measurement of the occurrence of nitrosamines in tissues treated with chemicals used in fisheries. What reagents are needed for each step?

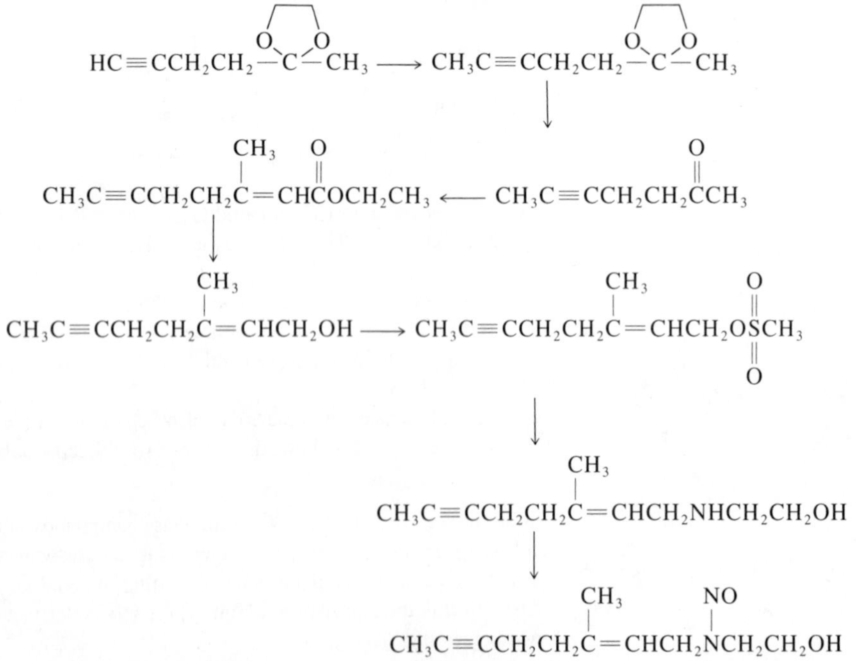

22.41 Write a mechanism that explains the following experimental result.

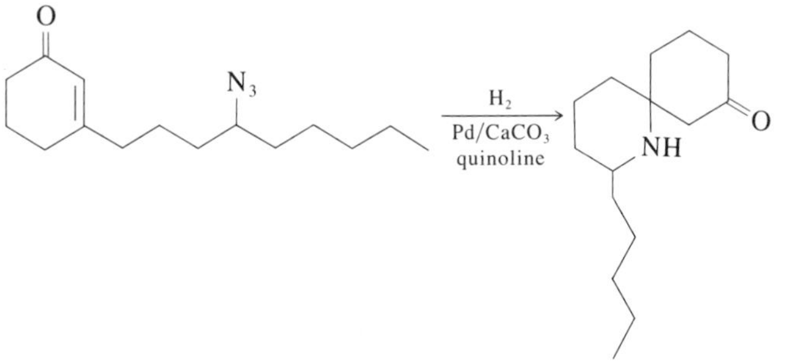

O
N3
H2
Pd/CaCO3
quinoline
O
NH

22.42 Compound A, $C_7H_7NO_3$, is converted into Compound B, C_7H_7NO, when treated with a reducing agent, such as tin in hydrochloric acid. The proton magnetic resonance spectrum of Compound B is shown in Figure 22.2. Assign structures to Compounds A and B.

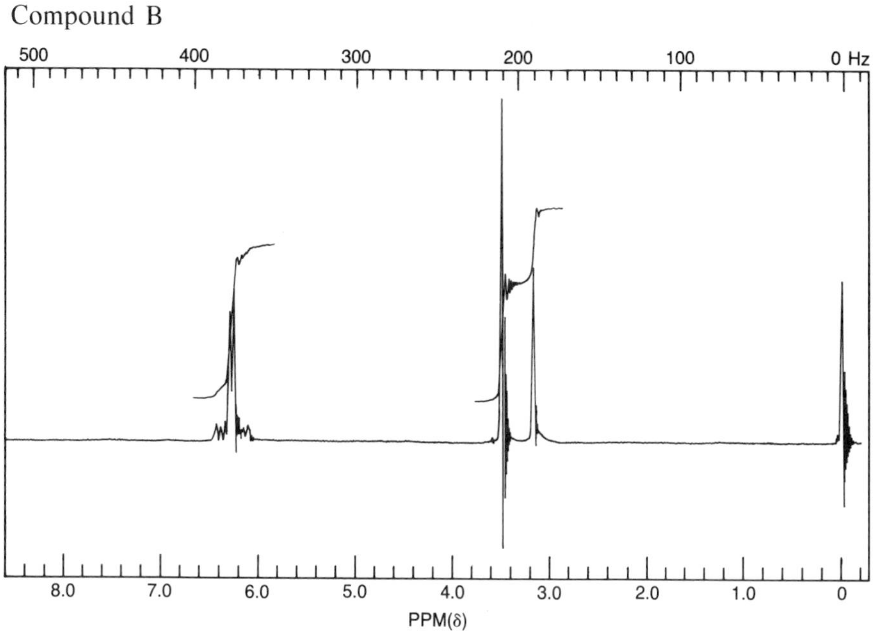

Compound B
500
400
300
200
100
0 Hz
8.0
7.0
6.0
5.0
4.0
3.0
2.0
1.0
0
PPM(δ)

FIGURE 22.2

22.43 Compound C, $C_{10}H_{15}N$, has the proton magnetic resonance spectrum shown in Figure 22.3 on the next page. Assign a structure to the compound.

22.44 The carbon-13 nuclear magnetic resonance spectrum of *N*-ethylacetamide has bands at 14.6, 22.8, 34.4, and 171.0 ppm. Assign these bands to the various carbon atoms in the compound. Table 11.4 (p. 413) may be useful.

22.45 Compound D, C_3H_9N, has two bands in its carbon-13 nuclear magnetic resonance spectrum, at 26.2 and 42.8 ppm. Assign a structure to Compound D.

22.46 Compound E, $C_8H_{11}N$, has bands in its carbon-13 nuclear magnetic resonance spectrum at 40.2, 112.7, 116.6, 129.0, and 150.7 ppm. Compound E is not soluble in water, but does dissolve in dilute hydrochloric acid. It does not react with benzenesulfonyl chloride in the presence of sodium hydroxide. Assign a structure to Compound E, and assign the bands that appear in its spectrum to the various carbon atoms.

Compound C

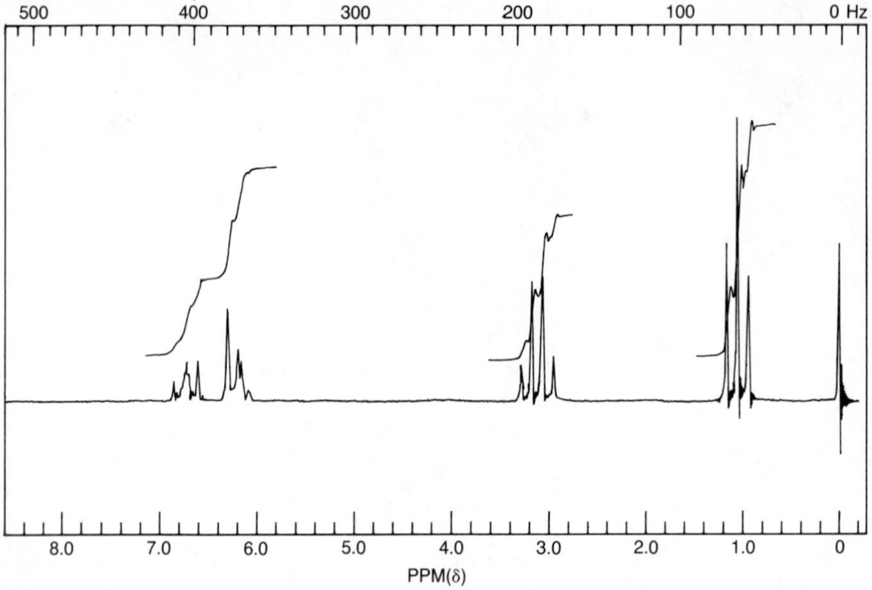
500
400
300
200
100
0 Hz
8.0
7.0
6.0
5.0
4.0
3.0
2.0
1.0
0
PPM(δ)

FIGURE 22.3

23

The Chemistry of Aromatic Components II. Synthetic Transformations

A • L O O K • A H E A D

Chapter 19 examined electrophilic substitution reactions of aromatic compounds. Chapter 22 described how to convert a nitro compound, the product of one important substitution reaction, to a primary aromatic amine and also how to convert the amine to a diazonium salt. The aryl diazonium ion has a good leaving group, molecular nitrogen, and undergoes substitution reactions in which the nitrogen is replaced regioselectively by a variety of substituents.

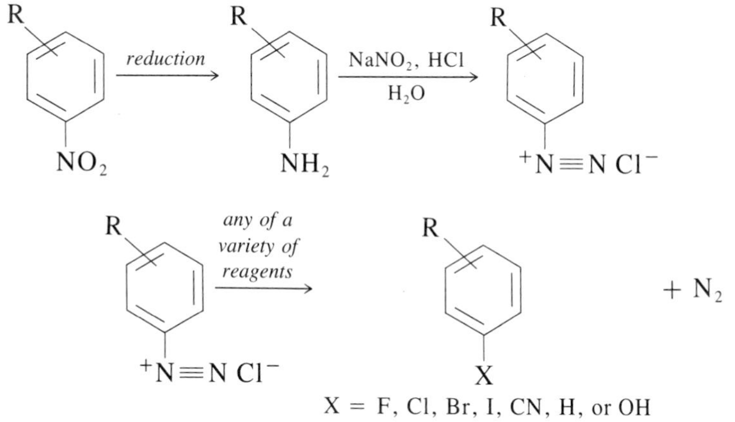

R
reduction
R
NaNO₂, HCl
H₂O
R
NO₂
NH₂
⁺N≡N Cl⁻
R
any of a
variety of
reagents
R
+ N₂
⁺N≡N Cl⁻
X

X = F, Cl, Br, I, CN, H, or OH

An aryl diazonium ion is also an electrophile and participates in electrophilic substitution reactions with activated aromatic rings to give compounds that are used as dyes.

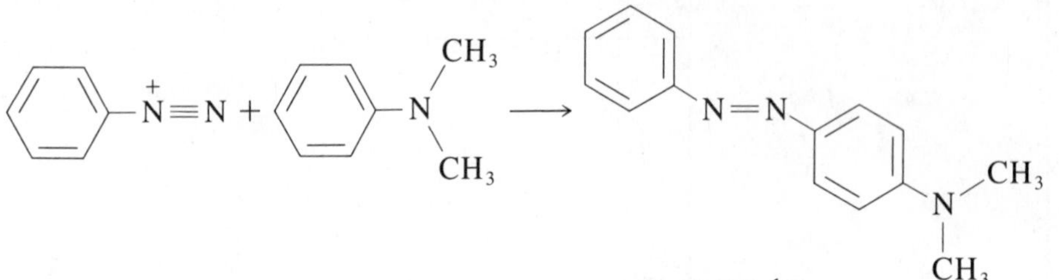
N≡N +
N
CH3
CH3
N=N
N
CH3
CH3

an azo dye

Aromatic rings do not react easily with nucleophiles. This chapter will show how substitution by electron-withdrawing groups on the aromatic ring or complexation of the ring with a transition metal makes the ring electrophilic enough to react with nucleophiles.

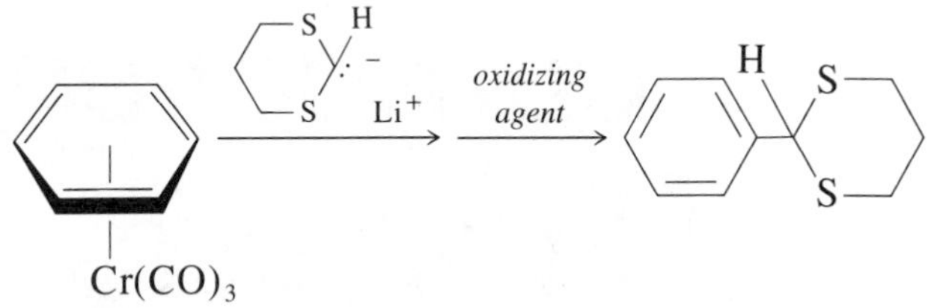
S
H
S
Li+
oxidizing
agent
H
S
S
Cr(CO)3

This reversal of the normal polarity of the aromatic ring allows compounds that are otherwise difficult to make to be synthesized directly.

23.1 SUBSTITUTION OF NITROGEN IN ARYL DIAZONIUM IONS

A. Sandmeyer Reactions

Treating aromatic primary amines with nitrous acid gives aryl diazonium ions, which are stable enough in solution at low temperatures (p. 913) to be used as reagents in syntheses. A molecule of nitrogen is the leaving group that is replaced in a variety of substitution reactions on aryl diazonium salts. Some of the reactions require catalysis by copper metal or copper(I) salts and are known as **Sandmeyer reactions.** The exact mechanism of the substitution reactions has been difficult to establish and probably involves radical intermediates in some cases and ionic substitution reactions in others. The important fact to remember is that in each case a primary amine group on an aromatic ring is replaced selectively by a halogen or hydrogen or a cyano or hydroxyl group. This type of reaction has wide usefulness.

For example, when toluene is brominated, a mixture of isomers is formed, including *o*-bromotoluene, *p*-bromotoluene, and even benzyl bromide when the reaction is carried out in the light (p. 832). If pure *o*-bromotoluene is needed, the best way to make it is by diazotization of *o*-toluidine and treatment of the resulting *o*-toluenediazonium bromide with copper metal or copper(I) bromide.

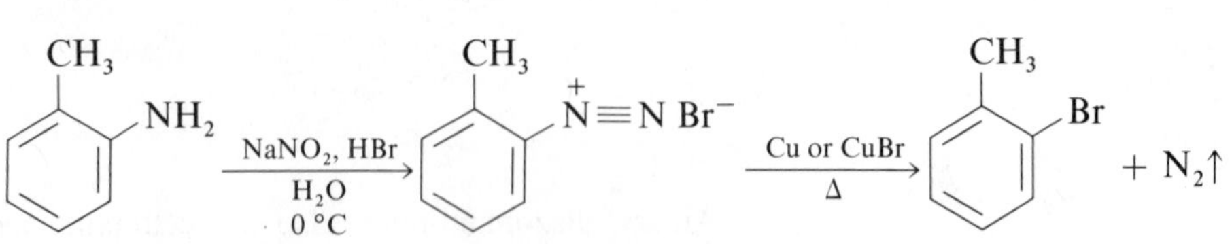
CH3
NH2
NaNO2, HBr
H2O
0 °C
CH3
N≡N Br−
Cu or CuBr
Δ
CH3
Br
+ N2↑

o-toluidine *o*-toluenediazonium bromide *o*-bromotoluene

Entirely analogous reactions are used to introduce chlorine regioselectively.

Elemental fluorine and iodine cannot be used for electrophilic aromatic substitution reactions. Aromatic fluoro and iodo compounds are prepared from the corresponding amines. Aniline, for example, is converted into fluorobenzene or iodobenzene.

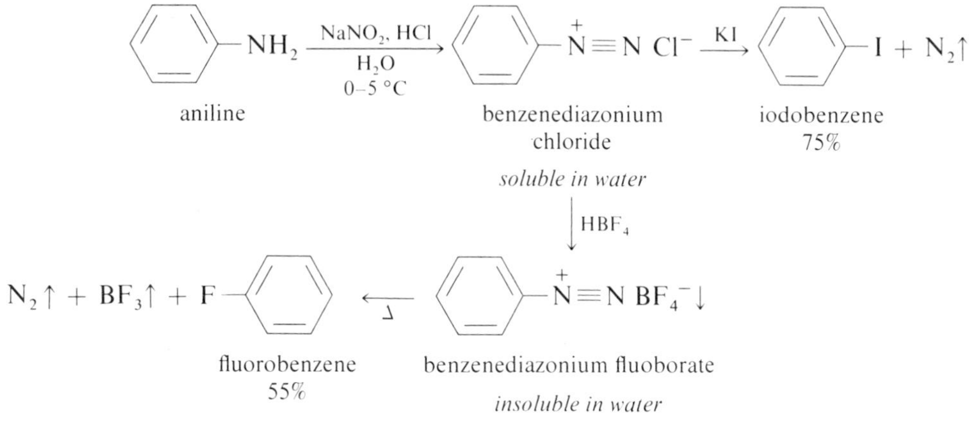

NH2
NaNO2, HCl
H2O
0–5 °C
N≡N Cl⁻
KI
I + N2↑
aniline
benzenediazonium chloride
soluble in water
iodobenzene 75%
HBF4
N2↑ + BF3↑ + F
Δ
N≡N BF4⁻↓
fluorobenzene 55%
benzenediazonium fluoborate
insoluble in water

Iodide ion directly replaces the nitrogen in benzenediazonium chloride to give iodobenzene. To make fluorobenzene, it is necessary to replace the chloride ion in benzenediazonium chloride with fluoborate anion, by treating the diazonium salt with fluoboric acid. The resulting diazonium fluoborate is much less soluble in water than the chloride is, and it precipitates. It is isolated, dried, and decomposed by heating to give fluorobenzene, boron trifluoride, and nitrogen.

A cyano group is another group that cannot be introduced directly onto an aromatic ring because aryl halides do not undergo nucleophilic substitution reactions easily (p. 956). Copper(I) cyanide is used to replace nitrogen in an aryl diazonium salt by a cyano group. Once again, the reaction proceeds with complete regioselectivity, as is illustrated by the conversion of *p*-toluidine to *p*-toluonitrile.

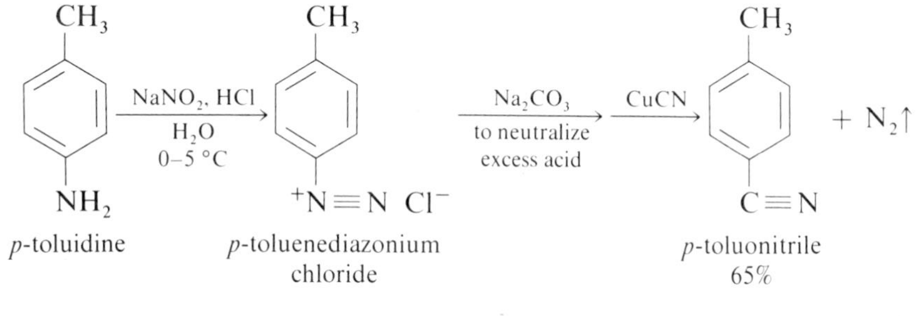

CH3
NaNO2, HCl
H2O
0–5 °C
NH2
p-toluidine
CH3
⁺N≡N Cl⁻
p-toluenediazonium chloride
Na2CO3
to neutralize excess acid
CuCN
CH3
C≡N
+ N2↑
p-toluonitrile
65%

In all the reactions described above, phenols are produced as products of a side reaction in which the diazonium ion reacts with water. The replacement of nitrogen by a hydroxyl group can be made the chief reaction by preparing the diazonium salt in an acid with a conjugate base that is a particularly poor nucleophile and then heating the reaction mixture. For example, *m*-nitrophenol, which cannot be prepared by the direct nitration of phenol (p. 785), is made in this way. Note the use of sulfuric acid in the diazotization reaction.

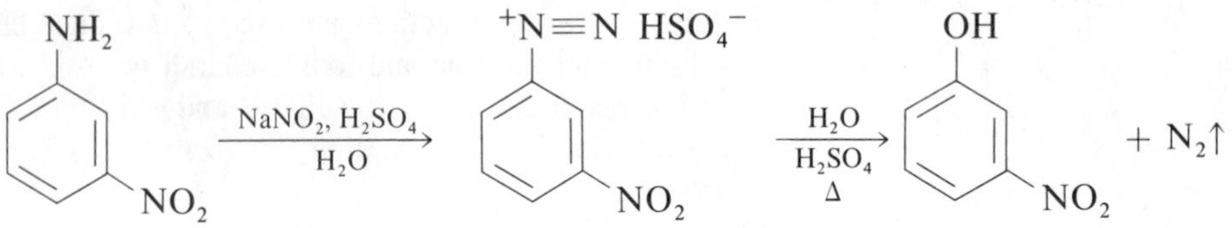
NH2
NaNO2, H2SO4
H2O
+N≡N HSO4−
H2O
H2SO4
Δ
OH
NO2
+ N2↑

m-nitroaniline — *m*-nitrobenzenediazonium hydrogen sulfate — *m*-nitrophenol 80%

Finally, the amino group can be used to direct electrophilic aromatic substitution on the ring and can then be removed through diazotization and replacement by hydrogen in a reduction reaction. The preparation of 2,4,6-tribromobenzoic acid illustrates this sequence of reactions. Bromination of benzoic acid is expected to give *m*-bromobenzoic acid as the chief product because the carboxyl group is ring-deactivating and meta-directing (p. 787). The amino group in *m*-aminobenzoic acid, however, will activate the ring and direct bromine atoms ortho and para to itself. The amino group can then be removed by diazotization followed by reduction of the diazonium ion with hypophosphorous acid, H_3PO_2, a reducing agent.

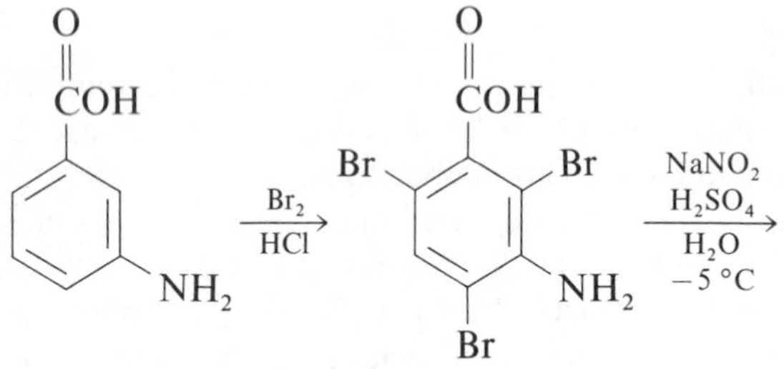
O
COH
NH2
Br2
HCl
Br
NaNO2
H2SO4
H2O
−5 °C

m-aminobenzoic acid

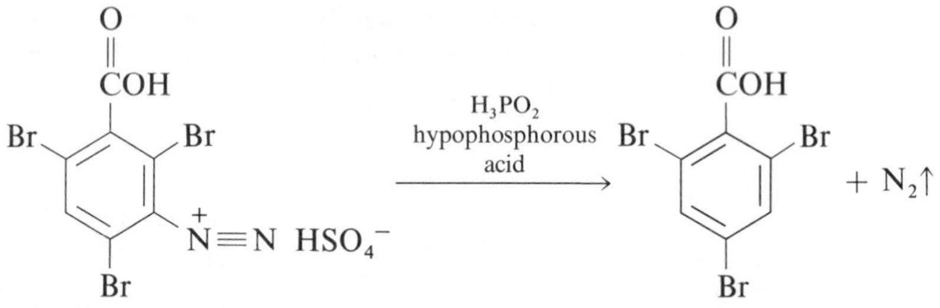
O
COH
Br
N≡N HSO4−
H3PO2
hypophosphorous
acid
+ N2↑

2,4,6-tribromobenzoic acid 75%

Study Guide Concept Map 23.1

The reactions outlined in the equations in this section represent ways in which functional groups can be positioned regioselectively on aromatic rings.

B. Problem-Solving Skills

Problem

How would you carry out the following conversion?

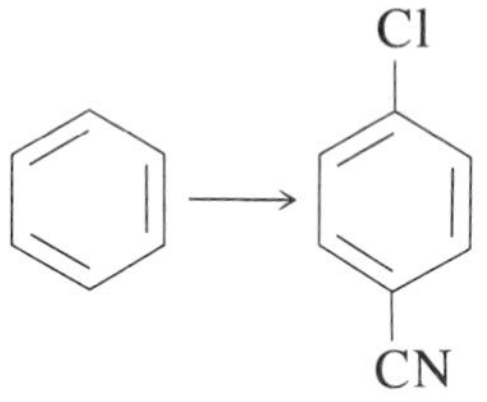

Solution

1. What functional groups are present in the starting material and the product?

   The starting material is benzene, an aromatic hydrocarbon. The product has a chlorine atom and a cyano group as substituents para to each other on the aromatic ring.

2. How do the carbon skeletons of the two compounds compare? How many carbon atoms does each contain? Are there any rings? What are the positions of branches and functional groups on the carbon skeletons?

   The product has one more carbon atom than the starting material; this carbon is part of the cyano group. A chlorine atom has also been substituted on the ring.

3. How do the functional groups change in going from starting material to product? Does the starting material have a good leaving group?

   The aromatic ring has undergone substitution at two positions.

4. Is it possible to dissect the structures of the starting material and product to see which bonds must be broken and which formed?

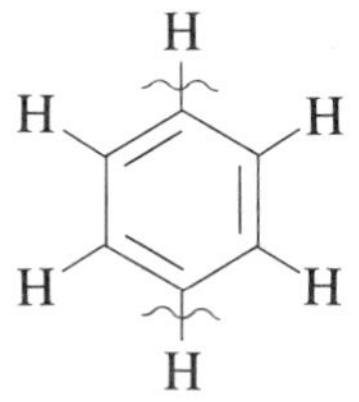

bonds to be broken

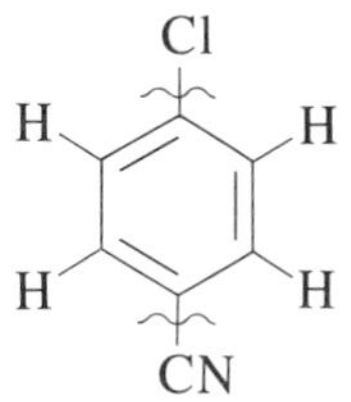

bonds to be formed

5. Do we recognize any part of the product molecule as coming from a good nucleophile or an electrophilic reaction?

   A chlorine atom can be introduced by direct electrophilic aromatic substitution. The cyano group must be introduced indirectly, by first producing a diazonium ion.

6. What type of compound would be a good precursor to the product?

   The amine corresponding to the cyano compound is needed.

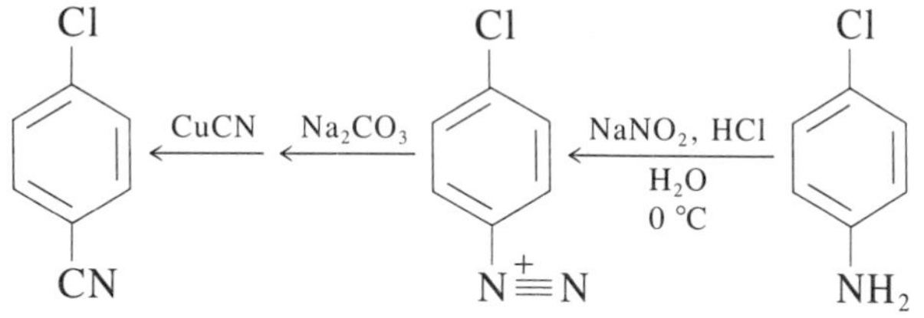

7. After this last step, do we see how to get from starting material to product? If not, we need to analyze the structure obtained in step 6 by applying questions 5 and 6 to it.

The amine is prepared by reduction of a nitro compound, which is the product of an electrophilic aromatic substitution. Because the chloro group is ortho, para-directing, it is introduced first.

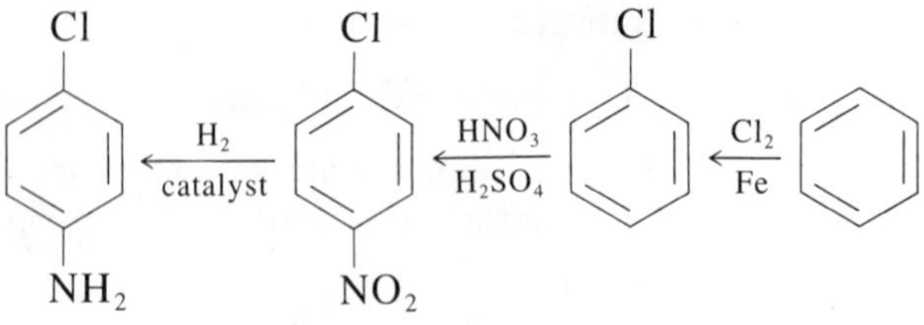
Cl
Cl
Cl
H₂
catalyst
HNO₃
H₂SO₄
Cl₂
Fe
NH₂
NO₂

To do this sequence the other way, that is, to make the amine first and then chlorinate it, would mean that the amino group would have to be protected as an amide and then deprotected later.

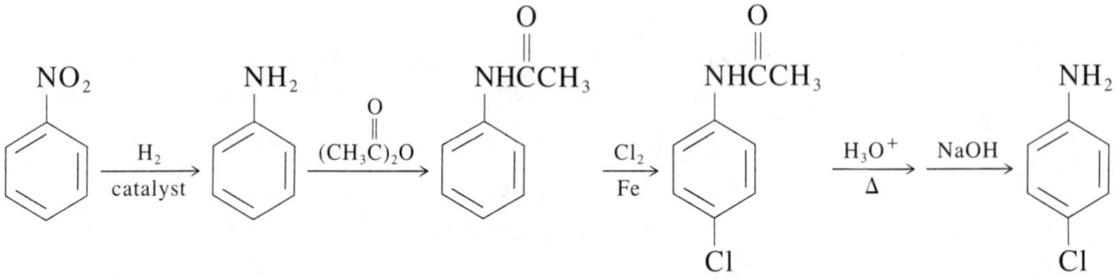
NO₂
NH₂
NHCCH₃
NHCCH₃
NH₂
H₂
catalyst
(CH₃C)₂O
Cl₂
Fe
H₃O⁺
Δ
NaOH
Cl
Cl

Direct chlorination of aniline to get the monochloro product is difficult. The amino group activates the ring strongly, so polychloro products are formed at first. Hydrogen chloride is a product of the substitution reaction and will protonate the amino group as the reaction progresses, making it meta-directing. Also, amines (nucleophiles) react directly with chlorine (an electrophile) at the nitrogen atom. All of these reasons make it advisable to introduce the chlorine atom first.

PROBLEM 23.1

The starting materials for the syntheses shown in Section 23.1A can be prepared by standard aromatic substitution reactions and further manipulations of the substituents. How would you prepare each of the following starting materials from benzene or toluene?

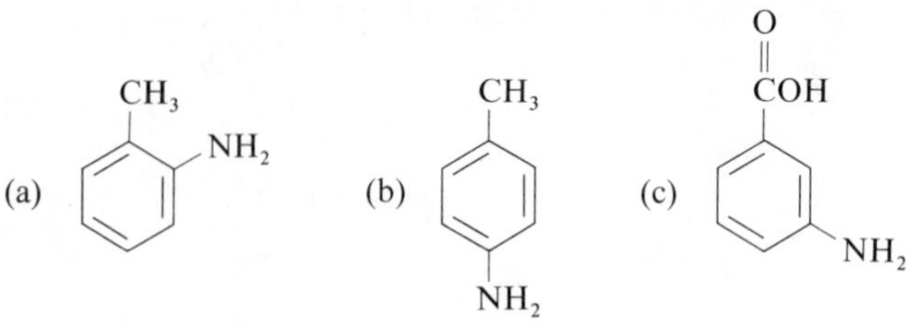
CH₃
NH₂
(a)
CH₃
(b)
NH₂
COH
(c)
NH₂

PROBLEM 23.2

Devise a synthesis for each of the following seven compounds, starting with benzene or toluene.

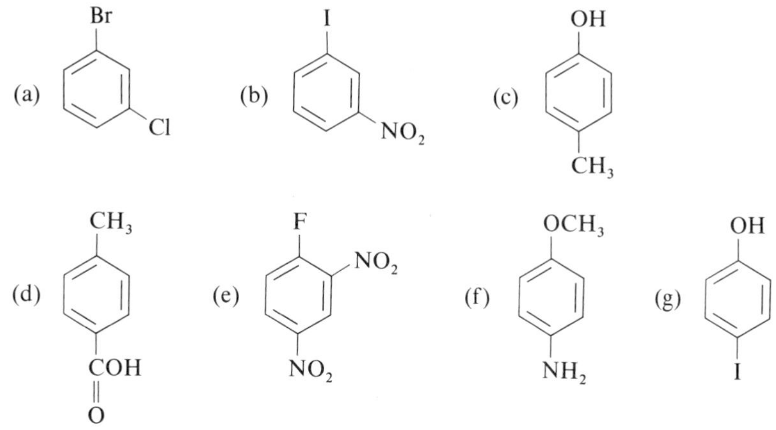
(a) Br Cl
(b) I NO2
(c) OH CH3
(d) CH3 COH O
(e) F NO2 NO2
(f) OCH3 NH2
(g) OH I

C. Benzyne and Its Reactions

An interesting phenomenon occurs when *o*-aminobenzoic acid, also known as anthranilic acid, is diazotized. Both nitrogen and carbon dioxide are lost from the molecule to give a very reactive intermediate. This intermediate, called **benzyne,** is a reactive dienophile (p. 691) and is also attacked by nucleophiles. The generation of benzyne by way of benzenediazonium-2-carboxylate is shown below. An alkyl nitrite is used with a strong acid to produce the diazonium salt of anthranilic acid. This diazonium chloride is then converted into the internal salt by removing the elements of hydrogen chloride with silver oxide.

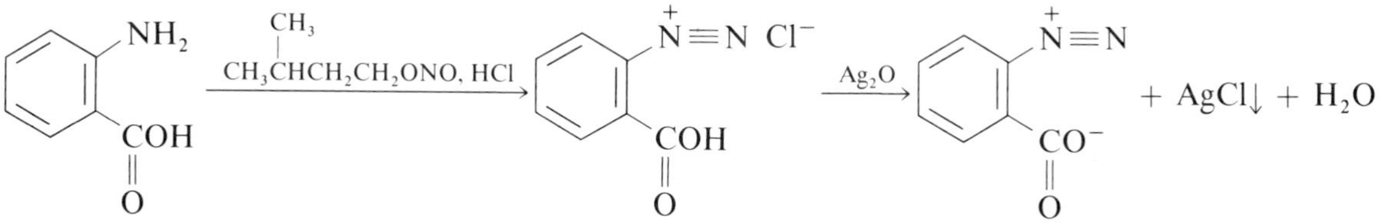
NH2
COH
O
CH3
CH3CHCH2CH2ONO, HCl
N≡N Cl−
COH
O
Ag2O
N≡N
CO−
O
+ AgCl↓ + H2O

o-aminobenzoic acid
anthranilic acid

benzenediazonium-
2-carboxylate
~60%

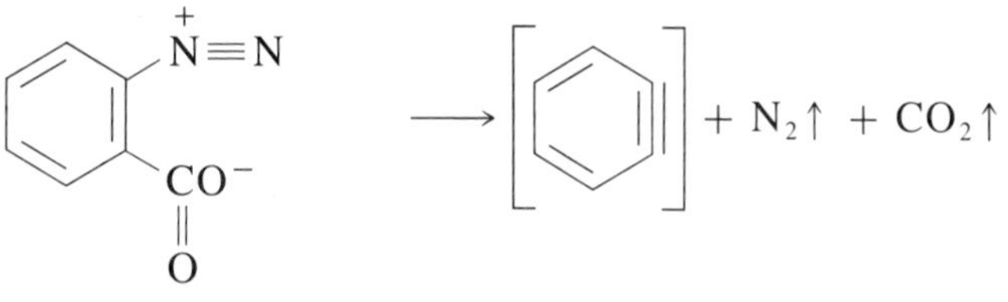
N≡N
CO−
O
+ N2↑ + CO2↑

benzenediazonium-2-carboxylate

benzyne

VISUALIZING THE REACTION

Formation and reactions of benzyne

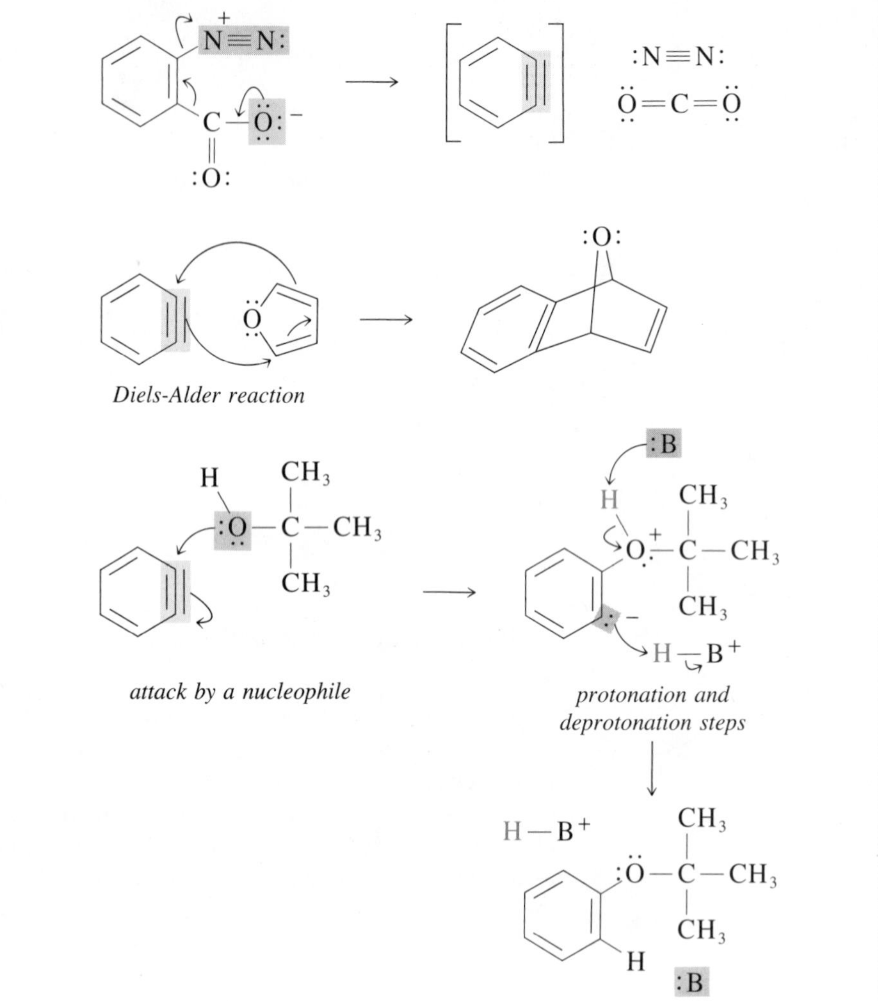

It is thought that benzyne has a triple bond in the benzene ring. The molecule is still an aromatic system; the aromatic sextet within the six-membered ring is undisturbed. Between two of the carbon atoms is an additional highly strained and, therefore, very reactive π bond, which can be represented as shown in Figure 23.1.

Other systems provide evidence for arynes as reaction intermediates. One of the most convincing experiments indicating such an intermediate uses iodobenzene, labeled with the radioactive isotope of carbon, ^{14}C, at the carbon atom bearing the iodine atom. The iodobenzene that is used is synthesized from aniline specifically labeled at the carbon atom bearing the amino group by diazotization and replacement of nitrogen by iodide ion. This sequence of reactions is a good example of an application of the syntheses discussed in Section 23.1A. The labeled iodobenzene is then subjected to a nucleophilic substitution reaction using the powerful nucleophile and strong base amide anion in liquid ammonia. The aniline formed in the reaction has the radioactivity distributed approximately equally at the carbon atom

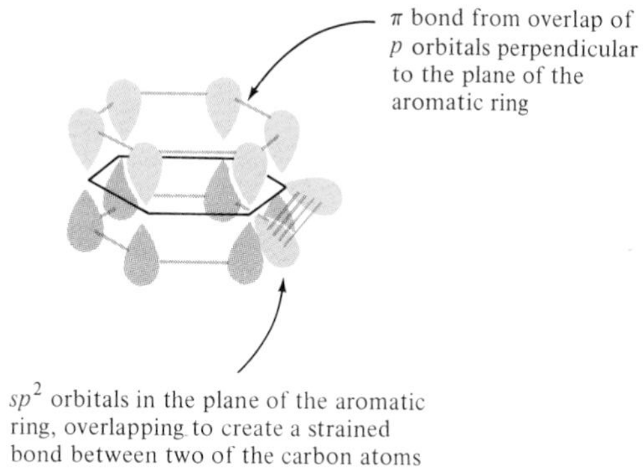

π bond from overlap of
p orbitals perpendicular
to the plane of the
aromatic ring
sp^2 orbitals in the plane of the aromatic
ring, overlapping to create a strained
bond between two of the carbon atoms

FIGURE 23.1 A representation of the orbitals involved in π bonding in benzyne.

originally bearing the amino group and the carbon atom adjacent to it. In the equation below, an asterisk is used to indicate the radioactive carbon atom.

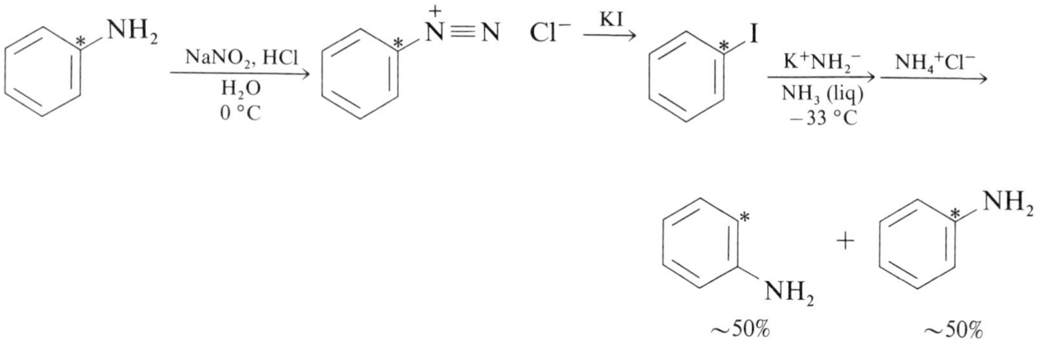

NH2
NaNO2, HCl
H2O
0 °C
N≡N Cl−
KI
I
K+NH2−
NH3 (liq)
−33 °C
NH4+Cl−
NH2
+
NH2
~50%
~50%

The experimental observation that radioactivity is distributed to two adjacent positions is explained by postulating a benzyne intermediate for the reaction. The strongly basic amide anion is thought to carry out an E_2 elimination reaction on iodobenzene to give benzyne, which then adds ammonia with roughly equal probability at the two adjacent carbon atoms.

VISUALIZING THE REACTION

Formation of benzyne and its reaction with a nucleophile

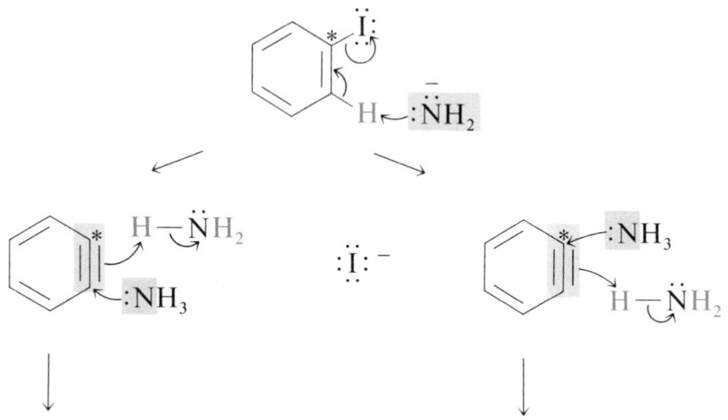

I
H
:NH2
H—NH2
:NH3
:I:−
:NH3
H—NH2

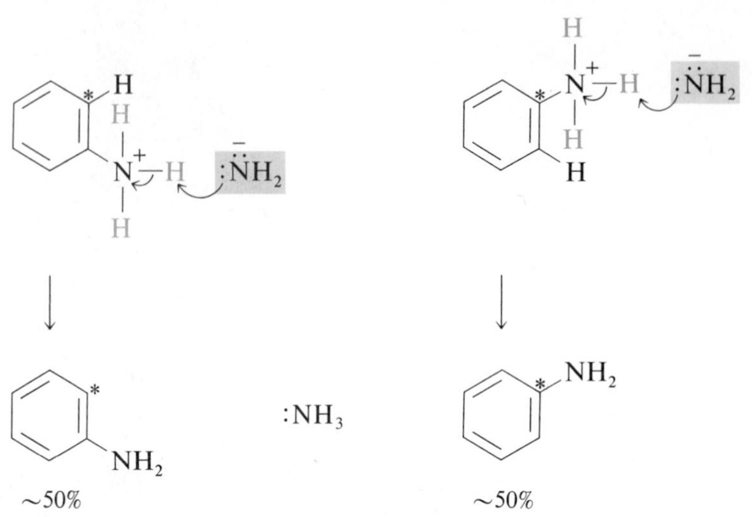
H
H
N
H
:NH₂
H
H
N
H
H
H
:NH₂
NH₂
:NH₃
NH₂
~50%
~50%

The reactions of aryl halides with potassium amide appear to be simple nucleophilic substitution reactions but are in fact **elimination-addition reactions.**

PROBLEM 23.3

When *o*-bromofluorobenzene is treated with magnesium metal in the presence of cyclopentadiene (p. 693), the product shown below is observed.

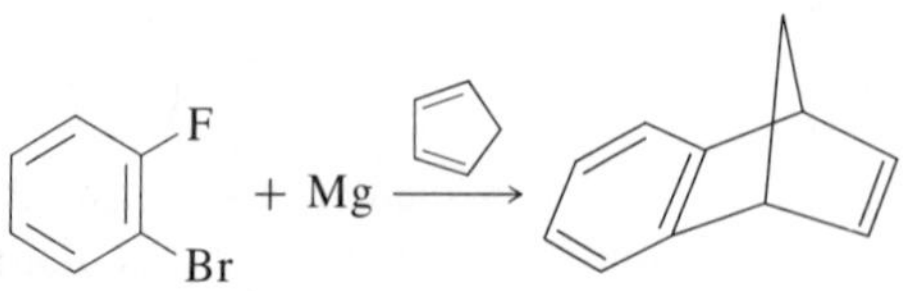
F
Br
+ Mg

Suggest a mechanism for this reaction.

23.2 THE DIAZONIUM ION AS ELECTROPHILE

A. Synthesis of Azo Compounds

One reaction of diazonium ions has important practical application in the dye industry. This is the coupling reaction in which the diazonium ion acts as an electrophile, substituting on the activated aromatic ring of either a phenol or an aromatic amine. A typical reaction is that of benzenediazonium chloride and phenol, which is carried out in a weakly basic solution. Electrophilic attack by the benzenediazonium ion takes place at one of the activated positions of the ring, ortho or para to the hydroxyl group.

VISUALIZING THE REACTION

The diazonium ion as electrophile

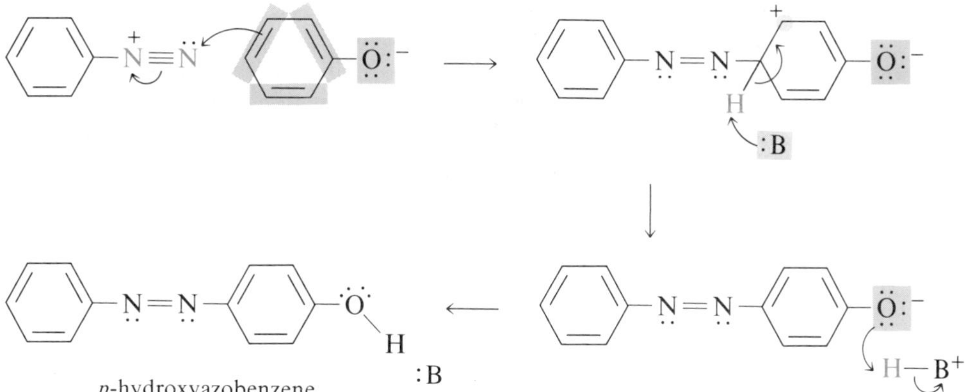

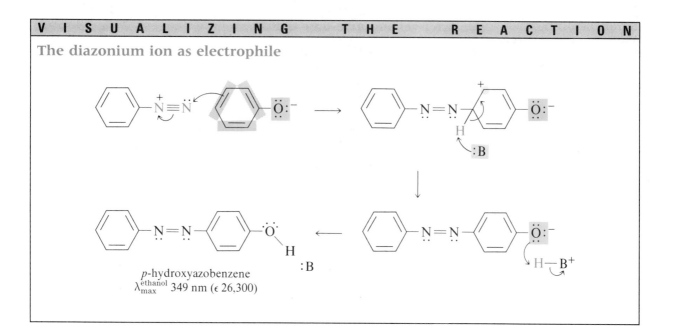

p-hydroxyazobenzene
$\lambda_{max}^{ethanol}$ 349 nm (ϵ 26,300)

The product of the above reaction is an **azo compound,** containing a nitrogen-nitrogen double bond that can have cis and trans isomers just as a carbon-carbon double bond can. The azo compound with a phenyl group at both ends of the azo linkage is azobenzene, of which both isomers are known.

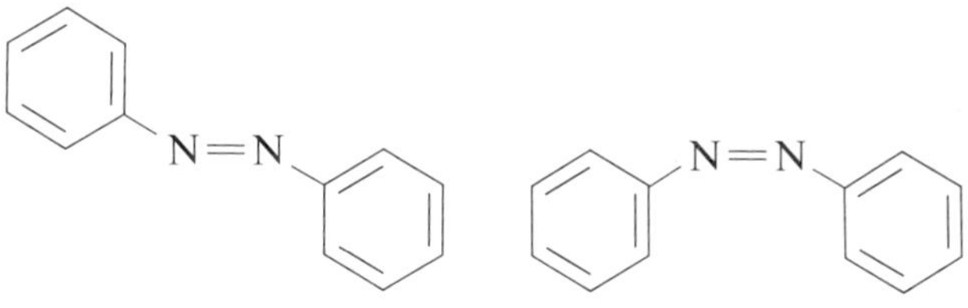

(E)-azobenzene (Z)-azobenzene

stereoisomers of azobenzene

Benzenediazonium chloride also couples with tertiary aromatic amines. For example, it reacts with N,N-dimethylaniline to give p-dimethylaminoazobenzene, a dye known as butter yellow, which was used to color margarine until it was discovered to be carcinogenic.

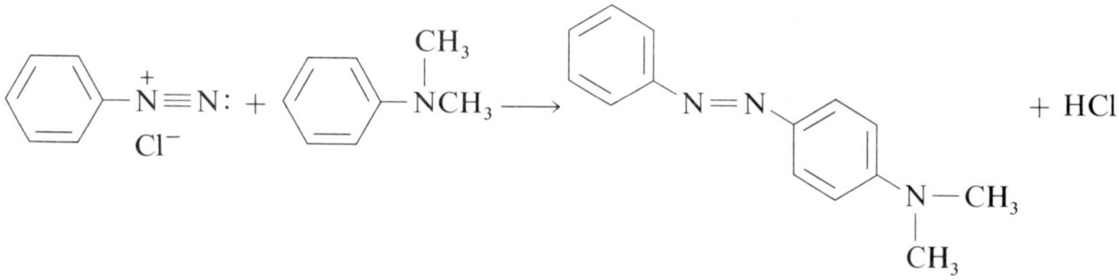

benzenediazonium chloride N,N-dimethylaniline p-dimethylaminoazobenzene butter yellow

B. Azo Dyes and Acid-Base Indicators

Compounds in which the azo linkage is between two aromatic rings are highly colored. These compounds have extended conjugated systems, and thus their absorption maxima are in the visible range of the electromagnetic spectrum (p. 719). (*E*)-Azobenzene, with $\lambda_{max}^{ethanol}$ 318 nm (ϵ 21,380), absorbs in the near-ultraviolet region and is visibly orange. This phenomenon is the basis for a qualitative analysis test for primary aromatic amines. The formation of red color when diazotized amine solution is added to 2-naphthol indicates the presence of an aromatic diazonium ion, and therefore of an aromatic primary amine.

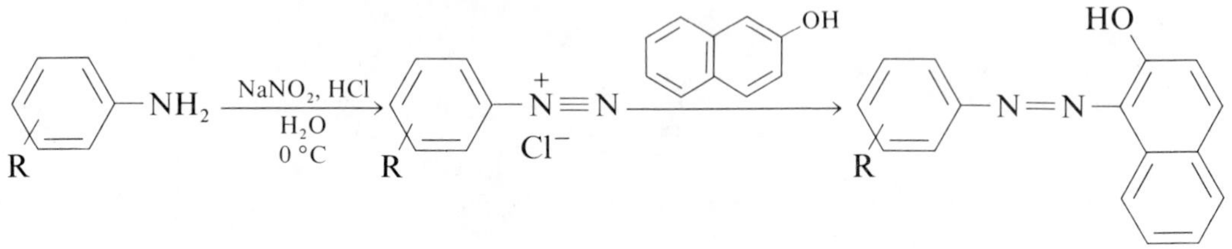
NH2
NaNO2, HCl
H2O
0 °C
N≡N
Cl⁻
OH
HO
N=N
R
R
R

primary aromatic amine — aryl diazonium ion — red dye

Hydroxyl and amino groups, especially if they are ortho or para to the azo bond, intensify the colors of azo compounds. For example, *p*-hydroxyazobenzene has $\lambda_{max}^{ethanol}$ 349 nm (ϵ 26,300), and *p*-dimethylaminoazobenzene has $\lambda_{max}^{ethanol}$ 408 nm (ϵ 27,540). Each of these compounds absorbs at a longer wavelength than azobenzene itself does, and, in each case, the molar absorptivity, ϵ, is also higher. Azo compounds in which an electron-donating group on one of the aromatic rings is conjugated with an electron-withdrawing group on the other ring have especially deep colors. A good example is the azobenzene in which a nitro group is substituted on one ring para to the azo linkage and a dimethylamino group is substituted on the other ring also para to the azo linkage.

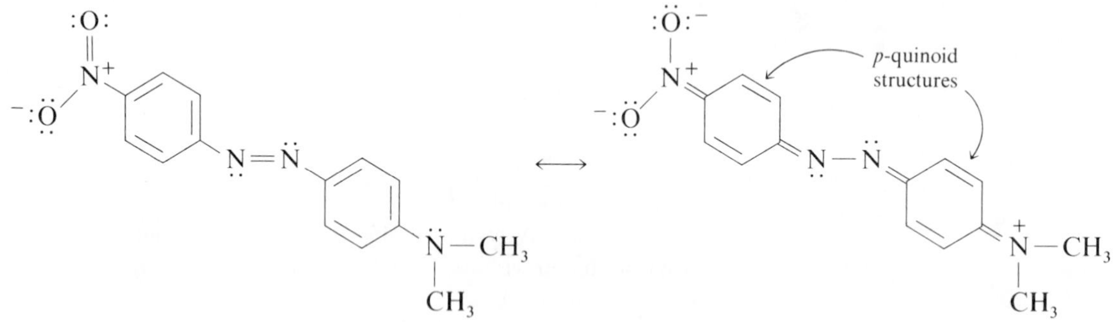
p-quinoid
structures
N=N
N−N
N−CH3
CH3
N−CH3
CH3

4-dimethylamino-4′-nitroazobenzene
$\lambda_{max}^{ethanol}$ 478 nm (ϵ 33,110)

One of the resonance contributors of the compound has quinoid structures, a feature that often leads to deep color in compounds (p. 849).

The aromatic rings of azo compounds allow the introduction of a variety of functional groups that can interact chemically with acidic, basic, or polar sites in the fibers used in making cloth or paper. A large variety of dyes, tailored to fit the chemical nature of the material to be dyed, have been created by extensions of the reactions shown above.

Many dye molecules have sulfonic acid groups as substituents, enabling them to adhere firmly to fibers such as silk and wool, which are composed chiefly of proteins and have basic functional groups on them. Methyl orange, synthesized from sulfanilic acid and *N,N*-dimethylaniline, is such a dye.

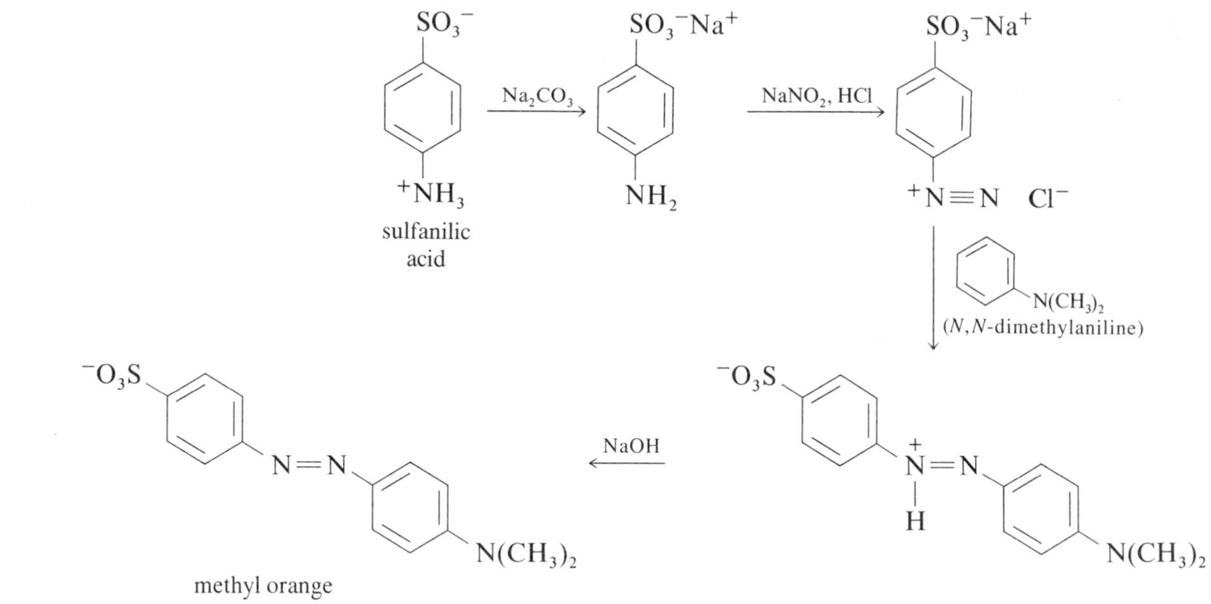

SO_3^-
$^+NH_3$
sulfanilic acid
Na_2CO_3
$SO_3^-Na^+$
NH_2
$NaNO_2$, HCl
$SO_3^-Na^+$
$^+N{\equiv}N\ \ Cl^-$
$N(CH_3)_2$
(*N,N*-dimethylaniline)
^-O_3S
N=N
$N(CH_3)_2$
H
NaOH
methyl orange

Methyl orange is also used as an acid-base indicator. In dilute solutions with a pH higher than 4.4, it is yellow; λ_{max} is 460 nm. When acid is added to the system, methyl orange is protonated, and the resulting dipolar ion predominates at pH values of 3.2 and lower. The protonated form has λ_{max} at 520 nm and appears red.

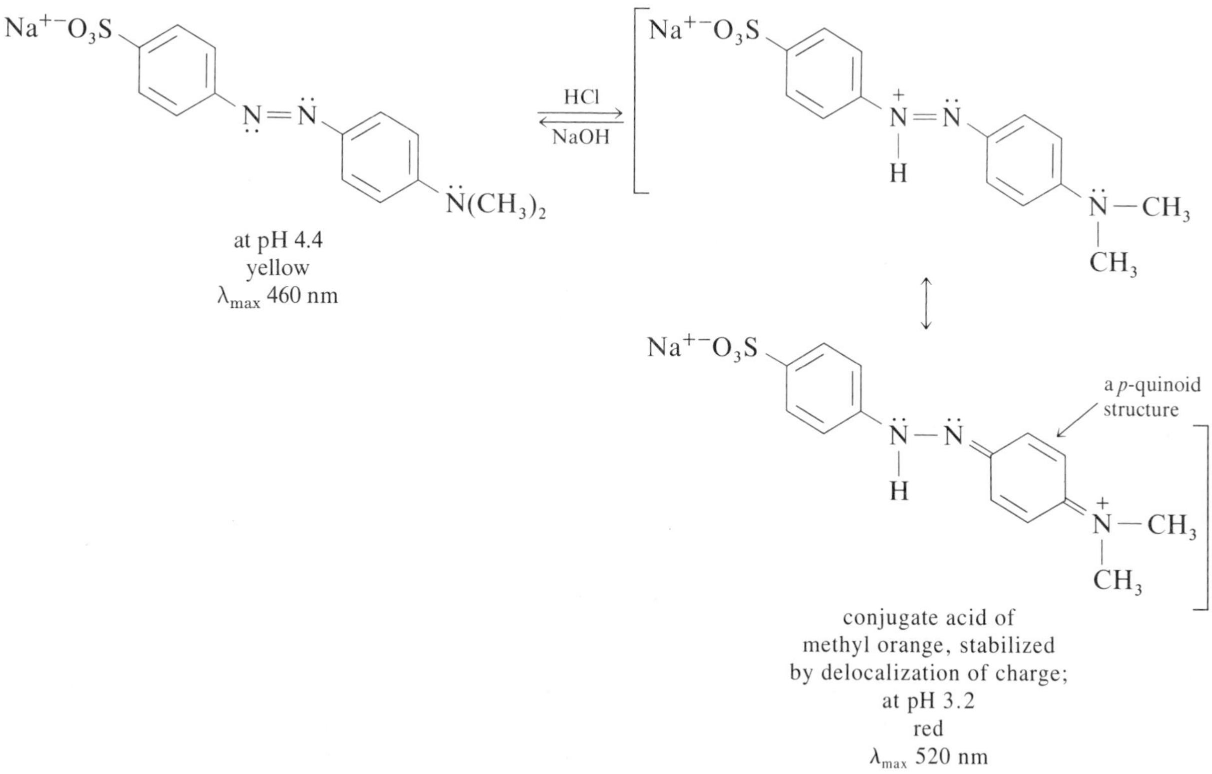

$Na^{+-}O_3S$
N=N
$N(CH_3)_2$
at pH 4.4
yellow
λ_{max} 460 nm
HCl
NaOH
H
N—CH_3
CH_3
a *p*-quinoid structure
conjugate acid of methyl orange, stabilized by delocalization of charge;
at pH 3.2
red
λ_{max} 520 nm

Methyl orange and its conjugate acid have different chromophores (p. 716) and thus absorb at different wavelengths in the range of visible light. The different colors that the compound shows allow it to be used to detect a change in the acidity of a system around the range of pH at which it is protonated and deprotonated.

PROBLEM 23.4

Study Guide Concept Map 23.2

There are three nitrogen atoms in methyl orange. Why is it protonated on the particular nitrogen atom shown in the equation above?

PROBLEM 23.5

Para red is widely used to dye cotton. The cotton fabric is soaked in one of the components of the dye and then the diazonium salt derived from the other component is added to the system so that the azo dye forms directly inside the fibers and is trapped there. Para red has the structure shown below. Write structures for the components that you would use to synthesize it.

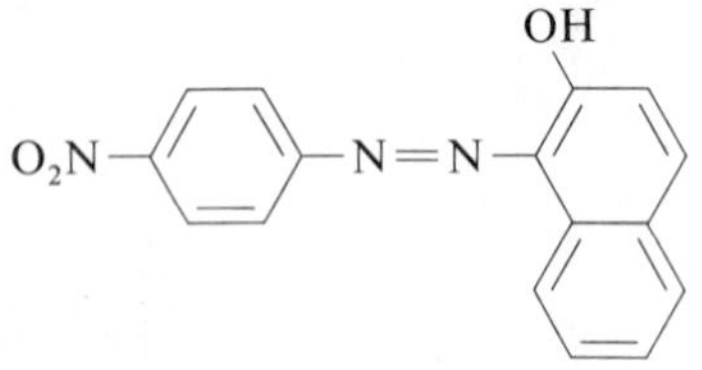
OH
O₂N
N=N

para red

PROBLEM 23.6

Extensive conjugation must be present in a molecule of an azo dye for it to have a blue color. Such a blue dye is synthesized by the following sequence of reactions. Supply structural formulas for Compounds A and B.

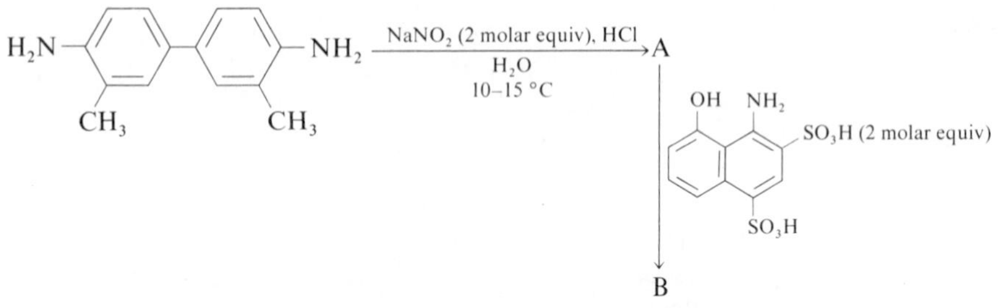
H₂N
NH₂
CH₃
CH₃
NaNO₂ (2 molar equiv), HCl
H₂O
10–15 °C
A
OH NH₂
SO₃H (2 molar equiv)
SO₃H
B

a blue dye

23.3 NUCLEOPHILIC AROMATIC SUBSTITUTION

A. Activation of Aryl Rings by Electron-Withdrawing Groups

Substitution of a nucleophile for a halogen in an aryl halide that does not also have electron-withdrawing groups as substituents is difficult. Such halides are inert to the reagents that react with alkyl halides under the usual conditions. When there are electron-withdrawing groups, especially nitro groups ortho or para to the halogen, however, nucleophilic substitution takes place with relative ease. 2,4-Dinitrophenylhydrazine, a reagent used to prepare derivatives of aldehydes and ketones in order to identify them, is synthesized by such a reaction.

Cl
NO_2
+ H_2NNH_2
(excess)
triethylene glycol
15–20 °C
20–30 min
$NHNH_2$
NO_2
+ $H_2NNH_3{}^+Cl^-$

2,4-dinitrochlorobenzene — hydrazine — 2,4-dinitrophenylhydrazine — hydrazine hydrochloride

Hydrazine, a base and a nucleophile, displaces chloride ion from the aromatic ring. One of the products of the reaction is hydrogen chloride, which reacts with excess hydrazine to form a salt.

This general reaction can involve any of a variety of pairs of nucleophiles and leaving groups. For example, 2,4-dinitrochlorobenzene can be converted into 2,4-dinitrophenol by heating with sodium carbonate and water.

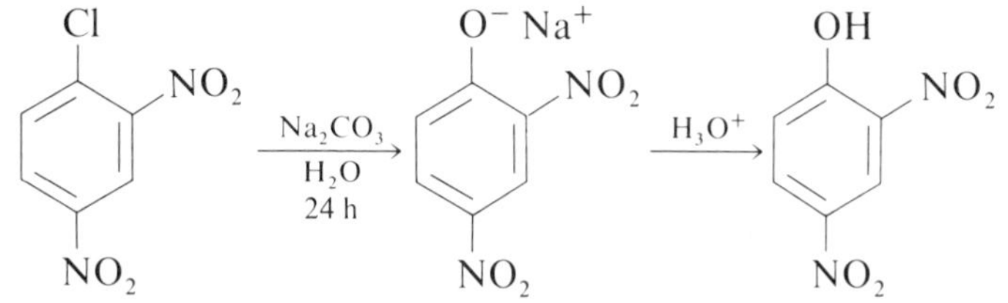

Cl
NO_2
Na_2CO_3
H_2O
24 h
O^- Na^+
H_3O^+
OH

2,4-dinitrochlorobenzene — sodium 2,4-dinitrophenolate — 2,4-dinitrophenol 91%

The nucleophile in this case is the hydroxide ion that is present in a solution of sodium carbonate in water.

A reagent that reacts very rapidly with amines is 2,4-dinitrofluorobenzene. It is used extensively in protein chemistry to label free amino groups on a protein or peptide chain (p. 1128). The reaction of this reagent with an amino acid, glycine, is illustrated below.

$$H_3\overset{+}{N}CH_2\overset{\overset{O}{\|}}{C}O^- \rightleftarrows H_2NCH_2\overset{\overset{O}{\|}}{C}OH$$

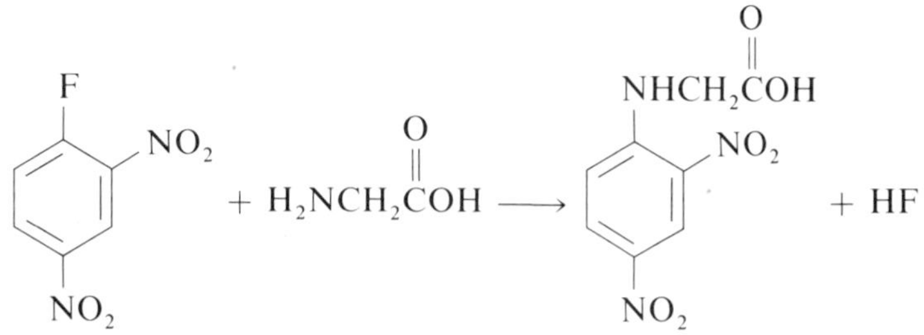

F
NO_2
+ H_2NCH_2COH
O
$NHCH_2COH$
+ HF

2,4-dinitrofluorobenzene — glycine — *N*-2,4-dinitrophenylglycine

Although fluoride ion is a poor leaving group in nucleophilic substitution reactions of alkyl halides, it is an excellent one in nucleophilic substitution reactions of aryl halides. In the following reaction, in which the cyclic amine piperidine is used to displace halide ion from a 2,4-dinitrohalobenzene, the fluoro compound reacts 3300 times faster than the iodo compound.

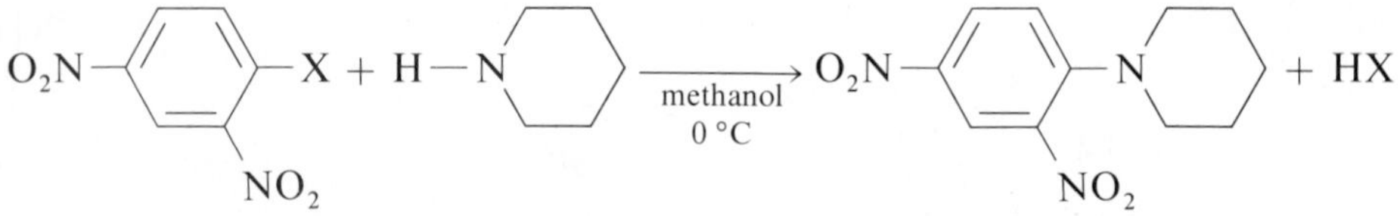
O2N
X + H—N
methanol
0 °C
O2N
N
+ HX
NO2
NO2
2,4-dinitrohalobenzene
piperidine
N-2,4-dinitrophenylpiperidine

X	k_{rel}
F	3300
I	1

Spectroscopic and kinetic evidence suggest that the reaction takes place by attack of the nucleophile on the aromatic ring at electron-deficient positions ortho and para to the electron-withdrawing nitro groups. An intermediate forms with a tetrahedral carbon atom, like the cation that is formed in electrophilic aromatic substitution. In this case, however, the intermediate is electron-rich rather than electron-deficient and is stabilized by delocalization of the negative charge to the nitro groups.

VISUALIZING THE REACTION

Nucleophilic aromatic substitution

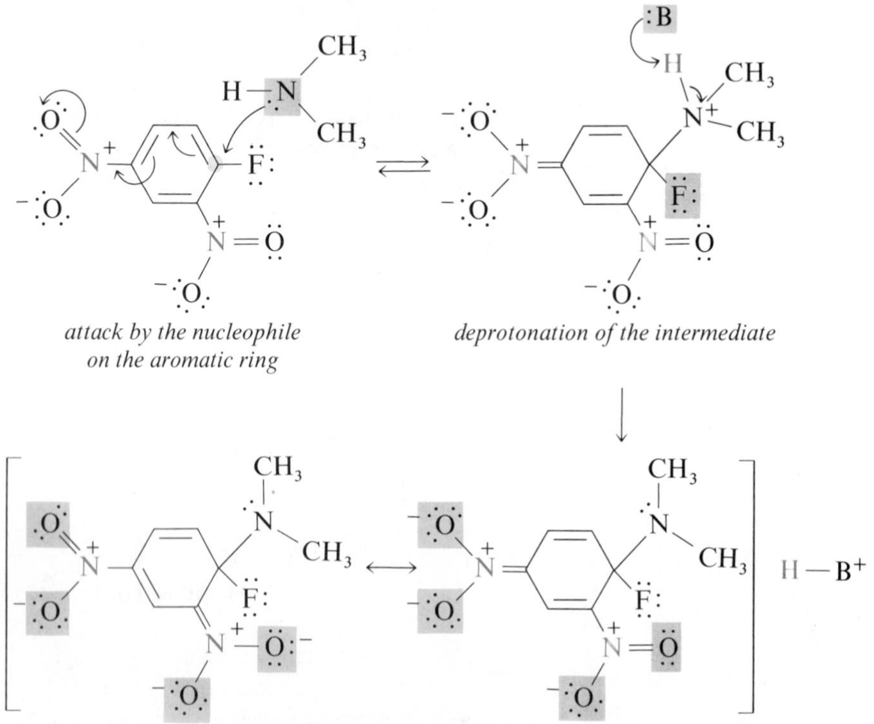
CH3
H—N
CH3
F
:B
H
CH3
N+
CH3
F
H—B+

attack by the nucleophile on the aromatic ring

deprotonation of the intermediate

tetrahedral anionic intermediate formed from the attack of nucleophile on aryl halide

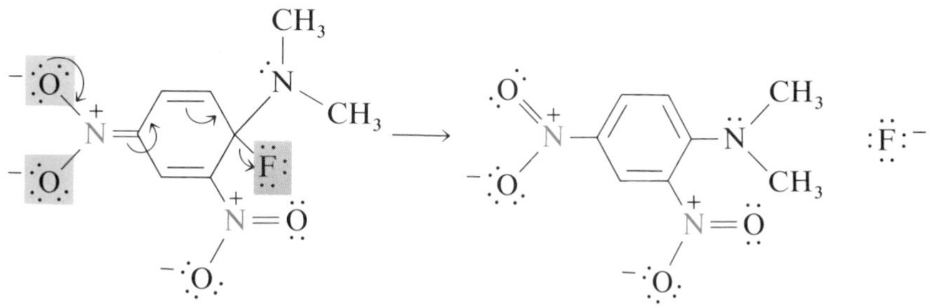
loss of fluoride ion,
the leaving group

The rate-determining step of the above reaction is the formation of the tetrahedral intermediate by nucleophilic attack of the amine at the carbon atom bearing the halogen atom. A fluoro compound reacts much faster than an iodo compound for two reasons. First, fluorine is much more electronegative than iodine. The carbon atom to which the halogen atom is bonded has a much larger positive charge in a fluorobenzene than in an iodobenzene. Second, a fluorine atom is much smaller than an iodine atom and, thus, offers less steric hindrance to an approaching nucleophile that will bond to the carbon atom to give the tetrahedral intermediate. Once the tetrahedral intermediate is formed, the amine loses a proton to a base, and the resulting anion is stabilized by delocalization of charge to the nitro groups. Departure of a fluoride ion reestablishes aromaticity and gives the product. This step is fast relative to the first step of the reaction.

In some cases, especially when the incoming nucleophile and the leaving group are both alkoxide ions, the intermediates are stable and can be isolated as salts, as shown below.

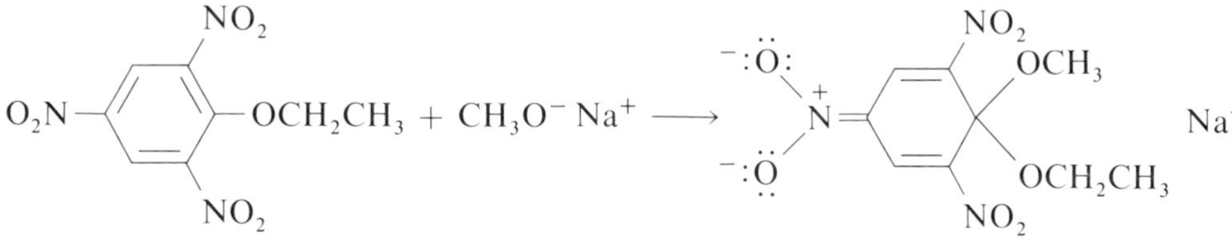
2,4,6-trinitroethoxybenzene
sodium methoxide
a Meisenheimer complex

These ionic compounds are called **Meisenheimer complexes** for the German chemist who first discovered them. The existence of such compounds is evidence that there is a tetrahedral anionic intermediate for nucleophilic substitution on aromatic rings.

Nucleophilic substitution on aryl halides, therefore, does not follow an S_N2 mechanism, in which no intermediate is formed. Instead such a reaction has two steps: addition of the nucleophile to give an intermediate, and elimination of the leaving group to restore the aromatic ring. The energy diagram for a nucleophilic aromatic substitution reaction resembles that for an electrophilic aromatic substitution reaction (Figure 19.4, p. 790).

PROBLEM 23.7

Complete the following equations.

(a) 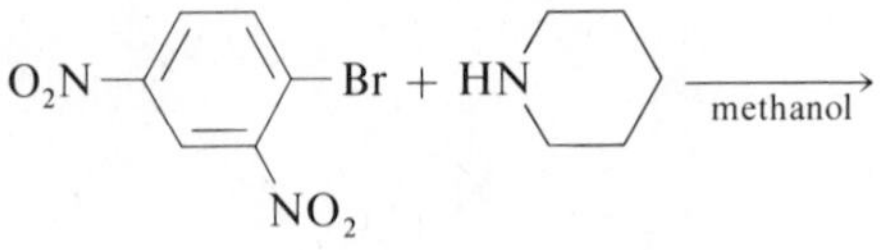
O2N
Br + HN
methanol
NO2

(b) 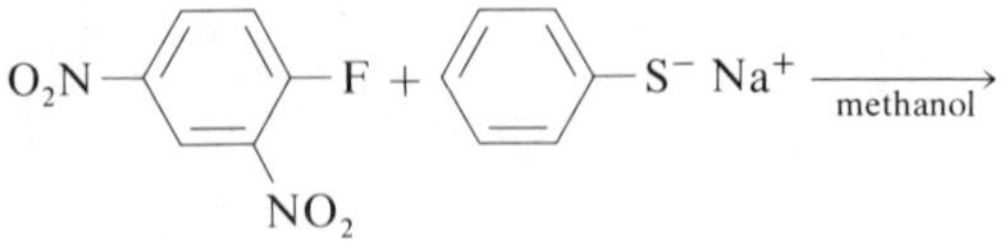
O2N
F +
S⁻ Na⁺
methanol
NO2

(c) 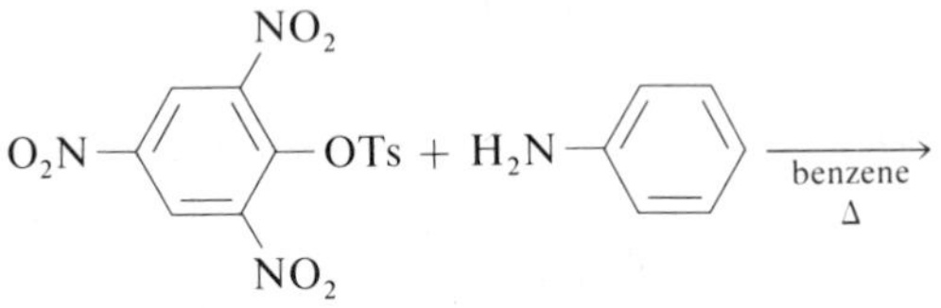
NO2
O2N
OTs + H2N
benzene
Δ
NO2

(d) 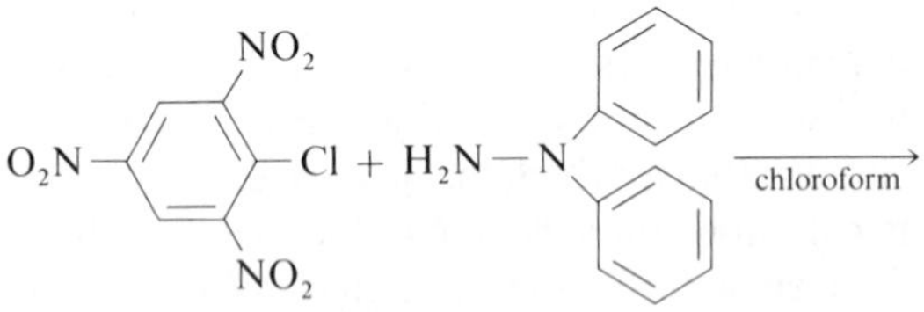
NO2
O2N
Cl + H2N—N
chloroform
NO2

(e) 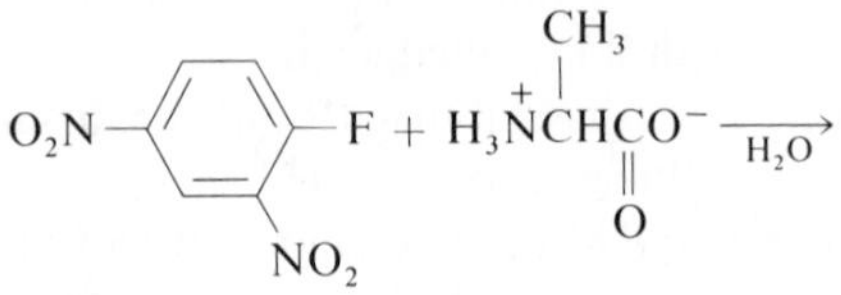
CH3
O2N
F + H3NCHCO⁻
H2O
O
NO2

B. Activation of Aryl Rings by Metal Complexes

Aromatic rings serve as ligands (p. 307) in organometallic compounds. For example, chromium hexacarbonyl reacts with benzene to give a complex of chromium with benzene.

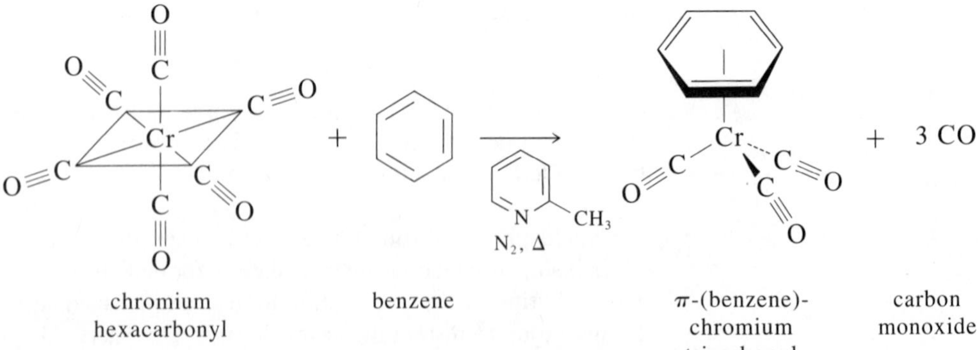
Cr
+
N
CH3
N2, Δ
Cr
+ 3 CO

chromium hexacarbonyl — benzene — π-(benzene)-chromium tricarbonyl 91% — carbon monoxide

eighteen electrons — *eighteen electrons*

Chromium is the fourth of the transition metals and has six electrons in its outermost shells, a $3d^5 4s^1$ electronic configuration. In chromium hexacarbonyl, six carbon monoxide molecules, each contributing two electrons, for a total of twelve electrons, give chromium an inert gas configuration of eighteen electrons. In the complex with benzene, the six π electrons of benzene replace the electrons of three of the carbon monoxide molecules. Because the transition metal atom has a variable valence, it serves to stabilize charge on or adjacent to an aromatic ring. Most interestingly, it activates the ring toward nucleophilic attack.

For example, π-(benzene)chromium tricarbonyl reacts with the 1,3-dithianyl anion (p. 756) to give an intermediate, which is decomposed by the oxidizing agent iodine.

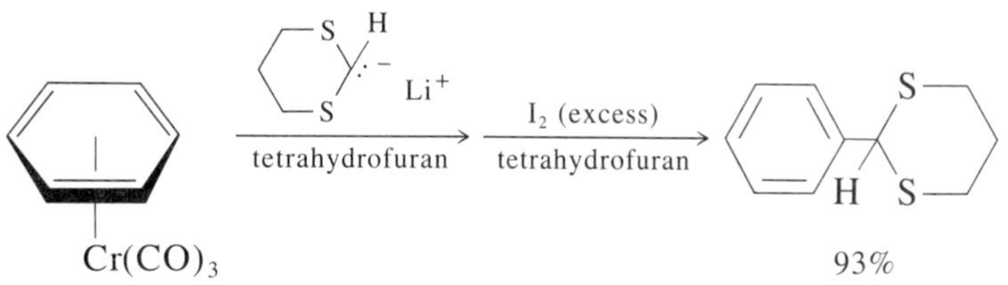
S H
Li+
S
I2 (excess)
tetrahydrofuran
tetrahydrofuran
S
H S
Cr(CO)3
93%

In this reaction, a carbon-carbon bond has been formed between the aromatic ring and a strongly basic and nucleophilic reagent.

The reaction proceeds by addition of the nucleophile to the aromatic ring on the face away from the bulky metal and carbonyl groups. The addition results in an anion that is stabilized by the transition metal atom, which can accept extra electrons.

VISUALIZING THE REACTION

Nucleophilic substitution on an aryl-metal complex

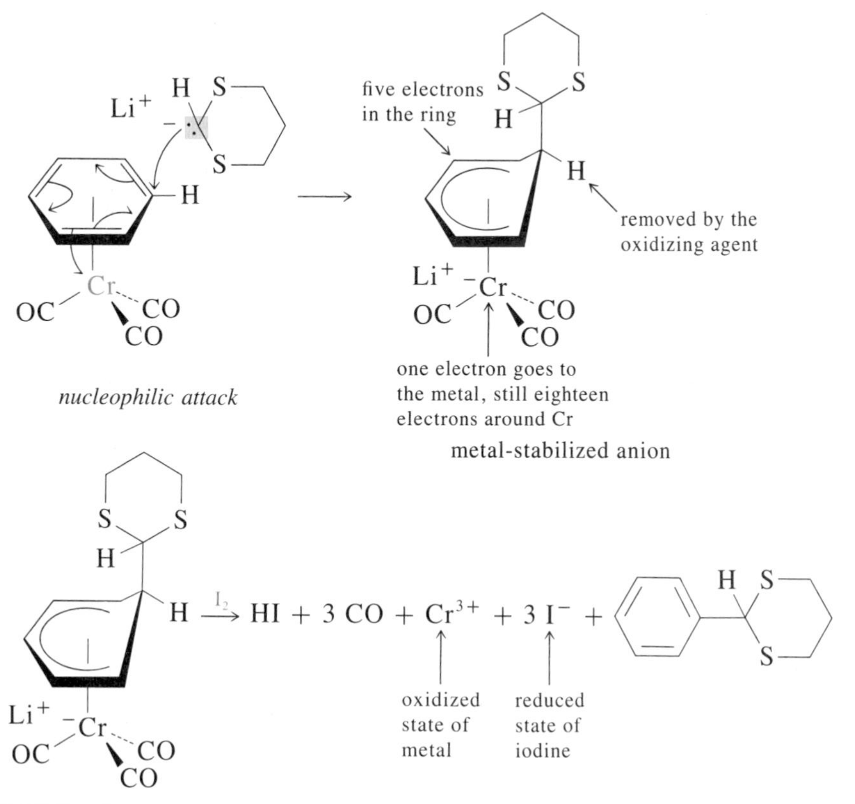
Li+
H S
S
H
Cr
OC
CO
CO
nucleophilic attack
five electrons in the ring
S
S
H
H
removed by the oxidizing agent
Li+
Cr
OC
CO
CO
one electron goes to the metal, still eighteen electrons around Cr
metal-stabilized anion
S
S
H
H
I2
HI + 3 CO + Cr3+ + 3 I− +
H S
S
Li+
Cr
OC
CO
CO
oxidized state of metal
reduced state of iodine

Such an aryl-metal complex reacts well with enolate anions from esters, with ions stabilized by nitriles, and with anions of carbon acids with pK_a values greater than 20. It does not react with stabilized enolates, such as that from diethyl malonate, or with Grignard reagents or organocuprates. Although *tert*-butyllithium and aryllithiums will react with such a complex, primary and secondary alkyllithiums do not.

Complexes of aryl halides with transition metals undergo nucleophilic substitution of the halogen just as aryl halides with nitro groups as substituents do (p. 956). The chromium complex of fluorobenzene, for example, undergoes nucleophilic substitution by the enolate anion of diethyl malonate to give diethyl phenylmalonate in high yield. Diethyl phenylmalonate is an intermediate in the synthesis of barbiturates and cannot be prepared directly from diethyl malonate in any other way (Problem 16.20, p. 670).

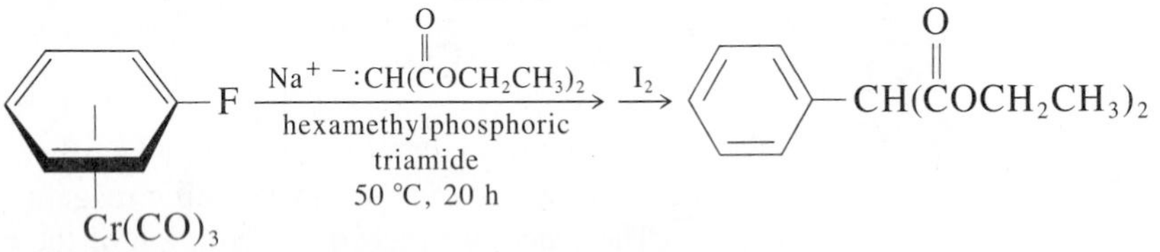
F
Na⁺ ⁻:CH(COCH₂CH₃)₂
hexamethylphosphoric
triamide
50 °C, 20 h
I₂
CH(COCH₂CH₃)₂
Cr(CO)₃

π-(fluorobenzene)-chromium tricarbonyl

diethyl phenylmalonate 95%

Studies have shown that at low temperatures the nucleophile attacks the ring carbon that is meta to the fluorine atom; only at higher temperatures does it migrate to the carbon bearing the halogen, at which point loss of the halogen occurs.

Study Guide Concept Map 23.3

PROBLEM 23.8

Give structures for the intermediates and products represented by letters in the following equations.

(a) 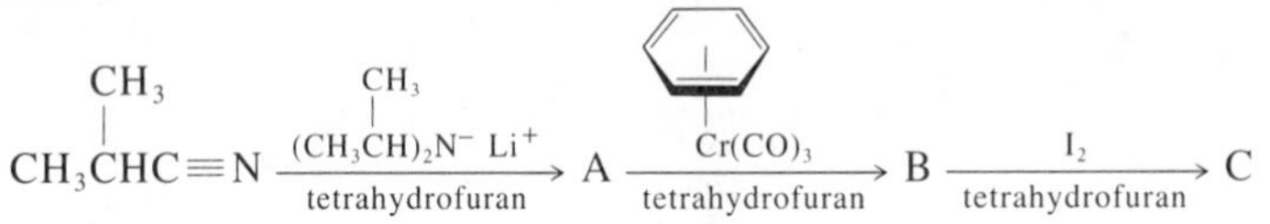
CH₃CHC≡N
(CH₃CH)₂N⁻ Li⁺
tetrahydrofuran
A
Cr(CO)₃
B
I₂
C

(b) 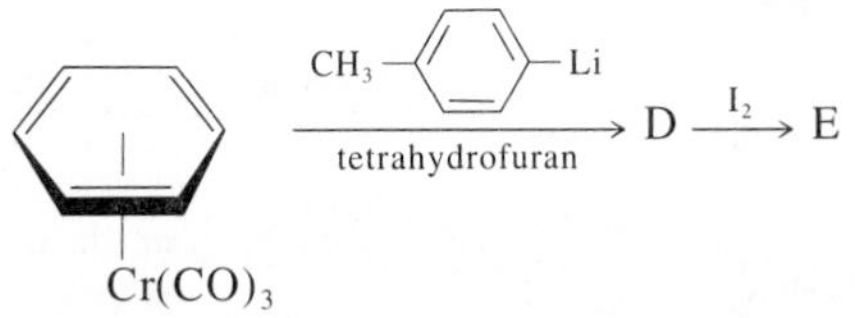
Cr(CO)₃
CH₃
Li
tetrahydrofuran
D
I₂
E

(c) 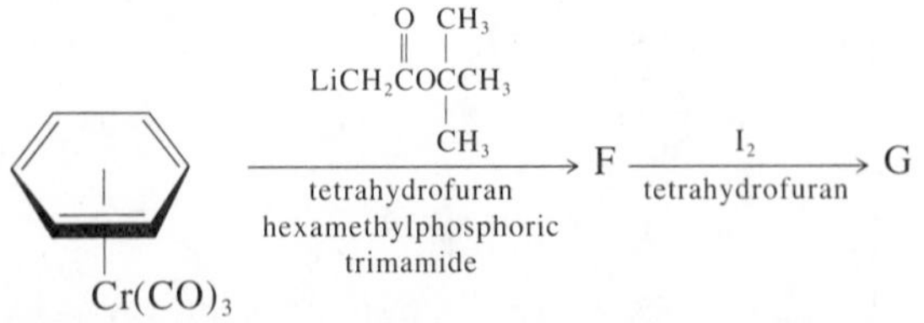
Cr(CO)₃
LiCH₂COCCH₃
tetrahydrofuran
hexamethylphosphoric
trimamide
F
I₂
G

(d) 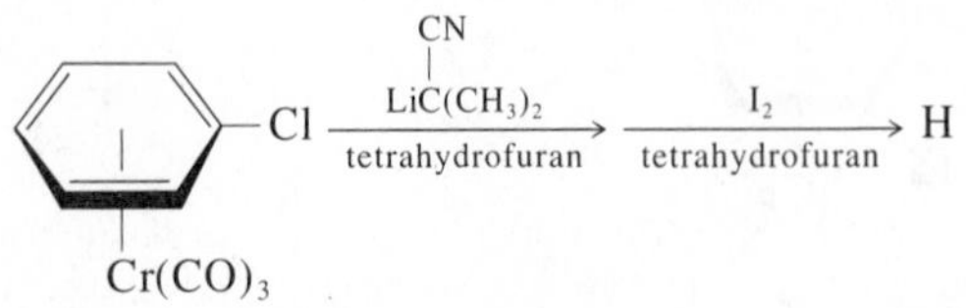
Cl
LiC(CH₃)₂
CN
tetrahydrofuran
I₂
H
Cr(CO)₃

PROBLEM 23.9

The painkiller Demerol is 4-carbethoxy-4-phenyl-*N*-methylpiperidine.

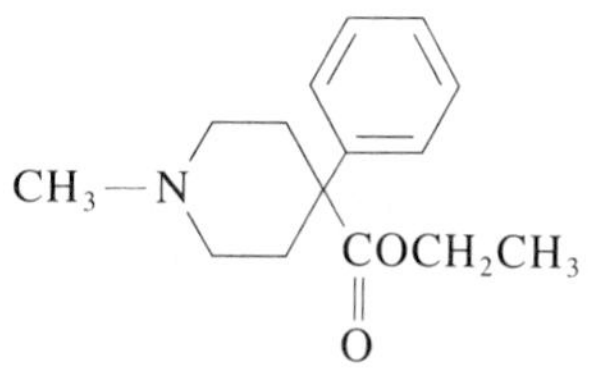
CH₃—N
COCH₂CH₃
O

It can be prepared from π-(fluorobenzene)chromium tricarbonyl and 4-carbethoxy-*N*-methylpiperidine. Propose a synthesis for Demerol.

23.4 OXIDATION OF SIDE CHAINS IN ARENES

The reactivity of the benzylic position of an alkylbenzene is reflected in the ease with which hydrocarbon side chains on aromatic rings undergo oxidation reactions. An alkane is normally inert to oxidation by a reagent such as potassium permanganate (pp. 314 and 557) or sodium dichromate (p. 439). But compounds having side chains on aromatic rings are oxidized to carboxylic acids by potassium permanganate or chromic acid. The aromatic ring itself is not affected. For example, *p*-nitrotoluene is converted to *p*-nitrobenzoic acid by heating it for 1 hour with an acidic sodium dichromate solution.

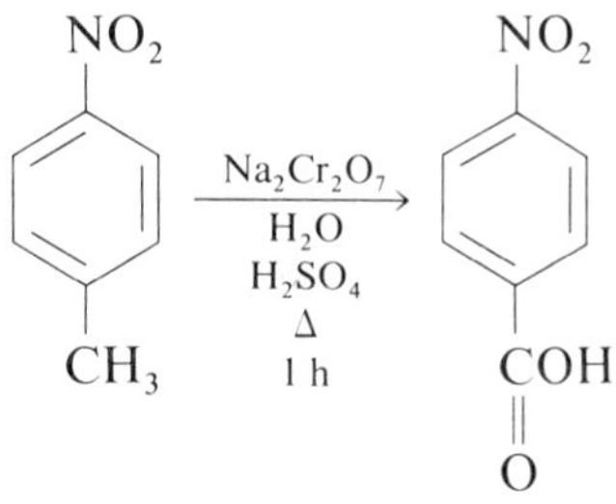
NO₂
CH₃
Na₂Cr₂O₇
H₂O
H₂SO₄
Δ
1 h
NO₂
COH
O

p-nitrotoluene — *p*-nitrobenzoic acid 86%

Under vigorous conditions, potassium permanganate oxides *p*-chlorotoluene to *p*-chlorobenzoic acid.

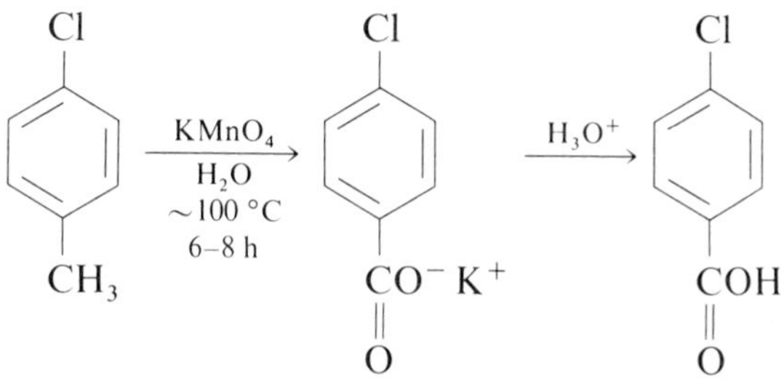
Cl
CH₃
KMnO₄
H₂O
~100 °C
6–8 h
Cl
CO⁻ K⁺
O
H₃O⁺
Cl
COH
O

p-chlorotoluene — potassium *p*-chlorobenzoate — *p*-chlorobenzoic acid 89%

Reduction of potassium permanganate in water produces a basic solution even when no other base is used for catalysis (p. 314). The substituted benzoic acid is formed as its potassium salt; the reaction mixture must be acidified to recover the free carboxylic acid.

Even if the side chain on the aromatic ring has more than one carbon atom, vigorous oxidation produces an aromatic acid. The other carbon atoms of the side chain are lost, as in the conversion of 1-phenylpropane to benzoic acid by aqueous potassium dichromate.

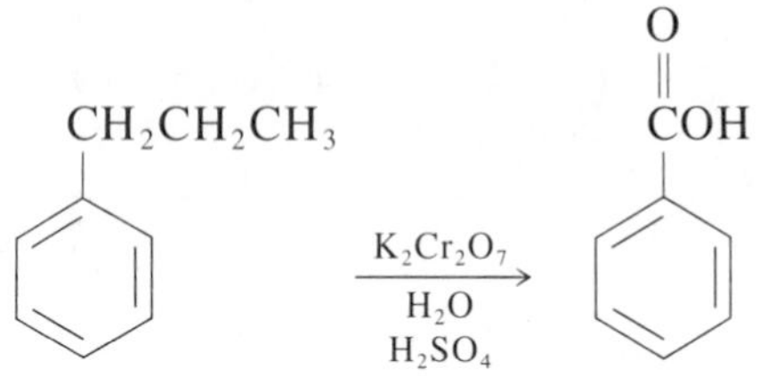

1-phenylpropane
propylbenzene

benzoic acid

All of these oxidation reactions are believed to start with the initial abstraction of a benzylic hydrogen atom by the oxidizing agent to give a stable benzylic radical (p. 833). A *tert*-butyl group is not oxidized, indicating the importance of this first step. The details of the mechanism from then on depend very much on the oxidizing agent and the reaction conditions.

Note that the reactions in which alkyl side chains are oxidized require that the starting compound be heated with the oxidizing agent for a period of time. If there is another functional group on the side chain that can be oxidized, it is possible to select conditions that will ensure that that group reacts and the side chain is not destroyed. For example, (*Z*)-1-phenylpropene is oxidized by cold dilute potassium permanganate to the corresponding diol.

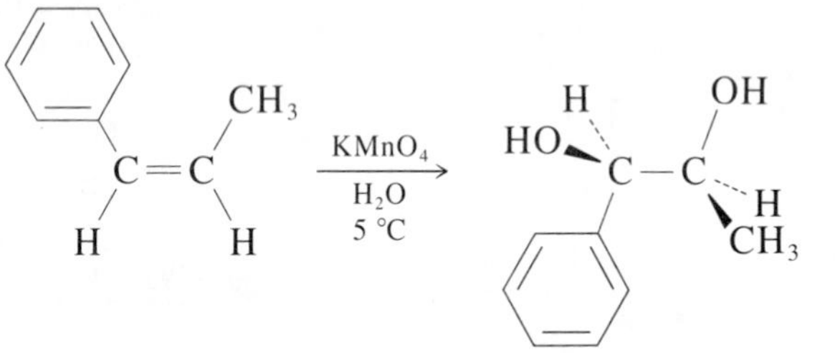

(*Z*)-1-phenylpropene

1-phenyl-1,2-propanediol
41%
racemic mixture

PROBLEM 23.10

Complete the following equations.

(a) 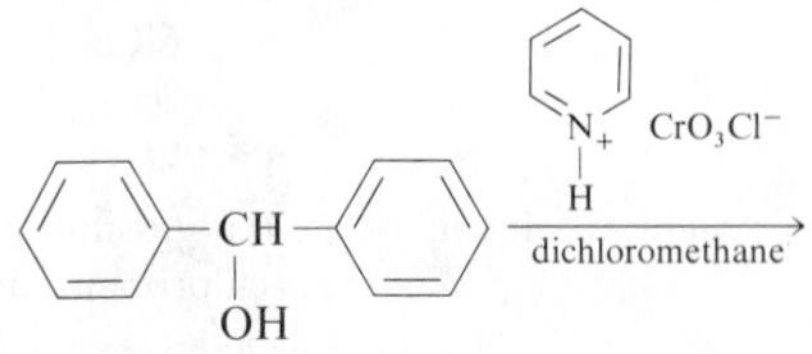

(b) 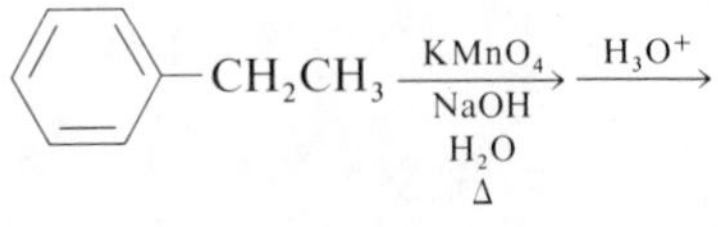

(c) 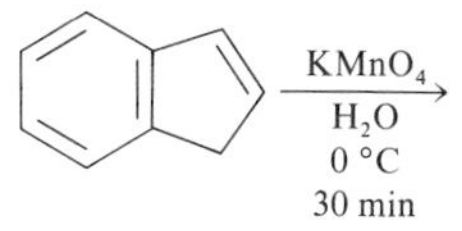

(d) 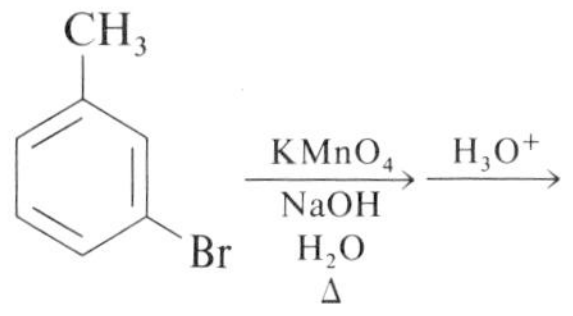

(e) 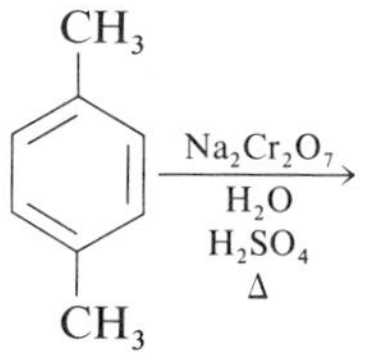

23.5
PHENOLS

A. Acidity of Phenols

Phenols are compounds in which a hydroxyl group is bonded directly to an sp^2-hybridized carbon atom of an aromatic ring. A typical phenol is much more acidic than an alcohol, but less so than a carboxylic acid. Phenol has pK_a 10.0; the pK_a of benzyl alcohol is approximately 16 and that of benzoic acid is 4.2. As was emphasized earlier (pp. 95 and 445), the alkoxide anion formed on the removal of a proton from an alcohol is strongly basic because the negative charge is localized on the oxygen atom. A carboxylate anion is much less basic than an alkoxide anion because the negative charge on the former is delocalized to two oxygen atoms. The anion of a phenol, a phenolate anion, also has delocalization of charge. Resonance contributors can be written that have the negative charge at the ortho and para positions of the aromatic ring.

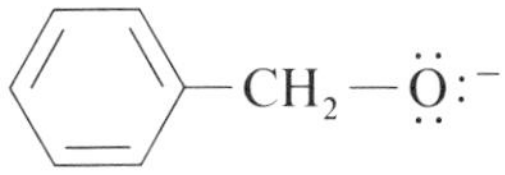

benzylate anion

no delocalization of charge
most basic

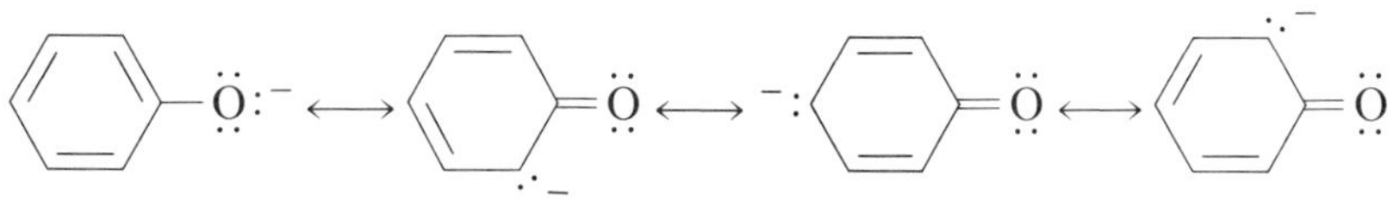

phenolate anion

delocalization of charge to aromatic ring

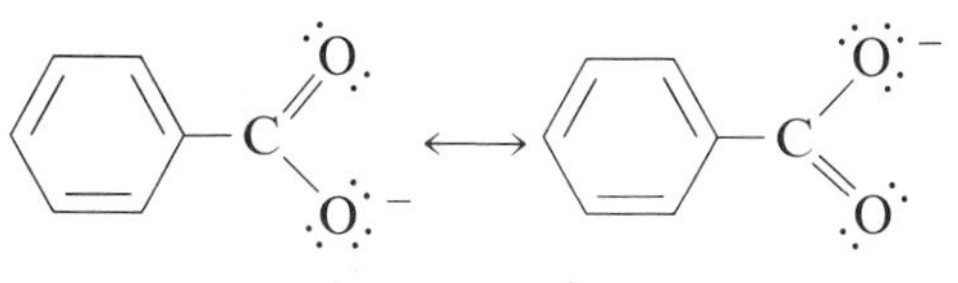

benzoate anion

delocalization of charge to 2 oxygen atoms
least basic

The benzoate anion has the greatest stability and the lowest basicity of the three anions shown above. The two resonance contributors for the benzoate anion are equivalent, and both have the negative charge on the most electronegative of the atoms present. These factors stabilize the anion.

There are more resonance contributors for the phenolate anion than there are for the benzoate anion. However, the four contributors for the phenolate anion are not all equivalent, and three of them put a negative charge on a carbon atom instead of on the more electronegative oxygen atom. Thus, the phenolate anion is stabilized more than the benzylate anion, in which no delocalization is possible, but not as much as the benzoate ion is.

The acidity of phenols has practical consequences in the laboratory. For example, a phenol can be separated from a carboxylic acid by taking advantage of the difference in acidity. Benzoic acid, a solid carboxylic acid, is insoluble in cold water but dissolves in a solution of sodium bicarbonate (p. 846). Its dissolving is accompanied by the evolution of bubbles of carbon dioxide, indicating that carbonic acid is being formed (p. 95). A phenol, such as 2-naphthol, is also insoluble in water, and does not react with aqueous sodium bicarbonate because it is less acidic than carbonic acid. It is not a strong enough acid to protonate the bicarbonate anion. A phenol is a stronger acid than water, however, and easily protonates hydroxide ion. Thus, 2-naphthol will dissolve in a dilute solution of sodium hydroxide in water.

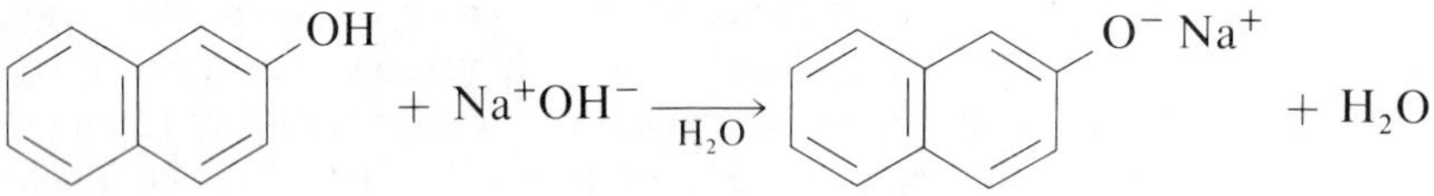
OH
+ Na⁺OH⁻
H₂O
O⁻ Na⁺
+ H₂O

2-naphthol
pK_a ~10
covalent compound
insoluble in water

sodium 2-naphtholate
ionic compound
soluble in water

water
pK_a 15.7

The acidity of phenols is affected by substituents on the aromatic ring.

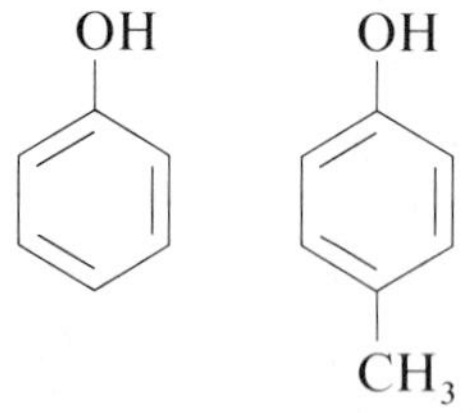
OH
OH
CH₃

phenol
pK_a 10.0

p-cresol
pK_a 10.2

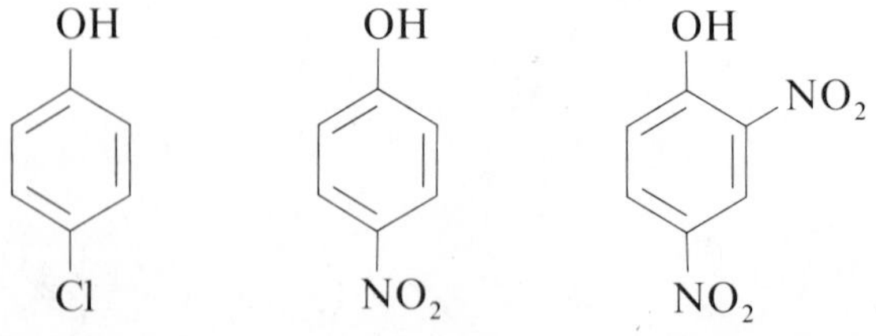
OH
Cl
OH
NO₂
OH
NO₂
NO₂

p-chlorophenol
pK_a 9.38

p-nitrophenol
pK_a 7.15

2,4-dinitrophenol
pK_a 4.02

Note that the effects of the above substituents on phenols parallel the effects that the same substituents have on the acidity of carboxylic acids (p. 555).

PROBLEM 23.11

(a) The pK_a of *m*-nitrophenol is 8.39, and that of *p*-nitrophenol is 7.15. How would you rationalize these facts? The pK_a of phenol is 10.0.

(b) 2,4,6-Trinitrophenol, also called picric acid, has a pK_a value of 0.25. Is picric acid soluble in sodium bicarbonate solution? Explain your answer.

PROBLEM 23.12

Assume that you are in a rather ill-equipped laboratory and you need to make a distinction between a carboxylic acid, say *p*-chlorobenzoic acid, and a phenol, *p*-chlorophenol. You have no instruments, but you do have a supply of simple inorganic reagents such as sodium hydroxide, hydrochloric acid, and sodium bicarbonate. What would you do? How would you decide which compound was which?

B. Herbicides from Phenols. Dioxin

Two compounds derived from polychlorinated phenols have been used extensively as **herbicides,** substances that kill plants. These compounds, (2,4,5-trichlorophenoxy)acetic acid abbreviated as (2,4,5-T) and (2,4-dichlorophenoxy)acetic acid (abbreviated as 2,4-D), mimic a plant hormone that controls the growth of cells. An application of the herbicide to a plant causes unchecked growth, which soon kills the plant.

OCH₂COH
Cl
Cl
Cl
OCH₂COH
Cl
Cl

(2,4,5-trichlorophenoxy)acetic acid
2,4,5-T

(2,4-dichlorophenoxy)acetic acid
2,4-D

These compounds are components of Agent Orange, which was used as a defoliant during the Vietnam War.

2,4,5-T is prepared from 1,2,4,5-tetrachlorobenzene by nucleophilic substitution reactions.

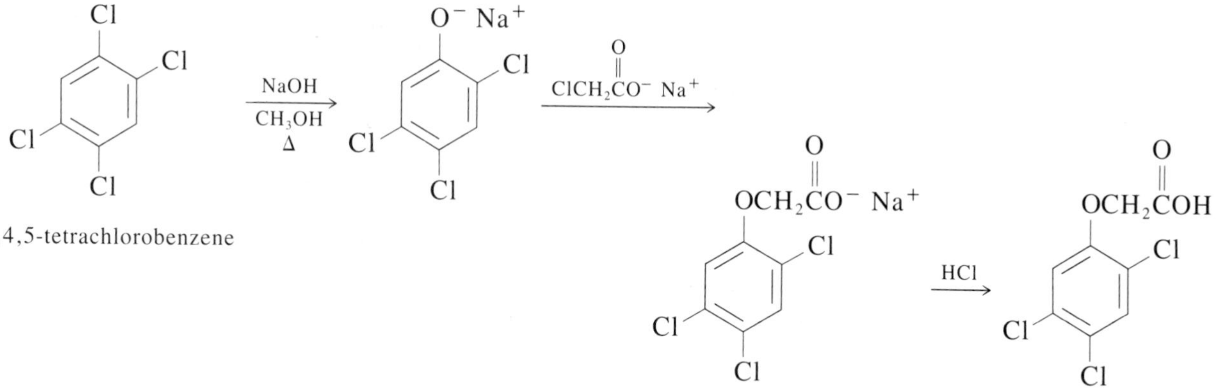

Cl
Cl
Cl
Cl
NaOH
CH₃OH
Δ
O⁻ Na⁺
Cl
Cl
Cl
ClCH₂CO⁻ Na⁺
OCH₂CO⁻ Na⁺
Cl
Cl
Cl
HCl
OCH₂COH
Cl
Cl
Cl

1,2,4,5-tetrachlorobenzene

2,4,5-T
85%

A side reaction of the conversion of 1,2,4,5-tetrachlorobenzene to 2,4,5-trichlorophenol is an internal nucleophilic substitution reaction that yields one of the most toxic organic chemicals known, 2,3,7,8-tetrachlorodibenzo[b,e][1,4]dioxin, usually just called dioxin. Small amounts of dioxin occur as a contaminant of 2,4,5-T.

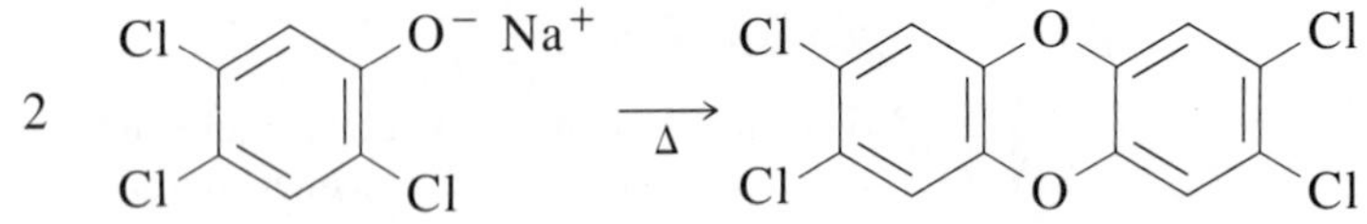
2
Cl
O⁻ Na⁺
Cl
Cl
Δ
Cl
O
Cl
Cl
O
Cl
dioxin

Dioxin is so toxic that it has an LD_{50} value of 0.6 μg/kg (μg is a microgram, 10^{-6} g) for guinea pigs. This means that a dose that small will kill half of the guinea pigs to which it is fed. Dioxin is also highly carcinogenic. The reasons for this high degree of toxicity are not well understood but are being studied intensively in many laboratories.

PROBLEM 23.13

The herbicide 2,4-D is made by mixing 2,4-dichlorophenol and chloroacetic acid with slightly more than 2 equivalents of sodium hydroxide in water solution and then acidifying the solution with hydrochloric acid. Write an equation showing all the species that are formed when the phenol and the acid are dissolved in base. Identify all of the nucleophiles and the leaving groups present in the solution. 2,4-D is formed in 87% yield. How do you explain the selectivity of the reaction?

23.6 ARENE OXIDES

The human body converts aromatic compounds into phenols in order to detoxify and excrete them. The totally nonpolar hydrocarbon is given polar functional groups that make it more soluble in the physiological solvent, water. The hydroxyl group on the hydrocarbon is used by the body to link the aromatic residue to other functionalities that increase the solubility of the molecule in water still further. For example, sulfate esters of a phenol may be formed, or the hydroxyl group may be linked to glucuronic acid (p. 1052) as an acetal. These functions serve to hold the hydrocarbon residue in solution and aid in its transport out of the body.

The liver is the body's major organ for detoxification. Drugs, pollutants, food additives, and even some natural components of food are chemical substances that cannot be usefully incorporated into the structure of the body and must be excreted somehow. Many substances are transformed in oxidation reactions catalyzed by enzymes found in abundance in the liver. Most important among these enzymes are a group of highly colored proteins known as **cytochromes.** These enzymes, which contain iron, are found in the microsomal fraction of liver cells. Their function is to convert the oxygen from the air into a highly reactive form that can attack systems normally resistant to oxidation by air.

When aromatic hydrocarbons undergo these oxidation reactions, the products are unstable oxiranes known as **arene oxides.** Naphthalene, for example, is converted by microsomal enzymes from the liver of the rat into naphthalene-1,2-oxide.

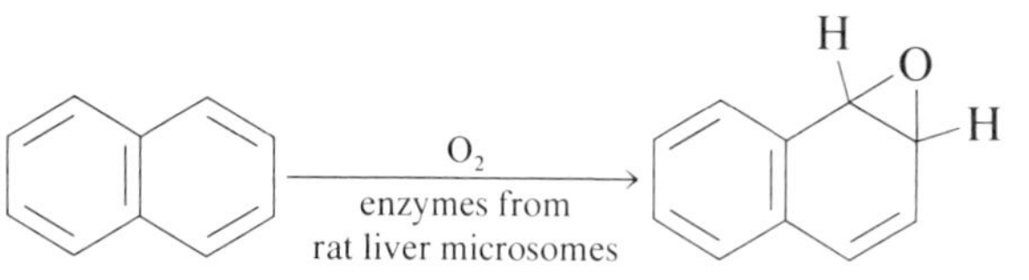

O$_2$
enzymes from
rat liver microsomes
H
O
H

naphthalene naphthalene-1,2-oxide

The reactive oxide is opened to a trans 1,2-diol by another enzyme. Experiments have shown that the oxygen atom in the benzylic hydroxyl group comes from molecular oxygen and the other one comes from water.

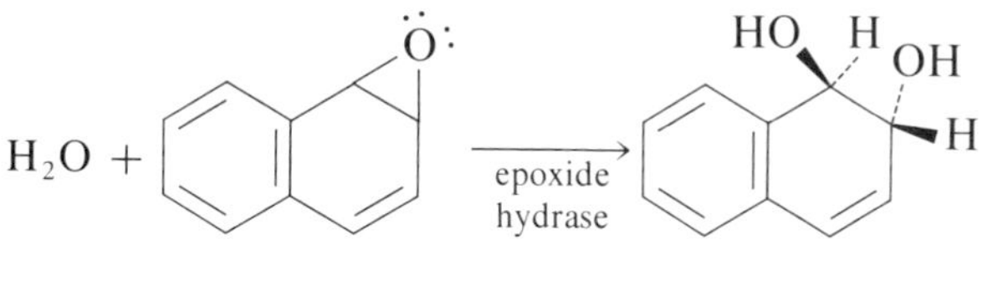

H$_2$O +
epoxide
hydrase
HO
H
OH
H

naphthalene-1,2-oxide *trans*-1,2-dihydro-1,2-dihydroxynaphthalene

The exact mechanism by which arene oxides react is of great interest because there is evidence that such oxides are responsible for the carcinogenicity of many aromatic hydrocarbons. It is ironic that the same oxidation processes that detoxify harmful chemical substances and rid the body of them may also create metabolites that are far more carcinogenic than the original hydrocarbons themselves.

Suspicion that substances in the residues from the burning of coal and wood cause cancer first arose in 1775 when a British doctor noticed that chimney sweeps were more likely to have cancer of the scrotum than men in other occupations. The harmful compounds, found widely in the environment in cigarette smoke, automobile exhausts, and industrial emissions, are polycyclic aromatic hydrocarbons. The one on which most interest has centered is benzo[a]pyrene (p. 970). About 40 of its oxygenated metabolites have been isolated and characterized. One of them in particular has been shown to be a potent mutagen. Compounds that are mutagens are suspected of being carcinogens as well because they disrupt the genetic processes in cells. A cancer cell is a cell that has been transformed so that the processes that normally control growth and reproduction no longer operate.

The reactions that are thought to be involved in the conversion of benzo[a]pyrene into its mutagenic metabolite are very much like those shown above for the biological conversion of naphthalene to *trans*-1,2-dihydro-1,2-dihydroxynaphthalene. A cytochrome catalyzes the conversion of benzo[a]pyrene into an arene oxide, which adds water enzymatically to give a trans diol. Another epoxidation takes place at the double bond in the same ring as the diol function, and two stereoisomeric diol-epoxides are formed (Figure 23.2). The new oxirane ring can have two orientations with respect to the hydroxyl groups that are already present. The 7,8-diol-9,10-epoxide of benzo[a]pyrene with the oxirane ring trans to the benzylic hydroxyl group, called the anti isomer, is more potent as a mutagen than is the other isomer.

Benzo[a]pyrene-7,8-diol-9,10-epoxide disrupts the genetic mechanism of the cell in several ways. The chemistry that is relevant to this discussion involves the reactivity of oxiranes toward nucleophiles. There are many nucleophilic sites on a DNA molecule. One such site is an amino group on a heterocyclic ring in deoxyguanosine (p. 970), which is part of the backbone of DNA (p. 1013).

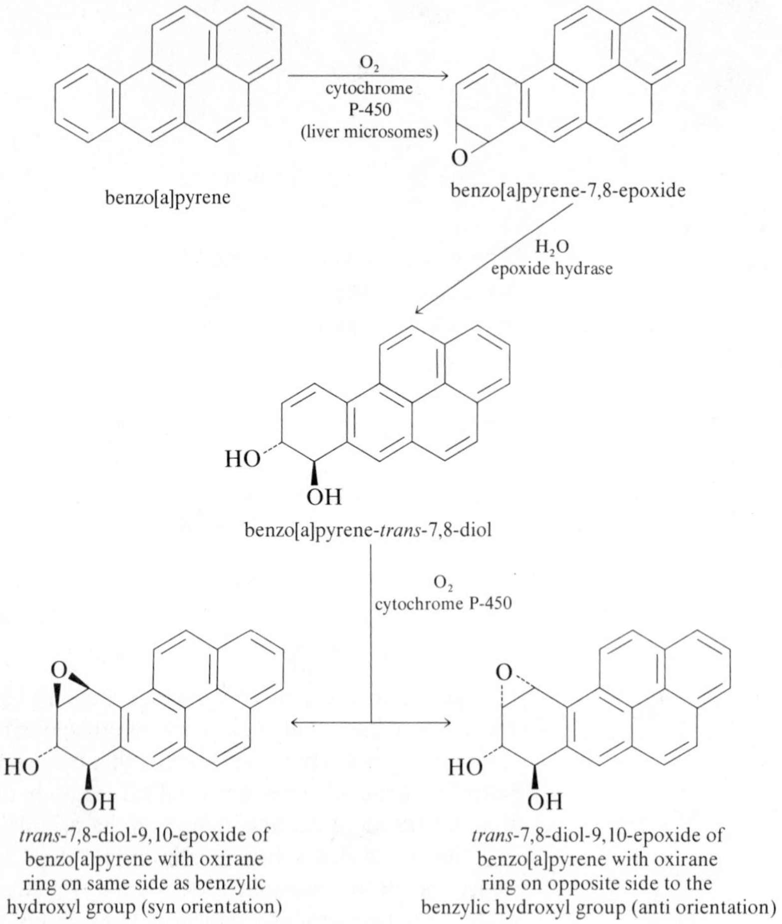
benzo[a]pyrene
O_2
cytochrome
P-450
(liver microsomes)
O
benzo[a]pyrene-7,8-epoxide
H_2O
epoxide hydrase
HO
OH
benzo[a]pyrene-*trans*-7,8-diol
O_2
cytochrome P-450
O
HO
OH
trans-7,8-diol-9,10-epoxide of benzo[a]pyrene with oxirane ring on same side as benzylic hydroxyl group (syn orientation)
O
HO
OH
trans-7,8-diol-9,10-epoxide of benzo[a]pyrene with oxirane ring on opposite side to the benzylic hydroxyl group (anti orientation)

FIGURE 23.2 The conversion of benzo[a]pyrene into its mutagenic metabolite.

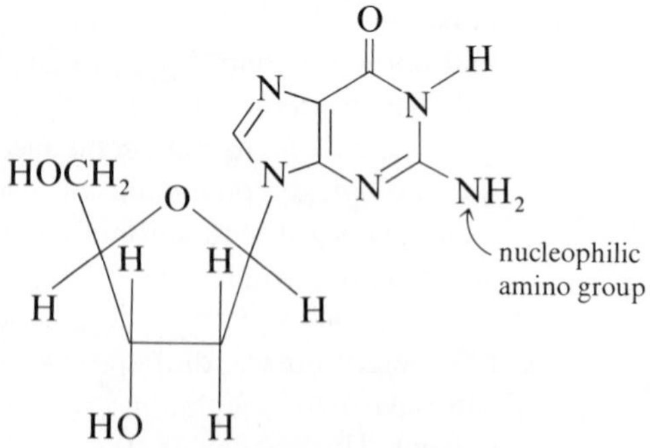
O
H
N
N
N
N
$HOCH_2$
O
H
H
H
H
NH_2
nucleophilic amino group
HO
H

deoxyguanosine
a component of DNA

Reaction of the benzo[a]pyrene-7,8-diol-9,10-epoxide with DNA, degradation of the giant molecule, and purification of the fractions containing the hydrocarbon residue lead to the isolation of the compound shown below.

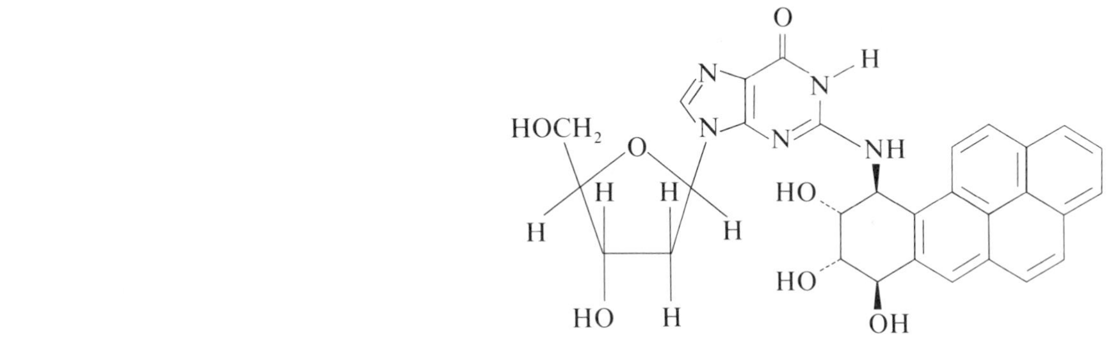

adduct of deoxyguanosine with benzo[a]pyrene-7,8-diol-9,10-epoxide

The amino group in deoxyguanosine reacts by a nucleophilic opening of the oxirane ring in benzo[a]pyrene-7,8-diol-9,10-epoxide. The same adduct is obtained when benzo[a]pyrene is incubated with mouse embryo cells, strongly indicating that the diol-epoxide is the metabolic intermediate responsible for the reaction with DNA.

If the aromatic hydrocarbon being metabolized contains an alkyl side chain, biological oxidations resembling the oxidation reaction discussed on p. 963 take place at the benzylic position. Benzene, with no alkyl side chain, is detoxified slowly by the body and is a cumulative poison because it cannot be excreted rapidly. It affects bone marrow and causes aplastic anemia and leukemia. In contrast, toluene is much less toxic. The major pathway for its detoxification by the body is oxidation of the side chain to benzyl alcohol.

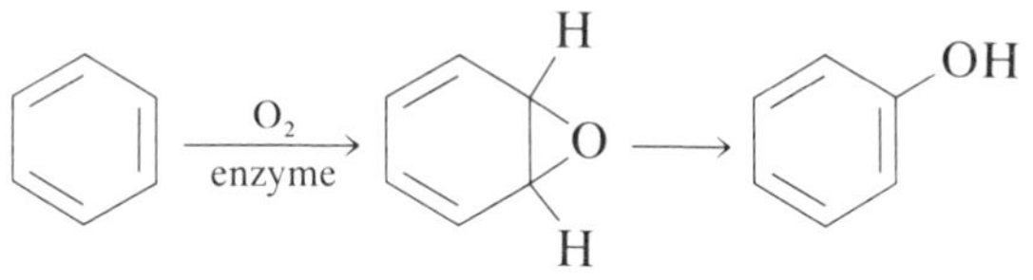

major oxidative route for the detoxification of benzene, slow

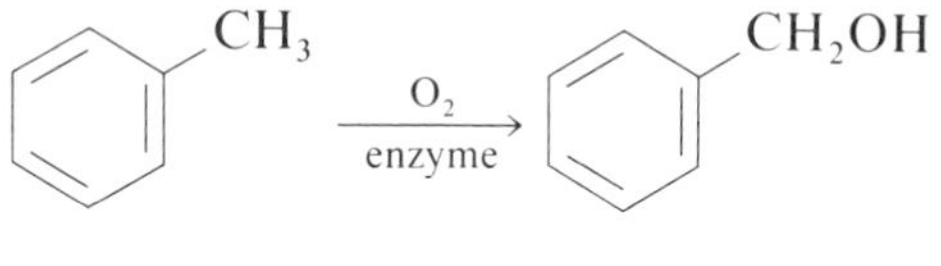

major oxidative route for the detoxification of toluene, fast

A realization of these facts has led to the substitution of toluene for benzene in the laboratory whenever possible.

PROBLEM 23.14

Two experiments provided part of the evidence for the formation of an arene oxide as an intermediate in the metabolism of naphthalene to *trans*-1,2-dihydro-1,2-dihydroxynaphthalene. In one experiment, naphthalene was incubated with air containing $^{18}O_2$ in ordinary water, $H_2{}^{16}O$. In the other, naphthalene was incubated with ordinary air in $H_2{}^{18}O$. Referring to the equations on p. 969, write mechanisms that predict the isotopic composition of the trans diol that was formed in each experiment.

ULTRAVIOLET SPECTROSCOPY OF AROMATIC COMPOUNDS

The aromatic ring is itself a chromophore. Benzene has major absorption bands in the ultraviolet region of the electromagnetic spectrum at 180 (ϵ 60,000), 200 (ϵ 8000), and 254 nm (ϵ 212) with other bands at 234, 239, 243, 249, 261, and 268 nm. Substitution of an alkyl group on the aromatic ring slightly increases the wavelengths of maximum absorptions. Toluene absorbs at λ_{max}^{hexane} 189 (ϵ 55,000) 208 (ϵ 7900) and 262 nm (ϵ 260) (Figure 23.3). The chromophore is an aromatic ring with one alkyl substituent on it.

For polycyclic hydrocarbons, the absorption bands in ultraviolet spectra shift to longer wavelengths and become more intense. For example, naphthalene has $\lambda_{max}^{methanol}$ 311 nm (ϵ 239) for the band corresponding to the absorption at 254 nm (ϵ 212) for benzene.

The interaction of the nonbonding electrons of an oxygen atom with the aromatic ring shifts the absorption bands observed for benzene to higher values for phenols and aromatic ethers. For this effect to appear, the oxygen atom must, of course, be directly bonded to the aromatic ring. A comparison of the spectrum of methoxybenzene (Figure 23.4) with that of toluene demonstrates this phenomenon. Methoxybenzene has $\lambda_{max}^{isooctane}$ 220 (ϵ 8100), 271 (ϵ 2200), and 278 nm (ϵ 2250).

The conversion of a phenol into a phenolate anion results in a shift of the absorption maxima in the ultraviolet spectrum to even longer wavelengths. This effect reflects the greater electron density in the aromatic ring that is possible when the phenol is deprotonated. The spectra of phenol taken in water and in aqueous sodium hydroxide (Figure 23.5) illustrate this effect. Both absorption bands for phenol (211 and 270 nm) move to longer wavelengths (235 and 287 nm) and increase in intensity for the phenolate anion. This change is typical of the spectra

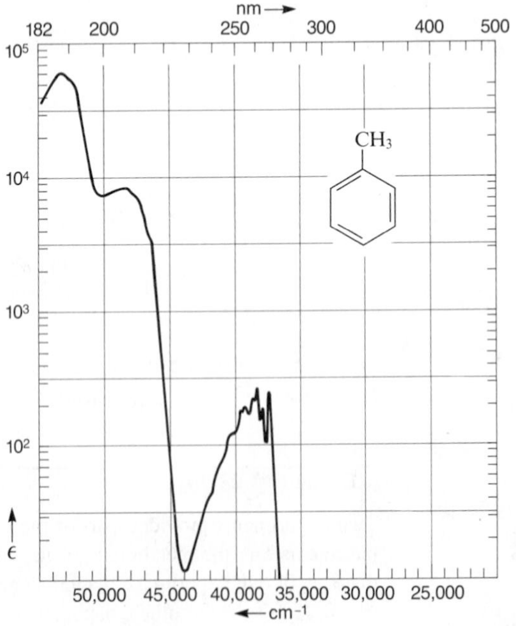
nm →
182
200
250
300
400
500
10⁵
10⁴
10³
10²
ε
CH₃
50,000
45,000
40,000
35,000
30,000
25,000
← cm⁻¹

FIGURE 23.3 Ultraviolet spectrum of toluene in hexane. (Adapted from *UV Atlas of Organic Compounds*)

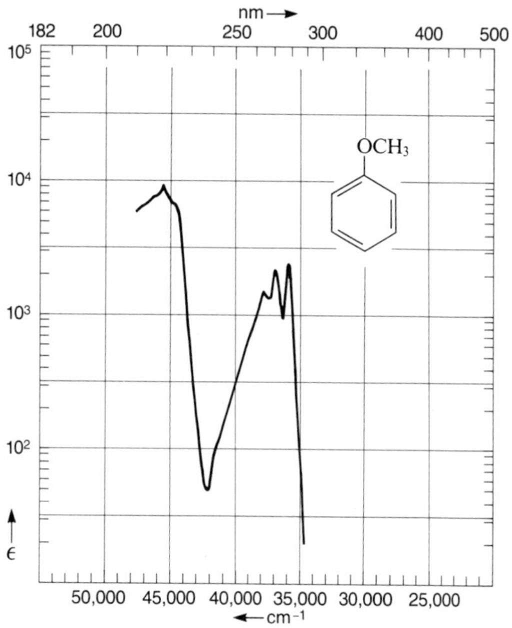
nm
182
200
250
300
400
500
10^5
10^4
10^3
10^2
ε
OCH_3
50,000
45,000
40,000
35,000
30,000
25,000
cm^{-1}

FIGURE 23.4 Ultraviolet spectrum of methoxybenzene in isooctane. (Adapted from *UV Atlas of Organic Compounds*)

of phenols and can be used as a diagnostic test that the aromatic compound containing oxygen is indeed a phenol.

For amines in which the nitrogen atom is bonded directly to an aromatic ring, interaction between the nonbonding electrons of the nitrogen atom and the π electrons of the ring is possible. Just as occurs with phenols and aryl ethers, the

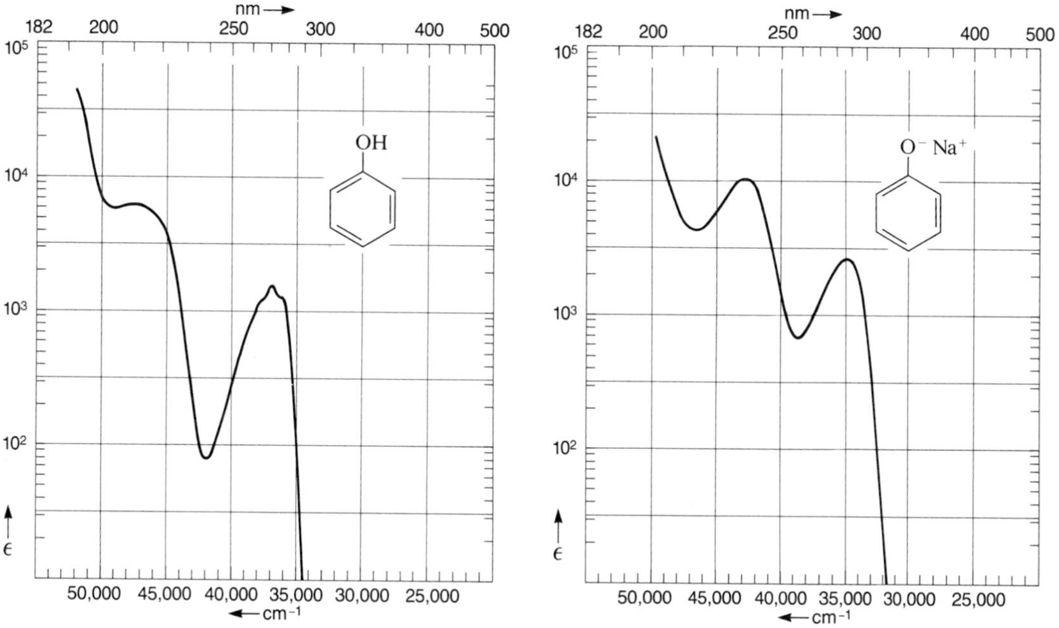
nm
182
200
250
300
400
500
10^5
10^4
10^3
10^2
ε
OH
50,000
45,000
40,000
35,000
30,000
25,000
cm^{-1}
nm
182
200
250
300
400
500
10^5
10^4
10^3
10^2
ε
$O^- Na^+$
50,000
45,000
40,000
35,000
30,000
25,000
cm^{-1}

FIGURE 23.5 Ultraviolet spectra of phenol taken in water and in aqueous sodium hydroxide. (Adapted from *UV Atlas of Organic Compounds*)

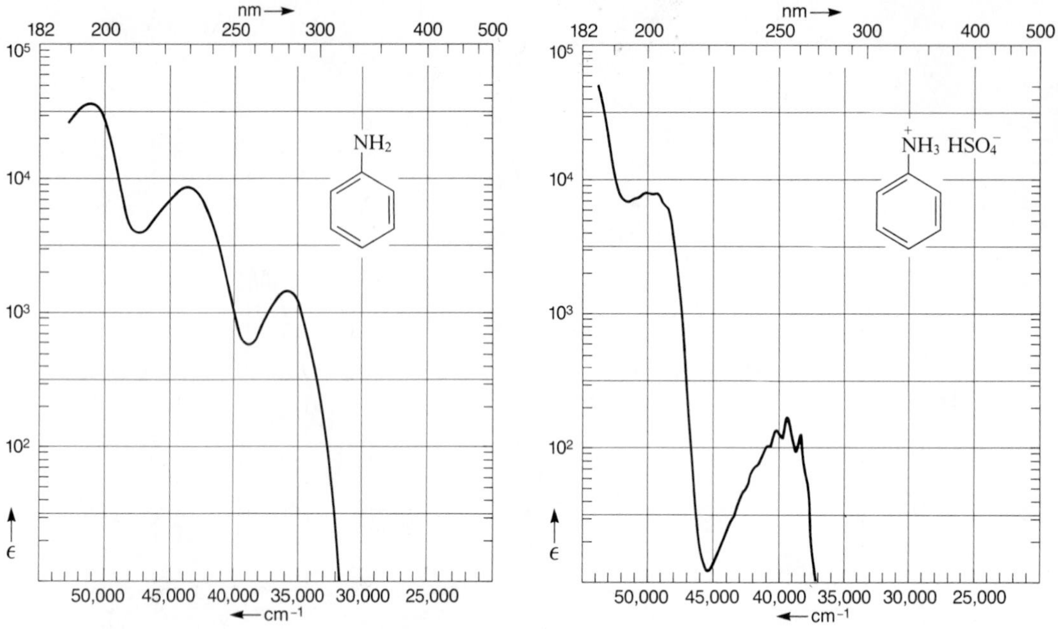
nm→
182 200 250 300 400 500
10⁵
10⁴
10³
10²
ε
NH2
50,000 45,000 40,000 35,000 30,000 25,000
←cm-1
nm→
182 200 250 300 400 500
10⁵
10⁴
10³
10²
ε
NH3 HSO4
50,000 45,000 40,000 35,000 30,000 25,000
←cm-1

FIGURE 23.6 Ultraviolet spectra of aniline taken in a buffer at pH 8.0 and in aqueous sulfuric acid. (Adapted from *UV Atlas of Organic Compounds*)

chromophore of the aromatic ring changes, and the absorption bands in the ultraviolet spectrum appear at longer wavelength and become more intense. The spectrum of an aromatic amine to which acid has been added resembles that of an alkylbenzene. Protonation of the nitrogen atom destroys the possibility of interaction between it and the ring. The spectra of aniline taken in a buffer at pH 8.0 and in aqueous sulfuric acid illustrate this phenomenon (Figure 23.6). Aniline has $\lambda_{max}^{H_2O}$ 196 (ϵ 34500), 230 (ϵ 8200), and 281 (ϵ 1400). The bands in the spectrum for the acidified solution at 200 (ϵ 7600) and 254 nm (ϵ 165) correspond to the bands at 230 and 281 nm in the spectrum for aniline in water. Thus, both the position of maximum absorption and the intensity of absorption decrease on protonation of the nitrogen atom in an aromatic amine. The spectrum of the anilinium ion looks very much like that of toluene (Figure 23.3).

PROBLEM 23.15

The ultraviolet spectrum shown in Figure 23.7 belongs to Compound X, $C_{10}H_{15}N$, and was taken in heptane.

Using the spectrum, distinguish between the following possible structures for Compound X.

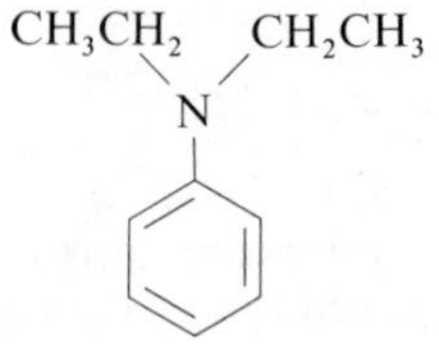
CH3CH2 CH2CH3
N

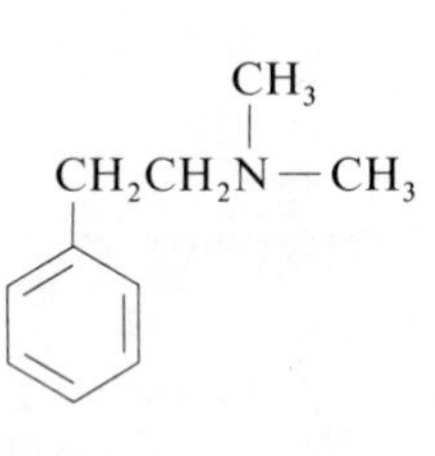
CH3
CH2CH2N—CH3

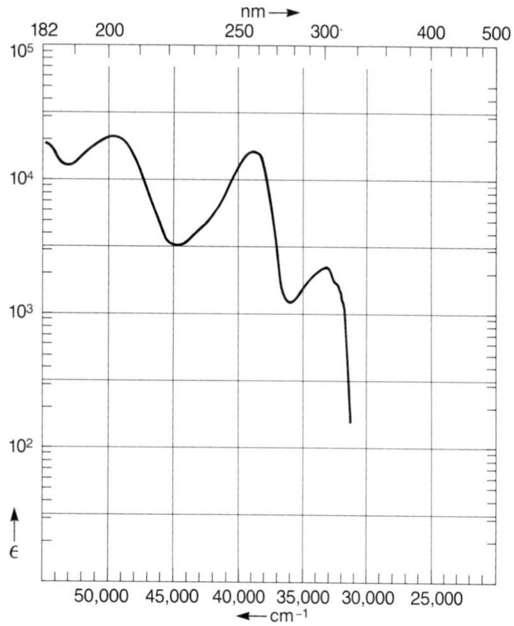
nm
182
200
250
300
400
500
10^5
10^4
10^3
10^2
ε
50,000
45,000
40,000
35,000
30,000
25,000
cm^{-1}

FIGURE 23.7 (Adapted from *UV Atlas of Organic Compounds*)

PROBLEM 23.16

Compound Y, C_7H_8O, has $\lambda_{max}^{methanol}$ 214 (ϵ 6030) and 273 nm (ϵ 1820). When Compound Y is put into methanol with some potassium hydroxide in it, the ultraviolet spectrum has bands at 238 (ϵ 5510) and 282 nm (ϵ 1990). What is a possible structure for Compound Y?

PROBLEM 23.17

Compound Z, C_7H_7BrO, has $\lambda_{max}^{methanol}$ 227 (ϵ 14,200), 281 (ϵ 1580), and 288 nm (ϵ 1280). The infrared spectrum of Compound Z has no absorption bands of any intensity between 4000 and 3200 cm^{-1} and between 2000 and 1600 cm^{-1}. Assign a structure to Compound Z that fits these data.

SUMMARY

Diazonium ions prepared by the reaction of primary aromatic amines with nitrous acid are important intermediates in many syntheses of aromatic compounds. The nitrogens of a diazonium ion can be directly replaced by a halogen, a cyano group, a hydroxyl group, or a hydrogen atom. Diazonium ions are also electrophiles that will react with ring-activated aromatic compounds to give azo compounds, useful as dyes and indicators. The reactions of diazonium ions are summarized in Table 23.1.

Elimination reactions of aryl halides with a strong base give benzyne, an electrophilic reactive intermediate that reacts with dienes and with nucleophiles. Benzyne is also formed when benzenediazonium 2-carboxylate loses nitrogen and carbon dioxide simultaneously.

Aryl halides that have strongly electron-withdrawing groups in the ortho and para positions undergo nucleophilic substitution reactions, following a mechanism in which the nucleophile adds to the aromatic ring to give an anionic intermediate,

TABLE 23.1 Reactions of Diazonium Ions

Ion	Reagent	Product
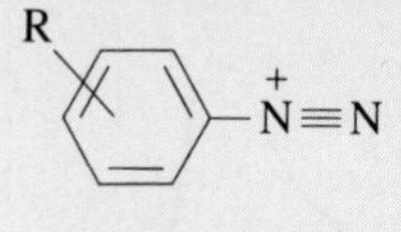	CuCl	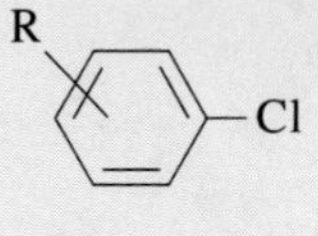
	CuBr	
	Na_2CO_3, CuCN	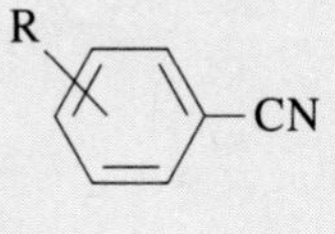
	KI	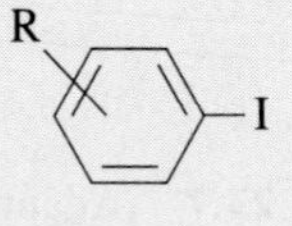
	BF_4^-	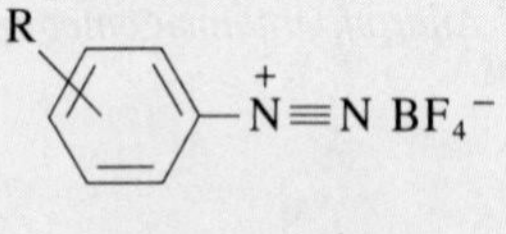
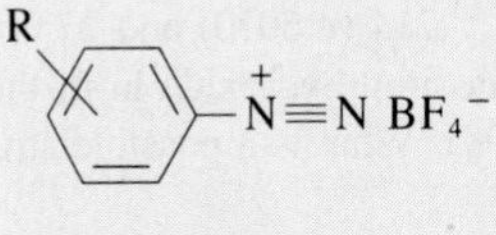	Δ	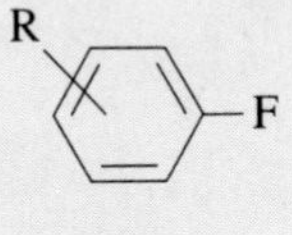
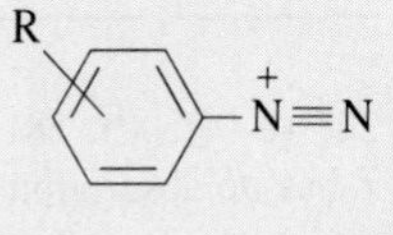	H_2O, Δ	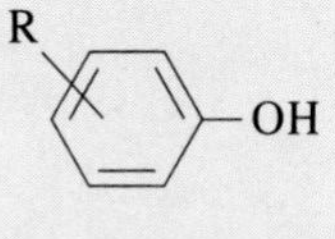
	H_3PO_2	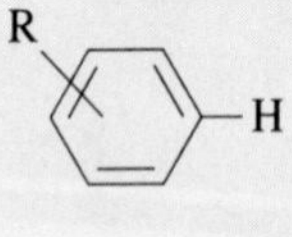
	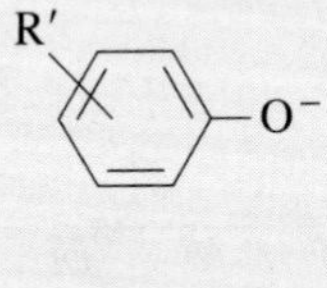	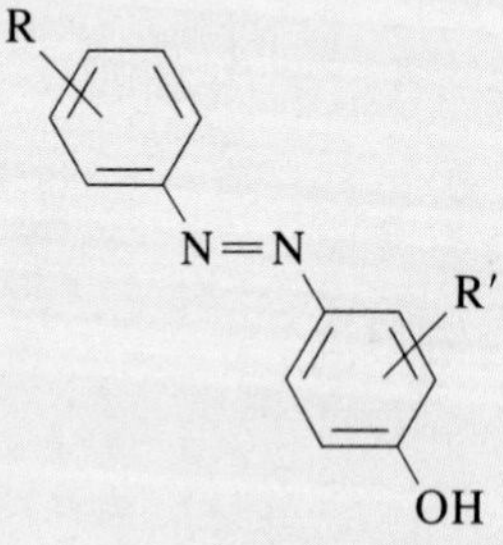
	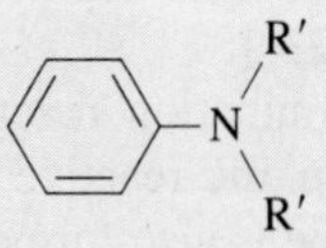	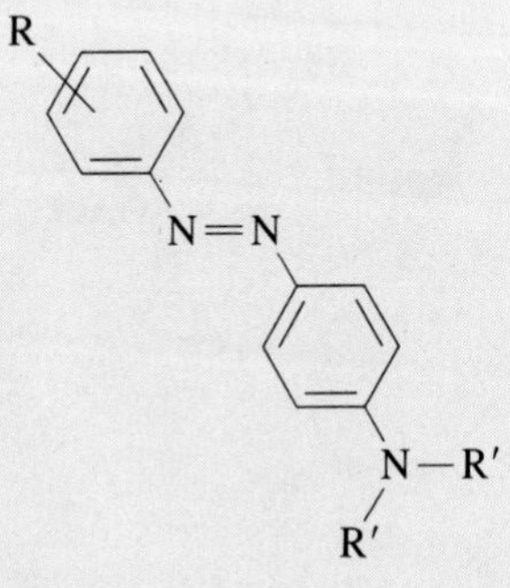

which then loses the leaving group. Benzene can be activated toward attack by a nucleophile by complexing it with a transition metal, such as chromium, that serves to stabilize negative charge.

The aromatic ring is not oxidized under conditions that result in the oxidation of an alkyl side chain to a carboxylic acid group. This is an important method for the preparation of aromatic carboxylic acids. Reactions of aromatic compounds other than electrophilic aromatic substitution are summarized in Table 23.2.

Phenols are compounds in which at least one hydroxyl group is bonded directly to an sp^2-hybridized carbon atom of an aromatic ring. Phenols, as a consequence, are much more acidic than alcohols, but less so than carboxylic acids.

Many aromatic compounds are important biologically. For example, dioxin, a side product from the synthesis of the herbicide (2,4,5-trichlorophenoxy)acetic acid is highly toxic. Arene oxides are products of the oxidation of polycyclic aromatic hydrocarbons (found in smoke) by enzymes in the liver and are implicated in the carcinogenicity of such compounds.

Aromatic compounds have distinctive ultraviolet spectra. Spectra of compounds that have nonbonding electrons on atoms adjacent to the aromatic ring that are available for conjugation with the ring differ characteristically from spectra of related compounds in which the electrons are not available. For example, the λ_{max} value for a phenol shifts to a longer wavelength if the phenol is deprotonated. A protonated amine has a λ_{max} value lower than that of the corresponding free amine.

TABLE 23.2 Reactions of Aromatic Compounds Other Than Electrophilic Substitution

Nucleophilic Substitution				
Aromatic Compound	**Nucleophile**	**Intermediate**	**Reagents for Second Step**	**Product**
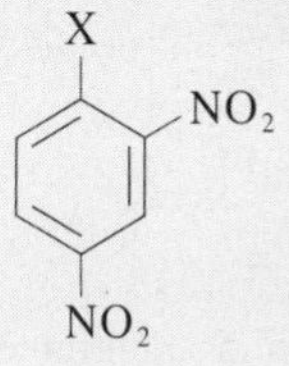	RNH_2	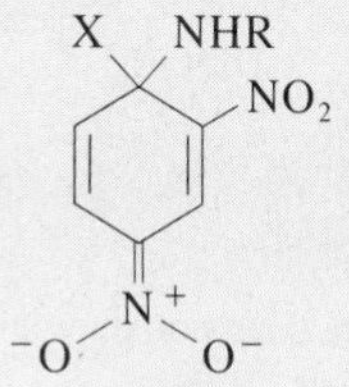	—	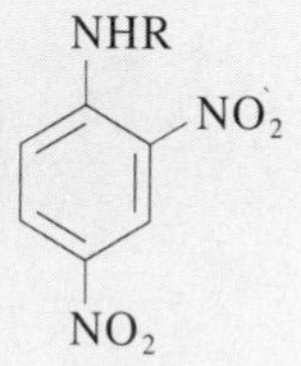
	RO^-	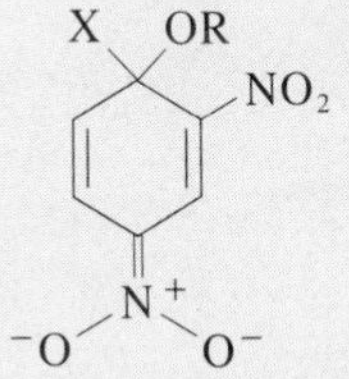	—	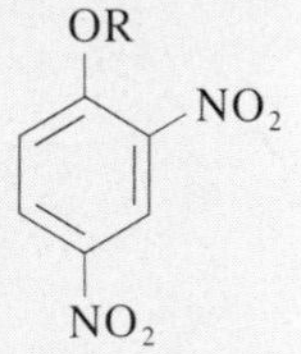
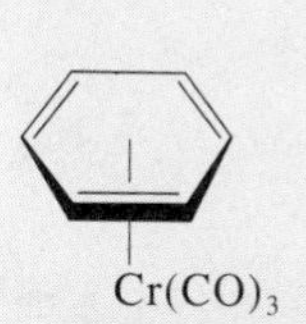	$R_3C:^-\ Li^+$ from R_3CH $pK_a > 20$	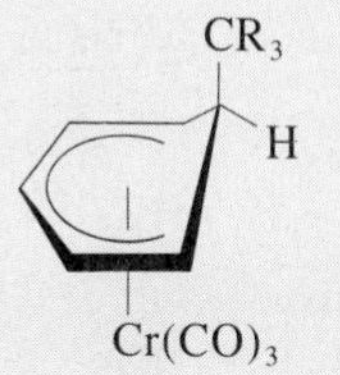	I_2	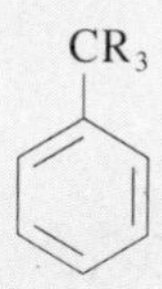
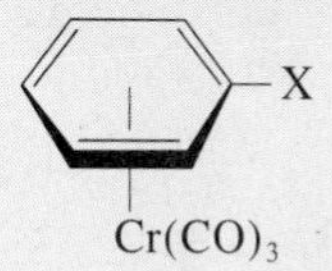	$R':^-\ M^+$ (may be $R_3C:^-\ Li^+$ or enolates)	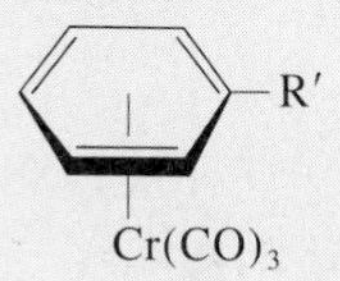	I_2	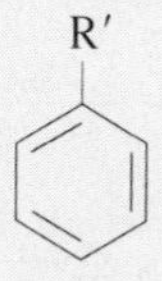

(*Continued*)

TABLE 23.2 *(Continued)*

Oxidations of Side Chains				
Aromatic Compound	**Oxidizing Agent**	**Intermediate**	**Reagents for Second Step**	**Product**
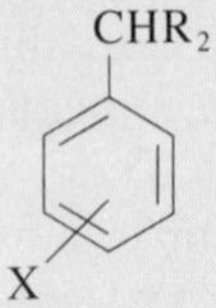	$Na_2Cr_2O_7$, H_2SO_4, H_2O, Δ	—	—	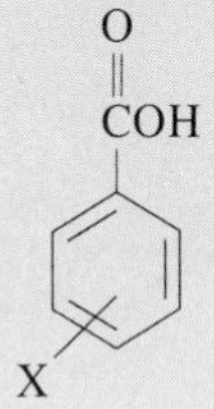
	$KMnO_4$, H_2O, Δ	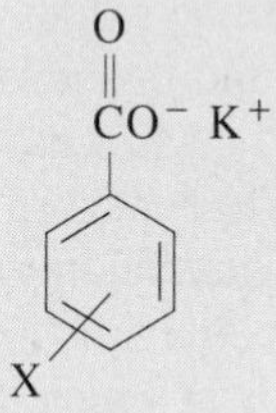	H_3O^+	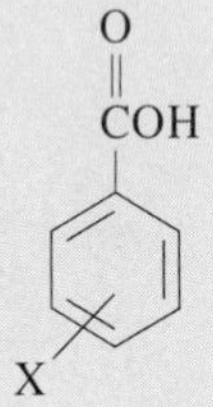

ADDITIONAL PROBLEMS

23.18 Give structural formulas for all intermediates and products designated by letters in the following equations.

(a) 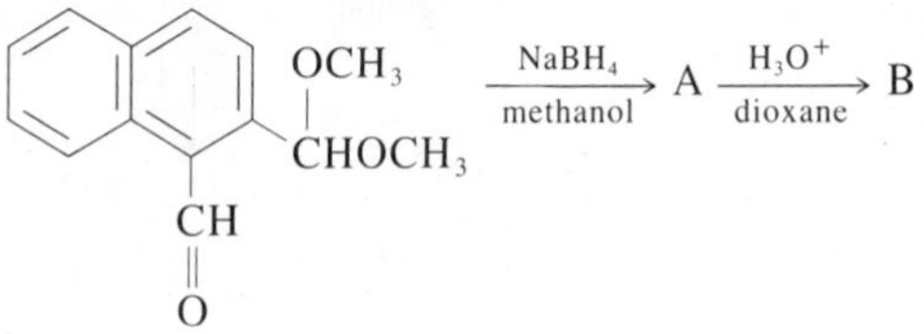

OCH₃
CHOCH₃
CH
O
NaBH₄
methanol
A
H₃O⁺
dioxane
B

(Hint: Compound B does not show an absorption band for a carbonyl group in its infrared spectrum.)

(b) 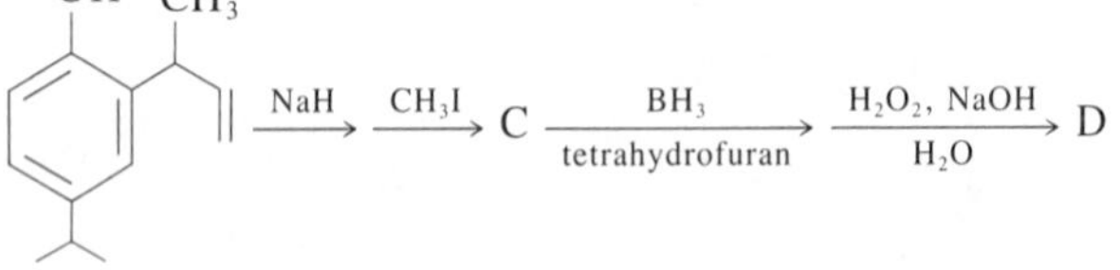

OH
CH₃
NaH
CH₃I
C
BH₃
tetrahydrofuran
H₂O₂, NaOH
H₂O
D

(c) $D \xrightarrow[H_2SO_4,\ H_2O,\ \Delta]{CrO_3} E \xrightarrow[\Delta]{H_3PO_4} F \xrightarrow{LiAlH_4} \xrightarrow{H_2O} G$

(d) 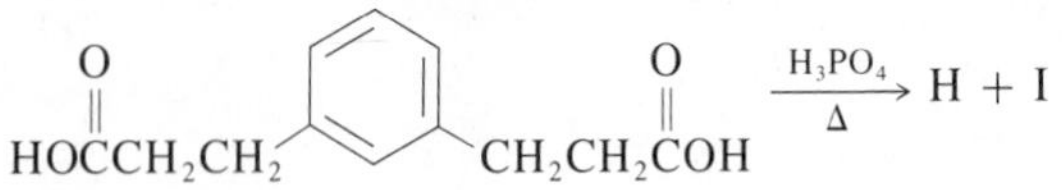

O
HOCCH₂CH₂
O
CH₂CH₂COH
H₃PO₄
Δ
H + I

(e) 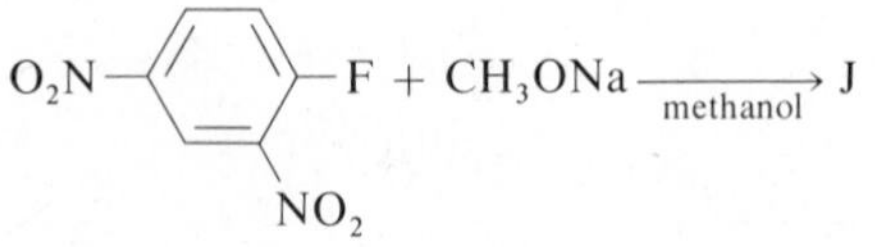

O₂N
F + CH₃ONa
methanol
J
NO₂

(f) 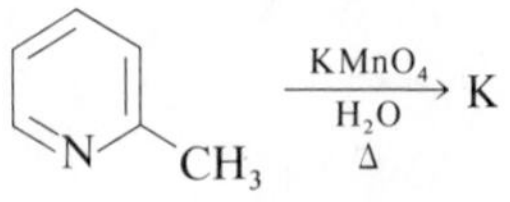

N
CH₃
KMnO₄
H₂O
Δ
K

(g) 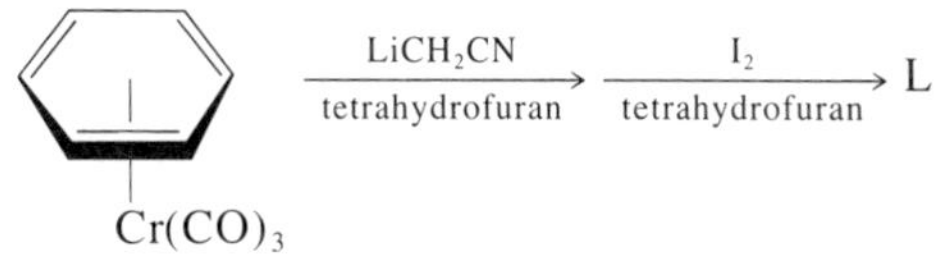
LiCH2CN
tetrahydrofuran
I2
tetrahydrofuran
L
Cr(CO)3

(h) 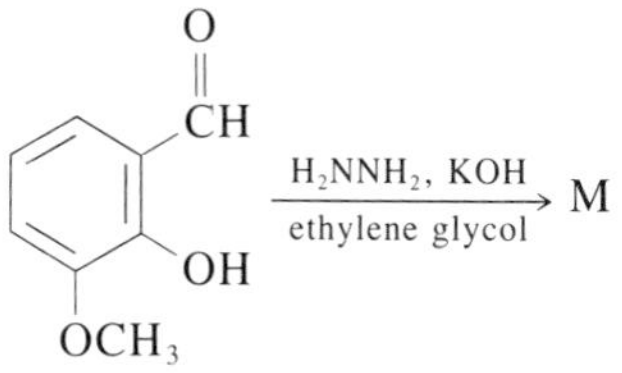
O
CH
OH
OCH3
H2NNH2, KOH
ethylene glycol
M

(i) 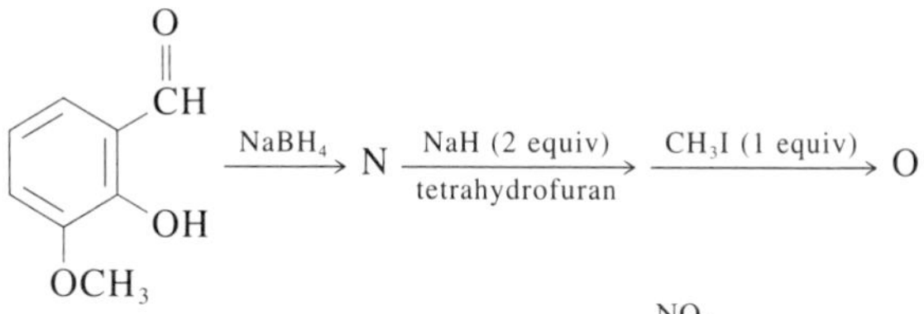
O
CH
OH
OCH3
NaBH4
N
NaH (2 equiv)
tetrahydrofuran
CH3I (1 equiv)
O

(j) 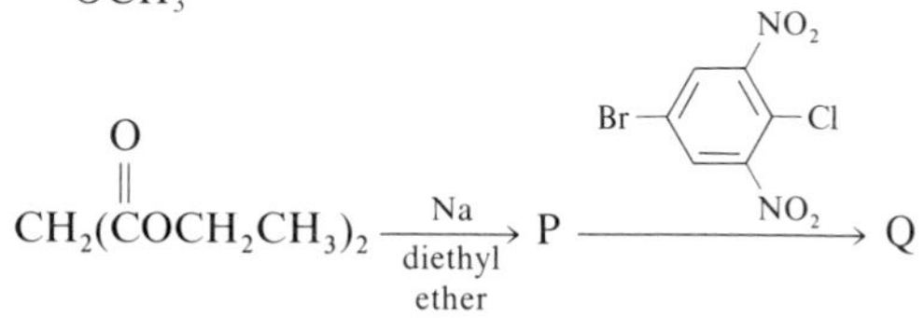
O
CH2(COCH2CH3)2
Na
diethyl
ether
P
NO2
Br
Cl
NO2
Q

(k) 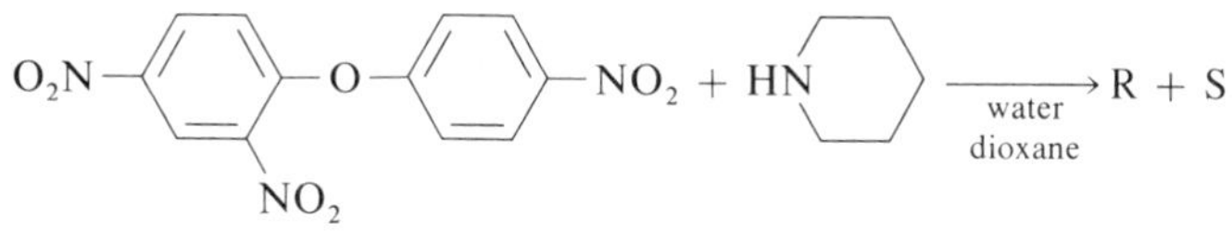
O2N
O
NO2
NO2 + HN
water
dioxane
R + S

(l) 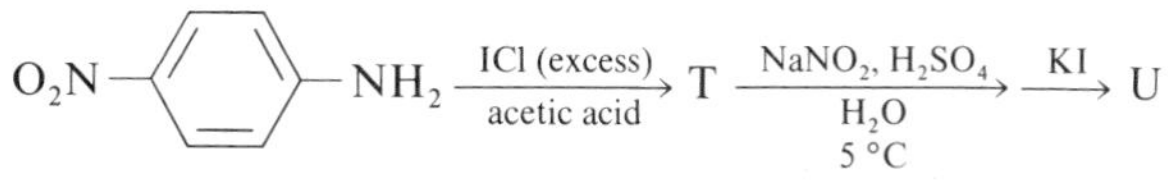
O2N
NH2
ICl (excess)
acetic acid
T
NaNO2, H2SO4
H2O
5 °C
KI
U

(m) 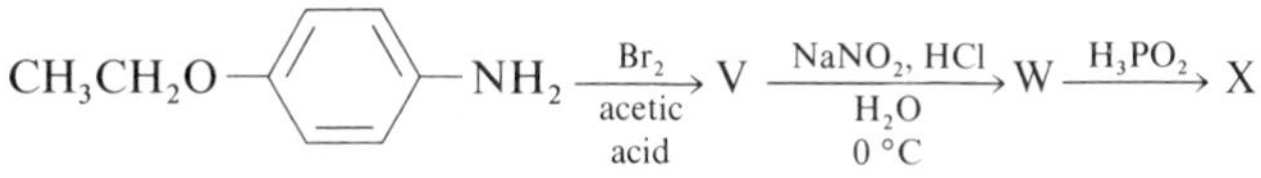
CH3CH2O
NH2
Br2
acetic
acid
V
NaNO2, HCl
H2O
0 °C
W
H3PO2
X

(n) 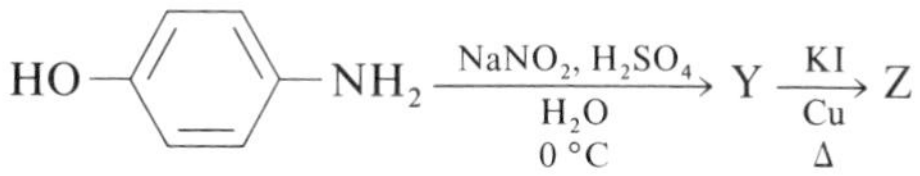
HO
NH2
NaNO2, H2SO4
H2O
0 °C
Y
KI
Cu
Δ
Z

(o) 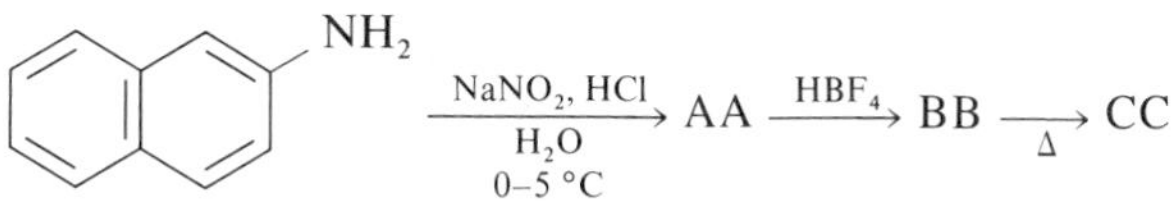
NH2
NaNO2, HCl
H2O
0–5 °C
AA
HBF4
BB
Δ
CC

(p) 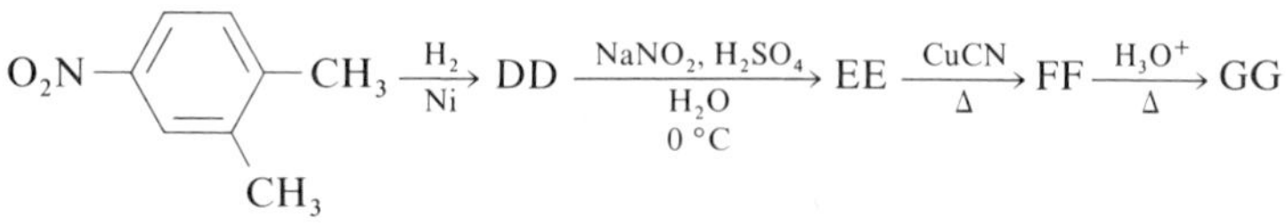
O2N
CH3
CH3
H2
Ni
DD
NaNO2, H2SO4
H2O
0 °C
EE
CuCN
Δ
FF
H3O+
Δ
GG

23.19 Write structural formulas for all products and intermediates designated by letters in the following equations.

(a) 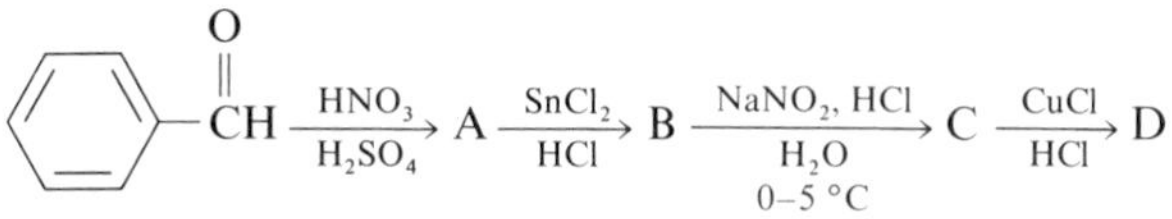
O
CH
HNO3
H2SO4
A
SnCl2
HCl
B
NaNO2, HCl
H2O
0–5 °C
C
CuCl
HCl
D

(b) 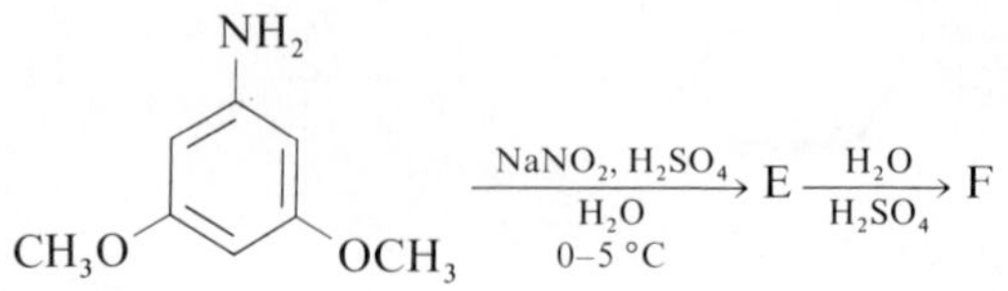

(c) 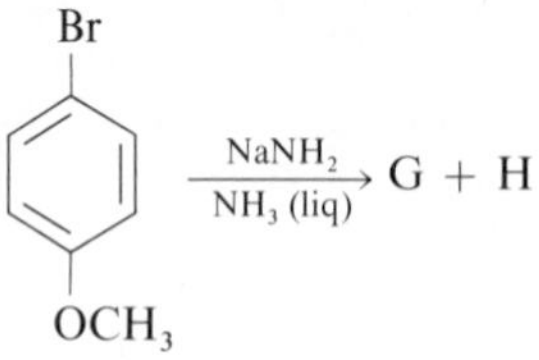

(d) 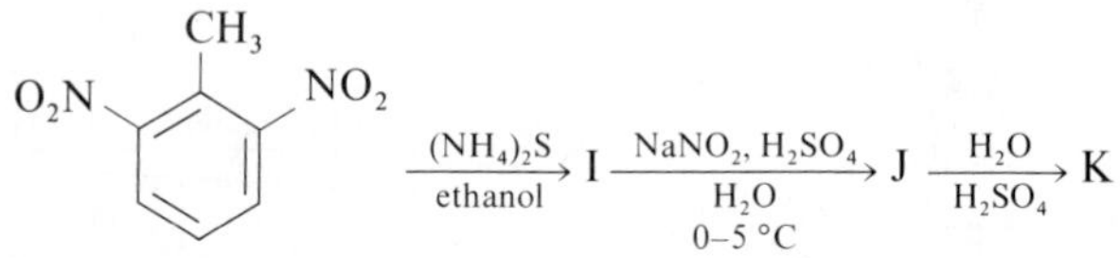

(e) 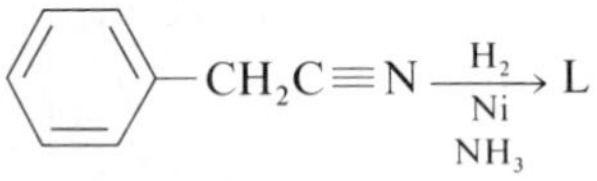

(f) 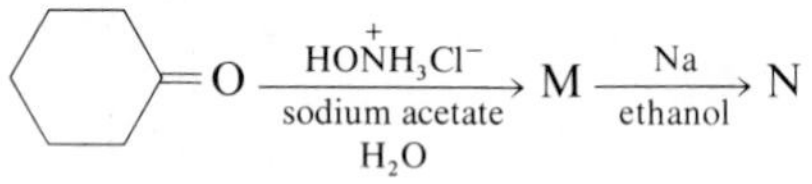

(g) 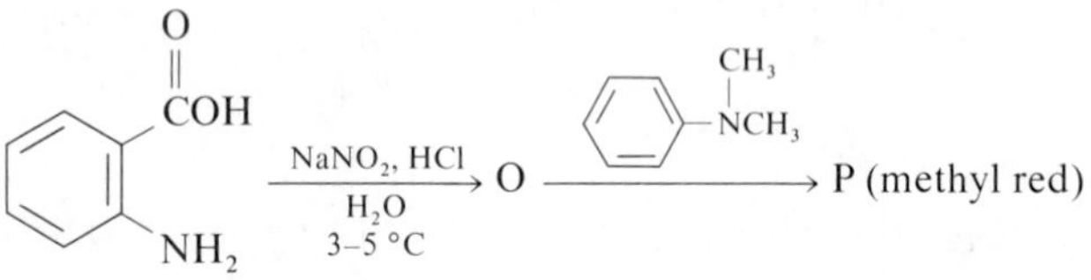

(h) 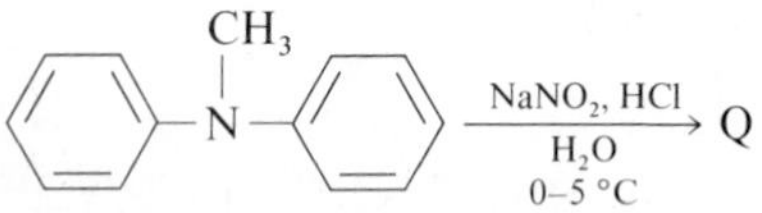

(i) 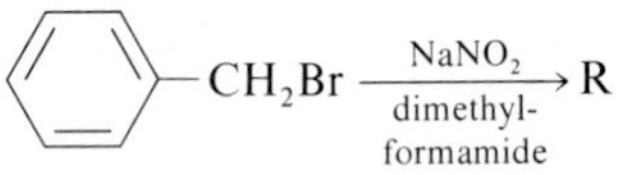

(j) 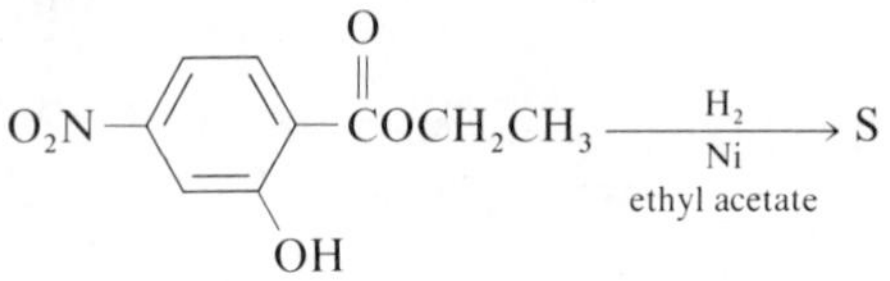

(k) 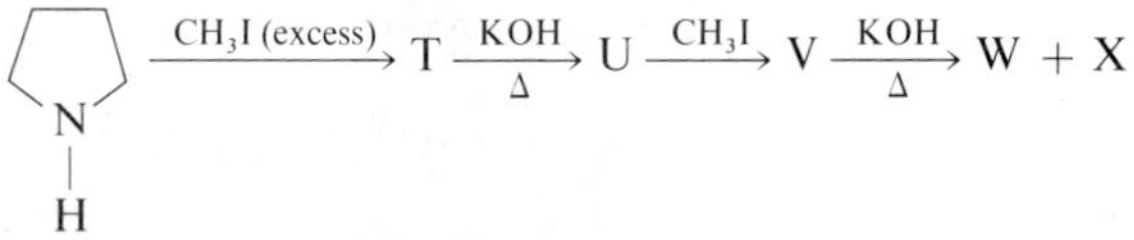

(l) 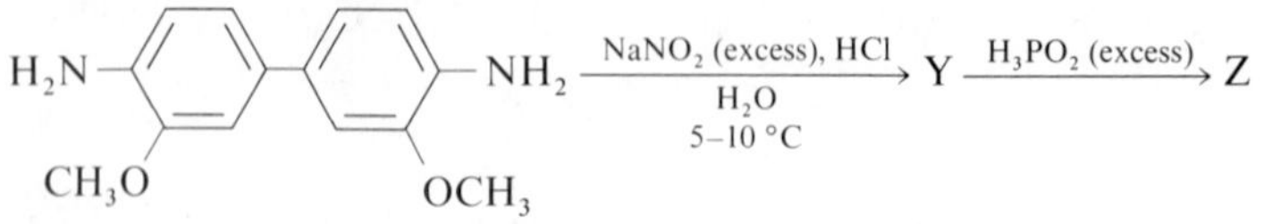

23.20 Write a mechanism that accounts for the following experimental observation.

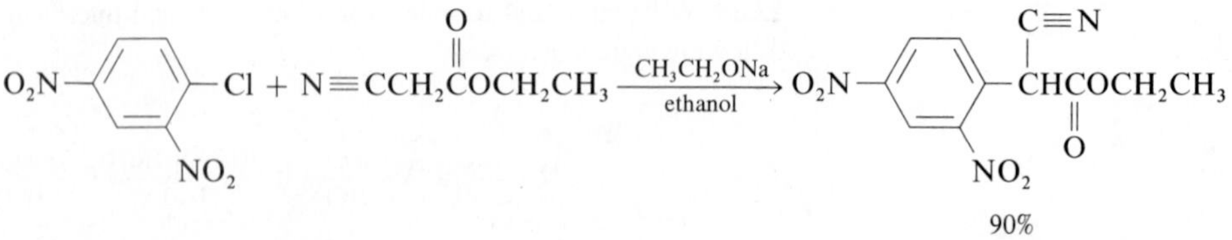

23.21 The following steps are used in the synthesis of the controversial sweetener saccharin. Give structures for the intermediate products, Compounds A and B.

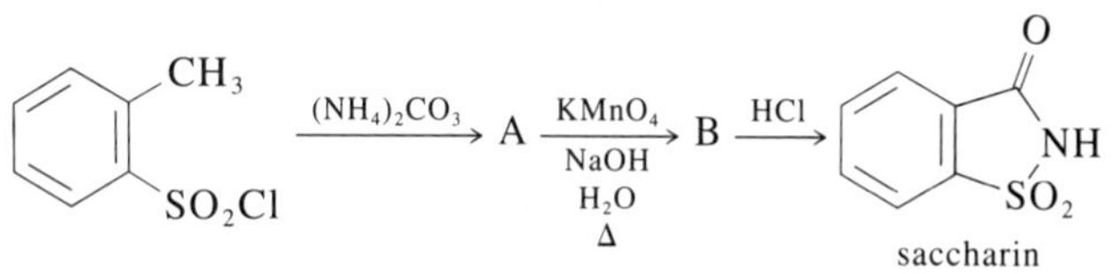
CH3
SO2Cl
(NH4)2CO3
A
KMnO4
NaOH
H2O
Δ
B
HCl
O
NH
SO2

saccharin

23.22 The following series of reactions was carried out to prepare compounds labeled with radioactive carbon, ^{14}C, in order to study the mechanism of solvolysis.

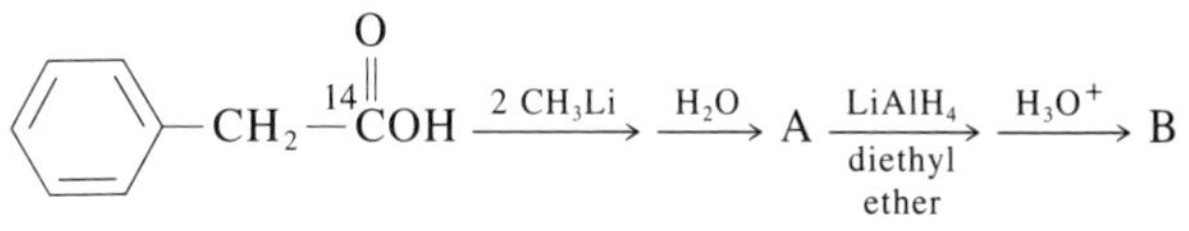
O
14
CH2—COH
2 CH3Li
H2O
A
LiAlH4
diethyl ether
H3O+
B

$$B \xrightarrow[\text{pyridine}]{\text{TsCl}} C \xrightarrow{\text{HCOOH}} D \xrightarrow[\text{diethyl ether}]{\text{LiAlH}_4} \xrightarrow{\text{H}_3\text{O}^+} B$$

$$B \xrightarrow{\text{HBr}} E$$

$$B \xrightarrow{\text{SOCl}_2} F$$

$$B \xrightarrow[\text{NaOH, H}_2\text{O, }\Delta]{\text{KMnO}_4} G + H$$

(a) What are the structures of Compounds A–H?
(b) Two readily available sources of ^{14}C are $^{14}CO_2$ and $Na^{14}CN$. How would you synthesize the radioactively labeled phenylacetic acid used as the starting material in the above series of reactions, assuming that bromobenzene is available?

23.23 4-Deuteriotoluene was needed for a study of the fate of an arene oxide derived from toluene. How would you synthesize 4-deuteriotoluene starting from toluene? Assume that heavy water, D_2O, is readily available.

23.24 Arrange each of the following groups of compounds in order of decreasing acidity.

(a)

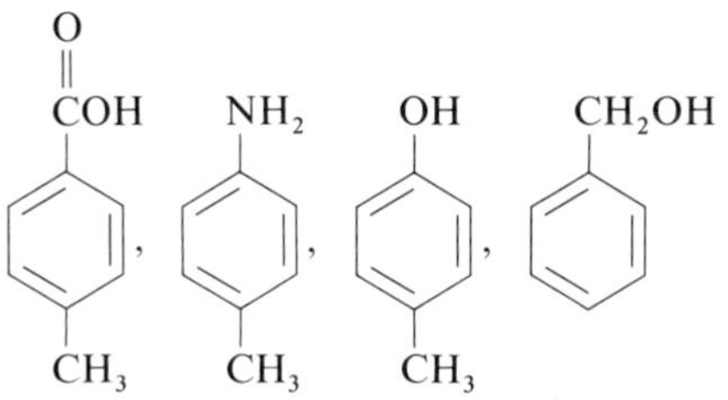
O
COH
NH2
OH
CH2OH
CH3
CH3
CH3

(b)

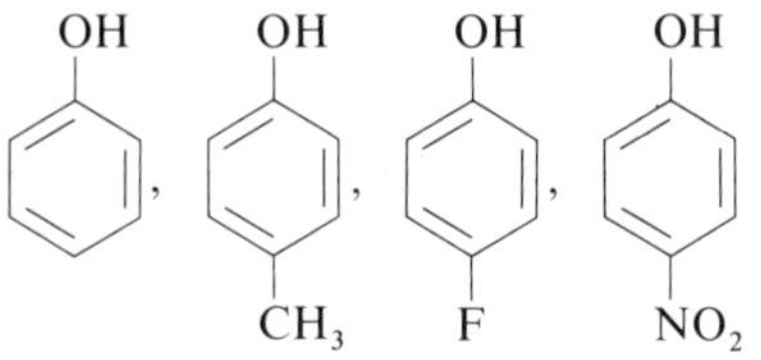
OH
OH
OH
OH
CH3
F
NO2

(c) 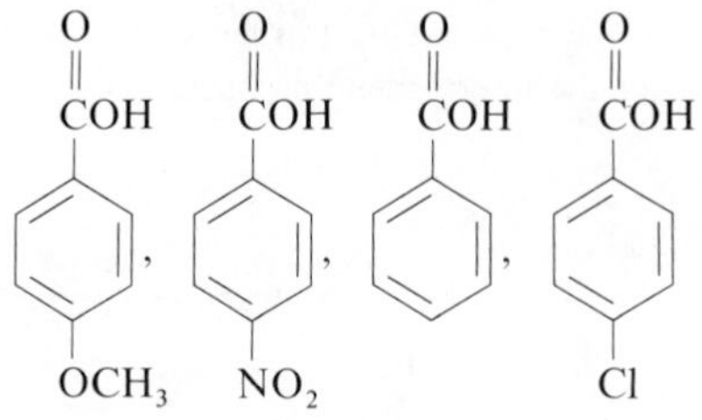

23.25 *p*-Nitrofluorobenzene reacts with a series of nucleophiles in methanol at 25 °C. The relative rates for these reactions are shown below. Explain the experimental observations. (Hint: Reviewing Section 7.2 may be helpful.)

Halide	**Nucleophiles**			
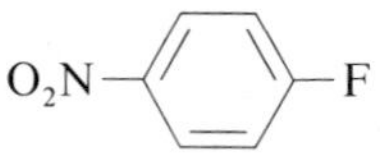	CH_3O^-	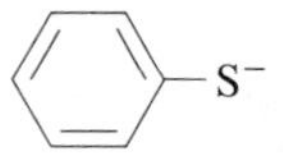	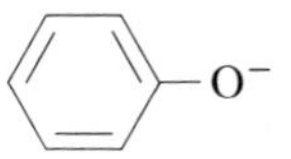	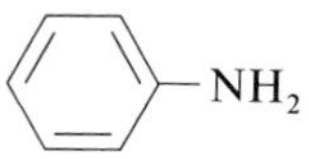
relative rates	11,600	10,800	65	1

23.26 The nucleophilic substitution reactions of 2,4-dinitrochlorobenzene with a series of substituted anilines were studied. The rate was found to depend on the substituent at the para position of the aniline. Write a general equation for the reaction described, and predict the order of reactivity for the following anilines, showing clearly which one would be expected to react the fastest and which the slowest.

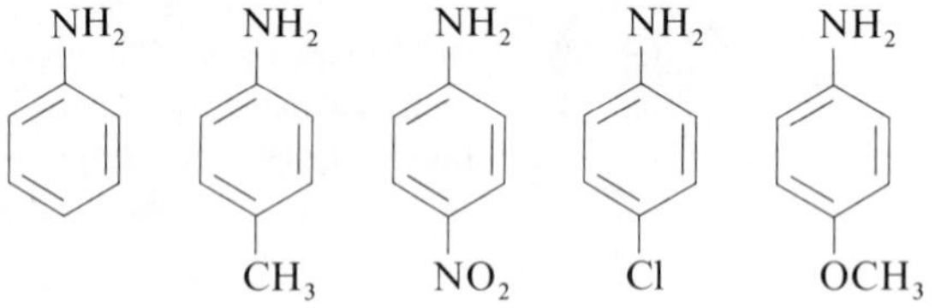

23.27 What reagents would be needed to carry out each of the following transformations? More than one step may be required in some cases.

(a) 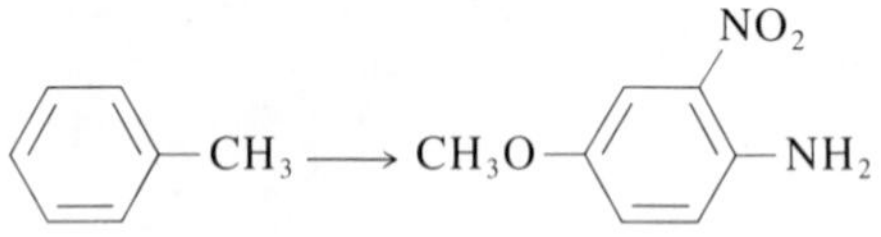

(b) 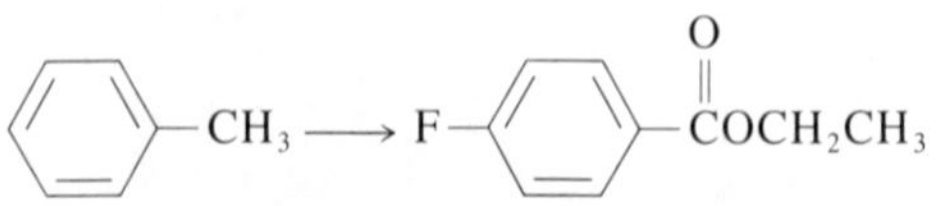

(c) 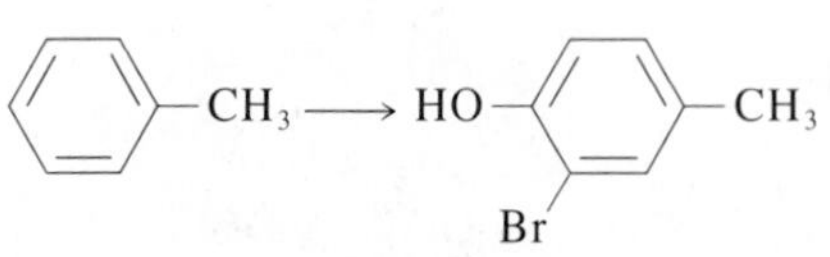

(d) 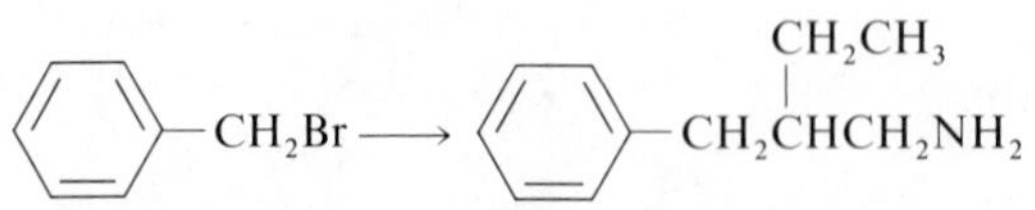

(e) 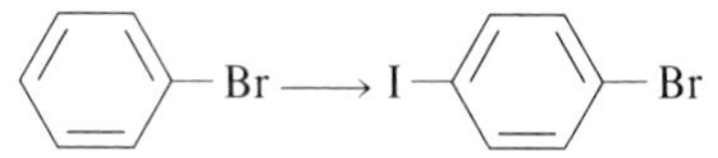
Br
I
Br

(f) 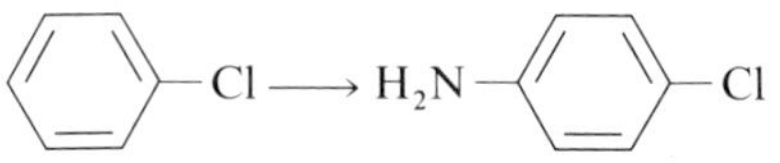
Cl
H2N
Cl

(g) 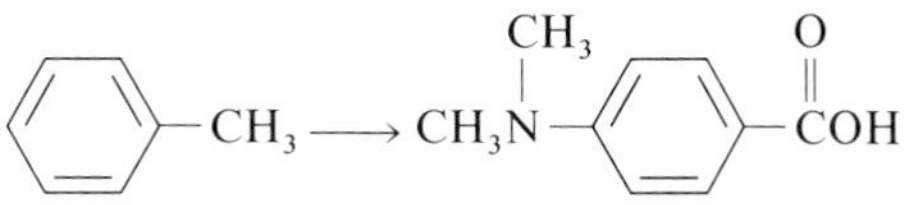
CH3
CH3
CH3N
O
COH

23.28 The reactions of several 2,4-dinitrochlorobenzenes with piperidine in ethanol as a solvent were studied. Structural formulas for the compounds and the relative rates of their reactions are shown below. What factor is influencing the rates in this series? How do you explain the effect? (Hint: A careful examination of the structural formula for the Meisenheimer complex on p. 959 will furnish a clue.)

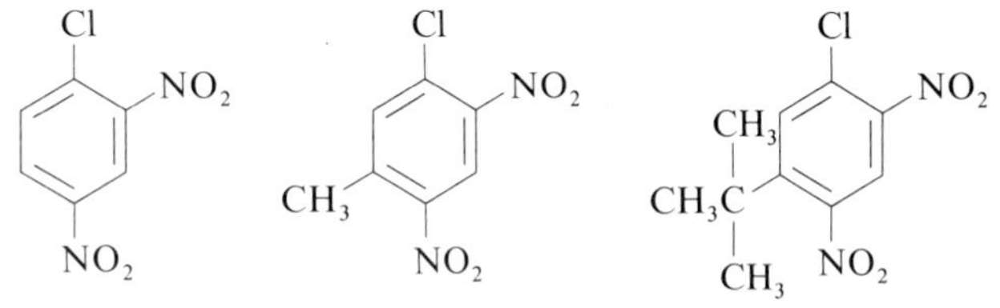
Cl
NO2
NO2
Cl
NO2
CH3
NO2
Cl
NO2
CH3
CH3C
CH3
NO2

k_{rel}	956	267	16.2

23.29 When the potassium salt of cyclopentadienyl anion is treated with iron(II) chloride, a stable compound in which an iron atom is sandwiched between two cyclopentadiene rings is formed. This compound, bis(cyclopentadienyl)iron, is also called ferrocene. Ferrocene undergoes aromatic substitution reactions. For example, heating ferrocene with acetic anhydride and phosphoric acid gives acetylferrocene. How do you explain the stability of ferrocene? [Hint: A similarly stable compound is not formed with iron(III). Looking at the periodic table may give you a clue.]

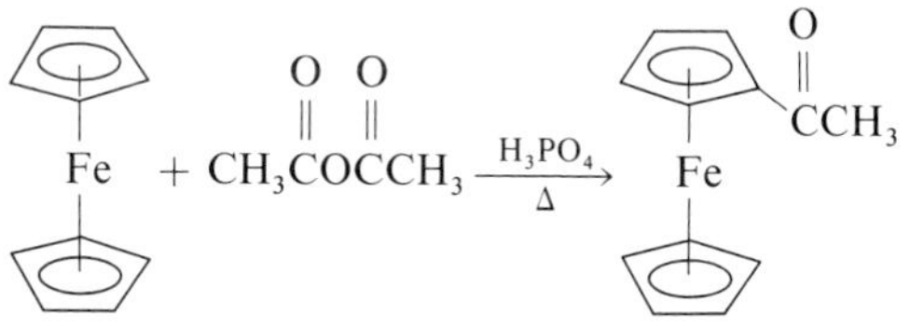
Fe
O O
CH3COCCH3
H3PO4
Δ
Fe
O
CCH3

ferrocene
all C—H *bonds on ferrocene are equivalent*

acetylferrocene

23.30 Compounds A and B are isomers with molecular ions at m/z 122. Their proton magnetic resonance spectra appear in Figure 23.8. Compound A has strong bands at 3329 and 1046 cm^{-1} in its infrared spectrum and a base peak at m/z 91 in its mass spectrum. The important bands in the infrared spectrum of Compound B are at 1245 and 1049 cm^{-1}, and the base peak for B is at m/z 94 in its mass spectrum. Assign structures to Compounds A and B. Show how the spectral data support your assignments. Write equations for the formation of the ion responsible for the base peak for each compound. (Hint: To figure out how the ion from B forms, decide on the structure of the neutral fragment that is lost.)

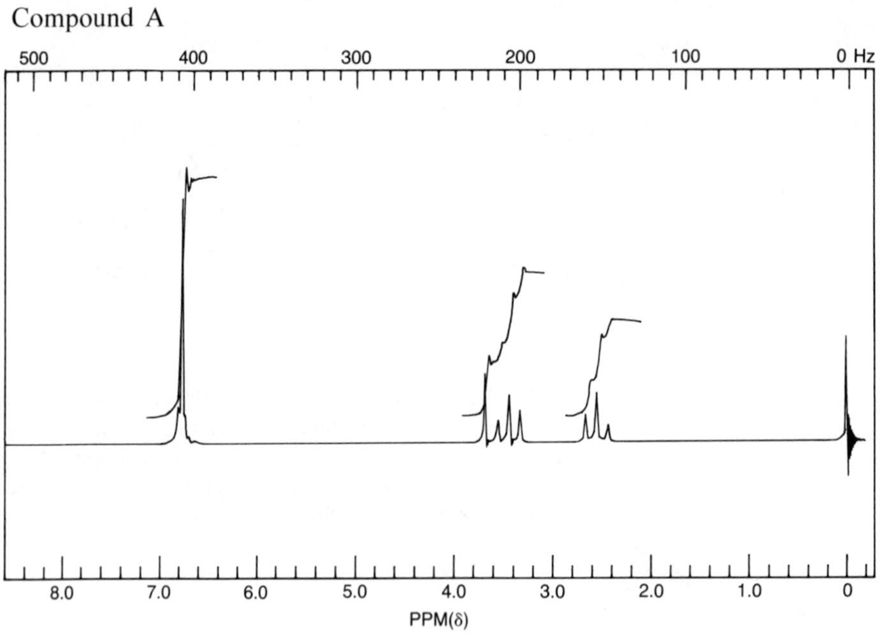
Compound A
500
400
300
200
100
0 Hz
8.0
7.0
6.0
5.0
4.0
3.0
2.0
1.0
0
PPM(δ)

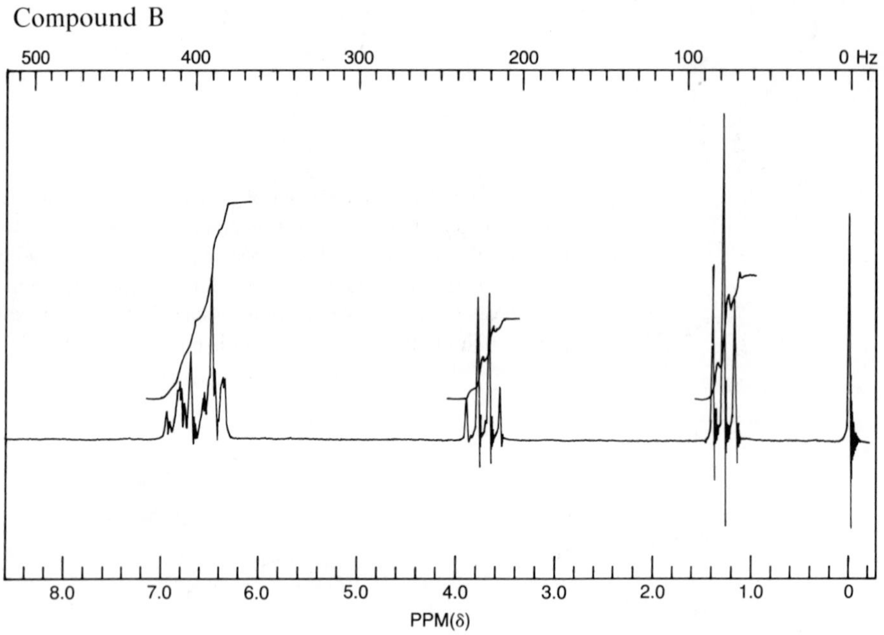
Compound B
500
400
300
200
100
0 Hz
8.0
7.0
6.0
5.0
4.0
3.0
2.0
1.0
0
PPM(δ)

FIGURE 23.8

23.31 Compound C has λ_{max} 225 (ϵ 8900) and 280 nm (ϵ 1600) in 0.1 N hydrochloric acid. In 1 N sodium hydroxide solution λ_{max} 244 (ϵ 11,700) and 298 nm (ϵ 2600) are observed. The infrared spectrum of Compound C has a broad absorption band at 3400 cm^{-1}. The proton magnetic resonance and mass spectra of C are shown in Figure 23.9. Its carbon-13 nuclear magnetic resonance spectrum has bands at 116.9 (d), 126.2 (s), 129.7 (d), and 153.6 (s) ppm. Assign a structure to Compound C. (Hint: The molecular ion is responsible for the base peak.) Discuss the different spectral data, showing how each piece supports your structural assignment.

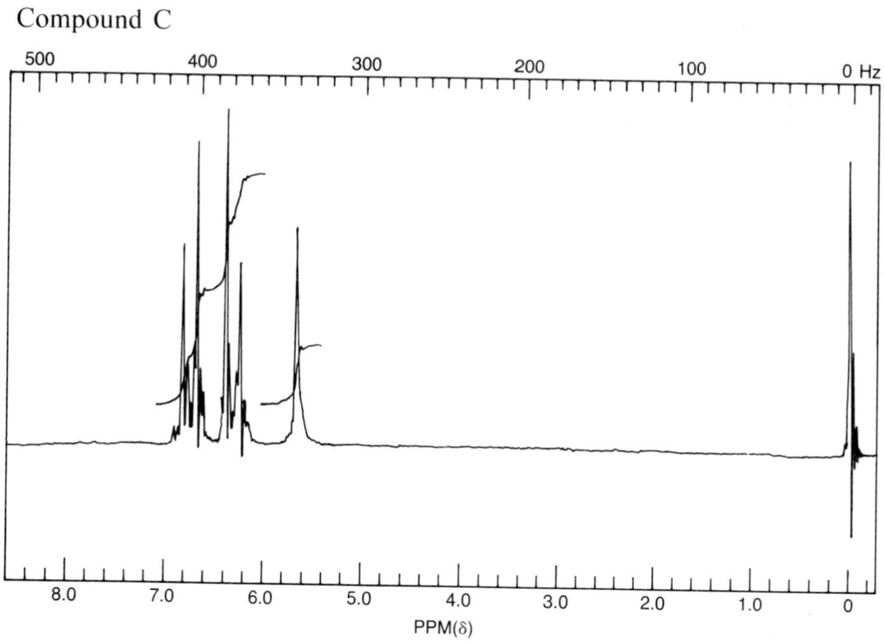
Compound C
500
400
300
200
100
0 Hz
8.0
7.0
6.0
5.0
4.0
3.0
2.0
1.0
0
PPM(δ)

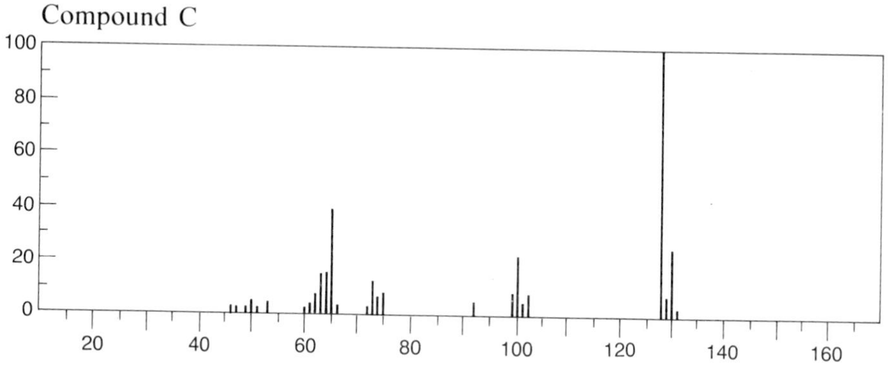
Compound C
100
80
60
40
20
0
20
40
60
80
100
120
140
160

FIGURE 23.9

23.32 The proton magnetic resonance spectrum of Compound D is shown in Figure 23.10 on the following page. After the sample of the compound is shaken with D_2O, the band at δ 3.2 disappears. Compound D has bands in its infrared spectrum at 3376, 3480, and 1621 cm^{-1}. Its carbon-13 nuclear magnetic resonance spectrum has bands at 13.0, 23.9, 115.4, 118.6, 126.7, 128.0, 128.3, and 144.3 ppm. When its ultraviolent spectrum is taken in 2 M hydrochloric acid, λ_{max} is 259 nm; at pH 9.2 λ_{max} is 280 nm. Assign a structure to Compound D, and show how the various pieces of spectral data support your assignment. Do the data allow you to make a definite assignment of structure?

23.33 Two isomeric compounds have the structural formulas shown below. One has $\lambda_{max}^{ethanol}$ 254 (log ϵ 2.34), 259 (log ϵ 2.38), 262 (log ϵ 2.37), 265 (log ϵ 2.27), and 268 (log ϵ 2.24) nm. The other has $\lambda_{max}^{ethanol}$ 246 (log ϵ 4.25), 284 (log ϵ 2.99), and 293 (log ϵ 2.84) nm. Which ultraviolet spectrum corresponds to which structure?

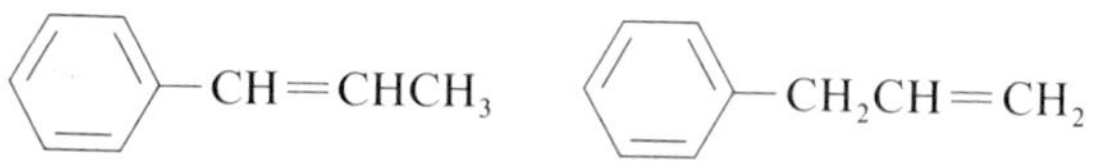
CH=CHCH3
CH2CH=CH2

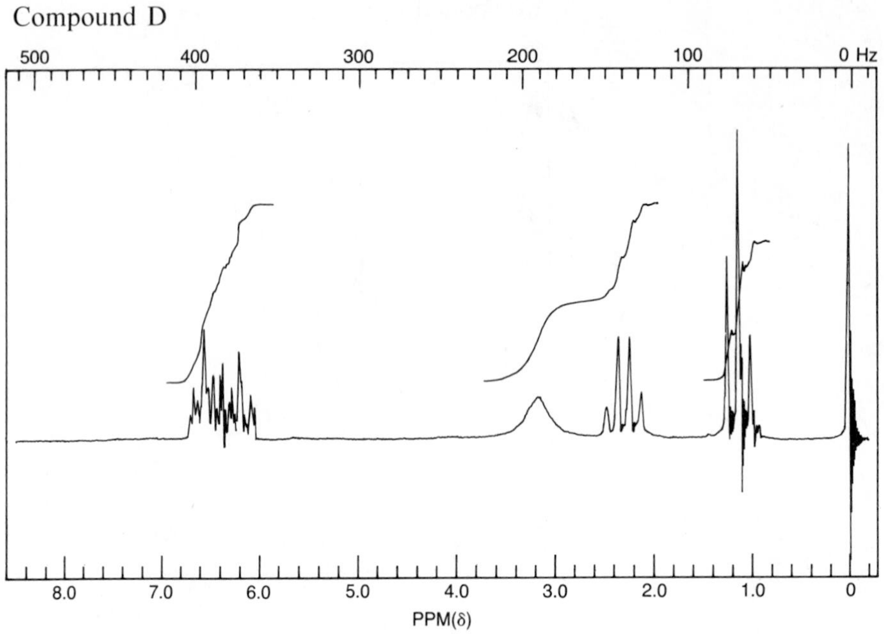
Compound D
500
400
300
200
100
0 Hz
8.0
7.0
6.0
5.0
4.0
3.0
2.0
1.0
0
PPM(δ)

FIGURE 23.10

23.34 Vitamin B_6, pyridoxine, has been isolated from rice bran and has the molecular formula $C_8H_{11}NO_3$. Its ultraviolet spectrum changes with pH. At pH 2.1, the spectrum shows $\lambda_{max}^{H_2O}$ 292 (ϵ 6950); at pH 10.2, $\lambda_{max}^{H_2O}$ 240 (ϵ 5500) and 315 nm (ϵ 5800) are observed.

(a) When the acidic proton in Vitamin B_6 is replaced by a methyl group, the spectrum of the resulting compound has a single band at 280 nm (ϵ 5800) that does not change with pH. Given this, what conclusions can you reach about the structure of Vitamin B_6?

(b) Oxidation of the methyl ether of Vitamin B_6 with barium permanganate in water at room temperature for 16 hours, followed by filtration of the manganese dioxide that was formed and acidification of the solution with sulfuric acid, gave two products, a lactone and a dicarboxylic acid. Their structures are shown below.

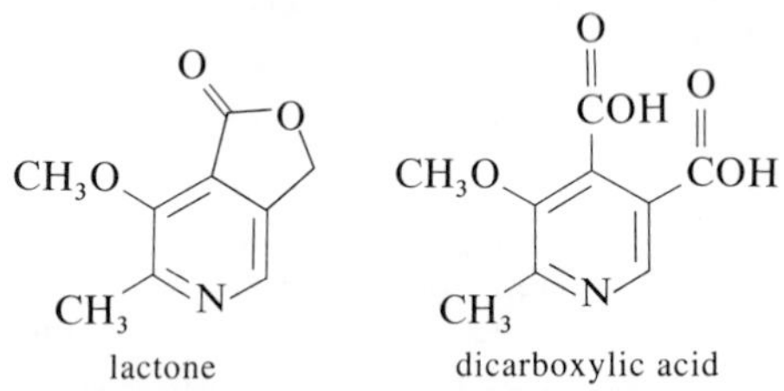
O
O
CH_3O
CH_3
N
O
COH
O
COH
CH_3O
CH_3
N

lactone from the oxidation of the methyl ether of Vitamin B_6 — dicarboxylic acid from the oxidation of the methyl ether of Vitamin B_6

What must be the structure of Vitamin B_6? Explain why two of the side chains on this aromatic ring were oxidized but the methyl group was untouched. Write equations for a reaction pathway to the lactone.

(c) If you were trying to confirm the structure of Vitamin B_6 using ultraviolet spectroscopy, what simpler compounds would you use as models? Draw structural formulas for some that would be useful. Under what conditions would you take the ultraviolet spectra of the models?

23.35 The dye butter yellow, *p*-dimethylaminoazobenzene (p. 953), has λ_{max} 408 nm in neutral solution. When acid is added to the solution, two different species are detected spectroscopically. One has λ_{max} 320 nm and the other λ_{max} 510 nm. Write structural formulas for the species that have chromophores giving rise to these absorption bands.

24

The Chemistry of Heterocyclic Compounds

A • L O O K • A H E A D

Cyclic organic compounds are divided into two large classes: those that have only carbon atoms in their rings are known as carbocyclic compounds, and those with atoms of elements other than carbon in their rings are called heterocyclic compounds. Heterocyclic compounds occur so widely in nature and are of such importance chemically that any discussion of organic chemistry will not get very far without mentioning them. For example, the heterocyclic compounds pyridine, tetrahydrofuran, and oxirane are already familiar to you.

pyridine *an aromatic heterocycle* tetrahydrofuran *a saturated heterocycle* oxirane *a saturated heterocycle*

Some heterocyclic compounds, such as pyridine, are aromatic. Others, such as tetrahydrofuran, are not. Aromatic heterocycles differ in reactivity. Some, such as thiophene, are much more reactive toward electrophilic substitution than benzene is. When they react, they are substituted at position 2 of the ring.

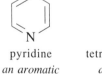

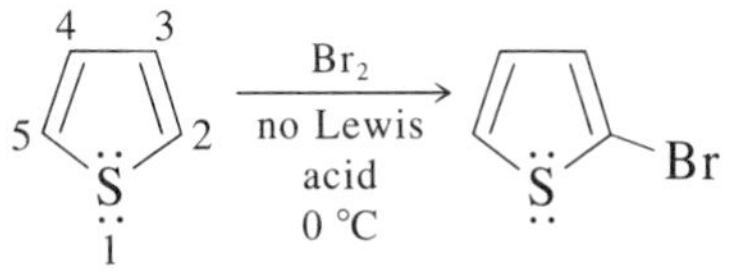

4
3
5
2
S
1
Br_2
no Lewis acid
0 °C
S
Br

Others, such as pyridine, need extreme conditions to react with electrophiles but react with nucleophiles with relative ease. Electrophilic substitution occurs at position 3.

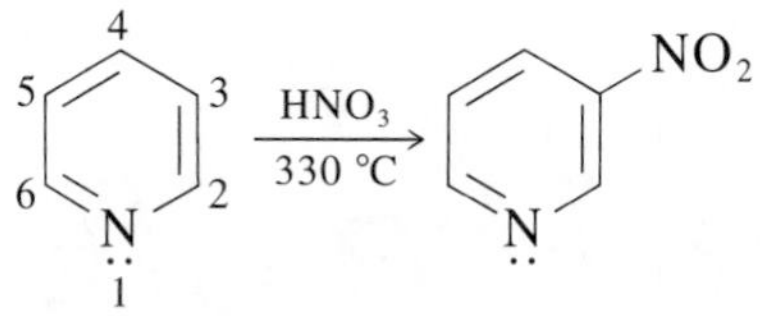

This chapter will explore these differences.

Heterocyclic compounds play important medicinal and biochemical roles. The purine and pyrimidine bases that are structural units in RNA and DNA are heterocyclic compounds, as are many drugs such as morphine, heroin, and cocaine. The chemistry of heterocyclic compounds is such an extensive field that this chapter is by necessity a highly selective look at some of it. The chapter emphasizes reactions that will expand and reinforce your understanding of the basic chemical principles that have been developed in the rest of this book.

24.1 NOMENCLATURE OF HETEROCYCLIC COMPOUNDS

Heterocyclic compounds may be classified in a number of ways: by the size of the ring, by the nature and number of **heteroatoms** (atoms other than carbon) in the ring, by the degree of unsaturation in the ring, and by whether or not the compound has aromatic character.

Aromatic heterocyclic compounds were included in Section 19.1 to show that nonbonding electrons on heteroatoms could be considered part of an aromatic sextet. The important aromatic heterocyclic systems that contain a single heteroatom are shown below. Nonbonding electrons shown inside a ring are part of the aromatic sextet for that system. Note that some aromatic heterocycles have additional nonbonding electrons on a nitrogen, oxygen, or sulfur atom that do not contribute to the aromaticity of the compound. Some of the chemical consequences of these structural properties are explored in Section 24.2.

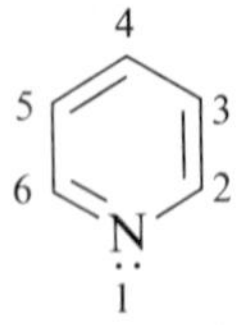
pyridine

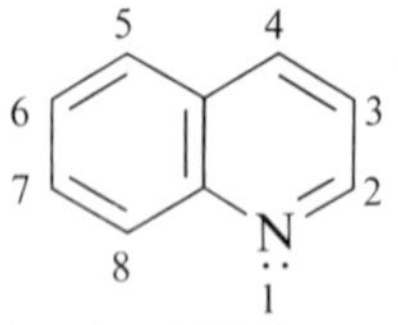
quinoline

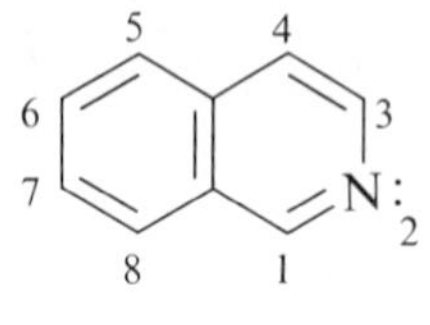
isoquinoline

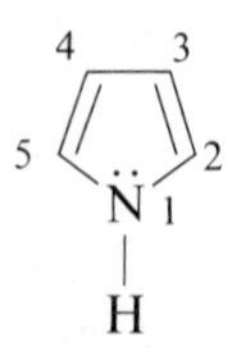
pyrrole

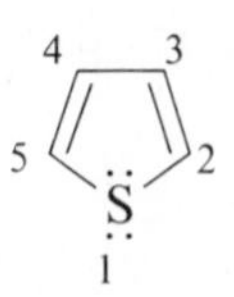
thiophene

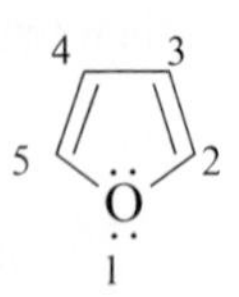
furan

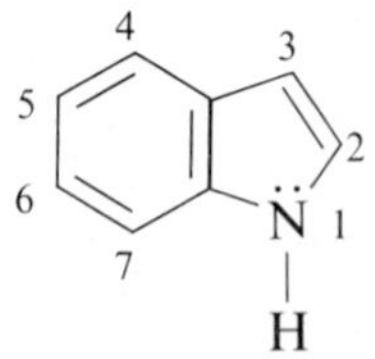
indole

The heteroatom in a heterocycle is given the lowest possible number consistent with an orderly progression around the ring system. In isoquinoline, a carbon atom has a lower number than the nitrogen atom to preserve an orderly sequence around the periphery of the two rings. Substituted heterocyclic compounds are named in the same way as other compounds, by giving the name of the substituent and a number to indicate its position on the ring system. The name of the substituent may appear as either a prefix or a suffix to the name of the heterocycle, depending on the rules for the precedence of functional groups (p. 480).

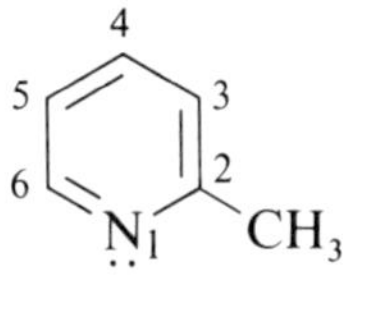
2-methylpyridine

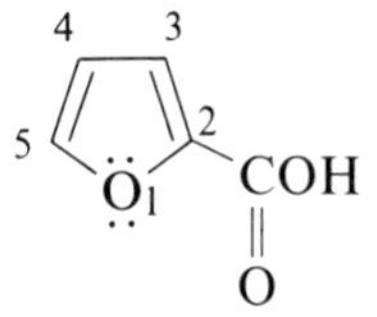
2-furancarboxylic acid
furoic acid

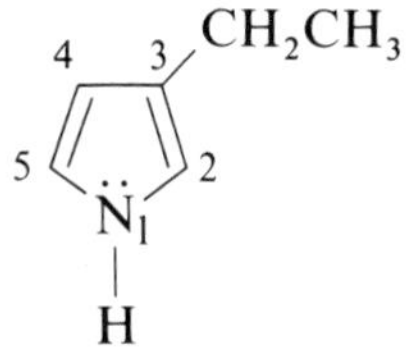
3-ethylpyrrole

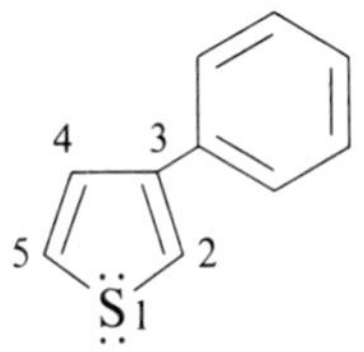
3-phenylthiophene

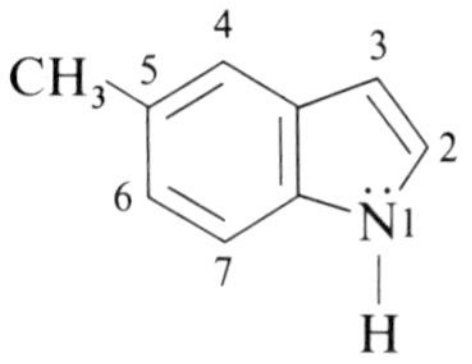
5-methylindole

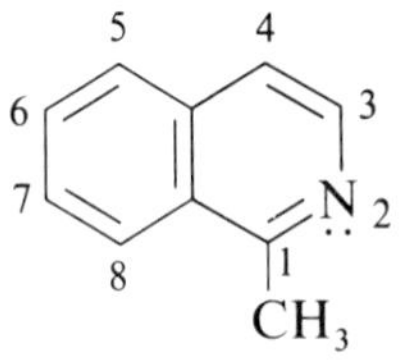
1-methylisoquinoline

The compounds shown above have only a single heteroatom in their rings. There are other heterocyclic compounds with some aromatic character that have two or more heteroatoms, one of which is nitrogen. When such compounds have a five-membered ring system, their names all end in **azole.** The rest of the name indicates what other heteroatoms are present.

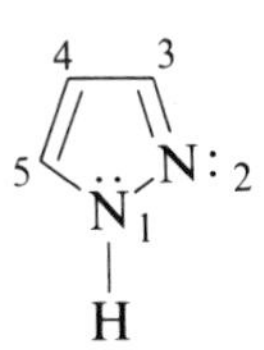
pyrazole

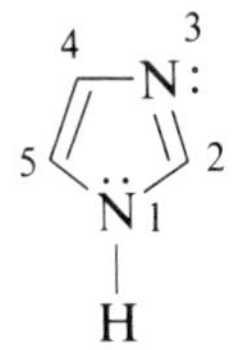
imidazole

thiazole

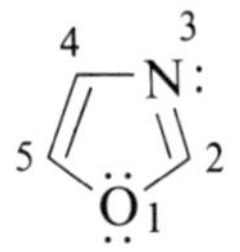
oxazole

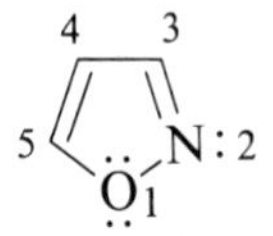
isoxazole

The names **pyrazole** and **imidazole** are given to the two isomeric compounds containing two nitrogen atoms in the ring. The name **thiazole** indicates that the ring has a sulfur atom and a nitrogen atom in it, and **oxazole** indicates the presence of oxygen and nitrogen. That name is reserved for the system in which the two heteroatoms are separated by a carbon atom; the compound in which oxygen and nitrogen are adjacent to each other is known as **isoxazole.**

The six-membered aromatic heterocyclic system with two nitrogens exists in three isomeric forms, the most important of which is pyrimidine. Purine is another heterocycle that is biologically important.

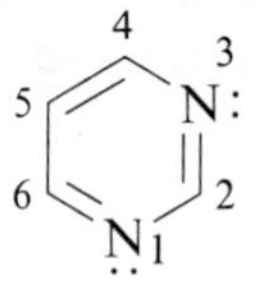

pyrimidine

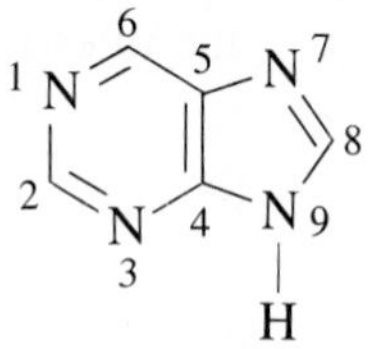

purine

The compounds that have been discussed so far in this section have the maximum degree of unsaturation. More saturated heterocycles also exist; for example, tetrahydrofuran is the fully saturated form of furan. Some other examples of heterocyclic compounds with varying degrees of unsaturation are shown in the structures below.

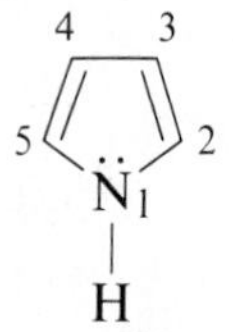

pyrrole

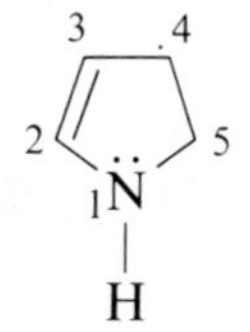

2-pyrroline

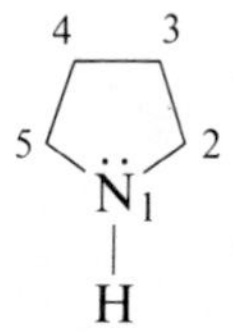

pyrrolidine

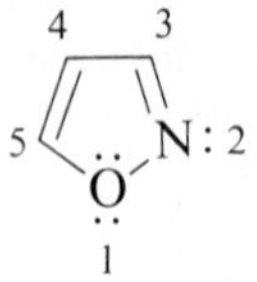

isoxazole

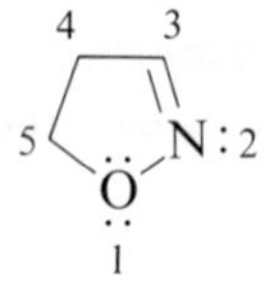

2-isoxazoline

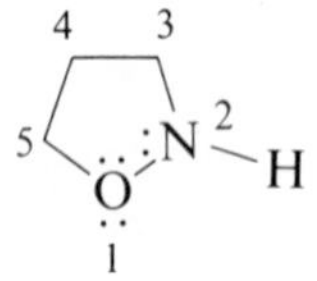

isoxazolidine

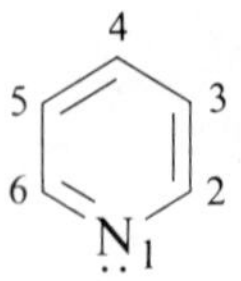

pyridine

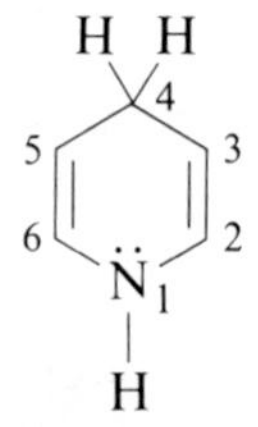

1,4-dihydropyridine

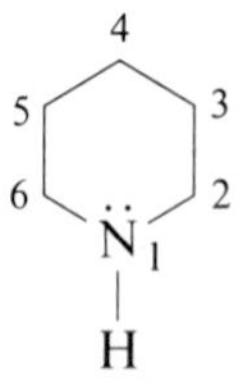

piperidine

a hexahydropyridine

⟵ increasing unsaturation

Certain heterocycles with four-membered rings are also important. For example, substituted **azetidinones** are components of a number of important antibiotics such as penicillin and cephalosporin.

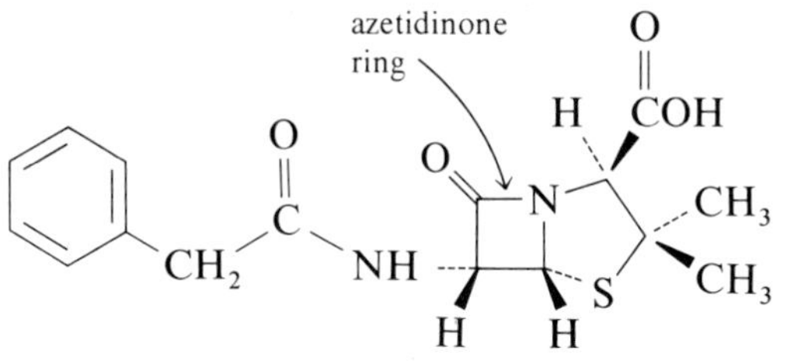
azetidinone ring
O
H COH
O
O
N
CH3
CH2
C
NH
S
CH3
H
H
penicillin G

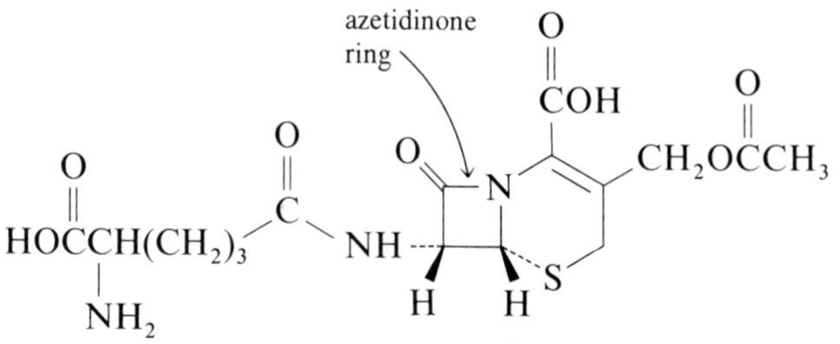
azetidinone ring
O
COH
O
O
O
O
CH2OCCH3
N
HOCCH(CH2)3
C
NH
S
NH2
H
H
cephalosporin C

The reactivity of the carbonyl group in the azetidinone, a strained four-membered amide ring (also called a **β-lactam**), in these antibiotics is responsible for their biological activity. The compounds act as acylating agents and disrupt the synthesis of bacterial membranes.

PROBLEM 24.1

Name the following compounds.

(a) 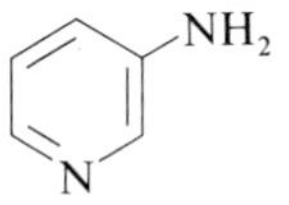
NH2
N
(b) 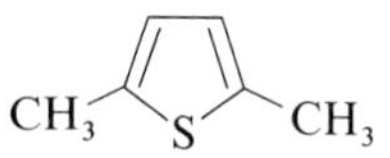
CH3
S
CH3
(c) 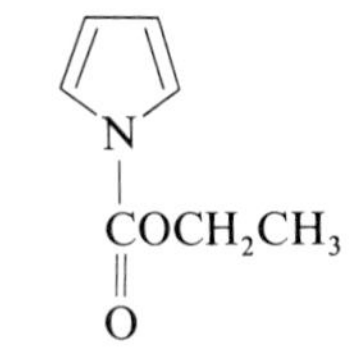
N
COCH2CH3
O

(d) 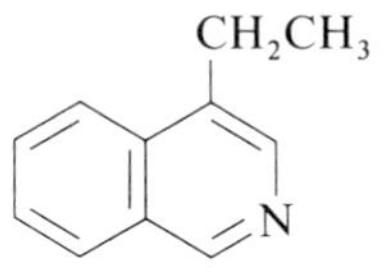
CH2CH3
N
(e) 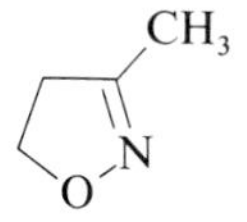
CH3
N
O
(f) 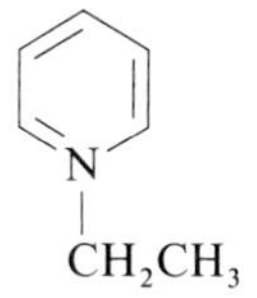
N
CH2CH3

(g) 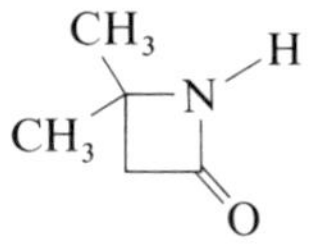
CH3
H
N
CH3
O
(h) 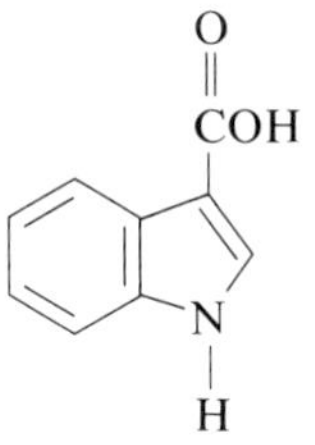
O
COH
N
H
(i) 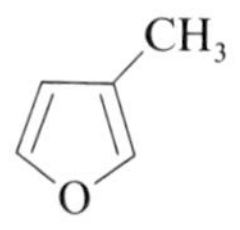
CH3
O
(j) 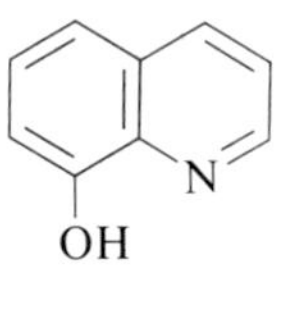
N
OH

(k) 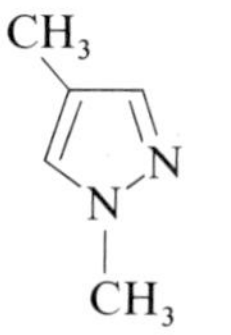
CH3
N
N
CH3
(l) 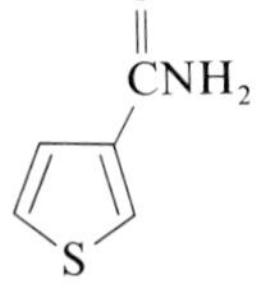
O
CNH2
S
(m) 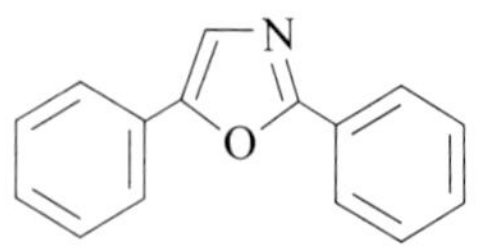
N
O

PROBLEM 24.2

Write structural formulas for the following compounds.

(a) 4-ethylindole (b) 6-aminoquinoline (c) 2-methyl-5-phenylthiophene
(d) 1,4-dimethylisoquinoline (e) 5-phenylisothiazole (f) 2,5-dimethylfuran
(g) 3-ethylisoxazole (h) 5-chloropyrimidine (i) 1,2-dihydropyridine
(j) 2,4-dimethyloxazole (k) 3-methyltetrahydrofuran (l) 3,5-dimethylpyrazole
(m) 3-ethyl-2-pyrrolecarboxylic acid

24.2 AROMATIC HETEROCYCLIC COMPOUNDS

Aromatic heterocyclic compounds resemble benzene in that each one can be shown experimentally to have resonance energy. In other words, each is more stable than would be expected of a similar compound with localized double bonds. Resonance energies determined by heats of combustion for different heterocycles, are shown in Table 24.1. Benzene and cyclopentadiene are included for comparison.

None of the aromatic heterocyclic compounds has as much resonance stabilization as benzene. All of the five-membered ring heterocycles, however, are much more stable than is expected for a cyclic diene, represented in Table 24.1 by cyclopentadiene. Thiophene, pyrazole, pyridine, pyrrole, and imidazole should have the chemical reactivity of aromatic compounds. Section 24.5 will examine electrophilic aromatic substitution reactions of some heterocycles to see to what extent this prediction is correct.

In Table 24.1, furan stands out as the compound with the least resonance stabilization of the simple heterocycles. The loss of aromaticity is not a large barrier to the reactions of furan. In general, it undergoes addition rather than substitution reactions with much greater ease than do the other heterocyclic compounds (p. 1001). The small degree of aromaticity of furan is also illustrated by the relative ease with which it serves as a diene in Diels-Alder reactions. It is much less reactive than cyclopentadiene, which as a cyclic diene reacts readily with all kinds of dienophiles (pp. 693 and 952), but much more reactive than thiophene or pyrrole. Furan reacts only with highly reactive dienophiles such as dimethyl acetylenedicarboxylate (p. 692) and benzyne (p. 950).

TABLE 24.1 Resonance Energies for Some Cyclic Compounds (Determined from Heats of Combustion)

Compound	Resonance Energy (kcal/mol)
benzene	36
thiophene	29
pyrazole	29
pyridine	28
pyrrole	22
imidazole	22
furan	16
cyclopentadiene	3

Study Guide Concept Map 24.1

24.3 REACTIONS OF HETEROCYCLES AS ACIDS AND BASES

An examination of the structural formulas for aromatic heterocycles raises interesting questions about the acidity and basicity of these compounds. For example, pyrrole, in which the nonbonding electrons on the nitrogen atom are part of the aromatic sextet, has a low basicity. When it accepts a proton, it does so on one of the carbon atoms adjacent to the nitrogen atom. On the other hand, the proton on the nitrogen atom can be removed by hydroxide ion to give the conjugate base of pyrrole. Salts containing the pyrrole anion are easily prepared in this way. The pK_a values for pyrrole and its conjugate acid are compared below with the values for ammonium ion and ammonia, taken from the table of pK_a values inside the cover of this book.

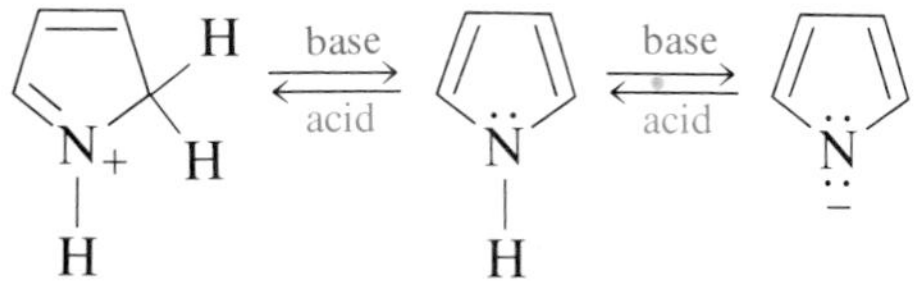

base
acid
base
acid

conjugate acid of pyrrole pK_a −3.80 | pyrrole pK_a ~15 | conjugate base of pyrrole

$$^{+}NH_4 \underset{\text{acid}}{\overset{\text{base}}{\rightleftharpoons}} :NH_3 \underset{\text{acid}}{\overset{\text{base}}{\rightleftharpoons}} :\ddot{N}H_2^{-}$$

ammonium ion, conjugate acid of ammonia, pK_a 9.4 | ammonia, pK_a 36 | amide anion, conjugate base of ammonia

The pair of nonbonding electrons on the nitrogen atom in pyrrole is much less available for protonation than the pair on ammonia. The conjugate acid of pyrrole is a strong acid, with pK_a −3.80. Pyrrole itself, with pK_a ~15, is a much stronger acid than ammonia. The conjugate base of pyrrole is still an aromatic species but has an extra pair of nonbonding electrons on the nitrogen atom. This anion is stabilized by delocalization of the negative charge. The negative charge on the amide anion, NH_2^-, in contrast, is localized on a single atom.

PROBLEM 24.3

Write resonance contributors for furan, thiophene, pyrrole, and the conjugate base of pyrrole. For the uncharged molecules, which resonance contributors are the major ones?

The reactions of imidazole as an acid or a base are important in many biological systems. Imidazole is a stronger base than pyrrole, because the second nitrogen atom has a pair of nonbonding electrons that is not part of the aromatic sextet. Imidazole also has a proton that is lost in base, and therefore it is an acid. The various species involved in the protonation and deprotonation reactions of imidazole are shown below. The conjugate acid of imidazole is stabilized by delocalization of the charge to both of the nitrogen atoms, giving two equivalent resonance contributors.

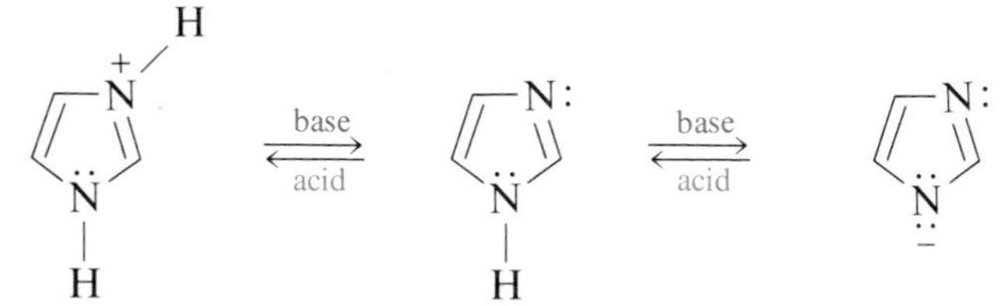

base
acid
base
acid

conjugate acid of imidazole pK_a 6.95 | imidazole pK_a 14.5 | conjugate base of imidazole

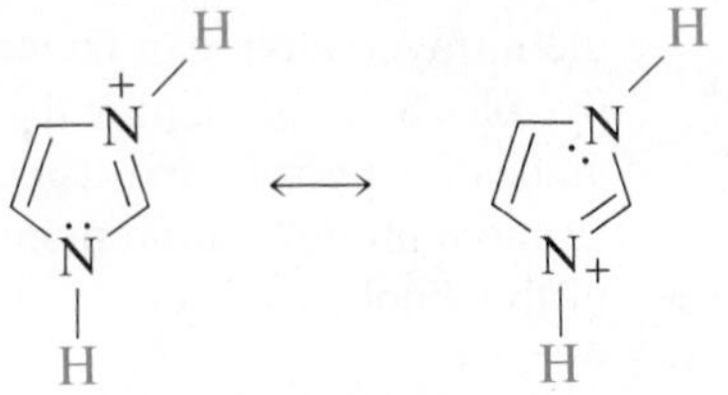

resonance contributors for the conjugate acid of imidazole

A comparison of the basicities of the six-membered ring heterocycles pyridine and pyrimidine is also interesting. Pyridine is a weak base but stronger than pyrrole. The conjugate acid has pK_a 5.2. The nonbonding electrons on the nitrogen atom are not part of the aromatic sextet but are present on a nitrogen atom that is sp^2 hybridized. It is postulated that the electrons are in an sp^2 hybrid orbital, rather than an sp^3 hybrid orbital as is the case for ammonia. Thus, they are held more closely to the nucleus of the nitrogen atom than is usual for amines because of the greater *s* character of the orbital (p. 56).

Introducing a second nitrogen atom into the pyridine ring lowers the basicity of the molecule still further. The conjugate acid of pyrimidine has pK_a 1.3. Nitrogen is more electronegative than carbon. The inductive effect of the second nitrogen atom makes the electrons on the first nitrogen less available for protonation.

Pyridine is the heterocycle that most resembles benzene in its structure and stability. Electrophilic substitution reactions of pyridine are discussed in Section 24.5B, where a closer comparison of the two aromatic compounds is made.

PROBLEM 24.4

For each of the following sets of bases, discuss the trend observed for the pK_a values given, which are for the conjugate acids of the compounds shown.

(a)

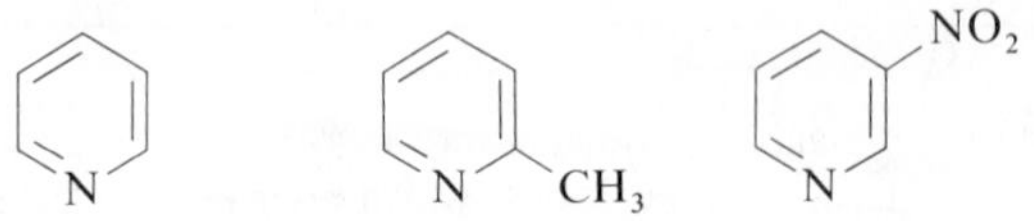

pyridine pK_a 5.2 | 2-methylpyridine pK_a 6.0 | 3-nitropyridine pK_a 0.8

(b)

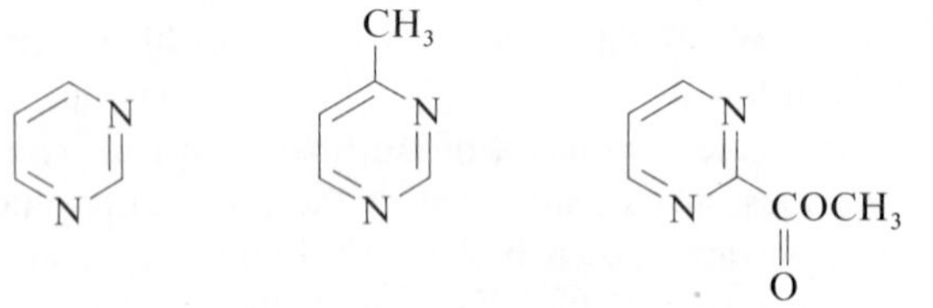

pyrimidine pK_a 1.3 | 4-methylpyrimidine pK_a 2.0 | methyl 2-pyrimidinecarboxylate pK_a −0.68

(c)

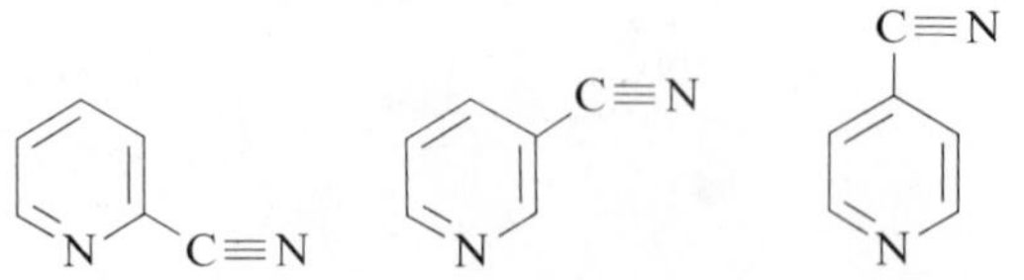

2-cyanopyridine pK_a −0.3 | 3-cyanopyridine pK_a 1.5 | 4-cyanopyridine pK_a 1.9

A. Five-Membered Heterocycles with One Heteroatom

A general way to synthesize heterocyclic compounds is by cyclization of a dicarbonyl compound using a nucleophilic reagent that introduces the desired heteroatom or atoms. An example of such a synthesis is the preparation of 2-methyl-5-phenylpyrrole from 1-phenyl-1,4-pentanedione and ammonia.

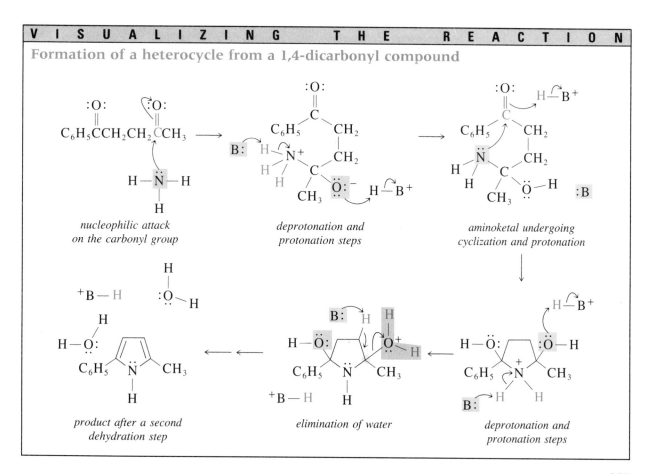

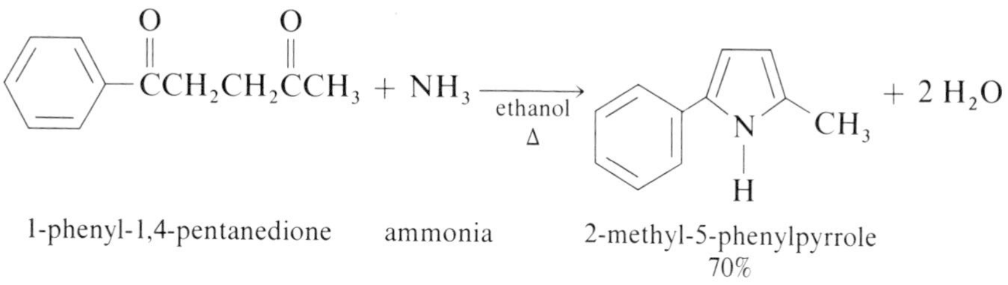
ethanol
Δ
1-phenyl-1,4-pentanedione
ammonia
2-methyl-5-phenylpyrrole
70%

The reaction may be pictured as involving an aminoketal intermediate resulting from nucleophilic attack of ammonia on a carbonyl group. The intermediate undergoes cyclization to a stable five-membered ring and dehydration. The last stage of the mechanism, shown below, represents two sequences of protonation and elimination reactions.

VISUALIZING THE REACTION

Formation of a heterocycle from a 1,4-dicarbonyl compound

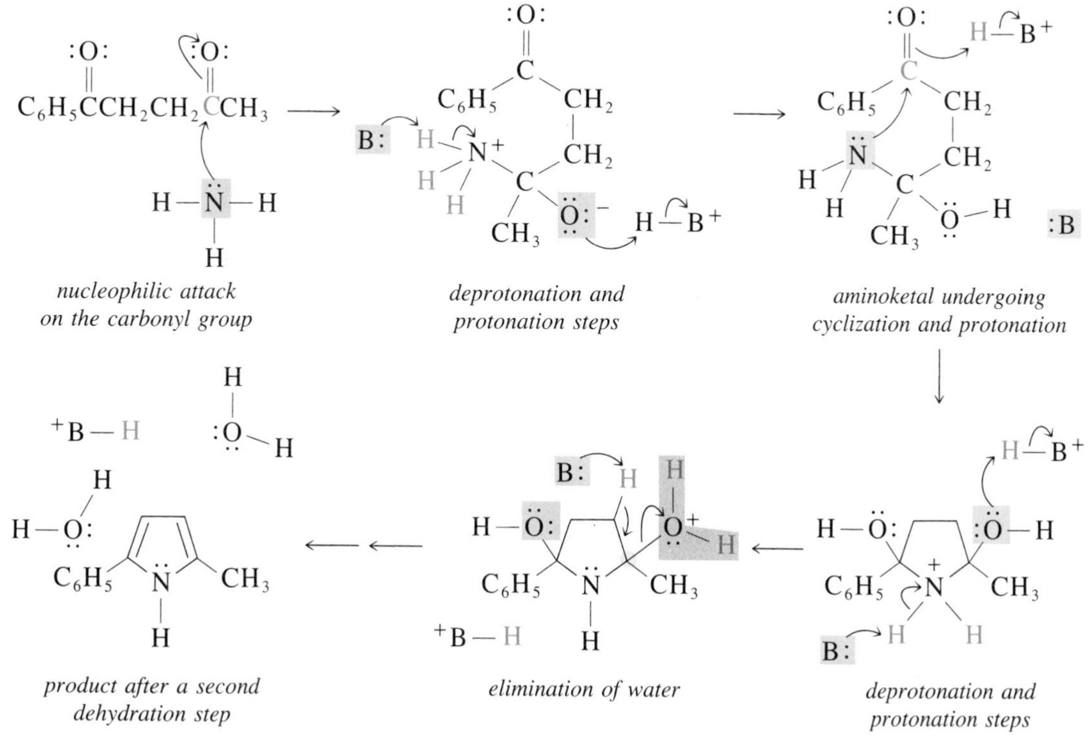
nucleophilic attack on the carbonyl group
deprotonation and protonation steps
aminoketal undergoing cyclization and protonation
deprotonation and protonation steps
elimination of water
product after a second dehydration step

In this and other mechanisms in this chapter, whenever protonation and deprotonation of the same species occur, the two consecutive steps are shown on a single structure.

If 1-phenyl-1,4-pentanedione is heated with phosphorous pentasulfide, which has the molecular formula P_4S_{10}, 2-methyl-5-phenylthiophene is formed.

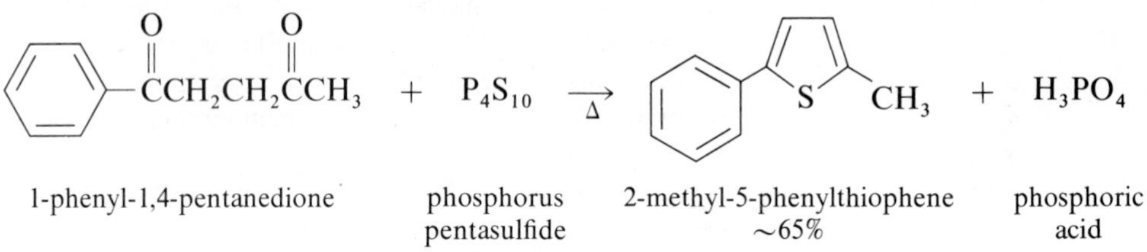
O
O
$CCH_2CH_2CCH_3$
+
P_4S_{10}
Δ
S
CH_3
+
H_3PO_4
1-phenyl-1,4-pentanedione
phosphorus pentasulfide
2-methyl-5-phenylthiophene ~65%
phosphoric acid

The structure of the phosphorus sulfide is too complex to allow a simple mechanism to be written, but the reagent acts by replacing carbonyl groups with thiocarbonyl functions, perhaps starting on the enol of the ketone. Phosphorus, which is pentavalent in the sulfide, is converted to phosphoric acid. The cyclization reaction may be represented as an attack by a nucleophilic sulfur atom on the other carbonyl group in the molecule.

VISUALIZING THE REACTION

Formation of a heterocycle from a thioenol

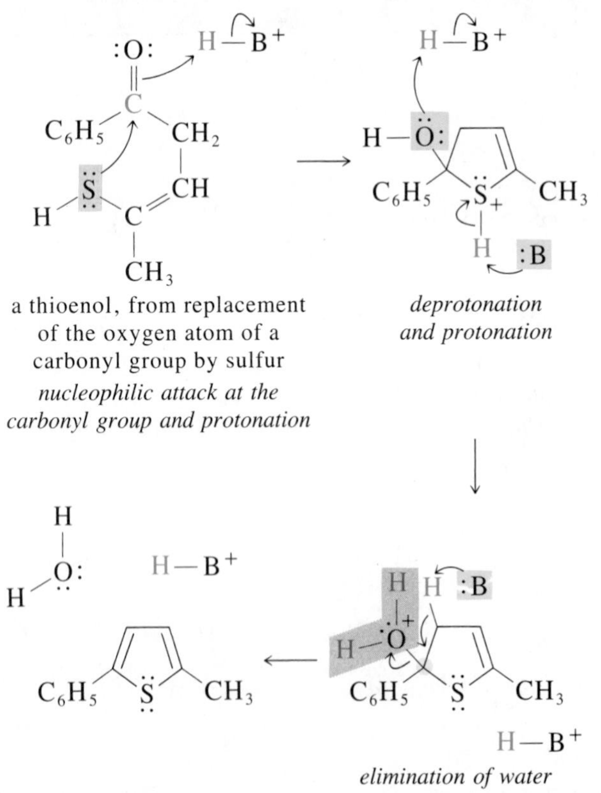
:O:
$H—B^+$
C
C_6H_5
CH_2
S
CH
H
C
CH_3
a thioenol, from replacement of the oxygen atom of a carbonyl group by sulfur
nucleophilic attack at the carbonyl group and protonation
H—O:
C_6H_5
S
CH_3
H
:B
deprotonation and protonation
H
H
:B
H—O
C_6H_5
S
CH_3
$H—B^+$
elimination of water
H
O:
H
$H—B^+$
C_6H_5
S
CH_3

In an analogous reaction, 1-phenyl-1,4-pentanedione cyclizes to 2-methyl-5-phenylfuran when it is heated in acid. If there is no nucleophilic reagent present that can supply a heteroatom other than oxygen, the furan ring is formed.

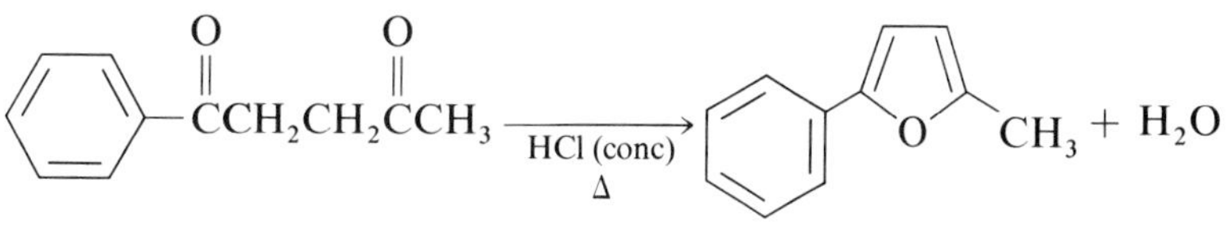

1-phenyl-1,4-pentanedione   2-methyl-5-phenylfuran

In that case, the enol form of one of the ketone functions serves as the nucleophile to form a cyclic hemiketal that then dehydrates.

In all of these reactions, the driving force is the formation of a stable five-membered ring that has aromaticity. Thiophene and pyrrole do not undergo reactions that lead to the opening of the ring. Furan, however, may be regarded as a cyclic hemiacetal that has been dehydrated, and it is hydrolyzed back to a dicarbonyl compound easily when heated with dilute acid.

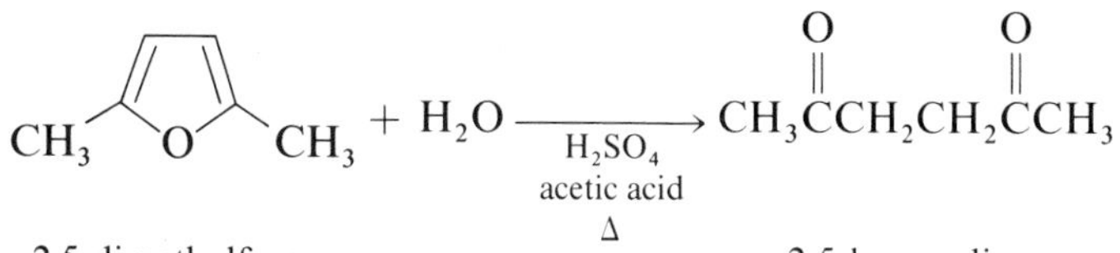

2,5-dimethylfuran   2,5-hexanedione
                                                                            86%

PROBLEM 24.5

Complete the following equations.

(a) $CH_3CCH_2CH_2CCH_3 \xrightarrow[\Delta]{P_4S_{10}}$ (b) $HCCH_2CH_2CH \xrightarrow{HCl}$

(c) $CH_3CCH_2CH_2CCH_3 \xrightarrow[\Delta]{P_4Se_{10}}$ (d) $CH_3CCH_2CH_2CCH_3 \xrightarrow[\Delta]{CH_3NH_2}$

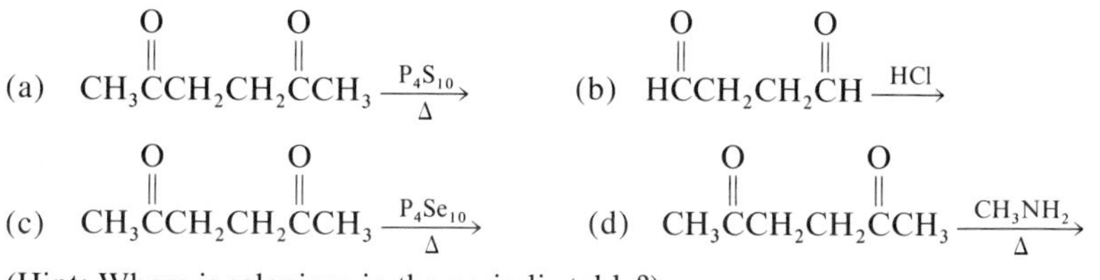

(Hint: Where is selenium in the periodic table?)

B. Five-Membered Heterocycles with Two Heteroatoms

Reagents with two adjacent heteroatoms, such as hydrazine and hydroxylamine, react with 1,3-dicarbonyl compounds to give pyrazoles and isoxazoles. For example, 2,4-pentanedione reacts with hydrazine to form 3,5-dimethylpyrazole and with hydroxylamine to give 3,5-dimethylisoxazole.

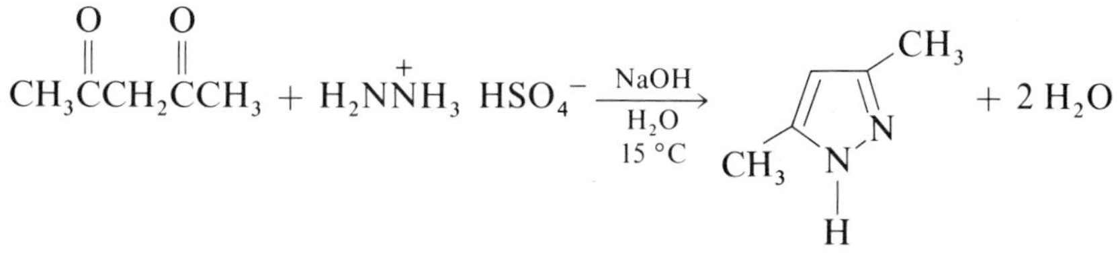

2,4-pentanedione   hydrazine sulfate   3,5-dimethylpyrazole
                                                                            ~80%

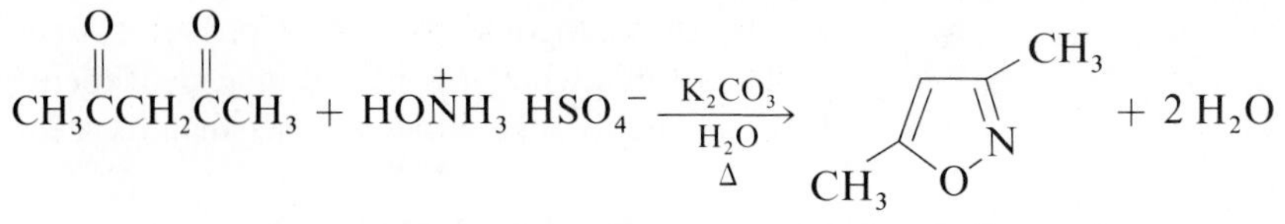

$$CH_3CCH_2CCH_3 + HO\overset{+}{N}H_3\ HSO_4^- \xrightarrow[\Delta]{\underset{H_2O}{K_2CO_3}} \text{3,5-dimethylisoxazole} + 2\ H_2O$$

2,4-pentanedione   hydroxylamine sulfate   3,5-dimethylisoxazole
84%

Hydrazine and hydroxylamine are both basic reagents that are most easily stored and handled as their salts. In the presence of bases such as hydroxide or carbonate ions, the free nucleophiles are generated and react with the carbonyl compounds. For example, 2,4-pentanedione reacts with hydroxylamine to give an oxime (p. 504), which cyclizes. The resulting intermediate readily dehydrates to give the aromatic ring.

PROBLEM 24.6

Write mechanisms for the formation of 3,5-dimethylpyrazole and 3,5-dimethyloxazole from 2,4-pentanedione.

Imidazoles are synthesized from two carbonyl compounds that are joined together with nitrogen atoms from ammonia. For example, when a mixture of 1,2-diphenyl-1,2-ethanedione (also called benzil) and benzaldehyde is heated with ammonium acetate in glacial acetic acid, 2,4,5-triphenylimidazole is obtained.

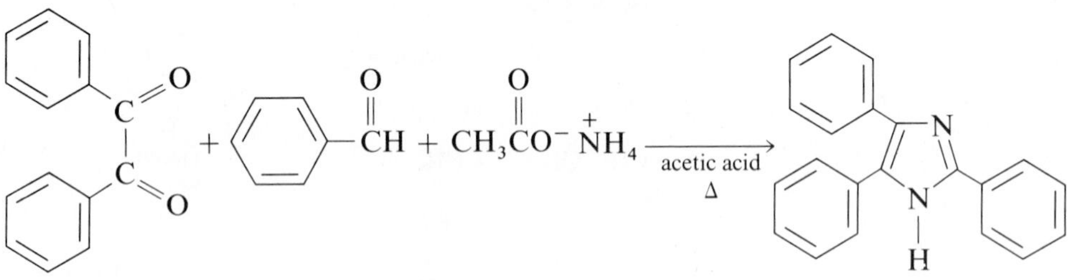

1,2-diphenyl-1,2-   benzaldehyde   ammonium   2,4,5-triphenylimidazole
ethanedione   acetate   90%
benzil

The three carbon atoms in the imidazole ring are the carbon atoms of the carbonyl groups in the organic reagents; the nitrogen atoms come from ammonia, which is in equilibrium with the ammonium ion from the salt in solution with the weak acid acetic acid.

PROBLEM 24.7

What is the major product of each of the following reactions?

(a) $\xrightarrow[\Delta]{\text{HONH}_3^+\ Cl^-,\ \text{NaOH}}$ 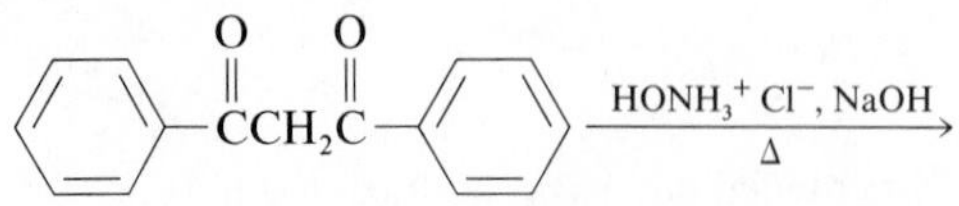

(b) 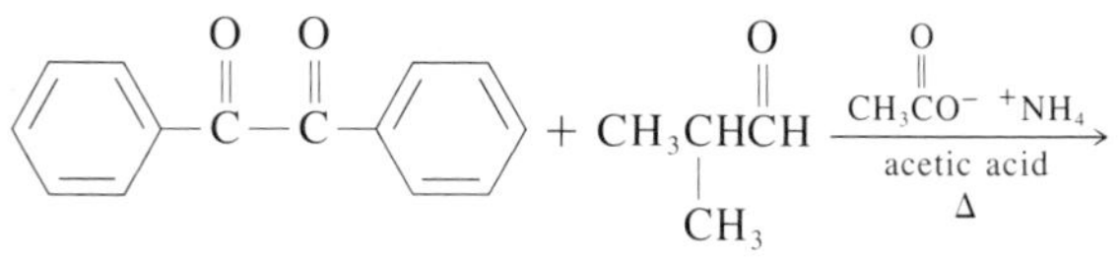

(c) 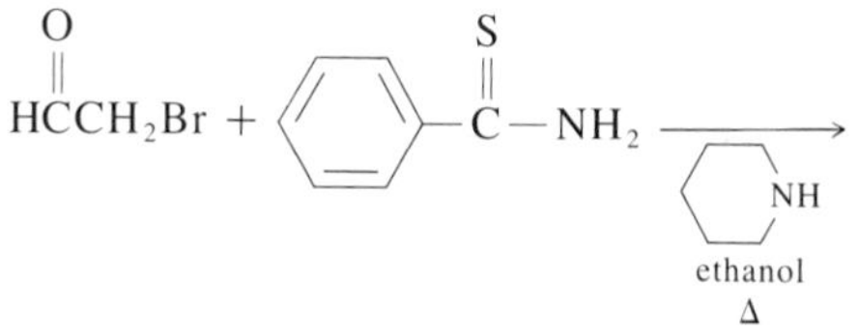

(d) 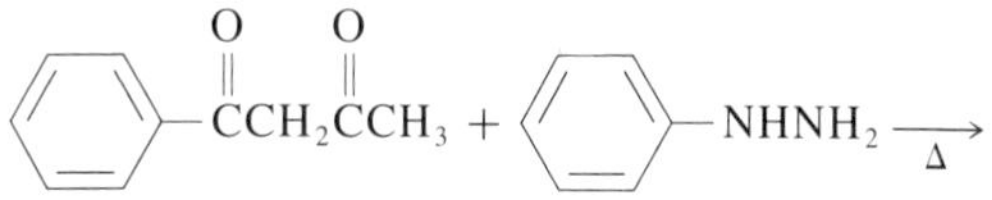

(Hint: Which carbonyl group is more likely to undergo nucleophilic attack first?)

(e) 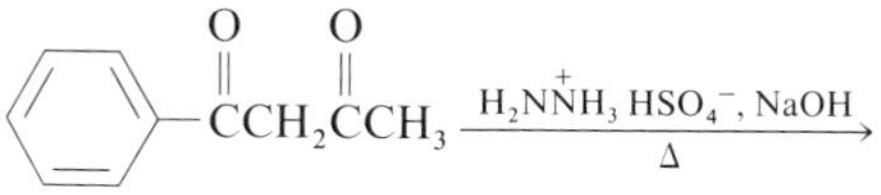

C. Six-Membered Heterocycles

Of the six-membered heterocycles, pyridine and various simple substituted pyridines can be obtained conveniently from natural sources. A great deal of work has been done, however, on the synthesis of pyrimidines because of their importance as drugs and as bases found in nucleic acids. The synthesis of barbiturates from derivatives of diethyl malonate and urea (p. 654) is one application of the most general way to create the pyrimidine ring. A 1,3-dicarbonyl compound is condensed with a reagent that is structurally related to urea. The products formed depend on the substituents present on each fragment. Two examples are given below.

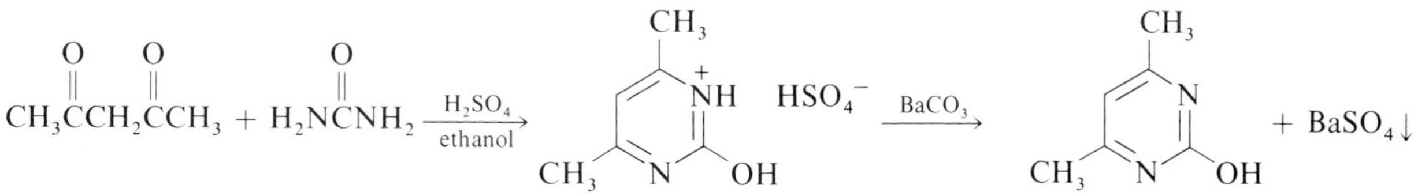

2,4-pentanedione urea 2-hydroxy-4,6-dimethylpyrimidine

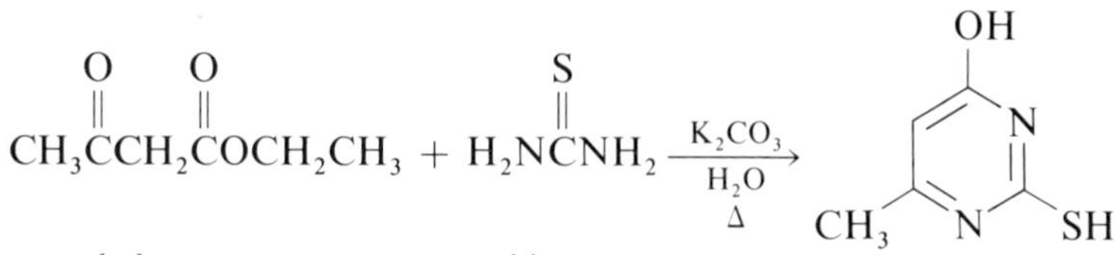

ethyl acetoacetate thiourea 4-hydroxy-2-mercapto-6-methylpyrimidine 95%

In 2,4-pentanedione, the carbonyl groups do not have good leaving groups bonded to them. The reaction with urea proceeds by condensation of the amino groups of urea with the carbonyl groups of the ketone and tautomerization to the aromatic system.

VISUALIZING THE REACTION

Formation of a pyrimidine from a 1,3-dicarbonyl compound and urea

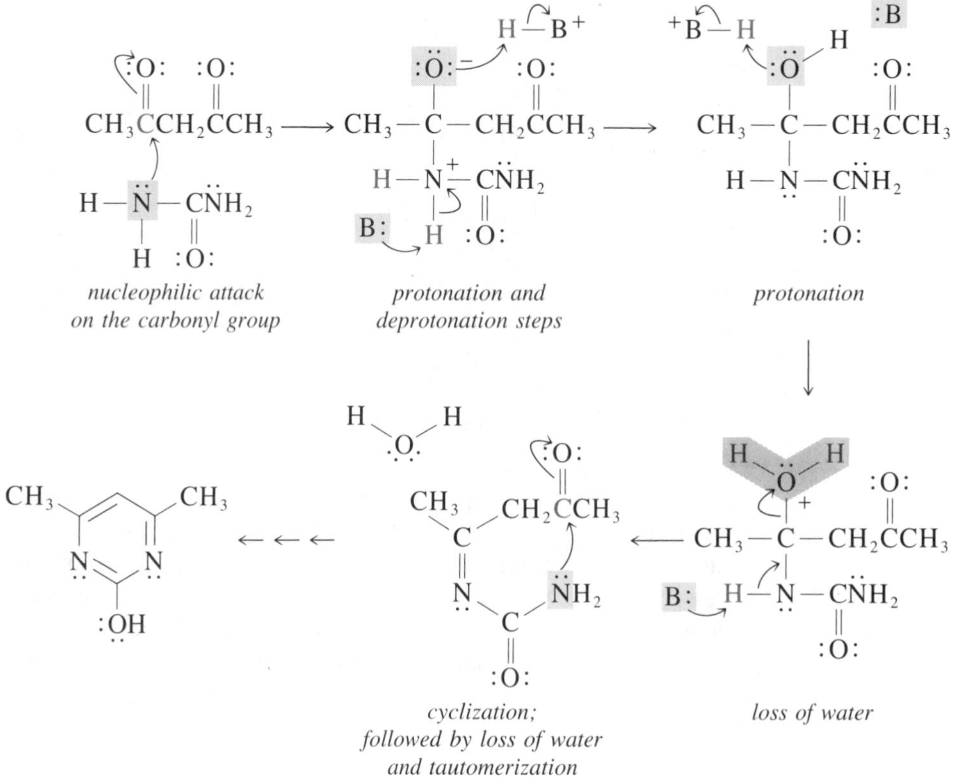
nucleophilic attack on the carbonyl group
protonation and deprotonation steps
protonation
loss of water
cyclization; followed by loss of water and tautomerization

When an ester group provides one of the carbonyl functions, as in diethyl malonate or ethyl acetoacetate, an alkoxide serves as a leaving group, providing a different pathway for the condensation reaction. The carbonyl group of the ester is retained in the pyrimidine. The carbon atoms that were part of the carbonyl groups of the ester or of urea (or thiourea) are marked by hydroxyl (or thiol) substituents on the fully aromatic form of the ring after tautomerization.

Study Guide Concept Map 24.2

PROBLEM 24.8

Complete the following equations.

(a) $$CH_3\overset{O}{\overset{\|}{C}}CH_2\overset{O}{\overset{\|}{C}}OCH_2CH_3 + H_2N\overset{O}{\overset{\|}{C}}NH_2 \xrightarrow[\text{ethanol}]{\text{HCl}}$$

(b) 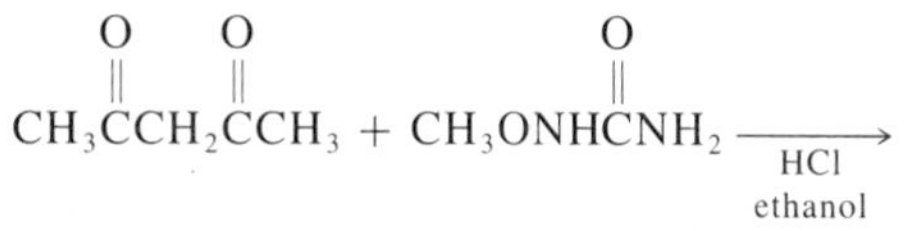
CH3CCH2CCH3 + CH3ONHCNH2
HCl
ethanol

(c) 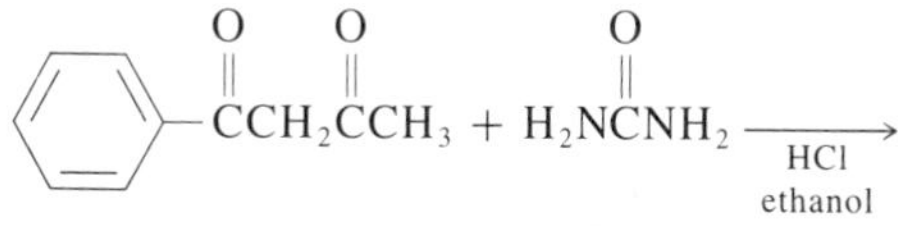
CCH2CCH3 + H2NCNH2
HCl
ethanol

24.5 SUBSTITUTION REACTIONS OF HETEROCYCLIC COMPOUNDS

A. Electrophilic Aromatic Substitution Reactions of Five-Membered Heterocycles

The five-membered aromatic heterocycles are all more reactive toward electrophiles than benzene is. The reactivity of the ring in these compounds resembles that of phenol in the ease with which substitution takes place. For example, thiophene reacts with bromine to give a mixture of bromothiophenes.

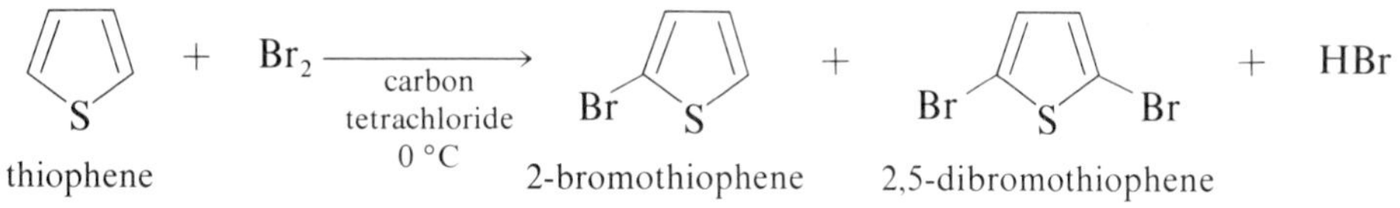
+ Br2
carbon
tetrachloride
0 °C
+
+ HBr
thiophene
2-bromothiophene
2,5-dibromothiophene

Furan, on the other hand, reacts with bromine by a 1,4-addition reaction, an indication of the relatively low aromaticity of this heterocycle. When the reaction is carried out in methanol, the product that is isolated is formed by solvolysis of the intermediate dibromide.

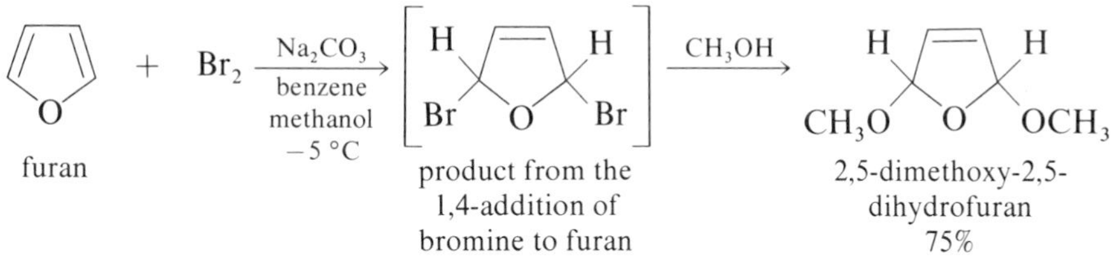
+ Br2
Na2CO3
benzene
methanol
−5 °C
CH3OH
furan
product from the
1,4-addition of
bromine to furan
2,5-dimethoxy-2,5-
dihydrofuran
75%

Five-membered aromatic heterocycles also undergo nitration reactions. However, because the mixture of nitric acid and sulfuric acid used for the nitration of benzene (p. 784) destroys the heterocycles, a milder nitrating agent prepared by dissolving nitric acid in acetic anhydride is used. Acetic anhydride acts as a dehydrating agent to create nitronium ions from nitric acid. Substitution in thiophene takes place chiefly at one of the carbon atoms adjacent to the heteroatom.

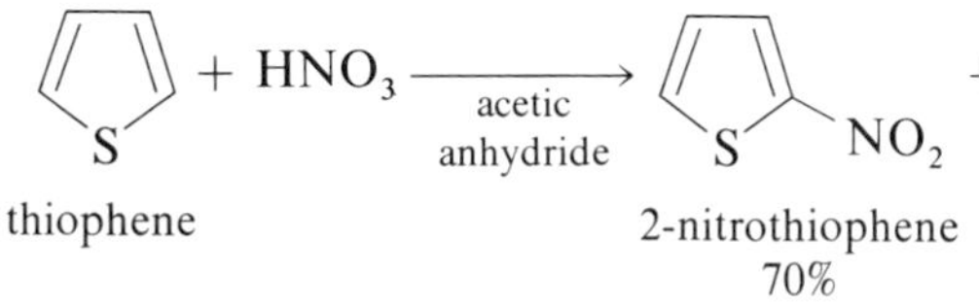
+ HNO3
acetic
anhydride
+ H2O
thiophene
2-nitrothiophene
70%

The regioselectivity of the substitution reactions of these heterocycles can be rationalized by the same kind of reasoning that was used to explain the directing effects of substituents on benzene (p. 792). The resonance contributors for the intermediates that would result from attack of the nitronium ion at carbon 2 and at carbon 3 of the thiophene ring are compared below.

VISUALIZING THE REACTION

Regioselectivity of electrophilic substitution on thiophene

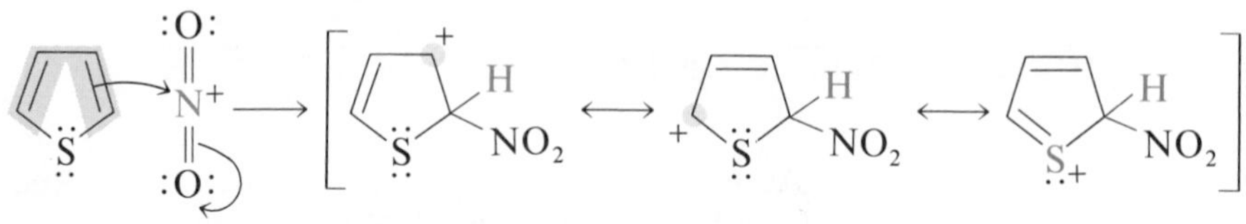
:O:
N+
:O:
S
H
NO2
S+

resonance contributors for the intermediate formed by attack of nitronium ion at carbon 2

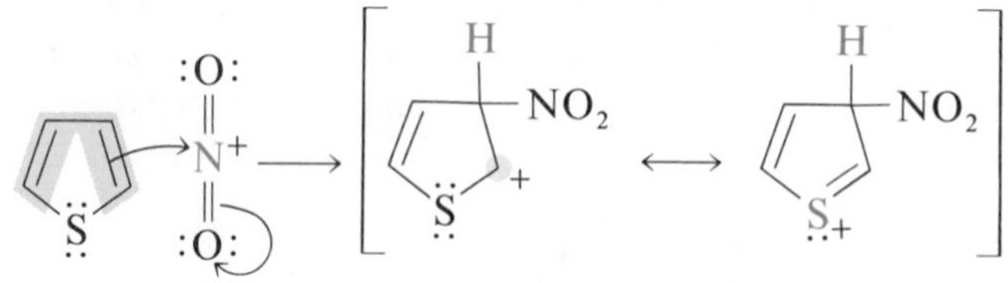
:O:
N+
:O:
S
H
NO2
S+

resonance contributors for the intermediate formed by attack of nitronium ion at carbon 3

The carbocation formed when nitronium ion attacks carbon 2 of thiophene is more stable than the other one because greater delocalization of charge is possible for it. The reaction thus follows the path leading through that intermediate and the lower-energy transition state corresponding to its formation. The intermediate carbocation loses a proton easily, and the product has the stable aromatic ring.

Friedel-Crafts acylation reactions are another type of aromatic substitution reaction (p. 802) that can be carried out with five-membered heterocycles. Thiophene, for example, reacts with benzoyl chloride in the presence of aluminum chloride to give phenyl 2-thienyl ketone.

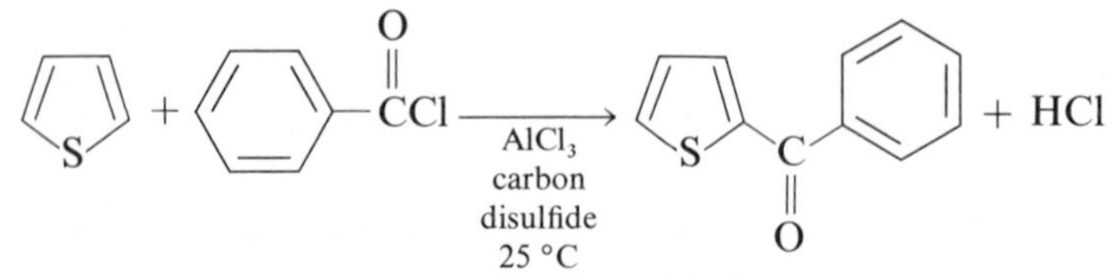
O
CCl
S
AlCl3
carbon
disulfide
25 °C
C
O
+ HCl

thiophene   benzoyl chloride   2-benzoylthiophene
phenyl 2-thienyl ketone
90%

Substituted aromatic heterocycles usually undergo the reactions typical of the functional groups that are present. For example, in a typical free-radical substitution reaction (p. 833), a hydrogen atom on the methyl group of 3-methylthiophene is replaced by bromine when *N*-bromosuccinimide is used.

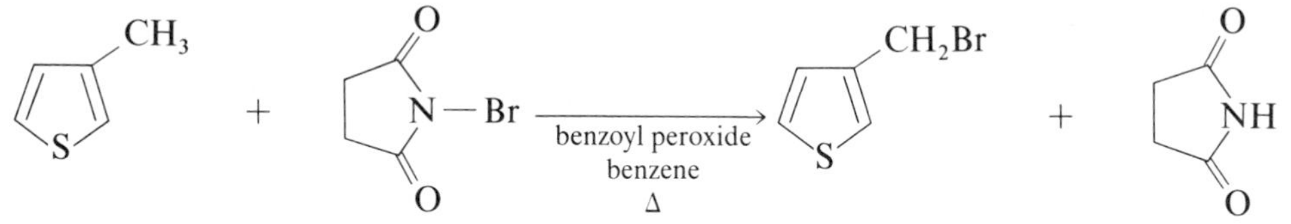

3-methylthiophene N-bromosuccinimide 3-(bromomethyl)thiophene 75% succinimide

The reactions shown above demonstrate that five-membered aromatic heterocycles behave much like benzene and its derivatives in electrophilic substitution reactions. Substitution occurs preferentially at carbon 2. The presence of the heteroatom makes the heterocyclic compound more reactive than benzene, so some of the reaction conditions must be modified. Substituents on a heterocyclic ring react in ways typical of their functional groups.

PROBLEM 24.9

Write structural formulas for all intermediates and products designated by letters in the following equations.

(a) 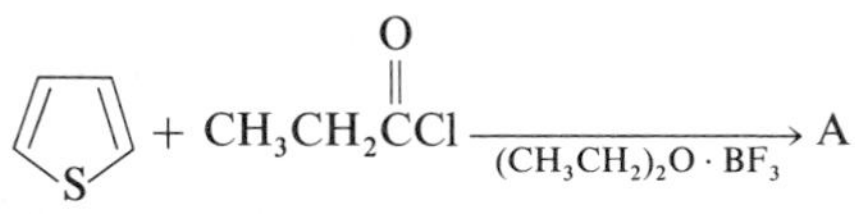

(b) 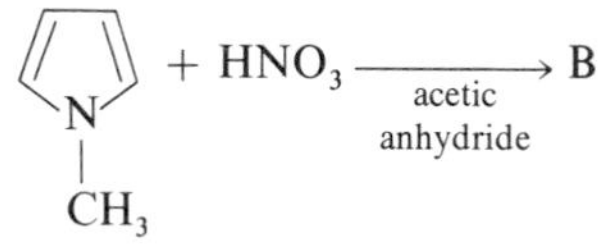

(c) 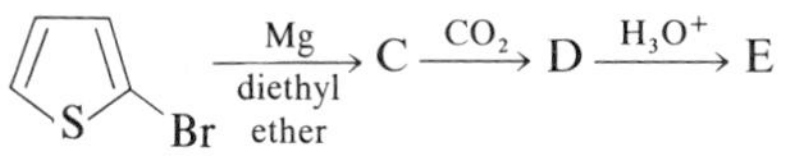

(d) 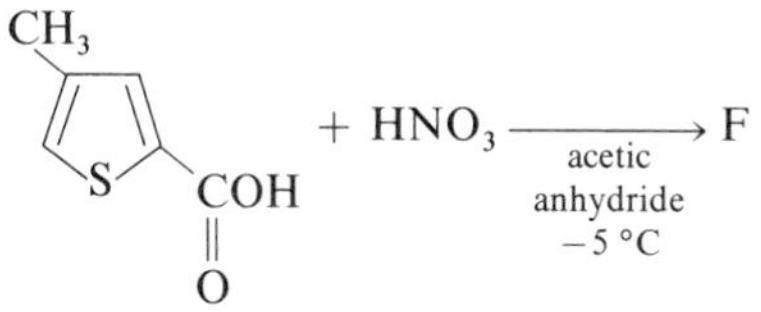

(e) 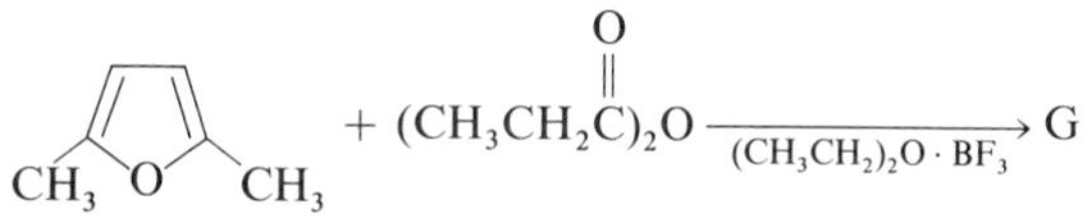

(f) 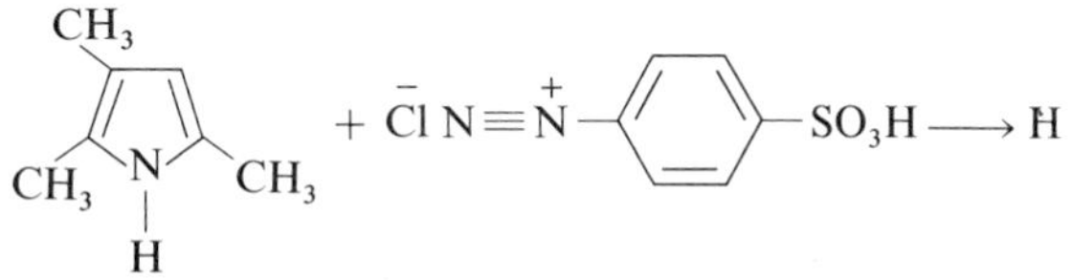

(g) 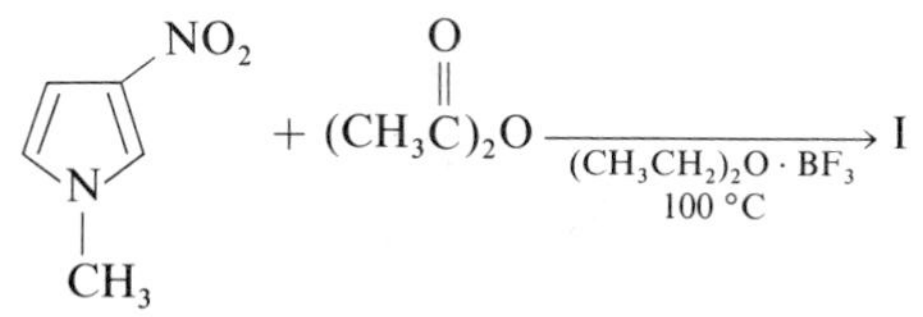

(h) 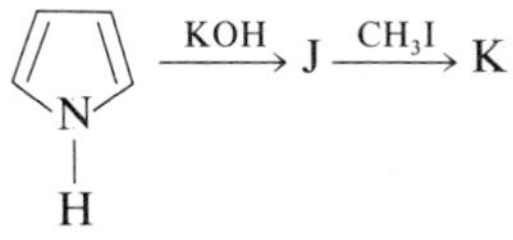

(i) 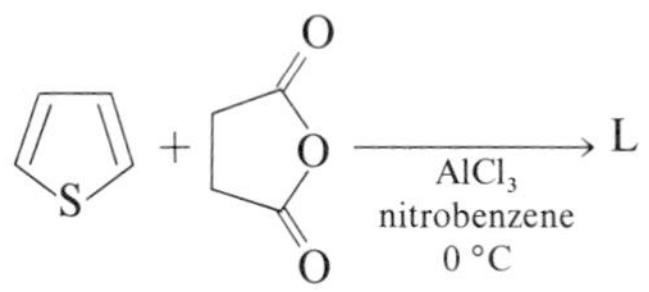

(j) 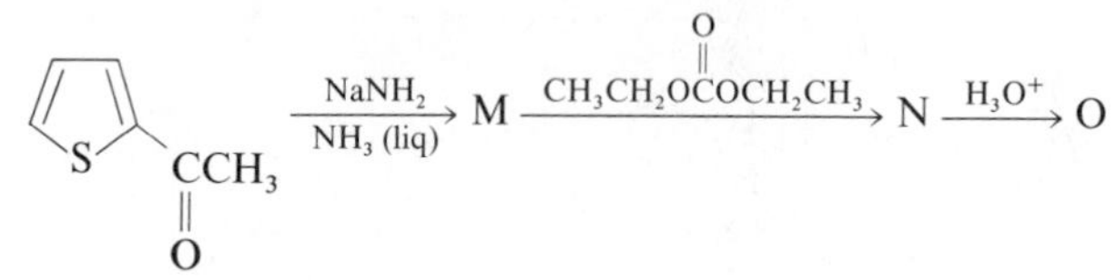

(k) 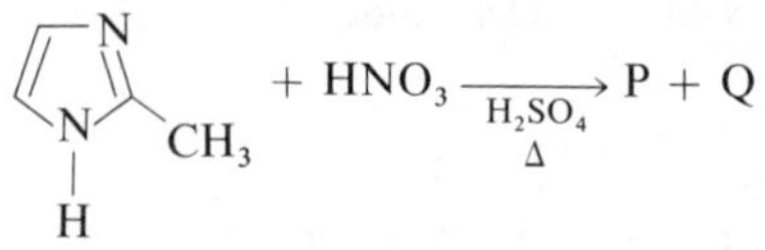

PROBLEM 24.10

When 2-acetyl-1-methylpyrrole is treated with nitric acid in acetic anhydride at 0 °C, two nitration products are obtained. Predict what their structures are by reasoning about the relative stabilities of the intermediates formed. Predict which isomer is the major product.

B. Aromatic Substitution Reactions of Pyridine

Although the five-membered heterocycles are much more reactive toward electrophilic substitution than benzene is, pyridine is much less reactive than benzene. Having a more electronegative nitrogen atom instead of one of the carbon atoms in the ring decreases the availability of electrons and makes it more difficult for an electrophilic attack to take place. The conditions that are necessary to bring about substitution on the pyridine ring are often more severe than those required to carry out multiple substitutions on nitrobenzene. An example is the nitration of pyridine, which takes place at 330 °C.

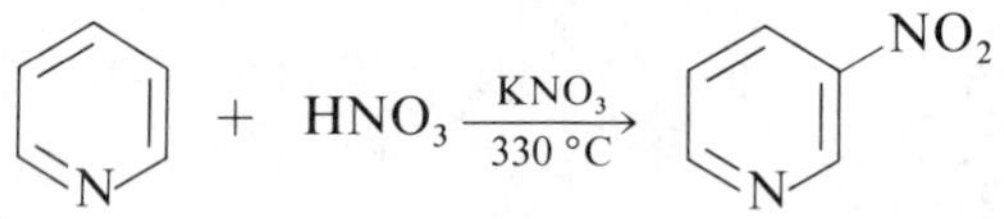

pyridine — 3-nitropyridine 15%

Even at this temperature only a fraction of the pyridine molecules are nitrated.

Substitution takes place preferentially at carbon 3 of the pyridine ring. The nitrogen atom in the ring deactivates the positions that are ortho and para to it more than it deactivates the meta positions. In this respect, it has the same effect as a nitro group on benzene. Part of the effect arises because pyridine is protonated or coordinates with Lewis acids under the conditions necessary for most substitution reactions. For example, in the nitration reaction, the species undergoing substitution is the pyridinium ion, not pyridine itself.

pyridine — nitric acid — pyridinium nitrate

The resonance contributors for the intermediates that arise when an electrophile attacks at carbon 2 or 3 of the pyridine ring are shown below.

VISUALIZING THE REACTION

Regioselectivity of electrophilic substitution on pyridine

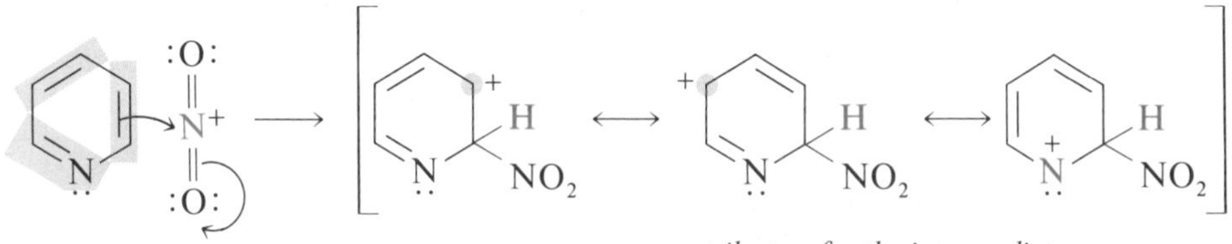

resonance contributors for the intermediate formed by electrophilic attack at carbon 2

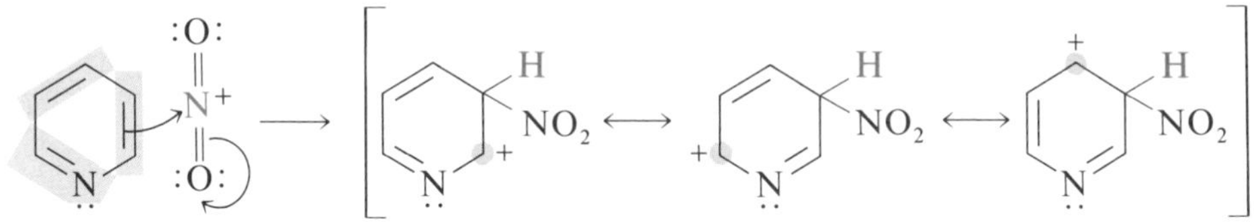

resonance contributors for the intermediate formed by electrophilic attack at carbon 3

Comparing the resonance contributors shows that the intermediate formed by electrophilic attack at carbon 3 is more stable than the other one. Delocalization of the positive charge in the intermediate from attack at carbon 2 puts the charge on the nitrogen atom, which will already bear a positive charge due to prior protonation or coordination with a Lewis acid. The resonance contributor with the positive charge on nitrogen also shows nitrogen, an element that is more electronegative than carbon, having only a sextet of electrons. The delocalization of charge for the intermediate from attack at carbon 3 does not require such a high-energy contributor.

Although pyridine itself is resistant to electrophilic attack, pyridines with electron-donating substituents react more easily. The amino group is particularly effective in activating the ring to substitution. 2-Aminopyridine, which is readily prepared from pyridine (p. 1006), is brominated easily.

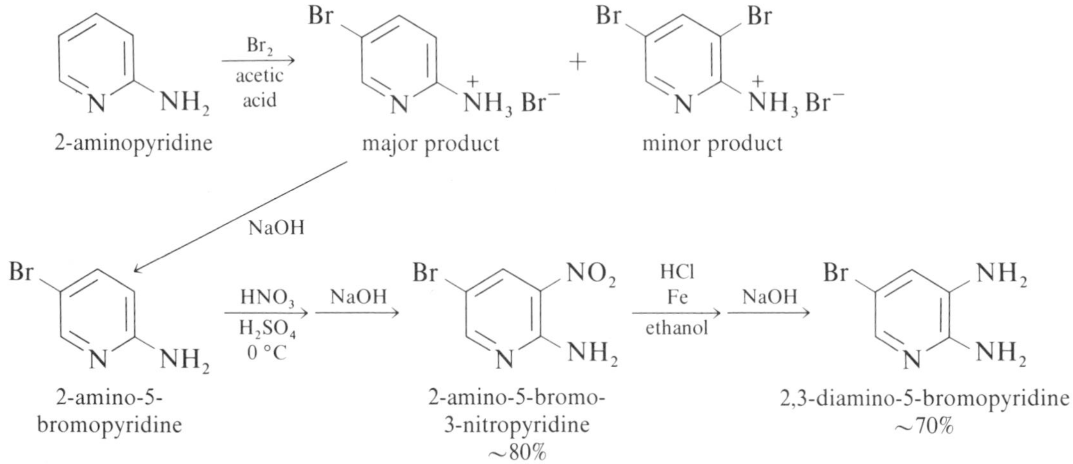
Br₂
acetic acid
2-aminopyridine
major product
minor product
NaOH
2-amino-5-bromopyridine
HNO₃
H₂SO₄
0 °C
NaOH
2-amino-5-bromo-3-nitropyridine
~80%
HCl
Fe
ethanol
NaOH
2,3-diamino-5-bromopyridine
~70%

The second substituent (the bromine) takes the position para to the activating group and meta to the deactivating nitrogen atom of the ring. The subsequent substitution reaction also occurs easily. (Compare these reaction conditions with those for the nitration of pyridine, p. 1004.) Substituents on the pyridine ring can be transformed into other functional groups by the same reactions used to make benzene derivatives. For example, the aromatic nitro compound is reduced to an aromatic amine with a metal and hydrochloric acid (p. 902).

PROBLEM 24.11

Write structural formulas for the products represented by letters in the following reactions.

(a)

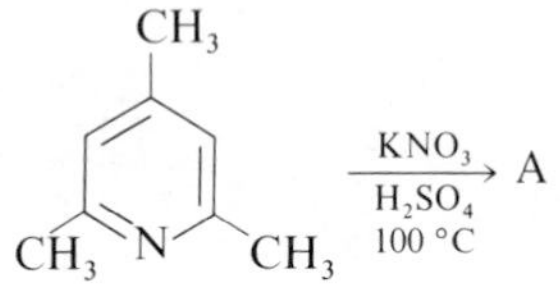

(b)

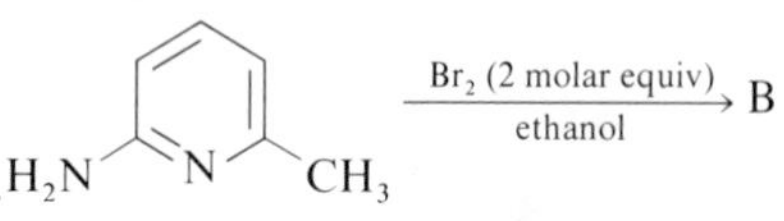

(c)

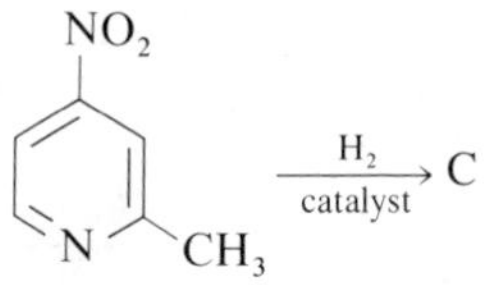

(d)

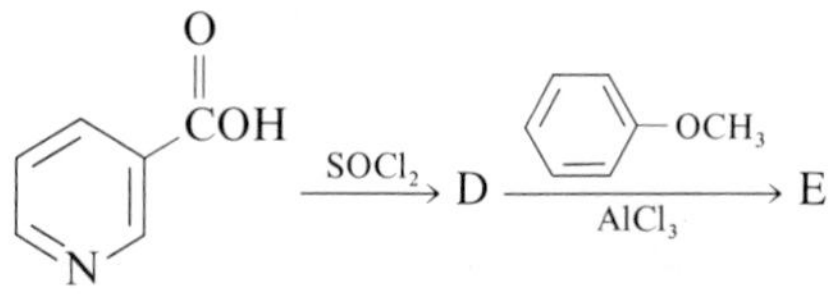

Pyridine derivatives differ most from benzene derivatives in the ease with which they react in nucleophilic aromatic substitution reactions. Pyridines with a halogen at position 2 or 4 have the kind of reactivity seen for 2,4-dinitrohalobenzenes (p. 957). For example, 2-bromopyridine reacts with aniline to give 2-(*N*-phenylamino)pyridine.

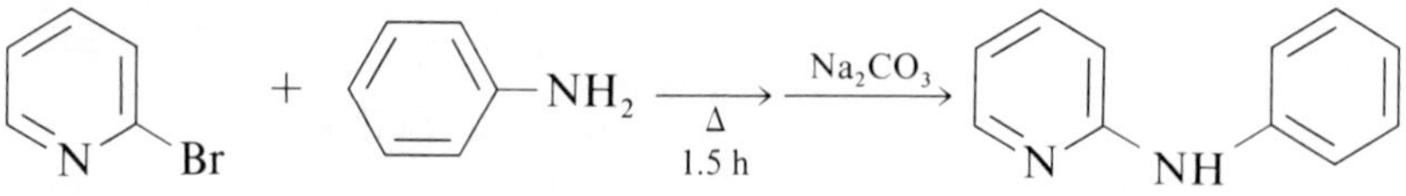

2-bromopyridine aniline 2-(*N*-phenylamino)pyridine

PROBLEM 24.12

Propose a detailed mechanism for the reaction of 2-bromopyridine with aniline.

Pyridine itself is vulnerable to nucleophilic attack. The most useful reaction of this type is the formation of 2-aminopyridine when pyridine is treated with sodium amide.

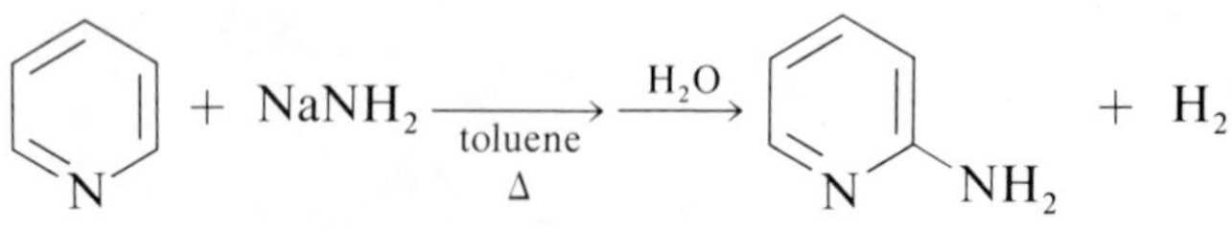

pyridine sodium amide 2-aminopyridine ~70% hydrogen

The reaction starts with an attack by the amide anion on carbon 2 of the ring, followed by loss of a hydride ion, which deprotonates the amino group. The product, 2-aminopyridine, is formed when water is added to the reaction mixture.

VISUALIZING THE REACTION

Nucleophilic substitution on pyridine

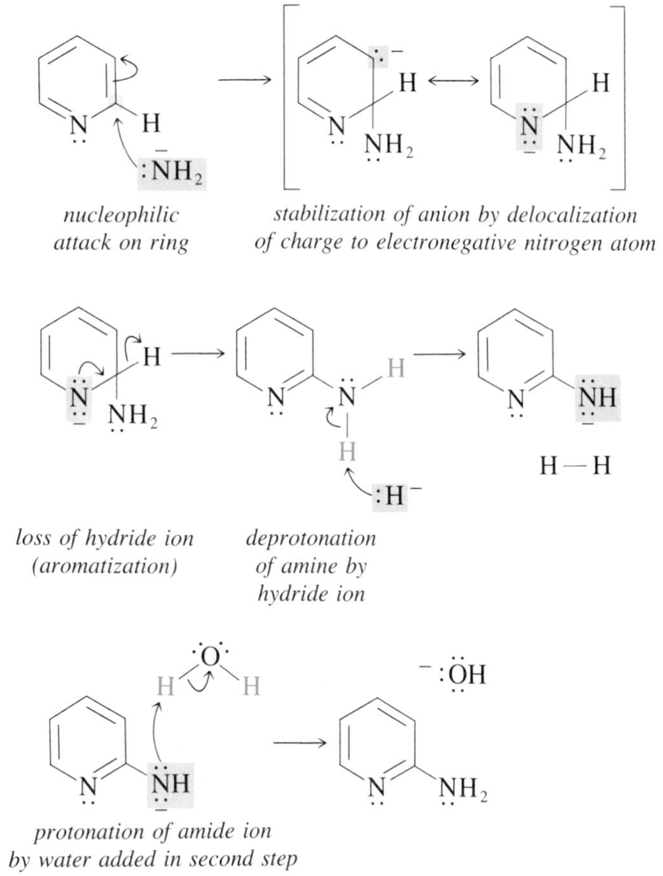

N
H
:NH2
H
N
NH2
H
N
NH2
nucleophilic attack on ring
stabilization of anion by delocalization of charge to electronegative nitrogen atom
H
N
NH2
N
N
H
H
:H−
N
NH
H—H
loss of hydride ion (aromatization)
deprotonation of amine by hydride ion
O
H
H
−:OH
N
NH
N
NH2
protonation of amide ion by water added in second step

Aminopyridines prepared in this way are valuable as reagents for synthesizing more highly substituted pyridines, as was shown above.

The reactivity of pyridine in electrophilic and nucleophilic aromatic substitution reactions is predictable on the basis of its electronic character. Electrophilic substitution occurs predominantly at carbon 3 and nucleophilic substitution at carbon 2. In most cases, the reactions of substituents on pyridine resemble those of the same substituents on benzene.

Study Guide Concept Map 24.3

PROBLEM 24.13

Pyridine reacts with phenyllithium to give 2-phenylpyridine and lithium hydride in a reaction analogous to the reaction of pyridine with amide anion.

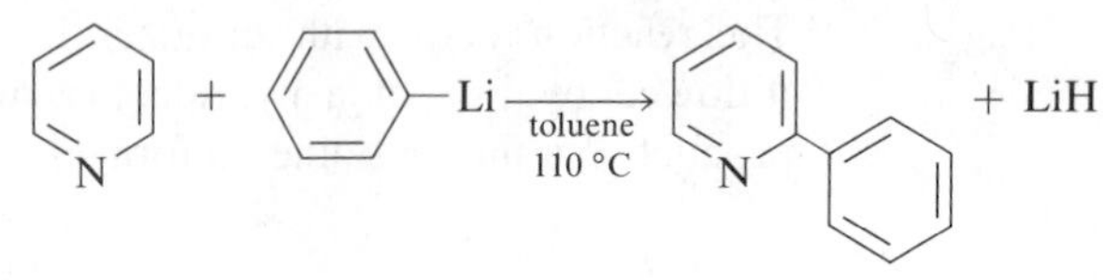

Suggest a mechanism for this reaction. How does it start? What is the leaving group? Why is such a leaving group feasible in this reaction and in the reaction of amide anion with pyridine?

C. Oxidation of Alkyl Side Chains on Pyridine

Pyridine is highly resistant to oxidation, unlike the five-membered aromatic heterocycles, which are destroyed by strong oxidizing agents. For example, pyridinium chlorochromate, a complex of chromium trioxide and hydrochloric acid with pyridine, is used as an oxidizing agent for alcohols that are sensitive to acid or to oxidize a primary alcohol to an aldehyde (p. 441). The ease with which an aromatic ring is oxidized is related to the electron density in the ring. The same factor influences the ease with which electrophilic aromatic substitution occurs. Thus, the five-membered aromatic heterocycles undergo substitution rapidly and are also susceptible to oxidation. Pyridine is not substituted easily and resists oxidation.

Methylpyridines, for example, are oxidized by hot potassium permanganate to the corresponding carboxylic acids (p. 963). The reaction is illustrated by the conversion of 4-methylpyridine to 4-pyridinecarboxylic acid.

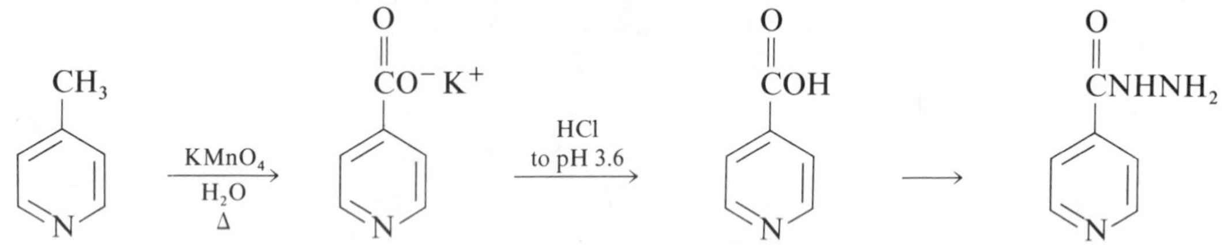

4-methylpyridine — potassium 4-pyridinecarboxylate — 4-pyridinecarboxylic acid, isonicotinic acid, 76% — isonicotinic acid hydrazide, isoniazid, *a drug used in the treatment of tuberculosis*

4-Pyridinecarboxylic acid, also known as isonicotinic acid, has medical usefulness; its hydrazide is a potent drug used in the treatment of tuberculosis.

PROBLEM 24.14

Write structural formulas for all compounds represented by letters in the following equations.

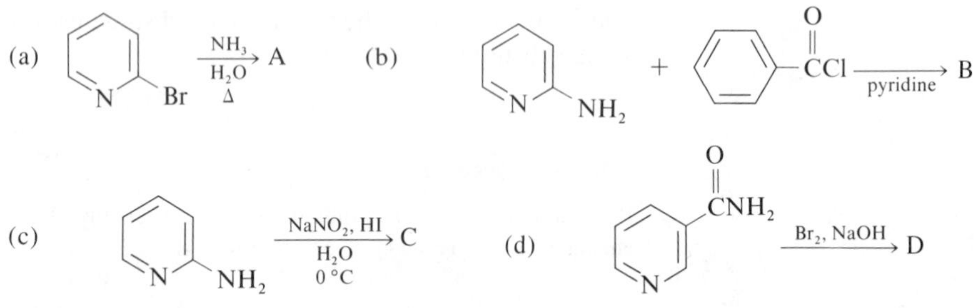

(e) 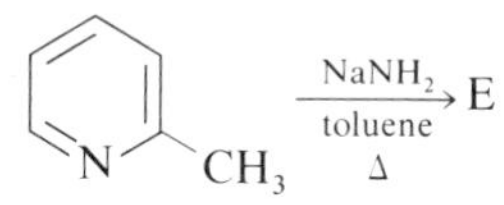

(f) 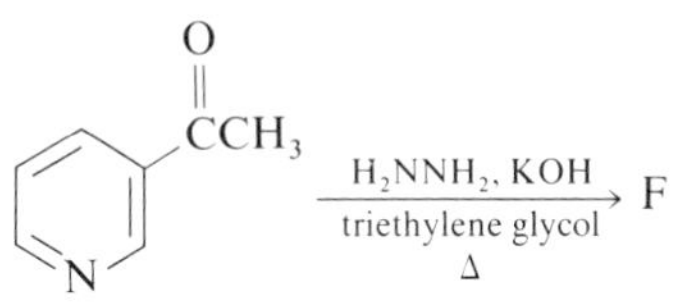

(g) 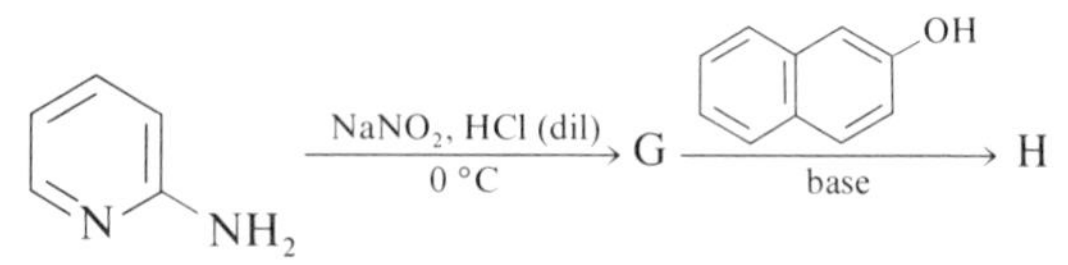

24.6
PYRIMIDINES, PURINES, AND PYRIDINES OF BIOLOGICAL SIGNIFICANCE

A. Pyrimidines and Purines

The chemistry of pyrimidines and purines is of interest because these heterocyclic rings are found in deoxyribonucleic acids (DNA) and ribonucleic acids (RNA), the complex molecules that transmit genetic information and mediate the synthesis of proteins in cells. The structures and chemical properties of several pyrimidines and purines determine what interactions are possible between different strands of DNA and between molecules of DNA and RNA. These interactions are believed to be largely responsible for the storage and transmission of genetic information in cells. The ways in which these heterocycles influence the structure and function of DNA and RNA will be discussed in greater detail in the next section.

The pyrimidine and purine bases that are found in DNA and RNA are shown below.

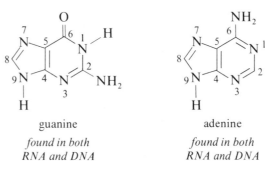

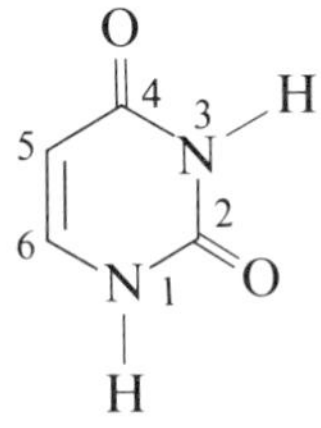

uracil

found in RNA

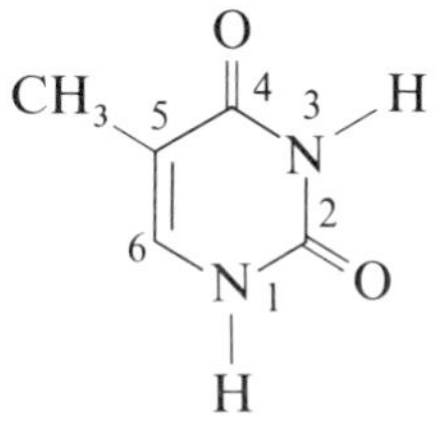

thymine

found in DNA

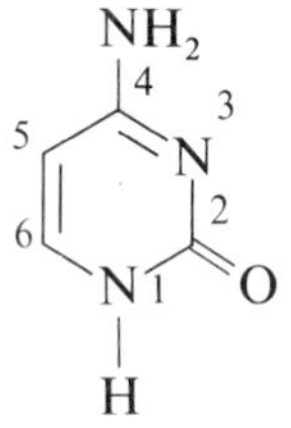

cytosine

found in both RNA and DNA

pyrimidine bases

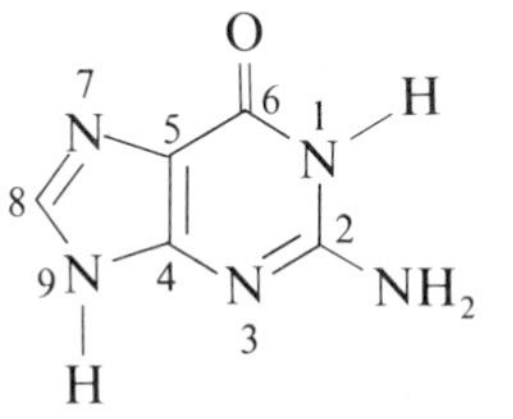

guanine

found in both RNA and DNA

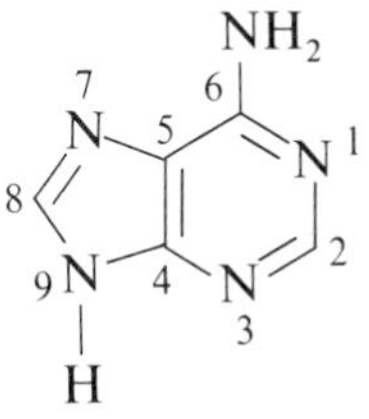

adenine

found in both RNA and DNA

purine bases

An examination of the structures of the purines shows that they contain an imidazole ring fused to a pyrimidine ring. The numbering system shown is the one commonly used for these compounds in biological systems. The tautomeric forms of the bases shown above are the ones that are important in water at pH 7, the conditions under which these bases are found in nucleic acids. The question of tautomerism is important because the exact location of hydrogen atoms on oxygen and nitrogen atoms determines the way the bases interact with each other by hydrogen bonding (p. 1015).

A number of other purines, shown in the structures below, occur in nature, including xanthine, hypoxanthine, and uric acid. Caffeine, found in coffee, tea, cola nuts, and cocoa, theobromine from cocoa, and theophylline from tea, are methylated xanthines.

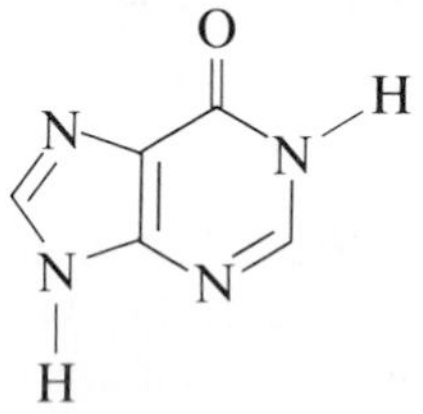
hypoxanthine

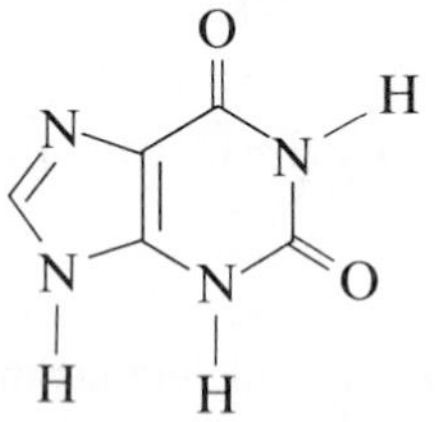
xanthine

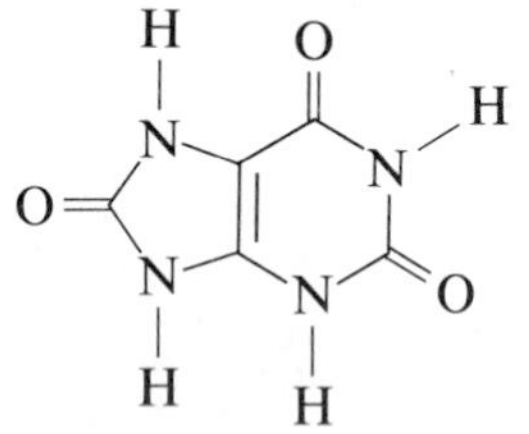
uric acid

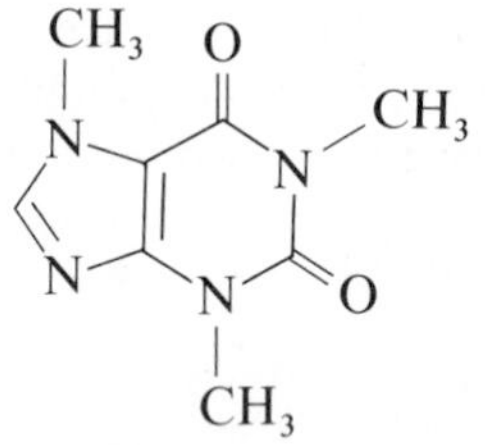
caffeine

found in tea, coffee, maté leaves, guarana paste, cola nuts

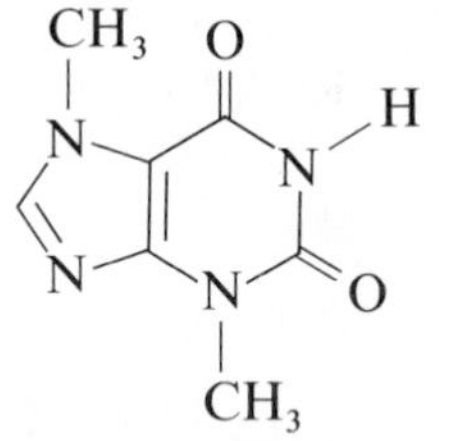
theobromine

principal alkaloid of cacao bean, also in cola nuts and tea

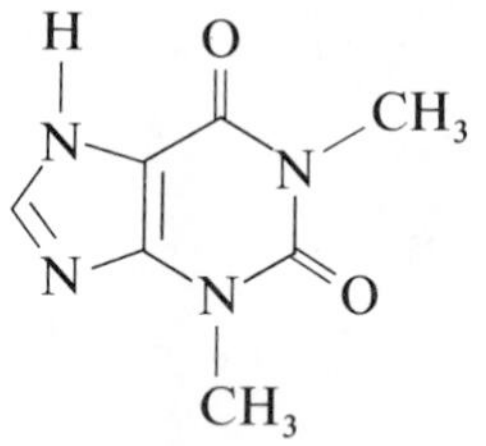
theophylline

small amounts in tea

Caffeine is a powerful stimulant of the central nervous system. Theophylline is a milder stimulant of the central nervous system and also a relaxant of smooth muscles. Theobromine does not have much activity as a stimulant. Hypoxanthine, xanthine, and uric acid are products of the metabolism of the purine bases adenine and guanine. The disease called gout results from the faulty metabolism and excretion of uric acid.

The purine and pyrimidine bases occupy a central place in the metabolic processes of cells because they are involved in regulating protein synthesis. Their importance has led chemists to design medications that incorporate their ring systems and mimic their structures. Researchers hope that such compounds will disrupt metabolic processes in cancer cells. Many compounds of this type have been synthesized and tested. Two that have been used in the treatment of cancer are 5-fluorouracil and 6-mercaptopurine.

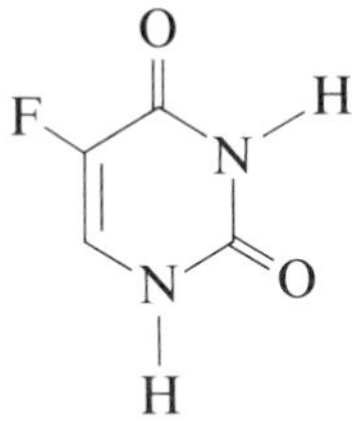

5-fluorouracil

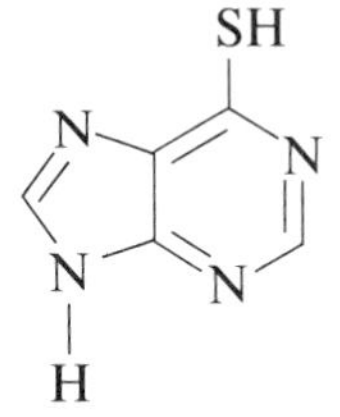

6-mercaptopurine

5-Fluorouracil was designed to resemble thymine structurally. It interacts with enzymes that function in the synthesis of RNA and DNA and prevents normal metabolic processes from taking place. 6-Mercaptopurine resembles adenine in structure, except that a nucleophilic sulfur atom replaces a nucleophilic nitrogen atom. This compound also acts in cells by blocking the synthesis of nucleic acids. Cancer cells grow in an uncontrolled way compared with normal cells. The designers of antitumor compounds hoped that the metabolic processes in cancer cells would therefore be more vulnerable to disruptive drugs than those of normal cells. To some extent this is true, but all the medications that are used in the treatment of cancer are also extremely toxic to normal body cells.

The roles that purine and pyrimidine bases play in biological systems give importance to their chemical reactions. The chemistry of pyrimidines resembles that of pyridines, except that the second nitrogen atom in the aromatic ring makes it even more resistant to electrophilic attack. For example, nitration of the ring in a pyrimidine occurs only when there are two ring-activating substituents. Shown below, uracil, which is 2,4-dihydroxypyrimidine in one of its tautomeric forms, can be nitrated.

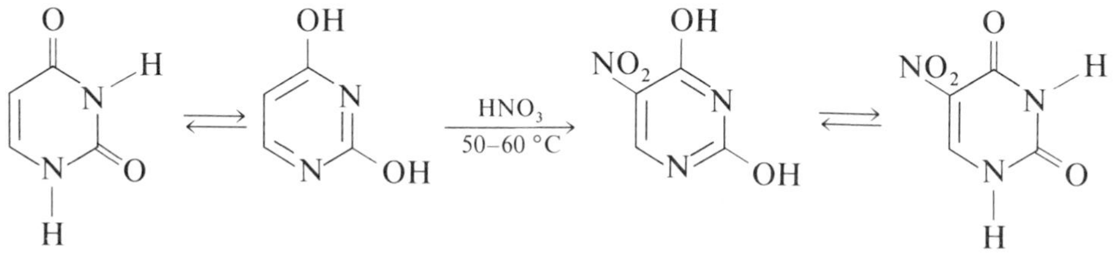
HNO₃
50–60 °C

uracil in its keto form — uracil in its enol form — 2,4-dihydroxy-5-nitropyrimidine 5-nitrouracil 99%

Note that the substituent attacks position 5 of the pyrimidine ring in an electrophilic substitution reaction. The basic arguments presented concerning substitution reactions on pyridine can be used to show that attack at the position meta to the two nitrogen atoms of the ring and ortho and para to the hydroxyl groups gives the most stable intermediates (p. 1005).

Aminopyrimidines and aminopurines react with dilute nitrous acid to give the corresponding hydroxy compounds (p. 945). This reaction is believed to be responsible for the damage that nitrous acid (and therefore nitrites) does to DNA, giving rise to mutations in organisms such as yeast. For example, cytosine is converted to uracil by the action of nitrous acid, as illustrated by the conversion of the nucleoside (p. 1013) cytidine to uridine.

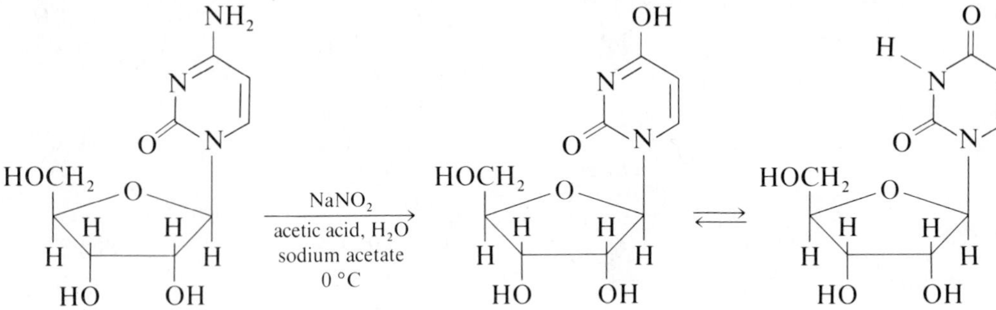
NaNO$_2$
acetic acid, H$_2$O
sodium acetate
0 °C

cytidine uridine

Cytidine and uridine, shown above, are ribonucleosides, but similar reactions occur with the aminopyrimidines and purines found in deoxyribonucleic acids. In the reactions that are involved in the transmission of genetic information, cytosine is paired with guanine by precise hydrogen-bonding interactions. Uracil interacts with adenine in a similar way. The conversion of cytosine to uracil by a chemical reaction changes the way the bases pair and therefore changes the genetic code (p. 1017).

Functional groups on pyrimidines and purines show essentially the same reactivity as they do on a benzene ring.

PROBLEM 24.15

Write structural formulas for all intermediates and products designated by letters in the following equations.

(a)

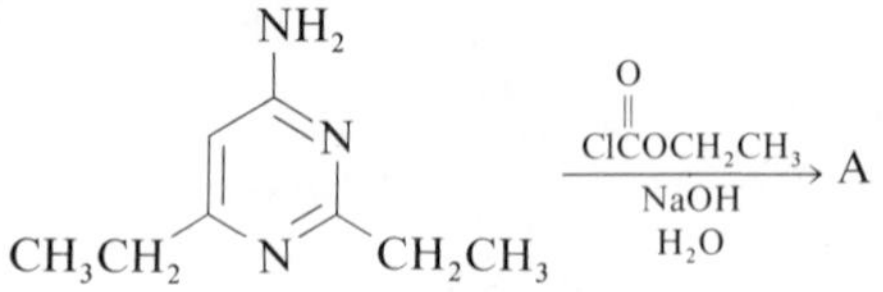

(b)

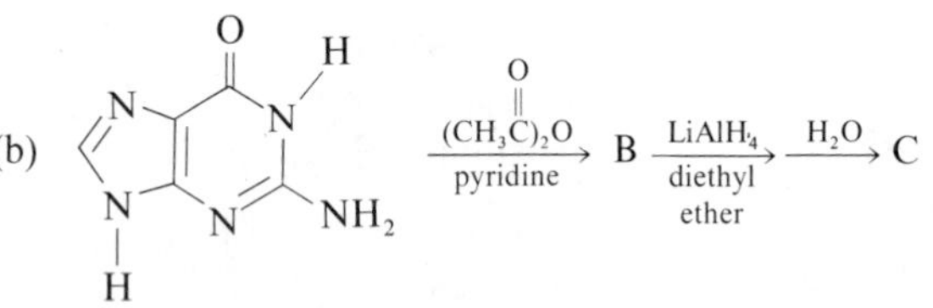

(c)

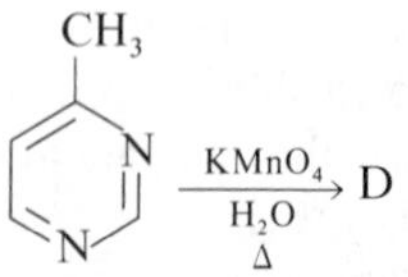

(d)

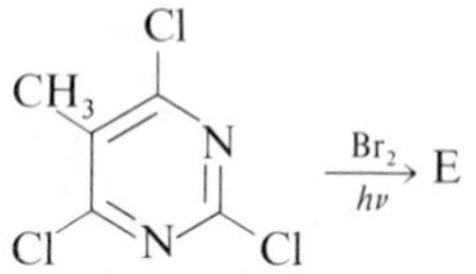

(e)

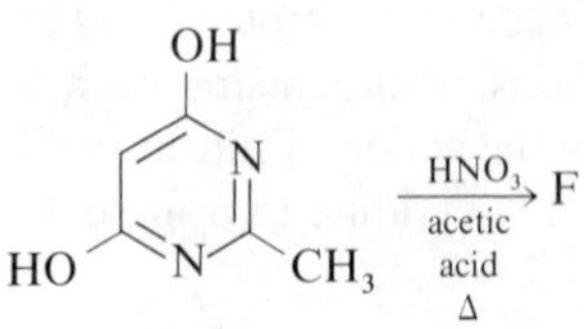

(f)

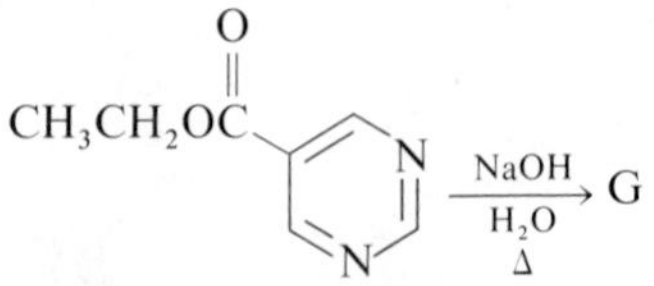

(g) 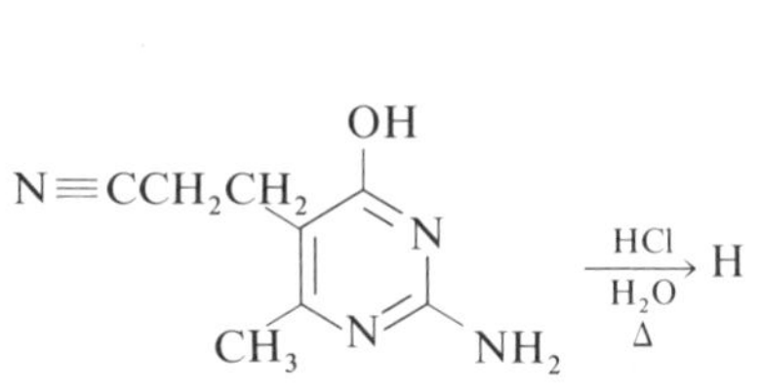

(h) 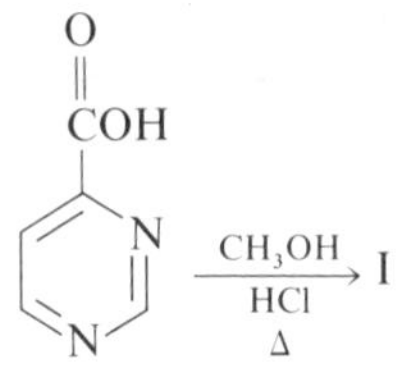

(i) 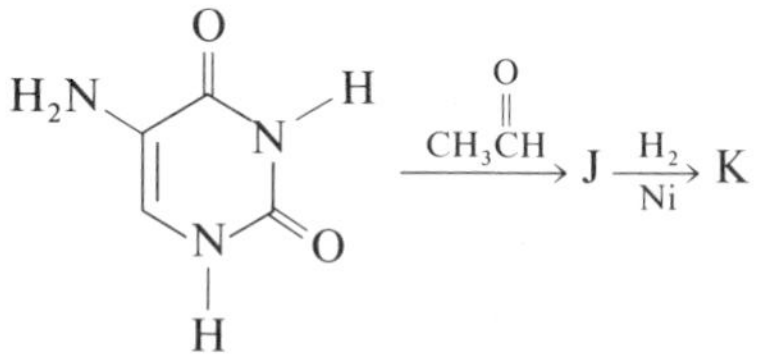

(j) 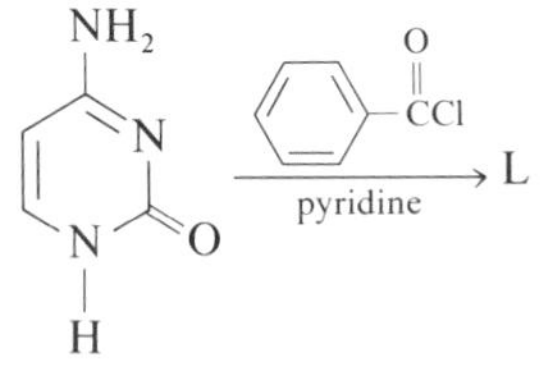

B. Ribonucleic Acids and Deoxyribonucleic Acids

Some of the largest molecules known, often having molecular weights higher than a million, are the nucleic acids found in the nuclei of cells. Nucleic acids have backbones made up of sugar units, which are five-membered rings. These rings are held together as the phosphate esters of the hydroxyl groups at carbons 3 and 5. In **ribonucleic acids** (abbreviated RNA), the sugar is D-ribose; in **deoxyribonucleic acids** (abbreviated DNA), it is D-2-deoxyribose (p. 1051).

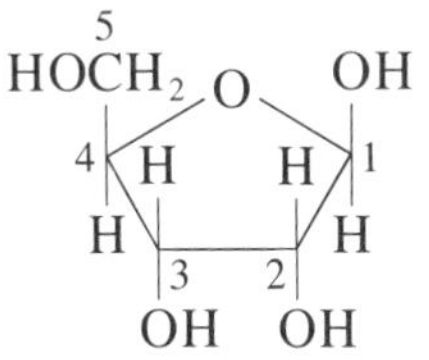

β-D-ribose

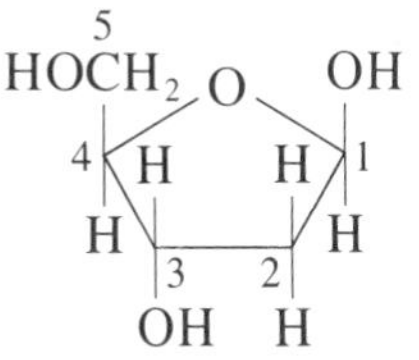

β-D-2-deoxyribose

Each sugar unit is also bonded at carbon 1 to a nitrogen heterocycle, resulting in compounds called **nucleosides.** The phosphate esters of nucleosides are known as **nucleotides.** RNA and DNA are polymers made up of nucleotide units and are therefore also called **polynucleotides.** The structural formulas and names of the nucleotides commonly found in RNA and DNA are shown below.

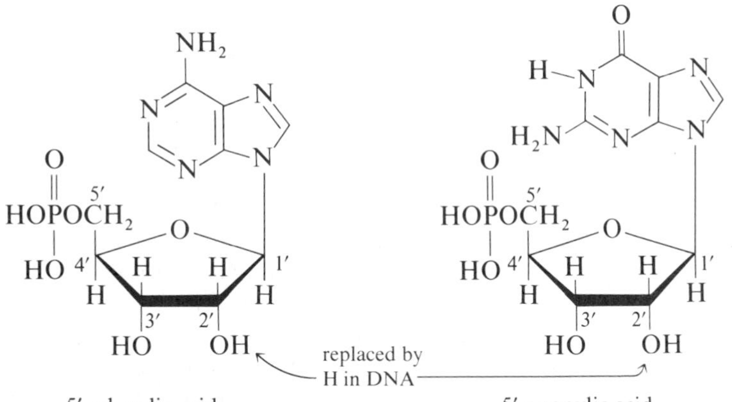
replaced by
H in DNA

5′-adenylic acid
adenosine 5′-phosphate
5′-AMP

5′-guanylic acid
guanosine 5′-phosphate
5′-GMP

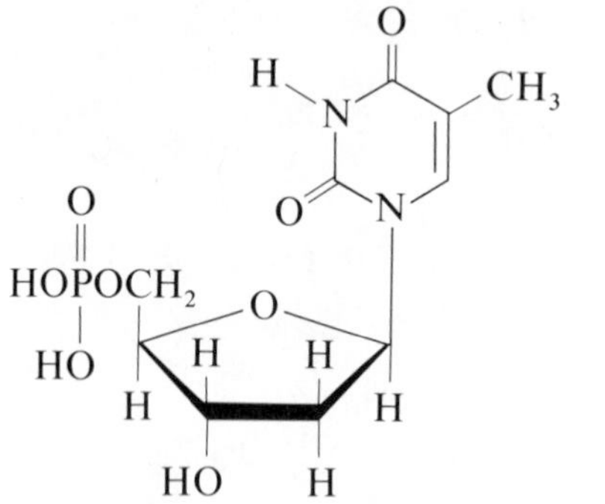

5′-thymidylic acid (DNA)
thymidine 5′-phosphate
5′-TMP

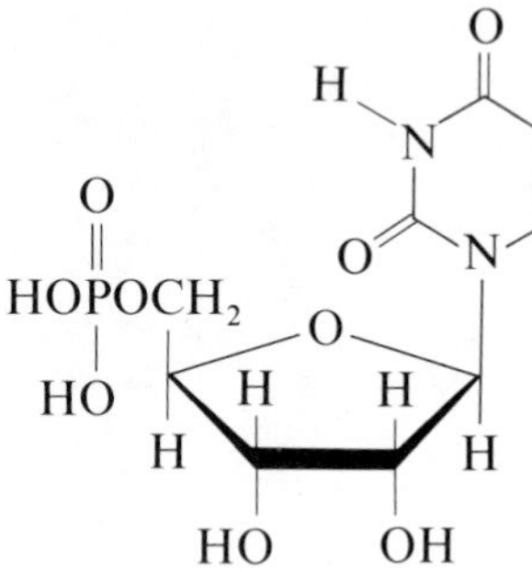

5′-uridylic acid (RNA)
uridine 5′-phosphate
5′-UMP

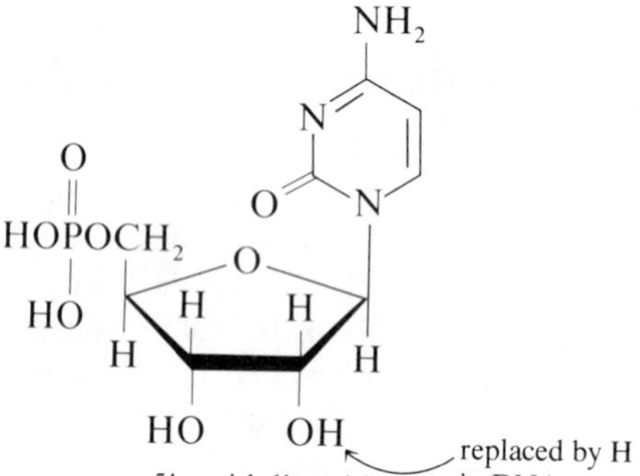
replaced by H in DNA

5′-cytidylic acid
cytidine 5′-phosphate
5′-CMP

The history of the determination of the structures and functions of RNA and DNA is a long one. It starts with the isolation of a material rich in phosphorus from the nuclei of pus cells and from the sperm of salmon by Friedrich Miescher in Germany in 1868. This material was first called nuclein and then nucleic acid. Early in this century, the components of nucleic acids—heterocyclic bases, sugars, and phosphoric acids—were identified. In the 1920s, the structures of the individual nucleotides were determined. Beginning in 1939, the British chemist Sir Alexander Todd investigated the structures of the nucleotides and showed that they are linked in polynucleotides as phosphoric acid esters at the hydroxyl groups of carbons 3 and 5 of the sugar units. For his contribution to the determination of the structures of RNA and DNA, he received the Nobel Prize in 1957.

Another large step was taken in the 1950s, when Erwin Chargaff of Columbia University investigated deoxyribonucleic acids from a large number of sources, including viruses, bacteria, molds, yeast, insects, plants, and mammals. He found that nucleic acids from different organisms contain differing amounts of the purine and pyrimidine bases, but in each case the molar ratio of adenine to thymine and of guanine to cytosine in deoxyribonucleic acid is close to 1.00. He reasoned that the constancy of this relationship in such a variety of organisms could not be accidental; it suggested some association of adenine with thymine and of guanine with cytosine.

In the 1950s, other researchers were making progress in developing techniques for determining the structures of complex organic molecules by x-ray diffraction. The patterns that develop on a photographic plate when a crystal is exposed to x rays can be interpreted in terms of the locations of the atoms that diffract the x rays, and thus give clues to molecular structure. Dorothy Crowfoot Hodgkins of the United Kingdom is one of the pioneers in this area of research. She received the Nobel Prize in 1964 for her work on the structure of Vitamin B_{12}. By the 1950s, the technique was advanced enough that it was being used to investigate the structures of proteins and nucleic acids. The best x-ray crystallographic data were obtained by Rosalind Franklin at King's College, London. She worked with the crystalline sodium salt of deoxyribonucleic acid from the thymus gland of a calf. The diffraction patterns that are seen for this DNA are best explained if it is assumed that the molecule has a helical structure with a distance of 3.4 Å between the different nucleotide units and a diameter of 20 Å.

James Watson and Francis Crick, working in the Cavendish Laboratory at Cambridge University, in England, recognized that a DNA molecule consisting of a single helical strand having these dimensions would not have the density that had already been determined for it. Very shortly after they saw the x-ray data obtained by Franklin, they worked out the idea that in the DNA molecule two strands are twisted around each other in a double helix, arranged so that the more hydrophobic nitrogen bases are inside the helix and the hydrophilic sugar and phosphate groups are on the outside. They proposed, in a paper published in 1953 along with papers by Maurice Wilkins, also at King's College, and Franklin describing the x-ray crystallographic data, that the two strands of the double helix were held together by hydrogen bonding between the bases adenine and thymine and the bases guanine and cytosine. Thus, although there is no regularity to the order in which the bases appear on the backbone of either chain, there is a one-to-one correspondence of the number of adenine units to thymine units and of guanine units to cytosine units for the two strands taken together. Adenine and thymine are said to constitute one **base pair,** and guanine and cytosine make up the other one. The hydrogen bonding responsible for the pairing of nucleotides containing these bases is shown schematically below.

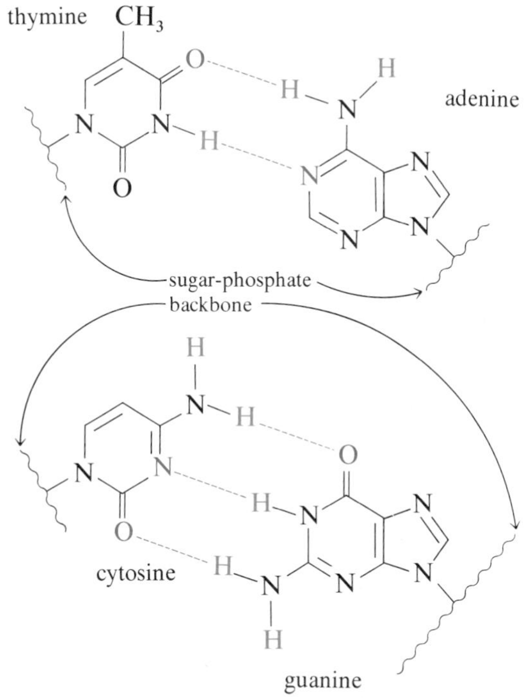
thymine
CH_3
adenine
sugar-phosphate
backbone
cytosine
guanine

the hydrogen bonding between nucleotides that is responsible for the pairing of the bases

The diameter of the double helix is such that a purine base on one strand must be matched with a pyrimidine base on the other. The flat rings of the bases lie parallel to each other in a stack up the axis of the helix. The sugar units to which the bases are attached and the phosphate linkages between the sugars together form a twisting ribbon on the outside of the helix. A polynucleotide has directionality. An unbonded phosphate group is present at carbon 5 of the final sugar unit at one end of the chain, and another is observed at carbon 3 of the analogous unit at the other end of the molecule. The two strands of the double helix are polynucleotide chains, headed in opposite directions. A fragment of a double helix and a view of its overall shape appear in Figure 24.1 to show some of these relationships.

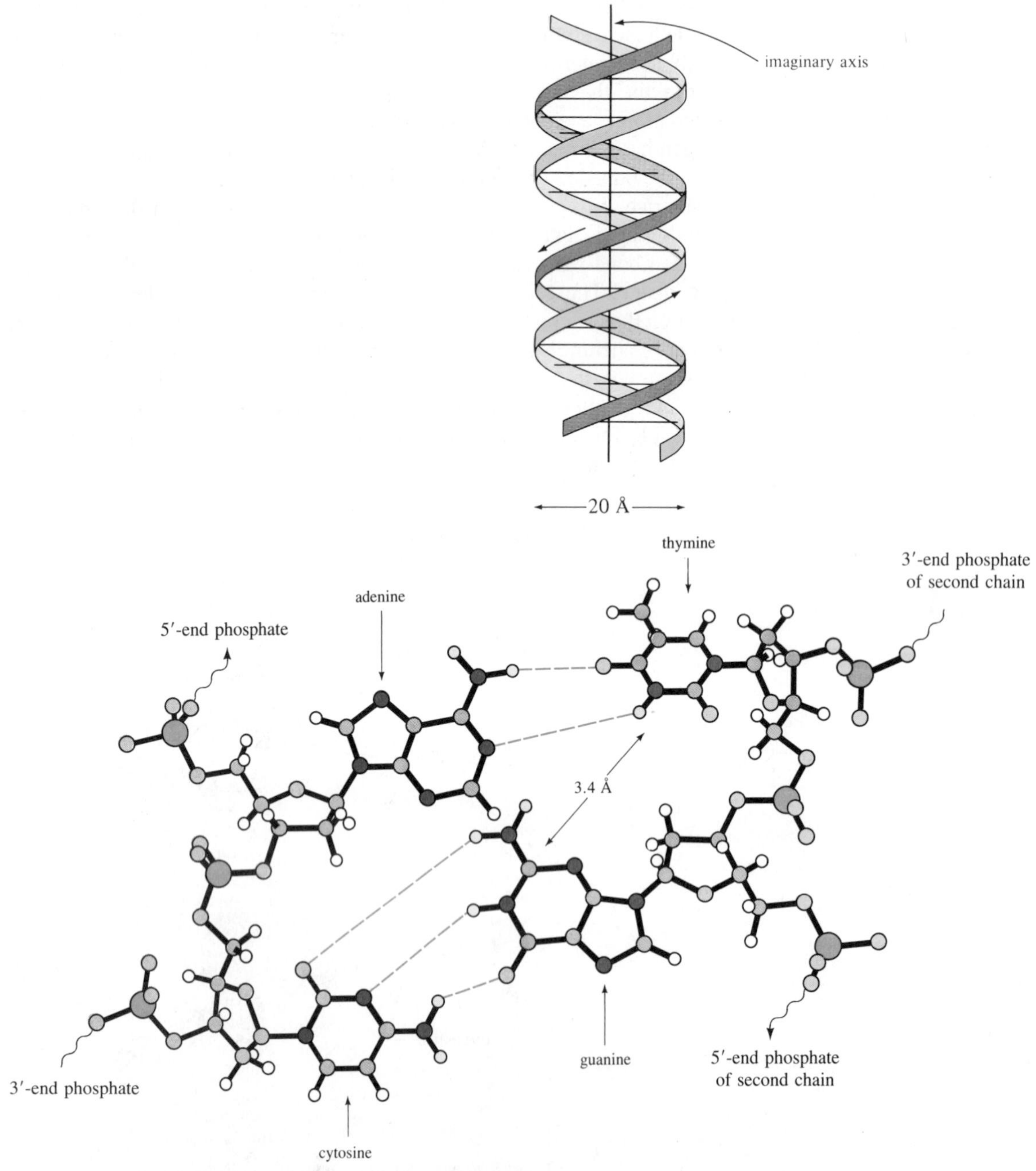
imaginary axis
20 Å
thymine
3′-end phosphate of second chain
adenine
5′-end phosphate
3.4 Å
guanine
5′-end phosphate of second chain
3′-end phosphate
cytosine

FIGURE 24.1 The double helix.

Even before a structure had been assigned to DNA, experimental evidence had indicated that DNA is involved in the storage and transmission of genetic information. Watson and Crick recognized that the pairing of the bases on one strand of the double helix with those on the other strand provided a mechanism by which genetic information could be transmitted. Either strand of a double helix could in principle, if separated from its partner, direct the synthesis of a new strand containing exactly the same purine and pyrimidine bases as were in the missing strand, as shown in Figure 24.2.

Watson and Crick proposed as the "central dogma" of the new science of molecular genetics that the bases are like the letters of an alphabet, that genes are in essence molecules of DNA, and that the order of the bases on the backbone of the polynucleotide chain constitutes a genetic code. The code is reproduced when DNA fosters the synthesis of molecules of ribonucleic acid. The order of bases in RNA chains is determined when these bases pair by means of hydrogen bonding with those on a strand of DNA. Ribonucleic acids, in turn, direct the synthesis of proteins.

Ribonucleic acid molecules are generally smaller than deoxyribonucleic acid molecules. Their molecular weights range from 30,000 to 2,500,000. A ribonucleic acid does not form a double helix because the hydroxyl group present at carbon 2 of the sugar unit ribose (unlike the hydrogen atom occupying that position in

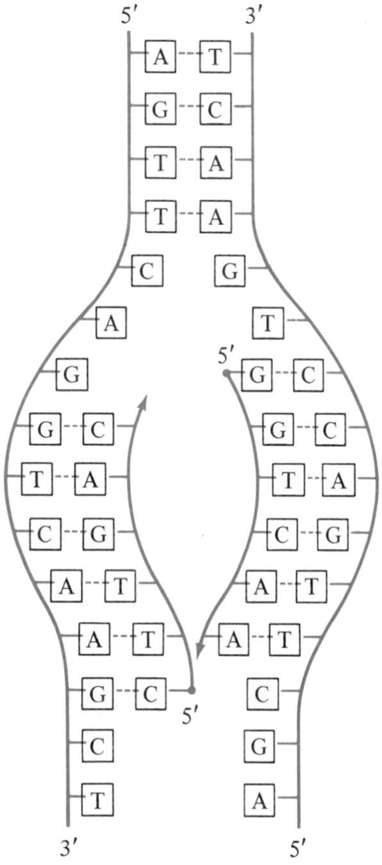
5′
3′
A
T
G
C
T
A
T
A
C
G
A
T
G
5′
G
C
G
C
G
C
T
A
T
A
C
G
C
G
A
T
A
T
A
T
A
T
G
C
5′
C
C
G
T
A
3′
5′

FIGURE 24.2 A schematic representation of the replication of DNA, the conversion of one double helix into two identical double helixes.

2-deoxyribose) is too bulky to allow two strands of RNA to come close enough to interact in that way. The RNA molecules exist as single strands that fold and loop with internal base pairing taking place at some points. In RNA, adenosine hydrogen-bonds with uracil, and guanine hydrogen-bonds with cytosine. Because RNA does not have a double helix in which every base on one strand is paired with a corresponding base on the opposite strand, analyses of samples of RNA do not show the regularity in molar ratios of purine bases to pyrimidine bases that is characteristic of DNA. This is an important and striking experimental difference between the two types of polynucleotides.

Experimentally, it has been found that three adjacent bases on one kind of ribonucleic acid, known as **messenger RNA,** are the code for a given amino acid. The first such three-nucleotide code was identified when it was found that polyuridylic acid directed the synthesis of the peptide polyphenylalanine. Thus, three uracil bases in a row constitute the code for the amino acid phenylalanine. An amino acid is often coded for by more than one sequence of three nucleotides. There are also portions of RNA molecules that do not code for anything, as far as is known. Other three-nucleotide sequences serve as punctuation marks, such as a signal to stop the synthesis of a polypeptide. Scientists have also found that some nucleotides are part of a code for two different amino acids. The code sequences for them overlap. The amino acids to be incorporated in a protein chain are transported to the messenger RNA by another type of RNA known as **transfer RNA.** Each amino acid has at least one transfer RNA that is specific for it.

Over the years, methods have been invented to determine the sequence of the bases in very small portions of DNA molecules. It is now possible, in principle, to compare the sequence of bases on a segment of DNA with the structure of the polypeptide that is synthesized from that gene. Complete analyses have been carried out for some reasonably complex proteins, such as the β-chain of mouse hemoglobin.

Watson, Crick, and Wilkins received the Nobel Prize in 1962 for their work on the structure of DNA. The model they developed has subsequently led to a tremendous amount of research into the chemistry of polynucleotides and the ways they interact with other constituents of cells. The deoxyribonucleic acid on which the x-ray crystallographic work was done was a highly purified and crystallized sample. Even so, it was observed to have a different structure depending on the humidity of its surroundings. The picture of the molecular structure of DNA, and of the way it may function, that emerges from the x-ray data is a simplified one. It must be modified when the activity of deoxyribonucleic acid in the living cell is considered.

Basic proteins known as **histones** are found with DNA in the nucleus of cells from a calf's thymus gland. Histones are positively charged at pH 7 and are held by ionic and hydrogen bonds to the phosphate groups on the double helix, stabilizing the molecule. A polypeptide chain lies along the groove in the helix structure and is bonded at several points to the polynucleotide strand. The combined polypeptide-polynucleotide structure is known as a **nucleoprotein.** In the nuclei of the cells of a calf's thymus gland, there is also a smaller fraction of acidic proteins. These are also found associated with DNA and with histones. The proteins constitute about 30% of the dry weight of the nucleus of a cell, suggesting that they have an important function there.

DNA has primary, secondary, and tertiary structure. The primary structure of DNA is the order of the nucleotides in the backbone of a single strand. The synthesis of a polynucleotide involves the orderly placement of one nucleotide unit after another in a chain held together by phosphate ester linkages. In the body, this process is catalyzed by an enzyme, DNA polymerase.

deoxyribonucleotide triphosphates + DNA template

$$\xrightarrow[\text{Mg}^{2+}]{\text{DNA polymerase}}$$

new DNA + pyrophosphate

The secondary structure of DNA is the right-handed helical shape that characterizes a strand of DNA, and the double helix that forms when the complementary strand is synthesized. In Figure 24.1 (p. 1016), the double helix is shown with its central axis as a straight line. It is seldom so. The double helix itself curves and twists in a variety of ways that are termed supercoiling, which constitutes the tertiary structure of DNA. Some forms of supercoiling are shown in Figure 24.3. DNA maintains the tertiary structure that is responsible for its biological activity only within a narrow range of pH, temperature, and solvent composition. When conditions are changed too drastically, the DNA is denatured (p. 1152). The most important cause of denaturation is the disruption of the ionic and hydrogen bonds that give the double helix its stability. For example, an increase in pH would result in the loss of protons from the basic proteins (p. 1121) associated with the phosphate groups on the double helix. The ionic bonds that hold them together would be disrupted, and the excess of negative charge on the helix would destabilize it.

PROBLEM 24.16

The points in a DNA helix at which adenine is paired with thymine are the first to pull apart when DNA is denatured. In fact, the stability of a particular form of DNA, as indicated by the temperature to which it can be heated before it melts is directly related to the relative amount of guanine (and cytosine) that it contains. How do you explain this experimental observation? (Hint: Another look at the diagram of the pairing of bases on p. 1015 may be helpful.)

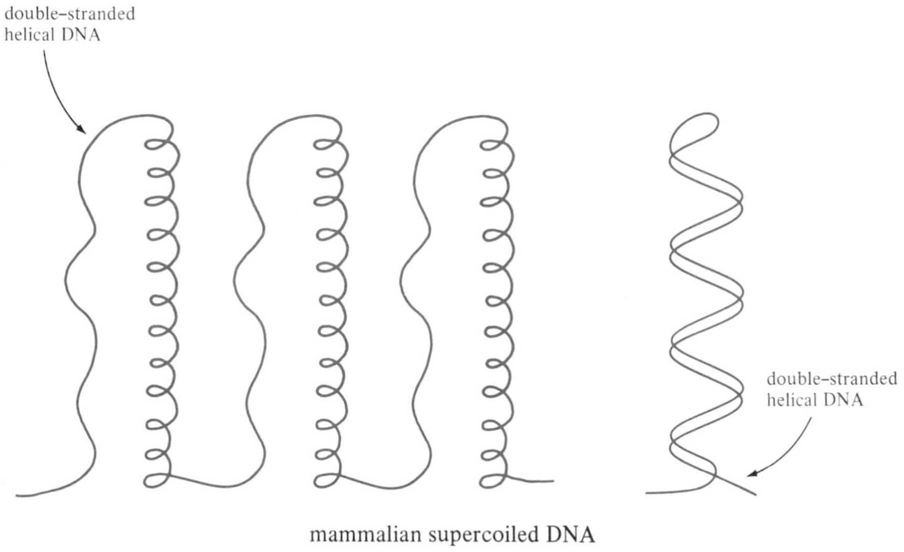
double-stranded
helical DNA
double-stranded
helical DNA
mammalian supercoiled DNA

FIGURE 24.3 Supercoiling of a DNA molecule.

Any reaction that interferes with the correct pairing of bases by hydrogen bonding changes the way in which the genetic code is transmitted. An incorrect transmission of genetic information by DNA may lead to a mutation in the species. Mutations take place all the time in nature, but some compounds, known as **mutagens,** greatly increase the rate at which mutations occur. Many mutagens are also **carcinogens,** compounds that increase the probability that a malignant tumor will develop in a living organism.

For a DNA molecule to serve as a template, for the synthesis either of other DNA molecules or of RNA molecules, which then direct protein synthesis, the double helix must unwind to some extent. In other words, a disruption of the most stable arrangement of the giant molecule must take place. The histones associated with DNA in multicelled organisms play an important part in the regulation of the process. They must somehow be detached to expose part of the polynucleotide before it can serve as a template. Exactly how this happens and what starts the process in living organisms is not yet understood.

The story of how genetic information is stored and transmitted is still unraveling. Recent advances in recombinant DNA technology allow DNA molecules to be cleaved selectively, then combined with fragments of DNA molecules from other organisms. These experiments have greatly increased public awareness of the potential for both useful and harmful manipulations of genes. In 1975, scientists who were working in this field stopped to consider the possible consequences of what they were doing. As a result of their deliberations, controls were put on the types of experiments that can be done and on the laboratory procedures and precautions that must be followed to minimize the risk. This is the first time that scientists ever joined together to impose formal controls on their own work.

Corporations now actively use transformed biological systems to synthesize commercially useful products such as human insulin for diabetics, growth hormones, and various immunological factors that may be useful in boosting resistance to diseases, including cancer.

PROBLEM 24.17

Write structural formulas for the following compounds. A review of the structures on p. 1009 may be helpful.

(a) adenosine 3′,5′-diphosphate (b) deoxyguanosine 5′-phosphate (c) cytidine
(d) thymidine 3′-phosphate (e) guanosine (f) uridine 2′-phosphate

PROBLEM 24.18

Some of the most potent mutagens are three-membered heterocycles, such as oxiranes or aziridines (three-membered rings containing nitrogen). Write an equation suggesting why such compounds are so destructive to DNA.

C. Biological Oxidation-Reduction Reactions

Many biological oxidation-reduction reactions are catalyzed by enzymes that are associated with coenzymes that have nicotinamide as part of their structures. In these coenzymes, nicotinamide, one of the B vitamins, is bonded to ribose at the nitrogen atom of the pyridine ring. The molecule also contains the nucleotide adenosine 5′-phosphate.

The oxidation-reduction reactions affect the nicotinamide portion of the molecule, so in equations the structural formula of nicotinamide adenine dinucleotide is abbreviated by using R to symbolize everything that is bonded to the nitrogen atom of the pyridine ring.

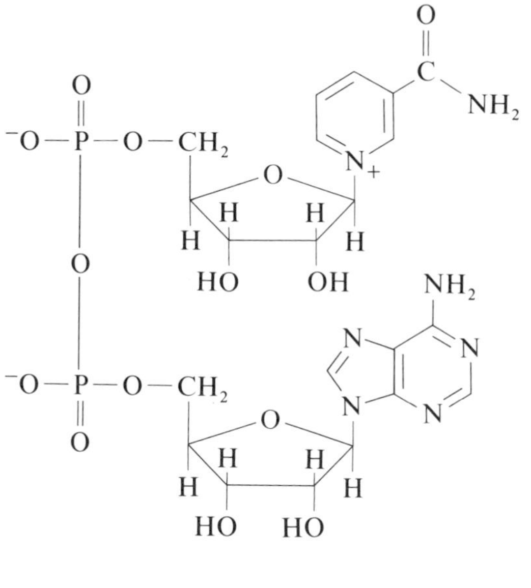

nicotinamide adenine dinucleotide
NAD^+

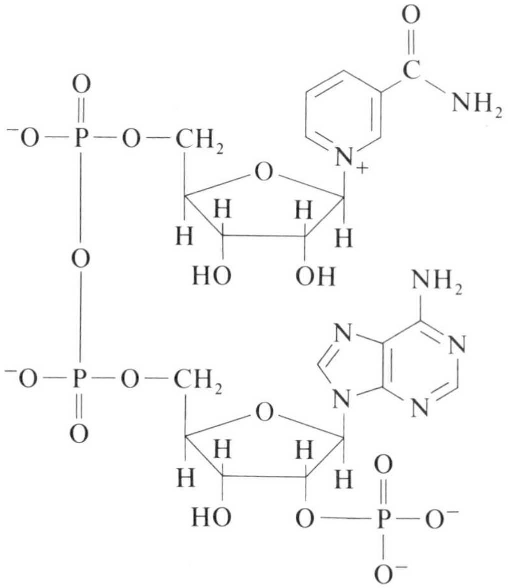

nicotinamide adenine dinucleotide phosphate
$NADP^+$

The presence of a positive charge on the pyridine ring is important. The nitrogen atom of the ring is in the form of a quaternary ammonium ion, and the ring is therefore highly activated toward nucleophilic attack, even more so than pyridine itself.

The coenzymes NAD^+ and $NADP^+$ are associated with a large number of enzymes known as dehydrogenases. A typical reaction catalyzed by a dehydrogenase found in the liver is the oxidation of ethanol to acetaldehyde.

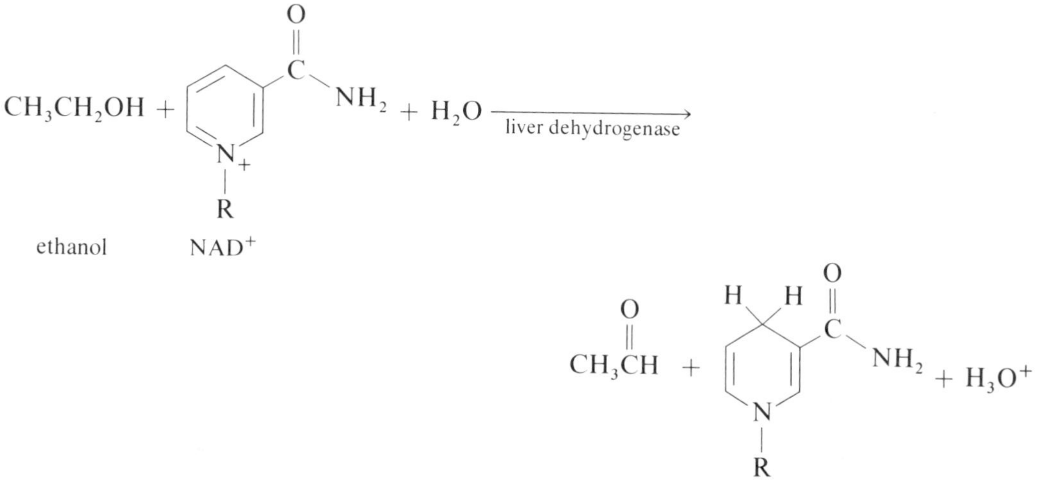
liver dehydrogenase

ethanol NAD^+ acetaldehyde NADH

In the process, NAD^+ is reduced to dihydronicotinamide adenine dinucleotide, abbreviated NADH, and a proton is also transferred to a water molecule.

The reaction starts with the transfer of a hydride ion from the alcohol to NAD^+. For example, if ethanol labeled with deuterium at carbon 1 is used, deuterium appears in the reduction product of NAD^+.

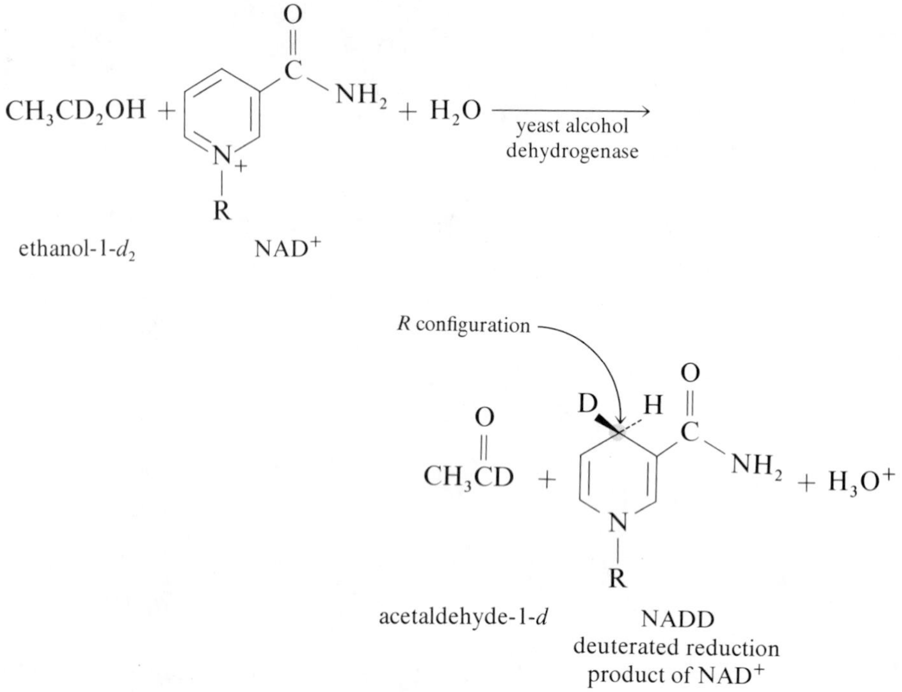
CH₃CD₂OH +
+ H₂O
yeast alcohol
dehydrogenase
ethanol-1-d₂
NAD⁺
R configuration
CH₃CD +
+ H₃O⁺
acetaldehyde-1-d
NADD
deuterated reduction
product of NAD⁺

The deuterium atom is transferred with high stereoselectivity to one face of the pyridine ring, so the reduction product with this enzyme and many others always has the *R* configuration at carbon 4. Not all enzymes have the same stereoselectivity. A number of enzymes catalyze the transfer of hydride ion to the other face of the nicotinamide residue.

Dehydrogenases catalyze reduction as well as oxidation reactions. The reduced nicotinamide serves as the reducing agent, again with high stereoselectivity. For example, if acetaldehyde-1-*d*, a product of the reaction shown above, is reduced with NADH, ethanol-1-*d* having the *S* configuration is produced.

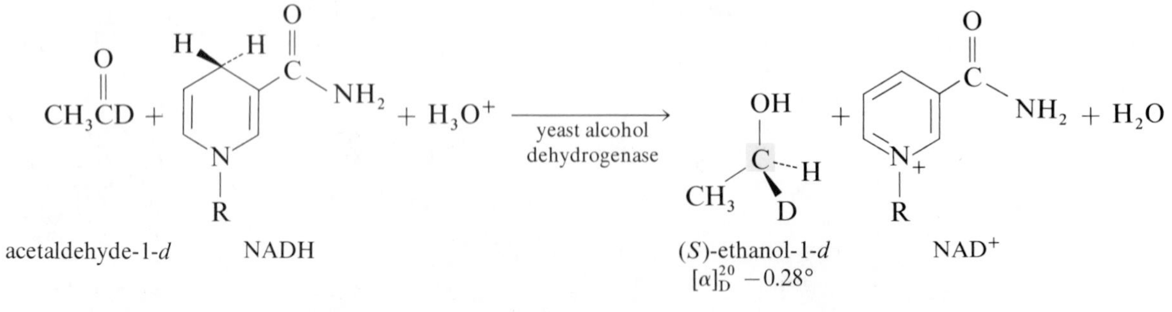
CH₃CD +
+ H₃O⁺
yeast alcohol
dehydrogenase
+ H₂O
acetaldehyde-1-d
NADH
(S)-ethanol-1-d
[α]D20 −0.28°
NAD⁺

If, on the other hand, NADD is used to reduce unlabeled acetaldehyde, the product is (*R*)-ethanol-1-*d*.

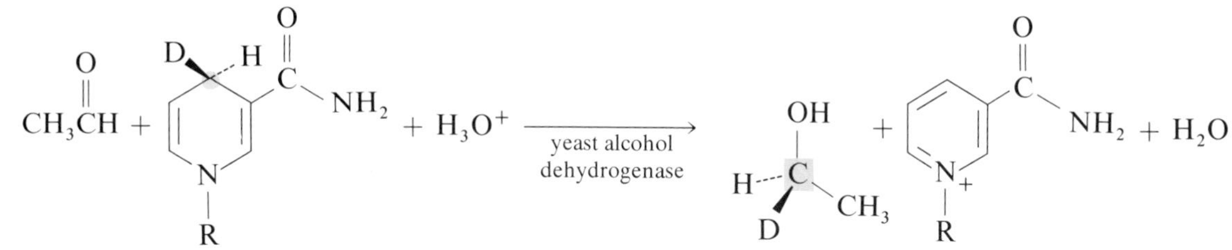
acetaldehyde
NADD
yeast alcohol dehydrogenase
(R)-ethanol-1-d
NAD+

In other words, the enzyme distinguishes between the two faces of the planar carbonyl group. The incoming hydride ion is always attached to the same side of the plane of the carbonyl group and is detached from only one face of the nicotinamide ring.

The reduction of a carbonyl compound by NADH is entirely analogous to reduction by a metal hydride such as sodium borohydride except that the NADH reaction shows stereoselectivity.

VISUALIZING THE REACTION

Reduction of a carbonyl group by NADH

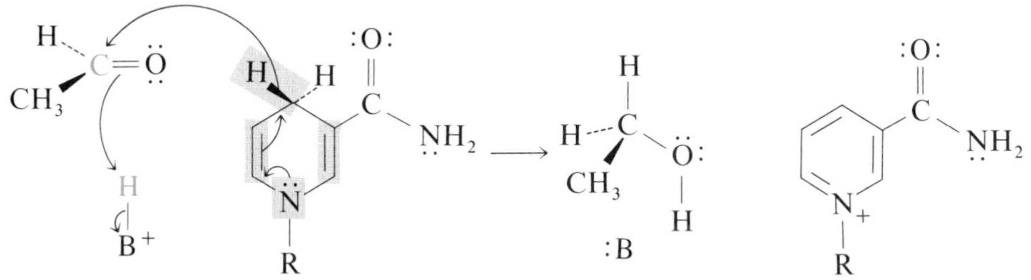

Ethanol that is not labeled with deuterium has no chirality. However, if one of the hydrogen atoms of the methylene group is replaced by another group, such as deuterium, one enantiomer of a pair is formed. Replacement of the other hydrogen atom by deuterium would give the mirror-image isomer of the first compound. The two hydrogen atoms on carbon 1 of ethanol are called **enantiotopic** because enantiomeric compounds are formed by replacing one or the other of them with another group. The two secondary hydrogen atoms on carbon 2 of butane are also enantiotopic. Replacing one of them with a chlorine atom gives (*R*)-2-chlorobutane, and replacement of the other forms the enantiomeric (*S*)-2-chlorobutane (p. 180). An achiral reagent does not distinguish between enantiotopic hydrogen atoms, but a chiral one does. Thus, when butane reacts with the achiral reagent, chlorine, there is an equal probability that each one of the two hydrogen atoms on carbon 2 will be replaced, and equal numbers of molecules of (*R*)- and (*S*)-2-chlorobutane are formed. When ethanol reacts with NAD^+ at the active site of the enzyme yeast alcohol dehydrogenase, however, only the hydrogen atom that occupies a certain position in space is transferred to NAD^+.

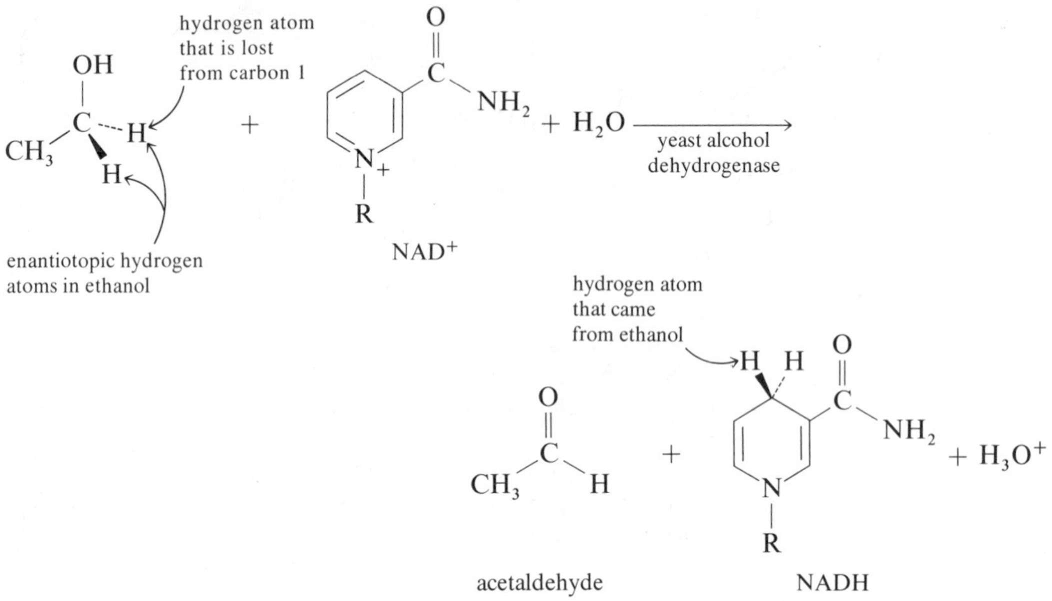
hydrogen atom
that is lost
from carbon 1
OH
C
H
CH3
H
+
O
C
NH2
N+
R
+ H2O
yeast alcohol
dehydrogenase
enantiotopic hydrogen
atoms in ethanol
NAD+
hydrogen atom
that came
from ethanol
H H
O
C
NH2
O
C
CH3
H
+
N
R
+ H3O+
acetaldehyde
NADH

This phenomenon is only detectable when the ethanol is labeled with deuterium, but it occurs whether the label is there or not.

PROBLEM 24.19

Write equations predicting the products of the reaction of (*R*)-ethanol-1-*d* with NAD^+ and of (*S*)-ethanol-1-*d* with NAD^+.

An enzyme can distinguish between the two enantiotopic hydrogen atoms of ethanol because the alcohol molecule fits the active site of the enzyme better in one orientation than it does in the mirror-image orientation. Ethanol fits the active site well in only one of two enantiomeric positions because the active site binds three parts of the molecule. There is no stereoselectivity if the active site interacts with only two parts of the molecule (Figure 24.4).

In a similar way, the two faces of nicotinamide in NAD^+ may be said to be enantiotopic. Depending on the nature of the enzyme, one face or the other receives the incoming hydride ion stereoselectively. Nicotinamide itself is achiral because it has a plane of symmetry that coincides with the plane of the ring, but the presence of ribose units in the coenzyme makes the molecule as a whole chiral. But even if it were not, the arguments that were made about ethanol could be used to show that the active site of an enzyme would interact differently with two enantiomeric orientations of a nicotinamide ring.

PROBLEM 24.20

Assume that the active site of a hypothetical enzyme interacts with the nitrogen atom in the pyridine ring of nicotinamide, with the carbonyl group of the amide function, and with carbon 5 of the ring as a hydrophobic site. Prove to yourself that the two faces of an achiral nicotinamide molecule are enantiotopic and thus distinguishable by such an enzyme.

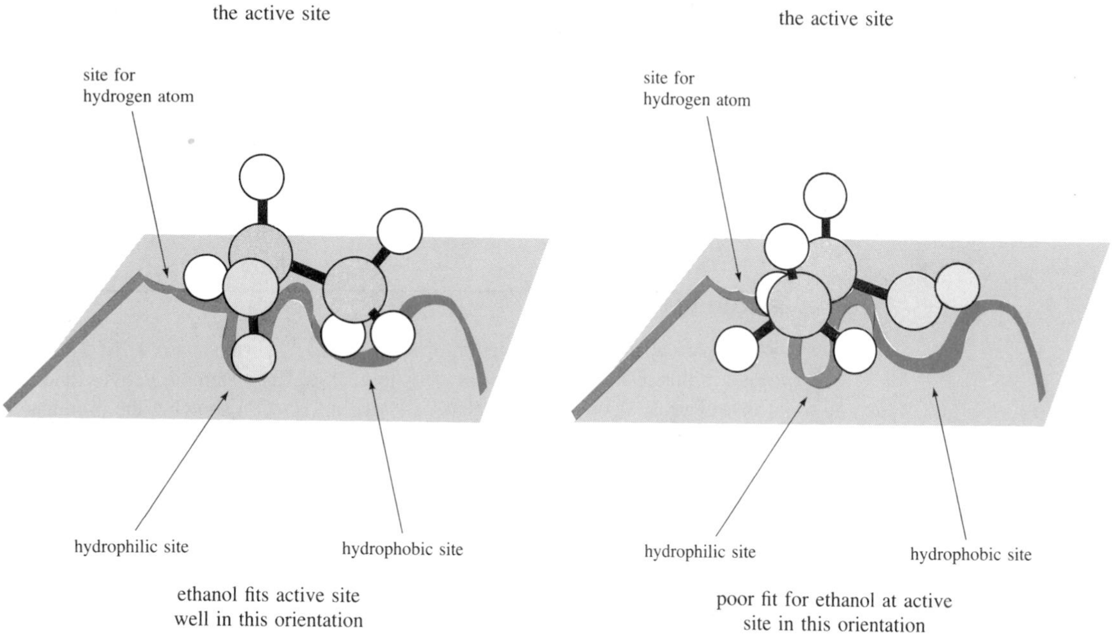

the active site
site for
hydrogen atom
hydrophilic site
hydrophobic site
ethanol fits active site
well in this orientation
the active site
site for
hydrogen atom
hydrophilic site
hydrophobic site
poor fit for ethanol at active
site in this orientation

FIGURE 24.4 Two possible ways in which ethanol can interact with the active site of an enzyme.

PROBLEM 24.21

Inspect the following structural formulas and decide which compounds contain enantiotopic hydrogen atoms.

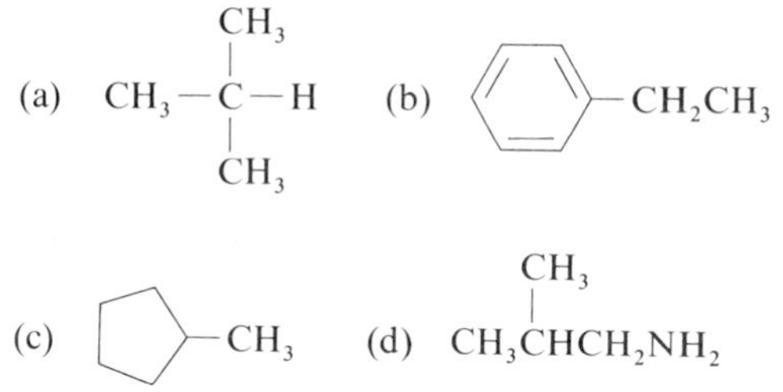

(a)
CH3
CH3—C—H
CH3
(b)
CH2CH3
(c)
CH3
(d)
CH3
CH3CHCH2NH2

PROBLEM 24.22

Hydrogen atoms are **diastereotopic** if replacing one or the other of them with another group gives rise to diastereomers. An example would be the hydrogen atoms shown in color in the structural formula below.

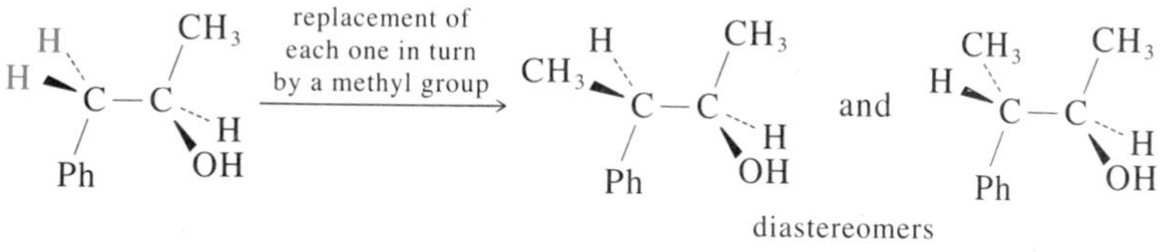

replacement of
each one in turn
by a methyl group
and
diastereomers

Are there diastereotopic hydrogen atoms in any of the compounds shown in Problem 24.21? How about the compounds shown below?

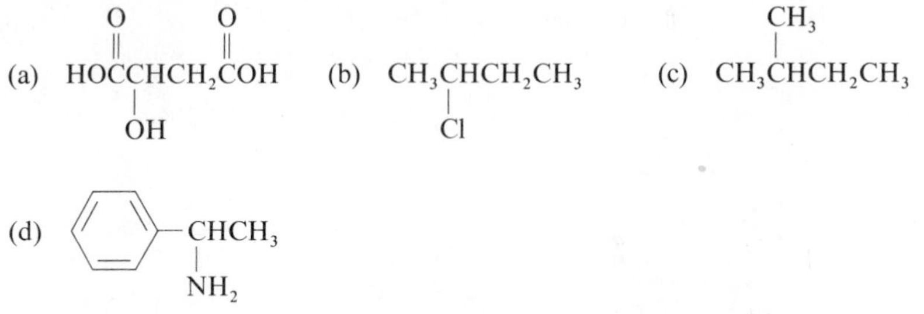

PROBLEM 24.23

NAD^+ participates in the oxidation of testosterone, a hormone (p. 477). The enzyme that catalyzes this reaction, testosterone dehydrogenase, has a stereoselectivity that is the opposite of that of yeast alcohol dehydrogenase. Write an equation showing the products that you expect from the reaction. Trace the fate of the hydrogen atoms removed from testosterone.

PROBLEM 24.24

Nicotinamide got its name because it is the amide of an acid derived from nicotine (p. 889). Write equations showing how you would synthesize nicotinamide from nicotine.

24.7 FIVE-MEMBERED HETEROCYCLES OF BIOLOGICAL SIGNIFICANCE

A. Sulfur Heterocycles

One of the first vitamins to be discovered was thiamine, Vitamin B_1, which is involved in essential metabolic processes. Thiamine pyrophosphate is found in every cell of the body and functions as a coenzyme in reactions that convert pyruvate ion, one of the products of glycolysis, to acetaldehyde and acetyl coenzyme A (p. 581). A deficiency of thiamine in human beings leads to a disease of the nervous system called beriberi.

In thiamine, a thiazole ring is joined to an aminopyrimidine ring by means of a bridging methylene group.

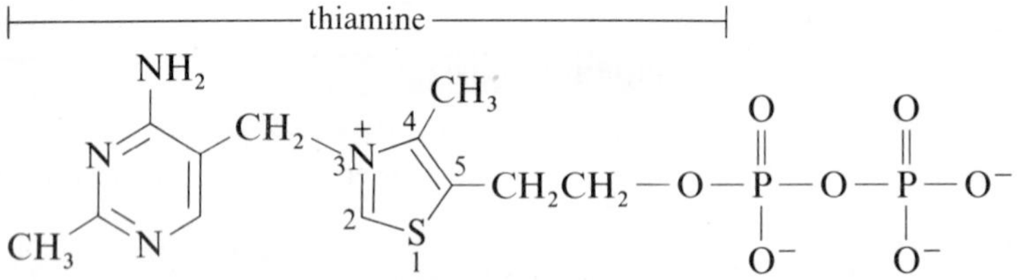

thiamine pyrophosphate

An interesting and important structural feature of the vitamin is the presence of a quaternary nitrogen atom in the thiazole ring. The hydrogen atom on carbon 2 of the thiazolinium ion, the carbon between the positively charged nitrogen atom and the sulfur atom, is extraordinarily acidic. In the 1950s, Ronald Breslow of Columbia University found that such a hydrogen atom in a thiamine analog exchanges for deuterium from deuterium oxide in the absence of either acid or base.

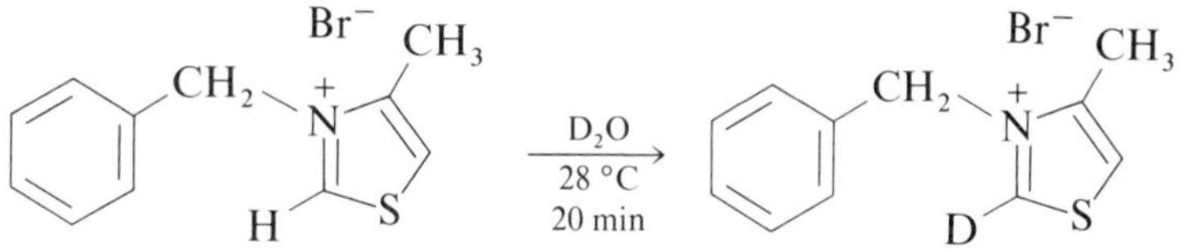
D₂O
28 °C
20 min
3-benzyl-4-methylthiazolinium bromide
3-benzyl-4-methylthiazolinium-2-*d* bromide

The course of this reaction was followed using nuclear magnetic resonance spectroscopy, and it was found that half of the hydrogen atoms at carbon 2 in a sample of the thiazolinium compound are replaced by deuterium atoms in 20 minutes at 28 °C. The rate of the reaction is extraordinarily fast for the breaking of a carbon-hydrogen bond, especially as no strong base is present.

The deuterium-exchange reaction starts with removal of a proton from the thiazolinium ion to give a carbanionic species that is an ylide.

VISUALIZING THE REACTION

The acidity of the thiazolinium ion

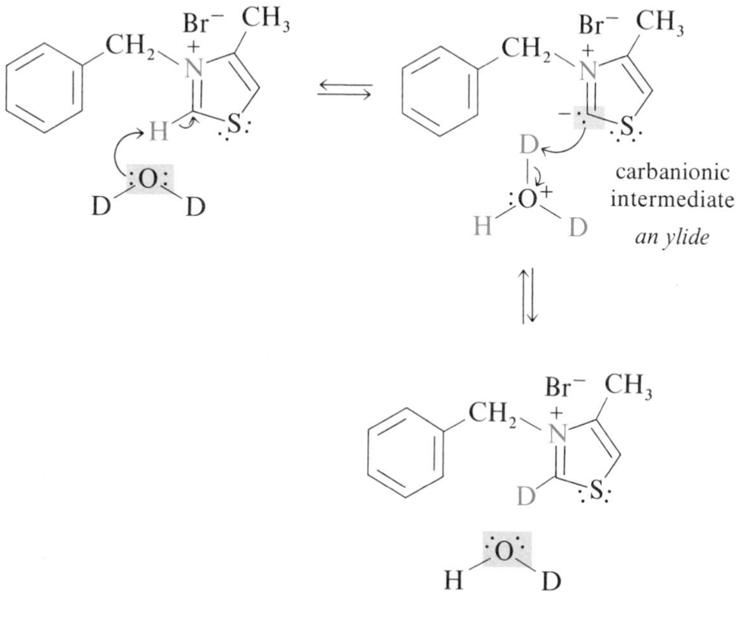
carbanionic
intermediate
an ylide

The negative charge is on a carbon atom adjacent to a positively charged atom (pp. 750 and 755), which stabilizes the carbanion. The sulfur atom of the thiazolinium ion can also stabilize the carbanion by accepting some electron density into its empty *d* orbitals. The combination of these two factors makes it possible for the thiazolinium ring of thiamine to ionize at neutral pH, in other words, under the conditions that exist in cells.

An important reaction that is catalyzed by thiamine pyrophosphate, in the presence of magnesium ions and an enzyme from brewer's yeast known as pyruvate decarboxylase, is the decarboxylation of an α-ketoacid to an aldehyde.

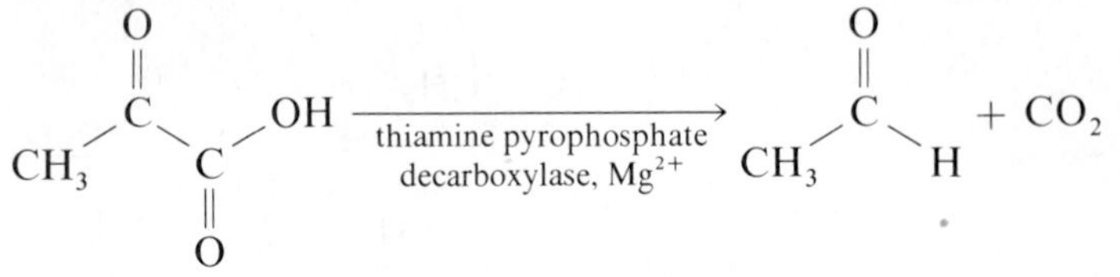
thiamine pyrophosphate
decarboxylase, Mg^{2+}

pyruvic acid — acetaldehyde — carbon dioxide

The reaction occurs by nucleophilic addition of the ylide from thiamine to the ketone function in pyruvic acid, giving an intermediate that loses carbon dioxide easily.

PROBLEM 24.25

Draw the structure of the intermediate for the above reaction. Suggest why it decarboxylates easily.

PROBLEM 24.26

How is acetaldehyde formed from the intermediate of Problem 24.25? To what class of reactions does this transformation belong?

PROBLEM 24.27

Thiamine is reasonably stable in acid. In pure water, it falls apart into a thiazole and a pyrimidine. In strong base, the thiazole ring opens. Suggest mechanisms for these two reactions, shown below.

(a)

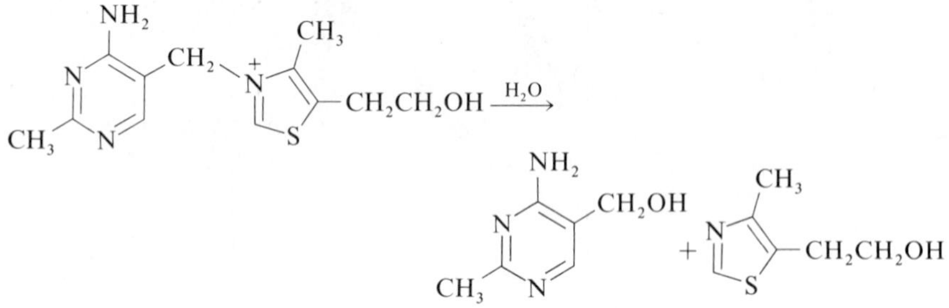

(b)

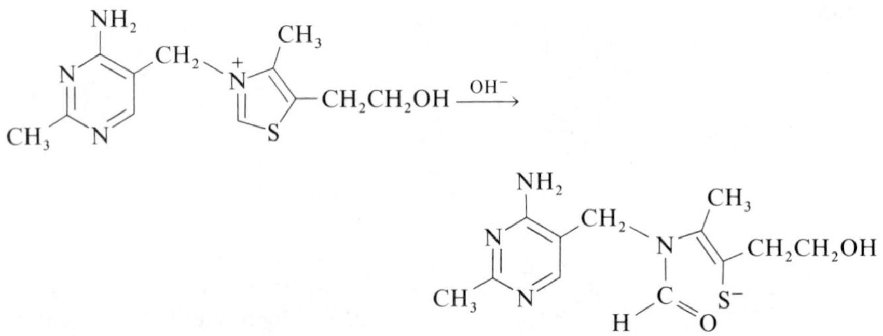

B. Nitrogen Heterocycles. Pyrrole

Pyrrole plays an important role in the chemistry of living organisms. The conversion of light energy from the sun to energy stored in the chemical bonds of the carbohydrates (p. 1050) synthesized by green plants is mediated by compounds known as **chlorophylls.** The essential structural feature of chlorophylls is a system of four pyrrole rings held together by bridges, each containing a single carbon atom. This ring system, known as **porphin,** also appears in heme, associated with the proteins hemoglobin (p. 1153) and myoglobin, which are responsible for the transport and storage of oxygen in the body tissues of warm-blooded animals.

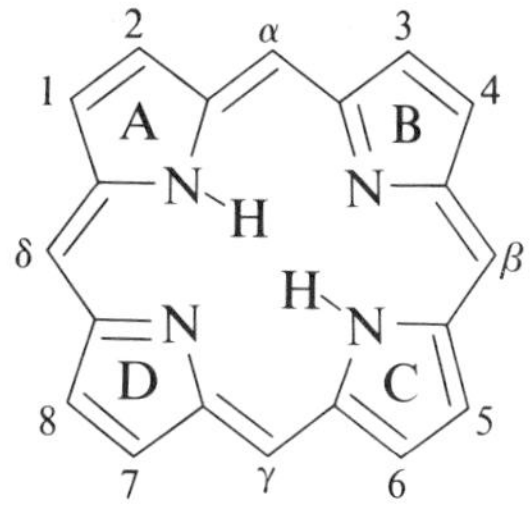

porphin

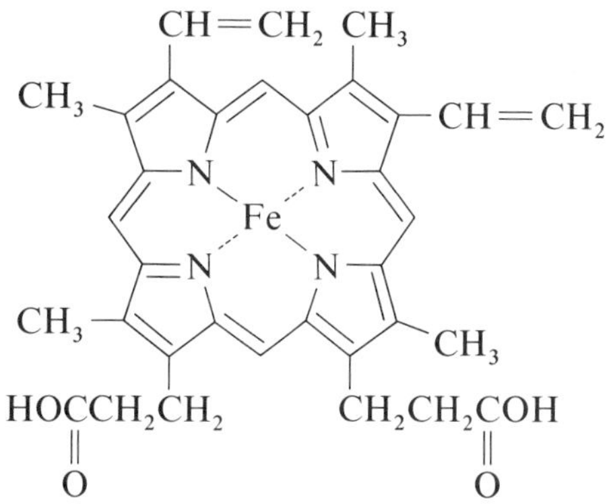

heme

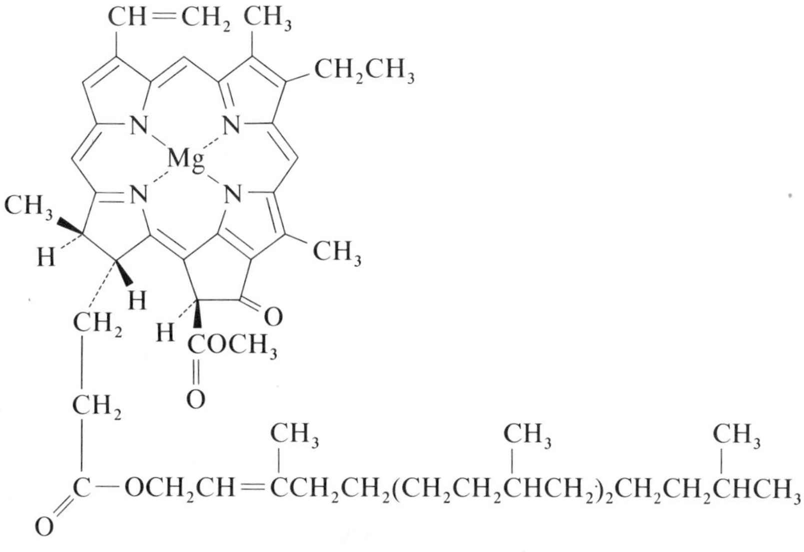

chlorophyll a

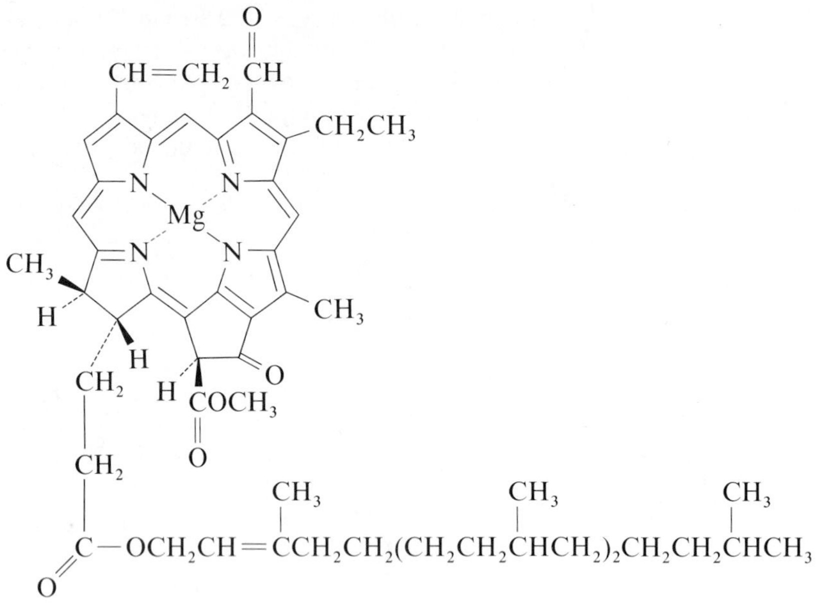

chlorophyll b

Porphin, with the four nitrogen atoms of the pyrrole rings pointing toward the center of its large ring system, complexes efficiently with metal ions. In heme, the ion is iron(II); in chlorophylls, it is magnesium(II). These metal ions may also complex with additional ligands above and below the plane of the heterocycle (p. 1153).

The porphin ring systems present in heme and chlorophylls have various substituents on the periphery. Substituted prophins are given the general name of **porphyrins.** The porphyrin in heme has the same degree of unsaturation as porphin does. In the chlorophylls, dihydroporphyrins, with one of the double bonds in the D ring of porphin reduced, are given the special name of **chlorins.** The compounds shown above are a few of many known porphyrins. The structural complexity means there are wide possibilities for isomerism in these systems.

The porphyrin ring system is very stable and has aromatic character. The extended conjugation present in porphyrins is responsible for the deep colors of these compounds (pp. 720 and 954). Porphin and heme have 22 π electrons, but only 18 of these are part of a cyclic array for which resonance contributors can be written. In this respect, a porphyrin resembles [18]annulene (p. 775), having $(4n + 2)$ π electrons, where n is 4.

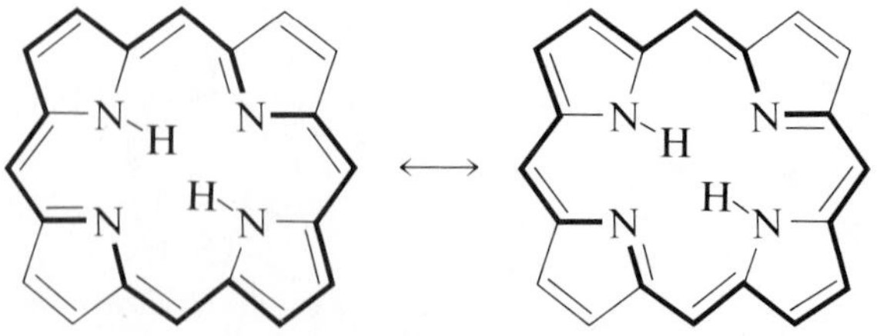

resonance contributors for porphin drawn in analogy to [18]annulene

Any attempt to include in the above resonance contributors the four π electrons that are outside the thick black lines gives structures that require that the protons on two of the nitrogen atoms be moved. Such a tautomeric transformation does occur in porphyrins. However, two structures that are related to each other by a change in the location of atoms as well as electrons are not resonance contributors.

Porphyrins are synthesized in nature with remarkable ease. The basic unit that combines with itself to give porphyrins substituted in a variety of patterns is a trisubstituted pyrrole called porphobilinogen. This compound is synthesized in living organisms by the condensation of two molecules of 5-amino-4-oxopentanoic acid (δ-aminolevulinic acid).

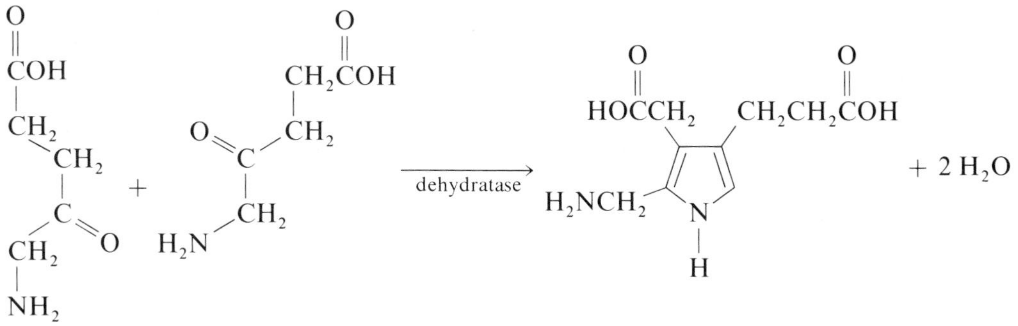
dehydratase
+ 2 H₂O

5-amino-4-oxopentanoic acid
δ-aminolevulinic acid

porphobilinogen

In the presence of the enzyme deaminase (or by simply heating with acid), porphobilinogen is converted into uroporphyrinogen I in high yield. A porphyrinogen is oxidized by air to the more stable aromatic porphyrin, unless special precautions are taken to exclude air.

VISUALIZING THE REACTION

Condensation of porphobilinogen units

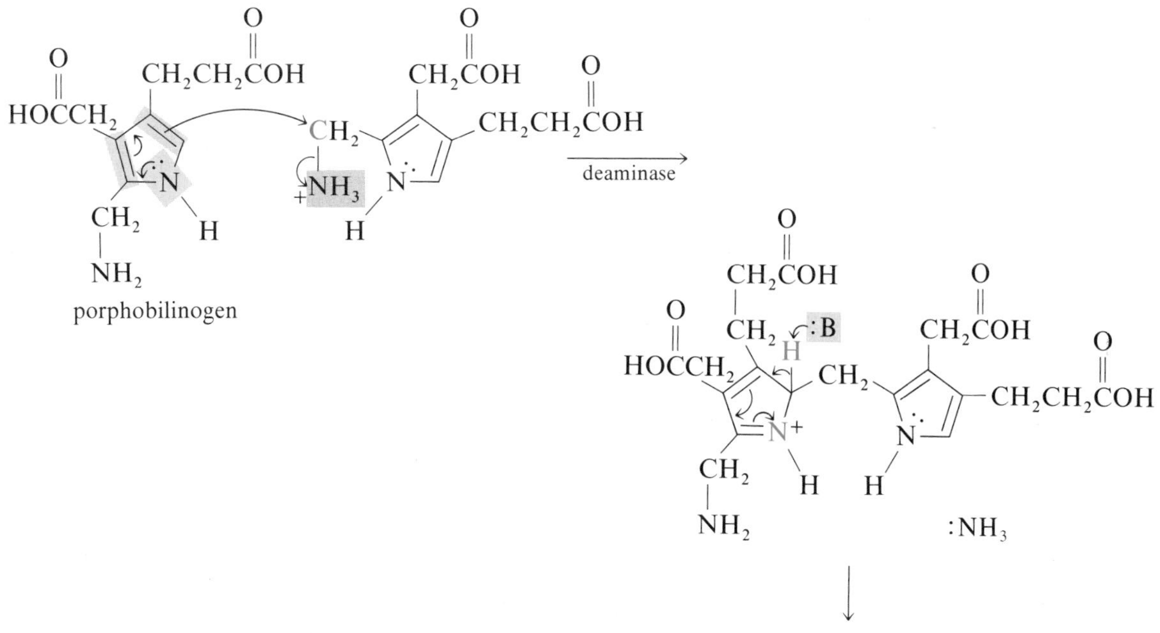
deaminase
porphobilinogen
:B
:NH₃

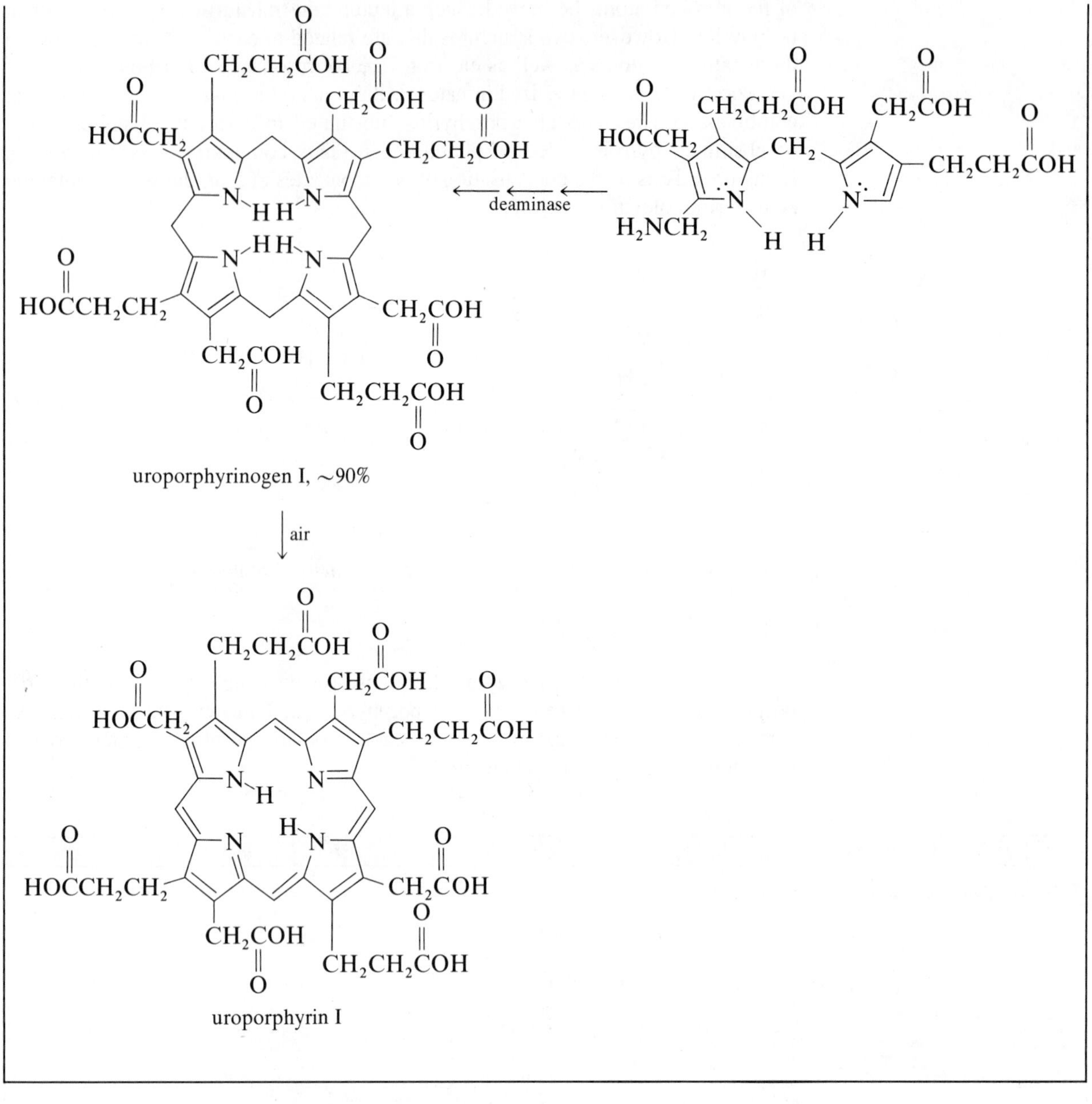
deaminase
uroporphyrinogen I, ~90%
air
uroporphyrin I

When two enzymes are present, the unsymmetrical porphyrin that is the structural precursor of heme and chlorophyll is formed from porphobilinogen.

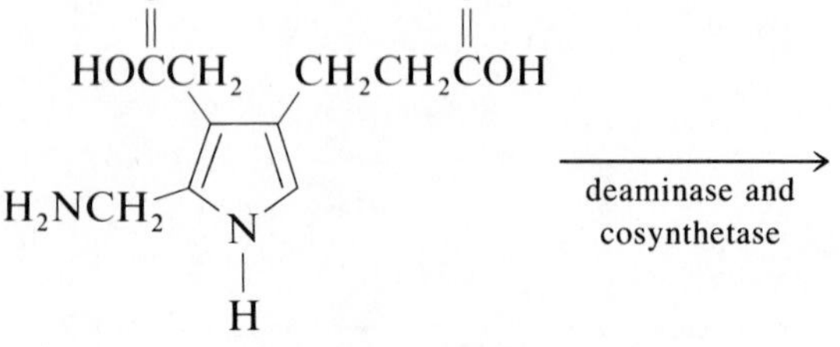
deaminase and
cosynthetase

porphobilinogen

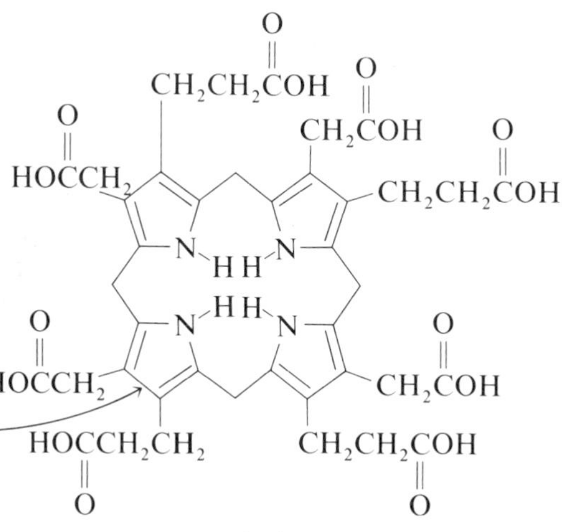
this pyrrole ring
inserted in a
reversed position

uroporphyrinogen III

PROBLEM 24.28

In one of the experiments designed to study the mechanism for the formation of uroporphyrinogen III, 5-amino-4-oxopentanoic acid labeled with ^{13}C at carbon 5 was synthesized and converted to porphobilinogen. Predict where the ^{13}C appears in the trisubstituted pyrrole.

24.8 ALKALOIDS

A. Tropane Alkaloids

Alkaloids, heterocyclic compounds containing nitrogen, occur most often in the seeds, leaves, and bark of plants. Alkaloids are bases, and many of them have a bitter taste and profound physiological effects.

A group of alkaloids containing a pyrrolidine ring that is bridged by three carbon atoms between the second and fifth carbons are known as **tropane alkaloids.** To this family belong cocaine, from the leaves of the coca shrub, and atropine, which is the racemic form of (−)-hyoscyamine and is obtained from henbane and the deadly nightshades.

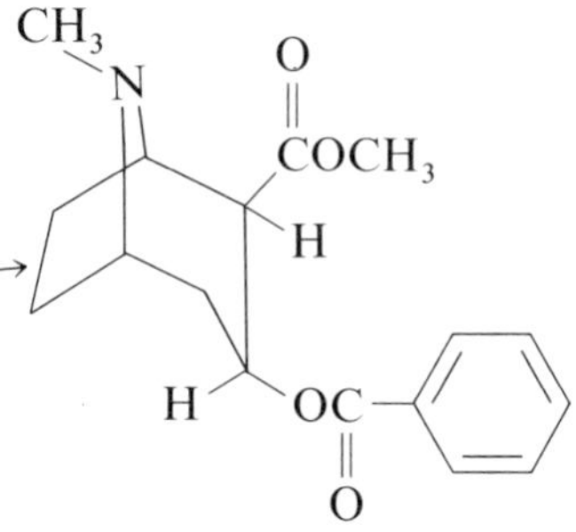
pyrrolidine
ring

(−)-cocaine

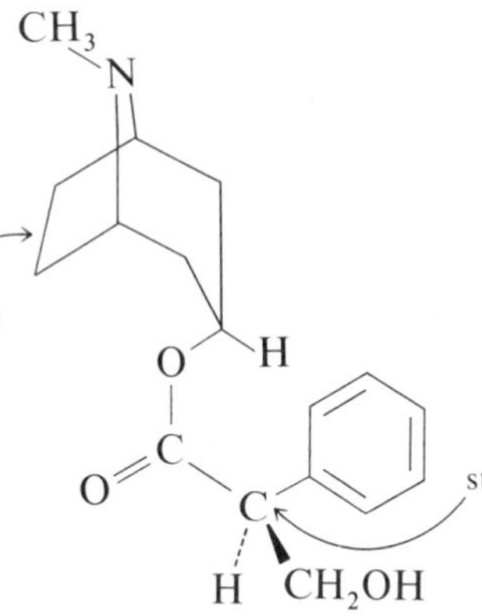
pyrrolidine
ring
stereocenter

(−)-hyoscyamine
atropine = (±)-hyoscyamine

Cocaine is a stimulant of the central nervous system and a local anesthetic because it blocks the transmission of nerve impulses. The drug is toxic and addictive and disrupts the rhythms of the heart. For this reason, a series of compounds that mimic the action of cocaine as a local anesthetic but lack its more harmful properties has been synthesized. Among them is Novocain (p. 903).

Atropine acts to relax the smooth muscles and thereby ease intestinal and bronchial spasms. Among other medical applications, atropine is used to dilate the pupil of the eye to allow examination of the retina. (−)-Hyoscyamine is the ester of a heterocyclic amino alcohol known as tropine with (*S*)-(−)-tropic acid, and it is hydrolyzed to these two components in cold water.

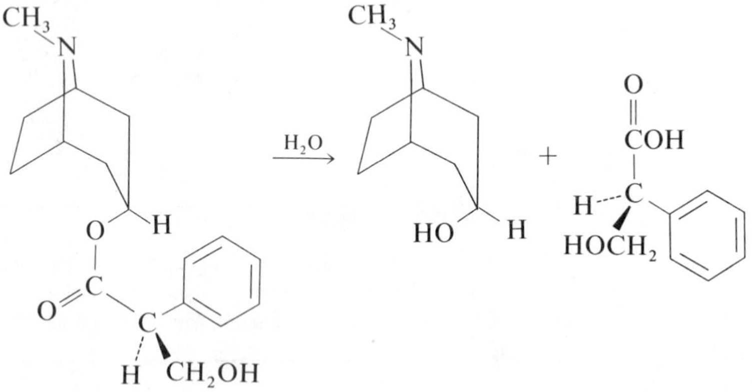
H_2O

(−)-hyoscyamine — tropine — (*S*)-(−)-tropic acid

Tropic acid is easily racemized, so (−)-hyoscyamine is converted into optically inactive atropine by warming with base and is hydrolyzed by basic solutions to tropine and racemic tropic acid.

PROBLEM 24.29

Write an equation outlining a mechanism for converting (−)-hyoscyamine to atropine with a base.

PROBLEM 24.30

Tropine is oxidized by chromic acid to a ketone, tropinone. When tropinone is reduced, tropine is not formed. Instead another alcohol, also $C_8H_{15}NO$, called ψ-tropine, is obtained. ψ-Tropine can be oxidized back to tropinone. Write equations showing what is happening.

Cocaine has essentially the same bicyclic ring structure as atropine, but the pattern of substitution is different. Hydrolysis of (−)-cocaine gives (−)-ecgonine, benzoic acid, and methanol.

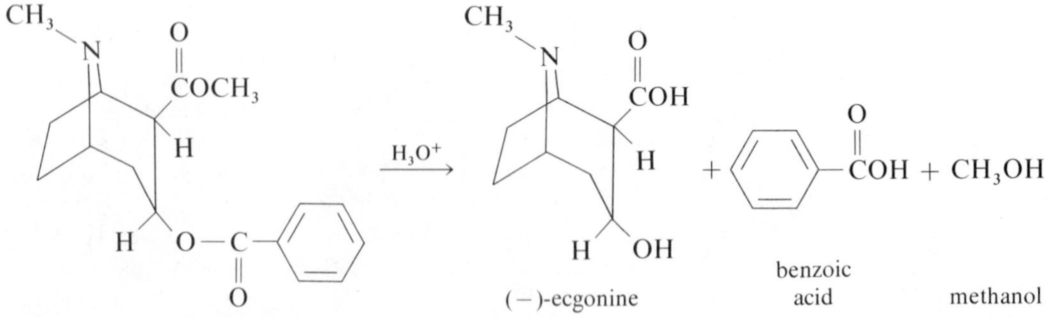
H_3O^+
+
+ CH_3OH

(−)-cocaine — (−)-ecgonine — benzoic acid — methanol

As shown below, the relationship between cocaine and atropine is clear when (−)-ecgonine is oxidized with chromic acid. Tropinone (Problem 24.30) is one of the products obtained, along with other compounds from oxidative cleavage of the ring.

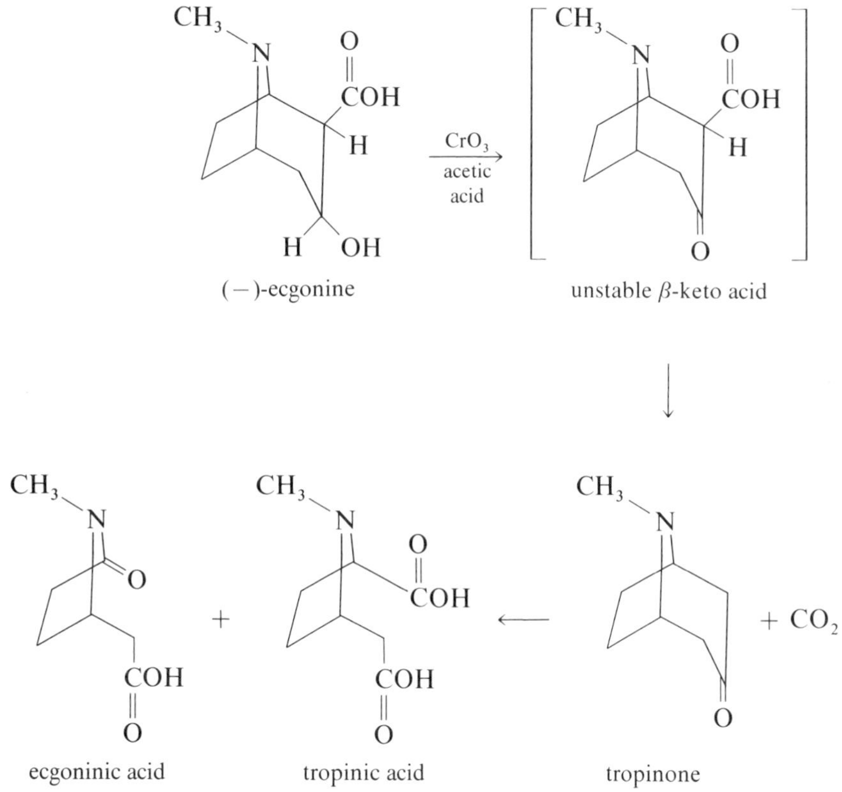
CH3
N
O
COH
H
H OH
CrO3
acetic
acid
(−)-ecgonine
unstable β-keto acid
ecgoninic acid
tropinic acid
tropinone
+ CO2

PROBLEM 24.31

A base isomeric with ecgonine has been synthesized from tropinone by treating it with hydrocyanic acid followed by hydrolysis of the resulting compound. Propose a structure for this base, known as α-ecgonine, and write equations for the reactions described. Is there any ambiguity about the structure you propose?

B. Indole Alkaloids

A large and important class of alkaloids is structurally related to the amino acid tryptophan and contains the aromatic indole ring system (pp. 988 and 1121). Among the indole alkaloids, one that appears to be of central importance in physiology is 5-hydroxytryptamine, also known as serotonin. This compound is widely distributed in nature and stimulates a variety of smooth muscles and nerves. It has an essential function in the central nervous system as a neurotransmitter. Several drugs that interfere with the metabolism of serotonin in the brain because they are structurally similar to it are known to induce mental changes, including symptoms that resemble those of schizophrenia. Structural formulas for serotonin and for three indole alkaloids that cause hallucinations, bufotenin, psilocene, and lysergic acid, are given on the following page.

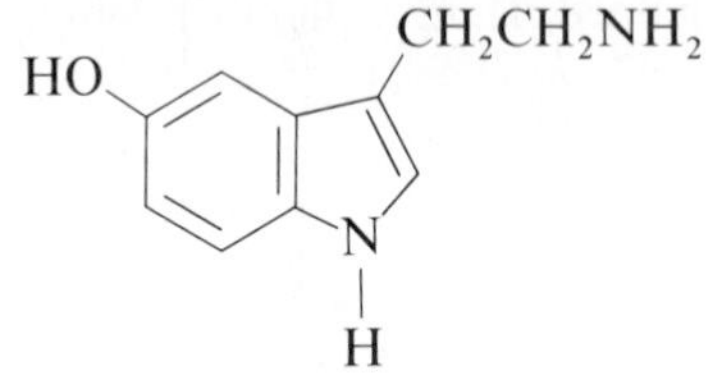

5-hydroxytryptamine
serotonin

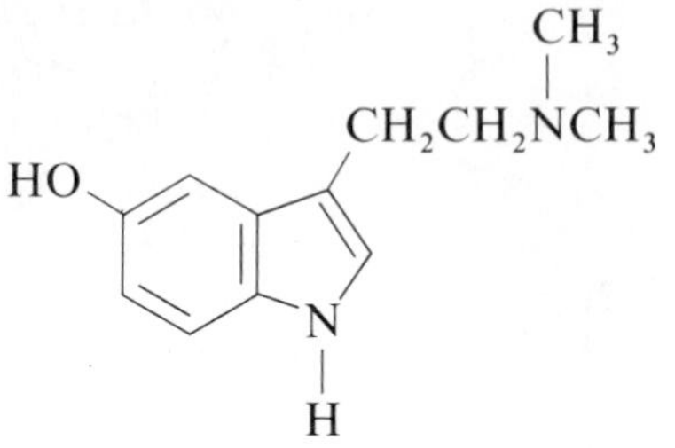

N,N-dimethyl-5-hydroxytryptamine
bufotenin

a psychoactive drug from the cahobe bean

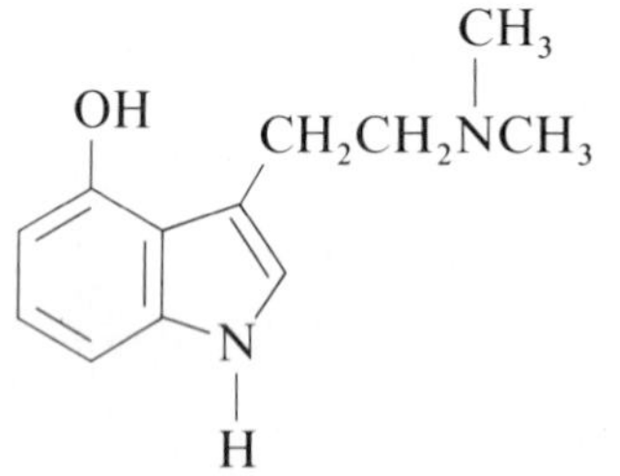

N,N-dimethyl-4-hydroxytryptamine
psilocine

active ingredient of hallucinogenic mushrooms

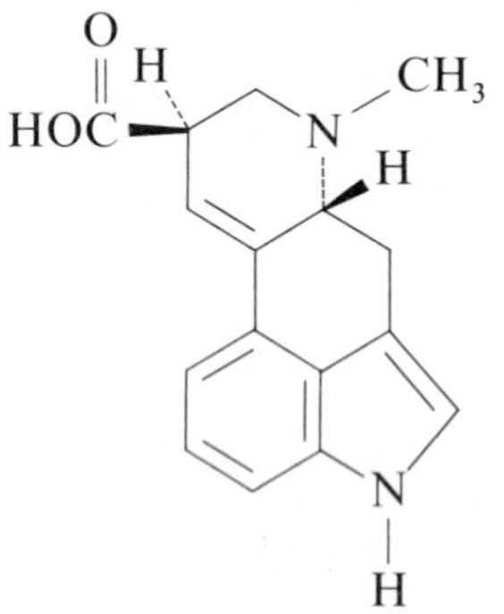

lysergic acid

an ergot alkaloid from a fungus of rye; LSD is the N,N-diethylamide of this compound

All of the compounds shown above have an indole ring substituted at carbon 3 by a two-carbon chain ending in an amino group. Serotonin is a primary amine, and bufotenin, psilocine, and lysergic acid are tertiary amines. In lysergic acid, the side chain forms part of two other rings. Serotonin, bufotenin, and psilocine also have phenolic hydroxyl groups on the indole ring. Serotonin is synthesized in mammals from the amino acid tryptophan, by hydroxylation of the aromatic ring (p. 968) and then decarboxylation.

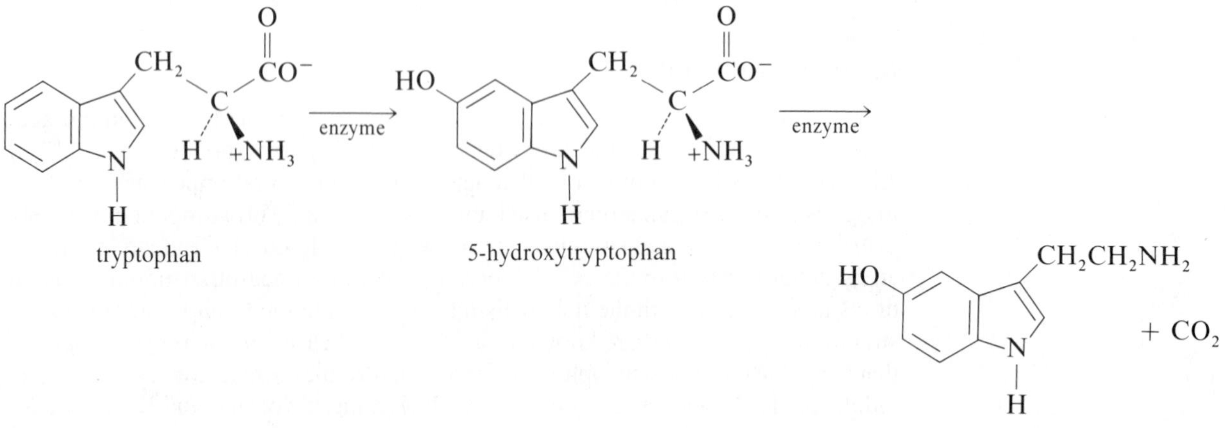

tryptophan → 5-hydroxytryptophan → serotonin + CO_2

PROBLEM 24.32

A synthesis of racemic lysergic acid is outlined below. Assign structures to the compounds designated by letters.

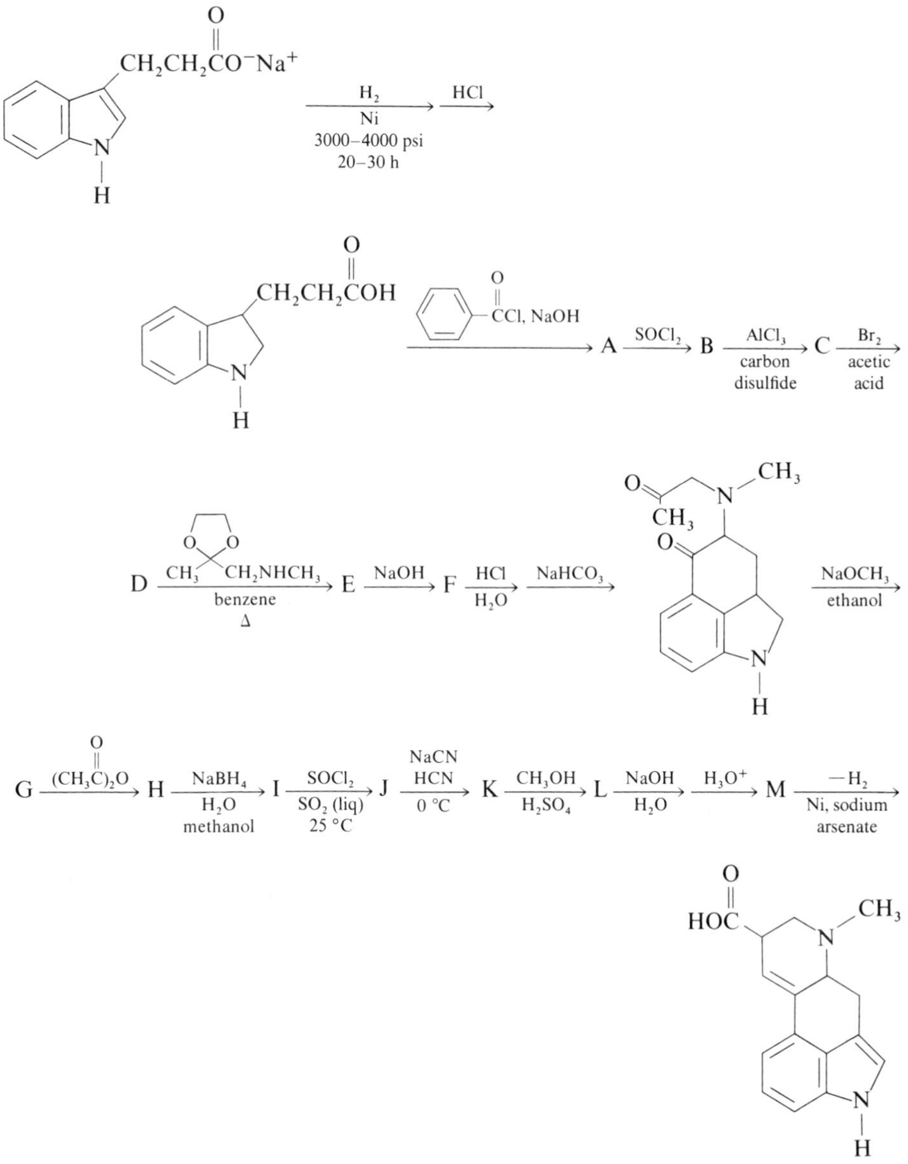

$CH_2CH_2CO^-Na^+$
H_2
Ni
3000–4000 psi
20–30 h
HCl
CH_2CH_2COH
CCl, NaOH
A
$SOCl_2$
B
$AlCl_3$
carbon disulfide
C
Br_2
acetic acid
D
CH_3
CH_2NHCH_3
benzene
Δ
E
NaOH
F
HCl
H_2O
$NaHCO_3$
CH_3
$NaOCH_3$
ethanol
G
$(CH_3C)_2O$
H
$NaBH_4$
H_2O
methanol
I
$SOCl_2$
SO_2 (liq)
25 °C
J
NaCN
HCN
0 °C
K
CH_3OH
H_2SO_4
L
NaOH
H_2O
H_3O^+
M
$-H_2$
Ni, sodium arsenate
HOC

(±)-lysergic acid

C. Isoquinoline Alkaloids

The most effective painkiller known is the alkaloid morphine, isolated from the juice obtained from unripe seed pods of the opium poppy, *Papaver somniferum*. Apparently morphine changes the perception of pain even when the pain itself is not much diminished. For this reason, the drug is valuable in medical practice. Unfortunately, morphine also has two severe disadvantages as a medication: it is addictive, and the body builds up a tolerance to it, so larger and larger doses may be necessary to provide the same relief from pain. The drug also depresses the function of the brain center that controls respiration; large doses of morphine (or of heroin, its synthetic diacetyl derivative) can kill by causing respiratory arrest. Another opium alkaloid is codeine, a monomethyl ether of morphine. Codeine is also a painkiller and is especially useful as a cough suppressant.

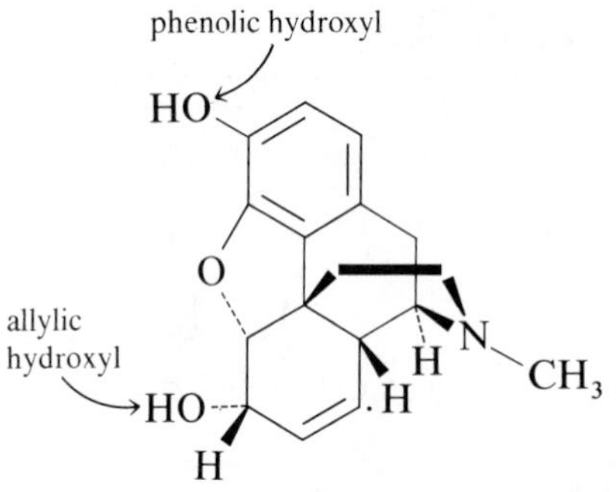
phenolic hydroxyl
allylic hydroxyl

morphine

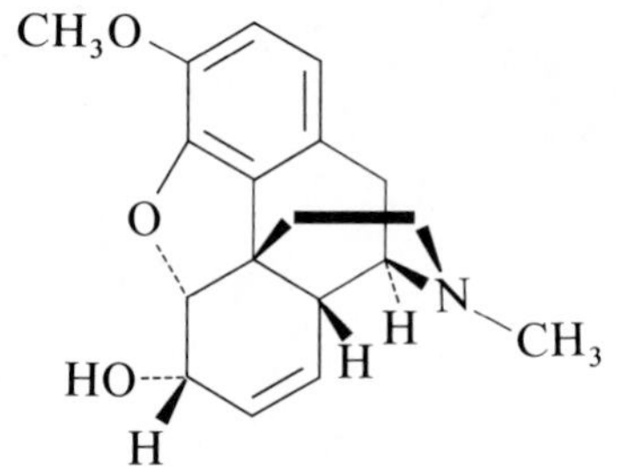

codeine

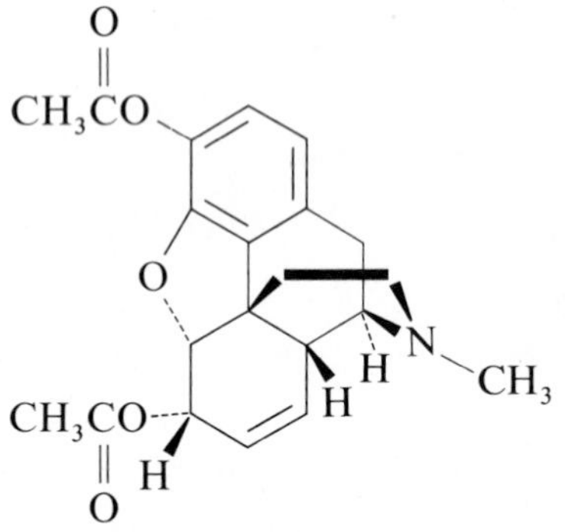

heroin

the opium alkaloids morphine and codeine, and the synthetic derivative heroin

All of these alkaloids have as part of their structure the benzylisoquinoline unit, which can be seen more easily in the structural formula of papaverine, another opium alkaloid.

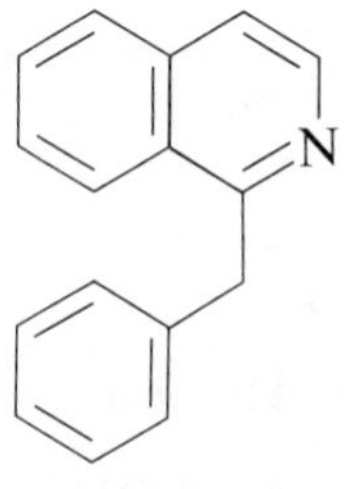

1-benzylisoquinone

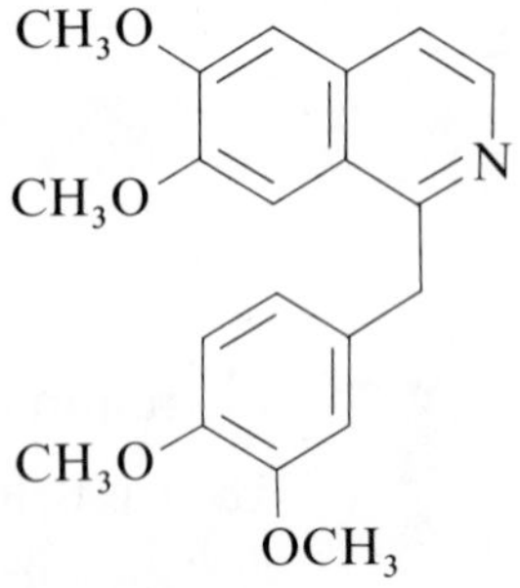

papaverine

These alkaloids, as well as some simpler ones, are synthesized in plants from the amino acid tyrosine. Tyrosine is first converted by oxidation of the aromatic ring (p. 968) into (3,4-dihydroxyphenyl)alanine (DOPA), a compound that is valuable in treating Parkinson's disease. (3,4-Dihydroxyphenyl)alanine is converted to β-(3,4-dihydroxyphenyl)ethylamine by decarboxylation or to (3,4-dihydroxyphenyl)pyruvic acid by transamination (p. 1123).

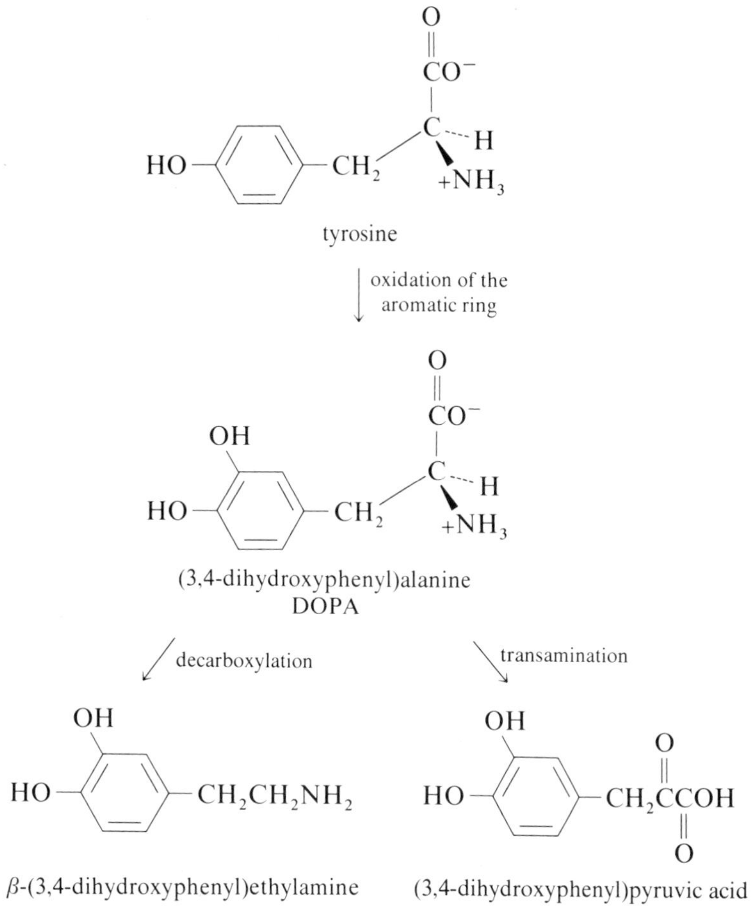
tyrosine
oxidation of the aromatic ring
(3,4-dihydroxyphenyl)alanine
DOPA
decarboxylation
transamination
β-(3,4-dihydroxyphenyl)ethylamine
(3,4-dihydroxyphenyl)pyruvic acid

The benzylisoquinoline skeleton of the opium alkaloids is created by condensation of β-(3,4-dihydroxyphenyl)ethylamine with (3,4-dihydroxyphenyl)pyruvic acid, followed by an electrophilic aromatic substitution reaction that forms the heterocyclic ring. The aromatic ring with two hydroxyl groups on it is highly activated, so a strong electrophile is not necessary for the reaction.

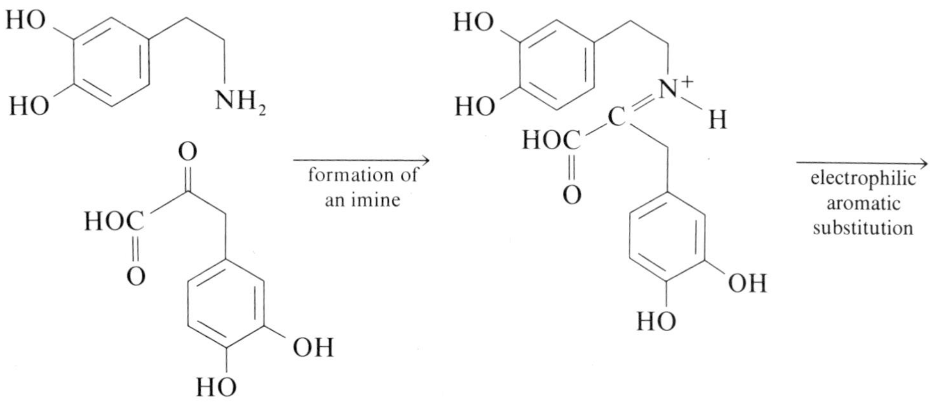
formation of an imine
electrophilic aromatic substitution

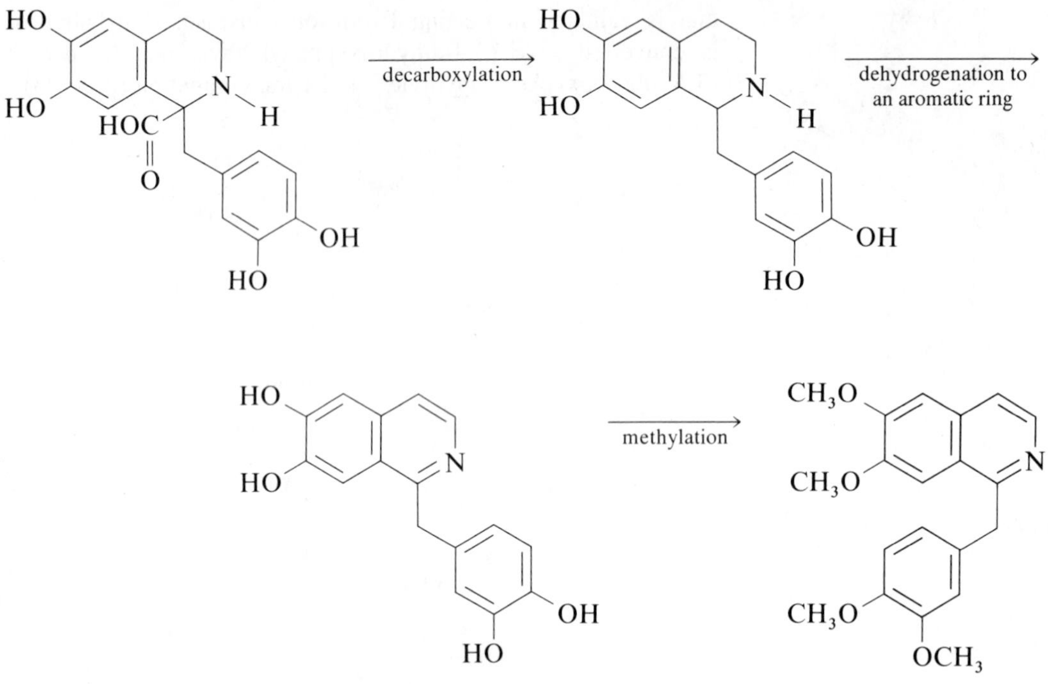
decarboxylation
dehydrogenation to an aromatic ring
methylation

papaverine

The intermediate formed by closure of the heterocyclic ring loses carbon dioxide and then hydrogen to form the stable aromatic isoquinoline compound. Methylation of the hydroxyl groups gives papaverine.

PROBLEM 24.33

Write mechanisms for the first three reactions in the sequence above. Use HB^+ and B: as the general acids and bases necessary in the reactions.

A key step in the biosynthesis of morphine is the oxidative coupling of the two phenolic rings (p. 850) in a benzylisoquinoline. In the following reactions, coupling occurs between the carbons ortho to one hydroxyl group and para to the other one.

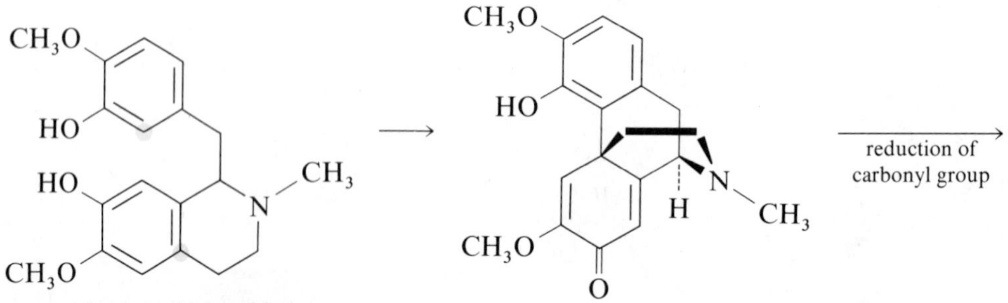
reduction of carbonyl group

positions at which oxidative coupling of the phenols takes place appear in color

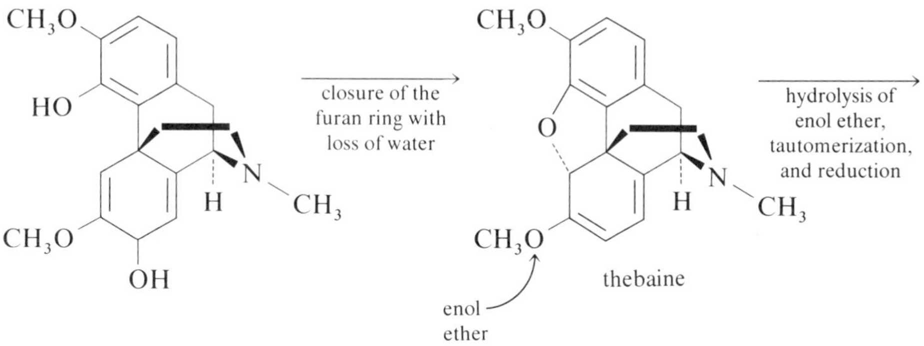
CH3O
HO
CH3O
OH
H
N
CH3
closure of the
furan ring with
loss of water
CH3O
O
CH3O
H
N
CH3
enol
ether
thebaine
hydrolysis of
enol ether,
tautomerization,
and reduction

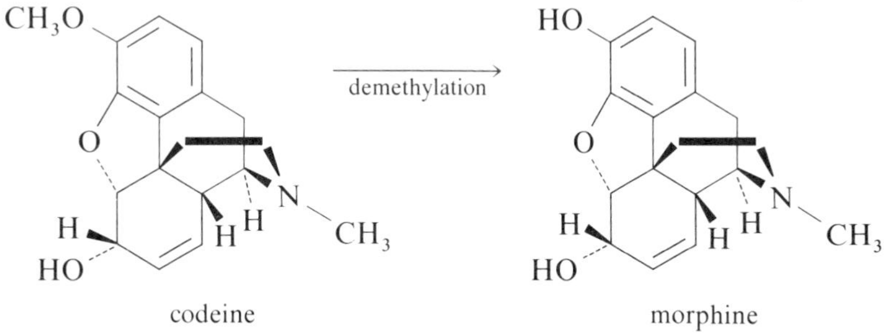
CH3O
O
H
HO
H
H
N
CH3
codeine
demethylation
HO
O
H
HO
H
H
N
CH3
morphine

Thebaine, an intermediate in the synthesis of codeine and morphine, is also an opium alkaloid.

PROBLEM 24.34

Using a 2-methoxy-5-methylphenol as a model for each of the two phenolic rings that undergo oxidative coupling in the precursor to morphine, write a mechanism for the coupling reaction. (Hint: A review of Section 20.7B may be helpful.)

PROBLEM 24.35

Suggest a mechanism for the formation of the furan ring in thebaine and for the hydrolysis of the enol ether in thebaine and the tautomerization of the product to give the α,β-unsaturated ketone that is the precursor to codeine.

PROBLEM 24.36

Tubocurarine, the potent curare alkaloid used by South American natives as a poison on the tips of their hunting arrows, is a bis(benzylisoquinoline) alkaloid. Dissect its structural formula, shown on p. 927, outlining the benzylisoquinoline units present.

PROBLEM 24.37

Write structural formulas for the intermediates and products indicated by letters in the following equations.

(a)

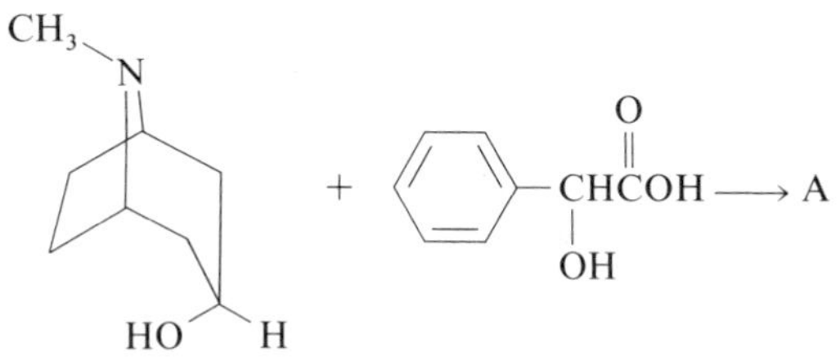
CH3
N
HO
H
+
O
CHCOH
OH
A

(b) 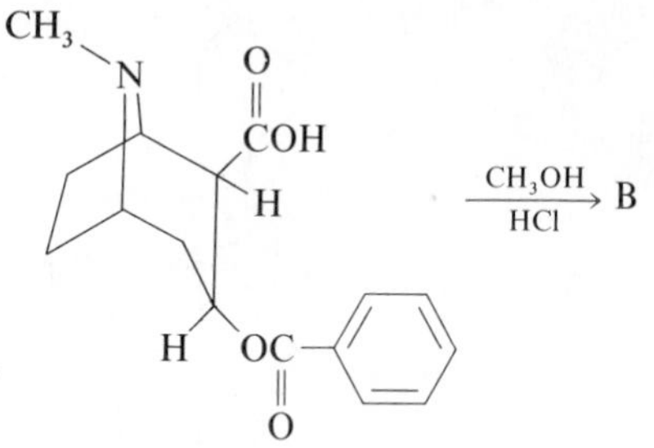

(c) 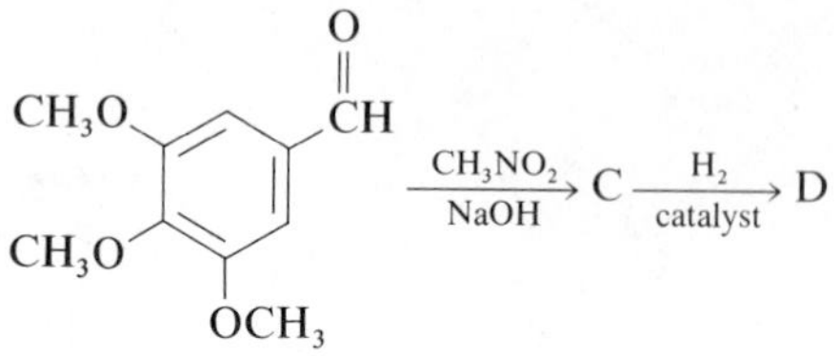

(d) 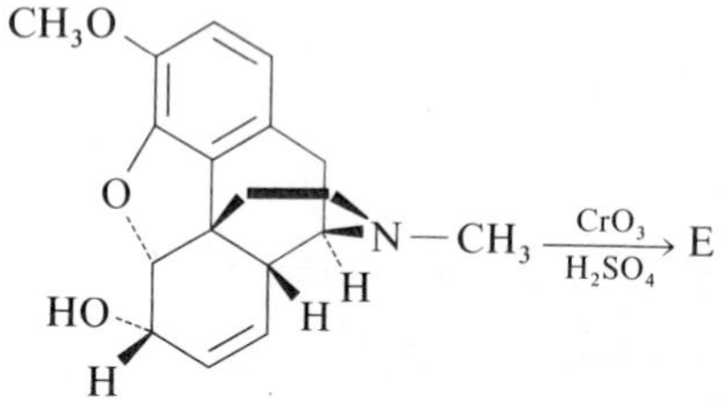

codeine

(e) 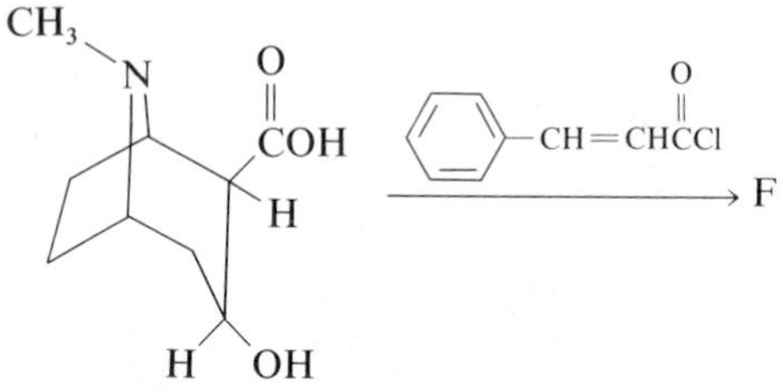

(f) 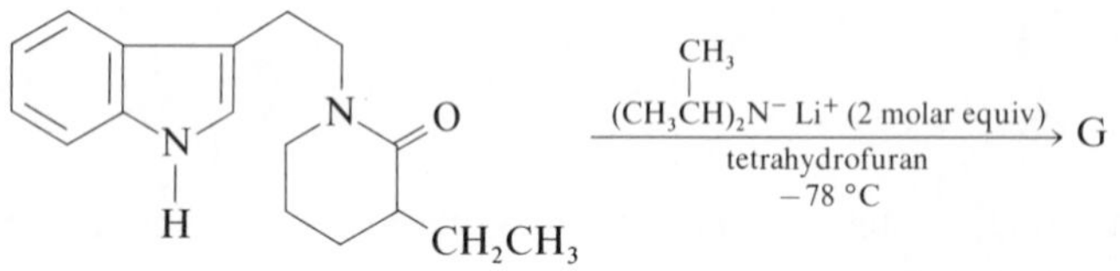

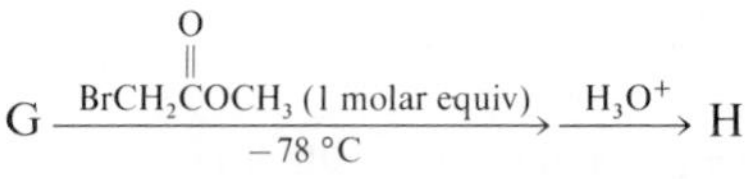

S U M M A R Y

Heterocyclic compounds are compounds that contain heteroatoms, atoms of elements other than carbon, in a ring. Heterocyclic compounds that have $(4n + 2)$ electrons (which may include nonbonding electrons from the heteroatom) in the ring are aromatic. Heterocyclic compounds may also be saturated or contain isolated double bonds. Because of the presence of nonbonding electrons on the heteroatom, heterocyclic compounds are bases. When the nonbonding electrons are part of the aromatic sextet, they are not as available for protonation. The basicity of such compounds (pyrrole is an example) is low.

An important way to synthesize heterocyclic compounds is to treat a diacarbonyl compound with a nucleophile containing the desired heteroatom. 1,4-Dicarbonyl compounds react with an external nucleophile containing a single heteroatom or by enolization of the carbonyl compound to give pyrroles, thiophenes, and furans, 1,3-Dicarbonyl compounds react with nucleophiles that have two adjacent heteroatoms to give five-membered heterocycles, such as isoxazoles from hydroxylamine and pyrazoles from hydrazine. 1,3-Dicarbonyl compounds react with urea to give pyrimidines. 1,2-Dicarbonyl compounds react with ammonia and aldehydes to give imidazoles. These methods for synthesizing heterocycles are summarized in Table 24.2.

TABLE 24.2 Synthesis of Heterocycles from Carbonyl Compounds and Nucleophiles

Carbonyl Compound	Reagent	Nucleophile or Intermediate	Product
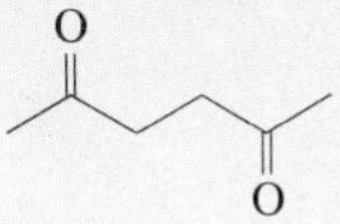	NH_3, Δ	NH_3	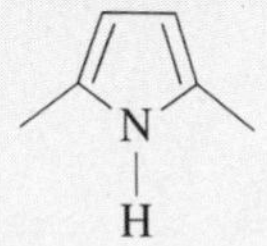
	HCl, Δ	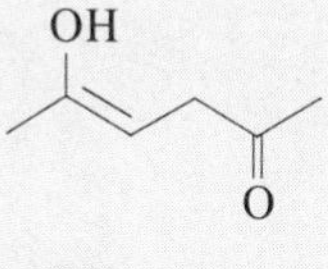	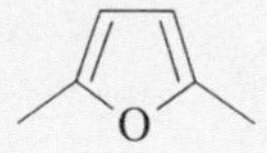
	P_4S_{10}, Δ	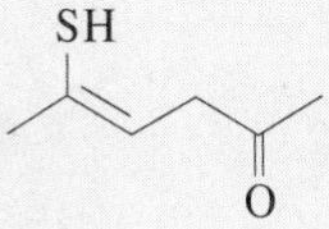	
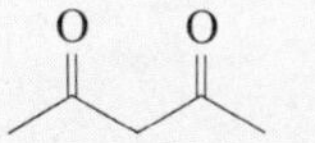	$H_2\overset{+}{N}H_3$ HSO_4^- NaOH	H_2NNH_2	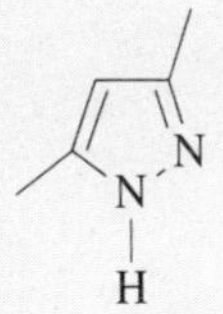
	$HO\overset{+}{N}H_3$ HSO_4^- K_2CO_3	$HONH_2$	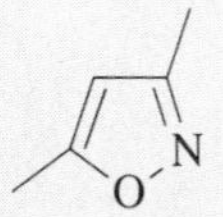
	$H_2N{-}C({=}O){-}NH_2$, Δ	$H_2N{-}C({=}O){-}NH_2$	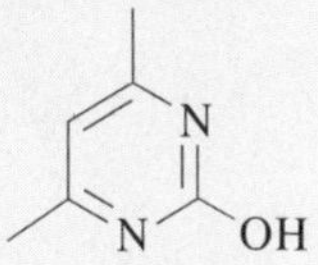
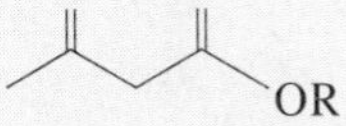	$H_2N{-}C({=}O){-}NH_2$, Δ	$H_2N{-}C({=}O){-}NH_2$	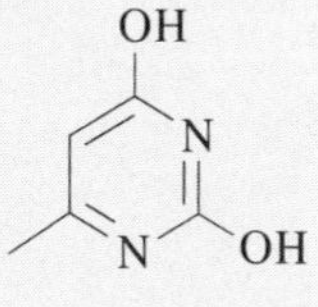
	$H_2N{-}C({=}S){-}NH_2$, Δ	$H_2N{-}C({=}S){-}NH_2$	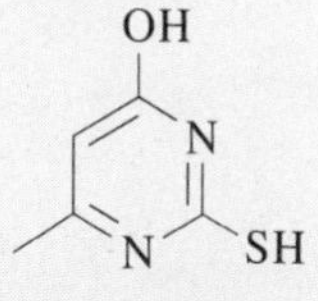
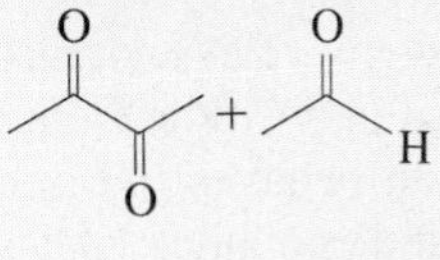	$CH_3\overset{O}{\overset{\Vert}{C}}O^-$ $\overset{+}{N}H_4$, Δ	NH_3	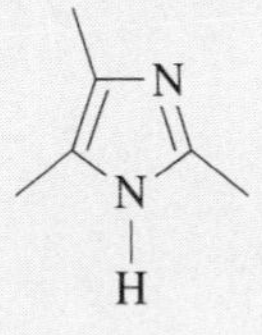

TABLE 24.3 Substitution Reactions of Heterocycles

Electrophilic Aromatic Substitution			
Heterocycle	**Reagent**	**Electrophile**	**Product**
 Y = S, NR	X_2	X_2	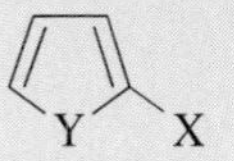
	HNO_3 $CH_3C(=O)OC(=O)CH_3$	$\overset{+}{N}O_2$	
Y = O, S, NR	$RC(=O)Cl$ or $RC(=O)OC(=O)R$ and BF_3 or $AlCl_3$	$RC{\equiv}O^+$	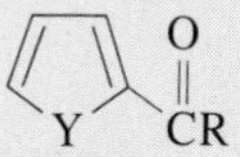
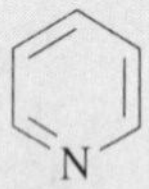	HNO_3, KNO_3 330 °C	$\overset{+}{N}O_2$	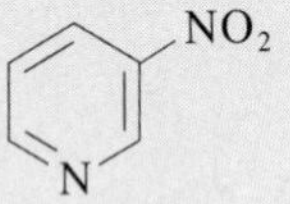 *low yield*
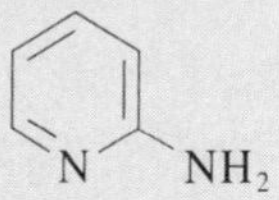	X_2	X_2	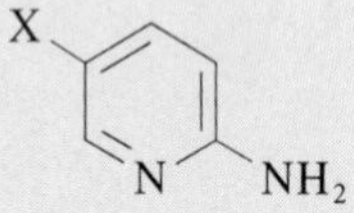
	HNO_3, H_2SO_4 then NaOH	$\overset{+}{N}O_2$	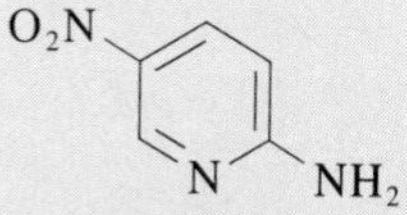
Nucleophilic Aromatic Substitution			
Heterocycle	**Reagent**	**Nucleophile**	**Product**
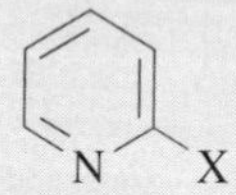	RNH_2 Δ	$R\ddot{N}H_2$	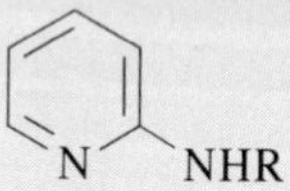
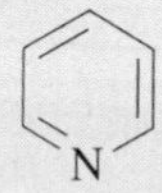	$Na^+\ NH_2^-$ Δ then H_2O	$:\ddot{N}H_2^-$	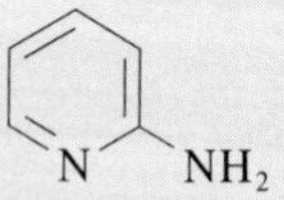
	RLi	$R:^-$	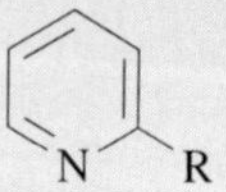

Five-membered aromatic heterocycles are reactive toward electrophilic substitution and are usually substituted in position 2. Furan is an exception. It has a low resonance stabilization and often behaves like a diene, undergoing 1,4-addition reactions with halogens or good dienophiles.

Six-membered aromatic heterocycles are deactivated toward electrophilic substitution, both because of the presence of an electronegative atom in the ring and because electrophiles complex with the heteroatom, making it even more electron-withdrawing. Under severe reaction conditions, substitution occurs at position 3. The presence of the electron-withdrawing heteroatom activates the ring toward nucleophilic substitution reactions, which occur at position 2. The inertness of the pyridine ring toward electrophilic attack also allows side chains on the ring to be oxidized without destruction of the ring. The substitution reactions of heterocycles are summarized in Table 24.3.

Heterocyclic compounds are important medicinally and biologically. Purine and pyrimidine bases bonded to the sugars ribose and 2-deoxyribose constitute the nucleosides that, as their phosphate esters, the nucleotides, are constituents of RNA and DNA. Alkaloids, of which morphine is an example, are an important class of natural products, many of which have biological activity. Nicotinamide and thiamine are among the many vitamins that are heterocycles. Porphyrins are part of heme, the carrier of oxygen in warm-blooded animals, and of chlorophyll, which mediates the conversion of light energy from the sun to the energy stored in carbohydrates.

ADDITIONAL PROBLEMS

24.38 Name the following compounds.

(a)
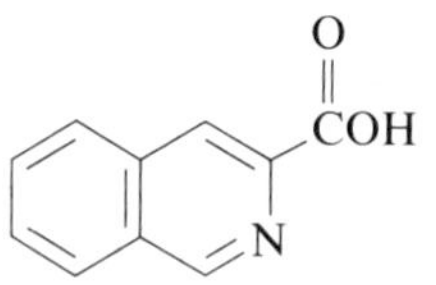
O
COH
N

(b)
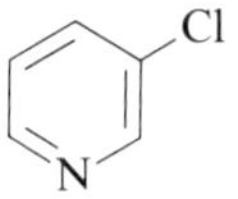
Cl
N

(c)
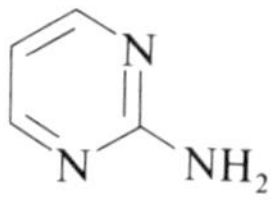
N
N
NH2

(d)
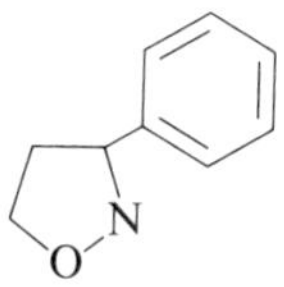
O
N

(e)
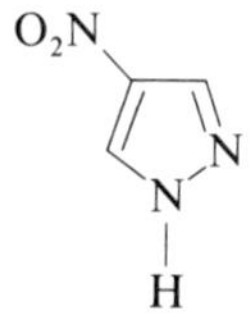
O2N
N
N
H

(f)
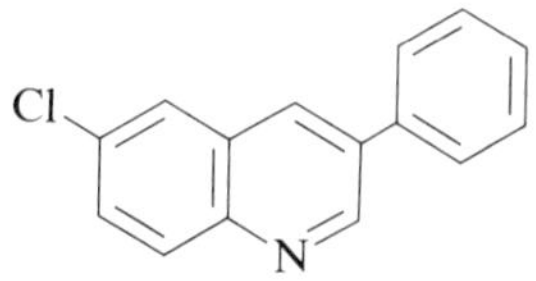
Cl
N

(g)
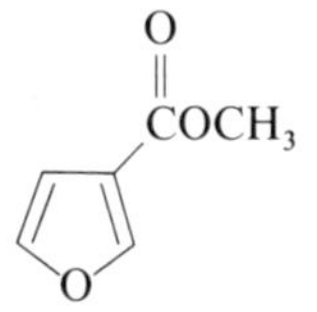
O
COCH3
O

(h)
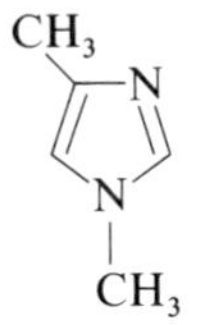
CH3
N
N
CH3

(i)
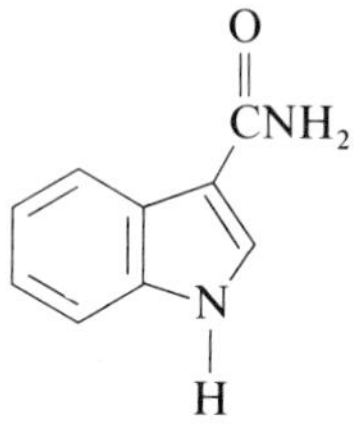
O
CNH2
N
H

(j)
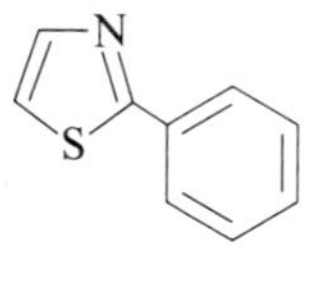
N
S

(k)
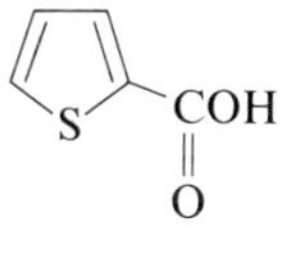
S
COH
O

24.39 Write structural formulas for all intermediates and products designated by letters in the following equations.

(a)
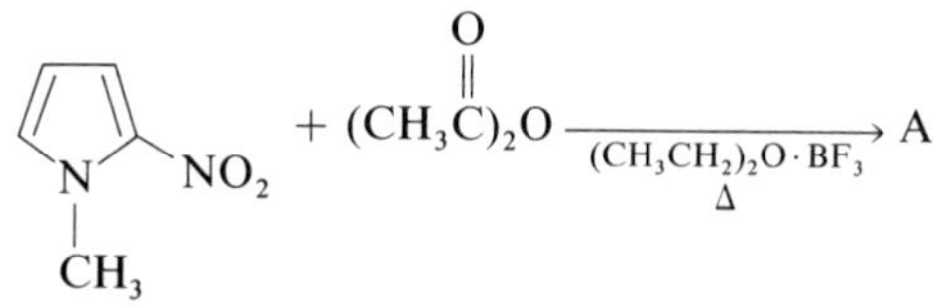
N
NO2
CH3
O
+ (CH3C)2O
(CH3CH2)2O · BF3
Δ
A

(b)
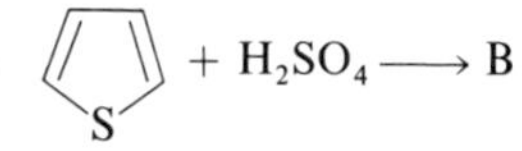
S
+ H2SO4
B

(c)
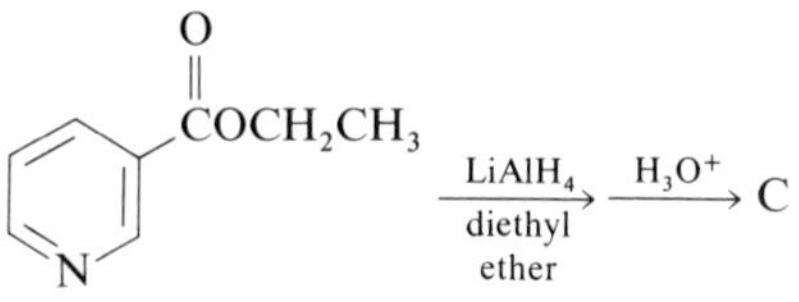
O
COCH2CH3
N
LiAlH4
diethyl
ether
H3O+
C

(d) 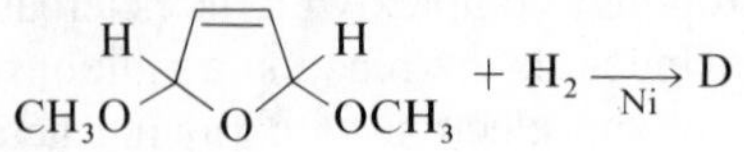

(e) 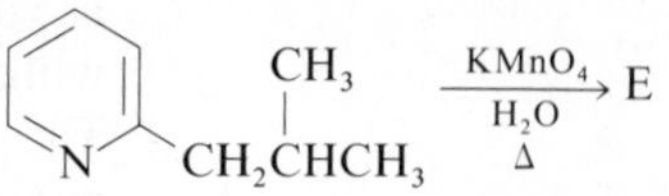

(f) 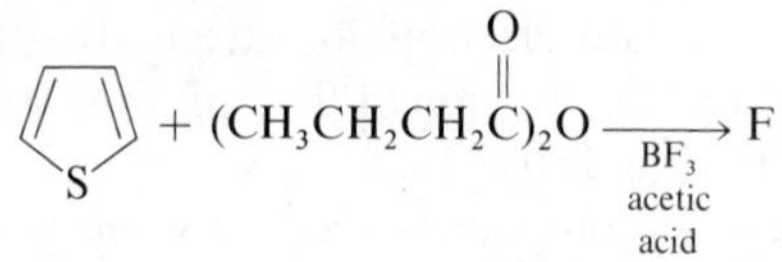

(g) 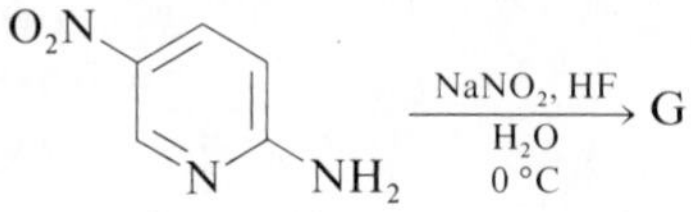

(h) 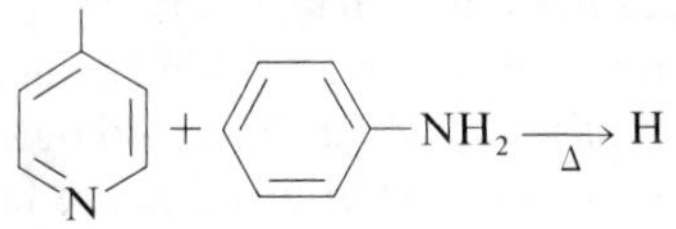

(i) 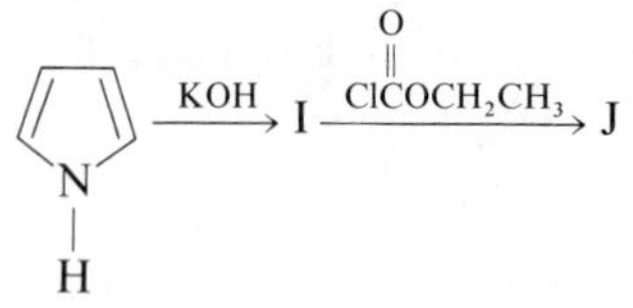

(j) 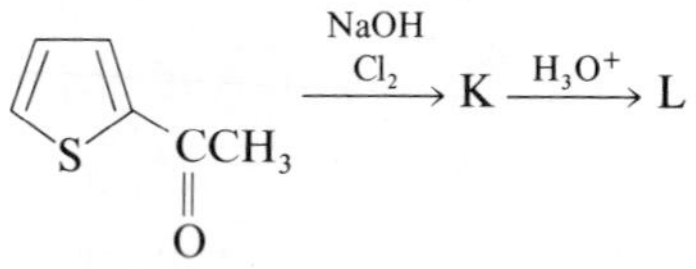

(k) 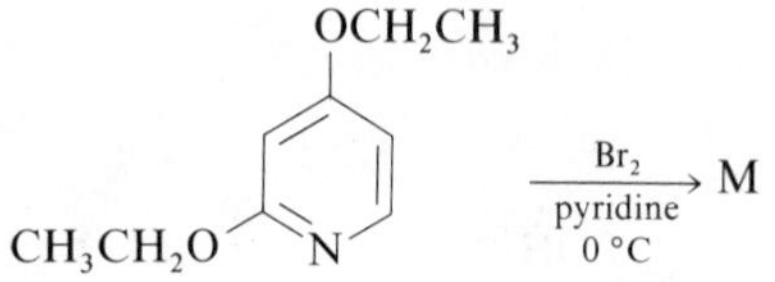

(l) 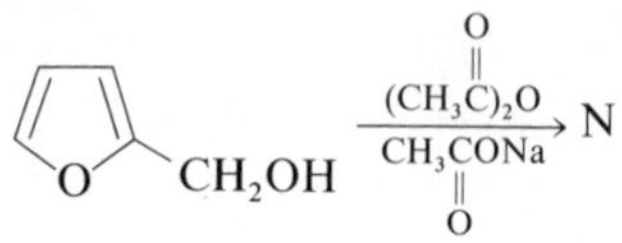

(m) 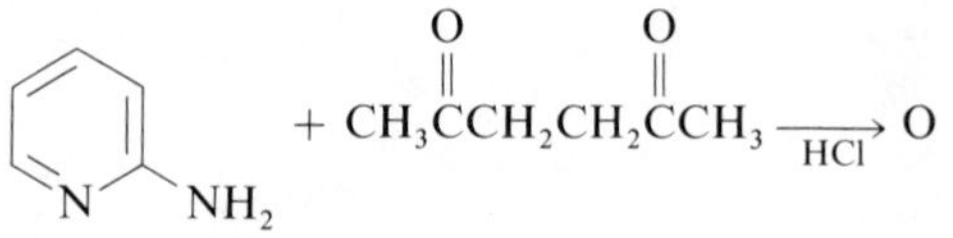

(n) 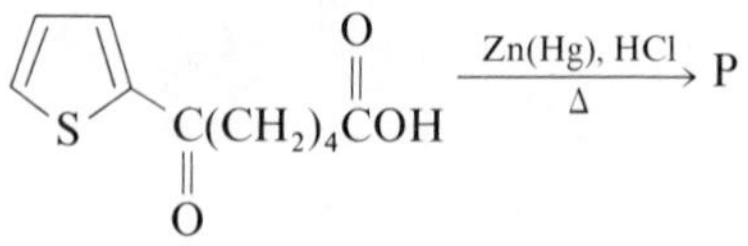

(o) 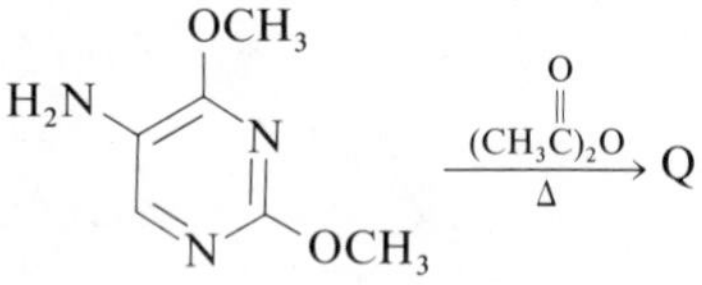

(p) 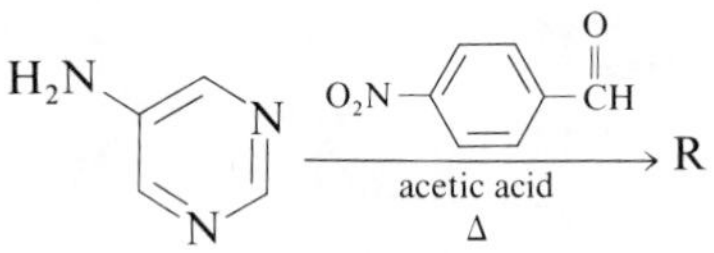

(q) 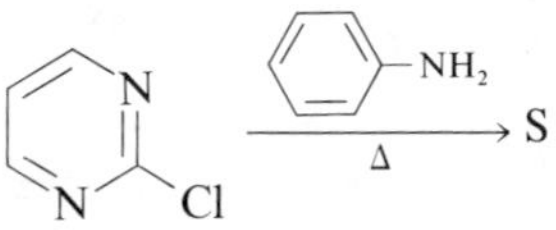

(r) 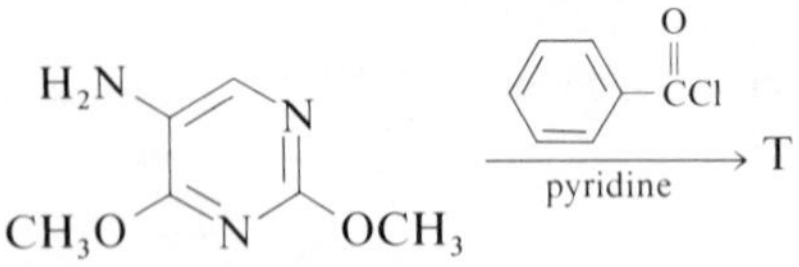

(s) 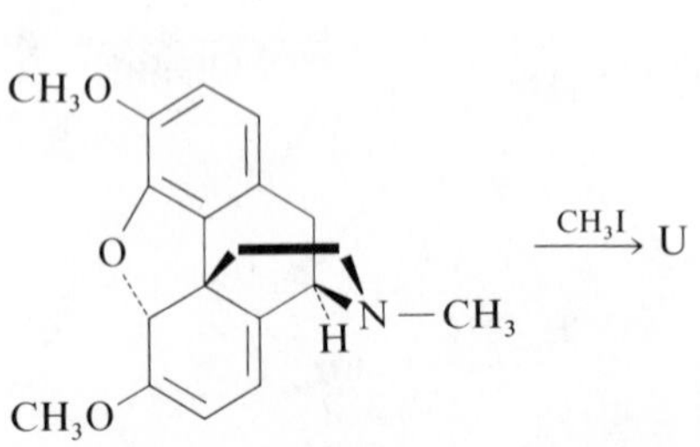

thebaine

(t) 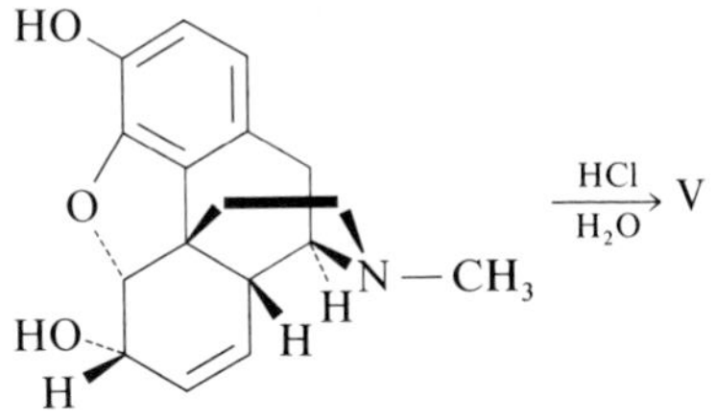

morphine

(u) 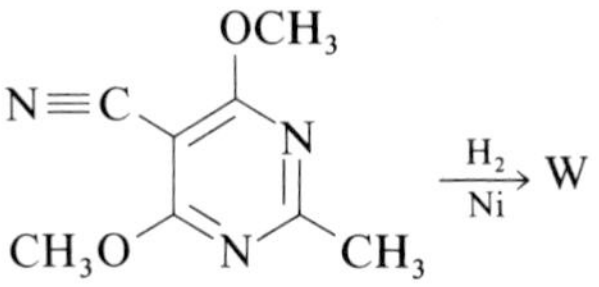

(v) 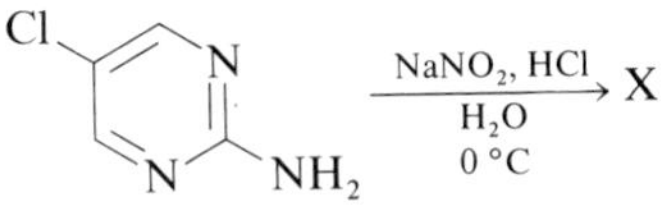

24.40 Give structural formulas for all intermediates and products designated by letters in the following equations.

(a) 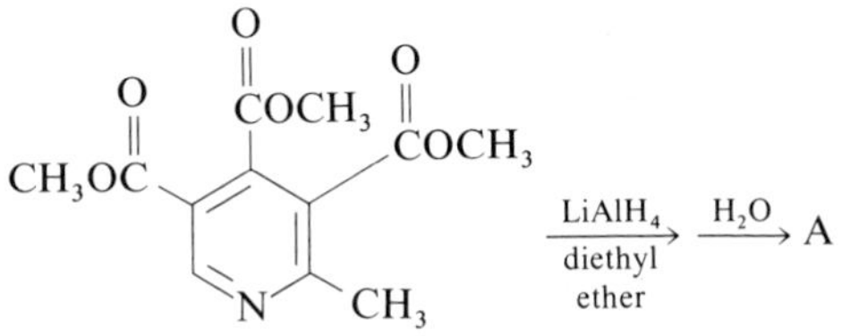

(b) 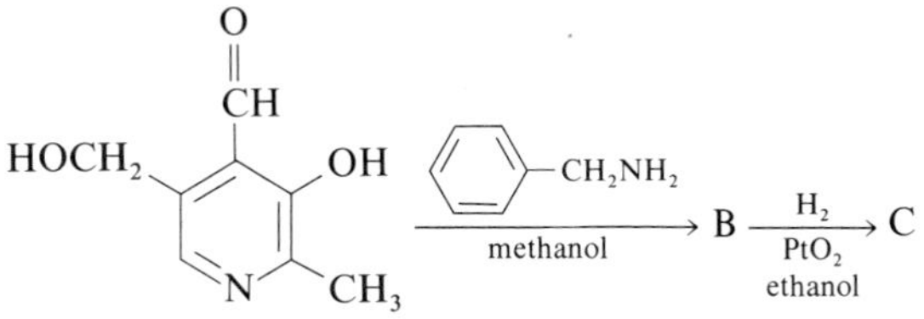

(c) 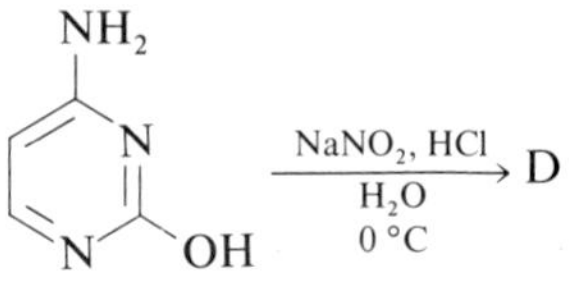

(d) 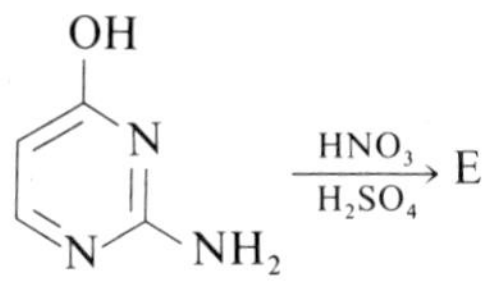

(e) 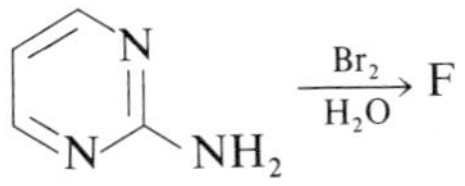

(f) 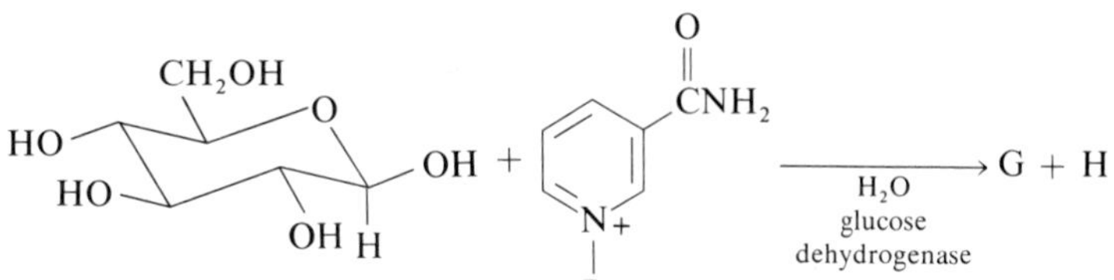

glucose (p. 1064)

(g) 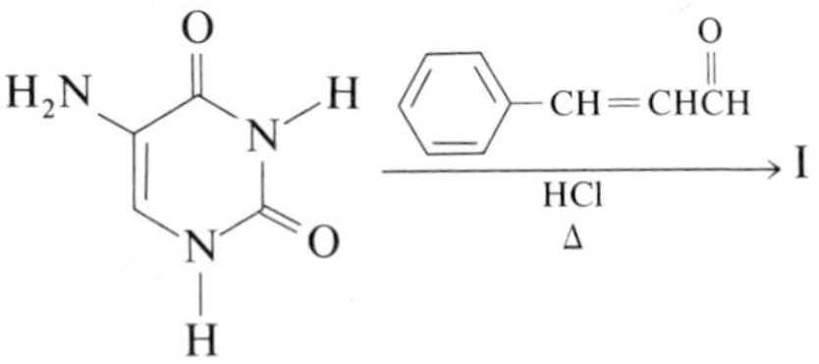

(h) 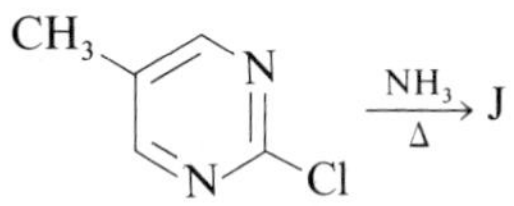

(i)

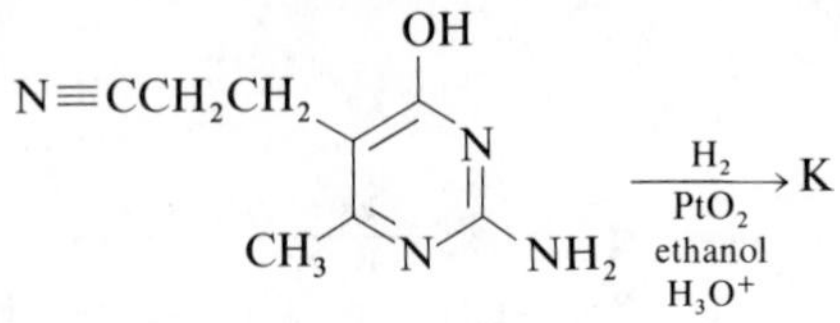

(j)

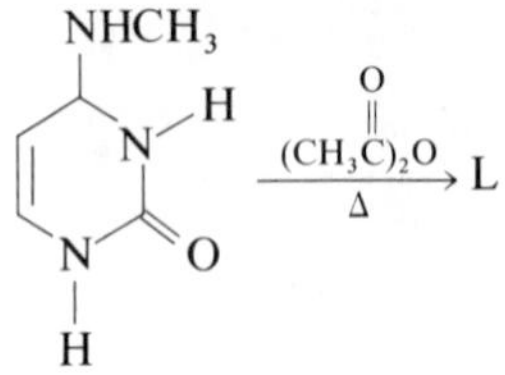

(k)

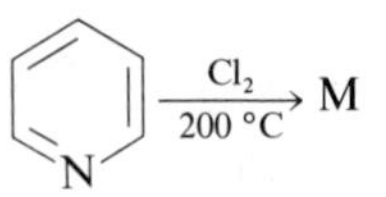

(l)

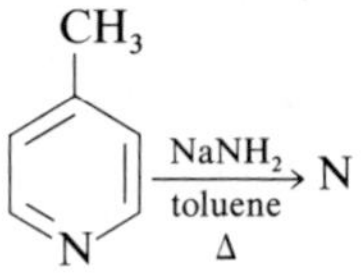

(m)

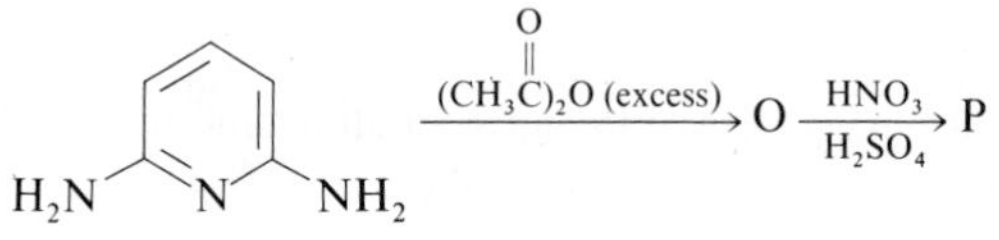

(n)

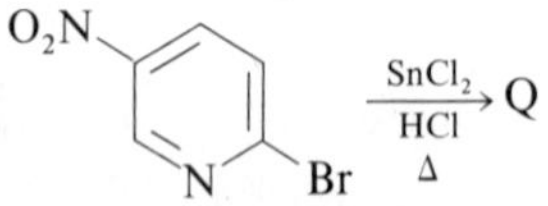

(o)

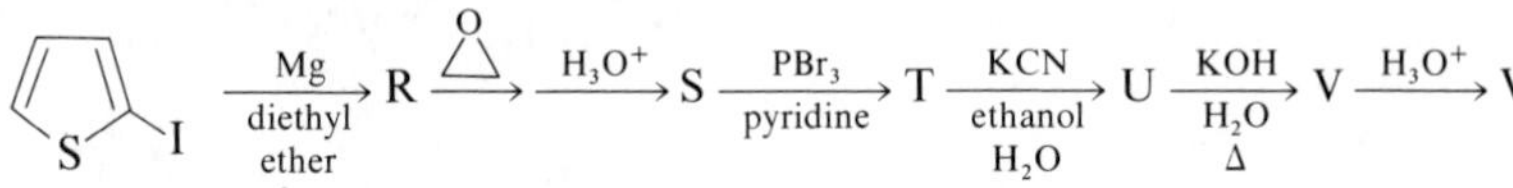

(p)

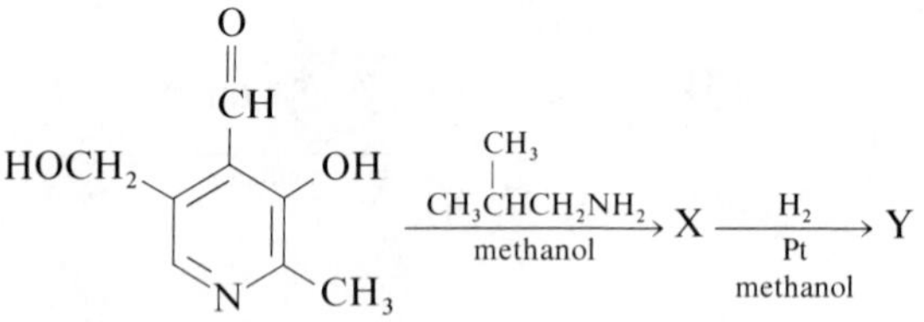

(q)

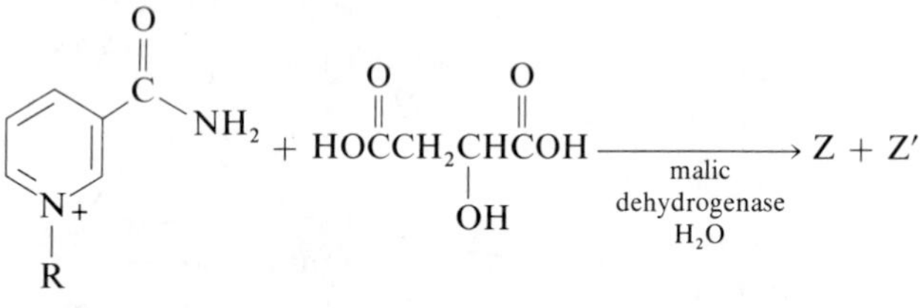

malic acid

24.41 Write detailed mechanisms showing how the following transformations could have taken place.

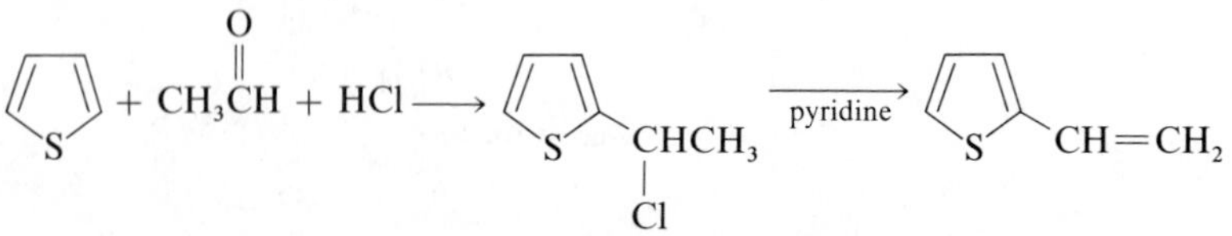

24.42 The two enantiomeric ethanol-1-*d* species obtained in the reactions shown on pp. 1022 and 1023 were collected at first in quantities too small to allow their optical rotations to be measured. The fact that they are enantiomers was originally proved by the following sequence of reactions.

$$CH_3\overset{\overset{\displaystyle O}{\|}}{C}D + NADH \longrightarrow \text{alcohol A} \xrightarrow[\text{pyridine}]{\text{TsCl}} B \xrightarrow[\substack{H_2O\\ \Delta}]{\text{NaOH}} C$$

$$C + NAD^+ \longrightarrow \underset{\substack{\text{containing}\\ \text{no deuterium}}}{\text{acetaldehyde}} + \underset{\substack{\text{molar equivalent}\\ \text{of deuterium}}}{\text{NADD containing 1}}$$

$$A + NAD^+ \longrightarrow \underset{\substack{\text{containing}\\ \text{deuterium}}}{\text{acetaldehyde}} + \underset{\text{no deuterium}}{\text{NADH containing}}$$

Write equations showing what is happening in this series of reactions. Be sure to use correct stereochemical representations of all the compounds involved. How do these results prove that the alcohols obtained in the original experiments are enantiomeric?

24.43 Arrange the following species in order of increasing acidity, and give your reasons for the order you choose.

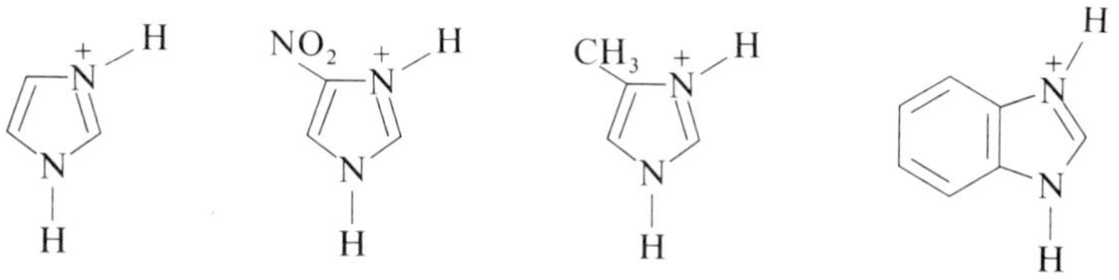

24.44 Propose a synthesis for each of the following compounds. You must start with thiophene, furan, pyrrole, or a methylpyridine as one of the reagents for each synthesis. The other organic reagents you use may not contain more than seven carbon atoms.

(a) 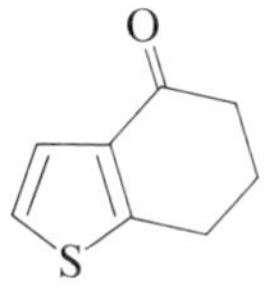 (b) 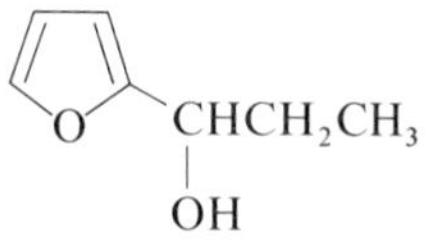 (c) 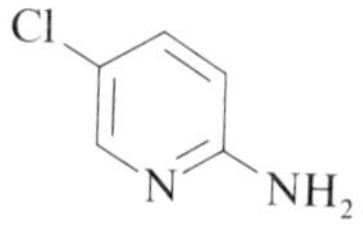

(d) 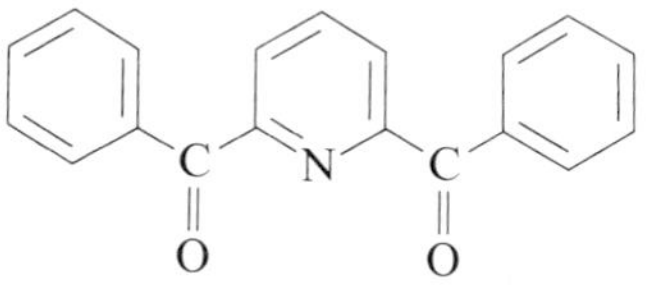 (e) 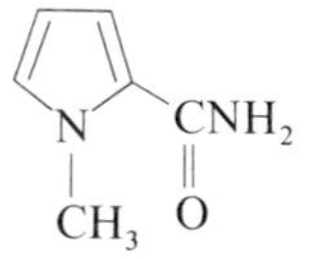 (f) 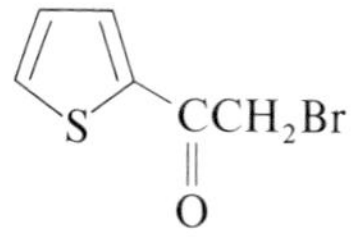

(g) 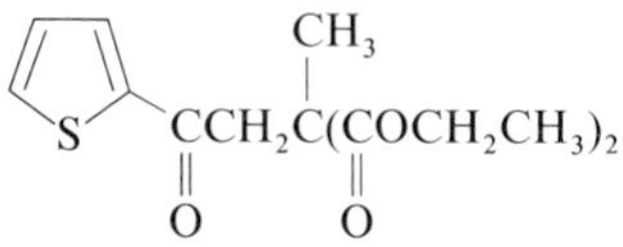 (h) 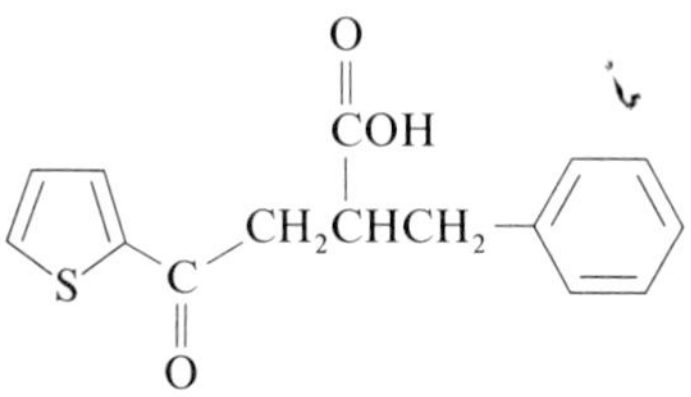

24.45 When phenylhydrazine is heated with 2-butenoic acid, water is evolved and a heterocyclic compound having the molecular formula $C_{10}H_{12}N_2O$ is formed. Propose a structure for this product. (Hint: A review of Section 18.2A may be helpful.) When phenylhydrazine is heated with ethyl acetoacetate, another heterocycle, $C_{10}H_{10}N_2O$, is formed. What is the structure of this compound?

25

Carbohydrates

A • L O O K • A H E A D

Carbohydrates are polyhydroxy aldehydes or ketones found abundantly in nature. Glucose, the primary source of energy in the human body, is a typical carbohydrate.

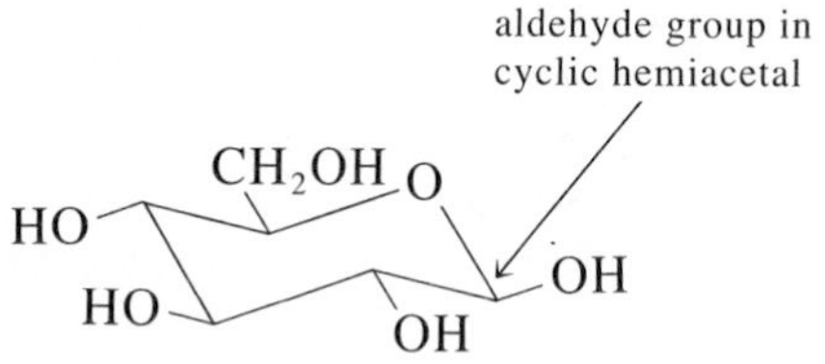
aldehyde group in
cyclic hemiacetal
CH₂OH
O
HO
HO
OH
OH

β-D-glucose

Carbohydrates react as carbonyl compounds and as alcohols. A study of their reactions is a review of those functional groups. The structures of carbohydrates range from the relatively simple, such as glucose, to the complex polymeric structures found in starches and cellulose.

Because compounds like glucose have several stereocenters, the history of the determination of the structure of carbohydrates is also the history of the development of the knowledge of stereochemistry. Fischer projection formulas, another way of representing stereochemistry, will be used in this chapter to clarify the stereochemical relationships among carbohydrates.

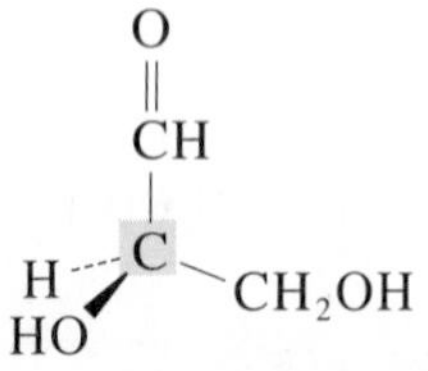
O
CH
C
H
HO
CH₂OH

(*R*)-(+)-glyceraldehyde
perspective formula

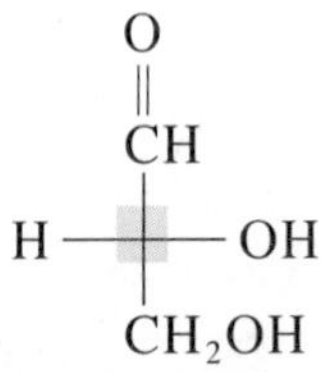
O
CH
H
OH
CH₂OH

(*R*)-(+)-glyceraldehyde
Fischer projection formula

This chapter will also show the reactions and reasoning processes that allowed early organic chemists to assign structures to these complex natural products.

Carbohydrates, important constituents of both plants and animals, are polyhydroxy aldehydes or ketones. Compounds classified as carbohydrates vary from those consisting of a few carbon atoms to gigantic polymeric molecules having molecular weights in the millions.

Carbohydrates that cannot be broken down into simpler units by hydrolysis reactions are known as **monosaccharides.** The most common monosaccharide is glucose, $C_6H_{12}O_6$, the chief form in which carbohydrates are metabolized in the human body. Fructose, $C_6H_{12}O_6$, is known as fruit sugar. Ribose, $C_5H_{10}O_5$, and 2-deoxyribose, $C_5H_{10}O_4$, are components of ribonucleic acids (RNA) and deoxyribonucleic acids (DNA), respectively, the giant molecules that play an important role in the storage and transmission of genetic information (p. 1013). Monosaccharides exist as cyclic hemiacetals or hemiketals (p. 499) in equilibrium with open-chain hydroxy aldehyde or ketone structures.

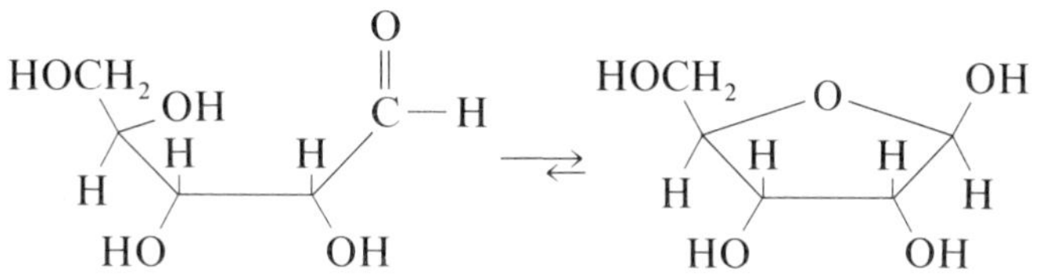

aldehyde form — hemiacetal form

ribose

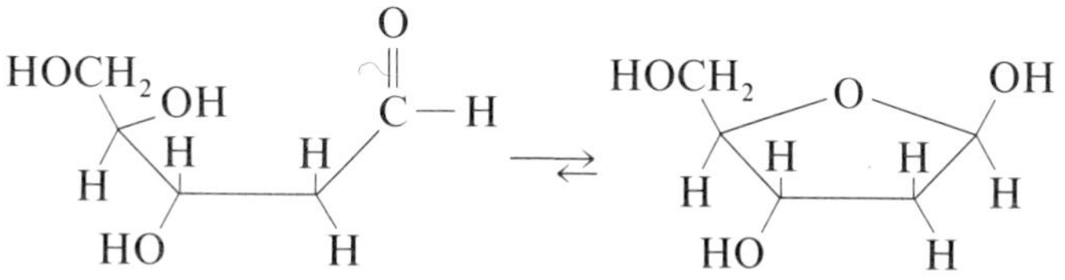

aldehyde form — hemiacetal form

2-deoxyribose

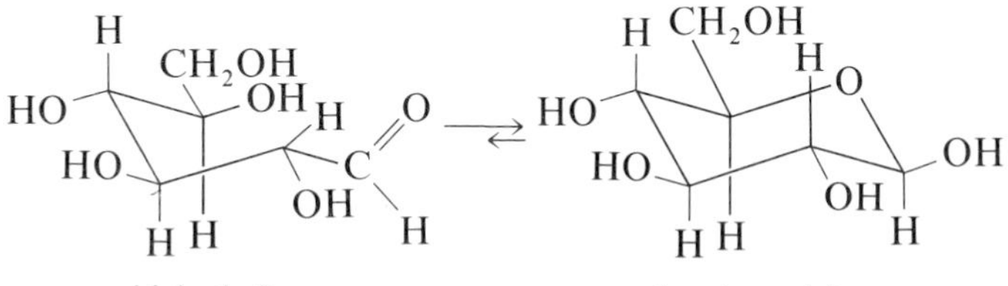

aldehyde form — hemiacetal form

glucose

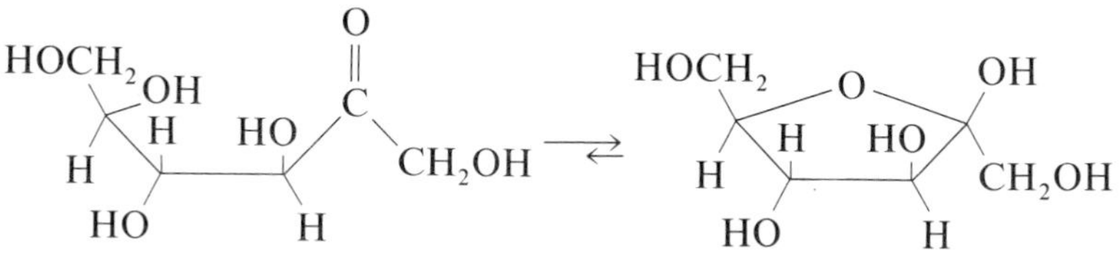

ketone form — hemiketal form

fructose

Monosaccharides can be further subdivided according to the number of carbon atoms they contain and whether they are aldehydes **(aldoses)** or ketones **(ketoses).** Ribose, with five carbon atoms, is a pentose. It has an aldehyde function, so it is an aldopentose. Glucose is an aldohexose, and fructose is a ketohexose.

PROBLEM 25.1

Besides pentoses and hexoses, there are trioses, with three carbon atoms, and tetroses, with four carbon atoms. Classify the following monosaccharides according to the number of carbon atoms and the nature of the carbonyl function they contain.

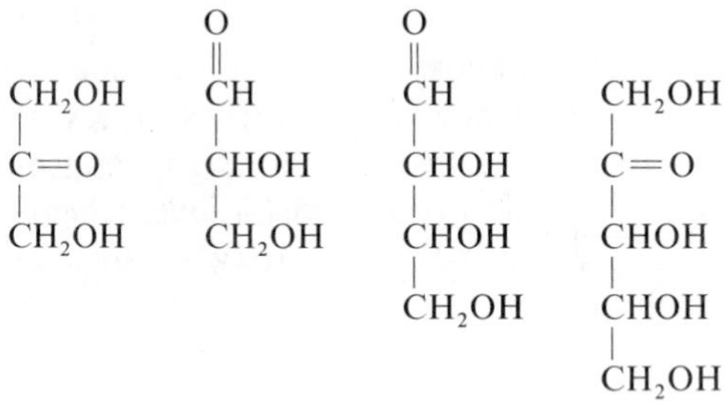

A B C D

Hydroxyl and carbonyl groups are not the only functional groups to appear in monosaccharides. Monosaccharides containing carboxyl groups and amino groups are common structural units in biologically important carbohydrates. Two such monosaccharides are 2-amino-2-deoxyglucose, $C_6H_{13}NO_5$, an amino sugar (also called glucosamine), and glucuronic acid, $C_6H_{10}O_7$, a sugar acid.

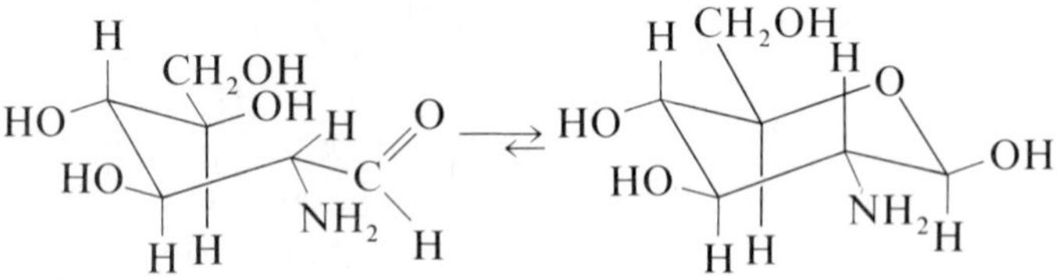

2-amino-2-deoxyglucose
glucosamine

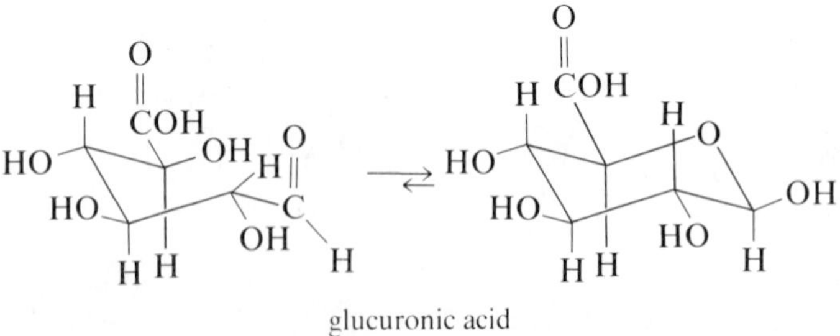

glucuronic acid

Common table sugar, sucrose, has the molecular formula $C_{12}H_{22}O_{11}$. When it is boiled with water with a trace of acid, as in candy making, it is converted into a mixture of glucose and fructose. Therefore, it is classified as a **disaccharide,** a compound made up of two monosaccharide units.

$$\underset{\text{sucrose}}{C_{12}H_{22}O_{11}} + H_2O \xrightarrow[\Delta]{H_3O^+} \underset{\text{glucose}}{C_6H_{12}O_6} + \underset{\text{fructose}}{C_6H_{12}O_6}$$

Maltose, another disaccharide, gives two molecules of glucose on hydrolysis.

$$\underset{\text{maltose}}{C_{12}H_{22}O_{11}} + H_2O \xrightarrow[\Delta]{H_3O^+} \underset{\text{glucose}}{2\ C_6H_{12}O_6}$$

Similarly, **trisaccharides** and **tetrasaccharides** give three and four monosaccharide units, respectively, on hydrolysis. Compounds containing from two to ten monosaccharide units are called **oligosaccharides.** Saccharides in this molecular weight range are individual, identifiable compounds with definite structures and molecular weights. They are crystalline compounds that are water-soluble and sweet-tasting. Oligosaccharides have many important physiological functions. For example, blood groups are determined by oligosaccharides combined with proteins on the surface of red blood cells. In fibrinogen, an important component in blood clotting, protein is also associated with oligosaccharides. Immunoglobulins, which are involved in the development of immunity to disease, are also composed of carbohydrates and proteins. The chemistry of carbohydrates is, thus, directly relevant to the processes of cell recognition and regulation of cell growth, so important to solving the problem of cancer.

When there are more than ten monosaccharide units in molecules, the compounds are defined as **polysaccharides.** Cellulose (p. 1103), the chief structural material of plants, is a high-molecular-weight polysaccharide made up of glucose units. Not all cellulose molecules have the same molecular weight. Instead, any sample contains molecules with a given range of molecular weights. Starch (p. 1101) is the polysaccharide form in which plants store glucose for their energy needs. Animals store glucose as another polysaccharide, glycogen.

Study Guide
Concept Map 25.1

25.2 STEREOCHEMISTRY OF SUGARS

A. Glyceraldehyde as the Standard for the Assignment of Relative Configurations

The chemistry of carbohydrates has been central to the development of chemists' understanding of stereochemistry and the assignment of absolute configuration (p. 195) to chiral compounds. In 1906, M. Rosanoff, an American chemist, suggested that (+)-glyceraldehyde be assigned a configuration and used as the standard for configurations of other sugars and, ultimately, other chiral compounds. The configuration that was chosen for (+)-glyceraldehyde was the *R* configuration, and, therefore, (−)-glyceraldehyde was assigned the *S* configuration.

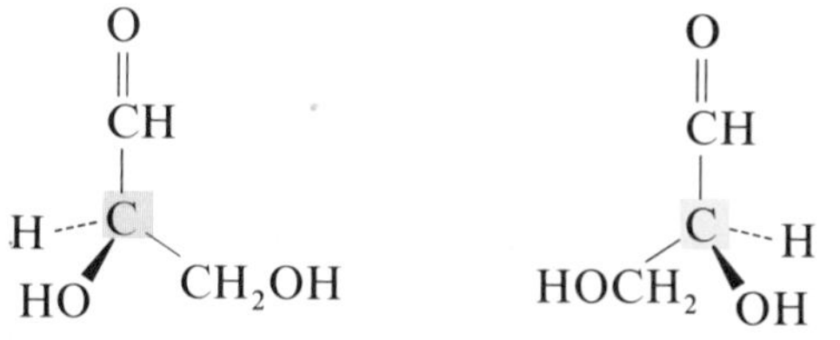
O
CH
H
C
HO
CH2OH
O
CH
C
H
HOCH2
OH

(*R*)-(+)-glyceraldehyde (*S*)-(−)-glyceraldehyde

configurations of (+)- and (−)-glyceraldehyde as assigned in 1906

This assignment was chosen for (+)-glyceraldehyde because the hydroxyl group and the hydrogen atom at carbon 5 of (+)-glucose had been assigned the corresponding configuration in 1891 by Emil Fischer. Because (+)-glyceraldehyde could be related by chemical transformations to (+)-glucose, the two compounds had to have the same stereochemistry at the stereocenter closest to the primary alcohol group in each molecule. Rosanoff and Fischer, of course, had a 50% chance that their assignments were correct. (+)-Glyceraldehyde had to have either the *R* configuration or the enantiomeric *S* configuration.

Rosanoff's suggestion was widely accepted. Over the years, many compounds were synthesized from or degraded to (+)-glyceraldehyde so that their relative configurations could be established. Each compound that was related to (+)-glyceraldehyde could then serve as a standard for other compounds, until the relative configurations of many compounds were known. The following reaction provides a simple example of the kinds of correlations that can be made. (*R*)-(+)-Glyceraldehyde is oxidized to (*R*)-(−)-glyceric acid (2,3-dihydroxypropanoic acid).

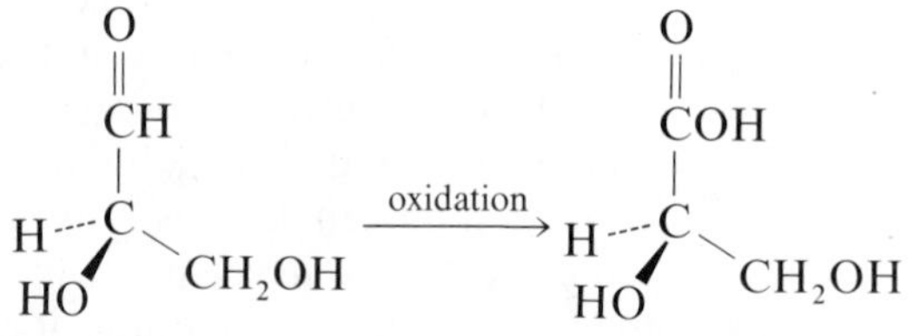
O
CH
H
HO
C
CH_2OH
oxidation
O
COH
H
HO
C
CH_2OH

(*R*)-(+)-glyceraldehyde (*R*)-(−)-glyceric acid

Because this reaction does not break bonds to the stereocenter, the relative configurations of the two compounds must be the same.

(*R*)-(+)Isoserine (3-amino-2-hydroxypropanoic acid), an amino compound, is used to relate (−)-lactic acid (2-hydroxypropanoic acid) from sour milk, to (+)-glyceraldehyde. Isoserine is converted to (*R*)-(−)-glyceric acid by one reaction and to (*S*)-(−)-3-bromo-2-hydroxypropanoic acid by another. The halogen-carbon bond in (*S*)-(−)-3-bromo-2-hydroxypropanoic acid is reduced to give (*R*)-(−)-lactic acid.

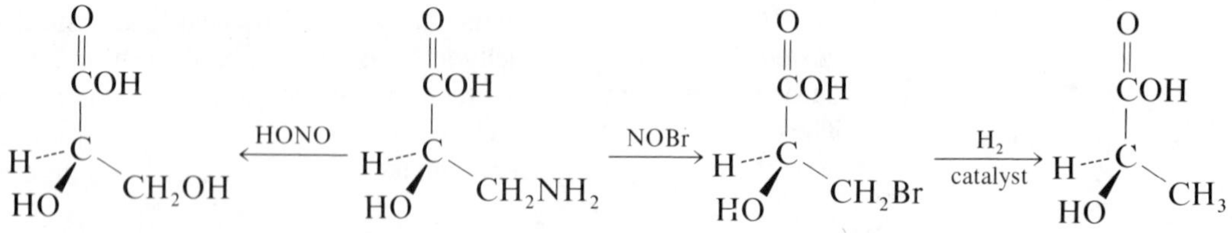
COH
CH_2OH
HONO
COH
CH_2NH_2
NOBr
COH
CH_2Br
H_2
catalyst
COH
CH_3

(*R*)-(−)-glyceric acid (*R*)-(+)-isoserine (*S*)-(−)-3-bromo-2-hydroxypropanoic acid (*R*)-(−)-lactic acid

Some of the reactions necessary for these conversions were discussed in Chapter 22. None of the transformations breaks bonds to the stereocenter. Therefore, the configurations of all the compounds involved relative to (+)-glyceraldehyde are established.

PROBLEM 25.2

(−)-But-1-en-3-ol can be hydrogenated to (−)-2-butanol. Ozonolysis of the butenol, followed by mild oxidation, gives (−)-lactic acid. Write structures showing the relative configurations of all the compounds mentioned. Give them the proper *R* or *S* designations.

B. Fischer Projections as Two-Dimensional Representations of Chiral Compounds

So far in this book three-dimensional molecules have been represented in two dimensions by perspective formulas (p. 18). Drawing these formulas is not too difficult for compounds containing one or two stereocenters, but becomes increasingly so for compounds having several stereocenters. Chemists needed a convention for representing three-dimensional molecules in the plane of paper in a consistent and simple way. The great German chemist Emil Fischer introduced such a convention. Fischer called his representations projection formulas because they essentially project a three-dimensional conformation of a molecule onto a two-dimensional surface. These representations are now called **Fischer projection formulas.**

Any two bonds at an sp^3-hybridized carbon atom are in a plane at right angles to the plane defined by the other two bonds. The Fischer projections at a given carbon atom represent a rendering in two dimensions of this view of the molecule. Lines intersecting at right angles represent the two planes. The projection formula for (*R*)-(+)-glyceraldehyde is given in Figure 25.1. In Fischer projection formulas, the longest carbon chain is usually written vertically.

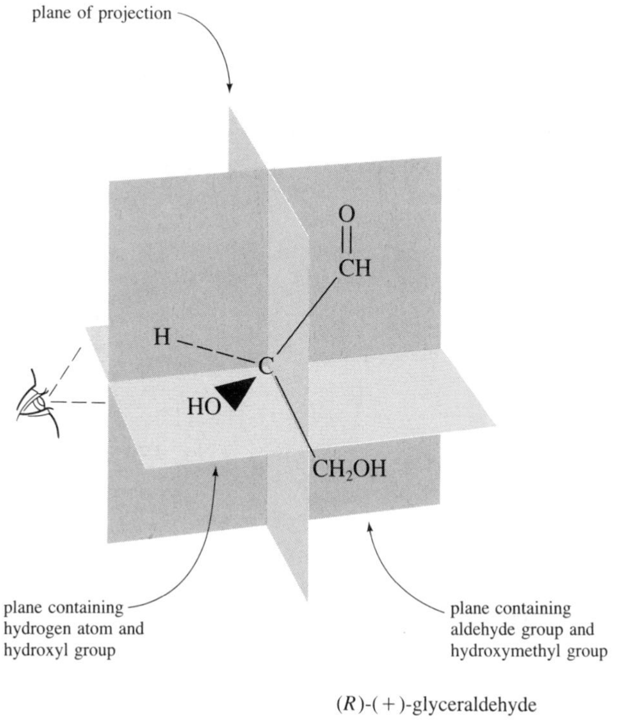
plane of projection
O
CH
H
C
HO
CH₂OH
plane containing hydrogen atom and hydroxyl group
plane containing aldehyde group and hydroxymethyl group
(R)-(+)-glyceraldehyde

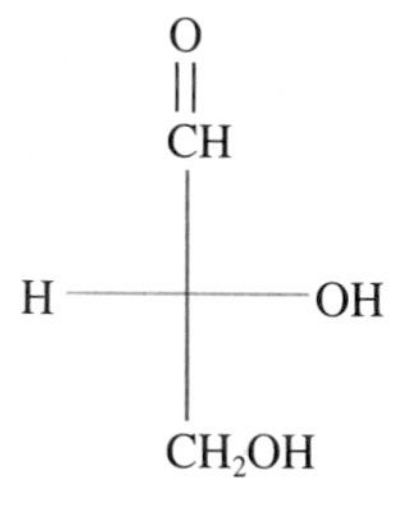
O
CH
H
OH
CH₂OH
Fischer projection formula

FIGURE 25.1 The planes that intersect at right angles at a tetrahedral carbon atom and the Fischer projection formula that indicates those planes.

The conformation that is drawn in a Fischer projection formula of any compound containing more than one stereocenter happens to be the least stable, fully eclipsed conformation of the molecule. Thus, Fischer projections do not represent the actual shape of the molecule as it exists in the crystalline state or in solution. They are a convenient way of comparing the configurations of various stereocenters. Fischer's original picture of the structure of glucose is shown below, along with a representation using wedges and dashes as a reminder of what the convention means. The sideways view of the glucose molecule shows that the first and last carbons are in reality quite close to each other.

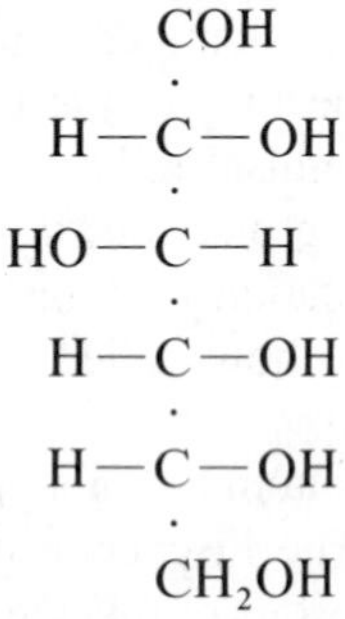
COH
H—C—OH
HO—C—H
H—C—OH
H—C—OH
CH2OH

(+)-glucose as Fischer represented it in his original projection formula

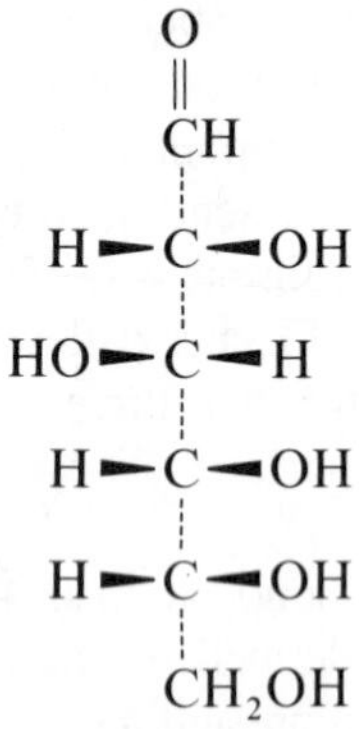
O
CH
H—C—OH
HO—C—H
H—C—OH
H—C—OH
CH2OH

an indication of which bonds project up and which down at every chiral center viewed individually in (+)-glucose

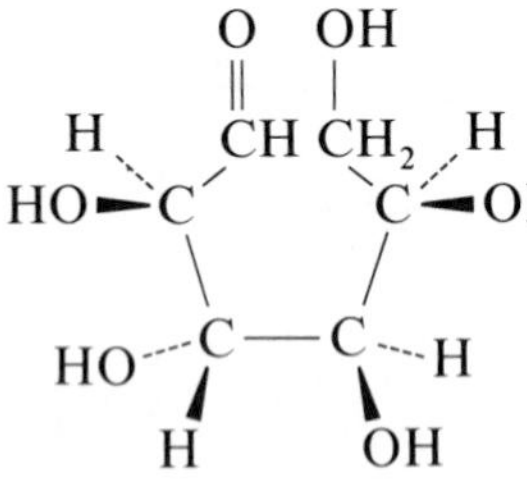
O OH
H CH CH2 H
HO—C C—OH
HO C—C H
H OH

sideways view

The original Fischer projections were later changed—the dots were replaced by lines that connected all the groups to the stereocenters. This is the form of Fischer projection currently used in many books. In 1893, Victor Meyer introduced another convention that has been widely used. He suggested that each stereocenter in a Fischer projection be represented as the crossing point of the bonds and that no carbon atom be shown at that point. These conventions are illustrated with various representations of (*R*)-(+)-glyceraldehyde.

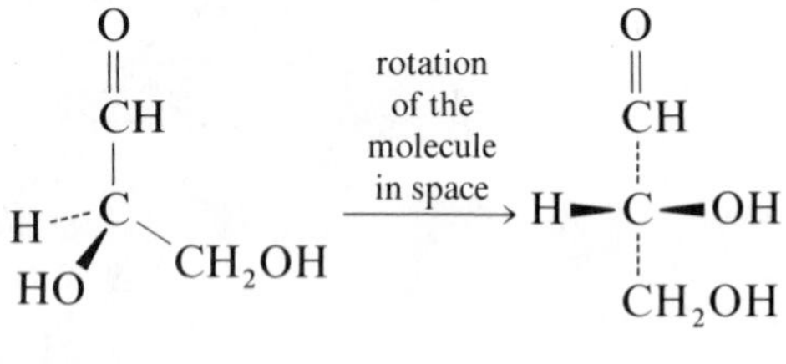
rotation
of the
molecule
in space

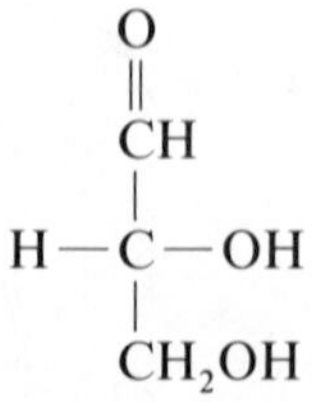

(R)-(+)-glyceraldehyde Fischer projection formula as often seen

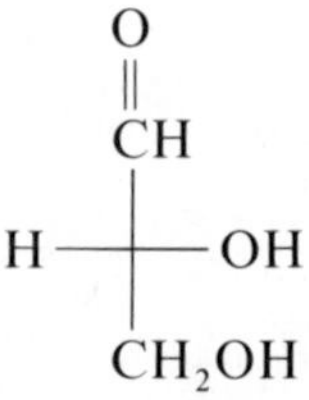

(R)-(+)-glyceraldehyde Victor Meyer's modification of the Fischer projection formula

This book will use Meyer's modification exclusively. The other form of the Fischer projection has nothing about it that clearly signals its stereochemical intent. Many times confusion arises because Lewis structures look very similar to this other form of Fischer projections. Lewis structures are not designed to give any stereochemical information. Unless the other form of a Fischer projection is identified as such every time it is used, there is nothing in its appearance to distinguish it from an ordinary representation of a tetrahedral carbon atom with no stereochemical information. The Meyer modification is clearly a different convention, the appearance of which is an immediate reminder that it is showing the specific directions in space for the four groups attached to the crossing point.

Projection formulas impose certain limitations. For example, exchanging the positions of any two substituents at a stereocenter converts the center into its enantiomeric configuration. The most obvious example of this is demonstrated below.

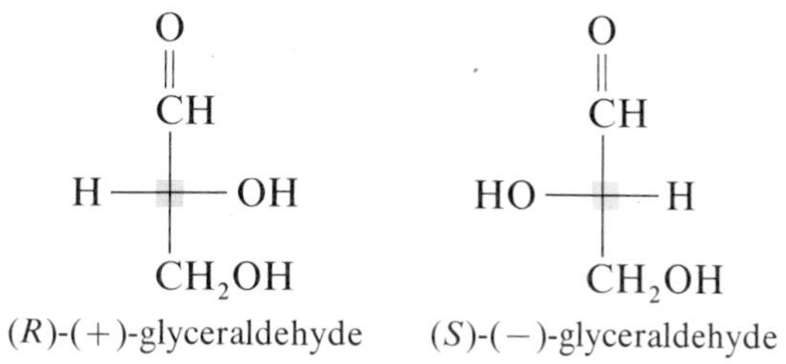

O
CH
H
OH
CH2OH
(R)-(+)-glyceraldehyde
O
CH
HO
H
CH2OH
(S)-(−)-glyceraldehyde

exchange of two substituents at the stereocenter converts the representation of one enantiomer into that of its mirror-image isomer

An interchange of the hydrogen atom and the hydroxyl group converts (*R*)-(+)-glyceraldehyde to (*S*)-(−)-glyceraldehyde. Any interchange of two substituents in a molecule containing a single stereocenter converts the representation to that of the enantiomer. All the structural formulas below are representations of (*S*)-(−)-glyceraldehyde arrived at by interchanging two substituents on the projection formula shown above for (*R*)-(+)-glyceraldehyde.

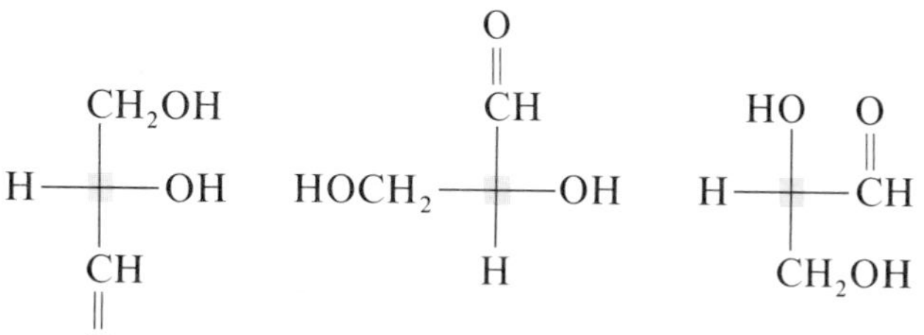

CH2OH
H
OH
CH
O
O
CH
HOCH2
OH
H
HO
O
H
CH
CH2OH

different Fischer projections of (S)-(−)-glyceraldehyde, all derived from the interchange of two substituents on the Fischer projection of (R)-(+)-glyceraldehyde

Two of the representations of (*S*)-(−)-glyceraldehyde illustrate another point. A Fischer projection formula can be rotated 180° *in* the plane of the paper and retain the same stereochemistry.

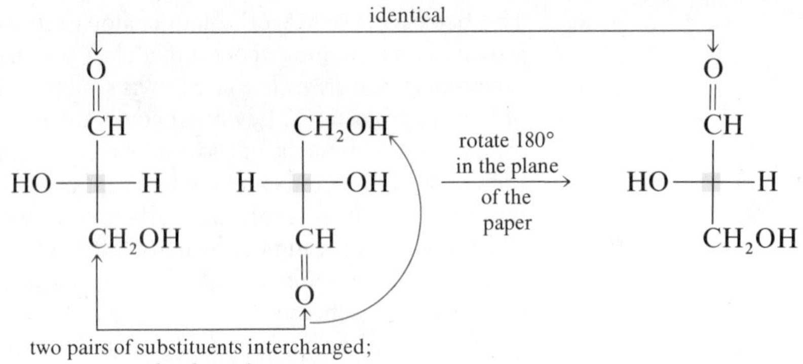
identical
O
CH
HO
H
CH2OH
CH2OH
H
OH
CH
O
rotate 180°
in the plane
of the
paper
O
CH
HO
H
CH2OH
two pairs of substituents interchanged;
molecule retains
the same stereochemistry

The maneuver shown above is the equivalent of interchanging two pairs of substituents at the stereocenter. The aldehyde function was exchanged with the hydroxymethyl group, and the hydroxyl group with the hydrogen atom. Such a set of two transformations converts the projection formula into another projection of the *same* compound.

Assignment of configuration to compounds shown as Fischer projections requires care. It is important to remember that the substituents on the vertical lines project back into the page and those on the horizontal lines project out of the page. In assigning configuration, priorities are assigned to each of the substituents at the stereocenter according to the Cahn-Ingold-Prelog rules (p. 196). If the substituent of lowest priority is on a vertical line, configuration can be assigned by letting the eye travel from the substituent of highest priority to those of second and third priority. A clockwise motion of the eye means that the center has the *R* configuration; a counterclockwise motion of the eye means that the center has the *S* configuration. This is just how configuration is assigned from a perspective formula of a molecule in which the group of lowest priority projects back into the page, for example:

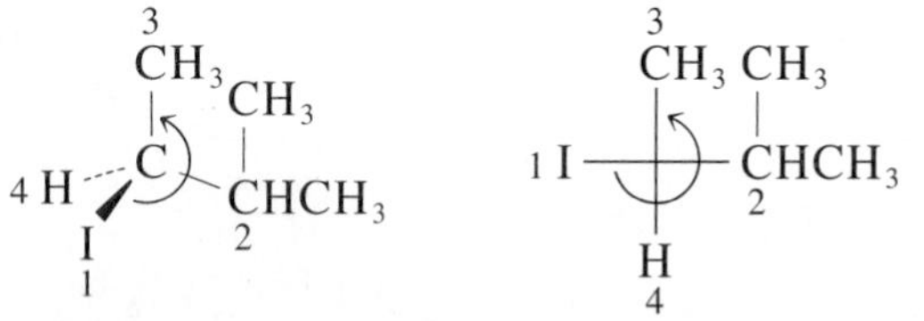
3
CH3
CH3
4 H
C
CHCH3
2
I
1
3
CH3 CH3
1 I
CHCH3
2
H
4

(*S*)-2-iodo-3-methylbutane (*S*)-2-iodo-3-methylbutane

If the Fischer projection is drawn so that the group of lowest priority is on the horizontal line, however, it is important to remember that that substituent is projecting *toward* the viewer, not away from the viewer, as is required for making an assignment of configuration. In such a case, the configuration of the molecule is the opposite of what it appears to be superficially. For example, with the following projection, the eye travels counterclockwise from the group of highest priority to that of the second and then the third priority; but because the lowest-priority group is on the horizontal line, though the configuration appears to be *S*, it is in fact *R*.

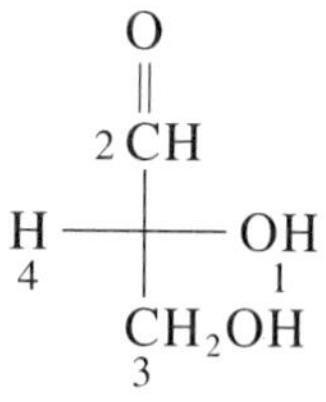

In looking at this Fischer projection to make the assignment, we are looking at the wrong face of the molecule, with the hydrogen atom projecting out of the plane of the page. If we were to view the molecule from behind the page, the eye would have to travel clockwise to go from the hydroxyl group to the aldehyde to the primary alcohol. Thus, any time the group of lowest priority is on the horizontal line in a Fischer projection, the correct configuration is the opposite of what it appears to be from looking at the representation as drawn.

PROBLEM 25.3

For each pair of Fischer projection formulas, decide whether they represent the same compound, enantiomers, or diastereomers. Check your conclusions by designating the configuration at each stereocenter as *R* or *S*.

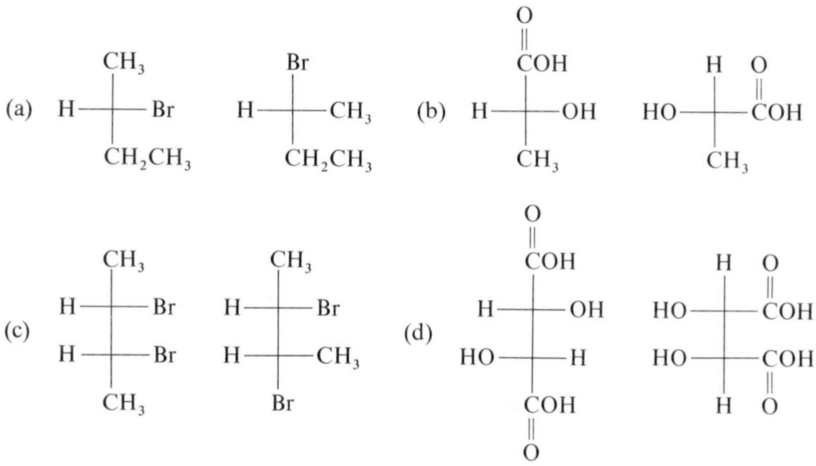

C. The Designation of Chiral Compounds as D or L

The *R* and *S* convention for the designation of configuration was not introduced until the 1950s (p. 196). Before that, chemists used another way of naming relative configurations. When Rosanoff suggested that a particular configuration be assigned to (+)-glyceraldehyde, he also proposed that that arrangement of atoms around the stereocenter be called the D configuration. All compounds having an arrangement of atoms similar to that at the stereocenter of (+)-glyceraldehyde at a comparable carbon atom are members of a D family. Those with the opposite configuration at such a carbon atom belong to an L family. Except for a few cases where relationships are quite easy to see, the system led to many complications and inconsistencies and has been largely abandoned except for carbohydrates and amino acids. This system is still used to some extent, however, especially in the biochemical literature, so you should understand it. Some examples of the D family are given below. The stereocenter that determines the family relationship is shaded in color in each compound.

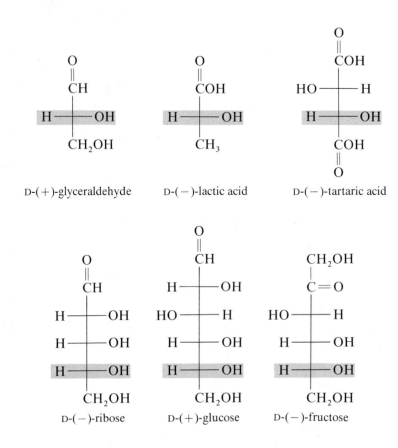

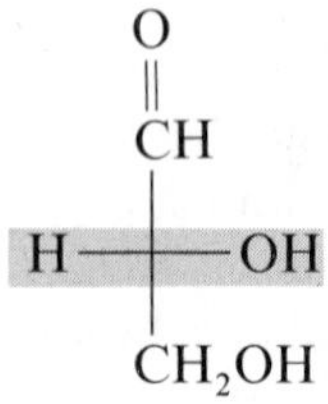
D-(+)-glyceraldehyde

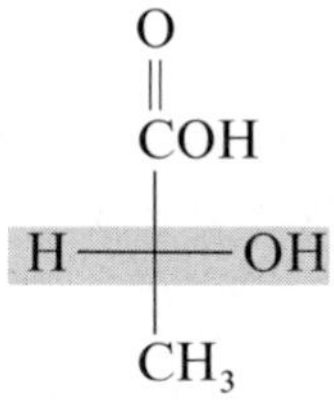
D-(−)-lactic acid

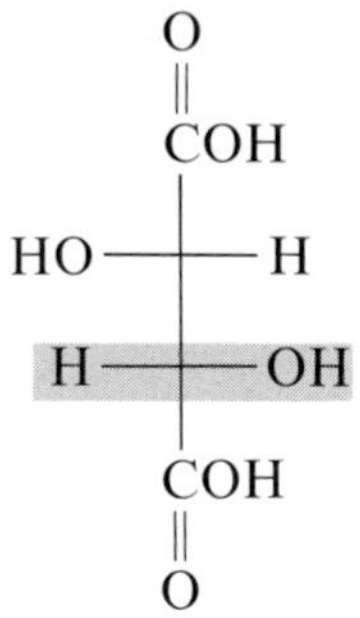
D-(−)-tartaric acid

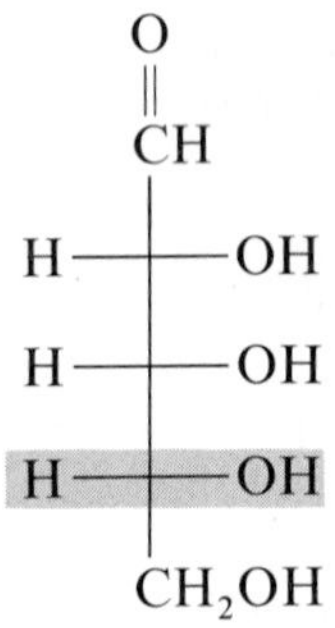
D-(−)-ribose

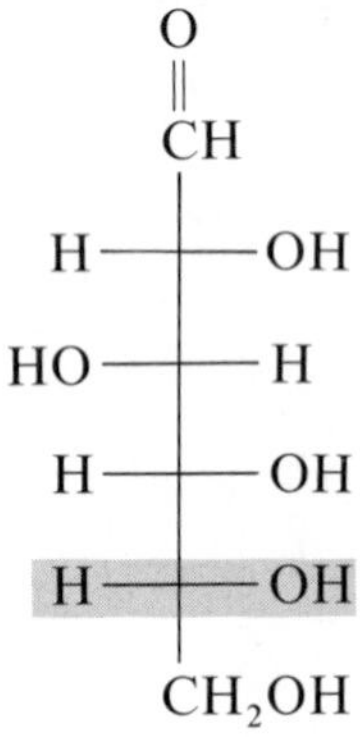
D-(+)-glucose

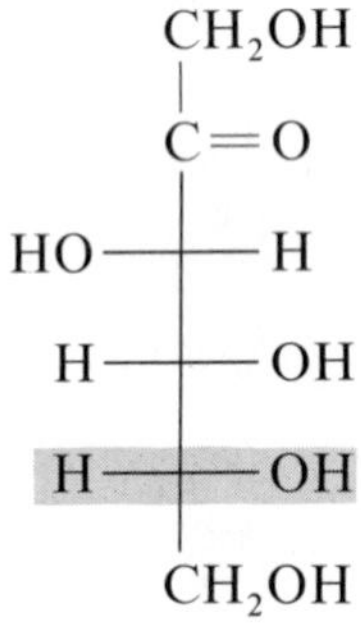
D-(−)-fructose

The symbol D has nothing to do with whether the compound is dextrorotatory or levorotatory, just as the designation *R* or *S* by itself does not give that information.

Some other compounds are members of the stereochemical family related to L-(−)-glyceraldehyde.

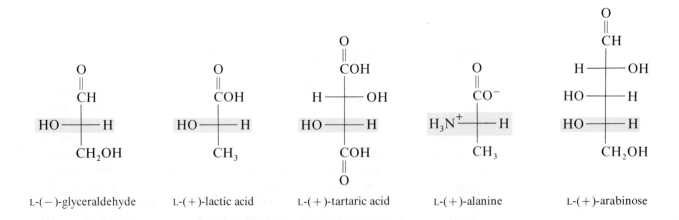

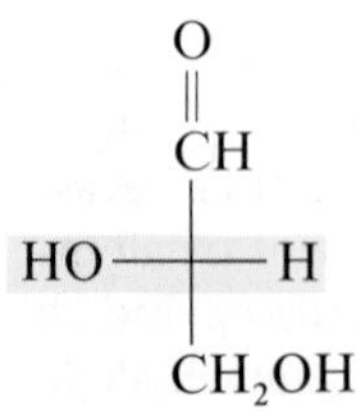
L-(−)-glyceraldehyde

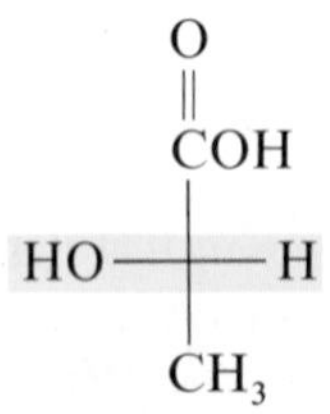
L-(+)-lactic acid

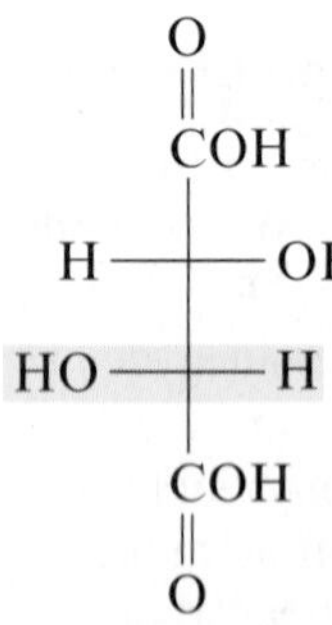
L-(+)-tartaric acid

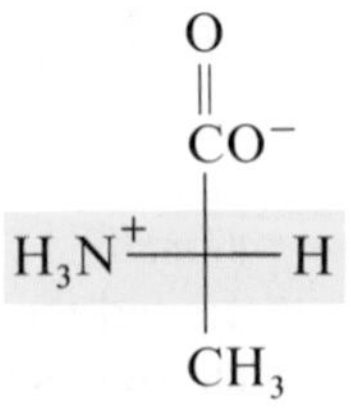
L-(+)-alanine

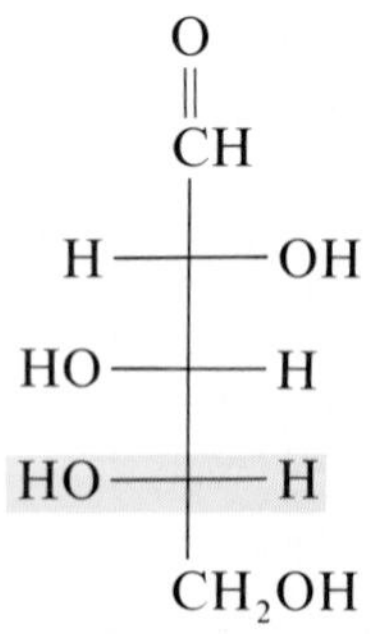
L-(+)-arabinose

The L designation, again, simply indicates that these compounds all have the same configuration at a certain stereocenter in their molecules.

PROBLEM 25.4

Designate as *R* or *S* the configuration at each stereocenter in the compounds of the D and L families given above.

D. The Relative Configurations of (+)-Glyceraldehyde and (−)-Tartaric Acid

(+)-Glyceraldehyde is converted by a series of chemical reactions into tartaric acid. The experiment was done in 1917 by A. Wohl in Germany. He started with the dimethyl acetal of (+)-glyceraldehyde and hydrolyzed it to the free aldehyde in the reaction mixture.

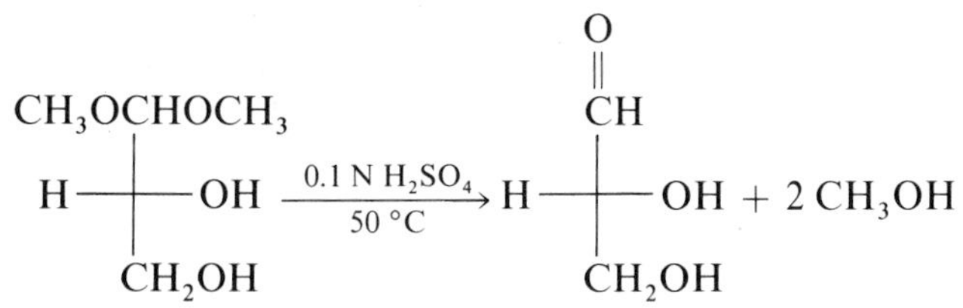

$CH_3OCHOCH_3$
H
OH
0.1 N H_2SO_4
50 °C
CH
O
OH + 2 CH_3OH
CH_2OH

dimethyl acetal of (+)-glyceraldehyde

(+)-glyceraldehyde

The aldehyde was treated with a mixture of hydrogen cyanide and aqueous ammonia to give cyanohydrins, which were not isolated but hydrolyzed to hydroxyacids in the same reaction mixture.

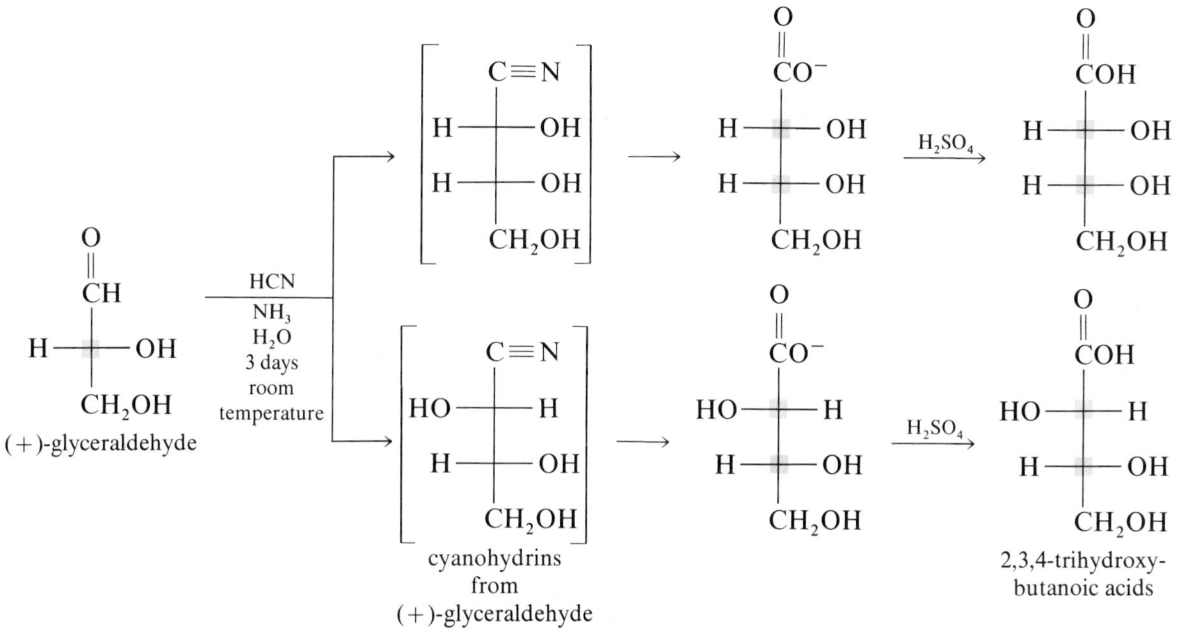

HCN
NH_3
H_2O
3 days
room
temperature
C≡N
CO^-
COH
H_2SO_4
CH_2OH
(+)-glyceraldehyde
cyanohydrins
from
(+)-glyceraldehyde
2,3,4-trihydroxy-
butanoic acids

In this process, a new stereocenter was formed in the molecule. The products are diastereomers that are formed in unequal amounts and can be separated from each other by recrystallization, because they have different physical properties, including solubilities. The trihydroxybutanoic acids were separated and then oxidized to tartaric acids.

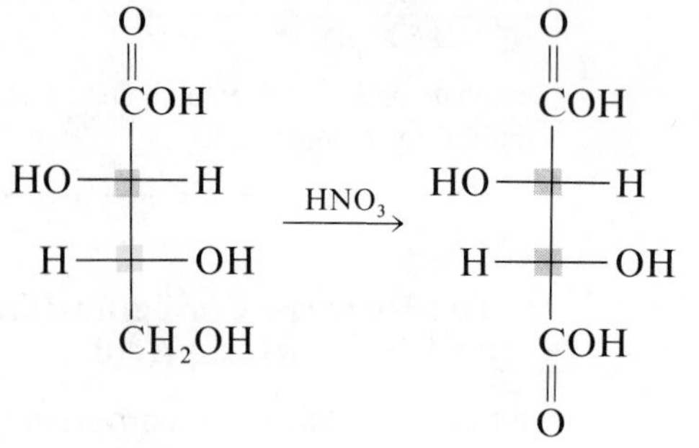

(2*S*,3*R*)-2,3,4-trihydroxybutanoic acid

(−)-tartaric acid
(2*S*,3*S*)-2,3-dihydroxybutanedioic acid

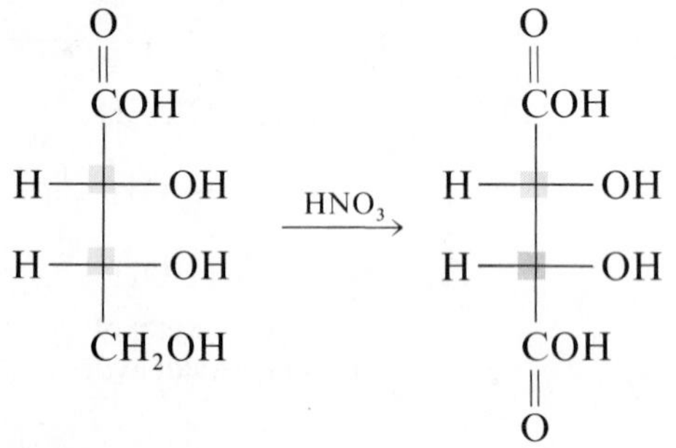

(2*R*,3*R*)-2,3,4-trihydroxybutanoic acid

meso-tartaric acid
(2*R*,3*S*)-2,3-dihydroxybutanedioic acid

The optically active tartaric acid formed in this synthesis is (−)-tartaric acid. The other tartaric acid is not optically active and cannot be resolved into optically active compounds. It is *meso*-tartaric acid, which is not optically active because the molecule has an internal plane of symmetry and is not chiral. This experiment established the relative configurations of (+)-glyceraldehyde and (−)-tartaric acid.

PROBLEM 25.5

Draw the perspective formula for *meso*-tartaric acid that corresponds to the Fischer projection given above. Be sure you see the plane of symmetry in the molecule.

PROBLEM 25.6

Using Fischer projections, write equations for the conversion of (−)-glyceraldehyde into tartaric acid. Show clearly the stereochemistry of the tartaric acids that would be formed, and indicate whether or not you expect them to be optically active.

E. Determination of Absolute Configuration

(+)-Tartaric acid, the enantiomer of (−)-tartaric acid, is the isomer that occurs most abundantly in nature. Bijvoet used it in 1951 for the determination of absolute configuration. This experiment, mentioned on p. 200, established the actual orientation in space of the atoms in (+)-tartaric acid.

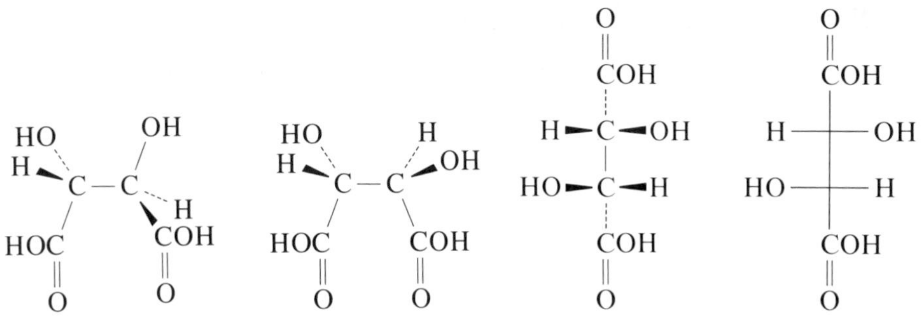

various representations of the absolute configuration of (+)-tartaric acid

The determination of the actual structure of (+)-tartaric acid also established the absolute configuration of its enantiomer, (−)-tartaric acid, which had been related structurally to (+)-glyceraldehyde by synthesis. This meant that the structure of (+)-glyceraldehyde did, indeed, correspond to the *R* configuration.

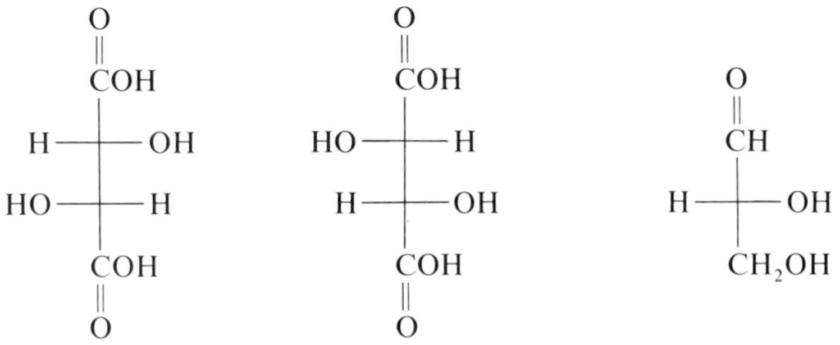

(+)-tartaric acid, x-ray structure determination

(−)-tartaric acid, structure known because it is the enantiomer of (+)-tartaric acid

(*R*)-(+)-glyceraldehyde, related to (−)-tartaric acid by synthesis

absolute configurations for (+)- and (−)-tartaric acids and for (+)-glyceraldehyde

With this one experiment, what had been known until that time as the relative configurations of hundreds of compounds suddenly became established as the absolute configurations. Assignments that had been made on the assumption that (+)-glyceraldehyde had the *R* configuration were all correct. The original assignment had been made by Fischer, who had decided arbitrarily that (+)-glucose had a certain configuration at the stereocenter farthest from the aldehyde group. He placed the hydroxyl group at that stereocenter on the right in his projection formula. He knew that his system would also be valid if the structure of glucose was actually the mirror image of the one that he had assigned it, but he had to choose one structure in order to be able to make comparisons with those of other carbohydrates. His selection proved to be correct.

Glucose, in its open-chain form, has four stereocenters and, therefore, 2^4, or 16, different stereoisomers. The stereoisomers exist as eight pairs of enantiomers. Fischer's assignment of stereochemistry at each of the stereocenters in glucose was the result of painstaking experimental work and logical thought about the implications of the experimental observations. He was given the Nobel Prize for this work in 1902. His experiments and arguments will be retraced in Section 25.7B.

PROBLEM 25.7

Draw the Fischer projection formula for the enantiomer of D-(+)-glucose (p. 1060).

25.3 THE STRUCTURE OF GLUCOSE

A. Glucose as a Pentahydroxyaldehyde

The open-chain form of glucose was assigned its structure on the basis of the following experimental data. Glucose has the molecular formula $C_6H_{12}O_6$. It reacts with mild oxidizing agents such as Tollens reagent (p. 557) and is therefore probably an aldehyde. When treated with acetic anhydride (p. 576), it forms a pentaacetate and thus has five hydroxyl functions. Structures that have two hydroxyl functions on the same carbon atom do not represent stable species but are the hydrates of carbonyl compounds (p. 497). Glucose must, therefore, have each hydroxyl group on a different carbon atom.

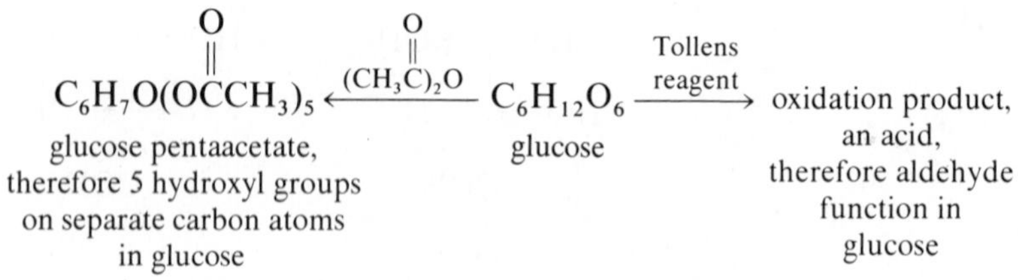
$C_6H_7O(OCCH_3)_5$
$(CH_3C)_2O$
$C_6H_{12}O_6$
Tollens reagent
oxidation product, an acid, therefore aldehyde function in glucose
glucose pentaacetate, therefore 5 hydroxyl groups on separate carbon atoms in glucose
glucose

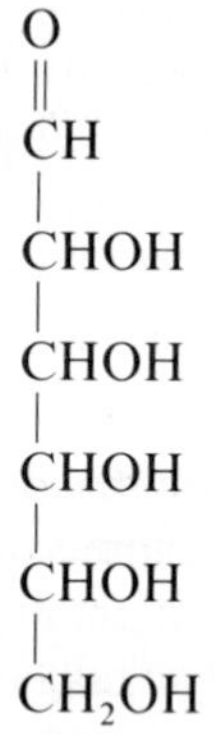
O
CH
CHOH
CHOH
CHOH
CHOH
CH_2OH

structure of glucose, excluding stereochemistry, determined from chemical data

The addition of hydrogen cyanide to an aldehyde to give a cyanohydrin (p. 487) that can be hydrolyzed to a carboxylic acid (p. 563) was applied to sugars in 1885. The German chemist Heinrich Kiliani used the reaction to prove that glucose had a straight chain of six carbon atoms. Adding hydrogen cyanide to glucose and hydrolyzing the nitrile gave a hydroxycarboxylic acid with one more carbon atom than glucose. This was reduced with hydrogen iodide to an acid identified as heptanoic acid.

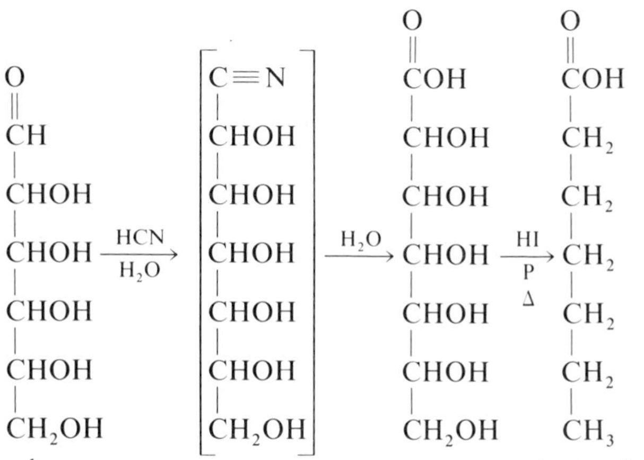

glucose heptanoic acid

The conversion of glucose, a six-carbon sugar, into a known, straight-chain, seven-carbon acid showed that the carbonyl group to which the hydrogen cyanide added was at the end of a straight chain of six carbon atoms. Thus, the structure of glucose as a 2,3,4,5,6-pentahydroxyhexanal was confirmed.

PROBLEM 25.8

When Kiliani tried the above sequence of reactions with fructose (p. 1060), he obtained 2-methylhexanoic acid. Write equations showing the results of the reactions for fructose. What do these reactions prove above the structure of fructose?

PROBLEM 25.9

(a) What do you think is the first step in the reduction of the polyhydroxyheptanoic acid to heptanoic acid with hydriodic acid?
(b) Phosphorus is not necessary for the reaction with hydriodic acid. When phosphorus is not used, molecular iodine forms in the reaction mixture. The reaction works on ordinary alcohols and on alkyl iodides.

$$ROH + 2\,HI \longrightarrow RH + H_2O + I_2$$

$$RI + HI \longrightarrow RH + I_2$$

The table of average bond energies (p. 61) reveals that the carbon-iodine bond is a relatively weak one. Use these facts to propose a mechanism for the conversion of an alcohol to an alkane by hydriodic acid.

B. Cyclic Structures of Monosaccharides

When a freshly prepared solution of a sample of glucose that has been recrystallized from methanol and has a melting point of 147 °C is put into the tube of a polarimeter, an initial specific rotation of +113° is observed. When the solution is left to stand in the polarimeter, the rotation falls until it reaches a value of +52.5°. If glucose is recrystallized from water at high temperatures, another crystalline form is obtained. This form has a melting point of 150 °C. A freshly prepared solution of these crystals placed in the tube of a polarimeter has an initial specific rotation of +19°. On standing, the optical rotation of this solution rises to +52.5°. Either solution can be evaporated and recrystallized under the conditions described above to give back the original form of glucose. Thus, the change in rotation is not a result of the decomposition of glucose in solution.

A change of optical rotation for a compound on standing in solution is called **mutarotation.** The phenomenon observed for the two forms of glucose, with both solutions arriving at the same final rotation, suggests an equilibrium between two stereochemically different forms of the compound.

These facts have been interpreted to mean that glucose normally exists as a cyclic hemiacetal (p. 499). The hydroxyl group on carbon 5 is close enough to the carbonyl group at carbon 1 for a six-membered ring to form.

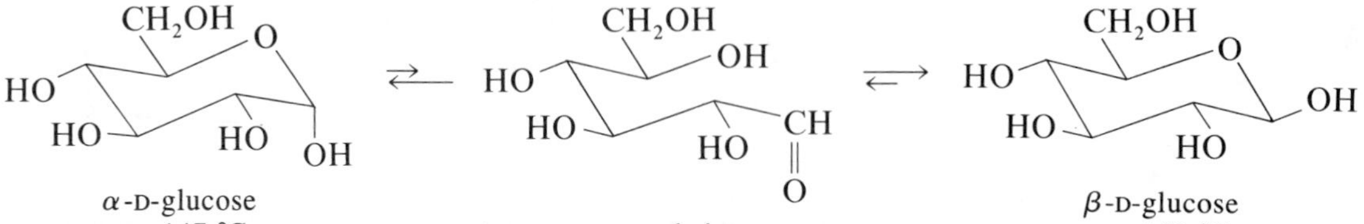
CH2OH
O
HO
HO
HO
OH
OH
CH
O

α-D-glucose
mp 147 °C
[α] +113°

open-chain
form of glucose

β-D-glucose
mp 150 °C
[α] +19°

When this cyclization occurs, a new stereocenter is created. The orientation of the hydroxyl group on carbon 1 can be axial or equatorial. The form in which it is axial is called ***α*-glucose,** which crystallizes at ordinary temperatures and has an initial rotation of +113°. The form in which the hydroxyl group at carbon 1 is equatorial is called ***β*-glucose.** It crystallized out of water at high temperatures and has an initial rotation of +19°. In aqueous solution, each of these two forms is in equilibrium with the open-chain form, which has the free aldehyde group.

The small concentration of this open-chain form in solution is responsible for the reactions of glucose that are typical of aldehydes. The equilibrium that exists among all three forms is responsible for the change in rotation from the initial values of +113° for α-glucose and +19° for β-glucose to the intermediate value of +52.5° for the equilibrium mixture. This value corresponds to a mixture consisting of 36% α-glucose and 64% β-glucose. β-Glucose has all of the large substituents in the equatorial positions in the chair conformation of the six-membered ring. It is more stable in solution than α-glucose, in which the hydroxyl group at carbon 1 is axial.

α-Glucose and β-glucose are stereoisomers that differ from each other at one stereocenter, carbon 1. Therefore, they are diastereomers. Diastereomers that differ from each other in stereochemistry at only one of many stereocenters are called **epimers.** α-Glucose and β-glucose are epimers at carbon 1. Other sugars are epimers of glucose at other carbon atoms.

Epimers in which the one difference in stereochemistry is at a potential carbonyl group in a cyclic hemiacetal are also called **anomers.** α-Glucose and β-glucose are, therefore, best defined as anomers of each other. Carbon 1 is the anomeric carbon atom, bearing an anomeric hydroxyl group.

A six-membered ring that includes an oxygen atom is related to the heterocyclic compound pyran.

O

pyran

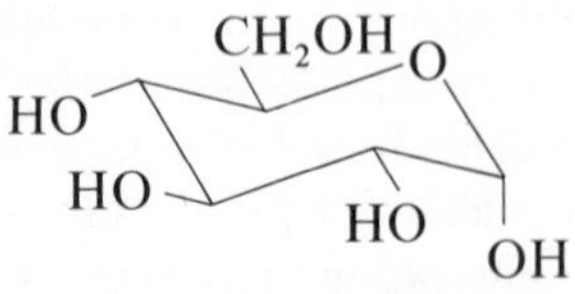
CH2OH
O
HO
HO
HO
OH

α-D-glucopyranose

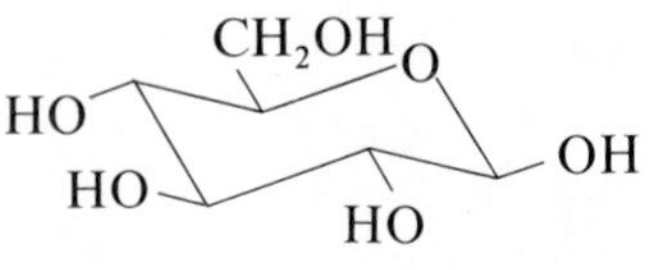
CH2OH
O
HO
HO
HO
OH

β-D-glucopyranose

A sugar in its six-membered cyclic form is called a **pyranose.** The names commonly used by carbohydrate chemists for the cyclic forms of glucose are α-D-glucopyranose and β-D-glucopyranose. The letter D defines the stereochemistry at carbon 5. The gluco portion of the name defines the stereochemistry relative to carbon 5 of the three stereocenters other than the anomeric one. Pyranose indicates a six-membered cyclic structure, and α and β define the stereochemistry at the anomeric carbon atom in the cyclic form.

The chair form of the six-membered ring is often used for representing the stereochemistry and reactions of pyranoses, but the six-membered ring is also widely represented in the planar form. This way of representing sugars is known as the **Haworth projection formula.** Such formulas are easily derived from the chair form.

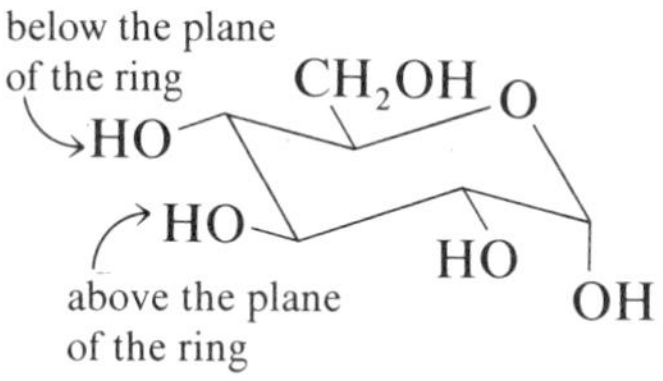
below the plane
of the ring
above the plane
of the ring

α-D-glucopyranose
the chair form

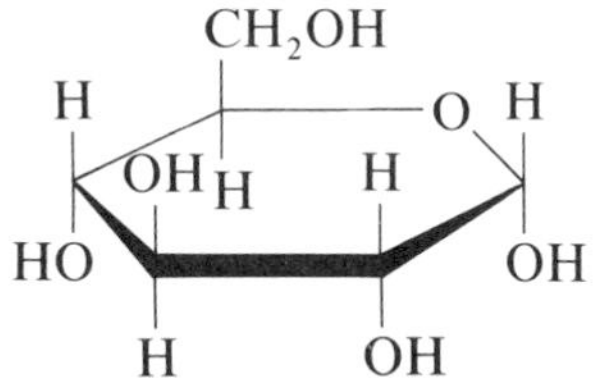

α-D-glucopyranose

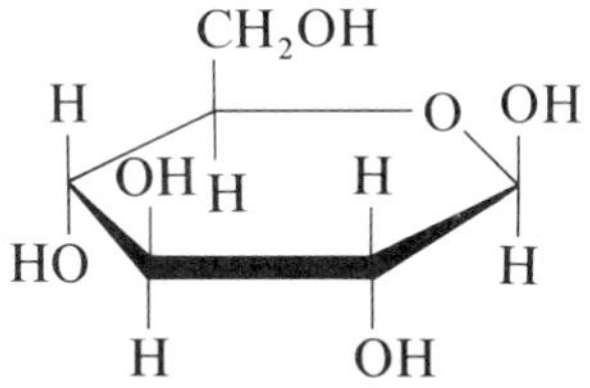

β-D-glucopyranose

Haworth projection formulas for glucose

In the Haworth representations, the lower edge of the ring is defined as projecting out of the plane of the paper toward the viewer. Hydroxyl groups, hydrogens, and other substituents are shown as being either above or below the plane of the ring, as shown above.

The ring in cyclic hemiacetals is not always six-membered. The hydroxyl group on carbon 4 in glucose is close enough to the carbonyl group to give a five-membered ring. Such a ring is related to the five-membered heterocycle furan. The sugar glucose is called a glucofuranose in this form. The pyranose form is more commonly observed for aldohexoses.

furan

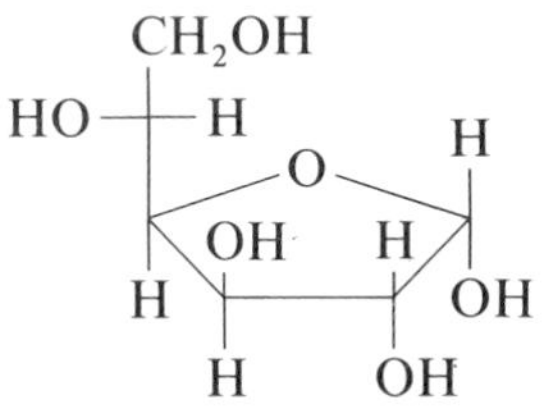

α-D-glucofuranose

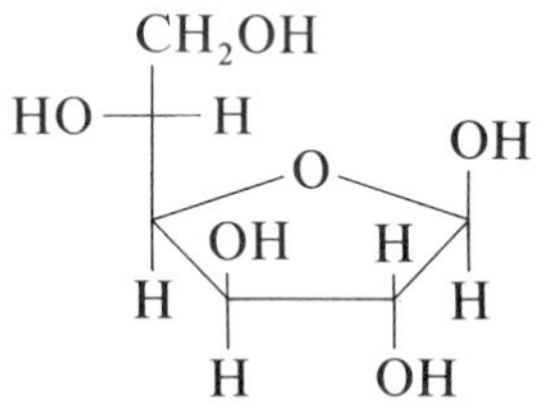

β-D-glucofuranose

Ketohexoses also exist in pyranose and furanose forms. The only crystalline form of fructose that is isolated is β-D-fructopyranose, which has an initial specific rotation of $-133.5°$. It undergoes rapid mutarotation to $-92°$. This mutarotation involves not only isomerization between the β- and α-pyranose forms, but also conversion to the furanose forms (Figure 25.2). The fructofuranose structure is important because fructose is present in sucrose and other oligosaccharides in that form.

Ribose, which is an aldopentose, exists as β-D-ribofuranose in ribonucleic acids, as shown on the next page.

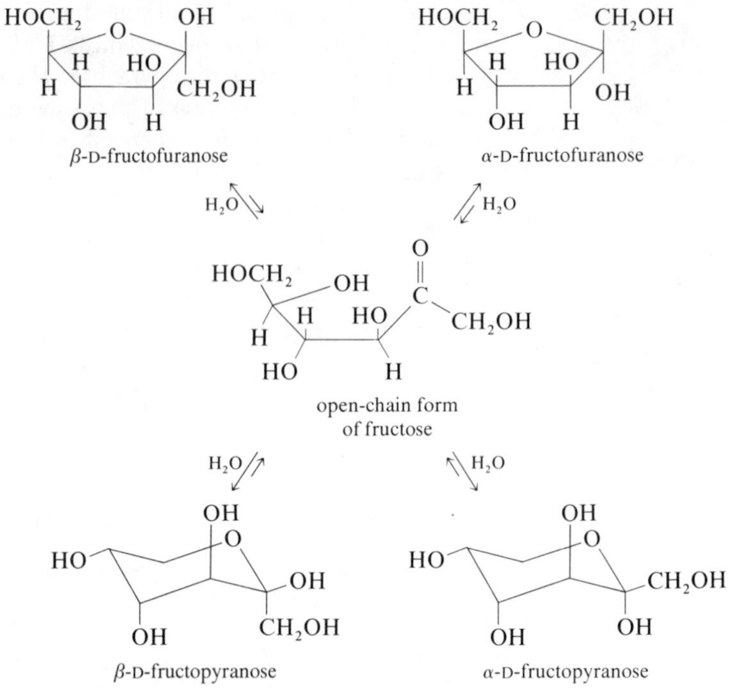
β-D-fructofuranose
α-D-fructofuranose
open-chain form of fructose
β-D-fructopyranose
α-D-fructopyranose

FIGURE 25.2 The mutarotation of fructose.

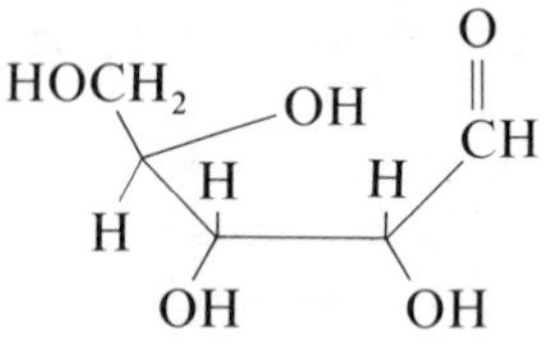

ribose in the open-chain form

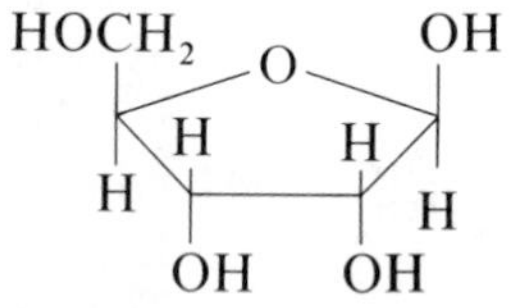

β-D-ribofuranose

Study Guide Concept Map 25.2

PROBLEM 25.10

In aqueous solution at equilibrium, 76% of ribose molecules are in a pyranose form. Draw a structural formula for β-D-ribopyranose.

25.4 REACTIONS OF MONOSACCHARIDES AS CARBONYL COMPOUNDS

A. Formation of Glycosides

When glucose is heated with methanol containing a little hydrogen chloride, two isomeric acetals are formed. In these compounds, the hemiacetal function has been converted into the cyclic monomethyl acetal. Of the two isomers, the product with the methoxy group axial, the α isomer, is about 98% of the mixture. This is true regardless of whether the starting compound is α- or β-glucose. This stereochemistry is an exception to the rule that larger groups occupy equatorial positions in the chair form of six-membered rings.

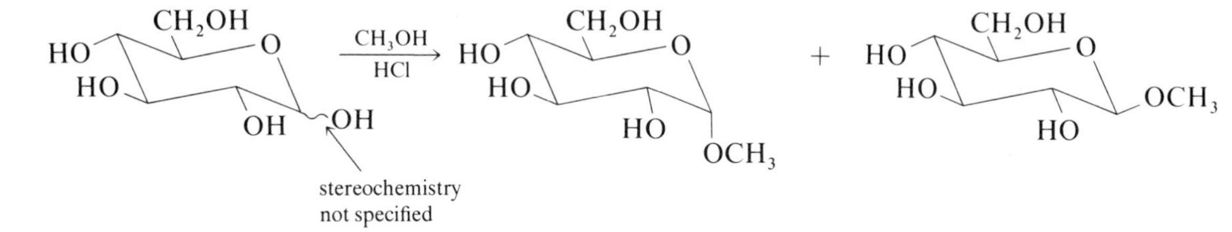

stereochemistry not specified

α- or β-D-glucopyranose

methyl α-D-glucopyranoside
mp 166 °C
[α] +158°
major product

methyl β-D-glucopyranoside
mp 105 °C
[α] −34°
minor product

As a class, carbohydrates with a full acetal linkage are known as **glycosides.** The compounds shown above are called methyl glucosides, or methyl glucopyranosides to reflect their cyclic nature.

PROBLEM 25.11

Write a mechanism for the formation of methyl α-D-glucopyranoside from α-D-glucopyranose. How does the mechanism that you propose explain the fact that both α- and β-glucopyranoses give the same mixture of methyl glucopyranosides? (Hint: Reviewing Section 13.7C may be helpful.)

Methyl glucopyranosides do not undergo mutarotation or show any of the aldehyde reactions that glucose exhibits. For example, they cannot be oxidized easily to carboxylic acids. As an acetal, the carbonyl group is effectively protected. The glucopyranosides are stable in basic solutions. They are hydrolyzed easily in acidic solutions to give an equilibrium mixture of α-D-glucopyranose and β-D-glucopyranose.

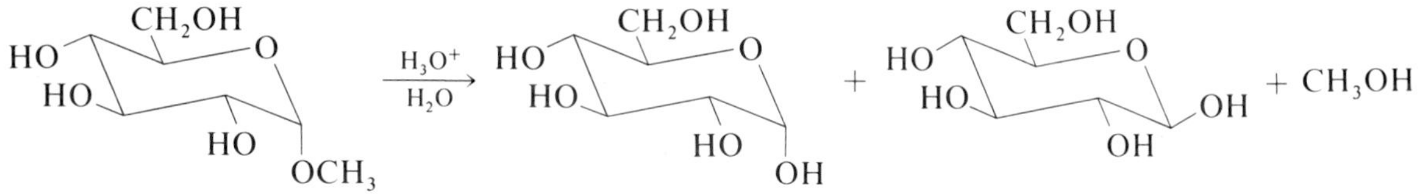

methyl α-D-glucopyranoside

α-D-glucopyranose

β-D-glucopyranose

Enzymes, which are stereoselective biological catalysts, can be used to hydrolyze glucosides selectively. The enzyme maltase will cleave only α-glucosides, and the enzyme emulsin cleaves only β-glucosides.

PROBLEM 25.12

Give a detailed mechanism for the hydrolysis of methyl α-D-glucopyranoside in dilute acid.

The glycosidic linkage occurs widely in nature. Such bonding between the anomeric carbon of one monosaccharide and a hydroxyl group of another is the way in which oligosaccharides and polysaccharides are formed. In plants, saccharides are also found bonded to any of a large number of alcohols and phenols, resulting in a variety of natural products that have medicinal and other practical uses. The hydroxyl compounds that are bonded to sugars in glycosides are called **aglycons.**

An example of a glycoside that has been used in medicine for some time is salicin, found in the bark of the willow tree. Salicin is the β-glycoside of *o*-(hydroxymethyl)phenol.

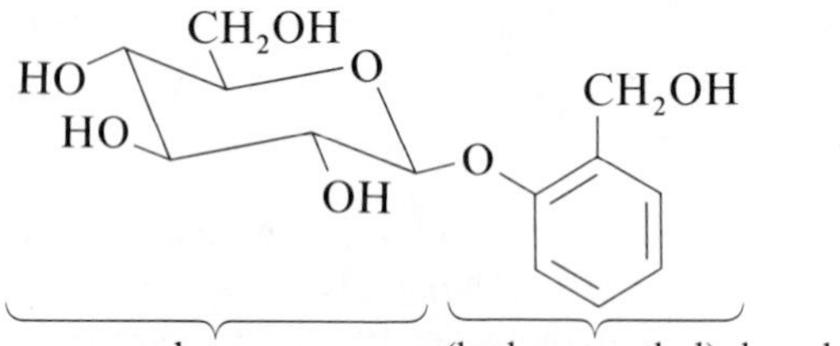
CH2OH
HO
HO
O
OH
O
CH2OH
glucose
a sugar
o-(hydroxymethyl)phenol
an aglycon

salicin, a naturally occurring glycoside

Willow bark preparations have been known since the time of the ancient Greeks as pain relievers. They were usually used externally because the willow juice is so bitter. Chemists sought to isolate the compound that gave willow its analgesic properties. The active component, salicin, was finally isolated from other plant sources and converted to salicylic acid. Salicylic acid has valuable medicinal properties, but it cannot be taken internally. In 1899, salicylic acid was converted into its acetyl derivative, acetylsalicylic acid, which is now commonly known as aspirin.

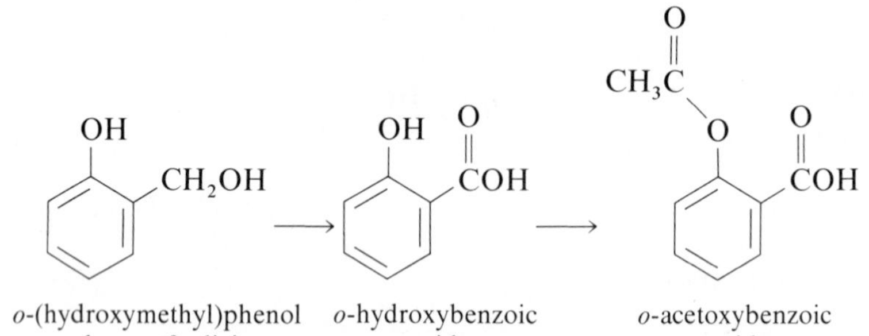
OH
CH2OH
OH
O
COH
CH3C
O
O
O
COH

o-(hydroxymethyl)phenol
aglycon of salicin

o-hydroxybenzoic acid
salicylic acid

o-acetoxybenzoic acid
acetylsalicylic acid
aspirin

Glycosidic linkages are also found in compounds such as digitoxin, called a **cardiac glycoside** (p. 1106) because it affects the action of the heart, and laetrile, the controversial compound that some people claim is active against cancer. Laetrile is one of a number of natural glycosides classified as **cyanogenic glycosides** because they release hydrogen cyanide when hydrolyzed either by acid or by enzymes and thus have considerable toxicity. The aglycon in laetrile is (*R*)-(−)-mandelonitrile.

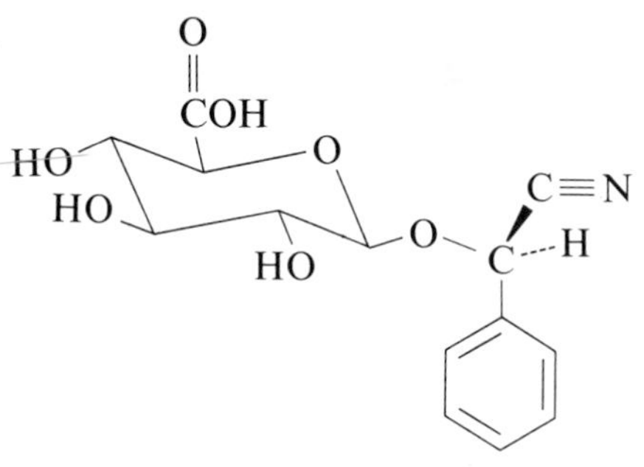

O
COH
HO
O
HO
C≡N
O
C
H
HO

glucuronic acid   (*R*)-(−)-mandelonitrile

laetrile

PROBLEM 25.13

Write equations for the transformations necessary to convert salicin to salicylic acid and salicylic acid to aspirin.

PROBLEM 25.14

The compound arbutin is a glycoside isolated from the bearberry. It has found use as a diuretic (a medication that increases the production of urine) and an antiseptic for the urinary tract. It can be hydrolyzed by emulsin (p. 1069) to glucose and hydroquinone (p. 847). What is the structure of arbutin?

B. Formation of Glycosylamines

If a sugar is heated with an amine in the presence of a trace of acid, a glycosidic linkage to nitrogen is formed. Such compounds are called **glycosylamines.** The reaction works particularly well with aromatic amines.

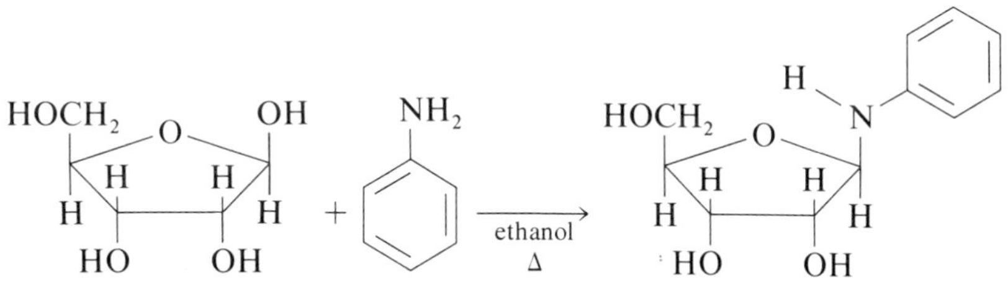

HOCH$_2$
O
OH
H
H
H
H
HO
OH
NH$_2$
+
ethanol
Δ
HOCH$_2$
O
H
N
H
H
H
H
HO
OH

β-D-ribofuranose   aniline   *N*-phenyl-β-D-ribofuranosylamine

This kind of linkage between an amine and a sugar is biologically significant. The structural units of ribonucleic acids (RNA) and deoxyribonucleic acids (DNA) are glycosylamines known as nucleosides (p. 1013).

C. Formation of Osazones. Configurational Relationships Among Monosaccharides

Sugars are highly soluble in water because they contain large numbers of hydroxyl groups that hydrogen-bond strongly with water molecules. Sugars have a tendency to form syrups, therefore crystallization of these compounds is difficult. However, Emil Fischer discovered early in his research that sugars react with phenylhydrazine to give beautiful yellow crystals. He named these compounds containing two phenylhydrazine residues **osazones,** from ose for sugar and azone for phenylhydrazone.

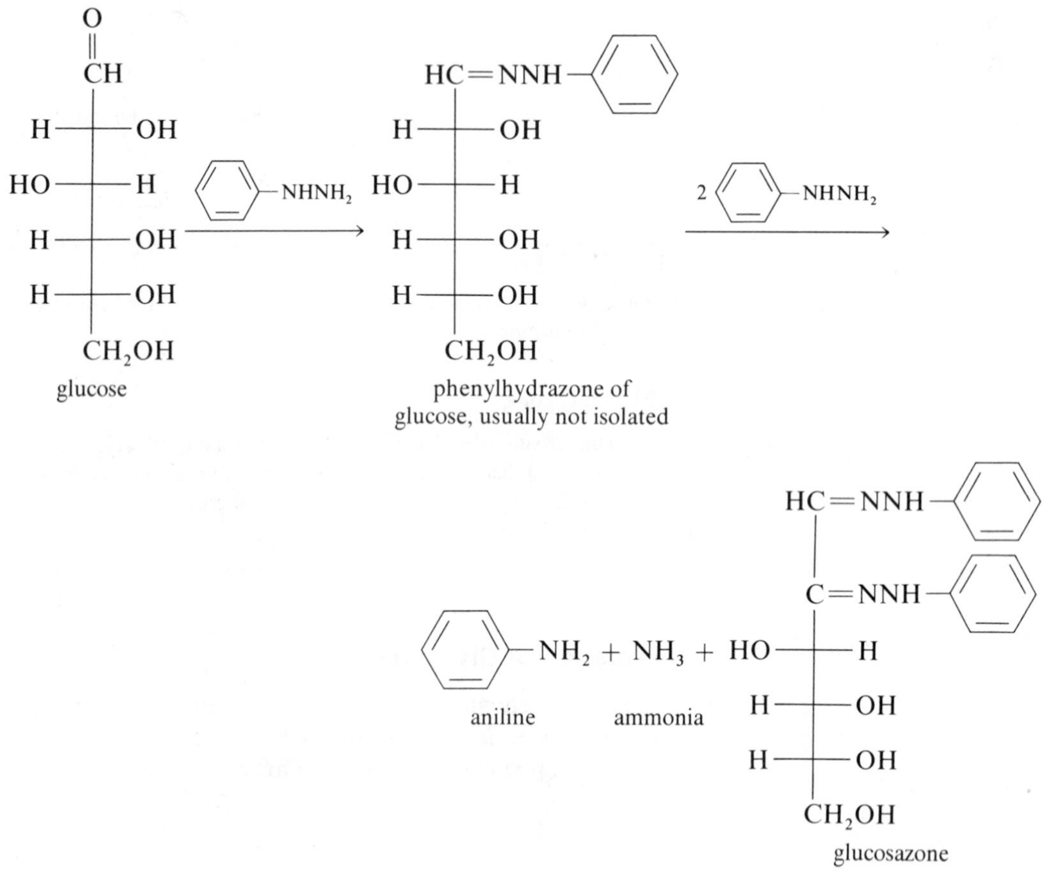
glucose
phenylhydrazone of glucose, usually not isolated
aniline
ammonia
glucosazone

The phenylhydrazone of glucose does form as an intermediate in the production of the osazone, but it is difficult to isolate unless the quantity of phenylhydrazine is carefully limited. The formation of the osazone involves an oxidation-reduction reaction; the second carbon of glucose is oxidized from a secondary alcohol to a carbonyl group and one molecule of phenylhydrazine is reduced to aniline and ammonia.

The forms in which different osazones crystallize are so distinctive that older biochemistry textbooks very often have colored photographs showing how they appear under the microscope. The preparation of an osazone and the examination of the crystals under a microscope were used to identify the common sugars. Fischer found that glucose, fructose, and mannose all gave the same osazone. He recognized that this meant that all three of these sugars had the same configuration at carbons 3, 4, and 5. Fructose, he knew, was a 2-ketohexose, so it did not have a stereocenter at carbon 2. This meant that glucose and mannose must be epimers at carbon 2, the carbon atom that goes from tetrahedral to trigonal planar when the osazone is formed and thus loses its chirality.

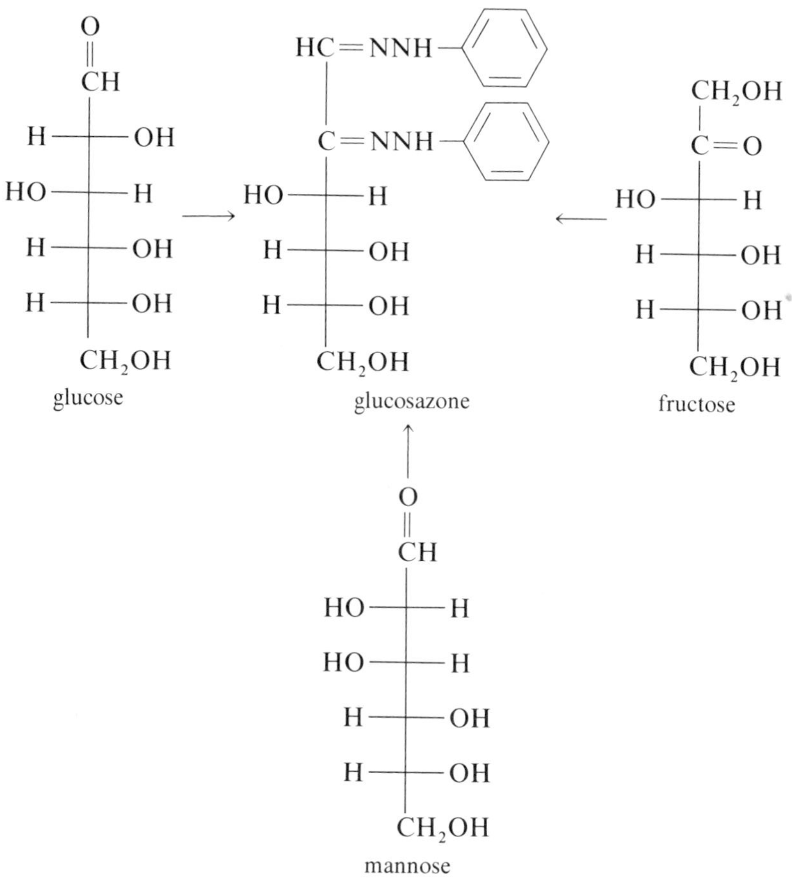
glucose
glucosazone
fructose
mannose

the configurational relationships of glucose, fructose, and mannose

PROBLEM 25.15

Besides glucose and mannose, there is one other isomeric aldohexose commonly found in nature. It is D-(+)-galactose, which is an epimer of glucose at carbon 4. What is the structure of galactose? Would its osazone have the same structure as glucosazone?

PROBLEM 25.16

Using the information that mannose is epimeric with glucose at carbon 2 and galactose is epimeric at carbon 4, write structural formulas for α-D-mannopyranose and β-D-galactopyranose.

D. The Interconversion of Glucose, Mannose, and Fructose

Glucose, mannose, and fructose are interconvertible in solution in the presence of a base. This transformation is called the **Lobry de Bruyn–Alberda van Ekenstein rearrangement,** named for the two Dutch chemists who discovered it at the end of the nineteenth century. If glucose is put into a solution of dilute sodium hydroxide, some of it is transformed into fructose, mannose, and decomposition products.

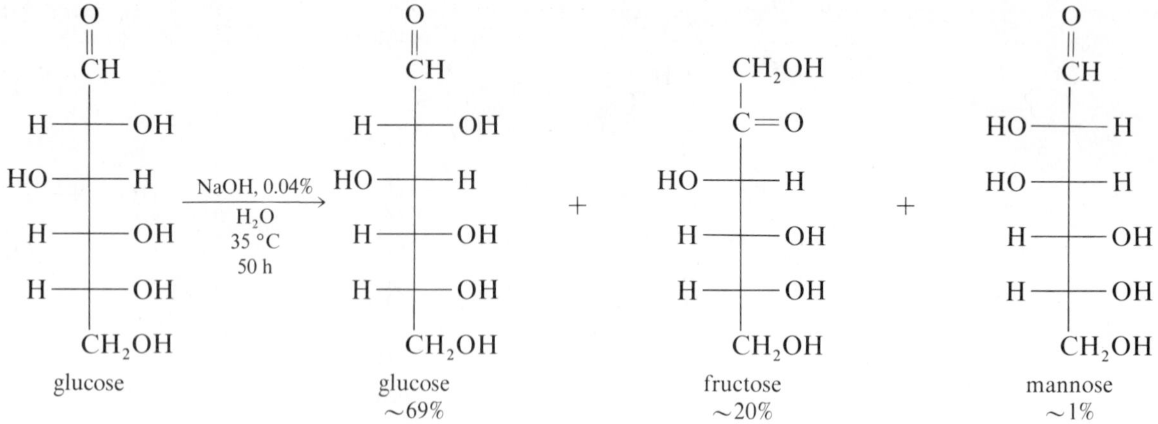
NaOH, 0.04%
H_2O
35 °C
50 h
glucose
glucose
~69%
fructose
~20%
mannose
~1%

If the reaction mixture stands longer, relatively more fructose and mannose are formed, but the total yield of hexoses falls to 70%.

The conversion of glucose to fructose is an important step in glycolysis, the process by which glucose is metabolized in the body. This transformation is believed to start with an enolization reaction. The hydrogen atom on the carbon adjacent to the carbonyl group is acidic enough to be removed to give an enolate ion, which is protonated by water. The intermediate is a 1,2-enediol, in which carbon atom 2 is now trigonal planar and has therefore lost its chirality.

VISUALIZING THE REACTION

Formation of an enediol from glucose

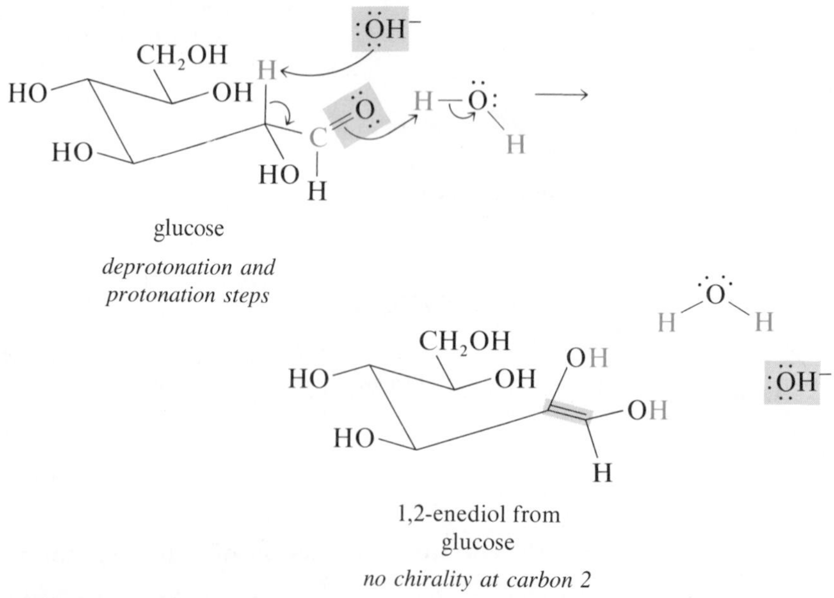
glucose
deprotonation and protonation steps
1,2-enediol from glucose
no chirality at carbon 2

The two protons on the enolic hydroxyl groups are acidic. Removal of the proton from the hydroxyl group on carbon 1 produces carbanion character at carbon 2. Protonation of the carbanion gives mannose or glucose, depending on which side of the molecule the reaction takes place.

VISUALIZING THE REACTION

Conversion of the enediol to mannose

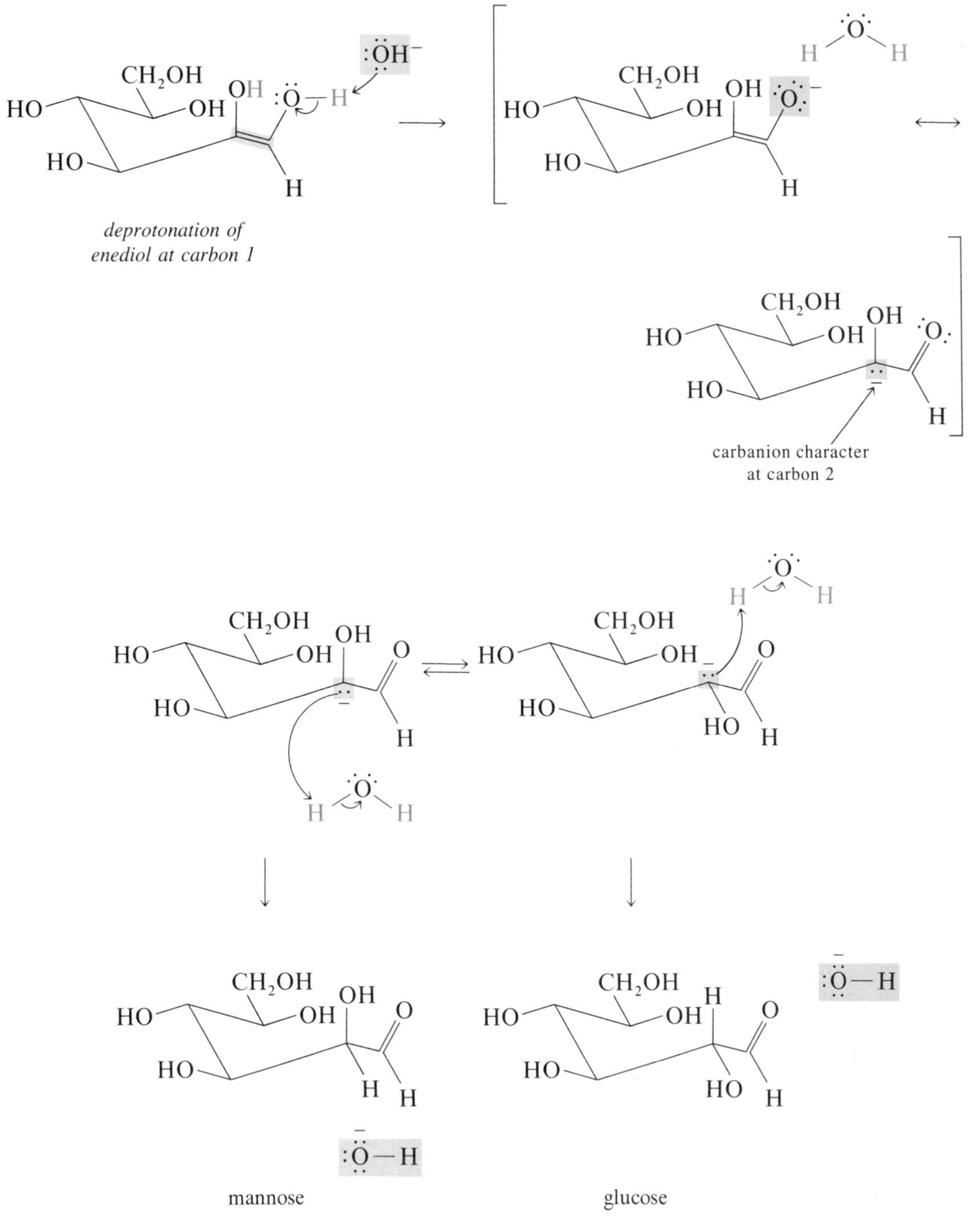
deprotonation of
enediol at carbon 1
carbanion character
at carbon 2
mannose
glucose

Removal of the proton from the hydroxyl group on carbon 2 yields the enolate ion that can be converted to fructose by protonation at carbon 1.

VISUALIZING THE REACTION

Conversion of the enediol to fructose

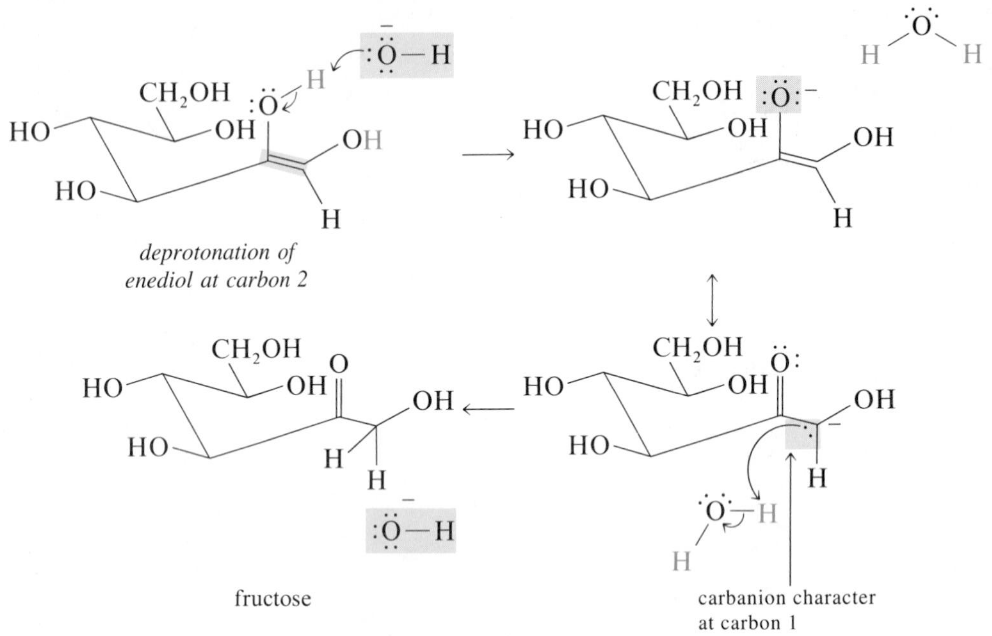

deprotonation of enediol at carbon 2
fructose
carbanion character at carbon 1

PROBLEM 25.17

(a) When the reaction described above was carried out in D_2O, it was found that mannose and glucose had deuterium attached to carbon 2, but in fructose deuterium was attached to carbon 1. Show how this evidence supports the mechanism proposed for the reaction.

(b) The mannose obtained as a product of the experiment in part a had 1.4 deuterium atoms attached to carbon per molecule. The fructose obtained had an average of 1.7 deuterium atoms attached to carbon. Referring to the mechanism, explain how fructose could have more than one deuterium atom per molecule. What does the deuterium content of mannose suggest?

25.5 REACTIONS OF MONOSACCHARIDES AS ALCOHOLS

A. Esterification

Carbohydrates contain many hydroxyl groups and thus react with acid anhydrides (p. 576) to give esters. The formation of a pentaacetate when glucose reacts with acetic anhydride (p. 1064) was taken as evidence that glucose is an open-chain aldehyde with a hydroxyl group on each of the other five carbon atoms. At present, the pentaacetate of glucose is assigned a cyclic structure with an acetyl group at carbon 1 and not at carbon 5. This is an interesting example of how an experimental observation remains valid, even though its interpretation changes with time and the accumulation of other experimental data.

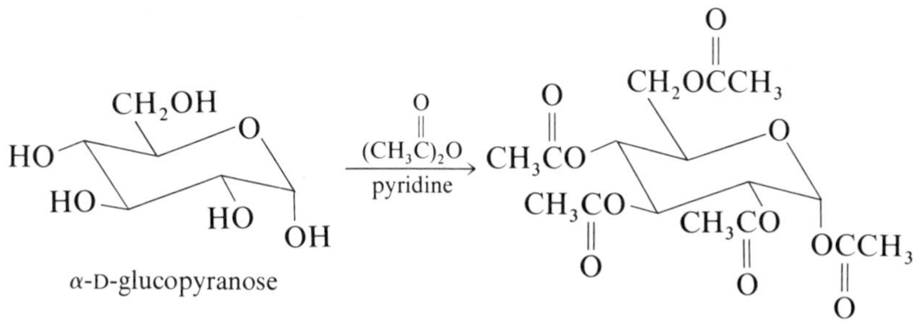
CH2OH
HO
HO
HO
OH
O
(CH3C)2O
pyridine
CH2OCCH3
CH3CO
OCCH3

α-D-glucopyranose

α-D-glucopyranose pentaacetate

The ester functions in glucopyranose pentaacetate undergo the typical ester reactions. For example, the acetyl groups can be hydrolyzed with dilute base to regenerate glucose.

PROBLEM 25.18

Write an equation for the hydrolysis of glucose pentaacetate with dilute aqueous base. Write the mechanism for the reaction using the structure below to represent glucose pentaacetate.

$$R{-}O{-}\overset{\overset{\displaystyle O}{\|}}{C}{-}CH_3$$

The acetoxy group at carbon 1 in the pentaacetate of glucose is more reactive than those at other carbons and can be replaced selectively. With hydrogen bromide in acetic acid, the α-bromide is formed.

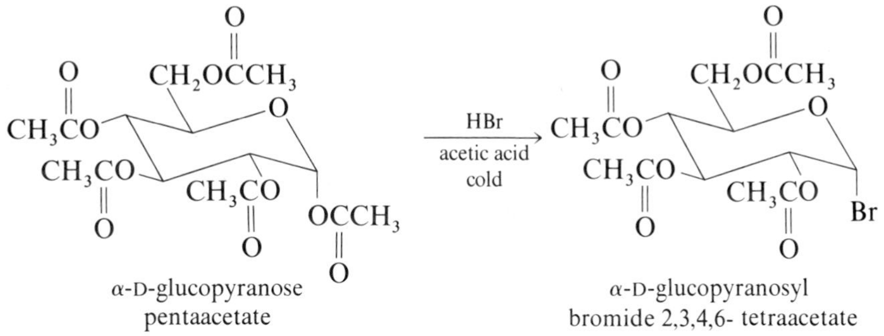
CH2OCCH3
CH3CO
OCCH3
HBr
acetic acid
cold
Br

α-D-glucopyranose pentaacetate

α-D-glucopyranosyl bromide 2,3,4,6- tetraacetate

Such bromides are important intermediates in the formation of glycosides. The synthesis of the β-glucoside salicin (p. 1070) is an example of such a reaction.

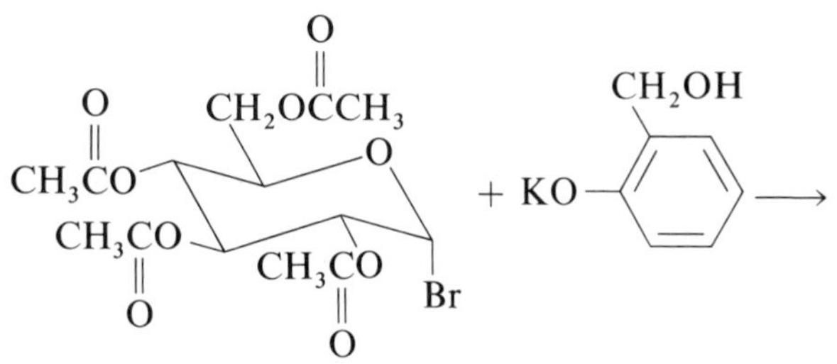
CH2OCCH3
CH3CO
Br
CH2OH
+ KO

α-D-glucopyranosyl bromide 2,3,4,6-tetraacetate

potassium o-(hydroxymethyl)phenolate

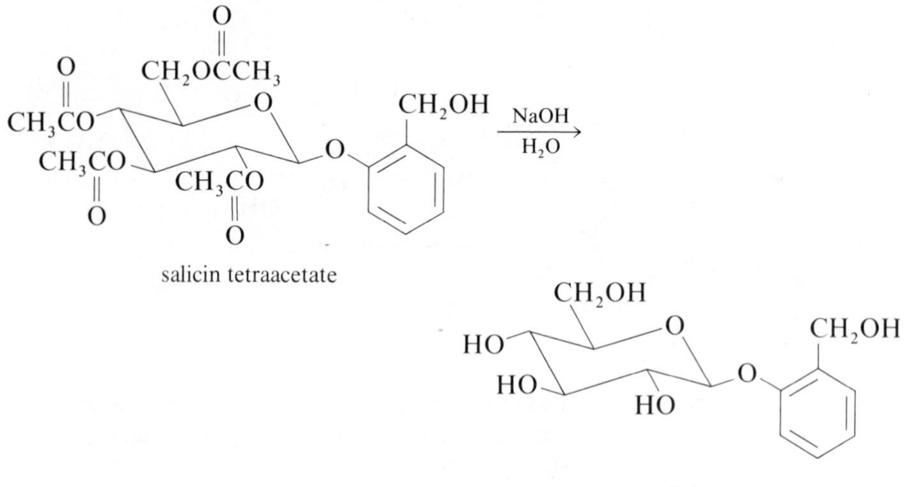
NaOH
H_2O

salicin tetraacetate

salicin

PROBLEM 25.19

Write a mechanism for the formation of the α-bromide from glucose pentaacetate.

PROBLEM 25.20

Why is it necessary to use basic hydrolysis to remove the acetate groups from salicin tetraacetate in the synthesis of salicin shown above?

PROBLEM 25.21

How would you classify the reaction shown as the first step of the preparation of salicin? What is the stereochemistry of that reaction?

B. Esters of Phosphoric Acid

In biological systems, the most important esters of sugars are the ones formed with phosphoric acid. There are three forms in which phosphoric acid appears in such systems. Two of these forms may be regarded as anhydrides of orthophosphoric acid, H_3PO_4.

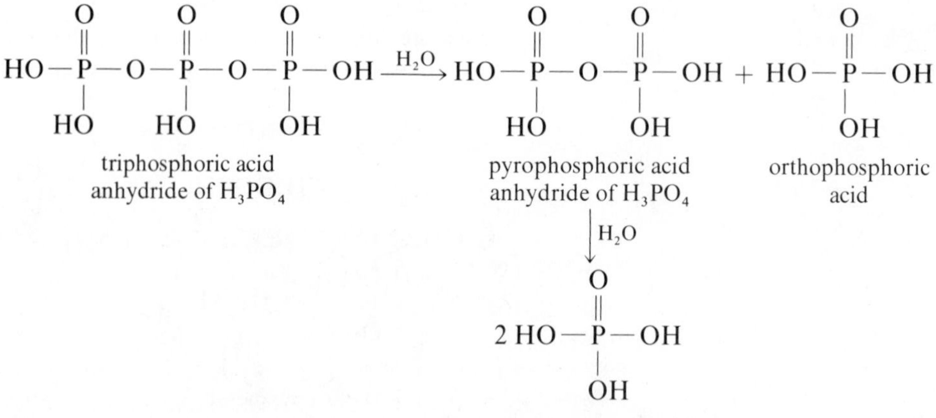
H_2O
H_2O

triphosphoric acid anhydride of H_3PO_4

pyrophosphoric acid anhydride of H_3PO_4

orthophosphoric acid

orthophosphoric acid

All of these forms of phosphoric acid are found in nucleotides, the phosphoric acid esters of nucleosides (p. 1013). Adenosine triphosphate, the triphosphate ester at carbon 5 of ribose in adenosine (p. 1013), is found extensively in living systems. Its presence is so synonymous with life on earth that when experiments are designed to search for life in outer space, it is one of the key compounds sought. Adenosine triphosphate stores chemical energy and makes it available to specific cell processes. It gives up energy by transferring a phosphate group to another molecule, which is converted into a reactive form as its phosphate ester. A specific example that starts off the process of glycolysis, the metabolism of glucose in the body, is the conversion of glucose to glucose 6-phosphate.

VISUALIZING THE REACTION

Biosynthesis of glucose-6-phosphate

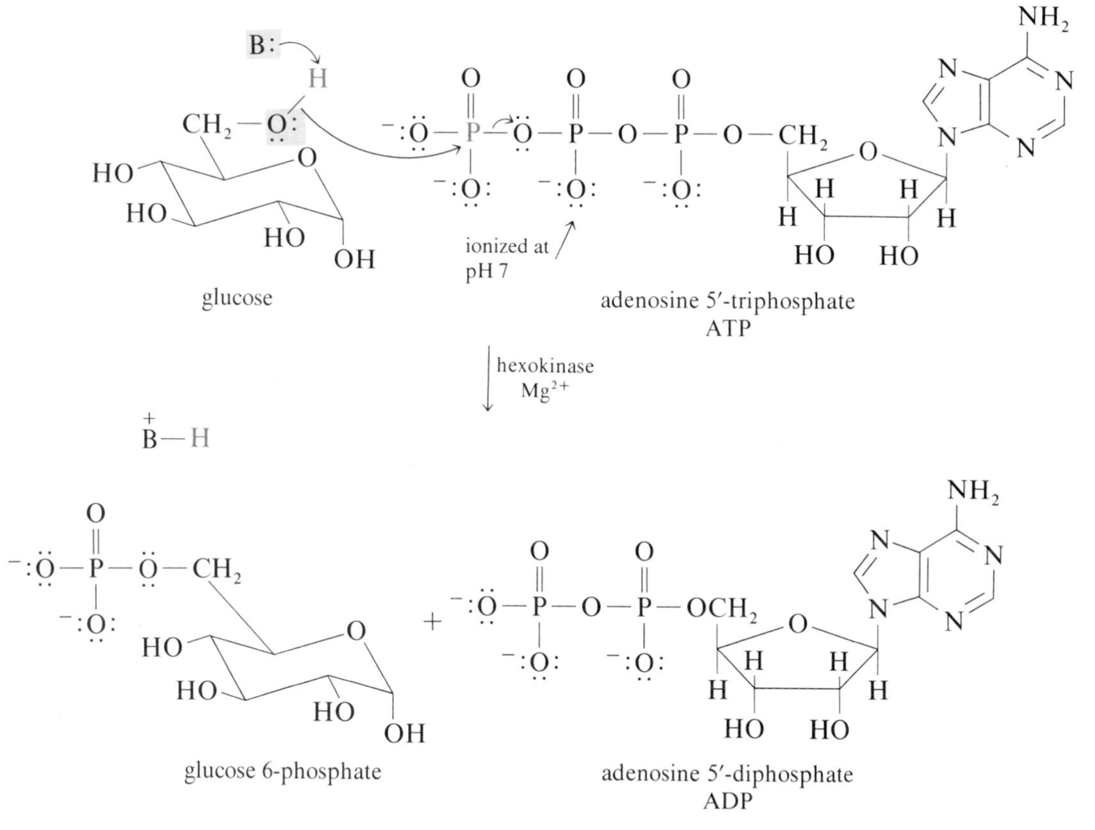

ionized at pH 7
glucose
adenosine 5′-triphosphate
ATP
hexokinase
Mg2+
glucose 6-phosphate
adenosine 5′-diphosphate
ADP

The pyrophosphate group is a leaving group in many biological substitution reactions, such as in the biosynthesis of terpenes (p. 706). It has the same function here. The reaction of glucose with adenosine triphosphate is a nucleophilic substitution reaction. The reaction is catalyzed by an enzyme, hexokinase, and magnesium ion is necessary. The magnesium ion forms a complex with the two terminal phosphate groups of adenosine triphosphate, reducing the negative charge on the ion at physiological pH and making it easier for the substitution reaction to take place.

Later in the process of glycolysis, a particularly reactive phosphate ester of an enol transfers a phosphate group back to the adenosine diphosphate and regenerates adenosine triphosphate. At this stage, some of the energy generated by the breakdown of glucose in the body is returned to storage in the triphosphate group. This reaction is also catalyzed by an enzyme.

VISUALIZING THE REACTION

Biosynthesis of ATP from ADP

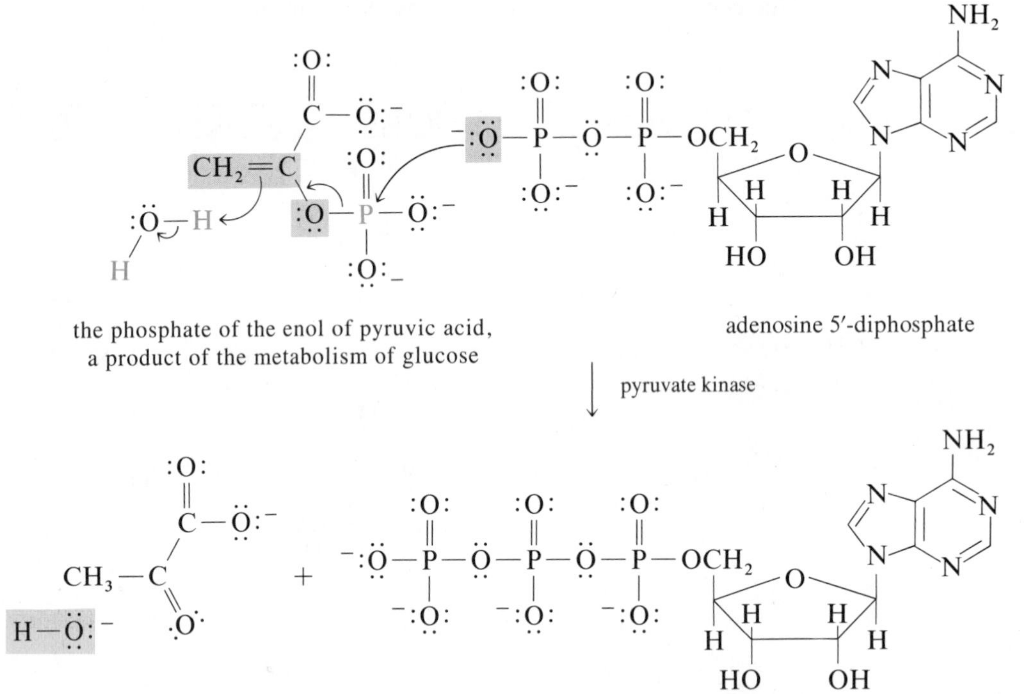
the phosphate of the enol of pyruvic acid, a product of the metabolism of glucose
adenosine 5′-diphosphate
pyruvate kinase

pyruvate anion

adenosine 5′-triphosphate

The loss of the phosphate group from the enol oxygen takes place easily because the stable carbonyl group in pyruvic acid is generated.

C. Ether Formation

If methyl α-D-glucopyranoside is treated with dimethyl sulfate in the presence of aqueous sodium hydroxide, the methyl ethers of the alcohol functions are formed. The base helps the hydroxyl groups to ionize, creating good nucleophiles, which then react by nucleophilic substitution with dimethyl sulfate.

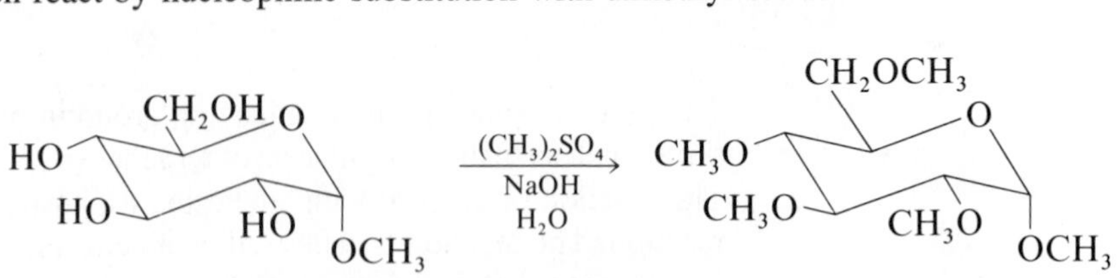
$(CH_3)_2SO_4$
NaOH
H_2O

methyl α-D-glucopyranoside

methyl 2,3,4,6-*O*-tetramethyl-α-D-glucopyranoside

The fully methylated sugar is named as a methyl glucopyranoside having four more methyl groups substituted on the oxygens at carbons 2, 3, 4, and 6. The name is therefore methyl 2,3,4,6-*O*-tetramethyl-α-D-glucopyranoside; the capital *O* indicates that the substituents named are on oxygen, not carbon.

The methyl ethers formed from carbohydrates are stable to bases and dilute acids and serve to protect the hydroxyl groups. The methoxy group at the anomeric carbon is different in its reactivity from the other ether groups in the molecule. It is part of an acetal linkage and is therefore sensitive to acid hydrolysis. That one methyl group can be removed, leaving the others in place.

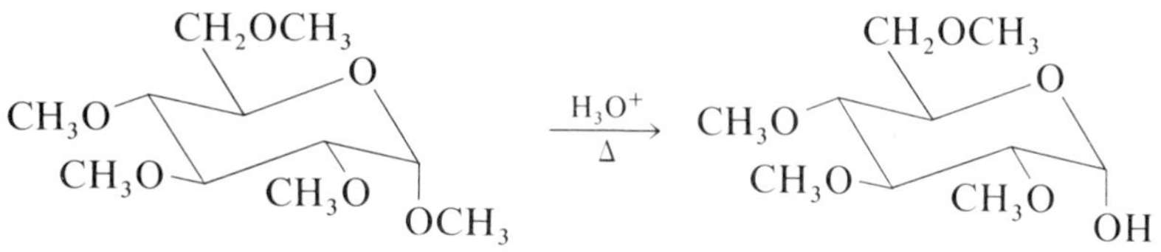

CH2OCH3
CH3O
CH3O
CH3O
OCH3
H3O+
Δ
OH

methyl 2,3,4,6-*O*-tetramethyl-α-D-glucopyranoside — 2,3,4,6-*O*-tetramethyl-α-D-glucopyranose

This is possible because protonation of the oxygen atom of the methoxy group and the loss of methanol at the anomeric carbon results in a carbocation that is particularly well stabilized by delocalization of charge to the adjacent oxygen atom (p. 501). Such stabilization is not possible for the cations that would result from the loss of any of the other methoxy groups, so much more drastic conditions are necessary to remove them.

PROBLEM 25.22

Write a mechanism for the hydrolysis in dilute acid of methyl tetramethylglucopyranoside. Convince yourself that no special stabilization is possible for carbocations formed by the loss of a methoxy group from any carbon atom other than the anomeric one.

PROBLEM 25.23

The methyl groups on the oxygens at carbons 2, 3, 4, and 6 of tetramethylglucopyranose can be removed by boiling with 57% hydriodic acid. One of the products of this reaction is iodomethane. Using $ROCH_3$ to represent a methyl ether, write a mechanism for the removal of the methyl group. Why is it possible to cleave an ether under these conditions but not with dilute hydrochloric acid, for example?

The removal of the methyl group at the anomeric carbon regenerates the hemiacetal linkage. The ring is once more in equilibrium with the open-chain form, and mutarotation takes place. Note that the hydroxyl group on carbon 5 is not methylated and retains its properties as a secondary alcohol function.

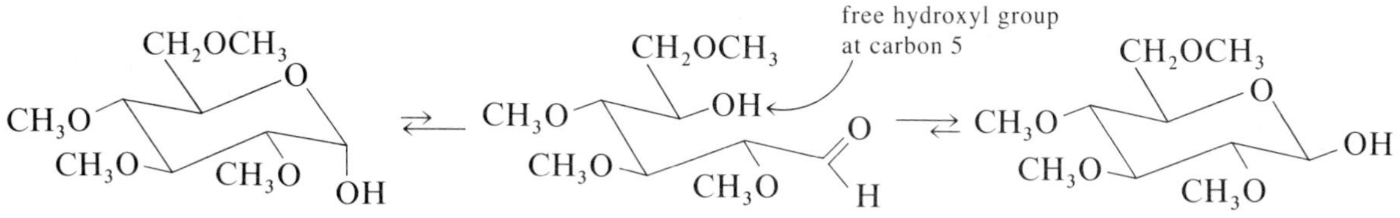

free hydroxyl group
at carbon 5
CH2OCH3
CH3O
OH
O
H

2,3,4,6-*O*-tetramethyl-α-D-glucopyranose — open-chain form — 2,3,4,6-*O*-tetramethyl-β-D-glucopyranose

Section 25.8A shows how reactions involving unmethylated hydroxyl groups can be used to determine the size of rings in sugars.

25.6 OXIDATION REACTIONS OF SUGARS

A. Reducing Sugars. Aldonic Acids

Sugars such as glucose and fructose are **reducing sugars.** This means that they are easily oxidized by mild oxidizing agents. The test for diabetes, in which urine is tested for glucose, is really a test for the presence of a reducing sugar. Sugars in which the carbonyl group is tied up in an acetal linkage are **nonreducing sugars.** Methyl glucopyranoside is an example of a nonreducing sugar. Disaccharides, such as sucrose, in which the anomeric carbon atoms of both monosaccharide units are bonded to each other so that both carbonyl groups are protected, are also non-reducing sugars.

The aldehyde function in glucose, like most aldehydes, is oxidized by very mild oxidizing agents such as **Benedict's reagent,** which is a solution in aqueous base of copper(II) sulfate and sodium citrate (used to complex with the copper ion). This reaction is the traditional one used for the diabetes test in the past.

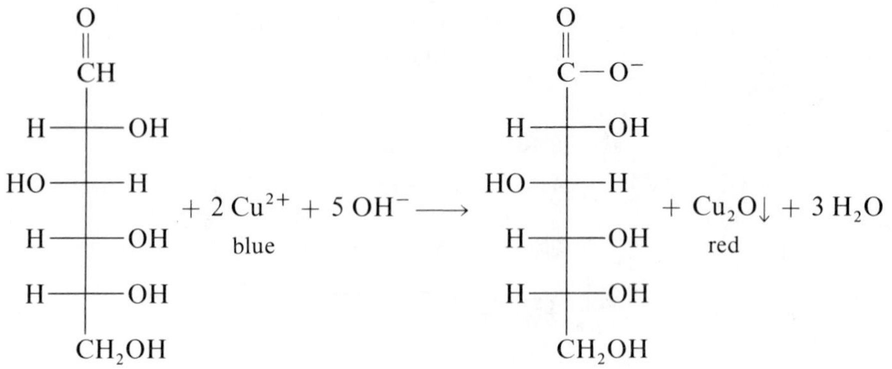
O
CH
H
OH
HO
H
H
OH
H
OH
CH2OH
+ 2 Cu2+ + 5 OH−
blue
O
C−O−
H
OH
HO
H
H
OH
H
OH
CH2OH
+ Cu2O↓ + 3 H2O
red

glucose   anion of gluconic acid

Silver ion in the presence of base is Tollens reagent (p. 557), which also serves as an oxidizing agent for reducing sugars. Glucose reduces silver ion to metallic silver.

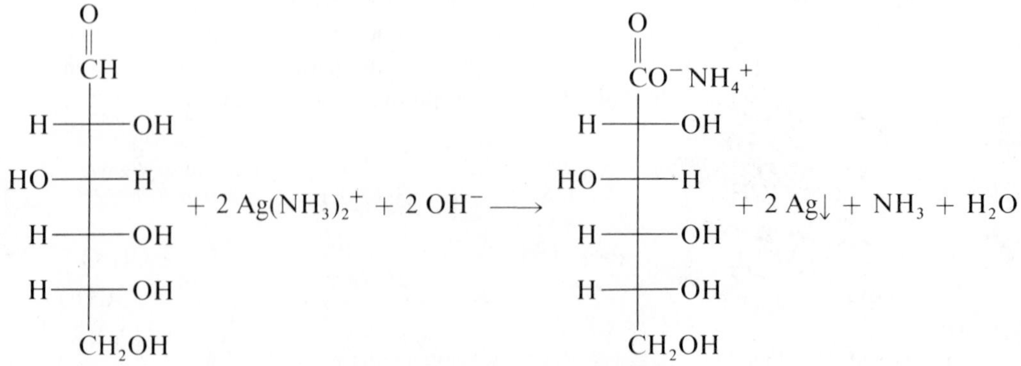
O
CH
H
OH
HO
H
H
OH
H
OH
CH2OH
+ 2 Ag(NH3)2+ + 2 OH−
O
CO− NH4+
H
OH
HO
H
H
OH
H
OH
CH2OH
+ 2 Ag↓ + NH3 + H2O

glucose   ammonium gluconate

Fructose also gives positive results with both of the above tests. α-Hydroxyketones in general are oxidized very easily to diketones and react with Benedict's reagent or in the Tollens test.

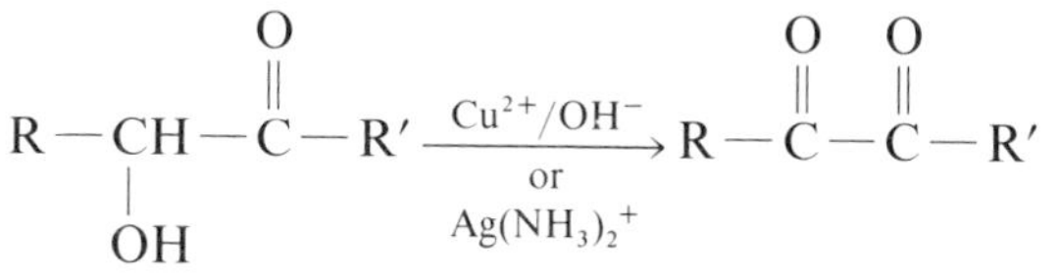

Cu^{2+}/OH^-
or
$Ag(NH_3)_2^+$

Bromine in water is a mild oxidizing agent that is used to distinguish a ketose from an aldose. Only the aldehyde function is oxidized by this reagent. Thus, mannose is converted into the corresponding carboxylic acid, mannonic acid.

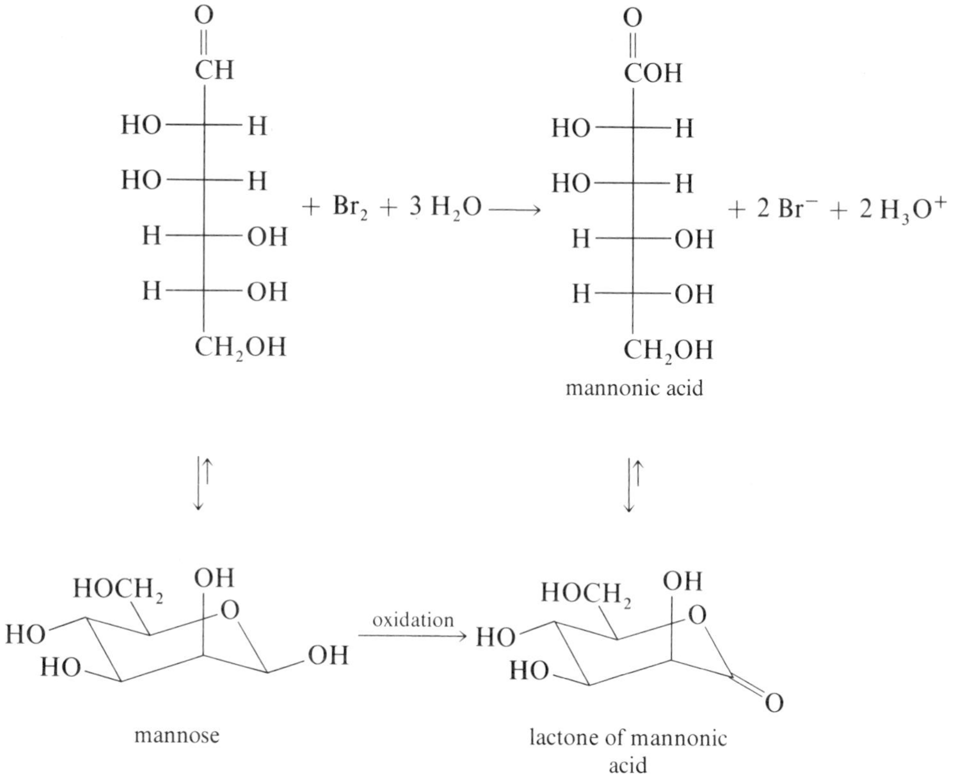

$+ Br_2 + 3\ H_2O \longrightarrow$
$+ 2\ Br^- + 2\ H_3O^+$
mannonic acid
oxidation
mannose
lactone of mannonic acid

In acidic solutions, the hydroxyacid exists as a lactone (p. 604).

Compounds in which carbon 1 of an aldose has been oxidized to a carboxylic acid are known as **aldonic acids.** In naming them, the **-ose** ending of the name of the sugar is converted into **-onic acid.** Thus, glucose gives gluconic acid, and mannose gives mannonic acid.

B. Nitric Acid as an Oxidizing Agent. Aldaric Acids

If a powerful oxidizing agent such as hot nitric acid is used, a sugar is oxidized at both ends of the chain to the dicarboxylic acid. Such acids are called **aldaric acids.** For example, glucose is oxidized by nitric acid to glucaric acid.

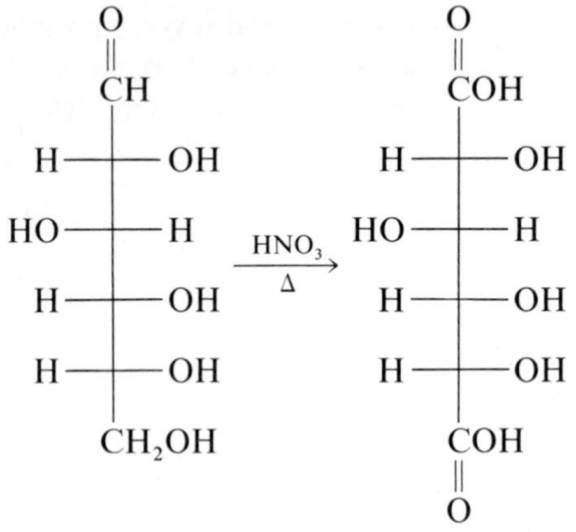
HNO3
Δ

glucose glucaric acid

This oxidation reaction played a large part in the determination of the stereochemistry of sugars. In this reaction, carbon 1 and carbon 6 of the sugar are converted into the same functional group. Thus, any symmetry present in the rest of the molecule becomes apparent. The discovery that both glucaric acid and mannaric acid show optical activity played an important part in the proof of the structure of glucose. Galactaric acid, derived from the other common natural aldohexose, has no optical activity.

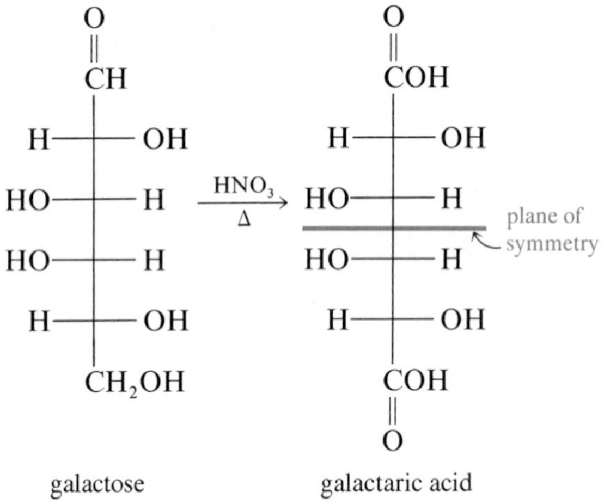
HNO3
Δ
plane of
symmetry

galactose galactaric acid
a meso compound

It is a meso form with a plane of symmetry between carbons 3 and 4. These experimental facts did not by themselves establish the structures of galactose, glucose, and mannose, of course, but they did limit the number of structures that were possible.

PROBLEM 25.24

Draw Fischer projection formulas for any other D-aldohexoses that would be oxidized to a meso aldaric acid.

C. Oxidation with Periodic Acid

Periodic acid (p. 459) is an oxidizing agent that is extremely useful in carbohydrate chemistry. It selectively cleaves the carbon-carbon bond in 1,2-diols, α-hydroxyaldehydes, and α-hydroxyketones. The reaction is carried out at or below room temperature in water solutions, an ideal solvent for carbohydrates, which are not very soluble in nonpolar organic solvents.

A polyhydroxy compound such as glucose gives many fragments when oxidized with periodic acid as shown below. Carbons 1 to 5 of glucose all end up in formic acid molecules, and carbon 6 appears as formaldehyde when degradation is complete.

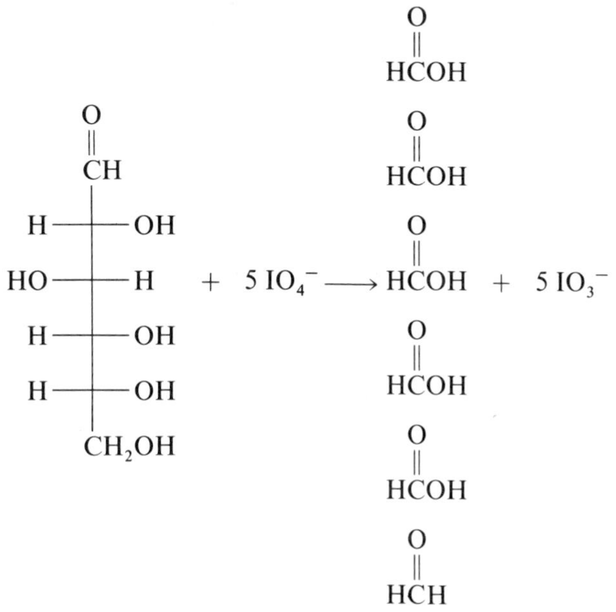

CH
H OH
HO H
H OH
H OH
CH2OH
+ 5 IO4−
HCOH
HCH
+ 5 IO3−

glucose

periodate oxidation products of glucose

Studies utilizing isotopic labeling have shown that periodate ion attacks the carbonyl group directly as a nucleophile. The carbonyl group is oxidized to a carboxylic acid. Cleavage of a diol gives two carbonyl groups, which undergo further oxidation to acids if hydroxyl groups are adjacent to them. Only the last carbon atom of glucose, carbon 6, remains as formaldehyde.

PROBLEM 25.25

Write a mechanism for the oxidation of a portion of the glucose molecule with periodic acid to make sure you understand why the observed products are formed.

α-Hydroxycarboxylic acids are oxidized only very slowly under the conditions used for the oxidation of carbohydrates. Glucaric acid, for example, reacts with only 3 equivalents of periodate ion.

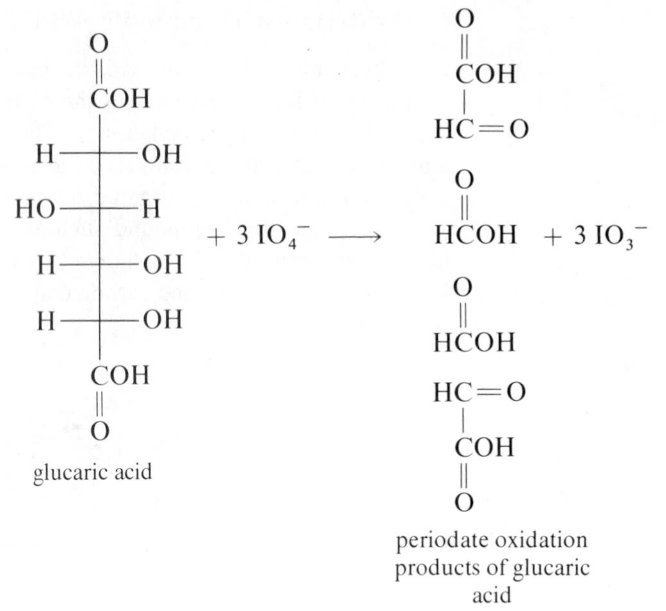
glucaric acid
periodate oxidation products of glucaric acid

D. Oxidation Reactions Used to Establish Relative Configuration and Ring Structure in Monosaccharides

The oxidation of methyl glycosides with periodic acid was used to establish the size of the ring in the cyclic structures of sugars and to correlate stereochemistry in different members of a series. Claude S. Hudson and his colleagues at the National Institutes of Health did elegant experiments in this area in the 1930s and 1940s. They showed that methyl α-D-glucopyranoside and methyl α-D-mannopyranoside were both oxidized to the same dialdehyde. Note that the ring structure in the glucopyranoside and mannopyranoside has only two adjacent diol units. Formic acid is formed from carbon 3 in both cases.

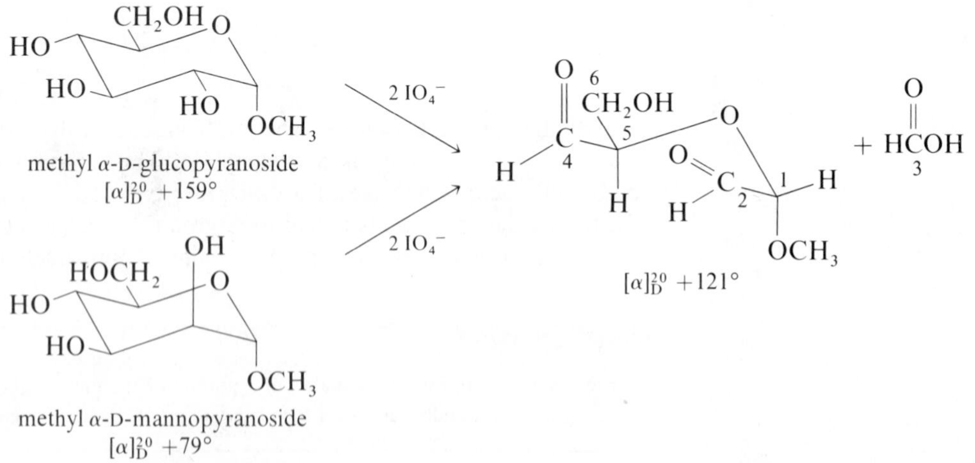
methyl α-D-glucopyranoside
[α]$_D^{20}$ +159°
methyl α-D-mannopyranoside
[α]$_D^{20}$ +79°
[α]$_D^{20}$ +121°

Stereochemistry at carbons 2, 3, and 4 is destroyed. The only stereocenters left in the dialdehyde are carbons 1 and 5. All the compounds assigned the methyl α-D-glycoside structure give the same dialdehyde, proving that they all have the same configuration at carbon 5, the stereocenter that determines their membership in the D family, and at carbon 1, which determines whether or not they are α-glycosides.

PROBLEM 25.26

Write an equation showing that methyl α-D-galactopyranoside gives the same dialdehyde as do the glucopyranoside and mannopyranoside.

The dialdehydes formed from these oxidation reactions are actually not good compounds for structure determinations. They are unstable and form hydrates reversibly with water. In many cases, they give cyclic hydrates, creating new stereocenters, which complicate the assignment of stereochemistry.

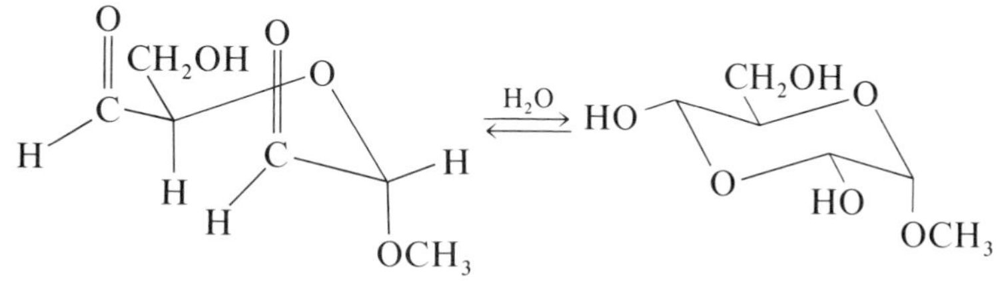

dialdehyde from periodate oxidation of α-D-glycosides

cyclic hydrate of dialdehyde

For these reasons, Hudson made further structural comparisons by oxidizing the dialdehydes to carboxylic acids with bromine water, in the presence of strontium carbonate to keep the reaction mixture from getting acidic.

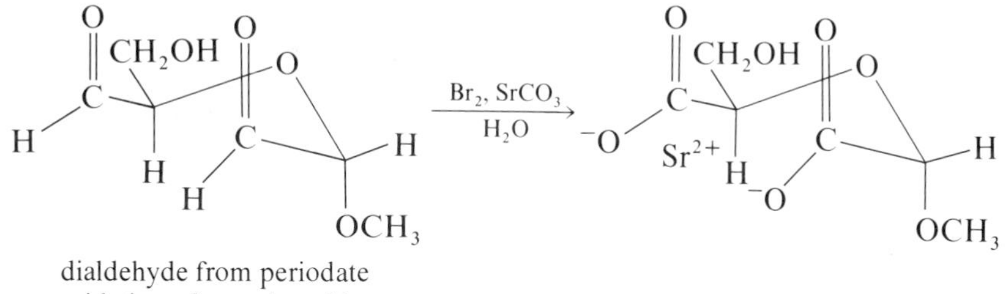

dialdehyde from periodate oxidation of α-D-glycosides

dicarboxylic acid from the oxidation of dialdehyde, isolated as the strontium salt

The same strontium salt was isolated whether methyl α-D-glucopyranoside or methyl α-D-mannopyranoside was the starting material, once again proving the identity of stereochemistry at carbons 1 and 5 of those compounds.

PROBLEM 25.27

Strontium carbonate has a low solubility in water. How does it keep the solution from turning acidic? Why is it necessary to keep the reaction mixture from getting acidic?

Acid hydrolysis of the strontium salt of the dicarboxylic acid cleaves the acetal linkage. The fragments formed are oxidized with bromine water. (−)-Glyceric acid and oxalic acid are isolated and identified, usually as their calcium or barium salts.

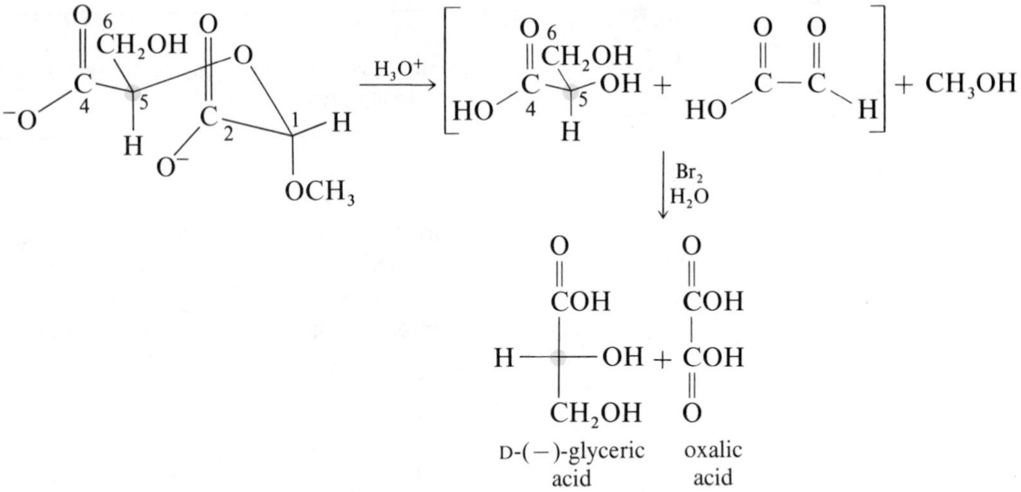
H3O+
+ CH3OH
Br2
H2O
D-(−)-glyceric acid
oxalic acid

Carbon 1 loses its chirality when the acetal group is hydrolyzed. The only stereocenter from the original sugar left untouched is carbon 5, which turns up in (−)-glyceric acid. (−)-Glyceric acid is related in configuration to (+)-glyceraldehyde and has the *R* or D configuration (p. 1054). These reactions, therefore, make a direct connection between the configuration at carbon 5 of hexoses and the configuration of (+)-glyceraldehyde.

PROBLEM 25.28

The aldopentoses also form pyranosides. When the degradative scheme outlined above was carried out on methyl β-D-arabinopyranoside, the strontium salt of the dicarboxylic acid had $[\alpha]_D^{20}$ + 55.7°. The same series of reactions on methyl α-D-arabinopyranoside gave a salt with $[\alpha]_D^{20}$ of −55.5°. Methyl β-D-arabinopyranoside has the following structure.

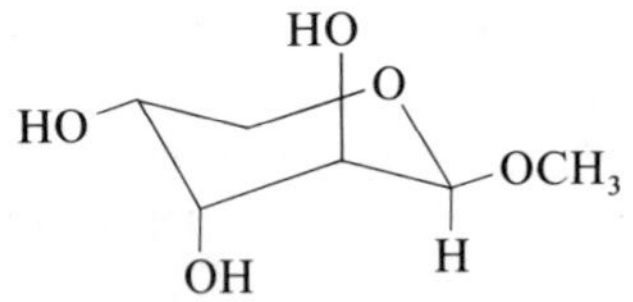

methyl β-D-arabinopyranoside

(a) Write equations for its degradation showing the intermediate dialdehyde and the salt of the diacid. Keep track of the stereocenters.
(b) What is the stereochemical relationship between the strontium salt obtained from the α-pyranoside and that from the β-pyranoside?

The formation of formic acid and a dialdehyde in the periodate oxidation of glucopyranosides is also proof of the size of the ring. If, for example, the ring were five-membered, no formic acid would be formed, and the major fragment would be a trialdehyde.

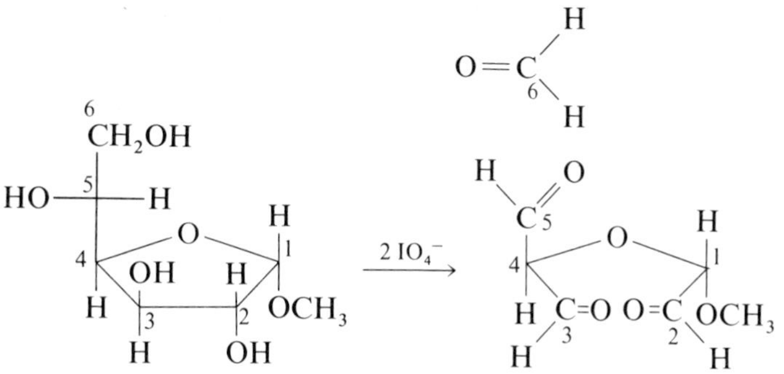

methyl α-D-glucofuranoside

periodate oxidation products expected from the methyl glucofuranoside

Study Guide Concept Map 25.3

This is not what is observed for the cyclic acetals formed by glucose under most reaction conditions.

25.7 SYNTHETIC TRANSFORMATIONS OF MONOSACCHARIDES

A. The Kiliani-Fischer Synthesis

The Kiliani reaction is the addition of hydrocyanic acid to a monosaccharide followed by the hydrolysis of the resulting cyanohydrin (p. 487) to obtain a polyhydroxycarboxylic acid. These carboxylic acids form lactones (p. 604). γ-Lactones have a five-membered ring and tend to be more stable than δ-lactones, which have a six-membered ring. Lactones are obtained whenever chemists attempt to isolate the free sugar acids.

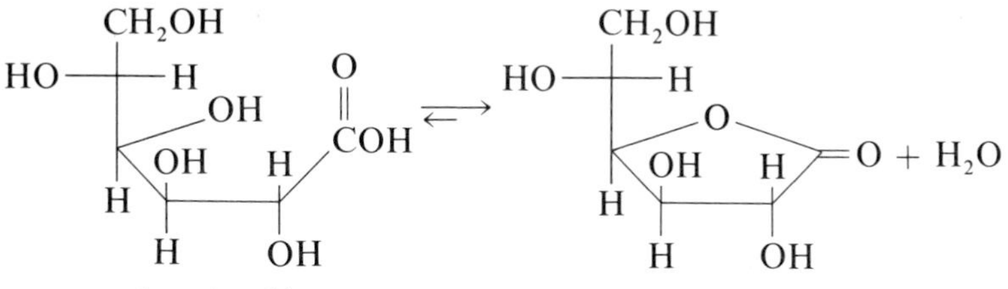

gluconic acid

γ-gluconolactone

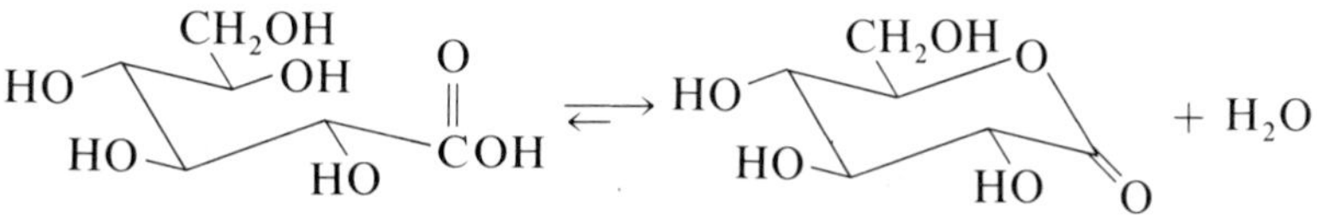

gluconic acid

δ-gluconolactone

The sugar lactones were known in the early days of carbohydrate chemistry. Fischer discovered in 1889 that these lactones could be reduced to aldoses. As his reducing agent, he used sodium amalgam, a small amount of sodium metal dissolved in mercury in order to moderate the reactivity of sodium in water solutions.

Sodium borohydride is now used for this reduction. For example, sodium borohydride is added to a solution of δ-gluconolactone carefully, so that the reducing agent is never present in excess. The lactone is reduced to the hemiacetal form of glucose.

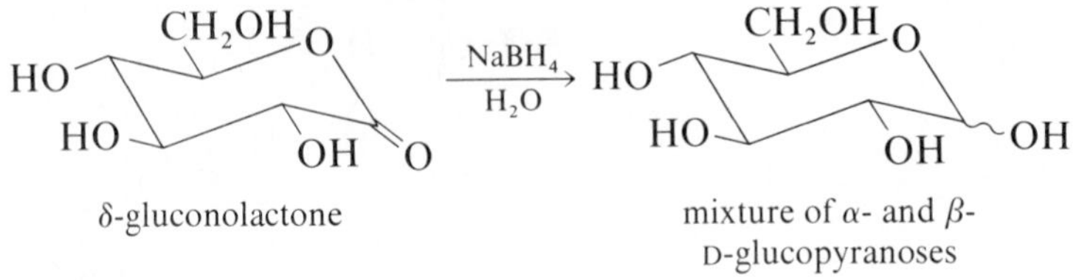
$NaBH_4$
H_2O
δ-gluconolactone
mixture of α- and β-D-glucopyranoses

PROBLEM 25.29

Why should an excess of sodium borohydride be avoided in the preceding reaction?

Fischer was able to reduce the lactone of the carboxylic acid resulting from the Kiliani reaction to an aldose. The combination of the two reactions, known as the **Kiliani-Fischer synthesis,** enables chemists to synthesize a longer-chain aldose from a shorter-chain one.

The synthetic sequence was first applied to L-arabinose, a natural aldopentose, and the products were L-mannose and L-glucose. Fischer painstakingly proved in 1890 that the L-glucose he had obtained was the enantiomer of natural glucose. Since then, D-arabinose has been converted to D-glucose and D-mannose. The diasteromeric nitriles, which are epimers at carbon 2, are not produced in equal amounts in the addition reaction.

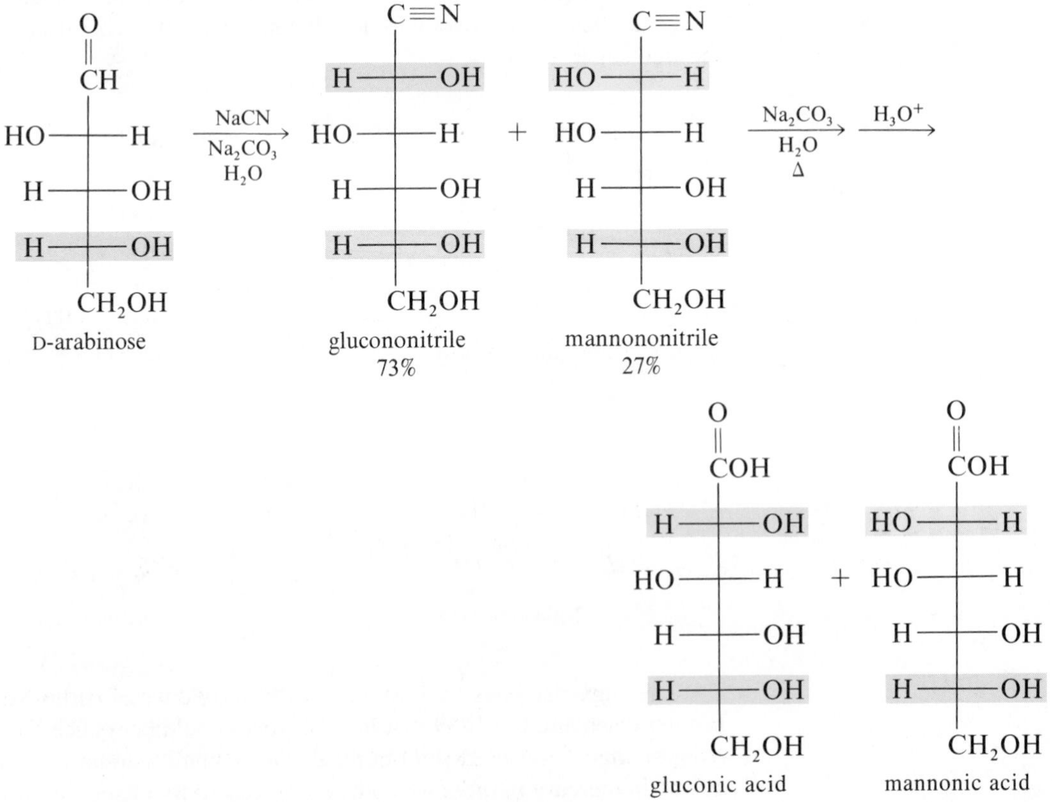
NaCN
Na_2CO_3
H_2O
D-arabinose
glucononitrile
73%
mannononitrile
27%
Na_2CO_3
H_2O
Δ
H_3O^+
gluconic acid
mannonic acid

In a basic solution, the nitrile with the same stereochemistry as glucose is formed in larger amounts. The nitriles, which are easily hydrolyzed, are usually not isolated, but are converted directly to the carboxylic acids.

The diastereomeric carboxylic acids can be separated. Their salts have different solubilities, and their lactones form with different degrees of ease. The separated acids, as their lactones, are reduced to the aldoses.

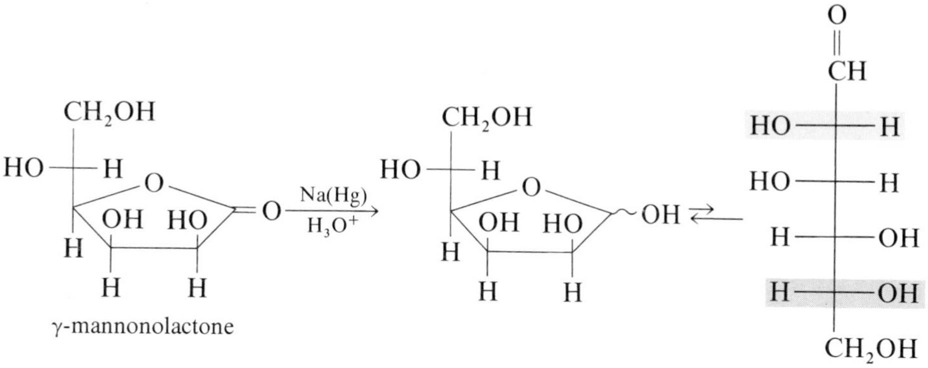

CH2OH
HO
H
O
OH HO
O
H
H H
Na(Hg)
H3O+
CH2OH
HO
H
O
OH HO
OH
H
H H
O
CH
HO
H
HO
H
H
OH
H
OH
CH2OH

γ-mannonolactone

D-mannose

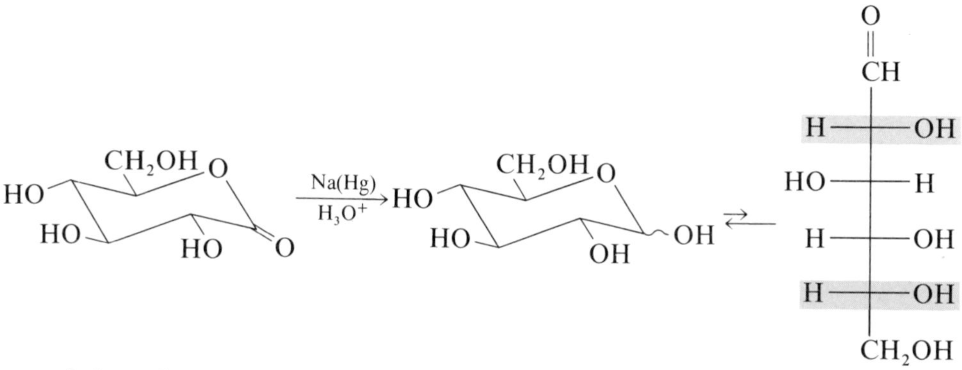

CH2OH
O
HO
HO
HO
O
Na(Hg)
H3O+
CH2OH
O
HO
HO
OH
OH
O
CH
H
OH
HO
H
H
OH
H
OH
CH2OH

δ-gluconolactone

D-glucose

Study Guide Concept Map 25.4

In this synthesis, mannonic acid gives a lactone with a five-membered ring, γ-mannonolactone; gluconic acid is isolated as a lactone with a six-membered ring, δ-gluconolactone. You should remember that when free carboxylic acids of sugars are isolated, they will form lactones, and that the lactones can be reduced to aldehydes.

PROBLEM 25.30

D-Glyceraldehyde can, in principle, be converted to two aldotetroses by the Kiliani-Fischer synthesis. On oxidation with nitric acid, one, D-erythrose, gives *meso*-tartaric acid. The other, D-threose, gives D-(−)-tartaric acid. Show the steps of the conversion and the oxidation to tartaric acids. Assign structures to D-erythrose and D-threose.

PROBLEM 25.31

By the Kiliani-Fischer synthesis, D-erythrose can be converted to D-arabinose and D-ribose. D-Arabinose, on oxidation with nitric acid, gives an optically active aldaric acid. The aldaric acid from D-ribose has no optical activity. Show how the reactions in Problem 25.30 and in this problem establish the stereochemistry of D-arabinose and D-ribose.

B. The Family of D-Aldoses

All the different chemical transformations necessary to establish the structures of the sixteen stereoisomers of glucose and of the pentoses and tetroses that connect them to D-(+)-glyceraldehyde constituted the compelling evidence for the stereochemistry of glucose and came from many sources. This section will retrace some of the reasoning that Fischer used to assign glucose its full stereochemical structure.

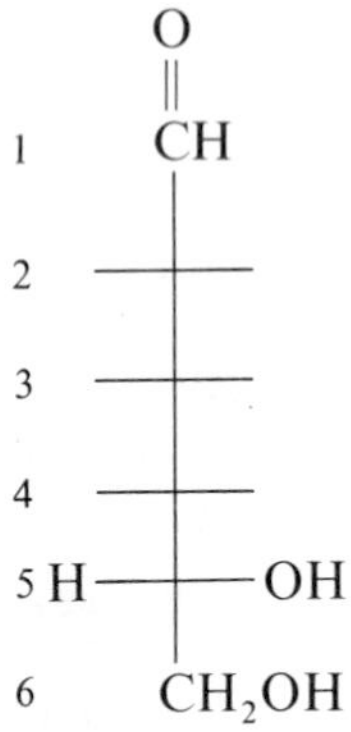
O
1 CH
2
3
4
5 H OH
6 CH₂OH

configuration at carbon 5 of glucose as assigned by Fischer

First of all, in his projection formula for glucose, Fischer arbitrarily assigned the hydroxyl group to a position to the right of carbon 5 (see the structure on the left).

Glucose and mannose could both be prepared from arabinose and were epimers at carbon 2 (by synthetic evidence and because they gave the same osazone). Arabinose gave an optically active aldaric acid and therefore had the hydroxyl group on the left on the carbon atom corresponding to carbon 3 in glucose (Figure 25.3). If that carbon atom had the opposite configuration in the aldopentose, the corresponding aldaric acid would have been a meso form with a plane of symmetry through the middle carbon atom.

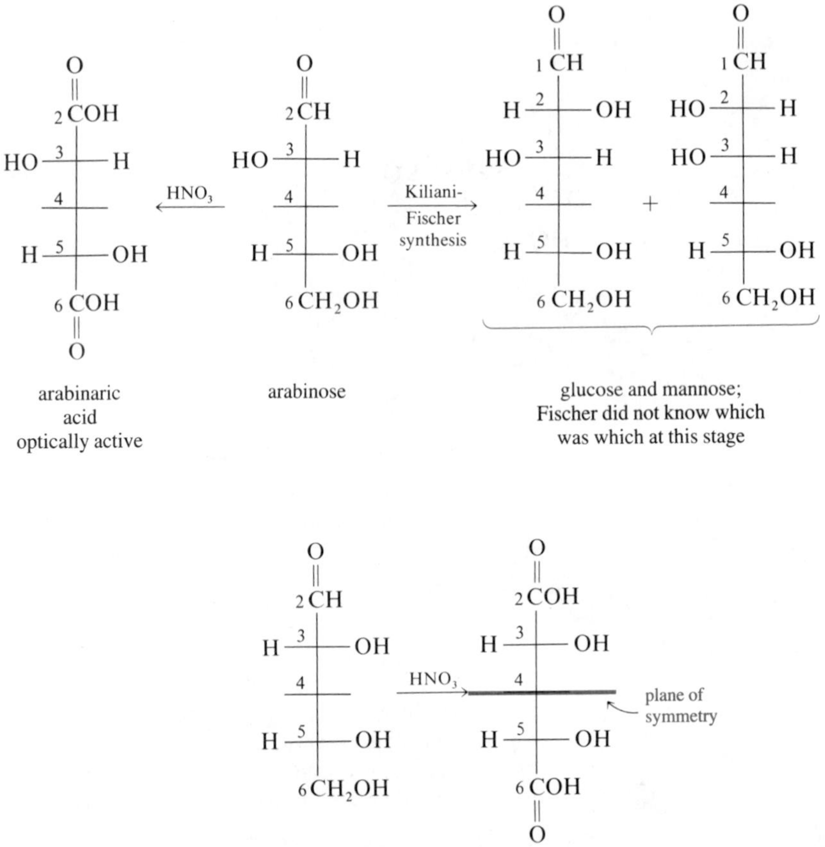
HNO₃
Kiliani-
Fischer
synthesis
arabinaric
acid
optically active
arabinose
glucose and mannose;
Fischer did not know which
was which at this stage
HNO₃
plane of
symmetry
aldopentose with
OH on carbon 3
on the right
meso acid,
not obtained
experimentally

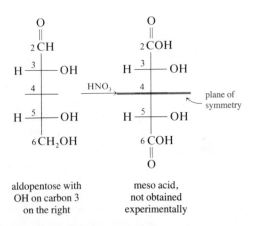

FIGURE 25.3 Reactions establishing the configuration at carbon 3 in glucose.

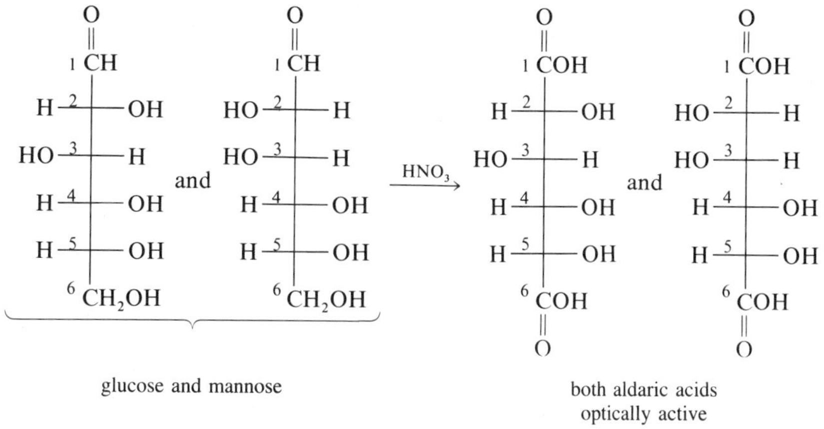

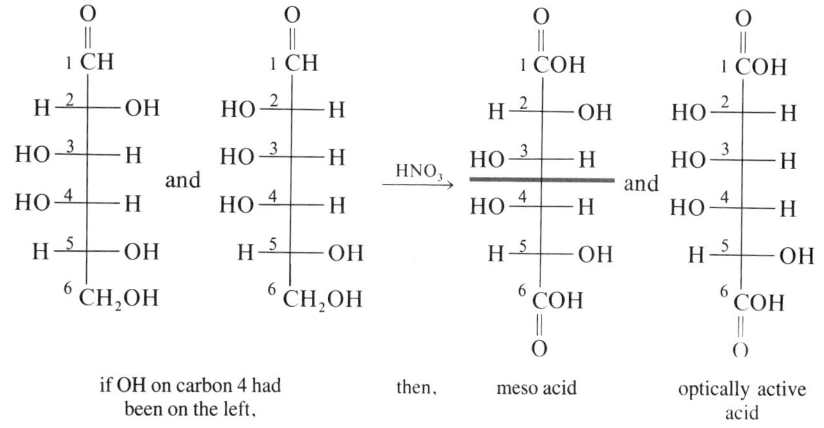

FIGURE 25.4 Evidence for the configuration at carbon 4 in glucose.

Both glucose and mannose are oxidized to optically active aldaric acids (Figure 25.4). This means that the hydroxyl group on carbon 4 is on the right. Otherwise, one of the two would have given a meso aldaric acid. This meso aldaric acid, in fact, is galactaric acid, which is from galactose, the epimer of glucose at carbon 4.

Fischer at this point had a pair of compounds to which he had assigned relative stereochemistry at every stereocenter. He knew that one structure was glucose and one was mannose, but he did not know which was which. He solved the problem by converting glucaric acid from glucose by a series of clever, selective reductions of a sugar lactone, of the aldehyde group, and of a lactone once more, into a new sugar, gulose (Figure 25.5 on page 1095). This sugar belongs to the L family and corresponds to glucose with aldehyde and hydroxymethyl groups interchanged. On oxidation with nitric acid, it gives the same dicarboxylic acid as glucose does.

An examination of the structures that are possible for glucose and mannose shows that only one of them would be transformed into a new compound by the interchange of the aldehyde and hydroxymethyl groups. The other structure is converted back into itself. This assertion can only be confirmed by a careful inspection of the structural formulas shown on the next page.

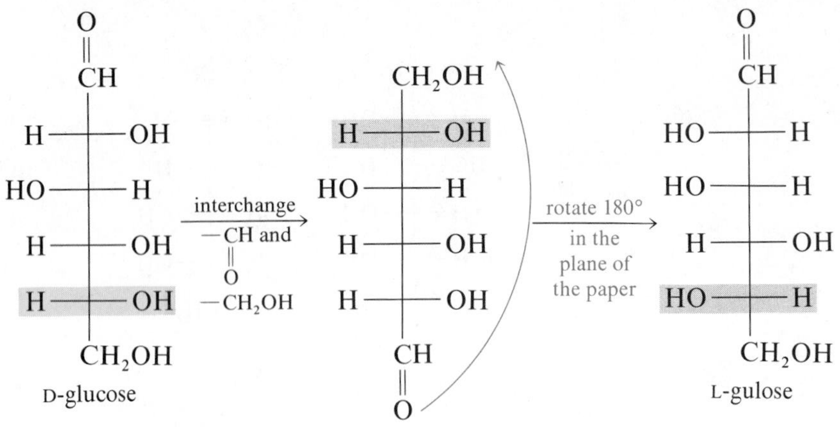
interchange
rotate 180° in the plane of the paper
D-glucose
L-gulose

glucose gives a new sugar when $\overset{O}{\overset{\|}{CH}}$ *and* CH_2OH *groups are interchanged*

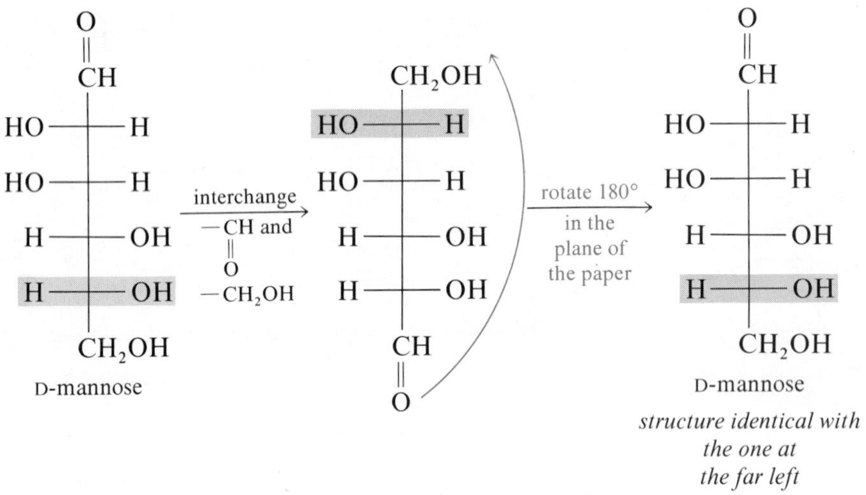
interchange
rotate 180° in the plane of the paper
D-mannose
D-mannose
structure identical with the one at the far left

mannose does not give a new sugar when $\overset{O}{\overset{\|}{CH}}$ *and* CH_2OH *groups are interchanged*

Glucose was the compound that was converted experimentally into a new sugar by Fischer's series of reactions. Therefore, glucose was assigned the structure that it now has, and mannose was given the structure that is epimeric at carbon 2, shown in the above reactions.

When Fischer started his work in carbohydrate chemistry in 1886, the only monosaccharides that were known were L-arabinose, D-glucose, D-galactose, D-fructose, and L-sorbose, which is another 2-ketohexose. He and his coworkers were responsible for the synthesis and the investigation of the chemical properties of twelve of the sixteen possible aldohexoses before his death in 1919. Two others were synthesized by that time in other laboratories. The last two were made in

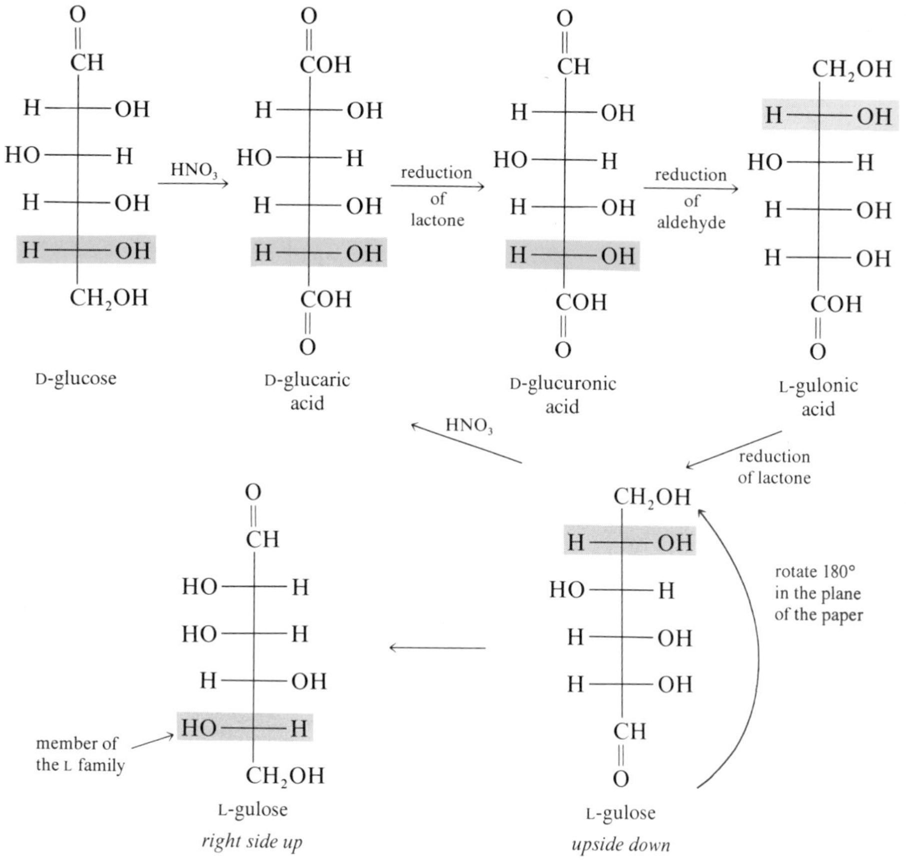
HNO₃
reduction of lactone
reduction of aldehyde
D-glucose
D-glucaric acid
D-glucuronic acid
L-gulonic acid
HNO₃
reduction of lactone
rotate 180° in the plane of the paper
member of the L family
L-gulose
right side up
L-gulose
upside down

FIGURE 25.5 Conversion of D-glucose to L-gulose.

1934. Stereochemical relationships in the family of monosaccharides from D-(+)-glyceraldehyde to the D-aldohexoses are shown in Figure 25.6 on the next page.

PROBLEM 25.32

The reduction of L-sorbose, a 2-ketohexose, gives as one of the products a sugar alcohol, D-glucitol (also called sorbitol). The same alcohol is obtained from the reduction of D-glucose. Assign a structure to L-sorbose.

25.8 DISACCHARIDES

A. Determination of the Structure of a Disaccharide. Lactose

Two disaccharides that occur in nature are sucrose and lactose. Sucrose is derived from plants and is prepared commercially from sugar beets and sugar cane. Lactose is found in the milk of animals and was known as milk sugar when it was first isolated. Other common disaccharides are prepared by breaking down poly-

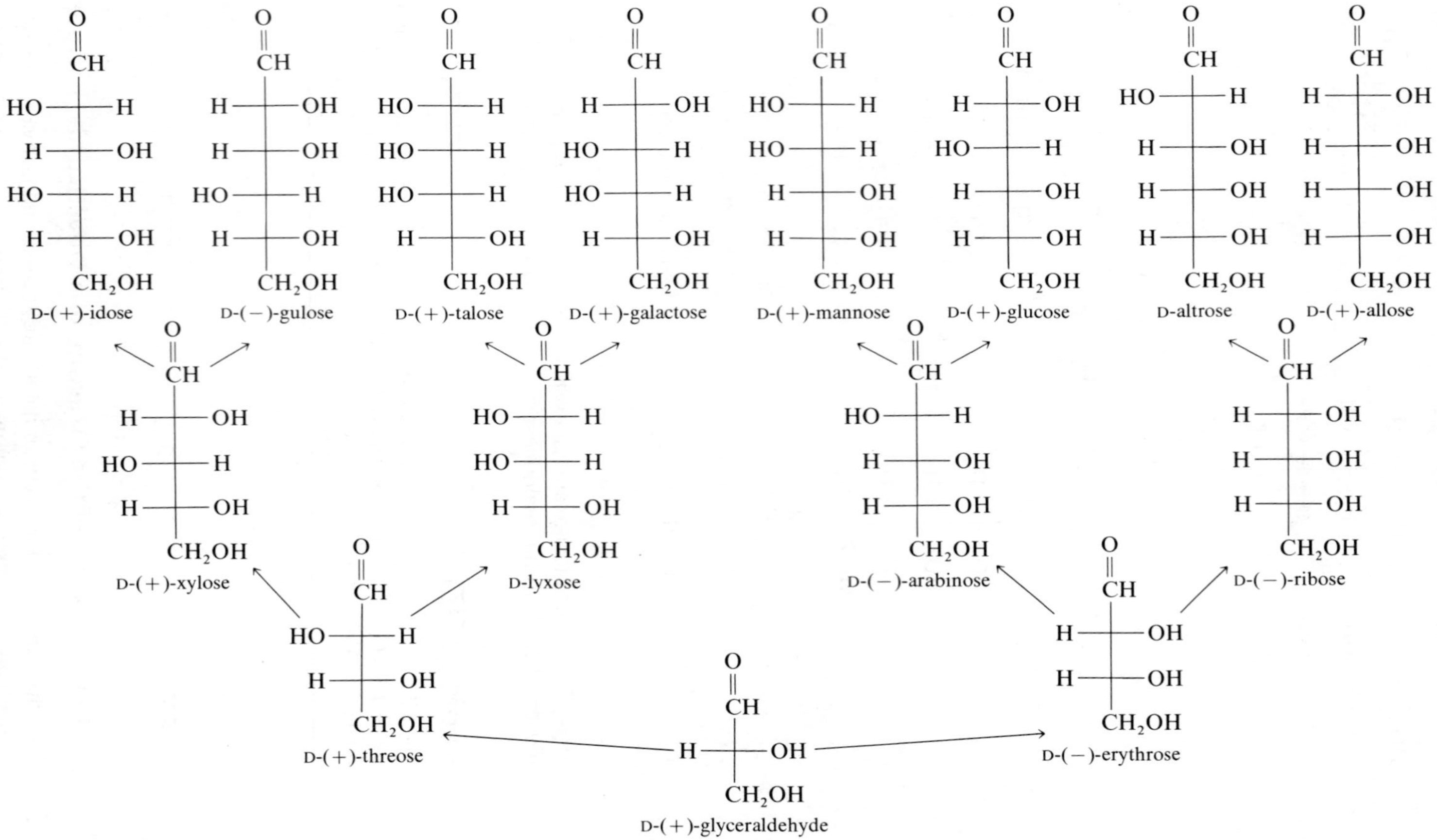
D-(+)-idose
D-(−)-gulose
D-(+)-talose
D-(+)-galactose
D-(+)-mannose
D-(+)-glucose
D-altrose
D-(+)-allose
D-(+)-xylose
D-lyxose
D-(−)-arabinose
D-(−)-ribose
D-(+)-threose
D-(−)-erythrose
D-(+)-glyceraldehyde

FIGURE 25.6 The family of D-aldoses.

saccharides. Maltose is formed from the enzymatic hydrolysis of starch. The partial hydrolysis of cellulose gives cellobiose.

In disaccharides, two monosaccharide units are held together by a glycosidic linkage. The stereochemistry and the position of the linkage and whether the monosaccharides are in the pyranose or the furnose form must be determined in order to establish the structure of a disaccharide. The determination of the structure of lactose, begun in the 1880s by Fischer and completed by the English chemist Sir Walter N. Haworth in the 1920s, serves as an example of the kinds of chemical transformations and reasoning that led to the assignment of structure.

One of the most important pieces of information is whether or not a disaccharide is a reducing sugar. Lactose is a reducing sugar, which means that one of the monosaccharide units has a potential carbonyl group in the hemiacetal form. Lactose gives an osazone, which can be hydrolyzed in dilute acid to galactose and glucosazone. Lactose gives a positive Benedict's test (p. 1082) and is oxidized by bromine water to a carboxylic acid (usually present as a lactone). It undergoes mutarotation, which means there may be an equilibrium between the open-chain aldehyde form and hemiacetal ring forms.

On hydrolysis with dilute acid, lactose gives galactose and glucose. If the hydrolysis is carried out after lactose has been oxidized with bromine water, the formation of gluconic acid indicates that the free aldehyde group is in the glucose residue. Carbon 1 of galactose must be tied up in a glycosidic linkage. The enzyme β-galactosidase cleaves lactose into galactose and glucose. This enzyme is specific for the β stereochemistry of a glycosidic linkage.

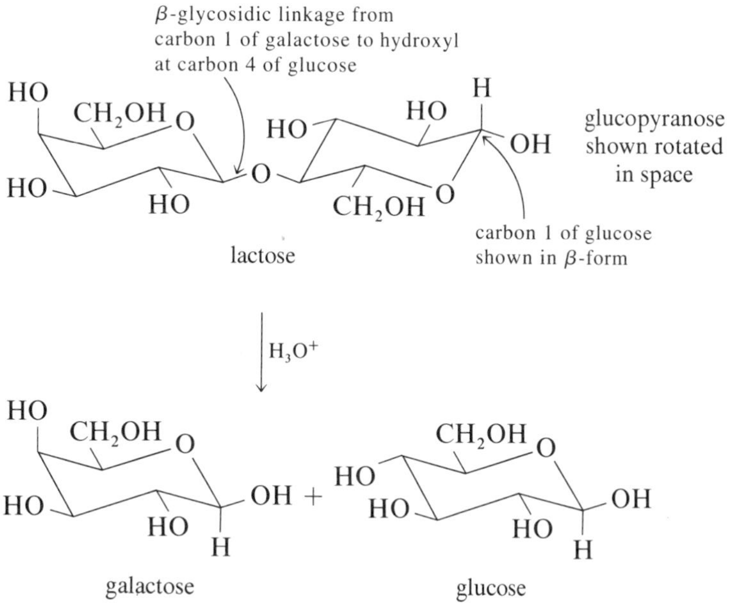

β-glycosidic linkage from carbon 1 of galactose to hydroxyl at carbon 4 of glucose
HO
CH₂OH
O
HO
HO
O
HO
H
OH
CH₂OH
O
glucopyranose shown rotated in space
carbon 1 of glucose shown in β-form
lactose
H₃O⁺
HO
CH₂OH
O
HO
HO
H
OH +
CH₂OH
O
HO
HO
HO
H
OH
galactose
glucose

Lactose is a substituted glucopyranose. The glucose unit is substituted on the oxygen bonded to carbon 4 by a galactopyranose substituent. This substituent is attached at carbon 1 with a β configuration, so it is a β-D-galactopyranosyl group. One systematic name for lactose is therefore 4-*O*-(β-D-galactopyranosyl)-β-D-glucopyranose. Another way of indicating the linkage between the galactose and glucose units is to name the compound *O*-β-D-galactopyranosyl-(1→4)-β-D-glycopyranose. This type of nomenclature is useful for more complex saccharides.

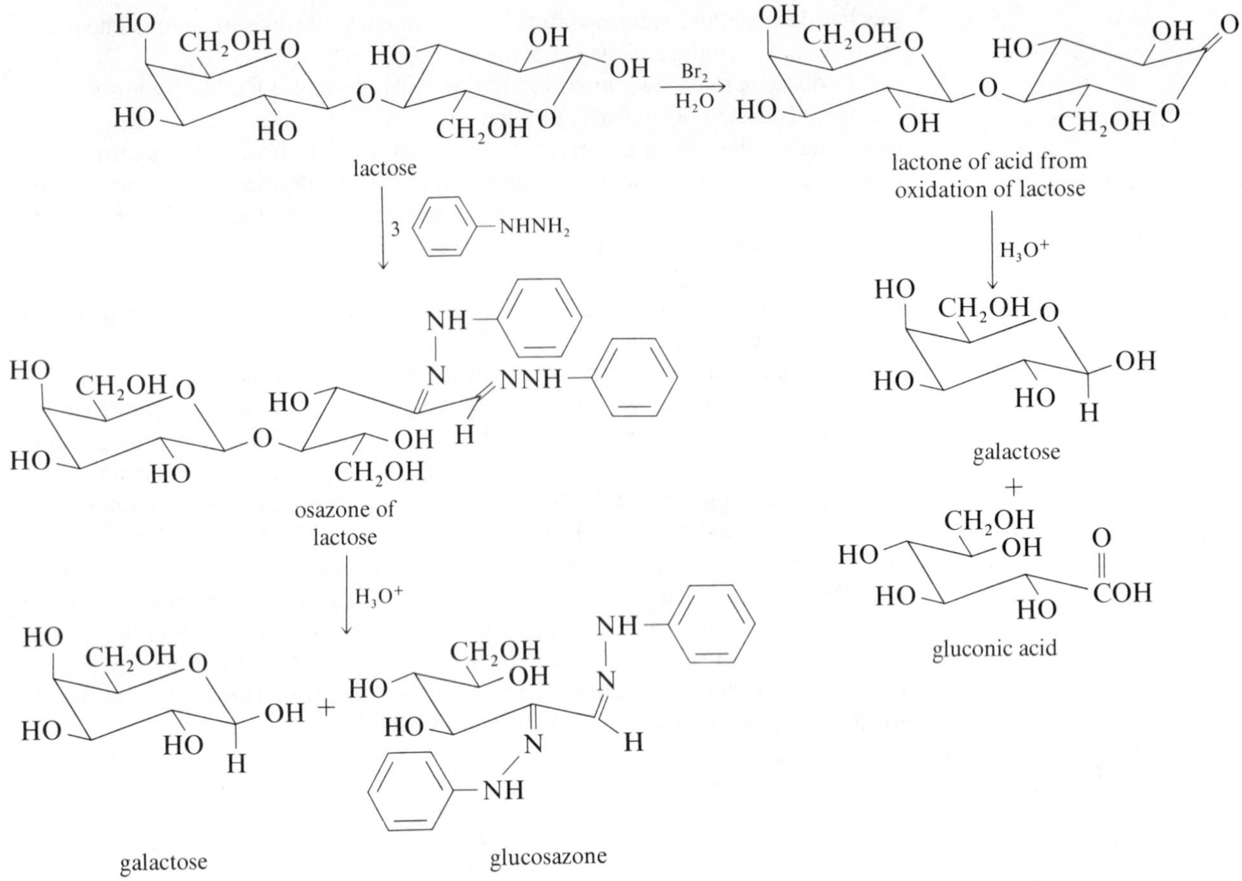
lactose
Br₂
H₂O
lactone of acid from
oxidation of lactose
3 NHNH₂
osazone of
lactose
H₃O⁺
H₃O⁺
galactose
+
gluconic acid
galactose
glucosazone

The size of the rings in the galactose and glucose units of lactose and the point of attachment of the galactose unit to glucose were determined by a series of methylation experiments. When lactose is methylated and then hydrolyzed, 2,3,4,6-*O*-tetramethyl-D-galactose and 2,3,6-*O*-trimethyl-D-glucose are formed.

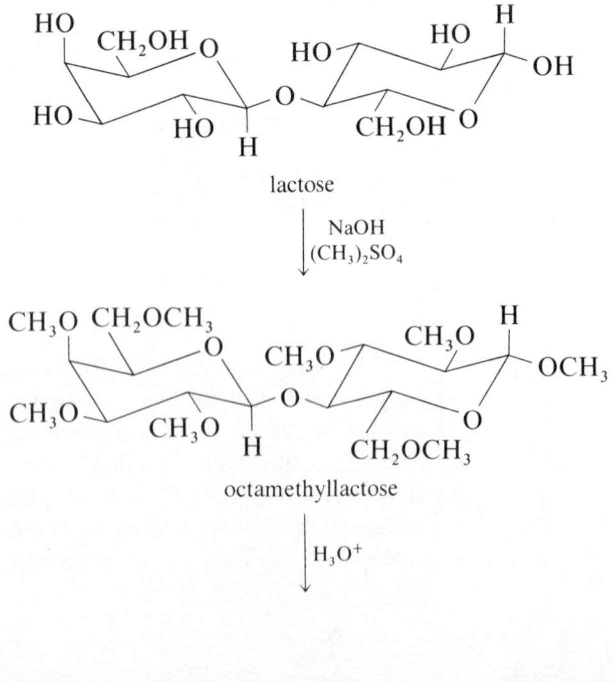
lactose
NaOH
(CH₃)₂SO₄
octamethyllactose
H₃O⁺

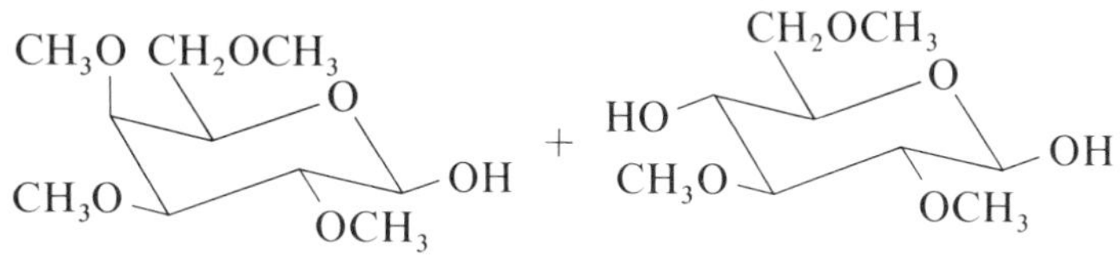

2,3,4,6-*O*-tetramethyl-D-galactose

2,3,6-*O*-trimethyl-D-glucose

assuming pyranose structure for glucose

The fact that the hydroxyl group at carbon 5 is the only one that is not methylated in galactose indicates that galactose exists as a pyranose ring in lactose. The hydroxyl groups at carbons 4 and 5 in the glucose unit are not methylated. This could mean that galactose is attached to the hydroxyl group at carbon 4 if glucose has the pyranose structure. On the other hand, if the glucose unit exists as a furanose, then the attachment must be at carbon 5.

This question was resolved by fully methylating the acid that resulted from the oxidation of lactose, then hydrolyzing. Under the conditions of the hydrolysis, the methyl ester function that had formed was also hydrolyzed. The glucose portion of the molecule was isolated as 2,3,5,6-*O*-tetramethyl-γ-gluconolactone. The formation of the lactone having a five-membered ring showed clearly that the hydroxyl group at carbon 4 is part of the glycosidic linkage with galactose. That hydroxyl group was the only one left unmethylated on the glucose unit once the pyranose ring was opened by the oxidation reaction.

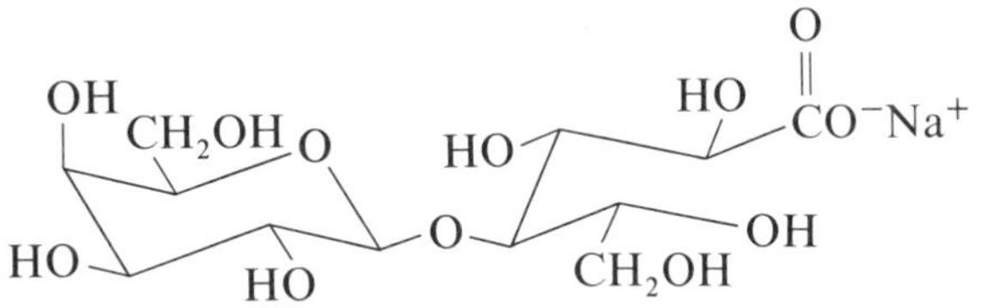

acid salt from the oxidation of lactose

$\downarrow$ NaOH, $(CH_3)_2SO_4$

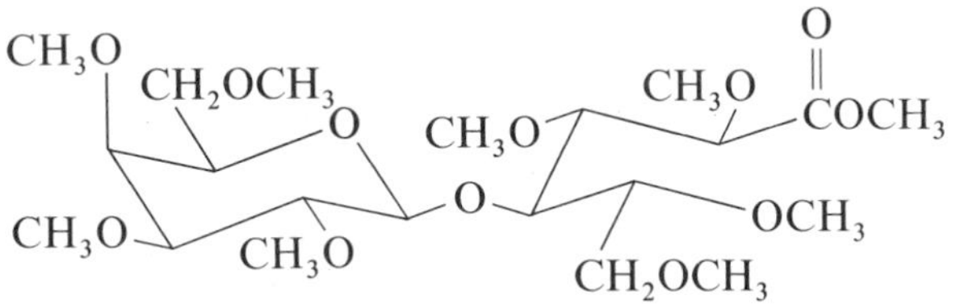

methyl groups at all free hydroxyl groups

$\downarrow$ H_3O^+, Δ, 6 h

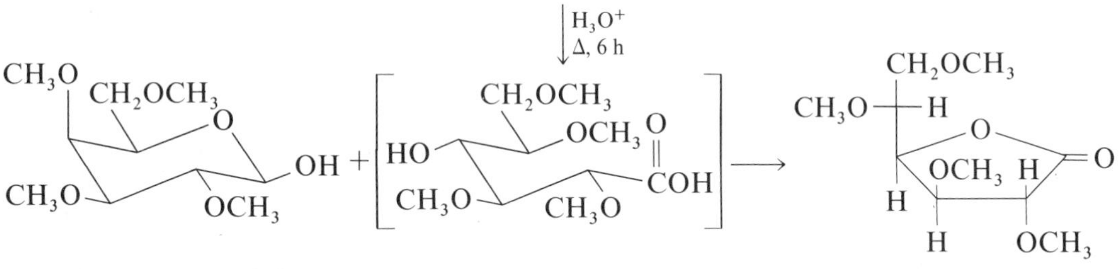

2,3,4,6-*O*-tetramethyl-β-D-galactose

hydroxyl group at carbon 4 left unmethylated; therefore, bonded to galactose in the disaccharide

2,3,5,6-*O*-tetramethyl-γ-gluconolactone; γ-lactone, because only the hydroxyl group at carbon 4 is available for reaction with the carboxylic acid

B. The Structures of Maltose and Cellobiose

Haworth did experiments in the 1920s that were similar to those described in the preceding section and established the structures of maltose and cellobiose. Maltose is 4-*O*-(α-D-glucopyranosyl)-D-glucopyranose, and cellobiose is 4-*O*-(β-D-glucopyranosyl)-D-glucopyranose. Both are reducing sugars that mutarotate. Both are hydrolyzed to two units of glucose. The difference between the two is in the stereochemistry of the glycosidic linkage. Maltose is hydrolyzed by the enzyme maltase, which is specific for α-glycosidic linkages, but emulsin, an enzyme specific for the cleavage of β-glycosidic linkages, is necessary for the hydrolysis of cellobiose.

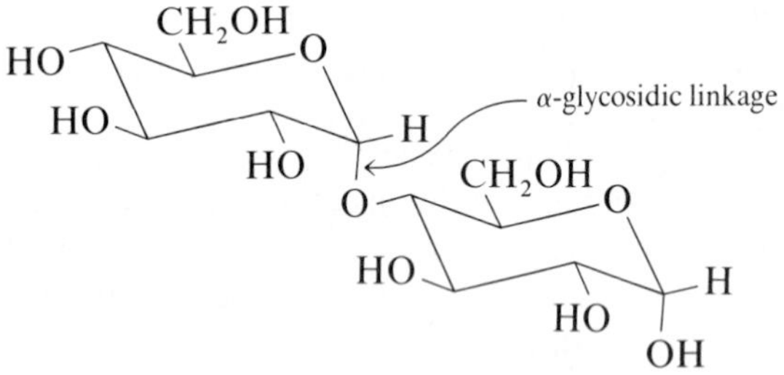
α-glycosidic linkage

maltose
product of the hydrolysis of starch

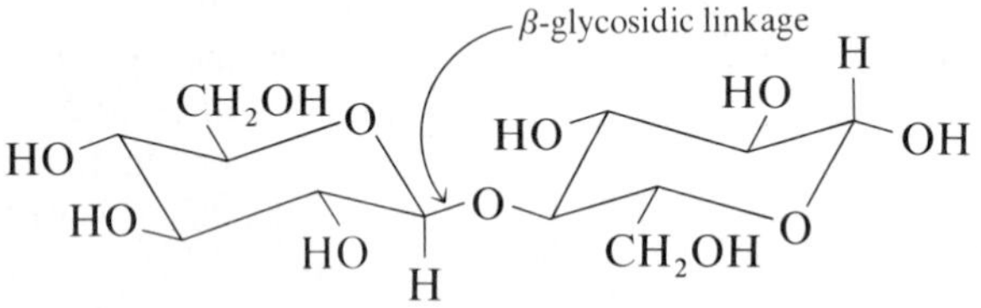
β-glycosidic linkage

cellobiose
product of the hydrolysis of cellulose

These structural differences are important because they suggest that similar differences exist in the polysaccharides starch and cellulose, from which these two disaccharides are derived by hydrolysis reactions.

C. The Structure of Sucrose

Sucrose is not a reducing sugar, and it does not exhibit mutarotation. It gives an osazone only after a long period of time. The osazone that is finally isolated is the osazone of glucose. Sucrose itself is dextrorotatory, $[\alpha]_D + 66°$. Hydrolysis of sucrose converts it into glucose, $[\alpha]_D + 52.5°$, and fructose, $[\alpha]_D -92°$. The resulting mixture is levorotatory. This phenomenon is known as the **inversion of sucrose.** The mixture of sugars that is formed is called **invert sugar.** Invert sugar is much sweeter than sucrose chiefly because fructose is eight times sweeter than sucrose. Invert sugar is the chief component of honey. The preparation of all sugar syrups and candies involves boiling table sugar with water and a bit of acid such as vinegar, lemon juice, or cream of tartar, which is the monopotassium salt of (+)-tartaric acid. The process described is nothing more than the preparation of invert sugar.

The fact that sucrose is not a reducing sugar means that there is no potential aldehyde or ketone function in the molecule in the hemiacetal form. Carbon 1 of glucose must be bonded to carbon 2 of fructose. Sucrose is hydrolyzed by maltase, so the glucosidic linkage must be α in configuration. The structure and stereochemistry of the fructose portion of the molecule was difficult to establish. Careful polarimetric studies by Hudson, started in 1908 and completed in 1934, and an x-ray structure determination in 1947 showed that fructose is present in sucrose as a β-fructofuranoside.

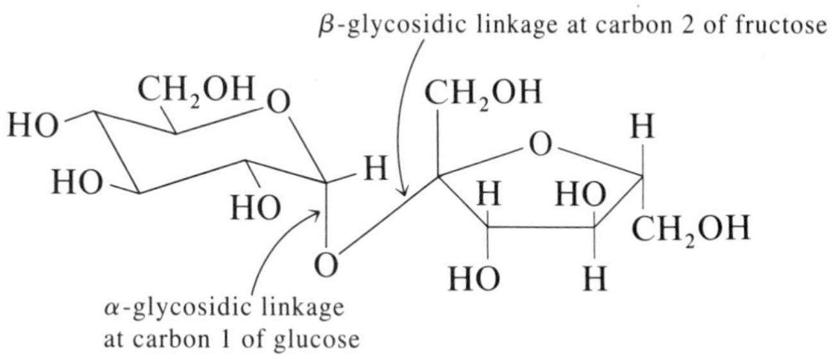
β-glycosidic linkage at carbon 2 of fructose
CH₂OH
HO
HO
HO
O
H
CH₂OH
O
H
H
HO
CH₂OH
HO
H
α-glycosidic linkage
at carbon 1 of glucose

sucrose

(α-D-glucopyranosyl)-β-D-fructofuranoside

or

(β-D-fructofuranosyl)-α-D-glucopyranoside

or

O-α-D-glucopyranosyl-(1→2)-β-D-fructofuranoside

PROBLEM 25.33

Write equations that account for the formation of glucosazone from sucrose.

PROBLEM 25.34

Gentiobiose is a disaccharide found in a number of natural products. It has the formula $C_{12}H_{22}O_{11}$ and is a reducing sugar. It is hydrolyzed by emulsin to glucose. 2,3,4,6-*O*-Tetramethyl-D-glucopyranose and 2,3,4-*O*-trimethyl-D-glucopyranose are obtained when gentiobiose is converted to its fully methylated form and then hydrolyzed. What is the structure of gentiobiose? Give it a systematic name.

25.9 POLYSACCHARIDES

A. Starch

The two polysaccharides of greatest biological and economic importance are starch and cellulose. Plants store their reserve carbohydrates in the form of starch. Humans get starch from roots, tubers, and seeds. Starch is a high-molecular-weight polysaccharide that can be broken down completely to glucose by acid hydrolysis or by the enzyme maltase, which establishes the presence of α-linkages. Starch can be fractionated into two different kinds of molecules. One is amylose, which has an average molecular weight of approximately 10^6. It is primarily a polymeric chain of glucopyranose units attached to each other by α-glycosidic linkages between carbon 1 of one glucopyranose and the hydroxyl group on carbon 4 of the next one. The structure of amylose is proved by methylation and hydrolysis, which gives almost exclusively 2,3,6-*O*-trimethylglucose.

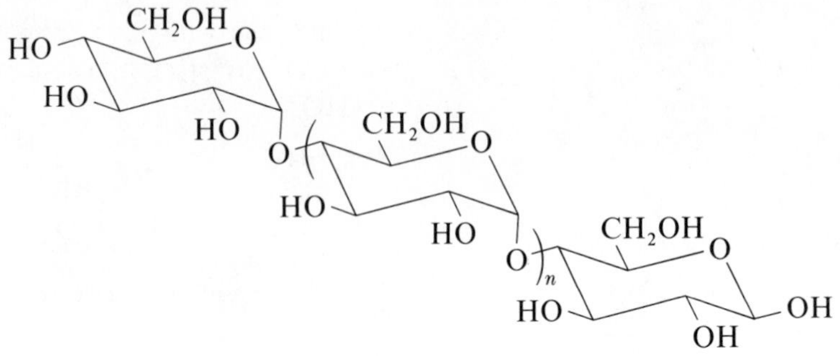
CH₂OH
O
HO
HO
HO
O
CH₂OH
O
HO
HO
O
n
CH₂OH
O
HO
OH
OH

amylose
$n \sim 1000–6000$

Amylose and iodine form a complex that is blue; this is the familiar color reaction of iodine in starch.

The chief fraction of starch is amylopectin, which has a more complicated structure. It is highly polymeric, with as many as a million glucose units in a single molecule. Complete methylation and hydrolysis of amylopectin gives about 3% of 2,3-*O*-dimethylglucose, suggesting that some glucose units are connected to others through the hydroxyl group at carbon 6 as well as through the oxygens at carbons 1 and 4. Enzymatic studies and hydrolysis reactions have shown that amylopectin has a randomly branched structure in which the hydroxyl groups at carbon 6 of some glucose units are indeed involved.

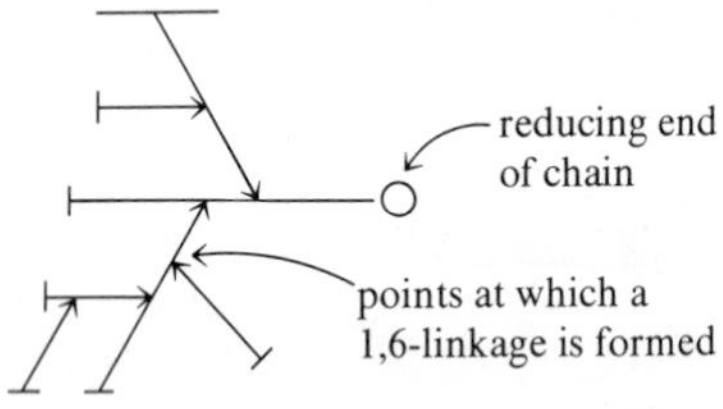
reducing end of chain
points at which a 1,6-linkage is formed

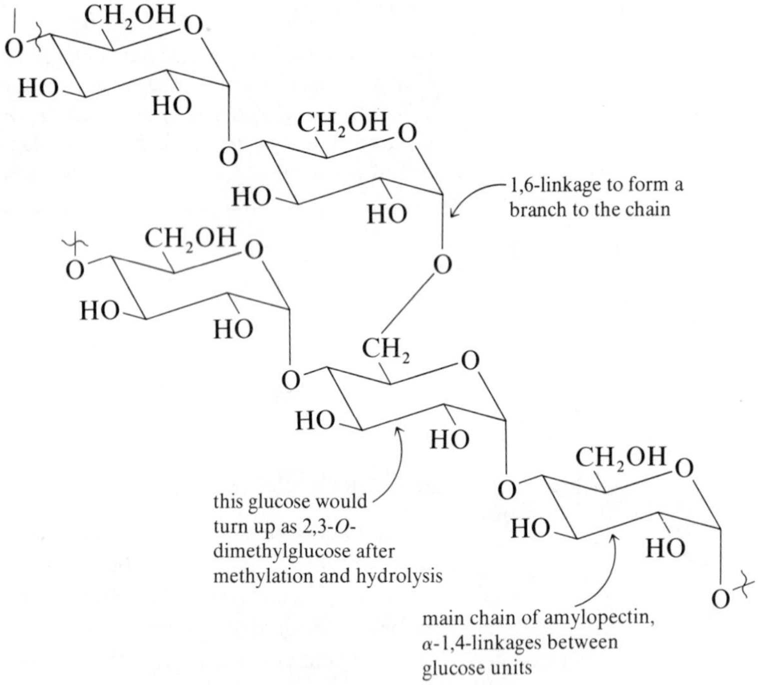
1,6-linkage to form a branch to the chain
this glucose would turn up as 2,3-O-dimethylglucose after methylation and hydrolysis
main chain of amylopectin, α-1,4-linkages between glucose units

amylopectin, showing the branching of the chain

The branching appears to occur once every 20 to 25 glucose units. The overall structure of amylopectin resembles the branching of a tree.

Glycogen, the form in which animals store carbohydrates, is also composed of glucose units and is similar to amylopectin in structure except that the branching occurs at shorter intervals, with about 12 glucose units in each branch. It is especially abundant in the liver of mammals and has also been isolated from kidneys, brains, and skeletal and cardiac muscles.

B. Cellulose

Cellulose is the organic substance that is most abundant in nature. It is the structural material of higher plants and is found in all of their parts. Wood is about 50% cellulose. Commercially important fibers such as cotton and flax consist almost completely of cellulose.

On hydrolysis, cellulose gives cellobiose and ultimately glucose. This establishes its structure as a linear chain of glucopyranose units attached to each other by β-glycosidic linkages from carbon 1 of one unit to the hydroxyl group on carbon 4 of another unit. The structure consists of long chains of six-membered rings in the most stable chair conformation, with all the larger substituents in the equatorial positions.

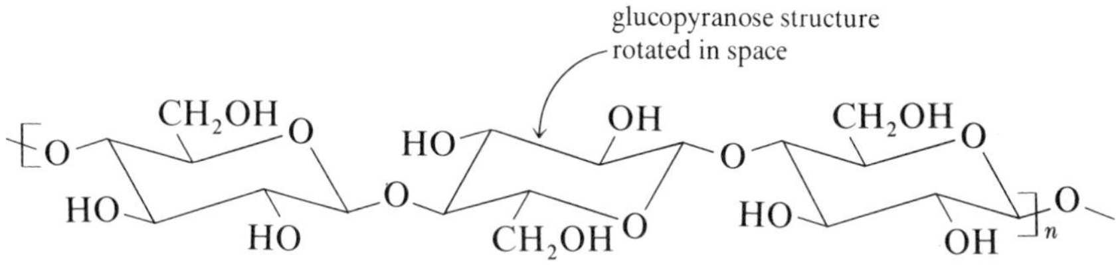
glucopyranose structure
rotated in space

cellulose
$n \sim 5000–10{,}000$

The individual molecules of cellulose are associated with each other in regular structures that have crystalline properties. Approximately 100 to 200 cellulose molecules are held together in each of these larger units. The exact nature of the interactions between the molecules has not been determined. Hydrogen bonding between adjacent strands of molecules does seem to be an important factor in determining the strength and rigidity of cellulose as a structural material.

Cellulose was the first polymeric material that was chemically modified to provide new materials useful to human beings. The polyhydroxy functions in cellulose participate in the typical reactions of alcohols. Thus, cellulose treated with acetic anhydride in acetic acid with a little sulfuric acid as a catalyst is converted into its acetate. Exactly how many acetyl groups are added to the molecule depends on the physical state of the cellulose at the beginning of the reaction, the exact reaction conditions, and how the material is treated afterwards. Some of the acetyl groups are hydrolyzed off by exposure to water during processing.

The most useful acetate is cellulose triacetate, which is soluble in mixtures of acetic anhydride and acetic acid.

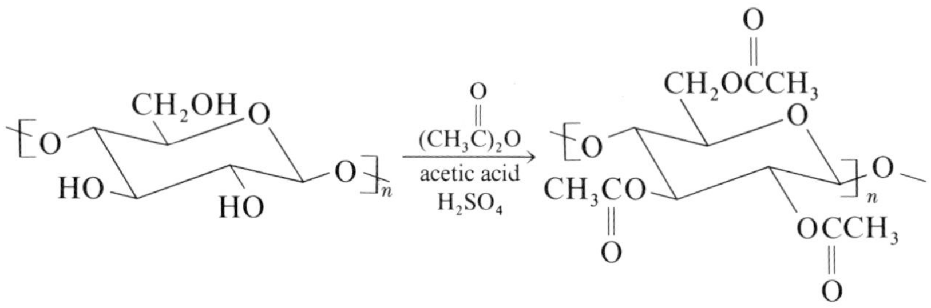

fragment of cellulose

fragment of cellulose triacetate

In this compound, all the free hydroxyl groups of cellulose have been converted into ester functions. When a solution of cellulose triacetate is forced through small holes into a solution of dilute acetic acid, the water precipitates it in the form of a continuous thread that can be used to weave fabrics in the textile industry. This product is known commercially as Arnel.

The use of cellulose in a variety of other products is based on similar modifications of structure. The hydroxyl groups are converted to other functional groups. In the process, the cellulose molecule becomes somewhat degraded and much more soluble in organic solvents. In this soluble state, it is manipulated by extruding the solutions as sheets or fine streams into other solutions that precipitate the compounds or reverse the original chemical reactions, regenerating cellulose in a new, more useful form. For example, the hydroxyl groups in cellulose are converted into alkoxide anions by base. These nucleophilic anions add to carbon disulfide to give compounds known as xanthate esters, which are stable as salts in basic solution but lose carbon disulfide in acid.

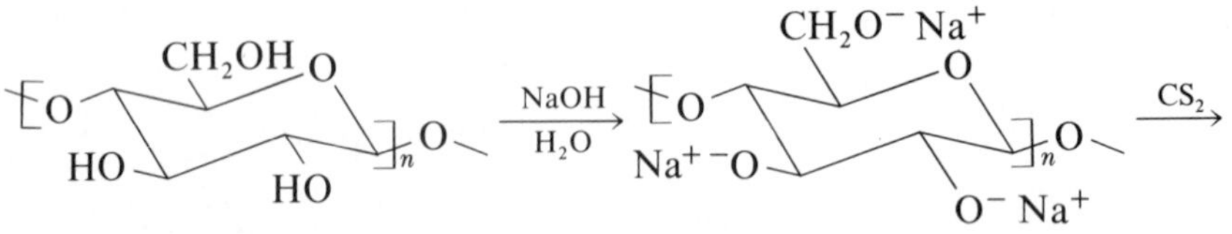
CH_2OH
O
HO
HO
NaOH
H_2O
$CH_2O^- Na^+$
$Na^+ {}^-O$
$O^- Na^+$
CS_2

cellulose — alkoxide anion of cellulose

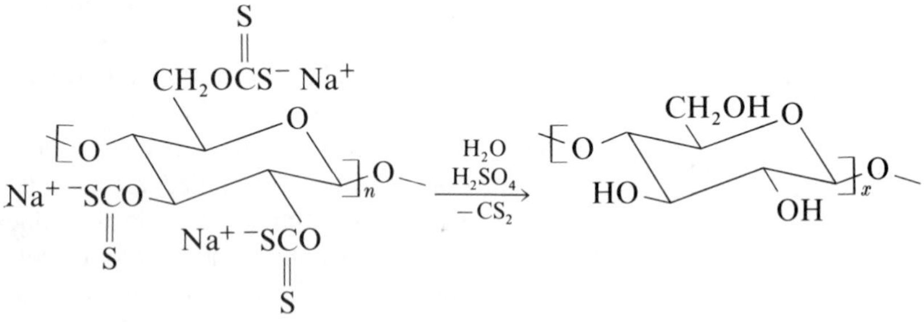
$CH_2OCS^- Na^+$
$Na^+ {}^-SCO$
$Na^+ {}^-SCO$
H_2O
H_2SO_4
$-CS_2$
CH_2OH
HO
OH

xanthate ester of cellulose — cellophane or rayon *transformed cellulose*

A basic solution of cellulose xanthate salts can be forced out of spinners into dilute sulfuric acid to form rayon threads. If thin slits are used, sheets of cellophane are formed. Both rayon and cellophane are essentially cellulose in a transformed physical state. The hydrogen bonding between the molecules of cellulose in its original state has been disrupted by the chemical process. When the hydroxyl groups reform after the reactions are over, the cellulose molecules have been physically forced by the extrusion process into new conformations, which have new physical properties.

PROBLEM 25.35

Cellulose is converted into a nitrate ester by treatment with a mixture of nitric and sulfuric acids. Cellulose trinitrate is called guncotton because of its explosive properties. Write an equation for the formation of cellulose trinitrate.

PROBLEM 25.36

Using ROH to represent cellulose, write a mechanism for the formation of cellulose xanthate.

PROBLEM 25.37

Among the hydrolysis products of chitin, the chief material in the shells of lobsters, is a disaccharide, chitobiose. It has the formula $C_{12}H_{24}O_9N_2$. Further hydrolysis of chitobiose gives *N*-acetylglucosamine (see p. 1052 for glucosamine). Assign a structure to chitobiose, which seems to be structurally analogous to cellobiose.

25.10 OTHER NATURAL PRODUCTS DERIVED FROM CARBOHYDRATES

A. Ascorbic Acid, Vitamin C

Vitamin C, ascorbic acid, is a sugar acid synthesized in plants and in the livers of most vertebrates, but not human beings. Human beings must constantly supplement their diets with the vitamin. The pathway by which the vitamin is synthesized in living organisms is reminiscent of the final steps in Fischer's proof of the structure of glucose, when he was converting D-glucose to L-gulose. The aldehyde group of D-glucuronic acid is reduced by an enzyme to give L-gulonic acid. L-Gulonic acid is converted to its γ-lactone by a different enzyme and finally oxidized to L-ascorbic acid by yet another enzyme system.

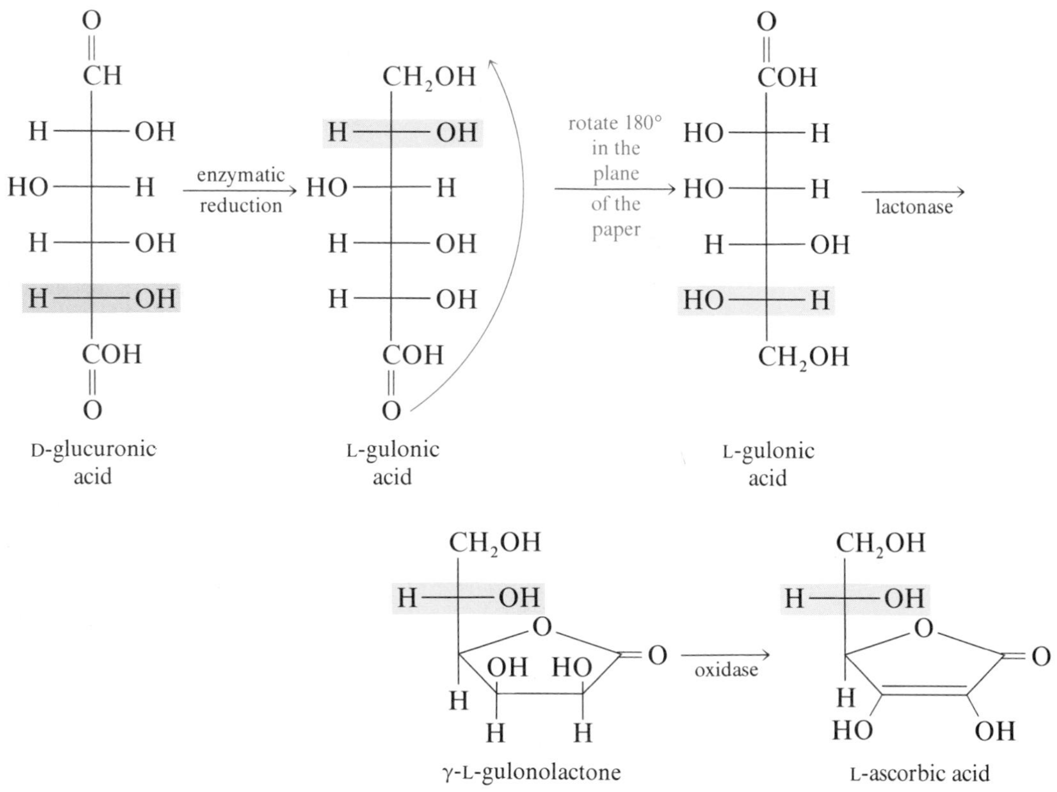

enzymatic
reduction
rotate 180° in the plane of the paper
lactonase
oxidase
D-glucuronic acid
L-gulonic acid
L-gulonic acid
γ-L-gulonolactone
L-ascorbic acid

Fischer reduced the same γ-gulonolactone to L-gulose (p. 1095).

Ascorbic acid is a lactone and an enediol. It is an unstable compound and a reducing agent. It easily undergoes further oxidation to L-dehydroascorbic acid, which also has some Vitamin C activity. The activity is lost if the lactone ring is opened by hydrolysis.

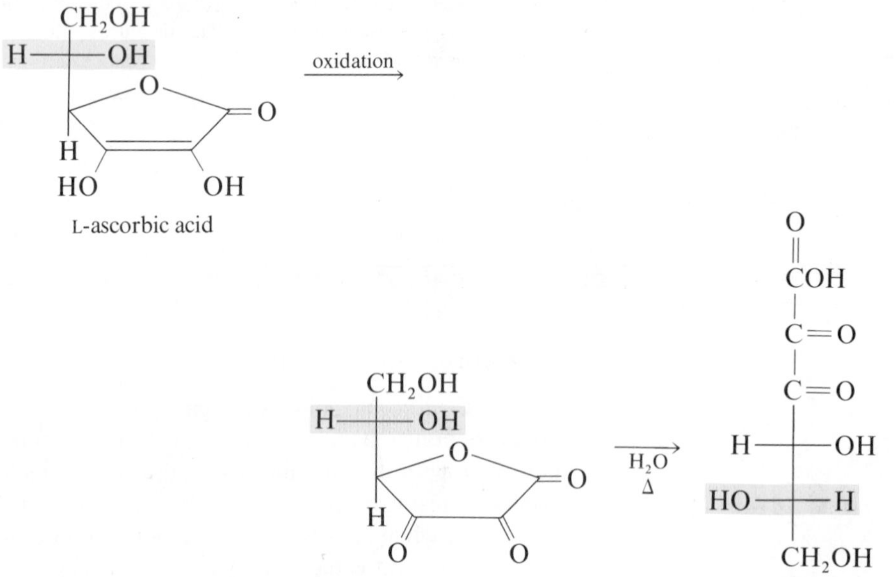
CH₂OH
H
OH
O
O
H
HO
OH
oxidation
CH₂OH
H
OH
O
O
H
O
O
H₂O
Δ
O
COH
C=O
C=O
H
OH
HO
H
CH₂OH

L-ascorbic acid

L-dehydroascorbic acid
has Vitamin C activity

L-diketogulonic acid
no Vitamin C activity

It is understood that ascorbic acid functions as a reducing agent in many biochemical reactions, but the exact physiological function of the vitamin is still unknown. It does appear to be essential to the maintenance of the integrity of cells and tissues.

PROBLEM 25.38

The acidic properties of ascorbic acid are due to the enol groups, not the opening of the lactone ring to generate a carboxylic acid. Which proton is the most acidic one in ascorbic acid? (Hint: Write the anions that would result from the removal of each proton and see which anion has the best delocalization of charge.)

B. Cardiac Glycosides

The cardiac glycosides have an effect on the actions of the heart, and many of them are highly toxic. A well-known example of a cardiac glycoside is digitoxin, isolated from the foxglove, *Digitalis purpurea*. This compound reduces the pulse rate, regularizes the rhythm of the heart, and strengthens the heart beat. It consists of a steroid aglycon attached to a trisaccharide that consists of three units of the sugar D-digitoxose.

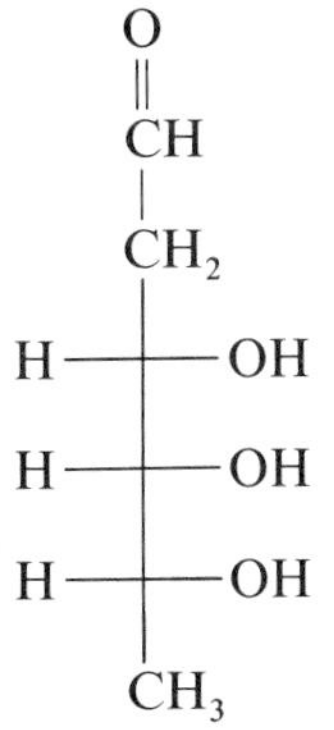

D-digitoxose
open-chain form

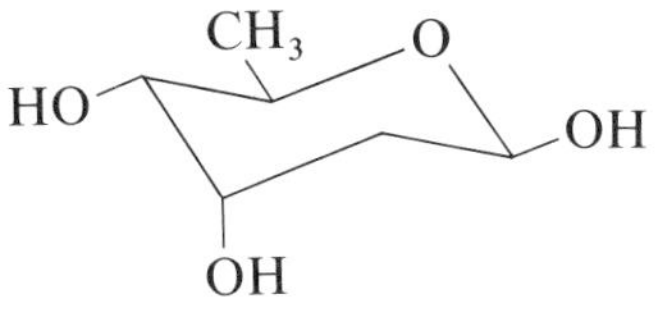

D-digitoxose
as hemiacetal

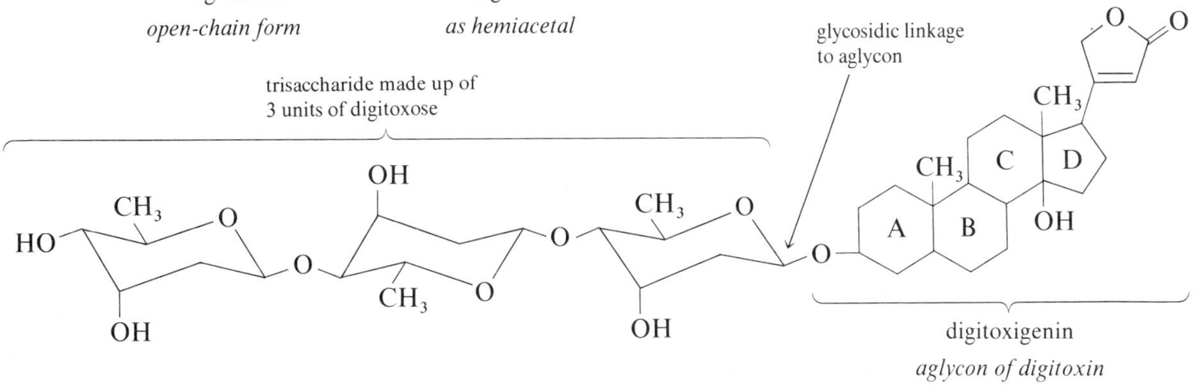

trisaccharide made up of 3 units of digitoxose
glycosidic linkage to aglycon
digitoxigenin
aglycon of digitoxin

digitoxin

Digitoxose is a 2,6-deoxyaldohexose with a configuration resembling that of ribose. The trisaccharide unit has β-linkages from carbon 1 of one digitoxose unit to the hydroxyl group of carbon 4 of another one. The aglycon in digitoxin is a steroid called digitoxigenin. The five-membered cyclic α,β-unsaturated lactone unit and the hydroxyl group at the junction of the C and D rings of the steroid are found in other compounds having similar effects on the heart. The sugars of other cardiac glycosides vary, however.

PROBLEM 25.39

Uzarin is a cardiac glycoside having the same aglycon as digitoxin does. The aglycon is attached by a β-glycosidic linkage to cellobiose. What is the structure of uzarin?

S U M M A R Y

Carbohydrates are polyhydroxy aldehydes (aldoses) or ketones (ketoses). A carbohydrate that cannot be broken down into simpler units by hydrolysis is a monosaccharide. Oligosaccharides contain from two to ten monosaccharide units. Polysaccharides are complex substances containing many monosaccharide units.

Glucose has four stereocenters in the open-chain form. Fischer projection formulas are used to depict molecules that have several stereocenters. One way carbohydrates are classified depends on whether the hydroxyl group at the stereocenter next to the primary alcohol group at the end of the chain is to the right (the D family) or to the left (the L family) in the Fischer projection.

Carbohydrates, because they have both hydroxyl and carbonyl groups within the same molecule, exist in solution in equilibrium with cyclic hemiacetals or hemiketals. The formation of such a hemiacetal or hemiketal introduces another stereocenter into the molecule. Glucose is isolated in two stereoisomeric forms: α-D-glucose in which the hydroxyl group of the hemiacetal is axial, and β-D-glucose, in which it is equatorial. When put in solution, each form equilibrates with the other. This phenomenon, observed as a change in the optical rotation of the solution, is called mutarotation. α-Glucose and β-glucose are anomers, stereoisomers that differ at the carbon atom derived from the carbonyl group. They are also epimers, a term used for diastereomers that differ at only one of a number of stereocenters.

Because they contain both hydroxyl and carbonyl groups, carbohydrates react as alcohols and as aldehydes or ketones. They are converted to esters by acid anhydrides and to ethers in the presence of a base and an alkyl halide or sulfate. The primary alcohol group at one end of a carbohydrate molecule and the aldehyde function at the other end are oxidized by nitric acid to carboxylic acid groups, giving dicarboxylic acids called aldaric acids. The carbon chain of a carbohydrate is cleaved by periodate ion between adjacent hydroxyl groups or between a carbonyl group and an adjacent hydroxyl group.

Aldoses are easily oxidized to monocarboxylic acids (aldonic acids) by mild oxidizing agents, such as bromine water, copper(II) ion, and silver(I) ion. These two metal ions are the oxidizing agents in Benedict's reagent and Tollens reagent, respectively, and are used to test for the presence of sugars. α-Hydroxyketones are also oxidized easily enough to give positive tests with these reagents.

The hemiacetal or hemiketal form in which a carbohydrate exists is converted to an acetal or ketal linkage when the sugar is treated with an alcohol in the presence of acid. Sugar acetals or ketals are called glycosides. The bonds between monosaccharide units in polysaccharides are all glycosidic linkages. Glycosidic linkages are found in many natural products joining alcohols or phenols, known as aglycons, to sugars. Sugar hemiacetals or hemiketals also bond with amines to give glycosylamines. Glycosylamines incorporating purines or pyrimidines are important as structural units of RNA and DNA.

Carbohydrates also react as carbonyl compounds with phenylhydrazine to give osazones. Treatment with base causes interconversion of carbohydrates by way of enediols. The reactions of carbohydrates are summarized in Table 25.1 on the facing page.

The Kiliani-Fischer synthesis converts an aldose to two epimeric aldoses that have one more stereocenter than the starting compound. Sodium cyanide is used to make epimeric cyanohydrins of the aldose. The cyanohydrins are hydrolyzed to epimeric carboxylic acids, which cyclize to two different lactones. Careful reduction of the lactones produces the hemiacetals of the two epimeric aldoses having lengthened chains.

Glucose is important as the principal source of energy in the human body. It is stored in plants as the polysaccharide starch and in animals as the polysaccharide glycogen. Cellulose, a polymer of glucose, is the principal structural material in plants and is converted industrially to fibers and films. Sucrose and lactose are important disaccharide components of foods.

TABLE 25.1 Reactions of Carbohydrates

Reactions as Carbonyl Compounds			
Carbohydrate	**Reagent**	**Intermediate**	**Product(s)**
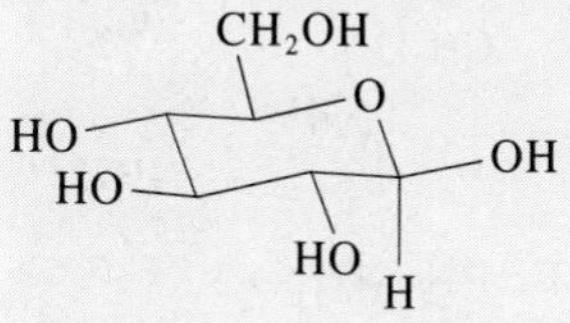	H_2O solution		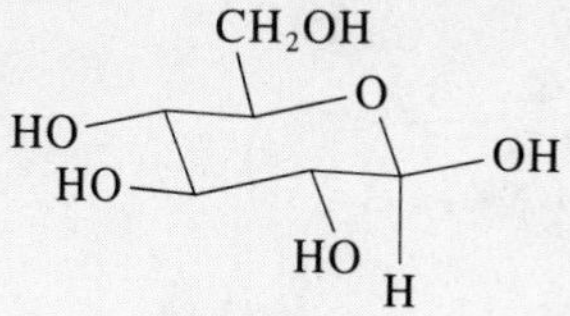 + 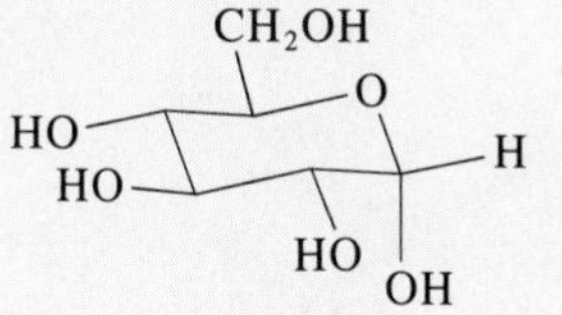
	ROH, HCl		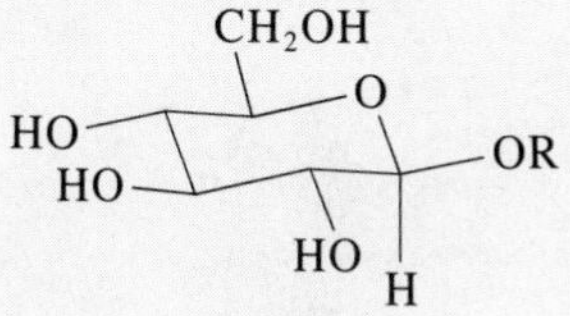 + 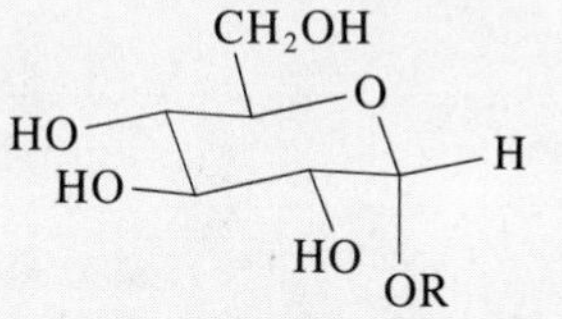
	R_2NH		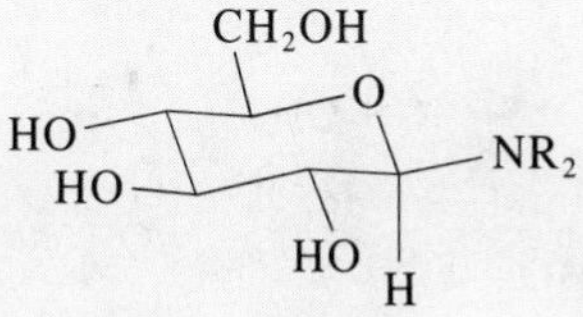
	Br_2, H_2O		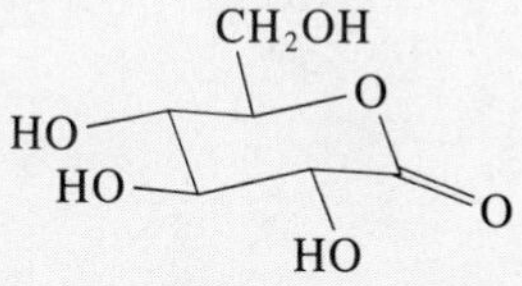
	Cu^{2+}, OH^-		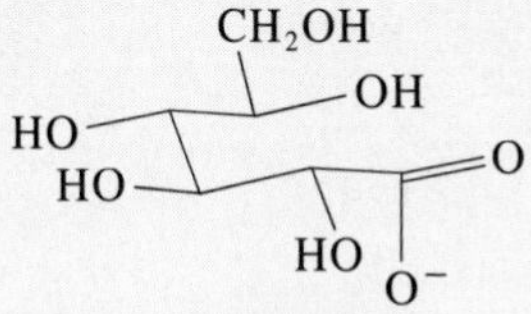 + Cu_2O
	$Ag(NH_3)_2{}^+$, OH^-		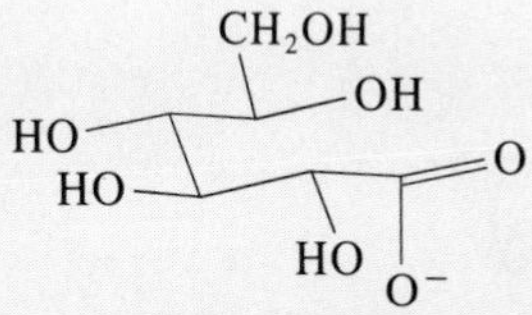 + Ag (The above two reactions are also seen with α-hydroxyketones.)

(*Continued*)

TABLE 25.1 *(Continued)*

Reactions as Carbonyl Compounds

Carbohydrate	Reagent	Intermediate	Product(s)
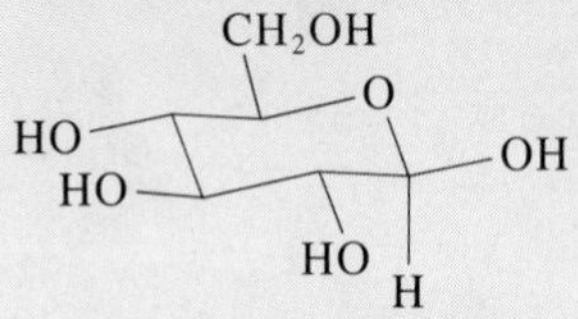	$C_6H_5NHNH_2$		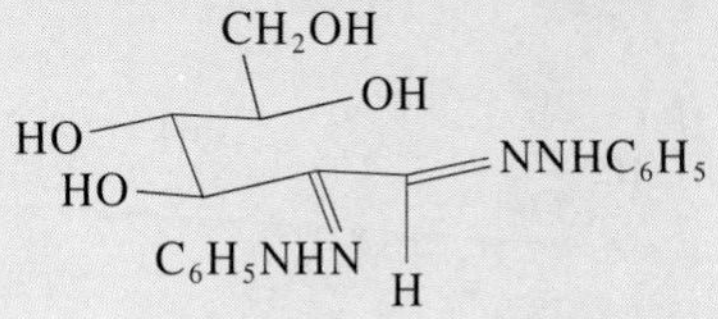
	NaOH, H_2O	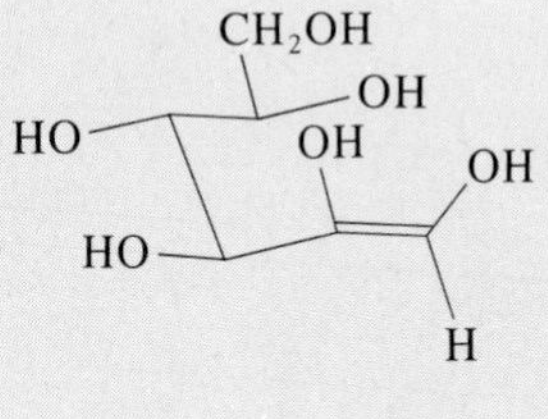	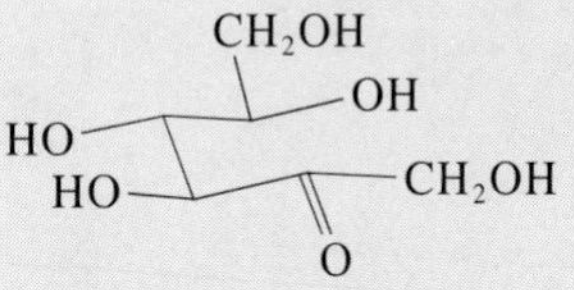 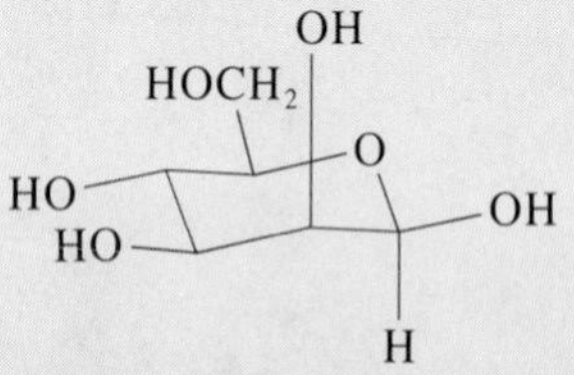

Reactions as Alcohols

Carbohydrate	Reagent	Product
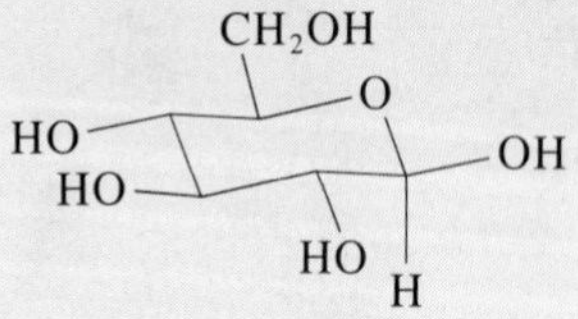	$\overset{O}{\overset{\|}{RC}}O\overset{O}{\overset{\|}{C}}R$	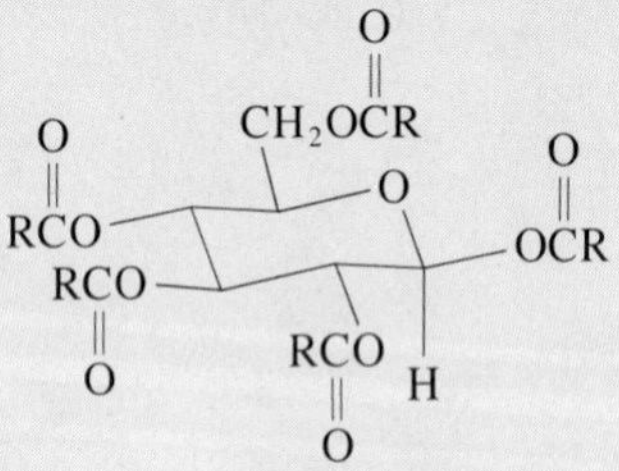
	NaOH, RX or R_2SO_4	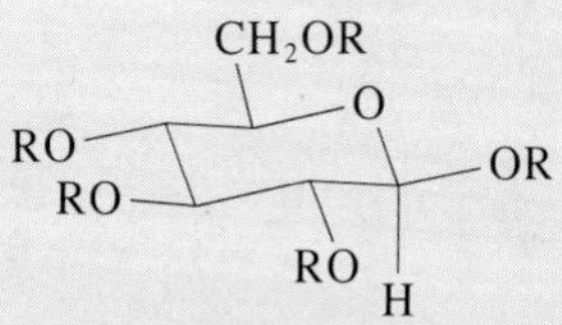
	HNO_3, Δ	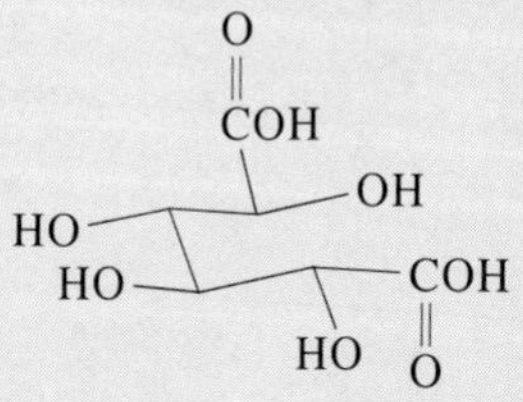
	IO_4^-	$\overset{O}{\overset{\|}{HCH}} + 5\ \overset{O}{\overset{\|}{HCOH}}$

ADDITIONAL PROBLEMS

25.40 Write structures for the following compounds, showing the stereochemistry.

(a) *N*-(α-D-glucopyranosyl)methylamine
(b) methyl 2,3,4,6-tetra-*O*-methyl-β-D-mannopyranoside
(c) D-mannaric acid (d) D-mannitol (e) α-D-fructofuranose-6-phosphoric acid
(f) 2-amino-2-deoxy-D-galactose
(g) *O*-α-D-galactopyranosyl-(1→6)-*O*-α-D-glucopyranosyl-(1→2)-β-D-fructofuranoside
(h) D-galacturonic acid

25.41 Name the following compounds.

(a) 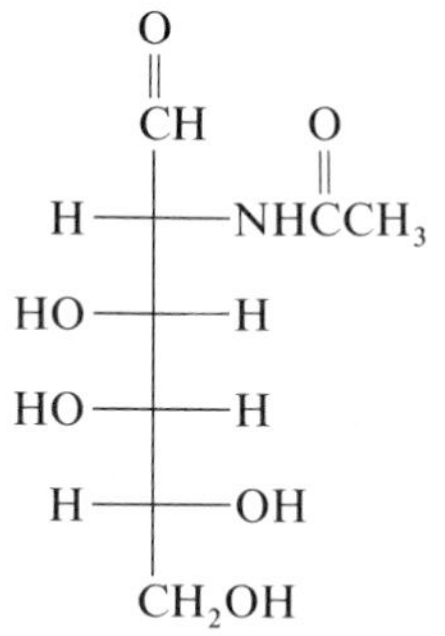

(b) 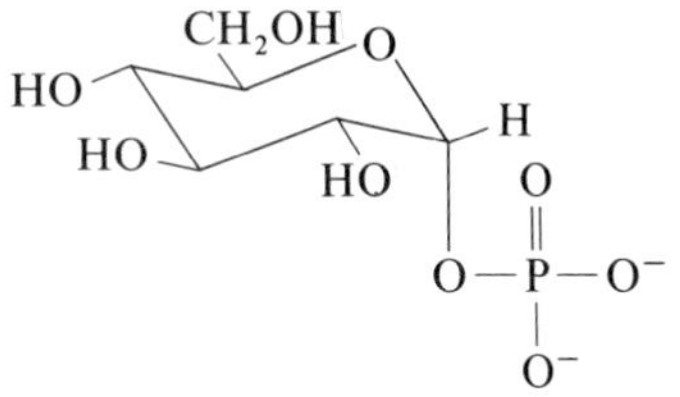

(c) 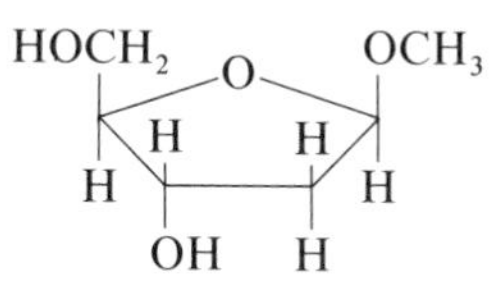

(d) 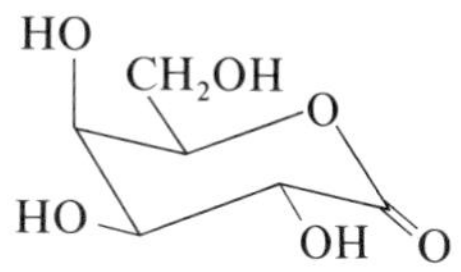

(e) 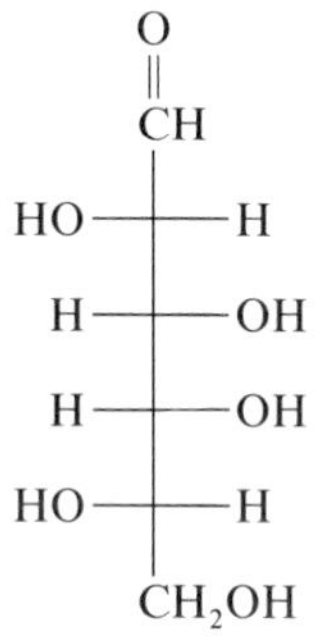

(f) 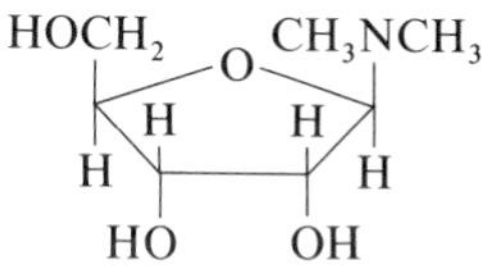

(g) 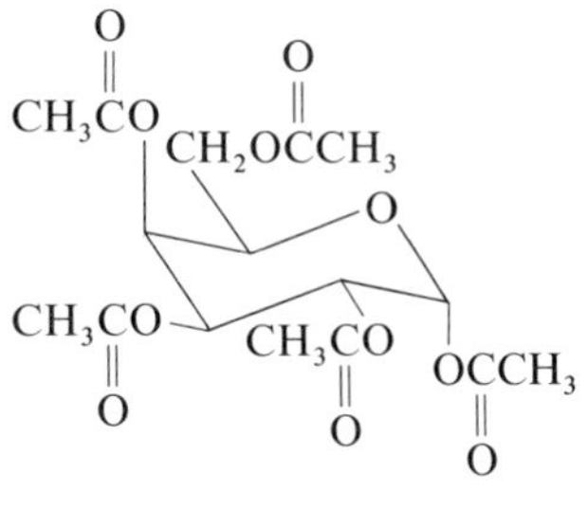

(h) 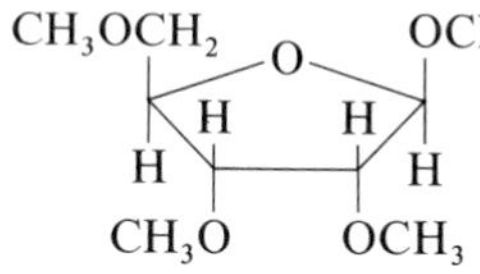

25.42 Write equations for the reactions of D-arabinose with the following reagents. Show stereochemistry where known.

(a) CH_3OH, HCl

(b) 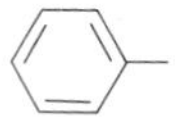 —$NHNH_2$ (excess), acetic acid (c) HNO_3, Δ

(d) HIO_4 (excess), H_2O
(e) Br_2, H_2O
(f) product of part a with NaOH, $(CH_3)_2SO_4$
(g) product of part f with dilute, aqueous HCl
(h) $CH_3\overset{O}{\overset{\|}{C}}O\overset{O}{\overset{\|}{C}}CH_3$ (excess), pyridine
(i) product of part h with cold HBr
(j) product of part i with sodium phenolate
(k) dilute, aqueous NaOH
(l) 
(m) NaCN, Na_2CO_3, H_2O

25.43 Write structural formulas for all compounds indicated by letters. Show the stereochemistry whenever it is known.

(a) 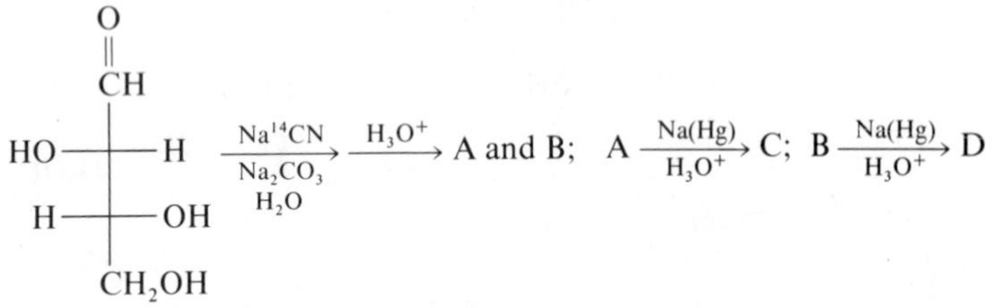

(b) 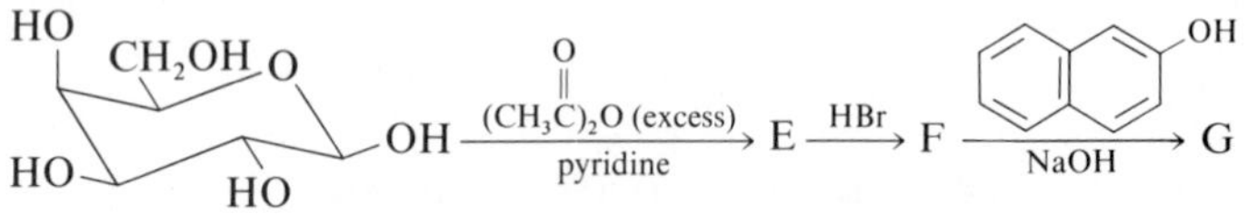

(c) 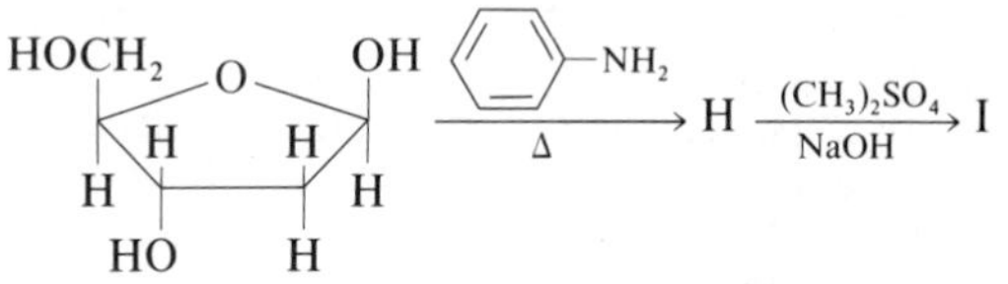

(d) 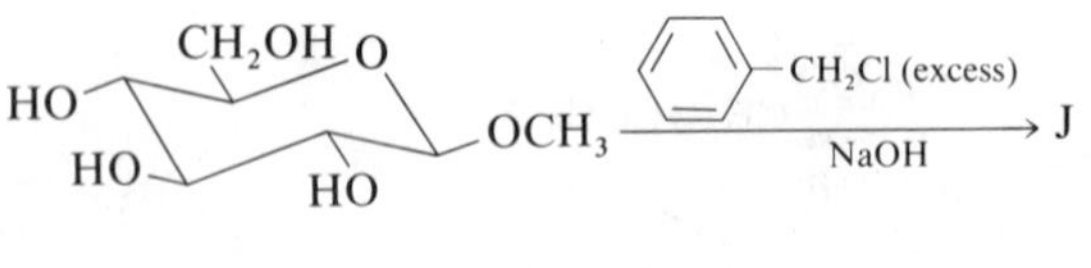

(e) 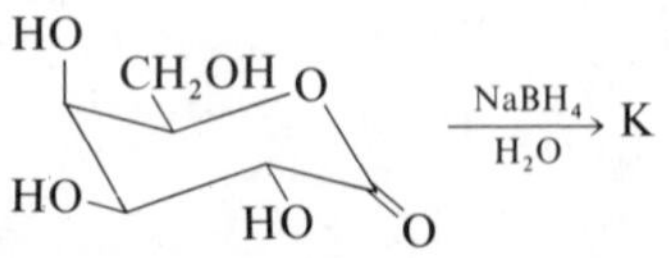

(f) 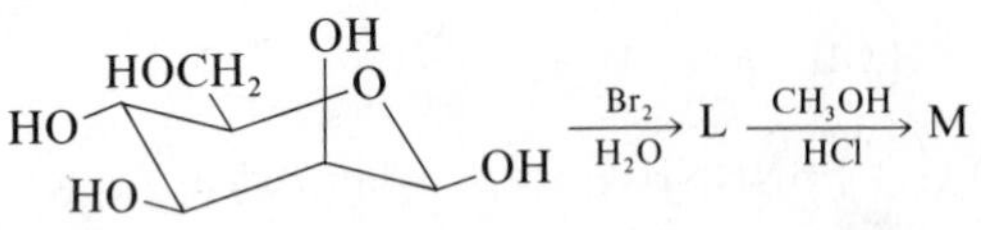

(g) 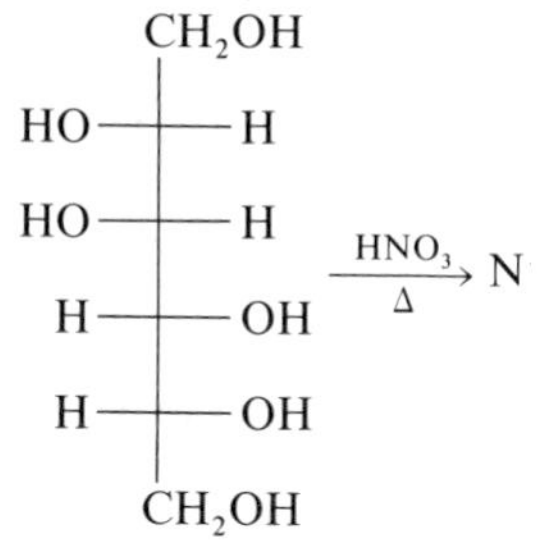

(h) 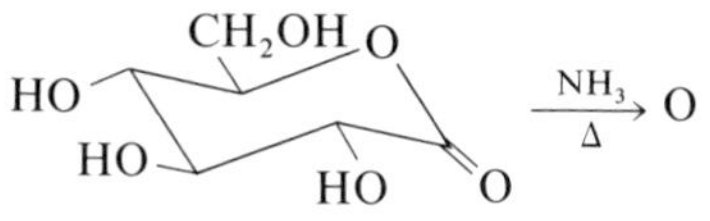

(i) 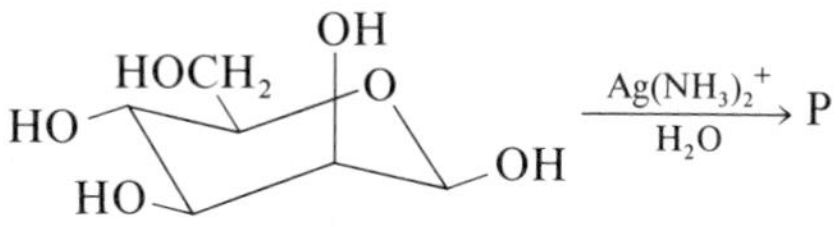

(j) 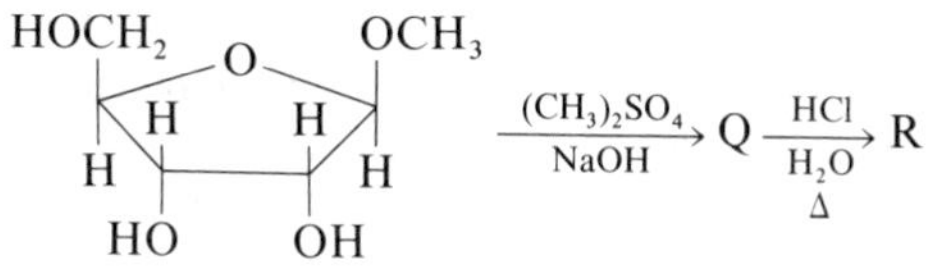

(k) 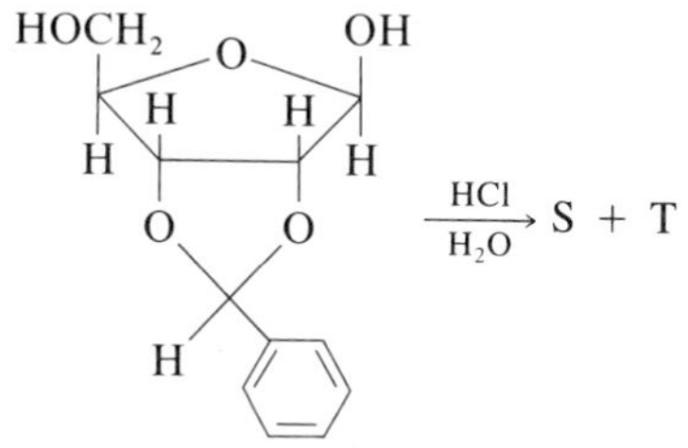

25.44 The absolute configuration of malic acid

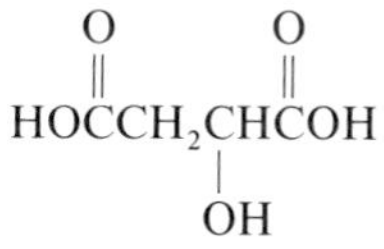

was determined by synthesizing it from (+)-tartaric acid (p. 1063). (+)-Tartaric acid can be converted to (+)-malic acid by the following chemical reactions. Give the correct stereochemical structures for the compounds present at the different stages of the conversion.

$$(+)\text{-tartaric acid} \xrightarrow[\text{HCl, }\Delta]{CH_3CH_2OH\text{ (excess)}} A \xrightarrow[\text{pyridine}]{SOCl_2} B$$ (the reaction of thionyl chloride in pyridine goes with inversion of configuration)

$$B \xrightarrow[CH_3CH_2OH]{\text{Zn (a reducing agent)}} C \xrightarrow[\text{NaOH, }\Delta]{H_2O} \xrightarrow[H_2O]{HCl} (+)\text{-malic acid}$$

25.45 An antiviral drug, marketed as Vira-A, has the scientific name 9-β-D-arabinofuranosyladenine. What is the structure of Vira-A? (The structure of adenine is found on p. 1009).

25.46

(a) When glucose is dissolved in D_2O, it is found that five deuterium atoms are rapidly incorporated into the molecule. Draw a cyclic structure for glucose and show which hydrogens have been rapidly replaced by deuterium atoms.
(b) When glucose is dissolved in $H_2{}^{18}O$, one oxygen atom in glucose slowly exchanges with the labeled water, so the glucose eventually contains one ^{18}O. Starting with a cyclic structure of glucose, write a mechanism that explains this experimental observation.

25.47 What products would you expect to get from sodium borohydride reduction of glucose, mannose, galactose, and fructose? In each case, comment on: (a) whether you expect a single product or a mixture to be formed; and (b) whether the products will be optically active. Is there a stereochemical relationship between the reduction products from fructose, glucose, and mannose?

25.48 α-D-Glucose reacts in its furanose form with 2 equivalents of acetone to give cyclic acetals with the hydroxyl groups at carbons 1 and 2 and carbons 5 and 6. Such cyclic acetals are known as isopropylidene derivatives of sugars. Draw the structural formula for the 1,2,5,6-di-*O*-isopropylidene-D-glucofuranose, and propose a mechanism for its formation.

25.49 Isomaltose is a disaccharide formed by the enzymatic hydrolysis of amylopectin (p. 1102). Isomaltose is a reducing sugar and undergoes mutarotation. It can be hydrolyzed by maltase or by dilute acid to two units of glucose. Isomaltose is oxidized by bromine water to isomaltobionic acid lactone. The lactone is converted by sodium hydroxide and dimethyl sulfate to the methyl ester of *O*-octamethylisomaltobionic acid, which in turn is hydrolyzed to 2,3,4,6-tetra-*O*-methyl-D-glucopyranose and 2,3,4,5-tetra-*O*-methyl-D-gluconic acid. Write equations for all the reactions described, showing how they can be used to assign a structure to isomaltose. What is the systematic name for isomaltose?

25.50 A glycoside called gaultherin is isolated from oil of wintergreen. Gaultherin is not a reducing sugar. It can be hydrolyzed by the enzyme primeverosidase into the disaccharide primeverose and the aglycon methyl salicylate. When hydrolyzed with dilute acid, primeverose, a reducing sugar, gives glucose and xylose.

(a) Xylose (p. 1096) is present in primeverose in the pyranose form. Draw the structure of β-D-xylopyranose.
(b) Primeverose is reduced by sodium borohydride. The reduction product is hydrolyzed by dilute acid to xylose and sorbitol (the reduction product of glucose). What does this prove about the structure of primeverose?
(c) Primeverose is synthesized from the reaction of α-D-xylopyranosyl bromide 2,3,4-triacetate with α-D-glucopyranosyl 1,2,3,4-tetraacetate in the presence of pyridine to give primeverose heptaacetate. Primeverose heptaacetate is then converted to primeverose by treatment with sodium methoxide in methanol. Write equations for these reactions, showing complete structures for the reactants, including their stereochemistry.
(d) The linkage between primeverose and the aglycon methyl salicylate is β. What is the complete structure of gaultherin?

25.51 Trisaccharide A, $C_{18}H_{32}O_{16}$, was isolated from a culture of *Betacoccus arabinosaceous*. Total acidic hydrolysis of the carbohydrate gave 2 units of glucose and one of galactose. Partial acidic hydrolysis of A gave glucose, galactose, lactose, and a disaccharide, B. Trisaccharide A is cleaved by a β-glycosidase to galactose and disaccharide B. The other glycosidic linkage was determined to be α from the optical rotation of the trisaccharide. Total methylation of A (eleven *O*-methyl groups) and then acid hydrolysis gave 2,3,4,6-tetra-*O*-methyl-D-glucopyranose; 2,3,4,6-tetra-*O*-methyl-D-galactopyranose; and 3,6-di-*O*-methyl-D-glucopyranose.

(a) Write structural formulas for trisaccharide A and disaccharide B.

(b) Trisaccharide A is a reducing sugar. Reduction of trisaccharide A with sodium borohydride gives an alcohol, C, $C_{18}H_{34}O_{16}$. Alcohol C reacts with 5 equivalents of periodate with the production of 2 equivalents of formic acid and 1 of formaldehyde. Write the structural formula of alcohol C. On the structural formula of alcohol C, indicate clearly the points of cleavage by periodate ion and which carbon atoms will appear as formic acid and as formaldehyde.

25.52 Patients who have severe bleeding or burns or who are undergoing surgery need to have huge volumes of blood plasma replaced. It is not always possible to provide as much human blood plasma as necessary, so artificial mixtures called plasma volume extenders have been developed to be used as part of the replacement. For example, a chemical modification of the starch fraction amylopectin has been synthesized and used as a plasma extender. The reaction used in the preparation of this plasma extender is shown below. Predict the structure of the modified polysaccharide formed, assuming that one hydroxyl group has been transformed for each glucose unit.

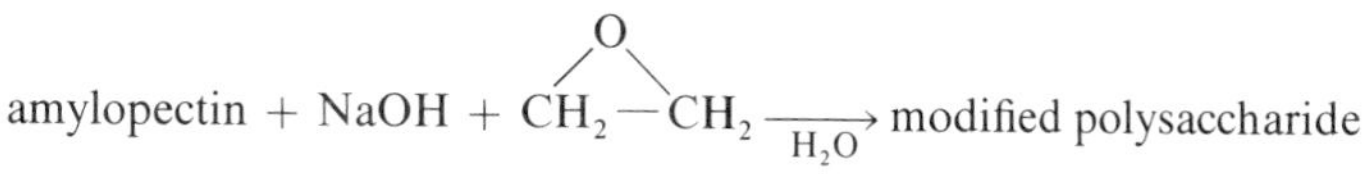

amylopectin + NaOH + CH2—CH2
O
H2O
modified polysaccharide

25.53 A cyanogenic glycoside called amygdalin has been substituted on the market for the compound originally patented as laetrile. Amygdalin has the same aglycon, (*R*)-(−)-mandelonitrile, as laetrile. The aglycon is linked by a β-glycosidic linkage to gentiobiose (Problem 25.34, p. 1101). What is the structure of amygdalin?

25.54 The amygdalin that is marketed contains considerable amounts of the isomer in which the stereocenter in mandelonitrile has the *S* configuration. Natural amygdalin, (*R*)-amygdalin, is easily epimerized to (*S*)-amygdalin by treatment with base. Write a mechanism for this epimerization. Why does it happen so easily?

25.55 D-(+)-Glyceraldehyde (p. 1096) is converted in basic or acidic solution to a mixture of D-fructose and D-sorbose, both shown in their open-chain forms below. The conversion proceeds faster if dihydroxyacetone is added to the solution. Write equations showing how D-(+)-glyceraldehyde is transformed into the mixture of hexoses.

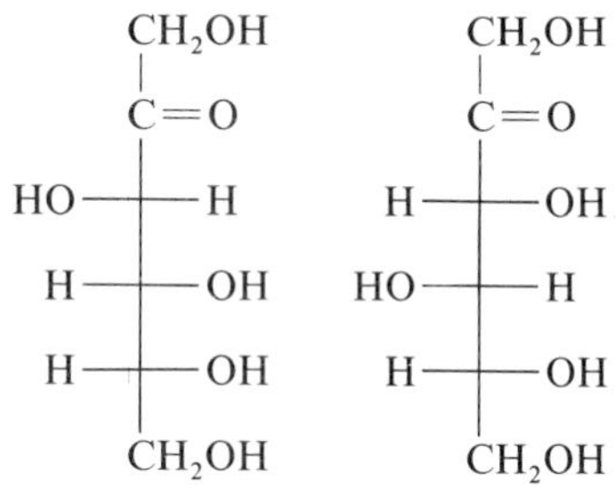

CH2OH
C=O
HO
H
OH
CH2OH
CH2OH
C=O
H
OH
HO
H
CH2OH

D-fructose D-sorbose

25.56 Shikimic acid is an important intermediate in the biosynthesis of amino acids containing aromatic rings. Shikimic acid itself is synthesized in the body from phosphoenolpyruvic acid, which is the phosphate ester of the enol of pyruvic acid. Various steps of this synthesis are given below. Write mechanisms for the different transformations, supplying acid or base (HB^+ and B:) catalysis as necessary. The named intermediates are known; the unnamed one has not been proven.

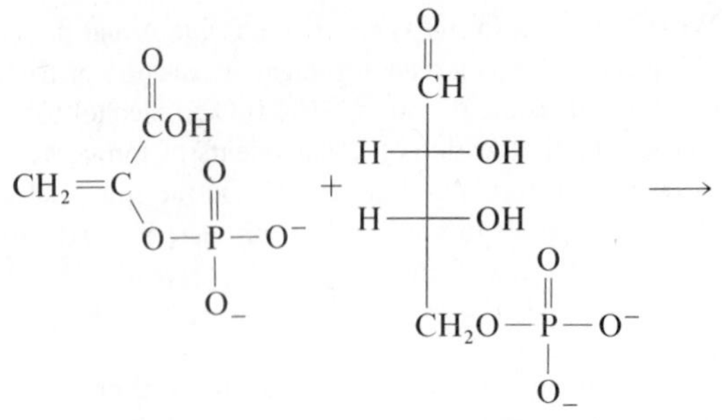

phosphoenolpyruvic acid

erythrose 4-phosphate

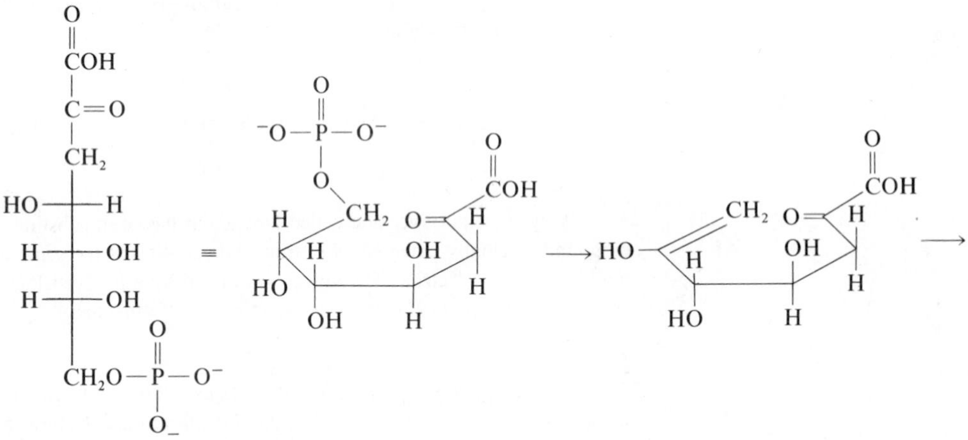

3-deoxy-α-arabinoheptulosonic acid 7-phosphate

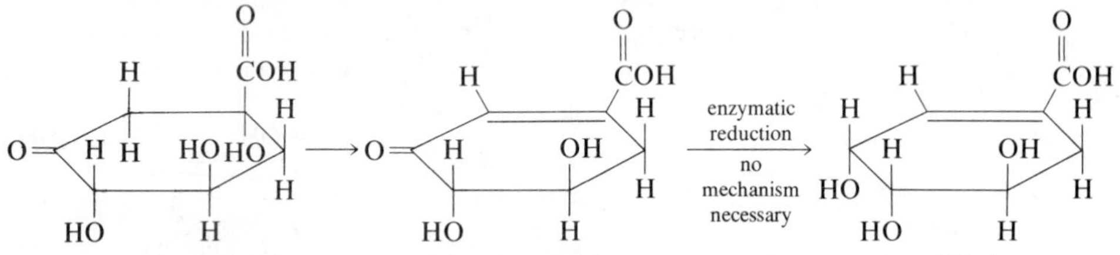
enzymatic
reduction
no
mechanism
necessary

5-dehydroquinic acid

5-dehydroshikimic acid

shikimic acid

26

Amino Acids, Peptides, and Proteins

A • L O O K • A H E A D

Proteins are one of the primary constituents of living matter. Even in plants, where carbohydrates are more abundant as structural materials, proteins are present in those parts that are responsible for growth and reproduction. The fundamental structure of proteins is simple. Protein molecules consist of long chains of amino acids bonded to each other by amide bonds, also known as peptide linkages, between the carboxylic acid group of one amino acid and the amino group in another (pp. 564 and 889). These chains are known as polypeptides and may contain from 50 to 300 amino acid units. A protein may consist of a single polypeptide chain or of several associated with one another. Molecular weights of proteins range from around 5000 to 1,000,000.

About twenty different amino acids are the building blocks of proteins. The number of ways in which these amino acids may be combined in sequence to give a protein is staggeringly large. The amino acids may be regarded as the letters of an alphabet from which an infinite variety of words can be created. Although the amino acid units of a polypeptide are linked to each other in a linear fashion, the polypeptide molecule assumes a folded conformation that is typical of that particular molecule and is to a great extent responsible for its biological activity.

This chapter will explore in greater detail the chemistry of amino acids and peptides and apply it to the complex proteins and their functions as structural units and catalysts in living organisms.

A. Structure and Stereochemistry of Amino Acids

The amino acids found in proteins derived from animals and higher plants are all α-aminocarboxylic acids of a particular stereochemical configuration. The older literature refers to them as L-amino acids, meaning that their configuration is related to that of L-(−)-glyceraldehyde (p. 1060).

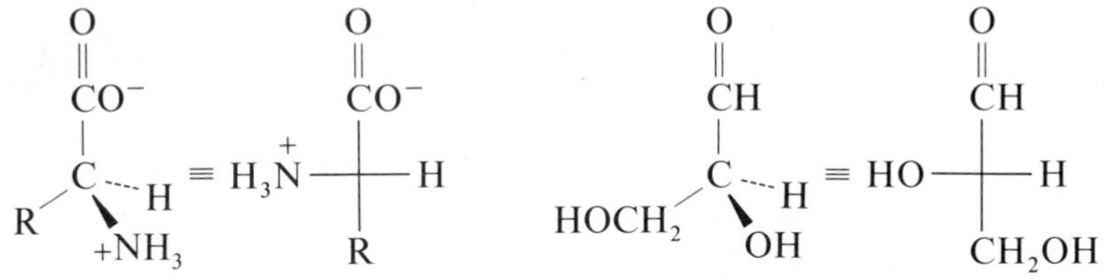

an L-amino acid L-(−)-glyceraldehyde

configurational relationship between L-amino acids and L-(−)-glyceraldehyde

The acidity and basicity of amino acids, derived from the presence of both a carboxyl and an amino group, are important properties. Additional basic or acidic functional groups in the portion of an amino acid molecule symbolized above as R also influence the chemistry. It is most useful to classify amino acids according to their ionic state at the **physiological pH,** the pH that exists in the cells and fluids of living organisms.

Study Guide Concept Map 26.1

The amino acids that are commonly found in proteins will be listed in Section 26.1C, but first the next section will discuss the reactions of amino acids as acids and bases.

B. Amino Acids as Acids and Bases

Amino acids have relatively high melting points, usually decomposing above 200 °C. They have high solubility in water and low solubility in nonpolar solvents. They have large dipole moments, and their solutions in water have high dielectric constants. All of these facts suggest that the units in the crystal lattice and in solution are charged species.

An examination of the relative basicities of a carboxylate anion and an amino group indicate that an amino acid should exist as a dipolar ion, also called a **zwitterion,** in which the amino group is protonated and the carboxyl group exists as a carboxylate anion (Problem 3.36, p. 107). In other words, the acidic group in an amino acid is a substituted ammonium ion and the basic group is the carboxylate anion.

The acid-base reactions of an amino acid are shown below.

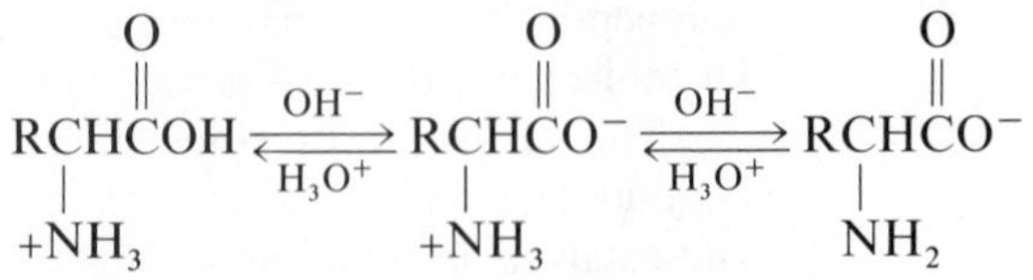

1 2 3

From this equation, it is clear that at very low pH, an amino acid exists in a form in which it is protonated at both the amino group and the carboxyl group (species 1). At such a pH, it bears a net positive charge and is a diprotic acid. At high pH, the amino acid bears a net negative charge and has two basic sites at which it can be protonated (species 3). Note that the equation indicates that it will be protonated first at the amino group, then at the carboxylate anion, which agrees with the relative basicities of the two groups. At some intermediate pH, the amino acid exists as a zwitterion with no net charge (species 2). The pH at which this occurs is known as the **isoelectric point,** pH_I, for the amino acid. At this pH, the amino acid is stationary in an electric field; it migrates neither to the negative pole nor to the positive pole because the charges on it are balanced. In contrast, at low pH, the amino acid bears a positive charge and migrates to the negative pole of an electric field, and at high pH, it migrates to the positive pole.

The behavior of a fully protonated amino acid as a diprotic acid can be seen clearly if a titration is carried out. For the typical amino acid, two pK_a values are determined. When half of an equivalent of base is added to the amino acid alanine ($R = CH_3$) in its fully protonated form, the pH of the solution is 2.34. At this point, the concentration of the zwitterionic form (species 2) equals the concentration of the diprotic acid (species 1). According to the Henderson-Hasselbach equation (Problem 3.38, p. 108), pK_a is equal to pH when the concentration of an acid and its conjugate base are equal. Therefore, the pK_a of the carboxylic acid group in alanine must be 2.34.

$$pK_a = pH + \log\frac{[HA]}{[A^-]}$$

when $\qquad [HA] = [A^-], \quad \log\frac{[HA]}{[A^-]} = 0$

and $\qquad pK_a = pH$

At pH 2.34, $\qquad \left[\mathrm{CH_3\underset{\underset{+NH_3}{|}}{C}H\overset{\overset{O}{\|}}{C}OH}\right] = \left[\mathrm{CH_3\underset{\underset{+NH_3}{|}}{C}H\overset{\overset{O}{\|}}{C}O^-}\right]$

therefore, $\qquad pK_a \text{ of } \mathrm{CH_3\underset{\underset{+NH_3}{|}}{C}H\overset{\overset{O}{\|}}{C}OH} = 2.34$

When an equivalent of base has been added, the zwitterion predominates. The pH at this point is 6.02, the isoelectric point of alanine. Finally, addition of another half of an equivalent of base gives a pH of 9.69, which corresponds to the pK_a of the ammonium ion derived from the amino acid. At this point, equal concentrations of the zwitterion and its conjugate base (species 3 above) exist.

The pH_I is the arithmetic mean of the pK_a values for the two acid groups in the compound.

$$pH_I = \tfrac{1}{2}(pK_{a_1} + pK_{a_2})$$

An isoelectric point of approximately 6 is typical for the amino acids that do not have additional acidic or basic groups on their side chains.

The pK_a value of 2.34 for the carboxylic acid group of alanine indicates that it is a stronger acid than a typical aliphatic carboxylic acid such as acetic acid, which has pK_a 4.76. The protonated amino group with its positive charge has an electron-withdrawing effect that strengthens the acid (p. 97). The amino group is also affected by the presence of the adjacent carboxylic acid group. It is less basic and its conjugate acid more acidic than is usually the case for an aliphatic amine. The conjugate acid of methylamine, for example, has pK_a 10.6, compared to pK_a 9.69 for the protonated amino group in alanine.

Amino acids that have additional amino or carboxyl groups in their molecules have isoelectric points that reflect their tendency either to be further protonated or to lose an additional proton. The amino acid lysine, for example, has three pK_a values: 2.18 for the carboxylic acid group, 8.95 for the α-amino group, and 10.53 for the amino group on carbon 6, the ϵ-amino group.

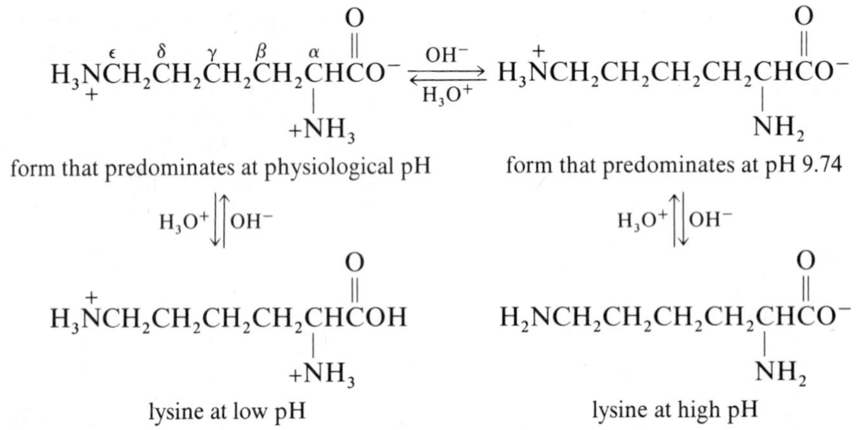
form that predominates at physiological pH
form that predominates at pH 9.74
lysine at low pH
lysine at high pH

The isoelectric point for lysine is 9.74. Note that the ϵ-amino group in lysine is more basic than the α-amino group, and thus the zwitterion for lysine is shown with the protonated amino group at carbon 6. The isoelectric point of lysine indicates that lysine would bear a net positive charge at pH 7, roughly the pH of cells. Only if the pH of the solution were raised to 9.74 would most of the lysine molecules be converted to the zwitterionic form having no net charge.

Glutamic acid, on the other hand, has an isoelectric point of 3.2. The carboxylic acid group on carbon 4 is like a normal aliphatic carboxylic acid group and has pK_a 4.25. The other carboxylic acid group has pK_a 2.19, and the protonated amino group has pK_a 9.67.

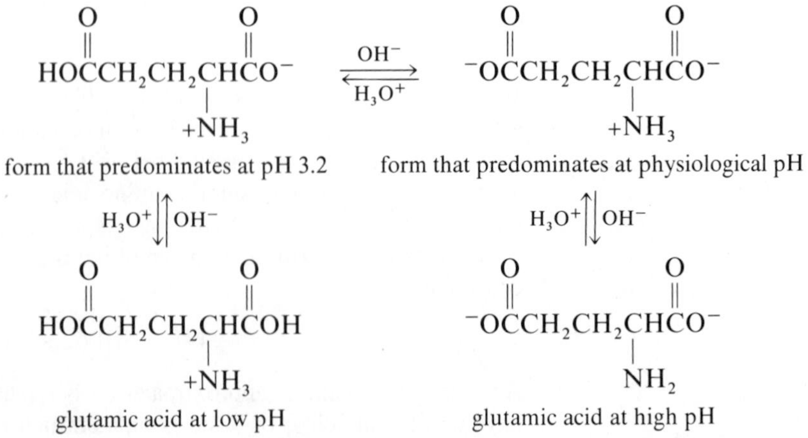
form that predominates at pH 3.2
form that predominates at physiological pH
glutamic acid at low pH
glutamic acid at high pH

At physiological pH, glutamic acid bears a net negative charge. Only when the pH is lowered to 3.2 does it exist as the zwitterion.

C. Classification of the Amino Acids Found in Proteins

The amino acids that are found in proteins can be classified into four groups on the basis of their properties at physiological pH. The first group contains the amino acids that have no polar substituents on their side chains. These all have isoelectric points close to 6 and have no net charge at pH 6–7. They are shown in Figure 26.1, along with the abbreviations of their names that will be used in writing formulas for peptides and proteins.

Other amino acids have polar functional groups that can participate in hydrogen bonding or function as nucleophiles in reactions, but have no net charge at physiological pH (Figure 26.2). Glycine is included among these because it does not have a large, hydrophobic side chain.

Two amino acids have extra carboxylic acid groups on them and therefore have a net negative charge at physiological pH (Figure 26.3).

Study Guide Concept Map 26.2

Finally, three amino acids have extra basic functional groups and are, therefore, positively charged at the pH of cellular fluids (Figure 26.4).

PROBLEM 26.1

(a) Would you expect the carboxyl group bonded to carbon 3 of aspartic acid to be more or less acidic than the carboxyl group bonded to carbon 4 of glutamic acid? Explain your answer.

(b) Would the isoelectric point for aspartic acid be higher or lower than that for glutamic acid?

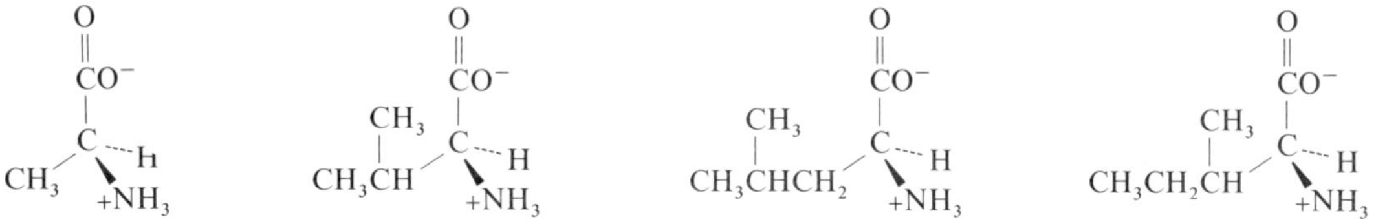

(+)-alanine Ala | (+)-valine Val | (–)-leucine Leu | (+)-isoleucine Ile

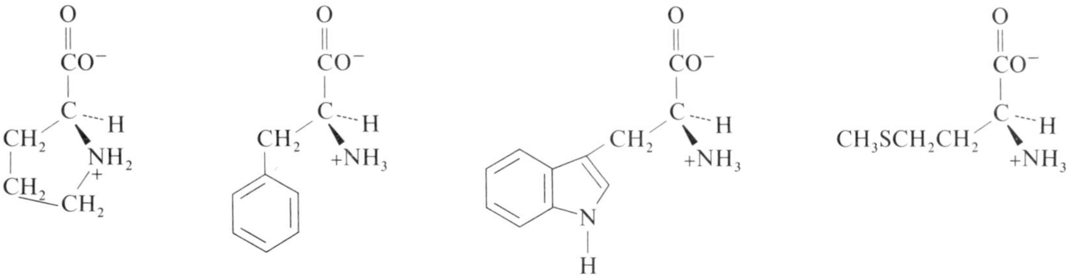

(–)-proline Pro | (–)-phenylalanine Phe | (–)-tryptophan Trp | (–)-methionine Met

FIGURE 26.1 Amino acids that have nonpolar R groups and no net charge at physiological pH.

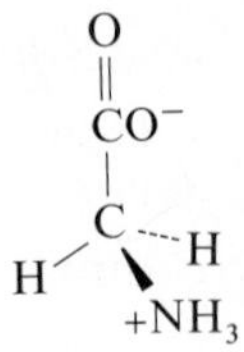

glycine
Gly

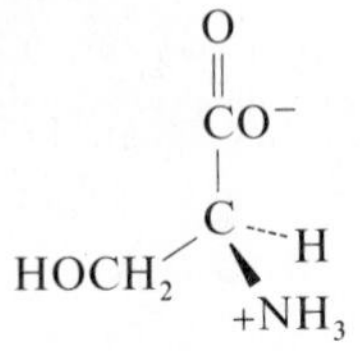

(−)-serine
Ser

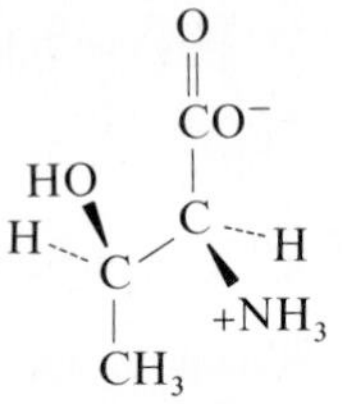

(−)-threonine
Thr

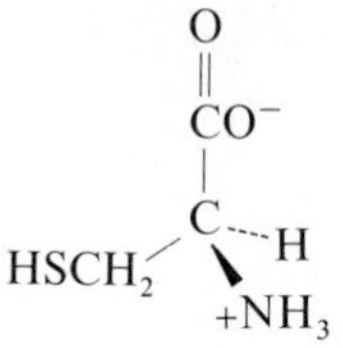

(−)-cysteine
Cys

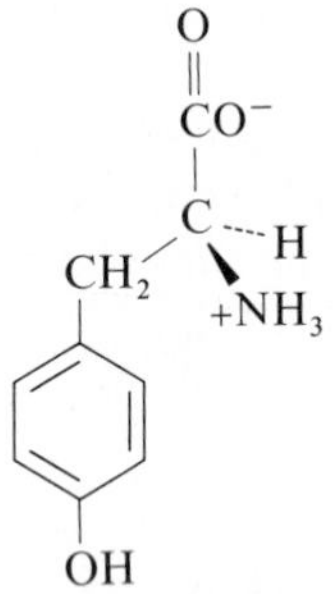

(−)-tyrosine
Tyr

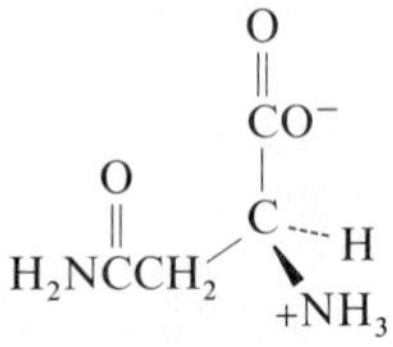

(−)-asparagine
Asn

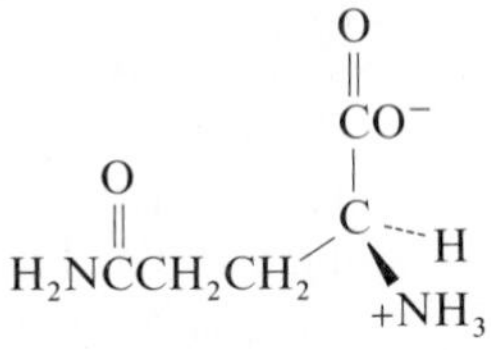

(+)-glutamine
Gln

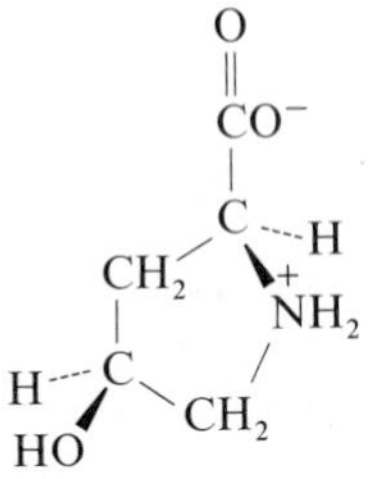

(−)-4-hydroxyproline
Hypro

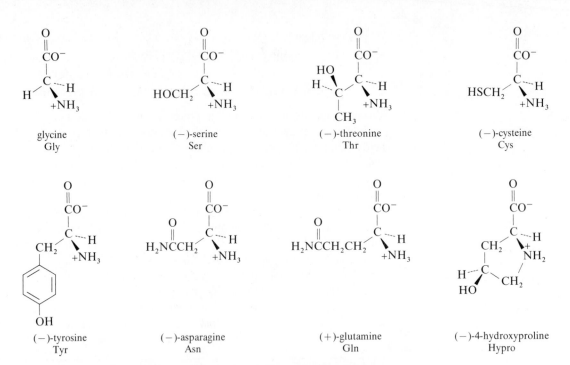

FIGURE 26.2 Amino acids that have polar R groups but no net charge at physiological pH.

FIGURE 26.3 Amino acids that are negatively charged at physiological pH.

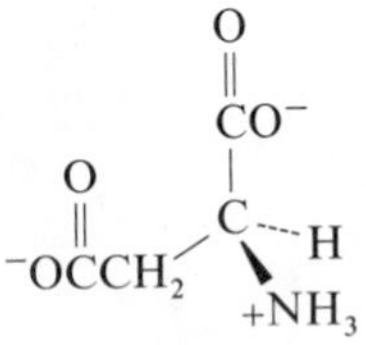

(+)-aspartic acid
Asp

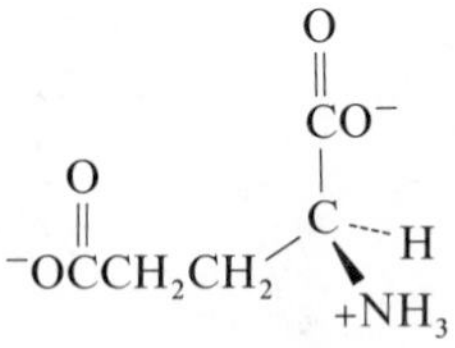

(+)-glutamic acid
Glu

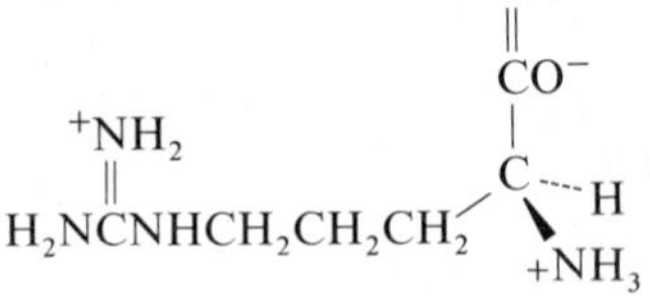

(+)-arginine
Arg

FIGURE 26.4 Amino acids that are positively charged at physiological pH.

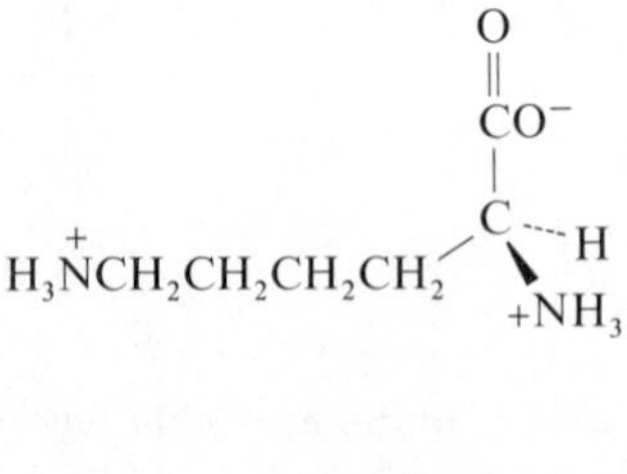

(+)-lysine
Lys

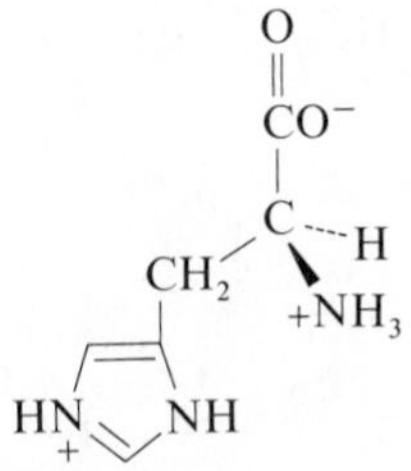

(−)-histidine
His

PROBLEM 26.2

The amino acids found in proteins are assigned to the L family according to their stereochemistry at carbon 2. Can they all be assigned the same configuration, *R* or *S*, at that carbon atom?

PROBLEM 26.3

Some of the amino acids shown in this section have more than one stereocenter. Pick them out. Assign configuration at the other stereocenters where possible.

PROBLEM 26.4

(a) Arginine is the most basic of the amino acids. The pK_a of the conjugate acid of the guanidinyl group on its side chain is 13.2. Why is the guanidinyl group so much more basic than the amino acid group in lysine (p. 1122)?

(b) Tryptophan is classified as a nonpolar amino acid even though it has a nitrogen atom in an indole ring (p. 1121) as part of its side chain. Why is tryptophan not basic? (Hint: A review of Section 19.1B and Problem 19.36, p. 820, may be helpful.)

D. Biosynthesis of Amino Acids

Many amino acids are synthesized in the body from ammonia and the degradation products of carbohydrates, such as α-ketocarboxylic acids, in the presence of reducing enzymes. The reaction, in fact, closely resembles the reductive amination reactions on p. 900. The synthesis of glutamic acid illustrates this reaction.

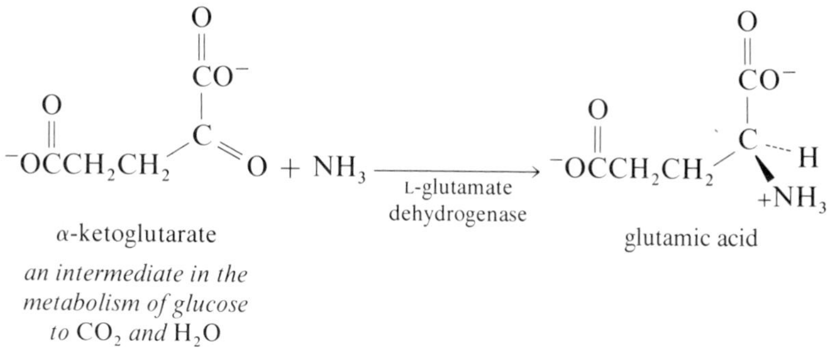

L-glutamate
dehydrogenase
α-ketoglutarate
glutamic acid
an intermediate in the metabolism of glucose to CO_2 and H_2O

The amino group in glutamic acid is transferred to other α-ketoacids in the body in another enzymatic reaction. A typical example is shown below.

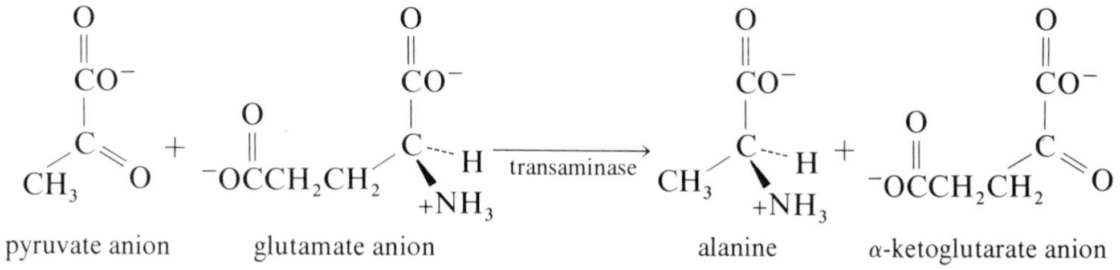

transaminase
pyruvate anion
glutamate anion
alanine
α-ketoglutarate anion

Such a transformation is called a **transamination** because an amino group is transferred from one species to another. A ketoacid is converted into the corresponding amino acid, and glutamic acid becomes α-ketoglutaric acid, which then reacts with ammonia to regenerate glutamic acid.

Because the human body cannot synthesize some of the amino acids necessary for protein synthesis, they must be obtained from food. These **essential amino acids** are arginine, histidine, isoleucine, leucine, lysine, methionine, phenylalanine, threonine, tryptophan, and valine.

PROBLEM 26.5

Amino acids undergo the reactions that are typical of primary alkyl amines and of carboxylic acids. In addition, other functional groups on the side chains can also react in typical ways. For the following equations, show the products that are expected. (Hint: For some of the reactions, it is necessary to remember that a small amount of the uncharged form of the amino acid is always in equilibrium with the zwitterionic form.)

(a)
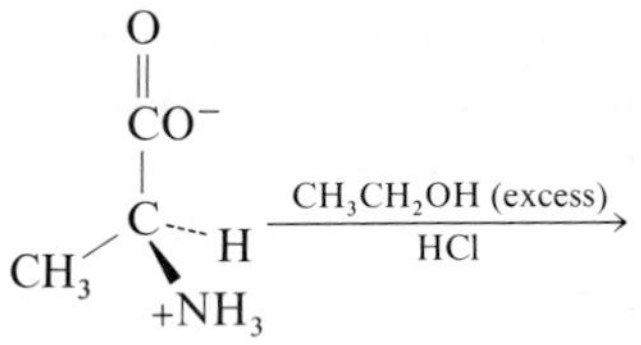

(b)
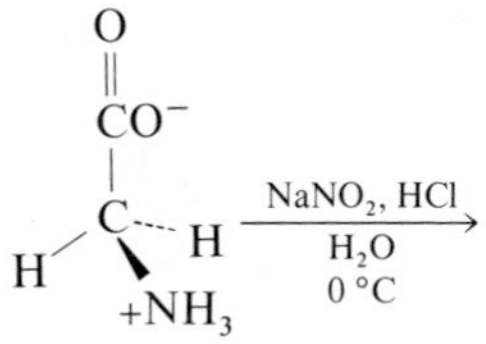

(c)
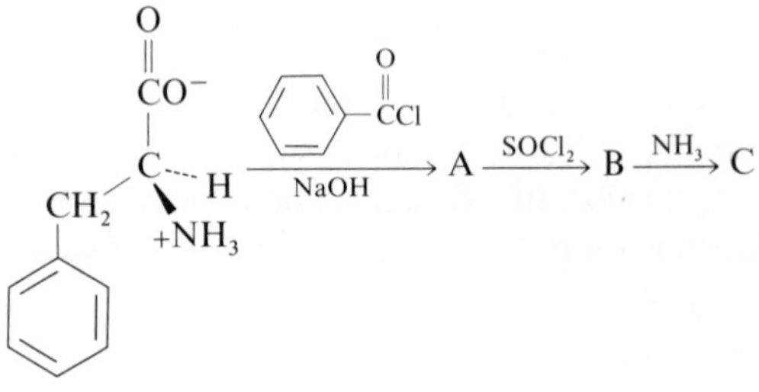

(d)
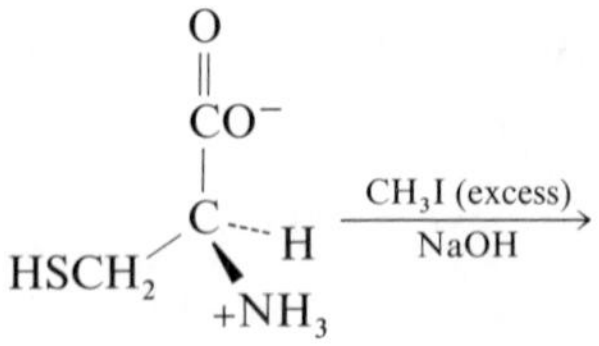

(e)
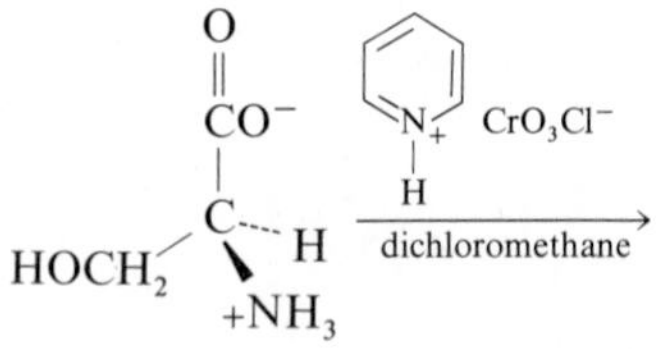

(f)
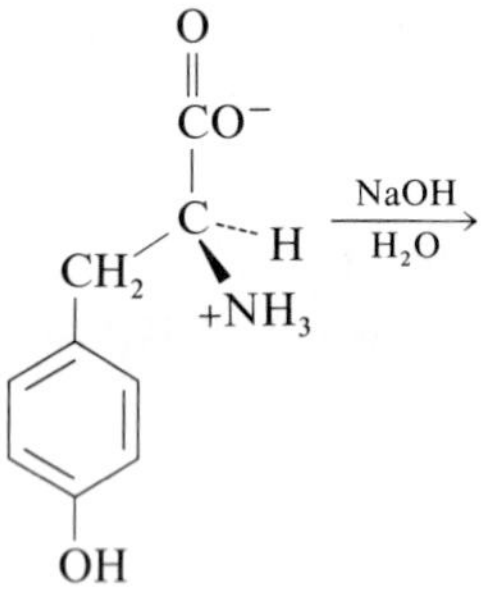

(g)
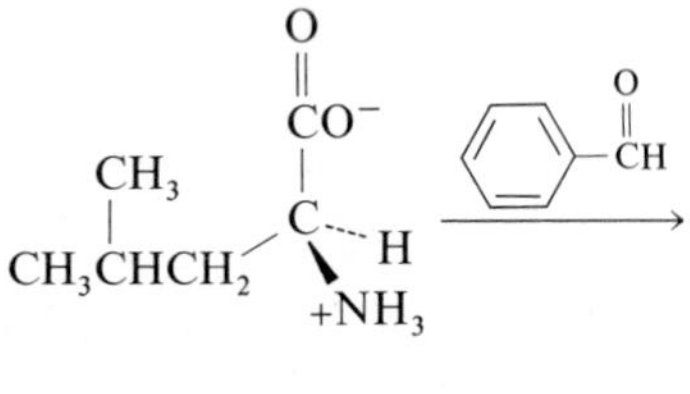

(h)
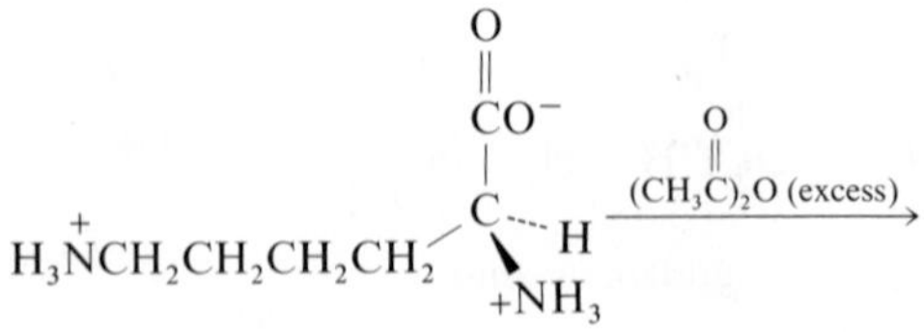

(i)
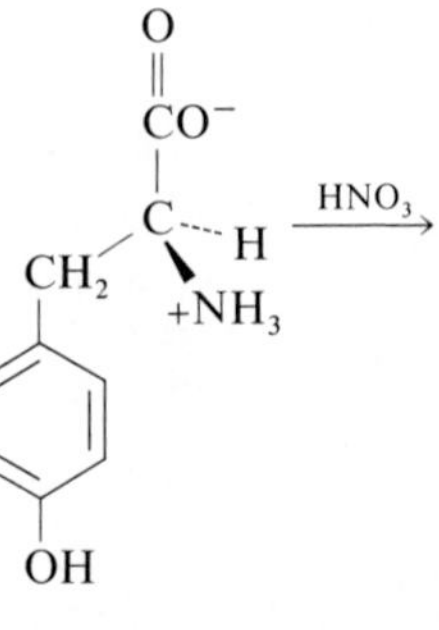

A. The Degradation of Proteins into Peptides and Amino Acids. Hydrolysis with Acid

In nature, amino acids are linked together by amide bonds, not only in the large molecules known as proteins but also in smaller molecules, many of which have hormonal activity. Such small molecules, containing only a few amino acid residues, are known as **peptides.** Proteins are cleaved into peptides in the process of digestion by specialized enzymes called **proteases.**

It is also possible to degrade proteins into peptides or amino acids by acidic hydrolysis of the amide bonds (p. 564). If a protein is subjected to concentrated hydrochloric acid at 37 °C for several hours, it is degraded into smaller peptide residues. If, on the other hand, it is heated with 20% hydrochloric acid for a period of days, all the amide bonds are cleaved and the protein is degraded completely into the individual amino acids; in fact, some amino acid molecules are destroyed by this treatment. Tryptophan does not survive vigorous acid hydrolysis, and serine and threonine, which contain hydroxyl groups, are partially destroyed. Asparagine and glutamine, which have amide groups on their side chains, are converted to aspartic acid and glutamic acid by this treatment, and ammonium chloride is formed as another product.

Rapid progress in the determination of the structure of proteins followed the development of chromatographic methods for the analysis of mixtures of peptides or amino acids. Many different techniques are used, and automated equipment can deliver a complete analysis of the amino acid content of a sample in a few hours. Some methods depend on the use of an electric field to separate the amino acids according to the charge they bear at a given pH. In paper chromatography, thin-layer chromatography, or column chromatography, amino acids of differing polarities are distributed differently between two phases, of which one is stationary and the other, mobile. The two phases may be two liquid phases of different polarities or a solid phase and a liquid phase. The more polar amino acids are held back by the more polar phase so that separations take place. At the end of the process, the amino acids are usually treated with ninhydrin, which reacts with amino acids other than proline and 4-hydroxyproline to give a purple dye.

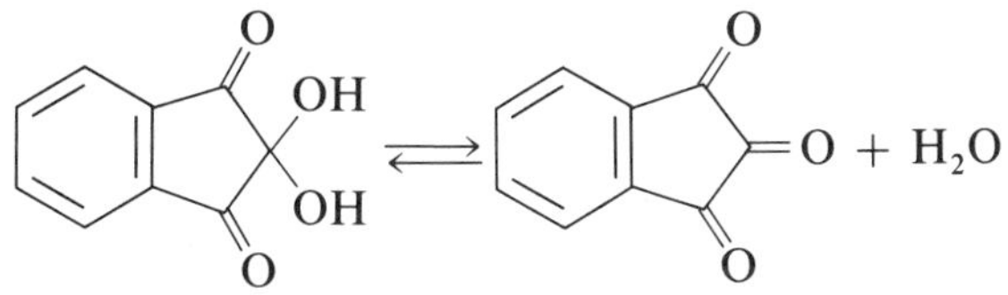
OH
OH
O
O + H₂O

ninhydrin

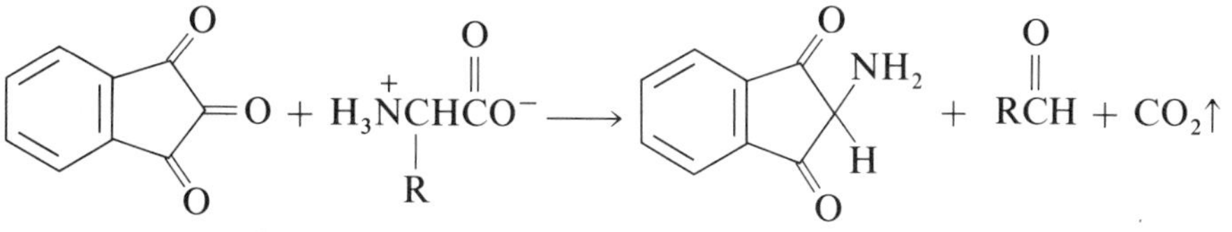
O
H₃NCHCO⁻
R
NH₂
H
RCH + CO₂

same product from all amino acids except proline and 4-hydroxyproline

different aldehyde from each amino acid

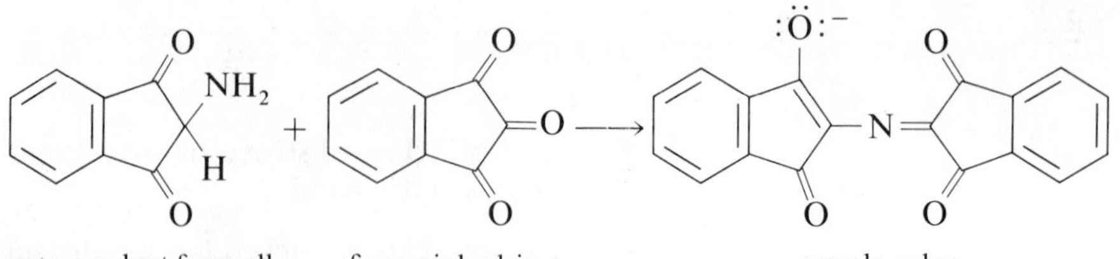
same product from all amino acids except proline and 4-hydroxyproline
from ninhydrin
purple color

If the experimental conditions are carefully standardized, the exact positions at which the dye spots appear on the paper sheet or thin-layer plate can be used to identify the amino acids present. The patterns of spots are sometimes called maps of amino acids. The same kinds of techniques can also be used to separate and identify peptides. Similarly, the absorption of the dye from ninhydrin can be detected spectroscopically and used to plot the appearance of the separated amino acids as they leave the column in column chromatography. Once again, careful standardization of conditions allows chemists to compare the experimental pattern of absorption peaks with known patterns to identify the amino acids present.

Separations may be carried out on the derivatives of amino acids rather than on the amino acids themselves. Automated systems that depend on chromatographic separation of the phenylthiohydantoins (p. 1129) of the amino acids in a protein are the latest and most rapid analytical tools.

PROBLEM 26.6

Write a mechanism for the formation of the purple dye from the reaction of excess ninhydrin with the amino derivative of ninhydrin shown above.

B. Nomenclature of Peptides

Peptides are named for the amino acids that constitute them, starting with the amino acid that has the free amino group. That end of the peptide chain is known as the ***N*-terminal amino acid,** and the amino acid that has the free carboxylic acid group is called the ***C*-terminal amino acid.** In naming peptides, starting with the *N*-terminal amino acid, each amino acid except the *C*-terminal one is treated as though it were an alkyl substituent on the next unit in the chain. This nomenclature is illustrated for a tetrapeptide.

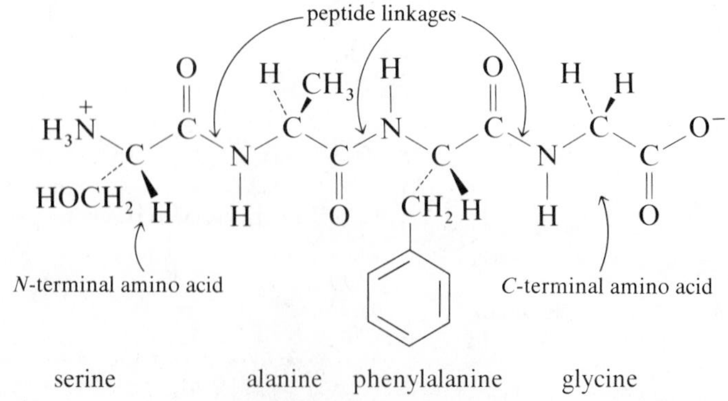
peptide linkages
N-terminal amino acid
C-terminal amino acid
serine
alanine
phenylalanine
glycine

serylalanylphenylalanylglycine

Ser-Ala-Phe-Gly

N-terminal amino acid (Ser) — *C*-terminal amino acid (Gly)

When written out in full, the name of a peptide containing more than three amino acids is cumbersome. Thus, three-letter abbreviations for the names of the amino acids are used to make the names of peptides more manageable. Separating the abbreviations with hyphens indicates that the amino acids are bonded to each other by peptide linkages in the order shown. The first amino acid listed is the *N*-terminal amino acid, and the last one is the *C*-terminal amino acid. This type of notation serves as a substitute for the full structural formula of the peptide. Separating the three-letter abbreviations for the amino acids with commas indicates that the amino acids are present in the peptide but the order of their bonding is not known. For example, complete hydrolysis of the tetrapeptide shown above would yield the information that it contains the amino acids alanine, glycine, phenylalanine, and serine (listed in alphabetical order) but no information on how they are bonded to each other.

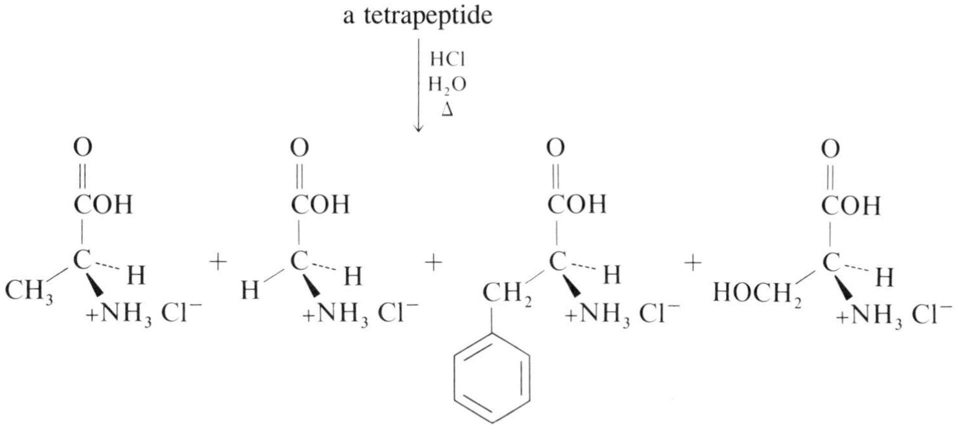

a tetrapeptide
HCl
H_2O
Δ

Ala Gly Phe Ser

$$\text{a tetrapeptide} \xrightarrow[H_2O,\ \Delta]{HCl} \text{Ala, Gly, Phe, Ser as their hydrochloride salts}$$

no information about the order in which the amino acids are bonded to each other

PROBLEM 26.7

Four amino acids can be combined in various ways to give twenty-four different peptides. Write complete structures for four peptides containing alanine, glycine, phenylalanine, and serine that are different from the one shown above. Show the stereochemistry, and label the *N*-terminal and *C*-terminal amino acids in each one.

PROBLEM 26.8

Draw the full structure of the peptide Leu-Asn-Lys-Tyr. Show the fragments you would expect to get from its complete hydrolysis with hydrochloric acid.

A peptide, like an amino acid, exists as a dipolar ion at physiological pH. Peptides also have typical isoelectric points that depend on the amino acids present in them. The carboxylic acid groups and amino groups that are part of the peptide linkages no longer have acidic or basic properties, so only the *N*-terminal amino group, the *C*-terminal carboxylic acid group, and any acidic or basic groups that are present in the side chains of the peptide will be ionized. The tetrapeptide Ser-Ala-Phe-Gly (p. 1126) has an isoelectric point around 6.0.

PROBLEM 26.9

For each of the following peptides, predict whether the isoelectric point will be below, above, or approximately at 6. Predict whether each peptide will move to the negative or positive pole of an electric field, or neither, if the pH of the solution is kept at 6.0.

(a) Gly-Phe-Thr-Lys
(b) Tyr-Ala-Val-Asn
(c) Trp-Glu-Leu
(d) Pro-Hypro-Gly

C. End-Group Analysis of Peptides and Proteins

Chemists have developed a series of reagents that react selectively with the *N*-terminal amino acid of a polypeptide, transforming it into some derivative that can be separated from all of the other amino acids in the chain and identified. One of the most useful of these is 2,4-dinitrofluorobenzene, **Sanger's reagent,** developed by Frederick Sanger during his determination of the structure of insulin. Sanger was given the Nobel Prize in 1958 for being the first to establish the sequence of amino acids in a protein.

2,4-Dinitrofluorobenzene, abbreviated as DNFB, is an aryl halide with nitro groups on the aromatic ring in positions that activate the halogen toward nucleophilic substitution (p. 957). The free amino group of the *N*-terminal amino acid displaces fluoride ion from the reagent, forming a dinitroarylamino group. Complete hydrolysis of the peptide gives free amino acids from all of the residues except the *N*-terminal one, which turns up labeled with the aryl group. The process is illustrated for the tripeptide Ser-Gly-Val, which is written in the form in which the amino group is not protonated.

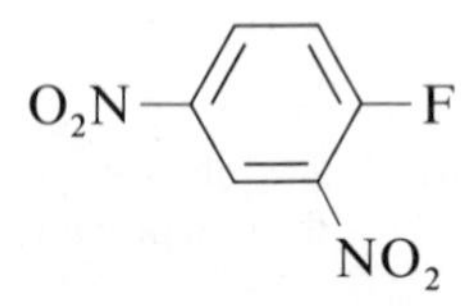
O_2N
F
NO_2

2,4-dinitrofluorobenzene

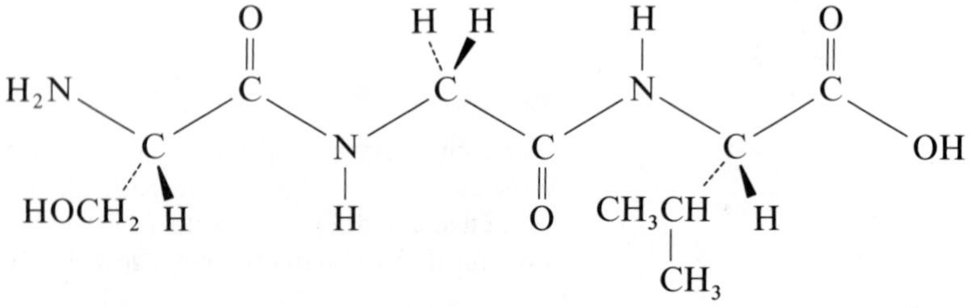
H_2N
$HOCH_2$
OH
CH_3CH
CH_3

Ser-Gly-Val

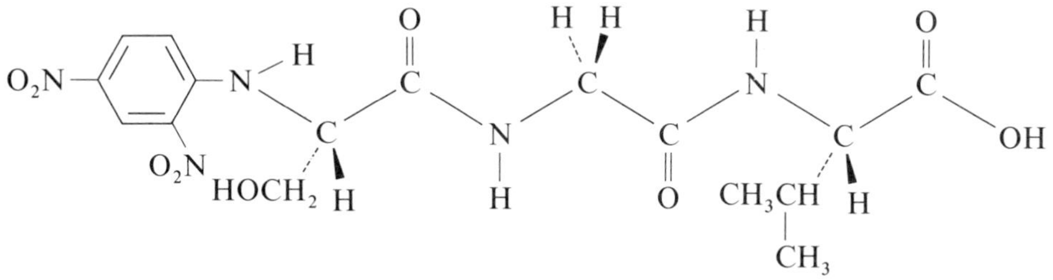
O₂N
H
O
H H
H
O
N
C
C
N
C
N
C
OH
O₂N
HOCH₂ H
H
O
CH₃CH H
CH₃

Ser-Gly-Val, labeled at serine with
a 2,4-dinitrophenyl group

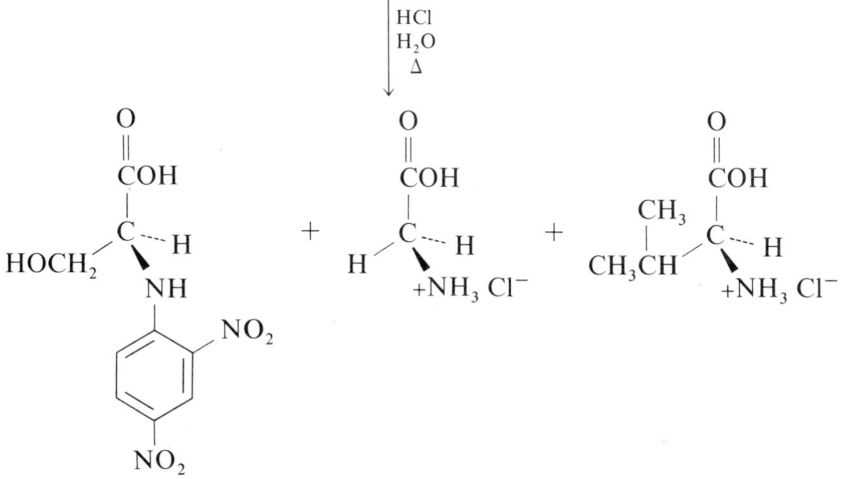
HCl
H₂O
Δ
O
COH
C
H
HOCH₂
NH
NO₂
NO₂
+
O
COH
C
H
H
+NH₃ Cl⁻
+
O
COH
CH₃
C
H
CH₃CH
+NH₃ Cl⁻

N-2,4-dinitrophenylserine
yellow color;
behavior unlike that of serine in chromatography or in an electric field

glycine

valine

The presence of the dinitrophenyl group on the nitrogen atom of serine converts the compound from an alkyl amine to an aryl amine and changes its acid-base properties drastically. The derivative of serine behaves differently than serine does with all the techniques used to separate amino acids and is easily identified by comparing it to an authentic sample of *N*-2,4-dinitrophenylserine.

PROBLEM 26.10

2,4-Dinitrofluorobenzene can be used to determine whether lysine is the *N*-terminal amino acid of a peptide or is somewhere in the middle of the peptide chain. Draw segments of hypothetical peptide chains with lysine in either position, and show with equations how the determination of the position of the lysine molecule could be carried out.

Another reagent used to analyze for *N*-terminal amino acids is phenyl isothiocyanate. This reagent, which is known as **Edman's reagent,** reacts with the free amino group to give a thiourea that is then cleaved off in anhydrous acid. The peptide is left with a new *N*-terminal amino acid that can be subjected to this degradation reaction and the amino acid that is lost becomes part of a substituted phenylthiohydantoin.

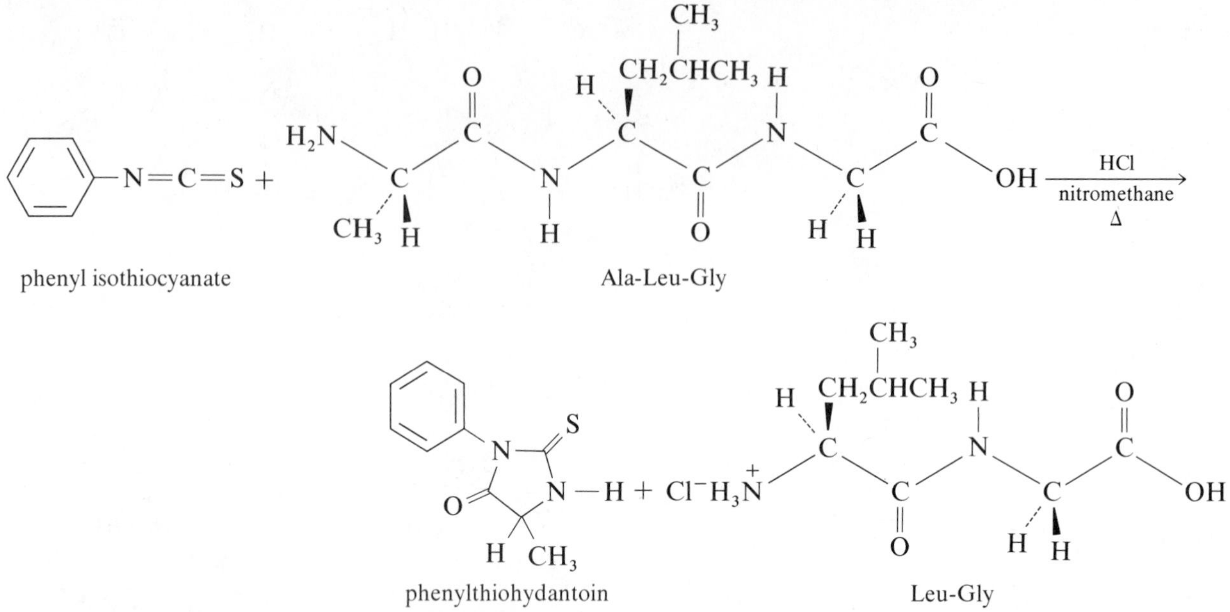

phenyl isothiocyanate

Ala-Leu-Gly

phenylthiohydantoin
derived from alanine, racemized

Leu-Gly

The substituted phenylthiohydantoin can be identified by comparing it with phenylthiohydantoins prepared from known amino acids. The reaction proceeds by way of an unstable intermediate that rearranges to give the phenylthiohydantoin that is isolated.

VISUALIZING THE REACTION

Degradation of a peptide by Edman's reagent

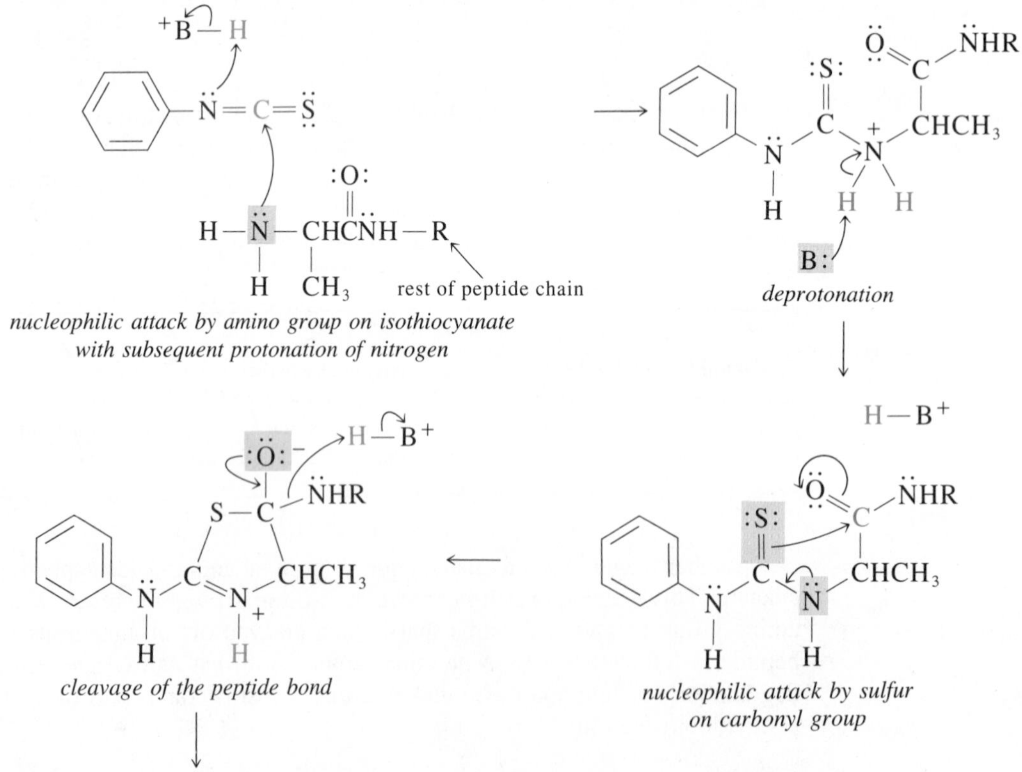

rest of peptide chain

nucleophilic attack by amino group on isothiocyanate with subsequent protonation of nitrogen

deprotonation

cleavage of the peptide bond

nucleophilic attack by sulfur on carbonyl group

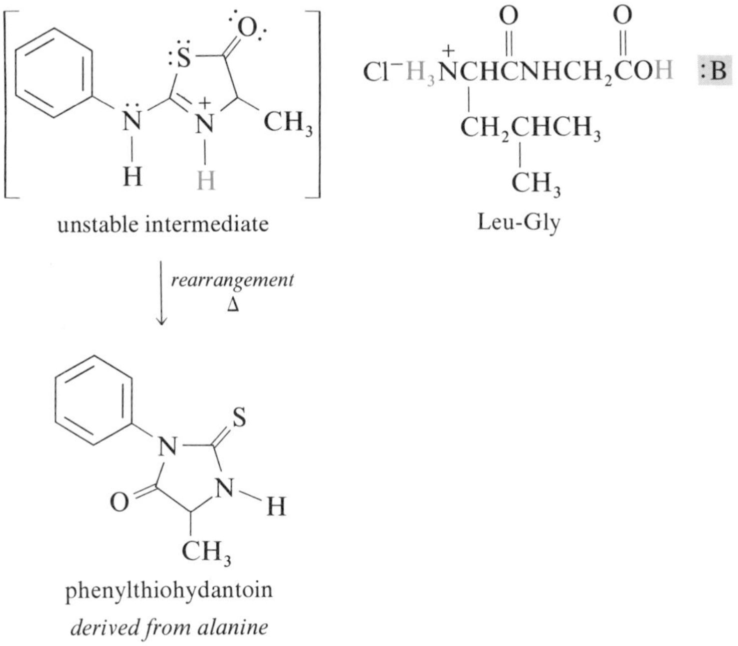
unstable intermediate
rearrangement
Δ
Leu-Gly
phenylthiohydantoin
derived from alanine

In principle, a large polypeptide could be degraded one amino acid at a time and each new *N*-terminal amino acid identified as its phenylthiohydantoin (see p. 1134). In general, though, the practical problems involved in handling small quantities of material and the side reactions that take place with some of the amino acids limit the number of times the degradation reaction can be used.

The enzyme carboxypeptidase selectively removes the *C*-terminal amino acid of a peptide. This process leaves a smaller peptide with a new *C*-terminal amino acid, which is then also attacked and removed by the enzyme. Therefore, this method must be used in connection with studies of the rate at which different free amino acids appear in the solution. For example, if the tripeptide Ala-Leu-Gly is treated with carboxypeptidase, the amino acid glycine is liberated into the solution first. After a while, leucine and alanine will also appear.

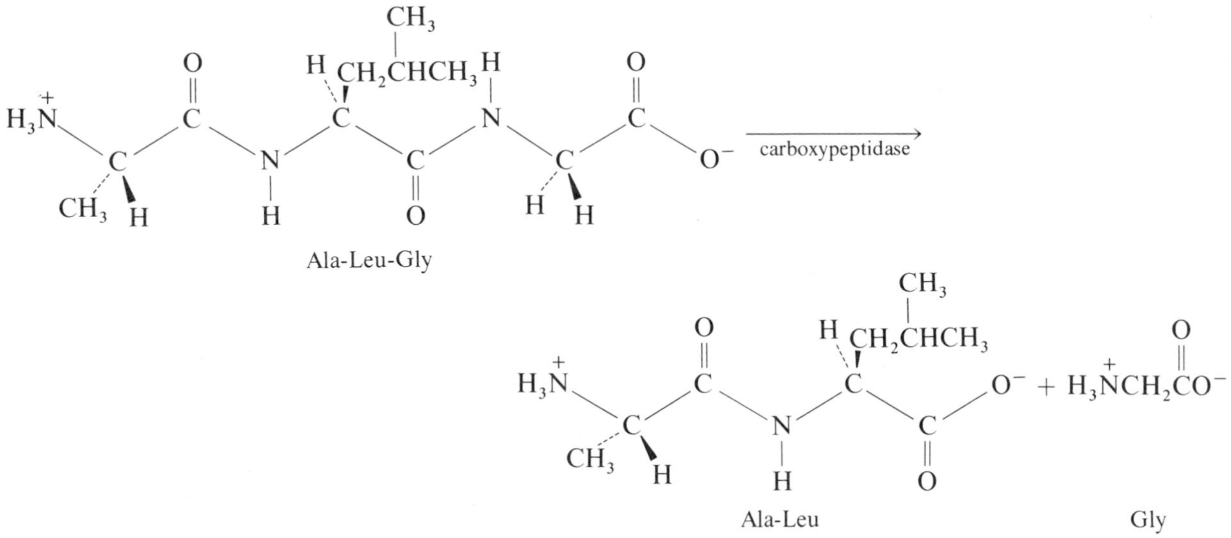
carboxypeptidase
Ala-Leu-Gly
Ala-Leu
Gly

PROBLEM 26.11

When the hydrolysis of a peptide by carboxypeptidase is carried out in water labeled with oxygen-18, the isotopic oxygen appears in the carboxylic acid groups of all the amino acids except the *C*-terminal one, allowing the *C*-terminal amino acid to be identified. Using Gly-Ala as an example, write a mechanism employing acid and base catalysis and $H_2{}^{18}O$ to show what happens in such a hydrolysis.

End-group analysis of many biologically interesting peptides is complicated by the fact that the *C*-terminal amino acid is in the form of an amide. Also, when glutamic acid is the *N*-terminal amino acid, it often forms a cyclic amide known as pyroglutamic acid, symbolized by a ring added to the three-letter abbreviation Glu. An important hormone secreted by the hypothalamus gland, the thyrotropin-releasing factor, has both of these features.

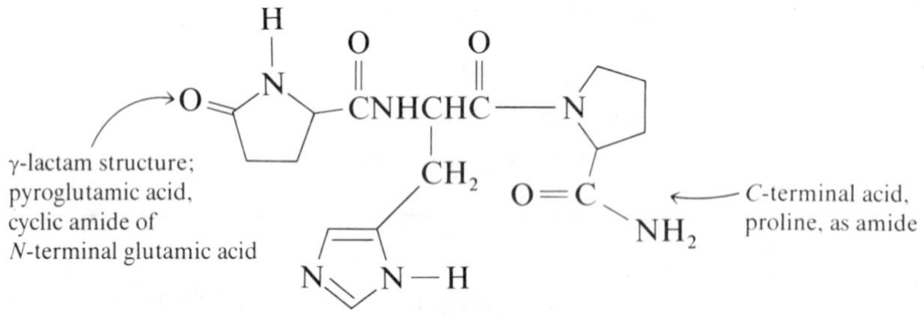
γ-lactam structure;
pyroglutamic acid,
cyclic amide of
N-terminal glutamic acid
C-terminal acid,
proline, as amide

└Glu-His-Pro-NH_2
thyrotropin-releasing factor
a tripeptide hormone

Such peptides do not react with reagents such as Sanger's reagent or Edman's reagent at the *N*-terminal amino acid, nor is the *C*-terminal amino acid cleaved off by carboxypeptidase. Hydrolysis of a molecule of the tripeptide hormone shown would give a molecule of ammonia, as well as the three amino acids glutamic acid, histidine, and proline.

D. The Degradation of Proteins and Peptides by Enzymes. The Disulfide Bridge

If hydrolysis of a protein is carried out with cold, concentrated hydrochloric acid for a relatively short period of time, a variety of peptide fragments are obtained. A protein may be more selectively cleaved at certain points of the peptide chain by enzymes. Trypsin, for example, attacks only the peptide bonds formed by the carboxyl groups of the basic amino acids lysine and arginine. The newly formed peptide fragments have either lysine or arginine as their *C*-terminal amino acid, except for the fragment that includes the original *C*-terminal end of the protein. Another enzyme, chymotrypsin, is less selective. It hydrolyzes the bonds formed by the carboxyl groups of tyrosine, phenylalanine, tryptophan, and methionine. The usefulness of these enzymes in selective cleavages of peptide bonds is illustrated by the determination of the structure of somatostatin, a tetradecapeptide that inhibits the secretion of pituitary growth hormone.

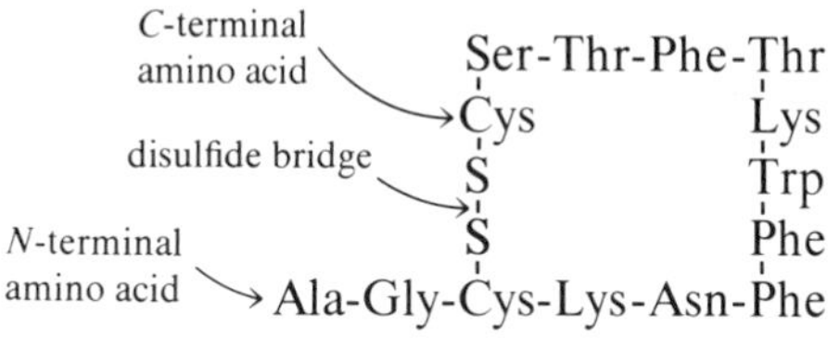
C-terminal amino acid
disulfide bridge
N-terminal amino acid
Ser-Thr-Phe-Thr
Cys
Lys
S
Trp
S
Phe
Ala-Gly-Cys-Lys-Asn-Phe

the structure of somatostatin, showing the disulfide bridge between two cysteine residues

The *C*-terminal amino acid of somatostatin and the amino acid third from the *N*-terminal one are cysteine. They are bonded to each other in oxidized form to give a disulfide bridge between the two ends of the peptide. Such bridges are important in stabilizing the tertiary structure of many proteins (p. 1150). The oxidized form of cysteine is known as cystine, a diamino acid with a disulfide bond. Cystine is reduced to give two molecules of cysteine.

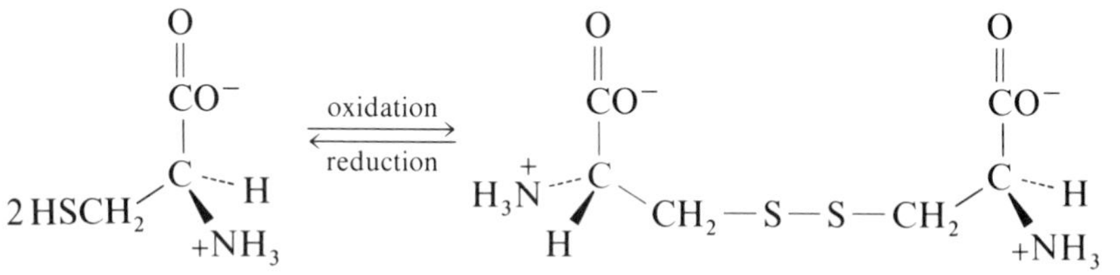
oxidation
reduction

cysteine — cystine

The proof of the structure of somatostatin starts with the reduction of the disulfide linkage with 2-mercaptoethanol, $HSCH_2CH_2OH$; cystine is thereby converted to cysteine. The thiol groups on the cysteine residues are then converted to inert groups by alkylation with iodoacetic acid.

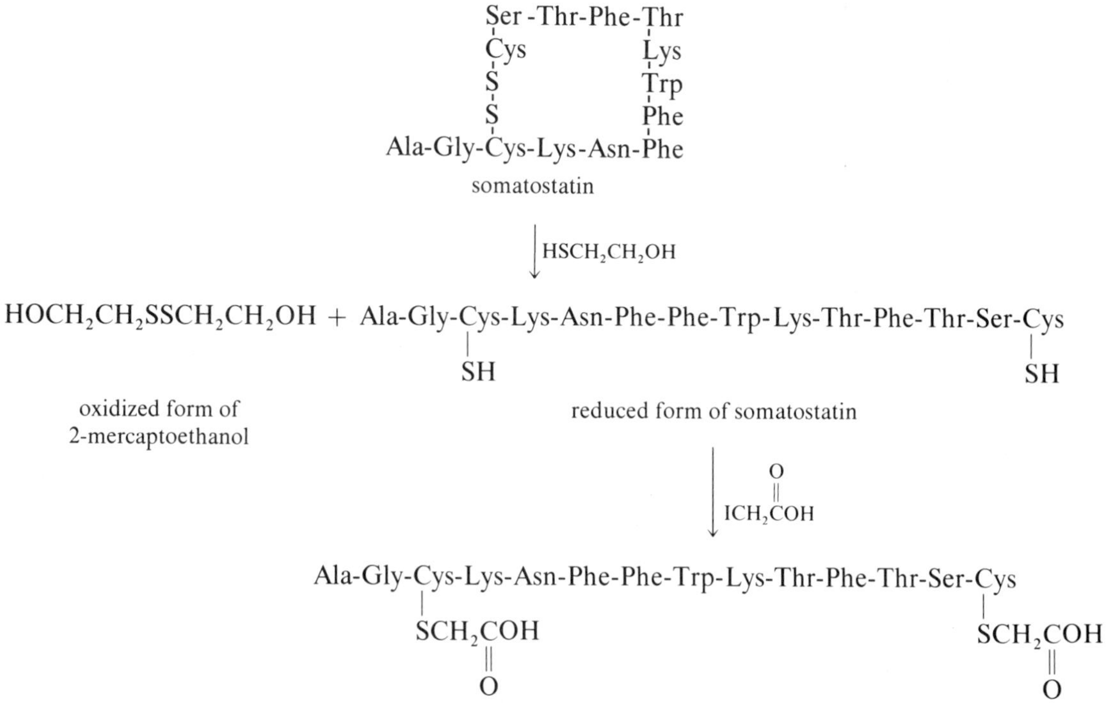
Ser-Thr-Phe-Thr
Cys
Lys
S
Trp
S
Phe
Ala-Gly-Cys-Lys-Asn-Phe
somatostatin
HSCH₂CH₂OH
HOCH₂CH₂SSCH₂CH₂OH + Ala-Gly-Cys-Lys-Asn-Phe-Phe-Trp-Lys-Thr-Phe-Thr-Ser-Cys
SH
SH
oxidized form of 2-mercaptoethanol
reduced form of somatostatin
ICH₂COH
Ala-Gly-Cys-Lys-Asn-Phe-Phe-Trp-Lys-Thr-Phe-Thr-Ser-Cys
SCH₂COH
SCH₂COH

the reduced form of somatostatin with the thiol groups of cysteine residues converted to acid residues

This step is necessary to prevent the cysteines from being oxidized back to cystine by air.

The peptide somatostatin is cleaved at two points by trypsin.

somatostatin
(reduced and alkylated)

↓ trypsin digestion

Ala-Gly-Cys(SCH_2COOH)-Lys + Asn-Phe-Phe-Trp-Lys + Thr-Phe-Thr-Ser-Cys(SCH_2COOH)

(Each Cys side chain is drawn as $SCH_2\overset{\|}{C}OH$ with =O, i.e. $SCH_2C(=O)OH$.)

Digestion of somatostatin with chymotrypsin gives more fragments than does digestion with trypsin.

somatostatin
(reduced and alkylated)

↓ chymotrypsin digestion

Ala-Gly-Cys($SCH_2C(=O)OH$)-Lys-Asn-Phe + Phe + Trp + Lys-Thr-Phe + Thr-Ser-Cys($SCH_2C(=O)OH$)

If the structures of the individual smaller peptides that result from acid or enzymatic hydrolyses can be determined, it is possible to reconstruct the structure of the original larger peptide by piecing together the different overlapping sequences of amino acids found in the fragments. The structures of small peptides are determined by reactions that selectively identify their *N*-terminal and *C*-terminal amino acids by end-group analysis (p. 1128). Thus, for example, if a determination is made that the *N*-terminal amino acid in somatostatin is alanine and the *C*-terminal amino acid is cysteine, the fragments obtained from the enzymatic cleavages can be combined to give the full structure of the molecule. The fragments from cleavage by trypsin and those from cleavage by chymotrypsin are lined up below to show how the different points of cleavage provide overlapping information about the order of the amino acids in the chain.

peptides from trypsin cleavage of somatostatin

Ala-Gly-Cys-Lys/Asn-Phe-Phe-Trp-Lys/Thr-Phe-Thr-Ser-Cys
Ala-Gly-Cys-Lys-Asn-Phe/Phe/Trp/Lys-Thr-Phe/Thr-Ser-Cys

peptides from chymotripsin cleavage of somatostatin

The total structure of somatostatin has also been determined by repeated use of Edman's reagent. The chain was degraded one amino acid at a time starting from the *N*-terminal end, and each amino acid was identified as its phenylthiohydantoin (p. 1129).

Study Guide
Concept Map 26.3

PROBLEM 26.12

A heptapeptide is isolated from milk protein that has been degraded by a protease from the bacterium *Bacillus subtilis*. The heptapeptide contains the amino acids arginine, glycine, isoleucine, phenylalanine, proline (2 residues), and valine.

(a) Treatment of the heptapeptide with trypsin gives arginine and a hexapeptide. What does this tell you about the structure of the original peptide?
(b) Some of the fragments that were obtained by partial hydrolyses of the heptapeptide are Pro-Phe-Ile, Arg-Gly-Pro, Pro-Phe-Ile-Val, and Pro-Pro. Assign a structure to the heptapeptide.

PROBLEM 26.13

The hormone arginine vasopressin, which is important in the regulation of water balance in the body, is a nonapeptide with one disulfide bridge. In the proof of its structure, the disulfide bridge was cleaved by oxidation with peroxyformic acid, which oxidizes cysteine residues to cysteic acid. During degradation, the cysteines appear as cysteic acid residues, symbolized below by Cys.

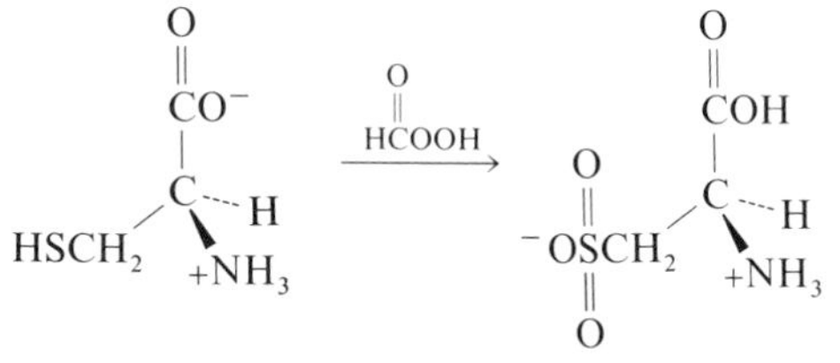

HSCH2
+NH3
HCOOH
-OSCH2
COH
cysteine
cysteic acid

Complete hydrolysis of the oxidized peptide with hydrochloric acid gave 3 equivalents of ammonia and the amino acids Arg, Asp, Cys (2), Glu, Gly, Phe, Pro, and Tyr. 2,4-Dinitrofluorobenzene reacted with one of the cysteic acid residues. Trypsin digestion gave an octapeptide with arginine as its *C*-terminal amino acid and glycinamide, Gly-NH_2. Partial acid hydrolysis gave the following fragments:

Asp-Cys Phe-Glu Cys-Tyr-Phe Glu-Asp-Cys Phe-Glu-Asp

In addition, a fragment having cysteine as its *N*-terminal amino acid and also containing arginine, glycine, and proline, in an undetermined order, was obtained.

(a) What is the order of amino acids derived from the degradation experiments?
(b) What is the full structure for arginine vasopressin?

26.3 SYNTHESIZING PEPTIDES AND PROTEINS IN THE LABORATORY

A. Protection of Reactive Functional Groups

The synthesis of a peptide requires that amide bonds be formed between the amino group of one amino acid and the carboxylic acid group of another one. To prepare a peptide with a particular structure, the amino acids must be added onto the chain in a precise order. The desired reaction involves a nucleophilic substitution by an amino group at a carbonyl group of a carboxylic acid, but there are nucleophilic groups such as other amino groups, thiols, and alcohols on the side chains of amino acids that are also expected to react with carboxylic acids. The reactions to form the backbone of a peptide chain must involve the amino group at carbon 2 and not any of the other nucleophilic functional groups. A free carboxylic acid group does not react readily with amines to give amides. Usually the acid chloride or acid anhydride is used so that a better leaving group than a hydroxyl group is present when an acid is converted into an amide (p. 582).

Thus, there are two major aspects to peptide synthesis. Functional groups must be selectively protected so that the reaction can be directed to the desired part of the molecule. The carboxylic acid group of an amino acid must be made reactive enough to form an amide bond under conditions that do not destroy peptide bonds and other functional groups that may be present in the molecule.

The problem that faces peptide chemists is complex; this section will discuss only reactions for the protection of the α-amino group in amino acids. The art of protecting different types of functional groups in a protein synthesis is the subject of lively research, and many different and highly selective protecting groups are being developed. In principle, the arguments concerning the α-amino group extend to the problems of protecting all functional groups.

The most important factor to consider when choosing a protecting group for the α-amino group is the ease with which the protecting group can be removed later in the synthesis. Amines have been protected as their amides, for example, when electrophilic aromatic substitution reactions were to be performed on aromatic amines (p. 948). An ordinary amide would not be a suitable protecting group in a peptide synthesis because there would be no way to remove the protecting group except by hydrolysis, which would also cleave peptide bonds.

A protecting group that can be cleaved under anhydrous acidic conditions that do not affect peptide bonds is a benzyl or *tert*-butyl carbamate (p. 921). The carbamate is prepared by reaction of the amino acid with a half ester of carbonic acid, the other half of which has a good leaving group on it. Benzyl chloroformate, for example, gives benzyl carbamates with amino acids.

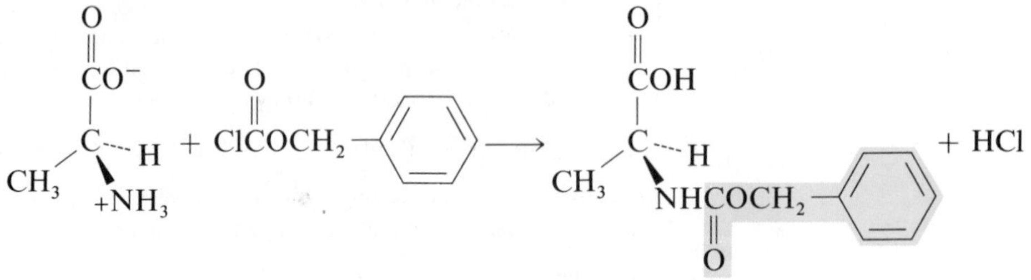

alanine — benzyl chloroformate — *N*-benzyloxycarbonylalanine, *N*-carbobenzoxyalanine, *N*-Cbz-Ala

The *N*-benzyloxycarbonyl protecting group is abbreviated as *N*-Cbz.

tert-Butyl chloroformate is not very stable, so *tert*-butyl azidoformate, in which the azide ion is the leaving group for the substitution reaction, is more commonly used as a reagent for protecting amino groups.

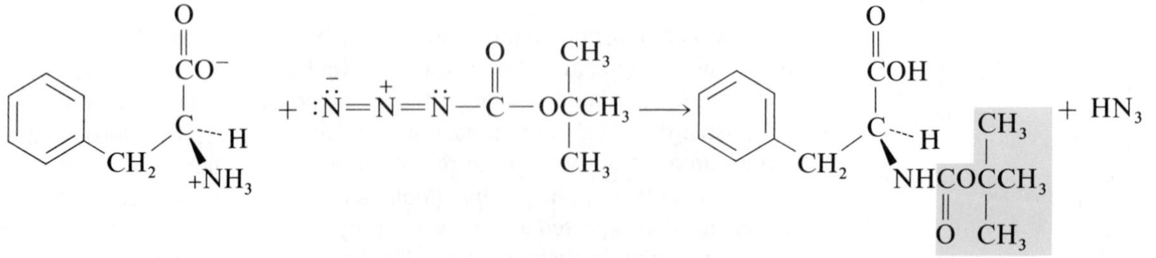

phenylalanine — *tert*-butyl azidoformate — *N*-*tert*-butoxycarbonylphenylalanine, Boc-Phe

The *tert*-butoxycarbonyl protecting group is abbreviated as Boc.

The benzyl and *tert*-butyl groups make good protecting groups because they both yield stable carbocations or radicals. Under acidic conditions, in the absence of water, they are both cleaved to give carbamic acids, which lose carbon dioxide and regenerate the amino group. The *tert*-butoxycarbonyl group comes off when treated with an acid such as dry trifluoroacetic acid in dichloromethane or dry hydrogen chloride in ether.

VISUALIZING THE REACTION

Removal of the *tert*-butoxycarbonyl protecting group

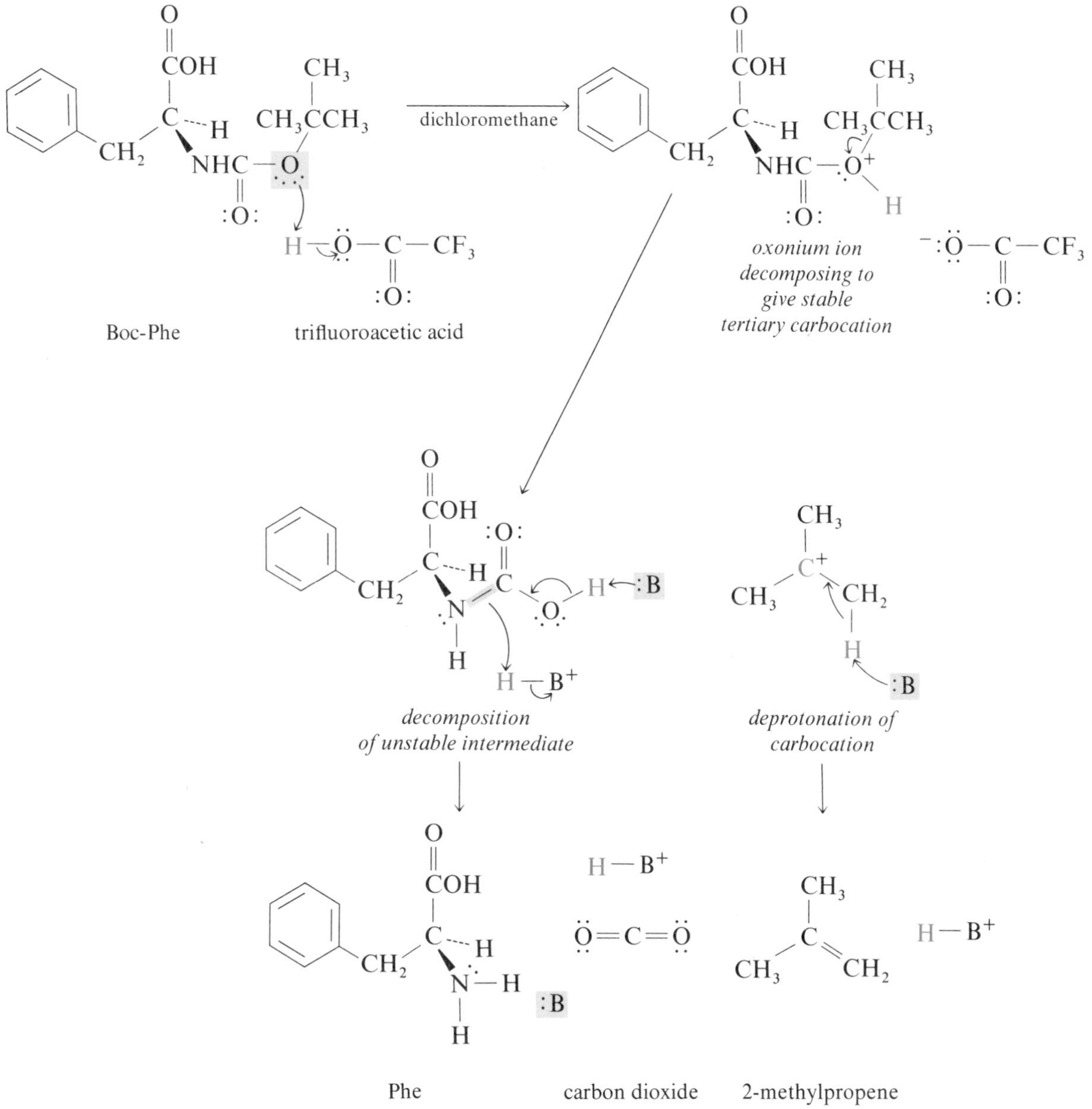

dichloromethane
Boc-Phe
trifluoroacetic acid
oxonium ion decomposing to give stable tertiary carbocation
decomposition of unstable intermediate
deprotonation of carbocation
Phe
carbon dioxide
2-methylpropene

The products, other than the amino acid, are the gases carbon dioxide and 2-methylpropene. The alkene comes from the *tert*-butyl cation.

The benzyloxycarbonyl group is removed by a stronger acid, such as dry hydrogen bromide dissolved in trifluoroacetic or acetic acid.

VISUALIZING THE REACTION

Removal of the benzyloxycarbonyl protecting group

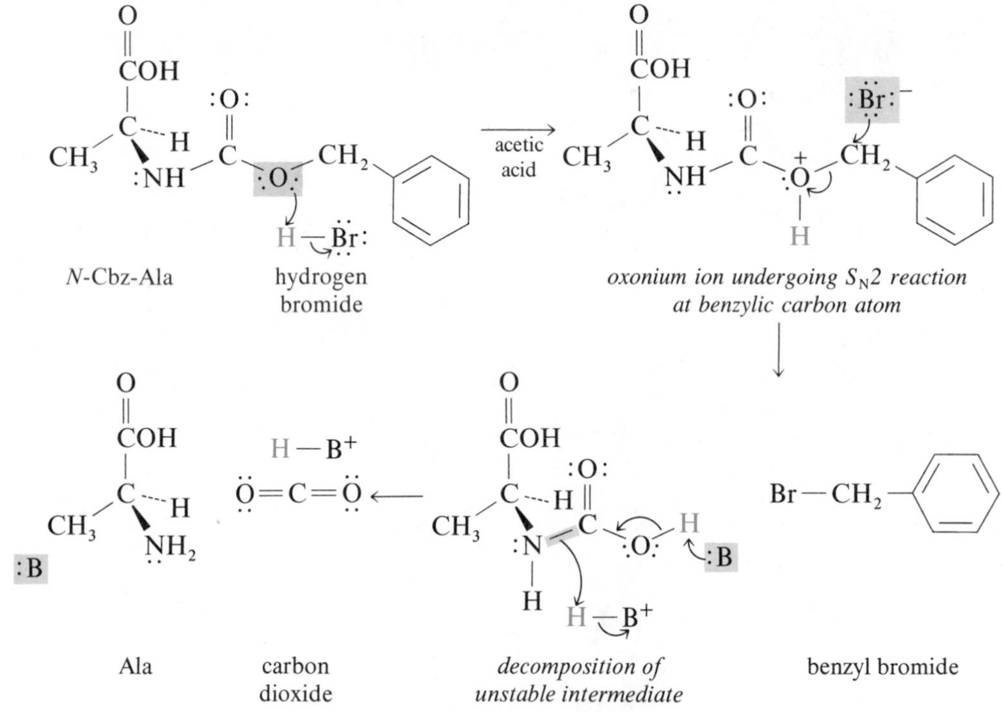
acetic acid
N-Cbz-Ala
hydrogen bromide
oxonium ion undergoing S_N2 reaction at benzylic carbon atom
Ala
carbon dioxide
decomposition of unstable intermediate
benzyl bromide

Alternatively, the bond between oxygen and the benzyl group is cleaved by hydrogenation.

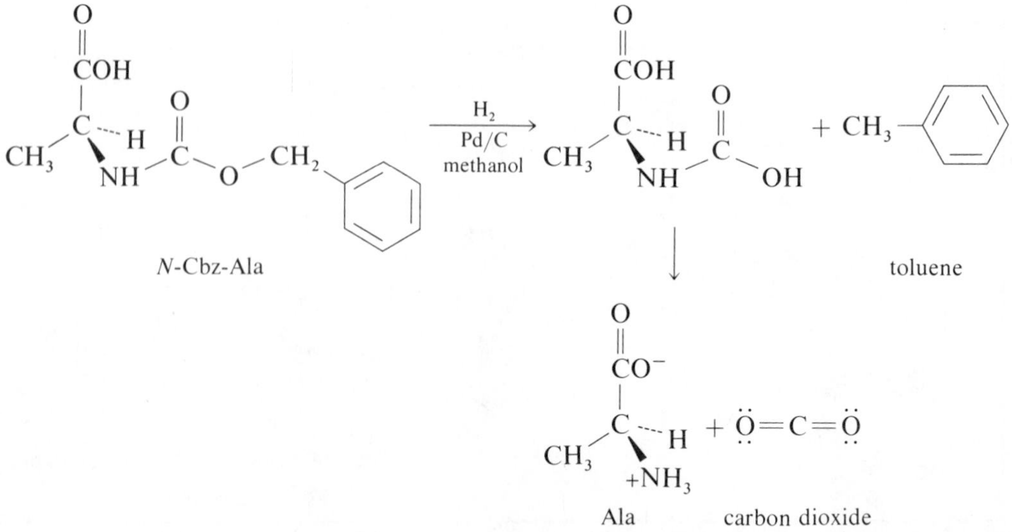
H_2
Pd/C
methanol
N-Cbz-Ala
toluene
Ala
carbon dioxide

The peptide bonds are stable under either of these sets of conditions. As carbamates, the amino groups are no longer basic and nucleophilic, and thus they are protected from reaction until the protecting group is removed at the appropriate time in the synthesis.

Study Guide Concept Map 26.4

The carboxyl group of the amino acid is usually protected as a relatively unreactive ester, such as the methyl or ethyl ester.

B. Activation of the Carboxyl Group and Formation of Peptide Bonds

For a peptide bond to be formed, it is necessary that the hydroxyl group present in the carboxylic acid function be converted into a good leaving group. This is done in one of two ways. The first was introduced on p. 922. In the presence of dicyclohexylcarbodiimide, an acid reacts with an amine to give an amide.

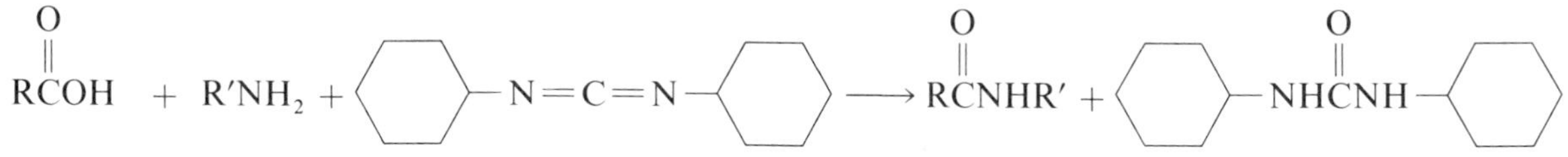

RCOH
R′NH2
N=C=N
RCNHR′
NHCNH

a carboxylic acid — an amine — dicyclohexylcarbodiimide (DCC) — an amide — dicyclohexylurea

Dicyclohexylcarbodiimide (often abbreviated as DCC) converts the carboxylic acid into an intermediate that has the properties of an acid anhydride.

VISUALIZING THE REACTION

Activation of a carboxylic acid by dicyclohexylcarbodiimide

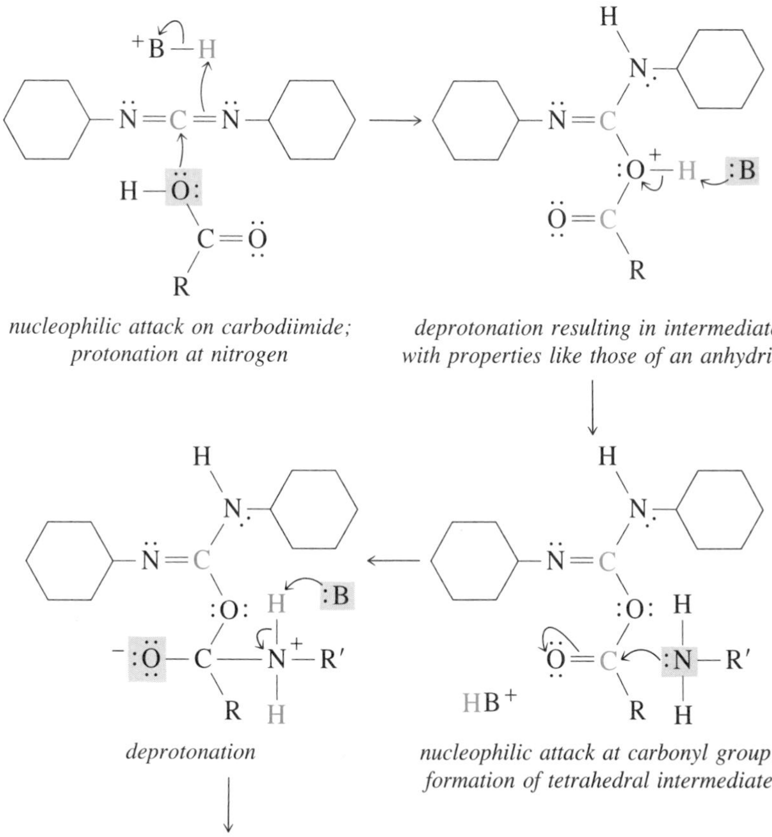

nucleophilic attack on carbodiimide; protonation at nitrogen

deprotonation resulting in intermediate with properties like those of an anhydride

nucleophilic attack at carbonyl group; formation of tetrahedral intermediate

deprotonation

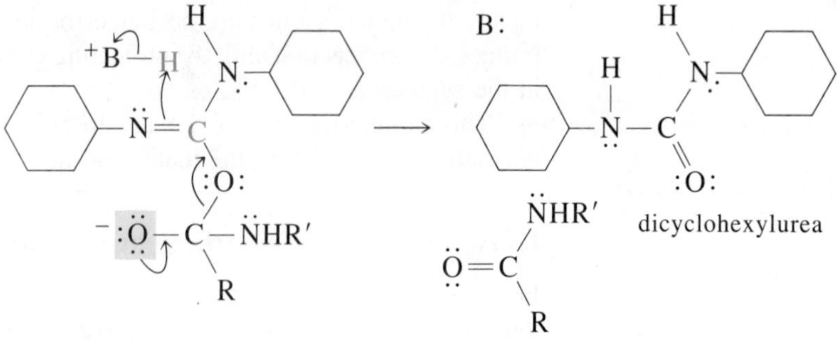
amide
dicyclohexylurea

regeneration of carbonyl group; protonation of nitrogen

The hydroxyl group of the carboxylic acid is converted into a good leaving group, so reaction with an amine to give an amide takes place.

The hydroxyl group can also be converted into a good leaving group by the preparation of an active ester of the amino acid. The *p*-nitrophenyl ester is often used. For example, the *p*-nitrophenyl ester of glycine, protected at the amino group by the benzyloxycarbonyl group, is prepared. The activated carboxyl group then reacts with the amino acid valine.

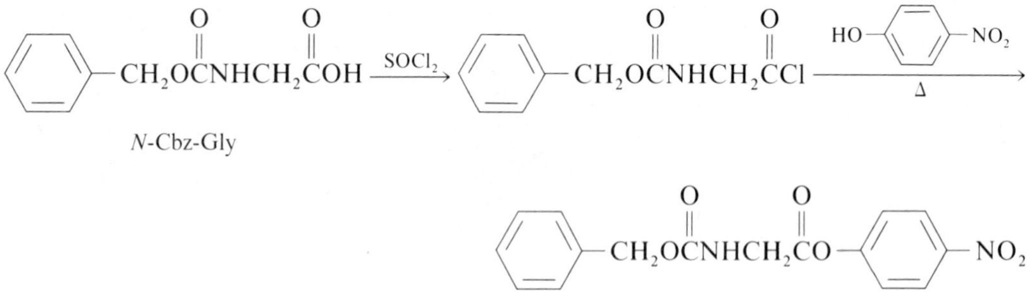
N-Cbz-Gly

p-nitrophenyl ester of *N*-benzyloxycarbonylglycine

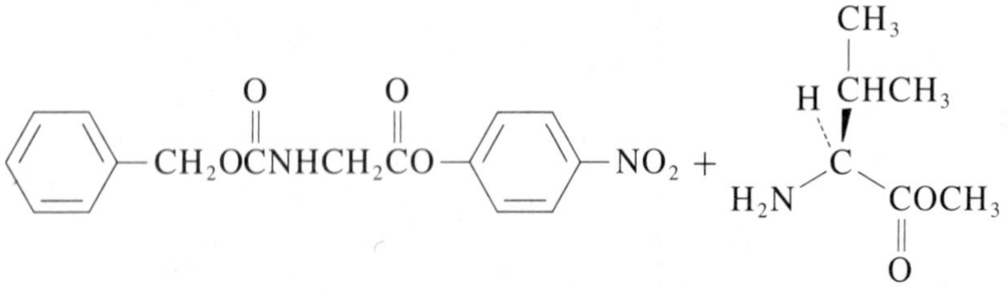

valine

protected at its carboxyl group as the methyl ester

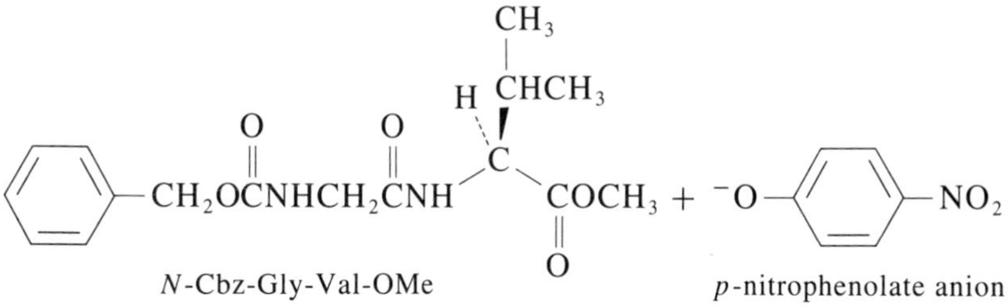

N-Cbz-Gly-Val-OMe
75%

p-nitrophenolate anion
good leaving group, stabilized by resonance and electron-withdrawing effect of nitro group

The conditions for the formation of the *p*-nitrophenyl ester are rather harsh and might not be suitable for activating the carboxyl group of a larger peptide. Other types of esters, formed in the presence of dicyclohexylcarbodiimide, are used in many peptide syntheses. Esters of *N*-hydroxysuccinimide are often used. Resonance stabilization of the *N*-oxysuccinimide anion makes it a good leaving group.

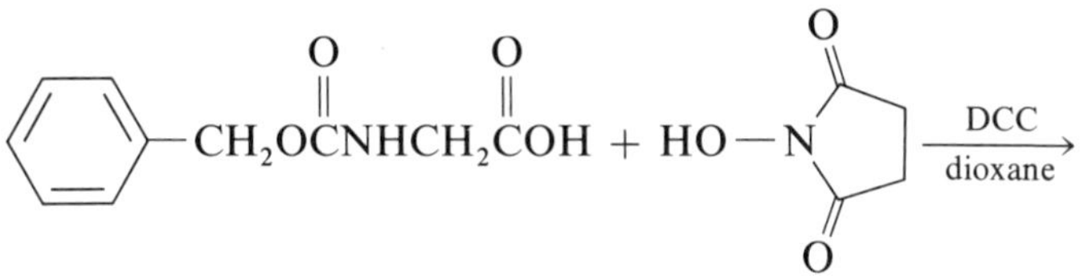
DCC
dioxane

N-Cbz-Gly

N-hydroxysuccinimide

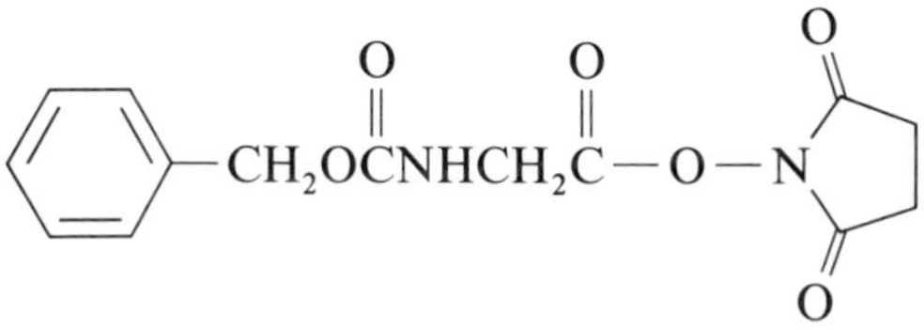

N-Cbz-Gly-*N*-Su

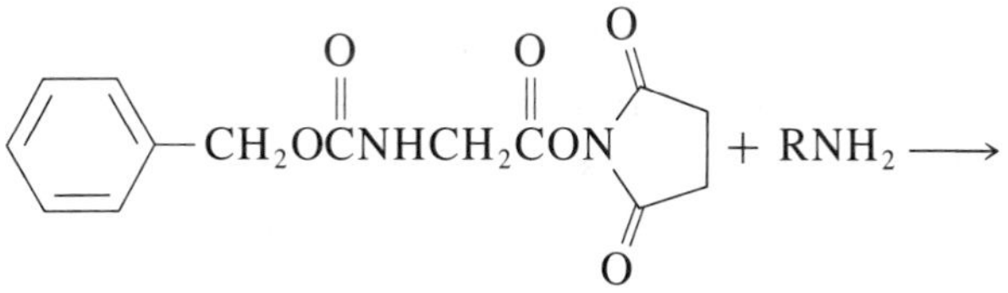

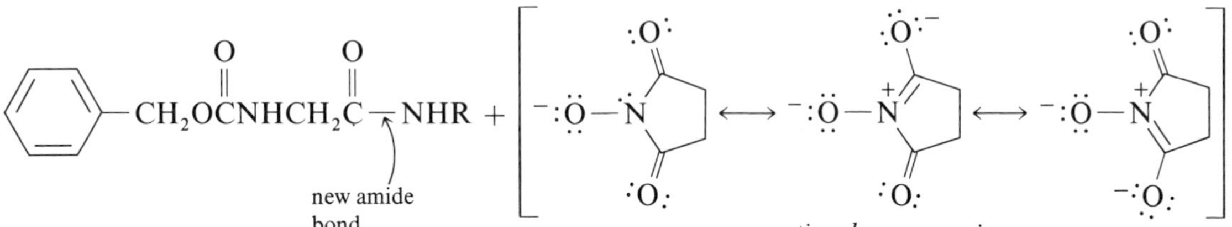
new amide bond

negative charge on anion stabilized by the electron-withdrawing effect of the adjacent nitrogen atom

Study Guide
Concept Map 26.5

The N-oxysuccinimide leaving group is abbreviated N-Su.

PROBLEM 26.14

Write structural formulas for the products and intermediates represented by letters in the following equations.

(a)
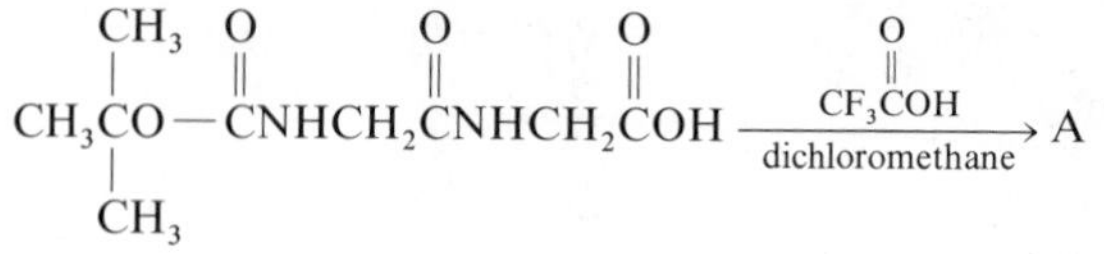

(b)
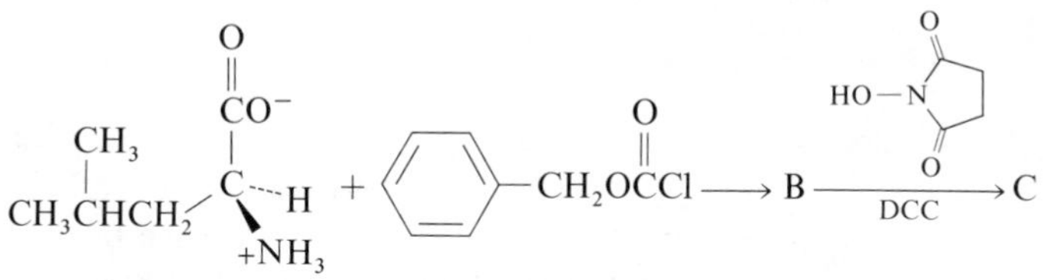

(c)
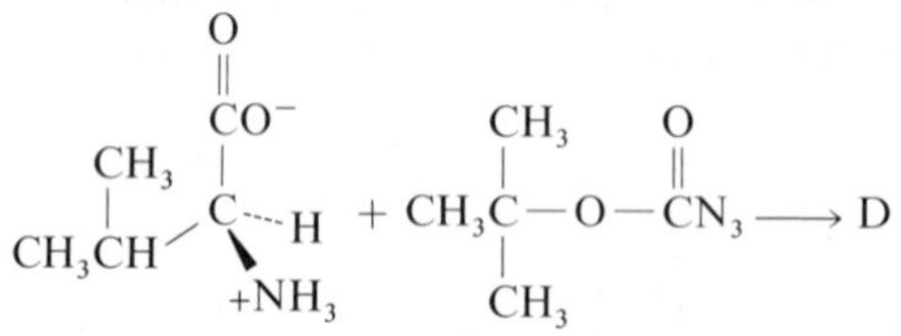

(d)
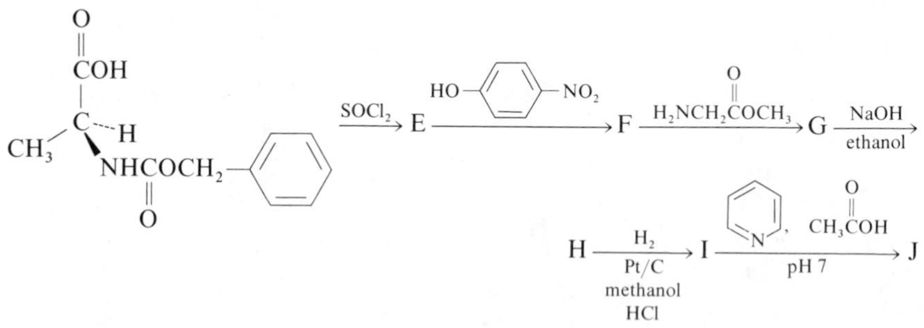

C. Synthesis of a Peptide

Methionine enkephalin is a pentapeptide found in the human brain. It is believed to function as a natural painkiller by interacting with the same sites in the brain with which potent opiates such as morphine (p. 1037) react. Its synthesis illustrates concretely some of the general points that have been made in preceding sections concerning the protection of functional groups and the activation of carboxyl groups. Methionine enkephalin has the structure Tyr-Gly-Gly-Phe-Met. It was synthesized in two fragments that were then joined. The synthesis of the smaller fragment Phe-Met is shown below.

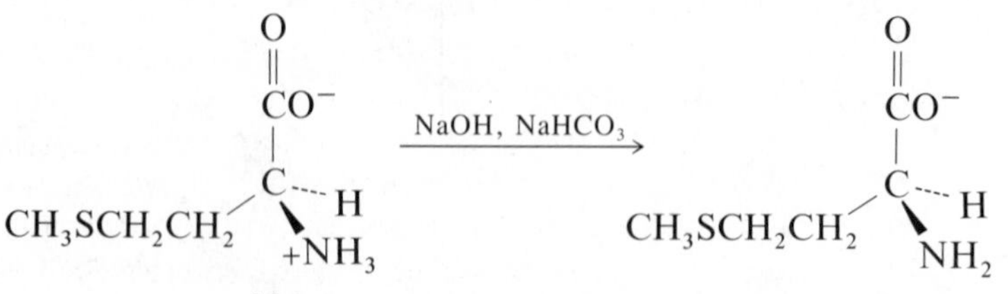

Met

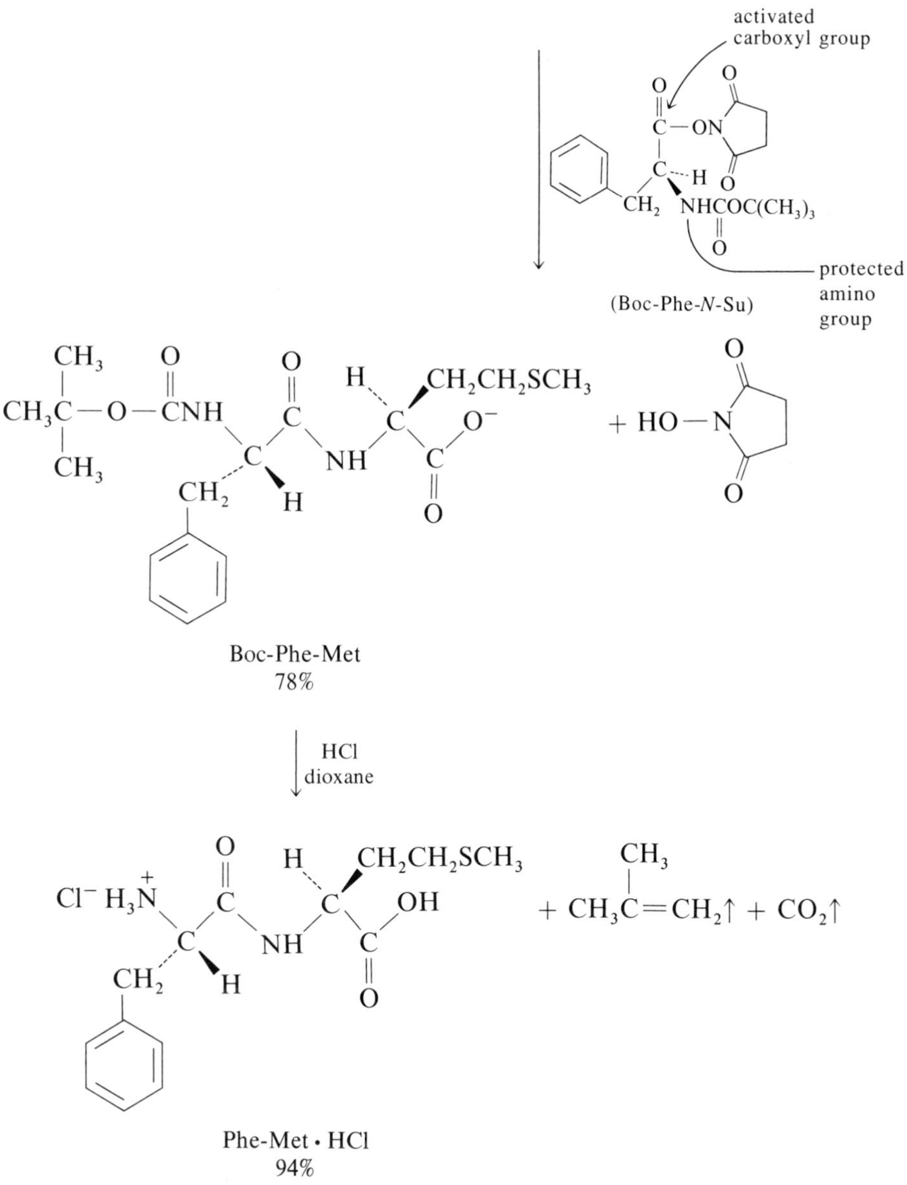

activated
carboxyl group
protected
amino
group
(Boc-Phe-N-Su)
Boc-Phe-Met
78%
HCl
dioxane
Phe-Met • HCl
94%

Phenylalanine, with its amino group protected by the *tert*-butoxycarbonyl group and its carboxyl group activated as the *N*-hydroxysuccinimide ester, reacts with methionine to give the protected dipeptide *N*-*tert*-butoxycarbonylphenylalanylmethionine, or Boc-Phe-Met. The protecting group is removed with dry hydrogen chloride in dioxane, and the amino group of the dipeptide is protected as its hydrochloride salt until it is used in the next step of the synthesis.

Glycylglycine is available from chemical supply houses. Tyrosine, protected as the *tert*-butyl ether at the phenolic oxygen and with a *tert*-butoxycarbonyl group on the amino group, reacts as its *N*-hydroxysuccinimide ester with glycylglycine to give a protected tripeptide. The carboxyl group of the tripeptide is activated by converting it to the *N*-hydroxysuccinimide ester with dicyclohexylcarbodiimide. The tripeptide and the dipeptide fragments are combined in the presence of base. Removal of the protecting groups followed by adjustment of the pH gives the pentapeptide methionine enkephalin.

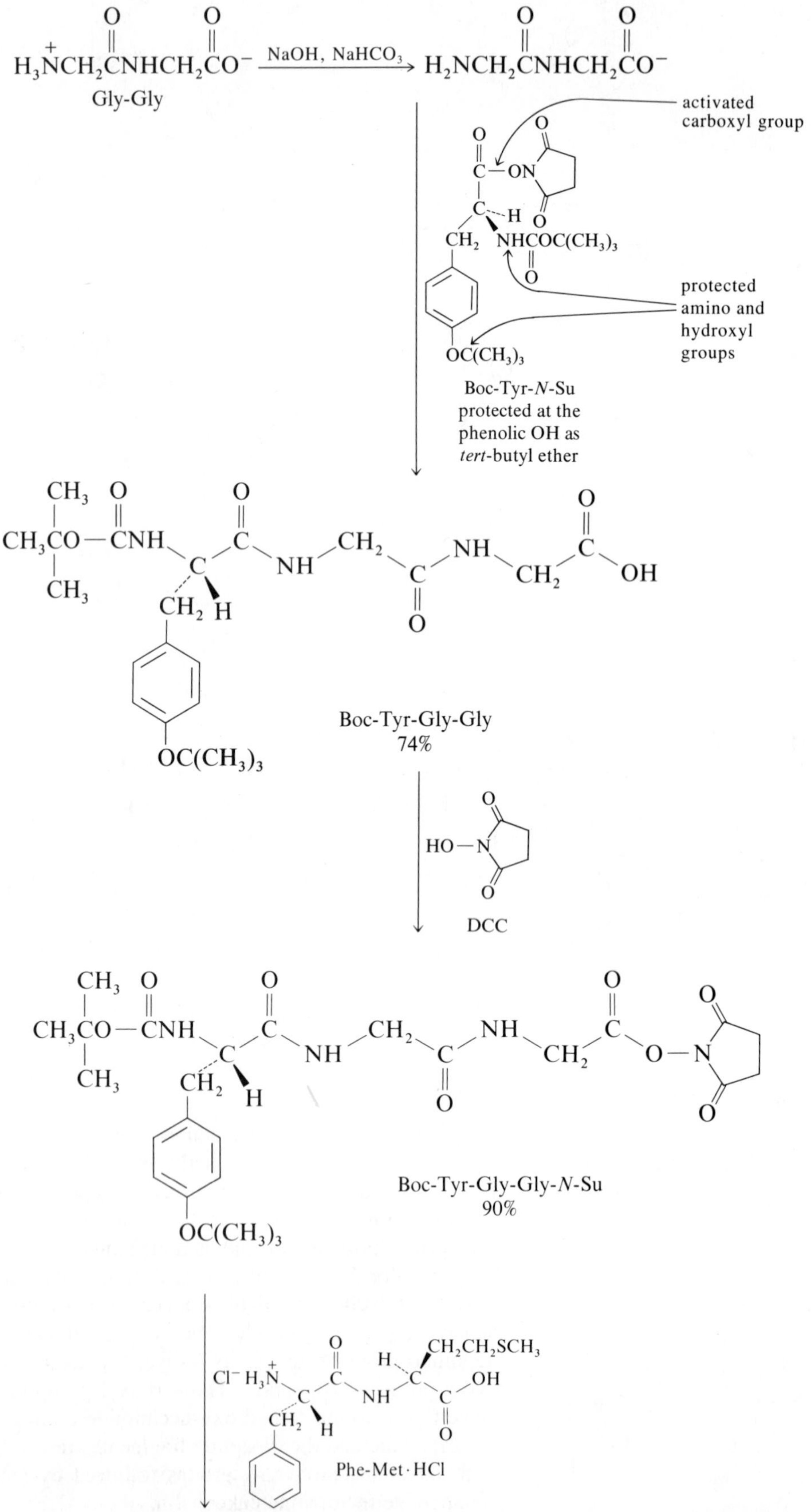
Gly-Gly
NaOH, NaHCO3
activated carboxyl group
protected amino and hydroxyl groups
Boc-Tyr-N-Su protected at the phenolic OH as tert-butyl ether
Boc-Tyr-Gly-Gly
74%
DCC
Boc-Tyr-Gly-Gly-N-Su
90%
Phe-Met · HCl

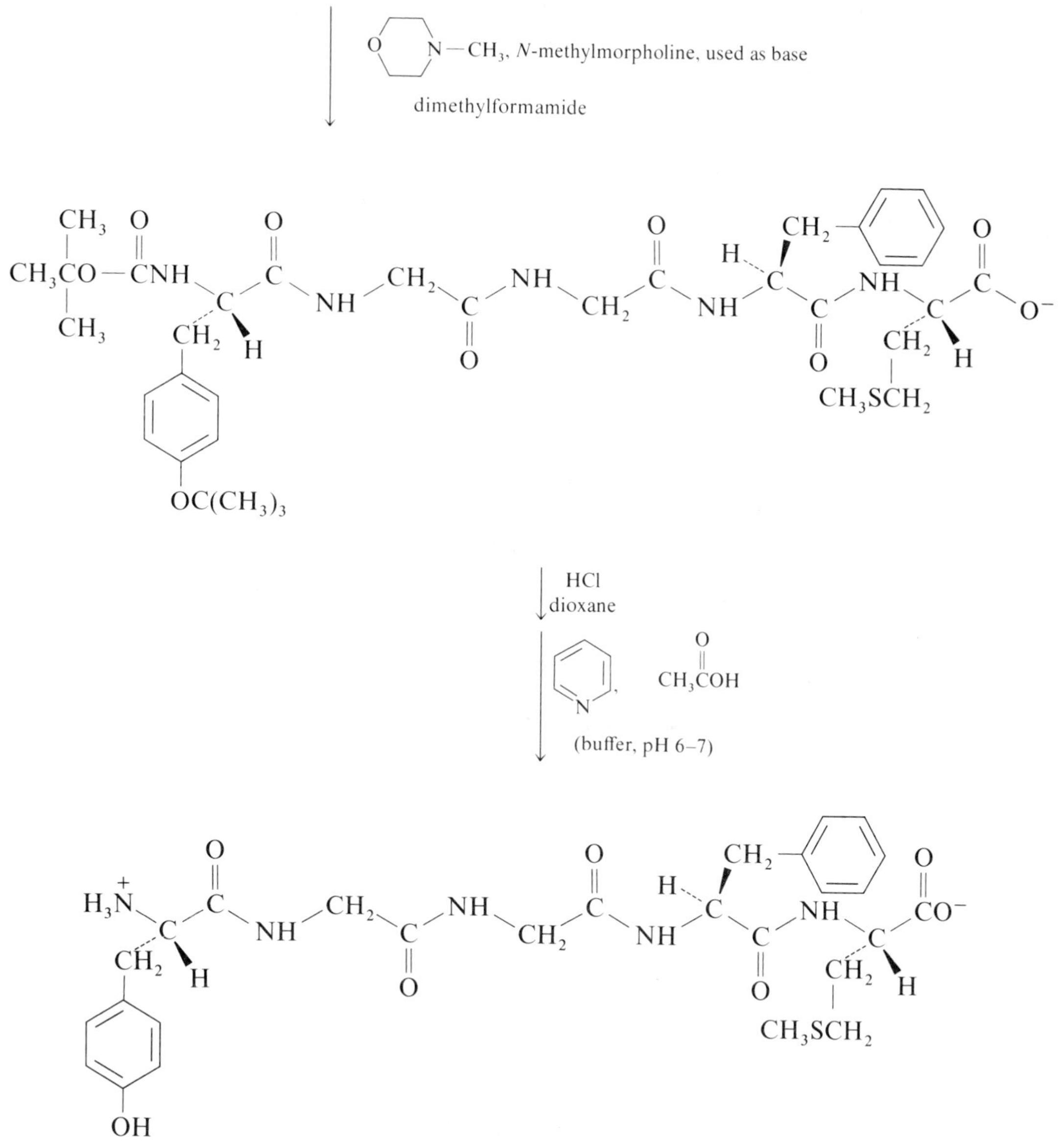
N—CH3, N-methylmorpholine, used as base
dimethylformamide
HCl
dioxane
(buffer, pH 6–7)

Tyr-Gly-Gly-Phe-Met
methionine enkephalin
73%

Even though methionine enkephalin is not a large peptide, its synthesis is subject to the limitations that have an increasing effect as the peptide to be synthesized becomes larger and the number of times that intermediates have to be isolated and purified increases. Each manipulation represents losses of material and time. For example, the overall yield of the enkephalin from the five steps shown above is

$$0.78 \times 0.94 \times 0.74 \times 0.90 \times 0.73 \times 100\% = 36\%$$

This yield does not reflect the additional losses incurred when making the protected and activated amino acids used in the synthesis.

Some fairly large peptides were synthesized by methods similar to those shown for the enkephalin, but real progress in the synthesis of larger peptides and proteins began in the early 1960s with the development of the technique of solid-phase synthesis, which is described in the next section.

PROBLEM 26.15

Why was the phenolic hydroxyl group of tyrosine protected in the synthesis of methionine enkephalin? Write an equation predicting what would happen if the hydroxyl group were not protected.

PROBLEM 26.16

Along with methionine enkephalin, another pentapeptide with similar physiological properties, leucine enkephalin, is found in the brain. It has the same structure as methionine enkephalin, except that leucine is substituted for methionine as the *C*-terminal amino acid. Leucine enkephalin has also been synthesized by a method similar to that used for methionine enkephalin. Propose a detailed synthesis for leucine enkephalin.

D. Solid-Phase Peptide Synthesis

Solid-phase peptide synthesis was introduced in 1963 by Robert B. Merrifield of Rockefeller University. This technique was designed to simplify the isolation and purification of peptides when many steps are necessary for a synthesis. Merrifield reasoned that if the growing peptide could be incorporated into a solid polymer, it could be handled as a solid and purified thoroughly at each stage of the synthesis by repeated washing and filtration steps, minimizing losses of peptides. The rates of the reactions for forming peptide bonds and the degree of completion of the formation of the peptide could be increased greatly by using large excesses of protected amino acids, which could then be washed away from the extended peptide chain on its solid support. Thus, it would be possible to add one amino acid at a time in precise order to build up large molecules of high purity in good yield. Ideally, the whole process would be automated so that the multiple cycles necessary to form a peptide bond, purify the extended protected peptide, remove the protecting group, and add a different amino acid to the newly exposed *N*-terminal amino acid could be repeated over and over again with minimal human attention.

It was necessary to establish a solid support system that could be covalently bonded to the *C*-terminal amino acid of the peptide by a linkage that would be stable to the reagents used in making peptide bonds and removing protecting groups. In addition, the support had to be insoluble in the solvents used in the reactions. It had to have structural stability and maintain its particle size so that it would survive the physical manipulations of repeated washings and filtrations. It had to be chemically inert except at the site where the original attachment of the peptide chain was made. Its structure had to allow the solutions of reagents to flow through it so that reactions could take place on a large portion of its surface area.

Much research focused on developing a polymeric material that would meet all these requirements. The material most widely used is a polystyrene that is cross-linked with divinylbenzene (p. 1181). The polystyrene is formed into beads approximately 5×10^{-3} cm in diameter. The cross-linking gives the polymer a high molecular weight and very low solubility in organic solvents. The amount of cross-linking is adjusted so that the polymer has some flexibility and the beads swell to double their dry size when they are placed in an organic solvent for the reactions. This allows reactions to take place inside the beads as well as on the surface. Even though the beads are so small that they can hardly be seen by the naked eye, each one has room for the growth of a trillion (10^{12}) peptide chains.

The polystyrene resin is converted in a Friedel-Crafts reaction with chloromethyl methyl ether into a polymer with chloromethyl groups on approximately 22% of the aromatic rings.

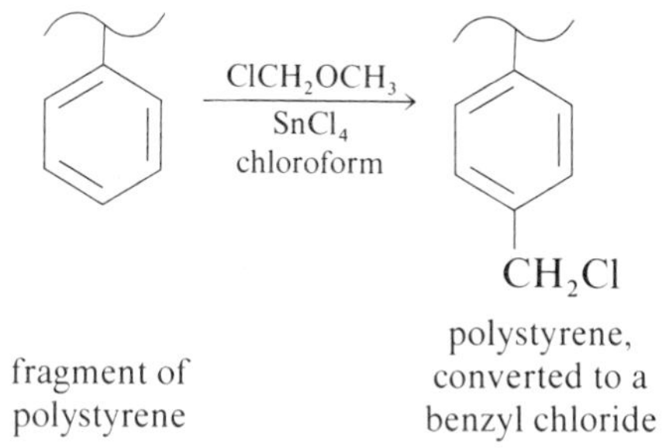

$ClCH_2OCH_3$
$SnCl_4$
chloroform
CH_2Cl
fragment of polystyrene
polystyrene, converted to a benzyl chloride

PROBLEM 26.17

Propose a mechanism for the reaction shown above.

The *C*-terminal amino acid of the peptide is attached to the polystyrene as a benzyl ester. The amino groups of the amino acids to be added to the chain are protected by *tert*-butoxycarbonyl groups, which can be removed under conditions that do not affect the benzylic ester bond (p. 1137). Dicyclohexylcarbodiimide (DCC) is used to activate the carboxylic acid group of the amino acid to be added to the chain (p. 1139). The different steps in a solid-phase synthesis of a pentapeptide, Gly-Val-Gly-Ala-Pro, are shown below and on the following page (abbreviations are used for the amino acids and protecting groups).

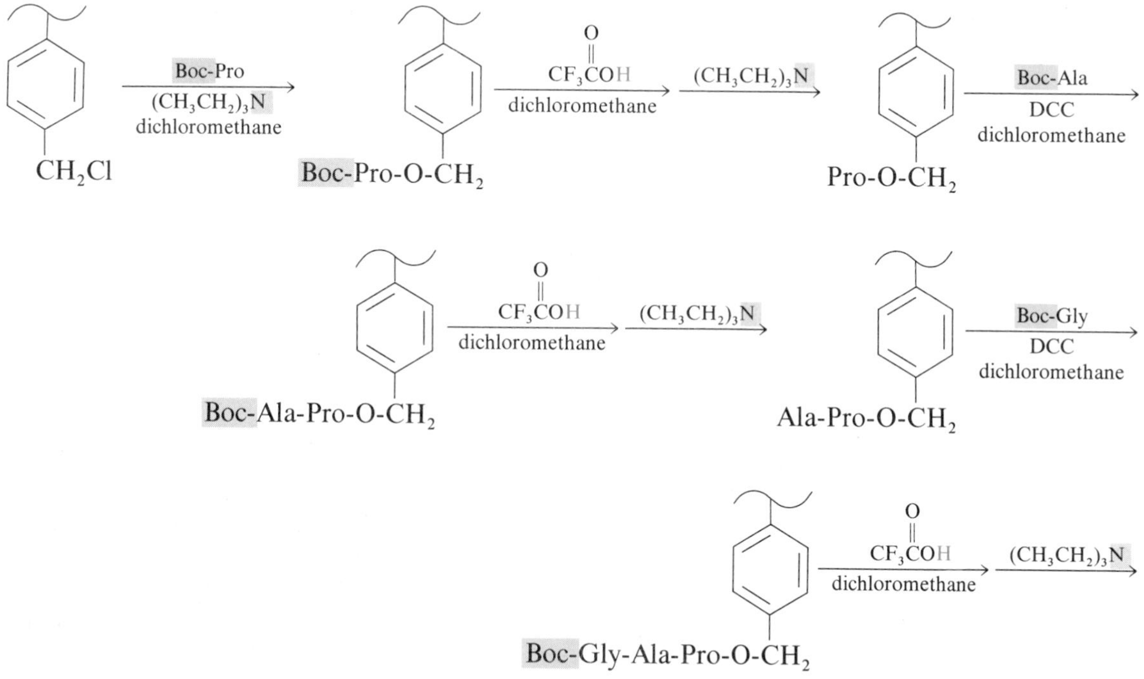

CH_2Cl
Boc-Pro
$(CH_3CH_2)_3N$
dichloromethane
Boc-Pro-O-CH_2
CF_3COH
dichloromethane
$(CH_3CH_2)_3N$
Pro-O-CH_2
Boc-Ala
DCC
dichloromethane
Boc-Ala-Pro-O-CH_2
CF_3COH
dichloromethane
$(CH_3CH_2)_3N$
Ala-Pro-O-CH_2
Boc-Gly
DCC
dichloromethane
Boc-Gly-Ala-Pro-O-CH_2
CF_3COH
dichloromethane
$(CH_3CH_2)_3N$

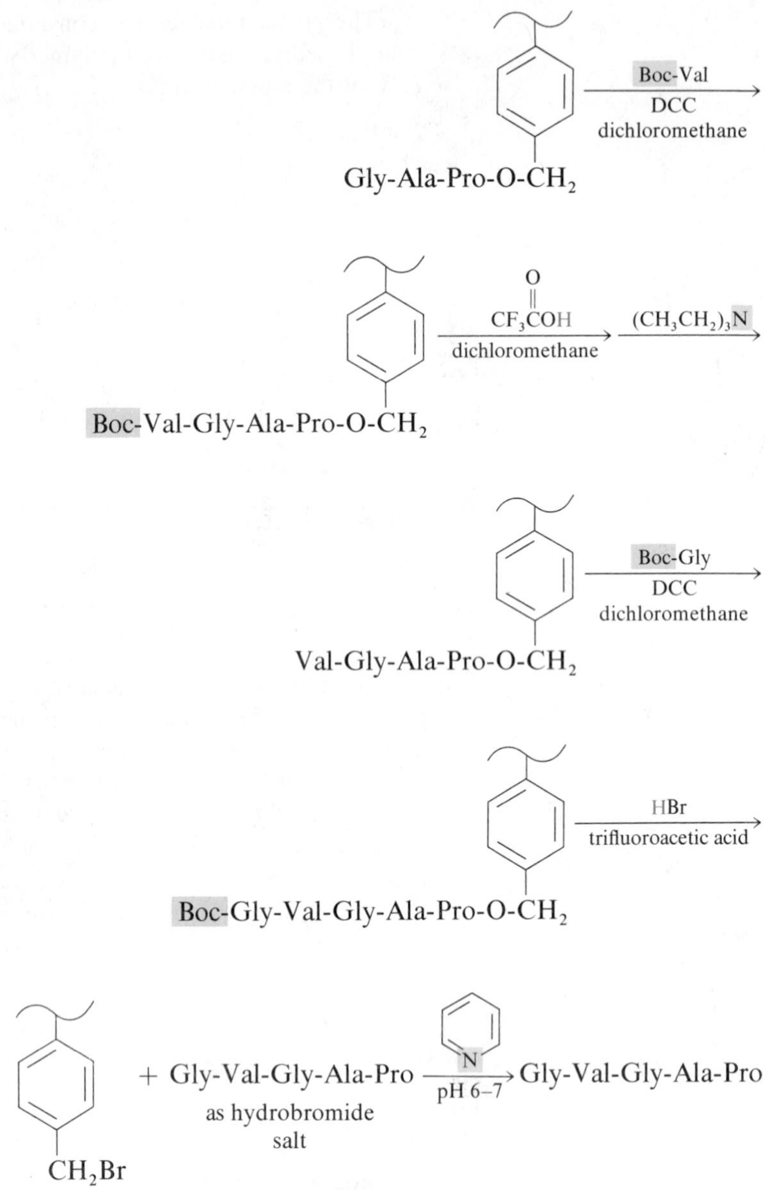
Gly-Ala-Pro-O-CH2
Boc-Val
DCC
dichloromethane
Boc-Val-Gly-Ala-Pro-O-CH2
O
CF3COH
dichloromethane
(CH3CH2)3N
Val-Gly-Ala-Pro-O-CH2
Boc-Gly
DCC
dichloromethane
Boc-Gly-Val-Gly-Ala-Pro-O-CH2
HBr
trifluoroacetic acid
CH2Br
+ Gly-Val-Gly-Ala-Pro
as hydrobromide
salt
N
pH 6–7
Gly-Val-Gly-Ala-Pro

The reagents for the peptide synthesis were all chosen so that the reaction sequences are repetitious and, therefore, can be automated. The *tert*-butoxycarbonyl group is removed with trifluoroacetic acid, which leaves behind a protonated amino group. Triethylamine removes the proton to give the free amino group, which reacts with the activated carboxylic acid group from another protected amino acid. Finally, the strong acid hydrogen bromide in trifluoroacetic acid cleaves both the protecting group and the benzylic ester bond. Note that aqueous acids, which would hydrolyze peptide bonds, are avoided at all stages. In addition to the reagents that are shown, there are a number of intermediate stages in which the polymer, with the attached peptide chains, is washed with solvents and filtered to purify it. Altogether, eleven separate steps are necessary in order to add one amino acid to the chain.

PROBLEM 26.18

Write equations for a solid-phase synthesis of Phe-Leu-Gly. Give the full structures of the benzylic ester, any protecting groups, and any reagents used to promote the formation of the peptide bond at least once. Show the full structure of the peptide at each stage of the synthesis.

Solid-phase peptide synthesis has been used to prepare a number of naturally occurring proteins, including insulin and the enzyme ribonuclease, which was synthesized in 1969 by Merrifield. Ribonuclease has 124 amino acid residues. Its synthesis required 369 chemical reactions and 11,931 steps in the automated process. The introduction of one amino acid took approximately four hours. This is a much more rapid and efficient process than was possible with classic solution-phase chemistry, but it does not begin to approach the efficiency with which proteins are synthesized in the living cell. A bacterial cell completes protein synthesis in seconds and keeps about 3000 different protein syntheses going simultaneously with no confusion of reagents and products.

26.4 CONFORMATION OF PEPTIDES AND THE STRUCTURES OF PROTEINS

Peptides consisting of a number of amino acids assume characteristic conformations. The carbonyl group and the nitrogen and hydrogen atoms around the peptide bond, as well as the two carbon atoms to which the carbonyl and amino groups are bonded, lie in a plane. Resonance interaction between the nitrogen atom and the carbonyl group gives a partial double-bond character to the carbon-nitrogen bond and prevents free rotation.

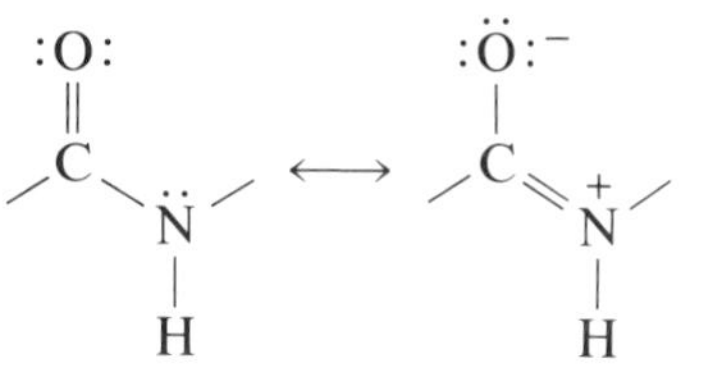

The oxygen atom of the carbonyl group and the hydrogen atom bonded to the nitrogen atom are trans to each other. The bond lengths and bond angles typical of the peptide group are illustrated in Figure 26.5.

The order of amino acids in the backbone of a polypeptide is known as the **primary structure** of the peptide. The conformation of the peptide chain gives it a **secondary structure.** Further bending and folding of peptide chains creates the **tertiary structure** of proteins. Finally, several polypeptide units may be associated with each other and with other simpler molecules such as sugars, inorganic residues, or coenzymes in what is known as the **quaternary structure** of proteins.

All the amino acids found in naturally occurring peptides and proteins have the same configuration, and this imparts a stereochemical regularity to peptide chains. They assume different conformations depending mainly on the types of amino acids that make up the chain. One common conformation is the **α-helix.** This structure was proposed in 1951 on the basis of x-ray diffraction experiments by L. Pauling and R. Corey. The peptide chain forms a right-handed coil having 3.7 amino acids

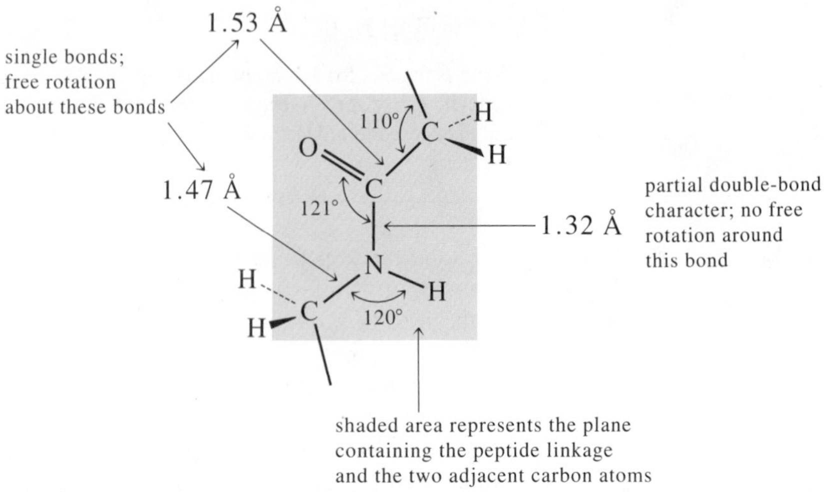
single bonds; free rotation about these bonds
1.53 Å
1.47 Å
110°
121°
120°
partial double-bond character; no free rotation around this bond
1.32 Å
shaded area represents the plane containing the peptide linkage and the two adjacent carbon atoms

FIGURE 26.5 Typical bond lengths and bond angles for the peptide group.

per full turn. The structure is given rigidity and stability by hydrogen bonding between the hydrogen atom bonded to a nitrogen atom and the carbonyl group of the amino acid four units down the chain at the next turn of the coil. The spacing of the amino acids is such that every peptide linkage is involved in hydrogen bonding (Figure 26.6).

The α-helix is particularly important in the structural proteins, such as the **α-keratins,** which make up the protein part of skin, nails, horns, hair, and feathers. Proteins that form structures such as hair and wool are rich in cystine, the oxidized form of cysteine (p. 1133). In them, several α-helixes are coiled around each other to give multiple strands that are held together by disulfide bridges. Giving a permanent wave, in fact, involves reduction of some of the disulfide linkages so that the hair loses its natural conformation, which is determined genetically. Then the hair is held in a new conformation on curlers and reoxidized to form new disulfide linkages.

Not all proteins have α-helical structures. The major constituent of silk is a protein called fibroin, which is a **β-keratin** or **pleated sheet.** The peptide chains are more extended in this conformation and lie side by side, with hydrogen bonding between the chains. In this structure, maximum hydrogen bonding is possible when adjacent polypeptide chains run in opposite directions (Figure 26.7). Such structures are most stable when the side chains of the amino acid residues are small and not charged. Silk fibroin consists mostly of glycine and alanine, for example.

Another important type of structural protein is the **collagens,** which make up connective tissues. When a collagen is boiled in water, gelatin is formed. Collagens are rich in glycine, alanine, and proline and contain 4-hydroxyproline, an amino acid that is seldom found in other proteins. The structure of collagens consists of a triple helix of polypeptide chains, each of which is a left-handed helix.

Keratins and collagens are known as **fibrous proteins.** The other large class of proteins is the **globular proteins,** in which the tertiary structure of the protein tends to be spherical or ellipsoidal, with much coiling of the polypeptide chain. Many proteins, including all enzymes, have this type of structure.

Parts of globular proteins have the α-helix conformation. Other parts are more randomly coiled. Not all side chains of amino acids fit well into the α-helix. Proline, in particular, causes a bend in the α-helix structure because the nitrogen

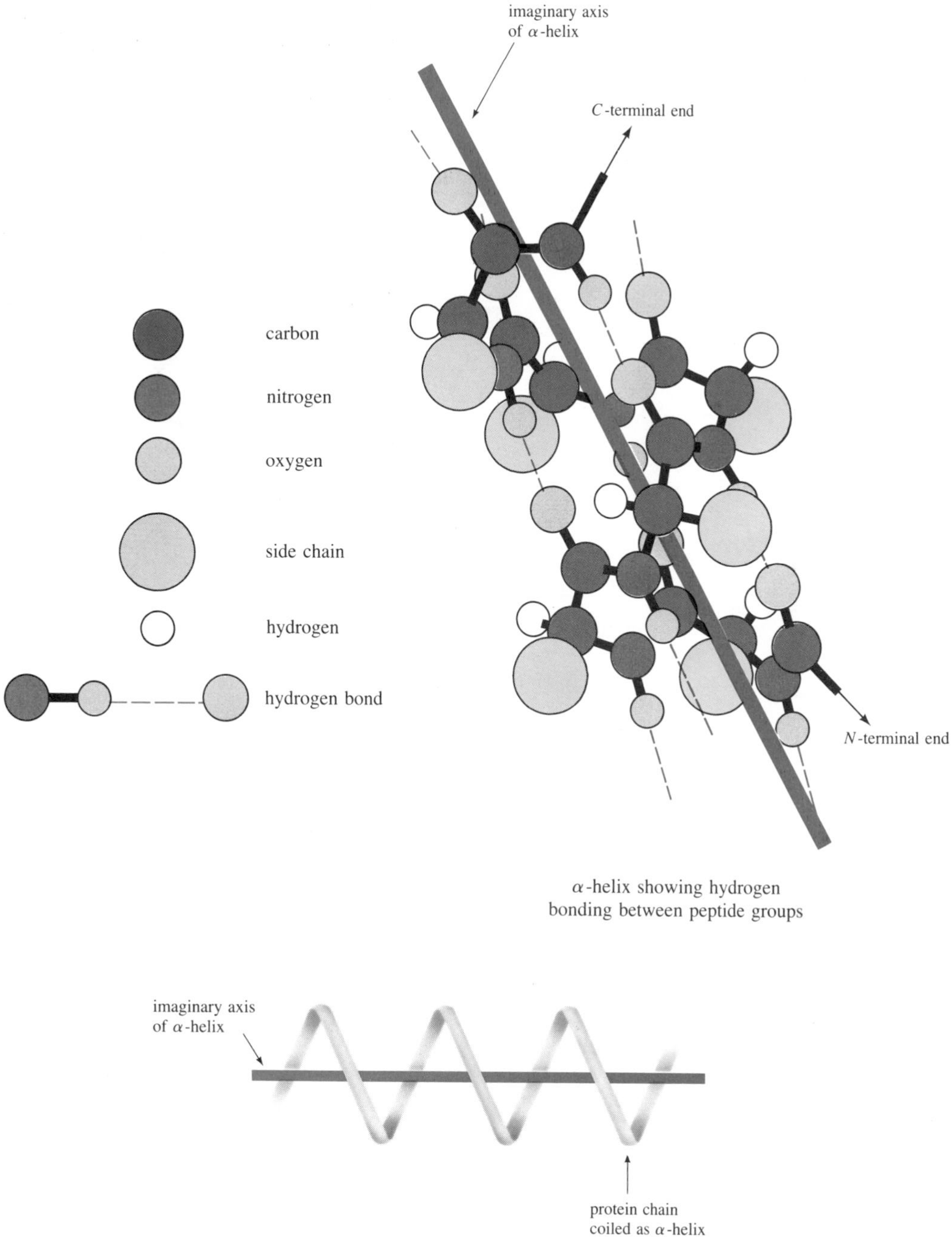
imaginary axis
of α-helix
C-terminal end
carbon
nitrogen
oxygen
side chain
hydrogen
hydrogen bond
N-terminal end
α-helix showing hydrogen
bonding between peptide groups
imaginary axis
of α-helix
protein chain
coiled as α-helix

FIGURE 26.6 The α-helix, showing hydrogen bonding between peptide groups and the coiled conformation of the peptide chain.

of the amino group is part of a five-membered ring. Furthermore, proline has no amide hydrogen atom to participate in hydrogen bonding. The side chains of lysine and arginine are positively charged at physiological pH and repel each other, disrupting the helical structure. Conversely, strong ionic interactions between carboxylate anions on the side chains of glutamic and aspartic acid residues and cationic centers on lysine and arginine residues may be more important than hydrogen bonding in determining conformation. Disulfide bridges between different parts

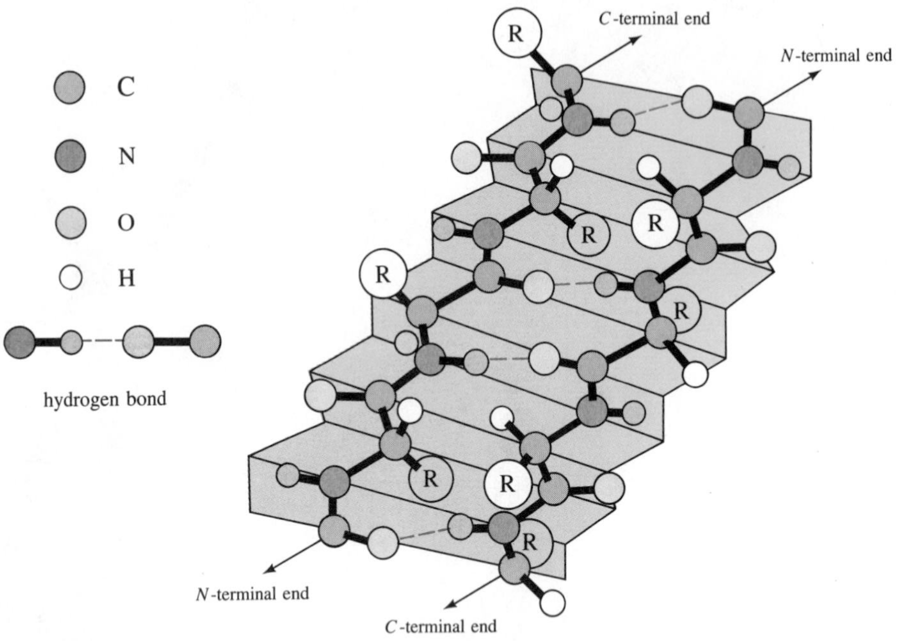
C
N
O
H
hydrogen bond
R
C-terminal end
N-terminal end
N-terminal end
C-terminal end

FIGURE 26.7 The pleated sheet conformation of some proteins.

of the peptide chain may also stabilize particular conformations of a protein. The whole molecule of a globular protein is arranged so that the hydrophobic side chains are protected on the inside, where they are stabilized by van der Waals interactions. The more polar and charged amino acid residues are on the surface of the protein in contact with the polar solvent water.

Thus, the overall structure of a protein in its natural state is determined by a multitude of interactions made possible by the specific order of amino acids in the chain. Among these are attractions between positively and negatively charged sites on the chain, repulsions between sites of similar charge, hydrogen bonding, van der Waals forces acting between hydrocarbon-like side chains on amino acids, and disulfide bridges. The next section looks closely at the structure of an enzyme, chymotrypsin, to see how its function is determined by these complex structural factors.

A protein retains its structure within a narrow range of pH, temperature, and ionic strength of the solution. Addition of acid, base, metal ions, or urea (which disrupts hydrogen bonding in proteins) to a solution of a protein **denatures** it, that is, changes its structure to a form in which biological activity is lost. An increase in temperature does the same thing—that is what happens when an egg is cooked. Sometimes a denatured protein will return to its original form and recover most of its biological activity if, for example, the optimum pH is restored. In other cases, the damage is irreversible. The particular structure that is native to a given protein is the favored one, and the protein will return to it if possible. The enzyme ribonuclease, for example, has four disulfide bridges between eight cysteine residues, widely spaced in the molecule. If the disulfide bridges are cleaved by reduction, the protein is easily denatured. Addition of urea, for example, causes it to lose all of its enzymatic activity. Removal of the urea allows the molecule to regain its activity, indicating that it folds back into the original tertiary structure and even reforms the same disulfide links by oxidation in air. A random reconnecting of eight

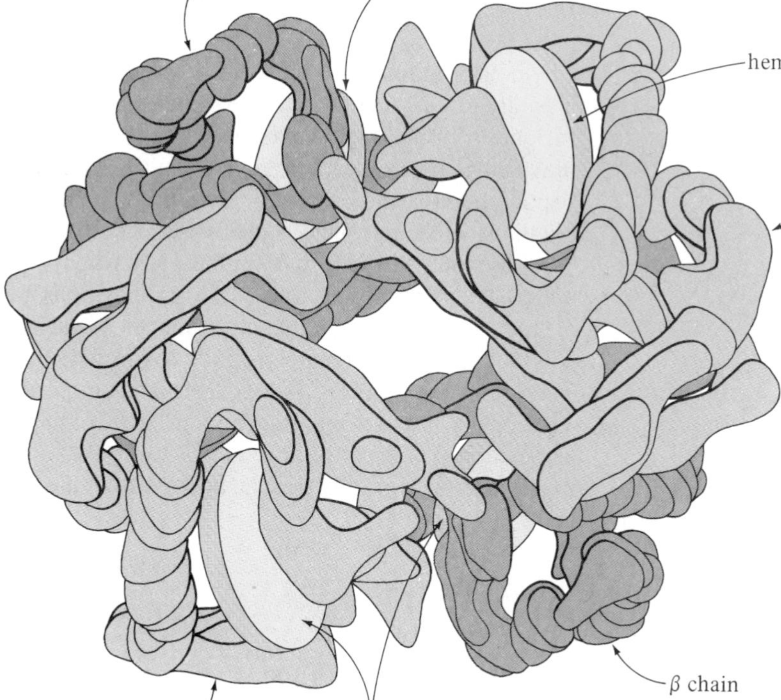

β chain
heme group
heme group
α chain
β chain
heme group
α chain

FIGURE 26.8 Schematic representation of the structure of hemoglobin, showing four polypeptide chains and four heme molecules. (Adapted from "The Hemoglobin Molecule," by M. F. Pertuz. Copyright © 1964 by Scientific American, Inc. All rights reserved.)

cysteine residues by four disulfide bridges could occur in more than 100 different ways, yet only one set of four disulfide bridges is actually formed.

Some proteins such as hemoglobin have quaternary structure. In hemoglobin, four polypeptide chains, each of which has tertiary structure, are associated with each other and with four molecules of heme. Heme is a heterocyclic ring (p. 1029) that coordinates with iron(II) ions to perform the oxygen-carrying function in the blood (Figure 26.8). In hemoglobin, the heme units are located in pockets of the protein that are lined with hydrophobic amino acid side chains. This placement protects iron(II) against being oxidized irreversibly by oxygen to iron(III), something that happens easily in ordinary solutions of iron(II) compounds. One ligand to the iron is a histidine residue, which also enables the iron to bind loosely and reversibly with oxygen. Carbon monoxide binds more strongly to hemoglobin than oxygen does. This can lead to carbon monoxide poisoning. Heavy smokers, who inhale large doses of carbon monoxide in tobacco smoke, have about 20% of their hemoglobin bound up in this nonfunctional form.

Study Guide
Concept Map 26.6

A protein such as hemoglobin, in which a relatively small molecule (heme) is associated with the protein, is called a **conjugated protein.** The small molecule associated with the protein is known as a **prosthetic group.** Many enzymes are proteins conjugated with coenzymes that are often compounds known to be vitamins.

Chymotrypsin, one of the digestive enzymes secreted by the pancreas, belongs to a family of enzymes that cleave proteins into smaller peptides. The group, which includes trypsin, is known as the serine proteases, because the side chain of serine plays an important part in their catalytic activity. Chymotrypsin hydrolyzes the peptide bond at the carboxylic acid group of amino acids that have large hydrophobic side chains, such as phenylalanine, tryptophan, and tyrosine (p. 1121).

Chymotrypsin is formed in the body when two dipeptide units, consisting of amino acids at positions 14 and 15 and positions 147 and 148, are removed from a precursor molecule, chymotrypsinogen, which has 245 amino acid residues. Chymotrypsin has three polypeptide chains held together by two disulfide bridges. There are three other disulfide bridges in the molecule, stabilizing its conformation. The enzyme has two important regions. The folding of the molecule brings histidine at position 57, aspartic acid at position 102, and serine at position 195 close together in what is known as the **active site** of the enzyme. Near this site is a region

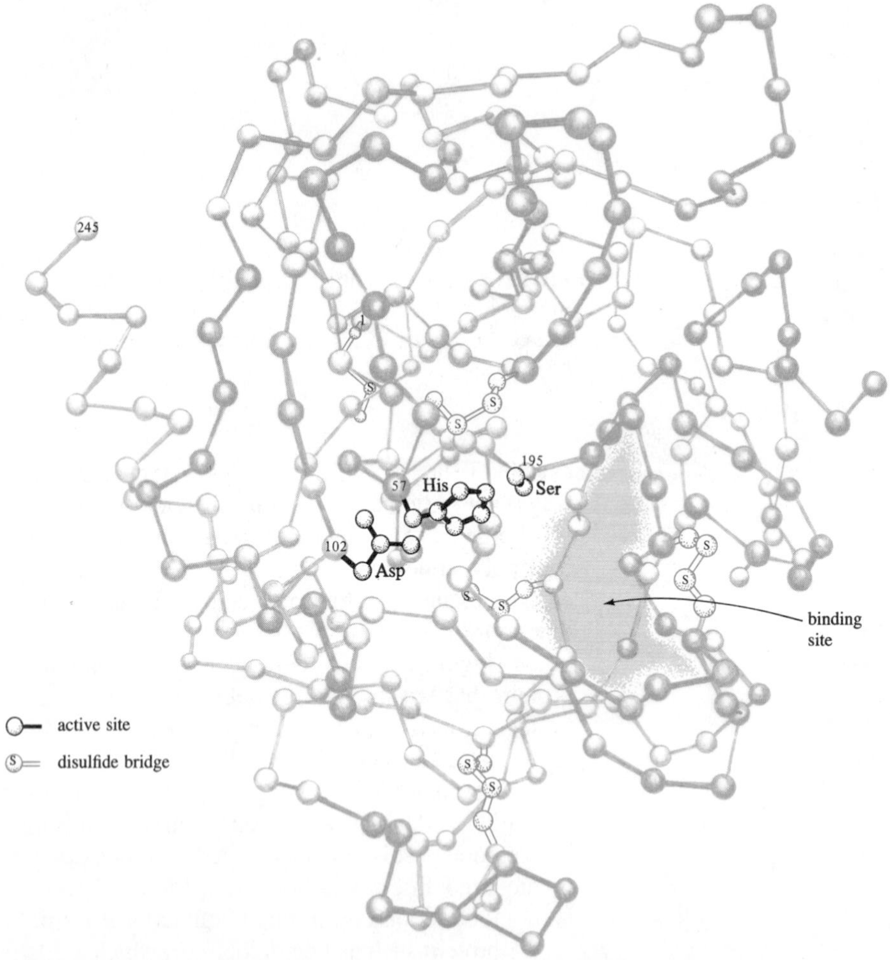
245
1
195
57 His
Ser
102
Asp
binding site
active site
disulfide bridge

FIGURE 26.9 A representation of the chymotrypsin molecule showing the placement of histidine, aspartic acid, and serine at the active site. The hydrophobic pocket that binds the peptide residue during cleavage is below and to the right of the serine at position 195. (Adapted from "A Family of Protein-Cutting Proteins," by Robert M. Stroud. Copyright © 1974 by Scientific American, Inc. All rights reserved.)

lined with hydrophobic groups where the proper portion of the peptide chain is positioned for cleavage. This region is known as the **binding site.** Figure 26.9 is a drawing of chymotrypsin in which each amino acid residue, except those involved in the active site and in the disulfide bridges, is represented only by carbon 2.

The way in which chymotrypsin acts has been determined by many experiments with different types of compounds that are hydrolyzed by the enzyme. Some reagents deactivate the enzyme, and a degradation of the molecule shows which amino acids have been affected. Spectroscopic methods have also been used to follow the course of protonation and deprotonation reactions. The mechanism for the hydrolysis of the peptide bond by the enzyme is an entirely familiar one. The only new aspects are the specific reagents that appear in the roles of acid, base, and nucleophile.

VISUALIZING THE REACTION

Acylation of the active site of chymotrypsin

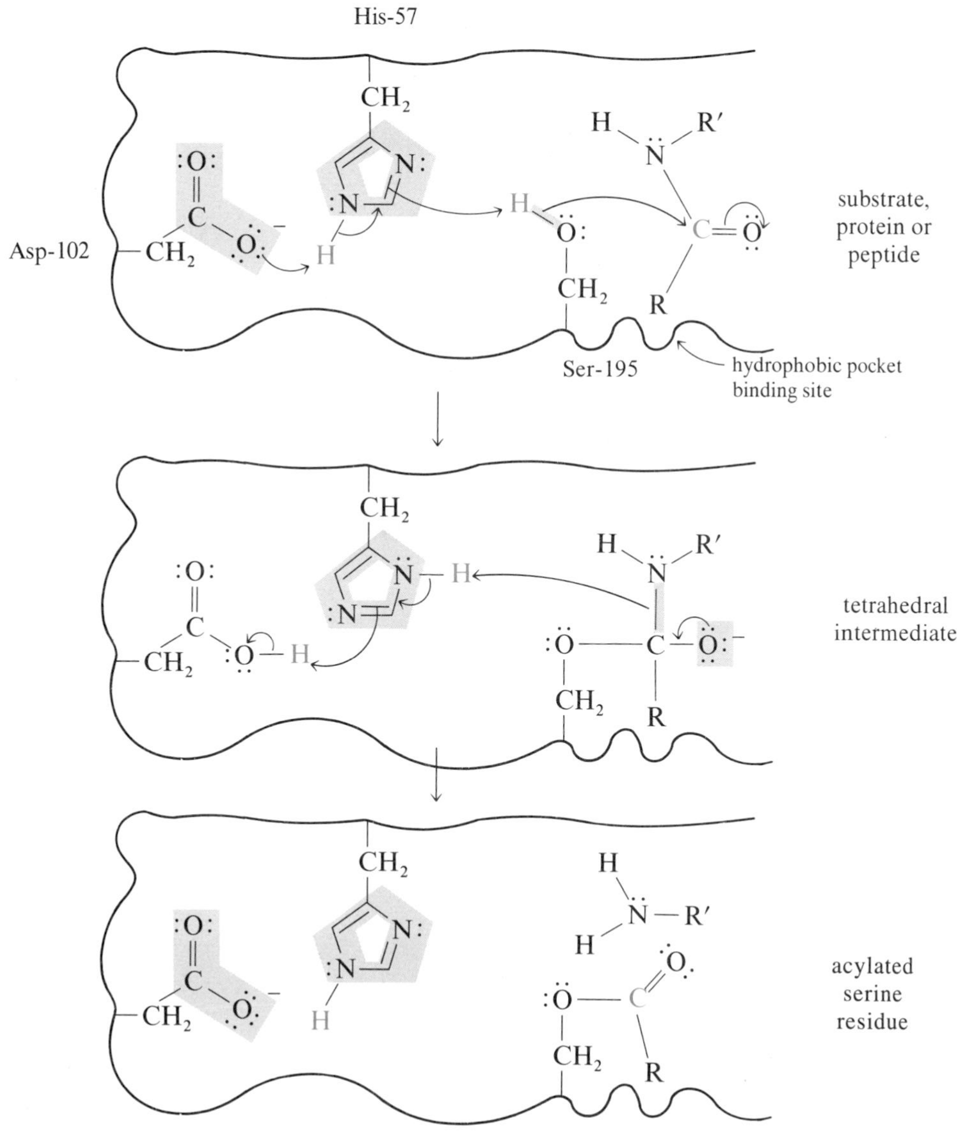
His-57
Asp-102
Ser-195
substrate, protein or peptide
hydrophobic pocket binding site
tetrahedral intermediate
acylated serine residue

The nucleophile, the hydroxyl group on the serine, attacks the carbonyl group of the peptide bond to give a tetrahedral intermediate (p. 568). Serine is made more nucleophilic by transferring its proton to histidine, which in turn is able to accept that proton because it can transfer a proton to a carboxylate anion on a conveniently placed aspartic acid residue. The tetrahedral intermediate breaks up after transfer of a proton from histidine to the amide nitrogen atom has created an amine as a leaving group. The histidine reclaims its proton from aspartic acid. At this stage, the enzyme is acylated at the serine residue. A transfer of an acyl group from the peptide to the enzyme has taken place. Serine is regenerated by a similar sequence of steps, with water as a participant.

VISUALIZING THE REACTION

Regeneration of the active site of chymotrypsin

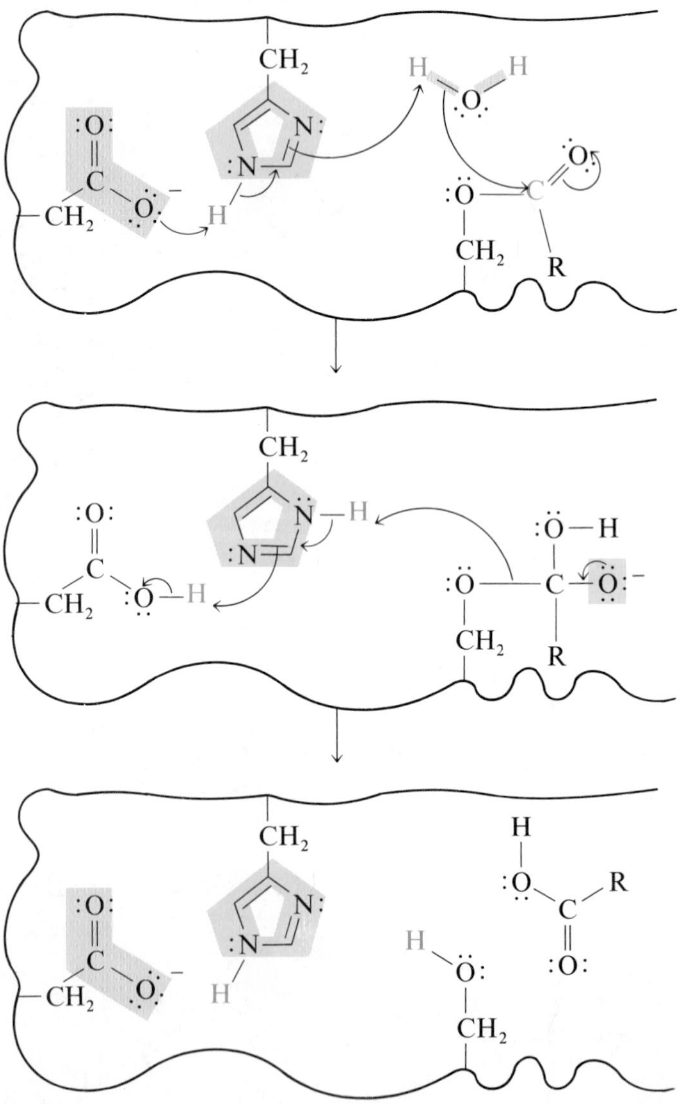
CH2
H
H
O
:O:
C
O
N:
:N
H
CH2
O
C
:O
CH2
R
N—H
:O—H
O—H
CH2
R

The ester bond to the serine hydroxyl group is hydrolyzed, and the peptide fragments diffuse away from the active site, which is then available for cleavage of another peptide bond. A process that requires strong acid or base in water solution and several hours of heating to 100 °C in the laboratory (p. 564) is achieved at pH 6–7, at 37 °C, in a fraction of a second in the body.

The activity of enzymes is a powerful demonstration of the effect that lowering the energy of the transition state can have on the overall rate of the reaction. The form of an enzyme molecule creates the maximum degree of coordination between the different steps necessary for the cleavage of the peptide bond, ensuring that the transition state for the reaction is achieved with a minimum energy of activation.

Chymotrypsin is only one of the thousands of enzymes that catalyze the chemical processes that support life. The structures of only a few of these enzymes are known well enough for scientists to begin to figure out how they function. Proteins that have similar functions in different organisms have remarkably similar, if not identical, structures. It almost seems as though nature has discovered certain solutions to chemical problems and uses them over and over again with minor modifications to accommodate the unique characteristics of different species.

S U M M A R Y

Proteins are high-molecular-weight compounds consisting of long chains of amino acids bonded together by peptide linkages (amide bonds). About twenty different amino acids are found in proteins. The order of amino acids in the peptide chain is considered to be the primary structure of a protein. The conformation of the peptide chain is the secondary structure. The major secondary structures are the α-helix and the pleated sheet. Further folding of the peptide chain creates the tertiary structure. Association of polypeptides with each other or with simpler units (such as heme in hemoglobin) gives rise to a quaternary structure.

Ionic interactions, hydrogen bonding, and disulfide bridges can be important in holding different parts of a protein molecule close together to give it its overall form. This form is essential to the biological function of the protein. When it is lost, the protein is denatured and inactive.

Amino acids contain both acidic and basic functional groups and can exist as zwitterions. The charge on an amino acid depends on the number of acidic and basic groups it contains and on the pH. The pH at which an amino acid has no net charge is called its isoelectric point.

The structure of a protein is determined by degradation reactions. Hydrolysis of the protein gives peptides or individual amino acids, which are identified by chromatographic methods. The *N*-terminal amino acid of a peptide may be determined by the reaction of that amino acid with 2,4-dinitrofluorobenzene (Sanger's reagent) before hydrolysis. Phenyl isothiocyanate (Edman's reagent) is used to degrade a peptide successively, starting with the *N*-terminal amino acid and removing amino acids one at a time and identifying them as their phenylthiohydantoins. The enzyme carboxypeptidase degrades peptides at the *C*-terminal end of the chain (Table 26.1).

In order to synthesize peptides in the laboratory, the carboxyl group of an amino acid must be activated and the amine group protected. Carbobenzoxy (*N*-Cbz) and *tert*-butoxycarbonyl (Boc) groups are common protecting groups. Both groups are easily removed by strong acids in dry solvents (Table 26.2). Carboxyl groups are activated by reaction with dicyclohexylcarbodiimide or by conversion to esters of *p*-nitrophenol or *N*-hydroxysuccinimide (Table 26.3). Most peptide syntheses are now carried out in the solid state.

TABLE 26.1 Degradation Reactions of Peptides

Peptide	Reagent and Reaction Conditions	Product(s)
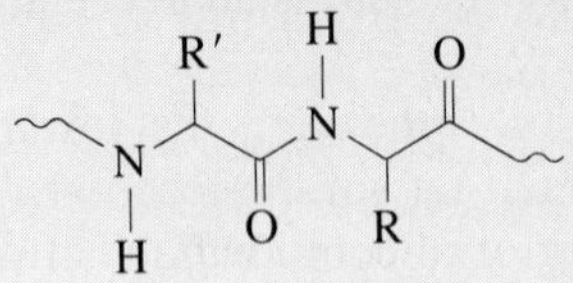	HCl, H_2O, Δ	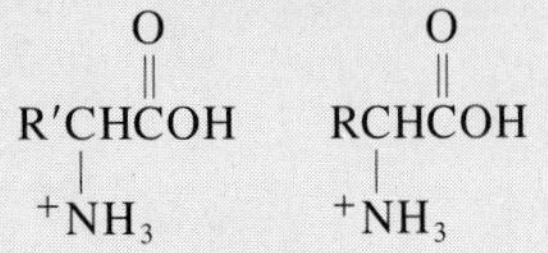
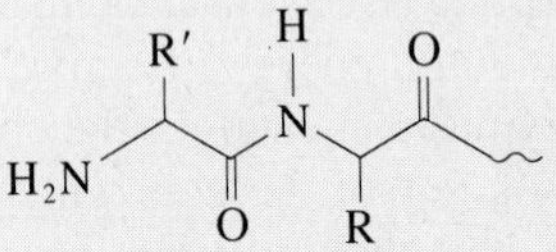	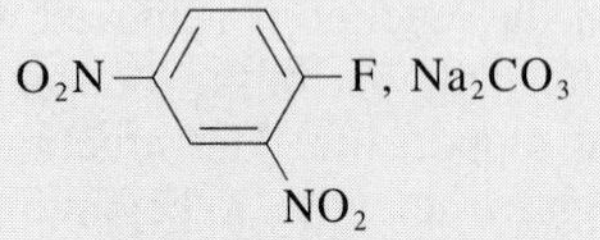 O_2N—⟨⟩—F, Na_2CO_3 NO_2 then HCl, H_2O, Δ	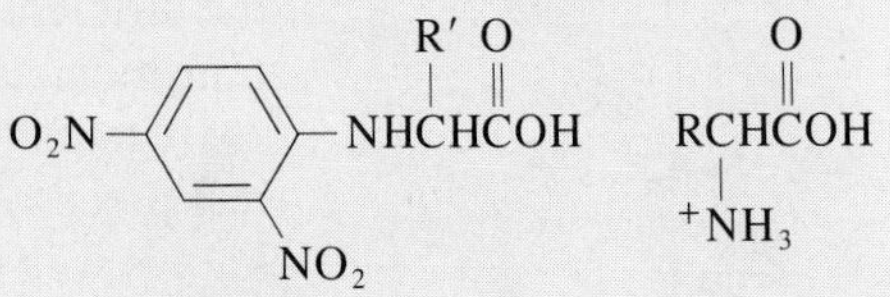
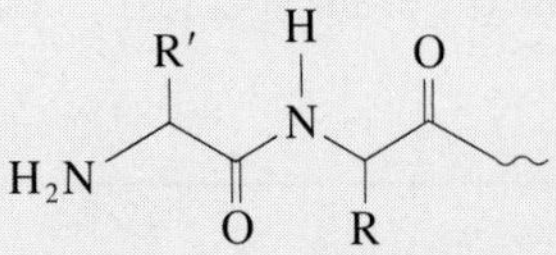	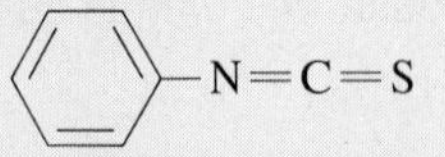 ⟨⟩—$N{=}C{=}S$ then HCl, nitromethane, Δ	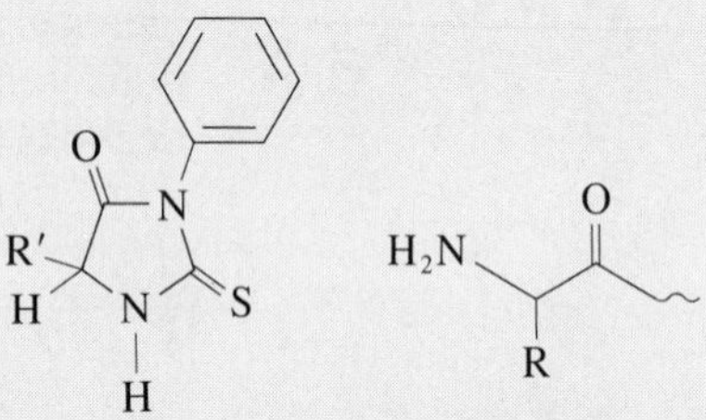
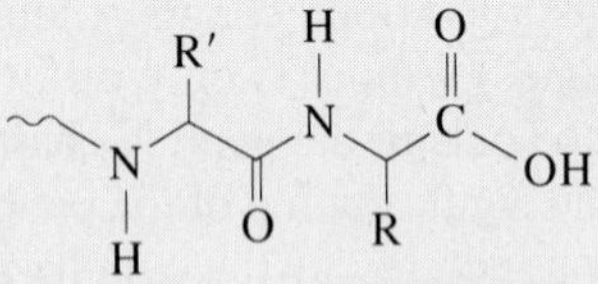	carboxypeptidase	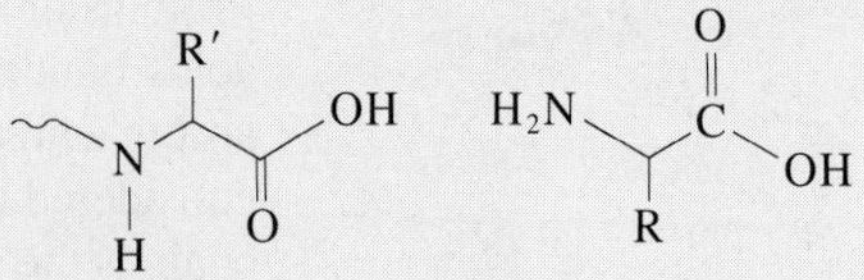
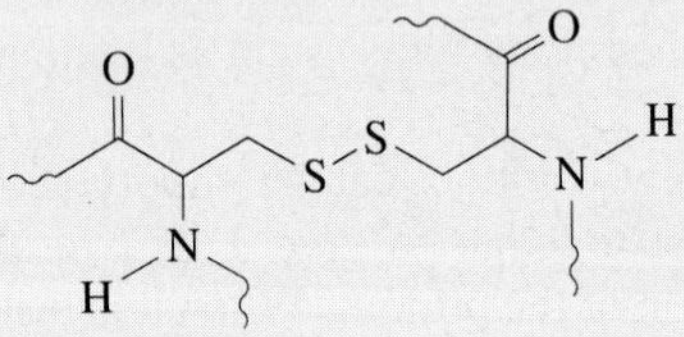	$HSCH_2CH_2OH$	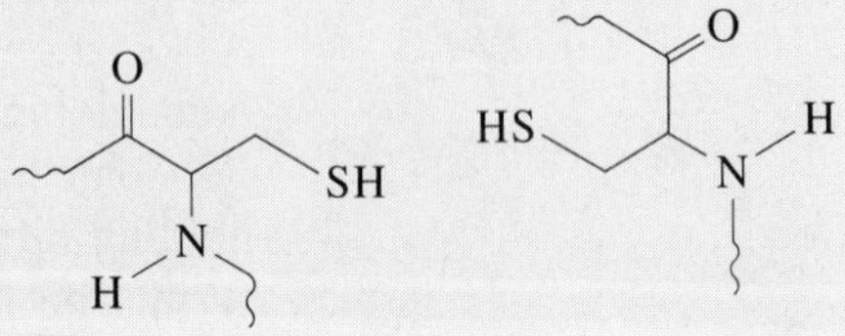
	then ICH_2COH	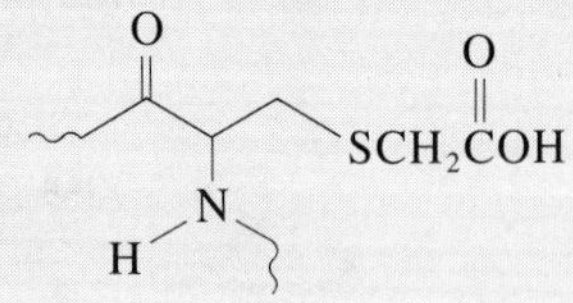

TABLE 26.2 Protection of Amino Group in Protein Synthesis

Functional Group	Reagent	Protecting Group	Reagent(s) for Removal	Products
RNH_2	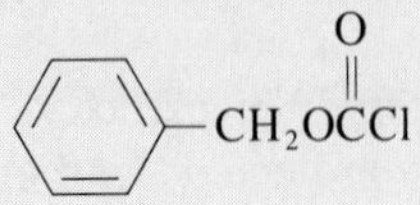 ⟨⟩—CH_2OCCl	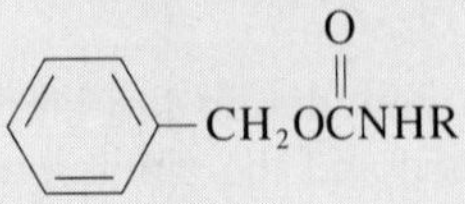 ⟨⟩—CH_2OCNHR *N*-Cbz	HBr, CH_3COH	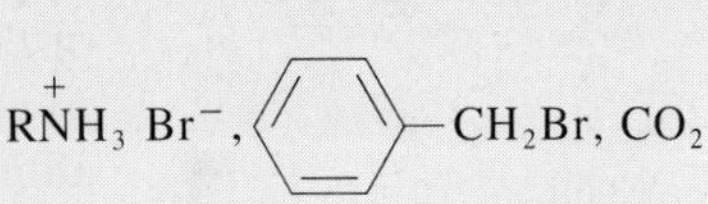 $\overset{+}{R}NH_3\ Br^-$, ⟨⟩—CH_2Br, CO_2
			H_2, Pd/C	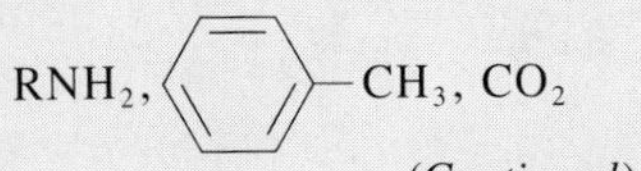 RNH_2, ⟨⟩—CH_3, CO_2

(Continued)

TABLE 26.2 (*Continued*)

Functional Group	Reagent	Protecting Group	Reagent(s) for Removal	Products
	$(CH_3)_3C-OC(=O)N_3$	$(CH_3)_3C-OC(=O)NHR$ Boc	HCl, ether or $CF_3C(=O)OH$, CH_2Cl_2	$R\overset{+}{N}H_3$, $CH_3C(CH_3)=CH_2$, CO_2

TABLE 26.3 Activation of Carboxyl Group in Protein Synthesis

Functional Group	Reagent	Activated Group
$RC(=O)OH$	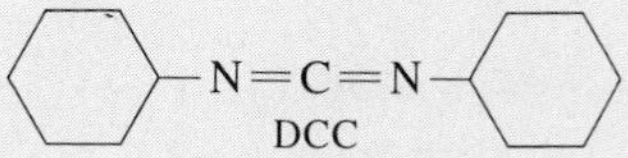 DCC	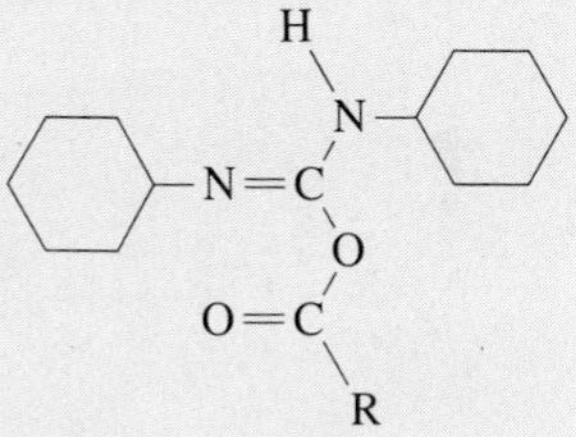
	$SOCl_2$, then 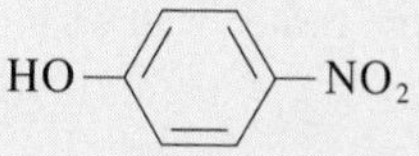	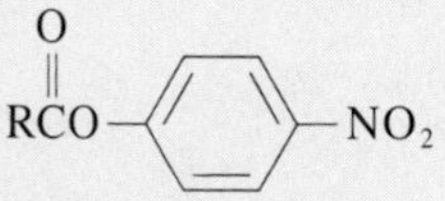
	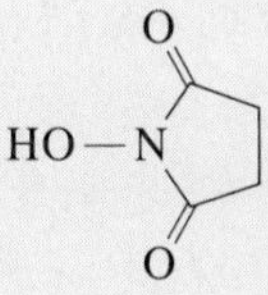 , DCC	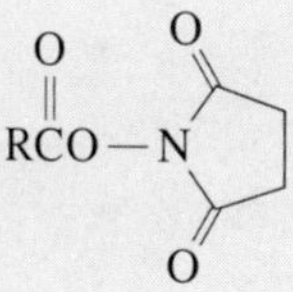

ADDITIONAL PROBLEMS

26.19 Write full structures for the following compounds.

(a) glycylalanyllysine (b) ethyl alaninate (c) *N*-benzyloxycarbonylmethionine
(d) *N*-*tert*-butoxycarbonylvaline (e) *N*-2,4-dinitrophenylleucine
(f) *p*-nitrophenyl tyrosinate

26.20 Name the following compounds.

(a) 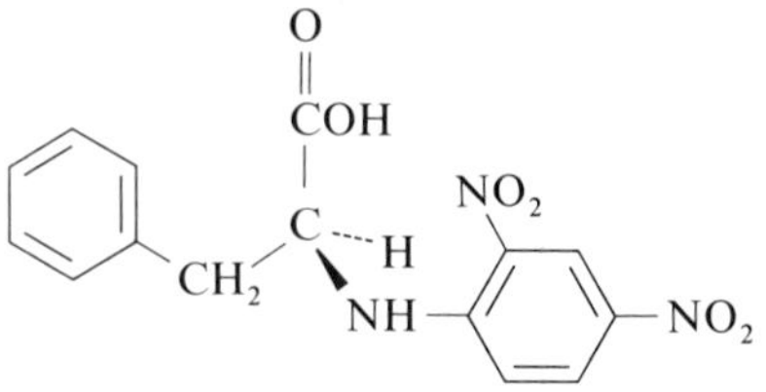

(b) 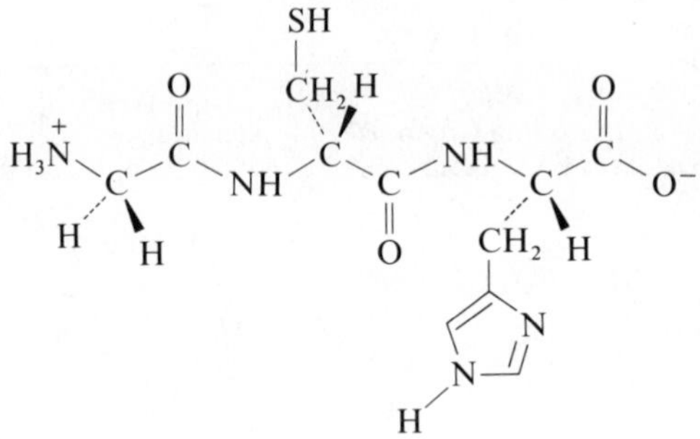

(c) 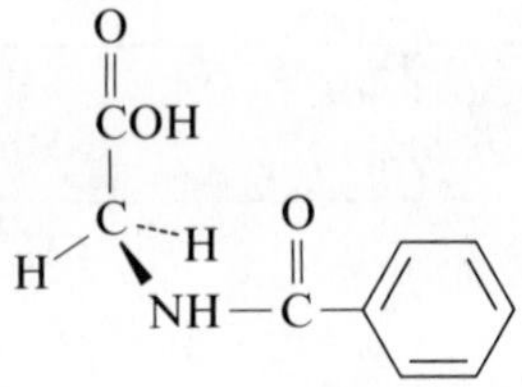

(d) 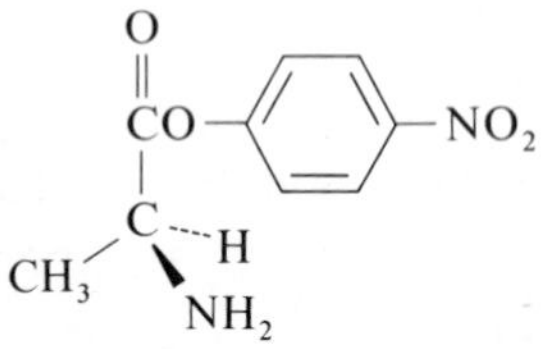

(e) 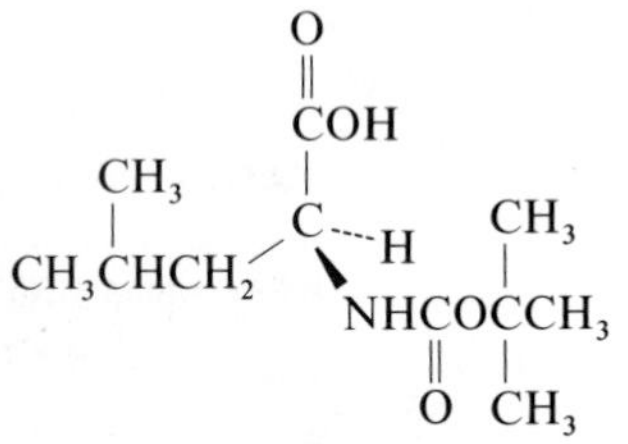

(f) 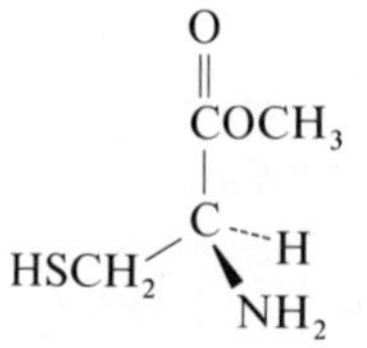

26.21 Write equations showing the reactions you predict would occur if the amino acid valine were treated with each of the following reagents.

(a) 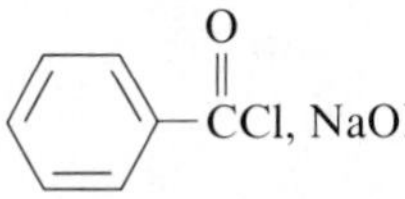, NaOH (b) CH_3OH, HCl (c) $NaNO_2$, HCl, H_2O, 0°C

(d) 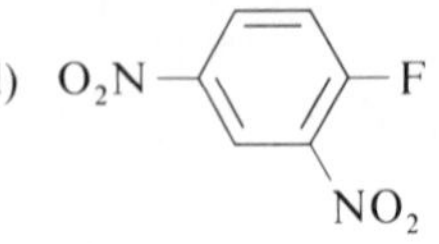 (e) 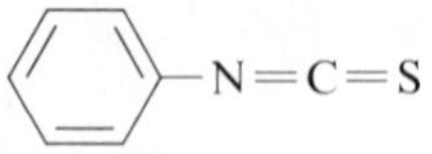 (f) product of (e), acid, Δ

(g) 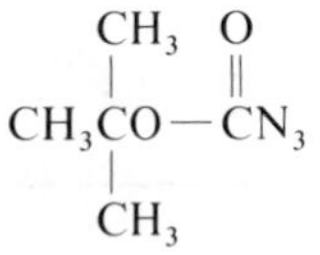 (h) product of (g) + 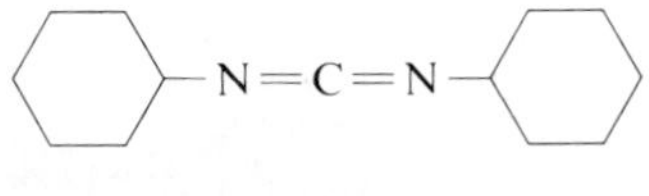

(i) product of (h) + 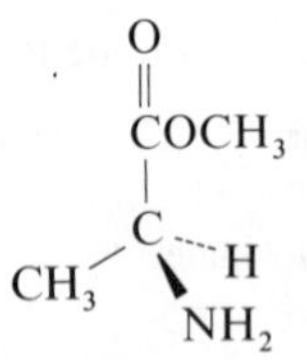 (j) 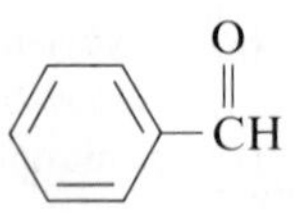

26.22 Complete the following equations.

(a) 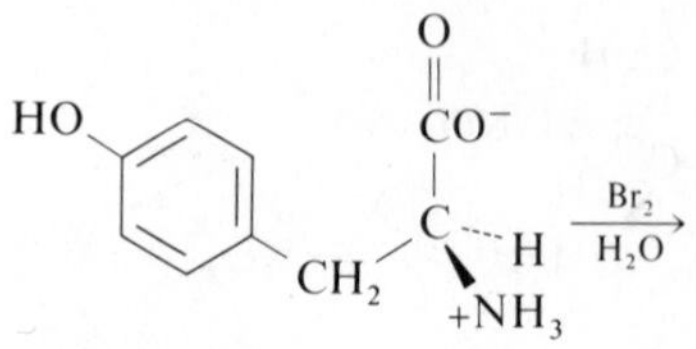

(b) 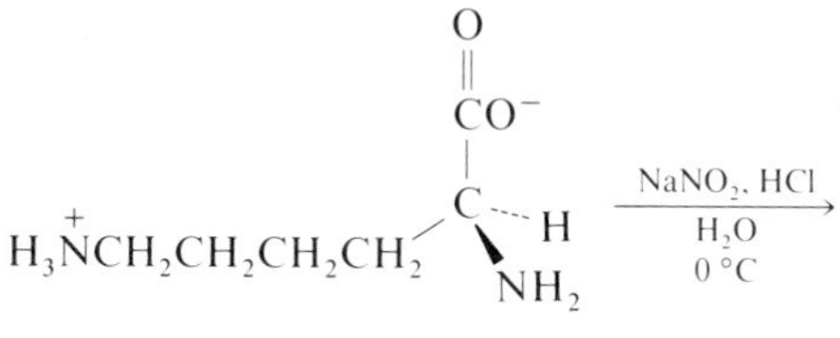

(c) 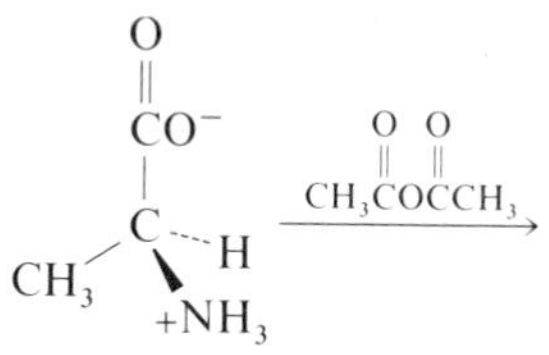

(d) 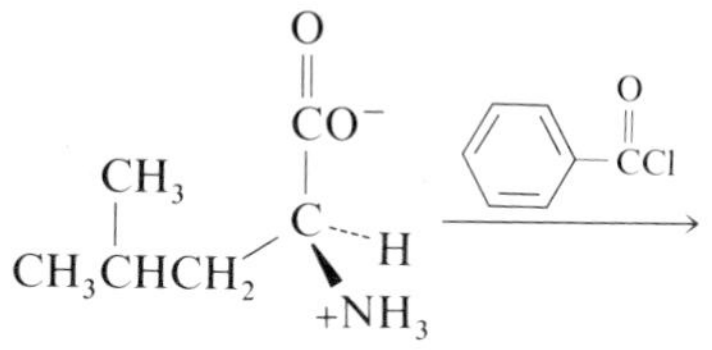

(e) 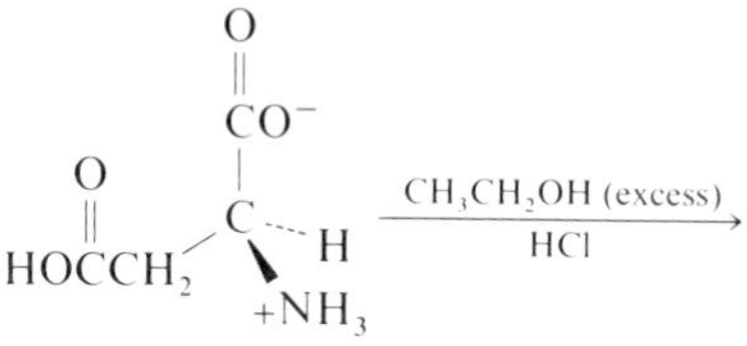

(f) 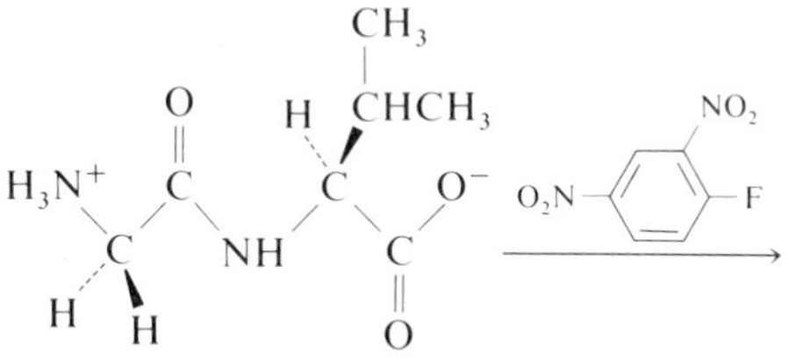

(g) product of (f) $\xrightarrow[\Delta]{H_2O,\ HCl}$

(h) 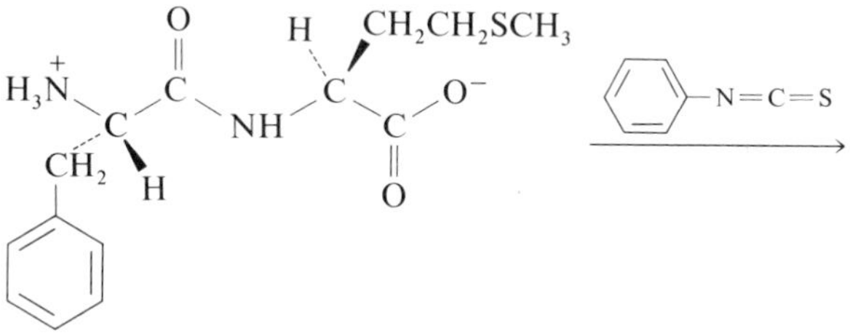

(i) product of (h) $\xrightarrow[\text{nitromethane},\ \Delta]{HCl}$

(j) 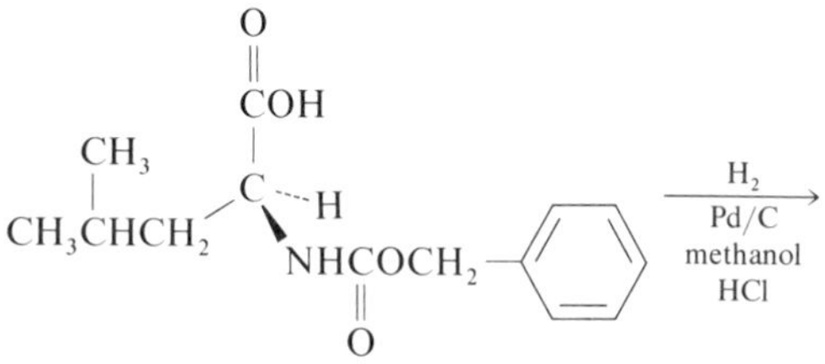

(k) 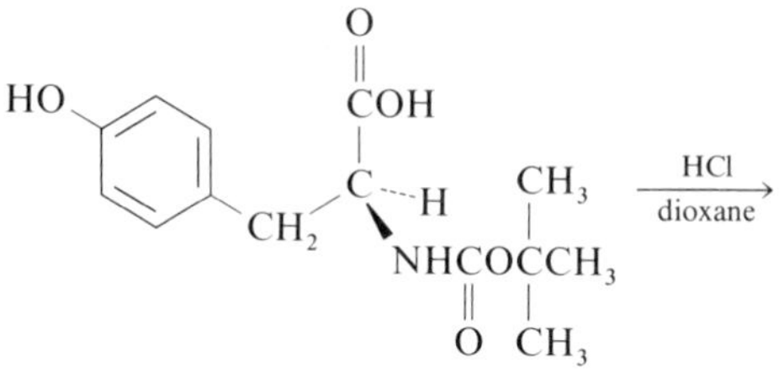

(l) 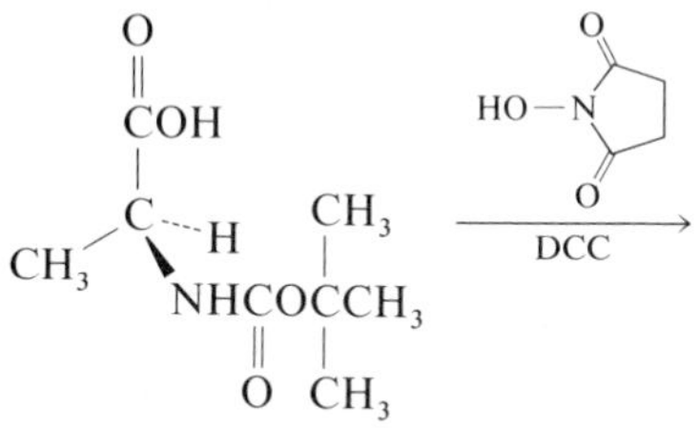

26.23 For each of the following peptides, draw the full structure and show the ionic state in which the compound would exist at pH 1, 6, and 11.

(a) Phe-Val-Asp (b) Lys-Ala-Gly (c) Leu-Tyr-Gly-NH_2 (d) Met-Gln-Ala

26.24 Since the discovery in 1975 that there are small peptides in the brain that have analgesic properties, chemists have been synthesizing molecules that are structurally analogous to these compounds. One such synthetic peptide that is 1500 times as active as an analgesic agent as the naturally occurring methionine enkephalin (p. 1142) has the following structure.

Tyr-D-Ala-Gly-Phe-Pro-NH_2

(a) Write the full structure for this peptide, showing correct stereochemistry at every stereocenter in the molecule.
(b) Synthesize a portion of the peptide consisting of three amino acids, starting from either end of the chain and showing the necessary protecting groups and reagents. Assume that all the necessary amino acids are available in the required stereochemical form. Also assume that any group you use to protect an amino group can also be used to protect a hydroxyl function and can be removed under the same experimental conditions from both groups.

26.25 Lysine vasopressin, also a naturally occurring hormone, has the same structure as arginine vasopressin (Problem 26.13, p. 1135) except that lysine is substituted for arginine. Lysine vasopressin was synthesized as outlined below. Write structures for the intermediates and products indicated by letters.

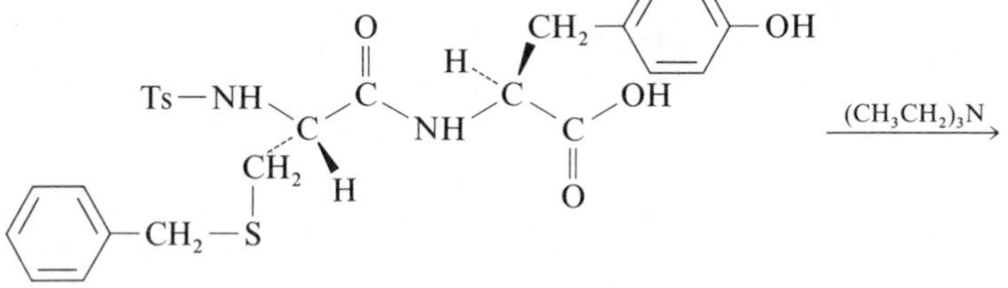

$$A \xrightarrow{CH_3\overset{CH_3}{|}{CH}CH_2O\overset{O}{\|}{C}Cl} B \xrightarrow{\text{Phe-Gln-Asn}} C$$

a protected pentapeptide

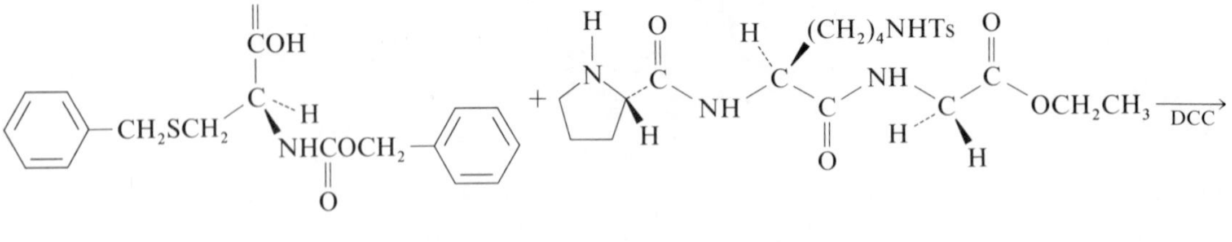

$$D \xrightarrow[\text{ethanol}]{NH_3} E \xrightarrow[\text{acetic acid}]{HBr} F$$

D: *a protected tetrapeptide, as the glycyl ester*; E: *a protected tetrapeptide, as glycyl amide*

$$F + C \xrightarrow{DCC} G \xrightarrow[NH_3\text{ (liq)}]{Na} H \xrightarrow{\text{air oxidation}} I$$

G: *a protected vasopressin*; I: lysine vasopressin

26.26 Synthetic, racemic leucine is resolved using an enzyme, hog renal acylase, which is isolated from hog kidneys. Hog renal acylase catalyzes the hydrolysis of the amides of L-amino acids only. The reactions are

$$\text{D,L-leucine} \xrightarrow{(CH_3\overset{O}{\overset{\|}{C}})_2O} N\text{-acetyl-D,L-leucine} \xrightarrow[\substack{\text{hog renal acylase} \\ H_2O,\ \text{pH 7 maintained} \\ \text{by buffer}}]{} \text{mixture}$$

Some useful information is that acetic acid has pK_a 4.8 and leucine has pH$_I$ 6.0. Also, leucine and N-acetylleucine are crystalline solids when bearing no net charge.

(a) Write equations for the reactions described, being sure to show the correct stereochemistry for the species present at each step. What are the components of the mixture formed at the end of the enzymatic hydrolysis?

(b) How would you separate the components of the mixture that results from the enzymatic treatment? Write equations showing what you would have to do to recover the individual compounds. (Hint: What ionic species are present in the solution when the reaction is over?)

(c) Write equations to describe what you would have to do to recover pure D-leucine.

26.27 In the human body, an amino acid is prepared for incorporation into a peptide chain by activation of its carboxyl group through the formation of an acylphosphate linkage. This reaction takes place with ATP (p. 1079) to give an acylphosphate group at carbon 5 of the ribose unit and a pyrophosphate anion. Outline a possible mechanism for the activation reaction using any amino acid.

26.28 The octapeptide xenopsin from the frog *Xenopus laevis* has a powerful contractile effect on muscles. The N-terminal amino acid of xenopsin is glutamic acid, which exists in the natural peptide as the cyclic amide pyroglutamic acid (p. 1132). Carboxypeptidase releases leucine first. Trypsin digestion yields ⌐Glu-Gly-Lys-Arg and ⌐Glu-Gly-Lys-Arg-Pro-Trp. Chymotrypsin gives the same hexapeptide as is obtained with trypsin and Ile-Leu. What is the structure of xenopsin?

26.29 An extract of the European mistletoe, *Viscum album,* contains a series of pharmacologically active peptides known as viscotoxins. The amino acid sequence of one of them, viscotoxin A_2, was determined using oxidation to break disulfide linkages, followed by digestion of the peptide with trypsin and chymotrypsin. The structures of the peptide fragments that were isolated from the two enzymatic cleavages are given below.

Peptides from trypsin digestion: Asn-Ile-Tyr-Asn-Thr-Cys-Arg,
Lys-Ser-Cys-Cys-Pro-Asn-Thr-Thr-Gly-Arg,
Ile-Ile-Ser-Ala-Ser-Thr-Cys-Pro-Ser-Tyr-Pro-Asp-Lys,
Phe-Gly-Gly-Gly-Ser-Arg,
Ser-Cys-Cys-Pro-Asn-Thr-Thr-Gly-Arg.

Peptides from chymotrypsin digestion: Asn-Thr-Cys-Arg-Phe,
Gly-Gly-Gly-Ser-Arg-Glu-Val-Cys-Ala-Ser-Leu,
Lys-Ser-Cys-Cys-Pro-Asn-Thr-Thr-Gly-Arg-Asn-Ile-Tyr,
Ser-Gly-Cys-Lys-Ile-Ile-Ser-Ala-Ser-Thr-Cys-Pro-Ser-Tyr-Pro-Asp-Lys.

Viscotoxin A_2 has 46 amino acid residues and a molecular weight of 4833. The N-terminal and C-terminal amino acids are both lysine. What is the order of the amino acids in viscotoxin A_2?

26.30 Fructose 1,6-diphosphate is converted into D-glyceraldehyde 3-phosphate and dihydroxyacetone phosphate (1,3-dihydroxy-2-propanone phosphate) by an enzyme known as aldolase. The enzymatic reaction proceeds through the formation of a Schiff base (an imine, p. 506) between the carbonyl group of fructose and the ϵ-amino group of lysine.

(a) Write a mechanism for the formation of the Schiff base at the active site of the enzyme, using HB^+ and B: to represent any acids and bases that are needed.
(b) Write a mechanism for the conversion of the enzyme-substrate complex to D-glyceraldehyde 3-phosphate and the Schiff base of dihydroxyacetone phosphate.
(c) If the imine function in the enzyme-substrate complex is reduced with sodium borohydride and the enzyme is hydrolyzed, a lysine derivative of dihydroxyacetone is isolated. Propose a structure for this compound.

26.31 Peptide P is widely distributed in the human body, especially in the nervous system. It is believed to mediate the body's response to pain. The following data were obtained in a determination of its structure:

1. Vigorous acidic hydrolysis of Peptide P gives Arg, Glu (2), Gly, Leu, Lys, Met, Phe (2), Pro (2). Enzymatic hydrolysis gives Arg, Gln (2), Gly, Leu, Lys, Met, Phe (2), Pro (2). Peptide P has eleven amino acid units in all.
2. When Peptide P is treated with phenyl isothiocyanate, the phenylthiohydantoins derived from arginine, proline, lysine, and proline can be obtained in that order.
3. Incubation of Peptide P with chymotrypsin gives Peptide A and Peptide B.
4. Peptide A contains Arg, Gln (2), Lys, Phe, Pro (2). Degradation of Peptide A with Edman's reagent gives the same phenylthiohydantoins derived from the intact Peptide P. Carboxypeptidase releases first Phe, then Gln to the solution.
5. Peptide B reacts with phenyl isothiocyanate to give the phenylthiohydantoins derived from phenylalanine, glycine, and leucine, in that order.
6. Peptide P is a strongly basic peptide with an isoelectric point above 8.9. No amino acid is released to the solution when the peptide is incubated with carboxypeptidase.
7. If Peptide P is incubated with 0.03 M HCl at 110 °C for 8–12 hours (a procedure developed to hydrolyze carboxylic acid amide bonds but leave most peptide bonds untouched), the resulting product is Peptide C, which reacts with carboxypeptidase. Gly, Met, Leu, Phe appear in the solution. The order in which these amino acids were released was not determined.

(a) What is the *N*-terminal amino acid of Peptide P?
(b) What is the *C*-terminal amino acid of Peptide P?
(c) What is the sequence of amino acids in Peptide A?
(d) What is the sequence of amino acids in Peptide B?
(e) What is the complete structure of Peptide B?
(f) What is the sequence of amino acids in Peptide C?
(g) What is the structure of Peptide P?

26.32 The solid-state synthesis of Peptide P (Problem 26.31) was carried out.

(a) Glutamine residues were protected at the α-amino group with *tert*-butoxycarbonyl groups and activated at the carboxylic acid as the *p*-nitrophenyl esters. Write equations, starting with glutamine and showing its conversion to Boc-Gln-*O*-*p*-nitrophenyl, giving the full structure of the protected and activated amino acid.
(b) Lysine was protected at the α-amino group by a *tert*-butoxycarbonyl group and at the ϵ-amino group by a benzyloxycarbonyl group. Write the full structure of lysine, as it would be used in the peptide synthesis.
(c) The peptide linkages other than those to glutamine were created by activating with dicyclohexylcarbodiimide the carboxylic acid group of the amino acid being added. Show the mechanism for the reaction of Boc-Phe (written out in full) with dicyclohexylcarbodiimide, and the addition of the *N*-terminal end of a peptide fragment (any one) to the activated amino acid to form a new peptide bond.
(d) The Boc groups in this synthesis were removed by the addition of 4 M HCl dissolved in dioxane. Write the equation for removing the protecting group from the peptide fragment synthesized in part c.
(e) It was discovered that methionine esters could not be formed efficiently with the benzylic chloride functions on the polymeric support because the side chain of methionine reacted with the benzylic chloride. What reaction might occur between the side chain of Boc-Met and benzyl chloride?

26.33 The amino acid proline is synthesized in the body from glutamic acid. This synthesis involves the following steps. Show a mechanism for the transformation in each step. ATP is adenosine triphosphate (p. 1079).

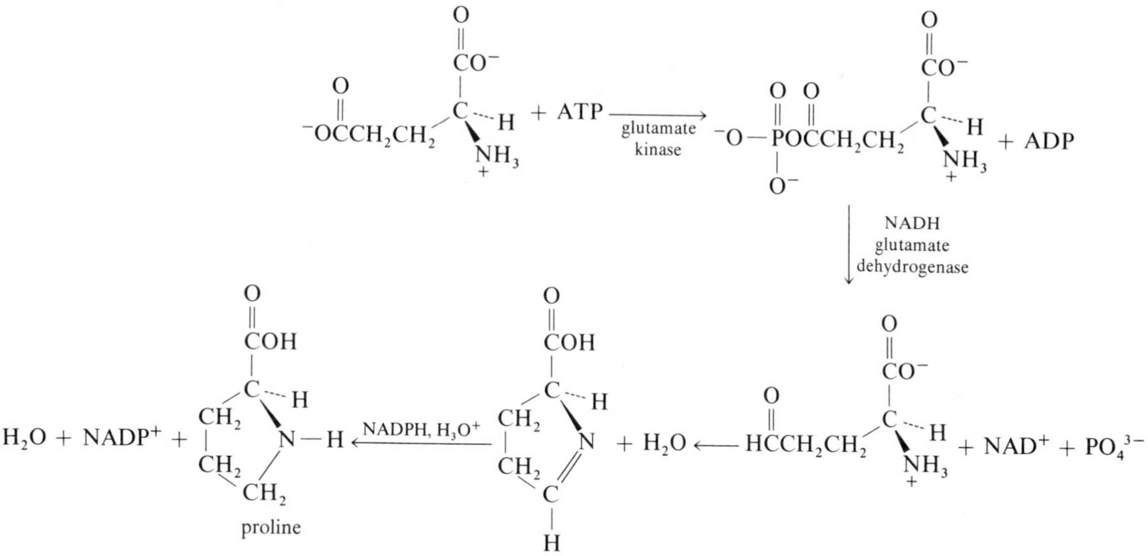

+ ATP
glutamate kinase
+ ADP
NADH
glutamate dehydrogenase
+ NAD⁺ + PO₄³⁻
+ H₂O
NADPH, H₃O⁺
H₂O + NADP⁺ +
proline

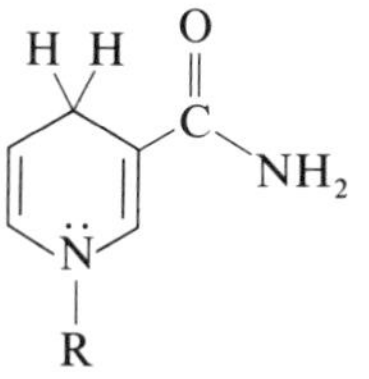

NADH, NADPH
(p. 1021)

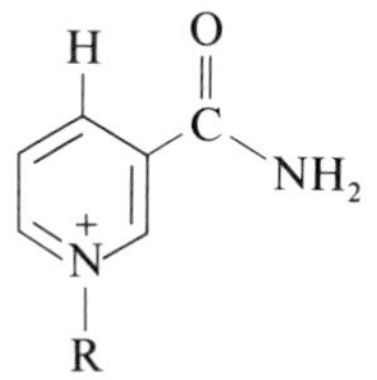

NAD^+, $NADP^+$
(p. 1021)

26.34 The human growth hormone isolated from the pituitary gland has 188 amino acid residues and has been synthesized by a solid-state technique similar to that of Merrifield (p. 1146). The first five amino acids in the chain, starting from the *C*-terminal end, are phenylalanine, glycine, cysteine, serine, and glycine, in that order. Devise a solid-state synthesis of this portion of the chain. Assume that the amino acids cysteine and serine are available with benzyl groups protecting the thiol and the hydroxyl group on their side chains. The benzyl group can be removed by reduction using sodium in ammonia.

26.35 A new amino acid, β-carboxyaspartic acid, was discovered in 1981 when researchers noticed that the acid hydrolysis of proteins associated with the ribosomes of the bacteria *Escherichia coli* gave more aspartic acid than did hydrolysis catalyzed by base. The amino acid has the structure shown below.

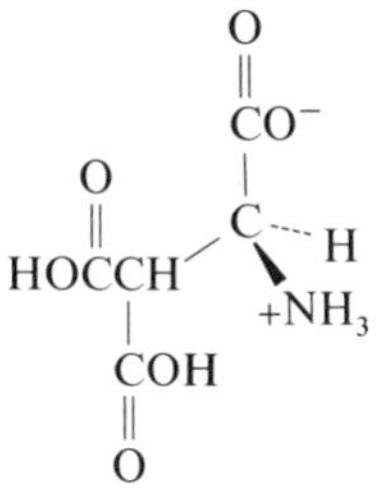

β-carboxyaspartic acid

(a) Why is there a difference in the products obtained from acid-catalyzed and base-catalyzed hydrolyses of proteins containing this amino acid?

(b) The new amino acid was synthesized in order to compare its spectral properties with those of products from the hydrolysis of the ribosomal proteins. The reactions used in the synthesis are shown below. Propose structures for the compounds designated by letters.

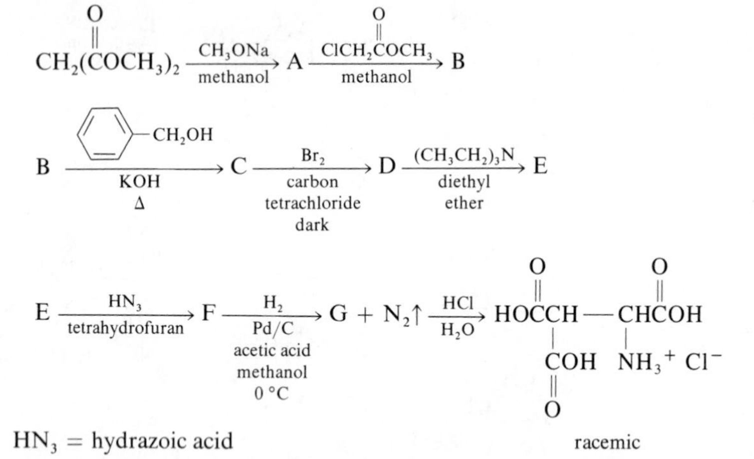
CH₂(COCH₃)₂ → CH₃ONa / methanol → A → ClCH₂COCH₃ / methanol → B
B → CH₂OH / KOH / Δ → C → Br₂ / carbon tetrachloride / dark → D → (CH₃CH₂)₃N / diethyl ether → E
E → HN₃ / tetrahydrofuran → F → H₂ / Pd/C / acetic acid / methanol / 0 °C → G + N₂↑ → HCl / H₂O → HOCCH—CHCOH / COH NH₃⁺ Cl⁻
HN₃ = hydrazoic acid
racemic

(c) The experimental directions specify that it is important to exclude light completely in the reaction of Compound C with bromine in carbon tetrachloride. What side reactions would be possible if the reaction mixture were exposed to light? (Hint: A review of Section 20.3 may be helpful.)

(d) Why does the reaction of Compound E with hydrazoic acid give Compound F in the orientation observed?

26.36 The amino acid proline has been used as the starting material for a number of chiral amines. One such synthesis is outlined below. Write structural formulas for the compounds designated by letters.

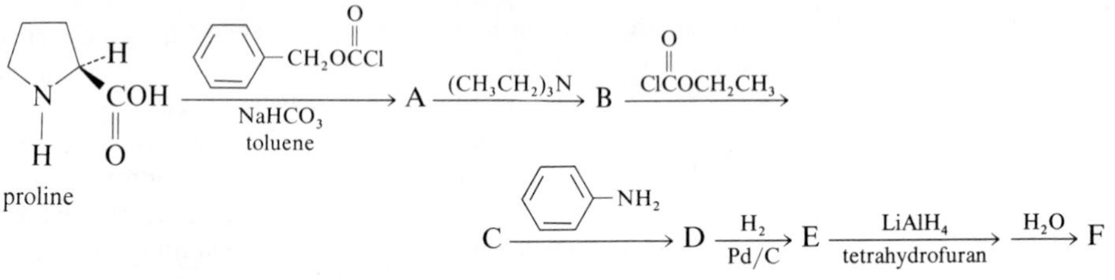
H
COH
N
H
proline
CH₂OCCl
NaHCO₃
toluene
A
(CH₃CH₂)₃N
B
ClCOCH₂CH₃
C
NH₂
D
H₂
Pd/C
E
LiAlH₄
tetrahydrofuran
H₂O
F

26.37 Distamycin A, a compound isolated from the fermentation broth of the microorganism *Streptomyces distallicus*, has antiviral and antitumor activity. The compound has the following structure.

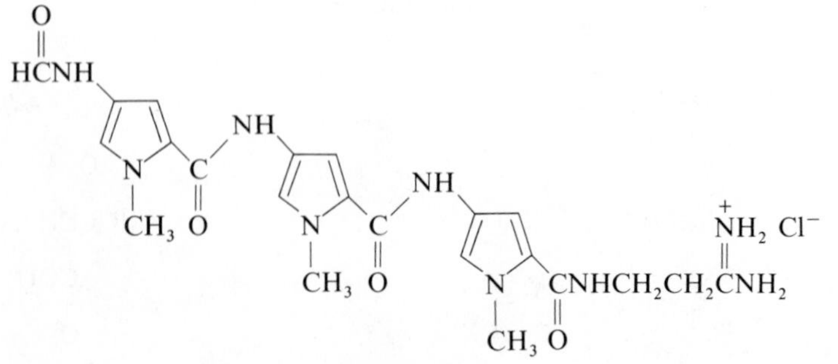
HCNH
NH
NH
CH₃
CH₃
CH₃
NH₂ Cl⁻
CNHCH₂CH₂CNH₂

distamycin A

The synthesis of the compound outlined below took advantage of the techniques developed for the synthesis of peptides. Give structural formulas for the compounds and intermediates represented by letters.

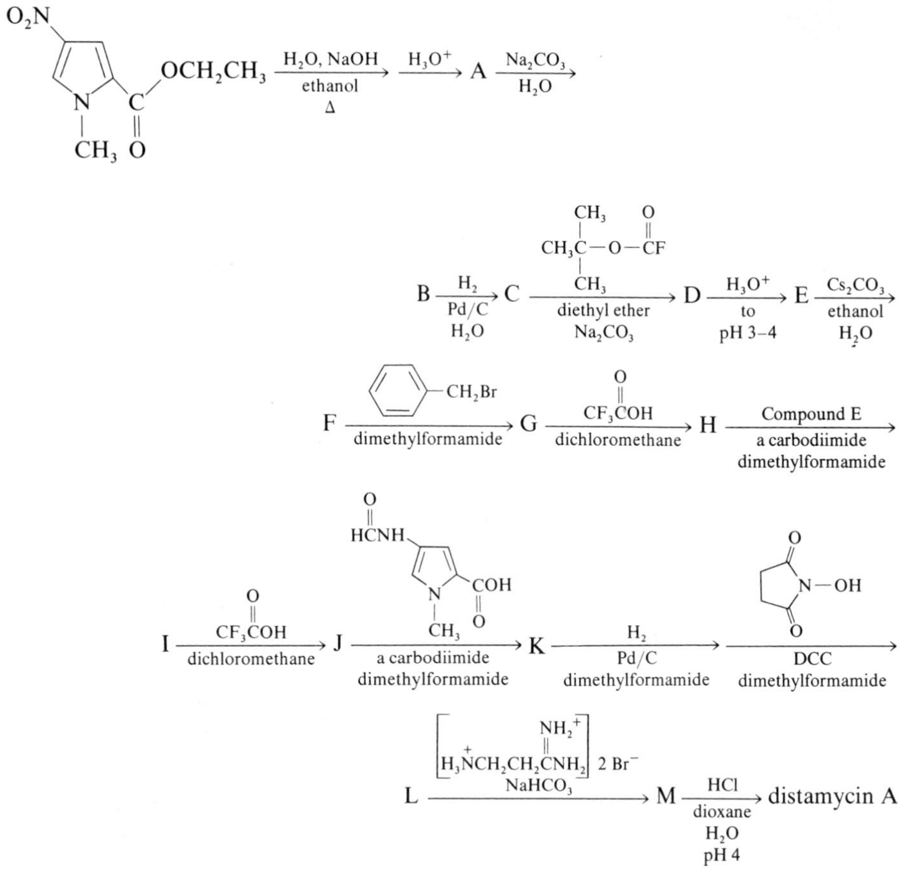
O2N
OCH2CH3
N
C
CH3 O
H2O, NaOH
ethanol
Δ
H3O+
A
Na2CO3
H2O
CH3 O
CH3C—O—CF
CH3
B
H2
Pd/C
H2O
C
diethyl ether
Na2CO3
D
H3O+
to
pH 3–4
E
Cs2CO3
ethanol
H2O
CH2Br
F
dimethylformamide
G
CF3COH
dichloromethane
H
Compound E
a carbodiimide
dimethylformamide
HCNH
COH
I
CF3COH
dichloromethane
J
a carbodiimide
dimethylformamide
K
H2
Pd/C
dimethylformamide
N—OH
DCC
dimethylformamide
NH2+
H3NCH2CH2CNH2
2 Br−
NaHCO3
L
M
HCl
dioxane
H2O
pH 4
distamycin A

27 Macromolecular Chemistry

A • L O O K • A H E A D

Polymers are large molecules created by repetitive reactions of simple molecular units. Important in modern industrial society, they are used to make everything from common household utensils to replacement parts for the human body. The small unit that appears many times in a polymer is a monomer (p. 285). The process by which a monomer is converted into a polymer is called polymerization. The growth of a polymer may occur through a steady unit-by-unit addition of monomer molecules to the end of the chain, in a chain-growth process. Alternatively, several monomer units may combine to give larger fragments, which then react to give the polymer. This is called a step-growth process. In some polymers, all of the monomer units are the same. Polymers that have more than one kind of repeating unit are called copolymers.

The size and stereochemistry of a polymer molecule are important in determining the properties of the material. This chapter will discuss the ways in which polymers are made and how their structures determine their properties.

27.1 INTRODUCTION TO MACROMOLECULES

A. Macromolecules of Biological Importance

The interactions that take place between different parts of a large molecule are important in determining its form and biological function. The stereoselectivity and the chemical specificity of reactions that take place at the active site of an enzyme, for example, are directly traceable to the precise form that the large protein molecule has in its natural state (p. 1152).

Proteins are only one category of large molecules that have biological functions. Others are the nucleic acids, DNA and RNA (p. 1013). The structural materials of plants, cellulose (p. 1103) and lignin, are also giant molecules. The spaces within the long fibers formed by cellulose are filled by lignins, which are complex molecules containing carbon-oxygen and carbon-carbon bonds between adjoining units of phenylpropanes with hydroxy and methoxy substituents on the aromatic rings.

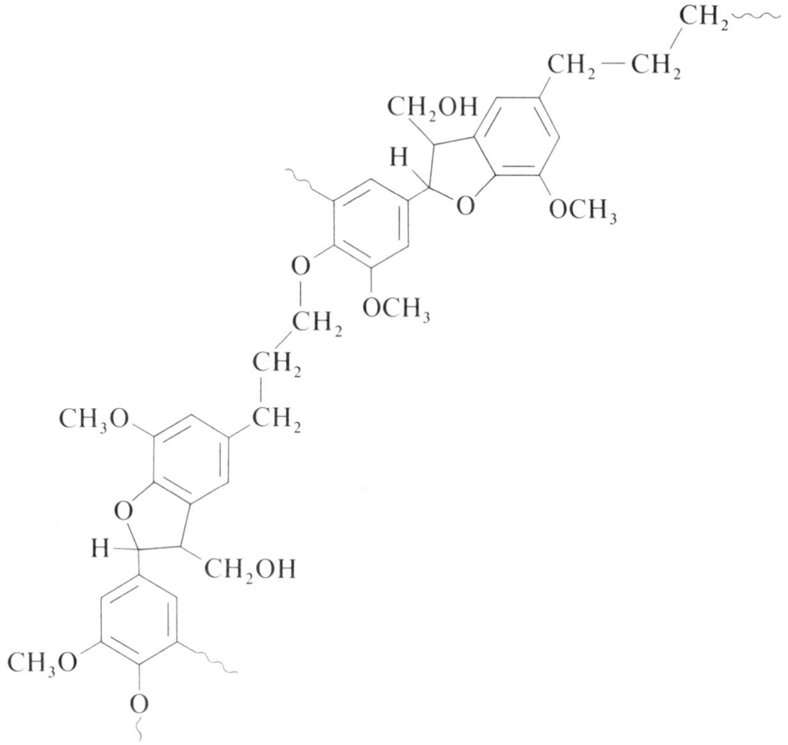

CH_2
CH_2-CH_2
CH_2OH
H
O
OCH_3
O
CH_2
OCH_3
CH_2
CH_3O
CH_2
O
H
CH_2OH
CH_3O
O

a fragment of lignin

The points at which the phenylpropane units are linked vary. The molecule is not linear but has many points of linkage, so it forms a three-dimensional network. The combination of the long crystalline fibers of cellulose with the network of lignin molecules penetrating them creates the strong, rigid structure of woody plants.

Rubber is another large molecule that owes its useful properties to its molecular size. When rubber is decomposed by heat, isoprene is formed (p. 705). The structure and chemistry of rubber is discussed in Section 27.5.

Proteins, cellulose, lignins, rubber, and the nucleic acids are naturally occurring representatives of a class of organic compounds of biological and, more recently, industrial importance. All these compounds have large molecular size in common. For this reason, they are known as **macromolecules,** or giant molecules. Their molecular structures can be dissected into smaller units. For example, proteins have units of amino acids, cellulose has units of glucose, and rubber has units of isoprene. The small unit that appears again and again in the macromolecule is called a **monomer,** and the large molecule that is composed of these units is known as a **polymer.** Isoprene is a monomer, and rubber is a polymer made up of many units of isoprene bonded together.

In some macromolecules, all the monomeric units are the same. Rubber contains only isoprene units. Cellulose is made up only of glucose units. In proteins or nucleic acids, on the other hand, the repeating monomeric units are not all identical. Chapter 26 showed that a large number of different protein molecules are possible because of the different order in which the twenty or so amino acids are incorporated into peptide chains. DNA and RNA are also made up of nucleotides with different structures, arranged in a chain with a varying order. Polymers in which there is more than one type of repeating unit are called **copolymers.** Rubber and cellulose are simple polymers, and proteins and nucleic acids are complex copolymers.

A polymer may also be classified according to the structure of the giant molecule. Rubber, cellulose, proteins, and nucleic acids are all **linear polymers.** The backbone of the macromolecule in each case consists of a long chain of monomer units held together by covalent bonds. The principal linkage between monomers in proteins is the peptide bond; in cellulose, it is the glycosidic linkage. Isoprene units are held together in rubber by covalent bonds between carbon atoms. In a linear polymer, once the chain is formed, other types of interactions, such as hydrogen bonding in proteins and van der Waals interactions between different parts of the hydrocarbon chain in rubber, contribute to the overall shape of the macromolecule.

Lignins represent another structural class of polymers, those in which there is additional covalent bonding between the monomer units. Not only do the monomers form long chains, but there are also covalent links between adjacent chains. Such polymers are said to be **cross-linked.** These links may be close together, as is the case in lignin, or they may be relatively far apart, as in some synthetic polymers. Such bonding affects the properties of a polymer and its potential uses. Later sections of this chapter will examine the relationship between structure and properties and show how the introduction of cross-linking affects these characteristics and thus changes the uses of a polymer.

Human beings have long made use of natural macromolecular substances. The structural rigidity of wood, which results from the properties of cellulose and lignin, makes it useful in construction. Many macromolecules have fibrous structures that can be spun into threads and then used to weave fabrics. Cellulose fibers in cotton and linen and protein fibers in silk and wool are used in clothing and household furnishings. Modern methods of transportation would not be possible without rubber. Human beings have also discovered ways to modify natural macromolecules to make them more useful. For example, a modification of cellulose led to the invention of paper, an important technological advance. Because only limited modifications could be made on natural molecules without destroying their essential structures, chemists started to think about creating new polymeric materials, starting with small reactive molecules. The remainder of this chapter is the story of their success.

Study Guide Concept Map 27.1

B. Macromolecules of Industrial Importance

By the middle of the nineteenth century, organic chemists had obtained as products in some of their experiments high-molecular-weight substances that were usually considered to be evidence of failed reactions. It was not until the early part of this century that chemists started to create polymers deliberately. To do this, they designed reactions that allowed them to control the average molecular weight, and therefore the properties, of the large molecules that were being formed. Hermann Staudinger of Germany, a pioneer in this field, was one of the first to recognize that controlling the conditions of polymerization was essential to the synthesis of useful substances. He received the Nobel Prize in 1953 for his work in this area of chemistry.

At first, chemists in this field tried to imitate nature. For example, the first really successful synthetic fiber was nylon, a polyamide created in the 1930s by the American chemist Wallace Carothers. The structure of nylon resembles a protein in having many amide bonds but is much more regular in its repeating units. Different kinds of nylons are shown below with the structural units that give rise to them.

$$\left[\overset{\displaystyle O}{\overset{\|}{C}}(CH_2)_4\overset{\displaystyle O}{\overset{\|}{C}}NH(CH_2)_6NH\right]_n \qquad H_2N(CH_2)_6NH_2 \qquad HO\overset{\displaystyle O}{\overset{\|}{C}}(CH_2)_4\overset{\displaystyle O}{\overset{\|}{C}}OH$$

nylon 66, a polyamide of 1,6-hexanediamine and hexanedioic acid (adipic acid)

$$\left[\overset{\displaystyle O}{\overset{\|}{C}}(CH_2)_{10}\overset{\displaystyle O}{\overset{\|}{C}}NH(CH_2)_6NH\right]_n \qquad H_2N(CH_2)_6NH_2 \qquad HO\overset{\displaystyle O}{\overset{\|}{C}}(CH_2)_{10}\overset{\displaystyle O}{\overset{\|}{C}}OH$$

nylon 612, a polyamide of 1,6-hexanediamine and dodecanedioic acid

$$\left[NH(CH_2)_5\overset{\displaystyle O}{\overset{\|}{C}}\right]_n \qquad H_2N(CH_2)_5\overset{\displaystyle O}{\overset{\|}{C}}OH$$

nylon 6, a polyamide of 6-aminohexanoic acid

n = number of repeating units in the polymer chain; for nylons that are useful as fibers, n = 50 to 120

The number in the name of a nylon indicates the structure of the polyamide. Nylon 66 is made from an amine and an acid, each having six carbon atoms; nylon 612 contains amine units with six carbon atoms and acid units with 12. Nylons of the appropriate molecular weight, where n is 50 to 120, can be formed into threads and used to make fabrics. In an early practical application, nylon was used to replace silk in women's stockings. It has since been used in a wide variety of fabrics, from carpets to parachutes to clothing of every description. The chemistry of the polymerization processes that give rise to nylons is discussed on pp. 1176 and 1185.

Since World War II, with the discovery that useful macromolecular compounds could be synthesized in the laboratory and then produced on a large scale in factories, much research has concentrated on developing new polymers. More industrial organic chemists work in polymer chemistry than in any other field. Objects made of polymers are so pervasive in everyday life that the names, especially the trade names, of polymers have become part of the language. News stories about the health-related and environmental effects of polymers and the monomers that go into their manufacture are common. Structural formulas for some familiar polymers, along with the monomers from which they are made, are given below.

$\left[CH_2CH_2\right]_n$	$CH_2{=}CH_2$
polyethylene	ethylene
$\left[CH_2\overset{\displaystyle Cl}{\overset{\vert}{C}}H\right]_n$	$CH_2{=}CHCl$
poly(vinyl chloride) PVC	vinyl chloride
$\left[CF_2CF_2\right]_n$	$CF_2{=}CF_2$
polytetrafluoroethylene Teflon	tetrafluoroethylene

$$\left[\mathrm{CH_2\underset{|}{C}H}\right]_n \ (\text{with } \mathrm{C{\equiv}N} \text{ on CH}) \qquad \mathrm{CH_2{=}CHC{\equiv}N}$$

a polyacrylonitrile
Orlon, Acrilan

acrylonitrile

poly(ethylene terephthalate)
a polyester
Dacron, Terylene

ethylene glycol

terephthalic acid

a polyurethane

ethylene glycol

m-phenylenediisocyanate
a diisocyanate

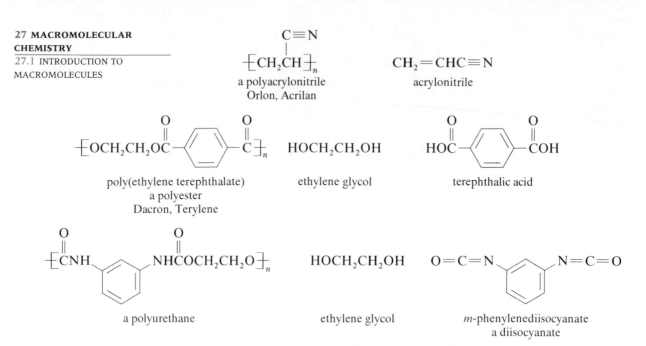

Polymers contain a wide variety of functional groups. These large molecules are made using essentially the same reactions as are applicable to low-molecular-weight compounds. Later sections in this chapter will concentrate on how these reactions generate large molecules having the physical properties that make them useful in practical applications.

C. Special Properties of Large Molecules

The literature of polymer chemistry reveals that researchers are intensely interested in the molecular weight of macromolecules. Every synthetic process is judged primarily by the range of molecular weights of the polymers formed. A synthesis of a polymer differs in this way from that of a low-molecular-weight organic compound. All the preparations in earlier chapters of this book have given compounds with well-defined structures. The synthesis of a polymer results in a mixture of compounds having a range of molecular weights. For example, the equation for the preparation of nylon 66 from hexanedioic acid and 1,6-hexanediamine can only indicate an approximate structure for the polymer that is formed.

$$n\ \mathrm{HO\overset{O}{\overset{\|}{C}}(CH_2)_4\overset{O}{\overset{\|}{C}}OH} + n\ \mathrm{H_2N(CH_2)_6NH_2}$$

hexanedioic acid — 1,6-hexanediamine, hexamethylenediamine

$$\downarrow$$

$$n\ {}^{-}\mathrm{O\overset{O}{\overset{\|}{C}}(CH_2)_4\overset{O}{\overset{\|}{C}}O^{-}} \qquad n\ \mathrm{H_3\overset{+}{N}(CH_2)_6\overset{+}{N}H_3}$$

salt of amine and acid

$$\downarrow \ \text{220 °C, } \sim\text{200–250 psi, 1–2 h}$$

$$\big\downarrow \text{270–280 °C}$$

$$-\!\!\!\!\left[\mathrm{NH(CH_2)_6NH\overset{\overset{\displaystyle O}{\|}}{C}(CH_2)_4\overset{\overset{\displaystyle O}{\|}}{C}}\right]\!\!\!\!-_n + 2n\ \mathrm{H_2O}\uparrow$$

nylon 66

The exact structure of the nylon molecule depends on how many of the repeating units become bonded together before the polymer chain stops growing. The repetitive structure of the polymer is indicated by the subscript n outside the brackets that enclose the repeating unit. The larger n is, the higher the molecular weight of the polymer. The molecules of a polymer formed in any reaction mixture do not have identical numbers of repeating units and thus differ in molecular weight. The product is a mixture of similar molecules that differ somewhat in size.

The concern about molecular weights is justified because the physical properties, and thus the practical uses, of a polymer depend heavily on its molecular weight. For example, nylon with a low molecular weight has no useful properties. It is a brittle solid. Only when the molecular weight reaches 10,000 does a nylon begin to show properties that make it useful as a fiber. Nylons that have molecular weights above 100,000 do not make good fibers but have high resistance to heat and high mechanical strength, so they are used in other industrial applications. One such nylon, reinforced with glass fibers, is used instead of steel to make valve covers for automobile engines.

Molecular weight is, of course, a reflection of molecular size. The molecules of a polymer must have a minimum size before the interactions that are important in determining the properties of the substance can take place. These interactions may be between different parts of the same molecule or between adjacent molecules. All the factors discussed in Chapter 26 in connection with the tertiary structure of proteins, such as hydrogen bonding and van der Waals interactions, are also important in determining the shapes of other macromolecules. The nature of the repeating unit in a polymer is, of course, of primary importance in determining which types of interactions are possible. For a nylon, which has a large number of amide linkages, hydrogen bonding may be important. In polyethylene, which is like a gigantic alkane molecule, van der Waals interactions between different parts of the molecule and between adjacent chains are most likely.

The shape of a polymer molecule and the types of interactions that it can have with neighboring molecules are determined by the regularity with which repeating units appear in the chain and by the stereochemistry at points where stereoisomerism is possible. For example, just like a low-molecular-weight alkene, a polymer may contain cis or trans double bonds. Rubber is an example of a polymer with cis double bonds. The trans isomer is a natural product known as gutta percha and has some properties that differ notably from those of rubber (p. 1204). If the repeating units in a polymer are chiral, as is the case for amino acids in proteins and glucose in cellulose, the macromolecule as a whole will exhibit chirality. Even a polymer formed from an achiral monomer such as propene has the possibility of stereoisomerism at the newly created tetrahedral carbon atoms along the backbone of the chain (p. 1190). In summary, polymers can exist as stereoisomers with differing stereochemistry at double bonds or at tetrahedral carbon atoms.

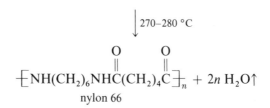

A given polymer can also exist in a large number of conformations arising from free rotations of the atoms that make up the backbone of the chain. Just as conformational isomerism is possible for butane with four carbon atoms in its chain (p. 150), the molecules in a given sample of a polymer exist in many different and constantly changing conformations. For polymers, too, anti and gauche arrangements of groups on adjacent carbon atoms are preferred over eclipsed conformations. The conformations that are favored and the range of motions possible for different parts of polymer chains are important in determining the properties of a polymer and how it will behave under different conditions.

One of the properties of a polymer that is very important for practical applications is how it behaves at different temperatures. The interactions between the large molecules of a polymer create solids that have a high degree of structural regularity. Linear polyethylene (p. 1191) is a highly crystalline substance with a melting point of approximately 135 °C. The polyethylene produced by some manufacturing processes has branching on the chain and does not form as highly crystalline a solid. Branched polyethylene has a lower melting point, about 120 °C. Some polymers, such as rubber, do not pack together well in regular structures and exist mostly as amorphous solids. Many polymers are partly crystalline, meaning that when the liquid polymer is cooled, parts of it form regions with a high degree of order and other parts solidify before such arrangement takes place. Nylon is such a partly crystalline polymer. These different kinds of interactions between polymer chains are represented in Figure 27.1.

Whether or not a polymer has crystalline regions determines properties such as its flexibility and mechanical strength. For example, the amorphous and coiled structure of rubber is responsible for its elasticity. Polymers with such elastic qualities are known as **elastomers.** Semicrystalline polymers are rather hard and can be drawn out into strong **fibers.** The process of drawing them out increases the alignment of the polymer chains and thus their crystallinity. Unless a polymer has

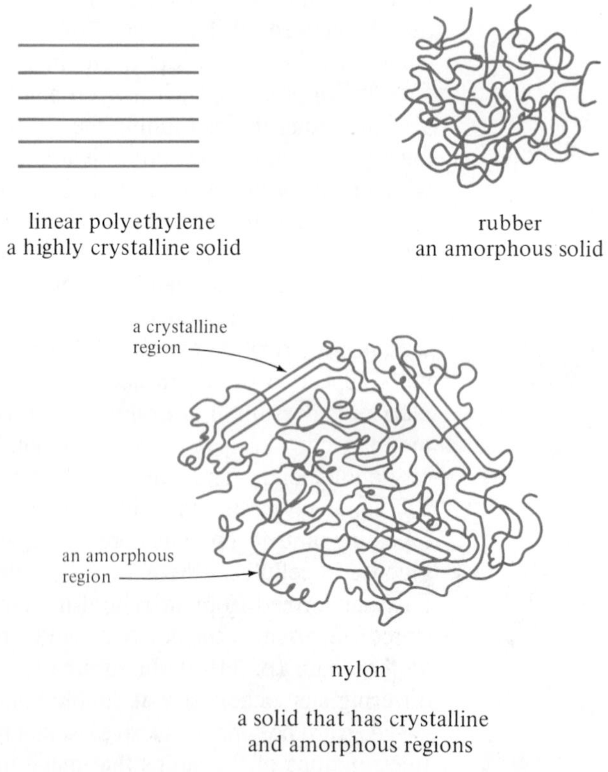
linear polyethylene
a highly crystalline solid
rubber
an amorphous solid
a crystalline region
an amorphous region
nylon
a solid that has crystalline and amorphous regions

FIGURE 27.1 Schematic representations of crystalline, amorphous, and semicrystalline polymers.

some crystalline regions, it cannot be formed into fibers; thus, an amorphous polymer does not make good fibers. Not all amorphous polymers have elasticity; those that do not are called **plastics** and are used in applications in which they can be molded under heat and pressure into objects such as toys and household goods. Polymers are classified as elastomers, fibers, or plastics according to their properties and their uses.

A polymer can exist in different physical forms at differing temperatures. At low temperatures, it exists as a solid. In the solid state, it may be partly or highly crystalline, or it may be amorphous. A solid that is not crystalline is called a **glass.** Transparency, for example, is typical of a solid that is a glass; a crystalline solid is opaque.

On being warmed to a given temperature, a polymer softens to a more pliable state but does not actually melt. This temperature varies a bit according to the method used to make the determination, but is typical for each compound. For rubber, this temperature is −70 °C; for nylon 66, it is 50 °C. Thus, rubber, as we know it, is in an intermediate state in which it is not a rigid solid but is not a liquid either. If a rubber ball is made very cold by putting it into liquid nitrogen (−196 °C), it loses all its bounce and shatters if thrown against a hard surface. This is typical of the brittle character of a glass.

At an even higher temperature, a polymer melts and becomes a liquid. This temperature is the melting point of the polymer, comparable to the melting point of a lower-molecular-weight organic compound. The melting point of raw rubber is 30 °C; for nylon 66, it is 265 °C. Knowledge of these properties of a polymer is clearly important. For example, raw rubber is useful at temperatures between −70 °C and +30 °C. Raw rubber, therefore, would be useless on a hot day. Similarly, the nylon that is used in valve covers for automobile engines must retain its shape and must therefore remain a solid at high temperatures. This nylon withstands exposure to temperatures of approximately 120 °C for long periods.

Study Guide
Concept Map 27.2

D. Types of Polymerization Reactions

Some of the polymers shown on pp. 1171–1172 contain all of the atoms of the monomeric unit within the polymer. For example, poly(vinyl chloride) consists of a large number of molecules of vinyl chloride bonded to each other by carbon-carbon bonds created at the doubly bonded carbon atoms in the monomer.

$$n\,CH_2{=}CHCl \xrightarrow[\text{water, gelatin, 50 °C, 14–18 h}]{(CH_3(CH_2)_{10}\overset{O}{\overset{\|}{C}}O)_2} \left[CH_2\underset{}{\overset{Cl}{\overset{|}{C}}}H\right]_n$$

vinyl chloride — *monomer*; poly(vinyl chloride) — *polymer*, ~90%

Such polymers are called **addition polymers** because they are created by the addition of one molecule of monomer to another one.

A polyamide, on the other hand, is formed when a diamine reacts with a diacid with the loss of molecules of water (p. 1172). Such polymers are called **condensation polymers** because they are formed by condensing together two different types of functional groups with the elimination of a small stable molecule, such as water. The water that is formed is driven off at high temperatures as steam to bring the reaction to completion.

A much more useful way to classify polymerization reactions is by their mechanisms. The types of functional groups present and their reactivity under different conditions determine the extent of polymerization and, therefore, the molecular weights of the products. Polymerization reactions, like other organic reactions, can have cationic, anionic, or radical intermediates. Polymers formed in reactions that proceed through these different intermediates often have different properties.

How a polymer chain grows is of great importance in determining how long the chain will be, which determines the molecular weight of the polymer. The reactions that give rise to polymers have been classified into two main types. One type is called **chain-growth polymerization.** In this type of polymerization, a reactive intermediate is formed and reacts rapidly with a monomer molecule to give a new reactive intermediate, which in turn reacts with yet another monomer molecule. The monomer is consumed rapidly and always adds onto a chain that gets longer and longer. A new reactive center is created at the end of the polymer chain, so the reaction perpetuates itself until all the monomer molecules have reacted or the reactive intermediate is destroyed by some end reaction. You probably recognize this description as corresponding to the steps of a free-radical chain reaction (pp. 823–825). There is an initiation step, many propagation steps in which each reactive intermediate generates a new reactive intermediate, and finally a termination step. Such a reaction often involves free radicals, but under the proper conditions chain reactions with cationic or anionic intermediates are also possible. Some of them will be examined in Sections 27.2B and 27.2C. The important factor in all of these reactions is the presence of a reactive intermediate that directs the course of the reaction, making it more probable at one molecular site than at others. In fact, in such a reaction the monomer units are usually incapable of reacting with each other until some reagent is added to create the reactive intermediate. The polymerization of an alkene, such as the reaction of vinyl chloride (p. 1175), is usually a chain-growth reaction.

The other type of polymerization is called a **step-growth reaction.** In such a reaction, the monomer units contain functional groups that are capable of reacting with each other without the formation of a reactive intermediate. The reactions are generally slower than those of chain-growth polymerizations, and the reaction sites are more random. The formation of a polyamide from a diamine and a diacid is an example of a step-growth reaction.

First step of polymerization

$$H_2N(CH_2)_6NH_2 + HO\overset{\overset{O}{\|}}{C}(CH_2)_4\overset{\overset{O}{\|}}{C}OH \longrightarrow H_2N(CH_2)_6NH\overset{\overset{O}{\|}}{C}(CH_2)_4\overset{\overset{O}{\|}}{C}OH + H_2O$$

1,6-hexanediamine hexanedioic acid
monomers

reactivity of amine function and carboxyl function not much different from reactivity of these functional groups in monomers

Second step of polymerization

$$2\ H_2N(CH_2)_6NH_2 + 2\ HO\overset{\overset{O}{\|}}{C}(CH_2)_4\overset{\overset{O}{\|}}{C}OH + 2\ H_2N(CH_2)_6NH\overset{\overset{O}{\|}}{C}(CH_2)_4\overset{\overset{O}{\|}}{C}OH$$

$$\downarrow$$

$$\underset{\text{monoamide}}{H_2N(CH_2)_6NH\overset{\overset{O}{\|}}{C}(CH_2)_4\overset{\overset{O}{\|}}{C}OH} + \underset{\text{diamide from 1 diamine and 2 diacid units}}{HO\overset{\overset{O}{\|}}{C}(CH_2)_4\overset{\overset{O}{\|}}{C}NH(CH_2)_6NH\overset{\overset{O}{\|}}{C}(CH_2)_4\overset{\overset{O}{\|}}{C}OH}$$

$$+ \underset{\text{diamide from 1 diacid and 2 diamine units}}{H_2N(CH_2)_6NH\overset{\overset{O}{\|}}{C}(CH_2)_4\overset{\overset{O}{\|}}{C}NH(CH_2)_6NH_2} + 3\,H_2O$$

A closer look at the individual steps of the above reaction reveals the problems associated with step-growth polymerization. The first step of the polymerization gives an amide formed from the diamine and the diacid. The amide retains amino and carboxylic acid groups, and the reactivities of these functional groups are not much different from the reactivities of the same functional groups in the monomers. The next stage is a random reaction of the carboxylic function of the monoamide with the diamine and of the amine function of the monoamide with another molecule of diacid. Instead of a steady and directed growth of a chain, a mixture of molecules results. Large increases in chain length will take place only when the monomer units have been used up and amides containing several units combine with each other. The polymerization process takes place in jumps rather than as a unit-by-unit addition to a continuously growth chain.

PROBLEM 27.1

Write equations showing the intermediate steps in the formation of poly(vinyl chloride), which is formed by a free-radical chain reaction (p. 823). The chain initiator is the dodecanoyloxy radical,

$$CH_3(CH_2)_{10}\overset{\overset{O}{\|}}{C}-O\cdot$$

PROBLEM 27.2

Draw structural formulas for the monomers that were used to produce the following polymers.

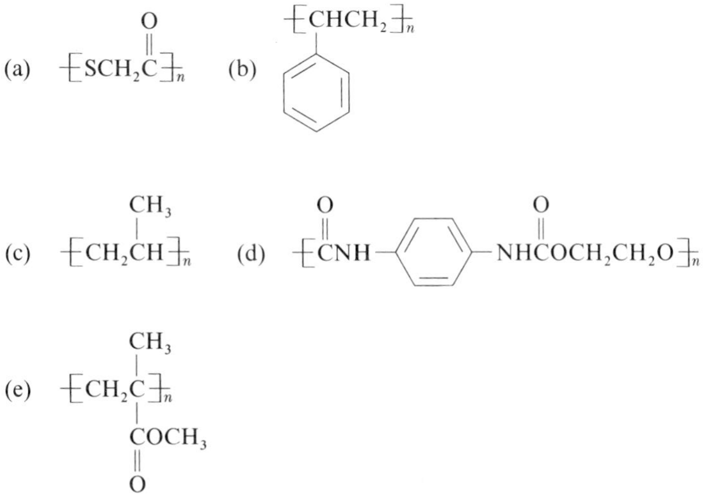

A. Free-Radical Reactions

The polymerization of styrene in the presence of a small amount of benzoyl peroxide is a typical example of a free-radical polymerization reaction.

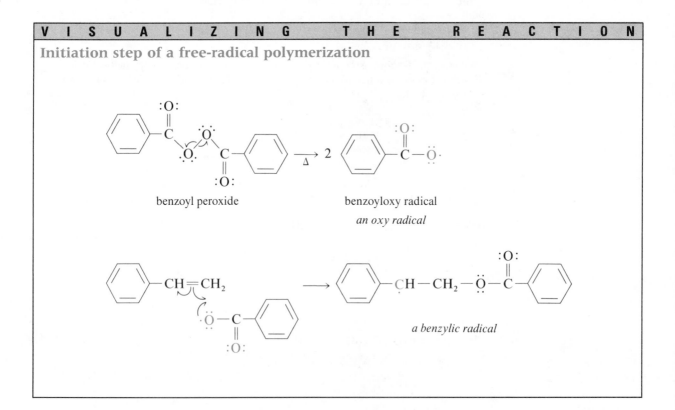

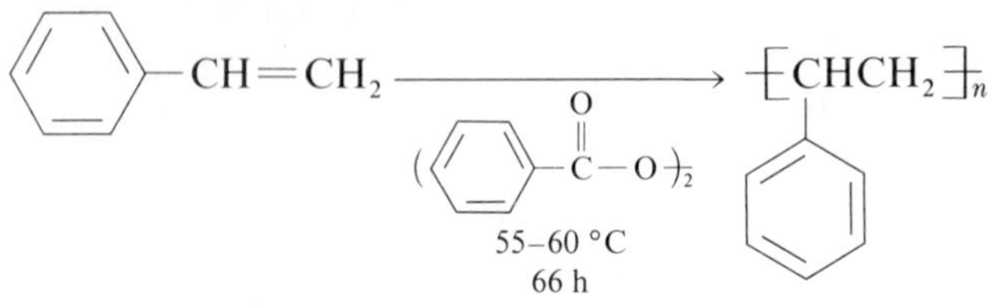
CH=CH₂
O
C—O
2
55–60 °C
66 h
CHCH₂
n
styrene
polystyrene
~100%

The reaction takes place in three stages. First, in the initiation step, benzoyl peroxide dissociates into two benzoyloxy radicals. A benzoyloxy radical, an electrophile, then reacts with the π electrons of the double bond in styrene to create a new radical, the stable benzylic radical.

VISUALIZING THE REACTION

Initiation step of a free-radical polymerization

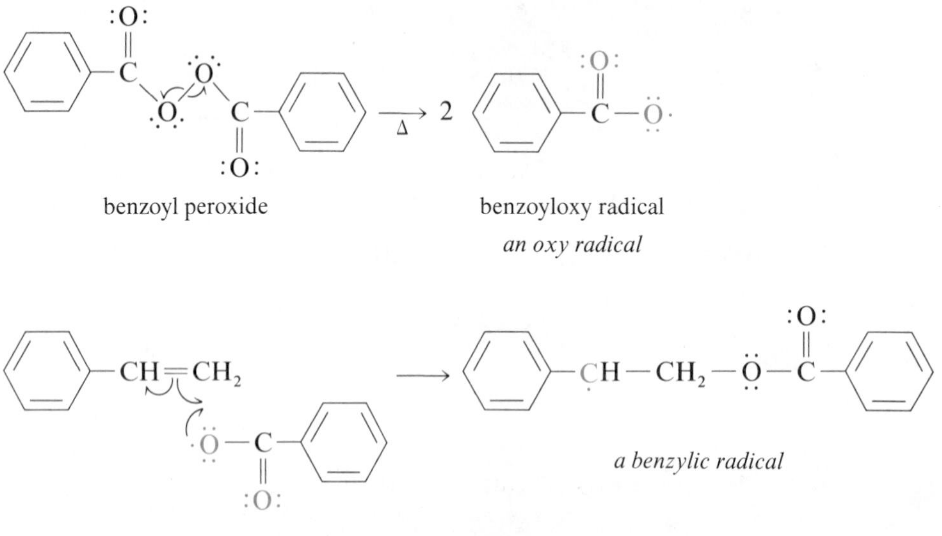
:O:
C
O
O
C
:O:
Δ
2
:O:
C—O·
benzoyl peroxide
benzoyloxy radical
an oxy radical
CH=CH₂
·O—C
:O:
CH—CH₂—O—C
:O:
a benzylic radical

The benzylic radical attacks another molecule of styrene, creating yet another radical intermediate.

VISUALIZING THE REACTION

Chain-propagation step of a free-radical polymerization

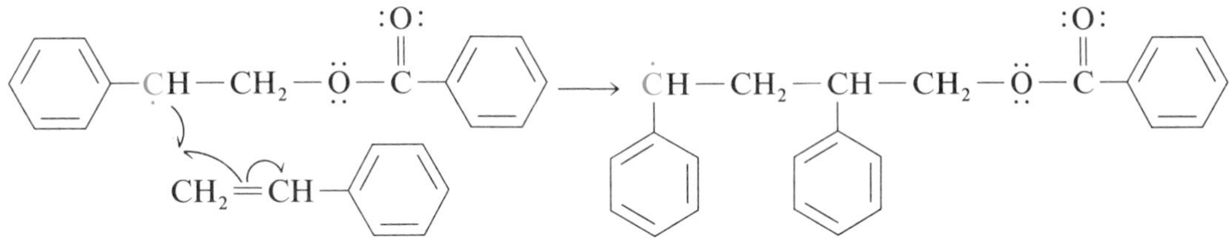

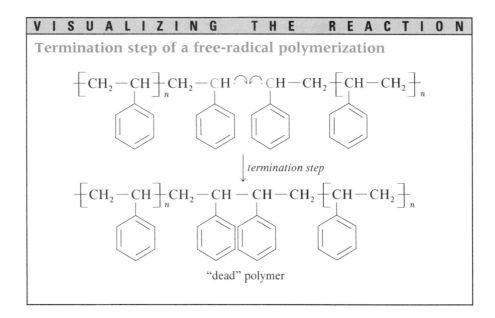

Each new radical continues to react with a molecule of the monomer in a series of chain-propagating steps. The growth of the chain of polystyrene is fast; it has been calculated that approximately 1500 monomer units are added to the chain each second. The reaction continues until all the monomer molecules have reacted or the radical intermediate is destroyed by one of a number of termination reactions. For example, a combination of two radicals will stop the polymerization process. Most polystyrene chains stop growing as a result of the combination of two polystyryl radicals, which gives a species with no reactive sites, called a **"dead" polymer.**

VISUALIZING THE REACTION

Termination step of a free-radical polymerization

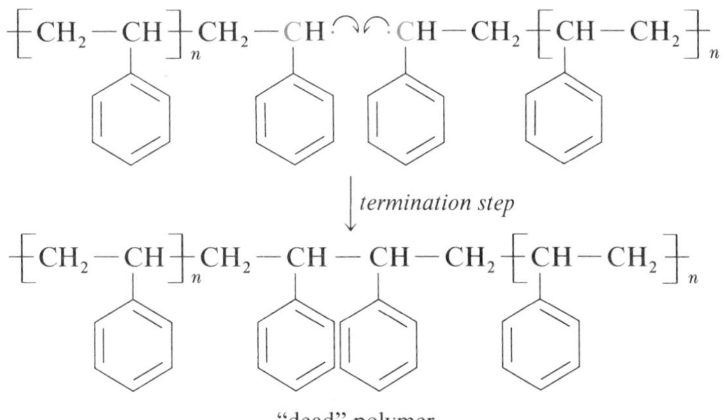
termination step
"dead" polymer

The polymer radical may also abstract a hydrogen atom from another chain, creating a new radical in what is known as a **chain-transfer reaction.**

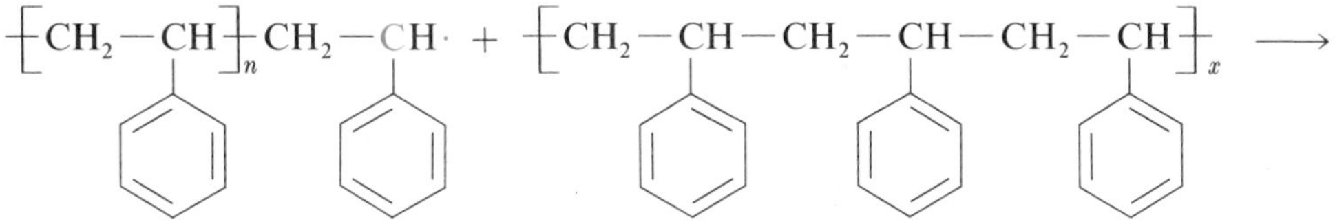

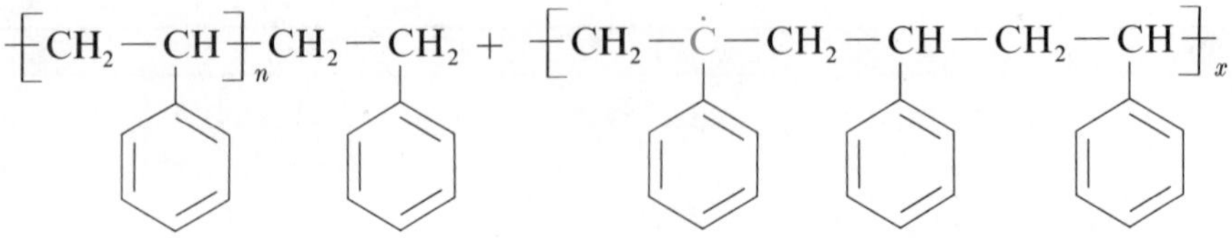

"dead" polymer

a new radical created by abstraction of a benzylic hydrogen atom from the middle of a polymer chain; a chain-transfer reaction

This new radical is in the middle of a chain, not at the end. It can react with a monomer unit, giving rise to branching of the chain.

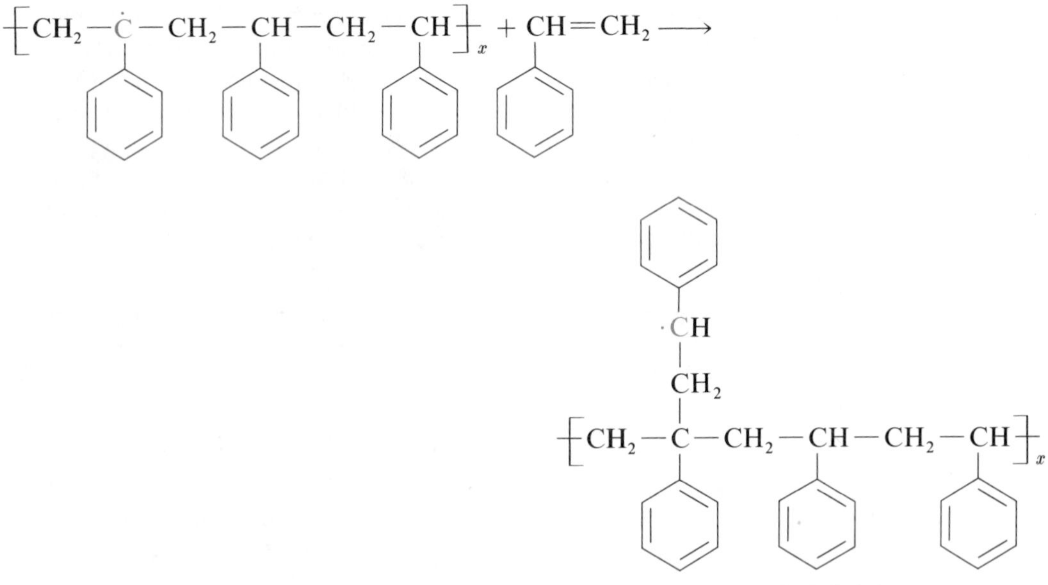

branched chain

Note that in writing the structure of the polymer, the ends of the molecule are left unspecified. The end groups are an insignificant portion of the total chain and are not necessarily identical for every molecule in the sample. The properties of the polymer are determined largely by the main body of the chain. Polystyrene, polymerized simply by adding a small amount of benzoyl peroxide to the monomer, reaches molecular weights of about 2,500,000 and is amorphous.

Polystyrene has many uses. The polymer may be molded into cases for radios and batteries. Toys and all kinds of containers are made from it. If a low-boiling hydrocarbon such as pentane or a chlorofluorocarbon (a Freon) is included in the polystyrene during processing, when the polymer softens as it is heated, the added compound vaporizes and creates bubbles that expand the polystyrene into a rigid but lightweight foam. This foam has very good insulating properties and is used as insulation in construction. Pellets of the foam are supplied to manufacturers, who use pressure and some heat to mold them into ice chests and disposable plastic cups for hot drinks. Egg cartons, which have to be rigid but cushiony at the same time, are made of polystyrene foam. The foam can be molded to the exact shape of an object that needs to be protected during transportation; thus, many delicate instruments and bottles of chemicals travel in their own polystyrene foam sheaths.

A multitude of copolymers involving styrene have also been designed to have particular properties. For example, a copolymer of styrene with acrylonitrile is superior to polystyrene in toughness, stability to light, and resistance to chemicals. The polymer, produced by polymerization of a mixture of the two monomers, is believed to have mostly an alternating arrangement of the two units in the chain.

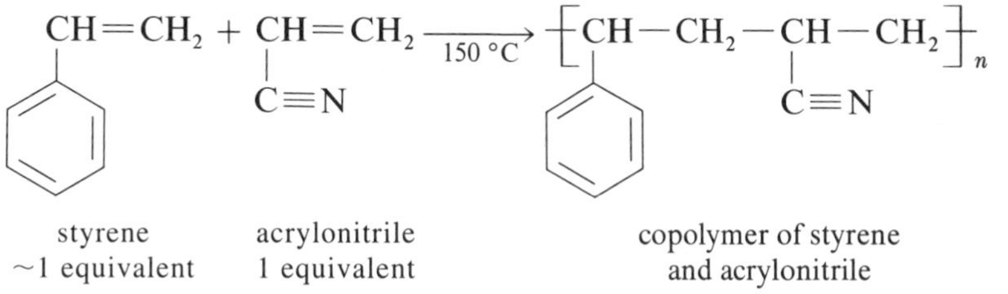

CH=CH2 + CH=CH2
C≡N
150 °C
CH—CH2—CH—CH2
n
C≡N
styrene
~1 equivalent
acrylonitrile
1 equivalent
copolymer of styrene
and acrylonitrile

The exact arrangement of monomer units in a copolymer depends on the relative reactivities and relative concentrations of the two monomers in the reaction mixture. Styrene-acrylonitrile copolymers have exceptional clarity and strength and have been made into a variety of objects, such as lenses for automobile headlights, household containers, and many devices for medical use, including disposable syringes and parts for artificial kidneys.

Another important copolymer of styrene is the cross-linked one it forms with *p*-divinylbenzene.

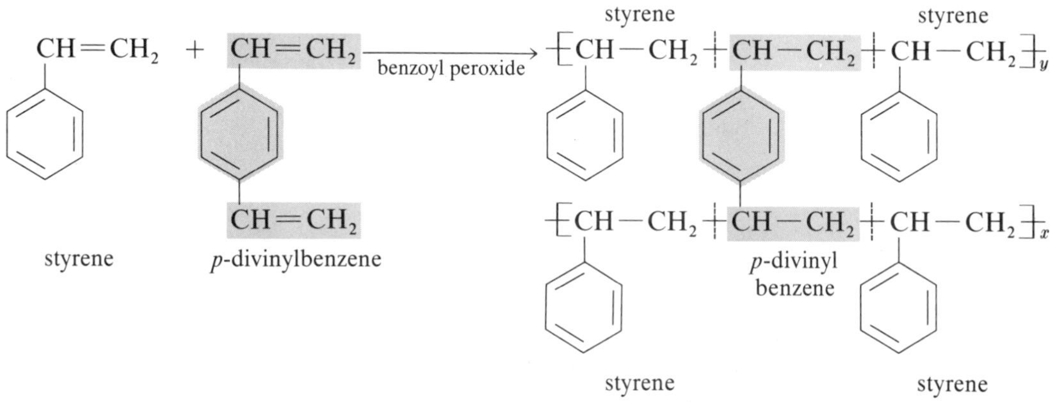

CH=CH2 + CH=CH2
CH=CH2
benzoyl peroxide
styrene
styrene
CH—CH2—CH—CH2—CH—CH2
y
CH—CH2—CH—CH2—CH—CH2
x
styrene
p-divinylbenzene
p-divinyl
benzene
styrene
styrene

copolymer of styrene and p-divinylbenzene showing how the p-divinylbenzene forms a cross-link between two polymer chains

Because *p*-divinylbenzene has two alkene functions, it becomes part of two polymer chains, forming a link between them. Depending on the relative amounts of styrene and *p*-divinylbenzene used, the links may be close together or widely separated.

In one practical application, the cross-linked copolymer of styrene and *p*-divinylbenzene, modified by incorporating polar functional groups such as sulfonic acid groups on the aromatic rings, is used as an ion-exchange resin. The whole polymer molecule, as its sodium salt, acts as a gigantic insoluble anion, which can exchange its cations with those in the solution that flows through it. Such resins soften water by exchanging sodium ions for the calcium and magnesium ions in hard water (p. 613) and are used in both water-treatment plants and water-softening units in homes. A styrene-divinylbenzene copolymer is also the support system for an automated solid-phase synthesis of peptides and proteins (p. 1146).

PROBLEM 27.3

For each of the following polymers, show the monomers that would be used to prepare it. Also, say whether each one is a copolymer or a simple polymer.

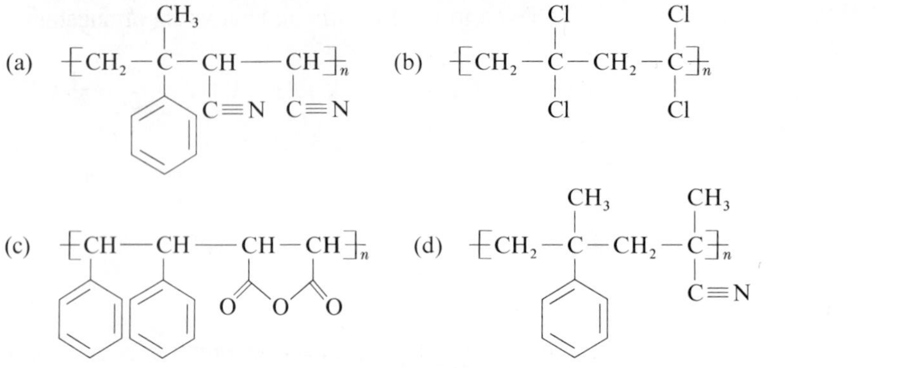

PROBLEM 27.4

The ion-exchange resin used for water softening is made by introducing sulfonic acid groups into the cross-linked copolymer of styrene and divinylbenzene. How would you carry out such a reaction?

B. Anionic Reactions

Some polymerization reactions are catalyzed by alkali metals or by organometallic compounds. The reactive species that participates in the growth of the chain in these cases is a carbanion. 2-Phenylpropene, commonly called α-methylstyrene, polymerizes by such a reaction.

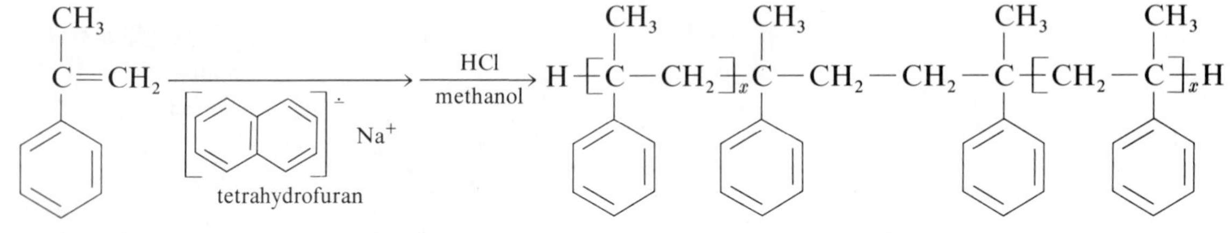
HCl
methanol
tetrahydrofuran

α-methylstyrene poly(α-methylstyrene)

The reaction is catalyzed by the radical-anion (p. 343) of naphthalene in tetrahydrofuran. The radical-anion of naphthalene transfers an electron (and the accompanying negative charge) to α-methylstyrene.

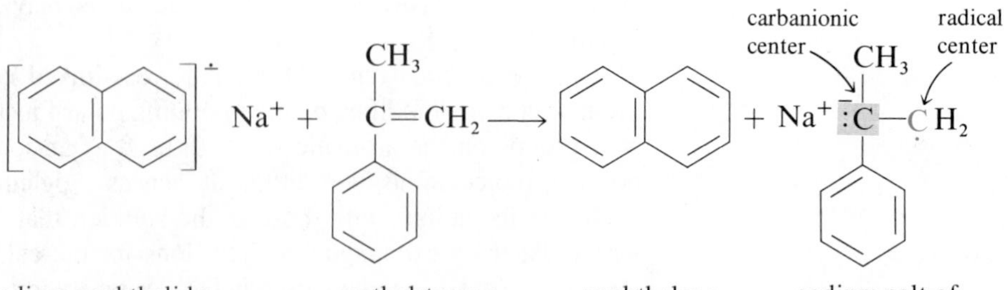
carbanionic center
radical center

sodium naphthalide; radical-anion of naphthalene

α-methylstyrene

naphthalene

sodium salt of the radical-anion of α-methylstyrene

Note that one electron has been added to the alkene. For the sake of clarity, this electron and the two electrons of the π bond are shown as if they were localized to give a radical center and a carbanionic center in the radical-anion. This is, of course, a highly simplified way of representing the reactive species, but it is useful in rationalizing the course of the reaction, which proceeds by a combination of two radicals to give a dianion.

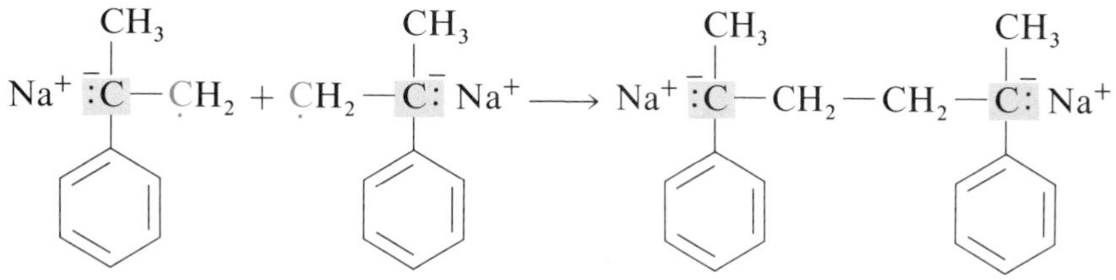

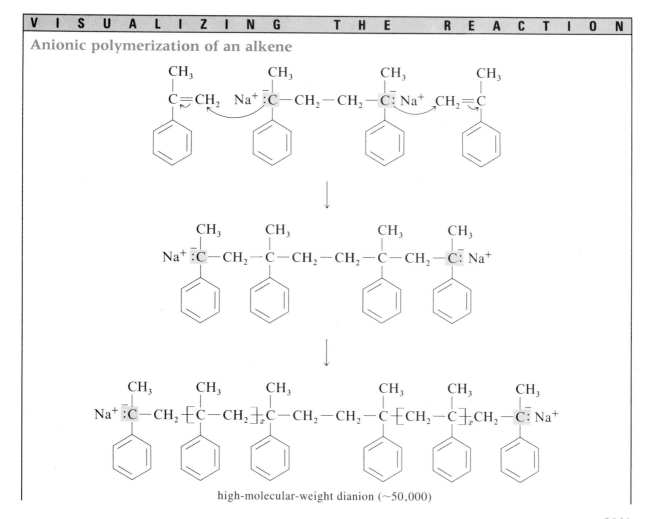

sodium salt of the radical-anion of α-methylstyrene

dianion
product of the reaction
of 2 radicals

The dianion is coordinated with sodium ions. It grows by adding monomer units at either end, creating new anions, which are stabilized by their association with cations.

VISUALIZING THE REACTION

Anionic polymerization of an alkene

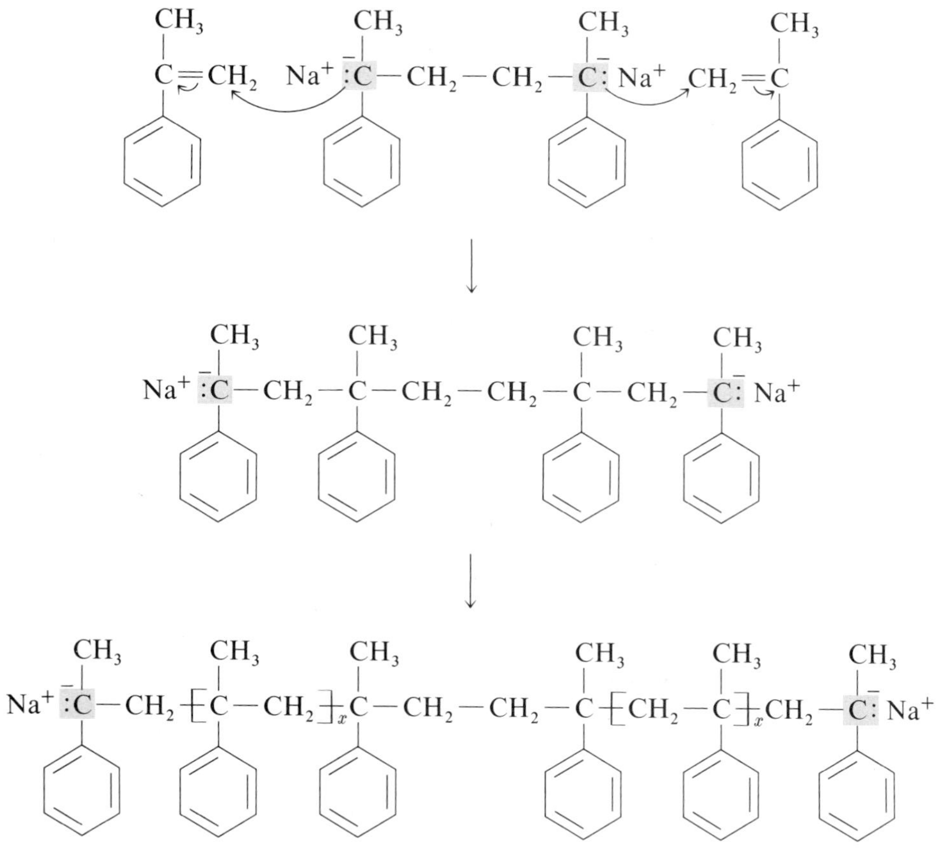

high-molecular-weight dianion (~50,000)

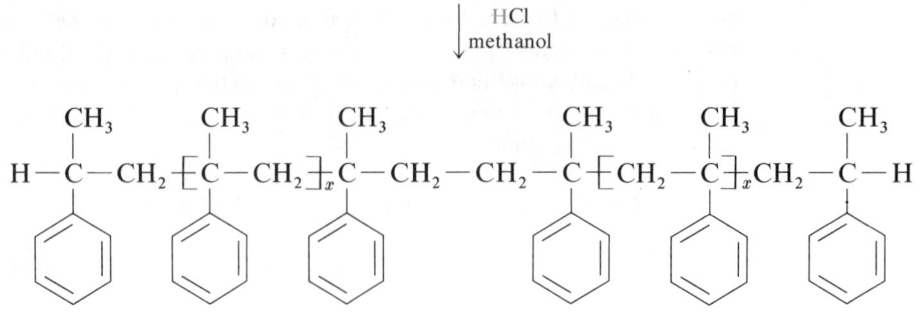

HCl
methanol

poly(α-methylstyrene)

The chain grows at both ends until some reagent that reacts with a carbanion is added to the reaction mixture. For example, the anions may be protonated by a dilute solution of hydrochloric acid in methanol.

This type of polymerization reaction was investigated extensively by Michael Szwarc at the State University of New York at Syracuse. He called the systems **living polymers** because the chains remain reactive until some reagent is deliberately added to stop the reaction. Note that two anionic centers have no tendency to react with each other. Thus, termination reactions involving the combination of two growing chains, an important reaction for free radicals, do not occur in anionic polymerizations. If the reagents are pure and the reaction mixture is protected from moisture, the polymer chains remain reactive. An anionic polymerization differs from a free-radical reaction in another way. Because all chains start at the same time, at the moment the anionic initiator is added to the reaction mixture, and the intermediates react with monomer at the same rate, polymers of remarkably uniform chain lengths and molecular weights are formed. In comparison, there is a much larger range of molecular weights in any sample of a polymer prepared by free-radical reactions.

Copolymerization of two monomers by an anionic mechanism can be done in such a way that large portions of the polymer chain consist of units of one monomer and other portions contain only units of the second one. Such a polymer is called a **block copolymer,** in contrast to a **random** or **alternating copolymer.** For example, the free-radical polymerization of a mixture of styrene and acrylonitrile gives an alternating copolymer (see p. 1181). It is also possible to create a block copolymer of these two monomers, as is illustrated in Figure 27.2. By controlling the concentration of the initiator and the amount of the first monomer used, the length of the chain in the first dianion can be determined. Addition of a known amount of a second monomer results in the growth of the chain to create a new dianion. Alternating additions of styrene and acrylonitrile give a long chain in which sections consisting of polystyrene are attached to sections that are polyacrylonitrile.

Nucleophilic opening of a ring to give an intermediate that will propagate a chain reaction is an important method of anionic polymerization. Nylon 6, for example, is prepared by the polymerization of the seven-membered cyclic lactam ϵ-caprolactam (p. 916). The reaction, shown on the following page, is catalyzed by adding any one of a number of amines to the reaction mixture as the salt of an organic acid.

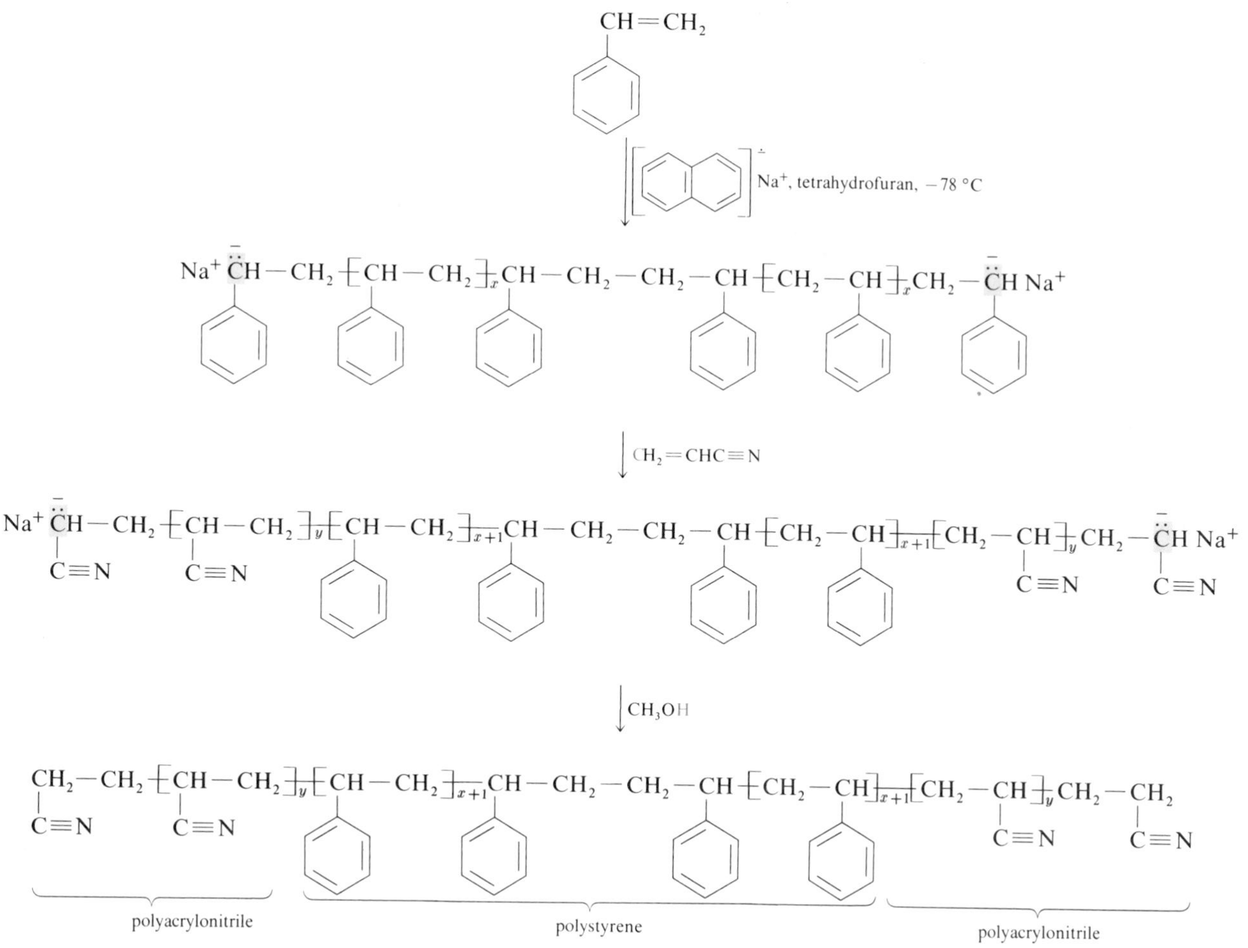

FIGURE 27.2 Formation of a block copolymer by the anionic polymerization of styrene and acrylonitrile.

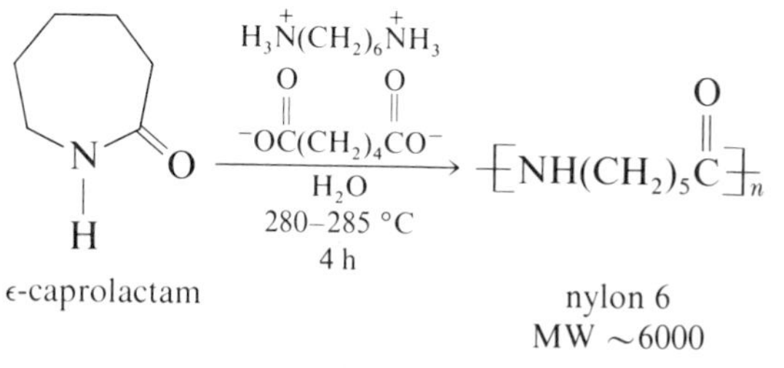

Caprolactam polymerizes so easily that water in catalytic amounts is enough to start the reaction at high temperatures. The reaction proceeds by attack of a nucleophile on the carbonyl group of the cyclic amide, giving rise to an amine, which then reacts with the carbonyl group of another molecule of the lactam. The reaction perpetuates itself because a new nucleophilic center is created every time a lactam ring is opened.

VISUALIZING THE REACTION

Anionic polymerization of a caprolactam

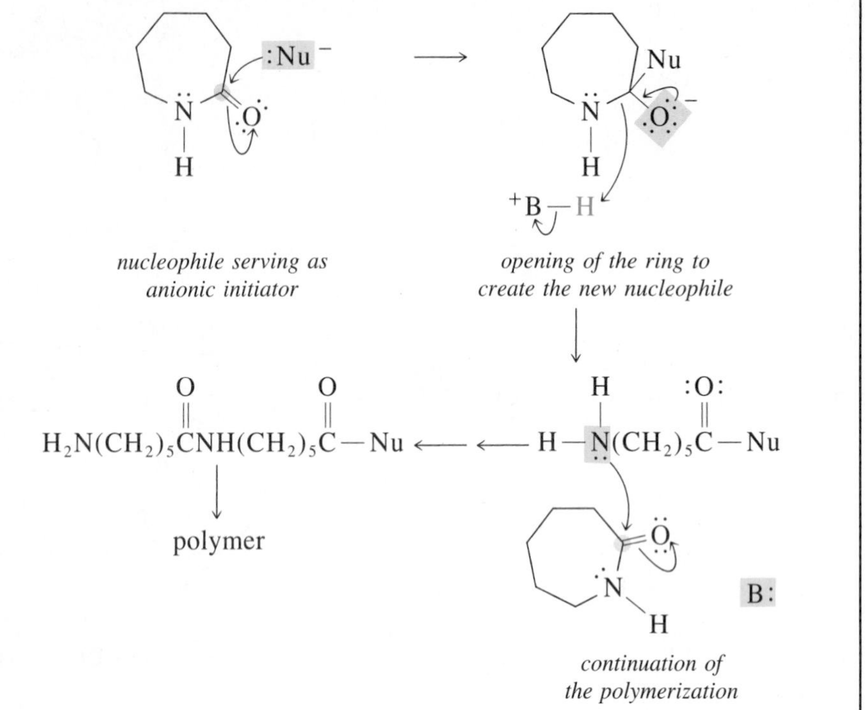
nucleophile serving as anionic initiator
opening of the ring to create the new nucleophile
polymer
continuation of the polymerization

The polymerization of caprolactam can be classified as a chain-growth reaction with anionic intermediates.

PROBLEM 27.5

Complete the following equations, showing the intermediates that would be responsible for the polymerization reactions.

(a) 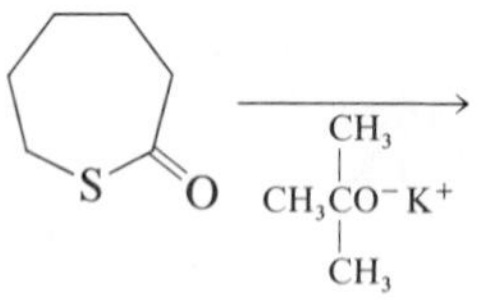

(b) 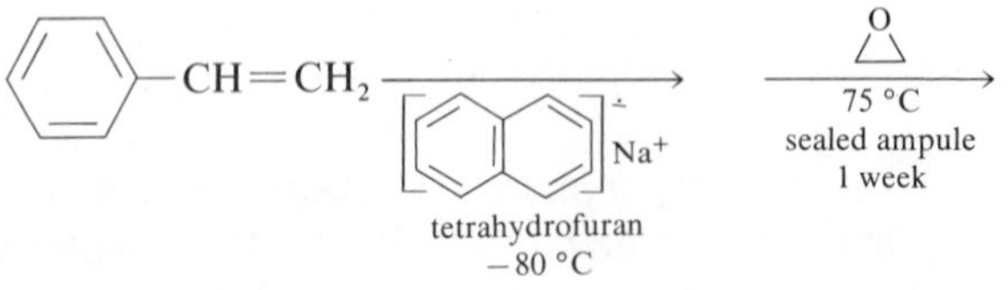
tetrahydrofuran
−80 °C
75 °C
sealed ampule
1 week

(c) $CH_2{=}CH\overset{O}{\overset{\|}{C}}OCH_3 + CH_2{=}CHCl \xrightarrow[\text{acetone, 50 °C}]{\text{benzoyl peroxide}}$

(d) 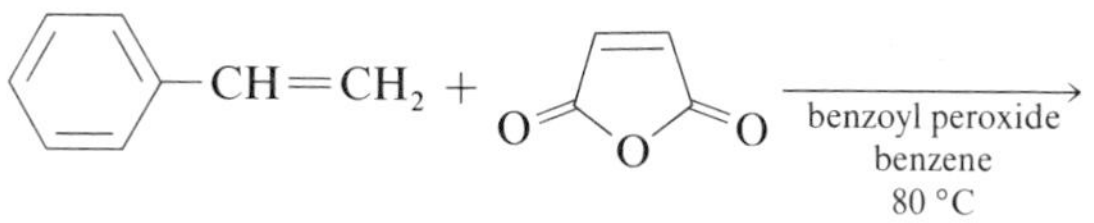

(e) 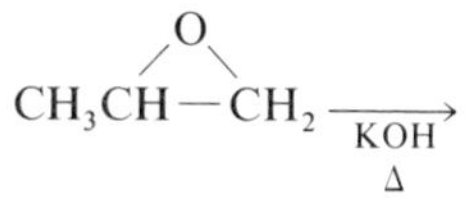

(Hint: At which carbon atom of the ring is the attack of the nucleophile more likely?)

PROBLEM 27.6

A block copolymer of methyl 2-methylpropenoate (methyl methacrylate) and isopropyl propenoate (isopropyl acrylate) has been prepared. Write equations showing how you would carry out such a preparation.

C. Cationic Reactions

The acid-catalyzed polymerization of an alkene was explored fairly early in this book (p. 285). Sulfuric acid, for example, will cause the polymerization of styrene, as will Lewis acids. When Lewis acids such as stannic chloride are used as polymerization catalysts, traces of water or a hydrogen halide are necessary for the reaction to take place, suggesting that some protic acid intermediate is involved. Carbocations are the chain-propagating species.

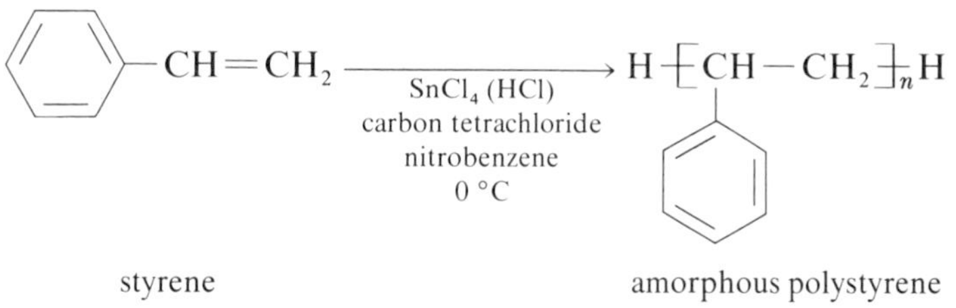
CH=CH$_2$
SnCl$_4$ (HCl)
carbon tetrachloride
nitrobenzene
0 °C
H⁅CH—CH$_2$⁆$_n$H
styrene
amorphous polystyrene

VISUALIZING THE REACTION

Cationic polymerization of styrene

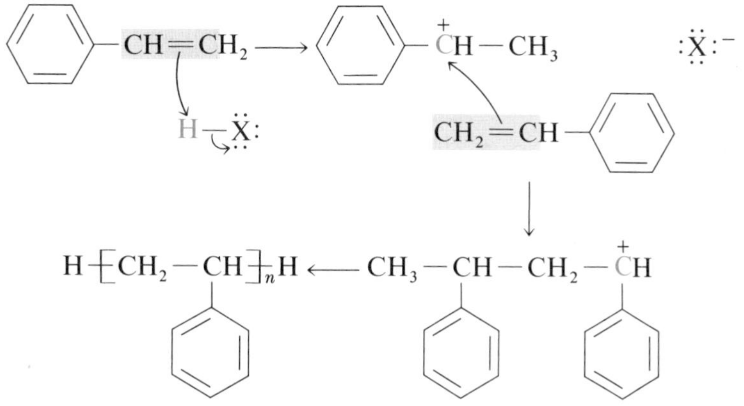

As in anionic polymerization, chain-termination reactions involving the combination of two reactive intermediates are not possible in cationic polymerization. The carbocation intermediates, however, are much more likely than the anions are to undergo other types of termination reactions. The reactions that terminate chains in cationic polymerizations are the familiar reactions of carbocations. Among them are reaction with a nucleophile, loss of a proton to give an alkene, and abstraction of a hydride ion from another molecule.

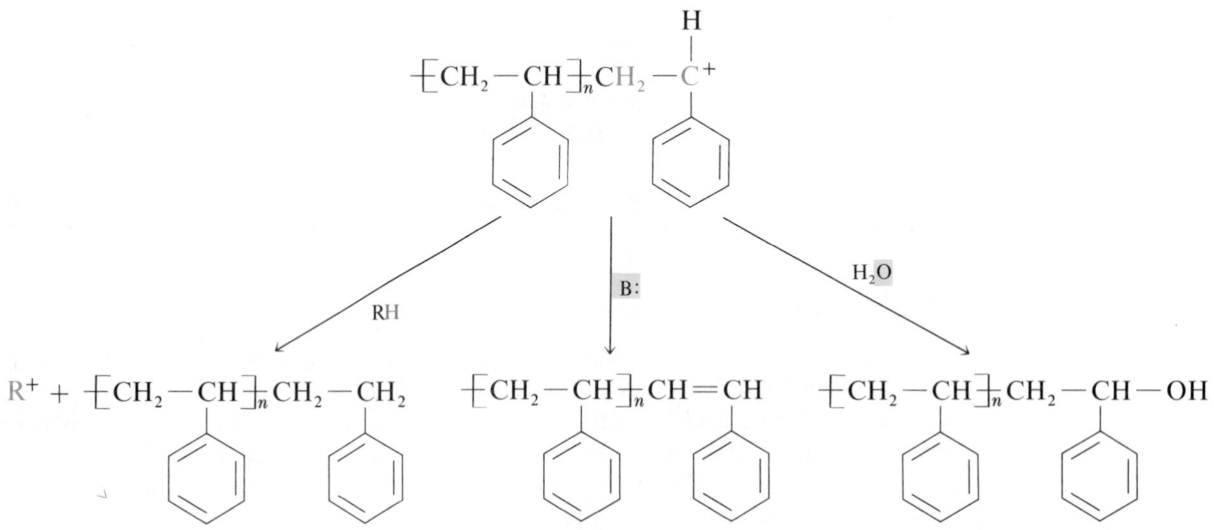
RH
B:
H_2O
R⁺ +

termination of the chain reaction by abstraction of hydride from another molecule

termination of the chain reaction by deprotonation of the cation

termination of the chain reaction by a combination with a nucleophile

Cyclic ethers also undergo polymerization reactions that follow cationic mechanisms. The most interesting of these are the reactions of oxetane and tetrahydrofuran.

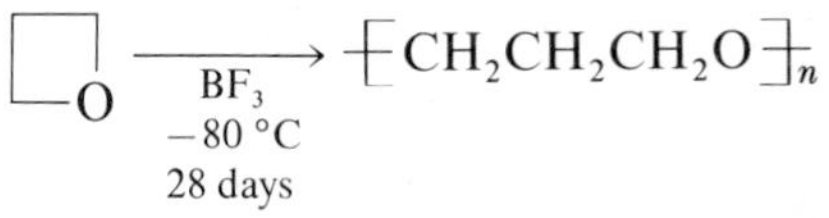
BF_3
−80 °C
28 days

oxetane

a polyether
95%

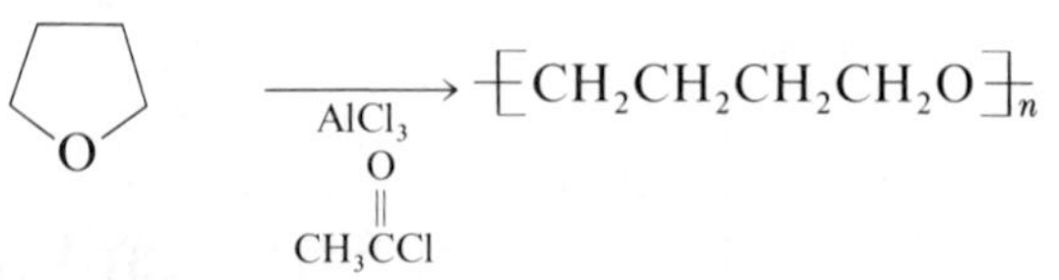
$AlCl_3$
CH_3CCl

tetrahydrofuran

a polyether

In each case, the oxygen atom of the cyclic ether is converted to a good leaving group by protonation or by reaction with a carbocation. The oxygen atoms of other ether molecules then serve as nucleophiles in S_N2 reactions and are converted in turn to reactive intermediates that propagate the chain reaction. The sequence is shown below for oxetane.

VISUALIZING THE REACTION

Cationic polymerization of a cyclic ether

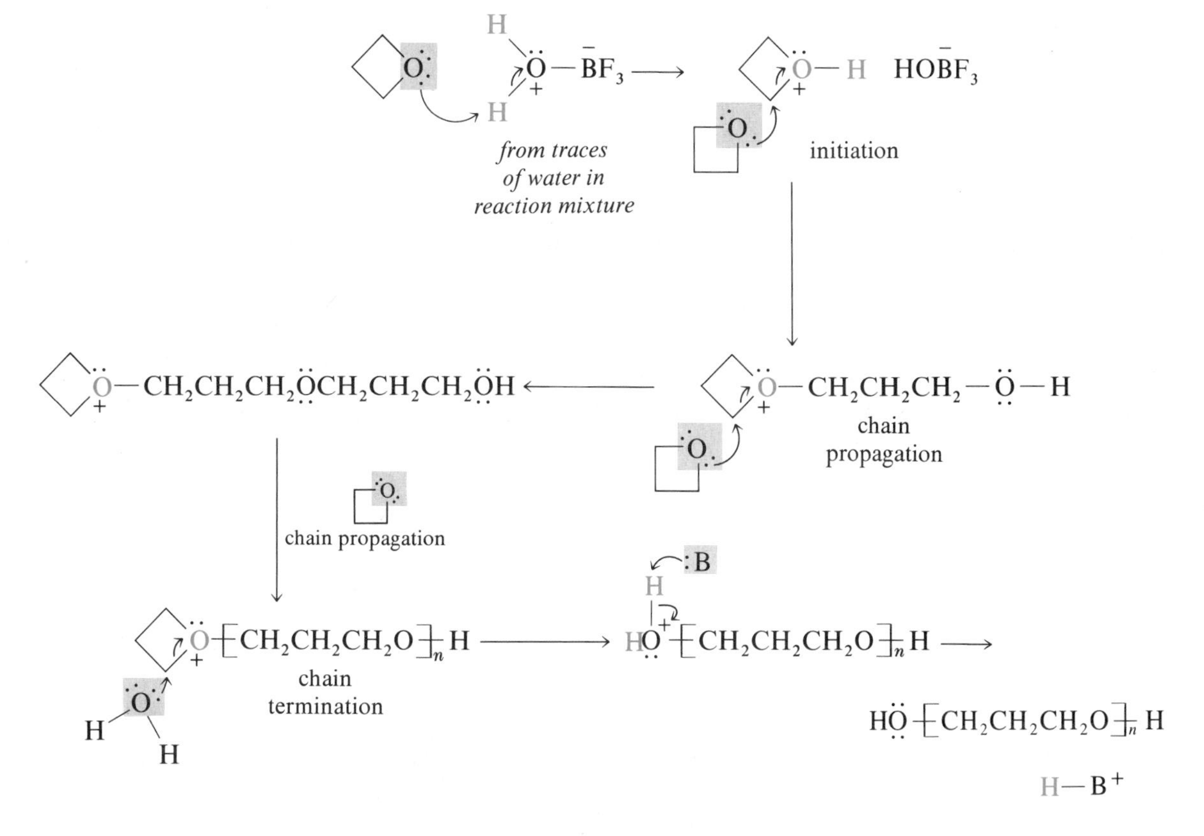

from traces
of water in
reaction mixture
initiation
chain
propagation
chain propagation
chain
termination

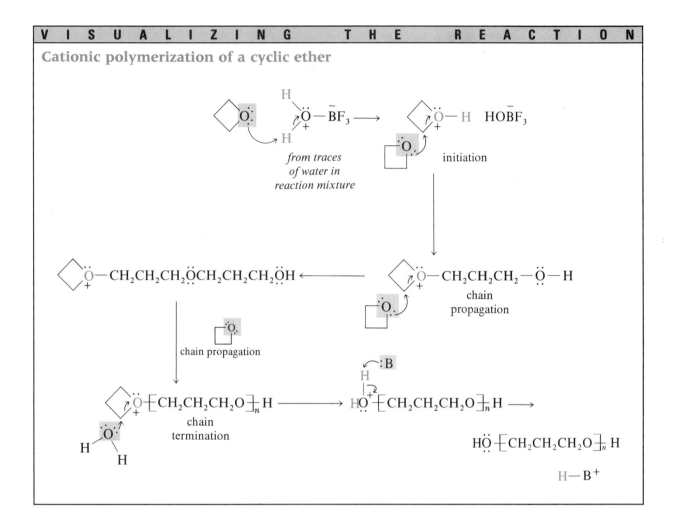

Termination of the reaction occurs when the growing chain reacts with some nucleophile, such as water.

PROBLEM 27.7

What is the reactive species that initiates the polymerization of tetrahydrofuran in the presence of aluminum chloride and acetyl chloride? Write equations showing the initiation and chain-propagating steps of the reaction.

PROBLEM 27.8

The following compounds polymerize under the conditions shown. Complete the equations, showing the intermediates that must be involved.

(a)  (b) 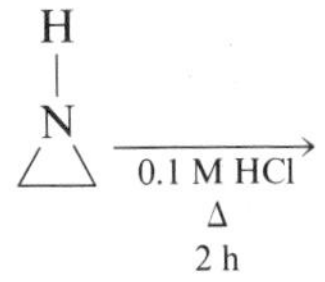

BF_3
0.1 M HCl
Δ
2 h

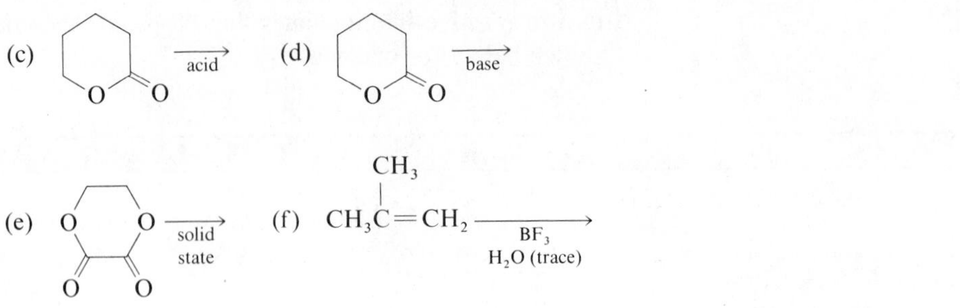
(c)
acid
(d)
base
(e)
solid state
(f)
BF3
H2O (trace)

27.3 POLYMERIZATION REACTIONS WITH CONTROLLED STEREOCHEMISTRY

A. Stereochemical Regularity in Polymer Structures

The repeating nature of the structure of a polymer chain allows three possible kinds of stereochemical relationship among the substituents on the chain. These are illustrated for polypropylene, formed from propene (propylene).

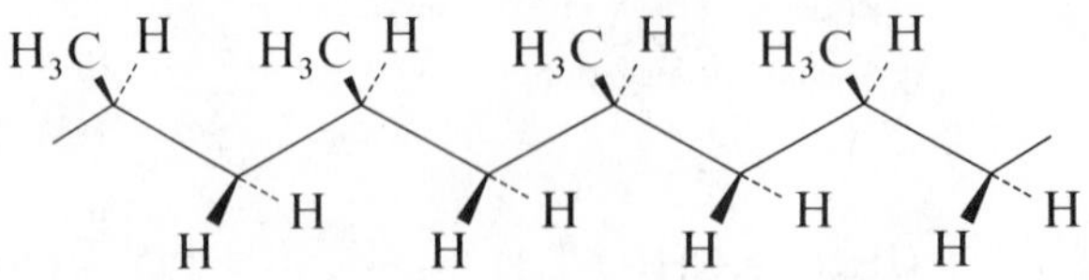

isotactic polypropylene

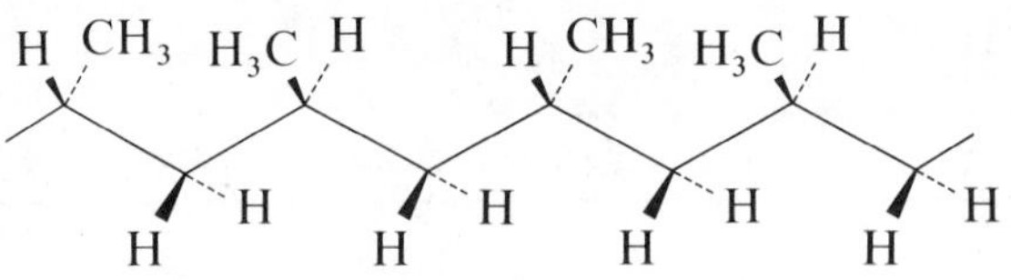

syndiotactic polypropylene

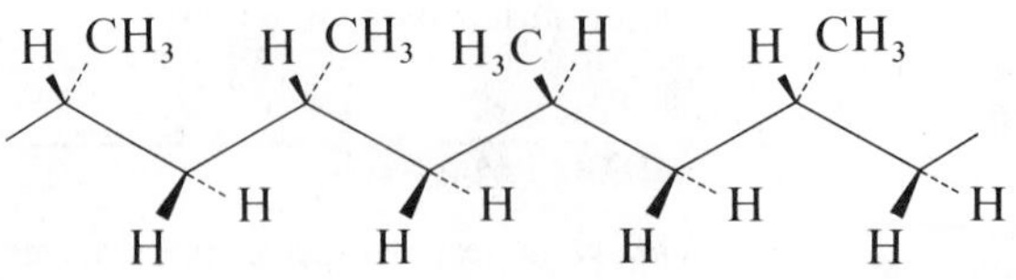

atactic polypropylene

For the first of these examples, shown with the chain stretched out into an extended form, all the methyl groups are on the same side of the chain. The configuration at each carbon of the chain is the same. Such a polymer is said to be an **isotactic polymer.** In a **syndiotactic polymer,** there is a regular alternation in the configuration at the tetrahedral carbon atoms bearing the substituents. In syndiotactic polypropylene, one methyl group is behind the plane of the paper, the next one in front, and the third one behind again. The third form of polypropylene is an **atactic**

polymer, one that has no stereochemical regularity. The arrangement of the methyl groups is completely random.

Whether a polymer is atactic, isotactic, or syndiotactic is important in determining its properties. Of hydrocarbon polymers, only isotactic and syndiotactic ones have the regularity of structure necessary for crystallinity, and thus the physical properties required for many applications. Indeed, early attempts to polymerize propylene using radical and cationic initiators resulted in soft polymers that were of no practical use. It was not until the development of new polymerization catalysts in the 1950s made it possible to make isotactic polypropylene (p. 1190) that the polymer was used industrially. Many household objects such as pails, dishpans, and plastic containers are made of polyethylene or polypropylene. Crystalline polypropylene, for example, has a melting point around 170 °C and can therefore be used in objects that will be exposed to boiling water. It can also be made into tough films such as those used in packaging food.

PROBLEM 27.9

Write structural formulas showing fragments of isotactic, syndiotactic, and atactic polystyrene.

PROBLEM 27.10

Why is isotactic or syndiotactic polypropylene more highly crystalline than the atactic polymer?

PROBLEM 27.11

A free-radical polymerization reaction of methyl methacrylate (Problem 27.6) produces atactic poly(methyl methacrylate). Give the structural formula for a fragment of such a polymer chain.

B. Heterogeneous Catalysis. Ziegler-Natta Catalysts

The polymerization reactions described in the first two sections of this chapter do not give rise to polymers having a high degree of stereochemical regularity. A careful choice of catalysts and solvent systems, however, makes it possible to control the stereochemistry of polymerization for some monomers. In the 1950s, Karl Ziegler of West Germany and Giulio Natta of Italy developed a catalyst system that polymerizes alkenes to give linear polymers with high stereoselectivity. These catalysts, which consist of mixtures of transition metal halides with organometallic compounds, mostly trialkylaluminums, are not soluble in the alkane solvents that are used for polymerizations of alkenes. The reaction mixture is, therefore, a heterogeneous one, and polymerization takes place at the surface of the catalyst. The use of these catalysts has revolutionized the production of alkene polymers. Polyethylene prepared in this way, for example, is almost completely linear and highly crystalline; other methods of polymerization result in a branched product of much lower strength and chemical resistance and thus less usefulness. Ziegler and Natta were jointly awarded the Nobel Prize in 1963 for the work that led to the discovery of these catalysts.

A typical reaction using a Ziegler-Natta catalyst is the preparation of crystalline polystyrene.

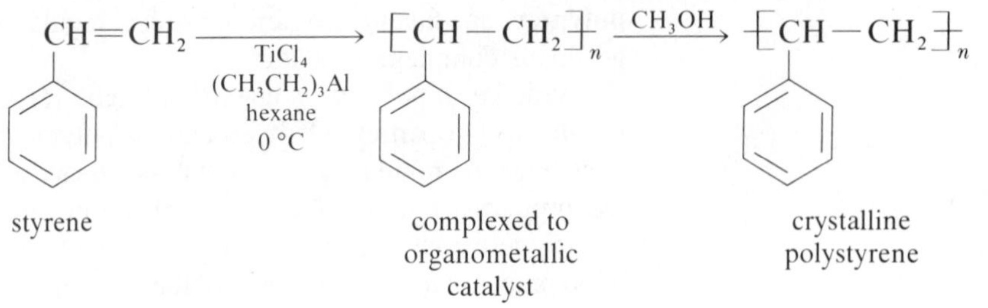
TiCl4
(CH3CH2)3Al
hexane
0 °C
CH3OH
styrene
complexed to organometallic catalyst
crystalline polystyrene

In a similar reaction, ethylene gives a polymer called high-density polyethylene to distinguish it from the softer, more highly branched polymer obtained by other methods.

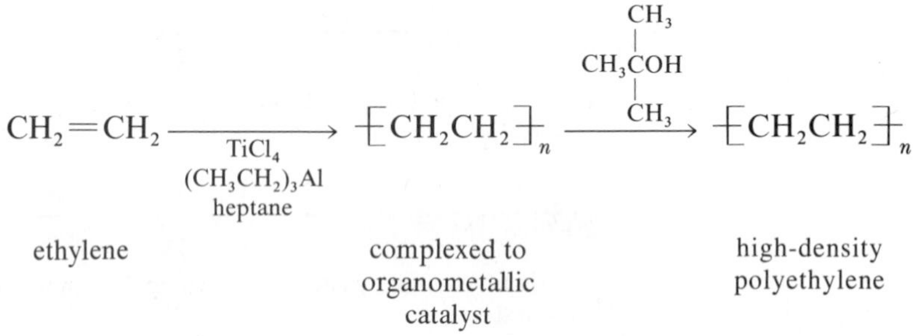
TiCl4
(CH3CH2)3Al
heptane
ethylene
complexed to organometallic catalyst
high-density polyethylene

Titanium tetrachloride is a typical transition metal halide, and triethylaluminum is the kind of organometallic compound used in the catalyst. They react with each other to give an organotitanium compound, which is believed to play the major role in the catalysis.

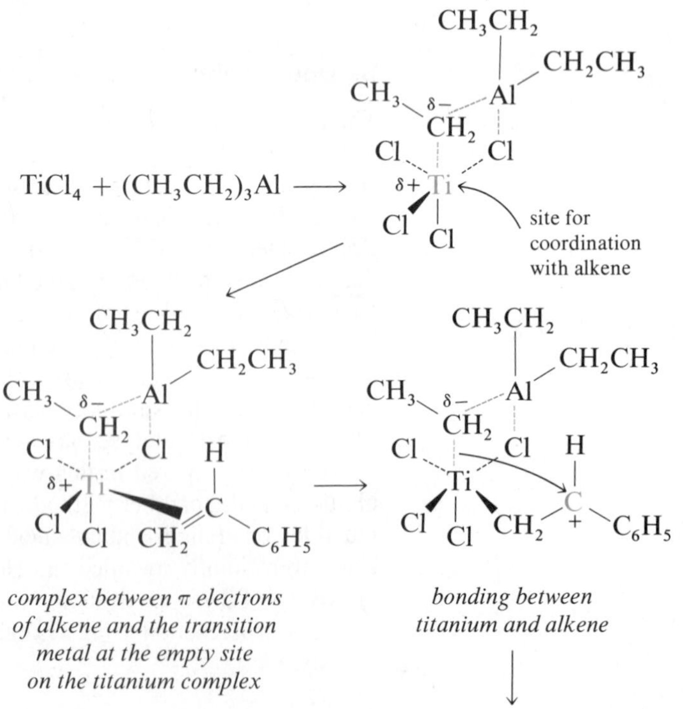
site for coordination with alkene
complex between π electrons of alkene and the transition metal at the empty site on the titanium complex
bonding between titanium and alkene

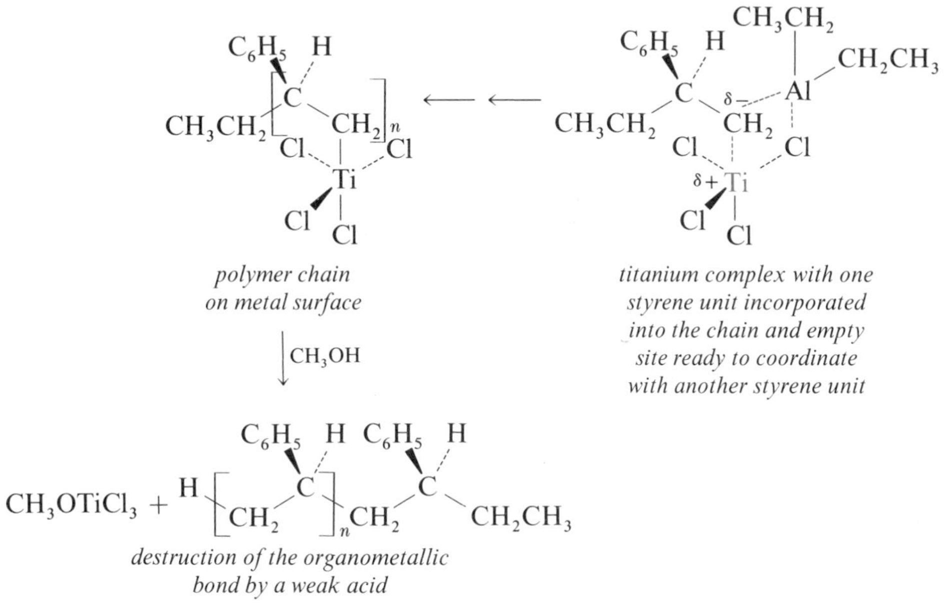
polymer chain
on metal surface
titanium complex with one
styrene unit incorporated
into the chain and empty
site ready to coordinate
with another styrene unit
destruction of the organometallic
bond by a weak acid

The reaction is pictured as proceeding through a π complex between the transition metal and the monomer. As bonding develops between the metal and the alkene, the metal complex acquires a partial negative charge while the benzylic carbon atom acquires a partial positive charge. At some point, the alkyl group bonded to the metal moves to the developing carbocation with a pair of electrons. The organotitanium compound that is formed is able to repeat this process, successively inserting monomer units between the metal atom and the rest of the growing polymer chain. The reaction takes place in a highly stereoselective way for reasons that are not well understood, partly because the exact nature of the catalyst is not known. A highly simplified representation was used above. Exactly what the organoaluminum compounds do during the reaction and whether it takes place at a single metal atom or with the cooperation of several are not known. The stereoselectivity is believed to arise from the stereochemistry imposed on the monomer as it is adsorbed onto the catalyst. There is no doubt that the nature of the metal catalyst somehow determines the stereochemistry of the reaction. For example, isoprene can be polymerized to give all *cis*- or all *trans*- 1,4-polyisoprene stereoselectively, depending on what transition metal is used in the catalyst (p. 1207).

Study Guide Concept Map 27.3

PROBLEM 27.12

Complete each of the following equations, showing the structure of the polymer you expect to be produced. Where it is known with certainty, the stereochemical nature of the polymer is indicated. Show this in the structural formula for the polymer. Assume that all copolymers have regular alternating structures.

(a) $CH_3CH_2\overset{\displaystyle CH_3}{\overset{|}{C}}HCH{=}CH_2 \xrightarrow[\substack{TiCl_4 \\ (CH_3CH_2)_2AlCl}]{} \xrightarrow{CH_3CH_2OH}$ isotactic

(b) $CH_2{=}CHOCH_3 \xrightarrow[\substack{VCl_3 \\ (CH_3CH(CH_3)CH_2)_3Al \\ \text{heptane}}]{} \xrightarrow{CH_3CH_2OH}$ isotactic

(c) $CH_2{=}CH_2 + (CH_3)HC{=}CH(CH_3) \xrightarrow[\substack{VCl_3 \\ (CH_3CH(CH_3)CH_2)_2AlCl \\ \text{heptane} \\ -30\ °C}]{} \xrightarrow{CH_3CH_2OH}$ syndiotactic

(d) $CH_2{=}C(CH_3)COCH_3$ (C=O) $\xrightarrow[\substack{VCl_3 \\ (CH_3CH(CH_3)CH_2)_3Al \\ \text{heptane}}]{} \xrightarrow{CH_3CH_2OH}$

(e) $CH_3CH_2CH{=}CH_2 \xrightarrow[\substack{FeCl_3 \\ (CH_3CH_2)_2AlCl}]{} \xrightarrow{CH_3CH_2OH}$

27.4 STEP-GROWTH POLYMERIZATION

A. Polyamides and Polyesters

The preparation of nylon 66 from a mixture of 1,6-hexanediamine and adipic acid (pp. 1171 and 1176) is an example of a step-growth polymerization. Nylon 66 is a typical polyamide. Some polyamides are formed in chain-growth processes, as was demonstrated for the anionic polymerization of ϵ-caprolactam to nylon 6 (p. 1185). Other polyamides are prepared by the reaction of an acid chloride with an amine in the presence of sodium hydroxide, the conditions of the Schotten-Baumann reaction (p. 576) for the preparation of simple amides. Nylon 610 is synthesized in this way in a two-phase system. The reaction is run in a blender to mix the water and organic layers thoroughly.

$$H_2N(CH_2)_6NH_2 + ClC(=O)(CH_2)_8C(=O)Cl \xrightarrow[\substack{H_2O \\ \text{tetrachloroethylene} \\ \text{blend} \\ 2\ \text{min}}]{NaOH} \left[NH(CH_2)_6NHC(=O)(CH_2)_8C(=O)\right]_n$$

1,6-hexanediamine; decanedioyl chloride (sebacyl chloride); poly(hexamethylenesebacamide), nylon 610, 85%, MW ~20,000

Poly(ethylene terephthalate), a typical polyester, on the other hand, is prepared by two transesterification reactions (p. 579) from dimethyl terephthalate and ethylene glycol. If the lower-boiling alcohol component that is formed at each stage is continuously removed from the reaction mixture, the polymerization is driven to completion. The first stage of this polymerization reaction is shown below.

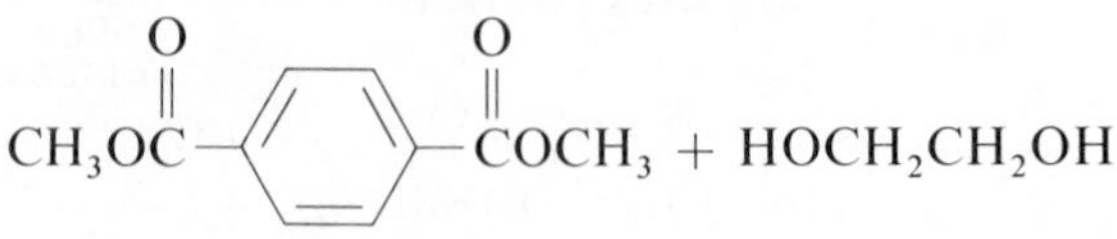

dimethyl terephthalate ethylene glycol

⇅ 150 °C

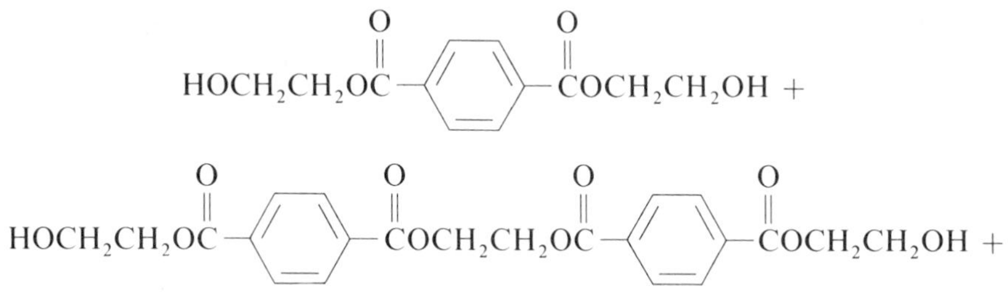

$$\text{HOCH}_2\text{CH}_2\overset{\overset{\displaystyle O}{\|}}{\text{OC}}\!-\!\!\bigcirc\!\!-\!\overset{\overset{\displaystyle O}{\|}}{\text{C}}\text{OCH}_2\text{CH}_2\text{OH} \;+$$

$$\text{HOCH}_2\text{CH}_2\overset{\overset{\displaystyle O}{\|}}{\text{OC}}\!-\!\!\bigcirc\!\!-\!\overset{\overset{\displaystyle O}{\|}}{\text{C}}\text{OCH}_2\text{CH}_2\overset{\overset{\displaystyle O}{\|}}{\text{OC}}\!-\!\!\bigcirc\!\!-\!\overset{\overset{\displaystyle O}{\|}}{\text{C}}\text{OCH}_2\text{CH}_2\text{OH} \;+$$

some higher polyesters $+$ CH$_3$OH$\uparrow$

Polyamides and polyesters have been put to many uses. Both are important in the synthetic fibers industry. For example, much cotton clothing contains at least some polyester fiber. Strong, relatively inflexible, but very light polyester films are used in sails for racing yachts and in the wings of human-powered planes.

Polyamides interact with water by hydrogen bonding at the amide bonds. The properties of a polyamide change somewhat, therefore, with changes in humidity. The chains of nylon 66, for example, become more mobile with increasing humidity as the hydrogen bonds between chains are replaced by hydrogen bonds to water molecules. Nylons that are particularly resistant to the absorption of moisture have been produced by making the hydrocarbon-like portions of the molecule larger.

Polyamides containing aromatic rings form highly crystalline structures of exceptional strength. Fibers made from these polyamides are stiffer than steel at a much lower density. Such polymers are being used to reinforce tires and to make bullet-proof vests and lightweight but very strong cords and cables.

$$\left[\text{NH}\!-\!\!\bigcirc\!\!-\!\overset{\overset{\displaystyle O}{\|}}{\text{C}}\right]_n \qquad \left[\text{NH}\!-\!\!\bigcirc\!\!-\!\text{NH}\overset{\overset{\displaystyle O}{\|}}{\text{C}}\!-\!\!\bigcirc\!\!-\!\overset{\overset{\displaystyle O}{\|}}{\text{C}}\right]_n$$

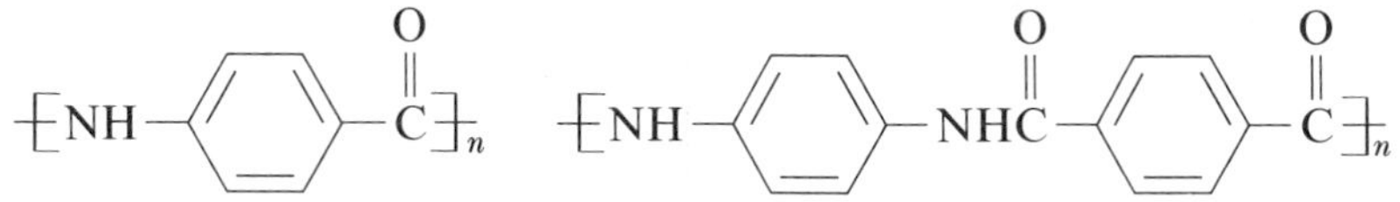

Kevlar

two polyamides that have high molecular rigidity

Polymers of every description are used in the health care industry. Not only are all kinds of disposable bottles, syringes, and laboratory ware made from different polymers, but replacements for parts of human bodies are being created. Every year, several million artificial parts are implanted into individuals who have suffered some loss as a result of either accident or disease. The development of biomaterials, artificial substances that are compatible with human tissues, is an important area of research. In some cases, the implant must become a permanent part of the body and not be rejected by it. Poly(ethylene terephthalate) mesh tubes, for example, are used to replace blood vessels; it is expected that the human tissue will grow into and around the mesh to make the implant part of the body's structure. In other cases, an implant is necessary for a period of time but should eventually be replaced by the body's own tissues. In these cases, a polyester such as poly(lactic acid) can be used so that the implant is absorbed by the body and leaves no permanent residue. This polyester is gradually hydrolyzed to lactic acid by the body, then metabolized to carbon dioxide and water, in the same way as natural lactic acid is.

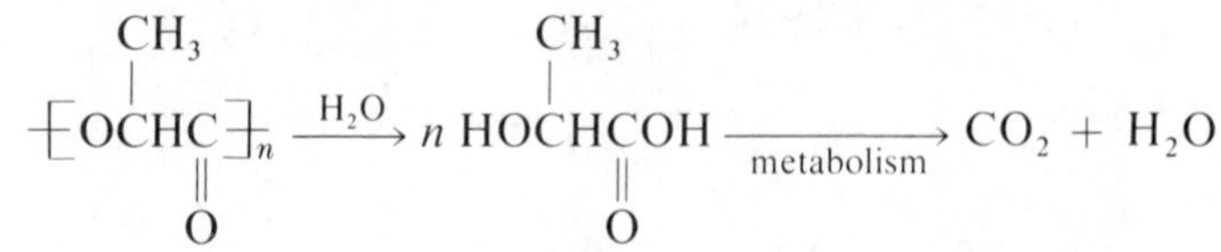

$$\begin{array}{c} CH_3 \\ | \\ \fbox{$-$}\,OCHC\,\fbox{$-$}_n \end{array} \xrightarrow{H_2O} n\ \begin{array}{c} CH_3 \\ | \\ HOCHCOH \end{array} \xrightarrow[\text{metabolism}]{} CO_2 + H_2O$$

poly(lactic acid)   lactic acid

a polyester

PROBLEM 27.13

Two important commercial products are the polycarbonate and polyester having the following structures.

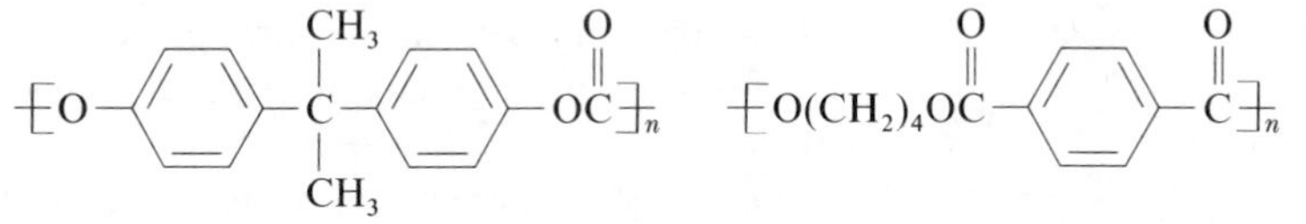

a polycarbonate   *a polyester*

Both are prepared by a transesterification reaction (p. 579). Dissect the structure of each polymer, decide what types of reagents are required, and write general equations for its preparation.

PROBLEM 27.14

Nylon 11 has the structure shown below. How would you synthesize it? Write an equation showing the necessary monomer and the general conditions for the polymerization reaction. Is the monomer likely to be available as a lactam?

$$\begin{array}{c} O \\ || \\ \fbox{$-$}\,NH(CH_2)_{10}C\,\fbox{$-$}_n \end{array}$$

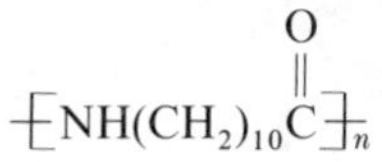

nylon 11

B. Polyurethanes

The reaction that forms the backbone of a polyurethane is the addition of an alcohol to an isocyanate (p. 921). For example, ethylene glycol reacts with 4,4′-diphenylmethane diisocyanate to give a polyurethane.

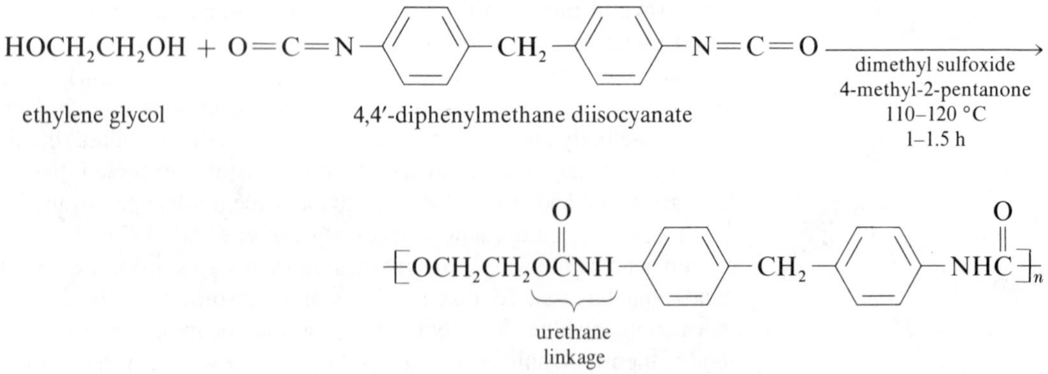

HOCH$_2$CH$_2$OH + O=C=N—⟨ ⟩—CH$_2$—⟨ ⟩—N=C=O $\xrightarrow[\substack{\text{4-methyl-2-pentanone} \\ 110\text{–}120\ °C \\ 1\text{–}1.5\ h}]{\text{dimethyl sulfoxide}}$

ethylene glycol   4,4′-diphenylmethane diisocyanate

$$\begin{array}{c} O \\ || \\ \fbox{$-$}\,OCH_2CH_2OCNH\!-\!\langle\ \rangle\!-\!CH_2\!-\!\langle\ \rangle\!-\!NHC\,\fbox{$-$}_n \end{array}$$

urethane
linkage

100%

This polymerization is carried out in a mixture of dimethyl sulfoxide and a ketone, which keeps the polymer in solution so that the reaction is not stopped by the precipitation of the high-molecular-weight product. Complete polymerization is possible under these conditions.

The chief application of polyurethanes is as foams that are used as cushioning material in furniture, pillows, mattresses, and automobile seats. For polyurethanes to fulfill this function, there must be a moderate amount of cross-linking between polymer chains and some method of creating bubbles in the melted polymer. Compounds that have rubbery properties usually have long, flexible polymer chains made from diol units that are themselves polymers. When cross-linking is desired, an excess of the diisocyanate is used so that some of the polymer chains end in unreacted isocyanate functions. This type of molecule is represented by using a wavy line to join the two isocyanate groups. The polymeric diisocyanate reacts with the urethane linkages in other polymer chains to link them.

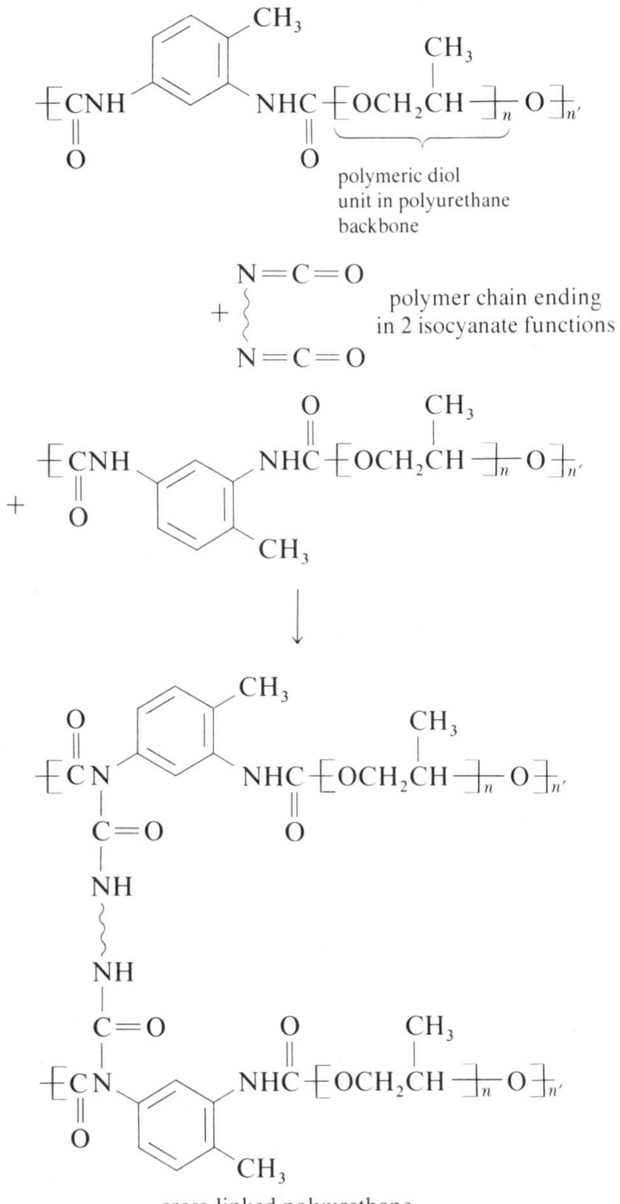
polymeric diol unit in polyurethane backbone
polymer chain ending in 2 isocyanate functions
cross-linked polyurethane

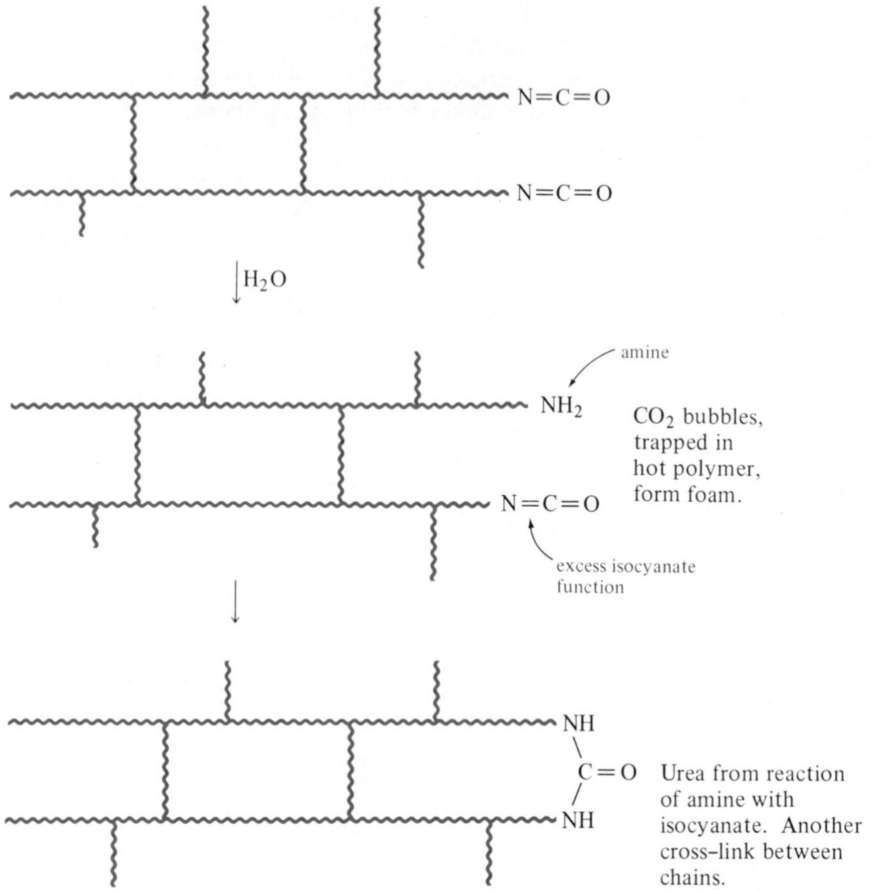
N=C=O
N=C=O
H₂O
amine
NH₂
CO₂ bubbles, trapped in hot polymer, form foam.
N=C=O
excess isocyanate function
NH
C=O
NH
Urea from reaction of amine with isocyanate. Another cross–link between chains.

FIGURE 27.3 Reaction of isocyanate end groups in polyurethane to give carbon dioxide and further cross-linking of the polymer.

If the cross-linking chains are long and flexible, the polyurethane network will also be flexible; that is, the polymer will be an elastomer. If a large number of cross-links are formed and the cross-linking chains are short and rigid, the whole polymer will be hard and inflexible.

An isocyanate reacts with water to give carbon dioxide and an amine (p. 918). This reaction is one method used to create polyurethane foam (Figure 27.3). A small amount of water is added to the hot polymer at a stage when many isocyanate end groups are present. The carbon dioxide forms bubbles that expand in the hot polyurethane, giving it a foamy texture that it retains when it cools. Meanwhile, the new amine groups that are created in the process react with the remaining excess isocyanate groups to give urea bonds. Toward the end of the polymerization process, no reactive end groups are left, and the structure of the foam is reinforced by further cross-linking reactions.

PROBLEM 27.15

Complete the following equations.

(a) $$\mathrm{Cl\overset{O}{\overset{\|}{C}}O(CH_2)_4O\overset{O}{\overset{\|}{C}}Cl + H_2N(CH_2)_6NH_2 \xrightarrow[H_2O]{NaOH}}$$

(b) 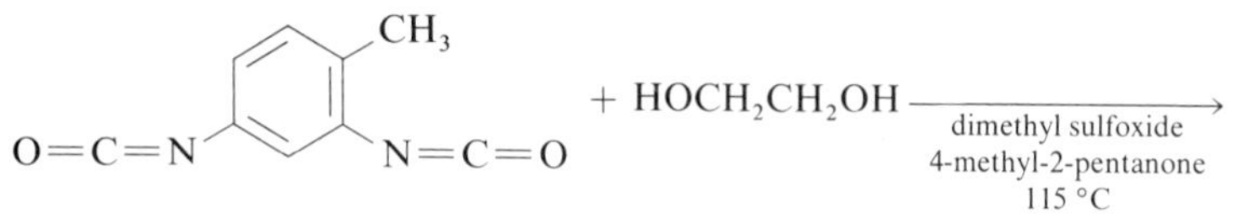

(c) $O{=}C{=}N(CH_2)_6N{=}C{=}O + HO(CH_2)_4OH \xrightarrow[185\text{–}195\ °C]{}$

C. Polymers Produced by Condensation Reactions of Formaldehyde

Polymers produced in reactions of formaldehyde with phenols have been known for over a hundred years. Leo Hendrik Baekeland of the United States was the first chemist to patent such a material at the beginning of this century. The polymer, Bakelite, is a stiff, three-dimensional network with very little solubility in organic solvents and a high resistance to electricity and heat. It is used in a wide variety of household objects and electrical fixtures.

Phenols condense with formaldehyde under either acidic or basic conditions. Under acidic conditions, polymerization gives a network of phenol rings held together by methylene groups at the ortho and para positions. These polymers exhibit a broad range of molecular weights in any one sample.

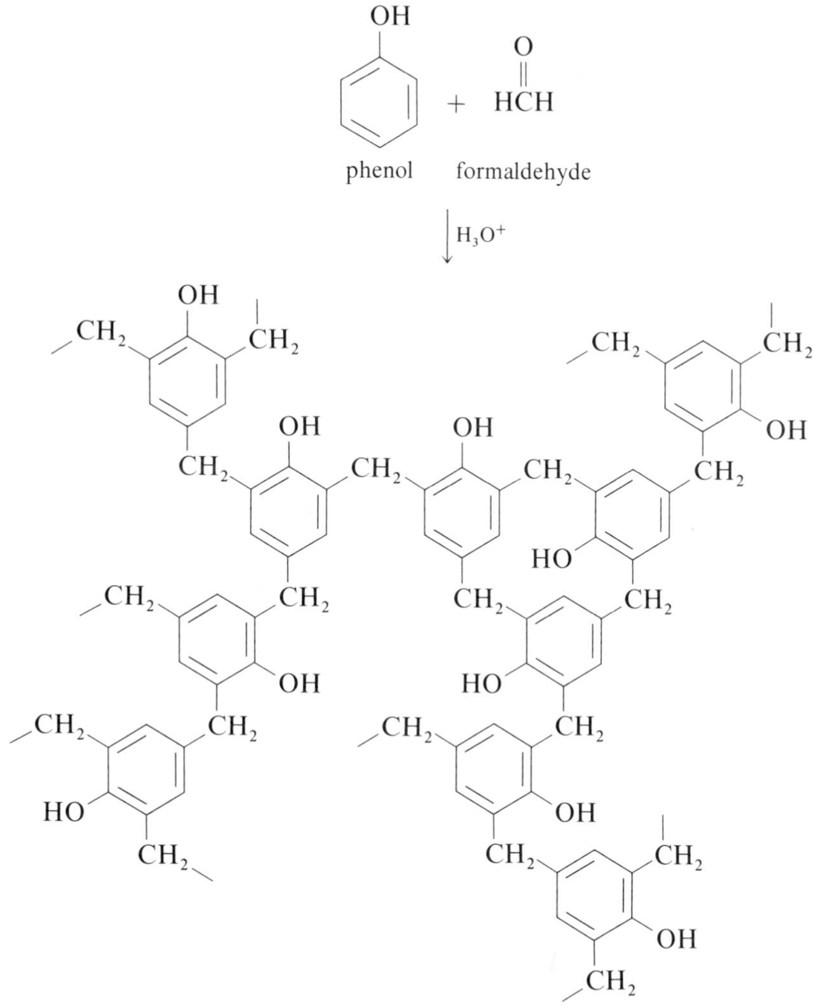
phenol
formaldehyde
H_3O^+

The reaction is thought to start by protonation of the carbonyl group of the aldehyde (or its hydrate) by the acid catalyst to give an electrophile that reacts with the phenol ring, which is highly activated toward electrophilic aromatic substitution at the ortho and para positions (p. 784). The resulting benzylic alcohols are converted into electrophilic cations by protonation and loss of water, and further electrophilic substitution of phenols takes place to create the network.

PROBLEM 27.16

Write a mechanism showing how formaldehyde condenses with phenol in the presence of acid. How is the product of that reaction converted into an electrophile that reacts with another unit of phenol?

Commercially important plastics are also obtained when urea and the heterocyclic triamine melamine react with formaldehyde. In both of these reactions, amino groups add to formaldehyde to give intermediates that have hydroxymethyl groups.

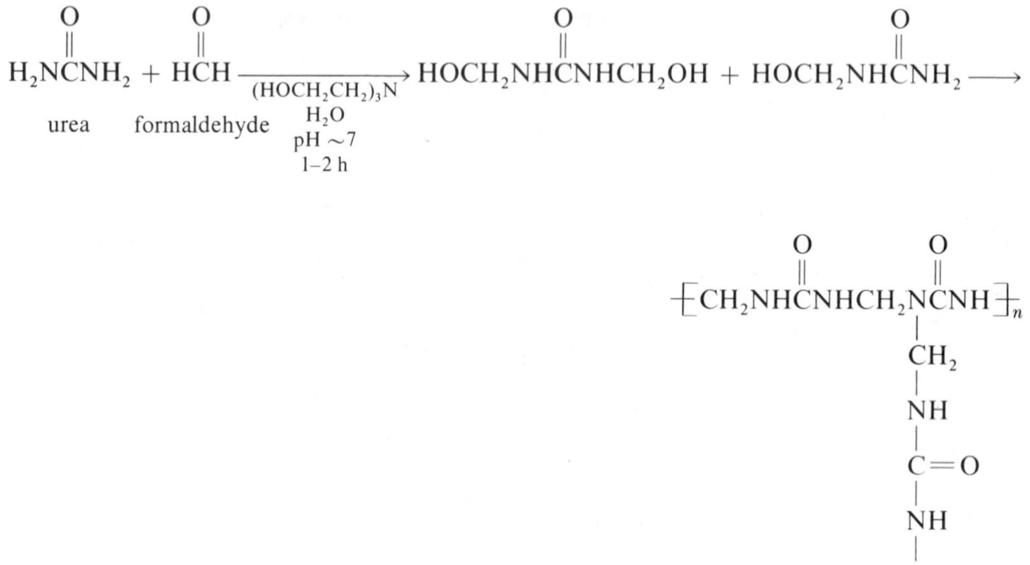
urea
formaldehyde
(HOCH₂CH₂)₃N
H₂O
pH ~7
1–2 h

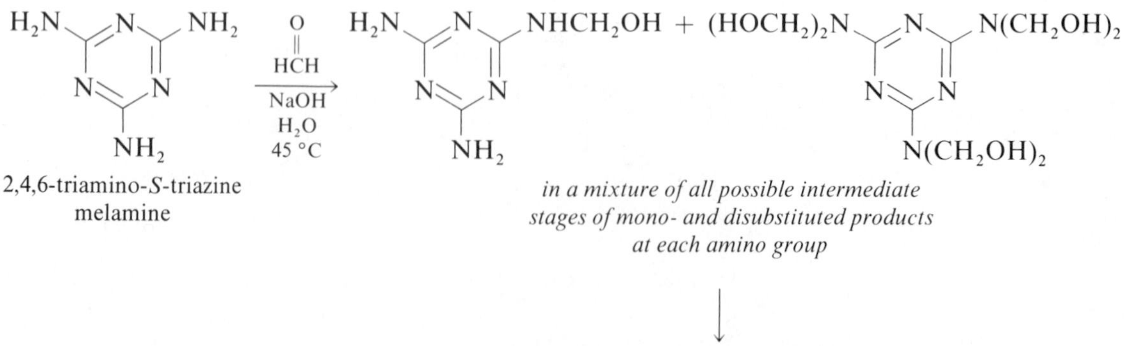
HCH
NaOH
H₂O
45 °C
2,4,6-triamino-S-triazine
melamine
in a mixture of all possible intermediate stages of mono- and disubstituted products at each amino group

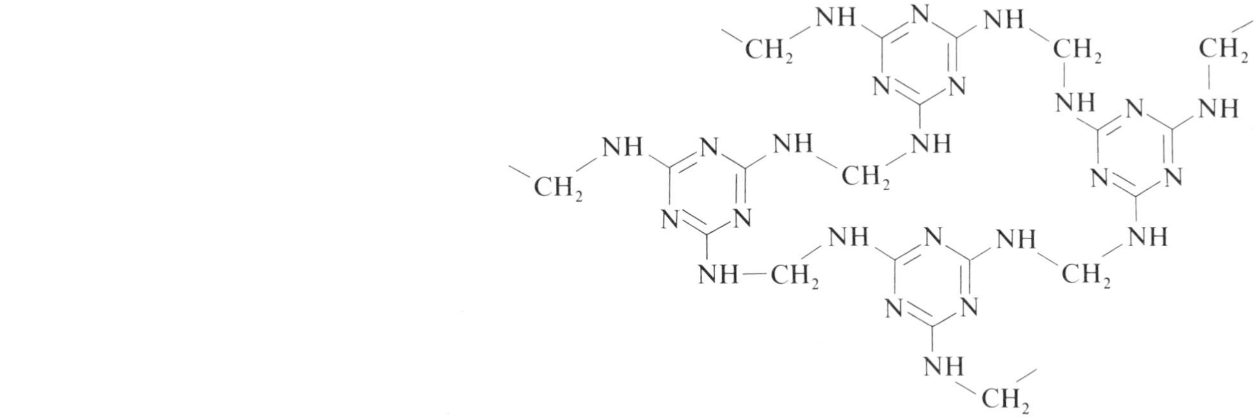

The resins obtained from the kinds of reactions shown above and on the previous page have many applications. A small amount of urea-formaldehyde resin applied to a cotton fabric gives it resistance to creasing. In fact, the predominant odor in a large fabric store is often that of formaldehyde. Boards are formed by blending sawdust or wood chips with such a resin and curing the mixture in a hot mold. Packaging paper is made more resistant to moisture by treating the paper pulp with urea-formaldehyde resin. Melamine resins are used in very much the same ways. Currently, however, there is some concern that exposure to formaldehyde may be a health hazard.

PROBLEM 27.17

Write a detailed mechanism for the reaction of urea with formaldehyde to produce a cross-linked polymer.

D. Epoxy Resins

Epoxy resins are in everyday use as adhesives. Such a resin is formed upon mixing two solutions: one of a polymer that contains oxirane rings that will form cross-links between polymer chains when treated with a nucleophile, and the other containing a polyamino compound, which is the nucleophilic reagent that starts the process and forms some of the cross-links itself. The polymer component contains hydroxyl groups, ether linkages, and oxirane rings. All these functional groups, along with the amino groups that are added during the cross-linking process, hydrogen-bond and coordinate strongly with surfaces such as glass, ceramics, and metal. Thus, the epoxy resin glues such surfaces tightly together.

Cross-linking takes place when an amine reacts with the oxirane end groups of the polymer. The linking groups may be amines or the alkoxide ions that are created when the oxirane ring is opened (Figure 27.4). Because the cross-linking takes place very fast, the amine and the linear polymer are mixed just before they are applied to the surfaces that they are to bond together. The bonding is so strong that mended objects frequently break elsewhere before they fall apart at the site of repair.

The most common epoxy resin is made from the oxirane derived from 3-chloro-1-propene and 2,2-bis(4′-hydroxyphenyl)propane, shown on the following page.

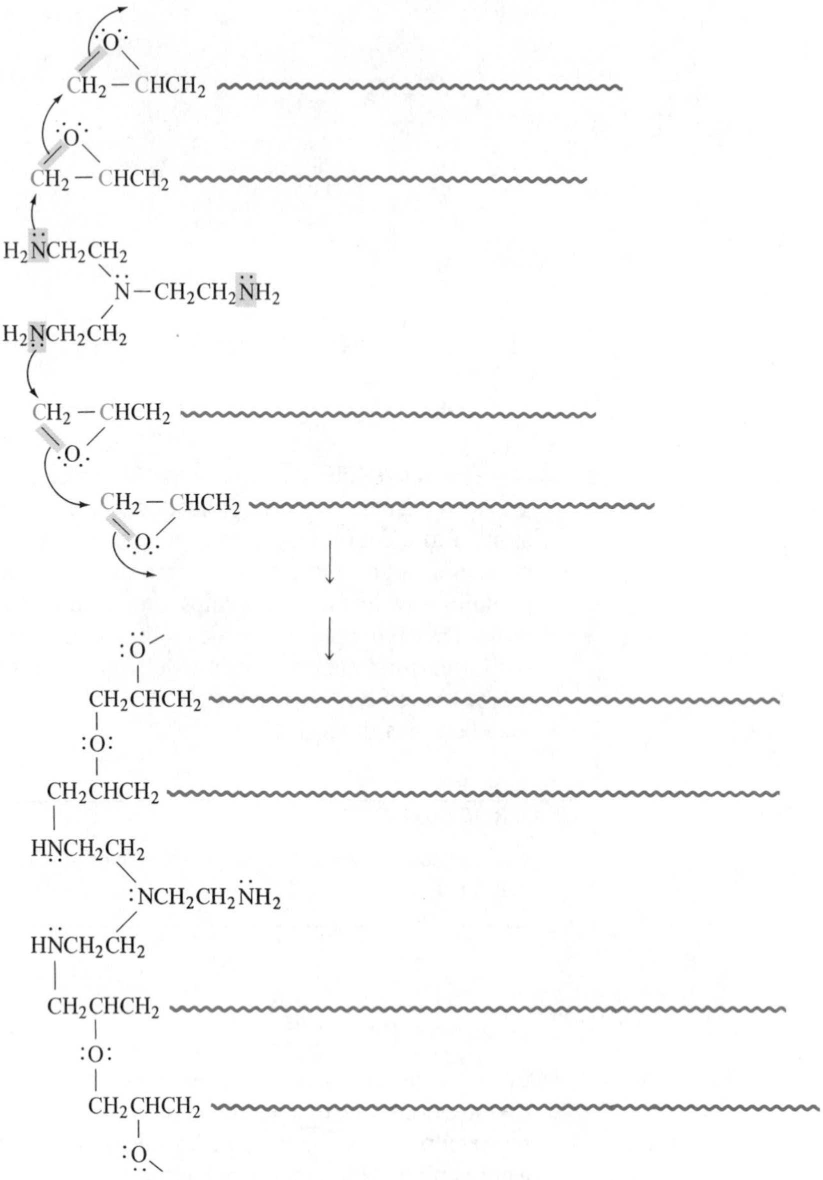

FIGURE 27.4 Cross-linking of an epoxy resin.

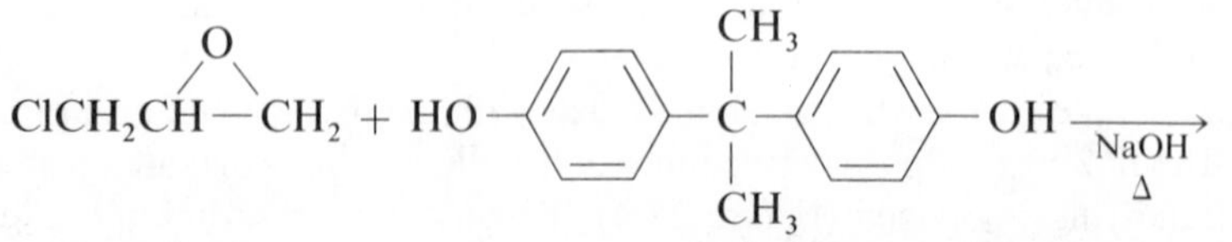

oxirane from 3-chloro-1-propene epichlorohydrin

2,2-bis(4′-hydroxyphenyl)propane

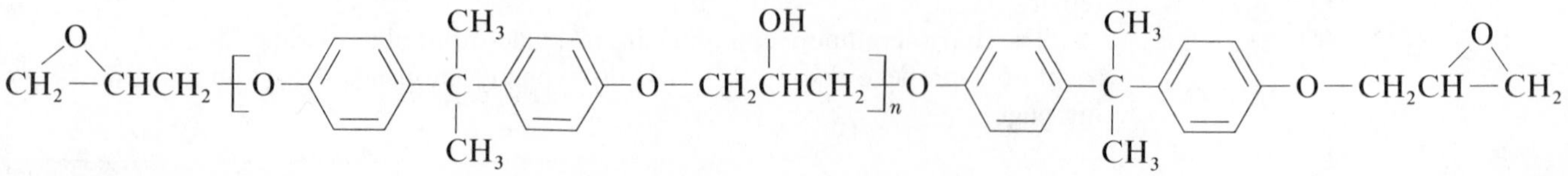

The oxirane is used in excess so that the end groups of the polymer chains are oxiranes.

The oxirane derived from 3-chloro-1-propene has the common name epichlorohydrin and is a toxic compound suspected of being a carcinogen. The reason for its toxicity is the ease with which it reacts with nucleophiles including biological nucleophiles, to link them together. When the oxirane ring is opened by the attack of one nucleophile, the alkoxide ion that is formed displaces chloride ion intramolecularly to give a new oxirane. This ring, in turn, opens in another nucleophilic substitution reaction.

V I S U A L I Z I N G   T H E   R E A C T I O N

Epichlorohydrin as a linking agent for nucleophiles

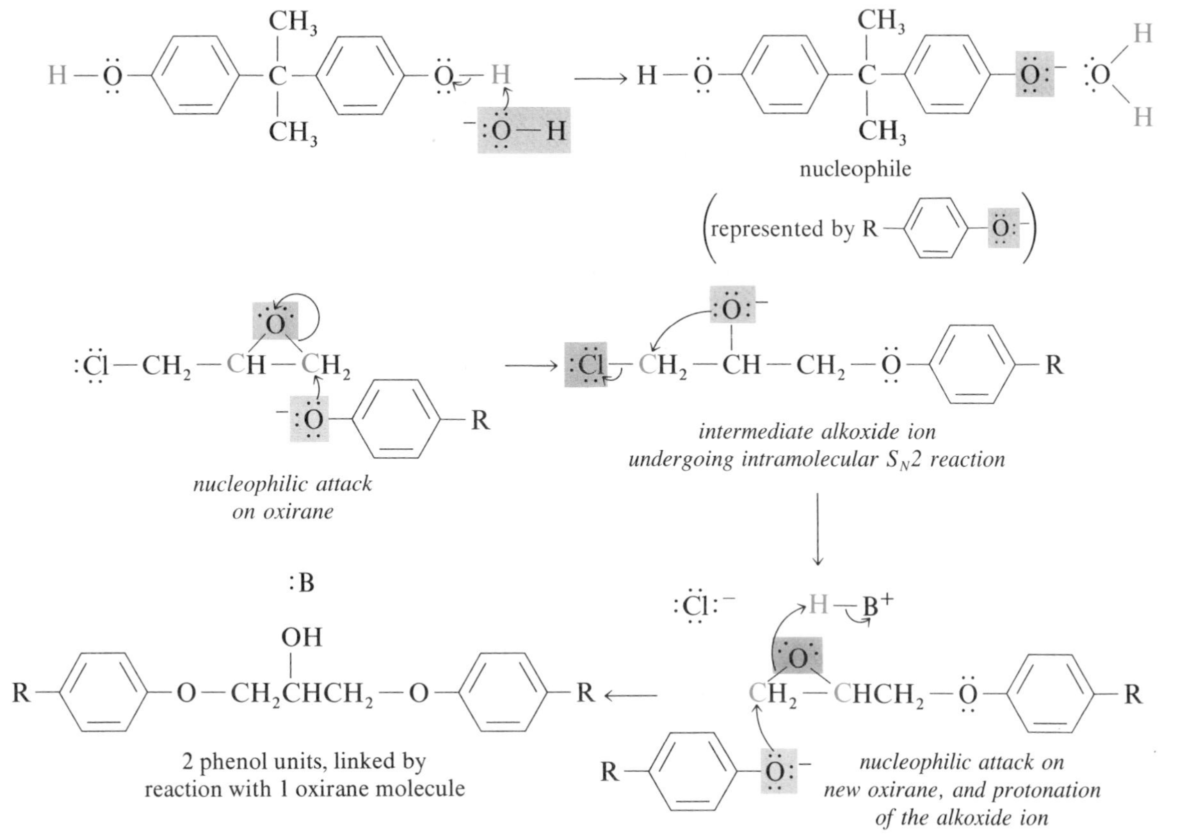

nucleophile
(represented by R
nucleophilic attack on oxirane
intermediate alkoxide ion undergoing intramolecular S_N2 reaction
2 phenol units, linked by reaction with 1 oxirane molecule
nucleophilic attack on new oxirane, and protonation of the alkoxide ion

Study Guide Concept Map 27.4

PROBLEM 27.18

The diphenol used in the preparation of polycarbonates and epoxy resin has the trivial name bisphenol A because it is synthesized from phenol and acetone. Propose a mechanism for its formation from these reagents.

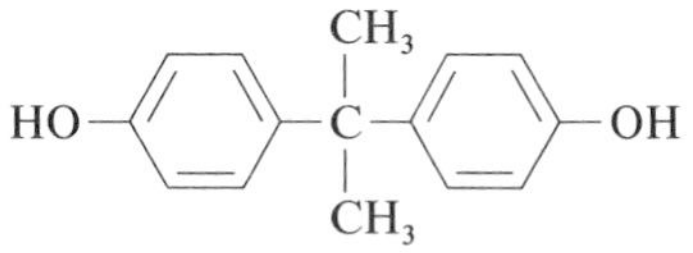

2,2-bis(4′-hydroxyphenyl)propane
bisphenol A

PROBLEM 27.19

The commercial synthesis of the oxirane derived from 3-chloropropene starts with propene and involves the use of gaseous chlorine at high temperatures, then chlorine with water at low temperatures, and finally calcium hydroxide. Write equations for these reactions showing how they lead to the formation of 2-(chloromethyl)oxirane. Classify each reaction according to its mechanistic type.

27.5 A NATURAL MACROMOLECULE, RUBBER

A. The Structure of Rubber and Gutta Percha

Rubber is produced by a number of plants that grow in tropical regions. The one that is most important commercially is a tree, *Hevea brasiliensis,* which was originally found in Brazil but now grows mostly in southeast Asia. The polymer is contained in a fluid known as latex, which is synthesized by cells under the bark of the tree. Latex is obtained by making cuts in the bark and collecting the liquid that flows out. *Hevea brasiliensis* synthesizes latex at a high rate; a tree may be tapped every other day to harvest rubber.

Rubber is a polymer made of isoprene units and containing cis double bonds. An isomeric polymer in which the double bonds are trans is obtained from trees of the genus *Dichopsis,* which are also found in southeast Asia. This polymeric material, which has properties quite different from those of rubber, is called **gutta percha.**

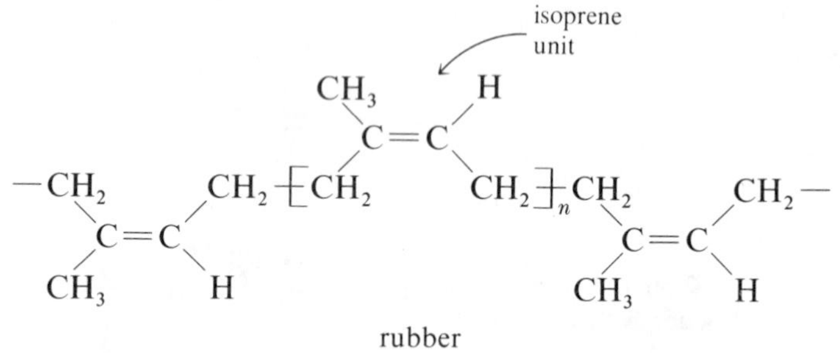
isoprene unit

rubber
cis double bonds
$n \sim 1500–15,000$
MW $\sim 100,000–1,000,000$

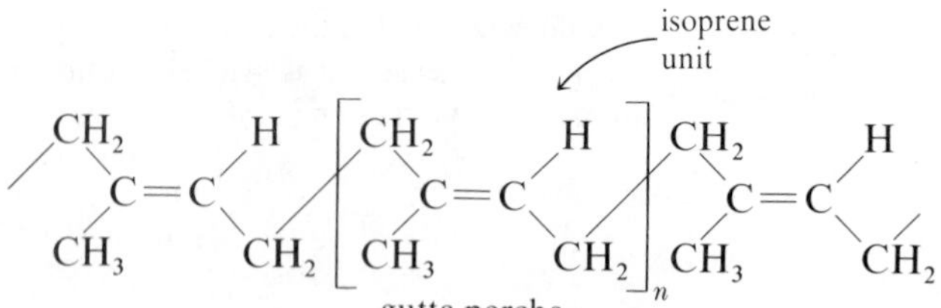
isoprene unit

gutta percha
trans double bonds
$n \sim 100$
MW ~ 7000

Both rubber and gutta percha are synthesized in plants from isopentenyl pyrophosphate. The reactions that give rise to gutta percha resemble those in the biosynthesis of terpenes such as geraniol and farnesol, which also have trans double bonds (p. 705). The molecular weight of gutta percha is approximately 7000, and that of rubber ranges from 100,000 up. The biosynthesis of rubber is obviously directed by a different type of enzyme than that involved in the production of gutta percha because of the stereoselective formation of cis double bonds in rubber.

The physical properties of rubber and gutta percha are quite different, as should be expected from the different shapes and sizes of their molecules. Rubber has a much more folded structure than gutta percha. The cis arrangement of the double bonds makes it harder for adjacent rubber molecules to fit close to each other in the ordered way that produces a crystalline structure. Thus, rubber is highly amorphous. Because of the random coiling of its large molecules, rubber can be stretched easily. When it is stretched, the molecules are forced into a more orderly arrangement, which is unstable for the polymer, so it snaps back into the more random coiled structure when the tension is released. The structure of gutta percha allows the polymer molecules to be packed very close to each other, so this polymer is much more crystalline in its natural state than rubber is. In general, gutta percha is harder and less flexible than rubber. For example, the relatively hard, inflexible covers of golf balls are made mostly of gutta percha. It is also a good electrical insulator and is used in many electrical applications, such as in casings for cables.

Raw rubber is affected by a number of environmental factors such as temperature (p. 1175) and the presence of oxygen. The allylic positions next to the double bonds, for example, are vulnerable to reaction with oxygen as a free radical (p. 842). The double bonds themselves react with ozone, which is always present but is especially abundant in air polluted by exhaust fumes from automobiles. Rubber is affected by organic solvents such as gasoline and oil; it tends to swell and dissolve in such hydrocarbons. Light also breaks down rubber. For these reasons, almost all rubber used industrially is treated to increase its stability; the process is known as **vulcanization.** Vulcanized rubber has greater strength, less stickiness, and greater elasticity than raw rubber. It is less soluble in most solvents and tends to retain its flexibility at lower temperatures.

Charles Goodyear of the United States discovered vulcanization in 1839 when he found that heating rubber with sulfur improved the properties of the rubber. Sulfur reacts irreversibly with rubber to link different chains and parts of the same chain together to give greater stability to the polymer. The chemistry is complicated, and cyclic structures containing sulfur are formed as well. The physical properties of rubber are changed even when only a very small percentage by weight of sulfur is used.

B. Synthetic Rubbers

Rubber is important to transportation as well as to many other aspects of modern technology. In the world wars, it became clear that industrial states could easily be cut off from the supply of natural rubber from tropical regions, such as southeast Asia. Therefore, research was started into ways of making synthetic rubber. One of the earliest replacements for natural rubber was neoprene, developed at DuPont in the United States. It is the polymer of 2-chloro-1,3-butadiene, a molecule that resembles isoprene except that a chlorine atom replaces the methyl group on the backbone of the chain, which gives rise to the common name of chloroprene.

A diene may polymerize by 1,2- or 1,4-addition to the conjugated double bonds (p. 686). 2-Chloro-1,3-butadiene polymerizes in a free-radical reaction by both types of addition.

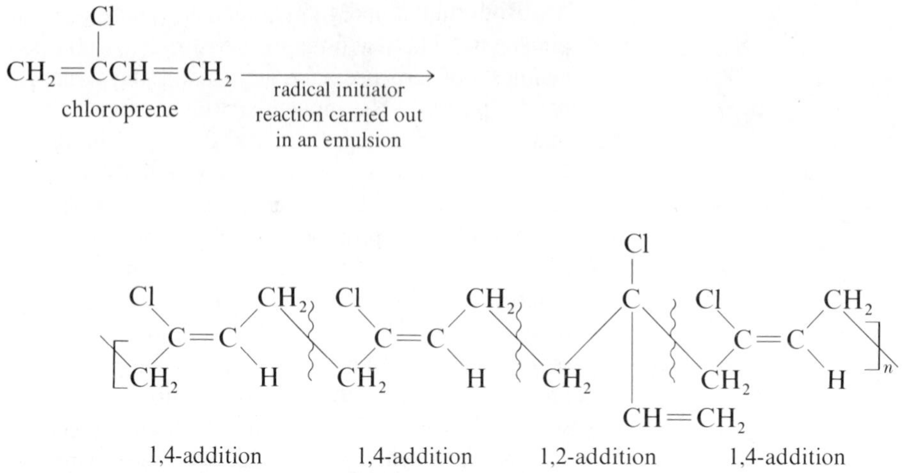
chloroprene
radical initiator
reaction carried out in an emulsion
1,4-addition
1,4-addition
1,2-addition
1,4-addition

a fragment of poly(chloroprene)
neoprene

The portions of the chain that result from 1,2-addition to the butadiene contain chlorine atoms bonded to carbon atoms that are tertiary and allylic. Heating causes isomerization of such an allylic halide group to the isomer in which the more stable internal double bond is present. Further heating with a metal oxide results in cross-linking of different polymer chains.

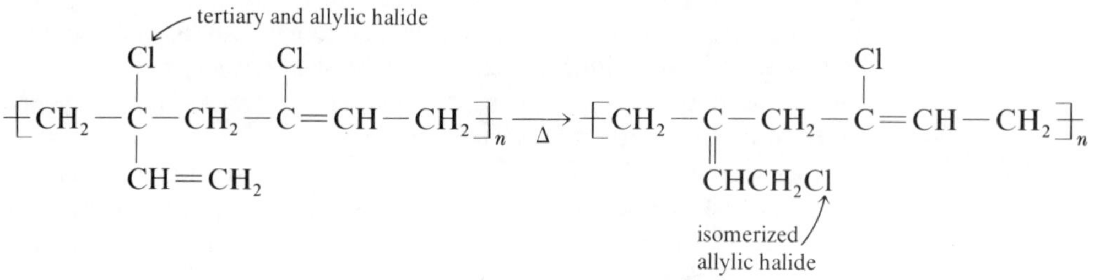
tertiary and allylic halide
isomerized allylic halide

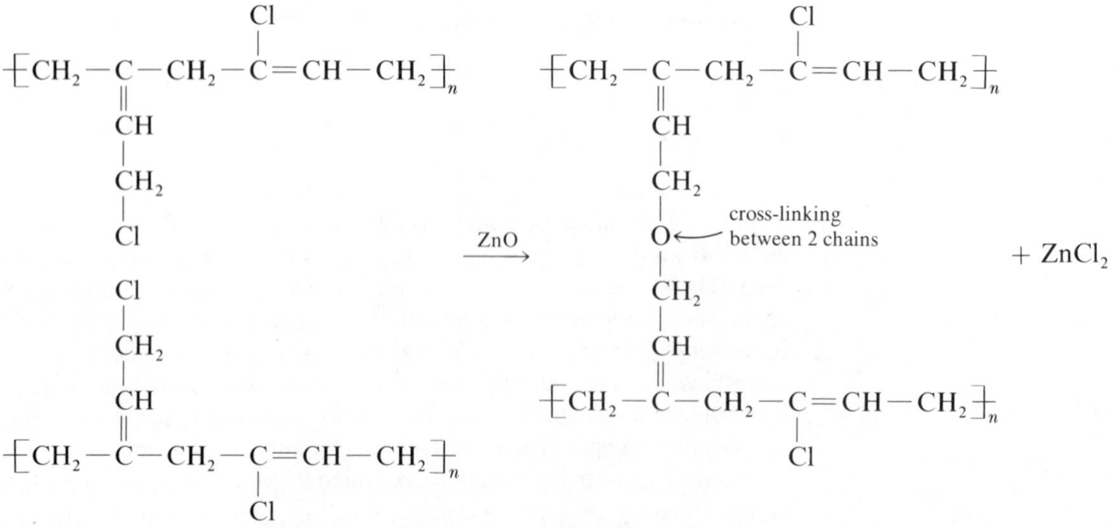
ZnO
cross-linking between 2 chains
+ ZnCl2

The cross-link is an ether group formed by nucleophilic displacement of the reactive allylic halide substituents on adjacent chains. Note that the unreactive vinylic chlorine atoms are not affected in either the isomerization step or the nucleophilic substitution reaction.

The stereochemistry of the double bond created between the second and third carbon atoms of a butadiene by 1,4-addition depends on the reaction conditions used. Neoprene, formed in emulsion by a free-radical polymerization, has mostly trans double bonds in the chain.

Neoprene is more resistant than natural rubber to oils and solvents and to reaction with ozone. It is tougher and resists wear better than rubber. It is used mostly in applications where its toughness and resistance to oil and grease are important, such as in gaskets, sealing rings, and engine mountings. It is also used in making protective gloves and aprons.

Many attempts had been made to polymerize isoprene to create a synthetic rubber that matched the properties of natural rubber. It was not until Ziegler-Natta catalysts became available in the 1950s that isoprene was polymerized stereoselectively to 1,4-polyisoprene with cis double bonds using one catalyst and 1,4-polyisoprene with trans double bonds using another one.

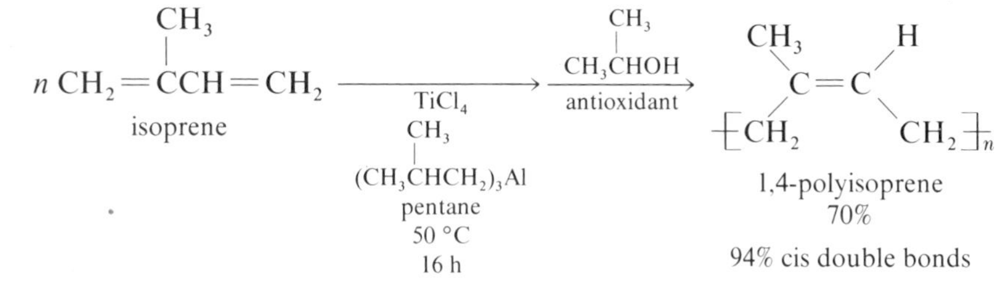
isoprene
TiCl4
(CH3CHCH2)3Al
pentane
50 °C
16 h
CH3CHOH
antioxidant
1,4-polyisoprene
70%
94% cis double bonds

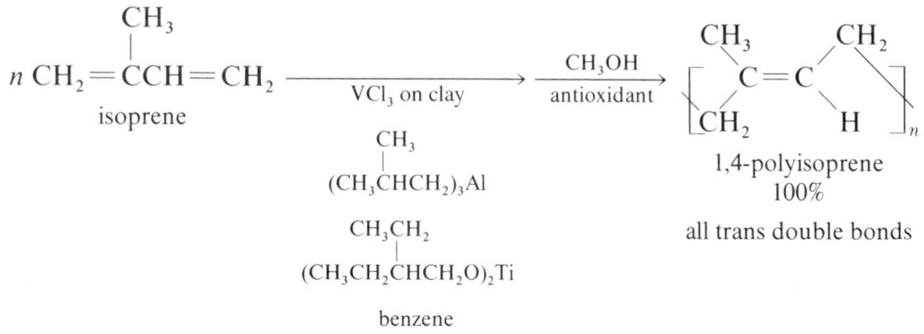
isoprene
VCl3 on clay
(CH3CHCH2)3Al
(CH3CH2CHCH2O)2Ti
benzene
CH3OH
antioxidant
1,4-polyisoprene
100%
all trans double bonds

Changing the transition metal used as the primary catalyst from titanium to vanadium changes the stereoselectivity of the polymerization reaction.

This chapter has explored some of the chemical reactions used to transform low-molecular-weight organic compounds into giant molecules that imitate natural molecules in their properties and usefulness. These syntheses are indications of the inventiveness and ingenuity of the human mind when faced with practical problems. Human beings have a great deal of curiosity about the limits of their understanding of natural phenomena. The experimental observations that are accessible to scientists spawn theories, which are followed by new experiments to test the theories. Polymer chemistry is an area where a questioning attitude, a constant asking of ''what if?'' has been particularly fruitful.

On the other hand, it must also be recognized that polymers have been a mixed blessing. Plastics are useful because they have good mechanical strength and are chemically unreactive. The same properties keep plastics from degrading naturally when they are discarded. Many of them, when they burn, depolymerize to give toxic degradation products; this creates potentially dangerous situations when such materials are used in carpeting and furniture for homes and institutions. Chlorofluorocarbons, used to make polystyrene foam (p. 1180), decompose when they absorb ultraviolet radiation to give chlorine atoms. Chlorine is implicated in chain reactions that destroy the ozone layer, which shields the earth from the portion of ultraviolet radiation from the sun that is most destructive to living organisms. In 1987, a worldwide conference agreed to start to limit the use of these compounds. The raw materials for polymers come from petroleum; a nonrenewable resource is thus being converted into plastics, many of which are put to only temporary use and are then discarded, creating problems for the solid-waste disposal systems of towns and cities. Each of us contributes to this waste. Perhaps the time has come to ask ourselves which applications of polymers represent a wise use of resources and which are, at best, a small convenience that we could do without.

The materials of the future must come increasingly from renewable sources and be less wasteful both of substance and of energy. Plants are especially versatile in converting the energy of the sun into a wide variety of natural products that can be transformed into other useful substances. Microorganisms, modified by genetic engineering, are already used in fermentation processes that convert readily available plant products such as glucose to both simpler and more complex chemicals. In the future organic chemists will be increasingly challenged to perfect methods for the transformation of common substances isolated from natural sources into other useful products. The stimulating interaction between experimental observation and careful thinking about the implications of the observed phenomenon that is the basis of the science of organic chemistry must continue if human beings are to use the resources of the earth wisely in meeting their needs.

PROBLEM 27.20

What products would you expect to get from the reaction of rubber with ozone? Does the process that takes place in the environment end up with a reductive or an oxidative decomposition of the ozonide?

PROBLEM 27.21

1,3-Butadiene, when polymerized in the presence of titanium tetrabromide and triethylaluminum, gives a 1,4-polybutadiene with mostly cis double bonds. Write an equation for this reaction.

PROBLEM 27.22

When 1-methoxy-1,3-butadiene is polymerized with a free-radical initiator, a low-molecular-weight, amorphous polymer is obtained. Polymerizing the same diene in the presence of vanadium trichloride and triisobutylaluminum gives a highly crystalline, high-molecular-weight polymer with trans double bonds. Write equations showing the differences in the structures of the two polymers.

PROBLEM 27.23

A copolymer made of acrylonitrile, butadiene, and styrene has a number of applications because it combines some of the properties of rubber with the toughness of an acrylonitrile-styrene copolymer (p. 1181). Write an equation for the preparation of the acrylonitrile-butadiene-styrene copolymer.

S U M M A R Y

Polymers are high-molecular-weight organic compounds with properties that make them useful as fibers, elastomers, and plastics. Because of their molecular size, polymers are also called macromolecules. Natural macromolecules include proteins, DNA, cellulose, and rubber. Synthetic macromolecules are made by subjecting monomers to polymerization reactions. Polymerization may occur by a chain reaction that involves free radicals, anions, or cations as reactive intermediates. In chain reactions, the monomer units add one by one to the end of the chain (Table 27.1). Some polymers, such as polyesters, polyamides, and polyurethanes, are formed by step-growth reactions. In these processes, monomer molecules first combine into larger fragments composed of several monomer units. These fragments then combine to give polymers (Table 27.2).

The monomer units in a simple polymer are all the same. A polymer made up of more than one kind of monomer is called a copolymer. Polymers are also classified as being linear, branched, or cross-linked.

The molecular weight and the stereochemistry of a polymer are important in determining its properties. Chiral monomer units will, of course, give rise to chiral polymers, but achiral monomers may also create stereocenters when they polymerize. The stereochemistry along a polymer chain is described as being isotactic when all carbon atoms in the chain have the same configuration, as syndiotactic when they have alternating *R* and *S* configurations, and as atactic when there is no regularity to the stereochemistry. Only isotactic and syndiotactic polymers can achieve the degree of crystallinity necessary to have a high enough melting point and enough mechanical strength to be used in practical applications. Such stereochemically regular polymers are prepared using special transition metal catalysts called Ziegler-Natta catalysts.

TABLE 27.1 Examples of Chain-Growth Polymerization

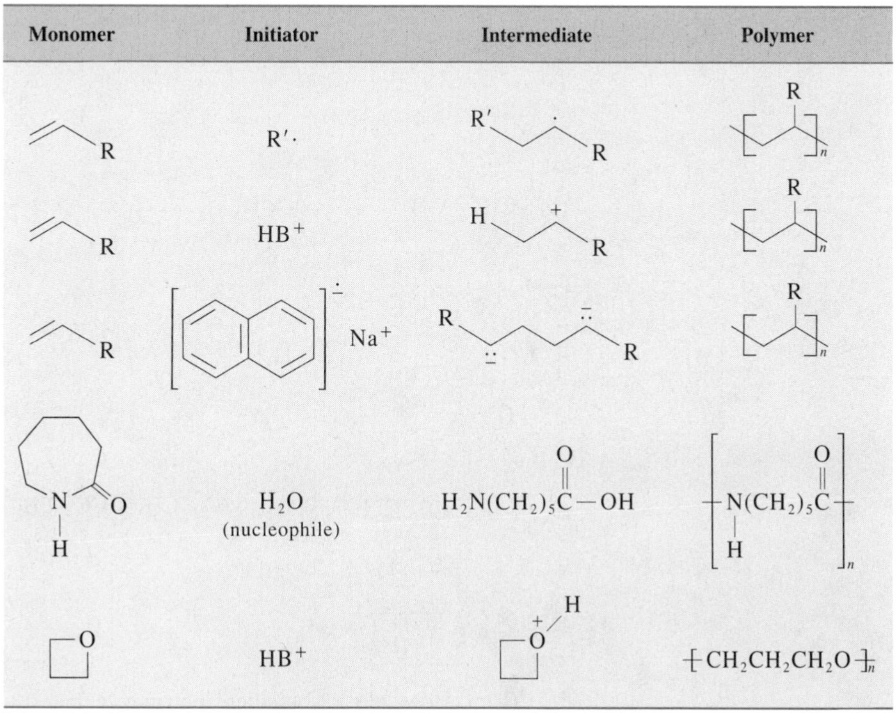
Monomer
Initiator
Intermediate
Polymer
R′·
HB⁺
Na⁺
H₂O
(nucleophile)
H₂N(CH₂)₅C—OH
N(CH₂)₅C
HB⁺
CH₂CH₂CH₂O

TABLE 27.2 Examples of Step-Growth Polymerization

Monomers	Smallest Intermediate Unit	Polymer
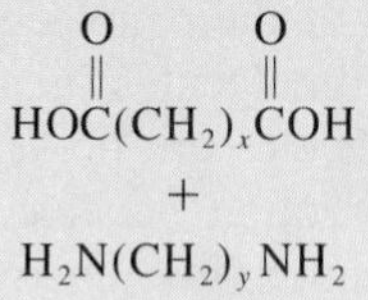	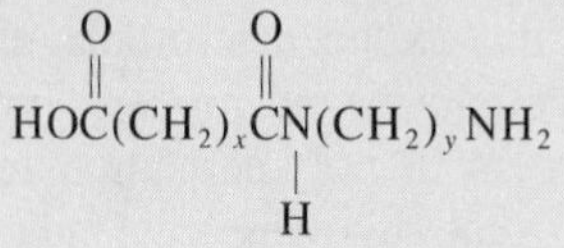	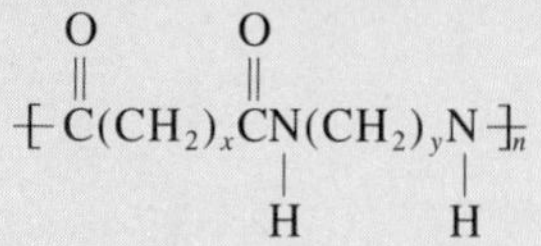
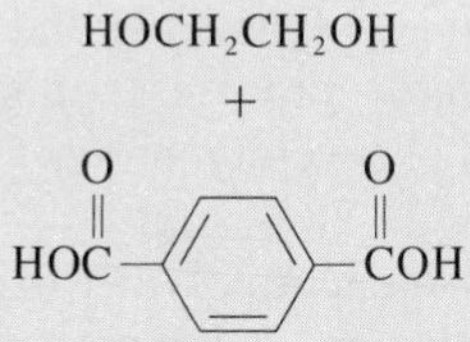	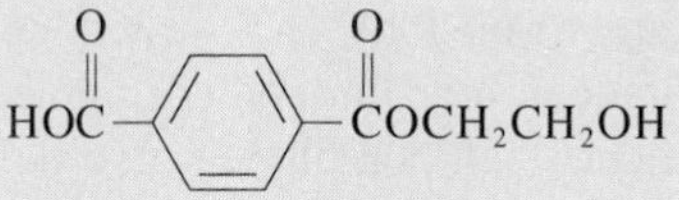	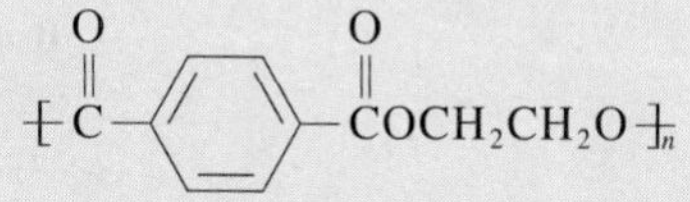
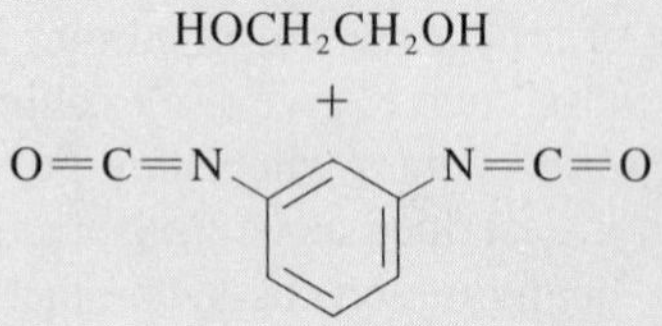	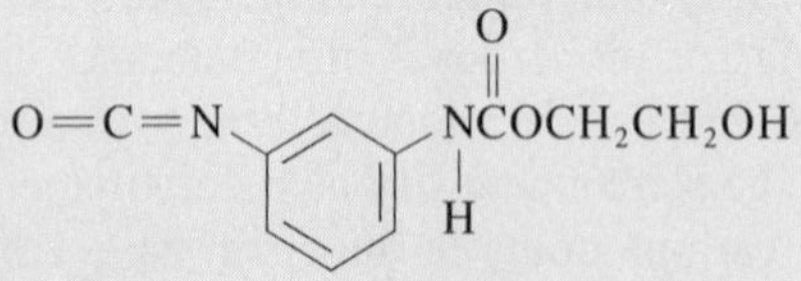	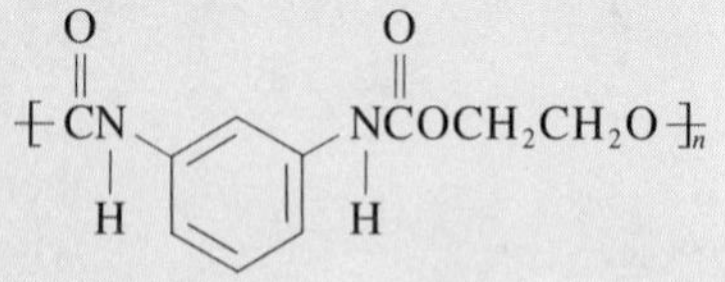

ADDITIONAL PROBLEMS

27.24 Predict the structures of the polymers that will result from the following reactions.

(a) 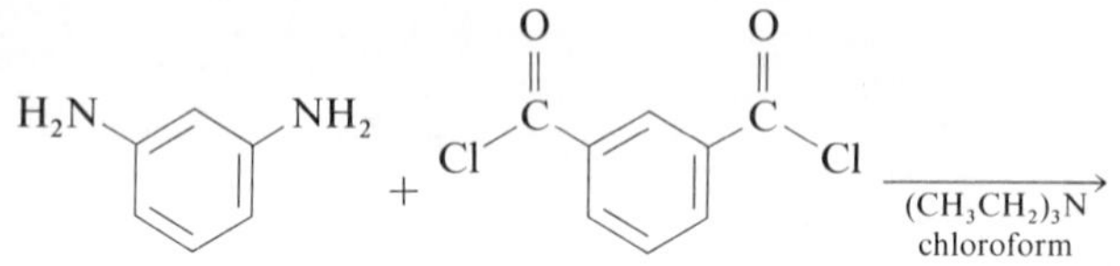

(b) 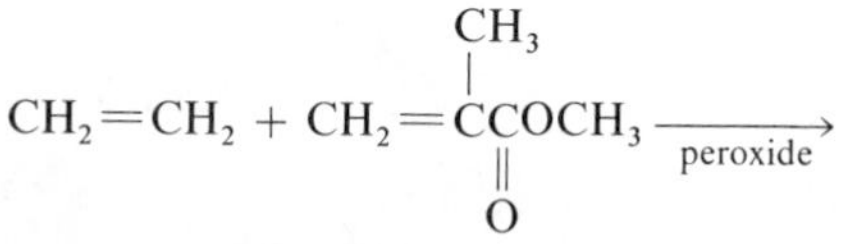

(c) 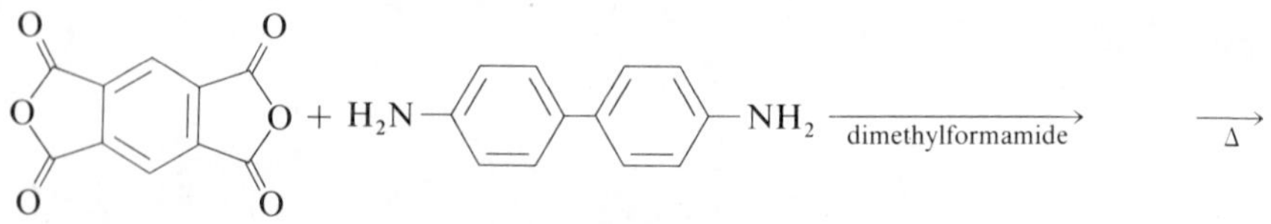

(d) 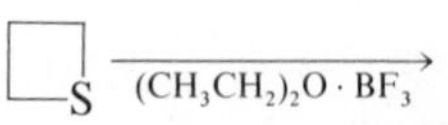

(e) 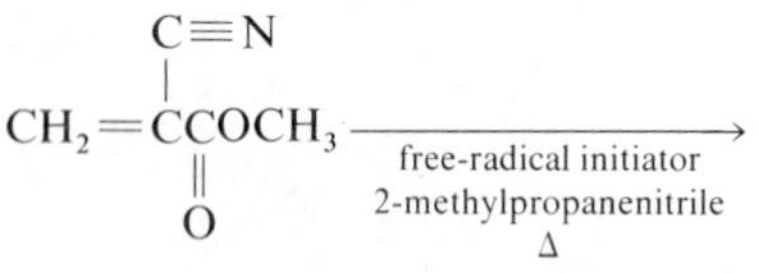

(f) 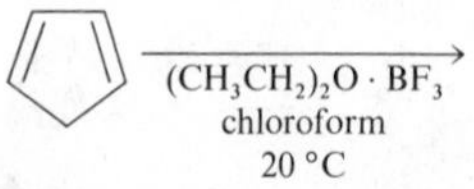

(Hint: How does 1,3-butadiene polymerize?)

(g) 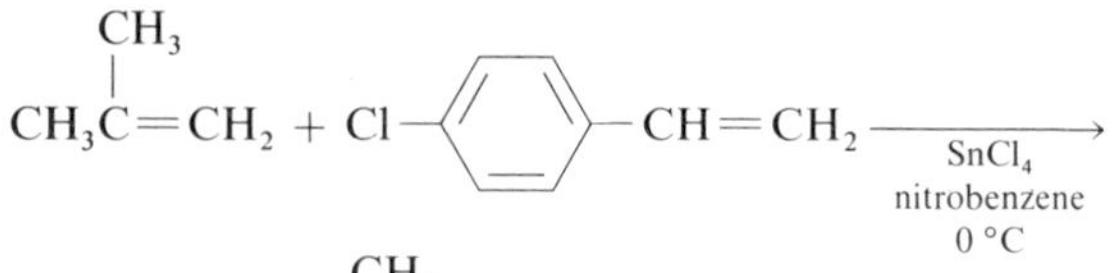
SnCl4
nitrobenzene
0 °C

(h) 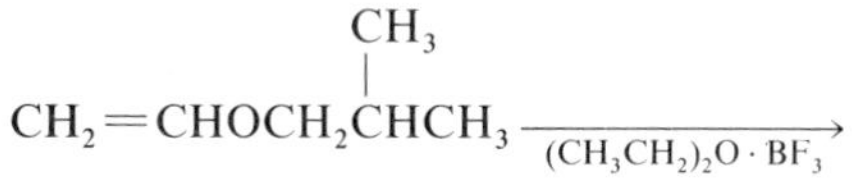
(CH3CH2)2O · BF3

(i) 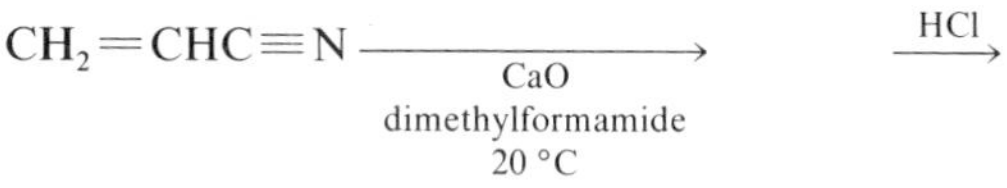
CaO
dimethylformamide
20 °C
HCl

(j) 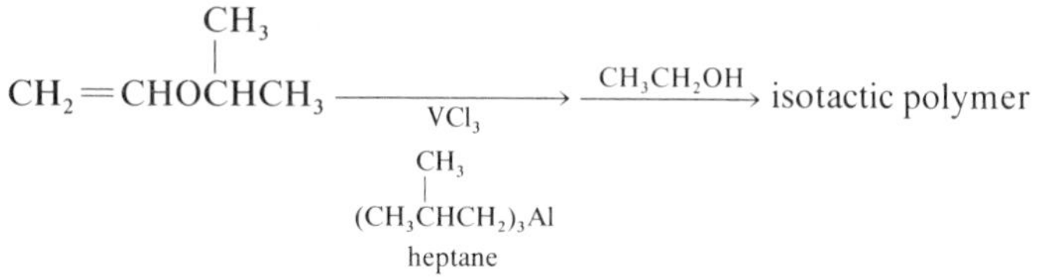
VCl3
(CH3CHCH2)3Al
heptane
CH3CH2OH
isotactic polymer

(k) 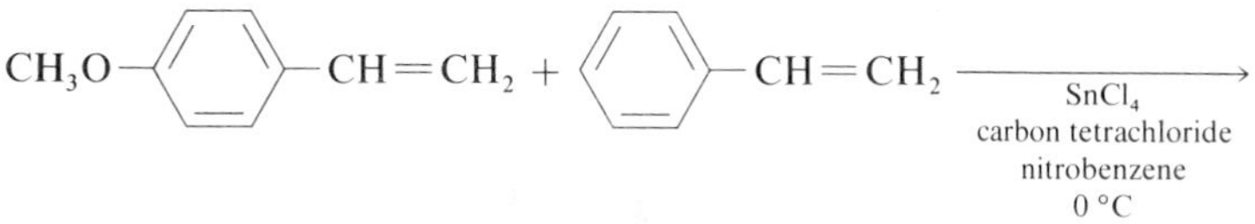
SnCl4
carbon tetrachloride
nitrobenzene
0 °C

(l) 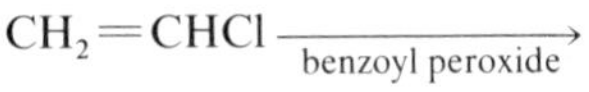
benzoyl peroxide

(m) 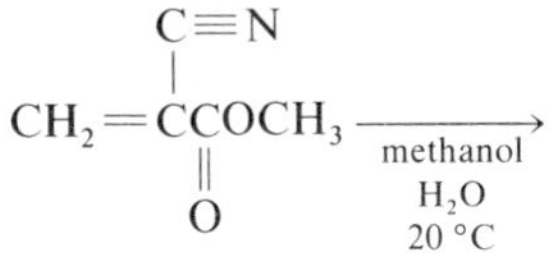
methanol
H2O
20 °C

(n) 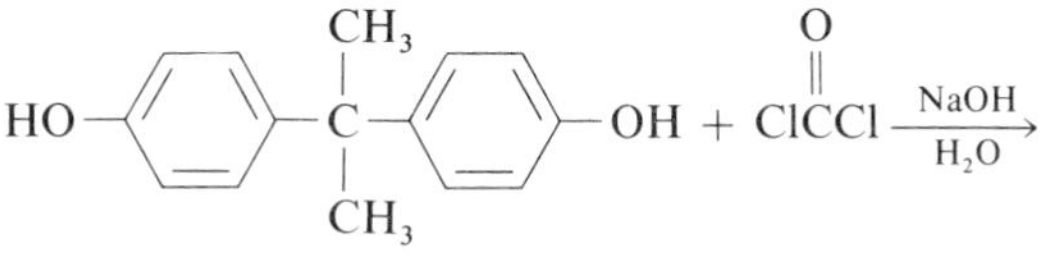
NaOH
H2O

(o) 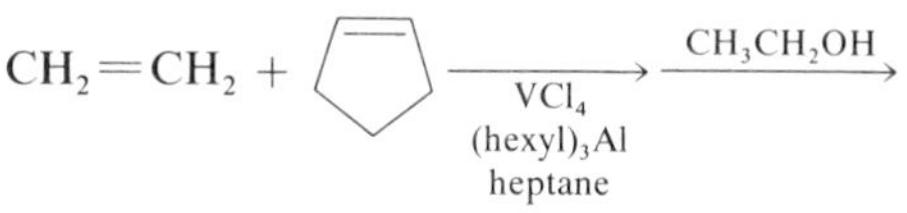
VCl4
(hexyl)3Al
heptane
CH3CH2OH

(p) 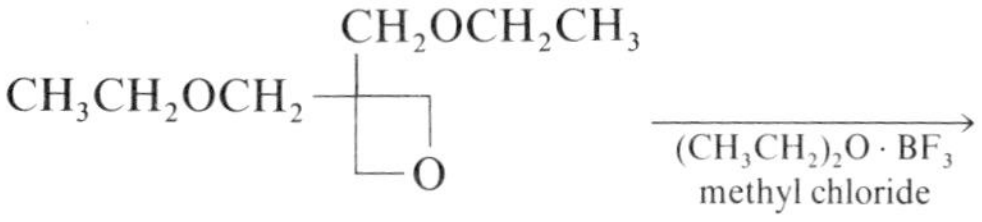
(CH3CH2)2O · BF3
methyl chloride

(q) 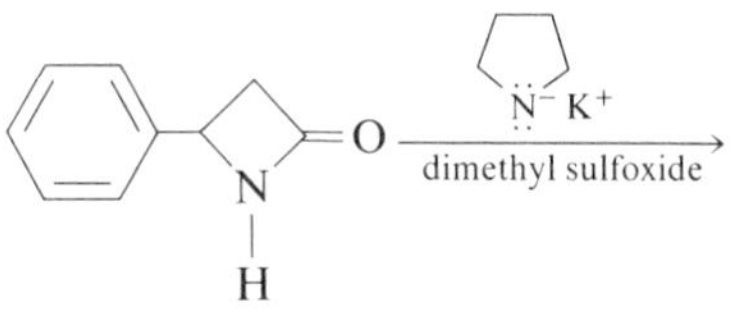
dimethyl sulfoxide

27.25 Unsaturated esters having the general formula $CH_2{=}CH\overset{O}{\overset{\|}{C}}OR$ and acrylonitrile, $CH_2{=}CHC{\equiv}N$, are successfully polymerized in chain reactions using free-radical or anionic initiators. Polymerization reactions do not take place when cationic initiators are used. How would you rationalize these experimental observations?

27.26 The reactivity of an alkene monomer in a free-radical polymerization reaction, as measured by the rate of the propagation step, depends on the vinylic substituent. The following order of reactivity has been observed.

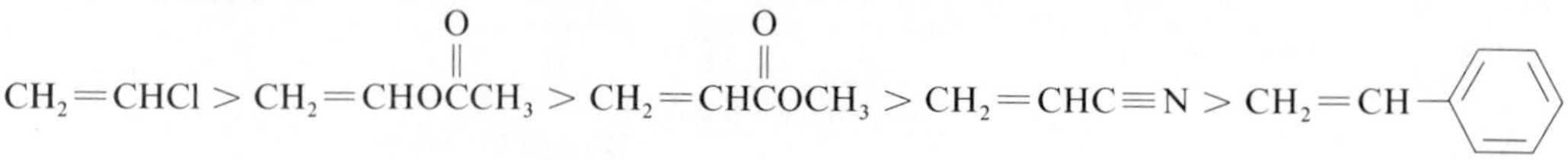

How would you rationalize this order of reactivity?

27.27

(a) The following polyimide has been used in a plastic film that retains its usefulness over a very wide range of temperatures. The repeating unit of the polymer is shown below. What monomeric units would you need to synthesize this polymer?

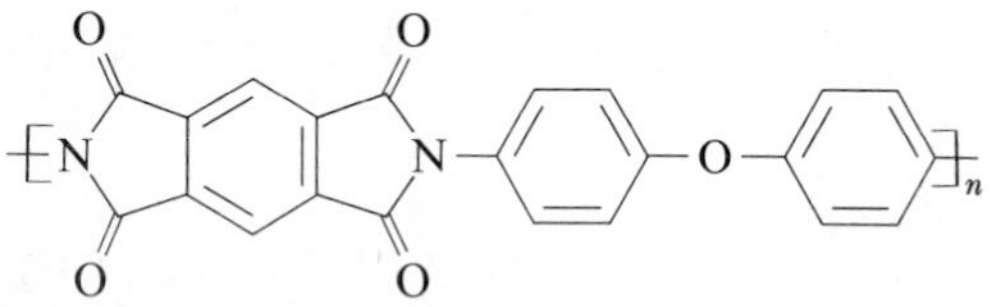

(b) The polymer shown above can be transformed into a more soluble polymer by reacting it with an amine. Write an equation predicting what will happen to the polymer if it is treated with 2-(*N*-methylamino)ethanol, $CH_3NHCH_2CH_2OH$.

27.28 Ion-exchange resins containing sulfonic acid groups are strongly acidic. Weakly acidic ion-exchange resins have also been made. One of them has the following partial structure. How would you make such a resin?

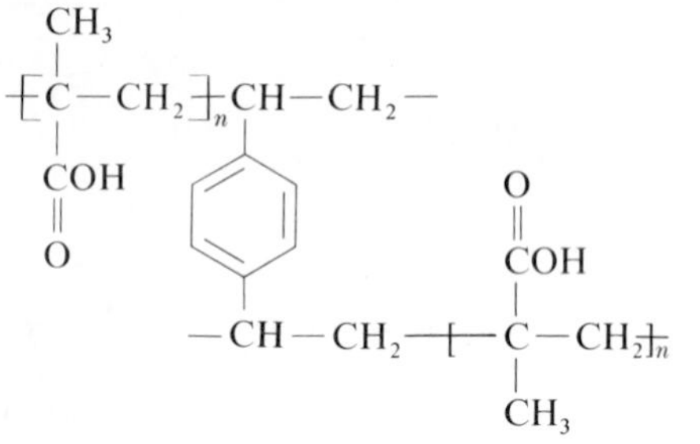

27.29 In the presence of a free-radical initiator, such as high-energy γ-radiation, acrylamide polymerizes at the carbon-carbon double bond to give a hydrocarbon chain with amide substituents on it. When a strong base is used to catalyze this polymerization, however, a polyamide that can be hydrolyzed to 3-aminopropanoic acid (β-alanine) is obtained. Write mechanisms that explain the different products obtained from these two reactions, shown below.

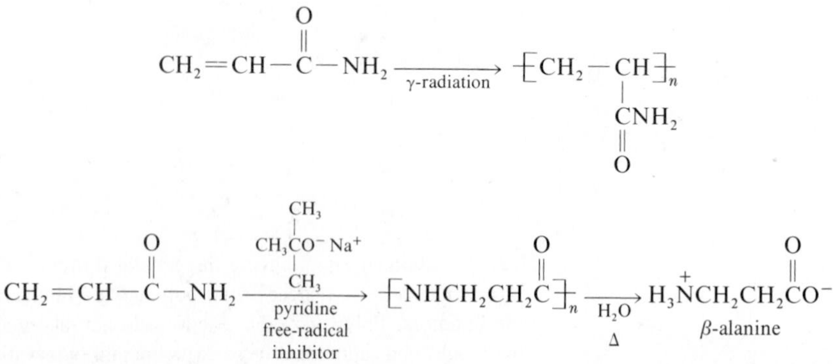

27.30 Fibers of a polyester, polypivalolactone, have a high degree of orientation and recover very quickly from deformation. This is a desirable property in fibers used in carpets; otherwise, the carpet shows footprints. The polyester is synthesized from either of the following monomers. Suggest a method for synthesizing the polymer from each monomer. For example, what type of catalyst would you use, and what reaction conditions would be necessary for each process?

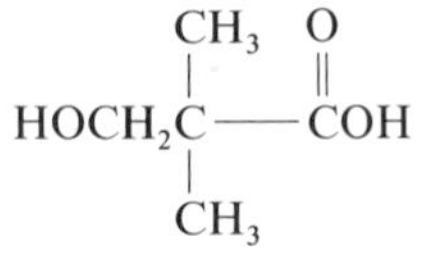

3-hydroxy-2,2-dimethylpropanoic acid

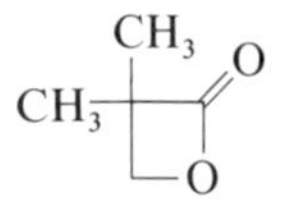

pivalolactone

27.31 When the polymerization of (*S*)-(−)-2-methyloxirane is catalyzed by solid potassium hydroxide, the product is a crystalline, optically active polymer. If the polymerization of the chiral oxirane is catalyzed by iron(III) chloride in ether, however, the resulting polymer is not optically active. Propose mechanisms for the two different types of polymerization that account for the different stereochemistry observed.

27.32 One of the polyamides having high molecular rigidity has the following repeating unit.

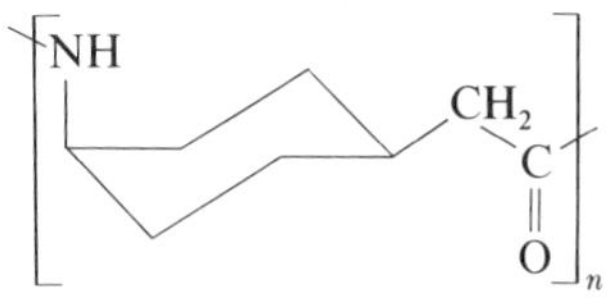

It is prepared by the polymerization of a bicyclic lactam.

(a) What is the structure of the lactam that will give rise to the polymer shown above?
(b) This lactam is synthesized in a series of steps starting with 1,3-cyclohexadiene and vinyl acetate and ending with a Beckmann rearrangement. Propose a synthesis for the lactam.

27.33 Under the proper conditions, the carbon-oxygen double bond of a carbonyl group takes part in polymerization reactions; for example, the two reactions shown below have been observed. Write a mechanism for each one, showing why the polymerization proceeds as it does.

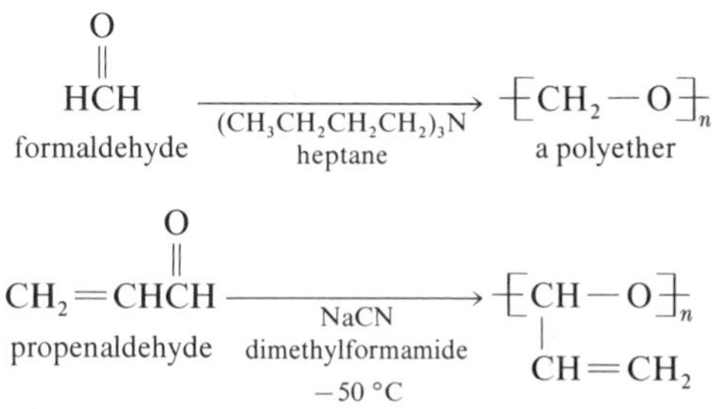
formaldehyde
(CH3CH2CH2CH2)3N
heptane
a polyether
propenaldehyde
NaCN
dimethylformamide
−50 °C

28

Concerted Reactions

A • L O O K • A H E A D

Concerted reactions take place in one step with high stereoselectivity and without the formation of a reactive intermediate. Bimolecular nucleophilic substitution reactions (p. 239) and Diels-Alder additions of dienes to dienophiles (p. 691) are familiar examples of concerted reactions. This chapter will explore a series of chemical transformations in which rearrangements of the positions of σ and π bonds take place by way of cyclic transition states. These cycloaddition reactions, electrocyclic reactions, and sigmatropic rearrangements are insensitive to reaction conditions such as solvent polarity and catalysis. They are believed to involve only a single step, in which reorganization of the bonding occurs simultaneously.

The reactions that are possible and the resulting stereochemistry depend in many cases on whether the reactions are thermal, meaning that the reactants are in their ground state, or photochemical, meaning that the excited state of a reactant is involved in the reaction. Chemists explain these differences by looking at the symmetry of the molecular orbitals involved in the changes in bonding taking place. An important theory states that the frontier orbitals, the highest-energy molecular orbital occupied by electrons of one reactant and the lowest-energy molecular orbital of the other reactant that is empty, are the ones that determine the course and the stereochemistry of the reaction. This chapter will show how to identify these orbitals and use their properties to make predictions about concerted reactions.

28.1 INTRODUCTION TO CONCERTED REACTIONS

A. Some Examples of Concerted Reactions

The Diels-Alder reaction (p. 691) is an example of a cycloaddition reaction, a reaction that forms a ring by addition to an alkene or an alkyne. Cycloaddition reactions are further classified according to the number of electrons involved in the transition state. A Diels-Alder reaction is called a **[4 + 2] cycloaddition reaction** because one component of the reaction, the diene, contributes four π electrons, and two π electrons come from the dienophile. A total of six electrons is involved in the transition state.

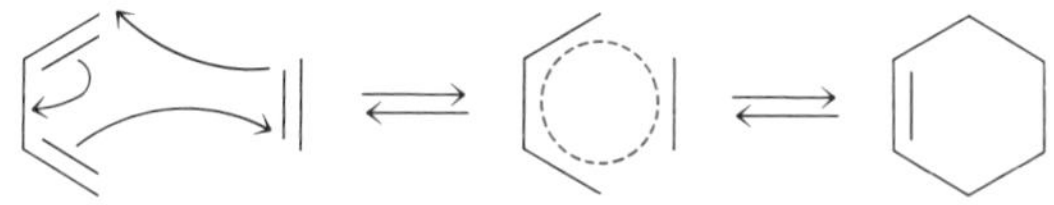

diene — dienophile — cyclic transition state — product

4 π electrons — *2 π electrons* — *6 π electrons*

The Diels-Alder reaction is an example of a **thermal cycloaddition reaction.** In contrast, some cycloaddition reactions occur only when one of the reactants absorbs light energy. For example, 2-butene does not add to itself to give a four-membered ring unless it is irradiated with ultraviolet radiation at a wavelength of 214 nm. The products obtained depend on the stereochemistry of the starting material.

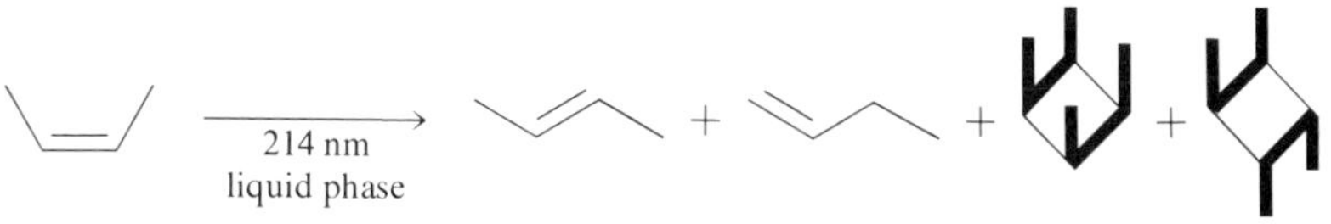
214 nm
liquid phase

(Z)-2-butene — (E)-2-butene *major product* — 1-butene — *methyl groups retain the cis orientation in each alkene unit*

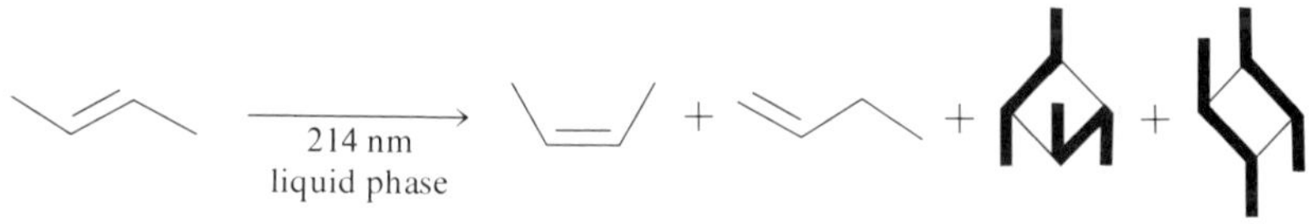
214 nm
liquid phase

(E)-2-butene — (Z)-2-butene — 1-butene — *methyl groups retain the trans orientation in each alkene unit*

Reactions that take place when molecules absorb ultraviolet or visible radiation are called **photochemical reactions.** The addition of 2-butene to itself upon absorption of ultraviolet radiation is a stereoselective photochemical [2 + 2] cycloaddition reaction, which is explored in greater detail in Section 28.2A.

The major photochemical reaction for both (E)- and (Z)-2-butene is the familiar stereochemical isomerization of the alkene (p. 715). A small amount of isomerization of the position of the double bond also takes place to give 1-butene. These reactions are not classified as concerted reactions.

Throughout the rest of this chapter, thermal reactions will be contrasted with photochemical reactions. The word thermal is used in this sense to mean that the reaction proceeds with the reactants in the ground state rather than the excited state achieved upon absorption of ultraviolet or visible radiation (p. 714). A thermal reaction may proceed at room temperature or even at very low temperatures. For example, the Diels-Alder reaction of cyclopentadiene with maleic anhydride, a thermal reaction, is carried out at 0 °C (p. 693).

Electrocyclic reactions take place when polyenes react intramolecularly to give cyclic alkenes or when cyclic alkenes undergo ring opening to become acyclic polyenes. These reactions, which are the reverse of each other, occur with high stereoselectivity. For example, when dimethyl *cis*-3-cyclobutene-1,2-dicarboxylate is heated to 140 °C, the product obtained is a dimethyl 2,4-hexadienedioate, which is almost entirely the (*Z*,*E*)-isomer.

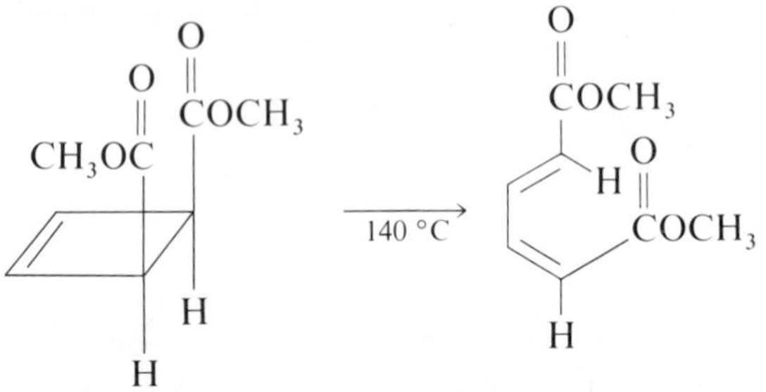
140 °C

dimethyl *cis*-3-cyclobutene-1,2-dicarboxylate

dimethyl (2*Z*,4*E*)-2,4-hexadienedioate

In principle, the above reaction is reversible, but the strain of the double bond in the small ring of cyclobutene and the conjugation of the double bonds with the carbonyl groups stabilize the acyclic diene with respect to the starting material. Thus, the reaction goes to completion at relatively low temperatures.

A conjugated triene, on the other hand, undergoes ring closure readily to give a stable six-membered ring. (2*E*,4*Z*,6*E*)-2,4,6-Octatriene gives *cis*-5,6-dimethyl-1,3-cyclohexadiene, but the (*Z*,*Z*,*E*) isomer closes to the corresponding trans compound.

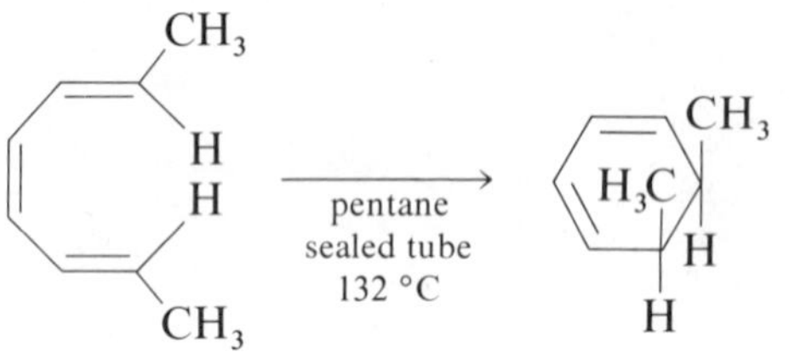
pentane
sealed tube
132 °C

(2*E*,4*Z*,6*E*)-2,4-6-octatriene

cis-5,6-dimethyl-1,3-cyclohexadiene

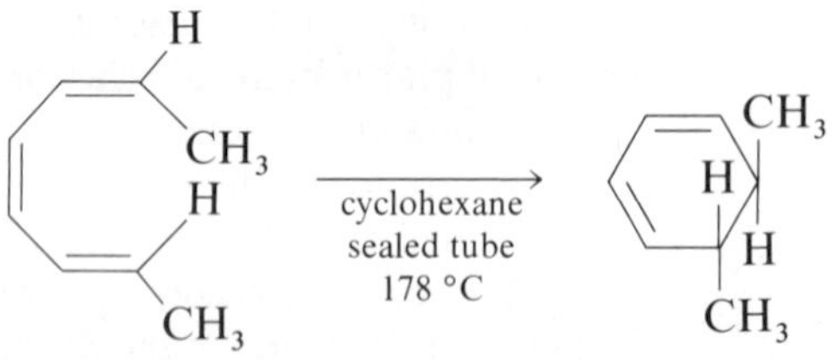
cyclohexane
sealed tube
178 °C

(2*Z*,4*Z*,6*E*)-2,4,6-octatriene

trans-5,6-dimethyl-1,3-cyclohexadiene

Thus, these reactions too are highly stereoselective.

In a **sigmatropic rearrangement,** a substituent and the pair of electrons that bond it to the rest of the molecule move along a conjugated system to a new position. The substituent that moves may be a hydrogen atom or an organic group. It may move from carbon to carbon or from some other element, such as oxygen, to carbon. In the simplest case, (Z)-1,3-pentadiene rearranges into itself, a reaction that can be detected only if the starting material is labeled with deuterium.

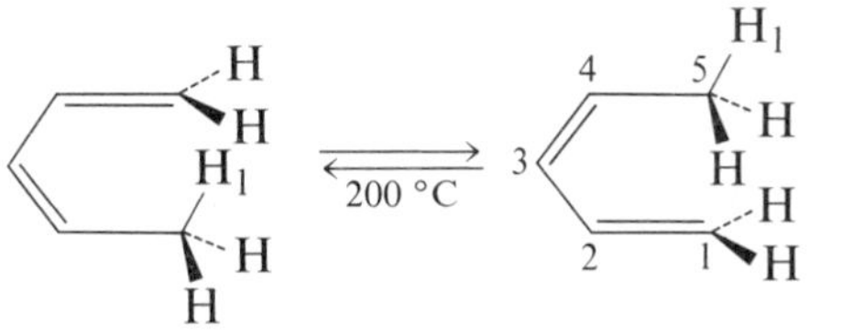
200 °C

(Z)-1,3-pentadiene (Z)-1,3-pentadiene

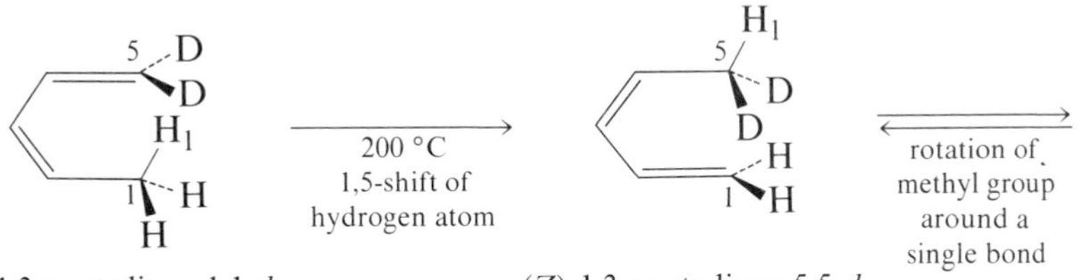
200 °C
1,5-shift of
hydrogen atom
rotation of
methyl group
around a
single bond

(Z)-1,3-pentadiene-1,1-d_2 (Z)-1,3-pentadiene-5,5-d_2

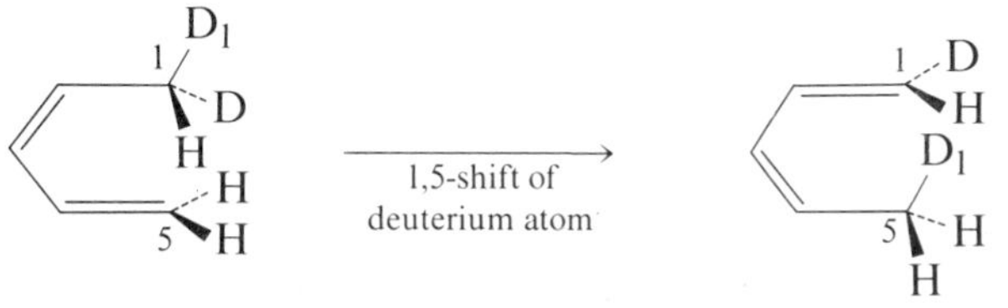
1,5-shift of
deuterium atom

(Z)-1,3-pentadiene-5,5-d_2 (1*E*,3*Z*)-1,3-pentadiene-1,5-d_2

In the above rearrangement, a hydrogen (or deuterium) atom, along with its σ bond, moves five atoms away along the carbon chain of a conjugated diene. The equation above illustrates [1,5] sigmatropic shifts of hydrogen and deuterium atoms. A σ bond is broken between the hydrogen atom and a carbon atom assigned the number 1 as the starting point of the rearrangement. A new σ bond is formed between the hydrogen atom, the first atom of the migrating group, and carbon 5 of the polyene system, thus the designation [1,5] for the shift. Other types of sigmatropic rearrangements will be examined in Sections 28.1B and 28.5, and the nomenclature will become clearer with practice. The important idea is that there has been a rearrangement involving the breaking of a σ bond in one portion of the molecule, the forming of one in another part of the molecule, and a new arrangement of the π bonds between these two points.

Sigmatropic rearrangements of hydrogen atoms are intramolecular reactions. They have been shown to be unimolecular and to be unaffected by solvent polarity. The reaction given above is stereoselective with the hydrogen (or deuterium) atom leaving the top side of the methyl group on one end of the pentadiene and arriving at the top side of the methylene group of the terminal double bond. According to all these experimental observations, sigmatropic rearrangements are concerted reactions.

This section has introduced some experimental facts about a variety of reactions that can be classified together. They are all concerted reactions involving the stereoselective reorganization of the electrons in π and σ bonds, usually through cyclic transition states. The next section will explore some of the attempts to correlate these experimental observations by means of theoretical explanations. The concepts and the language that have been developed during the evolution of these attempts have transformed the way chemists think and talk about a wide range of reactions.

PROBLEM 28.1

Study Guide Concept Map 28.1

To prove that the photochemical cycloaddition reactions of the 2-butenes shown on p. 1215 were indeed stereoselective, it was necessary to examine the products formed when a mixture of (*Z*)- and (*E*)-2-butenes were irradiated with ultraviolet radiation at a wavelength of 214 nm. Predict what the products observed in such experiments were.

PROBLEM 28.2

The following sequence of reactions was used to synthesize (*Z*)-1,3-pentadiene-1,1-d_2. Give structures for the compounds represented by letters.

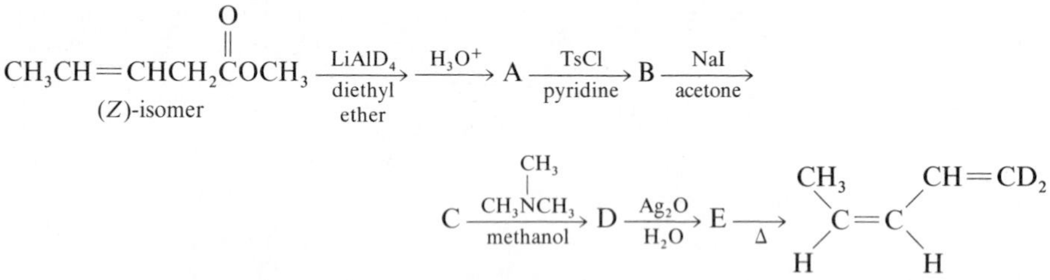
CH3CH=CHCH2COCH3
(Z)-isomer
LiAlD4
diethyl ether
H3O+
A
TsCl
pyridine
B
NaI
acetone
C
CH3NCH3
methanol
D
Ag2O
H2O
E
Δ
CH3
CH=CD2
C=C
H
H

B. The Evolution of the Theory of Concerted Reactions

In the 1960s, chemists began to develop theoretical explanations for the experimental facts described in the preceding section and many others observed in similar reactions. The need, not only for explanations but also for theories that would suggest new experiments, arose mostly out of greatly expanded research into the chemistry and synthesis of natural products. An illustration of the close relationship between such practical research and the development of theory in organic chemistry is the work done on Vitamin D by Egbert Havinga of the Netherlands. A form of Vitamin D is synthesized in the skin on exposure to sunlight.

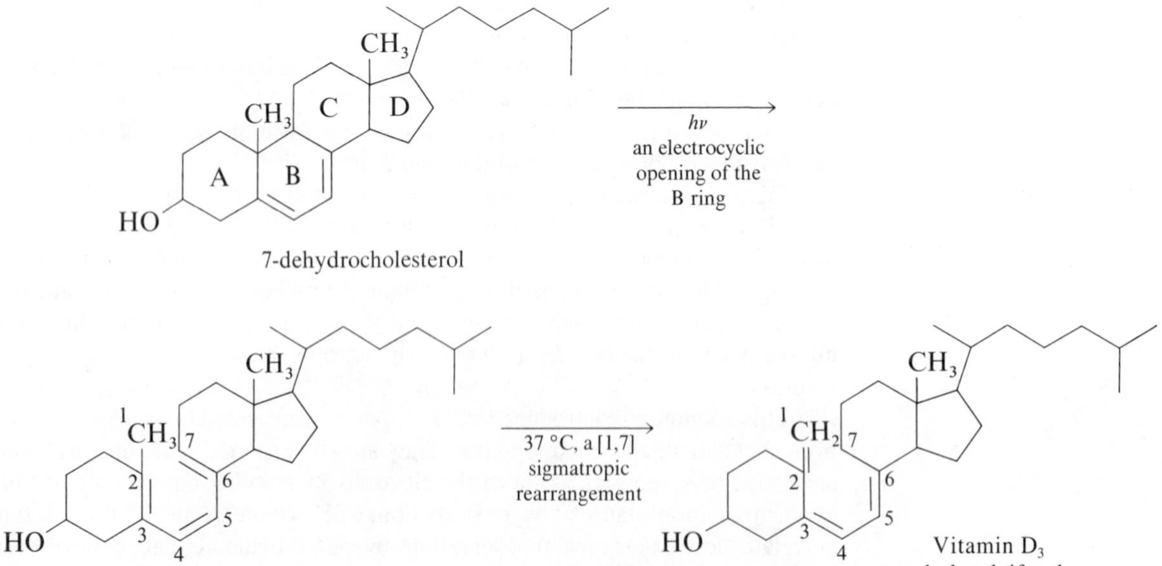
CH3
CH3
C
D
A
B
HO
hν
an electrocyclic opening of the B ring
CH3
CH3
HO
37 °C, a [1,7] sigmatropic rearrangement
CH3
CH2
HO
1
2
3
4
5
6
7

7-dehydrocholesterol

Vitamin D_3
cholecalciferol

The overall conversion may be classified as an electrocyclic opening of the B ring of 7-dehydrocholesterol to form a conjugated triene, followed by a sigmatropic rearrangement of a hydrogen atom from the methyl group to the other end of the conjugated system, a [1,7] sigmatropic shift. Recent research has shown that the sigmatropic rearrangement takes place slowly at body temperature, so Vitamin D_3 continues to be synthesized in the skin for as long as three days after exposure to the sun.

The reactions shown above are only two of many similar ones observed for the Vitamin D system. The reactions have differing stereochemistry depending on whether they occur photochemically or thermally, as will be seen in Section 28.3B. Havinga's very precise determination of the course of these reactions provided information that led his colleague Luitzen J. Oosterhoff to suggest that the stereochemistry characteristic of these reactions is determined by the symmetry of the molecular orbitals involved in the transformations.

Meanwhile, in the 1960s, Robert B. Woodward at Harvard University was doing research on the synthesis of Vitamin B_{12}. He observed the precise stereochemistry with which cyclohexadienes were formed from conjugated trienes and then converted back into trienes in thermal and photochemical reactions. At that time, the reactions in which cyclobutenes open to butadienes, such as the example given on p. 1216, were known. Woodward and his colleague Roald Hoffmann, in thinking about these observations, also arrived at the idea that the symmetry of the molecular orbitals that participate in the chemical reaction determines the course of the reaction. In a series of papers published in 1965, they applied their ideas systematically to electrocyclic, cycloaddition, and sigmatropic reactions. They proposed what they called the principle of the **conservation of orbital symmetry** in concerted reactions. In the most general terms, the principle means that in a concerted reaction, the molecular orbitals of the starting materials must be transformed into the molecular orbitals of the products in a smooth, continuous way. This is possible only if the orbitals have similar symmetry. The recognition of the generality and the significance of this way of examining organic reactions was only one of the many contributions made by Woodward, who received the Nobel Prize in 1965 for his work in organic chemistry.

C. A Review of π Molecular Orbitals

The bonding and antibonding π orbitals of ethylene (Figure 2.16, p. 50) and the electronic configuration of ethylene in the ground and the excited states are shown in Figure 28.1. The π molecular orbital has a nodal plane containing the two carbon atoms and the four hydrogen atoms. The π^* orbital has a second nodal plane in addition to the one that includes the σ-bonded backbone of the molecule and is of higher energy than the bonding π orbital. The more nodes there are in a molecular orbital (and in an atomic orbital), the higher the energy level of the orbital. Ethylene is a stable molecule in the ground state because the two electrons from the two *p* atomic orbitals are in the bonding π orbital. Absorption of ultraviolet radiation by ethylene raises the molecule to the excited state in which one electron has been promoted to the π^* orbital (p. 714).

The nodes and the relative energies of the π and π^* orbitals of ethylene are important because they represent the properties of all other alkenes with one double bond and two π electrons. This discussion assumes that substituents on the doubly bonded carbons do not change these properties of the molecular orbitals sufficiently to change the conclusions that can be drawn by looking at the simplified picture.

Just as the molecular orbital picture developed for ethylene can represent all compounds with a single π orbital and two electrons in it, a similar molecular

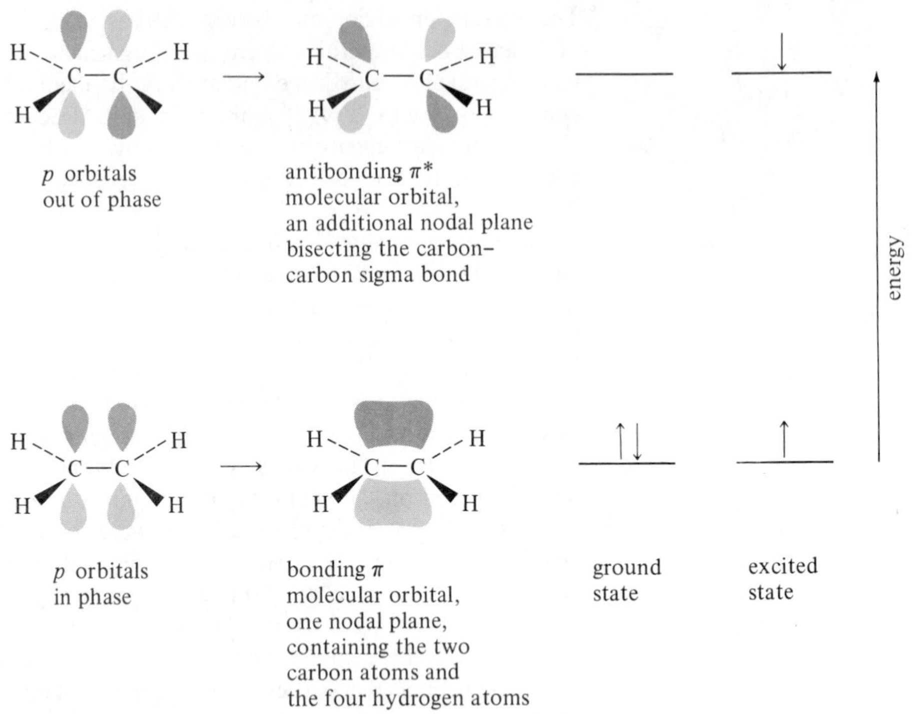
p orbitals
out of phase
antibonding π^*
molecular orbital,
an additional nodal plane
bisecting the carbon–
carbon sigma bond
energy
p orbitals
in phase
bonding π
molecular orbital,
one nodal plane,
containing the two
carbon atoms and
the four hydrogen atoms
ground
state
excited
state

FIGURE 28.1 The π and π^* orbitals of ethylene. The electronic configuration of ethylene in the ground state and after the absorption of a photon.

orbital picture for 1,3-butadiene (Figure 17.3) can be used to talk about a conjugated diene. The four molecular orbitals, designated as ψ_1, ψ_2, ψ_3, and ψ_4, are shown again in Figure 28.2. They are constructed by successively introducing additional nodes into the array of *p* atomic orbitals while retaining the overall symmetry of the system. Note how the energy level of an orbital is raised as the number of nodes in the orbital increases. In 1,3-butadiene, the symmetry of the molecule means that carbon 1 is equivalent to carbon 4 and that carbon 2 is equivalent to carbon 3 (and that carbon 1 is *not* equivalent to carbon 2, and carbon 3 is *not* equivalent to carbon 4). In constructing molecular orbitals, therefore, if a node is placed between carbon 1 and carbon 2, a node must also appear between carbon 4 and carbon 3.

The molecular orbitals shown in Figure 28.2 will be used to talk about any conjugated diene with four π electrons in the system. Once again, the assumption will be made that the substituents on the diene system do not change the nodal properties of the molecular orbitals sufficiently to affect the conclusions that can be drawn.

PROBLEM 28.3

How many nodes are there in each of the molecular orbitals, ψ_1, ψ_2, ψ_3, and ψ_4, of 1,3-butadiene?

What is the symmetry of a molecular orbital to which Woodward and Hoffmann refer in their principle? Each molecular orbital can be examined for symmetry in ways that are analogous to the method used earlier in this book to decide whether

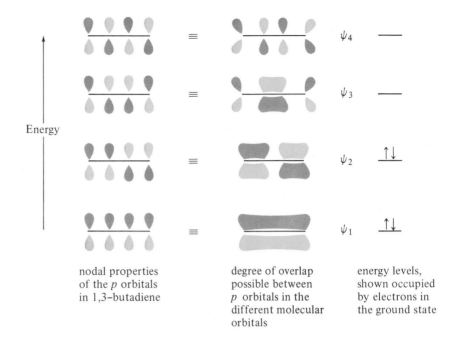

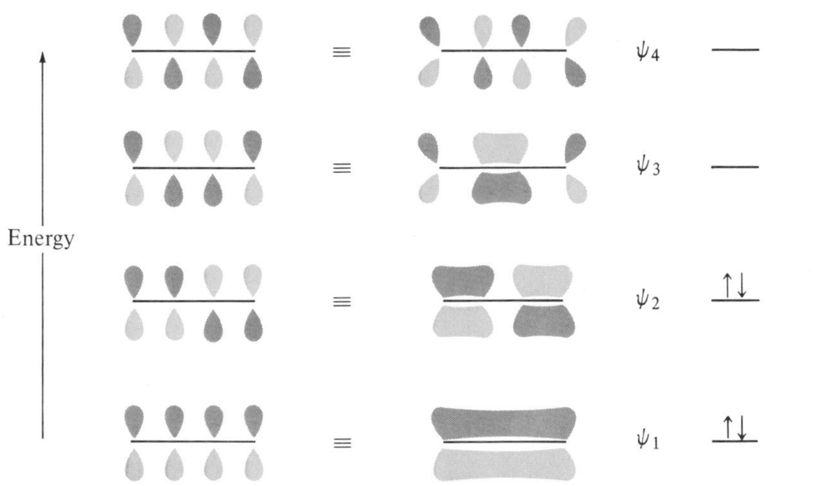
Energy
ψ4
ψ3
ψ2
ψ1
nodal properties of the p orbitals in 1,3-butadiene
degree of overlap possible between p orbitals in the different molecular orbitals
energy levels, shown occupied by electrons in the ground state

FIGURE 28.2 The four molecular orbitals of butadiene, created from four p atomic orbitals.

or not molecules are chiral. There are many such ways in which this can be done, but one example will be given to show that in principle each molecular orbital can be classified as symmetrical or antisymmetrical.

The two π molecular orbitals shown in Figure 28.1 can be described as symmetrical or antisymmetrical with respect to a plane bisecting the carbon-carbon bond. If you imagine such a plane, you will see that it cuts the π orbital into two halves. The left-hand side of the orbital is an exact mirror image of the right-hand side. Thus, the π orbital is **symmetrical** with respect to this plane. The π^* orbital is not. For example, to the left of the plane on the top there is a grey lobe, but to the right of the plane at the top and at the same distance away from the plane, there is a colored lobe. In other words, the sign of the wave function is reversed as it goes through the plane. The π^* orbital is thus **antisymmetrical** with respect to a plane bisecting the carbon-carbon bond.

PROBLEM 28.4

Of the molecular orbitals of 1,3-butadiene (Figure 28.2), decide which are symmetrical and which antisymmetrical with respect to a plane bisecting the bond between carbon 2 and carbon 3.

In a concerted reaction, the symmetry that is present in the reactants is maintained during the course of the reaction and is also present in the product. The Diels-Alder reaction provides a simple demonstration of this principle (Figure 28.3). The reactants, the diene and the dienophile, each have a plane of symmetry. They must approach each other in the transition state in such a way that the same plane of symmetry is maintained. The product, cyclohexene, also retains the same plane of symmetry.

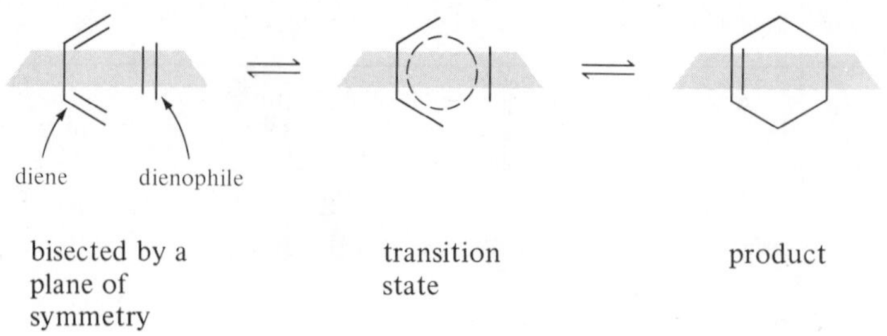
diene
dienophile
bisected by a plane of symmetry
transition state
product

FIGURE 28.3 A demonstration of the conservation of symmetry in the Diels-Alder reaction.

D. Interactions Between Molecular Orbitals

Interactions between the molecular orbitals of reactants and transformations of them into the molecular orbitals of products have been described in many different ways. One of the simplest and most useful is to say that a reaction takes place when the highest-energy molecular orbital that contains electrons of one reactant interacts with the lowest-energy molecular orbital that does not contain electrons of the other reactant. This is just like saying that a Lewis base, with electrons to donate, reacts with a Lewis acid, with an empty orbital ready to receive them, or that a nucleophile reacts with an electrophile. The only difference is that the orbitals performing these roles are examined very precisely in order to explain the highly specific chemistry and stereochemistry observed for concerted reactions.

The orbitals that interact have been called the **frontier orbitals** for the reaction by Kenichi Fukui of Japan. His contributions to the development of this way of looking at chemical reactions were recognized in 1981 by the award of the Nobel Prize in Chemistry jointly to him and Roald Hoffmann. The frontier orbitals are called the **highest occupied molecular orbital,** abbreviated **HOMO,** and the **lowest unoccupied molecular orbital,** or **LUMO.** For ethylene, for example, the π orbital is the HOMO, and the π^* orbital is the LUMO in the ground state. When ethylene is raised to its excited state in a photochemical reaction, the π^* orbital, with one electron in it, becomes the HOMO (Figure 28.1). In the ground state, ψ_2 is the HOMO and ψ_3 is the LUMO of 1,3-butadiene (Figure 28.2). These relationships will be taken up again in the next section and applied to reactions actually observed for compounds with molecular orbitals like those of ethylene and butadiene.

Why are these particular orbitals so important in determining the course of a concerted reaction? The electrons in the HOMO of a molecule are like the outer-shell electrons of an atom. They can be removed with the least expenditure of energy because they are already in a higher energy level than any of the other electrons in the molecule. For example, they are the electrons that are lost when the molecule is ionized in the gas phase. Spectroscopic techniques that cause this ionization can be used to determine the energy level of the HOMO for a molecule. The LUMO of a molecule is the orbital to which the electrons can be transferred with the least expenditure of energy.

The higher the energy of the HOMO of a molecule, the more easily electrons can be removed from it. The lower the energy of the LUMO of a molecule, the more easily electrons can be transferred into it. Therefore, the interaction between a molecule with a high HOMO and one with a low LUMO is particularly strong. In general, the smaller the difference in energy between the HOMO of one molecule and the LUMO of another with which it is reacting, the stronger is the interaction between the two molecules.

PROBLEM 28.5

Assume that an electron is raised from the HOMO to the LUMO of a molecule of 1,3-butadiene upon absorption of energy. Show the electronic configuration after the molecule absorbs ultraviolet radiation, and decide which molecular orbital will be the HOMO and which the LUMO of the diene in the excited state.

28.2 CYCLOADDITION REACTIONS OF CARBON COMPOUNDS

A. Photochemical Dimerization of Alkenes

The usefulness of the theoretical picture of the interactions between molecular orbitals developed in the preceding section is demonstrated by examining the cycloaddition reactions of alkenes. In Figure 28.4, the HOMO of one molecule of (*E*)-2-butene is shown approaching the LUMO of another molecule of (*E*)-2-butene in the ground state; that is, when ultraviolet radiation is not used. An examination of the way in which the two orbitals are approaching each other reveals that it will not be possible for a cyclobutane ring to form in a concerted manner. The bottom right-hand lobe of the HOMO is in phase with the top right-hand lobe of the LUMO and could interact to give a σ bond, but the lobes that are facing each other on the left-hand side are out of phase and would only interact in an antibonding way. The second bond necessary to complete the ring would not form.

If one of the molecules of (*E*)-2-butene is raised to the excited state, however, its HOMO will have a different symmetry, as shown in the energy diagram in Figure 28.1. If a sample of (*E*)-2-butene is irradiated with ultraviolet light, the concentration of molecules in the excited state is low; therefore, a molecule in the excited state is most likely to encounter and react with a molecule of (*E*)-2-butene in the ground state. The molecular orbital picture for such an interaction is shown in Figure 28.5. In this case, bonding interactions are possible at both ends of the alkene bonds as the two molecules approach each other face-to-face. Only one of the two possible stereochemical outcomes for two (*E*)-2-butene molecules reacting in this way has been drawn. The methyl groups on the left-hand carbon atom of each 2-butene are on the same side of the double bond and are cis to each other in the cyclobutane, and the same is true of the methyl groups on the other carbon atom of the reagent. The stereochemistry of the starting materials has been preserved in the product.

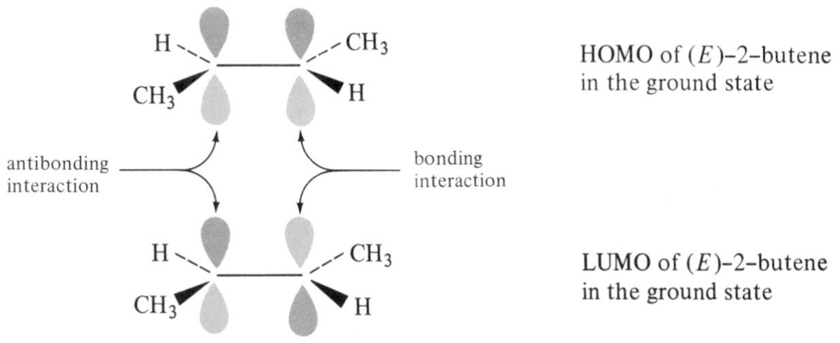
H
CH_3
CH_3
H
HOMO of (*E*)-2-butene in the ground state
antibonding interaction
bonding interaction
H
CH_3
CH_3
H
LUMO of (*E*)-2-butene in the ground state

FIGURE 28.4 A schematic representation of the interaction of the HOMO and LUMO of (*E*)-2-butene in the ground state.

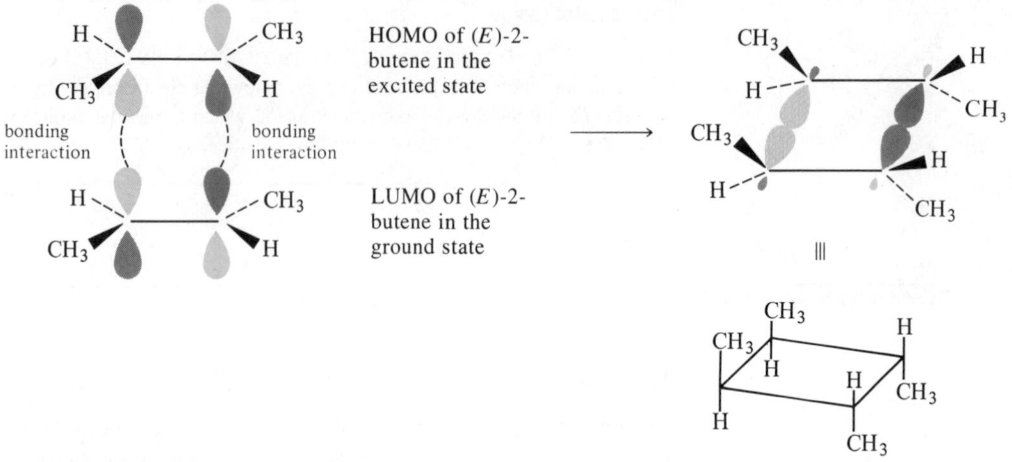

HOMO of (E)-2-butene in the excited state
bonding interaction
bonding interaction
LUMO of (E)-2-butene in the ground state

FIGURE 28.5 A schematic representation of the interaction of the HOMO of (*E*)-2-butene in the excited state with the LUMO of (*E*)-2-butene in the ground state.

The molecular orbital picture thus provides an explanation for the lack of reactivity of 2-butene in the ground state and for the stereoselective dimerization that occurs in the excited state.

PROBLEM 28.6

Another stereoisomer is also formed when (*E*)-2-butene is irradiated. Draw the molecular orbital picture showing how it arises.

PROBLEM 28.7

Draw molecular orbital pictures showing how the two cyclobutanes that are formed when (*Z*)-2-butene is irradiated with ultraviolet light come into being.

Two alkene functions react photochemically when DNA is damaged by ultraviolet radiation. This reaction is believed to be responsible for most of the harmful effects of such radiation on living organisms. For example, ultraviolet radiation kills bacteria quite efficiently and is used to sterilize equipment. Also, excessive ultraviolet radiation from sunlight (or tanning lamps) can cause skin cancer. Both of these effects arise from damage done to DNA so that it can no longer function properly to direct the synthesis of proteins in the cell.

Thymine residues that are adjacent to each other on a strand of DNA can undergo a photochemical [2 + 2] cycloaddition reaction to give a cyclobutane compound known as a thymine dimer.

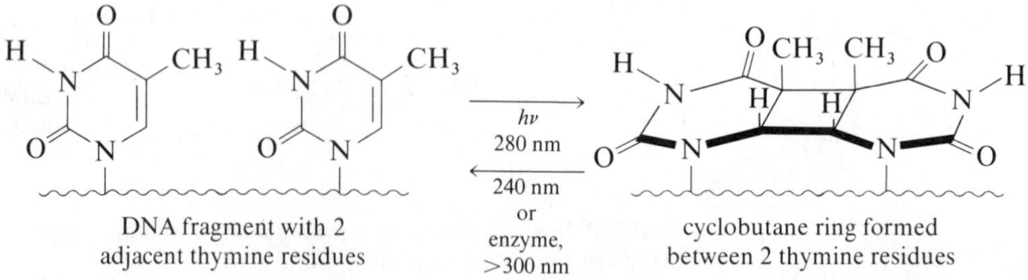

hν
280 nm
240 nm
or
enzyme,
>300 nm
DNA fragment with 2 adjacent thymine residues
cyclobutane ring formed between 2 thymine residues

This photochemical dimerization of thymine disrupts the normal functioning of DNA (p. 1020). Fortunately for creatures that must live under constant exposure to ultraviolet radiation, DNA that is damaged by the formation of dimers can be repaired. The cell has enzymes that repair DNA by cutting out the damaged parts and filling in the missing units. The dimer of thymine also absorbs ultraviolet radiation and undergoes an opening of the cyclobutane ring that converts it back to thymine units. These pathways provide a mechanism for DNA repair, but some damage always remains.

PROBLEM 28.8

The chromophore (p. 716) in thymine is an α,β-unsaturated carbonyl group. In the excited state, such a functional group adds to alkenes or alkynes in a [2 + 2] cycloaddition reaction. Given this information, predict the products of the following reactions.

(a) 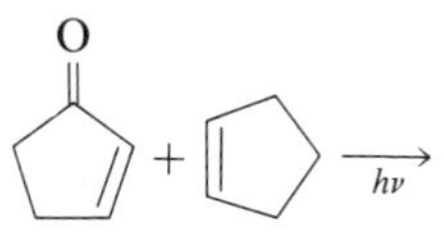

(b) 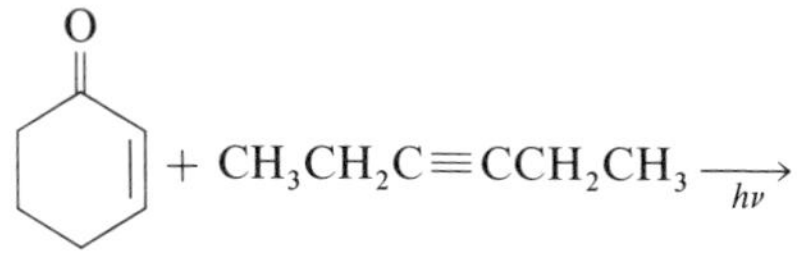

(c) 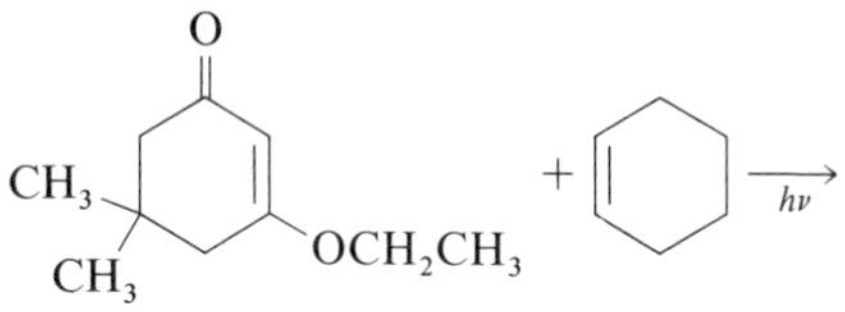

PROBLEM 28.9

The photochemical [2 + 2] cycloaddition reaction of alkenes has been used to create a cyclobutane ring in several syntheses of grandisol, a sex pheromone of the boll weevil. The cycloaddition products obtained from these syntheses are shown below. Write structures for the two reagents that must have been used in each synthesis. (Hint: In part a, one of them is *not* ethylene.)

(a) 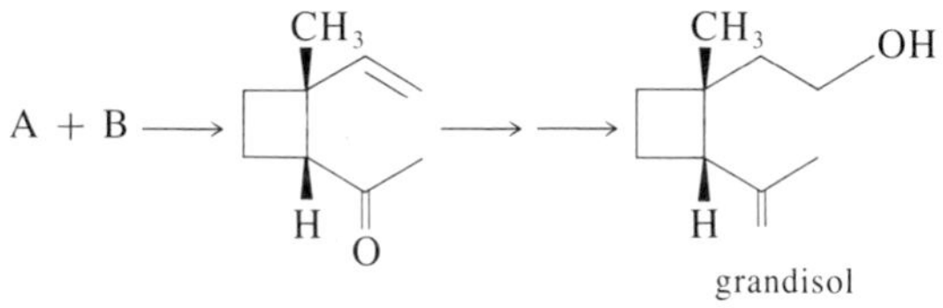

grandisol

(b) 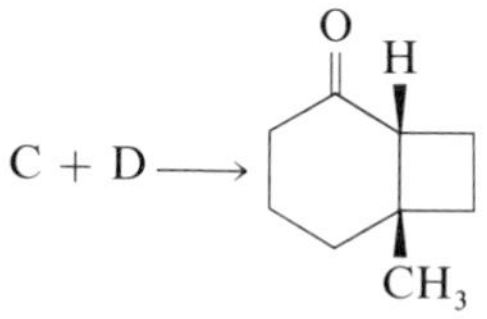

(c) 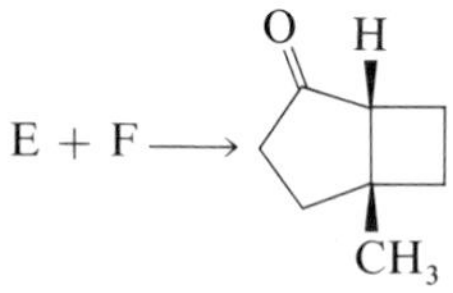

(d) 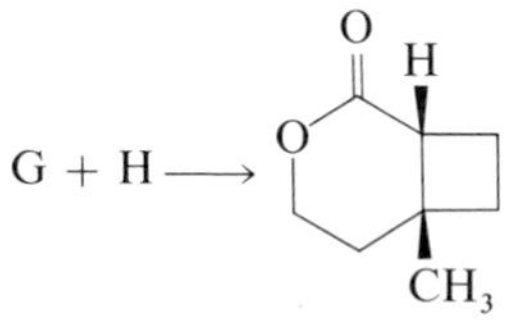

PROBLEM 28.10

When thymine is irradiated in solution, dimers other than the one shown on p. 1224 are produced. What other structures are possible for thymine dimers?

B. A Molecular Orbital Picture of the Diels-Alder Reaction

A molecular orbital picture of the Diels-Alder reaction shows the interaction of the HOMO of the diene with the LUMO of the dienophile (Figure 28.6). An examination of the HOMO of the diene shows that the lower lobes of the orbital at carbons 1 and 4 of the conjugated system have the same color as the corresponding upper lobes of the LUMO of the dienophile. Bonding interactions can develop between them if the dienophile approaches the ends of the diene at one face of the molecule. Note that it does not make any difference whether the dienophile is above or below the diene. Direct overlap of the lobes that are in phase gives rise to two bonding σ orbitals and the formation of a six-membered ring.

The substituents at the ends of the diene system and on the doubly bonded carbons of the dienophile retain the stereochemistry they had in the starting compounds because there is no twisting around single bonds during the reaction. The only part of the picture that remains puzzling is how the orbitals between carbons 2 and 3 of the diene system form the π bond in the product. At first glance, it looks as if the lobes of the *p* orbitals that are left over at carbons 2 and 3 will be antibonding. This is because only two of the orbitals that participate in the reaction are shown in the product; for the sake of simplicity, the others are ignored. As the σ bonds form, the interactions between the lobes of the *p* orbitals at carbons 1 and 2 and those at carbons 3 and 4, which are shown as bonding in the HOMO of the diene, change. The static picture in Figure 28.6 concentrates on only one part of the process and does not tell the whole story.

PROBLEM 28.11

Draw a molecular orbital picture of the Diels-Alder reaction showing the LUMO of the diene and the HOMO of the dienophile to prove to yourself that the stereochemistry of the reaction and the nature of the transition state are as described above.

Chemists have used a picture of the interactions of the molecular orbitals of cyclopentadiene and maleic anhydride to rationalize the observed stereochemistry (p. 692) in Diels-Alder reactions. The molecular orbital picture of the cycloaddition of cyclopentadiene with maleic anhydride is shown in Figure 28.7. The LUMO for maleic anhydride is shown to include the π bonds of the carbonyl groups that are conjugated with the double bond of the diene. The picture drawn is that of the LUMO of a 1,3,5-hexatriene system.

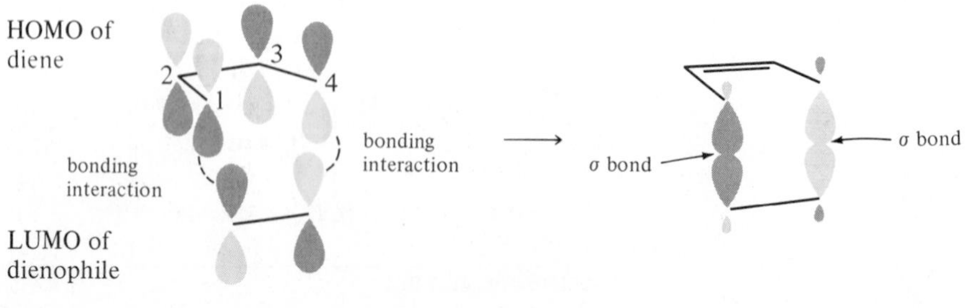
HOMO of diene
1
2
3
4
bonding interaction
bonding interaction
LUMO of dienophile
σ bond
σ bond

FIGURE 28.6 A schematic representation of the favorable interaction between the HOMO of a diene and the LUMO of a dienophile.

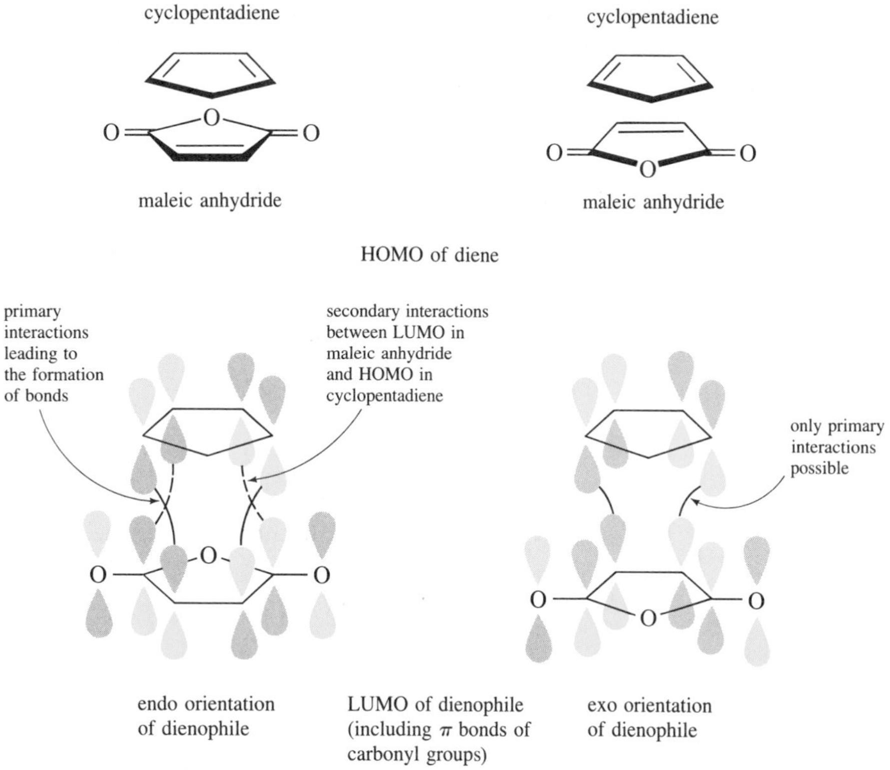
cyclopentadiene
cyclopentadiene
maleic anhydride
maleic anhydride
HOMO of diene
primary interactions leading to the formation of bonds
secondary interactions between LUMO in maleic anhydride and HOMO in cyclopentadiene
only primary interactions possible
endo orientation of dienophile
LUMO of dienophile (including π bonds of carbonyl groups)
exo orientation of dienophile

FIGURE 28.7 Schematic representation of the interactions between the HOMO of cyclopentadiene and the LUMO of maleic anhydride when the anhydride is endo to the cyclopentadiene ring and when it is exo.

PROBLEM 28.12

Start with six p orbitals in a row, showing the different signs for their lobes by different kinds of shading and construct the six π-type molecular orbitals of 1,3,5-hexatriene. Remember to increase the number of nodes in the orbitals systematically and to maintain the overall symmetry of the system. Show the relative energies for the molecular orbitals, decide which ones are bonding and which antibonding, where the six π electrons will be, and which is the HOMO for the system. The LUMO is already shown in Figure 28.7.

Chemists reason in the following way. If maleic anhydride approaches the diene so that the major part of the anhydride molecule lies underneath the cyclopentadiene ring, two kinds of interactions are possible. First, bonding interactions occur between the lobes at the ends of the diene system and the lobes at the double bond in the dienophile. These interactions give rise to the stable six-membered ring in the product. Second, additional weaker interactions are possible between the other lobes of the diene system and those on the carbon atoms of the carbonyl groups. These interactions do not lead to bonding because such a reaction would destroy the stable carbonyl groups and also result in highly strained systems with multiple rings. However, the interactions do stabilize the transition state and lower the energy of activation for the reaction. Thus, reaction with the endo orientation of the dienophile is faster than reaction with the exo orientation in which such secondary interactions are not possible (Figure 28.7). As a result, more of the product has the endo stereochemistry.

The regioselectivity observed for reactions of unsymmetrically substituted dienes and dienophiles was discussed in terms of resonance contributors (p. 696). Computer programs are available for calculating the molecular orbitals and other electronic properties of a wide range of molecules. The qualitative results that organic chemists get by drawing curved arrows are confirmed in many cases (and contradicted in a few) by theoretical calculations that describe molecular orbitals more precisely than has been done in this section. For example, all the lobes of the *p* orbitals making up the molecular orbitals have been shown as being the same size. Actually, the lobes are predicted by theory to be of different sizes, reflecting the different probabilities of finding an electron at any point along the conjugated system. Calculations for compounds such as 1-ethoxybutadiene and propenal (p. 696) confirm that the most important interaction for the HOMO of 1-ethoxybutadiene and the LUMO of propenal is between carbon 4 of the diene system and carbon 3 of propenal, thus determining the orientation observed in the product.

Study Guide Concept Map 28.2

28.3 ELECTROCYCLIC REACTIONS

A. Interconversion of Cyclobutenes and Conjugated Dienes

Cyclobutene rings open to conjugated dienes when heated (p. 1216). For example, the ring in *cis*-3,4-dimethylcyclobutene opens smoothly to form (2*Z*,4*E*)-2,4-hexadiene.

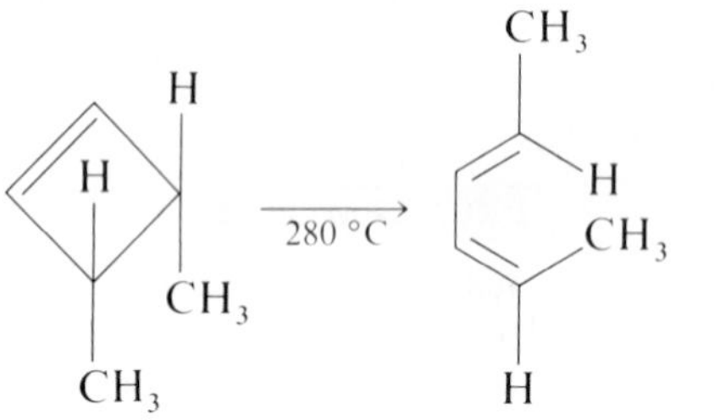
H
H
CH_3
CH_3
280 °C
CH_3
H
CH_3
H

cis-3,4-dimethylcyclobutene — (2*Z*,4*E*)-2,4-hexadiene >99% of the dienes formed

On the other hand, *trans*-3,4-dimethylcyclobutene gives (2*E*,4*E*)-2,4-hexadiene.

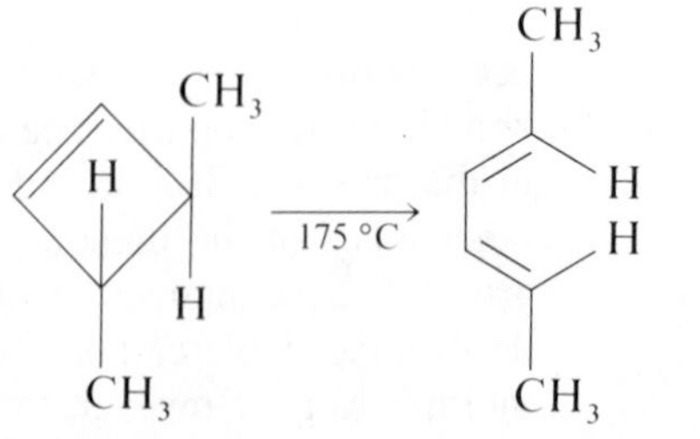
CH_3
H
H
CH_3
175 °C
CH_3
H
H
CH_3

trans-3,4-dimethylcyclobutene — (2*E*,4*E*)-2,4-hexadiene

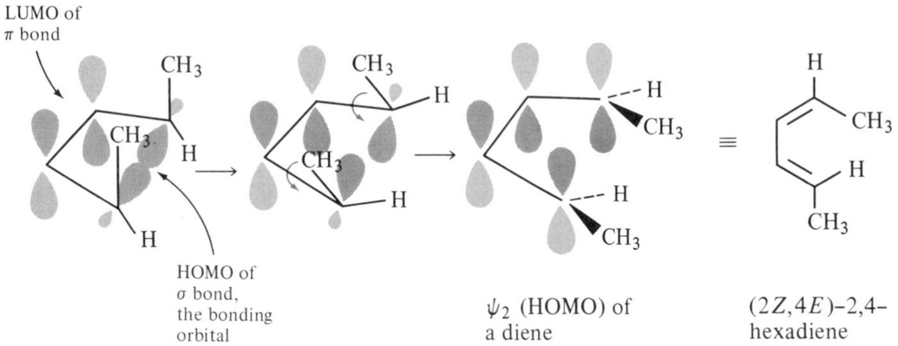
LUMO of
π bond
CH3
CH3
H
H
CH3
CH3
H
H
H
CH3
H
CH3
H
CH3
H
CH3
HOMO of
σ bond,
the bonding
orbital
ψ2 (HOMO) of
a diene
(2Z,4E)-2,4-
hexadiene

FIGURE 28.8 A schematic representation of the interactions of molecular orbitals that give rise to the stereoselective ring opening of *cis*-3,4-dimethylcyclobutene in a thermal reaction.

The ring-opening reaction consists of the breaking of a σ bond and the overlap of the lobes of that orbital with those of the adjacent π bond so that two new π orbitals are formed to accept the four bonding electrons of the diene system. Using the language of frontier orbitals (p. 1222), the reaction may be described as the addition of the HOMO of the σ orbital to the LUMO of the π bond in the cyclobutene (Figure 28.8).

The HOMO of the σ bond is the bonding orbital, shown with two lobes of the sp^3 hybrid orbitals on the carbon atoms overlapping. In order for the new π bonds to develop, bonding interactions must be possible between the lobes of the orbitals shown for the π bond in cyclobutene and the lobes that are freed from the σ bond as it breaks. For this to happen, the carbon atoms that are part of the breaking bond must rotate in order to bring the lobes of the σ orbital into line with those lobes of similar sign on the other two carbon atoms. As indicated in Figure 28.8, both carbon atoms must rotate in the same direction, either clockwise or counterclockwise, for the new π bonds to form smoothly. If you follow the motions of the two methyl groups during the reaction, you will see that they both move in the same direction that the carbon atoms to which they are bonded do, and the product is (2*Z*,4*E*)-2,4-hexadiene. This type of motion, in which the carbon atoms at the ends of the developing polyene system rotate in the same direction, is said to be **conrotatory.** Thus, *cis*-3,4-dimethylcyclobutene, when heated, undergoes conrotatory ring opening to (2*Z*,4*E*)-2,4-hexadiene.

PROBLEM 28.13

Draw the molecular orbital picture for the thermal opening of the ring in *trans*-3,4-dimethylcyclobutene.

PROBLEM 28.14

The following equilibrium has been observed. How do you account for this isomerization? Note that none of the other possible stereoisomers (what are they?) is formed.

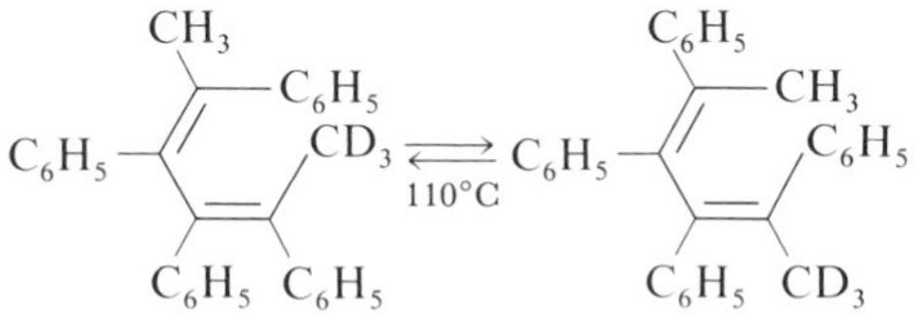
CH3
C6H5
CD3
C6H5
C6H5 C6H5
110°C
C6H5
CH3
C6H5
C6H5
C6H5 CD3

When *cis*-3,4-dimethylcyclobutene undergoes a photochemical ring opening, a mixture of products is obtained, among them (2*E*,4*E*)-2,4-hexadiene.

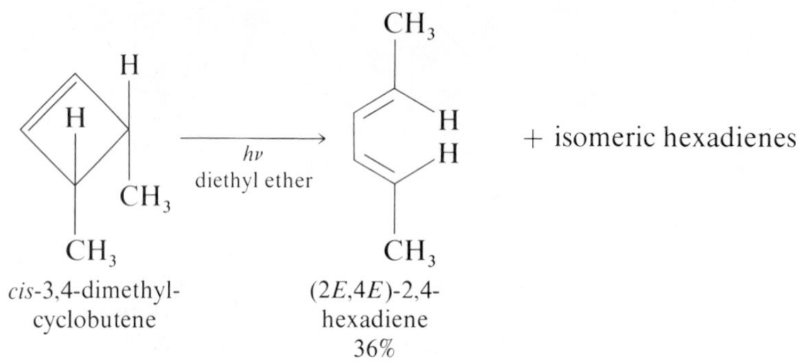
H
H
CH3
CH3
cis-3,4-dimethyl-cyclobutene
hν
diethyl ether
CH3
H
H
CH3
(2E,4E)-2,4-hexadiene
36%
+ isomeric hexadienes

Practically none of this product was obtained from the thermal ring opening of *cis*-3,4-dimethylcyclobutene (p. 1228). (2*E*,4*E*)-2,4-Hexadiene is believed to develop from the excited state of the cyclobutene. One molecular orbital picture that can be used to rationalize this experimental observation is shown in Figure 28.9.

In the photochemical reaction, the π bond of the cyclobutene is assumed to have absorbed ultraviolet energy which has promoted an electron to the π^* antibonding orbital. This antibonding orbital then plays the role of the HOMO for that portion of the molecule, and the antibonding σ orbital is the LUMO. As is seen in Figure 28.9, the new π bonds can form smoothly only if the two carbon atoms of the σ bond rotate in opposite directions. One carbon atom must rotate clockwise and the other one counterclockwise so that the lobes being freed from the σ bond can overlap with lobes of the same sign from the original π bond. When the carbon atoms at the ends of the developing polyene rotate in opposite directions with respect to each other, their motion is said to be **disrotatory.** The methyl groups move with the carbon atoms that are rotating, and (2*E*,4*E*)-2,4-hexadiene is formed.

(2*E*,4*E*)-2,4-Hexadiene undergoes a photochemical ring closure to give *cis*-3,4-dimethylcyclobutene.

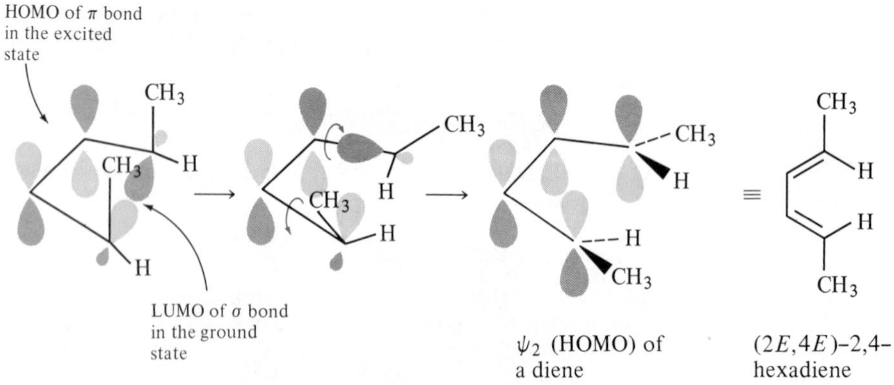
HOMO of π bond in the excited state
CH3
CH3
H
H
LUMO of σ bond in the ground state
CH3
CH3
H
H
CH3
CH3
H
H
ψ2 (HOMO) of a diene
CH3
H
H
CH3
(2E,4E)-2,4-hexadiene

FIGURE 28.9 A schematic representation of the interactions of molecular orbitals that give rise to the stereoselective ring opening of *cis*-3,4-dimethylcyclobutene in a photochemical reaction.

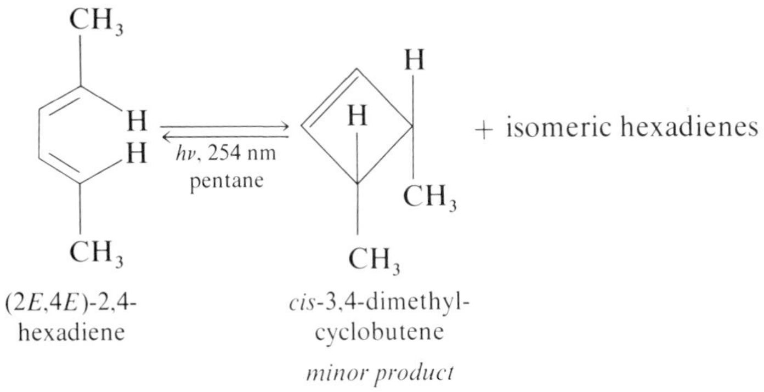
CH3
H
H
hν, 254 nm
pentane
H
H
CH3
CH3
CH3
+ isomeric hexadienes
(2E,4E)-2,4-
hexadiene
cis-3,4-dimethyl-
cyclobutene
minor product

The photochemical ring closure is a disrotatory process. The stereochemistry of a ring closure is determined by the symmetry of the HOMO of the polyene system undergoing the reaction. The HOMO for the diene in the excited state is ψ_3 (Figure 28.2, Problem 28.5). If the end carbon atoms of the diene system are to form a σ bond, they must rotate toward each other, one of them clockwise and the other counterclockwise, to bring the lobes of the same sign into contact so that overlap and bonding can take place (Figure 28.10).

In summary, cyclobutenes and conjugated dienes interconvert in a conrotatory way in the ground state. Photochemical interconversions of these systems occur in a disrotatory fashion. Note that, in all these reactions, the systems involved either start or end with four π electrons. The difference in the stereochemical course of the thermal and photochemical reactions can be rationalized by considering the difference in the symmetry of the molecular orbitals of the reacting molecules in the ground state and in the excited state.

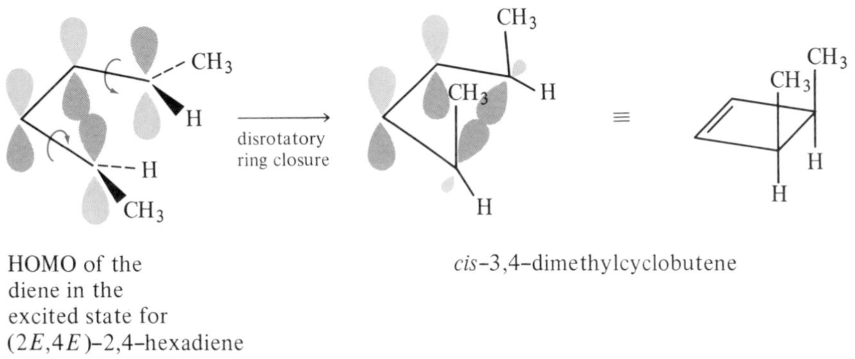
CH3
H
H
CH3
disrotatory
ring closure
CH3
CH3
H
H
CH3
CH3
H
H
HOMO of the
diene in the
excited state for
(2E,4E)-2,4-hexadiene
cis-3,4-dimethylcyclobutene

FIGURE 28.10 A schematic representation of the stereochemistry of photochemical ring closure of a diene.

PROBLEM 28.15

Predict what the product of the following reaction will be.

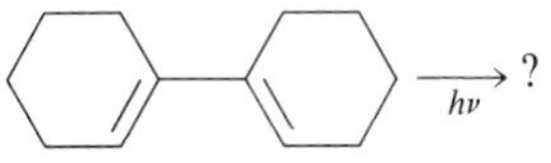
hν
?

PROBLEM 28.16

1,3-Butadiene selectively labeled with deuterium at carbons 1 and 4 has been prepared. For the isomer shown below, predict the structures of the products that would be obtained from a thermal and a photochemical cyclization reaction. Use molecular orbital pictures to explain your conclusions.

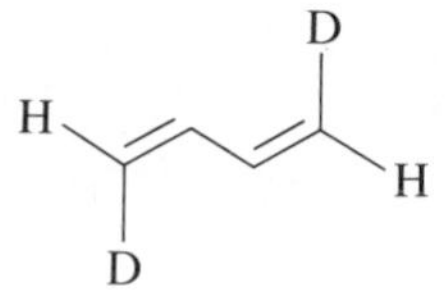
D
H
H
D

B. Interconversion of Cyclohexadienes and Trienes

The stereoselective ring closures of (2*E*,4*Z*,6*E*)- and (2*Z*,4*Z*,6*E*)-2,4,6-octatriene to *cis*- and *trans*-5,6-dimethyl-1,3-cyclohexadiene, respectively, have already been shown (p. 1216). An examination of the stereochemistry of the reactions indicates that the ring is closing in a disrotatory fashion. Thus, the thermal ring closure of a conjugated triene takes place with the opposite stereochemistry of that observed for a diene. This experimental observation can be understood by looking at the nature of the molecular orbital undergoing the transformation. The HOMO of (2*E*,4*Z*,6*E*)-2,4,6-octatriene is shown in Figure 28.11.

The HOMO of a triene system has two nodes besides the node in the plane of the molecule (Problem 28.12, p. 1227). The two end carbon atoms of the triene system must rotate toward each other if lobes having the same sign are to overlap to form a bond. The thermal ring closure of a triene is thus disrotatory.

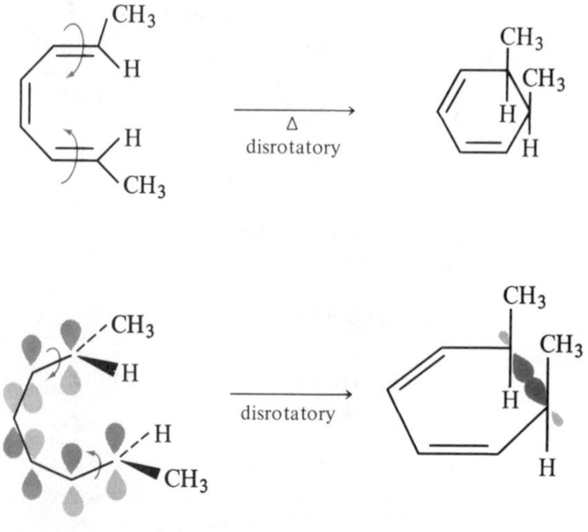
CH3
H
H
CH3
Δ
disrotatory
CH3
CH3
H
H
CH3
H
H
CH3
disrotatory
CH3
CH3
H
H

HOMO of the triene
in the ground state

(2*E*,4*Z*,6*E*)-2,4,6-
octatriene

cis-5,6-dimethyl-
1,3-cyclohexadiene

FIGURE 28.11 Schematic representation of the molecular orbital governing the stereochemistry of the thermal ring closure of a triene.

PROBLEM 28.17

Draw the molecular orbitals for (2*Z*,4*Z*,6*E*)-2,4,6-octatriene and demonstrate to yourself that *trans*-5,6-dimethyl-1,3-cyclohexadiene is the product of its disrotatory ring closure.

When (2*E*,4*Z*,6*E*)-2,4,6-octatriene is irradiated, *trans*-5,6-dimethyl-1,3-cyclohexadiene is formed, along with stereoisomers of the octatriene. Under the conditions of the reaction, an equilibrium is set up between the cyclohexadiene and the triene.

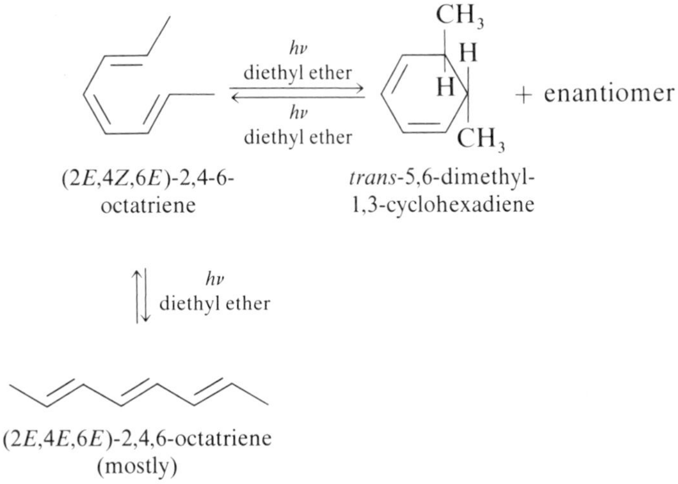
$h\nu$
diethyl ether
$h\nu$
diethyl ether
CH_3
H
H
CH_3
+ enantiomer
(2*E*,4*Z*,6*E*)-2,4-6-
octatriene
trans-5,6-dimethyl-
1,3-cyclohexadiene
$h\nu$
diethyl ether
(2*E*,4*E*,6*E*)-2,4,6-octatriene
(mostly)

The photochemical cyclization of a triene is conrotatory (Figure 28.12). When the triene absorbs ultraviolet radiation, an electron is promoted from the HOMO to the LUMO, which then becomes the HOMO of the excited state. It is the symmetry of this orbital that governs the stereochemistry of the photochemical ring closure of the triene. In order to cause the lobes of the same sign on the end carbon atoms of the polyene to overlap, the carbon atoms, and the methyl groups on them, must rotate in the same direction, either clockwise or counterclockwise. The photochemical conrotatory ring closure of (2*E*,4*Z*,6*E*)-2,4,6-octatriene gives *trans*-5,6-dimethyl-1,3-cyclohexadiene; in contrast, the thermal disrotatory ring closure of the triene gives rise to the cis isomer. In the case of the triene-cyclohexadiene transformations, just as for the diene-cyclobutene interconversions, thermal and photochemical reactions proceed with opposite stereochemistry.

PROBLEM 28.18

In Figure 28.11, only one of two possible disrotatory ring closures for the (2*E*,4*Z*,6*E*)-2,4,6-octatriene is shown. Similarly, in Figure 28.12, only one kind of conrotatory motion is indicated. Draw molecular orbital pictures to explore the consequences of the other possible disrotatory and conrotatory motions of the end carbon atoms of this polyene system. What is the stereochemical relationship of the new products to those in Figures 28.11 and 28.12?

Photochemical ring-opening reactions of cyclohexadienes are well known and have been most extensively investigated in connection with studies on Vitamin D. Vitamin D is a mixture of compounds that are effective in preventing and curing rickets, a disease in which bones become soft and easily deformed. The relationship of the disease to a dietary deficiency that can be remedied by cod liver oil has been recognized for over a hundred years. It was later recognized that certain foods

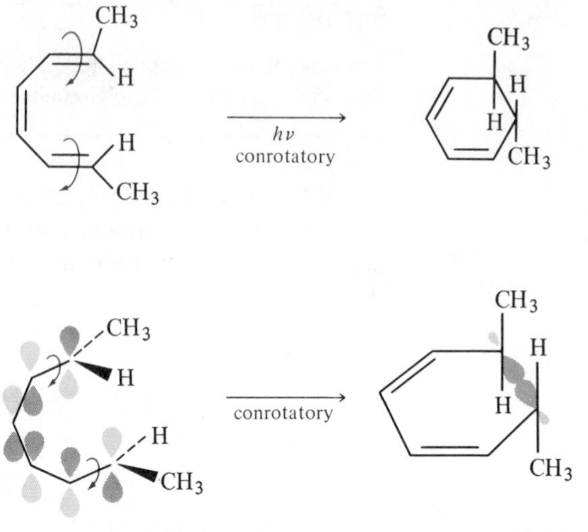
CH3
H
H
CH3
hν
conrotatory
CH3
H
H
CH3
CH3
H
H
CH3
conrotatory
CH3
H
H
CH3

HOMO of the triene
in the excited state

(2*E*,4*Z*,6*E*)-2,4,6-octatriene

trans-5,6-dimethyl-1,3-cyclohexadiene

FIGURE 28.12 Schematic representation of the molecular orbitals governing the stereochemistry of the photochemical ring closure of a triene.

contain compounds that can be transformed into Vitamin D by irradiation. These compounds, called **provitamins,** occur widely in both plants and animals. They are all steroids that have a diene system in the B ring. They have no Vitamin D activity until that ring is opened upon irradiation (p. 1218).

The most widely investigated provitamin D is ergosterol. Ergosterol undergoes a photochemical ring opening to give a compound called precalciferol. All the compounds that are provitamins for Vitamin D have the same structure except for variations in the side chain at carbon 17 of the steroid ring. The transformation of ergosterol to precalciferol is a conrotatory opening of the B ring. The stereochemical relationship between the methyl group at carbon 10 and the hydrogen atom at carbon 9 is particularly important, especially for the transformation of precalciferol into Vitamin D_2 in the next step of the reaction. The photochemical reaction is reversible. The conrotatory opening of the cyclohexadiene ring gives precalciferol in which the methyl group on carbon 10 points toward carbon 9 of the steroid system.

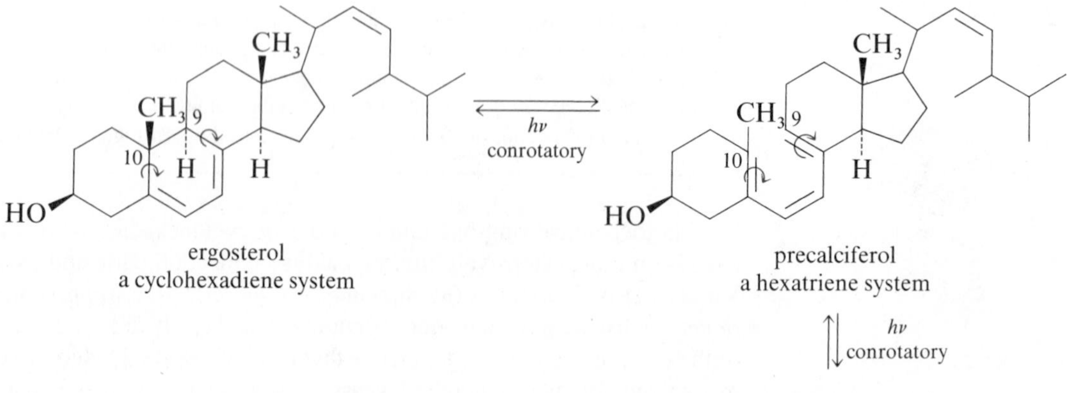
CH3
CH3
9
10
H
H
HO
hν
conrotatory
CH3
CH3
9
10
H
HO
ergosterol
a cyclohexadiene system
precalciferol
a hexatriene system
hν
conrotatory

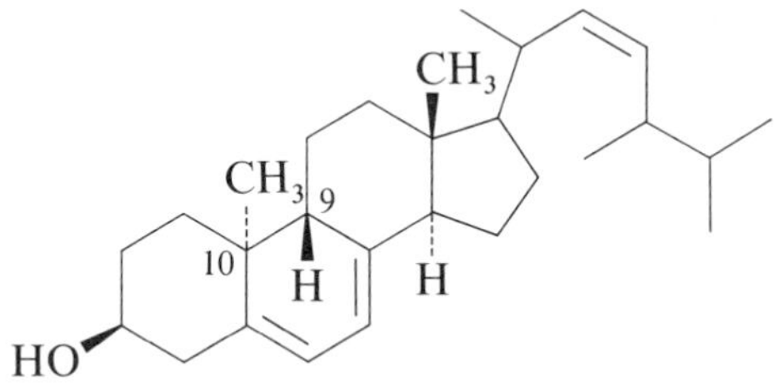

lumisterol
a cyclohexadiene system

Precalciferol is converted back to ergosterol by a reversal of this motion, but a new cyclohexadiene is formed if ring closure occurs with further rotation of the carbon atoms of the triene in the same direction. This cyclohexadiene is a stereoisomer of ergosterol called lumisterol. In lumisterol, the methyl group at carbon 10 lies below the plane of the steroid ring, and the hydrogen atom at carbon 9 lies above it.

The most important reaction of precalciferol, from a biological viewpoint, is its transformation into calciferol by means of a [1,7] sigmatropic rearrangement (p. 1247).

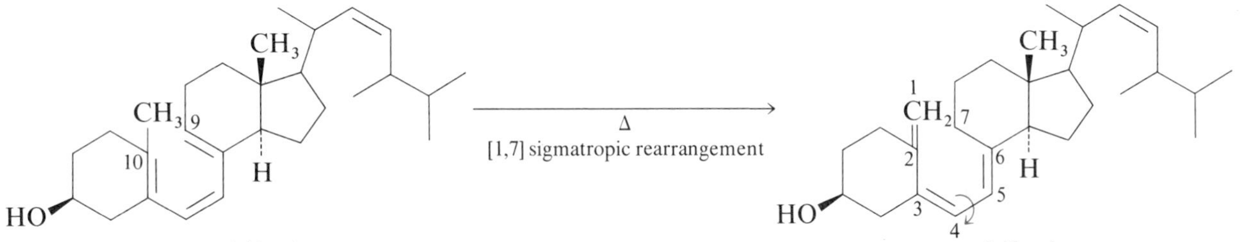
Δ
[1,7] sigmatropic rearrangement

precalciferol

calciferol
Vitamin D_2

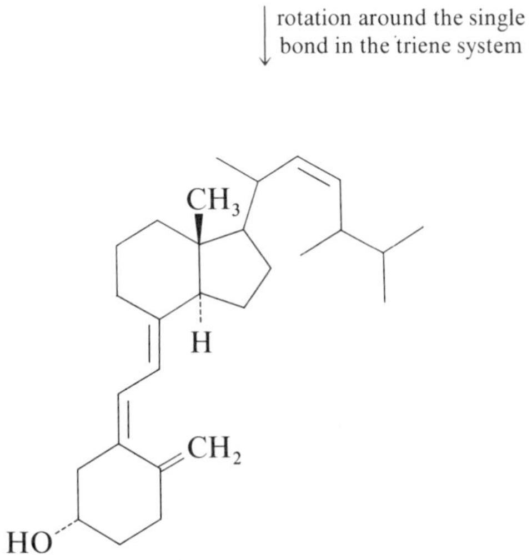
rotation around the single bond in the triene system

Vitamin D_2
in its most stable form,
the *s*-trans conformation

The shift of the hydrogen atom results in a change in the position of the double bonds in the triene, allowing the molecule to assume a more stable conformation through rotation around the central single bond of the system. The structure of Vitamin D_2 is usually drawn with the *s*-trans conformation at that bond.

When precalciferol is heated, two disrotatory ring closures occur. The photochemical and thermal electrocyclic reactions that have been observed, starting with the irradiation of ergosterol, are summarized below.

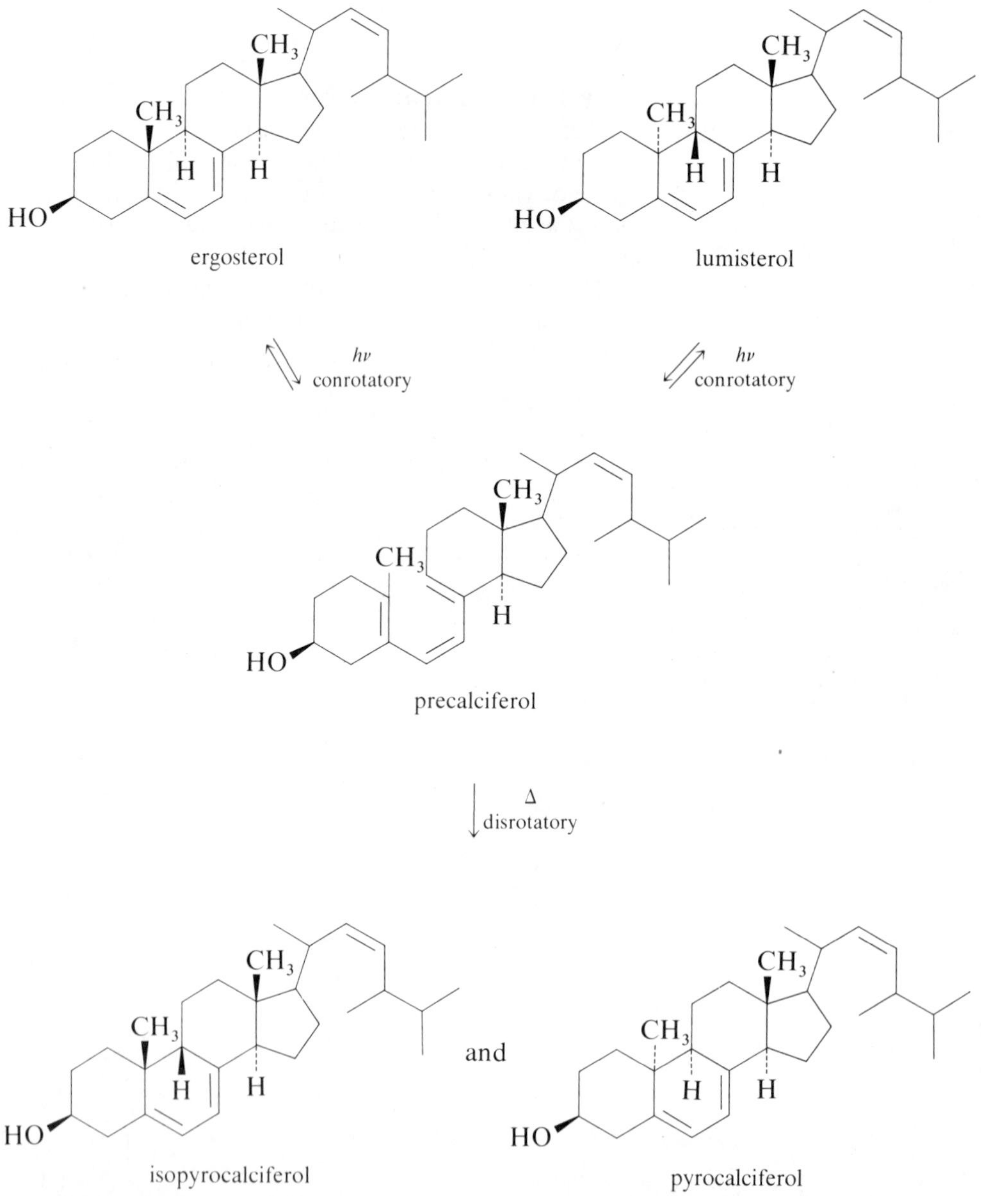
CH3
CH3
H
H
HO
ergosterol
CH3
CH3
H
H
HO
lumisterol
hν
conrotatory
hν
conrotatory
CH3
CH3
H
HO
precalciferol
Δ
disrotatory
CH3
CH3
H
H
HO
isopyrocalciferol
and
CH3
CH3
H
H
HO
pyrocalciferol

It is easy to understand why chemists were intrigued when they observed reactions in which the stereochemistry depended so precisely on reaction conditions. To explain these observations, they developed the theoretical models discussed in this section.

Study Guide Concept Map 28.3

PROBLEM 28.19

Using molecular orbital pictures, show how precalciferol is converted thermally into pyrocalciferol and isopyrocalciferol.

PROBLEM 28.20

Predict what the products will be when the terpene allo-ocimene is irradiated.

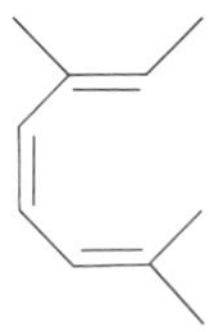

allo-ocimene

C. Some Generalizations about Cycloaddition and Electrocyclic Reactions

A cycloaddition reaction takes place in a concerted manner in the ground state only if $4n + 2$ electrons, where $n = 1, 2, \ldots$, are involved in the transition state. A cycloaddition reaction involving a total of $4n$ electrons participating in the transition state is concerted only in the excited state. These rules apply when the stereochemistry of the reaction has two unsaturated components approaching each other face to face and when bonding occurs on the same side of the molecule at each end of the diene or alkene system. This is the stereochemistry that is shown for the Diels-Alder reaction in Figures 28.6 and 28.7, for example.

These rules, known as the **Woodward-Hoffmann rules,** indicate how the total number of electrons involved in the transition state of a cycloaddition reaction governs the manner of the cycloaddition. Table 28.1 simply restates the generalizations formulated in the preceding paragraph. A concerted cycloaddition reaction is said to be **allowed** for a system with $4n$ electrons if one of the components of the reaction is raised to the excited state, but is **forbidden** if both of the components are in the ground state. Forbidden does not mean that no reaction can ever take place. It simply means that if a reaction is observed to occur under these conditions, it is not a concerted reaction.

Note that the generalization refers to the number of electrons and not to the number of orbitals. It is quite easy to remember the rules if you can keep firmly in mind that the Diels-Alder reaction is a $4n + 2$ system in which $n = 1$, and that it

TABLE 28.1 Rules for Concerted Cycloaddition Reactions in Which Both Components Retain Stereochemistry

Total Number of Electrons in the Transition State	Concerted Cycloaddition Reaction	
$4n$	allowed in the excited state	forbidden in the ground state
$4n + 2$	forbidden in the excited state	allowed in the ground state

TABLE 28.2 Rules for the Stereochemistry of Concerted Electrocyclic Reactions

Total Number of Electrons in the Transition State	Stereochemistry of the Allowed Concerted Reaction in the Ground State	Excited State
$4n$	conrotatory	disrotatory
$4n + 2$	disrotatory	conrotatory

is allowed in the ground state. What is allowed in the ground state is forbidden in the excited state, and any change in the number of electrons by two, either up or down, changes the nature of the allowed reaction.

The experimental data that have been given for electrocyclic reactions also lend themselves to generalizations about stereochemistry in terms of the number of electrons involved in the transition state. These Woodward-Hoffmann rules are summarized in Table 28.2 above. Once again, these rules do not mean that other types of reactions do not take place, but only that if a reaction is observed to disobey these rules, it is taking place by a different mechanism and is not a concerted reaction.

The two sets of rules given in Tables 28.1 and 28.2 summarize the experimental observations that have been described throughout this chapter. The basis for explaining these observations is molecular orbital theory. The theory allows predictions to be made about systems for which experimental data are not available. One of the predictions based on the rules is that systems containing four electrons on three atoms will add to alkenes in a concerted manner to give five-membered rings. The most fruitful experimental application of this prediction is described in the next section.

Study Guide Concept Map 28.4

PROBLEM 28.21

A useful synthesis of substituted phenanthrenes is by photochemical cyclization of 1,2-diarylethenes. The reaction gives a dihydrophenanthrene, which is then oxidized easily by oxygen or iodine to the phenanthrene. The reaction is shown below for 1,2-diphenylethene.

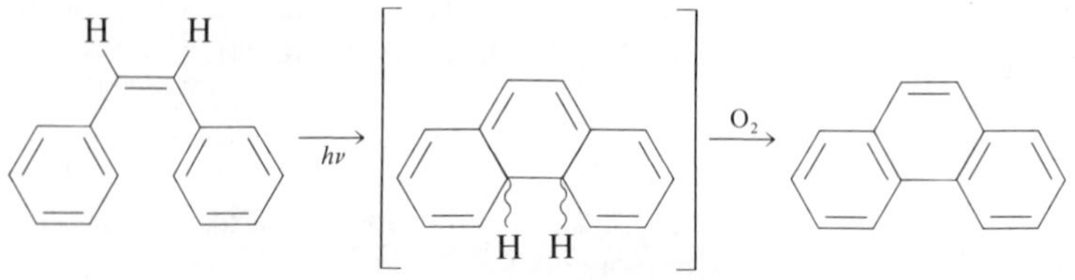
H
H
hν
H H
O_2

a dihydrophenanthrene intermediate

If all oxygen is removed from the reaction mixture before the irradiation, the dihydrophenanthrene can be isolated. How would you classify this reaction according to the Woodward-Hoffmann rules? What stereochemistry would you expect the dihydrophenanthrene to have?

PROBLEM 28.22

The Woodward-Hoffmann rules predict that ionic species should also undergo electrocyclic reactions. For example, the following reaction of a pentadienyl cation has been observed.

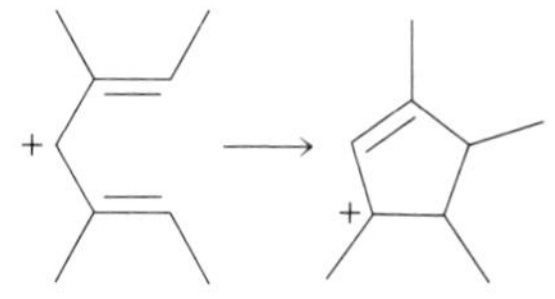

How would you classify this reaction? What stereochemistry do you expect the product cation to have? (Hint: How many electrons are involved in the transition state?)

PROBLEM 28.23

When cyclopentadiene and 3-iodo-2-methyl-1-propene are allowed to react in liquid sulfur dioxide in the presence of silver trifluoroacetate, the products that are formed can be rationalized as arising from the bicyclic tertiary cation shown below. How would you classify the reaction? What is a likely intermediate? How many electrons are involved in the transition state?

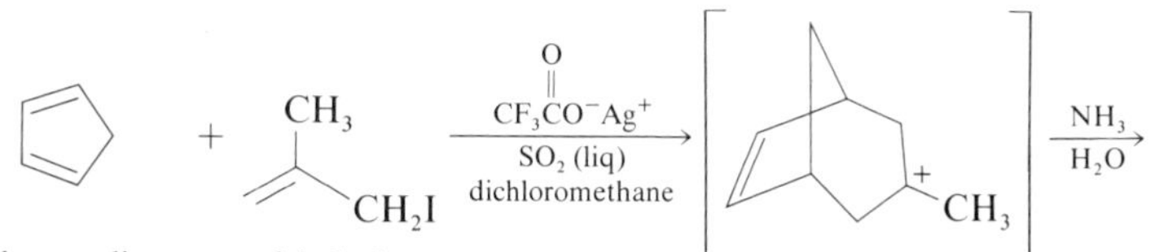

CH3
CH2I
O
CF3CO−Ag+
SO2 (liq)
dichloromethane
CH3
NH3
H2O

cyclopentadiene 3-iodo-2-methyl-1-propene

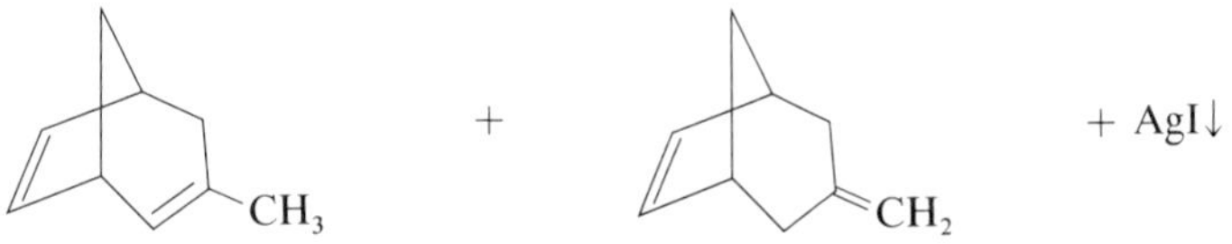

CH3
CH2
+ AgI↓

3-methylbicyclo[3.2.1]octa-2,6-diene 40% 3-methylenebicyclo[3.2.1]oct-6-ene 16%

28.4 1,3-DIPOLAR ADDITIONS. CYCLOADDITIONS OF COMPOUNDS CONTAINING NITROGEN AND OXYGEN

Molecular orbital theory is a powerful tool that can be applied to many different types of reactions. One of the most significant extensions of the theory was made by Rolf Huisgen of West Germany, who recognized that certain species containing oxygen or nitrogen could serve as sources of four electrons in [4 + 2] cycloaddition reactions and could therefore be used to create a variety of heterocyclic compounds. These reactions have all the characteristics of concerted reactions, including insensitivity to reaction conditions, such as solvent polarity, and high stereoselectivity.

Some species, though reasonably stable and isolable, have structures that can best be represented as a series of resonance contributors, one of which has a formal positive charge on one atom and a formal negative charge two atoms away. Structures characterized by this distribution of formal charge are called **1,3-dipolar species.** Ozone is a 1,3-dipolar species (p. 311) and so is phenyl azide. All the resonance contributors of phenyl azide have formal charges on various atoms. The 1,3-dipolar character of the compound is represented by the last structural formula shown below.

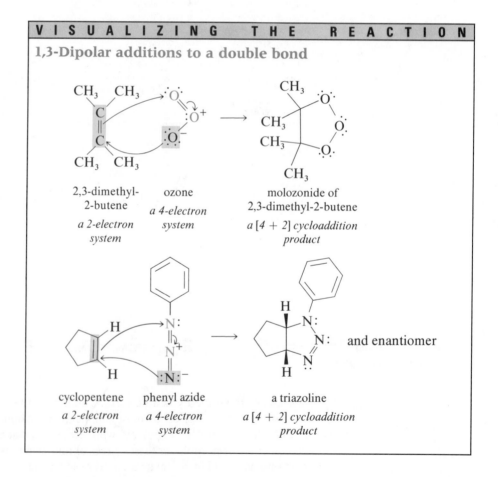

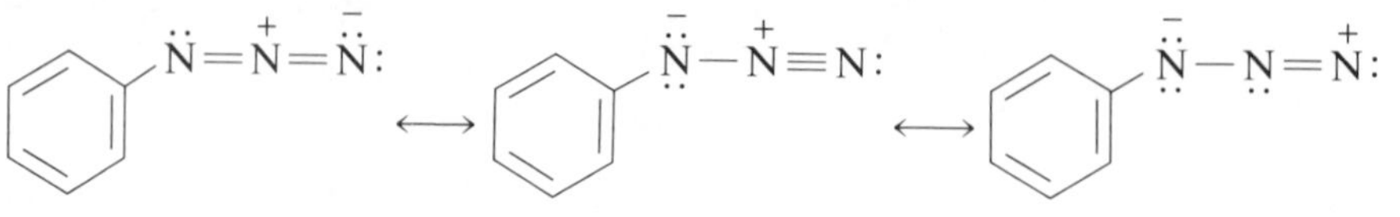

1,3-dipolar character

resonance contributors of phenyl azide

1,3-Dipolar species add to alkenes or alkynes (called **dipolarophiles**) in cycloaddition reactions. The addition of ozone or phenyl azide to an alkene involves the participation of a pair of π electrons and a pair of nonbonding electrons from the dipolar species and a pair of π electrons from the alkene.

VISUALIZING THE REACTION

1,3-Dipolar additions to a double bond

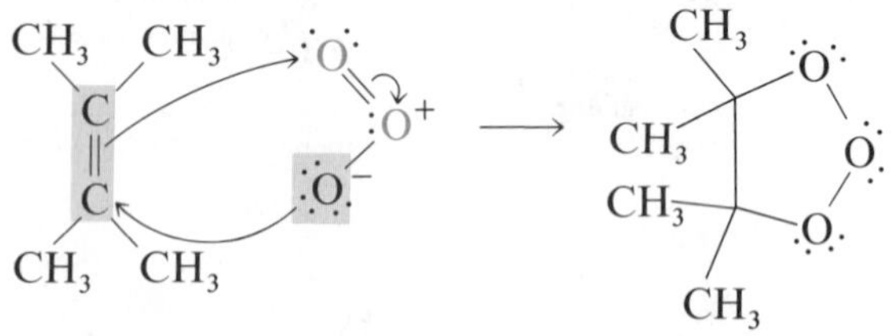

2,3-dimethyl-2-butene *a 2-electron system* — ozone *a 4-electron system* — molozonide of 2,3-dimethyl-2-butene *a [4 + 2] cycloaddition product*

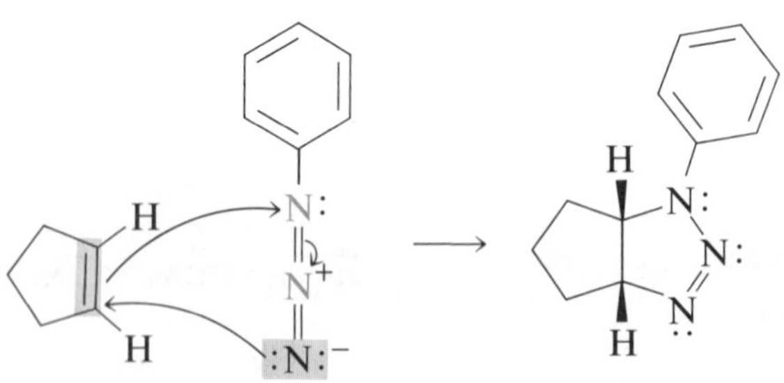

and enantiomer

cyclopentene *a 2-electron system* — phenyl azide *a 4-electron system* — a triazoline *a [4 + 2] cycloaddition product*

Each reaction shown above could be classified from an electronic point of view as a [4 + 2] cycloaddition reaction, which is allowed in the ground state. The dipolar species contributes four electrons to the reaction. These electrons are in three p orbitals. Two of the electrons are the electrons of the π bond; the other two are nonbonding electrons from an oxygen or a nitrogen atom. All three atoms in the dipolar species participate in the bonding. The product is, therefore, a five-membered ring containing the two doubly bonded carbon atoms of the alkene and three atoms from the dipolar species.

An important 1,3-dipolar species is diazomethane, CH_2N_2, a yellow gas that is explosive and highly toxic. Diazomethane is always handled in ether solution and protected from heat and light in order to decrease the chance of an explosive decomposition.

The structure of diazomethane, like those of ozone and phenyl azide, can only be described by writing a series of resonance contributors.

$$:\overset{-}{C}H_2-\overset{+}{N}\equiv N: \longleftrightarrow CH_2=\overset{+}{N}=\overset{-}{\ddot{N}}: \longleftrightarrow :\overset{-}{C}H_2-\underset{..}{N}=\overset{+}{N}: \longleftrightarrow \overset{+}{C}H_2-\ddot{N}=\overset{-}{\ddot{N}}:$$

major resonance contributors — 1,3-dipolar character

resonance contributors for diazomethane

The first two are the major resonance contributors for diazomethane because in them each atom is surrounded by an octet of electrons. The other two contributors show the 1,3-dipolar character of the molecule.

Diazomethane reacts with alkenes as a 1,3-dipolar species to give five-membered heterocyclic rings. For example, it reacts with stereoisomeric esters containing a carbon-carbon double bond to give diastereomeric products.

VISUALIZING THE REACTION

Stereoselectivity of a 1,3-dipolar addition

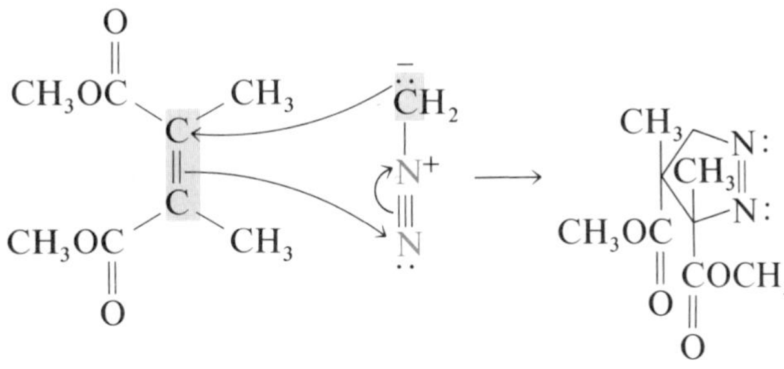

dimethyl dimethylmaleate — diazomethane — a pyrazoline *racemic mixture*

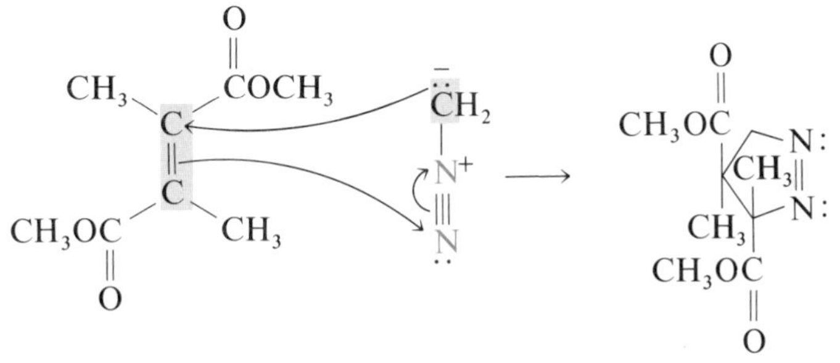

dimethyl dimethylfumarate — diazomethane — a pyrazoline *racemic mixture*

a 2-electron system — *a 4-electron system* — *a [4 + 2] cycloaddition product*

The retention of the stereochemical relationships of the substituents of the isomeric alkenes in the product pyrazolines clearly indicates that the reaction is stereoselective. Diazomethane adds to one face or the other of the alkene molecule in a concerted fashion.

Diazomethane adds to other alkenes as well. For example, it reacts with styrene to give a pyrazoline.

VISUALIZING THE REACTION

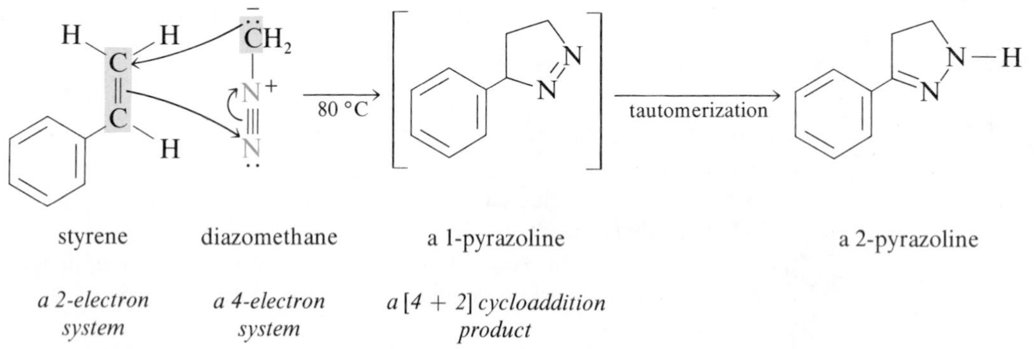
80 °C
tautomerization
styrene
diazomethane
a 1-pyrazoline
a 2-pyrazoline
a 2-electron system
a 4-electron system
a [4 + 2] cycloaddition product

The original products from the addition of diazomethane to alkenes are 1-pyrazolines, in which the double bond in the heterocyclic ring is between the two nitrogen atoms. Such compounds usually tautomerize easily to a more stable form, the 2-pyrazolines, which are the isolated products.

Another type of compound with 1,3-dipolar character is formed when an *N*-substituted hydroxylamine reacts with a carbonyl compound. Such compounds are called **nitrones.** For example, benzaldehyde reacts with hydroxylamine to give an oxime (p. 504). It also reacts with *N*-methylhydroxylamine to give a nitrone.

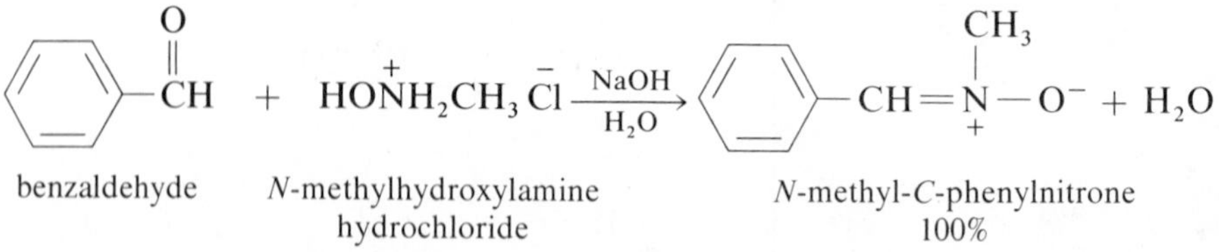
NaOH
H_2O
benzaldehyde
N-methylhydroxylamine hydrochloride
N-methyl-*C*-phenylnitrone
100%

The mechanism for the reaction of a substituted hydroxylamine with a carbonyl compound is similar to that for the formation of the ammonia derivatives of aldehydes and ketones (p. 505). The difference in products comes about from the lack of a second hydrogen atom on the nitrogen atom of the substituted hydroxylamine.

VISUALIZING THE REACTION

Formation of a nitrone

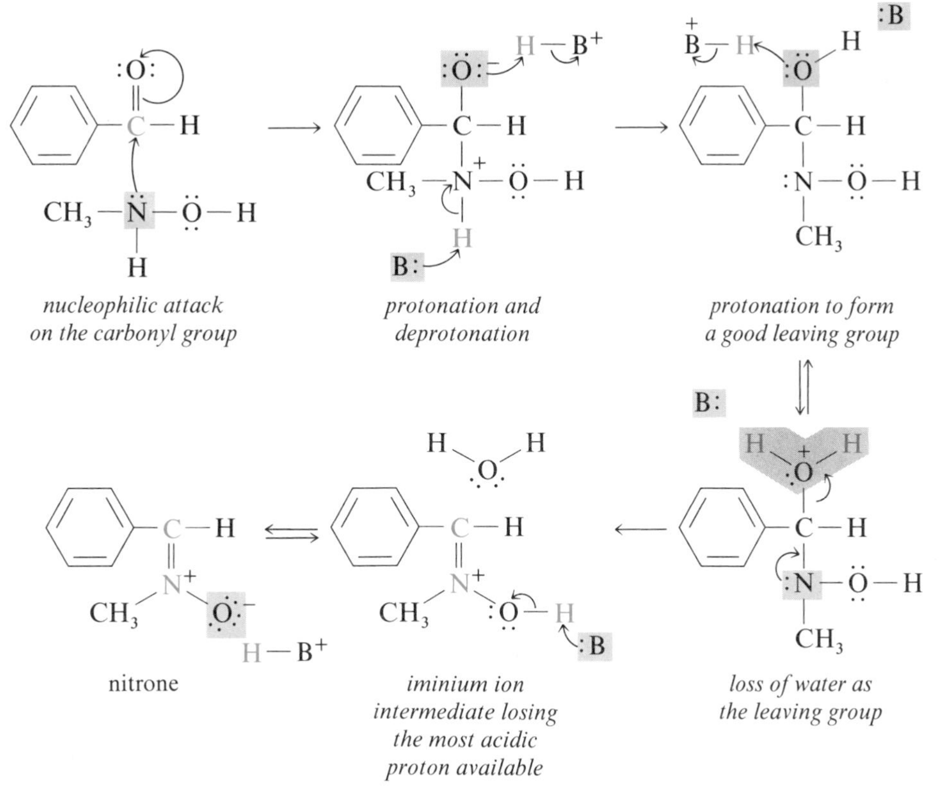

The iminium ion intermediate formed in the synthesis of an oxime has a hydrogen atom bonded to the positively charged nitrogen atom. That hydrogen atom is the most acidic one in the molecule and is lost in the last step of the reaction. In the reaction of an *N*-substituted hydroxylamine, the iminium ion intermediate has no hydrogen atom bonded to the quaternary nitrogen atom. In this case, the most acidic hydrogen atom is the one on the hydroxyl group. Removal of that proton by a base gives a nitrone.

A nitrone reacts with alkenes as a 1,3-dipolar compound. For example, it adds to styrene to give a five-membered heterocycle.

VISUALIZING THE REACTION

Reaction of a nitrone with an alkene

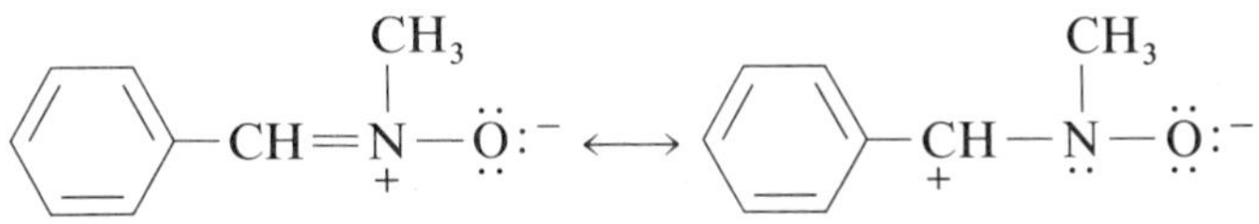

1,3-dipolar form of a nitrone

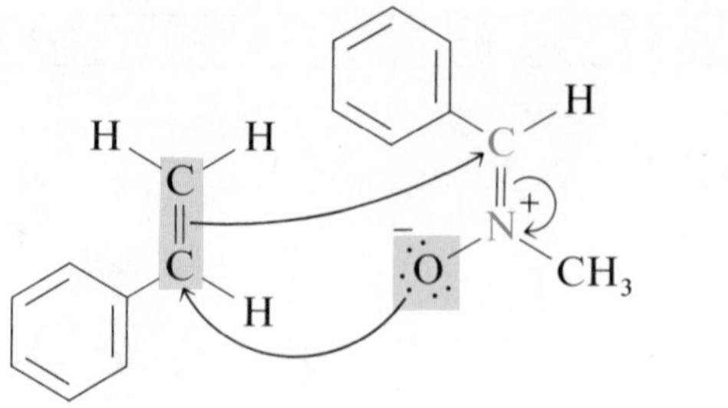

styrene — *a 2-electron system*

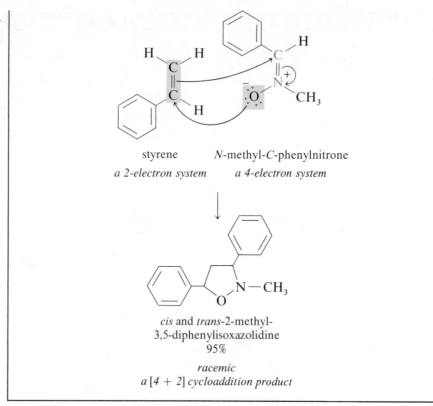

N-methyl-*C*-phenylnitrone — *a 4-electron system*

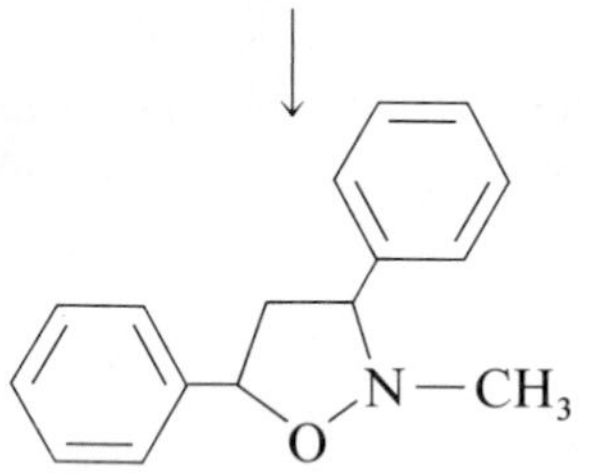

cis and *trans*-2-methyl-3,5-diphenylisoxazolidine
95%
racemic
a [4 + 2] cycloaddition product

A saturated five-membered heterocycle containing adjacent oxygen and nitrogen atoms is known as an isoxazolidine (p. 990). The formation of an isoxazolidine from a nitrone and an alkene is pictured as a [4 + 2] cycloaddition reaction. The nitrone contributes four electrons, two from a π bond and two nonbonding electrons from the oxygen atom, to the transition state.

The regioselectivity shown above for the addition of a nitrone to an alkene is quite general. Terminal alkenes give products in which the substituent is at carbon 5 of the isoxazolidine.

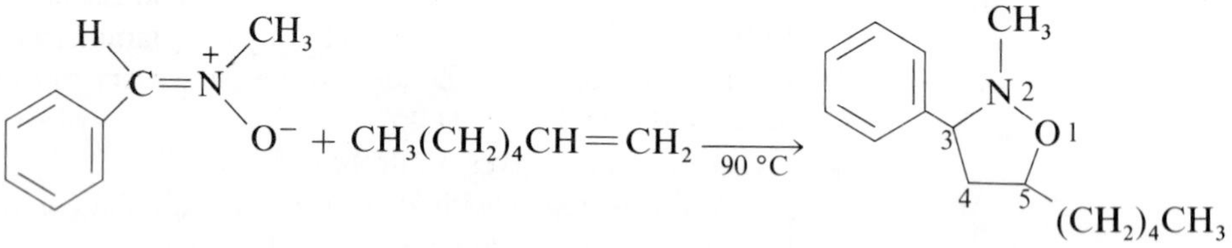

N-methyl-*C*-phenylnitrone — 1-heptene — 2-methyl-5-pentyl-3-phenylisoxazolidine 76%

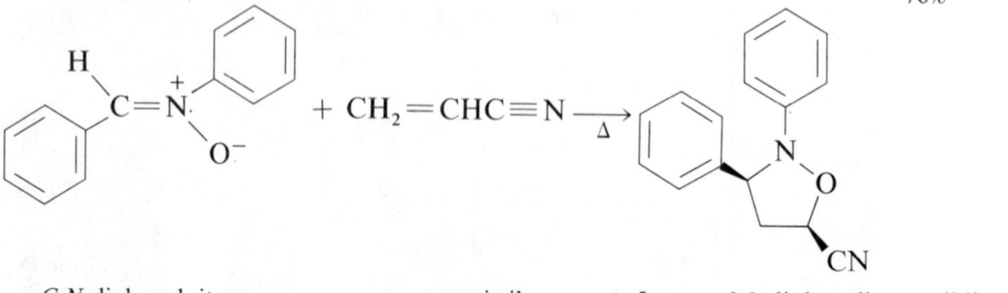

C,N-diphenylnitrone — propenenitrile — *cis*-5-cyano-2,3-diphenylisoxazolidine 100%

It is not necessary to synthesize and isolate a nitrone in order to perform the reaction that produces isoxazolidines. A mixture of a carbonyl compound, an *N*-substituted hydroxylamine, and an alkene heated together will give a high yield of an isoxazolidine. For example, butanal, *N*-phenylhydroxylamine, and styrene react to give 2,5-diphenyl-3-propylisoxazolidine.

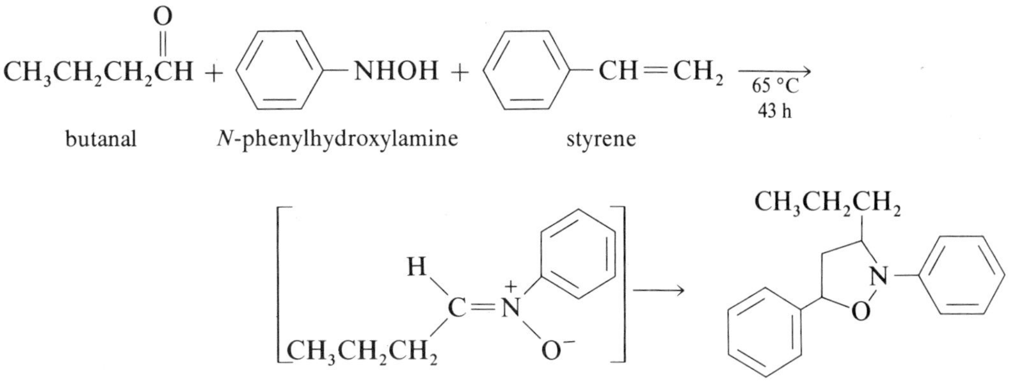

65 °C
43 h

butanal — *N*-phenylhydroxylamine — styrene

N-phenyl-*C*-propylnitrone

2,5-diphenyl-3-propylisoxazolidine
(stereochemistry not investigated)
99%

Study Guide Concept Map 28.2

The reactions described above are only a few examples of the kinds of heterocyclic systems that can be created by the addition of 1,3-dipolar species to alkenes or alkynes.

PROBLEM 28.24

Predict what the products of the following reactions will be.

(a) 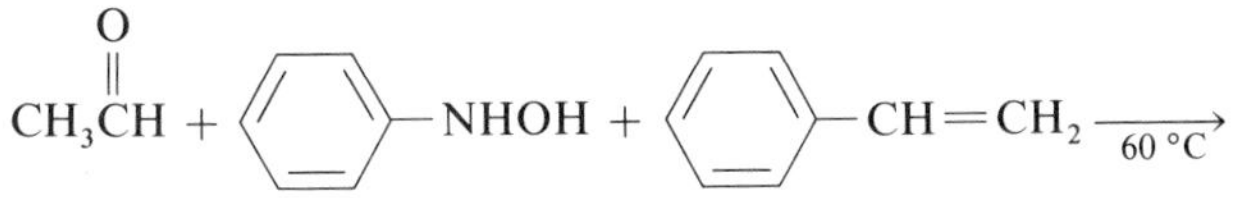

60 °C

(b) $CH_2{=}\overset{+}{N}{=}N^-$ (1 molar equiv) + $CH_2{=}CHCH{=}CH_2$ $\xrightarrow{\text{diethyl ether}}$

(c) 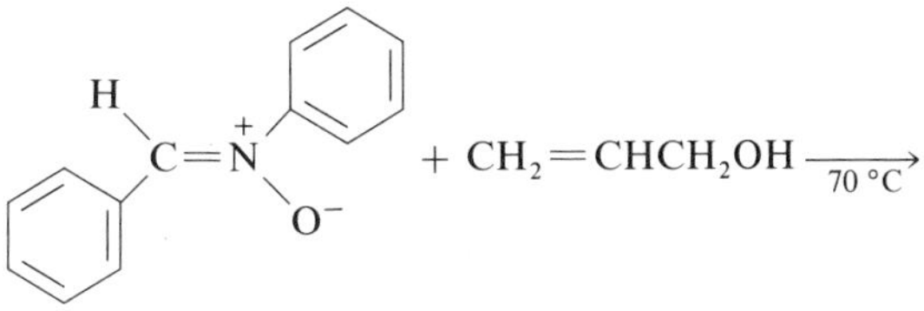

70 °C

(d) 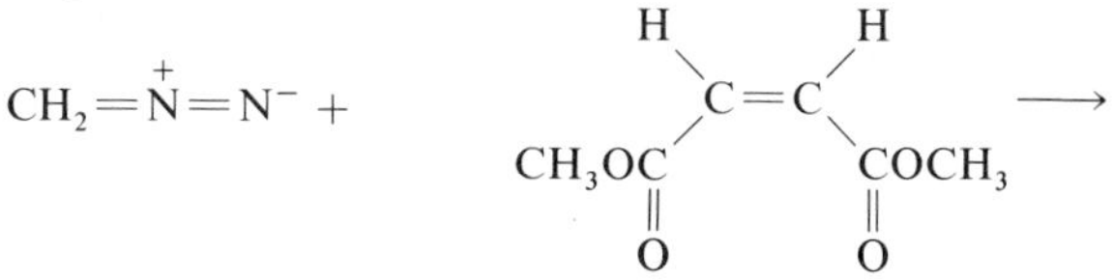

(e) 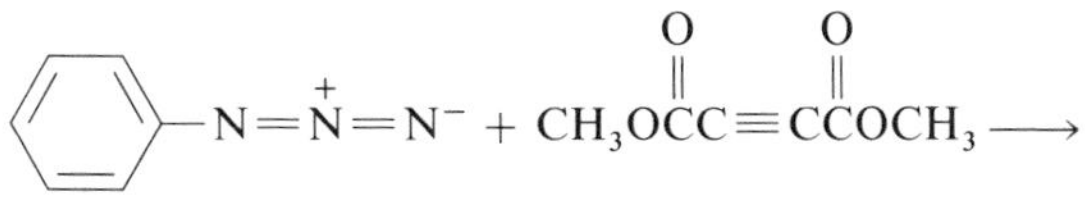

(f) $CH_2{=}\overset{+}{N}{=}N^- + HC{\equiv}CH \xrightarrow{20\ °C}$

(g) 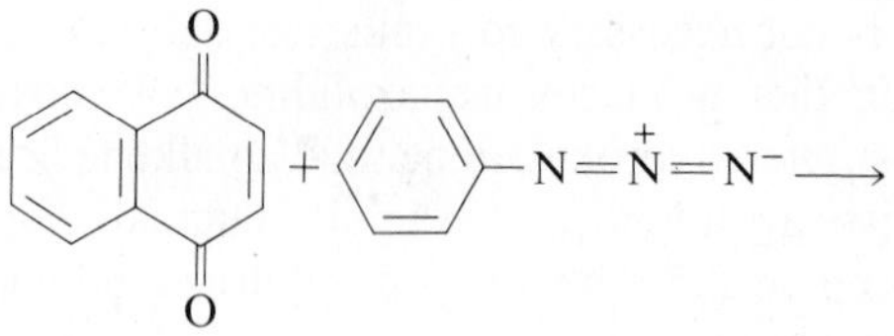

(h) 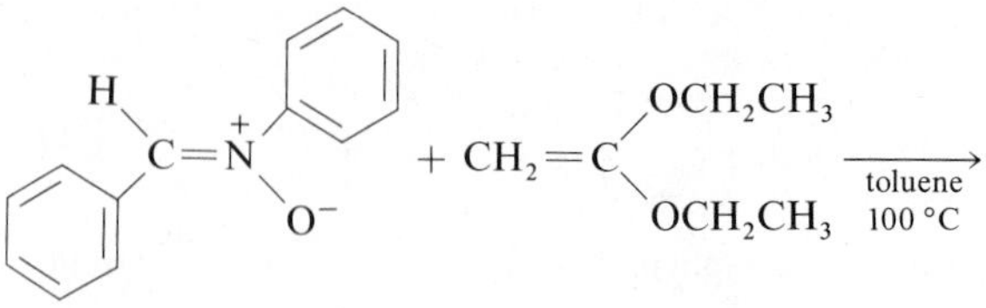

(i) 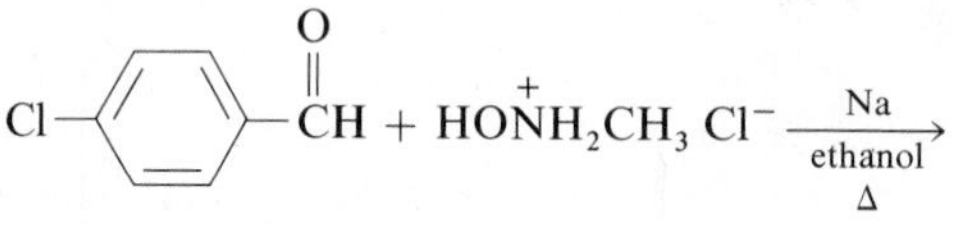

PROBLEM 28.25

The following heterocyclic compounds are obtained by 1,3-dipolar addition reactions. Draw the structure of the alkene or alkyne and the 1,3-dipolar species that would be required to synthesize each compound.

(a) 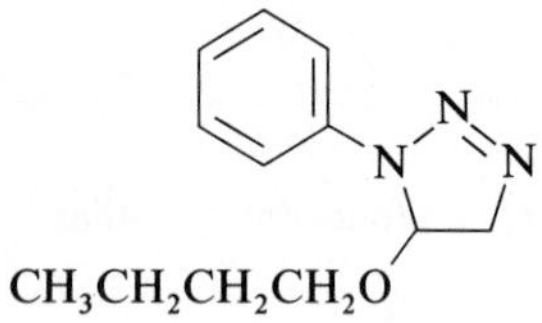

(b) 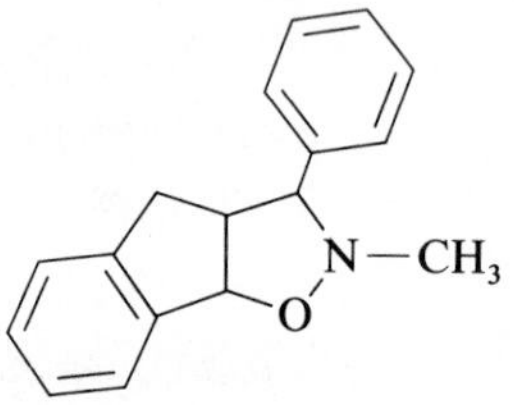

(c) 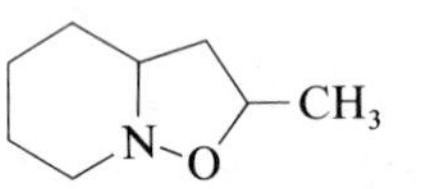

(d) 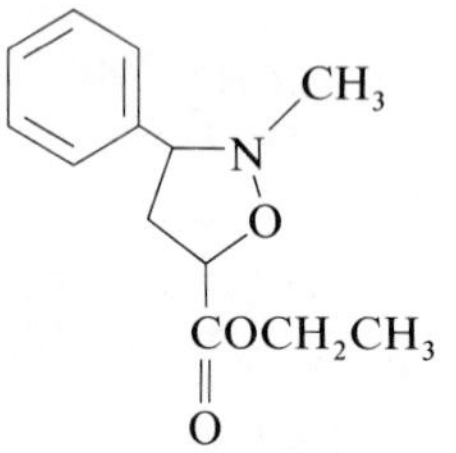

(e) 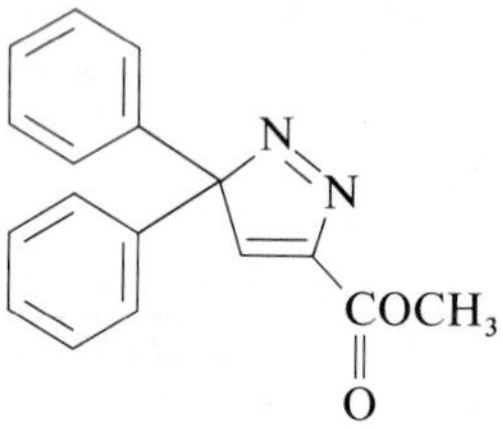

(f) 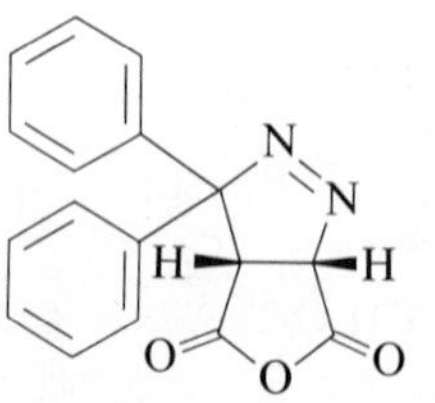

(g) 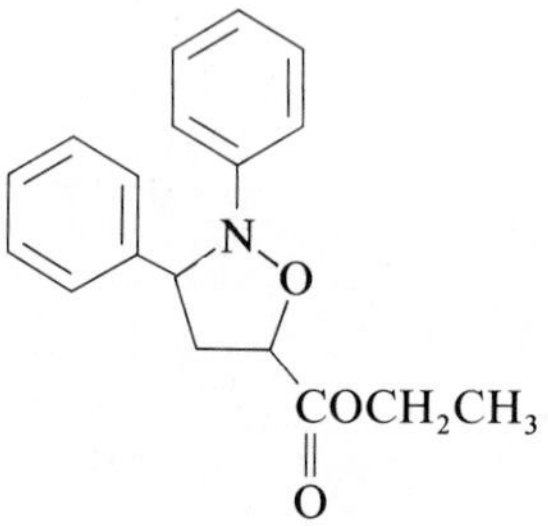

(h) 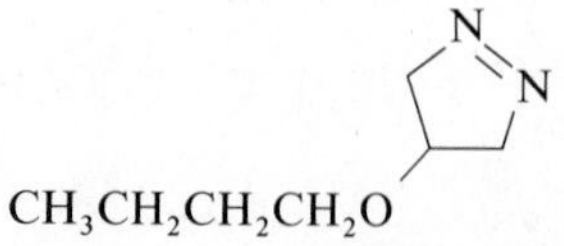

A. Hydrogen Shifts

A sigmatropic rearrangement involves the migration of a σ bond from one position to another one along a polyene chain with a simultaneous shift of π bonds. Such a migration of a hydrogen atom and its σ bond in (*Z*)-1,3-pentadiene was described on p. 1217. The migration occurs in a highly stereoselective way with the hydrogen atom moving along one surface of the polyene system.

The stereoselectivity of the reaction can be explained if a molecular orbital picture is drawn. In Figure 28.13, the process is depicted as the interaction of the HOMO of the carbon-hydrogen σ bond at the tetrahedral carbon atom with the LUMO of the diene that is the remainder of the molecule. A bonding interaction can take place between the *s* orbital of the hydrogen atom and the molecular orbital of the diene system at carbon 5 only if the hydrogen atom moves from the top of carbon 1 to the top of carbon 5. A new diene system with double bonds between carbons 1 and 2 and carbons 3 and 4 forms as the hydrogen atom migrates.

A similar look at the [1,7] sigmatropic rearrangement that occurs when pre-calciferol is converted to calciferol (p. 1235) suggests an interesting prediction (Figure 28.14). The reaction is pictured as an interaction between the HOMO of the σ bond that is migrating and the LUMO of the triene system. If you examine the symmetry of these orbitals, you will see that it is the bottom lobe of the orbital at

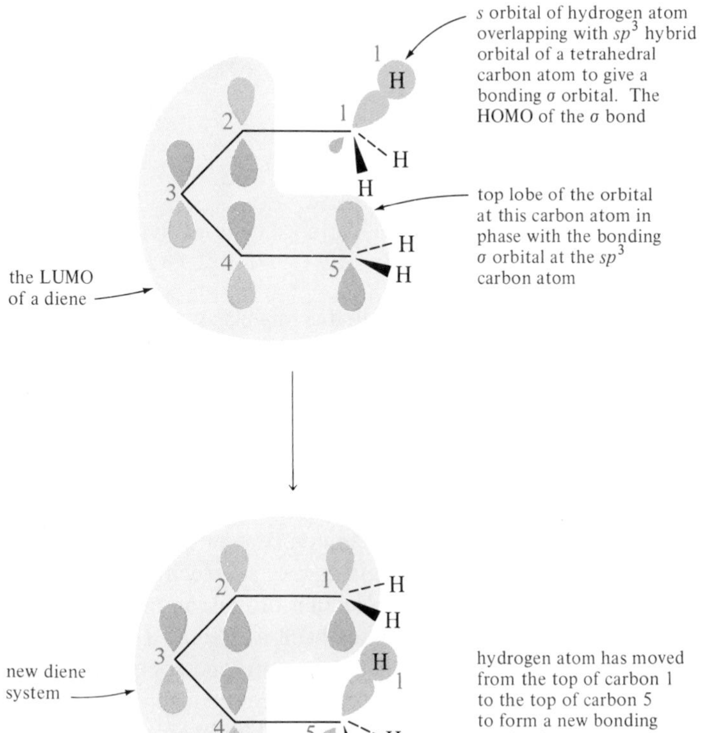
s orbital of hydrogen atom overlapping with sp^3 hybrid orbital of a tetrahedral carbon atom to give a bonding σ orbital. The HOMO of the σ bond
top lobe of the orbital at this carbon atom in phase with the bonding σ orbital at the sp^3 carbon atom
the LUMO of a diene
new diene system
hydrogen atom has moved from the top of carbon 1 to the top of carbon 5 to form a new bonding σ orbital

FIGURE 28.13 A molecular orbital picture of the stereochemistry of a [1,5] sigmatropic rearrangement involving hydrogen migration.

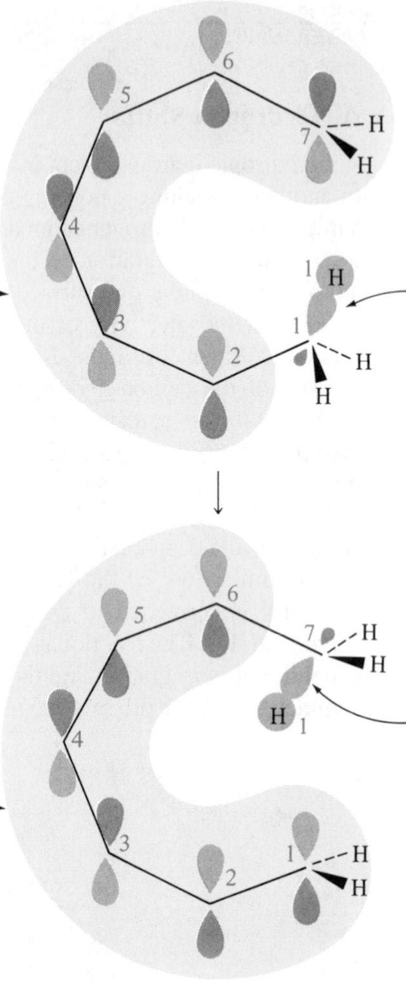
LUMO of the triene
HOMO of the σ bond between the hydrogen atom and the tetrahedral carbon atom
new triene system
hydrogen atom has moved from the top of carbon 1 to the bottom of carbon 7 to form a new bonding σ orbital

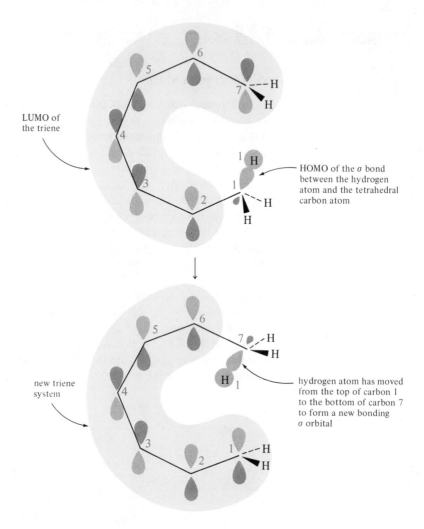

FIGURE 28.14 A molecular orbital picture of the stereochemistry of a [1,7] sigmatropic rearrangement involving hydrogen migration.

carbon 7 that is in phase with the σ bond at carbon 1. The hydrogen atom must move from the top of carbon 1 to the bottom face of carbon 7 for a bonding interaction to occur between the hydrogen atom and the carbon atom that is becoming tetrahedral. Such a migration is stereochemically possible in this kind of an open-chain triene because one end of the molecule lies above the other end in one conformation of the system.

A more general way of looking at these reactions is to consider the total number of electrons involved in the transition state. The [1,5] hydrogen shift occurs with a reorganization of six electrons, two from the σ bond and four in the diene system. In the molecular orbital interactions for the reaction as drawn in Figure 28.13, the reaction has been treated as a [4 + 2] interaction. A further generalization is that all sigmatropic rearrangements involving $4n + 2$ electrons in the transition state take place in a concerted fashion in the ground state with the stereochemistry shown in Figure 28.13. The hydrogen and alkyl shifts seen in the rearrangements of carbocations (p. 285) are concerted [1,2] sigmatropic rearrangements that take place along one face of the molecule. In these cases, only the two electrons of the migrating σ bond are involved in the transition state, as the bond shifts to a cationic carbon atom with an empty orbital. Such a case is a rearrangement of a system with $4n + 2$ electrons, where $n = 0$.

Using the same language, a [1,7] hydrogen shift is described as having six electrons from the triene system and two electrons from the σ bond involved in the transition state, for a total of eight electrons. Thus, such a reaction involves a $4n$ system, with $n = 2$. Rearrangements involving the $4n$ electrons are concerted in the ground state when the stereochemistry is that shown in Figure 28.14, with migration taking place from the top face of one part of the molecule to the bottom face of the other.

Just as in electrocyclic reactions and cycloaddition reactions, in sigmatropic rearrangements, the symmetry of the orbitals involved and the resulting stereochemistry of the reactions change in the excited state. The following problem provides an opportunity for you to prove this to yourself.

PROBLEM 28.26

When a hexatriene system undergoes a photochemical [1,7] hydrogen shift, the hydrogen atom migrates from the top of carbon 1 to the top of carbon 7. This observation can be rationalized by drawing a molecular orbital picture similar to the one shown in Figure 28.14, but using the HOMO that is appropriate for the excited state of the triene. Prove to yourself that this is so. You may want to refer to Problem 28.12 to see what the HOMO of the excited state of a triene looks like. You must now, of course, use the LUMO of the σ bond to the hydrogen atom.

B. The Cope Rearrangement

Not all sigmatropic rearrangements involve migration of hydrogen atoms. A large and important group of reactions takes place with the migration of a carbon atom and σ bond. A 1,2-alkyl shift in a carbocation is an example that you have already learned. A group of such reactions that do not have ionic intermediates is known as the **Cope rearrangement,** for Arthur Cope of the Massachusetts Institute of Technology who discovered and studied them.

A typical Cope rearrangement is the interconversion of 3-methyl-1,5-hexadiene and (*E*)-1,5-heptadiene.

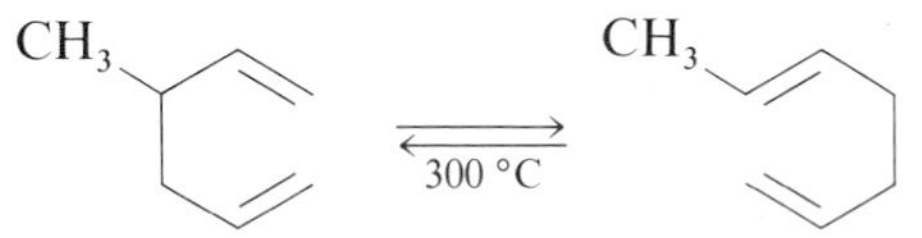

CH3
300 °C
CH3

3-methyl-1,5-hexadiene (*E*)-1,5-heptadiene

When either compound is heated to approximately 300 °C, it is converted into a mixture of the two. The two alkenes are similar to each other in energy, so an equilibrium is established, and the reaction does not go to completion in one direction or the other.

If the diene is chosen so that at least one of the double bonds in the rearranged diene is conjugated with an aryl group or a carbonyl group, then the reaction takes place at lower temperatures and gives good yields. An example of such a reaction is the rearrangement of 3-methyl-4-phenyl-1,5-hexadiene.

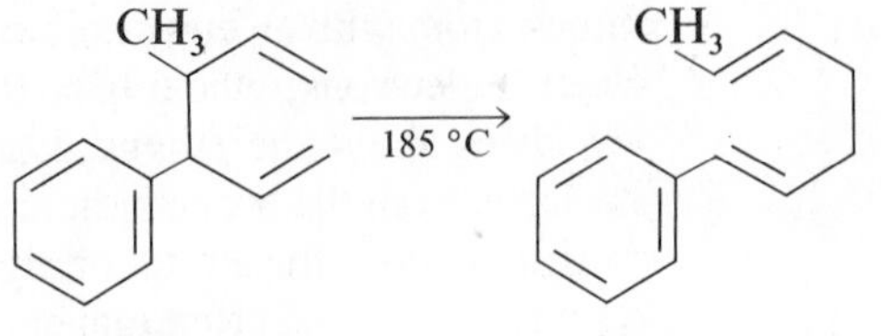

3-methyl-4-phenyl-1,5-hexadiene

(1*E*,5*E*)-1-phenyl-1,5-heptadiene 90%

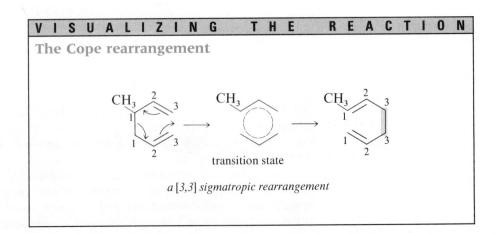

In the product, (1*E*,5*E*)-1-phenyl-1,5-heptadiene, one of the double bonds is conjugated with the aromatic ring, and the other one is an internal double bond. The product diene is more stable than 3-methyl-4-phenyl-1,5-hexadiene, in which both of the double bonds are in terminal positions.

Another example of a Cope rearrangement in which one of the double bonds is conjugated in the product diene is shown below.

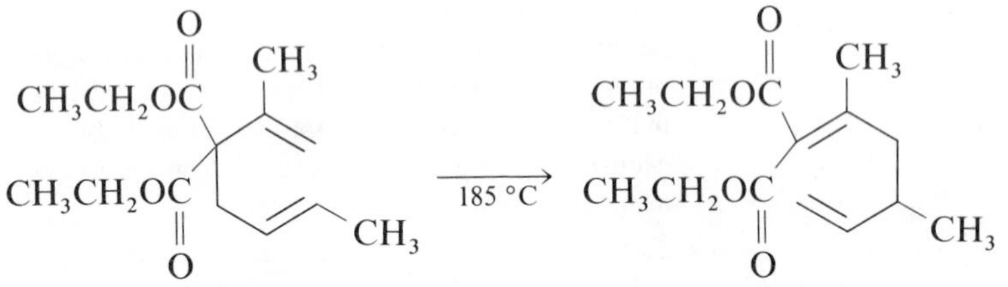

(*E*)-3,3-dicarboethoxy-2-methyl-1,5-heptadiene

ethyl 2-carboethoxy-3,5-dimethyl-2,6-heptadienoate

(*E*)-3,3-Dicarboethoxy-2-methyl-1,5-heptadiene is a derivative of diethyl malonate. After rearrangement, one of the double bonds is conjugated with both of the ester groups. The product diene is therefore more stable than the starting diene and predominates at equilibrium.

A reexamination of the rearrangement of 3-methyl-1,5-hexadiene to (*E*)-1,5-heptadiene (p. 1249) shows that the reaction involves breaking a σ bond at one part of the diene chain and forming a σ bond at another, while the π bonds in the molecule shift position.

VISUALIZING THE REACTION

The Cope rearrangement

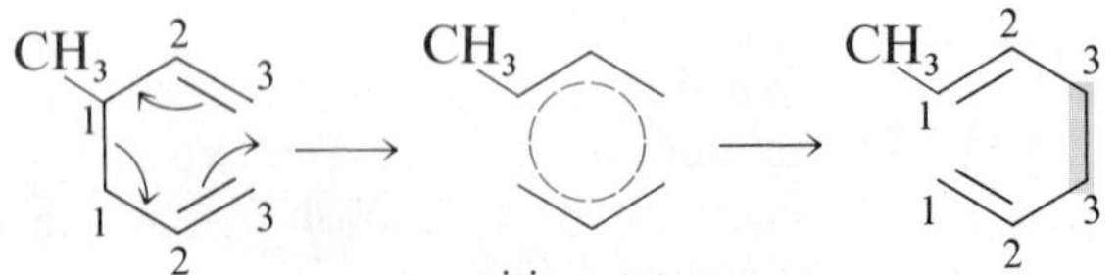

transition state

a [3,3] sigmatropic rearrangement

The arrows indicate that the Cope rearrangement has a transition state involving six electrons, two from the σ bond that is breaking and four from the two π bonds. The hexadiene is numbered with the carbon atoms of the breaking σ bond as the first carbon atoms of both halves of the chain. In this way of looking at the molecule, the new σ bond forms between the third atoms of each half of the chain. In other words, the σ bond has migrated three atoms for the top half of the molecule and also three atoms for the bottom half. For this reason, the reaction is called a [3,3] sigmatropic rearrangement.

PROBLEM 28.27

Complete the following equations.

(a) 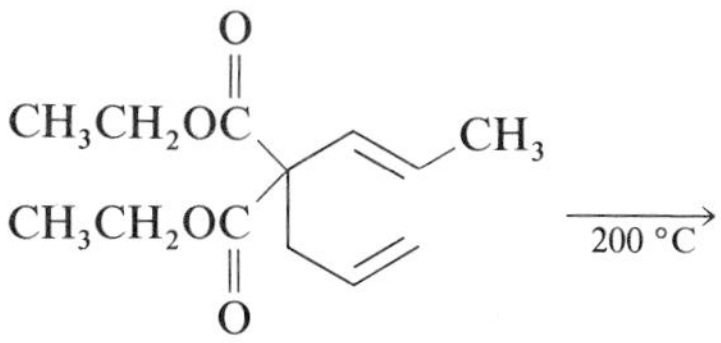
O
CH3CH2OC
CH3
CH3CH2OC
O
200 °C

(b) 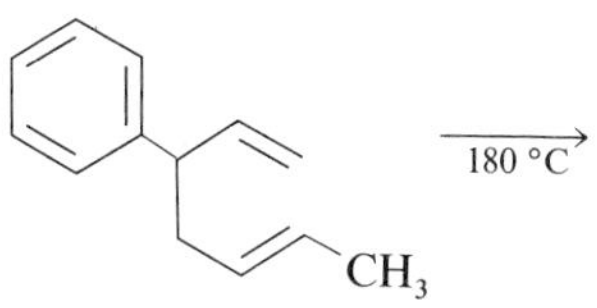
180 °C
CH3

(c) 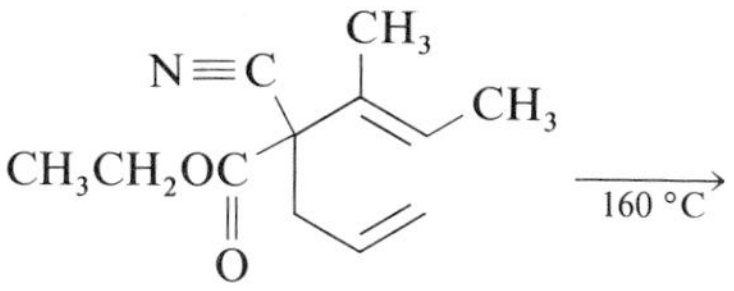
CH3
N≡C
CH3
CH3CH2OC
O
160 °C

(d) 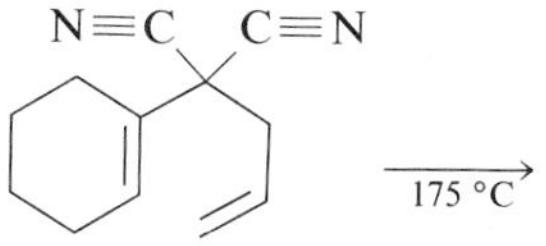
N≡C
C≡N
175 °C

PROBLEM 28.28

A synthesis of grandisol, a sex pheromone of the male boll weevil (Problem 28.9, p. 1225), started with a metal-catalyzed dimerization of isoprene to form what the researchers hoped would be a cyclobutane with the following structure. The product that was isolated when the reaction mixture was allowed to exceed room temperature was 1,5-dimethyl-1,5-cyclooctadiene. How would you rationalize this observation?

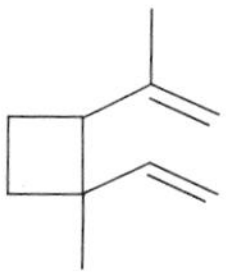

C. The Claisen Rearrangement

Sigmatropic rearrangements involving the cleavage of a σ bond at an oxygen atom are called **Claisen rearrangements.** The same Ludwig Claisen who discovered the condensation reaction of esters to form β-ketoesters (p. 662) also discovered that the allyl enol ether of ethyl acetoacetate undergoes a rearrangement to an acetoacetic ester substituted by an allyl group at the α-carbon atom.

VISUALIZING THE REACTION

Claisen rearrangement of an allyl ether

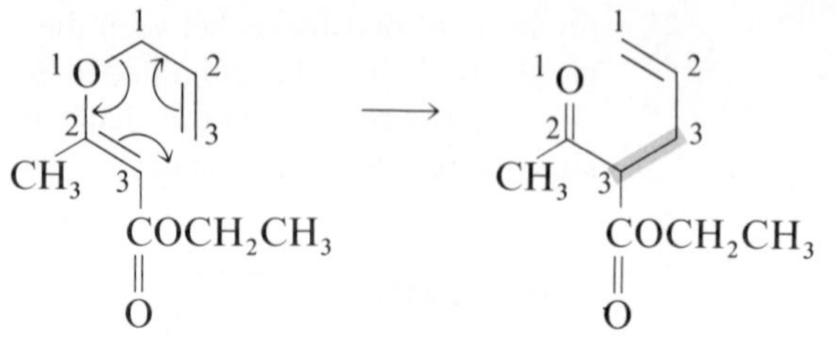
1
1
O
2
2
3
CH3
3
COCH2CH3
O
1
1
O
2
2
3
CH3
3
COCH2CH3
O

O-allyl ether of the enol of ethyl acetoacetate

ethyl 2-allylacetoacetate 85%

a [3, 3] sigmatropic rearrangement

A σ bond is broken between two atoms in this case, an oxygen atom and a carbon atom; another one is formed three atoms away on each portion of the chain. The σ bond migrates three atoms away from its starting position for both portions of the chain that participate in the rearrangement; thus, this reaction is also called a [3,3] sigmatropic rearrangement. The driving force for the reaction is the formation of the stable carbonyl group in the product.

The Claisen rearrangement is quite general. The rearrangement of the allyl ethers of phenols is another example. For example, allyl phenyl ether, easily prepared from phenol, rearranges to give *o*-allylphenol.

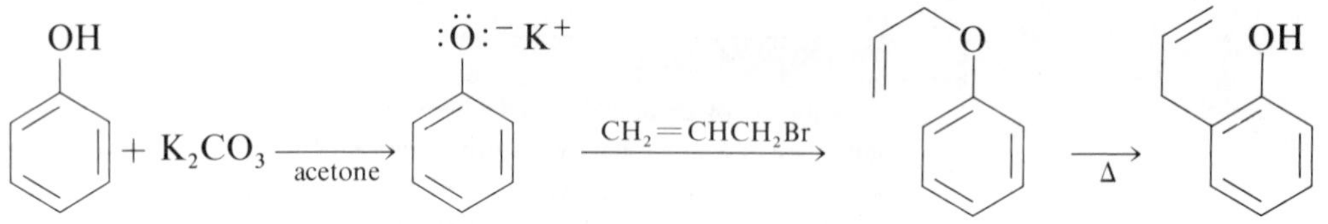
OH
+ K2CO3
acetone
:Ö:⁻ K⁺
CH2=CHCH2Br
O
Δ
OH

phenol — potassium carbonate — potassium phenolate — allyl phenyl ether 86% — *o*-allylphenol

The carbon atom attached to the aromatic ring is not the carbon atom that was bonded to the oxygen atom in the ether. If a 2-butenyl instead of an allyl group is present in a phenyl ether, a product phenol has a methyl branch on the side chain.

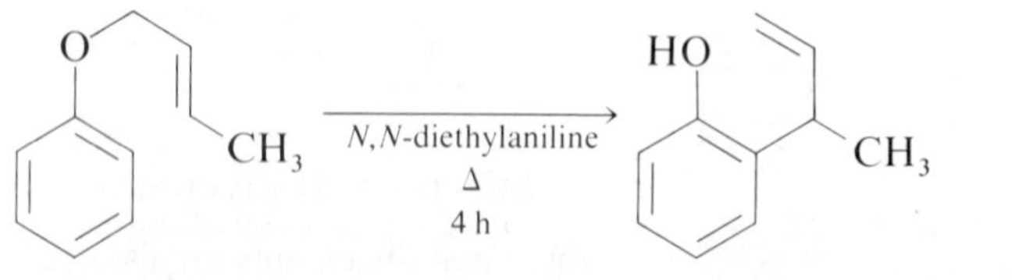
O
CH3
N,N-diethylaniline
Δ
4 h
HO
CH3

2-butenyl phenyl ether — 2-(1-methyl-2-propenyl)phenol 85%

This experiment, and others like it, show that attachment between the aromatic ring and the allyl side chain takes place three carbon atoms away from the oxygen atom.

The allyl group is not detached from the oxygen atom and reattached to the aromatic ring at the original point of bonding.

The rearrangement goes through a cyclic transition state.

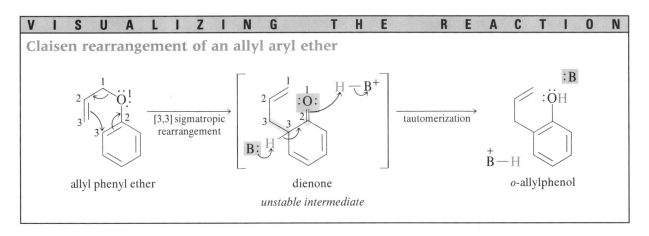

V I S U A L I Z I N G   T H E   R E A C T I O N

Claisen rearrangement of an allyl aryl ether

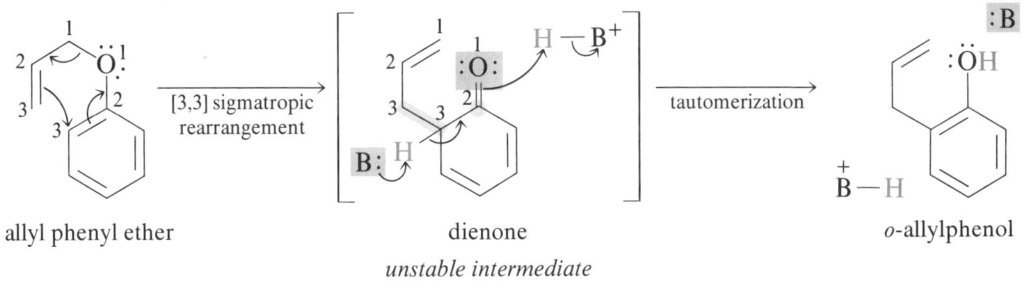

[3,3] sigmatropic rearrangement
tautomerization
:B
B:
allyl phenyl ether
dienone
unstable intermediate
o-allylphenol

Two of the π electrons of the aromatic ring participate in the reaction, which involves a total of six electrons in the transition state. The mechanism shown above requires an unstable dienone as an intermediate. Such a dienone is expected to tautomerize rapidly to the stable phenol by loss of a proton at the aromatic ring and protonation of the oxygen atom.

An allyl phenyl ether in which the ortho positions are blocked rearranges to put the allyl group on the para position of the ring. Thus, allyl 2,6-dimethylphenyl ether rearranges to 4-allyl-2,6-dimethylphenol.

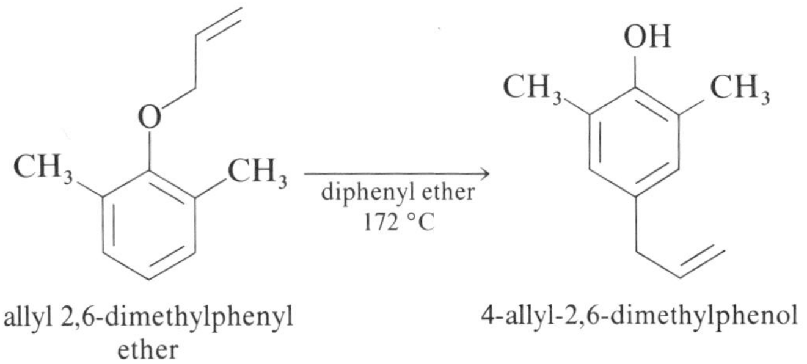

diphenyl ether
172 °C
allyl 2,6-dimethylphenyl ether
4-allyl-2,6-dimethylphenol

The reaction is thought to go through a dienone intermediate in which rearrangement to the ortho position has occurred, followed by a second rearrangement to the para position.

V I S U A L I Z I N G   T H E   R E A C T I O N

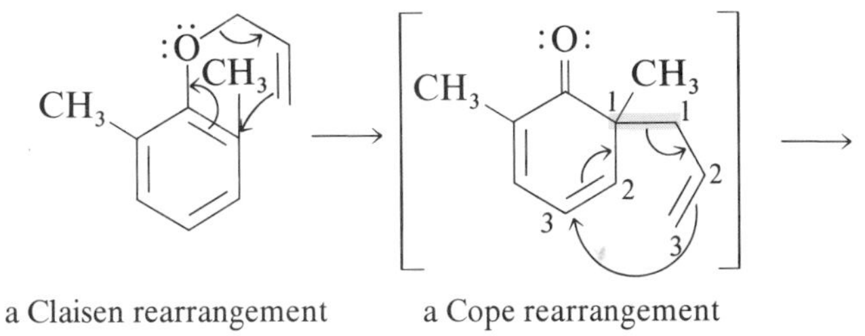

a Claisen rearrangement
a Cope rearrangement

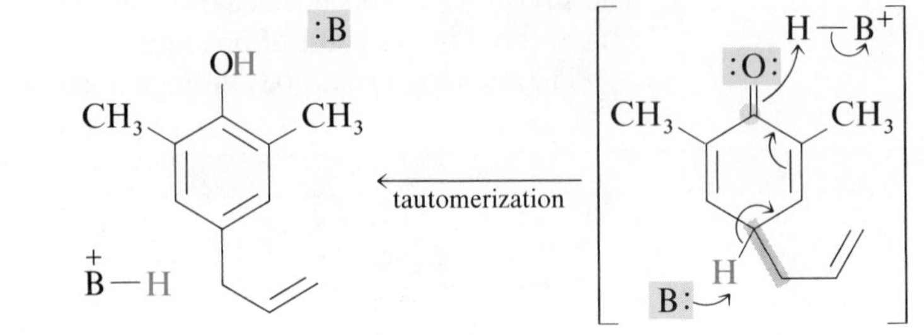
:B
OH
CH_3
CH_3
$\overset{+}{B}$—H
tautomerization
H—B+
:O:
CH_3
CH_3
H
B:

a *p*-dienone intermediate

Study Guide Concept Map 28.5

The [1,5] hydrogen shift (p. 1217), the Cope rearrangement (p. 1249), and the Claisen rearrangement (p. 1251) all take place by way of transition states involving six atoms and the reorganization of six electrons. This fact, kept firmly in mind, will help you to recognize these reactions when you see molecular structures that allow for such transformations.

PROBLEM 28.29

Complete the following equations.

(a) $CH_2{=}CHOCH_2CH{=}CH_2 \xrightarrow{255\ °C}$

(b)

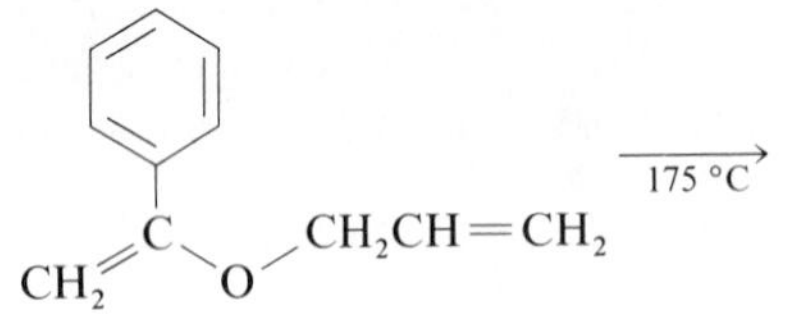
175 °C
C
$CH_2CH{=}CH_2$
CH_2
O

(c)

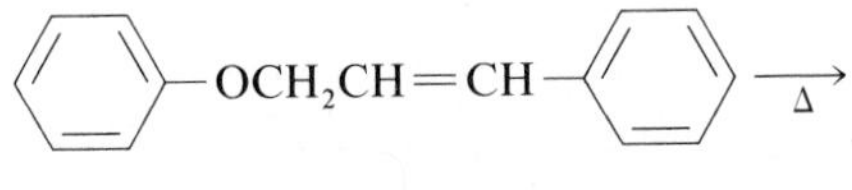
$OCH_2CH{=}CH$
Δ

(d)

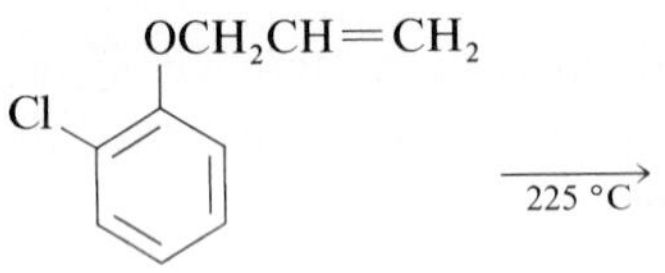
$OCH_2CH{=}CH_2$
Cl
225 °C

(e)

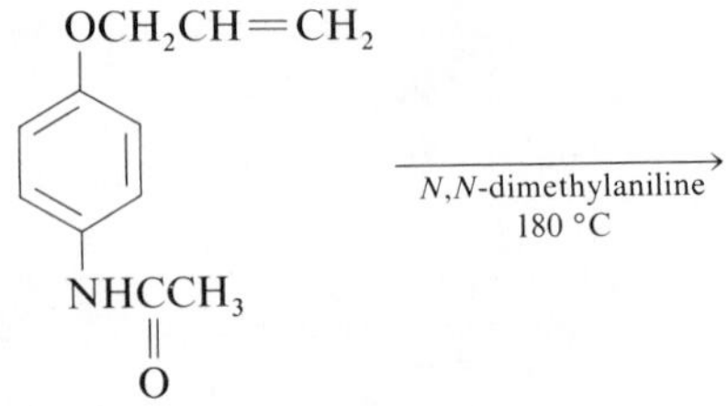
$OCH_2CH{=}CH_2$
$NHCCH_3$
O
N,N-dimethylaniline
180 °C

(f)

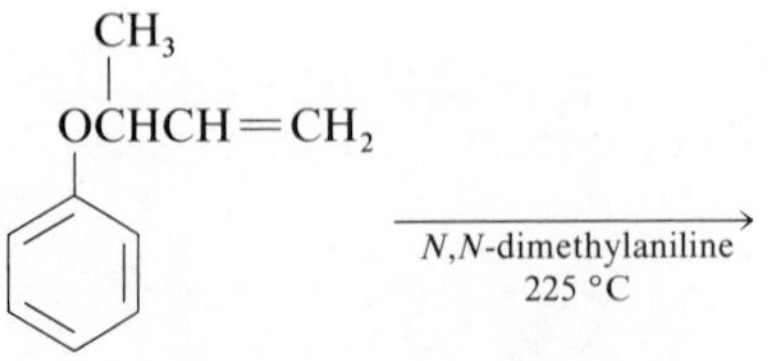
CH_3
$OCHCH{=}CH_2$
N,N-dimethylaniline
225 °C

(g)

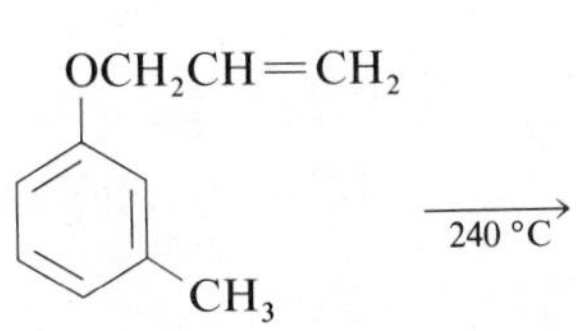
$OCH_2CH{=}CH_2$
CH_3
240 °C

28.6 REACTIONS OF CARBENES

Carbenes are reactive intermediates containing two single bonds and a pair of nonbonding electrons around a carbon atom. They are neutral species that must acquire another pair of electrons to complete a stable octet. Thus, they are electrophiles and add to double bonds to give cyclopropanes. Carbenes are formed when an alkyl halide that cannot undergo elimination to form an alkene reacts with a strong base. For example, tribromomethane reacts with potassium *tert*-butoxide to give dibromomethylene, a carbene.

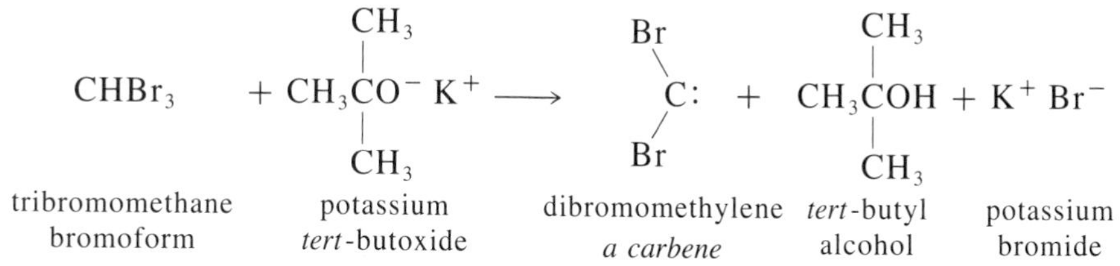
tribromomethane
bromoform
potassium
tert-butoxide
dibromomethylene
a carbene
tert-butyl
alcohol
potassium
bromide

The overall reaction is an α-elimination, taking place in two steps.

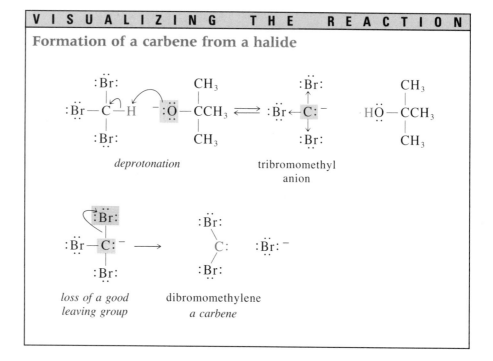

VISUALIZING THE REACTION

Formation of a carbene from a halide

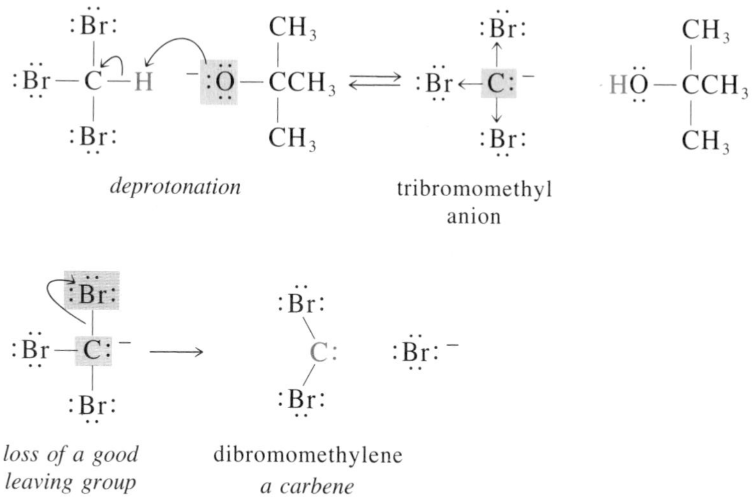
deprotonation
tribromomethyl
anion
loss of a good
leaving group
dibromomethylene
a carbene

The proton and the leaving group are lost from the *same* carbon atom of the alkyl halide rather than from adjacent carbon atoms as in β-elimination reactions that give rise to alkenes.

The reaction of a carbene such as dibromomethylene with an alkene is stereoselective. The carbene reacts with (Z)-2-butene to give *cis*-1,1-dibromo-2,3-dimethylcyclopropane and with (E)-2-butene to give the trans isomer.

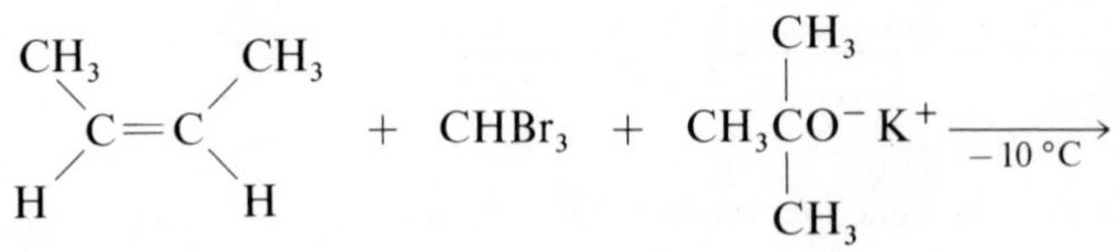

(*Z*)-2-butene — tribromomethane bromoform — potassium *tert*-butoxide

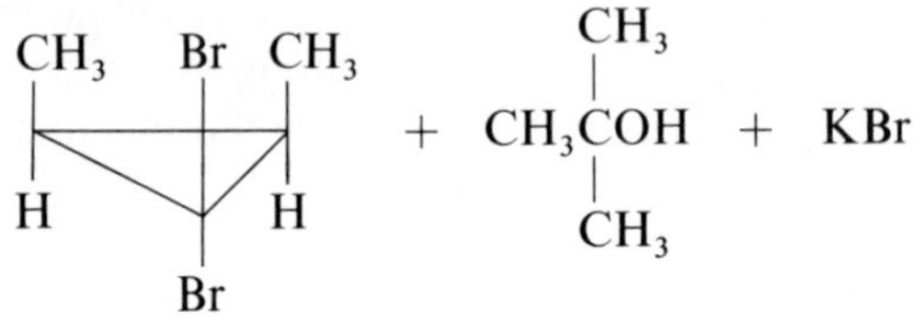

cis-1,1-dibromo-2,3-dimethylcyclopropane 80% — *tert*-butyl alcohol — potassium bromide

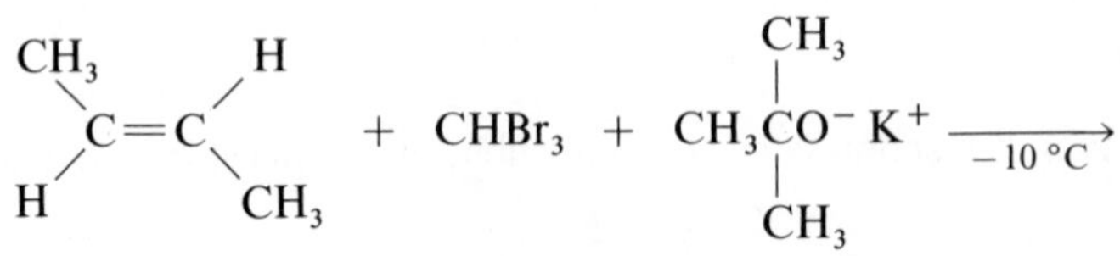

(*E*)-2-butene — tribromomethane bromoform — potassium *tert*-butoxide

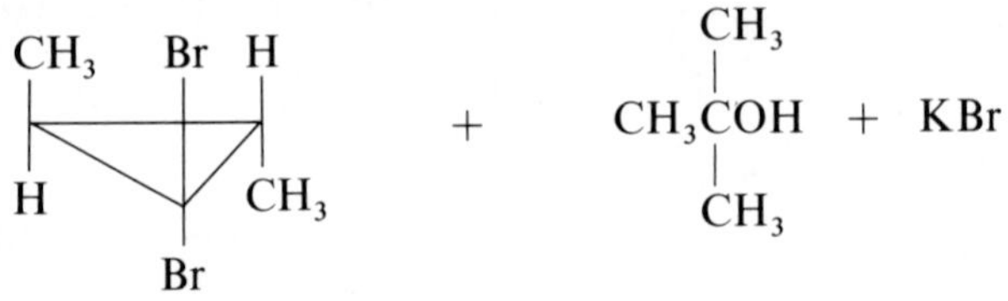

trans-1,1-dibromo-2,3-dimethylcyclopropane 68% *racemic* — *tert*-butyl alcohol — potassium bromide

Because the stereochemistry of the alkene is retained in the product cyclopropane, chemists think that the carbene adds to the double bond in a single step, in which both bonds to the doubly bonded carbon atoms of the alkene are formed simultaneously.

VISUALIZING THE REACTION

Addition of a carbene to an alkene

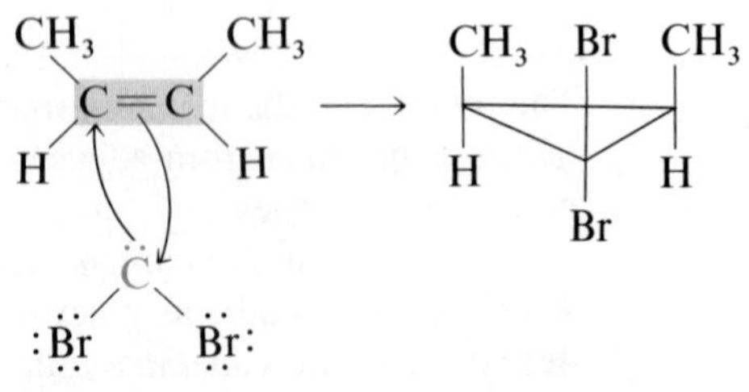

The number of electrons participating in the reaction, however, presents a problem at first glance. Four electrons are involved in the transition state, two from the π bond of the alkene and two from the carbene. This reaction resembles the concerted cycloaddition of two ethylene units, which is forbidden in the ground state (Table 28.1, p. 1237). The difference between the thermal cycloaddition of two alkene units to give cyclobutane and the cycloaddition of a carbene to an alkene lies in the stereochemistry that is possible for the two reactions. The four-carbon atom framework of a cyclobutane molecule requires that the two double bonds approach each other and react in a face-to-face manner (Figure 28.5, p. 1224). Otherwise, an impossibly twisted and strained four-membered ring would form. An examination of the orbitals of a carbene shows that a bonding overlap between the HOMO of a carbene and the LUMO of an alkene (or the other way around) is possible (Figure 28.15).

A carbene has three pairs of electrons around the central carbon atom. It is a trigonal carbon atom, with the three pairs of electrons each occupying an sp^2 hybrid orbital and an empty p orbital perpendicular to the plane defined by the carbon atom and the two substituents on it. The form of the carbene described above is called its **singlet state,** in which both nonbonding electrons occupy a single orbital and have opposing spins. For some carbenes, another electronic state, in which the nonbonding electrons become unpaired and each one occupies a different orbital, is more stable. Such carbenes are said to be in the **triplet state.** Triplet carbenes, because of their unpaired electrons, behave more like radicals than singlet carbenes do, and do not add to double bonds with retention of the stereochemistry of the alkene.

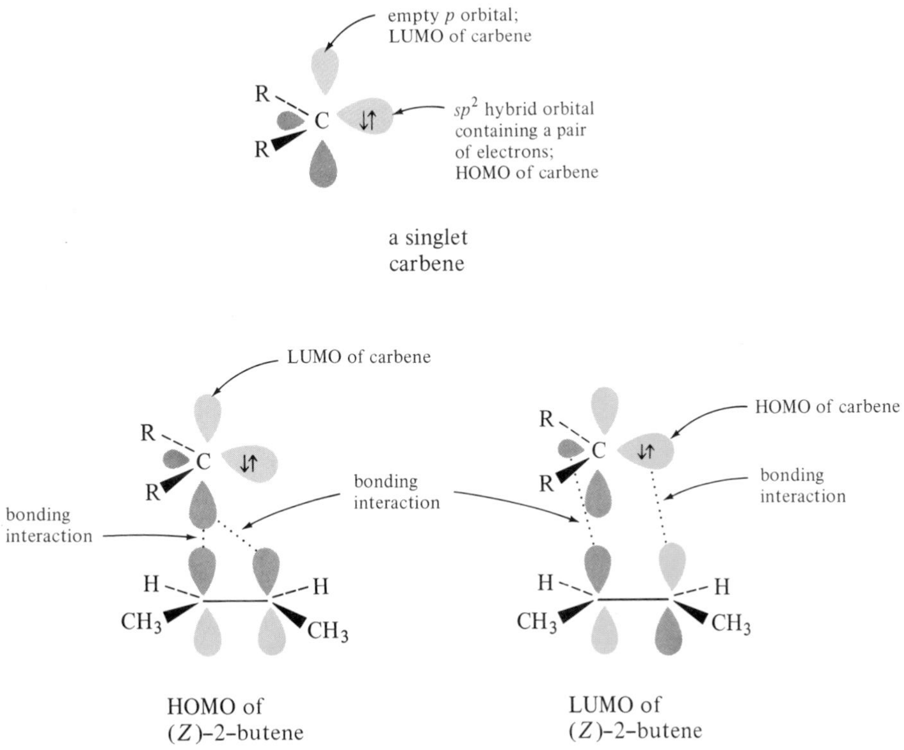
empty p orbital;
LUMO of carbene
R
C
R
sp² hybrid orbital
containing a pair
of electrons;
HOMO of carbene
a singlet
carbene
LUMO of carbene
HOMO of carbene
bonding
interaction
bonding
interaction
bonding
interaction
H
CH3
H
CH3
HOMO of
(Z)-2-butene
LUMO of
(Z)-2-butene

FIGURE 28.15 A representation of the interactions possible between the molecular orbitals of a carbene in its singlet state and an alkene.

The cycloaddition reactions of dibromomethylene are reactions of the singlet state of the carbene. Note that reaction is possible only if the carbene approaches the alkene sideways so that the plane defined by the carbon atom and its two substituents parallels the plane of the alkene. In this orientation, the empty *p* orbital of the carbene is pointing toward the electrons of the π bond. The carbene differs from an alkene in having this empty orbital, which makes a sideways approach of the carbene to the π bond and reaction to give a three-membered ring possible without the creation of an impossible amount of strain in the transition state.

PROBLEM 28.30

Prove to yourself that an approach of a carbene in which its nonbonding electrons are pointing toward the π orbital of the alkene cannot lead to a concerted cycloaddition reaction.

Addition of carbenes to double bonds is a useful way to synthesize cyclopropanes. One method often used for this reaction involves an organozinc reagent known as the **Simmons-Smith reagent,** for the two American chemists, Howard E. Simmons and Ronald D. Smith, who developed it. The reagent is prepared by treating metallic zinc with a salt of copper to make a zinc-copper couple, which reacts with diiodomethane to give the active species, usually formulated as $CH_2(ZnI_2)$. The exact structure of the reagent is not known, but it adds a methylene group to a wide variety of alkenes. The preparation of the Simmons-Smith reagent and its reaction with 2-cyclohexenone are shown below.

$$\underset{\text{diiodomethane}}{CH_2I_2} + \underset{\text{zinc-copper couple}}{Zn(Cu)} \xrightarrow[\text{diethyl ether}]{} \underset{\text{organozinc reagent, Simmons-Smith reagent}}{CH_2(ZnI_2)}$$

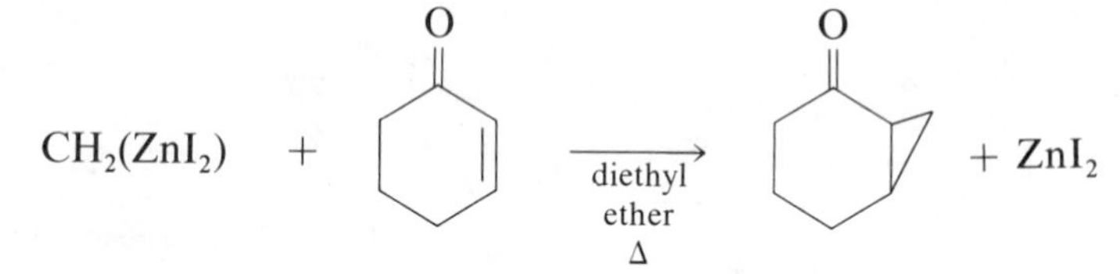

Simmons-Smith reagent — 2-cyclohexenone — bicyclo[4.1.0]heptan-2-one 90%

Study Guide Concept Map 28.6

PROBLEM 28.31

Complete the following equations.

(a) 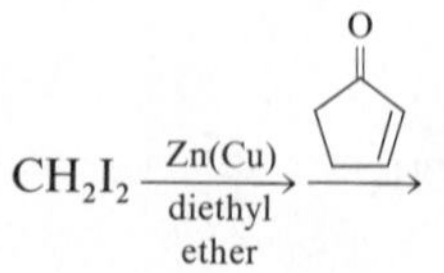

(b) 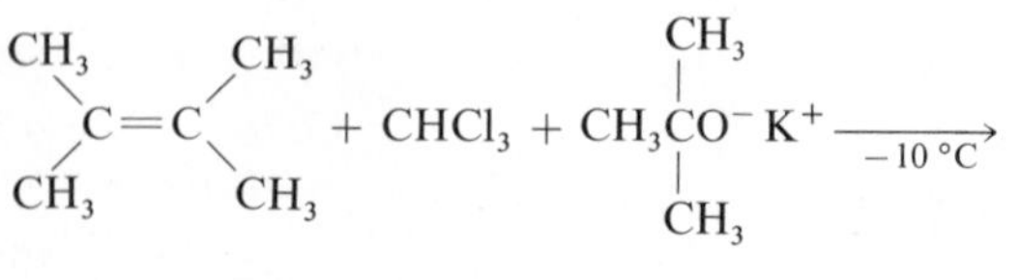

(c) 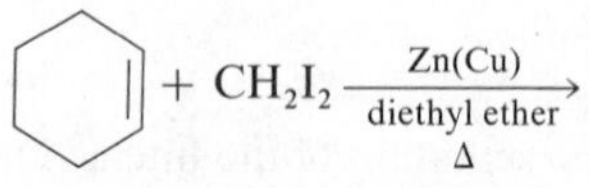

(d) 

(e) $CH_3(CH_2)_5CH{=}CH_2 + CH_2I_2 \xrightarrow[\text{diethyl ether},\ \Delta]{\text{Zn(Cu)}}$

(f) 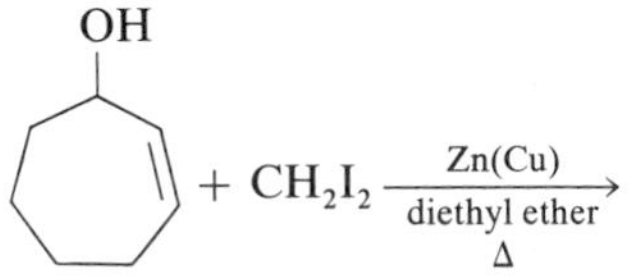

(g) 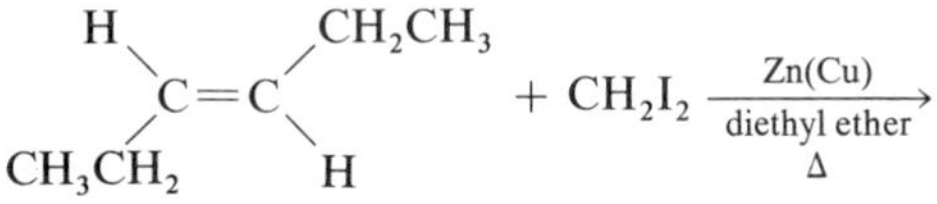

(h) 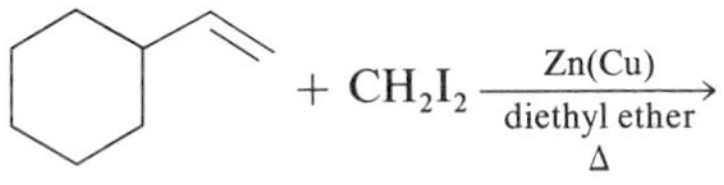

(i) 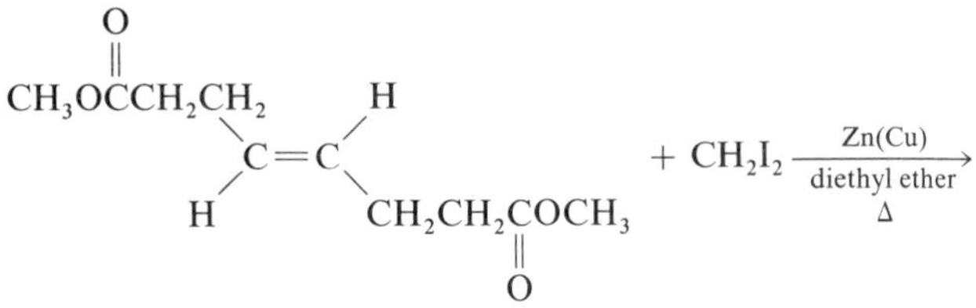

(j) 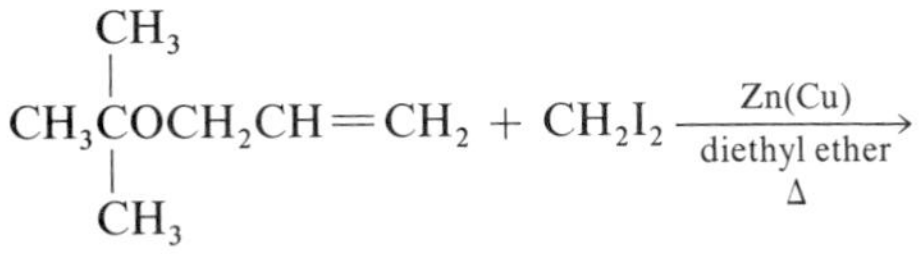

(k) 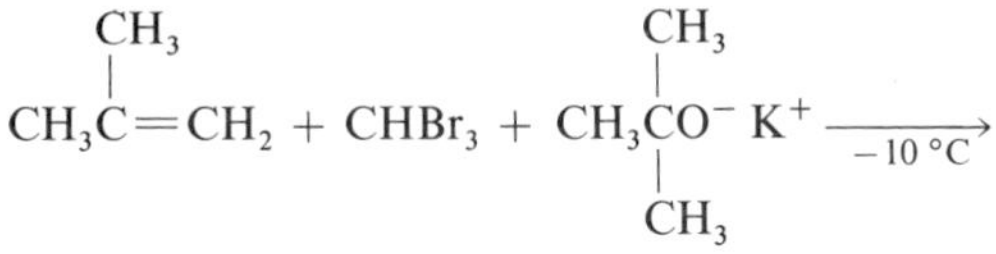

In summary, a consideration of the symmetry of molecular orbitals and a classification of a large number of organic reactions according to the number of electrons participating in the transition state have been used by chemists to explain experimental phenomena that seem unconnected at first glance. It is important to keep in mind, however, that in each case the theory is a rationalization of the experimental facts and an indication of new directions for experimentation. The concepts described in this chapter make it easier to correlate and remember facts about known reactions and to predict new reactions. The mere fact that reactions can be described in a certain way is no guarantee that they actually occur by those pathways. Experimental and theoretical work is still being done by chemists to improve their understanding of reactivity, regioselectivity, and stereoselectivity in the transformations of organic compounds.

S U M M A R Y

Reactions that take place in one step, without the formation of a reactive intermediate and with high stereoselectivity, are concerted reactions. Concerted reactions often occur by way of cyclic transition states and involve the reorganization of σ and π bonds.

In cycloaddition reactions, two compounds containing π bonds add to each other to give a cyclic product. If $(4n + 2)$ π electrons participate in the transition state, the reaction occurs in the ground state, or thermally. Diels-Alder reactions and additions of 1,3-dipolar species to π bonds to give five-membered heterocycles are examples of thermal cycloaddition reactions. When $4n$ π electrons participate in the transition state, the reaction is photochemical; one of the reactants must be in an excited state. Addition of one alkene to another to give a cyclobutane is this type of reaction (Table 28.3).

Chemists use molecular orbital theory to rationalize the reactivity patterns and stereochemistry observed in cycloaddition reactions. Such a reaction is pictured as occurring by the interaction of the frontier orbitals of the reactants, specifically, the highest occupied molecular orbital (HOMO) of one species and the lowest unoccu-

TABLE 28.3 Cycloaddition Reactions

Reactants	Conditions	Product(s)
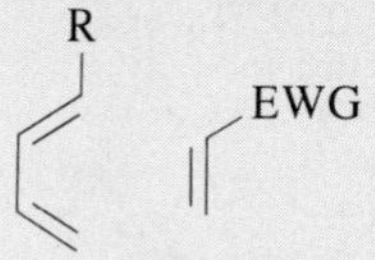	Δ	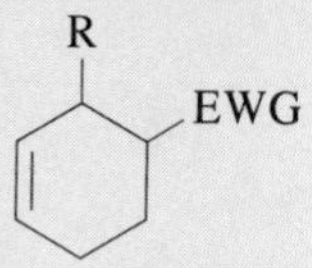
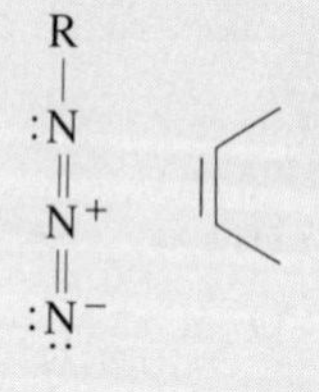	Δ	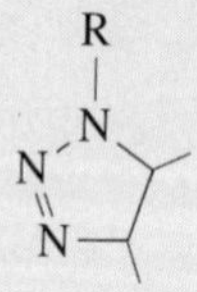
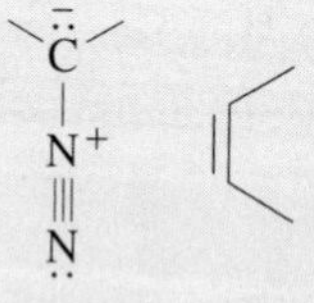	Δ	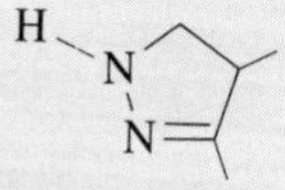
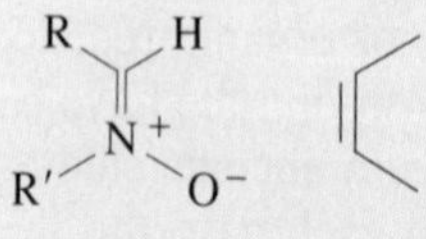	Δ	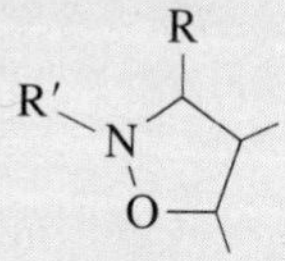
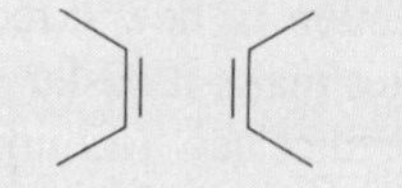	$h\nu$	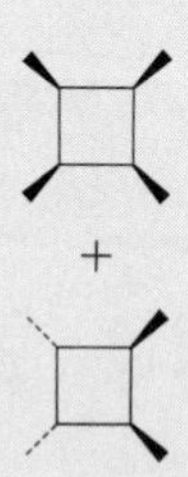

pied molecular orbital (LUMO) of the other. For example, the π molecular orbital of an alkene is its HOMO; the π^* molecular orbital is its LUMO. In a photochemical reaction of an alkene, an electron is promoted from the π to the π^* orbital, making the latter the new HOMO for the system. Bonds form when lobes of orbitals that are in phase with each other can overlap. Reactions and the stereochemistry of the reactions are, therefore, governed by the symmetry of the frontier orbitals.

Electrocyclic reactions involve the opening of cycloalkene rings to give polyenes or the conversion of polyenes to cycloalkenes. These reactions also show precise stereochemistry, which depends on the number of electrons in the transition state and on whether the reaction is proceeding in the ground state or the excited state. A system with $4n$ π electrons in the transition state undergoes reactions in the ground state by a conrotatory process. In such a process, the ends of the conjugated system undergoing the reaction rotate in the same direction to bring in-phase lobes of the HOMO of the system into a position in which they overlap. In a photochemical reaction, disrotatory motion (rotation of the ends of the conjugated system in opposite directions) is necessary to create new bonds. When $(4n + 2)$ π electrons are involved in the transition state, the symmetry of the HOMO of the polyene requires disrotatory motion for cyclization in the ground state and conrotatory motion for photochemical reactions (Table 28.4). The Woodward-Hoffmann rules summarize the experimental observations described for the reactions above and allow chemists to make predictions for other reactions involving those numbers of electrons in their transition states.

In sigmatropic rearrangements, an atom and a σ bond migrate from one position to another along a polyene chain with a simultaneous shift of π bonds. Sigmatropic rearrangements may involve shifts of hydrogen atoms, shifts of σ bonds

TABLE 28.4 Electrocyclic Reactions

	$4n$ π Electrons in Transition State	
Reactant	**Conditions**	**Product**
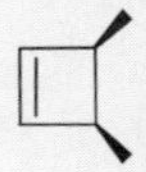	Δ, conrotatory	
	$h\nu$, disrotatory	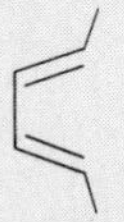
	$(4n + 2)$ π Electrons in Transition State	
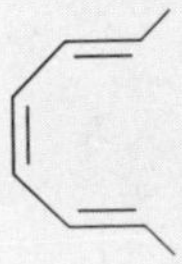	Δ, disrotatory	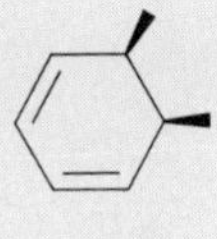
	$h\nu$, conrotatory	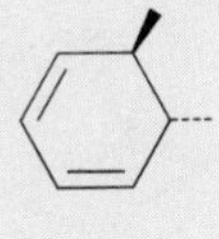

TABLE 28.5 Sigmatropic Rearrangements

Reactant	Conditions	Product
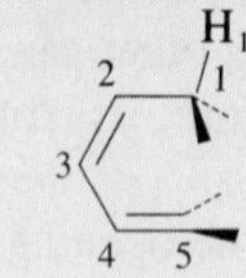	Δ	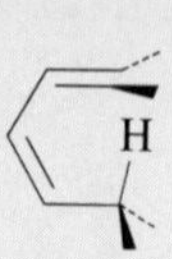 [1,5]
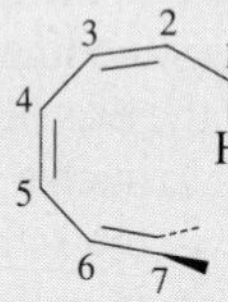	Δ	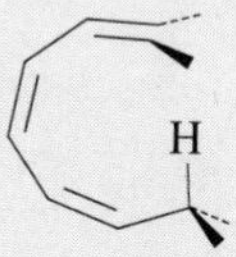 [1,7]
	hν	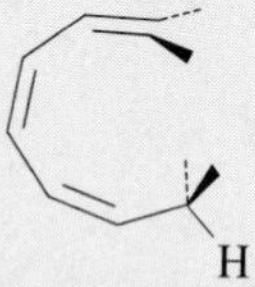
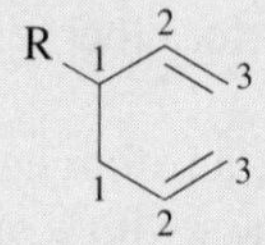	Δ	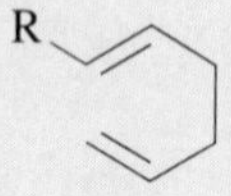 [3,3]
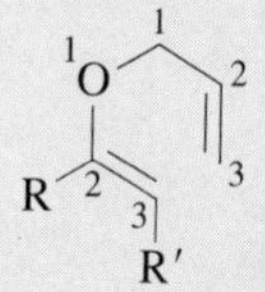 (R and R′ may be part of an aromatic ring.)	Δ	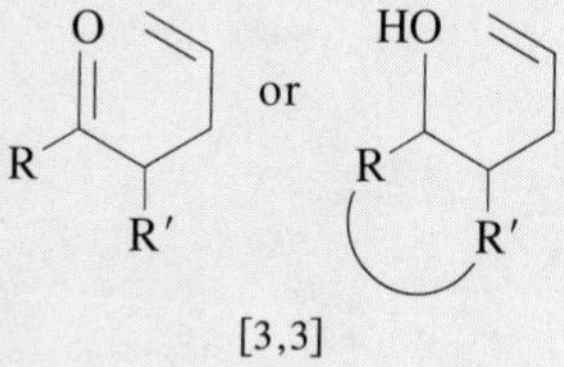 [3,3]

TABLE 28.6 Carbene Reactions

Reactant	Reaction Conditions	Intermediate	Reagent for Second Step	Product
CHX_3	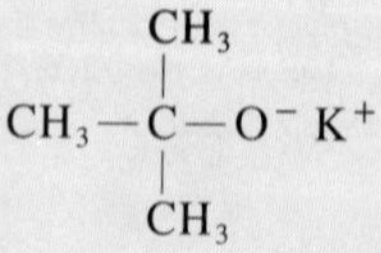	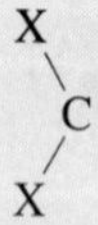		
			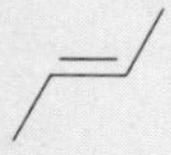	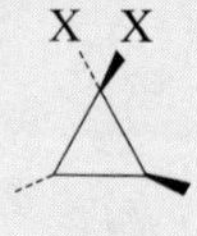
CH_2I_2	Zn(Cu)	$CH_2(ZnI_2)$		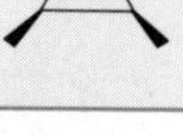

between carbon atoms (Cope rearrangements), or the breaking of a σ bond between carbon and oxygen atoms (Claisen rearrangements) (Table 28.5).

Carbenes are reactive intermediates containing two single bonds to a carbon atom bearing a pair of nonbonding electrons. A carbene is an electrophile and adds to double bonds to give cyclopropanes. A carbene may be in the singlet state; if it is, its reaction with an alkene is concerted, and the product cyclopropane retains the stereochemistry of the substituents on the doubly bonded carbons (Table 28.6). A triplet carbene behaves more like a radical and reacts with alkenes without retention of their stereochemistry.

ADDITIONAL PROBLEMS

28.32 Complete the following equations. Be sure to show the stereochemistry if it can be predicted.

(a) 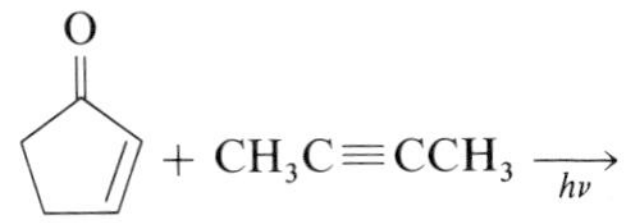

(b) 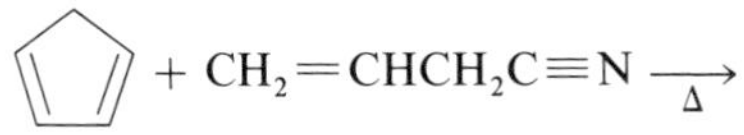

(c) 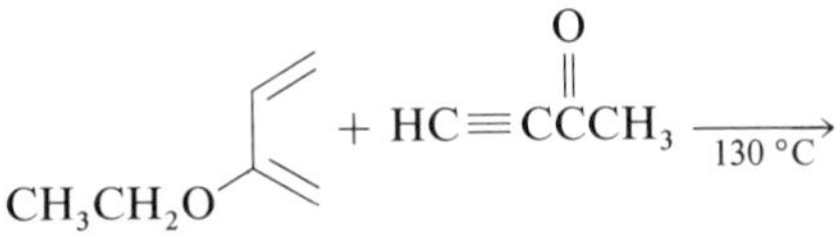

(d) 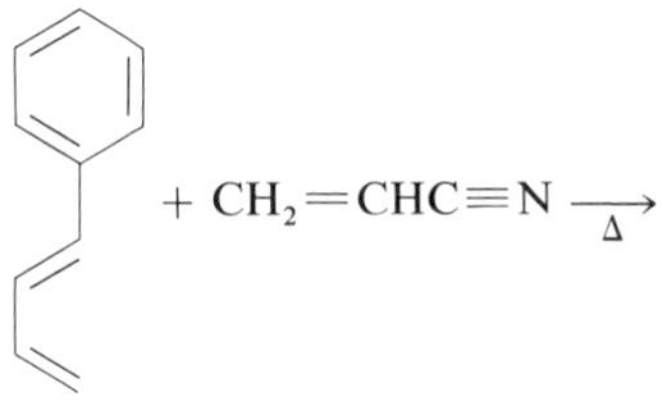

(e) 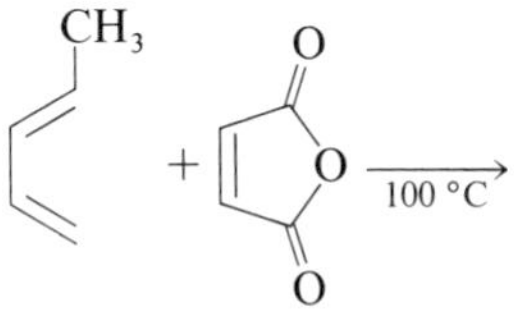

(f) 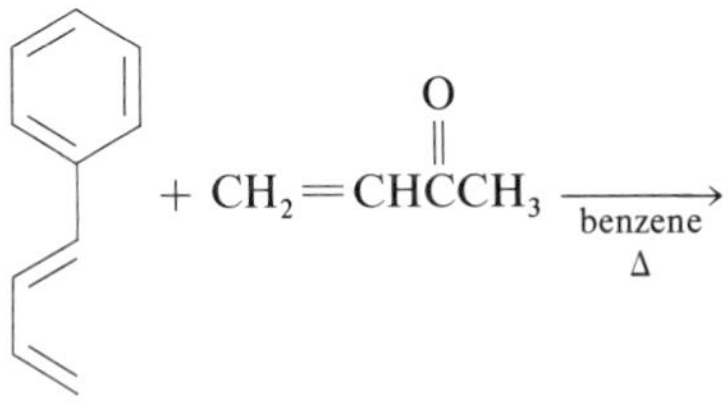

(g) 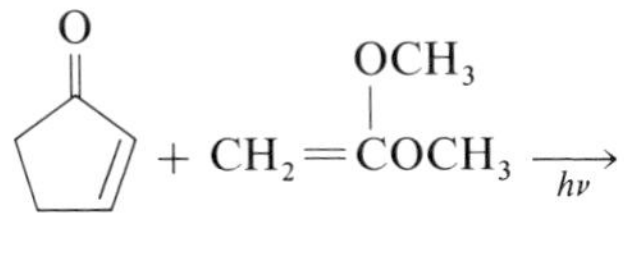

(h) 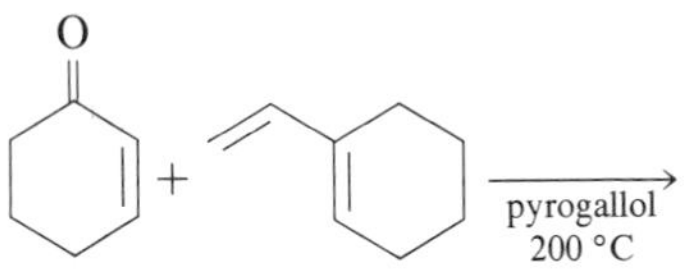

(i) 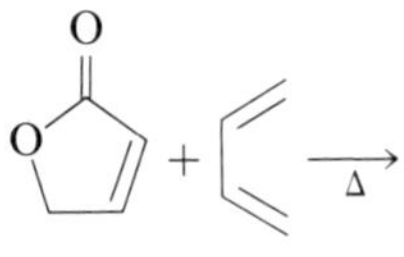

(j) 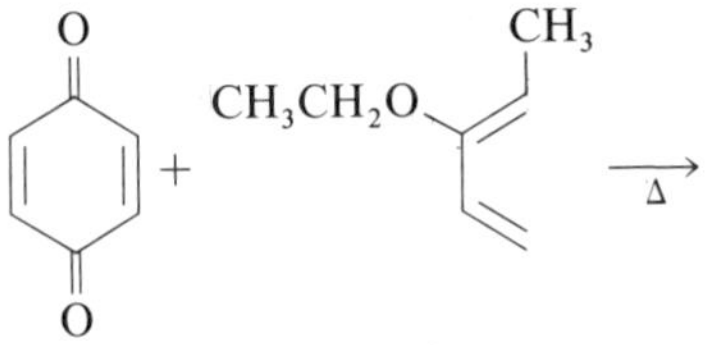

(k) 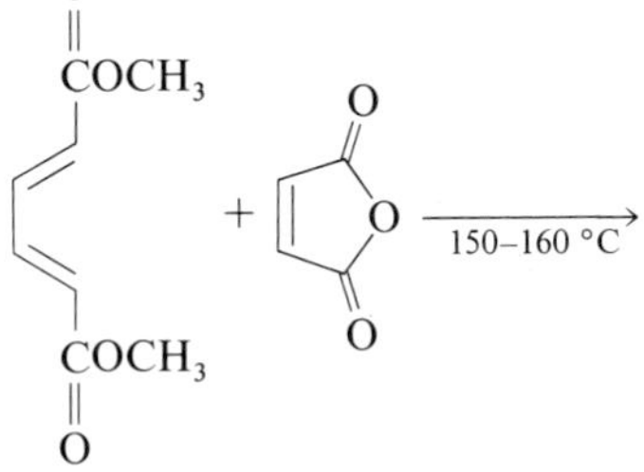

(l) 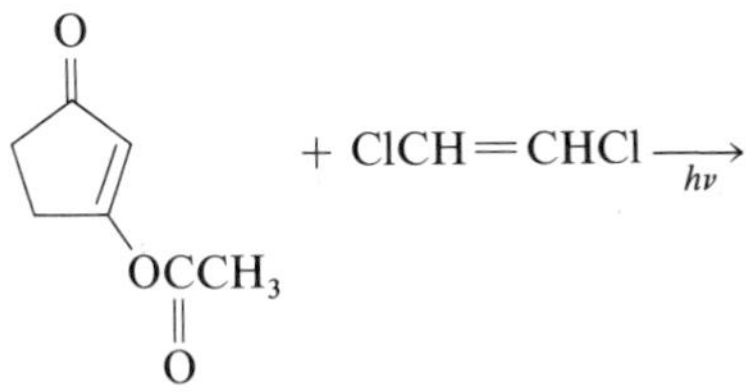

28.33 Complete the following equations. Show the stereochemistry if it is known.

(a) 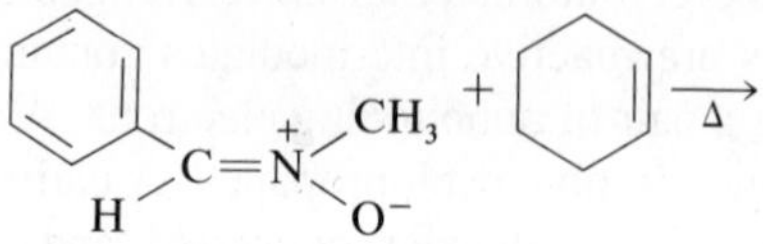

(b) 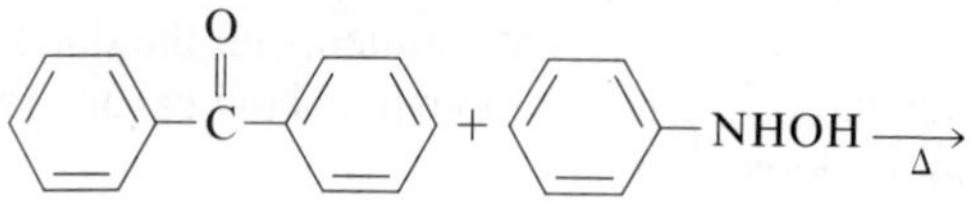

(c) 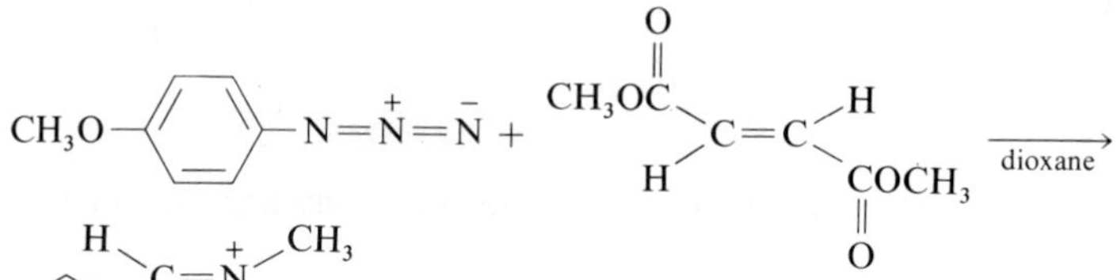

(d) 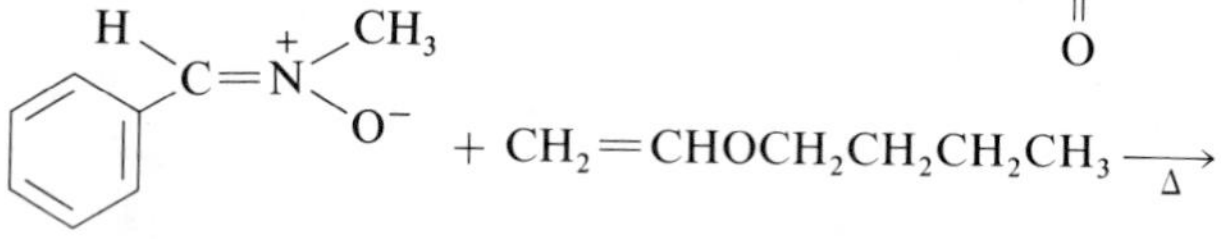

(e) 

(f) 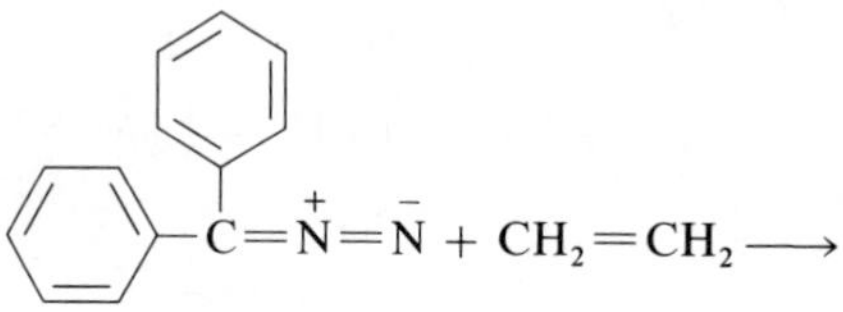

(g) 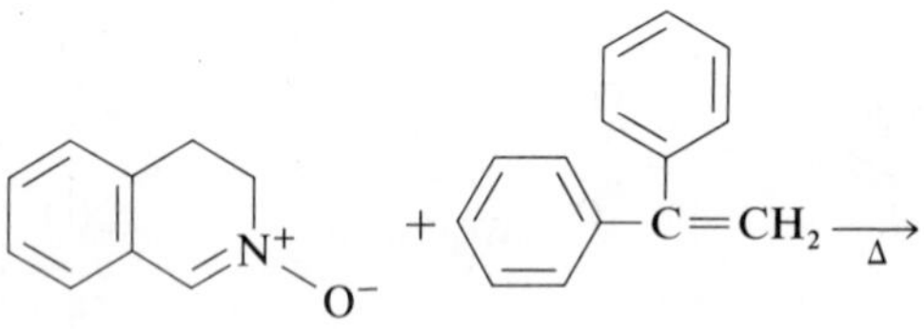

(h) 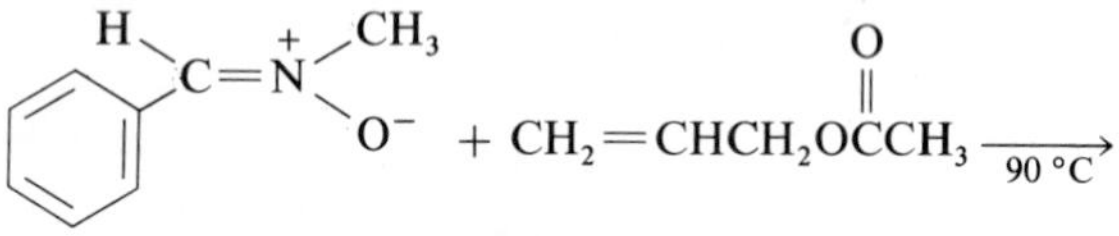

(i) 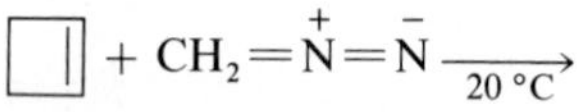

28.34 Complete the following equations.

(a) 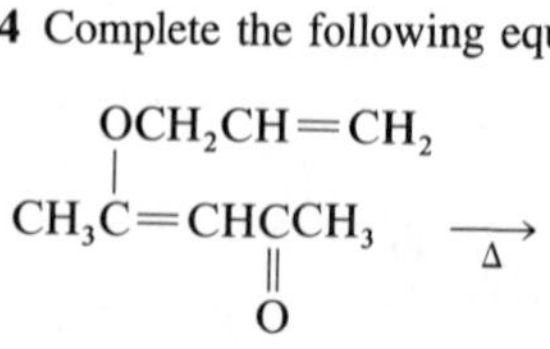

(b) 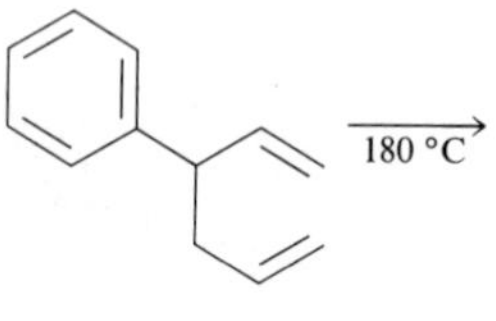

(c) 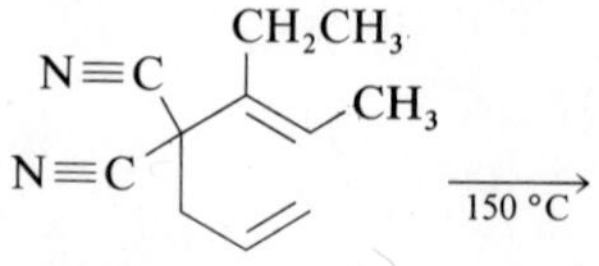

(d) 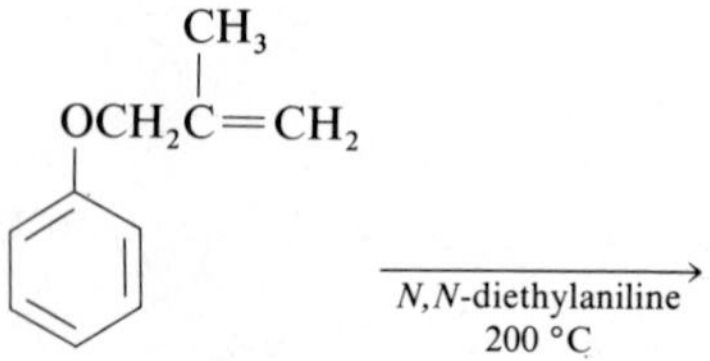

(e) 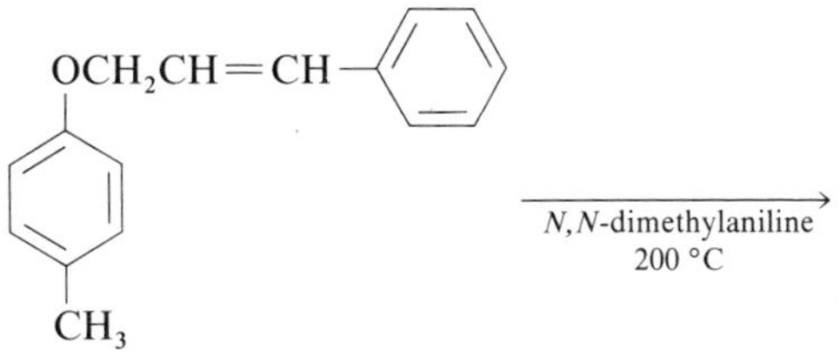
OCH$_2$CH=CH
CH$_3$
N,N-dimethylaniline
200 °C

(f) 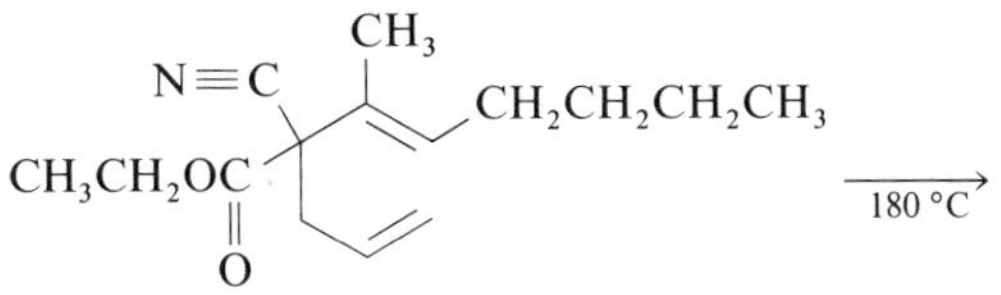
CH$_3$
N≡C
CH$_2$CH$_2$CH$_2$CH$_3$
CH$_3$CH$_2$OC
O
180 °C

(g) 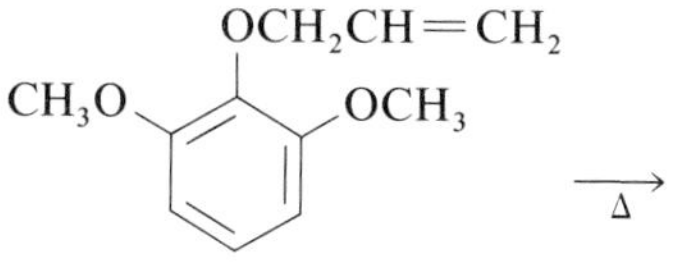
OCH$_2$CH=CH$_2$
CH$_3$O
OCH$_3$
Δ

28.35 Predict what the products of the following reactions will be.

(a) 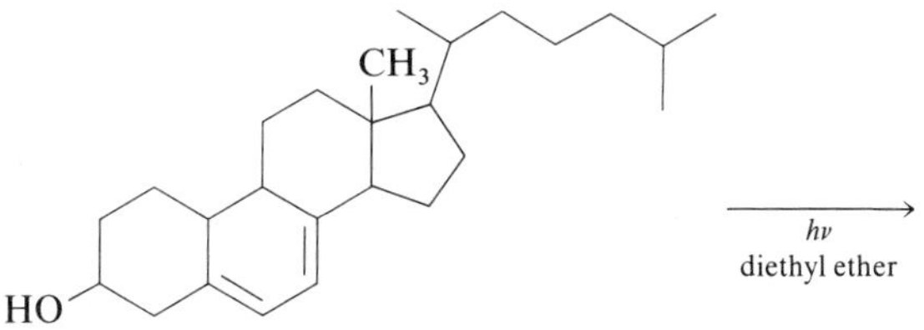
CH$_3$
HO
hν
diethyl ether

7-dehydro-19-norcholesterol
(7-dehydrocholesterol lacking the
methyl group numbered carbon 19)

(b) 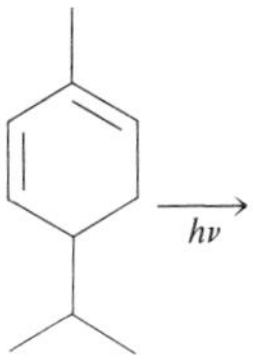
hν

(c) 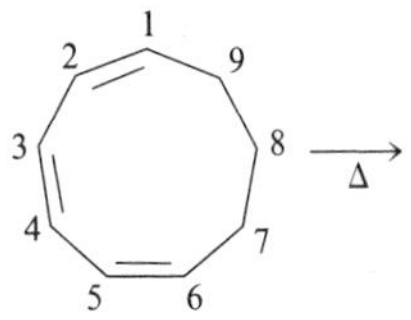
1
2
3
4
5
6
7
8
9
Δ

(d) 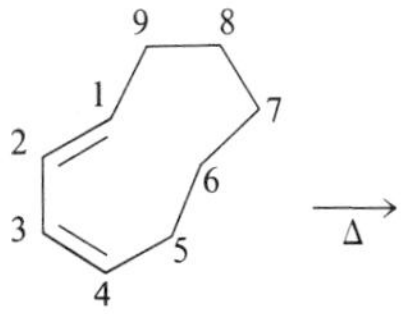
1
2
3
4
5
6
7
8
9
Δ

(e) 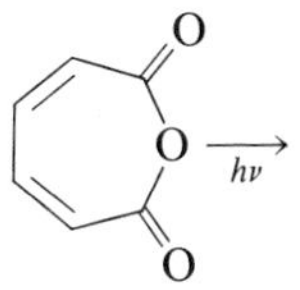
O
O
O
hν

(f) 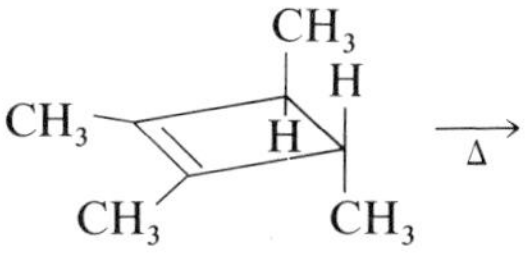
CH$_3$
H
CH$_3$
H
CH$_3$
CH$_3$
Δ

28.36 Complete the following equations.

(a) 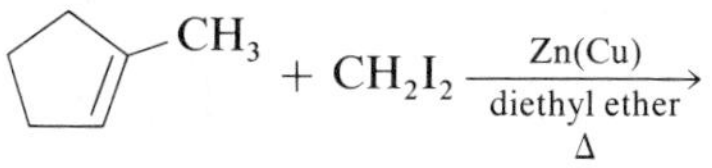
CH$_3$
+ CH$_2$I$_2$
Zn(Cu)
diethyl ether
Δ

(b) 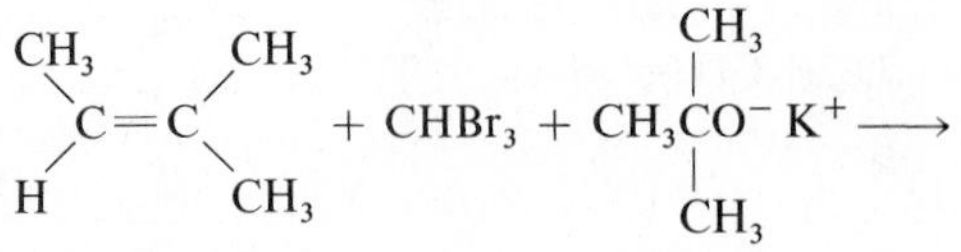

(c) $CH_2{=}CH\overset{\overset{\displaystyle O}{\|}}{C}CH_3 + CH_2I_2 \xrightarrow[\text{diethyl ether}\ \Delta]{\text{Zn(Cu)}}$

(d) 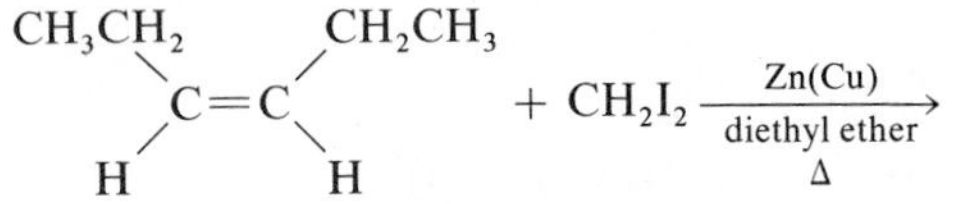

(e) 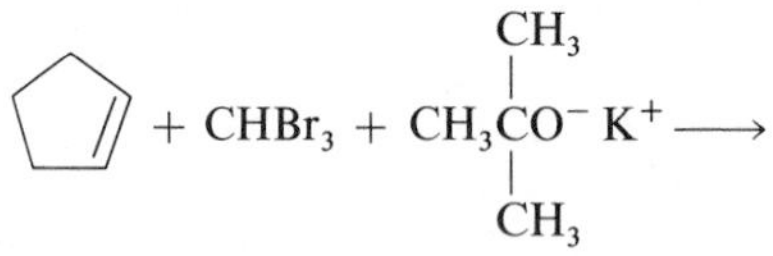

(f) 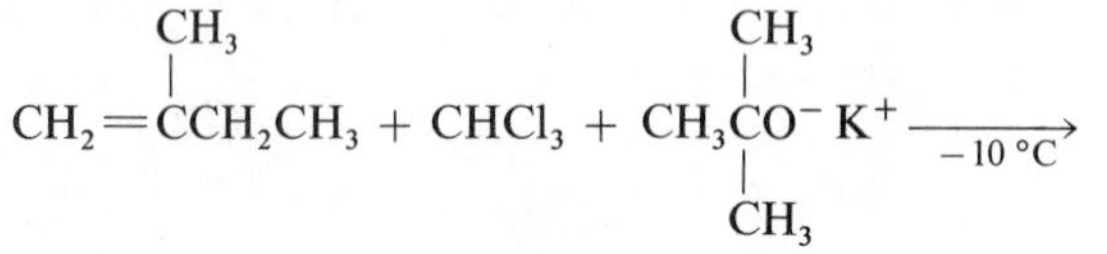

28.37 Allyl phenyl ether, specifically labeled with radioactive carbon-14 at the terminal carbon atom of the allyl group, was synthesized by the following sequence of reactions.

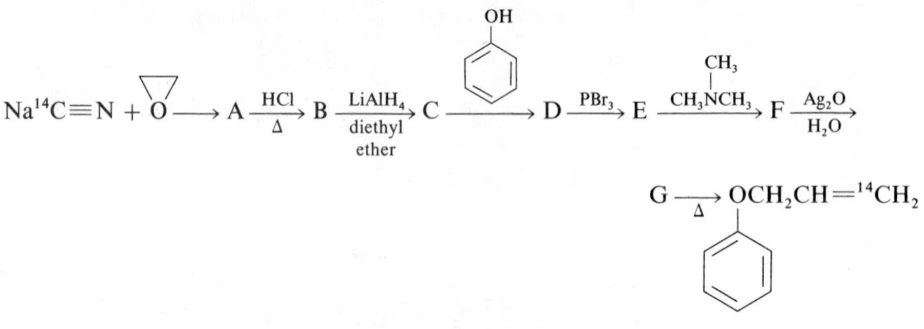

(a) Write structures for the compounds represented by letters.
(b) Why was this particular way of introducing the double bond used rather than another method, such as the dehydration of an alcohol?
(c) When the labeled allyl ether is heated, 2-allylphenol is formed. This phenol is subjected to the following reactions. Give structures for the compounds formed in this sequence of reactions.

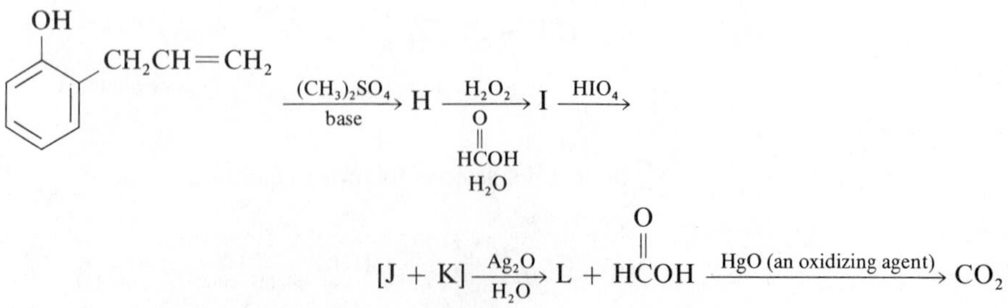

(d) Will the CO_2 that is obtained as a product of the sequence of reactions in part c have the radioactivity introduced into the allyl ether in the first synthesis?
(e) 2,6-Dimethylphenyl allyl ether, similarly labeled with carbon-14, was synthesized, rearranged, and degraded by the reactions shown in parts a and c. Will the CO_2 obtained from that experiment be radioactive?

28.38 The following experimental observations were made:

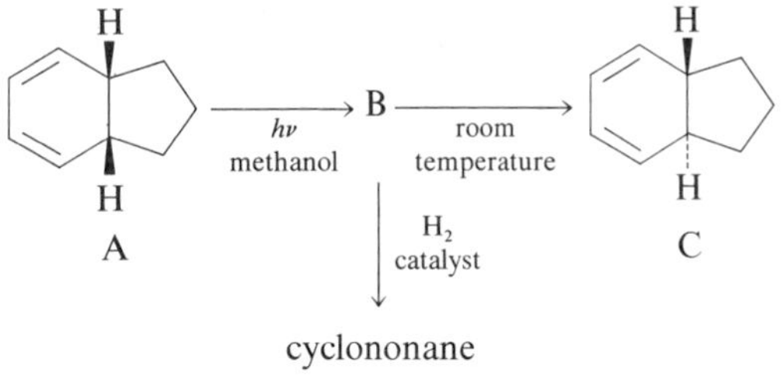
H
H
A
hν
methanol
B
room
temperature
H
H
C
H_2
catalyst
cyclononane

When *cis*-bicyclo[4.3.0]nona-2,4-diene (Compound A) is irradiated at -20 °C in methanol, Compound B is obtained. Compound B has λ_{max} 290 nm (ϵ 2050) and bands in the infrared at 1645, 1621, 1598, 975, 960, and 670 cm^{-1}. It has two broad multiplets in its nuclear magnetic resonance spectrum, six protons in the region 6.2–4.7 ppm and six protons at 2.6–0.6 ppm. If the temperature is kept below 0 °C, Compound B can be hydrogenated to cyclononane. At room temperature, it is converted to *trans*-bicyclo[4.3.0]nona-2,4-diene (Compound C). How would you explain these observations? What is the structure of Compound B?

28.39 When (1*Z*,3*E*)-1,3-Cyclooctadiene is heated to 80 °C, it converts to bicyclo[4.2.0]oct-7-ene (shown below). (1*Z*,3*Z*)-1,3-Cyclooctadiene undergoes a photochemical ring closure to give the same compound, with the same stereochemistry, but does not give a cyclobutene product when heated. How do you explain these observations?

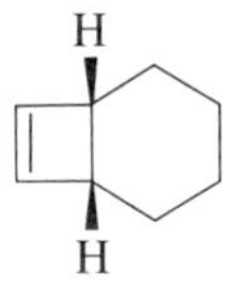
H
H

bicyclo[4.2.0]oct-7-ene

When bicyclo[4.2.0]oct-7-ene is heated above 300 °C, the cyclobutene ring opens and (1*Z*,3*Z*)-1,3-cyclooctadiene is formed. Is this opening of the ring a concerted reaction?

28.40 Cyclopentadiene reacts with (*E*)-1,2-dichloro-1,2-difluoroethylene to give two different products, shown below.

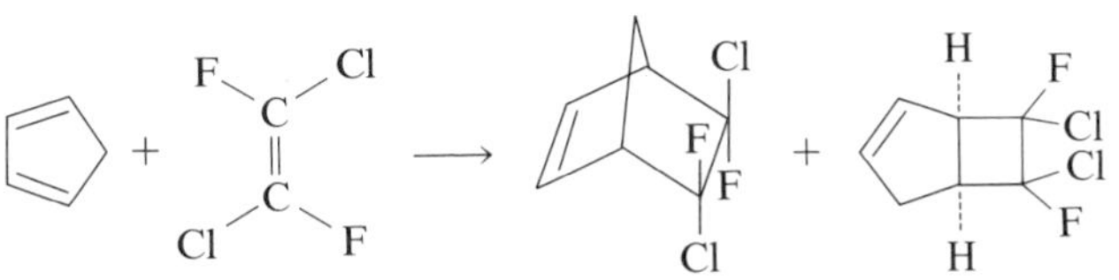
F
Cl
C
C
Cl
F
Cl
F
F
Cl
H
F
Cl
Cl
F
H
and enantiomer
98%
mixture of 4
possible diastereomers
2%

How do you account for the differing stereoselectivity of the reactions leading to the two types of products?

28.41 Write a mechanism for the following experimentally observed transformation.

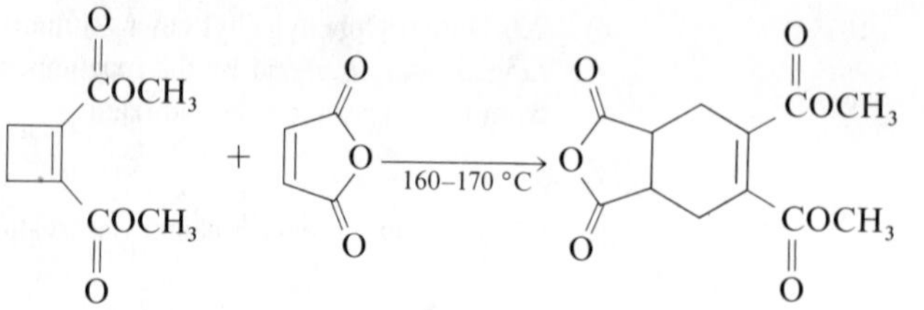

28.42 Of the two compounds shown below, one undergoes opening of the cyclobutene ring easily, whereas the other requires high temperatures. Predict which one reacts with ease, and explain why.

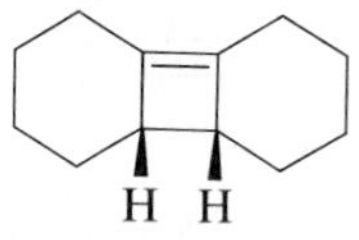

A

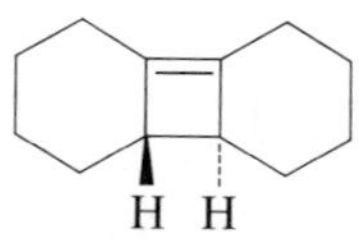

B

28.43 The cyclobutene diol shown below gives 3,4-dimethyl-2,5-dione, among other products, on heating. How would you explain the formation of this product?

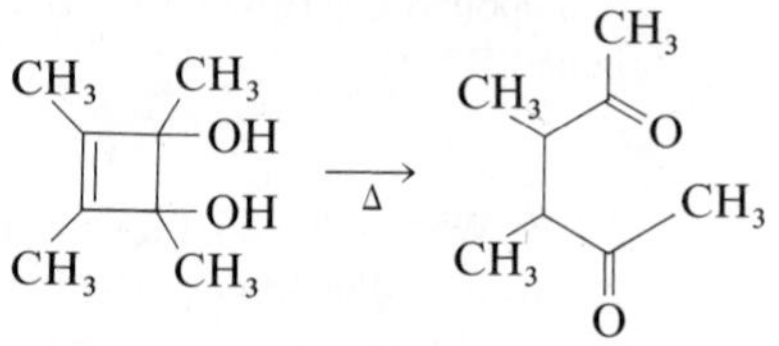

28.44 The following sequence of reactions was carried out.

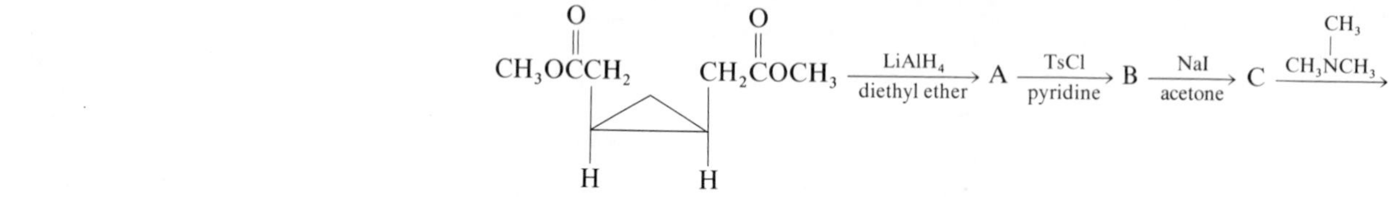

$$D \xrightarrow[H_2O]{Ag_2O} E \xrightarrow[\Delta]{} \underset{C_7H_{10}}{F} + \underset{C_7H_{10}}{G};\ F \xrightarrow[PtO_2]{H_2} \text{cycloheptane}$$

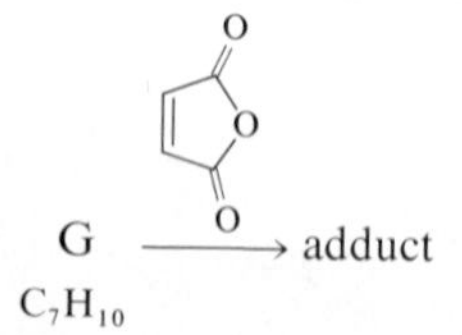

Compound F has no absorption in the ultraviolet region of the spectrum above 215 nm. It has no band in the infrared at 990 or 909 cm^{-1}, regions typical of C—H bending frequencies of terminal alkenes. Compound F has three multiplets in its nuclear magnetic resonance spectrum, with relative intensities of 4 : 4 : 2. Four of the protons lie in the region for vinylic protons; the other six are in the aliphatic region. Compound G has λ_{max}^{hexane} 248 nm. Give structures for Compounds A through G. How do you account for the formation of Compounds F and G?

28.45 The Hoffmann degradation of tropine, obtained from the hydrolysis of atropine (p. 925), gives 5-(*N*,*N*-dimethylamino)-1,3-cycloheptadiene, which isomerizes at 150 °C to 1-(*N*,*N*-dimethylamino)-1,3-cycloheptadiene. What kind of reaction is occurring?

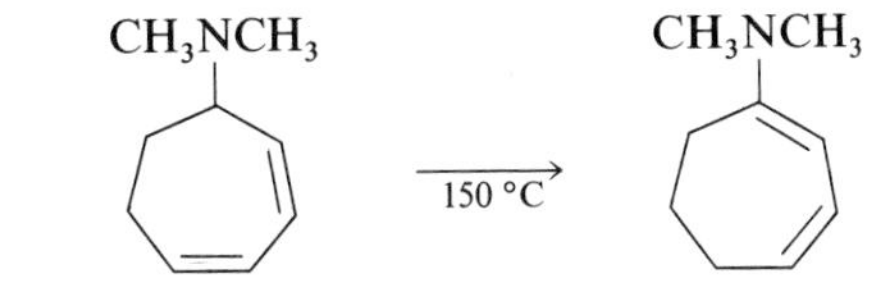

CH₃NCH₃
150 °C
CH₃NCH₃

5-(*N*,*N*-dimethylamino)-1,3-cycloheptadiene   1-(*N*,*N*-dimethylamino)-1,3-cycloheptadiene

28.46 The following sequence of reactions was carried out. Assign structures to the compounds designated by letters.

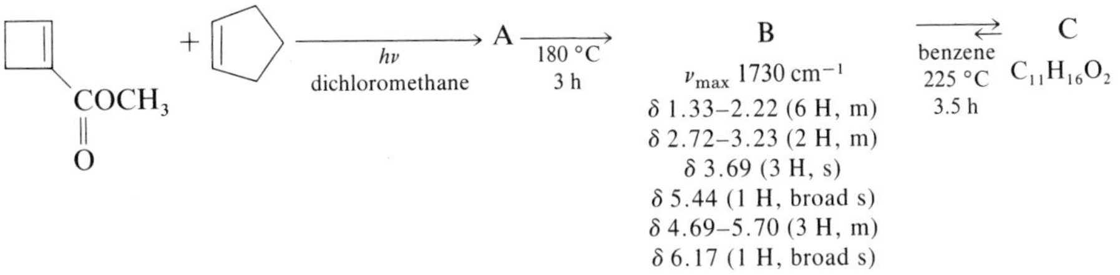

COCH₃
O
hν
dichloromethane
A
180 °C
3 h
B
νmax 1730 cm⁻¹
δ 1.33–2.22 (6 H, m)
δ 2.72–3.23 (2 H, m)
δ 3.69 (3 H, s)
δ 5.44 (1 H, broad s)
δ 4.69–5.70 (3 H, m)
δ 6.17 (1 H, broad s)
benzene
225 °C
3.5 h
C
C₁₁H₁₆O₂

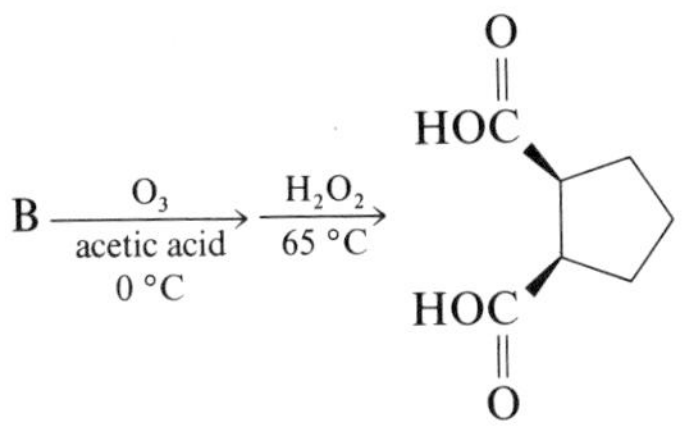

B
O₃
acetic acid
0 °C
H₂O₂
65 °C
HOC
O
HOC
O

Index

Boldface page number indicates text page on which indexed term is defined.